EIGHTEENTH EDITION

Zinsser Microbiology

EIGHTEENTH EDITION

Zinsser
Microbiology

EDITED BY

Wolfgang K. Joklik, D.Phil.
James B. Duke Distinguished Professor of Microbiology and Immunology
Chairman, Department of Microbiology and Immunology
Duke University Medical Center

Hilda P. Willett, Ph.D.
Professor of Microbiology
Duke University Medical Center

D. Bernard Amos, M.D.
James B. Duke Distinguished Professor of Immunology
and Professor of Experimental Surgery
Duke University Medical Center

APPLETON-CENTURY-CROFTS/Norwalk, Connecticut

Copyright © 1984 by Appleton-Century-Crofts
A Publishing Division of Prentice-Hall, Inc.

All rights reserved. This book, or any parts thereof, may not be used or reproduced in any manner without written permission. For information, address Appleton-Century-Crofts, 25 Van Zant Street, East Norwalk, Connecticut 06855.

84 85 86 87 88 89 / 10 9 8 7 6 5 4 3 2 1

Copyright © 1960, 1968 by Meredith Corporation
Under the title ZINSSER BACTERIOLOGY, this book was copyrighted as follows:
Copyright © 1957 by Meredith Corporation
Under the title ZINSSER'S TEXTBOOK OF BACTERIOLOGY, this book in part was copyrighted as follows: .
Copyright © 1948, 1952 by Meredith Corporation
Under the title A TEXTBOOK OF BACTERIOLOGY, this book in part was copyrighted as follows:
Copyright © 1910, 1914, 1916, 1918, 1922, 1928, 1934, 1935, 1937, 1939 by Meredith Corporation
Copyright © renewed 1962, 1963 by Mrs. Hans Zinsser
Copyright © renewed 1942, 1944, 1946, 1955, 1956 by Mrs. Ruby H. Zinsser
Copyright © 1950 by Mrs. Ruby H. Zinsser and Frederick F. Russell, M.D.
Copyright © renewed 1938 by Hans Zinsser

Prentice-Hall International, Inc., London
Prentice-Hall of Australia, Pty. Ltd., Sydney
Prentice-Hall Canada, Inc.
Prentice-Hall of India Private Limited, New Delhi
Prentice-Hall of Japan, Inc., Tokyo
Prentice-Hall of Southeast Asia (Pte.) Ltd., Singapore
Whitehall Books Ltd., Wellington, New Zealand
Editora Prentice-Hall do Brasil Ltda., Rio de Janeiro

Library of Congress Cataloging in Publication Data

Zinsser, Hans, 1878–1940.
  Microbiology.

  Includes bibliographies and index.
  1. Medical microbiology. I. Joklik, Wolfgang K.
II. Willett, Hilda P. III. Amos, Dennis Bernard,
1923- . IV. Title. [DNLM: 1. Microbiology. QW 4
Z783m]
QR46.Z5 1984   616'.01   83-25738
ISBN 0-8385-9978-8

Index compiled by Robert Zolnerzak
Design: Jean M. Sabato

PRINTED IN THE UNITED STATES OF AMERICA

To the memory of
Philip Hanson Hiss, Jr.,
Hans Zinsser,
Stanhope Bayne-Jones,
and David T. Smith

Contributors

Wolfgang K. Joklik, D.Phil.
James B. Duke Distinguished Professor of Microbiology
and Immunology; Chairman, Department of Microbiology
and Immunology, Duke University Medical Center

Hilda P. Willett, Ph.D.
Professor of Microbiology, Duke University Medical Center

D. Bernard Amos, M.D.
James B. Duke Distinguished Professor of Immunology
and Professor of Experimental Surgery, Duke University
Medical Center

David W. Barry, M.D.
Head, Department of Clinical Investigation and Virology,
Burroughs-Wellcome Company; Associate Professor of
Medicine, Duke University Medical Center

Dale L. Blazey, M.D., Ph.D.
Formerly Fellow in Microbiology and Immunology, Duke
University Medical Center; Presently Department of
Medicine, University of Washington School of Medicine,
Seattle

C. Edward Buckley, III, M.D.
Professor of Medicine and Assistant Professor of
Immunology, Duke University Medical Center

Rebecca H. Buckley, M.D.
James B. Sidbury Professor of Pediatrics and Professor of
Immunology, Duke University Medical Center

Richard O. Burns, Ph.D.†
Professor of Microbiology, Duke University Medical Center

James J. Crawford, Ph.D.
Professor of Oral Biology, University of North Carolina
School of Dentistry

Peter Cresswell, Ph.D.
Associate Professor of Immunology, Duke University
Medical Center

Jeffrey R. Dawson, Ph.D.
Associate Professor of Immunology, Duke University
Medical Center

Cornelia L. Dekker, M.D.
Formerly Assistant Clinical Professor of Pediatrics, Duke
University Medical Center; Presently Senior Clinical
Research Scientist, Burroughs-Wellcome Company

Thomas E. Frothingham, M.D.
Professor of Pediatrics and Professor of Community
Health Sciences, Duke University Medical Center

Harry A. Gallis, M.D.
Associate Professor of Medicine and Assistant Professor of
Microbiology, Duke University Medical Center

Laura T. Gutman, M.D.
Associate Professor of Pediatrics and Associate Professor
of Pharmacology, Duke University Medical Center

John D. Hamilton, M.D.
Associate Professor of Medicine, Duke University Medical
Center

Gale B. Hill, Ph.D.
Associate Professor of Obstetrics and Gynecology and
Associate Professor of Microbiology, Duke University
Medical Center

Joyce W. Jenzano, M.D.
Assistant Professor of Dental Ecology, University of North
Carolina School of Dentistry

Samuel L. Katz, M.D.
Wilburt C. Davison Professor of Pediatrics; Chairman,
Department of Pediatrics, Duke University Medical Center

Sandra N. Lehrman, M.D.
Formerly Clinical Assistant Professor of Pediatrics, Duke
University Medical Center; Presently Senior Virologist,
Burroughs-Wellcome Company

Thomas G. Mitchell, Ph.D.
Associate Professor of Mycology, Duke University Medical
Center

†Deceased.

vii

Suydam Osterhout, M.D., Ph.D.
Professor of Microbiology and Professor of Medicine,
Duke University Medical Center

Wendell F. Rosse, M.D.
Florence McAlister Professor of Medicine and Professor of
Immunology, Duke University Medical Center

Alfred Sanfilippo, M.D., Ph.D.
Assistant Professor of Pathology and Assistant Professor of
Experimental Surgery, Duke University Medical Center

David W. Scott, Ph.D.
Formerly Professor of Immunology, Duke University
Medical Center; Presently Dean's Professor of
Immunology, University of Rochester Medical Center

Daniel J. Sexton, M.D.
Formerly Fellow in Medicine, Duke University Medical
Center; Presently Oklahoma City Clinic, Oklahoma City

Ralph Snyderman, M.D.
Professor of Medicine and Professor of Immunology, Duke
University Medical Center

Norman F. Weatherly, Ph.D.
Professor of Parasitology, University of North Carolina
School of Public Health

Robert W. Wheat, Ph.D.
Professor of Microbiology and Assistant Professor of
Biochemistry, Duke University Medical Center

Catherine M. Wilfert, M.D.
Professor of Pediatrics and Professor of Clinical Virology,
Duke University Medical Center

Peter Zwadyk, Ph.D.
Associate Professor of Pathology and Associate Professor
of Microbiology, Duke University Medical Center

Contents

Section IV BASIC VIROLOGY

Section V CLINICAL VIROLOGY

Preface

With each passing year the term *Microbiology* becomes a less satisfactory umbrella for the many disciplines that it attempts to cover. Bacteriology, immunology, virology, mycology, and parasitology have each long since become separate and independent disciplines. They are together in a single text simply because they deal with the agents of infectious disease in humans and with the mechanisms employed by the host in his defense.

In spite of the undeniable triumphs of antibiotic chemotherapy, which has revolutionized the practice of medicine and very likely represents the greatest single triumph of biomedical science, "microbes" are by no means "conquered"; they continue to cause infections that demand a large amount of the physician's time. In fact, new infectious agents, unsuspected properties of known agents, additional mechanisms for the genesis and persistence of infections, and advances in our understanding of the behavior of infectious agents at the molecular, cellular, and organismal levels are constantly being reported. As a result, the scope and complexity of the material to be presented to students expand rapidly, and although the literature abounds with excellent papers and reviews, the compilation of a comprehensive textbook of manageable size becomes increasingly difficult.

This new edition of *Zinsser Microbiology*, the 18th, is designed for use by medical students experiencing their first exposure to medical microbiology. To that end, there is presented not only a description of the pathogenic infectious agents and the diseases that they cause, but also a discussion of the basic principles of bacterial physiology and genetics, of molecular and cellular immunology, and of molecular virology, the purpose of which is to provide a firm basis for growth with the field during the remainder of the student's future professional career. By the same token, the book will also fulfill the needs of advanced undergraduates who plan careers in medicine or biomedical research. The book is also designed as a reference source for instructors; to that end each chapter is supplemented with a selection of both reviews and important original papers that will permit a rapid entrée to any specialized topic that may require further study.

The 18th Edition represents a very extensive revision of the 17th Edition; all portions of the text that were not completely rewritten have been thoroughly updated. Section VII, Medical Parasitology, has been completely rewritten by a new contributor, Dr. Norman F. Weatherly; he replaces Dr. John E. Larsh, Jr., who contributed this section to the last six editions and who has now retired. Dr. Weatherly brings fresh approaches and new viewpoints without sacrificing authority. The Clinical Virology section also has several new contributors, who contributed completely new chapters on herpesviruses, papovaviruses, viruses in gastrointestinal tract infections, and hepatitis viruses. In the Basic Virology and Bacterial Physiology sections many chapters, particularly those on the molecular aspects of virus multiplication cycles, on tumor viruses, and on the molecular basis of genetics have been completely rewritten; these are areas in which exciting discoveries are constantly being made, many of them changing fundamentally our views of the arrangement of genetic material and how it is expressed. The same applies to the Immunology section where new chapters on immunopathology, immune responses to infection, and immunity to tumors and to pregnancy have been provided. This section provides a comprehensive account of both basic and clinical immunology, organized so as to highlight topics currently deemed of maximum relevance to medical students. Finally, the Medical Bacteriology and Medical Mycology sections have been brought up-to-date, with new material added on new diseases such as Legionnaire's disease, the toxic shock syndrome, the acquired immune deficiency syndrome (AIDS), and Lyme disease. In these sections, which like all other sections have been carefully edited by a single author so as to ensure a uniform format, emphasis is again placed on correlating the basic and clinical aspects of each infectious agent so that the student may acquire an appreciation of how fundamental research may be used in unravelling the complexities of host–parasite relationships. Each chapter consists of (1) an introduction to the important biologic properties of the organism, (2) a description of the clinical infection in humans, including a discussion of mechanisms of pathogenicity, (3) a section on laboratory diagnosis that provides information on modern culture and immunological procedures, and (4) a discussion of the currently recommended treatment.

With regard to the bibliography, we have again elected not to reference specific statements in the text but to append to each chapter a list of recent reviews and key original papers. The former will quickly guide the reader to any specific aspect of microbiology and immunology that he wishes to pursue; the latter makes available the detailed considerations and circumstances that have gone into the genesis of the most important discoveries. Many of the papers that are cited already are, or no doubt will soon become, "classics."

We have tried not to increase the size of the book—no easy task in view of the enormous amount of new information that has accumulated since publication of the last edition in 1980. Obviously, this has entailed the omission of a certain amount of older material; however, we are confident that there are no major gaps and that in our presentation of the newest advances we have not sacrificed careful and logical explanations of fundamental principles.

The list of individuals who have helped to produce this volume extends far beyond the circle of our colleagues who contributed textual material and to whom we are profoundly indebted. We would especially like to thank our many colleagues who permitted us to use illustrative material and who almost invariably supplied us with original photographs, and the many publishers who allowed us to reproduce previously published material. We would also like to thank the artists who did a superb job in drawing the innumerable charts and diagrams, and the many secretaries who cheerfully massaged the text innumerable times on their word processors. Finally, we wish to express our appreciation to the staff of Appleton-Century-Crofts for their efficient cooperation in producing this new edition.

Wolfgang K. Joklik
Hilda P. Willett
D. Bernard Amos

Preface to the First Edition

The volume here presented is primarily a treatise on the fundamental laws and technic of bacteriology, as illustrated by their application to the study of pathogenic bacteria.

So ubiquitous are the bacteria and so manifold their activities that bacteriology, although one of the youngest of sciences, has already been divided into special fields—medical, sanitary, agricultural, and industrial—having little in common, except problems of general bacterial physiology and certain fundamental technical procedures.

From no other point of approach, however, is such a breadth of conception attainable, as through the study of bacteria in their relation to disease processes in man and animals. Through such a study one must become familiar not only with the growth characteristics and products of the bacteria apart from the animal body, thus gaining a knowledge of methods and procedures common to the study of pathogenic and nonpathogenic organisms, but also with those complicated reactions taking place between the bacteria and their products on the one hand and the cells and fluids of the animal body on the other—reactions which often manifest themselves as symptoms and lesions of disease or by visible changes in the test tube.

Through a study and comprehension of the processes underlying these reactions, our knowledge of cell physiology has been broadened, and facts of inestimable value have been discovered, which have thrown light upon some of the most obscure problems of infection and immunity and have led to hitherto unsuspected methods of treatment and diagnosis. Thus, through medical bacteriology—that highly specialized offshoot of general biology and pathology—have been given back to the parent sciences and to medicine in general methods and knowledge of the widest application.

It has been our endeavor, therefore, to present this phase of our subject in as broad and critical a manner as possible in the sections dealing with infection and immunity and with methods of biological diagnosis and treatment of disease, so that the student and practitioner of medicine, by becoming familiar with underlying laws and principles, may not only be in a position to realize the meaning and scope of some of these newer discoveries and methods, but may be in a better position to decide for themselves their proper application and limitation.

We have not hesitated, whenever necessary for a proper understanding of processes of bacterial nutrition or physiology, or for breadth of view in considering problems of the relation of bacteria to our food supply and environment, to make free use of illustrations from the more special fields of agricultural and sanitary bacteriology, and some special methods of the bacteriology of sanitation are given in the last division of the book, dealing with the bacteria in relation to our food and environment.

In conclusion it may be said that the scope and arrangement of subjects treated in this book are the direct outcome of many years of experience in the instruction of students in medical and in advanced university courses in bacteriology, and that it is our hope that this volume may not only meet the needs of such students but may prove of value to the practitioner of medicine for whom it has also been written.

It is a pleasure to acknowledge the courtesy of those who furnished us with illustrations for use in the text, and our indebtedness to Dr. Gardner Hopkins and Professor Francis Carter Wood for a number of the photomicrographs taken especially for this work.

P. H. Hiss, Jr.
H. Zinsser

Zinsser
Microbiology

SECTION I
BACTERIAL PHYSIOLOGY

CHAPTER 1

The Historical Development of Medical Microbiology

The history of the many concepts now embodied in the doctrines of microbiology is an account of attempts to solve the problems of the origin of life, the putrefaction of dead organic materials, and the nature of communicable changes in the bodies of living humans and animals. The visible aspects of these phenomena were as apparent and interesting to ancient observers as they are to modern biologists. In the past, notions of ultimate causes were derived from the available factual knowledge colored by the theologic and philosophic tenets of the time. The early history of what has become the science of microbiology is to be found, therefore, in the writings of the priests, philosophers, and scientists who studied and pondered these basic biologic problems.

Infection and Contagion

Among ancient peoples, epidemic and even endemic diseases were regarded as supernatural in origin and sent by the gods as punishment for the sins of man. The treatment and, more important, the prevention of these diseases were sought by sacrifices and lustrations to appease the anger of the gods. Since man is willful, wanton, and sinful by nature, there was never any difficulty in finding a particular set of sins to justify a specific epidemic.

The concept of contagion and the practice of hygiene were, however, not entirely unknown to ancient man. The Old Testament is often quoted as indicating the belief that leprosy was contagious and could be transmitted by contact. The principle of contagion by invisible creatures was later recorded by Varro in the second century BC, and the concept was familiar to Greek, Roman, and Arabic writers. Roger Bacon, in the thirteenth century, more than a millennium later, postulated that invisible living creatures produced disease. The Venetian, Fracastorius, in 1546, wrote from a knowledge of syphilis that communicable disease was transmitted by living germs, "seminaria morbi," through direct contact or by intermediary inanimate fomites and through air "ad distans." Fracastorius expressed the opinion that the seeds of disease, passing from one infected individual to another, caused the same disease in the recipient as in the donor. This clear expression of the germ theory of disease was three centuries ahead of its time.

First Observation of Bacteria

Direct observation of microorganisms had to await the development of the microscope. The human eye cannot see objects smaller than 30 μm ($^1/_{1000}$ inch) in diameter, and although knowledge of magnifying lenses reaches back to the time of Archimedes, the science of optics was not clarified until the thirteenth century by the Franciscan monk, Roger Bacon. The telescope was invented by Galileo in 1608, followed by the microscope in the same century. The first person known to have made glass lenses powerful enough to observe and describe bacteria was the amateur lens grinder, Anton van Leeuwenhoek (1632–1723), of Delft, Holland. In letters to the experimentalist group, The Royal Society of London, Leeuwenhoek described many "animalcules," including the three major morphologic forms of bacteria (rod, sphere, and spiral), various free-living and parasitic protozoa from human and animal feces, filamentous fungi, and globular bodies we now know as yeasts, and he discovered spermatozoa. In his lifetime he made some 250 single-lensed microscopes. He searched everywhere in this new microcosmic world he had discovered. In letters from 1676 to 1683, he described the sizes, shapes, and even the motility of bacteria (Fig. 1-1), using simple single biconvex lens microscopes, such as illustrated in Figure 1-2. There is no doubt that he saw the most common forms of bacteria, the cocci, bacilli, and spirochetes. His observational reports were enthusiastic and accurate and developed some interest at the time, but unfortunately Leeuwenhoek did all this as a hobby and left no students to continue his work. However, in 1678, Robert Hooke, who developed the compound microscope, confirmed Leeuwenhoek's discoveries. Microorganisms were then occasionally studied by those primarily interested in classifying the various life forms observable with the microscope. These observations lay dormant and were not exploited by those interested in disease. The following 125 to 150 years witnessed the gradual development of knowledge and acceptance of the experimental method, which slowly disseminated throughout the expanding learned centers of the world. Improved microscopes became generally available only in the 1800s as a result of the Industrial Revolution, which allowed rapid technologic advances. Even then, no notable advance in microbiology was accomplished until after the attention of the scientific world was focused on microbes and their role in the controversies of the doctrine of spontaneous generation and the associated phenomenon of fermentation.

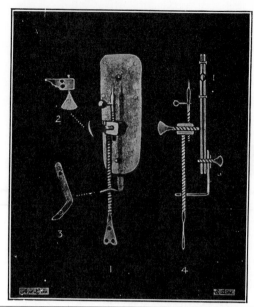

Figure 1-2. Leeuwenhoek's "Microscope." *(From Dobell: Anton van Leeuwenhoek and His "Little Animals," 1932. New York, Harcourt, Brace and Co.)*

Spontaneous Generation

The controversy over human ability to create life carried over from Greek mythology. Even Aristotle (384–322 BC) thought animals could originate from the soil. Samson, in the Bible (Old Testament), and again Virgil, about 40 BC, described recipes for producing bees from honey, and for centuries it was believed that maggots could be produced by exposing meat to warmth in the air. This was not refuted until Francesco Redi (1626–1697) proved that gauze placed over a jar containing meat prevented maggots forming in the meat. Redi also observed that adult flies, attracted by the odor of meat through the gauze, laid eggs on the cloth, and maggots developed from the eggs. Recipes for producing mice and other similar life forms in litter and refuse were gradually disproved and discarded in similar fashion. However, the question was not settled in all minds. When microbes were discovered, their association with putrefaction and fermentations again raised the question of spontaneous generation. John Needham, in 1749, observed the appearance of microorganisms in putrefying meat and interpreted this as spontaneous generation. Spallanzani, however, boiled beef broth for an hour, sealed the flasks, and observed no formation of microbes. Needham and, 100 years later, Pouchet (1859) argued that

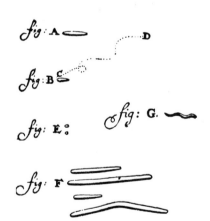

Figure 1-1. Leeuwenhoek's picture of bacteria from the human mouth. Dobell's identifications are as follows: **A.** A motile *Bacillus.* **B.** *Selenomonas sputigena.* **E.** Micrococci. **F.** *Leptothrix buccans.* **G.** Probably *Spirochaeta buccalis.* (From Dobell: Anton van Leeuwenhoek and His "Little Animals," 1932. New York, Harcourt, Brace and Co.)

access of air was necessary for the spontaneous generation of microscopic living beings. Disproof came from several lines of evidence. Franz Schulze (1815–1873) passed air through strong acids and then into boiled broth, while Theodor Schwann (1810–1882) passed air through red hot tubes and observed no growth. About 1850, Schroeder and von Dusch filtered air through cotton filters into broth and observed no growth. Pasteur was able to filter microorganisms from the air and concluded that this was the source of contamination. He developed an aseptic technique, using heat, in order to tranfer and work with his microbes and, finally, in 1859, in public controversy with Pouchet, prepared boiled broths in flasks with long narrow gooseneck tubes that were open to the air. Air could pass but microorganisms settled in the gooseneck, and no growth developed in any of the flasks. Finally, a British physicist, John Tyndall (1820–1893), proved that dust carried the germs, and the story was complete. Tyndall also found that bacterial spores could be killed by successive heating, a process now known as tyndallization.

The Germ Theory of Disease

Empirical Observations. A firm basis for the causal nature of infectious disease was established only in the latter half of the nineteenth century. One of the first proofs came from Agostino Bassi who, in the early 1800s, proved that a fungus, later named *Botrytis bassiana* in his honor, caused a disease of silkworms called "muscardine" in France and "mal segno" in Italy. In 1839, Schoenlein found the causative fungus in lesions of favus, and, in 1846, Eichstedt noted the contagiousness of pityriasis versicolor and discovered a fungus in skin scrapings from patients.

In the 1840s, the American poet-physician, Oliver Wendell Holmes, wrote "The Contagiousness of Puerperal Fever," in which he suggested that disease was caused by germs carried from one new mother to another. In 1861, Ignaz Semmelweis, who had drastically decreased childbirth deaths by antiseptic techniques and practices, published his extremely important "The Cause, Concept and Prophylaxis of Childbed Fever." However, the importance of antiseptics in reducing contagious disease was not fully realized until the late 1870s, when Joseph Lister demonstrated the value of spraying operating rooms with aqueous phenol.

Lessons Learned from Fermentations. Further emphasis on microbial activities came from the work of Louis Pasteur from the 1850s to the 1880s. In studies on the diseases of wine, Pasteur demonstrated that alcoholic fermentation of grapes, fruit, and grains was caused by microbes, then called "ferments." In good wine batches certain types of "ferments" existed in the vats, while in poor or bad fermentations other types of microbes were found, some of which Pasteur found to be capable of growing anaerobically. He suggested eliminating the bad types of "ferments" from fresh juices by heating at 63C for one-half hour, then

cooling and reinoculating with a culture from the satisfactory vats. Pasteur's success with the problems of the wine industry led the French government to request that he study a disease, pébrine, which was ruining the silkworm industry in Southern France. Pasteur struggled with this problem for several years before he isolated the causative organism and showed that farmers could eliminate the problem by using healthy, noninfected breeding stock.

Observations and Experiments with Animals. In 1850, Rayer and Davaine observed rod-shaped microorganisms in the blood of animals that had died of anthrax. Rayer recalled the experiments in 1825 of Barthélemy who had shown that anthrax was transmissible by inoculation in series in sheep, and by 1863, Davaine had experimentally transmitted anthrax by blood containing these rods but not by normal blood from which rods were absent. In 1872, Obermeier discovered the relationship of a spirillum to relapsing fever and demonstrated for the first time the presence of a pathogenic microorganism in the blood of a human being.

Importance of Pure Culture Techniques. Through all this time, etiologic research was not based on pure culture work. Pure cultures were obtained largely by accident, and investigators had no way, except by crude morphologic microscopic examination, of knowing when contaminants were present. This resulted in much equivocal thinking and work that hindered progress.

The first pure or axenic culture technique was developed by Joseph Lister in 1878. Lister used a syringe to make serial dilutions in liquid media to obtain pure cultures of a simple type of organism, which he named *Bacterium (Lactobacillus) lactis.* Meanwhile, Koch, as a student of Henle who insisted on proof that an organism caused disease, was also developing and refining techniques for the isolation of pure cultures. From the work of others, notably Ehrlich, Koch learned methods of staining bacteria on glass with aniline dyes for microscopic observation. In his early work on anthrax, Koch used sterile aqueous humor of the eyes of animals as a growth medium. But, having seen the advantages of older, solid but opaque media, such as potato, beets, starch, bread, egg white, and meat, Koch developed a transparent solid medium by mixing gelatin with Löffler's peptone solution. The gelatin mixture liquefied on warming, could be heat-sterilized and aseptically poured into plates, and upon cooling, it solidified. Microorganisms streaked upon it developed into macroscopic colonies as the result of the growth of a single invisible cell. However, gelatin liquefies at a relatively low temperature (26C), and Koch later switched to agar, the transparent red seaweed extract that solidifies below 43C.

Etiologic Proof of Infectious Agents. Koch was able to isolate the anthrax organism in pure culture by streaking on his solid media and found that even after many transfers, the organism could still cause the same symptoms and disease when inoculated into animals. On the basis of his experiences, Koch formulated criteria that provided proof that a specific bacterium caused a disease. We now call these Koch's postulates.

1. The organism must always be found in the diseased animal but not in healthy ones.
2. The organism must be isolated from diseased animals and grown in pure culture away from the animal.
3. The organism isolated in pure culture must initiate and reproduce the disease when reinoculated into susceptible animals.
4. The organism should be reisolated from the experimentally infected animals.

Koch's work thus provided impetus and means for proof of the germ theory of disease.

The 20-year period following Koch's work was the Golden Age of Bacteriology. By 1900, almost all the major bacterial disease organisms had been described. The list included anthrax (*Bacillus anthracis*), diphtheria (*Corynebacterium diphtheriae*), typhoid fever (*Salmonella typhi*), gonorrhea (*Neisseria gonorrhoeae*), gas gangrene (*Clostridium perfringens*), tetanus or lockjaw (*Clostridium tetani*), dysentery (*Shigella dysenteriae*), syphilis (*Treponema pallidum*), and others.

Viruses

Only with advances in technique and improvement in apparatus is it possible to make fundamental advances through new ideas and observations. The development of bacteriologic filters and the discovery of viruses is a case in point.

Bacteriologic Filters. As an alternate to heat sterilization, unsuccessful efforts to remove bacteria from solutions by filtration through paper and similar materials led Chamberland and Pasteur to test and develop unglazed porcelain as the first successful bacterial filter (1871–1884). The Berkefeld filter of Kieselguhr (diatomaceous earth) was developed shortly thereafter in 1891. Synthetic polymer filters of cellulose nitrate, cellulose acetate, polyester, and so forth have come into common use only in the last two or three decades because of technical advances allowing quality control of pore size. It is of interest to note that these are essentially space-age products developed in part for the rapid removal of microorganisms from jet and rocket fuels.

Discovery of Viruses. The tobacco mosaic disease agent was discovered by Iwanowski in 1892 in bacteria-free filtrates of diseased plant leaf juices. This finding, confirmed by Beijerinck in 1899, marked the beginning of studies on the so-called filterable agents.

The filterable agent causing foot-and-mouth disease in cattle, the first described animal virus, was discovered by Löffler and Frosch in 1898. The yellow fever virus of humans was discovered in 1900 by Walter Reed and his co-workers. Bacterial viruses, or bacteriophages, were discov-

ered in 1915 by Twort in England and d'Herelle in France.

Viruses could not be grown in artificial media, and Koch's criteria could not be specifically applied. Because these pathogens require a living host for propagation, rapid progress in their study developed only in recent years. Again, as in the Golden Age of Bacteriology, technology had to be developed. Outstanding were the development of the electron microscope, of the ultracentrifuge, and of tissue culture, and the application of sophisticated microchemical and biochemical techniques.

It is of interest to note that some filterable agents first thought to be viruses, such as the *Mycoplasma*, *Rickettsia*, and *Chlamydia*, all of which are extremely fastidious in their growth requirements and almost all of which require living host cells, were subsequently shown to be bacteria.

Immunity

Ancient peoples immunized themselves against venomous snakes by introducing small amounts of venom into scratches in the skin. The Chinese used variolization with dried material from dermic smallpox lesions for 20 centuries. This practice spread through Asia by trade routes and was well accepted in the Middle East. Later, Edward Jenner (1749–1823) noticed that milkmaids who developed cowpox were immune to smallpox and found that he was able to protect susceptible individuals by vaccinating them with cowpox. Pasteur developed a chicken cholera vaccine in 1877; he inoculated chickens with old attenuated cultures so that a mild disease rendered the chickens immune to virulent organisms. He called this "vaccination," after Jenner's procedure. Shortly afterward, in 1881, applying the same concept, Pasteur prepared temperature-attenuated anthrax grown at 42 to 43C and protected sheep by first injecting them with these bacteria before challenging them with virulent anthrax grown at lower temperatures. Salmon and Smith, in 1884–1886, used heat-killed cultures of hog cholera bacillus to develop resistance or immunity in swine against challenge by live virulent organisms. Pasteur developed rabies vaccine in 1886, again using the idea of injecting an attenuated living disease agent. In this case, Pasteur used dried animal spinal cords without, apparently, recognizing the viral form of the disease agent.

Two schools of thought arose in explanation of the increased resistance following vaccination. Metchnikoff developed, in the 1880s, the cellular theory of protection; Bordet and others proposed the humoral, or specific, antibody concept of immunity. There is now evidence that both theories are correct. The last two decades have resulted in the isolation and, in large measure, the structural description of the major humoral immune proteins, the immunoglobulins. These are now commonly referred to as IgA, IgG, IgD, IgM, and IgE. The functions of these various immunoglobulins are currently being intensively studied. Much work is also being devoted to the mechanisms of cellular interactions in immune reactions that occur not only in infectious disease caused by bacteria, viruses, fungi, and parasites but also in rejection reactions of tissue and organ transplants, and of cancer cells.

Antimetabolites

Many antimetabolites, which were pioneered in concept by Ehrlich in the mid to late 1800s, are now accepted household words, e.g., penicillin. The modern era of antibiotics developed only after Domagk reported in 1935 that Prontosil had a dramatic effect on streptococcal infections. It was soon discovered that Prontosil was converted in the body to sulfanilamide, the active chemical agent, which is an analog of the vitamin p-aminobenzoic acid. In the 1940s, as the result of the stimulus of World War II, Florey and Chain and their associates reinvestigated Fleming's penicillin, isolated and characterized it, and demonstrated its practical clinical value. As a result of millions of tests with thousands of organisms, we now have numerous other antibiotics active against almost all types of bacteria.

With the recognition of the metabolic and structural differences, at the molecular level, between pathogenic microorganisms and human or animal cells, the rationale for developing new chemotherapeutic compounds is now often based on exploiting these differences. There is every reason to believe that newer, more specific and potent drugs will be discovered. However, chemotherapy has created new problems. Many previously susceptible organisms are now resistant to therapeutic levels of many widely used drugs. In addition, drug sensitization reactions or allergies occasionally develop, clinical syndromes are modified, and the normal ecologic flora of the body is disturbed.

Impact of Microbiology on Genetics and Biochemistry, and the Development of Molecular Biology

The enormous advantages of the availability of homogeneous populations of cells for every conceivable type of investigation were soon realized. As a result, many of the epoch-making advances during the last century in cell physiology, biochemistry, and genetics have resulted from studies with microorganisms or materials isolated from them. During the last two or three decades, these advances have led to a precise way of investigating the structure and function of nucleic acids and proteins, which has become known as "molecular biology." For example, the demonstration of the central role of DNA as the repository of genetic information resulted from the studies of Griffith in the 1920s that pneumococci could be transformed from one capsular type to another, followed by the demonstration by Avery and associates during the 1940s that the transforming factor was DNA. Final proof beyond doubt was provided by the demonstration of Hershey and Chase in 1952 that viral nucleic acid itself contained all the information necessary for virus multiplication. At the same time, Watson and Crick developed the double-helix model of DNA, which led them to suggest that one of the complementary DNA strands could serve as the template for the synthesis of the other, thus providing a description of self-perpetuating gene replication and continuity.

Demonstration of the transcription from DNA of information in the form of messenger RNA synthesized in complementary sequence to DNA soon followed, again in a microbial system. Messenger RNA was then found to be translated into polypeptides on ribosomes. By the early 1960s Nirenberg, Ochoa, and others had worked out the nature of the triplet RNA base sequences corresponding to the codon signals for each amino acid.

More recently, attention has focused on the arrangement of genetic material, including the nature of genes, and on the mechanisms that control its expression. While much of this research work is still being carried out with microbial systems, cells of higher organisms, including mammalian cells, are also being used very extensively nowadays. A very important factor in this connection has been the development of the technique of tissue culture, which permits animal cells to be grown, cloned, and passaged like microorganisms. New concepts concerning the regulation of gene expression in mammalian cells are being developed rapidly. As a result, it should be possible to develop, within the foreseeable future, a rational system of antiviral chemotherapy, so that virus-caused diseases may be brought under effective control, just as antibiotics control bacteria-caused diseases. Further, there is every hope that the fundamental control mechanisms that operate in both normal and abnormal cell differentiation, including cancer, will become apparent before too long.

FURTHER READING

Bulloch W: The History of Bacteriology. London, Oxford University Press, 1938

Dobell C: Anton van Leeuwenhoek and His "Little Animals." New York, Harcourt, Brace, 1932

Dubos RJ: The Professor, the Institute and DNA. New York, Rockefeller University Press, 1976

Dubos RJ: Biochemical Determinants of Microbial Disease. Cambridge, Mass, Harvard University Press, 1954

Dubos RJ: Louis Pasteur, Free Lance of Science. Boston, Little, Brown, 1950

Florey HW: Antibiotics. London, Oxford University Press, 1949

Marquardt M: Paul Ehrlich. New York, Henry Schuman, 1951

Meleney FL: Treatise on Surgical Infections. London, Oxford University Press, 1948

Metchnikoff E: Immunity in Infective Diseases. Binnie FG (trans). Cambridge, Mass, Harvard University Press, 1905

Tyndall J: Essays on the Floating-Matter of the Air in Relation to Putrefaction and Infection. New York, D Appleton and Co, 1882

CHAPTER 2

The Classification and Identification of Bacteria

Procaryotes and Eucaryotes

By the 1960s it was clear that bacterial cells differ at the level of intracellular organization from all cells of higher life forms. Prior to 1961, bacteria were often classified as primitive plant cells or as protists. Protists were recognized by Haeckel in 1866 as mostly undifferentiated unicellular organisms that do not form the specialized tissues and organ systems so characteristic of higher plants and animals, i.e., the bacteria, algae, fungi, and protozoa. In the years following, through comparative cytologic staining observations with the light microscope, microbiologists gradually became aware that the internal structure and nucleoid or chromatin material of bacteria appeared to be different from that of other cells. However, details of the ultrastructure of bacteria were beyond the resolution of the light microscope. By the 1950s, application of the concurrently emerging techniques of electron microscopy and biochemical subcellular fractionation and identification led to the key discoveries that bacteria produce chemically unique envelope structures and lack the membrane-bound nucleus and intracellular organelles found in the cells of all other life forms. That is, bacterial cells differ in structure and intracellular organization from protists, plant and animal cells. The significance of these observations was first clearly enunciated in 1961 and 1962 by Stanier, van Niel, and Murray. Primarily for these reasons, in 1968 bacteria were placed in a new kingdom, the Kingdom Procaryotae, and are thus known as procaryotes (i.e., primitive nucleus). Because cells of all other living things produce membrane-bound nuclei, they are known as eucaryotes (i.e., true nucleus), a group that includes the protists, plants, and animals (Fig. 2-1). The concept of two patterns of intracellular organization led further to the endosymbiotic theory of evolution. That is, eucaryotic cells characteristically engulf (or endocytose) food particles, including bacteria, into their cytoplasm. In some cases, stable endosymbiotic relationships occur, as in the well-known case of the protozoan *Paramecium aurelia*. These ob-

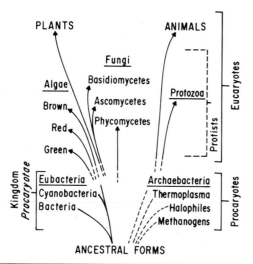

Figure 2-1. Suggested relationship of procaryotic and eucaryotic protists, based on increasing levels of intracellular and intercellular complexity.

servations led to the concept that primordial eucaryotic cells may have similarly evolved from procaryotic cells by endosymbiosis. There are now sufficient data to strongly indicate that the mitochondria and chloroplasts of eucaryotes originated by separate endosymbiotic events some billion years ago.

Bacteria

General Properties

The bacteria, or procaryotes, are single-celled organisms that reproduce by simple division, i.e., binary fission. Most are free living and contain the genetic information, energy-producing, and biosynthetic systems necessary for growth and reproduction. A few, such as the chlamydia and rickettsia, are obligate intracellular parasites and lack one or more of these attributes. Bacteria differ from the eucaryotes in a number of respects. They do not contain 80S ribosomes or membrane-bound organelles, such as the nucleus, mitochondria, lysosomes, endoplasmic reticulum, or golgi, and they lack the 9 + 2 fibril flagellum/cilia structure characteristic of eucaryotic cells. Bacteria have 70S ribosomes and a naked, single circular chromosome (nucleoid) composed of double-stranded deoxyribonucleic acid (DNA) that replicates amitotically. The cytoplasmic membrane carries out transport and energy-producing and specialized biosynthetic functions. In bacteria, motility is usually conferred by single-filament flagellar structures. Some bacteria produce external microfibrils (pili or fimbriae) that appear to have adhesive functions. Most bacteria produce envelope structures that contain a chemically unique rigid cell wall peptidoglycan. The chemical nature of their cell envelope structures imparts unique staining characteristics and permits bacteria to be arbitrarily divided into gram-positive, gram-negative, and acid-fast organisms. A few bacteria, such as the mycoplasma, do not possess cell walls. Morphologically, cell wall-producing bacteria occur as spheres (cocci), rods (bacilli), or as curved or spiral-shaped cells.

Bacterial Species, Species Biotype, and Strains
Description of Bacterial Species. Each kind of bacterium may be considered a species. A clearer definition is difficult until a genetic or phylogenetic description is available. Bacterial species are arbitrarily defined by a descriptive array of features encompassing a spectrum of phenotypic expressions that describe a sampling of the potential of a particular gene pool. Features used include, in sequence when possible, (1) structural traits, such as shape, size, mode of movement (i.e., flagella, axial filament, gliding, and so on), resting stage, gram stain reaction, and macroscopic growth, (2) biochemical and nutritional traits, such as the use of or action on various compounds (e.g., carbohydrates, protein), end-products, and other biochemical information on cell components and metabolites, (3) physiologic traits relative to oxygen, temperature, pH, and response to antibacterials, (4) ecologic traits, (5) genetic traits, and (6) the type reference strain where obtainable.

Species Biotype. Many bacterial groups of genera or species are comprised of slightly differing species or strains, respectively, that form clusters of interrelated organisms (in analogy with a cross-section in time of closely related, continually subdividing branches from a growing ancestral tree). The genus description includes the features of each of the species clusters within it. Within such a species cluster (not necessarily at dead center), a strain is chosen arbitrarily to best represent a species. This strain is known as the biotype of that species, and its characteristics will thereafter be used to describe that particular species. Biotype strains are sometimes referred to as "reference strains."

Strains. Biotype strains, even if well chosen, may not exhibit all the detailed features expressed by the totality of the strains within the species cluster. It is at this point that subspecies designations, such as serotypes (serovars) or phage types (phagevars), are used to indicate the mode of variation. Obviously, the borderlines of such groups are not always clear-cut, and many marginally interrelated strains may occur. An example of a group that forms a constellation of similar clusters is provided by the species, and even the genera, of the family Enterobacteriaceae. Its members share a common gene pool, as evidenced by their ability to undergo intergeneric transformation, conjugation, or transduction (Chap. 8). It should be noted that the genera and species of this family were designated years before their genetic relatedness by DNA hybridization was discovered.

The student will soon recognize that the lack of a definitive unifying criterion for species designation in bacteria has understandably resulted in variation in levels at which groups are divided, depending upon whether the describing investigator is a splitter or a lumper. In the splitter category are those who have designated each *Salmonella* serotype (serovar) as a species with its own name. The lumpers, on the other hand, have designated individual serotypes as numbered types within a single species of, for example, *Klebsiella* or *Streptococcus*.

Bacterial Taxonomy

According to the Linnaean scheme of classification used for higher plants and animals, bacteria may be classified as to Kingdom, Class (*-aceae*), Order (*-ales*), Family (*-aceae*), Tribe (*-ieae*), and a specific binomial name for genus and species designations. The specific name includes a latinized generic name that is capitalized and a trivial species name that is not. All latinized names are italicized in print or underlined in script. Beyond this point, however, the classification of bacteria is different from that of higher organisms, that is, concepts useful in describing relatedness in higher eucaryotic life forms have not been adequate for procaryotic cells. Attempts to set up traditionally ordered schemes of relationships among diverse bacterial groups have failed for lack of information. Without an accurate fossil record to trace the development of the various branches of the evolutionary family tree back more than three billion years, it is futile to construct presumed ancestrally related, i.e., arbitrarily phyletic, classifications based on comparisons of the extremely diversified present-day bacterial life forms. For this reason, the current trend is to describe groups of bacteria without implying degrees of relatedness among them. This approach is used in the most recent, Eighth Edition of *Bergey's Manual of Determinative Bacteriology*, published in 1974, which is the classification and nomenclature system most widely followed in the United States. For example, the Kingdom Procaryotae is recognized as containing two major divisions; I, the Cyanobacteria (or blue-green bacteria) and II, the Bacteria. The latter division includes some 19 parts (bacterial groups) that are given vernacular descriptions. Some 15 of these groups contain the three major classes of nonphotosynthetic bacteria that are of medical interest. These include (1) a diversity of cell-walled cocci, rods, and spiral bacteria, (2) rickettsiae (obligate intracellular parasites of eucaryotes), and (3) the cell wall-less mycoplasmas, but not the gliding, sheathed, budding, or appendaged bacteria. A modified presentation of the classification of medically important bacteria according to the Eighth Edition of *Bergey's Manual* is given in Table 2-1. The family, generic, and species names given in Table 2-1 have been validated. Since 1980, valid names of all bacterial species are published in the *International Journal of Systematic Bacteriology*.

In addition to generic and species names, well-known trivial names, such as tubercle bacillus (*Mycobacterium tuberculosis*) and typhoid bacillus (*Salmonella typhi*), often appear in medical literature. When a generic (genus) name is vernacularized in English, such as bacillus, salmonella, or pneumococcus, it is neither capitalized nor italicized.

Numerical Taxonomy

Because of the difficulties in constructing phylogenetic classifications based on only a few arbitrarily weighted characteristics, descriptive taxonomy has been revised for many strains in the form of computerized numerical comparisons of large numbers of diagnostic features. First devised by Michel Adanson in the 18th century, this system gives equal weight to all characteristics chosen for comparison, the argument being that if enough characteristics are examined, objective comparisons result that are unbiased by subjective judgments. The result of numerical taxonomy is a comparison of traits among several bacteria, expressed as a similarity coefficient (S), which is the percentage of the total number of characteristics or traits held in common between two organisms or groups of organisms. Designations of groups, or taxa, are then made on the basis of the ranges of degrees of similarity. These often fall within naturally defined groups.

In practice, some care based on general knowledge must be used in the selection of diagnostic features. Traits with a common basis in which one determines the other (if known) should be avoided. Again, use of large numbers of traits may overcome such problems. The similarity coefficient (S) is expressed as:

$$S = \frac{a}{a + b + c}$$

where:

a = number of features present in all strains
b = number of features present in one strain only
c = number of features present in strain two only, and so on.

Genetic Basis for Classification

Recent developments in comparative biochemistry indicate that gene-controlled stable metabolic patterns, cell polymers, and organelle structures have yielded information of possible evolutionary significance. An often quoted example is the observation that in a wide group of organisms, including bacteria, actinomycetes, blue-green bacteria, water molds, green algae, and vascular plants, the biosynthesis of the amino acid, L-lysine, occurs by decarboxylation of a common intermediate, α-ε-diaminopimelic acid.

TABLE 2-1. BACTERIA OF MAJOR MEDICAL IMPORTANCE

Family (-aceae)
  Genus
    Species
	Characteristic Properties

	GRAM-POSITIVE BACTERIA
Cocci	
Micrococcaceae	Spherical cells in clusters or packets, aerobic or facultatively anaerobic, catalase (+), cytochrome (+), nonmotile
Micrococcus	Cocci single, in clusters, or packets, aerobes, metabolism strictly respiratory
M. luteus	
M. roseus	
Staphylococcus	Cocci single, in pairs, and irregular clusters, facultatively anaerobic, metabolism respiratory and fermentative
S. aureus	
S. epidermidis	
Streptococcaceae	Cells spherical or ovoid, in pairs or chains, nonmotile, facultatively anaerobic, metabolism fermentative, complex nutritional requirements
Streptococcus	Homofermentors, catalase (−)
S. pyogenes	
S. pneumoniae	
S. agalactiae	
Peptococcaceae	Nonmotile, anaerobic, metabolism fermentative
Peptococcus	Spherical cells single, in pairs, tetrads, or irregular masses, catalase (±), coagulase (−)
Peptostreptococcus	Cells spherical to ovoid, in pairs or chains, catalase (−)
P. anaerobius	
Endospore-forming Rods	
Bacillaceae	Cells rod-shaped, often peritrichously flagellated
Bacillus	Strict aerobes or facultative anaerobes
B. anthracis	
Clostridium	Spores usually distend the organism, most strains strictly anaerobic
C. botulinum	
C. perfringens	
C. tetani	
Asporogenous Rods and Actinomycetes	
Lactobacillaceae	Rods, single, or in chains, complex nutritional requirements
Lactobacillus	Nonmotile, catalase (−), cytochrome (−), anaerobic or facultative, homolactic fermentation
L. acidophilus	
Listeria	Motile, aerobic, catalase (+)
L. monocytogenes	
Erysipelothrix	Nonmotile, often filamentous, catalase (−)
E. rhusiopathiae	
Propionibacteriaceae	Pleomorphic rods or filaments, motility (±)
Propionibacterium	Nonmotile, anaerobic to aerotolerant, propionic and acetic acids are major fermentation products
P. acnes	
Eubacterium	Motility (±), obligate anaerobes, usually fermentative, butyric and acetic acids produced
E. lentum	
Coryneform bacteria	Straight to slightly curved rods, irregular staining, generally aerobic
Corynebacterium	Nonmotile, catalase (+)
C. diphtheriae	
Mycobacteriaceae	Nonmotile, acid-fast, aerobic, slow growers
Mycobacterium	As for family
M. tuberculosis	
M. leprae	
Nocardiaceae	Mycelium produced, rudimentary or extensive, aerobic
Nocardia	Some species acid-fast
N. asteroides	
Actinomycetaceae	Nonmotile, nonacid-fast, predominantly diphtheroid in shape, tend to produce branching filaments

(continued)

TABLE 2-1. (cont.)

Family (-aceae) Genus Species	Characteristic Properties
Actinomyces	Anaerobic to facultatively anaerobic, catalase ($\pm$), major fermentation products are acetic, formic, lactic, and succinic acids, no gas
A. israelii *A. naeslundii* *Arachnia*	Facultatively anaerobic, catalase ($-$), major fermentation products are propionic and acetic acids, CO_2
A. propionica *Bifidobacterium*	Bifurcated forms common, anaerobic, catalase ($-$), major fermentation products are acetic and lactic acids, no gas
B. dentium *Bacterionema* *B. matruchotii*	Characteristic whip-handle morphology, facultative anaerobe
Rothia *R. dentocariosa*	Coccoid, diphtheroid, or filamentous, aerobic, catalase ($+$)
Streptomycetaceae	Aerobic soil organisms, form true mycelia, aerial spores, source of many antibiotics, mostly nonpathogens
Streptomyces *S. griseus* Micromonosporaceae *Micropolyspora* *M. faeni*	Primarily saprophytic soil forms. Spores on aerial and/or substrate mycelium Some facultative thermophiles

GRAM-NEGATIVE BACTERIA

Cocci and Coccobacilli

Neisseriaceae *Neisseria* *N. meningitidis* *N. gonorrhoeae*	Aerobic, coccal to rod-shaped, nonflagellated Biscuit-shaped diplococci, catalase ($+$), oxidase ($+$), few sugars fermented
Branhamella *B. catarrhalis*	Diplococci, catalase ($+$), oxidase ($+$), no acid from carbohydrates, nitrates reduced
Moraxella *M. lacunata*	Plump rods in pairs or short chains, strict aerobes, oxidase ($+$), catalase ($+$)
Acinetobacter Veillonellaceae	Rod-shaped to spherical, strict aerobes, oxidase ($+$), catalase ($-$) Anaerobic cocci, complex growth requirements, cytochrome oxidase ($-$), catalase ($-$), may produce volatile acids, parasitic
Veillonella *V. parvula* *Acidaminococcus* *A. fermentans*	Pyruvate utilized, carbohydrates not fermented Amino acids main energy source, pyruvate not utilized

Aerobic Bacilli and Cocci

Pseudomonadaceae *Pseudomonas* *P. aeruginosa* Genera of uncertain affiliation *Brucella*	Straight or curved rods, motile, metabolism respiratory, never fermentative Growth factors not required Coccobacilli, nonmotile, facultative intracellular parasites, fastidious growth requirements, strict aerobe, catalase ($+$)
B. abortus *B. melitensis* *B. suis* *Bordetella*	Minute coccobacilli, polytrichate if motile, strict aerobes, metabolism respiratory, mammalian parasites
B. pertussis *Francisella*	Pleomorphic single rods to cocci, capsules rare, require enriched media (usually cysteine) for growth
F. tularensis *Legionella* *L. pneumophila*	Stainable gram-negative with difficulty, catalase ($+$), rods

(continued)

TABLE 2-1. (cont.)

Family (-aceae) Genus Species	Characteristic Properties
Facultatively Anaerobic Bacilli	
Enterobacteriaceae	Small rods, facultative anaerobes, glucose fermented, metabolism respiratory and fermentative, catalase (+), oxidase (−), nitrates reduced to nitrites
Escherichia	Motile, mixed acid fermentation, H_2 and CO_2 produced, lactose fermented by most strains, H_2S (−), citrate not utilized
E. coli	
Edwardsiella	Usually motile, mixed acid fermentation, lactose not fermented, H_2S (+), indole (+)
E. tarda	
Citrobacter	Usually motile, citrate utilized, lysine not decarboxylated
C. freundii	
Salmonella	Usually motile, lactose not fermented, most strains aerogenic, indole (−), H_2S (+), lysine and ornithine decarboxylated
S. typhi	No gas from glucose
S. species	Acid and gas from glucose
Shigella	Nonmotile, lactose not fermented, no gas from glucose, H_2S (−)
S. dysenteriae	
S. flexneri	
S. sonnei	
Klebsiella	Nonmotile, encapsulated, 2,3-butanediol fermentation, citrate utilized, ornithine not decarboxylated
K. pneumoniae	Lactose fermented with acid and gas
Enterobacter	Motile, 2,3-butanediol fermentation, lactose fermented with acid and gas, citrate utilized, ornithine decarboxylated
E. aerogenes	
Serratia	Motile, lactose not fermented, H_2S (−), lysine decarboxylated, citrate utilized, pink to red pigment at R°, produce DNase
S. marcescens	
Proteus	Motile, lactose not fermented, urea hydrolyzed
P. mirabilis	
P. vulgaris	
Yersinia	Cells ovoid to rod-shaped, lactose not fermented, some species motile
Y. enterocolitica	
Y. pestis	
Vibrionaceae	Rigid rods, straight or curved, polar flagella, oxidase (+), catalase (+), metabolism respiratory and fermentative
Vibrio	
V. cholerae	
V. parahaemolyticus	
Aeromonas	
A. hydrophila	
Plesiomonas	
P. shigelloides	
Genera of uncertain affiliation	
Chromobacterium	Rods with rounded ends, motile, violet pigment produced at R°, metabolism respiratory or fermentative
C. violaceum	
Flavobacterium	Coccobacilli to slender rods, motile or nonmotile, pigmented, metabolism respiratory
F. meningosepticum	
Haemophilus	Coccobacillary to rod-shaped, pleomorphic, strict parasite, requires growth factors in blood
H. influenzae	
Pasteurella	Ovoid or rod-shaped, bipolar staining, catalase (+), metabolism fermentative
P. multocida	
Actinobacillus	Bacillary with interspersed coccal elements, nonmotile, fermentative metabolism
A. lignieresii	

(continued)

TABLE 2-1. (cont.)

Family (-aceae) Genus Species	Characteristic Properties
Cardiobacterium	Rods, nonmotile, fermentative metabolism, catalase ($-$), cytochrome oxidase ($+$)
C. hominis	
Streptobacillus	Rods in chains or filaments, nonmotile, may convert to L form, metabolism fermentative
S. moniliformis	
Calymmatobacterium	Pleomorphic rods, usually encapsulated, safety pin forms, nonmotile
C. granulomatis	
Anaerobic Bacilli	
Bacteroidaceae	Obligate anaerobic rods, non-sporeforming
Bacteroides	Fermentation products include succinic, acetic, formic, lactic, and propionic acids
B. fragilis	
Fusobacterium	Butyric acid a major fermentation product
F. nucleatum	
Leptotrichia	Straight or slightly curved, ends rounded or pointed, nonmotile, lactic acid is only major fermentation product
L. buccalis	
Helical Cells	
Spirochaetaceae	Slender, flexuous, helically coiled, produce axial fibrils, aerobic to anaerobic
Treponema	5–15 μm × 0.09–0.5 μm, anaerobic, motile, tight regular or irregular spirals
T. pallidum	
Borrelia	3–15 μm × 0.2–0.5 μm, anaerobic, 3–10 coarse, uneven coils, motile
B. recurrentis	
Leptospira	6–20 μm × 0.1 μm, aerobic, tightly coiled, one or both ends bent or hooked, motile
L. interrogans	
Spirillaceae	Rigid, helically curved rods, less than one to many turns, produce flagella, aerobic to anaerobic
Spirillum	0.25–1.7 μm × 2–60 μm, polytrichous polar flagella, aerobic to obligately microaerophilic
S. minor	
Campylobacter	0.2–0.8 μm × 0.5–5 μm, single polar flagellum, microaerophilic to anaerobic
C. fetus	
Rickettsias	
Rickettsiaceae	Small rods, coccoid and diplococcal, usually intracellular parasites associated with arthropods
Rickettsia	Grow in cytoplasm, not in vacuoles, sometimes in nucleus
R. rickettsii	
Rochalimaea	Grows extracellularly, can be grown in vitro on bacteriologic media
R. quintana	
Coxiella	Grows preferentially in vacuoles, resistant to extracellular drying or heat
C. burnetii	
Bartonellaceae	Rod, coccoid, ring- or disc-shaped, parasites of erythrocytes, cultivable on nonliving media
Bartonella	Characteristically in chains of several segmenting organisms, occur in human and arthropod vectors
B. bacilliformis	
Chlamydiaceae	Coccoid, obligately intracellular parasites, developmental cycle in cytoplasm, cultivable in yolk sac of chick embryo or tissue culture
Chlamydia	Nonmotile, limited metabolic activity, energy parasites
C. trachomatis	
	CELL WALL-LESS FORMS
Mycoplasmataceae	Cells highly pleomorphic, lack true cell wall, stain gram-negative, biphasic colony with fried egg appearance, penicillin-resistant, require sterol for growth
Mycoplasma	
M. pneumoniae	

Modified from Buchanan and Gibbons (eds): Bergey's Manual of Determinative Bacteriology, 8th ed. 1974. Philadelphia, Williams & Wilkins.

$$
\begin{array}{ccc}
\text{COOH} & & \text{COOH} \\
| & & | \\
\text{H}_2\text{NCH} & & \text{H}_2\text{NCH} \\
| & & | \\
(\text{CH}_2)_3 & \longrightarrow & (\text{CH}_2)_3 + \text{CO}_2 \\
| & & | \\
\text{HCNH}_2 & & \text{H}_2\text{CNH}_2 \\
| & & \\
\text{COOH} & &
\end{array}
$$

D, L-diaminopimelic   L-lysine
acid (*meso*-DAP)

By contrast, a different pathway, the aminoadipic acid pathway, is used for lysine biosynthesis by euglenoid algae, certain fungi including the Ascomycetes (for example, *Neurospora*), Basidiomycetes, and others. A second set of examples includes the observation that almost all procaryotes produce a similar rigid cell wall peptidoglycan structure (Chap. 6) and the fact that various microbial macromolecules of both cytoplasmic and surface origin can be used for comparison of bacterial relatedness by biochemical, serologic, genetic, and other means. Finally, comparison of homology, both of composition and of sequence, of DNA, RNA, and proteins allows examination of organisms at the genetic level.

Because these developments are still in progress, a complete correlation with the present empirical classification in Table 2-1 cannot yet be made. However, the older empirically developed relationships based on recognition of phenotypic expression of various traits and features are, after all, a reflection of gene pool expression and correlate in general with the results obtained by newer approaches that aim at defining relatedness on a molecular basis. Such information may or may not be of phylogenetic significance at this time. However, it is of use in assessing the similarity of present-day microorganisms.

Relatedness Based on Nucleic Acid Homology.

Genetic information is encoded in DNA base sequence. As organisms drift apart by mutation, recombination, transduction, and selection in different environments (i.e., evolution), their genomes change in size, nucleotide base composition, and nucleotide base sequence. The approximate sizes of the genomes of several bacteria are listed in Table 2-2. Analysis by both chemical and physicochemical methods shows that base composition is constant and characteristic for each organism. It is generally expressed in terms of the mole fraction of guanine plus cytosine (that is, [G + C]/[G + C + A + T] expressed as a percentage). Values of percent GC for different organisms vary from 25 to 75 (Table 2-3). At the level of descriptively defined species, it is to be noted that several genera exhibit similar DNA base compositions. However, similarity of base composition does not necessarily signify DNA homology, that is, similarity in base sequence. For example, the genomes of all vertebrates, including humans, contain approximately 44 mole percent GC, as do those of some mi-

croorganisms. Obviously, different organisms with similar DNA base compositions must have heterologous base sequences. Measurement of DNA homology has been quantified by several procedures that determine the extent of formation of molecular hybrids from two DNA strands of different origin. The generalized procedure is illustrated in

TABLE 2-2. APPROXIMATE DNA CONTENT OF VARIOUS ORGANISMS

Species	Daltons
Mammalian sperm	18×10^{11}
Salmonella typhimurium	28×10^8
Escherichia coli	25×10^8
Bacillus subtilis	20×10^8
Haemophilus influenzae	16×10^8
Neisseria gonorrhoeae	9.8×10^8
Rickettsia rickettsii	9.8×10^8
Chlamydia trachomatis	6.6×10^8
Mycoplasma species	5×10^8
Coliphage T4	1.3×10^8

Modified from Kingsbury: J Bacteriol 98:1400, 1969; Muller and Klotz: Biochim Biophys Acta 378:171, 1975; Sober (ed): Handbook of Biochemistry, 1968. Courtesy of Chemical Rubber Co.; Sorov and Becker: J Mol Biol 42:581, 1969.

TABLE 2-3. NUCLEOTIDE BASE COMPOSITION OF THE DNA OF VARIOUS BACTERIA

Organism	%GC
Clostridium perfringens, C. tetani	30–32
Staphylococcus aureus	32–34
Bacillus cereus, B. anthracis, B. thuringiensis, Mycoplasma gallisepticum, Streptococcus faecalis, Treponema pallidum, Veillonella parvula	34–36
Streptococcus (Diplococcus) pneumoniae, S. salivarius, S. pyogenes, Lactobacillus acidophilus, Listeria monocytogenes, Proteus vulgaris, P. mirabilis, Haemophilus influenzae	38–40
Moraxella bovis	40–42
Bacillus subtilis, Chlamydia trachomatis	42–44
Vibrio comma, Yersinia pestis, Propionibacterium acnes	46–48
Neisseria gonorrhoeae	48–50
Neisseria meningitidis, Bacteroides fragilis, Escherichia coli, Citrobacter freundii, Shigella dysenteriae. Salmonella typhi, S. typhimurium, S. enteritidis, S. arizonae, S. ballerup	50–52
Enterobacter aerogenes, Corynebacterium diphtheriae	52–54
Klebsiella pneumoniae, K. rhinoscleromatis, Brucella abortus	54–56
Lactobacillus bifidis	56–58
Serratia marcescens	58–60
Pseudomonas fluorescens	60–62
Pseudomonas aeruginosa, Mycobacterium tuberculosis	66–68
Micrococcus luteus (lysodeikticus), Nocardia sp.	70–80

Modified from Marmur et al.: Annu Rev Microbiol 17:329, 1963.

Figure 2-2. As indicated in Table 2-4, this approach has been useful in demonstrating the relative order and degree of DNA similarity of closely related groups of bacteria. In addition, the lack of DNA homology between dissimilar organisms is also clear, but this very specificity also limits the usefulness of this approach, since it cannot be used to study the relationship of dissimilar bacterial groups whose DNA do not hybridize.

On the other hand, ribosomal RNA exhibits more homology among widely dissimilar organisms than does DNA (e.g., compare Tables 2-4 and 2-5). This approach has been further refined by comparison of ribosomal RNA oligonucleotide sequences obtained by special nuclease digestion. Mutational change in RNA sequences has obviously been very low through the ages. When combined with other information, such as cell wall composition, metabolism, and physiology, such RNA homology studies can provide extremely interesting information. For example, most startling was the observation that the ribosomal RNA sequences, cell wall or envelope composition, modes of metabolism, and even coenzymes, of the anaerobic meth-

TABLE 2-4. DNA RELATEDNESS AMONG SOME BACTERIA

DNA Source	% Homology
	(to *E. coli*)
Escherichia coli	100
Shigella dysenteriae	89
Salmonella typhimurium	45
Enterobacter aerogenes	35
Serratia marcescens	20
Proteus vulgaris	10
Pseudomonas aeruginosa	1
Bacillus subtilis	1
Brucella neotamae	0
	(to *N. meningitidis*)
Neisseria meningitidis	100
Neisseria gonorrhoeae	80
Neisseria subflava	55
Branhamella catarrhalis	15
Mima sp., *Herellea*	5
Escherichia coli	3
Monkey kidney	0.1

Modified from Brenner: Int J Syst Bacteriol 23:298, 1973; Brenner et al.: J Bacteriol 94:486, 1967; Kingsbury: J Bacteriol 94:870, 1967; McCarthy and Bolton: Proc Natl Acad Sci USA 50:156, 1963.

anogens (which produce methane from carbon dioxide and hydrogen) appear to be very dissimilar to and apparently more primitive than other living bacteria. It would appear that the methanogens and other present-day procaryotes separated very early in evolutionary history. Because of their dissimilarities and apparent primitiveness in comparison to other bacteria, the methanogens, halophiles, and thermoplasma have been placed in the group Archaebacteria. To differentiate, all other bacteria are referred to as Eubacteria. Known pathogens appear only among the Eubacteria (Fig. 2-1).

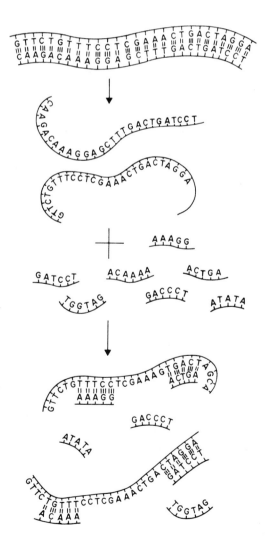

Figure 2-2. DNA-DNA hybridization for comparison of the nucleotide base sequences of the DNAs of two bacterial species. DNA from one species is denatured by heating or treatment with alkali, so as to separate the two strands. These are then entrapped on some supporting matrix (e.g., nitrocellulose membranes), so as to prevent their reassociation. To them small, broken, denatured (and therefore likewise single stranded) DNA oligonucleotide strands derived from the second species are added, which are radioactively labeled. The mixture is then heated and allowed to anneal by cooling slowly. During the annealing process, complementary oligonucleotides form hydrogen-bonded base-pair double stranded hybrid stretches that are resistant to solubilization by digestion with deoxyribonuclease, whereas unpaired single strands are susceptible. If, therefore, the two DNA samples are entirely homologous, all radioactively labeled DNA will hybridize and be converted to the DNase-resistant form; if they are completely dissimilar, all labeled DNA will remain enzyme-sensitive. The result is that radioactive hybrids will remain with the supporting matrix, whereas the DNase-sensitive radioactive single strands will be degraded and be removed by washing after the annealing process. Retention of radioactivity with the supporting matrix then becomes a measure of hybridization and DNA homology.

TABLE 2-5. RNA RELATEDNESS AMONG SOME BACTERIA

RNA Source	% Homology
	(to *E. coli*)
Escherichia coli	100
Proteus vulgaris	79
Enterobacter aerogenes	77
Serratia marcescens	76
Alcaligenes faecalis	36
Bacillus subtilis	16
Tetrahymena pyriformis	9
Saccharomyces chevalieri	6
Drosophila melanogaster	0

Modified from Pace and Campbell: J Bacteriol 107:543, 1971.

Identification of Bacteria

Among the primary concerns of the medical microbiologist are the isolation and rapid, accurate identification of disease-causing microorganisms, in order that adequate specific therapy can be initiated. This can be accomplished in most cases* by determinative procedures that, after isolation of the organisms in pure culture,† make use of knowledge of growth requirements, visible (colony) growth features, microscopic morphology and staining reactions, biochemical characteristics, serologic reactivity, animal pathogenicity, and antibiotic sensitivity.

Isolation of Organisms in Pure (Axenic) Culture

The approach used for the isolation of organisms depends upon the source of the clinical specimen. Blood, spinal fluid, and closed abscesses may yield almost pure bacterial cultures, while specimens of sputum, stool, skin, and body orifices usually contain mixtures of organisms. The specimen is generally streaked onto solid agar-containing medium so as to separate the bacterial population into individual cells. In some cases, two or more daughter cells of certain bacteria may adhere, and are known as a colony-forming unit (CFU). Pathogens present in small numbers in mixtures of organisms may be missed, since they may be overgrown by other bacteria, or they may be killed by metabolic acids or other products resulting from the growth of nonpathogens. Pathogens may also be missed if their growth requirements are not met. For this reason, selective culture techniques are employed in order to establish an environment in which the pathogen has a survival advantage; these include the use of selective media that

---

* Exceptions occur in the case of exotoxin-producing bacteria. Refer to discussions on botulism, tetanus, and staphylococcal food poisoning.
† Pure in the sense of axenic, that is, free of other organisms, rather than pure in the genetic sense.

are of specific pH, ionic strength, or chemical composition or that contain inhibitors or that lack nutrients for all but the organism in question. Control of gas phase and temperature of the growth environment must also be considered. Alternatively, for fastidious bacteria, enrichment media are used that contain nutrients ecologically favorable to the organism to be isolated. Such media include chocolate agar, blood agar, and nutrient agar and sometimes inhibitors, as discussed in later chapters under growth, nutrition, and specific organisms. The response of the unknown bacterium to such media is sometimes useful in identification. For example, *Haemophilus influenzae* requires for growth both pyridine nucleotide and hemin, the iron porphyrin derived from hemoglobin, and therefore will grow on chocolate agar (heated to lyse the cells) but not on blood agar. Unless the red cells are ruptured, the red cell pyridine nucleotide and hemin are not available to the bacterium.

Bacterial Colony Morphology

Bacteria multiply rapidly and form macroscopic, visible masses of growth when inoculated onto appropriate medium containing 2 percent agar and incubated 18 to 48 hours in a favorable atmosphere. For some organisms, such as the tubercle bacillus, *M. tuberculosis*, 2 to 8 weeks of incubation is required. Ideally, a colony is composed of the descendents of a single cell, i.e., a clone, but may also develop from two or more organisms or from a clump of cells. Such a colony may have the gross appearance characteristic of a certain type of bacterium, although it may contain contaminating organisms. The gross characteristics of colonies aid in identification, since colonies of different organisms vary in size, shape, color, odor, texture, and degree of adherence to the medium. Colony morphology is related in part to motility or to postfission bacterial movements, which depend on division planes formed by different species. Colonies have been described as loop-forming (wavy edges characteristic of long filaments, such as those of *Bacillus anthracis*), folding and snapping (serrated or crenated edges, such as those formed by *Yersinia pestis* and *Corynebacterium diphtheriae*), and slipping (smooth or lobate edges with spreading smooth growth films, characteristic of *Proteus vulgaris* or *Escherichia coli*).

The serologic characteristics of a bacterium are often correlated with mucoid (M), smooth (S), or rough (R) colony appearance (Fig. 2-3). M or S colonies are characteristic of bacteria recently isolated from natural habitats and are sometimes referred to as "wild-type."

The Mucoid Colony. M colonies exhibit a waterlike, glistening, confluent appearance and are characteristic of organisms that form slime or well-developed capsules. Notable examples are *Klebsiella pneumoniae*, *Streptococcus (Diplococcus) pneumoniae*, *H. influenzae*, and the pathogenic yeast, *Cryptococcus neoformans*. Capsular polymers may be group-, species-, or strain-specific and are usually antigenic for

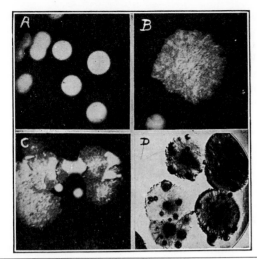

Figure 2-3. Types of bacterial colonies. **A.** Smooth. **B.** Rough. **C.** Smooth developing from rough colonies. **A, B,** and **C** are from a culture of an acid-fast bacterium. **D.** Colonies of the cholera vibrio with secondary daughter colonies.

mice and men. The capsule functions as a defense mechanism against phagocytosis, and among pathogenic bacteria, encapsulated organisms are usually more virulent than noncapsulated forms.

The Smooth Colony. S colonies give the appearance of homogeneity and uniform texture without appearing as liquid as mucoid colonies. S colony-forming bacteria are traditionally referred to as colonially or morphologically smooth (i.e., S forms). S forms are characteristic of freshly isolated wild-type organisms, such as the gram-negative enterobacteria (e.g., *Salmonella, Shigella, E. coli, Serratia,* and *Proteus* species), which produce a complete array of surface proteins and the well-known lipopolysaccharide somatic O antigens (Chaps. 3 and 6). For reasons discussed below, the identification of wild-type somatic antigens of S form organisms is usually confirmed by serologic agglutination tests.

The Rough Colony. R colonies are granulated and rough in appearance. R colonies are usually produced by mutant strains (R forms) that lack surface proteins or polysaccharides produced by freshly isolated wild-type parent organisms. Intermediate R mutants, detectable only biochemically, serologically, or genetically, may however, produce morphologically smooth colonies (Chap. 6).

Virulence is commonly, but not always, associated with the M or S colonies. Rough forms of enteric bacteria are usually avirulent and more easily killed by phagocytes, in contrast to the more resistant parent or wild-type S colony bacteria. However, with some organisms, such as the anthrax bacillus and the human and bovine types of tubercle bacilli, the R forms are more virulent.

The L Colony. L colonies were first associated with certain bacilli, notably *Streptobacillus moniliformis,* but have also been isolated from *H. influenzae* and various enteric bacteria. The rigid cell wall that is characteristic of bacteria is absent in L forms. Exposure to penicillin and other drugs and growth in osmotically supportive media facilitates L colony formation. Small coccoid and filamentous forms are observed, as well as large globoid forms, which also contain the minute forms. L forms normally resynthesize a cell wall once the penicillin or other drug is removed.

Microscopic Morphology and Staining Reactions

Light microscopic examination of gram-stained preparations is routinely used to determine the shape of bacteria. Common shapes are cocci (spherical), bacilli (rod-like), and spiral forms. The gram stain also allows a check on contamination by noting morphologic and staining homogeneity. Depending upon source and growth features, other differential stains may be used, as in the case of acid-fast staining of sputum smears or cultures. Special stains may also be used to detect capsules, flagella, spores, or intracellular inclusion bodies.

The Gram Stain. A heat-fixed bacterial smear on a glass slide is treated with the basic dye, crystal violet. All organisms take up the dye. The smear is then covered with Gram's iodine solution (3 percent I_2-KI in water or a weak buffer, pH 8.0, in order to neutralize acidity formed from iodine on standing). After a water rinse and decolorization with acetone, the preparation is washed thoroughly in water and counterstained with a red dye, usually safranin. The stained preparation is then rinsed with water, dried, and examined under the light microscope.

Bacteria can be differentiated into two groups by this stain, devised in the early 1800s by the Danish bacteriologist, Christian Gram. Gram-positive organisms stain blue, while about one third of the cocci, one half of the bacilli, and all spiral organisms stain red and are said to be gram-negative. Animal cells also stain gram-negative, and the gram stain procedure can therefore be used to detect large gram-positive bacteria, such as *Nocardia asteroides,* in tissues.

The mechanism of the gram stain appears to be generally related to the thickness of the cell wall, pore size, and permeability properties of the intact cell envelope. Gram-positive bacteria stain gram-negative if they lose osmotic integrity by rupture of the plasma membrane. Autolyzed, old, or dead gram-positive cells and isolated cell envelopes stain gram-negative. A unique bacterial cell wall chemical composition does not appear to explain the gram stain, since thick-walled yeast cells, which also stain gram-positive, have a chemical composition and structure different from gram-positive bacteria.

The Acid-fast Stain. The mycobacteria are lipophilic and difficult to stain, but once stained, they are resistant to destaining. Typically, a sputum or culture smear on a glass slide is stained with carbolfuchsin by steaming over a flame for several minutes. The phenol appears to aid penetration of the dye through the mycobacterial lipid. The slide is then rinsed with water, destained with acid alcohol, and counterstained briefly with methylene blue. When examined under the light microscope, mycobacteria appear as bright red bacilli against a light blue background.

Biochemical Characteristics

Various strains, species, and genera of organisms exhibit characteristic patterns of substrate utilization, metabolic product formation, and sugar fermentation. The latter are used routinely in the identification of enterobacteria: the *Salmonella* characteristically ferment sugars with production of acid and gas, in contrast to the anaerogenic (nongas-producing) *Shigella*.

Serologic Reactivity

The antigenic reactivity of an organism is often determined in order to identify the serogroup or serotype to which the organism belongs. These reactions are usually carried out on microscope slides by mixing a drop of antiserum containing specific antibody that reacts with a known cellular component. Visible reactions observed include flagellar clumping, cellular agglutination, and capsular reactions detected as an increase in the refractive index (i.e., swelling) around the cell. Extracts of cells also may be used for the detection of antigenic reactivity, a precipitate being formed on addition of specific antiserum.

Bacteriophage Typing

Epidemiologists have found bacteriophage typing to be of use in tracing the development and source of outbreaks of certain bacterial diseases, including for example, those caused by *Staphylococcus aureus*, *S. typhi*, and *Pseudomonas aeruginosa*. This is possible because different strains of a serologically or otherwise identical species of bacteria are susceptible to one or more different strains or types of species-specific bacterial viruses, known as bacteriophage or phage. Suspensions of each bacteriophage type are deposited on an agar plate newly inoculated with the suspected pathogen. Susceptible bacteria are lysed by the phage, leaving clear areas known as plaques (Chap. 63).

Pathogenicity for Animals

Identification of certain organisms is aided by inoculation into animals. Some organisms, such as the spirochete that causes syphilis, *Treponema pallidum*, cannot be grown in vitro but can be isolated by testicular inoculation and transfer in rabbits. In other instances, a pathogen that can be cultivated in vitro may be differentiated from similar pathogens by inoculation into susceptible animals that destroy the contaminant but yield to infection by the pathogen. For example, isolation of pneumococci present in only small numbers in contaminated sputum or in nasopharyngeal throat washings can be accomplished by inoculation into the mouse peritoneum, where the encapsulated pneumococcus thrives and from which it can be recovered in pure culture. Similarly, in the past, subcutaneous inoculation into the groin of the guinea pig was useful in differentiating pathogenic tubercle bacilli from morphologically similar, nonpathogenic acid-fast organisms.

Antibiotic Sensitivity

Antibiotics vary in their effect on different bacterial species and on strains of even the same species. Each pathogen must be tested for sensitivity to various concentrations of effective antimetabolites in order to determine the concentration level at which its growth is inhibited. The dosage necessary to give the blood level required for adequate therapy can then be established.

FURTHER READING

Books and Reviews

Balch WE, Fox GE, Magrum LJ, et al.: Methanogens: Reevaluation of a unique biological group. Microbiol Rev 43:260, 1979

Barghoorn ES: The oldest fossils. Sci Am May 1971, p 30

Barksdale L, Kim KS: Mycobacterium. Bacteriol Rev 41:217, 1977

Buchanan RE, Gibbons NE (eds): Bergey's Manual of Determinative Bacteriology, 8th ed. Baltimore, Williams & Wilkins, 1974

Edwards PR, Ewing WH: Identification of Enterobacteriaceae, 2nd ed. Minneapolis, Burgess, 1970

Gray MW, Doolittle WF: Has the endosymbiant hypothesis been proven? Microbiol Rev 46:1, 1982

Margulis L.: Origin of Eukaryotic Cells. New Haven, Yale University Press, 1970

Marmur J, Falkow S, Mandel M: New approaches to bacterial taxonomy. Annu Rev Microbiol 17:239, 1963

Meynell GG, Meynell E: Theory and Practice in Experimental Bacteriology. New York, Cambridge University Press, 1970

Raff RA, Mahler HR: The non-symbiotic origin of mitochondria. Science 177:575, 1972

Schopf JW: The evolution of the earliest cells. Sci Am 239:111, 1978

Wallace DC: Structure and evolution of organelle genomes. Microbiol Rev 46:208, 1982

Selected Papers

Brenner, DJ: Impact of modern taxonomy on clinical microbiology. ASM News 49:58, 1983

Cohen SS: Are/Were mitochondria and chloroplasts microorganisms? Am Sci 58:281, 1970

Gillis M, DeLey J, DeCleene M: The determination of molecular weight of bacterial genome DNA from renaturation rates. Eur J Biochem 12:143, 1970

Johnson JL: Genetic characterization. In Gerhardt P, Murray RGE, Costilow RN, et al. (eds): Manual of Methods for General Bacteriology. Washington, DC, American Society for Microbiology, 1981, p 450

Knoll AH, Barghoorn ES: Archean microfossils showing cell division from the Swaziland system of South Africa. Science 198:396, 1977

Margulis L: Evolutionary criteria in thallophytes: A radical alternative. Science 161:1020, 1968

Raven PH: A multiple origin for plastids and mitochondria. Science 169:641, 1970

Skerman VDB, McGowan V, Sneath PHA: Approved lists of bacterial names. Int J Syst Bacteriol 30:225, 1980

Whittaker RH: New concepts of kingdoms of organisms. Science 163:150, 1969

Woese CR, Fox GE: Phylogenetic structure of the prokaryotic domain: The primary kingdoms. Proc Natl Acad Sci USA 74:5088, 1977

Woese CR, Fox GE: The concept of cellular evolution. J Mol Evolution 10:1, 1977

CHAPTER 3

Bacterial Morphology and Ultrastructure

The Bacterial Cell

Most bacteria produce a layered cell envelope that includes the plasma membrane, cell wall, and associated proteins and polysaccharides. Some bacteria also produce external surface adherents, known as capsules or slimes. Filamentous appendages, flagella and pili, may occur. The cell wall is a semirigid structure that encloses and protects the protoplast from physical damage and conditions of low external osmotic pressure and generally allows bacteria to tolerate a wide range of environmental conditions. The protoplast is comprised of the naked cytoplasmic membrane and its contents. Internally, bacteria are relatively simple cells. Major cytoplasmic structures include a central fibrillar chromatin network or nucleoid surrounded by an amorphous cytoplasm that contains ribosomes. Cytoplasmic inclusion bodies, or energy storage granules, vary in chemical nature according to species and in amounts depending upon growth phase and environment. Some cytoplasmic structures, such as endospores, are limited to only a few bacteria. Typical gram-positive and gram-negative bacterial cells, which differ primarily in cell envelope organization, are shown in diagram form in Figure 3-1.

Bacterial Size and Form

Pathogenic bacteria vary widely in size (Fig. 3-2) and appear under the light or electron microscope as spheres (cocci), rods (bacilli), and spirals (Fig. 3-3). Cocci occur as single spherical cells, in pairs as diplococci, in chains as streptococci, or, depending upon division planes, in tetrads or in grape-like clusters. Bacilli may vary considerably in length, from very short rods (coccobacilli), to long rods that vary in length from 2 to 10 times their diameter. The ends of bacilli may be gently rounded, as in such enteric organisms as *Salmonella typhi*, or squared, as in *Bacillus anthracis*. Long threads of bacilli that have not separated into single cells are known as filaments. Fusiform bacilli, found in the oral and gut cavities, taper at both ends. Curved bacterial rods vary from small, comma-shaped or mildly helix-shaped organisms with only a single curve, such as *Vibrio cholerae*, to the longer sinuous spirochete forms, such as *Borrelia, Treponema*, and *Leptospira*, which have from 4 to 20 coils.

Microscopy

The Light Microscope
Brightfield Microscopy. The human eye cannot resolve objects much less than 0.1 to 0.2 mm nor detect an object less than about one-thousandth inch (30 μm) in di-

ameter. The light microscope must therefore be used to see bacteria, which often range from 0.5 to 2 μm in size. Detection of such small objects depends upon resolving power (R), which is the ability to distinguish between two adjacent points. Resolving power of the compound light microscope is determined by the wavelength of light (λ) and a characteristic of the objective lens, the numerical aperture (NA). This relationship, expressed as $R = \lambda/NA$, becomes $R = \lambda/2NA$ when the microscope is equipped with a condenser. Numerical aperture is determined by the product of the refractive index (n) of the medium, usually air (n=1) or oil (n $\simeq$ 1.5), between the lens and the object examined and the sine of half the angle (θ) formed by light rays entering the lens from a point on the object examined when centered under the lens. Numerical aperture is then calculated as $NA = n \sin \theta/2$. Because the range of visible light is fixed (0.4 to 0.7 μm), resolution with the oil immersion objective (NA usually 1.25) at best becomes 0.4 μm/2 $\times$ 1.25, or 0.16 μm. Because green light at about 0.55 μm is best detected by the human eye, the usual limit of light microscopy resolution is approximately 0.3 μm. In brightfield microscopy, objects appear dark in a brightly lighted field. Bacterial cells are not easily seen unless suspended in glycerol or nonaqueous solutions that enhance differences in refractive index or unless the cells are stained (to determine gross morphology). Capsules can be stained or visualized as halos around cells suspended in India ink. Flagella and pili are below the limit of resolution and must be specially stained with dyes that adhere to flagellar protein and augment their size.

Phase Contrast Microscopy. The phase contrast light microscope enhances small differences in refraction caused by cellular substructures and reveals some gross details of internal bacterial structure. Capsules, endospores, cytoplasmic particles, and cell wall can be observed.

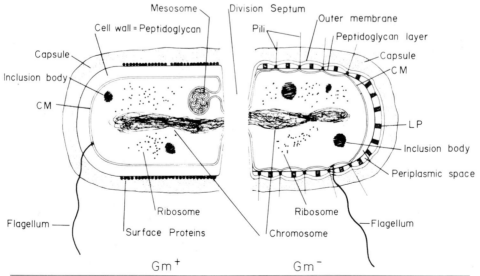

Figure 3-1. Diagram of idealized typical bacterial cells. Gram-positive (Gm$^+$) cell on left and gram-negative cell (Gm$^-$) on right, showing similarities and cell envelope differences. CM, cytoplasmic membrane; LP, lipoprotein.

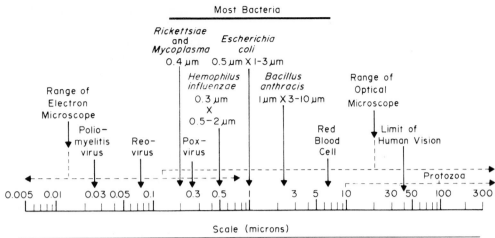

Figure 3-2. Relative sizes of bacteria.

Phase contrast is achieved by causing beams of partly out-of-phase light waves to interact (after passing through the object) to yield increases in light wave amplitude when in phase or decreases (darkness) when out of phase. The phase contrast system is essentially comprised of two annular rings, one in the condenser diaphragm and the other in the objective lens. The annular diaphragm allows only a ring of light to pass through the condenser and specimen on a glass slide to the objective lens. The objective lens contains a phase disc made of optically clear glass that has upon it a phase ring of dielectric material. Incident light from the phase condenser normally passes through this phase ring in the objective lens, resulting in a phase shift of one-quarter wavelength, i.e., $\frac{3}{4}$. However, when the incident light from the condenser passes through a specimen, substructures of different densities and thickness reflect, refract, or diffract light rays that pass through them. Refracted rays passing through the optically clear portion of the objective lens phase disc interact with light waves that passed through the phase rings. The result is a range of enhanced brightness (i.e., increased wave amplitude) where the two are in phase to enhanced darkness where the two waves are out of phase and interfere (i.e., decreased wave amplitude), hence the term "phase contrast."

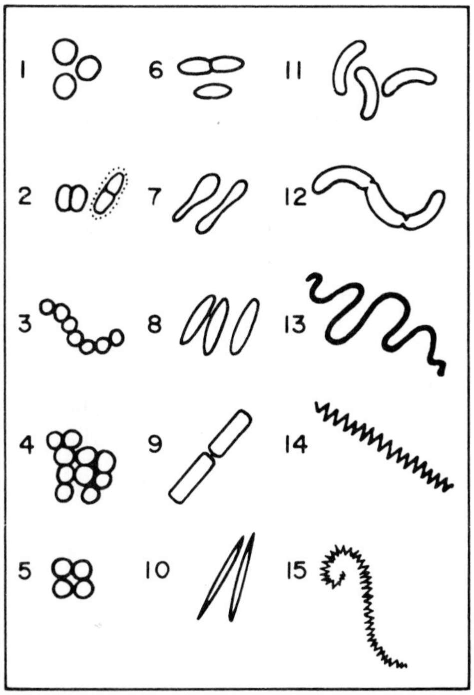

Interference Microscopy. Interference microscopes based on dual-beam optical systems generally achieve slightly less resolution but more depth and dimensional effect than the phase contrast microscope. They can also be used as an extremely accurate optical balance for the measurement of cellular components (usually of eucaryotes) down to 1×10^{-14} g, based on the well-known change in refractive index of 0.0018 per 1 percent change dry weight of substance in solution.

Darkfield Microscopy. This technique produces a black background against which objects appear brightly. A special condenser blocks direct illumination and directs light at an angle such that no incident light reaches the objective lens. If a specimen scatters light by reflection or refraction, light will enter the objective lens. The specimen appears brilliantly illuminated against the black background. This technique is valuable in examination of unstained organisms in fluid suspensions and has been effectively used to demonstrate the very thin spirochete that causes syphilis, *Treponema pallidum.*

Figure 3-3. Morphology. 1. Single cocci. 2. Cocci in pairs. 3. Cocci in chains. 4. Cocci in clusters. 5. Cocci in tetrads. 6. Coccobacilli. 7. Club-shaped bacilli. 8. Bacilli with rounded ends. 9. Bacilli with square ends. 10. Fusiform bacilli. 11. Vibrios. 12. *Spirillum.* 13. *Borrelia.* 14. *Treponema.* 15. *Leptospira.*

The Electron Microscope

Bacterial ultrastructure was demonstrated only after the development of the transmission electron microscope, sometimes abbreviated as TEM, which has a resolving power of about 0.001 μm, that is, some 200 times that of the light microscope. Transmission electron microscopy develops a two-dimensional image resulting from the variable electron density (stopping power) of the specimen interposed in the electron beam. Specimens must be fixed, stained, and dried. Improved techniques of fixation, embedding, and thin sectioning have gradually allowed better definition of structural components. The fine detail of flagella and pili, as well as of the cell envelope, membrane, and the internal cell fine structure, can be visualized by either shadow casting, (i.e., deposition of an electron-dense metal film at an angle) or negative staining procedures. The scanning electron microscope (SEM), has a practical limit of resolution of about 0.005 μm, or five-fold less than that of TEM. The scanning electron microscope produces a three-dimensional image by detection at a 90° angle of secondary electrons emitted from the specimen surface as a result of bombardment by the primary electron beam.

Negative Staining. This technique, which is used for ultrafine structure work, involves the application of solutions of salts containing heavy metal atoms, such as phosphotungstate, phosphosilicate, or ammonium molybdate. The electron-dense material deposits on the sample and forms a delicate, finely detailed outline of structural components.

Thin Sections. Thin sections of 0.1 μm or less, as compared to the usual 2 to 7 μm sections used in light microscopy, allow resolution and study of structures by electron microscopy without background interference from other components.

Freeze Etching. This technique involves freezing of specimens in situ with dry ice or liquid nitrogen. The specimen surface is then barely shaved or scratched with a microtome, followed by surface replication with carbon or metal. These preparations allow examination of surface layer and inner structures (Fig. 3-4).

Bacterial Ultrastructure

Appendages

Flagella

Flagella (singular, flagellum) are helical protein filaments of uniform length and diameter responsible for the rapid motility of bacteria (Fig. 3-5). The flagellum is composed of three parts: the filament, the hook, and the basal body (Figs. 3-6 and 3-7). The external filament connects to the

Figure 3-4. Structure of *Escherichia coli* after freeze etching. The cell on the left is dividing. Areas of intact cell surface can be seen on either side of the centrally exposed cytoplasmic membrane, which is studded with particles. Two distinct envelope layers can be seen at the cut surfaces above the plasma membranes and are especially visible on the cell on the right. At the upper left and right, the cells have been transected, showing granular cytoplasm and holes that appear to contain some fibrillar material, presumably DNA. *(From Bayer and Remsen: J Bacteriol 101:304, 1970.)*

hook at the cell surface. The hook is attached in the basal body, which is anchored in the plasma membrane. The basal body is comprised of a rod and two or more sets of encircling rings contiguous with the plasma membrane, peptidoglycan, and, in the case of gram-negative bacteria, the outer membrane of the cell envelope (Figs. 3-5, 3-6, and 3-7).

Free-swimming motility is a characteristic of flagellated vibrios, spirilla, and some bacilli. Pseudomonads may have either one polar flagellum (monotrichous), a tuft of several polar flagella (lophotrichous), or flagella at both poles (amphitrichous). By contrast, motile enterobacteria (e.g., *Salmonella*) or *Bacillus* species may have flagella distributed over the entire cell surface and are said to be "peritrichous." Flagella may vary in number from a few, as in some *Escherichia coli*, up to several hundred per cell, as is observed on *Proteus* species. *Proteus* species sometimes swarm as a thin film of growth on agar media surfaces. This phenomenon gave rise to the term "H antigen" (flagellar antigen), which derives from the German word *Hauch*, indicating a spreading film of growth, like breath condensing on a cold glass surface. In contrast, the term "O" or "somatic O antigen" of nonflagellated forms was

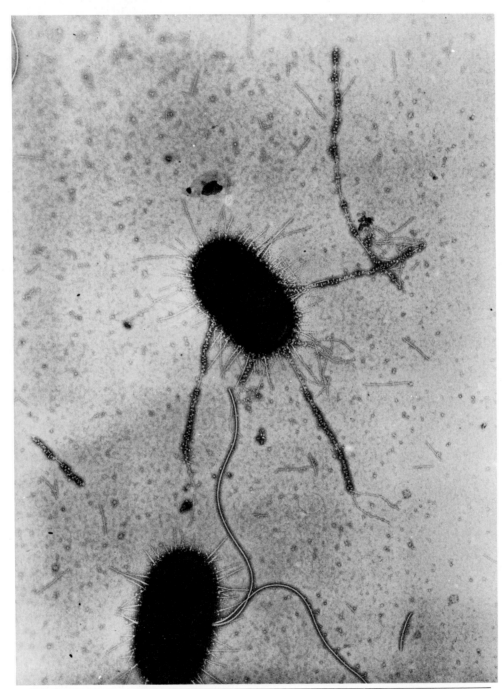

Figure 3-5. Negatively stained *E. coli* K12 Hfr Cavalli. Note two long large flagella, F-type sex pili with round RNA bacteriophage adsorbed to their sides and filamentous DNA bacteriophage adsorbed to their tips, and phage-free short common pili. *(Courtesy of Dr. Charles C. Brinton.)*

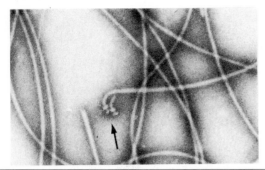

Figure 3-6. Isolated basal body–hook complex and flagellar filaments from *E. coli. (From DePamphlis and Adler: J Bacteriol 105: 384, 1971.)*

derived from the German term *ohne Hauch* (no film), indicating a nonspreading type of growth.

Bacterial motility can be observed microscopically in fluid suspensions (in a hanging drop or under a coverslip), by spread of bacterial growth as a film over agar, or as turbidity spreading through soft 0.4 to 0.5 percent (semisolid) agar. Flagella may be detected by darkfield or phase contrast microscopy and in stained preparations by light and electron microscopy. Flagellated cells react with specific antisera to give a typical, loose, flocculent agglutination, especially useful in diagnostic serologic identifications. Serologic agglutination is used in the identification of *Salmonella* species that alternately produce two flagellar filament antigens specified by separate structural genes (the H genes) by a process known as "phase variation" (Chap. 8).

Flagella are not required for viability. Flagella are removed by shaking with glass beads or by agitation in a

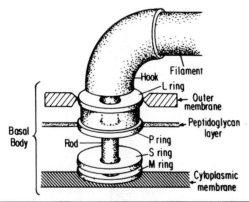

Figure 3-7. Idealized interpretive diagram of hook–basal body ring structures and integration into the cell envelope of *E. coli.* The top pair of rings are connected near their periphery and resemble a closed cylinder. The individual rings are referred to as the L ring (for attachment to outer-lipopolysaccharide-O-antigen complex membrane of the cell wall), the P ring (for its association with the peptidoglycan layer of the cell wall), the S ring (for supramembrane, which appears to be located just above the cytoplasmic membrane), and the M ring (for its attachment to the cytoplasmic membrane). *(Modified from DePamphlis and Adler: J Bacteriol 105:396, 1971.)*

blender. The cells remain viable and regain motility as the flagella regrow. Newly synthesized flagellin monomers appear at the distal tip of the growing flagellum. Flagella production is controlled by nutritional need or physiologic energy charge level. Cells of *E. coli* grown in the presence of glucose are inhibited from producing flagella by catabolite repression (Chap. 4). Flagellated cells, on the other hand, are able either to search for nutrients or to avoid poisons by following a gradient either toward a chemo-attractant or away from a repellent (p. 73).

Flagella propel the cell like a propeller, spinning around their long axis. Flagellar rotation is powered by proton current (Chap. 4). Flagellar function is governed by chemotactic responses, indicating a sensory feedback regulation system. Multiple flagella rotate counterclockwise to form a coordinated bundle and effect cell movement generally in the direction of a nutrient (i.e., positive chemotaxis). In the presence of a repellent, coordination is lost, the flagellar bundle becomes disorganized, and the cell tumbles and tends to move away from the repellent. Coordination of flagellar function involves chemoreceptors, known as "periplasmic binding proteins," that interact in membrane transport and at the methylation level of a specific plasma membrane protein. In the presence of chemoattractants, the methylation level of this protein is increased, whereas in the presence of a repellent, the methylation level is decreased or nil.

Axial Filaments. The spirochetes, i.e., the treponemes, leptospires, and borrelia, move by a traveling helical wave, a type of motion that allows penetration of viscous media. These bacteria produce flagellum-like axial filaments around which the cell is coiled (Fig. 3-8). These filaments occur in the periplasmic space between the inner and outer membranes of the cell (Fig. 3-9). *Treponema microdentium* produces two filaments per cell, *Treponema reiteri* produces six to eight, and some species produce many more. The filaments do not run from one pole of the cell to the other; instead, they originate at opposite poles of the cell and overlap at the center. No obvious connection holds the two overlapping ends together.

Treponeme axial filaments resemble flagella in length, diameter, and proximal hook at attachment sites. The axial filament is often surrounded by a sheath of unknown composition, often easily removed by distilled water. Axial filaments are suggested to mimic planetary drive gears by rotating against the cell body, imparting to it a twisting screwlike motion opposite in rotation to that of axial filament spin.

Microfibrils: Fimbriae and Sex Pili

Fimbriae, also called pili or common pili, are hair-like microfibrils of 0.004 to 0.008 μm, observable by electron microscopy on the surface of various bacteria. They are straighter, thinner, and shorter than flagella. Like flagella, fimbriae are composed of self-aggregating monomers,

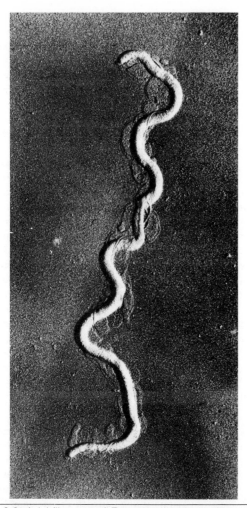

Figure 3-8. Axial filaments of *Treponema zuelzera* after treatment with distilled water to disrupt outer envelope. Shadowed with gold-palladium. A single axial filament originates at each pole. In the region of overlap, the two filaments have separated from each other and from the cell. × 15,000. *(From Barrier et al.: J Bacteriol 105:413, 1971.)*

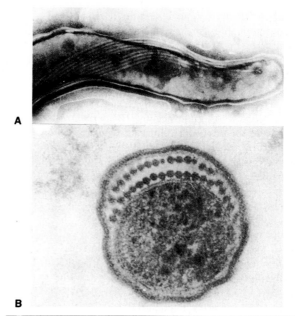

Figure 3-9. Axial filament of spirochetes from the mouth. **A.** End of a negatively stained cell showing multifibrillar axial filament. Insertion points of two fibrils are visible at the pole of the cell. × 60,000. **B.** Cross-section of a spirochete showing location of axial fibrils between outer membrane and cytoplasmic membrane. × 185,000. *(From Listgarten and Socransky: J Bacteriol 88:1087, 1964.)*

forming strands that originate in the plasma membrane. One of the best sources of richly piliated bacteria is the infected urinary tract.

Fimbriae are widespread among gram-negative bacteria. They also occur on some gram-positive bacteria. These include *Corynebacterium renale,* found in the urinary tract of cattle, *Actinomyces viscosus,* an organism present in the human oropharynx, *Actinomyces naeslundii, Streptococcus pyogenes, Streptococcus mutans,* and others.

Fimbriated cells adhere to interfaces, hydrophobic surfaces, and specific receptors. Fimbriae are, therefore, thought to afford bacteria survival advantages. They may be categorized according to functional activities or components as adhesins, lectins, evasins, and sex pili. The formation of pellicles on liquid media allows fimbriated bacteria to gain survival advantage access to oxygen in dense growth suspensions. More important in infectious diseases,

however, is the fact that fimbriae and other cell surface components act as specific adherence factors, adhesins, in the ecology of host–parasite interactions. That is, fimbrial adherence specificity may, in part, determine the ability of a bacterium to adhere to and colonize specific host tissue cells. For example, the 987P, K88, and K99 fimbriae of enteropathogenic (i.e., diarrheal) *E. coli* strains are important in the colonization by this organism of the intestine of piglets and of calves, and only fimbriated organisms of *Neisseria gonorrhoeae* (Kellogg types 1 and 2) are infectious.

The external location of the bacterial surface macromolecules is of importance in host–parasite interactions. For example, the streptococcal M protein, a known virulence factor, serves as an adherence factor (adhesin) in colonization of the pharynx and, unless neutralized with specific antiserum, prevents phagocytosis (acts as an evasin) and is leukocidal (acts as an aggressin or toxin).

Fimbriae fall into a category of proteins known as "lectins" that are also found in plants and animals. The adhesion of *Shigella flexneri* and *E. coli* fimbriae to red blood cells and tissues (e.g., intestinal cells) is specifically inhibited by D-mannose and α-D-methylmannoside. In a similar manner, fimbriae specific for α-methyl-D-galactose (*Pseudomonas aeruginosa*), D-galactose (*S. pyogenes*), L-fucose or D-mannose (*V. cholerae*), and a β-D-galactose-containing oligosaccharide (*N. gonorrhoeae*) have been reported.

Fimbriae of different strains of *N. gonorrhoeae* show great antigenic variation. This is apparently because of variation of fimbrial monomer units that are comprised of variable antigenic terminal peptide domains and a common nonantigenic, conserved, peptide domain. The latter is only antigenic when isolated by chemical means. The antigenic variability of gonococcal fimbriae thus appears to be another type of the phenomenon of host immune system evasion by parasite antigenic variation. On this basis, gonococcal fimbriae could be called "evasins."

Microfibrils of gram-negative bacteria are often referred to as common pili (fimbriae) or as sex pili, according to function. Both may occur independently or simultaneously on the same cell (Fig. 3-5). Some 100 to 200 ordinary fimbriae may be evenly distributed over the cell surface of *E. coli,* compared to only 1 to 4 sex pili found at random sites. Sex pili function in cell-to-cell adhesion in bacterial conjugation (Chap. 8). Sex pili are detected by the ability of cells to donate genes to recipients, by the presence of specific sex pilus antigen, or by the ability of the suspected pilus-bearing bacteria to inactivate certain bacteriophages, which attach specifically to sex pili (Chap. 8). Specific RNA phages attach along the sex pilus filament, whereas filamentous DNA phages attach to the pilus tips (Fig. 3-5).

Microfibrillar structures have also been implicated in the slow twitching motion of nonflagellated bacteria on surfaces, a process known as "surface translocation."

The Cell Envelope

The bacterial cell envelope includes the plasma membrane, the overlying cell wall, specialized proteins or polysaccharides, and any outer adherent materials. This multilayered organelle of the procaryotic cell comprises some 20 percent or more of the cell dry weight. The bacterial cell envelope contains transport sites for nutrients and receptor sites for bacterial viruses and bacteriocins (Chap. 8), influences host–parasite interactions, is the site of antibody and complement reactions (Chap. 13), and often contains components toxic to the host.

Topography and Visible Ultrastructure

SURFACE PATTERNS. Examination by electron microscopy of different bacteria after negative staining reveals a wide range of detailed surface patterns. The gram-positive staphylococcal, lactobacillal, and streptococcal cell surfaces appear smooth and devoid of regular patterns. Surfaces of *Clostridium botulinum* and *B. anthracis* (Sterne) reveal linear patterns of what are thought to be protein particles 0.006 to 0.008 μm in diameter. Tetragonal patterns are observed in *Bacillus polymyxa* and various other *Bacillus* species, and hexagonal patterns occur in still other bacilli and a few gram-negative species (Fig. 3-10). In contrast to the smooth or finely patterned surface of gram-positive bacteria, most gram-negative species exhibit a strikingly convoluted cell surface (Fig. 3-11).

Figure 3-10. Electron micrograph of a fragment of the envelope of *Lampropedia hyalina,* showing the surface pattern after tryptic digestion. × 160,000. *(Courtesy of Drs. J.A. Chapman and M.R.J. Salton.)*

Surface Adherents

Capsules and Slimes. Virulence of pathogens often correlates with capsule production. Virulent strains of pneumococci produce capsular polymers that protect the bacteria from phagocytosis. These bacteria form watery, mucoid (M), or smooth (S) colonies on solid media in contrast to noncapsule-forming rough (R) strains. Loss of cap-

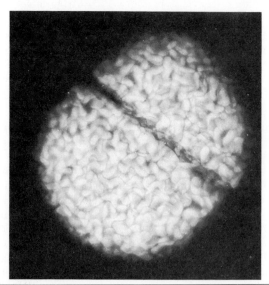

Figure 3-11. Convoluted surface pattern of whole cell of *Veillonella parvula,* a gram-negative bacterium that is a common inhabitant of the human oral cavity. Cell is in process of division. Negative stain, phosphotungstic acid. *(From Bladen and Mergenhagen: J Bacteriol 88:482, 1964.)*

sule-forming ability by S to R mutation correlates with loss of virulence and increased ease of destruction by phagocytes but does not affect viability. Capsules are dispensable.

Capsules form gels that tend to adhere to the cell, whereas slimes and extracellular polymers are more easily washed off. Capsules are easily visualized by negative staining. The capsule appears as a halo of light surrounding cells suspended in India ink (Fig. 3-12). Capsules may also be specially stained. If cells produce no visually demonstrable capsule and still react serologically with anticapsule sera, they are said to produce microcapsules (Fig. 3-13).

The Cell Wall

The cell wall, found in all free-living pathogenic bacteria except the cell wall-less mycoplasma, protects the cell from bursting in low osmotic pressure environments and maintains cell shape. The latter can be demonstrated by plasmolysis, by isolation of the particulate envelope after mechanical disruption of the bacterial cell, or by lysozyme digestion. The isolated cell wall retains the shape of the cell: envelopes of cocci resemble grape hulls, whereas envelopes of bacilli appear as long deflated balloons (Fig. 3-14). If whole cells or isolated envelopes are treated with lysozyme, the particulate cell wall of Eubacteria (but not Archaebacteria) characteristically dissolve. Chemically, the cell wall is composed of a peptidoglycan unique to bacteria.

Thickness. Cell envelopes of various organisms usually range from 0.150 to 0.500 μm in thickness as measured in thin sections but reach up to 0.8 μm in some strains of *Lactobacillus acidophilus*. Walls of young, rapidly growing bacterial cells are thinner than those of cells in old cultures or cells limited in protein synthesis by lack of a required amino acid or by antimetabolites. *Staphylococcus*

Figure 3-13. Electron micrograph of *Bacillus megaterium.* Treatment with specific antipolysaccharide serum reveals microcapsule-like cell wall polysaccharide, which is probably covalently linked to underlying murein sacculus. *(From Baumann-Grace and Tomcsik: Schweiz Z Pathol Bakteriol 21:906, 1958.)*

aureus cell walls, for example, may increase from 0.3 to 1.0 μm in thickness when protein synthesis is inhibited by chloramphenicol. The underlying plasma membrane remains approximately constant in thickness, about 0.0075 μm, in both gram-positive and gram-negative bacteria, and transverse sections exhibit typical trilaminar dark–light–dark unit membrane structure in electron micrographs. Pores of 0.001 to 0.01 μm in diameter can be demonstrated in isolated walls by molecular sieving; membrane pores appear to be about 1 nm in diameter.

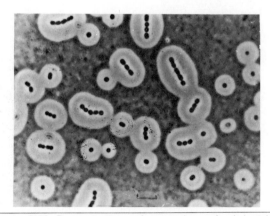

Figure   3-12. Demonstration of capsules of *Acinetobacter calcoaceticus* (*Bacterium anitratum*) cells suspended in India ink and observed by phase contrast microscopy. *(From Juni and Heym: J Bacteriol 97:461, 1964.)*

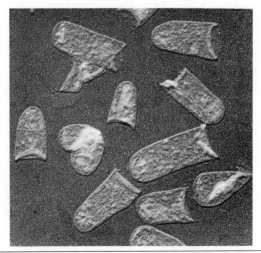

Figure 3-14. Electron micrograph of cell envelopes of *Mycobacterium tuberculosis.* × 22,000. *(From Ribi et al.: Proc Soc Exp Biol Med 100:647, 1959.)*

Differences Between Gram-positive and Gram-negative Cell Envelopes

The Gram-positive Cell Envelope. Gram-positive bacteria characteristically produce specific surface polysaccharides and proteins associated with the peptidoglycan. The better-known polysaccharides include teichoic acids, many of the pneumococcal capsular substances, and the streptococcal group polysaccharides. D-Polyglutamic acid polymers are produced by some *Bacillus* species, and the M protein of the group A streptococcus is a well-known virulence factor. The exact location and topographic arrangement of these substances is unknown. Thin cross-sections of gram-positive cells reveal a relatively thick, contiguous cell wall layer overlying the plasma membrane. This layer can be dissolved by lysozyme in most instances. Both protein and polysaccharide contribute to the layered wall substructure. Surface proteins of various bacilli and other organisms often form regular patterns (Fig. 3.10). The serologic type-specific M protein of the group A streptococcus forms a diffuse, thick, externally fimbriate wall layer, which can be removed by trypsin without destroying cell viability.

The Gram-negative Cell Envelope. Gram-negative bacteria exhibit three distinct, loosely arranged, envelope layers (Fig. 3-15). These include the convoluted, wrinkled, creviced, or undulating outer membrane (OM), which contains the somatic O antigen, a middle dense layer, and the inner plasma membrane. Both plasma and outer membranes are about 0.0075 μm (75 Å) thick. Both exhibit the typical bileaflet-trilayered sandwich structure seen by electron microscopy of membrane cross-sections: that is, two hydrophilic outer lamina (layers) of 25 Å each sandwich an inner hydrophobic 25 Å lamina comprised of fatty acid alkyl chains (Fig. 3-15). A lipoprotein, one third of which is covalently linked at one end to the outer surface of the peptidoglycan, inserts its lipid end into the outer membrane. The lipoprotein thus serves to anchor the outer membrane to the cell. The envelope can be isolated free of soluble cytoplasm by cell rupture and differential centrifugation. The inner membrane may be dissolved with mild nonionic detergent, leaving the outer membrane bound to insoluble peptidoglycan. The outer membrane can be disrupted by EDTA/detergent, aqueous phenol, or butanol extraction.

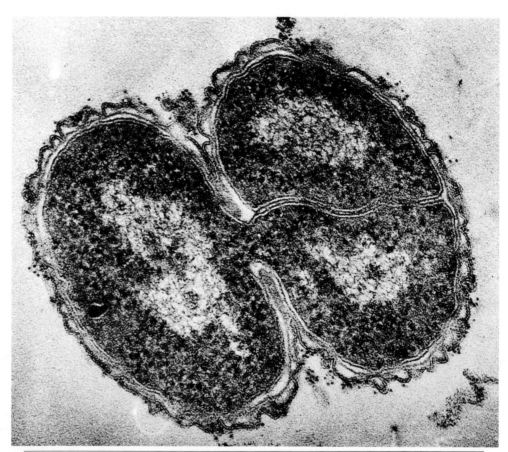

Figure 3-15. Transverse thin section of *Veillonella parvula*. The layered structure of the gram-negative cell envelope includes a convoluted outer unit membrane, a middle dense layer, and the plasma unit membrane. Note formation of division septa and fibrillar chromatin network in the cell interior. Dark spots of ferritin-labeled O-specific antibody are found primarily on the outside surface of the outer membrane. *(From Bladen and Mergenhagen: Ann NY Acad Sci 2:288, 1966.)*

OUTER MEMBRANE (OM). The outer membrane contains lipopolysaccharide (LPS), also known as somatic O antigen or endotoxin phospholipid, and unique proteins that differ from those of the plasma membrane. The inner and outer leaflets of the outer membrane are also uniquely asymmetrical. Phosphatidylethanolamine occurs almost entirely in the inner leaflet (lower or cell side and inner lamina), whereas lipopolysaccharide occurs only in the outer leaflet (top two lamina) of the outer membrane. Some proteins occur in the outer leaflet; others form transmembrane diffusion channels, known as porins, serve as nutrient or phage receptors, or perform other physiologic functions.

The outer membrane thus serves as a selective permeability barrier that keeps out hydrophobic substances and hydrophilic substances above a certain size range and retains periplasmic proteins.

Lipopolysaccharide (LPS). The lipopolysaccharide is responsible for many of the biologic activities associated with gram-negative bacteria. Isolated LPS is heat stable, is lethally toxic, is pyrogenic (i.e., it causes fever), is mitogenic for mouse but not human lymphocytes, stimulates bone marrow cell proliferation, activates the Hageman blood clotting factor in platelets, and activates complement. LPS also causes clotting of horseshoe crab amebocyte lysates at 10^{-12} dilution, providing a very sensitive assay for LPS.

Porin. These are major proteins uniquely found in the outer membrane and having molecular weights of around 35,000. They form transmembrane pores or diffusion channels that allow passage of small hydrophilic molecules through the outer membrane. Porins also serve as specific attachment sites for phage, vitamin B_{12}, and other nutrients. The occurrence of some porins is regulated environmentally by the presence or absence of certain substrates. Two classes of porins have been found. Enterobacterial porins exclude molecules greater than 600 daltons, whereas *P. aeruginosa* and perhaps gonococcal porins exhibit higher cutoff ranges (e.g., 6000 ± 3000). Porins are also known to be mitogenic, indicating that since both LPS and porins act similarly in this respect, the mammalian system is geared to detect either or both components of potentially infectious organisms.

BAYER JUNCTIONS. In gram-negative cells, points of connection between membrane and wall, known as adhesion sites or Bayer junctions, occur (Fig. 3-16). The Bayer junctions are physiologically active. Externally, they appear to be sites of bacteriophage DNA injection and complement-mediated lysis. Internally, they appear to be growth zones where they serve as sites for translocation of secretory protein, outer membrane proteins and lipopolysaccharides, and capsular polysaccharides. Sex pili as well as flagella also emerge from the cell at fusion sites between inner and outer membranes.

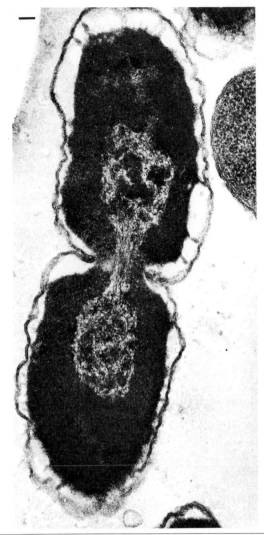

Figure 3-16. Plasmolyzed dividing cell of *Escherichia coli* B in an almost longitudinal section. Note adhesion sites (Bayer junctions) between outer and inner cytoplasmic membranes and division of nuclear material in the central portion. Bar equals 0.1 μm. *(From Bayer: J Gen Microbiol 53:395, 1968.)*

Protoplasts and Spheroplasts

Bacteria ordinarily lyse in water or serum when the rigid cell wall peptidoglycan layer of the cell envelope is dissolved by lysozyme or other agents. However, if stabilized by hypertonic solutions of sucrose or salts (0.2 to 0.5 M, depending upon the organism), a wall-less, osmotically sensitive spherical body called a "protoplast" is liberated. If envelope components are retained the osmotically sensitive body is called a "spheroplast." Gram-positive bacteria generally yield protoplasts, whereas gram-negative organisms yield spheroplasts since some outer membrane components inevitably are retained. Spheroplasts can also be produced by growth in hypertonic environments in the presence of cell wall synthesis inhibitors, such as penicillin (Fig. 3-17).

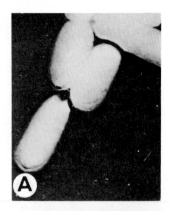

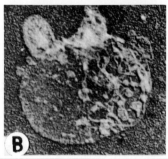

Figure 3-17. Spheroplast of *E. coli* W 173-25. **A.** Untreated cells. **B.** Cell treated for 90 minutes with 500 units of penicillin. × 9,000. *(From Schwarz et al.: J Mol Biol 41:419, 1969.)*

Periplasm

Periplasm, which occurs in the space between the plasma membrane and the outer membrane, may readily be observed in gram-negative bacteria (Fig. 3-18) but only with difficulty in gram-positive bacteria. This may be explained by the high internal osmotic pressures of gram-positive bacteria (8 to 20 atm) compared to those of gram-negative

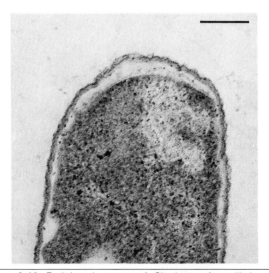

Figure 3-18. Periplasmic space of *Citrobacter freundii* shown in thin section. Bar equals 0.2 μm. *(Courtesy of Dr. Sara Miller.)*

bacteria (3 to 5 atm). The periplasmic space of gram-negative bacteria varies with growth conditions and among individual bacteria. In *E. coli,* the periplasmic space has been shown by cytochemical staining to contain various proteins, including alkaline phosphatase, acid hexosephosphatase, and cyclic phosphodiesterase. It is also thought to contain hydrolytic enzymes, such as acid phosphatase, DNase I, RNase I, and plasmid-controlled penicillinases, in addition to binding proteins that specifically bind sugars, amino acids, and inorganic ions (Chap. 5). These can be released from the cell by osmotic shock, i.e., rapid dilution of hypertonic (0.5 M sucrose) cell suspensions, after EDTA treatment. Other enzymes, such as the chromosomally controlled penicillinases of *E. coli,* may be membrane associated, since they are released from the cell only during spheroplast formation (Chaps. 8 and 9).

The Plasma Membrane

Beneath the rigid cell wall layer and in close association with it is the delicate cytoplasmic membrane, vitally important to the cell. In thin sections the plasma membrane shows a typical unit-membrane or trilaminar sandwich structure of dark–light–dark layers.

The Membrane as an Osmotic Barrier. Although bacteria are regarded as extremely tolerant to osmotic changes in their external environment, they undergo either plasmolysis or plasmoptysis when placed in media of varying salt concentrations. Placing cells in hypertonic solutions results in plasmolysis, that is, shrinkage of the membrane and cytoplasm from the cell wall. Gram-negative cells are more easily plasmolyzed than are gram-positive cells, which correlates with their relative internal osmotic pressures.

The osmotic barrier in bacteria is indicated by their ability to concentrate certain amino acids against concentration gradients. In gram-positive bacteria, a gradient of 300- to 400-fold may exist across the surface layers. Phosphate esters, amino acids, and other solutes contribute to the internal osmotic pressure. Osmotic activity is also indicated by the cells selective permeability toward various compounds.

Membrane-associated Structures. Membranes isolated after careful lysis account for some 30 percent or more of the cell weight. Up to 90 percent of the ribosomes may be isolated as a membrane-polyribosome-DNA aggregate.

Membrane Components. Membranes contain 60 to 70 percent protein, 30 to 40 percent lipid, and small amounts of carbohydrate. Phosphatidylethanolamines (75 percent), phosphatidylglycerol (20 percent), and glycolipids are found as major constituents. Choline, sphingolipids, polyunsaturated fatty acids, and steroids are rare. Pathogenic mycoplasma incorporate steroids from the environment into their plasma membranes. Glycolipids include diglycosyldiglycerides, found primarily in gram-positive bacterial membranes, which also contain lipoteichoic acids. A 55-carbon polyisoprenoid alcohol known as undecaprenol or bactoprenol, occurs in small amounts.

Various enzymic activities are associated with membrane proteins. These include the energy-producing bacterial cytochrome and oxidative phosphorylation system, the membrane permeability systems discussed in Chapter 5, and various polymer-synthesizing systems. An ATPase has been isolated from knob-like membrane structures similar to those found in eucaryotic mitochondria.

Mesosomes

These membrane-associated organelles are more easily demonstrated in gram-positive than in gram-negative bacteria. Mesosomes are usually seen as cytoplasmic sacs that contain whorled, lamellar, tubular, or vesicular structures and are often associated with division septa (Fig. 3-19). Attachment of mesosomes to both DNA chromatin and membrane has been demonstrated by thin-section electron microscopy. Formation of protoplasts (or spheroplasts) results in eversion of tubular or vesicular mesosomal components, which remain attached at one end to the outside of the membrane, whereas the enclosing mesosomal sacs disappear and are apparently pulled into the membrane by the stretched protoplast. Mesosomes have been reported to be artifacts of fixation procedures, but it is difficult to explain vesicular mesomal tubules as artifacts.

Cytoplasmic Structures

The Nuclear Body

Bacterial DNA can be detected as nucleoids or chromatin bodies by light microscopy using Feulgen staining. It is difficult to demonstrate chromatin bodies by direct staining because of the high concentration of ribonucleic acid. However, pretreatment with ribonuclease removes all or nearly all of the RNA, and chromatin bodies can then be seen at all stages of the growth cycle.

Electron microscopy of stained thin sections reveals nuclear material as an irregular, thin, fibrillar, DNA network, which frequently runs parallel to the axis of the cell (Fig. 3-16). A direct attachment to the membrane is sometimes obvious. Often the mesosome, itself a membrane-associated structure, appears to serve as a DNA-membrane attachment site. During multiplication, bacterial DNA remains as a diffuse chromatin network and never aggregates to form a well-defined chromosome during cell division, in contrast to the eucaryotic chromosomes. When bacterial cells are very gently lysed, the bacterial chromosome may be visualized by radioautography as a circular molecule. Although bacterial DNA represents only 2 to 3 percent of the cell weight, it occupies 10 percent or more of the cell volume. This loose arrangement allows for the ready diffusion of soluble materials to all parts of the nuclear structure.

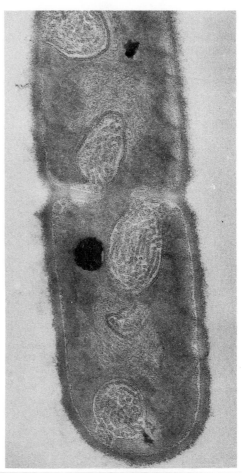

Figure 3-19. Dividing cell of *Lactobacillus plantarum,* showing prominent mesosomes associated with a newly forming crosswall. The immature type of mesosome seen at the bottom is not associated with nucleoplasm, in contrast to the upper mature mesosomes, which are surrounded by a triple-layered boundary and are continuous with the nucleoplasm. The black bodies are inclusion granules. × 94,300. *(From Kakefuda et al.: J Bacteriol 93: 474, 1967.)*

Ribosomes

Negatively stained thin sections allow resolution by electron microscopy of small cytoplasmic particles that correspond to the ribosomes present in pellets after lysed protoplasts or disrupted bacterial cells are centrifuged at 100,000 g. Ribosomes are composed of approximately 30 percent protein and 70 percent RNA and account for up to 40 percent of the protein and 90 percent of the ribonucleic acid of the cell. Gentle lysis of growing cells yields almost all ribosomes as polyribosome-membrane aggregates that contain all components of the protein-synthesizing mechanism; polyribosomes are chains of 70S ribosomes (monomers) attached to messenger RNA (Fig. 3-20). Ribosome numbers in the cell vary according to growth conditions: rapid-growing cells in rich medium contain many more ribosomes than do slow-growing cells in poor medium.

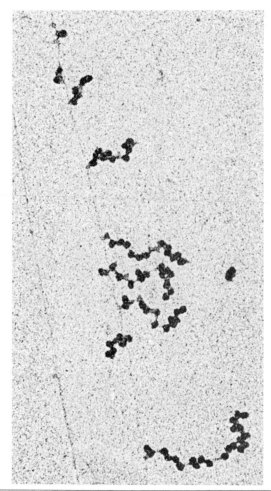

Figure 3-20. Electron photomicrograph of polyribosomes from *E. coli*. The central long vertical strand is DNA. The wavy lines extending from it are molecules of mRNA to which ribosomes are attached (i.e., polysomes). The gradual increase in length of mRNA from top to bottom indicates that transcription was proceeding in that direction. × 76,350. *(From Hamkalo and Miller: Annu Rev Biochem 42:379, 1973.)*

Subunits. The stability of ribosomes depends on the presence of Mg^{2+}. In low concentrations of magnesium (less than $10^{-4}M$ Mg^{2+}), 70S ribosomal monomers dissociate into 50S and 30S subunits. The 30S subunit measures approximately 0.007 by 0.016 μm and weighs about 800,000 daltons. It is composed of a single 16S RNA molecule with a molecular weight of about 500,000 and some 20 protein molecules. The 50S subunit measures about 0.014 by 0.016 μm and weighs 1.64×10^6 daltons. It is composed of one 23S RNA molecule with a molecular weight of 1.1×10^6, a 5S RNA molecule (40,000 daltons), and some 30 protein molecules.

Polyamines and Histone-like Proteins

Histone-like proteins have only recently been found in small amounts in association with *E. coli* DNA (Chap. 7), whereas the occurrence of polyamines in bacteria is well

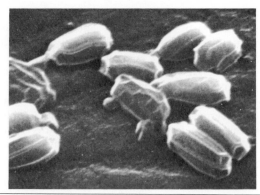

Figure 3-21. *Bacillus polymyxa* endospores examined in a scanning electron microscope. Both parallel ribbing and random reticulation of the surface structure are exhibited. × 4800. *(From Murphy and Campbell: J Bacteriol 98:727, 1969.)*

known. Polyamines are required for the growth of some *Haemophilus* species, but the lack of them does not retard the growth of other bacteria that are unable to make them. Polyamines are found associated with ribosomes and membranes. The principal ones are:

$$\text{Putrescine: } H_2N(CH_2)_4NH_2$$

$$\text{Spermidine: } H_2N(CH_2)_4NH(CH_2)_3NH_2$$

The precise function of these polyamines is not known. Polyamines exert an antimutagenic effect, they prevent dis-

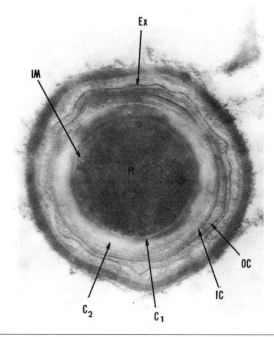

Figure 3-22. Cross-section of a *Bacillus cereus* endospore before release from parent cell, showing various spore coats and layers. Ex, exosporium; IM, inner membrane; R, ribosomal aggregates; OC, outer coat; IC, inner coat; C_1, inner dense cortical layer; C_2, less dense cortical layer. Note that parent cell still surrounds the exosporium. × 75,000. *(From Ellar and Lundgren: J Bacteriol 92:1748, 1966.)*

sociation of 70S ribosomes to 30S and 50S components, and they increase the resistance of protoplasts to osmotic lysis. The amount of spermidine varies inversely with the amount of Mg^{2+} in ribosomes, and 30S and 50S ribosomal subunits remain associated in the absence of Mg^{2+} if spermidine is present. From this it would appear that charge neutralization of polyanionic polymers, such as nucleic acid, may be at least partly nonspecific.

Cytoplasmic Granules

Granules, identified by appropriate staining procedures, indicate accumulation of food reserves, including polysaccharides, lipids, or polyphosphates. Granules vary with the type of medium and the functional state of the cells. Glycogen is the major storage material of enteric bacteria and may account for as much as 40 percent of the weight of some species. Similarly, some *Bacillus* and *Pseudomonas* species accumulate 30 percent or more of their weight as poly-β-hydroxybutyrate. Finally, polyphosphates are also known as the metachromatically staining Babes–Ernst or volutin granules found in abundance in *Corynebacterium diphtheriae*, the plague bacillus (*Yersinia pestis*), mycobacteria (e.g., *Mycobacterium tuberculosis*), and some other bacteria. Volutin granules stain in various colors, varying from red to blue, with toluidine blue and methylene blue.

Bacterial Endospores

Endospore formation is a distinguishing feature of organisms of the family Bacillaceae, which includes members of the aerobic genus, *Bacillus*, and the anaerobic genus, *Clostridium*. Endospores resist adverse environmental conditions of dryness, heat, and poor nutrient supply. If present in inadequately heat-sterilized canned foods, spores of the anaerobic *Clostridium* species germinate, grow, and cause spoilage. In the case of *C. botulinum*, this is especially dangerous because spoilage may not be obvious even though accompanied by the production of deadly exotoxin. Other sporeformers, such as the gas-gangrene bacillus, *Clostridium perfringens*, or the tetanus bacillus, *Clostridium tetani*, produce their toxic effects only when a medium of dead or injured tissue affords the spore a nidus for growth in the animal body.

The true endospore is a highly refractile body formed within the vegetative bacterial cell at a certain stage of growth. The size, shape, and position of the spore are relatively constant characteristics of a given species and are, therefore, of some value in distinguishing one kind of bacillus from another (Fig. 3-21). The position of the spore in the cell may be central, subterminal, or terminal. It may be the same diameter as the cell, smaller, or larger, causing a swelling of the cell. Various protective spore coats are formed within the vegetative cell before death and dissolution of the parent cell (Fig. 3-22). These include a rigid peptidoglycan layer, which differs in composition from that of the parent vegetative cell. Spore surface antigens are usually different from those of the parent vegetative bacilli.

FURTHER READING

Books and Reviews

Adler J: Chemotaxis in bacteria. Annu Rev Biochem 44:341, 1975

Archibald AR, Baddiley J, Blumson NL: The teichoic acids. Adv Enzymol 30:223, 1968

Balch WE, Fox GE, Magrum LJ, et al.: Methanogens: Reevaluation of a unique biological group. Microbiol Rev 43:260, 1979

Bayer ME: Ultrastructure and organization of the bacterial envelope. Ann NY Acad Sci 235:5, 1974

Beachey EH: Bacterial adherence: Adhesin-receptor interactions mediating the attachment of bacteria to mucosal surfaces. J Infect Dis 143:325, 1981

Berkeley RCW, Lynch JM, Melling J, et al. (eds): Microbiol Adhesion to Surfaces. Soc Chem Ind, London, E. Horwood Ltd, New York, Wiley, 1980

Brinton CC, et al.: Uses of pili in gonorrhea control: Role of bacterial pili in disease, purification and properties of gonococcal pili, and progress in the development of a gonococcal pilus vaccine for gonorrhea. In Brooks G, Gotschlich EC, Holmes KK, et al. (eds): Immunobiology of *Neisseria gonorrhoeae*. Washington, DC, American Society for Microbiology, 1978, p 155

Burchard RP: Gliding motility of prokaryotes: Ultrastructure, physiology, and genetics. Annu Rev Microbiol 35:497, 1981

Driel DV, Wicken AJ, Dickson MR, et al.: Cellular location of the lipoteichoic acids of *Lactobacillus fermenti* NCTC 6991 and *Lactobacillus casei* NCTC 6375. J Ultrastruct Res 43:483, 1973

Gaastra W, de Graaf FK: Host-specific fimbrial adhesins of noninvasive enterotoxigenic *Escherichia coli* strains. Microbiol Rev 46:129, 1982

Glauert AM, Thornley JJ: The topography of the bacterial cell wall. Annu Rev Microbiol 23:159, 1969

Goy MS, Springer MS, Adler J: Behaviour control mechanisms. Nature 280:279, 1979

Harold FM: Energy coupling: An overview. In Schlessinger D (ed): Microbiology 1979. Washington, DC, American Society for Microbiology, 1979, p 42

Henrichsen J: Bacterial surface translocation: A survey and a classification. Bacteriol Rev 35:478, 1972

Ino T: Genetics and chemistry of bacterial flagella. Bacteriol Rev 33:454, 1969

Inouye M, et al.: Mechanism of translocation of the outer membrane proteins of *Escherichia coli* through the cytoplasmic membrane. In Schlesinger D (ed): Microbiology 1979. Washington, DC, American Society for Microbiology, 1979, p 34

Kalckar HM: The periplasmic galactose binding protein of *Escherichia coli*. Science 174:557, 1971

Knox KW, Wicken AJ: Immunologic properties of teichoic acids. Bacteriol Rev 37:215, 1973

Koshland D: Sensory response in bacteria. Adv Neurochem 2:277, 1977

Kurland CG: Ribosome structure and function of the bacterial ribosome. Annu Rev Biochem 45:173, 1977

Lindberg AA: Specificity of fimbriae and fimbrial receptors. In Schlessinger D (ed): Microbiology 1982. Washington DC, American Society for Microbiology, 1982, p 317

Nikaido H: The role of outer membrane permeability in the sensitivity and resistance of gram-negative organisms to antibiotics. In Mitsuhashi S (ed): Drug Resistance in Bacteria. Tokyo, Japan Scientific Societies Press, 1982, p 317

Rogers H, Perkins, HR, Ward, JB (eds): Microbial Walls and Membranes. London, Chapman and Hall, 1980

Ryter A: Association of the nucleus and the membrane of bacteria: A morphological study. Bacteriol Rev 32:39, 1968

Salton MRJ: The Bacterial Cell Wall. Amsterdam, Elsevier, 1964

Salton MRJ: Bacterial membranes. CRC Crit Rev Microbiol 1:151, 1971

Schoolnik GK, Tal JYT, Gotschlich EC: Receptor binding and antigenic domains of gonococcal pili. In Schlessinger D (ed): Microbiology 1982. Washington DC, American Society for Microbiology, 1982, p 312

Shockman GD, Wicken AJ (eds): Chemistry and Biological Activities of Bacterial Surface Amphiphiles. New York, Academic Press, 1981

Wannamaker LW, Matsen JM (eds): Streptococci and Streptococcal Diseases. New York, Academic Press, 1972

Wicken AJ, Knox KW: Bacterial cell surface amphiphiles. Biochim Biophys Acta 604:1, 1980

Selected Papers

Berg HC: Dynamic properties of bacterial flagellar motors. Nature 249:77, 1974

Berg HC: How spirochetes may swim. J Theor Biol 56:269, 1976

Berg HC, Anderson RA: Bacteria swim by rotating their flagellar filaments. Nature 245:380, 1973

Boman HG, Nordstrom K, Normark S: Penicillin resistance in Escherichia coli K 12: Synergism between penicillinases and a barrier in the outer part of the envelope. Ann NY Acad Sci 235:569, 1974

Chen YU, Hancock REW, Mishell RI: Mitogenic effects of purified outer membrane proteins from Pseudomonas aeruginosa. Infect Immun 28:178, 1980

Davies JE, Benveniste RE: Enzymes that inactivate antibiotics in transit to their targets. Ann NY Acad Sci 235:130, 1974

Douglas JT, Lee MD, Nikaido H: Protein 1 of Neisseria gonorrhoeae outer membrane is a porin. FEMS Lett 12:305, 1981

Ebersold HR, Cordier, JL, Luthy P: Bacterial mesosomes: Method dependent artifacts. Arch Microbiol 130:19, 1981

Flessel CP, Ralph P, Rich A: Polyribosomes of growing bacteria. Science 158:658, 1967

Larsen SH, Reader RW, Kort EN, et al.: Change in direction of flagellar rotation is the basis of the chemotactic response in Escherichia coli. Nature 249:74, 1974

Morgan RL, Isaacson RE, Moon HW, et al.: Immunization of suckling pigs against enterotoxigenic Escherichia coli-induced diarrheal disease by vaccinating dams with purified 987 or K99 pili: Protection correlates with pilus homology of vaccine and challenge. Infect Immun 22:771, 1978

Nikaido H, Rosenberg EY: Porin channels in Escherichia coli: Studies with liposomes reconstituted from purified proteins. J Bacteriol 153:241, 1983

Nikaido H, Rosenberg EY, Foulds J: Porin channels in Escherichia coli: Studies with β-lactams in intact cells. J Bacteriol 153:232, 1983

Wadstrom T, Tylewska S: Glycoconjugates as possible receptors for Streptococcus pyogenes. Curr Microbiol 7:343, 1982

Wetzel BK, Spicer BK, Dvorak HF, et al.: Cytochemical localization of certain phosphatases in Escherichia coli. J Bacteriol 194:529, 1970

Yoshimura F, Zalman LS, Nikaido H: Purification and properties of Pseudomonas aeruginosa porin. J Biol Chem 258:2308, 1983

Zalman LS, Nikaido H, Kagawa Y: Mitochondrial outer membrane contains a protein-producing nonspecific diffusion channels. J Biol Chem 256:1771, 1980

CHAPTER 4
Energy Metabolism

Sources of Energy and Carbon

Bacterial cells, like the cells of all living organisms, accomplish work. For this they require a source of energy. Although the wide variety of compounds that serve as a source of energy for microorganisms is almost limitless, there is a remarkable simplicity in the basic metabolic patterns utilized to transform this energy into a useful form. Many of these systems are fundamentally similar to those found in the higher forms of life, but superimposed on these basic mechanisms are examples of differentiation unique to the bacterial world.

Bacteria can be divided into two large groups on the basis of their carbon requirement, the autotrophic (lithotrophic) bacteria and the heterotrophic (organotrophic) organisms (Table 4-1). The autotrophic bacteria can utilize carbon dioxide as the sole source of carbon and synthesize from it the carbon skeletons of all their organic metabolites. They require only water, inorganic salts, and CO_2 for growth. Their energy is derived either from light or from the oxidation of one or more inorganic materials. The photosynthetic autotrophs (photolithotrophs) obtain energy for their synthetic activities by the utilization of radiant energy. These are anaerobic organisms containing a magnesium porphyrin pigment closely related to chlorophyll a of green plants. Chemosynthetic autotrophs (chemolithotrophs) obtain their energy from oxidation-reduction reactions using simple inorganic electron donors, such as hydrogen, hydrogen sulfide, sulfur, or ammonia.

The heterotrophic bacteria are unable to utilize CO_2 as the sole source of carbon but require that it be supplied in an organic form. Most of these organisms require complex organic molecules, such as glucose, as electron donors. For the heterotrophic bacteria, a portion of the organic compound that serves the organism as an energy source invariably is used for the synthesis of many or all of the organic compounds required by the organism. This group contains all of those bacteria pathogenic for man.

Energy-yielding Metabolism

The systems in bacteria that transform chemical and radiant energy into a biologically useful form include respiration, fermentation, and photosynthesis. In respiration, molecular oxygen is the ultimate electron acceptor, while in fermentation, the foodstuff molecule is usually broken down into two fragments, one of which is then oxidized by the other. In photosynthesis, light energy is converted into chemical energy. In all types of cells, however, and regardless of the mechanism used to extract useful energy, the reaction is accompanied by the formation of adenosine triphosphate (ATP). ATP is a common intermediate of both energy-mobilizing and energy-requiring reactions, and its formation provides a mechanism by which the available energy may be channeled into the energy-requiring biosynthetic reactions of the cell. The study of energy metabolism is the study of ATP production.

The metabolic activity of bacteria is very high. This is manifested both in a very rapid rate of cell division and in a high rate of catabolism. Associated with these processes is a very noticeable evolution of heat, much greater than for other organisms. Since the heat produced during metabolism represents that fraction of the total free energy change that is unavailable to the organisms for the performance of work, bacteria in general are less efficient as converters of free energy than are organisms with a slower metabolic rate.

Bioenergetics

Principles of Thermodynamics

Fundamentally, the bacterial cell is a physicochemical system whose activities occur in large part by the flow of chemical energy. The same laws of thermodynamics that

TABLE 4-1. CLASSIFICATION OF BACTERIA ACCORDING TO SOURCE OF CARBON AND ENERGY

Type	Carbon Sources	Energy Sources	Electron Donors	Examples
Photolithotrophs	CO_2	Light	Inorganic compounds (H_2S, S)	Green and purple sulfur bacteria
Photoorganotrophs	Organic compounds (in addition to CO_2)	Light	Organic compounds	Purple nonsulfur bacteria
Chemolithotrophs	CO_2	Oxidation-reduction reactions	Inorganic compounds (H_2, S, H_2S, Fe, NH_3)	Hydrogen, sulfur, iron, and denitrifying bacteria
Chemoorganotrophs	Organic compounds	Oxidation-reduction reactions	Organic compounds (glucose)	Most bacteria

deal with energy and its transformation also hold in the biologic world. The most useful of these concepts is that of free energy. Knowledge of the free energy change (ΔG) of a reaction tells us whether a reaction may proceed spontaneously or whether it must be driven by other reactions.

The standard free energy change $\Delta G°$ for any reaction may be calculated as the difference between the standard free energy of the products and the standard free energy of the reactants:

$$\Delta G° = \Sigma G°_{products} - \Sigma G°_{reactants}$$

For the reaction to proceed spontaneously as written, $\Delta G°$ must be negative, that is, the products must be lower on the free energy scale than the reactants. Thus, reactions only go downhill energetically, from compounds of higher to those of lower free energy. Reactions that have a positive $\Delta G°$ do not occur spontaneously but must be supplied with free energy greater than $\Delta G°$ from another source if they are to proceed.

In the bacterial cell where work is performed under isothermal conditions, a large fraction of the system's energy cannot be made to perform in the manner required of work. This energy is regarded as unavailable or lost to entropy. Entropy refers to the degree of randomness or disorientation of a system. The driving force of all processes is the tendency to seek the position of maximum entropy, and heat is either given up or absorbed from the surroundings by the system to reach this state. A large positive entropy means death, whereas negative entropy and order characterize life.

The relationship between the entropy and free energy change in a reacting system is expressed in the equation:

$$\Delta G = \Delta H - T\Delta S$$

in which ΔG is the free energy change of the system, ΔH is the change in total energy of the system, T is the absolute temperature, and ΔS is the change in entropy. This relationship is useful in predicting quantitatively the direction of a chemical reaction and whether that reaction will or will not proceed.

The free energy change of chemical reactions can in principle be quite accurately measured. Since the equilibrium reached in a chemical reaction is a function of the drive toward minimum free energy of the reaction components, the equilibrium constant is a mathematical function of the free energy change of the components of the reaction. Thus, for the reaction:

$$A + B \rightleftharpoons C + D$$

the free energy change is:

$$\Delta G = G°' + RT \ln \frac{[C][D]}{[A][B]}$$

where R is the gas constant, T is the absolute temperature, and the brackets denote initial molar concentrations of reactants and products. The symbol $\Delta G°'$ designates the standard free energy change of the reaction. It is a fixed

TABLE 4-2. RELATIONSHIP BETWEEN THE EQUILIBRIUM CONSTANT AND THE STANDARD FREE ENERGY CHANGE AT 25C

K_{eq}	$\Delta G°'$ cal/mole
0.001	+4089
0.01	+2726
0.1	+1363
1.0	0
10.0	−1363
100.0	−2726
1000.0	−4089

constant for any given chemical reaction and is a measure of the decrease in free energy of the reaction at 25C, at a pH of 7.0, as 1 mole of reactant is converted to 1 mole of product.*

At equilibrium, there is no free energy change, and $\Delta G = 0$. Also:

$$K'_{eq} = \frac{[C][D]}{[A][B]}$$

By substituting the equilibrium constant K'_{eq} in the above reaction:

$$\Delta G°' = - RT \ln K'_{eq}$$

The standard free energy change of any chemical reaction can thus be calculated from its equilibrium constant. Table 4-2 shows this relationship between the equilibrium constant and the standard free energy change. When the equilibrium constant is high, the reaction tends to go to completion, and the standard free energy change is negative. Such a reaction is an exergonic or downhill reaction and proceeds with a decline of free energy. When the equilibrium constant is low, the reaction does not go far in the direction of completion, the free energy change is positive, and energy must be put into the system. Such processes are endergonic or uphill reactions.

Energy from Organic Compounds

The heterotrophic bacterial cell ultimately derives its energy from the chemical energy stored in the molecules of its carbon substrate. The complete oxidation of glucose releases 686,000 calories/mole:

$$C_6H_{12}O_2 \rightleftharpoons 6\ CO_2 + 6\ H_2O$$

$$\Delta G° = -686,000 \text{ calories}$$

In *Escherichia coli* about 50 percent of the glucose is oxidized to CO_2. This results in enough ATP to convert the remaining 50 percent of the substrate into cell material.

---

* The standard free energy change at pH 7.0 is designated by $\Delta G°'$; that at pH 0.0 by $\Delta G°$.

In biologic oxidations as well as in chemical ones, the essential characteristic is the removal of electrons from the substance being oxidized. Since the majority of biologic oxidations involve a dehydrogenation, biologic oxidations may be expressed more simply in terms of the transfer of hydrogen if it is remembered that such a transfer actually involves a loss of electrons:

$$AH_2 + B \rightleftharpoons A + BH_2$$

Here the substrate AH_2 is oxidized to A and the substance B acts as the hydrogen acceptor and is reduced to BH_2.

In the transfer of hydrogen from the substrate to a final hydrogen acceptor, reducing equivalents are removed, two at a time, and passed via a graded series of oxidation-reduction systems such that the derivatives en route are alternately reduced and oxidized. The occurrence of several reversible oxidation-reduction reactions between the initial substrate and final oxidant makes for a smoother release of energy, providing a system whereby oxidations involving large amounts of energy resulting from the complete oxidation of a carbohydrate are split into several integrated partial reactions, and the energy is stored or liberated in smaller packets. In this sequence of reactions, chemical energy is transferred from one reaction to another by means of a common intermediate or coupling factor. An example of this common intermediate principle is the coupling between the oxidation of one substrate (AH_2) and the reduction of another (B) by the appropriate AH_2 and BH_2 dehydrogenases in the presence of the specific coupling factor (NAD), which is alternately reduced and oxidized:

The most important common intermediate, however, in the transfer of chemical energy is ATP. It effectively links or couples enzymatic reactions involving the transfer of phosphate groups (Fig. 4-1).

Free Energy Changes of Oxidation-Reduction Reactions

The calculation of free energy changes for oxidation-reduction reactions is based on the oxidation-reduction potential, a quantitative measure of the ability of the system to accept or donate electrons reversibly with reference to the standard hydrogen electrode. In biologic systems, the normal potential of any two oxidation-reduction systems enables one to predict the direction of interaction. A system with a more positive normal oxidation-reduction potential than another system has a greater tendency to take up electrons, i.e., it is a stronger oxidizing agent. The standard free energy change of oxidative reactions in calories per mole may be calculated from equilibrium data. Some electrode potentials of biochemical interest are shown in Table 4-3.

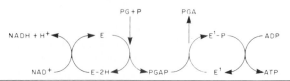

Figure 4-1. Primary chemical coupling between the phosphorylation of ADP and the NAD-linked oxidation of 3-phosphoglyceraldehyde (PG) to 3-phosphoglycerate (PGA) via 1,3-diphosphoglycerate (PGAP). E/E-2H and E′/E′-P stand for 3-phosphoglyceraldehyde dehydrogenase and 3-phosphoglycerate kinase, respectively.

Key Position of Adenosine Triphosphate

In both aerobic and anaerobic cells, all of the usable energy released by oxidation is transformed to ATP for use in driving the various energy-requiring reactions involved in the biosynthesis of cell material. The amount of ATP available from a particular substrate depends upon whether the organism employs a fermentative type of metabolism or whether the compound is completely oxidized to CO_2 and H_2O (p. 59).

ATP occurs in all types of cells. In the intact cell at pH 7.0, the molecule is completely ionized and exists as a complex with Mg^{2+} (Fig. 4-2). Its free energy of hydrolysis is significantly higher than that of simple esters, glycosides, and many phosphorylated compounds. Molecules, such as ATP, that are characterized by a free energy of hydrolysis at pH 7 more negative than 7 kcal per mole are classified as high-energy compounds. These energy-rich compounds include a number of other important molecules: acetylphosphate, aminoacyladenylates, phosphoenolpyruvate, and the esters of coenzyme A and lipoic acid, all of which serve as a driving force for the various endergonic reactions of the cell.

The standard free energies of hydrolysis of various phosphate compounds are shown in Table 4-4. Compounds with the more negative values have a higher equi-

TABLE 4-3. STANDARD OXIDATION-REDUCTION POTENTIALS OF SOME CONJUGATE REDOX PAIRS

System (pH 7.0)	E'_0, V
$\frac{1}{2}O_2/H_2O$	0.82
NO_3^-/NO_2^-	0.42
Fe^{3+} cytochrome a/Fe^{2+}	0.29
Fe^{3+} cytochrome c/Fe^{2+}	0.22
Ubiquinone ox/red (pH 7.4)	0.10
Fe^{3+} cytochrome b/Fe^{2+} (pH 7.4)	0.07
Fumaric acid/succinic acid	0.03
FMN old yellow enzyme/$FMNH_2$	−0.12
Oxaloacetic acid/malic acid	−0.17
Pyruvic acid/lactic acid	−0.19
Acetaldehyde/ethanol	−0.20
$NAD^+/NADH + H^+$	−0.32
Ferredoxin ox/red (*Clostridium pasteurianum*) (pH 7.5)	−0.42
$H^+/\frac{1}{2}H_2$	−0.42
Acetic acid/acetaldehyde	−0.60

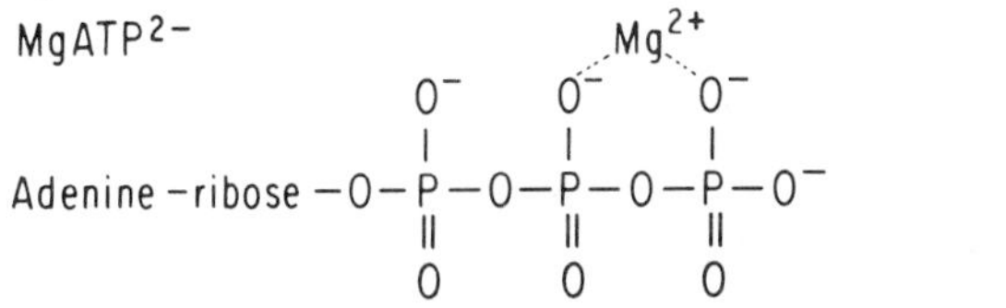

Figure 4-2. Magnesium complex of ATP.

librium constant than those lower in the scale. This scale is thus a quantitative measure of the affinity of the compound for its phosphoryl group. Those high in the scale tend to lose their phosphate groups, and those lower in the scale tend to hold on to their phosphate groups. ATP is unique because its free energy of hydrolysis occupies the midpoint of this thermodynamic scale of phosphorylated compounds. The direction of enzymatic phosphate group transfer is specified by this thermodynamic scale. Phosphate groups are transferred only from compounds of high potential to acceptors of low potential, i.e., down the scale.

The ATP–ADP system functions as an intermediate carrier of phosphate groups. ADP serves as a specific acceptor of phosphate groups from cellular phosphate compounds of very high potential formed during the oxidation of substrate by the cell:

$$\text{Phosphoenolpyruvate} + \text{ADP} \rightleftharpoons \text{pyruvate} + \text{ATP}$$

The ATP so formed then donates its terminal phosphate group enzymatically to phosphate acceptor molecules, such as glucose, transforming them to phosphate derivatives with a higher energy content:

$$\text{ATP} + \text{D-glucose} \rightleftharpoons \text{ADP} + \text{D-glucose 6-phosphate}$$

Generation of ATP. Bacteria utilize two fundamentally different classes of reactions to make energy available. One class consists of those reactions that generate ATP and other energy-rich compounds by substrate-level phosphorylation. Included in this group are ATP-yielding reactions of the glycolytic pathway, arginine fermentation, and a number of bizarre ATP-yielding processes characteristic of clostridia. In these reactions, a part of the energy that is released is initially conserved in energy-rich compounds formed in dehydrogenase (or lyase) reactions and then

transferred to the ATP system by kinase reactions (Fig. 4-1). The second general class of reactions for the synthesis of ATP in bacteria is oxidative phosphorylation or photophosphorylation. In these reactions, during the flow of electrons from the first electron carrier to the final electron acceptor in a catabolic redox sequence, ATP is generated via the mechanism of electronic transport phosphorylation.

Energy-yielding Heterotrophic Metabolism

Fermentation and Respiration

Although all heterotrophic microorganisms ultimately obtain their energy from oxidation-reduction reactions, the amount of energy obtained and the mechanisms by which they extract it vary. There are two basic mechanisms employed, fermentation and respiration.

In fermentation processes, electrons are passed from the electron donor, an intermediate formed in the breakdown of the substrate molecule, to an electron acceptor, which is some other organic intermediate in the fermentation process. Fermentation results in the accumulation of a mixture of end products, some more oxidized and some more reduced than the substrate. The average oxidation level of the end products in a fermentation, however, is always identical to that of the initial substrate. Fermentations are carried out by both obligate and facultative anaerobes (p. 68).

Respiration is a process in which molecular oxygen usually serves as the ultimate electron acceptor. When oxygen is the ultimate acceptor the process is referred to as "aerobic respiration" to distinguish it from anaerobic respiration, in which an inorganic compound, such as nitrate, sulfate, or carbonate, is used.

Fermentation is a less efficient mechanism than respiration for extracting energy from the substrate molecule. When organisms ferment glucose, only a small amount of the energy potentially available in the glucose molecule is released. Most of the energy is still locked up in the product of the reaction, e.g., lactate. When organisms oxidize glucose completely to CO_2 and H_2O, all of the available energy of the glucose molecule is released:

$$\text{Glucose} \rightarrow 2 \text{ lactate} \qquad \Delta G^{\circ\prime} = -47.0 \text{ kcal mole}^{-1}$$

$$\text{Glucose} + 6 \text{ O}_2 \rightarrow 6 \text{ CO}_2 + 6 \text{ H}_2\text{O} \qquad \Delta G^{\circ\prime} = -686.0 \text{ kcal mole}^{-1}$$

Among the microorganisms that carry out aerobic respiration are the obligate aerobes and facultative anaerobes. In addition, some of the facultative anaerobes can also employ nitrate as their terminal electron acceptor. The organisms, however, that use sulfate or carbonate as electron acceptors in anaerobic respiration are obligate anaerobes. A more complete discussion of aerobes and anaerobes is found on page 68.

TABLE 4-4. STANDARD FREE ENERGY OF HYDROLYSIS OF PHOSPHORYLATED COMPOUNDS

	$\Delta G^{\circ\prime}$ kcal	Direction of Phosphate Group Transfer
Phosphoenolpyruvate	−14.8	
1,3-Diphosphoglycerate	−11.8	
Acetylphosphate	−10.1	
ATP	−7.3	
Glucose 1-phosphate	−5.0	
Fructose 6-phosphate	−3.8	
Glucose 6-phosphate	−3.3	

Dissimilation of Glucose. Glucose occupies an important position in the metabolism of most biologic forms, and its dissimilation provides a metabolic pathway common to most forms of life. The ability to utilize a sugar or related compound of a configuration different from glucose is the result of the organism's ability to convert the substrate to intermediates common to the pathways for glucose fermentation.

Entry into Cell. The utilization of a specific monosaccharide is also dependent upon the presence in the organism of specific carrier systems for the transport of the sugar across the cell membrane. A diversity of systems of this type occurs. Some of these utilize ATP generated by electron transport. In *E. coli*, the phosphotransferase system derives its energy directly from phosphoenolpyruvate (PEP) rather than from ATP, and phosphorylation of the sugar occurs during transport. Transport systems present in bacteria are discussed in Chapter 5 on page 70.

Glycolytic Pathway

The major route of glucose catabolism in most cells is glycolysis, in which the glucose molecule is degraded to two molecules of lactic acid without the intervention of molecular oxygen. The basic concepts of glycolysis are incorporated in the 11 enzymatic reactions of the Embden–Meyerhof–Parnas (EMP) scheme, shown in Figure 4-3. Although the basic pathway is the same for all cell types, the properties of certain of the enzymes in the pathway are not uniform in all species or cell types. Such variations apparently are introduced for purposes of cellular differentiation and control of specific steps in the pathway.

Glycolysis consists basically of two major phases. In the first, glucose is phosphorylated either by ATP or PEP,

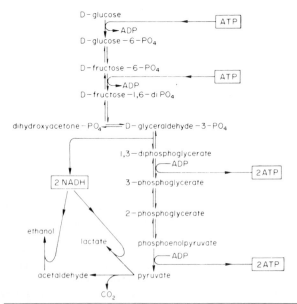

Figure 4-3. The Embden–Meyerhof–Parnas glycolytic scheme.

depending on the organism, and cleaved to form glyceraldehyde 3-PO_4. In the second phase this 3-carbon intermediate is converted to lactic acid in a series of oxidoreduction reactions that are coupled to the phosphorylation of ADP. A mechanism is thus provided for the conservation of the energy originally present in the glucose molecule.

Phase I of Glycolysis

The conversion of glucose to glyceraldehyde 3-PO_4 proceeds as follows:

(1) glucose + ATP → glucose 6-PO_4 + ADP

(2) glucose 6-PO_4 $\rightleftharpoons$ fructose 6-PO_4

(3) fructose 6-PO_4 + ATP $\rightleftharpoons$ fructose 1,6-diPO_4 + ADP

(4) fructose 1,6-diPO_4 $\rightleftharpoons$ glyceraldehyde 3-PO_4 + dihydroxyacetone PO_4

Reaction 3, the phosphorylation of D-fructose to fructose 1,6-diphosphate, occupies a very strategic position in the glycolytic pathway. Alternate pathways of hexose metabolism diverge from the other hexose phosphates in the earlier part of the pathway. This reaction may be regarded as the first one characteristic of the glycolytic sequence proper and thus constitutes a very important branch and control point, subject to strong metabolic regulation. Phosphofructokinase, the enzyme catalyzing this pathway, is an allosteric enzyme responding to fluctuations in adenine nucleotide levels (p. 61). Control at this point ensures that when an abundant supply of ATP is available, as occurs when lactate and pyruvate are oxidized to CO_2 via the citric acid cycle, glycolysis will be essentially blocked and glucose synthesis will be favored. The reverse is also true. When glycolysis is absolutely required for energy generation, glycolysis will be favored and carbohydrate synthesis turned off.

Reaction 4, the cleavage of fructose 1,6-diphosphate to glyceraldehyde 3-PO_4 and dihydroxyacetone PO_4, is catalyzed by aldolase. Different types of aldolases are produced by different cell types. In bacteria, fungi, and blue-green algae, the aldolases are of class II and differ from the animal class I enzyme in a number of their properties. The products of this reaction are interconvertible by an enzyme, triose phosphate isomerase, that directs most of the dihydroxyacetone phosphate into the central stream of glycolysis. Whereas most of this compound is metabolized via glyceraldehyde 3-phosphate, it has an alternate fate that is essential to lipid metabolism, the formation of glycerol phosphate.

Phase II of Glycolysis

During the second stage of glycolysis, the two molecules of glyceraldehyde 3-PO_4 formed from one molecule of glucose are oxidized in a two-step reaction that leads to the synthesis of ATP.

(5) glyceraldehyde 3-PO$_4$ + NAD$^+$ + P$_i$ $\rightleftharpoons$
1,3-diphosphoglycerate + NADH + H$^+$

(6) 1,3-diphosphoglycerate + ADP $\rightleftharpoons$ 3-phospho-
glycerate + ATP

In the first of these reactions, the aldehyde group of glyceraldehyde 3-phosphate is oxidized to the oxidation level of a carboxyl group. The other important component of the reaction is the oxidizing agent nicotinamide adenine dinucleotide (NAD) that accepts electrons from the aldehyde group of glyceraldehyde 3-PO$_4$. The electrons are then carried to pyruvate that is formed later in the glycolytic pathway.

In the second reaction, the 1,3-diphosphoglycerate that was formed in reaction 5 transfers a phosphate group to ADP, with the resultant formation of 3-phosphoglycerate. As a result of these two reactions the energy derived from the oxidation of an aldehyde group has been conserved as the phosphate bond energy of ATP.

These two reactions are a prototype example of substrate-level oxidative phosphorylation. In these reactions the phosphorylation of ADP is coupled to the NAD-linked oxidation of 3-phosphoglyceraldehyde, as shown in Figure 4-4. In this type of coupling, hydrogen is transferred from an initial donor to a final acceptor via transitional intermediates and via intermediate carrier compounds. The intermediate, 1,3-diphosphoglycerate, is the common covalent intermediate in the above reactions.

The dehydration of 2-phosphoglycerate to phosphoenolpyruvate, as shown in Figure 4-3, is the second reaction of the glycolytic sequence in which a high-energy phosphate bond is generated. The formation of this bond involves an internal rearrangement of a phosphorylated molecule, leading to the conversion of a phosphoryl group of low energy into one of high energy. In the subsequent reaction the phosphate group from phosphoenolpyruvate is transferred to ADP, yielding ATP and pyruvate.

Energy Yield

In the glycolytic pathway, a total of 4 moles of ATP are formed per mole of glucose used. Since 2 moles of ATP are used in the initial steps, the net ATP yield is 2 moles per mole of glucose fermented. The stoichiometry observed in the production of pyruvate from hexoses is:

$$C_6H_{12}O_6 + 2\ NAD^+ + 2\ ADP + 2\ P_i \rightarrow$$

$$2\ CH_3COCOO^- + 2\ NADH + 2\ ATP$$

$$\Delta G^{\circ\prime} = 15\ kcal/mole$$

Only a very small proportion of the total free energy potentially derivable from the breakdown of a hexose molecule is actually made available via this pathway. This is because of the inherent inefficiency of the system and because the reaction products are compounds in which carbon is still at a relatively reduced level. The ultimate fate of the key metabolite, pyruvate, depends upon the means employed for the regeneration of NAD$^+$ from NADH. For

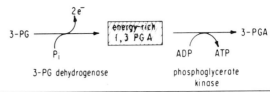

Figure 4-4. Substrate-level phosphorylation during glycolysis. These two consecutive reactions comprise one of the most important sequences of the glycolytic pathway. In these reactions, the energy of oxidation of the aldehyde group in 3-phosphoglyceraldehyde (3-PG) to the carboxylate group in 3-phosphoglycerate (3-PGA) is conserved in the form of ATP. 1,3-Diphosphoglycerate is an energy-rich intermediate in the sequence.

this purpose microorganisms have evolved a variety of pathways (p. 46).

Phosphogluconate Pathway (Pentose Phosphate Pathway)

Whereas the EMP scheme is the major pathway in many microorganisms, as well as in animal and plant tissues, it does not represent the only available pathway for carbohydrate metabolism. The phosphogluconate pathway, also known as the pentose phosphate pathway or the hexosemonophosphate shunt, is a multifunctional pathway that may be used in the fermentation of hexoses, pentoses, and other carbohydrates (Fig. 4-5). For some organisms, such as the heterolactic fermentors, this pathway is their major energy-yielding pathway. For most organisms, however, its principal use is (1) to generate reducing power in the form of NADPH for biosynthetic reaction, (2) to provide pentoses for nucleotide synthesis, and (3) to provide a mechanism for the oxidation of pentoses by the glycolytic sequence. Because of its multifunctional use, in contrast to glycolysis, it cannot be visualized as a consecutive set of reactions leading directly from glucose and always ending in its complete oxidation of six molecules of CO$_2$.

The point of departure of this route from the EMP system is the oxidation of glucose 6-phosphate to 6-phosphogluconate, which is in turn decarboxylated and further oxidized to D-ribulose 5-phosphate. The dehydrogenases catalyzing these reactions, glucose 6-phosphate dehydrogenase and 6-phosphogluconate dehydrogenase, require NADP$^+$ as an electron acceptor. The overall equation for the pathway to this stage is:

$$\text{Glucose 6-phosphate} + 2\ NADP^+ + H_2O \rightleftharpoons$$

$$\text{D-ribose 5-phosphate} + CO_2 + 2\ NADPH + 2\ H^+$$

The D-ribose 5-phosphate produced results from the reversible isomerization of D-ribulose 5-phosphate.

In some organisms or metabolic circumstances, the phosphogluconate pathway proceeds no further. In others, the pool of ribose and ribulose 5-phosphates is converted to xylulose 5-phosphate, which is the starting point for a series of transketolase and transaldolase reactions, leading

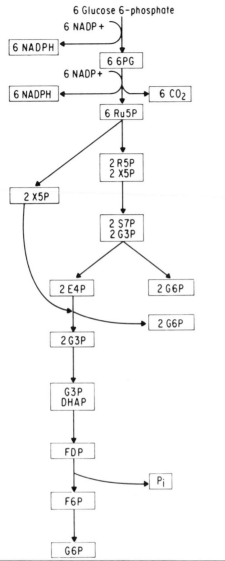

Figure 4-5. Phosphogluconate pathway. Complete oxidation of glucose 6-phosphate to CO_2.

ultimately to the initial compound of the pathway, glucose 6-phosphate. By a complex sequence of reactions in which six molecules of glucose 6-phosphate are oxidized to six molecules each of ribulose 5-phosphate and CO_2, the phosphogluconate pathway can also bring about the complete oxidation of glucose 6-phosphate to CO_2:

$$\text{Glucose 6-phosphate} + 12\ NADP^+ + 7\ H_2O \rightarrow$$

$$6\ CO_2 + 12\ NADPH + 12\ H^+ + P_i$$

The complete oxidative pathway is operative only when the demand for NADPH is high, such as occurs in organisms actively engaged in lipid biosynthesis.

Heterolactic fermenting microorganisms utilize this pathway instead of glycolysis for the fermentation of glucose and the pentoses. These organisms lack the glycolytic enzymes, phosphofructokinase, aldolase, and triose phosphate isomerase, but possess the key enzyme, phosphoketolase, that cleaves xylulose 5-phosphate to acetylphosphate and glyceraldehyde 3-phosphate. The use of this pathway provides an explanation for the source of ethanol in these organisms (Fig. 4-6). In addition to the heterolactic fermenters, other organisms also using this pathway include *Brucella abortus* and species of *Acetobacter*.

When glucose is fermented through the phosphogluconate pathway, the net yield of ATP is half of that characteristic of the EMP pathway. This lower energy yield is characteristic of a pathway of dehydrogenation before cleavage.

Entner–Doudoroff Pathway

Although this pathway has been established as a functional system only for the genus *Pseudomonas*, enzymes characteristic of the pathway have been detected in other organisms. The pathway diverges at 6-phosphogluconate from the phosphogluconate pathway. In this sequence, 6-phosphogluconate is dehydrated and then cleaved to yield one molecule of glyceraldehyde 3-PO_4 and one molecule of pyruvate, from which ethanol and CO_2 are formed via the same series of reactions as in alcoholic fermentation by yeast (Fig. 4-6). Like the phosphogluconate pathway, only one molecule of ATP is produced per molecule of glucose fermented.

Fate of Pyruvate Under Anaerobic Conditions

The fermentation of glucose is always initiated by a phosphorylation at the expense of ATP, to yield glucose 6-PO_4. The pyruvic acid to which glucose 6-PO_4 is converted is a key intermediate in the fermentative metabolism of all carbohydrates. In its formation, NAD is reduced and must be reoxidized in order to achieve a final oxidation-reduction balance. This reoxidation characteristically occurs in the terminal step reactions and is accompanied by the reduction of a product derived from pyruvic acid.

Bacteria differ markedly from animal tissues in the manner in which they dispose of pyruvic acid. In mammalian physiology, the main course of respiration is such that substrates are oxidized to CO_2 and H_2O, oxygen being the ultimate hydrogen acceptor. Among the bacteria, however, incomplete oxidation is the rule rather than the exception, and the products of fermentation may accumulate to an extraordinary degree. The final product in certain organisms is either alcohol or lactic acid. In others, the pyruvic acid is further metabolized to such products as butyric acid, butyl alcohol, acetone, and propionic acid. Bacterial fermentations are of practical importance because they provide products of industrial value and are useful in the identification of bacterial species (Fig. 4-7).

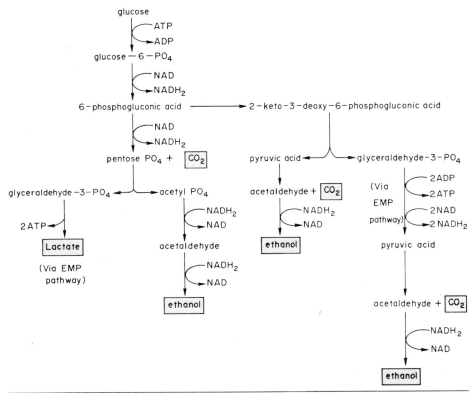

Figure 4-6. Alternate pathways of glucose fermentation in certain bacteria. The pentose phosphate pathway used in heterolactic fermentation is shown on the left. A variation of this pathway, the Entner–Doudoroff pathway, on the right, results in the alcoholic fermentation of glucose.

Alcoholic Fermentation

The oldest known type of fermentation is the production of ethanol from glucose. In yeasts that carry out an almost pure alcoholic fermentation, the alcohol arises from the decarboxylation of pyruvic acid by pyruvate decarboxylase, the key enzyme of alcoholic fermentation. The free acetaldehyde formed is then reduced to ethanol by alcohol dehydrogenase, and the NADH is reoxidized. Although a number of bacteria produce alcohol, it is produced via other pathways (Fig. 4-6).

Homolactic Fermentation

All members of the genera *Streptococcus* and *Pediococcus* and many species of *Lactobacillus* ferment glucose predominantly to lactic acid with no more than a trace accumulation of other products. In the dissimilation of glucose by the homofermenters, pyruvate is reduced to lactic acid by the enzyme lactic dehydrogenase, with NADH acting as the hydrogen donor. The homofermentative mechanism owes its characteristically high yields of lactic acid to the action of aldolase, which cleaves the hexose diphosphate into two

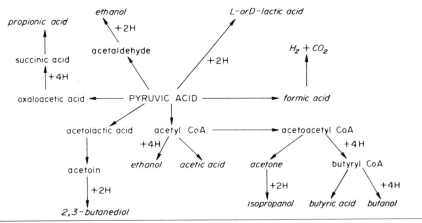

Figure 4-7. Fate of pyruvate in major fermentations by microorganisms.

equal parts, both of which form pyruvate and, hence, lactate. The same fermentation occurs in animal muscle (Fig. 4-3).

Heterolactic Fermentation

In addition to the production of lactic acid, some of the lactic acid bacteria (*Leuconostoc* and certain *Lactobacillus* species) produce a mixed fermentation in which only about one-half of the glucose is converted to lactic acid, the remainder appearing as CO_2, alcohol, formic acid, or acetic acid. The heterolactic fermentation differs fundamentally from the homolactic type in that the pentose phosphate pathway rather than the EMP scheme is employed. The release of carbon 1 of glucose as CO_2 is characteristic of glucose fermentations by all heterolactic organisms. Also of significance is the finding that the energy yield as measured by growth is one-third lower per mole of glucose fermented than observed for homolactic organisms.

Propionic Acid Fermentation

Propionate is a major end product of fermentations carried out by a variety of anaerobic bacteria (p. 681). This fermentation pattern is characteristic of the genus *Propionibacterium*, anaerobic gram-positive nonsporeforming rods closely related to the lactobacilli. The propionic acid that they produce from glucose or from lactic acid contributes to the characteristic taste and smell of Swiss cheese. The ability of these organisms to ferment lactic acid, an end product of other fermentations, is significant in that it allows the organisms to net an additional ATP.

In the fermentation of hexose, the early stages of the glycolytic pathway are employed. Part of the pyruvate, derived either from hexose or from lactic acid is further oxidized to CO_2 and acetyl-CoA, thus accounting for the presence of acetate and CO_2 among the reaction products. The energy-rich bond of acetyl-CoA is used for the synthesis of ATP.

$$3 \ CH_3CHOHCOOH$$
$$\downarrow -6H$$
$$3 \ CH_3COCOOH$$
$$ADP + P_i \ \rceil$$
$$\qquad +6H$$
$$ATP \ \lfloor$$
$$\downarrow$$
$$2 \ CH_3CH_2COOH + CH_3COOH + CO_2$$

The conversion of pyruvate or lactic acid to propionic acid is a complex cyclic process consisting of the following series of reactions:

(1) pyruvate + methylmalonyl-CoA $\xrightleftharpoons{\text{biotinyl enzyme}}$ oxaloacetate + propionyl-CoA

(2) oxaloacetate + 4 H → succinate

(3) succinate + propionyl-CoA $\rightleftharpoons$ succinyl-CoA + propionate

(4) succinyl-CoA $\rightleftharpoons$ methylmalonyl-CoA

The first step in the sequence is a unique CO_2 transfer reaction, catalyzed by a biotin-containing enzyme that acts as both CO_2 acceptor and donor. Propionyl-CoA, one of the products of this reaction, gives rise to propionate.

Mixed Acid Fermentation

This type of fermentation is characteristic of most of the Enterobacteriaceae. Organisms within the genera *Escherichia*, *Salmonella*, and *Shigella* ferment sugars via pyruvate to lactic, acetic, succinic, and formic acids. In addition, CO_2, H_2, and ethanol are produced. The nature and quantitative relationships of these products vary with the organism. All of the enterobacteria produce formate, which either accumulates or, under acid conditions, is converted by formic hydrogenlyase to molecular hydrogen and carbon dioxide. The formic acid produced in this fermentation is derived from pyruvate in a cleavage involving coenzyme A to yield acetyl-CoA and formate. The acetyl-CoA is rapidly converted to acetyl PO_4. The combined reaction whereby formate and acetate are produced from pyruvate is known as the phosphoroclastic split.

$$Pyruvate + CoA \rightleftharpoons acetyl\text{-}CoA + formate$$

$$Acetyl\text{-}CoA + P_i \rightleftharpoons CoA + acetyl \ PO_4$$

The conversion of formate to CO_2 and H_2 is catalyzed by formic hydrogenlyase, an inducible enzyme complex whose formation is inhibited by aerobiosis. Fermentations by *E. coli* and most *Salmonella* are characterized by CO_2 and H_2 production, but in *Shigella* and *Salmonella typhi*, no CO_2 and H_2 are produced and an equivalent amount of formic acid accumulates. The inability of *S. typhi* and *Shigella* to cleave formate is useful in the diagnostic bacteriology laboratory:

$$HCOOH \rightarrow H_2 + CO_2$$

The overall fermentation of glucose by *E. coli* is as follows:

$$2 \ glucose + H_2O \rightarrow 2 \ lactate + acetate$$
$$+ \ ethanol + 2 \ CO_2 + 2 \ H_2$$

The ethanol formed by *E. coli* comes from acetyl-CoA via acetaldehyde and its subsequent reduction. The lactic acid is produced via the EMP scheme.

Butanediol Fermentation

Several groups of organisms, including *Enterobacter*, *Bacillus*, and *Serratia*, produce 2,3-butanediol in fermentations that are otherwise of the mixed acid type. Two molecules of pyruvate, the precursor of acetoin (acetylmethylcarbinol), are decarboxylated in the formation of one molecule of the neutral acetoin, which is then reduced to 2,3-butanediol:

$$2 \text{ CH}_3\text{COCOOH} \rightarrow \text{CH}_3\text{-COHCOOH} + \text{CO}_2$$

pyruvic acid · · · · · · · · · · · · · · · COCH₃

acetolactic acid

$$\text{CH}_3\text{CHOHCHOHCH}_3 \xleftarrow{(2\text{H})} \text{CH}_3\text{CHOHCOCH}_3 + \text{CO}_2$$

2,3-butanediol · · · · · · · · · · acetoin

This reduction is slowly reversible in air and, when made strongly alkaline, is the basis for the Voges-Proskauer reaction, a test for acetoin.

The diversion of part of the pyruvate to 2,3-butanediol greatly reduces the amount of acid produced relative to the mixed acid fermentation and is responsible for the positive methyl red reaction often used in the differentiation of *Escherichia* and *Enterobacter*.

Butyric Acid Fermentation

Among the primary characteristic products of carbohydrate fermentation by many organisms in the genus *Clostridium* are butyric acid, acetic acid, CO_2, and H_2. The clostridia employ phosphotransferase systems for sugar uptake and the EMP pathway for degradation of hexose phosphates to pyruvate. The conversion of pyruvate to acetyl-CoA is catalyzed by the pyruvate-ferredoxin oxidoreductase system. In this reaction sequence, the two hydrogens are not transferred to NAD as in the pyruvate dehydrogenase reaction (p. 50) but are used to reduce ferredoxin. Ferredoxin has a very low redox potential, which at pH 7.0 is about the same as that of the hydrogen electrode. Therefore, when clostridia ferment carbohydrates, reduced ferredoxin can transfer electrons to hydrogenase and hydrogen can be evolved.

The key reaction in the butyric acid fermentation is the formation of acetoacetyl-CoA by the condensation of two molecules of acetyl-CoA derived from acetate or from pyruvate:

$$2 \text{ CH}_3\text{CO-SCoA} \rightleftharpoons \text{CH}_3\text{COCH}_2\text{CO-SCoA} + \text{HSCoA}$$

This C_4 compound is the key to all of the C_4-forming reactions of the clostridia. Its subsequent reduction and conversion to butyric acid permit ATP formation. In some organisms, the primary acidic products of the fermentation are reduced, resulting in the accumulation of neutral end products: butanol, acetone, isopropanol, and ethanol. The end products of clostridial fermentations can thus be very numerous and vary with the species.

Generally, only obligate anaerobes form butyrate as a primary fermentation product. In addition to several species of *Clostridium*, members of the genera *Fusobacterium*, *Butyrivibrio*, and *Eubacterium* produce butyric acid.

Fermentation of Nitrogenous Organic Compounds

Some anaerobes are not primarily butyric acid producers. Amino acids, formed from proteins by extracellular proteases, and purine and pyrimidine bases are fermented by a variety of organisms. Single amino acids can serve as major energy sources for selected species of anaerobic bacteria. For proteolytic clostridia, however, such as *Clostridium sporogenes*, *Clostridium difficile*, and *Clostridium botulinum* types A and B, the most characteristic type of amino acid fermentation is the Stickland reaction, a coupled oxidation-reduction involving a pair of amino acids, one of which serves as the electron donor and the other as the electron acceptor. An example of this type of reaction is the fermentation of alanine and glycine:

$$\text{Alanine} + 2 \text{ glycine} + 2 \text{ H}_2\text{O} \rightarrow$$

$$3 \text{ acetic acid} + 3 \text{ NH}_3 + \text{CO}_2$$

The degradation of amino acids in a protein hydrolysate is very complex. The process always involves oxidation and reduction reactions between one or more amino acids or nonnitrogenous compounds derived from amino acids. The oxidation reactions are usually similar to corresponding reactions in aerobic organisms, i.e., oxidative deaminations, transaminations, and α-keto acid oxidations. The reduction reactions are more distinctive and utilize such electron acceptors as amino acids, α- and β-keto acids, α and β unsaturated acids, or their coenzyme A thiolesters, and protons. The ultimate reduction products include a variety of short-chain fatty acids, succinic acid, δ-aminovaleric acid, and molecular hydrogen.

Aerobic Respiration

In aerobic cells, energy is obtained from the complete oxidation of the substrate, with molecular oxygen usually serving as the ultimate hydrogen acceptor. In respiration the large amount of energy set free in the formation of water is made available to the process. The pathways of aerobic dissimilation are exceedingly complex. They consist of many enzymes and a large number of biochemical reactions. The most important respiratory mechanism for terminal oxidation is the tricarboxylic acid cycle of Krebs, which, together with the known reactions of glycolysis, can account for the complete oxidation of glucose. This cycle is unique in that it provides the cell not only with an energy source but also with carbon skeletons for the synthesis of cellular material.

Tricarboxylic Acid Cycle

Oxidative Decarboxylation of Pyruvate. In aerobic cells, the pyruvate formed from the glycolytic pathway is enzymatically oxidized by the pyruvate dehydrogenase complex to a 2-carbon compound, acetyl-CoA:

$$\text{CH}_3\text{COCOOH} + \text{NAD} + \text{CoA-SH} \rightarrow$$

$$\text{CH}_3\text{CO-S-CoA} + \text{CO}_2 + \text{NADH}$$

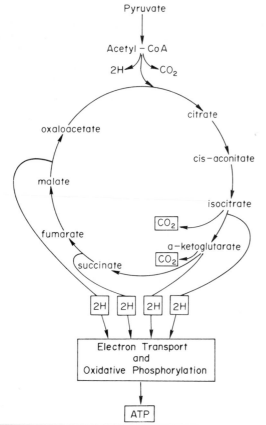

Figure 4-8. The tricarboxylic acid cycle. The four pairs of H atoms liberated are fed into the respiratory chain.

The electrons accepted from pyruvate by NAD are carried in the form of NADH to the respiratory chain.

Oxidation of Acetyl-CoA. The TCA cycle carries out the oxidation of the acetyl moiety of acetyl-CoA to CO_2 with transfer of the reducing equivalents to NAD, NADP, and FAD. Acetyl-CoA enters the cycle via the citrate synthase reaction, in which oxaloacetate and acetyl-CoA are condensed to form citric acid (Fig. 4-8). In one turn of the cycle this 6-carbon molecule is then decarboxylated and oxidized to regenerate the 4-carbon oxaloacetate and liberate two carbon atoms as CO_2. In so doing, four pairs of electrons are enzymatically extracted from the intermediates of the cycle. Anything capable of generating acetyl-CoA can be oxidized via the cycle. Important synthetic mechanisms utilize reactants of the cycle to provide a common meeting ground for carbohydrate, lipid, and protein metabolism.

Anaplerotic Reactions

Under normal conditions, the reactions by which the TCA cycle intermediates are formed and drained away remain in balance. This is made possible by enzymatic mechanisms for replenishing the TCA cycle intermediates as they are diverted to biosynthetic pathways.

Phosphoenolpyruvate (PEP)-Carboxylase Reaction. The most important of these anaplerotic (filling-up) reactions is the enzymatic carboxylation of pyruvate to oxaloacetate.

$$\text{Phosphoenolpyruvate} + CO_2 + P_i \rightarrow \text{oxaloacetate} + P_i$$

The reaction is catalyzed by pyruvate carboxylase, an allosteric enzyme containing biotin. The rate of the forward reaction is very low unless acetyl-CoA, the fuel of the TCA cycle, is present in excess.

Glyoxylate Cycle. When microorganisms are grown on fatty acids or acetate as a sole carbon source, acetyl-CoA is formed without the intermediate formation of pyruvic acid. Under such circumstances there is no mechanism for the generation of oxaloacetate from pyruvate by the PEP carboxylase reaction. Growth on acetate, however, induces the synthesis of two enzymes, isocitrate lyase and malate synthase, which together with some of the enzymes of the TCA cycle carry out a modification of the TCA cycle, the glyoxylate cycle. The glyoxylate cycle bypasses the CO_2-evolving steps of the TCA cycle. The succinate formed can be converted by reactions of the TCA cycle to oxaloacetate, which may then condense with acetyl-CoA to start another turn of the cycle, or alternately the oxaloacetate may be used for biosynthetic purposes (Fig. 4-9).

Electron Transport

Components of System

In procaryotic organisms the redox carriers and enzymes involved in electron transport are located in the plasma membrane. Although the same basic types of carriers are present as occur in the mitochondria of higher organisms,

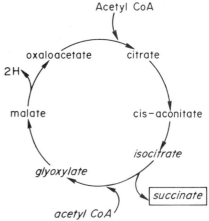

Figure 4-9. The glyoxylate cycle. This cycle provides both energy and 4-carbon intermediates for biosynthetic purposes. In each turn of the cycle, two molecules of acetyl-CoA enter, and one molecule of succinate is formed. The reactions in the pathway between isocitrate and malate are catalyzed by auxiliary enzymes. All of the others are reactions of the TCA cycle.

considerable diversity also exists, especially in the number and properties of the cytochrome components.

The major components that participate in the transport of electrons from an organic substrate to oxygen are pyridine and flavin-linked dehydrogenases, iron-sulfur proteins, quinones, and cytochromes.

Pyridine-linked Dehydrogenases. Nicotinamide adenine dinucleotide (NAD^+) is the coenzyme most frequently employed as an acceptor of electrons from the substrate (Fig. 4-10). It functions in five of the six oxidative steps in the oxidation of glucose. NADP-linked dehydrogenases serve primarily to transfer electrons from intermediates of catabolism to intermediates of biosynthesis. The pyridine nucleotides are relatively loosely bound to the enzyme protein by noncovalent bonds. They are, therefore, not fixed prosthetic groups but substrates that can serve as dissociable carriers of electrons.

Flavin-linked Dehydrogenases. These enzymes contain the vitamin riboflavin, either as flavin mononucleotide (FMN) or flavin adenine dinucleotide (FAD) (Fig. 4-11). Unlike the pyridine-linked dehydrogenases, in the flavin-linked dehydrogenases, the flavin nucleotide is tightly bound. Among the most important of the flavoproteins is NADH dehydrogenase, which catalyzes the transfer of electrons from NADH to the next component of the electron transport chain. Some of the flavoproteins, such as succinic dehydrogenase, are active in primary dehydrogenations. Another class of flavin-linked enzymes, the flavin oxidases, are reoxidized by molecular oxygen to yield hydrogen peroxide (p. 55). Among the numerous enzymes in this group are D-amino acid oxidase and xanthine oxidase. In addition to flavin nucleotide, some of the flavoproteins contain metals (iron and molybdenum) that are essential for catalytic activity.

Iron–Sulfur Proteins. These proteins contain iron and an equimolar amount of acid-labile sulfur. They apparently function as electron carriers by undergoing reversible Fe(II)-Fe(III) transitions. A number of different iron–sulfur proteins have been described from a variety of sources. These vary in the number of iron–sulfur centers per molecule and oxidation-reduction potential. Among the best studied are the ferredoxins from the anaerobic nitrogen-fixing *Clostridrium pasteurianum* and from the photosynthetic bacterium *Chromatium*. Their precise role in electron transport is unknown.

Quinones (Coenzyme Q). Also participating in electron transport is the lipid-soluble ubiquinone, which functions with its hydroquinone as a redox couple (Fig. 4-12). It is not bound to specific proteins but is present as a small pool in the liquid phase of the membrane, where it serves as an electron acceptor for one group of enzymes and an electron donor to the next component of the chain. As a mobile liquid-soluble substrate, it is available to enzymes more rigidly locked into position in the membrane.

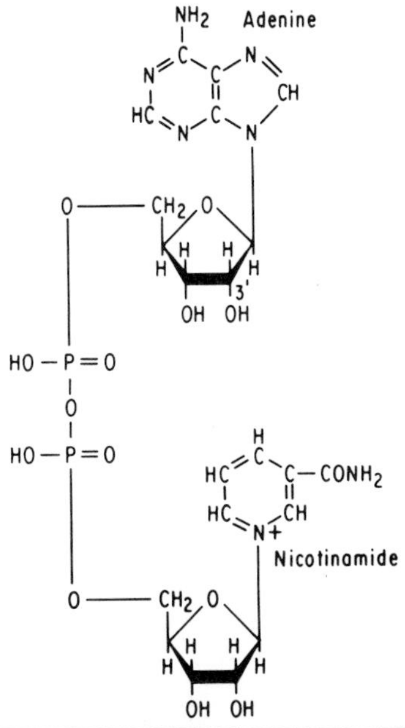

Figure 4-10. Nicotinamide adenine dinucleotide (NAD). In nicotinamide adenine dinucleotide phosphate (NADP) the 3′ hydroxyl group is esterified with phosphate.

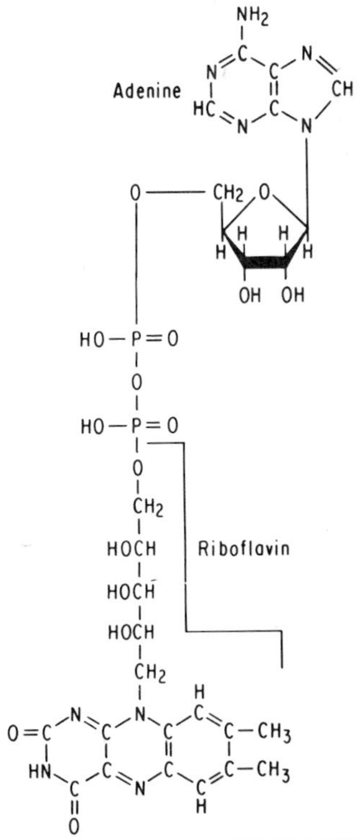

Figure 4-11. Flavin adenine dinucleotide (FAD).

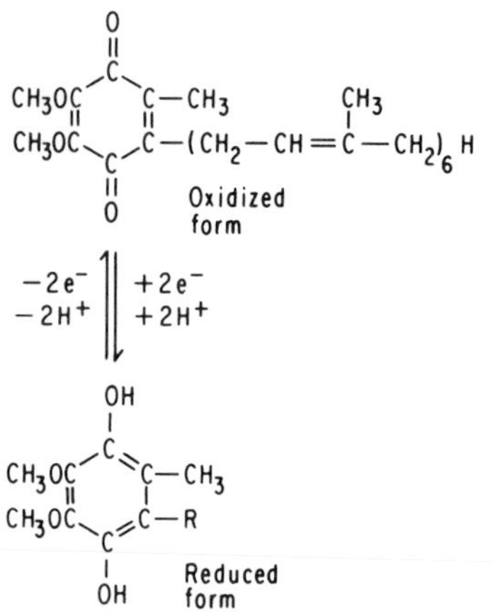

Figure 4-12. Ubiquinone (coenzyme Q), a carrier of electrons in gram-negative microorganisms. In most gram-positive species it is replaced by menaquinone.

Although ubiquinone is ubiquitous in its occurrence and is the major quinone of the electron transport system of mitochondria, additional quinones play a role in bacterial electron transport. In most gram-positive organisms, menaquinone replaces ubiquinone, and in many of the Enterobacteriaceae, both quinones are present. In *Mycobacterium phlei* the naphthoquinone vitamin K_9 occurs and has been shown to act between the flavoprotein and cytochrome *b*.

Cytochromes. The cytochromes are iron–porphyrin components of the electron transport chain. During their catalytic cycle, they undergo reversible Fe(II)-Fe(III) valence changes and act sequentially to carry electrons toward molecular oxygen. Although basically similar to the mammalian cytochromes, the bacterial cytochromes are more diverse and possess properties not encountered in mammalian systems. Four broad classes of bacterial cytochromes have been identified on the basis of their characteristic absorption spectra. Among the various species, there is considerable diversity in the classes of cytochromes present, as well as in their structure, functions, and conditions for existence.

Some bacteria have more than one autooxidizable cytochrome of the *a* class. Many contain cytochrome *o*, a widely distributed heme protein resembling cytochrome *c* that appears to serve as a terminal oxidase. The conditions under which a bacterium is grown markedly affect both the total and relative amounts of the cytochrome components in the organism. Oxygen deprivation tends to cause the replacement of cytochrome oxidase aa_3 by *o* in *Paracoccus denitrificans* and the enhanced synthesis of cytochrome oxidase *d* relative to *o* (*E. coli, Haemophilus parainfluenzae*). Such changes reflect attempts by the organism to compensate for oxygen deficiency by the increased synthesis of higher concentrations of alternate oxidases that have an increased affinity for oxygen or that exhibit higher turnover numbers. Whereas mammalian cytochromes function primarily as members of a respiratory electron transport chain, bacterial cytochromes also transport electrons to nonoxygen acceptors (Fig. 4-13). In *E. coli*, for example, cytochromes function as part of the nitrate reductase system (p. 56). They also play a role in photosynthesis (p. 58).

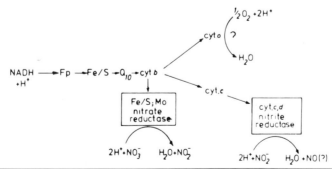

Figure 4-13. Electron transport chains in *P. denitrificans*, illustrating the linear sequence of redox carriers associated with aerobically grown cells as compared with anaerobically grown cells in the presence of NO_3^-. *(From Haddock and Jones: Bacteriol Rev 41:47, 1977.)*

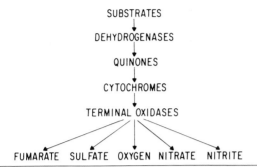

SUBSTRATES

↓

DEHYDROGENASES

↓

QUINONES

↓

CYTOCHROMES

↓

TERMINAL OXIDASES

FUMARATE SULFATE OXYGEN NITRATE NITRITE

Figure 4-14. Generalized scheme of electron transport systems in bacteria.

Electron Transport Chain

In each revolution of the TCA cycle, there are four dehydrogenations. In three of these, NAD serves as the electron acceptor, and in the fourth, the electron acceptor is FAD. The reoxidation of these reduced coenzymes is accomplished by passing of the electrons through a series of intermediate carriers, capable of undergoing freely reversible oxidation and reduction. The last carrier in the series reacts with oxygen in a reaction mediated by a terminal oxidase (Fig. 4-14). The series of carriers that link the dehydrogenation of an oxidizable substrate with the reduction of molecular oxygen to water is termed the "electron transport chain." The carriers participate in a series of reactions of gradually increasing E'_o values. Thus, electrons will tend to pass from the more negative carrier NAD in Figure 4-15 to the more positive carrier above it on the

scale. A decline in free energy is associated with each electron transfer and is directly related to the magnitude of the drop in electron pressure. The decline in free energy during the passage of the pair of electrons from NADH to molecular oxygen is large enough to make possible the synthesis of ATP from ADP and P_i.

The typical electron transport sequence of:

$$\text{Flavoprotein} \rightarrow \text{cyt } b \rightarrow \text{cyt } c \rightarrow \text{cyt } a \rightarrow O_2$$

found in mammals is also found in bacteria. However, the structure of the electron transport chain in bacteria appears to be more complex, in that a number of transport chain systems may exist. The flavoprotein dehydrogenases may be inputs to a number of the chain systems, and the chains may be more branched in structure than is found in mammalian systems. A bacterial respiratory chain should be conceptualized as a three-dimensional model, with each member having possibly more than one input and output, rather than as a two-dimensional linear and almost unbranched chain, as is often done in representations of the mammalian respiratory chain (Fig. 4-16).

Oxidative Phosphorylation

The movement of electrons down the respiratory chain of carriers is coupled with the production of energy-rich phosphate bonds as a result of oxidative phosphorylation. The multiplicity of catalysts in the chain provides a device for bleeding off the energy in convenient packets. Oxidative phosphorylation is, thus, a process whereby the large amount of free energy liberated during the complete oxidation of metabolites via the citric acid cycle can be utilized to drive the synthesis of ATP.

The search for the mechanism by which respiratory chain or photosynthetic oxidoreductions are coupled to ATP synthesis has been a very challenging one, and numerous hypotheses have been advanced. At present, the most widely accepted of these is the chemiosmotic hypothesis, which is based on the assumption that a proton-motive force consisting of a pH gradient (ΔpH) and an electrical potential difference ($\Delta\psi$) can be generated by the redox reactions of electron transport and that this proton-motive force drives the synthesis of ATP. The essence of this hypothesis, as shown in Figure 4-17, emphasizes the role of the membrane as the site in which electron transport and phosphorylation are coupled and which is impermeable to protons and hydroxyl ions.

According to the chemiosmotic mechanism, the electron carriers are so arranged in the membrane that electron transfer from centers of low to high redox potential is coupled obligatorily with the transport of protons across the membrane from the inside to the outside of the cell (Fig. 4-18). As a consequence of this electrogenic proton transport, protons are delivered to the outside at high electrochemical potential and return to their starting side by traveling through a proton channel in the membrane leading to an ATPase. The ATPase system couples the hydrolysis of ATP with the translocation of protons through

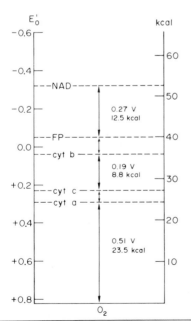

Figure 4-15. Release of free energy as a pair of electrons passes down the respiratory chain to oxygen. Sufficient energy is generated in three of the segments for the formation of a molecule of ATP: between NAD and FP, between cyt b and cyt c, and between cyt a and O_2.

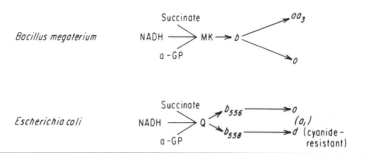

Figure 4-16. Bacterial respiratory chains with terminal pathways that are branched either before or at the cytochrome oxidase level. *(From Jones: In Haddock and Hamilton (eds): Microbial Energetics, 1977. London, Cambridge University Press.)*

the channel away from the ATPase. Since this reaction is reversible, the passage of protons toward the ATPase is coupled with the synthesis of ATP, and the ATPase functions as an ATP synthase.

Microbial ATPases may be visualized in photomicrographs of membranes as knob-shaped structures that stud the inner surface of the membrane and project into the cytoplasm. The molecular structure and composition of the bacterial ATPase complex appear to be similar to that of mitochondria and chloroplasts. Two distinct regions comprise the enzyme complex: the headpiece F_1 and the membrane portion or basepiece F_o. The purified ATPase of *E. coli* consists of five distinct subunits.

Comparison with Mammalian Systems. Although many bacteria can carry out oxidative phosphorylation, bacterial systems differ from intact mammalian systems in a number of ways. They have low P:O ratios and are quite variable in their properties even within a particular group of bacteria. Especially variable are such properties as the requirement for cofactors and sensitivity to various inhibitors. In mitochondrial systems, the number of moles of ATP formed relative to the gram atoms of oxygen consumed, i.e., the P:O ratio, approaches integral values for different substrates undergoing one-step oxidation. When a single pair of electrons travels from NADH to oxygen along the respiratory chain in mitochondria, three molecules of ATP are formed from ADP and phosphate. Since the formation of 3 moles of ATP requires the input of at least 3 times 7000 calories and the oxidation of NADH delivers 52,000 calories (Fig. 4-15), the oxidative phosphorylation of 3 moles of ADP conserves 3 (7000/52,000) or about 40 percent of the total energy

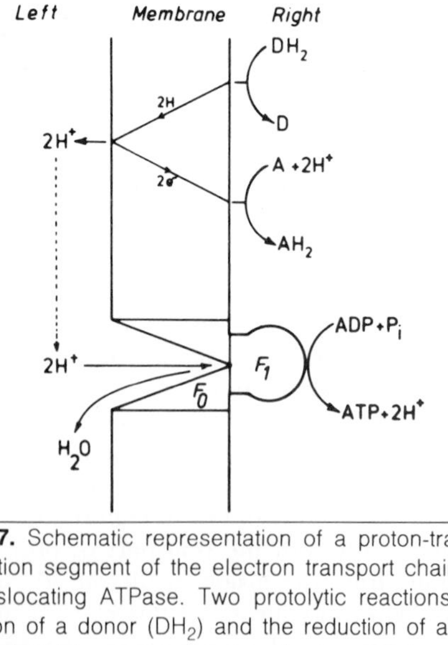

Figure 4-17. Schematic representation of a proton-translocating oxidoreduction segment of the electron transport chain and of a proton-translocating ATPase. Two protolytic reactions, involving the oxidation of a donor (DH_2) and the reduction of an acceptor (A), are catalyzed by an enzyme complex, comprising an alternating sequence of a hydrogen carrier and an electron carrier, arranged across the membrane to form a proton-translocating oxidoreduction loop. *(From Haddock and Jones: Bacteriol Rev 41:47, 1977.)*

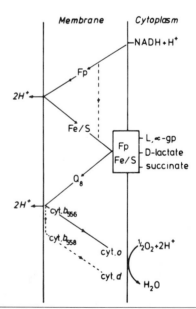

Figure 4-18. Proposed functional organization of the redox carriers responsible for aerobic electron transport in *E. coli*. The scheme includes the various routes for aerobic electron transport in *E. coli*, with the dashed lines indicating alternative pathways for reducing equivalents. L-α-gp, L-α-glycerophosphate. *(From Haddock and Jones: Bacteriol Rev 41:47, 1977.)*

yield when 1 mole of NADH is oxidized by oxygen. In the respiratory chain of mammalian systems, there are three segments in which there is a relatively large free energy drop: from NAD to flavoprotein, from cytochrome *b* to cytochrome *c*, and from cytochrome *a* to oxygen. It is at these points in the respiratory chain that high-energy intermediates are generated during electron transfer. In many bacteria, it appears that there are only one or two of these energy conservation sites. Loss of these sites and the presence of nonphosphorylative electron transport bypass reactions, both of which have been demonstrated, may account for the lower P:O ratios observed in most bacteria. In certain bacteria, however, such as *M. phlei*, 3 moles of ATP are synthesized per mole of oxidized substrate by three distinct respiratory chains.

Mitochondrial systems are characterized by a tight coupling of oxidation to phosphorylation, thus providing a means by which the rate of oxidation of foodstuffs is regulated by the requirements of the cell for useful energy. The utilization of ATP to drive the various energy-requiring processes of the cell automatically increases the available supply of ADP and inorganic phosphate. This, in turn, becomes available to react in the coupling mechanism and to permit respiration to proceed. Classic respiratory control has also been demonstrated in several bacterial systems, as evidenced by the ability of ADP, uncoupling agents, and ionophorous antibiotics to stimulate respiration by collapsing the transmembrane proton-motive force. This is probably the major control process in linear respiratory systems under highly aerobic conditions. The major function of a branched respiratory system is to allow some flexibility in the exact route of electron transfer. Thus, in the presence of low concentrations of molecular oxygen, electron transfer would be routed via the terminal branch that is most capable of maintaining a high potential rate of respiration. The presence of multiple cytochrome oxidases allows this rerouting to take place immediately (Fig. 4-16).

Pyridine-Nucleotide Transhydrogenase. Transhydrogenases have been detected in bacteria that catalyze the following reaction:

$$NADPH + NAD^+ \rightleftharpoons NADP^+ + NADH$$

The reaction makes possible the utilization of the reducing equivalents of NADPH by the electron transport chain. In aerobic organisms, however, when excess ATP is present, the reverse reaction allows reduction of $NADP^+$ for biosynthetic purposes. Although the molecular mechanism of the transhydrogenase reaction is unclear, it has the earmarks of a process dependent upon the energized state of the membrane.

Flavin-mediated Reactions

Although electron transport accounts for the great bulk of oxygen utilization in microbial systems, a number of other enzymes also catalyze reactions with O_2. Many of these are inducible enzymes produced by the organism in large quantities when it is grown on various aromatic compounds as the sole carbon source. Some of these enzymes, such as the D- and L-amino acid oxidases, are flavoproteins, autooxidizable by molecular oxygen. The reaction with oxygen is accompanied by the formation of hydrogen peroxide, which is highly toxic:

$$FPH_2 + O_2 \rightarrow FP + H_2O_2$$

Aerobic organisms produce catalase, an enzyme that breaks down the hydrogen peroxide to water and oxygen:

$$2\ H_2O_2 \rightarrow 2\ H_2O + O_2$$

The streptococci are facultative anaerobes that grow readily in the presence of air although they lack cytochromes and catalase. These organisms contain a peroxidase that destroys any peroxide produced by flavoprotein enzymes. The destruction of hydrogen peroxide by this mechanism requires the action of two enzymes, a flavoprotein NADH oxidase and a peroxidase:

$$NADH + H^+ + O_2 \xrightarrow{\text{NADH oxidase}} NAD^+ + H_2O_2$$

$$NADH + H^+ + H_2O_2 \xrightarrow{\text{peroxidase}} NAD^+ + 2\ H_2O$$

$$\text{Sum: } 2\ NADH + 2\ H^+ + O_2 \rightarrow 2\ NAD^+ + 2\ H_2O$$

Most anaerobic organisms lack both catalase and peroxidase.

Superoxide Dismutase

The reduction of oxygen results in the production of free radical intermediates that are very toxic to the bacterial cell. Among the most important of these is the superoxide anion O_2^-. Superoxide is generated during electron transport to molecular oxygen as well as in the autooxidation of hydroquinones, leukoflavins, ferredoxins, and flavoproteins. The catalytic actions of several enzymes (e.g., xanthine oxidase) also evolve O_2^-.

All aerobic and aerotolerant bacteria possess the enzyme superoxide dismutase, which scavenges the superoxide radical in a very bizarre enzymatic reaction:

$$O_2^- + O_2^- + 2H^+ \xrightarrow{\text{superoxide dismutase}} H_2O_2 + O_2$$

Catalase prevents the accumulation of the noxious H_2O_2 formed from O_2^-.

Superoxide dismutases are metalloenzymes. An interesting finding is that the same basic type of enzyme is found in bacteria and in eukaryotic mitochondria. Both contain Mn^{2+} and have many homologies of their amino acid sequence. A second type of enzyme found in the cytoplasm of eukaryotes contains Cu^{2+} and Zn^{2+} and has a significantly different structure.

The absence of catalase, peroxidase, and superoxide dismutase in anaerobic organisms provides at least a partial explanation for the oxygen toxicity in these organisms. The inhibitory role of other highly toxic radicals, such as singlet oxygen and the hydroxyl radical generated by the reduction of oxygen, remains to be explored.

Anaerobic Electron Transport

In obligately aerobic bacteria, oxygen is the only terminal electron acceptor, and cytochromes of the *a, d,* and *o* types can function as terminal oxidases (Fig. 4-16). In facultatively anaerobic organisms, however, a wide variety of other electron acceptors may be used. When these organisms are grown under aerobic conditions, they contain a functional respiratory chain, whereas under anaerobic conditions, their electron transport systems may be coupled to electron acceptors other than oxygen (Fig. 4-14). Anaerobic electron transport systems have also been demonstrated in some obligate anaerobes.

ATP generation can be coupled to many of these electron-accepting, hydrogen-consuming reactions in both facultatively and obligately anaerobic organisms. There is also evidence that some of the systems are coupled to active transport of metabolites in a manner comparable to that of the aerobic electron transport chain. Systems in which the reductive processes are coupled with phosphorylation are shown in Table 4-5.

Nitrate and Nitrite Reduction

The best characterized of the anaerobic electron systems is nitrate respiration, in which nitrate is used as the terminal electron acceptor. Nitrate respiration occurs in a wide range of bacterial species, including strictly aerobic, anaerobic, and facultatively anaerobic organisms:

$$NO_3^- + H_2 \rightarrow NO_2^- + H_2O$$

The reduction of nitrate to nitrite is catalyzed by a membrane-associated electron transport system consisting of dehydrogenases, electron carriers, and nitrate reductase (Fig. 4-19). In general, the dehydrogenases are inducible

TABLE 4-5. REDUCTIVE PROCESSES COUPLED WITH PHOSPHORYLATION

Reaction*	kcal/Electron Equivalent from H_2
CO_2 reduction to methane	3.9
Sulfate reduction to sulfide	4.5
Fumarate reduction to succinate	10.3
Nitrate reduction to nitrite	19.5
Nitrite reduction to N_2	31.7
O_2 reduction to H_2O (for comparison)	28.3

From Thauer et al.: Bacteriol Rev 41:118, 1977.
*CO_2, CH_4, H_2, N_2, NO, and N_2O in the gaseous state; all other substances in aqueous solution.

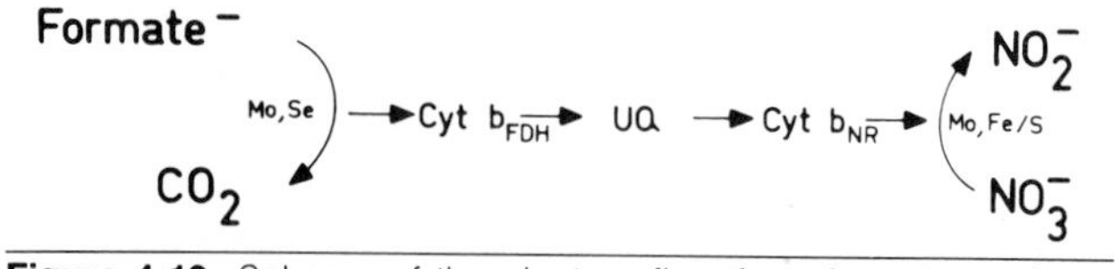

Figure 4-19. Scheme of the electron flow from formate to nitrate in *E. coli. (From Thauer et al.: Bacteriol Rev 41:100, 1977.)*

enzymes. In *E. coli* grown anaerobically in the presence of nitrate, formate is the most effective electron donor. Other substrates that may also function as electron donors for nitrate reduction include lactate, succinate, and NADH.

The reduction product of nitrate respiration, nitrite, is highly toxic, and growth of most organisms is limited. In a few organisms, however, such as *Bacillus* and *Pseudomonas,* nitrate can be reduced beyond the level of nitrite to molecular nitrogen by a series of anaerobic respiratory processes that in sum are called "denitrification." The one physiologic property characteristic of most denitrifying organisms is their ability to produce nitrogen gas by respiratory nitrate reduction:

$$2 NO_2^- + 3 [H_2] + 2 H^+ \rightarrow N_2 + 4 H_2O$$

Nitric oxide (NO) and nitrous oxide (N_2O) are intermediates in the reduction process (Fig. 4-20).

Fumarate Reduction

A number of bacteria, including both strict and facultative anaerobes, use fumarate as an electron acceptor (Fig. 4-21):

$$Fumarate^{2-} + 2 H_2 \rightarrow succinate^{2-}$$

Fumarate can easily be formed from a wide range of carbon sources, such as malate, aspartate, and pyruvate, and thus is readily available to the organisms as an electron acceptor. The standard redox potential of the fumarate/succinate couple ($E'_0 = +33$ mV) is greater than that of most of the other redox couples of metabolism, which makes it useful in the oxidation of various hydrogen donors (e.g., NADH, lactate, formate).

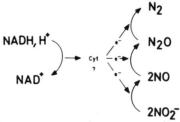

Figure 4-20. Scheme of the electron transport system involved in nitrite reduction to N_2. *(According to Payne: Bacteriol Rev 37:409, 1973; from Thauer et al.: Bacteriol Rev 41:100, 1977.)*

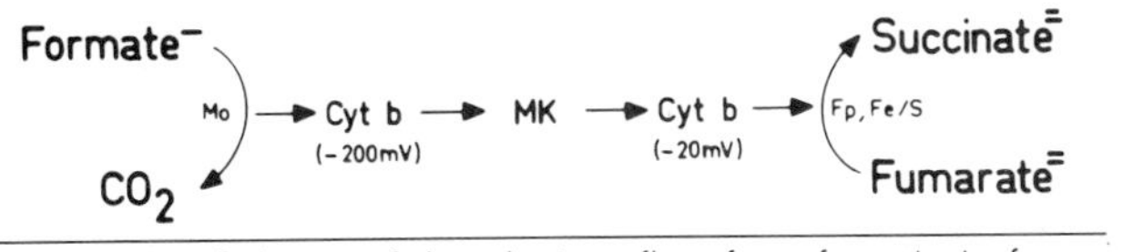

Figure 4-21. Scheme of the electron flow from formate to fumarate in *Vibrio succinogenes*. (*According to Kröger: 27th Symposium Society General Microbiology; from Thauer et al.: Bacteriol Rev 41:100, 1977.*)

Energy-yielding Autotrophic Metabolism

During the past decade, our concept of autotrophy has become increasingly blurred as we have gained a better understanding of the biochemistry of the organisms previously classified unequivocally as autotrophs or heterotrophs. At present, the unique property that may be considered to be common to all autotrophs is their ability to obtain the major part of their biosynthetic carbon from carbon dioxide or the metabolism of a 1-carbon compound. They obtain their energy from light (phototrophs), from the oxidation of inorganic compounds (chemolithotrophs), or from the oxidation of methyl groups attached to atoms other than carbon (methylotrophs) (Table 4-6).

Chemolithotrophs

These organisms are widely distributed in nature, where they play an important role in the maintenance of the nitrogen, carbon, and sulfur cycles. A variety of inorganic compounds can serve as their energy source. There is,

however, no shared mechanism of inorganic chemical oxidation among the members of the group. The different substrates (H_2, S^{2-}, NH_4^+, NO_2^-, Fe^{2+}) are all oxidized by different enzyme complexes and pathways, and the oxidation of a reduced inorganic compound is not a unique property restricted to autotrophs (Table 4-6).

Hydrogen Bacteria. These are aerobic organisms, most of which can utilize compounds in addition to hydrogen as an energy source. They possess the enzyme hydrogenase, which activates molecular hydrogen:

$$H_2 \xrightleftharpoons{\text{hydrogenase}} 2\,H^+ + 2e$$

The acceptors that function subsequent to the primary step vary with the kinds of coupling reactions that exist between the hydrogenase and the final electron acceptor. In some members of the group, a coupling with pyridine nucleotides and the electron transport chain occurs, with oxygen serving as the ultimate electron acceptor. An example of this type of coupling is found in *P. denitrificans*, which contains a membrane-bound respiratory chain very similar to that of mitochondria (Fig. 4-13). It has been speculated that the mitochondrion probably evolved from the plasma membrane of an ancestor of *P. denitrificans* via endosymbiosis.

Nitrifying Bacteria. In the nitrifying bacteria *Nitrosomonas* and *Nitrobacter*, the E'_0 values for the oxidations involved do not permit a coupling with the reduction of NAD. In *Nitrobacter*, the electrons enter the transport chain at the level of cytochrome a_1:

$$\textit{Nitrosomas}\quad NH_3 + 1\tfrac{1}{2}\,O_2 \rightarrow NO_2^- + H_2O + H^+$$

$$\textit{Nitrobacter}\quad NO_2^- + \tfrac{1}{2}\,O_2 \rightarrow NO_3^-$$

TABLE 4-6. SOME ORGANISMS EXPLOITING UNCONVENTIONAL SOURCES OF ENERGY

Group	Energy Source	Heterotrophic Growth	
		+*	−†
Phototrophs	Light	*Rhodospirillum rubrum*	*Chromatium okenii*
Lithotrophs	H_2	*Alcaligenes eutrophus*	*Methanobacterium thermoautrophicum*
		Paracoccus denitrificans	
	S^{2-}	*Thiobacillus acidophila*	*Thiobacillus denitrificans*
		Thiobacillus intermedius	*Thiobacillus thiooxidans*
	NH_4^+	None	*Nitrosomonas europaea*
			Nitrosospira briensis
	NO_2^-	*Nitrobacter agilis*	*Nitrobacter* sp.
			Nitrococcus mobilis
Methylotrophs	CH_4 and other $[CH_3-]$ compounds	Methylotroph strain XX	*Methylobacter* sp.
			Methylomonas methanooxidans
	$[CH_3-]$ compounds other than CH_4	*Arthrobacter* 2B2	Methylotroph 4B6
		Pseudomonas 3A2	Methylotroph C2A1

Modified from Smith and Hoare: Bacteriol Rev 41:419, 1977.
*Growth on organic compounds in the absence of the specific energy source (versatile strains).
†No growth on organic compounds in the absence of the specific energy source (specialist strains).

Most of the nitrifying bacteria are obligate anaerobes, incapable of using organic substrates as an energy source.

Methanogenic Bacteria. Methane is the most reduced organic compound, and its formation is the terminal step in an anaerobic food chain. Methane is generated mainly from acetate, CO_2, and H_2 by a specialized group of anaerobic bacteria, the methanogens, that inhabit anaerobic environments where organic matter is being decomposed. Methanogenic bacteria have the most stringent anaerobic requirements among anaerobes and carry out methanogenesis only where the redox potential is lower than -330 mV. They thus occupy a very narrow ecologic niche.

The biodegradation of organic compounds to methane in anaerobic habitats involves a microbial metabolic food chain, the complexity of which depends on the habitat. In the absence of nitrate, sulfate, or elemental sulfur, carbon dioxide becomes a major electron sink for anaerobic respiration, permitting obligate proton-reducing bacteria to carry out the oxidation of fatty acids and alcohols in such environments. By rapidly oxidizing and removing hydrogen from the anaerobic habitat, conditions thermodynamically favorable for the more complete anaerobic oxidation of carbon skeletons are produced.

The reduction of CO_2 to CH_4 proceeds stepwise, but the intermediates (formate, formaldehyde, and methanol) remain firmly bound to carriers that have not been completely identified. One carrier, coenzyme M, has been identified as 2-mercaptoethanesulfonic acid. Methylcoenzyme M is probably the direct precursor of methane:

$$HCO_3^- + H_2 \rightarrow HCOO^- + H_2O$$

$$HCOO^- + H_2 + H_2 \rightarrow CH_2O + H_2O$$

$$CH_2O + H_2 \rightarrow CH_3OH$$

$$CH_3OH + H_2 \rightarrow CH_4 + H_2O$$

$$HCO_3^- + H^+ + 4H_2 \rightarrow CH_4 + 3H_2O$$

$$\Delta G'_o = -32.4 \text{ kcal}$$

The formation of methane from CO_2 and H_2 cannot be coupled to ATP synthesis by substrate-level phosphorylation. It thus appears that methanogenic bacteria gain ATP by electron transport phosphorylation.

Methylotrophs

This group of organisms is characterized by the ability to fulfill their energy requirement by the oxidation of methyl groups attached to atoms other than carbon. Some of these are obligate methylotrophs, growing only at the expense of compounds containing no carbon–carbon bonds (methane, methanol). Others are facultative methylotrophs, capable of growing on a variety of carbon sources including C_1-compounds.

The oxidation of methane to carbon dioxide proceeds via a series of two-electron oxidation steps. In the metabolism of methane, formaldehyde occupies a key position, since it is at this level that the carbon is both assimilated into biomass and dissimilated to carbon dioxide to provide energy.

Phototrophs

These organisms derive their energy for growth from light by the process of photosynthesis. Mechanistically this is the most complex mode of energy-yielding metabolism. Although the overall reaction is basically the same in all photosynthetic organisms, bacteria possess an evolutionarily more primitive mechanism.

Photosynthesis. Photosynthesis consists of an oxidation-reduction sequence in which carbon dioxide is reduced to the level of carbohydrate at the expense of a variety of hydrogen donors activated by light reactions.

$$2 \text{ H}_2\text{A} + CO_2 \rightarrow (CH_2O) + H_2O + 2 \text{ A}$$

The nature of the compound H_2A varies with the organism, and it is this property that distinguishes bacterial photosynthesis from that present in evolutionarily higher forms. In plants and algae that can grow aerobically in the light, H_2A can be water, and oxygen is liberated in the reaction. In bacteria, however, photosynthesis proceeds only under anaerobic conditions, no oxygen is evolved, and H_2A must be supplied as reduced sulfur or organic compounds. The photosynthetic bacteria are typically aquatic species and include the green and purple sulfur bacteria and the purple nonsulfur bacteria.

PHOTOSYNTHETIC APPARATUS. Photosynthesis occurs in a deeply invaginated membrane system containing pigments, electron carriers, lipids, and proteins. In bacteria this apparatus is much simpler than the chloroplasts of plants. The basic membranous unit is an enclosed sac or thylakoid, which contains the chlorophyll and carotenoid pigments that function in the absorption of light energy.

PHOTOSYNTHETIC ELECTRON TRANSPORT. Whereas plants have two distinct photochemical reaction centers, bacteria have only one. The primary photochemical event is initiated when a molecule of chlorophyll absorbs a quantum of light and transfers it to a reaction center buried within the membrane. The chlorophyll serves in some manner to effect a photochemical separation of oxidizing and reducing power, resulting in a flow along two transport systems. One of these systems accepts the electron delivered to the acceptor, and the other replaces it (Fig. 4-22).

In a photosynthetic bacterium, such as *Rhodospirillum rubrum*, a nonheme-iron center complexed with quinone serves as the immediate electron acceptor and a cytochrome *c*-like protein as the electron donor to the activated chlorophyll. The flow of electrons is accompanied by the generation of ATP. The coupling mechanism is similar in principle to the coupling mechanisms employed in respiratory chain phosphorylation.

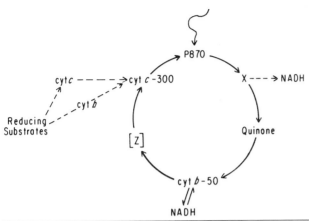

Figure 4-22. A generalized scheme for photosynthetic electron flow in photosynthetic bacteria. Electrons lost from the cyclic electron flow system during the formation of NADH are replaced by electron transport from reduced substrates. X is probably an iron protein or an iron-quinone complex; Z is an intermediate suggested by kinetic data. *(From Jones OTG: In Haddock and Hamilton (eds): Microbial Energetics, 1977. London, Cambridge University Press.)*

The fixation of the CO_2 into carbohydrate requires a supply of both NADPH and ATP. This relationship may be summarized in the following general statement:

$$H_2A + NAD(P)^+ + yADP + yP_i \xrightarrow{mhv}$$

$$A + NAD(P)H + H^+ + yATP$$

For photosynthetic bacteria, the H_2A may be an inorganic substance, such as H_2S, or an organic compound, such as succinate. The energy of absorbed photons drives this reaction. In this event, the flow of electrons is open or noncyclic. In the absence of oxidizable substrate, however, light-induced electron flow occurs along a circular path (Fig. 4-22). The electrons that come from the excited chlorophyll molecule may quite simply return to it again after they have traveled a circular route around the closed chain of electron carriers. This circuitous flow of electrons is a device to conserve some of the energy of the high-energy electrons that leave the excited chlorophyll. Cyclic photophosphorylation may represent the most primitive form of photosynthesis, useful to organisms in an environment rich in organic compounds and requiring only ATP.

DARK PHASE OF PHOTOSYNTHESIS. The formation of glucose in photosynthesis is a dark process that begins with the reduced NADP and ATP generated by light. The mechanism by which CO_2 fixation occurs in all photosynthesizing organisms is cyclic in nature and occurs in both eucaryotic and procaryotic organisms. This complex series of reactions is known as the "Calvin cycle," the initial reaction of which is the synthesis of ribulose 1,5-diphosphate from ribulose 5-phosphate.

Ribulose 5-PO_4 + ATP → ribulose 1,5-PO_4 + ADP

Ribulose 1,5-PO_4 + CO_2 → 2 3-phosphoglyceric acid

These two reactions are specific for organisms that use CO_2 as a sole carbon source and are not found in organisms that have a heterotrophic metabolism. The reduction of the two molecules of 3-phosphoglycerate occurs at the expense of NADPH and ATP formed in the light reaction. The two molecules of 3-phosphoglyceraldehyde thus formed are then converted into glucose essentially by reversal of the reactions of glycolysis.

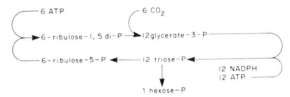

Energetics of Bacterial Metabolism

Energy Conservation

When facultative organisms are grown under aerobic conditions, growth is always more vigorous than that obtained under anaerobic conditions. This is because of the greater amount of phosphate bond energy made available by respiration.

The anaerobic breakdown of the glucose molecule to lactate is accompanied by the phosphorylation of 2 moles of ADP and proceeds with a free energy decline of −38,000 calories/mole. This is 16,000 calories less than the calculated −56,000 calories free energy change, which would be expected from the simple breakdown of glucose to lactate. In the intact cell, a significantly large part of this free energy loss is conserved in the form of ATP. In the breakdown of 1 mole of glucose to the level of lactate, 16/56, or approximately 28 percent, of the energy is thus conserved in the 2 moles of ATP formed.

In growth under aerobic conditions, the combined processes of glycolysis and oxidative phosphorylation provide a total of 38 moles of ATP. Since the calculated $\Delta F'$ for the complete combustion of glucose is −686,000 calories/mole and since an approximate input of 8000 calories/mole of ATP is required, the efficiency of energy conservation under aerobic conditions is (38 × 8000)/686,000, or approximately 45 percent (Table 4-7).

This energy that is conserved in the form of ATP is used by the cell to perform its various activities. The most important forms of work carried out by the bacterial cell at the expense of ATP are active transport and the biosynthesis of cellular components from small precursor molecules. Almost all of the energy of bacterial cells is put into biosynthetic work. Their sole mission is to multiply. Since bacteria normally live in natural environments over which they have no control and from which they cannot escape, the ability to multiply rapidly fits them to survive. For this task, a large amount of energy is required.

TABLE 4-7. ENERGETICS OF GLUCOSE METABOLISM

	ATP Yield	$\Delta G'$ (calories)
Glucose $\to$ 2 lactic acid	2	56,000
Glucose + $6O_2 \to 6\ CO_2 + H_2O$	38	686,000
Sequence		
Glucose $\to$ fructose 1,6-diP	-2	
2 Triose P $\to$ 2 3-phosphoglyceric acid	$+2$	
$2\ NAD^+ \to 2\ NADH \to 2\ NAD^+$	$+6$	
2 Phosphoenolpyruvic acid $\to$ 2 pyruvic acid	$+2$	
2 Pyruvic acid $\to$ 2 acetyl-CoA + 2 CO_2		
$2\ NAD^+ \to 2\ NADH \to 2\ NAD^+$	$+6$	
2 Acetyl-CoA $\to$ 4 CO_2	$+24$	
Net $C_6H_{12}O_6 + 6O_2 + 6H_2O$	$+38$	

Growth Yield

One of the approaches used to evaluate the efficiency of bacterial energy conservation is the measurement of molar growth yields. When growth is limited by the energy source, the total growth obtained in a culture is proportional to the amount of carbohydrate added. When anaerobic organisms employing a fermentative metabolism are grown in a complex medium, the substrate is used almost exclusively for the generation of ATP. Since the amount of ATP produced by various fermentations can be calculated, the growth yield as a function of ATP (Y_{ATP}) provides a means of determining the efficiency with which ATP is used by different organisms. The growth yields for different organisms utilizing a wide variety of substrates and employing different pathways are constant: values of approximately 10 g of cell material per gram-mole of ATP have been obtained under batch conditions of growth. Although significantly higher values have been obtained under energy-limited, continuous culture conditions (12.4 and 14.0 for *E. coli* and *Klebsiella pneumoniae*, respectively), the values are still considerably less than would be expected if all of the ATP used in cellular growth was coupled to biosynthesis. It would thus appear that bacteria are inefficient converters of free energy, and there is a large outflow of entropy from the cell. Perhaps this is the price they must pay to cope with the vicissitudes of life outside the laboratory. In order to compete effectively in their natural, nutrient-limited environment, they must be able to react rapidly when nutritional constraints are occasionally relieved.

Control of Energy Metabolism

If a microorganism is to function efficiently, the rate of metabolism along the various branching and anastomosing metabolic pathways must be regulated in such a way that optimal use is made of the available substrates. For this purpose, the microbial cell has evolved an extremely so-phisticated system of controls, some of which regulate its energy-supplying processes. Regulation of the energy-yielding metabolism takes place on different levels: (1) regulation of enzyme production, (2) end product inhibition of enzyme activities, and (3) general metabolic regulation of enzymatic activities by substrate and product levels.

Genetic Regulation

In its natural habitat, the cell is confronted with a variety of potential energy sources. The survival of a particular species in its highly competitive environment has resulted from the ability of that species to adapt to new experiences in its environment. In so doing, the enzymatic machinery for the degradation of a wide variety of organic compounds is produced. Although the potential for the dissimilation of different substrates is great, the enzymes for such activities are produced (induced) only when needed. Controls of this type, induction and repression, are exceedingly common in microorganisms and are examples of genetic regulatory mechanisms. In general, induction exerts effective control of catabolic sequences involving carbon and energy sources, where the synthesis of enzymes catalyzing a particular sequence is turned on or off, depending on the demands for that specific sequence. The classic example of genetic regulation is the utilization of the disaccharide lactose by *E. coli*. The molecular biology of the lactose operon is discussed on page 141.

Catabolite Repression

This type of control is frequently observed when organisms are grown on glucose or some other rapidly metabolizable energy source. Often referred to as the "glucose effect," catabolite repression results in a repression of synthesis of enzymes that would metabolize the added substrate less rapidly than glucose. When the *lac* system is induced, the rate of synthesis of β-galactosidase is considerably reduced in cultures growing upon glucose, compared with cells for which some other metabolite is provided as the carbon source. Glucose elicits catabolite repression by depressing the level of 3′-5′-cyclic AMP (cAMP) in the cell. The addition of cAMP to cultures overcomes glucose repression by stimulating transcription of the inducible enzyme, β-galactosidase. The level of cAMP in the cell varies with conditions of growth and reflects the energetic needs of the cell. The level is low when the available energy exceeds the biosynthetic requirement for energy, and the level of cAMP rises when the organisms's carbon supply is depleted. The molecular aspects of catabolite repression are presented in Chapter 8.

Metabolite Regulation

Adenylate Energy Charge

The adenine nucleotides (ATP, ADP, and AMP) are metabolic energy modulators strategically placed to regulate the entire metabolic economy of the cell. In general, catabolic sequences contain regulatory enzymes that are activated by ADP or AMP or inhibited by ATP. Degradation of the substrate, therefore, proceeds maximally only when there is a need for ATP.

In its role as a primary metabolic coupling agent, the adenylate system has been compared with a storage battery in its ability to accept, store, and supply chemical energy. The term "adenylate energy charge" defines the relative amount of energy stored in the system and may be expressed on a linear scale by the equation:

$$\text{Energy charge} = \frac{\text{ATP} + \frac{1}{2}\text{ADP}}{\text{ATP} + \text{ADP} + \text{AMP}}$$

The catalytic properties of a number of enzymes are modified by changes in the energy charge. As the energy charge increases, adenylate-regulated enzymes in ATP-regenerating sequences decrease in activity. The adenylate system is poised to run optimally in a steady state in which the energy charge is between 0.8 and 0.9 and strongly resists deviations from this range (Fig. 4-23). In view of the very rapid turnover rate of ATP in growing bacteria (1 to

10 S^{-1}), the tight stabilization of the energy charge is indicative of very sensitive and fast controls.

Pasteur Effect

In facultative organisms, the fermentative capacity of the cell is blocked in the presence of oxygen, and the energy is supplied almost exclusively by respiration. As a result, less glucose is consumed, and the accumulation of lactate is decreased. This phenomenon, first recognized by Pasteur in fermenting yeast, is known as the "Pasteur effect." The benefits of this effect are obvious in terms of the energy gain realized in switching from an anaerobic metabolism to an aerobic one. Anaerobic glycolysis releases only about 8 percent of the energy that is obtained from the complete breakdown of glucose. Therefore, if oxygen is available and the glucose is oxidized to CO_2 and H_2O without the accumulation of lactic acid, the energy needs of the cell can be met by the utilization of less glucose.

Several factors may be responsible for the Pasteur effect, but the major determinant is the key enzyme phosphofructokinase, which plays a central role in the regulation of glycolysis (p. 44). In the generation of ATP via the respiratory pathways, increased levels of ATP relative to ADP inhibit phosphofructokinase, thereby decreasing the flow of glucose into the glycolytic pathway.

The inhibition of phosphofructokinase activity by ATP is illustrative of regulation of an allosteric enzyme by a negative effector. The reaction catalyzed by phosphofructokinase, the formation of fructose 1,6-diphosphate from fructose 6-phosphate, is a critical control point subject to strong metabolic regulation. It is exquisitely sensitive to changes in the adenylate energy charge (Fig. 4-23), ensuring that whenever a plentiful supply of ATP is available, as when pyruvate is metabolized aerobically to CO_2 via the TCA cycle, glycolysis will be essentially blocked. The reverse is also true. When glycolysis is absolutely required for the generation of energy, i.e., when the ATP drops to low levels and ADP or AMP accumulates, glycolysis will be favored.

In addition to ATP, other products of respiration can modulate glycolytic activity. The need for additional controls stems from the dual function of glycolysis as an amphibolic pathway. The glycolytic rate must respond to the need for supplying synthetic intermediates as well as the need for regenerating ATP. Phosphofructokinases from various bacterial species exhibit many similar, though not identical, regulatory properties. In general, activators of phosphofructokinase, such as AMP, ADP, P_i, and fructose 1,6-diphosphate, are those that tend to accumulate during anaerobic conditions, whereas inhibitors of the enzyme, such as ATP and citrate, are those that tend to increase on transition from anaerobic to aerobic conditions.

Other regulatory inputs that are important either generally or in specific cases are the charges of the nicotinamide adenine nucleotides (NADH/NADH + NAD$^+$) and (NADPH/NADPH + NADP$^+$) and positive feed-forward effects (stimulation of a reaction by precursors one or more steps earlier in the sequence).

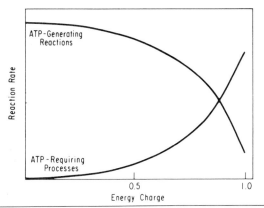

Figure 4-23. Generalized response to the adenylate energy charge expected from enzymes involved in regulation of ATP-regenerating and ATP-utilizing sequences. *(From Atkinson: Biochemistry 7:4030, 1968.)*

Localization of Enzymic Activities

In eucaryotic organisms, the mitochondrion and the chloroplast contain the units that transform oxidative energy into the bond energy of ATP. No mitochondria are found in bacteria, but a functional equivalent is present. In such cells, either the cell membrane itself or extensions of the membrane contain the subunits that carry out energy transductions. Although bacteria have essentially a single membrane system, this system is a composite of almost all the membrane systems found in the more complex forms of life.

The main components of the respiratory chain—succinic dehydrogenase, NADH dehydrogenase, cytochromes of groups *a*, *b*, and *c*, ubiquinones, and naphthoquinones—are found in membranes or in particles obtained by homogenization and subsequent differential centrifugation. The membranes and their fragments also contain several enzymes of the TCA cycle and firmly bound dehydrogenases.

Because of the small size of the bacterial cell and the nature of the open membrane system of which the respiratory apparatus is an integral part, there is in bacteria direct interaction between the membranes carrying the respiratory chain and the cytoplasm, which is the source of coenzymes, substrates, and ADP. The metabolites that accumulate in the cytoplasm as a result of glycolysis are rapidly oxidized by the enzymes in the membranes.

FURTHER READING

Books and Reviews

Atkinson DE: Adenine nucleotides as stoichiometric coupling agents in metabolism and as regulatory modifiers: The adenylate energy charge. In Vogel HJ (ed): Metabolic Regulation. Metabolic Pathways. New York, Academic Press, 1971, vol 5, pp 1–21

Atkinson DE: Cellular Energy Metabolism and Its Regulation. New York, Academic Press, 1977

Barker HA: Amino acid degradation by anaerobic bacteria. Annu Rev Biochem 50:23, 1981

Brown CM, Macdonald-Brown DS, Meers JL: Physiological aspects of microbial inorganic nitrogen metabolism. Adv Microb Physiol 15:1, 1977

Chance B: Electron transfer: Pathways, mechanisms, and controls. Annu Rev Biochem 46:967, 1977

Chapman AG, Atkinson DE: Adenine nucleotide concentrations and turnover rates. Their correlation with biological activity in bacteria and yeast. Adv Microb Physiol 15:253, 1977

Colby J, Dalton H, Whittenbury R: Biological and biochemical aspects of microbial growth on C_1 compounds. Annu Rev Microbiol 33:481, 1979

Fridovich I: Superoxide dismutases. Annu Rev Biochem 44:147, 1975

Gottschalk G: Bacterial Metabolism. New York, Springer-Verlag, 1979

Gottschalk G, Andreesen JR: Energy metabolism in anaerobes. Int Rev Biochem 21:85, 1979

Gunsalus IC, Stanier RY (eds): The Bacteria. New York, Academic Press, 1961, vol 2

Haddock BA, Hamilton WA (eds): Microbial Energetics. London, Cambridge University Press, 1977

Haddock BA, Jones CW: Bacterial respiration. Bacteriol Rev 41:47, 1977

Jones CW: Aerobic respiratory systems in bacteria. In Haddock BA, Hamilton WA (eds): Microbial Energetics. London, Cambridge University Press, 1977, pp 23–59

Jones CW: Energy metabolism in aerobes. Internat Rev Biochem 21:49, 1979

Jones OTG: Electron transport and ATP synthesis in the photosynthetic bacteria. In Haddock BA, Hamilton WA (eds): Microbial Energetics. London, Cambridge University Press, 1977, pp 151–183

Kaback HR: Membrane vesicles, electrochemical ion gradients, and active transport. Curr Top Memb Transp 16:393, 1982

Klotz IM: Energy Changes in Biochemical Reactions. New York, Academic Press, 1967

Kondratieva EN: Interrelation between modes of carbon assimilation and energy production in phototrophic purple and green bacteria. Int Rev Biochem 21:117, 1979

Konings WN, Boonstra J: Anaerobic electron transfer and active transport in bacteria. In Bronner F, Kleinzeller A (eds): Curr Top Memb Transp 9:177, 1977

Lehninger AL: Bioenergetics, 2nd ed. New York, Benjamin, 1971

Lehninger AL: Biochemistry, 2nd ed. New York, Worth Publishers, 1975

Mah RA, Ward DM, Baresi L, et al.: Biogenesis of methane. Annu Rev Microbiol 31:309, 1977

Mitchell P: Vectorial chemiosmotic processes. Annu Rev Biochem 46:996, 1977

Moloney PC: Coupling between H^+ entry and ATP synthesis in bacteria. Curr Top Memb Transp 16:175, 1982

Pastan I, Adhya S: Cyclic adenosine 5′-monophosphate in *Escherichia coli.* Bacteriol Rev 40:527, 1976

Quayle JR, Ferenci T: Evolutionary aspects of autotrophy. Microbiol Rev 42:251, 1978

Ramaiah A: Pasteur effect and phosphofructokinase. Curr Top Cell Regul 8:297, 1974

Richenberg HV: Cyclic AMP in prokaryotes. Annu Rev Microbiol 8:353, 1974

Sanwal BD: Allosteric controls of amphibolic pathways in bacteria. Bacteriol Rev 34:20, 1970

Smith AJ, Hoare DS: Specialist phototrophs, lithotrophs, and methylotrophs: A unity among a diversity of procaryotes? Bacteriol Rev 41:419, 1977

Stanier RY, Doudoroff M, Adelberg EA: The Microbial World, 4th ed. Englewood Cliffs, NJ, Prentice-Hall, 1976

Stouthamer AH: The search for correlation between theoretical and experimental growth yields. Int Rev Biochem 21:1, 1979

Thauer RK, Jungermann K, Decker K: Energy conservation in chemotrophic anaerobic bacteria. Bacteriol Rev 41:100, 1977

Wolfe RS, Higgins IJ: Microbial biochemistry of methane—a study in contrasts. Int Rev Biochem 21:267, 1979

Zeikus JG: The biology of methanogenic bacteria. Bacteriol Rev 41:514, 1977

Zeikus JG: Chemical and fuel production by anaerobic bacteria. Annu Rev Microbiol 34:423, 1980

Selected Papers

Anderson KB, von Meyenburg K: Charges of nicotinamide adenine nucleotides and adenylate energy charge as regulatory parameters of the metabolism in *Escherichia coli.* J Biol Chem 252:4151, 1977

Archibald FS, Fridovich I: Manganese, superoxide dismutase, and oxygen tolerance in some lactic acid bacteria. J Bacteriol 146:928, 1981

Cross AR, Anthony C: The electron-transport chains of the obligate methylotroph *Methylophilus methylotrophus.* Biochem J 192:429, 1980

DiGuiseppi J, Fridovich I: Oxygen toxicity in *Streptococcus sanguis.* The relative importance of superoxide and hydroxyl radicals. J Biol Chem 257:4046, 1982

Gest H: Evolution of the citric acid cycle and respiratory energy conversion in prokaryotes. FEMS Microbiol Lett 12:209, 1981

Graham A, Boxer DH: The organization of formate dehydrogenase in the cytoplasmic membrane of *Escherichia coli.* Biochem J 195:627, 1981

Kashket ER: Effects of aerobiosis and nitrogen source on the proton motive force in growing *Escherichia coli* and *Klebsiella pneumoniae* cells. J Bacteriol 146:377, 1981

Krasna AI: Regulation of hydrogenase activity in enterobacteria. J Bacteriol 144:1094, 1980

Krinsky NI: Singlet oxygen in biological systems. Trends Biochem Sci 2:35, 1977

McCord JM, Keele BB Jr, Fridovich I: An enzyme-based theory of obligate anaerobiosis: The physiological function of superoxide dismutase. Proc Natl Acad Sci USA 68:1024, 1971

Mitchell CG, Dawes EA: The role of oxygen in the regulation of glucose metabolism, transport and the tricarboxylic acid cycle in *Pseudomonas aeruginosa.* J Gen Microbiol 128:49, 1982

Mitchell P: Chemiosmotic coupling in energy transduction: a logical development of biochemical knowledge. Bioenergetics 3:5, 1972

Moody CS, Hassan HM: Mutagenicity of oxygen free radicals. Proc Natl Acad Sci USA 79:2855, 1982

CHAPTER 5

Physiology of Bacterial Growth

Requirements for Growth

Growth may be defined as the orderly increase of all of the chemical constituents of the cell. It is a process that entails the replication of all cellular structures, organelles, and protoplasmic components from the nutrients present in the surrounding environment. In order for bacteria to grow, they must be provided with all of the substances essential for the synthesis and maintenance of their protoplasm, a source of energy, and suitable environmental conditions.

As a group, bacteria are extremely versatile organisms. They exhibit tremendous capabilities for the utilization of quite diverse food materials, ranging from completely inorganic substrates to very complex organic compounds. Many species also have learned to grow in a wide diversity of ecologic niches with extremes in temperature, acidity, or oxygen tensions. The ability of bacteria to exist under such circumstances is proof of their tremendous adaptability and reflects their capacity to respond successfully to a stimulus that is completely foreign to their past history.

Nutrient Requirements

Carbon

Two basic patterns characterize the nutritional requirements of bacteria and reflect their metabolic potentialities (Table 4-1). At one end of the spectrum are the autotrophic bacteria (lithotrophs) that require only water, inorganic salts, and CO_2 for growth. These organisms synthesize a major portion of their essential organic metabolites from CO_2. At the other end of the spectrum are the heterotrophic bacteria (organotrophs) that require an organic form of carbon for growth. Although glucose is quite extensively used as the organic source of carbon in routine laboratory practice, a wide variety of other substances also can be used as an exclusive or partial source of carbon by different species of bacteria. Among the most versatile bacteria are the pseudomonads, some of which can utilize over 100 different organic compounds as the sole carbon and energy source.

Nitrogen

The nitrogen atoms of amino acids, purines, pyrimidines, and other biomolecules come from NH_4^+. The flow of nitrogen into these compounds starts with the reduction of atmospheric N_2 to NH_4^+ by nitrogen-fixing microorganisms. NH_4^+ is then assimilated into amino acids by way of glutamate and glutamine, the two pivotal molecules in nitrogen metabolism (Fig. 5-1).

Nitrogen Fixation. Higher organisms depend upon certain nitrogen-fixing species of bacteria and blue-green algae to convert N_2 into the organic form. Because of the strength of the $N{\equiv}N$ bond, nitrogen-fixation is very energy demanding, requiring ATP and a powerful reductant. The process is catalyzed by a complex enzyme, the nitrogenase complex, composed of two kinds of iron–sulfur proteins. In most nitrogen-fixing organisms, reduced ferredoxin is the source of electrons:

$$N_2 + 6e^- + 12\ ATP + 12\ H_2O \rightarrow$$
$$2\ NH_4^+ + 12\ ADP + 12\ P_i + 4\ H^+$$

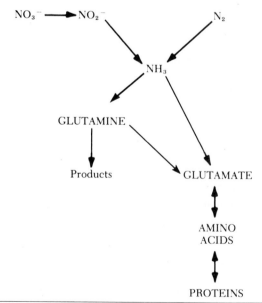

Figure 5-1. Inorganic nitrogen assimilation in microorganisms.

Among the systems that have been extensively studied are those of the anaerobe *Clostridium pasteurianum* and the aerobes *Azobacter vinelandii* and *Klebsiella pneumoniae*.

Nitrate Reduction. Nitrate reduction may be accomplished by two distinct physiologic mechanisms: (1) assimilatory nitrate reduction, a process in which nitrate is reduced via nitrite and probably hydroxylamine to ammonia, which is then assimilated, and (2) dissimilatory nitrate reduction in which nitrate serves as an alternative electron acceptor to oxygen (i.e., anaerobic respiration, p. 56), with N_2 and NO_2 being the usual products. Nitrate assimilation is quite widespread in microorganisms, but dissimilatory nitrate reduction is common only in anaerobic bacteria and in facultatively anaerobic organisms growing at low oxygen tensions. Assimilatory nitrate reduction is catalyzed by two enzymes, nitrate reductase and nitrite reductase, both of which are distinct from the dissimilatory reductases.

Ammonia Assimilation. Ammonia occupies a central position in the metabolism of organisms grown on organic sources of nitrogen. Its assimilation by prokaryotic organisms is accomplished by three major pathways (Fig. 5-2).

The first of these, the formation of glutamic acid from ammonia and α-ketoglutaric acid by 1-glutamate dehydrogenase, is the primary pathway for the formation of α-amino acids directly from ammonia. The versatility of glutamic acid as amino group donor in a number of transamination reactions permits the introduction of α-amino groups into most of the other amino acids.

$$NH_3 + \text{α-ketoglutarate} + NADPH + H^+ \xrightarrow{\text{glutamate dehydrogenase}} \text{glutamate} + NADP^+ + H_2O$$

Glutamate dehydrogenase has a high K_m for ammonia and functions efficiently only when the environmental ammonia concentration is high. When the concentration of ammonia is low, the two enzymes, glutamine synthetase and glutamate synthase, together provide an alternate route for the incorporation of ammonia.

$$\text{Glutamate} + NH_3 + ATP \xrightarrow{\text{glutamine synthetase}} \text{glutamine} + ADP + P_i$$

$$\text{Glutamine} + \text{α-ketoglutarate} + NADPH + H^+ \xrightarrow{\text{glutamate synthase}} 2 \text{ glutamate} + NADP^+$$

Glutamine synthetase exerts strong regulatory control over the synthesis of enzymes responsible for the formation of glutamate from ammonia (Chap. 7).

Growth Factors

Many of the heterotrophic bacteria are unable to grow unless supplied with one or more specific growth factors. These substances, usually provided in the culture medium in the form of yeast extract or whole blood, include the B complex vitamins, amino acids, purines, and pyrimidines. The B complex vitamins play a catalytic role within the cell either as components of coenzymes or as prosthetic groups of enzymes. Organisms that do not require an exogenous source of a given growth factor are capable of synthesizing their own. They are referred to as prototrophic in respect to that trait in order to distinguish them from auxotrophic mutants that require the growth factor for growth.

Inorganic Ions

Small amounts of a number of inorganic ions are required by all bacteria. In addition to nitrogen, sulfur, and phosphorus, which are present as constituents of important biologic compounds, potassium, magnesium, and calcium occur in bacteria functionally associated with certain anionic polymers. Magnesium functions to stabilize ribosomes, cell membranes, and nucleic acids and is required for the activity of many enzymes. Potassium also is required for the activity of a number of enzymes, and in gram-positive organisms its concentration in the cell is influenced by the teichoic acid content of the cell wall. A requirement for iron, manganese, zinc, copper, and cobalt has also been shown in most organisms, and for others, molybdenum and selenium are essential. The need for others is more difficult to access because of the minute amounts required and the presence of trace amounts as contaminants in constituents of media.

Trace elements play an important role in host–parasite interactions. In the animal host, powerful iron-binding proteins in the body fluids function to withhold iron from microbial invaders. The successful microbial invader has developed its own powerful iron chelators that vigorously extract iron from a variety of environments. A number of these iron compounds (siderophores) have been recog-

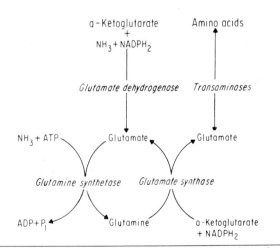

Figure 5-2. Pathways of ammonia assimilation in prokaryotic organisms. *(After Brown et al.: Adv Microb Physiol 15:1, 1977.)*

nized in different bacterial species. Their existence emphasizes the essential role of iron for the organism as well as the evolutionary significance of systems that insure that the organism can compete successfully with the host for essential nutrients that may be present in limited amounts. For a further discussion of iron-binding proteins, see page 72.

Oxygen

The oxygen requirement of a particular bacterium reflects the mechanism employed for satisfying its energy needs. On the basis of their oxygen requirements, bacteria may be divided into five groups:

1. Obligate anaerobes that grow only under conditions of high reducing intensity and for which oxygen is toxic
2. Aerotolerant anaerobes that are not killed by exposure to oxygen
3. Facultative anaerobes that are capable of growth under both aerobic and anaerobic conditions
4. Obligate aerobes that require oxygen for growth
5. Microaerophilic organisms that grow best at low oxygen tensions, high tensions being inhibitory

In the obligate and aerotolerant anaerobes, the metabolism is strictly fermentative. In the facultative anaerobe, however, a respiratory mode of metabolism is employed when oxygen is available, but in its absence fermentation occurs. The requirement of the microaerophilic aerobic organisms for a reduced oxygen tension is probably indicative of the presence in these organisms of enzymes that are inactivated under strongly oxidizing conditions.

When organisms are grown in the presence of air, a number of enzymatic reactions occur that result in the production of hydrogen peroxide and the superoxide radical. In aerobes and the aerotolerant and facultative anaerobes, the enzyme superoxide dismutase prevents the accumulation of the superoxide ion, but in the obligate anaerobe this enzyme is absent:

$$2\ O_2^- + 2\ H^+ \xrightarrow[\text{dismutase}]{\text{superoxide}} O_2 + H_2O_2$$

The hydrogen peroxide formed in the dismutase reaction is rapidly destroyed by the enzyme catalase, which is present in the aerobic and facultative anaerobes. Although some aerotolerant organisms, such as the lactic acid bacteria, lack catalase, they possess peroxidases that catalyze the destruction of H_2O_2, thereby enabling the organism to grow in the presence of oxygen (p. 55).

Carbon Dioxide

In addition to the chemolithotrophic and photolithotrophic bacteria that use CO_2 as the principal source of cellular carbon, chemoorganotrophs also have a requirement

for an adequate supply of CO_2 for heterotrophic CO_2 fixation and for the synthesis of fatty acids. Since carbon dioxide is normally produced during the catabolism of organic compounds, usually it does not become a limiting factor. Some organisms, however, such as *Neisseria* and *Brucella*, presumably have one or more enzymes with a low affinity for CO_2 and require for growth a higher concentration (10 percent) of CO_2 than is usually present in the atmosphere (0.03 percent). This need must be considered in the isolation and culture of these organisms.

Physical Requirements

Oxidation-Reduction Potential

The oxidation-reduction potential (E_h) of the culture medium is a critical factor in determining whether growth of an inoculum will occur when transferred to a fresh medium. For most media in contact with air, the E_h is about +0.2 to 0.4 volt at pH 7. Obligate anaerobes are unable to grow unless the E_h is at least as low as −0.2 volt. In order to establish anaerobic conditions in a culture, oxygen may be excluded by the use of anaerobic culture systems or by the addition of sulfhydryl-containing compounds, such as sodium thioglycollate (mercaptoacetate). During growth of both aerobic and anaerobic bacteria, there is a progressive decrease in the oxidation-reduction potential of the environment, an observation that is of extreme clinical importance in wound infections where a mixed population of aerobic and anaerobic organisms is capable of setting up an infection in an initially aerobic setting (Chap. 46).

Temperature

For each bacterium, there is an optimal temperature at which the organism grows most rapidly and a range of temperatures over which growth can occur. Cellular division is especially sensitive to the damaging effects of high temperature; very large and bizarre forms are often observed in cultures grown at a temperature higher than that supporting the most rapid division rate.

Bacteria are divided into three groups on the basis of the temperature ranges through which they grow: psychrophilic, −5 to 30C, optimum at 10 to 20C; mesophilic, 10 to 45C, optimum at 20 to 40C; and thermophilic, 25 to 80C, optimum at 50 to 60C. The optimum temperature is usually a reflection of the normal environment of the organism. Thus, bacteria pathogenic for man usually grow best at 37C. One very practical example of the importance of temperature on the growth of microorganisms in vivo is found in studies with *Mycobacterium leprae*. Growth of this organism in vivo is temperature-dependent, as reflected by the distribution of lesions in clinical cases of leprosy. The skin usually shows the most obvious lesions, whereas the internal organs are not involved. The usual laboratory animals are not susceptible to infection with leprosy bacilli, but by inoculation of the foot pads of mice, a site with a reduced body temperature, successful passage of the organism can be obtained.

Hydrogen Ion Concentration

The pH of the culture medium also affects the growth rate, and here also there is an optimal pH with a wider range over which growth can occur. For most pathogenic bacteria the optimal pH is 7.2 to 7.6. Although a given medium may be initially suitable for growth, subsequent growth may be severely limited by metabolic products of the organisms themselves. This is especially pronounced in bacteria exhibiting a fermentative type of metabolism where large amounts of inhibitory organic acids accumulate.

Osmotic Conditions

The concentration of osmotically active solutes inside a bacterial cell is, in general, higher than the concentration outside the cell. Except for the mycoplasmas and other cell wall-defective organisms, the majority of bacteria are unusually osmotically tolerant and have evolved complex transport systems and osmotic sensor-regulating devices for the maintenance of constant osmotic conditions within the cell.

A hitherto unrecognized class of cell constituents, the membrane-derived oligosaccharides (MDO), has recently been discovered in *Escherichia coli*. In *E. coli* and other gram-negative bacteria, there are two distinct aqueous compartments, the cytoplasm contained within the inner membrane and the periplasmic space contained between the inner and outer membranes. When the organisms are grown in medium of low osmolarity, the cytoplasmic membrane, which has little mechanical rigidity, will swell unless prevented from doing so by an osmolarity of the periplasmic space, similar to that found in the cytoplasm. In cells grown in medium of low osmolarity, MDO is the principal source of fixed anion in the periplasmic space and thus acts to maintain the high osmotic pressure and Donnan membrane potential of the periplasmic compartment. These unique oligosaccharides are structurally well suited for their regulatory role. They have molecular weights in the range of 2200 to 2600 and are thus impermeable to the outer membrane, a property essential for their specific function. They consist of 8 to 10 glucose units in a highly branched structure, multiply substituted with membrane-derived phosphoglycerol and succinate residues, which allow them to maintain a high anionic charge. Cells grown in medium of low osmolarity synthesize MDO at a maximum rate, the rate of synthesis apparently being regulated at the genetic level in response to changes in osmolarity of the medium.

Uptake of Nutrients

Extracellular Enzymes

Bacteria, in common with all living organisms, are surrounded by a semipermeable membrane that restricts the entry of most molecules into the cell. Highly specialized systems have evolved for the transport of small molecules across the membrane barrier. The large molecules found in the organism's natural environment, however, cannot be used unless the organism produces exoenzymes that are liberated outside the cell. The development of such systems has permitted certain bacteria to occupy specific ecologic niches and has played an important role in the conservation and recycling of carbon, nitrogen, and other elements.

Enrichment Culture. If one looks in the right place and in the proper way, one will always find some organism that can break down any selected naturally occurring substance. This hypothesis is the basis for the enrichment culture technique used in isolating organisms capable of breaking down macromolecules of high molecular weight. In attempting to find an organism that can degrade a particular substrate, the substrate is added to a medium containing essential inorganic salts and inoculated with a mixture of organisms derived from a source in which it is probable that destruction of that substrate has been occurring. Soils and muds are the most useful sources of inoculum, and it was from such a source that organisms were isolated capable of degrading the pneumococcal polysaccharide capsule.

Location. Whereas gram-negative procaryotes can retain high concentrations of soluble secretory proteins in their periplasmic space, gram-positive procaryotes may compensate for the absence of an equivalent storage compartment by anchorage of these proteins to the plasma membrane. They may also be able to achieve effective local concentrations in the surrounding medium by regulating the amount of membrane-bound or secreted molecules in response to various stimuli. Distinction between membrane-bound and periplasmic enzymes may be made by determining whether the enzyme is extracted by procedures that remove the cell wall and retain the intact cytoplasmic membrane of the protoplast.

Properties and Synthesis. Among the substances attacked by exoenzymes are polysaccharides, such as cellulose, starch, and pectins, mucopolysaccharides, including chitin, hyaluronic acid, and chondroitin sulfate, proteins, lipids, and nucleic acid. A number of the more invasive pathogenic bacteria, such as *Streptococcus pyogenes*, *Staphylococcus aureus*, and certain *Clostridium* species, elaborate a variety of exoenzymes that destroy vital components of the body tissues and thus contribute to the overall pathogenesis of infection.

Exoenzymes may be constitutive or inducible, and in most cases the rate of synthesis appears to be regulated by end product inhibition and catabolite repression. No universal statement can be made concerning the stage of the growth cycle during which exoenzymes are produced. Although many are formed toward the end of the logarithmic phase of growth, such as the lecithinase of *Clostridium perfringens* and the hyaluronidase of *S. aureus*, others, including the nicotinamide adenine dinucleotidase of *S. pyogenes* and the proteinase of *Clostridium botulinum*, are formed in approximately equivalent proportions during most of the growth cycle.

Only recently has there been any understanding of how exoenzymes are synthesized and secreted. For many of these studies the penicillinase of *Bacillus licheniformis* has served as an ideal model. It is a monomeric protein of known primary sequence coded for by a single structural gene, according to the signal hypothesis. In its synthesis the precursor penicillinase with an additional hydrophobic NH_2-terminal signal segment is formed on a membrane-bound ribosome and extruded into the membrane, with folding occurring at the outer surface. A lipid residue is attached during cotranslational secretion. The membrane enzyme can be cleaved by a protease to form exopenicillinase, which then passes through the cell wall and into the external medium.

Membrane Transport

Growth and survival of an organism depend upon its ability to transfer solutes from the external milieu into the cytoplasm. This transfer is not a matter of selective permeability, as is often inferred in the definition of the cytoplasmic membrane, but of transport. With only a few exceptions, such as water and ammonia, which enter the cell by passive diffusion in response to a concentration gradient, the passage of metabolites is accomplished by specific transport or carrier systems. During the course of bacterial evolution, a large number of very diverse transport systems have evolved for the capture of nutrients. The substrates for these systems range from the trace metals, vitamins, and major nutrients to the precursors of extracellular macromolecules. For these nutrients, translocation across the membrane constitutes the first step in metabolism.

Transport Systems
Porin and Maltose Channels. Although most of the specific transport systems of bacteria are energy-dependent, a few do not require metabolic energy. In this latter category are two systems located in the outer membrane of *E. coli*, the porin channel (p. 33) and the maltose channel (lambda receptor). The porin channel is quite nonspecific and admits all compounds with a molecular weight of 600 or less. Interaction of the maltose channel protein with the periplasmic maltose-binding protein confers specificity on this outer membrane pore (p. 72).

In the cytoplasmic membrane of *E. coli*, there is also at least one energy-independent transport system, the glycerol facilitator. Except for this, most bacterial transport systems are geared to the performance of osmotic work and allot a considerable fraction of their energy supply to the work of transport.

Facilitated Diffusion. This simple mode of transport is typified by the process of glycerol uptake. In *E. coli* a single membrane-associated protein facilitates the rapid equilibrium of substrate across the cell membrane (Fig. 5-3). Functioning in tandem with a cytoplasmic ATP-dependent kinase, it effects the capture of glycerol from the medium. Once phosphorylated the glycerol is trapped inside the cell. Glycerol kinase is an effective scavenger and pacemaker of carbon source consumption and has properties suitable for this role, a high affinity for substrate and susceptibility to remote feedback inhibition. The glycerol facilitator behaves in many ways as a membrane channel.

Phosphoenolpyruvate: Sugar Phosphotransferase System (PTS). This transport system mediates group translocation, a vectorial pathway in which metabolic sequences are oriented across the membrane so as to catalyze transport and chemical transformation concurrently. The sugars are apparently phosphorylated as they are translocated across the cell membrane.

COMPONENTS OF THE SYSTEM. The phosphotransferase system is a multiprotein complex consisting of four proteins, as shown in Figure 5-4. Two of these proteins are general proteins of the system required for the phosphorylation of all sugar substrates, whereas the other two proteins are sugar specific, a given pair being required for the transport of a particular sugar.

The two nonspecific components of the system, Enzyme I and HPr, are soluble proteins and are produced constitutively. Enzyme I catalyzes the transfer of the energy-rich phosphoryl group from phosphoenolpyruvate to HPr, a low molecular weight histidine-containing protein. Enzyme I itself appears to be phosphorylated during the transfer reaction (Reaction 1).

(1) Phosphoenolpyruvate + Enzyme I $\underset{\phantom{Mg^{2+}}}{\overset{Mg^{2+}}{\rightleftharpoons}}$ pyruvate
$+ \ P \sim$ Enzyme I

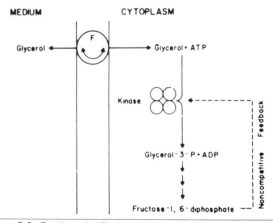

Figure 5-3. Facilitated diffusion of glycerol. A membrane-associated facilitator protein functions in tandem with a cytoplasmic kinase for the capture of glycerol from the medium. *(From Andrews and Lin: Fed Proc 35:2185, 1976.)*

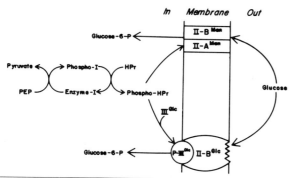

Figure 5-4. Schematic representation of the phosphotransferase system of *Salmonella typhimurium*. The phosphoryl group is sequentially transferred from PEP to Enzyme I, HPr, and then to one of the sugar-specific proteins, II-A^{Man} (an integral membrane protein) or III^{Glc} (a soluble and/or peripheral membrane protein). II-B^{Man} and II-B^{Glc} catalyze the transfer of the phosphoryl group from II-A^{Man} and III^{Glc} to the sugar (glucose) concomitant with the translocation of the sugar across the membrane. *(From Weigel et al.: J Biol Chem 257:14461, 1982.)*

$$(2) \quad P \sim \text{Enzyme I} + \text{HPr} \rightleftharpoons \text{Enzyme I} + P \sim \text{HPr}$$

$$(3) \quad \text{Phosphoenolpyruvate} + \text{HPr} \xrightarrow{\text{Enzyme I, Mg}^{2+}}$$
$$\text{pyruvate} + P \sim \text{HPr}$$

$$(4) \quad P \sim \text{HPr} + \text{sugar} \xrightarrow{\text{Enzyme II, Mg}^{2+}} \text{sugar-P} + \text{HPr}$$

The phosphorylated form of HPr serves as a phosphoryl group donor to a family of membrane-bound substrate-specific Enzyme II complexes, each of which consists of two distinct proteins. One of these receives the phosphoryl group in covalent form before its final transfer to the sugar.

At least one of the two sugar-specific proteins, Enzyme II-B, is an integral component of the plasma membrane and is responsible for recognizing and binding the sugar substrate. The other protein component of the sugar-specific pair is either membrane-bound (II-A proteins) or, in some organisms, is found in the soluble fraction (Factor III). Most of the sugar-specific protein pairs are inducible.

The PTS is widespread among microorganisms, being generally present in facultative aerobes and anaerobic organisms but not in the obligate aerobes. The intracellular accumulation of the corresponding sugar phosphate concomitant to translocation and phosphorylation is an energetically favorable pathway, permitting the uptake of an external sugar and its conversion to the first catabolic product in a single step. For sugars such as galactose that are not taken up via the PTS, two energy-requiring steps are required. The sugar phosphate product of PTS translocation is trapped and accumulates within the cell.

In addition to catalyzing the transport and phosphorylation of its sugar substrates, PTS is involved in a number of other processes. The proteins of the PTS function as a chemoreception system, permitting the organisms to recognize sugar substrates of the PTS in the extracellular en-

vironment and to swim up concentration gradients of these compounds. Another major function of PTS is to regulate the utilization of certain non-PTS substrates, such as glycerol, maltose, and melibiose in *S. typhimurium* and also lactose in *E. coli*. In culture media containing both a PTS and a non-PTS sugar, the PTS sugar is utilized before induction of the catabolic systems for the non-PTS sugar. This regulatory function, known as "the glucose effect" or "diauxic" growth, has been intensely studied for many years. It involves both the inhibition of uptake of the non-PTS sugar (inducer exclusion) and regulation of adenylate cyclase. The inhibition occurs directly at the level of the functional permease and is effected by a direct interaction between that transport protein and one or more PTS components.

In addition to the PTS, bacteria also contain other transport systems that operate via group translocation. Adenine is glycosylated to adenosine monophosphate in a vectorial reaction catalyzed by a membrane-bound enzyme. During transport fatty acids are converted to acyl-CoA by the acyl-CoA synthetase.

Active Transport (Substrate Translocation). Active transport resembles facilitated diffusion in that it also requires the participation of specific membrane-associated transport proteins. Active transport of solute, however, occurs at the expense of metabolic energy. A source of energy is required because the cell must do work in moving the substance through the cell membrane against a concentration gradient. Within the bacterial cell, concentrations of solutes may be several thousand times as great as those outside the cell. The substrate molecule is not altered during transport but appears in the cytoplasm in a chemically unchanged form.

Substrate translocation in bacteria is mediated by two distinct mechanisms. One of these requires only membrane-associated components. The second class of mechanisms, found in gram-negative organisms and utilized for the transport of a wide range of substrates, requires the participation of soluble binding proteins present in the periplasmic space.

β-GALACTOSIDE PERMEASE. This is the archetypal substrate translocation system (Fig. 5-5). It is a system by which lactose is transported by a carrier into the cell and accumulated to a concentration many times that in the medium. A single membrane protein or permease (M protein) coded for by the *y* gene of the *lac* operon, is responsible for the recognition and translocation of substrate. The carrier demonstrates typical Michaelis-Menten kinetics. Under conditions of energy abundance the binding site of the M protein has a high affinity for the substrate when it is oriented externally and a low affinity when it is internally oriented. This results in the energetically uphill movement of substrate into the cell. When the cell is deprived of energy, the function of the M protein is reduced to the role of facilitating diffusion of the substrate to a transmembrane equilibrium.

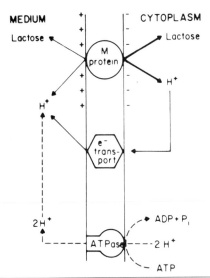

Figure 5-5. β-Galactoside permease system. A single membrane protein (M protein) is responsible for the recognition and translocation of substrate. *(From Andrews and Lin: Fed Proc 35:2185, 1976.)*

The energy-coupling mechanisms employed in active transport are best interpreted by the chemiosmotic theory of transport and metabolism proposed by Mitchell (p. 53). According to this concept the accumulation of solutes is brought about by coupling solute movement to proton movement down the electrochemical gradient. The electrochemical gradient or proton motive force can be generated by either oxidation or ATP hydrolysis.

PERIPLASMIC BINDING PROTEIN. In gram-negative organisms, a number of active transport systems are associated with binding proteins localized in the periplasmic space. These may be released by cold osmotic shock treatment, which damages the outer layer of the cell wall and releases proteins in the periplasmic space. Following osmotic shock and loss of these proteins, the uptake of a number of metabolites is impaired. The binding proteins are low molecular weight, water-soluble proteins that bind the metabolite with high affinity and specificity but have no known enzymatic activity. They are thus believed not to carry the substrate across the membrane but to function as very efficient scavengers in the recognition of substrate, which is then transported by a membrane-bound transport system. Unlike the β-galactoside permease system, shockable transport systems requiring the water-soluble binding protein specifically require ATP as an energy source.

Transport systems for which binding proteins are essential include those for galactose, maltose, sulfate, glutamine, and several amino acids. One of the best characterized of these is the maltose transport system, the only system in *E. coli* that is able to transport maltose across the cytoplasmic membrane. In this organism, the accumulation of maltose requires five proteins, including a periplasmic maltose-binding protein, an outer membrane protein, and

three plasma membrane proteins. The outer membrane protein also serves as the receptor for bacteriophage λ. The membrane-bound proteins appear to consist of both integral and peripheral membrane proteins that are present in smaller amounts than the corresponding periplasmic component. According to the model depicted in Figure 5-6, accumulation of substrate occurs as a result of a cycle of conformational changes of membrane proteins.

In the galactose system, the galactose-binding protein also acts as a signal receptor for chemotaxis, but there does not appear to be a general correlation between the presence of periplasmic binding proteins and chemotaxis.

Iron Uptake and Transport

Siderophores. Iron is a nutrient probably universally required by living cells. In aerobic environments and at neutral pH, however, the concentration of soluble iron is too low to achieve maximum growth rates. Under these conditions it is present in its ferric form (Fe^{3+}) and occurs primarily as highly insoluble hydroxides, carbonates, and phosphates. Since iron is involved in several critical stages in metabolism, microorganisms have evolved multiple systems for its acquisition. A low affinity system is present in most bacteria, permitting the organism to utilize the polymeric forms of iron in spite of the profound insolubility of Fe^{3+}. For this pathway, relatively high levels of iron are required to achieve maximum growth rates, and no specific solubilizing and transporting compounds or membrane receptors are required.

The high affinity iron assimilation system is comprised of two parts, the siderophore and the matching mem-

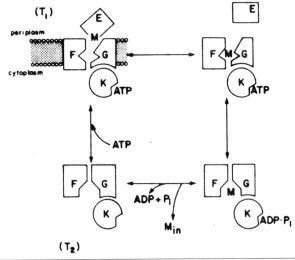

Figure 5-6. Model for the accumulation of maltose across the cytoplasmic membrane. In this model, the *mal F, mal G,* and *mal K* gene products form a complex in the cytoplasmic membrane. The *mal E* gene product, maltose-binding protein, interacts with this complex from the external periplasmic side of the membrane. *M* indicates a molecule of substrate, maltose, or longer maltodextrin.*(From Shuman: J Biol Chem 257:5455, 1982.)*

brane-associated receptors and transport apparatus. Siderophores are relatively low molecular weight (500 to 1000), virtually ferric-specific ligands whose function is to supply iron to the cell. They are viewed as the evolutionary response to the appearance of O_2 in the atmosphere and concomitant oxidation of Fe^{2+} to Fe^{3+}. Although considerable structural variation exists among the many siderophores that have been characterized, most are of two general structural types: catechols, of which enterobactin (Fig. 5-7) is the best characterized, and hydroxamates (Fig. 5-8), typified by ferrichrome, which are most common in fungi. Citrate also functions in some organisms as a high affinity carrier, but its receptor on the outer membrane is induced only by growth of the organism in the presence of the substrate.

Membrane Receptors. Enterobactin is a very powerful chelating agent produced very rapidly by *E. coli* under conditions of iron stress and secreted into the medium. The organism also has developed a scavenger ability to utilize siderophores, such as ferrichrome produced by other species, and possesses efficient uptake systems for the purpose. At least three system-specific receptors have been identified in the outer membrane of the cell envelope. The synthesis of these receptors and of enterobactin appears to be regulated coordinately by the intracellular iron concentration. Assimilation of the siderophore-iron complex implicates reduction of the metal iron and release into the cell after enzymatic hydrolysis of the ligand moiety. Uptake is strongly dependent upon an energized membrane state.

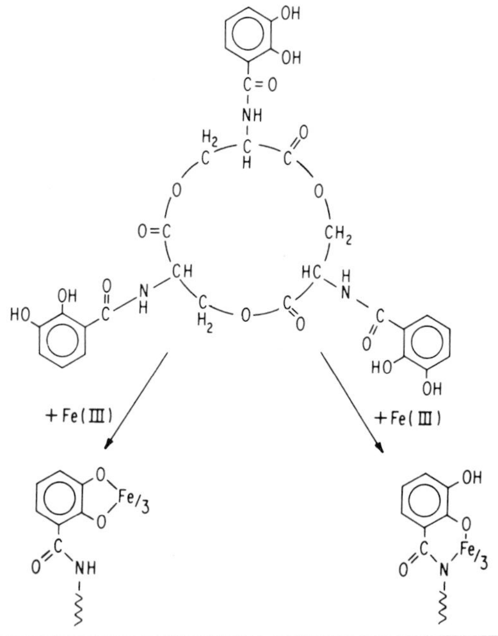

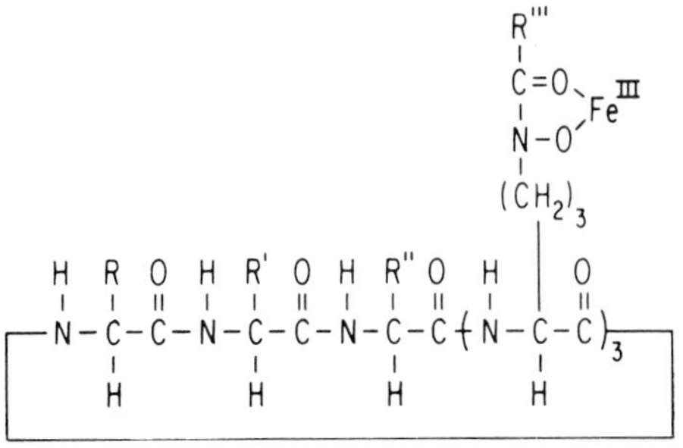

Figure 5-8. General structure of a hydroxamate siderophore. In ferrichrome, a hydroxamate produced by fungi, R = R^1 = R^2 = H; R^3 = CH_3.

In addition to their role in iron transport, the iron-siderophore receptors serve as receptors for certain bacteriophages and colicins. Mutants lacking a specific receptor are phage resistant and insensitive to the specific colicins and are simultaneously defective in iron transport. The system is in many ways analogous to that used in the transport of vitamin B_{12} in *E. coli.* In both systems, the large size of the molecule to be transported necessitates the requirement for outer membrane transport components. These transport components also function in the receptor systems for various bacteriophages and colicins.

Bacterial Chemotaxis

Motile bacteria have a well-developed sensory system that allows them to compete successfully in their natural environment. The system enables the organism to detect changes in concentration of certain chemicals and to move either toward (positive chemotaxis) or away (negative chemotaxis) from the substance, depending on its nature. Bacteria are attracted to many different kinds of chemicals, most of which can serve as nutrients. There is, however, no correlation between the metabolism of a substance and its ability to attract bacteria. Although most of the repellents that cause negative chemotaxis are toxic, toxicity is not essential for a negative response.

Chemotactic Response. Elegant quantitative methods have been used to follow the motion of bacteria by microscopic and photomicrographic techniques. In the absence of a stimulus a bacterium swims in a straight line for a few seconds and then turns abruptly, appearing to tumble head over tail for a fraction of a second before swimming off in a new direction. The bacterium responds to chemical stimuli by modification of this normal pattern of swimming. Bacteria tumble less frequently when they encounter increasing concentrations of attractant, and they tumble more frequently when the concentration decreases. This sensing of change in concentration is temporal, i.e., the organism has some kind of memory that allows it to compare the environment of its past with that of its present and to interpret this signal.

Figure 5-7. Enterobactin, the cyclic triester of 2,3-dihydroxy-*N*-benzoyl-1-serine, a phenolic iron transport compound. Two possible coordination geometries with Fe^{3+} are given. *(After Anderson et al.: Nature 262:722, 1976.)*

Sensory Apparatus

CHEMORECEPTORS. The component of the sensory system that recognizes the chemical and measures the change in concentration is a chemoreceptor located either in the plasma membrane or in the periplasmic space. Receptors are protein molecules specifically designed to receive signals only from those molecules or physical conditions that the apparatus needs to sense. In *E. coli,* there are approximately 20 attractant receptors and 10 repellent receptors. Most of the receptors are specific for one or two chemicals at high affinities, but they usually exhibit a limited range of substances with which they will react, some with appreciably lower affinity. The total environment sensed by the bacterium is, thus, a product of the specificity of each individual receptor multiplied by the repertoire of receptors present on its surface. A few of the receptors, such as those for aspartate and serine, are constitutive, but most, especially for the sugars, are induced by growth on a particular substrate. These receptors are present in substantial concentration. There are about 10,000 molecules of periplasmic galactose, ribose, and maltose receptors per cell when fully induced and about 5000 molecules of aspartate and serine receptors per cell. For such sugars as maltose, ribose, and galactose, the chemoreceptor is a small soluble protein located in the periplasmic space. These are the same binding proteins active in the uptake of the corresponding sugar, although uptake is not necessary for taxis. Other chemoreceptors are integral membrane proteins, as in the case of amino acids and those sugars transported into the cell via the phosphotransferase system. Transport and chemotaxis are thus very closely related.

TRANSDUCER PROTEINS. Three transducer proteins, or methyl-accepting chemotaxis proteins (MCPs), play central roles in the processing of transmembrane signals, acting as the comparator in the sensory system and relaying information to the flagellar apparatus about changes in the concentration of chemoeffectors (Fig. 5-9). These integral membrane proteins are products of the *tsr* (MCP I), *tar* (MCP II), and *trg* (MCP III) genes, and each is specific for mediating signals from a different set of stimuli. The transducer proteins receive signals from the chemoreceptors, which presumably induce a conformational change in the transducer proteins. As a result, posttranslational methylation of a glutamyl residue by methyltransferase and the methyl donor, *S*-adenosylmethionine, occurs. The degree of methylation reflects the cell's environment and increases until it reaches a plateau level that is a function of the receptor occupancy. Adaptation to the stimulus is complete, and prestimulus behavior is resumed when methylation reaches the plateau, and protein methyltransferase activity is balanced by the activity of a protein methylesterase. There is thus a dynamic process of methylation and demethylation occurring constantly. Control of these processes is the mechanism that permits response and adaptation.

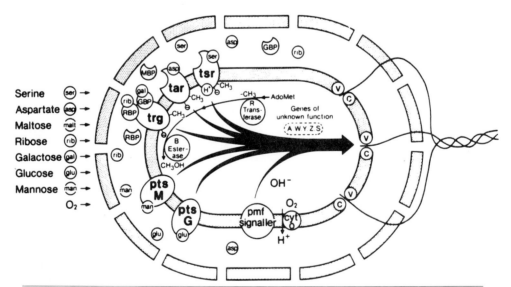

Figure 5-9. Schematic representation of sensory transduction in chemotaxis in *E. coli* and *S. typhi-murium.*The chemoeffectors (e.g., serine, aspartate) can cross the outer membrane and bind to receptors in the periplasmic space (RBP, GBP, MBP) or in the membrane (e.g., pts M, trg). For signals processed through MCPs (tsr, tar, trg), adaptation is dependent on methylation or on demethylation catalyzed by protein methyltransferase (R) or protein methylester (B), respectively. For signals processed through other signaling proteins (pts M, pts G, pmf signaler), adaptation may be independent of the activity of R. All sensory transduction pathways converge prior to C and V, which represent the switch on the flagellar motor. A, W, Y, Z, S are chemotaxis gene products with unassigned functions in signal processing. *(From Niwano and Taylor: J Biol Chem 79:11, 1982.)*

Information from the three transducer proteins converges on the central components of the sensory system, producing an immediate effect upon flagellar rotation. However, the nature of this signal from the transducer to the flagellar structure remains obscure. Flagellar rotation is thought to be driven by a rotary motor located in the basal structure of the flagellum (p. 26). In a free-swimming cell, all of the flagella come together to form a synchronously rotating bundle of filaments that drives the cell through the medium. During smooth swimming, the flagella are all rotated counterclockwise. A reversal of rotation of one or more filaments disrupts the flagellar bundle and leads to tumbling. The chemotactic response that stems from the regulation of tumble frequency thus arises as a result of the regulation of flagellar reversal. Addition of an attractant results in suppression of tumbling by causing counterclockwise rotation of the bacterial flagella, whereas addition of a repellent causes increased tumbling, due to a clockwise rotation.

Growth of Bacterial Populations

Measurement of Bacterial Growth

Unlike the higher forms of life, bacteria do not have an obligatory life cycle. When placed in a nutritionally complete medium, a bacterial cell grows larger and eventually divides to form two cells. This continues, with the production of a population of vegetative, undifferentiated cells.

In the development of a bacterial culture there is an increase both in cell mass and in number of organisms, but there is no constant relationship between the two parameters. In quantitative studies dealing with cell growth, it is therefore necessary to distinguish between cell concentration, or the number of cells per unit volume of culture, and bacterial density, defined as total protoplasm per unit volume. In most biochemical studies of bacteria, the significant variable is bacterial density. However, in problems concerned with the genetics or infectivity of bacteria, it is necessary to know the actual cell concentration.

Determination of Bacterial Mass. Although various techniques are in general use for estimating bacterial densities and cell concentrations, no single method determines mass and number in a single operation. Cell mass can be determined directly in terms of dry weight. This method, although time consuming, is especially useful for reference in isolation and purification work and in the basic calibration of other methods.

The most widely used method for estimating total microbiologic material in suspension is the measurement of optical density of a broth culture in a spectrophotometer. Turbidimetric techniques are especially useful in determining mass of cells during growth, as in the evaluation of the effects of drugs on bacteria. Other methods, such as nitrogen determination and measurement of cell volume after centrifugation, are useful when problems are encountered with the clumping of cells or light absorption by colored materials in the turbidimetric assay. In permeability studies, data on packed cell volume have proved useful.

Determination of Cell Number. The number of organisms in a culture may be determined either by total direct count or by indirect viable count. Total count of both living and dead organisms may be made by use of a bacterial counting chamber, such as the Petroff–Hauser counter, or more conveniently by the Coulter counter, an electronic particle counter that measures both the distribution of sizes and the numbers in bacterial suspensions.

For determination of viable numbers, it is necessary to plate out a sample of the culture. The microbial population is diluted in a nontoxic diluent, and an aliquot of the diluted population is dispersed in or on a suitable solid medium, such that after incubation, each viable unit forms a colony. The number of viable individuals or clusters originally present is determined from the colony count and the dilution. Samples containing fewer than 100 microorganisms per milliliter, such as urine or clear natural water, often require concentration rather than dilution before counting. This is done by passing the sample through a sterile membrane filter of pore size capable of retaining all the bacteria, then transferring the membrane to an absorbent pad saturated with a nutrient broth.

Bacterial Culture Systems

Closed Systems

As bacteria are usually grown in the laboratory in batch culture, the conditions approximate that of a closed system. If a suitable medium is inoculated with bacteria and small samples are taken at regular intervals, a plotting of the data will yield a characteristic growth curve (Fig. 5-10). The changes of slope on such a graph indicate the transition from one phase of development to another. Usually logarithmic values of the number of cells are plotted rather than arithmetic values. Logarithms to the base 2 are the most useful, since each unit on the ordinate represents a doubling of population. The bacterial growth curve can be divided into four major phases: lag phase, exponential

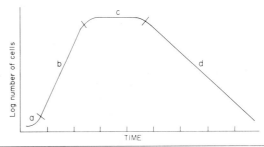

Figure 5-10. Bacterial growth curve, showing the four phases of growth: a, the lag phase; b, the exponential phase; c, the stationary phase, and d, the phase of decline.

growth phase, stationary phase, and phase of decline. These phases reflect the metabolic state of the organisms in the culture at that particular time.

Lag Phase. The growth curve of a bacterial culture based on numbers of organisms fails to reflect the dynamic state of the culture during this period. Following inoculation, there is an increase in cell size at a time when little or no cell division is occurring. There is a marked increase in macromolecular components, metabolic activity, and susceptibility to physical and chemical agents. Often referred to as a phase of rejuvenescence or physiologic youth, the lag phase is a period of adjustment necessary for the replenishment of the cell's pool of metabolites to a level commensurate with cell synthesis at a maximum rate. By taking a large inoculum from a logarithmic phase culture, the lag phase may be essentially eliminated. When inocula are taken from the period of decline, however, hours may elapse before growth is established.

Exponential Growth Phase. In the exponential or logarithmic phase, the cells are in a state of balanced growth. During this state, the mass and volume of the cell increase by the same factor in such a manner that the average composition of the cells and the relative concentrations of metabolites remain constant. During this period of balanced growth, the rate of increase can be expressed by a natural exponential function. The cells are dividing at a constant rate determined both by the intrinsic nature of the organism and environmental conditions. There is a wide diversity in the rate of growth of the various microorganisms. The doubling time for a broth culture of *E. coli* at 37C is approximately 20 minutes, as compared with a minimum doubling time of approximately 10 hours for mammalian cells at the same temperature.

Stationary Phase. When routine culture conditions are used, the accumulation of waste products, exhaustion of nutrients, change in pH, and other obscure factors exert a deleterious effect on the culture, resulting in a decreased growth rate. During the stationary phase, the viable count remains constant for a variable period, depending upon the organism, but eventually gives way to a period of decreasing population. In some cases the cells in dying cultures become quite elongated, abnormally swollen, or distorted, a manifestation of unbalanced growth.

Growth Rate and Doubling Time

Knowledge of the growth rate is important in determining the state of the culture as a whole. If one assumes the doubling of the initial mass M_i in time g, the final concentration of microorganisms M is:

$$(1) \qquad M = M_i 2^n$$

where n is the number of cell divisions in time t. The equation:

$$(2) \qquad g = \frac{t}{n}$$

expresses the doubling time or mean generation time and is the reciprocal of the growth rate constant ρ, which is usually expressed as the number of doublings per hour. The term "doubling time" represents the average generation time of the culture as a whole, usually determined by doubling of the microbial mass in the culture. The doubling time g is best determined by calculation. To accomplish this, the increase of cell mass is determined in a known time interval, and the generation time is calculated from the values obtained. Rearranging equation (2) to the form

$$n = \frac{t}{g}$$

which is substituted into equation (1), we have:

$$(3) \qquad M = M_i 2^{\frac{t}{g}}$$

By conversion to the logarithmic form and rearranging we obtain:

$$(4) \qquad g = \frac{\ln 2\, t}{\ln M - \ln M_i} = \frac{0.69t}{\ln M - \ln M_i}$$

Equation (4) is the formula for calculating the doubling time from two measurements that give the increase of the mass in time t. Measurements must be performed under constant conditions, and the amount of microorganisms is best determined as dry weight. From equation (4) may be derived the growth rate, ρ:

$$\rho = \frac{\ln M - \ln M_i}{0.69t}$$

For calculating the specific growth rate or exponential growth rate of an organism, the logarithmic form of equation (3) is used:

$$(5) \qquad \ln M = t \frac{\ln 2}{g} + \ln M_i$$

For the exponential phase of growth, the expression $\frac{\ln 2}{g}$ is a constant. Therefore, in equation (5) we can substitute μ. The resulting equation:

$$(6) \qquad \ln M = \mu t + \ln M_i$$

expressing the increase of mass within a certain time, is the equation of a straight line. When the values of t are plotted on the abscissa and the values of $\ln M$ on the ordinate, a straight line is obtained, and the constant μ is the slope of this straight line. It determines the growth rate of the bacterial mass as a function of time. It is, therefore, called the "specific growth rate" or "instantaneous growth rate" constant. Its value can be determined either graphically, by calculation:

$$(7) \qquad \mu = \frac{\ln 2}{g} = \frac{0.69}{g}$$

or it can be calculated directly from equation (6):

$$(8) \qquad \mu = \frac{\ln M - \ln M_i}{t}$$

where the time t is the interval $t_1 - t_2$ during which the bacterial mass M_i increases to the value of M.

The instantaneous growth rate μ is specific for every organism and culture medium. It is governed primarily by such factors as the growth capacity of the organism but also is affected by the environment. In order to express the real maximum value, the value corresponding to the logarithmic phase of the growth curve, the culture must grow in an unrestricted medium with substrate and growth factors in excess, so that the growth rate is independent of these.

Continuous Culture

In many different types of investigative work, it is advantageous to use organisms that are in their exponential phase of growth. With use of the routine batch culture techniques, however, it is impossible to maintain the culture at a steady state over a long period of time. Continuous culture techniques circumvent this problem and permit a bacterial population of fixed size to be grown for many generations under constant conditions and at a selected growth rate.

One of the most widely used continuous system cultivators is the chemostat (Fig. 5-11). This is an open-system apparatus consisting of a culture tube in which the population of organisms is maintained at a constant size by continuous dilution. This is accomplished by the admission of fresh nutrient medium into the culture tube at a defined and constant rate, with simultaneous removal by overflow of an equal volume of the bacterial suspension. In the chemostat, the most important factor controlling growth of the organisms is the rate at which fresh medium is added to the culture tube. The ratio of the rate at which fresh medium is added to the culture (f) to the operating volume of the culture (V) is referred to as the dilution rate (D):

$$D = f/V$$

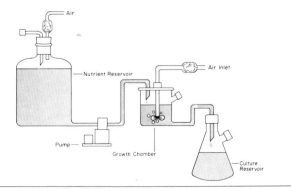

Figure 5-11. Simplified diagram of a chemostat.

The dilution rate is the number of volumes of medium that pass through the culture vessel in 1 hour. Its reciprocal, $1/D$, is the mean residence time of an organism in the culture tube.

In a chemostat, the organisms are growing, but they are also being washed out of the culture tube. The net change in the number of organisms with time is therefore determined by the relative rates of growth and washout:

$$(9) \qquad \frac{dN}{dt} = \mu N - \frac{f}{V}$$

Thus:

$$(10) \qquad \frac{dN}{dt} = \mu N - DN$$

where μ is the instantaneous growth rate constant and N is the number of organisms per milliliter. For the size of the population to remain constant in the culture tube, it is necessary that:

$$(11) \qquad \frac{dN}{dt} = 0, \text{ and}$$

$$(12) \qquad \mu = D$$

If the dilution rate is constant, the concentration of all of the components of the cell becomes constant, and a steady state is established that may be maintained for an extended period of time.

The key to an understanding of the mode of action of the chemostat lies in the way in which the growth rate depends on the concentration of a limiting growth factor in the culture medium. In a batch culture, substrate is consumed as the organisms grow. As a result, the substrate concentration continually decreases, accompanied by a parallel decrease in the growth rate. In a chemostat, however, the continued addition of fresh medium fixes the substrate concentration and the growth rate at some predetermined value that is less than the maximum growth rate. The nutrient medium is so constituted that it contains a large excess of all except a single required nutrient, which now becomes the growth-limiting factor. As long as the growth rate exceeds the dilution rate, the number of organisms increases with time. Since, however, an increase in the number of organisms in the culture tube results in a continuous decrease in the concentration of the growth-limiting factor in the tube, a decrease in the bacterial growth rate eventually results. The growth rate continues to fall until it is equal to the dilution rate, at which time the concentration of organisms in the culture tube stabilizes and remains at a constant level.

At very low nutrient concentrations, the specific growth rate is directly proportional to the concentration (c) of the nutrient. This relationship may be expressed in equation (10). so that:

$$(13) \qquad \frac{dN}{dt} = \mu(c) \, N - D$$

and in the steady state where:

$$(14) \qquad \frac{dN}{dt} = 0$$

$$(15) \qquad \mu(c) = D$$

The chemostat has been especially useful in the study of population genetics and of regulatory mechanisms that control the flow of material into the major classes of macromolecules.

Effect of Growth Rate on Bacterial Structure and Composition.

The structure and chemical composition of bacteria vary markedly with changes in the growth rate. Some of these changes are shown in Table 5-1, in which the growth rate of *S. typhimurium* was controlled by providing different carbon and nitrogen sources. The size of the cell and the average number of nuclear bodies per organism are both directly related to the organism's growth rate. The most sensitive indicator, however, of a change in growth rate is the ribosomal RNA. Except at very low growth rates, the number of ribosomes per milligram of protein is proportional to the growth rate.

The response of RNA to changes in growth rate can be demonstrated by the use of shift experiments, either a shift up if the change of medium leads to an increase in growth rate or a shift down if there is a decrease (Fig. 5-12). In both types of experiments, the first process to respond to the change in environment is the synthesis of RNA. When bacteria growing in one medium at 37C under conditions of balanced growth are transferred to another medium that will support a higher growth rate, the transition is characterized by an instantaneous increase in the rate of RNA synthesis, followed more slowly by corresponding changes in the rates of protein and DNA synthesis. The slower increase in the rate of protein synthesis after a shift to a richer medium is closely correlated with an increase in total RNA per cell, which regulates the capacity of the translational apparatus. According to current ideas, one of the early physiologic effects of such a nutritional shift up is a greater extent of tRNA charging with amino acids, which reduces ribosome idling and thus synthesis of guanosine tetraphosphate. A low concentration of this nucleotide is presumed to increase the expression of rRNA and ribosomal protein genes, thereby stimulating ribosome synthesis (p. 147).

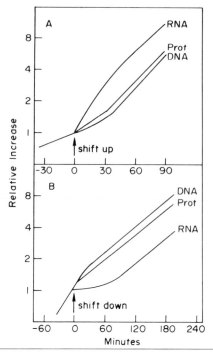

Figure 5-12. Relative rate of RNA, DNA, and protein synthesis in *E. coli* following a shift in medium. **A.** Cells are shifted up from a glucose-salts medium to a nutrient broth. **B.** Cells growing in nutrient broth are shifted down to a glucose-salts medium. The value of each component is normalized to 1.0 at 0 time.

Synchronous Growth

As bacteria are usually grown, cell division occurs at random, giving a mixture of cells representing all phases of the division cycle. Studies on such cultures yield average values only. Techniques are available, however, that synchronize division and induce all the cells in a bacterial culture to divide simultaneously. By withdrawing cells representing a single age class, a sample is obtained large enough for analysis by standard biochemical techniques. Two fundamentally different approaches have been employed. In the first, the cells are sorted out according to age or size, while in the second, synchrony is induced by manipulating environmental conditions. When applicable, the former procedures are preferable, since they are less likely to introduce distortions in the physiologic state of the cells.

TABLE 5-1. EFFECT OF GROWTH RATE ON CELL SIZE AND COMPOSITION OF *SALMONELLA TYPHIMURIUM*

Medium	Growth Rate (μ*)	Dry Weight per Cell (g $\times$ 10^{-15})	Nuclei per Cell	RNA per Nucleus (g $\times$ 10^{-15})
Lysine salts	0.6	240	1.1	22
Glucose salts	1.2	360	1.5	31
Nutrient broth	2.4	840	2.4	65
Heart infusion	2.8	1090	2.9	84

From Schaechter et al.: J Gen Microbiol 19:608, 1958.
*Growth rate expressed as generations/hour.

Selection by Size and Age

FILTRATION. One useful technique for synchronization of bacteria is based on size selection by filtration. This method takes advantage of the cyclic changes in size that accompany the division cycle. A culture is filtered through a stack of filter papers that retains the larger cells near the top of the pile and allows the small cells to pass through and to be collected in the filtrate. The cells in the filtrate, or those obtained by eluting selected papers, grow synchronously during subsequent incubation.

MEMBRANE ELUTION. One of the most widely used methods for obtaining synchronously dividing populations is that of Helmstetter, in which cells of uniform age are selected from a growing population. A population of cells is passed through a membrane filter of pore size small enough that the bacteria remain irreversibly bound on the surface of the membrane. When the filter is then inverted and medium is passed through, the only cells appearing in the eluted medium are the unbound sister cells, the youngest cells in an exponential phase culture. Under the proper conditions, new daughter cells can be removed from the population growing on the surface and, when incubated, will grow in a synchronous manner (Fig. 5-13). This technique has been very useful in studying the biochemical events accompanying the bacterial division cycle.

VELOCITY SEDIMENTATION. The use of velocity sedimentation for the separation of exponentially growing cell populations into different age classes on the basis of size and density differences is at present a very popular method, albeit a hazardous approach for selection synchrony. Its use is based upon the principle that the rate of sedimentation of an ideal spherical particle is proportional to the square of its radius and to the density difference between the particle and the suspending fluid. In addition to

sucrose density gradients, gradients employing lactose, glucose, dextran, Percoll, and others have been used.

Selection by Induction Techniques. Some techniques for obtaining synchronized cells have used shock treatments, such as temperature, starvation, and illumination. Although synchronized growth is induced by these techniques, and the degree of synchrony is usually as good as or better than that achieved by selection techniques, induction techniques often introduce physiologic abnormalities.

One of the earliest techniques utilized temperature and was based on the assumption that the processes that occur during the division cycle are differentially sensitive to temperature. If the temperature of an exponentially growing culture is reduced from 37C to 25C for 15 minutes and then returned to 37C, cell division of most of the cells is synchronized. The most successful methods for synchronizing bacterial growth with temperature shifts have involved repeated changes in temperature at intervals of a generation time.

Starvation treatments of various forms have also been used extensively. A phasing of cell division can be obtained in cultures of a thymine-requiring mutant following withdrawal and readdition of thymine to the cultures. The inclusion of rifampicin in this system yields populations of cells that are synchronized with respect to a single round of chromosome replication.

Bacterial Cell Cycle

Cells that are growing are destined to divide. Their growth rate, and thus the frequency at which they divide, depends on the species and environmental conditions. Within a short period, often as short as 20 minutes, a bacterium can create a complete duplicate of itself, which then in turn is capable of duplicating. In exponentially growing cultures, the organisms divide after each doubling of their cell volume. In rod-shaped organisms, the doubling of cell volume between successive divisions takes place entirely by a doubling of cell length.

Bacteria do not exhibit the characteristic cell cycle observed in eucaryotic organisms. Whereas in eucaryotic cells DNA synthesis is confined to the S phase of the cell cycle, in exponentially growing bacteria DNA synthesis occurs virtually throughout the entire division cycle (Fig. 5-14). In bacteria the duplication sequences do not necessarily follow one after the other but overlap, the amount of overlapping depending on the culture medium.

Bacterial Cell Division

The number of chromosomes per cell and the extent of their replication are determined by the physiologic state and age of the cell. Chromosomes of very slow-growing

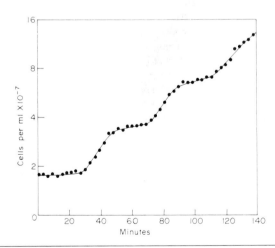

Figure 5-13. Cells of *E. coli* synchronized by the selection of newly formed cells eluted from membrane filter. *(After Helmstetter: In Norris and Ribbons (eds): Methods in Microbiology 1969, Vol 1. New York, Academic Press.)*

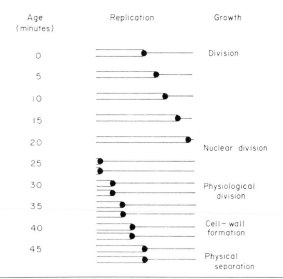

Figure 5-14. DNA replication in *E. coli* growing with a doubling time of 45 minutes. Newborn cells at 0 time possess half-replicated chromosomes. The end of a round occurs at 20 to 25 minutes. Nuclear division occurs, and a new round of replication begins. At the time of division, cells contain two half-replicated chromosomes. *(After Clark: J Bacteriol 96:1223, 1968.)*

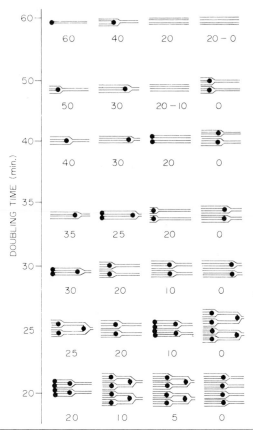

Figure 5-15. Relationship between chromosome replication and the division cycle in *E. coli* B/r growing at different rates. The time required for a replication point to proceed from the original to the terminus of the genome is 40 minutes; the time between the end of a round of chromosome replication and cell division is 20 minutes. The dot indicates a replication point, and the numbers indicate the time in minutes prior to cell division. *(After Cooper and Helmstetter: J Mol Biol 31:521, 1968.)*

cells have a single replication point. In such cells, replication occupies only part of the cell cycle, and gaps in DNA synthesis occur. Cells in rapidly growing cultures, however, contain multiple replication points. Because replication occurs simultaneously at several points along these chromosomes, the entire chromosome can be replicated in a fraction of the time that would be required if there were but a single replication fork. Genetic material can thus be duplicated at rates that otherwise could not be attained.

In bacteria, cell division and DNA replication are closely coordinated in order to insure that each daughter cell receives an equal portion of the chromosome. The control of DNA replication in bacteria is concerned primarily with the control of initiation of rounds of replication. The precise times at which initiation takes place are correlated with the mass of the cell.

Chromosome Replication

In bacteria, chromosome replication is a bidirectional process involving two replication forks that move in opposite directions around the chromosome. In *E. coli* with a generation time of 60 minutes, approximately 40 minutes is required for one of these replication forks to travel one-half the length of the chromosome. Because of the simultaneous movement of the two forks, a complete doubling of the DNA molecule is accomplished during this 40-minute period. Cell division takes place about 20 minutes after the completion of each round of replication. During this 20-minute period there is no further replication of the chromosome. Replication is initiated at 60 minutes when the cell has attained twice its initial unit mass, i.e., when the ratio of number of unit mass equivalents to number of chromosome origins reaches 1 (Fig. 5-15).

An explanation of these observations is provided in the model of Cooper and Helmstetter, which explains the relationship between DNA replication and the cell division cycle of *E. coli*. In this model, if the generation time is 60 minutes or less, there are two events of fixed duration: *C*, which equals 40 minutes and is the time required for one round of DNA replication to be completed after initiation at the origin, and *D*, which equals 20 minutes and is the time interval between completion of a round of replication and cell division. The *C* and *D* values of 40 and 20 minutes, respectively, apply to cells with generation times of 20 to 60 minutes. In the faster-growing cells, which are obtained by the use of a richer medium, growth is twice as fast, with a doubling of mass every 30 minutes. In such cells, the first round of chromosome replication is initiated at zero time, and the first replication forks meet at the terminus 40 minutes later. By 30 minutes, however, the cell will have doubled its mass and will start new rounds of replication at the two copies of the chromosome origin that were formed as soon as the first round of replication began. The result is that between 30 and 40 minutes, there are three pairs of replicating forks on the chromosome. Therefore, in media where the mass doubling time is less than 40 minutes, successive rounds of chromosome replication will overlap to give a dichotomously replicating chromosome (Fig. 5-15).

In summary, as presently conceptualized, a duplication sequence begins with the initiation of chromosome synthesis at a unique site, the replicator site of the chromosome. Initiation of chromosome replication requires protein synthesis and takes place when the ratio of cell mass to the number of initiation sites reaches a critical value. Chromosome replication begins when a fixed amount of initiator protein substances has accumulated regardless of the position of other replication points on the chromosome. Thus, in a given culture medium, the rate of cell division is determined by the time required for accumulation of the initiator substances. Initiation is not influenced by the presence, absence, position, or rate of movement of replication points on the chromosome.

Role of Cell Membrane in Chromosome Separation

Electron microscopy has failed to show any bacterial structure equivalent to the mitotic apparatus of eukaryotic cells. It is evident, however, that there must be some mechanism for ensuring equitable separation of chromosomes after each replication. At present, the replicon model of Jacob and associates provides the best explanation of how this might be achieved. According to this model, the circular DNA is attached to the cytoplasmic membrane, where replication occurs by passage of the DNA through an enzyme complex in the membrane. Following replication the newly formed DNA copy becomes attached to an adjacent site on the membrane. Separation and equipartition of the DNA copies result from the synthesis of new cell membrane in an annular region between the two attachment points.

Terminal Steps in Cell Division

In the commonly invoked model for the timing of some of the events leading to cell division in *E. coli*, two parallel sequences of events, initiated at the same time in the cell cycle, are required. It is assumed that both sequences occur through most of the cell cycle but that completion of DNA synthesis is required to produce a positive signal (a protein) that permits the final stages of septation. As shown in Figure 5-16, a 40-minute period of protein synthesis is followed by a final process that takes about 20 minutes to complete. Since neither RNA, DNA, nor protein synthesis is required during this period, it is assumed that it involves some sort of assembly of preformed components of the septum. The final stages of septation, however, require the participation of a terminal protein that is synthesized immediately after the completion of chromosome duplication at 40 to 55 minutes. According to this model, the rate-limiting process for division in the bacterial cell cycle is the completion of events in the temporal division sequence and not those dependent on chromosome replication.

Septum Formation. Compartmentalization of physiologic division of the cell occurs prior to the actual physical separation of the daughter cells. This is the time at which the cytoplasmic events at one end of the cell become independent of cytoplasmic events occurring at the opposite end. The end of a round of replication occurs well in advance of this compartmentalization. This stage of growth is characterized by the development of a weak septum followed by the formation of a strong crosswall.

Cell Wall Growth. During balanced growth, a greater surface area is normally required to cover a greater volume. In gram-positive cocci, cell wall growth appears to be limited to discrete areas at the equator, as demonstrated by staining the cell surface with fluorescent antibodies and observing the location of the label during subsequent

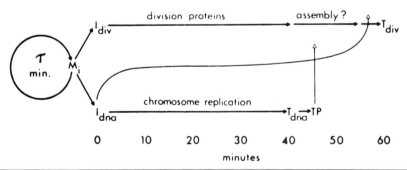

Figure 5-16. Model for the control of the timing of cell division in *E. coli*. The initiation mass (M$_i$) is attained every mass doubling time (T min), and at this time two parallel sequences of events are initiated. I$_{div}$ indicates the initiation of a sequence of events, required for division, which takes 60 minutes to complete and consists of at least three sequential processes. Initiation of chromosome replication (I$_{dna}$) sets up conditions that block the completion of a late step in the division sequence. However, termination of chromosome replication (T$_{dna}$) allows the synthesis of termination protein (TP), which relieves this inhibition, so that division will normally occur at the termination of the division sequence at 60 minutes. *(From Donachie: In Kolber and Kohiyama (eds): Mechanism and Regulation of DNA Replication, 1974. New York, Plenum Press.)*

growth (Fig. 5-17). A very detailed description of wall growth in *Streptococcus faecalis* has also provided evidence for a single, central growth zone (Fig. 5-18). In these studies, naturally occurring wall bands detectable by electron microscopy were used as surface markers to define the junction between old and newly synthesized wall material. According to the model proposed for cell wall growth in this organism, new peripheral wall is generated by the peeling apart of the crosswall at its base. Both peripheral wall elongation and centripetal crosswall extension apparently result from biosynthetic activity in the vicinity of the leading edge of the crosswall. An interplay of murein-hydrolyzing and murein-synthesizing enzymes has been invoked to explain the sequence of interrelated events associated with the enlargement of the cell wall. A closed molecule like the murein sacculus can be modified only when covalent bonds are split. In *S. faecalis*, murein hydrolase (*N*-acetylmuramidase) has been detected in zones of active murein synthesis.

In gram-negative organisms, there is no clearcut picture of how the surface of the cell grows. There is evidence for zonal growth of the mucopeptide layer of the cell envelope, but the lipopolysaccharide is added by diffuse intercalation during cell elongation.

Separation of Bacteria. Separation of the two daughter cells takes place some time after the completion of the transverse septum. The separation begins with a constriction between the daughter cells and a thickening of the cell wall in the constriction region. Further thickening of the wall accompanies constriction between the cells until the pole of the cell assumes its characteristic rounded shape. The normal thickness of the cell wall is achieved only after constriction is complete and the two daughter cells have separated. Separation of the daughter cells is

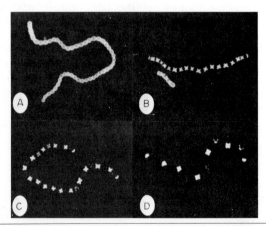

Figure 5-17. Growth of the wall of *S. pyogenes,* followed by ultraviolet photomicrography of growing chains of cells after initial labeling with fluorescent antibody. **A.** Immediately after antibody treatment, showing even fluorescence of cells. **B.** Cells after 15 minutes of growth. **C.** After 30 minutes. **D.** After 60 minutes. New (nonfluorescent) wall material is formed around the equator of each cell. *(From Cole: Bacteriol Rev 29:326, 1965.)*

mediated by murein hydrolases that are localized at the site of septum formation and are latent in nondividing cells.

Asymmetrical Cell Division

Regulation of the site at which cell division occurs is under genetic control. In rod-shaped organisms, division sites normally arise in an equatorial position, partitioning the daughter genomes into cells of approximately the same length. A number of bacterial mutants have been described, however, in which the division site location is abnormal (Fig. 5-19). The most extensively studied of these are the minicells of *E. coli*. These are tiny, spherical cells produced from the ends of the parent, rod-shaped cells. *E. coli* minicells lack host chromosomes although they may contain plasmid DNA. Minicells lacking DNA are unable to synthesize any polymer whose synthesis is DNA-dependent.

Minicell formation also occurs in mutants of *Bacillus subtilis* as a result of abnormal division site location. Motility is retained in the *B. subtilis* minicells, an indication of functional energy metabolism. Such cells are produced by a structurally normal division mechanism and contain a normal cell surface. Minicells have been very useful in many areas of experimentation, although the mechanism responsible for abnormal division site location is unknown.

Differentiation in Bacterial Cells

Sporulation

One of the most unique properties of certain bacteria is their ability to form endospores. At some point in the vegetative cell cycle of sporeforming organisms, growth is arrested and the cell undergoes progressive changes that result in the formation of an endospore (Fig. 5-20). A spore is a dormant structure capable of surviving for prolonged periods and endowed with the capacity to reestablish the vegetative stage of growth under appropriate environmental conditions. The process involved in sporulation, as well as the breaking of the spore's dormancy and subsequent emergence of a vegetative cell, represents a primitive example of unicellular differentiation.

Properties of Endospores

Endospores are formed by species of *Bacillus, Clostridium,* and *Sporosarcina* during their stationary phase of growth after depletion of certain nutrients in the culture medium. A single spore is produced within a vegetative cell and differs from the parent cell in its morphology and composition, increased resistance to adverse environments, and absence of detectable metabolic activity. Although the thermal resistance of spores has been of primary concern

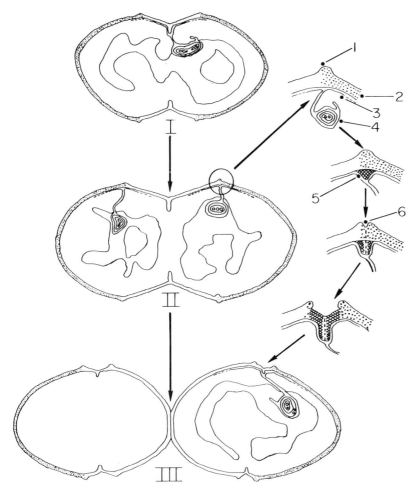

Figure 5-18. Diagrammatic representation of cell wall division model for *S. faecalis*. This simulated time-lapse sequence was reconstructed from a large number of electron micrographs. **1.** Wall band. **2.** Cell wall. **3.** Cell membrane. **4.** Mesosome. **5.** New wall synthesis. **6.** Wall notch. *(Adapted from Higgins and Shockman: Crit Rev Microbiol 1:29, 1971.)*

to the medical microbiologist in his efforts to control infection caused by spore-producing organisms, the increased resistance of spores to the lethal effects of desiccation, freezing, radiation, and deleterious chemicals is of greater importance in their natural environment. The primary selective value of the spore lies in its longevity in the soil coupled with its ability to germinate under the proper environmental conditions.

Basis of Spore Resistance. At the present time, no single theory provides an adequate explanation of the mechanism of spore resistance. Although one mechanism may be of primary importance, other mechanisms also may be involved in achieving and maintaining the resistant state.

In the sporulating cell, resistance to various chemical and physical agents appears at different stages, concomitant with changes in the physicochemical composition of the cell. Resistance to radiation, drying, and toxic chemicals appears after the cell becomes refractile and depends, at least in part, upon the properties of the cystine-rich spore coat proteins. Thermal resistance is probably due to the very low content of water, which renders the proteins and nucleic acids more resistant to denaturation. This re-

duction in water content occurs late in sporulation at the time of cortex formation and at the time that the spore first appears as a refractile object. The massive synthesis during this stage of dipicolinic acid, a spore-specific component, may be responsible for the dehydration. Dipicolinic acid is a chelating agent present in high concentrations in all bacterial endospores (Fig. 5-21). It is present as the calcium salt in the core of the mature spore and accounts for as much as 10 percent of the dry weight of the spore. Both calcium and dipicolinic acid are required for heat resistance.

Metabolic Differences Between Vegetative Cells and Spores

As a bacterial cell passes from the vegetative to the sporulating state, dramatic changes occur in its morphology and physiology. During sporulation, certain storage polymers, such as poly-β-hydroxybutyrate, accumulate and are later utilized. Extensive turnover of macromolecules occurs, and some enzyme levels change drastically. New and characteristic spore structures are synthesized, while previously existing structures are degraded.

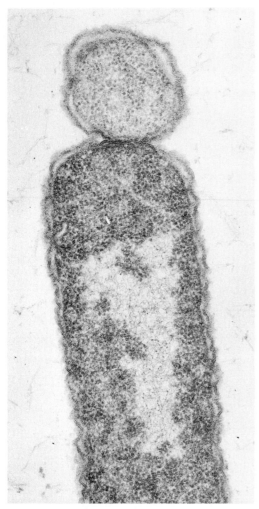

Figure 5-19. Thin section of an *E. coli* minicell-producing cell (X 925) dividing to yield a minicell. × 49,140. Electron micrograph taken by D. P. Allison. *(From Frazer and Curtiss: Curr Top Microbiol Immunol 69:1, 1975.)*

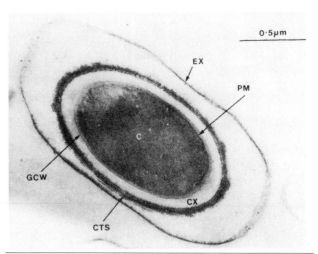

Figure 5-20. Electron micrograph of spore of *Bacillus cereus* showing the core (C), plasma membrane (PM), germ cell wall (GCW), cortex (CX), coats (CTS), and exosporium (EX). *(From Warth: Adv Microb Physiol 17:1, 1978.)*

The pools of small molecules found in spores are distinctly different from those of vegetative cells. There is an accumulation of dipicolinic acid, divalent cations, and high levels of L-glutamic acid. The predominant component of the drastically reduced acid-soluble phosphate pool is 3-phosphoglyceric acid, which constitutes as much as 75 percent of the acid-soluble phosphate in the spore as compared with only 5 percent in the vegetative cell. Conversely, spores contain very low levels of ATP, whereas in vegetative cells, ATP is a major component.

During sporulation many differences may be observed in the pattern of enzyme activities. Some of these are associated with mechanisms related to spore formation, whereas others are specific components of the spore itself. A heat-stable catalase found in spores is immunologically distinct from the vegetative cell enzyme, and certain enzymes, such as glucose dehydrogenase, a ribonuclease, and spore lytic enzymes are present solely in the spore. These en-zymes as well as the spore coat proteins represent spore-specific gene products.

Spore-specific Structures

The developing spore (forespore) is unique in being surrounded by two membranes of opposite polarity relative to each other. The outer forespore membrane has reversed surface polarity relative to both the inner forespore membrane and the mother cell plasma membrane. As a result of this orientation, transport from the mother cell into the spore can occur only by passive or facilitated diffusion. The accumulation of high levels of calcium in the developing spore can be accounted for by this mechanism. In the mother cell, calcium enters by an active transport mechanism, is concentrated, and moves into the forespore compartment by facilitated diffusion, where it chelates with dipicolinic acid. The parallel synthesis of dipicolinic acid permits the forespore to continue withdrawing Ca^{2+} from the surrounding cytoplasm by chelation of the incoming cation (Fig. 5-22).

An important spore-specific structure is the cortex, the major component of which is peptidoglycan. Cortical peptidoglycan, however, possesses some unique structural properties not found in bacterial cell walls. Unlike the bacterial wall in which most of the muramic acid residues of the glycan chain are substituted with a tetrapeptide and 60 percent of these are then cross-linked, most of the muram-

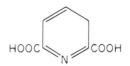

Figure 5-21. Dipicolinic acid (pyridine 2,6-dicarboxylic acid.)

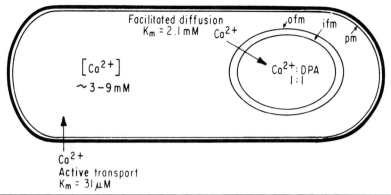

Figure 5-22. Model proposed for Ca^{2+} accumulation during sporulation. pm, plasma membrane of mother cell; ofm, outer forespore membrane; ifm, inner forespore membrane; DPA, dipicolinic acid. *(From Ellar: In Stanier et al. (eds): Relations Between Structure and Function in the Prokaryotic Cell, 1978. London, Cambridge University Press.)*

ic acid residues in cortical peptidoglycan are either present as the spore-specific muramic acid lactam or are linked to single L-alanine residues, and only 3 percent crosslinking occurs. This reduced degree of crosslinking may play a significant role in the compressive contraction of the cortex and dehydration during sporulation. In some spores, a layer of peptidoglycan close to the forespore inner membrane is of the vegetative peptidoglycan type and becomes the cell wall of the germinating spore.

The spore coat consists of layers of spore-specific keratin-like proteins that give to the spore properties of survival value. Their formation is initiated with the synthesis and accumulation of large amounts of a precursor that is processed by specific proteolytic enzymes to produce spore coat protein monomers, which are then integrated into the layers of the spore coats. Posttranslational changes lead to the incorporation of half-cysteine-rich polypeptides.

Initiation of Sporulation

Sporulation is a response to nutritional deprivation. Although limitation of any one of a variety of nutrients can initiate sporulation, the most pronounced effect is exerted by the available carbon and nitrogen sources. The regulation of spore formation is negative: the cell makes a repressor from some ingredient in the medium that prevents the initiation of sporulation. When this ingredient is exhausted, inhibition is released, and sporulation begins. A metabolizable supply of both carbon and nitrogen is required to inhibit sporulation. If there is a deficiency in the supply of either of these, inhibition will be relieved and sporulation initiated.

The specific factor that regulates the initiation of sporulation is guanosine triphosphate (GTP). A decrease in the GTP pool of growing cells suffices to initiate sporulation in *B. subtilis*. All conditions of nutritional deficiency known to initiate sporulation cause a GTP decrease at the time at which sporulation begins. Two types of nutritional deprivation are responsible for GTP decreases under conditions of nutrient limitation: (1) a decrease in the purine precursor, P-ribosyl-PP, caused by a limited carbon supply and (2) the stringent response to amino acid deprivation which is correlated with an increase in the concentration of the highly phosphorylated guanine nucleotides, ppGpp and pppGpp (p. 147). Under sporulation conditions, ppGpp inhibits the function of inosine 5'-monophosphate (IMP) dehydrogenase, the first enzyme in the synthesis, thus explaining the decrease of guanine nucleotides.

Biochemistry of Sporulation

At some period in the development of a cell, the metabolism is irreversibly channeled in the direction of sporulation. There is no single point of commitment for the sporulation process as a whole but a separate point of commitment for each new specific macromolecule. Accompanying these physiologic changes is an ordered series of cytologic and structural changes.

The morphologic changes observed in *B. subtilis* are presented in Figure 5-23 as a time scale divided into seven sequentially appearing stages. The process is timed from the end of exponential growth at t_0 and at hourly periods thereafter.

Stage 0. This represents the state of the cell at the end of exponential growth at a time when each vegetative cell contains two chromosomes.

Stage I. This stage is characterized by a broad axial chromatin filament that occupies the center of the cell. The most characteristic events of this stage are the production and secretion of antibiotics and various exoenzymes, especially proteases. Proteases play an important role in the intracellular turnover of protein, but the correlation between sporulation and antibiotic production is not known.

SECTION I   BACTERIAL PHYSIOLOGY

STAGE	MORPHOLOGIC EVENT	BIOCHEMICAL EVENT
	Vegetative cell	
	Chromatin filament	Exoenzymes Antibiotic
	Spore septum	Alanine dehydrogenase
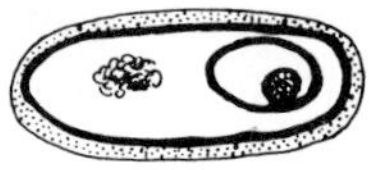	Spore protoplast	Alkaline phosphatase Glucose dehydrogenase Aconitase Heat-resistant catalase
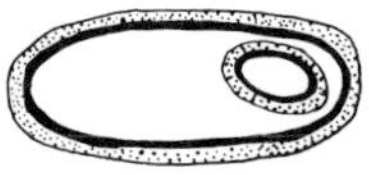	Cortex formation (refractility)	Ribosidase Adenosine deaminase Dipicolinic acid
	Coat formation	Cysteine incorporation Chemical resistance
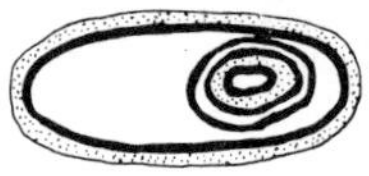	Maturation	Alanine racemase Heat resistance

Figure 5-23. Morphologic and biochemical events associated with sporulation. A composite diagram of data obtained with different species of *Bacillus. (After Mandelstam: Microbial Growth, 1969. London, Cambridge University Press.)*

Stage II. This stage is characterized by the formation of a forespore septum near one pole of the cell, resulting in the segregation of the nuclear material into two compartments, the mother cell and the forespore units. The forespore septum is formed de novo. During this period there is a marked increase of alanine dehydrogenase, an enzyme important in germination.

Stage III. During this stage the spore protoplast is formed as a consequence of the unidirectional growth of the cytoplasmic membrane of the mother cell around the forespore (Fig. 5-24). The complete engulfment of the forespore results in a cell bounded by two membranes, the inner one of which becomes the cytoplasmic membrane of the germinating spore. During this stage, the activity of a number of enzymes is greatly elevated. One of these, aconitase, is an example of an enzyme appearing during

the early stages of sporogenesis, which, like sporulation itself, is subject to catabolite repression. With the exhaustion of the carbon and nitrogen supply at the end of vegetative growth, repression is relieved, aconitase is formed, and functional tricarboxylic acid and glyoxylic acid cycles are produced.

Stage IV. A spore-specific peptidoglycan is laid down between the inner and outer forespore membranes and constitutes the major component of the cortex layer. The spore appears for the first time as a refractile body. During this stage, the synthesis of dipicolinic acid is initiated, and calcium begins to accumulate. At the end of this stage, metabolic activity of the forespore is markedly reduced and, subsequently, probably contributes little to its own development.

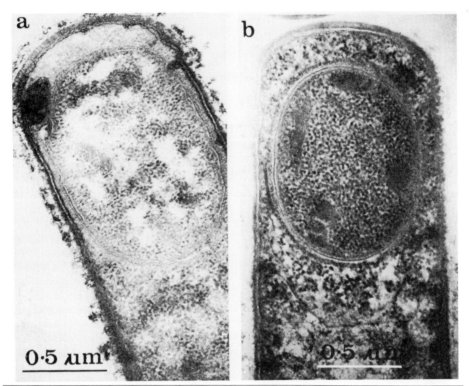

Figure 5-24. Electron micrographs of sporulating cells of *B. cereus*. **A.** After 6 hours of spore formation. **B.** One hour later, the volume is reduced 51 percent, the cytoplasm is more tightly packed, and the DNA has a fibrous structure and peripheral location. The cortex is just beginning to form. *(By D.F. Ohyn. From Warth: Adv Microb Physiol 17:1, 1978.)*

Stage V. During this period, the deposition of spore coat protein and the incorporation of cystine lead to a completed coat structure.

Stage VI. Spore maturation occurs, during which the cytoplasm of the spore protoplast becomes more homogeneous and electron-dense. There is continued synthesis of or modification of cortical peptidoglycan and uptake of dipicolinic acid and calcium. Alanine racemase, an enzyme important in spore germination, is formed, and a number of ill-defined changes occur that make the spore resistant to organic solvents and to heat.

Stage VII. The mature endospore is liberated from the mother cell by a lytic enzyme synthesized or activated subsequent to maturation of the endospore.

Genetic Regulation of Sporulation

Most of our current knowledge of the molecular genetics of sporulation has been obtained from studies on *B. subtilis*. By use of asporogenous mutants blocked at different stages of the developmental cycle and transformation and transduction techniques, a linkage map has been constructed that shows the position of many genetic loci for sporulation. The location of spore genes on the linkage map reveals a complex mechanism of expression, since the sequence of spore genes on the chromosome does not correspond with the time of gene expression.

Sporulation and germination genetic loci are not confined to one segment of the genome but are clustered in distinct and widely dispersed regions of the chromosome. The clustering of developmental genes occurs in about five chromosomal segments, in regions distinct from essential functions. There are between 40 and 60 of these clusters (operons), but the number of genes within each is unknown.

Gene Expression During Sporulation. With the appearance of the forespore, the cell becomes a compartmentalized structure. In the compartment developing into the forespore, DNA replication is inhibited, and the substances synthesized include some spore-specific products. In the second compartment, the sporangium, replication of vegetative DNA continues, and products of the vegetative cell continue to be produced. Thus, in the sporulating cell both the vegetative and spore genomes are transcribed. The mRNA contains some of the same mRNA molecules as derepressed cells transferred from a rich medium to a poor one but, in addition, molecules specific to sporulation. There is no evidence that the mRNA produced late in sporulation is stable and retained in the spore for protein synthesis during outgrowth.

Role of RNA Polymerases. The changes that occur during sporulation result from the turnoff of certain vegetative genes and the expression of new classes of genes. This turnoff is thought to be caused by a change in the template specificity of the RNA polymerase. The change occurs not in the subunit structure of the core enzyme that is conserved but in the diminished activity of the major vegetative 55,000-dalton sigma factor. Inhibition of sigma factor activity may be attributed to a protein inhibitor that prevents its binding at the promoter site.

Although deficient in sigma factor, sporulating cells produce a modified form of RNA polymerase that exhibits distinctive transcriptional specificity. This altered specificity has been attributed to the presence during sporulation of multiple sigma-like factors that interact with *B. subtilis* core RNA polymerase to impart multiple transcriptional specificities at the level of promoter recognition (p. 127). One of the best characterized of these sigma-like factors is a 37,000-dalton subunit isolated from a modified form of RNA polymerase. The modified enzyme, which lacks the usual sigma factor, selectively transcribes two cloned genes whose expression is under early sporulation control. Neither of these genes is transcribed by the usual form of *B. subtilis* RNA polymerase containing a sigma factor (55,000 daltons).

Germination and Outgrowth

Simultaneous structural and physiologic changes also are manifested during the transformation of a dormant spore into a vegetative cell. The process of spore germination consists of three sequential phases: (1) an activation state that conditions the spore to germinate in a suitable environment, (2) a germination stage, during which the characteristic properties of the dormant spore are lost, and (3) an outgrowth stage during which the spore is converted into a new vegetative cell.

Activation is a reversible process that is essential for spore germination. Spores do not germinate or germinate very slowly unless activated either by heat or various chemical treatments. Activation probably involves the reversible denaturation of a specific but presently undefined macromolecule. Germination is an irreversible process triggered by the exposure of activated spores to specific stimulants, such as amino acids, nucleosides, and glucose. This is the stage during which the dormant stage is ended. During the early stages of germination there is a loss of refractility, swelling of the cortex, and appearance of fine nuclear fibrils. Accompanying these changes is a loss of resistance to deleterious physical and chemical agents, increase in the sulfhydryl level of the spore, a release of spore components, and an increase in metabolic activity. The germination of spores is not inhibited by antibiotics that perturb protein and nucleic acid synthesis, indicating that the enzymes responsible for germination are already present in the spore.

During outgrowth there is de novo synthesis of proteins and structural components that are characteristic of vegetative cells. During this stage the spore core membrane develops into the cell wall of the vegetative cell. Outgrowth is a period of active biosynthetic activity and is markedly inhibited by interference with the energy supply or by antibiotics that inhibit cell wall, protein, or nucleic acid synthesis.

If heat-activated spores are germinated under suitable conditions, a high degree of synchrony is obtained. Such a synchronous population provides a model system for the study of the initiation of transcriptional and translational events occurring during differentiation. Since the dormant spore is devoid of functional mRNA, the conversion of the spore to a vegetative cell during outgrowth requires transcription and de novo synthesis of gene products. The appearance of these gene products is ordered, determined by the time of transcription of particular genes. Ribosome synthesis starts early, vegetative cell wall synthesis begins later, and DNA synthesis begins just before division of the cell. During a stepwise doubling of cell number for several generations, the initiation of certain enzymes occurs at a specific time during each division cycle and results in a doubling of each enzyme during only a fraction of the total cycle.

FURTHER READING

Books and Reviews

Adler J: Chemotaxis in bacteria. Annu Rev Biochem 41:341, 1975

Brown CM, Macdonald-Brown DS, Meers JL: Physiological aspects of microbial inorganic nitrogen metabolism. Adv Microb Physiol 15:1, 1977

Bulla LA Jr, Bechtel DB, Kramer KJ, et al.: Ultrastructure, physiology and biochemistry of *Bacillus thuringiensis*. CRC Crit Rev Microbiol 8:147, 1980

Carlberg DM: Principles of optical and electrical methods for the determination of bacterial populations. In Lorian V (ed): Antibiotics in Laboratory Medicine. Baltimore, Williams & Wilkins, 1980, pp 55–72

Coleman R: Membrane-bound enzymes and membrane ultrastructure. Biochim Biophys Acta 300:1, 1973

Dalton H: Utilization of inorganic nitrogen by microbial cells. Int Rev Biochem 21:227, 1979

Dills SS, Apperson A, Schmidt MR, et al.: Carbohydrate transport in bacteria. Microbiol Rev 44:385, 1980

Doi RH: Genetic control of sporulation. Annu Rev Genet 11:29, 1977

Donachie WD, Masters M: Temporal control of gene expression in bacteria. In Padilla EM, Whitson, GL, Cameron IL (eds): The Cell Cycle. New York, Academic Press, 1969, pp 37–76

Ellar DJ: Spore-specific structures and their function. In Stanier RY, Rogers HJ, Ward JB (eds): Relations Between Structure and Function in the Prokaryotic Cell. London, Cambridge University Press, 1978, pp 295–325

Fridovich I: Superoxide dismutases. Annu Rev Biochem 44:147, 1975

Glenn AR: Production of extracellular proteins by bacteria. Annu Rev Microbiol 30:41, 1976

Guirard BM, Snell EE: Nutritional requirements of microorganisms. In Gunsalus IC, Stanier RY (eds): The Bacteria. New York, Academic Press, 1962, vol IV, pp 33–93

Harold FM: Membrane and energy transduction in bacteria. Curr Top Bioenergetics 6:83, 1977

Helmstetter CE: Methods for studying the microbial division cycle. In Norris JR, Ribbons DW (eds): Methods in Microbiology. London, Academic Press, 1969, vol I, pp 327–363

Helmstetter CE: Regulation of chromosome replication and cell division in *Escherichia coli*. In Padilla EM, Whitson GL, Cameron IL (eds): The Cell Cycle. New York, Academic Press, 1969, pp 15–35

Hengstenberg W: Enzymology of carbohydrate transport in bacteria. Curr Top Microbiol Immunol 77:97, 1977

Higgins IJ, Best DJ, Hammond RC, et al.: Methane-oxidizing microorganisms. Microbiol Rev 45:556, 1981

Higgins ML, Shockman GD: Prokaryotic cell division with respect to wall and membranes. CRC Crit Rev Microbiol 1:29, 1971

Konings WN: Active transport of solutes in bacterial membrane vesicles. Adv Microbiol Physiol 15:175, 1977

Koshland DE Jr: Bacterial chemotaxis. In Sokatch JR, Ornston LN (eds): The Bacteria. Mechanisms of Adaptation. New York, Academic Press, 1979, vol VII

Lampen JO: Phospholipoproteins in enzyme excretion by bacteria. In Stanier RY, Rogers HJ, Ward JB (eds): Relations Between Structure and Function in the Prokaryotic Cell. London, Cambridge University Press, 1978, pp 231–247

Lloyd D, Poole RK, Edwards SW: The Cell Division Cycle. Temporal Organization and Control of Cellular Growth and Reproduction. New York, Academic Press, 1982

Losick R, Pero J: *Bacillus subtilis* RNA polymerase and its modification in sporulating and phage-infected bacteria. Adv Enzymol 44:165, 1976

Maaløe O, Kjeldgaard NO: Control of Macromolecular Synthesis. New York, Benjamin, 1966

Malek I, Fenel Z: Theoretical and Methodological Basis of Continuous Culture of Microorganisms. New York, Academic Press, 1966

Mallette MF: Evaluation of growth by physical and chemical means. In Norris JR, Ribbons DW (eds): Methods in Microbiology. London, Academic Press, 1969, vol I, pp 521–566

Mandelstam J: Regulation of bacterial spore formation. In Microbiol Growth. London, Cambridge University Press, 1969, pp 377–402

Maurizi MR, Switzer RL: Proteolysis in bacterial sporulation. Curr Top Cell Regul 16:163, 1980

Neilands JB: Iron absorption and transport in microorganisms. Annu Rev Nutr 1:27, 1981

Neilands JB: Microbial envelope proteins related to iron. Annu Rev Microbiol 36:285, 1982

Oxender DL: Amino acid transport in microorganisms. In Hokin LE (ed): Metabolic Pathways. Metabolic Transport. New York, Academic Press, 1972, vol VI, pp 133–185

Postma PW, Roseman S: The bacterial phosphoenolpyruvate: Sugar phosphotransferase system. Biochim Biophys Acta 457:213, 1976

Priest FG: Extracellular enzyme synthesis in the genus *Bacillus*. Bacteriol Rev 41:711, 1977

Rogers HJ: Bacterial iron metabolism and host resistance. In Schlessinger D (ed): Microbiology—1974. Washington DC, American Society for Microbiology, 1975, pp 289–298

Rogers HJ, Ward JB, Burdett IDJ: Structure and growth of the walls of gram-positive bacteria. In Stanier RY, Rogers HJ, Ward JB (eds): Relations Between Structure and Function in the Prokaryotic Cell. London, Cambridge University Press, 1978, pp 139–178

Saier MH: Bacterial phosphoenolpyruvate: sugar phosphotransferase systems: Structural, functional, and evolutionary interrelationships. Bacteriol Rev 41:856, 1977

Schaeffer P: Sporulation and the production of antibiotics, exoenzymes, and exotoxins. Bacteriol Rev 33:48, 1969

Schlessinger D (ed): Bacilli: Biochemical genetics, physiology, and industrial applications. In Microbiology—1976. Washington DC, American Society for Microbiology, 1976, pp 1–449

Schlessinger D (ed): Cell envelope and cell division in bacilli. In Microbiology—1977. Washington DC, American Society for Microbiology, 1977, pp 5–103

Shockman GD, Daneo-Moore L, Higgins ML: Problems of cell wall and membrane growth, enlargement and division. Ann NY Acad Sci 235:161, 1974

Simon M, Silverman M, Matsumura P, et al.: Structure and function of bacterial flagella. In Stanier RY, Rogers, HJ, Ward JB (eds): Relationship Between Structure and Function in the Prokaryotic Cell. London, Cambridge University Press, 1978 pp 271–283

Slepecky RA: Synchrony and the formation and germination of bacterial spores. In Padilla EM, Whitson GL, Cameron IL (eds): The Cell Cycle. New York, Academic Press, 1969, pp 77–100

Smith AJ, Hoare DS: Specialist phototrophs, lithotrophs, and methylotrophs: A unity among a diversity of procaryotes. Bacteriol Rev 41:419, 1977

Snow GA: Mycobactins: Iron-chelating growth factors from mycobacteria. Bacteriol Rev 34:99, 1970

Sonenshein AL, Campbell KM: Control of gene expression during sporulation. In Chambliss G, Vary JC (eds): Spores VII. Washington DC, American Society for Microbiology, 1978

Stanier RY, Adelberg EA, Ingraham JL: The Microbial World, 4th ed. Englewood Cliffs, NJ, Prentice Hall, 1976

Stolp H, Starr MP: Principles of isolation, cultivation, and conservation of bacteria. In Starr MP, Stolp H, Trüper HC, Balows A, Schlegel HG (eds): The Prokaryotes. New York, Springer-Verlag, 1981, vol I, pp 135–175

Tipper DJ, Pratt I, Guinand M, et al.: Control of peptidoglycan synthesis during sporulation in *Bacillus sphaericus*. In Schlessinger D (ed): Microbiology—1977. Washington DC, American Society for Microbiology, 1977, pp 50–68

Selected Papers

Adler J, Hazelbauer GL, Dahl MM: Chemotaxis toward sugars in *Escherichia coli*. J Bacteriol 115:824, 1973

Andrews KJ, Lin ECC: Selective advantages of various bacterial carbohydrate transport mechanisms. Fed Proc 35:2185, 1976

Bechtel DB, Bulla LA Jr: Ultrastructural analysis of membrane development during *Bacillus thuringiensis* sporulation. J Ultrastruct Res 79:121, 1982

Chang CN, Nielsen JBK, Izui K, et al.: Identification of the signal peptidase cleavage site in *Bacillus licheniformis* prepenicillinase. J Biol Chem 257:4340, 1982

Churchward GG, Holland IB: Envelope synthesis during the cell cycle in *Escherichia coli* B/r. J Mol Biol 105:245, 1976

Clark DJ: Regulation of deoxyribonucleic acid replication and cell division in *Escherichia coli* B/r. J Bacteriol 96:1214, 1968

Cooper S, Helmstetter CE: Chromosome replication and the division cycle of *Escherichia coli* B/r. J Mol Biol 31:519, 1968

Cox CD, Rinehart KL Jr, Moore ML, et al.: Pyochelin: Novel structure of an iron-chelating growth promoter for *Pseudomonas aeruginosa*. Proc Natl Acad Sci USA 78:4256, 1981

Cozzone AJ: How do bacteria synthesize proteins during amino acid starvation? Trends Biochem Sci 6:108, 1981

Dignam SS, Setlow P: *Bacillus megaterium* spore protease. J Biol Chem 255:8408, 1980

Dix DD, Helmstetter CE: Coupling between chromosome completion and cell division in *Escherichia coli*. J Bacteriol 115:786, 1973

Donachie WD, Begg KJ, Vincente M: Cell length, cell growth and cell division. Nature 264:328, 1976

Ellar DJ, Lundgren DG, Slepecky RA: Fine structure of *Bacillus megaterium* during synchronous growth. J Bacteriol 94:1189, 1967

Frost GE, Rosenberg H: Relationship between the *ton B* locus and iron transport in *Escherichia coli*. J Bacteriol 124:704, 1975

Haldenwang WG, Losick R: Novel RNA polymerase σ factor from *Bacillus subtilis*. Proc Natl Acad Sci USA 77:7000, 1980

Haldenwang WG, Losick R: A modified RNA polymerase transcribes a cloned gene under sporulation control in *Bacillus subtilis*. Nature 282:256, 1979

Heefner DL, Harold FM: ATP-linked sodium transport in *Streptococcus faecalis*. J Biol Chem 255:11396, 1980

Heller KB, Lin ECC, Wilson TH: Substrate specificity and transport properties of the glycerol facilitator of *Escherichia coli*. J Bacteriol 144:274, 1980

Helmstetter CE, Krajewski CA: Initiation of chromosome replication in *dna* A and *dna* C mutants of *Escherichia coli* B/rF. J Bacteriol 149:685, 1982

Helmstetter CE, Pierucci O: DNA synthesis during the division cycle of three substrains of *Escherichia coli* B/r. J Mol Biol 102:477, 1976

Herbert D, Elsworth R, Telling RC: The continuous culture of bacteria; a theoretical and experimental study. J Gen Microbiol 14:601, 1956

Hirota Y, Suzuki H, Nishimura Y, et al.: On the process of cellular division in *Escherichia coli*: A mutant of *E. coli* lacking a murein-lipoprotein. Proc Natl Acad Sci USA 74:1417, 1977

Jacob F, Brenner S, Cuzin F: On the regulation of DNA replication in bacteria. Cold Spring Harbor Symp Quant Biol 28:329, 1963

Jones THD, Kennedy EP: Characterization of the membrane protein component of the lactose transport system of *Escherichia coli*. J Biol Chem 244:5981, 1969

Kennedy EP: Osmotic regulation and the biosynthesis of membrane-derived oligosaccharides in *Escherichia coli*. Proc Natl Acad Sci USA 79:1092, 1982

Lai JS, Sarvas M, Brammar WJ, et al.: *Bacillus licheniformis* penicillinase synthesized in *Escherichia coli* contains covalently linked fatty acid and glyceride. Proc Natl Acad Sci USA 78:3506, 1981

Lopez JM, Dromerick A, Freese E: Response of guanosine 5′-triphosphate concentration to nutritional changes and its significance for *Bacillus subtilis* sporulation. J Bacteriol 146:605, 1981

Marriott ID, Dawes EA, Rowley BI: Effect of growth rate and nutrient limitation on the adenine nucleotide content, energy

charge and enzymes of adenylate metabolism in *Azotobacter beijerinckii*. J Gen Microbiol 125:375, 1981

McIntosh MA, Earhart CF: Coordinate regulation by iron of the synthesis of phenolate compounds and three outer membrane proteins in *Escherichia coli*. J Bacteriol 131:331, 1977

Mesibov R, Adler J: Chemotaxis toward amino acids in *Escherichia coli*. J Bacteriol 112:315, 1972

Mitchell WM, Misko TP, Roseman S: Sugar transport by the bacterial phosphotransferase system. Regulation of other transport systems (lactose and melibiose). J Biol Chem 257:14553, 1982

Nelson SO, Scholte BJ, Postma PW: Phosphoenolpyruvate:sugar phosphotransferase system-mediated regulation of carbohydrate metabolism in *Salmonella typhimurium*. J Bacteriol 150:604, 1982

Norris TE, Koch AL: Effect of growth rate on the relative rates of synthesis of messenger, ribosomal and transfer RNA in *Escherichia coli*. J Mol Biol 64:633, 1972

Ochi K, Kandala J, Freese E: Evidence that *Bacillus subtilis* sporulation induced by the stringent response is caused by the decrease in GTP or GDP. J Bacteriol 151:1062, 1982

Parks LC, Rigney D, Daneo-Moore L, et al.: Membrane-DNA attachment sites in *Streptococcus faecalis* cells grown at different rates. J Bacteriol 152:191, 1982

Philson SB, Llinas M: Siderochromes from *Pseudomonas fluorescens*. Isolation and characterization. J Biol Chem 257:8081, 1982

Pierucci O: Dimensions of *Escherichia coli* at various growth rates: Model for envelope growth. J Bacteriol 135:559, 1978

Reizer J, Panos C: Regulation of β-galactoside phosphate accumulation in *Streptococcus pyogenes* by an expulsion mechanism. Proc Natl Acad Sci USA 77:5497, 1980

Ryals J, Little R, Bremer H: Control of RNA synthesis in *Escherichia coli* after a shift to higher temperature. J Bacteriol 151:1425, 1982

Ryter A: Association of the nucleus and the membrane of bacteria: A morphological study. Bacteriol Rev 32:39, 1968

Setlow P, Kornberg A: Biochemical studies of bacterial sporulation and germination. J Biol Chem 245:3637, 1970

Shepherd N, Churchward G, Bremer H: Synthesis and function of ribonucleic acid polymerase and ribosomes in *Escherichia coli* B/r after a nutritional shift-up. J Bacteriol 143:1332, 1980

Shuman HA: Active transport of maltose in *Escherichia coli*. J Biol Chem 257:5455, 1982

Smit J, Nikaido H: Outer membrane of gram-negative bacteria XVIII. Electron microscopic studies on porin insertion sites and growth of cell surface of *Salmonella typhimurium*. J Bacteriol 135:687, 1978

Smith WP, Tai P-C, Davis BD: *Bacillus licheniformis penicillinase:* Cleavages and attachment of lipid during cotranslational secretion. Proc Natl Acad Sci USA 78:3501, 1981

Springer MS, Goy MF, Adler J: Sensory transduction in *Escherichia coli:* Two complementary pathways of information processing that involve methylated proteins. Proc Natl Acad Sci USA 74:3312, 1977

Springer WR, Koshland DE Jr: Identification of a protein methyltransferase as the *che R* gene product in the bacterial sensing system. Proc Natl Acad Sci USA 74:533, 1977

Stock JB, Waygood EB, Meadow ND, et al.: Sugar transport by the bacterial phosphotransferase system. The glucose receptors of the *Salmonella typhimurium* phosphotransferase system. J Biol Chem 257:14543, 1982

Tso W-W, Adler J: Negative chemotaxis in *Escherichia coli*. J Bacteriol 118:560, 1974

Uratani-Wong B, Lopez JM, Freese E: Induction of citric acid cycle enzymes during initiation of sporulation by guanine nucleotide deprivation. J Bacteriol 146:337, 1981

Warth AD: Molecular structure of the bacterial spore. Adv Microb Physiol 17:1, 1978

Weigand RA, Vinci KD, Rothfield LI: Morphogenesis of the bacterial division septum: A new class of septation-defective mutants. Proc Natl Acad Sci USA 73:1882, 1976

Weigel N, Waygood EB, Kukuruzinska MA, et al.: Sugar transport by the bacterial phosphotransferase system. Isolation and characterization of enzyme I from *Salmonella typhimurium.* J Biol Chem 257:14461, 1982

Willsky GR, Malamy MH: Characterization of two genetically separable inorganic phosphate transport systems in *Escherichia coli.* J Bacteriol 144:356, 1980

Wong S-L, Doi RH: Peptide mapping of *Bacillus subtilis* RNA polymerase σ factors and core-associated polypeptides. J Biol Chem 257:11932, 1982

CHAPTER 6

Composition, Structure, and Biosynthesis of the Bacterial Cell Envelope and Energy Storage Polymers

The Bacterial Envelope

The bacterial cell surface and envelope components are of medical interest for many reasons. There was early recognition that cell surface antigens may play a role in the virulence of pathogens and that these antigens were useful for serologic identification and vaccine prophylaxis. The biosynthesis of the cell wall structure unique to bacteria later was found to be inhibited by various antibiotics, a fact that established a strong rationale for the study of the cell envelope as an approach to selective chemotherapy. It is now known that bacteria are probably the most important natural modulators of the host immune system. A body of knowledge is therefore developing that attempts to relate the structure of specific microbial products with their biologic activities in the mammalian host. It is for this reason that, in many instances, microbial envelope components are being intensively studied.

Envelope components provide survival value to microorganisms by serving many functions. Acidic polymers act as water and divalent metal (e.g., Mg^{2+} and Ca^{2+}) sequestering agents, and additionally, the lipophilic nature of the integuments of the acid-fast and gram-negative bacteria provides barriers to the entrance of various noxious chemicals and perhaps prevent loss of needed metabolites. The outer membrane of gram-negative bacteria may also serve to position periplasmic nutrient gathering and protective hydrolases outside the cytoplasmic membrane where they are most needed. In addition, the chemical diversity of bacterial surface components is overwhelming. This is important in host–parasite interactions, such as opsonization and phagocytosis, wherein it may be necessary to coat the parasite surface antigens with specific antibody and complement in order to promote bactericidal phagocytosis (Chaps. 14 and 20). It follows that bacterial pathogens that produce copious amounts of various capsular polysaccharides or other envelope polymers will be difficult to phagocytose, especially in the early stages of infection before sufficient specific antibodies are produced. Though little specific information is yet available, it would appear that human and animal systems have difficulty in destroying enzymatically some of the more unusual components of bacterial integuments.

Gross Composition

The structure and chemical nature of bacterial envelopes fall into three categories that correlate with whether a bacterium stains gram-positive, gram-negative, or acid-fast (Chap. 2). Other than in cell wall peptidoglycans, which are of remarkably similar general structure, the three groups of organisms differ considerably in the types of envelope lipids, polysaccharides, and proteins produced and in their ultrastructural arrangement. These differences are most notable in the occurrence of an outer membrane on gram-negative bacteria that is not found on gram-positive and acid-fast bacteria and in the presence in both gram-positive and acid-fast bacteria of polysaccharides linked to peptidoglycan, a pattern not found in gram-negative cells. In addition, acid-fast and related organisms produce a variety of complex lipids not found in gram-positive or gram-negative organisms. The different types of envelope components produced by the three groups of organisms are summarized in Table 6-1. The chemical composition, structure, biosynthesis, and, as indicated, the general location of many of these components within the various layers of the cell envelope can be described. However, the exact topologic, three-dimensional orientation, physical interrelationships, and function of many of these polymers as yet can only be inferred.

The Cell Wall Peptidoglycan

The rigid cell wall is a single, giant, bag-shaped macromolecule composed of a network of crosslinked peptidoglycan, sometimes also called murein or mucopeptide. This supermolecule and associated components may account for 2 to 40 percent of the cell dry weight. The glycan component is invariably constituted of the two amino sugars, glucosamine and muramic acid. These occur as alternating β-1,4-linked N-acetyl-D-glucosamine (GlcNAc) and N-acetylmuramic acid [3-0-(1'-D-carboxyethyl)-N-acetyl-D- glucosamine] (i.e., MurNAc) residues (Fig. 6-1). Chains vary from <10 to >170 disaccharide units. The glycan and peptide units are linked through the lactic acid carboxyl group of MurNAc to the amino terminus of a tetrapeptide. These glycotetrapeptides are crosslinked through the tetrapeptide

TABLE 6-1. COMPARISON OF CHARACTERISTIC TYPES OF BACTERIAL ENVELOPE COMPONENTS

Gram-positive Bacteria	Acid-fast Bacteria	Gram-negative Bacteria
Peptidoglycan 0.02 to 0.06 μm (multilayer)	Peptidoglycan 0.01 μm (monolayer?)	Peptidoglycan 0.01 μm (monolayer?)
Proteins	Polypeptides	Lipoproteins
Lipoteichoic acids	Mycolic acid-glycolipids	Outer membrane
Teichoic acids	Arabinogalactans	Lipopolysaccharide
Teichuronic acids	Wax D	Proteins
Polysaccharides	Cord factor	Phospholipid
	Sulfolipids	Polysaccharides
	Mycosides	

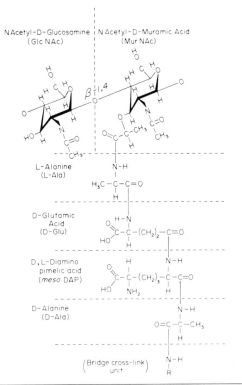

Figure 6-1. Structure of the peptidoglycan repeating unit. (See Fig. 6-2, where L-Ala-D-Glu-DAP-D-Ala are represented by a, b, c, d, respectively.)

units, forming a continuous sheet, which is the cell wall peptidoglycan bag-shaped sacculus. The invariant feature of the tetrapeptide component is the presence of D-alanine, which is always the linkage unit between peptidoglycan chains.

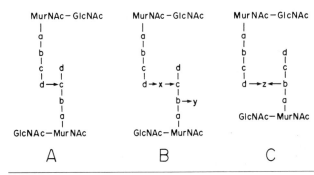

Figure 6-2. General patterns of bacterial peptidoglycan cross-linkage structures. Symbols used are MurNAc, N-acetylmuramic acid; GlcNAc, N-acetylglucosamine; a, b, c, d are peptide units; amino acids (see Figs. 6-1, 6-3); y, amide substituent linked to α-carboxyl of D-isoglutamic or 3-hydroxyglutamic acid; x and z are bridge peptide amino acids. Type C bridge units, z, contain two -NH-termini, i.e., -NH-R-NH-.

Peptide Structure and Variations

As shown in Figure 6-1, the muramic acid-linked peptide component of many bacteria is the tetrapeptide, -L-Ala-D-iso-Glu-meso DAP (or L-Lys)-D-Ala, which may be generalized as a-b-c-d. Crosslinkage between two peptidoglycan chains may be established directly or through an interposed peptide bridge, as illustrated in Figure 6-2. The peptidoglycans of most organisms conform to the two types, A and B, illustrated in Figure 6-2, type C being rarely found. For example, Escherichia coli and all gram-negative bacteria, Corynebacterium diphtheriae, the nocardia, mycobacteria, and Bacillus species are all of type A (Fig. 6-2A and Fig. 6-3A), whereas Staphylococcus aureus, the micrococci, and the streptococci accomplish interchain crosslinkage through an interposed peptide bridge, as shown in Figure 6-2B and Figure 6-3B.

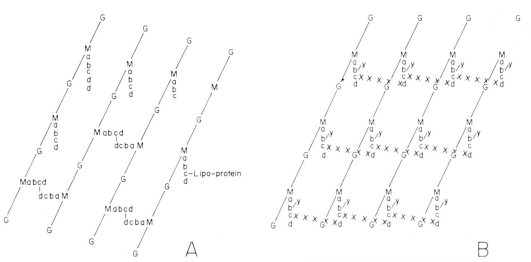

Figure 6-3. A. Schematic generalization of E. coli peptidoglycan structure. B. Schematic generalization of S. aureus H peptidoglycan structure. M (N-acetylmuramic acid); G (N-acetylglucosamine); a (L-alanine); b (D-glutamic acid); c (either meso-diaminopimelic acid, in A, or L-lysine, in B); d (D-alanine); X (pentaglycine bridge); Y (-NH₂).

Further modifications of peptide structures include the in vivo removal of terminal D-alanine residues from tetrapeptides (e.g., Fig. 6-3A) as in *E. coli*, or the removal of whole peptide units from the glycan chain. This happens in *E. coli*, which loses some 30 percent, and in *Micrococcus luteus*, where half or more of the glycan chains are free of tetrapeptide units. Crosslinkage of peptidoglycan in these organisms, therefore, may approach only 30 to 70 percent. In contrast, *S. aureus* glycan retains all of its tetrapeptide units, which are completely crosslinked (Fig. 6-3B). In addition, peptidoglycan-bound polypeptides or proteins occur in gram-positive, gram-negative, and acid-fast organisms. For example, in *E. coli*, other enterobacteria, and apparently most gram-negative bacteria, a lipoprotein is attached covalently to the ε-amino group of diaminopimelic acid (Fig. 6-3A). The lipoprotein is thought to aid in anchoring the outer membrane to the cell through lipophilic interactions of the N-terminal covalently bound lipoprotein fatty acids.

Strength and Shape. The alternating L- and D-amino acid sequence of the peptidoglycan tetrapeptide adds structural strength (D-, L-heteropolymers are known to be stronger than L- or D-homopolymers) and allows amino acid R-groups to align on one side of the peptide chains: intrapeptidoglycan hydrogen bonding (as shown with models) can occur between juxtaposed chains aligned in this way. Further, the equatorial or planar β-1, 4-linkage and the pyranose ring forms of the MurNAc-GlcNAc components of the glycan chains in peptidoglycan conform to the most thermodynamically stable of polysaccharide structures. The same type structure is also found in chitin and cellulose.

Cell shape is not determined by peptidoglycan chemical structure, since rod-shaped and spherical *E. coli* yield chemically similar peptidoglycan. The exact three-dimensional or spatial organization of the peptidoglycan chains is unknown. X-ray crystallography indicates an amorphous (i.e., noncrystalline) arrangement. However, it is thought that the glycan chains tend to be aligned perpendicular to the long axis of the cell, in contrast to a more or less parallel (i.e., axial) orientation of the crosslinked tetrapeptide units (Fig. 6-4). For *E. coli*, which contains approximately 10^6 repeating disaccharide–tetrapeptide units, the amount of peptidoglycan allows only one to three layers of peptidoglycan. In a gram-positive cell, which may contain 20 times as much or more peptidoglycan, there could be as many as 40 layers.

Glycan Variations

These include (1) the occurrence of *N*-glycolylmuramic acid in *Mycobacterium* and *Nocardia kirovani*, (2) the partial substitution of muramic acid lactam in spore peptidoglycan, (3) the occurrence of small amounts of the D-mannosamine muramic acid derivative (2 percent of total muramic acid) in *M. luteus*, (4) the occurrence of free amino

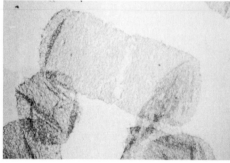

Figure 6-4. Orientation of *E. coli* cell wall peptidoglycan. **Top.** *E. coli* cell wall after SDS and trypsin treatment. **Bottom.** After limited endopeptidase digestion of peptide crosslinkages, the glycan components appear to be oriented perpendicular to the long axis of the cell, indicating that the dissolved peptide crosslinks were oriented with the long axis of the cell. *(From Verwer et al.: J Bacteriol 136:723, 1978.)*

groups on the glucosamine and muramic acid groups of peptidoglycans, and (5) the occurrence of O-acetyl groups on muramic acid or glucosamine. In addition, phosphodiester groups, which are substituted on the C-6 hydroxyl groups of muramic acid or glucosamine of gram-positive organisms, may serve as links to such polymers as teichoic acid, teichuronic acid, and other polysaccharides.

Cell Wall Lytic Enzymes

Because of their D-amino acid content and lack of aromatic amino acids, peptidoglycans are not susceptible to L-proteases, such as trypsin and chymotrypsin, which often are used to remove nonpeptidoglycan proteins from cell wall preparations. However, bacteriolytic enzymes that act on peptidoglycans occur widely and have been very useful in elucidating peptidoglycan structures. These enzymes fall into three major groups: (1) endo-β-1, 4-*N*-acetylhexosaminidases, which cleave the glycan strands either between *N*-acetylmuramic acid and *N*-acetylglucosamine (for example, the muramidases or lysozymes of egg white, tears, and white blood cells) or between the alternative acetylglucosamine-muramic acid glycoside linkages (for example, the β-*N*-acetylhexosaminidases), (2) endopeptidases, many of which attack D-alanine at bridge peptide crosslinkages, whereas others may specifically hydrolyze the interpeptide bridge linkages, as in the case of lysostaphin, which splits glycylglycine bonds in the pentaglycine bridge of *S. aureus* (Fig. 6-3B), and (3) amidases, which cleave the glycan-peptide junction between *N*-acetylmuramic acid and L-alanine, thereby separating the glycan strands from the interlocking peptides.

Many of the lytic enzymes described above are found as bacterial autolysins. One of the most notable autolysins is the *N*-acetylmuramyl-L-alanine amidase found in *Streptococcus pneumoniae*, which is activated by low pH or bile salts.

Lysozyme Resistance. Lysozyme is a very important first line of host resistance to bacterial infection (6 to 15 μg/ml human serum). Susceptibility to lysozyme varies from one bacterium to another, but gram-positive organisms are generally more sensitive than gram-negative cells because of the outer membrane barrier of the latter. In either case, organisms usually are more sensitive in the log phase of growth (minimal peptidoglycan) than at the stationary phase of growth (e.g., continuance of peptidoglycan synthesis, after protein synthesis stops, yields thicker cell walls). Lysozyme resistance of some organisms can be attributed to modification of peptidoglycan structure through O-acetylation of *N*-acetylmuramic acid residues (e.g., *S. aureus*) or by removal of *N*-acetyl groups from *N*-acetylglucosamine, which leaves lysozyme-resistant free amino groups on the peptidoglycan chain glucosamine residues (e.g., *Bacillus* species and the group A *Streptococcus pyogenes*).

Biologic Activities of Muramylpeptides

Adjuvants enhance the immune response in a nonspecific manner, thus making it possible to boost responses to specific immunogens. Long studied because of this property, mycobacterial cell wall products were initially thought to be active because of their peculiar lipids. Investigation of the mycobacterial components, however, led to the unexpected discovery that water-soluble peptidoglycan components common to many bacteria are responsible for the adjuvant effect. The smallest active fragment is *N*-acetylmuramyl-L-alanyl-D-isoglutamine, known as muramyldipeptide or MDP (Fig. 6-1). MDP was recently discovered also to have the peculiar properties of being a pyrogen (fever-causing agent) and a somnagen (sleep-inducing factor) in small animals. The full significance of these recent discoveries is yet to be realized.

Nonpeptidoglycan Components

Envelope Proteins

Both gram-positive and gram-negative cells produce envelope proteins that may influence host–parasite interaction. These proteins include the M proteins of the group A streptococci, the *S. aureus* protein A (pA) (Table 6-2), and the lipoprotein and porin of *E. coli* and other gram-negative organisms.

Envelope Polysaccharides: Capsular Polysaccharides

A variety of chemically diverse capsule and surface polymers are produced by both gram-positive and gram-negative bacteria. Among the best known capsular polymers are the soluble specific substances (SSS) or acidic polysaccharides produced by the gram-positive pneumonia-causing organism, *S. pneumoniae*, and the gram-negative *Klebsiella pneumoniae*, and hyaluronic acid, which is produced by group A *S. pyogenes*. These polymers are composed of repeating oligosaccharide units of two to four monosaccharides, one of which is usually a uronic acid. The *Klebsiella* polymers often contain substituents, such as acetic and pyruvic acids, and sometimes the methyl ethers of hexoses. Examples of medically important types of capsules are listed in Table 6-2.

Capsules are dispensable. Loss of the ability to produce capsules has no effect on growth rate or viability. In addition to genetic capability, environment and cultural conditions also markedly influence capsule production. Polysaccharide capsule production is enhanced by large amounts of carbohydrate, the restrictive growth conditions of low nitrogen, sulfur, or phosphorus, low temperature, or high salt concentration. In contrast, the *Bacillus anthracis* polypeptide capsule, which is composed of D-glutamic acid linked through gamma carboxyl groups, is produced in large amounts only when the organism is grown in an environment containing a high CO_2 concentration. Other factors may also affect capsule production. For example, the hyaluronic acid capsule of groups A and C of *S. pyogenes* can be demonstrated only very early in the growth of hyaluronidase-producing strains.

Several different types of genetically determined capsular polysaccharides may be produced by subgroups of a single species. These can often be differentiated serologically, hence the term "serogroup" or "serotype." For example, over 70 immunologically distinct capsular polysaccharide serotypes of *S. pneumoniae* have been defined. Other organisms, such as *Klebsiella*, *Escherichia*, and *Haemophilus*, also produce several capsular-type polysaccharides. Differences in capsular types are detected by Neufeld's quellung reaction, in which cells are mixed with specific anticapsule serum. If the appropriate capsular substance is present, a swelling appears around the cell.

Although capsular polysaccharides may be washed off bacterial cells, small amounts often remain cell bound and are detectable by sensitive serologic procedures. In the case of the pneumococci, many of the capsular polysaccharides have the chemical structure of teichoic acid-like polymers, which are thought to be cell wall-bound. However, because of the ease with which the pneumococci autolyze, it is difficult to know whether the easily removed capsular materials are in fact excreted as extracellular polysaccha-

TABLE 6-2. VARIOUS KINDS OF BACTERIAL SURFACE POLYMERS

Bacterium	Surface Polymer	Components
1. *Bacillus anthracis*	Polypeptide capsule	$[-\alpha\text{-D-(-)-glutamic acid}]_n$
	Teichoic acid	Polyolphosphate, glucose
2. *Yersinia (Pasteurella) pestis*	Protein	Protein
3. *Streptococcus pyogenes*	Hyaluronic acid capsule	$[-N\text{-acetyl-D-glucosamine-}\beta\text{-1,4D-glucuronic acid-}\beta\text{-1,3-}]1_n$
	M (virulence) antigens	Proteins
	Group A polysaccharide	Polyrhamnan, N-acetyl-D-glucosamine
4. *Staphylococcus aureus*		
Smith strain	Polysaccharide	$[2\text{-}N\text{-(-}N\text{-acetyl-alanyl)-D-glucosamine uronic acid}]_n$
Copenhagen strain H	Teichoic acid	Ribitol-1,5-phosphodiester, N-acetylglucosamine, D-alanine
	Protein A	Protein
5. *Staphylococcus epidermidis*	Teichoic acid	Glycerol-1,3-phosphodiester, D-glucose
6. *Streptococcus (Diplococcus) pneumoniae*	Type II	D-Glucose, L-rhamnose, D-glucuronic acid
	Type III	D-Glucose, D-glucuronic acid
	Type V	D-Glucose, D-glucuronic acid, N-acetyl-L-pneumosamine (2-acetamido-2,6-dideoxy-L-talose), N-acetyl-L-fucosamine (2-acetamido-2,6-dideoxy-L-galactose)
	Type VI (teichoic acid)	$[\text{Galactose, glucose, L-rhamnose, ribitol-PO}_4]_n$
	Type XVIII (teichoic acid)	$[\text{D-Glucose, D-galactose, L-rhamnose, glycerol-PO}_4, \text{O-acetyl}]_n$
7. *Escherichia coli*	Vi antigen capsule	$[N\text{-Acetyl-D-galactosamine uronic acid}]_n$
	Colominic acid	$[N\text{-Acetylneuraminic acid}]_n$
	Polysaccharides	Acidic polymers of various uronic acids, hexoses, and amino sugars
	Common antigen	$(N\text{-Acetyl-D-glucosamine-1,4-}N\text{-Acetyl-D-mannosamine uronic acid, fatty acids})_n$
	K88, K99 antigens	Protein (pili)
	Mucous antigen (colanic acid)	Glucuronic acid, galactose, fucose
8. *Haemophilus influenzae*	Type a capsules	Glucose, phosphate
	b	Polyribose-ribitol phosphate
	c	Hexose, phosphate
	d	N-Acetyl-D-glucosamine uronic acid
	e	Hexose, N-acetyl-D-glucosamine
	f	Galactosamine, phosphate
9. *Klebsiella pneumoniae*	Type I capsule	D-Glucose, fucose, glucuronic acid, pyruvic acid
10. *Neisseria meningitidis*	Capsular polysaccharides	
	Serogroup A	$[N\text{-Acetyl-,O-acetyl-mannosamine}]_n$ phosphate
	Serogroup B	$[N\text{-Acetylneuraminic acid}]_n$
	Serogroup C	$[N\text{-Acetyl,O-acetylneuraminic acid}]_n$
11. *Pseudomonas aeruginosa*	Capsules	Alginic acid-like polymer (mannuronic and guluronic acids) DNA, various other polysaccharides

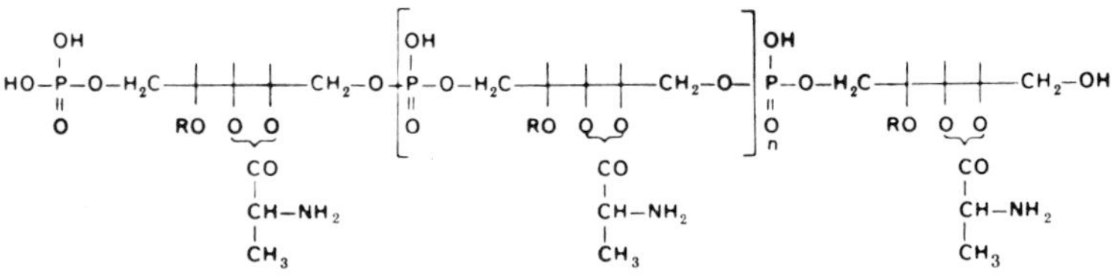

Figure 6-5. Ribitol teichoic acids. *Bacillus subtilis*: R = β-glucosyl, n = 7; *S. aureus* H: R = α- and β-N-acetylglucosaminyl, n = 6; *Lactobacillus arabinosus* 17-5: R = α-glucosyl, alternate ribitol residues also have α-glucosyl at the 3-position, n = 4.5. *(From Baddiley: Endeavour 23:33, 1964.)*

rides, as opposed to cell wall-bound autolytically released teichoic acids, or whether they are lipoteichoic acids pulled from the cell by the long water-soluble polymer chains after or before loss of a lipophilic cytoplasmic membrane anchor. In the case of gram-negative cells, covalent linkage of polysaccharides to peptidoglycan is unknown. However, capsular-like acid polymer chains linked to outer membrane lipopolysaccharides are found in some gram-negative bacteria.

Capsular polysaccharides have been shown, in some cases, to be bacteriophage (i.e., bacterial viruses) receptors. These bacteriophages produce specific depolymerases, which apparently serve to dissolve the capsule at the site of attachment in order to allow the virus access to the cell below the capsule.

The Envelope of Gram-positive Bacteria

Cell wall polysaccharides of gram-positive bacteria contribute 10 to 50 percent of the mass of the cell wall, and, in many instances, these polymers appear to be covalently linked to peptidoglycan. Cell surface polysaccharides produced include lipoteichoic, teichoic, teichoic-like, and teichuronic acids on the one hand, and nonteichoic acidic and neutral polysaccharides on the other.

Teichoic Acids

Teichoic acids (wall, from the Greek *teichos*) are polymers of phosphodiester-linked polyols (Fig. 6-5), which in several cases have been shown to be linked to the cell wall through muramic acid 6-phosphate. Up to 50 percent of the peptidoglycan may be combined with teichoic acids in the bacilli and staphylococci. Cell wall teichoic acids usually contain ribitol, or occasionally glycerol, and appear to be covalently linked to peptidoglycan through substituted phosphodiester groups on the C-6-hydroxyl of *N*-acetylmuramic acid residues. Membrane or lipoteichoic acids (LTA) that are not cell wall-bound are glycerophosphate polymers, which terminate in glycolipid (Fig. 6-6). Lipoteichoic acid, which remains membrane-associated in protoplasts, appears to be anchored in the cytoplasmic membrane through its lipophilic groups (Fig. 6-7).

Teichoic acids are specifically modified in different bacteria by addition to the polyol units of ester-linked D-alanine, D-lysine, or O-glycoside-linked glucose, galactose, or *N*-acetylhexosamines. The substituted teichoic acids are important as specific cell surface antigens of *Staphylococcus*,

Streptococcus, Lactobacillus, and *Bacillus* species. For example, almost all (96 percent) human strains of *S. aureus* produce glucosamine-substituted ribitol cell wall teichoic acids, whereas *Staphylococcus epidermidis* produces glucose-substituted glycerol teichoic acids.

Many teichoic acid-like polymers also occur in a variety of bacteria. These polymers are composed of phosphodiester-linked sugar repeating units, repeating units built of oligosaccharide-phosphate only, or oligosaccharides linked to a polyolphosphate. A notable example is the teichoic acid-like C polysaccharide of *S. pneumoniae* (composed of phosphate, *N*-acetyl-D-galactosamine, D-glucose, *N*-acetyl-2,4-diamino-2,4,6-trideoxyhexose, ribitol, and choline), which functions in cell wall division. If cells are grown on ethanolamine instead of choline, the C polysaccharide contains ethanolamine, and the cells divide but do not separate into individual cells, forming long filaments instead. In contrast to choline-grown cells, the ethanolamine-grown cells contain only nonactive monomers of *N*-acetylmuramyl-L-alanine amidase (MurNAc-L-Ala). The isolated C polysaccharide nonenzymatically aggregates with these nonactive MurNAc-L-Ala amidase monomers to form higher molecular weight, active enzyme aggregates. On the other hand, the lipoteichoic acid of this organism (a Forssman antigen) inhibits both the aggregation phenomenon and the enzymatic activity of the aggregate. These observations have led to the concept that bacterial cell envelope polymers may be part of feedback regulation systems that control cell wall growth.

Teichuronic Acid

This polymer is produced by various *Bacillus* species and is composed of *N*-acetylgalactosamine (GalNAc) and glucuronic acid (GlcUA), linked as the disaccharide repeating unit (GlcUA → 1,3 → GalNAc)$_n$, but contains no phos-

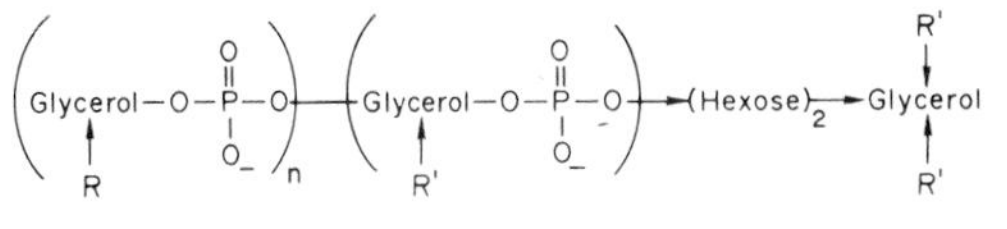

LIPOTEICHOIC ACID

Figure 6-6. Postulated structure of lipoteichoic (membrane teichoic) acid includes possibility of fatty acid (R′) substituted glycerophosphate interposed between glycerol teichoic acid chain and glycolipid. R-H or glycosyl; R′-H or esterified fatty acid residue; hexose disaccharide may be either α-1,2- or β-1,6-glucosylglucose or α-1,2-galactosylglucose; n > 28. *(After Knox and Wicken: Bacteriol Rev 37:215, 1973.)*

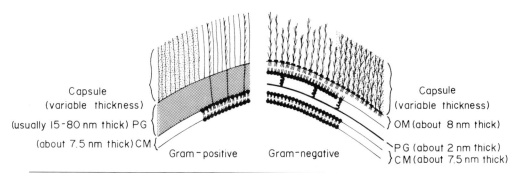

Figure 6-7. Diagram comparing some major envelope structures of gram-positive and gram-negative bacterial cells. PG, peptidoglycan; CM, cytoplasmic membrane; OM, outer membrane. The peptidoglycan layer of typical gram-positive cells is much thicker than that of gram-negative cells. Gram-positive cells often have polysaccharides covalently linked to peptidoglycan (represented by straight lines ending at peptidoglycan layer), as well as lipoteichoic acids that penetrate the peptidoglycan layer from the cytoplasmic membrane (represented by singly feathered lines). In contrast, gram-negative cells often exhibit a periplasmic space between the cytoplasmic membrane and the outer membrane, in which is found the relatively thin peptidoglycan layer. Helical lipoproteins, covalently linked to the peptidoglycan, are thought to anchor the outer membrane to the cell. No polysaccharides are bound to the peptidoglycan of gram-negative bacteria, but lipopolysaccharides (doubly feathered lines) are found in the outer leaflet of the outer membrane.

phate. Teichuronic acid may be found in the same cell together with teichoic acid. Teichuronic acid is apparently also covalently linked to cell wall and is synthesized in large amounts by cells deprived of phosphate, which cannot, therefore, make teichoic acids.

Nonteichoic Acid Cell Wall Polysaccharides

A characteristic, highly branched, group-specific, and cell wall-bound L-rhamnose polymer occurs as such in the group B and group G streptococci, whereas it is substituted with N-acetyl-D-glucosamine in group A streptococci and with N-acetyl-D-galactosamine in group C organisms.

The Envelope of Acid-fast and Related Bacteria

Members of the genus *Mycobacterium* and some *Nocardia* species, which characteristically stain red with carbolfuchsin and resist decolorization with acid–alcohol, are said to be "acid-fast." This staining property appears to correlate with the presence of cell wall-bound mycolic acids in the intact bacterium. Mycolic acids occur principally as esters bound to cell wall polysaccharides and as components of extractable (i.e., free) glycolipids known as "cord factors." The latter occur uniformly in mycobacteria and nocardia and also in some members of the nonacid-fast genus, *Corynebacterium* (the human and animal parasites and pathogens but not the plant-associated corynebacteria). The corynebacteria, nocardia, and mycobacteria (i.e., the CNM group of bacteria) also share a similar cell wall peptidoglycan structure and crosslinkage pattern in addition to a common antigen, which is probably the peptidoglycan-bound

arabinogalactan. Arabinomannans and galactosamine-containing polymers also occur, but the relationship of these to the arabinogalactans is obscure. Only the nocardia and mycobacteria produce cell wall-bound mycolic acids. The CNM group of organisms is considered to be gram-positive but is not reported to produce teichoic acid.

Mycolic Acids

The general structure and hypothetical biosynthesis of mycolic acids as α-substituted, β-hydroxy fatty acids is shown in Figure 6-8. The biosynthesis of the complex, branched-chain fatty acids is thought to proceed by condensation of the carboxyl groups of one long-chain fatty acid to the α position of another. The corynebacteria, nocardia, and mycobacteria each produce characteristic types of mycolic acids, which are presumably synthesized by similar mechanisms from similar precursors. As might be expected, the chain lengths and complexity of mycolic acids increase from the corynemycolic acid (about C_{32-36}) through the nocardic acids (about C_{50}) and mycolic acids (up to C_{90}) (Fig. 6-9). The synthesis of mycolic acid in *Mycobacterium tuberculosis* is reported to be inhibited by isoniazid (Chap. 9) by an unknown mechanism.

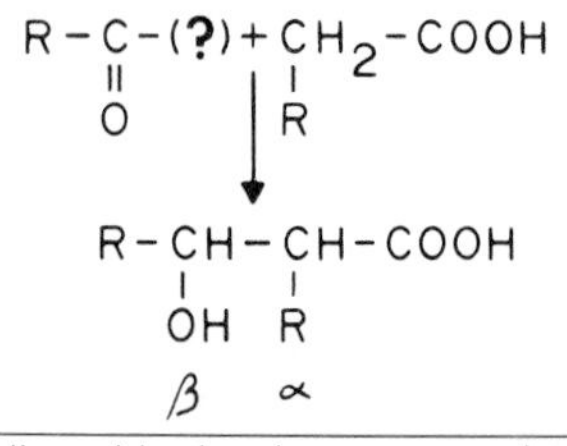

$$R-C-(?)+CH_2-COOH$$
$$O \qquad R$$
$$\downarrow$$
$$R-CH-CH-COOH$$
$$OH \quad R$$
$$\beta \qquad \alpha$$

Figure 6-8. Mycolic acid, showing presumed route of biosynthesis and general structure as α-substituted, β-hydroxy fatty acids.

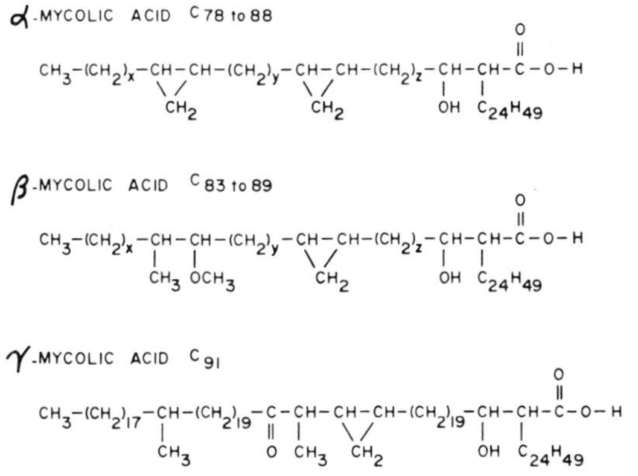

Figure 6-9. Structures of selected α, β, and γ mycolic acids. *(After Toubiana et al.: Cancer Immunol Immunother 2:189, 1977; Ribi et al.: Cancer Immunol Immunother 3:171, 1978.)*

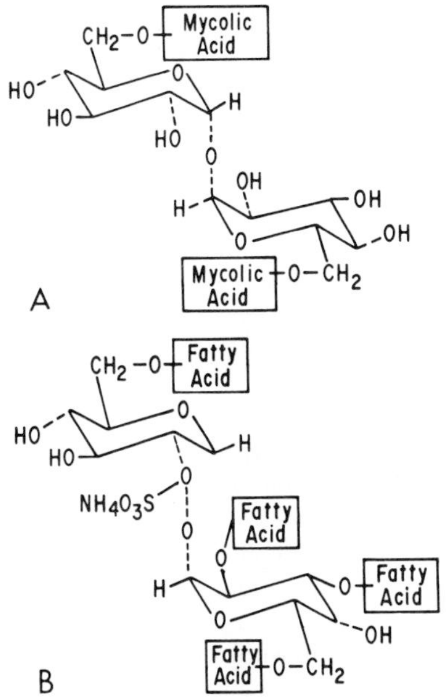

Figure 6-10. A. Cord factors, 6,6'-dimycolyltrehalose. **B.** Sulfolipids, 2,3,6,6'-tetraacyltrehalose-2-sulfate, where R groups are different fatty acids. *(After Goren: Bacteriol Rev 36:33, 1972.)*

Cell Wall

The cell wall of *M. tuberculosis* contains approximately equal amounts of peptidoglycan, arabinogalactan, and lipid. Greater than 50 percent of the lipid components are esterified mycolic acids, whereas some 25 percent appear to be normal fatty acids. Peptidoglycan-linked poly-L-glutamic acid also occurs in *M. tuberculosis*. The common peptidoglycan structure of *C. diphtheriae*, *Nocardia*, and *M. tuberculosis* is essentially represented in Figure 6-3A. An alternative *meso*-diaminopimelic acid-peptide bridge may occur in the mycobacteria, and *N*-glycolylmuramic acid occurs in peptidoglycans of the mycobacteria and at least some nocardia but not in the corynebacteria. Muramic acid phosphate is thought to serve as a linkage between cell wall-bound arabinogalactans and peptidoglycan, although glycosidic linkage of arabinogalactan may occur, possibly through *N*-acetylgalactosamine. Mycolic acids are bound to cell wall uniformly through the C-5-hydroxyl of D-arabinose residues of the arabinogalactan.

Glycolipids

Several unusual glycolipids occur in the acid-fast and related bacteria that are not cell wall-bound. These include cord factors, sulfolipids, mycosides, and lipopolysaccharides.

Trehalose Mycolates: Cord Factors. Cord factors were first discovered as toxic petroleum ether-soluble glycolipids in virulent *M. tuberculosis*, which grew in culture as serpentine cords. Cord factors are 6,6'-dimycolyl esters of the α,α'-1,1'-linked glucose disaccharide, trehalose (Fig. 6-10A), found throughout the corynebacteria, nocardia, and mycobacteria. *C. diphtheriae* produces trehalose esterified with the C_{32} acids, corynemycolic acid, and corynemycolenic acid. In some *Nocardia* species, only nocardic acids are found, but in *Nocardia asteroides*, cord factors contain a mixture of C_{28} to C_{36} corynemycolic acids, whereas the cell wall is esterified with C_{50} to C_{56} nocardic acids. The cord factor of *M. tuberculosis* contains a series of mycolates ranging in size from C_{78} to C_{90}. These mycolic acids characteristically contain a methyl group, a methoxyl group, and cyclopropane rings (Fig. 6-9).

Interest in cord factors centers on their toxic properties. For example, mycobacteria or isolated cord factor increases the susceptibility of experimental animals to gram-negative endotoxin. More generally, the multiple injection of cord factor, mixed in oil, into mice intraperitoneally induces wasting and ultimate death. Similar injection intravenously into mice can cause granulomatous responses in lungs, with the appearance of tubercles indistinguishable from those caused by infection with live *M. tuberculosis*. In addition, some immunity may be conferred. At the cellular level, cord factors appear to disrupt mitochondria, decreasing respiration and oxidative phosphorylation. Finally, a specific cord factor isolated from *M. tuberculosis* exhibits antitumor activity in mice when administered in a mixture (Freund's adjuvant) with the cell wall skeleton of BCG.

Sulfolipids. Sulfolipids appear to be peripherally located within the envelope and seem to be responsible for the neutral-red staining properties of cord-forming mycobacteria. Sulfolipids appear to be nontoxic but potentiate the toxicity of cord factor. More importantly, the production of sulfolipids appears to correlate with virulence in *M. tuberculosis*. Virulence may be related to the ability of these sulfolipids to act as evasins in preventing phagosome–lysosome fusion in macrophages, which may explain the success of *M. tuberculosis* as a facultative intracellular parasite. The principle sulfatide of *M. tuberculosis* has been identified as 2,3,6,6'-tetraacyltrehalose-2-sulfate (Fig. 6-10B).

Mycosides. These nonimmunogenic and apparently nontoxic compounds are presumed to be peripherally located, since the production of specific mycosides often correlates with type of colonial growth and sometimes corresponds to susceptibility to specific bacteriophages. Chemically, the mycosides are of interest because they contain 6-deoxytalose and O-methyl ether derivatives of deoxytalose, fucose, and rhamnose. The mycosides A and B are phenolic glycolipids (Fig. 6-11), whereas mycosides C are a group of peptidoglycolipids (Fig. 6-12). Mycosides C appear in some instances to be strain-specific phage receptors.

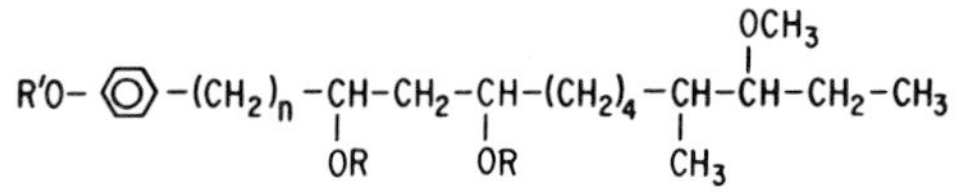

Phospholipids and Lipopolysaccharides. These components are probably membrane-associated, although their exact location within the cell is unknown. Mycobacterial phospholipids include dephosphatidylglycerol (cardiolipin) and phosphatidylethanolamine, which are found in common with other bacteria. In addition, the mycobacteria produce phosphatidylinositol mono- and oligosaccharides, such as tetracylated phosphatidylinositol pentamannosides.

Lipoglycans (lipopolysaccharides) have also been described in the mycobacteria. *M. tuberculosis* and *Mycobacterium phlei* produce a branched-chain polymer composed of eleven 6-O-methyl-D-glucose and seven D-glucose residues in α-1,4 linkage. The reducing terminus is linked to acylated glycerol. The polymer occurs acylated with 0 to 3 succinyl ester groups, 3 acetyl, and 1 each propionate, isobutyrate, and octanoate. An unusual α-1,4-linked 3-O-methyl-D-mannose polymer has also been described, which is reported to stimulate the fatty acid synthetase of *M. phlei*.

The Envelope of Gram-negative Bacteria

Gram-negative bacteria produce a unique three-layered envelope. This structure consists of an outer membrane overlying a thin peptidoglycan or murein layer, which may be separated by a periplasmic space from the inner cytoplasmic membrane (Fig. 6-13; see also Figs. 3-1, 3-17, and 6-7).

The outer membrane contains phospholipids, the characteristic lipopolysaccharide (LPS) of gram-negative bacteria, and matrix proteins that appear to be unique to the outer membrane. The cell surface area can be completely accounted for in terms of protein (60 percent) and

Figure 6-11. Mycosides A and B. Mycoside A: R' = trisaccharide composed of 2-O-methylfucose, 2-O-methylrhamnose, 2,4-D-O-methylrhamnose; n = 16 to 20. Mycoside B: R' = 2-O-methylrhamnose; n = 4 to 18. R = palmitic or mycocerosic acids. *(After Goren: Bacteriol Rev 36:33, 1972.)*

LPS (40 percent). The outer membrane is asymmetrical, in that all the LPS occurs in the outer leaflet, in contrast to phospholipid, which is found almost entirely in the inner leaflet of the outer membrane. Phospholipids of the outer membrane appear to be similar to those of the plasma membrane.

The Barrier Effect. The tightly fitting hydrophilic LPS, metal ligands, and porins that form the outer membrane surface serve as a barrier to small lipophilic molecules. These include many antibiotics, chemicals, and detergents. Treatment with EDTA decreases the effectiveness of this barrier to many antibiotics and detergents. EDTA treatment of cells results in loss of some 30 percent of outer membrane components. The EDTA-caused lesion can be quickly repaired by the cell if the cell is placed in a nutrient medium that contains divalent metals, e.g., Mg^{2+}. In the presence of even low Mg^{2+} concentrations, the major proteins tightly complex with peptidoglycan, LPS, and lipoprotein (see below). Mutations that result in loss of anionic LPS core components and outer membrane proteins decrease the outer membrane barrier effect.

Outer Membrane Attachment to Peptidoglycan. The mode of linkage between the outer membrane components and peptidoglycan is not completely known. Outer membrane is released from some cells during growth, indicating a transitional or loose linkage to the cell. Current evidence indicates that divalent metal ligands and lipophilic binding are involved in the attachment of the outer membrane to the cell wall peptidoglycan layer. The lipophilic attachment is established through interaction of the outer membrane–inner leaflet components, i.e., phospholipid, free lipoprotein, and the peptidoglycan-bound lipoprotein. Matrix proteins may also be involved.

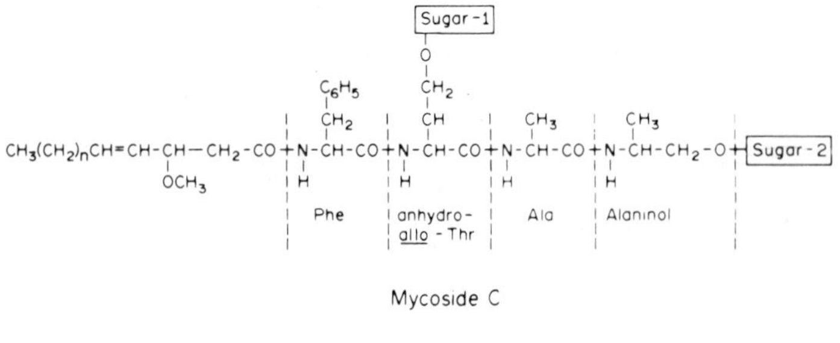

Figure 6-12. Mycoside C. Sugar 1, diacetyl-5-deoxytalose; sugar 2, mono-O-methylrhamnose to tri-O-methylrhamnose derivatives. *(After Goren: Bacteriol Rev 36:33, 1972.)*

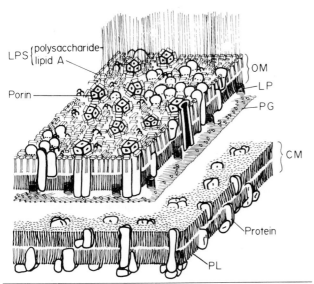

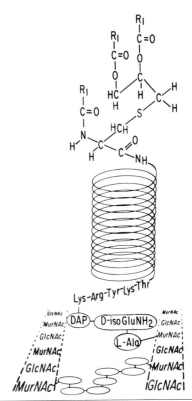

Figure 6-13. Interpretive diagram of gram-negative bacterial envelope structure. LPS, lipopolysaccharide; OM, outer membrane; LP, lipoprotein; PG, peptidoglycan; CM, cytoplasmic membrane; PL, phospholipid. *(Modified from DiRienzo et al.: Annu Rev Biochem 47:481, 1978.)*

Lipoprotein and Peptidoglycan

Lipoprotein contributes up to 40 percent of the mass of the isolated 4 percent SDS-insoluble cell wall sacculus of some enterobacteria. Lipoprotein probably occurs in all gram-negative bacteria. The peptidoglycan structure of all gram-negative bacteria so far examined appears to be identical to the type A structure illustrated in Figure 6-2. The insoluble peptidoglycan, after trypsin digestion to remove lipoprotein, seldom accounts for more than a small percentage of the cell dry weight. The lipoprotein is covalently linked to peptidoglycan through the trypsin-sensitive peptidoglycan sequence, Lys-Tyr-Arg-Lys, to *meso*-diaminopimelic acid (DAP) units of the peptidoglycan (Fig. 6-14). The lipoprotein molecules appear by electron microscopy to be spaced every 10 to 12 nm on the outer surface of the peptidoglycan structure. This corresponds to 1 lipoprotein for every 10 to 12 peptidoglycan disaccharide units 1.03 nm long, or some 10^5 bound lipoprotein molecules per cell.

A free form of the lipoprotein of molecular weight approximately 7500 occurs in the outer membrane in twice the concentration of the bound form. The lipoprotein, which self-aggregates and is strongly lipophilic, exists in helix form (Fig. 6-14). Because it binds avidly to lipid and protein, it is thought to aid in anchoring the outer membrane complex to the cell wall (Fig. 6-13). This view is supported by the fact that over 90 percent of the lipoprotein may be isolated with the outer membrane after lysozyme digestion of the peptidoglycan sacculus.

The lipoprotein contains no histidine, proline, phenylalanine, or tryptophan. The lipoprotein appears to be composed of repetitive sequences with nonpolar amino

Figure 6-14. Diagram of lipoprotein-peptidoglycan structure. *(Modified from Braun and Hantke: Annu Rev Biochem 43:89, 1974.)*

acids every 3.5 residues. This arrangement would allow the alignment of the nonpolar amino acid side-chain groups on one side of the helical structure. Such a structure could be of importance in binding with other outer membrane components. The lipoprotein N-terminal amide fatty acids include palmitic acid (65 percent), palmitoleic acid (11 percent), and *cis*-vaccenic acid (11 percent). Ester-bound fatty acids include palmitic acid (45 percent), palmitoleic acid (11 percent), and *cis*-vaccenic acid (24 percent).

Lipopolysaccharide (LPS)

Lipopolysaccharides are responsible for many of the biologic properties of gram-negative bacteria. Serologic specificity resides primarily in the variable polysaccharide portion of LPS, which corresponds to the somatic O antigen of gram-negative organisms. O antigens also act as specific receptor sites for certain bacteriophages. LPS is also known as endotoxin, a term used before discovery of LPS and its location in the outer membrane (Figs. 6-7, 6-13). The term "endotoxin" was meant to indicate heat-stable, cell-bound, polysaccharide-like toxin found in gram-negative bacteria in order to contrast and differentiate this type of toxin from heat-labile, protein toxins found in culture filtrate outside the cell, i.e., exotoxins. Heat-stable endotoxic properties of LPS are largely due to the glycolipid portion called "lipid A." Endotoxic properties include fever production (pyrogenicity), lethality, tissue necrosis activity, complement activation, B cell mitogenicity in mice, immunoadjuvant activity, and antitumor activity. It is not surprising, therefore, that endotoxins have been subjected to intensive investigation.

Lipopolysaccharide Structure. The LPS is composed of three regions: O-specific polysaccharide (region I), core polysaccharide (region II), lipid A (region III). Serologic specificity resides in region I, toxicity in region III.

For serologic and biochemical studies, regions I and II can be separated from the water-insoluble region III by mild acid hydrolysis (pH 3). The complete structure of the lipopolysaccharide of *Salmonella typhimurium* is indicated in Figure 6-15.

O-SPECIFIC POLYSACCHARIDE (REGION I). The O antigen polymer is composed of repeating oligosaccharide units of three to four monosaccharides. A variety of monosaccharides, including various pentoses, a 4-aminopentose, hexoses, 2-aminohexoses, 6-deoxy- and 3,6-dideoxyhexoses, 6-deoxyamino sugars with amino groups at the C-2, C-3, and C-4 position on the hexose carbon skeleton, and aminohexose uronic acids have been isolated from the region I polymers of various bacteria.

The LPS of *Salmonella, Escherichia, Shigella, Citrobacter,* and related genera often contain similar O antigen monosaccharides and can be grouped into a limited number of similar chemotypes on the basis of O antigen sugar composition. Sugar composition, sequence, linkage groups, and additional substituents (e.g., acetyl group) of the re-

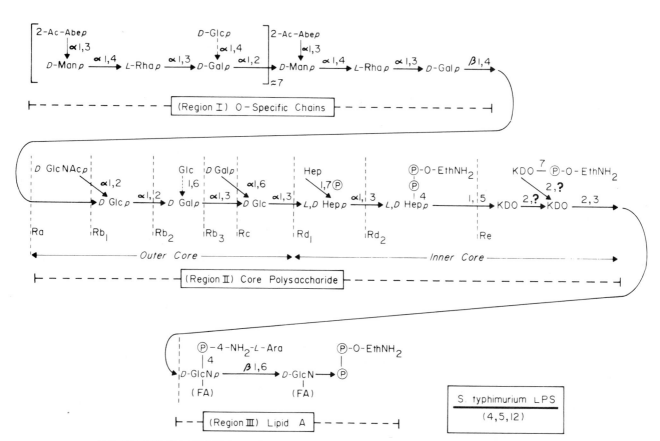

Figure 6-15. Diagram of lipopolysaccharide structure of *S. typhimurium.* Note three regions indicated as O side chain, core (outer and inner), and lipid A. Reading right to left, vertical dashed lines indicate limits of structures produced by Re to Ra core polysaccharide mutants. Roman numerals indicate order of addition of core sugars during biosynthesis. Glycosidic bonds are indicated where known. Question marks indicate uncertainty about bonds. Abe, abequose (3,6-dideoxy-D-galactose); Ac, acetyl; 4-NH$_2$-L-Ara, 4-amino-L-arabinose; EthNH$_2$; ethanolamine; F.A., fatty acid; Gal, galactose; Glc, glucose; GlcN, glucosamine; Hep, L-glycero-D-mannoheptose; (β-OH) MA, β-hydroxymyristic acid; KDO, 2-keto-3-deoxymannooctulosonic acid; Man, mannose; Rha, rhamnose. *(After Lüderitz et al.: In Weinbaum et al. (eds): Microbial Toxins, 1971, vol 4. New York, Academic Press; Rietschel et al.: Eur J Biochem 28:166, 1972; Galanos et al.: Int Rev Biochem 14: 239, 1977.)*

peating oligosaccharide units determine the antigenic (serologic) and sometimes the bacteriophage specificity of the particular O antigen. Bacteriophage attachment to O antigens is determined by the specificity of phage depolymerases. Over a thousand different serologic combinations have been found in *Salmonella*.

CORE POLYSACCHARIDE (REGION II). This region is arbitrarily separated into inner and outer core areas. Inner core contains 2-keto-3-deoxyoctonic acid (KDO) and heptose, both of which are unique to bacteria, as well as phosphate and pyrophosphate-bound ethanolamine. The KDO is bound to the glucosamine of lipid A by an acid-labile (pH 3) ketosidic linkage. The outer core is composed of the hexoses glucose, galactose, and *N*-acetylglucosamine. All *Salmonella* are believed to share a common core polysaccharide which differs from that of *Shigella* or *Escherichia* species. *E. coli*, however, may produce several different core structures.

The core polysaccharide of *Salmonella*, as shown in Table 6-3, is a linear polysaccharide. Mutants incapable of completing the O antigen polysaccharide and the various steps in the biosynthesis of core polysaccharide are generally referred to as "R mutants" because they tend to form rough colonies, especially the deep rough mutants that lack the ability to complete the inner core polysaccharide. This is in contrast to the S colony produced by wild-type organisms that make complete LPS molecules. For these reasons, it has been possible to follow the concept of colony morphologies in making chemotype designations of the mutants in accordance with the stage, i.e., Ra to Re, to which they can carry out the synthesis of the LPS core polysaccharide (Table 6-3). Outer core mutants are sometimes referred to as "superficial R mutants," since in the presence of high salt concentration (e.g., 1 percent NaCl), they sometimes produce rough colonies that can be differentiated from S colony forms. Additionally, the differences between the various LPS chemotypes and the organisms that produce them can be detected chemically, serologically, or by susceptibility to specific bacteriophages.

Of clinical interest is the interesting correlation that gram-negative bacteremic patients with high antibody titers to Re LPS appear to recover better (less shock and death) than those with low or no Re LPS antibody.

LIPID A (REGION III). Lipid A of the Re mutant of *Salmonella* (Fig. 6-16) is composed of a β-1,6-linked D-glucosamine disaccharide substituted at positions 4′ and 1 by phosphomonoester groups. The hydroxyl group at C6′ serves as the attachment site of KDO and the core polysaccharide. Long-chain fatty acids linked to the hydroxyl and amino groups of the glucosamine disaccharide confer lipophilic properties to the lipid A molecule. Amide-linked D-3-hydroxy fatty acids consisting of 14 carbons are characteristically found in the lipid A of various gram-negative bacteria. In *Salmonella*, the 4′ phosphoglucosamine disaccharide position is substituted with 4-amino-L-arabinose, whereas in some instances the 1 position phospho group is substituted with phosphoethanolamine: neither amino nor hydroxyl groups of these substituents are acylated. Microheterogeneity occurs, in that the 4-amino sugar is not found in *E. coli* or *Shigella*.

Hydrolysis of LPS by mild acid cleaves the ketosidic KDO–lipid A linkage. Because of its high fatty acid content, lipid A behaves as a difficultly water-soluble phosphoglycolipid, which is separable from the water-soluble, O antigen, core KDO product. LPS and free lipid A tightly bind divalent metals in complexes with phospholipids and outer membrane proteins. LPS and lipid A also form biologically active, self-aggregating, high molecular weight complexes, or micelles, of variable molecular weight, ranging from 10^5 to 10^8 or more. Deacylation of LPS or free lipid A results in loss of major biologic activities, including lethal toxicity, pyrogenicity, mitogenicity (in mice), and complement activation activity. However, removal of only the α-1-phosphate group (i.e., R, in Fig. 6-16) by mild acid hydrolysis results in loss of endotoxicity without loss of adjuvant activity.

Biosynthesis of Extracellular Polysaccharides and Intracellular Storage Polymers

Extracellular Polymers. The extracellular slimes of some microorganisms, such as the lactic acid bacteria, are produced by a single transglycosylating enzyme from a single substrate in such quantity that viscous solutions result. As an example, *Leuconostoc mesenteroides* produces a predominantly α-1,6-linked polyglucan slime, or dextran, which is sometimes used as a blood plasma substitute. It is produced only when the proper substrate, sucrose (glucose 1,2-fructose), is supplied in the growth medium. The reaction requires a specific transglucosidase and proceeds by utilization of the energy inherent in the sucrose-glycoside bond (Fig. 6-17).

Glycogen. Bacterial glycogen, an intracellular α-1,4-linked polyglucan with α-1,6 branch points, is synthesized from adenosine-diphosphateglucose by a variety of organisms, including *E. coli*, *Enterobacter aerogenes*, *M. luteus*, *Rhodopseudomonas* species, and the mycobacteria. It is produced at a rate that varies inversely with the growth rate and accumulates most readily in media rich in carbohydrate and poor in nitrogen or sulfur, which restrict growth but not metabolism. Glycogen synthesis occurs by transfer of glucose from ADP-glucose to the terminal nonreducing end groups of the glycogen polymer (Fig. 6-18). As in the animal system, branching is achieved by the action of an amylo-α-1,4 to α-1,6-transglucosylase.

Poly-β-Hydroxybutyric Acid (PHB). Some *Bacillus* species are capable of producing this chloroform-soluble polymer in amounts ranging from 7 to 40 percent of the

TABLE 6-3. STRUCTURES OF LIPOPOLYSACCHARIDES OF WILD-TYPE S FORM AND OF DEFECTIVE O ANTIGEN POLYMER AND CORE POLYSACCHARIDE R MUTANTS (R FORMS) OF *SALMONELLA*

Source*	O Antigen Polymer	Outer Core	Inner Core	Lipid A	Chemotype

Smooth or S form (parent wild-type) — O Antigen Polymer: [Multiple repeating unit oligosaccharide]$_n$ — Chemotype: S

```
                          P
                          ↓
→ Glc → Gal → Glc → Hep → Hep → KDO → KDO → Lipid A
    ↑      ↑     ↑     ↑                  ↑
  GlcNac  Gal   Hep  PPOEthN            KDO
                                         ↑
                                       POEthN
```

Semirough mutants — O Antigen Polymer: [Single repeating unit oligosaccharide]$_1$ — Chemotype: SR

```
                          P
                          ↓
→ Glc → Gal → Glc → Hep → Hep → KDO → KDO → Lipid A
    ↑      ↑     ↑     ↑                  ↑
  GlcNAc  Gal   Hep  PPOEthN            KDO
                                         ↑
                                       POEthN
```

Rough (R form) mutants — O Antigen Polymer: None

Chemotype: Ra
```
                          P
                          ↓
→ Glc → Gal → Glc → Hep → Hep → KDO → KDO → Lipid A
    ↑      ↑     ↑     ↑                  ↑
  GlcNAc  Gal   Hep  PPOEthN            KDO
                                         ↑
                                       POEthN
```

Chemotype: Rb
```
                        P
                        ↓
  Glc → Gal → Glc → Hep → Hep → KDO → KDO → Lipid A
               ↑     ↑     ↑                ↑
              Gal   Hep  PPOEthN          KDO
                                           ↑
                                         POEthN
```

Chemotype: RcP$^+$
```
                  P
                  ↓
  Glc → Hep → Hep → KDO → KDO → Lipid A
         ↑     ↑                ↑
        Hep  PPOEthN           KDO
                                ↑
                              POEthN
```

Chemotype: Rd$_1$P$^+$
```
  Hep → Hep → KDO → KDO → Lipid A
         ↑                ↑
       PPOEthN          KDO
                          ↑
                        POEthN
```

Chemotype: Rd$_1$P$^-$
```
  Hep → Hep → KDO → KDO → Lipid A
                     ↑
                    KDO
                     ↑
                   POEthN
```

Chemotype: Rd$_2$
```
  Hep → KDO → KDO → Lipid A
               ↑
              KDO
               ↑
             POEthN
```

Chemotype: Re
```
        KDO → KDO → Lipid A
               ↑
              KDO
               ↑
             POEthN
```

Adapted from Lüderitz: Angew Chem 9:20, 1970; Muhlradt: Eur J Biochem 18:20, 1971; Nikaido: In Leive (ed): Bacterial Membranes and Walls, 1973, Vol 1. New York, Marcel Dekker.

*Chemotype designations follow the concept of colony morphologies. That is, S or smooth colony forms make wild-type complete O antigen, whereas SR and Ra to Re mutants that lose (at some stage in the sequence of biosynthesis) the ability to complete either the O antigen polymer or the core polysaccharide tend to form R or rough colonies. KDO, 2-keto-2-deoxyoctonic acid; Hep, heptose; Glc, glucose; Gal, galactose; GlcNAc, *N*-acetylglucosamine; P, phosphate; PPOEthN, pyrophosphorylethanolamine; POEthN, phosphoethanolamine.

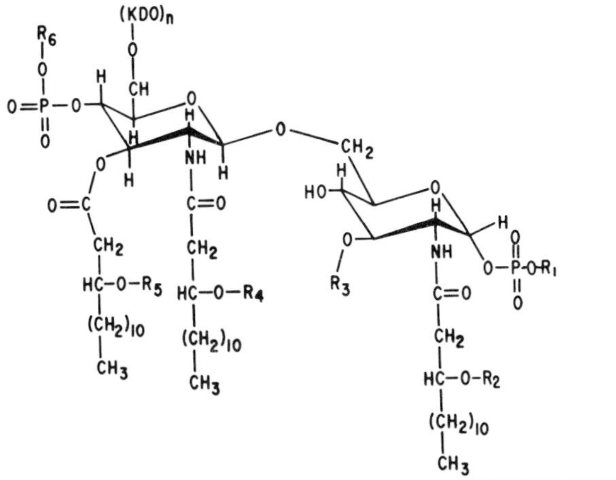

Figure 6-16. Structure of naturally occurring KDO-lipid A, from the Re mutant of *Salmonella typhimurium*. The glucosamine disaccharide is linked $\beta(1-6)$, and both are *N*-acyl substituted with β-hydroxymyristic acid groups. R_1 = H, phosphate, or phosphorylethanolamine; R_2 = H (may be palmitic acid in *S. minnesota*); R_3 = β-hydroxymyristic acid; R_4 = lauric acid; R_5 = double ester, two β-hydroxy myristic acids; R_6 = H, or 4 amino-4 deoxy-L-arabinose in α-1-phosphate linkage; n = 2 or 3. *(After Takayama et al.: J Biol Chem 258:7379, 1983.)*

dry cell weight. The polymer appears to be encased in a thin layer of protein, which is assumed to contain the enzymes involved in its metabolism. Biosynthesis and degradation of PHB are apparently carried out by the same enzymes (Fig. 6-19). The polymer is depleted during growth and again accumulates during the stationary phase or restrictive conditions of growth.

Polyphosphate. Babes–Ernst or volutin granules, thought to be polymetaphosphates, have been identified in various organisms, including *E. coli*, *E. aerogenes*, *C. diphtheriae*, *Mycobacterium*, and *Saccharomyces cerevisiae*. During active growth, only minute amounts of polyphosphates are detectable, whereas amounts equivalent to 1 to 2 percent of the dry cell weight accumulate under conditions of restrictive growth in media poor in N or S. Synthesis of polyphosphate in *E. coli* proceeds as shown in Figure 6-20.

Biosynthesis of Cell Envelope Polymers

Bacterial envelope polymers are synthesized by membrane-bound enzymes from nucleotide-sugar precursors that are synthesized in the cytoplasm. A membrane-bound cofac-

$$n\ \text{Sucrose} \rightleftharpoons (-\text{Glucose} \xrightarrow{\beta-1,6})_n + n\ \text{fructose}$$

Figure 6-17. Dextran biosynthesis.

$$\text{ADP}-\text{Glucose} + (\text{Glycogen})_n \longrightarrow$$

$$(\text{Glycogen})_{n+1} + \text{ADP}$$

Figure 6-18. Glycogen synthesis in bacteria.

tor, generally known as glycosylphosphate lipid carrier, is involved in the biosynthesis of peptidoglycan and the O-specific polysaccharide chain of lipopolysaccharides. The carrier lipid has been identified as the phosphomonoester of C_{55}-polyisoprenoid alcohol, undecaprenol (Fig. 6-21). This substance was first named "bactoprenol" before it was known that analogous lipid cofactors are also found in eucaryotic cells, including those of mammalian systems. The membrane-bound carrier forms glycosylpyrophospholipid oligosaccharide intermediates, which are then transferred to form polymers. Similar membrane lipid-linked oligosaccharide derivatives are involved in the biosynthesis of membrane-bound mannan in *M. luteus (lysodeikticus)* and capsular polysaccharides in *E. aerogenes*.

Teichoic Acid and Teichuronic Acid Biosynthesis

Polyolphosphodiester polymers, or teichoic acids, are formed by particulate enzyme systems from CDP-glycerol or CDP-ribitol (Fig. 6-22). The newly made teichoic acid chains are not transferred to preexisting peptidoglycan. The polyolphosphate chains appear to be added to nascent peptidoglycan chains during the synthesis of both polymer units in the membrane. The teichoic acid-bound peptidoglycan unit is then transferred as a unit to the growing cell wall peptidoglycan. The membrane-associated teichoic acid carrier appears to be undecaprenol. The synthesis of teichuronic acid appears to occur by a similar mechanism.

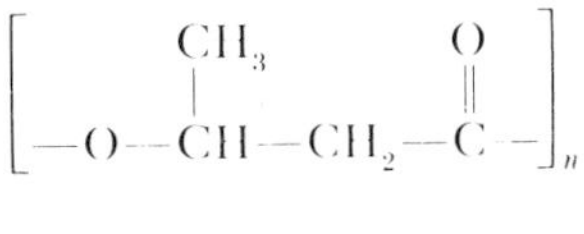

$$\left[-O-\underset{\underset{CH_3}{|}}{CH}-CH_2-\overset{\overset{O}{\|}}{C}- \right]_n$$

1. $2\ \text{Acetyl}\sim\text{SCoA} \rightleftharpoons$

    $\text{Acetoacetyl}\sim\text{SCoA} + \text{HSCoA}$

2. $\text{Acetoacetyl}\sim\text{SCoA} + \text{NADH} \rightleftharpoons$

    $\beta-\text{OH}-\text{butyryl}\sim\text{SCoA} + \text{NAD}^+$

3. $(\text{PHB})_n + \text{CoA}-S-\overset{\overset{O}{\|}}{C}-CH_2-CHOH-CH_3 \rightleftharpoons$

    $(\text{PHB})_{n+1} + \text{HSCoA}$

Figure 6-19. Poly-β-hydroxybutyric acid biosynthesis.

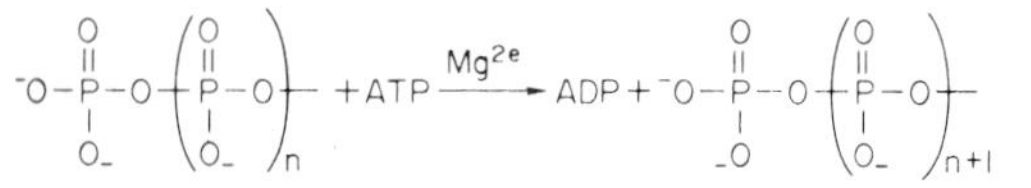

$$\text{}^-O\!-\!\overset{\overset{O}{\|}}{\underset{\underset{O_-}{|}}{P}}\!-\!O\!\left(\!\overset{\overset{O}{\|}}{\underset{\underset{O_-}{|}}{P}}\!-\!O\right)_n\!\!+\text{ATP}\xrightarrow{\text{Mg}^{2\oplus}}\text{ADP}+\text{}^-O\!-\!\overset{\overset{O}{\|}}{\underset{\underset{O_-}{|}}{P}}\!-\!O\!\left(\!\overset{\overset{O}{\|}}{\underset{\underset{O_-}{|}}{P}}\!-\!O\right)_{n+1}$$

Figure 6-20. Polyphosphate biosynthesis.

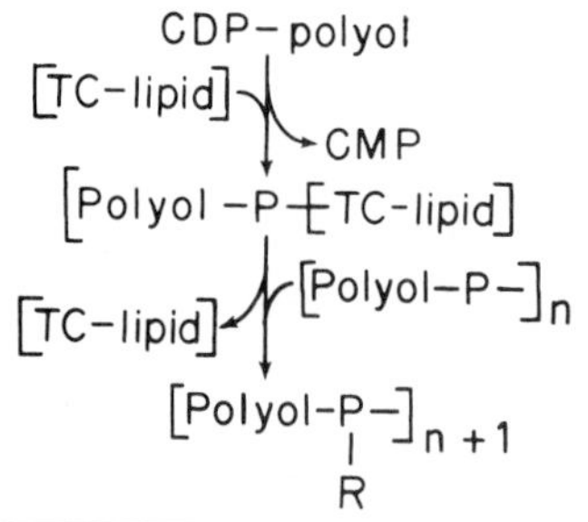

Peptidoglycan Biosynthesis

There are five stages in the biosynthesis of bacterial peptidoglycan: (1) the biosynthesis of soluble precursors in the cytoplasm, (2) transfer of precursors to membrane-bound carrier lipid (phosphopolyisoprenol) and the formation of disaccharide pentapeptide units, (3) transfer of the disaccharide pentapeptide to the cell wall, thereby extending the peptidoglycan backbone polymer, (4) formation of crosslinks between peptidoglycan polymers, and (5) regeneration of monophosphocarrier lipid.

1. The first stage of peptidoglycan synthesis involves formation of uridinediphospho-*N*-acetylmuramic acid (UDP-MurNAc) from uridinediphospho-*N*-acetylglucosamine (UDP-GlcNAc) by soluble cytoplasmic enzymes. First, the enolpyruvic acid group is transferred from phosphoenolpyruvic acid to the acetylglucosamine carbon-3-hydroxyl group, which is followed by its enzymic reduction by means of NADPH to a 3-0-lactic acid group. The amino acids of the pentapeptide are then added to the lactyl carboxyl group of UDP-MurNAc in stepwise fashion (Fig. 6-23), each addition except the last being catalyzed by a separate soluble enzyme and requiring ATP and a divalent cation, either Mg^{2+} or Mn^{2+}. The last two amino acids are added to the MurNAc-tripeptide as the dipeptide, D-alanyl-D-alanine, by a reaction that also requires ATP and divalent cation. The D-alanine is produced from L-alanine by a racemase.

2. A series of steps then occurs in the particulate membrane fraction (Fig. 6-24). First, the phospho-acetylmuramyl-pentapeptide group is transferred to the membrane-bound carrier lipid, with formation of a pyrophosphate bridge and release of UMP (Fig. 6-24, step 1). Then β-1,4-linked disaccharide-pentapeptide-pyrophospho carrier lipid is formed by addition of acetylglucosamine from UDPGlcNAc to the C-4 hydroxyl of the muramic acid component (Fig. 6-24, step 2). Various modification steps may follow, depending upon the species; for example, *S. aureus* amidates the α-carboxyl group of glutamic acid and also forms a pentaglycine bridge peptide by stepwise addition of glycine to the

$$\text{}^-O\!-\!\overset{\overset{O}{\|}}{\underset{\underset{O_-}{|}}{P}}\!-\!O\!\left(\!CH_2\!-\!CH=\!\overset{\overset{CH_3}{|}}{\underset{}{C}}\!-\!CH_2\!\right)_{11}\!\!H$$

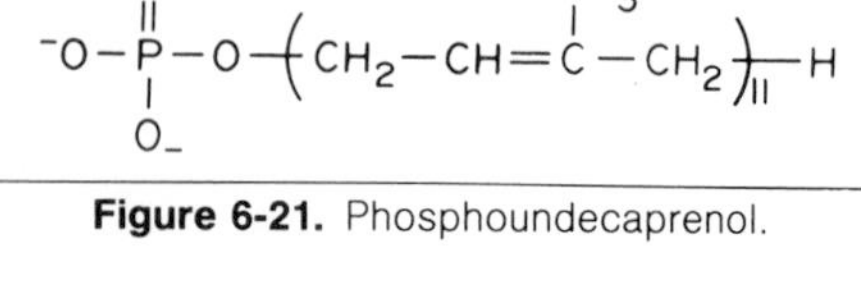

Figure 6-21. Phosphoundecaprenol.

Figure 6-22. Postulated sequence of polyolphosphate (teichoic acid) biosynthesis in gram-positive bacteria. CDP-polyol, cytidine diphosphate polyol; CMP, cytidine monophosphate; TC-lipid, teichoic acid carrier lipid; R, peptidoglycan biosynthetic intermediate.*(After Fiedler et al.: Ann NY Acad Sci 235:198, 1974.)*

ε-amino group of the muropeptide L-lysine residues by mediation of a special 4-thiouridine-containing tRNA, which differs from the glycine-tRNA involved in protein biosynthesis (Fig. 6-24, steps 3 and 4).

3. The third stage, elongation of the peptidoglycan backbone, occurs by transglycosylation (translocation) of the disaccharide-pentapeptide unit from the carrier lipid to the cell wall acceptor peptidoglycan backbone (Fig. 6-24, step 5), thereby releasing the pyrophospho carrier lipid, from which phosphatase regenerates the monophospho carrier lipid by removal of phosphate (Fig. 6-24, step 6).

4. The cell wall polymer formed at this stage is a noncrosslinked peptidoglycan with pentapeptide units terminating in D-Ala-D-Ala·COOH. Closure of the peptide bridge linkages by transpeptidation to form the crosslinked murein polymer finishes the biosynthetic sequence. In *S. aureus* this is accomplished by a reaction (Fig. 6-25) in which the penultimate D-alanine carboxyl group is linked to the free amino group of the pentaglycine bridge peptide of a neighboring peptidoglycan chain with concomitant release of the terminal D-alanine.

This sequence of reactions is essentially similar in all bacteria studied so far. Differences known include the species-specific substitution of alternative amino acids during pentapeptide synthesis (for example, *meso*-diaminopimelic acid instead of L-lysine) and various peptide and bridge peptide modification reactions. In addition, some bacteria, including *E. coli*, *Bacillus subtilis*, *Lactobacillus casei*, and others, appear to control the extent of interpeptidoglycan crosslinkage. Removal of whole peptides may occur by action of the autolysin, MurNAc-L-Ala amidase, as in the case of *M. luteus*, in which 30 to 70 percent of the glycan chains are not crosslinked. Additionally, one or both D-alanines may be removed from soluble UDP-MurNAc-pentapeptide precursors or noncrosslinked pentapeptidoglycan by carboxypeptidase. This explains the occurrence of tri- and tetrapeptides, as in *E. coli* (Figs. 6-3 and 6-25).

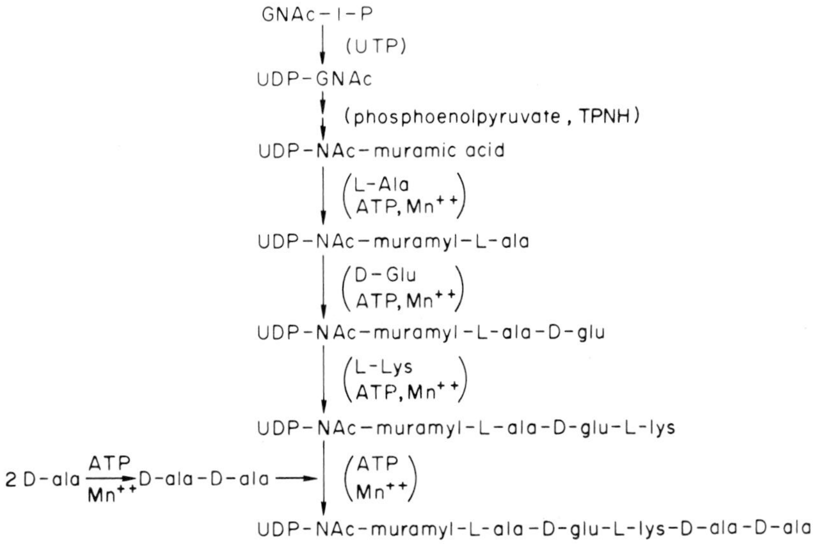

GNAc−I−P
↓ (UTP)
UDP−GNAc
↓ (phosphoenolpyruvate , TPNH)
UDP−NAc−muramic acid
↓ $\begin{pmatrix} L-Ala \\ ATP, Mn^{++} \end{pmatrix}$
UDP−NAc−muramyl−L−ala
↓ $\begin{pmatrix} D-Glu \\ ATP, Mn^{++} \end{pmatrix}$
UDP−NAc−muramyl−L−ala−D−glu
↓ $\begin{pmatrix} L-Lys \\ ATP, Mn^{++} \end{pmatrix}$
UDP−NAc−muramyl−L−ala−D−glu−L−lys

$2\,D-ala \xrightarrow[Mn^{++}]{ATP} D-ala-D-ala \longrightarrow$ ↓ $\begin{pmatrix} ATP \\ Mn^{++} \end{pmatrix}$

UDP−NAc−muramyl−L−ala−D−glu−L−lys−D−ala−D−ala

Figure 6-23. Biosynthesis of the UDP-*N*-acetyl-glucosamine-*N*-acetylmuramic acid pentapeptide precursor of cell wall peptidoglycan.

Lipopolysaccharide Biosynthesis

Lipid A, LPS core, and side chain are synthesized in the membrane in several stages. Lipid A (region III) serves as an acceptor for the stepwise addition of core polysaccharide units (region II), followed by the addition of preassembled O-specific polysaccharide units (region I). The completed LPS is then rapidly and irreversibly translocated into the outer membrane. Translocation may occur at discrete export sites, seen as adhesion sites between inner and outer membranes in plasmolyzed cells

(Fig. 3-20). Randomization of newly synthesized lipopolysaccharide in outer membrane occurs within several minutes of growth.

Lipid A Biosynthesis (Region III). The earliest precursor of lipid A is a monomer of diacyl-glucosamine-α-1-phosphate. Both acyl groups are β-OH-myristic acids, one being esterified at C-3, the other in acyl-amide linkage on the C-2 amino group. The lipid A disaccharide is next synthesized from two of these monomers by a reaction in

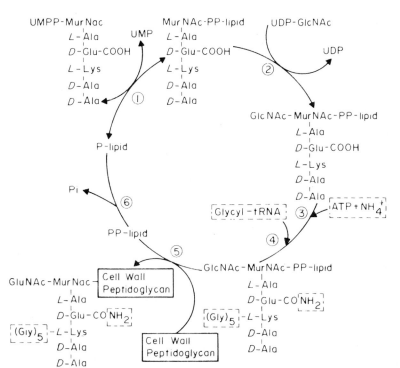

Figure 6-24. Biosynthesis of peptidoglycan: membrane-involved carrier lipid reactions.

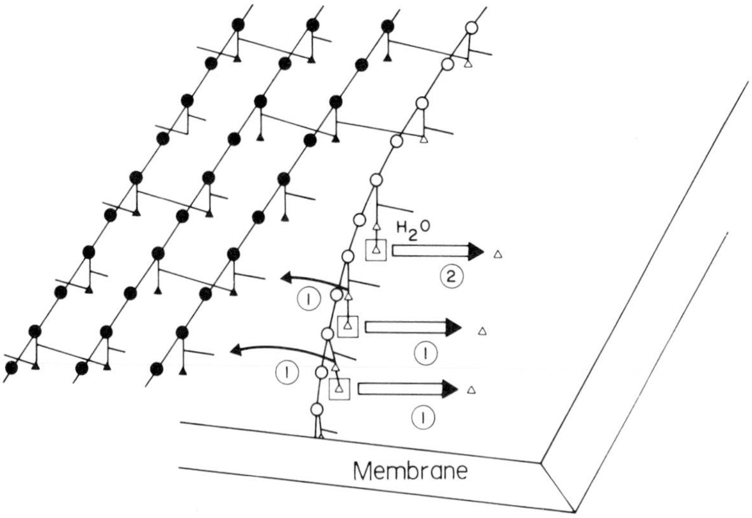

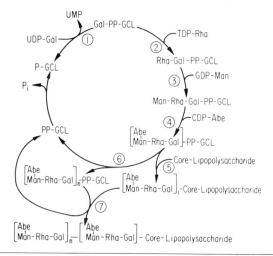

Figure 6-25. 1. Process of peptidoglycan crosslinkage by transpeptidation and release of D-alanine (i.e., D-alanine = Δ). Newly synthesized peptide glycan strand (open circles) donates carbonyl group of penultimate D-alanine of D-ala-D-ala dipeptide (i.e., -Δ-Δ) to ε-amino group (e.g., of diaminopimelic acid or lysine) of acceptor strand, resulting in transpeptidation (i.e., crosslinkage) and release of terminal D-alanine (Δ). 2. DD-Carboxypeptidase action can hydrolyze D-alanyl-D-alanine, resulting in loss of terminal D-alanine and concomitant loss of crosslinkage of 30 percent or more of the peptidoglycan. *(Modified from Ghuysen: J Gen Microbiol 101:13, 1977.)*

which one of them is transferred to the other via a UDP-monomer intermediate. The next step involves addition of 4-amino-L-arabinose to the C4'-phosphate and of phosphoethanolamine to the 1-phospho group. Three molecules of KDO are then added to the C3' position, followed by esterification of available hydroxl groups with various fatty acids, to yield the Re LPS, sometimes also called the "Re glycolipid" (Fig. 6-16). The Re glycolipid is the first of the various lipid A products to be found in the outer membrane. Apparently, the inability to synthesize at least the Re glycolipid is lethal to the bacterium.

Core Polysaccharide Biosynthesis (Region II). The basic mechanism of core polysaccharide synthesis involves membrane-associated enzymes that catalyze the stepwise addition of monosaccharides from nucleotide sugar intermediates to the nonreducing terminus of the growing core polymer (Table 6-3). Inner core units (KDO, heptose, phosphate, and ethanolamine) are added stepwise, followed by the outer core hexose units (glucose, galactose, and *N*-acetyl-D-glucosamine). The sequence of addition is ordered by specific enzyme recognition of both acceptor and activated nucleotide-sugar donor.

Much of the knowledge of core polymer structure and biosynthesis has been made possible by the availability of a series of *Salmonella* R or rough form mutants that are defective in LPS biosynthesis (Table 6-3, Fig. 6-15). Such mutants are defective in their ability to synthesize nucleotide sugar precursors, such as UDP-galactose, or they lack one or more of the necessary nucleotide sugar transferases or O polymer transferases. By using cell envelope preparations from appropriate mutant organisms, the stepwise addition of each core component in sequence can be demonstrated.

O Antigen Biosynthesis (Region I). The synthesis of the O antigen polysaccharide hapten is basically similar to that of peptidoglycan, in that soluble nucleotide sugars are utilized by membrane enzymes to synthesize carrier lipid-O antigen repeating oligosaccharide unit intermediates,

which are then polymerized. The biosynthetic sequence in particulate cell envelope membrane preparations occurs in four stages: (1) preassembly of O antigen repeating units as oligosaccharide-pyrophosphate carrier lipid intermediates, (2) polymerization of the preassembled repeating units, (3) transfer of the O-chain polymer to LPS core polysaccharide, and (4) regeneration of monophospho carrier lipid (Fig. 6-26).

1. In *S. typhimurium* the first step in synthesis of the oligosaccharide repeating unit involves the transfer of galactose-1-phosphate from UDP-galactose (UDPGal) to membrane-bound phospho carrier lipid to form galactosyl-pyrophospho carrier lipid

Figure 6-26. Biosynthesis of lipopolysaccharide: cytoplasmic membrane-involved carrier lipid reactions. P-CL, membrane phospho carrier lipid; UDP-, uridine diphosphate; UMP-, uridine monophosphate; TDP-, thymidine diphospate; CDP-, cytidine diphosphate; PP-CL, diphospho carrier lipid; Pi, phosphate; Gal-, galactosyl; Rha-, rhamnosyl; Man-, mannosyl. Note that different transferases add the initial single and subsequent polymerized oligosaccharide repeating units to the growing lipopolysaccharide chain.

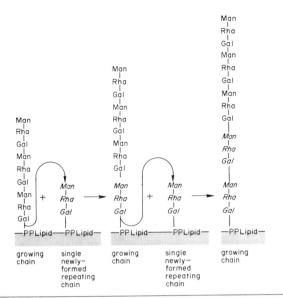

Figure 6-27. Polymerization of preassembled repeating units during the biosynthesis of O antigen. Mechanism allows extension of growing polysaccharide side chain into external environment without requiring movement of biosynthesis machinery.

with release of UMP. Transfer of L-rhamnose from thymidine diphosphorhamnose (TDPRha), mannose from guanosine-diphosphomannose (GDP-Man), and abequose from cytidine diphospho-abequose (CDPAbe) follow to complete the tetra-saccharide repeating unit intermediate (Fig. 6-26).

2. This step involves the transfer of the proximal ends of growing chains from membrane-attached PP-lipid to single, newly formed oligosaccharide repeating units, which are themselves linked to membrane-bound PP-lipid (Fig. 6-27). This is exactly the opposite of the mechanisms of intracellular biosynthesis of cytoplasmic polymers, such as glycogen, in which each new repeating unit is added to the distal ends of growing polymer chains.

3. Transfer of O antigen oligosaccharide repeating units to the core polysaccharide involves two steps. First, a single repeating unit is transferred, and then a different enzyme transfers the polymerized O antigen polysaccharide (Fig. 6-26).

4. Monophosphopolyisoprenol, the P-lipid carrier, is regenerated from the corresponding pyrophosphate by phosphatase action, as described above for peptidoglycan synthesis (Fig. 6-26).

Of interest is the ability of bacteriophage to change the composition and structure of the O antigen side chain repeating units (Chap. 63).

FURTHER READING

Books and Reviews

Agarwal MK (ed): Bacterial Endotoxins and Host Response. Amsterdam, Elsevier/North Holland, 1980

Ashwell G, Hickman J: The chemistry of the unique carbohydrates of bacterial lipopolysaccharides. In Weinbaum G, Kadis S, Ajl SJ (eds): Microbial Toxins. New York, Academic Press, 1971, vol 4, pp 235–266

Asselineau C, Asselineau J: Trehalose-containing glycolipids. Prog Chem Fats Other Lipids 16:59, 1978

Barksdale L: *Corynebacterium diphtheriae* and its relatives. Bacteriol Rev 34:278, 1970

Barksdale L, Kim K-S: Mycobacterium. Bacteriol Rev 41:217, 1977

Braun V: Lipoprotein from the outer membrane of *Escherichia coli* as an antigen, immunogen, and mitogen. In Schlessinger D (ed): Microbiology—1977. Washington, D.C., American Society for Microbiology, 1977, p 257

Braun V, Hantke K: Biochemistry of bacterial cell envelopes. Annu Rev Biochem 73:89, 1974

Dinarello CA: Production of endogenous pyrogen. Fed Proc 38:52, 1979

DiRienzo JM, Nakamura K, Inouye M: The outer membrane proteins of gram-negative bacteria: Biosynthesis, assembly, and functions. Annu Rev Biochem 47:481, 1978

Ghuysen JM: Use of bacteriolytic enzymes in determination of wall structure and their role in cell metabolism. Bacteriol Rev 32:425, 1968

Ghuysen JM: The concept of the penicillin target from 1965 until today. J Gen Microbiol 101:13, 1977

Goren MB: Mycobacterial lipids. Bacteriol Rev 34:33, 1972

Goren MB: Phagocyte lysosomes: Interactions with infectious agents, phagosomes, and experimental perturbations in function. Annu Rev Microbiol 31:507, 1977

Grov A: Biological aspects of protein A. In Schlessinger D (ed): Microbiology—1977. Washington, D.C., American Society for Microbiology, 1977, p 350

Inouye M (ed): Bacterial Outer Membranes Biogenesis and Functions. New York, Wiley, 1979

Kaplan MH: Nature of the streptococcal and myocardial antigens involved in the immunologic cross-reaction between group A streptococcus and heart. In Nowotny A (ed): Cellular Antigens. New York, Springer-Verlag, 1972, p 70

Lederer E: Natural and synthetic immuno-stimulants related to mycobacterial cell wall. Med Chem 5:257, 1977

Lüderitz O, Jann K, Wheat R: Somatic and capsular antigens of gram-negative bacteria. In Florkin M, Stotz EH (eds): Comprehensive Biochemistry. Amsterdam, Elsevier, 1968, vol 26A, pp 105–228

Lüderitz O, Staub AM, Westphal O: Immunochemistry of O and R antigens of *Salmonella* and related Enterobacteriaceae. Bacteriol Rev 30:192, 1966

Lüderitz O, Wesphal O, Staub HM, et al.: Isolation and chemical and immunological characterization of bacterial lipopolysaccharides. In Weinbaum G, Kadis S, Ajl SJ (eds): Microbial Toxins, New York, Academic Press, 1971, vol 4, p 145

Lüderitz O, Gallanos C, Lehman V, et al.: Chemical structure and biological activities of lipid A's from various bacterial families. Naturwissenchaften 65:578, 1978

Morrison DC, Rudbach JA: Endotoxin-cell-membrane interactions leading to transmembrane signaling. Contemp Top Mol Immunol 8:187, 1981

Nikaido H, Nakae T: Outer membrane of gram-negative bacteria. Adv Microb Physiol 19, 1980

Osborn MJ: Structure and biosynthesis of the bacterial cell wall. Annu Rev Biochem 38:501, 1969

Osborn MJ, Rick PD, Lehmann V, et al.: Structure and biogenesis of the cell envelope of gram-negative bacteria. Ann NY Acad Sci 235:52, 1974

Preiss J: The regulation of the biosynthesis of α-1,4-glucans in bacteria and plants. Curr Top Cell Regul 1:125, 1969

Schleifer KH, Kandler O: Peptidoglycan types of bacterial cell walls and their taxonomic implications. Bacteriol Rev 34:407, 1972

Schleifer KH, Krause RM: The immunochemistry of peptidoglycan. Separation and characterization of antibodies to the glycan and to the peptide subunit. Eur J Biochem 19:471, 1971

Schwab JH: Suppression of the immune response by microorganisms. Bacteriol Rev 39:121, 1975

Tipper DJ: Mode of action of β-lactam antibiotics. Rev Inf Dis 1:39, 1979

Tipper DJ, Wright A: The structure and biosynthesis of bacterial cell walls. In Sokatch J (ed): The Bacteria. New York, Academic Press, 1979, vol 7, chap 6

Ward JB: Teichoic and teichuronic acids: Biosynthesis, assembly, and location. Microbiol Rev 45:211, 1981

Watson SW, Levin J, Novitsky TJ (eds): Endotoxins and their detection with the limulus amebocyte lysate assay. Progress in Clinical and Biological Research. New York, Alan R. Liss, 1982, vol 93

Westphal O, Westphal U, Sommer T: The history of pyrogen research. In Schlessinger D (ed): Microbiology—1977. Washington, D.C., American Society for Microbiology, 1978, p 221

Wicken AJ, Knox KW: Biological properties of lipoteichoic acids. In Schlessinger D (ed): Microbiology—1977. Washington, D.C., American Society for Microbiology, 1978, p 360

Yamamura Y, Kotani S, Azuma I, et al. (eds): Immunomodulation by Microbial Products and Related Synthetic Compounds. Amsterdam, Excerpta Medica, 1982

Yotis WW (ed): Recent advances in staphylococcal research. Ann NY Acad Sci 236:520, 1974

Selected Papers

Antoine A, Tepper BS: Characterization of glycogens from mycobacteria. Arch Biochem 134:207, 1969

Baird-Parker AC: The basis for the present classification of staphylococci and micrococci. Ann NY Acad Sci 236:7, 1974

Barrow WW, Ullom BP, Brennan PJ: Peptidoglycolipid nature of the superficial cell wall sheath of smooth-colony-forming Mycobacteria. J Bacteriol 144:814, 1980

Bettinger GE, Young FE: Reversible formation of undecaprenyl glucosaminyl lipids by isolated Bacillus subtilis membranes. In Schlessinger D (ed): Microbiology—1977. Washington, D.C, American Society for Microbiology, 1978, p 69

Bracha R, Davidson R, Mirelman D: Defect in biosynthesis of the linkage unit between peptidoglycan and teichoic acid in a bacteriophage-resistant mutant of Staphylococcus aureus. J Bacteriol 134:412, 1978

Chetty C, Klapper DG, Schwab JH: Soluble peptidoglycan-polysaccharide fragments of the bacterial cell wall induce acute inflammation. Infect Immun 38:1010, 1982

Decad GM, Nikaido H: Outer membrane of gram-negative bacteria, XII. Molecular-sieving function of cell wall. J Bacteriol 128:325, 1976

Glaser L, Lindsay B: The synthesis of lipoteichoic acid carrier. Biochem Biophys Res Commun 59:1131, 1974

Heptinstall S, Archibald AR, Baddiley J: Teichoic acids and membrane functions in bacteria. Nature 225:519, 1970

Inouye M, Hirashima A, Lee N: Biosynthesis and assembly of a structural lipoprotein in the envelope of Escherichia coli. Ann NY Acad Sci 235:83, 1974

Iwanaga S, Morita T, Harada T, et al.: Chromogenic substrates for horseshoe crab clotting enzyme. Haemostasis 7:183, 1978

Kamio Y, Nikaido H: Outer membrane of Salmonella typhimurium: Accessibility of phospholipid head groups to phospholipase C and cyanogen bromide activated dextran in the external medium. Biochemistry 15:2561, 1976

Krueger JM, Pappenheimer JR, Karnovsky ML: The composition of sleep-promoting factor isolated from human urine. J Biol Chem 257:1664, 1982

Krueger JM, Pappenheimer JR, Karnovsky ML: Sleep-promoting effects of muramyl peptides. Proc Natl Acad Sci USA 79:6102, 1982

Lambert PA, Hancock IC, Baddiley J: Occurrence and function of membrane teichoic acids. Biochim Biophys Acta 472:1, 1977

Leloir LF: Two decades of research on the biosynthesis of saccharides. Science 172:1299, 1971

Mannel D, Mayer H: Isolation and chemical characterizations of the enterobacterial common antigen. Eur J Biochem 86:361, 1978

McCabe WR, Kreger BE, Johns MI: Type-specific and cross-reactive antibodies in gram-negative bacteremia. N Engl J Med 287:261, 1972

Misaki A, Seto N, Azuma I: Structure and immunological properties of D-arabino-D-galactans isolated from cell walls of Mycobacterium species. J Biochem 76:15, 1974

Nielsen JBK, Lampen J: Glyceride-cysteine lipoproteins and secretion by gram-positive bacteria. J Bacteriol 152:315, 1982

Oeding P: Cellular antigens of staphylococci. Ann NY Acad Sci 236:15, 1974

Orskov F, Orskov I, Jann B, et al.: Immunochemistry of Escherichia coli O antigens. Acta Pathol Microbiol Scand 71:339, 1967

Qureshi N, Takayama K, Ribi E: Purification and structural determination of nontoxic lipid A obtained from the lipopolysaccharide of Salmonella typhimurium. J Biol Chem 257:11808, 1982

Rick PD, Osborn MJ: Isolation of a mutant of Salmonella typhimurium dependent on D-arabinose-5-phosphate for growth and synthesis of 3-deoxy-D-mannooctulosonate (ketodeoxyoctonate). Proc Natl Acad Sci USA 69:3756, 1972

Scherrer R, Gerhardt P: Molecular sieving by the Bacillus megaterium cell wall and protoplast. J Bacteriol 107:718, 1971

Schnaitman CA: Effect of ethylenediaminetetraacetic acid, triton X-100, and lysozyme on the morphology and chemical composition of isolated cell walls of Escherichia coli. J Bacteriol 108:553, 1971

Takayama K: Selective action of isoniazid on the synthesis of cell wall mycolates in mycobacteria. Ann NY Acad Sci 235:426, 1974

Takayama K, Qureshi N, Mascagni P, et al.: Fatty acyl derivatives of glucosamine-1-phosphate in Escherichia coli and their relation to lipid A. J Biol Chem 258:7379, 1983

Yin ET, Galanos C, Kinsky S, et al.: Picogram-sensitive assay for endotoxin: gelation of Limulus polyphemus blood cell lysate induced by purified lipopolysaccharides and lipid A from gram-negative bacteria. Biochim Biophys Acta 261:284, 1972

Zinner SH, McCabe WR: Effects of IgM and IgG antibody in patients with bacteremia due to gram-negative bacilli. J Infect Dis 133:37, 1976

CHAPTER 7

The Molecular Basis of Genetics and Metabolic Regulation

The foregoing chapters describing the metabolic and structural properties of bacterial cells reveal the magnitude of the genetic information contained in these organisms. The relatively ordered growth and division of single bacterial cells described in Chapter 5 shows that the information contained in the bacterial chromosome is expressed in a manner that insures the production of progeny fully equipped to repeat the performance of the progenitors.

Bacteria vary widely with respect to their genetic potential. Some species grow and divide within environments of restricted quality, whereas others are remarkable in accomplishing the process of cell replication within widely variant environments. These metabolically flexible species demonstrate one of the basic features of the genetic apparatus of bacteria, that it is not differentiated.

The bacterial chromosome, which is equivalent to a covalently closed, circular, double-stranded DNA molecule, must be organized in a way to allow immediate expression of all of its encoded information together with its own duplication and segregation during cell division. The following sections of this chapter deal with the structural organization of the bacterial chromosome, its replication, the mechanisms that result in expression of its content, and examples of how this expression is regulated.

Structure of the Bacterial Chromosome

Early visualization of the stained bacterial nucleoid revealed a highly condensed structure not unlike the smaller chromosomes of eucaryotic cells. These elegant pictures, although giving a misleading impression of the structural nature of the bacterial chromosome, nonetheless revealed what appeared to be an orderly segregation of the genetic apparatus, in that elongated cells in the prefission state contained appropriately situated chromatinic bodies (Fig. 7-1). Electron microscopic examination of fixed thin sections of cells containing active DNA, whether in bacteria or in the interphase nucleus of eucaryotic cells, shows the DNA as compacted, twisted filaments that occupy a large volume of the space enclosed by the delineating membrane, that is, the cytoplasmic membrane in bacteria (Fig. 7-2) and the nuclear membrane in eucaryotic cells. The extent of compaction of the bacterial chromosome is evident by comparing the extended length of the duplex DNA molecule and the dimensions of the cell in which it is packaged. For example, the chromosome of *Escherichia coli* is 1400 μm in length and is contained in a rod-shaped cell of about 1 by 2 μm. The structure of DNA must accommodate an orderly compaction while maintaining its ability to preside over cellular functions, including its own replication.

The basic arrangement in duplex DNA is two strands of deoxyribonucleotide polymers in an antiparallel orientation where the purines (adenine and guanine) on one

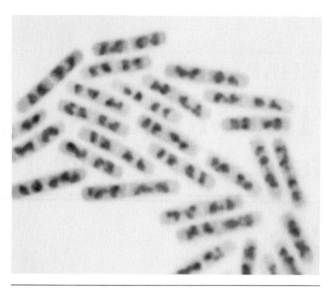

Figure 7-1. The stained nucleoid of *Bacillus cereus.* The staining procedure requires prior fixation of the cells, usually by exposure to osmic acid fumes, and hydrolysis of the abundant cellular RNA that would otherwise mask the DNA. *(Courtesy of Dr. C. Robinow.)*

strand form hydrogen bonds with pyrimidines (thymine and cytosine) on the other strand (Fig. 7-3). The resulting stacking of these planar bases creates a hydrophobic environment and effectively decreases the water concentration in the vicinity of the hydrogen bonds. This results in an effective increase in the strength of these bonds because the participating elements cannot be transferred to water. These interactions and the relatively free rotation about the five single bonds of the sugar-phosphate backbone, the alternative conformations of the deoxyribose, and the free rotation of the N-glycosidic bond between the base and the sugar (Fig. 7-4) cause the duplex DNA molecules to assume a helical conformation or the structure of lowest free energy level. Environmental conditions, such as ionic strength, alter the interactions mentioned above, which, because of the degree of freedom in bond rotation and sugar conformations, allow duplex DNA to assume different types of helical conformations. Table 7-1 lists some of these forms and their properties. The B structure is the one deduced by Watson and Crick from the x-ray diffraction data of Franklin and Wilkins using fibers of the sodium salt of DNA at 92 percent humidity. In real life, DNA closely corresponds to the right-handed double helical B configuration over most of its length. It is possible, however, that segments of DNA are in other helical forms. The most extreme departure from the B form of DNA is the left-handed helical Z DNA. It is known that sequences of alternating purine-pyrimidine nucleotides can assume the Z configuration under physiologic conditions and that this is a natural form, since antibodies prepared against synthetic Z DNA have been shown to react with normal, cellular DNA. The helical conformation of duplex DNA represents a first order compression of the molecule in the

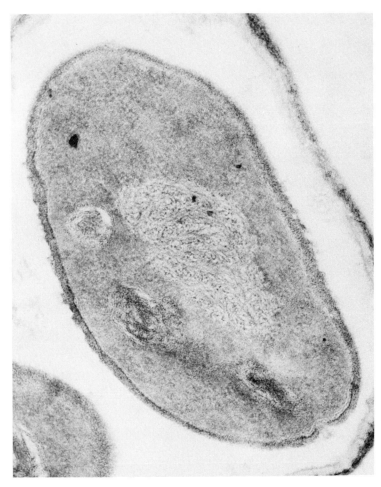

Figure 7-2. Germinating spore of *B. subtilis*. Kellenberger fixation. Vestopal embedding. The fibrillar material inside the spore is the nucleoid which is becoming activated. Following total maturation the DNA will occupy a large portion of the vegetative cell. X 100,000. *(From a study of bacterial nuclei by Carl Robinow and John Marak, 1962. Courtesy of Dr. C. Robinow.)*

sense that a helix is shorter in length than an extended polymer. The helix also is required for a higher order structure, superhelicity, that is vital for the survival of all bacteria and that represents a further compaction of the molecule. The establishment of this suprastructure will be described later.

The function of DNA whether in replication, transcription, gene transfer, or genetic recombination is based upon proteins that specifically bind to the DNA to initiate their action. The double helix per se contributes to specific protein-DNA interaction. The helical arrangement of antiparallel base-paired strands of DNA results in the formation of sterically distinct grooves on the surface of the duplex molecule. Figure 7-5 illustrates the major and minor grooves of B DNA. These topographic features provide for both general and specific recognition by proteins. General recognition is caused by the complementarity of the periodic arrangement of the constituents of the polydeoxyribonucleotide chains in the double helix and the periodicity inherent in the standard β-pleated sheet and α-helix structure in proteins. For example, a β-ribbon structure in a protein consists of a pair of antiparallel peptide chains that can have a right-handed twist and a 7 Å period between alternate α carbon atoms. The distance between adjacent phosphate groups on the B-helix backbone also is 7 Å. If the alternating α carbons are substituted with basic substituents (such as found in the amino acids, lysine, histidine, and arginine), this β-ribbon not only fits into the minor groove of the DNA helix but would be held there by interactions between the basic groups on the protein and the negatively charged phosphates. Similar considerations can be extended to regions of α-helix structure in protein and the major groove of the B-helix. Specific protein-DNA interaction involves components in addition to those involved in general recognition and is related to specific nucleotide sequences in the DNA. The stacked bases of the four nucleotides are oriented so that their edges are consistently located in the major groove or in the minor groove. Figure 7-3 shows the AT and GC hydrogen-bonded base pairs and indicates the respective major and minor groove orientation of the bases. Some proteins involved in DNA metabolism act within the major groove, others in the minor groove, and still others in both grooves.

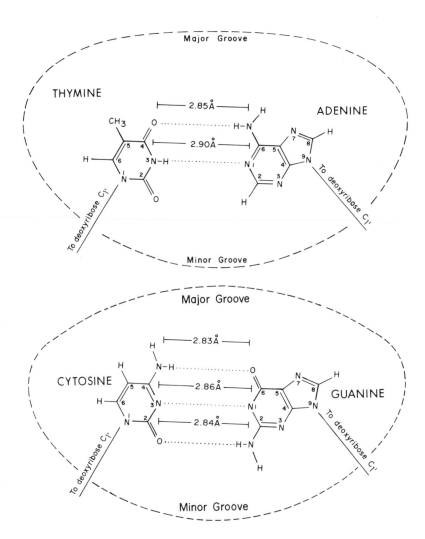

Figure 7-3. Purine-pyrimidine base pairs. The two arcs represent the minor and major groove orientation of the bases in B DNA.

DNA Suprastructure—Superhelicity

Bacterial chromosomes as well as ancillary genetic elements (plasmids) exist and function as covalently closed circular molecules. This circularity, however, is not a sine qua non for the biologic function of all DNA molecules, as certain viral DNAs exist at least for part of their life cycles as linear molecules. Circularity aids in establishing an essential suprastructure in DNA, namely, supercoils or superhelicity. Supercoiling of the double helix occurs when it writhes in space to form a new, high order helix caused by curvature of the original helix axis. Superhelicity can occur in two ways: The double helical molecule can wrap around a cylindrical structure to form a coil, or covalently closed circular duplex DNA can coil in an interwound structure. The nucleosomes of the chromatin of eucaryotic organisms where the DNA is wrapped around a complex of histones to form a left-handed toroid exemplify the first type of supercoil. The second type of supercoiling occurs in bacteria and is loosely analogous to what occurs when a stretched rubber band is twisted and then allowed to relax while the twist is maintained. In the

bacterial cell the significant supercoiling occurs by a right-handed superhelical twist. It is of interest to note that although the handedness of the coil in the nucleosome is opposite that found in the coiled bacterial DNA, these structures are topologically identical and represent negative superhelicity (see below). In eucaryotic organisms the superhelical character of the DNA is maintained by the nucleosomes, whereas bacteria depend upon enzymes to maintain a proper balance of suprastructure. It is the existence of these enzymes, known as topoisomerases, that underscores the importance of superhelicity for the proper functioning of DNA. In order to understand how topoisomerases catalyze morphologic changes in DNA, we first have to consider the structural features that allow covalently closed circular duplex DNA molecules to exist as supercoils. The B form of DNA is a double helix that rotates in a right-handed way about an axis perpendicular to the plane of the stacked bases. A complete (360°) turn occurs for approximately each 10 base pairs along the double helix. Inspection of the linear representation of B DNA reveals that the sugar-phosphate backbones when turning about this axis cross at alternate sides of one another for

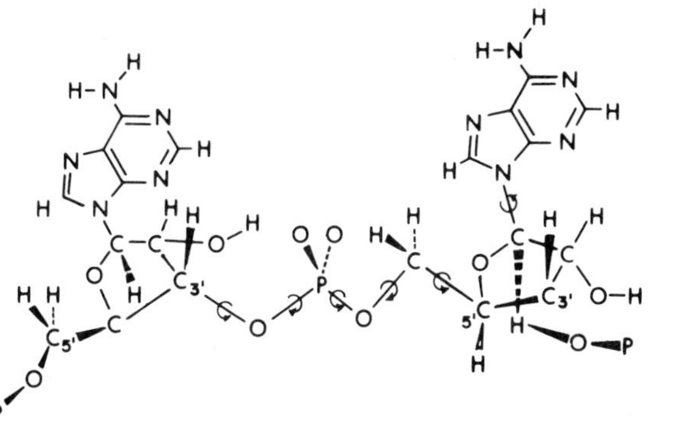

Figure 7-4. Five rotatable bonds in a segment of a polynucleotide chain. These and free rotation at N-C glycosidic bond together with variations in conformation in the pyranose ring of deoxyribose allow DNA to exist in different forms.

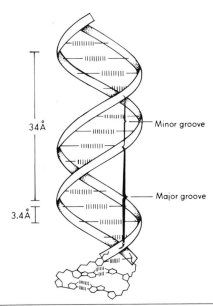

Figure 7-5. A model of B form DNA. *(From Kornberg: DNA Replication, 1980. San Francisco, W. H. Freeman and Co.)*

each turn of the helix (Fig. 7-6). When this linear structure is closed to form a double-stranded circle, the two strands of DNA become topologically linked to one another. The covalently closed, inextricably locked circles still contain 10 base pairs for each turn of the helix, since this arrangement represents the lowest energy level of this structure. If the number of times the two antiparallel circles are mutually linked is decreased, the tendency toward establishment of appropriate base pairing causes curvature along the helix axis in a right-handed sense, and the circular duplex molecule becomes supercoiled. When the DNA molecule is in the superhelical form the axis of the double helix is pictured as writhing in space, and the writhing number (Wr) is a measure of axis deformation and the degree of superhelicity. The value of Wr is related to the degree of interlocking of the two strands of DNA (expressed as the linking number, Ln) and to the number of turns of the helix (the twisting number, Tw). The expression:

$$Wr = Ln - Tw$$

describes this relationship and recapitulates what was previously said, that is, when Ln and Tw are equal, $Wr = 0$. The covalently closed circular DNA molecules isolated

from bacterial cells contain a value of Ln that is usually less than the potential value of Tw, and therefore the value of Wr is negative; these molecules are negatively supercoiled. If the value of Tw is decreased relative to the value of Ln a positive supercoil will result. This condition, $Tw < Ln$, can be brought about by the intercalation of planar dyes between the stacked bases. These relationships are depicted in Figure 7-6.

Topoisomerases

Topoisomerases control the superhelical density of DNA molecules by altering the linking number. Most topoisomerases increase the linking number and consequently are involved in decreasing superhelicity. All bacteria, however, contain a topoisomerase known as DNA gyrase, which is capable of decreasing the linking number and consequently catalyzes the formation of negative supercoils. A change in the linking number in a covalently closed circular duplex DNA molecule requires breakage of phosphodiester bonds. Topoisomerases are generally classified on the basis of their ability to catalyze breaks in one or in two strands of the duplex. Type I topoisomerases produce single-stranded breaks, whereas type II topoisomerases produce double-stranded breaks. Topoisomerases have been described in both eucaryotic and procaryotic cells. Topoisomerase I (ω protein), the product of the *topA* gene in *E. coli* and *Salmonella typhimurium*, is capable of relaxing supercoiling in steps of one; it does not possess gyrase activity, that is, the ability to introduce negative superhelical turns. The basic operation of this enzyme is to cause a break in one strand of DNA, pass the complementary strand through the break, and then reform the phosphodiester bond. These events lead to an increase in

TABLE 7-1. PROPERTIES OF THREE HELICAL FORMS OF DNA

	A DNA*	B DNA*	Z DNA*
Helix (sense)	Right-handed	Right-handed	Left-handed
Nucleotides/turn	11	10	12 (6 dimers)
Rise/nucleotide	2.55 Å	3.4 Å	3.7 Å
Rotation/nucleotide	33°	36°	−60°/dimer
Helix pitch	28 Å	34 Å	45 Å
Base pair tilt	20°	6°	7°
Diameter	23 Å	19 Å	18 Å

Table constructed in consultation with Dr. T-S Hsieh.

*A plane of symmetry perpendicular to the helix axis exists between each base pair in A and B DNA and between alternating purine-pyrimidine dimers in Z DNA. The repeating unit in Z DNA, therefore, is considered to be the dimer.

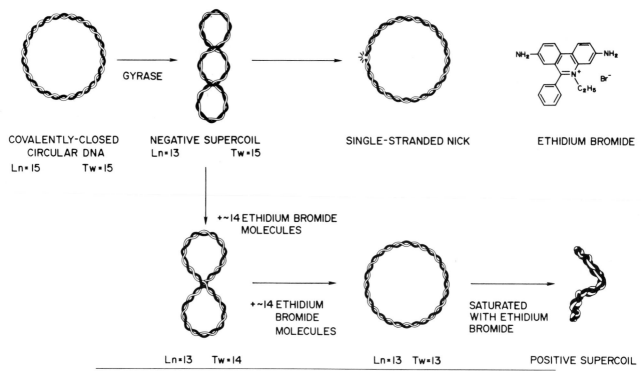

Figure 7-6. Circular DNA. A small, covalently closed, circular DNA molecule composed of 150 base pairs is depicted at the left of the figure. Each turn (twist) of the helix (10 base pairs) results in a single linking of the interwound strands: the 15 linkages in the double-stranded molecule are seen in the structure at the lower left, where one of the strands is constrained so as to lie in a plane. Two negative supercoils are introduced into the molecule by the action of DNA gyrase, which changes the linking number in steps of two. The negative supercoiling is relieved by introducing a single-stranded break in the molecule or by reducing the twist (helix winding number) by intercalating the dye ethidium bromide between the stacked bases. Each intercalated ethidium bromide molecule causes a 26° left-handed rotation of the helix; approximately 14 molecules of the dye must be intercalated for a net rotation of 360° (a reduction in the twisting number, *Tw*, by 1). Continued addition of the dye causes further unwinding of the helix and a compensatory positive supercoiling of the molecule. Ethidium bromide is useful in the analytic and preparative separation of various forms of DNA. The use of this and related dyes is based on the following: ethidium bromide is less dense than DNA, and linear DNA will bind more of the dye than will supercoiled DNA. Therefore, in isopycnic centrifugation in the presence of moderate concentrations of dye supercoiled DNA will band at a higher density than will linear DNA.

the linking number by one and a consequent removal of a single negative supercoil. The breakage-reunion mechanism requires no additional energy source because a 5'-phosphoryl end of the nucleotide chain is transferred to the enzyme to cause the initial scission and then transferred back to the 3'-hydroxyl to effect sealing. Topoisomerase I is not required for the viability of such bacteria as *E. coli*, although removal of the *topA* gene leads to wide-ranging pleiotropic effects, most of which are related to DNA function. The viability of *topA* mutant cells is probably maintained by the activity of topoisomerase II, which in bacteria is the DNA gyrase. This enzyme performs all the functions of topoisomerase I but in addition is capable of catalyzing the formation of negative superhelicity. It is the gyration activity of this enzyme that distinguishes it

from the type II topoisomerases found in the nucleus of eucaryotic cells. DNA gyrase is required for the viability of bacteria; antibiotics that inhibit the activity of this enzyme prevent their growth. DNA gyrase from *E. coli* is oligomeric, composed of two α subunits and two β subunits. The α and β subunits, respectively, are the products of the *gyrA* and *gyrB* genes, which were formerly termed *nalA* and *cou* because mutation in these genes rendered the organism resistant to nalidixic acid and coumermycin (and novobiocin). The mode of action of DNA gyrase is different from that of topoisomerase I. The enzyme changes the linking number of DNA by transiently altering the writhe of the molecule. This feat is accomplished by the simultaneous stabilization of a positive and negative coiling (a situation that is permitted because there is no net change in the superhelicity). The enzyme produces a double-stranded break at one of the supercoiled nodes to allow passage of the unbroken double-stranded segment (Fig. 7-7). DNA gyrase, unlike type I topoisomerase, hydrolyzes ATP in the course of its action. This activity is associated with the β subunit and is probably required for conformational changes to allow rearrangement of the superhelical structure. Figure 7-8 depicts a model for the activity of gyrase. Additional aspects of topoisomerase activity are described under DNA replication.

Supercoiling and the Organization of the Bacterial Nucleoid

A variety of techniques have been employed to study chromosome organization in *E. coli.* The results of these analyses show that the chromosome is organized into 20 to 70 superhelical domains. It has been speculated that this arrangement is not completely random but represents a means whereby the cell can differentially regulate the expression of functions in the individual domains. This view is supported by the observation that preferred sites for the action of DNA gyrase are situated along the chromosome at intervals of 100,000 nucleotides, which would be sufficient for maintaining 30 to 40 domains. In addition to functional considerations this domained arrangement serves to sensibly compact the chromosome.

As previously mentioned, nucleosomes are an important organizational and functional feature in eucaryotic chromatin. Similar structures can be seen in appropriately prepared bacterial DNA, and it is tempting to speculate that some functional analogy can be made between these structures and the eucaryotic nucleosomes. The eucaryotic nucleosome core is composed of dimers of the four major histones H2A, H2B, H3, and H4. These proteins are arranged to form an octomeric cylindrical core about which is wrapped a segment of DNA in a left-handed toroidal superhelix. The periodicity of nucleosomes along the DNA of eucaryotic cells, although variant, is about 1 every 200 nucleotide pairs. A similar periodicity occurs with the nucleosome-like structures in bacteria. The nucleosome-like nature of these structures is further supported by the isolation of four distinct histone-like proteins from the

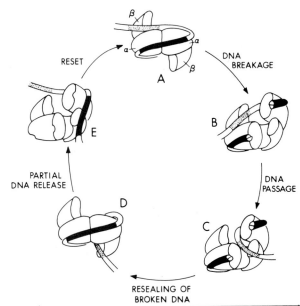

Figure 7-8. Model for action of DNA gyrase. **A.** The α_2 β_2 gyrase molecule with a section of DNA wrapped in a right-handed superhelix. The solid DNA represents approximately a 50 base pair region within which a double-stranded scission occurs. The transiently formed gap is a passageway for the stippled section of DNA. Steps **B** and **C** result in conversion of a right-handed loop (**A**) to a left-handed loop (**D**). The conformational changes in the enzyme that allow these gymnastics are apparently caused by the binding of ATP to the β subunit. *(From Morrison and Cozzarelli: Proc Natl Acad Sci USA 78:1416, 1981.)*

DNA of *E. coli.* These have been designated HLP11a, HLP11b, HLPI, and H protein. HLP11a and HLP11b have amino acid compositions similar to that of histone H2B, and H protein cross-reacts with antibodies prepared against histone H2A from calf thymus. As mentioned previously, one of the primary roles of nucleosomes of eucaryotic cells is to maintain the essential negative supercoiled structure in the DNA. Bacteria, on the other hand, have DNA gyrase to carry out this function. This fact together with the observation that bacteria contain histone-like proteins sufficient for interacting with only 20 percent of the DNA suggest that if the nucleosome-like particles serve to restrain superhelicity they do so as an auxiliary to gyrase.

The Cell Cycle and Chromosome Replication

The cell cycle of procaryotic cells forms an interesting contrast with the cycle of eucaryotic cells. The greater part of the cell cycle in eucaryotic cells is a period cytologically

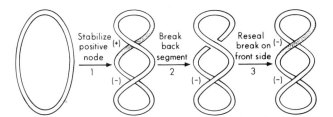

Figure 7-7. The sign-inversion mechanism for DNA gyrase. See text for details. *(From Brown and Cozzarelli: Science 206:1082, 1979.)*

characterized as the interphase, that is, a time during which the genetic apparatus is delineated by a membrane from the remainder of the protoplasm. The interphase period is initiated with the G1 phase; during this period the cell enlarges, owing to the synthesis of proteins and ribosomes and to the assembly of organelles. The various assembly processes continue during interphase, but it is the length of the G1 phase that generally governs the duration of the cell cycle. Events that occur during the G1 phase commit the cells to subsequent DNA replication (S phase) after which they proceed through terminal interphase (G2 phase) and into mitosis with subsequent cell division. The timing of events that lead to cell division in bacteria is different. Cells like *E. coli,* when growing at other than exceedingly slow rates, synthesize DNA during the entire division cycle. This means that other functions of the chromosome, such as transcription and genetic recombination, must be accommodated by an orderly process of replication. The orderly replication of the bacterial chromosome was first suggested by the demonstration that daughter chromosomes are semiconserved replicas of the parental DNA molecule, that is, composed of one complete parental strand and one newly synthesized strand of DNA. The classic experiment that demonstrated the principle of semiconservative replication is depicted in Figure 7-9. This and other observations relating to DNA replication fostered the concept of the *replicon,* or the unit of replication. Accordingly, the bacterial chromosome is a replicon, as are ancillary genetic elements, such as plasmids. Replicons are now considered to be units of replication defined by an origin and a terminus. This expands the replicon concept to include not only DNA molecules but also the units

of replication within eucaryotic chromosomes. The replicon model of DNA replication has been verified innumerable times since its inception. The most significant extension of this view of replication is the demonstration that the point or origin of initiation of replication for a given replicon is always the same, composed of a heritable nucleotide sequence. This arrangement is not inconsequential for the bacterial cell as it is interaction of this sequence with specific proteins that regulates the rate at which replication is initiated. This arrangement maintains the remaining parts of the DNA molecule, including the various genes, in the same order relative to the dynamic replication apparatus and, therefore, provides for a means of primitive gene amplification. It was pointed out in Chapter 5 that rapidly dividing bacteria often inherit chromosomes that are already undergoing replication. Genes located closest to the origin are present in higher copy number because a partially replicated chromosome is diploid over the replicated portion and haploid over the remaining portion. Perhaps it is not fortuitous that genes whose products are required in high levels by the cell are located in the amplifiable region.

Replication of many replicons, including the bacterial chromosome, occurs in a bidirectional manner. In *E. coli* two replication forks are established at the origin of replication and proceed counter to one another until they meet at a terminator 180° from the starting point; this termination area also is heritable. One obvious advantage of divergent replication is that it halves the time required for a single replication fork to travel around the entire chromosome. For example, in the case of the *E. coli* chromosome, which contains approximately 4000 kilobase pairs, a repli-

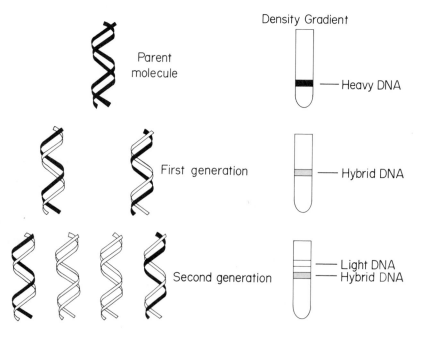

Figure 7-9. The semiconservative replication of the bacterial chromosome; the classic Meselson and Stahl experiment. Cells of *E. coli* that had grown for several generations in medium containing the heavy isotope ^{15}N were transferred to medium containing ^{14}N, and the distribution of the isotopes in the DNA was monitored after subsequent generations. The left side of the figure depicts the distribution of uniformly labeled heavy (^{15}N) DNA following 0, 1, and 2 generations, respectively, in light (^{14}N) medium. The right side of the figure depicts the location at equilibrium in a cesium chloride gradient of heavy, hybrid, and light DNA. These results are consistent with the semiconservative replication of DNA. The results also illustrate another very significant aspect of DNA replication, namely, that it occurs in a continuous and sequential fashion, as shown by the fact that all of the heavy DNA becomes hybrid prior to the appearance of light DNA.

cation fork travels at 45 kilobase pairs per minute at 37C. A single replication fork, therefore, would require 90 minutes to travel around the chromosome, while the same net synthesis is accomplished during the 45-minute trek of divergently moving forks.

Replication of the Bacterial Chromosome

The simplicity of the concept of semiconservative replication of DNA as applied to the bacterial chromosome is inversely related to the complexity of the mechanisms used to insure that this process occurs in a regulated and faithful manner. Chromosome replication can be separated into three phases: initiation, elongation, and termination. Early studies of DNA replication were devoted to defining the elements involved in elongation, because of the obvious requirement for DNA synthesis in the replication process. Later studies were aimed at defining the elements involved in the initiation process because it was shown that the regulation of the rate of replication is physiologically related to the rate at which this process is initiated. The termination process, conceptually the simplest of the three, has not been extensively studied, and it is unclear whether this is a passive or an active event.

A number of techniques have been employed to identify and characterize the various proteins that participate in chromosome replication. A fruitful approach employs mutant methodology, and a number of temperature-sensitive mutations affecting replication have been identified. These mutations can be classified on the basis of their effect on DNA replication at the nonpermissive temperature. Organisms containing certain ones of these mutations immediately cease DNA sythesis when shifted to nonpermissive temperatures, whereas others exhibit a delay before synthesis is arrested. These responses have identified elements involved in elongation (fast stop) and initiation (slow stop). The delayed response is explained by completion of ongoing rounds of replication and inability to reinitiate new rounds of replication. Extracts of these mutant strains have been used in complementation assays to follow the purification of the replication proteins from normal strains. Unfortunately, the bacterial chromosome, owing to its large size, cannot be readily prepared in a cell-free form suitable for use as a template in these assays. This problem has been circumvented by using smaller replicons, such as phage and plasmid DNA, for these studies. Many of the proteins of interest initially were intransigent to purification because of their low intracellular concentration. This problem has largely been overcome by the use of recombinant DNA technology. Chimeric, multicopy plasmids that contain the structural genes for these replication proteins have been constructed; these proteins are readily purified from strains bearing these plasmids. The results from studies using these purified proteins and the small replicons have led to an extrapolated view of bacterial chromosome replication.

The centerpiece of replication is DNA polymerase, the most extensively studied replication protein. Bacteria generally contain three types of polymerases, which, by analogy to those found in *E. coli*, are known as polymerases I, II, and III. Polymerases I and III are directly involved in the replication process; no definite role in replication has been described for polymerase II. All DNA polymerases act by the stepwise addition of a single deoxyribonucleotide to the 3'-OH of the sugar moiety of a primer molecule that is precisely hydrogen bonded to the template strand. As explained below, either DNA or RNA serves as primer.

The product of a single step in the polymerization is a template-associated primer extended by a properly base paired nucleotide and a pyrophosphate that is released from the nucleotide triphosphate substrate. The polymerases possess an additional activity that guards against 3'-OH extension where the substrate is improperly base paired with the template strand. This activity is a 3'→5' exonuclease that serves a proofreading function by reducing errors in replication. DNA polymerase when encountering a mispaired potential primer nucleotide will remove it and use the next encountered, precisely matched nucleotide. Table 7-2 shows the properties of the DNA polymerases of *E. coli*.

The continuous semiconservative replication of DNA from a single point of synthesis has two basic requirements. Since all DNA polymerases extend DNA in a 5'→3' direction the two antiparallel template strands must be copied counter to one another. In addition, the helix must be unwound at the replication point so that the template strands become available to the polymerase. The 3'→5'-oriented template could be copied continuously because the unwound helix and 3'-OH group of the complementary primer are appropriately oriented. However, the 5'→3' template must be copied discontinuously. The reason for the discontinuous synthesis is that the progressive unwinding of the helix exposes template that can be copied only after a primer is synthesized. The primer employed here must be RNA because its synthesis is independent of primer. The manner in which the two strands of DNA are synthesized defines the potentially continuously synthesized one as the leading strand and the discontinuously synthesized one as the lagging strand. The coordination of the synthesis of the leading and lagging strands requires the participation of several different proteins. The unwinding of the helix is caused by proteins known as helicases. One of these, the *E. coli rep* protein, is known to bind to the 3'→5'-oriented template and to utilize the energy of ATP to melt the DNA; this protein acts in conjunction with other helicases that bind to the 5'→3'-oriented strand at the replication fork. The unwound section of the helix formed by the action of the helicases is maintained in the single-stranded form by single-strand binding proteins. Leading strand synthesis is effected by polymerase III holoenzyme, which is composed of a core structure consisting of three polypeptides (α, ϵ, θ) and at least four easily separable

TABLE 7-2. PROPERTIES OF DNA POLYMERASES FROM E. COLI

Features of DNA polymerases I, II and III of *E. coli.*

	Pol I	Pol II	Pol III
Turnover number [a,b]	~600	~30	~9000
Activity/cell [a,b,c]	X	<X/20	<X/20
Molecular weight	105,000[c] one subunit	120,000 one subunit	140,000 one subunit[d]
In vivo requirement	+	?	+
Exonuclease:			
$5' \rightarrow 3'$	+	?	+
$3' \rightarrow 5'$	+	+	+
Polymerization:			
$3' \rightarrow 5'$	+	+	+
$5' \rightarrow 3'$	−	−	−

	Pol I	Pol II	Pol III
Template and primer requirement			
Duplex DNA	−	−	−
DNA with short gaps	+	+	+
Denatured DNA	+	−	−
"Primed" single strands	+	−	−
Nicked DNA	+	−	−

Table constructed in consultation with Dr. Paul Modrich.

[a]Nucleotides polymerized/min/molecule of enzyme at 37C.

[b]Activity/cell vs. turnover number shows that cells contain many more molecules of pol I than pol III; this probably reflects the multiple roles of pol I in DNA replication, repair, and recombination.

[c]Pol I can be separated into a large and small fragment by proteolysis. The large fragment contains the polymerase and the $3' \rightarrow 5'$ exonucleolytic activity and the small fragment the $5' \rightarrow 3'$ exonucleolytic activity. The large fragment, sometimes called the "Klenow" fragment, is frequently employed in genetic engineering procedures.

[d]Pol III is the α subunit of polymerase III holoenzyme (Table 7-3).

polypeptides (τ, γ, δ, β). The lagging strand synthesis requires prior formation of primer molecule, which is in the form of short stretches of polyribonucleotides. Primer formation requires prior assembly of a primosome on the lagging strand template. Once the short RNA primer is synthesized, DNA polymerase III holoenzyme extends the

primer by adding deoxyribonucleotides to its 3' end. Synthesis continues until a 1000 to 2000 nucleotide extension occurs. These short molecules are known as Okazaki fragments after their discoverer. During this synthesis the helix continues to unwind, leading strand synthesis occurs, and the primosome moves in the direction of helix, unwinding to ready itself for synthesis of another primer. The primosome is composed of a number of proteins including the primase, the product of the *dnaG* gene. The role of RNA as primer points to a basic difference in RNA polymerase and DNA polymerases. RNA polymerases initiate complementary RNA synthesis in the absence of a primer, whereas all DNA polymerases require a primer. The discontinuous segments of DNA on the lagging strand template are the sites of action of polymerase I in replication. This enzyme uses its polymerase and $5' \rightarrow 3'$ exonuclease functions in a coordinated manner to remove the RNA primer segments and to fill the gaps between the disjointed DNA segments. The complete continuity of the lagging strand is established by the activity of DNA ligase (see below). Table 7-3 lists several of the proteins involved in DNA replication, and Figure 7-10 depicts how some of these are proposed to fit into the described scheme of replication.

TABLE 7-3. REPLICATION PROTEINS OF E. COLI

Protein	Native Mass kdal	Subunits	Function
SSB	74	4	Single-strand binding
Protein i	66	3	Primosome assembly and function
Protein n	28	2	
Protein n'	76	1	
Protein n"	17	1	
dnaC	29	1	
dnaB	300	6	
Primase	60	1	Primer synthesis
pol III holoenzyme	(760)	(2)	
α	140	1	Processive chain elongation ×2
ϵ	25	1	
θ	10	1	
β	37	1	
γ	52	1	
δ	32	1	
τ	83	1	
pol I	102	1	Gap filling, primer excision
Ligase	74	1	Ligation
Gyrase	400	4	Supercoiling
gyrA	210	2	
gyrB	190	2	
rep	65	1	Helicase
Helicase II	75	1	Helicase
dnaA	48		origin of replication

Modified from Kornberg: 1982 Supplement to DNA Replication, 1982. San Francisco, W.H. Freeman & Co.

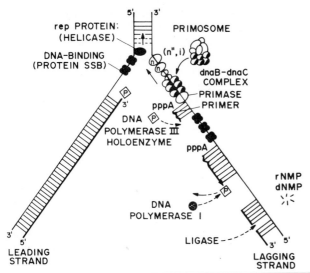

Figure 7-10. A scheme for DNA chain growth at a replication fork of the *E. coli* chromosome. The role of the various proteins is explained in the text and Table 7-3. *(From Kornberg: 1982 Supplement to DNA Replication, 1982. San Francisco, W.H. Freeman and Co.)*

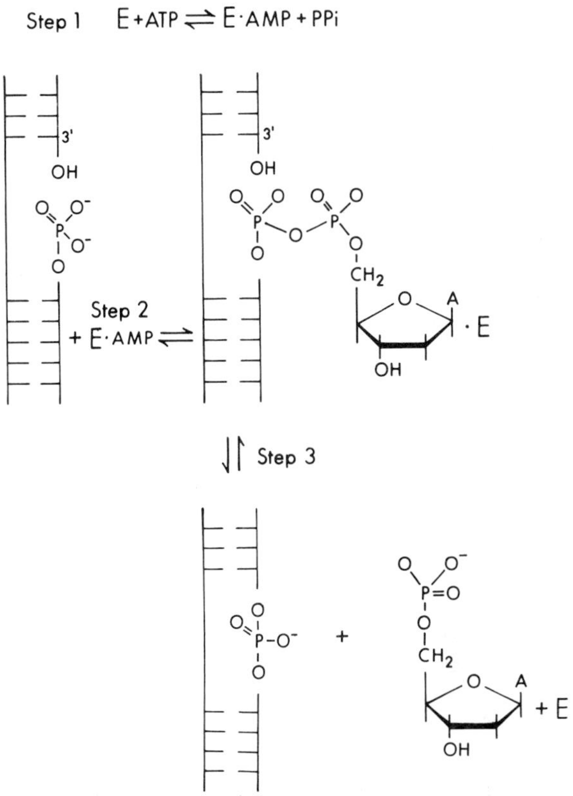

DNA Ligases

Establishment of the continuity of polydeoxyribonucleotide strands of DNA is a terminal event not only in DNA replication but also in genetic recombination and in repair of damaged DNA. The specificity of this class of DNA-joining enzyme insures that aberrant DNA molecules are not repaired; they normally use as substrate only single-stranded breaks that are bordered by properly hydrogen-bonded nucleotides containing a 3'-OH and 5'-phosphoryl group. DNA ligases require a source of energy to catalyze the formation of phosphodiester bonds. Bacterial DNA ligases, such as those from *E. coli* and *Bacillus subtilis*, use nicotinamide adenine dinucleotide (NAD), whereas those from mammalian sources and those encoded by certain bacteriophage use ATP. These nucleotide cosubstrates serve as a source of the activation energy for phosphodiester bond formation. In either case, AMP is transferred to a lysine residue on the ligase to form a phosphoramidate. This is followed by activation of the 5'-phosphoryl group, closure of the nick, and release of AMP. Figure 7-11 illustrates this process. All DNA ligases carry out a conservative sealing reaction, in that base pair complementarity is maintained in the repaired region. Some ligases, under certain conditions, carry out radical joining reactions, so that a nucleotide is joined by an entirely new neighbor. The ligase specified by bacteriophage T4 acts in either a conservative or a radical manner. Its radical activity, which is manifested at high enzyme concentration, is frequently used for blunt end ligations in recombinant DNA endeavors (Chap. 8).

Step 1 $E + ATP \rightleftharpoons E \cdot AMP + PPi$

Step 2 $+ E \cdot AMP \rightleftharpoons$

Step 3

Figure 7-11. DNA ligase. ATP is shown as the source of the activating adenylyl group. Bacterial ligases generally use NAD (nicotinamide adenine dinucleotide) as the source of AMP, in which case NMN (nicotinamide mononucleotide) is a product of the reaction.

Topoisomerases and Chromosome Replication

Topoisomerases play a number of roles in DNA replication. It is known that chromosome replication fails to occur when the negative superhelical density of the DNA falls below a critical value. This is probably so because the potential energy in the superhelical structure is used to drive replication processes. It is known also that unwound helix resulting from negative superhelicity is required for the entry of certain replication proteins. Topoisomerases, in addition to regulating the superhelical density of the chromosome, possess another activity that expands their role in replication. Both of the topoisomerases in *E. coli* are able to resolve catenanes of circular DNA. These interlocked rings are the products of the replication of covalently closed, circular, duplex DNA. DNA gyrase is active in resolution by virtue of its ability to produce double-stranded breaks. Topoisomerase I, in spite of the fact that it produces only single-stranded breaks, is also able to resolve catenanes. This enzyme acts preferentially at sites opposite single-stranded breaks, which results in the pro-

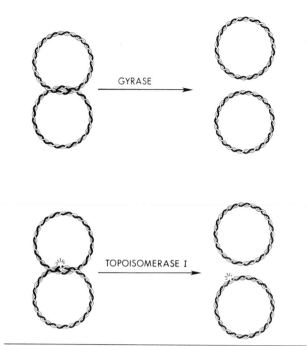

Figure 7-12. The resolution of catenanes by DNA gyrase and topoisomerase I. See text for details.

duction of a passageway for the intact circle (Fig. 7-12). DNA gyrase can be isolated in a modified form that lacks gyration, negative coil relaxation, and ATPase activities. This enzyme, which is called topoisomerase II′, has the ability to relax positive superhelicity. It is speculated that this activity is significant in removing positive supercoils that accumulate in advance of the replication fork.

Regulation of Chromosome Replication

The physiologically significant regulation of the rate of chromosome replication in bacteria is equivalent to the modulation of the frequency with which this process is initiated. The nucleotide sequence corresponding to the origin of replication specifies not only the point on the chromosome where this process begins but also how frequently it occurs. It is clear that the events that occur to initiate chromosome replication are in some way connected to the growth rate of the cell. The chromosomes of cells undergoing rapid growth and division contain more replication forks than do slow-growing cells. Early observations suggested that a constant unit of cytoplasmic mass exists for each replication fork. It is clear that de novo protein synthesis is required for the initiation of each round of replication, which strongly suggests that at least one of the proteins involved in the regulatory process is consumed, inactivated, or sequestered during the replication cycle. These observations tend to associate general RNA and protein synthesis with the regulation of initiation of replication, but the precise manner in which this connection is made is unclear.

The regulation of chromosome replication is probably related to those proteins that are known to be mechanistically involved in the initiation process. Several of these proteins have been identified by studying in vitro initiation in plasmids into which the origin of replication of the *E. coli* chromosome *oriC* region has been introduced by recombinant DNA methods. Among these proteins are the *dnaA* gene product, RNA polymerase, DNA gyrase, and *dnaB* and *dnaC* proteins. Other uncharacterized proteins also are involved in this process.

Information Content of the Bacterial Genome

The genome of bacteria consists of the chromosome and all other DNA molecules that are stably established in the cell. The properties of a bacterial cell are largely governed by the information encoded within the chromosome, although plasmids and prophages (Chap. 8), which usually range in size from less than 1 percent to several percent of the size of the chromosome, frequently contain determinants for significant properties. For example, antibiotic resistance determinants frequently are borne on plasmids, and some toxins are produced by prophage.

One of the striking features of the gene content of bacterial cells is that it is generally maintained at a minimum level. This is a conservative arrangement and aids the cell in its apparent sole purpose, growth and division. If multiple copies of a given gene are present in the chromosome, the increased dosage usually offers the organism an obvious advantage. For example, there are seven highly homologous regions in the *E. coli* chromosome that encode the sequences for 23S, 16S, and 5S ribosomal RNA (rRNA). However, even these are not totally homologous, since most of them also contain sequences for different transfer RNA (tRNA) molecules. Owing to the active role ribosomes play in protein synthesis, metabolically active cells require a high concentration of rRNA. Yeast cells, for example, contain 140 rRNA genes and amphibians, such as *Xenopus*, contain 500 to 600 rRNA genes, which are amplified 1000-fold in their very active oocytes.

Bacterial genomes are generally distinguished from those of higher organisms by the lack of repetitive sequences, that is, the DNA of bacteria contains unique sequence over most of its length. The DNA of higher organisms contains varying proportions of their DNA as moderately or highly repetitive components. In yeasts, for example, 20 percent of the DNA exists as moderately repetitive sequences, whereas in mammals highly repeated short (500 base pair) and long (5000 base pair) interspersed nucleotide sequences occur. A given mammalian genome may contain several families of these repeated sequences, each containing several thousand copies. The role of these sequences is not known, but their presence in eucaryotic chromosomes and absence in procaryotic chromosomes mirror the relative complexity of the two types of genetic apparatus.

A convenient method for assessing the complexity of any genome is to measure the rate of reassociation of single-stranded DNA produced by melting sheared fragments (500 base pairs long) of total cellular DNA. The reassociation of the complementary strands requires random collision, a bimolecular reaction that, because the participants are in equal concentration, will follow second order kinetics. If the original duplex DNA is composed of totally unique sequences, reassociation of the separated components will occur uniformly following the differential equation:

$$dC/dt = -kC^2$$

where C is the concentration of DNA that is single stranded at time t and k is the single second order rate constant that describes the reaction. If, on the other hand, the original DNA contains repeated sequences, the sheared DNA when melted will contain a disproportionately higher concentration of single-stranded molecules that can participate in reassociation mixed with those derived from the unique sequence. The unique single strands must search out their single complementary partner by random collision, whereas a single strand derived from repeated sequence can choose from among many complementary strands. In this case the reassociation kinetics will be composed of a rate constant to describe the rapid reassociation of the repeated sequences and a rate constant to describe the slower reaction. Figure 7-13 depicts the typical C_0t plotting form for a reassociation or renaturation analysis of DNA sample of varying complexity.

Transcription

Primary Decoding Process

The three major types of RNA encoded in the DNA of bacteria are used in protein synthesis. Ribosomal RNA (rRNA) and transfer RNA (tRNA) are involved in the synthesis of proteins encoded within messenger RNA (mRNA). Owing to the requirement for a variety of proteins for proper cellular function a large portion of the chromosome is dedicated to encodement of an equally large variety of mRNA molecules and much less for rRNA and tRNA. In spite of this, rRNA and tRNA are the relatively abundant species comprising, respectively, 85 percent and 5 percent of the total cellular RNA. The reason for this bias in concentration is the stability of rRNA and tRNA and the lability of mRNA. The stability of rRNA and tRNA is conservative in that ribosomes and tRNA are recycled for use in many rounds of translation. The instability of mRNA, on the other hand, accommodates differential and regulated gene expression by preventing engorgement of the cell with needless and unused mRNA.

There are three distinct species of RNA in bacterial ribosomes. These are characterized by sedimentation coefficient: 23S (3100 nucleotides), 16S (1520 nucleotides), and 5S (120 nucleotides). There are, however, many types of

tRNA species. All of these stable polyribonucleotides are synthesized as transcripts containing extraneous nucleotides. The functional forms of rRNA and tRNA are processed from these larger transcripts by specific ribonucleases (see RNA Processing, below).

Bacterial genomes also encode highly specialized transcripts in addition to the three major ones described above. The *oriC* primer RNA used for initiation of chromosome replication is one such example. Another example is the 362 nucleotide M1 RNA, which is a specific cofactor for ribonuclease P, one of the enzymes involved in processing rRNA. This RNA is absolutely required for the activity of this enzyme. *Staphylococcus aureus* produces a tRNA-like molecule that is specifically involved in cell wall biosynthesis. In this case a glycyl-tRNA is the activated form of glycine used for the nontemplate-directed synthesis of the pentaglycine bridge found in the peptidoglycan layer of this organism (Chap. 6).

RNA Polymerase, Promoters, and Terminators

The synthesis of all types of RNA present in bacterial cells is mediated by DNA-dependent RNA polymerase. The

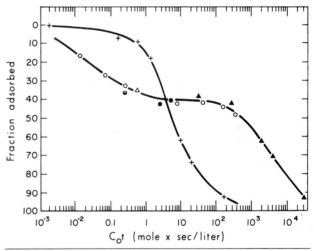

Figure 7-13. The kinetics of reassociation of calf thymus DNA (○, △) and *E. coli* DNA (x). The rate of reassociation of sheared DNA at several concentrations was determined. C_0t is expressed in moles of nucleotides in the single-stranded form at time zero times seconds per liter. Fraction adsorbed is a measure of the amount of DNA in the double-stranded form; this quantity is determined by passing the DNA through a column of hydroxyapatite. Double-stranded DNA is adsorbed, single-stranded DNA is not adsorbed. The curve shows that the calf thymus DNA is composed of a rapidly reassociating fraction and a slowly reassociating fraction. The rapidly reassociating fraction represents repetitive sequences in the calf thymus DNA sample, and the slowly reassociating fraction represents unique sequences. The curve for the *E. coli* DNA is typical of one described by a single, second order rate constant and indicates that the DNA contains unique sequences. *(From Britten and Kohne: Science 161: 529, 1968.)*

polymerase carries out three discrete processes: (1) selection of sites for the initiation of synthesis of meaningful transcripts, (2) polymerization of ribonucleotides to form a polyribonucleotide that is complementary to one strand of the DNA duplex, and (3) termination of the polymerization reaction at appropriate positions of the DNA template. The RNA polymerases from a number of bacteria are typical of the one found in *E. coli.* This enzyme is an oligomer composed of three types of protein subunits, α (36.5 kdal), β (150 kdal), and β' (160 kdal). The enzyme can be reconstituted from its components by the following pathway:

$$2\alpha + \beta \rightarrow \alpha_2\beta \overset{\beta'}{\rightarrow} \alpha_2\beta\beta'$$

The enzyme in this form is capable of using a single-stranded DNA template plus ATP, GTP, CTP, and UTP to synthesize randomly initiated polyribonucleotides that are complementary to the DNA. This core form of the enzyme, however, is incapable of using as template double-stranded DNA, as found in the chromosome. Addition of another protein, sigma (σ, 70 kdal), to the core polymerase endows it with two properties: (1) ability to use double-stranded DNA as template and (2) initiation of RNA synthesis at specific locations on a genophore (i.e., chromosome or plasmid). These specific regions are known as promoters. They are integral parts of units to be transcribed in that they are conjoined to appropriate nucleotide sequences to insure proper initiation of transcription. Neither the core polymerase nor the holoenzyme (i.e., with sigma) requires a primer for initiation of RNA synthesis.

The sigma subunit when complexed with the core enzyme is known to contact two sets of nucleotides within the promoter region. These are centered at the -35 and -10 positions, that is, at positions 35 and 10 nucleotides upstream or before the point of transcription initiation. DNA sequence analysis of a large number of promoter regions shows a close similarity in these structures; several of these are shown in Table 7-4. Two heptameric consensus sequences can be derived from these analyses, -TTGACA- at the -35 region and -TATAAT- at the -10 region. The region of DNA occupied by holopolymerase is about 60 nucleotide pairs, i.e., from -40 to $+20$. This is demonstrated by nuclease protection experiments that show that this region, when bound by RNA polymerase, is protected from digestion with DNase. Points of contact between the DNA and the polymerase (or any DNA-binding protein) can be established more precisely by determining the comparative pattern of methylation of the DNA by dimethylsulfate in the presence and absence of polymerase. This alkylating agent normally methylates N^7 of guanines in the major groove and N^3 of adenine in the minor groove of duplex DNA. Similar experiments can be performed with the agent ethylnitrosourea, an alkylating agent for backbone phosphates. Using these reagents it can be shown that RNA polymerase protects bases in the vicinity of the consensus sequences. Interestingly, the bound polymerase causes enhancement of methylation at the N^1 of adenine and N^3 of cytosine in a region from -9 to $+3$. Enhancement of methylation at these positions is an indication of an unwinding of the helix. These ring nitrogens are normally protected from alkylation by their participation in hydrogen bonding in the double helix (Fig. 7-3). The 12 base pair separation that occurs upon polymerase binding is thought to correspond to establishment of the open complex that in vivo signals the initiation of transcription.

Upon formation of the transcriptionally active, open complex a nucleotide triphosphate is positioned by hydrogen bonding to the appropriate template deoxyribonucleotide. This ternary complex of template, polymerase, and nucleotide triphosphate is the final initiation complex from which extension of the RNA chain proceeds. Initiation usually occurs with a purine nucleotide, although some transcripts can be initiated with pyrimidine nucleotides. The polymerization process is initiated by transfer of a 5' phosphoryl nucleotide to the 3'-OH of the in-place nucleotide. This and all subsequent steps in polymerization involve release of PP_i from the substrate nucleotide triphosphate. Polymerization proceeds and obeys the rules of base pairing so that the finished RNA transcript is aligned in an antiparallel manner with the template strand of DNA. Once polymerization is initiated the sigma factor is no longer required. This subunit dissociates from the polymerization complex following a 6 to 7 nucleotide extension and is used to form more holoenzyme. Polymerization continues until a termination signal is encountered. Termination involves the cessation of RNA elongation and release of the polymerase. The signal for termination is encoded in the DNA. This signal, however, probably is not effective as DNA but rather as the homologous RNA product. It is known that the terminal portions of transcriptional units contain a sequence of dyad symmetry that provides for the potential formation of a stem and loop structure. This secondary structure is conjectured to cause

---

TABLE 7-4. PROMOTER SEQUENCES SHOWING SIMILARITY AT THE −35 AND −10 REGIONS

	−35 Region	−10 Region	RNA
lac	A C C C C A G G C T T T A C A C T T T A T G C T T C C G G C T C G|T A|T G T|T|T|G T G T G G(G)(A)A T T G T G A G C G G
galP$_2$	A T T T A T T C C A T G T C A C A C T T T T C G C A T C T T T G T|T A|T G C|T|T A T G G T T(A)T T T C A T A C C A T
araBAD	G G A T C C T A C C T G A C G C T T T T T A T C G C A A C T C T C|T A|C T G|T|T T C T C C A T(A)C C G T T T T T
trp	A A A T G A G C T G T T G A C A A T T A A T C A T C G A A C T A G|T T|A A C|T|A G T A C G C(A)A G T T C A C G T A
tRNA$_{tyr}$	C A A C G T A A C A C T T T A C A G C G G C G C G T C A T T T G A|T A|T G A|T|G C G C C C C(G)C T T C C C G A T A
rrn D$_1$	C A A A A A A A T A C T T G T G C A A A A A A T T G G G A T C C C|T A|T A A|T|G C G C C T C C(G)T T G A G A C G A

From Rosenberg and Court: Annu Rev Genet 13:323, 1979.

the RNA polymerase to pause and to undergo a conformational change to effect termination. The efficient termination of certain transcripts requires an additional factor, rho. This factor, as purified from *E. coli,* is a hexameric protein composed of 50 kdal subunits. It contains a nucleotide triphosphatase activity that is stimulated by polyribonucleotides to form nucleoside diphosphates and inorganic phosphate. The mode of action of the termination protein is unknown. There is a correlation, however, between the terminal structure of those transcripts that are rho-independent and those that are rho-dependent. Those transcripts that do not require rho for termination contain a sequence of 6 to 8 uridylate nucleotides at their ends, whereas no such sequence is found in those that require rho. The poly U sequence and rho could serve the same function, namely, protracting the stalling time of RNA polymerase. In any case, each type of transcript possesses the terminating stem and loop structure that persists in the completed transcript. Figure 7-14 summarizes the transcription process.

Sigma Factors and Endospore Formation.

The core RNA polymerase-sigma factor relationship is complex in *B. subtilis,* which is able, under appropriate conditions, to undergo a primitive developmental process, endospore formation (Chap. 5). This developmental pathway is usually triggered by nitrogen, carbon, or phosphorus limitation and is under the control of a special set of genes. Three types of sigma factors exist in vegetatively growing *B. subtilis,* a predominant one σ^{55} and two minor ones σ^{37} and σ^{28} (the superscripts are the molecular weights in kdal). The activity of σ^{37} is under the control of sporulation regulatory genes. This sigma factor is involved in the production of an additional factor σ^{29}, which is specifically involved in the sporulation process. The discriminatory transcription directed by the sigma factors in *B. subtilis* is based in the promoters recognized. The σ^{55} is similar to the sigma factor in *E. coli* and recognizes promoters with the same consensus sequence. The σ^{37} and σ^{28} direct the binding of RNA polymerase to different sequences, which

TABLE 7-5. CONSERVED PROMOTER NUCLEOTIDE SEQUENCES RECOGNIZED BY THE SIGMA FACTORS OF *BACILLUS SUBTILIS*

−35 Region	−10 Region	Sigma Factor
T T G A C A	T A T A A T	σ^{55}
G G − T − A A A	T A T T G T T T	σ^{37}
C T A A A	C C G A T A T	σ^{28}

From Losick and Pero: Cell 25:582, 1981.

forms the basis for a cascade of events that result in endospore formation. Table 7-5 compares the promoter recognition sequences for these σ factors.

Other Aspects of Transcription Process

Although sigma-liganded RNA polymerase recognizes promoters containing similar nucleotide sequences, not all promoters act with equal efficiency. In other words, the nature of promoters provides the cell with an intrinsic means for regulating the rate of transcription. Variation in promoter efficiency is caused not only by variation in the structure of the promoter per se but also by bordering nucleotide sequences. Two events that obviously are significant in consideration of transcriptional efficiency are the initial binding of RNA polymerase to the promoter and the ease with which the open complex is established. The initial binding of RNA polymerase (closed promoter complex) to promoters is rapid if unimpeded by regulatory proteins. The rate of formation of the open complex is slower and represents the rate-limiting step in the initiation of transcription. However, once the open complex is established, a process that involves a conformational change in the polymerase, association of the holoenzyme with the DNA is strengthened, and transcription is initiated. Several factors can influence the rate at which the productive open complex is formed. As pointed out above, sequences bordering the promoter can influence this pro-

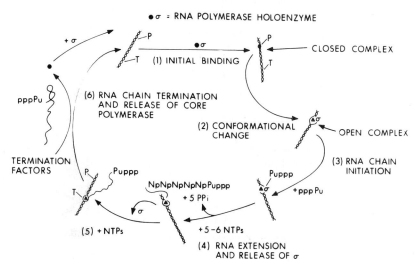

Figure 7-14. General scheme for transcription in bacteria. P, promoter; T, terminator.

cess, as can regulatory proteins that may aid or impede helix unwinding. Negative superhelical density is also important for efficient transcription initiation. This DNA suprastructure contains regions of underwound helix. In vitro analysis of transcription has shown that whereas certain templates are virtually unable to support transcription when in linear form, they are very efficient in doing so when in the form of highly negatively supercoiled, covalently closed circles.

The elongation process in transcription does not occur in a totally passive fashion. In vitro transcription studies using defined templates and homogeneous holopolymerase show that pausing and termination occur at a high frequency prior to complete transcript synthesis. Although the bases for these pauses are not understood, they may be related to the vagaries of local sequence in the DNA or in the RNA. Since RNA can form intramolecular hydrogen bonds between complementary bases, it is possible that formation of secondary structures that mimic the normal terminator structures is unavoidable. Spurious termination of transcription apparently is prevented, in vivo, by the antitermination activity of a number of protein factors, some of which may be elements that normally participate in transcription and translation. The most extensively studied antitermination process is one that occurs in bacteriophage λ. This process is highly specific for λ transcription, but it demonstrates that antitermination of transcription can occur, and it forms a model for how host-specified factors possibly prevent termination at the weak, adventitiously formed terminators that arise in the course of chromosome-directed transcription. The phage λ antitermination process requires the phage-specified N protein (11.8 kdal), which acts to relieve termination at two early termination signals distal to the promoters P_L and P_R (Chap. 8). The RNA polymerase complexes that embark from P_L and P_R encounter the sequences nut_L and nut_R at which they acquire the N protein. This modified transcription complex fails to cease transcription at all subsequent termination signals. It is of interest that the activity of the N protein requires a number of host-specified factors, including the NusA protein that normally participates in all termination events, i.e., whether or not they require the rho factor. The activity of the N protein also forms part of the basis for the phenomenon of retroregulation, which is described later in this chapter (refer to Fig. 7-25).

Processing of RNA

The functional rRNA and tRNA molecules of bacterial cells are processed from larger transcripts by specific ribonucleases (RNases). The large RNA transcripts (30S) that contain 23S, 16S, 5S, and tRNA molecules are substrates for RNase III, which recognizes specific intrastrand duplex regions in RNA molecules. These regions of double strandedness, caused by the complementarity of adjacent nucleotide sequences, are positioned in the molecule so that their cleavage by RNase III releases the smaller precursors of 23S, 16S, 5S, and tRNA. Other RNases are required for final maturation of the functional molecules. For example, the 5S RNA released by the action of RNase III is initially contained in a 9S RNA molecule. This larger molecule is processed by RNase E to form the 5S RNA. The pathway for processing the precursors of 23S and 16S RNA is different. The precursors of these ribosomal RNAs, released from the 30S transcript by RNase III, are immediately sequestered into ribosome assembly intermediates. The 5' and 3' ends are then trimmed by RNases to produce the mature 23S and 16S molecules. The tRNA precursors that are released from the rRNA-containing transcripts are further processed by RNase P, a unique ribonucleoprotein enzyme (see transcription, above). Other tRNA molecules also are processed from larger transcripts. These transcripts may contain a single tRNA molecule, in which case extraneous nucleotides must be trimmed away from the 3' and 5' ends, or the transcripts may contain more than one tRNA molecule. One transcript in *E. coli* contains seven tRNA molecules. The tRNAs are released from these transcripts by the action of RNase P, and final trimming is accomplished by enzymes, such as RNase D. Many of the tRNA molecules that are released from larger transcripts lack the terminal 3' sequence-CCA, which is required for function of all tRNA molecules. This sequence is added in a stepwise manner by the enzyme nucleotidyl transferase.

The processing of mRNA also occurs in bacteria. Many of the mRNAs of bacterial cells contain the information for more than one protein, i.e., they are polygenic. Cases are known where the differential expression of these genes is modulated by processing the mRNA. Examples of this are given later in this chapter in the section dealing with the regulation of gene expression.

Processing of stable RNA in procaryotic cells does not contrast to any great extent with that process in eucaryotic cells. Eucaryotic cells contain an additional RNA molecule (5.8S) in the large subunit. This RNA is in the spacer region occupied by tRNA in procaryotic cells; tRNA sequences are not found as part of the large transcript. The 5S RNA forms a separate transcriptional unit. Although the pathways of rRNA and tRNA processing vary among eucaryotic organisms, the general principles applied to procaryotic and eucaryotic stable RNA processing are similar. The type of mRNA processing in eucaryotic cells strongly contrasts with the restricted processing of this molecule in procaryotic cells. Many eucaryotic mRNA molecules are formed by a unique processing mechanism. The sequences for eucaryotic mRNA are frequently synthesized as contained within a large transcript in which the segments of the functional mRNA are separated by blocks of noncoding nucleotide sequences. The functional mRNA is formed by splicing together the appropriate segments. The segments that are ultimately expressed as mRNA are known as exons, and the intervening sequences are known as introns (Chap. 8). The mRNA synthesized in the nucleus of eucaryotic cells is further modified by the addition of a cap structure at the 5' end of the molecule. This cap structure, 7-methyl-guanosine (5′) ppp (5′)Np, facilitates translation and stabilizes the mRNA. Many nuclear-mRNA contain a poly A tail 40 to 200 residues in length. This sequence, which is added posttranscriptionally, may aid in stabilizing the mRNA. Neither capping nor poly A tailing occurs in bacteria.

Translation

Translation is the process whereby the triplet nucleotide code is deciphered to the primary amino acid sequence in proteins. Translation and transcription are thought to be tightly coupled in procaryotic cells so that mRNA synthesis occurs while its early sequences are directing peptide synthesis. This condition contrasts with that in eucaryotic cells, where mRNA must be transported from the nucleus for translation in the cytoplasm. The translation process is complex and requires a number of participating factors, including mRNA, ribosomes, aminoacyl tRNA, several protein factors, and GTP.

Ribosomes

The most complex structure involved in the translation process is the ribosome, which can be considered the workbench of protein synthesis. Ribosomes are designated according to their sedimentation coefficient; the functional ribosome in procaryotes is 70S and is composed of two subparticles, 30S and 50S. Its mass is two-thirds RNA and one-third protein. The composition of the large (50S) and small (30S) subunits reflect their complexity. The 50S subunit is composed of one copy each of 23S and 5S RNA molecules complexed with 32 different proteins designated L1 to L32. Each of these proteins is present, with the exception of L7 and L12 of which two copies of each are present; these proteins (L7 and L12) differ only by an acetyl group on the amino terminus of L7. The 30S subunit consists of a single 16S RNA complexed with 21 different proteins (S1 to S21). Many of the ribosomal proteins are modified by methylation or acetylation at the amino terminal position or by methylation of internal lysine residues. The role of these modifications is in the assembly of the ribosome and in modulation of protein-RNA and protein-protein interaction in the assembled particle. Each of the modified ribosomal proteins is required for normal cell growth. A number of mutant strains of *E. coli* have been described that lack a specific ribosomal protein modification or that completely lack one of the ribosomal proteins. These cells are impaired, usually displaying cold or heat sensitivity.

Ribosomes can be easily prepared and their subunits separated. These particles can be disassembled in vitro and the RNA and protein components purified. This has allowed study of the reassembly of these particles. These studies show that the secondary structure of the ribosomal RNAs provides a matrix for an ordered addition of the proteins. For example, the assembly of the 30S particle is initiated by the binding of proteins S4, S8, S15, S17, and S20, with the binding of S20 being facilitated by the previously bound S4 and S8. The binding of the remaining proteins also occurs in an ordered sequence. Figure 7-15 shows an assembly map for reconstitution of the functional 30S ribosome. An ordered pathway also is involved in the assembly of the 50S subunits. A principle that derives from these studies is that it is the RNA that directs the assembly of the ribosome subparticles, i.e., these serve, at least in part, as scaffolds of assembly. It is significant in this context that the shape of the 23S RNA, established by intramolecular hydrogen bonding, is roughly the same shape as the assembled 50S subunit.

The two types of ribosome subunits represent a division of labor in the translation process. The 30S subunit is primarily involved with aminoacyl tRNA as a decoder of mRNA, and the 50S subunit is involved in formation of peptide bonds.

The three-dimensional shapes of the ribosomal subunits have been studied by electron microscopy, and the orientation of the protein components has been revealed by the use of immunoelectron microscopy and by neutron diffraction. Figure 7-16 shows the shapes of the subunits and orientation of some of the proteins deduced from these studies.

tRNA and tRNA Synthetases

The role of tRNA in protein synthesis is to carry amino acids in an activated form to the translation complex and to faithfully read the triplet code in the mRNA. In this sense the amino acids are passive participants in the translation process. Once they are esterified to the tRNA their proper placement in the polypeptide chain depends upon tRNA function. This adaptor role for tRNA was initially demonstrated in an in vitro protein-synthesizing system

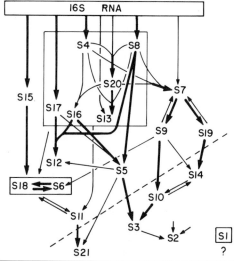

Figure 7-15. The assembly map of the 30S ribosome subunit of *E. coli.* (From Nomura and Held: In Nomura et al. (eds): Ribosomes, New York, Cold Spring Harbor Press, 1974.)

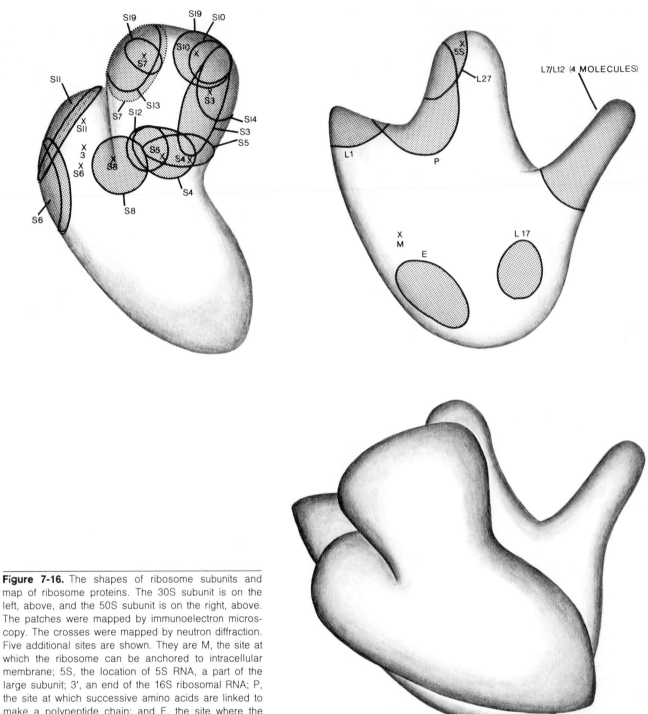

Figure 7-16. The shapes of ribosome subunits and map of ribosome proteins. The 30S subunit is on the left, above, and the 50S subunit is on the right, above. The patches were mapped by immunoelectron microscopy. The crosses were mapped by neutron diffraction. Five additional sites are shown. They are M, the site at which the ribosome can be anchored to intracellular membrane; 5S, the location of 5S RNA, a part of the large subunit; 3', an end of the 16S ribosomal RNA; P, the site at which successive amino acids are linked to make a polypeptide chain; and E, the site where the newly synthesized polypeptide chain emerges from the ribosome. *(Adapted from Lake: Sci Am 245:84, 1981.)*

where cysteinyl-tRNA was modified by catalytically removing the —SH group to produce alanyl-tRNA, with the result that alanine was incorporated into the polypeptide chain at positions normally occupied by cysteine.

The adaptor role for tRNA reveals a point in protein synthesis that must operate at a level of fidelity to match the high degree of accuracy in the translation process, that is, the amino acids must be joined to only their cognate tRNAs. This fidelity is based in enzymes, aminoacyl-tRNA synthetases, that synthesize aminoacyl-tRNA by stereospecifically recognizing the amino acid and the appropriate tRNA. Stereospecific recognition of the amino acids by aminoacyl-tRNA synthetases produces no conceptual problems because of the differences in the structures of these compounds. A conceptual problem does arise, however, when considering stereospecific recognition of tRNA molecules. The structure of tRNA molecules must be similar enough to participate in a unitary fashion in the translation process but yet be sufficiently different for specific recognition by the cognate aminoacyl-tRNA synthetase. As expected, tRNA molecules have similar structural features. They range in size from 73 to 93 nucleotides, and all are able to fold by virtue of intramolecular hydrogen bonding to produce domains. These are most clearly seen in the

cloverleaf structure depicted in Figure 7-17. The results of x-ray diffraction analysis of tRNA crystals show that the tRNA molecules are actually L-shaped (Fig. 7-17). There are two common features in all tRNA molecules: they all terminate with a —CCA sequence at their 3' ends, and they all have an anticodon triplet sequence at a constant geometric region. The remaining parts of the molecule, although similar in conformation, vary in detail. Additional variation in tRNA molecules is caused by posttranscriptional modifications; some of these modifying groups are on the bases, whereas others are on the ribose. Table 7-6 presents a list of some of the tRNA-modifying groups.

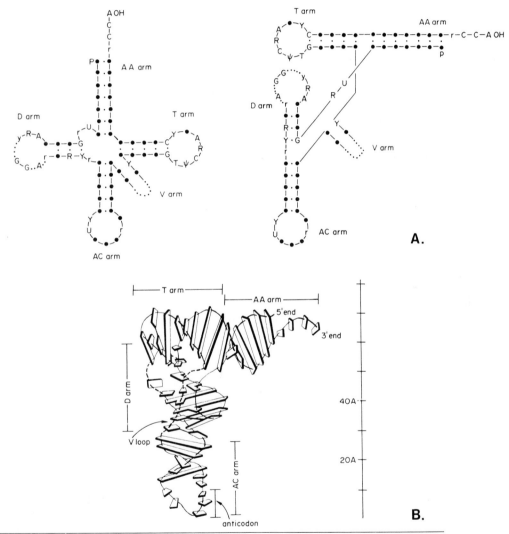

Figure 7-17. The structure of transfer RNA. **A.** Sequence analysis of tRNA shows that the molecule can possess secondary structure maintained by intramolecular hydrogen bonding; the cloverleaf configuration is depicted above. The solid circles in the figure represent nucleotides that are generally variable among various tRNA molecules; the letters r and y indicate positions that are usually occupied, respectively, by purine and pyrimidine; R and Y indicate positions invariably occupied, respectively, by purine and pyrimidine; H indicates the position usually occupied by a modified nucleoside. **B.** X-ray crystallographic studies reveal additional structure and suggest the rearrangement of the relative position of the domains as depicted at the upper left. A three-dimensional representation of the tRNA molecule is shown. *(Courtesy of Dr S.H. Kim.)*

The role of most of these modifications is unknown, but they are speculated to serve a function in the interaction of tRNA with aminoacyl-tRNA synthetases, ribosomes, and mRNA. As shown in Table 7-7, more than a single codon exists for each amino acid except tryptophan. In order for the cell to take full advantage of this genetic code, more than a single tRNA molecule is required for each amino acid. This isoaccepting species of tRNA are recognized by single aminoacyl-tRNA synthetases. In other words, bacteria contain a specific aminoacyl-tRNA synthetase for each amino acid, not for each tRNA. This fact adds to the remarkable discriminatory properties of these enzymes.

The reactions catalyzed by aminoacyl-tRNA synthetases are essentially the same, but there are relatively subtle differences that may reflect the evolutionary development of their specificities. All aminoacyl-tRNA synthetases carry out a two-step reaction in joining amino acid and tRNA. The carboxyl group of the amino acid is first activated by formation of the aminoacyladenylate (Fig. 7-18), and then the amino acid is transferred to the tRNA to form an ester linkage with one of the free hydroxyl groups of the 3'-terminal adenine nucleotide. The reaction is as follows:

$$\text{Enz} + \text{ATP} + \text{amino acid} \rightleftharpoons \text{Enz-aminoacyl-AMP} + \text{PP}_i$$

$$\text{Enz-aminoacyl-AMP} + \text{tRNA} \rightleftharpoons \text{aminoacyl-tRNA} + \text{AMP} + \text{Enz}$$

Most aminoacyl-tRNA synthetases carry out the amino acid activation reaction in the absence of tRNA, whereas others, such as glutaminyl-tRNA synthetase, must bind their cognate tRNA prior to carrying out this reaction. There is an additional degree of freedom in the development of specificity that is offered by the mechanism of peptide bond synthesis during the translation process; that is, this process requires tautomerization of the esterified amino

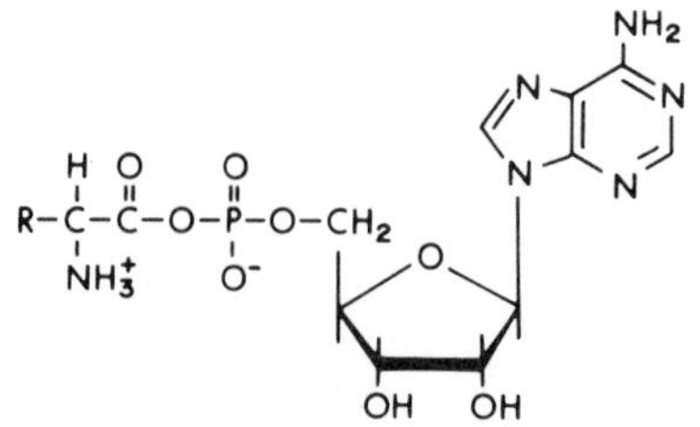

Figure 7-18. Aminoacyladenylate, the primary activation of amino acids for protein biosynthesis.

acid between the 3'-OH and 2'-OH of the terminal adenine nucleotide of the tRNA. Accordingly, it would make no difference for subsequent peptide bond formation if the amino acid was initially placed at either the 3'-OH or 2'-OH of the tRNA. It would appear that advantage has been taken of this situation, since some aminoacyl-tRNA synthetases esterify the 3'-OH, others the 2'-OH, and still others either the 3'-OH or 2'-OH. In spite of the high fidelity of transfer of amino acids to the proper tRNA, mistakes sometimes occur. There is, however, an additional activity of aminoacyl-tRNA synthetases that helps prevent the subsequent misincorporation of amino acids caused by this illicit union. A deacylase activity has been described that hydrolyzes mischarged tRNA molecules. For example, isoleucyl-tRNA synthetase rapidly deacylates a valine esterified to tRNA^{ile} to yield free valine and tRNA^{ile}. This reaction is not a simple reversal of the synthetase reactions because it occurs in the absence of AMP and PP$_i$. It probably is not surprising that the evolution of enzymes with the remarkable discriminatory abilities of the aminoacyl-tRNA synthetases resulted in the development of a group of enzymes with widely varying gross anatomy. Table 7-8 shows examples of the structural variation among the aminoacyl-tRNA synthetases found in *E. coli*.

TABLE 7-6. MODIFIED NUCLEOSIDES*

I	Inosine	s^{4}U	4-thiouridine
m^{1}I	1-methyl inosine	s^{2}m^{5}U	2-thio-5-methyl uridine
m^{1}A	1-methyl adenosine	V	5-oxyacetic acid uridine
		s^{2}am^{5}U	2-thio-5-acetic acid methyl ester uridine
m^{2}A	2-methyl adenosine		
m^{6}A	N^6-methyl adenosine	s^{2}cm^{5}U	2-thio-5-carboxymethyl uridine
i^{6}A	N^6-isopentenyl adenosine	cmm^{5}U	5-carboxymethyl uridine methyl ester
ms^{2}i^{6}A	2-methylthio-N^6-isopentenyl adenosine	cm^{5}U	5-carboxymethyl uridine
t^{6}A	N^6-(N-threonylcarbonyl) adenosine	Um	2'-O-methyl uridine
m^{1}G	1-methyl guanosine	ψm	2'-O-methyl pseudouridine
		mam^{5}s^{2}U	5-methylaminomethyl-2-thiouridine
m^{2}G	N^2-methyl guanosine	X	3-(3-amino-3-carboxypropyl) uridine
m$_2^2$G	N^2-dimethyl guanosine	s^{2}m^{5}C	2-thio-5-methyl cytidine
m^{7}G	N^7-methyl guanosine	s^{2}C	2-thiocytidine
Gm	2'-O-methyl guanosine	Cm	2'-O-methyl cytidine
T	(ribo) Thymidine	ac^{4}C	N^4-acetyl cytidine
ψ	Pseudouridine (5-ribofuranosyl uracil)	m^{3}C	N^3-methyl cytidine
D or hU	5,6-dihydrouridine	m^{5}C	5-methyl cytidine

After Kim: Prog Nucleic Acid Res Mol Biol 17:181, 1976.

*Modified nucleosides found as minor components in tRNA molecules. The biologic role of most of these modifications is unknown.

TABLE 7-7. NUCLEOSIDE SEQUENCES OF RNA CODONS AND CORRESPONDING AMINO ACIDS

1st Base	2nd Base				3rd Base
	U	C	A	G	
U	Phe	Ser	Tyr	Cys	U
	Phe	Ser	Tyr	Cys	C
	Leu	Ser	Ochre*	Opal*	A
	Leu	Ser	Amber*	Trp	G
C	Leu	Pro	His	Arg	U
	Leu	Pro	His	Arg	C
	Leu	Pro	Gln	Arg	A
	Leu	Pro	Gln	Arg	G
A	Ileu	Thr	Asn	Ser	U
	Ileu	Thr	Asn	Ser	C
	Ileu	Thr	Lys	Arg	A
	Met	Thr	Lys	Arg	G
G	Val	Ala	Asp	Gly	U
	Val	Ala	Asp	Gly	C
	Val	Ala	Glu	Gly	A
	Val	Ala	Glu	Gly	G

From Crick: Cold Spring Harbor Symp Quant Biol 31:1, 1966.
*Nonsense codon.

The Three Phases of Translation

The translation process proceeds in three distinct phases: initiation, elongation, and termination of polypeptide synthesis.

Initiation. The essential task accomplished by the initiation process is the orientation of the initial aminoacyl-tRNA in such a way that the triplet code (Table 7-7) in the mRNA is read in the proper frame. Polypeptide synthesis in bacteria is uniformly initiated with N-formylmethionine, which is specified by the codons AUG or GUG. The tRNA used in this initial step accepts methionine, which is then N-formylated by a specific transformylase that uses N^{10}-formyltetrahydrofolate as a cosubstrate. The selective reading of the initial AUG or GUG is directed by information in the mRNA and the 30S ribosomal subunit, acting in conjunction with three initiation factors, IF1, IF2, and IF3, and the nucleotide triphosphate, GTP.

TABLE 7-8. SUBSTRUCTURE OF REPRESENTATIVE AMINOACYL-tRNA SYNTHETASES OF E. COLI

Substructure	Synthetase	Molecular Weight of Constituent Polypeptide Chains	
α	Glutaminyl	69,000	
	Isoleucyl	112,000	
α_2	Histidinyl	84,000	
	Tryptophanyl	37,000	
α_4	Alanyl	95,000	
$\alpha\beta$	Glutamyl	(56,000)	(46,000)
$\alpha_2\beta_2$	Phenylalanyl	(39,000)	(94,000)

The translation initiation complex is formed in a stepwise manner. First, 30S ribosomes complexed with IF3 form a binary complex with a region of the mRNA at a position on the 5′side of the initiating codon. Most bacterial mRNA molecules contain a sequence typified by 5′—AAAGGAGGU—3′ situated no further than nine and no less than five nucleotides upstream (toward the 5′ end) from the initiating AUG or GUG triplet. This sequence is the centerpiece of the 30S subunit binding site and is complementary with a sequence in the 3′-terminal region of the 16S ribosomal RNA. This sequence in the mRNA is not the sole determinant in directing 30S binding, as sequences more distal to the initiating codon can influence the strength and efficiency of binding. Once the initial binary complex is established, it accepts the N-formylmethionyl-tRNA. This union is directed by IF2 complexed with GTP. The IF2-GTP complex joins with N-formylmethionyl-tRNA to form a ternary complex that binds to the mRNA-30S complex. This event is aided by IF1 and is accompanied by the release of IF3. The release of IF3 at this stage is essential because its persistence would preclude completion of the initiation complex, which involves addition of the 50S subparticle. This is so because in addition to its role in directing proper union of mRNA and 30S subunits, it also possesses a ribosome dissociation function. This activity insures a pool of 30S particles for use in translation initiation by neutralizing their tendency to spontaneously join with 50S subunits to form initiation-inert 70S ribosomes. Formation of the initiation complex is completed by addition of the 50S subparticle. This is accompanied by the hydrolysis of GTP to GDP and P_i and the release of IF2 and IF1.

Elongation. Once the initiation complex is formed, the growth of the polypeptide chain occurs by the repetitive addition of amino acids. Each step in this elongation process requires (1) the codon-directed binding of aminoacyl-tRNA to a binding site, which at the first step is adjacent to the site occupied by fmet-tRNA but which at each successive site is occupied by peptidyl-tRNA, (2) peptide bond formation that results from a peptidyl transfer from the fmet-tRNA or peptidyl-tRNA to the amino acid of the newly bound aminoacyl-tRNA, and (3) translocation of the mRNA and newly synthesized peptidyl-tRNA to the site occupied by the discharged tRNA (i.e., the tRNA that no longer contains formylmethionine or peptide). This final step results in expulsion of the discharged tRNA and orientation of the ribosome in such a way that the next codon in the mRNA is able to direct the binding of the pertinent aminoacyl-tRNA. This elongation process identifies two tRNA-binding sites on the ribosome: the site from which fmet-tRNA or peptidyl-tRNA transfers the esterified carboxyl group is known as the "peptidyl" or "donor" site, and the site to which the incoming aminoacyl-tRNA binds is called the "acceptor" site. (These are frequently referred to as the P and A sites, respectively.) Historically, the donor site was initially called the P site because aminoacyl-tRNAs at this site were able to transfer their amino acids to puromycin, an antibiotic that is a structural analog of the 3'-aminoacyl portion of the tRNA and that binds to the 50S portion of the A site (Chap. 9).

At least three protein factors and GTP, as well as peptidyl transferase, which is an integral part of the 50S ribosome, are involved in polypeptide chain elongation. The three elongation factors are Tu, Ts, and G. The T designation derives from the role these factors play in transferring aminoacyl-tRNA from the cytoplasm to the mRNA-ribosome complex. The Tu and Ts designations derive from the heat stability of the factors (unstable and stable, respectively). The G designation stems from a GTPase activity of the G factor. The Tu (mol wt 40,000) and Ts (mol wt 19,000) factors associate avidly. They purify and crystallize as a unit. In vitro studies have shown that the Tu-Ts complex reacts with GTP and aminoacyl-tRNA to form a Tu-aminoacyl-tRNA-GTP complex, which in turn reacts with either the initiation complex or a complex undergoing chain elongation. At this point the aminoacyl-tRNA binds to the acceptor site on the mRNA-ribosome complex, resulting in the release of Tu-GDP and iP. The Tu-Ts complex is regenerated by the displacement of GDP by Ts (Fig. 7-19, step 4). Interaction of aminoacyl-tRNA with the elongation complex apparently is facilitated by the complementarity that exists between the thymidine-pseudouridine-cytosine loop of the tRNA and the 5S RNA component of the 50S ribosome subparticle.

Addition of the aminoacyl-tRNA to the acceptor site is immediately followed by peptide bond formation between the esterified carboxyl group of the formylmethionine or the growing peptide chain of the tRNA at the donor site and the amino group of the amino acid bound by the tRNA at the acceptor site. This process results in occupancy of the donor site by a discharged tRNA and peptidyl-tRNA at the acceptor site. As mentioned above, the enzyme (peptidyl transferase) that effects peptide bond formation is part of the 50S ribosome. It should be emphasized that peptide bond formation at this stage does not require the expenditure of energy, since the esterified carboxyl group at the donor site is at a sufficiently high energy level to participate in group transfer.

Following peptide bond formation and prior to the addition of the next aminoacyl-tRNA to the protein-synthesizing complex, the aminoacyl-tRNA acceptor site must be vacated and properly oriented with respect to the next codon in the mRNA. These events are carried out by a simultaneous translocation of both the mRNA and the peptidyl-tRNA by a process that requires G factor and GTP. G factor possesses a potent GTPase activity, that is manifested only in the presence of ribosomes. The energy of hydrolysis of GTP, in some unknown manner, is coupled to the movement of ribosomes over the mRNA or movement of mRNA through the ribosome. The translation mechanism results in the reframing of the ribosome-mRNA complex, with expulsion of tRNA from the P site. GDP dissociates from the complex, which then becomes capable of participating in the next elongation step.

Termination. The process of polypeptide chain elongation described above results in a constant association of tRNA-attached growing polypeptide chain with the mRNA-ribosome complex. This process demands a mechanism for termination that results in the cessation of polypeptide chain growth, as well as in release of the tRNA from the carboxyl terminus of the polypeptide chain. It is known that termination is caused by information encoded in the mRNA and by protein-release factors. Three codons, UAA, UAG, and UGA, for which there normally are no tRNA species with corresponding anticodons and which consequently do not serve to specify amino acids in polypeptide chains, serve as efficient termination signals. Two release factors (RF1 and RF2) act in conjunction with nonsense codons to effect release of the polypeptide chain. RF1 acts with either UAA or UAG, and RF2 with UAA or UGA. An additional factor, RF3, acts in conjunction with RF1 and RF2 to facilitate release. This factor has no release activity by itself but increases the affinity of the terminating codons in RF1 and RF2 in the release process. Peptidyl transferase is thought to be involved in the release of the peptide from the terminal tRNA. One view of the role of peptidyl transferase in termination is that RF factor converts peptidyl transferase into a hydrolase, which then transfers the peptidyl group of peptidyl-tRNA to water. This event serves to simultaneously stop protein synthesis and to release the protein. Figure 7-19 summarizes the translation process.

Other Aspects of Translation

Wobble. As previously pointed out, a triplet code utilizing four different nucleotides generates 64 (4^3) different codons. Three of these codons (UAA, UGA, and UAG) generally do not have complementary tRNAs in bacteria and normally serve as signals for translation termination. In spite of the fact that the array of tRNAs found in any cell can read the remaining 61 triplets in a sensible way, these cells do not contain 61 precisely cognate tRNA molecules. The lack of a requirement for 61 cognate tRNA molecules is explained by the ability of some tRNAs to read more than one codon. This is accomplished by the wobble that is permitted in pairing of the base that occupies the third position (3') in the codon and the base at the first position (5') in the anticodon (recall that codon-anticodon recognition requires antiparallel alignment of the paired triplets). This third position redundancy in the genetic code is limited to interaction of only some bases. Table 7-9 shows the base pairing that is allowed at the third position in the codons. Included in the table are the modified bases inosine, produced from the deamination of adenine, and 5-oxyacetic acid uridine; these bases, when found at the 5' position of anticodons, expand the wobble recognition that would otherwise be precluded by the nonmodified bases. Employing the wobble rules, a cell would require a minimum of 32 tRNAs to read 61 codons.

Reading of the genetic code by wobble does not lead to mistakes during translation because of the precise pairing that is required in the first two positions of the code. Inspection of the codon-amino acid assignments shown in Table 7-7 shows that evolution of the genetic code has accommodated this situation. The 17 amino acids that have four codons use the same bases in the first two positions, and the three amino acids that have six codons use only two sequences in these positions.

Codon Usage. The codon-tRNA arrangement described above apparently has significant use in the translation process because it allows an intrinsic regulation of rates of translation. This is so because of the degeneracy in the code for each amino acid and of the ability of a tRNA to recognize more than one code word by third position wobble. Because of this, the codons within a given reading frame can be selected to provide weak, intermediate, or strong interaction with the anticodon. Comparison of the nucleotide sequence specifying proteins that are produced at high levels show that the 61 codons are not used randomly but rather are biased toward those that provide intermediate strength interaction with the available anticodon. The bias in codon usage for synthesis of proteins found in low abundance is toward those providing weak and strong interaction with anticodons. It is reasoned that weak interactions cause slowed translation by impeding initial codon-anticodon interaction, whereas strong interactions cause a sticky union that would impede rapid progression of the translation process. Interactions of intermediate strength are reasoned to be consistent with a rapid rate of translation. Table 7-10 shows examples of how the strengths of codon-anticodon interaction can be modulated by use of third position degeneracy. These considerations illustrate how the intrinsic rate of synthesis of individual polypeptides governed by promoter efficiency and strength of ribosome binding sites can be further modulated by codon usage.

Localization of Translation in Bacteria. Several of the proteins synthesized by bacterial cells are ultimately located in the cytoplasmic membrane, the periplasmic space, the outer membrane in the case of gram-negative cells, or extracellularly. These noncytoplasmic locations require a mechanism whereby these generally hydrophilic proteins gain entrance to, or passage through, the hydrophobic membrane. Many of these proteins, as initially synthesized, contain a leader or signal sequence at the amino terminal portion of the molecule. This sequence, which contains many hydrophobic amino acid residues, plays a role in transport of these proteins into or through the membrane. These hydrophobic regions of the protein as they are initially synthesized cause their mRNA-ribosome translation complexes to associate with the cytoplasmic membrane. This association results in a partitioning of translation complexes between the membrane and the cytoplasm. This partitioning results in an incidental establishment of a qualitative difference between the two types of translation complexes, i.e., they display differential sensitivity to certain antibiotics. For example, the membrane-bound complexes in *E. coli* are more sensitive to tetracycline and less sensitive to chloramphenicol than are the cytoplasmic complexes.

Multiple Use of Translational Reading Frames within a Nucleotide Sequence. Proteins are generally thought of as being specified by nucleotide sequences that are *uniquely* used to govern their primary structure. Cases exist, however, where a given region of a genome can be used to specify two or three proteins, each of which possesses a unique amino acid sequence. This is accomplished by reading a given nucleotide sequence in a different frame. For example, the nucleotide sequence shown below could ultimately be read in the three frames as depicted:

$$\overline{\text{ATGATGATG}}$$

$$--\overline{\text{ATGATGATG}}-$$

$$-\overline{\text{ATGATGATG}}--$$

Such a situation has been described with coliphage ϕX174. The genome of this virus codes for at least 10 proteins. The B protein, which contains 120 amino acids, is completely encoded within the *A* gene, which also codes for the A protein, which contains 513 amino acids. The E protein, which contains 91 amino acids, is encoded within the *D* gene, which also codes for the D protein, which contains 152 amino acids. This gene-protein relationship also is true for the related coliphage G4, where, in addition, it has been shown that the K protein (56 amino acids) is encoded within a nucleotide sequence that overlaps genes *A* and *C*. Furthermore, the region of gene *A*, which encodes the K protein, is within the region that encodes protein B. This observation means that this region of the *A* gene is being read in three different frames. Although this type of nucleotide sequence conservation is not widespread, it does show what can be accomplished by evolution and the selection of compatible codons.

Shared nucleotide sequence, to the extent described above, while compatible with encodement of structural proteins, apparently does not occur to any great extent in such highly functional proteins as enzymes. Relatively more minor cases of ambivalent use of nucleotide sequence have been described, however. For example, it is known that the third nucleotide of termination codons in some polygenic mRNAs serves as the first nucleotide of an adjacent AUG initiating codon.

Variations in Protein Synthesis

The general principles governing transcription and translation in various types of organisms are similar, but as was

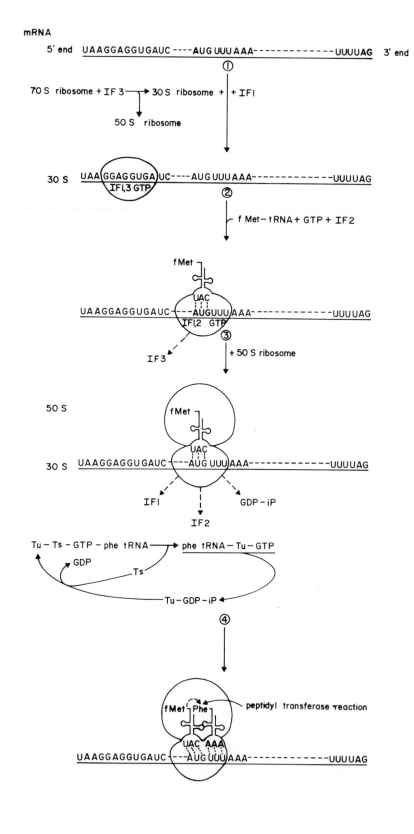

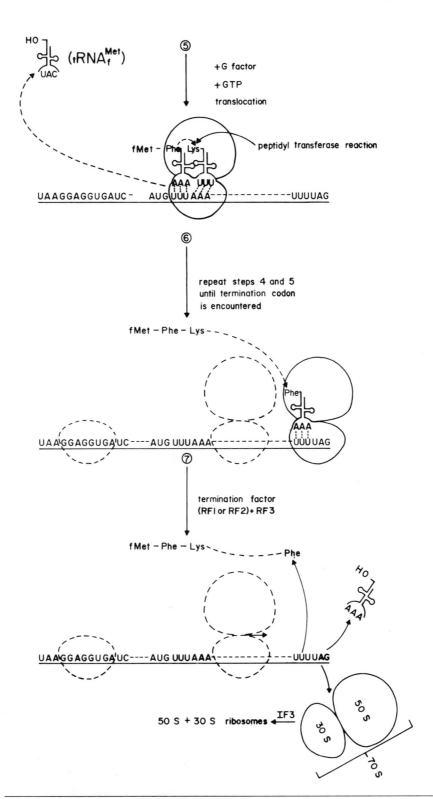

Figure 7-19. Messenger RNA-directed protein synthesis. The process is initiated by interaction of the 30S ribosome with a binding site at the 5'-triphosphoryl end of the RNA molecule. The nucleotide sequence representing the ribosome-binding site is complementary to the 3'-OH sequence of 16S rRNA. Ribosome-binding sites usually contain sequences of 4 to 6 nucleotides complementary to this region. Steps 1, 2, and 3 represent formation of the initiation complex, steps 4 and 5 represent elongation, and steps 6 and 7 represent termination.

TABLE 7-9. CODON-ANTICODON BASE PAIRING PERMITTED BY WOBBLE AND EFFECT OF BASE MODIFICATION ON WOBBLE

5′ Base of Anticodon	3′ Base of Codon
A	U
C	G
U	A or G
G	C or U
I	U, C, or A
O⁵U (5-oxyacetic acid uridine)	A, G or U

previously described there are variations in the pathways whereby genetic programs are expressed. Although the various elements involved in protein synthesis accomplish similar results in different types of organisms, the structures of these elements vary. This point is strikingly revealed by the differential effect of antibiotics. Several of the clinically useful antibiotics have the procaryotic ribosome as their site of action. However, the cytoplasmic ribosomes of eucaryotic organisms are generally unaffected by these bacteriostatic and bactericidal agents (Chap. 9). Differences among procaryotic ribosomes also are revealed by antibiotic susceptibility. For example, the 50S ribosomal subparticle of gram-positive cells is generally susceptible to the lincosamide antibiotics (lincomycin, clindamycin), whereas those of gram-negative cells are insensitive. This isolated example also reveals heterogeneity in ribosomes of gram-negative cells, as *Bacteroides fragilis*, a gram-negative organism, is very sensitive to clindamycin.

There are other significant manifestations that may reflect differences in ribosomes from gram-negative and gram-positive cells. In gram-negative cells AUG and GUG are initiation codons. These serve the same purpose in gram-positive cells, but in some mRNAs of gram-positive cells, UUG serves as an initiation codon. This may reflect a difference in the ribosomes as well as in the ribosome-binding sites of the mRNA. It is known that the formation of translation initiation complexes of gram-positive cells is less dependent on initiation factors than in gram-negative cells. Gram-positive-derived 30S subunits can discriminate between mRNA from gram-positive and gram-negative cells.

TABLE 7-10. EXAMPLE OF CODON SELECTION THAT GOVERNS STRENGTH OF INTERACTION WITH ANTICODON

Amino Acid	Codon	Number of Hydrogen Bonds	Relative Strength of Interaction with Anticodon*
Isoleucine	AUU	6	Weak
	AUC	7	Intermediate
Glycine	GGU	8	Intermediate
	GGC	9	Strong

*Relative strength applies to the codons possible for each amino acid, in this case, AUU and AUC for ileu, GGU and GGC for gly.

Protein Synthesis in Mitochondria. The transcription and translation processes in mitochondria consist of a blend of features found in procaryotes, the eucaryotic nucleus, and cytoplasm, along with features unique to these plastids. This arrangement is frequently cited to support the endosymbiont theory, i.e., that these organelles have procaryotic origins (Chap. 2). The RNA polymerase found in most mitochondria appears to be similar to nuclear polymerases and is unaffected by such antibiotics as rifampin that inhibit bacterial RNA polymerases. Some of the mRNAs of mitochondria are synthesized, containing introns and exons, and require processing typical of nuclear mRNA. On the other hand, all translation in mitochondria is initiated with N-formylmethionyl-tRNA, and the ribosomes are similar in structure to those of bacteria. This similarity is revealed by the fact that mitochondrial ribosomes are inhibited by many of the antibiotics that act on bacterial ribosomes; among these are erythromycin, lincomycin, chloramphenicol, and the aminoglycosides. Most of the proteins found within the mitochondria are imported from the cytoplasm. The coding capacity of the genome of the mitochondrion is reserved for eight or nine proteins, including the respiratory proteins, and rRNA and tRNA. The limited protein repertoire of the mitochondrial genome has allowed the evolution of a fascinating method for reading the genetic code, which has resulted in a departure from the universal manner in which the genetic code is read by procaryotic cells and by the translation apparatus in the cytoplasm of eucaryotic cells. The mitochondria of organisms as diverse as yeast and humans contain 24 distinct species of tRNA, all encoded by the mitochondrial genome. This number is fewer than in other translation systems and is even fewer than the 32 required for reading all possible codons by the third position wobble. The conservation in production of tRNAs by mitochondria is made possible by invoking different rules for codon-anticodon recognition. This involves a two-out-of-three recognition scheme that is made possible by grouping codons into families where the first two nucleotides are the same and read by a single tRNA. This is allowed by invoking different wobble rules so that the base in the first position (5′) of the anticodon can read all four bases in the third position (3′) of the codon. This is accomplished by placing U in the wobble position of the anticodon; 7 of the 16 codon families are read by this mechanism, which means that only 7 tRNAs are required to read these 28 codons. Normal wobble also is used to decrease the number of tRNAs required to read the remaining codons. Six tRNA molecules have a modified U at the wobble position. This modification prevents complete third position degeneracy and restricts reading to A and G. Eight of the remaining nine tRNAs have G in the wobble position and therefore can read C or U in the third position of the codon. Table 7-11 shows the genetic code of the yeast, *Saccharomyces cerevisiae*. The evolution of this reading scheme resulted in some variation in the universal code in mitochondria. UGA, a normal translation termination codon in procaryotes, is a tryptophan codon and CUA

TABLE 7-11. CODONS AND ANTICODONS OF THE YEAST MITOCHONDRIAL GENETIC CODE*

Codon	AA	Anti	Codon	AA	Anti	Codon	AA	Anti	Codon	AA	Anti
UUU	Phe	AAG	UCU			UAU	Tyr	AUG	UGU	Cys	ACG
UUC			UCC	Ser	AGU	UAC			UGC		
UUA	Leu	AAU	UCA			UAA	Ter		UGA	Trp	ACU*
UUG			UCG			UAG			UGG		
CUU			CCU			CAU	His	GUG	CGU		
CUC	Thr	GAU	CCC	Pro	GGU	CAC			CGC	Arg	GCA
CUA			CCA			CAA	Gln	GUU	CGA		
CUG			CCG			CAG			CGG		
AUU			ACU			AAU	Asn	UUG	AGU	Ser	UCG
AUC	Ile	UAG	ACC	Thr	UGU	AAC			AGC		
AUA			ACA			AAA	Lys	UUU	AGA	Arg	UCU
AUG	Met	UAC	ACG			AAG			AGG		
GUU			GCU			GAU	Asp	CUG	GGU		
GUC	Val	CAU	GCC	Ala	CGU	GAC			GGC	Gly	CCU
GUA			GCA			GAA	Glu	CUU	GGA		
GUG			GCG			GAG			GGG		

After Bonitz et al.: Proc Natl Acad Sci USA 77:3167, 1980.
*The codons (5' → 3') are at the left, and the anticodons (3' → 5') are at the right. The wobble nucleotides of the anticodons are underlined.

and its family members are threonine rather than leucine codons.

Regulation of Gene Expression

A wide variety of mechanisms serve to integrate the expression of the genetic program of bacteria. These mechanisms are directly or indirectly responsive to the environment and prevent unnecessary synthesis of nucleic acids and/or proteins. A quantitatively significant control acts at the level of transcription initiation and involves elements that modulate the accessibility of RNA polymerase to transcription initiation sites. There are two general types of transcriptional regulatory elements: (1) nucleotide sequences that are conjoined to the genes they modulate and (2) proteins that interact with these conjoined sequences to retard or enhance transcription. The conjoined sequences are *cis*-active, owing to their orientation relative to the genes they affect. The regulatory proteins are *trans*-active because regardless of where they are produced they reach their *cis*-active targets by diffusion. The activity of these regulatory proteins is modulated by low molecular weight compounds that signal the physiologic or environ-

mental state of the organism. Some regulatory proteins act as repressors and others as activators of transcription, and the low molecular weight effector molecules modulate the activity of these proteins in an appropriate manner. For example, enzymes involved in catabolism of certain compounds may be synthesized only when the compound is present. The induction of these enzymes could be accomplished in one of two ways, i.e., by the compound neutralizing the normal repressor function or by activating the activator function of a regulatory protein. In either case, the inducing molecule binds to stereospecific sites on the regulatory protein to decrease or increase the affinity for its *cis*-active target.

Regulation of transcription initiation events is not the only manner by which bacteria modulate gene expression. Regulation is known to act at most levels of informational macromolecule synthesis. For example, mechanisms are known that modulate protein synthesis at the level of translation. In fact, the modulation of protein synthesis by regulating translation is a significant control in bacteria. This type of control occurs in three general ways: (1) intrinsically, by the secondary structure that exists by virtue of the ability of mRNA to form intramolecular hydrogen bonds between regions containing sequences of complementary bases, (2) by proteins that bind to regions of mRNA and impede translation, (3) by the processing and controlled degradation of mRNA. Other, intrinsic types of regulatory mechanisms have evolved that adventitiously use the principles and elements of information transfer. The foremost example of this is the translational control of transcription termination that is involved in the attenuation mechanism concerned in the control of amino acid biosynthetic systems. This process is explained later in this chapter.

Genetic regulatory systems in bacteria vary widely in scope. *Cis* and *trans* acting elements have evolved that are highly specific in controlling the synthesis of one RNA molecule, whereas other regulatory elements exert more global effects on gene expression. Some genes are under control of both types of regulation, with the less specific element being epistatic over the specific regulation. An example of this situation is the regulation of the *lac* operon, the expression of which is inducible but also catabolite repressible. The basis for this is explained below.

Operons and Regulons

The term "operon" was initially coined by Jacob and Monod to describe the relationship between a *cis*-active regulatory element, the operator, which serves as the target for a regulatory protein, the product of the *lacI* gene, which is responsible for the inducible expression of the structural genes for β-galactosidase, galactoside permease, and galactoside transacetylase. This arrangement provides for the unit control of three distinct proteins by the regulated synthesis of a single polygenic mRNA molecule. The term "operon" is now applied to structural genes together with their *cis*-active regulatory regions regardless of whether or not these conjoined nucleotide sequences serve as repressor targets. The concept of the operon has been extended to include monogenic and polygenic transcripts, including rRNA-tRNA gene clusters that are considered polygenic. Unit control provided by polygenic mRNA is common in bacteria, where the genes involved in related function are frequently contained in a single operon. The coordinate control provided by this arrangement aids in the synthesis of single intermediary metabolites, in the initial catabolism of potential carbon, nitrogen, and energy sources, or in the balanced formation of critical elements used, such as those involved in DNA, RNA, and protein synthesis. Figure 7-20 shows examples of proteins in *E. coli* that are encoded by polygenic mRNA. The benefit of this type of gene organization is obvious in some cases and less so in other cases. For example, the clustering of the genes encoding the nine enzymes involved in histidine biosynthesis obviously provides a useful unit control. A similar rationale can be applied to the arabinose operon, which encodes the three enzymes required for the conversion of L-arabinose to D-xylulose-5-P. The organization of the genes for several ribosomal proteins together with the β and β′ subunits of RNA polymerase would seem to provide some coordination of transcription and translation. One of the most curious operons in *E. coli* is one encoding

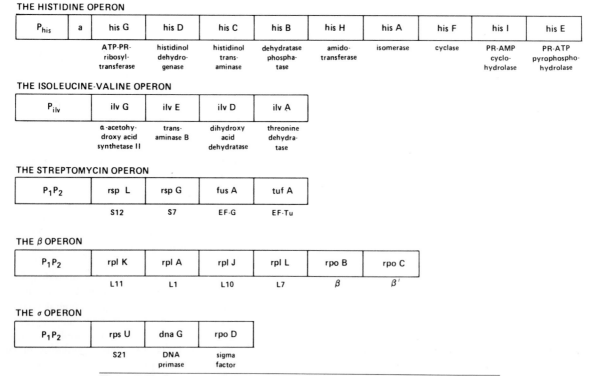

Figure 7-20. Some examples of operons in *E. coli*. The genes are not drawn to scale.

the ribosomal protein S21, DNA primase, and the sigma factor of RNA polymerase. This organization tends to link the synthesis of protein, DNA, and RNA. It is of interest to note that each of these proteins acts at the initial steps of the polymerizations in which they are involved. The roles of primase and sigma factor were previously described in this chapter; protein S21 is required for the proper interaction of the 3′ nucleotide sequence of 16S RNA with mRNA ribosome-binding sites. It should be emphasized at this point that unit control of polygenic mRNA does not always result in unit production of the encoded proteins, as the translational controls previously alluded to can provide differential rates of protein synthesis.

Genes involved in related function are not always organized into polygenic operons but rather are dispersed on the chromosome. Frequently, however, these dispersed genes respond to a common regulatory element. When this occurs these genes are collectively referred to as a "regulon." An example of such an arrangement is the arginine biosynthetic system, which consists of eight structural genes, only three of which are in a single operon. These six units of regulation are under the control of the product of the *argR* gene. The seven rRNA operons of the *E. coli* chromosome also are, sensu strictu, a regulon, as they respond to a common regulatory signal. It is frequently found that physiologic connections between cellular processes are maintained or enhanced by common regulation. The operon organization of the mixture of elements involved in informational macromolecule synthesis is an example of this. An additional example from intermediary metabolism where multiple operons constitute a meaningful regulon is the common regulation of the operons encoding the enzymes required for the β-oxidation of fatty acids and the enzymes of the glyoxylate shunt. In this case the simultaneous induction of these systems insures the anaplerotic use of acetyl-CoA by bypassing the two CO_2 evolving steps of the TCA cycle (Chap. 5).

Control of Transcription Initiation

The Lactose Operon

The lactose operon in *E. coli* is among the most completely understood regulatory systems in bacteria. The expression of this operon, as previously described, is subject to a specific as well as to a general type of control. The elements involved in the specific control are designed to sense the presence of the substrate for β-galactosidase, lactose, while the general control senses the nutritional quality of the prevailing environment. The model for the regulation of this operon, initially proposed by Jacob and Monod and based on the results of physiologic and genetic experiments, stated simply that β-galactosidase was induced by lactose or by structurally related compounds by interacting with the product of the *lac I* gene to prevent its repressive activity exerted at the operator site. The results of a seminal experiment that shows that the inducibility of the *lac* operon is under the control of the *lac I* gene is

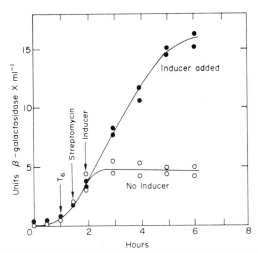

Figure 7-21. The delayed expression of inducibility of β-galactosidase synthesis in merozygotes formed by conjugation between inducible β-galactosidase-positive males (Hfr i^+z^+ T_8^s Sm^r) and constitutive β-galactosidase-negative females ($F^-i^-z^-T_8^r$ Sm^r). The abscissa represents time after formation of the merozygote. Coliphage T6 and streptomycin are added to kill the male cells; the female cells and the zygotes are resistant. The female cell contains neither β-galactosidase (z^-) nor repressor (i^-). Introduction of the i^+, z^+ alleles from the male cell allows immediate and constitutive expression of the *z* gene. However, at later times, the i^+ gene product, the repressor, accumulates to a level sufficient to repress the operon, and the synthesis of β-galactosidase then requires the presence of inducer. *(From Pardee et al.: J Mol Biol 1:165, 1959.)*

shown in Figure 7-21. This model has been verified by the isolation and structural analysis of the elements involved. The *lac I* gene is the structural gene for the *lac* repressor, a polypeptide of 37,500 daltons that associates to form a tetramer. This tetrameric protein contains two types of sites, one for binding inducer molecules and another that binds to a defined nucleotide sequence, the operator. The symmetry of the repressor and of the *lac* operator are complementary. The repressor tetramer appears as dumbbell-shaped with molecular dimensions of 45 by 60 by 120Å and possesses three two-fold axes of symmetry. The operator contains a region of two-fold rotational symmetry in its nucleotide sequence. The repressor can be oriented so that its long axis is aligned to present a two-fold symmetry to match that of the operator (Fig. 7-22). Although the general shape of the *lac* repressor and the nucleotide sequence of the *lac* operator suggest a complete, complementary, divalent interaction between the protein and the DNA, other lines of evidence suggest that the repressor-operator domains interact with differing affinities. This inequality is probably influenced by the nature of neighboring nucleotide sequences. In spite of this, the quasi-symmetrical interaction contributes to the very avid repressor-operator interaction. In vitro measurements of repressor binding to *lac* operator DNA yield an equilibrium constant of about 10^{13}, which reflects a slow rate of dissociation of the operator-repressor complex (the half-

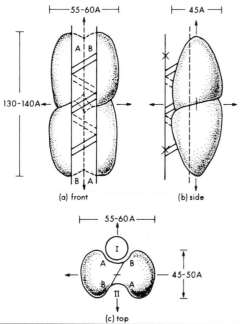

Figure 7-22. Diagrammatic representation of the *lac* repressor showing the possible interaction with the operator. Each of the four subunits has two different surfaces, A and B, that can interact with the DNA, resulting in the possibility of two operator binding sites, I and II. *(From Steitz et al.: Proc Natl Acad Sci USA 71:593, 1974.)*

life of the complex is about 30 minutes). A number of compounds structurally related to lactose are known to bind to the repressor to decrease its affinity for the operator. The physiologic role of inducer is to drive the repressor from the operator to allow RNA polymerase to proceed into the structural genes to transcribe the corresponding trigenic mRNA. It is curious that lactose itself has a very low affinity for the repressor and, therefore, is an ineffective inducer of the *lac* operon. The natural inducer is allolactose (6-0-β-D-galactopyranosyl-D-glucose), which is formed from lactose (4-0-β-D-galactopyranosyl-D-glucose) by the action of the low uninduced level of β-galactosidase. This interesting relationship between the inducing molecule (allolactose) and the molecule to be metabolized (lactose) gives the *lac* operon a self-regulated character.

Catabolite Repression

The expression of the *lac* operon is also regulated by catabolite repression, the previously mentioned more general type of control. The molecular basis for catabolite repression is a nucleotide sequence that lies before the promoter and that binds the 3′,5′-cyclic AMP-liganded catabolite activator protein. This type of regulation is exerted on a variety of transcriptional units and delineates promoters into two broad classes, catabolite-sensitive and catabolite-insensitive. In general, the synthesis of those enzymes that are used to metabolize carbon and energy sources to forms that enter the primary energy-yielding pathways is under the control of catabolite repression.

The catabolite activator protein (CAP) is dimeric, consisting of identical 22,500 dalton subunits. The protein contains sites for 3′,5′-cyclic AMP and DNA. Analysis of a number of CAP binding sites identifies —TGTN$_{10}$CAC— as a possible consensus sequence. Effective CAP binding requires that the protein be liganded with 3′,5′-cyclic AMP. The role of the cyclic nucleotide-protein complex in enhancing transcription is to render the promoter site more accessible to RNA polymerase or to facilitate formation of the open initiation complex (see Transcription, p. 125). The cyclic AMP-CAP complex apparently does not directly interact with RNA polymerase as CAP-binding sites range in distance from immediately before to about 80 nucleotides from polymerase binding sites. The cyclic AMP-CAP complex probably aids in helix unwinding. This view is supported by the observation that transcription from catabolite-sensitive promoters is more sensitive to the antibiotic nalidixic acid than is that from catabolite-insensitive promoters. Nalidixic acid decreases the negative superhelical density of DNA, which results in a concomitant decrease in the amount of underwound helix. An increase in helix stability in the vicinity of catabolite-sensitive promoters would be expected to impede the intrusion of RNA polymerase at these sites.

The physiologic connection between transcription and cyclic AMP-CAP lies in the regulation of the intracellular levels of the cyclic nucleotide. It is the fluctuation in these levels that, at least in part, explains catabolite repression in *E. coli.* As previously pointed out, catabolite repression is a conservative mechanism that prevents cells like *E. coli* from synthesizing proteins of no use to the cell in the prevailing environment. The case of β-galactosidase is an example of this situation; i.e., if glucose or other rapidly metabolizable carbon sources are present in the medium, the endogenous level of cyclic AMP is low and transcription of the *lac* operon is curtailed regardless of the presence of lactose. The decrease in the level of the cyclic nucleotide is caused by a modulation of the activity of adenylcyclase, the enzyme that catalyzes the conversion of ATP to 3′,5′-cyclic AMP. Figure 7-23 depicts the *lac* operon with its regulatory elements.

Other Aspects of Catabolite Repression. The physiologic phenomenon of catabolite repression is widespread among microorganisms. In some organisms, however, the phenomenon does not involve cyclic AMP. *Bacillus megatherium,* for example, displays the physiologic characteristics of catabolite repression but does not synthesize cyclic AMP. The eucaryotic microbe, *S. cerevisiae,* synthesizes cyclic AMP and shows a catabolite sensitivity in the synthesis of certain proteins, but the cyclic nucleotide apparently is not involved in this response. Cyclic AMP plays a more central role in this lower eucaryote as it is essential in phasing the cell cycle. Mutants of *S. cerevisiae* that lack adenyl cyclase require exogenous cyclic AMP for completion

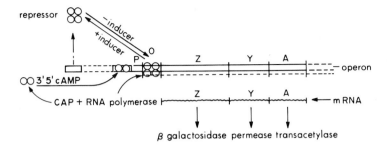

Figure 7-23. The lactose operon. The structural genes for β-galactosidase, galactoside permease, and galactoside transacetylase are *z*, *y*, and *a*, respectively. P is the promoter, the binding site for RNA polymerase. O is the operator region, the binding site for the repressor, which is specified by the *lac i* gene. CAP is catabolite activator protein, which in the 3′,5′-cyclic AMP-liganded form is required for the maximum expression of the *lac* operon.

of the cell cycle. Removal of cyclic AMP from cultures of these mutant strains causes an accumulation of non-budding cells, presumably in the G1 phase of growth. This effect probably reflects a lack of phosphorylation of essential proteins by cyclic nucleotide-dependent protein kinases because the lack of adenyl cyclase can be overcome by an additional mutation that renders the kinase independent of cyclic AMP. Cyclic AMP plays a similar role in most eucaryotes. Mammalian cells, for example, employ cyclic AMP to modulate the activity of protein kinase, which regulates the functions of key proteins by setting the level of their reversible phosphorylation. Although as yet unproven, this epigenetic modification is probably central to gene expression or to cell division. There is no evidence, as in procaryotic cells, for the direct involvement of cyclic AMP in promoting gene expression in eucaryotic cells, but there is an analogous transcription-promoting mechanism that is based in steroid hormones and specific receptor proteins. For example, it is known that specific receptors in the cytoplasm of chicken oviduct cells bind progesterone and that the resulting complexes enter the nucleus to induce the production of a number of egg white proteins. A 19 base pair consensus sequence preceding the genes for ovalbumin, conalbumin, and ovomucoid has been tentatively identified as the binding site for the hormone-protein complex. Interaction of the complex at these sites apparently renders the conjoined promoters more accessible to RNA polymerase in much the same manner as the cyclic AMP-CAP complex functions in *E. coli.*

Positive Genetic Regulatory Proteins

The cyclic AMP-CAP complex described above is a positive regulatory system that acts on a large number of systems. Other positive-acting regulatory proteins are known that are highly specialized in that they are required for RNA polymerase to initiate transcription of specific operons. In these cases the activity of the regulatory proteins is modulated by inducing molecules, which cause them to bind at sites in the vicinity of the pertinent promoters to effect transcription. For example, the *malT* gene of *E. coli* encodes a protein that is required to activate the genes for maltose and maltodextrin utilization. Among these are the *malPQ* operon, which encodes maltodextrin phosphorylase and amylomaltase, and the *malB* gene cluster, which con-

tains five genes whose products are required for the transport of maltose and the larger maltodextrins. An additional example of positive regulation is the arabinose operon of *E. coli*, where the product of the *araC* gene functions in the unliganded form as a repressor and in the arabinose-bound form as an activator of transcription. The *araC* gene product also acts as an autogenous regulator in that it represses its own synthesis.

How Do Regulatory Proteins Find their *cis*-Active Targets?

Most of the genetic regulatory proteins that bind to specific nucleotide sequences with a high affinity are able to bind at other nucleotide sequences but with a generally lower affinity. In other words, these proteins are endowed with structural features to allow general as well as specific recognition of DNA. This type of protein is found largely associated with DNA rather than free in the cytoplasm of the bacterial cell. For example, more than 90 percent of the *lac* repressor is nonspecifically bound to the DNA of *E. coli*. It would appear that the nonspecific binding of these proteins would shortcircuit the diffusion pathway by which they reach their specific targets. Experimental evidence suggests that the converse is true—nonspecific binding actually aids these proteins in finding their specific targets. It has been shown that the rate constant for association of the *lac* repressor with the *lac* operator contained within a small (∼ 200 bp) DNA molecule is ∼ 10^8 mole^{-1}sec^{-1}, that is, the rate expected of a diffusion-limited reaction. If the *lac* operator is contained in a much larger DNA molecule, the rate of binding to the specific sequence increases to 10^{10} to 10^{11} mole^{-1}sec^{-1}. This value approaches the in vivo rate (10^{12} mole^{-1}sec^{-1}) deduced from induction kinetic experiments. A current view of how the extraneous DNA facilitates specific binding pictures the regulatory protein as binding to the DNA molecule anywhere along its length and then sliding or diffusing along the double helix for some distance before dissociating from the DNA. If the operator sequence is encountered in one of these random walks, avid association occurs to establish proper repressor function. An alternative view of how nonspecific binding facilitates specific binding proposes that randomly bound protein is transferred directly to specific binding sites. This mechanism when considered along with the flexibility of DNA and its domained structure could be re-garded as a means of reducing the volume in which the specific binding occurs. This would increase the rate of as-sociation of the protein and its specific target.

Attenuation

The differential rate of synthesis of the enzymes involved in the formation of low molecular weight cellular compo-nents is generally repressed by a signal that corresponds to an excess of the pertinent end-product. A repressor protein activated by the end-product could adequately ex-plain this response, and this type of regulation does occur in some biosynthetic systems. An additional mode of regu-lation prevalent in amino acid biosynthetic systems occurs by a translational control of transcription termination. This process senses the level of specific aminoacyl-tRNA rather than free amino acid. This type of attenuation con-trol is able to occur in bacteria because translation can be tightly coupled to transcription, i.e., translation is able to closely follow transcription so that the initially synthesized RNA is translated as the mRNA molecule is extended by RNA polymerase. The mRNA encoding the biosynthetic enzymes for several of the amino acids contains leader se-quences that precede the enzyme-encoding portion of the transcript. These leader regions contain relatively short open reading frames (translational units) that are rich in codons for the regulating amino acid. For example, the leader region of the histidine operon in *E. coli* and *S. typhi-murium* contains a translational unit for a 16 amino acid peptide that includes 7 consecutive histidine residues. These leader regions contain additional nucleotide se-quences that lie between the short peptide coding se-quence and the protein-encoding region. These sequences are capable of generating secondary structure by hydrogen bonding. One of these structures is a transcriptional ter-mination signal (terminator), which, if present, causes tran-scriptional termination to occur, thereby preventing forma-tion of the productive region of the mRNA. The frequency of this termination or attenuation event is controlled by regulating the formation of the terminator structure. This is accomplished by an additional sequence of nucleotides that is able to form a secondary structure (the preemptor) with the downstream sequence, which potentially can be used to form the secondary structure that is the terminator (Fig. 7-24). In other words, the preemptor and terminator are constructed from a common nucleotide sequence so that only one of the structures can exist at any given time. The choice of forming the preemptor or the terminator is a function of the rate of the coupled translation of the spe-cific codon-rich portion of the mRNA. If rapid translation occurs, the ribosomes involved in this process cover the earliest encountered sequence to prevent its hybridization to its complement so that preemptor formation is prevent-ed. This frees the common sequence so that it can hybrid-ize with its downstream complement to form the terminator, and transcription ceases at this site. If, howev-er, translation is impeded, the first possible secondary

structure, the preemptor, will form and preclude forma-tion of the terminator. In this case, RNA polymerase will continue to synthesize the complete mRNA. The physio-logic connection in this regulatory scheme is the concen-tration of the aminoacyl-tRNA cognate with the abundant codons in the attenuator peptide. For example, in the case of the *his* operon, if the level of histidyl-tRNA is sufficient to synthesize the histidine-rich peptide at a maximum rate, translation will remain tightly coupled to transcription. In this case the participating ribosomes will mask the proxi-mal segment of polyribonucleotide used to form the pre-emptor structure, and the distal polyribonucleotide segment can participate with its alternate complementary sequence to form the terminator. The attenuation that oc-curs will be manifest as a repression in the rate of synthe-sis of the *his* operon-encoded enzymes. The converse occurs when the level of histidyl-tRNA is low. When trans-lation fails to occur for one reason or another, an addi-tional stem and loop structure is potentially able to form in the leader region of the mRNA. This secondary struc-ture precludes formation of the preemptor because it is composed of the upstream sequence of this potential structure and a promoter proximal sequence, termed the "protector." Protector formation precludes formation of the preemptor, which allows the terminator to form and to cause attenuation. Figure 7-24 depicts the potential stem and loop structures in the leader sequence of the trypto-phan operon and the hypothetical condition of these struc-tures under the three relevant physiologic conditions. Mutations that affect the charging level of tRNA or that af-fect the energy of formation of the secondary structures in the leader sequence result in partial constitutive synthesis of the pertinent amino acid biosynthetic system. Mutations affecting the Km of tRNA synthetases for the tRNA or the specific amino acid lead to uncontrolled enzyme synthesis as do mutations that stabilize the preemptor or destabilize the terminator structure. Mutations that affect the struc-ture of tRNA molecules also alter their participation in at-tenuation. For example, *hisT* mutants lack an enzyme that converts a uridine to a pseudouridine adjacent to the anti-codon of tRNA[his]. Apparently the lack of this modification causes inefficient reading of the *his* codons in the attenua-tor peptide, which leads to derepressed expression of this operon. This modifying enzyme is not totally specific for tRNA[his], as failure to modify other tRNAs also occurs, for example, *hisT* mutants lead to derepressed rates of synthe-sis of isoleucine-valine and leucine biosynthetic enzymes, which also are under attenuation control. An additional modifying enzyme, the product of the *trpX* locus, is in-volved in adding an isopentenyl group on the N^6 position of an adenine in tRNA[trp]. The attenuation control of the tryptophan operon is decreased owing to *trpX* mutations.

Regulation of Translation

Although procaryotic organisms primarily regulate protein synthesis at the initial steps in mRNA synthesis, significant controls are known to operate at the level of translation.

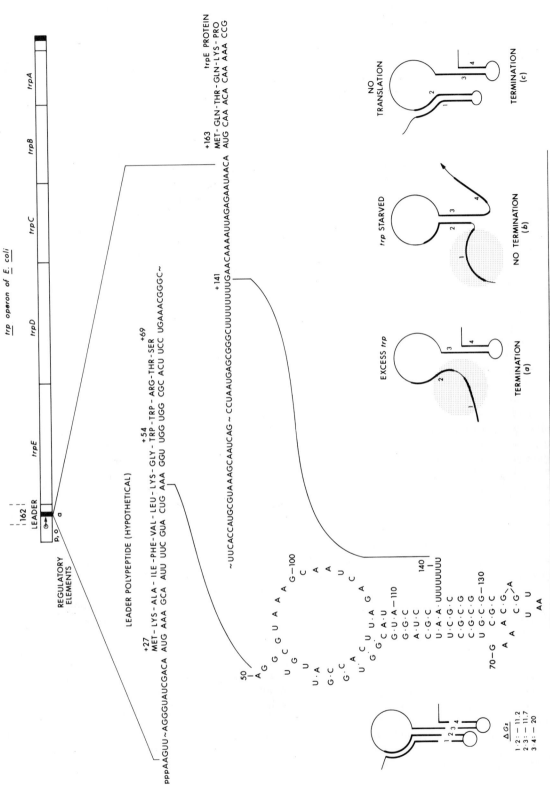

Figure 7-24. The attenuator segment of *trp* operon mRNA. The proposed secondary structure is shown. Note the tandem tryptophan codons in the leader polypeptide. The frequency of translation of these codons governs the formation of the secondary structures as shown; see text for details of attenuation process.

145

Intrinsic controls based in the strength of ribosome-binding sites and codon usage were previously described. The significance of secondary structure in mRNA also was previously mentioned. This secondary structure potentially plays a number of roles in the regulation of translation. Ribosome-binding sites within polygenic mRNA can be sequestered within secondary structure; cases are known where translation of upstream genes is required to make these sites available. Intramolecular hydrogen bonding in mRNA can also lead to the formation of nuclease-specific sites, such as those acted upon by RNase III. Cleavage at these sites could lead to positive or negative effects with respect to differential expression of the genes within a polygenic mRNA. On the one hand, this type of activity could unmask potentially efficient ribosome-binding sites or, on the other hand, could result in a 3' terminus that is more susceptible to the many 3' exonucleases of the cell. Removal of the stem and loop transcriptional termination structure generally renders a mRNA more susceptible to 3' exonucleolytic digestion. Messenger RNA metabolism has been invoked to explain differential gene expression from some polygenic molecules. For example, the steady state levels of the products of the *rpsU, dnaG, rpoD* operon (Fig. 7-20), protein S-21, primase, and sigma factor are produced, respectively, in concentrations of 50,000, 50, and 3000 copies per cell. At least part of this striking differential expression is explained in terms of mRNA metabolism.

The intrinsic stability of mRNA also is considered a significant control of translation. As pointed out above, mRNA molecules are susceptible to ribonucleases and generally undergo rapid turnover. A mRNA molecule that is not participating in translation is particularly sensitive to degradation owing to lack of protection by bound ribosomes. The average half-life of mRNA in bacteria is about 2 minutes. Some mRNA molecules, however, are long-lived, perhaps reflecting the extent to which the cells use their encoded proteins. For example, the mRNA encoding the protein of the lipoprotein of the outer membrane of *E. coli* (Chap. 6) has a half-life of 11.5 minutes. This protein is the most abundant one in *E. coli.*

Retroregulation

Retroregulation is defined as the control of gene expression from sites distal (i.e., after, not before) to the gene. This situation forms a stark contrast to regulatory sites (promoters, operators) that precede the gene. This phenomenon has been described in bacteriophage λ and is based upon a quasi-programmed degradation of mRNA. Chapter 63 describes the lytic vs. lysogenic life style of phage λ. Each of these modes of existence follows from complicated regulatory circuits that involve many of the genetic regulatory principles previously described. The retroregulation described here apparently coordinates events that are important for effective lysogenic integration of the prophage into the host chromosome. Among the functions required for integrative lysogeny are the int protein required for site-specific recombination and the cI

repressor for curtailment of the phage's vegetative program. Efficient integration of λ would logically be expected to require that sufficient cI repressor be available prior to integration, since expression of vegetative functions, including replication, by the integrated prophage would be detrimental to the cell. Retroregulation apparently solves this problem. This phenomenon is explained on the basis of alternative secondary structures in the mRNA distal to the *int* gene: one of these structures is a normal transcription termination signal and the other, which requires additional 3'-oriented nucleotides, is an RNase III site. The nuclease site forms only if synthesis of the mRNA is not terminated at the potential transcription termination site. Early in phage infection, this transcript originates from the P_L promoter and is transcribed by an RNA polymerase that has acquired the antiterminator N protein. This polymerase complex eventually transcribes through the terminator distal to the *int* gene, which results in formation of the RNase III site. Cleavage of this site renders the mRNA subject to digestion by 3' exonucleases with the resulting destruction of the *int* coding sequences. Later in infection, transcription from the PI promoter is activated by a regulatory protein. The RNA polymerase involved in this transcription does not acquire the N protein, and therefore mRNA synthesis stops prior to acquisition of the nucleotides required for formation of the RNase III site. At this time the int protein is produced, the cI repressor has accumulated, and successful integrative lysogeny occurs. Figure 7-25 depicts this regulatory circuit.

Feedback Regulation of Ribosomal Protein Synthesis

A significant extrinsic regulatory mechanism has been described for the modulation of ribosomal protein synthesis in *E. coli.* Approximately 52 different proteins are distributed between the 30S and 50S ribosomal subunits. The structural genes for these are organized into a series of operons the expression of which is coordinated with the synthesis of rRNA. Part of this coordination is based upon the ability of one of the proteins from each operon to bind to its respective mRNA to cause autogenous regulation of its own synthesis as well as that of other proteins encoded within the polygenic mRNA. There is a correlation between the site on the mRNA recognized by the dual function protein and the cistrons whose translation is inhibited (Fig. 7-26). The proteins within each operon that function in autogenous regulation are those that make first contact with ribosomal RNA in the course of ribosome assembly (p. 129). It appears that the cells have evolved a regulatory system that adventitiously makes use of an essential property of these proteins, namely, nucleic acid recognition. This regulatory scheme connects the availability of rRNA and the proteins with which it participates to form mature ribosomes. The rRNA and mRNA compete for the protein —when rRNA is in excess, the dual function protein is used for ribosome assembly, but when the rRNA is saturated with the protein, it feedback inhibits further transla-

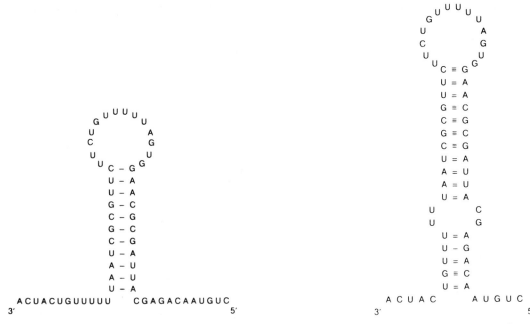

If the transcription complex does not contain the antitermination factor the mRNA terminates in a normal, exonuclease protective stem and loop.

If the transcription complex contains the antitermination factor transcription continues so that a potential RNAse III site forms. Cleavage at this site destroys the normal exonuclease protective stem and loop so that 3' terminal mRNA is destroyed.

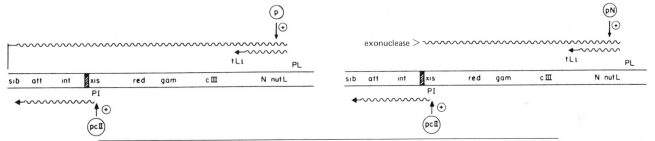

Figure 7-25. The process of retroregulation as described for bacteriophage λ. The expression of the *int* gene is controlled by degradation of its mRNA. Transcription beyond the terminator immediately following the *int* gene causes production of a RNase III site. Cleavage at this site renders the mRNA subject to 3' exonucleases, which rapidly destroy the *int* coding sequences. Transcription beyond the terminator occurs only with RNA polymerase complexes that have acquired an antiterminator factor. See text for details. *(From Guarneros et al.: Proc Natl Acad Sci USA 79:238, 1982.)*

tion of its mRNA. This view is supported by comparative sequence analysis, which shows a striking similarity between the site on the rRNA and the segment of mRNA beyond which translation is blocked (Fig. 7-26).

Regulation of Stable RNA Synthesis

The rate of synthesis of tRNA and rRNA in bacteria is coordinated with the growth rate. As pointed out in Chapter 5 the concentration of these stable RNA molecules is greatest in rapidly growing cells. The rate of synthesis of other elements involved in protein and RNA synthesis is influenced by the growth rate. For example, the rate of transcription of the ribosomal protein-encoding operons is greatest in rapidly growing cells. In both cases an inverse relationship exists between the level of the nucleotides 3'-diphosphate 5'-diphosphate guanosine and 3'-triphosphate 5'-diphosphate guanosine (ppGpp and pppGpp) and the rate of RNA synthesis. The level of these nucleotides is controlled by a ribosome-associated enzyme (stringent factor) that transfers the pyrophosphoryl group from ATP to the 3'-OH of GDP or GTP. The activity of this enzyme responds to the level of amino acids available for protein synthesis in that its activity is triggered by binding an uncharged tRNA to a translation complex. This is the basis for the so-called stringent and relaxed response of stable RNA synthesis to amino acid starvation. The stringent re-

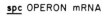

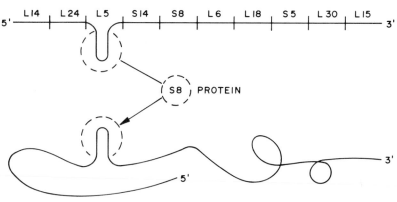

Figure 7-26. Feedback regulation of ribosomal protein synthesis in *E. coli*. The competition of mRNA and 16S RNA for protein S8 governs the rate of translation of the *spc* operon as shown.

sponse is an immediate decrease in the rate of stable RNA synthesis upon amino acid starvation. If, however, the ribosome-associated stringent factor (the product of the *rel* gene) is inactivated by mutation, stable RNA synthesis will continue following amino acid starvation; this is the relaxed response. An additional enzyme, the product of the *spoT* gene, also governs the level of ppGpp and pppGpp and is involved in degrading these nucleotides.

It is not entirely clear whether or not the growth rate control and the amino acid control of stable RNA synthesis are part of the same phenomenon. All ribosomal RNA operons in *E. coli* contain tandem promoters, which raises the possibility that they could be used for transcriptional events responsive to different signals. The manner in which the guanosine nucleotides regulate the activity of RNA polymerase is unknown. It is interesting to note, however, that these nucleotides have a reciprocal effect on the expression of many biosynthetic systems, i.e., they are stimulatory rather than inhibitory. Those systems that are stimulated, such as amino acid biosynthetic systems and proteases, supply the protein synthetic machinery with necessary substrates.

Posttranslational Regulatory Events

The genetic regulatory mechanisms described above represent the first line of control for integrating cellular functions. Many other types of regulatory phenomena must be exerted on the informational macromolecules of the cell in order that their roles be played in an orderly and productive fashion. Examples of how the activities of RNA molecules are modulated by modification, processing, and interacting with protein have previously been described. The following section deals with the mechanisms employed in the posttranslational management of the proteins synthesized by the cell. These processes include events as varied as controlled proteolysis, transmembrane transport, epigenetic modification, and in the case of en-

zymes, the modulation of their activity. These are not mutually exclusive events, and in some cases more than one of them may affect the function of a single protein.

Protein Degradation

Proteolysis of individual proteins in bacteria can be extreme, so that the entire molecule is degraded to free amino acids, or it can be limited, so that some specific function of the protein is affected. In general, normal proteins in bacteria turn over at a very slow rate. Proteins that are deformed as the result of incorporation of amino acid analogs or as a consequence of missense mutations are rapidly destroyed. This housekeeping activity also acts on truncated proteins that form as a result of premature termination of translation caused by nonsense mutations. Proteins that are damaged by free radicals or other forces are cleared from cytoplasm by proteolysis. The rate of proteolysis of all proteins, including the normal ones, is increased when the cells are starved for nitrogen and carbon. This response is under the control of guanosine tetraphosphate, the accumulation of which, as we previously described, signals the lack of amino acids. The amino acids produced from the degradation of proteins are used for protein synthesis, and some of them serve as sources of carbon and energy. Protein turnover of this type represents an extreme attempt by the cell to establish a collection of new proteins better suited for survival of the cell under the conditions of starvation. An example in which the use of the products of proteolysis are used in a highly directed fashion is during endospore formation by members of the genus *Bacillus*. In this case massive protein breakdown occurs, with the products being used for synthesis of the components of the endospore (Chap. 5).

The manner in which the activity of the intracellular proteases is regulated is largely unknown. *E. coli* contains at least eight different proteases. One of these, protease La, is interesting in that it requires ATP for activity. This protease is large, containing four identical 120 kdal polypeptides and is involved in the breakdown of abnormal proteins with the concomitant hydrolysis of ATP. This protease is the basis for earlier observations that showed that the breakdown of abnormal proteins is energy dependent in bacteria. How this unusually large protease recognizes abnormal proteins and spares normal proteins remains a mystery.

There is some evidence that proteins in bacteria can be marked for proteolysis, i.e., they probably are covalently modified so that they can be distinguished from later synthesized molecules. The molecular basis for this sensitization is unknown. It is of interest in this context that a small polypeptide, ubiquitin, which has been found in a wide variety of eucaryotic cells, has been shown to be involved in protein degradation in reticulocytes. Ubiquitin is added to designated proteins by formation of a peptide bond between its terminal carboxyl group and amino groups on lysine of the doomed protein. This peptide adduct identifies the protein as a substrate for proteases. The coupling of ubiquitin to the protein is an energy-dependent process. The carboxyl group of the ubiquitin is activated by formation of an adenylate in a reaction that is analogous to the one carried out by aminoacyl-tRNA synthetases, i.e., ATP is used and iPP is generated. Ubiquitin, in spite of its name, has not been found in bacteria, and it remains to be seen whether or not comparable mechanisms exist in these organisms.

Many types of bacteria produce extracellular proteases. The activity of these enzymes is largely uncontrolled, as they are digestive and break down otherwise impermeable proteins to more freely diffusible peptides and amino acids. Still other proteases are located in the periplasmic space; this location serves to separate the proteolytic enzymes from the cytoplasm but yet keeps them from diffusing away from the cell. This segregation of proteases in the cytoplasm and in the periplasmic space is roughly analogous to the organization seen in eucaryotic cells where a number of acidic proteases (active at acidic pH) are contained within the lysosome while an alkaline proteolytic system is free in the cytoplasm.

Specific proteases also exist in bacteria. For example, the signal sequence (p. 70) that is frequently found on proteins to be secreted is cleaved off by a peptidase as these proteins traverse the membrane. This signal peptidase is membrane-associated. Another particularly notorious protease that has been described in E. coli is derived from the recA gene product, a protein that plays a vital role in general genetic recombination (Chap. 8). This DNA-binding protein is converted to a protease by agents that damage DNA. Its proteolytic activity is specifically involved in inactivating certain repressor proteins. The consequences of this activity are described in Chapter 8.

Export of Proteins

As described above, many of the proteins destined for transmembrane transport contain a signal sequence that associates with the cytoplasmic membrane as the protein is synthesized. This results in a loose association of the translation complex with the membrane. The question of transmembrane transport of protein in bacteria has been approached by recombinant DNA technology. Hybrid genes have been constructed by fusing the leader sequence coding information of normally transported proteins with the coding sequence for normally soluble cytoplasmic proteins. The lamB gene encodes a protein that acts as the receptor for bacteriophage λ, serves in the transport of maltose and maltodextrins, and is located in the outer membrane of E. coli. This protein traverses the cytoplasmic (i.e., inner) membrane by way of a signal peptide. Fusion of a sufficient portion of the coding sequence for the signal peptide with the coding sequence for β-galactosidase leads to export of β-galactosidase to the outer membrane. If, however, a smaller portion of the leader sequence coding region is fused to the β-galactosidase sequence, the enzyme remains associated with the inner membrane. These observations suggest that mere association with the cytoplasmic membrane does not lead to export. Additional studies on the export of other proteins suggest that structure other than the signal peptide is important for transport. For example, removal of the coding sequence for about 10 percent of the carboxy-terminus portion of β-lactamase, which is normally found in the periplasmic space of S. typhimurium, renders the truncated protein nontransportable and locked in the cytoplasm. This observation and those with other normally transported proteins suggest that, in addition to the signal sequence, other structures within the transported protein serve as anchors that are required for proper transmembrane portage. Many proteins destined for export do not contain signal peptides and are synthesized in the cytoplasm. The manner in which these proteins traverse the membrane is unknown.

Control of Enzyme Activity—Integration of Metabolic Pathways

The discussion, in Chapter 5, of the biosynthetic capabilities of bacteria pointed out that many bacterial species are able to exist by a relatively autotrophic metabolism. Autotrophic metabolism implies the existence of complex enzyme systems that participate in the synthesis of the various types of small molecules required for cell replication. An organism, such as E. coli or S. typhimurium, that can effect its own replication using a compound such as glucose as the sole source of carbon would be expected to possess mechanisms whereby an integration of the various pathways is maintained to allow the efficient use of carbon and energy.

The accrued benefit of efficient regulation of carbon flow through a biosynthetic pathway is apparent when the amphibolic nature of the pathways using primary carbon sources is considered. Figure 7-27 represents the glycolytic pathway and tricarboxylic acid cycle for the aerobic utilization of glucose and illustrates some of the points at which intermediates in this pathway are shunted out of the strict-

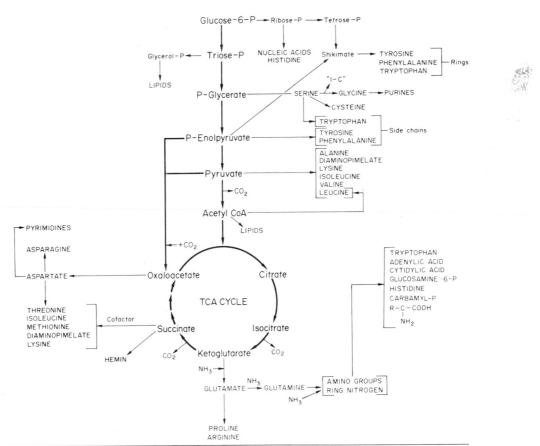

Figure 7-27. The amphibolic nature of the glycolytic and tricarboxylic acid pathways. Amphibolic pathways are those that serve as an energy source and as a source of biosynthetic intermediates.

ly energy-yielding pathway into biosynthetic pathways. It is apparent that if a balanced supply of these various end-products is to be produced by the cell, some regulatory mechanisms must exist, viz., if an unbridled utilization of the glycolytic intermediates occurred, the TCA cycle would cease to function. One manner in which regulation could possibly be accomplished is by the evolutionary selection of enzymes with appropriate substrate affinities and maximum reaction velocities. Without a doubt, this type of regulation does occur to some extent, but it does not represent the quantitatively significant metabolic regulatory mechanism in bacteria. It is apparent that an organism that depended solely upon this type of regulation would be suited for efficient growth in a limited number of environments. The demand for specific end-products differs depending upon the environment, and the synthesis of compounds that are already present represents a waste of carbon and energy.

The genetic regulatory mechanisms described above are also involved to some degree in regulating carbon flow through various pathways, for the differential rate of synthesis of enzymes responds to metabolic regulatory signals. However, genetic regulation is more properly viewed as a regulation of macromolecule synthesis and less as a regulation of intermediary metabolism. For example, when a derepressed biosynthetic pathway receives a signal indicating excess end-product, the synthesis of the pertinent enzymes may be curtailed, but carbon will continue to flow through the pathway. In this case, the activity of the pathway will decrease only when the concentrations of the enzymes constituting the pathway are diluted by growth of the cells. This obviously would constitute a sluggish and inefficient manner for controlling the flow of carbon. However, inhibition of the activity of the pathway would result in an immediate response to the end-product and an efficient utilization of metabolites.

Efficient and highly specific mechanisms have evolved whereby the flow of carbon through the various biosynthetic pathways is regulated. The negative feedback inhibition that end-products exert on their own biosynthesis represents the foremost, and conceptually the simplest, example of metabolic regulation. It is known that in virtually all biosynthetic pathways that have been studied in bacteria, one of the enzymes of the pathway is effectively inhibited by the end-product. The enzyme upon which this control is exerted is invariably the first enzyme specific for the pathway. This provides efficient control of the entire pathway, since inhibition of an enzyme other than the first would result in accumulation of the pertinent biosynthetic intermediates.

Simple End-product Inhibition

Physiologically, the simplest case of end-product inhibition is that exerted on a pathway specifically involved in the synthesis of a single end-product. There are numerous examples of this pattern of end-product inhibition in bacteria, plants, and animals. Historically, one of the most significant examples of this type of control is the inhibition by isoleucine of threonine deaminase, the first enzyme in the isoleucine biosynthetic pathway.

End-product Inhibition in Multifunctional Pathways

Other, physiologically more complex patterns of end-product inhibition are found in multifunctional pathways, i.e., pathways whose intermediates serve as the source of more than one end-product. An example of this type of pathway is the synthesis of lysine, methionine, and threonine from aspartic acid (Fig. 7-28). It is clear that if the first enzymatic reaction in this pathway, catalyzed by aspartokinase, was inhibited exclusively by one of the three end-products, starvation for the remaining two amino acids would occur.

This problem is obviated in *E. coli* by the presence of three distinct aspartokinases: aspartokinase I is inhibited by threonine, aspartokinase III is inhibited by lysine, and aspartokinase II is not inhibited. The presence of independently regulated enzymes provides for a fractional inhibition of the total activity. Interestingly, this regulatory problem is not solved uniformly by all bacteria. For example, in *Bacillus polymyxa* and *Rhodopseudomonas capsulatus*, a single aspartokinase is present, and in these organisms, lysine, threonine, or methionine alone cannot cause enzyme inhibition, but efficient inhibition is observed in the presence of both lysine and threonine. This type of end-product inhibition is known as "concerted" or "multivalent." Figure 7-28 shows the pathway for the synthesis of the aspartate family of amino acids in *E. coli*. It is noteworthy that regulation of metabolite flow occurs at points other than the first enzyme in the multifunctional portion of the pathway. Each branch point in the pathway is controlled by the ultimate, specific end-product. Another interesting feature of this pathway, which illustrates the high degree of integration of the various catalytic steps, is that two homoserine dehydrogenases (aspartate semialdehyde ⇌ homoserine), I and II, are present. I, like aspartokinase I, is inhibited by threonine, whereas II, like aspartokinase II, is not inhibited. The ultimate in metabolic integration is demonstrated by these pairs of enzymes. It has been

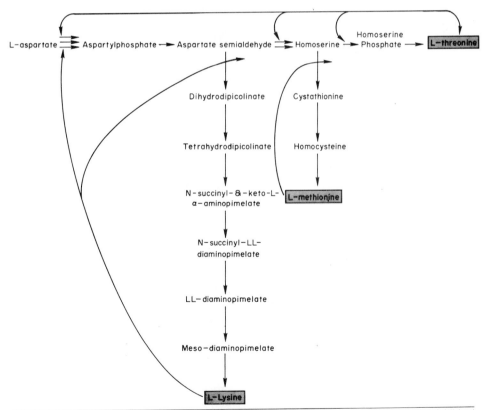

Figure 7-28. Pathways for synthesis of the aspartate family of amino acids. The enzymes inhibited by the end-products are indicated. End-product inhibition provides for efficient control of carbon flow through this highly branched pathway.

shown that the activity of aspartokinase I and homoserine dehydrogenase I resides in the same protein and that aspartokinase II and homoserine dehydrogenase II are also a single protein.

COOPERATIVE END-PRODUCT INHIBITION. Other patterns for regulating metabolic flow in multifunctional pathways are known. For example, cooperative end-product inhibition is exerted on the first enzyme in purine biosynthesis, glutamine phosphoribosyl pyrophosphate amidotransferase. The products of the purine biosynthetic pathway are 6-hydroxypurine ribonucleotides and 6-aminopurine ribonucleotides. The transferase is inhibited to a greater degree by pairs consisting of hydroxynucleotides and aminonucleotides than by pairs of hydroxynucleotides or pairs of aminonucleotides.

CUMULATIVE END-PRODUCT INHIBITION. Another pattern of regulation for multifunctional enzymes has been reported for glutamine synthetase of *E. coli.* Glutamine serves as an amino donor in the biosynthesis of tryptophan, adenylic acid, cytidylic acid, glucosamine 6-phosphate, histidine, and carbamylphosphate. Each of the end-products is able to partially inhibit the activity of the synthetase. The degree of inhibition exerted by combinations of the end-products is cumulative in that the residual activity is the product of the fractional activities in the presence of the individual inhibitors. For example, when tested independently, tryptophan, cytidylic acid, carbamylphosphate, and adenylic acid show respective fractional activities of 0.84, 0.86, 0.87, and 0.59. In combination, the total residual activity is:

$$0.84 \times 0.86 \times 0.87 \times 0.59 = 0.37.$$

This type of cumulative effect provides another means whereby the multiple products of a multifunctional enzyme can effectively regulate metabolite flow by modulating the activity of a key enzyme.

The various patterns of end-product inhibition described above endow the cell with obvious physiologic advantages. However, this type of modulation of enzyme activity is not reserved strictly to biosynthetic systems. The activity of catabolic enzymes can be regulated in similar fashion. In these cases, however, the physiologic relationship of the inhibiting molecule to the affected enzyme is not always as obvious as it is in cases of end-product inhibition. A case where the relationship is apparent is inhibition of the key glycolytic enzyme, phosphofructokinase, by ATP. The relationship between effector molecule and enzyme in this case constitutes, at least partially, the molecular basis of the Pasteur effect, the inhibition of glycolysis by respiration (Chap. 4).

Metabolite Activation

Metabolic pathways are integrated by the ability of certain metabolites to stimulate enzymes whose products ensure the efficient use of the stimulating molecule. This type of activation occurs in both biosynthetic and catabolic pathways. Excellent examples of this type of activation occur with the biosynthetic and biodegradative threonine deaminases of *E. coli.* Biosynthetic threonine deaminase, as previously mentioned, is specifically involved in the biosynthesis of, and is inhibited by, isoleucine. The synthesis of isoleucine is coupled to the synthesis of valine by virtue of shared enzymes. It is known that valine stimulates the activity of threonine deaminase in the presence of a low concentration of isoleucine. It would seem that valine exerts a positive control in order to balance isoleucine formation with its own biosynthesis, so that both may be efficiently used for protein biosynthesis. Biodegradative threonine deaminase is a catabolic enzyme produced by *E. coli* when it is grown under relatively anaerobic conditions. Its role is distinct from the biosynthetic enzyme, and it is not inhibited by isoleucine. However, its activity is stimulated by adenylic acid. Since the ultimate function of this enzyme is to provide energy by the degradation of threonine, the role of AMP stimulation of its activity may be regarded as a regulatory signal to ensure metabolite flow through a pathway that will ultimately provide for the utilization of the activator molecule, i.e., the formation of ATP from AMP. Although a number of examples of metabolite activation exist, care must be exercised in attributing physiologic significance to them. In many instances, the observed activations may be fortuitous.

Allosteric Mechanisms

One of the most striking features of the types of inhibitions and activations described above is that the effector molecules, i.e., the inhibitor or activator ligands, bear little if any structural resemblance to the substrates of the enzymes whose activity they modulate. This fact, together with the observation that various treatments, such as exposure to mercurials, high pH, high temperatures, and proteolytic enzymes, can render these enzymes refractory to inhibition by the specific ligand without affecting their catalytic function, indicates that the effector molecule binds to a site other than the active site of the enzyme. Mutant enzymes can be obtained that are no longer subject to the action of effector ligands. The term "allosteric" is used to describe this property. The activity of allosteric enzymes is modulated by the binding of low molecular weight ligands to stereospecific sites distinct from the active site. Therefore, the basis of this property is different from the strict competitive inhibition by a ligand that is isosteric with respect to the substrate.

Another property of allosteric enzymes that in most cases may be visualized as physiologically significant is that the velocity of the reaction catalyzed increases with a high-order concentration of substrate. In other words, the reaction kinetics bear a sigmoidal relationship with increasing substrate or effector concentration. Not all allosteric enzymes exhibit high-order kinetics. Some yield typical Michaelis-Menten kinetics, in which a plot of reaction velocity

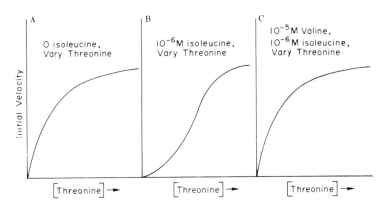

Figure 7-29. Kinetic analysis of L-threonine deaminase in the presence of allosteric effectors. The initial reaction velocity in the presence of various concentrations of the substrate L-threonine is shown (**A**). The negative allosteric effector, L-isoleucine, induces a cooperativity (**B**), which is removed by the positive effector L-valine (**C**).

versus substrate or effector concentration generates a rectangular hyperbola.

Allosteric effects are usually separated into two categories: homotropic effects, caused by interactions between identical ligands, and heterotropic effects, caused by interactions between different ligands. The sigmoid curves obtained for some allosteric enzymes when substrate concentration is plotted against initial reaction velocity are a positive homotropic effect. Inhibition of the enzyme by a stereospecific ligand is a heterotropic effect. Heterotropic effects need not always be antagonistic but can also be positive. An example of this is shown in Figure 7-29, which illustrates the effects of isoleucine and valine on the activity of biosynthetic threonine deaminase. Native threonine deaminase exhibits cooperative homotropic effects with respect to substrate only in the presence of low concentrations of the inhibitor, isoleucine. The induction of this homotropic effect by the end-product inhibitor is, in itself, a heterotropic effect and is negative. Higher concentrations of isoleucine will completely inhibit the enzyme at the substrate concentrations shown in Figure 7-29. The homotropic cooperatively induced by isoleucine can be reversed by valine, and this action may be considered a positive heterotropic effect with respect to threonine binding and a negative heterotropic effect with respect to isoleucine binding.

Heterotropic effects are usually reciprocal. For example, in the case cited previously, if isoleucine induces cooperativity with respect to substrate binding, the substrate will induce cooperative isoleucine binding. This relationship is illustrated in Figure 7-30.

The high-order kinetics exhibited by many allosteric enzymes probably is of some survival value to the organism. Basically, such a kinetic pattern allows the enzyme to respond in a maximum way to small changes in substrate concentration and permits maintenance of a finite pool of the substrate. It is evident in the case of a substrate, such as threonine, which serves as a source of protein as well as a source of isoleucine, that unrestricted threonine deaminase activity would deplete the threonine required for protein synthesis. Similar physiologic significance can be attached to other allosteric enzymes that are endowed with this type of kinetic property.

Several models have been offered to explain how the binding of allosteric ligands modulates the activity of enzymes. All these models are predicated on the observations that (1) effector-ligand binding sites, as mentioned above, are different from catalytic sites, and (2) all allosteric proteins are oligomeric, i.e., composed of identical protomers. A protomer is defined as the smallest subunit possessing, at least potentially, all of the structure necessary for accommodating the stereospecific ligands known to affect the native enzyme. The unifying principle of the various models is that all invoke subunit interaction mediated by conformational changes within the protomers. The question is whether the conformational changes are induced by the binding of the allosteric ligand or whether they occur spontaneously and are stabilized by the binding

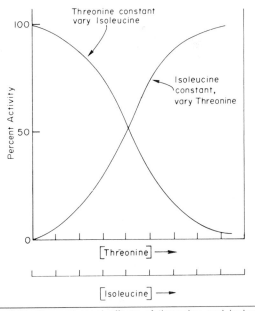

Figure 7-30. The reciprocal effects of threonine and isoleucine on the activity of threonine deaminase. Threonine is the substrate for the enzyme, and isoleucine is an allosteric inhibitor of enzyme activity. Isoleucine induces cooperativity with respect to enzyme activity, and threonine induces cooperativity with respect to isoleucine inhibition.

of the ligands. These two conditions form the basis for the sequential and concerted mechanisms of allosteric transitions, respectively. In a sequential mechanism, ligand binding to one of the enzyme's protomers is attended by a conformational change in the protomer, which because of its contact with adjacent protomers may change their conformation so as to affect their affinity for the ligand. The concerted transition model, on the other hand, supposes that symmetry among the protomers is constantly preserved and that a spontaneous isomerization between different conformational forms exists. According to this model, in the simplest case (such as that for a simple end-product-inhibited enzyme), two isomers of the enzyme would exist, one catalytically active and the other inactive. The inhibiting ligand, which binds only to the inactive form of the enzyme, pulls the equilibrium toward the inactive species, and the substrate or positive effector ligand binds to the active form to pull the equilibrium toward the active species. These models, which are shown in Figure 7-31, are consistent with some of the experimental results obtained with allosteric enzymes. Both models allow effective enzyme activity modulation by either affecting the apparent dissociation constant (K_m) for substrate or the maximum reaction velocity (V_{max}). Systems in which ligand binding affects the K_m are known as K systems, those in which the V_{max} is affected are known as V systems.

The two models outlined previously represent the mechanistic extremes with respect to allosteric transition. Each of these mechanisms, as well as mechanisms that combine features of both, function in certain systems, and, undoubtedly, other mechanisms also exist.

It would be surprising if regulation of all enzymes occurred by identical mechanisms. It seems more probable that a variety of effective regulatory mechanisms exists. The evolutionary selection of regulatory properties would, of course, be secondary to the selection of the enzymatically active protein per se. The added structure that endows an enzyme with the ability to respond to regulatory signals must be compatible with the primitive catalytic structure. Any structure that allows efficient regulation ought to be preserved. Aspartate transcarbamylase is composed of two different types of polypeptide chains, of which there are six of each in the native enzyme. These are organized into two different types of subunits, one of which contains the catalytically active site, while the other contains the inhibitor (CTP) binding site. There are two trimeric catalytic subunits and three dimeric regulatory subunits in the native enzyme. The role of the two types of subunits can be demonstrated by exposing the native enzyme either to the sulfhydryl reagent, *p*-hydroxymercuribenzoate, or to high temperatures, in which case the enzyme dissociates into the two types of subunits. The catalytic subunit, when freed of the regulatory subunit, is no longer inhibited by CTP and demonstrates normal (as opposed to sigmoidal) reaction kinetics. Also, the isolated regulatory subunit binds CTP and is able to restore the regulatory property to the catalytic subunit. As opposed to aspartate transcarbamylase, threonine deaminase is composed of four identical polypeptide chains. Nonetheless, this enzyme possesses stereospecific sites for substrate (threonine), isoleucine, and valine. In contradistinction to aspartate transcarbamylase, the primary structure of the single type of polypeptide chains in threonine deaminase provides for the ultimate formation of different stereospecific binding sites. The comparative anatomy of these two enzymes serves to support the contention that effective

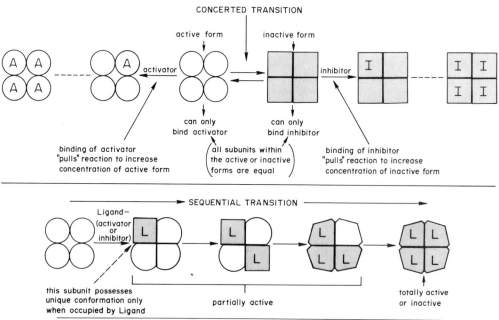

Figure 7-31. Models for concerted and sequential transitions in allosteric proteins.

regulatory properties are able to reside in grossly different types of structures.

Allosteric proteins do not function solely in regulating intermediary metabolism. Most of the genetic regulatory proteins that were previously described contain separate sites with linked function based on ligand-induced or ligand-stabilized conformational changes. The *lac* repressor, for example, possesses two distinct types of binding sites, one for the *lac* operator and the other for inducer molecules. It was previously stated that the role of the inducer is to drive the repressor from the operator. This is caused by an induced conformational change that results in a decrease in the affinity of the repressor protein for the operator. Other genetic regulatory proteins are similarly constructed, i.e., their DNA binding sites and effector binding sites are separate.

Epigenetic Modification and Protein Function

The physiologic properties of proteins, whether enzymatic, structural, or regulatory, depend upon secondary, tertiary, and quaternary structure. The folding of polypeptide chain occurs, for the most part, as a spontaneous process and is governed by the primary sequence of amino acids. Folding occurs during translation so that frequently active enzymes are associated with the translation complex (ribosome bound). This spontaneous maturation is not always a dead-end process because many times the covalent structure of the protein must be altered to allow proper function. These covalent alterations are considered as epigenetic modifications. In general, these modifications involve peptide bonds, substitutions on amino and carboxyl groups, or alteration of individual amino acid side chains. Examples of structural changes by proteolysis were described earlier in this chapter, and the extensive epigenetic modification of ribosomal proteins also was described. Several other types of modification also occur in bacteria. For example, pilin and flagellin, the protein components, respectively, of the appendage-like pili and the motive flagellae, may be phosphorylated and glycosylated (Chap. 2). Some epigenetic modifications are more critical than others. For example, the unique modification diacyl glycerol S-(sn-1-2,3-diacyl glyceryl) in thiol ether linkage with the terminal cysteine residue of the major outer membrane protein (i.e., lipoprotein) of gram-negative cells is essential for the integrity of the outer membrane of these organisms (Chap. 6).

All of these epigenetic modifications are based in the activity of enzymes produced by the bacteria. It is significant that some modifying enzymes produced by bacteria are involved in covalent modification of proteins of nonbacterial origin. The toxins produced by *Corynebacterium diphtheriae* and *Vibrio cholerae* are examples of these. Diphtheria toxin contains an enzyme activity that specifically ADP ribosylates the elongation factor (EF-2) of mammalian cells (Chap. 32), and cholera toxin specifically ADP

ribosylates the regulatory subunit of adenylcyclase in the mucosal cells of the small intestine (Chap. 38).

Regulation of Enzyme Activity by Epigenetic Modification

The role in metabolic regulation of the modification of enzyme structure by covalently associated adducts is exemplified by the adenylylation of the glutamine synthetase of *E. coli*. The central role of glutamine in the general nitrogen metabolism of the cell is summarized in Figure 7-32 and in Chapter 5. The previously described cumulative inhibition of the activity of glutamine synthetase exerted by some of the end-products for which glutamine serves as an amide donor apparently is, by itself, insufficient for the proper integration of this activity into the general metabolism of the organism. As can be seen in Figure 7-32, one of the functions of glutamine is to serve as a source of nitrogen in the conversion of α-ketoglutarate to glutamic acid. This reaction is catalyzed by glutamate synthase, another key enzyme in nitrogen assimilation. *E. coli* and many other gram-negative cells have evolved an elaborate mechanism that allows glutamine synthetase activity to respond to the levels of α-ketoglutarate and glutamine present in the cells. This mechanism, which involves the adenylylation alluded to above, also causes the catalytic potential of glutamine synthetase to respond to UTP and ATP.

Glutamine synthetase exists in two forms, adenylylated and unadenylylated. The adenylylated form of the enzyme is less active than is the unadenylylated form and also is more sensitive to inhibition by histidine, tryptophan, CTP, and AMP but less sensitive to inhibition by alanine and glycine. Adenylylation and deadenylylation of glutamine synthetase is catalyzed by a single enzyme (P_I). Adenylylation involves the transfer of AMP from ATP to form a phosphodiester bond with a tyrosyl residue in each of the 12 subunits that constitute the enzyme. Deadenylylation involves phosphorolysis of this bond to form ADP and unmodified enzyme. It is the modulation of these two ac-

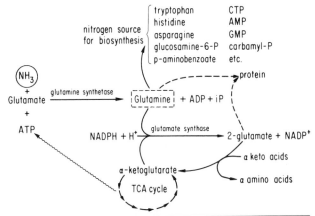

Figure 7-32. The central role of glutamine in ammonia assimilation by the Enterobacteriaceae.

tivities that provides the linkage between glutamine synthetase activity and the metabolic signals mentioned above. The adenylylation activity is stimulated by glutamine and is inhibited by α-ketoglutarate and UTP, so that when a physiologically sufficient quantity of glutamine is present, a less active glutamine synthetase is produced, but when the level of α-ketoglutarate is high, the more active form of synthetase is maintained. (Recall that α-ketoglutarate requires glutamine for its conversion to glutamate via the glutamate synthase reaction.) The effect of UTP may also be significant in that it allows glutamine synthetase to respond to a signal of excess concentration of this nucleotide. (Recall that glutamine is a source of amide nitrogen in this nucleotide.) The deadenylylating activity is stimulated by α-ketoglutarate, UTP, and ATP and is inhibited by glutamine. The physiologic significance of these effects is as explained above, but an additional feature, stimulation of the conversion of the synthetase to a more active form by ATP, may also have significance. It would provide a signal to ensure that α-ketoglutarate participates in the tricarboxylic acid cycle rather than being shunted out of this energy-yielding pathway. The adenylylation-deadenylylation activities of P_I are further regulated by an additional protein, P_{II}. This regulatory protein exists in two forms, unuridylylated and uridylylated, which are required, respectively, for the adenylylation and deadenylylation activity of P_I. The uridylylation-deuridylyation of P_{II} is catalyzed by a single protein, uridylyl transferase-uridylyl-removing (UT-UR) enzyme. The activity of this enzyme is metabolically regulated. The UT activity requires α-ketoglutarate and ATP and is inhibited by glutamine and inorganic phosphate, while the UR activity requires manganous ions alone or magnesium ions plus ATP and α-ketoglutarate. These coupled protein modifications (Fig. 7-33), each of which responds to metabolic signals, represent a cascade mechanism whereby the signals are amplified to effect the activity of a key metabolic reaction. The significance of the proper integration of this key enzyme, glutamine synthetase, is further underscored by the large amount of energy expended in effecting this regulation.

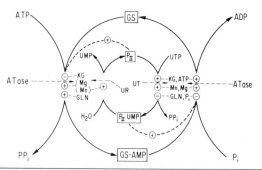

Figure 7-33. The interrelationship between the adenylylation-deadenylylation of glutamine synthase and the uridylylation-deuridylylation of the P_{II} regulatory protein and the control of these covalent modifications by various metabolites. UT, uridylyl transferase; UR, uridylyl-removing enzyme; GS, glutamine synthetase; KG, α-ketoglutarate; +, stimulation; −, inhibition. *(From Stadtman and Ginsburg: In Boyer (ed): The Enzymes, 3rd ed. 1974. New York, Academic Press.)*

FURTHER READING

Books and Reviews

Adhya S, Garges S: How cyclic AMP and its receptor protein act in *Escherichia coli.* Cell 29:287, 1982

Altman S: Biosynthesis of transfer RNA in *Escherichia coli.* Cell 4:21, 1975

Bauer W, Vinograd J: Circular DNA. J Nucleic Acid Chem 2:265, 1974

Berg OG, Winter RB, vonHippel PH: How do genome-regulatory proteins locate their DNA target sites? Trends Biochem Sci 7:52, 1982

Champoux JJ: Proteins that affect DNA conformation. Annu Rev Biochem 47:449, 1978

Cozzarelli NR: DNA gyrase and the supercoiling of DNA. Science 207:953, 1980

Englesberg E, Wilcox G: Regulation: Positive control. Annu Rev Genet 8:219, 1974

Gottesman M, Oppenheim A, Court D: Retroregulation: Control of gene expression from sites distal to the gene. Cell 29:727, 1982

Grosjean H, Fiers W: Preferential codon usage in prokaryotic genes: The optimal codon-anticodon interaction energy and the selective codon usage in efficiently expressed genes. Gene 18:199, 1982

Holmes WM, Platt T, Rosenberg M: Termination of transcription in *E. coli.* Cell 32:1029, 1983

Jacob F, Monod J: Genetic regulatory mechanisms in the synthesis of protein. J Mol Biol 3:318, 1961

Kim SH: Three-dimensional structure of transfer RNA. Prog Nucleic Acid Res Mol Biol 17:181, 1976

Kolter R, Yanofsky C: Attenuation in amino acid biosynthetic operons. Annu Rev Genet 16:113, 1982

Kornberg A: DNA Replication. San Francisco, Freeman, 1980

Kornberg A: 1982 Supplement to DNA Replication. San Francisco, Freeman, 1982

Koshland DE Jr: Conformational aspects of enzyme regulation. Curr Top Cell Regul 1:1, 1969

Kozak M: Comparison of initiation of protein synthesis in procaryotes, eucaryotes, and organelles. Microbiol Rev 47:1, 1983

Kreil G: Transfer of proteins across membranes. Annu Rev Biochem 50:317, 1981

Lake JA: The ribosome. Sci Am 245:84, 1981

Lewin B: Gene Expression. Eucaryotic Chromosomes, 2nd ed. New York, Wiley, 1980, vol 2

Losick R, Chamberlin M (eds): RNA Polymerase. Cold Spring Harbor, NY, Cold Spring Harbor Laboratory, 1976

Losick R, Pero J: Cascades of sigma factors. Cell 25:582, 1981

Miller JH, Reznikoff WS (eds): The Operon. Cold Spring Harbor, NY, Cold Spring Harbor Laboratory, 1978

Monod J, Wyman J, Changeux JP: On the nature of allosteric transitions: A plausible model. J Mol Biol 12:88, 1965

Mount DW: The genetics of protein degradation in bacteria. Annu Rev Genet 14:279, 1980

Nomura M, Dean D, Yates JL: Feedback regulation of ribosomal protein synthesis in *Escherichia coli.* Trends Biochem Sci 7:92, 1932

Pettijohn DE: Structure and properties of the bacterial nucleoid. Cell 30:667, 1982

Rickenberg HV: Cyclic AMP in prokaryotes. Annu Rev Microbiol 28:353, 1974

Schimmel PR, Söll D: Aminoacyl-tRNA synthetases: General features and recognition of transfer RNAs. Annu Rev Biochem 48:601, 1979

Siebenlist U, Simpson RB, Gilbert W: *E. coli* RNA polymerase interacts homologously with two different promoters. Cell 20:269, 1980

Stadtman ER, Ginsburg A: The glutamine synthetase of *Escherichia coli*: Structure and control. In Boyer PD (ed): The Enzymes, 3rd ed. New York, Academic Press, vol 10, p 755

Umbarger HE: Amino acid biosynthesis and its regulation. Annu Rev Biochem 47:533, 1978

Watson JD: Molecular Biology of the Gene, 3rd ed. New York, Benjamin, 1976

Weissbach H, Pestka S (eds): Molecular Mechanisms of Protein Biosynthesis. New York, Academic Press, 1977

Selected Papers

Adams JM, Capecchi MR: N-formylmethioninyl-sRNA as the initiator of protein synthesis. Proc Natl Acad Sci USA 55:147, 1966

Barry G, Squires C, Squires CL: Attenuation and processing of RNA from the *rplJL-rpoBC* transcription unit of *Escherichia coli.* Proc Natl Acad Sci USA 77:3331, 1980

Bird RE, Louarn J, Mortascelli J, et al.: Origin and sequence of chromosome replication in *Escherichia coli.* J Mol Biol 70:549, 1972

Bonitz SG, Berlani R, Coruzzi G, et al.: Codon recognition rules in yeast mitochondria. Proc Natl Acad Sci USA 77:3167, 1980

Burgess RR, Travers AA, Dunn JJ, et al.: Factor stimulating transcription by RNA polymerase. Nature 22:43, 1969

Burton ZF, Gross CA, Watanabe KK, et al.: The operon that encodes the sigma subunit of RNA polymerase also encodes ribosomal protein S21 and DNA primase in *E. coli* K12. Cell 32:335, 1983

Caskey CT, Tomkins R, Scolnick E, et al.: Sequential translation of trinucleotide codons for the initiation and termination of protein synthesis. Science 162:135, 1968

Crick FHC: Codon-anti-codon pairing: The wobble hypothesis. J Mol Biol 19:548, 1966

Crick FHC, Barnett L, Brenner S, et al.: General nature of the genetic code for proteins. Nature 192:1227, 1961

DeCrombrugghe B, Chem B, Gottesman M, et al.: Regulation of lac mRNA synthesis in a soluble cell-free system. Nature [New Biol] 230:37, 1971

Gilbert W, Muller-Hill B: The *lac* operator is DNA. Proc Natl Acad Sci USA 58:2415, 1967

Glaser G, Sarmientos P, Cashel M: Functional interrelationship between two tandem *E. coli* ribosomal RNA promoters. Nature 302:75, 1983

Godson GN, Fiddes JC, Barrell BG, et al.: Comparative DNA sequence analysis of the G4 and φX 174 genomes. In Denhardt D, Dressler D, Ray D (eds): The Single-stranded Phages. Cold Spring Harbor, NY, Cold Spring Harbor Laboratories, 1978

Guarneros G, Montanez C, Hernandez T, et al.: Posttranscriptional control of bacteriophage λ*int* gene expression from a site distal to the gene. Proc Natl Acad Sci USA 79:238, 1982

Haseltine WA, Block R: Synthesis of guanosine tetra- and pentaphosphate requires the presence of a codon-specific, uncharged transfer ribonucleic acid in the acceptor site of ribosomes. Proc Natl Acad Sci USA 70:1564, 1973

Ippen K, Miller JH, Scaife F, et al.: New controlling element in the *lac* operon of *E. coli.* Nature 217:825, 1968

Jacob F, Brenner S, Cuzin F: On the regulation of DNA replication in bacteria. Cold Spring Harbor Symp Quant Biol 28:329, 1963

McLaughlin JR, Murray CL, Rabinowitz JC: Initiation factor-independent translation of mRNAs from gram-positive bacteria. Proc Natl Acad Sci USA 78:4912, 1981

McLaughlin JR, Murray CL, Rabinowitz JC: Unique features in the ribosome binding site sequence of the gram-positive *Staphylococcus aureus* β-lactamase gene. J Biol Chem 256:11283, 1981

Meselson M, Stahl FW: The replication of DNA in *Escherichia coli.* Proc Natl Acad Sci USA 44:671, 1958

Nakajima N, Ozeki H, Shimura Y: Organization and structure of an *E. coli* tRNA operon containing seven tRNA genes. Cell 23:239, 1981

Nichols JL: Nucleotide sequence from the polypeptide chain termination region of the coat protein cistron in bacteriophage R17 RNA. Nature 225:147, 1970

Nirenberg MW, Leder P: RNA and protein synthesis: The effect of trinucleotides upon the binding of sRNA to ribosomes. Science 145:1399, 1964

Nirenberg MW, Matthei JH: The dependence of cell-free protein synthesis in *E. coli* naturally occurring or synthetic polyribonucleotides. Proc Natl Acad Sci USA 47:1588, 1961

Nordheim A, Rich A: The sequence $(dC-dA)_n \cdot (dG-dT)_n$ forms left-handed Z DNA in negatively supercoiled plasmids. Proc Natl Acad Sci USA 80:1821, 1983

Pardee AB, Jacob F, Monod J: The genetic control and cytoplasmic expression of "inducibility" in the synthesis of β-galactosidase by *E. coli.* J Mol Biol 1:165, 1959

Roberts JW: Termination factor for RNA synthesis. Nature 224:1168, 1969

Silhavy TJ, Shuman HA, Beckwith J, et al.: Use of gene fusions to study outer membrane protein localization in *Escherichia coli.* Proc Natl Acad Sci USA 74:5411, 1977

Silverstone AE, Magasanik B, Reznikoff WS, et al.: Catabolite sensitive site of the *lac* operon. Nature 221:1012, 1969

Steitz JA, Jakes K: How ribosomes select initiator regions in mRNA: Base pair formation between the 3′ terminus of 16S rRNA and the mRNA during initiation of protein synthesis in *Escherichia coli.* Proc Natl Acad Sci USA 72:4734, 1975

von Ehrenstein G, Weisblum B, Benzer S: The function of sRNA as amino acid adapter in the synthesis of hemoglobin. Proc Natl Acad Sci USA 49:669, 1963

Warren SG, Edwards BFP, Evans DR, et al.: Aspartate transcarbamylase from *Escherichia coli*: Electron density at 5.5Å resolution. Proc Natl Acad Sci USA 70:117, 1973

Watson JD, Crick FHC: A structure for deoxyribose nucleic acid. Nature 171:737, 1953

Yanofsky C, Carlton BC, Guest JR, et al.: On the collinearity of gene structure and protein structure. Proc Natl Acad Sci USA 51:266, 1964

CHAPTER 8

Genetic Variation and Gene Transfer

Although the hydrogen-bonded, helical structure of DNA imparts an extraordinary degree of stability to such a large molecule, this structure can be altered. Alterations, regardless of how subtle they may be, are passed, by virtue of the semiconservative mode of DNA replication, from generation to generation. Alterations or mutations in the DNA, if they persist, are usually reflected in a concomitant change in some property of the cell. Common types of mutations encountered in microbiology affect easily recognizable properties, such as nutritional requirements, morphology, and susceptibility to antibiotics and bacteriophage. Mutation in bacteria is not necessarily a dead-end process. Many well-described systems exist that provide for gene transfer among bacteria. These systems allow the assortment of mutationally derived characters.

Mutation and Variation

Mutations may occur spontaneously, or they may be induced by agents (mutagens) that directly or indirectly cause alterations in the structure of DNA. Spontaneous mutation frequencies are generally low and occur with a probability of from 10^{-7} to 10^{-12} per organism. Events leading to potential mutations occur at a somewhat higher frequency, but many of these alterations do not persist because of repair mechanisms present in the organism. These mutation-mitigating mechanisms will be discussed later.

Organisms with altered characteristics, because of the low frequency of their occurrence, are difficult to detect in the absence of environments that allow them to multiply in preference to the overwhelming number of nonmutated cells. An organism that acquires a nutritional requirement by mutation would be difficult to recognize against an overwhelming background of nonmutated cells. This is so regardless of the presence or absence of the required nutrilite. An antibiotic-resistant organism, on the other hand, will rapidly replace a population of susceptible cells when the inhibitor is present in the medium. When both the wild-type organism and the mutant can multiply, the population will be enriched for the faster growing organism, with the result that the original population will eventually be displaced by the progeny of the more fit organism. In this case, the conversion of the population will not occur as rapidly as in the case of the antibiotic resistance previously mentioned because the selective pressure is not as strong. Nonetheless, because of the usually short bacterial generation time, a relatively short period will allow the mutation-imparted growth advantage to be compounded through many generations, with the result that the progeny of the mutated cell will prevail. This population conversion represents an example of darwinism, i.e., the appearance of a different organism by mutation and selection or genetic adaptation.

Mutagenesis

As was mentioned above, mutation frequencies can generally be increased by a variety of agents. It may not always be clear, however, whether a given agent is acting as a mutagen or providing an environmental background against which a mutant phenotype is preferentially expressed (selection). In order to answer this question, the fluctuation test of Luria and Delbrück may be used. This test was originally used to determine whether the origin of bacteriophage resistance in *Escherichia coli* was darwinian or lamarckian, that is, by spontaneous mutation and selection or by acquired immunity. This procedure was designed to determine whether bacteriophage-resistant organisms appeared prior to or following exposure to phage. A typical fluctuation test works as follows: When a series of Petri plates seeded with bacteriophage are spread with approximately 10^9 cells from a single culture, a narrow fluctuation occurs in the number of resistant colonies per plate. However, when the original culture is used as the source of a small inoculum (approximately 1000 cells) for a series of cultures (100) and allowed to grow, subsequent plating of 10^9 cells from the individual cultures shows a wide fluctua-

tion in the number of resistant colonies per plate. These results are explained by the occurrence of random mutations in the individual cultures, the resistant mutants in some arising early and producing large clones within the cultures, and others arising late and therefore forming smaller clones. This experiment is depicted in Figure 8-1.

In general, there are four types of alterations that occur in the nucleotide sequence of DNA: (1) deletions, the loss of one or more nucleotides, (2) additions, the acquisition of one or more nucleotides, (3) transversions, the substitution of a purine for a pyrimidine and vice versa, and (4) transitions, purine/purine or pyrimidine/pyrimidine substitutions.

The effects of mutation are potentially unlimited. They may affect the properties of DNA with respect to its replicative function as well as information transfer. Most of the DNA is involved in specifying the primary structure of proteins, and, therefore, the probability is high that alterations in the nucleotide sequence will be expressed as alterations in proteins. It is now clear that aberrations produced by mutation can have multiple effects on the process of protein synthesis. Consider, for example, a deletion of nucleotides within a gene that, in addition to inactivating the gene product, may cause premature polypeptide chain termination by generating a nonsense codon. The fusion of the internal 3'-OH and 5'-OH ends created by the deletion may result in the production of a nonsense codon or in alteration of the reading frame that ultimately generates nonsense codons. Nonsense codons, when located in polygenic mRNA molecules, have a pleiotropic* effect on the expression of cistrons that are located distal to the nonsense codon and the initiating codon, i.e., in the downstream direction of translation. These effects are polar, in that they cause decreased expression of distal genes. The reason for this perturbed expression is at least partially explained in terms of chain termination and ribosome dissociation, which cause a decrease in the rate of the initiation of translation of the next gene in the polygenic mRNA. Additions of nucleotides have similar effects. The most commonly encountered consequence of additions is alteration of the reading frame. It should be noted that additions and deletions can also correct previously generated frameshift mutations.

Transversions and transitions also may cause nonsense mutations, but the probability of creating missense mutations is greater. A missense mutation is one in which the triplet code is altered so as to specify an amino acid different from that normally located at a particular position in a protein. The potential consequences of missense mutations are many. For example, the alterations can range from total inactivation of an enzyme to more subtle changes in catalytic ability expressed in such parameters as

---

* A pleiotropic mutation is one that affects more than one phenotype. In the case stated above, the nonsense mutation causes (1) inactivation of the gene in which it is located and (2) decreased expression of genes located between it and the 3'-OH end of the polycistronic mRNA.

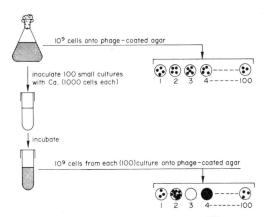

Figure 8-1. A fluctuation test. The results of this experiment show that phage-resistant mutants appear spontaneously, prior to exposure to phage. The number of resistant mutants in each sample from a single culture is relatively constant. When several cultures are sampled, however, the number of mutants varies over a wide range, indicating that the mutants arise at different times in the course of the growth of the individual cultures. The time between the appearance of the mutant and sampling governs the number of resistant organisms per culture.

enzyme-substrate apparent dissociation constants (Km) or maximum catalytic ability (V_{max}). However, even in cases where enzyme activity is altered, gross topologic changes do not occur, and these inactive enzymes are detectable with antibodies prepared against the native enzyme. Because of their serologic cross-reactivity these enzymatically inactive proteins have been termed "cross-reactive material" (CRM).

Missense, or partial missense, mutations also may give rise to temperature-sensitive gene products. A missense mutation may, for example, alter the internal bonding of a protein in such a way that its secondary or tertiary structure is compatible with biologic function at low, but not high, temperatures. The elevated temperature induces conformational changes that are incompatible with biologic function, and these mutations are termed "conditional." Many types of conditional mutants exist in which the function of a gene is normal or sufficient for survival under one set of circumstances but nonfunctional under different circumstances.

Mutagens

Many physical and chemical agents are mutagenic, and although their modes of action may differ, they possess the common ability to alter the nucleotide sequence in DNA. Mutagens exert their effect by promoting errors in replication or in repair of DNA.

The foremost physical mutagenic agents are ultraviolet light and high-energy ionizing radiation. The primary effect of ultraviolet light on DNA is the production of pyrimidine dimers caused by linking the 5,6 unsaturated bonds of adjacent pyrimidines to form a cyclobutane ring (Fig. 8-2). These dimers are stable and can be quantified

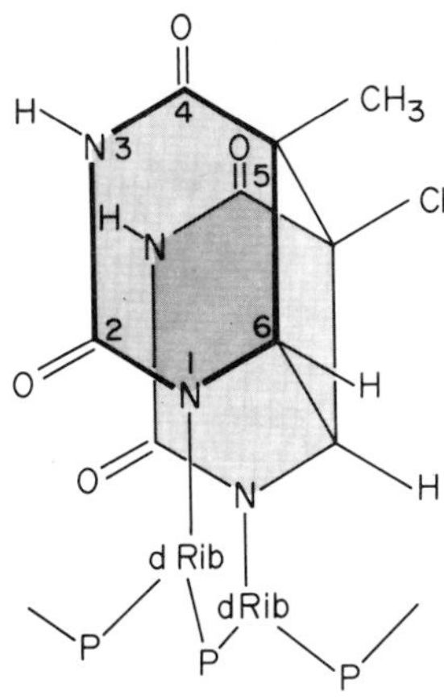

Figure 8-2. Pyrimidine dimer.

following irradiation of a bacterial culture and extraction of the DNA. The relative abundance of the three types of pyrimidine dimers is 50 percent thymine-thymine, 40 percent thymine-cytosine, and 10 percent cytosine-cytosine. The production of pyrimidine dimers within a DNA strand is by far the greatest, although dimers formed between pyrimidines on adjacent strands occur at a frequency of about 10 percent.

The mutations that are caused by ultraviolet irradiation of bacterial cells are not directly caused by the presence of pyrimidine dimers but rather result from an error-prone process of postreplication repair of dimer-distorted regions on the DNA. The various functions that constitute this repair process are normally repressed in cells containing undamaged DNA but are induced by damaged DNA. Because of the inducible nature of these functions, this type of repair has been termed "SOS repair" (after the international distress signal). This process and other mechanisms of DNA repair are discussed below.

Ionizing radiations have greater penetrance than ultraviolet radiation and are effective by virtue of their ability to produce free radicals that tend to labilize molecules. This type of radiation is particularly efficient in causing single-stranded breaks in the DNA molecule and produces a high incidence of multisite (deletion) mutations.

The discovery of mutagenic effects produced by various types of radiation provided the initial tool for experimentally altering genetic material. However, owing, at that time, to the relatively ill-defined manner by which radiation causes genetic aberration, agents with more precisely defined functions were sought so that more precise modifications could be promoted. A wide variety of chemicals is now known to be mutagenic, and a great deal of experimental evidence has accumulated to provide insight into their mode of action.

The chemical mutagens can be classified according to the manner in which they alter nucleotide sequences. A general classification of chemical mutagens is as follows: (1) agents that alter the pyrimidines or purines so as to cause errors in base pairing or that labilize the bases to spontaneous chemical modification, (2) agents that interact with the DNA and its secondary structure, producing local distortions in the helix and, therefore, promoting replication and recombination errors, and (3) base analogs that are incorporated into the DNA and cause replication errors. Examples of the first class are nitrous acid and alkylating agents. Nitrous acid deaminates adenine to form hypoxanthine. Adenine pairs with thymine during replication, but hypoxanthine pairs with cytosine, therefore causing a transition from an AT to a GC pair. Such agents as ethylethane sulfate or nitrogen and sulfur mustards cause alkylation at the 1 and 7 position of guanine. The alkylation may cause pairing errors or induce the slow hydrolysis of the purine, which results in depurination and gap formation in the DNA.

The foremost representatives of the second class of chemical mutagens are the acridine dyes, such as proflavine and acridine orange. These, as well as other planartype compounds, are able to intercalate between the stacked bases that form the DNA helix and distort this structure in such a way that recombination between homologous chromosomes is attended by occasional insertions or deletions of bases. Additions and deletions cause a shift in the reading frame and, therefore, have profound effects on the gene product.

The mutagenic activity of intercalating dyes is greatly enhanced by adding an alkylating function to the molecule. These compounds, which are known as nitrogen half-mustards, are typified by the compound ICR 191 (Fig. 8-3 and Table 8-1).

An example of a widely used mutagenic base analog is 5-bromouracil, which, when in the keto form, functions as an analog of thymine and pairs with adenine. However, rare tautomerization to the enol form occurs, and the analog is then more apt to pair with guanine. This copy error results in a transition from an AT pair to a GC pair (Fig. 8-4). Table 8-1 lists the action of several mutagens that are widely used in studying bacteria and their viruses.

Intrinsic Mutagenesis and Insertion Sequences

Apparent spontaneous mutation frequencies can be increased by alteration of so-called mutator genes (*mut*). For example, *E. coli* contains at least six different genetic loci within which mutation increases mutation frequencies. Although the genetic aberrations enhanced by some of these *mut* genes are known (viz., *mutT* causes AT to CG transversions), the responsible elements have not been identified. Mutations of *mut* lead to the loss of functions

TABLE 8-1. MUTAGENS USED FOR BACTERIA

Mutagen	Apparent in vivo Specificity	Additional Advantages	Disadvantages
2-AP (2-aminopurine)	Transitions AT $\rightleftharpoons$ GC	—	In some cases a weak mutagen
5-BU (5-bromouracil)	Transitions AT $\rightleftharpoons$ GC	—	Must depress normal thymine incorporation; weak mutagen
NH$_2$OH	Transitions AT $\rightleftharpoons$ GC	—	In some cases a weak mutagen
Sodium bisulfite	Specific transition GC $\rightarrow$ AT	—	Weak mutagen
Mutator gene (*mutT*)	Specific transversion AT $\rightarrow$ CG	No treatment required	Genetic construction required
EMS (ethylmethane sulfonate)	Specific transition GC $\rightarrow$ AT	Powerful mutagen	Very dangerous to handle
NG (nitrosoguanidine)	Specific transition GC $\rightarrow$ AT induces small deletions at very low rate	Very powerful mutagen	Very dangerous to handle, frequent secondary mutations
ICR 191	Frameshifts, small insertions and deletions	Powerful mutagen	Compound difficult to obtain
Nitrous acid	Transitions and probably transversions, deletions	—	High amount of killing required for good mutagenesis
UV (ultraviolet radiation)	Transitions and transversions, deletions, possibly stimulates insertions and chromosomal rearrangements	—	High amount of killing required for good mutagenesis, certain strains too sensitive
Mu-1 phage	Insertions, some deletions	Random induction of nonleaky, polar, nonreverting mutations	—
Spontaneous (no mutagen)	Transitions, transversions, insertions, deletions (frameshifts)	Wide spectrum of mutations, no complications due to secondary mutations	Low level of mutants, many siblings in each culture

Provided by J.H. Miller, Modified from Miller: Experiments in Molecular Genetics, 1972. Cold Spring Harbor, NY, Cold Spring Harbor Laboratory.

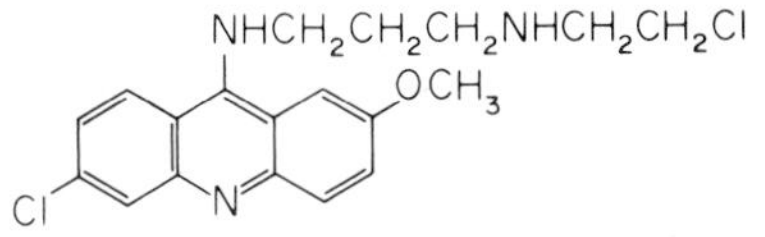

NHCH₂CH₂CH₂NHCH₂CH₂Cl

OCH₃

Cl

Figure 8-3. The frameshift mutagen ICR 191 (Institute for Cancer Research, compound 191). The planar, acridine-type compounds are not usually mutagenic in bacteria. However, when an alkylating function is introduced into these intercalating agents, they acquire the ability to cause the addition and deletion of bases and consequent frameshifts.

that normally participate in the repair of genetic damage. Most of the *mut* loci in *E. coli* are thought to be normally involved in correcting mispaired bases (mismatch repair). A mutator function specified by coliphage T4 has been described. Alteration in the DNA polymerase specified by this phage causes transversions of GC to TA. The error-prone nature of DNA replication by this enzyme is caused by the mutational loss of its 3′-OH to 5′OH exonuclease activity, which normally provides a proofreading role (p. 121).

An additional mechanism for bringing about gene inactivation is by the insertion of extraneous nucleotide sequences within the gene. It is known that a large number of so-called spontaneous mutations are accounted for by the translocation of insertion sequences from one region of a genophore (chromosome, plasmids) to another region of the same or to a different genophore. When the recipient region for these 800 to 1400 base pair long insertion

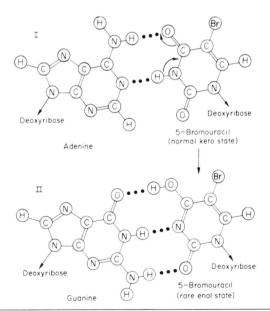

Figure 8-4. Proper and spurious base pairing with the tautomers of 5-bromouracil. 5-Bromouracil is a thymine analog, which, when in its normal keto state, pairs with adenine but, when in its rarer enol state, pairs with guanine. Replication of 5-bromouracil-containing DNA is occasionally accompanied by the mispairing of the analog with guanine, with the result that mutant daughter DNA molecules are produced.

sequences is the nucleotide sequence of a functional gene, inactivation of the gene occurs. Several types of insertion sequences, each with a unique nucleotide arrangement, have been described. Among these are IS1, IS2, IS3, and IS4, all of which are active in *E. coli*. These elements are normal constituents of chromosomes and plasmids. It is known, for example, that the *E. coli* chromosome contains approximately 10, 5, and 3 copies, respectively, of IS1, IS2, and IS3. The insertion of IS elements within a gene has pleiotropic effects. If an IS element is present in a gene within an operon, in addition to inactivating the gene, it will exert a strong polarity effect on the expression of distally located genes within the operon. The basis for this effect varies depending upon the IS element present. IS2, for example, possesses a rho-dependent transcription termination site and, therefore, causes premature termination of polygenic mRNA synthesis. Other IS elements exert polarity effects by virtue of the fact that their nucleotide sequence is transcribed as a continuum with the polygenic mRNA of the operon in which they exist, with the result that frameshift-generated termination codons will be encountered upon subsequent translation. The IS2 element exerts polar effects only if inserted within an operon in one of its two possible orientations. The reason for this orientation-specific effect is that IS2 contains a promoter sequence that is active in one orientation (i.e., the nonpolarigenic one). This property of IS2 endows it with the ability to turn on otherwise dormant genes when it is appropriately transposed. As might be expected, IS elements can disintegrate from genophores. This loss of the IS is frequently accompanied by deletion as well as rearrangement of neighboring nucleotide sequences.

The type of gene splitting effected by IS elements can also occur by the integration of episomes or parts of episomes within genes. These processes are described more fully later in this chapter.

Bacteria as Indicators of Mutagenic/Carcinogenic Compounds

The ease and economy with which bacterial systems serve as experimental tools have made them ideal for use as indicators of the mutagenic activity of various chemicals. Bacterial systems have been used to demonstrate the mutagenic effect of a number of known carcinogens. Ames and his collaborators, employing a series of well-defined histidine auxotrophs of *Salmonella typhimurium*, have shown that a number of carcinogenic compounds are capable of increasing the frequency of reversion to histidine sufficiency. In practice, these tests are performed simply by spreading an auxotrophic strain of the bacterium on an agar plate containing minimal medium to which the suspected mutagen has been added. An increase in the spontaneous reversion rate is assessed by counting the revertant colonies that subsequently appear on the plate. A pharmacologic aspect can be introduced into this system by incorporating into the medium sterile microsomal preparations derived from various tissues. For example, it has been shown that cigarette smoke condensates display powerful mutagenic activity only if lung or liver microsomes are present in the test medium. The results of this approach to testing the mutagenic effects of carcinogens, together with other results, lend support to the theory that cancer can be caused by somatic mutations.

Repair of Genetic Damage

The bacterial cell is not defenseless against genetic damage, since a number of enzymes are present that are able to participate in the repair of at least some types of aberrant gene structure. The type of DNA repair that has received the most attention is reversal of damage induced by ultraviolet irradiation or, more specifically, reversal of pyrimidine dimer-imparted distortions.

Pyrimidine dimers can be removed by two different processes. One process occurs only in the light and is known as "photoreactivation" or "light repair," and the other, which does not require light, is known as "dark repair." Photoreactivation occurs by separation of pyrimidine dimers by an enzyme (photolyase) that splits the cyclobutane ring that joins the bases to regenerate the original pyrimidines. The enzyme performing this function binds to the pyrimidine dimers but is inactive unless irradiated with long ultraviolet or visible light.

The mechanism of dark repair, unlike photoreactivation, requires the participation of many enzymes. Two distinct processes are involved: (1) removal of pyrimidine dimer and (2) reestablishment of the continuity of the DNA molecule. Removal of pyrimidine dimers occurs in two phases, the first of which is an incision by an endonuclease to produce a single-stranded nick bordered by a 3'-OH nucleotide and the 5'-phosphoryl pyrimidine dimer. The enzyme, correndonuclease II, that effects incision is specified by the *uvrA* and *uvrB* loci. Inactivation of these functions by mutation prevents ultimate excision of pyrimidine dimers and renders the cell extrasensitive to UV irradiation. The initial single-strand scission exposes the pyrimidine dimer to an exonucleolytic activity that ultimately removes the dimer along with other nucleotides. This single-stranded nick also is a substrate for polynucleotide ligase and may be sealed prior to removal of the dimer. However, another protein, which is the product of the *uvrC* locus, is conjectured to function in repair by specifically impeding ligase-mediated repair of the original scission and, therefore, allowing the competing exonucleolytic activity to function. Mutations in *uvrC* also cause increased sensitivity to UV irradiation. Several exonucleolytic activities have been described that are potentially capable of removing the pyrimidine dimer exposed by the incision process described above. Among these is polymerase I, which possesses a 5' to 3' exonucle-

ase activity. Repair is completed by filling the postincision gap by the action of DNA polymerase, using the bordering 3'-OH nucleotide as primer and the intact strand as template. Once the gap is filled, the remaining single-stranded nick is sealed by the action of polynucleotide ligase.

Analysis of DNA that has undergone pyrimidine dimer excision and repair reveals that the patched regions of the chromosome may contain either short (10 to 30 nucleotides) or long (1000 to 3000 nucleotides), single-stranded stretches of newly synthesized DNA. The process of short patch repair predominates and results from the concerted polymerizing and 5' to 3' exonucleolytic activities of polymerase I. These concerted activities result in nick translocation, which is a well-documented property of polymerase I.

The less frequently occurring long patch pathway of repair results from extensive gap widening brought about by combined activities, including a single-strand nuclease activity specified by the *recBC* genes. This enzyme, exonuclease V, is also involved in the process of genetic recombination (p. 181). The gap is then filled by the action of polymerase II or III, which have been shown to initiate synthesis using the 3'-OH border of gaps as primer. DNA-binding protein (p. 122) is also involved in this process. Its role is two-fold—protecting the long, single-stranded regions from nucleases and preventing secondary folding of these regions. This latter function is necessary for processive* polymerization over long stretches of template. Other functions, including that of the genetic recombination *recA* gene product, are also involved in this process. Figure 8-5 illustrates some features of these excision repair processes.

Excision repair of pyrimidine dimers is not found exclusively in microorganisms, and many of the enzymes described above are found in eucaryotic organisms, including humans. It is noteworthy that patients with xeroderma pigmentosum, a heritable disease that imparts extreme ultraviolet sensitivity to the skin, are defective in their ability to repair pyrimidine dimer-containing DNA.

Although the above discussion of repair mechanisms dealt with reversal of damage to DNA caused by ultraviolet irradiation, other types of damage can similarly be reversed. It is known, for example, that other classes of excision enzymes are capable of initiating removal of nucleotides containing monoadducts, such as alkylated bases. The repair of these lesions apparently involves the postincision mechanisms described above for pyrimidine dimers.

Bacteria, as well as other types of cells, possess additional DNA repair systems based on N-glycohydrolases and apurinic and apyrimidinic endonucleases (AP-endonucleases). The glycohydrolases serve to remove aberrant

---

* A processive polymerase is one that effects continuous elongation of the polymer without being released from its primer-template complex. A processive nuclease is one that effects sequential hydrolysis of phosphodiester bonds without dissociating from its polymeric substrate.

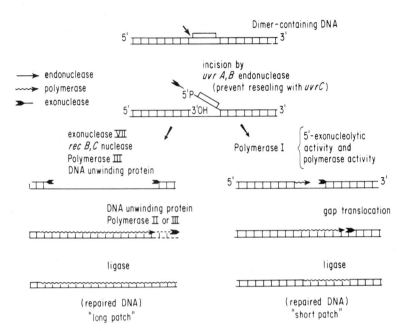

Figure 8-5. The enzymatic repair of pyrimidine dimer-containing DNA. The pathway on the left represents short patch repair. The pathway on the right represents long patch repair.

bases from the DNA and the AP-endonucleases remove the debased sugars. The glycohydrolases are generally specific in their action. For example, one of these removes uracil that results from the misincorporation of deoxyuridine into DNA in place of deoxythymidine or that arises from the deamination of cytosine. A separate enzyme functions to remove hypoxanthine that results from dITP incorporation or from deamination of adenine. Other enzymes for removing alkylated bases also have been described.

SOS Functions and DNA Repair

Mutant bacteria that lack the *uvr* pathways of repair, although being more sensitive to UV irradiation than their nonmutated counterparts, carry out another type of repair process that contributes to their survival. This process is known as "postreplication repair." It essentially removes pyrimidine dimers by recombination, is error prone, and is the basis for UV mutagenesis. Postreplication repair is one of many processes that is influenced by the SOS regulatory system; among others are excision repair, filamentation (cell division), prophage induction, cessation of respiration, and alleviation of host-controlled restriction. The participating genetic loci for these phenomena are directly or indirectly under the control of the *lexA* protein, which functions as a repressor. The SOS regulatory system is activated by agents that damage DNA, including UV irradiation and alkylating agents. Damaged DNA or its products convert the *recA* protein into a protease that cleaves the *lexA* protein to destroy its repressor function. This results in activation of the genes (including *recA*) that are under the control of *lexA*. Other aspects of this system are described under the section on genetic recombination.

Genetic Suppression

The previously described repair mechanisms reverse genetic aberration and restore the original phenotypic expression of the affected gene. Other mechanisms are available that do not correct the original mutation per se but that are capable of restoring the original phenotype. This occurs by the action of suppressor mutations. Suppression is defined as the reversal of a mutant phenotype by another mutation at a position on the DNA distinct from that of the original mutation. The presence of suppressor mutations is ascertained by demonstrating, with appropriate genetic analysis, that the original mutation can be recovered independently of the suppressor mutation, and vice versa.

Suppression is a generic term, in that its molecular basis may be varied. Suppressor mutations may be categorized as follows:

1. The suppressor mutation may open an alternate pathway for the production of a product, the synthesis of which is prevented by the primary mutation. A corollary to this type of suppression is that in which the second mutation augments a low residual activity caused by the primary mutation.
2. The suppressor mutation may be located in a gene whose mutated product may replace that of the affected gene.
3. The suppressor mutation may alter the internal environment of the cell so that the function of the product of the mutated gene is restored to normal.
4. The suppressor mutation may introduce an additional alteration that completely or partially ne-

gates the effect of the primary mutation. This latter case includes not only interactions between missense mutations but also corrections of frameshift mutations by subsequent additions or deletions.

5. The suppressor mutations may alter the properties of one of the factors that participate in the transfer of information from DNA to protein.

The first three mechanisms represent indirect suppression in which the consequence of the primary mutation is circumvented rather than corrected. The last two mechanisms provide for a correction within the product of the mutated gene. This is direct suppression.

Indirect Suppression

Several examples of indirect suppression exist. Examples of 1, 2, and 3 listed above are as follows:

1. A *Neurospora* mutant with an effective decrease in synthesis of carbamylphosphate requires pyrimidines for growth. Carbamylphosphate is also used for arginine biosynthesis. If a mutation that lowers the activity of ornithine transcarbamylase, the enzyme utilizing carbamylphosphate for arginine biosynthesis, is introduced into the former mutant, the pyrimidine requirement is suppressed. Apparently, the second mutation introduces a sparing effect on carbamylphosphate that allows this compound to be diverted from arginine to pyrimidine biosynthesis.

2. A mutation in one of two genes that code for the two polypeptide chains that form isopropylmalate isomerase, an enzyme specifically involved in leucine biosynthesis in *S. typhimurium*, can be suppressed by mutations far removed from the leucine locus. It is thought that this suppressor mutation allows a surrogate polypeptide chain to replace the mutated, nonfunctional polypeptide chain in the isomerase.

3. A mutation in *Neurospora* produces a tryptophan synthetase that is extremely sensitive to zinc and is therefore inactive. (All culture media contain traces of zinc as an impurity.) This mutation is suppressed by a mutation that prevents the organisms from accumulating the toxic metal, thus providing an environment in which the zinc-sensitive enzyme is active.

Direct Suppression

As pointed out above, direct suppression results in an alteration of the mutated gene product so that it is again functional. Direct suppression may be intragenic or intergenic. Both types have been extensively investigated.

Intragenic Suppression. Intragenic suppression requires two distinct mutations within a single gene, and the effect of one mutation cancels the other. An example of

this type of suppression is a mutation in the tryptophan synthetase A protein of *E. coli*, which causes a substitution of a glycine with a glutamic acid residue in position 120 of the polypeptide chain and inactivation of the protein. Another mutation causes a replacement of a tyrosine with a cysteine residue in position 175 of the polypeptide chain. This second mutation cancels the effect of the first mutation and restores activity. In this particular case the amino acid substitutions have reciprocal effects, i.e., the glutamic acid or cysteine substitution alone produces an inactive protein.

Intergenic Suppression. The most extensively studied case of intergenic suppression is that involving the nonsense codons UAA, UAG, and UGA. These triplets, when present in mRNA, cause termination of polypeptide chain extension. This characteristic is suppressible by mutations that alter elements that participate in the translation process. The molecular basis of this suppression is relatively well understood and involves mutational alteration of tRNA.

The most straightforward way of suppressing nonsense codons would be by altering the anticodon of a tRNA molecule so as to cause reading of the otherwise undecipherable triplet. This mechanism would seem to require that the cell contain duplicate tRNA molecules, each of which reads the same triplet. Otherwise, any benefit derived by an anticodon alteration allowing nonsense codon translation would be negated by inability to read the normal, cognate codon. The *supIII* gene is an example of how this problem is solved. A mutation in the *supIII* gene, which is one of two tandem genes for a minor species of tRNA^{tyr}, causes suppression of UAG (amber) nonsense codons. Tyrosine-accepting tRNA from *supIII*$^-$ and *supIII*$^+$ strains has been sequenced.* The results of these analyses reveal that the only alteration in the *supIII* product is in the anticodon loop of the molecule, where AUG in the wild type is changed to AUC in the mutant. The anticodon AUC reads the nonsense codon UAG. (Note that, by convention, nucleotide sequences are expressed in a 5′ to 3′ sequence, and therefore one of the triplets must be reversed in order to display the proper base pairing.) The result of the recognition of the UAG codon by the mutated tyrosine-accepting tRNA is the insertion of tyrosine into the polypeptide chains at the position of the nonsense codon-mediated chain termination.

Suppression of nonsense codons does not always require alterations in the anticodon of suppressor tRNA molecules. The nonsense codon UGA is suppressed by mutations in the gene that specifies tryptophan-accepting tRNA. This suppressor mutation results in an A to G sub-

---

* Note that the suppressor allele designation is the reverse of genetic convention. Classically, wild-type alleles are designated with a (+) and mutated alleles with a (−). Suppressor alleles are designated as *sup*$^+$ when active in suppression and *sup*$^-$ when inactive. This is so regardless of their wild-type or mutated character.

stitution in the dihydrouracil arm, far removed from the anticodon. It is thought that this alteration in tRNA[trp] increases wobble (p. 134) and allows C to pair with A while also maintaining normal CG pairing. This type of change in the tRNA[trp] molecule probably is the only one that is acceptable for viability of the organism because there is a single gene for tRNA[trp]. This situation provides an interesting contrast with the case of the tRNA[tyr] described above.

Figure 8-6 shows how the nonsense codons arising from sense codons are suppressed by mutant tRNAs. It is evident that suppression by this pathway results either in complete restoration of the amino acid sequence of the polypeptide or in substitution of an amino acid that is compatible with the biologic function of the protein. Similar suppression mechanisms exist for the remaining nonsense codon, UAA, as well as for missense mutations. Mechanisms for suppressing frameshift mutations also are available. One of these involves an additional nucleotide in the anticodon for phenylalanine tRNA, in which case the anticodon reads UUUC rather than UUU as the codon for phenylalanine. This results in restoration of the reading frame that was shifted by the insertion of an extra base.

Phenotypic Suppression

Mutations may also be suppressed by nonmutational alteration of the components of information transfer. For example, sublethal concentrations of streptomycin and other related antibiotics that cause misreading of the genetic code may allow missense or nonsense codons to be translated as sense codons. This phenomenon is described in detail in Chapter 9.

Most agents capable of causing phenotypic suppression are effective at low concentrations. Higher concentrations are bactericidal or bacteriostatic. The ambiguities introduced by suppressing compounds, while being sufficient to return the function of a mutated gene, do not seriously detract from the transfer of information from normal genes. Bacterial systems, in most cases, possess supernumerary quantities of gene products, and a low level of ambiguity is sufficient to produce an effective quantity of functional gene product without effectively lowering the level of the functional product produced from a non-mutated gene.

Gene Transfer

Zygote formation, in eucaryotic cells, results from the fusion of the gametes of sexually compatible members of the same species. This sexual union contributes to invigoration of the species and speciation by allowing reassortment of traits acquired by mutation. Zygotes are formed in bacteria by more primitive means of gene transfer. Three general mechanisms are known whereby the genetic information of one bacterial cell can be introduced into a recipient bacterial cell. These are characterized by the vectors of transfer. The most primitive mode of transfer occurs by the process of transformation, in which naked DNA derived from one cell is taken up by another cell. Another type of relatively primitive transfer is accomplished by virus infection; this process is termed "transduction." A third mode of transfer, which is probably most analogous to the fusion of gametes, is conjugation, where genetic transfer is preceded by cell-to-cell contact. It should be emphasized that this latter process does not involve cell fusion. Incomplete, rather than full, diploids result from gene transfer in bacteria. These partial diploids are known as "merozygotes" and consist of the entire genome (endogenote) derived from the recipient cell and the incomplete genome (exogenote) derived from the donor cell. Regardless of the mode of gene transfer, merozygotes

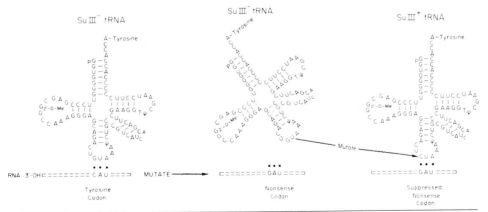

Figure 8-6. Reading of a nonsense codon by a suppressor tRNA. The product of the *suIII* gene of *E. coli* is a tyrosine-accepting tRNA. A mutation affecting the anticodon region of this tRNA allows it to read the nonsense codon UAG.

may support genetic recombination in which portions of the DNA of the exogenote effectively replace segments of the DNA of the endogenote.

Transformation

Historically, identification of the active transforming principle permitted the equating of genetic material with DNA. Avery, MacLeod, and McCarty showed that DNA purified from an encapsulated strain of pneumococcus was capable of imparting onto a nonencapsulated strain the ability to form capsules and that the resulting strain was genetically stable.

Bacterial transformation is the basis for several experimental determinations of the properties of DNA. Gene transfer by soluble DNA provides a test for the biologic activity of DNA that has been subjected to various in vitro treatments (such as with mutagens). Transformation may be used to assay the gene dosage in a given DNA solution, since the number of transformants (cells that have participated in recombination with externally added DNA) is directly proportional to the concentration of the pertinent genetic regions. In addition, transformation has been extensively used in in vitro recombinant DNA studies that have resulted in the cloning of many genetic regions (see Fig. 8-11, p. 182).

Transformation occurs naturally in a number of gram-positive and gram-negative species and can be effected by artificial means with certain other species. In brief, the process of transformation involves the binding and uptake of naked DNA molecules from the medium by the bacterial cell. The ability to perform this process is referred to as "competence." Some variation in the mechanism of DNA uptake can be expected a priori in view of the different structures of gram-positive and gram-negative cell envelopes. Upon entry into the cytoplasm, the DNA molecule must be processed in order that it can be stably inherited. In the case of transforming chromosomal DNA, this involves recombination with the host chromosomal DNA. When the transforming molecule is plasmid DNA, the molecule must be restored to its native form as a self-replicating genetic entity. A specific type of transformation with the DNA of bacteriophage is termed "transfection" and may result in replication of the viral DNA and formation of mature virus particles.

Genetic transformation has been most studied in gram-positive bacteria, including *Streptococcus pneumoniae*, *Streptococcus sanguis*, and to a lesser extent, *Bacillus subtilis*. For both *S. pneumoniae* and *S. sanguis*, cultures become competent only when a certain critical cell density is reached. This acquisition of competence actually depends upon a low molecular weight protein, termed "competence factor" (CF), that is continuously synthesized and secreted into the medium. CF reaches an effective concentration only at higher cell densities and, at this concentration, rapidly induces competence in all cells in the culture. The

mechanism of CF action is not well described, although it does appear to involve interaction with a cell–surface receptor to induce the synthesis of certain proteins that are presumably involved in transformation. Competence factors have also been described in *Bacillus,* although the acquisition of competence is limited to only 10 to 20 percent of the cell culture.

Adsorption of transforming DNA by competent cells involves first reversible then irreversible binding. Irreversibly bound DNA is still present on the cell surface, since it is sensitive to DNase, but it cannot be removed by washing or by adding excess carrier DNA. The binding reaction is specific for duplex DNA, since RNA-DNA hybrids and RNA are not bound. Denatured DNA may be bound, but this is much less efficient than for duplex DNA; for example, the transformation rate of pneumococcus with denatured DNA is 200-fold lower than with duplex DNA. Adsorption of the DNA occurs independently of the sequence of the DNA, since foreign (i.e., heterologous) DNAs are bound and taken up as efficiently as homologous DNAs. Only homologous DNAs, however, participate in subsequent general recombination with the recipient chromosome. Uptake of DNA refers to conversion of the DNA to a DNase-resistant form and may reflect entrance into the cytoplasm (as in pneumococcus) or sequestration in the periplasmic space (as in *S. sanguis* and *B. subtilis*). In the pneumococcus, uptake is initiated by a membrane-bound endonuclease that cleaves bound DNA at sites opposite from nicks that are introduced during the course of binding. One strand of the bound DNA is then passed into the cytoplasm, while the other strand is degraded into oligonucleotides that are released into the medium. The single-stranded DNA is then said to be in eclipse because if it is isolated and added to other competent cells it fails to effect transformation.

The transforming ability of the DNA is only recovered after subsequent integration into the host chromosomal DNA and restoration to duplex form. The single-stranded DNA in eclipse is noncovalently associated with a protein that appears to be formed only during the state of competence. The DNA-protein eclipse complex is quite stable, and the protein appears to protect the DNA from cellular nucleases and may also facilitate recombination with the chromosome. Similar processes occur in *S. sanguis* and *B. subtilis* except that the eclipse complex remains in the periplasmic space. At least for *S. sanguis*, integration of the transforming DNA into the chromosome occurs directly without a definite cytoplasmic intermediate, suggesting that a structure may be formed between the eclipse complex, cytoplasmic membrane, and chromosome prior to integration.

Transforming homologous DNA is integrated into the host chromosome by recombination that involves the single-stranded donor DNA displacing a homologous strand from the recipient chromosome and base pairing with its complementary strand. The integrity of the duplex chromosome must be restored by repair functions. Heterozygosity as reflected in mismatched bases may undergo correction by mechanisms that show some degree of preference for donor or recipient markers depending upon the nature of the mutated DNA sequence and possibly other factors, including the methylation patterns of the two strands. A heterozygous duplex that escapes correction and undergoes replication will generate progeny that are now homozygous for the markers in question. When the transforming DNA is plasmid DNA and bears no homology with the chromosome, a special problem exists in restoring the molecule to its closed circular form, since only a linearized single strand of the plasmid DNA is transferred to the cytoplasm. Successful transformation with plasmid DNA requires two molecules so that complementary single strands can be taken up, and the closed circular duplex is formed by homologous pairing and DNA replication. Transformation of pneumococcus with monomeric plasmid DNA accordingly shows two-hit kinetics, and, as expected, dimeric plasmids show one-hit kinetics.

Several gram-negative bacteria have natural transformation systems, with that of *Haemophilus influenzae* being the best studied. Competence is induced in *H. influenzae* by certain cultural conditions that prevent cell division but allow protein synthesis. No competence factor has been shown. Although competent *H. influenzae* can bind foreign DNAs, only donor DNA from that species or closely related *Haemophilus* species is actually taken up into the cytoplasm. This discrimination, which is not shown by the gram-positive bacteria previously described, depends upon the presence of an 11 base pair sequence, 5'-AAGTGCGGTCA-3' residing within the donor DNA molecule that is apparently recognized by recipient cell surface proteins. DNA molecules containing the recognition sequence are taken up into the cytoplasm in the duplex form. No strand degradation occurs, and no eclipse complex has been described. Incorporation of the donor DNA involves strand displacement of its homolog and general recombination. The significance of this species-specific transformation process is not known but could serve to limit the uptake of exogenous DNA to those molecules that are then capable of stable inheritance by homologous recombination with the host chromosome. A very similar system of transformation has been described in *Neisseria gonorrhoeae*, although a specific DNA recognition sequence has not yet been demonstrated.

Certain species of bacteria, such as *E. coli* K-12 and *S. typhimurium* LT2, are not capable of natural genetic transformation but can be induced to a competent state by artificial means. The acquisition of competence in these species generally involves placing the cells in a $CaCl_2$ solution that by an unknown mechanism permits the adsorption and uptake of exogenous DNA. Artificially induced transformation, unlike most forms of natural transformation, is much more efficient with plasmid DNA than with chromosomal DNA. Artificial transformation is an important phenomenon in that it is the basis of gene transfer in recombinant DNA experimentation (see below).

Conjugation

Bacterial cells frequently harbor extrachromosomal, autonomously replicating DNA molecules known as "plasmids." Two classes of plasmids exist: (1) those that replicate autonomously and (2) those that replicate autonomously but also can integrate into the chromosome and replicate as any other chromosomal character. The latter type of plasmid is usually referred to as an "episome." Some types of episomes are capable of promoting gene transfer to compatible cells and, therefore, confer sexual fertility to cells that harbor them. This episome-imparted fertility is the basis for bacterial conjugation. Although the most extensively studied conjugal system is that found in *E. coli*, it is known that interchange of genetic material by conjugation can occur in other types of gram-negative cells, and conjugal systems of gene transfer have also been described for a limited number of gram-positive cells.

The episome (fertility or F factor) that confers fertility to the *E. coli* cell is a double-stranded, circular DNA molecule with a molecular weight equivalent to about 63 million daltons. The F factor is the vector for bacterial gene transfer, and as will be explained below, only genetic information that is physically integrated with this factor will be transferred in conjugation (although, as will subsequently be explained, certain functions of the transfer apparatus can serve as helper for transmission of otherwise nonconjugal plasmids). Figure 8-7 depicts the genetic map of the F factor: some of the loci are involved in the autonomous replication of the episome, some in the process of conjugal transfer, and others in more specialized functions that contribute to persistence of the episome in its host cell. The earliest described product of F factor expression was the F pilus, which is usually present in one to three copies on the male cell surface. A large portion of the F factor *tra* operon (Fig. 8-7) codes for the assembly of the F pilus, which is actually a multimeric protein complex that appears to be arranged as a hollow cylinder. F pilus is composed of a single subunit, F pilin, that has a molecular weight of 12,000 and contains two phosphate residues, one glucose residue, and perhaps some other sugars. F pili are responsible for aggregation of male (F-bearing) and female (non-F-bearing) cells; it is within these mating aggregates that gene transfer occurs. While it has been suggested that the F pilus provides a conduit through which conjugal transfer of DNA proceeds, this hypothesis is weakened by the observations that upon mating pair formation, the F pilus undergoes rapid retraction and that most DNA transfer occurs among cells with wall-to-wall rather than pilus-to-wall contacts. Retraction of F pilus may promote formation of conjugally useful wall-to-wall contacts.

The F factor is able to exist in three states: (1) as an autonomously replicating episome, (2) as an integral part of the chromosome, and (3) as an autonomously replicating episome containing relatively small segments of material derived from the chromosome. The terminology

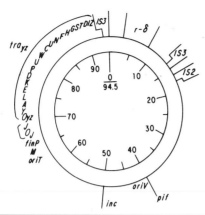

Figure 8-7. Genetic map of *E. coli* fertility (F) factor. The inner circle represents physical distance in kilobases. *traA, L, E, K, B, V, W, C, U, F, H,* and *G* are involved in the synthesis of pilin and pilus assembly. *finP* is a fertility inhibition function. *traN* and *traG* are involved in stabilization of mating pairs. *traMYGDIZ* are involved in conjugal DNA metabolism. *traS* and *T* are part of a surface exclusion system involving membrane proteins that reduce the ability of F-bearing cells to act as recipients in conjugation. *pif* renders F-bearing cells resistant to certain phages (viz., coliphage T-7, which consequently is a female-specific phage). *inc* governs incompatibility, i.e., the inability of related plasmids to coexist with F. *oriV* is the origin of vegetative replication of F, and *oriT* is the origin of replication for conjugal transfer of F. γ-δ, IS2 and IS3 are insertion sequences through which F is integrated into the chromosome to form Hfr cells.

applied to cells bearing the sex or fertility factor in these states are F$^+$, Hfr, and F′, respectively. Cells not containing the fertility factor are termed F$^-$. The types of male cells are characterized by the manner in which they effect gene transfer. The F$^+$ cell is able to transfer only the genetic information contained in the fertility factor. The only genetic alteration effected by introduction of the fertility factor into an F$^-$ cell is conversion of the F$^-$ cell to an F$^+$. This transfer does not involve loss of maleness of the F$^+$ cell. The reason for this will become obvious when the mechanism of transfer is considered. Hfr cells are capable of transferring the entire chromosome of the cell, and although this occurs rarely, relatively large portions of the chromosome are transferred at a high frequency. However, matings between Hfr and F$^-$ cells rarely result in acquisition of maleness by the F$^-$ recipients. This is one of the original observations that led to an understanding of some of the characteristics of Hfr-mediated gene transfer.

The conjugal transfer of the F plasmid is triggered by formation of a stable mating pair, perhaps through the function of the *traM*-encoded membrane protein. Conjugation involves endonuclease-mediated nicking of one strand of F plasmid DNA at the origin of transfer replication (*oriT*) and passage of the 5′-terminal single-strand DNA into the recipient. Only the 5′-terminal single strand is transmitted, and this invariant polarity is the basis of the direction of transfer of Hfr strains, as described below. The single strand is converted to duplex DNA by discontinuous DNA

replication in the recipient cell, and upon complete transfer and replication of the F sequence, circularization occurs by an unknown mechanism. As the single strand of DNA is passed into the recipient, the donor F plasmid is restored to duplex form by DNA replication. This explains why the male cell remains F$^+$ after the mating process. Donor replication of the F plasmid, however, is not the driving force for transfer of the single strand because inhibition of donor DNA synthesis does not affect F conjugation. The energy source for conjugal DNA transfer may involve supercoiling of DNA and the action of DNA gyrase, since nalidixic acid, a potent inhibitor of gyrase, also prevents conjugation.

F-mediated transfer of the Hfr chromosome proceeds in the same polarized fashion that is initiated at *oriT*. However, since the F factor is integrated into the bacterial chromosome, the transfer of the initial F sequence is followed by contiguous chromosomal sequences. As conjugation proceeds, more distal chromosomal sequences are transferred in a continuous linear manner. Because of random shearing of the mating aggregate and interruption of conjugation, a gradient of transfer frequency exists so that the probability of transmission of a given marker decreases with its distance from the leading terminus of DNA. In the rare event that the entire Hfr chromosome with the distal F sequence is transmitted to the F$^-$ recipient, conversion to Hfr-determined maleness may occur.

Hfr-mediated gene transfer provides a convenient method for locating the relative position of various genes on the chromosome. This is accomplished by performing crosses between appropriate genetically marked strains and interrupting transfer of the chromosome at various times by subjecting the conjugates to moderate shear forces and plating the male/female mixture on a medium that will kill the male cells but that is selective for the desired recombinant cells. Viability of the zygote but death of the male cell is usually accomplished by using phage-resistant or antibiotic-resistant female cells and sensitive male cells. Application of this technique with a series of Hfr strains with the origin of transfer at different positions on the chromosome facilitated construction of the rather extensive genetic map of *E. coli* shown in Figure 8-8. The results of a typical interrupted mating experiment are shown in Figure 8-9.

The Formation of Hfr Cells From F$^+$ Cells

Hfr cells arise from F$^+$ cells. The molecular event involved in this conversion is the integration of the fertility factor into the chromosome. This integration is not a particularly rare event and occurs normally in cultures of F$^+$ cells. This fact accounts for the originally observed fertility of a mating between F$^+$ and F$^-$ cells (fertility is defined here as the ability of markers from the chromosome of the F$^+$ cell to recombine with those of the chromosome of F$^-$ cells) and provided for the discovery of bacterial sexuality. The introduction of alleles into F$^-$ cells from an F$^+$ population occurs with a frequency of approximately 10^{-4} to 10^{-5},

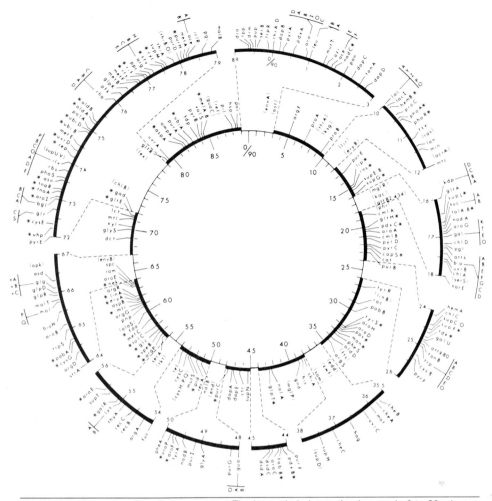

Figure 8-8. Map of the *E. coli* chromosome. The inner circle bears the time scale 0 to 90 minutes, or the time required for transfer of the entire chromosome from a male to female cell. The positioning of 0 time is arbitrary, since different Hfr strains possess different origins of transfer. The meaning of the gene symbols may be found in the article credited below. The contour length of the *E. coli* chromosome is 1000 μm, or about 3 times 10^6 nucleotide pairs. Assuming that the average gene contains 1000 nucleotide pairs, the *E. coli* chromosome contains enough information for 3000 genes. It is possible, therefore, that the map depicted above is only 10 percent saturated with respect to gene content (310 genes are shown). *(From Taylor: Bacteriol Rev 34:155, 1970. For a newer, recalibrated version see Bachman and Low: Microbiol Rev 44:1, 1980.)*

and virtually any chromosomal marker can be introduced. This apparent fertility of the F⁺ population is explained by a relatively random insertion of the F factor into the chromosome and its subsequent mobilization into the F⁻ cell. In other words, the fertility is caused by the presence of Hfr cells. A number of Hfr types exist that are distinguished by the region of the chromosome in which the F factor has integrated as well as the direction of Hfr-mediated transfer of the chromosome (see below). The enhanced ability of a population of cells, each one with the F factor integrated at the same position of the chromosome, to transfer chromosomal material is the basis for the Hfr designation, i.e., high-frequency recombination.

Close examination of normal, recombination-sufficient (*recA*⁺) Hfr cells reveals that integration of the fertility factor is not totally random but that there are favored regions for integration of the F factor into the chromosome. In *recA*⁺ cells, integration occurs by a mechanism of general recombination, where base sequence homology specifies the initial interaction between the participating genophores. The F factor and the chromosome contain common nucleotide sequences, which are the insertion sequences (IS) that were discussed in the section of this chapter dealing with mutagens. The F factor is known to contain three types of IS, known as IS2, IS3, and γ-δ (see Fig. 8-7). The same elements have been found among the

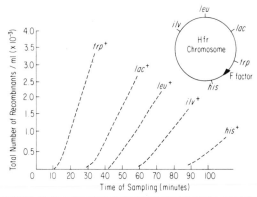

Figure 8-9. Interrupted mating experiment. Hfr strain of *E. coli* with the fertility factor integrated at the position shown is mixed with an F strain that is auxotrophic for the indicated amino acids. Samples are removed from the mating mixture at the times shown, subjected to moderate shear forces, and plated on medium that is selective for the desired recombinants and that prevents growth of the Hfr. The rate of appearance of recombinants equals the time of entry of the competent region of the Hfr chromosome and, therefore, represents the position of the various loci relative to the fertility factor. The decrease in the slopes of the lines representing the frequency of recombinants for late entering loci is caused by the smaller number of zygotes formed that contain the relevant genetic loci. This smaller number is the result of accidental separation of mating pairs during the course of the linear, polarized transfer of the Hfr chromosome.

IS sequences contained in the *E. coli* K-12 chromosome, and it is their location that governs the predominating sites of recombination between the F factor and the chromosome. The F factor is also able to insert into the chromosome of *recA⁻* cells using the illegitimate recombinational capacity of its IS elements. This type of integration, which occurs less frequently than when the IS elements serve as portable regions of homology, is more fully described later in this chapter.

Regardless of the mechanism of F integration into the bacterial chromosome, the integration event imposes a specific orientation of the F sequence, in particular *oriT,* with regard to the flanking chromosomal markers. Because of the polarized nature of F-mediated DNA transfer, this orientation defines the direction of transfer of the chromosome for a given Hfr, with markers to one side of F being transferred early and those markers to the other side being transferred late. The conjugal characteristics of a given Hfr strain, therefore, include both the site of F integration and the direction of transfer from that site.

Formation of F′ Factors

Precise excision of the F factor from the Hfr chromosome may occur by recombination between those sequences that participated in the original insertion event. Occasionally, the F factor excises from the chromosome in an imprecise manner so that neighboring chromosomal sequences are incorporated into the autonomous replicon. These altered

F plasmids are called F′ factors and are of two types. Type I factors contain a chromosomal fragment that is joined at one end to a site within the original F sequence and so do not contain an entire F factor. In some type I factors, the chromosome segment is fused to one end of an IS element contained within F, and it is thought that these F′ plasmids arise by *recA*-independent recombination promoted by the IS sequence. This type of excision leaves a portion of the F factor in the bacterial chromosome designated a sex factor affinity locus (*sfa*), which because of its homology with F is a preferred site for subsequent reintegration of the F factor. Type I F′ plasmids can also result from aberrant recircularization of conjugally transmitted Hfr DNA, particularly when recombination in the recipient cell is prevented by a *recA* mutation. Generation of type I factors by this mechanism does not result in an *sfa* locus in the donor cell, since the plasmids are formed from transferred DNA in the recipient. Type II F′ factors result from a recombination event occurring at sites located in the chromosomal sequences that flank the integrated F factor and therefore contain the intact F sequences. Excision of some type II factors occurs by *recA*-dependent recombination between repeated sequences normally found in the bacterial chromosome, such as IS elements and ribosomal RNA operons (Chap. 7).

F′ factor-bearing cells are capable not only of transferring the conjoined genes at a high frequency but also, in contrast to Hfr cells, of conferring maleness to female cells at an equally high frequency. In other words, the shorter distance between the point of entry and the distally located portion of the fertility factor decreases the probability that the DNA will be sheared during transfer. Genetic transfer mediated by F′ cells is also known as "sexduction." F′ cells have been particularly useful in genetic and biochemical analysis because virtually entire populations of recipient cells can be converted to a homogeneous merodiploid state. The F⁺-Hfr-F′ pathway is depicted in Figure 8-10.

Transduction

The type of gene transfer in which the DNA of one bacterial cell is introduced into another bacterial cell by virus infection is known as "transduction." The transfer of genetic material by bacteriophage can be classified according to the scope of potential transfer. Some types of bacteriophage are able to transfer genes derived from their previous host cells in a relatively indiscriminate manner, so that virtually any chromosomal marker can be transferred. This process is termed "generalized transduction." However, other types of bacteriophage are capable of transferring only certain genes from the previously infected host. This process is termed "specialized transduction."

The ability of a phage to mediate specialized or generalized gene transfer depends upon its relationship not only with the organism that is to serve as the source of genetic information to be transferred but also with the recip-

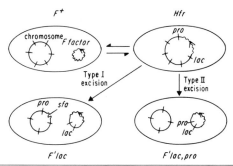

Figure 8-10. The formation of Hfr and F' cells. The integration of the fertility factor occurs by a reciprocal recombination event between the circular factor and the circular chromosome. Disintegration of the fertility factor also occurs by reciprocal recombination. The site of recombination governs the chromosome-derived material that will be present in the F'. Type I excision, as shown, frequently results in a portion of the fertility factor remaining in the chromosome, creating a sex factor affinity locus.

ient organism. An understanding of the events that permit phage to mediate gene transfer depends upon a knowledge of the physiology and molecular biology of phage infection (Chap. 63). An abbreviated and generalized outline of the infectious process is as follows: adsorption of the bacteriophage to stereospecific receptor sites on the bacterial surface is followed by injection into the cell of the DNA contained within the phage particle. If the phage is of the virulent type, the genetic program of the DNA will be expressed, in which case an abundance of new phage particles will be produced and released by subsequent cell lysis. However, if the infecting phage is of the temperate type, alternative pathways are available following injection of the DNA. The vegetative program of the DNA may be expressed, in which case the lytic response ensues, or the vegetative program may be prevented by the production of a phage-specified repressor, in which case the phage DNA may enter into a benign relationship with the host cell. The presence of the phage repressor establishes a lysogenic state, in which the DNA of the phage may be incorporated into the host chromosome in such a way that the two genomes are linearly contiguous. The phage genome with its vegetative functions repressed, whether or not integrated into the chromosome, is known as a "prophage." The cell acquires a significant new property as a consequence of lysogeny, namely, that it is immune to infection by homologous phage. This immunity is important in demonstrating transduction.

Generalized Transducing Particles

If, during phage maturation, host DNA rather than phage DNA is packaged into the phage capsid, a generalized transducing particle will be formed. This aberrant packaging occurs randomly with respect to the price of host DNA that is occluded in the phage particle, and it occurs at a relatively low frequency, and only a minor fraction of the population of a phage lysate contains host rather than viral

DNA. The size of the host DNA contained in generalized transducing particles is similar to the size of the genome that is normally carried by the phage. This is reflected in the amount of genetic information that the particle subsequently is able to transfer. For example, the DNA molecular weights of coliphage P1 and *Salmonella* phage PLT 22 are 60 million and 26 million, respectively, and genetic analysis has revealed that transducing fragments introduced into recipient cells by P1 are twice the size of those introduced by PLT 22. The production of generalized transducing particles is not a totally passive process. Mutant strains of *Salmonella* phage PLT 22 exist in which 50 percent of the particles released from an infected cell contain host-derived DNA. These lysates transduce, at a high frequency, the same random assortment of host characters as do normal lysates.

Specialized Transducing Particles

The production of specialized transducing particles resides in the ability of certain phage genomes to integrate reversibly into the genome of the cells that they infect. As alluded to above, integration of the phage genome into the host chromosome occurs by a process similar to that described for fertility factor integration (p. 170). It is thought that production of specialized transducing DNA occurs by a process similar to that which produces F's, i.e., when prophages are released from the chromosome, they frequently carry a portion of host DNA. There are significant quantitative and qualitative differences between the host DNA carried by F's and released prophage. For example, relatively large F's can be produced by the reciprocal crossover mechanism and still be recovered, owing to their mode of transfer. Only a limited amount of host DNA can be detected in the released prophage because of the limited size of DNA that can be packaged into the phage capsid. The F factor-Hfr-F' route potentially produces episomes carrying a variety of host markers, but the release of prophage results in episomes containing but a few host determinants. The reason for this latter limitation is that prophage integrates at one or, at best, a few positions on the host chromosome (recall that F integration is not so restricted). The basis for specialized transduction is, therefore, the location of the prophage on the host chromosome (i.e., which bacterial genes are contiguous with it) and the manner in which the prophage is released. Specialized transducing lysates are produced by destroying the lysogenic condition with ultraviolet radiation or with radiomimetic treatments (viz., mitomycin). The aberrant excision process that produces the specialized transducing DNA occurs with a frequency of 10^{-5} to 10^{-6}, which is similar to the frequency with which generalized transducing particles are produced. It is possible, however, to obtain lysogenized strains of bacteria that produce specialized transducing particles at a very high frequency. This will subsequently be described.

The process just described for the formation of specialized transducing DNA results in exclusion of a portion of phage DNA from the transducing particle. The extent of the exclusion will depend upon how much of the host DNA is included in the excised prophage—the greater the amount of host DNA, the greater the exclusion of phage DNA. This principle has obvious consequences with respect to the viability of the particle that is subsequently released from the cell. For example, in the case of coliphage λ, which integrates on the host chromosome between the galactose and biotin loci, λgal-transducing particles may be formed that are still capable of growing vegetatively. These particles are termed $\lambda pgal$ (p for plaque-forming), and they lack nonessential portions of the λ genome. However, other λ transducing particles that also carry galactose genes may be formed that are incapable of vegetative growth because they lack essential λ genes. These defective particles are termed $\lambda dgal$(defective λ-carrying galactose genes), and they require helper phage if they are to grow. The role of the helper phage is to provide the function lacking in the defective particle.

Transmission of Genetic Information by Transducing Particles

Specialized transducing particles are formed because of the special relationship existing between the viral and the host genomes, and it is, therefore, obvious that only temperate phages demonstrate this ability. The reason that generalized transduction also requires a temperate phage vector is not quite so obvious and is probably more directly related to the events that occur at the time of DNA transfer to the recipient organism rather than to the manner in which generalized transducing particles are formed. Because of the low relative abundance of transducing particles, a high multiplicity of infection must be used to demonstrate transduction. Therefore, the probability is high that a cell receiving a transducing particle also will be infected by normal phage. If the nontransducing particles are of a virulent nature, death of the cell ensues, thus preventing recovery of recombinant cells. However, with temperate phage, the immunity that attends lysogeny is able to prevent lysis, and the DNA introduced by the transducing particle is able to participate in recombination with the host chromosome. It is of interest to note that transduction with certain virulent coliphages can be demonstrated by employing genetic variants with reduced virulence. When T1 phage containing an amber mutation is grown in a strain of *E. coli* containing an amber suppressor mutation, a productive lysis ensues. If this lysate is then used to infect another strain of *E. coli* that does not contain the amber suppressor, the phage cannot grow, and transduction can be demonstrated.

The DNA contained in generalized transducing particles, upon injection into the host cell, is apparently available for recombination by the usual mechanisms, but the fate of the DNA of a specialized transducing particle is somewhat different. For example, when transducing a galactose-negative strain to a galactose-positive with λ, two pathways of recombination can occur: (1) the region of the phage DNA bearing the gal^+ gene may participate in generalized recombination with the gal^- region of the host, or (2) the λgal may integrate into the normal λ site so as to retain the gal genes. The latter type of recombination, which occurs with the greater frequency, results in the production of a cell that contains both alleles of gal. The transductant cells that arise by lysogenization and integration of λgal, upon induction with UV, will yield a lysate that contains predominantly transducing particles if the original transducing phage was of the $\lambda pgal$ type. If the original transducing particle was of the $\lambda dgal$ type, subsequent recovery of transducing particles requires complementation by a normal λ, in which case the secondary lysate will contain approximately equal numbers of $\lambda dgal$ and λ. In either case the phage preparations are known as Hft (high-frequency-transducing) lysates. Hft lysates provide a useful source of DNA enriched for chromosomal genes.

Episomes and Plasmids

As was pointed out above, episomes are extrachromosomal genetic units that are capable of replicating either autonomously or as integral parts of the host chromosome. Both the F factor and the DNA of temperate phage fit this description. Other types of extrachromosomal genetic units, plasmids, are unable to integrate into the host chromosome, and some of these elements are transferred by an infectious process similar to F transfer. Those plasmids that are incapable of autonomous transfer frequently gain entrance to other cells by transduction.

Bacteriocinogenic Factors

A variety of bacteria harbor extrachromosomal elements (bacteriocinogenic factors) that produce bactericidal proteins, known as "bacteriocins." Bacteriocins may be relatively low molecular weight, heat-stable, nonsedimentable proteins, or they may be large, sedimentable material that in electron micrographs appears as phage components. Bacteriocins bind to stereospecific sites on the bacterial surface and either manifest their activity while bound or gain entrance to the cell. A single surface-bound molecule of bacteriocin frequently is bactericidal. The metabolic perturbation wrought differs among the various types of bacteriocins. Some of these effects are cessation of DNA, RNA, and protein synthesis, cessation of respiration, cellular leakage, and impairment of general metabolism.

The bacteriocins that have received the most attention are the colicins (those produced by *E. coli*), of which a large number exist. The mode of action of some of the colicins is known in some detail. Colicin E1 causes an impairment of ATP formation, which is reflected in a cessation of protein synthesis, a perturbation of DNA and RNA synthesis, and an inability to accumulate certain compounds from the medium. Colicin E2 causes single-stranded nicks in DNA and, therefore, interferes with DNA synthesis. Colicin E3 inactivates 30S ribosomes by removing 50 terminal nucleotides from 16S RNA.

The presence of bacteriocinogenic factors in bacterial cells frequently is cryptic, owing to repression of bacteriocin production. However, as in the case of lysogenic phage, the repression can be destroyed by irradiation with ultraviolet light as well as by other treatments. Unlike induction of prophage, the derepression of bacteriocins does not always result in death of the cell. The bacteriocins consequently are analogous to defective prophage in this respect, although the basis for lack of killing may be different. It is known, for example, that the colicinogenic factor (Col E3) that contains the structural gene for colicin E3 also contains a structural gene for a colicin E3 immunity factor. This immunity factor, which is a protein of 10,000 molecular weight, reacts stoichiometrically with the colicin E3 molecule to prevent its activity. It is highly probable that other bacteriocinogenic factors specify immunity substances that prevent self-destruction of the cells that harbor them.

Some colicinogenic factors appear to be analogous to F factors, whereas others act more like nonintegrating prophages. Colicinogenic factor I, for example, specifies I-pilus formation and consequently promotes its own transfer by cell-to-cell contact. It has also been reported that other colicinogenic factors can promote host gene transfer. The responsible factor, like the fertility factor, may integrate into the host chromosome. Other types of colicinogenic factors that are incapable of autonomous transfer are able to be transferred with a helper F factor. Regardless of the episomal versus plasmid existence of bacteriocinogenic factors, most are inducible, as explained above, by treatment with ultraviolet irradiation or mitomycin C. The bacteriocinogenic state in this respect appears analogous to the lysogenic state. As release from lysogeny results in expression of vegetative phage functions, so release from bacteriocinogeny results in expression of the bacteriocin gene. The reason for the existence of bacteriocinogenic factors is not entirely clear: some may provide survival value by preventing colonization of ecologic niches by intruding organisms (such as in the case of Col E1, E2, and E3), and others may represent evolutionary intermediates between prophage and sexual fertility factors (granting the obvious advantages of sexuality).

Transmissible Drug Resistance

The practical aspects of infectious gene transfer are underscored by the existence among members of the Enterobacteriaceae of a genetic capability that allows the passage of antibiotic resistance from one organism to another. The potential hazard of this transfer lies in the fact that resistance to multiple antibiotics is involved. This phenomenon of transmissible drug resistance was discovered in Japan in 1959 when a strain of *Shigella flexneri* resistant to four drugs, chloramphenicol, tetracycline, streptomycin, and sulfonamides, was isolated from a case of dysentery. The occurrence of multiple-resistant strains of shigellae,

salmonellae, *E. coli*, *Enterobacter*, *Proteus*, and related inhabitants of the animal or human intestine, as well as of other gram-negative cells, is now common throughout the world. The scope of resistance has expanded to include most types of antibiotics.

Multiple antibiotic resistance occurs by acquisition of an extrachromosomal element known as a "resistance (R) factor." Many of the R factors are similar to F′ factors in that they contain genetic information for autonomous replication and conjugal transfer as well as genetic determinants for antibiotic resistance. The conjugal transfer of antibiotic resistance occurs by a mechanism similar to F′ sexduction. These two classes of genetic information are usually contained on two functionally distinct units within the R factor: one is a sex factor unit (frequently termed "resistance transfer factor" or RTF) that contains the information for autonomous replication and conjugal transfer, and the other is a unit that specifies antibiotic resistance (frequently termed the "r determinant"). The r determinant of most R factors is strictly dependent upon the RTF for its autonomous replication, although some R factors, may dissociate to form two independent replicons. The RTF-specified conjugal transfer of R factors is mediated through characteristic sex pili. Some RTFs produce F pili, others I pili (i.e., the type produced by colicinogenic factor I), and still others produce unique N pili.

r Determinants. The mechanism of antibiotic resistance imparted by r determinants is usually specific for each antibiotic. Examples of the antibiotic-inactivating enzymes as well as other resistance principles elicited by r determinants are presented in Chapter 9. It should be noted here, however, that even for a single mode of resistance, variations may exist depending upon the R factor involved. An excellent example of this is the R factor-determined penicillinases. Penicillinase (β-lactamase) hydrolyzes the β-lactam ring of penicillin to form penicilloic acid. The β-lactam ring is the functional group for inhibition of cell wall biosynthesis (Chap. 9), and its hydrolysis destroys the bactericidal action of penicillin. The β-lactamases specified by R factors can be grouped into at least four main types based upon their substrate specificities (i.e., against the β-lactam type antibiotics, benzylpenicillin, ampicillin, cephaloridine, cephalexin, carbenicillin, and cloxacillin), immunologic cross-activity, and electrophoretic mobility.

Transposons. The r determinant portion of an R factor may be a multipartite unit consisting of genetic determinants for resistance to many antibiotics. These determinants are transferred en bloc from one organism to another so that the recipient organism acquires multiple antibiotic resistance. The genes that encode the antibiotic detoxifying principles (e.g., enzymes) are frequently contained within genetic elements known as "transposons" (Tn). These nucleotide sequences derive their name from the ability to be transferred from one position to another within a replicon or to be transferred to a different replicon. This process occurs by an illegitimate recombinational event independent of the normal (recA), generalized, recombination mechanism. A salient feature of this recombinational event is that it occurs by replication so that the transposon is not lost from its original site. The mechanism of transposition is more fully described later in this chapter under Recombination. Several different types of transposons have been characterized, and in each case it has been shown that the expressed genetic information is flanked by identical nucleotide sequences containing from 140 to 1400 base pairs. These nucleotide sequences are thought to function in a manner similar to the insertion sequences (IS) that were previously described in this chapter.

As in the case of insertion sequences, the nucleotide sequences that flank the resistant determinants probably serve as substrate for specialized recombination enzymes. The existence of these types of translocatable elements, without considering their origin, provides at least one of the mechanisms for development of R factors that bear determinants for resistance to several antibiotics.

Translocatable antibiotic resistance elements compound the potential hazard of transmissible resistance. As explained below, the promiscuous conjugal relationship that exists among the members of the Enterobacteriaceae is tempered by the inability of certain plasmids to be stably established in cells that harbor certain other plasmids and by the ability of cells to hydrolyze foreign DNA. The efficient operation of both of these barriers in preventing acquisition of antibiotic resistance characters can potentially be overcome by the translocation of transposons from the invading plasmid to a replicon that had previously become established in the recipient cell.

The mobility of transposons and their apparent ability to breach genetic barriers are demonstrated by cases of plasmid-specified ampicillin resistance in the Enterobacteriaceae, H. influenzae, and N. gonorrhoeae. Owing to the widespread use of ampicillin in the treatment of infections caused by enteric organisms, a large reservoir of these organisms resistant to this antibiotic has developed. The accumulation of these resistant organisms was soon followed by the development of ampicillin resistance among strains of H. influenzae, an organism that was originally sensitive to this antibiotic. More recently, ampicillin-resistant strains of N. gonorrhoeae have been isolated from clinical sources. DNA hybridization analysis of the Haemophilus plasmid that confers ampicillin resistance shows that it contains the Tn3 transposon that was originally identified in the Enterobacteriaceae. A similar analysis of the resistance plasmid from N. gonorrhoeae shows that it contains 40 percent of the Tn3 transposon. The peril of the transposon acquisition by the Neisseria is amplified by the fact that the ampicillin-resistance character is able to be conjugally transferred. This latter capability raises the possibility that other Neisseria, such as Neisseria meningitidis, could sexually participate with the gonococcal strain to eventually acquire ampicillin resistance.

It should be emphasized that chromosomal mutation also leads to antibiotic resistance in the Enterobacter-

iaceae, which is again exemplified by penicillinase production (viz., the ampA locus of the E. coli chromosome codes for a type I β-lactamase). The significant difference between chromosome-mediated and R factor-mediated resistance is that the latter imparts multiple resistance, whereas the former imparts resistance to a single class of drug. Episome-specified multiple resistance is brought about by a single event, namely, acquisition of the R factor. The development of multiple resistance based in the chromosome requires several mutational events, so that accumulation of multiple antibiotic resistance of the chromosomal type within a given strain requires sequential selection in the presence of a series of antibiotics. This fact forms one of the bases for using a combination of unrelated antibiotics in treatment of infectious disease.

The potential hazard of transmissible antibiotic resistance is further amplified by the observation that transfer can occur among the various members of the Enterobacteriaceae. A strain of E. coli inhabiting the intestine of a healthy individual could possess an R factor. A shigella or other pathogen entering this setting could acquire multiple antibiotic resistance by transfer from the E. coli, with obvious repercussions upon attempting to clear the newly acquired pathogen from this site.

Antibiotic Resistance Plasmids and Determinants of Gram-positive Bacteria

Antibiotic resistance in gram-positive bacteria, as in gram-negative bacteria, can be plasmid-specified. The archetypical plasmidal antibiotic resistance in gram-positive cells is the so-called penicillinase plasmids of Staphylococcus aureus. Unlike the RTFs of gram-negative cells, the penicillinase plasmids of S. aureus are not transferred by a conjugal process but instead rely upon a viral vector (transduction) for their transfer.

The β-lactamases of plasmid-containing strains of S. aureus are usually produced at a maximal rate only in the presence of β-lactam antibiotics (i.e., penicillin, ampicillin, and so on), and the enzyme is found both associated with the cell and in the surrounding medium. These features are in contrast with the regulation and localization of the β-lactamases produced by gram-negative cells, where the enzyme may be produced constitutively or be induced but where it usually remains cell associated.

At least four types of β-lactamases are produced by the plasmids of S. aureus. These have been designated types A, B, C, and D and are distinguished from one another on the basis of turnover number, immunologic cross-reactivity, substrate specificity, and amino acid sequence. The types of penicillinase produced by plasmid-containing S. aureus correlate with the typing of the particular strain on the basis of bacteriophage sensitivity (Chap. 25). Penicillinases types A and C are produced by members of phage group I or III but never II, and type B is produced by members of group II. Type D β-lactamase is relatively rare. It is immunologically different from types

A, B, and C, is produced constitutively, and is found in *S. aureus* cells of phage groups I, II, and III.

Penicillinase plasmids frequently contain other characteristics, such as resistance to erythromycin (or, in the case of the rare plasmid mentioned above, resistance to fusidic acid) and resistance to the inorganic ions, mercury, arsenate, cadmium, lead, and bismuth. Plasmids bearing determinants for resistance to other antibiotics may be present in *S. aureus* and in other gram-positive organisms. For example, resistance to streptomycin, tetracycline, neomycin, and fusidic acid are plasmid-borne in *S. aureus*.

Two transposons have been described in *S. aureus* encoding erythromycin and erythromycin-spectinomycin resistances (Tn551 and Tn554, respectively). It is likely that, as in gram-negative organisms, a number of transposons will be found in *S. aureus* that can promote formation of multiple resistance determinants.

Antibiotic resistance in *S. aureus* can also occur by mutation of the chromosome. Table 8-2 presents the relative frequency of the location of resistance determinants for a number of antibiotics. The appearance of methicillin resistance on this list is noteworthy, because this drug was primarily developed for its property of resistance to hydrolysis by β-lactamase. Methicillin resistance also exemplifies the apparent genetic complexities that accrue as a consequence of the selective pressures introduced by antibiotics. In methicillin resistance, it is uncertain whether the responsible genetic determinant is chromosomal or plasmidal or both because most of the strains isolated as methicillin-resistant are nonetheless bearers of plasmids that produce a methicillin-inactive β-lactamase.

Plasmid-specified antibiotic resistance is also found in other types of gram-positive organisms. The various streptococci, including members of groups A, B, C, D, G, and H, have been shown to contain plasmids conferring resistance to macrolide antibiotics (erythromycin), lincosamide antibiotics (lincomycin, clindamycin), and streptogramin B. Resistance to these antibiotics in streptococci occurs by way of an inducible enzyme that causes dimethylation of the 23S ribosomal RNA, and simultaneous resistance to

these three antibiotics, which have 50S ribosome subunits as their site of action (Chap. 9). Other plasmids described in the streptococci confer resistance to chloramphenicol, streptomycin, kanamycin, sulfonamides, and tetracycline. The resistance principles in these cases are similar to those specified by the R factors of gram-negative cells (viz., antibiotic-modifying enzymes). There are two additional noteworthy observations regarding antibiotic resistance in the streptococci. The first is that, to date, β-lactamase has not been described, and although this raises interesting questions relative to the epidemiology of antibiotic resistance, it also is clinically fortunate owing to the wide use of the β-lactam antibiotics in the treatment of streptococcal infections. The other observation, which represents a therapeutically unfortunate property, is that antibiotic resistance in many of the streptococci is transferred to normally sensitive streptococcal cells by a conjugal process.

Conjugal transfer of plasmids in *Streptococcus faecalis* (group D) appears to occur by two different mechanisms. Some plasmids, as exemplified by the erythromycin-resistance plasmid pAMβ1, transfer inefficiently in liquid cultures (approximately 1 erythromycin-resistant transconjugant per 10^6 donor cells) but can transfer much more efficiently when matings are performed on filter membranes (1 transconjugant per 10^2 to 10^4 donor cells). The filter membrane acts as a matrix upon which conjugally productive cell-cell contacts are supported. Another group of plasmids, such as pAMγ1, which encodes hemolysin and bacteriocin functions, transfer at high frequency in liquid cultures (1 transconjugant per 10^1 to 10^2 donor cells). This efficient transfer system involves secretion by recipient cells of clumping-inducing agent (CIA) that causes donor cells to become adherent and form stable mating pair aggregates with recipient cells. CIA is a chymotrypsin-sensitive peptide with a molecular weight of 1000, and it interacts with a plasmid-containing donor cell through a cell surface receptor to induce synthesis of aggregation substance. This cell surface protein is found only on donor cells that are induced by CIA and are capable of clumping. Upon acquisition of a plasmid like pAMγ1, the synthesis of CIA is repressed, and the cells are no longer able to induce clumping in other cells containing pAMγ1. Several CIAs exist, however, and each appears to be specific for a given plasmid and is synthesized in the absence of that plasmid. Accordingly, a cell containing pAMγ1 still secretes CIAs that affect donor cells containing other plasmids of this type. Because the CIAs effect communication between organisms that leads to conjugation, they have been referred to as "sex pheromones."

Antibiotic resistance in streptococci can also reside within chromosomal determinants that, in some cases, effect conjugally mediated resistance transfer in the absence of a plasmid vector. In *S. faecalis*, chromosomal tetracycline resistance is encoded by a 10 megadalton sequence designated Tn916. Tn916 is a transposable element and can move from chromosomal sites to plasmids and back to the chromosome independently of general recombination.

TABLE 8-2. GENETIC LOCI OF RESISTANCE TO VARIOUS ANTIBIOTICS IN *S. AUREUS*

Antibiotic Resistance in Clinical Strains	Genetic Locus
Penicillinase	Plasmid in >95% of strains
Streptomycin	Probably plasmid in 30% of strains, chromosomal in 70%
Tetracycline	Generally plasmid
Erythromycin	Generally plasmid
Neomycin	Plasmid
Fusidic acid	Plasmid in 70% of strains, chromosomal in about 30%
Trimethoprim	Probably chromosomal in a few strains
Novobiocin	Probably chromosomal in one strain
Methicillin	Conflicting (see text)

Modified from Lacey and Richmond: Ann NY Acad Sci 236:395, 1974.

Tn916 is also capable of transferring itself to tetracycline-sensitive cells via a DNase-resistant conjugal process that requires cell-cell contact and occurs without participation of plasmid functions. In the tetracycline-resistant transconjugants, Tn916 is found to be inserted into the chromosome at several different sites. The Tn916-containing transconjugants are themselves capable of conjugal transfer of tetracycline resistance. It is important to note that the conjugal process effects transfer only of Tn916, and that no other chromosomal markers are known to be concomitantly transmitted. The mechanism by which Tn916 effects self-transfer may be related to the transposition process, since derivatives of Tn916 that show enhanced transposition frequencies also transfer at similarly enhanced frequencies. Because of this unique combination of transposition and self-transfer functions Tn916 is the prototype of a new class of genetic elements termed "conjugative transposons." Chromosomal drug resistance determinants active in conjugation have also been described in *S. pneumoniae*, encoding tetracycline and chloramphenicol resistances, and in *Streptococcus agalactiae*, encoding tetracyline, chloramphenicol, and erythromycin resistances. Although these elements have not yet been shown to transpose independently of general recombination, they do effect en bloc transfer of drug resistance and demonstrate significant DNA homology with Tn916. These elements could have originated by transposition into Tn916 of other transposons encoding chloramphenicol or erythromycin resistance. Conjugative transposons may be a common means for dissemination of drug resistance in pathogenic bacteria with recent studies suggesting their existence in *Clostridium difficile* (determining tetracycline resistance) and in the gram-negative organism *Bacterioides fragilis* (encoding both clindamycin and tetracycline resistances).

Regulation of Antibiotic Resistance

Considerable variation exists in the regulatory mechanisms governing expression of antibiotic resistance, which is due in part to the diverse origins of resistance determinants. For instance, the chloramphenicol acetyltransferase (*cat*) gene found in gram-negative bacteria is constitutively expressed with regard to chloramphenicol and is subject only to more general regulation by catabolite repression. The *cat* gene of *S. aureus*, however, is induced by chloramphenicol through a mechanism involving autogenous regulation of the *cat* gene by the encoded gene product, chloramphenicol acetyltransferase. Similarly, the β-lactamase genes of gram-negative plasmids are constitutively expressed, while those of *S. aureus* are induced by β-lactam antibiotics. It is not known whether the attendant lag in appearance of an induced antibiotic resistance function has any clinical significance with regard to antimicrobial chemotherapy.

The regulatory mechanisms affecting expression of antibiotic resistance are in general similar to the classic systems of gene expression described in *E. coli* K-12

(Chap. 7). Tetracycline resistance (*tet^r*), encoded by the gram-negative plasmid R222, for example, is subject to negative control by a repressor protein that curtails transcription of the *tet^r* genes in an analogous manner to the *lac* repressor-operator interaction. The presence of tetracycline antagonizes the action of the *tet* repressor and effects derepression of *tet^r*. Sequence analysis of the promoter region of the *S. aureus cat* gene has demonstrated an inverted repeat in a similar position to that of the inverted repeat in the *lac* operator. Autogenous regulation of *cat* could involve binding of chloramphenicol acetyltransferase to this sequence that represses transcription of *cat*. Derepression could occur through chloramphenicol binding to the protein and releasing the protein from the regulatory sequence just as allolactose binds to the *lac* repressor, reducing its affinity for the *lac* operator.

Regulation of resistance to other inhibitors of protein biosynthesis may involve posttranscriptional control of translation. The *S. aureus* plasmid pE194 encodes resistance to the MLS antibiotics (macrolides, lincosamide, streptogramin) that is brought about by enzymatic dimethylation of the 23S ribosomal RNA. The DNA sequence preceding the resistance gene contains a promoter, a ribosome-binding site, and a leader peptide sequence. There are four complementary inverted repeats within the controlling region, and alternative stem and loop secondary structures in the messenger RNA are possible. The controlling region thus resembles the attenuator structures of amino acid biosynthetic operons, such as *trp* and *his*, except for the lack of a transcriptional terminator following the leader peptide sequence. It is hypothesized that transcription of the resistance determinant occurs constitutively and that the messenger RNA assumes a secondary structure so that the leader peptide sequence is freely translated. A downstream stem and loop structure persists that sequesters the ribosome-binding site for the resistance gene and prevents translation of the coding sequence. When erythromycin is present in low concentrations, effective translation of the leader sequence is curtailed and ribosomes stall on the messenger RNA. The stalled ribosomes disrupt the predominant secondary structure and permit an alternative stem and loop to form that exposes the ribosome-binding site of the structural gene. Translation of this sequence proceeds (probably by already resistant ribosomes normally found in small numbers in the cell), and induction of erythromycin resistance occurs. The similarity of this mechanism to attenuation lies in the ability of ribosomes to affect the secondary structure of messenger RNA. However, in the case of attenuation, alternative secondary structures influence transcriptional termination, whereas these structures in the erythromycin resistance gene affect actual translation of the formed messenger RNA.

The maximum level of antibiotic resistance conferred by a given determinant is in part related to the amount of gene products formed, which is in turn dependent upon the permissible rates of transcription and translation. Increase in drug resistance beyond the maximum level of a single determinant can occur by gene amplification, that is, an increase in the actual number of resistance determinants present per cell. The *S. faecalis* plasmid pAMα1 is a small nonconjugative plasmid that encodes tetracycline resistance within a 2.65 megadalton DNA sequence. This sequence is bounded by short direct repeats. When a culture of *S. faecalis* containing pAMα1 is grown in the presence of tetracyline, it is found that the maximum level of tetracycline resistance (as reflected in the minimal inhibitory concentration) increases with time of exposure to tetracycline. Examination of the plasmid DNA from cells with high level tetracycline resistance reveals that the plasmids have increased in size and that this increase reflects the presence of tandem repeats of the 2.65 megadalton tetracycline resistance determinant. The amplification process is reversible, since growth in the absence of tetracycline results in cells with pAMα1 containing a single copy of the 2.65 megadalton sequence. Amplification of tetracycline resistance is dependent upon general homologous recombination and may involve recombination between the directly repeated sequences flanking the tetracycline resistance determinant. Similar amplification of plasmid-borne drug resistance has been seen in *Proteus mirabilis* and *E. coli* K-12. An analogous process exists in some eucaryotic cells; for example, resistance to methotrexate involves amplification of the dihydrofolate reductase gene.

Plasmid Incompatibility Groups

The various plasmids and episomes discussed above can be classified according to their ability to cohabitate. Plasmids that are distinguishable and that cannot coexist stably in the same cell belong to the same incompatibility group. Conversely, plasmids within one incompatibility group can coexist with those of another group.

The episomes of the Enterobacteriaceae have been placed into several incompatibility groups (group FI, FII, Iα, Iε, N, C, O, T, W, P, L, x). The relatedness of various functionally distinct episomes is underscored by the members of these incompatiblity groups. For example, the fertility factor F, the colicinogenic factors Col v2 and Col v3, as well as resistance factor R386, are members of the same group. DNA hybridization studies indicate extensive homology among the DNA of plasmids within incompatibility groups and little homology between groups.

Incompatibility grouping potentially could be of epidemiologic value, for it provides a method of partially identifying R factors. For example, both of the resistance factors R444 and R390 carry resistant determinants for ampicillin, streptomycin, tetracycline, chloramphenicol, and sulfonamides, but R444 belongs to incompatibility group FII, whereas R390 belongs to group N. The epidemiologic sleuth finding these R factors in a case of dysentery might conclude that they arose from different origins, and he would be right, for R444 was originally identified in *Proteus morganii* and R390 in *Proteus rettgeri*.

Plasmids found in other organisms can also be classified according to incompatibility groups. There are, for example, at least two incompatibility groups of penicillinase plasmids in *S. aureus*.

The molecular basis for incompatibility is obscure but may be related to the replicon properties of plasmids. It is known that plasmid replication is at least partially self-regulated, and, therefore, a cytoplasmic, plasmid-specified repressor could possess ambivalent recognition of an incoming plasmid to prevent its replication. Maintenance of the plasmid at a membrane site may also be involved, since it is known that replication and segregation involve DNA membrane association, and competition for membrane sites may cause incompatibility. The number and types of membrane sites may also explain why cells contain one or two copies of one type but ten or more copies of a different type of plasmid.

Restriction and Modification of DNA

The foregoing outline of gene transfer suggests that the genotypes of bacteria are susceptible to alteration by the genophores with which they coexist in nature. This appears especially true with members of the Enterobacteriaceae, owing to shared mechanisms of gene transfer as well as to common ecology. However, a mechanism exists that tempers this potentially promiscuous interchange of genetic material. It is known that in many bacteria, endonucleases are present that protect the cell from invasion by DNA derived from other strains of bacteria, and these endonucleases are the basis for the phenomenon of restriction. These restricting endonucleases can be active against all types of foreign DNA, and the restriction is manifest regardless of the mode of infection with the foreign DNA, i.e., by phage infection, conjugation (including sexduction), or transformation. Restriction systems are highly specific and are accompanied by equally specific modification systems. Modification resides in an enzyme that acts on the same site of the DNA that is recognized by the restriction system and alters it in such a way as to prevent subsequent cleaving of the DNA by restricting endonucleases. Restriction and modification enzymes exist as companions and constitute highly specific epigenetic systems. The genetic information for restriction-modification systems may be located on the chromosome or on plasmids, such as prophage, R factors, or bacteriocinogenic factors. It is tempting to speculate that the presence of restriction systems on plasmids adds a new dimension to the parasitism of this type of element, since it endows them with a mechanism for excluding other extraneous DNA.

The phenomenon of modification and restriction has been most extensively studied using DNA-containing bacteriophages as indicators for the presence of modifying and restricting activities. The degree of modification and restriction can be assessed by determining plating efficiency of the phage on a given host. (Plating efficiency equals the number of infective centers produced per phage particle.)

The specificity of modification and restriction can be exemplified by phage λ replicating in two different hosts, *E. coli* strains K12 and B. If a lysate of λ is prepared from *E. coli* K12 and then used to infect *E. coli* B, the plating efficiency is about 10^{-4}. However, the plating efficiency on strain K12 is 1. Conversely, if strain B is the original host, the plating efficiency on strain K12 is about 4×10^{-4}, and the efficiency of plating on strain B is 1. These results, explained in terms of modification and restriction, show that by growing in one strain, the phage acquires a strain-specific modification so that upon infecting the same strain, it is protected from restriction. However, when grown on another strain with a different modification-restriction specificity, the phage is restricted. That the restriction is not absolute is shown by the fact that although the plating efficiency on a restricting host is decreased, some of the phage escapes the restriction. This fact reflects the epigenetic nature of modification in that it proceeds without DNA synthesis and is in fact a modification of existing DNA structure. Similarly, restriction does not require DNA synthesis. This is evident from the observation that a host that is lysogenically homoimmune for λ (i.e., carrying a λ prophage) and that, therefore, does not allow replication of λ DNA, will restrict incoming, unmodified λ. The degree of restriction is reflected in the number of restriction sites present on the DNA molecule, i.e., the number of sites subject to the endonucleolytic activity of the restriction enzyme. This can be demonstrated by determining the extent of restriction when a single DNA molecule is acted on by two restriction systems with different specificities. For example, if *E. coli* K12 is lysogenized with phage PI, which carries its own restriction system, the restriction of *E. coli* B-grown λ is greater (plating efficiency equals 10^{-7}) on the lysogen (KP1) than it is on strain K (plating efficiency equals 4×10^{-4}) itself.

The first description of a molecular basis for modification was glucosylation of the DNA of the T-even coliphages. When T-even coliphages are grown through the cycle in a mutant of *E. coli* that is unable to synthesize uridinediphosphoglucose, an intermediate in the phage DNA glucosylation reaction (Chap. 63), phages are produced that lack glucose on their DNA. These phages are unable to grow in *E. coli* containing a membrane-associated nuclease that is active on the nonglucosylated phage DNA, i.e., the phages are restricted. These phages will grow in *Shigella*, which lacks the nuclease, and because *Shigella* possesses the glucosylating system, unrestricted phage will be produced, i.e., the *Shigella* modifies the DNA. The *Shigella*-derived phage can now grow on the previously restrictive strain of *E. coli* because the phage DNA is once again glucosylated.

It is evident from the description of restriction and modification systems that their specificity resides in the host. It is for this reason that the term "host-controlled modification and restriction" is frequently applied to this phenomenon.

The quantitatively important modification mechanism involves methylation of specific sites on the DNA, which results in protection against the activity of restricting endonuclease. Two types of restriction-modification systems involving methylation have been described. Type I systems, represented by those of *E. coli* B and *E. coli* K12, are complex and require a number of participating factors for both the restricting and modifying activities. Restriction requires a divalent metal, S-adenosylmethionine ATP, and double-stranded DNA as substrate for nucleolytic cleavage. The modification activity requires a divalent metal and S-adenosylmethionine, which serves as a methyl donor. Modification of the DNA involves transfer of the methyl group from S-adenosylmethionine to produce either 5-methylcytosine or N-6-methyladenine in the DNA, depending upon the specificity of the methylase. Genetic analysis has revealed that three loci are involved in type I restriction-modification systems, *hss*, *hsr*, and *hsm* (*hs* stands for host specific, *s*, *r*, and *m* stand for specificity, restriction, and modification, respectively.)

Mutant strains have been selected that are phenotypically restrictionless (R⁻), and 50 percent of these are also modificationless (M⁻). Strains that are solely M⁻ have not been obtained for obvious reasons: this would be a lethal situation, since the presence of active restriction in the absence of modification would result in destruction of the cellular DNA. The frequency of the phenotypically R⁻M⁺ and R⁻M⁻ (50/50) strains and the three genetic loci involved in the phenomenon of restriction and modification are now understood in terms of the subunit structure of the enzyme involved. The restriction and modification activities reside in a single functional unit, together with a protein subunit specified by the *hss* locus and that stereospecifically recognizes certain nucleotide sequences. The nuclease and methylase subunits are produced, respectively, by the *hsr* and *hsm* loci.

The DNA specificity site for type I modification is the substrate for the methylation reaction, but it is not the substrate site for the endonuclease. The enzyme must bind to the specificity site and then migrate to other sites where endonucleolytic cleavage occurs. It is known that an excessive hydrolysis of ATP is associated with the endonuclease subunit. This expenditure of energy might be involved in the migration of the enzyme along the DNA.

It is known, for example, that the endonuclease-methylase complex must migrate a minimum distance before the cleavage of the DNA occurs. Type I nucleases are peculiar in that they do not turn over, i.e., an enzyme molecule is capable of performing a single catalytic event. The modifying enzyme methylates both strands of DNA at the recognition site, although methylation of a single strand is sufficient to prevent restriction (migration of the nuclease from the site of initial binding). The modification of both strands of DNA means that upon semiconservative replication one of the daughter strands is methylated. This feature serves to protect rapidly replicating DNA from the action of endogenous restriction. It is known that the rate of methylation of a totally unmodified duplex DNA site is slow compared to the rate of replication and restriction and that the rate of methylation of a hemimodified site is very rapid. The relative rates of nuclease and methylase activity on DNA with unmodified versus hemimodified specificity sites provide the organism with appropriate protection against incoming foreign DNA that lacks the specific methylation, while protecting its own DNA where potential restriction sites are fully or hemimethylated.

The type II restriction and modification systems are also composed of endonucleases and methylases, but they are different in many aspects from the type I systems. The type II endonucleases and methylases exist as separate proteins and are active without the mediation of a common site-specific subunit. The sole requirement for the restricting nucleolytic activity is Mg^{2+}. Like type I methylases, the corresponding type II activity also uses S-adenosylmethionine as the methyl group donor. The most significant property of the type II systems is that the recognition site on the DNA is the substrate not only for the methylase but also for the endonuclease, i.e., the cleavage occurs at the unmodified recognition site.

Because the nucleotide sequence at the recognition site governs the specificity of the restriction-modification system and also is the site of action of the restricting nuclease, all the fragments produced from an unmethylated DNA molecule by a type II endonuclease contain the same termini. This has permitted sequence analysis of these recognition sites. The sequences of the recognition sites for many of the type II restriction enzymes have been determined. Table 8-3 lists several of these sequences, together with the positions at which the double-stranded break occurs in the unmodified DNA and the positions of methylation in the modified DNA. An interesting common feature of all these sequences is the presence of dyad symmetry. This type of symmetrical structure appears to be prevalent in regions of DNA that are designed specifically to bind proteins. Two types of scissions are made by type II restriction endonucleases: some cleave the double-stranded DNA to produce nonoverlapping ends, while others cleave on a bias to produce fragments with single-stranded complementary ends. Most site-specific nucleases produce 3' overlapping ends, and some others produce 5' overlapping ends. More than 150 site-specific nucleases have been obtained from microorganisms, and in the case of many of these, the companion methylases have also been described. The specificity shown by type II restriction nucleases raises the question as to whether these systems serve a function different from the one of protection involved for the type I restriction system. It has been shown, for example, that these site-specific nucleases can be used in vivo for rare, site-specific, genetic recombinational events. Regardless of their in vivo role, these endonucleases have been used extensively for DNA sequence analysis and for study of gene structure and function (Chap. 64). In addition to these analytic procedures, these site-specific nucleases have been extensively employed for various feats of genetic engineering and for the cloning of desirable genes, using such organisms as *E. coli* as the recipient. An outline of a cloning procedure is given in Figure 8-11. The contribution that recombinant DNA

TABLE 8-3. TYPE II RESTRICTION-MODIFICATION RECOGNITION SITES

Specificity Designation	Enzyme	Substrate ↓ = Hydrolysis • = Methyl
*Eco*RI	Endonuclease	↓ GAATTC CTTAAG ↑
	Methylase	• GAATTC CTTAAG •
*Eco*RII	Endonuclease	↓ NCC(A)GGN NGG(T)CCN ↑
	Methylase	• NCC(A)GGN NGG(T)CCN •
*Hin*dIII	Endonuclease	↓ NAAGCTTN NTTCGAAN ↑
	Methylase	• NAAGCTTN NTTCGAAN •
*Hin*dII	Endonuclease	↓ NGTPyPuACN NCAPuPyTGN ↑
	Methylase	• NGTPyPuACN NCAPuPyTGN •

*The nomenclature for restriction-modification systems uses the first letter of the genus designation, the first two letters of the species designation, a strain or type designation (if appropriate), and the identification of the genophore that codes for the system, if other than the chromosome. *Eco*RI signifies the restriction-modification system of *E. coli* antibiotic resistance factor RI. A suggested nomenclature for restriction-modification systems is given in Smith and Nathans: J Mol Biol 81:419, 1973.

technology has made to the solution of many problems in biology is well known.

Recombination

A primary significance of the various types of gene transfer previously described is the formation of merozygotes (partial or incomplete zygotes) within which genetic recombination may occur. Recombination simply defined is the process by which a genetic linkage group acquires information from another linkage group.

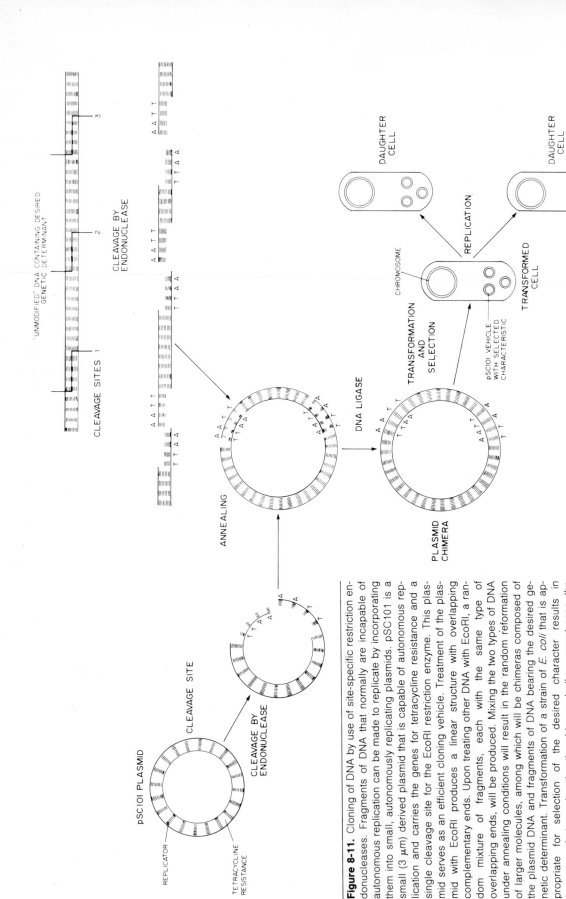

Figure 8-11. Cloning of DNA by use of site-specific restriction endonucleases. Fragments of DNA that normally are incapable of autonomous replication can be made to replicate by incorporating them into small, autonomously replicating plasmids. pSC101 is a small (3 μm) derived plasmid that is capable of autonomous replication and carries the genes for tetracycline resistance and a single cleavage site for the EcoRI restriction enzyme. This plasmid serves as an efficient cloning vehicle. Treatment of the plasmid with EcoRI produces a linear structure with overlapping complementary ends. Upon treating other DNA with EcoRI, a random mixture of fragments, each with the same type of overlapping ends, will be produced. Mixing the two types of DNA under annealing conditions will result in the random reformation of larger molecules, among which will be chimeras composed of the plasmid DNA and fragments of DNA bearing the desired genetic determinant. Transformation of a strain of E. coli that is appropriate for selection of the desired character results in formation of clones bearing the chimera. In the present case, the selective pressure can be enhanced by the simultaneous selection of the tetracycline resistance and of the phenotypic expression of the pertinent character.

Three types of recombination can be distinguished in prokaryotic systems: (1) general recombination, occurring between regions of extensive sequence homology (such as *recA* promoted recombination in *E. coli*), (2) site-specific recombination, which occurs at unique nucleotide sequences on both donor and recipient molecules (exemplified by the integration of phage λ into the *E. coli* chromosome through interaction between identical phage and bacterial sequences), and (3) illegitimate recombination, which is site-specific for the donor molecule but occurs at random on relatively nonspecific sequences on the recipient molecule (the movement of transposable genetic elements such as phage Mu and the transposon Tn3 are examples of this type of recombination). The three types of recombination differ, therefore, in the amount of sequence homology required between donor and recipient molecules (general recombination requires the greatest amount of homology, while illegitimate recombination requires the least). General recombination promotes efficient exchange of genetic information between highly related linkage groups to generate products of a very similar nature, while site-specific and illegitimate recombination permit interaction between disparate linkage groups that can result in products of novel form.

Mechanisms of General Recombination

Plausible models of genetic recombination have been fostered by extensive studies of the reassortment of heritable traits among bacteria and bacteriophage, by description of enzyme-DNA interactions, by mutant methodology, and by knowledge of the structural and mechanical properties of DNA.

Bacteriophage have served as powerful tools for the study of genetic recombination. Genetic crosses with phage are readily performed by mixedly infecting bacterial cells with appropriately genetically marked phage and noting the properties of the progeny by easily recognized traits, such as plaque morphology. Phage chromosomes are small and can be extracted intact from either infected cells or phage particles. This feature allows physical and

biochemical analyses of whole molecules of DNA. The results of transformation experiments also have revealed interesting properties of recombinant DNA molecules. An advantage of this type of gene transfer is that well-characterized DNA can be employed as the exogenote and its fate followed during the process of recombination. Both of these genetic tools have shown that incipient recombinant DNA molecules are heterozygous for the genetic markers under investigation. Heterozygosity is generally considered a property of the diploid state and, in classic genetic terms, is defined as the presence of allelic forms of the same gene. The heterozygosity mentioned above, which disappears upon one round of replication of the DNA molecule within which it is found, represents a region of heteroduplex DNA, a region within the double-stranded DNA molecule composed of single strands from each parent.

Current efforts to explain recombination are aimed at defining the molecular events involved in the formation of heteroduplex (heterozygous) DNA flanked by recombinant markers (derived from each parent). Models of genetic recombination hold that heteroduplex DNA is formed by the assimilation of a single strand of DNA from one chromosome by another chromosome. The essential propinquity for this assimilation is complementarity of the bases within the participating regions of the chromosomes and, therefore, may be restricted in application to the process of generalized recombination. Some of the processes described below, however, should also apply to site-specific and illegitimate genetic recombination.

Single-strand assimilation is most probably initiated by a nuclease that produces a single-stranded nick in the donor duplex DNA molecule. This initial scission is followed by displacement of the 5′-terminal bordering strand by the action of DNA polymerase, which initiates DNA synthesis at the 3′-OH border of the scission. (This polymerization may, in fact, be preceded by gap formation.) The single strand that is locally freed from the duplex structure is then available for displacing a like strand from a homologous region of a recipient DNA molecule (Fig. 8-12). Although the precise mechanism of assimilation is obscure,

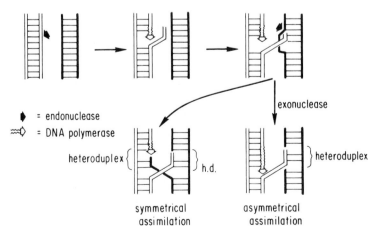

= endonuclease

= DNA polymerase

Figure 8-12. Initiation of recombination by asymmetrical and symmetrical strand assimilation. Asymmetrical strand assimilation results in the formation of the heteroduplex DNA in one of the participating molecules, whereas symmetrical exchange results in formation of heteroduplex DNA on both molecules. The arms of the chromosomes that flank the heteroduplex region are not recombinant in either case.

results are available that show that superhelical, but not relaxed circular, phage DNA is capable of assimilating homologous, single-stranded DNA fragments in vitro and that the tripartite structure formed is relatively stable. These observations as applied to genetic recombination are consistent with the view that in the assimilation process the resident strand of the recipient DNA is removed by the action of nuclease following formation of a tripartite structure. Strand assimilation could be propagated by the simultaneous action of polymerase on the donor molecule and nuclease activity on the recipient molecule. This model (Fig. 8-12) adequately explains heteroduplex formation but does not reveal how the rearrangement of the flanking arms of the participating chromosomes occurs so as to produce the recombinant configuration.

Two extremes can be visualized to explain the formation of recombinant molecules. One involves continued interaction of the type of structure shown in Figure 8-12 with recombination enzymes, and the other involves an isomerization of this putative recombinant intermediate. In the first process, the stepwise, or concerted, interaction of the initial recombinant intermediate with recombination enzymes results in a symmetrical cross-strand exchange (Fig. 8-12), which, unlike the asymmetrical assimilations described above, results in the formation of heteroduplex DNA in each of the participating molecules. The formation of recombinant molecules is completed by an additional symmetrical assimilation involving the DNA strands of the participating molecules that did not undergo the first cross-strand exchange. This process, depicted in Figure 8-13, requires participation of all four strands of DNA and would require a high degree of synchrony of the participating enzymes. An alternative mechanism for generating recombinant molecules following the asymmetrical strand assimilation (Fig. 8-12) is by an isomerization brought about by rotating one pair of homologous arms 180 degrees about an axis between and parallel to them. The recombinational event is terminated by nucleolytic cleavage of the cross strands, as depicted in the figure. The isomerization model results in the formation of heteroduplex DNA in only one of the chromatids. This would seem to rule out its operation in vivo as a significant mechanism for genetic recombination, because genetic segregation data suggest that heteroduplex DNA frequently forms on both chromatids. This problem is readily solved, and plau-

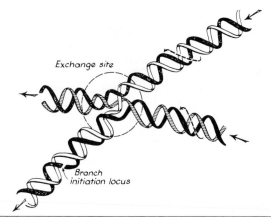

Figure 8-14. Migration of a cross-strand exchange by rotary diffusion. Heteroduplex DNA can be formed on both DNA molecules by rotating each participating duplex in the same sense. *(Courtesy of Dr. T. Broker.)*

sibility of the model is maintained, by the process of rotary diffusion. Rotary diffusion of the cross strand occurs by rotating the individual DNA duplexes in the same sense. This results in migration of the cross strand and formation of heteroduplex DNA on both strands (Fig. 8-14).

It should be emphasized that the foregoing models for genetic recombination are purely speculative. The first model is based on extensive interaction of the participating molecules with various enzymes, whereas the second model requires a minimal interaction with these same enzymes, followed by molecular gymnastics that may appear restricted by the nature of the DNA molecule. However, theoretical calculations suggest that isomerizations of the type invoked above are possible, as is the rotary diffusion. It is of interest to note that complexities of structure, such as superhelicity, rather than impeding these events, may in fact facilitate them. The models above, both of which involve an intimate union of participating molecules, are supported by the isolation of branched DNA molecules of coliphage T4 extracted from cells in which permissive conditions for initiation of recombination exist and failure to observe these forms when these conditions are absent

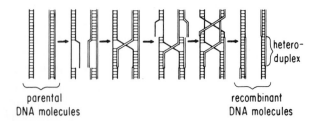

Figure 8-13. The formation of recombinant DNA molecules. Recombinant DNA molecules can be formed by the participation of all four strands of DNA in single-strand assimilation reactions. The enzymes involved are described in the text. *(Modified from Holliday: Genet Res 5:282, 1964.)*

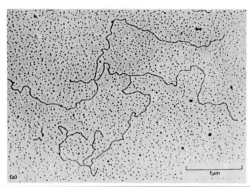

Figure 8-15. Joint molecules of coliphage T4 DNA radioautographs of DNA molecules isolated from T4-infected *E. coli. (Courtesy of Dr. T. Broker.)*

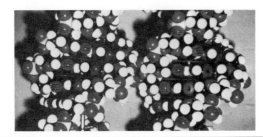

Figure 8-16. Space-filling model of a cross-strand exchange between two molecules of DNA. The cross-strand exchange resulting from the symmetrical assimilation of single strands of DNA does not involve gross distortion of the participating molecules. Base stacking and bond angles are preserved. *(Courtesy of Dr. B. Alberts.)*

(Fig. 8-15). Space-filling models of the unions specified by the models show a remarkable preservation of base stacking and bond angles at cross-strand regions (Fig. 8-16).

The Genetics and Enzymology of Genetic Recombination

Several genetic loci thought to be involved in genetic recombination in *E. coli* have been identified, and the effects of mutation of these loci have been extensively investigated. A central role in general recombination is accorded the *recA* gene, since mutations inactivating *recA* cause almost total cessation of homologous recombination. The *recA* mutations have pleiotropic effects, such as enhanced sensitivity to UV and x-irradiation, as well as enhanced DNA degradation following irradiation. The *recA* gene product is a 37,800 dalton protein that is both an ATP-dependent recombinase and protease. The cofactor for all *recA* protein reactions, such as binding of duplex DNA and duplex unwinding, is single-stranded DNA. The *recA* protein catalyzes single-stranded DNA assimilation by a duplex DNA molecule in a reaction that requires ATP hydrolysis and homologous base pairing. This reaction results in formation of the Holliday structure as described above and thus leads to formation of heteroduplex DNA. The *recA* protein can utilize single-stranded DNA as such or can exploit nicked duplex DNA or partially denatured DNA as a source of single-stranded regions to effect homologous pairing with the target molecule. In the case of a partially denatured donor DNA molecule, the action of topoisomerase would be required to effect intertwining of the recombining molecules.

Two other genes, *recB,C*, are also prominent in recombination, since mutations *recB,C* decrease recombination to about 1 percent of that seen in wild-type cells. This effect, however, is less profound than that of *recA* mutations, and *recB,C* mutants also are less sensitive to UV light and show less DNA degradation after irradiation. The *recB,C* genes encode the subunits of an ATP-dependent exo- and endonuclease (also known as "exonuclease V"). When *recB,C* enzyme is incubated with duplex DNA in the presence of ATP and single-stranded DNA binding protein, the nuclease activity of the enzyme is suppressed, and instead the enzyme effects an ATP-dependent unwinding of the duplex. Therefore, by either its nuclease activity (perhaps working from existing nicks in the DNA) or by its duplex unwinding activity, the *recB,C* enzyme can provide single-stranded DNA that is a potential *recA* substrate.

It has been recognized from genetic studies that the probability of recombination within a given length *E. coli* chromosomal DNA is not constant and that recombination is increased in certain regions. The basis for this is the presence of a specific DNA sequence, designated *chi*, that is capable of enhancing recombination in an orientation-specific manner from one end of the sequence. The frequency of recombination decreases with increasing distance from a given *chi* sequence. Analysis of the *E. coli* K-12 chromosome suggests that the eight base pair *chi* sequence is present once in every 5 to 15 kilobases. Recent evidence suggests that the *recB,C* enzyme may recognize the *chi* sequence and effect unidirectional duplex unwinding from that point.

Mutant methology has revealed more than one pathway for recombination in *E. coli*. The decreased recombination and viability of *recB⁻*, *recC⁻ E. coli* can be suppressed by mutations in the *sbcB* locus, which bring about a lack of exonuclease I. Although the basis of this suppression is obscure, *recB⁻ recC⁻ sbcB⁻* cells are recombination competent. However, recombination in *recB⁻ recC⁻ sbcB⁻* strains can be decreased to 0.05 to 0.5 percent of normal by mutation in the so-called *recF* or *recL* loci. These observations suggest that two recombination pathways normally are operative, a major one, the *RecBC* pathway, and a minor one, the *RecF* pathway, and both of these pathways require an intact *recA* gene.

Several other proteins, in addition to those mentioned above, have also been implicated in genetic recombination. For example, the DNA polymerases I, II, and III and their associated nucleolytic activities might participate in the strand assimilation reactions depicted in Figures 8-12 and 8-13, and polynucleotide ligase would be required for sealing the nicks that are generated during recombination. It is known that at least some other proteins that participate in DNA replication are also involved in recombination. Notable among these are the DNA-binding proteins that facilitate the melting as well as the reannealing of DNA by preventing it from assuming secondary structure. This latter type of protein is indispensable for the displacements and assimilations depicted in Figures 8-12 and 8-13. In addition, the activity of topoisomerases is also likely to be important in providing for intertwining of duplex molecules by nicking-closing reactions (type I activity) and by providing driving energy for strand assimilation from negative supercoiling of the DNA molecule (DNA gyrase activity).

The *recA* gene product is involved in not only general recombination but also in DNA repair processes, including the error-prone SOS system. Following damage to DNA by such agents as UV radiation or the radiomimetic drugs, e.g., mitomycin C, the expression of the *recA* gene is induced so that the *recA* protein levels increase to over 100-fold greater than the noninduced level. This induction depends upon the protease function of the *recA* protein. The required cofactors for the protease are ATP and single-stranded DNA, which is presumably generated after damage to the chromosomal DNA. One substrate of the activated *recA* protease is the *lexA* gene product, which is the repressor of the *recA* gene. Proteolysis of the *lexA* protein effects derepression of the *recA* gene and accounts for the enhanced level of *recA* expression seen after DNA damage. The *lexA* gene product is also the repressor of the error-prone repair functions so that these are also induced by *recA* protease-mediated cleavage of the *lexA* protein. Therefore, the diverse functions of the *recA* gene product are essential for general recombination and also serve to integrate the recombination and DNA repair processes. Coincidentally, the λ repressor is also cleaved by the *recA* protease, which explains the ability of UV light to induce the λ prophage. This may serve a teleologic role for the phage to permit escape from a damaged genetic background.

Site-Specific Recombination

The integration of bacteriophage λ into *E. coli* K-12 occurs by recombination at specific attachment sites (*att*) on both the phage and bacterial chromosomes. These sites contain identical 15 base pair core sequences. Integration of phage λ depends upon the function of the λ *int* gene product, the integrase, which is a type I isomerase and can catalyze DNA nicking-closing reactions in the absence of ATP. It is thought that integrase nicks both the phage and bacterial core regions, and homologous base pairing ensues followed by strand closure to unite the molecules. This process is strongly dependent upon the energy associated with negative supercoiling of the λ chromosome, since relaxed λ molecules are incapable of integration in in vitro systems, and the DNA gyrase inhibitor, coumermycin, can prevent λ integration in vivo (in fact, the identification of an activity permitting relaxed λ molecules to participate in integrative recombination led to the discovery of DNA gyrase). Integration of λ also requires the integration host factor, which is encoded by the chromosomal *himA* and *himD* genes, but the role of these host factors is unknown.

Excision of prophage λ from the bacterial chromosome is an exact reversal of the integration process and requires both the integrase and an additional phage protein encoded by the *xis* gene. Since excision involves strand breaking and joining within the core sequences, both λ and bacterial chromosomes are in general restored precisely. Occasionally, imprecise excision of λ occurs that can generate specialized transducing phage derivatives. This process may occur at bacterial sequences bearing similarity to the core region of the normal bacterial *att* site.

The primary *E. coli* K-12 *att* site is located between *gal* and *bio*. Integration of λ can occur into other sites on the chromosome if the *att* locus is deleted by mutation. These secondary sites bear some homology to the core sequence but are recognized much less efficiently by the integrase

protein (about 0.5 percent of the frequency of normal integration at the primary *att* site locus). Integration of λ at secondary attachment sites and subsequent imprecise excision has been exploited as a genetic tool to isolate specializing transducing phages-bearing genes from many regions of the *E. coli* K-12 chromosome. Figure 8-17 depicts the λ integration-excision pathway.

Illegitimate Recombination

This form of recombination is site-specific only for the donor molecule in the sense that the ends of the element directing recombination are the invariant junction sites in the recombinant molecule. Illegitimate recombination is best exemplified by the process of transposition, whereby certain genetic elements can move into various target sequences in either an intramolecular or intermolecular fashion in the absence of extended homology or apparent site-specificity. Transposition is exhibited by the simple insertion elements, such as ISI, by small drug resistance transposons encoding single antibiotic resistances, such as Tn3 (ampicillin resistance), and by longer drug resistances elements such as Tn10 (tetracycline resistance). The bacteriophage Mu is perhaps the largest transposable element (38 kilobases). Despite the diversity of transposable elements, two features of transposition appear to be general. First, insertion of a transposable element creates a short

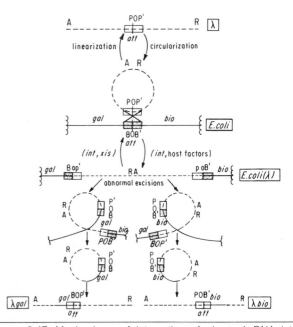

Figure 8-17. Mechanisms of integration of phage λ DNA into and out of the *E. coli* chromosome. The site specific recombination is directed by the sequence —GCTTTTTTTATACGAA— —CGAAAAAAATATGCTT— which is present at the *att* site and on the phage. Abnormal excisions, as shown, lead to the formation of particles containing adjacent host chromosome DNA. *(From Kornberg: DNA Replication, 1980. San Francisco, W. H. Freeman & Co.)*

duplication of the target molecule about either end of the inserted element. Second, transposition does not involve actual excision of the transposable element but involves in situ replication of the transposing sequence in such a manner that a new copy of the element is formed within the target sequence while the original copy is conserved.

The mechanism of transposition has been best studied for Tn3. Tn3 is a 5 kilobase transposable element that encodes β-lactamase and two other protein products, a transposase (*tnpA*) and a resolvase (*tnpR*). Tn3 is bounded by 38 base pair inverted repeats that appear to be essential for transposition and also contains a *cis*-active internal resolution site that is required for completion of the transposition process. Transposition of Tn3 occurs in two distinct steps. First, replicative fusion of the Tn3-containing replicon to the target molecule occurs by joint action of the transposase, which may recognize the target sequence and some site within the Tn3 element (perhaps one of the inverted repeats) and replication of the Tn3 element to form a cointegrate structure (Fig. 8-18). The temporal relationship between transposase action and replication is not known, but the replication function is clearly provided by the host cell and extends only over the Tn3 sequence. The second step in transposition involves resolution of the cointegrate by the action of the recombinase. The recombinase effects recombination between the Tn3 copies within the AT-rich internal resolution site; deletion of this site absolutely prevents resolution of the cointegrate. Although the transposition sequence shown in Figure 8-18 occurs between two molecules, Tn3 can also transpose between sites in the same molecule by essentially the same process. These intramolecular transposition events are associated with inversions or deletions of the intervening chromosome according to the manner in which an intramolecular cointegrate is resolved. Many of the details of Tn3 transposition, such as how the transposase recognizes specific sequences and effects fusion of donor and target molecules, are not known. Even less is known about other transposition mechanisms of other transposable elements, but it does not appear that cointegrates are common intermediates in transposition. Very different processes may occur to effect transposition of the larger sequences, such as bacteriophage Mu.

The importance of transposable elements in the spread of antibiotic resistance determinants has been well recognized. The recombination functions of transposons may also be exploited by the bacterial cell to effect regulation of chromosomal gene expression. In *Salmonella*, the flagellar antigens may be of two types, H1 and H2. These antigens are alternatively expressed by a given cell (or clonal population derived from a single cell); this alternating expression is called phase variation. Molecular analysis of the phase variation system has shown that the H2 flagellin gene is adjacent to a 970 base pair sequence that is capable of existing in either orientation with respect to H2 (Fig. 8-19). In one orientation, a promoter located at one end of this sequence directs transcription of the adjacent H-2 gene and a contiguous gene, *rhl*, which encodes

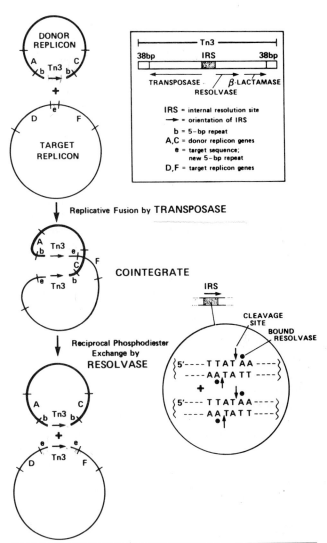

Figure 8-18. A scheme for replicative transposition of Tn3. Additional details are given in the text. *(From Kornberg: 1982 Supplement of DNA Replication, 1982. San Francisco, W. H. Freeman & Co.)*

the repressor of the H-1 gene. This orientation is correlated with the H2 flagellar phenotype. The opposite orientation of the 970 base pair sequences permits no transcription of H-1 and *rhl*, and the cell expresses the alternating H1 flagellar antigen. Inversion of the controlling element is effected by a site-specific recombinase encoded by the *hin* gene, which is contained within the 970 base pair sequence. The *hin* gene product appears to recognize the inverted repeats found at either end of the controlling sequence. These repeats are essential for phase variation, since deletion of either repeat prevents inversion of the element and fixes flagellar antigenic types. The similarities between the phase variation system to transposable elements is all the more striking with the finding that the *hin* gene product shares significant protein sequence homology with the *tnpR* gene product of Tn3. The *Salmonella*

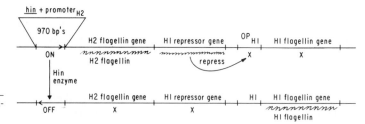

Figure 8-19. The basis for the control of flagellar phase variation in *S. typhimurium*. Other details are given in the text.

phase variation system may represent the acquisition by the bacterial cell of a portion of a transposable element that retains the site-specific recombinase function and constitutes a useful regulatory mechanism. Recent work in *N. gonorrhoeae* suggests that pilus expression may also be controlled by an invertible regulating element analogous to the *hin* sequence.

FURTHER READING

Books and Reviews

Boyer HW: DNA restriction and modification mechanisms in bacteria. Annu Rev Microbiol 25:153, 1971

Boyer HW: Restriction and modification of DNA: Enzymes and substrates. Fed Proc 33:1125, 1974

Broker TA, Doermann M: Molecular and genetic recombination of bacteriophage T4. Annu Rev Genet 9:213, 1975

Cambell A: Episomes. New York, Harper, 1969

Clark AJ: Recombination deficient mutants of *E. coli* and other bacteria. Annu Rev Genet 7:67, 1973

Clewell DB: Plasmids, drug resistance, and gene transfer in the genus *Streptococcus.* Microbiol Rev 45:409, 1981

Drake JW: The Molecular Basis of Mutation. San Francisco, Holden Day, 1970

Dressler D, Potter H: Molecular mechanisms in genetic recombination. Annu Rev Biochem 51:727, 1982

Goodgal SH: DNA uptake in *Haemophilus* transformation. Annu Rev Genet 16:169, 1982

Gorini L: Informational suppression. Annu Rev Genet 4:107, 1970

Gorini L, Beckwith JR: Suppression. Annu Rev Microbiol 20:401, 1966

Gottesman S: Lambda site-specific recombination: The *att* site. Cell 25:585, 1981

Grossman L, Braun A, Feldberg R, et al.: Enzymatic repair of DNA. Annu Rev Biochem 44:14, 1975

Hayes W: The Genetics of Bacteria and their Viruses, 2nd ed. New York, Wiley, 1968

Heffron F: Tn3 and its relatives. In Shapiro JA (ed): Mobile Genetic Elements. New York, Academic Press, 1983, p 223

Helinski DR: Plasmid-determined resistance to antibiotics: Molecular properties of R factors. Annu Rev Microbiol 27:437, 1973

Kleckner, N. Translocatable elements in procaryotes. Cell 11:11, 1977

Kleckner N: Transposable elements in prokaryotes. Annu Rev Genet 15:341, 1981

Levy SB, Clowes RC, Koenig EL (eds): Molecular Biology, Patho-genicity and Ecology of Bacterial Plasmids. New York, Plenum Press, 1981

Lewin B: Gene Expression. Plasmids and Phages. New York, Wiley, 1977, vol 3

Linn S, Lautenberger JA, Barnet E, et al.: Host-controlled restriction and modification enzymes of *Escherichia coli* B. Fed Proc 33:1128, 1974

Miller JH: Experiment in Molecular Genetics. Cold Spring Harbor, NY, Cold Spring Harbor Laboratory, 1972

Radding CM: Genetic recombination: Strand transfer and mismatch repair. Annu Rev Biochem 47:847, 1978

Rowbury RJ: Bacterial plasmids with special reference to their replication and transfer properties. Prog Biophys Mol Biol 31:271, 1977

Smith HO, Danner DB, Deich R: Genetic transformation. Annu Rev Biochem 50:41, 1981

Stahl FW: Special sites in generalized recombination. Annu Rev Genet 13:7, 1979

Starlinger P, Saedler H: IS-elements in microorganisms. Curr Top Microbiol Immunol 75:111, 1976

Strickberger MW: Genetics. New York, Macmillan, 1968

Watson JD: Molecular Biology of the Gene, 3rd ed. New York, Benjamin, 1976

Willetts N: The genetics of transmissible plasmids. Annu Rev Genet 6:257, 1972

Willetts N, Skurray R: The conjugation system of F-like plasmids. Annu Rev Genet 14:41, 1980

Witkin EW: Ultraviolet mutagenesis and inducible DNA repair in *Escherichia coli.* Bacteriol Rev 40:869, 1976

Selected Papers

Alberts BM, Frey L: T4 bacteriophage gene 32: A structural protein in the replication and recombination of DNA. Nature 227:1313, 1970

Avery OT, MacLeod CM, McCarty M: Induction of transformation by a deoxyribonucleic acid fraction isolated from pneumococcus type III. J Exp Med 79:137, 1944

Broker TR, Lehman IR: Branched DNA molecules: Intermediates in T4 recombination. J Mol Biol 60:131, 1971

Cavalli S, Sforza LL, Lederberg J, et al.: An infective factor controlling sex compatibility in *Bacterium coli.* J Gen Microbiol 8:89, 1953

Franke A, Clewell DB: Evidence for conjugal transfer of a *Streptococcus faecalis* transposon (Tn916) from a chromosomal site in the absence of plasmid DNA. Cold Spring Harbor Symp Quant Biol 45:77, 1980

Goodman HM, Abelson J, Landy A, et al.: Amber suppression: A nucleotide change in the anticodon of tyrosin tranfer RNA. Nature 217:1019, 1968

Gottesman S, Beckwith JR: Directed transposition of the arabinose operon: A technique for the isolation of specialized transducing bacteriophages for any *Escherichia coli* gene. J Mol Biol 44:117, 1969

Hedges RW, Datta N, Kontomichalou P, et al.: Molecular specificities of R factor-determined β-lactamases: Correlation with plasmid compatibility. J Bacteriol 117:56, 1974

Hirsh D: Tryptophan transfer RNA as the UGA suppressor. J Mol Biol 58:439, 1971

Holloman WK, Wiegand R, Hoessli C, et al.: Uptake of homologous single-stranded fragments by superhelical DNA: A possible mechanism for initiation of genetic recombination. Proc Natl Acad Sci USA 72:2394, 1975

Horiuchi K, Zinder ND: Cleavage of bacteriophage f1 DNA by the restriction enzyme of *Escherichia coli* B. Proc Natl Acad Sci USA 69:3220, 1972

Horinouchi S, Weisblum B: Posttranscriptional modification of mRNA conformation: Mechanism that regulates erythromycin-induced resistance. Proc Natl Acad Sci USA 77:7079, 1980

Jakes KS, Zinder ND: Highly purified colicin E3 contains immunity protein. Proc Natl Acad Sci USA 71:3380, 1974

Kelly TJ, Smith HO: A restriction enzyme from *Haemophilus influenzae*. II. Base sequence of the recognition site. J Mol Biol 51:393, 1970

Kier LD, Yamasaki E, Ames BN: Detection of mutagenic activity in cigarette smoke condensates. Proc Natl Acad Sci USA 71:4159, 1974

Lacey RW, Richmond MH: The genetic basis of antibiotic resistance in *S. aureus*: The importance of gene transfer in the evolution of this organism in the hospital environment. In Yotis WW (ed): Recent Advances in Staphylococcal Research. Ann NY Acad Sci 236:395, 1974

Landy A, Ross W: Viral integration and excision: Structure of the lambda *att* sites. Science 197:1147, 1977

Lederberg J, Tatum EL: Gene recombination in *Escherichia coli*. Nature 158:558, 1946

Luria S, Delbruck M: Mutation of bacteria from virus sensitivity to virus resistance. Genetics 28:491, 1943

Meselson MS: Formation of hybrid DNA by rotary diffusion during genetic recombination. J Mol Biol 71:795, 1971

Meselson MS, Radding CM: A general model for genetic recombination. Proc Natl Acad Sci USA 72:358, 1975

Meselson MS, Yuan R: DNA restriction enzyme from *E. coli*. Nature 217:110, 1968

Meyer TF, Mlawer N, So M: Pilus expression in *Neisseria gonorrhoeae* involves chromosomal rearrangement. Cell 30:45, 1982

Morrow JS, Cohen S, Chang A, et al.: Replication and transcription of eukaryotic DNA in *Escherichia coli*. Proc Natl Acad Sci USA 71:1743, 1974

Novick RP, Brodsky R: Studies on plasmid replication. I. Plasmid incompatibility and establishment in *Staphylococcus aureus*. J Mol Biol 68:285, 1972

Novick RP, Edelman I, Schwesinger MD, et al.: Genetic translocation in *Staphylococcus aureus*. Proc Natl Acad Sci USA 76:400, 1979

Novick RP, Khan SA, Murphy E, et al.: Hitchhiking transposons and other mobile genetic elements and site-specific recombination systems in *Staphylococcus aureus*. Cold Spring Harbor Symp Quant Biol 45:67, 1980

Setlow RB, Carrier WL: The disappearance of thymine dimers from DNA: An error correcting mechanism. Proc Natl Acad Sci USA 51:226, 1964

Sigal N, Alberts B: Genetic recombination: The nature of a crossed strand exchange between two homologous DNA molecules. J Mol Biol 71:789, 1972

Simon M, Zieg J, Silverman M, et al.: Phase variation: Evolution of a controlling element. Science 209:1370, 1980

Smith GR, Schultz DW, Taylor AF, et al.: Chi sites, *recB,C* enzyme and generalized recombination. Stadler Symp 13:25, 1981

Smith HO, Wilcox KW: A restriction enzyme from *Haemophilus influenzae*. I. Purification and general properties. J Mol Biol 51:379, 1970

Vovis GF, Horiuchi K, Zinder ND: Kinetics of methylation of DNA by a restriction endonuclease from *Escherichia coli* B. Proc Natl Acad Sci USA 71:3810, 1974

Yuan R, Hamilton DL, Burckhardt J: DNA translocation by the restriction enzyme from *E. coli* K. Cell 20:237, 1980

Zinder ND, Lederberg J: Genetic exchange in *Salmonella*. J Bacteriol 64:679, 1952

CHAPTER 9

Antimicrobial Agents

One of the major triumphs of medical science in the twentieth century has been the virtual eradication of many infectious diseases by the use of specific chemotherapeutic agents. Two important discoveries heralded a new era in chemotherapy and revolutionized the therapy of infectious diseases. The first was the discovery in 1935 of the curative effect of the red dye Prontosil on streptococcal infections. Prontosil was the forerunner of the sulfonamides. Although it has no antibacterial activity in vitro, in the body it releases its active component, *p*-aminobenzenesulfonamide (sulfanilamide). The second discovery, and the discovery that ushered in the golden age of antibiotic therapy, was the account published in 1940 of the properties of an extract of cultures of *Penicillium notatum*. Although the discovery of penicillin had been made in 1929 by Fleming, it was Florey and Chain and their associates at Oxford University who demonstrated its unequalled potency and the feasibility of its extraction from culture fluids. A few of the useful antibiotics, like penicillin, were entirely fortuitous discoveries, but from the discovery of streptomycin in 1944 to the present, the search for such agents has been a highly planned, scientifically designed effort.

they are used, whereas bacteriostatic agents exert only an inhibitory effect and rely on the cellular and humoral defense mechanisms of the host for the final eradication of the infection.

The ideal chemotherapeutic agent is one to which susceptible organisms do not become genetically or phenotypically resistant. Although it is desirable that the agent be effective against a broad range of microorganisms, problems often arise as a consequence of the use of such broad-spectrum drugs (p. 457). The ideal agent should not be allergenic, nor should continued administration of large doses cause adverse side effects. It should remain active in the presence of plasma, body fluids, or exudates. It is desirable that the agent be water-soluble and stable and that bactericidal levels in the body be rapidly reached and maintained for prolonged periods.

Although the search for new drugs to fill the gaps in therapy for infectious diseases has been most successful with antimicrobial agents of microbial origin, important compounds of synthetic origin have also been introduced. The synthetic chemist has added immeasurably to our therapeutic armamentarium, especially in the area of mycobacterial chemotherapy.

Chemotherapeutic Agents

Desirable Properties

Selective toxicity is an essential property of a chemotherapeutic agent; it must inhibit or destroy the pathogen without injury to the host. The ideal chemotherapeutic agent is one that is bactericidal rather than bacteriostatic in its effects. Bactericidal agents kill the organisms against which

Antibiotics

As originally defined, an antibiotic was a chemical substance produced by various species of microorganisms that was capable in small concentrations of inhibiting the growth of other microorganisms. The advent of synthetic methods, however, has resulted in a modification of this definition. The term "antibiotic" now refers to a substance produced by a microorganism or to a similar substance produced wholly or partially by chemical synthesis that, in low concentrations, inhibits the growth of other microorganisms.

Antibiotics are widely distributed in nature, where they play an important role in regulating the microbial population of soil, water, sewage, and compost. They differ markedly both chemically and in their mechanism of action. There is thus little or no relation among the antibiotics other than their ability to affect adversely the life processes of certain microorganisms. Of the several hundred naturally produced antibiotics that have been purified, only a few have been sufficiently nontoxic to be of use in medical practice. Those that are currently of greatest use have been derived from a relatively small group of microorganisms belonging to the genera *Penicillium*, *Streptomyces*, *Cephalosporium*, *Micromonospora*, and *Bacillus*.

Although originally obtained from *Streptomyces venezuelae*, chloramphenicol is now produced completely by a synthetic process. The penicillins and cephalosporins are the prime examples of the use of chemical modifications of a core nucleus to alter their pharmacokinetics, antibacterial spectrum, and potential for influencing the production of inactivating enzymes or for resisting their action.

Mechanisms of Action

Chemotherapeutic agents interfere at a number of vulnerable sites in the cell. They interfere with (1) cell wall synthesis, (2) membrane function, (3) protein synthesis, (4) nucleic acid metabolism, and (5) intermediary metabolism. One must keep in mind, however, that there may be a number of stages between the initial or primary effect of the drug and the eventual death of the cell that results. In addition, some agents may have more than one primary site of attack or mechanism of action.

Cell Wall Inhibitors

The bacterial cell is surrounded by a rigid cell wall that protects the fragile protoplasmic membrane beneath from osmotic and mechanical trauma. The assimilation in the cell of low molecular weight, soluble substances from the external environment creates an osmotic pressure within the cell that is many times that of the surrounding medium. Any substance that destroys the wall or prevents the synthesis or incorporation of the wall polymers in growing cells leads to the development of osmotically sensitive cells and death. Since the bacterial cell wall is unique, and the mechanism for its biosynthesis is completely lacking in eucaryotic cells, agents that interfere at this site are highly specific and more likely to be low in toxicity.

The component of the wall that confers rigidity is the peptidoglycan layer. This substance consists of polysaccharide chains composed of alternating units of N-acetylglucosamine and N-acetylmuramic acid. Short pep-

tides linked to the carboxyl group of muramic acid are covalently crosslinked with peptides of neighboring polysaccharide chains (p. 94).

The biosynthesis of the bacterial cell wall consists of three stages, each of which occurs at a different site in the cell (Fig. 9-1). In the first stage, which occurs in the cytoplasm, the recurring units of the backbone structure of murein, N-acetylglucosamine and N-acetylmuramyl-pentapeptide, are sythesized in the form of their UDP derivatives. The second stage takes place in the membrane where N-acetylmuramylpentapeptide is transferred from UDP to a membrane-bound lipid, undecaprenyl phosphate. The other monosaccharide unit, N-acetylglucosamine, is added to yield the recurring disaccharide of the peptidoglycan backbone attached to the undecaprenyl phosphate. These lipid intermediates are then modified, the nature of the modification depending upon the organism. In *Staphylococcus aureus* the modification consists of the amidation of the α-carboxyl of glutamic acid and the addition of five glycine residues to form an open pentaglycine chain. The entire disaccharide unit with its crosslinking chain is transferred to the growing end of the peptidoglycan. Undecaprenol remains in the membrane as its pyrophosphate, from which the orthophosphate ester is regenerated by enzymatic hydrolysis. The third stage of cell wall synthesis takes place outside the cell membrane and consists of the crosslinkage between parallel peptidoglycan chains through a transpeptidation reaction.

At subinhibitory concentrations, antibiotics that interfere with the biosynthesis of peptidoglycan often cause an accumulation in the culture medium of uridine nucleotide intermediates. The nature of these intermediates provides information on the specific reaction site of the drug. Protec-

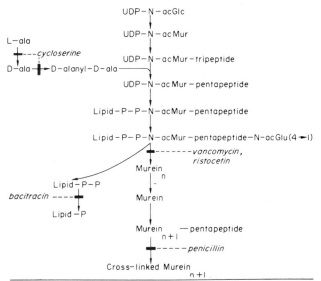

Figure 9-1. Site of action of antibiotics that interfere with cell wall synthesis.

TABLE 9-1. ANTIBIOTICS AFFECTING CELL WALL SYNTHESIS

Antibiotic	Source	Antibacterial Activity	Major Therapeutic Applications
Bacitracin	*Bacillus licheniformis*	Many gram-positive species, *Neisseria*	Sterilization of gut before surgery, topical application
Cycloserine	*Streptomyces orchidaceus*	Many gram-positive and gram-negative bacteria, *Mycobacterium*	Tuberculosis caused by drug-resistant bacilli
Fosfomycin	*Streptomyces fradiae*	Many gram-positive and gram-negative species	Not licensed in US
Penicillin	*Penicillium chrysogenum*	Many gram-positive organisms, *Neisseria*, spirochetes, actinomycetes	Gram-positive cocci infections, syphilis, gonorrhea, meningococcal meningitis
Vancomycin	*Streptomyces orientalis*	Gram-positive bacteria, spirochetes	Severe drug-resistant staphylococcal infections

tion against a concentration of the antibiotic that normally causes lysis and death can be obtained using a medium with a high osmotic pressure. Under such conditions, cells become spherical and are converted to spheroplasts (p. 33). Species of bacteria that lack cell walls, such as the mycoplasmas and halophilic bacteria, are not inhibited by antibiotics that interfere with cell wall synthesis.

Among the antibiotics whose primary action is on cell wall biosynthesis are the β-lactam antibiotics (penicillins, cephalosporins), fosfomycin, cycloserine, vancomycin, and bacitracin (Table 9-1).

β-Lactam Antibiotics

Penicillins

Penicillin is still the most widely used antibiotic for general therapeutic use. The only major problem associated with its use is its ability to provoke hypersensitivity reactions in a small percentage of individuals. In the years since the

first crude product was obtained from *P. notatum*, the penicillin molecule has been chemically manipulated, and numerous natural and semisynthetic congeners have been produced, several of which are very useful therapeutically. By greater attention to cultural conditions that provide the chemical precursors of the antibiotic and by substitution of a high-yielding mutant of *Penicillium chrysogenum*, penicillin production has been enhanced manyfold. The term "penicillin" is generic for the entire group of natural and semisynthetic penicillins. The basic structure consists of a thiazolidine ring joined to a β-lactam ring, to which is attached a side chain that determines many of the antibacterial and pharmacologic properties of a particular type of penicillin (Fig. 9-2).

Natural Penicillins. Of the natural penicillins, benzylpenicillin or penicillin G is clinically the most useful. Its almost exclusive formation is ensured by the addition to the culture medium of the appropriate precursor, phenylacetic acid. Penicillin G is effective against most gram-positive

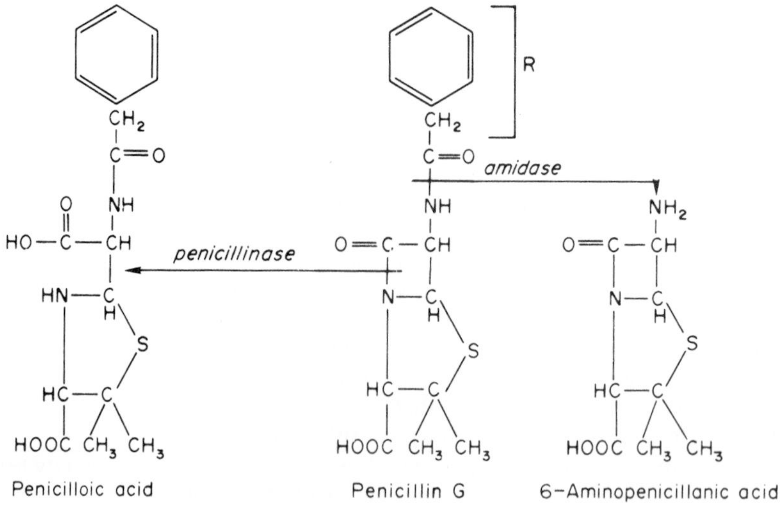

Figure 9-2. Benzylpenicillin (penicillin G) and products of its enzymatic hydrolysis. Penicilloic acid is inactive. 6-Aminopenicillanic acid is the starting point for semisynthetic penicillins.

TABLE 9-2. ORGANISMS FOR WHICH PENICILLINS ARE ANTIMICROBIAL AGENTS OF CHOICE

Organism	Type of Penicillin	Alternative Therapy in Case of Penicillin Allergy
Gram-positive Cocci		
Staphylococcus aureus		
Nonpenicillinase-producing	Penicillin G	A cephalosporin, vancomycin, erythromycin
Penicillinase-producing	Methicillin or nafcillin	As above
Streptococcus pneumoniae	Penicillin G	As above
Streptococcus pyogenes	Penicillin G	A cephalosporin, erythromycin
Streptococcus faecalis	Penicillin G (or ampicillin) plus streptomycin or gentamicin	Vancomycin plus streptomycin, gentamicin
Streptococcus (viridans group)	Penicillin G	A cephalosporin, vancomycin, erythromycin
Gram-positive Bacilli		
Bacillus anthracis	Penicillin G	A tetracycline or erythromycin
Corynebacterium diphtheriae	Penicillin G	Erythromycin
Listeria monocytogenes	Ampicillin	A tetracycline
Gram-negative Cocci		
Neisseria meningitidis	Penicillin G	Chloramphenicol, erythromycin
Neisseria gonorrhoeae	Penicillin G	Spectinomycin, tetracycline, cefoxitin, ampicillin
Gram-negative Bacilli		
Haemophilus influenzae	Ampicillin	Chloramphenicol
Pasteurella multocida	Penicillin G	A tetracycline
Proteus mirabilis	Ampicillin	A cephalosporin, gentamicin, tobramycin
Salmonella (non-*typhi* species)	Ampicillin	Chloramphenicol
Shigella	Ampicillin	
Anaerobic Organisms		
Actinomyces israelii	Penicillin G	A tetracycline, erythromycin, clindamycin
Anaerobic streptococci	Penicillin G	A tetracycline, erythromycin, clindamycin
Bacteroides (non-*fragilis* species)	Penicillin G	Clindamycin, chloramphenicol
Clostridium	Penicillin G	Tetracycline, chloramphenicol

organisms and gram-negative cocci (Table 9-2). Among the least susceptible of the gram-positive cocci are the enterococci and penicillinase-producing strains of *S. aureus*. In general, gram-positive bacteria inhibited by the natural penicillins are more susceptible to these penicillins than to the semisynthetic congeners.

The chemical modification of the penicillin molecule has extended its range of activity and has counteracted some of its undesirable properties. The major disadvantages of penicillin G are that (1) it is inactivated by the acid pH of the gastric juice, (2) it is destroyed by penicillinases, bacterial enzymes that split the β-lactam ring, and (3) its use is sometimes associated with hypersensitivity reactions, which range in severity from rash to immediate anaphylaxis.

Penicillin V is a side chain variant of penicillin G with a similar antimicrobial spectrum. Its sole advantage is that it is more stable in an acid medium and thus better absorbed from the gastrointestinal tract.

Semisynthetic Penicillins. The penicillin nucleus itself, 6-aminopenicillanic acid, is the primary structural requirement for biologic activity. It can be prepared in quantity from benzyl or other natural penicillins by the action of an amidase derived from a number of microbial species (Fig. 9-2). Once this penicillin nucleus is available, various side chains can be attached, and an almost unlimited number of semisynthetic penicillins can be produced (Table 9-3).

TABLE 9-3. CLASSIFICATION OF PENICILLINS

Type and Generic Name	Acid Resistance
Natural penicillins	
Benzylpenicillin (G)	−
Phenoxymethyl penicillin (V)	+
Semisynthetic penicillins	
Penicillinase-resistant	
Methicillin	−
Nafcillin	+
Isoxazolyl penicillins	
Cloxacillin	+
Dicloxacillin	+
Oxacillin	+
Extended spectrum penicillins	
Ampicillin	+
Amoxicillin	+
Antipseudomonas penicillins	
Carbenicillin	*
Azlocillin	*
Mezlocillin	*
Piperacillin	*
Ticarcillin	*

*Not well absorbed from gut, must be given parenterally.

PENICILLINASE-RESISTANT PENICILLINS. Among the most useful of the semisynthetic penicillins is the penicillinase-resistant group, in which an alteration of the side chain has provided protection for the β-lactam ring from the action of penicillinase without removing its antibacterial activity. Such penicillins include methicillin and nafcillin, which are acid-labile, and the isoxazolyl penicillins (cloxacillin, dicloxacillin, and oxacillin), which combine resistance to penicillinase with resistance to acid (Fig. 9-3). Semisynthetic penicillinase-resistant penicillins are the drugs of choice only for penicillin-resistant *S. aureus* and *Staphylococcus epidermidis*.

GRAM-NEGATIVE SPECTRUM PENICILLINS. The most striking change brought about by chemical manipulation of the penicillin side chain is an increase of activity against gram-negative organisms. Among the most clinically useful of these broad-spectrum compounds are ampicillin and amoxicillin, which are acid-stable but penicillinase-sensitive (Fig. 9-4). These drugs, however, are not effective for a number of gram-negative species commonly associated with nosocomial infections. For these organisms, especially *Pseudomonas aeruginosa*, carbenicillin, ticarcillin, and piperacillin are especially useful (Fig. 9-5). These agents are also often effective against certain strains of *Proteus* that are not susceptible to other penicillins or to the cephalosporins (Table 9-4).

Mecillinam is the most active of a new group of amidinopenicillanic derivatives. It has unusual antibacterial properties for a penicillin. It is extremely active against *Escherichia coli* but requires more than 60 times the concentration for an equal effect on gram-positive organisms. In contrast to ampicillin, it is active against many strains of

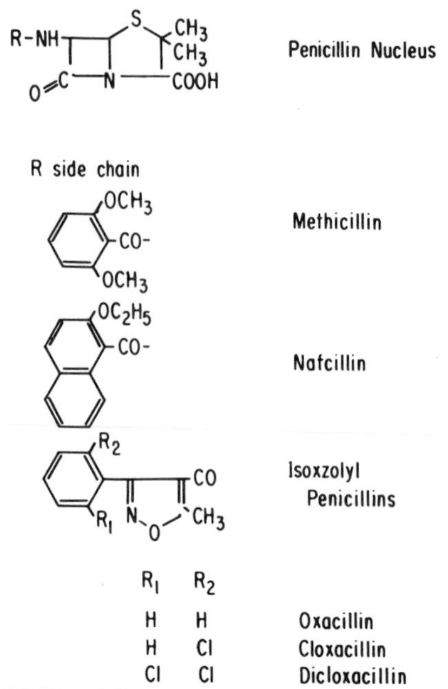

Figure 9-3. Penicillinase-resistant penicillins. These semisynthetic penicillins are especially useful in the treatment of penicillin G-resistant staphylococcal infections. They are inactive against gram-negative bacilli.

Klebsiella and *Enterobacter*. *Pseudomonas* is not inhibited by mecillinam, and its action on *Proteus* is variable.

Cephalosporins

Antibiotics of this group are produced by the fungus *Cephalosporium*. One of the natural compounds, cephalosporin C, resembles penicillin in structure and mode of action. It possesses a β-lactam ring that is fused with a six-member

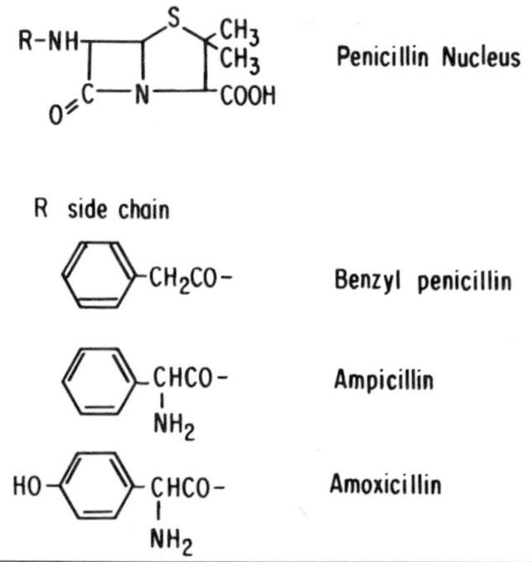

Figure 9-4. Benzyl penicillin (penicillin G) and the aminopenicillins. The aminopenicillins are more active than penicillin G against group D streptococci and a number of gram-negative species. They are not stable to β-lactamases of either gram-positive or gram-negative organisms.

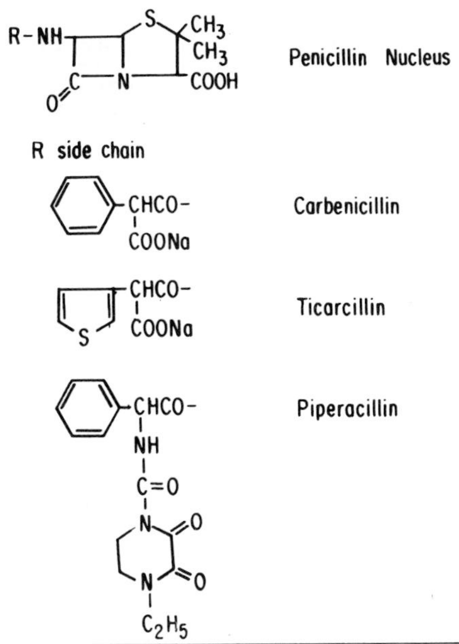

Figure 9-5. Antipseudomonas penicillins. The principal advantage of carbenicillin and ticarcillin is in their activity against *P. aeruginosa* and indole-positive *Proteus* species. Piperacillin has an extended spectrum, similar to ampicillin for gram-positive species and to carbenicillin for gram-negative species. Its activity against *Pseudomonas* is greater than that of carbenicillin.

dihydrothiazine ring instead of the five-member thiazolidine ring characteristic of the penicillins. The cephem nucleus (7-aminocephalosporanic acid), obtained by acid hydrolysis, lends itself to modification that alters both microbiologic activity and pharmacologic properties (Fig. 9-6). All of the cephalosporins possess a sulfur atom at position 1 of the dihydrothiazine ring, with the exception of moxalactam, which has an oxygen atom. Substitutions at position 7 or nearby affect stability against β-lactamases, and further changes in the acyl side chain can alter both antibacterial and pharmacologic properties (Fig. 9-7). Sub-

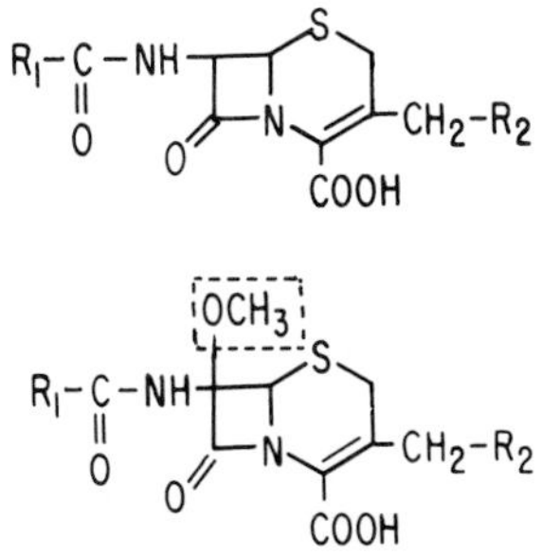

Figure 9-6. Basic structure of the cephalosporins (top) and cephamycins (bottom). In the cephamycins a methoxyl group replaces the hydrogen at the 7 position of 7-aminocephalosporanic acid. This confers unusually high resistance to the β-lactamases. Substitutions at the R_1 and R_2 positions have yielded clinically useful derivatives.

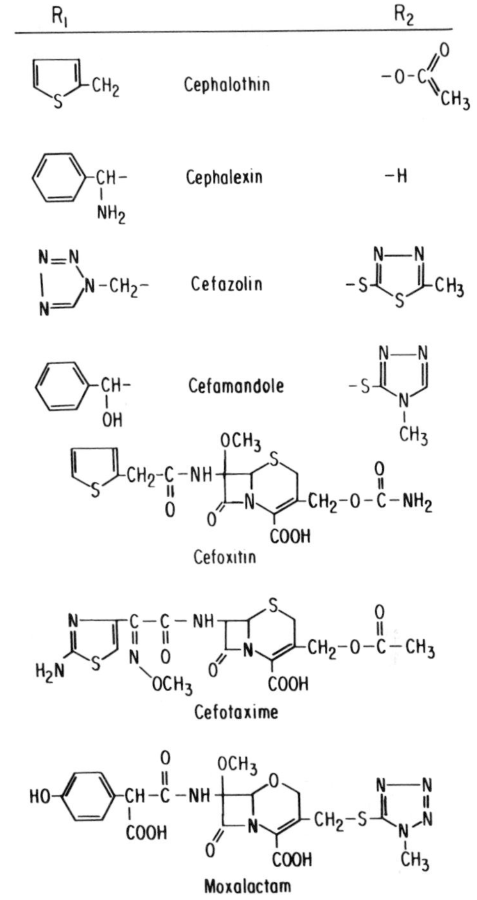

Figure 9-7. Cephalosporins and cephamycins. All are derivatives of the active nucleus of the natural product, cephalosporin C, except cefoxitin, which is a cephamycin derivative (see Fig. 9-6).

stitutions at position 3 of the ring usually affect pharmacologic properties to a greater degree than they affect microbiologic activity.

The cephamycins are similar in structure to the cephalosporins, but they are derived from species of *Streptomyces*. The best-known of the cephamycin derivatives is cefoxitin.

Spectrum of Activity. The cephalosporins are widely used antibiotics. They are active against most organisms susceptible to the penicillins and have been useful alternatives in patients who are allergic to penicillin. The older, first generation cephalosporins all have very similar antimicrobial activity in vitro. They are bactericidal against most gram-positive cocci and many of the common gram-negative bacilli of clinical importance. However, *Enterobacter* species, indole-positive *Proteus* strains, and *Pseudomonas* species are resistant to these older agents. Of the first generation cephalosporins currently marketed in the United States, cephalothin, cefazolin, and cephalexin are the most useful (Table 9-5).

The newer second and third generation cephalosporins are more resistant to the action of cephalosporinases

TABLE 9-4. INFECTIOUS ORGANISMS FOR WHICH PENICILLINS ARE NOT THE ANTIMICROBIAL AGENTS OF CHOICE

Organism	Initial Therapy	Definitive Therapy	Alternate Therapy if Strain is Susceptible
Gram-positive Bacilli			
Corynebacterium diphtheriae	Erythromycin	Erythromycin	
Anaerobic Gram-negative Bacilli			
Bacteroides fragilis	Clindamycin	Same	Chloramphenicol, carbenicillin, ticarcillin
Enterobacteriaceae			
Enterobacter species	Gentamicin or tobramycin*	Same as initial therapy or carbenicillin	Amikacin, tetracycline
Escherichia coli	As above†	Ampicillin or a cephalosporin	Carbenicillin or ticarcillin, kanamycin, amikacin
Klebsiella pneumoniae	As above	A cephalosporin	Amikacin, chloramphenicol
Proteus mirabilis	As above	Ampicillin or a cephalosporin	
Other *Proteus* species	As above	Same as initial therapy	Carbenicillin or ticarcillin, amikacin
Providencia species	As above	Same as initial therapy	As above
Salmonella typhi	Chloramphenicol	Chloramphenicol	Ampicillin or trimethoprim-sulfamethoxazole
Serratia	Gentamicin or tobramycin	Same	Amikacin, carbenicillin or ticarcillin
Shigella species	Ampicillin	Same	Trimethoprim-sulfamethoxazole
Other Gram-negative Bacilli			
Acinetobacter	Gentamicin or tobramycin (±carbenicillin)	Same	Trimethoprim-sulfamethoxazole, minocycline
Brucella species	Tetracycline (±streptomycin)	Same	Chloramphenicol (+streptomycin)
Francisella tularensis	Streptomycin	Same	Tetracycline, chloramphenicol
Haemophilus influenzae‡	Chloramphenicol	Ampicillin	
Pseudomonas aeruginosa	Tobramycin or gentamicin (±ticarcillin or carbenicillin)	Same	Amikacin, polymyxin
Vibrio cholerae	Tetracycline	Same	Trimethoprim-sulfamethoxazole
Yersinia pestis	Streptomycin	Same	Tetracycline, chloramphenicol
Miscellaneous Bacteria			
Legionella species	Erythromycin	Same	Rifampin
Nocardia species	A sulfonamide with ampicillin	Same	Trimethoprim-sulfamethoxazole
Mycoplasma pneumoniae	Erythromycin	Same	Tetracycline
Rickettsia species	Tetracycline	Same	Chloramphenicol

*Before speciation of organism, an aminoglycoside should be given. If susceptibility data permit, a shift to a less toxic agent is recommended.
†Treatment for systemic infections. For uncomplicated urinary tract infections, ampicillin or sulfisoxazole should be used.
‡Treatment for meningitis. For other *Haemophilus* infections ampicillin is the agent of choice.

TABLE 9-5. CLASSIFICATION OF CEPHALOSPORINS AND CEPHAMYCINS

First Generation	Second Generation	Third Generation
Cephalothin	Cefoxitin[†]	Cefotaxime
Cephaloridine	Cefamandole	Ceftizoxime
Cephapirin	Cefuroxime	Ceftriaxone
Cephalexin*	Cefaclor*	Cefmenoxime
Cephradine*		Ceftazidime
Cefazolin		Cefsulodin
Cefadroxil*		Cefoperazone
		Moxalactam[†]

*Oral agents.
[†]Cephamycins.

produced by gram-negative organisms. Cefamandole's major advantage is its increased activity for *Enterobacter, Proteus,* and *Haemophilus.* Cefoxitin has an increased activity for *Proteus, Haemophilus,* and *Bacteroides fragilis.* In the third generation cephalosporins (moxalactam, cefotaxime), the antibacterial spectrum has been expanded to include *P. aeruginosa* and the more resistant gram-negative organisms (Table 9-6). At present, the precise clinical use of these new, highly active β-lactam agents has not been defined.

The cephalosporins are relatively nontoxic, but they may elicit hypersensitivity reactions in a small proportion of individuals.

Mechanism of Action

The β-lactams are bactericidal agents whose killing action is attributable to an interference with cell wall synthesis. The bacterial response following exposure to these agents is very complex and varies with respect to the organism and to the particular antibiotic. This is a reflection of the

TABLE 9-6. IN VITRO ACTIVITY OF REPRESENTATIVE CEPHALOSPORINS AGAINST GRAM-NEGATIVE BACTERIA

Organism	MIC (μg/ml)		
	Cephalothin*	Cefamandole*	Moxalactam*
Escherichia coli	4	0.5	0.06
Enterobacter aerogenes	4	1	0.125
Enterobacter cloacae	> 128	8	0.25
Proteus morganii	> 128	8	1
Serratia marcescens	> 128	64	1
Pseudomonas aeruginosa[†]	> 128	> 128	16 (range 8–32)

From Webber and Yoshida: Rev Infect Dis 4 [Suppl]:S496, 1982.
*Cephalothin and cefamandole are first- and second-generation cephalosporins. Moxalactam is a third-generation agent.
[†]Mean MIC values for three strains.

interaction with specific targets in the membrane, the penicillin-binding proteins, and events occurring during normal cell wall growth. The precise molecular mechanism, however, by which these agents exert their lethal effect remains elusive.

The last stage in cell wall synthesis has been identified as the phase during which inhibition occurs. The peptidoglycan crosslinking enzyme system, considered as a whole, is the target specifically inhibited by penicillin. During this stage the linear peptidoglycan strands are crosslinked by a transpeptidation step in which a peptide bridge is formed between two adjacent strands with the elimination of the terminal D-alanine. When bacteria are grown in the presence of penicillin, uncrosslinked uridine nucleotide intermediates of cell wall synthesis accumulate, and new walls cannot be formed. Electron micrographs of thin sections of penicillin-treated organisms show an accumulation of fibrous material at the growing point of the septum (Fig. 9-8).

Penicillin inhibits not only the transpeptidases responsible for the crosslinking but also, and reversibly, D-alanine carboxypeptidases, which specifically remove D-alanine from a pentapeptide side chain (Chap. 6, p. 108). These enzymes are especially important in the final stages of cell wall synthesis in gram-negative bacilli where there is usually only a single crosslink between one peptide side chain and another.

A structural analogy between penicillin and the D-alanyl-D-alanine end of the pentapeptide in the uncrosslinked precursor of the cell wall has been invoked to explain the molecular basis for the antibacterial action of penicillin (Fig. 9-9). The labile CO-N bond in the β-lactam ring of

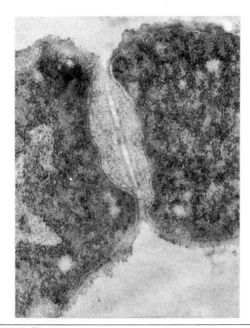

Figure 9-8. Fibrous material accumulated at the growing point of cell of *Bacillus megaterium* induced by penicillin. *(From Fitz-James and Hancock: J Cell Biol 26:657, 1965.)*

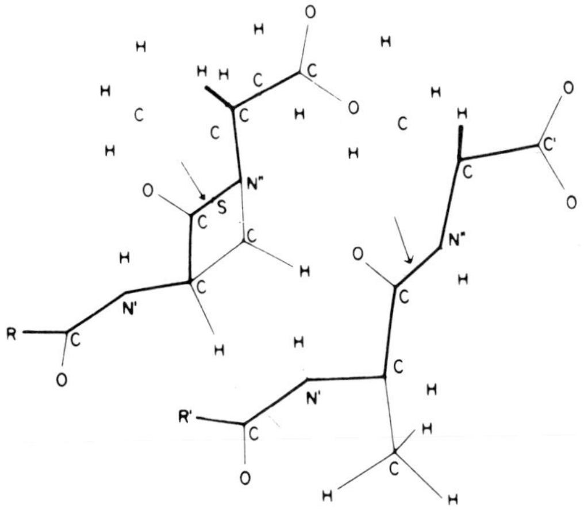

Figure 9-9. Stereomodels of penicillin (left) and of the D-alanyl-D-alanine end of the peptidoglycan strand (right). Arrows indicate the position of the CO-N bond in the β-lactam ring of penicillin and of the CO-N bond in D-alanyl-D-alanine at the end of the peptidoglycan strand. *(From Blumberg and Strominger: Bacteriol Rev 38:291, 1974.)*

penicillin lies in the same position as the peptide bond involved in the transpeptidation. It has thus been proposed that penicillin, acting as a substrate analog of the normal transpeptidation substrate, combines with the transpeptidase and thereby irreversibly inactivates it. Antibiotics containing a β-lactam ring behave chemically as acylating agents, reacting to produce penicilloyl derivatives. The action of penicillin on these enzymes might, therefore, involve acylation of the enzymically active site, with formation of a rather stable inactive complex (Fig. 9-10).

The inhibition of biosynthetic reactions by β-lactam drugs is accompanied by distinctive morphologic changes. The nature of these changes depends on the specific organism, the antibiotic used, and the concentration. The morphologic differences observed are thought to be due to the particular penicillin-binding protein that is affected. These proteins, which are present in the membranes of all bacteria except *Mycoplasma*, bind covalently to β-lactam antibiotics and vary in molecular weight, enzymic activity, and apparent function. In *E. coli* there are at least seven penicillin-binding proteins, some of which have transpeptidase activity and are lethal targets for the action of β-lactam antibiotics. Each appears to have a unique function, including extension of the peripheral cell wall, maintenance of the rod shape, and formation of the septum. Some of the β-lactam antibiotics have binding affinities toward only one or two specific penicillin-binding proteins of *E. coli*. Each of these agents elicits a distinct morphologic response in the organism, such as lysis, filamentation, or ovoid cell formation. For example, cephalothin and benzylpenicillin produce lytic effects on *S. aureus*, cephalexin causes filamentation of *E. coli*, and mecillinam tends to produce osmotically fragile round forms.

The effects of the β-lactams are observed only in growing organisms. If growth is prevented by the omission of a nutrient or by the addition of a bacteriostatic agent, penicillin is without effect. The specific response also depends on the concentration. If staphylococci are exposed to concentrations of penicillin above the minimal inhibitory concentration (MIC), the septum loses its density, the cell wall becomes thinner, and there is rapid loss of viability and cellular lysis. At sub-MICs, however, the cell wall remains normal, but the septum becomes much thicker. Gram-negative bacilli exposed to sub-MICs of β-lactam antibiotics become elongated and, in the absence of septation, form long filamentous cells. Elongation is inhibited by high concentrations of these agents, and the bacilli form large osmotically fragile globular forms that lyse unless the medium is osmotically stabilized.

The lytic effect of penicillin is viewed as a result of unbalanced metabolism and involves endogenous peptidoglycan hydrolyzing enzymes, autolysins, which normally participate in cell wall turnover and separation of bacteria after cell division. In gram-positive organisms, penicillin also causes the release of lipids and teichoic acids, an event that has been linked with the onset of autolytic activity, which is normally regulated by these compounds.

Cells that are not undergoing multiplication can survive in the presence of penicillin because their peptidoglycan is unbroken and there is no reparative crosslinking activity for the penicillin to block. These organisms are persisters and may be responsible for recurrence of infection after penicillin treatment has been terminated.

Resistance to β-Lactam Antibiotics

Biochemical resistance to the β-lactam antibiotics may be attributed to three distinct mechanisms: (1) inactivation of the drug, (2) alteration of the target site, and (3) blocking transport of the drug into the cell. The major basis for bacterial resistance to penicillin, whether occurring naturally or acquired in vivo, is the inactivation of the drug by β-lactamases. These enzymes split the β-lactam ring of the penicillins and cephalosporins between the C and N atoms to form inactive compounds. Penicillin amidase also inactivates penicillin by removing the acyl side chain. These amidases, however, are not widespread among bacteria and are not an important cause of penicillin resistance in pathogenic species. Their primary importance is in the preparation of semisynthetic penicillins and cephalosporins.

β-Lactamases are found in both gram-positive and gram-negative organisms and are responsible for many highly resistant strains, including penicillin-resistant staphylococci obtained from patients. In the development of semisynthetic penicillins and cephalosporins, one of the primary goals was to find compounds that are insensitive to β-lactamase activity.

The β-lactamases of gram-positive bacteria are inducible and have a very high affinity for their substrates. They are extracellular and are produced in relatively large amounts. The production of β-lactamases in gram-positive

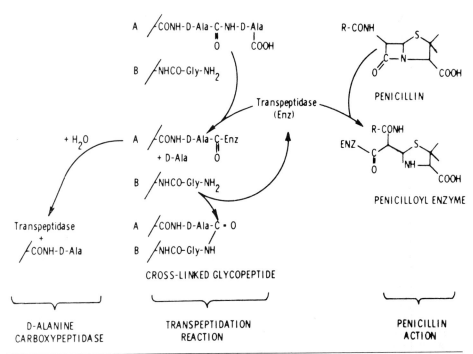

Figure 9-10. Proposed mechanism of transpeptidation and its relationship to penicilloylation and D-alanine carboxypeptidase activity. A. The end of the main peptide chain of the glycan strand. B. The end of the pentaglycine substituent from an adjacent strand. If the acyl enzyme intermediate can react with water instead of the acceptor (left), the enzyme would be regenerated and the substrate released. The overall reaction would be the hydrolysis of the terminal D-alanine residue of the substrate (D-alanine carboxypeptidase activity). *(From Blumberg and Strominger: Bacteriol Rev 38:291, 1974.)*

organisms is mediated by both chromosomal and plasmid genes. However, as in the case of *S. aureus*, the gene on the plasmid can be derived from that on the chromosome via a transposon. In *S. aureus*, penicillinase production is mediated mainly by a plasmid gene that can be easily transferred to other susceptible organisms by bacteriophages.

The β-lactamases from gram-negative organisms are different in many respects from the enzymes of gram-positive species. They are usually constitutive and are cell bound whether they are mediated by R factors or chromosomes. They are produced in smaller amounts and have a much lower affinity for their substrates. The release by gram-positive species of large amounts of β-lactamase into the immediate environs results in a population effect. By contrast, localization of the enzyme in the periplasmic space of gram-negative species restricts access of the drug to the membrane target sites only after the drug has penetrated the cell wall layer. In these organisms, penicillinase resistance cannot be attributed to β-lactamase activity alone but is due to a complex interaction between β-lactamase and a permeability barrier in the outer membrane.

The second mechanism of resistance to the β-lactam drugs is due to an alteration either in the amounts or the affinities of the penicillin-binding proteins for the antibiot-

ic. The most important examples of resistance of this type are those occurring in *Streptococcus pneumoniae* in South Africa and in resistant strains of *Neisseria gonorrhoeae*. Methicillin resistance in *S. aureus* has been attributed to altered penicillin-binding proteins, and there are indications that in some gram-negative organisms, resistance to certain cephalosporins is related to reduced affinity for the target.

Before the β-lactam antibiotics can inhibit growth, they must be able to reach the susceptible target site(s) on the membrane. In gram-positive organisms the cell wall is not a major barrier to entry of these drugs. Gram-negative bacteria, however, have a complex outer membrane that retards the entry of the anionic β-lactam antibiotics. Penicillin G has more difficulty penetrating the outer membrane of gram-negative bacilli than do ampicillin and the cephalosporins. Most gram-negative organisms are therefore relatively resistant to penicillin G. *P. aeruginosa* provides a good example of the role of permeability and other mechanisms in resistance. Wild-type strains of *Pseudomonas* are intrinsically resistant to most β-lactam antibiotics through (1) the ability of the outer membrane to restrict entry of the antibiotic molecules and (2) the inactivation of those that do enter by periplasmically located β-lactamases. Altered penicillin-binding proteins have also been described in these organisms. Acquired penicillin resistance due to changes in permeability and outer membrane protein has been described in *N. gonorrhoeae*. Further discussion of penicillin resistance, its genetics, epidemiology, and clinical importance is found on page 225 of this chapter and in Chapters 8 and 25.

Cycloserine

Cycloserine is a broad-spectrum antibiotic but has found clinical use only in the treatment of tuberculosis. Even here because of central nervous system toxicity its use is limited to the retreatment of drug-resistant cases, where it is administered as a member of a three or more drug combination.

Cycloserine is an inhibitor of peptidoglycan synthesis. When susceptible organisms are grown in the presence of minimal inhibitory concentrations of the drug, there is an accumulation of an incomplete cell wall precursor lacking the terminal D-alanyl-D-alanine dipeptide. The accumulation of this precursor may be reduced by the addition of D-alanine to the culture medium.

The molecular basis for the bactericidal activity of cycloserine lies in its structural similarity to D-alanine (Fig. 9-11). Cycloserine is a competitive inhibitor of two sequential reactions in the synthesis of peptidoglycan in which D-alanine is incorporated, alanine racemase and D-alanyl-D-alanine synthetase. The synthetase has two binding sites for cycloserine, the donor and acceptor sites. The donor site is believed to be the primary site of antibiotic action. The concentration of cycloserine required for 50 percent saturation is lowest for this site and correlates directly with the minimal inhibitory concentration for growth. The effectiveness of cycloserine as a competitive enzyme inhibitor can be attributed to a significantly higher affinity than the natural substrate D-alanine for both of the D-alanine reacting sites of the D-alanyl-D-alanine synthetase. The ratio of K_m to K_i is about 100. This high affinity may be related to the rigid planar ring structure of cycloserine, which effectively holds the molecule in the proper conformation for binding to the active sites on the inhibited enzymes.

Resistance of cycloserine may be attributed to two different mechanisms. In certain organisms, such as *E. coli* and *Streptococcus faecalis*, the antibiotic is effective only if a transport system for D-alanine is present. Loss of some component of the alanine transport system can protect the organism against D-cycloserine activity. In other mutants, resistance is attributed to elevated levels of both alanine racemase and D-alanyl-D-alanine synthetase.

Fosfomycin (Phosphonomycin)

This antibiotic is unusual in that it is a small molecule that shows certain homologies with the metabolite phosphoenolpyruvate (PEP). It is a broad-spectrum antibiotic with favorable pharmacologic characteristics. It has been used clinically in Spain and several other countries but has not been licensed in the United States.

Fosfomycin enters the bacterial cell on permeases that normally transport L-α-glycerophosphate or D-glucose 6-phosphate. It inhibits the first stage of peptidoglycan biosynthesis by interfering with the condensation step of UDP-N-acetylglucosamine with PEP. Fosfomycin binds covalently to the transferase effecting this enzymatic condensation, thereby inactivating it. Forsfomycin does not interfere with PEP-requiring reactions in animals because only in bacteria is a central carbon of PEP activated to form an ether linkage with N-acetylglucosamine (Fig. 9-12). In animal enzymes, the catalytic attack is at the phosphorus atom of PEP to cleave the phosphorus-oxygen bond. There is no P-O bond in fosfomycin.

The development of resistant strains belonging to different bacterial genera has been observed among hospital isolates. Two mechanisms for fosfomycin resistance have been described: an altered transport system and an altered target enzyme with decreased affinity for fosfomycin.

Vancomycin

Vancomycin is a narrow-spectrum, bactericidal antibiotic active against many species of gram-positive cocci. Its toxicity has limited its clinical usefulness, but it remains valuable alternative therapy for a number of serious infections, especially in patients allergic to the β-lactam drugs and in serious infections caused by methicillin-resistant *S. aureus* or multiple-resistant pneumococci.

Vancomycin is a complex glycopeptide with a molecular weight of about 1450 (Fig. 9-13). It interferes with the biosynthesis of peptidoglycan by binding very rapidly and

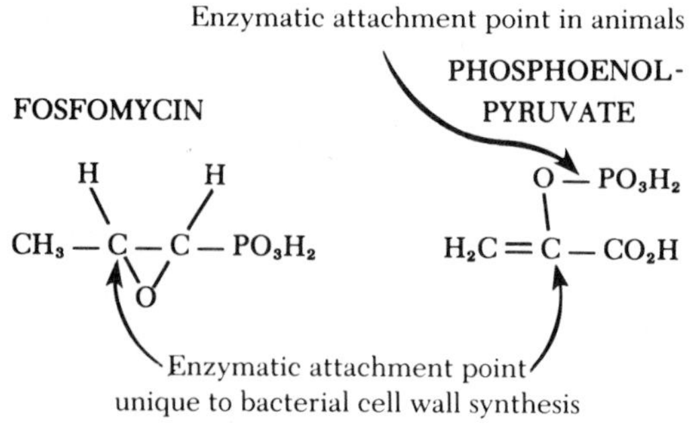

Figure 9-12. Structural relationship between fosfomycin and phosphoenolpyruvate and the contrast between phosphoenolpyruvate utilization in bacterial cell wall synthesis and in enzymatic transformations common to animals. *(Adapted from Woodruff et al.: Chemotherapy 23 [Suppl I]:1, 1977.)*

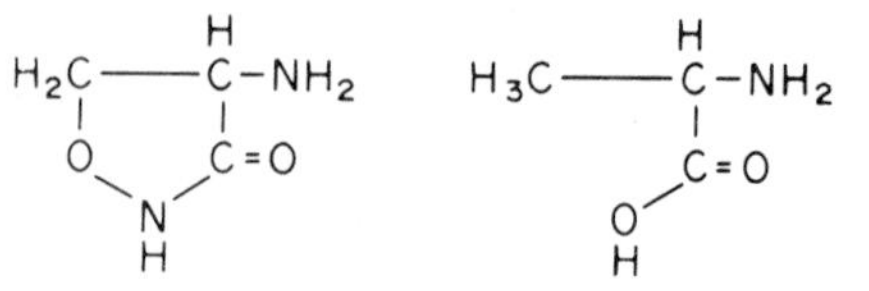

Figure 9-11. Structural relationship between cycloserine (left) and D-alanine (right).

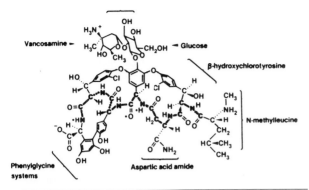

Figure 9-13. Vancomycin, a glycopeptide. The structure consists of a disaccharide (vancosamine and glucose), two β-hydroxychlorotyrosine units, three substituted phenylglycine systems, N-methylleucine, and aspartic acid amide. The peptide backbone is shown in heavier type. The atoms involved in hydrogen bonding with acetyl-D-alanyl-D-alanine are indicated by asterisks. *(From Pfeiffer: Rev Infect Dis 3 [Suppl]:S205, 1981.)*

irreversibly to the acyl-D-alanyl-D-alanine terminus of one of the membrane-bound peptidoglycan precursor molecules. In the presence of sufficient vancomycin one would expect that both glycan chain extension and incorporation of new chain by transpeptidation would be curtailed. Growth in the presence of low concentrations of vancomycin results in the accumulation of membrane-bound lipid intermediates.

The development of resistance during clinical use of vancomycin is not a problem.

Bacitracin

Bacitracin is a polypeptide antibiotic that is bactericidal for many gram-positive organisms and pathogenic *Neisseria* (Fig. 9-14). Its toxicity limits its clinical usefulness to topical administration only. Resistance to the drug does not readily develop.

Bacitracin interferes with the third stage of peptidoglycan biosynthesis in which the linear peptidoglycan is synthesized via a membrane-bound phospholipid intermediate and transferred to an endogenous acceptor. In the last step of this stage of cell wall biosynthesis, the pyrophosphate form of the phospholipid (undecaprenyl pyro-

phosphate) is dephosphorylated to yield inorganic phosphate and regenerated phospholipid.

$$C_{55} \text{ isoprenyl-PP} \rightleftharpoons C_{55} \text{ isoprenyl-P} + P_i$$

Bacitracin binds to the pyrophosphate and specifically blocks this reaction. Bacitracin thus inhibits cell wall synthesis by preventing reentry of the lipid carrier into the reaction cycle of peptidoglycan biosynthesis.

Cell Membrane Inhibitors

The cell membrane plays a vital role in the cell. It poses an osmotic barrier to free diffusion between the internal and external environment. It effects the concentration of metabolites and nutrients within the cell and serves as a site for respiratory and certain biosynthetic activities. Several antibiotics have been shown to impair one or more of these functions, resulting in major disturbances in the viability of the cell. The action of agents whose primary target of attack is the cell membrane is independent of growth and begins immediately when cells and antibiotic come together. Unlike inhibitors that interfere with cell wall biosynthesis and that are relatively innocuous for mammalian tissues, antibiotics that attack the cell membrane distinguish less successfully between microorganisms and host tissues. Few have found a place in clinical medicine because of their toxicity (Table 9-7).

TABLE 9-7. ANTIMICROBIAL AGENTS THAT AFFECT THE CELL MEMBRANE

Agent	Source	Antimicrobial Activity	Major Therapeutic Applications
Polymyxin B	*Bacillus polymyxa*	Gram-negative bacteria	Serious drug-resistant *Pseudomonas* infections
Nystatin	*Streptomyces noursei*	Many fungi	Cutaneous or oral candidiasis
Amphotericin B	*Streptomyces niveus*	Many fungi	Systemic fungous infections
Miconazole	Synthetic	Many fungi	Vaginal candidiasis, tinea pedis or cruris
Ketoconazole	Synthetic	Many fungi	Many cutaneous and systemic fungous infections

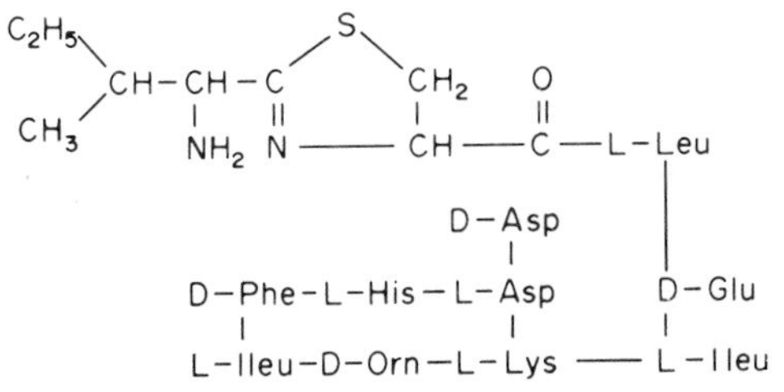

Figure 9-14. Bacitracin A. One of a group of polypeptide antibiotics containing a thiazoline ring structure.

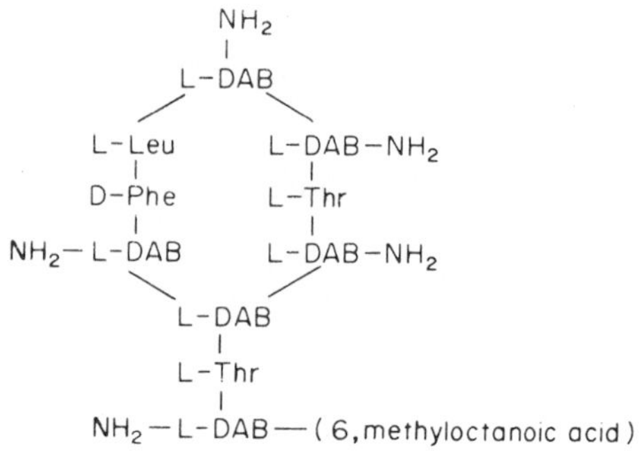

Figure 9-15. Polymyxin B. DAB is α, γ-diaminobutyric acid, a component that is present together with L-threonine and D-6-methyloctanoic acid in all polymyxins.

Polymyxins

The polymyxins are a family of relatively simple polypeptides characterized by poor diffusibility and significant toxicity. The members of this group are designated by the letters A, B, C, D, and E, but of these only polymyxin B and polymyxin E (colistin) are currently of clinical use (Fig. 9-15). The polymyxins are decapeptides that contain a high percentage of 2,4-diaminobutyric acid (DAB), a fatty acid, and a mixture of D- and L-amino acids. The in vitro and in vivo activity of polymyxin is restricted to gram-negative organisms. Because of severe adverse reactions, polymyxin is usually reserved for serious *P. aeruginosa* infections when the organism is resistant to other antibiotics.

Polymyxin binds specifically to the outer surface of cell membranes, altering their structure and osmotic properties. There is a leakage of metabolites and inhibition of a number of biochemical processes secondary to membrane damage. Damage to the membrane is attributed to an electrostatic binding and disruption of the structure of the phospholipid and lipopolysaccharide components, probably by competitively displacing Mg^{2+} or Ca^{2+} from negatively charged phosphate groups on membrane lipids (Fig. 9-16).

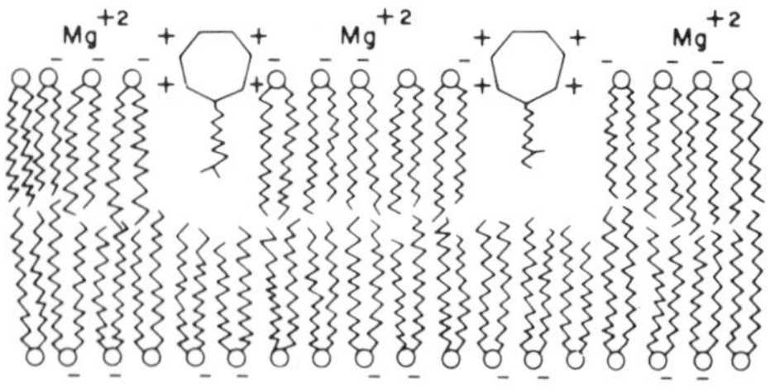

Figure 9-16. Hypothetical model for interaction of polymyxin with a phospholipid bilayer. It is proposed that the fatty acid tail of the peptide penetrates the hydrophobic domain of the bilayer, with the peptide amino groups interacting electrostatically with phospholipid phosphates. *(From Storm et al.: Annu Rev Biochem 46:723, 1977.)*

Polyenes

The polyenes are macrolide antibiotics, the two most important of which are the antifungal agents, amphotericin B and nystatin. All of the polyene agents possess a large lactone ring containing a flexible hydroxylated portion and a rigid hydrophobic system of unsubstituted conjugated double bonds. In amphotericin B and nystatin there is also an amino sugar mycosamine (Figs. 9-17, 9-18).

Amphotericin B remains the cornerstone of antifungal therapy. It is active against most fungi that can cause deep-seated infections. It does, however, produce a number of side effects, of which nephrotoxicity is the most serious. Nystatin is also a clinically useful polyene, but because of its toxicity it cannot be administered parenterally. Its most important use is in the treatment of topical or superficial infections, especially those caused by *Candida*.

The polyenes selectively inhibit organisms whose membranes contain sterols. They are active against yeast, fungi, and other eucaryotic cells but are not inhibitory for the procaryotic bacteria because of the absence of sterols in their membranes. The antifungal activity of the polyenes is referable to changes in the permeability of the membrane produced by antibiotic-sterol interaction, resulting in cells becoming selectively permeable especially to potassium ions and small vital constituents up to the size of glucose. Exogenous sterol antagonizes this effect. Approximately one molecule of the antibiotic binds to each sterol molecule in the membrane. The selective action of both amphotericin B and nystatin for fungi is attributable to their preferential binding to ergosterol, which all fungi contain in their membranes rather than cholesterol, which is present in mammalian membranes. Since some interaction with cholesterol does occur, however, these antifungal agents are often associated with severe adverse reactions in the host that are dose-related. Interaction of the polyenes with ergosterol molecules results in the formation of lethal membrane pores across the membrane. In molecular models that have been constructed to explain the mechanism of pore formation, approximately 10 molecules of amphotericin B are assembled with their long axes parallel to form a cylinder. All of the hydrophilic groups are on the inside and the hydrophobic groups are on the outside, with one sterol molecule tucked in between each pair of molecules on the hydrophobic surface. Such a structure would provide a pore 0.7 nm in diameter and, when two

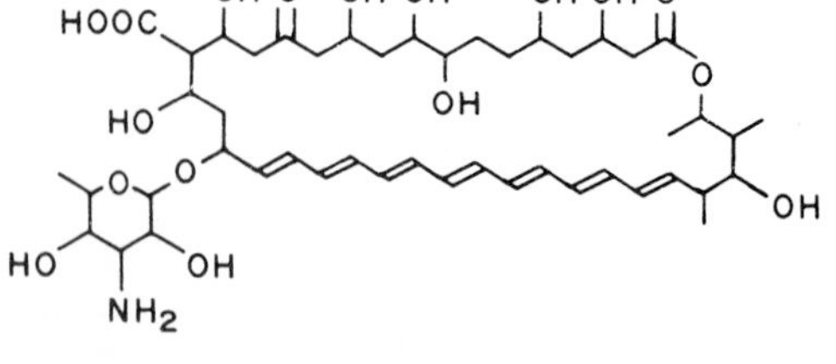

Figure 9-17. Amphotericin B, a polyene antibiotic with selective activity for fungi.

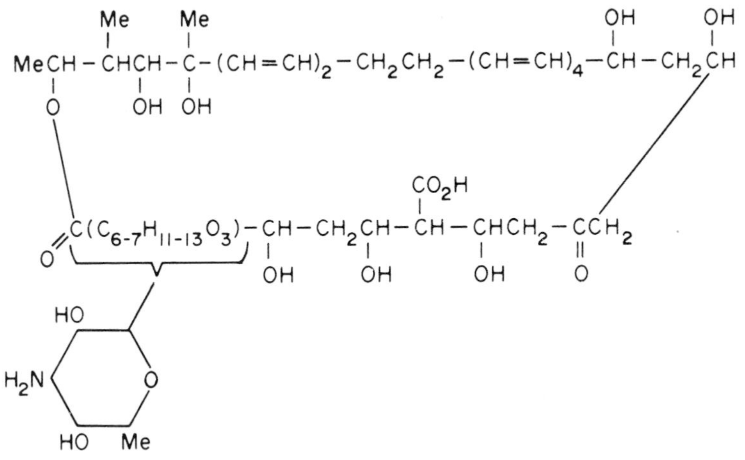

Figure 9-18. Nystatin, an antifungal agent.

of these are placed end to end, would span the membrane and provide the permeability observed. The pores appear to be transient in existence, a small fraction of the antibiotic being in the form of pores at any one time.

Mechanisms for drug-induced resistance to the polyenes include (1) decrease in the membrane ergosterol content and (2) modification of membrane sterols to ones that bind less efficiently to polyenes.

Imidazoles

Miconazole and Ketoconazole. Of the imidazole derivatives currently available, these agents are clinically the most useful (Fig. 9-19). They are broad-spectrum antifungal agents active against dermatophytes, dimorphic fungi, and yeasts, as well as some bacteria and protozoa. Use of miconazole is limited to topical and intravenous use, but ketoconazole is efficacious by the oral route in a broad spectrum of superficial and deep mycotic infections in hu-

mans. Unlike miconazole, which has side effects in man when administered parenterally, ketoconazole causes very few adverse reactions. Both drugs alter the permeability of the cell membrane, resulting in a leakage of potassium ions and phosphorus-containing compounds. These changes in the cell membrane are the result of an interference with the synthesis of ergosterol in the fungal cell. Since inhibition of ergosterol synthesis coincides with the accumulation of lanosterol-like sterols, inhibition of ergosterol synthesis is attributed to an inhibition of one of the metabolic steps involved in the dimethylation at the C14 site of the precursor of ergosterol, lanosterol. Miconazole also influences the nature of fatty acids, free or esterified, by inducing a shift from unsaturated to saturated acyl moieties. This latter effect enhances membrane disturbances, decreases growth, and leads to decreased activity of membrane-bound enzymes.

Morphologic changes observed in *Candida albicans* following miconazole treatment primarily involve alterations of the cell membrane, changes in cell volume, and defective cell division.

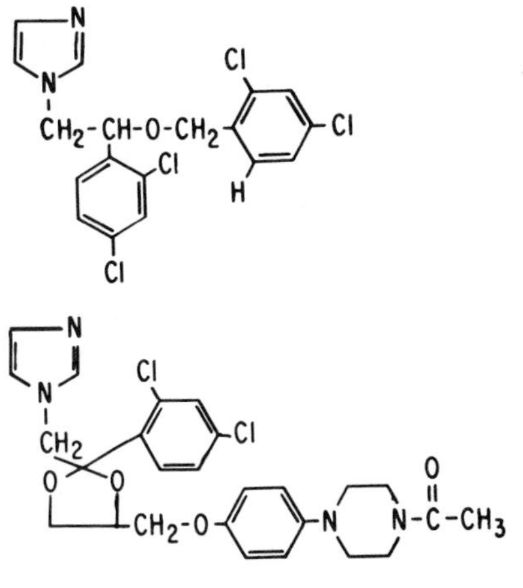

Inhibitors of DNA Function

A number of antimicrobial drugs specifically interfere with the structure and function of DNA, but few of these agents have shown a selective toxicity acceptable for clinical use. In spite of their limited clinical use, these drugs have been very useful as biochemical tools and have contributed significantly to the study of molecular biology (Table 9-8). The structure of the DNA molecule is intimately related to its two primary roles, duplication and transcription. Any agent that disturbs the structure of the organized double helix of DNA is potentially capable of causing profound effects on all phases of cell growth and metabolism. Among the mechanisms employed by the drugs for altering the structure or function of DNA are crosslinking and intercalation between the stacked bases of the double helix.

Figure 9-19. Chemical structure of miconazole (top) and ketoconazole (bottom), imidazole derivatives useful in antifungal chemotherapy.

TABLE 9-8. ANTIMICROBIAL AGENTS THAT INTERFERE WITH DNA FUNCTION

Agent	Source	Antimicrobial Activity	Clinical Use
Mitomycin	*Streptomyces species*	Many species	None*
Nalidixic acid	Synthetic	Gram-negative organisms	Urinary tract infections caused by *Escherichia coli, Proteus, Enterobacter**
Novobiocin	*Streptomyces niveus*	Gram-positive organisms	Penicillin-resistant staphylococcal infections*†

*Useful biochemical tools.
†No valid indications for its use at present.

Mitomycin

The addition of mitomycin to growing bacterial cells results in inhibition of cell division with the formation of long filamentous forms, bacteriostasis, and death (Fig. 9-20). The bactericidal effect coincides with inhibition of DNA synthesis and usually is accompanied by massive degradation of the preexisting DNA.

Before its inhibitory effects are expressed in vivo, mitomycin is converted enzymatically to a highly reactive hydroquinone derivative that acts as a bifunctional alkylating agent. This reactive, short-lived species readily crosslinks with DNA by bonding to two sites, one on each of the complementary strands. Guanine residues in DNA are the most probable sites for alkylation. The formation of covalent crosslinks in DNA prevents separation of the complementary strands, thereby inhibiting progress of the replicating fork and causing a blockage of DNA synthesis. Apparently, however, crosslink formation is not the only mechanism by which mitomycin damages the cell. Monoalkylated sites on a single strand of DNA are found with a 10 times greater frequency than crosslinks. The DNA degradation that follows treatment with mitomycin is due to the excision of the crosslinked zones and to endonucleolytic breaks in the damaged DNA at the monoalkylated sites. The appearance of nucleases is associated with lysogenic phages induced by mitomycin.

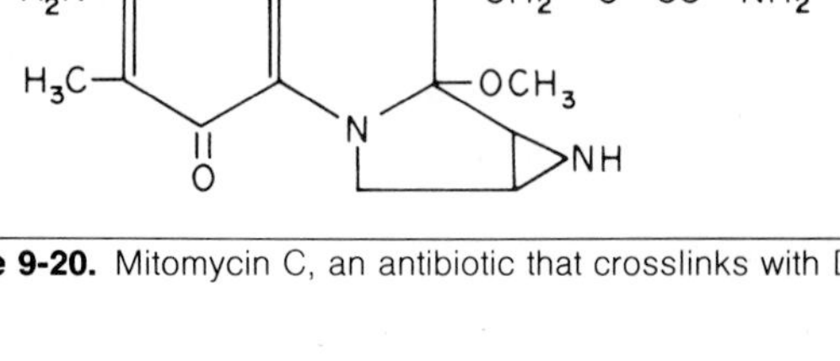

Mitomycin exhibits some selectivity in the crosslinking of DNA. Under certain conditions, it blocks the synthesis of host cell DNA but permits viral DNA synthesis. The reason for the relative resistance of viral DNA is not clear, but the finding has been useful in studies of viral DNA synthesis in the absence of host cell DNA synthesis.

Since mitomycin fails to distinguish between the DNA of the infecting organism and that of the host, its toxicity prohibits clinical use.

Nalidixic Acid

Nalidixic acid, a synthetic derivative of 1,8-naphthyridine, has been used successfully in the clinical management of uncomplicated urinary tract infections caused by gram-negative organisms (Fig. 9-21).

In addition to its clinical value, it has been a useful tool in studies on the regulation of bacterial cell division. It selectively and reversibly blocks DNA replication in susceptible bacteria but exhibits no demonstrable mutagenicity. Nalidixic acid inhibits the A subunit of DNA gyrase and induces the formation of a relaxation complex analog. Complex formation, unlike the introduction of supertwists into closed circular DNA, is insensitive to novobiocin, another drug that also inhibits gyrase activity. Nalidixic acid, but not novobiocin, inhibits the nicking-closing activity of the swivelase component of DNA gyrase that relieves the positive winding stress on the supercoiled DNA. In nalidixic acid mutants, mutation at the *nal A* gene locus confers resistance to high levels of nalidixic acid but not to novobiocin, whose target gene is widely separated from the *nal A* gene.

Novobiocin

This antibiotic is bactericidal for a variety of bacteria, especially gram-positive organisms (Fig. 9-22). Although a number of biosynthetic processes in vivo are inhibited by novobiocin, its primary inhibitory effect is on the replication of DNA. It inhibits the supercoiling of DNA by DNA gyrase (Chap. 7, p. 123). The target site for novobiocin is the subunit B component of DNA gyrase, rather than subunit A, as in the case of nalidixic acid. This accounts for

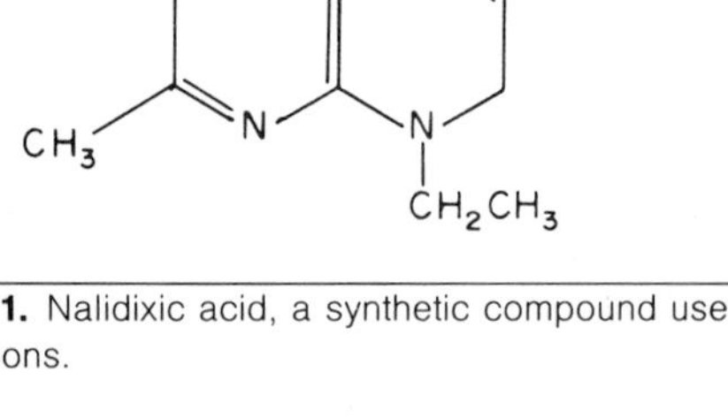

Figure 9-20. Mitomycin C, an antibiotic that crosslinks with DNA.

Figure 9-21. Nalidixic acid, a synthetic compound used in urinary tract infections.

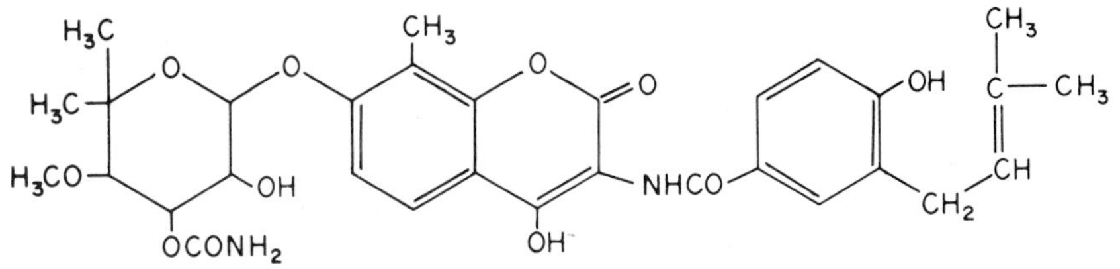

Figure 9-22. Novobiocin. Its antibacterial spectrum resembles that of penicillin.

some of the observed differences between the action of these two agents.

At present there are no valid indications for the therapeutic use of novobiocin.

Inhibitors of Protein Synthesis and Assembly

Protein synthesis is the end result of two major processes: (1) DNA-dependent ribonucleic acid synthesis, or transcription, and (2) RNA-dependent protein synthesis, or translation. An antibiotic that inhibits either of these processes will inhibit protein synthesis. Although the antibiotics that primarily inhibit translation are the ones that have been most useful clinically, the agents that inhibit transcription have been useful in characterizing the steps involved in protein synthesis (Table 9-9).

Inhibitors of Transcription

During transcription, the genetic information in DNA is transferred to a complementary sequence of RNA nucleotides by the enzyme RNA polymerase, a complex protein composed of four subunits, β, β', α, α. Associated with

TABLE 9-9. INHIBITORS OF PROTEIN SYNTHESIS AND ASSEMBLY

Antibiotic	Source	Antimicrobial Spectrum	Major Therapeutic Applications
Actinomycin	*Streptomyces* species	Many gram-positive and gram-negative bacteria	None
Chloramphenicol	*Streptomyces venezuelae*	Many gram-positive and gram-negative bacteria, rickettsiae, chlamydiae	Typhoid fever, *Haemophilus influenzae* meningitis, anaerobic infections
Clindamycin	Semisynthetic derivative from *Streptomyces lincolnensis*	Gram-positive organisms, anaerobic species	*Bacteroides* and other anaerobic infections
Erythromycin	*Streptomyces erythreus*	Similar to penicillin	Diphtheria, *Mycoplasma pneumoniae*, *Legionella* and penicillin-resistant streptococcal and pneumococcal infections
Fucidin	*Fusidium coccineum*	Gram-positive organisms, *Neisseria*	Penicillin-resistant staphylococcal infections*
Griseofulvin	*Penicillium griseofulvum*	Dermatophytes and other fungi with chitin-containing walls	Dermatophytosis
Nitrofurantoin	Synthetic	Many gram-positive and gram-negative organisms	Uncomplicated urinary tract infections
Rifampin	Semisynthetic derivative from *Streptomyces mediterranei*	Gram-positive bacteria, *Mycobacterium tuberculosis*	Tuberculosis
Streptomycin	*Streptomyces griseus*	Many gram-positive and gram-negative bacteria, *Mycobacterium tuberculosis*	Tuberculosis, plague, tularemia
Tetracyclines	*Streptomyces* species†	Many gram-positive and gram-negative bacteria, rickettsiae, mycoplasma, chlamydia	Brucellosis, cholera, infections by chlamydiae, rickettsiae, mycoplasma

*No clinical indication for its use at present.
†Tetracycline is produced by *Streptomyces viridifaciens*, chlortetracycline and demethylchlortetracycline by *Streptomyces aureofaciens*, and oxytetracycline by *Streptomyces rimosus*. Doxycycline and minocycline are semisynthetic derivatives.

this core polymerase, which forms the internucleotide linkages, is the sigma factor, a dissociable component that acts catalytically in the accurate initiation of RNA chains. Antibiotics that either alter the structure of the template DNA or inhibit the RNA polymerase will interfere with the synthesis of RNA and, consequently, with protein synthesis.

Actinomycin

Actinomycin D is a bright-red oligopeptide that is active against many gram-positive and gram-negative organisms as well as mammalian cells (Fig. 9-23). Actinomycin forms complexes specifically with DNA, thereby impairing DNA function. This binding is dependent upon guanine residues and helical secondary structure. The action of actinomycin is attributed to the planar character of its chromophoric ring, which permits it to intercalate between the adjacent stacked base-pairs of the double helix. To permit this insertion, there is a preliminary local unwinding of the double helix to produce spaces into which the planar chromophore can move. The actinomycin molecule is stabilized internally by hydrogen bonding between its two cyclic pentapeptides that lie in the minor groove of the double helix and by bonding of guanine residues with the L-threonines of one of the peptide rings. In the duplex molecule, hydrogen bonding between guanine-cytosine base-pairs remains undisturbed although there is some distortion of the smooth coil of the sugar phosphate backbone.

Whereas actinomycin inhibits both DNA synthesis and DNA-dependent RNA synthesis, RNA synthesis is inhibited at much lower concentrations. Actinomycin blocks RNA synthesis by preventing the progression of RNA polymerase along the DNA template. Since normal progression of the polymerase is along the minor groove, the presence of cyclic pentapeptide rings in the groove blocks polymerase movement.

Although its toxicity prevents clinical use, actinomycin has been extensively used as a tool for specifically shutting off DNA-directed RNA synthesis. It has been helpful in following the decay of mRNA and for showing the absence of DNA-dependent steps in the growth of many RNA viruses.

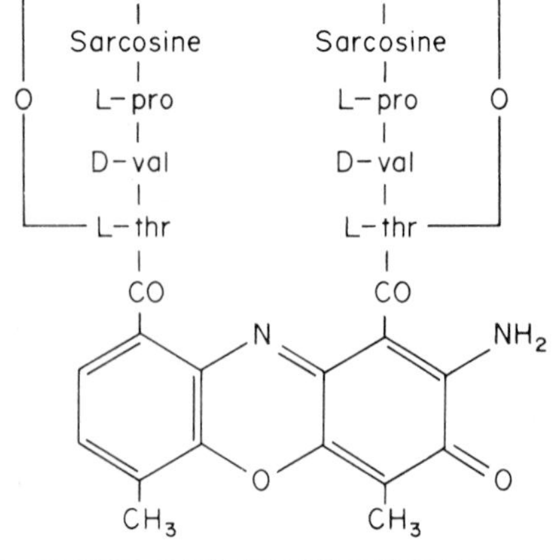

Rifampin (Rifampicin)

The rifamycins are ansa compounds, i.e., compounds that contain an aromatic ring system spanned by a long aliphatic bridge. Some of the members of the group occur in nature, but most are semisynthetic derivatives (Fig. 9-24). One of the most useful members of the group is rifampin. Rifampin has a wide antibacterial spectrum and is especially effective against gram-positive organisms and mycobacteria. It is a major drug for the treatment of tuberculosis and leprosy and for meningococcal prophylaxis.

Rifampin inhibits protein synthesis by selectively inactivating the DNA-dependent RNA polymerase. In the first step of transcription, the polymerase is bound to a specific initiation site on the DNA template, followed by the binding to the enzyme of the first nucleoside 5'-triphosphate. To this initiation complex, the second nucleoside 5'-triphosphate is added, and the first phosphodiester bond is formed. Rifampin binds noncovalently to the β subunit of RNA polymerase and blocks RNA chain initiation after the formation of the first phosphodiester bond. The binding by rifampin apparently induces an isomerization or conformational change of the enzyme, which is responsible for its inactivation. Only the initiation of RNA synthesis is arrested. There is no effect on chain elongation. Rifampin is inactive on DNA-directed RNA polymerases from eucaryotic nuclei. Its effect on mammalian DNA viruses and oncogenic viruses is discussed in Chapter 61.

Inhibitors of Translation

In bacterial cells, the translation of mRNA into protein can be divided into three major phases: initiation, elongation, and termination of the peptide chain. Protein synthesis starts with the association of mRNA, a 30S ribosomal subunit, and formylmethionyl-tRNA$_f$ (fMet-tRNA$_f$) to form a 30S initiation complex. The formation of this complex also requires GTP and three protein initiation factors. The codon, AUG, is the initiation signal in mRNA and is recognized by the anticodon of fMet-tRNA$_f$. A 50S ribosomal subunit is subsequently added to form a 70S initiation complex, and the bound GTP is hydrolyzed.

There are two distinct sites on the ribosome, the P (peptidyl) and A (aminoacyl) sites. At the end of the initiation stage, the fMet-tRNA$_f$ molecule occupies the P site, and the other site for a tRNA molecule is empty. In the first step of the elongation cycle, an aminoacyl-tRNA is inserted into the vacant A site on the ribosome. The particular species inserted depends on the mRNA codon that is positioned in the A site. Protein elongation factors and GTP are required for polypeptide chain elongation.

In the next step of the elongation phase, the formyl-

Figure 9-23. Actinomycin D. This agent forms complexes with DNA by binding to deoxyguanosine residues.

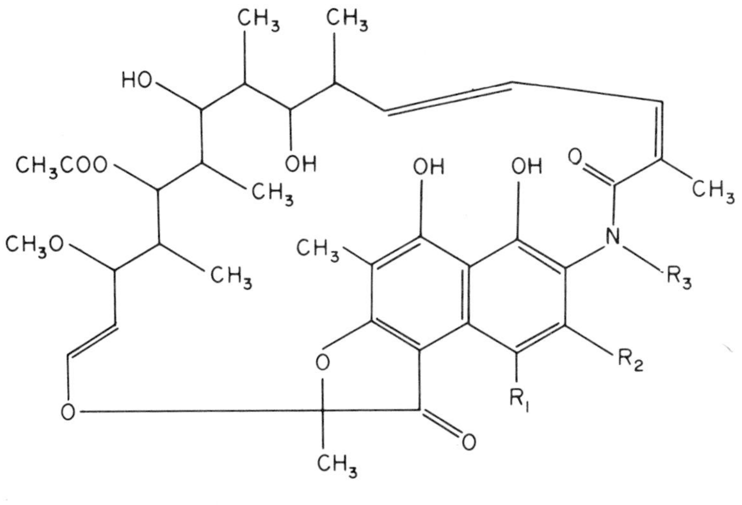

Figure 9-24. Basic structure of the rifamycins. Various derivatives are substituted in the R_1, R_2, and R_3 positions. In rifampicin, R_1 is —OH; R_2 is —CH = N—N ⬭ N— CH_3; R_3 is H.

methionyl residue of the fMet-tRNA$_f$ located at the peptidyl donor site is released from its linkage to tRNA$_f$ and is joined with a peptide bond to the α-amino group of the aminoacyl-tRNA in the acceptor site to form a dipeptidyl-tRNA. The enzyme catalyzing this peptide formation is peptidyl transferase, which is part of the 50S ribosomal subunit.

Following the formation of a peptide bond, an uncharged tRNA occupies the P site, whereas a dipeptidyl tRNA occupies the A site. The final phase of the elongation cycle is translocation, catalyzed by elongation factor EF-G and requiring GTP. It consists of three movements: (1) the removal of the discharged tRNA from the P site, (2) the movement of fMet-aminoacyl-tRNA from the acceptor site to the peptidyl donor site, and (3) the movement or translocation of the ribosome along the mRNA from the 5′ toward the 3′ terminus by the length of three nucleotides. After translocation, the stage is prepared for the binding of the next aminoacyl residue to the fMet-AA-tRNA, each addition requiring aminoacyl-tRNA binding, peptide bond formation, and translocation. Peptidyl-tRNAs replace the fMet-tRNA in the second and in all subsequent cycles.

The polypeptide chain grows from the amino terminal toward the carboxyl terminal amino acid and remains linked to tRNA and bound to the mRNA-ribosome complex during elongation of the chain. When completed, it is released during chain termination. Termination is triggered when a chain termination signal (UAA, UAG, or UGA) is encountered at the A site of the ribosome. Protein release factors bind to the terminator codons, triggering hydrolysis by the peptidyl transferase. The polypeptide is released, and the messenger-ribosome-tRNA complex dissociates.

It is the elongation phase of protein synthesis that is most susceptible to inhibitors (Fig. 9-25). For a number of these, the ribosome is the site of action, and a complex array of inhibitory effects has been observed. Several medically important antibiotics owe their selective antimicrobial action to a specific attack on the 70S ribosomes of bacteria, while mammalian 80S ribosomes are unaffected.

A subdivision of antibiotics into major classes has been made on the basis of their binding to the 30S or 50S ribosomal subunits, with the assumption that the site of fixation provides presumptive evidence for its site of action.

Inhibitors of the 30S Ribosomal Subunit

As noted above, 30S ribosomal subunits provide attachment sites for mRNA, and move relative to it during translation. The 30S subunit also provides a binding site for fMet-tRNA$_f$ and aminoacyl-tRNA. Inhibition of protein synthesis on the 30S subunit could result if (1) mRNA is prevented from attaching, (2) movement of mRNA relative to the 30S subunit is impaired, or (3) if the aminoacyl acceptor site is blocked. Numerous antibiotics act at the level of the 30S ribosomal subunit. Among these are streptomycin, other aminoglycosidic aminocyclitol antibiotics, and the tetracyclines.

Aminoglycosidic Aminocyclitol Antibiotics. This group contains many important antimicrobial agents, such as streptomycin, neomycin, kanamycin, gentamicin, tobramycin, and amikacin (Fig. 9-26). The unique aminocyclitol ring structure that forms the backbone of each of these compounds is a derivative of inositol in which various hydroxyl groups have been replaced with amino groups or substituted amino groups (Fig. 9-27). Kasugamycin is the simplest of these compounds and consists of an amino-containing sugar joined by a glycosidic linkage to inositol itself. Streptidine is the aminocyclitol component of streptomycin, and 2-deoxystreptamine provides the backbone for kanamycin, gentamicin, tobramycin, and amikacin (Fig. 9-28). To this aminocyclitol ring, two amino sugars are attached by a glycosidic linkage, hence the name aminoglycosidic aminocyclitol. Spectinomycin, an antibiotic usually included in discussions of this group, contains an

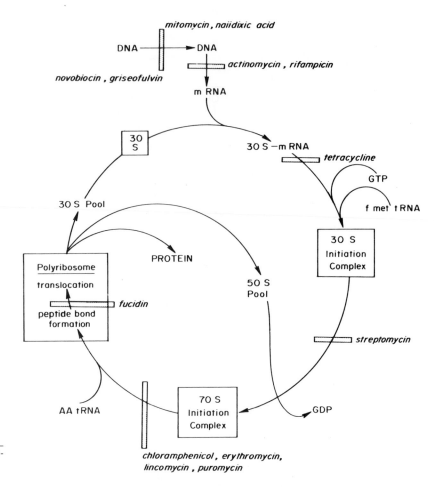

Figure 9-25. Site of action of antibiotics that inhibit DNA, RNA, and protein synthesis.

aminocyclitol ring structure but lacks an amino sugar residue (Fig. 9-29). Unlike the other members of this group, it is bacteriostatic rather than bactericidal in action.

The pharmacokinetic properties of the group are attributable to their polarity as polycations. They are poorly

absorbed orally, penetrate poorly into the cerebrospinal fluid, and are rapidly excreted by the kidney. A number of adverse reactions are associated with their use. Among the most serious are damage to the vestibular portion of the eighth nerve, auditory toxicity, and renal damage.

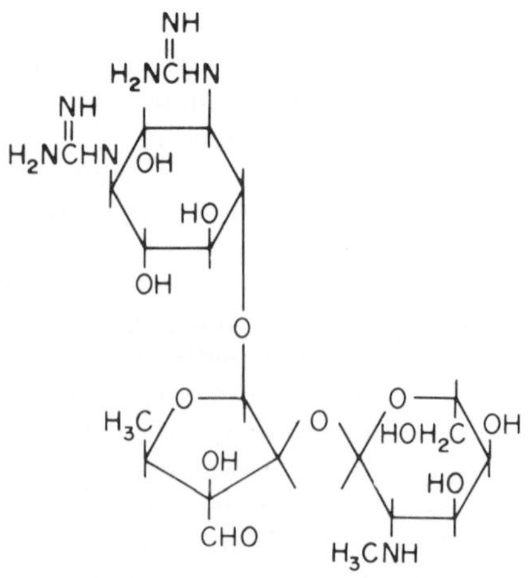

Figure 9-26. Streptomycin. The streptidine moiety (upper ring structure) is an inositol substituted with two guanido groups linked to the streptobiosamine moiety by a glycosidic bond.

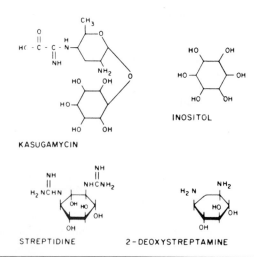

Figure 9-27. Structural formulae of components of various aminoglycosidic aminocyclitol antibiotics. *(From Moellering: Med J Aust [Spec Suppl] 2:4, 1977.)*

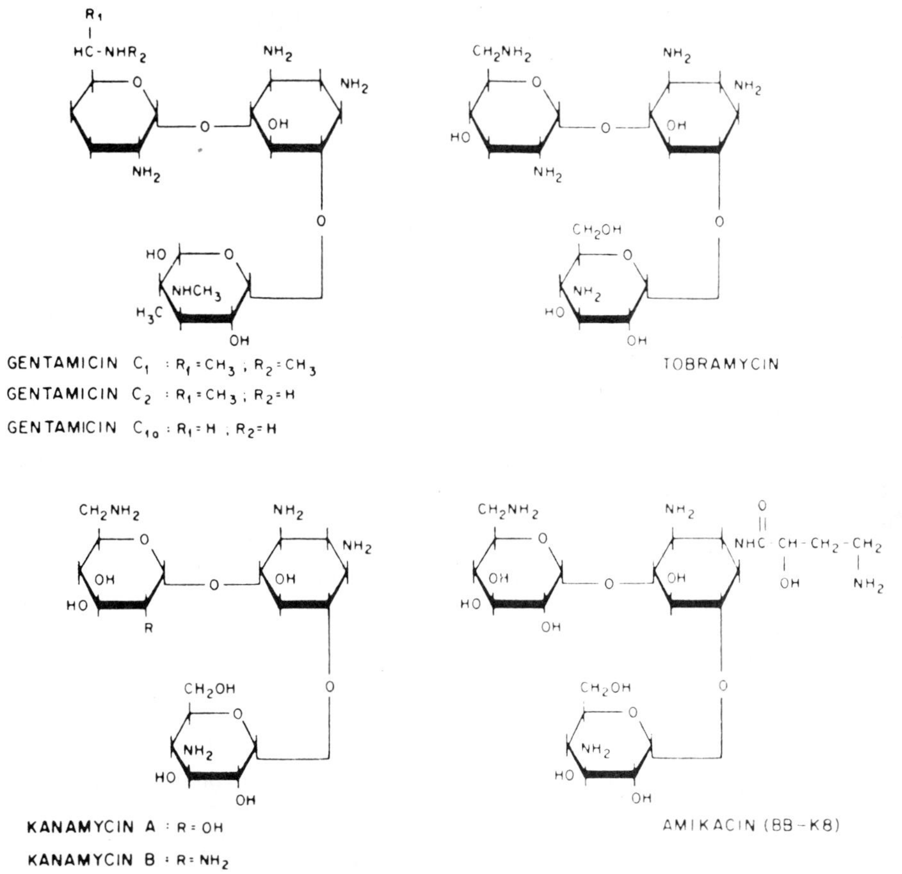

Figure 9-28. Structural formulae of the gentamicins, tobramycin, the kanamycins, and amikacin. *(From Moellering: Med J Aust [Spec Suppl] 2:4, 1977.)*

STREPTOMYCIN. Streptomycin is bactericidal for a wide variety of gram-positive and gram-negative species and for *Mycobacterium tuberculosis.* Clinically, the chief merit of streptomycin lies in its ability to attack certain organisms that are not affected by penicillin (Table 9-4). However, the introduction of newer agents with a broader antibacterial spectrum has limited the current indications for streptomycin. It remains the drug of choice for tularemia and may be used in the therapy of plague and brucellosis. It is also useful in combination with penicillin in the treatment of

endocarditis caused by enterococci or viridans streptococci and in combination with other agents in the management of tuberculosis.

The rapid development of resistance to high levels of streptomycin has been a major factor in limiting its clinical usefulness. Resistance to the aminoglycoside antibiotics transferred by R factors is frequently seen in clinical strains and is probably responsible for most of the resistance to this group of antibiotics. The biochemical mechanisms of streptomycin resistance are discussed on page 228.

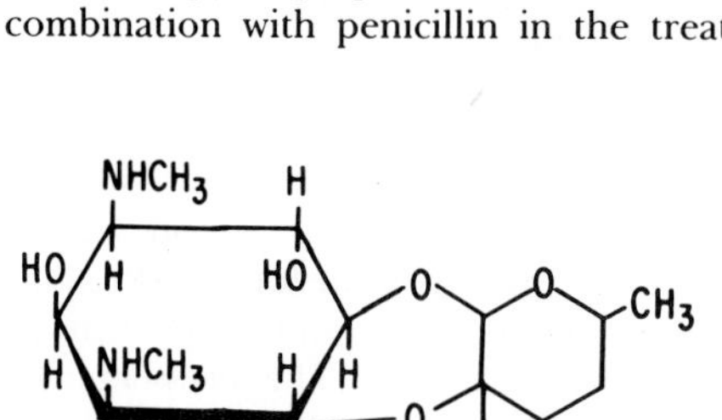

Mechanism of Action. Although streptomycin induces many effects when added to growing bacterial cells, the lethal effect of streptomycin results from irreversible binding of the drug to ribosomes and the subsequent interference with protein synthesis.

When streptomycin is added to in vitro polypeptide-synthesizing systems, it has two major effects: (1) it markedly inhibits polypeptide synthesis, and (2) it causes misreading of the synthetic polynucleotide messengers. If ribosomes from streptomycin-resistant strains are used, both of these effects are abolished or greatly reduced.

Figure 9-29. Spectinomycin. A bacteriostatic aminocyclitol antibiotic that differs from the aminoglycoside group in that it lacks an amino sugar residue.

Inhibition of Protein Synthesis. Streptomycin binds irreversibly to the 30S ribosomal subunit, drastically interrupting the ribosome cycle at the initiation of protein synthesis. It does not inhibit the formation of the initiation complex but inhibits the initiation of peptide chains on the complex.

In vitro polypeptide-synthesizing systems employing the natural messenger, viral RNA, have been useful in delineating the nature of the inhibition by streptomycin. When streptomycin is added to such a system, polypeptide synthesis in progress on polysomes is slowed down, but ribosomes are allowed to leave mRNA either prematurely or at the termination signal. Upon release they dissociate into subunits but subsequently reassociate at the normal initiation sites on mRNA to form streptomycin monosomes, irreversibly inactivated initiation complexes that contain stabilized mRNA. These modified complexes cannot form peptide bonds. Their movement along the mRNA is impeded, thereby blocking the ribosome cycle at an early stage in the initiation of protein synthesis. These effects on initiation and elongation are attributed to conformational changes at both acceptor and donor sites by streptomycin.

A single protein in the 30S ribosomal subunit, S12, is the ultimate target of streptomycin, but it cannot bind the antibiotic by itself. Interaction of streptomycin with the 30S subunit is complex, involving at least four additional ribosomal proteins, S3, S5, S9, and S14. Binding to 16S RNA may also be involved.

Streptomycin-induced Misreading. One of the effects produced by streptomycin when added to an in vitro polypeptide-synthesizing system is misreading of messenger. Streptomycin can inhibit the incorporation of certain synthetic polymer-directed incorporations (e.g., poly U-phenylalanine) and can stimulate the incorporation of amino acids not coded for by certain synthetic polymers. The extensive misreading in in vitro homonucleotide systems provides an explanation for the ability of streptomycin to suppress mutations in a class of mutants that are conditionally streptomycin-dependent (CSD). Such mutants can grow in the absence of their specific growth requirement provided streptomycin is present in the medium. This behavior apparently is due to a misreading of the genetic message by streptomycin. The wrong message of the mutant is not corrected physically but is read differently in the presence of streptomycin. The net result of this misreading of an incorrect message is the synthesis of a functional protein. In strains that permit correction by streptomycin, streptomycin mimics the action of a suppressor, a mutation at a second distinct site that reverses the mutant phenotype. Unlike the genetic suppression, however, the correction produced by streptomycin is phenotypic and is not heritable.

Although it is difficult to assess the in vivo significance of misreading of the code, an apparent relationship does exist among misreading, killing action, and phenotypic suppression. The streptamine or deoxystreptamine moiety of the aminoglycoside is the chemical structure responsible for misreading. Aminoglycoside drugs, such as spectinomycin, that lack this moiety cause no misreading and are bacteriostatic only.

The chemical basis for all of the in vivo effects of streptomycin is not known, but the major effects observed, i.e., inhibition of protein synthesis, misreading, and phenotypic suppression, have been attributed to the irreversible binding of streptomycin to the 30S ribosomal subunit. This binding produces a conformational change at the aminoacyl-tRNA binding site, resulting in an interference both with the binding of aminoacyl-tRNA and the fidelity of translation. One of the key steps in ensuring fidelity of protein biosynthesis, the selection of aminoacyl-tRNA by the ribosome, involves two recognition steps, one involving the ternary complex of aminoacyl-tRNA, EFTu, and GTP, the other occurring after GTP hydrolysis and involving only the aminoacyl-tRNA. Streptomycin exclusively inhibits the latter recognition step, known as proofreading. By impairing the ability to reject an error, streptomycin-induced misreading results.

Streptomycin Resistance. In a bacterium, three phenotypically distinct responses to streptomycin are possible: sensitivity, resistance, or dependence. These responses are determined by multiple alleles of a single genetic locus, the *str* locus that codes for protein S12 of the 30S subunit. This protein determines the sensitivity of the entire 30S ribosomal particle to streptomycin. A mutation to resistance would result if this site were eliminated or if it were altered in such a manner that the bound drug could no longer exert its effect. There is evidence for the existence of single-step, high-level mutations to resistance of both of these types. Resistance conferred by mutations in protein S12 can be suppressed by mutations in protein S4 and S5 of the 30S subunits.

The amount of misreading observed in a cell-free polypeptide synthesizing system containing ribosomes from a *str*^{r} mutant is 10-fold less than that observed with ribosomes from *str*^{s} strains, even in the absence of streptomycin. This observation may explain the finding that mutations to streptomycin resistance often introduce other phenotypic effects. They may simultaneously confer auxotrophy, change host-prophage relationships, alter control of inducible or repressible enzymes, or alter patterns of genetic suppression.

Streptomycin Dependence. When a population of streptomycin-sensitive organisms is plated on a medium containing a high level of streptomycin, a number of the survivors require streptomycin for growth. These mutants are streptomycin-dependent and arise from the *str*^{s} wild-type strain in a single step. The *str*^{d} phenotype is determined by a mutation allelic with or very close to the *str*^{r} locus. The genetic basis for streptomycin dependence in these mutants is thus quite different from the conditionally streptomycin-dependent (CSD) phenotype discussed above (p. 167).

As with streptomycin resistance, mutation to dependence also exhibits pleiotropic effects. Unlike streptomycin resistance, however, where phenotypic suppression is restricted, in str^d strains phenotypic suppression is enhanced. In these strains, most of the nonsense and missense mutations that normally occur are weakly suppressed, and ambiguity in translation is introduced.

The absolute requirement for streptomycin for growth by str^d mutants can be satisfied by certain other agents that also cause misreading. This would be expected if, as proposed, the str^d mutants represent a special class of str^r mutants in which a change in the configuration of the str protein of the 30S subunit prevents correct reading of codons in the cell's mRNA, and certain specific requirements become apparent unless streptomycin or another agent that produces misreading is present. It is thus hypothesized that the dependence on streptomycin represents the requirement for an agent able to overcome ribosomal restriction by introducing selective translational ambiguity.

Bacterial resistance that emerges following the clinical use of streptomycin and the other aminoglycosides is usually mediated by plasmids that code for aminoglycoside-modifying enzymes. This type of resistance is discussed on page 228.

OTHER AMINOGLYCOSIDE ANTIBIOTICS. In addition to streptomycin, the aminoglycoside antibiotics that have been used clinically are neomycin, kanamycin, gentamicin, tobramycin, and amikacin (Table 9-10). Compared with the penicillins and cephalosporins, the margin between toxic and therapeutic doses is very narrow for all of the aminoglycoside antibiotics. They are, however, important therapeutic tools, especially in the treatment of aerobic gram-negative infections. Their successful clinical application has been threatened by the increasing frequency with which resistant bacteria are isolated. At present, gentamicin, tobramycin, and amikacin are clinically the most useful. They are active against a broad spectrum of gram-negative bacteria including *Pseudomonas*. With the exception of *S. aureus*, gram-positive organisms are generally resistant to all of these drugs. Gentamicin is widely used for serious gram-negative bacillary infections, but its use in nosocomial infections is becoming more limited because of an increase in gentamicin-resistant organisms in the hospital environment. Tobramycin is the drug of choice for *P. aeruginosa* infections because it is two to six times more active than gentamicin for this organism. Amikacin is a semisynthetic aminoglycoside that is resistant to most of the bacterial enzymes that acetylate, adenylate, or phosphorylate the other aminoglycosides. It should be reserved for the treatment of infections caused by gentamicin- and tobramycin-resistant organisms.

Because of its toxicity, the current clinical use of neomycin should be restricted to the preoperative suppression of intestinal flora. Kanamycin is active against most gram-negative bacteria with the exception of *Pseudomonas*. It is less toxic than neomycin but has limited clinical usefulness at present because of the emergence of bacterial resistance, especially in the hospital environment.

Mechanism of Action. All of the aminoglycosides act on protein synthesis at the level of the 30S ribosomal subunit, and most of them also produce ambiguity in translation. The level of misreading produced by neomycin, kanamycin, and gentamicin is much greater than that produced by streptomycin. As expected, those aminoglycosides that cause translational errors are all capable of phenotypic suppression of nonsense and missense mutations. In contrast to streptomycin, neomycin, kanamycin, and gentamicin exhibit concentration-dependent multiphasic

TABLE 9-10. CLINICALLY USEFUL AMINOCYCLITOL ANTIBIOTICS

Agent*	Source	Major Clinical Use
Streptomycin	*Streptomyces griseus*	Tularemia, plague, brucellosis
Neomycin	*Streptomyces fradiae*	Superficial skin infections, bacterial conjunctivitis, preoperative suppression of bowel flora
Kanamycin	*Streptomyces kanamyceticus*	Preoperative suppression of bowel flora
Gentamicin	*Micromonospora purpurea*	Infections by susceptible gram-negative rods; in combination with penicillin for enterococcal infections
Tobramycin	*Streptomyces tenebrarius*	Systemic *Pseudomonas* infections; alternative to gentamicin for gram-negative rod infections
Amikacin	Semisynthetic	Reserved for treatment of severe gram-negative rod infections resistant to other aminoglycosides
Sisomicin	*Micromonospora myosensis*	Gram-negative rod infections, especially *Pseudomonas* and indole-positive *Proteus*
Netilmicin	Derivative of sisomicin	Gram-negative rod infections (not *Pseudomonas, Serratia*), especially gentamicin-resistant strains that produce adenylylating enzymes
Spectinomycin	*Streptomyces spectabilis*	Gonorrhea in patients with penicillin allergy

*All of the agents listed are aminoglycosidic aminocyclitols except spectinomycin. Their activity is primarily directed against aerobic gram-negative bacilli and *Staphylococcus aureus*.

effects on isolated ribosomes, indicating the existence of at least two binding sites for each agent rather than a single site. These sites appear to be different from the streptomycin-binding site.

AMINOCYCLITOL ANTIBIOTICS

Spectinomycin. This antibiotic lacks an amino sugar residue that characterizes the aminoglycoside group. Unlike the aminoglycosides, it is bacteriostatic rather than bactericidal in action. Its clinical use is restricted to the treatment of uncomplicated gonorrheae in patients allergic to penicillin. There is no cross-resistance between spectinomycin and penicillin.

Spectinomycin inhibits protein synthesis at the level of the messenger-ribosome interaction but causes no misreading. Since in spectinomycin the streptamine nucleus, which is the structure responsible for misreading, is replaced by a stereoisomer, not only is the misreading property lost, but the agent is bacteriostatic and not bactericidal in its action.

Tetracyclines.

The tetracyclines are a family of closely related broad-spectrum antibiotics with activity against a wide range of gram-positive and gram-negative species, mycoplasmas, rickettsiae, and chlamydia. Included in the group are the parent compound tetracycline, several natural products (chlortetracycline, oxytetracycline, demethylchlortetracycline), and a number of semisynthetic derivatives (doxycycline, minocycline) (Fig. 9-30). The antibacterial spectrum of the tetracycline group is very broad and overlaps that of penicillin, streptomycin, and chloramphenicol. The tetracyclines inhibit only rapidly multiplying organisms and are bacteriostatic.

The antimicrobial, pharmacologic, and therapeutic properties of the older fermentation-derived tetracyclines are similar. The semisynthetic derivatives have significantly improved therapeutic properties due to improved pharmacokinetics. One of the most undesirable side effects associated with the administration of the natural tetracyclines is superinfections, caused by incomplete oral absorption and resulting in inhibition of the normal flora of the intestine. Such infections are caused by the outgrowth of drug-resistant indigenous bacteria or fungi that normally are kept in check by the drug-sensitive members of the intestinal flora. The two semisynthetics, doxycycline and minocycline, are more lipophilic than the natural antibiotics. They are almost completely absorbed from the intestine and, therefore, less inhibitory to the normal gut flora. Another major adverse reaction associated with the use of the tetracyclines results from their deposition in calcified tissue, causing staining and impairment of the structure of bone and teeth. This effect also is diminished with the use of the more lipophilic semisynthetic derivatives.

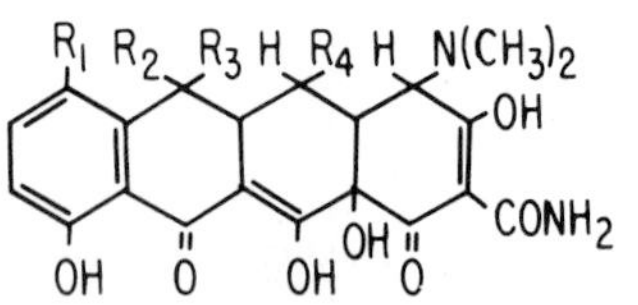

Unfortunately, with overusage of the natural tetracyclines, a high level of drug resistance has developed, especially in the hospital environment. Organisms that have become insensitive to one tetracycline also exhibit approximately the same level of resistance to other members of the group. The exception is minocycline, which remains active against organisms resistant to other tetracyclines.

Although the tetracyclines are broad-spectrum antibiotics with relatively low host toxicity, at the present time they are clinically secondary in importance to the broad-spectrum penicillins and cephalosporins. They are reserved primarily for specific indications, such as the treatment of brucellosis, cholera, infections caused by chlamydiae, rickettsiae, and *Mycoplasma pneumoniae* and of urinary tract infections by sensitive gram-negative species. They also remain a very important group of reserve drugs for ampicillin- and cephalosporin-resistant organisms. Tetracycline, the least expensive of the group, is generally the preferred drug.

MECHANISM OF ACTION. Unlike the aminoglycoside antibiotics, the tetracyclines inhibit protein synthesis of both procaryotic and eucaryotic cells. They are, however, much more effective inhibitors of protein synthesis in intact procaryotic cells than in eucaryotic cells. This selective activity is primarily due to the suicidal ability of bacteria to accumulate these drugs by an energy-dependent transport system that is not present in eucaryotic cells. The tetracyclines are transported across the membrane as a complex with magnesium ions. Within the cell, they bind to phosphate residues of the 30S ribosomal subunit via chelation with magnesium. This binding of tetracycline interferes with the binding of aminoacyl-tRNA to the acceptor site on the ribosome and, thereby, inhibits protein synthesis. Tetracycline binds specifically to the S4 and S18 proteins of intact 70S ribosomes, with secondary binding to proteins S7, S13, and S14.

Bacteria develop resistance to the tetracyclines predominantly by becoming less permeable to the antibiotic. The biochemical, genetic, and epidemiologic aspects of tetracycline resistance are discussed on page 226.

	R$_1$	R$_2$	R$_3$	R$_4$
Tetracycline	H	OH	CH$_3$	H
Oxytetracycline	H	OH	CH$_3$	OH
Chlortetracycline	Cl	OH	CH$_3$	H
Demethylchlortetracycline	Cl	OH	H	H
Doxycycline	H	H	CH$_3$	OH
Minocycline	N(CH$_3$)$_2$	H	H	H

Figure 9-30. Structural formulae for tetracycline and some of its analogs.

Nitrofurans.

One of the best known members of this group of synthetic antibacterials is nitrofurantoin, which has been used clinically in the treatment of urinary tract infections, especially in patients unable to tolerate sulfon-

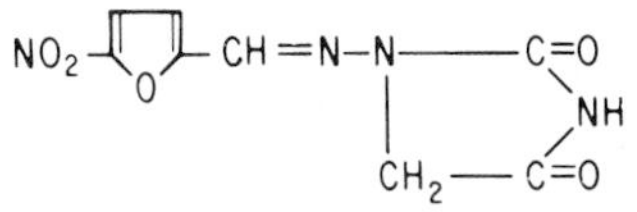

Figure 9-31. Furadantin, a nitrofuran, is a clinically useful synthetic antimicrobial agent. Nitrofurans are derivatives of the furans, 5-membered ring sugars, and possess a nitro group in the 5-position.

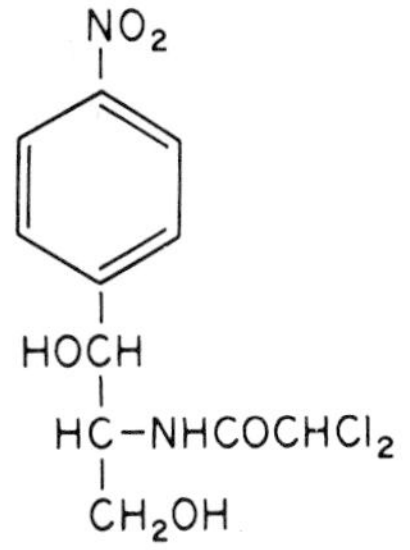

Figure 9-32. Chloramphenicol, an antibiotic that inhibits peptide bond formation.

amides. All of the nitrofurans possess a nitro group in the 5 position of the furan ring. The activity spectrum is wide and includes both gram-positive and gram-negative organisms (Fig. 9-31). It is most active against *E. coli, Klebsiella-Enterobacter* species, and some strains of enterococci. At present it is recommended only for uncomplicated urinary tract infections.

The nitrofurans inhibit protein synthesis both in vitro and in vivo. They appear to inhibit preferentially the synthesis of inducible enzymes by blocking the initiation of translation. The unique feature of this inhibition is that it apparently discriminates among various mRNAs. At concentrations sufficient to inhibit the translation of mRNAs of inducible enzymes, such as lactose operon and galactose operon enzymes, it has almost no effect on the translation of other type mRNAs, such as mRNA from the tryptophan operon or coliphage RNA.

Inhibitors of the 50S Ribosomal Subunit

In the synthesis of protein, the 50S ribosomal subunit provides an attachment site for peptidyl-tRNA, the donor site. It also contains the active center for catalyzing the peptide bond-forming reaction of protein synthesis. Inhibition of protein synthesis at the 50S ribosomal subunit could result (1) if attachment of peptidyl-tRNA is prevented, (2) if there is an interference of peptide bond formation, or (3) if the translocation step is inhibited, i.e., the movement of peptidyl-tRNA and the ribosome relative to each other. Among the antibiotics that act at the level of the 50S ribosomal subunit are chloramphenicol, lincomycin, and the macrolide group of agents.

Chloramphenicol. Chloramphenicol is a bacteriostatic agent, active against a wide range of gram-positive and gram-negative bacteria, rickettsiae, and chlamydiae (Fig. 9-32). It is rapidly and completely absorbed from the gastrointestinal tract and penetrates well into all tissues, including the brain and cerebrospinal fluid. Its clinical use, however, demands caution because of its associated toxic effects, the most important of which is bone marrow suppression. Currently its primary use is in the treatment of anaerobic infections, *Haemophilus influenzae* meningitis, and infections due to *Salmonella typhi* (Table 9-4).

MECHANISM OF ACTION. Chloramphenicol inhibits growth of bacteria by interfering with protein synthesis. It binds exclusively to the 50S subunit. The binding is ste-

reospecific, and a 1:1 equivalence exists between the number of ribosomes present and the number of chloramphenicol molecules bound. Other antibiotic inhibitors of the 50S subunit, such as erythromycin and lincomycin, compete with chloramphenicol for this binding.

Since polyribosome formation continues in the absence of protein synthesis, the drug does not interfere with the initiation of protein synthesis. Chloramphenicol inhibits peptide bond formation. Information on the nature of this inhibition has come from the use of a model system, the "puromycin reaction," in which peptidyl transfer is uncoupled from other reactions of protein synthesis, making possible the identification of inhibitors that act specifically on the peptidyl transferase. When puromycin is added to a system synthesizing peptides, it is incorporated into the growing peptide chain instead of the next incoming amino acid, and peptidyl puromycin, i.e., an incomplete peptide chain terminated by puromycin, is released from the ribosome (see also p. 217). Chloramphenicol prevents the formation of peptidyl puromycin at concentrations that inhibit protein synthesis in vivo. This inhibition results from the binding of chloramphenicol to a site on the 50S subunit in the vicinity of but not identical with the site that binds the aminoacyl end of aminoacyl-tRNA in the peptidyl transferase catalytic center.

Several proteins have been identified that are thought to form the binding site(s) of chloramphenicol. The most important of these appears to be L16, which is concerned in the peptidyl transferase activity and also forms part of the acceptor site. The interaction of chloramphenicol with 70S ribosomes is rapidly reversible and is highly specific. Chloramphenicol is completely inactive against 80S ribosomes.

RESISTANCE. Chloramphenicol and the macrolide antibiotics, erythromycin and lincomycin, compete for binding at the chloramphenicol binding site on the 50S subunit. Since they bind to or near the same unique site on the ribosome in a 1:1 ratio and are mutually exclusive, a ribosome to which erythromycin is bound is resistant to chloramphenicol. This type of resistance is important in the effective clinical utilization of antibiotics whose activity is related to inhibition of 50S subunit function. Since the simultaneous use of a competitive pair of these agents results in the therapeutic effects of a single agent, there would be no advantage to the simultaneous administration of two or more of the 50S inhibitors.

Resistance to chloramphenicol may be mutational (chromosomal) or mediated by plasmids. In some organisms, mutation to chloramphenicol resistance may be attributed to an alteration of the 50S ribosomal subunit resulting in a decreased affinity for the drug. Ribosomes from such mutants have low activity in cell-free protein synthesis and are able to bind labeled chloramphenicol, erythromycin, or lincomycin.

Resistance to chloramphenicol is not infrequently accompanied by resistance to the tetracyclines. This cross-resistance phenomenon is largely confined to enteric bacteria and is mediated by plasmids. The phenomenon involves the intra- or interspecies transfer of plasmids that carry, among other genetic properties, genes determining multiple drug resistance, including resistance to chloramphenicol (p. 228). In this type of resistance, the biochemical basis is either acquisition of the ability to degrade the antibiotic or altered membrane permeability. High-level resistance is usually associated with the presence of the enzyme chloramphenicol acetyltransferase, which inactivates the drug by catalyzing its O-acetylation in the presence of acetyl-CoA. In gram-negative organisms, the enzyme is constitutive. R factor-mediated resistance by this mechanism has been responsible for widespread epidemics of chloramphenicol-resistant typhoid fever and *Shigella* dysentery in Central and South America and in Vietnam.

The second mechanism of resistance, altered permeability, is less often encountered and is inducible. It has been detected both in enteric bacteria and in *P. aeruginosa.*

Erythromycin. Erythromycin is the most important member of the macrolide antibiotics, a group characterized chemically by a macrocyclic lactone ring of 12 to 22 carbon atoms to which one or more sugars are attached (Fig. 9-33, Table 9-4). Although primarily bacteriostatic, it may be bactericidal for some organisms if sufficiently high concentrations are used and conditions of testing are appropriate.

Erythromycin may be used as the primary drug in *M. pneumoniae* and *Legionella pneumophila* infections and in diphtheria. It is also useful in patients allergic to penicillin who have infections caused by group A streptococci or pneumococci. Bacteria develop resistance readily to erythromycin both in vitro and in vivo, but at present most clinical isolates are sensitive to this agent.

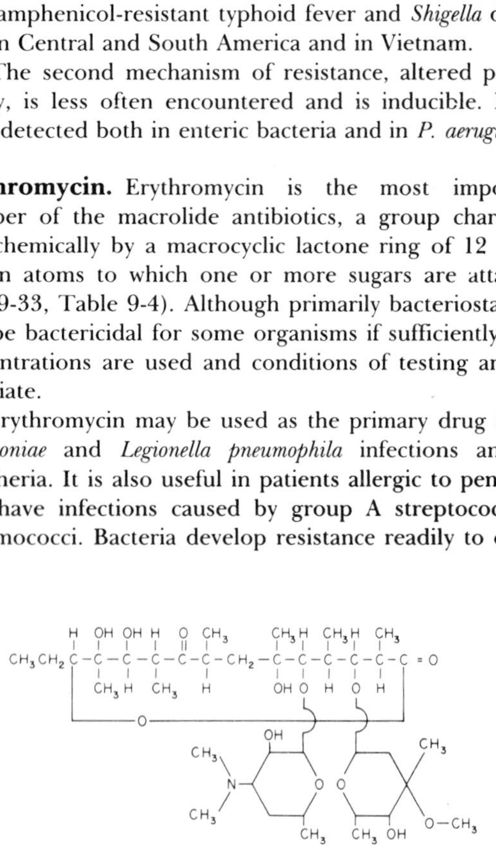

MECHANISM OF ACTION. The 50S ribosomal subunit is the site of action of erythromycin, but the mechanistic details of its action on protein synthesis remain ill defined. Although chloramphenicol and erythromycin compete for a binding site on the 50S subunit, they appear to bind to different but interacting sites. In the 70S ribosome, L15 protein possesses both erythromycin binding and peptidyl transferase activity, both functions being modulated by interaction with protein L16, to which chloramphenicol binds. Evidence obtained by use of the puromycin fragment reaction (p. 217) also suggests differences between erythromycin and chloramphenicol in their precise effects on peptidyl transfer. Erythromycin does not inhibit the fragment reaction, but chloramphenicol and lincomycin do.

Experimental evidence supports the hypothesis that in intact bacteria erythromycin blocks the translocation step. It appears to block specifically the release of the charged tRNA bound to the donor site (P site) of the ribosome after peptide bond formation. Persistence of deacylated tRNA at this site interferes with the translocation of peptidyl tRNA from the acceptor site back to the donor site. It is unclear whether the peptidyl tRNA is immobilized at the acceptor site or whether it binds to a site closely adjacent to the donor site that does not allow proper positioning for peptide bond formation.

RESISTANCE. Resistance to erythromycin is a genetically controlled property of ribosomes and may be either mutational or plasmid-mediated. The ribosomes from a number of mutants highly resistant to erythromycin have a reduced ability to bind erythromycin and are less active in in vitro polypeptide synthesis than ribosomes from erythromycin-sensitive strains. These differences can be correlated with an altered 50S ribosomal subunit protein component, either protein L4 or L12, the conformation of which is less favorable for erythromycin binding. This one-step high-level resistance is the result of chromosomal mutation. A second mechanism of resistance that has been well documented in clinical isolates involves an alteration in the 23S RNA of the 50S subunit. Resistance is plasmid-mediated and is due to the dimethylation of a specific adenine residue in the 23S subunit, thereby reducing its affinity for erythromycin. This type of resistance may be either constitutive or inducible. The modified ribosomes are cross-resistant to lincomycin and to other macrolide antibiotics.

Lincomycin and Clindamycin. The activity spectrum of these antibiotics is similar to that of erythromycin, but they are chemically unrelated (Fig. 9-34). They are especially active against group A streptococci, pneumococci, and penicillinase-producing staphylococci.

Clindamycin is the chloroderivative of lincomycin and is superior to the parent compound in its activity and absorption properties. It shows significantly greater activity against most clinically significant anaerobic bacteria, espe-

Figure 9-33. Erythromycin, a macrolide antibiotic.

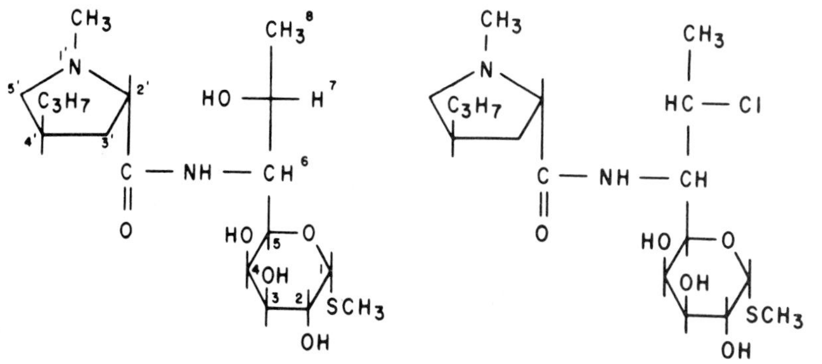

Figure 9-34. Structural formulas for lincomycin (left) and clindamycin (right). The replacement of the 7-hydroxy group in lincomycin by chlorine, with inversion of configuration, yields a drug with significantly more activity and very efficient absorption. *(From Lewis: Fed Proc 33:2303, 1974.)*

cially *B. fragilis*. Clindamycin is usually administered clinically as its phosphate ester, which is inactive in vitro but in vivo is hydrolyzed to the parent compound by lipases present in the intestinal tract. Its antibacterial range, low toxicity, and clinical efficacy make it a suitable substitute for penicillin in the treatment of infections where penicillin is contraindicated. It should be reserved primarily, however, for the treatment of infections caused by *B. fragilis* or for other anaerobic infections in persons allergic to penicillin.

MECHANISM OF ACTION. Lincomycin binds exclusively to the 50S ribosomal subunit and competes with chloramphenicol for its binding site on the ribosome. It does not bind to the 60S subunit of mammalian 80S ribosomes. Qualitatively, at least, the mode of action of lincomycin is similar to that of chloramphenicol. Lincomycin, like chloramphenicol, blocks the peptide bond-forming step, as indicated by its inhibition of the puromycin fragment reaction. Unlike chloramphenicol, however, whose polyribosomes are completely preserved in the presence of high concentrations of the drug, in the presence of lincomycin an extensive and rapid breakdown of polyribosomes occurs at all drug concentrations, and the majority of the ribosomes dissociate to 50S and 30S ribosomal subunits. This breakdown of polyribosomes by lincomycin could result from selective inhibition of some very early phase of polypeptide synthesis. The current evidence indicates that inhibition of transpeptidation by lincomycin is due to an interference with the correct positioning of aminoacyl-tRNA and peptidyl-tRNA at the acceptor and donor sites.

RESISTANCE. Lincomycin-resistant strains emerge during the course of therapy. Some of these strains are resistant only to lincomycin and erythromycin, while others show the dissociated type of resistance characteristic of the macrolides, i.e., strains sensitive to lincomycin when tested in the absence of erythromycin are resistant to lincomycin when tested in its presence. This type of resistance involving inhibitors of the 50S ribosomal subunit is due to the induction by erythromycin of a RNA methylase that dimethylates an adenine moiety on the 23S RNA. Binding of lincomycin to the altered subunit is inhibited (p. 228).

Puromycin. This antibiotic provides an excellent example of the structural analog concept of antimetabolite action (Fig. 9-35). It is structurally analogous to the terminal aminoacyl adenosine portion of tRNA and, therefore, inhibits protein synthesis by terminating the growth of polypeptide chains. As a result, growth of the cells is prevented.

Since puromycin inhibits protein synthesis at a step that is present in all living cells, it inhibits growth of both procaryotic and eucaryotic species. As a result, it is not clinically useful but is included in this discussion because of its usefulness in elucidating the reactions involved in peptide bond formation.

MECHANISM OF ACTION. As mentioned previously, there are two binding sites on the ribosome, the peptidyl-tRNA donor site (P site), which is the site where the peptidyl group of peptidyl-tRNA is donated to the incoming tRNA or to puromycin, and the aminoacyl-tRNA acceptor site (A site), which accepts aminoacyl-tRNA during chain elongation. When peptidyl-tRNA is on the P site and the A site is vacant, puromycin can bind at the A site with subsequent formation of a covalent peptidyl-puromycin derivative. The peptidyl-puromycin then dissociates from the ribosome. Growth of the peptide chain is reinitiated at frequent intervals as the ribosome moves along the mRNA molecule, resulting in the abortive synthesis of a collection of oligopeptides with random N-terminal amino acids. Puromycin also causes an increase in the rate of subunit exchange, suggesting that the release of peptide chains is accompanied by the release of 50S subunits bearing tRNA, while only 30S subunits remain associated with mRNA, which subsequently recombines with 50S subunits to form initiation complexes (Fig. 9-36).

PUROMYCIN REACTION. The finding that puromycin reacts to form a peptide with the C terminus of the growing peptide chain on the ribosome has led to the use of this reaction as a model system for the study of peptide bond formation in protein synthesis (Fig. 9-36B). In a simplified peptide bond-synthesizing system, the N-formyl-methionyl hexanucleotide fragment CAACCA-fMet, obtained by the digestion of N-formyl-methionyl-tRNA with ribonuclease, undergoes a ribosome-catalyzed reaction with puromycin in the presence of methanol to give N-formyl-methionyl-puromycin, a reaction that is analogous to peptidyl transfer in protein synthesis. Supplementation of the system

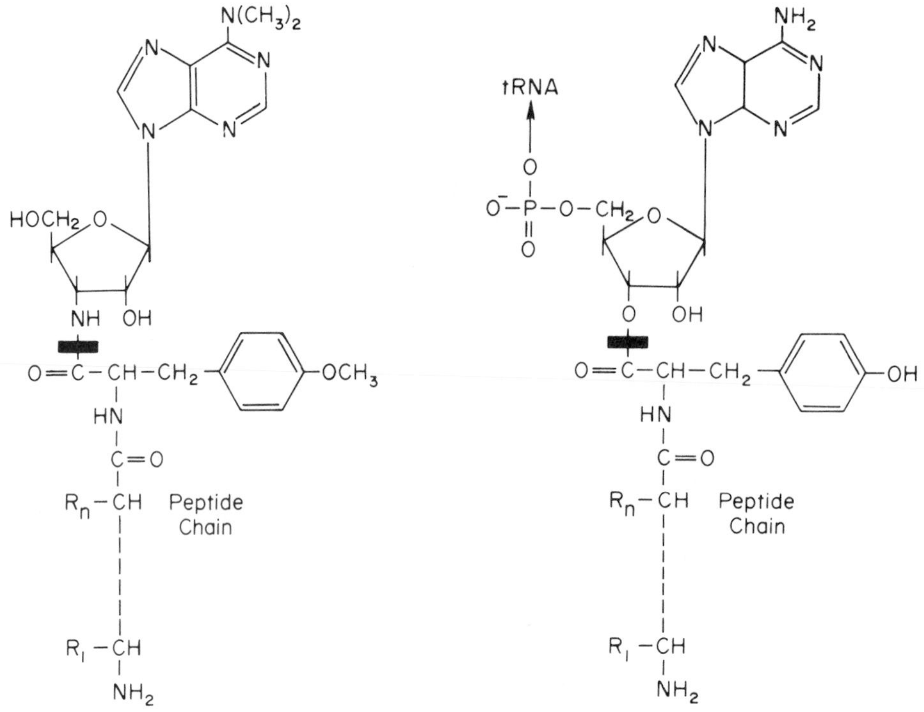

Figure 9-35. Mechanism of action of puromycin, showing its structural similarity to the terminal AMP residue of aminoacyl-tRNA. I is peptidyl-puromycin, II is peptidyl-tRNA. The crossbar marks the bond that is normally cleaved during extension of the peptide chain but that cannot be cleaved in peptidyl-puromycin.

with methanol permits rapid catalysis by isolated 50S subunits and eliminates the requirement for 30S subunits normally required for peptidyl transferase activity. The reduced size of both substrate analogs eliminates other reactions that are normally interlinked with peptidyl transfer, and confines the reaction to the immediate vicinity of a catalytic center on the 50S subunit.

Fusidic Acid. This is a steroidal antibiotic with a rather narrow antibacterial spectrum (Fig. 9-37). It inhibits the growth of gram-positive bacteria but has no significant activity against gram-negative species. Fucidin, the sodium salt of fusidic acid and the clinically useful form, has been used successfully in the treatment of a variety of serious staphylococcal infections (Table 9-9). Although it is little used at present, fucidin provides a useful addition to the antistaphylococcal armamentarium.

Mechanism of Action. Fusidic acid inhibits protein synthesis in whole cells and in cell-free extracts. Unlike 50S ribosomal subunit inhibitors, it neither binds to the ribosomes nor inhibits chloramphenicol binding. The specific site of attack of fusidic acid is the translocation reaction, the last composite step in peptide bond formation. In this step, the discharged tRNA in the donor P site is released from the ribosome, the peptidyl-tRNA is shifted from the acceptor A site to the donor site, and the ribosome moves

the length of one codon along the mRNA. These reactions are mediated by elongation factor EF-G and GTP, which form a labile complex with the ribosome. In the course of translocation, the bound GTP is cleaved to GDP and P_i, and the EF-G factor, GDP, and P_i are released from the ribosome. In the presence of fusidic acid, EF-G and GTP form a stable complex with ribosomes. The antibiotic freezes the EF-G $\sim$ GDP $\sim$ ribosome complex and prevents the release of EF-G from the ribosome, a process required for the next round of translocation and GTP hydrolysis (Fig. 9-38). Fusidic acid also blocks protein synthesis in systems containing 80S ribosomes by the same mechanism.

Inhibitor of Protein Assembly

Griseofulvin

This is a fungistatic agent specific for fungi whose walls contain chitin (Fig. 9-39). It has no effect on fungi with cellulose cell walls or on bacteria, yeasts, or yeast protoplasts. Its clinical use is limited to the management of dermatophyte infections, for which it is standard therapy. Following oral administration, griseofulvin is delivered to the stratum corneum via the sweat or by deposition in keratinocytes.

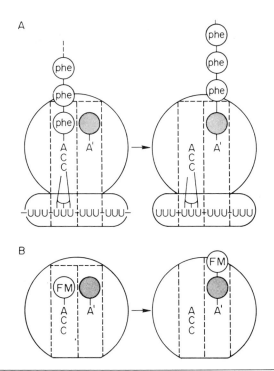

Figure 9-36. A. Reaction of puromycin (dark sphere) with polyphenylalanyl-tRNA-charged 70S ribosomes. The puromycin occupies the aminoacyl acceptor site. B. Fragment reaction employed in the study of peptide bond formation. The system contains 50S ribosomal subunits, an fMet-oligonucleotide, puromycin, and alcohol. *(Adapted from Monro et al.: Cold Spring Harbor Symp Quant Biol 34:357, 1969.)*

Treatment of growing cells with griseofulvin causes morphologic abnormalities, such as swelling and branching in the growing tip, while old cells distant from the growing point are not affected. It inhibits mitosis in the metaphase, causing multipolar mitosis and abnormal nuclei. The molecular basis for the antimitotic action of griseofulvin has

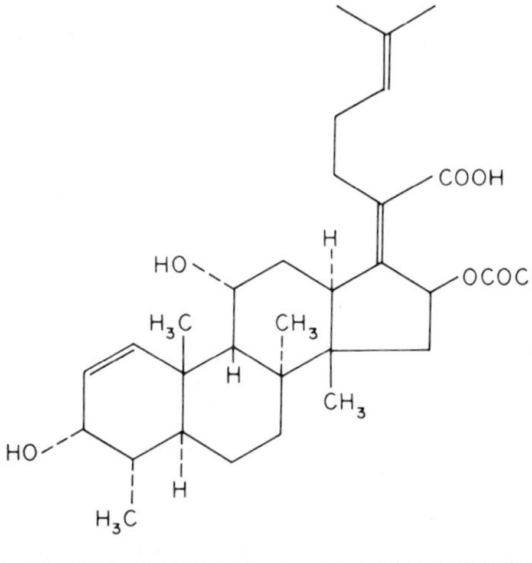

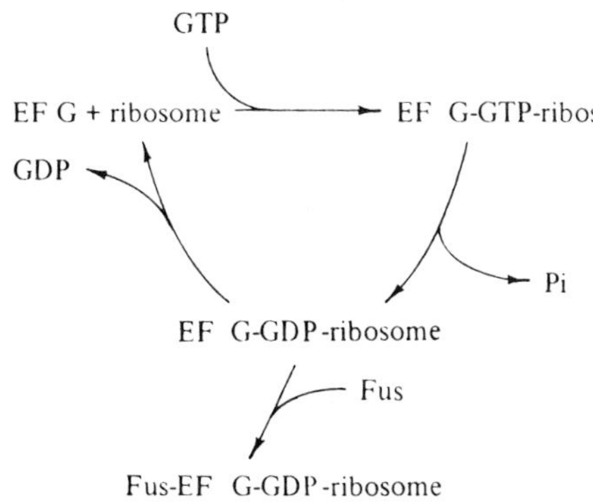

Figure 9-38. Schematic representation of mechanism of fusidic acid (Fus) inhibition of the GTPase reaction.

been attributed to an interference with the assembly process of tubulin into microtubules. Griseofulvin binds to proteins involved in tubulin assembly. Cell processes that depend on microtubule function, such as the movement of chromosomes during mitosis, are thus inhibited by the drug.

Metabolite Analogs

Enzymes are often inhibited by compounds possessing a structure similar to the natural substrate. Such inhibitors combine with the enzyme in such a manner as to prevent the normal substrate-enzyme combination and subsequent catalytic reaction. Many inhibitors of this type are analogs of the bacterial growth factors, organic factors required by all bacteria for growth. Such growth factors include the B-complex vitamins, amino acids, purines, and pyrimidines. Enzymes essential in the synthesis and utilization of certain of these factors can be inhibited by various antimetabolites, compounds structurally related to the metabolites.

Competitive versus Noncompetitive Inhibition. The antimetabolites that cause inhibition of enzymatic reactions are of two major types, competitive and noncompetitive. Competitive inhibition can be overcome by

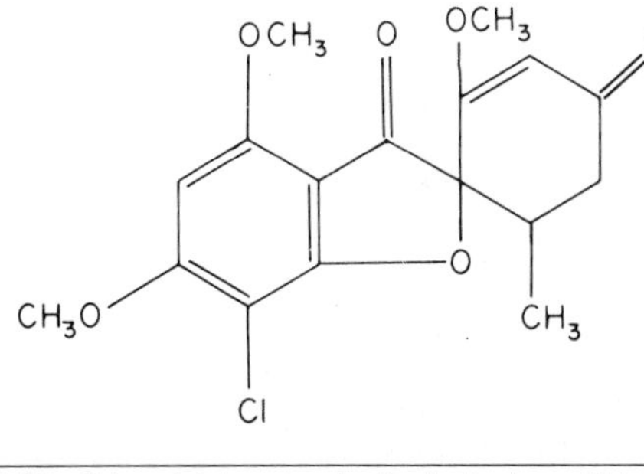

Figure 9-37. Fusidic acid, which is used clinically as its sodium salt, fucidin.

Figure 9-39. Griseofulvin, a selectively toxic antibiotic for fungi whose walls contain chitin.

increasing the substrate concentration, whereas noncompetitive inhibition cannot be reversed by the substrate.

In the competitive type of inhibition, both inhibitor (I) and substrate (S) compete for the same enzyme site:

$$E + S \rightleftharpoons ES \rightarrow E + P$$

$$E + I \rightleftharpoons EI$$

The EI complex yields no reaction products, and although the formation of EI is reversible, the continuing competition with substrate reduces the effective free enzyme concentration. In inhibitions of this type, the percentage of inhibition of the enzyme is a function of the ratio of the concentrations of inhibitor and substrate rather than a function of the absolute concentration of the inhibitor alone. This relationship may be treated quantitatively by use of the Michaelis-Menten equation, which defines the relationship between the enzyme reaction rate (v) and the substrate (S) concentration. Competitive inhibition is most easily recognized by using Lineweaver-Burk plots, i.e., plots of 1/v versus 1/S at varying concentrations of inhibitor (Fig. 9-40). In competitive inhibition, the plot is characterized by straight lines of differing slope that intersect at a common intercept on the 1/v axis. Thus, at any inhibitor concentration, there is a substrate concentration that can evoke full activity of the enzyme.

In noncompetitive inhibition, inhibition depends only on the concentration of the inhibitor and is not reversed by increasing the substrate concentration. In contrast to the competitive type of inhibition, the inhibitor binds at a locus on the enzyme other than the substrate binding site. It may bind to the free enzyme, to the ES complex, or to both, resulting in the formation of inactive EI and ESI complexes:

$$E + I \rightleftharpoons EI$$

$$ES + I \rightleftharpoons ESI$$

The rate of conversion of S→P is slowed but not stopped. The effect exerted may be on the affinity of the enzyme for substrate or on the rate of the reaction. In a Lineweaver-Burk plot of noncompetitive inhibition, the plots differ in slope and do not share a common intercept on the 1/v axis. The intercept on the 1/v axis is greater for the inhibited than for the uninhibited reaction, indicating that the enzyme activity cannot be restored regardless of the substrate concentration.

Usefulness of Competitive Inhibitors in Chemotherapy.

With the introduction of the therapeutic agent, sulfanilamide, attention was focused on the potential value of metabolite analogs in the designing of new chemotherapeutic agents. Although thousands of analogs of the essential metabolites have been designed, many of which have been effective inhibitors in vitro, with the exception of analogs of *p*-aminobenzoic acid (PABA), most of them have lacked the requisite selectivity necessary for clinical use. The basis for this lack of selectivity lies in the similarity among most of the enzymatic reactions present in bacterial and mammalian cells. Although a compound may strongly inhibit an isolated, purified enzyme in vitro, such conditions are nonexistent in vivo. In the living cell, which is analogous to an open system, substrate is continuously supplied to the enzyme by the previous enzyme of the metabolic pathway, and its product is removed by the next enzyme. In this system of balanced growth, a steady state is reached in which the level of substrate is often insufficient to saturate the enzymes. If a competitive inhibitor is added to this system, the competitor is bound by the enzyme, and inhibition of activity results. In biosynthetic pathways, however, reactions that precede the inhibited reaction continue to supply the natural substrate. In time, the concentration of substrate is sufficiently high to reverse the inhibition, and the reaction is resumed at a rate characteristic of the higher level. Therefore, structural analogs, in order to be successful antimicrobial agents, either must have a much higher affinity for the enzyme than has the natural substrate or must function as something more than a simple competitive inhibitor. Most of the competitive inhibitors that have been studied have a lower affinity for the target enzyme. An exception is the antibiotic cycloserine, which is an analog of D-alanine (p. 202). This drug binds to D-alanyl-D-alanine synthetase 100 times more effectively than the natural substrate, which may account for its clinical efficacy.

Whereas structural analogs that inhibit by competing at the substrate binding site of an enzyme are usually only transient inhibitors, these inhibitors may be quite effective

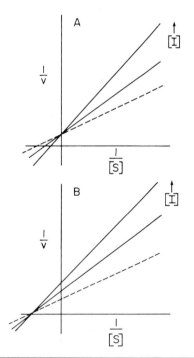

Figure 9-40. Lineweaver-Burk plots of competitive (A) and noncompetitive (B) inhibition. The reaction in the absence of inhibitor is represented by the dotted line.

when they function as end-product inhibitors or repressors. In such cases, they mimic the effect of the essential metabolite and inhibit the activity or the synthesis of new enzymes. Their action is thus to reduce the availability of the natural analogs.

Inhibitors of Tetrahydrofolate Synthesis

Sulfonamides

The term "sulfonamide" is a generic name for derivatives of p-aminobenzenesulfonamide or sulfanilamide. First administered in 1935 by Domagk as the red dye Prontosil, sulfanilamide was the first effective chemotherapeutic agent to be used systemically for the prevention and cure of bacterial infections in humans. In vitro, Prontosil is inactive against bacteria, but in the body it is broken down to p-aminobenzenesulfonamide, the chemotherapeutic moiety of the molecule.

Since the introduction of sulfanilamide, numerous derivatives have been synthesized and tested for their clinical value in various types of infections. The minimal structural requirement for antibacterial action is that the sulfur be linked directly to the benzene ring and that the NH_2 group in the para position be either retained as such or replaced only by radicals that can be converted in the tissues to a free amino group (Fig. 9-41). Of the thousands of derivatives synthesized, less than 25 are clinically useful. Although the advent of the antibiotics detracted from the popularity and usefulness of the sulfonamides, they continue to play an important but smaller role in the control of infectious diseases.

Antibacterial Spectrum. Sulfonamides exhibit inhibitory activity against a broad spectrum of gram-positive and gram-negative species as well as *Nocardia, Chlamydia,* and certain protozoa, such as *Pneumocystis* and *Plasmodium.* Although there is usually a direct correlation between in vitro and in vivo efficacy of the sulfonamides, the presence in tissues of such substances as p-aminobenzoic acid, methionine, and the purines can neutralize their inhibitory activity. Since bacteria are impermeable to folic acid, its presence in tissues does not interfere with the efficacy of the drug. In vivo usefulness is also conditioned by the extent of protein binding of the drug after absorption, since the conjugate is therapeutically inactive.

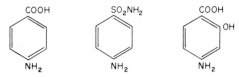

Figure 9-41. Structural relationship between p-aminobenzoic acid (left), sulfanilamide (center), and p-aminosalicylic acid (right).

Adverse Effects. A number of toxic reactions have been attributed to the various sulfonamides, the most serious of which are renal blockage from crystalluria, some systemic disorders that are probably allergic in origin, and blood dyscrasias.

Current Clinical Use. The major current indications for the use of sulfonamides include (1) nocardiosis, (2) suppressive therapy for paracoccidioidomycosis, (3) prophylaxis in sulfadiazine-sensitive meningococcal infections, (4) alternative prophylaxis for acute rheumatic fever in patients allergic to penicillin, and (5) uncomplicated urinary tract infections caused by susceptible *E. coli* strains. For presurgical sterilization of the gut, some of the poorly absorbed sulfonamide derivatives in combination with neomycin have been used extensively. Some of the clinically useful sulfonamides and their major uses are shown in Table 9-11.

Sulfonamides are also used in combination with antifolate drugs in an attempt to potentiate their activity and to prevent the development of resistance. Since the antifolates block the same metabolic pathway as the sulfonamides but at a different site, the combination of a sulfonamide and an antifolate drug, such as trimethoprim, is synergistic.

TABLE 9-11. SULFONAMIDES OF CLINICAL IMPORTANCE

Drug	Properties	Clinical Use
Sulfadiazine	Rapidly absorbed and excreted	Meningitis
Sulfafurazole (Gantrisin)	Rapidly absorbed and excreted	Urinary tract infections
Sulfamethoxazole (Gantanol)	Rapidly absorbed and excreted	Urinary tract infections
Sulfamethoxypyridazine (Midicel, Kynex)	Rapidly absorbed, slowly excreted	Urinary tract infections, respiratory infections
Sulfadimethoxine (Madribon)	Rapidly absorbed, slowly excreted	Urinary tract infections, respiratory infections
Succinylsulfathiazole (Sulfasuxidine)	Poorly absorbed	Treatment of *Salmonella* and *Shigella* carriers, suppression of intestinal flora before surgery
Phthalylsulfathiazole	Poorly absorbed	Treatment of *Salmonella* and *Shigella* carriers, suppression of intestinal flora before surgery

Mechanism of Action. The sulfonamides are structural analogs of *p*-aminobenzoic acid (PABA), a precursor of folic acid (pteroylglutamic acid). The biologically active form of folic acid is tetrahydrofolic acid (FH_4), a coenzyme important in the transfer and reduction of 1-carbon fragments. Tetrahydrofolate serves as the acceptor of the β-carbon atom of serine when it is cleaved to yield glycine. This reaction is of special significance as a source of active 1-carbon units required in the synthesis of methionine, thymine, and the purines.

Sulfonamides interfere with the synthesis of folic acid by inhibiting the condensation of PABA with 2-amino-4-hydroxy-6-dihydropteridinylmethyl pyrophosphate to form dihydropteroic acid. The sulfonamides compete with PABA in this reaction not simply by occupying the active site on the enzyme but by acting as alternative substrates for dihydropteroate synthetase (Fig. 9-42).

Organisms that synthesize folic acid are sensitive to sulfonamides, whereas those that have a requirement for preformed folic acid are insensitive because of the absence of the sulfonamide-inhibited reaction. The addition of PABA to a system in which growth has been inhibited by the sulfonamides neutralizes the inhibitory effect. Certain metabolites involved in folic acid coenzyme-requiring reactions (i.e., methionine, serine, thymine, and the purines) also overcome inhibition produced by these drugs.

Man, like certain microorganisms, requires preformed folic acid for growth and cannot synthesize it from PABA. The successful use of the sulfonamide drugs in therapy in spite of the presence of folic acid in human tissues, is due to the impermeability of the bacterial cell to folic acid as it occurs in the tissues.

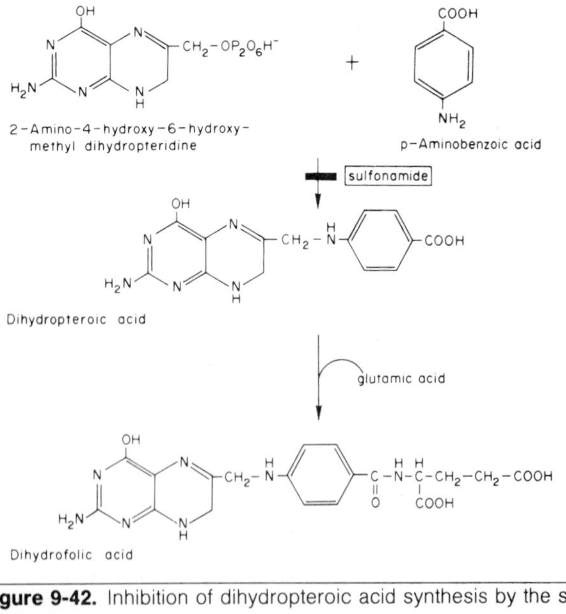

Resistance. The emergence of drug-resistant strains has limited the clinical usefulness of the sulfonamides. This has been a serious handicap, especially in the treatment of bacillary dysentery and meningococcal meningitis, infections for which the sulfonamides were previously very useful. Resistance to one sulfonamide results in cross-resistance to other members of the group. The major mechanism responsible for this increased resistance is R factor-mediated and is due to the production of a target enzyme with diminished affinity for the drug. Other mechanisms of resistance are listed in Table 9-12, on p. 227, and on pages 228–229.

Other Analogs of *p*-Aminobenzoic Acid

Sulfones. Derivatives of 4,4'-diaminodiphenylsulfone (dapsone) form a group of agents that display marked specificity, primarily against the genus *Mycobacterium* (Fig. 9-43). Although previously used in the treatment of tuberculosis, their present use is limited to the management of leprosy. Dapsone is the agent most useful clinically and, when employed in the early stages of leprosy, is successful in halting progression of the disease. A number of toxic reactions accompany its use, including hemolytic anemia, peripheral neuropathy, dermatitis, and erythema nodosum. Sulfones interfere with the metabolism of PABA, which in vitro neutralizes the drug's activity.

_p_-Aminosalicylic Acid (PAS). The antimicrobial activity of PAS is highly specific for *M. tuberculosis*. It is a bacteriostatic agent structurally similar to PABA and is antagonized by PABA in vitro (Fig. 9-41). At present, it is used primarily as a second-line drug in the chemotherapy of tuberculosis.

Inhibitors of Dihydrofolate Reductase

Trimethoprim. This antifolic acid agent is a very potent and selective inhibitor of bacterial dihydrofolate reductase (Fig. 9-44). Its spectrum of activity is similar to that of the sulfonamides and includes most gram-positive cocci and gram-negative rods. Although sulfonamides are bacteriostatic even at high concentrations, trimethoprim may be bactericidal. For clinical use, trimethoprim is combined with the sulfonamide sulfamethoxazole, with which it acts synergistically. The trimethoprim-sulfamethoxazole combination (co-trimoxazole) has proved especially useful in the management of chronic urinary tract infections and *Pneumocystis* infections. Clinical trials suggest that it may also be useful in patients with bronchitis, shigellosis, and typhoid fever resistant to both chloramphenicol and ampicillin. Adverse reactions are the same as for the sulfonamides.

MECHANISM OF ACTION. The inhibitory action of trimethoprim is based on its ability to reduce the pool of tetrahydrofolate cofactors in the bacterial cell to a level

Figure 9-42. Inhibition of dihydropteroic acid synthesis by the sulfonamides.

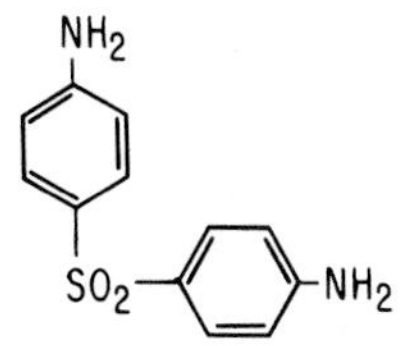

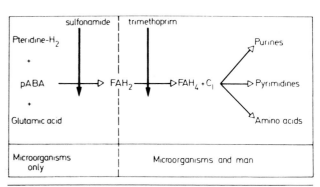

Figure 9-43. Dapsone (4,4'-diaminodiphenyl sulfone), the basic therapeutic agent for *Mycobacterium leprae* infections.

Figure 9-45. The folic acid pathway: loci of trimethoprim and sulfonamide inhibition. *(From Burchall: In Corcoran and Hahn (eds): Antibiotics III. Mechanism of Action of Antimicrobial and Antitumor Agents, 1975. New York, Springer-Verlag.)*

tetrahydrofolate cofactors in the bacterial cell to a level that is inadequate for growth (Fig. 9-45). The tetrahydrofolates function in a battery of biosynthetic reactions in which they serve as carriers of 1-carbon fragments. In one of these reactions, the synthesis of thymine by thymidylate synthetase, tetrahydrofolate reverts to the dihydro state. This must be converted to the more reduced form in order for growth to continue. The enzyme that catalyzes this reaction is dihydrofolate reductase, the enzyme that is inhibited by trimethoprim. Depletion of the tetrahydrofolate pool interferes with the synthesis of purines, pyrimidines, several amino acids, pantothenate, and N-formylmethionyl-tRNA, resulting in cessation of growth and ultimately death. The highly selective activity of trimethoprim for bacterial reductases and low toxicity for mammals are due to its extremely high affinity for bacterial enzymes and weak binding to mammalian reductases.

The use of trimethoprim in combination with sulfamethoxazole exploits the biochemical differences between humans and bacteria in order to selectively damage the parasite. Since both drugs block the folic acid pathway but at different points, the double blockage is effective in cutting off completely the supply of tetrahydrofolate to the bacteria.

At the present time, the resistance levels of clinical isolates to trimethoprim or to trimethoprim-sulfamethoxazole are still fairly low (4 to 11 percent), although there is concern that plasmid-encoded resistance may eventually compromise the clinical effectiveness of the agents. Several mechanisms of trimethoprim resistance have been described, but the major cause of significant resistance among clinical isolates is the production of a plasmid-encoded trimethoprim-resistant form of dihydrofolate reductase.

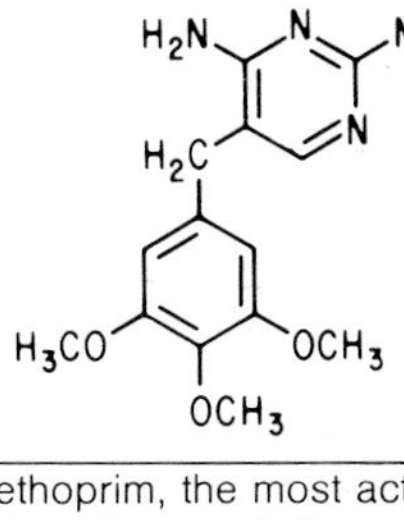

Figure 9-44. Trimethoprim, the most active and selective agent of a series of synthetic inhibitors of dihydrofolate reductase.

Other Metabolite Analogs

Isoniazid

Isoniazid, the hydrazide of isonicotonic acid, is highly specific for *M. tuberculosis* (Fig. 9-46). It is effective in very low concentrations and is bactericidal only for actively growing organisms. Isoniazid does not immediately inhibit growth but does so only after the organisms undergo one or two divisions. Tubercle bacilli exposed to the drug lose their acid-fast staining property. Isoniazid penetrates cells with ease and, unlike streptomycin, is as effective against bacilli within monocytes as against extracellular organisms.

Isoniazid remains the keystone of initial treatment of pulmonary tuberculosis and is highly effective, well absorbed, and of low toxicity. Drug resistance, however, may emerge in the presence of very large bacterial populations, such as are associated with pulmonary or renal cavities. This can be largely circumvented by the simultaneous use of one or more companion drugs (ethambutol or rifampin). At present, this combination therapy represents the mainstay in the treatment of tuberculosis. The most important side effects of isoniazid are related to its hepatotoxicity and toxicity for the peripheral and central nervous systems. Of these, hepatitis is the most significant and is thought to be secondary to the conversion of isoniazid to acetylhydrazine. Symptoms of vitamin B_6 deficiency, such as peripheral neuritis and seizures, may also occur.

Mechanism of Action. In spite of extensive studies, there is no convincing evidence that pinpoints any single mechanism as the primary site of attack by isoniazid. Its structural similarity to both niacin and pyridoxal suggests that the drug might act as an antimetabolite against either of these vitamins. Interference with either would have pleiotropic effects because of their importance in many aspects of metabolism. Although pyridoxal reverses competitively the inhibitory action of isoniazid on tubercle bacilli and on various pyridoxal-requiring reactions, the signifi-

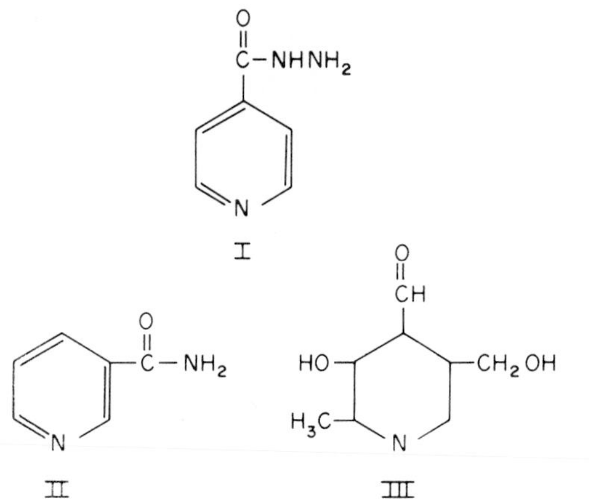

Figure 9-46. Structural relationships among I, isoniazid (isonicotinic acid hydrazide), II, nicotinamide, and III, pyridoxal.

cance of these effects is difficult to evaluate because of the formation of a pyridoxal-isonicotinyl hydrazone.

Also difficult to evaluate at present is the significance of the structural relationship of isoniazid to nicotinamide adenine dinucleotide (NAD). Isoniazid reduces the NAD supply of the cell by activating NADase, the enzyme that breaks down NAD. In the intact cell the enzyme is associated with the membrane where it normally occurs in an inactive form. Isoniazid activates NADase by altering the conformation of a protein inhibitor with which the enzyme is normally associated. This results in the rapid breakdown of NAD and depletion of the cell's supply.

Any acceptable explanation of the mechanism of action of isoniazid must also explain the reason for its exquisite sensitivity for mycobacteria. The only observation to date that may provide this explanation is the finding that mycolic acid synthesis by sensitive strains of *M. tuberculosis* is inhibited by isoniazid. It is unclear, however, whether limitations on the synthesis of this lipid would in itself be lethal to the organisms.

Flucytosine (5-Fluorocytosine)

Flucytosine, the fluorine analog of cytosine, is the only antifungal agent in clinical use that is a true antimetabolite (Fig. 9-47). It is useful in the systemic treatment of some deep-seated fungal infections in humans, in particular candidiasis, cryptococcosis, and chromomycosis. Only in chro-

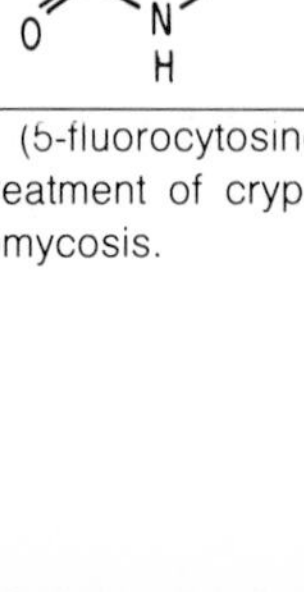

Figure 9-47. Flucytosine (5-fluorocytosine), a fluorine analog of cytosine, useful in the treatment of cryptococcosis, candidiasis, torulopsosis, and chromomycosis.

momycosis, however, is it the drug of choice. Since flucytosine-resistant strains may emerge rapidly if the drug is used alone, it is used primarily in combination with amphotericin B, especially in cryptococcal meningitis.

Mechanism of Action. Flucytosine enters the cell via the cytosine permease. It is first deaminated to 5-fluorouracil by cytosine deaminase and subsequently phosphorylated and incorporated into RNA. Another pathway for 5-fluorouracil involves the formation of 5-fluorodeoxyuridine monophosphate, a noncompetitive inhibitor of thymidylate synthetase. This interferes with DNA synthesis and leads to defective cell division. At present, no definitive statement can be made as to whether impaired DNA synthesis or the dysfunction of fungal RNA leading to disturbed protein synthesis is the primary cause of the drug's antifungal activity.

Resistance to flucytosine can be attributed to many mechanisms. Resistant mutants may have either deficient enzyme systems involved with the metabolism of the drug, increased de novo synthesis of competing pyrimidines, or compensating mechanisms for the abnormal RNA function. The plurality of mechanisms of resistance to flucytosine, most of which are independent of each other and result from one-step mutation, explains the high frequency of resistance to this agent.

Drug Resistance

The introduction of the sulfonamides and penicillin opened a new era in clinical medicine and stimulated a wave of optimism in the fight against infectious diseases. Early in the use of these drugs, however, it was realized that even though devastating epidemics had been curbed, disease caused by infectious organisms remained a serious problem. One of the major factors contributing to the persistence of infectious diseases is the tremendous capacity of microorganisms for circumventing the action of inhibitory agents.

The ability of many microorganisms to develop resistance to the chemotherapeutic drugs offers a serious threat to their future usefulness and demands both resourcefulness and ingenuity in meeting and counteracting this problem. If the use of these agents is to be successful, we must abandon the notion that they affect only those organisms against which they are directed at any particular time and that regardless of how recklessly they are used, the organisms are powerless to respond.

Origin of Drug-resistant Strains

There are two major mechanisms by which increased resistance to antibiotics and other drugs used in clinical practice may arise: (1) by mutation (chromosomal) and (2) by genetic exchange.

Selection of Drug-resistant Mutants

In the past, the origin of drug-resistant strains of microorganisms has aroused much controversy. Considerable effort has been directed toward determining whether resistant cells represent the products of phenotypic adaptation induced by some interaction of the drug with the organism or whether they are mutants arising independently of the antibiotic by mutation. It is now firmly established that drug resistance arises by the latter mechanism, i.e., by a random mutation that results in an altered susceptibility to the drug, the drug serving only as a selective agent favoring the survival of resistant over sensitive organisms once the genetic alteration has taken place and has been expressed phenotypically.

Mutations generally occur at a frequency of about 1 in 10^5 to 10^{10} cell divisions. Knowledge of the mutation rate for a particular organism as well as the site of attack of a specific drug is important for a rational approach to chemotherapy. The successful use of combination treatment in the management of tuberculosis provides such an example. The exceedingly large numbers of tubercle bacilli present within the tuberculous lesion provide an opportunity for the rapid emergence of resistant strains if a single drug is administered. By the use of combined therapy, however, the likelihood of an organism mutating to resistance to two drugs administered simultaneously is very low, about 1 in 10^{15} cell divisions.

Resistance Mediated by Genetic Exchange

Genetic information that controls bacterial drug resistance occurs both in the bacterial chromosome and in the DNA of extrachromosomal plasmids. The resistance trait may be transmitted from these loci by the transfer of genetic material from a resistant cell to a sensitive one via transformation, conjugation, or transduction. Bacteria rarely, if ever, exchange their chromosomal genes. Plasmids, however, constitute a very efficient and powerful means for the dissemination and rearrangement of genetic information. The host range of some plasmids is quite broad, e.g., R (resistance) plasmids residing in certain *Pseudomonas* species that can be transmitted by conjugation into certain soil bacteria, photosynthetic bacteria, and *Neisseria* (Fig. 9-48).

Plasmids may be classified into two major types: conjugative or nonconjugative. Conjugative plasmids are self-transmissible from one cell to another and have a region concerned with conjugation and the synthesis of a sex pilus. The plasmids that are associated with the transference of drug-resistance markers by conjugation are referred to as R factors. They consist of two distinct components: (1) the resistance transfer factor (RTF), which initiates and controls the conjugation process, and (2) the r determinant, a series of one or more linked genes that confer resistance to specific antimicrobial agents. The r determinants and RTFs are independent replicons, each of which is capable of replicating and operating on its own in the bacterial cell. The r determinants confer resistance only in the cell possessing them unless the RTF is also present. When a resistance determinant and its RTF are present in the same cell, association of the two occurs and an R factor is produced. For transfer of R factors, conjugation is required. RTFs code for the formation of specific pili through which the transfer of R factors is accomplished.

Infectious drug resistance of this kind has far-reaching epidemiologic implications. It was first recognized in Japan in 1957 during an epidemic of bacillary dysentery, when strains of *Shigella dysenteriae* became simultaneously resistant to chloramphenicol, streptomycin, sulfanilamide, and tetracycline. Multiple drug resistance is now very common in many countries throughout the world, complicating and in some cases precluding the successful treatment of many bacterial infections. The drug-resistant determinants found thus far to be transmitted in this manner include those for

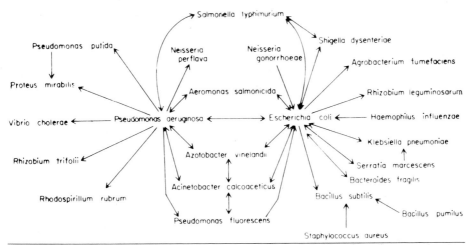

Figure 9-48. Flow of plasmids among organisms by DNA-mediated transformation or conjugation. The arrows indicate the flow of genetic information from donor to recipient. *(From Young and Mayer: Rev Infect Dis 1:55, 1979.)*

streptomycin, sulfonamides, chloramphenicol, tetracycline, neomycin, kanamycin, ampicillin, and furazolidone. Environmental exposure of the normal intestinal flora to antibiotics favors the growth of organisms that carry R factors. Following infections with pathogenic species, these drug-resistant saprophytes can transmit resistance to cells of the sensitive pathogen, which may then, if antibiotics are used in treatment, completely replace the initially drug-sensitive organism. There is clear-cut clinical evidence that transfer of drug resistance of this type occurs within the human intestinal tract.

Nonconjugative plasmids are unable to initiate self-transfer and do not encode for a sex pilus. They are smaller in mass (about 5×10^6 daltons) than conjugative plasmids (40×10^6 to 200×10^6 daltons) and rarely encode for more than two antibiotic resistance genes. Their transfer is mediated by coresident conjugative plasmids by the process of mobilization. In *N. gonorrhoeae* this is the mechanism by which nonconjugative R plasmids are transmitted from gonococcus to gonococcus.

Nonconjugative as well as conjugative plasmids can also be transmitted via transduction or transformation. In staphylococci, all plasmids are nonconjugative, including the penicillinase plasmid, and dissemination is solely by transduction. The staphylococcal penicillinase plasmids contain the determinants for the enzyme β-lactamase (p. 227). Genetic markers, such as resistance to erythromycin and to a number of inorganic ions, may also be located on this plasmid, but resistance to streptomycin, tetracycline, chloramphenicol, and the macrolides, which can also be transduced, are located on different plasmids.

The acquisition of genetic material by plasmids and chromosomes is not limited by the classic recombination processes. Many drug resistance genes reside on transposons, DNA sequences that have the capacity for excising themselves from one genome and inserting themselves into another. The resistance genes can transpose from plasmid to plasmid, from plasmid to chromosome, or plasmid to bacteriophage. Genes that specify resistance to a number of our most useful antibiotics, including ampicillin, chloramphenicol, tetracycline, kanamycin, streptomycin, and trimethoprim, are found on transposons, thus providing a possible explanation for the rapid evolution of R plasmids that possess a wide variety of antibiotic resistance determinants.

Biochemical Mechanisms of Drug Resistance

Resistance is due to genetically controlled peculiarities of the metabolism or structure of the cell that enable it to escape the action of the drug. Among the biochemical mechanisms by which microorganisms resist the inhibitory effect of an antimicrobial agent are (1) decreased permeability of the organism to the drug, (2) inactivation of the inhibitor by enzymes produced by the resistant organism, (3) modification of the properties of the drug receptor site, and (4) increased synthesis of an essential metabolite that is antagonistic for the drug (Table 9-12).

Among clinical isolates, chromosomal drug resistance usually causes changes in cellular structures that make the organism impermeable to the antibiotic or that render the specific target site indifferent to the presence of the drug. R plasmid-mediated resistance usually involves either a decrease in the permeability of the cell or enzymatic inactivation of the inhibitor.

Decreased Cell Permeability

Altered permeability to antimicrobial agents may involve changes in specific receptors for the drug, loss of capacity for active transport through the cell membrane, or structural changes in one or more components of the cell envelope that influence permeability in a relatively nonspecific manner.

This is the most commonly encountered mode of resistance to the tetracyclines and is a cause in some cases for sulfonamide resistance. The impermeability mechanism is specific for each drug, since loss of either sulfonamide or tetracycline resistance in strains resistant to both drugs does not impair resistance to the other drug. Bacterial resistance to tetracycline is caused by at least four different determinants carried on plasmids in the host bacterial cell. In most cases, resistance is inducible by subinhibitory concentrations of tetracycline, and for some plasmids it can reach 200 times the resistance of sensitive cells. The most common tetracycline resistance is that borne on transposon Tn10. Coincident with the appearance of induced resistance in *E. coli* is the synthesis of three different proteins, which appear to be located in the cell envelope and which are thought to be components of a new transport system that limits tetracycline uptake and accumulation by the cell. In tetracycline-sensitive cells, there are two transport systems for tetracycline, only one of which is energy dependent. Both of these systems are altered in resistant organisms. The energy-dependent component of uptake is replaced in resistant cells by a non-energy-requiring uptake at a lower rate, and the energy-independent rate is decreased. This decrease in accumulation of tetracycline has been attributed to an active efflux of the drug.

In gram-negative organisms, the genes that determine tetracycline resistance are often, but not always, found on transposon 10, which is a component of several R factors. The plasmid location of the determinant for tetracycline resistance in *S. aureus* probably also involves a transposon.

Enzymatic Inactivation of the Drug

This type of resistance, commonly observed among clinical isolates of resistant organisms, is the primary mechanism of resistance to penicillin, chloramphenicol, and the aminoglycoside antibiotics. Specific inactivating enzymes for these agents occur in a number of bacteria carrying R factors and other plasmids (Table 9-12).

TABLE 9-12. MECHANISMS OF ANTIMICROBIAL RESISTANCE

Mechanism	Agents	Organisms
Failure to enter cell*	β-Lactams	*Pseudomonas aeruginosa, Enterobacter*
	Aminoglycosides	*Pseudomonas aeruginosa, Serratia, Streptococcus faecalis*
	Chloramphenicol, trimethoprim	*Pseudomonas aeruginosa*
Alteration in transport system, cell wall, cell membrane		
Reduced uptake or increased removal	Tetracyclines	Enterobacteriaceae
Membrane not energized	Aminoglycosides	Anaerobes
Enzymatically modified drug not transported	Aminoglycosides	Enterobacteriaceae, *Pseudomonas*
Enzymatically modified drug poorly transported	Chloramphenicol	*Pseudomonas*
Altered sugar transport system	Fosfomycin	Enterobacteriaceae
Enzymatic inactivation of drug	β-Lactams	*Staphylococcus aureus,*
β-Lactamase		Enterobacteriaceae, *Pseudomonas, Haemophilus influenzae*
Chloramphenicol acetyltransferase	Chloramphenicol	*Staphylococcus aureus,* Enterobacteriaceae
Acetylation, phosphorylation, nucleotidylation	Aminoglycosides	*Staphylococcus aureus, Streptococcus,* Enterobacteriaceae, *Pseudomonas*
Alteration of target		
Methylation of 23S RNA	Erythromycin, clindamycin	*Staphylococcus aureus*
DNA gyrase	Nalidixic acid	Enterobacteriaceae
RNA polymerase	Rifampin	Enterobacteriaceae
Penicillin-binding proteins	Penicillin	*Neisseria gonorrhoeae, Streptococcus pneumoniae, Streptococcus faecalis, Staphylococcus aureus,* Enterobacteriaceae,
30S ribosome	Streptomycin	Enterobacteriaceae
Synthesis of resistant pathway		
Dihydrofolate reductase	Trimethoprim	Enterobacteriaceae
Dihydropteroate synthetase	Sulfonamides	Enterobacteriaceae, *Staphylococcus aureus*

Adapted from Neu: Rev Infect Dis 5[Suppl 1]:S9, 1983.
*Intrinsic resistance.

Inactivation of β-Lactam Antibiotics. From a clinical standpoint, the nost important and widespread of the degradative enzymes that attack the β-lactam antibiotics are the β-lactamases. These enzymes hydrolyze the β-lactam ring of the penicillin and cephalosporin antibiotics, converting it to the inactive derivative, penicilloic acid.

The complex cell envelope of gram-negative organisms makes them intrinsically less sensitive than gram-positive species to many of the β-lactam antibiotics, and significant differences exist between the properties of the lactamases of the two groups. The lactamases of gram-positive organisms can be divided serologically into four distinct types, all of which are predominantly active against penicillin and show little activity against cephalosporins. In contrast, gram-negative bacteria produce a plethora of lactamases that exhibit a range of hydrolytic activities against penicillins and cephalosporins. In gram-positive organisms, the lactamases are inducible enzymes synthesized within the bacterial cell and then secreted into the surrounding medium. In most gram-positive species, the gene for β-lactamase production is located mainly on a plasmid. Penicillin resistance in gram-negative bacteria can be determined either by chromosomal or plasmid genes, but in clinical isolates, resistance is usually mediated by plasmid genes on R factors. In the enteric bacteria, the β-lactamases are produced constitutively in small amounts and remain bound to the cells. They prevent access of β-lactam antibiotics to the membrane-associated target sites by destroying the antibiotics as they pass through the cell envelope. The most important lactamase (TEM-1), which has a broad activity range against penicillins and cephalosporins, is carried on a transposon (Tn4), which undoubtedly accounts for its wide distribution.

Inactivation of Aminoglycoside Antibiotics. In clinical isolates of gram-negative organisms, resistance to the aminoglycoside antibiotics is due to the production of enzymes that specifically modify the antibiotic so that it can no longer gain entry into the cell. The genes for the aminoglycoside-modifying enzymes are carried on R factors, and several of the genes have been found on transposons.

These enzymes inactivate the drug by acetylation of amino groups, phosphorylation of hydroxyl groups, or adenylylation of hydroxl groups (Table 9-13). Twelve enzymes have been identified that inactivate the aminoglycoside antibiotics. Five of these are phosphorylating enzymes, three are acetylating enzymes, and four adenylylate some of the antibiotics. They are produced constitutively and are located near the cell surface, probably in the periplasmic space. Except for the streptomycin and spectinomycin-inactivating enzymes, one enzyme can inactivate a number of different aminoglycosides, and one antibiotic can be inactivated by more than one enzyme or mechanism. It is believed that the primary effect of the enzymatic modification of the drug is to interfere with the transport of the antibiotic into the cell. The modified compound is apparently unable to induce the transport system needed for entry of the drug.

A number of semisynthetic derivatives of the aminoglycoside antibiotics have been synthesized in an attempt to find agents that are resistant to the aminoglycoside-modifying enzymes. One of these derivatives, amikacin, is resistant to all but one of the enzymes capable of inactivating the aminoglycosidic aminocyclitol antibiotics. This explains the enhanced activity spectrum of amikacin against organisms resistant to kanamycin, gentamicin, and tobramycin.

Inactivation of Chloramphenicol. In the majority of drug-resistant clinical isolates of gram-positive and gram-negative species, resistance is mediated by a plasmid coding for an inactivating enzyme, chloramphenicol acetyltransferase. The enzyme is found intracellularly and is synthesized constitutively in gram-negative organisms. In *S. aureus*, however, the enzyme is induced by the presence of the drug. Like the penicillinases, the chloramphenicol

acetyltransferases of gram-positive and gram-negative species appear to constitute a family of immunologically related but electrophoretically distinguishable proteins.

Modification of Drug Receptor Site

Streptomycin Resistance. Resistance controlled by chromosomal genes is usually due to changes in enzymes or active sites involved in essential metabolic reactions in the cell. An example of this mechanism is resistance to streptomycin where differences exist between the ribosomes of streptomycin-resistant and streptomycin-sensitive organisms. As discussed above, streptomycin binds to a specific site on the ribosome, thereby deranging protein synthesis. Any mutation that deletes this site or alters it in such a manner that the drug cannot exert its effect results in streptomycin resistance. The binding site that is modified in streptomycin mutants involves a single amino acid replacement in the S12 protein on the 30S ribosomal subunit coded for by the *strA* gene. Following exposure of cultures of *E. coli* to high levels of streptomycin, survivors generally are all mutants in this gene. This mechanism of resistance, however, is less significant clinically than is plasmid-mediated enzymatic inactivation.

Kasugamycin Resistance. Kasugamycin is an aminoglycoside antibiotic that acts on the 30S subunit of 70S ribosomes. It inhibits protein synthesis but does not cause misreading or phenotypic suppression. Mutation to resistance causes an alteration, not in the ribosomal protein but in the 16S ribosomal RNA. Kasugamycin resistance is associated with a failure to methylate two adenine residues in the sequence AACCUG near the 3' end of the 16S RNA. This alteration prevents binding of the drug to the ribosome.

Erythromycin Resistance. Resistance to erythromycin is associated with an altered 50S ribosomal subunit. In *E. coli* and a number of other species, the alteration is in a specific protein of the 50S subunit (L4 or L12), resulting in reduced affinity of the ribosome for erythromycin. In *S.*

TABLE 9-13. AMINOCYCLITOL-INACTIVATING ENZYMES

Antibiotic	Acetyltransferases (AAC)			Phosphotransferases (APD)					Adenylyltransferases (AAD)				
	2'	6'	3	3'	2"	3"	6	5"	2"	4'	3"(a)	6	9
Kanamycins	+	+	+	+	(+)	−	−	−	+	+	−	−	−
Tobramycin	+	+	+	−	(+)	−	−	−	+	+	−	−	−
Amikacin	−	+	−	(+)	(+)	−	−	−	−	+	−	−	−
Gentamicin	+	+	+	−	+	−	−	−	+	−	−	−	−
Sisomicin	+	+	+	−	+	−	−	−	+	−	−	−	−
Netilmicin	+	+	(+)	−	+	−	−	−	−	−	−	−	−
Neomycin	+	+	+	+	−	−	−	(+)	−	+	−	−	−
Streptomycin	−	−	−	−	−	+	+	−	−	−	+	−	−
Spectinomycin	−	−	−	−	−	−	−	−	−	−	+	−	+

+, normal substrate; (+), substrate for some forms of the enzyme; −, nonsubstrate for enzyme.

aureus, however, binding of the drug in resistant strains is blocked by dimethylation of a specific adenine sequence in the 23S ribosomal RNA. A plasmid-mediated inducible ribosomal RNA methylase is apparently responsible for erythromycin resistance in *S. aureus.*

Rifampin Resistance. Mutants resistant to rifampin have an RNA polymerase with an altered β subunit. Alteration of the β subunit is accompanied by failure of the core enzyme to bind the antibiotic and is the result of a chromosomal mutation.

Penicillin Resistance. The most important examples of resistance caused by altered target sites are those resulting from the alteration of penicillin-binding proteins (PBPs) in *N. gonorrhoeae* and of resistant strains of *S. pneumoniae* in South Africa. Methicillin resistance in *S. aureus* also appears to be due to altered PBPs.

Synthesis of Resistant Pathway
Sulfonamide and Trimethoprim Resistance. Resistance to the sulfonamides may be mutational or plasmid-mediated and may involve more than one mechanism. The major cause of significant sulfonamide resistance among clinical isolates, however, is the plasmid-mediated production of an altered dihydropteroate synthetase, which is 1000 times less sensitive to the drug than the wild-type enzyme. Similarly, trimethoprim resistance is mediated by R plasmids that code for a trimethoprim-resistant dihydrofolate reductase (DHFR). Two types of plasmid-encoded DHFRs are known, both of which are several thousand times more resistant to trimethoprim than the chromosomal enzyme of the wild type. The synthesis of a plasmid-encoded replacement enzyme that is selectively refractory to the antimicrobial agent provides a mechanism for bypass of the blocked reaction.

FURTHER READING

Books and Reviews

Atkinson BA, Amaral L: Sublethal concentrations of antibiotics, effects on bacteria and the immune system. CRC Crit Rev Microbiol 1982, pp 101–138

Benveniste R, Davies J: Mechanisms of antibiotic resistance in bacteria. Annu Rev Biochem 43:471, 1973

Blumberg PM, Strominger JL: Interaction of penicillin with the bacterial cell: penicillin-binding proteins and penicillin-sensitive enzymes. Bacteriol Rev 38:291, 1974

Brown JR, Ireland DS: Structural requirements for tetracycline activity. Adv Pharmacol Chemother 15:161, 1978

Cherubin CE, Neu HC, Turck M: Current status of cefotaxime sodium: A new cephalosporin. Rev Infect Dis 4 [Suppl]:S281, 1982

Chopra I, Howe TGB: Bacterial resistance to the tetracyclines. Bacteriol Rev 42:707, 1978

Chopra I, Howe TGB, Linton AH, et al.: The tetracyclines: prospects at the beginning of the 1980s. J Antimicrob Chemotherap 8:5, 1981

Corcoran JW, Hahn FE (eds): Antibiotics III. Mechanism of Action of Antimicrobial and Antitumor Agents. New York, Springer-Verlag, 1975

Cozzarelli NR: The mechanism of action of inhibitors of DNA synthesis. Annu Rev Biochem 46:641, 1977

Davies J: Aminoglycoside-aminocyclitol antibiotics and their modifying enzymes. In Lorian V (ed): Antibiotics in Laboratory Medicine. Baltimore, Williams & Wilkins, 1980

Davies J, Smith DI: Plasmid-determined resistance to antimicrobial agents. Annu Rev Microbiol 32:469, 1978

Dhawan VK, Thadepalli H: Clindamycin: A review of fifteen years of experience. Rev Infect Dis 4:1133, 1982

Elwell LP, Falkow S: The characterization of plasmids that carry antibiotic resistance genes. In Lorian V (ed): Antibiotics in Laboratory Medicine. Baltimore, Williams & Wilkins, 1980, pp 433–453

Feder HM Jr, Osier C, Maderazo EG: Chloramphenicol: A review of its use in clinical practice. Rev Infect Dis 3:479, 1981

Finland M, Kass EH, Platt R: Trimethoprim-sulfamethoxazole revisited. Rev Infect Dis 4:185, 1982

Franklin TJ, Snow GA: Biochemistry of Antimicrobial Action, 3rd ed. New York, Chapman & Hall, 1981

Gale EF, Cundliffe E, Reynolds PE, et al.: The Molecular Basis of Antibiotic Action. New York, Wiley, 1972

Garrod LP, Lambert HP, O'Grady F: Antibiotics and Chemotherapy, 5th ed. Edinburgh, Churchill Livingstone, 1981

Gottlieb D, Shaw PD (eds): Antibiotics I. New York, Springer-Verlag, 1967

Hamilton-Miller JMT: Chemistry and biology of the polyene macrolide antibiotics. Bacteriol Rev 37:166, 1973

Hancock REW: Aminoglycoside uptake and mode of action—with reference to streptomycin and gentamicin. I. Antagonists and mutants. J Antimicrob Chemother 8:249, 1981

Hancock REW: Aminoglycoside uptake and mode of action—with reference to streptomycin and gentamicin. II. Effects of aminoglycosides on cells. J Antimicrob Chemother 8:429, 1981

Hewitt WL: The cephalosporins and cephamycins: a perspective. In Remington JS, Swartz MN (eds): Current Clinical Topics in Infectious Diseases. New York, McGraw-Hill, 1981, vol 2, pp 234–258

Kass EH, Evans DA (eds): Future prospects and past problems in antimicrobial therapy: The role of cefoxitin. Rev Infect Dis 1:1, 1979

Katz E, Demain AL: The peptide antibiotics of *Bacillus:* chemistry, biogenesis, and possible functions. Bacteriol Rev 41:449, 1977

Kobayashi GS, Medoff G: Antifungal agents: recent developments. Annu Rev Microbiol 31:291, 1977

Lacy RW: Antibiotic resistance plasmids of *Staphylococcus aureus* and their clinical importance. Bacteriol Rev 39:1, 1975

Mandell GL, Douglas RG Jr, Bennett JE: Principles and Practice of Infectious Diseases. New York, Wiley, 1979, vol 1

Mandell GL, Klastersky J, Finegold SM, et al.: Special Issue on Problem Pathogens. Arch Intern Med 142:1983, 1982

Mitsuhashi S (ed): Drug Action and Drug Resistance. 2. Aminoglycoside Antibiotics. Baltimore, University Park Press, 1975

Moellering RC Jr (ed): Symposium on Cefamandole. J Infect Dis 137 [Supp]:S1-S190, 1978

Moellering RC Jr, Young LS (eds): Moxalactam International Symposium. Rev Infect Dis 4 [Suppl]:S489-S726, 1982

Murray BE, Moellering Jr: Cephalosporins. Annu Rev Med 32:559, 1981

Neu HC: The in vitro activity, human pharmacology, and clinical effectiveness of new β-lactam antibiotics. Ann Rev Pharmacol Toxicol 22:599, 1982

Ninet L, Bost PE, Bouanchaud DH, et al.: The Future of Antibiotherapy and Antibiotic Research. New York, Academic Press, 1981

Ogawara H: Antibiotic resistance in pathogenic and producing bacteria, with special reference to lactam antibiotics. Microbiol Rev 45:591, 1981

Pestka S: Insights into protein biosynthesis and ribosome function through inhibitors. Prog Nucleic Acid Res Mol Biol 17:217, 1976

Restrepo A, Stevens DA, Utz JP (eds): First International Symposium on Ketoconazole. Rev Infect Dis 2:519, 1980

Rogers HJ, Perkins HR, Ward JB: Microbial Cell Walls and Membranes. New York, Chapman & Hall, 1980, Chap 9

Salton MRJ, Shockman GD (eds): β-Lactam Antibiotics. Mode of Action, New Developments, and Future Prospects. New York, Academic Press, 1981

Schlessinger D: Genetic and antibiotic modification of protein synthesis. Annu Rev Biochem 43:135, 1974

Sherris JC, Minshew BH: Mutational antibiotic resistance. In Lorian V (ed): Antibiotics in Laboratory Medicine. Baltimore, Williams & Wilkins, 1980, pp 418–432

Storm DR, Rosenthal KS, Swanson PE: Polymyxin and related peptide antibiotics. Annu Rev Biochem 46:723, 1977

Tanaka N: Fusidic acid. In Corcoran JW, Hahn FE (eds): Antibiotics III. Mechanism of Action of Antimicrobial and Antitumor Agents. New York, Springer-Verlag, 1975, pp 436–447

Tomasz A: The mechanism of the irreversible antimicrobial effects of penicillins. Annu Rev Microbiol 33:113, 1979

Vasquez D: Inhibitors of protein synthesis. FEBS Letters 40 [Suppl]:S63-S84, 1974

Washington JA II: The effects and significance of subminimal inhibitory concentrations of antibiotics. Rev Infect Dis 1:781, 1979

Wehrli W, Staehelin M: Actions of the rifamycins. Bacteriol Rev 35:290, 1971

Wise RI, Kory M (eds): Reassessments of vancomycin—a potentially useful antibiotic. Rev Infect Dis 3 [Suppl]:S200-S300, 1981

Woodruff HB, Mata JM, Hernandez S, et al.: Fosfomycin: laboratory studies. Chemotherapy 23 [Suppl 1]:1, 1977

Selected Papers

Brewer NS: The aminoglycosides. Streptomycin, kanamycin, gentamicin, tobramycin, amikacin, neomycin. Mayo Clinic Proc 52:675, 1977

Brufani M, Cerrini S, Fedeli W, et al.: Rifamycins: An insight into biological activity based on structural investigations. J Mol Biol 87:409, 1974

Bryant DW, McCalla DR: Nitrofuran-induced mutagenesis and error prone repair in Escherichia coli. Chem Biol Interact 31:151, 1980

Courvalin P, Weisblum B, Davies J: Aminoglycoside-modifying enzyme of an antibiotic-producing bacterium acts as a determinant of antibiotic resistance in Escherichia coli. Proc Natl Acad Sci USA 74:999, 1977

Garvin RT, Biswas DK, Gorini L: The effects of streptomycin or dihydrostreptomycin binding to 16S RNA or to 30S ribosomal subunits. Proc Natl Acad Sci USA 71:3814, 1974

Gellert M, O'Dea MH, Itoh T, et al.: Novobiocin and coumermycin inhibit DNA supercoiling catalyzed by DNA gyrase. Proc Natl Acad Sci USA 73:4474, 1976

Gellert M, Mizuuchi K, O'Dea MH, et al.: Nalidixic acid resistance: A second genetic character involved in DNA gyrase activity. Proc Natl Acad Sci USA 74:4772, 1977

Hancock REW, Raffle VJ, Nicas TI: Involvement of the outer membrane in gentamicin and streptomycin uptake and killing in Pseudomonas aeruginosa. Antimicrob Agents Chemother 19:777, 1981

Hartman B, Tomasz A: Altered penicillin-binding proteins in methicillin-resistant strains of Staphylococcus aureus. Antimcrob Agents Chemother 19:726, 1981

Helser TC, Davies JE, Dahlberg JE: Changes in methylation of 16S ribosomal RNA associated with mutation to kasugamycin resistance in Escherichia coli. Nature [New Biol] 233:13, 1971

Hermans PE: General principles of antimicrobial therapy. Mayo Clin Proc 52:603, 1977

Herrlich P, Schweiger M: Nitrofurans, a group of synthetic antibiotics, with a new mode of action: Discrimination of specific messenger RNA classes. Proc Natl Acad Sci USA 73:3386, 1976

Horinouchi S, Weisblum B: Posttranscriptional modification of mRNA conformation: Mechanism that regulates erythromycin-induced resistance. Proc Natl Acad Sci USA 77:7079, 1980

Kawaguchi H: Discovery, chemistry, and activity of amikacin. J Infect Dis 134:S242, 1976

Kelly JA, Moews PC, Knox JR, et al.: Penicillin target enzyme and the antibiotic binding site. Science 218:479, 1982

Keys TF: Antimicrobials commonly used for urinary tract infection. Sulfonamides, trimethoprim-sulfamethoxazole, nitrofurantoin, nalidixic acid. Mayo Clin Proc 52:680, 1977

Kozarich JW, Strominger JL: A membrane enzyme from Staphylococcus aureus which catalyzes transpeptidase, carboxypeptidase, and penicillinase activities. J Biol Chem 253:1272, 1978

Langlois R, Cantor CR, Vince R, et al.: Interaction between the erythromycin and chloramphenicol binding sites on the Escherichia coli ribosome. Biochemistry 16:2349, 1977

Langlois R, Lee CC, Cantor CR: The distance between two functionally significant regions of the 50S Escherichia coli ribosome: the erythromycin binding site and proteins L7/L12. J Mol Biol 106:297, 1976

Levy SB, McMurray L: Plasmid-determined tetracycline resistance involves new transport systems for tetracycline. Nature 276:90, 1978

Lewis C: Clinically useful antibiotics obtained by directed chemical modification—lincomycins. Fed Proc 33:2303, 1974

Luzzatto L, Apirion D, Schlessinger D: Polyribosome depletion and blockage of the ribosome cycle by streptomycin in Escherichia coli. J Mol Biol 42:315, 1969

McCoy EC, Petrullo LA, Rosenkranz HS: Non-mutagenic genotoxicants: novobiocin and nalidixic acid, 2 inhibitors of DNA gyrase. Mutat Res 79:33, 1980

McMurray L, Petrucci RE Jr, Levy SB: Active efflux of tetracycline encoded by four genetically different tetracycline resistance determinants in Escherichia coli. Proc Natl Acad Sci USA 77:3974, 1980

Modolell J, Davis BD: Rapid inhibition of polypeptide chain extension by streptomycin. Proc Natl Acad Sci USA 61:1279, 1968

Moellering RC: Microbiological consideration in the use of

tobramycin and related aminoglycosidic aminocyclitol antibiotics. Med J Aust [Spec Suppl] 2:4, 1977

Monro RE, Staehelin T, Celma ML, et al.: The peptidyl transferase activity of ribosomes. Cold Spring Harbor Symp Quant Biol 34:357, 1969

Neu HC: Tobramycin. J Infect Dis 134:S3, 1976

Nozawa Y, Kitajima Y, Sekiya T, et al.: Ultrastructural alterations induced by amphotericin B in the plasma membrane of *Epidermophyton floccosum* as revealed by freeze-etch electron microscopy. Biochim Biophys Acta 367:32, 1974

Ozaki M, Mizushima S, Nomura M: Identification and functional characterization of protein controlled by the streptomycin-resistant locus in *E. coli*. Nature 222:333, 1969

Pattishal KH, Acar J, Burchall JJ, et al.: Two distinct types of trimethoprim-resistant dihydrofolate reductase specified by R plasmids of different compatibility groups. J Biol Chem 252:2319, 1977

Pestka S: Translocation, aminoacyl-oligonucleotides, and antibiotic action. Cold Spring Harbor Symp Quant Biol 34:395, 1969

Pfeiffer RR: Structural features of vancomycin. Rev Infect Dis 3 [Suppl]: S205, 1981

Polak A, Scholer HJ: Mode of action of 5-fluorocytosine and mechanisms of resistance. Chemotherapy 21:113, 1975

Sheldrick GM, Jones PG, Kennard O, et al.: Structure of vancomycin and its complex with acetyl-D-alanyl-D-alanine. Nature 271:223, 1978

Siewert G, Strominger JL: Bacitracin: an inhibitor of the dephosphorylation of lipid pyrophosphate, an intermediate in biosynthesis of the peptidoglycan of bacterial cell walls. Proc Natl Acad Sci USA 57:767, 1967

Small GD, Setlow JK, Kooistra J, et al.: Lethal effect of mitomycin C on *Haemophilus influenzae*. J Bacteriol 125:643, 1976

Sugino A, Peebles CL, Kreuzer KN, et al.: Mechanism of action of nalidixic acid: Purification of *Escherichia coli nal A* gene product and its relationship to DNA gyrase and a novel nicking-closing enzyme. Proc Natl Acad Sci USA 74:4767, 1977

Tai P-C, Wallace BJ, Davis BD: Selective action of erythromycin on initiating ribosomes. Biochemistry 13:4653, 1974

Tai P-C, Wallace BJ, David BD: Streptomycin causes misreading of natural messenger by interacting with ribosomes after initiation. Proc Natl Acad Sci USA 75:275, 1978

Tomasz A, Waks S: Mechanism of action of penicillin: Triggering of the pneumococcal autolytic enzyme by inhibitors of cell wall synthesis. Proc Natl Acad Sci USA 72:4162, 1975

Uno J, Shigematsu ML, Arai T: Primary site of action of ketoconazole on *Candida albicans*. Antimicrob Agents Chemother 21:912, 1982

Wallace BJ, Davis BD: Cyclic blockade of initiation sites by streptomycin-damaged ribosomes in *Escherichia coli*: an explanation for dominance of sensitivity. J Mol Biol 75:377, 1973

Waring MJ: Drugs which affect the structure and function of DNA. Nature 219:1320, 1968

Wehland J, Herzog W, Weber K: Interaction of griseofulvin with microtubules, microtubule protein and tubulin. J Mol Biol 111:329, 1977

Wilkowske CJ: The penicillins. Mayo Clinic Proc 52:616, 1977

Wilson WR: Tetracyclines, chloramphenicol, erythromycin, and clindamycin. Mayo Clin Proc 52:635, 1977

Yarbrough LR, Wu Y-H, Wu C-W: Molecular mechanism of the rifampicin-RNA polymerase. Biochemistry 15:2669, 1976

CHAPTER 10

Sterilization and Disinfection

An understanding of the basic principles of sterilization and disinfection is fundamental to the intelligent practice of medicine. Although new techniques of sterilization and disinfection are continually being introduced, we still use some of the same agents and procedures that were introduced centuries before there was any concept of infection. Most of the simple chemical agents once used in therapy have been replaced by more specific chemotherapeutic agents, but many of the group have retained their importance as effective antiseptics or disinfectants in the destruction of microorganisms in the nonliving environment. At the present time no group of chemical agents is more widely used.

Definitions. The following terms are useful in describing the damaging effects of certain chemical and physical agents on microorganisms. The term *sterilization* is an absolute one that implies the total inactivation of all forms of microbial life in terms of the organism's ability to reproduce. The suffix *-cide* is added when a killing action is implied, while *-stasis* is added when the organism is merely inhibited in growth or prevented from multiplying. A *bactericide* destroys bacteria; a *germicide* or *disinfectant* is an agent that kills microorganisms capable of producing an infection. A *bacteriostatic* agent is a substance that prevents the growth of bacteria. An *antiseptic* opposes sepsis or putrefaction either by killing microorganisms or by preventing their growth. This term is commonly used for agents that are applied topically to living tissues.

The selection of an appropriate procedure or agent is determined by the specific situation and whether it is necessary to kill all microorganisms or only certain species. For example, the complete destruction of all microorgan-

isms present in or on any material is essential in surgical procedures, in the preparation of all media and glassware used in the microbiology laboratory, and in the canning of nonacid high-protein foods. In the care of cases of communicable diseases, the destruction of the pathogen is necessary in order to prevent spread of the infection to susceptible individuals. For this purpose a disinfectant is adequate and is usually employed as a cleaning-up process (Table 10-1).

Dynamics of Sterilization and Disinfection

Death Rate of Microorganisms. Knowledge of the kinetics of death of a bacterial population is useful in understanding the basis of sterilization by lethal agents. In the case of a microorganism, the only valid criterion of death is the irreversible loss of the ability to reproduce. This is usually determined by plating techniques that quantitate by colony count the number of survivors.

When a bacterial population is exposed to a lethal agent, there is with time a progressive reduction in the number of survivors. The kinetics of death of a microbial population are usually exponential: the number of survivors decreases with time. If the logarithm of the number of survivors is plotted as a function of the time of exposure, a straight line is obtained (Fig. 10-1), the negative slope of which defines the death rate. The death rate, however, tells only what fraction of the initial population survives a given period of exposure to the antibacterial

TABLE 10-1. TYPES OF ANTIMICROBIAL AGENTS

Type of Agent	Agents	Applications
Physical	Dry heat (160–180C)	Sterilization
	Moist heat (115–150C)	Sterilization
	Moist heat (65–100C)	Disinfection
	Ionizing radiation (gamma, electrons)	Sterilization
	Ultraviolet radiation	Disinfection
Chemical (vapors)	Ethylene oxide	Sterilization
	Formaldehyde	Sterilization or disinfection
Chemical (low selectivity)	Alcohols, aldehydes, halogens, phenols, ammonium compounds	Disinfection or preservation
Chemical (moderate selectivity)	Antibiotics (bacitracin, polymyxins)	Topical chemotherapy
	Dyes (acridines, triphenylmethanes)	Antisepsis
	Metal chelate complexes	Antisepsis
	Organic arsenic compounds	Chemotherapy
	Organic mercury compounds	Preservation or antisepsis
Chemical (high selectivity)	Synthetic (*p*-aminosalicylic acid, isonicotinic acid hydrazide, sulfonamides, trimethoprim)	Chemotherapy
	Antibiotics (aminoglycosides, amphotericin, cephalosporins, chloramphenicol, erythromycin, griseofulvin, lincomycin, nystatin, penicillins, rifamycins, tetracyclines)	Chemotherapy

From Gardner: In Block (ed): Disinfection, Sterilization, and Preservation, 2nd ed, 1977. Philadelphia, Lea & Febiger.

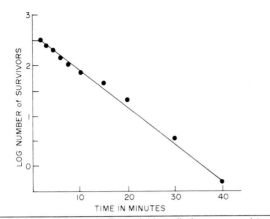

Figure 10-1. Death rate of *Escherichia coli* when exposed to 0.5 percent phenol at 20C. *(From Chick: J Hyg (Camb) 10:237, 1910.)*

agent. In order to determine the actual number of survivors, one must also know the initial population size. This relationship is expressed mathematically by the formula:

$$K = 1/t \log B/b$$

where B is the initial number of organisms and b is the number remaining after time t.

Although the logarithmic curve is mathematically convenient and approximately correct when relatively high concentrations of a disinfectant are used, with lower concentrations the disinfection curve is sigmoidal, the rate being slow in the early stages, then proceeding rapidly for most of the disinfection process, and finally slowing down at the end. The flattening of the slope toward the end of the process is extremely important from the standpoint of sterilization, resulting in the requirement for a more prolonged or intense treatment in order to destroy the resistant survivors that are more likely to be present in an initially large microbial population. Practical experience has shown that under no circumstances can one extrapolate the exponential death rate to zero and assume that the time of exposure so indicated will guarantee sterility.

Because of the exponential form of the survivor-time curve, the larger the initial number of cells to be killed, the more intense or prolonged is the treatment required for sterilization. In addition, as might be expected, the rate of disinfection varies with the concentration of disinfectant. The effect of concentration on rate, however, is not constant but varies with the different disinfectants, as discussed below.

Antimicrobial Chemical Agents

Factors Affecting Disinfectant Potency. In contrast to chemotherapeutic agents that exhibit a high degree of selectivity for certain bacterial species, disinfectants are

highly toxic for all types of cells. The effectiveness of a particular agent is determined to a great extent by the conditions under which it operates.

CONCENTRATION OF AGENT. Many agents are lethal for bacteria only when used in extremely high concentrations. Others may stimulate, retard, or even kill the organism in very low concentrations. The concentration required to produce a given effect, however, as well as the range of concentrations over which a given effect is demonstrable, varies with the disinfectant, the organism, and method of testing. A close relationship exists between the concentration of drug employed and the time required to kill a given fraction of the population. This relationship is shown in the expression:

$$C^n t = K$$

where C is the drug concentration, t is the time required to kill a given fraction of the cells, and n and K are constants. With phenolic compounds, a change in the concentration of the disinfectant has a pronounced effect on the disinfection rate, e.g., reducing the concentration by one half increases approximately 64-fold the time required for sterilization. With most disinfectants, however, the effect is much less dramatic.

TIME. When bacteria are exposed to a specific concentration of a bactericidal agent, even in excess, not all of the organisms die at the same time, but rather there is a gradual decrease in the number of living cells. Disinfection is usually considered as a process in which bacteria are killed in a reasonable length of time, but there are varying opinions about what this should be (Fig. 10-1).

pH. The hydrogen ion concentration influences bactericidal action by affecting both the organism and the chemical agent. When suspended in a culture medium of pH 7, bacteria are negatively charged. An increase of pH increases the charge and may alter the effective concentration of the chemical agent at the surface of the cell. The pH also determines the degree of ionization of the chemical. In general, the nonionized form of a dissociable agent passes through the cell membrane more readily than the relatively inactive ionic forms.

TEMPERATURE. The killing of bacteria by chemical agents increases with an increase in temperature. At low temperatures, for each 10C temperature increment, there is a doubling of the death rate. With some agents, such as phenol, the rate is increased five to eight times, suggesting a more complex reaction and the interplay of additional factors.

NATURE OF THE ORGANISM. The efficacy of a particular agent depends upon properties of the organism against which it is tested. The most important of these are the species of organism, the growth phase of the culture, the presence of special structures, such as spores or capsules, the previous history of the culture, and the number of organisms in the test system (Fig. 10-2).

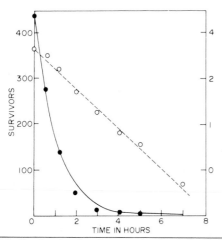

Figure 10-2. Death rate of anthrax spores treated with 5 percent phenol at 33.3C. Number of surviving spores is plotted on an arithmetic scale and on a logarithmic scale. *(From Chick: J Hyg (Camb) 8:92, 1908.)*

PRESENCE OF EXTRANEOUS MATERIALS. The presence of organic matter, such as serum, blood, or pus, influences the activity of many disinfectants and renders inert substances that are highly active in their absence. These foreign materials alter disinfectant activity in a number of ways: surface absorption of the disinfectant by protein colloids, formation of a chemically inert or less active compound, and binding of the disinfectant by active groups of the foreign protein. Among the disinfectants whose inhibitory activity is greatly diminished by organic material with a high protein content are the aniline dyes, mercurials, and cationic detergents. The mercurials are markedly inhibited by compounds containing sulfhydryl groups and the quaternary ammonium compounds are inhibited by soaps and lipids.

Evaluation of Disinfectants. At present, the primary method for the evaluation of disinfectants employs phenol as the standard reference material. The method is based on a tube dilution procedure designed to determine the highest dilution of a germicidal agent that will kill the test organism within a series of time intervals under specified conditions. The results are expressed in terms of the phenol coefficient number, which is a ratio of the highest dilution of disinfectant killing the test organism within a specified time to the greatest dilution of phenol showing the same result. The official quantitative test for disinfectants is that of the Food and Drug Administration of the US Department of Agriculture. This official method specifies standardized conditions for the testing of disinfectants against strains of *Salmonella typhi*, *Staphylococcus aureus*, and *Pseudomonas aeruginosa* of known susceptibility to phenol. Criticism has been directed against the phenol coefficient test because of the limited information that it provides. It has been recommended that the test be supplemented with a series of additional methods designed to determine various properties of the disinfectant: the effect of variations in time and temperature on the bactericidal power of the disinfectant against a variety of organisms including spores, its activity both in the absence and in the presence of organic matter, penetrability, the extent of bacteriostasis, tissue toxicity, and in vivo testing.

Mechanisms of Antimicrobial Action. The mechanisms by which drugs kill or inhibit the growth of microorganisms are varied and complex. Sequential or simultaneous changes often occur that make it difficult to differentiate primary from secondary effects. In general, however, all of the observable effects of chemical agents on bacteria are the result of changes in its macromolecular components. Some of these changes damage the cell membrane, some irreversibly inactivate proteins, while others induce extensive nucleic acid damage.

Agents that Damage the Cell Membrane

The structural integrity of the membrane depends upon the orderly arrangement of the proteins and lipids of which it is composed. Exposure of bacteria to organic solvents and detergents results in a structural disorganization of the membrane and interference with normal function. The net effect is the release of small metabolites from the cell and interference with active transport and energy metabolism.

Surface-active Disinfectants. Substances that alter the energy relationships at interfaces, producing a reduction of surface or interfacial tension are referred to as surface-active agents. They have wide application both in industry and in the home as wetting agents, detergents, and emulsifiers. Surface-active agents are compounds that possess both water-attracting (hydrophilic) and water-repelling (hydrophobic) groups. The interface between the lipid-containing membrane of a bacterial cell and the surrounding aqueous medium provides a susceptible target site for agents of this type. The hydrophobic portion of the molecule is a fat-soluble, long-chain hydrocarbon, while the hydrophilic portion may be either an ionizable group or a nonionic but highly polar structure. Included in the surface-active agents are cationic, anionic, nonionic, and amphoteric substances (Table 10-2).

CATIONIC AGENTS

Quaternary Ammonium Compounds. The most important antibacterial surface-active agents are the cationic compounds in which a hydrophobic residue is balanced by a positively charged hydrophilic group, such as a quaternary ammonium nucleus (Fig. 10-3). When bacteria are exposed to agents of this type, the positively charged group associates with phosphate groups of the membrane phospholipids, while the nonpolar portion penetrates into the hydrophobic interior of the membrane. The resulting distortion causes a loss of membrane semipermeability and leakage from the cell of nitrogen and phosphorus-containing compounds. The agent itself may then enter the cell

TABLE 10-2. SURFACE-ACTIVE AGENTS

Trade Name	Type of Compound	Structure
Zephiran	Cationic	Alkyldimethylbenzyl ammonium chloride
Triton K-12	Cationic	Cetyldimethylbenzyl ammonium chloride
Ceepryn chloride	Cationic	Cetylpyridinium chloride
Duponol LS	Anionic	Sodium oleyl sulfate
Triton W-30	Anionic	Sodium salt of alkylphenoxyethyl sulfonate
Carbowax 1500 dioleate	Nonionic	Oleic acid ester of polymerized polyethylene glycol
Tween 80	Nonionic	Sorbitan monooleate polyoxyalkylene derivative

and denature its proteins. The activity of quaternary ammonium compounds is greatest at an alkaline pH. Although these compounds are bactericidal for a wide range of organisms, gram-positive species are more susceptible (Table 10-3). Antibacterial activity is reduced in the presence of organic matter.

ANIONIC AGENTS. Among the anionic detergents are soaps and fatty acids that dissociate to yield a negatively charged ion. These agents, most active at an acid pH, are effective against gram-positive organisms but are relatively ineffective against gram-negative species because of their lipopolysaccharide outer membrane. By combining an anionic agent with acid, very effective acid-anionic surfactant sanitizers have been devised that are synergistic and display very rapid bactericidal action (within 30 seconds). They have been especially useful in the food and dairy processing industries for disinfecting equipment and utensils.

The anionic detergents cause gross disruption of the lipoprotein framework of the cell membrane. The primary injury of the bile salts, long used by microbiologists to lyse pneumococci, is dissociation of the cell membrane, permitting autolytic enzymes to act upon substrates from which they are restricted in the intact cell. When used together, the cationic and anionic detergents neutralize each other.

NONIONIC AGENTS. As a group, the nonionic detergents show little or no antimicrobial activity, and a few promote bacterial growth. For example, Tween 80 facilitates a diffuse submerged growth of *Mycobacterium tuberculosis* and

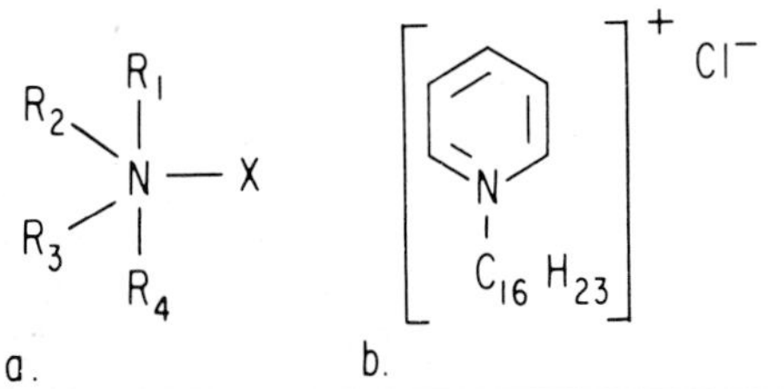

provides the organism with a source of oleic acid, which is stimulatory. Triton X-100, however, another nonionic detergent, has a specific solubilizing effect on the cytoplasmic membrane and selectively separates the proteins of the cell wall and membrane.

AMPHOTERIC AGENTS. Chemically, these agents consist of an amino acid, usually glycine, substituted with a long-chain alkyl amine group. Marketed in Europe under the trade name Tego, the effectiveness of this group of agents has continued to be a controversial subject in spite of the reported claims that they are equally as effective as the cationic quaternaries but somewhat less toxic. Among the uses that have been recommended for these agents are surgical handwashing, floor disinfection in hospitals, and disinfective cleanup in dairies, soft drink bottling plants, and slaughter houses.

Phenolic Compounds. These compounds cause membrane damage with leakage of cell contents and lysis. At low concentrations that are rapidly bactericidal, membrane-bound oxidases and dehydrogenases are irreversibly inactivated.

PHENOL. At present, phenol (carbolic acid) is no longer used as a major disinfectant, its use being limited primarily to the testing of new bactericidal agents (p. 236). It has been replaced as a practical disinfectant by less caustic and toxic phenol derivatives.

The antibacterial activity of phenol is greatly increased by various substitutions in the phenol nucleus; the compounds of greatest importance are the alkyl- and chloro-derivatives and the diphenyls. Not only do many of these derivatives have a very high antibacterial activity, but they are considerably less toxic than phenol. Since most phenolic disinfectants have a low solubility in water, they are formulated with emulsifying agents, such as soaps, which also increase their antibacterial action.

CRESOLS. The simplest of the alkyl phenols are the cresols. Ortho-, meta-, and paracresol are appreciably more active than phenol and are usually employed as a mixture, tricresol. Cresols, obtained industrially by the distillation of coal tar, are emulsified with green soap and sold under the trade names of Lysol and Creolin.

Figure 10-3. a. General formula of the quaternary ammonium compound. R_1, R_2, R_3, R_4 are alkyl groups that may be alike or different. The nitrogen atom has a valency of 5, and X is usually a halogen. For marked antibacterial activity, one of the four radicals must have 8 to 18 carbon atoms. **b.** Cetylpyridinium chloride (Ceepryn), a quaternary.

TABLE 10-3. INHIBITING CONCENTRATIONS* OF QUATERNARY AMMONIUM COMPOUNDS

Compounds	Gram-negative Bacteria		Gram-positive Bacteria	
	Escherichia coli	Pseudomonas fluorescens	Bacillus subtilis	Staphylococcus aureus
Benzethonium chloride	1,000	300	3	3
Benzalkonium chloride	200	300	3	4
Dodecyltrimethyl ammonium chloride	500	500	5	5
Dodecylbenzyldimethyl ammonium chloride	750	750	2	2
Cocobenzyldimethyl ammonium chloride	225	225	2	2
Didecyldimethyl ammonium chloride	225	750	0.7	7

Adapted from Petrocci: In Block (ed): Disinfection, Sterilization, and Preservation, 2nd ed, 1977. Philadelphia, Lea & Febiger.
*Concentration in ppm.

DIPHENYL COMPOUNDS

Hexachlorophene. The halogenated diphenyl compounds exhibit unique antibacterial properties. Of these compounds, the most important is the chlorinated derivative, hexachlorophene, which is highly effective against gram-positive organisms, especially staphylococci and streptococci. Hexachlorophene is bactericidal if used in sufficiently high concentrations and, unlike many disinfectants, retains its antimicrobial potency when mixed with soaps or when added to various cosmetic preparations. It has been used in a wide variety of products, such as germicidal soaps, antiperspirants, toothpastes, and furnace filters.

During the period 1961–1971, routine daily hexachlorophene bathing of newborn infants was an accepted procedure in many nurseries in order to reduce colonization of the umbilical stump with *S. aureus* (p. 454). Numerous studies attest to the effectiveness of this procedure in decreasing the incidence of severe staphylococcal disease and hospital-based epidemics. In 1971, however, prompted by evidence of neurotoxicity following dermal absorption of the agent, the Food and Drug Administration placed strict controls on the use of hexachlorophene and curtailed its use in the newborn nursery.

Alcohols. Alcohols provide an insight into the interaction of organic solvents with lipid membranes. They disorganize lipid structure by penetrating into the hydrocarbon region. Short-chain alcohols produce quantitatively greater changes in membrane organization than do the higher homologs. In addition to their effect on the cell membrane, alcohols and other organic solvents also denature cellular proteins.

The aliphatic alcohols, especially ethanol, have been widely employed as skin disinfectants because of their bactericidal action and ability to remove lipids from skin surfaces. Their action as disinfectants, however, is severely restricted by their inability at normal temperatures to kill spores, and for this reason they should not be relied upon for the sterilization of instruments.

Ethanol is used extensively to sterilize the skin prior to cutaneous injections (Fig. 10-4). It is also used for the disinfection of clinical thermometers and is very effective provided a sufficient period of contact is allowed (Table 10-4). It is active against gram-positive, gram-negative, and acid-fast organisms and is most effective at a concentration of 50 to 70 percent.

The bactericidal activity of isopropyl alcohol is slightly greater than that of ethanol, and it is less volatile. For these reasons, it has been recommended that it replace ethanol for the sterilization of thermometers. The toxic effects of isopropyl alcohol, however, are greater and longer lasting than those produced by ethanol. Toxic reactions have been reported in children who received alcohol sponge baths to reduce fever. Isopropyl alcohol vapors may be absorbed through the lungs and produce narcosis.

Agents that Denature Proteins

Proteins are the most abundant organic molecules in a bacterial cell and are fundamental to all aspects of cell structure and function. In its native state, each protein possesses a characteristic conformation that is required for its proper functioning. Agents that alter the conformation of the protein by denaturation cause an unfolding of the polypeptide chain so that the chains become randomly and irregularly looped or coiled. Among the chemical agents

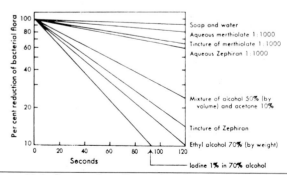

Figure 10-4. Comparative effects of various antiseptics on the resident bacterial flora of the hands and arms. *(From Altemeier: In Block (ed): Disinfection, Sterilization, and Preservation, 2nd ed, 1977. Philadelphia, Lea & Febiger.)*

TABLE 10-4. COMPARISON OF DISINFECTANTS AGAINST VARIOUS MICROORGANISMS IN THE STERILIZATION OF CLINICAL THERMOMETERS

Time of Immersion of Thermometer in Disinfectant	Disinfectant						
	Solution Iodine 2%	Tincture Iodine 2%	Ethyl Alcohol wt/vol 95%	Ethyl Alcohol wt/vol 70%	Ethyl Alcohol wt/vol 50%	Isopropyl Alcohol wt/vol 70%	Isopropyl Alcohol wt/vol 50%
Streptococcus hemolyticus							
Control	+	+	+	+	+	+	+
20 sec	0	0	+	0	0	0	0
60 sec	0	0	0	0	0	0	0
80 sec	0	0	0	0	0	0	0
Streptococcus faecalis							
Control	+	+	+	+	+	+	+
20 sec	+	+	+	+	+	+	+
40 sec	+	0	+	+	+	+	+
80 sec	+	0	+	+	+	+	+
100 sec	0	0	+	+	+	+	+
Escherichia coli							
Control	+	+	+	+	+	+	+
80 sec	+	+	+	+	+	+	+
100 sec	0	0	+	+	+	+	+
120 sec	0	0	+	+	0	0	0
Staphylococcus aureus							
Control	+	+	+	+	+	+	+
60 sec	+	+	+	+	+	+	+
80 sec	+	0	+	+	+	+	+
120 sec	0	0	+	+	+	+	+
3 min	0	0	+	+	+	+	+
4 min	0	0	+	+	+	0	+
5 min	0	0	+	+	+	0	+
10 min	0	0	+	0	0	0	0

Data from Gershenfeld, et al., 1951. Adapted from Morton: In Block (ed): Disinfection, Sterilization, and Preservation, 2nd ed, 1977. Philadelphia, Lea & Febiger.
+, growth in 48 hours; 0, No growth in 48 hours.

that denature cellular proteins are the acids, alkalies, alcohols, acetone, and other organic solvents. The organic solvents have already been discussed in the previous section, since their primary target appears to be the cell membrane.

Acids and Alkalies. These agents exert their antibacterial activity through their free H^+ and OH^- ions, through the undissociated molecules, or by altering the pH of the organism's environment. The strong mineral acids and strong alkalies have disinfectant properties proportional to the extent of their dissociation in solution. Some hydroxides, however, are more effective than their degree of dissociation would indicate, suggesting that the metallic cation also exerts a direct toxic action on the organism.

The intact molecule of the organic acids is responsible for their antibacterial activity. Although the extent of their dissociation in solution is less than that of mineral acids, they are sometimes more potent disinfectants. Benzoic acid, widely used as a food preservative, is approximately seven times as effective as hydrochloric acid, showing that both the whole molecule and the organic radical possess disinfectant activity. Other organic acids that have been used extensively as food preservatives to extend the storage life of food products include lactic, acetic, citric, and propionic acids (Table 10-5). The use of food preservatives in the United States is subject to strict regulation by the Food and Drug Administration of the federal government.

TABLE 10-5. THE USE OF ACID PRESERVATIVES

Acid	Examples of Specified Foods
Propionic	Flour, confectionery, bread
Sulfurous	Fruit juices
Benzoic	Fruit juices, liquid coffee extract
p-Hydroxybenzoic acid esters	Pickles, tomato puree
Sorbic	Cheese, flour, confectionery

From Kimble: In Block (ed): Disinfection, Sterilization, and Preservation, 2nd ed, 1977. Philadelphia, Lea & Febiger.

Agents that Modify Functional Groups of Proteins and Nucleic Acids

The catalytic site of an enzyme contains specific functional groups that bind the substrate and initiate the catalytic events. Inhibition of enzyme activity results if one or more of these functional groups is altered or destroyed. Important functional groups of the cell wall, membrane, and nucleic acids are also susceptible to inactivation.

Compounds containing mercury or arsenic combine with sulfhydryl groups; formaldehyde, anionic detergents, and acid dyes react with amino and imidazole groups; basic dyes, quaternary ammonium compounds, and cationic detergents react with acidic groups, such as hydroxyl or phosphoric acid residues. The presence of organic matter and other substances containing free reactive groups markedly reduces the effectiveness of agents whose toxicity results from combination with reactive groups of cell components.

Heavy Metals. Soluble salts of mercury, arsenic, silver, and other heavy metals poison enzyme activity by forming mercaptides with the sulfhydryl groups of cysteine residues. The initial reaction is reversible, and if extraneous —SH groups are provided in the form of glutathione or sodium thioglycollate, most of the cells recover. The binding ability of the mercurials extends to a broad range of ligands other than SH-containing groups, e.g., carboxylates, phosphates, and amines.

MERCURIALS. Various forms of mercury have been employed in medicine for many years. Mercuric chloride, once popular as a disinfectant, is very toxic and at present has limited use. Organic mercurials, such as Metaphen, Merthiolate, and Mercurochrome, are less toxic and, although unreliable as skin disinfectants, are useful antiseptic agents (Table 10-6). The phenylmercury salts are among the most efficient inhibitors of gram-positive and gram-negative bacteria, fungi, yeasts, and algae. They have been especially useful in the control of pseudomonads and other microbial contaminants in pharmaceutical, ophthalmic, and cosmetic preparations (Table 10-6).

SILVER COMPOUNDS. Silver compounds are widely used as antiseptics, either as soluble silver salts or as colloidal preparations. The inorganic silver salts are efficient bactericidal agents, but their practical value is restricted by their irritant and caustic effects. The most commonly employed of the silver salts is silver nitrate, which is highly bactericidal for the gonococcus and is routinely used, as legally required by state law, for the prophylaxis of ophthalmia neonatorum in newborn infants. Colloidal silver compounds in which silver is combined with protein and from which silver ions are slowly released have been extensively used as antiseptics, especially in ophthalmology. These compounds are primarily bacteriostatic, however, and relatively poor disinfectants. The most recent application of silver compounds has been in the handling of burn patients. Topical application of silver nitrate or silver sulfadiazine in cream has significantly reduced the mortality in these patients.

Oxidizing Agents. The most useful antimicrobial agents in this group are the halogens and hydrogen peroxide. They inactivate enzymes by converting functional —SH groups to the oxidized S—S form. The stronger agents also attack amino groups, indole groups, and the phenolic hydroxyl group of tyrosine.

HALOGENS. Chlorine and iodine are among our most useful disinfectants. For certain purposes—iodine as a skin disinfectant and chlorine as a water disinfectant—they are unequaled. They are unique among disinfectants in that their activity is almost exclusively bactericidal and that they are effective against sporulating organisms.

Iodine. Iodine exists principally in the form of I_2 at pH values below 6, where maximal bactericidal action is manifested. The rate of killing decreases as the pH is increased above 7.5. The iodide ion, I^-, formed as a result of iodine hydrolysis in aqueous solutions, has no significant bactericidal effect; the triiodide ion, I_3^-, also present in aqueous solutions, has minimal activity. Iodine tincture USP contains 2 percent iodine and 2 percent sodium io-

TABLE 10-6. HIGHEST DILUTION OF DISINFECTANT KILLING MICROORGANISMS AFTER 10 MINUTES BUT NOT AFTER 5 MINUTES AT 37C

Microorganism	Phenylmercuric Nitrate	Merthiolate	Metaphen	Mercurochrome	Mercuric Chloride	Phenol
Staphylococcus aureus	1:192,000	1:120,000	1:140,000	1:160	1:16,000	1:85
Streptococcus pyogenes	1:144,000	1:112,000	1:110,000	1:320	1:32,000	1:100
Streptococcus pneumoniae (Type 1)	1:96,000	1:64,000	1:72,000	1:240	1:20,000	1:90
Escherichia coli	1:48,000	1:32,000	1:32,000	1:180	1:10,000	1:75
Neisseria gonorrhoeae	1:80,000	1:48,000	1:48,000	1:240	1:20,000	1:90
Bacillus subtilis	1:24,000	1:36,000	1:16,000	1:300	1:12,000	1:80

Data from Birkhaug, 1933, and Brewer, 1968. From Grier: In Block (ed): Disinfection, Sterilization, and Preservation, 2nd ed, 1977. Philadelphia, Lea & Febiger.

dide in dilute alcohol. The principal use of iodine is in the disinfection of the skin, and for this purpose it is probably superior to any other agent (Table 10-4). Mixtures of iodine with various surface-active agents that act as carriers for the iodine are known as iodophors. They have been widely used for the sterilization of dairy equipment.

Chlorine. In addition to chlorine itself, there are three types of chlorine compounds, the hypochlorites and the inorganic and organic chloramines. The disinfectant action of all chlorine compounds is due to the liberation of free chlorine. When elemental chlorine or hypochlorites are added to water, the chlorine reacts with water to form hypochlorous acid, which in neutral or acidic solution is a strong oxidizing agent and an effective disinfectant.

$$Cl_2 + H_2O \rightarrow HOCl + H^+ + Cl^-$$

$$Ca(OCl)_2 + H_2O \rightarrow CA^{2+} + H_2O + 2\ OCl^-$$

$$Ca(OCl)_2 + 2\ H_2O \rightarrow Ca(OH)_2 + 2\ HOCl$$

$$HOCl \rightleftharpoons H^+ + OCl^-$$

The dissociation of hypochlorous acid depends on pH, which determines the disinfection efficiency.

Although chlorine is one of the most potent bactericidal agents, its activity is markedly influenced by the presence of organic matter. For example, in the disinfection of water it is first necessary to determine its chlorine demand. This is due to the possible presence in water of substances capable of combining with chlorine. It is customary to add sufficient chlorine to the water supply to satisfy the chlorine demand of the water and, at the same time, to provide enough residual for complete disinfection. In the case of swimming-pool water, however, a wide spectrum of organisms is being constantly introduced, and the contact time with chlorine may be very short. A concentration of 0.6 to 1.0 ppm of free chlorine residual should be maintained to assure rapid kill (15 to 30 seconds).

Hypochlorites are the most useful of the chlorine compounds. They are available in liquid or powder form as salts of calcium, lithium, and sodium. Today, hypochlorites are widely used in the food and dairy industries for sanitizing dairy and food-processing equipment. They are employed as sanitizers in most households, hospitals, restaurants, and public buildings and are marketed under such popular labels as Clorox and Purex bleach.

HYDROGEN PEROXIDE. In a 3 percent solution, hydrogen peroxide is a harmless but very weak antiseptic whose primary clinical use is in the cleansing of wounds. When hydrogen peroxide is applied to tissues, oxygen is rapidly released by the tissue catalases, and the germicidal action is brief. Although the antibacterial action of hydrogen peroxide is usually attributed to its oxidizing ability, it is probable that the formation of free hydroxyl radicals from the peroxide accounts for most of this activity. As a disinfectant of inanimate materials, hydrogen peroxide is a very useful and effective agent. It has been used increasingly in the last 10 years, especially for the disinfection of medical-surgical devices and soft plastic contact lenses (Table 10-7).

Dyes. Some of the coal-tar dyes, especially the triphenylmethanes and the acridines, not only stain bacteria but are inhibitory at very high dilutions. Within the usual pH range, the basic dyes are the most effective. They exhibit a marked affinity for the acidic phosphate groups of nucleoproteins and other cell components, and are inactivated by serum and other proteins. Although previously used as antiseptics for wound and skin infections, their current medical use is limited primarily to the treatment of dermatologic lesions. The selective activity of the triphenylmethane dyes for gram-positive organisms has been utilized in the laboratory in the formulation of selective culture media.

TRIPHENYLMETHANE DYES. Of the aniline dyes, derivatives of triphenylmethane, especially brilliant green, malachite green, and crystal violet, have been used for many purposes. They are highly selective for gram-positive organisms (Table 10-8). The activity of triphenylmethane dyes is a property of the pseudobase formed upon ionization of the dye (Fig. 10-5). The pseudobase is more lipid-soluble than the cation, and it is probable that it is in this form that it gains access to the interior of the cell.

The specific mode of action of most of the dyes in this group remains ill defined. The action of crystal violet is attributed to its interference with the synthesis of the peptidoglycan component of the cell wall, where it blocks the conversion of UDP-acetylmuramic acid to UDP-acetylmuramylpeptide. In gram-negative organisms, lipo-

TABLE 10-7. LENS DISINFECTION WITH 3 PERCENT HYDROGEN PEROXIDE SOLUTION

	Time Required for Disinfection*
Neisseria gonorrhoeae	0.3
Haemophilus influenzae	1.5
Pseudomonas aeruginosa	2.2
Bacillus subtilis (vegetative)	2.7
Propionibacterium acnes	3.0
Escherichia coli	3.0
Proteus vulgaris	3.1
Bacillus cereus (vegetative)	5.5
Proteus mirabilis	6.0
Streptococcus pyogenes	8.0
Staphylococcus epidermidis	9.7
Staphylococcus aureus	12.5
Herpes simplex	12.8
Serratia marcescens	20.5
Candida albicans	21.2
Fusarium solani	26.1
Aspergillus niger	45.3
Candida parapsilosis	96.9

From Spaulding et al.: In Block (ed): Disinfection, Sterilization, and Preservation, 2nd ed, 1977. Philadelphia, Lea & Febiger.
*Minutes of exposure required to reduce inoculum to 0.5 organisms/ml.

TABLE 10-8. MAXIMUM DILUTIONS OF TRIPHENYLMETHANE DYES INHIBITING GROWTH FOR 24 HOURS

Organism	Brilliant Green	Malachite Green	Crystal Violet
Escherichia coli	1:675,000	1:40,000	1:85,000
Salmonella typhi	1:500,000	1:30,000	1:85,000
Shigella dysenteriae	1:1,500,000	1:250,000	1:400,000
Bacillus subtilis	1:15,000,000	1:4,000,000	1:4,000,000
Staphylococcus aureus	1:4,000,000	1:1,000,000	1:1,000,000

From Kliger: J Exp Med 27:463, 1918.

polysaccharide in the outer membrane provides a major penetration barrier for the uptake of the dye and accounts for the selectivity of gentian violet for gram-positive organisms.

ACRIDINE DYES. The acridine dyes, often referred to as "flavines" because of their yellow color, exert a bactericidal and bacteriostatic effect upon a number of organisms. Among the compounds of clinical use are proflavine and acriflavine, which have been employed in wound antisepsis. Unlike the aniline dyes, antimicrobial activity is retained in the presence of serum or pus.

The acridine dyes interfere with the synthesis of nucleic acids and proteins in both bacterial and mammalian cells. They are planar heterocyclic molecules that interact with double-stranded helical DNA by intercalation (Fig. 10-6). Because of its flat hydrophobic structure, acridine is inserted between two successive bases in DNA, separating them physically (Fig. 10-7). When the chain is replicated, an extra base is inserted into the complementary chain opposite the intercalated drug. When the latter chain is then replicated, the new chain will also contain an extra base.

Alkylating Agents. The lethal effects of formaldehyde, ethylene oxide, and glutaraldehyde result from their alkylating action on proteins. Inhibitions produced by such agents are irreversible, resulting in enzyme modification and inhibition of enzyme activity.

ALDEHYDES

Formaldehyde. Formaldehyde (Fig. 10-8) is one of the least selective agents acting on proteins. Carboxyl, hydroxyl, or sulfhydryl groups of protein are alkylated by direct replacement of a hydrogen atom with a hydroxymethyl group. Its reaction with a sulfydryl group of an enzyme protein is as follows:

$$E-SH \ + \ H-C=O \rightarrow E-S-\underset{\underset{H}{|}}{\overset{\overset{H}{|}}{C}}-OH$$

Formaldehyde is commercially available in aqueous solutions containing 37 percent formaldehyde (formalin) or as paraformaldehyde, a solid polymer that contains 91 to 99 percent formaldehyde. Formalin is used for preserving fresh tissues and is the major component of embalming fluids. When used at a sufficiently high concentration, it destroys all organisms, including spores. Formalin has been used extensively to inactivate viruses in the preparation of vaccines, since it has little affect on their antigenic properties. Generally, from 0.2 to 0.4 percent formalin has been used for this purpose. As a gas, formaldehyde has been used for years to decontaminate rooms, buildings, fabrics, and instruments.

Glutaraldehyde. Within the last few years, glutaraldehyde has been used increasingly as a cold sterilant for surgical instruments (Table 10-9). At present, it is the only available highly effective cold chemical sterilant recommended by the United States Center for Disease Control for use on respiratory therapy equipment. Glutaraldehyde is 10 times more effective than formaldehyde as a bactericidal and sporicidal agent and is considerably less toxic. Its bactericidal effectiveness is not diminished by protein-containing materials.

The mode of action of glutaraldehyde has been attributed to its binding of sulfhydryl or amino groups, but the specific target in the cell has not been defined (Fig. 10-8).

ETHYLENE OXIDE. Ethylene oxide is an alkylating agent extensively employed in gaseous sterilization. It is active

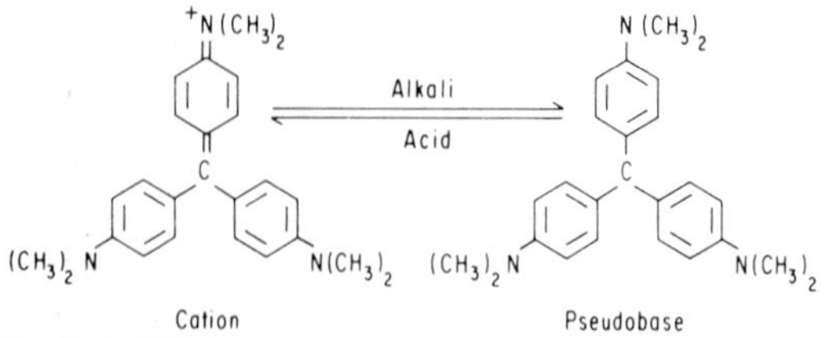

Figure 10-5. Crystal violet (gentian violet), a triphenylmethane dye. Activity of these dyes is a property of the pseudobase.

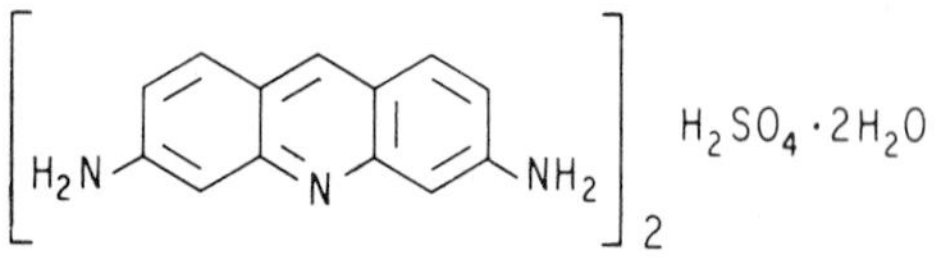

Figure 10-6. Proflavine (3,6-diaminoacridine hemisulfate), an acridine dye.

Figure 10-7. Diagrammatic representation of the secondary structure of DNA containing intercalated proflavine molecules. *(From Lerman: J Cell Comp Physiol 64 [Suppl 1] 1964.)*

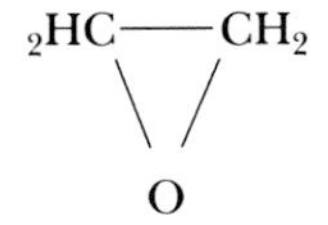

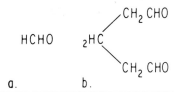

Figure 10-8. Aldehyde disinfectants. **a.** Formaldehyde. **b.** Glutaraldehyde.

against all types of bacteria, including spores and tubercle bacilli, but its action is slow. It can be used for sterilizing a wide range of materials, but its greatest applicability lies in the sterilization of materials that would be damaged by heat, such as polyethylene tubing, electronic and medical equipment, biologicals, and drugs. It has been of special value in the sterilization of heart-lung machines.

The alkylating action of ethylene oxide is responsible for its bactericidal activity. It is an epoxy compound with the formula:

$$_2HC \overset{\displaystyle \diagup \diagdown}{\underset{O}{\rule{0pt}{0pt}}} CH_2$$

The ethylene oxide ring opens in the presence of a labile hydrogen and forms a hydroxy ethyl radical (CH_2CH_2 OH), which then attaches to the position in the protein formerly occupied by the hydrogen. A labile hydrogen is available in carboxyl, amino, sulfhydryl, hydroxyl, and phenolic groups of proteins. Irreversible death of the cell results from blockage of these reactive groups. In addition to its action on protein, ethylene oxide also reacts with DNA and RNA, possibly by esterification of phosphate groups or reaction with the ring nitrogen of purines and pyrimidines. It is mutagenic for bacteria. Its use as a disinfectant presents some hazard of potential toxicity for humans, including mutagenicity and carcinogenicity.

Antimicrobial Physical Agents

A wide variety of bacterial species have adapted to the very diverse physical conditions that exist in the various ecologic niches on earth. Bacteria are found in the ocean waters at all latitudes from 20 degrees south to 90 degrees north and at the bottom of the seas at hydrostatic pressures as great as 13,000 pounds per square inch. Some

TABLE 10-9. DISINFECTION LEVELS OF SELECTED GERMICIDES

Class	Use Concentration	Activity Level
Gas		
Ethylene oxide	450–800 mg/liter*	High
Liquid		
Glutaraldehyde, aqueous	2%	High
Formaldehyde + alcohol	8% + 70%	High
Formaldehyde, aqueous	3 to 8%	High to intermediate
Iodine + alcohol	0.5% + 70%	Intermediate
Alcohols	70%	Intermediate
Chlorine compounds	0.1 to 0.5%[†]	Intermediate
Phenolic compounds, aqueous	0.5 to 3%[‡]	Intermediate to low
Iodine, aqueous	1%	Intermediate
Iodophors	0.007 to 0.015%[§]	Intermediate to low
Quaternary ammonium compounds	0.1 to 0.2% aqueous	Low
Hexachlorophene	1%	Low
Mercurial compounds	0.1 to 0.2%	Low

From Spaulding et al.: In Block (ed): Disinfection, Sterilization, and Preservation, 2nd ed, 1977. Philadelphia, Lea & Febiger.
*In autoclave-type equipment at 55 to 60C.
[†]Chlorine.
[‡]Dilution of concentrate.
[§]Available iodine.

live in natural hot springs at temperatures near boiling, while others grow at temperatures near freezing. Perhaps the one expansive area that most closely approaches sterility is the surface of a sandy desert. Although the dry heat found there is destructive to most bacteria, the greatest sterilizing effect is exerted by rays from the sun in the near ultraviolet spectrum.

Most pathogenic bacteria have limited tolerance to extreme variations in their physical environment and have little survival ability outside the living body. Others, however, produce spores that are highly resistant to deleterious physical conditions in the environment and endow the organism with an increased survival value. In the following discussion, attention will be focused primarily on those physical agents that are useful as sterilizing agents.

Heat

Heat is the most reliable and universally applicable method of sterilization and, whenever possible, should be the method of choice. As with other types of disinfection, the sterilization of a bacterial population by heat is a gradual process, and the kinetics of death are exponential. The first order inactivation by heat means that a constant fraction of the organisms undergoes an inactivating chemical change in each unit of time and that one such change is sufficient to inactivate an organism.

The time required for sterilization is inversely related to the temperature of exposure. This relationship may be expressed by the term *thermal death time,* which refers to the minimum time required to kill a suspension of organisms at a predetermined temperature in a specified environment. Because of the high temperature coefficients involved in heat sterilization, a minimal change in temperature significantly alters the thermal death time. In accordance with the law of mass action, the sterilization time is directly related to the number of organisms in the suspension.

MECHANISM OF THERMAL INJURY. Heat inactivation of bacteria cannot be defined in simple biochemical terms. Although the lethal effect of moist heat above a particular temperature is usually attributed to the denaturation and coagulation of protein, the pattern of thermal damage is quite complex, and coagulation undoubtedly masks other more subtle changes induced in the cell before coagulation becomes apparent. Since every part of the cell is damaged by some specific temperature, the problem is that of defining which site essential to viability is most sensitive to heat.

The production of single-strand breaks in DNA may be the primary lethal event. The loss of viability of cells exposed to mild heat can be correlated with the introduction of these breaks. Damage to the DNA appears to be enzymatic, as a result of an activation or release of a nuclease. The ability of the cell to repair this damage and to recover viability depends upon the physiologic state and genetic makeup of the organism.

Heat also causes a loss of functional integrity of the membrane and leakage of small molecules and 260 nm absorbing material. This material is of ribosomal origin and apparently results from degradation of the ribosomes by ribonucleases activated by the heat treatment. There appears to be a correlation between the degradation of ribosomal RNA and the loss of viability of cells exposed to high temperatures.

The mechanism by which organisms are destroyed by dry heat is different from that of moist heat. The lethal effects of dry heat, or desiccation in general, are usually ascribed to protein denaturation, oxidative damage, and toxic effects of elevated levels of electrolytes. In the absence of water, the number of polar groups on the peptide chain decreases, and more energy is required to open the molecules, hence the apparent increased stability of the organism.

Moist Heat. Objects may be sterilized either by dry heat applied in an oven, or by moist heat provided as steam (Table 10-10). Of the two methods, moist heat is preferred because of its more rapid killing. Exposure of most mesophilic nonsporeforming bacteria to moist heat at 60C for 30 minutes is sufficient for sterilization. Among the exceptions are *S. aureus* and *Streptococcus faecalis,* which require an exposure time of 60 minutes at 60C. A temperature of 80C for 5 to 10 minutes destroys the vegetative form of all bacteria, yeast, and fungi. Among the most heat-resistant cells are the spores of *Clostridium botulinum,* an anaerobic organism that causes food poisoning. Spores of this organism are destroyed in 4 minutes at 120C, but at 100C, 5.5 hours are required.

The application of moist heat in the destruction of bacteria may take several forms: boiling, live steam, and steam under pressure. Of these, steam under pressure is the most efficient because it makes possible temperatures above the boiling point of water. Such temperatures are necessary because of the extremely high thermal resistance of bacterial spores.

An autoclave is a chamber in which steam sterilization is carried out. The basic essential in this type of steriliza-

TABLE 10-10. MINIMUM TIMES REQUIRED FOR STERILIZATION BY MOIST AND DRY HEAT AT VARIOUS TEMPERATURES

Temperature	Moist Heat		Dry Heat Time (min)
	Time (min)	Pressure	
121C	15	15	—
126C	10	20	—
134C	3	30	—
140C	—	—	180
150C	—	—	150
160C	—	—	120
170C	—	—	60

From Meynell and Meynell: Theory and Practice in Experimental Bacteriology, 1970. Cambridge, University Press.

tion is that the whole of the material to be sterilized shall be in contact with saturated steam at the required temperature for the necessary period of time. For sterilizing small objects, an exposure of 20 minutes at 121C (15 pounds steam pressure per square inch) is used and provides a substantial margin of safety. Attention must be given, however, to the need for sufficient time to allow for the load to reach the required temperature before the actual sterilizing period begins.

For the sterilization of certain liquids or semisolid materials that are easily destroyed by heat, a fractional method of sterilization is employed. This process, often called tyndallization, consists of heating the material at 80 or 100C for 30 minutes on three consecutive days. The rationale for this fractional type of sterilization is that vegetative cells and some spores are killed during the first heating and that the more resistant spores subsequently germinate and are killed during either the second or third heating. The method is useful in sterilizing heat-sensitive culture media containing such materials as carbohydrates, egg, or serum.

PASTEURIZATION. As indicated above, most vegetative bacteria can be killed by relatively short exposures to temperatures of 60 to 65C. The most important application of temperatures in this range is in the pasteurization of milk and preparation of bacterial vaccines. Although originally devised by Pasteur as a means of destroying microorganisms that cause spoilage of wine and beer, the pasteurization process is now primarily used to make beverages and foods safe for consumption. The most widespread application of this treatment is the pasteurization of milk, which consists of heating to a temperature of 62C for 30 minutes, followed by rapid cooling. This temperature does not sterilize the milk, but it does kill all disease-producing bacteria commonly transmitted by milk.

Dry Heat.

Sterilization by dry heat requires higher temperatures and a longer period of heating than does sterilization with steam. Its use is limited primarily to the sterilization of glassware and such materials as oils, jellies, and powders that are impervious to steam. The lethal action results from the heat conveyed from the material with which the organisms are in contact and not from the hot air that surrounds them, emphasizing the importance of uniform heating of the entire object to be sterilized. The most widely used type of dry heat is the hot air oven. Sterilizing times of 2 hours at 180C are required for the killing of all organisms, including the sporeformers.

Other useful forms of dry heat include incineration of objects to be destroyed and flaming by passage of transfer needles, coverslips, or small instruments through the flame of a Bunsen burner.

Freezing

Although many bacteria are killed by exposure to cold, freezing is not a reliable method for sterilization. Its primary use has been in the preservation of bacterial cul-

tures. Repeated freezing and thawing are much more destructive to bacteria than is prolonged storage at freezing temperatures. Although it was previously believed that the lethal effect of freezing resulted from damage of the membrane by ice crystals, such a mechanism plays only a minor role in the death of frozen organisms. In the freezing of bacteria, the formation of ice crystals outside the cell causes the withdrawal of water from the cell interior, resulting in an increased intracellular electrolyte concentration and a denaturation of proteins. The cell membrane is damaged, and a leakage of intracellular organic compounds ensues. The leakage material contains inorganic phosphorus, ribose, peptides, and nucleotides that arise as a consequence of the activation of latent ribonucleases and peptidases.

When bacteria are frozen rapidly to temperatures below −35C, ice crystals form within the cell and produce a lethal effect during defreezing. If cultures are dried in vacuo from the frozen state by the process of lyophilization or freeze drying, the initial mortality is greatly diminished. This method is widely used for the preservation of bacterial cultures.

Radiation

Sunlight possesses appreciable bactericidal activity and plays an important role in the spontaneous sterilization that occurs under natural conditions. Its disinfectant action is due primarily to its content of ultraviolet rays, most of which, however, are screened out by glass and by the presence of ozone in the outer regions of the atmosphere. Other electromagnetic rays of shorter wavelength, such as x-rays and gamma-rays, as well as rays produced by radioactive decay and by ion accelerators, also exert a pronounced effect when absorbed by bacteria (Fig. 10-9).

Effects of Radiation. Only absorbed light promotes photochemical reactions. As a molecule absorbs light, it receives energy in the form of discrete units termed "quanta." The energy of a quantum is inversely related to its wavelength. In the primary reaction, only 1 quantum of light is absorbed by each molecule of absorbing substance. The number of quanta absorbed by a biologic system is proportional to the product of the duration and intensity of the radiation as well as to the absorption coefficient of the irradiated material. The absorption of a quantum by an electron in an atom results in activation of the molecule, which then either uses the extra energy for chemical changes, such as decomposition and internal rearrangements, or loses it entirely as heat or fluorescence.

Radiation may have sufficient energy to remove an electron completely from an atom and produce an electrical charge (ionization) or only enough energy to raise electrons to states of higher energy (excitation). Energy equivalent to 10 electron volts is required to pull an electron completely out of an atom. This is provided by x-rays and gamma-rays that ionize atoms by the ejection of electrons from any atom through which the radiation passes.

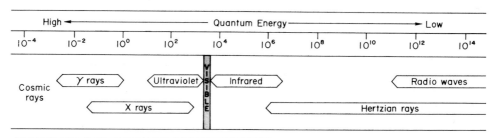

Figure 10-9. The electromagnetic spectrum. The wavelength in nanometers is plotted on an exponential scale.

In the visible and ultraviolet range, although the quantum energy absorbed by the molecule cannot remove an electron completely, the excitation produced often leads to photochemical changes. In the infrared range of the spectrum, the energy is inadequate to initiate a chemical change in biologic material, and the absorbed energy is dissipated as heat.

Ultraviolet Radiation

MECHANISM OF ULTRAVIOLET RADIATION INJURY. The effectiveness of ultraviolet light as a lethal and mutagenic agent is closely correlated with its wavelength. The most effective bactericidal wavelength is in the 240 to 280 nm range with the optimum at about 260 nm, which corresponds with the absorption maximum of DNA. The major mechanism of the lethal effect of ultraviolet light on bacteria is attributed to this absorption by, and resultant damage to, the DNA. Ultraviolet radiation leads to the formation of covalent bonds between pyrimidine residues adjacent to each other in the same strand, resulting in the formation of cyclobutane-type pyrimidine dimers. These dimers distort the shape of the DNA and interfere with normal base pairing. This results in an inhibition of DNA synthesis and secondary effects, such as inhibition of growth and respiration. Although other effects have been shown to be produced by ultraviolet radiation, such as photohydration of cytosine and crosslinkage of complementary strands of DNA, the extremely high dose required rules them out as a possible mechanism for ultraviolet damage to cells.

Repair Mechanisms. If, following treatment with ultraviolet light, cells are immediately irradiated with visible light in the 300 to 400 nm range, both the mutation frequency and the bactericidal effect of ultraviolet light are greatly reduced. This phenomenon is known as "photoreactivation" and results from the activation by light of an enzyme that hydrolyzes the pyrimidine dimers. Another mechanism also exists in some cells for the repair of light-damaged DNA. This is a dark-repair mechanism that invokes the activity of an elaborate set of enzymes for the excision of the ultraviolet-induced dimers and repair of the damaged strand. Excision-repair is probably the primary cause of an enhanced resistance to ultraviolet light by some strains of bacteria.

PRACTICAL APPLICATIONS. Ultraviolet radiation can be produced artificially by mercury vapor lamps. The unit of radiation energy is measured in terms of microwatts per unit area per unit of time. A 15-watt ultraviolet light delivers 38 μW/cm^2/second of radiation at a distance of 1 meter. Ultraviolet radiation is equally effective against grampositive and gram-negative organisms. The lethal dose for most of the common, nonsporeforming bacteria varies from 1800 μW/cm^2 to 6500 μW/cm^2. Bacterial spores require up to 10 times this dose.

Although the bactericidal properties of ultraviolet radiation are indisputable, it cannot be properly classified as a sterilizing agent because of the many uncertainties surrounding its use. Unlike ionizing radiation, the energy of ultraviolet radiation is low, and its power of penetration is very poor. It does not penetrate into solids and penetrates into liquids very slightly. For this reason, ultraviolet rays have no effect on organisms shielded or protected from the incident beams.

The primary application of ultraviolet irradiation is in the control of airborne infection, where it is used for the disinfection of enclosed areas, such as entryways, hospital wards, and operating rooms. Whereas the application of ultraviolet irradiation has failed to produce uniform results in the control of airborne infections in public places, it has provided a significant reduction in the incidence of secondary infections following surgery.

Ionizing Radiations

PROPERTIES OF IONIZING RADIATIONS. Ionizing radiations are classified according to their physical properties and fall into two major categories: (1) those that have mass and that may be either charged or uncharged and (2) those that are energy only. Some of the ionizing radiations are products of radioactive decay (α-, β-, γ-rays), and others are produced in an x-ray machine, by particle bombardment, or by nuclear reactors. Primary cosmic rays from outer space that bombard the earth and its atmosphere are composed of protons, alpha particles, and heavy atomic nuclei. Few of these rays, however, reach the earth at altitudes of sea level because of the protective blanket provided by the atmosphere. The ionizing radiations that are of greatest practical value for purposes of sterilization are the electromagnetic x-rays and gammarays and the particulate cathode rays (artificially accelerated electrons). These radiations have a much higher energy content than ultraviolet radiation and consequently have a much greater capacity to produce lethal effects.

The penetrating power of ionizing radiations contributes to their effectiveness as sterilizing agents. Although cathode rays, because of their particulate nature, have a greater intrinsic energy and, consequently, the greater power of penetration, x-rays and gamma-rays have a relatively greater penetrating ability. Because of the nature of the mechanisms involved, optimum activity never occurs at the surface of the material being treated. With gamma-rays it occurs just below or inside the surface; with cathode rays it occurs a few centimeters deeper.

RADIATION UNITS. The effect of ionizing radiations upon living systems depends upon the amount of energy absorbed. The amount of radiation to which a system is exposed is the exposure dose. For x-rays or gamma-rays the exposure dose is measured in roentgens and is equivalent to an energy absorption of about 83 erg/g of air. That portion of the exposure dose that is actually absorbed by the biologic system, the absorbed dose, is the biologically effective dose, the unit of which is the rad. The rad is based on an energy absorption of 100 ergs/g of air. For practical purposes the Mrad, equal to 10^6 rad, is used because this is the dose required for sterilization.

LINEAR ENERGY TRANSFER. In the passage of ionizing radiations through matter, the energy of the protons is transferred by collisions with an orbital electron in an atom of the absorbing medium. Following such a collision, an electron is ejected from the atom with high energy and at great speed. As this electron moves through the medium, it will ionize and excite atoms with which it interacts. The energy given to it will be dissipated as it moves through the medium (Fig. 10-10).

RADIOLYSIS OF WATER. Ionizing radiation has both direct and indirect effects on the macromolecules of the cell. The direct effect is exerted by the initial transfer of energy within a limited vital or sensitive area, according to the target theory. However, since biologic systems are composed of large amounts of water, the primary mechanism for the lethal effect of ionizing radiation is an indirect one. As ionizing radiations pass through water, they cause the water molecules to ionize:

$$H_2O \xrightarrow[\text{quantum}]{\text{energy}} H_2O^+ + e^-$$

The resulting positive ion reacts with unionized water to yield another species of charged water molecule and the free hydroxyl radical:

$$H_2O^+ + H_2O \rightarrow H_3O^+ + \cdot OH$$

The ejected electron can also react with unionized water to give an OH^- ion and a free hydrogen radical:

$$e^- + H_2O \rightarrow OH^- + H\cdot$$

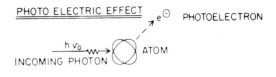

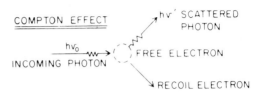

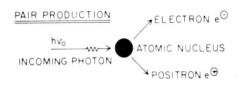

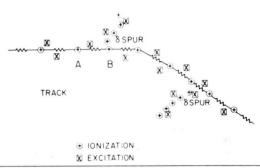

⊕ IONIZATION
⊠ EXCITATION

Figure 10-10. Photon energy interaction with matter and a diagrammatic depiction of events along an ionization track. *(From Silverman and Sinskey: In Block (ed): Disinfection, Sterilization, and Preservation, 2nd ed, 1977. Philadelphia, Lea & Febiger.)*

The hydroxyl radical is a strong oxidizing agent, and the hydrogen radical is a powerful reducing agent. The hydroxyl radical is highly reactive with macromolecules, especially DNA, and ruptures it by effecting a break in both of the constituent chains.

The presence of oxygen in cells at the time of irradiation enhances the magnitude of the irradiation effect. The increased effect results from the interaction of oxygen with radiation-produced free radicals, drawing these radicals into destructive auto-oxidative chain reactions. The presence of oxygen during irradiation also promotes the formation of hydrogen peroxide and organic peroxides:

$$H\cdot + O_2 \rightarrow \cdot HO_2$$

$$2 \cdot HO_2 \rightarrow H_2O_2 + O_2$$

Certain chemical compounds, such as those that contain sulfhydryl groups, protect biologic systems from the damaging effects of ionizing radiation by diverting the absorbed energy.

LETHAL EFFECTS. Most pathogenic nonsporeforming bacteria are relatively sensitive to ionizing radiation, but spores are among the most radiation-resistant microorganisms known. Even among the nonsporeformers, however, considerable variation exists. Gram-positive nonsporeformers are generally more resistant than are the gram-negative organisms, and within the latter group the pseudomonads are among those most readily killed.

The rate or intensity of irradiation is of little importance in determining the fraction of organisms killed. It is the total dose administered that is important. The death of microorganisms exposed to ionizing radiation is usually exponential throughout the sterilization period, although in some cases it tends to be sigmoidal. The slope of the time-survivor curve is determined by the intensity of the irradiation, but in terms of dose against percentage killed the relationship is always exponential. Toward the end of the process, however, a tailing effect may become prominent, emphasizing the importance of ensuring that the full dose has been given.

PRACTICAL APPLICATIONS. Although the sterilizing dose is dependent upon the initial level of contamination, a dose of 2.5 Mrad of ionizing radiation has been accepted as the sterilizing dose. This dose is sufficient to kill the most resistant microorganisms and also provides an adequate margin of safety under conditions of practical use.

The main areas in which ionizing radiations have been applied for purposes of sterilization are in pharmacy and medicine. They are especially suitable for sterilizing such articles as catgut and nylon sutures and disposable items, such as plastic syringes, needles, surgical blades, catheters, prostheses, surgical gloves, plastic tubing, plastic dishes, and blood transfusion sets.

Ultrasonic and Sonic Vibrations

Sound vibrations at high frequency, in the upper audible and ultrasonic range (20 to 1000 kc), provide a useful technique for disrupting cells, especially for the extraction of enzymes. The soundwave generators that are widely employed for disruption of cells operate in the frequency range of 9 to 100 kc per second. No specific frequency has been found to be uniquely effective, but generally, ultrasonic waves are more effective as the frequency is increased.

The passage of sound through a liquid produces alternating pressure changes, which, if the sound intensity is sufficiently great, cause cavities to form in the liquid. The cavities, which are about 10 μ in diameter, grow in size until they collapse violently with the production of high local velocities and pressures of the order of 1,000 atmospheres. During this violent collapse stage, the cell

disintegrates. In addition to disintegrating cells, cavitation produces a number of chemical and physical changes in the suspending medium, which may be deleterious to certain enzymes. Among the most important of these is the formation of hydrogen peroxide when cavitation takes place in a liquid containing dissolved oxygen. Ultrasonic vibrations have also been shown to cause a depolymerization of macromolecules and intramolecular regroupings. Double-strand breaks are produced in transforming DNA by sonic vibration, and integration into the host genome is inhibited.

Microorganisms vary markedly in their sensitivity to sonic and ultrasonic vibrations. The most susceptible are the gram-negative rods, and among the most resistant are the staphylococci, which require long periods of exposure. Although sonic vibrations are lethal to many members of the exposed bacterial population, there are many survivors. Consequently, treatment with sonic vibrations is of no practical value in sterilization and disinfection.

Filtration

The principal method used in the laboratory for the sterilization of heat-labile materials is filtration. Although mechanical sieving plays a role in all filtration processes, electrostatic and absorption phenomena and the physical construction of the filter also exert a pronounced effect.

A number of different types of filters are employed for purposes of sterilization. These include (1) Seitz filters, consisting of discs of an asbestos-cellulose mixture that is discarded after a single filtration, (2) sintered glass filters, prepared by fusing together fine glass fragments, (3) candle filters, which are thick-walled tubes made of diatomaceous earth (Berkefeld filter) or unglazed ceramic (Selas and Chamberland filters), and (4) membrane filters consisting of porous discs of cellulose esters. Candle and membrane filters are available in a wide range of pore sizes. They absorb very little of the fluid being filtered and thus are very useful for the sterilization of certain materials that cannot tolerate, without deterioration, the high temperatures used in heat sterilization.

Membrane Filters. Membrane filters, composed of biologically inert cellulose esters, are available in pore sizes of 14 μm to 0.023 μm. The 0.22-μm filter is the most widely used for sterilization filtration purposes because the pore size is smaller than bacteria. It should always be used with solutions containing serum, plasma, or trypsin where species of *Pseudomonas* or other small bacteria may be present.

Membrane filters are also useful in microbiology for purposes other than sterilization. Because of their inertness, they provide an optimum method for the collection of microorganisms, since any bacteriostatic agent present in the suspending medium may be easily flushed from the filter. This technique has been invaluable in the sterility testing of disinfectants and antibiotics where it is necessary to overcome the inhibitory effects of the drug. In clinical microbiology it has been useful for the culture of organisms present at a low concentration in a large volume of fluid.

Membrane filters act essentially as two-dimensional screens, retaining all particles that exceed the pore size. In liquid filtration, a large number of particles somewhat smaller than the pore size are retained by van der Waals forces, by random entrapment in the pores, and by build-up on previously retained particles. However, the important characteristic of the membrane filter is that all particles larger than the pore size are positively retained on the filter surface.

FURTHER READING

Books and Reviews

Allwood MC, Russell AD: Mechanisms of thermal injury in nonsporulating bacteria. Adv Appl Microbiol 12:89, 1970

Bennett EO: Factors affecting the antimicrobial activity of phenols. Adv Appl Microbiol 1:123, 1959

Block SS (ed): Disinfection, Sterilization, and Preservation, 2nd ed. Philadelphia, Lea & Febiger, 1977

Chambers CW, Clarke NA: Control of bacteria in nondomestic water supplies. Adv Appl Microbiol 8:105, 1966

Chatigny MA: Protection against infection in the microbiological laboratory: devices and procedures. Adv Appl Microbiol 3:131, 1961

Él'piner IE: Action of ultrasonic waves on microorganisms, viruses, and bacteriophages. In Ultrasound: Its Physical, Chemical, and Biological Effects. New York, Consultants Bureau, 1964, chap 9

Farrell J, Rose AH: Low-temperature microbiology. Adv Appl Microbiol 7:335, 1965

Gardner JF: Principles of Antimicrobial Activity. In Block SS (ed): Disinfection, Sterilization, and Preservation, 2nd ed. Philadelphia, Lea & Febiger, 1977, pp 883–911

Grier N: Mercurials—inorganic and organic. In Block SS (ed): Disinfection Sterilization, and Preservation, 2nd ed. Philadelphia, Lea & Febiger, 1977, pp 361–394

Gray TRG, Postgate JR (eds): The Survival of Vegetative Microbes. 26th Symp. Soc Gen Microbiol. Cambridge, University Press, 1976

Harold FM: Antimicrobial agents and membrane function. Adv Microb Physiol 4:45, 1970

Heckly RJ: Preservation of bacteria by lyophilization. Adv Appl Microbiol 3:1, 1961

Hugo WB (ed): Inhibition and Destruction of the Microbial Cell. New York, Academic Press, 1971

Kimble CE: Chemical food preservatives. In Block SS (ed): Disinfection, Sterilization, and Preservation, 2nd ed. Philadelphia, Lea & Febiger, 1977, pp 834–858

MacLeod RA, Calcott PH: Cold shock and freezing damage to microbes. In Gray TRG, Postgate JR (eds): The Survival of Vegetative Microbes. 26th Symp Soc Gen Microbiol. Cambridge, University Press, 1976, pp 81–109

McDade JJ, Philips GB, Sivinski HD, et al.: Principles and applications of laminar-flow devices. In Norris JR, Ribbons DW (eds): Methods in Microbiology. New York, Academic Press, 1969, vol 1, pp 137–168

Meynell GG, Meynell E (eds): Theory and Practice in Experimental Bacteriology, 2nd ed. Cambridge, University Press, 1970

Morton HE: Alcohols. In Block SS (ed): Disinfection, Sterilization, and Preservation, 2nd ed. Philadelphia, Lea & Febiger, 1977, pp 301–318

Mulvany JG: Membrane filter techniques and microbiology. In Norris JR, Ribbons DW (eds): Methods in Microbiology. New York, Academic Press, 1969, vol 1, pp 205–253

Petrocci AN: Quaternary ammonium compounds. In Block SS (ed): Disinfection, Sterilization, and Preservation, 2nd ed. Philadelphia, Lea & Febiger, 1977, pp 325–347

Pizzarello DJ, Witcofski RL: Basic Radiation Biology. Philadelphia, Lea & Febiger, 1967

Reddish GF (ed): Antiseptics, Disinfectants, Fungicides, and Chemical and Physical Sterilization, 2nd ed. Philadelphia, Lea & Febiger, 1957

Russell AD, Jenkins J, Harrison IH: The inclusion of antimicrobial agents in pharmaceutical products. Adv Appl Microbiol 9:1, 1967

Spaulding EH, Cundy KR, Turner FJ: Chemical disinfection of medical and surgical materials. In Block SS (ed): Disinfection, Sterilization, and Preservation, 2nd ed. Philadelphia, Lea & Febiger, 1977, pp 654–684

Sykes G: Methods and equipment for sterilization of laboratory apparatus and media. In Norris JR, Ribbons DW (eds): Methods in Microbiology. New York, Academic Press, 1969, vol 1, pp 77–121

Selected Papers

Birkhaug KE: Phenylmercuric nitrate. J Infect Dis 53:250, 1933

Brewer JH: Reduction of infectivity of certain pathogenic bacteria by "Mercurochrome." JAMA 137:858, 1948

Chick H: An investigation of the laws of disinfection. J Hyg (Camb) 8:92, 1908

Chick H: The process of disinfection by chemical agencies and hot water. J Hyg (Camb) 10:237, 1910

Fried VA, Novick A: Organic solvents as probes for the structure and function of the bacterial membrane: Effects of ethanol on the wild type and an ethanol-resistant mutant of Escherichia coli K12. J Bacteriol 114:239, 1973

Gershenfield L, Greene A, Witlin B: Disinfection of clinical thermometers. J Am Pharm Assoc, Sci Ed 40:457, 1951

Gezon HM, Thompson DJ, Rogers KD, et al.: Control of staphylococcal infections and disease in the newborn through the use of hexachlorophene bathing. Pediatrics 51:331, 1973

Gustafsson P, Nordström K, Normark S: Outer penetration barrier of Escherichia coli K12: Kinetics of the uptake of gentian violet by wild-type and envelope mutants. J Bacteriol 116:893, 1973

Laipis PJ, Ganesan AT: In vitro repair of x-irradiated DNA extracted from Bacillus subtilis deficient in polymerase I. Proc Natl Acad Sci USA 69:3211, 1972

Schnaitman CA: Solubilization of the cytoplasmic membrane of Escherichia coli by Triton X-100. J Bacteriol 108:545, 1971

Shew CW, Freese E: Lipopolysaccharide layer protection of gram-negative bacteria against inhibition by long-chain fatty acids. J Bacteriol 115:869, 1973

Silvernale JN, Joswick HL, Corner TR, et al.: Antimicrobial actions of hexachlorophene: Cytological manifestations. J Bacteriol 108:482, 1971

Waldstein EA, Sharon R, Ben-Ishai R: Role of ATP in excision repair of ultraviolet radiation damage in Escherichia coli. Proc Natl Acad Sci USA 71:2651, 1974

SECTION II
IMMUNOLOGY

CHAPTER 11

Introduction to Immunity

Components of the Immune System
    Antibodies
    Antigens
    Consequences of the Combination of Antibody
        with Antigen
    The Complement Cascade
    Cellular Aspects of Immunity
        The Thymus and Other Lymphoid Organs
        T Cell Maturation and Proliferation
        Antigen Presentation and T Cell Function
        **Cell–Cell Collaboration and Major Histocompati-
          bility Complex Restriction**

T Cell Function and Immunopathologic States
States of Immunity
Natural Immunity

Immunopathology and Clinical Immunology
    Immunodeficiency
    Autoimmunity
    Reactions Against Red Cells
    Reactions Against White Cells, Tissues, and
        Platelets
    Summary

When an unimmunized child becomes infected with an infectious agent such as *Corynebacterium diphtheriae*, there is a high probability of serious morbidity and a smaller, but still definite, probability of death. When a child previously vaccinated against diphtheria encounters the organism, however, the chances of a clinically apparent infection are slight, and the probability of death is negligible.

This example represents some of the most obvious attributes of the immune system. The protection afforded by vaccination is specific, i.e., the child immune to diphtheria will have no increased immunity to measles, and the protected (or memory) state will persist for years. Any subsequent exposure will not only reinforce the immunity, but the capacity to respond will be faster and more vigorous.

To determine if an individual has been successfully immunized, it is convenient and customary to demonstrate the presence of antibody in serum (Chap. 12). Antibodies are globular proteins produced by the lymph nodes and spleen and there are numerous ways of detecting them. Serum from an immune individual will, for example, agglutinate bacteria, or neutralize the adverse effects of many toxins. Whereas many of the procedures used to measure antibodies were introduced nearly 100 years ago, we now know that antibody is only one of several different effectors of immunity. Of at least equal importance is the cellular immune response (Chap. 14) mediated by small

mononuclear cells circulating in the bloodstream and lymphatic vessels and also migrating through the intracellular spaces. A well-known measure of the cellular response is the reaction of a previously immunized subject to intracutaneously administered extract of tubercle bacilli, the tuberculin reaction (Chap. 21), also called delayed hypersensitivity. The most prominent feature of tuberculin hypersensitivity is the effusion of small mononuclear cells or lymphocytes into the extravascular spaces and especially into the areas adjacent to small vessels—the perivascular cuffing so typical of cellular responses. Tuberculin is nonirritant and nontoxic and the inflammatory changes are not seen in unimmunized subjects.

The response to infection is the most obvious function of the immune system and patients in whom immunity is impaired characteristically suffer, and often die from, chronic infection. Defense against infection is by no means the only function of the immune system. Some immunobiologists are concerned with such diverse functions of immunity as aging (Chap. 15), autoimmunity (Chap. 18), and the defense against tumors (Chap. 20).

The antecedents of the immune system are probably to be found at least as far back in evolution as the protozoa. Sponges of one colonial type resist the encroachment of members of another colony (even of the same species) by mechanisms closely analogous to those of the cellular immune responses of fishes, birds, and mammals. As animals evolved the primitive responses became modified to become the highly complex systems of higher forms.

The student will probably have more difficulty in understanding immunologic responses than in mastering any of the other systems. Part of the difficulty has already been alluded to above; whereas some of our understanding of immunity dates back over a century, much of our knowledge is extremely recent and indeed is still in a state of flux. However, if the student is to have an understanding of the many diseases caused by subtle or not so subtle malfunctions of the immune system, and of the immunologic factors in manipulations such as transplantation and blood transfusion, familiarity with the topics dealt with in this section is essential.

Components of the Immune System

Antibodies

The antibodies are globular proteins, also referred to as immunoglobulins (Ig). Their basic structure is built around the union of a heavy (H) chain and a light (L) chain, which are often, but not invariably, covalently bound. There are five main classes of H chain—α, γ, μ, ϵ, δ—which give the designations of IgA, IgG, IgM, IgE, and IgD to the corresponding antibody. There are two main classes of L chain, κ and λ, and the basic unit consists of an H chain with either a κ or a λ chain. Each H and each L chain is made of an assemblage of peptides. Extensive amino acid substitution is a special feature of the peptides adjacent to the amino (N) terminus. This part of the H or L chain is called the variable (V) region, and short sequences within the V region, called hypervariable sequences, allow the antibody to bind specifically with antigen. They also confer a special feature called the idiotype. Thousands of different hypervariable sequences are represented in the antibody population of every individual. In contast, that portion of the molecule closest to the carboxyl (C) terminus is constant for each class of H or L chain from a given individual, so this portion of the molecule is called the constant region. Substitutions at individual amino acids in the constant region are characteristic of each class or subclass (e.g., γ_1, γ_2) of chain, whereas substitutions at other positions are characteristic of an individual. These latter characteristic configurations are the basis of what are known as allotype differences between individuals. Km are markers of human L chains and Gm are markers of human H chain (Chap. 17). Just as the V regions control antibody specificity, the constant regions, especially of the H chains, confer different functional properties on the immunoglobulin molecule.

Immunoglobulins are produced by lymphocytes of a restricted lineage, called B lymphocytes (B cells), and their direct descendants, known as plasma cells. Approximately 10 percent of the blood lymphocytes and 5 percent of thoracic duct lymphocytes are B cells. The majority of the B

cells and almost all plasma cells, however, reside in the peripheral lymphoid organs. These are the spleen, lymph nodes, and lymphoid tissues such as tonsils, adenoids, Peyer's patches, and appendix. B cell precursors are present in the yolk sac, fetal liver, and bone marrow. Birds possess a unique central lymphoid organ, called the bursa, in which B cell differentiation occurs. In mammals bursal function has been decentralized and the bone marrow appears to have taken over many of its functions in regulating differentiation.

Immature or pre-B lymphocytes do not secrete immunoglobulin. As the cells mature, immunoglobulin is formed and becomes inserted into the plasma membrane (surface immunoglobulin, or SIg). The first SIg to appear is IgM. With increasing maturation, IgD is also formed. At about this stage of differentiation the mature B cell can be stimulated to secrete immunoglobulin into the environment. Even though IgM remains the predominant SIg of a B cell, the antibody it secretes may be of another class. Up to this stage the B cell is capable of mitosis. However, when it reaches the final stage of differentiation as a plasma cell, it carries very little SIg and there are changes (losses and gains) in the production of several glycoproteins. These membrane-associated (or membrane-inserted) proteins are referred to as differentiation antigens. The fully differentiated plasma cell rarely divides or enters the bloodstream. Exceptions occur in the rare plasma cell malignancy, myeloma (Chap. 20).

The immunoglobulin H and L chains are capable of combining with a wide variety of organic molecules known collectively as antigens (see below and Chap. 12). The combining site formed by apposition of parts of the H and L chains closes around the antigen and expels water so that extremely tight or high affinity binding can result. The H chain genes of humans are on the 14th chromosome, with the λ gene on the 22nd and the κ chain gene on the second chromosome. Included in the genome is a large family of genes for hypervariable regions and a more limited series of genes for the constant regions of the α, γ, μ, ϵ, and δ chains. The number of possible immunoglobulins that can be formed by an individual is increased by the existence of two additional genetic elements called D and J genes (Chap. 12). The complete immunoglobulin has a quarternary structure and at least four domains can be distinguished—three for the constant region and one for the V region.

Each complete immunoglobulin molecule has two or more H chains, each with its complementary L chain. Each molecule of IgG, IgD, and IgE typically has two H homologous chains and two L homologous chains. IgA occurs in two forms: A lower molecular weight form with two H and two L chains is present in the bloodstream; and a more complex form with four H and L chains formed by the union of two molecules of IgA through an additional J chain is found in secretions. Secretory IgA molecules are not multimeric at their site of formation, but become so when passing through the endothelium of the mucous membrane. During their passage they acquire a glycoprotein secretory piece that provides protection against proteolytic digestion. IgM has multiple chains, typically 10 H and 10 L chains. In every case, the H and L chains are paired. The H chains (with their attendant L chains) are also paired by the protein called a J chain; thus IgM is usually referred to as a pentamer, even though it has 10 H and 10 L chains.

The size of the molecule is one of the factors determining the tissue distribution of an immunoglobulin. IgG is found in the blood and tissue spaces. Being larger, IgM is unable to pass through capillary walls easily and is mainly restricted to the bloodstream. It does not cross the placenta. The normal newborn infant has no IgM unless the placenta has become inflamed, as in congenital syphilis (Chap. 19). IgA, as has been mentioned, is in its multimeric form the immunoglobulin of mucous surfaces, and in its dimeric form is a minor immunoglobulin in the blood. IgG, IgM, and IgA all have protective functions against microorganisms or their products. IgE is present in very low concentration in the bloodstream and tends to be sequestered on the surface of a highly specialized granular cell known as a mast cell. When IgE bound to a mast cell reacts with its appropriate antigen, the mast cell releases histamine and other permeases from its granules; thus IgE is the immunoglobulin responsible for allergic or anaphylactic reactions (Chap. 21). No function is yet known for IgD, and this immunoglobulin might have remained unknown but for the very rare uncontrolled proliferation of IgD-producing plasma cells in myelomas.

A typical immune response results in the stimulation of many different lymphocytes, some of which form antibody. Even those that form antibody have different lineages. The descendants of a single lymphocyte are known as a clone. Conventional antibody responses are polyclonal and include many antibody molecules differing somewhat in their binding affinities. Of particular importance are those antibodies formed by a single clone of antibody-producing cells. These are called monoclonal antibodies. Such antibodies are rare in the natural state but are produced in the laboratory by the progeny of the fusion of a single antibody-producing lymphocyte with a myeloma cell. For convenience, the myeloma cell selected is usually a special mutant that does not secrete antibody. Antibody production is then controlled by the B cell partner of the fusion, whereas the myeloma partner contributes an unlimited capacity to proliferate. A conventional antibody can be likened to the white light of a searchlight and a monoclonal antibody to the monochromatic light of a laser beam. Antibodies formed after immunization are focused to a considerable extent but do not approach the exquisite specificity of monoclonal antibody.

Antigens

Antigen is the traditional name given to a molecule that can be bound by the combining site of an antibody; diphtheria toxin is an antigen in the classic sense. However, because the combining site of an antibody molecule forms

a relative small cavity that can only encompass a structure not much larger than a pentapeptide, it is obvious that the whole protein is not an antigen. Various terms have been coined to give a closer definition of that part of an antigenic molecule that is actually bound. Antigenic site is one term that has been used. Furthermore, although an antigen is defined as a structure capable of being recognized by an antibody, some antigenic substances can stimulate immunity whereas others cannot. A substance that can be bound by an antibody but is not capable of eliciting immunity is called a hapten. A hapten that is not immunogenic by itself can become an immunogen when combined with another molecule called a carrier. Haptens are small and can be of any chemical class (from azobenzene arsonate, to nucleotides, to amino acid sequences). Carriers are usually large proteins, commonly with a molecular weight of over 10,000 daltons, although polysaccharides can be carriers. Experimentally, poly-L-lysine has been extensively used as an effective carrier. Clinically, a chemical may bind to lysine or cysteine of a protein from the skin; thus, the body's own proteins can become carriers. This feature becomes especially important in the contact hypersensitivity to allergens such as the urushiols of poison ivy.

Certain portions of a protein or glycoprotein molecule may be more highly charged or may be more accessible than others. These portions are sometimes called the immunodominant loops or sites, with this term also being replaced by the term epitope. Myoglobin is known to have five separate oligopeptide epitopes, and albumin has a similar number; the epitope of a blood group antigen may consist of four or five monosaccharides or amino sugars complexed to a protein and more than one such epitope may be carried by the same protein (Chap. 16). The amino acids of an epitope may be part of a linear array, but in a protein with tertiary structure they need only be contiguous in their spatial arrangement and need not be linear.

Consequences of the Combination of Antibody with Antigen

Antibody–antigen reactions are exothermic and occur almost instantaneously. The rate of reaction is influenced by temperature and concentration and by the binding constant for each antibody, according to the laws of mass action. However, the gross manifestation of antibody–antigen reactions involve steps other than simple binding, and take minutes to hours to appear and may involve many intermediates. One example already cited is the release of histamine from a tissue mast cell following the reaction of IgE with antigen. Another example, which is explained below, is the activation of a complex series of blood proteins known collectively as the complement cascade.

The increased uptake by phagocytic cells of microorganisms in the presence of antibody is called opsonization. This process, one of the first activities of antibodies to be detected, is of fundamental importance in the defense against infection. Only complement and low concentrations of antibodies are needed and the rate of uptake of bacteria by granular cells in the blood (monocytes and polymorphonuclear neutrophils, or PMN) can be increased 100-fold.

Another important feature of antibody is its ability to form large complexes with soluble antigen. This occurs where both antibody and antigen is multivalent and where neither is present in great excess. Because even the smallest of the antibodies, e.g., IgG, is divalent and the same epitope may be present several times in a large molecule of antigen, a lattice forms. In vitro, the lattice becomes so large that the complexes precipitate from solution. Antibodies with this property are therefore known as precipitins. In vivo, the complexes can be trapped during filtration through the glomerulus. The trapped complexes can fix complement, and they can also be chemotactic for PMN. In either event, damage to the glomerular basement membrane triggers that form of glomerulonephritis found in serum sickness (Chap. 18).

The Complement Cascade

Complement is the collective term used to describe a family of proteins and glycoproteins with nearly 20 members. It has two main branches, the classic and the alternative (or alternate) pathways. The classic pathway is usually activated when two or more molcules of certain classes of antibody lie close to each other as they bind to antigen. The antigens can be identical repeating units, or dissimilar epitopes of a complex molecule. In the presence of calcium, three proteins combine to form the first component, C1, which binds to the adjacent constant portions of the bound antibodies. The bound C1 has enzymatic activity and cleaves the next component, unfortunately named C4, (since the components were named in the order of their discovery, and not in their sequence of activation) to a fragment, C4b. This binds to the cell surface if the antigen is on a red cell membrane, and can then cleave C2 to reveal its enzymatically active site. C2 then cleaves and attaches C3. C3 is the focal point at which the two pathways meet. When C3 is activated, a C567 complex soon forms, and sets the stage for C8 and C9, which are together known as the attack complex because they are capable of forming a lesion in the surface of a cell.

The critical third component can also be activated by many other agents, including cobra venom, yeast cell walls, and bacterial endotoxin. The alternative pathway became notorious, not only because it could lead to nonantibody-mediated cytolysis, but because during investigations of this initially controversial process it was found that several molecules with biologically active properties were released. Best known of these are C5a and C3a, which are chemotactic and inflammatory. More recently it has been found that C4a fixed during activation of the classic pathway has some similar but weaker biologic effects. The intricate enzymatic processes, the varied activities of the released peptides, and the system of checks and balances that normally inhibit the indiscriminate activation of these powerful agents are described in Chapters 14 and 18.

Cellular Aspects of Immunity

Up to this point we have considered antibodies formed by the B lymphocytes and the plasma cells. These topics were dealt with first because they are straightforward. The reactions of the other great lineage of cells, the thymus-derived or T lymphocytes, can now be described. T cells, one of the most primitive and most fundamental of the body defenses, also interact with B cells for the production of antibody against the majority of antigens. The antigens that need the intervention of T cells to trigger antibody production by B cells are called T-dependent antigens. The cellular and humoral defense systems interdigitate and regulate each other in an intricate manner.

The Thymus and Other Lymphoid Organs
During development, the thymus develops as an outgrowth from the third and fourth pharyngeal pouches. Although the anatomy of the thymus differs considerably in different species, it basically consists of a mass of tightly packed mononuclear cells forming a cortex around a looser medulla, which is rich in epithelial elements. The epithelial cells are responsible for the differentiation and maturation of lymphocytes. The thymus is regarded as a central lymphoid organ, as opposed to the spleen and lymph nodes, which are regarded as peripheral lymphoid organs.

T cell precursors enter the thymic rudiment from primitive centers in the yolk sac, liver, and bone marrow. They divide rapidly and differentiate to form mature T cells. The maturation stages can be analyzed through changes in cell surface markers. Unlike B cells, mature T cells do not carry surface immunoglobulins, but appear to express a receptor with an idiotype that can be recognized exactly as the combining site of an antibody and is recognized by an antiidiotype. T cells also recognize antigens only in the context of certain self antigens, products of the major histocompatibility complex (Chap. 17), whereas B cells can "see" antigen per se.

T Cell Maturation and Proliferation
Two well-known markers of T cells are the T1a and Thy 1 markers. These are integral components of the plasma membranes of T lymphocytes. Many thymocytes die without leaving the thymus, but about 5 percent enter the circulation and may be found in the bloodstream or in the T-dependent paracortical sites of lymph nodes and spleen. A large number of subclasses of mature T cells have been described, including suppressor T cells (T_S), helper T cells (T_H), and effector or cytotoxic T cells (T_E). Suppressors and helpers are conspicuously marked by other cell surface antigens (Chap. 14).

The thymus is regarded as one of several privileged sites. These are locations that are shielded from antigen under normal conditions. (The anterior chamber of the eye, the brain, and the testes are also privileged sites.) Thus, the immature T cell rarely, if ever, encounters environmental antigens. When it reaches the periphery it meets antigen for the first time, usually in the paracortical areas of the lymph node, where proliferation and final maturation occur.

Antigen Presentation and T Cell Function

As mentioned earlier, those antibody responses requiring the intervention of T cells are called T-dependent. The sequence of events is as follows: Antigen, e.g., a product released from a microorganism, enters the intracellular space and is rapidly collected in the draining lymphatics in solution or after being entrapped by a macrophage. This is the afferent arc of the immune response. Antigen or antigen-loaded macrophages then enter the draining proximal lymph node and come into contact with dendritic cells, where they undergo processing. As part of the central arc of the immune response, the dendritic cell presents the processed antigen to B cells or T cells. Two subsets of T cells usually proliferate rapidly—T_H cells, which facilitate the production of antibody and the differentiation of T_E; and T_S, which damp down immune responses. Following these interactions, B cells proliferate, antibody is formed, and/or T_E are released to take part in the efferent arc of the immune response.

Cell–Cell Collaboration and Major Histocompatibility Complex Restriction

To unravel the complexities of the maturation of T and B cells, cell transfer experiments have been carried out in mice. Mice can be rendered athymic at birth for studies of T cell deficiency, or can be lethally irradiated and reconstituted with thymic lymphocytes to produce B cell deficiencies; T cells can be mixed with B cells to provide more complete reconstitution. These experiments are successful only when the T cells and B cells come from the same inbred strain or from strains that are identical at what is known as the major histocompatibility complex (MHC). The MHC of humans is called the HLA system, and the MHC of the mouse is H-2. The MHC consists of an array of genes of very different structure and function that together are essential for immunoregulation. Some of the products of the MHC are cell surface antigens. One type of molecule (class 1) is found on lymphocytes, macrophages, and the majority of tissues. These antigens are involved in the recognition and rejection of grafted kidneys and other forms of organ transplant. Class 1 antigens may also become modified in certain viral infections or following exposure to chemical agent. The virally infected cell is recognized as modified self, and can be destroyed by T_E. Recognition of altered self class 1 antigens by T cells appears to represent a major component of the defense against virus that has entered a tissue cell.

Class 2 molecules have a more restricted distribution and are detectable on B cells, on some macrophages and activated T cells, but rarely on resting T cells. Class 2 antigens (sometimes called Ia antigens) are essential for T–B interactions, and probably for macrophage–lymphocyte interactions. Because the Ia antigens of each unrelated individual are rarely identical, B cells from one individual are unable to collaborate with T cells from another, so as a corollary it can be inferred that the Ia antigens are of great importance in normal cellular interactions leading to the maturation of an immune response. Some of the complement components, notably C4, are also a frequent or perhaps constant component of the MHC. A more complete description of the intricacies of the MHC is given in Chapter 17.

T Cell Function and Immunopathologic States

The suppressor and helper functions of the corresponding subsets of T cells have been clearly defined in clinical as well as in experimental situations. A number of lymphocytic leukemias, for example, appear to be clones of T cells frozen in their differentiation by the oncogenic process. Szezary cell leukemia can provide a source of pure suppressor cells. The role of T effector cells and of another subset of lymphocyte, the null cell, has been determined experimentally; the demonstration of their role of T cells in killing tumor cells in vivo has been inferred by observations of patients with depressed T cell function resulting from immunosuppressive drugs used to prevent rejection of kidney or liver transplants. Lymphoid tumors arise if immunosuppression is too intense, but interestingly have, on occasion, disappeared when the drug was withheld.

Infants with congenital absence of the thymus (DiGeorge syndrome) have normal immunoglobulin levels but impaired T_E function (Chap. 15). They reject skin grafts slowly, have absent or weak cutaneous hypersensitivity responses to *Candida* antigen or dinitrochlorobenzene (both of which measure T cell function), and they are unduly susceptible to infection from a wide range of infectious agents. Transplant patients are treated with immunosuppressive agents to minimize the risk of rejection. One price such patients may pay for retaining their transplanted kidney or heart is lymphoma or leukemia, as mentioned above; a more frequent threat is chronic infection (Chap. 19). Cytomegalovirus is a common invader of the lungs of kidney transplant recipients, and can also be a cause of infection in pregnant women. Pregnancy itself often results in a subtle impairment of T cell function (Chap. 20). Thus, infection is common in all subjects with impaired T cell reactivity and, in the DiGeorge syndrome, resistance to infection develops when the infant is successfully reconstituted with thymic tissue. Although the exact manner in which T cells protect against microorganisms is not known, three functions that are easily demonstrated in vitro may operate in vivo: antibody-dependent cell-mediated cytotoxicity (ADCC), natural killing (NK), and cell-mediated lympholysis (CML) (Chap. 19). ADCC depends upon the ability of mononuclear cells from nonimmune subjects to lyse sensitive target cells in the presence of even minute traces of antibody; NK is the lytic capacity of a special subset of nonimmune lymphocytes to lyse other target cells in the absence of antibody; and CML is a cytolytic function of T_E from previously immunized subjects against antigens (usually of the MHC) of the donor. ADCC has been described as a mechanism for eliminating effete red cells (Chap. 16) and, as previously stated, NK is thought to have a role in the elimination of virally infected cells or of some tumor cells.

States of Immunity

Immunity can be active, adoptive, or passive. Active immunity follows exposure to antigen. Adoptive immunity is a term used to describe immunity that results when lymphoid cells are transferred from an immune to a nonimmune host, and passive immunity results from the transfer of serum or colostrum (the early secretions of the mammary gland following parturition). The human infant is passively immunized by the transplacental passage of immunoglobulins, mainly IgG, in utero. Many domestic animals, notably pig, goat, and sheep, are born agammaglobulinemic because placentation is such that globulins cannot pass to the fetal circulation. The newborn is achlorhydric, and antibodies from the colostrum pass through the gut within a few hours of birth, raising immunoglobulin to maternal levels. The importance of passive immunity can be demonstrated by depriving piglets of colostrum; such piglets die of overwhelming infections within 48 hours if not protected in a germfree environment. Active immunity develops as the individual is exposed to environmental antigens. It is the method of choice for the protection of an individual. The therapeutic transfer of passive immunity, e.g., to tetanus, was once an important and sometimes lifesaving procedure. It is now little used because (1) it is of short duration because the transferred antibody is soon metabolized and is not replaced; (2) the serum used was frequently from immune animals, and the recipient soon produced antibody to the animal serum; and (3) much more is known about the production of highly effective vaccines. Very little is known about adoptive immunity except in the experimental animal, where it has been invaluable for analyzing cellular immune processes.

Natural Immunity

Active immunity is acquired through environmental exposure, e.g., to *Escherichia coli* ingested orally or to a derivative of the toxin of *Clostridium tetani* by injection. Although the immunity developed is specific for the immunogen, the individual often produces antibodies that also react with antigens from very different sources. A person with group O or A blood (Chap. 16) who is exposed to *E. coli* will not only develop antibodies to components of the coliform or-

ganisms (species-specific antigens), but will also have antibodies reacting with red cells from a donor with group B blood because some of the carbohydrates of *E. coli* variants resemble those found on the group B red cell. This is called cross-reactivity. Cross-reactivity amplifies the range of immunity. It has also been exploited for purposes of diagnosis. A subject exposed to typhus develops antibodies that cross-react with antigens of a subline (X19) of *Proteus vulgaris;* it has been easier to use *Proteus* than *Rickettsia* in laboratory diagnostic procedures. Cross-reactivity is also in part responsible for the important phenomenon of herd immunity. When a population of people or animals is exposed to a pathogen causing an epidemic, some individuals will become infected while others will escape. Differences in genetic control of immune responsiveness as well as previous environmental exposure (natural immunity) act to protect certain individuals who have not been specifically immunized. The genetics of regulation of the immune response are discussed in Chapter 17.

Immunopathology and Clinical Immunology

Immunodeficiency

An imaginative immunologist named Robert Good drew attention to the value of what he called "experiments of nature" in aiding our knowledge of normal functions. Good was referring specifically to the wealth of information derived from studies of abnormal function in the major immunodeficiency states—DiGeorge with impaired T cell, and Bruton with impaired B cell activities. At one time only the major forms of immunodeficiency were well understood. With increasing sophistication has come the awareness that there are multitudinous specific functional abnormalities. These can result in too little or too much T cell help or suppression, lack of a T cell subset, selective deficiencies of a single class or subclass of immunoglobulin, failure to provide a functioning component of the complement cascade, or even malfunctioning of the proteolytic function of the granulocytes and macrophages. All of these conditions may be temporary or permanent, idiopathic or iatrogenic (Chap. 18).

Autoimmunity

Excessive or uncontrolled reactivity of any of the components of the immune system has grave consequences. Whereas low level of responsiveness to self antigens is constantly required for self surveillance, augmented levels lead to autoaggressive disease. The list of diseases from known or suspected immunologic reactivity against self antigens is steadily expanding. Hashimoto's thyroiditis, lupus erythematosis, and rheumatoid arthritis are but a few of the autoaggressive diseases discussed in Chap. 18. The list of suspected autoimmune diseases is much longer. In many instances it is possible to produce the disease experimentally. The experimental disease, produced by injecting tissue or a tissue extract usually with an adjuvant to intensify the response, tends to be self curing and this represents a major distinction from the natural disease, which is more frequently progressive. A series of postulates drawn up by Witebsky and comparable to Koch's postulates for microbial infection (Chap. 33) set rules for the determination of autoimmunity. (It is frequently difficult to satisfy Witebsky's postulates for human disease; hence, many diseases for which immune cells, antibodies, complement components and their inhibitors are probably responsible must still be regarded as being of doubtful etiology.) This is one of the major areas in which our knowledge of diseases will develop a surer foundation in future.

Reactions Against Red Cells

Red blood cells carry a rich variety of individual specific antigens. Some of them, although highly immunogenic, are so frequent (public) or so rare (private) that they are seldom involved in untoward reactions, although they can be responsible for immunizing against a fetus or against transfused cells. Severe transfusion reactions are usually due to incompatibility for the antigens of the ABO blood group system. These can be regarded as oligosaccharides, usually carried by a protein core. Individuals of blood group O have fucose as the terminal sugar; blood group A persons have an added N-acetyl-galastosamine; and those of group B, galactose. The glycosyl transferases are expressed in a codominant manner, so persons with glycosyl transferases for galactosamine and galactose have both A and B antigens. Because these sugars are commonly found in bacteria and in foodstuffs, natural antibodies, usually IgM, are frequently formed. Persons of group O usually have natural anti-A and anti-B, persons of group A have anti-B, and persons of group B have anti-A. Transfusion of incompatible blood into recipients with preformed antibodies leads to agglutination, rapid destruction of the transfused cells, and to the excretion of breakdown products in the urine. Oliguria is a serious sequal (Chap. 16).

Antigens of the Rh system are sometimes responsible for less severe transfusion reactions. The Rh antigens are proteins and are rarely encountered in the environment, so preformed antibodies are rare. The Rh antigens appear to be the product of a complex of structural genes. One allelic form frequently designated Rh D is present in about 85 percent of whites and can induce transfusion reactions in the 15 percent of Rh D negative individuals. Formerly of great concern, the inheritance of Rh D from the father when the mother was Rh negative could lead to sensitization of the mother and to the transplacental transfer of antibodies leading to the condition of hemolytic disease of the newborn. A triumph of immunoprophylaxis, the passive administration of high titered anti-Rh antiserum to the mother during pregnancy before active immunization can occur has almost abolished this complication.

Reactions Against White Cells, Tissues, and Platelets

A cause of minor transfusion reactions when there is no red cell incompatibility may be due to reactions against the leukocytes of transfused whole blood. This occurs in some parous women and also in patients who have been previously transfused with whole blood. The reactions are usually against antigens of the HLA antigens of leukocytes and tissues. The HLA incompatibility may also lead to the rapid destruction of transfused platelets in patients with aplastic anemia or leukemia and who have required frequent transfusions of blood or platelets. The risk can be diminished by giving buffy coat poor blood, or in the case of platelet transfusion, by matching donor and recipient for HLA-A and HLA-B antigens.

Matching for organ transplantation is more complex than matching for blood transfusion. HLA-A and HLA-B are members of an antigenic complex that has some points of resemblance to the Rh system, although the alleles are more numerous and the genes not as tightly clustered. Unlike Rh, HLA-A, HLA-B, and HLA-C antigens are found on most tissues, as well as on white cells. They are not present on mature red cells. A second class of antigens, HLA-DR, is also coded for by structural genes of the complex, and these are important not only for cell–cell collaboration (Chap. 17), but are implicated in transplant rejection, especially of bone marrow and kidney. The HLA-DR antigens are not expressed on platelets and no significant amount of any HLA antigen is found on the mature red cell.

Summary

In this brief presentation, the major components of the immune system, the central and peripheral lymphoid organs, the cells, accessory cells, and their products, have been introduced. Mention has been made of the wonderfully intricate way in which the system is self regulated and of the grave consequences that result from imbalances in immune function. Some of the ways in which knowledge of the immune system has been used to prevent and correct a variety of diseases have been discussed. The stage is now set for a more complete understanding of the topics outlined here, and a projection for the future is given in the Overview (Chap. 22).

CHAPTER 12

Immunogens (Antigens) and Antibodies and their Determination

In the first part of this chapter, discussion will center on the types of molecules that initiate an immune response (immunogens) and the nature of those distinct sites (antigenic determinants) that are reactive with products of an immune response (e.g., antigen-specific T and B lymphocytes and humoral antibodies). The central sections of this chapter will cover the molecular structure and function of antibodies, their in vitro reactions with antigenic determinants, and their synthesis by B cells and plasma cells. The final section is devoted to outlines of how the determinations are performed.

General Properties of Immunogens

An immunogen is a substance that an exposed individual can specifically recognize and respond to. The same substance is called an antigen when the reaction with it of preformed products of an immune response (e.g., antibody) is being measured. Strong immunogens, such as bacteria and certain viruses, tend to be very complex, presenting an array of distinct macromolecular components, including proteins, glycoproteins, phospholipids, glycolipids, polysaccharides (carbohydrates), and nucleic acids. Many of the isolated proteins, glycoproteins, and large complex polysaccharides are strongly immunogenic, while the phospholipids, glycolipids, and smaller, less complex carbohydrates are not. Immune responses to nucleic acids are generally difficult to demonstrate, but antibodies are frequently present in the serum of patients and animals with autoimmune diseases, and experimentally they can be elicited to heat-denatured, single-stranded DNA and RNA.

One reason that large macromolecules are strong immunogens is that the immune system is capable of recognizing several distinct antigenic determinants on a single molecule (Fig. 12-1). There are also examples of immunogens that contain many repeating copies of the same antigenic determinant. For example, the flagellar organelles of *Salmonella adelaide* consist of replicates of a single polypeptide unit, or monomer. Each monomer expresses a single strong (immunodominant) antigenic determinant and several weak determinants, and the organelle represents a repeating array of these determinants. Macromolecules expressing several nonidentical antigenic determinants (Fig. 12-1) may also possess one or more immunodominant determinants. As a consequence of multiple recognition of distinct antigenic determinants, the immune response to macromolecular immunogens is necessarily complex, consisting of distinct subsets of lymphocytes and antibodies that interact with many nonidentical determinants on a single molecule.

One approach to designing suitable experimental systems for studying specific aspects of an immune response to a given antigenic determinant is to isolate the individual antigenic determinants from the macromolecular immunogen and to characterize them chemically and even

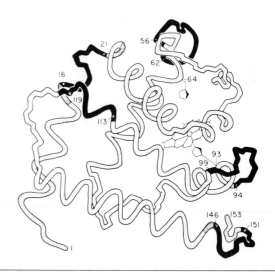

Figure 12-1. The antigenic structure of sperm whale myoglobin. The polypeptide backbone structure for sperm whale myoglobin as determined by Kendrew et al. (Nature 190:666, 1961) is shown schematically. Five distinct antigenic determinants (solid black areas) consisting of six to seven amino acid residues have been described by Atassi. Some rabbit antisera recognize a few additional amino acid residues (striped areas) in the first and final antigenic determinants. The stippled areas have not been shown to be antigenic. The numbers refer to amino acid residue positions. *(From Atassi: Immunochemistry 12:423, 1975.)*

to synthesize them. This approach has been used by Atassi and co-workers for the immunogens sperm whale myoglobin (Fig. 12-1) and eggwhite lysozyme, but it is a very time-consuming and laborious procedure. More often, experimental immunologists resort to designing a new antigenic determinant from a small, chemically characterized molecule (hapten) and chemically complexing it to a suitable macromolecular carrier (usually a protein or a synthetic polypeptide). The hapten, e.g., 2,4-dinitrophenol (DNP), is not immunogenic alone but functions as an immunodominant determinant when presented to the immune system as a hapten-carrier conjugate. The antibodies specifically produced to the haptenic determinant can be readily purified from immune serum (antiserum) by affinity chromatography or adsorption methods (immunoadsorption), utilizing hapten coupled to an inert matrix as a solid phase adsorbent. The hapten-carrier technique can be used to produce antibodies specific for a variety of small molecules, e.g., penicillin, certain steroids, hormones, lipids, and individual nucleic acid bases. The carrier macromolecule itself may or may not be immunogenic; the important feature is that the overall conjugate is immunogenic in the animal species and strain to be utilized.

A number of small molecular weight components of complex organisms, like glycolipids, which are nonimmunogenic alone, function as haptens when they are administered as part of an immunogenic complex. Thus the Forssman hapten, a glycolipid consisting of *N*-acetylgalactosamine-galactose-galactose-glucose joined to a ceramide lipid moiety, is nonimmunogenic in purified form. In its native plasma membrane environment it is a strong immunogen. Since it is a membrane component of wide and diverse tissue and species distribution (a heterophilic antigen), immunization of a Forssman-negative species with Forssman-positive tissue or cells results in the production of Forssman-specific antibodies.

Requirements for Immunogenicity

Foreignness
The immune system possesses the ability to distinguish between self and nonself. For example, rabbits do not usually respond to their own serum albumins, but they do respond to human serum albumins. It has been demonstrated that during ontogeny the immune system learns to distinguish self from nonself. In rare instances, components of sites in the body that are poorly drained by the lymphatic system or are anatomically sequestered from the immune system are treated as foreign if injury renders the sites accessible to immunocompetent cells later in life (for example, the lens protein of the eye and certain thyroid proteins).

Size and Shape
Molecular shape does not seem to influence immunogenicity. Proteins and polypeptides with globular, rod-like, and random coil configurations can be immunogenic. On the other hand, molecular size does affect immunogenicity; the smaller the molecule, the less immunogenic it is likely to be. For example, bovine and porcine insulin (mol wt about 6000) are poor immunogens, but if insulin is polymerized by chemical methods or is heat-aggregated, it may become immunogenic. Similarly, the H antigen of *S. adelaide* (the flagellar protein antigen described earlier and in Chaps. 3 and 37) can be prepared in various stages of polymerization: as intact organelles (flagella), as 38,500 dalton monomers (flagellin or MON), or as polymerized flagellin (POL) (prepared by aggregating monomers). In terms of immunogenicity, the following relationship holds:

Flagella > POL > MON

The minimum size for immunogenicity was thought to be about 1000 daltons from several experimental studies. Angiotensin II (1031 daltons), a native polypeptide hormone, and α-dinitrophenol-hepta-L-lysine (1200 daltons), a hapten-synthetic polypeptide conjugate, are immunogenic in guinea pigs under certain conditions. However, a synthetic compound of 450 daltons (*p*-azobenzene arsonate-*N*-acetyltyrosineamide) has been shown to be immunogenic in guinea pigs. Smaller compounds that are apparently immunogenic, like the urushiols from poison ivy (120 to 320 daltons) and picryl chloride (trinitrophenyl chloride), are not truly immunogenic by themselves but become so by virtue of their chemical reactivity with self skin proteins, thereby creating larger conjugates. The small catechol molecule is an example of a highly reactive hapten, and many skin proteins and membranes function as natural carriers. As a general rule, however, the number of antigenic determinants that a molecule expresses is directly related to its size; this may be a critical factor determining immunogenicity.

Complexity and Composition
Synthetic polypeptides have been effectively used to study the effect of complexity and composition on immunogenicity. Linear homopolymers, such as poly-L-lys or poly-L-glu, are usually not immunogenic alone. If they are used as conjugated haptens with a suitable protein carrier, specific antibody responses can be detected (Table 12-1). Linear random copolymers are immunogenic. For example, poly-L-$(glu_{60}lys_{40})$, where the subscripts denote the mole percent concentrations of initial reactants, is immunogenic in rabbits but not in humans, while poly-L-$(glu_{50}lys_{30}ala_{20})$ is immunogenic in both. Increasing complexity as well as host genetic factors contribute to a molecule's immunogenicity. The amino acid composition of the copolymers is also important in that the addition of aromatic amino acids, such as tyrosine, leads to dramatic increases in immunogenicity. Incorporation of amino acids, such as cysteine, lysine, alanine, and glutamic acid, leads to more modest increases in immunogenicity.

Branched copolymers, such as poly-L-(tyr,glu)-poly-DL-

TABLE 12-1. ESTIMATED SIZE OF ANTIGENIC DETERMINANTS

Immunogen	Test Component (Hapten) Showing Maximal Inhibition*	Size in Most Extended Form[†]
Dextran	Isomaltohexaose (6 units of α(1-6) linked D-glucose)	3.4 × 1.2 × 0.7 nm
Polyalanyl bovine serum albumin (polyalanine = hapten; BSA = carrier)	Pentaalanine	2.5 × 1.1 × 0.65 nm
Polylysyl rabbit serum albumin (polylysine = hapten; RSA = carrier)	Penta- (or hexa-) lysine	2.7 × 1.7 × 0.65 nm

Data from Kabat: J Immunol 97:1, 1966.
*Test haptens of increasing size were used. Maximal inhibition of the antibody reactions was observed with the compounds listed in the table. Haptens of greater size showed no further increase in inhibitory capacity.
[†]Estimated maximal size. The actual size may, in fact, be smaller.

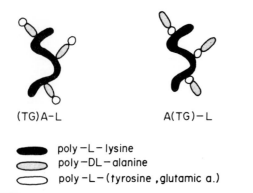

(TG)A-L   A(TG)-L

▬▬▬ poly – L – lysine
◗◗◗ poly – DL – alanine
◯◯◯ poly – L – (tyrosine , glutamic a.)

Figure 12-2. Representative structures for two branched copolymers. In both polymers a poly-L-lysine (L) backbone has been used. In one (T,G)A-L, a poly-DL-alanine (A) branch has been added to available ε-amino groups in the poly-L-lysine backbone, followed by a short copolymer of tyrosine and glutamic acid (T,G). In the other, A(TG)-L, the short copolymer (T,G) was added first, followed by a poly-DL-alanine homopolymer. Several model studies have shown that (T,G)A-L is immunogenic, whereas A(T,G)-L is not.

ala-poly-L-lys, or (T,G)A-L, are extremely potent immunogens (Fig. 12-2). The (T,G)A-L polymer has also been used to demonstrate that the immunodominant groups (T,G) must be readily accessible in order for recognition to take place. Similar branched polymers have been used extensively in studies elucidating genetic control over a number of immune responses (Chaps. 15 and 19).

Recognition by Two Lymphocyte Types

Certain immunogens require only stimulation of specific B cells in order for humoral antibodies to be formed (T cell-independent immunogens) (Chap. 15). More often, however, an immunogen will require specific recognition by both B and T cells. While the specific T cells do not produce any detectable antibody, they do function as helper cells in the overall specific activation of B cells (the precursor to the antibody-forming cell, or plasma cell). Additional details may be found in Chapter 15.

Other Parameters

Additional important factors contribute to a molecule's immunogenicity. First, as noted earlier, the immune response to some immunogens may show species dependence: for example, dextran (a glucose polymer) can be immunogenic in humans and mice but not in rabbits and guinea pigs. Second, within a species there are differences in reactivity between individuals. This is most easily demonstrated in inbred strains of mice or guinea pigs (Chaps. 15 and 17). Some guinea pigs are responders, and some are non-responders to the hapten-polypeptide conjugate α-dinitrophenol-poly-L-lys. Responders have a specific immune response (Ir) gene to the poly-L-lys carrier, and nonresponders lack this gene. Both responders and nonresponders are capable of responding to the hapten (DNP) on a different carrier. Inbred guinea pigs of strain 2 carry

the appropriate Ir gene, and all strain 2 animals respond. No member of a second inbred line, strain 13, responds. The F₁ hybrid between strains 2 and 13 does respond, so the capacity to respond is dominant. Third, the dose administered, the route of injection, and the timing between injections may influence the quantity and quality of the immune response an animal makes. Indeed, one can induce an immune paralysis or a tolerant state, as opposed to an immune response, by varying these conditions (Chap. 15). Finally, the use of adjuvants can result in increased immunogenicity for a previously demonstrated weak immunogen. Adjuvants include water-in-oil emulsions, such as Freund's complete adjuvant (mineral oil, water, and an emulsifying agent), finely particulate suspensions (alum precipitates), and silica or bentonite particles. Their mechanisms of action vary, but all adjuvants are thought to protect immunogens from rapid, nonspecific elimination and to give rise to prolonged responses by slow release of the trapped (or adsorbed) immunogen. Complete Freund's adjuvant contains, in addition to the emulsion, heat-killed mycobacteria; their inclusion results in an intense local inflammatory reaction and augmentation of the immune response.

Antigens and Antigenic Determinants

Antigenic determinants are the discrete sites on a molecule to which antibodies or immune cells are specifically directed. An immunogen generally expresses two or more distinct antigenic determinants.

Size of Antigenic Determinants

The size of an antigenic determinant can be determined by estimating the minimum size of a fragment of an immunogen that can combine with a specific antibody. The assay commonly employed is inhibition of precipitate formation between the intact immunogen and antibody by fragments of the original immunogen. For example, a dextran-antidextran precipitin reaction can be inhibited by glucose polymers, while polypeptide-antipolypeptide precipitation can be inhibited by oligopeptides. The results of a few such studies are summarized in Table 12-1. The extended size of antigenic determinants appears to be about 3.0 by 1.5 by 0.7 nm.

Conformation

It was previously stated that molecular conformation does not affect immunogenicity. This was a simplification to emphasize the concept that any shape of molecule can be immunogenic. However, the form of the molecule does affect the presentation of the antigen and, thus, the specificity of the antibody. For example, the antibodies elicited with the linear tripeptide copolymer poly-L-(tyr,ala,glu) do not interact with the tripeptide tyrosylalanylglutamic acid. Under physiologic conditions the tripeptide copolymer exists as an α-helix, and it is this conformation that is recognized by the immune system and not the linear sequence of tyr,ala,glu. Conversely, antibodies to the linear tripeptide unit (presented as a hapten-carrier complex) do not react with the α-helical copolymer. This system and other similar ones indicate that the immune system is capable of recognizing and responding to linear sequences and/or specific conformations of amino acids.

Accessibility

Antigenic determinants must be accessible to antibodies. For example, antibodies to the branched copolymer (T,G)A-L (Fig. 12-2) do not react with A(T,G)-L because these antibodies are directed largely at the T,G determinants, which are sterically blocked in A(T,G)-L by the linear alanine polymer. Also note that the antigenic determinants expressed by myoglobin (Fig. 12-1) occur on exposed surfaces.

Specificity and Cross-reactivity

A corollary of the immune system's ability to discriminate self from nonself is that both the recognition phase and the response phase are specific for the challenging immunogen. The antibodies elicited to a single antigenic determinant are complementary to that particular determinant. (The classic contributions of Landsteiner and his contemporaries demonstrated the specificity of antibodies produced to hapten-protein conjugates.) One often reproduced example of these exquisite studies is shown in Figure 12-3. The antibody was produced in rabbits to a metasulfonyl amino benzene hapten conjugated to horse serum proteins. A change in position of the sulfonyl group from meta to para or ortho in the test hapten resulted in complete or partial loss of antibody recognition. Further, slight and extremely subtle changes in the nature of the side group from sulfonyl to arsenyl or carboxyl also had a dramatic effect on antibody binding.

Cross-reactivity has two primary causes. The first is chemical similarity. This is illustrated by the example just given, in which antibody to the metasulfonyl aminobenzene hapten reacts, but less strongly (cross-reacts), with the ortho- and parasulfonyl haptens. This particular antibody also cross-reacts with meta-arsenyl or carboxyl substitutions of the amino benzene group. The second cause of cross-reactivity is the presence of the same antigenic determinant in two otherwise dissimilar antigens. For example, the red blood cell membranes of sheep and certain other animal species and the membranes of certain bacteria cross-react with Forssman antigen-specific antiserum because all contain Forssman antigen. Cross-reactivity of this type is often quite unpredictable unless the molecular structures are known.

Carrier-specific Determinants

The function of the protein carrier in conferring immunogenicity upon the coupled haptens has been studied intensively. The carrier moiety is not necessarily immuno-

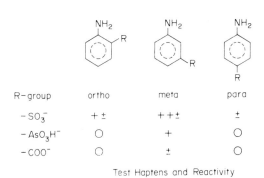

Test Haptens and Reactivity

Figure 12-3. Specificity of antibodies produced to the meta-aminobenzene sulfonate hapten. The hapten was coupled through an azo intermediate (via the available amino group) to a protein carrier. The resulting antibodies were tested with the homologous hapten and with several structural analogs coupled to different protein carriers. The extent of precipitation was scored 0, ±, +, +±, ++, in order of increasing reaction. (*Adapted from Landsteiner: The Specificity of Serological Reactions, 1962, p 169. New York, Dover Publications.*)

logically inert. Haptens like 2,4-dinitrophenol (DNP) have been conjugated to a wide variety of proteins, such as bovine serum albumin (BSA), bovine gamma globulin (BGG), ovalbumin (OVA), and keyhole limpet hemocyanin (KLH). Animals primed with DNP-BGG conjugates and challenged later with DNP-BGG respond vigorously with antibody specific for DNP (as well as components on the carrier molecule). If, instead, DNP-BGG-primed animals are challenged with DNP on a different carrier, like DNP-KLH, only a small DNP-specific-antibody response is elicited. The recognition of carrier-specific determinants in an immune response to a hapten is a reflection of cooperation at the cellular level. The carrier specificity resides in the T lymphocyte, and the specificity of the antibody produced to the hapten is a function of a B lymphocyte. Both are required for an effective immune response. Further details may be found in Chapter 15.

Isolation of Antigenic Determinants

Antigenic activity from complex biologically important immunogens can sometimes be recovered following controlled proteolytic digestion. A 6600 dalton fragment derived from human serum albumin contains a single antigenic determinant capable of binding a specific subpopulation of antibodies from an antiserum raised to the intact albumin. Much of the antibody raised against tobacco mosaic virus coat protein (TMVP) subunits is directed against a single oligopeptide in TMVP.

The polypeptide hormone glucagon, which is 29 amino acids in length, has been split into several fragments by trypsin digestion. Antibodies raised to intact glucagon (with adjuvant) in guinea pigs are directed primarily at the amino terminal end of the hormone, while cell-mediated immune recognition sites are confined largely to the carboxyl terminal end. Other high molecular weight proteins, like myoglobin (Fig. 12-1) and lysozyme, owe their immunogenicity to short, continuous, linear sequences of amino acids (myoglobin) or, more interestingly, to groups of amino acids that are far apart in linear sequence but lie in proximity on the surface of the molecule (lysozyme).

Since the antigenic determinants of myoglobin have been isolated, chemically characterized, and even synthesized, it has been possible to determine how some of the parameters discussed above contribute to immune recognition and response. For example, the immune response in mice to myoglobin is under genetic control. A given inbred strain of mouse will respond to a defined subset of the five possible determinants; a second inbred strain may respond to a different (often overlapping) subset. In addition, the conformation and exact amino acid composition of each isolated determinant is critical in recognition and response by antibody (B lymphocytes) and T lymphocytes. The determinants recognized by T lymphocytes may or may not be distinct. Clinical application of these techniques (preparation of vaccines) is directed at splitting immunogens into immunodominant fragments that may increase the immune response or give the specific response (cellular or antibody) desired. It may also be inferred that certain individuals will give more complete responses to a vaccine than will others.

Antibodies

General Properties of Immunoglobulins

Antibodies as Globular Proteins

Early experiments showed that animals challenged with killed organisms could withstand later challenge with live, virulent organisms or with toxins produced from these organisms. Serum from these animals could be transferred to normal animals and would protect them from challenge. To characterize the substances in serum having protective capacity (antibodies), several approaches were used. One was to sequentially precipitate proteins from serum with salts, especially with ammonium or sodium sulfate. High molecular weight proteins, such as fibrinogen, precipitated readily with no interference with the protective activity of the serum. Increasing the salt concentration precipitated both the serum globulins and the protective activity but left the major serum protein (albumin) in solution. A more sophisticated approach became possible when Tiselius found that different serum proteins moved at characteristic velocities in an electric field. Kabat and Tiselius then electrophoresed a potent precipitating antiserum and compared the patterns given by whole antiserum to serum that had been exposed to (absorbed by) antigen (Fig.

12-4). After absorption the slow moving γ-globulin peaks showed a marked depletion indicating that the precipitating antibody was predominantly of γ mobility. Other antibody activity was also found in the alpha and beta peaks. While such methods are no longer used, the designations of IgA and IgG derive from them. More modern methods have given greater precision to the definition of the immunoglobulins, have permitted their subdivision into subclasses (e.g., IgG1), and have identified additional classes of immunoglobulin (Table 12-2).

Immunoglobulin Classes

In humans, five major classes of immunoglobulins have now been described: IgG, the major serum component, IgM, or macroimmunoglobulin, IgA, the predominant immunoglobulin in extracellular secretions, IgD, a minor serum component but an important surface marker for a class of B lymphocytes (Chap. 15), and IgE, the immunoglobulin implicated in anaphylactic hypersensitivity (Chap. 21). The existence of IgG, IgM, IgA, and IgE has been documented in most other mammals; the existence of IgD in these mammals is accepted on the basis of its important functional role in human and mouse B lymphocytes. The physicochemical, biologic, and physiologic properties of the immunoglobulins are listed in Table 12-2.

Although Tiselius analysis could distinguish between some immunoglobulins, even an isolated class exhibits electrophoretic heterogeneity. For example, IgG purified from immune serum shows a broad range of mobilities when separated by ammonium sulfate precipitation and ion exchange chromatography and then subjected to electrophoresis (Table 12-2). This electrophoretic heterogeneity is related to further subclassification of IgG and to distinct biologic activities. Human serum IgG comprises a population of four subclasses (Table 12-3) and, within

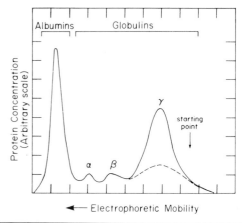

Figure 12-4. Free-zone electrophoresis of rabbit serum following hyperimmunization with ovalbumin. Data were obtained before (——) and after (---) absorption of the serum with the immunizing antigen. Albumin is the most prominent and fastest moving serum protein. Subsequent serum protein peaks (α, β, and γ) were named for their characteristic electrophoretic mobilities. *(Adapted from Tiselius and Kabat: J Exp Med 69:119, 1939.)*

TABLE 12-2. PHYSICAL, PHYSIOLOGIC, AND BIOLOGIC PROPERTIES OF THE HUMAN IMMUNOGLOBULINS

	IgG	IgA	IgM	IgD	IgE
Physical properties					
Sedimentation coefficient, $S_{20\omega}$	6.8–7.0	6.6–14.0	18.0–19.0	7.0	7.9
Molecular weight, daltons	143,000–160,000	159,000–447,000	900,000	177,000–185,000	188,000–200,000
Electrophoretic mobility*	γ^2–α^1	γ^2–β^2	γ^1–β^1	γ^1	γ^1
Carbohydrate content, %	2.3	7.5	7–11	13	11–12
Number of 4-chain units per molecule	1	1–3	5	1	1
Heavy chains	γ	α	μ	δ	ϵ
Light chains	κ or λ	κ or λ	κ or λ	κ or λ	κ or λ
Heavy chain allotypes	Gm	Am	—	—	—
Light chain allotypes	Km(κ)	Km(κ)	Km(κ)	Km(κ)	Km(κ)
Heavy chain subclasses	γ1, γ2, γ3, γ4	α1, α2	—	—	—
Heavy chain molecular weight, daltons	50,000–55,000	62,000	65,000	70,000	75,000
Light chain molecular weight, daltons	23,000	23,000	23,000	23,000	23,000
Physiologic properties					
Normal adult serum concentration, mg/ml	8–16	1.4–4.0	0.4–2.0	0.03	ng amounts
Percent total immunoglobulin	80	13	6·	1	0.002
Synthetic rate, mg/kg/d	26	27	5.7	0.4	0.003
Functional catabolic rate in serum, %/day	6	25	15	37	70–90
(or half-life, day)	(23)	(6)	(5)	(3)	(<3)
Intravascular distribution, %	48	41	76	75	51
Biologic properties					
Agglutinating capacity	±	++	++++	—	—
Complement-fixing capacity	+	—	++++	—	—
Anaphylactic hypersensitivity (homologous)	—	—	—	—	++++
Guinea pig anaphylaxis (heterologous skin)	+	—	—	—	—
Fixation to homologous mast cells	—	—	—	—	++++
Placental transport to fetus	+	—	—	—	—
Rheumatoid factor-binding activity	+	—	—	—	—
Tumor-blocking activity	+	?	±	?	?
Present in external secretions	+	++++	±	—	++

*Various classes of immunoglobulins exhibit from α^1 to γ^2 electrophoretic mobilities (see Fig. 12-4 and Fig. 12-24).

each subclass, individual antibodies specific for unique antigenic determinants. Subclasses have also been described for IgA (Table 12-2). The subdivisions are based upon slight amino acid composition differences in the heavy chains that can often be distinguished serologically.

The next important breakthrough with respect to understanding immunoglobulin structure was the discovery of the significance of serum paraproteins.

Paraproteins

Neoplastic and proliferative diseases of antibody-forming cells (plasma cell tumors, myeloma) produce an excess quantity of a serum immunoglobulin (paraprotein). The serum paraprotein from a single patient is homogeneous upon electrophoresis, implying a single component. The plasma cell tumor in a myeloma patient produces both the characteristic serum protein (paraprotein) and, often, a low molecular weight protein, excreted in urine and called Bence Jones protein. Myeloma paraproteins of every immunoglobulin class (IgG, IgA, IgD, IgE) and subclass have been described in the literature. In Waldenstroms' macroglobulinemia the plasma cell tumors produce an excess of an IgM serum protein. No two paraproteins of the same class and subclass from different patients have ever been shown to be completely identical, and, with few exceptions, a paraprotein shows none of the electrophoretic heterogeneity of conventional immunoglobulins. This electro-

TABLE 12-3. PROPERTIES ASSOCIATED WITH SUBCLASSES OF HUMAN IgG

Combined Properties of IgG Subclasses	IgG1	IgG2	IgG3	IgG4
Approximate occurrence in total serum IgG, %	65–75	15–23	7	3
Synthetic rate, mg/kg/d in serum	25	?	3.4	?
Fractional catabolic rate, %/d, in serum (half-life, d)	8 (23)	6.9 (23)	16.8 (7)	6.9 (23)
Intravascular distribution, %	51	53	64	54
Allotypic markers (Gm types)	1,2,3,4,17	23	5,6,10,11,13,14,21	?
Complement-fixing capacity	+	+	+	−
Heterologous skin-binding capacity	+	−	+	+
Placental transport to fetus	+	±	+	+
Polypeptide structure and location of interchain disulfide bonds				

KEY:

___ = Light chain

_____ = Heavy chain (2× the length of the light chain)

| or / = Interchain disulfide bonds

---

phoretic homogeneity led to the supposition that a single cell had become neoplastic and had formed a clone of cells all producing an identical immunoglobulin with a single amino acid sequence.

The paraproteins, because of their high concentrations, are readily purified in quantity from patient sera. They have been important in elucidating both structural and functional attributes of immunoglobulins (Tables 12-2 and 12-3). Antisera to myeloma proteins produced in goats or rabbits can distinguish the various classes and subclasses of human immunoglobulins.

Plasmacytomas may be experimentally produced in two inbred strains of mice, BALB/c and NZB. Intraperitoneal injections of mineral oil frequently induce the formation of these tumors. As in the human disease, mouse myelomas produce a characteristic serum paraprotein and may produce a urinary Bence Jones protein. These tumors may be passaged in mice of the same inbred strain or may be adapted to tissue culture. In addition, Köhler and Milstein and Sharff and co-workers have described methods by which cultured murine myeloma tumor cell lines (producing an immunoglobulin of unknown specificity) can be fused with splenic lymphocytes from mice immunized with a defined antigen. They further described techniques whereby hybrid cell lines (hybridomas) can be selected that secrete a monoclonal immunoglobulin with distinct antigenic specificity (determined by the immune spleen cell population) while retaining the growth characteristics of the parent tumor cell line. Essentially, an antibody with any desired antigen specificity can be immortalized.

Characterization of Antibodies

Most of the early physical–chemical studies were done with heterogeneous IgG purified from rabbit, horse, and human sera. Several important structural features of IgG were predicted from these studies long before amino acid sequences and x-ray diffraction studies were reported. For example, the molecular weight of IgG was calculated from sedimentation and diffusion studies, the molecule was predicted to be asymmetrical from its viscosity parameters, and its unique susceptibility to proteolytic agents suggested that the molecule consisted of enzyme-resistant globular domains covalently linked by enzyme-sensitive regions. Hapten-binding studies with hapten-specific antibodies demonstrated the existence of two antigen-binding sites per IgG molecule. The existence of a third domain was suggested for the binding of complement components once hapten or antigen had bound. Thus, these early studies predicted a minimum of three functional and structural domains. This concept has been abundantly verified.

Polypeptide Structure of IgG

Classic studies of Edelman and his co-workers showed by zone electrophoresis under denaturing and reducing conditions that human and rabbit IgG consisted of two types of polypeptide chains. Porter and co-workers also demonstrated that rabbit IgG consists of two sizes of polypeptide chains by gel filtration under similar denaturing and reducing conditions (Fig. 12-5). The heavy (H) chains

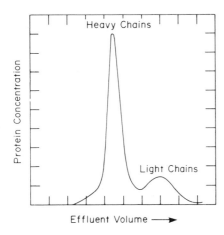

Figure 12-5. Separation of the heavy (H) and light (L) chains of rabbit IgG by gel filtration column chromatography in the presence of 1 N acetic acid (denaturing conditions). *(Adapted from Fleischman et al.: Arch Biochem Biophys [Suppl 1]:174, 1962.)*

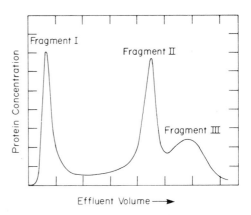

Figure 12-6. Separation of peptides following digestion of rabbit IgG with the enzyme, papain. Separation was achieved by chromatography over the cation-exchange resin, carboxymethyl cellulose, and elution with an increasing ionic strength gradient. The peaks were labeled I, II, and III, in order of their appearance. *(Adapted from Porter: Biochem J 73:119, 1959.)*

(50,000 to 53,000 daltons) contain a small amount of covalently coupled carbohydrate; the light (L) chains (22,000 to 25,000 daltons) generally lack carbohydrate. From the mole/mole and wt/wt ratios and the molecular weight of the intact IgG, a 4-chain structure (2H plus 2L) was postulated.

Porter and co-workers digested rabbit IgG with the proteolytic enzyme papain and chromatographed the resulting subunits on cation-exchange columns (Fig. 12-6). The peaks I and II (Fab fragments) contained univalent binding sites for hapten or antigen, while peak III contained no hapten-binding site and was readily crystallizable (Fc fragment) in low ionic strength buffer. The weakly bound Fab fragments (peak I) were subsequently shown to derive from more anionic immunoglobulins. The intact IgG was characterized by a sedimentation coefficient of 7S, while both the Fab and Fc sedimented with 3.5S.

Digestion of rabbit IgG with pepsin resulted in a 5S component that precipitated with antigen and was, therefore, a bivalent fragment F(ab')$_2$. The analogous Fc' fragment consists of a small pFc' subunit and several small peptides. If reducing agents, such as β-mercaptoethanol, were added to the 5S fragment, disulfide bonds were broken and two Fab' fragments (3.5S) resulted. Extensive reduction of Fab' or Fab with β-mercaptoethanol yielded an Fd' or Fd fragment, respectively, and free light chain. Taken together, Edelman and Porter's contributions resulted in the proposed structure for rabbit IgG shown in Figure 12-7. Subsequent studies with the different human immunoglobulin classes and subclasses have shown that they all possess the same fundamental 4-chain structure. The differences reside in the number of 4-chain units per immunoglobulin molecule and the additional appearance of accessory polypeptides. For example, IgM and often IgA (Fig. 12-8) contain a j chain (15,000 daltons) that may stabilize the polymeric structure of these immunoglobulins with disulfide bonds. IgA (and sometimes IgM) can also

contain an associated T piece (transport piece, or secretory component) (60,000 daltons) that appears to enhance their secretion into extracellular fluids. A summary of the different structures is illustrated in Figure 12-8, and differences in their physiologic and biologic properties are summarized in Table 12-2.

Interchain and Intrachain Disulfide Bonds

From Figures 12-7 and 12-8 and Table 12-3, it is clear that the various subclasses of IgG contain several interchain HL and HH disulfide bonds. These bonds play a role in stabilizing the structure, but they are not essential, since reduction of the disulfide bonds in the absence of denaturing agents does not dissociate the IgG molecule. In addition, some IgA molecules do not possess HL disulfide bonds.

The intrachain disulfide bonds are critical for conferring the domain structure upon immunoglobulin molecules. If the polypeptide chains are stretched out and the positions of cysteine residues involved in intrachain disulfide bonds are marked, one can see that by forming the disulfide bonds, loops of some 60 to 80 amino acids are generated (Fig. 12-9). For human IgG (440 amino acids in length), each 110 amino acid segment contains a disulfide-bonded loop of 60 to 80 amino acids. There are two loop sections per light chain and four loop sections per heavy chain. The disulfide-binding pattern, critical amino acid residues, and overall lengths for various paraproteins are also shown in Figure 12-9.

The differences in the molecular weights of the major immunoglobulin classes are due to differences in the H chains: The H chains for IgM (μ) and IgE (ϵ) are longer than the γ chain by about 110 amino acids and contain one additional loop segment. The L chains κ or λ are shared by all classes of heavy chains (γ, μ, α, δ, ϵ).

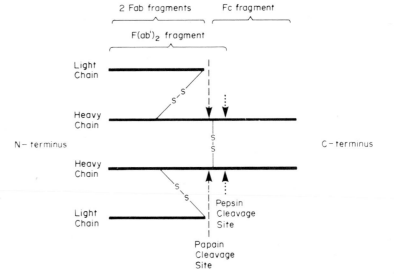

Figure 12-7. Proposed structure for rabbit IgG showing the 4-chain basic structure, important disulfide bonds, and probable sites of enzymatic cleavage by papain or pepsin. The enzyme papain cleaves the heavy chains on the amino terminal side of the single interheavy chain disulfide bond, while multiple pepsin cleavage sites are to the carboxyl terminal side of this disulfide bond. Reduction of the light-heavy chain disulfide bonds in the presence of denaturants results in the separation of light chains and heavy chains or heavy chain fragments.

Peptide Mapping and Sequence Studies

Bence Jones proteins are now known to be homogeneous L chains and represent a catabolic product of the characteristic serum paraprotein. They are chemically identical from a given individual, and are, therefore, ideal for chemical structure determination. They can be readily purified, characterized by tryptic peptide mapping, and sequenced. When several different L chains had been characterized in this manner, it became apparent that light chains possessed two distinct regions, a variable (V) amino (N) terminal end (no two Bence Jones proteins have identical V regions), and a constant, carboxyl (C) terminal end (with only two major variants, κ or λ). The constant regions (C$_κ$) of all κ Bence Jones proteins possess 95 to 100 percent sequence homology in residues 111 to 214 and little homology with C$_λ$ constant regions. There are also two allelic forms of human kappa chains: some possess a leucine (Km 1,2) at position 191, while others contain a valine at this

position (Km 3). This genetic marker, or allotype, which can also be detected by specific antisera, follows classic mendelian inheritance patterns. Individuals and families can by typed with respect to the Km marker. In a heterozygote (Km 1,2/Km 3) a single immunoglobulin molecule contains a single allotype and never both. No allotypic markers have been demonstrated for human λ chains.

The κ chains terminate in Cys at residue 214, and λ chains terminate in Ser at residue 215, with a penultimate Cys residue. These cysteine residues are involved in HL interchain disulfide bonds. Amino acid sequence studies of many Bence Jones proteins have shown the N-terminal half of the molecule (V region) to be variable (in selected areas) and have implicated this region as the antigen-binding site, the variability reflecting individual antibody specificities. However, the first 18 to 20 amino acids of the N-terminal end exhibit less heterogeneity, and based upon these sequences, kappa chains can be subtyped into at

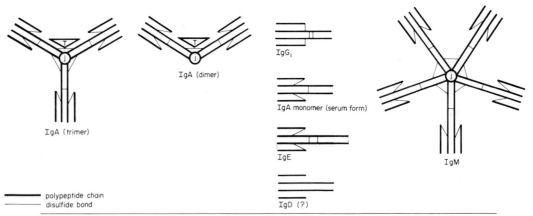

Figure 12-8. Representative structures for the major human immunoglobulin classes. Each contains the basic 4-chain polypeptide unit, and some may exist as polymers (e.g., IgM and IgA). The accessory polypeptides j and T may also be present in the polymeric forms. Differences in the IgG subclasses are noted in Table 12-3.

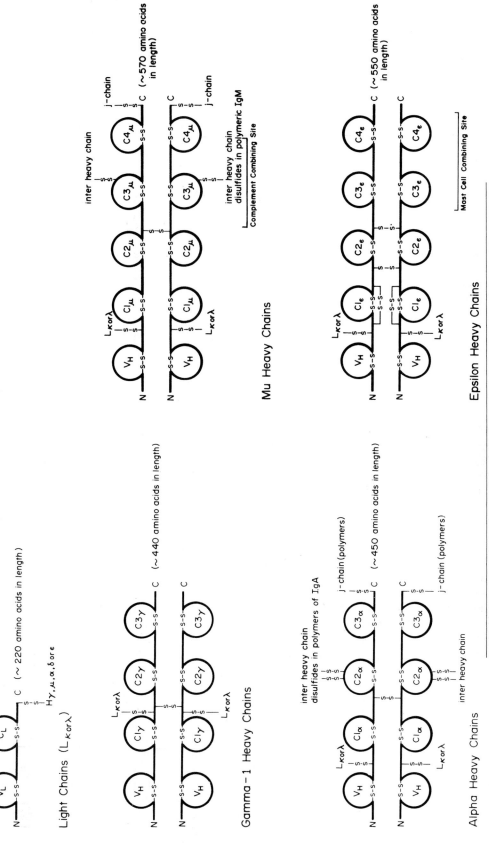

Figure 12-9. Detailed description of the light and heavy chains for various human immunoglobulins. A single schematic representation of a light chain is shown, illustrating variable (V_L) and constant (C_L) region domains of 60 to 80 amino acids each and the position of the light-to-heavy chain disulfide bond. Heavy chain dimers are illustrated for $\gamma 1$, α, μ, and ϵ chains, respectively. The variable and constant region domains in each are marked, and positions of the light-to-heavy disulfide bonds and various other stabilizing bonds are shown. The carboxyl terminal disulfide bonds in α and μ chains may be covalently linked to j chain in the polymeric forms. Physiologic properties of the immunoglobulins (Tables 12-2 and 12-3) are largely controlled by the constant region domains. *(For further details see Winkelhake: Immunochemistry 15:695, 1978.)*

Light Chains ($L_{\kappa \text{ or } \lambda}$)

Gamma–1 Heavy Chains

Alpha Heavy Chains

Mu Heavy Chains

Epsilon Heavy Chains

AMINO ACID RESIDUE POSITION FROM AMINO TERMINAL END

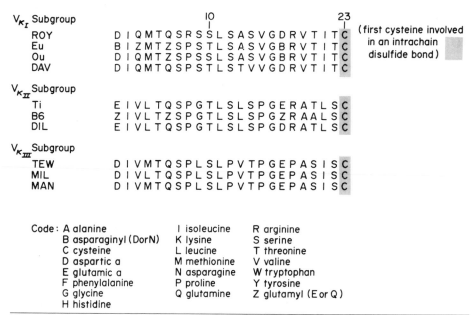

Figure 12-10. The subtyping of human κ chains upon amino acid sequence analyses. Five groups of κ chains can be distinguished in this manner by examination of the amino acids in positions 9 and 13. Three are illustrated here. For example, the V$_{\kappa II}$ subgroup is characterized by glycine at position 9 and leucine at 13. *(Data and the one-letter amino acid code adapted from Hood and Prahl: Adv Immunol 14: 291, 1971.)*

least three to five groups (e.g., κ$_I$, κ$_{II}$, κ$_{III}$) (Fig. 12-10), and the lambda light chains can be subtyped into five groups (λ$_I$, λ$_{II}$, λ$_{III}$, λ$_{IV}$, λ$_V$). Therefore, sequencing an unknown Bence Jones protein through the first 18 to 20 amino acids is often sufficient for typing and subtyping, since V$_\kappa$ is always found with C$_\kappa$ (and V$_\lambda$ with C$_\lambda$). Other constant or conserved residues that fall within the confines of the variable region include the cysteine residues responsible for the loop structure, which seems to be structurally important.

When the H chains from homogeneous myeloma proteins were similarly characterized, an analogous structural relationship became apparent (Fig. 12-9). In all cases the N-terminal 110 amino acid segment was found to contain variable sequences (V$_H$), and the remaining segments except for a few changes that account for heavy chain allotypic markers (*Gm* markers) were found to be constant (C$_H$).

TABLE 12-4. RELATED SEQUENCE CHANGES IN THE *Gm1* (OR α) ALLOTYPIC MARKER IN HUMAN IgG1 HEAVY CHAINS

Gm Type	Amino Acid Residues Involved	Residue Positions
1+	-Asn-Glu-Leu-	356-357-358
1−	-Glu-Glu-Met-	

Data adapted from Waxdal et al.; Biochemistry 7:1967, 1968; Frangione et al.; Nature 221:145, 1969; Burton and Deutsch: Immunochemistry 7:145, 1970.

For example, all IgG1 myeloma H chains (γ1) share a largely identical amino acid sequence from residue 111 to the C-terminal end. Like the *Km* allotypes of κ light chains, *Gm* genetic differences can be distinguished serologically. IgG1 heavy chain constant regions show very distinct amino acid sequence differences from the H chains of other IgG subclasses (γ2, γ3, γ4). These, in turn, have their own unique constant regions and allotypic markers (Table 12-3). The *Gm* allotype markers can, in some cases, be defined by sequence studies, and in the case of *Gm1* (or *a*) of IgG1 heavy chains this amounts to amino acid changes in two residues (Table 12-4). Allotypic markers on other classes of immunoglobulins have only been described for IgA$_2$. *Am2*$^+$ has no HL disulfide bonds, and *Am2*$^-$ has the usual HL disulfide bonds.

The N-terminal 18 to 20 amino acids in the variable regions of the H chains (γ, μ, α, δ, and ε) of several human paraproteins have been compared, and an interesting relationship has been found. By analogy with the light chains, the N-terminal 18 to 20 amino acids fall into four subtypes (V$_{HI}$, V$_{HII}$, V$_{HIII}$, and V$_{HIV}$), but in contrast to the class (κ or λ) association for L chains, the H chain subtypes appear to be shared by all H chain classes and subclasses. In addition, in one case, an individual produced both an IgG and an IgM paraprotein that shared identical amino acid sequences in the variable regions of both γ and μ chains; the light chains (κ) in both paraproteins were identical. Additional reports have appeared that document the appearance of three monoclonal paraproteins in a single patient (IgM, IgG, IgA). Preliminary evidence suggests that they may share an identical V region as well. These observations are frequently cited as evidence for separate genes controlling the synthesis of variable and constant regions of immunoglobulins. In the first example, the identical variable gene product appears to be associated with two distinct types of constant region gene products (C_γ and C_μ).

Hydrodynamic studies suggest that the Fab fragments of rabbit and human immunoglobulins are linked to the Fc domain through a flexible polypeptide region (the hinge region). Chemical studies have confirmed this model. The hinge region contains two relatively inflexible proline-rich sequences connected by a highly flexible glycine-rich region (Fig. 12-11). This primary structure is entirely compatible with the flexible hinge concept.

Control of the effector activities of immunoglobulin molecules appears to center exclusively in the Fc regions. For example, fixation of C1q, the first component of complement (Chap. 13), depends upon the availability of defined Fc subunit sites in IgM, IgG1, IgG2, and IgG3. Indeed, with the availability of several enzyme-produced fragments of the Fc region (like Fc′, pFc′ and TpFc) and available sequence data for these regions, it has been possible to assign the C1q recognition site to the C_{H2} domain of IgG and the C_{H3}-C_{H4} domain of IgM. A variety of effector cells contain Fc receptors for IgG (lymphocytes, macrophage, PMNs and eosinophils), IgM (lymphocytes), and IgE (mast cells and basophils). Finally, certain physiologic properties of immunoglobulin molecules, including biologic half-life, placental transport, seromucosal secretion, and intestinal absorption are, in large part, dictated by the nature of the Fc region.

The Antibody-binding Site

Structural studies have centered on the N-terminal 110 amino acids that comprise the variable region in both L and H chains. Since the information for antigenic determinant complementarity must be built into the primary sequence of each chain, and a given antigen initiates the biosynthesis of antibody molecules having complementary receptor sites, these variable regions have all the requirements that one would impose on hypothetical binding sites.

Several lines of evidence indicate that the variable regions are directly involved in binding. Rabbit antisera directed at the binding sites of hapten-specific antibodies produced in other rabbits (anti-idiotype antibodies) have been highly informative. Donor and recipient animals must be matched for rabbit immunoglobulin allotypes, so that the specificity detected can only be ascribed to differences in the variable regions (idiotypes). Anti-idiotypic antibodies may block the hapten-specific binding of the donor antibody, and anti-idiotypic immunoglobulins only react with intact donor immunoglobulin or the isolated Fab or F(ab′)₂ fragments. Therefore, the idiotypic specificity must reside in the region of the binding site and, most likely, in the variable regions of the H and L chains.

It has been stated earlier that proteolytic fragments Fab and F(ab′)₂ contain antigen-specific binding sites. More extensive pepsin digestion of the Fab fragments from a mouse IgA myeloma protein (with significant DNP-lysine specificity) yielded an F_v fragment containing only the V_H region of the heavy chain and the V_L region of the light chain. The F_v fragment was also capable of binding DNP-lysine.

Further evidence that the variable region comprises the antigen-binding site has resulted from the studies of Wu and Kabat and of Capra and Kehoe. The amino acid sequences of variable regions of both paraprotein L and H chains were analyzed for the extent of variation in each residue. Wu and Kabat studied the variability in human κ chains irrespective of their subtype ($V_{\kappa I}$, $V_{\kappa II}$, $V_{\kappa III}$) and described three hypervariable regions (residues 24 to 34, 50 to 56, and 89 to 97) (Fig. 12-12). Certain residues were invariant and, therefore, presumably important structurally (for example, Cys 23 and 88, Try 35, Gly 99, and Gly 101). By comparing the sequences of only $V_{\kappa I}$ light chains, the first hypervariable region could be confined to residues 30 to 32. The hypervariable regions are believed to

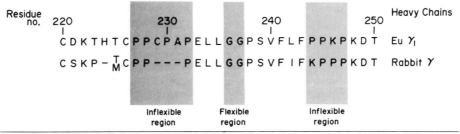

Figure 12-11. Structural analysis of the immunoglobulin hinge region. Partial sequences in the region of the proposed hinge region are shown for rabbit IgG and the human IgG1 paraprotein Eu. The hinge regions for the two immunoglobulins are not equal in length; residues were matched for best fit, and the dashes represent the insertion of artificial gaps in the rabbit γ chain. Note that relatively stiff, proline-rich, conserved sequences are found to either side of a flexible, double-glycine region. The one-letter amino acid code of Figure 12-10 has been used. *(Adapted from Day: Advanced Immunochemistry, 1972. Baltimore, Williams & Wilkins.)*

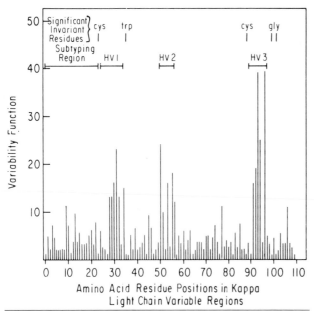

Figure 12-12. An analysis of human κ chain variable region sequences. Variability in a given residue position is defined as the number of different amino acids found in that position divided by the frequency of the most common amino acid at that position. Variability can theoretically range from a ratio of 1.0 (conserved residue) to 400 (maximal variation). Illustrated at the top of the figure are the subtyping region, the three hypervariable (HV) regions, and important invariant residues. *(Adapted from Wu and Kabat: J Exp Med 132:211, 1979.)*

be direct contact residues with antigen, while areas of minimal variability in the V region are necessary for a certain basic tertiary structure and for intersubunit interactions (such as V_L and V_H contact.)

The H chain variable regions have been similarly studied by Capra and Kehoe. Regions of hypervariability were assigned to four regions (residues 31 to 35, 50 to 56, 81 to 85, and 95 to 102). Again the Cys residues 22 and 92, responsible for the loop, were invariant, as were Trp 36, Arg 38, Pro 41, Gly 42, Trp 47, Ala 88, Tyr 90, Gly 106, Val 111, Ser 112.

Several of the murine paraproteins from BALB/c and NZB mice have been extensively studied by sequence analysis throughout the light and heavy chain variable regions. These studies suggest that not only is the variation of amino acids in the hypervariable regions important in determining antibody specificity, but the size of hypervariable regions may vary. This has the effect of changing the structural shape of the antibody-binding site to form pockets of varying depths and, indeed, long crevices.

Largely due to the work of Singer and co-workers, important residues in the region of the binding site were defined by the affinity labeling technique. In this technique, antibody with affinity for a hapten is reacted with an analog of the hapten containing a chemically reactive group. As a result of recognition and binding, a covalent bond may be formed between the modified hapten and an amino acid at or near the binding site (usually Tyr or Lys).

The reactive residue can then be identified by limited sequence analysis of tryptic digests. These studies have implicated Tyr residues at positions 24 to 34, 50 to 75, and 90 to 114 and Lys residues at positions 54 of H chains as being important for binding. Significant labeling of Try 33, 34, 86 and Lys 54 was noted for light chains.

In 1974 the three-dimensional structure for the paraprotein IgG New, which has affinity for the γ-hydroxyl derivative of vitamin K, was reported by Amzel and co-workers. Crystals of the Fab fragments were prepared, and by a combination of amino acid sequencing and x-ray crystallography, the structure was determined at 0.35 nm resolution. X-ray crystallography of the Fab–vitamin K complex is shown in Figure 12-13. The results of these studies are consistent with all the direct and indirect evidence that has accumulated concerning the nature of antibody-binding sites. The size of the site (1.6 by 0.7 by 0.6 nm) has been largely predicted. The positions or roles of the H and L hypervariable regions in the site are now obvious, and the positions of potentially reactive Tyr and Lys residues are also apparent.

Summary of Antibody Structure

Antibody molecules in man are comprised of a basic 4-chain unit of 2L and 2H chains. The class and subclass of immunoglobulin are defined by the H chain (γ1, γ2, γ3, γ4, α1, α2, μ, δ, or ε). All immunoglobulins share either κ or λ light chains. Polymeric forms of IgM contain five basic units linked by inter H chain disulfide bonds and a cysteine-rich j chain. Polymeric forms of IgA (two or more basic units) also contain a j chain and may contain an additional secretory component (T) if found in extracellular secretions. The domain structure of immunoglobulins is intimately related to its function. There is an N-terminal variable region domain responsible for antigen binding, and there are specialized constant region domains implicated in the binding of complement components and important for recognition by certain types of cells. Antibody function is thought to be mediated through recognition of antigen and subsequent activation of an auxiliary system (e.g., complement). A complete x-ray crystallographic analysis for the human paraprotein Dob is illustrated in Figure 12-14, where many of the features discussed above can best be appreciated.

Antigen–Antibody Reactions

Thermodynamics

Theoretically, if an antigen–antibody reaction is truly reversible, it should obey the law of mass action:

(1) $Ag + Ab \rightleftharpoons AgAb$ complex

It follows that at equilibrium, the following relationship should apply:

(2) $$K = \frac{[AgAb]}{[Ag][Ab]}$$

where the values in brackets reflect the molar concentrations of reactants and product at equilibrium. Once the equilibrium constant has been determined, the change in the standard free energy, $-\Delta G°$, can be calculated (the free energy of a reaction taking place under standard conditions of 1M concentrations of reactants):

$$(3) \qquad -\Delta G° = RT\ln K$$

where R is the universal gas constant (1.986 times 10^{-3} kcal/K/mole), and T is the absolute temperature (in degrees Kelvin). If the equilibrium constant K is measured at two or more temperatures, the enthalpy, $\Delta H°$, can be calculated:

$$(4) \qquad \frac{d(\ln K)}{dT} = \frac{\Delta H°}{RT^2}$$

Alternatively, the $\Delta H°$ can be measured directly by microcalorimetry. The entropy change, $\Delta S°$, can be calculated with the measured values of $-\Delta G°$ and $\Delta H°$.

$$(5) \qquad -\Delta G° = -\Delta H° + T\Delta S°$$

Since antigens may be multideterminant and antibodies are at least bivalent, it is difficult to obtain quantitative data for most antigen–antibody reactions. For example, if the antigen–antibody complex precipitates, the reactants and product are not in a true state of equilibrium. However, if the reactions are carried out in extreme antigen excess, where the complexes are soluble, valid data can be obtained.

Hapten–Antibody Reactions

It is relatively easy to treat hapten–antibody reactions quantitatively when the hapten is univalent and the antibody is at least bivalent (polymeric forms of immunoglobulin possessing additional binding sites, e.g., IgA and IgM). The reaction of antibody site (S) with free hapten (H) can be written as follows:

$$(6) \qquad H + S \underset{k_r}{\overset{k_f}{\rightleftharpoons}} HS$$

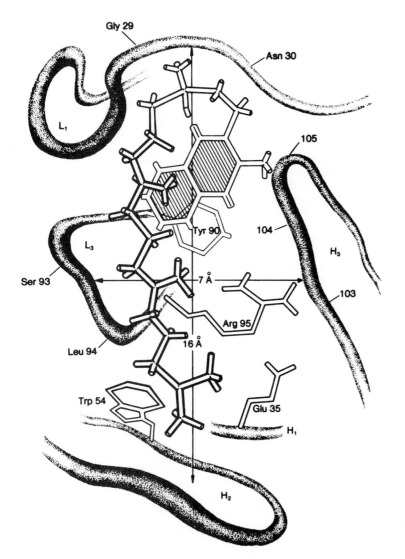

Figure 12-13. The antibody-combining site of the human IgG1 paraprotein New as determined by x-ray diffraction studies. The position of the hapten, a derivative of vitamin K, has also been determined. Two of the light chain hypervariable regions (L_1, L_3) and three of the heavy chain hypervariable regions (H_1, H_2, H_3) can be seen in this representation. The aromatic rings of vitamin K_1 (striped centers) are in proximity to light chain tyrosine residue 90. The aliphatic chain of vitamin K_1 extends from one side of the antibody binding site (L_1, Gly 29) to the other (H_2, Trp 54). *(From Amzel et al.: Proc Natl Acad Sci USA 71:1427, 1974.)*

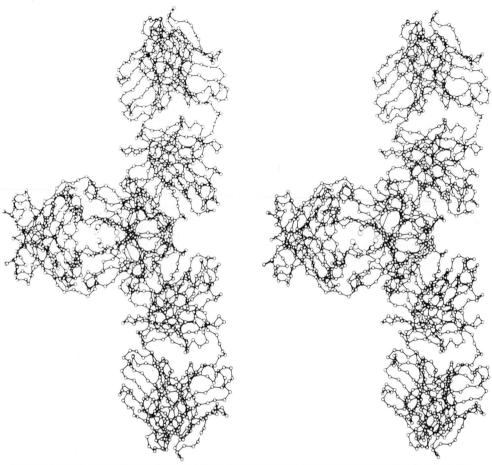

Figure 12-14. Stereoview of the three-dimensional structure of a human IgG1 (κ) protein Dob. The small circles represent the α-carbon atoms of the polypeptide backbones, and the large circles in the Fc region of the molecule denote carbohydrate units. The Fc region is oriented to the left of center, while the Fab regions are at the top and bottom of the figure. The V regions consist of two domains (H and L), showing β-pleated sheet structures in two dimension; they are connected by open stretches of linear polypeptide to the C_L- C_{HI} domains. The Fc domains show a prominent central groove in which much of the covalently coupled carbohydrate resides. The antibody-binding sites lie at the very top and bottom ends of the molecule. (Many individuals can visualize three-dimensional stereograms by staring at the two figures. A third image will appear in the center field in three dimensions.) *(From Silverton et al.: Proc Natl Acad Sci USA 74:5140, 1977.)*

where the equilibrium constant, K, is defined by the ratio of the forward to reverse rate constants ($K = k_f/k_r$). Although the rate constants can, in some cases, be measured, more often K is measured directly by equilibrium dialysis. In principle, the technique involves the dialysis of a mixture of free hapten and specific antibody, where only the free hapten can pass easily through dialysis tubing. By knowing the initial concentration of hapten and the concentration of free hapten at equilibrium, one can calculate the amount of bound hapten, the association constant, K, and the valence of the antibody (the number of binding sites per molecule of antibody). If:

r = moles of hapten-bound/total moles of antibody
    (calculated)
c = moles of free hapten at equilibrium (measured)

n = valence of antibody (or number of sites/antibody
    molecule) (calculated)
K = equilibrium constant (or affinity constant) (calculated)

the mass action formula (equation 6) can be rewritten and rearranged according to the method of Scatchard, where:

$$(7) \qquad\qquad r/c = nK - rK$$

By plotting r/c against r, K can be obtained from the slope and n from the intercept on the abscissa. In practice, several different concentrations of hapten are dialyzed in individual experiments with a fixed concentration of antibody to generate the points in Figure 12-15. Curvature in this type of data presentation reflects heterogeneity in the binding affinities of a population of antibodies in the serum. Antisera generally contain antibodies of varying affin-

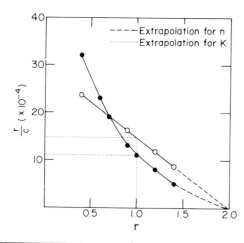

Figure 12-15. Scatchard plots for two hypothetical hapten–antibody systems. The linear relationship reflects a homogeneous antibody response to the haptenic determinant. The nonlinear relationship reflects a heterogeneous antibody response. Both sets of data can be used to obtain values for n and K (or K_o) by the proper extrapolations.

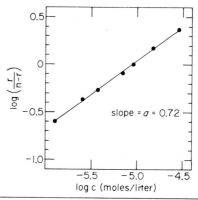

Figure 12-16. Sips analysis of the nonlinear data illustrated in Figure 12-15. The value of the Sips constant (a) can be obtained, and an average affinity constant can be recalculated from equation 9.

ity for the antigens to which they are produced. The average association constant may be obtained by measuring the value of r/c at which half of the sites are filled (Fig. 12-15). The extent of heterogeneity can be estimated by several methods, one of which assumes a Sips distribution of equilibrium constants; thus:

$$(8) \qquad (cK_o)a = \frac{r}{n-r}$$

where a is a power function reflecting the dispersion of equilibrium constants about an average constant, K_o. If $a = 1$, the expression reduces to that of equation 7. By plotting the log of equation 8:

$$(9) \qquad \log \frac{r}{n-r} = a \log c + a \log K_o$$

one usually obtains a straight line (Fig. 12-16), the slope of which can be used to calculate a. For homogeneous binding (a single affinity constant) $a = 1$, and for a heterogeneous mixture of hapten-binding antibodies $a < 1$. When a is determined, the average K_o can be calculated.

Reactions of Antibodies with Multivalent Antigens

In principle there are several methods for measuring in vitro antigen–antibody reactions, such as precipitation, complement fixation, agglutination, and radioimmunoassay. These techniques are described later in this chapter. For quantitative measurement of association constants (or affinity), antigen–antibody reactions must be studied in the region of antigen excess (where the complexes are soluble), or the binding of univalent Fab fragments with multivalent antigens can be studied. Both techniques can be used to obtain association constants.

Avidity vs. Affinity

To this point affinity has been defined as having a strict relationship to the experimentally determined equilibrium constant. The term "avidity" is often used semiquantitatively to describe the overall combining properties of an antiserum with a multivalent antigen. The avidity of a particular antiserum depends upon the number of different antibodies directed at distinct antigenic determinants on the antigen and their intrinsic association constants. If, in addition, an antigen possesses two or more identical antigenic determinants in close proximity (e.g., viral capsid proteins), the binding of a single antibody to two adjacent determinants results in a large increase in the observed association constant. For example, the primary immune response to an infectious agent, such as a bacterium or a virus, is characterized by IgM. Although the intrinsic binding affinity of an isolated IgM Fab subunit might be characterized as weak, multideterminant binding by pentameric IgM with a multivalent infectious agent(s) would result in very efficient recognition at a critical time in the infection. This latter phenomenon is known as "monogamous bivalent binding." A combination of some or all of these factors determines the avidity of a particular antiserum and also complicates the exact determination of intrinsic affinity constants.

Implications of a Heterogeneous Antibody Response

One of the peculiar aspects of the humoral immune response is that it is usually heterogeneous, even to simple chemically defined haptenic determinants. For example, if an animal is immunized with the hapten–protein conjugate DNP-BSA and the resulting antibodies specific for the hapten (DNP) are isolated by immunoadsorption techniques, they will demonstrate a range of intrinsic affinity constants from very low to very high affinity. In other words, in this population of DNP-specific antibodies, there are subpopulations with unique intrinsic affinities for the hapten. Each one of these probably has a unique antibody-binding site and, therefore, a unique variable region amino acid sequence (or single idiotype). These observations are consistent with the hypothesis that individual subpopulations of antibodies are products of a clone of cells derived from a single parent cell. Implicit in this hypothesis is that a single antibody-forming cell (and its precursor) is capable of synthesizing an antibody of a single restricted specificity. Upon antigenic challenge, cells with both high and low affinities for the antigen may be stimulated to proliferate and transform to antibody-producing plasma cells. Heterogeneous antibodies, therefore, reflect a heterogeneous response. It is also possible to manipulate this response. If very small amounts of antigen are used in the immunization schedule, only precursor cells (B cells) with high affinity receptor molecules (cell surface immunoglobulins) will respond, and antibodies of high affinity will be produced. In more conventional immunizations when excess antigen is administered repeatedly, the affinity of the antibodies rises as clones of high affinity precursor cells compete for antigen and expand in preference to those of lower affinity cells.

Antibody Formation

Kinetics of Antibody Synthesis

The primary response to an immunogen is both quantitatively and qualitatively different from all subsequent, or secondary, responses. Following initial exposure, there is a lag period of a few days to a few weeks before antibody can be detected, with the length often dependent upon the nature of the immunogen. With few exceptions, the first antibodies detected are IgM, and these can be distinguished from IgG by their sensitivity to reduction with β-mercaptoethanol and their sedimentation coefficients (Table 12-2).

If the immunogen is first administered with adjuvant, the response is modified, and it is often possible to detect an early IgG and an IgM response that persists for several days. In contrast, a primary IgM response elicited in the absence of adjuvant generally peaks and rapidly wanes (Fig. 12-17). Note that the response curve consists of a lag period, an exponential increase in antibody concentration, a peak or steady state of antibody concentration, and an exponential decay (if no adjuvant is used).

Repeated immunizations are usually characterized by an enhanced (anamnestic) response (Fig. 12-17). After the second or subsequent exposure to antigen, there is a substantially shorter lag period, a higher level of specific antibody is produced, and the steady state level persists over longer periods of time. The persistent response is predominantly IgG, but there is an IgM response that often resembles a primary response after each booster injection. The quantitative difference between a primary response and a secondary response is due to an increase in the number of potentially reactive cells. The primary stimulus causes a clonal expansion of specific memory cells that further expand upon secondary challenge. The capacity for

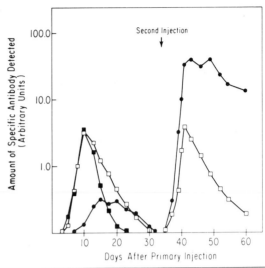

Figure 12-17. Primary and secondary responses to administered immunogen. The ordinate represents detectable serum antibody in arbitrary units (log scale); the abscissa, time in days following the first injection of immunogen. IgM (□) and IgG (●) responses are shown for immunogen given with adjuvant. Note the lag period before the primary response can be detected and the presence of a small, early IgG response. When adjuvant is not used, the IgM response decays rapidly (■), and there is often little or no early detectable IgG response.

eliciting a secondary response can persist for several years, long after circulating specific antibody can be detected. However, the magnitude of the secondary response does depend on the interval between primary and secondary challenge. If the interval is too short and circulating antibody is present, the secondary response will be reduced. If the interval is too long, antigen-reactive cells gradually die out, and a reduced response is therefore observed. These are, of course, generalizations. Sometimes antibody persists for 20 years; in another individual, or with a different antigen, it may disappear in a few months. Differences in regulatory mechanisms are discussed in Chapter 15.

The shift in production of IgM to IgG appears to be a very important feature of the immune response to certain types of antigens. It has been postulated that antigens that require T cell cooperation for an antibody response by B cells also give rise to the switch in immunoglobulin class production upon second exposure. Occasionally, it has been possible to demonstrate that plasma cells and certain myeloma tumor cells produce both IgM and IgG, but with apparently identical binding specificity. Hence, the general rule that one cell produces antibody of a single specificity still holds.

Synthesis and Secretion

Several murine monoclonal plasmacytomas have been either adapted to in vitro culture or maintained by in vivo transfer. These plasma cell tumors and human myelomas, in which over 40 percent of the protein that is synthesized is immunoglobulin, have proved invaluable in studying the biosynthesis of IgG, IgM, and IgA. Most of them produce a homogeneous immunoglobulin, while a few produce multiple classes of immunoglobulin (IgG and IgM, or IgG and IgA, and so on). These classes possess an identical V_H sequence and an identical L chain (V_κ, C_κ or V_λ, C_λ) and differ only in the C_H region (C_γ, C_μ, C_α, and so forth). This suggested that more than one gene controls the synthesis of a heavy chain, minimally, a gene that determines the V_H sequence and one that determines the C_H sequence. These and other data support the translocation hypothesis, which proposes that the variable regions of both H and L chains are controlled by a set of V genes that are distinct and separate from a set of genes that code for the constant regions of immunoglobulin molecules. This hypothesis conflicts with the general rule of one gene–one polypeptide, since it proposes that at least two genes code for one polypeptide in the case of immunoglobulin chains.

More recent data suggest that immunoglobulin messenger RNA (mRNA) is transcribed from three gene segments, V, J, and C in the case of L chains, and from four gene segments V, D, J, and C, in the case of H chains (see below). The cytoplasmic RNA consists of a 5′-end segment coding for a leader polypeptide important in signaling the secretion of the polypeptide, the entire V, D, J, and C message, and a poly-A-rich 3′ terminus. Both L and H chains are synthesized on membrane-bound polyribosomes of the rough endoplasmic reticulum and are secreted into the cisternae, the leader polypeptide being cleaved during this process. Assembly of H and L chains into the basic 4-chain immunoglobulin structure and the formation of the stabilizing interchain disulfide bonds occur within the cisternae prior to extracellular secretion. Carbohydrate is added covalently to the H chains by a series of glycosyl transferases during transport through the Golgi complex.

Polymerization of IgM or IgA monomers (the predominant intracellular precursor form) and addition of j chains occur just prior to secretion. Secretory component is synthesized by certain epithelial cells and is added to IgA polymers (and some IgM) after secretion by plasma cells as these molecules are transported into the glandular lumen. It has been suggested that addition of secretory component is mandatory for secretion of IgA.

A small proportion of immunoglobulin molecules is tightly bound to the cell membrane by way of a hydrophobic sequence of amino acids at the C terminus of the heavy chains. The hydrophobic sequence is absent from secreted immunoglobulins (see below). In precursor B cells, this type of receptor immunoglobulin may in fact be the major form of immunoglobulin synthesized. After transformation to a plasma cell, however, most of the immunoglobulin synthesized is secreted.

The Origin of Antibody Diversity

Since an individual may have the capacity to respond to as many as perhaps 10^6 different antigenic determinants, the mechansim for generating this huge number of unique antibodies has received special attention. The requirements are simplified in one respect, since it is assumed that any given L chain can combine with any given H chain; 10^3 different L chain variable regions and 10^3 different H chain variable regions could therefore generate 10^6 different antibodies.

Two popular theories were originally proposed to account for the diversity of variable regions: the germ line theory and the somatic mutation theory. The germ line theory as originally proposed simply states that an individual inherits 10^3 genes coding for L chains and 10^3 genes coding for H chains and that each V region gene is physically adjacent to its respective C region gene. This model predicts that the genes coding for C region classes (κ, λ, γ, μ, α, δ, ϵ), subclasses (e.g., $\gamma 1$, $\gamma 2$, $\gamma 3$, $\gamma 4$), and any existing allotypes (e.g., Gm markers) are repeated many times in the genome. This theory had difficulty in accounting for switches in the immunoglobulin class produced while an identical V region sequence is maintained. In addition, the commitment of a single cell to the production of a single antibody with a single specificity would require a mechanism for excluding the transcription of all other immunoglobulin genes, and this restriction would have to be passed down to all daughter cells.

The somatic mutation theory proposes that an individual inherits a limited number of V and C region genes and that new V region sequences are generated by somatic mutation. In its initial form this theory could not account for the various subtypes of V region sequences for L and H chains (e.g., $V_{\kappa I}$, $V_{\kappa II}$, $V_{\kappa III}$, and so on), but it has been modified to propose that an individual inherits V region genes for each L and H chain subtype and C region genes for the major classes, subclasses, and allotypes described thus far. In substance, it proposes that the total genetic information required to generate 10^6 antibodies is limited. This theory has its own set of problems, such as an explanation of the mechanism involved in the selection of an important somatic mutation occurring in the V region genome (and not the C region) and its maintenance through cell division.

In recent years speculation as to the mechanism of the generation of diversity has been rendered largely redundant by rapid and spectacular advances in the understanding of immunoglobulin gene structure. Relatively accurate assessments of the number of copies of immunoglobulin variable and constant region genes have been made by molecular hybridization. The specific messenger RNA for a given unique L or H chain isolated from a plasmacytoma is transcribed into radioactively labeled DNA (cDNA) using reverse transcriptase (Chap. 62). The rate with which this labeled DNA probe hybridizes with the genome of the plasmacytoma cell or embryonic DNA is then measured. The faster it hybridizes, the more DNA that specify L or H chains is present in the cell, that is, the more gene copies it contains. The results indicate that in the mouse there are very few C region genes and the number of V_λ region genes has been estimated at 1 to 5 copies per genome, while a similar estimate of V_κ gene copies in the mouse runs as high as 350.

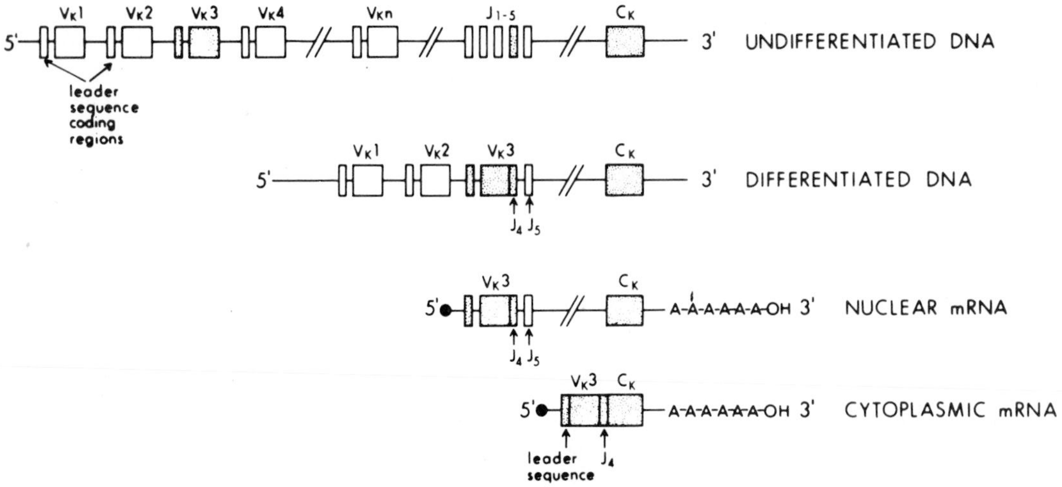

Figure 12-18. Construction of a murine κ light chain mRNA. In this schematic representation of the genome the boxes represent coding regions (exons), and lines separating the boxes are intervening sequences. Multiple V$_\kappa$ genes exist, each consisting of two gene segments coding for the leader sequence and the N-terminal sequence of the mature κ chain, separated by an intervening sequence. In undifferentiated cells V$_\kappa$ genes, J genes, and the C$_\kappa$ gene are well separated on the chromosome. Plasma cell differentiation and commitment to the production of a single κ chain involve translocation of a V$_\kappa$ gene (V$_{\kappa 3}$ above) to a position 5′ to a selected J gene (J$_4$ above). Nuclear mRNA is transcribed from this configuration. Processing of the precursor mRNA involves loss of intervening sequences (including the J$_5$ sequence above), modification of the 5′ terminus, and addition of polyadenylic acid to the 3′ terminus. The precursor form of the κ light chain is translated from the cytoplasmic mRNA on membrane-associated ribosomes. The leader sequence of the protein is removed in the rough endoplasmic reticulum to give mature κ chains.

In the case of murine L chain variable regions there are similarities and differences in genetic organization. In both cases V regions are constructed from two initially separate genetic elements, V and J, on the same chromosome. V$_\kappa$ genes (Fig. 12-18) code for a 22-amino acid leader sequence, initially translated but ultimately cleaved from the κ chain prior to secretion, and amino acids 1 to 95 of the completed κ chain. The coding region for the leader sequence is split by an intervening sequence of 175 base pairs. Amino acids 96 to 108 of the classically defined V region are encoded by one of four functional J$_\kappa$ genes. One V$_\kappa$ gene is translocated by an unknown mechanism to a position 5′ to the selected J$_\kappa$ gene to make a functional V region gene. The two remain separated by an intervening sequence of 2.5 to 4 kilobases. The entire V region gene together with the C$_\kappa$ gene is transcribed into RNA, the intervening sequences are removed by the splicing processes common in eucaryotes (Chap. 58), and the result is a polyadenylated mRNA coding for an intact κ chain.

In the case of λ genes there are only two V$_\lambda$ genes, similar in construction to V$_\kappa$ genes, each of which can translocate to a position where it is transcribed together with one of two J$_\lambda$C$_\lambda$ sets. In the case of V$_{\lambda 2}$ one of the alternative J$_\lambda$C$_\lambda$ sets is inactive. Subsequent to the translocation event, the transcription of the V$_\lambda$J$_\lambda$C$_\lambda$ complex and subsequent RNA processing and translation are identical to the steps described for κ chains.

The construction of murine heavy chain V regions is more like that of the V$_\kappa$ region than the V$_\lambda$ region, in that multiple sequentially ordered V$_H$ genes exist (Fig. 12-19), each coding for a leader sequence and amino acids 1 to 97 of the V$_H$ region. An additional order of variability is introduced by additional genes, the D genes. To form a functionally active V$_H$ region, one of a number (greater than 100) of V$_H$ genes is translocated to a position adjacent to one of 10 to 15 D gene segments. A second translocational event juxtaposes the VD segment of one of four J$_H$ segments to code for a complete V$_H$ region. An intervening sequence of 6.5 kilobases separates the J$_H$ cluster from C$_\mu$, the μ chain constant region gene. Transcription of the VDJ segment, together with C$_\mu$, gives rise to a precursor RNA that is ultimately spliced and processed to yield an mRNA coding for an intact μ chain.

An additional property of the H chain gene cluster not shared by the L chain systems is the potential for class switching. Following assembly of the VDJ gene set, it can be translocated to a position immediately 5′ to any of the H chain constant region genes (δ, γ$_3$, γ$_1$, γ$_{2b}$, γ$_{2a}$, ε, or α), forming a functional set that can code for an intact chain of any class or subclass and that shares the V$_H$ region initially present in the μ chain. Thus, the phenomenon of two immunoglobulins of different classes sharing a common binding site and idiotype can now be explained in the mouse at the molecular level.

The precise mechanism of the class switch has not been elucidated, but particular regions of the DNA between the constant region genes appear to be involved. These are known as switch regions, or S regions, and con-

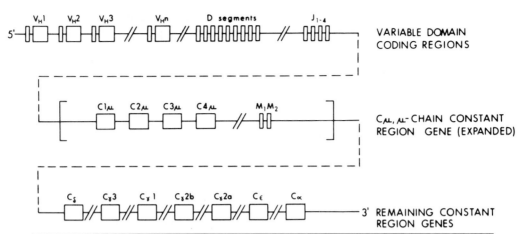

Figure 12-19. Schematic representation of the heavy chain gene complex of the mouse in undifferentiated cells. During differentiation to an IgM-producing plasma cell one of the V_H genes is translocated to a position adjacent to one of 10 to 15 D segments. A further translocation associates the V_HD coding sequence with one of the four J genes to yield a complete variable region gene. This gene is coexpressed with the C_μ domains (Fig. 12-9), and the M exons coding for the membrane integration sequence at the C-terminus of membrane-associated μ chains. Further differentiation, to an IgG1 secreting cell for example, would involve the translocation of the V_HDJ transcription unit to a position immediately 5' to the $C_{\gamma 1}$ gene. Individual C region genes are organized in a similar way to the C_μ gene represented above, with individual exons coding for separate domains.

sist of tandem repetitive sequences, which appear to provide recognition sites for the translocation process, postulated to be a homologous recombination, i.e., recombination between genes on the same chromosome. The S_μ region for example, which lies between the J_H cluster and the C_μ gene, has the nucleotide sequence $[(GAGCT)_n (GGGGT)]_m$ where n varies between 1 and 7, and m is of the order of 150.

The individual H chain, C region genes are organized in a fashion remarkably reminiscent of the organization of the ultimately translated protein. Each domain is encoded by an exon that is separated by an intervening sequence from an exon encoding the adjacent domain. The hinge region of C_γ genes is encoded by a small exon between those coding for C_{H1} and C_{H2} of the γ chain. Intervening sequences are removed from transcribed RNA prior to translation, as described above.

Membrane IgM, which acts as the antigen receptor of B lymphocytes, is attached to the membrane by virtue of a sequence of 26 hydrophobic amino acids at the C-terminal end of the μ chain. The hydrophobic sequence is encoded by the region M, which is 3' to C_μ (Fig. 12-19). The synthesis of the membrane form of the μ chain appears to be regulated at the RNA-splicing level. For soluble μ chains the C_μ DNA region up to the 3'-terminus of the exon coding for the fourth C domain is transcribed. For membrane μ chains an entire transcript, which includes the M region, is modified at the splicing level, such that RNA coding for the C-terminal 20 residues of the soluble μ chains is removed, and the M region, coding for 41 amino acids, which include the 26 amino acid hydrophobic stretch, is retained. This mechanism is consistent with the fact that B

cells can simultaneously express surface IgM and also secrete IgM.

Methods of Detecting and Quantitating Antigen and Antibody Reactions

Since it is important for the clinician to understand the processes used in the clinical immunology laboratory, this section is concerned with various in vitro methods of detecting and quantitating antigen–antibody interactions. The techniques used for detecting antigens differ depending on whether the antigen in question is soluble or cell-bound. Accordingly, the following is divided into two subsections covering these areas. This section is not intended to be a detailed exposition of the methods but to give an explanation of the principles used. The appropriate detailed sources are indicated in the Further Reading section.

Soluble Antigens

Quantitative Precipitation
Most techniques examining the reactions of soluble antigens with their antibodies depend upon precipitation. This occurs when divalent IgG antibody reacts with a multivalent antigen. Precipitation depends upon the formation of

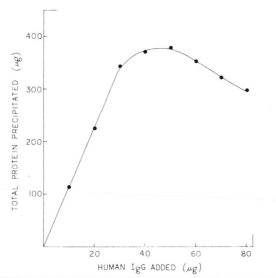

Figure 12-20. A quantitative precipitin assay of goat antiserum of human IgG with human IgG as antigen.

large, insoluble lattices of interconnected antigen and antibody molecules. A typical quantitative precipitin curve is shown in Figure 12-20. The quantitative assay represented involves adding increasing amounts of antigen to constant volumes of antiserum. The precipitates formed are washed, and the total protein in the precipitates is determined for each point, commonly by using the Kjeldahl nitrogen estimation method. The amount of antibody precipitated with increasing antigen is linear initially, eventually reaching a peak (the equivalence point). When antigen is added in excess, the amount of antibody precipitated usually falls, since adequate crosslinking for the formation of large complexes no longer occurs. An antiserum characterized in this way can subsequently be used to determine the amount of antigen in unknown samples, always working in the linear portion of the quantitative precipitin curve, i.e., in antibody excess.

Ouchterlony Analysis

This is a powerful yet simple technique that allows qualitative detection of antigens in solution and permits the determination of antigenic relationships between different antigens. A 2 percent (w/v) solution of hot agar in isotonic saline is poured into a Petri dish or onto a microscope slide to a depth of 1 to 2 mm. Using a special punch, various patterns of circular holes are cut in the agar layer after cooling. A commonly used pattern consists of six holes in a hexagonal arrangement surrounding a central hole that contains antiserum. When an antiserum and its corresponding antigen in solution are placed in adjacent wells, both diffuse through the agar layer and eventually meet and combine. A band of precipitated antigen forms between the wells. Typical precipitation patterns seen using related and unrelated antigens with a polyspecific serum are shown in Figure 12-21. If the two antigens, A and B, are identical (Fig. 12-21a), a continuous precipitin line is formed between the two antigen wells and the antiserum well. Two unrelated antigens produce the result shown in Figure 12-21b, where the line produced with one antigen completely crosses that produced with the other. This also implies that the antiserum contains two unrelated populations of antibodies, specific for antigen A and for antigen B, respectively. An interesting pattern is seen when A and B have some antigenic determinants in common. In Figure 12-21c, the antiserum primarily reacts with antigen B, but some of the antigenic determinants are shared by A. The result is a spur of precipitation with antigen B, resulting from diffusion of antibodies that fail to react with A through the zone of precipitation given by A, and subsequent reaction with B.

Radial Immunodiffusion

Molten agar is first premixed with antiserum and poured onto a plate as in the Ouchterlony method, and wells are cut, usually as two lines of holes. Dilutions of an antigen solution are introduced into each well. Circular zones of precipitation form around the well as the antigen diffuses

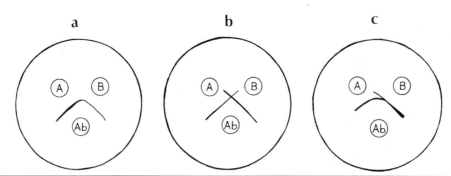

Figure 12-21. Ouchterlony analysis of a precipitating antiserum (Ab) to two antigens, A and B. **a.** A and B are antigenically identical, and the antiserum reacts equally well with both, showing a line of identity. **b.** A and B are antigenically unrelated, and separate crossed lines of precipitation are produced with the antiserum. **c.** A and B are antigenically related, with antigen A possessing only a portion of the antigenic determinants of B, resulting in a spur of precipitation with B.

into the agar and reacts with the antibody. The greater the amount of antigen, the larger the zone of precipitation. In fact, for a fixed volume of antigen solution, the square of the diameter of the zone of precipitation is proportional to the antigen concentration. Radial immunodiffusion plates are commercially available for the quantitation of levels of various human serum proteins. Standards containing known amounts of the protein in question are placed in the same plate as the unknown sample, e.g., patient's serum, and a standard curve is constructed from them. Figure 12-22 shows a radial immunodiffusion plate for the determination of human IgG.

Electroimmunodiffusion

In this quantitative technique of immunoprecipitation devised by Laurell, antiserum is also incorporated into the agar layer as above. However, instead of relying upon diffusion to establish zones of precipitation, antigen migration is induced electrophoretically. Agarose (ion-free agar) is substituted for agar to reduce the cathodal migration of incorporated IgG. The buffer commonly used is one first described by Michaelis, which contains sodium diethylbarbiturate and sodium acetate, and has a pH of 8.2 and an ionic strength of 0.1. The samples (1 to 3 μl) are introduced into wells cut into the plate, and electrophoresis is performed with a potential gradient of 3 to 4 volts/cm. Cooling the plate during electrophoresis is usually necessary. Rocket-shaped precipitin lines are formed, the heights of which are proportional to the antigen concentration under standard conditions. Figure 12-23 shows examples of the kinds of patterns produced.

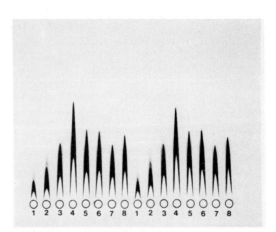

Figure 12-23. Electroimmunodiffusion used for albumin determination. Duplicate samples (1 to 8) containing varying amounts of human albumin were subjected to electrophoresis through agarose containing goat antiserum to human albumin. *(Courtesy of A.G. Hoechst.)*

Immunoelectrophoresis

The immunoprecipitation method of Ouchterlony gives a confusion of lines when a complex antiserum and a complex antigen mixture (e.g., human serum and rabbit or goat anti-whole human serum) are placed in adjacent wells. The immunoelectrophoresis technique of Grabar and Williams overcomes this problem by separating the complex antigen mixture electrophoretically before antiserum is added. Figure 12-24 shows the result of a typical immunoelectrophoretic analysis.

A 2 percent agar solution containing Michaelis buffer is allowed to solidify on a microscope slide. A special die then cuts the pattern shown in Figure 12-24. Wells for the sample are adjacent to a trough for antiserum. A sample (2 to 5 μl) is then introduced into the sample well, and electrophoresis is performed for about 45 minutes at 6 volts per cm to separate the different components. Subsequently, the antiserum (40 to 50 μl) is introduced into the trough. Diffusion of separated antigens and the antiserum gives rise to the arcs of precipitation visible in Figure 12-24. Some 30 different human serum proteins can be differentiated by this method. By using antisera to individual proteins any arc of questionable identity can usually be unambiguously identified. Immunoelectrophoresis is a frequently used diagnostic tool in medicine, useful in the detection of monoclonal gammopathies (paraproteins), for example, by an enormously enlarged and distorted IgM precipitin line in cases of Waldenström's macroglobulinemia, or in the detection of Bence Jones proteins (light chain dimers) in serum or urine, as well as identifying many other serum protein deficiencies or disorders.

Figure 12-22. Radial immunodiffusion used to determine human IgG levels. The outer annulus of the plate contains agarose with incorporated goat antiserum to human IgG. In the wells are samples containing various amounts of IgG. Circular areas of precipitation can be seen around each well. *(Courtesy of A.G. Hoechst.)*

Two-dimensional Immunoelectrophoresis

This technique is a combination of the electrophoretic separation method used in conventional immunoelectrophoresis and electroimmunodiffusion. Initially, the antigen

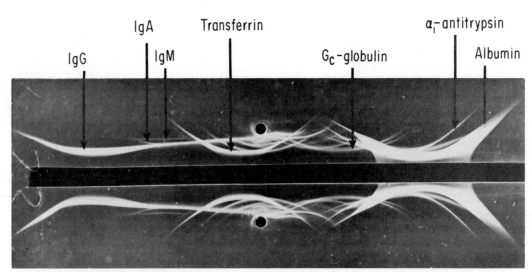

Figure 12-24. Immunoelectrophoresis of normal human serum. The electrophoretic separation and precipitation with a goat antiserum to whole human serum were performed as described in the text. The precipitin arcs corresponding to some serum proteins are indicated. *(Courtesy of A.G. Hoechst.)*

mixture is separated in agarose electrophoretically as described for immunoelectrophoresis. A strip of agarose containing the separated antigens is then placed on a second slide and an antibody-containing agarose solution is allowed to solidify adjacent to it. Electrophoresis as described for electroimmunodiffusion is then performed at right angles to the original electrophoretic separation, giving rise to peaks such as those shown in Figure 12-25. Quantitation is possible with this technique by comparing the surface area of the precipitated arcs with those given by known amounts of standard antigens.

Radioimmunoassays and Enzyme-linked Immunosorbent Assays

Some of the most sensitive techniques used in clinical and basic science laboratories are the various kinds of radioimmunoassays for the detection and quantitation of diverse substances, from hormones to immunoglobulin allotypes. The principle of quantitation used is that of inhibition. Binding of a radioactively labeled antigen to its antibody is inhibited by known amounts of unlabeled antigen to generate a standard curve; unknown samples are compared to this.

The quantitation of bound labeled antigen can be accomplished in various ways. Systems have been devised using electrophoretic separation of antibody-bound and unbound labeled antigen, separation of the immune complex from free antigen on the basis of size by gel filtration, and precipitation of the antigen–antibody complex by an anti-immunoglobulin serum or by the addition of saturated ammonium sulfate. However, the most convenient is the solid-phase radioimmunoassay. In this technique, antibody is covalently attached to an insoluble matrix, commonly agarose beads that have been activated to bind protein

amino groups using cyanogen bromide. A mixture of labeled standard antigen and unlabeled test material is then added. Antibody-bound radioactively labeled antigen can then be separated from free antigen by centrifugation or filtration. The greater the recovery of free label, the higher the concentration of the unknown. An example of the kind of data obtained in a solid-phase radioimmunoassay is shown in Figure 12-26.

Technical problems with radioimmunoassay techniques are the decay of isotope used to label the antigen (commonly ^{125}I, with a half-life of 60 days) and the general dangers inherent in handling radioactive substances. An alternative method that circumvents these problems is the

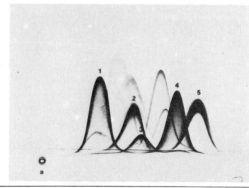

Figure 12-25. Two-dimensional immunoelectrophoresis. The application well (a) of the first dimension (electrophoresis from left to right) contained normal human serum. The agarose gel of the second dimension (from bottom to top) contained an oligospecific antiserum. The numbered precipitates belong to the following proteins: (1) transferrin, (2) α_2-macroglobulin, (3) ceruloplasmin, (4) α_1-antitrypsin, (5) α_1-acid glycoprotein. *(Courtesy of A.G. Hoechst.)*

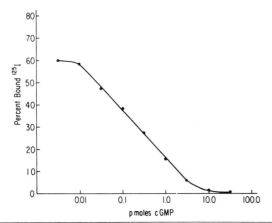

Figure 12-26. A solid-phase radioimmunoassay for guanosine 3′-5′-cyclic monophosphate (cyclic-GMP). Shown is a standard curve obtained by measurement of the binding of a radioactive derivative of cyclic-GMP (^{125}I-succinyl-cyclic-GMP-tyrosine methyl ester) to agarose beads with covalently attached rabbit antibody to cyclic-GMP, in the presence of known amounts of unlabeled cyclic-GMP. Unknown samples are quantitated by comparison with this standard curve.

enzyme-linked immunosorbent assay, often abbreviated ELISA. In this approach the antigen standard in question is conjugated to an enzyme, such as alkaline phosphatase or horseradish peroxidase. Glutaraldehyde is often used for the coupling. In the solid-phase variation of the technique, antibody is again attached to an insoluble matrix, and binding of enzyme-linked antigen to the antibody is inhibited by standard amounts of unlabeled antigen or by the unknown samples. After washing, the amount of enzyme-linked antigen coupled to the insoluble antibody is quantitated using a chromogenic substrate for the enzyme, such as *p*-nitrophenyl phosphate in the case of alkaline phosphatase.

Complement-fixation Tests

Complexes of antibodies and antigens, both soluble and cell-bound, have the ability to fix serum complement (Chap. 14). Thus, measurements of complement fixation can be used to determine the presence of immune complexes and therefore antigens or antibodies.

Complement levels are usually measured by adding sheep erythrocytes presensitized with rabbit antisheep erythrocyte antibodies to aliquots of the complement (normally from guinea pigs). Upon incubation at 37C, hemolysis occurs and can be quantitated by centrifuging the samples and estimating the hemoglobin released into the supernatant by visible absorption spectroscopy.

The complement-fixation protocol for a soluble antigen involves incubating dilutions of antigen with its antiserum in the presence of an amount of guinea pig complement that will give 90 percent hemolysis of subsequently added, sensitized sheep erythrocytes if no fixation occurs. After incubation, usually at 4C for periods of up to

18 hours, the sensitized sheep erythrocytes are added and hemolysis measured after 1 hour at 37C. If complement fixation by antigen–antibody complexes has occurred, this is manifested in reduced hemolysis. The protocol for cellular antigens is the same, except the antigen–antibody–complement complexes (i.e., cells) can be removed by centrifugation before the sensitized sheep erythrocytes are introduced. By comparing them with known standard amounts of antigens, unknown samples of antigen can be quantitated.

Cellular Antigens

Agglutination Reactions

The adherence of red blood cells to each other in the direct and indirect agglutination procedures used in immunohematology and blood banking are described in Chapter 16. However, the process of agglutination can be modified to detect other antigens. One such modification is called "passive agglutination." Erythrocyte membranes are modified by soluble antigens bound to their surface. The method of attachment depends upon the nature of the antigen. Bacterial lipopolysaccharides bind to red cells upon simple incubation, as do some proteins. Reactive haptens, such as trinitrobenzene sulfonic acid, will covalently attach to red cell membranes. Red cells treated with tannic acid will bind protein antigens and are more easily agglutinable than are untreated red cells. Bifunctional crosslinking agents, such as bisdiazobenzidine, or 1,3-difluoro-4,6-dinitrobenzene can also be used to bind a variety of antigens to erythrocytes. Cells with bound antigen form a convenient method of detecting and assaying antibodies to almost any hapten or soluble antigen. Inhibition of passive agglutination can also be used to quantitate haptens or antigens in solution.

Other assays based on the general principles of hemagglutination include mixed hemabsorption and leukoagglutination. Both of these assays were at one time used for the detection of antigens of leukocyte membranes (Chap. 17), but both present many technical difficulties and leukocyte antigens are now almost always determined by the cytotoxicity test.

Cytotoxicity Assays

Lymphocytes for HLA typing can be easily prepared from human blood by Ficoll–Hypaque gradient centrifugation. Ficoll is a commercial high molecular weight polysaccharide that induces spontaneous red cell agglutination. Hypaque is a radiopaque dye that is used to vary the density of the Ficoll–Hypaque solution in water. The density of the mixture is adjusted to 1.078 g/cm³, a density at which agglutinated erythrocytes sink, while lymphocytes float. By simply layering diluted defibrinated blood on the Ficoll–Hypaque solution and centrifuging, lymphocytes can be isolated from the interface.

For HLA typing, the lymphocytes are incubated in small typing trays with human alloantisera specific for the different HLA antigens. Following a wash step to remove unbound antibody, rabbit complement is added to effect lysis. To differentiate between live and dead cells, a vital dye, commonly trypan blue, is added. Such dyes penetrate the membrane of cells lysed by complement but fail to stain living cells. By testing lymphocytes with a battery of sera, unambiguous HLA typing can usually be achieved. Similar procedures are used for detecting the presence of antigens present only in certain stages of development of animal and human cells. These differentiation antigens are discussed in Chapters 15, 17, and 19.

A more quantitative measure of cytotoxicity is the chromium-51 release assay. Lymphocytes are prelabeled by incubating with $Na_2{}^{51}CrO_4$, which is rapidly incorporated but released from a viable cell at a rate of approximately only 5 percent per hour. A cell that has suffered membrane damage due to antibody and complement will release the ^{51}Cr label very rapidly. Figure 12-27 shows a titration of an anti-HLA antiserum upon human peripheral blood lymphocytes labeled with ^{51}Cr. Dilutions of the serum were added to the cells, followed by the addition of rabbit complement.

Yet another assay of cell death by antibody and complement involves adding fluorescein diacetate (FDA) to the cells. FDA is itself membrane-permeable and nonfluorescent. On entering a cell, natural esterases convert FDA into fluorescein, thus causing the cells to become visibly fluorescent under ultraviolet light. Fluorescein itself can leave the cell only slowly, and most is retained until lysis occurs. Thus, after a cytotoxicity assay, dead cells are non-fluorescent and live cells are fluorescent.

Immunofluorescence

This is a very sensitive technique for the detection of cellular antigens, whether they are cytoplasmic, nuclear, or plasma membrane-bound. Direct immunofluorescence re-

quires an antibody that is conjugated to a fluorescent dye, such as fluorescein or rhodamine. The isothiocyanate derivatives of the dyes are used to bind them covalently to IgG preparations of the antisera. Intact or sectioned cells are then incubated with the fluorescent antibodies, washed, and examined under a microscope with an ultraviolet light source. Cells that have bound the antibody are brightly fluorescent and easily visible. The site of intracellular localization can be determined in tissue sections.

The sensitivity of the technique can be increased dramatically by utilizing indirect immunofluorescence. In this case the fluorescent dye is conjugated to IgG from an anti-immunoglobulin serum. The cells under study are first incubated with antigen-specific antiserum, washed, and then incubated with the fluorescent-conjugated anti-immunoglobulin serum. Stained cells are again detected by observation under an ultraviolet microscope. Examples are shown in Figure 12-28.

The availability of fluorescent dyes of different colors greatly increases the flexibility of the technique, particularly for examining membrane antigens using the technique of capping. Capping is an energy-requiring process whereby certain membrane molecules accumulate over one pole of a cell (particularly a lymphoid cell) when a specific antibody binds to them. When a fluorescent antibody is used to induce capping, a brightly fluorescent cap can be observed on the cell under ultraviolet light. If an antiserum conjugated with fluorescein is used to induce capping, the cap is green. A second antibody, conjugated with rhodamine (which emits a red fluorescence) can be added to the capped cells at 0C, where capping no longer occurs, thus introducing a second stain. If the two antisera react with the same membrane antigenic complex, the rhodamine-labeled antibody will be visible in the cap. If the two antibodies react with different membrane markers, the rhodamine fluorescence will be visible on the rest of the cell and not isolated in the cap. This technique has been

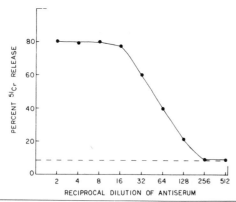

Figure 12-27. A titration of a cytotoxic anti-HLA-A2 alloantiserum against human peripheral blood lymphocytes from an HLA-A2-positive individual, using the ^{51}Cr-release assay. The dashed line indicates the background release of ^{51}Cr by cells in the absence of antiserum but with added complement.

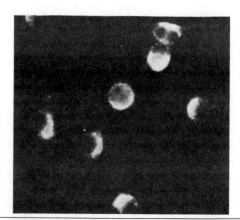

Figure 12-28. Cytoplasmic staining of a tumor-associated embryonic antigen in acetone-fixed mouse L cells, using indirect immunofluorescence. The specific antibody is a rabbit antiserum to a mouse teratoma, and the developing serum is a goat antiserum to rabbit IgG conjugated with fluorescein. *(Courtesy of Dr. Linda R. Gooding.)*

used to prove that membrane immunoglobulin on mouse B cells is not associated with H-2 antigens, and that H-2 antigens coded by different genes are not associated with each other on the membrane.

The technology has been developed for separating cells within a population that bind fluorescent antibodies from those that do not. The method uses a machine called a fluorescence-activated cell sorter (FACS). A suspension of cells is passed through the FACS as a stream of positively charged microdroplets. Illumination of the droplets with laser light allows discrimination of those containing fluorescent cells from those containing nonfluorescent cells. Microdroplets containing fluorescent cells can be selectively deviated by briefly exposing them to a negative charge, and they can be collected separately. Cell populations that differ in intensity of fluorescence, rather than being absolutely positive and negative respectively, can also be separated by the FACS.

Radioactive Binding Assays

Immunofluorescent detection of antibodies bound to cell surfaces is not a quantitative method. Enumeration of particular antigenic determinants on cells can be achieved by binding assays using ^{125}I-labeled anti-immunoglobulin reagents or ^{125}I-protein A. Protein A, a molecule isolated from the cell wall of *Staphylococcus aureus,* has the capacity to bind to the Fc region of most subclasses of the IgG of many mammals. Either protein A or an anti-immunoglobulin (frequently an affinity-purified Fab$'_2$ preparation of, for example, rabbit antimouse IgG) can be labeled with ^{125}I without losing its binding capacity and then used in a manner analogous to indirect immunofluorescence. Cells are first incubated with specific antibody, washed, and then incubated with the ^{125}I-labeled reagent. After further washes the amount of ^{125}I bound to the cells is determined. Using known numbers of cells and ^{125}I reagents of known specific activity estimates of the number of antigenic determinants per cell can be obtained. Fab$'_2$ of rabbit antimouse IgG is particularly useful in binding assays using murine monoclonal antibodies.

Immunoferritin Techniques

The use of specific antibodies to detect cellular antigens by electron microscopy is a technique of increasing value. The principle is similar to that of immunofluorescence, except that the antibody is conjugated to an electron-dense molecule, such as the iron-containing molecule, ferritin. Cells are incubated with ferritin-conjugated antibody, and sections are examined by electron microscopy. Ferritin molecules can be observed on the cell surface where antibodies are bound. Other markers, such as viruses, have been used in place of ferritin and can be distinguished from ferritin morphologically, allowing the detection of two antigens on the same cell using antisera conjugated to ferritin and, for example, to tobacco mosaic virus.

Hemolytic Plaque Assays

These techniques differ from the others described in that, rather than seeking to detect antibodies or antigens, they serve to detect and enumerate antibody-producing cells.

In the plaque assays, cells producing hemolytic antibodies are most readily detected. Antibody-secreting lymphoid cells, for example, spleen cells from a mouse immunized against sheep red cells, are mixed with target erythrocytes in a warm (46C) isotonic, 0.6 percent (w/v) agarose solution. This cell-laden agarose is then overlayered on a preformed 1.2 percent (w/v) agarose layer in a Petri dish. The plates are incubated at 37C in a humid atmosphere for 2 hours, during which time IgM and IgG secreted by the antibody-producing cells diffuse and bind to the target erythrocytes. On addition of complement and further incubation, erythrocytes that have bound IgM antibody will lyse, causing visible, clear plaques in the otherwise red agarose, with a plasma cell or B lymphocyte at the center of each plaque (direct plaques). IgG antibodies do not cause lysis because of their lower lytic and complement-fixing activity, but the inclusion of an anti-immunoglobulin serum in the top agarose layer enhances complement binding and hemolysis (indirect plaques).

The flexibility of the hemolytic plaque assay can be increased to include antigens other than erythrocytes by modification of erythrocytes with haptens or protein antigens, as described for passive hemagglutination.

Monoclonal Hybridoma Antibodies

As mentioned earlier in the section describing paraproteins, techniques have recently been developed that allow the production of monoclonal antibodies by cell lines to virtually any designated antigens. The basic technique involves the polyethylene glycol-induced fusion of splenic B lymphocytes from an immunized mouse or rat with cells from a mutant plasmacytoma lacking the gene for hypoxanthine-guanine-phosphoribosyl-transferase (HGPRT). The parent cell line is unable to grow in medium containing hypoxanthine, aminopterin, and thymidine (HAT medium) as a result of its enzyme deficiency, whereas hybrid cells containing the *HGPRT* gene donated by the splenocyte grow continuously. The unfused splenocytes die off in this culture system. In addition to the expression of the *HGPRT* gene, the fused cells also express the products of the immunoglobulin genes of the B lymphocytes, giving rise to hybridoma cell lines that produce antibodies to the immunizing antigen. The use of a nonimmunoglobulin-secreting plasmacytoma in the fusion simplifies the pattern of immunoglobulin production by eliminating the light and heavy paraprotein chains. Cloning the hybrid cells following fusion results in cell lines producing monoclonal antibodies. Such cell lines can be grown in vitro or adapted to growth in animals for the production of large amounts of antibody of extremely high titer and with great specificity.

Many monoclonal antibodies of diagnostic and clinical relevance have been produced and are commercially available. These include blood group-specific reagents and antibodies that discriminate between subsets of human lymphocytes. The potential for monoclonal antibodies in diagnostic microbiology and clinical chemistry as well as in biomedical research appears almost limitless.

FURTHER READING

GENERAL

Kabat EA: Structural Concepts in Immunology and Immunochemistry, 2nd ed. New York, Holt, Rinehart, 1976

Mandel TE, Cheers C, Hasking CS, et al. (eds): Progress in Immunology III. Amsterdam, North Holland, 1977

Nisonoff A: Introduction to Molecular Immunology. Sunderland, Mass, Sinauer, 1982

Nisonoff A, Hopper JE, Spring SB: The Antibody Molecule. New York, Academic Press, 1975

Nossal GVJ, Ada GL: Antigens, Lymphoid Cells and the Immune Response. New York, Academic Press, 1971

ANTIGENS AND ANTIGENIC DETERMINANTS

Atassi MZ: Precise determination of the entire antigenic structure of lysozyme. Immunochemistry 15:909, 1978

Karush F: Immunological specificity and molecular structure. Adv Immunol 2:1, 1960

Landsteiner K: The Specificity of Serological Reactions. New York, Dover Publications, 1962

Senyk G, Williams GB, Nitecki DE, et al.: The functional dissection of an antigen molecule: Specificity of humoral and cellular responses to glucagon. J Exp Med 133:1294, 1971

IMMUNOGENICITY

Sela M: Antigenicity: Some molecular aspects. Science 166:1365, 1969

IMMUNOGLOBULIN STRUCTURE

Heremans JF: Immunoglobulin A. In Sela M (ed): The Antigens. New York, Academic Press, 1974, vol II, chap 6

Koshland ME: Structure and function of the j-chain. Adv Immunol 20:41, 1975

Melchers F, Potter M, Warner N (eds): Lymphocyte hybridomas. Curr Top Microbiol Immunol 1978, vol 81

Moller G (ed): Immunoglobulin D. Immunol Rev 1977, vol 37

Moller G (ed): Immunoglobulin E. Immunol Rev 1978, vol 41

Potter M: Antigen-binding myeloma proteins of mice. Adv Immunol 25:141, 1977

ANTIBODY-COMBINING SITE

Capra JD, Kehoe JM: Hypervariable regions, idiotype, and the antibody-combining site. Adv Immunol 20:1, 1975

Wu TT, Kabat EA: An analysis of the sequences of the variable regions of Bence-Jones proteins and myeloma light chains and their implications for antibody complementarity. J Exp Med 132:211, 1970

ANTIBODY EFFECTOR FUNCTIONS

Metzgar H: Effect of antigen binding on the properties of antibody. Adv Immunol 18:169, 1974

Winkelhake JL: Immunoglobulin structure and effector functions. Immunochemistry 15:695, 1978

ORIGIN OF ANTIBODY DIVERSITY

BJack C, Hirama M, Lenhard-Schuller R, et al.: A complete immunoglobulin gene is created by somatic recombination. Cell 15:1, 1978

Seidman JG, Leder, P: The arrangement and rearrangement of antibody genes. Nature 276:790, 1978

Valbuena O, Marcu KB, Weigert M, et al.: Multiplicity of germline genes specifying a group of related mouse kappa chains with implications for the generation of immunoglobulin diversity. Nature 276:780, 1978

Blomberg B, Traunecker A, Eisen H, et al.: Organization of four mouse light chain immunoglobulin genes. Proc Natl Acad Sci USA 78:3765, 1981

Early P, Huang H, Davis M, et al.: An immunoglobulin heavy chain variable region gene is generated from three segments of DNA: V_H, D and J. Cell 19:981, 1980

Early P, Rogers J, Davis M, et al.: Two mRNAs can be produced from a single immunoglobulin μ gene by alternative RNA processing pathways. Cell 20:313, 1980

Honjo T: The molecular mechanisms of the immunoglobulin class switch. Immunology Today 3:214, 1982

Rogers J, Early P, Carter C, et al.: Two mRNAs with different 3' ends encode membrane-bound and secreted forms of immunoglobulinμ chain. Cell 20:303, 1980

SPECIFIC METHODS USED IN THE QUANTITATION OF ANTIGEN AND ANTIBODY REACTIONS

Dorval G, Welsh KI, Wigzell H: Labeled staphylococcal protein A as a immunological probe in the analysis of cell surface markers. Scand J Immunol 3:405, 1971

Engvall E, Jonsson K, Perlmann P: Enzyme-linked immunoabsorbent assay. Biochim Biophys Acta 251:427, 1971

Jerne NK, Nordinn AA, Henry C: The agar plaque technique for recognizing antibody-producing cells. In Amos DB, Koprowski H (eds): Cell Bound Antibodies. Philadelphia, Wistar Institute, 1973

Laurell CB: Electrophoretic and electroimmunochemical analysis of proteins. Scand J Clin Invest 29:124, 1972

Mancini G, Carbonara AO, Heremans JF: Immunochemical quantitation of antigens by single radial immunodiffusion. Immunochemistry 2:235, 1965

Mittal KK, Mickey MR, Singal DP, et al.: Serotyping for homotransplantation. XVIII. Refinement of microdroplet lymphocyte cytotoxicity test. Transplantation 6:913, 1968

Ressler N: Two-dimensional electrophoresis of serum protein antigens in an antibody-containing buffer. Clin Chem Acta 5:795, 1960

Rose NR, Friedman H: Manual of Clinical Immunology. Washington, D.C., American Society for Microbiology, 1976

CHAPTER 13

The Complement System

During the latter part of the nineteenth century it became apparent that microbial invasion produced a number of human diseases and that specific immunization could provide an effective means of preventing many of them. Studies of immunologic mechanisms of host defense against bacteria demonstrated that microorganisms injected into the peritoneal cavities of immune animals underwent rapid dissolution. Bacteria were similarly lysed in vitro when added to the cell-free serum of immunized animals. Serum that had been aged for a few weeks or had been heated at 56C no longer supported a bactericidal reaction despite the fact that it still contained antibacterial antibody. The bactericidal reaction therefore required antibody and a heat-labile serum factor initially termed "alexin" (Greek, to ward off) and now called "complement."

Complement is a series of proteins that interact as an enzyme cascade and function as an immune effector of the acute inflammatory response (Table 13-1). Activation of the complement system on the surface of cells results in the production of structural and functional membrane alterations that lead to cell death. During sequential activation of complement (C), a number of biologic events are initiated that facilitate the localization and destruction of foreign material by immune effector cells. The C system can be activated by two synergistic pathways. The classic pathway, which consists of three components, C1, C4, and C2, is activated by immune complexes of antigen and IgG or IgM antibodies. The first component, C1, functions as a naturally occurring antiglobulin and, under appropriate conditions, binds to the Fc portion of IgG or IgM molecules. Following binding, C1 acquires enzymatic activity and can cleave its natural substrates, C4 and C2, which bind to the immune complex and have proteolytic activity capable of cleaving C3. The second pathway, the properdin or alternative pathway, bypasses C1, C4, and C2 and enters the reaction sequence at the level of C3.

Biochemistry of the Complement System

The realization that complement involved a number of components came with the original definitions based on differential inactivation by physical or chemical means. The order of interaction was not at first appreciated, and it is for historical reasons that the order of interactions in the classic pathway is C1, C4, C2, and C3. The late components, C5, C6, C7, C8, and C9, were numbered in order of their activation. At least 16 proteins are engaged in the complement cascade.

The proteins circulate in the plasma in inactive form. When activated, many of the proteins are converted from proenzymes to enzymes. Some fragments become adherent to other proteins or to surfaces; other fragments are released into the fluid phase, where they may have inflam-

TABLE 13-1. THE HUMAN COMPLEMENT SYSTEM

Component	Concentration in Serum (μg/ml)	Molecular Weight	Sedimentation Coefficient	Relative Electrophoretic Mobility
Classic pathway				
C1q	180	400,000	11.1	γ_2
C1r	—	180,000	7.5	β
C1s	110	86,000	4.5	α
C4	640	206,000	10.0	β_1
C2	25	117,000	4.5	β_1
Alternative (properdin) pathway				
Properdin	25	184,000	5.4	γ_2
Factor D(C3PAse, GBGase)	—	24,000	3.0	α_2
Factor B(C3PA, GBG)	200	93,000	5.6	β_2
C3b	—	171,000	9.0	α_2
Terminal components				
C3	1500	180,000	9.5	β_2
C5	80	95,000	8.7	β_1
C6	75	110,000	5.5	β_2
C7	55	163,000	6.0	β_2
C8	80	79,000	8.0	γ_1
C9	230		4.5	α
Inhibitory components				
C1 INH	180	90,000	—	α_2
β_1H	133	150,000	5.6	β_1
C3b INA (KAF)	25	100,000	5.5	β_2
C6 INA	—	—	—	β_1
Anaphylatoxin INA	—	310,000	—	α

matory properties. Activation occurs at a specific site. In the classic pathway this is typically where there is multiple binding of antibody to antigen in such a way that several of the constant domains of immunoglobulin are closely apposed. This results in the binding and activation of C1. Despite very extensive investigation no conformational changes have been detected in the antibody molecules. In the properdin or alternative pathway antibody is not required and triggering of this pathway is usually dependent on less specific surface characteristics for the binding of its initial component, C3. Microbial products are potent activators of the alternative pathway, and as shown below, this is probably the more important of the two pathways in immunologic defense.

Classic Pathway of Complement Activation

As stated above, the initial complex in the classic pathway of complement activation consists of antibody and the first component of complement. C1 is in itself a complex of three molecules, C1q, C1r, and C1s, which are held together by calcium. C1q is a very complex molecule that appears to have six areas capable of binding the Fc portion of certain antibodies. These binding areas are held together in a chain structure. Only IgM and IgG of the IgG1 and IgG3 subclasses are capable of binding to C1q. Further, C1q must be simultaneously affixed to immunoglobulins by at least two of its binding areas. Thus, one IgM

molecule or two IgG molecules are capable of binding C1q, provided they are affixed to antigens that are sufficiently close together to allow the C1q molecule to span the doublet. The binding of C1q activates (probably autoactivates) the attached C1r molecule, converting it from a proenzyme to an enzyme; this enzyme in turn activates C1s to an enzymatic form (Fig. 13-1).

C1 complex formation is modified by several considerations: (1) The necessity for a specific geometric ar-

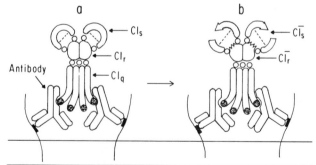

Figure 13-1. The formation of the antibody C1 complex of the classic pathway. Antibody bound by antigen exposes a binding area that binds C1q; two such interactions must occur for firm binding. This binding results in the activation of C1r to an enzyme, which, in turn, cleaves C1s, resulting in a proteolytic enzyme.

rangement of antibody molecules limits the conditions for activation. (2) Interactions between antibody and antigen and between antibody and C1q are unstable and, hence, reversible. (3) A specific inhibitor of C1r and C1s is the C1q esterase inhibitor present in the serum. This molecule also inhibits other plasma enzymatic activities; its deficiency results in hereditary angioendema (see below). The second complex of the classic pathway, C4b2a, is activated by the enzymatic action of C1s (Fig. 13-2). Native C4 is cleaved, and the larger part of the molecule (C4b) can covalently bind to proteins, e.g., those on cell surfaces. As many as 20 C4 molecules are scattered randomly about the first complex, providing a significant degree of amplification. Native C2, in the presence of magnesium ions, binds to C4 and is likewise cleaved by C1s. By this cleavage, C2 is converted from a proenzyme to an enzyme, the C3 convertase of the classic pathway. This enzyme reacts with two natural substrates (Fig. 13-3). In its first reaction it binds native C3 from plasma onto the C4 previously bound to the cell surface and cleaves it into two fragments. C3a is released into the fluid phase and has anaphylatoxin-like activity. C3b is a sticky protein capable of binding by a thiolester linkage to polysaccharide structures on the cell surface. C3b molecules that bind near the enzyme complex serve as the binding site of C5. C5 is the second substrate for C3 convertase and is cleaved into C5a and C5b. C5a is a potent chemotactic factor and anaphylatoxin; C5b is the first step of the terminal complex.

For IgM antibody to sheep red cells only one complement lesion is necessary for cell lysis. This hypothesis is known as the "one-hit" theory. As proof, it was necessary to determine whether the number of cells destroyed by complement activation was directly proportional to the amount of C present. By varying the number of available effective molecules of C1, C4, and C2, a direct correlation between the number of cells destroyed and the amount of

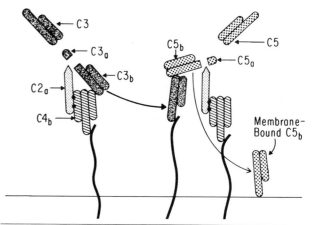

Figure 13-3. The cleavage of C3 and C5 by the classic pathway. Native C3 is bound by C4b and cleaved by C2a. The resultant molecule, C3b, is bound to membrane components. It, in turn, binds C5 and presents it for cleavage by a neighboring C42 complex. The resulting C5b is then bound to the plasma membrane.

C utilized was demonstrated, thereby supporting a one-hit theory of complement lysis.

Alternative (Properdin) Pathway

Early in the 1950s Pillemer described a factor in serum that combined with zymosan, a yeast cell wall polysaccharide, and selectively consumed C3 while apparently not requiring C1, C4, and C2. Pillemer termed the factor that bound to zymosan "properdin" (Latin *perdere*: to destroy). Other workers subsequently demonstrated a similar bypass of the early acting C components when serum was incubated with a factor derived from cobra venom or with bacterial lipopolysaccharides (endotoxins). The action of the properdin pathway required magnesium, a hydrazine-sensitive factor called "factor A," and a heat-labile factor termed "factor B." Since these factors bore striking chemical similarities to C4 and C2, the properdin system was at first discounted and thought to be an interaction between natural antibody and C1, C4, and C2. Subsequent investigation has clearly established the existence of an alternative (properdin) C pathway. In recent years, components of the properdin system have been isolated and purified and are distinct from the early acting proteins of the classic complement sequence. The potential biologic significance of the alternative pathway is that it may allow the activation of C by microbial products or altered self substances prior to the development or in the absence of specific IgG antibodies.

Seven proteins have been identified as participating in the activation or regulation of the alternative C pathway, these being C3 and the additional components, factor B, factor D, C3b, properdin, factor H, and factor I (Fig. 13-4). It has been proposed that the initial activation of this pathway begins as a random event of C3 and factor B forming a reversible complex that may be activated by fac-

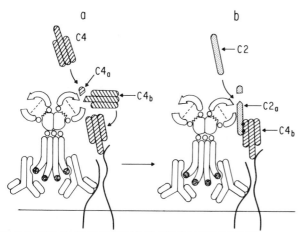

Figure 13-2. Formation of the C42 complex. Native C4 is cleaved by C1s, and C4b, the larger part, binds to components of the membrane. C2, bound to C4b by magnesium, is also cleaved by C1s, resulting in an enzyme, C2a.

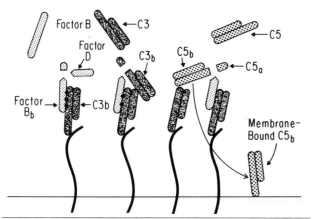

Figure 13-4. Formation and activity of the alternative pathway complex. C3b bound to membrane components binds factor B from plasma in the presence of magnesium. Factor B is cleaved by a serum enzyme, factor D, resulting in a proteolytic enzyme, factor Bb. This $\overline{C3bBb}$ complex is stabilized by another serum protein, properdin. This complex binds native C3 and cleaves it to C3b and C3a. C3b is bound by nearby membrane components and is capable of binding C5 from plasma. C5b bound in this way may be presented for enzymatic cleavage by neighboring C3bBb complex into C5a and C5b. C3b initiates the terminal complex.

tor D to form $\overline{C3Bb}$. This labile fluid phase enzyme is the initial C3 cleaving factor of the alternative pathway. It is capable of producing small amounts of C3b, which unless protected (see below) would be rapidly destroyed by factors H and I. This process of C3b formation and destruction is probably constantly ongoing at very low levels in serum. Certain substances are alternative pathway activators in the sense that they permit the binding of C3b on their surfaces in a way that allows the formation of more stable intermediates. These alternative pathway intermediates allow the C sequence to go to completion in the absence of immunoglobulin, C1, C4, or C2. Alternative pathway activators are generally complex polysaccharides or lipopolysacchrides, such as zymosan (yeast cell wall substance), inulin (plant polysaccharide), or endotoxin. The mechanism by which an activator facilitates the alternative pathway is by binding C3b in such a way as to make C3b less accessible to degradation by factors H and I (Fig. 13-2). Removal of sialic acid from membranes depresses the binding of factor H to any membrane-bound C3b and favors the activation of the alternative pathway by factor B. Thus membranous substances poor in sialic acid are good activators of the alternative pathway. The site of C3b deposition on the activator is designated as S, and S-C3b then may bind factor B, which upon binding becomes susceptible to cleavage by activated factor D. Once activated, factor B (Bb) together with bound C3b forms a C3/C5 convertase (S-$\overline{C3bBb}$). This enzyme undergoes a spontaneous rapid decay, with a half-life of approximately 2 minutes at 37C.

S-$\overline{C3bBb}$, in addition to its C3/C5 cleaving activity, is capable of activating precursor properdin to activated P (P). P can bind to S-$\overline{C3bBb}$ to form the properdin-stabilized C3/C5 convertase, S-$\overline{C3bPBb}$. In the presence of P, the half-life of the C3/C5 convertase is prolonged to 8 minutes at 37C.

The two types of alternative pathway C3/C5 convertases, the labile S-$\overline{C3bBb}$, and the properdin-stabilized S-$\overline{C3bPBb}$ are capable of fixing C5 through C9 to the activating agent, thus completing fixation of the complement sequence. The formation and degradation of the two C3/C5 convertases are regulated by factor I and factor H. These factors degrade C3b to C3b$_i$, which may then be further degraded by proteases to C3c and C3d. Factors I and H rapidly degrade fluid phase C3b but also are capable of slower degradation of C3b on S-C3b, S-$\overline{C3bBb}$, and S-$\overline{C3bPBb}$. In the absence of factor I, alternative pathway intermediates form a positive feedback loop and continue to convert all available C3 to C3b.

An endogenous activator of the alternative pathway has been found in some humans with hypocomplementemic glomerulonephritis. This substance has been termed "C3 nephritic factor" (NeF). C3NeF has been identified as an immunoglobulin that binds to the C3 convertase ($\overline{C3bBb}$) and forms $\overline{C3bBbNeF}$. NeF apparently stabilizes the $\overline{C3bBb}$ complex in a way similar to properdin and promulgates C activation by the alternative pathway. It has been postulated that NeF is an autoantibody directed toward $\overline{C3bBb}$.

It is apparent that the classic and alternative pathways act synergistically in whole serum, since selective blockage of the classic pathway depresses the rate of C3 cleavage by such properdin pathway activators as endotoxin, zymosan, and inulin. This synergism may reflect the role of $\overline{C4b2a}$ in producing the C3b required for the alternative pathway function.

Attack Complex

The $\overline{C423b}$ complex of the classic pathway and the $\overline{C3bBbC3b}$ complex of the alternative pathway are able to activate native C5 by enzymatic cleavage. The C5b that results is capable of binding to the membrane in a relatively loose fashion. Binding is increased by the presence of other components, particularly C3b, on the membrane and is markedly increased by the binding of the next component, C6. This forms the nidus for the formation of the terminal complex or the membrane attack complex (MAC), which evolves when the final components, C7, C8, and C9, are assembled (Fig. 13-5). Although enzymatic activity may be part of this assembly, the major effect of the combination is to create a complex of hydrophobic protein within the lipid bilayer. During these interactions C9 self-associates to complete the attack complex. The final assemblages can be seen in electron microscopy as tubular structures of approximately 9.0 to 11.0 nm (Fig. 13-6). It is clear that the mere presence of such a complex is not a sufficient condition for penetration of the lipid bilayer and that some,

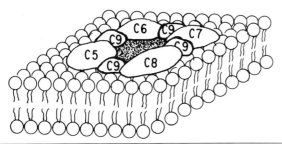

Figure 13-5. The membrane attack complex. The binding of C5 to the membrane is capable of binding C6. C7, C8, and C9 are subsequently bound. Under appropriate but presently unknown conditions, this intermembrane complex forms a lesion that permits the passage of internal contents, and the cell is lysed.

perhaps two or more, MACs must be assembled for penetration to occur. This process is relatively inefficient, particularly when the source of the complement is from the same species as the cells under attack. Further, in some bacteria that have capsules, this mechanism of membrane attack is totally ineffective, since the membrane complex is assembled at some distance from the cell. Nevertheless, when membrane attack is effective, as typified by the reaction against red cells and lysable bacteria, a water-soluble passage is made in the lipid bilayer, and the cell is destroyed.

Modulation of the Complement System

The classic and alternative pathways of complement activation are modulated by the enzymatic instability of certain C intermediates and by specific inhibitors of certain components. The unstable enzyme intermediates of the classic pathway include the decay of C2a from the $\overline{C4b2a}$ enzyme (half-life about 7 minutes) and of C3b from the $\overline{C14b2a3b}$ complex. The convertases of the alternative pathway are also unstable. Fluid phase C3Bb rapidly dissociates, and S-$\overline{C3bBb}$ and $\overline{S-C3bPBb}$ also decay rapidly (half-lives about 2 minutes and 8 minutes, respectively) due to the loss of factor B. Specific inhibitors of the complement sequence also regulate C enzyme activation. C1 esterase inhibitor, a normal serum constituent, combines stoichiometrically with C1, resulting in inactivation of its esterase activity. An inherited deficiency of this inhibitor is associated with hereditary angioedema (see below). Factor I in concert with factor H degrades C3b to C3b$_i$ by cleaving the β-chain of the protein into two fragments that remain held together by disulfide linkages. Proteolytic enzymes present in serum then degrade C3b$_i$ into C3c and C3d. Factor H alone accelerates the rates of decay of both C3/C5 convertases of the alternative pathway. Additionally, factor H interferes with the interaction of C3b with C5 in the classic pathway. In addition to the inhibitors regulating the complement sequence, a carboxypeptidase anaphylatoxin inactivator has been described, which cleaves the C-terminal arginine from the C3a and C5a fragments, rendering them less biologically active.

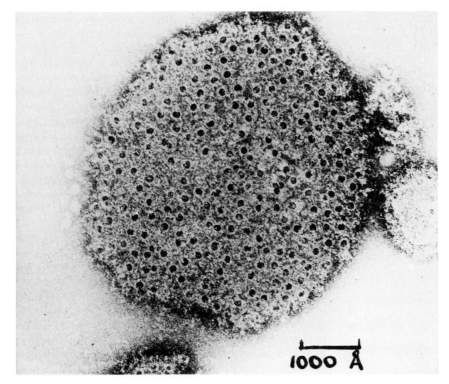

Figure 13-6. The intermembrane lesions of the membrane attack complex (C5b-9) may be seen on electron microscopy. The majority of these do not penetrate through the membrane. Some channels can be seen in cross section at the edge of the membrane. × 117,600.

Biologic Activities Mediated by the Complement System

Virus Neutralization

The binding of antibodies to certain viruses results in the formation of infectious virus–antibody complexes (VA). The addition of C1 and C4 to VA produces virus neutralization. Sequential addition of the later acting C components produces additional virus neutralization (Table 13-2). One mechanism of virus neutralization by C may be to prevent the attachment of the virus to cells by steric hindrance produced by virus-bound C.

Another mechanism by which C may neutralize particular types of viruses does not require antibody. RNA tumor viruses (Chap. 62) are lysed by human serum but not by serum from many nonprimates. Although C-mediated lysis of oncornaviruses does not require antibody to the virus it does require an intact classic C pathway. The C sequence is activated by the direct interaction of C1q with the P15E component of the viral envelope. This leads to formation of $\overline{\text{C1}}$ and subsequently to lysis of the viral envelope by the membrane attack complex. The ability of human serum to lyse oncornaviruses may provide a natural defense against RNA tumor viruses.

Immune Adherence

Nelson, in 1953, demonstrated that immune complexes that had reacted with C1, C4, C2, and C3 adhered to cells that possessed an immune adherence (IA) receptor. Cells that possess the IA receptor include primate erythrocytes, polymorphonuclear leukocytes, macrophages, some B lymphocytes, and nonprimate platelets. The IA phenomenon may have biologic significance in enhancing phagocytosis (opsonization) by leading to the binding of antigens to phagocytes. IA may also be a mechanism for rapid clearance of immune complexes from the circulation by fixed phagocytic cells (macrophages) in the liver and spleen, and it may permit localization of nonphagocytized complexes at B lymphocyte-rich areas of the spleen and lymph nodes, facilitating additional antibody production. The C fragment most responsible for the IA reaction is C3b or its cleavage product $C3b_i$.

Opsonization

Antigens that have bound antibody and C1, C4, C2, and C3 are far more easily phagocytized than antigen alone. This opsonization is facilitated by the binding of antigen-fixed C3b and C3bi to receptors on the plasma membranes of phagocytic cells. The addition of C5 further enhances the phagocytosis of immune complexes. Complement-dependent opsonization may be particularly important in limiting bacterial infections by bacteria, such as pneumococci, that are poorly opsonized in the absence of C.

Anaphylatoxin Activity

In 1910, Friedberger found that guinea pigs developed severe anaphylaxis when injected intravenously with serum that had been incubated with immune precipitates. Removal of the immune precipitates before injection did not prevent anaphylaxis, and he therefore proposed that a toxic factor, anaphylatoxin, was produced by the interaction of serum with immune complexes. It is now realized that anaphylaxis can be produced by IgE-mediated reactions (Chap. 21) as well as by C activation, but the term "anaphylatoxin" has persisted. It is used to describe a pharmacologically active substance derived from serum that (1) causes hypotension when injected intravenously, (2) causes abrupt contraction of guinea pig ileum, with subsequent tachyphylaxis, (3) is blocked from producing ileum contraction by antihistamines, (4) fails to contract estrous rat uterus, (5) increases vascular permeability in guinea pig skin, and (6) degranulates mast cells from guinea pigs but not from rat mesenteric preparations. It is now clear that the production of anaphylatoxin activity in serum requires C activation. Two complement-derived peptides, C3a and C5a, with molecular weights of 9040 and 11,200, respectively, have activities in common with classic anaphylatoxin, that is, anaphylatoxin purified from complement-activated serum. C5a is 100 to 1000 times more potent than is C3a in contracting guinea pig ileum, while C3a produces degranulation of rat mesenteric mast cells, a property not shared by classic anaphylatoxin. C5a therefore appears to be identical to classic anaphylatoxin, while C3a, although less potent, produces similar biologic effects. C4a, an 8740 dalton peptide cleaved from C4 also has anaphylatoxin activity but is less potent even than C3a. The primary amino acid structures of C5a, C3a, and C4a share considerable homology, and it is likely that the parent molecules (C5, C3, and C4) arose by gene duplication of a common ancestor. A naturally occurring carboxypeptidase termed "anaphylatoxin inactivator" removes the C-terminus residue from the anaphylatoxins and diminishes their biologic activities considerably. The biologic function of anaphylatoxins in vivo may be to induce vascular per-

TABLE 13-2. BIOLOGIC ACTIVITIES OF COMPLEMENT

Activity	Components Involved
Virus neutralization	Virus + Ab, $\overline{\text{C14b}}$ and entire C pathway
	Retroviral P15E + C1q and entire C pathway
Immune adherence	C3b, $C3b_i$
Opsonization	C3b, $C3b_i$
Anaphylatoxin activity	C4a, C3a, C5a
Chemotactic activity	C5a
Induction of respiratory burst and secretion by phagocytes	C5a
Cytolysis	C5b-9(membrane attack complex)

meability and stasis, thus allowing the efflux of serum proteins, including antibodies, into surrounding tissues and permitting establishment of stable chemotactic gradients necessary for recruitment of phagocytic cells.

Chemotactic Activity

Chemotaxis is the unidirectional migration of cells toward an increasing concentration of a chemical attractant (Chap. 18). Leukocytes, such as polymorphonuclear leukocytes and macrophages, are capable of chemotactic migration. Small lymphocytes may also migrate chemotactically. Leukocyte chemotaxis can be quantified in vitro using a chamber with two compartments separated by a porous filter. The cells are placed on one side of the filter, and substances to be tested for chemotactic activity are placed on the other. The number of cells migrating through the filter

can then be counted. Activation of C in serum by inflammatory agents, such as immune complexes or endotoxins, produces chemotactic activity for polymorphonuclear leukocytes and macrophages (Fig. 13-7). Most of the chemotactic activity in activated serum is due to C5a. C5a and C3a have been highly purified and their primary structure determined. C5a is a glycoprotein composed of 74 amino acids. The carboxyl terminal amino acid is arginine, and this amino acid may be cleaved from C5a by a carboxypeptidase normally present in serum. Loss of the terminal arginine reduces but does not abrogate the chemotactic potency of C5a. Although C3a has amino acid sequence similarities to C5a, it lacks a methionyl residue present in C5a at the fifth position from the carboxy terminus. Purified C3a is not chemotactic, and it is noteworthy that human polymorphonuclear leukocytes contain a specific receptor for C5a but not C3a. Evidence against the signifi-

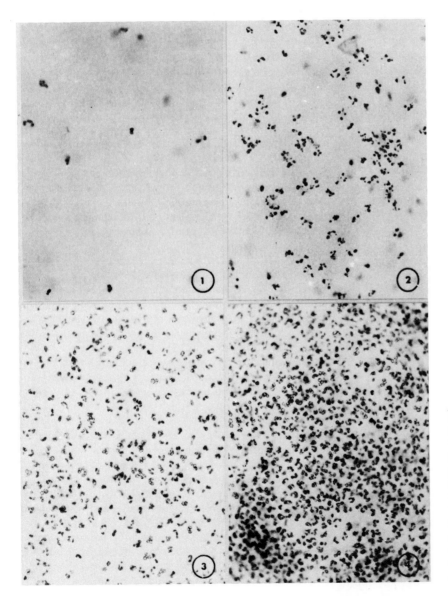

Figure 13-7. Neutrophil response to chemotactic factor generated by graded doses of *Veillonella alcalescens* endotoxins incubated in guinea pig serum. The neutrophils are placed in one compartment of a double chamber closed by a membrane. The endotoxin is placed in the second compartment. The cells migrate into the endotoxin as a function of concentration. **1.** Serum alone. **2.** Serum plus 0.05 μg endotoxin. **3.** Serum plus 0.5 μg endotoxin. **4.** Serum plus 5.0 μg endotoxin. × 325. *(From Snyderman et al.: J Exp Med 128:259, 1968.)*

cance of a formerly reported C567 chemotactic factor is that human and animal sera congenitally devoid of C6 or C7 generate normal levels of chemotactic activity upon C activation. Naturally C5-deficient individuals develop only low levels of activity when their C is activated.

Injection of C5a into the skin produces vasodilatation, vascular stasis, and local accumulation of polymorphonuclear leukocytes and macrophages. The histologic picture looks quite similar to an Arthus reaction, and indeed C5a is an important mediator of the acute inflammatory response initiated by immune complexes. At higher concentrations than that required for chemotaxis, C5a initiates a respiratory burst and secretion of lysosomal enzymes by neutrophils and mononuclear phagocytes (Chap. 18).

Deficiencies of Complement System and Their Relationship to Human Diseases

Complete or partial deficiencies of all the classic C components, with the exception of some C inhibitors, have been described in humans or animals (Table 13-3). Some of these complement deficiencies are associated with severe diseases, while in others, clinical manifestations are sporadic. Defects in the C system could result in impaired

TABLE 13-3. COMPLEMENT COMPONENT DEFICIENCIES AND DISEASE ASSOCIATIONS

Component	Clinical Status
C1	SLE, glomerulonephritis
C4	SLE
C2	SLE, recurrent infections, inflammatory bowel disease, asymptomatic
C3	Pyogenic infections (incl. *Neisseria meningitidis*), fever, and rashes
C5	SLE and recurrent infections, pneumococcal pneumonia, recurrent disseminated gonococcal infections (DGI)
C5 dysfunction	Eczema and infections
C6	*Neisseria gonorrhoeae* arthritis, DGI, recurrent *N. meningitidis* infections, *Streptococcus pneumoniae* and *N. meningitidis* infections, primary biliary cirrhosis
C7	Raynaud's phenomenon, recurrent *Neisseria* infections, chronic pyelonephritis, asymptomatic
C8	Xeroderma pigmentosum, SLE, *N. gonorrhoeae* infection, *N. gonorrhoeae* and staphylococcal endocarditis, *N. meningitidis* infection, and IgA deficiency, asymptomatic
C9	Asymptomatic

elimination of microbial antigens or circulating immune complexes. Indeed, C deficiencies have been associated with recurrent bacterial and fungal infections as well as with collagen-vascular inflammatory diseases.

A deficiency of dysfunction of C1 esterase inhibitor results in hereditary angioedema, an autosomal dominant heritable disease. It is characterized by acute and transitory local accumulations of edema fluid, which, when localized in the larynx, can become life-threatening by obstructing the tracheal airways. During attacks, hemolytic C activity in these individuals is markedly depressed due to consumption of C4 and C2 by unregulated C1 esterase. The edema may be produced by a kinin cleaved from C2 by the action of C1s. Attacks appear to be triggered by activation of Hageman factor (factor XII, the plasma protein that initiates clotting), which leads to the formation of plasmin and kallikrein. These proteases cleave and activate C1, resulting in activation of the C system. The most promising approach to treating hereditary angioedema has been the use of androgen therapy with the attenuated sex hormones, such as fluoxymesterone and particularly danazol. These agents are quite effective in preventing attacks and appear to do so by increasing the synthesis of C1 esterase inhibitor.

Partial C1q deficiencies have been found in several patients with combined immunodeficiency disease. Normal levels of C1q were restored upon bone marrow transplantation. Selective deficiency of C1r has been found in a few patients with glomerulonephritis and polyarthritis. C2 deficiency is the most common of the heritable C deficiencies. Approximately half of the individuals with C2 deficiency enjoy normal health, while the remainder may have lupus-like diseases. Deficiencies of the fourth and second component of complement have been found in a number of patients with collagen-vascular diseases, particularly systemic and discoid lupus erythematosus. Genes, both regulatory or structural for C2 and C4, as well as for factor B, are closely linked to the major histocompatibility complex (HLA) of humans, and HLA-linked genes are involved in many diseases. Therefore, the lupus-like diseases may not have been directly due to lack of C2 or C4 but to some other gene of this complex (Chap. 18).

Individuals with defects of C3 suffer from severe recurrent bacterial infections. Sera from these patients generate less chemotactic activity and support less phagocytic activity than do normal sera. It should be noted that an almost complete deficiency of these components is required to produce clinically apparent disease. A defect of the C3b inactivator has also been described. A patient with this deficiency had very low levels of C3 because of hypercatabolism of this component, presumably because of constant activation of the alternative C pathway by C3b. This patient suffered from bacterial infections, and serum from this individual was not able to generate chemotactic activity, probably because C3 is required for the activation of C5. The depression of C3 could be reversed in vivo by the administration of C3b inacativator.

Isolated deficiencies of each of the terminal C components have been discovered in humans. Several families with heritable C5 deficiency have been found. In one, the proband had systemic lupus erythematosus. A sister with very low but detectable levels of C5 had several bouts of pneumococcal pneumonia. In another family, the proband had low but measurable levels of C5 (about 0.5% of normal hemolytic activity). She suffered from repeated episodes of disseminated gonococcal infections. While two other family members with similar levels of C5 were healthy, their risk factor may have differed from that of the proband. However, individuals with isolated deficiencies of C5, C6, C7, or C8 do appear to be generally more susceptible to recurrent infection with neisserial organisms. Of 31 individuals found to be deficient in either C5, C6, C7, or C8, 15 have had infections with *Neisseria gonorrhoeae* or *Neisseria meningitidis*. Thus, the bactericidal activity of C may be important for protection against neisserial infection. Some individuals with terminal C deficiencies have had rheumatic disorders.

The association of C defects with infectious diseases could certainly be anticipated, but the high frequency of collagen-vascular inflammatory diseases with deficiencies of C, in particular C1r, C4, and C2, was less predictable. This latter association suggests that patients with isolated deficiencies of the early acting classic C components are more susceptible to subtle infections that cause systemic inflammatory diseases, or that elimination of antigens and immune complexes in general by these individuals is suboptimal, thus producing chronic stimulation of the immune response. In any case, understanding the intriguing relationship of C defects with inflammatory diseases will lead to a better understanding of the pathophysiology of these diseases, as well as of the biologic role of the C system.

FURTHER READING

Books and Reviews

Hugli TE: The structural basis for anaphylatoxin and chemotactic functions of C3a, C4 and C5a. Crit Rev Immunol 4:321, 1981

Porter RR, Reid KBM: The biochemistry of complement. Nature 275:699, 1978

Selected Papers

Chenoweth DC, Hugli TE: Demonstration of specific C5a receptor on intact human polymorphonuclear leukocytes. Proc Natl Acad Sci USA 75:3943, 1978

Cooper NR, Jensen FC, Welsh RM Jr, et al.: Lysis of RNA tumor viruses by human serum: Direct antibody-independent triggering of the classical complement pathway. J Exp Med 144:970, 1976

Jensen J: Anaphylatoxin in its relation to the complement system. Science 155:1222, 1967

Lachman PJ, Rosen FS: Genetic defects of complement in man. Springer Semin Immunopathol 1:399, 1978

Müller-Eberhard HJ, Schreiber RD: Molecular biology and chemistry of the alternative pathway of complement. Adv Immunol 29:1, 1980

Shin HS, Snyderman R, Friedman E, et al.: Chemotactic and anaphylatoxic fragment cleaved from the fifth component of guinea pig complement. Science 162:361, 1968

Snyderman R, Phillips JK, Mergenhagen SE: Biological activity of complement in vivo: Role of C5 in the accumulation of polymorphonuclear leukocytes in inflammatory exudates. J Exp Med 134:1131, 1971

Tschopp J, Muller-Eberhard HJ, Podack ER: Ultrastructure of the membrane attack complex of complement: Detection of the tetramolecular C9-polymerizing complex C5b-8. Proc Natl Acad Sci USA 79:7474, 1982

CHAPTER 14

Cellular Basis of the Immune System and Immunoregulation

Cellular Aspects of the Immune System

The immune system is primarily concerned with distinguishing between nonself (usually microbial invaders) and self components. The central role of lymphocytes in this process and the subdivision of lymphocytes involved in various immune responses was not appreciated until the 1960s. Since then a great wealth of information, not only about the reactions involving lymphocytes but also about the regulation of antibody responses, has been gained. In this chapter, the organization of this immune system and the properties and interactions of its various components are considered.

Components of the Immune System

Barriers and Drainage

As stated in the introduction, the immune system per se is really a second line of defense that has evolved in multicellular, differentiated organisms. All vertebrates possess primary barriers that prevent the invasion of most foreign organisms (Chaps. 11 and 19). Once past these "walls," invaders drain into a web of lymphatics. The lymphatic system, schematically shown in Figure 14-1A, underlies the skin and extends throughout the body. Thin-walled lymphatic vessels carry foreign matter (and cells leaving the tissue spaces) into larger lymphatics and ultimately, via the thoracic duct, into the bloodstream. Interposed along this network of lymphatics are the lymph nodes, in which foreign materials are filtered out and processed, and in which the immune responses are initiated. In addition, lymphocytes circulate through the nodes so that antigens entering the lymph are exposed to a broad spectrum of antigen-specific lymphocytes in a relatively brief time.

Filtration can be demonstrated by injecting carbon particles into the footpad of an experimental animal. Within minutes, the popliteal lymph node (behind the knee) becomes "black"; later, any particles that escape the popliteal node pass through efferent lymphatics to the inguinal node, and ultimately may be dispersed throughout the body.

Figure 14-2 shows typical immune elimination curves to illustrate further both the nonspecific and specific aspects of the entire immune system. An antigen labeled with a radioactive tracer molecule is introduced intravenously. The first phase is a brief period of equilibration between intra- and extravascular space (line 1), occurring primarily through diffusion. Nonimmune catabolism of antigen, during which time antigen may be degraded, follows

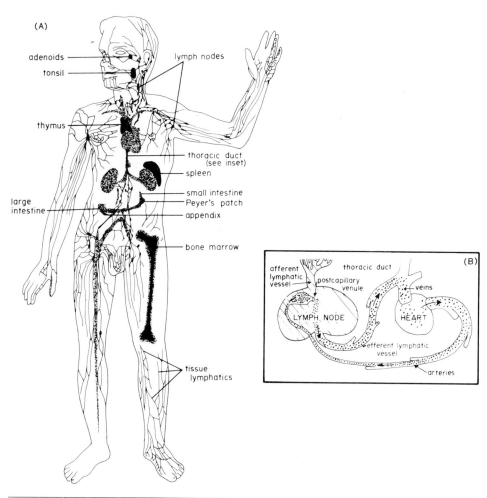

Figure 14-1. A. The human lymphoid system, which includes a network of lymphatic vessels (*thin lines*) along which are lymph nodes (*black dots*). The spleen, blood lymphocytes, adenoids, tonsils, appendix, and Peyer's patches compose the remainder of this system. The bone marrow (e.g., in the femur shown here) and the thymus are primary sites of lymphocyte production. The tissue lymphatics carry antigen, antibodies, and lymphocytes around the body. **B.** The lymphocyte recirculation pathway involves lymphocytes entering the nodes via the postcapillary venules (see Fig. 14-4C) and percolating through the node before exiting in the efferent lymphatics. These drain into larger vessels (e.g., the thoracic duct) and eventually enter the neck veins to join the blood circulation, and the process is repeated. (*A*, *adapted from Jerne: Sci Am 229:52, 1973. B, adapted from Gowans: Hosp Prac 3:34, 1968.*)

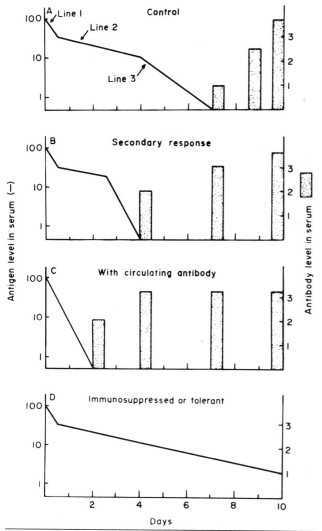

Figure 14-2. Nonimmune and immune elimination of antigen. Radioactively labeled antigen is injected intravenously on day 0, and samples of blood serum are measured for remaining radioactivity (*solid lines*) or titrated for free antibody activity (*shaded bars,* arbitrary scale). Immune complexes are often detectable before free antibody. **A.** Normal animals not previously exposed to antigen. **B.** Previously immunized animals, but with no antibody detectable on day 0 (secondary stimulus). Note shorter lag period. **C.** Immunized animals with circulating antibody still present show accelerated (immune) elimination. **D.** Immunosuppressed or tolerant animals. Note extended nonimmune elimination (catabolism) and no antibody formation. *(Adapted from Talmage et al.: J Immunol 67:243, 1951.)*

(line 2). This occurs with both self and nonself antigens. After a lag period, which usually lasts for several days and is dependent on the half-life of the antigen, there is an immune or rapid elimination of the residual antigen (line 3). This occurs after the immune system has started to produce antibodies that complex with the antigen. Complexes are very rapidly taken up by phagocytic cells and eliminated. Part B of Figure 14-2 shows the pattern in an animal

that has previously seen this antigen. Line 1 is identical to that seen in a normal animal. However, line 2 is of shorter duration before the secondary immune response occurs. This typifies the difference between a primary immune response and an anamnestic (meaning: not to forget) memory response, as described in Chapter 13. Part C of Figure 14-2 depicts the situation when antigen is introduced into an organism that still has circulating antibody. No lag period is observed and the antigen is eliminated in an immune fashion. Part D of Figure 14-2 shows an antigen to which the host is tolerant, or an irradiated or otherwise immunologically incompetent host.

Macrophages and the Phagocytic (Reticuloendothelial) System

Initial nonspecific aspects of the immune response are carried out by macrophages and other phagocytic cells of the reticulendothelial system. These include monocytes and polymorphonuclear neutrophils (PMN) of the blood; macrophages in the lymphoid tissues; Kupffer's cells in the liver; Langerhans' cells in the skin; and alveolar macrophages in the lungs. Many of these cells function primarily as scavenger cells, ingesting foreign or effete autologous materials, but also participate in inflammatory reactions. Some cells, such as medullary and dendritic macrophages in the lymph nodes, also function to present antigens to lymphocytes (see below), initiating the immune response.

Macrophages are derived ultimately from a stem cell in the bone marrow. The blood monocyte is the intermediate cell in this maturation process. Elegant studies have demonstrated the differentiation of rapidly dividing marrow stem cells to blood monocytes to tissue macrophages. During this process, the cells change morphologically and biochemically, notably in the accumulation of lysosomal enzymes that ultimately destroy ingested particulate matter through a process of endocytosis, phagolysomic localization, and enzyme release.

The Lymphocytes and Recirculation

The cells that carry out a specific immune response belong to the two major classes of lymphocytes: T cells and B cells. The involvement of antibody in immune responsiveness has been known since the last century, but the source of antibody was unknown until its production by plasma cells—the end stage of differentiation of the B lymphocytes—was demonstrated. The importance of small lymphocytes as the central defenders in the immune system was only realized 25 years ago from the work of Gowans. He cannulated the major lymphatic vessel of the rat, the thoracic duct, and noted that the thoracic duct lymphocytes could transfer the entire immune potential of the host animal. Moreover, when reinjected intravenously into a rat with an indwelling cannula, thoracic duct lymphocytes (TDL) from the first animal came back through the cannula after a brief sojourn in the host; that is, they recirculated from blood to lymph and back again (Figure 14-1B). During this process, lymphocytes in the blood attach to specialized endothelial cells in the postcapillary venules (PCV) of the lymph nodes. They migrate between the endothelial cells lining the PCV and begin to percolate through the node.

Small lymphocytes (6 to 8 μm in diameter), which make up more than 95 percent of all cells traversing the lymphatics and 20 percent of the nucleated cells in the blood, also bear specific antigen-recognition units or receptors on their membranes. These receptors give the lymphocytes their specificity in recognizing different antigens. When they pass through a lymph node in which antigen is trapped (see below), they bind to the antigen via these receptors. They are thus held out of the circulation and remain in the node to join in an immune response pattern. This process is called recruitment.

Four other major discoveries led to much of what we now know of lymphocyte function:

1. The finding that lymphocytes were not, as had been previously accepted, end cells; they could be stimulated to divide by a plant lectin, phytohemagglutinin (PHA). We also know that, contrary to earlier suppositions, at least a proportion of lymphocytes live almost indefinitely. Lymphocytes from subjects receiving high levels of irradiation therapeutically (for example, for ankylosing spondylitis) show chromosomal breaks when stimulated by PHA to divide years later. These cells are so damaged that they cannot further divide; thus, they must have persisted in the resting state for years.

2. The observation that mice thymectomized at birth lose much of their immune response capacity in adult life. This led to the discovery of the dichotomy of immune function in terms of thymus-dependent or -independent reactions and T and B cells (see below).

3. Further recirculation experiments involving TDL injected into like (syngeneic) or unlike (allogeneic) rats. In syngeneic rats, as described above, the cells recirculate and can be recovered from the thoracic duct. In allogeneic rats, the injected cells tend to colonize the spleen and other lymphoid organs and then to divide and attack the host in what is called a graft-vs.-host reaction.

4. The in vitro experiments demonstrating that lymphocytes could attack foreign target cells opened up studies of lymphoid function in tissue culture.

These observations were important in laying the foundations for what we now know as modern immunology.

Structure of Lymphoid Organs

The Thymus—A T Cell Factory

Not all lymphoid organs are built for filtration; some are factories for making more lymphocytes. These so-called central or primary lymphoid organs are exemplified by the thymus, where T cells differentiate. Figure 14-3A shows a cross-section of the mammalian thymus, a bilobed organ in the anterior mediastinum. The thymus, the first lymphoid organ to appear in ontogeny, is derived from the third and fourth pharyngeal pouches. Although thymic lymphocytes (thymocytes) were originally believed to be formed directly from epithelial cells, cell transfer and parabiosis experiments established that stem cells migrate into the thymus via the bloodstream. Stem cells are first identifiable in the yolk sac, then in the fetal liver, and still later in the bone marrow. Migration continues from the bone marrow in adults (Chap. 15). Shortly after their arrival in the thymus, precursor (T) cells organize into a thin layer in the outer cortex and begin dividing. Gradually, these cells migrate into the medulla, where they become smaller, acquire new cell surface antigenic markers, and, most importantly, develop immunocompetence. Though the exact differentiation scheme(s) is still being investigated, evidence to date indicates that differentiation and maturation are under the influence of hormones such as thymopoietin and thymosin, produced by the epithelial elements in the thymus. The cells producing these hormones can now be identified with monoclonal antibodies, as shown in Figure 14-3B. A proportion of the medullary thymocytes exit from the thymus and are continuously replaced through cell division. There is also a great deal of intrathymic cell death. It is believed that the diversity of the immune system is generated in the thymus and that many of the cells, especially in the embryo, are mutants with differing specificities. It is also postulated that thymocytes expressing antiself reactivity are treated as lethal mutants and destroyed. Though there is no proof for this long-held notion, recent data suggest that the thymus is where T cells become tolerant to self antigens and learn what self is (Chap. 17).

There is no known equivalent of the thymus for B cell development in mammals, although B cells also go through maturational and developmental processes (Chap. 15). In avian species, the bursa of Fabricius (a lymphoid organ in communication with the cloaca) plays a major role in B cell development. In mammals, B cells probably differentiate in the microenvironment of the bone marrow or fetal liver. B cell maturation may also occur in other lymphoid organs including the spleen, tonsils, and possibly in the Peyer's patches of the gut. The first identifiable poststem cell stage has been called the pre-B cell. These cells have immunoglobulin (Ig) in their cytoplasm but not on their surface. Subsequently, the surface membrane immunoglobulin they express is IgM; this is later joined by membrane IgD in most B cells. Other surface markers have also been identified sporadically on developing B cells, but surface IgM (with or without IgD) is the hallmark of B cells. Further discussion of B and T cell markers appears later in this chapter.

Peripheral Lymphoid Organs

The lymph nodes (and to a certain extent, the spleen) are intricate filters where immune recognition of foreign materials occurs. Figure 14-4, a lymph node, can be contrasted with the thymus shown in Figure 14-3. Lymph nodes have

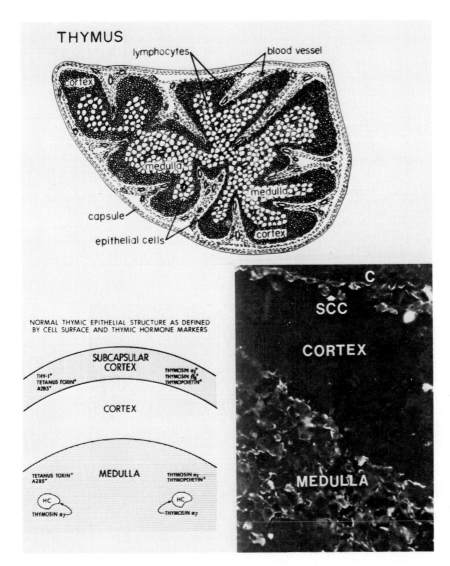

Figure 14-3. Thymus architecture. Dense accumulation of thymic lymphocytes (thymocytes) in the cortex surrounding a lighter medullary area, composed primarily of epithelial cells and more mature thymocytes. Lower right: Reactivity pattern of normal human thymus labeled with monoclonal antibody A2B5. Scheme at lower left depicts this pattern, which matches the staining observed with several thymic hormones. *(Photomicrographs courtesy of Dr. Barton F. Haynes; reprinted from J Immunol 130:1182, 1983.)*

two main areas: a central medullary area and a peripheral cortex without discrete borders. Lymph fluid enters via afferent lymphatics carrying cells, debris, and foreign particles from the extracellular spaces, and passes through a series of sinusoids lined with phagocytic macrophages.

Within the superficial cortex, which is predominantly composed of B cells (Fig. 14-4A), are structures called primary follicles. These oval-shaped areas include a dense accumulation of lymphocytes in a mesh of macrophage-like cells called dendritic reticulum cells. During an immune response, new centers—called secondary follicles—are formed, in which the developing germinal center is partially covered by a cap of phagocytic cells. The germinal center contains many lymphocytes and large, actively metabolizing blastoid cells as well as dendritic reticulum cells and tingible body macrophages actively involved in the immune response.

Surrounding the follicles in the paracortical region (or deep cortex) are further accumulations of lymphocytes, primarily T cells. Interspersed in the deep cortex are the

postcapillary venules. Between the endothelial walls of the postcapillary venules, lymphocytes enter the node from the blood. The role of the postcapillary venules in the recirculation of lymphocytes can be seen when radioactively labeled lymphocytes (obtained from the thoracic duct lymph) are injected intravenously. Within minutes, labeled cells traverse the postcapillary venules into the deep cortex T cell area and later into the superficial cortex (B cell area). After percolating through the rest of the lymph node, most of these cells exit via efferent lymphatics and recirculate (Fig. 14-1B).

Like the lymph nodes, the spleen acts not only as a filter but also as a site of immune response. The spleen has no lymphatic vessels and is supplied exclusively by the bloodstream. The spleen is divided into a lymphoid white pulp and an erythroid red pulp. The latter area removes effete red cells. It also serves as a site of hematopoiesis and, in some species with a contractile spleen (e.g., dog), as a reservoir of red cells. The white pulp, surrounding the splenic arterioles, is analogous to the lymph node cor-

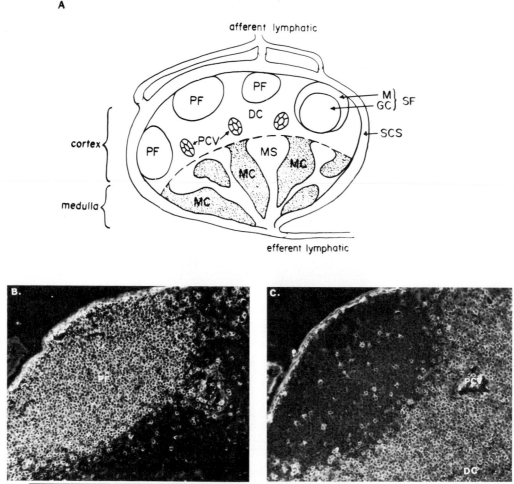

Figure 14-4. Lymph node structure. **A.** Schematic drawing of a lymph node. PF, primary follicle; SF, secondary follicle containing a germinal center (GC) and mantle (M); DC, diffuse cortex containing the postcapillary venules (PCV); MS, medullary sinus; MC, medullary cords. **B.** and **C.** T and B cell localization. Frozen sections of a mouse lymph node were stained with fluorescein-labeled antibodies to mouse immunoglobulin (**B**), or with fluorescein-tagged anti-T cell reagent (**C**). The antiimmunoglobulin stains B cells because of their surface immunoglobulin; they are localized in the primary follicle (light area in **B**). T cells in the diffuse cortex and a few in the PCV are labeled in **C**. There is little overlap of the two cell types. *(Photomicrographs courtesy of Dr. G. Gutman.)*

tex (the so-called periarteriolar lymphocyte sheath) and contains accumulations of less densely packed lymphocytes and numerous follicles. Interposed between the sheath and the red pulp is the marginal zone, which is somewhat analogous to the lymph node medulla. In addition, mature plasma cells can be found in parts of the red pulp. Plasma cells are rarely found outside the organized lymphoid tissues.

Other lymphoid tissues, such as the Peyer's patches, tonsils, and adenoids are also somewhat specialized. The most obvious differences among them are the varying proportions of T and B cells and the class of antibody produced in different types of lymphoid tissue. (Peyer's patches, for example, contain many IgA-secreting cells.) Interestingly, the ratio of T and B cells within an organ is

determined by receptor site interaction with the endothelial cells of the PCV. That is, Peyer's patch PCV bind more B than T cells. Hence, this organ possesses nearly twice as many B as T cells, in contrast with the peripheral nodes.

Properties of T and B Cells

The most notable property of the lymphocytes, aside from their ability to recirculate, is their specificity endowed by surface receptors which to bind specific antigen. The receptor on B cells is immunoglobulin. Although all mature B cells are surface Ig-positive, only a small proportion (less than 0.1 percent) can interact with any given antigen. For example, one small fraction binds to tetanus toxoid, another fraction to polio virus, and so on. Thus, lymphocytes are believed to belong to many small clones, together representing every possible specificity in the repertoire of the host animal. Its surface receptors enable a given lymphocyte to respond to an antigen encountered in a lymph node. Upon interacting with a multivalent antigen, the receptors on B cells coalesce into a "cap" (Fig. 14-5). The cap is then endocytosed so that the cell loses its receptors, which are resynthesized within 6 to 12 hours. Capping involves actin in the cytoskeletal structure beneath the membrane, and leads to the stimulation of lymphocyte migration. The T cell receptor is a specialized membrane-bound molecule or molecules with some characteristics of immunoglobulins (e.g., idiotypes), but lacking conventional light and heavy chains. It appears to be a heterodimer of disulfide-linked chains of 40,000 to 45,000 daltons.

Numerous other markers help distinguish subsets of T and B lymphocytes (Table 14-1). Some are shared by both T and B cells, whereas others (like Thy-1 in the mouse) are restricted to T cells or denote B cells (surface Ig). The function of certain markers is apparent: e.g., the surface immunoglobulins or the receptors for the constant domain of immunoglobulin (Fc receptors). Markers such as the histocompatibility (H) antigens (Chap. 17), including so-called murine Ia antigens (D-region related in humans), are essential for collaboration and communication between cells. Stages of differentiation are exemplified by the murine Tla and Lyt markers of T cells. The latter antigens

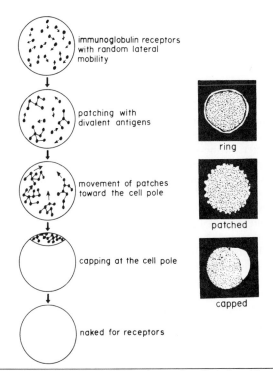

Figure 14-5. The process of capping on B cells induced by multivalent antigen or by bivalent antiimmunoglobulin reagents. Arrows denote movement of crosslinked membrane molecules. *(Adapted from Hood et al. (eds): Immunology, 1978. Sunderland, Mass, Benjamin/Cummings Publishing Co.)*

also reflect functionally different T cell subsets: $Lyt1^+$ cells generally function as helper cells, whereas $Lyt2^+$ cells are active as cytotoxic and suppressor cells. In humans, the analog of Lyt1 is defined by a monoclonal antibody called T4; monoclonal T5,8 reacts with the $Lyt2^+$ equivalent in humans (see also Chap. 15).

In addition to these qualitative changes, there may be fluctuations in expression. For example, the Ia antigen concentration increases when B cells are stimulated to proliferate; the surface Ig concentration falls to undetectable levels when B cells mature to plasma cells. Proliferation, induced by substances called polyclonal activators or mitogens (lipopolysaccharide, a bacterial product, is frequently used for this), precedes differentiation and greatly changes the profile of cell surface markers. Receptors for many physiologically important molecules, such as insulin or transferrin, also increase in number following stimulation of the lymphocyte.

Results of an Antigenic Encounter

Role of Macrophages

Within minutes after an antigen enters a lymph node, it is taken up (phagocytized) by macrophages in the subcapsular sinus, the cortex, and the medulla. Similar events occur in the spleen if the antigen is blood-borne, or in the tonsils and adenoidal lymphoid tissues if the portal of entry is nasal or oral. These sessile or tissue-bound macrophages comprise part of the reticuloendothelial system.

After uptake by macrophages, antigen enters cytoplasmic vacuoles called phagosomes, which fuse with lysosomes, organelles laden with hydrolytic enzymes. The number of lysosomes can increase during phagocytosis or after antigenic stimulation, the latter mediated by products of T cells. Such a cell is called an activated macrophage. Lysosomal enzymes degrade most of the antigen. Some antigenic material is preserved in a recognizable (or processed) form, either intracellularly or on the macrophage surface, often in association with Ia antigens. Processed antigen has been shown to be highly immunogenic for lymphocytes. Some antigen, presumably in a highly immunogenic form, is also maintained on the surface of the nonphagocytic follicular dendritic reticulum cells of the lymph nodes.

Studies in vitro have shown that macrophages are critically important in triggering the activation of specific lymphocytes. For example, spleen cells can be separated by exposure to a glass surface into an adherent, macrophage-rich fraction and a nonadherent lymphocytic fraction. Antigen incubated with either fraction does not stimulate an immune response. However, antigen added to the adherent macrophages followed by the addition of nonadherent cells will stimulate the latter to yield an antibody response in vitro. This accessory function is relatively resistant to γ- or x-irradiation, whereas most lymphocytes are radiosensitive.

Our detailed knowledge of the localization of antigen in the lymph nodes is based largely on studies involving

TABLE 14-1. T AND B LYMPHOCYTE AND MACROPHAGE PROPERTIES

	T Cells	B Cells	Macrophages
Differentiation site	Thymus	Marrow (bursal equivalent)	Marrow
Recirculation	Fast	Slower	
Life span	Long-lived	Short to long-lived	
Specificity	Antigen + MHC	Antigen (hapten)	
Memory	+	+	−
Antigen receptor	?	Immunoglobulin	
Fc receptors	Some (Tγ, Tμ)	Most	Many
C3b receptors	−	+	+
SRBC-rosette	+(Human)	−	−
Major histocompatibility complex antigens (K/D-like)	+	+	+
Ia antigens	Minority	+	Some
Thy-1, brain associated antigen	+	−	−
T4 marker	Helpers	−	−
T8 marker	Cytotoxic/suppressors	−	−
Localization (percent)			
Peripheral blood	70 to 80	20	+
Thoracic duct	90	10	−(Early)
Lymph node	75 (Deep cortex)	25 (Follicles)	+
Spleen	40 to 50	40 to 50	+
Thymus	100 (90 percent immature)	−	Few
Bone marrow	Few mature	Some mature B	+
Mitogen response			
PHA	+	±	−
ConA	+	−	−
Pokeweed	+	+	−
Endotoxin (lipopolysaccharide)	−	+ (Mouse, some human B)	± (Factor release)
Sensitivity to:			
Corticosteroids	+	+	−
Irradiation	+	+ +	−
Antilymphocyte serum	+	+	±
Immunosuppressive drugs	+	+	−

radioiodinated flagellar proteins from *Salmonella* organisms, as well as other antigens. Classic studies visualized the localization of these antigens during the primary response in medullary macrophages and in the marginal sinus around the primary follicles, close to many B cells. In contrast, during a secondary response, antigen was localized predominantly within the follicles on the dentritic cells. Follicular localization could be mimicked in unimmunized animals by the administration of preformed antigen–antibody complexes. This suggests that antigen combines with circulating antibody in immune animals and is trapped quickly in the follicles, where it can interact with T and B memory cells.

Morphologic Changes in Lymphoid Tissues During the Immune Response

As stated above, antigen processed by macrophages stimulates specific lymphocytes. This results in the trapping of antigen-specific circulating small T and B cells in the responding lymph node and their transformation into large blast cells, which stain intensively with the RNA-stain pyronin (pyroninophilic cells) (Fig. 14-6). Blastogenesis is first seen in the diffuse cortex, the T cell area, and shortly thereafter, in the superficial cortical region of B cells. The blast cells continue division to yield daughter cells, or clones, of differentiated T and B effector cells, whereas some daughter cells persist as memory cells. Within days, some of the educated cells migrate to form secondary follicles and germinal centers; later, plasma cells may be seen along the medullary cords.

The extent of the changes in the various lymphoid tissue regions depends on the nature of the antigen as well as the route of injection. The picture described above is typical for the great majority of protein antigens, the so-called T-dependent antigens, which elicit both antibody formation and delayed hypersensitivity (see below).

Some antigens, like dinitrochlorobenzene or the urushiols of poison ivy, stimulate a contact sensitivity (T cell) reaction. This form of stimulation is sharply localized to the paracortical (T cell) region. Other antigens, primarily polysaccharides, elicit a pure antibody response in the absence of a significant T cell reaction. The response to

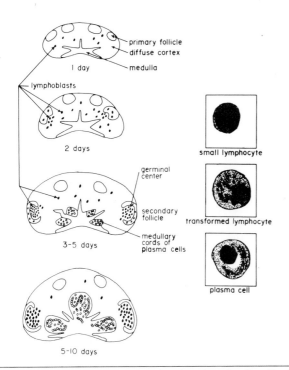

Figure 14-6. Morphologic changes in a lymph node after stimulation with a thymus-dependent antigen. Lymphoblastoid cells (*dark dots*) increase with time after immunization to form the germinal center (by 3 to 5 days). Insets show morphologic stages of an individual lymphocyte, stimulated to eventually produce antibody as a plasma cell. *(Adapted from Hood et al. (eds): Immunology, 1978. Sunderland, Mass, Benjamin/Cummings Publishing Co.)*

these antigens is largely in the superficial cortex and follicles.

Clonal Nature of the Immune System

Each of the B and T lymphocytes in the immune system has specificity for a single antigen. That is, a given B cell once stimulated will produce antibody of the same specificity as its IgM receptor, and of no other specificity. The frequency of lymphocytes with a specificity for one foreign antigen is less than 1 in 1000. Therefore, shortly after local antigenic stimulation, the lymphatics are relatively depleted of lymphocytes specifically reactive with that particular antigen, whereas the total numbers of circulating lymphocytes appear unchanged because the reaction responding to a single antigen is small. Those antigen-specific cells proliferate and differentiate. This process is called clonal selection or recruitment. After the initial stage of activation is over, antigen-specific T and B cells re-enter the circulating pool and disseminate themselves throughout the body. The entire scenario of lymphocyte recruitment, stimulation, and release results in the temporary enlargement of lymph nodes. Part of the enlargement is from nonspecific vascular effects secondary to the release of lymphokines, monokines, and prostaglandins following the activation of T cells and macrophages.

Enlargement lasts until the antigenic response (to a bacterial invader, for example) wanes.

T and B Cell Collaboration

B cells generally require the presence of T cells to be stimulated by antigen toward antibody formation. This process of T cell help was discovered in the late 1960s. Animals depleted of T cells (e.g., by neonatal thymectomy) fail to produce antibody to most antigens tested. When thymus cells are injected into these mice, the capacity to form antibody is restored (Fig. 14-7). Studies using labeled lymphocytes demonstrated that the antibody was produced by the B cells, not the added T cells. The capacity for T cell help is mediated by a subpopulation of T cells, which bear distinct markers such as Lyt1 or T4 (Table 14-1).

The mechanism by which helper T (T_H) cells and B cells collaborate is a subject of intense investigation. An absolute requirement for direct contact via an antigen bridge between T_H and B cells seems to be unlikely because an in vitro response to antigen can occur when T and B cells are separated by a cell-impermeable membrane. It is clear that T cells are stimulated by antigen presented on accessory cells such as macrophages (or the dendritic reticulum cells), which produce a factor called interleukin 1 (IL-1). Once activated, T cells produce a large number of growth and differentiation factors that promote their own continued proliferation as well as that of B cells (Fig. 14-8). Some of the helper T cell factors are specific for a single antigen, but most are nonspecific. Continuous T cell lines that produce one or more of these factors have recently been described, thus permitting their biochemical characterization as well as a delineation of their role in T–B collaboration.

T-Independent Immune Responses

As noted earlier some antigens, notably polysaccharides and lipopolysaccharides from bacteria, seem to be able to stimulate B cells without T cell help. These T-independent antigens possess several characteristics that enable them to trigger B cells directly: (1) They are usually large polymers with repeating antigenic determinants; (2) they are often mitogens that can cause B cell proliferation; and (3) many can activate the complement cascade (Chap. 13). Each of these properties has independently led to hypotheses of B cell activation. For example, repeating determinants can simultaneously interact with several immunoglobulin receptors on B cells, resulting in a highly avid interaction and receptor capping. The latter is also observed after B cells are incubated with bivalent F(ab')$_2$, but not with univalent Fab, anti-immunoglobulin reagents (Fig. 14-5). Only the former can induce B lymphocyte proliferation. Capping of receptors is followed by their loss from the cell surface. They are soon resynthesized, often with an increased number of expressed receptors—hence, capping is considered to be a critical event that is necessary, but not sufficient alone for triggering B cells. As many B cells possess recep-

BONE MARROW–THYMUS RECONSTITUTION EXPERIMENT.

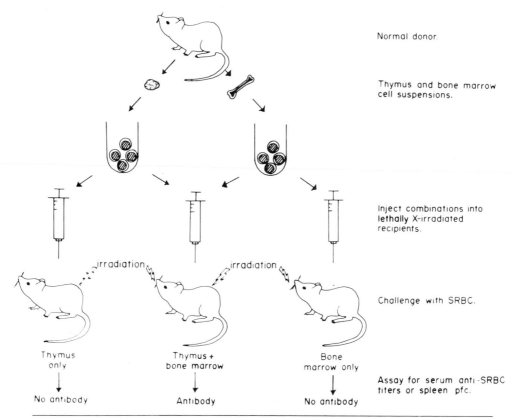

Normal donor.

Thymus and bone marrow cell suspensions.

Inject combinations into lethally X-irradiated recipients.

Challenge with SRBC.

Thymus only   Thymus + bone marrow   Bone marrow only

No antibody   Antibody   No antibody

Assay for serum anti-SRBC titers or spleen pfc.

Figure 14-7. Protocol to demonstrate that both T and B cells are needed to collaborate in antibody formation to certain antigens. Protocol: Lethally irradiated mice received either thymocytes or bone marrow cells or mixtures of both types of cells. Upon immunization with a T-dependent antigen, only recipients of both thymus and marrow cells responded with antibody production. Subsequent experiments demonstrated that the antibody was made by the marrow-derived (B) cells, whereas the thymocytes provided carrier or helper cell recognition to these B cells. *(Adapted from Golub: The Cellular Basis of the Immune Response, 1977. Menlo Park, Calif, Sinauer Associates.)*

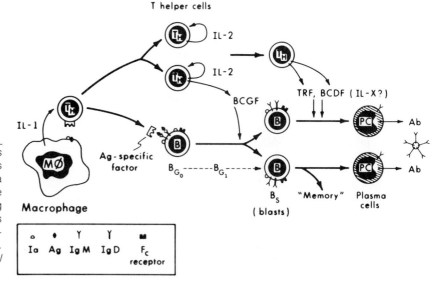

Figure 14-8. Scheme for cellular interactions leading to antibody formation. T helper cells recognizing antigen in association with self Ia antigens (on macrophage surfaces) produce several products (lymphokines), including specific helper factor. These allow B cells specific for the antigen to proliferate and differentiate into actively secreting plasma cells. *(Adapted from Scott: Crit Rev Immunol 4(3) 1983.)*

tors for activated complement components (Table 14-1), the ability of T-independent antigens to activate the complement sequence is considered important.

Interestingly, T-independent antigens primarily stimulate an IgM response and little B cell memory, whereas T-dependent antigens elicit IgM, IgG responses, and memory. The IgM response is relatively T cell independent, while the IgG response is highly T cell dependent. An IgG response to T-independent antigens can occur, but different IgG subclasses appear to be synthesized. Some data suggest that B cells responding to different antigens differ in buoyant density, cell surface immunoglobulin isotype, and other surface markers. Regardless of subpopulation differences, any minimal hypothesis still requires B cell triggering to occur via surface immunoglobulin receptor focusing of antigenic determinants. Events subsequent to this binding depend on the properties of the antigen and its ability to elicit T cell helper effects.

Memory and Affinity

The typical immune response illustrated by Figs. 12-17 and 14-2 reflects one property that is the hallmark of immunity and the basis of prophylactic immunization: immunologic memory. Once an individual has been exposed to an antigen, the subsequent response to that antigen occurs more quickly and is usually of a different class—IgG rather than IgM. Memory for an anamnestic (secondary) response may persist in humans for decades. Memory is a property of both B and T cells and takes some time to develop adequately. For example, if a second antigen injection is given too soon after the first, a poor secondary response will be observed. In addition, memory for some antigens, e.g., tetanus toxoids, can last for years; however, memory to other antigens can wane unless booster immunizations are given periodically. These properties explain the critical time course and booster immunization protocols for prophylactic vaccination with certain antigens.

Memory is still poorly understood. Presumably, it results from an expansion of specific T and B cell clones upon initial stimulation by antigen X. Some of these progeny differentiate into effector cells, such as T helper cells and plasma cells, whereas others remain as putative memory cells. When antigen is seen a second time, these cells divide rapidly and presumably differentiate to yield the expanded clones of specific effector T and B cells that typify the anamnestic response.

Another feature of the secondary antibody response is a change in the affinity of the antibodies produced. This can be explained by the realization that in an individual's repertoire of anti-X lymphocytes are cells that vary in the affinity of their surface immunoglobulin receptors for antigen X. Upon stimulation with a sufficient quantity of antigen X, anti-X antibodies of heterogeneous affinity are produced. The average affinity seen in this primary response is relatively low. By a combination of immune elimination (Fig. 14-2) and the ability of anti-X clones of higher affinity to combine with and be stimulated by de-

creasing amounts of antigen X, there is an increase in the average affinity with time after immunization. This is called the maturation of the immune response. Subsequent administration of antigen X leads to the immediate stimulation of these higher affinity clones of memory cells (which may have survived longer due to their better ability to compete for and bind small amounts of remaining antigen) and the typical high affinity secondary response. This enables the host to be better protected from subsequent infection.

Immunologic tolerance, the process that prevents individuals from responding to their own antigens, may be considered the opposite of memory. Consistent with the above clonal interpretation of antibody affinity in secondary antibody responses, in experimental tolerance the highest affinity B cells would be the most readily rendered tolerant. This would lead to a relatively low affinity response (if any) upon immunization of tolerant animals. Not only is this observed in experimental tolerance, but it is not uncommon to detect low affinity autoantibodies in clinically normal individuals (see below and Chap. 18).

Effector Mechanisms in Cellular Immunity

Stimulation with antigen leads not only to antibody responses of the type described above and in Chapters 12 and 21, but to the activation of T cells other than T helpers. Other cell types activated include delayed hypersensitivity T cells, cytotoxic or killer T cells, and regulatory suppressor T cells. The reactions of these subpopulations are collectively termed cellular immunity because the immune state cannot be transferred with serum (humoral immunity), but can be transferred by lymph node or spleen cells from an immune donor.

Delayed Hypersensitivity

Delayed-type hypersensitivity was originally described in the late 19th century in studies on immunity to tuberculosis (see also Chap. 21). The inflammatory response observed when guinea pigs exposed to tubercle bacilli were later inoculated with boiled bacterial culture filtrate (so-called old tuberculin, or OT) were the classic manifestation of a delayed hypersensitivity reaction. To this day, skin testing with purified protein derivative (PPD) of OT from these organisms is used as a standard assay for prior exposure to tubercle. The delayed reaction, which is described below and in Chapter 33, is also used as a diagnostic aid with a variety of bacterial, fungal (e.g., coccidiomycosis), plant (e.g., poison ivy), and protozoan (e.g., leishmaniasis) antigens.* Delayed reactions can also be elicited by hapten carrier conjugates or even by haptens that spontaneously bind to tissue cells or products. Thus, a patient's T cell competence can be tested by the

---

* Some of the complexities of these tests, especially in the sick, are described in Chapter 21.

ability to be sensitized against such simple compounds as dinitrochlorobenzene.

Delayed hypersensitivity, by definition, differs from immediate-type hypersensitivity, not only in the time of appearance of the inflammatory response, but also in its histologic character and specificity. Little change is noted at the site of a delayed hypersensitivity skin reaction during the first 12 hours. Erythema and induration gradually progress during the next 24 to 48 hours. The induration is due to the influx of large numbers of mononuclear cells, notably macrophages, as well as a few lymphocytes. The macrophages are of hematogenous origin from monocytes originally derived from the marrow.

Mechanistically, delayed reactions depend upon the release of lymphokines from a minority population of activated T cells that both attract and hold inflammatory mononuclear cells at the skin test site. One lymphokine is a macrophage–monocyte chemotactic factor measured by the migration of mononuclear cells through the pores of a membrane. Another important one is macrophage migration inhibitory factor (MIF). When packed into a capillary tube by centrifugation, macrophages will begin to migrate out after a few hours. Migration is inhibited if antigen is present and if the suspension contains as few as 1 to 2 percent of T lymphocytes from a sensitized individual. Other factors produced by activated T cells affect vasocular dilation and vascular permeability.

T Cell Cytotoxicity

Cell-mediated reactions of a substantially different nature are manifested by another subpopulation of T cells. These are called cytotoxic or effector T cells because of their ability to destroy target cells bearing specific antigens. They are typified by the cytotoxic cells generated during allograft or tumor rejection (Chaps. 17 and 20). Such cells can be generated in mixed lymphocyte culture (the MLC reaction), in which leukocytes from two unrelated (allogeneic) individuals are mixed in vitro. After 5 to 7 days, substantial proliferation can be measured due to the recognition of foreign histocompatibility antigens on the leukocytes. Though the majority of the proliferating cells may be T helper or T suppressor cells, another cell type also participates in this reaction. These are the cytotoxic cells, which cause damage to target cells. Target cells, for example, mitogen-stimulated lymphocytes from the allogeneic donor, are labeled (Chap. 12) with chromium in the form of chromate. Damage is measured by determining the radioactivity released.

Cytotoxic reactions against targets carrying tumor or viral antigens (in addition to alloantigens) can also be readily detected by this method. In this case, the cytotoxic cells are often generated in vivo by inducing an animal to reject a transplanted, virally induced tumor. The effector cells generated preferentially destroy target cells simultaneously displaying viral antigens along with self histocompatibility antigens (Chap. 17). They may fail to react to the same virus on cells from a different individual. This restriction is also involved in T–B cell collaboration, as well as in the transfer of delayed hypersensitivity reactions (see below).

Graft-vs.-Host Reactions

Another category of cell-mediated reaction is the so-called graft-vs.-host (GVH) response. The opposite of GVH is host-vs.-graft (HVG). The GVH reaction occurs when immunocompetent T lymphocytes are injected into a host that is unable to respond. Failure to respond may be for genetic reasons (i.e., the donor has no markers the host can recognize), or because the host is immunoincompetent. This last situation is of critical importance in the use of bone marrow transplantation to repair immunodeficiencies or in the treatment of leukemia. Avoidance of GVH reaction by lectin purification of marrow cells is described in Chapter 16. The GVH response seems to require the collaboration of two types of T lymphocytes—perhaps the same as delayed hypersensitivity or helper T and cytotoxic T cell precursors. The HVG reaction can occur in any transplant situation, but the term is mainly restricted to situations in which an incompletely immunosuppressed host reacts against the cells from a bone marrow transplant.

Suppressor T Cells

A final category of cell-mediated reactions is manifested by suppressor T cells. These cells resemble cytotoxic T cells in some, but not all, surface markers (e.g., in the mouse, suppressor and cytotoxic cells are Lyt2$^+$, but the suppressor cells also carry the H-2J marker). The involvement of suppressor cells in the regulation of antibody responses is described below. They also play a role in modulating other cell-mediated reactions. Gershon, the discoverer of suppressor cells, suggested that these T cells are the true "conductors of the immunologic orchestra," damping down or preventing unnecessary immunologic activity, including antiself reactivity.

Immunoregulation

Specific self–nonself discrimination characterizes the immune system. Not only must subsets of interacting cells be organized in such a manner as to deal with pathogens (and other environmental agents), but they must do so in an optimal fashion. Once the hazards of infection are dealt with, this system must shut itself off. Moreover, to prevent autoimmunity, there must be methods of ensuring that serious antiself reactions are not induced. Several levels of regulation are therefore necessary in the immune response; these are described in the following sections.

Regulation by Antigen: Immunologic Tolerance

Ehrlich realized the importance of lack of responsiveness to self when he coined the phrase "horror autotoxicus" to connote the possible consequences of antiself reactivity. An experimental basis for the development of self tolerance is provided by observations on dizygotic cattle twins. These fraternal twins—unique in that they share hematopoietic stem cells in utero—fail to reject a skin graft from the twin after attaining immunologic competence. These animals maintain a stable chimeric state with both their own and their twin's blood cells; moreover, though they do not reject each other's skin grafts, they can reject grafts from unrelated cattle. These data led to the suggestion that during prenatal exposure to their twin's cells (and, therefore, their allogeneic antigens), these dizygotic animals learn to accept the alloantigens as self. A corollary of Burnet's clonal selection hypothesis, this process occurs by the elimination (or repression) of the specific reactive lymphocyte clones when the immune system is immature. Experimental induction of unresponsiveness to foreign alloantigens in mice and chickens, respectively, by pre- or perinatal injection of allogeneic cells verified this hypothesis.

Since that time, numerous systems have been developed to experimentally trick the immune system of even adult animals into specifically accepting foreign antigens as self. The factors affecting this process include the dose, form, and chemistry of the antigen, and the age of the recipient.

Role of Antigen Dose, Form, and Chemistry

In the 1960s, it was reported that mice treated repeatedly with either very high (milligram) or very low (nanogram or lower) doses of antigen failed to make an immune response to an optimal immunizing challenge with the same antigen. These mice responded normally to unrelated antigens. Thus, pretreatment led to specific unresponsiveness to subsequent antigenic challenge, which is the minimal definition of tolerance. It was later shown that high dose tolerance somehow inactivated both T and B cells, whereas low dose tolerance only affected T cells (see below).

The importance of antigen form is exemplified by the fact that gammaglobulin and other serum protein antigens, which normally are weakly immunogenic, become tolerogenic (i.e., would induce tolerance) after high-speed centrifugation to remove aggregates. The soluble supernatant preparations were tolerogens, whereas the aggregated proteins in the pellets were immunogens. This difference is probably due to the ease with which aggregates are taken up by phagocytic cells in the immune network.

Haptens coupled to self serum proteins (again notably gammaglobulins) or even self lymphocytes are very tolerogenic. This property probably reflects the ability of isologous gammaglobulins and cells to pass unchecked through or even localize in certain areas of lymphoid organs. Hence, tolerance to the penicilloyl group has been induced by coupling it to self IgG carriers to prevent or even reverse hypersensitivity to penicillin. Similarly, the anti-DNA autoimmunity syndrome of New Zealand black mice, which resembles lupus erythematosus (Chap. 18), has been prevented or ameliorated by coupling nucleosides to isologous gammaglobulins.

Further efforts at tolerance induction have involved chemical modification of the antigen so that antibodies no longer recognize it; nonetheless, some cells do. Interestingly, some antigens can be chemically modified to become powerful tolerogens by stimulating suppressor T cells in another form of functional tolerance (see below). Efforts in these directions may eventually be applied to reversing other hypersensitivities, autoimmune states, and even providing experimental tolerance for transplantation purposes.

Role of Age in Tolerance

As stated above, experimental tolerance as originally described is induced in the perinatal period when the immune system is immature. The manipulations reviewed above permit tolerance induction in adults, which is more clinically desirable. Compared to adults, however, neonatal animals are often more susceptible to tolerance induction. This is particularly true in rats and mice, which are very immunoincompetent at birth; in contrast, the human neonate is fairly mature (see Chap. 15). The susceptibility of young animals to tolerance induction can be attributed to a combination of at least two factors: (1) a poor antigen-trapping network in lymphoid organs, which may enable antigens to contact many lymphocytes directly rather than be processed by macrophages; and (2) the presence of larger proportions of immature lymphocytes. For example, in the mouse, the first neonatal B cells possess only surface IgM receptors; these cells are more sensitive to in vitro tolerance induction than the more mature IgM$^+$, IgD$^+$ bearing B cells, which appear later in ontogeny. It has also been suggested that adult B cells that have had their IgD removed revert to an immature B cell behavior, especially in terms of tolerance induction. While this is clearly an oversimplification, it is possible that these receptors (IgM vs. IgD) provide different (negative and positive) signals to the B cell bearing term. The results of possible tolerogenic signals are discussed later in this chapter.

Kinetics and Loss of Tolerance

The immune response to a T-dependent antigen requires collaboration between both T and B cells. Hence, tolerance to such antigens could be apparent if either T or B cells are unresponsive. Adult mice rendered tolerant to soluble human gammaglobulin possess both unresponsive T and B cells. T and B cells differ in terms of the kinetics of tolerance induction, the doses required, and the waning of tolerance. Unresponsiveness is seen in the recipients of T cells taken from donors 1 day after injection of tolerogen mixed with normal B cells, whereas B cells do not become unresponsive for at least 1 week. As shown in Figure 14-9, tolerance in T cells lasts longer than tolerance in B cells. *Tolerant donors remain unresponsive as long as their T cells are tolerant.* In addition, the dose of tolerogen required to induce unresponsiveness in B cells was at least 100-fold that required for T cell tolerance.

The waning of tolerance is presumably due to the gradual elimination of the tolerogen (below a tolerogenic threshold), and the generation of new antigen-reactive cells in the thymus and marrow. If tolerogen is reinjected before this waning occurs, tolerance can be induced in these newly generated cells and unresponsiveness persists. The implications of this sequence of events for understanding self tolerance and autoimmunity are obvious. Some self antigens present in high concentrations may produce tolerance in both T and B cells. Other antigens, which may only occur at lower concentrations in the body, only produce T cell unresponsiveness. In the steady state, both kinds of self antigens are tolerated by the host. Moreover, maintenance of minimal amounts of self antigen is sufficient to prevent loss of tolerance and autoimmunity. Under some circumstances, the host can be tricked by altered self antigens (e.g., tissue antigens altered during viral infection). Helper T cells recognizing virally modified self antigens can cooperate with nontolerant B cells, if present. This could circumvent self tolerance and lead to an autoimmune response.

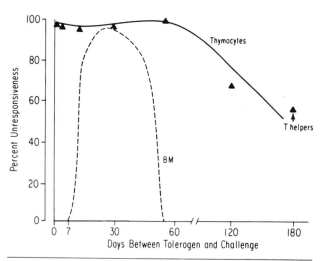

Figure 14-9. Kinetics and waning of tolerance in T and B cells. Tolerance in helper T cells *(triangles)* was measured by the decrease in response to a hapten coupled to a carrier protein to which the donor was tolerant. *(From Scott: Immunologic Tolerance: Mechanisms and Potential Therapeutic Applications, 1974. New York, Academic Press.)* This pattern was identical in kinetics to tolerance observed in thymocytes *(solid line, redrawn from Weigle et al.: Prog Immunol 1:312, 1971)* and contrasts with bone marrow or B cell tolerance *(dotted line).*

Mechanisms of Tolerance

It is apparent that there is no single unifying mechanism of tolerance. Rather, it is likely that several mechanisms all occur concurrently to maintain self tolerance in T cells, B cells, or both. It was assumed from Burnet's hypothesis that self tolerance is due to the absence (i.e., elimination) of autoreactive clones as they develop in the maturing immune system. Recent data, however, indicate the presence of some degree of autoreactivity, especially low affinity antibody. Cells with high affinity receptors are most likely to interact with and be triggered by an antigen to produce high affinity antibody. Paradoxically, these cells are the very ones most likely to become tolerant under different conditions. Low avidity autoantibody producing cells (to DNA) have been revealed in vitro in normal mouse strains and low avidity autoantibodies can be detected in clinically normal human sera.

Because of these observations, three mechanisms—other than clonal deletion—have received support: receptor blockade by free antigen, control by antibody feedback, and active suppression. Receptor blockade has been observed with tolerogens such as haptenated isologous gammaglobulins. Thus, tolerogen-bearing lymphocytes persist in the host. They cannot be triggered by antigen, presumably because their antigen receptors are blocked, but they can be activated to antibody production by mitogens (which act via nonimmunoglobulin receptors). This process of receptor blockade may explain self tolerance to antigens continuously present in relatively high concentrations. Functional self tolerance could also be due to the presence of so-called blocking antibodies, or even suppressor cells specific for certain self antigens (see below).

Finally, a lack of responsiveness exists to many antigens that are physically sequestered from the immune system. These never induce tolerance while the immune system is developing. Thus, sperm antigens—if exposed to the immune system, e.g., following vasectomy—can stimulate autoantibody production.

In summary, self tolerance can be maintained at many levels. No single mechanism is sufficient to prevent autoreactivity to self antigens, which differ greatly in concentration and chemistry.

Regulation by Antibody

As early as 1909, immunologists realized that specific antibody could influence the response to antigen. Preformed antigen–antibody complexes (especially in antibody excess) induced poor antibody responses, whereas free antigen or complexes in antigen excess were often potent immunogens. Thus, the presence of antibody could modulate immune responsiveness. This has enormous practical consequences because the placental transfer of maternal antibodies against polio, etc., could possibly prevent adequate immunization of human infants. Hence, vaccination is delayed until the infants are 3 to 9 months old, at a time when maternal antibody is no longer present in appreciable concentrations in the baby's circulation (see also Chap. 19).

Effect of Antibody on the Humoral Response: Parameters

When passively administered, antibodies of the IgG class were more efficient at inhibiting the immune response than was IgM to the same antigen. However, recent results show murine IgG$_1$ is quite effective at suppressing immunity, whereas IgG$_2$ antibodies have little effect and may actually augment responsiveness. Intact IgG seems to be better than F(ab')$_2$ fragments, which are still superior to univalent Fab antibody fragments. This suggests that bivalency and the Fc portion of the antibody play important roles in antibody feedback (perhaps acting through Fc receptors on B cells and macrophages) and that feedback is not simply due to masking antigenic determinants.

Passive IgG antibody can be effective even when administered 24 hours after antigen. This observation has important clinical applications because an immune response could be aborted after antigen has already entered the body. In the case of hemolytic disease of the newborn (erythroblastosis fetalis), an Rh negative mother delivering an Rh positive infant can be treated with anti-Rh antibodies soon after parturition and soon after the mother has received a transfusion of baby's cells during placental separation. This prevents sensitization to the baby's erythrocytes and has virtually eliminated the occurrence of this disease upon subsequent Rh positive pregnancies (Chap. 16).

The primary response is the response most readily inhibited by passive antibody, though one can partially affect the secondary response if enough high affinity antibody is given with antigen. Inhibition of helper T cell priming is not generally affected greatly by antibody under conditions that would ablate primary B cell response to antigen. This may reflect the greater sensitivity of T cells to priming by low doses of antigen.

The affinity of passively administered antibody plays a critical role in the effectiveness of this regulation. High affinity antibody—(typified by the IgG antibody found late in the immune response)—is more suppressive. Unlike tolerance, in antibody feedback the only antibody precursors that seem to escape regulation are those of high affinity (i.e., higher than the passive antibody).

As the systems described above are artificial, it was important to show that antibody might in fact regulate its own formation homeostatically in vivo. Cyclical antibody formation, in which antibody levels fall after a single injection of antigen and then abruptly rise again, has been repeatedly reported as possibly due to antibody feedback acting on a reservoir of remaining antigen. In fact, when rabbits immunized to a given antigen were plasmapherized and their plasma replaced by normal plasma, the antibody titers, after an initial sharp drop, began to rise and plateaued at control values. This effect is due to de novo protein synthesis initiated by the drop in antibody titer

leading to stimulation by retained antigen. Thus, the antibody made in response to antigen X was regulating its own synthesis and level.

Effect of Antibody on Cell-Mediated Immunity: Enhancement

Antibodies to cell surface antigens (e.g., histocompatibility antigens or tumor antigens) can suppress the immune response to an allograft or tumor, respectively. This phenomenon, called immunologic enhancement, is also discussed in Chapter 22, and may be demonstrated in two forms: active and passive. Active enhancement results from deliberate immunization with killed tumor cells, and is presumably due to an ongoing antibody response to the tumor cells because killed cells are poor immunogens for a cellular immune response. The antibodies made actively by the host can thus prevent sensitization to a subsequent viable tumor cell challenge immunization. These antibodies can be transferred to a normal recipient to suppress tumor graft rejection; this is called passive enhancement.

Several mechanisms of suppression of cellular (and of humoral) immunity are possible. Antibody can bind to and block cell surface antigens to prevent sensitization. Also, modulation (loss through capping) of cellular antigens could occur. Finally, immune elimination of the injected cells would prevent sensitization. These are all afferent mechanisms of regulation.

Antibodies could allow tumor/graft and survival by efferent suppression of induced immune effector function; e.g., by blocking target tumor antigens so killer T cells cannot destroy tumor cells (high dose enhancement). Finally, antibody, especially immune complexes, may directly inactivate immunocompetent cells, a "central mechanism." Evidence for and against all of these mechanisms exists; each seems to operate in certain clinical situations and in some experimental tolerance systems.

Idiotype Regulation

Idiotypes (i.e., antigenic determinants of the antibody combining site, Chap. 13) may also play an important role in immune regulation. Idiotypic determinants may induce the formation of antiidiotypic antibodies, which can regulate the formation of the antibody carrying the initial idiotype. (Though idiotypes are antigenic determinants on one's own specific immunoglobulins, tolerance is presumably not induced to these antigens because a sufficient concentration of each idiotype is not available to tolerize the developing immune system.) Jerne proposed that the immune system is a network of idiotypes that is perturbed by antigenic exposure. That is, an idiotype formed in response to antigen X could activate antianti-X (antiidiotype) formation and so on. Such antiidiotypes can both suppress or augment the immune response. Heterologous IgG$_1$ antiidiotype will suppress the formation of a given idiotype, but not reduce total antibody against antigen X. This is because the immune response to most antigens is idiotypically heterogeneous. Heterologous IgG$_2$ anti-idiotype has been shown to augment idiotype (anti-X) production. The latter occurs as a result of stimulation of idiotype positive helper cells (expressing idiotype X on their surface). Therefore, both T and B cells vs. a given antigen (X) may share idiotypic determinants. It was reported several years ago that major histocompatibility complex (MHC) reactive T cells possess idiotypes that are similar, if not identical, to the idiotypes found on B cells (vs. MHC antigens) and on anti-MHC antibody (Chap. 16). The production of this kind of antiidiotype has been used successfully to prolong allograft (MHC different) survival, a potentially exciting clinical application derived from a basic research tool.

Suppressor T Cell Regulation

Subclasses of T cells (primarily of the Lyt2$^+$ and Ly1$^+$,2$^+$ phenotype in the mouse, and bearing T5,8 markers in humans) can actively suppress the immune response to an antigen specifically or nonspecifically. These cells produce factors (see below) that can interfere with a variety of immunologic activities such as T cell helper activity, IgE responses, delayed hypersensitivity, or even T-independent triggering of B cells. Their action can be direct or indirect, i.e., interference with helper factor activity or macrophage function. Suppressor T cell activity is itself regulated in a feedback loop in such a way that a homeostatic control of responsiveness can be maintained. Recently, this loop has been shown to involve subsets of T suppressor cells that recognize antigen and others that recognize idiotypes. They communicate with each other via soluble factors containing an antigen- or idiotype-specific binding site and other elements that restrict or focus their activity on appropriate acceptor/target cells.

Nonantigen-specific suppressor T cells may also be involved in at least two clinical situations: immunodeficiency and autoimmunity. Patients with an acquired hypogammaglobulinemia may have normal B cell levels, but their T cells suppress the differentiation of these cells to plasma cells. A loss of suppressor activity in New Zealand black mice prior to the onset of autoimmune symptoms reminiscent of lupus has also been described. Regulation of immune reactivity by the induction of specific suppressor T cell clones has become an important goal for the future.

Genetic Basis of Immune Responses: Gene Regulation

In mice, rats, guinea pigs, and humans, genes have been mapped that regulate the immune response of an individual to a given antigen. Most of these genes are mapped in the MHC region of the 17th chromosome in the mouse or the sixth chromosome in the human, and are called immune response (Ir) genes.

The Ir genes may affect the immune response in either a qualitative (all or none) or quantitative manner, as exemplified in two prototype models. The first is the poly-L-lysine (PLL) gene of guinea pigs, described by Levine and Benacerraf in 1963. Animals lacking this gene can make neither antibody nor delayed hypersensitivity responses to PLL or to hapten-conjugated PLL (see below). The second model, reported by McDevitt and Sela, involves the response of mice to the branched polymer, (T, G)-A--L (see Fig. 12-3). "Nonresponders" to (T, G)-A--L are actually low responders, capable of developing an IgM response but unable to switch to IgG synthesis.

Use of these synthetic polypeptides and their hapten-conjugates has greatly advanced our understanding of Ir genes in the regulation of the immune response to many natural antigens, especially when the latter are injected at low doses that are only immunogenic in a few individuals.

Ir Genes and Ia Antigens

The original studies in the PLL gene were done in outbred guinea pigs immunized with dinitrophenylated (DNP-) PLL. Some of these animals were responders and some nonresponders. Breeding experiments established that responder × nonresponder crosses gave responder offspring if these animals were of a certain histocompatibility type, and that responsiveness was a function of a dominant autosomal gene. The two inbred strains of guinea pigs, strain 2 and strain 13, were found to be responder and nonresponder, respectively. Hence, all 2 × 13 F$_1$ animals were responders because they codominantly expressed strain 2 antigens.

Interestingly, strain 13 (nonresponder) guinea pigs could make an anti-DNP antibody response to DNP-PLL only if it were coupled to methylated bovine serum albumin (mBSA); that is, nonresponders could respond if they were given an immunogenic carrier. Nonresponder animals, previously tolerized to BSA, failed to give this response. The DNP-PLL-BSA immunized nonresponder animals never developed delayed hypersensitivity to DNP-PLL skin test, though they were positive when challenged with BSA. Thus, the PLL gene seems to control T cell functions, such as helper activity and delayed hypersensitivity.

Linkage to the histocompatibility complex (Chap. 17) was also observed in the murine response to (T, G)-A--L: H-2^b mice are good IgG responders and H-2^k mice are poor responders; F$_1$ (H-2^b × H-2^k) mice are all responders. Subsequent studies mapped the Ir genes to a set of loci called the I-region. Because neonatally thymectomized responders behave like nonresponders, i.e., they make IgM but no IgG anti-(T, G)-A--L, the Ir gene defect is reflected in T cell activity involving the IgM–IgG switch in antibody synthesis. If the amino acid histidine is substituted for tyrosine, the resulting polymer, (H, G)-A--L, is immunogenic in H-2^k mice but not in H-2^b mice. The Ir genes show exquisite specificity, and responsiveness depends on the expression of the appropriate I-region associated (Ia) antigens by macrophages and other antigen-presenting cells (see Chap. 17).

Another locus seems to control responsiveness to the terpolymer, glutamic acid, alanine, and tyrosine (GAT). In this case, nonresponders fail to make antibody to GAT because they possess T suppressor cells that actively interfere with the immune response. This is shown by immunizing nonresponder mice with GAT-mBSA (to which they respond). If T cells from GAT-treated mice are mixed with the GAT-mBSA immune spleen cells, no response is generated. T cells in nonresponder mice make an antigen-specific suppressor factor that can inhibit the anti-GAT response in lieu of T cells. Though these studies have been with model systems, the immune response to naturally occurring molecules such as ovalbumin and myoglobin may be regulated by the same genetic systems of Ia-restricted antigen presentation and possible suppressor cell recognition.

Complementing Ir Genes

Recently, it has been noted that some immune responses are controlled by two Ir genes. For example, H-2^b mice and H-2^a mice are nonresponders to GLT, but their F_1 hybrids are responders. This form of gene complementation is explained by one strain possessing an appropriate I-A gene product and the other providing an I-E gene product. This allows permissive Ia antigens to be displayed on macrophage surfaces to trigger an immune response. This form of complementation is relatively easy to demonstrate in rodents because crosses of the appropriate strains can be made at will; however, its existence and the complexities thereof in humans are much more difficult to document.

Ir genes in humans have been described from family studies of disease susceptibility. These are more fully discussed in Chapter 17.

Summary

In summary, the immune system is built of tissues that make and educate lymphocytes to provide a network for antigen capture, T and B cell recirculation, and interaction with antigen to generate an immune response. Each cell in this system has a specific function, and these functions are regulated at many different levels. Both the quantity and quality of antigen control the nature of immune response, and can induce tolerance. The antibody produced can feed back to regulate its own formation. Suppressor T cells, induced by antigen or antiidiotype, also control the level of the immune response. Finally, Ir genes determine the quantity and quality of responsiveness. All these control mechanisms ensure the immune response to foreign pathogens or autochthonous tumor cells and prevent overproduction or autoimmunity.

FURTHER READING

Books and Reviews

Benacerraf B, Germain RN: The immune response genes of the major histocompatibility complex. Immunol Rev 38:70, 1978

Cooper MD, Lawton AR: The development of the immune system. Sci Am 231:59, 1974

Ford WF: Lymphocyte migration and the immune response. Prog Allergy 19:1, 1975

Golub ES: The Cellular Basis of the Immune Response. Sunderland, Mass, Sinauer Assoc, 1982

Hood LE, Weissman IL, Wood WB: Immunology. Menlo Park, Ca, Benjamin/Cummings, 1978

Metcalf D, Moore MAS: Haematopoietic Cells. Amsterdam, North-Holland, 1971

Samter M (ed): Immunological Diseases, 3rd ed. Boston, Little, Brown, 1978

Scott DW: Mechanisms of tolerance. Crit Rev Immunol 4(3) 1983

Selected Papers

Eardley DD, Hugenberger J, McVay-Bordeaux L, et al.: Immunoregulatory circuits among T-cell sets. I. T helper cells induce other T cells to exert feedback supression. J Exp Med 147:1106, 1978

Eichmann K, Rajewsky K: Induction of B and T cell immunity by anti-idiotypic antibody. Eur J Immunol 5:661, 1975

Gowans JL, Uhr JW: The carriage of immunological memory by small lymphocytes in the rat. J Exp Med 124:1017, 1966

Jerne N: The somatic generation of immune recognition. Eur J Immunol 1:1, 1971

CHAPTER 15

Normal and Abnormal Development of the Immune System

Development of Normal Immune Function

Certain well known attributes of the human fetus and new-born suggest abnormal immune function. These include diminished lymphoid tissue, an increased incidence of and often poor response to infection, and low to absent immune responses to certain types of antigens. These features led to the former misconception that the human fetus and newborn are immunologically null. Although very little is known yet about the ontogeny of human specific immune responsiveness, information is rapidly accumulating that the fetus is immunologically competent from a very early age.

The most ambitious and successful work relating to the ontogeny of specific immune responsiveness has been accomplished through hysterotomy and immunization of sheep and monkeys in utero. Sheep can respond to antigens such as bacteriophage ϕX 174 when immunized as early as the 35th day of their 149-day gestation period—a time of minimal lymphoid development, when only an epithelial thymus is evident. Later in development, sheep embryos respond to other antigens such as ferritin and ovalbumin, but they never achieve the adult sheep's ability to respond to some other antigens, such as *Salmonella typhosa.* Similar types of observations were made in the monkey. These findings support the concept that the capacity for specific immunologic responsiveness does not develop as an all-or-none phenomenon, but rather that it appears in stepwise fashion for different types of antigens.

Although the time course of immunologic responsiveness appears to be genetically programmed for any given species, the question has arisen whether various morphologic and phenomenologic changes that occur in the lymphoid system of developing embryos might be related to antigen exposure. Studies in newborn piglets tend to support such a concept. In this species there is a six-layered placenta and no placental transfer of immunoglobulin (Ig) takes place. In germ-free colostrum-deprived piglets, no surface immunoglobulin bearing B lymphocytes, plasma cells, natural antibodies, or germinal centers can be found, and the total lymphoid mass is extremely small. Nevertheless, newborn piglets are fully immunologically competent, as antibody production can be detected within 48 hours after immunization. In studies with the chick embryo, a somewhat different picture has emerged. Cells containing IgM were detected on the 13th day of embryonation and only 24 hours after yolk sac stem cells enter the bursa of Fabricius, when the embryo presumably has not encountered antigen. IgG-containing cells could first be found at 21 days of embryonic life, and they appeared in follicles that had prior to that time been engaged only in IgM synthesis. Therefore, because both IgM- and IgG-producing cells were normally found in the chick prior to hatching, these workers postulated that the first stage of B cell maturation, e.g., development of the capacity to become anti-

gen reactive and to bear surface immunoglobulin, is a clonal developmental state and is thus antigen independent. The second stage they refer to as the clonal proliferation stage, which begins when the B cells encounter antigen. They also hypothesized that IgG-producing cells are derived from IgM-producing cells through a genetic switch mechanism controlling immunoglobulin heavy chain synthesis; this thesis was supported by the fact that injection of anti-μ chain antiserum into the chick embryo at the 13th day consistently led to agammaglobulinemia involving all classes of immunoglobulins.

Studies in the fetal mouse (and now also in the human fetus) have shown that first fetal liver and then bone marrow contain lymphoid cells with a relatively large nucleus and a narrow rim of cytoplasm that stain weakly with fluoresceinated anti-IgM, but do not bear surface IgM (Fig. 15-1). These cells are referred to as pre-B cells. It has been postulated that they are precursors of multiple clones of B lymphocytes. Immature B cells that lack cytoplasmic IgM but bear surface IgM are thought to derive from pre-B cells. In contrast to pre-B cells, which are not tolerizable by anti-μ antibody antigen, immature B cells are highly susceptible to tolerance induction. More mature B cells bearing single, double, or triple isotypes develop from these cells and from them come memory B cells and plasma cells specific for one particular isotype (Fig. 15-1). It is thought by some workers that all stages except the last may occur without antigen exposure.

Ontogeny in the Human Fetus

The human immune system, as with that of other species, arises in the embryo from gut-associated tissue. The thymus first appears as a proliferation of epithelial cells lining the third and fourth branchial pouches at about the sixth or seventh week of embryonic life. Lymphoid tissue first appears in the thymus at about the eighth week of gestation, presumably as a result of a thymic microenvironmental influence on pluripotential lymphoid stem cells that have migrated there from the yolk sac, fetal liver, or spleen. The earliest lymphoid cells within the thymus are large subcapsular cells that lack mature T cell antigens but express antigens recognized by monoclonal antibodies to the T10 antigen and to the transferrin receptor (T9 or 5E9 antigen), both of which are present on many types of rapidly dividing cells (Table 15-1). During maturation in the cortex, these human prothymocytes acquire a thymocyte-distinct antigen, T6 (also known as human thymocyte antigen 1, or HTA-1), probably analogous to the murine thymus leukemia (TL) antigen. They also acquire antigens T4 and T5 (or T8). The latter are simultaneously coexpressed with each other and with T10 and T6 on approximately 70 percent of thymocytes. Fetal cortical thymocytes are among the most rapidly dividing cells in the body, with a mean generation time of 6 to 8 hours. Many of these very immature cells die in situ; the remainder migrate to the medulla, where they acquire the T3 antigen, present

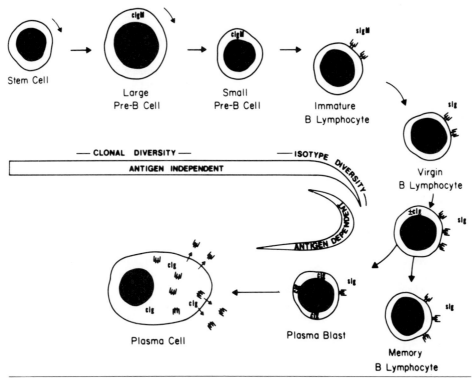

Figure 15-1. Outline of developmental sequence of B cells. *(From Lawton et al.: In Goldberger (ed): Biological Regulation and Development, vol II, 1979. New York, Plenum.)*

on 95 percent of peripheral blood T lymphocytes and 25 percent of thymocytes. A few medullary thymocytes coexpress T6 and T3, but most lose T6 and acquire T1 and T3 antigens (present on 95 to 100 percent of peripheral blood T cells) and segregate into T4$^+$ and T5$^+$ (or T8$^+$) subsets (Table 15-1). Functions are acquired concomitantly with the latter antigenic changes, but are not fully developed until the cells emigrate from the thymus. Such med-

ullary cells are resistant to the lytic effects of corticosteroids. They leave the thymus via the bloodstream and are distributed throughout the body, with heaviest concentrations in the paracortical areas of the lymph nodes, the periarteriolar areas of the spleen, and the thoracic duct lymph. Thymic cells forming spontaneous sheep erythrocyte rosettes (E) have been noted at as early as 8 weeks of gestation. Lymphocytes with membrane markers character-

TABLE 15-1. ACQUISITION OF LYMPHOCYTE SURFACE MARKERS AND FUNCTION DURING INTRATHYMIC DEVELOPMENT

Cell Type	Antigens	Function
Large subcapsular cells (prothymocytes)	T10, T9 (5E9)	?None
Cortical thymocytes	T10, T6, T11, T4, T5 (T8)	?PHA-responsive
Medullary thymocytes		
Majority	T1, T3, T4, T10, T11	PHA, Con A, MLR- and antigen-responsive; helper function
Minority	T1, T3, T5 (T8), T10, T11	Con A, MLR-responsive; cytotoxic suppressor function
Peripheral blood T cells		
Majority	T1, T3, T4, T11	PHA, Con A, MLR- and antigen-responsive; helper function
Minority	T1, T3, T5 (T8), T11	Con A, MLR-responsive; cytotoxic suppressor function

PHA, phytohemagglutinin; Con A, concanavalin A; MLR, mixed leukocyte response.

istic of T cells comprise 65 to 100 percent of thymus cells by 18 weeks of gestation, but comprise only 5 percent of spleen cells at this age. By 20 to 22 weeks of gestation, T cells represent 10 to 30 percent of the fetal splenic lymphocyte population. Reactivity to phytohemagglutinin first appears in the thymus at 10 weeks of gestation, in the spleen at 13 weeks of gestation, and in the peripheral blood at 14.5 weeks of gestation. Mixed leukocyte reactivity was first detected in the thymus at 12.5 weeks of gestation and antigen-binding cells at 20 weeks of gestation. Evidence of specific cellular immune responsiveness appears even earlier in the fetal liver; cells from this organ respond in mixed leukocyte culture as early as 7.5 weeks of gestation, and mild but definite graft-vs.-host disease followed the infusion of liver cells from 8- to 9-week fetuses into patients with severe combined immunodeficiency disease. Precursors of phagocytes and macrophages are also present in the liver at 8 weeks of gestation, and natural killer cell function has been detected in fetal liver cells as early as 9 to 11 weeks of gestation.

The first evidence of development of the humoral limb of the human immune system is found in the appearance of pre-B cells (Fig. 15-1) in the fetal liver as early as 5 to 7 weeks of gestation. These cells have the Epstein–Barr virus receptor, but are Ia and C3b receptor negative. From these develop surface IgM^+ and Ia^+ early, or immature B cells at about 8 to 9 weeks of gestation. Virgin B cells, positive for surface immunoglobulin of multiple isotypes and for the C3b receptor, are found in the liver by 9.5 weeks of gestation, and in the peripheral blood, bone marrow, and spleen by 11.5 weeks of gestation. By 14 weeks of gestation the percentages of blood lymphocytes bearing surface IgM and IgD have been reported to be similar to those found in normal adult blood. The synthesis and secretion of IgM and IgE may occur as early as

10.5 weeks of gestation, and IgG as early as 12 weeks of gestation, as shown by the incorporation by cultured fetal lymphoid cells of radiolabeled amino acids into immunoglobulins. No evidence of significant IgA synthesis was found in normal noninfected fetuses; free secretory piece has been found in the urine of premature infants, however, indicating the normal fetal synthesis of this protein. Despite the capacity of fetal B lymphocytes to differentiate into immunoglobulin-synthesizing and -secreting B plasma cells, plasma cells are not normally found in lymphoid tissues of the fetus until about 20 weeks of gestation, and then only rarely. Peyer's patches have been found in significant numbers by the fifth intrauterine month, and plasma cells have been seen in the lamina propria by 25 weeks of gestation. At birth there may be primary lymphoid nodules, but usually secondary follicles are not present and typical plasma cells are extremely few in number. These follicles appear shortly after birth as a result of antigenic stimulation afforded by extrauterine life. Intrauterine infection with *Treponema pallidum*, cytomegalovirus, rubella virus, toxoplasma, or other agents does, however, result in abundant fetal plasma cell formation and in development of mature centers in the lymphoid tissues. Nevertheless, peripheral blood B lymphocytes from even full-term normal newborn infants are capable of undergoing only limited differentiation into mature plasma cells when stimulated in vitro with the T cell-dependent polyclonal B cell activator, pokeweed mitogen, and then primarily only into IgM-producing cells.

The human fetus begins to receive significant quantities of maternal IgG transplacentally at around 12 weeks of gestation, although small amounts have been detected in the fetus at as early as 7.5 weeks of gestation. The quantity increases steadily until at birth cord serum contains a concentration of IgG comparable to or greater than

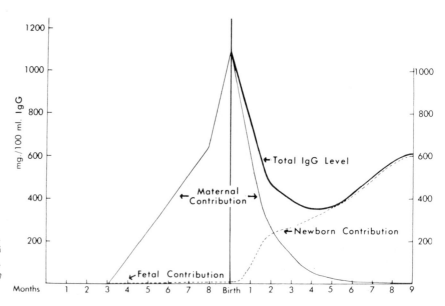

Figure 15-2. Schematic representation of IgG concentrations in the fetus and newborn. *(From Allansmith: In Falkner (ed): Human Development, 1966. Philadelphia, Saunders.)*

that of maternal serum (Fig. 15-2). IgG is the only class to cross the placenta to any significant degree, and all four of its subclasses do this, but IgG_2 does so least well. A small amount of IgM (10 percent of adult levels) and a few nanograms of IgA, IgD, and IgE are normally found in cord serum; as none of these proteins cross the placenta in any significant quantity, they are presumed to be of fetal origin. These observations raise the possibility that certain antigenic stimuli normally cross the placenta to provoke responses even in noninfected fetuses. Indeed, some atopic infants frequently have reaginic antibodies to antigens (such as egg) to which they have had no known exposure during postnatal life, suggesting that synthesis of these IgE antibodies could have been induced in the fetus by placentally transmitted antigens ingested by the mother. Complement components do not cross the placenta; levels of all components in cord serum are approximately 50 percent of adult values.

Ontogeny in the Infant and Child

T cells are present in cord blood in roughly the same quantity as in adults, although the percentage of E rosette-forming cells is somewhat less. The percentage of T cells of the helper phenotype ($T4^+$) is normal, but the percentage of those of the suppressor phenotype ($T8^+$) is slightly but statistically lower than in adult blood. Cord blood T cells have the capacity to respond normally to the two T cell mitogens, phytohemagglutinin (PHA) and concanavalin A (Con A), and they are capable of mounting a normal mixed leukocyte response (a response considered to be mediated only by T cells). Thus, the absence of these responses in tests of cord blood lymphocytes is a priori evidence of profound primary dysfunction of the T cell system. The newborn infant has the capacity to develop delayed hypersensitivity in vivo at birth, as demonstrated by experimentally induced rhus sensitivity and by bacillus of Calmette and Guerin-induced tuberculin reactivity. Premature and full-term infants were not, however, sensitized as regularly by the topical application of dinitrofluorobenzene as were older infants. The rejection times of first set allografts were somewhat prolonged in newborns—12 to 96 days, as opposed to a normal adult rejection time of 10 to 11 days. The number of infants studied was small, however, and all of the skin grafts were either from parents to their infants or from the donors of fresh blood used in exchange transfusions of the infants, factors that could have accounted for the prolonged rejection times.

Newborn infants are quite susceptible to infections with gram-negative organisms because they have not received IgM antibodies (e.g., heat-stable opsonins) to these organisms from their mothers. Quantities of the heat-labile opsonin, C3b, are also lower in newborn serum than in adults. These factors probably account for the finding of impaired phagocytosis of some organisms by newborn polymorphonuclear cells, as these cells phagocytize and kill such bacteria normally in the presence of normal adult serum. Maternally transmitted IgG antibodies serve quite adequately as heat-stable opsonins for most gram-positive bacteria, and IgG antibodies to viruses afford adequate protection against those agents. However, because there is a relative deficiency of the IgG_2 subclass, antibodies to capsular polysaccharide antigens may be deficient. Because premature infants have received less maternal IgG at the time of birth than full-term infants, their serum opsonic activity is low for all types of organisms. B lymphocytes are present in cord blood in percentages and numbers equal to those in adult blood. As in adults, a majority of these bear both surface IgM and IgD; however, the percentage of those coexpressing IgM, IgD, and another isotype (i.e., triple isotype-bearers) is much higher than in adults. The capacity of cord blood large granular lymphocytes (LGLs) to act as effectors in antibody-dependent and natural killer assays is roughly two thirds that of adults.

The neonatal human begins to synthesize antibodies of the IgM class at an increased rate very soon after birth, in response to the immense antigenic stimulation of the new environment. Premature infants appear to mature immunologically at about the same rate as full-term infants. At about 6 days after birth, the serum concentration of IgM rises sharply. This rise continues until adult levels are achieved by approximately 1 year of age (Fig. 15-3). Cord serum usually does not contain IgA detectable by single radial diffusion; serum IgA normally first becomes detectable by this method at around the 13th day of postnatal life. Serum IgA gradually increases during early childhood until adult levels are achieved and preserved between the sixth and seventh years of life (Fig. 15-3). Cord serum contains an IgG concentration comparable to or greater than that of maternal serum. Maternal IgG gradually disappears during the first 6 to 8 months of life (Fig. 15-2), whereas the rate of infant IgG synthesis increases (with IgG_1 and IgG_3 synthesized faster than IgG_2 and IgG_4 during first year), until adult concentrations of total IgG are reached and maintained by 7 to 8 years of age (Fig. 15-3). However, IgG_1 and IgG_4 reach adult levels first, followed by IgG_3 at 10 years and IgG_2 at 12 years of age. The total immunoglobulin levels in the infant usually reaches a low point at approximately the fourth to fifth month of postnatal life. IgD is detectable in nanogram concentrations in the sera of premature infants, and adult concentrations are achieved by 1 year of age. The rate of development of IgE has generally been found to follow that of IgA. After adult concentrations of each of the three major immunoglobulins are reached, these levels remain remarkably constant for a given individual. Studies of serum immunoglobulin levels in identical twins suggest that hereditary factors play a major role in maintaining the constancy of serum immunoglobulin concentrations in normal individuals.

Lymphoid tissue is proportionally small but rather well developed at birth and matures rapidly in the postnatal period. The thymus is largest relative to body size during fetal life and at birth is ordinarily two thirds its mature

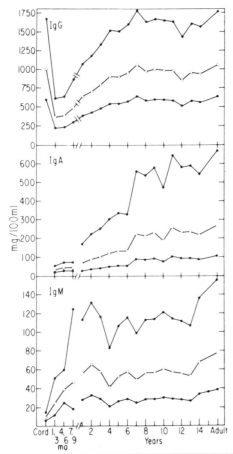

Figure 15-3. Immunoglobulin concentrations in 201 normal subjects from infancy to adulthood. The lines connecting the open
circles represent the geometric means. The boundaries are
obtained by taking the antilogs of the mean logs ± two pooled
standard deviations of the logs. *(From Buckley et al.: Pediatrics
41:600, 1968.)*

weight, which is attained during the first year of life. It
reaches its peak mass, however, just before puberty, and
then gradually involutes thereafter. By 1 year of age all
lymphoid structures are mature histologically. Absolute
lymphocyte counts in the peripheral blood also reach a
peak during the first year of life (Fig. 15-4). Peripheral
lymphoid tissue (including lymph nodes, tonsils, adenoids,
appendix, and the lamina propria plasma cell system) increases rapidly in mass during infancy and early childhood.
It reaches adult size by approximately 6 years of age, exceeds those dimensions during the prepubertal years, and
then undergoes involution coincident with puberty. It may
be pertinent that the adult-sized peripheral lymphoid mass
is achieved at essentially the same age that adult concentrations of serum IgG and IgA are reached. The spleen,
however, gradually accrues its mass during maturation and
does not reach full weight until adulthood. The mean
number of Peyer's patches is one half the adult number at
birth, and gradually increases until the adult mean number
is exceeded during adolescent years.

Primary Immunodeficiency Diseases

Since the first example of a human host deficit was described in 1952, reports of clinical immunodeficiency syndromes have appeared in the literature at an almost
exponential rate. In 1970 the World Health Organization
formulated a classification to facilitate recognition and understanding of these disorders. In it the possible cellular
deficit (T, B, or stem cell) was given for each category of
immunodeficiency. This classification was drawn up before
information had been accumulated regarding the numbers
of T and B cells usually present in the peripheral blood of
these patients, as initial methods for identifying and enumerating these cells were not developed until the early
1970s. Studies with these techniques have since shown
that cells having surface markers characteristic of B or T
cells or their subsets are rarely completely lacking in any
of these conditions. The classification in Table 15-2, which
now includes several more recently identified primary
immunodeficiency disorders, groups them according to
their functional deficiences and according to the currently
presumed cellular level at which the defects occur. It is important to keep in mind that, with one or two possible exceptions (e.g., failure of development of the thymus gland
in the DiGeorge syndrome and profound deficiencies of
key enzymes in the purine salvage pathway in some forms
of severe combined immunodeficiency or in Nezelof's syndrome), the primary biologic error is not known for any of
these defects. Moreover, none of the defects studied thus
far has been found to have associated deficiencies of particular HLA antigens; therefore, they are unlikely to represent defects involving HLA-linked immune response
genes.

Although the true incidence of these syndromes is unknown, well-defined immunodeficiency is thought to be
rare. From data in the report of the Working Party on Hypogammaglobulinemia of the British Medical Research
Council, it has been estimated that the incidence of agammaglobulinemia is roughly 1 in 50,000. Selective absence
of serum and secretory IgA is the most common defect,
with reported incidences ranging from 0.03 to 0.97 percent. The primary immunodeficiencies occur more frequently in children (60 percent) than in adults (40
percent) and there is a 5:1 male to female ratio in childhood, changing to a 1:1.4 ratio in adults.

Antibody Deficiency Disorders

Antibody deficiency may occur clinically either as an apparent congenital or as an acquired abnormality, with deficiencies in all immunoglobulin classes (agammaglobulinemia or hypogammaglobulinemia) or in one or more but
not all classes (dysgammaglobulinemia). In addition, antibody deficiency may occur in the presence of normal or
near normal serum immunoglobulin concentrations. Most
patients with these disorders are recognized because they

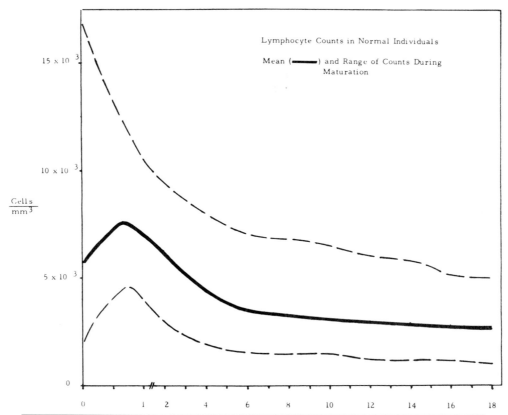

Figure 15-4. Absolute lymphocyte counts in normal individuals during maturation. *(From Altman: Blood and other body fluids. In Dittner DS (ed): Biological Handbook. Washington, D.C., Fed Am Soc Exp Biol 1961, p 125.)*

have recurrent infections or a history of treatment failure, but some individuals with selective IgA deficiency or infants with transient hypogammaglobulinema may have few or no infections. Table 15-3 lists some of the general features of these disorders.

X-Linked Agammaglobulinemia

Commonly referred to as Bruton's agammaglobulinemia (a misnomer because a small amount of immunoglobulin can usually be detected), this was the first described of the immunodeficiency states. A majority of boys afflicted with this malady remain well during the first 6 to 9 months of life, presumably by virtue of maternally transmitted immunoglobulin. Thereafter, they repeatedly acquire infections with high-grade extracellular pyogenic organisms such as pneumococci, streptococci, and *Haemophilus* unless given prophylactic antibiotics or gammaglobulin therapy. Infections with other organisms such as meningococci, staphylococci, and *Pseudomonas* occur less frequently. The most common types of infections include sinusitis, pneumonia, otitis, furunculosis, meningitis, and septicemia. Despite these chronic or recurrent infections, patients with this disorder usually grow normally unless they develop bronchiectasis or persistent enterovirus infections. Chronic fungal

infections are not usually present and *Pneumocystis carinii* pneumonia rarely occurs unless there is an associated neutropenia. Virus infections and live virus vaccines are also usually handled normally, with the notable exceptions of hepatitis and enterovirus infections. Several examples of paralysis after polio vaccine administration have occurred, presumably due to mutation of the vaccine virus to a more neutrotropic form. In addition, chronic progressive eventually fatal central nervous system infections with various echoviruses have occurred in more than 20 such patients. These observations suggest a primary role for antibody, particularly secretory IgA, in host defense against this group of viruses, because normal T cell function has been present in all x-linked agammaglobulinemics with persistent enterovirus infections reported thus far. Several have presented with a dermatomyositis-like picture before neurologic abnormalities were apparent, raising the possibility of a viral etiology for other collagen-like diseases, such as rheumatoid arthritis, which occurs in approximately 20 percent of these patients.

The diagnosis of x-linked agammaglobulinemia is suspected if serum concentrations of IgG, IgA, and IgM are far below the 95 percent confidence limits for appropriate age- and race-matched controls (i.e., usually less than 100 mg/dl total immunoglobulin). The demonstration of anti-

TABLE 15-2. CLASSIFICATION OF PRIMARY IMMUNODEFICIENCY DISORDERS

Disorder	Functional Deficiencies	Level of Defect Presumed Cellular
X-linked agammaglobulinemia	Antibody	Pre-B Cell
Common variable (B lymphocyte) hypo-gammaglobulinemia	Antibody	B lymphocyte
Selective IgA deficiency	IgA antibody	IgA B lymphocyte
Secretory component deficiency	Secretory IgA	Mucosal epithelium
Selective IgM deficiency	IgM antibody	T helper cells
Immunodeficiency with elevated IgM	IgG and IgA antibodies	IgG, IgA B lymphocytes
Transient hypogammaglobulinemia of infancy	None: immunoglobulins low but antibodies present	Unknown
Antibody deficiency with near-normal immunoglobulins	Antibody	Unknown; ?B cell
X-linked lymphoproliferative disease	Anti-EBNA antibody	B cell; ?also T cell
DiGeorge's syndrome	T cellular; some antibody	Dysmorphogenesis of third and fourth branchial pouches
Nezelof's syndrome (including with PNP or ADA deficiency)	T cellular, some antibody	Unknown; ?thymus; ?T cell; metabolic defects
Severe combined immunodeficiency syndromes (autosomal recessive; ADA deficiency; x-linked recessive; bare lymphocyte syndrome; reticular dysgenesis)	Antibody and T cellular; phagocytic in reticular dysgenesis	Unknown; metabolic defects(s); ?T cell; ?stem cell; ?thymus
Wiskott–Aldrich syndrome	Antibody, T cellular	Unknown
Ataxia telangiectasia	Antibody; T cellular	B lymphocyte; helper T lymphocyte
Immunodeficiency with short-limbed dwarfism	T cellular	G1 cycle of many cells
Immunodeficiency with thymoma	Antibody; some T cellular	B lymphocyte; excessive T suppressor cells
Hyperimmunoglobulinemia E	Specific immune responses; excessive IgE	Unknown
Chronic mucocutaneous candidiasis	Variable cellular	?antigen overload

EBNA, Epstein–Barr virus nuclear antigen; ADA, adenosine deaminase, PNP, purine nucleoside phosphorylase.

body deficiency in serum and in external secretions is of great importance in distinguishing this disorder from transient hypogammaglobulinemia of infancy. Tests for natural antibodies to blood group substances; for antibodies to antigens given during standard courses of immunization,

TABLE 15-3. CLINICAL CHARACTERISTICS OF ANTIBODY DEFICIENCY DISORDERS

1. Recurrent infections with high grade extracellular encapsulated pathogens
2. Few problems with fungal or viral (except enterovirus) infections
3. Chronic sinopulmonary disease
4. Growth retardation not striking
5. Antibody deficiency in serum and secretions
6. May or may not lack B lymphocytes with surface immunoglobulins or complement receptors
7. Absence of cortical follicles in lymph node and spleen in x-linked agammaglobulinemia
8. Paucity of palpable lymphoid and nasopharyngeal tissue in x-linked agammaglobulinemia
9. Compatible with survival to adulthood or for several years after onset except for those with persistent enterovirus infections, autoimmune disorders, or malignancy

e.g., diphtheria, tetanus, or pneumococcal; and for antibodies to and ability to clear bacteriophage ΦX 174 are useful in this regard. Polymorphonuclear functions are usually normal if heat-stable opsonins (e.g., IgG antibodies) are provided, but some patients with this condition have had transient, persistent, or cyclic neutropenia.

The number of total T cells is usually increased and the percentages of T cell subsets have been found to be normal in most of these patients. In contrast, blood lymphocytes bearing surface Ig, Ia-like antigens, the Epstein–Barr virus receptor, or reacting with a specific anti-B cell serum are absent or present in very low number. Cells with another B cell marker, the receptor for C3b and C3d, are usually present in normal number, and normal numbers of pre-B cells are found in the bone marrow of such patients. Evidence has been presented to suggest that this defect results from a maturation arrest at the stage of the earliest pre-B cell, which produces μ constant region without variable region. In vitro studies of pokeweed mitogen-induced B cell differentiation have shown little or no synthesis of immunoglobulin. Hypoplasia of adenoids, tonsils, and peripheral lymph nodes is the rule. Germinal centers are not found in these tissues, and plasma cells are lacking or rarely found. Mixed lymphocyte responsiveness and lymphocyte proliferative responses to antigens and mitogens are normal. Cell-mediated immune responses can be detected in vivo and the capacity to reject allografts is intact. The thymus has appeared morphologically normal in all autopsied cases, Hassall's corpuscles are present, and lymphoid cells are abundant in thymus-dependent areas of peripheral lymphoid tissues.

Except in those unfortunate patients who develop polio, persistent enterovirus infection, or lymphoreticular malignancy (an incidence as high as 6 percent has been reported), the overall prognosis is reasonably good if humoral replacement therapy is instituted early. Systemic infection can be prevented by intramuscular or intravenous immune serum globulin (primarily IgG) or plasma infusions (loading dose of IgG, 200 mg/kg; maintenance dose, 100 mg/kg every 3 to 4 weeks). However, many patients go on to develop crippling sinopulmonary disease despite this, as presently no effective means exists for replacing secretory IgA at the mucosal surface.

Agammaglobulinemia with Immunoglobulin-bearing B Lymphocytes

This condition, also known as common variable agammaglobulinemia or idiopathic late onset immunoglobulin deficiency, may appear similar clinically in many respects to x-linked agammaglobulinemia. Although there are well-documented examples of this disorder in infants and young children, most patients present with a history of recurrent infection beginning several years after birth, with little or no clinical evidence of susceptibility to infection prior to that time. While these defects are generally considered to be acquired, there are few documentations of true acquisitions of immunoglobulin deficiency. Moreover, the high incidences of abnormal immunoglobulin concentrations, autoantibodies and autoimmune disease, and malignancy in families of such patients suggest a hereditary influence. The principal differences between this disorder and x-linked agammaglobulinemia are the generally later age of onset, the somewhat less severe infections, and an almost equal sex distribution. In contrast to patients with the x-linked form, patients with common variable agammaglobulinemia may have normal-sized or enlarged tonsils and lymph nodes, and the latter may have cortical follicles. Splenomegaly is relatively common in this disorder. Additionally, patients often have normal or near-normal numbers of circulating immunoglobulin-bearing B lymphocytes. Nevertheless, the serum immunoglobulin and antibody deficiency are usually just as profound as in the x-linked disorder, and the kinds of infections experienced and bacterial etiologic agents involved are generally the same for the two defects. Thus far, no documented example of fatal echovirus meningoencephalitis has occurred in patients with common variable agammaglobulinemia; however, their susceptibility to hepatitis virus infection appears to be as great as those of x-linked patients.

This condition has been variably associated with a sprue-like syndrome, with or without nodular follicular lymphoid hyperplasia of the intestine; thymoma; alopecia areata; hemolytic anemia; gastric atrophy; achlorhydria;

and pernicious anemia. Frequent complications include giardiasis (seen far more often here than in x-linked agammaglobulinemia), bronchiectasis, gastric carcinoma, lymphoreticular malignancy, and cholelithiasis. Lymphoid interstitial pneumonia, pseudolymphoma, amyloidosis, and noncaseating granulomata of the lungs, spleen, skin, and liver have also been seen.

Despite the normal numbers of circulating immunoglobulin-bearing B lymphocytes and the presence of lymphoid cortical follicles, the lymphocytes do not differentiate in vivo into immunoglobulin-producing plasma cells or in vitro even in the presence of the polyclonal B cell activator, pokeweed mitogen. Thus, although the primary biologic error responsible for this defect is unknown, it appears in most patients to be due to abnormal terminal differentiation of the B cell line. Because this disorder occurs in first-degree relatives of patients with selective IgA deficiency and some patients with hyper-IgM have later become panhypogammaglobulinemic (or vice versa), it is possible that these diseases all belong to the same spectrum of B cell maturation arrests. The inconstancy of excessive suppressor T cell activity and the quantitatively poor plasma cell production even in the presence of optimal numbers of T helper cells suggest that excessive T suppressor cell activity is probably not basic to these disorders.

The treatment of patients with common variable hypogammaglobulinemia is essentially the same as for the x-linked disorder.

Selective IgA Deficiency

An isolated absence or near-absence (i.e., less than 10 mg/dl) of serum and secretory IgA is thought to be the most common well-defined immunodeficiency disorder. In the only large United States study, a frequency of 1:886 was reported for IgA deficiency in a series of 73,569 blood donors. Though this disorder has been observed in apparently healthy individuals, it is commonly associated with ill health. The kinds of health problems experienced by such patients often reflect the type of clinic from which they are drawn. Among 75 such patients from an allergy–immunology clinic, there were high frequencies of chronic or recurrent respiratory tract infection and atopic diseases. In contrast, 30 IgA-deficient patients drawn primarily from a rheumatology clinic had a high frequency of autoimmune and/or collagen vascular diseases. This disorder is also detected frequently in neurology clinics, where at least some of the patients have had phenytoin-induced IgA deficiency.

As would be expected when there is a deficiency of the major immunoglobulin of external secretions, infections occur predominantly in the respiratory, gastrointestinal, and urogenital tracts. Bacterial agents responsible are essentially the same as in other types of antibody deficiency syndromes. Although a high incidence of viral hepatitis was noted in one group of IgA deficient patients, there is no clear evidence that patients with this disorder have an undue susceptibility to other viral agents. In a prospective study of influenza infection in 90 such subjects, the rate of infection was no higher than in simultaneously studied normal controls. Children with IgA deficiency vaccinated with killed polio virus intranasally produced local IgM and IgG antibodies. While the children were being followed, several contracted rubella, and IgM and IgG antirubella antibodies were found in their secretions during convalescence. Recent studies suggest that the compensating IgM is locally synthesized and capable of combining with secretory piece for local secretion similar to IgA. With regard to antibacterial antibodies, secretory IgM antibodies to the cariogenic microorganism *Streptococcus mutans* have been detected in patients with selective IgA deficiency. Serum concentrations of other immunoglobulins are usually normal in patients with selective IgA deficiency, although an IgG_2 subclass deficiency has been reported in some, and IgM (usually elevated) may be of the low molecular weight variety.

In addition to limiting the attachment of infectious agents to mucosal surfaces, secretory IgA antibodies probably act to prevent absorption of other foreign antigens, such as those in the diet. The latter could stimulate either enhanced IgE antibody formation or complement-fixing IgG antibodies cross-reactive with organ- or tissue-specific antigens, resulting in allergic, immune complex, or autoimmune phenomena. In support of this possibility is the finding of high incidences of allergy and of IgG antibodies against cow's milk and ruminant serum proteins in patients with IgA deficiency; circulating immune complexes have also been found following milk ingestion in some such patients. The antiruminant antibodies often present technical problems in radial diffusion measurement of IgA, because most antibody-agar plates are made with goat antisera. A very high incidence of autoantibodies was noted in one group of IgA-deficient subjects; this may have bearing on the frequent association of IgA deficiency with collagen vascular and autoimmune diseases. Intestinal nodular hypoplasia has been seen in a few such patients, and there is a high incidence of a sprue-like syndrome in adults with selective IgA deficiency; the latter may or may not respond to a gluten-free diet.

The basic defect leading to selective IgA deficiency is unknown. In 10 of 11 IgA deficient individuals studied, more than 80 percent of the IgA-bearing B cells also coexpressed surface IgM and IgD (similar to cord blood B cells), as compared with less than 10 percent triple isotype-bearing IgA B cells in controls. Other studies demonstrated that blood lymphocytes from 11 of 14 such patients synthesized cytoplasmic IgA when stimulated by pokeweed mitogen. However, none of the 14 patients' cells secreted IgA in any appreciable quantity into the culture supernatants. B lymphocytes from 3 of the 14 patients failed to synthesize IgA at all. Cocultures of cells from the former three patients with those from normal subjects resulted in selective suppression of IgA synthesis by normal B cells; no suppression was seen in cocultures with cells from the 11 patients who synthesized but did not secrete IgA. Studies of T cell function have been normal in most patients with selective IgA deficiency. That this defect may not always be permanent is supported by observations of the author who, in following over 120 such patients, has seen the spontaneous development of normal serum IgA concentrations in 8 children documented for several years to have absent or extremely low IgA. The occurrence of IgA deficiency in both males and females and in families suggests autosomal inheritance. In some families this appears to be a recessive trait, and in others it appears to be dominant with variable expressivity.

Of possible etiologic and great clinical significance is the presence of antibodies to IgA in the sera of as high as 44 percent of patients with selective IgA deficiency. Anti-IgA antibodies can fix complement and remove IgA from the circulation 4 to 20 times faster than the normal catabolic rate for IgA. At least eight IgA deficient patients have been described as having severe or fatal anaphylactic reactions after intravenous administration of blood products, and anti-IgA antibodies appeared to be clearly incriminated in four of these. For this reason, only multiply washed erythrocytes or blood products from other IgA absent individuals should be administered to these patients, and immune serum globulin (which contains a small amount of IgA) is contraindicated.

Currently there is no treatment for IgA deficiency beyond the vigorous treatment of specific infections with appropriate antimicrobial agents. Even if serum IgA could be replaced (in the face of anti-IgA antibodies), it would not be transported into the external secretions, as the latter is an active process involving only locally and endogenously produced IgA.

Secretory Component Deficiency

A patient with chronic intestinal candidiasis and diarrhea was found to lack IgA in his external secretions, despite having a normal concentration of serum IgA. This was ultimately traced to a lack of secretory piece, which prevented the normal secretion of locally produced IgA onto his mucous membrane surfaces.

Selective IgM Deficiency

If one uses a strict definition of a serum concentration less than 10 percent of the normal mean or less than 10 mg/dl, there are very few well-documented cases of this entity. Septicemia due to meningococci and other gram-negative organisms, pneumococcal meningitis, tuberculosis, recurrent staphylococcal pyoderma, periorbital cellulitis, bronchiectasis, recurrent otitis, and other respiratory infections have all been reported in this disorder. In vivo production of IgM antibodies is markedly impaired. However, in vitro studies of B lymphocyte function in two such patients revealed normal production of immunoglobulins of all isotypes when cocultured with normal T cells. This was not due to excessive autologous suppressor T cell activity, but apparently to a lack of adequate T helper cell function.

There is no specific treatment for this disorder; early and vigorous treatment with antibiotics is recommended to avoid fatal septicemia.

Immunodeficiency with Elevated IgM

This disorder is characterized by very low serum concentrations of IgG and IgA, but either a normal or, more frequently, a markedly elevated concentration of polyclonal IgM. Some of these patients have low molecular weight IgM molecules that give falsely high IgM values on radial diffusion. High titers of IgM antibodies to blood group substances and to *Salmonella* O antigen have been found in some patients, but very low titers or no IgM antibody has been noted in others.

Similar to patients with x-linked agammaglobulinemia, patients with this defect may become symptomatic during the first or second year of life with recurrent pyogenic infections, including otitis media, sinusitis, pneumonia, and tonsillitis. In contrast to patients with x-linked agammaglobulinemia, however, the frequent presence of lymphoid hyperplasia in this condition often leads the physician away from a diagnosis of immunodeficiency. Even more than with some of the other antibody deficiency syndromes, there is an increased frequency of autoimmune disorders in the hyper-IgM syndrome. Hemolytic anemia and thrombocytopenia have been seen in several patients, and transient, persistent, or cyclic neutropenia is a common feature. The neutropenia is considered a possible explanation for the unusually high frequencies of *P. carinii* pneumonia and extensive verruca vulgaris lesions seen. Thymic-dependent lymphoid tissues and T cell functions are usually normal, but several have had partial T cell deficiencies. A sex-linked mode of inheritance has been proposed, but several examples in females now seem to make this less certain.

The precise defect in the hyper-IgM syndrome has yet to be elucidated. Normal or only slightly reduced numbers of IgM and/or IgD bearing B lymphocytes have been found in the blood of these patients. In vitro studies have demonstrated either no immunoglobulin synthesis or only IgM production; deficient T cell help and increased T suppressor cell activity have each also been reported. However, cultured B cell lines from such patients have shown the capacity to synthesize only IgM, suggesting a primary B cell defect. Histologic studies of peripheral lymphoid tissues have revealed the presence of primary follicles in some, but not all, of these patients. When plasma cells have been present in the perifollicular areas or medullary cords of lymph nodes, they have usually contained only IgM.

Because these patients have an inability to make IgG antibodies, the treatment for this condition is the same as for agammaglobulinemia, i.e., immune serum globulin replacement therapy.

Transient Hypogammaglobulinemia of Infancy

This condition has been described as a prolongation and accentuation of the physiologic decline in serum Ig concentrations normally seen during the first 3 to 7 months of life. Until recently there had been scant documentation of this disorder. Eleven patients with this defect were evaluated with various tests of immunologic function in the author's laboratory over a 12-year period. Two groups were identified: six patients who were found by screening relatives of patients with other types of immunodeficiency; and five patients whose sera were sent for study because the patients were having frequent or unusual infections. Those in the first group had no significant health problems during early infancy or childhood. In the second, recurrent infection was a problem during early infancy but not in later years. All 11 patients could synthesize antibodies to human type A and B erythrocytes and to diphtheria and tetanus toxoids, usually by 6 to 11 months of age, and well before immunoglobulin concentrations became normal. Lymphocyte studies in vitro showed no abnormalities in the percentages of cells in the different subpopulations or in their responses to the mitogens. The finding of only 11 cases of transient hypogammaglobulinemia of infancy among over 10,000 patients whose sera underwent immunoglobulin studies over a 12-year period suggests that this is not a common entity, contrary to earlier opinions.

Although the etiology or etiologies of transient hypogammaglobulinemia of infancy remain unknown, two possible pathogenetic mechanisms have been suggested. First, it was hypothesized that maternal alloantibodies to fetal Gm antigens might cross the placenta and suppress fetal immunoglobulin production. A prospective study in which serial serum immunoglobulin determinations were made from birth throughout infancy in offspring of mothers found during pregnancy to have specific Gm agglutinins, however, failed to demonstrate any evidence of depressed immunoglobulin synthesis in the infants. A second hypothesis was that the condition may be a manifestation of genetic heterozygosity for some other immunodeficient state. Considering the variety of defects reported in the relatives of patients with transient hypogammaglobulinemia of infancy, it is difficult to envision how transient hypogammaglobulinemia of infancy could be a manifestation of heterozygosity for so many different conditions unless the primary biologic error underlying all of them proves to be similar.

Gammaglobulin replacement therapy is not indicated in this condition. In addition to the known risk of inducing antiallotype antibodies, passively administered IgG antibodies could suppress endogenous antibody formation in the same manner that RhoGAM suppresses anti-D antibodies in Rh negative mothers delivering Rh positive infants.

Antibody Deficiency with Near-Normal Immunoglobulins

Only scattered reports have appeared in the literature describing patients with apparently normal T cell function and normal or near-normal immunoglobulin concentrations but with deficient antibody responses. Recently, the author and her associates have studied the antibody-forming capacities of 12 such patients. Blood group antibody titers were absent in all but two of these patients, diphtheria titers were low in all, and tetanus titers were low in 10 patients. Geometric mean antibody titers to 13 pneumococcal serotypes were significantly lower than those of 27 normal controls before and after immunization with tridecavalent pneumococcal polysaccharide vaccine. All patients cleared bacteriophage ϕX 174 normally, but all primary immune responses were far below the normal range. Secondary responses to ϕX 174 were also below the normal range in all but two, however, in both cases most of the secondary response was IgM rather than IgG. This type of immune problem may be far more common than generally thought. It would not be detected unless functional tests of antibody-forming capacity are regularly conducted in the assessment of humoral immunity. Because these patients do not have the ability to produce antibodies normally, they are candidates for immune serum globulin replacement therapy.

Possibly related to disorders of specific antibody production in patients with normal or near-normal immunoglobulins are deficiencies of immunoglobulin subclasses or κ or λ chains. Descriptions of patients with κ or λ chain deficiencies have reported only moderately low concentrations of all three major immunoglobulin classes, but clearly deficient antibody responses, closely resembling the patients mentioned above with antibody deficiency but near-normal concentrations of immunoglobulins.

X-Linked Lymphoproliferative Disease

This disorder, also referred to as Duncan's disease (after the original kindred in which it was described), is a recessive trait characterized by an inadequate immune reaction to infection with Epstein–Barr virus. Affected males are apparently healthy until they experience infectious mononucleosis. Two thirds of the 100 patients studied thus far died of overwhelming Epstein–Barr virus-induced B cell proliferation during mononucleosis. A majority of those patients surviving the primary infection developed hypogammaglobulnemia and/or B cell lymphomas. There is a marked impairment in production of antibodies to the Epstein–Barr virus nuclear antigen (EBNA), whereas titers of antibodies to the viral capsid antigen have ranged from zero to markedly elevated. Antibody-dependent cell-mediated cytotoxicity (ADCC) against Epstein–Barr virus-infected cells has been low in many, and natural killer (NK) function is also depressed. There is also a deficiency in long-lived T cell immunity to Epstein–Barr virus. Patients with this disorder usually have normal numbers of total B and T cells, but studies of their lymphocyte subpopulations with monoclonal antibodies have frequently revealed elevated percentages of cells of the suppressor (T8) phenotype. Lymphocytes from survivors usually give normal proliferative responses to T and B cell mitogens. However, immunoglobulin synthesis in response to polyclonal B cell mitogen stimulation in vitro is markedly depressed. Thus, both Epstein–Barr virus-specific and nonspecific immunologic abnormalities occur in these patients.

Cellular Immunodeficiency Disorders

The two conditions described below are often referred to as examples of isolated defects of the T cell system, as serum immunoglobulin concentrations are usually normal or elevated. However, B cell function is necessarily also compromised due to a deficiency of T helper cells; hence, the defects are actually combined. Some important clinical characteristics of cellular immunodeficiency disorders are listed in Table 15-4. In general, patients with partial or absolute defects in T cell function more often have infections or other clinical problems for which there is no effective treatment and/or that are of a more severe nature than do those with antibody deficiency disorders. It is also rare that such individuals survive beyond infancy or childhood.

Thymic Hypoplasia (DiGeorge's Syndrome)

This syndrome results from dysmorphogenesis of the third and fourth pharyngeal pouches during early embryogenesis, leading to hypoplasia or aplasia of the thymus and parathyroid glands. Other structures forming at the same age are also frequently affected, resulting in anomalies of the great vessels (right-sided aortic arch), esophageal atresia, bifid uvula, congenital heart disease (atrial and ventricular septal defects), a short philtrum of the upper lip, hypertelorism, an anti-Mongoloid slant to the eyes, mandibular hypoplasia, and low set often notched ears. The diagnosis is usually first suggested by the presence of hypocalcemic seizures during the neonatal period. This condition has occurred in both males and females and there is little evidence that it is heritable.

Since the original description of the syndrome, it has become apparent that a variable degree of hypoplasia is more frequent than total aplasia of the thymus and parathyroid glands. Some children have little trouble with life-threatening infections and grow normally, even though recurrent respiratory tract infections caused by common pathogenic bacteria and chronic oral candidiasis do occur; such patients are often referred to as having partial DiGeorge's syndrome. Other patients who have more marked thymic hypoplasia may resemble patients with severe combined immunodeficiency in their susceptibility to infection with low-grade or opportunistic pathogens (i.e.,

TABLE 15-4. CLINICAL CHARACTERISTICS OF CELLULAR IMMUNODEFICIENCY DISORDERS

1. Recurrent infections with low-grade or opportunistic infectious agents such as fungi, viruses, or *Pneumocystis carinii*
2. Delayed cutaneous anergy
3. Accompanied by growth retardation, short life span, wasting, and diarrhea
4. Susceptible to graft-vs.-host disease if given fresh blood, plasma, or unmatched allogeneic bone marrow
5. Fatal reactions from live virus or bacillus of Calmette and Guerin vaccination
6. High incidence of malignancy

fungi, viruses, and *P. carinii*) and to graft-vs.-host disease from nonirradiated blood transfusions.

Concentrations of serum immunoglobulins are usually near normal for age, but some fractions, particularly IgA, may be diminished and IgE may be elevated. T cell numbers are decreased and there is an increased number of B cells. Analyses of T cells using monoclonal antibodies have demonstrated that, despite decreased numbers of total T cells, there are normal proportions of those with the helper and suppressor phenotypes. The DNA synthetic response of peripheral blood lymphocytes following mitogen stimulation, like the intradermal delayed hypersensitivity response, has been absent, reduced, or normal, depending upon the degree of thymic deficiency. Serial sections through the anterior mediastinums of DiGeorge patients more often than not have revealed tiny nests of thymic tissue of variable quantity. The tissue contained Hassall's corpuscles and a normal density of thymocytes, and corticomedullary distinction was present. Lymphoid follicles usually appear normal, but lymph node paracortical areas and thymus-dependent regions of the spleen show variable degrees of depletion, depending upon the degree of thymic hypoplasia. Because of variability in the severity of the immunodeficiency and the fact that it may become much less severe with time, it is difficult to evaluate the previously claimed benefits of fetal thymus transplantation. However, in the complete DiGeorge's syndrome, the latter is still considered the treatment of choice.

Cellular Immunodeficiency with Immunoglobulins (Nezelof's Syndrome)

This syndrome is characterized by lymphopenia, diminished lymphoid tissue, abnormal thymus architecture, and the presence of normal or increased levels of most of the five immunoglobulin classes. Unlike DiGeorge patients, these children have no endocrine or cardiovascular anomalies. They often present with recurrent or chronic pulmonary infections, failure to thrive, oral or cutaneous candidiasis, chronic diarrhea, recurrent skin infections, gram-negative sepsis, urinary tract infections, severe varicella, and/or progressive vaccinia. An autosomal recessive pattern of inheritance has been suggested in some cases, but an x-linked mode seemed more likely in others. Other findings include lymphopenia, neutropenia, and eosinophilia. Serum immunoglobulins may be normal or elevated for all classes, but selective IgA deficiency, marked elevation of IgE, and elevated IgD levels occur not infrequently. Restricted heterogeneity of molecules of the IgG class, as well as IgG subclass imbalance, were reported in three female siblings with this disorder.

Studies of cellular immune function have shown delayed cutaneous anergy to ubiquitous antigens in all such patients and low to absent in vitro lymphocyte responses to mitogens and allogeneic cells. Analyses of blood lymphocytes with monoclonal antibodies have revealed profound deficiencies of total T cells and T cell subsets, with usually a normal helper (T4$^+$) to suppressor

(T8$^+$) cell ratio, in contrast to the acquired immune deficiency syndrome (AIDS), where there is a characteristic inversion of the T4 to T8 ratio. Peripheral lymphoid tissues are hypoplastic and demonstrate paracortical lymphocyte depletion, whereas AIDS patients usually have lymphadenopathy. The thymuses are very small, have poor corticomedullary distinction, a paucity of thymocytes, and usually no Hassall's corpuscles; however, again in contrast to AIDS, thymic epithelium is present. These could all be useful features to distinguish Nezelof's syndrome from AIDS in the pediatric age group, as this is the primary immunodeficiency disorder most like to be confused with it. Despite the profound cellular immunodeficiency, however, Nezelof patients usually survive longer than do infants with severe combined immunodeficiency. The fatal or serious infections in both syndromes are very similar and include varicella, vaccinia, *P. carinii*, cytomegalovirus, rubeola, *Pseudomonas*, and *Mycobacterium kansasii.* Though antibody-forming capacity has been impaired in a majority, it has not been absent and has been apparently normal in roughly one third of the reported cases. Moreover, plasma cells are usually abundant in the lamina propria and lymph nodes. Although patients with this disorder have been successfully reconstituted by matched sibling bone marrow transplants, most other forms of therapy have been unsuccessful.

With Nucleoside Phosphorylase Deficiency. In 1975, an absence of the purine salvage pathway enzyme, purine nucleoside phosphorylase (PNP), was noted in the erythrocytes of a female child who had profound lymphopenia and normal to elevated serum immunoglobulins, similar to that described for Nezelof patients. This observation was of great interest, as adenosine deaminase (ADA)—found deficient in a number of infants with autosomal recessive severe combined immunodeficiency—and PNP act sequentially in that pathway. Red blood cell lysates from her clinically normal parents (a second cousin marriage) contained only 50 percent of the mean PNP activity found in normal erythrocytes, consistent with an autosomal recessive mode of inheritance for the enzyme deficiency. Subsequently, 11 additional patients with cellular immunodeficiency with immunoglobulins have been found to have PNP deficiency. In contrast to patients with ADA deficiency, serum and urinary uric acid are markedly deficient, and no characteristic physical or skeletal abnormalities have been noted. Two patients have suffered from a progressive neurologic disorder with spastic tetraplegia, two developed an autoimmune hemolytic anemia, and one has idiopathic thrombocytopenic purpura. Deaths have occurred from generalized vaccinia, varicella, lymphosarcoma, and graft-vs.-host disease following blood transfusions. In contrast to a majority of Nezelof patients, the thymuses of PNP-deficient patients have had some Hassall's corpuscles at postmortem examination, reminiscent of some patients with ADA deficiency. Analyses of lymphocyte subpopulations with monoclonal antibodies in two such patients revealed a marked deficiency of T cells and T cell subsets, but an increased number of cells with NK phenotype and function. Attempts to correct the immunologic and enzymatic deficiencies of PNP-deficient patients by enzyme replacement therapy using normal erythrocyte transfusions have been successful in only one case. Deoxycytidine therapy has also been unsuccessful.

Severe Combined Immunodeficiency Disorders

The syndrome(s) of severe combined immunodeficiency (SCID) are distinguished by their apparent congenital absence of all adaptive immune function. For some time it has been assumed that an absence or a failure of proliferation and/or differentiation of the primordial stem cell is the basis of these syndromes. Unfortunately, this theory does not explain the great diversity of genetic, enzymatic, hematologic, and immunologic features observed. This disorder is the most severe of all of the recognized immunodeficiencies. Unless immunologic reconstitution can be achieved through immunocompetent tissue transplants or enzyme replacement therapy, or gnotobiotic isolation can be carried out, death usually occurs before the patient's first birthday and almost invariably before the second. The major subcategories of this disorder are discussed below.

Autosomal Recessive Severe Combined Immunodeficiency Disease

This was the first described of the SCID syndromes, reported initially by Swiss workers in 1958. In this and all other forms of SCID, there is an apparent congenital absence of both cellular and humoral immune function. Affected infants present within the first few months of life with frequent episodes of otitis, pneumonia, sepsis, diarrhea, and cutaneous infections. Growth may appear normal initially but extreme wasting usually develops after infections and diarrhea begin. Infections with opportunistic organisms such as *Candida albicans*, *P. carinii*, vaccinia, varicella, measles, cytomegalovirus, and bacillus of Calmette and Guerin frequently lead to death because of the difficulties encountered in diagnosis and adequate treatment. In addition to their undue susceptibility to infection, these infants also lack the ability to reject foreign tissue and are, therefore, at risk for graft-vs.-host disease. Immunocompetent cells capable of producing graft-vs.-host reactions can derive from maternal lymphocytes that cross the placenta while the infant is in utero, or from the inadvertent administration of blood products containing viable histoincompatible lymphocytes.

Immunologic evaluations in virtually all SCID patients have shown a near total lack of cellular immune function, with profound lymphopenia and absent or extremely low lymphocyte proliferative responses to mitogens and allogeneic cells in vitro, delayed cutaneous anergy in vivo, and

an inability to reject transplants. Serum immunoglobulin concentrations are usually diminished to absent and no antibody formation occurs following immunization. Occasionally normal or elevated concentrations of individual immunoglobulin classes have been observed. Analyses of lymphocyte populations and subpopulations have demonstrated marked heterogeneity among SCID patients, even among those with similar inheritance patterns or with ADA deficiency. Despite the uniformly profound lack of T or B cell function, some patients have had low numbers of both B and T lymphocytes, whereas others have had elevated numbers of B cells, and occasionally, even normal numbers of both T and B cells have been found. Cytofluorographic studies with monoclonal antibodies to mature T cells and subsets have generally revealed some, albeit low, percentages of cells reacting with all such reagents; however, none have had blood cells reacting with the monoclonal antibody to the T6 antigen, present on immature cortical thymocytes. Thus, the lymphocytes present appear to have acquired surface markers characteristic of mature T cells. In contrast to similarly lymphopenic patients with AIDS, SCID patients rarely have an inverted ratio of helper (T4$^+$) to suppressor (T8$^+$) cells. Recently, a new phenotype of SCID was characterized by the author in which virtually all of the lymphocytes of two infants with SCID were large granular lymphocytes with NK cell phenotype and function. NK function has been totally lacking in other SCID patients, again illustrating the striking heterogeneity at a cellular level. Typically, SCID patients have very small thymuses (less than 2 g), which usually fail to descend from the neck, contain few thymic lymphocytes, lack corticomedullary distinction, and usually lack Hassall's corpuscles (see exceptions below). Recently, in studies of postmortem thymic tissue from two infants with SCID and one with Nezelof's syndrome (none ADA or PNP deficient), Haynes and Buckley found abnormalities in thymic epithelial surface antigens. The SCID patients' thymic epithelium failed to bind tetanus toxoid (a marker for GD and GT gangliosides) and the Nezelof patient's thymus failed to react with a monoclonal antibody to GQ ganglioside; age-matched normal thymuses bound both tetanus toxin and the antibody to GQ ganglioside. Whether these abnormalities were primary or secondary to some other basic defect in these patients is unknown. It should be noted that, despite the profound thymocyte depletion in SCID patients, thymic epithelium is present—in contrast to the situation in AIDS, where there is marked epithelial atrophy. Both the follicular and paracortical areas of the peripheral lymph nodes are depleted of lymphocytes in SCID patients; tonsils, adenoids, and Peyer's patches are absent or extremely underdeveloped.

Immune serum globulin injections fail to halt the progressively downhill course of SCID. Transplantation of bone marrow cells from HLA genotypically identical or D locus compatible donors has resulted in apparent complete correction of the immunologic defect in a number of these patients, with over 30 known long-term survivors. Fetal tissue transplants have been less effective, and unfractionated bone marrow cell infusions from D locus incompatible donors have invariably resulted in fatal graft-vs.-host disease. An important new development, i.e., use of soybean agglutinin to deplete postthymic T cells from donor marrow (see Treatment), has recently allowed haplo-identical bone marrow to be used for correction of the immunologic defect without graft-vs.-host disease in nearly 30 more of these infants.

With Adenosine Deaminase Deficiency. More recently, an absence of the enzyme ADA has been observed in some, but not all, patients with the autosomal recessive form of SCID; approximately 30 families have now been identified in which the enzyme deficiency was associated with severe immunodeficiency. Although most such patients have had profound lymphopenia at birth or from the earliest age studied, a few ADA-deficient SCID patients have had normal or fluctuating lymphocyte counts early on, only to decline by 6 weeks to 2 years of life. As in the case of SCID with normal ADA, cells with differentiation markers characteristic of mature T and B lymphocytes have been found in varying quantities in these patients, providing evidence against this group of disorders being secondary to a stem cell defect. In marked contrast to the classic SCID, some ADA-deficient patients have been found to have a few Hassall's corpuscles in their thymuses and changes suggestive of early differentiation.

Other distinguishing features of ADA-deficient SCID patients have included the presence of rib cage abnormalities similar to a rachitic rosary and multiple skeletal abnormalities of chondroosseous dysplasia on radiographic examination; these occur predominantly at the costochondral junctions, at the apophyses of the iliac bones, and in the vertebral bodies.

Matched sibling bone marrow transplants have resulted in lymphocyte chimerism and partial or complete correction of the immunologic defect in ADA-deficient SCID. The in vitro addition of exogenous ADA to lymphocytes from one ADA-deficient SCID patient conferred an ability to give DNA synthetic responses to mitogenic stimuli. Applying this observation clinically, enzyme replacement therapy was developed consisting of the administration of 15 ml/kg glycerol-frozen, irradiated, packed normal erythrocytes every 2 to 4 weeks. Only a few patients so treated have shown immunologic and/or clinical improvement. The enzyme remains within the transfused erythrocytes in the recipient and gradually disappears as those cells are destroyed; hence, repeated therapy is necessary.

X-Linked Recessive Severe Combined Immunodeficiency Disease

This is thought to be the most common form of SCID in the United States. In pedigrees where there has been a proven or putative x-linked mode of inheritance, there have been no examples of deficiencies of the purine salvage pathway enzymes ADA or PNP. Clinically, immunologically, and histopathologically, patients with the x-linked

form usually appear similar if not identical to those with the autosomal recessive form.

Bare Lymphocyte Syndrome

In this form of combined immunodeficiency there is a lack of expression of HLA-A, HLA-B, and HLA-C antigens and the absence of B2 microglobin on lymphocytes. Nine examples of this defect have now been reported, all from the Mediterranean area. The immune defect is similar to that in the other forms of SCID except that it was only partial in five of the cases; in those, the expression of HLA antigens was reduced on both lymphocytes and platelets but not completely suppressed. The associated defects of immunity and of HLA expression support the concept of a biologic role of HLA determinants in the development of functional T lymphocytes.

Severe Combined Immunodeficiency with Leukopenia (Reticular Dysgenesis)

In 1959, identical twin male infants were described who exhibited a total lack of both lymphocytes and granulocytes in their peripheral blood and bone marrow. Seven of the eight infants thus far reported with this defect died between 3 and 119 days of age from overwhelming infections; the eighth underwent complete immunologic reconstitution from a bone marrow transplant. The organisms responsible for demise of patients with this disorder have been both bacterial and viral, including cytomegalovirus, *Pseudomonas*, *Klebsiella*, and pyogenic cocci. Mature, normal-appearing granulocytes (though markedly reduced in number) were noted in three patients and a normal percentage of E-rosetting T cells in the cord blood of a fourth patient, arguing against a total failure of stem cell differentiation in this defect. However, despite the normal percentage of T cells in the latter patient's cord blood, the cells failed to give an in vitro proliferative response to mitogens. Serum immunoglobulins were very low and there was no lymphocyte proliferative response to mitogens in the few patients in whom immunologic evaluations were conducted. Lymphoid tissues have been difficult to identify grossly at postmortem examinations. The thymus glands have all weighed less than 1 g, no Hassall's corpuscles have been present, and few or no thymocytes were seen. A genetic influence seems likely from reports of familial occurrences; an autosomal mode of inheritance seems most likely.

Partial Combined Immunodeficiency Disorders

In addition to the severe combined immunodeficiency disorders discussed above, there are other conditions that manifest varying degrees of both humoral and cellular immunodeficiency.

Immunodeficiency with Thrombocytopenia and Eczema (Wiskott–Aldrich Syndrome)

This x-linked recessive syndrome is characterized clinically by the triad of eczema, thrombocytopenic purpura with normal-appearing megakaryocytes, and undue susceptibility to infection. Often there is prolonged bleeding from the circumcision site or bloody diarrhea during infancy. Atopic dermatitis and recurrent infections also usually develop during the first year of life. In younger patients, infections are commonly those produced by pneumococci and other bacteria having polysaccharide capsules, resulting in episodes of otitis media, pneumonia, meningitis, and sepsis. Later, as cellular immune function wanes, infections with agents such as *P. carinii* and herpesviruses become more frequent. Survival beyond the teens is rare; infections or bleeding are major causes of death, but there is also a 12 percent incidence of fatal malignancy in this condition. A papovavirus has been recovered from a reticulum cell sarcoma of the brain and from the urine of patients with this syndrome.

The earliest demonstrable evidence of immunodeficiency is an impaired humoral immune response to polysaccharide antigens. Absent or markedly diminished isohemagglutinin titers are found uniformly, and poor or no responses are seen following immunization with polysaccharide antigens. Antibody titers to proteins also fall with time, and anamnestic responses are often poor or absent. Studies of immunoglobulin metabolism have shown an accelerated rate of synthesis as well as hypercatabolism of albumin, IgG, IgA, and IgM, resulting in highly variable immunoglobulin concentrations, even within the same patient. The predominant pattern is a low serum IgM, elevated IgA and IgE, and a normal or slightly low IgG concentration. Restricted heterogeneity of the immunoglobulins and the frequent appearance of transient paraproteins have also been reported. While lymphocytes from these patients have sometimes responded normally to mitogen stimulation in vitro, more often responses are moderately depressed and cutaneous anergy is a frequent finding. Analyses of blood lymphocytes with monoclonal reagents have revealed low percentages of cells reacting with antibodies to all T cells and to the helper (T4$^+$) and suppressor (T8$^+$) subsets. However, as with the other primary cellular immunodeficiencies, there is usually no imbalance in the T4 to T8 ratio. Biopsy and autopsy observations have shown a gradual loss of lymphoid elements in both the thymus and thymus-dependent portions of the peripheral lymphoid organs with age.

The thrombocytopenia appears to be due to an intrinsic platelet abnormality, as antiplatelet antibodies have not been demonstrated and survival times of homologous but not autologous ^{51}Cr-labeled platelets have been normal in these patients.

Initially, treatment was directed toward control of bleeding with platelet transfusions and of infections by appropriate antibiotic therapy. More recently, antibody replacement has been possible due to the availability of immune serum globulin preparations suitable for intravenous use. Several patients who required splenectomy for uncontrollable bleeding have had impressive rises in their platelet counts and have done well clinically while on chronic antibiotic and antibody replacement therapy. Most encouraging, however, are several recent complete corrections of both the platelet and immunologic abnormalities in patients with this disorder by matched sibling bone marrow transplants after conditioning with irradiation or busulfan and cyclophosphamide.

Ataxia Telangiectasia

This is a complex syndrome with neurologic, immunologic, endocrinologic, hepatic, and cutaneous abnormalities. The most prominent clinical features are progressive cerebellar ataxia, oculocutaneous telangiectasias, chronic sinopulmonary disease, a high incidence of malignancy, and variable humoral and cellular immunodeficiency. Ataxia typically becomes evident soon after the child begins to walk and progresses until he or she is confined to a wheelchair, usually by the age of 10 to 12 years. The telangiectasias develop between 3 and 6 years of age. Recurrent, usually bacterial, sinopulmonary infections occur in approximately 80 percent of these patients, but common viral exanthems and smallpox vaccination have not usually resulted in untoward sequelae. However, fatal varicella occurred in one of the author's patients. Gonadal agenesis, abnormal liver function, elevated serum α-fetoprotein concentrations, nonketotic hyperglycemia, and insulin-resistant diabetes have been noted in some patients.

The malignancies reported in this condition have usually been of the lymphoreticular type, but adenocarcinoma of the stomach, basal cell carcinoma, dysgerminoma, and medulloblastoma have also been seen. There is also an increased incidence of malignancy in unaffected relatives of such patients. Recently their cells as well as those of heterozygous carriers of the defect have been reported to have increased sensitivity to ionizing radiation, defective DNA repair, and frequent chromosomal abnormalities. An autosomal recessive mode of inheritance seems operative. The most frequent humoral immunologic abnormality is the selective absence of IgA, found in from 50 to 80 percent of these patients; hypercatabolism of IgA is also known to occur. IgE concentrations are usually low, and the IgM may be of the low molecular weight variety, IgG$_2$ or total IgG may be decreased. Specific antibody titers may be decreased or normal. In vivo there is impaired but not absent cell-mediated immunity, as evidenced by delayed cutaneous anergy and prolonged allograft survival. Death from graft-vs.-host disease has not been reported, and skin allografts are usually eventually rejected. In vitro tests of lymphocyte function have generally shown moderately depressed proliferative responses to T and B cell mitogens. Enumeration of blood T cells and subsets in five patients with this disorder revealed reduced percentages of total T cells and T cells of the helper (T4) phenotype, with normal or increased percentages of cells of the suppressor (T8) phenotype. Studies of immunoglobulin synthesis in 20 patients revealed depressed synthesis of all three of the major isotypes. The latter was not due to excessive suppression, but could be partially reversed by coculturing with normal T cells, suggesting a helper T cell defect. However, the failure of selectively IgA-deficient patient B cell IgA synthesis to be increased by normal T helper cells speaks for an intrinsic B cell defect as well. The thymus is very hypoplastic, exhibits poor organization, and is lacking in Hassall's corpuscles. No satisfactory treatment has been found.

Immunodeficiency with Short-limbed Dwarfism

In 1964 an unusual form of short-limbed dwarfism with frequent and severe infections was reported among the Amish; non-Amish cases have since been described. This form of dwarfism is distinct from achrondroplasia, and some affected individuals have also had cartilage–hair hypoplasia. Features include short and pudgy hands; redundant skin; hyperextensible joints of hands and feet but an inability to completely extend the elbows; and fine, sparse light hair and eyebrows. Radiographically the bones show scalloping and sclerotic or cystic changes in the metaphyses and flaring of the costochondral junctions of the ribs. Though recurrent sinopulmonary infections due to various viral and bacterial agents are seen, severe and often fatal varicella infections appear to be a particular hazard. Progressive vaccinia and vaccine associated poliomyelitis have also been observed.

The severity of the immunodeficiency varies; in one series, 11 of 77 patients died before age 20, but 2 were still alive at age 76. Three patterns of immune dysfunction have emerged: defective antibody mediated immunity, defective cellular immunity (most common form), and severe combined immunodeficiency. In vitro studies of lymphocyte numbers and function have shown decreased numbers of total T cells, but normal proportions of those of the helper (T4$^+$) and suppressor (T8$^+$) phenotypes. The most striking abnormality appears to be one of defective cell proliferation due to an intrinsic defect related to the G 1 phase, resulting in a longer cell cycle for individual cells. This abnormality also occurs in fibroblasts from these patients. The trait appears to be autosomal recessive with variable penetrance. One patient with cartilage–hair hypoplasia had correction of his immune defect by matched sibling bone marrow transplantation.

Immunodeficiency with Thymoma

The patients are adults who almost simultaneously develop recurrent infections, panhypogammaglobulinemia, deficits in cell-mediated immunity, and benign thymoma. They may also have eosinophilia or eosinopenia, aregenerative or hemolytic anemia, agranulocytosis, thrombocytopenia, or pancytopenia. Antibody formation is poor, and progressive lymphopenia develops, although percentages of immunoglobulin-bearing B lymphocytes are normal. Several patients with this disorder have been shown to have excessive suppressor T cell activity. The thymomas are predominantly of the spindle cell variety, although other types of benign and malignant thymic tumors have also been seen.

Hyperimmunoglobulinemia E Syndrome

The hyper-IgE syndrome is a primary immunodeficiency characterized by recurrent severe staphylococcal abscesses and markedly elevated levels of serum IgE. The disorder was first reported by the author and her co-workers in two young boys in 1972; since then she has evaluated a total of 21 patients with the condition and several other examples have been reported. These patients all have histories of staphylococcal abscesses involving the skin, lungs, joints, and other sites from infancy; persistent pneumatocoeles develop as a result of their recurrent pneumonias. Seven of the author's patients required thoracic surgery because of infection in these chronic lung cysts. The only sites that have not been foci of infections include the urinary and gastrointestinal tracts and the bones (except for mastoids). The pruritic dermatitis that occurs is not typical atopic eczema and does not always persist; respiratory allergic symptoms are usually absent.

Laboratory features include exceptionally high serum IgE concentrations; elevated serum IgD concentrations; usually normal concentrations of IgG, IgA, and IgM; pronounced blood and sputum eosinophilia; abnormally low anamnestic antibody responses to booster immunizations; and poor antibody and cell-mediated responses to neoantigens. In vitro studies have shown normal percentages of E rosette-forming, T3, T4, and T8 positive lymphocytes, and there is no increase in the percentage of IgE-bearing B lymphocytes. In addition, all but two of the author's patients had normal lymphocyte proliferative responses to mitogens. In contrast, lymphocyte proliferative responses to antigens or allogeneic cells from family members were absent or very low. Histologic sections of lymph nodes, spleen, and lung cysts have shown striking eosinophilia. Hassall's corpuscles and normal thymic architecture were observed at postmortem examination of one patient.

Phagocytic cell ingestion, metabolism and killing, and total hemolytic complement activity have been normal in all patients. Defects of mononuclear and/or polymorphonuclear chemotaxis have been present in some but not all patients, thus they are not the basic problem in these patients.

The fact that both males and females have been affected, as have members of succeeding generations, suggests an autosomal dominant form of inheritance with incomplete penetrance. The most effective management for this condition consists of chronic administration of a penicillinase-resistant penicillin, with the addition of other antibiotics or antifungal agents as required for specific infections, and appropriate thoracic surgery for superinfected pneumatocoeles or those persisting beyond 6 months.

Chronic Mucocutaneous Candidiasis

This is a clinical syndrome, probably due to multiple etiologies, in which the host experiences chronic candidal infection of the skin and mucous membranes, but only rarely experiences life-threatening systemic infections of the types seen in patients with severe T cell dysfunction. Often *Candida* is the single agent to which these patients are unduly susceptible. When conventional therapy fails, the infection may spread to involve the skin of the face, scalp, trunk, and extremities, including the nails. Some patients have endocrinopathies involving the parathyroid, thyroid, adrenal, and/or pancreatic glands. Autoantibodies to endocrine, gastric, and other tissue specific antigens, commonly found both in these patients and in asymptomatic members of their families, may be of pathogenetic significance in these endocrinopathies.

Many patients with chronic mucocutaneous candidiasis do not have either associated endocrinopathy or any demonstrable immunologic abnormality. Serum immunoglobulins are generally normal or increased, but IgA deficiency has been reported. Precipitating or agglutinating antibodies to *Candida* are usually present. Even in those patients who have had in vivo and/or in vitro evidence of deficient cell-mediated immunity, it is not clear whether that was the primary problem or if it was secondary to the extensive fungal disease (e.g., an antigen-overload mechanism). Reversal of delayed cutaneous anergy has occurred following high-dose intravenous amphotericin B and clearing of the fungal lesions.

Recently, ketoconazole has been found to be an extremely effective form of therapy, with a majority of patients with this condition showing dramatic clearing on this drug. Intravenous amphotericin B may still be required if *Candida* meningitis or septicemia develop. Transfer factor has been claimed to be of benefit, but this has not been proven by controlled trials.

Acquired Immune Deficiency Syndrome

This disorder, first noted in male homosexuals, is a new highly lethal epidemic immunodeficiency disease of obscure pathogenesis affecting primarily the T cell system. Careful retrospective reviews have failed to detect cases diagnosed prior to 1979, but over 2200 cases from 33 states and 13 foreign countries had been reported to the U.S. Communicable Diseases Center by September of 1983. The case acquisition rate of approximately 50 per week appears to be rapidly increasing. Patients with this syndrome present with life-threatening opportunistic infections and/or Kaposi's sarcoma. The opportunistic infectious agents have now included most of the bacterial, fungal, and parasitic agents customarily associated with cellular immunodeficiency, with *P. carinii, C. albicans, Myco-*

bacterium avium intracellare, Herpes simplex, Toxoplasma gondii, hepatitis b, cytomegalovirus, and cryptococcus being most commonly reported. Indeed, the spectrum of infectious agents found in individual AIDS patients far exceeds that ever seen in infants with the most severe primary combined immunodeficiency disorders. Until recently, Kaposi's sarcoma was a tumor rarely seen in North America or Europe, occurring only in persons aged 50 years or older and responding well to chemotherapy or irradiation. A rapidly fatal form of it has been reported to be endemic in equatorial Africa, predominantly in black boys and young men. However, since 1979 several hundred cases of the latter type of Kaposi's sarcoma have been reported in the United States in young homosexual males. All of the available epidemiologic data suggest a transmissable agent as the cause of the severe alterations in immune function observed in individuals so affected. The overall mortality from AIDS is over 40 percent and will probably approach 100 percent. Many patients who recover initially die subsequently from malignant disease or overwhelming infection; there are very few reports of complete remission.

At least four major groups of individuals appear to be at risk for AIDS: homosexual males (representing about 75 percent of reported cases), intravenous drug abusers with no history of homosexuality (13 percent), Haitian immigrants who are not homosexual and do not abuse drugs (6 percent), and hemophiliacs (0.7 percent). In addition, the disorder has also been reported in female sexual partners of affected males and in a growing number of children born to affected or susceptible individuals. Other apparent contributing factors include promiscuity among homosexuals, use of lyophilized preparations of concentrated factor VIII from thousands of blood donors by hemophiliacs, and, possibly genetic factors. The frequency of HLA-DR5 is reported increased in patients with Kaposi's sarcoma. In addition to the classic presentation with either severe opportunistic infections or Kaposi's sarcoma, a prodromal phase characterized by generalized adenopathy, recurrent fever, weight loss, and leukopenia is becoming recognized with high frequency among susceptible individuals. The fact that the prodromal phase can last for many months, together with other epidemiologic data, suggests a prolonged incubation period for the causative agent (i.e., 6 to 48 months).

The immunologic abnormalities found in AIDS are those of a severe and profound cellular immunodeficiency. These include an absence of delayed hypersensitivity; an absolute lymphopenia due to an absence of phenotypic T helper (OKT4$^+$ or Leu 3a$^+$) cells reversal of the usual ratio of T helper to T suppressor (OKT8$^+$ or Leu 2a$^+$) cells, whereby the former are severely reduced in number and the latter come to predominate; depressed lymphocyte responses to mitogens; and impaired natural killer cell function in vitro. In contrast, humoral immunity generally remains intact: hypergammaglobulinemia is the rule, antibody titers to a wide range of antigens, including those of the agents that infect them, are often very high. Complement components are normal. Thymus glands from patients dying from this disorder show marked epithelial atrophy, despite persistence of lymphoid elements; this picture is somewhat the opposite of that seen in thymuses of SCID patients, where epithelium is preserved but thymocytes are markedly depleted.

Although the epidemiologic data suggest an infectious, possibly viral, etiology for AIDS, no agent has been linked etiologically to the disease. Because many such patients are chronically infected with cytomegalovirus, hepatitis, or herpesviruses, these were initially thought to be leading candidates. However, not all AIDS patients have been infected with any one of these agents, and many homosexuals who do not have AIDS are chronically infected with these viruses. The pronounced depression of cellular immunity and the quantitative modifications of subpopulations of T lymphocytes suggest either that T cells or a subset might be a preferential target for the putative causative infectious agent, or that the alterations may occur as a consequence of the subsequent infections they experience. An attractive hypothesis is that AIDS may be caused by a human retrovirus related to the human T cell leukemia virus (HTLV). Retrovirus infections are known to cause lymphoid malignancies in several species, including T cell leukemia and mycosis fungoides (a T cell lymphoma) in humans (Chap. 20). Increased rates of the latter, as well as increased isolation of HTLV, have been reported in populations inhabiting the Caribbean islands and Southern Japan. HTLV is a lymphotrophic retrovirus that preferentially infects T helper cells. Paradoxically, such infected lymphoma cells, though of the helper phenotype, actually show suppressor function when studied in vitro. Moreover, naturally occurring feline leukemia virus (also a retrovirus) causes thymic atrophy, lymphopenia, and profound immunosuppression. In support of the hypothesis that HTLV or a related retrovirus may have an etiologic role in AIDS is the fact that antibodies specifically reactive with internal structural proteins of HTLV have recently been found more frequently in AIDS patients' sera than in normal sera, and HTLV has now been isolated from peripheral blood lymphocytes of several AIDS patients. Because HTLV can also infect T cells of other primates, it may soon be possible to develop an animal model to test this hypothesis directly.

For the present, there is no known effective treatment for AIDS. Prevention based on the above epidemiologic information and treatment of those opportunistic infections for which therapies are available are the principal approaches to management. Bone marrow transplantation has not been successful in correcting the profound immunodeficiency.

Phagocytic Cell Disorders

The essential role of phagocytic cells in host defense is seen in those patients who, although fully endowed with all of the necessary components and functions needed for specific immune responsiveness, have either an insufficient number or inadequate function of polymorphonuclear cells and/or macrophages. Infections experienced by such individuals are similar in many respects to those of patients with antibody deficiency, e.g., primarily infections with high-grade encapsulated pathogens; their severity can, and often does, equal or surpass those in the agammaglobulinemic host. Similar types of infections are seen in patients with hereditary deficiencies of components of inhibitors of the complement system (see Chap. 13).

Disorders of Production

Hereditary Neutropenia
This is an autosomal trait that occurs in both recessive and dominant forms and is characterized by the absence of granulocytes from the peripheral blood from birth. Arrested myeloid differentiation is the apparent cause of this condition, for the bone marrows of these patients contain an increased number of granulocyte precursors. The recessive form is invariably fatal in the first year of life, but the dominant form is a benign condition in which the life span is normal and the abnormality is detected only as an incidental finding. No explanation for the latter paradox is apparent.

Cyclic Neutropenia
Cyclic neutropenia, or periodic myelodysplasia, is an entity characterized by a periodic diminution in the number of peripheral blood polymorphonuclear leukocytes. The rhythmicity of the disorder is remarkably uniform, with the neutropenia occurring every 21 days in the majority of instances. Fever, malaise, and ulcers on the oral mucous membranes and occasionally arthritis, abdominal pain, sore throat, lymphadenitis, and cutaneous ulceration accompany the neutropenia. The patients are typically sick for 1 week, then well for 2 weeks, before the cycle repeats. Serial bone marrow studies have shown an intermittent failure of bone marrow neutrophil maturation at the promyelocyte stage. All other aspects of immunity are usually normal. Interestingly, splenectomy has relieved the clinical symptoms of some patients, even though it did not alter the neutropenic episodes. The condition occurs in both sexes with approximately equal frequency; in females the rhythmicity ordinarily has no relation to menstruation. A few instances of familial cyclic neutropenia are known, but there is no clear evidence for a genetic mechanism.

Neutropenia with Immunodeficiency
As noted earlier in this chapter, neutropenia may occur in association with several of the primary immunodeficiency disorders, most often in reticular dysgenesis, infantile x-linked agammaglobulinemia, and x-linked immunodeficiency with hyper-IgM. When neutropenia occurs, it

may be transient, persistent, or cyclic in nature, with the transient variety being most common. In the transient type, neutropenia may occur at the onset of a severe infection and give way to leukocytosis later in the course of the illness. The cyclic variety is different from the periodic myelodysplasia described above in that the periodicity is usually not uniform.

Acquired Neutropenia

This may occur secondary to toxic effects of drugs, pollutants, or irradiation or as a consequence of autoimmune reactions involving leukocytes. The latter may be seen as a transient phenomenon in the newborn as a consequence of maternal isoimmunization by fetal neutrophils during pregnancy. Neutropenia has also been noted in patients with neoplasms, overwhelming infections, or endotoxemia. Neutropenia is frequent in hypersplenism, in aplastic anemia, and in certain forms and/or stages of leukemia.

Disorders of Function

In these disorders, the number of peripheral blood leukocytes is usually normal (a few exceptions are noted below), but their function may be impaired at one or more of the following phases: (1) chemotaxis; (2) opsonization; (3) ingestion of particles; or (4) intracellular killing.

Chemotactic Defects

These occur in two forms: (1) those in which the blood chemotactic activity is diminished; and (2) those in which the cellular response is impaired. In the first instance, diminished chemotactic activity has been attributed to the presence of plasma or serum inhibitors of chemotaxis in children with recurrent staphylococcal infections, in alcoholics with advanced liver disease, in Hodgkin's disease, and in agammaglobulinemia. Controversy exists as to whether the inhibitors are abnormal constituents or a normal factor that is unopposed by an absence of a normal antagonist to the inhibitor because the inhibitor could be neutralized by normal plasma in some instances. Drugs, such as colchicine and steroids, can also act as inhibitors of chemotaxis. Chemotactic factors per se can be diminished in conditions in which C3 and/or C5 are diminished or dysfunctional (see congenital deficiency, Chap. 13). Other examples of such deficiencies are seen in the normal newborn, where both components are low; and in acute glomerulonephritis, where C3 is markedly reduced.

The newborn's polymorphonuclear cells have also been noted to have an intrinsic impairment in their ability to respond to normal chemotactic stimuli. Impairment has also been found transiently in patients with diabetes mellitus, in terminal shock, in rheumatoid arthritis, and in alcoholics.

Depressed responsiveness of polymorphonuclear cells to normal chemotactic stimuli has also been noted in the Chediak–Higashi–Steinbrink syndrome. This rare disorder of humans, Aleutian mink, and cattle is characterized by gigantism of cytoplasmic lysosomes in white cells, melanocytes, Schwann cells, and possibly other tissues. Clinically, affected individuals have partial (oculocutaneous) albinism with resultant photophobia; undue susceptibility to viral and enteric bacterial infections; hepatosplenomegaly, lymphadenopathy, anemia, and leukopenia; cutaneous ulcers; and neurologic changes. Peripheral blood smears show abnormally large peroxidase-positive granules in the neutrophils and eosinophils. An intracellular microbicidal defect has also been demonstrated in the granulocytes of such patients. This appears to be due to microtubular abnormalities that can be partially corrected by ascorbic acid in vitro. All aspects of adaptive immunity are normal, but very low natural killer cell activity has been found recently in these patients. Death from infection generally occurs before the fifth year of life, but a few patients have survived until the late teens, and some deaths have been attributed to lymphoma. A simple autosomal recessive mode of inheritance has been postulated. Heterozygous carriers have been identified by the presence of the granulation anomaly in leukocytes.

Another type of intrinsic polymorphonuclear abnormality is that seen in the lazy leukocyte syndrome where, not only do the neutrophils fail to respond normally to chemotactic stimuli, but they have reduced random mobility. Patients afflicted with this disorder have mild but recurrent infections characterized by gingivitis, stomatitis, otitis, and low-grade fever associated with severe neutropenia and poor mobilization of granulocytes from the marrow. The serum does not contain inhibitors of chemotaxis, and the phagocytic and killing capacities of the polymorphonuclear cell are normal. A somewhat similar disorder has been described in two kindreds in which afflicted members had congenital ichthyosis and recurrent fungal infections, but without the neutropenia and abnormal random mobility.

Opsonic Defects

These may be due to deficiencies of the heat-labile (complement-derived C3b) or heat-stable (specific IgM and IgG antibodies) opsonins that alter the surfaces of bacteria to facilitate phagocytosis. As pointed out earlier, newborn serum is deficient in C3 and C5; thus, it is not only incapable of generating normal quantities of C3a and C5a chemotactic factors, but also cannot provide adequate quantities of the opsonin, C3b. In addition to this deficiency of heat-labile opsonin, cord serum is virtually devoid of specific IgM antibodies to gram-negative bacteria. Patients with congenital deficiencies of C3 or C3b inactivator or with dysfunction of C5 (Leiner's syndrome, see Chap. 13) produce inadequate quantities of heat-labile opsonins and may have recurrent often severe infections. Patients with B cell functional deficiencies are the classic examples of patients deficient in heat-stable opsonins, accounting to a large extent for their special tendency to develop pyogenic infections.

Ingestion Defects

The capacity of the phagocytic cell to ingest bacteria is determined by the maturity of the cell, the presence of specific membrane receptors, and the energy potential of the cell. In acute myelocytic leukemia, the ingestion capacity of the cells is severely limited. Phagocytic cell membrane receptors for C3b and the Fc portion of IgG are often saturated in conditions such as systemic lupus erythematosus, rheumatoid arthritis, multiple myeloma, and macroglobulinemia where circulating immune complexes or immunoglobulin aggregates interact continuously with the cells. A familial deficiency of a phagocytosis promoting tetrapeptide, tuftsin, has also been reported to result in diminished ingestion of staphylococcal organism by polymorphonuclear cells. The negative surface charge of the phagocyte membrane can be altered by drugs such as levorphanol, a morphine analog, or by viruses such as influenza that bind to the membrane. Any condition in which there is an associated inhibition of glycolysis, such as that due to hypophosphatemia of malnutrition, can produce a phagocytic defect on the basis of diminished cellular energy potential.

Killing Defects

A variety of disorders can alter the phagocytic cell's ability to kill bacteria, even though the cells may have responded normally to chemotactic stimuli and the bacteria have been well opsonized and ingested normally. Among the best known of these conditions is the syndrome chronic granulomatous disease of childhood (CGD). This is a fatal group of disorders characterized by chronic suppurative infections, draining adenopathy, pneumonia, hepatomegaly with liver abscess, osteomyelitis, splenomegaly, hypergammaglobulinemia, and dermatitis, with onset of symptoms usually before 1 year of age. Adaptive immunity is usually entirely normal, although rarely IgA deficiency or hypogammaglobulinemia have been seen. Neutrophils from such patients are defective in their ability to kill catalase-positive bacteria (such as *Staphylococcus aureus*, *Klebsiella*, *Aerobacter*, and *Proteus* organisms, and *Serratia marcescens*) and some fungi (*Candida* and *Aspergillus*), despite a normal ability to phagocytize these organisms. Moreover, these intracellular bacteria or fungi enjoy protection from antibiotics and can later seed out into the body fluids. These intracellular organisms evoke granulomatous reactions in the liver and spleen and other organs that they invade; pigmented histiocytes are found in such reactions. Catalase-negative organisms that generate their own peroxide, such as group A streptococci and pneumococci, are killed normally by these cells. Leukocytes from these patients have normal increments of glucose consumption, lactate production, Kreb's cycle activity, and lipid turnover during phagocytosis of latex particles. In contrast, their leukocytes fail to show normal oxygen consumption, direct oxidation of glucose, and hydrogen peroxide formation. The inability of leukocytes from patients with chronic granulomatous disease to lyse bacteria is related to their inability to stimulate the direct pathway of glucose metabolism to form hydrogen peroxide during phagocytosis. The myeloperoxidase, iodide (or other halide), hydrogen peroxide system is an important bactericidal system of human polymorphonuclear cells; hence, the failure of CGD cells to generate peroxide, the superoxide anion radical, and/or singlet oxygen provides an explanation for their impaired microbicidal activity.

In the syndrome of CGD there appear to be several different basic biochemical defects and more than one mechanism of genetic transmission. When normal granulocyte membranes are stimulated during phagocytosis, there is activation of the membrane-bound electron transport chain, reduced nicotinamide adenine dinucleotide phosphate (NADPH) oxidase, which acts on NADPH to reduce molecular oxygen to the superoxide anion. The NADPH-oxidase system is known to consist of at least two components, a flavoprotein and cytochrome-b. Failure of oxidase activity could be due to defective transport chain activation, absence of any of the chain components, or lack of substrate. Probable examples of each of these types of defects have been reported. In a recent multicenter study involving 27 CGD patients from Europe, the heme-containing protein, cytochrome b-245, was undetectable in all 19 males in whom the defect appeared to be located on the X chromosome; female carriers had reduced cytochrome b-245 concentrations and variable proportions of cells that were unable to generate superoxide. In all eight patients with probable autosomal recessive inheritance, cytochrome b-245 was present but nonfunctional. The reason for the latter is not clear, but a defective mechanism for adding electrons or hydrogen to the cytochrome is one possibility. Further study may identify as yet unknown defective enzymatic components, controls, or effectors. The failure of polymorphonuclear cells from affected individuals to reduce the redox dye nitroblue tetrazolium (NBT) to purple formazan during phagocytosis serves as a simple screening test for this disorder.

In vitro fungicidal and bactericidal activity is also markedly diminished in the neutrophils and monocytes of patients with congenital absence of the lysosomal enzyme myeloperoxidase. Only one of the five individuals (from three pedigrees) known to have this defect had increased susceptibility to infection, however. In contrast to the situation in CGD, leukocyte metabolism during phagocytosis was normal, as was NBT dye reduction, but the inactivation of bacteria was markedly diminished. As noted above, intracellular killing is also defective in Chediak–Higashi leukocytes, presumably on the basis of delayed degranulation and decreased deposition of lysosomal enzymes into the phagocytic vesicles. Bactericidal capacity may be depressed in patients with severe (e.g., less than 1 percent of normal) deficiencies of glucose-6-phosphate dehydrogenase G-6-PD, but most individuals with this hereditary defect (that involves both erythrocytes and leukocytes) have between 20 and 50 percent of the normal amount of G-6-PD, and hence no clinical problems with infection. Acquired but transient defects in phagocytic killing have been noted in patients with severe thermal injuries and in children undergoing craniospinal irradiation for treatment of acute lymphoblastic leukemia. Large doses of corticosteroids can depress reduced nicotinamide adenine dinucleotide- (NADH) oxidase activity and leukocytes from such patients show diminished NBT dye reduction. Finally, newborn leukocytes, despite their unresponsiveness to chemotactic factors and inadequate heat-labile and heat-stable opsonins, have no intrinsic microbicidal defects; they do, however, have increased resting rates of oxygen consumption, hexose monophosphate (HMP) shunt activity, and NBT dye reduction.

Treatment

The principal modes of therapy for the primary immunodeficiency disorders include protective isolation, use of antibiotics for the eradication or prevention of bacterial and fungal infections, and attempted replacement of missing humoral or cellular immunologic functions. The complexities of both the immunodeficiency diseases and their treatment emphasize the need for all such patients to be evaluated in centers where detailed studies of immune function can be conducted before therapy is selected or begun.

Antibody Deficiency Disorders

Judicious use of antibiotics and regular administration of antibodies are the only treatments that have been shown effective for this group of disorders. Patients with agammaglobulinemia, x-linked immunodeficiency with hyper-IgM, antibody deficiency with near-normal immunoglobulins, or the Wiskott–Aldrich syndrome are candidates for humoral replacement therapy. The latter is contraindicated in patients with selective absence of serum and secretory IgA or transient hypogammaglobulinemia of infancy. Currently, the only commercially available forms of replacement therapy are immune serum globulin for intramuscular use or modified immune serum globulin for intravenous use. Both consist primarily of IgG antibodies, with only traces of the other classes. The recommended dose of 100 mg/kg per month of IgG is strictly arbitrary; studies are underway in the author's laboratory to establish guidelines for selection of optimal doses for individual patients. All five immunoglobulin classes can be provided through periodic infusions of plasma from a known safe donor previously immunized with vaccines such as tetanus toxoid, influenza, and pneumococcal antigens to provide high titers of specific antibodies. For patients with severe defects in cellular immunity, there is a risk of graft-vs.-host disease, unless the plasma is irradiated to incapacitate any viable lymphocytes of donor origin. Therapy is planned to provide the recipient 100 mg/kg of IgG every 3 to 4 weeks, so a typical dose of plasma from a normal donor would be 10 ml/kg. As with immune serum globulin thera-

py, an initial loading dose of two to three times the maintenance dose is given. Systemic reactions may occur but, except in patients with selective IgA deficiency, these are less common than with either the intramuscular or intravenous forms of immune serum globulin.

Cellular Immunodeficiency

The only adequate therapy for patients with severe forms of cellular immunodeficiency is immunologic reconstitution by means of an immunocompetent tissue transplant. To date, only three types of immunocompetent tissue have been used successfully for this purpose. Mature bone marrow that is major histocompatibility complex-compatible or haplo-identical with the recipient is the tissue of choice, except in the case of the complete DiGeorge's syndrome, where fetal thymic tissue appears satisfactory. Fetal liver or thymic tissues have been much less effective in the treatment of SCID. The major risk to the recipient from transplants of bone marrow or fetal tissues is that of a graft-vs.-host reaction. Although this reaction occurs even in HLA-identical transplants, it is usually transient and milder than the severe and usually fatal variety seen in recipients of incompatible marrow. Fetal tissue transplants, even though incompatible, have a low risk of fatal graft-vs.-host reaction, for reasons not fully understood. Recently, a major advance has allowed the use of haplo-identical (half-matched) bone marrow cells for correction of the immunologic defect in SCID. In this technique, which takes advantage of the affinity of human T cells for soybean lectin and for sheep erythrocytes, postthymic T cells are selectively and completely removed, leaving the stem cells intact for transplantation. To date, nearly 30 infants with SCID who would have otherwise died due to a lack of an HLA-identical donor have been treated successfully with this approach, with virtually no signs of graft-vs.-host reaction. In the past, the main indication for an immunocompetent tissue transplant has been a severe deficit in cellular immunity. Patients with less severe forms of cellular immunodeficiency will reject such grafts unless they are treated with immunosuppressive agents prior to transplantation. However, several patients with Wiskott–Aldrich syndrome and other forms of partial cellular immunodeficiency have been treated successfully with bone marrow transplants following immunosuppression. It is entirely possible that a number of genetic defects, including the syndromes of CGD, will be correctable in the future using this approach.

Agents that have been advocated as augmentors of existing cellular immunity include transfer factor, thymosin, and levamisole. Although some clinical and less impressive immunologic improvements have been claimed in individual case reports, the natural variability of the less severe forms of cellular immunodeficiency in which these substances have reportedly given benefit precludes any rational assessment. Controlled clinical trials are needed for each of these agents before their usefulness can be established.

FURTHER READING

Books and Reviews

Babior BM, Crowly CA: Chronic granulomatous disease and other disorders of oxidative killing by phagocytes. In Stanbury JB, Wyngaarden JB, Fredrickson DS, et al. (eds): The Metabolic Basis of Inherited Disease. New York, McGraw-Hill, 1983, pp 1956–1985

Bergsma D, Good RA, Finstad J, et al. (eds): Immunodeficiency in Man and Animals. Birth Defects: Original Article Series. Sunderland, Mass., Sinauer Associates, 1975, vol 11

Bortin MM, Rimm AA: Severe combined immunodeficiency disease. Characterization of the disease and results of transplantation. JAMA 238:591, 1977

Buckley RH: Immunodeficiency diseases. In Kelley WN, Harris ED, Ruddy S, et al. (eds): Textbook of Rheumatology, Philadelphia, Saunders, 1981, vol II, pp 1351–1377

Buckley RH: Primary immunodeficiency diseases. In Wyngaarden JB, Smith LG (eds): Cecil Textbook of Medicine, 16th ed. Philadelphia, Saunders, 1982, pp 1789–1796

Buckley RH, Sampson HA: The hyperimmunoglobulinemia E syndrome. In Franklin EC (ed): Clinical Immunology Update. New York, Elsevier/North-Holland, 1981, pp 147–167

Kauder E, Mauer AM: Neutropenias of childhood. J Pediatr 69:147, 1966

Lawlor GJ, Ammann AJ, Wright WC Jr, et al.: The syndrome of cellular immunodeficiency with immunoglobulins. J Pediatr 84:183, 1974

Lawton AR, Cooper MD: B cell ontogeny: Immunoglobulin genes and their expression. Pediatr 64(Suppl—Host Defenses):750, 1979

McKusick VA, Elderidge R, Hostetler JA, et al.: Dwarfism in the Amish. II. Cartilage-hair hypoplasia. Bull Johns Hopkins Hosp 116:285, 1964

Moller G (ed): Ontogeny of Human Lymphocyte Function. Immunol Rev 57:1–161 1981

Reimann HA: Cyclic neutropenia. In: Periodic Diseases, Philadelphia, FA Davis 1963, p 94

Salvaggio JE (ed): Primer on allergic and immunologic diseases. JAMA 248:2579–2772 1982

Spector BD, Perry GS, Kersey JH: Genetically determined immunodeficiency diseases (GDID) and malignancy: Report from the Immunodeficiency-Cancer Registry. Clin Immunol Immunopathol 11:12, 1978

Stiehm ER, Fulginiti VA (eds): Immunologic Disorders in Infants and Children, 2nd ed. Philadelphia, Saunders, 1980

Waldmann TA, Broder S: Polyclonal B cell activators in the study of the regulation of immunoglobulin synthesis in the human system. Adv Immunol 32:1, 1982

Working Party on Hypogammaglobulinemia: Hypogammaglobulinemia in the United Kingdom. Medical Research Council Special Report Series 310. London, Her Majesty's Stationary Office, 1971

Webster ADB: Metabolic defects in immunodeficiency diseases. Clin Exp Immunol 49:1, 1982

Selected Papers

Buckley RH, Gard S, Schiff RI, et al.: T cells and T cell subsets in a large population of patients with primary immunodeficiency.

In Wedgwood R, Rosen FS (eds): Primary Immunodeficiencies. New York, Alan R. Liss, 1983

Buckley RH, Gilbertsen RB, Schiff RI, et al.: Heterogeneity of lymphocyte subpopulations in severe combined immunodeficiency: Evidence against a stem cell defect. J Clin Invest 58:130, 1976

Buckley RH, McQueen JM, Ward FE: HLA antigens in primary immunodeficiency diseases. Clin Immunol Immunopathol 7:305, 1977

Centers for Disease Control Task Force on Kaposi's Sarcoma and Opportunistic Infections: Epidemic aspects of the current outbreak of Kaposi's sarcoma and opportunistic infections. N Engl J Med 306:248, 1982

Gallo RC, Sarin PS, Gelmann EP, et al.: Isolation of human T cell leukemia virus in acquired immune deficiency syndrome (AIDS). Science 220:865, 1983

Giblett ER, Ammann AJ, Wara DW, et al.: Nucleoside phosphorylase deficiency in a child with severely defective T cell immunity and normal B cell immunity. Lancet 1:1010, 1975

Gottlieb MS, Schroff R, Schauker HM: *Pneumocystis carinii* pneumonia and mucosal candidiasis in previously healthy homosexual men: Evidence of a new acquired cellular immunodeficiency. N Engl J Med 305:1425, 1981

Harada S, Sakamoto K, Seeley JK, et al.: Immune deficiency in the X-linked lymphoproliferative syndrome. I. Epstein–Barr virus specific defects. J Immunol 129:2532, 1982

Haynes BF, Warren RW, Buckley RH, et al.: Demonstration of abnormalities in expression of thymic epithelial surface antigens in severe cellular immunodeficiency diseases. J Immunol 130:1182, 1983

Hayward AR, Ezer G: Development of lymphocyte populations in the human fetal thymus and spleen. Clin Exp Immunol 17:169, 1974

Lindsten T, Seeley JK, Ballow M, et al.: Immune deficiency in the X-linked lymphoproliferative syndrome. II. Immunoregulatory T cell defects. J Immunol 129:2536, 1982

Maugh TH: Singlet oxygen: A unique microbicidal agent in cells. Science 182:44, 1973

Miller ME, Oski FA, Harris MD: Lazy leukocyte syndrome. Lancet 1:665, 1971

Ownby DR, Pizzo S, Blackmon L, et al.: Severe combined immunodeficiency with leukopenia (reticular dysgenesis) in siblings: Immunologic and histopathologic findings. J Pediatr 89:382, 1976

Pierce GF, Polmar SH: Lymphocyte dysfunction in cartilage hair hypoplasia. II. Evidence for a cell cycle specific defect in T cell growth. Clin Exp Immunol 50:621, 1982

Polmar SH, Stern RC, Schwartz AL, et al.: Enzyme replacement therapy for adenosine deaminase deficiency and severe combined immunodeficiency disease. N Engl J Med 295:1337, 1976

Reisner Y, Kapoor N, Kirkpatrick D, et al.: Transplantation for severe combined immunodeficiency with HLA-A, B, D, DR incompatible parental marrow cells fractionated by soybean agglutinin and sheep red blood cells. Blood 61:341, 1983

Segal AW, Cross AR, Garcia RC, et al.: Absence of cytochrome b-245 in chronic granulomatous disease. A multicenter European evaluation of its incidence and relevance. N Engl J Med 308:245, 1983

Tiller TL, Buckley RH: Transient hypogammaglobulinemia of infancy: Review of the literature, clinical and immunologic features of 11 new cases and long term follow-up. J Pediatr 92:347, 1978

Touraine J-L: The bare-lymphocyte syndrome: Report of the Registry. Lancet 1:319, 1981

Waldmann TA, Broder S, Goldman CK, et al.: Disorders of B cells and helper T cells in the pathogenesis of the immunoglobulin deficiency of patients with ataxia telangiectasia. J Clin Invest 71:282, 1983

Wilfert CM, Buckley RH, Mohanakumar T, et al.: Persistent and fatal central nervous system echovirus infections in patients with agammaglobulinemia. N Engl J Med 296:1485, 1977

CHAPTER 16

Immunohematology

Blood and tissue cells are distinguished by cell surface molecules, usually glycoproteins that can function as antigens when the cells are transferred to another individual. These markers of self are usually referred to as "alloantigens" or, in the older literature, as "isoantigens." The transfer of blood, blood transfusion, can be regarded as the oldest and most frequent form of transplant. The presence of alloantigens on transfused red cells can lead, rarely, to intravascular hemolysis, renal damage, and even death if the recipient has high-titered preformed antibodies in the circulation. In the abscence of preformed antibodies the reactions are less severe. The white blood cells may possess some antigens in common with red cells but also carry others that are undetectable on red cells. Incompatibility for these can cause febrile but nonhemolytic reactions in the presensitized patient. Lymphocytes, polymorphonuclear neutrophils (PMN), and platelets are all richly endowed with alloantigens. Immunoglobulins may also carry alloantigenic markers on the light chain (Km) or on the heavy chain (Gm). While these rarely if ever lead to clinical manifestations, patients may develop antibodies to Km or Gm after transfusion.

This chapter concentrates on the alloantigens of red blood cells and of blood platelets. Antigens of other tissues are described in Chapter 17. Many of the antigens are unique to red cells and are of special importance in transfusion and in hemolytic diseases, especially hemolytic disease of the newborn. Knowledge of autoimmune reactions, especially of the autoimmune hemolytic reactions to altered self antigens, is dramatically illustrated by the autoimmune hemolytic anemias.

Attempts at blood transfusion were first undertaken seriously in the middle of the 17th century, but they were not successful. Several of the recipients died, and the practice was outlawed in both France and England. It was revived early in the 19th century by Blundell, a Scottish obstetrician, for saving patients with postpartum bleeding. Transfusion was sometimes successful, but often the patient became hypotensive, passed dark urine, then became oliguric and frequently died.

The reason for these adverse reactions remained mysterious until 1904, when Landsteiner discovered that the serum of some normal human donors agglutinated the red cells of other donors. From an analysis of the pattern of reactions, as described later, he defined the major blood groups (A, B, and O). Blood could usually be transfused safely when the groups of the donor and recipient were the same. Serious reactions occurred when the serum of the recipient agglutinated the donated cells. Since that time, many other blood groups have been described; the red cell alloantigens number more than 300. Antibodies to these antigens (isoantibodies or alloantibodies) are often encountered in two clinical situations: in pregnancy, when the fetus has different antigenic components than the mother, and following blood transfusion, when the transfused blood has different antigens than the recipient.

The blood group antigens may be used as individual identifying markers, since they are present at birth and remain intact throughout the life of the individual. Antigens that are characteristic of certain ethnic groups have been used to trace migrations of peoples and intermixture of races. Since the blood group antigens are passed on by genetic mechanisms, they may be used to test the laws of heredity in human beings in a way available by almost no other means. Finally, the chemical characteristics of some of these antigens have been elucidated and give important information concerning the nature of the antigen–antibody interaction.

Antigens

Antigens on the surfaces of blood cells are initially defined by their interaction with antibodies. Cells reacting with a given antibody are said to possess the antigen against which that antibody was made. For example, cells reacting with an antibody called anti-X are said to be X-positive, those not reacting to be X-negative. When a new antibody is found, it is tested on several samples of X-positve and X-negative cells. If it gives the same pattern of reactions as the first antibody with these cells (i.e., if it reacts only with X-positive cells), it also is called anti-X. If it gives different reactions, it is clearly not anti-X and has specificity for another antigen. The reactions of this new antibody with a panel of red cells containing many different antigens are then compared with those of other defined antibodies, for example, anti-Y and anti-Z, until it is seen that its reactions are identical to those of some previously defined antibody. If an identical pattern of reaction is not found, the

antibody is said to define a new blood group antigen (Table 16-1).

Chemistry and Genetics

Chemical determinants of the blood group antigens are either polysaccharide or protein. The polysaccharide antigens occur on either glycoproteins or glycolipids, and the determining structure is usually complex, consisting of more than one sugar moiety. The sugars making up these structures are glucose, galactose, fucose, N-acetylglucosamine, N-acetylgalactosamine, and N-acetylneuraminic acid (sialic acid), which imparts most of the charge to the cell surface.

The polysaccharide antigens are generated by enzymes (glycosyl transferases), usually acting in sequence to construct complex polysaccharides. These complexes are attached to either proteins or lipids, which are then attached to the membrane. The structure of the antigen is ultimately genetically determined by the presence or absence of a gene(s) coding for the appropriate enzymes.

The protein antigens of the red cell are usually found on integral proteins that are fixed to the hydrophobic lipid layers of the membrane. The antigenic structure is either due to particular peptide sequences of 1–6 amino acids or to such sequences modified by attached glycosidic residues. Since antigenicity in this case depends upon amino acid sequence, the presence or absence of the antigen depends upon the presence or absence of appropriate DNA sequences in the genome.

Most of the antigens of blood cells are expressed when a single copy of the gene responsible for their expression is present in the genome; they are thus said to be dominant or codominant. In the case of protein antigens, slight variations in the DNA sequence of an exon yields one or more variations in antigenic protein. Antigens resulting from such variations are called "alleles." Since the gene locus occurs on both homologous chromosomes, an individual may express at most two of these alleles. Such allelism can also occur with polysaccharide antigens.

TABLE 16-1. THE ASSIGNMENT OF SPECIFICITY TO NEW ANTIBODIES BY AGGLUTINATION REACTIONS WITH CELLS OF SEVERAL DONORS

	Donor 1	Donor 2	Donor 3	Donor 4	Donor 5
Anti-X	+	+	−	−	+
Anti-Y	−	−	+	−	+
Anti-Z	+	−	+	−	−
Unknown antisera					
1	+	+	−	−	+
2	−	−	−	+	−

The cells of five donors were reacted with serum previously denoted anti-X, anti-Y, and anti-Z (hypothetical designations). New antibody 1 has the same reactions as anti-X; if no differences were found when other cells were tested, this antibody would be also called anti-X. New antibody 2 does not give the same reactions as anti-X, anti-Y, or anti-Z, and, therefore, it does not react with the X, Y, or Z antigen. If its reactions are not the same as any known antibody, it is said to define a new specificity, and a name, for example, anti-Q, is assigned. The antigen is then called the Q antigen, and cells are said to be Q-positive or Q-negative.

In this case, however, the alternative genes at the gene locus must code for alternative glycosyl transferases (e.g., A and B).

Genes coding for different nonallelic antigens may occur near one another on the same chromosome. In this case, they are said to be linked. Alleles at two (or more) linked loci will usually be inherited together and will be found together in a nonrandom distribution in the population.

Antigens may be lacking or appear to be lacking for a variety of reasons.

In the case of protein antigens:

1. The gene may code for an antigen for which no antiserum is available for detection.
2. There may be localized deletion of the DNA segment coding for the antigen.
3. There may be unequal crossing over of similar genes, which are closely linked on a chromosome, resulting in a hybrid protein molecule lacking some of the antigens of either parent protein.

In the case of polysaccharide antigens:

1. The gene coding for an enzyme may be deleted.
2. The variant enzyme that is coded for may not be able to glycosylate the substrate to form an antigen.
3. The polysaccharide substrate upon which the appropriate enzyme is supposed to act to produce the antigen is missing; this is usually because an enzyme necessary to the construction of that substrate is missing.
4. If the polysaccharide is attached to a protein (i.e., a glycoprotein), the entire protein or a part of it may be missing.

ABO, Lewis, and Secretor Systems

In 1902, Landsteiner discovered the ABO blood group by mixing the red cells of each of his laboratory assistants with the serum of all the others. From these reactions, he defined the preeminent blood groups A, B, O, and (later) AB. The cause of the agglutination in these experiments was found to be naturally occurring antibodies to the antigens on the red cells. These antibodies accounted for many of the serious reactions that had occurred when blood had been transfused (Table 16-2).

The antibodies regularly occur in the serum of people who lack the antigen on the red cells even though the donor has been neither transfused nor pregnant. They are not present at birth and gradually develop during early childhood. Their presence has been attributed to antigenic stimulation from oligosaccharides of the same chemical conformation as the A or B antigens. Such oligosaccharides abound in nature. The quantity of the isoantibody may be markedly increased, however, by the antigenic stimulation of incompatible blood or by immunization with blood group substance.

Antigens of the ABO blood groups are inherited characteristics. According to the theory of their transmission proposed in 1924, the allelic *A* and *B* genes give rise to their respective gene products, the A and B antigens, whereas the allelic *O* gene gives rise to no gene product, i.e., it is an amorph. Thus, the phenotype (detectable antigens on the red cell) of a person whose genetic makeup is *AO* will be group A, but he may pass the *O* gene to his offspring.

It later became apparent that O red cells did contain an antigen. These cells were more strongly agglutinated than were A, B, or AB cells by certain animal sera and plant products (lectins), as well as by certain alloantibody-containing sera. The antigen responsible for these reactions was called the "H antigen," and it was expressed on the cell surface in inverse proportion to the amount of A and/or B antigen expressed on the cell.

Rarely, the A, B, and H antigens may all be lacking. This rare blood type is called "Bombay" and is due to the absence of the gene responsible for the formation of the H antigen, which in turn serves as a precursor for the A and B antigens. Patients with this blood type have anti-A, anti-B, and anti-H in their serum.

TABLE 16-2. THE ABO BLOOD GROUPS

Group	Specific Antigens on Cell	Quantity of H Antigen	Antibodies in Serum	May Receive Blood from*	May Give Blood to*
O	None	High	Anti-A Anti-B	O	O, A, B, AB
A	A, A_1	Low	Anti-B	A, O	A, A_2, AB
A_2	A	More than A	Anti-B, occ. anti-A_1	A (usually), A_2, O	A, A_2, AB
B	B	Like A	Anti-A	B, O	B, AB
AB	A, A_1, B	Less than A or B	None	AB, A, B, O	AB
A_2B	A, B	More than AB	None	AB, A, B, O	AB

*In most instances, type-specific blood is given (e.g., type A to type A patient, and so on). Occasionally this rule may be broken in special circumstances, but the groups must be selected from these columns. When, for instance, type-O blood containing anti-A is infused into a person whose cells are type A, the antibodies are normally diluted so that no reaction takes place.

Subgroups of A and B have been described. By cross-absorption experiments, it was found that the usual A cells had two closely related antigens, designated A and A_1. Most anti-A antisera contained antibodies to both antigens, but absorption of such antisera with cells from a weak variant called A_2 left only anti-A_1. Red cells of subgroup A_2 contain only the A antigen and frequently react less strongly than do cells of most (90 percent) group A donors. This is important, since the presence of A_2 may be missed, especially in subjects in group A_2B, since the B antigen is readily detected while the A_2 reaction is weak.

The antigens of the ABO(H) system are present on many other tissues in addition to red blood cells. They may appear in water-soluble form in the secretions of the body (with the exception of cerebrospinal fluid). Their presence in water-soluble form appears to be under separate genetic control by the so-called secretory gene (*Se*), which is inherited as a mendelian dominant. People who lack the gene have the cellular ABO(H) antigens on their red cells but do not have the water-soluble forms in their saliva or in other secretions.

The antigens of another blood group system that is closely related, the Lewis system, may also be found in the secretions. There are two antigens in the Lewis system, Lewis a (Le^a) and Lewis b (Le^b). Individuals may have either antigen or neither in a complex genetic pattern that will be explained below. These antigens occur only in water-soluble form but are adsorbed onto the red cell surface from the serum.

The relationship among the ABH and Lewis systems has been suggested by studies on the chemical nature of the antigens involved. The basic structure of the antigens of the ABH and Lewis systems is a four-sugar oligosaccharide called "paragloboside," which may be attached to protein component or to a lipid component (Fig. 16-1).

The *H* gene produces an enzyme (a fucosyl transferase), which is able to place a fucose on the terminal galactose of paragloboside. Persons lacking the *H* gene (Bombay) make no H substance either on cells or in water-soluble form. The *A* gene codes for a transferase that places an *N*-acetylgalactosamine molecule on the terminal sugar of the H substances, and the *B* gene codes for an enzyme that places a galactose there, thus producing, respectively, the A and B antigens.

The Lewis gene (*Le*) produces an enzyme that can place a fucose on the penultimate sugar of paragloboside in the *water-soluble* form. If this is the only addition to that tetrasaccharide, the resulting antigen is Le^a. On the water-soluble antigens, the *H* gene is active only if the secretor *Se* gene is present. If both *H* and *Se* genes are present, a fucose molecule will be attached to the terminal sugar as well, resulting in the Le^b antigen. Thus, Le^a antigen occurs in the absence of either the *Se* gene (most commonly) or the *H* gene (very rarely).

The polysaccharide structure upon which these occur may be simple or complex. The backbone may be lengthened by the insertion of *N*-acetylglucosamine-galactose moieties (Fig. 16-2) or may be branched either simply or more complexly. New antigens are thus generated. The

ANTIGENS OF THE ABH AND LEWIS GROUPS

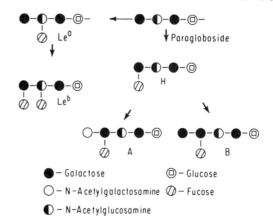

Figure 16-1. The formation of blood groups from paragloboside. Paragloboside or more complex molecules having the same terminal sugars (see Fig. 16-4) may have added a fucose on the terminal sugar, resulting in H substance. In the soluble antigens, this occurs only when both the secretor gene and the *H* gene are present. H substance may be subsequently glycosylated to A substance or B substance. Paragloboside may also be fucosylated on the subterminal sugar; this results in the Lewis a (Le^a) antigen. This reaction occurs only on soluble antigens. If both *H* gene and secretor gene are present, a second fucose is added on the terminal sugar, resulting the the Lewis b (Le^b) gene.

antigen reacting with anti-i cold agglutinin resides in the lengthened but straight backbone (II). This form is more common in the cells of the infant. The antigen reacting with the anti-I cold agglutinins resides in the branched backbone (III and IV), more common in the adult.

BLOOD GROUP GLYCOLIPIDS

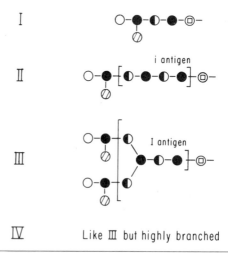

Figure 16-2. Variations in the polysaccharide backbone of the blood group substances. The simplest form is paragloboside (type I backbone). If additional *N*-acetylglucosamine-galactose residues are added without branching, the i antigen is generated (type II). If these are branched as in type III or type IV backbones, the I antigen is generated.

The A$_2$ antigens appears to be formed by a variant glycosyl transferase enzyme that is not readily able to add the determinant sugar to complexes with branched backbones (III and IV). This results in a weakened expression of the antigen and in an antigen that is sufficiently different from the normal antigen to be distinguished by specific antibody, anti-A$_2$.

P Antigens

The complex polysaccharides of the red cell surface are involved in another system of antigens, the P system. This system is biochemically similar but not genetically related to the ABO system. The two common antigens of the P system are called P and P$_1$. The P antigen is a relatively simple polysaccharide that evolves by the sequential linear addition of sugar moities to galactosyl-ceramide (Fig. 16-3). The enzymes necessary for the addition of these sugars are almost universally present. If the first enzyme is missing, no antigens of the P system are made, and the phenotype is pp. If the second enzyme is missing, the antigen P^k results; this phenotype is rare and is found only in homozygotes, i.e., when the gene coding for the enzyme is missing from both homologous chromosomes. The addition of a terminal galactose results in the almost universal P antigen. The P antigen is the antigen with which the autoimmune antibody, the Donath–Landsteiner antibody, reacts.

The other common antigen in the P system, the P$_1$ antigen, is derived from paragloboside by the addition of a terminal galactose without the addition of a fucose, as in the case of the formation of the H antigen (Fig. 16-3). About 80 percent of people in the population have the enzyme that makes this antigen. These people, thus, have both the P$_1$ and P antigen, and the phenotype is the P$_1$ phenotype. The red cells of those lacking this enzyme have only the P antigen; this phenotype is the P$_2$ phenotype. These individuals may make antibody against the P$_1$ antigen (anti-P$_1$).

The P$_1$ and P antigens are genetically linked in some way, since those persons lacking the enzyme that converts galactosyl-ceramide to the P^k antigen by the addition of a galactose also are unable to convert paragloboside to the P$_1$ antigen. The reason for this is not, at the present time, apparent.

MNSs System

For over 20 years, the A and B antigens were the only two known antigens of human red cells. In 1927, Landsteiner and Levine injected rabbits with human red cells. From these experiments, the MN system emerged. The MN system was thought to consist of two allelic antigens, M and N. People were divided into three groups: (1) those reacting only with the antiserum containing anti-M (genotype MM), (2) those reacting only with anti-N (genotype NN), and (3) those reacting with both (genotype MN). Later, a genetically linked allelic antigen system (S and s) was identified, and presently many related antigens are known.

ANTIGENS OF THE P BLOOD GROUP

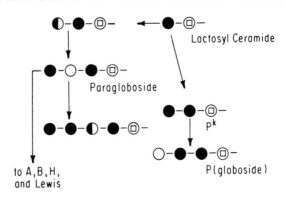

Figure 16-3. Formation of the P groups. Sugars are sequentially added to lactosyl-ceramide to form either the P^k and subsequently the P antigens (globoside) or paragloboside. Paragloboside, which is not fucosylated, may be changed into the P$_1$ antigen by the addition of a terminal galactose.

The antigens for these systems appear to be present on glycoprotein molecules called "glycophorin A" (glycophorin α) and "glycophorin B" (glycophorin δ). The amino terminus of the glycophorin α (the one present in greatest concentration) is polymorphic, and variation in the first and fifth amino acids yields the M or N antigen (Table 16-3). Glycosyl residues terminating in sialic acid must be present on the three subterminal amino acids for antigenicity. Other variations in the amino acids yield less common allelic antigens (Mg, Mc, and so on).

Glycophorin δ is very similar to glycophorin α for the first part of the molecule. However, the amino acid sequence characteristic of the N antigen is always seen at the terminus. The S and s antigens are due to an alternative amino acid sequence within the portion of glycophorin δ that differs from glycophorin α. The genes for the α and δ glycophorin molecules are probably near one another on the chromosome, thus accounting for the linkage of the MN blood groups on glycophorin α and Ss groups on glycophorin δ.

In about 2 percent of the black population, both the S and s antigens are missing; this is due to a deletion of a portion of the glycophorin δ, which is either partial (in which case an antigen, U, is retained) or complete (when U is also missing). The glycophorin α is missing in rare patients called En^a (−). Such patients lack M and N anti-

TABLE 16-3. CHEMICAL BASIS OF THE M AND N ANTIGENS

Antigens		Glycophorins
M	$\begin{array}{cccccc} & CHO & CHO & CHO \\ & | & | & | \\ ser & - ser - thr - thr - gly - val - \end{array}$	α
N	$\begin{array}{cccccc} & CHO & CHO & CHO \\ & | & | & | \\ leu & - ser - thr - thr - glu - val - \end{array}$	α and δ

gens (except for the N antigen on the δ glycophorin) and have a markedly diminished content of sialic acid in the membrane. These studies of the glycoprotein antigens illustrate many biochemical correlates of serologically determined phenotypes.

Rh System

For practical purposes, the most important of the blood group systems, other than the ABO system, is the Rh system, since it is within this system that the major difficulties of transfusion and of fetomaternal incompatibility occur. The Rh system was first discovered by Levine and Stetson when transfusion of the mother of a baby with hemolytic disease of the newborn (erythroblastosis fetalis) with the blood of the father resulted in an acute hemolytic reaction despite the fact that the father was the same ABO group as the mother. The authors suggested that the baby had on his cells an antigen lacking in the mother that had passed from the baby to the mother and had caused antibody to be formed. Since that baby had inherited the antigen from the father, the father's cells injected into the mother also were rapidly destroyed in an acute hemolytic reaction. This untoward reaction was due to an antibody to a new antigen, soon known as the "Rh" antigen. Eighty-five percent of the white population possesses the antigen, and it was found to be related to four other antigens—two linked loci with two common alleles each.

This antigen (called somewhat erroneously the "Rh antigen," since a similar antigen was detected by serum produced in animals immunized with rhesus monkey red cells) was soon found to be part of a complex system of antigens that were identified with antibodies occurring as the result of sensitization by transfusion. Considerable confusion occurred because of the variant nomenclature evolved by different investigative groups. According to Wiener, eight different gene complexes were possible at a single locus, and the five antigens occurred in these complexes in different combinations (R^1, R^2, and so on). According to Race and Fisher, three tightly linked loci, each with two alleles, were responsible for the expression of the genetic information. The alleles were named C and c, E and e, D and d. Since the antibody for the d antigen was never identified, the existence of this allele cannot be affirmed. The different names used in each classification system are shown in Table 16-4.

Each person has two of these gene complexes, one inherited from each parent, and, in turn, one or the other is passed to each of his children. Careful family studies or the use of special antibodies is needed to determine the gene complex from the antigens present on the cell surface, e.g., a person DCe/DcE (R_1R_2) could not be distinguished readily from one DCE/Dce (R_2R_0), since the same antigens would be present on the cells of both. However, the combination could be readily deduced from reactions with cells from the children if the spouse was cde/cde. Similar problems arise in HLA typing (Chap. 17).

The antigens of the Rh system vary considerably in their capacity to immunize. The D (Rh_o) antigen is the most potent and, in common parlance, is the antigen referred to when speaking of Rh-positive or Rh-negative blood. D(Rh_o) is the antigen most commonly involved in sensitization by transfusion or pregnancy. Nevertheless, sensitization occurs in only 70 percent of Rh-negative recipients even when Rh-positive red cells are repeatedly injected. The other antigens of the Rh system are very much less potent and, consequently, cause less difficulty in transfusion or fetomaternal incompatibility. Antibodies to these antigens are sometimes found as a cause of autoimmune hemolytic anemia.

The chemical nature and organization of these antigens are not entirely known, but assumptions can be made on the basis of unusual variants of the system. These variants are of several apparent types:

TABLE 16-4. ALTERNATIVE NOMENCLATURE OF ANTIBODIES AND GENE COMPLEXES OF THE RH SYSTEM

	Wiener →		Antibodies					Frequency in Whites
			Anti-Rh$_0$	Anti-rh′	Anti-rh″	Anti-hr′	Anti-hr″	
	↓	Fisher →	Anti-D	Anti-C	Anti-E	Anti-$\bar{c}$	Anti-e	
		↓						
Genes or Gene Complexes	R^1	DCe	+	+	−	−	+	0.408
	r	__ce	−	−	−	+	+	0.39
	R^2	DcE	+	−	+	+	−	0.14
	R^0	Dce	+	−	−	+	+	0.025
	r″	__cE	−	−	+	+	−	0.011
	r′	__Ce	−	+	−	−	+	0.009
	R^z	DCE	+	+	+	−	−	0.001
	r^y	__CE	−	+	+	−	−	Very rare

1. Structural alleles. Unusual variants like the C^w antigen appear to be due to changes (perhaps a single amino acid) in the structure of the more common antigen.

2. Repeated sequences. The anti-G antibody reacts with all cells that have either the D or the C antigen, suggesting that these two antigens contain repeated segments with which the antibody reacts.

3. Compound antigens. Some antibodies (such as anti-f) react with antigenic sequences present only when two of the common antigens of the system are present and are coded on the same chromosome (in this case, both the c and e antigens). This suggests that the antibody reacts with parts of each antigen.

4. Suppressions. One of more of the antigens may be suppressed. Expression of the D antigen appears to be suppressed in D(Rh_o)-negative (usually -ce) antigen complexes. Partial or complete suppression of the expression of the antigens of the E site is also common, particularly in blacks. The antigens of both the C and E sites may be suppressed (D--/D--); this results in a stronger expression of the D antigen. Finally, all antigens of the Rh complex may be lacking (Rh null); the reason for the absence is probably complex but may involve the loss of the entire protein upon which the Rh antigens are carried. Such cells are leaky to cations and do not survive well in the circulation.

Other Blood Group Systems

Kell Group. In 1946, an antibody not related to the Rh system was found in a case of hemolytic disease of the newborn. The antibody was called anti-Kell or anti-K. The allelic antigen was described by another antibody called anti-Cellano or anti-k. Anti-K is frequently involved in crossmatching difficulties. The *Kk* genetic locus is closely linked to two other loci, *Kp* and *Js*. The alleles of *Kp* are Kp^a (rare) and Kp^b (very common). The alleles of *Js* are Js^a (Sutter), which is found in about 20 percent of blacks and very rarely in whites, and Js^b, which occurs in almost all whites and most blacks. These three closely linked genetic loci are similar in some respects to the three loci present in the Rh complex and are presumably present in a sequence on a protein molecule. Just as in the case of the Rh antigens, deletions may occur, and rare examples of cells lacking antigens of the Kell system have given valuable information about the genetics of the system. A phenotype with weak expression of Kell antigens (the McLeod phenotype) may also be associated with a morphologic abnormality of red cells and chronic granulomatous disease, an immune deficiency related to poorly functioning granulocytes.

Duffy Group. An antibody describing a new antigen system was found in a patient named Duffy and was called anti-Duffy^a (anti-Fy^a). The allele was named Fy^b. The cells of all whites are either Fy(a+, b−), Fy(a−, b+), or Fy (a+, b+). About 60 percent of all blacks are Fy(a−, b−) (Duffy null). Such cells lack a receptor for the malarial parasite, *Plasmodium vivax*. Selection is believed to be responsible for the preponderance of Duffy null blood type in African blacks.

Kidd (Jk) System. Two antigens have been described belonging to this system: Jk^a and Jk^b. The antibodies to these antigens are particularly unstable on storage and do not remain in the serum of sensitized patients.

Other Systems. Many other antigens have been defined on the red cell surface by the fact that antibodies have been produced to them, usually presenting as an unexpected transfusion reaction. In some instances, the antigens are rare, and in others, the antigens are found in over 99.9 percent of the population. Because of their frequency in distribution, these so-called private and public blood groups do not often cause trouble in the clinical setting. They do, however, add to the antigenic complexity of the red cell surface, and when reactions to them occur, they may be severe. Interestingly, the otherwise ubiquitous HLA antigens are not detectable on red cells by conventional techniques. The Bennett Godspeed (Bg^a) system of antigens, which is related to some HLA-A and HLA-B series antigens, are present on red cells but present few clinical difficulties.

Antigens of Platelet Surface

Although platelets and red cells are derived from the same precursor cells, they possess quite different glycoproteins on the surface. Platelets do not have Rh, Duffy, Kell, and MNSs antigens on their surface. They do possess the ABO, Lewis, I, i, and HLA antigens. Many of these antigens may not be integral parts of the membrane but may have been adsorbed from the serum. The platelet surface is capable of adsorbing a number of proteins. The Lewis antigens and perhaps the ABH and HLA antigens may be adsorbed and not endogenous to the platelet itself.

HLA antigens are variably present on platelets. However, it is clear from clinical facts that such antigens are important causes of allosensitization in platelet transfusions, since patients allosensitized by transfusion of platelets from random donors will produce antibodies that cause the destruction of all platelets except those identical in their HLA antigenic constitution. As the techniques have evolved, more precise knowledge about the role of the HLA antigens in the platelet will become evident.

Relatively little is known about the specific alloantigens of the platelet. The most important system is the so-called Pl^{Al} (Zw^a) system, which is responsible for two clinical syndromes. This antigen is located on glycoprotein III of the platelet membrane, and 98 percent of the population have this antigen. Anti-Pl^{Al} antibody formation may occur either by transfusion or during pregnancy because of transference of platelets bearing the antigen to the mother's circulation. When this occurs, the antibody formed in the Pl^{A1}-negative mother may recross the placenta, destroying the platelets of the infant and resulting in neonatal purpura. Rarely, the transfusion of Pl^{A1}-positive platelets to a Pl^{A1}-negative person may result in an unusual reaction in which the recipient's Pl^{A1}-negative platelets are destroyed. Apparently, the glycoprotein upon which the antigen is resident is able to transfer from the transfused platelets to the endogenous platelets, resulting in fixation of anti-Pl^{A1} antibody to the adsorbed antigen. This syndrome is called "post-transfusion purpura."

Other alloantigen systems must exist for the platelet, but because of the difficulty in techniques in identifying the binding of antibodies to platelets, relatively little is known about them. These may become important as they provide a source of sensitization that may make platelet transfusion impossible and may limit transplantation, particularly bone marrow transplantation.

Granulocytes

Granulocytes have a membrane antigen constitution that differs from both platelets and red cells. HLA antigens appear to be well expressed on the granulocyte surface, and antibodies to these antigens may result in destruction of the white cells, sometimes with adverse physiologic effects. In addition, the ABO, I, and i antigens are probably also expressed. Whether other blood groups are expressed is not clear.

In addition to antigens shared by other components, there is a series of specific granulocyte antigens called NA1, NB1 . . . NF1. Each of these antigens appears to have allelic expression. Linkage among themselves and other blood groups is not clear. These antigens may elicit antibodies that can cause neonatal neutropenia and may result in adverse reactons when granulocytes are transfused.

Antibodies

Antibodies to blood group antigens are usually IgG1, IgG3, or IgM. IgA antibodies are less common, and to date no IgD or IgE antibodies to human red cells have been found.

In general, the antibodies directed against the polysaccharide antigens have the following characteristics: (1) they are usually, but not always, IgM, (2) they usually react better at temperatures below 37C (cold-reacting antibodies), and (3) although they may occur as a consequence of transfusion or pregnancy, they may also occur without previous sensitization by blood or blood products (naturally occurring antibodies). Antibodies directed against protein antigens have the following characteristics: (1) they are usually IgG, (2) they react equally well or better at 37C than at lower temperatures, and (3) they do not occur without prior sensitization (except in the case of autoimmune disease).

Tests used to detect the presence of antibody depend in large part upon the immunoglobulin type of the antibody. The simplest test system detects agglutination. Agglutination occurs when one antigen-binding site of the antibody is attached to an antigen site on one cell and the other to an antigen site on another cell (Fig. 16-4). As more antibodies bind to the cell surface, a lattice of cells is bound together as an agglutinate. In forming this lattice, the antibody must overcome or bridge the repulsive forces that keep red cells apart. Red cells in saline suspension are mutually repelled by two forces: (1) the negative charge at the red cell surface and (2) the cloud of positive ions that accompanies the red cell when it is suspended in saline solution, the so-called zeta potential. Of these, the latter is the more important force. When an antibody reacts with a red cell, the negative surface charge is reduced. IgM antibodies are sufficiently large to bridge the gap imposed by the zeta potential, and agglutination usually occurs (Fig. 16-4a). Some IgG antibodies may also be able to bring about agglutination, particularly if the antigen is present in high density. More frequently, however, IgG molecules are too small to bridge the gap between the mutually repelled cells, and no agglutination occurs (Fig. 16-4b).

In these circumstances, IgG antibodies may be demonstrated in one of three ways: (1) the force of the zeta potential may be overcome by centrifugal force, (2) the zeta potential may be dissipated by suspending the cells in some anisotropic medium, such as 22 percent albumin or polyvinylpyrrolidone (PVP), after which the cell-to-cell gap is small enough to be spanned by IgG molecules (Fig. 16-4c), or (3) antibody molecules on each cell may be joined together by antiglobulins made in rabbits or goats (Fig. 16-4d). This antiglobulin reaction is sometimes known as the "Coombs reaction."

Antibody reacting with the antigen but incapable of bringing about agglutination may be detected by the blocking reaction. In this case, the cells are coated with the IgG blocking antibody and then treated with the agglutinating (IgM) antibody of the same specificity. No agglutination occurs because the nonagglutinating antibody has covered or blocked the antigen sites, thus preventing any reaction with agglutinating IgM antibody.

The presence of antibody can sometimes be detected by its ability to bring about lysis of the cell through the activation of hemolytic serum complement (Chap. 14). For this to occur, the first component of complement must be fixed and activated, a process that requires the presence of only one IgM molecule on the red cell surface but two IgG antibody molecules in some geometric juxtaposition (a doublet). Hence, IgM molecules are generally more efficient in the fixation of complement than are IgG molecules. In some instances, however, complement may not be fixed under any circumstances, as with the antibodies to antigens of the Rh system. This is thought to be because the spacing of antigen sites on the red cell surface is such that the doublet cannot be formed.

Human red cells are resistant to the hemolytic action of human complement. Thus, uncompleted sequences may

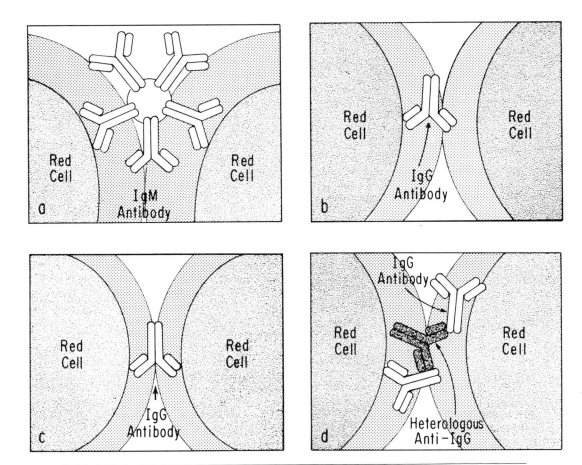

Figure 16-4. The repulsive forces between red cells. The combination of a negative surface charge and a surrounding ionic cloud (the zeta potential) tends to keep red cells in saline suspension apart. **a.** Agglutination by an IgM molecule that can span the zeta potential. **b.** An IgG molecule cannot interact with two cells because of their mutual repulsion. **c.** Reduction in the mutual repulsion by dissipation of the zeta potential allows an IgG molecule to react with two cells and agglutinate them. **d.** An antiantibody reacts with two IgG antibodies, one on each cell, resulting in agglutination. The ionic cloud is indicated by stippling.

accumulate on the surface and can be detected by the use of antibody against the components of complement. Since the third component (C3) tends to accumulate in the largest amount, anti-C3 may be used in an agglutination test, such as the direct Coombs test, to detect the immunologic reaction on the red cell surface.

As noted above, the reaction of the antibody and the antigen on the red cell may be strongly dependent on the temperature. Thus, some reactions may not occur at all at 37C but may be very strong at 0C (cold-reacting antibodies). Since others may occur equally well at the two temperatures, it is important to test for activity at both temperatures.

A highly sensitive, temperature-dependent (37C), lytic process that does not require complement is the antibody-dependent cellular cytotoxicity assay (ADCC). Macrophages and some (Fc-positive) lymphocytes can readily lyse the red cells of many species, including humans, in the presence of minute quantities of specific IgG antibody, including anti-Rh.

Clinical Applications of Blood Groups

Transfusion of Red Cells

The antigens of the blood cells and their antibodies are important in transfusion of blood, which may be regarded as a form of temporary transplantation. Fortunately, most of the antigens of the red cell are poorly immunogenic, and naturally occurring antibodies are uncommon. This means that, with the exception of the ABO group (where naturally occurring antibodies are common) and the $D(Rh_o)$ antigen (which is quite immunogenic), blood may usually be transfused without regard to matching the red cell antigens of donor and recipient. Nevertheless, great care is taken to be certain that antibodies are not present in the serum of the recipient that might react with the antigens on the red cells of the donor, which would result in the immune destruction of the transfused cells.

The development of alloantibodies following transfusion is relatively uncommon and occurs in 1 to 3 percent of all transfusions given when patients are matched for ABO and D blood groups. On the other hand, some patients appear to develop antibodies very readily to relatively nonimmunogenic antigens, and some chronically transfused patients may develop many antibodies. This renders transfusion difficult in that blood lacking all the antigens is difficult to find.

Transfusion reactions that result in the immunologic destruction of red cells are, in general, of three types: major hemolytic reactions, minor hemolytic reactions, and delayed reactions.

Major Hemolytic Reactions. When the destruction of the transfused blood is sudden and massive, several reactions take place that may result in serious morbidity or death of the patient. (1) Hemoglobin is released into the blood and filtered by the kidneys, resulting in hemoglobinuria. (2) The blood pressure may fall, and this combined with the massive hemoglobinuria may result in renal failure. (3) The coagulation system may be activated so that it no longer functions, resulting in hemorrhage. Reactions of this type are usually due to incompatibility within the ABO system.

Minor Hemolytic Reactions. When the rate of destruction is less rapid, the result may be only a falling hematocrit as the transfused blood is removed by the spleen and the appearance of jaundice as the hemoglobin is metabolized. Little needs to be done except to find compatible blood for further transfusion if necessary. Patients requiring frequent blood platelet or granulocyte transfusion may also develop febrile reactions. These are usually due to the reactions of antibodies to leukocytes (HLA).

Delayed Transfusion Reactions. Certain blood group antibodies do not persist in the serum (particularly the Kidd and Duffy antibodies). Thus, at the time of compatibility testing before transfusion, no incompatibility is found even though the patient has been previously sensitized. When the blood is transfused, it elicits an anamnestic response, and the concentration of antibody rises rapidly. This results in the destruction of the transfused blood 5 to 10 days later. Frequently the destruction is manifest by hemoglobinuria. The reaction is usually not serious.

Transfusion of Platelets and Granulocytes
The philosophy used in the transfusion of platelets is somewhat different. It is recognized that the histocompatibility antigens, particularly the HLA antigens, present on the platelet surface are immunogenic. Nevertheless, plate-

lets are transfused for the most part without regard to the antigenic constitution of donor or recipient. Unfortunately, no simple in vitro compatibility test has been devised. This means that when antibodies develop, their presence is best signaled by the fact that the infusion of platelets does not result in the expected rise in numbers of circulating platelets. When this occurs, platelets from persons of the same HLA type (see below) are substituted and, in general, are more compatible. However, it is clear that antibodies may also develop to antigens other than the HLA antigens, with the result that the platelets may be destroyed even though HLA antigens are compatible.

Transfusion of granulocytes is even less well understood than the transfusion of platelets. Granulocyte transfusions are usually performed only to combat overwhelming infections, and, therefore, the opportunity to study antigranulocyte reactions is limited. The granulocytes present in stored blood would not be expected to survive for any appreciable length of time. Consequently, although antibodies may develop against granulocytes and can produce febrile reactions, there has been little systemic study of immune responses to granulocytes.

Hemolytic Disease of the Newborn

In the preceding discussion, many allusions have been made to hemolytic disease of the newborn (erythroblastosis fetalis or HDN), a disease first recognized in the early 1930s. As detailed above, its explanation came with the discovery of the Rh groups, and although they are by no means the only cause of the problem, they account for a vast majority of cases.

The problem occurs when antigens are present on the red cells of the fetus that are lacking on those of the mother. If cells from the baby cross the placenta, antibodies are produced in the mother. These antibodies recross the placenta and cause the destruction of the cells of the infant. The effects may be mild or may be severe enough to cause death, even before birth.

Sensitization does not generally occur during the first pregnancy, but a rise in antibody titer is seen shortly after the birth of the child. With the next pregnancy, the titer of antibody may increase slowly during the pregnancy but again may rise sharply after delivery. It is thought that a transfusion of sufficient cells from the infant to the mother to initiate antibody production occurs at delivery.

Once the mother is immunized, small transplacental leaks of cells during subsequent pregnancies may increase the antibody titer. Thus, the disease is usually manifest only during the second and subsequent incompatible pregnancies.

Even when the infant is Rh positive and the mother is Rh negative, sensitization to the Rh_o (D) antigen is much less common when the mother's serum contains antibodies to A or B antigens on the infant's cells. Thus, an O Rh (negative) mother is less likely to become sensitized to the Rh_o antigen than is a mother who is A, B, or AB and Rh negative. In the latter case, it is thought that the immunizing dose of fetal cells is cleared by the anti-A or anti-B in the mother's plasma before the Rh incompatibility can be recognized.

Only IgG antibody molecules are able to cross the placenta. Therefore, if antibodies of other immunoglobulin types are made, they cannot reach the infant. Further, it has been postulated that those IgG antibodies with the greater affinity for the antigen are more likely to be responsible for the onset of the hemolytic reactions.

Hemolytic disease of the newborn due to the D (Rh$_o$) antigen is preventable. If the mother has not been previously sensitized, she is given a large dose of anti-D shortly after delivery. This prevents the initiation of an immune response, sensitization does not occur, and the next pregnancy does not result in an erythroblastotic infant.

The antigens of the ABO (H) system are poorly expressed on fetal cells. This may account in part for the rarity of hemolytic problems in cases where the mother's serum contains IgG antibodies to either the A or B antigens that may be on the baby's cells. Whatever the reason, less than 1 in 3000 of such incompatible mother–infant combinations result in serious hemolytic anemia.

Autoimmune Disease of the Hematopoietic System

The foregoing discussion is mainly concerned with reactions due to isoantibodies or alloantibodies. However, in certain instances, immunologic tolerance is apparently broken, and the patient begins to make antibodies to antigens present on his own (autologous) blood cells.

Classically, this process has involved the red cells, in the form of autoimmune hemolytic anemia, but, more recently, the occurrence of thrombocytopenia (lack of platelets) and leukopenia (lack of white cells) has been ascribed to the same process.

Several different mechanisms may account for the apparent loss of self tolerance.

1. The antibody may be produced by a benign or malignant clone of cells, as in the case of the cold agglutinins. When this occurs, the antibody is monoclonal, and each molecule is identical in structure and specificity to every other.
2. The antibody may be produced as a result of viral invasion of the immune system, as when Epstein–Barr virus invades B cells in infectious mononucleosis.
3. The antibody may be produced that is part of a normal immune reaction against invading organisms but that cross-reacts with normal antigens.
4. Antibodies may be produced to antigens that are adsorbed to the surface of the cell. Various drugs, such as quinine and sulfa, may elicit such antibodies.

5. In alterations of the immune system by disease, the balance of controlling cells, particularly lymphocytes, may be lost so that more helper function is manifest than suppressor function. In this setting, antibodies against blood cell antigens are part of a generalized reaction, as in systemic lupus erythematosus.

In most instances, the reason for the loss of tolerance is not known and can only be surmised.

Autoimmune Hemolytic Anemia

The antibodies causing autoimmune hemolytic anemia are usually of either the IgM or IgG immunoglobulin class. They are further divided into those that react only at temperatures below body temperature (cold-reacting antibodies) and those that react at body temperature (warm-reacting antibodies). In general, the warm-reacting antibodies are IgG, while the cold-reacting ones are IgM. The mechanism of hemolysis and the resultant clinical syndrome are related to both the immunoglobulin class and the temperature of reactivity of the antibody molecule.

The antigens present on the red cells that react with autoantibody may or may not belong to groups defined by isoantibodies (alloantibodies). Thus, some warm-reacting IgG autoantibodies may react with cells of only specific Rh phenotype. Most autoantibodies are directed against very public antigens, antigens possessed by all but a few individuals in the population. These antigens are frequently in the non- (or pauci-) polymorphic backbone molecules that may bear allomorphic blood group antigens, e.g., the protein bearing the Rh antigens but lacking in Rh$_{null}$ cells, glycophorin δ (U), glycophorin α (Wr^b), and so on.

The cold-reacting IgM antibodies (cold agglutinins) appear to react with polysaccharide antigens on red cells. All cold agglutinins react with all red cells, under appropriate conditions. The most common of these antigens (I and i) are part of the polysaccharide backbone of the ABO antigens (see above). The i antigens are more markedly expressed on fetal cells, the I antigen on adult cells. Uncommonly, the cold agglutinin may react with polysaccharide antigens on glycophorin α, the Pr antigens.

Immune Thrombocytopenia

Autoimmune destruction of the platelets by warm-reacting IgG antibody occurs in a syndrome called "idiopathic (immune) thrombocytopenic purpura" (ITP). The antibody-mediated nature of the destruction has been hard to define, but newer techniques have shown the presence of IgG on the platelet surface. This syndrome appears to be analogous to warm-reacting, IgG-mediated hemolytic anemia, since (1) in both instances, the major destruction appears to occur in the spleen, and both syndromes may be benefited by splenectomy, (2) in both instances, remission may often be induced by corticosteroids, which may be due in part to suppression of antibody or suppression of sequestration, and (3) both are commonly seen in patients with systemic lupus erythematosus, chronic lymphocytic leukemia, and other diseases of the immune system.

Leukopenia

Although neutropenia due to autoimmune destruction has often been suspected, especially in patients with systemic lupus erythematosus, the demonstration of the antibodies or of cellular destruction has been singularly difficult to obtain. Recently, techniques similar to those used for platelets have demonstrated such antibodies, particularly in Felty's syndrome (rheumatoid arthritis, neutropenia, and splenomegaly).

FURTHER READING

Books and Reviews

Mollison PL: Blood Transfusion in Clinical Medicine, 6th ed. Philadelphia, Davis, 1977

Petz LD, Garratty G: Acquired Immune Hemolytic Anemias. New York, Churchill Livingstone, 1980

Race RR, Sanger R: Blood Groups in Man, 6th ed. Philadelphia, Davis, 1977

Zmijewski CM: Immunohematology, 3rd ed. New York, Appleton, 1978

Selected Papers

Anstee DJ: The blood group MNSs—active sialoglycoproteins. Semin Hematol 18:13, 1981

Hakomori S: ABH and I-i antigens of human erythrocytes: Chemistry, polymorphism and their developmental change. Semin Hematol 18:39, 1981

Landsteiner K: Über Agglutinationserscheinungen normalen menschlichen Blutes. Wien Klin Wochenschr 14:1132, 1901

Landsteiner K, Levine P: On the inheritance of agglutinogens of human blood demonstrable by immune agglutinins. J Exp Med 48:731, 1928

Levine P, Stetson RE: An unusual case of intragroup agglutination. JAMA 113:126, 1939

Marcus DM, Kundu SK, Suzuki H: The P blood group system: Recent progress in immunochemistry and genetics. Semin Hematol 18:63, 1981

Pollack W, Hager HJ, Reckel R, et al.: A study of the forces involved in the second stage of hemagglutination. Transfusion 5:158, 1965

Rosse WF: Interactions of complement with the red-cell membrane. Semin Hematol 16(2):128, 1979

Rosse WF: The lysis of erythrocytes by incomplete antibodies. Am J Clin Pathol 77(1):1, 1982

CHAPTER 17

Immunogenetics of Tissue Antigens

In the beginning of Chapter 16, Immunohematology, it is stated that the study of the alloantigens of red blood cells is one of the major branches of immunogenetics. In this chapter we will describe another major component of immunogenetics, structural and functional studies of the antigenic markers of leukocytes and tissues. With the exception of the ABO(H) antigens, few of the red cell antigens are intrinsic to tissues, although they may, like Lewis substance, be adsorbed onto them. Instead, other antigenic markers are found; some are detected by serologic procedures, while others are detected by their ability to stimulate cellular reactivity or to bring about the rejection of a transplant. Thus these are called "transplantation" or "histocompatibility" (H) antigens. Other markers arise as a cell differentiates. Differentiation markers have been most intensively studied in rodents, where they are often distinguished by the prefix Ly. Lyt antigens are differentiation markers of T lymphocytes, and Lyb are markers of B cells. Since different individuals have variant forms (alleles) of these antigens, they are also called differentiation alloantigens.

Some tissue antigens that are constant for all members of a species are detected by the reactions of antisera (xenoantisera) raised in other species, and the antigens recognized are xenoantigens. The Forsmann antigen is usually recognized by xenoantisera, and such antibodies are also useful in the detection of tumor antigens (Chap. 20). Most of our information about tissue antigens has been gained by studying antibodies produced by the immunization of one individual with cells or tissues from another member of the same species. These are called "alloantibodies," and the corresponding targets are alloantigens. While there are probably more antigens on the surface of nucleated cells than on red cells, one set of antigens is much more important than all of the others. These are the antigens of the major histocompatibility complex (MHC), which form the principal focus of this chapter.

Histocompatibility Antigens

When skin or other tissues are transplanted back to the original individual or to an identical twin, the graft survives for the life of the individual. Skin or kidney transplanted to another individual, even a fraternal twin, in the absence of immunosuppressive treatment is almost certain to be lost through the process of rejection. Rejection itself is a complex process first studied systematically in the rabbit by Medawar in England and by Shinoi in Japan, both of whom noted that skin allografts (from other rabbits) at first healed in as if they were autografts and became vascularized by the sixth day. By the ninth day perivascular cuffing by lymphocytes could be observed, the small blood vessels became dilated, and tortuous, and the blood flow became sluggish. Allografts then died and were sloughed off, autografts continued healing-in and became indistinguishable from nontransplanted skin. Gorer observed similar changes with transplanted sarcomas in mice, but infiltration with macrophages was a prominent feature.

The importance of lymphoid cells in graft rejection was shown by Mitchison and later by Winn. Immunity could be transferred by lymph node cells from an animal rejecting a graft (adoptive transfer) but could not be transferred by serum (passive transfer). Graft rejection is generally regarded as an example of delayed-type hypersensitivity, but it also has some features of cutaneous basophilic hypersensitivity (Chap. 21). Nude mice (which are athymic) or mice thymectomized at birth cannot reject skin grafts, emphasizing the role of the T cell in transplant rejection. Antibody is believed to play a significant part in the rejection of second-set grafts applied to a previously immunized recipient. The target of the rejection process is usually, but not invariably, the vascular endothelium of the graft, and the presence of passenger lymphocytes in the transplant helps to trigger the host response. Rejection may be slow or rapid. Rapid rejection occurs when donor and recipient differ by what is called a "major histocompatibility (H) difference." Slow rejection is more typical of differences of so-called minor H antigens. Relatively little is known about properties of the minor H loci, since antibodies to minor H loci are difficult to demonstrate. From skin graft experiments in mice the existence of over 300 minor H loci has been postulated, and approximately 100 have been described in more detail. Individually, minor locus differences cause only slow and indolent rejection, but differences at multiple minor H loci can lead to more rapid rejection. In humans the immune response to minor H antigens seems to be weaker than in most experimental animals and is easily controllable by corticosteroids and other immunosuppressive agents. By contrast, rejection due to major H differences is usually difficult to control. The ABO(H) antigens can act as H antigens in humans and appear to be of intermediate strength. Since the A and B antigens are so easy to determine, transplantation from a donor of blood group A or B to an incompatible recipient can be avoided easily. The Lewis antigens adsorbed onto leukocytes and tissues can serve as targets for antibody-mediated lysis.

Systematic study of the major H antigens began in the 1940s, but progress was relatively slow until the mid 1960s, when attempts to predict kidney graft survival were stimulated by the variable outcome of the first transplants. While the most informative studies were with mice, and though much of the present research is restricted to mice or humans, enough is known of various other mammalian and even avian species to support the generalization that the major H system of all higher animals is amazingly conserved. Structural studies have confirmed significant sequence homologies of the glycoprotein products of the major histocompatibility complex (MHC) of several species and corresponding success in hybridizing cDNA between species.

Major Histocompatibility Complex

The general features of the MHC of every species are as follows. The MHC includes members of two classes of structural genes lying close to each other on the chromosome (Fig. 17-1). The first class of genes codes for H antigens found on every tissue (except possibly brain). In humans these are the HLA-A, B, and C antigens; in mouse they are called the H-2K, D, L, and R antigens. With the exception of H-2L, the other genetic determinants are known to be at different sites on the chromosome, since they can, on occasion, be separated by crossing-over or recombination. The frequency with which crossing-over occurs is a function of the distance between genes. Between HLA-A and HLA-B the frequency is about 0.5 percent. Crossing-over between HLA-B and C is rare so it is thought the genes are very close. In the mouse, H-2L is even closer to H-2D.

The second class of structural gene codes for antigens that are expressed only on certain special cells, and the distribution, at present, seems somewhat arbitrary. These antigens are found on B lymphocytes, on macrophages and vascular endothelial cells, on sperm, on precursors of the myeloid cells of the bone marrow, on some cells of the thymic epithelium, and on the Langerhans cells of the skin. These have been termed "B cell antigens," since they were discovered on B cells first and because they are either absent from or present in very low concentration on T cells and somatic cells. In the mouse, the B cell antigens are known collectively as "Ia antigens." The phrase Ia-like is now being used with increasing frequency for the multiplicity of class II antigens of humans. The best known of the human Ia-like antigens is HLA-DR. Until recently it was generally belived that DR was the only class II antigen, but cDNA probes have revealed the presence of many genes, and immunochemical and serologic tests have revealed many gene products, all of which are very similar.

The MHC of humans also includes genes controlling the expression of complement components C2 and C4 and of Bf, the proactivator of C3. Structural and regulatory genes both appear to be represented. Interestingly, two variants of C4 distinguishable by electrophoretic mobility are expressed on human red cells as the Chido and Rogers blood group antigens. Complement component regulation is also known to be MHC related in other species. The MHC is known to have essential functions in the collaboration between cells of the immune system. This

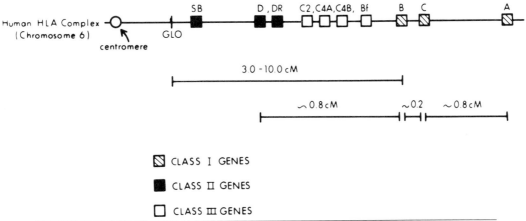

Figure 17-1. A representation of the sixth chromosome of humans, which carries genes for the HLA complex. *HLA-A, B,* and *C* are genes for the heavy chains of class I antigens found on somatic cells. *HLA-DR, DC/DS,* and *SB* are compound. In each case a relatively constant α chain gene is accompanied by one or more genes for β chains; and α and β chains are tightly but not covalently bound. *C4a* and *C4b* are genes for alleles of the fourth components of complement, and the products are also present on red cells as the Chido and Rogers blood group antigens. The H-2 complex of the mouse is very similar to HLA. Both contain an array of genes or pseudogenes as yet not known to be translated.

was first realized when it was found that lymphocytes from one individual could stimulate those of another when mixed together in culture, the so-called "MLC reaction" or MLR. The ability to stimulate is most potent when the stimulator carries Ia-like antigens, while T cells are the most active responders. The ability to stimulate was originally believed to be the property of a discrete locus called HLA-D in humans and MLR-S in mice, but no gene product has been isolated and the ability to stimulate is now believed to be a function of the array of Ia antigens on the B cell, macrophage, or other Ia-positive cell. In the mouse, T-dependent antibody production is triggered only when the antigen presenting cell has the same Ia antigens as the T cell donor (see below).

Other attributes of the MHC are the ability to generate cytotoxic T lymphocytes that can kill suitable target cells in vitro, to generate suppressor cells and suppressor substances, and to attack an immunoincompetent recipient in what is called a graft-vs.-host (GVH) reaction (Chap. 20). Immune response genes, regulating qualitative differences in response (which may be all or none), are found in several subregions of H-2I of the mouse MHC and have been located to the MHC in other species. Evidence for human immune response genes is increasing, but as in the mouse, numerous other genes also appear to have regulatory roles. *H-2* is implicated in susceptibility to many viruses including tumor viruses, and *HLA* is linked or associated with susceptibility to a wide variety of diseases.

T, H-2, Qa, and Tlª

Four genetic systems located on the same (seventeenth) chromosome of the mouse are involved in development and differentiation. The T region is located close to the

centromere. At a distance from T are clustered the genes of *H-2, Qa,* and *Tl*ª. *T* and *H-2* are highly polymorphic, and many mutants have been detected. *T* is the symbol for Brachyury, short tail, and the allele *T* is dominant. Most of the T region mutants (*t*) have been detected in crosses of wild mice to stocks carrying *T*. *T/t* hybrids are viable tailless mice. *T* itself is a recessive lethal, as are most *t* mutants. The cross *T/t*ˣ × *T/t*ˣ yields only *T/t*ˣ offspring. Five separate genetic loci are clustered in the T region, although each is called a complementation group rather than a locus. The hybrid *t*ˣ/*t*ʸ is viable if *t*ˣ and *t*ʸ are from different complementation groups. The embryo dies at a specific point in development if *t*ˣ and *t*ʸ are different members of the same complementation group. Male sterility and abnormal male transmission ratio are but two of many other features associated with this system. Two studies have suggested that the human sixth chromosome also has *T*-like and *H-2*-like genes. It has also been suggested that *T* is as important for cell-cell cooperation during embryogenesis as *H-2* is for cell-cell cooperation in immunologic responsiveness.

Serologic and functional studies of *H-2* have allowed the description of the K I S and D regions (Fig. 17-2). The K region has only one known product, the class 1 glycopeptide H-2K. The H-2K locus is unique in its mutability with a mutational frequency of 10^{-4}. An associated I region abnormality has been found to occur with some H-2K mutations, and not infrequently strong MLR responses are given in mixes of lymphocytes of the ancestral and the mutant. H-2K itself has not been implicated in MLR responses.

The I region has been divided into IA, IB, IC, IE, and IJ. The distinction between IC and IE has become blurred, and the antigenic product, a class 2 glycoprotein is usually

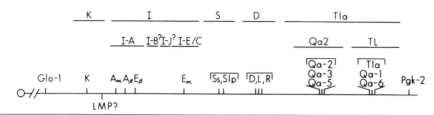

Figure 17-2. A representation of the seventeenth chromosome of the mouse, which carries many genes that influence development and differentiation. The T complex has been mapped close to the centromere, but chromosonal inversion may affect the location of genes of some T chromosomes. Extensive duplication of genes has been revealed by molecular genetic procedures, and not all are transcribed. In the H-2 region, H-2D, K, and L code for class I genes, and H-2, IA, and E for the class II genes that are responsible for cell–cell interactions. Qa and Tl resemble class I genes; their products have a restricted distribution.

referred to as IE-C. Both IA and IC-E have genes coding for an α chain and a β chain. IA combines with IAβ and IE-Cα with IE-Cβ, but pairing between homologous products of the two chromosomes of an F_1 hybrid has been reported. An invariant chain Ei is always present during the intracellular processing of the α and β chains to form the transmembrane form of a class 2 antigen, but the gene for the invariant chain, like the gene for the β_2 microglobulin product of a class 1 antigen is located on a different chromosome. The immune response function of IA depends upon the IA antigens and is codominant. Recessive responses have been localized to the IE-C region. No product is known for the I-B region, but this region reportedly regulates the response to IgG. *IJ* is the controller of immunosuppression. Certain studies had placed *IJ* within the I region, but gene cloning has failed to identify an *IJ* gene. S is a genetic region coding for the C4 proteins and for Bf, which is sex-limited. Probing with cDNA probes and with cosmids has revealed an extreme complexity in the D region and in the interval between D and Tl^a. H-2L and H-2R have been placed close to H-2D, and there is tentative evidence for another class 1 product H-2M. Little is known of the 30 additional genes or pseudogenes mapped close to D. It is possible they play a role in tissue differentiation.

Telomeric to D are genes for Qa2. Several antigens have been described, some may be alleles at the same locus. Qa1, at one time regarded as a product of a separate locus, has not been separated from Tl^a through recombination. Like Qa2, Tl^a is expressed on subpopulations of T lymphocytes. Qa is found on splenic and other lymphocytes while Tl is restricted to thymic cells and some thymic leukemias.

Biochemistry

As mentioned earlier, two classes of structural genes have been identified in the MHC. One codes for a 44,000 dalton glycoprotein that binds noncovalently to β_2-microglobulin (β_2M) (Fig. 17-3). In humans, the three examples of this gene are called *HLA-A, B,* and *C.* In the

mouse two loci are known, *H-2D* and *H-2K,* and two others, *H-2L* and *H-2R,* are probably separate from *D* but lie close to it. Other genetic loci have been recognized by recombinant DNA techniques, one lies close to *K,* the others closer to *D,* but the gene products have not yet been found.

The second class of structural gene products of the MHC are the Ia and HLA-DR antigens. These antigens are composed of two glycoprotein units of approximately 33,000 (α) and 28,000 (β) daltons (Fig. 17-4), and in both HLA and H-2 the two proteins are noncovalently bound.

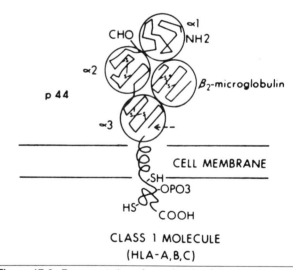

Figure 17-3. Representation of a molecule of the HLA-A or HLA-B series. The 44,000 dalton heavy chain is transmembrane with the C terminus inside the cell. The arrow marks a papain-sensitive site situated close to the insertion into the membrane. Two loops are formed by internal disulfide bonds. These represent functionally distinct domains. The conformation of the molecule is, in part, stabilized by the light chain, which is β_2-microglobulin. Neither the carbohydrate side chain nor β_2-microglobulin appears to contribute to the alloantigenic specificity. The H-2D and H-2K molecules of the mouse show considerable homology to human HLA-A and HLA-B. *(Modified from Owen and Crumpton: Immunology Today 1:117, 1980.)*

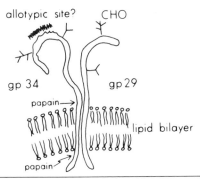

Figure 17-4. Representation of the HLA-DR molecule. The 34,000 dalton α-chain and the 29,000 dalton β-chain are noncovalently bound but difficult to dissociate. Like the HLA-A and HLA-B molecules, HLA-DR is transmembrane, with a papain-sensitive site close to the insertion. During intracellular processing (glycosylation) the dimer is bound to a third, invariant, chain, which is lost as the molecules are inserted into the cell membrane. The I-E-C molecules of the H-2 system of the mouse resemble HLA-DR.

These protein chains are, however, so tightly bound that reduction and boiling are necessary to separate them. A third, invariant chain (I) is loosely bound to the α and β chains and, like them, is processed in the Golgi. The I chain is shed before the complex is inserted into the plasma membrane.

The antigens of both classes have the amino terminal ends exposed. These are transmembrane proteins with cytoplasmic carboxyl terminal ends. The molecule is retained in the lipid membrane of the cell by a hydrophobic region near the carboxy terminus. Cleavage with papain leaves a fragment of about 10,000 daltons bound to the membrane and releases a soluble product that can still be bound by antibody. Although HLA-A, B, and C and H-2D and K are glycoproteins, the carbohydrate does not contribute to antigenic specificity. With rare exceptions the β_2-microglobulin chain is invariate within a species. It shows homology between species and also with a portion of the Fc domain of immunoglobulin. It too does not contribute to specificity except in the sense that it influences the conformation of the whole molecule and, hence, its antigenicity. The α chain of the class 2 molecules shows little variability, and most of the polymorphism is determined by the β chain. As with class 1 antigens, the carbohydrate is not thought to be responsible for the polymorphism.

The HLA-A, B, and C and the H-2D, K, and L gene products all exist in a rich variety of variant (allelic) forms; at least 20 alleles each of HLA-A, HLA-B, H-2K, and H-2 D can be identified (Table 17-1). The HLA-A and B molecules each carry at least two epitopes. This has been formally proved by allowing a lectin to bind to the carbohydrate of the HLA-A molecule. The bound lectin will sterically hinder the binding of certain monoclonal antibodies. Similarly, antibody to β₂ microglobulin will attach and sterically block the binding of other antibodies to HLA-A. At least two sites are also present on HLA-B and

TABLE 17-1. CURRENT HLA ANTIGENS

HLA-A Locus	HLA-B Locus	HLA-C Locus	HLA-DR Locus	HLA-D Locus
A1	B5	Cw1	DR1	Dw1
A2	Bw51	Cw2	DR2	Dw2
A3	Bw52	Cw3	DR3	Dw3
A9	B7	Cw4	DR4	Dw4
Aw23*	B8	Cw5	DR5	Dw5
Aw24	B12	Cw6	DRw6	Dw6
A10	Bw44	Cw7	DR7	Dw7
A25	Bw45	Cw8	DRw8	Dw8
A26	B13		DRw9	Dw9
A11	B14		DRw10	Dw10
Aw19	B15			Dw11
A29	Bw62			Dw12
Aw30	Bw63			
Aw31	Bw16			
Aw32	Bw38			
Aw33	Bw39			
A28	B17			
Aw34	Bw57			
Aw36	Bw58			
Aw43	B18			
	Bw21			
	Bw49			
	Bw50			
	Bw22			
	Bw54			
	Bw55			
	Bw56			
	B27			
	Bw35			
	B37			
	B40			
	Bw60			
	Bw61			
	Bw41			
	Bw42			
	Bw46			
	Bw47			
	Bw48			
	Bw53			
	Bw59			
	Bw4			
	Bw6			

*Splits or variants of an antigen are listed as indentations below the original antigen. The small "w" (for workshop) indicates an antigen that has been accepted provisionally at an international workshop as a specific HLA antigen. When the specificity of such an HLA antigen has been established, the w is deleted.

also on H-2D and H-2K. The proof of their existence depends upon the ability of one antibody to block the binding of some but not all antibodies directed against the same molecule. This effect will be discussed below in the section on subtypic and supertypic specificities.

So far, we have treated HLA and H-2 as being closely analogous systems. Indeed it is believed that they are really homologous and that the MHC evolved very far back in vertebrate evolution, since birds, frogs, and reptiles share so many of the same attributes. However, inevitably there are differences. Some are major. In humans, HLA-DR, which is responsible for MLR stimulation, is centromeric to both HLA-A and HLA-B, whereas in the mouse the 1 region lies between K and D (Fig. 17-2). The Rhesus monkey MHC appear to resemble that of humans while the rat appears to have only one class 1 antigen. Gene duplication during evolutionary diversification can readily account for the observed differences and for the clustering of so many genes and pseudogenes in the same segment of chromosome to form a gene family.

MHC Determinants and T Lymphocyte Function

Earlier in this chapter it was pointed out that most if not all higher vertebrates have an MHC. This observation raises an intriguing question. What physiologic role does the MHC play that makes it so important that it has been conserved through evolution in the more complex vertebrates? Recent studies demonstrate that most (or all) T cell functions are integrally associated with recognition of self-MHC antigens. Interestingly, different T cell functions are restricted by one or the other of the two different classes of MHC molecules (see below).

Helper T Cells Are Restricted for Ia Antigens

The importance of MHC determinants in controlling T cell function was shown by studies in which helper T cell function was measured in congenitally athymic (nude) mice that had been reconstituted with T cells from a variety of inbred strains. Inbred nude mice could only be efficiently reconstituted by thymocytes from mice of the same H-2 type. H-2-compatible T and B cells have been found to cooperate in antibody responses to a variety of antigens. Using mice that bear intra-H-2 recombinant haplotypes, it has been shown that the H-2 gene product that controls this effect maps in the I-A subregion. No other H-2I region product influences the cooperative event, and, similarly, H-2K and H-2D region products do not normally appear to be involved in restricting helper T cell function.

Immunogenetic analysis of T cell-macrophage interaction has provided similar results. Macrophages play an essential role in antigen presentation to T cells (Chaps. 12 and 14). When T cells are sensitized on antigen-pulsed macrophages, only H-2-compatible macrophages will efficiently present antigen to T cells. This can be measured as proliferation of T cells to homologous antigen in culture and as helper cell activity of the sensitized T cells.

Two other important features of these systems are worth noting. First, T cells from F_1 hybrids between two inbred strains ($P_1 \times P_2$) that have been sensitized on antigen-pulsed macrophages from one parent (P_1) will proliferate in response to a second challenge of the antigen only if F_1 macrophages or macrophages from P_1 are present. These T cells will provide helper cell function only for B cells from the F_1 or the parent that served as a source of antigen presenting cells. Thus, T cell specificity is not for antigen per se but for antigen *and* the appropriate H-2 specificity. Whether the T cell recognition structure is composed of one receptor for an antigen-modified H determinant or is two separate receptors, one for antigen, the other for the appropriate H-2 molecule, is controversial and is of considerable concern in immunity to viruses (Chap. 19). However, these studies and those in which cytotoxic T lymphocyte specificity has been examined (see below) clearly demonstrate that the receptor specificity of T cells, in contrast to B cells, involves a self-H determinant recognition unit.

The second important point from these studies derives from experiments measuring responses to antigens under *IR* gene control (Chap. 15) using F_1 hybrids between responder and nonresponder strains of mice. T lymphocytes from F_1 mice primed to such an antigen are often restricted in their ability to interact with antigen-pulsed parental macrophages or parental B cells. Primed F_1 T cells may interact with antigen-pulsed macrophages from high responder but not low responder macrophages or, alternatively, can be found to interact only with high responder B cells to produce antigen. Such selectivity is not observed using antigens that are not demonstrably under *IR* gene control. The model of immune responsiveness originally proposed by Medawear envisaged three components that corresponded to the three areas of response of the nervous system, afferent, central, and efferent. Using this analogy, the absence of the appropriate *IR* allele may result in a deficiency in cell-to-cell cooperation at the afferent (antigen presentation) stage, at the central (T-B interaction) stage, or both.

Implications for Clinical Management of Human Diseases

Studies of T cell-macrophage interaction in humans, measured by proliferative responses to antigen-pulsed macrophages, and of cytotoxic T cell responses to minor H or viral antigens suggest that human T cell function is restricted by H determinants. Cytotoxic T lymphocyte responses would be restricted by self-HLA antigens, predominantly HLA-A and HLA-B. HLA-DR and other class 2 molecules may also be implicated. The immunogenetics of helper T cell function in humans have not been fully delineated. However, the data accumulated thus far are in accord with those from laboratory animals.

The need for compatibility for HLA-D for the avoidance of severe GVH disease in clinical bone marrow transplantation will be presented later. Here it is appropriate to discuss the use of fetal liver for the reconstitution of immunologic deficiency. From studies of Zinkernagel and others, it is clear that the presence of thymic tissue of the right MHC type is necessary for the optimal maturation of fetal T cells. When fetal liver or other immature cells are used for reconstitution and where HLA matching is impracticable, the simultaneous transplantation of thymus from the same fetus should help prevent some of the late sequelae of lymphocyte precursor transplantation, namely, chronic infection.

The HLA System of Humans

HLA in Families

The human MHC is located on the short arm of the sixth chromosome. Other markers of the chromosome are shown in Figure 17-1. Since the MHC occupies only a relatively short segment of the chromosome, crossing-over within the MHC is infrequent, and the whole complex is usually inherited as a single unit. This unit is called a "haplotype." Because most human populations are heterogeneous and because gene expression is codominant, the two HLA haplotypes carried by an individual can usually be distinguished. Their unit character can be traced back within a family. The haplotype frequently remains unchanged for many generations. Within a nuclear family (father, mother, children) haplotypes are inheritable from the father as gametic units. For convenience let us call these gametic units a and b. Two haplotypes are inheritable from the mother, c and d. The children must then be ac, ad, bc, or bd. If there are five children in a sibship, at least two of them inherit the same pair of haplotypes, e.g., ad, ad. These are called "HLA-identical sibs." If two sibs share one haplotype (and differ at the other) they are called "haploidentical," e.g., ad, ac. If they share neither, they are "HLA different," e.g., ad, bc. Skin grafts exchanged between HLA-identical sibs persist for several weeks (mean 24 days, range 14 to 40 days), and rejection, when it does occur, is attributable to minor H differences. Grafts between HLA-different sibs are rapidly rejected (mean 11 days, range 6 to 14 days). Grafts between haploidentical sibs show a wide variation in rejection time (mean 13 days, range 6 to 40 days). The same immunogenetic relationship holds true for kidney grafts. HLA-identical kidney grafts are readily acceptable and require only moderate amounts of immunosuppressive drugs to prevent rejection, while haploidentical grafts are somewhat unpredictable; some are very well accepted, and some are irreversibly rejected. HLA-different sibling grafts are rarely performed, since they offer few if any advantages over ca-

daveric grafts, and there is little ethical justification for using a live donor where there is no clear advantage to the recipient. The MLR response follows the same genetic pattern: HLA-identical pairs do not stimulate each other, HLA-different pairs stimulate strongly, and responses between haploidentical sibs are highly variable.

Testing for HLA within a family in the clinical laboratory is relatively straightforward, even though there are very many HLA antigens. Highly specific antibodies to most of the HLA antigens are available, and the test, measurement of cell damage produced by complement after sensitization of lymphocytes with antibody (Chap. 13), is in theory simple. Typing for HLA-DR antigens is more complicated. B lymphocytes comprise only about 15 percent of human peripheral blood lymphocytes, i.e., about 3 percent of the blood leukocytes, so special procedures are required. Usually this involves the initial removal of T cells and all nonlymphoid cells. The test is then basically the same as for A, B, and C typing, but B cells are technically more difficult to manipulate, are more easily damaged during preparation, and are more sensitive to the natural antibodies in the rabbit serum used as complement. Assignment of HLA-A, B, C, and DR specificities to a haplotype is best achieved by studying the segregation of positive reactions given by the specific antisera within a family. Monoclonal antibodies are being used increasingly in experimental laboratories. They give highly reproducible results and often have exquisite specificity. Unfortunately, only a limited repertoire of monoclonal antibodies is currently available, and some do not fix complement and must be tested in binding assays. Interestingly, many of the monoclonal antibodies bind to determinants not previously recognized.

Typing for HLA-D is quite different and depends upon results of stimulation in MLR. The MLR is a biologic assay in which lymphocytes from two individuals are cultured together for six days, and their proliferative response is measured by their capacity to incorporate radioactive thymidine. To make the response unidirectional and therefore more specific, cells from one of the individuals (the stimulator) are x-irradiated or treated with mitomycin C (both procedures inhibit replication) before mixing. The stimulator cells used for typing are carefully selected to carry two identical copies of the same HLA-D specificity and are called "homozygous typing cells" (HTC). HTC can only stimulate cells from responders having a different D allele. Some of the most highly discriminating HTC come from children of a consanguineous (usually first cousin) marriage, in which two copies of exactly the same haplotype are inherited from a common great grandparent. As an example, HTC of specificity HLA-Dw1 will stimulate a responder lymphocyte from a Dw2/Dw3 subject but will not stimulate Dw1/Dw2 or Dw1/Dw3 cells.

A preferred test for HLA-D is the primed lymphocyte test (PLT) in which a responder, e.g., Dw1/Dw1 is stimulated with cells from a donor, Dw1/Dw2 or Dw2/Dw2.

The responding cells are allowed to proliferate for 7 to 10 days; after this time in culture, the priming response has ended. Stimulator cells from subjects to be typed are then added to the primed cells, and a known Dw2 cell is used as a positive standard. The Dw2 cell stimulates a rapid, anamnestic type of response. If the unknown carries Dw2, the response will also be rapid. If it carries other specificities, e.g., Dw1/Dw4, proliferation will still occur, but the kinetics are different and more like those of a primary MLR.

Cytotoxic lymphocytes (CTL) are generated during an MLR from precursor cells. CTL will kill cells from the sensitizing donor and also cells from other people. The specificity of CTL is different from that of the PLT, since HLA-B antigens as well as other HLA products can be detected by CTL.

HLA in Caucasian Populations

Since HLA was first described in Caucasians and the early investigations were all carried out in Europe or North America, most of our information is from Caucasians. As will become apparent, different ethnic groups may be very dissimilar. Even within Caucasians there are differences depending upon varying ancestral racial admixture.

The first HLA-A and B alleles received their designations before the loci could be distinguished and are numbered in order of their discovery. Separate sequences of numbers are used for the other loci. Since DR was originally believed to be the serologic manifestation of the functional specificity HLA-D, there is a correspondence between many of the D and DR alleles. The HLA-D/DR region is, however, known to be more complex than was originally suspected, and there appear to be many more genes for class II antigens than are present in the mouse. Therefore, the current terminology does not reflect the true complexity of these polymorphic loci. Genes for at least three distinct α chains and for at least six β chains have been localized to the segment of the haplotype between GLO and HLA-B. The products of three loci have been isolated and partially characterized; these are DR, SB, and DC/DS. It is likely that additional products of class II genes will be recovered, since there is presumptive evidence for a fourth set of α and β chain genes. All α chains and all β chains have considerable homologies within their class but also show differences, especially in the hydrophilic intracytoplasmic domains. There are also extensive differences between the uncoded flanking sequences of β chains from different loci.

New HLA specificities, not yet formally recognized, are given a provisional or workshop (w) designation. A complication has arisen with HLA-C alleles. Many C locus alleles have been adequately characterized, but the w is retained to avoid confusion with complement components, especially C4, which is coded for by genes on the same haplotype as HLA-Cw4. There are approximately 40 al-

leles of HLA-B, 20 of HLA-A, and 12 of HLA-C, so there should be nearly 10,000 different haplotypes distinguishable by HLA-A,B, and C typing and more than 120,000 if HLA-DR is included. In practice some haplotypes are found more frequently than would be expected by chance, and others are exceedingly rare.

The nonrandom association of alleles is called "linkage disequilibrium" (ld). The haplotypes A1B8 and A3B7 offer convenient examples. The alleles A1 and A3 and B7 and B8 have approximately the same frequencies in northern Europeans and, therefore, either A1 or A3 would be expected to be found with B7 or B8. In practice A1 is found with B8 and A3 with B7 many times more often than in the opposite configuration. The ld includes HLA-C and the C4 alleles and extends centromerically to include DR and even GLO several centimorgans away. Two explanations, which are not mutually exclusive, for ld have been advanced. One is that ld is a consequence of the population explosion in former centuries and that certain combinations confer a selective advantage. Modern populations originated in relatively small groups in which limited mobility must have imposed a degree of inbreeding, and as the populations expanded the original haplotypes must have proliferated. The second is that some combinations of MHC alleles confer disease resistance or increased fertility and thus have selective advantages. The converse also seems to be true, that certain haplotypes are frequently associated with specific diseases, but disease association would not necessarily exclude other compensatory (unknown) advantages. One practical consequence of ld is that it is often possible to trace the inheritance of a genetic trait within a family even where the gene cannot be identified in the heterozygous state, provided the gene is in strong C-d. This is most clearly illustrated by the metabolic abnormality hemochromatosis (see below) and to a varying extent with numerous other diseases.

HLA in Non-Caucasians

African and American blacks, American Indians, Orientals, and Polynesians have very different frequencies for many of alleles of HLA, and some antigens are restricted to these populations (Table 17-2). HLA-A1 and B8 are both common in Caucasians and blacks but are extremely rare in Orientals. There are numerous examples. The specificity Aw42 (a variant of A10) is almost exclusively confined to blacks. American Indians have high frequencies of A9 and Bw35, while populations from New Guinea lack A2, the most common A locus specificity of Caucasians and of most other populations. The strong correlations between D and DR specificities so characteristic of Caucasians are also different in other populations, and the disease associations also differ. Some of the differences are believed to be due to founder effects and to gene drift. These are common phenomena in small populations, especially when there is a degree of inbreeding. They may also reflect the evolutionary pressures of different environments.

TABLE 17-2. HLA-A AND -B GENE FREQUENCIES IN DIFFERENT ETHNIC GOUPS

	European Caucasoids	North American Caucasoids	American Blacks	African Blacks	Japanese	American Indians
HLA-A	(228)*	(290)	(128)	(102)	(195)	(89)
A1	15.8	16.1	8.1	3.9	1.2	2.5
A2	27.0	28.0	16.3	9.4	25.3	45.3
A3	12.6	14.1	7.0	6.4	0.7	0.6
Aw23	2.4	1.9	10.6	10.8		
Aw24} A9	8.8	7.3	5.1	2.4	37.2	23.2
A25	2.0	2.6	0.4	3.5		
A26} A10	3.9	3.4	2.3	4.5	12.7	0.6
A11	5.1	5.1	2.8	—	6.7	—
A28	4.4	4.2	5.8	8.9	—	2.8
A29	5.8	3.6	2.3	6.4	0.2	0.6
Aw30	3.9	2.9	13.0	22.1	0.5	1.1
Aw31	2.3	4.5	2.8	4.2	8.7	19.9
Aw32	2.9	3.7	1.9	1.5	0.5	1.1
Aw33	0.7	1.2	5.1	1.0	2.0	0.6
Aw43	—	—	—	4.0	—	—
Blank	2.2	1.3	16.5	11.0	4.2	1.8
HLA-B	(228)	(290)	(128)	(102)	(195)	(89)
B5	5.9	5.9	4.9	3.0	20.9	14.0
B7	10.4	10.5	12.6	7.3	7.1	0.6
B8	9.2	10.4	5.5	7.1	0.2	1.7
B12	16.6	13.8	14.0	12.7	6.5	1.7
B13	3.2	2.6	0.4	1.5	0.8	—
B14	2.4	5.1	4.6	3.6	0.5	—
B18	6.2	3.1	3.6	2.0	—	0.6
B27	4.6	5.6	0.8	—	0.3	6.2
B15	4.8	5.9	4.7	3.0	9.3	13.7
Bw38	2.0	2.5	0.4		1.8	
Bw39} Bw16	3.5	1.4	0.4	1.5	4.7	14.5
B17	5.7	4.9	11.2	16.1	0.6	—
Bw21	2.2	3.8	4.4	1.5	1.5	—
Bw22	3.6	2.3	3.9	—	6.5	0.6
Bw35	9.9	8.6	12.5	7.2	9.4	22.1
B37	1.1	1.7	1.2	—	0.8	—
B40	8.1	9.2	3.9	2.0	21.8	16.6
Bw41	—	—	—	1.5	—	—
Bw42	—	—	—	12.3	—	—
Blank	3.6	2.8	11.0	17.9	7.6	7.8

Adapted from Amos and Kostyu: Adv Human Genet 10:137, 1980.
*Number tested.

HLA and Disease

Since mice of certain H-2 type are more susceptible to leukemogenesis or oncogenesis by specific tumor viruses, it seemed logical to see if a similar association existed between HLA and leukemia or Hodgkin's disease. Associations were indeed found but were not very strong, and studies of HLA and disease might have been discontinued had not extraordinarily strong associations been found between the specificity HLA-B27 and a disease of the spine called "ankylosing spondylitis" (AS). Two independent studies reporting that over 90 percent of patients with AS carried the antigen B27 immediately spurred a host of studies of HLA antigen frequencies in other diseases. Many endocrine diseases, rheumatoid syndromes, and

autoaggressive diseases are now known to be associated with a particular HLA antigen (Fig. 17-5).

Disease studies are conducted at two levels: family versus population. Association studies are conducted at the population level. The number of individuals with disease having the particular antigen is compared with the number of individuals with the antigen in a control group. From this an index, the relative risk, is calculated. Since human families are usually small, only one member may be affected. It is often difficult to test unaffected family members. Therefore, most studies of HLA and disease are designed to demonstrate an association at the population level and to establish the magnitude of the relative risk. This can be as high as 90 for B27 and AS, but more typically it is between 3 and 10 for the autoimmune diseases.

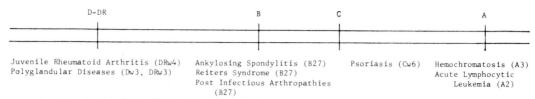

Figure 17-5. Many diseases are associated with different regions of the HLA haplotype. Diseases of endocrine glands, including Graves' disease, are most frequent in people carrying specific HLA-D or DR alleles. Over 90 percent of Caucasians and Japanese suffering from ankylosing spondylitis carry the HLA-B allele B27. Disease associations with HLA-C and HLA-A are less common, but the recessive gene (*h*) for the iron storage disease, hemochromatosis, has been tightly linked to HLA-A. The A3 allele is most frequently the marker, but the gene *h* can be linked to other alleles, including A29.

Information of a different kind can be gained from family studies, especially when large kindreds having more than one affected member are available. The genetic locus responsible for disease susceptibility may not be in strong linkage disequilibrium with a specific HLA allele. This would be the case if the gene was on the sixth chromosome but at some distance from it. Under these conditions, an HLA allele and a disease gene could be transmitted from an ancestor as a haplotype to some family members. However, recombination can occur within an extended family. This is observed when the two genes are not very closely linked. An apparent example has been reported for some forms of the developmental abnormality, spina bifida. Linkage rather than association might also be expected where other, unlinked genes modify the disease phenotype. In some diseases the association with HLA is first noticed in population surveys, and proof that a disease susceptibility gene is present on the sixth chromosome is obtained through linkage studies. Genes for ankylosing spondylitis, 21-hydroxylase deficiency, C2 and C4 deficiencies, and hemochromatosis are examples of linkage first established in population screens.

Two interesting and potentially important findings have not been fully explored, differences in recombination frequency and polygenic effects. While the average frequency of recombination between HLA D/DR and HLA-B or HLA-B and HLA is of the order of 0.5 percent, some exceptional families have been found. The frequency of recombination was over 50 percent in a large family with multiple cases of arthritis, 20 percent recombination has been found in some series of acute myelocytic leukemia, and several families with two or three recombinants have been reported in other conditions. While these are usually families in which a disease gene appears to segregate, it is notable that recombination is frequent in unaffected members. Polygenic interactions are difficult to unravel in human families. Immune responsiveness is almost certainly influenced by several unlinked systems.

The reasons that so many diseases are HLA linked are speculative. The X chromosome carries a number of genes for diseases or congenital abnormalities, including hemophilia, X-linked immunodeficiency, and color blindness.

Other autosomes must also determine arrays of abnormalities, and these are beginning to be known as the human chromosomes become better mapped. Many of the HLA-associated or HLA-linked diseases are frequent, X linked are rare, and the relative risks between other markers and a disease, e.g., blood group A and peptic ulcer, are usually very low as compared to the HLA-linked diseases. The implication is that many of the HLA-linked diseases are related to malfunctions of regulation and that the MHC is critically important for many regulatory functions. Perhaps from a better understanding of HLA we will derive a better understanding of endocrine interactions, including endocrine-immune and neurologic-immune interactions. Other possibilities are that the rich array of polymorphic glycoproteins of the MHC provide receptors for viruses and many reactive compounds or that the disease-related gene may have become trapped by translocation within the MHC, the so called hitchhiker effect. The HLA system will continue to provide valuable information not only on diseases but also on physiologic function.

HLA and Paternity Exclusion

HLA is being increasingly relied upon for paternity exclusion because it is so highly polymorphic. Its resolving power is greatest where the putative father has a different ethnic background from the mother. In combination with blood group markers and enzyme polymorphisms, it can provide a high degree of resolution provided that the testing is carried out by a highly competent laboratory.

Immunogenetics of Transplantation

The laws of transplantation were formulated in mice, since inbred strains of different H-2 types were readily available for testing the requisite genetic crosses. Tissue from one inbred strain was freely transplantable to another member of the same inbred stain, whereas it was usually rejected promptly by a mouse of a different strain. The F_1 hybrid would accept a graft from either parent but could not donate to either. Either parent could donate to a proportion of the progeny of a mating between two F_1 hybrids (the F_2). After much investigation it was proved that the H-2 system was uniquely immunogenic in transplant rejection and that a graft was accepted only if there were no H-2 antigens on the donor tissue that were not also present in the recipient. If the donor had H-2 antigens lacking from the recipient, antibodies to H-2 were formed, and cytotoxic lymphocytes were generated.

Since mice were inbred, they are homozygous for H-2. Humans are rarely homozygous for HLA, so grafts from parent to child may be promptly rejected, H-2 and HLA both show gene-dose effects. The reaction against a donor sharing one haplotype is often less strong than that against a donor of a completely different type. Reference has already been made to the finding that skin or kidney from a sibling who carries the same pair of HLA haplotypes (HLA identical) stimulates only a weak immune response due to non-HLA antigen differences. Within a family the order of preference for donation of a kidney is (1) identical sibling, (2) haploidentical sibling or parent–child, and (3) haplodistinct sibling (Fig. 17-6).

For various reasons only a minority of recipients receive an HLA-identical transplant and most donor recipient combinations are mismatched at one or both haplotypes. Since the majority of organs, including many kidneys and all hearts, lungs, and livers, are from cadaveric donors there have been numerous attempts at reducing the antigenic disparity by selecting the most compatible recipient from a pool awaiting transplantation.

European transplant centers have consistently found better graft survival when donor and recipient are matched for two A locus and two B locus antigens. Results from the United States have not always been in agreement with these finding. The difference can be explained in part by the greater heterogeneity of the U.S. urban population, since antigens that cannot yet be characterized may be in linkage disequilibrium with the known antigens. As knowledge of class II antigens developed, an increasing number of centers have begun to match for DR, and positive correlations between DR match and graft survival have been reported. Since few laboratories can presently type for alleles of the newer loci (SB and DC/DS) and since there is probably more than one DR molecule per haplotype, it will be several years before the full significance of the new loci is known.

Avoidance of Transplant Reactions

Since the number of antigens in the HLA system is so large and there are still new antigens to be detected, the chance of matching an unrelated donor with a waiting recipient is slim unless both possess very common haplotypes in which linkage disequilibrium is very strong (e.g., A1 B8 DR3 or A3 B7 DR2). Therefore, emphasis is placed upon reducing the severity of the immune response against the graft. The first measure is to crossmatch donor and recipient just as blood for transfusion is tested (Chap. 16), using donor lymphocytes as targets. Recipients with preformed antibodies to HLA-A or B antigens may reject a kidney from a donor giving a positive crossmatch in hyperacute fashion. The next measure is to transfuse the recipient with whole blood several weeks prior to transplant. While this procedure increases the risk of sensitization and the formation of cytotoxic antibodies, a large proportion of transfused recipients do not form antibodies and accept a graft more freely. Improved graft survival following transfusion appears to be in part due to elimination of the most immunoreactive recipients and in part to the triggering of an as yet unknown state of immunosuppression. Finally, mention must be made of immunosuppressive drugs and antithymocyte serum. Traditionally, corticosteroids and azathioprine have been the drugs found empirically to be most effective. The cost of nonspecific immunosuppression can be an increased risk of infection by cytomegalovirus and other infectious agents (Chap. 19) and an increased frequency of lymphoma. Cyclosporin A appears to be more effective than older immunosuppressive agents but is nephrotoxic and hepatotoxic in high doses. Antithymocyte serum (ATS) is often given. Unfortunately there is no reliable in vitro assay for the potency of different batches of ATS as immunosuppressive agents in vivo. Monoclonal antibodies to T cell subsets have been used to reduce the severity of immune reactions against the transplant. Many of the patients seem to have developed antibodies to the monoclonal antibody, but it is not clear if these were to mouse proteins, in which case recipients could be tolerized to mouse immunoglobulins, or to the idiotype, in which case the monoclonal antibodies would have only temporary effects.

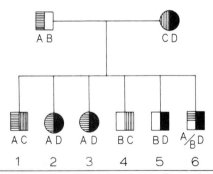

Figure 17-6. Inheritance of HLA in a family. The code letters A B C D represent the haplotypes, which could be formally written, for example, HLA-A1 B8 Cw1 Dw3, A2 B12 Cw1 Dw5, A9 B13 Cw3 Dw1, and A10 Bw15 Cw3 Dw4. Sibs 2 and 3 are HLA identical; 1 and 2, 3 and 5, and so on are haploidentical; 3 and 4, 1 and 5 are nonidentical. Sib 6 is a recombinant, having inherited part of the A haplotype (for example, A1) and part of the B (for example, B12 Cw1 Dw5).

Transplantation of Bone Marrow

Patients with immunodeficiency disease, bone marrow aplasias, metabolic deficiencies, leukemias, and other malignancies are candidates for bone marrow transplantation. The special problems faced in bone marrow transplantation are (1) reactions of lymphoid cells in the transplanted marrow against the host (GVH reactions), (2) rejection or failure of engraftment, (3) infection from the cumulative effects of drugs and GVH reactions, (4) recurrence of leukemia, and (5) lack of adequate stimulus for differentiation (Chaps. 14 and 20).

Mention has been made of the HLA-D locus and its effects in MLR stimulation. In the immunodepressed recipient, whether the immunodepression is drug induced or idiopathic, marrow from a donor who gives a positive MLC reaction with the recipient is more likely to cause a fatal GVH reaction in vivo, although GVH can occur even when the donor and recipient are HLA-identical siblings. The GVH reaction includes a desquamating skin eruption, diarrhea, and progressive weight loss.

The three major categories of disease for which replacement therapy by bone marrow or fetal thymus or liver is indicated, namely, severe immunodeficiency states, aplastic anemia, and leukemia, present different problems. In congenital immunodeficiency states, failure of engraftment is common, and even when chimerism (i.e., colonization with cells from the transplant) is achieved, immune function may still be inadequate. These problems are discussed earlier in this chapter and also in Chapter 15. Failure of engraftment or rejection of the graft are fairly frequent problems in aplastic anemia, and fatal GVH reactions have been observed. Fatal reactions are, however, most common when the recipient is leukemic. Failure of engraftment and rejection of the transplant are minimized by pretreatment of the recipient with cytotoxic drugs, such as Cytoxan, even though such treatment may in itself be hazardous. Despite these complications, about 50 percent of transplants are successful. In leukemia, the loss from GVH disease or interstitial pneumonitis (possibly secondary to GVH) may exceed 50 percent, and many of the remaining patients die from a recurrence of leukemia or die from various causes before the transplanted marrow has had an adequate opportunity to colonize the recipient. Incompatibility for minor H loci, sex differences between donor and recipient, and recognition of what may be regarded as altered self (viral or chemical) (Chap. 14) are all possible complicating factors. In mice and rats, the effects of minor H locus differences are most marked when the donor has been sensitized, and cross-reactivity with environmental agents may serve as a sensitizing stimulus. Germ-free recipients rarely develop GVH, especially if the donor is also germ free, even when MHC barriers are crossed. Newer modalities of therapy offering considerable promise include total lymphoid irradiation and treatment of the donor marrow with lectins. Whole body irradiation was used to reduce the load of leukemic cells and to create an aplastic marrow that would not crowd out the transplanted marrow. X or γ radiation-induced damage to lungs and gut presented difficult problems in management, and a new form of lead shielding that protects the viscera while leaving the long bones, sternum, and lymphoid tissues exposed (total lymphoid irradiation) has improved survival. The precursor cells for GVH reactions appear to have a high concentration of receptors for the lectin soybean agglutinin. Removal of cells with an affinity for the insoluble lectin greatly reduces the risk of GVH even when donor and recipient are HLA-D/DR incompatible and have significant responses in MLR.

Transplantation of Cornea

Successful corneal transplants have been performed without recourse to tissue typing or immunosuppression. This is not because the cornea lacks antigens but because the anterior chamber of the eye is one of a limited number of what are known as immunologically privileged sites. Other such sites include poorly vascularized areas of the brain and the testes. These sites lack lymphatic drainage, and the small amount of antigen released into the bloodstream from experimental grafts in these sites in not sufficient to trigger a cellular response. Interestingly, it has recently been found that HLA may be very important in corneal transplantation in those cases where the eye of the recipient is chronically but not severely inflamed. In a large series of corneal transplants, failure was rare where the recipient site was not inflamed but quite common when there was even a moderate state of inflammation. In this situation, a direct correlation was observed between the degree of incompatibility and the incidence of failure of the transplant. HLA matching is therefore indicated in corneal transplants when the recipient cornea is moderately vascularized or when a previous cornea has been rejected.

FURTHER READING

Books and Reviews

Amos DB, Kostyu DD: HLA—A central intelligence system of man. Adv Hum Genet 10:137, 1980

Bodmer WF, Batchelor JR, Bodmer JG, et al. (eds): Histocompatibility Testing 1977. Copenhagen, Munksgaard, 1978

Diczfalusy E (ed): Immunological Approaches to Fertility Control. Stockholm, Karolinska Institute, 1974

Goetz D (ed): The Major Histocompatibility System in Man and Animals. New York, Springer-Verlag, 1977

Klein J: Biology of the Mouse Histocompatibility-2 Complex. New York, Springer-Verlag, 1975

Lengerova A, Vojtiskova M (eds): Immunogenetics of the H-2 System. Basel, Karger, 1971

Najarian J, Symonds R: Transplantation. Philadelphia, Lea & Febiger, 1972

Terasaki PI (ed): Histocompatibility Testing 1980. Los Angeles, UCLA Tissue Typing Laboratory, 1980

Selected Papers

Artzt K, Shin H-S, Bennett D: Gene mapping within the T/t complex of the mouse II. Cell 28:471, 1982

Awdeh ZL, et al.: Extended HLA complement allele haplotypes evidence for T/t-like complex in man. Proc Natl Acad Sci USA 80:259, 1983

Carrole MC, Poter RR: Cloning of a human complement C4 gene. Proc Natl Acad Sci USA 80:264, 1983

Gottesman SRS, Hall KY, Walford RL: A thesis of genetic linkage of immune regulation and aging: The major histocompatibility complex as a supergene system. In Cooper EL, Brazier MAB, (eds): Developmental Immunology: Clinical Problems and Aging. New York, Academic Press, 1982

Guy K, van Heyningen V: Further intricacy of HLA-DR antigens. Immunol Today 3:237, 1982

Monaco JJ, McDevitt HO: Identification of a fourth class of proteins linked to the murine major histocompatibility complex. Proc Natl Acad Sci USA 79:3001, 1982

Ploegh HL, Orr HI, Strominger JL: Molecular cloning of a human histocompatibility antigen cDNA fragment. Proc Natl Acad Sci USA 77:6081, 1980

Porter KA, Andres GA, Calder MW, et al.: Human renal transplantation. II. Immunofluorescent and immunoferritin studies. Lab Invest 18:159, 1968

Reyes AA, Schold M, Itakura K, et al.: Isolation of a cDNA clone for the murine transplantation antigen H-2K^b. Proc Natl Acad Sci USA 79:3270, 1982

Seigler HF, Gunnells JC Jr, Robinson RR, et al.: Renal transplantation between HLA-identical donor-recipient pairs: Functional and morphological evaluation. J Clin Invest 51:3200, 1972

Shearer GM, Rehn TG, Garbarino CA: Cell-mediated lympholysis of triphenyl-modified autologous lymphocytes. J Exp Med 141:1348, 1975

Steinmetz M, et al.: Clusters of genes encoding mouse transplantation antigens. Cell 28:489, 1982

Woods DE, et al.: Isolation of cDNA clones for the human complement protein factor B, a class III major histocompatibility complex gene product. Proc Natl Acad Sci USA 79:5661, 1982

Zinkernagel RM, Doherty PC: H-2 compatibility requirement for T cell-mediated lysis of target cells infected with lymphocytic choriomeningitis virus. J Exp Med 141:1427, 1975

CHAPTER 18

Immunopathology

Introduction

Among the primary functions of the immune system are protection against microbial invasion, development and spread of neoplasms, and removal of effete or denatured cells or substances. To subserve these functions, components of the immune system must discriminate self from nonself or altered self, and then efficiently localize and destroy material recognized as nonself. The discrimination of self from nonself requires the recognition of foreign substances. Recognition of specific epitopes on antigens is mediated in part by immunoglobulins and by lymphoid cells containing specific surface receptors.

Foreign substances and altered self components can also be recognized following nonspecific interaction of the material with, for example, factors from the alternative pathway of the complement system (see Chap. 13). The interaction of host recognition factors with foreign materials leads to the production or release of biologically active molecules. These products enhance vascular permeability locally, produce vascular stasis, and chemotactically attract phagocytic wandering cells—such as polymorphonuclear leukocytes and macrophages—to the local sites of the immune reaction (Fig. 18-1). The generation of biologically active phlogistic products amplifies the initial reaction and permits the rapid influx of inflammatory cells capable of localizing, ingesting, and degrading the inciting agent. The production of biologically active molecules and the attraction and activation by these molecules of phagocytic wandering cells are termed an immune effector function. The inflammatory process is a mechanism by which the immune system mediates the actual localization and destruction of antigens.

The sequence of events leading to an immunologically mediated inflammatory reaction is: (1) binding of a recognition component to an antigen, (2) modification of the recognition component, which results in the activation or release of immune effector molecules that (3) produce a local inflammatory reaction by altering vascular permeability, producing vascular stasis, and chemotactically attracting and activating immune effector cells. Whereas these processes must be considered as primarily defensive against microbial and parasitic invasion, they are of necessity highly destructive not only for the organism but sometimes for the host harboring the infectious agent. There is also a high probability that the immune system acts to eliminate dead, neoplastic, or injured cells. It is therefore not surprising that immune mechanisms can damage and destroy other tissues and cause a number of important human diseases such as rheumatoid arthritis, systemic lupus erythematosis, and hemolytic anemia. Understanding how inflammation is initiated and regulated is thus essential not only for understanding immunologically mediated host defense, but also for comprehending immunopathologic processes.

Mechanisms of Inflammatory Cell Accumulation and Phagocytosis

The accumulation of inflammatory cells at sites of antigen and their biologic functions at these sites are central events in immune function. Polymorphonuclear leukocytes and macrophages are motile cells that have many common physiologic characteristics. Both cell types can perceive

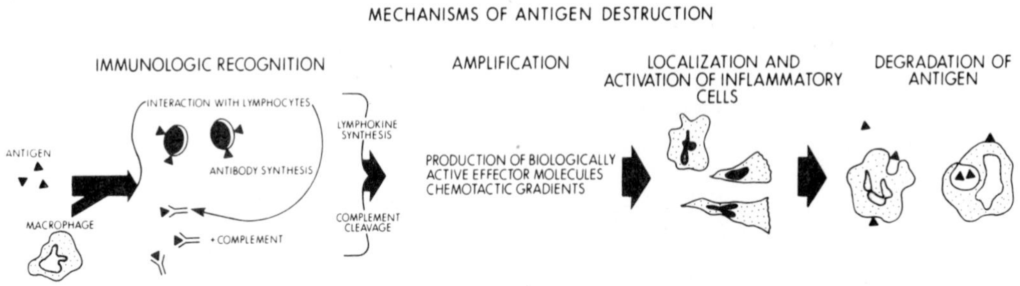

MECHANISMS OF ANTIGEN DESTRUCTION

Figure 18-1. Immunologic mechanism of antigen destruction. Immune recognition of antigen leads to the production of effector molecules, which mediate inflammatory cell influx and activation. Antigen is degraded by phagocytic cells that accumulate.

gradients of chemoattractant molecules and can migrate directionally along such gradients. They also perform endocytosis (ingestion), secrete lysosomal enzymes, and generate superoxide anions. Polymorphonuclear leukocytes and macrophages perceive chemotactic factors by means of specific surface receptors. These cells have receptors for bacterial products and for synthetic polypeptide chemotactic factors that may be analogous to them. They also have receptors for C5a, for a chemoattractant produced by polymorphonuclear leukocytes that have engulfed particulate matter, and for certain leukotrienes.

Binding of chemoattractants to the surface of phagocytes results in orientation of the cells toward the source of the chemotactic gradient. Upon orientation, the cells lose their round configuration and become polarized in shape. The configuration of motile cells is triangular, with the base of the triangle facing toward the chemoattractant gradient (Fig. 18-2). Changes in cell shape require rearrangement of intracellular cytoskeletal elements. Microtubules provide a front-to-back polarization, and actin filaments accumulate at the front and rear of the cells to provide the contractile forces required for movement. Chemotactic factors also initiate other responses by leukocytes such as superoxide anion production and lysosomal enzyme secretion. The concentration of chemotactic factors required to initiate these latter processes is 10-fold

REQUIREMENTS FOR CHEMOTAXIS

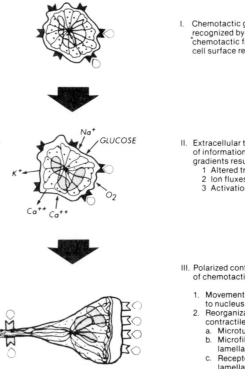

I. Chemotactic gradients are recognized by interaction of chemotactic factors with cell surface receptors.

II. Extracellular to intracellular translation of information concerning chemotactic gradients resulting in:
1 Altered transmembrane potential
2 Ion fluxes
3 Activation of metabolic processes

III. Polarized contraction of cell in direction of chemotactic gradients induced by:
1. Movement of centriole posterior to nucleus
2. Reorganization of cytoskeletal and contractile elements
a. Microtubules orient from centriole
b. Microfilaments polymerize in lamellapod and uropod
c. Receptors bind chemoattractant at lamellapod and sweep posteriorly to uropod

Figure 18-2. Requirements for leukocyte chemotaxis. The interaction of chemotactic factor receptors (*shaded*) with chemotactic factors (*hollow*) triggers the indicated cellular responses.

that required for induction of chemotaxis. The concentration gradient effect may delay the release of potentially toxic products from the cells until they arrive at the inflammatory site where the concentration of chemoattractants is greatest.

Chemotactic factors induce transmembrane fluxes of calcium, sodium, and potassium and thus alter cellular transmembrane potential. Following exposure to chemotactic factors, a transient intracellular elevation of cyclic AMP (cAMP) can be observed. The oligopeptide chemoattractant receptor appears to be modulated by a guanine nucleotide regulatory unit similar to other receptors, such as those for neurotransmitters. Arachidonic acid is released from the phospholipids of the activated phagocytic cells. Metabolism of arachidonic acid via enzymatic pathways involving cyclooxygenase and lipoxygenase results in the production of prostaglandins and leukotrienes. Transmethylation reactions mediated by S-adenosyl-methionine are required for chemotaxis and regulate the affinity of the chemoattractant receptor. One such reaction, methylation of membrane phospholipids, is inhibited by chemotactic factors in macrophages. Inhibition of the formation of methylated phospholipids results in changes in the composition of membrane phospholipids cells exposed to chemoattractants. Alterations in the biophysical properties of the membrane at the leading edge of the chemotactically migrating cell is important for motility.

Lymphocytes are also highly motile cells, but do not respond to the same chemotactic factors as polymorphonuclear leukocytes and macrophages. The migration of lymphocytes is stimulated by specific antigens, mitogens, and by as yet undefined factors produced by lymphocytes and by macrophages.

Phagocytosis is initiated by binding of antigen to the surface of phagocytic cells. Antigens carrying bound immunoglobulin or the complement fragment C3b are more readily phagocytosed because they bind to Fc and C3b membrane receptors of polymorphonuclear leukocytes and macrophages. However, phagocytosis can occur in the absence of receptor involvement. Particle ingestion results from the envelopment and fusion of phagocytic cell membrane around the foreign material (Fig. 18-3) (see Chap. 19). Intracellular lysosomes migrate to the phagocytic vesicle, fuse with it, and empty their contents, thus forming a phagolysosome. Within the phagolysosome, antigenic digestion and microbial killing generally occur. During the process of phagocytosis, lysosomal hydrolases and superoxide anions may be released extracellularly.

Immunopathologic Processes

The nature of immunologically mediated inflammatory responses depends upon the immunologic recognition component that identifies the antigen. Inflammatory responses can produce adverse reactions in the host, ranging from minor local tissue irritation to selective destruction of organs or even sudden death. Immunologically mediated inflammatory reactions can be divided into four general types (Table 18-1). The response is often mixed and sometimes all of these reactions operate simultaneously.

MECHANISMS OF PHAGOCYTOSIS

I. Recognition of Foreign Material by:

   1 Fc receptors

   2 C3b receptors

   3 Nonspecific membrane alterations

II. Transduction of Extracellular Information From Particle to Cell Interior Resulting in:

   1 Polymerization of actin

   2 ? Myosin Mg^{++} ATPase dependent contraction

   3 Activation of respiratory burst and movement of lysosomal granules to area of particle injestion

III. Particle Engulfment Leading to:

   1 Formation of phagolysosome

   2 Respiratory burst

   3 Killing and digestion

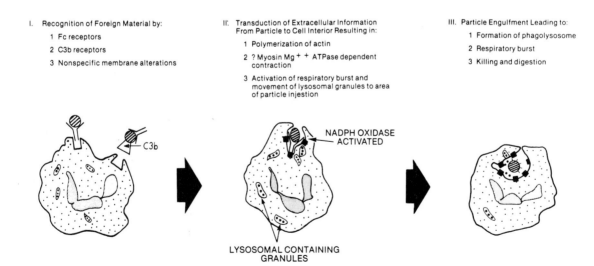

NADPH OXIDASE ACTIVATED

LYSOSOMAL CONTAINING GRANULES

Figure 18-3. Requirements for phagocytosis. Particulate antigen binding to the membrane of phagocytic cells initiates cellular responses that lead to envelopment of the antigen. This process is enhanced when the antigens have bound immunoglobulins and/or C3b.

TABLE 18-1. IMMUNOPATHOLOGICAL PROCESSES

Type of Inflammation	Immune Recognition Component	Soluble Mediator	Inflammatory Response	Disease Example
Reagenic, allergic	IgE	Basophil and mast cell products (i.e., Histamine, ECF)	Immediate flare and wheal, smooth muscle constriction	Atopy, anaphylaxis
Cytotoxic antibody	IgG, IgM	Complement	Lysis or phagocytosis of circulating antigens, acute inflammation in tissues	Autoimmune hemolytic anemia, thrombocytopenia associated with systemic lupus erythematosus
Immune complex	IgG, IgM	Complement	Accumulation of polymorphonuclear leukocytes and macrophages	Rheumatoid arthritis, lupus erythematosus
Delayed hypersensitivity	T lymphocytes	Cytokines	Mononuclear cell infiltrate	Tuberculosis, sarcoidosis polymyositis, granulomatosis, vasculitis

Inflammation Initiated by Reagenic (IgE) Antibodies

The Fc domains of IgE antibodies bind to receptors on mast cells and basophils. This Fc binding frees the Fab portion of the molecule for binding to specific antigen. Shortly after the appropriate antigen is bound, the cells degranulate and secrete their intracellular products, which include histamine, eosinophil chemotactic factors (ECF), and heparin. Release of mediators from basophils or mast cells causes an increase in local vascular permeability within seconds, and produces vascular stasis and smooth muscle contraction. Reagenic reactions are responsible for such allergic phenomena as urticaria, seasonal rhinitis, asthma, and systemic anaphylaxis.

Tissue Destruction Mediated by Cytotoxic Antibody

The development of antibody to antigens on the surface of a host's own cells can lead to tissue destruction. Injury results from the binding of complement-fixing antibodies to host tissue cells. Activation of the complement cascade ensues, leading to the release of inflammatory mediators and the accumulation of inflammatory cells (Chap. 13). Secretion of lysosomal enzymes and toxic oxygen radicals by inflammatory cells and direct cytolysis of target cells through complement action all contribute to tissue destruction. An example of a disease resulting from the development of antibody directed toward self tissues is Goodpasture's syndrome. It is the result of antibody binding to basement membrane antigens in the lungs and kidneys, causing the explosive onset of hemorrhagic pneumonitis and rapidly progressive glomerulonephritis. Antibodies to cytoplasmic membranes and to other tissue antigen are frequently encountered in certain rheumatologic disorders, particularly systemic lupus erythematosus. Autoimmune hemolytic anemia occurs following the deposition of complement-fixing

antibodies plus C3 cleavage products on circulating red blood cells. Consequently, the cells are rapidly destroyed either by macrophages in the reticuloendothelial system (particularly in the spleen) or, less commonly, by intravascular hemolysis mediated by complement. Also common in individuals with systemic lupus erythematosus is idiopathic thrombocytopenic purpura, in which antibody develops against platelet antigens. This leads to thrombocytopenia following the rapid clearance of platelets by the reticuloendothelial system.

Immune Complex Diseases

Formation or deposition of immune complexes in tissues produces an inflammatory response called, the Arthus reaction. The response to immune complexes is characterized by accumulation of polymorphonuclear leukocytes, followed by influx of macrophages. Several mechanisms exist by which immune complexes initiate this reaction. Antibody molecules cluster tightly as they bind to the array of antigenic determinants. If the immunoglobulin (Ig) is either IgM or IgG (subclass 1, 2, or 3), a site between the second and third common region of the H chain binds and activates C1. As a result of this activation, C4 and C2 are cleaved and activated, and C3 is cleaved into two fragments (see Chap. 14). The large fragment (C3b) binds to the immune complex; the small fragment (C3a) diffuses into the surrounding tissues to enhance vascular permeability, to contract venular smooth muscle, and to degranulate mast cells and basophils. Cleavage of C5 releases the extremely potent inflammatory polypeptide C5a. Diffusion of C5a from the site of immunologic reactions establishes a gradient of this chemoattractant, with the highest concentrations being at the site of the immune complex itself. Polymorphonuclear leukocytes and macrophages detect this gradient and migrate to the site of immune complex deposition. Upon arrival at the immune complex, Fc and

C3b receptors on the phagocyte bind to the immune complexes and phagocytosis ensues (Fig. 18-3).

If a small amount of immune complex is deposited, the material can be taken up and digested by phagocytes without causing tissue destruction. If the deposition of immune complex is extensive or if significant portions of the immune complexes are lodged in vessel walls, permanent tissue destruction can ensue. Tissue damage is most severe when the complexes are not easily internalized because the cells attracted to them secrete lysosomal enzymes and toxic oxygen radicals externally. Death of phagocytes and of tissue cells releases additional proteolytic enzymes. The released enzymes then cleave additional C5, thereby producing more C5a and thus increasing further the accumulation of phagocytes at the inflammatory site.

Antigen–antibody complexes formed in the circulation may evoke serum sickness reactions. Serum sickness occurs after an individual develops complement-fixing antibody to a circulating antigen (Fig. 18-4). As antibody is produced, antigen–antibody complexes form in the circulation. During the early phase of antibody synthesis, the amount of antibody available for binding is small and the complexes are formed in a setting of great antigen excess. Such complexes are not pathogenic. As antibody production increases, usually within several days of antigenic exposure,

the immune complexes become larger and the ratio of antigen to antibody decreases. Complexes in slight antigen excess tend to be deposited in the walls of small blood vessels, where they initiate inflammatory lesions of vasculitis. If the reaction is severe it leads to hemmorrhagic necrosis and local tissue destruction. Palpable purpuric skin lesions (leukocytoclastic vasculitis), arthritis, glomerulitis (Fig. 18-5), and fever, as well as depressed serum complement levels, are common clinical manifestations. In the absence of additional antigen, and as antibody production continues, the remaining circulating immune complexes increase in size. They are then rapidly cleared by the reticuloendothelial system and the illness resolves. Examples of both types of inflammation initiated by immune complexes are common in rheumatic diseases such as rheumatoid arthritis and systemic lupus erythematosus (SLE).

Certain animals develop immune complex diseases that bear striking similarities to human and experimental serum sickness, and to human SLE. By studying such animals, a great deal has been learned about the immunopathology of human autoimmune diseases. Studies on two strains of mice from New Zealand—one white (NZW) and one black (NZB)—have been particularly informative. The NZB mice develop severe autoimmune hemolytic disease, and the F₁ hybrid (NZB/W) is subject to very severe autoimmune diseases. At about 4 months of age, antibodies to nuclear protein appear in the circulation. Glomerulonephritis, autoimmune hemolytic anemia, and generalized vasculitis develop and the animals die young.

Disease in NZB/W mice illustrates the additive contribution of genetic, immunologic, and infectious factors to the production of a spontaneous immune complex disease. Genetically, multiple autosomal genes appear to be involved. Immunologically, the hybrids have heightened B cell responsiveness and depressed T cell suppressor function. The mice produce excessive levels of antibody to many experimental antigens and have a depressed ability to reject skin grafts. The disease itself appears when the NZB/W mice begin to produce unusually large amounts of antibody to Gross leukemia virus, an agent that infects many normal mouse strains but frequently produces no disease. In the NZB/W mice, however, antigen–antibody complexes develop and deposit in a granular "lumpy-bumpy" immunofluorescent pattern in the renal glomerulus, leading to immune complex-induced glomerulonephritis (Fig. 18-5). The formation of circulating immune complexes causes a systemic vasculitis and a fall in the serum complement titer.

The requirement for a genetic predisposition along with exposure to the appropriate infectious agent or other environmental factor is also likely in human SLE. Studies have demonstrated that humans with SLE usually have the same set of DR (class II) histocompatibility antigens (Chap. 17). Moreover, abnormal T cell suppressor function is seen not only in patients with this disease, but also in family members. Numerous other human illnesses appear to occur secondary to the development of either circulating or localized immune complexes. These include

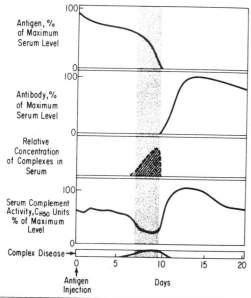

Figure 18-4. Relationships between the serologic events and experimental immune complex disease. Injected antigen is equilibrated with body fluid and undergoes gradual catabolic attrition until day 6, when antibody production begins. Rapid attrition of antigen characteristic of immune elimination occurs between day 7 and day 10. During this interval, antibody production results in the presence of identifiable antigen–antibody complexes in the serum, depression of complement activity, and tissue changes in kidney glomeruli and blood vessel walls characteristic of immune complex disease. Free antibody appears after antigen elimination and is followed by a rebound in serum complement activity.

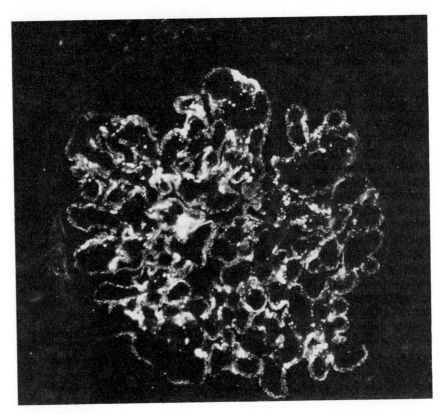

Figure 18-5. Immune complex glomerulone-phritis from a human with serum sickness following treatment with horse antithymocyte globulin. Staining is with fluoresceinated antihorse gammaglobulin. (*Photograph generously provided by Dr. Alfred P. Sanfilippo, Duke University Medical Center.*)

certain adverse reactions to drugs, hypersensitivity pneumonitis, and reactions to viruses such as hepatitis B virus.

Inflammatory Reactions Initiated by Mononuclear Leukocytes

Lymphocyte-initiated inflammatory reactions are termed delayed hypersensitivity because maximal inflammatory cell accumulation does not appear for 48 to 72 hours after secondary antigenic exposure. For example, if an individual previously sensitized to the tubercle bacillus is injected with antigen from this organism, a delayed type of inflammatory response ensues. The foreign material is encountered first by macrophages, which partially digest and alter the antigen. This altered form of antigen is recognized by small lymphocytes that contain specific surface receptors for the antigen. Exposure to the antigen initiates synthesis and release of lymphokines from lymphocytes and monokines from macrophages which diffuse from the inflammatory site to areas of the vessel wall closest to the immunologic event. Increased vascular permeability follows chemotactic gradients that attract macrophages and other lymphocytes, resulting in inflammatory cell accumulation. Small numbers of polymorphonuclear leukocytes precede the mononuclear cell influx, but the number of these cells is far less than seen in inflammatory response mediated by immune complexes. Lymphokines also activate the macrophages, which become more metabolically active, develop higher levels of hydrolytic enzymes, and

are better able to bind to and destroy tumor cells and many intracellular parasites.

Lymphocytes at the inflammatory site undergo blastogenesis and release lymphokines that recruit other nonsensitized lymphocytes, thus expanding the clones of cells capable of recognizing and responding to the specific antigen. If successful in complete destruction of the inciting organism and its antigens, the inflammatory response resolves and produces no tissue necrosis. However, the inflammatory process continues if the antigen is large in quantity or difficult to digest—as are the waxes of the tubercle bacillus—or if the antigen is an organism resistant to phagocytic destruction. New cells arrive to replace the dying cells already present at the site, resulting in the release of proteolytic enzymes and toxic oxygen radicals. Lesions typical of delayed hypersensitivity reaction or of the related Jones–Mote–Dines cutaneous basophilic anaphylaxis are seen in mycobacterial and fungal diseases, sarcoidosis, and a number of rheumatologic disorders, including polymyositis and the granulomatous vasculitides.

FURTHER READING

Books and Reviews

Cochrane C, Koffler D: Immune complex diseases in experimental animals and man. Adv Immunol 16:186, 1973

Samter M (ed): Immunological Diseases. Boston, Little, Brown, 3rd ed., 1978

Snyderman, R: Mechanisms of inflammation and tissue destruction in the rheumatic diseases. In Wyngaarden JB, Smith LH (eds): Cecil Textbook of Medicine. Philadelphia, Saunders, 1982

Theofilipoulos AW, Dixon FJ: The biology and detection of immune complexes. Adv Immunol 22:89, 1979

Selected Papers

Dixon FJ, Feldman JP, Vazquez JJ: Experimental glomerulonephritis: The pathogenesis of a laboratory model resembling the spectrum of human glomerulonephritis. J Exp Med 113:899, 1961

Dixon FJ, Feldman JP, Vazquex JJ, et al.: Pathogenesis of serum sickness. Arch Pathol 65:18, 1958

Lohr KM, Snyderman R: Disorders of human leukocyte chemotaxis. Clin Immunol Rev 1:67, 1981

Snyderman R, Goetzl EJ: Molecular and cellular mechanisms of leukocyte chemotaxis. Science 213:830, 1981

Steinberg AD, Huston DP, Taurog JD, et al.: The cellular and genetic basis of murine lupus. Immunol Rev 55:121, 1981

Talal N: Disordered immunologic regulation and autoimmunity birth defects. In Bergsma D, Goldstein AL (eds): Neurochemical and Immunologic Components in Schizophrenia. New York, Alan R Liss, 1978

Theofilipoulos AN, Dixon FJ: Etiopathogenesis of murine SLE. Immunol Rev 55:179, 1981

Waldman TA, et al.: NIH conference. Disorders of suppressor immunoregulatory cells in the pathogenesis of immunodeficiency and auto-immunity. Annu Intern Med 88:226, 1978

CHAPTER 19

Immune Responses to Infection

The study of immune reactions to pathogenic organisms has provided much of the framework for our current understanding of basic immunology. Indeed, many of the fundamental concepts of immunology were developed by microbiologists studying natural host reactions to specific infectious organisms (Table 19-1). Although recent progress has been based largely on systems using unusual or artificial antigens, manipulated hosts, and purely in vitro studies, the concepts and possible mechanisms obtained from such models should ultimately relate to natural phenomena. In clinical practice, the interface between immunology and microbiology becomes most evident with the infected patient; indeed, a major component of clinical immunology involves serologic evaluation of the infected patient. Responsiveness to infectious organisms is clearly the major function of host immunity. This chapter examines immune reactions to infection, considering the influence of both the host and offending organism.

Infectious Organisms: Basis for Host Immune Diversity

The complexity of the immune system in humans reflects the variety of infectious organisms encountered. In evolutionary terms, the development of new pathogenic organ-

TABLE 19-1. EARLY MICROBIOLOGISTS–IMMUNOLOGISTS AND THEIR CONTRIBUTIONS

Scientist	Discovery (Year)	Organism/Model
Jenner	Vaccination, cross-reactivity (1796)	Cowpox, smallpox
Pasteur	Attenuated vaccine (1878)	Cholera
Metchnikoff	Phagocytosis (1882–84)	Bacteria, fungi
Richet & Hericourt	Serologic immunity (1888)	*Staphylococcus*
Charrin & Roger	Serum agglutinins (1889)	*Pseudomonas aeruginosa*
Pfeiffer	Serologic specificity (1889)	Cholera
Behring & Kitasato	Serologic antitoxins, neutralization (1890)	*Clostridium tetani*
Pfeiffer & Isaeff	Bacteriolysis (1894)	Cholera
Bordet	Complement (1895)	Cholera
Kraus	Serum precipitins (1897)	Culture supernates
Bordet & Gengou	Complement fixation (1901)	Cholera
Neufeld	Quellung reaction (1902)	*Streptococcus pneumoniae*
Richet & Porter	Anaphylaxis (1902)	Actinaria
Wright & Douglas	Opsonization (1903)	Bacteria

isms has presumably been associated with concomitant changes in the defense mechanisms of otherwise susceptible hosts to ensure survival of the species. Thus one might expect that humans, as members of the most mobile and ubiquitous of species, would have highly sophisticated means of immunologic defense for protection against the wide range of pathogens to which they are exposed. The diversity of immunologic protection is reflected at several distinct levels: (1) mechanistic variety; (2) broad specificity of recognition; and (3) polymorphism of histocompatibility antigen expression.

Mechanistic Variety

The human host has developed a variety of immunologic mechanisms for coping with the many different characteristics of infectious organisms. Some of the major factors that influence the effectiveness of specific immune mechanisms include the modes of invasion and intrahost spread, and structural features that protect the organism from certain types of immune-mediated damage. Therefore, depending on the organism and the route of invasion optimal immune responsiveness might be systemic or local (or both), and might involve cellular or humoral components or a combination of both.

Range of Specificity

The human host has the capacity to recognize 10^6 to 10^8 different antigenic determinants (epitopes). It would seem that this high degree of diversity is essential when considering the tremendous number of microbes that may be encountered and the variety of antigens expressed by each organism. Driven by this diversity of potentially destructive organisms and their multitudinous (toxic) products, the individual has evolved two interlocking, interdependent (but also to some extent independent) systems of defense. The humoral immune defense is given its versatility by recombination within the DNA coding for the immunoglobulin hypervariable regions and by the ability of hypervariable sequences to join onto different constant regions (Chap. 12). Variability in binding is thus combined with variability

of function. Cellular immune responses are also richly variable although this variability is achieved by processes as yet unknown. Because the processes are different and the epitopes recognised are often different, the net result is the potential ability of a host to respond to virtually any immunogenic combination of antigenic determinants expressed by an organism.

Histocompatibility Polymorphism

The discovery of histocompatibility (transplantation) antigens in mice and humans during the 1940s and 1950s raised fundamental questions about their true function in natural immunity, as tissue transplantation cannot be considered a frequent natural event. Moreover, the high degree of polymorphism associated with these antigens was difficult to comprehend until it was shown that cell-mediated lysis of virus-infected cells is restricted on the basis of both viral and histocompatibility antigen expression by the sensitizing cell (Table 19-2). Thus, polymorphism may provide a selective advantage for host cytotoxic responses against certain viruses due to differences in immunogenicity of particular viral-histocompatibility antigen combinations.

Mechanisms of Host Defense

Natural vs. Adaptive Immunity to Infection

Natural immunity may be considered as immune responsiveness present prior to initial exposure with a specific organism. Adaptive immunity, on the other hand, relates to the host immune response against a specific organism following infection. However, "natural" immunity is actually an adaptive response by the host following exposure to commensal and other nonpathogenic organisms or substances that share antigens with the pathogen in question. Thus, "germ-free" animals that have not been exposed to microbial or other specific cross-reacting antigens show essentially no natural or preformed immunity, and quickly

TABLE 19-2. MAJOR HISTOCOMPATIBILITY COMPLEX RESTRICTION OF T-KILLER CELLS OF STRAIN H-2K^{k}D^k INFECTED BY VIRUS A

Virus	Target Cell Strain	Killing
A	K^{k}D^k	+
A	K^{k}D^d	+
A	K^{d}D^k	+
A	K^{d}D^d	−
B	K^{k}D^k	−
B	K^{d}D^d	−

succumb to sepsis when taken from their sterile environment. The importance of induced immunity as provided by vaccination cannot be overemphasized in terms of clinical impact. Immunization against formerly epidemic diseases such as smallpox and diphtheria has greatly reduced worldwide mortality and morbidity from these and other infectious diseases.

Nonimmunologic Factors

A major factor in the prevention of infection is the elimination of pathogens at the site of initial contact, usually the skin, respiratory, or gastrointestinal tracts. In addition, secretory and excretory functions play a crucial role in providing chemical and biologic protection:

1. Physical defenses. These include surface epithelial cell turnover (desquamation); secretion of cleansing solutions such as saliva, tears, and mucus; action of surface cilia; and excretion of organisms with bodily wastes.
2. Chemical defenses. These include a large variety of agents that may act as disinfectants or by inhibiting microbial attachment. Lysozyme (muramidase) in tears, saliva, and nasal secretions, lactic acid in sweat, lactoperoxidases in saliva, unsaturated fatty acids in sebum, and gastric acids in the gastrointestinal tract can all directly inhibit bacterial growth. Alternatively, glycolipids and fibronectin can act to inhibit bacterial attachment to certain epithelial surfaces.
3. Biological defenses. These may be considered as the nonimmunologic interference of a specific infection by a previous or simultaneous infection. As a result of derepression, host cells infected by virus can produce interferon, which inhibits virus replication in neighboring cells. Likewise, normally nonpathogenic bacteria can inhibit the growth of pathogenic strains. For example, fatty acids produced by enteric anaerobes in normal gut flora can inhibit the growth of some species of *Salmonella* and *Shigella*, whereas *Staphylococcus aureus* growth on the skin is inhibited by *Propionibacterium acnes*.

The importance of these nonimmune types of defense is evident from the increased susceptibility to infection seen when they are altered or impaired. Overwhelming infection is a major life-threatening risk to burn victims who have lost the protective effect of their skin. Damage to the cilia in respiratory epithelium from smoking or other insults increases the risk of pneumonia. Urinary retention as a result of obstructive urologic disorders is associated with increased urinary tract infections. Neutralization of gastric acids can also increase the risk of enteric infections.

Immune Defense at the Epithelial Surface

In addition to the many nonimmunologic factors that are protective at the point of initial contact with pathogens, immunologic inhibition of attachment or binding by organisms to the host provides another effective defense. When carefully studied, most microorganisms are found to have binding sites for attachment to specific cells. Perhaps the best characterized situation involves the influenza virus where the viral hemagglutinin (binding site) reacts with a neuraminic acid receptor on the respiratory epithelial cell surface.

A major immune defense at the body surface is mediated by the predominant immunoglobulin (Ig) found in secretions, preformed secretory IgA; it is therefore, the primary means of humoral protection at the site of microorganism attachment. IgA activity is directed primarily at neutralizing viruses and preventing bacterial attachment to epithelial surfaces. Because bound secretory IgA can activate complement by the alternative pathway, opsonization and, in some cases, actual lysis may be induced. The IgA molecule is protected against some proteolytic enzymes by the secretory piece, which is synthesized in epithelial cells. Two subclasses of IgA (IgA1 and IgA2) in secretions have different susceptibilities to specific proteolytic products of various bacteria, thus providing a greater potential for protection. Other immunoglobulins can also supplement secretory IgA activity; for example, individuals with selective IgA deficiencies appear to be adequately protected by an increased level of IgM in their secretions.

Immune Defenses Against Extracellular Organisms

The wide range of potentially invasive microorganisms, many of which produce toxic products, is matched by the diversity of host immune responsiveness. The principal reactions that appear effective against extracellular organisms are phagocytosis, neutralization, and humorally mediated lysis.

Phagocytosis

Phagocytosis is the principal mechanism for destruction of extracellular bacterial pathogens as well as several viral and fungal organisms. Both polymorphonuclear leukocytes and macrophages have phagocytic capacity, but differ in several respects. The polymorphonuclear leukocyte is an end-stage myeloid cell (e.g., neutrophil) that has a functional circulating half-life of only 6 to 7 hours, followed by a variable, but brief (less than 1 week) survival in tissues.

Macrophages, which are derived from premonocytic cells in the bone marrow, have a circulating half-life of less than a day as monocytes, and then pass into various tissues where they mature into functional histiocytes. Macrophages may also proliferate locally, such as Kupffer's cells in the liver and alveolar macrophages in the lung. The predominant function of neutrophils is phagocytosis of extracellular bacteria. Macrophages, however, perform a variety of complex functions in addition to simple phagocytosis (see Chap. 16), and may be stimulated to produce various factors that can effectively regulate the functions of other cells (e.g., T lymphocytes). From the standpoint of phagocytosis alone, however, there are many similarities between neutrophil and macrophage function. In both cases, four stages of activity can be identified (Fig. 19-1).

Chemotaxis. The attraction of phagocytic cells to the site of infection can be mediated by several different mechanisms. Products from the coagulation, fibrinolytic, and kinin pathways can act as chemotactic factors primarily for neutrophils. Bacterial production of *N*-formylmethionine in initiating protein synthesis attracts both neutrophils and macrophages directly, as do some other formylated oligopeptides. The production of chemotactic factors from the activation of the classic and alternative complement pathways (C5a, C3a, and C3bB) probably provides the predominant chemotactic influence (see Chaps. 14 and 18). Finally, various cell products are chemotactic: eosinophil chemotactic factor (ECF-A) from mast cells can attract eosinophils; histamine from mast cells and basophils can attract neutrophils; and lymphokines from activated T cells can attract macrophages. Although not truly chemotactic themselves, some factors that influence vascular permeability can result in increased adherence of phagocytes to vascular endothelial cells.

Attachment. As most pathogenic bacteria have surface components that resist phagocyte attachment, several opsonins (from the Greek word *opsonein*, to prepare food for) exist that bind to organisms and increase their phagocytic susceptibility. Both macrophages and neutrophils have surface receptors for the Fc portion of IgG1, IgG3, or C3b. Thus, opsonization may be accomplished by three different types of humoral responses: (1) when sufficient IgG1 or IgG3 is produced and binds to the organisms; (2) when the antibody response itself is insufficient for opsonization, but can fix enough complement on the surface of the organism; and (3) when the alternative complement pathway is activated directly by components of the organism (e.g., bacterial endotoxin and fungal polysaccharides).

Ingestion. After attachment, pseudopodia are extended by the phagocyte to envelope the organism in a sequential interaction of surface receptors and binding (opsonin) sites. They then fuse to form a phagocytic vacuole lined by cell surface membrane. This phagosome subsequently fuses with primary lysosomes to destroy the organism. The motion of pseudopodia for engulfment, phagosome migra-

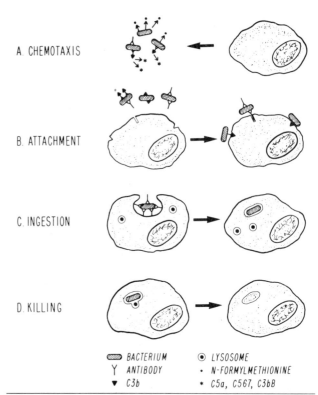

A. CHEMOTAXIS

B. ATTACHMENT

C. INGESTION

D. KILLING

▭▭▭ BACTERIUM   ⊙ LYSOSOME
Υ ANTIBODY   · N-FORMYLMETHIONINE
▼ C3b   * C5a, C567, C3bB

Figure 19-1. Phagocytosis. Chemotaxis of phagocytes is mediated by bacterial products and complement activation. Attachment of phagocytes is enhanced by opsonins bound to the microorganism, such as the Fc portion of immunoglobulin and C3b. Engulfment of the microorganism results in the formation of a phagosome, which then fuses with cytoplasmic lysosomes, destroying the microbe by oxygen-dependent or oxygen-independent reactions (see text).

tion, and lysosome fusion appears to be mediated by a complex cytoplasmic microfilament and microtubular system.

Killing. Both oxygen-dependent and oxygen-independent systems can be involved in the destruction of phagocytized organisms:

1. Oxygen-dependent mechanisms. A significant increase in oxygen consumption is seen within minutes of contact with an opsonized target. The activation of reduced nicotinamide adenine dinucleotide phosphate (NADPH) oxidase catalyzes the reduction of oxygen to superoxide (O_2-). Two superoxide anion radicals may then combine in a reaction catalyzed by superoxide dismutase to form hydrogen peroxide (H_2O_2). The most important function of H_2O_2 in microbial killing appears to be in the oxidation of halide ions (predominantly chloride) to form microbicidal halide radicals. This reaction is catalyzed by myeloperoxidase (MPO), the major component of the azurophilic granules of neutrophils. Hydrogen peroxide at high concen-

trations can also kill susceptible microorganisms directly, and can interact with superoxide anion radicals to form toxic hydroxyl radicals.

2. Oxygen-independent mechanisms. Other microbicidal products produced by phagocytes are not dependent on oxidative mechanisms. These include enzymes such as lysozyme, phospholipase A2, serine esterases, lipases, and acid phosphatase. Substances such as histones, cationic proteins, and lactoferrin also can exert antimicrobial activity under anaerobic conditions.

The importance of phagocytosis for defense against many organisms is reflected by the increased susceptibility to infection, especially by the pyogenic bacteria, in patients with disorders of phagocyte function. These disorders may represent primary or secondary impairment of phagocyte activity. Primary disorders include chronic granulomatous disease (CGD), MPO deficiency, Chédiak–Higashi syndrome, and glucose-6-phosphate dehydrogenase (G-6-PD) deficiency (see Chap. 22). The most serious of these primary phagocyte disorders is chronic granulomatous disease, where susceptibility to bacteria and fungi of normally low virulence is seen within the first years of life. Secondary disorders include side effects of immunosuppressive drugs and corticosteroids on phagocyte function, and deficiencies of complement or immunoglobulin.

Neutralization

Neutralization represents a humoral response that inhibits the infectivity of microorganisms and blocks the action of their toxic products. It is an important mechanism in host defense against extracellular viruses, many motile bacteria, and microbial toxins.

Virus Neutralization. Upon initial contact, viruses are susceptible to neutralization by antibodies present in the extracellular spaces of the host. Later, when infection has taken hold, antiviral antibodies may be important in preventing extracellular dissemination (Fig. 19-2) through several mechanisms. In the most direct case, antibody bound to viral coat proteins results in a critical conformational change preventing cellular adsorption, penetration, or uncoating. Antibody may also aggregate virus particles, or may be bound in such quantity on each virus as to sterically inhibit infectivity. Finally, complement fixation can augment antibody-mediated neutralization, presumably by amplifying the steric or conformational changes caused by antibody.

Bacterial Neutralization. Although phagocytosis is the primary defense against most extracellular bacteria, neutralizing antibody can prevent initial attachment (see above). It can also impede dissemination of some strains by blocking certain binding sites, such as on pili, or by causing physical aggregation to reduce infectivity. Moreover, antibodies to flagella or pili can inhibit motility and spread.

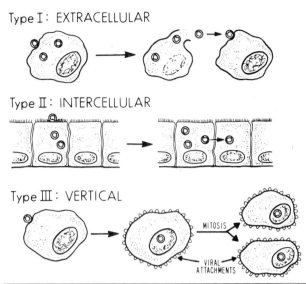

Figure 19-2. Viral dissemination. The three types of viral spread are not mutually exclusive; for example, many viruses that spread by intercellular means (type II) also shed infective virions (type I).

Neutralization of Toxins. Toxic products of various bacterial and fungal organisms can induce severe damage to the affected host. Most exotoxins and many endotoxins can stimulate neutralizing antitoxins. The apparent mechanism of many antitoxins is to inhibit sterically the active region of the toxin by binding to an adjacent part of the molecule. Direct competition between antibody and toxin binding site does not seem necessary for most antitoxin activity.

Lysis

In some cases antibody can directly mediate complement dependent lysis of a microbial organism. Various gram-negative bacteria (*Neisseria* species, *Haemophilus influenza*, and some enterobacteria) can be directly lysed in vitro by IgM and, to a lesser extent, by IgG. In addition, certain viruses (influenza and herpes simplex), and perhaps protozoa (*Trypanosoma*) are also susceptible to in vitro lysis. The endotoxin in gram-negative bacteria may induce lysis indirectly by activating complement via the alternative pathway. The potential in vivo importance of lysis is suggested by the increased incidence of *Neisseria* infections in patients with deficiencies in those terminal complement components that are not potent opsonins, but are critical for lysis.

Immune Defenses Against Intracellular Organisms

Several immune mechanisms can fight off organisms that have gained entrance to cells by attacking the infected cell directly or by providing a potentially lethal cellular environment.

Complement-dependent Cytolysis. Infected cells expressing viral antigens can by lysed in vitro by immune sera and complement activation via the alternative pathway. Paradoxically, complement activation via the classic pathway seems less effective (except where noted) in lysing infected targets.

Lymphocyte-mediated Cytotoxicity. Cell-mediated cytotoxicity clearly represents a most important defense against intracellular infections. The complex interaction of T cytotoxic cells, macrophages, and T regulatory cells is presented in Chapter 15. Cytotoxic T cells specifically sensitized to viral antigens associated with histocompatibility antigens (major histocompatibility complex restriction) appear soon after infection, and have been demonstrated in vivo as well as in vitro (Table 19-2). Other lymphoid cells that have cytotoxic activity against virus-infected targets include natural killer (NK) cells and killer (K) cells that mediate antibody-dependent cellular cytotoxicity (ADCC). Although NK cells lyse a variety of targets in vitro, virus-infected cells appear to be especially susceptible. In addition, interferon enhances NK activity, suggesting an in vivo role for NK cells against virus infection. Likewise, K cells have been shown to effect lysis of virus-infected target cells with very small amounts of antibody, but no definitive in vivo role for these cells has been established.

Activated Macrophages. In situations where enhanced phagocytosis appears critical for handling organisms that can persist in host cells (including the macrophage itself), sensitized T lymphocytes can augment cytotoxic activity (Fig. 19-3). Macrophages can ingest organisms such as the tubercle bacillus that, if virulent enough to grow, will subsequently infect other macrophages attracted to the reaction site. Macrophage activation by lymphokines produced by sensitized T cells leads to increased microbial destruction due to release of various proteinases, which may also result in host tissue damage.

Immune Defense Against Protozoa and Metazoa

Protozoa and metazoa are complex in terms of structure, antigenic diversity, and life cycle, thus presenting the host immune system with multiple moving targets. In addition, these parasites often have the ability to evade or modify the immune system (as discussed below), providing an even more difficult task for the infected host. It should, therefore, not be surprising that the outcome of a majority of parasitic infestations is concomitant immunity, whereby the parasite can provoke a response that is able to prevent overwhelming sepsis or reinfection, but cannot completely

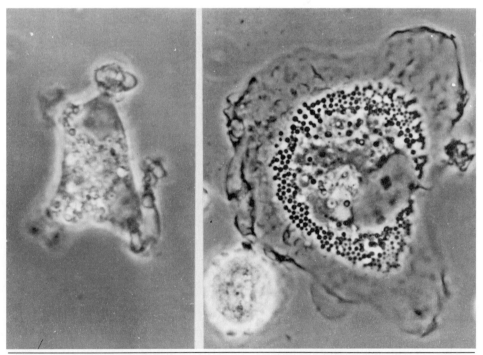

Figure 19-3. Phase photomicrograph of macrophage activation. **Left**. Unactivated macrophage with well-developed pseudopodia but few lysosomes. **Right**. Macrophage activated by incubation for 1 day with immune lymphocytes and antigen. The cell is much enlarged and shows increased spreading, prominent nucleoli, and a large ring of phase-dense lysosomes. Both × 1560. *(Courtesy of Dr. D.O. Adams.)*

eliminate the organism. Because of this homeostatic situation, the morbidity caused worldwide by parasitic disease is enormous. Parasites can be primarily extracellular or intracellular, and in many cases have stages where both types of infection are present simultaneously. A concerted humoral and cellular immune response is usually required for protection, which is often only partial.

Humoral Mechanisms. The best demonstration of a direct effect of antibody against protozoa is provided by the observation that individuals with IgA deficiency are more susceptible to giardiasis. Antibody responses, in themselves insufficient to control an infection, may inhibit certain parasitic functions. *Schistosomula* appear more susceptible to lysis with complement activated by the classic rather than the alternative pathway, and complement-dependent antibody lysis may play a partially protective role against toxoplasma and trypanosomes. Some metazoan forms, notably *Schistosomula,* can activate complement directly by the alternative pathway, resulting in lysis. Antibodies to parasites may also act via neutralization: IgG antibodies to the surface coat glycoproteins of the merozoite in malarial infections can block its invasion of red cells; specific antibodies in *Taenia* infections can inactivate the migrating oncosphere. Antibodies can also function as opsonins in parasite diseases; IgG antibodies in toxoplasmosis can enhance macrophage mediated phagocytosis, and IgG antibodies to *Schistosomula* attract eosinophils via surface receptors for IgG. In some situations, antibody production may be nonbeneficial or even detrimental. The high immunoglobulin levels associated with malaria and trypanosomiasis represent polyclonal B cell activation with only a small percent of the antibody being reactive with the organism. Delayed antibody production to oncospheres in *Taenia* infections may actually coat the organism and protect it from host cellular immune attack.

Cellular Mechanisms. Cellular immune responses, especially to intracellular parasites, have been demonstrated in many parasitic diseases. Delayed type hypersensitivity skin responses can be elicited from patients infected with some protozoa (e.g., *Entamoeba, Leishmania, Toxoplasma,* and *Trypanosoma*) or metazoa (*Echinococcus, Schistosoma,* and *Trichinella*). Cytotoxic T cells directed against host cells infected with intracellular parasites have been detected in models for several protozoan infections, as well as against the extracellular schistosome, which can itself acquire host histocompatibility antigens. From in vitro studies, ADCC reactions mediated by K cells have been thought to play a role in the destruction of erythrocytes infected with plasmodia. *Pneumocystis carinii* and *Toxaplasma gondii* infection in AIDS suggests a major contribution against protozoa is normally supplied by T cells. Activated macrophages also appear to play an important role in many protozoan infections, though most of the evidence for this is indirect.

Allergic Mechanisms. An important feature of immune defense to parasitic infections involves IgE-mediated re-

sponses. Elevated IgE levels are commonly associated with metazoan infections, as is eosinophelia. The stimulus for the relative increase in IgE production upon metazoan infection involves regulation of B cell activity at the helper T cell level. A major effect of IgE is to activate (via Fc binding) a variety of cells that can release pharmacologically active products. In response to infection, mast cells and basophils are the most important target cells, and their products act to directly impair infestation by damaging the parasite or inhibiting attachment. In addition, other effector cells, especially eosinophils, are attracted by release of chemotactic factors (ECF-A) from mast cells and basophils. Antiparasite IgE antibodies may also act in a complement-independent manner to destroy parasites by an ADCC response mediated by macrophages. Attracted eosinophils can also effect an ADCC response directed by IgG antibodies. Eosinophils have been shown to damage larval forms in schistosomal and trichinella infestations, apparently by deposition of their granule products—especially those containing major basic protein—directly on the parasite.

Characteristics of Microorganisms that Influence Host Responsiveness

Infections remain a common occurrence in humans despite the development of antibiotics and widespread immunization. Organisms have apparently adapted well to many host defenses, and have even developed mechanisms that circumvent or directly interfere with potentially detrimental responses. Several characteristics have been examined that influence host responsiveness in a manner usually beneficial to the offending organism.

Antigen Expression
Clearly, the nature and expression of antigens by an organism are critical factors in determining the character and effectiveness of host immune responses. Some infectious organisms have evolved means of thwarting host immunity at the afferent, efferent, or central levels. Such mechanisms include avoiding detection by expressing nonimmunogenic antigens; by shedding, modifying, or altering antigen expression; by interfering with host antigen processing; and even by mimicking or assuming host antigens.

Diminished Immunogenicity. The slow viruses such as Kuru and Creutzfeldt–Jacob may persist and spread unchecked because they are nonimmunogenic and do not stimulate an immune host response. This characteristic also makes the immunologic detection and study of such organisms extremely difficult. Antigens expressed on the host cell surface by some organisms (e.g., measles, mumps, influenza A, and herpesviruses) can be modulated off by the presence of an immune (usually antibody) response, whereas others (e.g., *Leishmania*) may induce the loss of associated surface histocompatibility antigens, thus impeding cytotoxic T cell responses.

Cross-Reactivity. Organisms may attempt to reduce their immunogenicity by mimicking normal host antigens, as host responses to self components are normally suppressed. For example, the hyaluronic acid in the capsule of group A streptococci is identical to that in human joint fluid, and streptococcal antigens on the cell membrane and in the cell wall associated with the M protein cross react with antigenic determinants on the mycardial sarcolemma. A similar effect occurs when an organism is masked by acquisition of host antigens. Shistosomes have been shown to coat their surface with host blood group and histocompatibility antigens within days after infection, but the added protection this imparts is unclear. In many cases, expression of antigens cross-reactive with host components results in the development of autoimmune disease, such as seen between *Klebsiella pneumoniae* and HLA-B27 in patients with ankylosing spondylitis (Chap. 17).

Induced Unresponsiveness. A common microbial mechanism for inhibiting host responsiveness involves the elaboration of large amounts of antigen, which can often overload the host and cause anergy. Nonspecific cellular unresponsiveness can result from several types of fungal, viral, protozoan, metazoan, and severe bacterial infections. Mechanistically, the effect of a high antigen burden on the host is to cause interference with virtually all limbs of the immune system, and in a simplistic sense, competitively exhaust the capacity for normal responsiveness. A variety of nonimmune effects on the host from the heavy antigen (organism) burden can also have significant secondary inhibitory effects on immunity, which can contribute to the anergic state. A more specific form of host immune inhibition can result from the release of soluble antigens by organisms, often leading to specific tolerance induction. Such tolerance can be the result of effector cell blockade, T helper cell inhibition, or T suppressor cell activation.

Numerous viruses can induce unresponsiveness by mechanisms only partially understood. Measles virus can cause lympholysis. Other viruses remain quiescent in T lymphocytes but replicate and cause cytolysis when the T cell is activated. Cytomegalovirus (CMV), which is especially likely to colonize the immunocompromised host, itself appears to contribute to the immunosuppression. The most potent aquired unresponsive state is in acquired immune deficiency syndrome (AIDS). At the time of this writing, AIDS remains an incurable, highly lethal syndrome of unclear etiology, although epidemiologic studies have indicated that it is transmissible by parenteral contact, especially among certain cohorts, such as promiscuous homosexuals, intravenous drug users, and patients receiving frequent blood transfusions (e.g., hemophiliacs). To date, casual contact with AIDS victims has not been shown to provide risk of disease transmission. Other preliminary studies have suggested that AIDS victims suffer a

number of immunologic abnormalities such as depressed T helper cell levels, reduced specific cytotoxic T cell responsiveness to alloantigens and mitogens, elevated levels of alpha-interferon, beta-2 microglobulin, and thymosin, and atypical lymph node morphology. These and other indirect data have led to an assumption by many that AIDS is caused by a viral agent such as a variant form of human T cell leukemia virus (HTLV). However, other transmissible viral infections (e.g., herpes, hepatitis, Epstein-Barr virus, CMV) are more common to the population at high risk for AIDS so it remains to be seen whether these or other organisms are associated with AIDS as either true etiologic agents, as additional risk factors for potential victims, or simply as manifestations of severe immunosuppression in AIDS victims.

Variation. Perhaps the most dramatic influence of microbial antigenicity on host immune responsiveness is that of antigenic variation. In the most extreme case, that of infection with *Trypanosoma brucei,* there is a sequential change of the glycoprotein coat antigens that coincides with the IgM peak antibody response to each microbe population (Fig. 19-4). Because this intravascular parasite is relatively susceptible to antibody mediated lysis, the persistance of infection is facilitated by the replacement with an antigenically new population before the organism can be completely eliminated. This cyclic change in coat antigens is controlled by the trypanosome genome, which codes for each variant separately.

A similar type of antigenic variation is also seen with *Borrelia recurrentis* ("relapsing fever"). Many viruses and bacteria show a type of antigenic variation that is more subtle and occurs over a much longer period of time. This type of variation more accurately represents antigenic drift (e.g., influenza B). In other cases an intermediate degree of variation truly representing new genetic variants is seen (antigenic shift). With influenza A in humans, a complex series of antigenic changes have taken place over years, as evidenced by the diverse specificity of serum samples obtained from different generations.

Mode of Infection

Organisms that enter the body in an unusual fashion often bypass the defenses encountered at the normal site of entry. Organisms transmitted by penetrating wounds, insect vectors, iatrogenic inoculation, or through maternal–fetal transfer often have this type of advantage. Certain organisms evade host responses by becoming physically inaccessible, as with intracellular viral infections where spread is not dependent on an extracellular phase (Fig. 19-2), or where the extracellular phase involves shedding in secretions or excretions (e.g., herpes simplex in saliva and polyoma in urine). Microbial growth may also occur in areas that are poorly accessible to the immune system, as in the lumen of sebaceous, salivary, or mammary glands, or in the renal tubules. Obligate anaerobes such as *C. tetani* or *C. perfringens* thrive unchecked in necrotic tissues. Com-

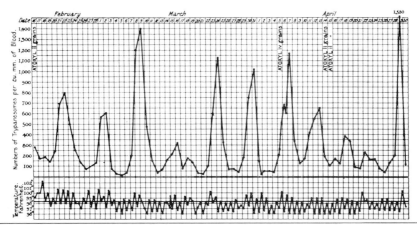

Figure 19-4. Cyclic reappearace of trypanosomes in the peripheral blood of a patient with African sleeping sickness. Each peak corresponds to the appearance of a new antigenic variant. A number of variants are present in the infecting inoculum, and specific antibody probably plays a passive selective role in the appearance of new variants. *(From Ross and Thomas: Proc R Soc Lond [Biol] 82:411, 1910.)*

mensal bacteria and some cestodes that reside within the lumen of the intestinal tract are also well protected from host defenses, as are certain fungi (dermatophytes) and viruses (warts) that primarily infect the corneal layer of the skin.

Structural characteristics of some organisms can also impede the effectiveness of host reactions. Specialized features for epithelial penetration (e.g., cercarial stage of schistosomes) and attachment (e.g., scolex of cestodes) can physically combat host attempts at expulsion of these metazoa. Organisms susceptible to host immune reactions can also be protected by the formation of cystic capsules around them, either within host cells (*Toxoplasma*) or in extracellular sites (*Echinoccus*). In a manner analogous to antigen variation, the ability of some organisms to express different antigens during their life cycle may also help in the evasion of host responses. Many metazoa and some protozoa have antigenically distinct determinants at different stages during their development.

Inhibitory Products

Many organisms produce products that can directly or indirectly interfere with host immunity. The exotoxin produced by cholera is associated with a reduced number of circulating T cells. *Trichinella* and trypanosomes may also produce lymphocytotoxic toxins. The endotoxin (lipopolysaccharide) of gram-negative bacteria also appears to produce a nonspecific inhibition of T cells, partly as a result of increased prostaglandin production from macrophages stimulated by this product. The effects of exotoxins are, however, very complex and may actually stimulate some components of the immune system. Thus, a nontoxic derivative of endotoxin is of possible value in enhancing immunity, e.g., against cancer.

Microbial enzymatic products may also directly suppress host reactions. Some bacteria (e.g., *Neisseria, Strepto-*

coccus) can produce proteases that cleave immunoglobulins. Complement components may be directly inactivated by enzymatic products of strains such as *Pseudomonas aeruginosa*. Hemolysins released by streptococci and staphylococci can damage phagocytic lysosomes and, in effect, promote the phagocyte to kill itself by autodigestion, whereas some *Shigella* have been found to kill phagocytes after their ingestion. Organisms such as *Mycobacterium tuberculosis* and *Toxoplasma gondii* may inhibit lysosomal fusion with the phagosome in macrophages, thus permitting their growth and eventual destruction of the phagocyte. Some viruses are also known to infect lymphocytes with a resulting impairment of function. The preferential infection of B cells by Epstein–Barr virus usually stimulates a cytotoxic T cell response, resulting in a B cell lymphopenia.

Finally, some microbial structural products may act to inhibit immune effector function. Most notable are a variety of bacterial antigens that resist phagocytosis or killing. The polysaccharide capsule of some bacteria and fungi inhibits phagocytosis and digestion. The cell wall structures of other organisms can also resist phagocytic killing and digestion by mechanical means. Numerous specific surface component antigens are effective in resisting phagocytosis or digestion (e.g., O antigen and K antigen of *Escherichia coli*, Vi antigens of *Salmonella*, and M substance of *Streptococcus*).

Host Characteristics that Influence Immune Responses to Infection

Age

Primary changes that occur in the immune system during fetal, neonatal, and adult development can predispose the host to different types of infection. During gestation, circulating T and B cells develop by 3 to 4 months, but well-defined lymph node development is not seen until the fifth month. Placental transfer of maternal IgG plays an important role in providing passive fetal immunity, but some synthesis of IgM and IgG can be elicited by the second and third trimesters. During fetal development, susceptibility is generally limited to those organisms that are blood-borne and can be transmitted by the maternal circulation (e.g., CMV, rubella, toxoplasma, and syphilis). At birth, neonatal levels of maternally transferred IgG are relatively high but slowly begin to decrease, whereas IgM levels are low but slowly rise. Milk and colostrum provide immunoglobulins (especially IgA) and leukocytes; indeed, the number of leukocytes in a daily feeding of milk is roughly equivalent to those already in the neonate. Epidemiologic studies suggest that this transferred immunity is important because breast fed infants are less susceptible to viral and bacterial infections. Cell-mediated immunity is relatively intact in the neonate, with an apparent elevation of T suppressor cell activity for 12 to 48 months that, to some extent, may suppress B cell activation. This combination of factors results in the increased susceptibility seen during the first months of life to gram-negative bacteria because of relatively low IgM levels. From about 1 to 12 months of age, the relative decrease in IgG results in an increased risk from bacteria that are more susceptible to IgG-mediated immunity (*Staphylococcus, Streptococcus,* and *Haemophilus*). Following thymic involution, regulatory T cell function and activation gradually declines, which probably contributes to the increased susceptibility to infections seen in the elderly.

Genetic Factors

The most overt examples of genetic effects on the ability for the host to combat infections are seen with the primary immunodeficiency diseases (Chap. 22). The nature of an immunologic defect is often first diagnosed by the pattern of predisposing infections. Defects in humoral immunity frequently correlate with increased infections by extracellular microbes such as pyogenic bacteria, whereas depressed cellular immunity is usually associated with increased viral and fungal (intracellular) infections.

The inheritance of genes that control the type and magnitude of immune responses to specific antigens (immune response genes) (see Chap. 18) appear to influence host responsiveness to infection. For example, the level of host immunity appears related to the clinical nature of hepatitis B virus infection. Vaccinated hosts are generally protected, whereas nonvaccinated hosts develop different types of disease depending in part upon their ability to respond. Strong responders generally experience acute (transient) hepatitis, and weak responders have a mild indolent infection with little liver damage. Intermediate responders, unable to eliminate infection, develop persistent and often severe immune-mediated hepatic damage (chronic active hepatitis). Similarly, weak responders to *Mycobacterium leprae* develop lepromatous leprosy, as op-posed to the tuberculoid leprosy seen in strong responders. Certain HLA phenotypes have been associated with these as well as other infection-related, autoaggressive diseases, such as type 1 diabetes, ankylosing spondylitis, and anterior uveitis, which suggests a genetic influence at the level of immune recognition or responsiveness (Chap. 18).

Genetic factors may also indirectly affect host susceptibility to certain organisms. Individuals who do not have the Duffy a and b erythrocyte antigens (Fy a$^-$b$^-$) are resistant to *Plasmodium vivax* infection, correlating with the inability of the organism to bind and penetrate these red cells. Likewise, individuals who have sickle cell trait (HbS) have increased resistance to *Plasmodium falciparum*. In the homozygous condition (S-S), this is probably as a result of decreased survival of infected red cells that become sickled by oxygen metabolism of the organism within the red cell. In the heterozygous condition (S-A), growth is arrested at the large ring stage.

Nutritional Factors

Vitamin A is important for the function of lysosomal membranes: deficiencies affect normal phagocytic function and are associated with an increased incidence of infections that are normally destroyed by activated macrophages (e.g., *M. tuberculosis*). Vitamin C acts on lysosomal membranes, can potentiate neutrophil chemotaxis, and reverse suppression of T cell activity caused by influenza A virus. Deficiency states have been associated with increased bacterial infections, and high doses may reduce the severity (but probably not the incidence) of common colds. Exogenous iron is a requirement for bacterial replication and depressed host iron binding capacity is associated with increased gram-negative bacterial infections. In vitro mixed lymphocyte culture responses are reduced in the absence of zinc. In vivo, it binds to a bacteriocidal polypeptide in amniotic fluid. In African mothers with zinc deficiency, a significant increase in amniotic fluid infections and fetal death has been found. Patients with protein deficiency (kwashiorkor) have decreased circulating T cells, and depleted T-dependent areas in their lymph nodes and spleen. Clinically, they show depressed delayed-type hypersensitivity reactions (anergy), produce antibodies with lower affinity, and have an increased susceptibility to infection.

Immunopathologic Reactions to Infection

An infection can have three different pathologic effects: (1) no host damage; (2) damage caused by the organism; and (3) damage caused by the host response to the organism (i.e., immunopathologic). Normally, an inflammatory response benefits the host by resolving the infection. However, in some cases, host damage from inflammation ex-

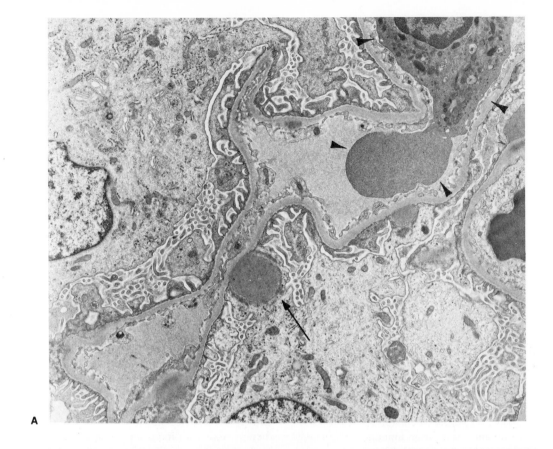

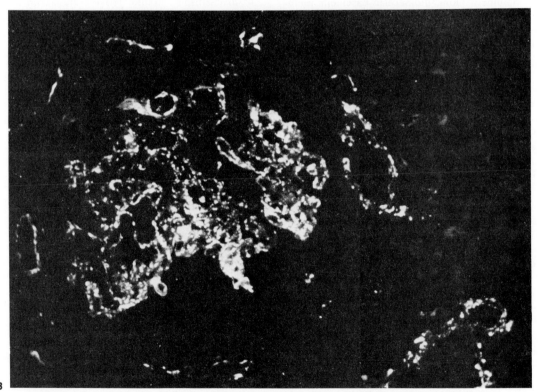

Figure 19-5. Postinfectious glomerulonephritis. **A**. The typical appearance of electron dense, subepithelial immune complex deposit (arrow) associated with postinfectious glomerulonephritis (× 5000). An attracted neutrophil (arrowhead) is seen nearby in the glomerular capillary loop. **B**. Immunofluorescent staining shows abundant C3 deposits in these complexes (× 250).

ceeds damage caused by the organism, as with hepatitis B virus, tuberculosis, and leprosy. In these diseases, the organism itself causes minimal damage, whereas damage from the host response can lead to significant morbidity and death.

Essentially, all four major types of immunopathologic reactions (Chap. 18) may take place in response to infection. Type I reactions (IgE mediated), which occur in response to many metazoan infections, Type IV reactions (cell mediated), which are seen with intracellular microbes, and some aspects of Type II reactions (cytotoxic–antibody) have been discussed earlier in this chapter. However, infections may also induce other kinds of immunopathologic responses, such as Type III (immune complex) reactions, as well as various autoimmune diseases.

Immune Complex Diseases

Immune complexes commonly form during infections, but infrequently cause significant damage. Nevertheless, certain organisms can induce severe immune complex diseases, such as streptococcal infection in children, which can result in acute glomerulonephritis. Typically, 1 to 2 weeks after a pharyngeal or skin infection with a nephritogenic strain of *Streptococcus pyogenes,* hematuria and proteinuria develop concomitantly with large immune complexes that bind complement along glomerular basement membranes (Fig. 19-5). These lesions will usually heal with minimal impairment, but acute complications may result. Other organisms associated with immune complex glomerulonephritis include bacteria (e.g., staphylococci and enterococci), fungi (e.g., candida), protozoa (e.g., malaria), and viruses (e.g., hepatitis B virus [HBV]). Other clinical manifestations of immune complex diseases associated with infection include vasculitis (e.g., polyarteritis nodosa with HBV), dermatitis (e.g., erythema nodosa with leprosy), and pneumonitis (e.g., allergic alveolitis with actinomyces). Although identification of specific microbial antigens within the immune complexes is often difficult, the characteristic temporal, clinical, and morphologic features associated with the formation of these complexes provides presumptive evidence of their etiology (Chap. 18).

Autoimmune Diseases

Infectious organisms are often involved in the pathogenesis of autoimmune diseases. Some microbes or their products may induce nonspecific immune reactions: B cells undergo polyclonal (nonspecific) activation in response to certain organisms (e.g., mycoplasma, trypanosoma, plasmodia, measles, and Epstein–Barr virus) and bacterial products (e.g., lipopolysaccharide, *S. aureus* protein A, and purified protein derivative of tuberculin), which may detract from the specific response required for protection. Animal models have implicated infectious organisms for autoimmune diseases such as systemic lupus erythematosis and rheumatoid arthritis. Several rheumatic, endocrinologic, and neurologic diseases with autoimmune features have

been associated with microorganisms, especially in individuals with certain HLA phenotypes. As mentioned previously, some infections cause secondary disease as a result of autodestructive immune mechanisms (e.g., hepatitis and leprosy). Other organisms appear to stimulate host responses to cross-reactive self antigens (e.g., rheumatic fever with streptococcal antigens). Some autoimmune hemolytic anemias are also associated with specific infectious organisms: a cold agglutinin to the I blood group can be seen in patients infected with *Mycoplasma pneumoniae.* Interestingly, some microbial enzymes can convert the A_1 blood group to B, resulting in a hemolytic anemia from circulating anti-B isohemagglutinins.

As discussed in this chapter, infectious organisms are largely responsible for the complex nature of host immunity, and over the years have provided practical models for studying immunologic mechanisms. Despite tremendous advances in our understanding of host–microbial interactions, however, infection still remains the greatest cause of human morbidity and mortality. Our ability to intervene and amplify host immunity, especially by immunization, has already had a dramatic effect in reducing and even eradicating previously widespread diseases such as smallpox, but such developments are only a beginning. Continued progress over the next few years in prevention and therapy, based on increased understanding of immunologic mechanisms, should provide the basis for competing with infectious organisms on a more equitable basis.

FURTHER READING

Books and Reviews

Amos DB, Janicki B, Schwarts R (eds): Interface between Infection and Immunity. New York, Academic Press, 1979

Bloom BR: Games parasites play: How parasites evade immune surveillance. Nature 279:21, 1979

Cohen S, Warren KS (eds): Immunology of Parasitic Infections. Oxford, Blackwell, 1982

Damian RT: Molecular mimicry in biological adaptation. In Nickol BB (ed): Host–Parasite Interface: At Population, Individual and Molecular Levels. New York, Academic Press, 1978

Dick G (ed): Immunological Aspects of Infectious Disease. Baltimore, University Park, 1979

Dixon FJ (ed): Viral immunopathology. Springer Semin Immunopathol 2:233, 1979

Gadebusch HH (ed): Phagocytosis and Cellular Immunity. Boca Raton, Fla, CRC Press, 1979

Klebanoff SJ, Clark RA: The Neutrophil: Function and Clinical Disorders. Amsterdam, North Holland, 1978

Kraus R (ed): Immunopathology of parasitic diseases. Springer Semin Immunopathol 2:355, 1980

McGregor DD, Kostiala AAI: Role of lymphocytes in cellular resistance to infection. Contemp Top Immunobiol 5:237, 1976

Mims CA: The Pathogenesis of Infectious Disease. New York, Academic Press, 1982

Moller G (ed): The immune response to infectious diseases. Transplant Rev 19:1, 1974

Moller G (ed): MHC restriction of anti-viral immunity. Immunol Rev 58:1, 1981

Moeller G (ed): Immunoparasitology. Immunol Rev 61:1, 1982

Notkins AL: Viral infections: Mechanisms of immunologic defense and injury. Hosp Practice 9:65, 1974

Theofilopoulos AN, Dixon FJ: Autoimmune diseases. Immunopathology and etiopathogenesis. Am J Pathol 108:321, 1982

Zinkernagel RM, Doherty PC: MHC-restricted cytotoxic T cells: Studies on the biological role of polymorphic major transplantation antigen determining T cell restriction-specificity, function and responsiveness. Adv Immunol 27:51, 1979

Selected Papers

Ambrusco DR, et al.: Lactoferrin enhances hydroxyl radical production by human neutrophils, neutrophil particulate fractions, and an enzymatic generating system. J Clin Invest 67:352, 1981

Babior BM: Oxygen-dependent microbial killing by phagocytes. N Engl J Med 298:659, 721, 1978

Benacerraf B: Role of MHC gene products in immune regulation. Science 212:1229, 1981

Densen P, et al.: Phagocyte strategy vs microbial tactics. Rev Infect Dis 2:817, 1981

DeVries RRP, et al.: HLA-linked control of susceptibility to tuberculoid leprosy and association with HLA-DR types. Tissue Antigens 16:294, 1980

Diamond RD, et al.: Damage to pseudohyphal forms of *Candida albicans* by neutrophils in the absence of serum in vitro. J Clin Invest 61:349, 1978

Feizi T, Taylor-Robinson D, Shields MD, et al.: Production of cold agglutinins in rabbits immunized with human erythrocytes treated with Mycoplasma pneumoniae. Nature 222:1253, 1969

Kornfeld SJ, et al.: Secretory immunity and the bacterial IgA proteases. Rev Infect Dis 3:521, 1981

Kress Y, et al.: Resistance of *Trypanosoma cruzi* to killing by macrophages. Nature 257:394, 1975

Masson SJ, et al.: The Duffy blood group determinants: Their role in the susceptibility of human and animal erythrocytes to *P. knowlesi* malaria. Br J Haematol 36:327, 1977

Ogra SS, Ogra PL: Immunologic aspects of human colostrum and milk. J Pediatr 92:546, 1978

Petersen BH, et al.: *Neisseria meningitidis* and *Neisseria gonorrhoeae* bacteremia associated with C6, C7, or C8 deficiency. Ann Intern Med 90:917, 1979

Ratzan KR: The role of surface factors in the pathogenesis of infection. In Weinstein L, Fields BN (eds): Seminars in Infectious Disease. New York, Stratton, 1979, vol 2, p 145

Root RK, et al.: The microbicidal mechanisms of human neutrophils and eosinophils. Rev Infect Dis 3:565, 1981

Ross R, Thomas D: A case of sleeping sickness studied by precise enumerative methods: Regular, periodical increase of the parasites disclosed. Proc R Soc London (Biol) 82:411, 1910

Sissons JGP, Oldstone MBA: Killing of virus-infected cells by cytotoxic lymphocytes. J Infect Dis 142:114, 1980

Vickerman K: Antigenic variation in trypanosomes. Nature 273:613, 1978

Wright LW, Levy NL: Generation on infected fibroblasts of human T and non-T lymphocytes with specific cytotoxicity, influenced by histocompatibility, against measles virus-infected cells. J Immunol 122:2379, 1979

Zinkernagel RM, Doherty PC: H-2 compatibility requirement for T-cell-mediated lysis of target cells infected with lymphocytic choriomeningitis virus. Different cytotoxic T-cell specificities are associated with structures coded for in H-2K or H-2D. J Exp Med 141:1427, 1975

Zinkernagel RM, Callahan GN, Klein J, et al.: Cytotoxic T cells learn specificity for self H-2 during differentiation in the thymus. Nature 271:251, 1978

CHAPTER 20

Immunity to Tumors and Pregnancy

Introduction

A tumor or neoplasm can be defined for purposes of this chapter as the uncontrolled replication of some cells of a tissue. Neoplasms are usually classified as malignant or benign. Malignant tumors frequently spread to distant sites called metastases. Even tumors that do not metastasize, such as gliomas in the brain and basal cell carcinomas of skin, invade and destroy surrounding tissues. Benign tumors, by definition do not metastasize and they usually remain encapsulated, so there is a clear distinction between new growth and surrounding normal tissues. The three major classes of tumors in humans are carcinoma, derived from epithelial structures; sarcoma, affecting mesothelial structures; and tumors of the reticuloendothelial system, including the lymphomas and leukemias. Carcinoma is the

most frequent malignancy in humans, but lymphoma/leukemia has been more intensively studied.

The majority of tumors in humans are of unknown etiology, but cellular oncogenes related to viral transforming elements (Chap. 62) are being demonstrated in tumors other than those of the lymphoid system, where they were first described. The induction of tumors may be attributable to carcinogens in food or in the environment (chemically induced tumors), to physical agents including ionizing radiation and ultraviolet light, to spontaneous mutations, or to viruses. A tumor may be conveniently regarded as the progeny of a cell that has escaped regulation. This is often associated with a chromosomal rearrangement. As the cells divide they may also mutate. Some tumors, especially of the lymphoid system, appear to be clonal with little variation between cells, whereas others show a diversity, either because they were polyclonal from the onset or through subsequent mutation. Teratocarcinomas show the greatest diversity, with many different tissue types present in a single tumor. Individual tumor cells can sometimes be induced to revert to their original state. For example, some provocative studies in which tumor cells having distinctive chromosomal and other markers were mixed with cells from a developing embryo and reimplanted in the uterus resulted in a chimeric fetus. Some of the cells of the fetus displayed the tumor-cell markers.

Because a tumor arises from the tissues of the individual, it is not surprising that most tumors are feebly immunogenic to the original host. There is more than a slight resemblance between immunity to tumors and immunity to parasites (Chap. 19), wherein the invader (tumor or parasite) escapes detection or destruction by any of several processes.

Evidence for immunity to human tumors is incomplete and much is inferential. Certain highly malignant types of tumor may regress spontaneously, e.g., melanoma or choriocarcinoma, and tumor-reactive antibodies can sometimes be demonstrated in convalescent serum. Cutaneous hypersensitivity to tumor extracts has frequently been demonstrated, but the interpretation of the data is complex because the tumor and the extract used for testing are frequently contaminated with bacteria or bacterial products. This chapter first reviews the nature of antigens known to be tumor specific or tumor associated, and then discusses the immune responses. Of necessity, much of the information will be from the experimental animal, and even so, there are difficulties. An autochthonous tumor (one growing in its original site) is not the same as a transplanted tumor, even when the transplant is to a member of the same inbred strain (syngeneic or genetically identical to the autochthonous host). Tumor transfer can greatly modify the biochemistry the antigenic expression and the vascularization of a tumor. Nevertheless, certain

generalizations can be made as a result of decades of careful investigation.

Tumor Antigens

Neoantigens

Antigens that stimulate an effective, tumor-specific rejection (Table 20-1) are designated tumor-specific transplantation antigens (TSTA) (Table 20-2). They most likely represent neoantigens and are encoded by altered genes (single point mutations, deletions and/or translocations, viruses). These antigens are specific for that tumor, i.e., two methylcholanthrene tumors induced in the same inbred strain or even at different sites on the same animal express unique TSTA. Therefore, despite histologic similarities, the tumors are antigenically disparate and reflect distinct genetic alterations.

Oncogenic viruses may induce the expression of several neoantigens. Some of these are attributed to viral-encoded components used either in packaging the virus or in its synthesis. Viral-encoded neoantigens may be group specific (common to all viruses of a group, e.g., herpes viruses) or type specific (e.g., herpes simplex) (Table 20-2). In contrast to the unique TSTA of chemically induced tumors, group-specific neoantigens are common to all tumors induced by viruses in the same group and express common neoantigens. Retroviruses (oncogenic RNA viruses) possessing acute transforming properties may also express neoantigens encoded by linked transforming genes (see below). These neoantigens should be common to all tumors induced by all retroviruses carrying the same transforming element (e.g., ras). Finally, retroviruses lacking a transforming element (therefore, less oncogenic) may insert next to or within a cellular oncogene (see below) and

TABLE 20-1. IMMUNITY TO METHYLCHOLANTHRENE-INDUCED TUMORS—LIVE TUMOR CELL CHALLENGE OF EITHER NON-IMMUNE MICE SYNGENEIC TO THE TUMOR CELL DONOR OR SYNGENEIC MICE PREIMMUNIZED WITH IRRADIATED TUMOR CELL VACCINES

Tumor Used for Immunization*	Lethal Dose in Nonimmune Animals (Cell No.)	Consecutive Doses Given Preimmunized Animals Resulting in No Tumor Growth (Cell No.)	Lethal Dose in Preimmunized Animals (Cell No.)	Comments
MDAD	10^3	10^3, 10^4, 10^5, 10^6, 10^7	NT	Mice preimmunized with irradiated MDAD resist a challenge of 10^7 live MDAD tumor cells, but succumb to 10^5 live MDAQ tumor cells.
MDAQ	10^4	10^4, 10^5, 10^6	10^7	

Adapted from Klein et al.: Cancer Res 20:1561, 1960.

*Tumors were induced on the right hind leg of mice with methylcholanthrene. The mice were cured by amputation at 3 months. A tumor cell suspension was used either live for challenge or x-irradiated for immunization.

TABLE 20-2. ANTIGENIC CHANGES ASSOCIATED WITH ONCOGENESIS*

Change Detected	Virally Induced Tumors	Chemically Induced Tumors	Spontaneous Tumors
Expression of neoantigens			
Tumor-specific transplantation antigens	+	+ +	±
Individually specific	±	+ +	?
Common neoantigens	+	±	?
Oncofetal/developmental antigens reexpresed	+	+	+
Decreased expression of normal surface components (e.g., histocompatibility antigens, blood group antigens)	+	+	+

*Tumor-specific transplantation antigens detected by tumor transplantation; others detected serologically.
+, occasionally detected; + +, frequently detected; ±, seldom; ?, unknown.

modify it.* If the viral promoters and transcriptional elements influence only the level of cellular oncogene expression, no new antigen should be expressed. If the cellular oncogene affected is genetically altered, a new antigen may also be expressed. Because the particular cellular oncogene affected and the mode of alteration may vary from one viral-induced tumor to another, it is unlikely that these neoantigens will be the same.

Oncofetal and Developmental Antigens

In addition to the tumor-associated antigens described above, chemically induced, virally induced, and spontaneous tumors sometimes express components identical to those detected on fetal tissue and/or tissues in intermediate stages of differentiation. Oncogenesis is characterized by relatively stable genetic alterations that result in altered gene products and/or changes in the level of gene product expression. Genes active during embryogenesis and early development may be directly affected by the transforming event or indirectly through alteration of a regulatory element. Few of the oncofetal antigens encoded by these genes are very immunogenic; none appear to act as TSTA. Many are designated antigens because they can be detected and characterized with xenogeneic antibodies raised against either tumor or fetal tissues. Some oncofetal antigens are cell associated and sometimes shed (e.g., carcinoembryonic antigen, or CEA), while others are actively secreted (e.g., α-fetoprotein and human chorionic gonadotropin, or HCG). Some oncofetal antigens are common to tumors of a given histologic type; others are common to a variety of tumors of distinct histologic type. Oncofetal antigens are often expressed at low levels on normal tissue, and thus their increased expression in tu-

mor tissue appears to reflect enhanced synthesis or decreased catabolism.

Changes in the Expression of Normal Surface Components (Antigens)

Cell surface histocompatibility antigens (Chap. 17) and/or blood group antigens (Chap. 16) are sometimes absent in tumor tissue, though they are expressed in adjacent normal tissue. In the case of blood group antigens (notably ABH), the terminal carbohydrates determining blood group specificity are missing; the core oligosaccharide structure is still expressed and may be modified by addition of more fucose or sialic acid residues. When histocompatibility antigens are lost (as detected serologically) the entire glycoprotein may be missing; sometimes this loss is accompanied by reexpression of an early differentiation antigen loosely linked to the major histocompatibility complex. Therefore, one cause for reduced expression of a normal cell surface component might be loss or reduced expression of a particular glycosyl transferase, or increased activity of fucosyl or sialyl transferases. Alternatively, the change in expression may be due to the function of different regulatory elements.

Sometimes it is possible to detect a host response associated with these antigenic changes. For example, serum anti-I/i blood group immunoglobulins may be detected in cancer patients, and their appearance correlates with the loss of blood group ABH activity (Chap. 16). These antigens are rarely strong enough to act as TSTA.

The Oncogene Hypothesis: Probable Origin of Tumor Antigens

Oncogenesis ultimately depends on a fundamental, stable change in the cellular genome. Its consequences are manifest in altered cell properties including growth patterns, responses to tissue-specific chalones (growth regulatory molecules), state of differentiation, cell metabolism, as well

---

* As the name implies, oncogenes are implicated in tumorigenesis. They appear to subserve essential but unknown functions in the normal cell and have been highly conserved in evolution. Identical or closely similar genes are found in Drosophila melanogaster.

as possible biochemical and antigenic changes discussed above.

Cell transformation may result from infection with an acute transforming retrovirus such as Rous sarcoma virus (chickens), simian sarcoma virus (woolly monkeys), or Harvey sarcoma virus (rats, mice). Each of these viruses contains a linked transforming element (src, sis, and ras, respectively). The transforming properties of these genes can be assayed biologically by transfection of murine fibroblasts with DNA restriction fragments containing the transforming genes. Presence of the gene results in transformation (oncogenesis) of these fibroblasts and the loss of contact inhibition. The transformed cells overlay each other in a haphazard fashion, whereas the original fibroblast grows to form a confluent sheet, but the cells stop dividing when they come into contact.

Alternatively, oncogenesis may result from the activation/alteration of one of several normal cellular oncogenes. These cellular genes may normally function at different stages of cell differentiation and their expression is strictly regulated. For example, the different stages of B lymphocyte differentiation may represent the phenotypic expression of distinct cellular oncogene subsets. B cell lymphomas of early, middle, or late stages of B cell differentiation may clonally express aberrant function of one of the subsets.

Activation or alteration of a cellular oncogene may occur following (1) viral activation; (2) chromosome breaks and translocation of the cellular oncogene; or (3) deletions or single point mutations. Viruses implicated in causing human cancer such as the DNA containing Epstein–Barr virus in Burkitt's lymphoma and human T cell leukemia virus (HTLV), an RNA retrovirus, may facilitate activation/alteration of normal cell oncogenes. This may be through proximal effects (insertion close to a cellular oncogene) or induction of chromosome breaks and translocation of a cellular oncogene. For Burkitt's lymphoma, translocations involving chromosomes 8 and 14, 8 and 2, and 8 and 22 have been characterized.

Altered cellular oncogenes have been reported. The most complete data (for T24 bladder carcinoma) suggest that a single base change (thymidine for guanine) results in coding for a valine residue (vs. a glycine residue) at position 12 of the first exon (Fig. 20-1), encoding a protein, p21. Protein p21 in the bladder tumor cells differs from the p21 product of the normal cell oncogene by a single amino acid. This change has a dramatic effect on the structure of p21, and is associated with cell transformation. DNA restriction fragments of the T24 altered oncogene (encoding the altered p21) transform suitable target fibroblasts; corresponding fragments from normal cells do not. Tumors from human colon and lung appear to carry very similar altered cellular oncogenes. Whereas some bladder cancers show activation/alteration of the oncogene c-Ha-ras (with sequence homology to the transforming element of Harvey sarcoma virus), others do not; the implication is that different cellular oncogenes may be activated in other bladder tumors.

The oncogene hypothesis provides a basis for understanding the antigenic changes that accompany transformation. It predicts what types of antigens might be detected and, hopefully, whether these altered components, or neoantigens, might function as TSTA.

Tumor Immunity

General Observations

It is easier to demonstrate immunogens (Table 20-1), to characterize TSTA and other tumor-associated antigens, and to characterize the nature and extent of the immune response in inbred rodents than it is in humans. For human cancer, the importance of an associated immune response must be inferred. For example, there are confirmed reports of spontaneous regressions of human cancers, albeit at a very low frequency; however, a significant number of these regressions involve choriocarcinomas where reactivity against paternal histocompatibility antigens is suspected. In patients with T cell or combined immunodeficiencies (Chap. 15), there is also a higher frequency of malignancies, notably lymphosarcomas. The same pattern of increased frequency of lymphosarcomas is also observed in transplant patients receiving immunosuppressive therapy (Chap. 17). However, in none of the above situations is there an increased frequency of carcinomas other than those of the epidermis, and as carcinomas constitute almost 90 percent of all human cancers, this observation may argue against a significant antitumor immune response leading to rejection. This may be a fallacious argument, however, because many of the abnormal states of immunity favor proliferation of subsets of lymphocytes and it has been a frequent observation that viruses replicate well in dividing lymphocytes. Hyperplasia of other tissues is frequently premalignant. Other in vivo and in vitro illustrations of an immune response associated with malignancy will be reviewed in subsequent sections.

Evidence for Tumor Immune Responses In Vivo

The role of the immune system in regulating the rate of growth, the spread and the spontaneous regression of human tumors (immune surveillance) has been hotly debated ever since the earliest observations of the spread and regression of experimental and spontaneous tumors in animals. For the most concrete evidence we must refer to the experimental animal, and there is abundant information from inbred strains of rats and mice. Lewis, Foley, and others showed that if a chemically induced tumor was completely removed (by surgery or through vascular ligation), the animal became resistant to challenge with an inoculum of the same tumor (transplanted or stored frozen), but remained fully susceptible to inoculation with other tumors. Viral tumors caused by the Maloney sarcoma virus

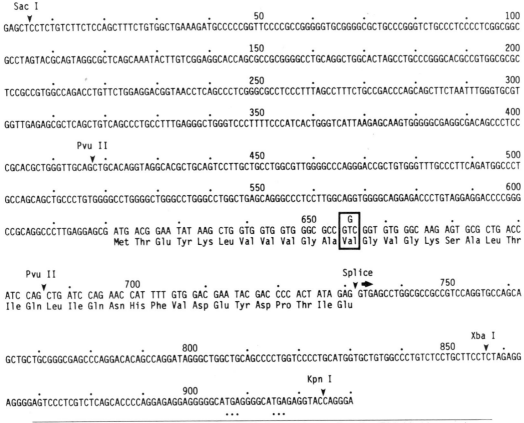

Figure 20-1. A comparison between the DNA fragment encoding the T24 bladder carcinoma oncogene (the first exon of the p21 protein) and the normal human oncogene homologue. The DNA sequence reads 5' to 3' from left to right. *Sac I, Pva II, Xba I,* and *Kpa I* indicate restriction enzyme cleavage sites (arrow). *Splice* indicates where the first exon terminates. The sequences are identical save a single base change in the codon originally coding for glycine (*GGC*) in the normal cell oncogene homologue to a codon coding for valine (*GTC*) (broad arrow). *(Adapted from Reddy et al.: Nature 300:149–152, 1982.)*

or by inoculation of newborn mice with Rous sarcoma tissue may regress spontaneously or regression can be induced. Animals bearing viral tumors are resistant to challenge with the same type of tumor. The frequency with which virally induced tumors are rejected is determined by the strain of mouse, and this argues directly in favor of genetic influences and indirectly in favor of a role for immune response genes in determining the response (Chap. 17).

Tumor-specific immunity can frequently be transferred from an immune animal to a nonimmune, syngeneic animal by the transfer of lymphocytes (adoptive transfer of immunity). Draining lymph node lymphocytes from an animal that has rejected a tumor will transfer adoptive immunity to naive, syngeneic animals. The strongest protection is given by lymphocytes from the node draining the site of the original tumor (Mitchison assay), especially if the node cells are mixed with the tumor cell inoculum (Winn assay). If these procedures are repeated using draining lymph node B cells alone or T cell subsets (Chap. 14), the optimum protection is afforded by T cells of the subset con-

ferring delayed type responses. Many manifestations of cellular immunity are lost as the tumor becomes enlarged and especially when it becomes metastatic. This is clearly reflected in skin tests with antigen and in the normal lymphocyte transfer test (NLT) of Brent and Medawar. This test, which is in many ways an in vivo correlate of the mixed lymphocyte response (MLR) (Chap. 17), is performed by injecting lymphocytes into the dermis of another individual. An inflammatory reaction develops at the injection site in 24 to 48 hours, and is accompanied by the release of angiotensin. Not only do patients with advanced disease fail to react to many environmental antigens including tuberculin, *Candida,* tricophytin, or mumps, but they frequently give markedly impaired responses to allogeneic lymphocytes in the NLT, especially if the cells transferred were from another patient with advanced malignancy. Unfortunately, lack of availability of highly purified tumor-associated antigen for skin testing, immunosuppression related to chemotherapy, dietary deficiencies, and many other variables also interfere with many of the clinical tests that have been tried; therefore, tests of tumor-associated cell-mediated responses are often determined in vitro in a more controlled environment.

In Vitro Evidence for Tumor Immunity

The response of patient lymphocytes to autochthonous tumor antigens (tumor cells or extracts) or to an allogeneic tumor of the same histologic type is compared to the response of lymphocytes from a normal individual. The response can be quantified by enumerating T lymphoblasts microscopically, by incorporation of ^{3}[H] thymidine into newly synthesized DNA or by the elaboration of several lymphokines including macrophage migration inhibitory/activating factor (MIF/MAF), T cell growth factor (TCGF, or interleukin II), and immune interferon. The proliferating cells (in mice) are largely Lyt1$^+$, whereas in human systems the cells are T3- and T4-positive. Later in the mixed cultures it is sometimes possible to detect tumor cytotoxic T cells (Lyt2, 3$^+$ in mice, T8$^+$ in humans). The appearance of cytotoxic cells is dependent on the earlier Lyt1$^+$ or T3/T4 cell response, but because of the great variability in reactivity between different subjects the reactions are difficult to interpret.

T suppressor cells (Lyt2, 3$^+$) isolated from solid tumors or malignant effusions in mice can also be detected by in vitro assays. Some of these have been shown to be tumor (antigen) specific, while others are nonspecifically suppressive. It is thought that the soluble factor produced in specific suppression contains a polypeptide encoded by the I–J subregion of the murine major histocompatibility complex (Chap. 17) and a polypeptide(s) that is antigen specific. Cells of the appropriate suppressor phenotype (T8$^+$) have been recovered from human solid tumors and malignant effusions but have been difficult to characterize functionally.

Humoral responses to tumors can often be demonstrated, but, again, interpretation is difficult. Tumor-reactive antibodies in serum and malignant effusions have been shown by immunofluorescence, immunohistochemical, and radioisotopic assays. The serum level of these tumor-reactive antibodies may change with tumor reduction/growth, i.e., increasing following cytoreductive therapy and decreasing with increased tumor growth. There may also be a reciprocal relationship between the serum level of free, tumor-reactive antibody and the level of immune complex detected. The significance of malignancy-associated immune complexes is demonstrated by increased incidence of complex-induced glomerular nephritis (about 10 percent of all cancer patients).

Controversy concerns the specificity of these antibodies. Early claims of tumor specificity were supported by the apparent specificity of (1) inhibition of cell-mediated cytotoxicity in vitro (blocking antibodies); or (2) enhancement of animal tumor growth in vivo (enchancing antibodies). Extensive, subsequent analyses have demonstrated that though the antibodies are tumor reactive, the initial hopes that they are tumor specific have not been realized. They may detect autoantigens and oncofetal anti-

gens and, not infrequently, histocompatibility antigens of the tumor. The lack of absolute tumor specificity for the majority of tumor-reactive antibodies does not rule out a significant role in control of neoplastic growth in vivo, but the clinician is rightly concerned that any induced antibody might enhance, rather than suppress, tumor growth (see later sections). In a few cases, notably murine leukemias, it is possible to induce a tumor-reactive antibody of exquisite specificity and significant protective activity. In this case, the antibody is specific for an idiotypic determinant (Chap. 12) expressed by the surface immunoglobulin of the malignant B cell, and has been used therapeutically.

Evidence for Natural Resistance to Tumor Growth and Metastasis

Though it is possible that the serum of normal subjects contains natural antitumor antibodies, their significance is not clear. There is excellent evidence for a natural cyto toxic cell, or natural killer (NK) cell subpopulation, which may be effective during the early stages of tumor growth and prevent, to a limited extent, the metastatic spread of tumor cells. Approximately 1 to 5 percent of all blood mononuclear cells in humans and rodents function as NK cells. These cells bind to and kill a variety of transformed cell types, as well as a subpopulation of immature, normal cells in the thymus and bone marrow. This latter activity suggests that their primary role may be that of a regulatory cell. The NK activity is strongly associated with cells that can be isolated because of their low buoyant density or their distinct cytoplasmic granules (Fig. 20-2). They are large lymphocytes with a greater cytoplasmic to nuclear ratio than most peripheral blood lymphocytes and contain cytoplasmic, azurophilic granules. These cytoplasmic granules, specialized lysozomes, are important in the NK cell's activity.

The NK cell is of debatable cell lineage, as several different types of cells appear to have comparable activities. The NK cells express cell surface phenotypic markers common to both T cells and monocytes, yet lack other markers typical of either. The NK cells may express Fc receptors, but the ability to bind tumor cells (targets) does not depend on the presence of Fc receptors or, therefore, on natural or specific antibodies. Human NK cells weakly bind to sheep erythrocytes and can be separated from strong E-rosetting cells readily. The binding of NK cells to tumor target cells most likely occurs through recognition of target cell glycolipids/glycoproteins and perhaps differentiation antigens.

The in vivo relevance of NK cells to control of malignancy has been demonstrated by experiments in which a cell suspension has been depleted/enriched of NK activity in vitro and transferred to NK deficient mice that are subsequently challenged with a murine leukemia or melanoma. Cloned NK cells have also been reported to afford NK deficient mice protection against NK sensitive tumors. Beige mutant mice (bg/bg) are NK deficient and succumb to murine leukemias and melanomas, whereas heterozygous

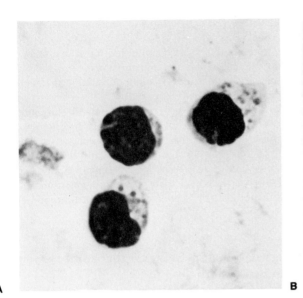 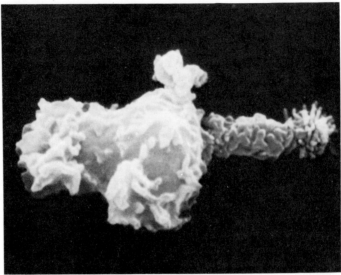

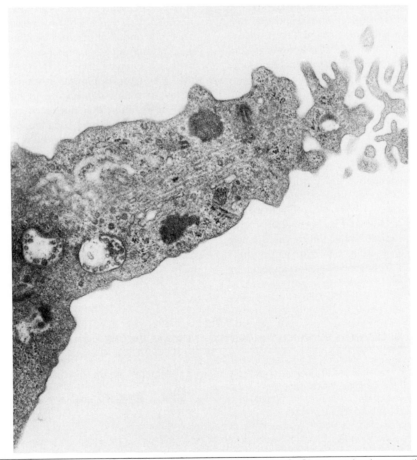

Figure 20-2. Large granular lymphocytes or natural killer cells. **A**, Giemsa stained preparation of large granular lymphocytes (LGL) from rat spleen; **B**, scanning electon microscope; and **C**, transmission electon microscope photographs of a similar cell from human peripheral blood. The elongated process shown in **B**, and sectioned in **C** can often be seen in time-lapse photographs to be in close contact with the target before cytolysis. The membrane lesions produced have been described as cylindrical structures resembling those produced by the attack complex of complement. *(From JR Dawson, H. Koren, and K. Muse, unpublished observations.)*

littermates (bg/+) are resistant. Like the beige mouse, the Chediak–Higashi syndrome in humans (Chap. 15) is characterized by a cellular deficiency with large, aberrant cytoplasmic granules. These patients are susceptible to many infections despite their ability to mount normal delayed hypersensitivity and antibody responses. No NK activity can be demonstrated in vitro with their lymphocytes and 85 percent of them enter a lymphoma-like phase before they die. While the evidence suggests an in vivo role for NK cells, again most of the activity is directed at lymphosarcomas. NK activity against freshly isolated carcinomas is variable.

Interrelationship Between Tumor-Specific Immune Responses and Nonspecific, Natural Mechanisms

Data obtained from Winn assays and adoptive transfer of immunity suggest that the T cell subpopulation responsible for delayed hypersensitivity is essential for the rejection of syngeneic tumor grafts. Rejection is thought to be mediated through activated, nonspecific macrophages and, possibly, NK cells because antigen-activated T cells elaborate (1) chemotactic factors that attract macrophage and monocytes to the site of the reaction; (2) macrophage-activating lymphokines (MIF/MAF); and (3) macrophage arming factors, possibly with tumor-specific recognition components. Resting macrophages show very low tumoricidal activity, whereas activated macrophages show greatly enhanced tumoricidal activity. Potent proteases and reactive oxygen intermediates are elaborated by activated macrophages.

The NK cells express receptors for T cell growth factor (interleukin II) and interferon, both of which may be elaborated by antigen-activated T cells. NK cell activity can be increased in vitro two- to five-fold by pretreatment with interferons. In phase I clinical trials, interferon therapy also results in a transient increase in cancer patient peripheral blood NK activity. T cell growth factor appears to stimulate NK cell activity and to promote proliferation of cells with NK activity.

Another subclass of lymphocyte, the K cell, may overlap the NK subpopulation and functions as a killer cell in conjunction with specific antibody (antibody-dependent cell-mediated cytotoxicity, or ADCC). K cells in humans express a cell surface antigen phenotype in common with NK cells, and are also identified with large granular lymphocytes (LGL), but their activity depends on expression of an Fc receptor for IgG (and perhaps IgM). This cell also responds to interferon pretreatment, giving two- to five-fold increases in activity. Activated macrophage may also function as ADCC effector cells against antibody-coated targets.

Interrelationship Between the Tumor and the Immune System

There are several reasons why tumors could escape immune elimination (Table 20-3). First, the tumor immunity that can be demonstrated early in tumor growth gradually declines with time. This suggests the induction of tolerance. Second, tumors that become clinically apparent are weakly immunogenic, implying that immunogenic tumor cells are eliminated at early stages. Alternatively, tumors may adapt to an immune response by developing mechanisms to repair damage or to circumvent immune lysis. The observation that nucleated cells are more difficult to lyse with antibody and complement than red cells suggests that membrane defects may be repaired to varying degrees by nucleated cells. Drugs that inhibit protein synthesis increase the sensitivity to lysis of some, but not all, tumors. Third, tumor cell surface components may be modulated, or down-regulated, in the presence of an immune effector cell or molecule. This has been most clearly documented with thymic leukemias in the presence of antibody. Loss of histocompatibility antigens would, of necessity, interrupt reactivity against viral tumors in which major histocompatibility complex restriction operates (Chap. 17).

TABLE 20-3. MECHANISMS BY WHICH TUMORS MIGHT ESCAPE IMMUNE ELIMINATION

1. The host becomes tolerized to his/her tumor
2. Tumors may be selected (by interaction with the immune system) that are weakly immunogenic or that have developed mechanisms for resisting immune-specified lysis
3. Immunogenic components may be modulated in the presence of an immune response
4. Suppression of the immune response:
   a. Nonspecific factors elaborated by the tumor cells
   b. Nonspecific suppressor macrophage induced by either the tumor or the immune response
   c. Nonspecific, T suppressor cells induced by the immune response
   d. Specific, T suppressor cells
   e. Therapy-related immunosuppression
5. Immune stimulation or enhancement of tumor growth by products of the immune response
6. An imbalance between the mass of tumor cells and specific and nonspecific components of the immune system

Suppression of the immune response may be induced by several mechanisms:

1. Tumor cells may suppress immunity by synthesis and release of nonspecific immunoregulatory proteins and polypeptides, prostaglandins, and corticosteroids.
2. Macrophages recovered from tumor infiltrates may nonspecifically suppress NK and T cell proliferative responses, both specific and nonspecific.
3. T suppressor cells have been demonstrated in some animal tumor inflammatory cell infiltrates.
4. Antineoplastic chemotherapy or radiation therapy is strongly suppressive of the immune system and of NK activity.

Antibodies and/or blocking factors may result in increased growth of the tumor. Blocking factors may combine with antigen and interfere with effector function; under certain conditions they may also down-regulate the immune response. In animals, enhancing antibodies may directly stimulate tumor growth or inhibit the immune system. As with ADCC, very small amounts of antibody are required. Finally, just as a patient may succumb to overwhelming infection in the face of demonstrable immunity to a pathogen, the mass of a tumor may also exceed the capacity of the immune system to bring about its rejection.

Immunotherapy

Active Specific Immunotherapy

Active immunization with tumor cell vaccines has been attempted but has had no conspicuous success in humans. Clinicians are concerned that the use of crude tumor cell extracts or irradiated tumor cells as vaccines could result in active enhancement of tumor growth, as opposed to immunoprophylaxis. If purified tumor antigens are to be utilized, suitable immunogenic components must be identified and purified, and this is not yet possible in humans. For ethical reasons, immunotherapy of humans is only attempted in the late stages of disease, i.e., at a stage when the patient is already anergic.

Passive Cellular Immunotherapy

Cloned cytotoxic T lymphocytes specific for murine tumors have been expanded in vitro and adoptively transferred to nonimmune, syngeneic animals before challenge with live tumor cells. The results of these experiments have been disappointing. In contrast, transfer of cloned NK effector cells to NK deficient mice, e.g., bg/bg mice, does protect the recipients from challenge with live, NK-sensitive tumor cell lines. The transfer of cloned NK effector cells also protects C57 B1/6 mice from developing x-irradiation-induced leukemia, if given soon after the final

x-irradiation. Transfer at later stages fails to suppress development of leukemia.

Passive Antibody Immunotherapy

Tumor-reactive antibodies have been effective in preventing the growth of feline and murine leukemias if given simultaneously with or shortly after challenging the animal with live tumor cells. However, with few exceptions, this technique is not successful against established leukemias/lymphomas and not at all for carcinomas. Tumor-reactive antibodies cannot easily be delivered to the center of a large tumor mass; much of the antibody is trapped and/or eliminated nonspecifically in liver, lungs, and kidney. In addition, the role of antibody- and complement-mediated cytolysis of tumor cells in vivo is minimal despite quite effective in vitro activity; the contribution of ADCC in vivo is unknown.

Early in this century, Ehrlich thought that antibody could be used to deliver a "magic bullet" in the form of a cytotoxic drug conjugated to an antibody. Later it was thought that if the antibody could be made radioactive (with ^{125}I or ^{131}I, for example), the radioactivity would be tumoricidal once the antibody had homed to the tumor. Unfortunately, most conjugates are also rapidly eliminated from circulation, highly radioactive antibodies lose appreciable binding capacity and the recipient may become sensitized to the foreign antibody and develop anaphylaxis.

Recent attempts have concentrated on the use of monoclonal, tumor-reactive antibodies. Their homogeneity reduces some of the problems associated with administration of complex, polyclonal reagents, but does not eliminate them. Conjugates under consideration include monoclonal antibodies conjugated to ricin, to potent chemotherapeutic drugs, or even to interferon (see below). In addition, new techniques for labeling antibodies including chelation of ^{111}In are promising in terms of preserving antibody binding capacity and in vivo localization. It remains possible that monoclonal reagents and new methods of labeling will allow the radiologist to detect the location of otherwise unsuspected metastases through the use of antibody sufficiently radioactive to be detected, though not in itself therapeutic.

Biologic Response Modifiers

A number of bacteria and bacterial cell wall products have been used to increase the activity of nonspecific effector cells such as macrophage and NK cells. Attenuated M. tuberculosis (BCG) therapy has been used in the management of melanoma. The organism is used live and is given intralesionally. Corynebacterium parvum therapy has been used in conjunction with chemotherapy. Its effect has been equivocal; its advantage is that a formalin-fixed vaccine is used. In many of the C. parvum trials, the reagent was administered intravenously; recent reports suggest that intratumor inoculation may prove more beneficial. Significant activation of macrophage and natural killer cells that correlated with a clinical response was observed. A lyophilized vaccine of heat and penicillin treated *Streptococcus pyogenes* A3, Su strain (OK432), also augments NK activity in vivo and in vitro and has significant antitumor activity.

Synthetic muramyl dipeptides and tripeptides have been shown to activate macrophage in vivo when incorporated in liposomes. A corresponding reduction in the ability of B16 murine melanoma to metastasize to the lung was also obtained by this therapy. The list of natural cell wall components and synthetic compounds under investigation is impressive. Most result in macrophage activation and augmentation of NK; some may achieve this through stimulation of interleukin II and/or interferons. Endotoxins have also been used to stimulate immunity.

Lymphokine therapy has been under active consideration since interferons were shown to augment NK activity in vitro. With the advent of recombinant sources of interferons (α, β, γ, and subtypes) it has been possible to show that the major types and subtypes differ in their ability to augment NK activity, to protect tumor cells from immune lysis, and to inhibit tumor cell proliferation. Phase I trials with systemic or intramuscular administration of interferons have shown an increase in the NK activity of peripheral blood lymphocytes, although changes in NK activity have not always correlated with clinical regression/progression.

Interleukin II has recently been produced through recombinant DNA techniques, and based on its ability to promote T cell differentiation and augment natural killing, passive immunotherapy of animal tumor models is being attempted. Previous attempts with interleukin II purified from the supernatant fluid of mitogen-stimulated T lymphocytes were equivocal, but the amount of lymphokine was undoubtedly limiting.

Based on the supposition that an ineffective antitumor response reflects an immunodeficiency (which might be imposed by either the tumor or the early immune response), attempts to reconstitute the immune system with hormones known to stimulate T cell differentiation have been contemplated. Thymosin $\alpha1$, thymopoietin fraction 5, thymic hormone factor (THF), and facteur thymique serique (FTS) have been isolated and shown to exhibit thymic hormone-like activity. Patients with immunosuppressive malignancies, such as Hodgkin's lymphoma, treated with THF have shown increases in T cell number, proliferative responses to mitogens, and cutaneous delayed hypersensitivity to recall antigens. Peripheral blood NK activity was also augmented, but the studies are not far enough along to assess the effect on tumor growth.

Other Procedures

Considerable interest was aroused by reports that the filtration of tumor-bearer plasma through a column containing formalin-fixed Cowan strain staphylococci (or protein A derived from this strain and conjugated to an inert matrix) removed immune complexes (blocking factors).

Reinfusion of the column-treated plasma correlated with rapid necrosis of mammary and other tumors. Unfortunately, this treatment often results in toxicity, presumably by leaching of toxins from staphylococci.

Lawrence and co-workers discovered that cell-free extracts of human peripheral blood leukocytes from tuberculin-positive patients could transfer tuberculin-specific reactivity to unreactive patients. Transfer factor was subsequently found to be heat-stable, dialyzable, and of low molecular weight. Transfer factor has been used to achieve conversions of bacterial and fungal reactivities. Patients with severe fungal infections, e.g., candidiasis, have benefitted from this treatment. Transfer factor therapy has been used in malignant melanoma; prolonged remissions were sometimes noted.

Human Tumor Markers

The ectopic production of certain enzymes and hormones has been utilized to confirm the presence of certain tumors and to monitor the response of these tumors to therapy. The classic example of ectopic production of a hormone by a tumor is the synthesis of HCG by choriocarcinoma and trophoblastic tumors. HCG can be detected in serum by radioimmunoassay or enzyme immunoassay (Chap. 12) and monoclonal anti-HCG antibodies have been developed for this purpose. Serum levels provide an excellent assessment of tumor response to therapy. Antibodies to α-fetoprotein can be used in a similar way to monitor the response of patients with hepatomas and germinal yolk sac tumors to therapy. The association between a rising serum α-fetoprotein level and the growth of α-fetoprotein-producing tumors is excellent. Pregnancy must be excluded, because α-fetoprotein levels rise in pregnancy and α-fetoprotein is concentrated in amniotic fluid.

Antibodies to β_2 microglobulin, a polypeptide associated with certain products of major histocompatibility gene complex (Chap. 17), have been used in radioimmunoassays and enzyme-linked immunoassays to determine serum concentrations. Changes in the serum level of β_2 microglobulin are useful in determining the prognosis of Hodgkin's disease, non-Hodgkin's lymphoma, chronic lymphocytic leukemia, and multiple myeloma. β_2 Microglobulin levels are also raised in AIDS (Chap. 15).

Antibodies to carcinoembryonic antigen (CEA) were at first thought to be specific for colon carcinoma and fetal tissue. With the advent of sensitive radioimmunoassays, this strong association failed. Several other tumors express and shed CEA, and normal tissues express a low level of CEA. Serum detection of CEA is subject to a number of variables, for example, smokers and noncancer patients with liver disease may also show elevated serum CEA. Nonetheless, for some patients with CEA-producing tumors, serum CEA monitoring has been used as a valuable adjunctive assay. Futhermore, xenogeneic anti-CEA and, recently, monoclonal anti-CEA have been radiolabeled with ^{131}I, administered intravenously and used to detect small primary and metastatic CEA-producing tumors by radiographic imaging techniques.

Figure 20-3 illustrates the radioimmunodetection of a human cancer with a radiolabeled antitumor monoclonal antibody. The monoclonal antibody, 791T/36, was produced against an osteosarcoma cell line (791T) but cross-reacts with some malignant lung, cervix, and bladder cell lines. It may identify an antigen of the oncofetal/developmental type, but nevertheless, can be used very effectively in applications of this type.

Thus the available polyclonal antibodies and monoclonal reagents can be used in assays to monitor the serum level of a variety of tumor-associated molecules, and, therefore, the response of the tumor to therapy. They can be used to detect the metastatic spread of tumor cells in lymph node biopsies by immunohistochemical techniques, for example. They have found use in the in vivo detection of occult metastases and small primary tumors by radiographic techniques. They might be used in the form of reagent conjugates for passive immunotherapy if tumor localization can be improved, and the expression of the tumor reactive epitope is limited on normal tissues.

Maternofetal Relationship

Without an efficient defense system, the individual must die; the immune system provides for specific defense and to do so it must monitor self and nonself. The embryo and fetus are nonself, but if the immune system destroyed the fetus, the species would die. Thus, to the complexities of immunoregulation is added other prerequisites; not only the fetus but also the male gamete and the fertilized ovum must also avoid immunologic attack. This section examines some of the ways in which avoidance is accomplished.

In his considerations of the relationship between mother and fetus, Medawar advanced several possibilities. These have been added to and tested by many others. The embryo and fetus might not express antigens that could be recognized as foreign. The uterus might provide an especially privileged site where immune reactions did not occur. The mother's immune system might be paralyzed for the period of gestation, either specifically by becoming tolerized, or nonspecifically. The mother could subvert the immune attack by providing blocking antibodies. The placenta might provide a barrier between the mother and the fetus. These points will now be analyzed.

Lack of Antigenicity

It is possible to examine embryonic antigens in early cleavage embryo's in vitro and later during fetal development. By direct cytotoxicity or binding assays it has been established that the early embryo does express at least some of the paternal antigens. The amount of tissue available for testing is extremely small and these studies are difficult to carry out. An ingenious innovation by Simmons and Russel was to implant the developing embryo of a mouse under the kidney capsule of an allogenic recipient. They then observed the growth of the transplant. Extraembryonic tissue—the ectoplacental cone—survived. The em-

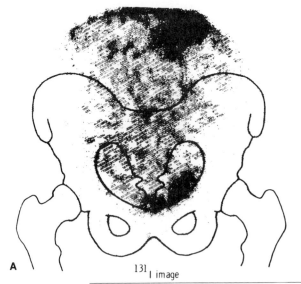

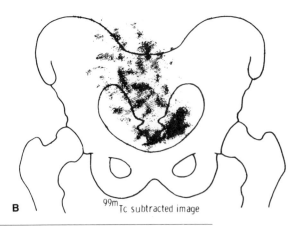

^{131}I image ^{99m}Tc subtracted image

Figure 20-3. An anterior radionuclide image of the pelvis of a colorectal cancer patient showing localization of the monoclonal, ^{135}I-labeled 791T/36 antibody. ^{99m}Tc was used to label red blood cells and ^{113m}In was used to label blood transferrin. **A** shows the radiographic analysis before image enhancement. **B** shows the analysis after image enhancement by subtraction of ^{99m}Tc and ^{113m}In backgrounds. *(From Farrands et al.: Lancet 21:399, 1982.)*

bryo itself did not, thus it was inferred that the very early embryo did carry immunogenic forms of transplantation antigens, although the trophoblast apparently did not. There have been subsequently numerous transplant studies using fetal skin and other tissues (including hearts). Though there may be some prolongation of survival, the eventual outcome is rejection, confirming that the fetus is immunogenic. In humans, the HLA antigens are well expressed by the 12th week of gestation, and in the mouse, H-2 can be detected beginning at about the 12th day. From these and other studies we can reject the hypothesis that the fetus lacks antigens, although the tissues of the early embryo are probably less immunogenic than those of the mature fetus.

The Uterus as Privileged Site

A privileged site can be defined as one in which a transplant can grow without being destroyed by an immunologic reaction. From studies with allografts and xenografts, e.g., of human tumors to the check pouch of the Syrian hamster or to a raised skin pedicle, it has been learned that a tissue will often not immunize the recipient unless it can establish connections with lymphatic vessels. The anterior chamber of the eye and the surface of the brain are, like the cheek pouch, also privileged sites. Because the decidua of the uterus was, for a long time, believed to be devoid of lymphatic vessels, it was logical to examine the possibility that the uterine cavity was a privileged site. The hypothesis was tested by Schlesinger. Fragments of tumor placed on the decidua were rejected normally. Because tumor might be more immunogenic than normal tissues, other cell types have also been grafted to the uterus. The most ingenious experiments were those of Billingham, who cut a longitudinal slit in one horn of the bifid uterus of a rat and inserted a graft made by everting a cylinder of rat tail skin. Not ony was the tube of skin rejected, the draining lymph nodes also became hypertrophic. Other studies have also shown hyperplasia of the regional nodes in pregnancy. Thus the fetus is recognized and the uterine cavity is not a privileged site.

Impairment of Maternal Immune Responses

This remains one of the more complex aspects of the immune response to the fetus. There are subtle changes in immune responsiveness during pregnancy. Isoagglutinin levels are unchanged in the human. Humans and mice both can make antibodies to paternal antigens. However, grafts of tumor or of skin may persist longer than they would on the nonpregnant animal, and secondary antibody responses may be reduced. Infections are more frequent during pregnancy, especially those with cytomegalovirus, suggesting an impairment of the effector cells of cellular immunity. It has been mentioned above that the lymph nodes draining the uterus may become enlarged in pregnancy as they are in graft rejection. In pregnancy, T cells from the nodes can produce migration inhibitory factor, but do not generate cytotoxic effectors again supporting

the concept of selective immunologic unresponses. Pharmacologic doses of steroid hormones have been reported to depress T cell responses, but this is disputed and neither estrogen nor progesterone appear to prolong skin graft survival. The protein hormones, especially HCG can decrease spleen and lymph node weights and depress immune responses. The mixed lymphocyte response (MLR) in humans and the graft-vs.-host response in experimental animals is depressed by HCG, whereas phytohemagglutin (PHA) responses are impaired in a dose-related manner. HCG is placentally derived; thus, the most profound effects on the immune system may be caused by placental metabolites. The α_2 macroglobulin levels also rise to a variable extent in pregnancy, and fast or activated α_2 macroglobulin is known to be an effective down-regulator of cellular immune responses. It is possible that the large number of histocytic cells in the placenta are responsible for the activation of the α_2 macroglobulin.

The Placenta as Immunologic Barrier

Although there has been widespread support for the idea that the placenta, by separating maternal and fetal circulations, acts as a barrier, there was for many years a controversy as to how the barrier functioned. Kirby and others believed that an inert fiberinoid substance constituted a barrier. More recently, Faulk has shown that a most important factor is the absence of HLA-A and HLA-B antigens from trophoblastic cells. Thus the placenta does function as a barrier, but it is antigenically inert rather than being coated with an insulating substance. Trophoblast cells not infrequently pass into the maternal circulation. Lymphocytes and granulocytes may also enter the maternal circulation during parturition and may contribute to the development of anti-HLA, anti-Lewis, and other antibodies by the mother.

There is still some uncertainty about the antigenicity of nontrophoblastic cells in the placenta. Immunoglobulins including antibodies to HLA are recovered commercially from human placentas. Palm proposed that graft-vs.-host reactions occurring in the placentas of rats were protective. The maternal cells that might otherwise pass into the fetus and cause injury were by this hypothesis immobilized by the graft-vs.-host reaction. In support of this suggestion are the observations that: (1) the placentas of hybrid rodents are heavier and more densely cellular than the placentas of inbred strains; and (2) male rats died from a runting syndrome when the parents were compatible for Rt-1, the major histocompatibility complex of the rat, but incompatible for minor antigens. The mother would of course lack the male-associated antigen of the male offspring and could respond to it.

Protective Antibodies

Antibodies can depress cellular responses. This is called immunologic enhancement in tumor host relationships, and blocking in most other situations. Barrett and Breyere found that female mice mated to a male of another strain would form enhancing antibodies as well as antibodies to H-2, the amount of antibody formed being a function of the number of pregnancies. Rocklyn has demonstrated the formation of blocking antibodies by healthy gravid humans and found these antibodies to be absent from the serum of women with a history of repeated abortion. Buckley and her colleagues used the MLR-blocking antibodies present in a multiparous woman to suppress graft-vs.-host reactions that might otherwise have developed when her bone marrow was transplanted to her immunodeficient child. Not all pregnancy sera have this effect on the MLR or on graft-vs.-host reactions.

Immunization and Abortion

Spontaneous abortion is quite frequent in first pregnancies. Some women however repeatedly abort and may lack the protection to the fetus afforded by blocking antibodies. Skin grafts from the father or a third party donor have been reported to reduce this tendency. Beer, Faulk, and their colleagues have immunized women with histories of repeated abortion with injections of lymphocytes and have reported the subsequent delivery of healthy infants at term. Although it might appear that immunization could lead to adverse reactions against the fetus, these have not been reported. Care is taken of course to avoid sensitization against ABH or Rh antigens. Anti-HLA antibodies reactive with the infant probably do not cross the placenta, and there are no known adverse effects to the fetus from maternal immunity to HLA.

FURTHER READING

GENERAL

Beer AE, Billingham RE: Immunobiology of mammalian reproduction. Adv Immunol 14:1, 1971

Benacerraf B, Unanue ER: Tumor immunology. In Benacerraf B. Unanue ER (eds): Textbook of Immunology. Baltimore, Williams & Wilkins, 1979, pp 196–217

Hokama Y, Nakamura RM: Immunology of cancer. In Hokama Y, Nakamura RM (eds): Immunology and Immunopathology. Boston, Little, Brown, 1982, pp 415–456

Sell S: Immunology, Immunopathology and Immunity, 3 ed. Hagerstown, Harper & Row, 1980

THE EXPRESSION OF TUMOR ANTIGENS:
THE ONCOGENE HYPOTHESIS

Cooper GM: Cellular transforming genes. Science 218:801, 1982

Muschel RJ, Khoury G, Lebowitz P, et al.: The human c-ras[H]
oncogene: A mutation in normal and neoplastic tissue from the same patient. Science 219:853, 1983

Popovic M, Sarin PS, Robert-Gurroff M, et al.: Isolation and transmission of human retrovirus (human T-cell leukemia virus). Science 219:856, 1983

Premkumar Reddy E, Reynolds RK, Santos E, et al.: A point mutation is responsible for the acquisition of transforming properties by the T24 human bladder carcinoma oncogene. Nature 300:149, 1982

Tabin CJ, Bradley SM, Bargmann CI, et al.: Mechanism of activation of a human oncogene. Nature 300:143, 1982

TUMOR IMMUNITY

Foley EJ: Antigenic properties of methylcholanthrene-induced tumors in mice of strain of origin. Cancer Res 13:853, 1953

Herberman RB (ed): Natural Cell-Mediated Immunity Against Tumors. New York, Academic Press, 1980

Kaliss N: Immunological enhancement of tumor homografts in mice: A review. Cancer Res 18:992, 1958

Kawase I, Urdal DL, Brooks CG, et al.: Selective depletion of NK cell activity *in vivo* and its effect on the growth of NK-sensitive and NK-resistant tumor cell variants. Int J Cancer 29:567, 1982

Klein G, Sjogren HO, Klein E, et al.: Demonstration of resistance against methylcholanthrene-induced sarcomas in the primary autochthonous host. Cancer Res 20:1561, 1960

Lawrence HS, Valentine FT: Transfer factor and other mediators of cellular immunity. Am J Pathol 60:437, 1970

Mitchison MA: Passive transfer of transplantation immunity. Proc R Soc Lond [Biol] 143:72, 1954

Moller, G (ed): Experiments and the concept of immunological surveillance. Transplant Rev 28:1, 1976

Prehn RT, Main JM: Immunity to methylcholanthrene induced sarcomas. J Natl Cancer Inst 18:769, 1957

Winn HJ: Immune mechanisms in homotransplantation. II. Quantitative assay of the immunologic activity of lymphoid cells stimulated by tumor homografts. J Immunol 86:228, 1961

IMMUNOTHERAPY

Terry WD, Rosenberg SA (eds): Immunotherapy of Human Cancer. New York, Excerpta Medica, 1982

Warner JF, Dennert G: Effects of a cloned cell line with NK activity on bone marrow transplants, tumor development and metastasis in vivo. Nature 300:31, 1982

HUMAN TUMOR MARKERS

Hellstrom KE, Hellstrom I, Brown JP: Human tumor associated antigens identified by monoclonal antibodies. In Baldwin RW, Miescher PA, Muller-Eberhard HJ (eds): Springer Seminars in Immunopathology, New York, Springer-Verlag, 1982, vol 5, p 127

Lapis K, Jeney A, Price MR (eds): Tumor Progression and Markers. Amsterdam, Kugler, 1982

CHAPTER 21

Allergy and Atopy

Allergy

Atopy was defined in 1923 by Coca and Cooke as a familial predisposition toward increased sensitivity to common environmental agents. The unusual sensitivity of atopic individuals causes altered reactivity to antigens. This form of altered reactivity is called an allergy. One or more of several related diseases are typical of the atopic individual. Cooke included anaphylaxis, asthma, angioedema, urticaria, atopic dermatitis, certain types of acute gastroenteritis, and allergy to infectious agents as illnesses characteristic of atopic individuals. Careful clinical observations revealed a tendency for atopic diseases to cluster in families. Cooke suggested that genetic influences contribute to the atopic individual's predisposition toward allergic diseases. This astute observation has been abundantly confirmed. The allergic diseases of atopic individuals are the most prevalent of all human immune disorders. Collectively, allergic disorders account for continuing morbidity in a large number of patients and are a major concern of the practicing physician. Atopic diseases such as asthma and drug allergy cause significant impairment of health. Another atopic disease, anaphylaxis, can cause death. Exposure to otherwise harmless environmental antigens is the principle cause of the atopic allergic disease. Prevention is often possible. With adequate therapy, atopic allergic diseases can be controlled.

Hypersensitivity

Allergic reactions are also called hypersensitivity reactions. There are two general types of hypersensitivity reactions. Both are classified and named with respect to the time interval between antigen challenge and the appearance of the reaction. Immediate hypersensitivity reactions occur shortly after antigen exposure and are initiated by the reaction of antigen with antibodies. Delayed hypersensitivity reactions occur several days after antigen challenge and are initiated by the reaction of antigen and the cells of the immune system. The immune mechanisms responsible for immediate hypersensitivity provide host protection against infections. The immune mechanisms responsible for delayed hypersensitivity provide host resistance to infection.

The allergic diseases of atopic individuals are caused by a nondestructive form of immediate hypersensitivity called anaphylactic hypersensitivity. A second form of immediate hypersensitivity, Arthus hypersensitivity, produces cell death and extensive local destruction of normal tissue structures (Chap. 18). An understanding of the mechanisms responsible for hypersensitivity reactions and the ability to distinguish between them is clinically important.

Hypersensitivity to an antigen can be distinguished from normal sensitivity by evaluating (1) the dose dependence of the size of the tissue reaction caused by the antigen, and (2) estimating the chance of a similar sized reaction in normal individuals. This assessment is usually made with a skin test; in vivo provocative tests of other tissues are also possible. Large reactions are more likely to cause symptoms and a hypersensitivity disease. Allergic patients exhibit severe hypersensitivity to antigen doses that provoke little or no reaction in nonallergic persons. Patients with the atopic allergic diseases differ from healthy individuals in their exquisite anaphylactic hypersensitivity to many common environmental antigens.

Quantitation of Hypersensitivity

Hypersensitivity reactions can be evaluated in several ways. Over an optimal range of antigen doses, a quantitative sigmoid dose relationship exists between (1) the size of the hypersensitivity reaction, and (2) the antigen dose. Figure 21-1 illustrates this relationship and the variability expected in reactions to the same hypothetic antigen among a large number of individuals. There are several important attributes of the relationship shown in Figure 21-1. First, neither allergic nor normal subjects react to very low antigen doses. Second, both allergic and normal subjects can exhibit equivalent reactions to high antigen doses. Third, differences between allergic and normal subjects are detectable only over an optimal dose range along the sigmoid part of the curve.

The classic approach to quantifying a hypersensitivity reaction, an end point titration, stems directly from conventional immunoserologic techniques. This method tests

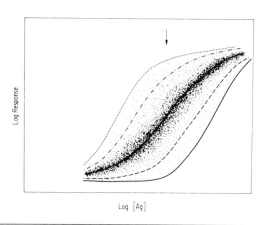

Figure 21-1. The relationship between the size of a hypersensitivity reaction and the antigen dose used to provoke the reaction is complex. At extremely low antigen doses, all subjects fail to react to the antigen. At extremely high doses of the antigen, all subjects exhibit similar nonspecific inflammatory reactions to the antigen. Extremely low and extremely high antigen doses fail to detect the maximum variability among the reactions needed to distinguish between allergic and nonallergic subjects. Maximum antigen induced reaction size variation can be detected in a population of subjects only with intermediate doses of antigen. This illustration shows the expected range distribution of reactions to varied doses of a hypothetic antigen in a test population. Hypersensitivity is defined as a reaction that occurs in certain individuals but not in others. Similarly, an allergic reaction to an antigen occurs in certain individuals who have disease provoked by antigen exposure, but does not occur in other healthy individuals. Differences among individuals related to hypersensitivity and/or allergy are detectable only over the intermediate antigen dose range. Within this population, the reactivity of a single subject can be quantified in terms of the antigen dose needed to elicit an arbitrarily selected threshold reaction or in terms of the actual size of the reaction at some optimal antigen dose. Neither assessment provides a clear estimate of sensitivity nor responsiveness (see text). In the hypothetic population shown here, the optimal antigen dose would be the antigen concentration beneath the arrow.

the reaction to serial dilutions of antigen beginning with low antigen doses and proceeding through sequential higher doses toward the range of maximum response variation. Tests are made at several antigen doses until an arbitrarily selected significant reaction size is obtained and taken as the end point of the titration. This dose is used as an estimate of the subject's sensitivity to the antigen. Using this method, 1000-fold or greater differences can be detected in the antigen doses needed to elicit similar sized reactions among healthy and allergic subjects. A second method for measurement and evaluation of hypersensitivity reactions can be based on an a priori selection of an antigen dose that yields the largest variation among the reactions in the individuals tested. This dose usually lies at or above the center of the optimal range shown in Figure 21-1. The size of the observed reaction is used as an estimate of the subject's responsiveness to the antigen.

Both traditional methods have serious conceptual and technical limitations in quantitative studies of hypersensitivity. The best estimates of an individual's responsiveness and sensitivity to an antigen lie at the midpoint of the subject's own particular sigmoid dose response curve to the specific antigen being tested. Among subjects, estimates of both parameters vary independently. The independence of responsiveness and sensitivity contributes a confounding source of variation when either traditional method is used to quantify hypersensitivity reactions. Measurements of reactions to several antigen doses across the middle portion of the sigmoid dose response curve should be performed to obtain a quantitative estimate of both responsiveness and sensitivity.

Antigens and Allergens

The antigens responsible for the atopic allergic diseases are often called allergens in order to identify the expectation of a harmful but reversible immediate hypersensitivity reaction. Known allergens include: (1) airborne products of plant, animal, and insect origin; (2) products of microorganisms; (3) drugs and other chemicals, and (4) certain foods. Pollen allergens represent examples of this kind of antigen. Biochemically, this classification of allergens includes polysaccharides, proteins, and small hapten molecules of natural and synthetic origin. Useful classifications of allergens can also be based on: (1) the very frequent association of a particular antigen with a specific disease; (2) the ecologic probability of harmful exposure; and (3) the immunogenic properties of the particular allergen.

Other antigens are clinically important because of their value in tests of exposure and sensitization to infectious and noninfectious microorganisms prevalent in the natural environment. Tuberculin, a product of the tubercle bacillus, is an example of this kind of antigen. Delayed cutaneous hypersensitivity reactions to tuberculin and related antigens are used to demonstrate evidence of intact cellular immunity and prior sensitization to microorganisms. Because of the role of prior exposure and immune memory in these assessments of immunity and sensitization, delayed hypersensitivity skin test antigens are also called recall antigens.

Responders and Anergy

Healthy subjects and most patients become sensitized to many common antigens. Healthy persons usually exhibit immediate and delayed cutaneous hypersensitivity to antigens prepared from streptococci, staphylococci, mumps virus, and certain common fungi. Current immunization practices also produce cutaneous hypersensitivity to diphtheria and tetanus antigens. Patients with immune disorders frequently do not have delayed skin test reactions to recall antigen doses that elicit reactions in healthy individuals. This inability to exhibit a delayed cutaneous reaction to a recall antigen can occur despite known prior exposure, the presence of immediate hypersensitivity and detectable circulating antibody to the antigen, and cutaneous reactions to other antigens. Individuals who exhibit specif-

ic impairment of delayed reactivity to an antigen in the presence of antibody are said to exhibit an immune deviation. Individuals who lack cell and antibody reactivity to a specific antigen despite prior exposure are considered nonresponders to the particular antigen. Care must be taken to distinguish patients who are nonresponders to a particular antigen from patients who are unresponsive because of infection, disease activity, drug therapy, or a nonspecific defect in immunity. The presence of skin test reactions to other recall antigens is used to distinguish between nonresponders to a specific antigen and general unresponsiveness, or anergy. Skin tests with multiple recall antigens are necessary to detect patients with anergy.

Immediate Hypersensitivity

There are two kinds of immediate hypersensitivity reactions, anaphylactic hypersensitivity and Arthus hypersensitivity. Anaphylactic hypersensitivity occurs within seconds to minutes after antigen challenge and is initiated by IgE antibodies. Anaphylactic hypersensitivity is characteristic of the atopic diseases. Arthus hypersensitivity occurs within hours after antigen challenge and is initiated by complement fixing antibodies (Chap. 18).

Anaphylactic Hypersensitivity

This form of immediate hypersensitivity is named for anaphylaxis, the most dangerous of all allergic diseases. Anaphylaxis was first clearly described in 1839 by Magendie, who reported the sudden death of rabbits following several injections of egg albumin. Richet (1902) identified the reaction as the opposite of protection (phylaxis) and originated the term anaphylaxis. Prausnitz (1921) was able to adoptively sensitize himself, but not guinea pigs, for anaphylactic hypersensitivity to fish proteins with serum from Kustner, his allergic colleague. The Prausnitz–Kustner (P–K) skin test remained the only direct evidence of the existence of anaphylactic antibodies until the Ishizaka's elegant biochemical identification of IgE as the antibody class responsible for anaphylactic hypersensitivity in 1967.

IgE antibodies cause hay fever, asthma, angioedema, urticaria, atopic dermatitis, certain types of gastrointestinal reactions, and the allergic component of infectious diseases. Anaphylactic hypersensitivity causes the precipitious onset of local skin and mucosal edema, changes in smooth muscle tone, and increased secretions from mucosal surfaces, but does not cause irreversible tissue changes or destruction. Despite the nondestructive nature of anaphylactic hypersensitivity, death can occur when the tissue reaction involves the air passages or other vital structures essential for physiologic function.

Experimental Anaphylaxis. Systemic anaphylaxis in experimental animals occurs within seconds after antigen challenge. Death occurs when the allergic reaction impairs the function of a physiologic system necessary for life.

Anaphylactic death can be experimentally induced in monkeys, guinea pigs, rabbits, mice, and many other animals. Differences in the tissues involved in anaphylactic hypersensitivity reactions among animal species accounts for observed differences in the physiologic consequences of experimentally induced anaphylaxis. If life can be sustained and the reaction controlled, the tissue changes induced by anaphylaxis can be reversed and complete recovery is possible.

Guinea pigs can be sensitized for anaphylaxis with as little as 0.1 mg of protein. Within 30 to 60 seconds after challenge with the same dose of antigen 3 weeks later, the sensitized animal exhibits agitation, piloerection, sneezing, increased secretions from the respiratory passages, often defecates and/or urinates, and makes a few spasmodic jumps. Respiration becomes slowed and forceful and the animal becomes cyanotic and appears preoccupied with breathing. Convulsions usually occur and respiratory death follows within minutes. Necropsy reveals distention of the lungs with air. Bronchospasm is the primary cause of anaphylactic death in the guinea pig. In the rabbit, death results from pulmonary arteriolar constriction and circulatory failure. Mice die of circulatory shock due to hemoconcentration and death can be prevented by the use of plasma expanders.

Immunoglobulin E. The antibodies responsible for initiating anaphylaxis have been called skin-sensitizing or homocytotrophic antibodies (Fig. 21-2). Skin sensitization is an example of homocytotrophism, which means the capacity of IgE antibodies to bind preferentially to the tissues of the animal species in which the antibodies were produced. The avid binding of IgE to tissue mast cells and circulating basophils is a unique attribute of this immunoglobulin class. At the present time, anaphylaxis and related allergic diseases are the only known important biologic consequence of IgE antibodies; thus, anaphylactic antibodies and IgE antibodies can be used as synonyms.

CHEMICAL PROPERTIES OF IgE. IgE is one of the five major isotypic classes of human antibody globulins (Chap. 12). Similar antibodies have been detected in other species. IgE does not pass the placental barrier to the fetus. IgE binds tightly to specific receptors on tissue mast cells. The 10- to 12-day half-life of cell-bound IgE and the short 2- to 3-day serum half-life suggests the cell membrane may protect bound IgE against extracellular proteases. Myeloma proteins have been used for physicochemical studies of the molecular properties of IgE because of the limited nanogram quantities of IgE normally present in serum and other body fluids. Human IgE has a sedimentation coefficient ($S_{20,w}$) of 7.9 and weighs approximately 188,000 daltons. IgE migrates electrophoretically as an α-1 globulin, contains 12 percent carbohydrate, and has five intrachain disulfide bonds per ϵ chain and three interchain disulfide bonds per molecule. Anaphylactic antibodies do not fix complement through the classic pathway and are sensitive to sulfhydryl reducing agents. Heating at 56C for 30 minutes results in irreversible changes in the conformation of the Fc portion of IgE, and continued thermal denaturation

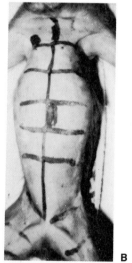 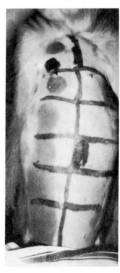 

A   B   C

Figure 21-2. Monkey passive cutaneous anaphylaxis can be elicited with human anaphylactic antibody. **A.** Unreacted skin test sites sensitized with serum from patients sensitive to two different pollen antigens. **B.** Intravenous challenge with ragweed antigen results in bluing of sites passively sensitized with anaphylactic antibodies to ragweed pollen, but not at sites sensitized with orchard grass on the left side of the animal's abdomen. **C.** Challenge with orchard grass antigen shown results in bluing of sites sensitized with anaphylactic antibody to orchard grass. This sequence of challenges demonstrates the specificity of the reaction. *(Adapted from Buckley, Metzgar: J Allergy 36:382, 1965.)*

for 4 hours results in the loss of the skin sensitizing property of the molecule. Similar heat treatment of the Fab portion of the molecule causes no appreciable conformation change and does not decrease the antigen binding activity of the Fab fragment.

SITES OF IgE PRODUCTION. Gastrointestinal and respiratory tract mucosal tissues secretions contain measurable quantities of IgE. Adenoidal and tonsillar tissues, peribronchial and peritoneal lymph nodes, and the lamina propria of the gastrointestinal and respiratory tract mucosa contain IgE laden plasma cells. In contrast, relatively few IgE containing cells are found in the peripheral lymph nodes and spleen. Although IgE molecules do not contain a secretory piece, the distribution of IgE producing cells close to protective mucosal surfaces identifies an area of potential importance of IgE function.

BINDING TO MAST CELLS. IgE molecules bind reversibly through the Fc portion of the molecule to receptor sites on the surface of mast cells and basophils. Fc binding is specific for receptors on cells from the homologous and closely related species. The basophil IgE glycoprotein receptor weighs 80,000 daltons and contains 10 to 15 percent carbohydrate. The average number of receptors on basophils from different subjects varies between 30,000 to 100,000 per cell. The number of IgE molecules bound per cell varies between 10,000 to 40,000. The basophil receptor site for IgE has an extremely avid association constant (K_a 10^{-8}) comparable to the specific binding of antigen and antibody. Polyvalent allergens bridge two or more receptor bound IgE antibodies and initiate the release of chemical mediators responsible for anaphylactic hypersensitivity reactions.

Mast Cell Mediators.   The chemical compounds that cause anaphylactic hypersensitivity are produced by tissue mast cells or circulating basophils. Mast cells produce large quantities of the chemical mediators of anaphylaxis or permeases responsible for the loss of fluid from small blood vessels and tissue edema associated with allergic reactions. The chemical mediators include: (1) histamine, (2) serotonin, (3) leukotrienes, (4) bradykinin, (5) heparin, and, possibly, (6) acetylcholine. The leukotrienes, a group of closely related compounds, are responsible to the biologic activity formerly attributed to the slow reacting substances of anaphylaxis. Mediator release is also accompanied by the release of eosinophil chemotactic factor, which attracts eosinophils into the local area of allergic injury. Differences in the tissue distribution of mast cells among different species and variation in their mediator content contribute to the diversity of local changes associated with anaphylaxis in different animal species. At least two types of functionally different mast cells exist. For example, lung mast cells contain rich quantities of heparin, whereas mast cells proximate to mucosal surfaces produce large quantities of leukotrienes.

Figure 21-3 summarizes the mechanisms that initiate and modify mediator release and illustrates how pharma-

cologic agents intervene in anaphylaxis. Antigen challenge of mast cell bound IgE activates a series of enzymatic mechanisms leading to the release of leukotrienes, histamine, and other mediators from mast cell granules. The process is modulated by the relative intracellular concentrations of cyclic adenosine monophosphate (cAMP) and cyclic guanosine monophosphate (cGMP). An increase in the ratio of cAMP to cGMP retards mediator release. Locally active hormones such as prostaglandin E_1 and E_2, corticosteroids, and stimulators of β-adrenergic receptors increase intracellular cAMP. Infection and medications such as propranolol block the β-adrenergic receptors and interfere with β agonist therapy of anaphylactic hypersensitivity. Stimulation of α-adrenergic and cholinergic receptors increases modulation by cGMP and facilitates mediator release. Theophylline, a methylxanthine, inhibits the activity of phosphodiesterase, an enzyme which inactivates cAMP. This increases the ratio of cAMP to cGMP and retards mediator release. Predictably, the combined effects of theophylline and β agonist therapy is additive. Disodium chromoglycate has a direct stabilizing effect on cell membranes and prevents permease release from the mast cell granules. Calcium entry blockers can also stabilize the mobilization of extracellular and intracellular calcium and prevent the mast cell and smooth muscle changes associated with allergic reactions. Antihistamines block the action of histamine on the type one (H_1) and type two (H_2) receptors of the cells involved in the inflammatory process.

HISTAMINE. The quantity of histamine produced during IgE initiated reactions varies between different cells and tissues and between species. In humans, for example, histamine appears most important in anaphylactic hypersensitivity reactions in the skin and upper airway. The effect of histamine on target cells depends upon the kind of histamine receptor possessed by the cell. Species differences also exist in the distribution of H_1 and H_2 receptors and experimental studies of antihistamines in animals do not apply directly to humans. Table 21-1 summarizes differences in the physiologic activities of histamine receptors. Of special importance is the possible role of the H_2 receptor in the intrinsic modulation of interactions among monocytes, lymphocytes, basophils, and eosinophils in the inflammatory processes. Methylation, acetylation, and oxidation by monoamine oxidase inactivate and degrade histamine.

SEROTONIN. This mediator is especially important in rat and mouse anaphylaxis. For example, the mouse uterus is 1000 times more responsive to serotonin than to histamine. Serotonin, 5-hydroxytryptamine, is structurally related to the amino acid tryptophane and is degraded by monoamine oxidase. Serotonin, histamine, and heparin are released from mast cells and intestinal chromaffin cells, and are present in platelets, spleen, brain stem, and lung tissue. The physiologic effects of serotonin are increased intestinal peristalsis, respiratory rate, smooth muscle contraction, and decreased central nervous system activity. Serotonin increases the intracellular level of cGMP in human

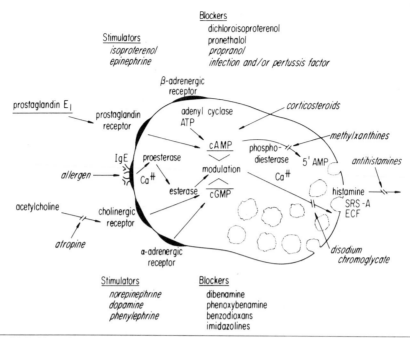

Figure 21-3. Many factors modulate the release of permease by mast cells and basophils. An awareness of the cell mechanisms modulating permease release provides a pharmacologic and clinical basis for intervention in anaphylactic hypersensitivity reactions. Clinically important drugs and compounds that facilitate or block permease release are shown in italics. Permease release is diminished by increased cell levels of cyclic AMP (cAMP) and is augmented by increased levels of cyclic GMP (cGMP). Permease release is initiated by the reaction of allergen and cell bound IgE. β agonists (isoproterenol, epinephrine) stimulate cAMP production. Methylxanthines retard the cAMP degradation by mast cell phosphodiesterase. Both effects increase intracellular cAMP and decrease permease release. Corticosteroids augment the number of available beta receptors. Prostaglandins E_1 and E_2 augment cell cAMP levels. α receptor agonists and acetylcholine increase intracellular cGMP levels and facilitate permease release. Disodium chromoglycate and possibly calcium entry blockers stabilize cell membranes and block permease release. Antihistamines block the permease activity of histamine at sites distant from the mast cell or basophil.

peripheral blood mononuclear cells and enhances macrophage activity. Injection of serotonin into the skin produces local pain and erythema and provokes histamine release.

LEUKOTRIENES. The leukotrienes were formerly called the slow reacting substances of anaphylaxis, or SRS-A. Elegant studies by Samuelsson and many others have revealed that these compounds, like the prostaglandins and thromboxanes, are oxygenated products of arachidonic acid metabolism. The diverse biologic activities formerly attributed to SRS-A are caused by leukotrienes C_4, D_4, and E_4. These cysteinyl-containing compounds are released from allergen-challenged lung tissues of asthmatic patients and produce bronchial smooth muscle spasm, increased permeability and fluid loss from postcapillary venules, and mucus secretion. Endobronchial allergen challenge provokes leukotriene release and these mediators are primarily responsible for the bronchial constriction of asthma in humans. The leukotrienes are also produced in quantity by polymorphonuclear leukocytes. Leukotriene B_4 stimulates

leukocyte chemotaxis, provokes cell adhesion and aggregation, and promotes superoxide generation and enzyme release by leukocytes.

BRADYKININ. This small peptide is produced by the action of peptidases on serum globulin and other proteins and represents a second class of slow reacting substances. The tissue changes induced by bradykinin occur more slowly than changes caused by histamine. Bradykinin causes smooth muscle contraction, enhances capillary permeability, and may also enhance the migration of leukocytes through blood vessel walls. The injection of bradykinin into the skin of humans causes pain, but does not produce a pruritic flare and wheal.

HEPARIN. Lung but not bronchial mast cells contain large quantities of this acid, mucopolysaccharide. Heparin decreases the coagulability of blood by altering the surface charge of formed elements of blood and blood vessel walls. Prolonged clotting times are observed in anaphylaxis and may be a consequence of the release of heparin from lung mast cells.

TABLE 21-1. PHYSIOLOGIC EFFECTS OF HISTAMINE

H$_1$ Histamine Receptor	H$_2$ Histamine Receptor
Initiates and/or facilitates inflammation; stimulates cyclic GMP and AMP production	Inhibits inflammation—impairs histamine release from basophils, neutrophil chemotaxis, and lysosome enzyme release; retards lymphocyte cytolysis and lymphokine production; stimulates eosinophil chemotaxis
Increases vascular permeability by loosening tight junctions between postcapillary venule epithelial cells	Increases vascular permeability, inhibits adrenergic induced vasoconstriction
Stimulates smooth muscle contraction—guinea pig ileum, guinea pig and human bronchi	Stimulates smooth muscle relaxation—rat uterus, sheep and rabbit bronchi
Stimulates salivary and lacrimal gland secretion	Stimulates gastric acid secretion
Slows atrioventricular node conduction in the guinea pig and rat	Stimulates contractile force of the heart and sinoatrial node rate in the guinea pig

ACETYLCHOLINE. The possible role of acetylcholine in cholinergic urticaria is one basis for its inclusion among the chemical mediators. Acetylcholine is primarily localized at neural synaptic connections and causes peripheral vasodilation. Cholinergic agonists are used to detect airway hyperreactivity in patients with asthma. Acetylcholine is inactivated by cholinesterase.

Arthus Hypersensitivity

A second kind of immediate hypersensitivity reaction occurs within hours after antigen challenge. This allergic reaction, Arthus hypersensitivity, is usually initiated by complement-fixing IgG antibodies (Chap. 18). Arthus-type hypersensitivity produces cell death and extensive local destruction of normal tissue structures. The injury is not reversible and is usually localized at the site of the hypersensitivity reaction. Cutaneous Arthus type reactions can usually be distinguished from the preceding IgE-mediated allergic reactions. It is equally important but often more difficult to distinguish between this form of immediate cutaneous hypersensitivity and delayed cutaneous reactions caused by cellular immunity.

The typical cutaneous Arthus hypersensitivity reaction is initiated by intradermal antigen challenge in the presence of a large excess of circulating complement-fixing (IgG and IgM) antibodies. Complexes of antigen and antibody are formed in the zone of the antigen excess at the site of antigen challenge. The immune complexes activate the complement system and initiate the local release of permeases and chemotactic factors for polymorphonuclear leukocytes. The skin test site becomes tender, warm, and swollen and exhibits central bruising surrounded by petechiae. The local reaction reaches maximum intensity in 24 to 36 hours and can progress to local necrosis. Histopathologic examination of the site reveals evidence of edema, an intense polymorphonuclear leukocyte infiltrate, thrombosis of small blood vessels, local bleeding, cell death, and necrosis with disruption of the normal anatomic structures.

Delayed Hypersensitivity

Delayed hypersensitivity is the observable outcome of a cell-mediated immune reaction (Chap. 14) in the tissues of a sensitized individual. The reaction can involve any sensitized tissue, but is most readily observed in the skin. Much clinically useful information about cellular immunity to antigens in different diseases is based on studies of the delayed skin test reaction. The intracutaneous deposition of antigen in a sensitive individual initiates the local accumulation of lymphocytes and other mononuclear inflammatory cells at the site of antigen challenge. The resulting inflammation is called delayed cutaneous hypersensitivity.

There are two types of delayed cutaneous hypersensitivity. The first, cutaneous basophilic hypersensitivity, is less well known than the form of delayed hypersensitivity identified by Koch as tuberculin hypersensitivity. Zinsser later showed that the tissue response to most microorganisms is similar to the delayed cutaneous tuberculin reaction. Both types of delayed hypersensitivity are clinically important.

Cutaneous Basophilic Hypersensitivity

First Dienes, and then Jones and Mote, described a transient, flat, erythematous cutaneous response provoked by the repetitive daily injection of soluble protein antigens. The reaction characteristically appears 3 to 6 days after beginning the process of sensitization. The local tissue reaction begins 3 to 4 hours after antigen challenge and reaches maximum size after 24 to 48 hours. Histologically, local edema and perivascular mononuclear cell infiltrates develop at 6 hours and are followed by a diffuse infiltrate containing lymphocytes and basophils. The response persists and increases with repetitive antigen challenges until the development of a typical flare and wheal anaphylactic hypersensitivity response heralds the generation of antibodies. Subsequent challenges with antigen provoke an Arthus hypersensitivity reaction.

A similar but persisting reaction can be elicited by exposure of the skin to contact allergens. These simple chemicals form immunogenic complexes with skin proteins and induce cell-mediated immunity to the chemical. Contact allergens rarely provoke an antibody response. In experimental animals, basophils may account for 20 to 60 percent of the infiltrating cells. In humans, the basophilic component of the reaction often goes unrecognized because of the need for special stains to detect the human basophil. Poison ivy dermatitis is an example of this form of contact allergy. Drug hypersensitivity reactions can exhibit similar histopathologic findings. Cutaneous basophilic hypersensitivity reactions are sensitive to the therapeutic effects of corticosteroids.

The cutaneous basophilic hypersensitivity reaction can be passively transferred with lymphocytes from sensitized animals. The use of syngeneic animals may facilitate, but is not a strict requirement for, adoptive sensitization. It is important to note that cells from tuberculin-sensitized guinea pigs elicit a cutaneous basophilic reaction instead of tuberculin hypersensitivity in adoptively sensitized recipients. Indirect evidence implicates a specifically sensitized T lymphocyte in the initiation of cutaneous basophilic hypersensitivity.

Similar histopathologic reaction can be passively transferred with antibody. Hapten-specific cutaneous basophilic hypersensitivity can be transferred with serum from guinea pigs immunized with a protein conjugate of the hapten in adjuvant. Allergen-specific late cutaneous basophilic reactions can be passively transferred with IgE antibodies. As with antibody-dependent, cell-mediated cytotoxicity, it is difficult to exclude a role of cytophilic antibodies in this form of cellular immunity. The importance of these observations stems from the possibility that several immune mechanisms converge and recruit the augmented expression of essentially the same form of cellular immunity.

Tuberculin Hypersensitivity

Tuberculin is prepared from the aqueous ultrafiltrate of a broth culture of *Mycobacterium tuberculosis*. The delayed cutaneous hypersensitivity reaction to tuberculin is classically elicited with a Mantoux-type intradermal injection of 0.1 ml of a standardized solution of a purified protein derivative (PPD). The antigen dose is selected to elicit a measurable response in individuals previously sensitized to the human tubercle bacillus. In epidemiologic studies in healthy populations previously unexposed to soil mycobacteria and other mycobacterial infections, such as Alaskan Eskimos, the tuberculin skin test provides an unambiguous assessment of sensitization to the human tubercle bacillus. Epidemiologic studies in healthy populations exposed to cross-reacting soil mycobacteria suggest an assessment of the size of the tuberculin skin test reaction is important. Small tuberculin skin test reactions can be caused by sensitization to the cross-reacting soil mycobacteria. Only healthy subjects with tuberculin skin test reactions equal or

greater than 10 mm at 48 to 72 hours have a 90 percent or greater chance of sensitization and risk of subsequent infection with *M. tuberculosis*.

Unfortunately, epidemiologic criteria based on healthy subjects are not applicable to sick patients. The immune function of sick patients is frequently impaired by age, disease, or the treatment of the disease. Tuberculin skin test responses less than 10 mm in diameter are prevalent among hospitalized patients. A falsely negative tuberculin skin test should be anticipated in immunocompromised patients. The use of clinical judgment has been suggested in the interpretation of small tuberculin skin test reactions in patients. A simultaneous assessment of the skin test reactions to positive control antigens and use of an appropriate specificity control, such as an avian tuberculin, can detect specific sensitization to the human tubercle bacillus in sick patients.

In its simplest form, little or no immediate hypersensitivity reaction occurs at the tuberculin skin test site. The cutaneous reaction reaches maximum size between 48 and 72 hours after antigen challenge, and the involved skin feels firm or indurated to the touch. Histologically, the typical delayed tuberculin response reveals a mononuclear cell infiltrate composed of lymphocytes, large mononuclear cells, a few plasma cells, and a small but variable number of polymorphonuclear leukocytes (Fig. 21-4). Although the histopathologic changes of tuberculin hypersensitivity begin 4 to 6 hours after antigen challenge and may initially resemble the perivascular mononuclear cell infiltrate of the cutaneous basophilic reactions, basophils are not prominent and the reaction can persist for more than a week. Tuberculin hypersensitivity can cause cell death and subsequent fibrosis, but does not cause extensive anatomic destruction similar to the Arthus hypersensitivity reaction.

An unambiguous interpretation of the tuberculin skin test 48 to 72 hours after antigen challenge is not always possible. Immediate hypersensitivity reactions can occur at the skin test site. Occasionally, an anaphylactic hypersensitivity reaction appears within seconds to minutes and leads to intense local itching followed by a typical flare and wheal reaction. Histologic examination of the skin test site during this interval reveals only vascular congestion and edema, and no evidence of cell death or inflammatory cells. The tuberculin skin test more often provokes local redness and swelling within 6 to 12 hours, followed by increasing warmth, tenderness, local hemorrhage, and central necrosis—changes consistent with an Arthus reaction. These changes reach maximum after 24 to 36 hours, and can obscure or mimic the subsequent delayed reaction. When the complexity of the initial phase of the tuberculin reaction is ignored, the contribution of immediate hypersensitivity to the delayed response can be falsely interpreted as a positive skin test.

As observed by Zinsser, many microbiologic antigens provoke hypersensitivity reactions. Many of these reactions are similar to the classic tuberculin reaction. With other microbiologic antigens, the early phases of the Mantoux skin test reaction is more prominent. For example, histo-

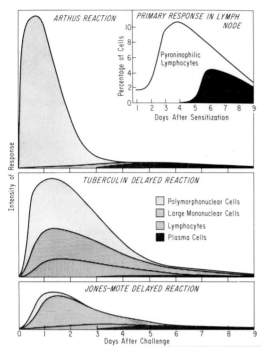

Figure 21-4. The time course of changes in the cell composition differs among the several kinds of cutaneous hypersensitivity reactions elicited by tuberculin. Note the similarity of the primary response to antigen in the lymph node and the prominence of lymphocytes and mononuclear cells in the typical Jones–Mote basophilic hypersensitivity reaction. This form of hypersensitivity contrasts with the predominance of polymorphonuclear leukocytes in the Arthus reaction. The typical cutaneous response to tuberculin exhibits a cell composition that lies between these two extremes. Differences in the characteristics of the cutaneous response to microbiologic antigens reflect differences in the kind of hypersensitivity responsible for the skin test reaction.

plasmin and coccidioidin, antigens prepared from the pathogenic fungi *Histoplasma capsulatum* and *Coccidioides immitis,* usually produce a tuberculin-type reaction. With other microbiologic antigens, the skin test reaction may not persist for 48 to 72 hours, and the reaction detected at 24 hours may correlate best with exposure and infection. The reaction to blastomycin, an antigen from the pathogenic fungus *Blastomyces dermatitides,* typically reaches maximum size at 24 hours and exhibits different clinical and histopathologic changes than the tuberculin reaction. The time of interpretation and the meaning of diagnostic skin tests must be determined separately for each microbiologic antigen.

Atopic Diseases

Atopy occurs with varying severity in 10 to 30 percent of the population in the United States. The total magnitude and cost of impaired health due to atopy is considerable, but not clearly known. It is estimated that 1 percent of total health care costs are expended on allergy-related diseases. In atopic patients, allergen exposure can cause anaphylaxis, asthma, bronchitis, conjunctivitis, rhinitis, sinusitis, angioedema, urticaria, and atopic dermatitis. Food allergy can occur early in life in atopic subjects, but is rare in adults. Drug allergy is a significant clinical problem.

Clinical Epidemiology

Conjunctivitis, rhinitis, asthma, and drug and food hypersensitivity reactions are the most prevalent atopic diseases. It is estimated that 37 million patients have allergic diseases of varying severity. Approximately 19 million patients have allergic rhinitis. Approximately 9 million patients have bronchial asthma, and asthma is the cause of 4000 deaths per year. Almost a million patients are at risk of anaphylaxis from sting insect venom, a less common but well publicized cause of allergic death. From 1 to 4 percent of hospitalized patients develop problems with adverse drug reactions attributed to drug allergy. Allergy and infection are prevalent in patients with frequent or chronic respiratory diseases. Because of the relatively mild impairment and reversibility of most allergic disorders, atopic patients are most often managed in the ambulatory practice of their physicians.

Because of the ambulatory prevalence of allergic disorders, hospitalized patients in medical training institutions do not adequately represent the clinical problems of allergic patients. This same problem is reflected in the composition of the faculty of many medical schools. Allergy is a component of infectious processes and, when augmented in atopic patients, can lead to hospitalization. Patients with severe impairment due to anaphylaxis, asthma, sinusitis, and bronchitis, food allergy, and drug hypersensitivity frequently require hospitalization, but with proper treatment enjoy a good prognosis for recovery.

The clinical identification of patients with an atopic predisposition is not difficult. Older patients often have a past history of unusual reactions to other environmental allergens. A past diagnosis of a prior atopic disease can provide a clue to the nature of the current illness. The familial prevalence of atopy is so common that the family history can help corroborate the diagnosis in young patients.

Genetic Factors
Although shared exposure could account for the familial prevalence of atopic diseases, much evidence suggests inherited factors contribute to the clustering of closely related illnesses within families. Approximately 25 percent of atopic patients have at least one atopic parent. This incidence increases to more than 40 percent when grandparents are included, and to more than 50 percent when the patient's siblings are also included. The particular atopic diseases exhibited by different family members are varied.

Environmental factors clearly contribute to the varied clinical features of each atopic disease. In sharp contrast to much evidence of the inheritance of atopic reactivity, studies of atopic monozygotic twins reveal a 70 percent discordance in the occurrence of the specific diagnosis in allergic diseases. Moreover, healthy family members can exhibit allergen reactivity equivalent to that of atopic patients. The clinical diagnosis of an allergic disease is not an adequate genetic phenotype for atopy.

Evidence of a genetic component of atopy stems from recent studies of the associated segregation of the specific immune response to specific allergens and genetic markers in human families; and from the heritability of serum IgE concentrations and augmented reactions to nonspecific pharmacologic agents, respectively. In certain families, the size of the skin test response to pollen allergens appears associated with a particular HLA haplotype (Chap. 19). Family studies of serum IgE concentrations suggest the presence of a major non-HLA-linked gene and a small polygenic effect. Mechanisms responsible for the neurophysiologic control of reactivity, such as the α-adrenergic hyperreactivity and β-adrenergic hyporeactivity of asthmatic patients, may also be under genetic control. Studies in monozygotic and dizygotic twins suggest a high degree of concordance in the altered neurophysiologic control of exercise induced asthma. Twin studies also reveal a genetic component in the responsiveness to inhaled methacholine, an agonist for the acetylcholine receptor of mast cells. Collectively, these observations suggest the genetic control of the reactivity of allergic patients is complex.

Natural Sensitization

The occurrence of in utero sensitization and IgE antibody production is suggested by differences in the IgE antibody concentrations in cord and maternal serum and by the inability of IgE to cross the placenta. During the first year of life, IgE antibodies are directed predominantly toward foods such as egg albumin and cow's milk. IgE antibodies to inhalent allergens (dust, animal danders, etc.) become detectable during the second year of life and antibody activity to foods decrease as the child matures. Exposure of older susceptible individuals to airborne allergens provokes sensitization and production of IgE antibodies to pollen proteins. The mechanisms responsible for sensitization to low concentrations of airborne pollens appear complex. Many naturally occurring allergens have biologic properties in addition to their immunogenicity. Certain pollen allergens contain components with lectin-like or lectin-binding activity and may bind to the intracellular substances between mucosal cells. Other allergen components are mitogenic for certain lymphocyte populations and may contribute a regulatory effect to the process of sensitization. A glycoside called rutin is present on a tobacco glycoprotein allergen and appears experimentally capable of producing an exclusive IgE response in susceptible mice.

Although the mechanisms responsible for pollen sensitization are not completely understood, the route of allergen exposure and quantity of allergen responsible for sensitization are important. The immediate burden of low-dose airborne allergen exposure falls on the mucosal surfaces of the respiratory tract. Mucoiliary clearance mechanisms in the nasal airway, trachea, and bronchi carry the inhaled allergens to the back of the pharynx where the antigen is swallowed. The secretory immune system of the gastrointestinal tract bears the primary burden of exposure to sensitizing doses of both ingested and inhaled allergens. The route of sensitization can lead to the subsequent sensitization of other mucosal surfaces. Sensitized lymphoid cells from gastrointestinal mucosal lymphoid aggregates reach the lymphatic circulation and home preferentially to the submucosal lymphoid structures of other secretory surfaces. Specifically sensitized cells generated in the gut can home to breast and peribronchial lymphoid aggregates and produce antibodies in the colostrum and respiratory tract secretions, respectively, without generating serum antibodies.

Protection and Disease

The normal protective function of anaphylactic hypersensitivity is not known. The preservation of the ability to generate IgE antibodies during the evolution of humans and animals suggests anaphylactic hypersensitivity has intrinsic survival value. Circumstantial evidence suggests IgE could have an important role in the protection of secretory surfaces against parasites and related pathogens. This evidence includes:

1. the route of naturally occurring sensitization
2. the developmental sequence in which IgE antibodies are generated to specific environmental allergens
3. the localization of the IgE producing cells close to mucosal surfaces
4. the nondestructive nature of IgE initiated reactions.

The reaction of IgE antibodies and allergens initiates increased local mucus production, smooth muscle contraction, augmented blood flow, increased vascular permeability, and local edema. These changes have been viewed as a gate-keeping activity in which IgE antibodies modulate an intermediate level of nondestructive responsiveness between antigen exposure and the more destructive components of the central immune system.

An important hypothetic concept can be deduced from this circumstantial evidence. The importance of this concept lies in its potential application to the pathophysiology of many clinical problems. The failure to retard the excessive growth of gut microflora, neutralize local toxic effects, and hasten elimination is known to cause injury to the absorptive surface of the gastrointestinal tract and maladsorption. Reliance on the destructive components of the central immune system for protection against gut microflora would also destroy the absorptive surface, impair nutrition, and lead to disease or death. The nondestructive neutralizing capacity of IgA coproantibodies are ideally suited to retard the rapid growth phase of ingested pathogens and help sequester gut microflora. For those gut flora antigens that reach the lamina propria of the bowel, the nondestructive tissue changes initiated by IgE antibodies provide an additional protective mechanism. The IgE initiated production of increased local edema, mucus secretion, and smooth muscle contractions in the nonrigid bowel wall retards antigen exposure without tissue destruction, causes diarrhea, and hastens elimination of the offending antigen and associated microflora. The adaptive significance of this postulated mechanism stems from the way IgE initiated reactions decrease gut transit time and augment the protective effects of IgA coproantibodies. The maladaptive significance of this hypothetic mechanism stems from the way the inappropriate initiation of a similar IgE reaction in a semirigid structure, such as the nasal airway, larynx, trachea, or the bronchi, would impair respiration and causes disease. An inappropriate localization of IgE-initiated protective mechanisms could account for allergic respiratory diseases. The route and kind of allergen exposure, the normal physiologic function of the site of the reaction, and the potential penalty of alternative forms of immune protection are important considerations in hypotheses about possible protective function of IgE antibodies.

Parasitism

A protective role for anaphylactic antibodies has been identified in experimental studies of certain parasitic diseases. These experiments identify mechanisms of possible importance in the maintenance of hemostasis between the host and gut microflora. Much evidence suggests these protective mechanisms are complex. For example, the subcutaneously injected larvae of the rat hookworm (*Nipponstrongylus brasiliensis*) migrate through the tissues of the infected animal to the lungs and moults, and then continues to the small intestine via the trachea and the esophagus. The nematode matures in the small intestine and mates. Egg production begins approximately 1 week after infection. A "self cure" of the infestation begins 2 weeks later, and results in a sharp reduction in egg production and in the expulsion of the adult worms from the intestine. This self cure occurs coincident with the development of anaphylactic hypersensitivity to worm allergens. The self cure can be blocked by antihistamines, antiserotonin compounds, and corticosteroids. Sensitized cells and antibody are both needed to passively transfer this capacity for self cure. Active immunization with the nematode allergen in Freund's complete adjuvant and generation of nonanaphylactic antibodies fails to protect against worm infestation.

Parasite induced potentiation of IgE antibody production to other allergens may be a part of this protective mechanism. When mice are experimentally infected with *Nematospiroides dubius*, immunity is induced more slowly and is primary directed toward the larval stage of the infection. Following oral ingestion, the parasite larvae burrow into the wall of the small intestine, mature, and return as adults to the gut lumen in 7 to 8 days. Larvae infected mice develop leukocytosis without eosinophilia. The intestinal nodules at the site of larval implantation become infiltrated with neutrophils and ultimately disappear. The adult worms usually persist in the gut lumen for months, but expulsion can be induced within 2 to 4 hours by a second larval infection. The natural history of infection in conventional mice contrasts sharply with the outcome in germ-free mice. In the absence of bacteria, fewer *N. dubius* larvae develop into adult worms, and both worm survival and egg production is reduced. Germ-free mice develop a marked eosinophilia during the larval stage of the worm infection. The intestinal nodules become infiltrated with mononuclear cells and persist at the site of the larval implantation in the small intestine. Fewer worms are returned to the gut. Bacterial monocontamination of the germ-free mouse reduces the number of persisting intestinal nodules and facilitates worm development and survival; this trend is most pronounced with species of bacteria that normally infest the mouse small intestine. These observations suggest concurrent exposure to the bacterial components of the gut flora are an important component of the natural history of worm infestation. The potentiation of nondestructive anaphylactic hypersensitivity to other antigens may be an important part of this postulated mechanism. Similar mechanisms could contribute to the anaphylactic hypersensitivity to many common allergens.

Diagnostic Tests

The direct allergen skin test is the single most important diagnostic aid in the evaluation of allergic patients. Currently available in vitro tests for IgE antibodies to allergens are less reliable and more costly than the direct allergen skin test. Epicutaneous (scratch or prick) and intradermal (intracutaneous) skin tests are most often used to evaluate atopic patients. The direct allergen skin test elicits a local flare reaction caused by vasodilation and a central wheal due to the edema formed by fluid lost from blood vessels. An example of an epicutaneous scratch test is illustrated in Figure 21-5. Two bisecting perpendicular diameters of the flare or wheal are measured and used to estimate the area of the reaction. Using this simple technique, replicate measurements of the flare response provoked by the same allergen dose in the same exquisitely sensitive individuals yield coefficients of variation of 3 to 7 percent. This compares favorably with the 17 to 24 percent coefficients of variation observed with in vitro measurements of IgE antibodies. Epicutaneous tests with 100 to 400 μg/ml allergen concentrations yield very few false positive reactions and rarely cause anaphylaxis. Less sensitive patients can be tested with an intradermal injection of 0.03 ml of a sterile $^1/_{1000}$ dilution of the allergen concentration used for epicutaneous tests. Intradermal skin tests are more sensitive than epicutaneous tests, but elicit an appreciable incidence of false positive reactions and are more likely to cause anaphylaxis. With both types of skin tests, a positive reaction begins within seconds with local itching

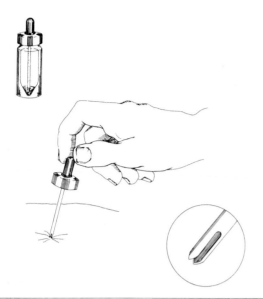

Figure 21-5. This form of the direct epicutaneous skin test uses a two-tined solid needle to make a minute circular scratch on the outer surface of the skin. The small hollow space between the barbs carries a microdrop of allergen to the scratched skin surface. Measurements of the flare response induced by this technique yield reproducible estimates of allergen responsiveness. Measureable differences detected by this technique in atopic patients are 10- to 20-fold better than the conventional 1 to 4 + method of scoring skin tests. The direct skin test is clinically useful and can be used as a quantitative measure of anaphylactic hypersensitivity.

followed by the development of erythematous flare. Intense reactions exhibit a pale edematous, elevated irregular central wheal at the test site. Positive reactions reach maximum size in 15 to 30 minutes and usually subside in 2 to 6 hours.

Oral and inhalation tests with allergens are also clinically important. These provocative in vivo tests elicit local anaphylactic reactions in the mucous membranes of the respiratory and gastrointestinal tracts. A diagnosis of food allergy or occupational respiratory impairment made on the basis of positive skin tests may be misleading because of differences in the reactivity of the skin and the mucous membranes. Oral ingestion or inhalation of the allergen is used to provoke the characteristic symptoms and physiologic changes associated with the disease. Oral provocation tests are the only reliable way to identify the cause of food allergy. Allergen inhalation challenge is necessary to unambiguously establish a diagnosis of occupational respiratory allergy. The initial allergen dose selected for oral and inhalation challenges should be less than the least dose capable of reliably producing a cutaneous reaction in the patient. The dose is increased until a reaction is produced or the test dose exceeds the level of environmental exposure. The risk of anaphylaxis during oral and inhalation tests increases with the dose of allergen. Oral and inhalation provocative tests are not without risk and should not be attempted without continued observation and the availability of drugs to treat anaphylaxis.

The P–K test is the classic method for measuring serum anaphylactic antibodies. The least dilution of donor serum giving a positive reaction 24 hours or more after adoptively sensitizing skin test sites in a nonatopic recipient is taken as the titer of anaphylactic antibodies in the serum. The P–K test has been used to detect IgE antibodies in patients who cannot be subjected to direct skin tests. Experimental in vitro immunochemical assays of IgE antibodies have replaced this use of the P–K test because of the risk of transfer of hepatitis to the serum recipient.

Experimental Methods

Components of the allergic reaction initiated by IgE antibodies can be measured by in vitro methods such as the Schultz–Dale bioassay of mediator release (Fig. 21-6), chemical measurements of histamine release from passively sensitized peripheral blood leukocytes (Fig. 21-7), and isotope (Fig. 21-8) and enzyme immunoassays. A modified form of the Schultz–Dale method using passively sensitized human or monkey smooth muscle has been used to detect human anaphylactic antibodies.

The low concentrations of IgE in serum and body secretions can be measured by immunoassay using radioactive isotopes or an enzyme as a ligand on antibody or antigen, as described in Chapter 12. Radioisotope-labeled IgE is allowed to bind to anti-IgE antibodies covalently linked to an insoluble polymer. The unlabeled IgE in an unknown serum competes with the isotope-labeled IgE for binding sites on the polymer. The resulting decrease in binding of the isotope-labeled IgE is proportional to the quantity of IgE in the unknown serum. This assay has been used to evaluate IgE concentrations in the serum, secretions, and the site of IgE production. A modification of the assay has been used to assay the in vitro production of IgE by cultured B cells.

Radioimmunoassay is also used to measure the concentration of specific IgE antibodies. Methods for measurement of IgE concentrations make use of an antiglobulin reaction similar to the Coombs test (Fig. 21-8). A specific allergen is covalently linked to an insoluble polymer, incubated with test serum, and then washed free of unbound antibodies and other serum proteins. The antigen-polymer complex and bound IgE antibodies are incubated with immunoadsorbent purified radioisotope-labeled anti-IgE And again washed. The amount of isotope retained by the polymer is measured to determine the concentration of specific IgE in the unknown serum.

IgE and Human Disease

Serum IgE averages a seven-fold increase between birth and early adult life and does not change appreciably with advancing age. The distribution of IgE concentrations in healthy persons appears multimodal and ranges between 5 and 1200 nanograms per milliliter of serum. Rarely, exceptionally high concentrations (10,000 ng/ml) can be detected in apparently healthy individuals. Serum IgE concentrations in atopic patients overlap the range observed

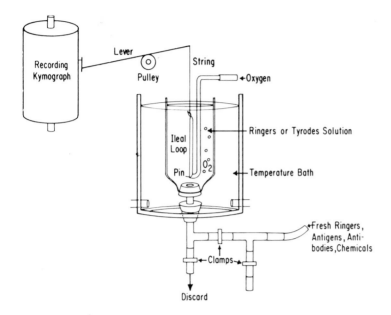

Figure 21-6. The Schultz–Dale type in vitro reaction provides a measure of permease released by antibody sensitized tissue. Introduction of antigen into the water bath results in the release of histamine, which causes contraction of the smooth muscle of the ileal loop, and displacement of the lever and scribe on the revolving recording kymograph drum.

in normal nonatopic controls. Although serum IgE concentrations are not helpful in evaluating the individual patient, averaged concentrations in atopic patients are higher than in healthy controls. Serum IgE concentrations in patients with infected atopic dermatitis are generally higher than in patients with uncomplicated dermatitis, asthma, or allergic rhinitis.

Diagnostically useful concentrations of IgE are found in atopic patients with allergic bronchopulmonary aspergillosis, who have asthma, peripheral blood eosinophilia, transient x-ray lung infiltrates, *Aspergillus fumigatis* in the sputum, positive immediate and late skin test reactions, precipitating serum antibodies to the aspergillus antigen, and extreme (more than 6000 ng/ml) elevations of serum

IgE. Elevated IgE concentrations have been reported in patients with eosinophilic gastritis and in certain forms of polyarteritis. Parasitic infections cause marked elevation of serum IgE; this trend is most marked during the tissue phase of infection. Average serum IgE concentrations are increased in smokers and in workers with byssinosis. Among the immunodeficiency diseases, patients with the Wiskott–Aldrich syndrome, advanced Hodgkin's disease and other disorders of T cell-dependent immune function frequently have elevated serum IgE concentrations. Reduced serum IgE concentrations occur in patients with generalized defects in immunoglobulin synthesis, in certain patients with IgA deficiency, and in patients with ataxia telangiectasia.

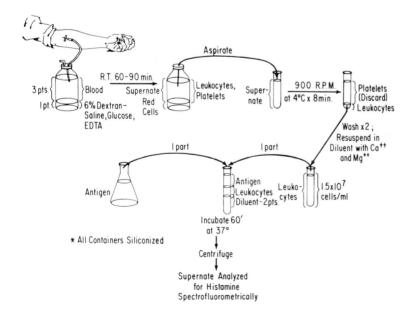

Figure 21-7. The quantity of histamine released by antibody sensitized leukocytes can be measured directly. Careful assays with this complex procedure have provided direct evidence of seasonal change and the effect of treatment on anaphylactic antibody in human disease.

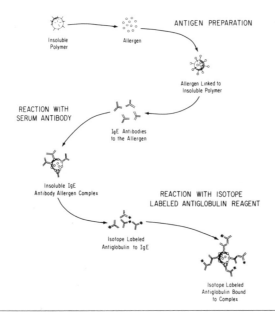

ANTIGEN PREPARATION

Figure 21-8. Direct measurements of IgE antibody to allergens is possible. An allergen is chemically linking to an insoluble polymer. The washed insolubilized allergen is mixed with serum containing IgE antibodies to the allergen. The insoluble allergen-polymer is washed free of serum proteins not specifically bound to the allergen, and the complex is mixed with a solution containing an isotope-labeled antiglobulin (antibodies to IgE). The polymer-allergen-IgE complex binds isotope-labeled antiglobulin. The insoluble reactants are washed and counted. The isotope counts bound to the insoluble polymer provides a measure of serum IgE antibodies to the allergen. *(Adapted from Aas, Johansson: Allergy Clin Immunol 48:143, 1971.)*

Allergic Rhinitis

The airborne pollutant content of the entire inspired air volume normally passes through the nasal airways. The nasal turbinates produce turbulence, which promotes mixing, warming, and humidification of the inspired air, and promotes contact between the airborne particles and the mucous membranes. Particulate pollutants adhere to the nasal mucus blanket and are carried by ciliary activity to the pharynx and swallowed. The inhalation of allergens by sensitized individuals causes the release of chemical mediators that disrupt the dynamic balance between local tissue factors and autonomic nervous system control of nasal airway patency. Acute allergen exposure causes sneezing and a watery nasal discharge, and the nasal airway exhibits mucosal erythema and strands of watery mucus bridging between enlarged turbinates and the nasal septum. Chronic allergen exposure produces hypertrophy of the nasal turbinates and mucosal pallor caused by the proliferation of mucus-secreting goblet cells.

The symptoms of pollen-induced allergic rhinitis are seasonal, whereas those caused by perennial or persisting allergen exposure continue throughout the year. Serum anaphylactic antibodies concentrations increase after sea-

sonal pollen exposure. Allergic rhinitis patients often develop nasal polyps. The IgE concentrations in fluids expressed from these polyps can exceed serum IgE concentrations, suggesting that synthesis occurs in or near the mucosal surface. Seasonal allergen exposure also provokes an enrichment of IgE in the nasal secretions as well as the serum of pollen sensitive patients.

Asthma

The wheezing and shortness of breath of asthma can be a symptom of many pathologic states. Care must be taken to exclude asthmatic symptoms that are unrelated to exposure to inhaled substances. Even when asthma is exposure related, the etiology can be complex. Symptoms can be caused by one or more combinations of several major mechanisms. First, bronchospasm and excessive airway mucus secretion can be induced by allergen exposure. Transient allergen induced asthma is probably the most prevalent cause of reversible airways disease. Second, inhalation exposure of organic dusts, including particles of tobacco smoke and noninfectious components of fungal and bacterial allergens, can induce bronchospasm and mucus secretion, and can inflame the bronchi. Finally, oxidant gas exposure, chemical or drug exposure, and certain infections predispose to subsequent specific and nonspecific provocation of bronchospasm and mucus secretion.

The nonspecific causes of asthma include temperature change, irritants, exercise, virus infections, drugs, and other chemicals. Airways hyperreactivity can be detected by bronchial challenge with histamine, methacholine, and allergens. Methacholine is an agonist of the mast cell acetylcholine receptor. Airways challenge with methacholine provides an assessment of inducible permease release as opposed to an assessment of the direct effect of inhaled histamine on bronchial smooth muscle. Nonspecific airways reactivity to cholinergic agonists can be separated from reactivity to specific allergens. The allergen-induced airways reactivity persists, whereas the reactivity to inhaled methacholine can be blocked by pretreatment with atropine.

Chemicals such as aspirin and yellow food dyes, which have a direct effect on prostaglandin production, cause nasal polyps and asthma in susceptible individuals. Food and drug additives, such as the sulfite compounds used in medications, beverages, and on seafoods and vegetables to prevent oxidation, are also a cause of asthma and severe reactions. These chemical and drug-induced reactions mimic hypersensitivity reactions, but are not caused by immune mechanisms. Industrial chemicals can act as an occupational allergens and also induce nonspecific hyperreactivity.

The inhalation of organic dusts composed of noninfectious components of bacteria and fungi is likely to become an increasingly important cause of asthma. Industrial fermentation technology is currently used to prepare many complex chemicals as well as beverages and antibiotics. Currently evolving industrial applications of gene technology, which make use of genetic information inserted into microorganisms (Chap. 20) to produce large quantities of complex proteins, are likely to use similar fermentation technology. In the absence of adequate industrial hygiene, an increased prevalence of occupational asthma can be anticipated. In addition to acting as allergens, the products of certain microorganisms may have other important effects. These include the ability to directly activate the complement properden pathway and cause inflammation, and the ability to produce lectin-mitogen activity. Interestingly, simultaneous exposure to inhaled mitogen and protein antigen are needed to induce the tissue changes characteristic of organic dust-induced hypersensitivity lung disease. Inhalation challenge with antigen or mitogen alone fails to provoke significant tissue reactions.

Atopic Dermatitis

This heritable chronic pruritic form of eczema is frequently associated with asthma and allergic rhinitis. The infantile form of the dermatitis presents as an acute dermatitis over the cheeks and usually remits during the first 6 years of life. Half of the patients afflicted during childhood develop allergic rhinitis later in life, and 20 percent have continuing symptoms and/or develop asthma. In adults, the chronic pruritic dermatitis involves the anticubital and popliteal fossae and other flexural folds of the skin. Thickening and augmentation of normal skin lines (lichenification) and dryness of the skin are characteristic of the chronic form of atopic dermatitis.

A scratch on the skin of most normal individuals produces red dermatographism, an erythematous line and subsequent flare and wheal about the scratch. This response is known as the triple response of Lewis, and implies the local release of histamine. Patients with atopic dermatitis replace the initial erythematous line with a white blanch extending outward from the line of the scratch. This response is known as white dermatographism. An injection of acetylcholine beneath the skin of nonatopic individuals produces an erythematous persisting flare beyond the injection wheal. Patients with atopic dermatitis exhibit an initial flare, but replace the flare with blanching within 2 to 3 minutes. The blanching reaction characteristic of patients with atopic dermatitis and other atopic individuals has been interpreted as evidence of vasoconstriction.

Histopathologic examination of the dermatitis reveals evidence of intercellular edema and the migration of lymphocytes, eosinophils, and neutrophils into the epidermis and into intraepithelial vesicles. Acanthosis occurs, the stratum corneum epidermal cell nuclei are retained, and the upper portion of the dermis exhibits edema, vascular dilatation, thickening of the dermal capillaries and perivascular infiltrates composed of neutrophils, lymphocytes, and eosinophils. A predominantly lymphocytic infiltrate is typical of the dermis in the chronic form of atopic dermatitis. In many respects, the histopathology of chronic atopic dermatitis resembles contact dermatitis.

Although an etiologic relationship between the dermatitis and hypersensitivity to specific allergens has not been demonstrated, positive allergen skin tests are usually found in those patients with atopic dermatitis who have allergic rhinitis or asthma. Eight out of 10 patients with active dermatitis have elevated serum IgE levels, which become normal during periods of remission. Patients with atopic dermatitis exhibit decreased delayed cutaneous hypersensitivity to recall antigens and decreased reactivity to poison ivy and tuberculin. Impaired in vitro correlates of cell-mediated immunity have been detected in patients with severe atopic dermatitis. The skin and upper airway of patients with atopic dermatitis is usually infested with *Staphylococcus aureus*. The pruritic excoriated dermatitis can become complicated by staphylococcal pyoderma and/or a generalized exfoliative erythroderma.

Food Allergy

Food allergy occurs in infants and children, but is uncommon in adults. Certain foods, such as peanuts and seafoods, are common causes of allergic reactions in atopic subjects. An allergic reaction is most suspect when ingestion of the food causes gastrointestinal symptoms accompanied by typical hives, wheezing, or upper airways congestion. Gastrointestinal symptoms alone or other atypical allergic symptoms following food ingestion are much more difficult to interpret in relation to possible food allergy. Unexplained gastrointestinal symptoms are not uncommon in nonatopic adult patients and are usually attributed to an overactive bowel. The occurrence of similar symptoms in an atopic adult patient can erroneously suggest a diagnosis of food allergy. This suspicion can be heightened by food allergen skin tests, which are often falsely positive. It is important to establish a correct diagnosis; patients with unexplained symptoms and positive skin tests often avoid important nutrients because of presumed food allergy.

Unlike the respiratory and cutaneous allergic disorders, a positive direct allergen food skin test is of little value in patients with food allergy. A negative food allergy skin test provides more reliable information. Tests with elimination diets and the cautious ingestion of suspect foods are handicapped by patient apprehension and the chance of occurrence of atypical emotional symptoms. Because of these problems, the reproducible provocation of the patient's symptoms following blinded oral food challenge is necessary to establish a diagnosis of food allergy. When the reaction attributed to the food is life-threatening, food challenge tests should not be attempted without adequate measures to control serious allergic reactions.

Drug Allergy

Adverse drug reactions are not rare. Drug hypersensitivity is one cause of adverse drug reactions and, if present, limits the ability to intervene in the disease process. This handicap becomes a clinical problem when therapeutic al-

TABLE 21-2. POSSIBLE CAUSES OF ADVERSE DRUG REACTIONS

Classification	Reaction Mechanism	Examples of Clinically Important Adverse Reactions
Pharmacologic—adverse reactions caused by the expected drug action	Excessive drug dose Primary alteration Secondary metabolic alteration	Medication error Constitutional difference in the activity of an enzyme Disease induced failure of liver metabolism or renal excretion
Idiosyncratic—adverse reactions caused by an unusual drug action	Constitutional metabolic difference Constitutional immunologic difference Immunity altered by the disease Immunity altered by the drug	G-6-PD deficiency, dilantin-induced megaloblastic anemia, isoniazid-induced polyneuritis Antibiotic induced changes in host microflora; colitis, vaginitis; Jarish–Herxheimer endotoxin reaction Epstein–Barr virus induced reaction; airways hyperreactivity induced by respiratory syncytical virus X-ray or immunosuppressive therapy, hydralazine, dilantin, penicillamine, and gold-induced immune alterations
Allergic—adverse reactions caused by hypersensitivity to the drug	Reactions caused by responsive cells Reactions caused by serum antibodies and other macromolecules	Antigen induced drug reactions analogous to contact reactions and the cutaneous basophil response Mitogen reactions Noncomplement dependent IgE reactions—penicillin allergy; anaphylaxis; angioedema; urticaria; asthma Complement-dependent antibody induced reactions —serum sickness; nephritis; arthritis
Pseudo-allergic—mimics a hypersensitivity reaction	Biochemical	Nonspecific activation of inflammation and/or complement by chemicals (aspirin, food dyes, metabisulfite) or by aggregates (x-ray contrast media, insulin, polyanions)

ternatives are not available. Fortunately, many instances of presumed drug allergy are actually caused by other kinds of adverse drug reactions. The diagnosis of penicillin allergy can be corroborated by a skin test. Effective methods exist for desensitizing some penicillin-sensitive patients when the use of the antibiotic is highly desirable. Adequate clinical test procedures do not exist for other drugs capable of causing drug allergy. Provocative in vivo tests are occasionally used in patients with drug fever and rash, but are not without hazard.

Table 21-2 presents a useful formulation of the many possible causes of adverse drug reactions. A determination of (1) the circumstances in which the reaction occurred; (2) the temporal sequence between drug exposure and the onset of specific symptoms; and (3) the patient's past susceptibility to other allergic disorders can provide the basis for a clinical assessment of the problem. An awareness of the possible occurrence of idiosyncratic interactions between the drug and innate attributes of the patient; between two or more specific drugs; and between the drug and microorganisms is important in the clinical evaluation of patients with possible drug allergy.

Anaphylaxis

The clinical recognition of pruritis as a first symptom of anaphylaxis in patients is important. Anaphylaxis usually begins in patients with intense itching about the ears, scalp, and throat. Prompt parenteral treatment with 0.1 to 0.3 ml of $1/1000$ adrenalin can prevent death and modify the severity of the reaction. If untreated, the pruritis is followed rapidly by angioedema, urticaria, asthma, malaise, and weakness. Syncope, cyanosis, shock, and death can occur. Fatal anaphylaxis is rare and usually results from respiratory failure. Anaphylaxis occurs most often in atopic patients who have a constitutional predisposition to severe allergic reactions. An awareness of the reversibility of this violent form of hypersensitivity is clinically important.

Adoptive sensitization with human anaphylactic antibodies can cause atopic disease in transfused nonatopic patients. Following transfusion of blood from an atopic donor, the serum P–K titer of anaphylactic antibodies in the serum of the recipient decreases within a few hours and allergen challenge does not cause a reaction. Anaphylactic antibodies become fixed to the recipient's tissues during a 4-hour latent period, after which the direct allergen skin tests become positive and reach maximum reactivity within 24 to 48 hours. Passively transferred anaphylactic antibodies can be detected in the normal recipient for 6 to 8 weeks, and allergen inhalation during this period causes hay fever and asthma.

The incidence of human anaphylaxis decreased following the substitution of antibiotic therapy for heterologous serum therapy of infectious diseases. Allergic reactions to penicillin and other drugs then became prevalent causes of anaphylactic hypersensitivity in patients. Wasp and bee stings also cause unexpected anaphylaxis. Certain forms of experimental therapy, such as the suppression of immunity in transplant patients with heterologous antihuman lymphocyte and thymocyte serums can cause anaphylaxis and serum sickness.

Hyperimmune human sera against certain toxins, including snake venoms and infectious agents, are available and can be used to prevent or treat tetanus, mumps, measles, hepatitis, and other infectious agents that cannot be treated with antibiotics. The production of human monoclonal antibodies (Chap. 12) to tetanus toxin has been reported. Extension of this novel technology could make possible passive immunotherapy with homologous antibodies to a large number of infections that cannot be effectively treated at the present time. The risk of anaphylaxis from passive immunotherapy with homologous serum or antibodies is remote.

Hypersensitivity and Infection

Figure 21-9 presents a simplification of the response of allergic patient to microorganisms. Immunity, hypersensitivity, and unresponsiveness to different antigens of the same infectious agents can occur simultaneously. For example, a patient who is sensitized and partially protected to toxins and other antigenic components of an infectious agent can be allergic to harmless components of the same organism. Sufficient exposure to the infectious agent may cause an allergic as well as an infectious disease. An allergic component of an infectious disease is suspected when the symptoms are unusually severe, protracted, or occur in a known

allergic patient. Modification of the hypersensitivity reactions becomes an important part of the treatment of the infection.

Treatment of Atopic Diseases

There are three main approaches to the treatment of allergic diseases: (1) preventive avoidance of allergens, (2) pharmacologic intervention in the reaction, and (3) immunologic modification of the ability to react to allergens. Preventive avoidance is the most effective form of therapy. Unfortunately, freedom from allergen exposure is frequently not consistent with a satisfactory quality of life. Pharmacologic intervention becomes necessary. Drug therapy is most effective in patients with mild transient forms of allergic disease that pose no risk of continuing impairment or death. Allergen immunotherapy is indicated in patients who can not be managed by more conservative measures.

Allergen immunotherapy is a form of repetitive immunization. This treatment has also been called desensitization, a term that is probably inaccurate. Injection therapy is initiated with an extremely small dose of allergen. The allergen dose is progressively increased in small increments as tolerated by the patient until it exceeds the quantity of environmental allergen usually encountered by the patient. The speed of the dose progression is limited by local reactions at the injection site. Properly formulated allergen mixtures and treatment plans do not cause significant local or systemic reactions. The effect of therapy is proportional to the total quantity of allergen administered over time. Repeated high allergen doses generally yield maximum benefit. This form of treatment is indicated in:

1. patients who are at risk of anaphylaxis
2. patients who are incapacitated by the pharmacologic effects of medications ordinarily used to control their symptoms
3. patients with persisting symptoms and impairment despite adequate pharmacologic management.

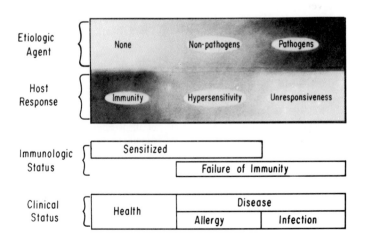

Figure 21-9. The relationships among etiologic agents, host responsiveness, and the immunologic and clinical status of patients can be represented schematically. Sensitization for allergic or hypersensitivity reactions probably represents an intermediate form of host protection.

Continuing impairment due to a hypersensitivity disease and/or the risk of death are the primary indications for allergen immunotherapy.

Convincing evidence of the effectiveness of immunotherapy has been obtained with pollen allergens in patients with allergic rhinitis, with stinging insect venom allergens in patients with anaphylaxis, and with cat dander in patients with asthma. Allergen immunotherapy is a form of physiologic conditioning analogous to exercise and not a form of pharmacologic intervention. The precise mechanisms by which immunotherapy alters the natural history of allergic disease are not clearly known. Evidence exists for three potentially important mechanisms. First, treatment causes a nonspecific effect on the overall allergen reactivity of the patient. This effect may be due to a physiologic change in reactivity or a nonspecific effect on the production of IgE antibodies. Second, immunotherapy induces a change in specific reactivity. Experimental methods for the in vitro measurement of IgE antibodies have been used to follow the seasonal variation in IgE antibodies to ragweed allergen. The usual seasonal increase in immunotherapy patients is decreased or does not occur in treated patients. The magnitude of this induced change is generally proportionate to the total allergen dose used for treatment. The third potentially important effect is the generation of blocking antibodies. Parenteral immunization with allergens produces serum IgG antibodies that block passive anaphylactic sensitization. IgG antibodies are produced in greater quantity than IgE antibodies. Serum and extracellular fluid antibodies compete with cell-bound IgE antibodies for allergen, thereby decreasing the chance of mediator release. The serum antibodies of the IgG class that block the release of the chemical mediators of anaphylaxis are called blocking antibodies.

FURTHER READING

Books and Reviews

Bienenstock J: The physiology of the local immune response and the gastrointestinal tract. In Brent L, Holborow J (eds): Progress in Immunology II. New York, American Elsevier, 1974, vol 4

Boushey HA, Holtzman MJ, Sheller JR, et al.: Bronchial hyper-reactivity. Am Rev Resp Dis 121:389, 1980

Buckley CE: Genetics of allergy. In Gupta S, Good RA (eds): Cellular, Molecular and Clinical Aspects of Allergic Disorders. New York, Plenum, 1979

Kaliner M, Shelhamer JH, Davis PB, et al.: Autonomic nervous system abnormalities and allergy. Ann Intern Med 96:349, 1982

Middleton E, Reed CE, Ellis EF (eds): Allergy, principles and practice, 2nd ed. St. Louis, Mosby, 1983

Ogilvie M: Immunity to parasites. In Brent L, Holborow J (eds): Progress in Immunology II. New York, American Elsevier, 1974, vol 4

Salvaggio J (ed): Primer on allergie and immunologic diseases. JAMA 248:2579, 1982

Samuelsson B: Leukotrienes: Mediators of immediate hypersensitivity reactions and inflammation. Science 220:567, 1983

Tada T, Taniguchi M, Takemor T: Properties of primed suppressor T cells and their products. Transplant Rev 26:106, 1975

Selected Papers

GENERAL

Adkinson NF: The radioallergosorbent test in 1981—Limitations and refinements. J Allergy Clin Immunol 67:87, 1981

Austen KF: Biologic implications of the structural and functional characteristics of the chemical mediators of immediate hypersensitivity. Harvey Lect 73:93, 1977–78

Barbee RA, Halonen M, Lebowitz R, et al.: Distribution of IgE in a community population sample: Correlations with age, sex, and allergen skin test reactivity. J Allergy Clin Immunol 68:106, 1981

Buckley CE, Lee KL, Burdick DS: Methacholine-induced cutaneous flare response: Bivariate analysis of responsiveness and sensitivity. J Allergy Clin Immunol 69:25, 1982

Buckley RH, Seymour F, Sanal SO, et al.: Lymphocyte responses to purified ragweed allergens in vitro. J Allergy Clin Immunol 59:70, 1977

Fish JE, Ankin MG, Adkinson NF, et al.: Indomethacin modification of immediate-type immunologic airway responses in allergic asthmatic and non-asthmatic subjects: Evidence for altered arachidonic acid metabolism in asthma. Am Rev Resp Dis 123:609, 1981

Francus T, Siskind GW, Becker CG: Role of antigen structure in the regulation of IgE isotype expression. Proc Natl Acad Sci 80:3430, 1983

Hunt KJ, Valentine MD, Sobotka AK, et al.: A controlled trial of immunotherapy in insect hypersensitivity. N Engl J Med 299:157, 1978

Ishizaka K, Dayton DH Jr (eds): The biological role of the immunoglobulin E system. Washington, DC, US Government Printing Office, 1973

Ishizaka K, Ishizaka T: Identification of E antibodies as a carrier of reaginic activity. J Immunol 99:1187, 1967

Knox BA, Clarke A, Harrison S, et al.: Cell recognition in plants. Determinants of the stigma surface and their pollen interactions. Proc Natl Acad Sci USA 73:2788, 1976

MacGlashan DW, Schleimer RP, Peters SP, et al.: Comparative studies of human basophils and mast cells. Fed Proc 42:2504, 1983

Ohman JL, Marsh DG, Goldman M: Antibody responses following immunotherapy with cat pelt extract. J Allergy Clin Immunol 69:319, 1982

Patterson R, Greenberger PA, Radin RC, et al.: Allergic bronchopulmonary aspergillosis: Staging as an aid to management. Annu Intern Med 96:286, 1982

Pence HL, Mitchell DQ, Greely RL, et al.: Immunotherapy for mountain cedar pollinosis. J Allergy Clin Immunol 58:39, 1976

Reddy PM, Magaya H, Pascual HC, et al.: Reappraisal of intracutaneous tests in the diagnosis of reaginic allergy. J Allergy Clin Immunol 61:36, 1978

Siriganian RP, Hook WA, Levine BB: Specific in vitro histamine release from basophils by bivalent haptens. Evidence for activation by simple bridging of membrane bound antibody. Immunochemistry 12:149, 1975

Spiegelberg HL, Boltz-Nitulescu G, Plummer JM, et al.: Characterization of the IgE Fc receptors on monocytes and macrophages. Fed Proc 42:124, 1983

Warren CPW, Holford-Stevens V, Wong C, et al.: The relationship between smoking and total immunoglobulin levels. J Allergy Clin Immunol 69:370, 1982

Watson L, Knox RB, Creaser EH: Con A differentiates among grass pollens by binding specifically to wall glycoproteins and carbohydrates. Nature 249:574, 1974

PARASITISM

Larrick JW, Buckley CE III, Schlagel GD, et al.: Does hyper-immunoglobulinemia-E protect tropical populations from allergic disease? J Allergy Clin Immunol 71:184, 1983

Wescott RB: Metazoa-protozoa-bacteria interrelationships. Am J Clin Nutr 23:1502, 1970

DELAYED HYPERSENSITIVITY

American Thoracic Society: Diagnostic standards and classification of tuberculosis and other mycobacterial diseases. Am Rev Resp Dis 123:343, 1981

Dorsch W, Ring J: Induction of late cutaneous reactions by skin-blister fluid from allergen-tested and normal skin. J Allergy Clin Immunol 67:117, 1981

Dvorak HF, Mihm MC, Dvorak AM, et al.: Morphology of delayed hypersensitivity reactions in man. Lab Invest 31:111, 1974

Gleich GJ: The late phase of immunoglobulin E mediated reaction: A link between anaphylaxis and common allergic disease. J Allergy Clin Immunol 70:160, 1982

Snider DE: The tuberculin skin tests. Am Rev Resp Dis 125:108, 1982

ANAPHYLAXIS

Smith PL, Kagey-Sobotka A, Bleeker ER, et al.: Physiologic manifestations of human anaphylaxis. J Clin Invest 66:1072, 1980

ANGIOEDEMA AND URTICARIA

Harvey RP, Wegs J, Schocket AL: A controlled trial of therapy in chronic urticaria. J Allergy Clin Immunol 68:2762, 1981

Kaplan AP: Mediators of urticaria and angioedema. J Allergy Clin Immunol 60:324, 1977

Mathews KP: Management of urticaria and angioedema. J Allergy Clin Immunol 66:347, 1980

ALLERGIC RHINITIS AND ASTHMA

Ahmed T, Fernandez RJ, Wanner A: Airway responses to antigen challenge in allergic rhinitis and allergic asthma. J Allergy Clin Immunol 67:135, 1981

Griffin MP, McFadden ER, Ingram RH: Airway cooling in asthmatic and nonasthmatic subjects during nasal and oral breathing. J Allergy Clin Immunol 69:345, 1982

Hall WJ, Hall WB, Speers DM: Respiratory syncytial virus infection in adults. Clinical, virologic and serial pulmonary function studies. Ann Intern Med 88:203, 1978

Henderson AF, Heaton RW, Dunlop LS, et al.: Effects of nifedipine on antigen-induced bronchoconstriction. Am Rev Resp Dis 127:549, 1983

Kamat SR, Kamat GR, Salpekar VY, et al.: Distinguishing byssinosis from chronic obstructive pulmonary disease: Results of a prospective five-year study of cotton mill workers in India. Am Rev Resp Dis 124:31, 1981

Karr RM, Davies RJ, Butcher BT, et al.: Occupational asthma. J Allergy Clin Immunol 61:54, 1978

Kulczycki A: A role of immunoglobulin E and immunoglobulin E receptors in bronchial asthma. J Allergy Clin Immunol 68:5, 1981

Stevenson DD, Simon RA: Sensitivity to ingested metabisulfites in asthmatic subjects. J Allergy Clin Immunol 68:16, 1981

Willoughby WF, Willoughby JB, Cantrell BB, et al.: In vivo responses to inhaled proteins. Lab Invest 40:399, 1979

Zeiss CR, Patterson R, Pruzansky JJ, et al.: Trimellitic anhydride-induced airway syndromes: Clinical and immunologic studies. J Allergy Clin Immunol 60:96, 1977

ATOPIC DERMATITIS

Fiser PM, Buckley RH: Human IgE biosynthesis in vitro: Studies with atopic and normal blood mononuclear cells and subpopulations. J Immunol 123:1788, 1979

Sampson HA: Role of immediate food hypersensitivity in the pathogenesis of atopic dermatitis. J Allergy Clin Immunol 71:473, 1983

DRUG HYPERSENSITIVITY

Arndt KA, Jick H: Rates of cutaneous reactions to drugs: A report from the Boston Collaborative Drug Surveillance Program. JAMA 235:918, 1976

Greenberger PA, Patterson R, Simon R, et al.: Pretreatment of high risk patients requiring radiographic contrast media studies. J Allergy Clin Immunol 67:185, 1981

Latkin JD, Strong DM, Sell KW: Polymyxin B reactions, IgE antibody and T cell deficiency: Immunochemical studies in a patient after bone marrow transplantation. Ann Intern Med 893:204, 1975

Rosenow ED: The spectrum of drug induced pulmonary disease. Annu Intern Med 77:977, 1972

Sullivan TJ, Yecies LD, Shatz GS, et al.: Desensitization of patients allergic to penicillin using orally administered beta-lactam antibiotics. J Allergy Clin Immunol 69:274, 1983

FOOD ALLERGY

Aas K: The diagnosis of hypersensitivity to ingested foods. Clin Allergy 8:39, 1978

Bock SA: Food sensitivity: A critical review and practical approach. Am J Dis Child 134:973, 1980

CHAPTER 22

Overview: Future Trends in Immunology

Introduction

This chapter is an innovation, an attempt to help the reader focus on significant new trends. It is possible, in a limited sense, to forecast some of the future developments in immunology because the number of quantum jumps in knowledge likely to occur at any one time is relatively small. Each major discovery is followed by a period of relative quiescence, in which that discovery is confirmed and then extended by numerous other investigations that fill in the missing details. By identifying some of the major conceptual or technologic advances that have occurred recently, and by explaining how these are likely to influence further developments, it is possible to provide a framework that can serve as a guide to future reading. Several developments in molecular biology and immunology are so recent that they could not be fully described in the preceding chapters. Several years will elapse before some of the advances now being developed can be exploited clinically. A few of these are highlighted in this chapter, and may serve as a guide for subsequent reading.

Vaccines Constructed with Cloned Genes

One of the most exciting new concepts is leading to the production of a new generation of vaccines radically different from any previously available vaccines. Vaccination or immunization against infectious diseases has already progressed through several stages. Live vaccinia virus or attenuated rabies virus were among the first of the immunogens used. Bacterial toxins, modified to the form of toxoid, provided another set of protective substances. In this way, the infectious diseases of childhood and some of the diseases that become endemic in wartime or following massive civil disasters have been controlled. The successful cultivation of polio and flu viruses has provided other advances. In the past 80 to 100 years, most of the life-threatening diseases of western cultures have been eliminated or regulated by a combination of immunization, antibiotics, and improvements in nutrition and hygiene. Despite this, the venereal diseases have steadily increased in frequency, and many microorganisms have become or are becoming antibiotic resistant. (Often this is because an inadequate course of medication is taken, thus encouraging the emergence of resistant strains.) Tuberculosis and parasitic diseases are major causes of morbidity and contribute to early death in developing countries, and cancer is a leading cause of death in countries with a well established health delivery system. Conventional methods of immunization have been ineffective or of doubtful value against these diseases.

The reasons that vaccines for syphilis, gonorrhea, giardiasis, leprosy, and other diseases have been ineffective are varied. Sometimes it is not known why a whole organismal product is ineffective, as under the appropriate conditions, antigenicty of products of the organism can be demonstrated. Limiting concentrations of immunogen in highly complex organisms, suppression, and subversion of immunity are responsible for at least some of the failures. One of the new and exciting developments promising a new generation of vaccines comes from molecular biology.

The other arises from the demonstration that one antigen can regulate the form of immune response made to another antigen on the same molecular complex.

Chromatography, affinity chromatography, gel electrophoresis, and high performance liquid chromatography have all facilitated the preparation of relatively pure antigen. The purified antigen can be used for the production of polyclonal antibodies of high specificity, or of monoclonal antibodies. The antigen can be sequenced or the antibody can be used to recover ribosomal RNA. In either case, a probe can be contructed and the gene isolated. The isolated gene can be used to transfect a microorganism, which is then used for the production of large quantities of purified lymphokine or other modifier of the immune response. The isolated gene can also be transduced into a poxvirus.

The introduction of new genes into poxviruses offers one of the most versatile procedures for immunization imaginable. One example of the manner in which the principle can be exploited is provided through the incorporation of a gene for the protective antigen from the virus of hoof and mouth disease. Introduced into one of the many poxviruses nontoxic for cattle, it has the potential of a living replicating vaccine with a carrier-adjuvant effect made possible by the large size of the poxvirus. Such a vaccine has the potential for active immunization, but with none of the risk of hoof and mouth disease. Genes coding for antigens of hepatitis, cytomegalovirus, and malaria are among those presently being inserted into viruses of low pathogenicty. This list is likely to grow explosively in the near future.

Modification of the Immune Response

Gene transfection not only offers a means of presenting an individual with substantial quantities of antigen without exposure to the original infectious agent, but allows the presentation of an immunogen without competition from other epitopes or molecules of the original material. The advantage gained by presenting single molecules has been perceived for some time, but there have been relatively few concrete examples. One has recently been provided by studies of a tobacco glycoprotein related to ruten. Rutens are bioflavenoids present in many plant extracts. Polyphenol-containing tobacco glycoprotein was conjugated to albumin. The molecular complex elicited an IgE reponse, whereas albumin alone elicited high titers of hemagglutinating antibodies in addition to IgE. Thus, one molecule (rutin) appears to have suppressed part of the immune response to albumin. Another and more familiar example is provided by concanavalin A, which provokes a powerful immunosuppressive response in addition to being a polyclonal activator for B cells. Obviously there are numerous ways in which different components of the immune response can be preferentially stimulated, and these are likely to be intensively studied. The strategy of incorporating genes into poxviruses is likely to stimulate a search for immunogenic molecules that can be used to vaccinate against organisms that have not by themselves stimulated a protective response, as well as molecules that can turn off a response or elicit a highly specific response by a single component of the immune system. This type of research would not have been practicable in the past because of the limited potency and high cost of preparing large quantities of highly purified antigens. Both difficulties are overcome by incorporating the desired gene into a replicating highly immunogenic virus that provokes a strong cellular response. The new vaccines will be especially valuable for use in developing countries, where cost is a major consideration and where, because of less efficient delivery systems, conventional vaccines have deteriorated beyond usefulness before they reached the field station. As a by-product of this research, knowledge of molecules that provoke other types of immunity—i.e., suppression rather than responsiveness, IgE rather than IgG—is anticipated. Of greatest value would be a preparation that selectively deletes the memory cell for antibody or the cytotoxic effector memory cell.

Regulation and the T Cell Receptor

Whereas recent developments have largely elucidated the genetic rearrangements that precede immunoglobulin synthesis and the mechanisms whereby antibody diversity is achieved, research on the molecular genetics of T cells has been delayed because it has not been possible to identify the T cell receptor or to explain how T cell reactions are mediated. It now appears certain that the number of T cell receptors is low as compared to the number of immunoglobulin molecules on the surface of a B cell. Isolation of the T cell receptor was impracticable until monoclonal antibody technology was developed. Preliminary data on the first T cell receptors are now being published. Many laboratories are competing to isolate the receptors of different classes of T cell. Helper T cell receptors appear to consist of a dimer, with each chain approximately 44,000 daltons. Once the nature of the receptors is confirmed, it can be anticipated that the structural genes will be incorporated into plasmids and used to transfect L cells. An understanding of the mechanism of the generation of diversity of T cells will soon follow. This information will lead to a series of new insights into cellular reactivity, and will ultimately have great clinical relevance.

Although it is known that the individual with a defect in delayed type responsiveness is at risk for a variety of diseases, arguments still rage as to the ways in which delayed responses are protective and over the relative importance of the different subclasses of T cells, e.g., cytotoxic effectors and effectors of cutaneous basophilic anaphylaxis, pure delayed responsiveness, antibody dependent cellular cytotoxicity (ADCC), and natural killing (NK). It is not even agreed that the effectors of ADCC and NK are members of the T cell lineage. One school of thought regards all of the effectors of cellular immunity as belonging to a single lineage and representing different stages of differentiation. This hypothesis is supported by observations of fluctuations of T4 to T8 ratios in the progeny of cloned T cells, by the capacity of cells from neonatal piglets to participate in ADCC but not NK reactions, and by observations that the lesions produced by NK cells and cytotoxic T cells are morphologically identical and both have a resemblance to the lesions produced by complement. If this hypothesis is confirmed, immunotherapy using NK cells derived from cloned cytotoxic T cells should be easily accomplished.

The regulation of immune responsiveness is currently limited to procedures such as the use of adjuvants (in many forms) to augment immune responses; administration of immunosuppressive drugs and antilymphoid sera to suppress immunity; and, most recently, total lymphoid irradiation and lectin treatment of bone marrow, as preparation of the recipient of bone marrow replacement therapy. Bone marrow transplantation is being used with increasing frequency as an adjunct to chemotherapy of solid tumors. There are additional regulatory mechanisms not yet tapped. Physiologically, the nervous system and the endocrine organs can regulate immune responses in ways that have not yet been explored because we do not know what questions to ask or how to formulate them. The subtle immunosuppression of pregnancy, and the use of conditioned reflexes to modify immune responsiveness are examples of this.

Organ Transplantation

An intricate form of toleration (which is not tolerance in its usually accepted form) develops toward a long transplanted organ. In some instances it has been possible to stop all immunosuppression without any impairment of the graft. In other instances an immunization against tetanus or the onset of a minor respiratory infection appears to have triggered a violent rejection episode. The basis for this toleration is being explored.

In the mouse, the Ia antigen is now recognized as critical for the presentation of T-dependent antigen. The DR, DC/DS, and SB antigens probably serve the same function in humans. Because the mouse has a limited repertoire of Ia antigens, its capacity to respond to T-dependent antigens is also limited. An ingenuous approach to transplantation suggests that if the donor organ is denuded of all antigen-presenting cells and transplanted to a totally allogeneic host, the recipient (being unable to provide the appropriate Ia antigens) is unable to process donor antigen and cannot respond against the graft. This hypothesis is supported by several lines of experimental evidence. If substantiated for humans, we may find HLA typing for incompatibility of cadaveric organs, together with depletion of all donor macrophages and other antigen-presenting cells, the next approach to transplantation. Conversely, as an approach to the immunotherapy of cancer, bone marrow from a donor matched for some, but not all, of the

rich complement of Ia-like antigens may allow immune recognition of an otherwise antigenically undetectable tumor. In any event, more precise knowledge of antigen processing is needed.

Antigen Processing

Current dogma that antigen processing is by the dendritic cell is being challenged. Processing appears to require recognition and brief intracytoplasmic processing by a macrophage. The manner of presentation probably determines whether B cells or T cells will be triggered. These investigations are difficult because the amount of processed antigen required appears to be small and because many of the antigens are readily degraded outside the environment of the presenting cell. Understanding of presentation and processing and of the cell types that have this capability will further permit refinement of the type of immune response produced. It also appears probable that the presenting macrophage determines one path to immunity and simultaneously depresses other forms of responsiveness. This knowledge will also allow us to control the immune response more precisely.

Tumor Immunity

The knowledge that onc genes are present in all eucaryotic organisms so far examined and that translocations affecting onc and mic genes are associated with, and probably causative of, oncogenesis is reinvigorating tumor immunology because a specific product of the individual's own tumor can be sought.

An urgent question to be asked concerns the effects of blood transfusion on the cancer patient. (This is different from the possible relationship between transfusion and acquired immune deficiency syndrome [AIDS].) If blood transfusion decreases the ability of the kidney recipient to reject a transplant, it should have a comparable effect on the patient in immunostasis with the tumor.

Many immunologists hope to control immune responses by administering monoclonal antibodies to T cell subsets. Results of administering monoclonal antibodies to patient rejecting kidneys were at first highly encouraging. Because the patient with advanced malignancy is in a state of immunodepression and, perhaps, immunosuppression, it was hoped that the administration of antibodies to suppressor cells would allow immune attack to proceed against the tumor. This aspect of immunotherapy should be carefully watched, but little of clinical significance may emerge until human monoclonal antibodies become available. Much additional information about the regulation of immunity will come from studies based on the etiologic agent of AIDS, once this is known. AIDS appears to be the ultimate in acquired cellular unresponsiveness but with a compensatory polyclonal activation of B cells. Similar polyclonal activation is known for other conditions, e.g., malaria but its basis and significance have not been understood. In understanding AIDS we will thus learn about disturbances in immunoregulation in other conditions.

Concluding Remarks

Monoclonal antibodies; DNA sequencing; gene cloning and transfection; onc genes; resolution of the identity and specificity of major histocompatibility complex-linked and other immune response genes; T cell receptors; the biochemistry of antigen processing; the control of memory cells; the interrelationship among the central and autonomic nervous systems, the endocrine system, and the immune system; the integration of the effectors of immunity with those of blood coagulation; complement; and the monokine regulators of cell growth and function are all fruitful subjects for future study. This list reflects some of the newer technologies and perspectives that are beginning to expand even further our rapidly changing concepts of the immune system and of immune surveillance.

SECTION III
MEDICAL BACTERIOLOGY

CHAPTER 23

Host-Parasite Relationships

Infectious disease is the result of an unsuccessful relationship between parasite and host. It is the summation of the vectors of both agents and ranges in severity from human rabies, which is almost invariably fatal, to the common cold, which is almost invariably nonfatal. In between lie the common infectious diseases, the etiologic agents of which are described in this book. Although new agents of infectious disease continue to be described, i.e., Lyme disease, the rate of new discovery has slowed remarkably. Increasing attention is being paid to perturbations in the host defense mechanisms that open the doors to infections with our own microbial flora, a subject that is only now beginning to be understood. The advances in public health services, vaccines, antibiotics, and improved nutrition have tremendously reduced infection as one of the common causes of death in advanced societies. It has been the advances in therapeutics and consequent alteration of host defense mechanisms that have tremendously increased the incidence of infectious diseases of medical progress. Although there are many general principles regarding host defense mechanisms and determinants of microbial pathogenicity, each specific patient and each specific pathogen must be carefully studied for clues as to why that particular combination produced a particular illness. In many instances, however, the specific reasons why a particular combination results in a given clinical picture are not clearly discernible. This chapter discusses a few of the host factors that may modify infection, the problem of the compromised host, clinical manifestations of infectious diseases, and a few of the microbial attributes that are sited as the explanations for what is seen clinically.

The Host

To a considerable extent, chance plays a role in any infectious disease. An individual must be in the wrong place at the wrong time—historically, using a blanket previously used by a patient with smallpox, being in the center of a typhoid epidemic, living with someone who has active tuberculosis, or being a member of a family in which there is a case of meningococcal meningitis. What then ensues depends upon a number of host factors. All infectious diseases begin at the surface of the host, with the exception of certain intrauterine transmitted infections. The first barrier, therefore, is the gross surface area, which is usually skin, respiratory tract, gastrointestinal tract, or genitourinary tract. The specific surface involved relates to how the pathogenic microorganism reaches the host, i.e., puncture of the skin by a needle, inhalation of an aerosolized droplet, swallowing of an infected substance, or venereal transmission of a microbe. Starting with the skin, there is an obvious physical barrier aided by frequent desquamation. The spraying of a suspension of bacteria on the skin usually produces no untoward effect. Simple drying is sufficient to eliminate most bacteria, and the fatty acids on the surface layers and the skin pH exert an antibacterial effect. Direct intradermal inoculation of volunteers with virulent bacteria may require up to 5 to 10 million viable organisms. However, there are exceptions, such as *Francisella tularensis*, the cause of tularemia, which in human volunteers may require a very small number of viable organisms to produce a serious infection. Any break in the integrity of the skin will, of course, remove the physical barrier. This could range from a simple abrasion to a severe burn or drug-related exfoliative dermatitis. More commonly, a puncture wound with a foreign body, such as a splinter, a suture, or an indwelling intravenous needle and/or plastic catheter, can provide a nidus for initiating infection.

Besides the integrity of the mucosal lining, the gastrointestinal tract is protected to a great extent by the acid secreted in the stomach, which is inimical to most microbial forms of life. This acid condition, however, can be modified by the ingestion of food or other substances that neutralize the acid or by diseases and/or surgery of the stomach. Overcoming this barrier can be accomplished by large numbers of bacteria—feeding a million organisms of *Salmonella,* a common cause of food poisoning, will provoke disease in volunteers. Another feature protecting the gastrointestinal tract is its fairly rapid motility, resulting in a very rapid transit time for microbes that survive the stomach acid. In addition, the lower parts of the gastrointestinal tract have an enormous bacterial flora of their own that can play a competitive role in the establishment of an exogenous pathogenic organism but that can also be altered by the use of antibiotics, bringing new pathogens to light.

The respiratory surface has as its physical protection a mucous coating that can entrap most microbial forms of life. This mucous coating is swept up from the respiratory tree by cilia, where it is coughed up and swallowed reflexly, thereby depositing the potentially infectious material into the sterilizing environment of the stomach. Finally, the surface of the urinary tract—the ureter, bladder, urethra—has as its protective barrier the simple flow of urine. Instillation of organisms capable of causing urinary tract infections into the healthy bladder is followed by their prompt excretion in the urine, with no untoward effects. Many of these cellular surfaces, however, have receptors to which pathogenic organisms can attach and thereby lessen the effectiveness of the simple physical means of eliminating these agents. Obstruction in any of these areas—the gastrointestinal tract, the respiratory tract, the genitourinary tract—either from a physical agent or from disease, can alter these bacteria-removing barriers in such a way that either resident microbial flora or invasive organisms can replicate and produce disease.

Once microbial forms of life breach these gross physical barriers, they encounter an extremely hostile environment. At this particular point, they encounter a very complex cellular response called "inflammation." This response is as old as recorded history—*rubror et tumor cum calore et dolore.* The type of cellular response varies with both time and the specific type of microorganism. In very general terms, the polymorphonuclear leukocyte is primarily effective against bacteria. This effectiveness is greatly enhanced by the presence of complement. The later appearance of specific antibody tremendously improves the efficiency of these cells. Preexisting antibody usually prevents the appearance of disease. The macrophage and the subsequent multinucleated giant cell and epithelioid cell

are most effective against mycobacterial and mycotic infections. Inteferon, lymphocytes, and specific antibody globulin are the most effective deterrents to progressive viral infection. Lastly, the eosinophil plays a role in certain parasitic infections.

It should also be pointed out, however, that the inflammatory response may be triggered by many other events, i.e., physical, chemical, immunologic, in addition to microbial invasion, resulting in a clinical picture that at times is very difficult to separate from infection. In bacterial infections, the first 20 minutes of the host response usually determines whether or not a given invading organism will produce disease. At the cellular level, one of the earliest responses is that of polymorphonuclear leukocyte migration. Chemotaxis, the attraction of phagocytic cells to the site of infection or tissue injury, is an early event in response to infection. When the polymorphonuclear leukocyte arrives at the site of the invading microorganism, a change occurs in the plasma membrane of the phagocytic cell, resulting in invagination around the bacterial cell with subsequent fusion to produce a phagocytic vacuole. Biochemical changes within the leukocyte, at the time of phagocytosis, may kill the ingested organism. In addition, fusion of lysosomes with the phagocytic vacuole to produce a phagolysosome may result in digestion of the organism by the enzymatic contents of the lysosome.

In general, the antimicrobial systems of the polymorphonuclear leukocyte are of two types, the oxygen-dependent and the oxygen-independent systems. The oxygen-dependent system can be further subdivided into myeloperoxidase (MPO)-mediated systems, which when combined with hydrogen peroxide and appropriate oxidizable cofactor (iodide, bromide, or chloride) have marked antimicrobial activity. Myeloperoxidase-independent systems include the production of hydrogen peroxide, superoxide anion, hydroxyl radicals, and singlet oxygen. Oxygen-independent microbial systems include the acid pH within the phagocytic vacuole, the release of lysozyme, lactoferrin, and granular cationic proteins. Antimicrobial systems similar to these have been described in macrophage cell cultures. Complement possesses a dual role in the phagocytic process. Fission of fragments C3 and C5a generates chemotactic factors, while bound C3b has primarily the function of opsonization. This opsonic function is based on the ability of these fragments to bind bacteria or other microbial forms to sites that can specifically interact with receptors on the surface of phagocytic cells. The presence of preformed antibody to the invading agent tremendously improves the efficiency of phagocytosis. This preformed antibody may be from previous infection or from stimulation by a related but not necessarily microbiologic source of antigen. Any event that interferes with this classic inflammatory response obviously will allow microbes to become established. Interfering effects may range from simple foreign bodies and impaired blood supply preventing adequate polymorphonuclear leukocyte response to blatant leukopenia, due either to natural disease or the effects of drug therapy. More subtle changes in chemotaxis, phagocytosis, and intracellular killing are also now recognized as being responsible for certain infectious diseases.

In addition to the classic defense mechanisms of the polymorphonuclear leukocyte, specific antibody globulin and complement, most microbial forms, upon invasion of the host, generate a form of incompletely understood cell-mediated immunity. This form of immunity results from an interaction of lymphocytes and macrophages, with the result that both cellular types can play a role in resistance to certain types of infection, including representatives of the bacterial, mycotic, viral, and parasitic groups of pathogenic microorganisms. Unfortunately, a number of diseases, as well as certain therapeutic agents, can interfere with this form of immunity, which permits the replication of the above group of agents, resulting in disease.

Should the invading organism survive the immediate localizing effect of inflammation, there exist simple gross physical barriers, such as fascial planes, muscles, serosal cavities, and bone. More effective is the extensive barrier of the lymphatic system, with its filtering lymph nodes and phagocytic potential. It is possible, however, for microorganisms to overcome this barrier either through gross numbers or through a remarkable ability to resist phagocytic mechanisms, with the result that the bloodstream is invaded. Bacteremia (also fungemia, viremia, parasitemia) signals a failure of these local barriers and usually portends more significant illness. However, even here there is an incredibly effective clearance mechanism scattered throughout the body, characterized as the reticuloendothelial system, but concentrated in the liver, spleen, marrow, and lung. This system is comprised of a remarkably efficient phagocytic macrophage system capable of clearing the blood of a wide range of foreign agents. Despite this effective clearance mechanism, some organisms can survive to set up sites of infection distal to the site of primary invasion, with the result that patients may present with metastatic disease involving the brain, heart, joints, and other organ sites.

The Compromised Host

Every patient with an infectious disease is to some extent a compromised host. In some instances, we know enough about the defense mechanisms to pinpoint the area compromised. In many instances, however, we are unable to define the defect. With continued observations of infectious diseases, however, it has been possible to define certain settings in which increased infection can be predicted. Among the most common compromised hosts are the very young and the very old. Certain prevalent basic diseases, such as alcoholism and diabetes, are often associated with infection. Less commonly, the uremic patient is a source of host–parasite problems. Natural diseases involving the hematopoietic system, Hodgkin's disease, and other lymphomas, leukemias, and multiple myeloma are notoriously associated with high attack rates of infection. Various solid tumors, such as carcinomas and sarcomas, also present

problems with increased infection attack rates. Inherited and acquired primary immunodeficiency diseases present problems. Certain viral infections have left the host predisposed to bacterial diseases.

In addition to these basic illnesses, there exist superimposed conditions, which in some respects are the diseases of medical progress. For this reason, they are commonly found in hospitals and are associated with hospital-acquired infection or nosocomial infection. Physical barriers may be altered by the use of indwelling intravenous and urethral catheters. One can change the normal indigenous flora by the use of broad-spectrum antibiotics, thereby leaving a void to be filled by drug-resistant microorganisms, which in themselves have a low degree of pathogenicity but in the compromised host are capable of producing significant illness. A decrease in the total circulating pool of phagocytic cells can be caused by cytotoxic chemotherapy of cancers and by irradiation. Recently, altered leukocyte responses, such as decreased migration, diminished phagocytosis and decreased bactericidal capacity, have presented problems. Finally, interference with the classic immune globulin production or impairment of cell-mediated immunity by therapy with steroids or cytotoxic drugs, irradiation, or antilymphocytic globulins can lead to infection. In the compromised host, one does not necessarily have problems with the usual pyogenic cocci. Instead, one may see organisms usually selected by prior chemotherapy, such as *Pseudomonas* species, Enterobacteriaceae, *Mycobacterium,* fungi, herpes viruses, and such parasites as nematodes, protozoa, and *Pneumocystis.*

Clinical Manifestations of Infectious Diseases

Most infectious diseases start with exposure to the infectious agent, followed by an incubation period and then the manifestations of the disease. Patients ill enough to enter the health care system are usually evaluated by obtaining a history from the patient, doing a complete physical examination, and then employing selected laboratory procedures. Fever is almost always an accompaniment of clinical infectious diseases, although it may also accompany a wide variety of illnesses totally unrelated to infection. The fever may be subtle or abrupt in its onset, and it may move with great speed or at a very slow tempo. It is usually the inflammatory response that brings a patient to a physician because of the attendant pain and loss of function. It is generally possible by history and/or physical examination to localize the site of the infection. By virtue of the tempo and clinical appearance of the infection, differentiation among the usual subdivisions of microbiology is made, i.e., bacterial, virus, fungal/mycobacterial, or parasitic. However, the specific etiology of any given infectious disease can only be proven by demonstrating the infectious agent in smears or cultures or by demonstrating an appropriate antibody response. Because the number of antimicrobial drugs is so great and the spectrum so varied, it is necessary that some attempt be made to identify specifically the etiology of any infectious disease. Additional laboratory procedures, such as x-rays, have been very helpful in defining infectious illnesses more precisely. The more recent advent of computed axial tomography (CAT) scan has revolutionized the localization of abscesses. In addition, the host response to the infectious agent may be roughly gauged by measuring the number and types of peripheral white blood cells. Classically, an excessive production of polymorphonuclear leukocytes with the appearance of early forms suggests bacterial infection. Normal or low white blood cell counts with prevalent lymphocytes suggest viral infection. Eosinophilia has been associated with some types of parasitic infections. Nonspecific parameters of inflammation, such as the sedimentation rate and C-reactive protein, can be measured, and the lack of an increase in such measurements may suggest a viral infection. The least reliable, but nonetheless commonly used, parameter is the response of the patient to antimicrobial therapy.

Most infectious diseases run a reasonably predictable course, with spontaneous resolution. How antimicrobial therapy changes this clinical course is an extremely important element of infectious diseases. Finally, it should be stressed that certain infections, although they may resolve either spontaneously or with therapy, can persist in a latent form to reappear at a later date. This problem has become of increasing significance, as therapeutic immunosuppression may allow these diseases to change from a latent host–parasite relationship to an overt disease-producing relationship.

The Parasite

Of the many thousands of known microorganisms, less than 300 have acquired the ability to produce disease in humans and animals. Organisms capable of producing disease are usually referred to as "pathogens." One of the most fundamental properties of such a microorganism is its ability to multiply within the human host. However, even here, there are a few exceptions wherein organisms multiply on the surface of the host. In addition, there are not uncommon intoxications as opposed to infections resulting from products produced outside the host but coming in contact with the host. "Virulence" is a term usually used to designate the degree of pathogenicity of a given type of microorganism or of a specific strain. Virulence is usually measured by the numbers of microorganisms necessary to kill or alter a particular animal species or test system under standardized conditions. The importance of genetic variability in both pathogenicity and virulence has been repeatedly demonstrated. "Communicability" may be defined as the ability of the organism to spread under natural conditions from human to human or from animal to human or human to animal. The mechanisms of such communicability cover a remarkable range, such as simple respiratory droplet nuclei, anal/oral ingestion, contact, insect or animal bite.

Surface Factors

Most infectious diseases require that the parasitic agent initiate infection on the body surfaces—the skin or the mucous membranes of the respiratory, gastrointestinal, or genitourinary tract. The initial contact resulting in adherence of the microorganism to the underlying cell is an important determinant of pathogenicity. There is evidence to suggest that there are receptor sites on both bacteria and surface cells that would allow the microorganism to attach to the host and thereby gain pathogenic properties over organisms that do not possess these receptors. This has been abundantly documented in the area of virology and more recently extended to chlamydia and mycoplasma. Bacteria and protozoa have also been shown to take advantage of this mechanism of pathogenicity. *Escherichia coli* both in the urinary tract and in the enterotoxic relationship in the gastrointestinal tract has utilized this mechanism through the fimbriae. In addition, *Vibrio cholerae* and *Shigella dysenteriae* also adhere to gastrointestinal epithelium, and *Plasmodium* species, the cause of malaria, utilize this mechanism for initial attachment to the susceptible erythrocyte.

In addition to attachment to a susceptible host cell, the microbial surface can play a very significant role in resisting phagocytosis, thereby adding another dimension of pathogenicity. These surface characteristics are usually the result of a slime layer or capsule of either polysaccharide or polypeptide structure. Examples of such surface structures are the polysaccharide capsules of *Streptococcus pneumoniae*, *Neisseria meningitidis*, and *Haemophilus influenzae*. Patients with defective reticuloendothelial systems (e.g., as a result of splenectomy) are particularly susceptible to bacteremic infections with these pathogens. In addition, *Bacillus anthracis* presents an example of a surface capsule of polypeptide composition. Although the surface capsule can explain why the organism can multiply within the human host, it does not explain why disease is produced. Antibodies directed specifically against these surface antigens dramatically reverse the resistance to phagocytosis. Such antibodies are called opsonins and, in the presence of complement and an intact polymorphonuclear leukocyte, present the most fundamental mechanism of resistance against bacteria that have crossed the surface barriers of the host. However, there are some microorganisms that have developed an ability to survive within phagocytic cells as a partial explanation of their pathogenicity.

Toxins

A very large number of substances have been described in association with bacteria that can either damage host cells or interfere with defense mechanisms. In relatively few instances do these substances explain what is seen clinically or provide a rationale for treatment and/or prevention. In some instances, the mechanisms of action of these substances have been clearly described at the biochemical level. These substances are called toxins and can be divided into those that are easily separated from the bacterial cell and those that are intimately connected with the bacterial cell.

Extracellular Toxins

These substances, called exotoxins, are usually protein in nature and possess enzymatic activity. They can be roughly divided into two-component protein toxins and the bacterial cytolysins.

Two-component Protein Toxins. The isolation, purification, and chemical characterization of the two-component toxins have revealed a fascinating thread of similarity despite the startling diversity of the organisms and the diseases from which they were isolated. Usually one component is associated with absorption to the surface of a susceptible cell (implying specificity) and in some way is related to the transfer of the other component across the cell membrane. One mechanism suggested for this transfer is the making of a tunnel with hydrophilic channels. Once inside the cell, the toxic component exerts an enzyme-like effect that alters the normal function of the susceptible cell. Cholera toxin has been divided into an A and a B component. The A component is subdivided into two units, while the B component has five units. Although many B components can attach to ganglioside receptors on the susceptible cell surface, only a few A units are required to get across the cell membrane in order to produce the toxic effect. The A unit of this toxin activates adenylate cyclase in a remarkably wide spectrum of cells. The ultimate effect of this toxin is the production of a profuse outpouring of fluid and electrolytes, resulting in a profound life-threatening diarrhea. Interestingly, some strains of *E. coli* capable of producing enteritis have been found to produce a similar heat-labile toxin.

Diphtheria toxin consists of a single polypeptide chain that can be cleaved proteolytically into an A and a B fragment. The B fragment is the binding element, and the cell receptor site involves a glycoprotein-nucleotide interaction. The A fragment is then transported across the cell membrane to exert its toxic effect. It catalyzes the ADP ribosylation of elongation factor-2 (EF2), thereby interfering with protein synthesis. Although many types of cells can be damaged by this toxin, the major clinical manifestations besides the pharyngitis are carditis and peripheral neuropathy. A similar toxin is produced by *Pseudomonas aeruginosa*, a gram-negative rod commonly associated with nosocomial infections because of its antibiotic resistance. However, the role of this toxin in the pathogenesis of disease caused by this organism is less clear.

Tetanus toxin is less well understood, since the molecular site of action of this protein is not known. The protein is released from the bacterial cell as a single-chain polypeptide. Nicking of this chain proteolytically produces a heavy (H) and a light (L) chain. The H chain binds to a neuroreceptor ganglioside. Determination of whether or not the L chain is responsible for the toxicity will have to await the discovery of the molecular site of toxin activity. Clinically, the toxin releases inhibitory impulses, with the production of trismus (lockjaw) and tetanic convulsions. Botulinum toxin, the most powerful of all biologic toxins, is also secreted as a single polypeptide chain. Nicking this chain proteolytically produces an H and an L chain. Evidence suggests that the H chain is associated with binding. Here again, the receptor site and the molecular nature of the activity of the toxin remain to be elucidated, although ultimately the release of acetylcholine is inhibited with the clinical result that profound paralysis of a frequently fatal nature develops.

Bacterial Cytolysins. These substances were initially discovered because they damage the membranes of erythrocytes with an easily observed result—hemolysis. Much of the earlier work refers to these toxins as hemolysins. However, some of these toxins have a much broader range of target cell membranes, and some that are quite toxic to intact cells have no effect on red blood cells. Although hemolysis is occasionally seen as a clinical complication of infectious diseases, the fact that these toxins can produce tissue necrosis and could be lethal upon intravenous administration to experimental animals has resulted in their receiving a great deal of experimental attention as a possible explanation for clinical syndromes. Although some of these cytolysins probably explain what is seen clinically, there are as yet few therapeutic or preventive applications. Examples of such cytolysins are (1) those that hydrolyze membrane phospholipids (phospholipases), seen with *Clostridium* and *Staphylococcus* pathogens, (2) thiol-activated cytolysins (also called oxygen-labile) that bind to cholesterol and alter membrane permeability and are seen with *Streptococcus* and *Clostridium* pathogens, and (3) cytolysins that exert a detergent-like activity on cell membranes (i.e., *Staphylococcus* pathogens) and have a remarkable range of activity in membrane systems with a very rapid rate of lysis. There also remain cytolysins isolated from bacteria, for which there is no explanation at present of their mechanism of action.

A few of the above toxins can be altered by exposure to dilute concentrations of formaldehyde, with the production of toxoids—substances that have retained antitoxic antigenicity but have lost physiologic toxicity. These have provided powerful tools for the prevention of disease (e.g., tetanus and diphtheria).

Endotoxins

The second large category of toxins, those that are intimately associated with the bacterial cell, are called endotoxins. They are classically associated with gram-negative bacteria, and considerable chemical manipulation is required to extract them from bacterial cell walls. Although protein-polysaccharide-lipid complexes can be isolated that have considerable immunologic significance, most of the toxic activity has been ascribed to the lipid moiety. Although these substances are potent pharmacologic agents, they differ from the above described exotoxins in that they are very stable to heat, do not form toxoids, are less specific in their action, and are considerably less toxic on a weight basis. A fairly large number of toxic effects have been described following the administration of endotoxins to experimental animals. However, the relationships of these effects to the clinical manifestations of disease are not clear. The most obvious effects, such as pyrogenicity, the ability to produce shock, and their effect on nonspecific immunity, are probably the most important.

Host Response

There are a number of microbial agents for which no toxic substance has been implicated in the pathogenesis of the disease. In some instances, much of the clinical picture is the result of the host response to infection. This may range from the sudden outpouring of polymorphonuclear leukocytes producing consolidation of the lung in *S. pneumoniae* infections to the cell-mediated immunity resulting in activated macrophages, granuloma reaction, and caseous tissue necrosis seen with *Mycobacterium* infections.

FURTHER READING

Books and Reviews

Ciba Foundation Symposium 80: Adhesion and Microorganism Pathogenicity. London, Pitman Medical, 1981

Cohen P, van Heyningen S (eds): Molecular Action of Toxins and Viruses. New York, Elsevier Biomed Press, 1982

Costerton JW, Irvin RT, Cheng KJ: The role of bacterial surface structures in pathogenesis. CRC Crit Rev Microbiol 8:303, 1981

Dixon RE (ed): Nosocomial Infections. New York, Yorke Medical Books, 1981

Gill DM: Bacterial toxins: a table of lethal amounts. Microbiol Rev 46:86, 1982

Grieco MH (ed): Infections in the abnormal host. New York, Yorke Medical Books, 1980

Hahn H: The role of cell-mediated immunity in bacterial infections. Rev Infect Dis 3:1221, 1981

Hanson JM, Rumjanek VM, Morley J: Mediators of cellular immune reactions. Pharmacol Ther 17:165, 1982

Hirsch RL: The complement system: its importance in the host response to viral infections. Microbiol Rev 46:71, 1982

Horwitz MA: Phagocytosis of microorganisms. Rev Infect Dis 4:104, 1982

Jeljaszewicz J, Wadstrom T (eds): Bacterial Toxins and Cell Membranes. New York, Academic Press, 1978

Mizuro D, Cohn ZA, Takeya K, et al. (eds): Self-defense Mechanisms—Role of Macrophages. Tokyo, University of Tokyo Press/Elsevier Biomed Press, 1982

O'Grady F, Smith H (eds): Microbial Perturbation of Host Defenses. New York, Academic Press, 1981

Powanda MC, Canonico PG (eds): Infection—The Physiologic and Metabolic Responses of the Host. New York, Elsevier Biomed Press, 1981

Root RK, Cohen MS: The microbiocidal mechanisms of human neutrophiles and eosinophiles. Rev Infect Dis 3:565, 1981

Rosenstreich DL, Weinblatt AC, O'Brien AD: Genetic control of resistance to infection in mice. CRC Crit Rev Immunol 3:263, 1982

Smith H, Skehel JJ, Turner MJ: The Molecular Basis of Microbial Pathogenicity. Deerfield Beach, Fla, Verlag Chemie, 1980

Taylor PW: Bactericidal and bacteriolytic activity against gram-negative bacteria. Microbiol Rev 47:46, 1983.

Wolbach B, Baehner RL, Boxer LA: Review: clinical and laboratory approach to the management of neutrophil dysfunction. Isr J Med Sci 18:897, 1982

Selected Papers

Miles AA, Miles EM, Burke J: The value and duration of defense reactions of the skin to the primary lodgement of bacteria. Br J Exp Pathol 38:79, 1957

Steere AC, Grodzicki RL, Kornblatt AN, et al.: The spirochetal etiology of Lyme disease. N Engl J Med 308:733, 1983

CHAPTER 24

Normal Flora and Opportunistic Infections

Ecology is the study of the interactions between living organisms and their environment. Disease in humans and animals caused by microorganisms is only a small segment of this constantly changing balance of forces. Microorganisms are ubiquitous and are present under all conditions that permit the existence of any form of life. They are constantly engaged in the synthesis of new organic compounds and in the degradation of complex animal and plant tissues. Here, as among the higher forms of life, there is a constant struggle for survival between individuals and various species.

Ecologic Relationships. The complex relationships between the different microbial species may be neutral, antagonistic, or synergistic. Few species are strictly neutral in their reactions because they interfere in a passive manner by utilizing the available food supply or excreting toxic materials. The majority of microbial species exhibit a positive antagonism arising from alterations in the physical environment or from the elaboration of antibiotics and bacteriocins that are specifically inhibitory for certain organisms. These factors are important in soil, water, and animal ecology. In the upper respiratory tract, antagonism between the gram-positive cocci and gram-negative bacilli and their independent and mutual antagonism to fungi

was recognized only after the introduction of antibiotics into clinical medicine. For example, following treatment with penicillin, the normal flora (predominantly streptococci, *Neisseria,* and *Haemophilus* species) are replaced by gram-negative enteric bacilli or *Pseudomonas.* On the other hand, broader spectrum antibiosis (as achieved by chloramphenicol and tetracyclines or combinations of penicillins, cephalosporins, and aminoglycosides) may result in the emergence of resistant bacteria or overgrowth by *Candida* species.

Synergism may be described as a cooperation of two or more microbial species that produces a result that could not be achieved individually. Although this phenomenon may be relatively frequent in nature, it is uncommon in disease states. Examples in disease states are the synergistic gangrene described by Meleney, and some forms of anaerobic lung abscess caused by organisms that are ordinarily normal mouth flora (see Chap. 45).

The various ecologic relationships that exist between microorganisms and the human host may be subdivided into commensalism, symbiosis, and parasitism or opportunism. Commensalism refers to the mutual but almost inconsequential association between bacteria and higher organisms. Symbiosis refers to a mutually beneficial relationship between two species. Parasitism is that complex spectrum of relationships whereby one organism derives benefits at the expense of another.

These interactions were analyzed in a monograph by Theobald Smith in 1934. Smith emphasized the concept that the phenomenon of disease caused by infectious agents is largely a by-product of evolving parasitism; that is, violent reactions between host and parasite tend to lessen as parasitism approaches a biologic equilibrium. The rapid and destructive actions of some microorganisms are expressions of bungling parasitism. The skillful or well-adapted parasite enters its host with ease and may produce lesions only as a means of securing exit in order to infect a new host. In a sense, commensalism represents an ideal form of parasitism.

The terms "pathogen" and "opportunist" require definition. In general, a pathogen is a microorganism that is capable of infecting or parasitizing "normal" individuals. As the field of immunology becomes more sophisticated, we may find that no one is "normal," and that such terminology is artificial. At the present time, however, the terms are useful. Certain organisms appear to represent bona fide pathogens in that their hosts are sufficiently large numbers of the general population who lack demonstrable underlying disease. Among these are *Staphylococcus aureus*, *Streptococcus pyogenes*, and *Streptococcus pneumoniae*, among the bacteria; *Histoplasma capsulatum* and *Coccidioides immitis* among the fungi; and the plethora of common cold viruses and viruses that cause childhood diseases (measles, mumps, varicella, and rubella). On the other hand, organisms such as *Pseudomonas aeruginosa*, *Serratia marcescens*, *Candida albicans*, *Pneumocystis carinii*, and *Nocardia asteroides*, uncommonly cause de novo disease but are almost always encountered under unusual circumstances—either in abnormal hosts or in situations where the normal flora have been supplanted. The factors relating to the pathogenic potential of various microorganisms are discussed in the chapters that follow.

Natural Habitats

The diversity of physical and chemical conditions present in different environments results in the segregation of microorganisms into different physical niches, depending on available nutrients, temperature, moisture, and other conditions. Knowledge of the flora of various natural habitats is important in understanding human acquisition of disease.

Soil. The soil is a great reservoir of microorganisms, the majority of which are nonpathogenic. Some microorganisms reach the soil in the excreta or cadavers of animals; others, such as the autotrophic bacteria, actinomycetes, and fungi, are indigenous. Among the pathogens that may be present in soil are *Clostridium tetani* and *Clostridium perfringens*, the etiologic agents of tetanus and gas gangrene.

These organisms have been cultured from uncontaminated soil and thus are able to grow in this environment, as well as being introduced via animal and human feces. Another species, *Clostridium botulinum*, whose toxin is responsible for the symptoms of botulism, is also present in soil and from this source may find its way into improperly processed foods or contaminated wounds. *Bacillus anthracis*, the causative agent of anthrax, is deposited in the soil when animals die of the disease. It infects herbivorous animals and occasionally humans by entering the body through the skin or mucous membranes. *Clostridrium* and *Bacillus* species produce endospores that are of survival advantage to the bacterial cell in that they confer resistance to adverse environmental conditions. Certain of the pathogenic fungi, *C. immitis*, *H. capsulatum*, *Cryptococcus neoformans*, and *Blastomyces dermatitidis*, have been grown from the soil. Inhalation of spores aerosolized from soil results in entry of the organism into the respiratory tract of humans and other animals.

Water. Most bodies of salt and fresh water contain microorganisms, many of which are adapted to extremely adverse conditions (e.g., psychrophilic, halophilic, and thermophilic bacteria). Pathogenic bacteria, however, are usually present only in water that is directly contaminated by human and animal urine and feces. Among the pathogenic organisms that often reach water used for drinking or recreational purposes are *Salmonella* and *Shigella* species, the cholera vibrio, hepatitis virus, polio, and other enteroviruses. These organisms, however, are infrequently isolated directly from water. Isolation of *Escherichia coli*, which is hardier and persists in water for longer periods, thus serves as an index of fecal contamination. The bacillus responsible for certain epidemics of legionnaires' disease, *Legionella pneumophila*, may also be isolated from water. Aerosolization of contaminated water may therefore result in epidemics of this illness.

Air. Although microorganisms are frequently found in air, they do not multiply in this medium. The outdoor air rarely contains pathogens, probably because of the bacteridical effects of desiccation, ozone, and ultraviolet radiation. Indoor air, however, may contain pathogenic viruses and bacteria that are shed by humans from the skin, hands, clothing, and, especially, the upper respiratory tract.

Talking, coughing, and sneezing produce progressively larger numbers of respiratory droplets, many of which contain bacteria and viruses. A sneeze (Fig. 24-1) may produce as many as 10^6 particles from 10 μm to 2 mm in diameter. The larger droplets may travel a distance of 1 to 3 m before reaching the ground. These larger droplets rapidly settle to the floor and dry, leaving organisms attached to dust particles. The smaller droplets remain suspended in air and evaporate rapidly leaving behind droplet nuclei a few micrometers in diameter, which may or may not contain organisms. These droplet nuclei settle very slowly, and, in an ordinary room filled with people, they are wafted about in air currents and remain suspended almost

Figure 24-1. Droplet dispersal following a sneeze by a patient with a cold; note strings of mucus. *(From Jennison: Aerobiology 17:106, 1947.)*

indefinitely. Under such conditions great accumulations of potentially infective particles may occur.

Animals and Animal Products. Animals are hosts for many of the microorganisms that produce disease in humans (e.g., tularemia, brucellosis, psittacosis, salmonellosis, plague, anthrax, insect-borne viral and rickettsial diseases, and parasitic diseases). These diseases may be transmitted directly, by vectors, by contamination of soil, water, or other materials, or by ingestion of meat or dairy products.

Milk from normal cows, even when drawn under aseptic conditions, usually contains from 100 to 1000 nonpathogenic organisms per milliliter. Other organisms may be present when cows are diseased or they may be added during collection. Diseases that may be transmitted by diseased cows or milk handlers include tuberculosis, salmonellosis, streptococcal infections, diphtheria, shigellosis, brucellosis, and staphylococcal food poisoning. Pasteurization processes and the destruction of diseased cattle have decreased the incidence of milk-borne infections.

Microbial Flora of the Normal Human Body

Humans are constantly bombarded by the myriad of microorganisms that occupy their environment. Fortunately, however, humans do not provide a favorable habitat for most of these saprophytes, as they must compete with the commensal flora that are already adapted to the human environment. Those organisms that comprise the normal flora must overcome barriers to colonization produced by flow of body juices, mucociliary clearance, and local immune mechanisms. In addition, the persistence of various organisms in their specific niches may depend upon attachment to receptors on host cells in these areas (Chap. 23).

Skin

Human skin normally contains a varied microbial population (Table 24-1). One animal, the mite *Demodex folliculorum*, is commonly found in sebacious glands and hair follicles, especially on the face. Two lipophilic yeasts, *Pityrosporum ovale* and *Pityrosporum orbiculare*, are present on the scalp or chest and back, respectively. Nonlipophilic yeasts, *Torulopsis glabrata* and *C. albicans* are variably present. The bacterial flora are predominantly *Staphylococcus epidermidis*, *Micrococcus*, aerobic and anaerobic diphtheroids, and sarcinae. *S. aureus* regularly inhabits only the nose and perhaps the perineum, but transient colonization by this and other bacteria such as α- and nonhemolytic streptococci may occur at any site. Saprophytic mycobacteria occasionally are found on the skin of the external auditory canal and the genital and axillary regions.

Most of these organisms inhabit the stratum corneum and the upper parts of hair follicles. A small number, however, are present deeper within follicles and serve as a reservoir for replenishing flora after washing. Washing may decrease skin counts by 90 percent, but normal numbers are found again within 8 hours. Abstinence from washing does not lead to an increase in numbers of bacteria on the skin. Normally 10^3 to 10^4 organisms are found per square centimeter; however, counts may increase to 10^6 per cm^2

TABLE 24-1. NORMAL FLORA AT VARIOUS BODY SITES

Skin	*Staphylococcus, Micrococcus, Corynebacterium, Propionibacterium*, yeasts
Conjunctiva	*Staphylococcus, Corynebacterium, Haemophilus*
Oropharynx	*Staphylococcus, Streptococcus, Bacteroides, Fusobacterium, Actinomyces*
Intestinal tract	*Bacteriodes, Bifidobacterium, Eubacterium, Lactobacillus, Clostridium*, streptococci, coliforms, yeasts
Genitourinary tract	*Lactobacillus* (vagina), *Staphylococcus*, streptococci, *Corynebacterium*

in more humid areas such as the groin and axilla. Small numbers of bacteria are dispersed from the skin to the environment; however, certain individuals may shed up to 10^6 organisms in 30 minutes of exercise. Many of the fatty acids found on the skin may be bacterial products that inhibit colonization by other species. The flora of hair are similar to those of the skin.

The relative freedom of the normal conjunctiva from infections may be explained by the mechanical action of the eyelids, the washing effect of the normal secretions that contain the bacteriolytic enzyme lysozyme, and the production of inhibitors by the normal flora of the eye. *S. epidermidis* and various aerobic and anaerobic diphtheroids are frequently isolated from the conjunctival sac, presumably arising from the flora of the eyelids.

Nose, Nasopharynx, and Accessory Sinuses

Innumerable bacteria are filtered from the air as it passes through the nasopharynx, trachea, and bronchi. The majority of these organisms are trapped in mucous secretions and are swallowed; thus the sinuses, trachea, bronchi, and lungs are usually sterile. The nasopharynx is the natural habitat of the common pathogenic bacteria and viruses that cause infections in the nose, throat, bronchi, and lungs. Humans are the primary host for these bacteria. Thus, individuals with active infection or convalescent and asymptomatic carriers maintain the reservoir from which others become infected. Certain individuals become nasal carriers for streptococci and staphylococci and discharge these organisms in enormous numbers from the nose into the air. Efforts to eradicate *S. aureus* from the nares of such individuals by the use of antibiotics have met with only limited success.

The pharynx usually contains a mixture of viridans (α-) and nonhemolytic streptococci, *Neisseria* species, and *S. epidermidis*. These organisms are inhibitory to *S. aureus* and *Neisseria meningitidis*. Many strains of viridans streptococci are inhibitory to *S. pyogenes*. Children infected with *S. pyogenes* may have fewer inhibitory strains than those who are not infected. Also, colonization with inhibitory flora increases with age. The normal flora of the pharynx may be eradicated by high doses of penicillin, resulting in colonization and overgrowth with gram-negative organisms such as *E. coli, Klebsiella, Proteus,* and *Pseudomonas*. If the viridans streptococci, however, are made resistant to penicillin by stepwise increases in dosage, no abnormal colonization occurs.

The nasopharynx of the newborn infant is sterile, but within 2 to 3 days the infant acquires the common commensal flora and the pathogenic flora carried by the mother and nursing staff. The carrier rate of pathogens (such as group A streptococci, *Haemophilus influenzae,* and pneumococci) may be almost 100 percent in infants, and is higher in children than in adults.

Mouth

Streptococcal species comprise 30 to 60 percent of the bacterial flora of the surfaces within the mouth. These are primarily viridans streptococci: *Streptococcus salivarius, Strep-*

tococcus mitior, Streptococcus mutans, and *Streptococcus sanguis. S. mitior* is primarily found on the buccal mucosa, *S. salivarius* on the tongue, and *S. mutans* and *S. sanguis* on the teeth and dental plaque. Specific binding to mucosal cells or to tooth enamel has been demonstrated with these organisms. Bacterial plaques developing on teeth may contain as many as 10^{11} streptococci per gram in addition to actinomycetes, *Veillonella,* and *Bacteroides* species. Anaerobic flora such as *Bacteroides melaninogenicus,* treponemes, fusobacteria, clostridia, and peptostreptococci are present in gingival cervices where the oxygen concentration is less than 0.5 percent. Many of these organisms are obligate anaerobes and are killed by higher oxygen concentrations. These organisms, when aspirated into the tracheobronchial tree, may play a role in the pathogenesis of anaerobic pneumonia and lung abscess. The natural habitat of the pathogenic species *Actinomyces israelii* is the gums. Among the fungi, species of *Candida* and *Geotrichum* are found in 10 to 15 percent of individuals.

The newborn infant's mouth is not sterile, but, in general, contains the same types of organisms as in the mother's vagina. This usually consists of a mixture of lactobacilli, corynebacteria, staphylococci, micrococci, coliforms, yeast, and streptococci. Among the streptococci are enterococci, microaerophilic and anaerobic species, and, of specific importance in neonatal sepsis and meningitis, group B streptococci. These organisms diminish in number during the first 2 to 5 days after birth, and are replaced by the types of bacteria present in the mouth of the mother or nurse. Anaerobic flora appear after the eruption of teeth and the production of gingival crevices.

Intestinal Tract

In normal fasting individuals, the stomach is usually sterile or contains less than 10^3 organisms per milliliter. Organisms swallowed from the mouth are either killed by the hydrochloric acid and enzymes in gastric secretions, or passed quickly into the small intestine, where forward peristalsis is of primary importance in maintaining sterility or keeping the number of organisms less than 10^3 per milliliter. When organisms are present in the duodenum or jejunum, they usually consist of small numbers of streptococci, lactobacilli, and yeasts, especially *C. albicans*. Larger numbers of bacteria are normally found in the terminal ileum. Under abnormal conditions, such as gastric achlorhydria, abnormal peristalsis (scleroderma and diabetes), or blind loops following gastric surgery, bacterial overgrowth may occur. This may result in megaloblastic anemia due to consumption of vitamin B_{12}, or fat malabsorption and diarrhea secondary to deconjugation of bile salts.

In contrast to the small intestine, the colon is heavily colonized and is the major reservoir of microorganisms in the body. Approximately 20 percent of the fecal mass consists of bacteria (10^{11} organisms/g wet weight). Distribution of bacteria in the colon is as follows: *Bacteroides,* 10^{10-11}; *Bifidobacterium,* 10^{10-11}; *Eubacterium,* 10^{10}; *Lactobacillus,* 10^{7-8}; coliforms, 10^{6-8}; streptococci, 10^{7-8}; *Clostridium,* 10^6; and variable numbers of yeasts. Thus, more than 90 percent of the fecal flora consist of *Bacteroides* and *Bifidobacterium,* both of which are obligate anaerobes. Factors that determine the balance of these bacterial species are poorly understood. Enterobacteria produce colicins, substances that are inhibitory to a small number of similar strains of bacteria, perhaps accounting for the predominance of a few serotypes of these species.

Flora may be altered by the administration of antibiotics; e.g., cephalosporins decrease aerobic and anaerobic streptococci and lactobacilli, whereas clindamycin may totally eradicate most of the anaerobic species (*Bacteroides*, bifidobacteria, and lactobacilli). Eradication of normal flora and overgrowth of enterotoxin producing *Clostridium difficile* is the mechanism of induction of some forms of antibiotic induced colitis. Colonic flora may play a role in disease when the integrity of the gut is compromised, as in appendicitis, diverticulitis, intestinal perforation, and postoperative infections, and may supply the flora (coliforms) that are the major cause of urinary tract infection and gram-negative bacteremia.

The intestinal tract of the newborn child is usually sterile, although a few organisms may be acquired during delivery. Under normal circumstances, intestinal flora are established within the first 24 hours after delivery, primarily from the above listed organisms. The stool of the breast-fed infant is soft, light yellowish-brown in color, and has a faintly acid odor. *Lactobacillus bifidus* is the prominent organism in these stools, others being enterococci, coliforms, and staphylococci. In contrast, artificially fed infants have hard, dark brown, foul-smelling stools and contain *Lactobacillus acidophilus*, coliforms, enterococci, and anaerobic bacilli, including clostridial species. *L. bifidus* may return following addition of 12 percent lactose to cow's milk or other formulas.

Since the time of Pasteur, the significance of the intestinal flora has been a controversial issue. Are the microorganisms essential for life, a natural but inevitable handicap, or a nonessential asset? Pasteur's studies on microbial fermentations suggested that the intestinal organisms might play an essential role in the metabolism of foodstuffs analogous to that of the protozoa in the guts of termites, but subsequent studies in germ-free animals have shown that these flora are not essential. Intestinal flora do, however, have complex effects on the rate of maturation of intestinal epithelial cells, as well as on the levels of various cytoplasmic enzymes in these cells.

The bacteria of the intestinal tract possess a variety of constitutive and inducible enzyme systems. Among these are the glycosidases that are active against dietary sugars or glucuronide metabolites excreted by the liver. When the small intestine is deficient in lactase (as occurs in over 80 percent of some populations), lactose is metabolized by colonic bacteria, resulting in abdominal discomfort, flatulence, and diarrhea due to increased water retention and lowered pH. Also, many bacteria produce urease, resulting in the hydrolysis of urea to CO_2 and NH_3. Other reactions occurring with nitrogen compounds include deamidation of amino acids, N-esterification, dealkylation, and hydrolysis and reduction of diazo compounds. Bile acids (cholanic

acids conjugated with taurine or glycine) are dehydroxylated or hydrolyzed to free bile acids by intestinal bacteria. Bilirubin is metabolized to urobilinogen, which is reabsorbed by the intestine and excreted in the urine or bile.

The role of bacteria in the metabolism of drugs is poorly studied. An interesting phenomenon, however, is known to occur with salicylazosulfapyridine, a drug used in the treatment of inflammatory bowel diseases. This compound is metabolized to sulfapyridine and 5-aminosalicylate by the intestinal microflora, but not in germ-free animals. It is possible that the antiinflammatory effects of this drug are related to this intraluminal cleavage and high local levels of 5-aminosalicylate. Many antibiotics are inactivated by the intestinal flora, which serve as the major reservoir for the generation of antibiotic resistance transfer factors. In addition, intestinal bacteria manufacture a number of vitamins, including niacin, thiamine, riboflavin, pyridoxine, folic acid, pantothenic acid, biotin, and vitamin K. Broad spectrum or poorly absorbable oral antibiotics may greatly reduce or alter the normal flora, thereby inducing vitamin deficiencies in individuals with poor nutrition.

Genitourinary Tract

The secretions around the urethra of the female and the uncircumsized male frequently contain *Mycobacterium smegmatis*, a harmless commensal, which when found in voided specimens of urine may be confused with *Mycobacterium tuberculosis*. The outermost portion of both the male and female urethra may contain many bacteria, especially diphtheroids, nonhemolytic streptococci, and *S. epidermidis*. In addition, the female flora contain a large number of anaerobic lactobacilli (Döderlein's bacillus). The sterility of the internal urethra is generally maintained primarily by the normal flow of urine and evacuation of the bladder. Urine aspirated from the bladder with a needle is normally sterile.

The vulva of the newborn child is sterile, but after the first 24 hours of life it gradually acquires a rich and varied flora of nonpathogenic organisms, such as diphtheroids, micrococci, and nonhemolytic streptococci. After 2 to 3 days, estrogen from the maternal circulation induces the deposition of glycogen in the vaginal epithelium, which facilitates the growth of lactobacilli. These organisms produce acid from glycogen, and a flora develops that resembles that of the adult female. After the passively transferred estrogen is excreted, the glycogen disappears, the lactobacilli are lost, and the pH again becomes alkaline. At puberty, the glycogen reappears and an adult flora again returns. On both aerobic and anaerobic cultures, these flora usually consist of diphtheroids, lactobacilli, micrococci, *S. epidermidis*, *Streptococcus fecalis*, microaerophilic and anaerobic streptococci, ureaplasmas, and yeasts. The presence of *Haemophilus vaginale* or chlamydia is usually associated with symptomatic vaginitis.

Despite the close proximity of the anus, the vaginal flora of normal women only rarely shows even small numbers of coliforms. It has been shown, however, that women who are prone to recurrent urinary tract infection generally demonstrate vaginal and urethral colonization with coliforms prior to the invasion of the bladder by these organisms. During pregnancy, 15 to 20 percent of women demonstrate the presence of group B streptococci (*Streptococcus agalactiae*), an agent assuming increased importance in the etiology of neonatal sepsis and meningitis. After the menopause the flora resemble those found before puberty.

Bacteria in the Blood and Tissues

Occasionally commensals from the normal flora of the mouth, nasopharynx, and intestinal tract are carried into the blood and to tissues. Under normal circumstances, they are eliminated by normal defense mechanisms, particularly phagocytosis by reticuloendothelial cells. A few organisms may remain viable for a time in lymph nodes and may be cultured from biopsies of such tissues. Any unusual organisms of questionable pathogenicity that appear in only one of a series of blood cultures should be regarded as a contaminant from the skin or a stray transient. Simple manipulations such as chewing, toothbrushing, dental work, genitourinary catheterization or instrumentation, and proctosigmoidoscopy may also be associated with transient bacteremia. This phenomenon is generally of little consequence in the normal host. However, in the presence of abnormal heart valves, prosthetic heart valves, or other prosthetic devices made of foreign materials, these bacteremias may lead to colonization and infection by pathogenic organisms or saprophytes of low pathogenicity.

Acquisition of Infectious Agents

Infections may be acquired via a variety of mechanisms, including foods and water, respiratory secretions, venereal contact, inoculation via foreign materials, disruption of normal skin and mucosal barriers, animal vectors, or invasion of normal flora. These pathogenic mechanisms are discussed in the following chapters dealing with bacterial, fungal, and viral diseases.

Nosocomial Infections

Of particular interest to the medical and nursing professions are hospital-acquired (nosocomial) infections. These infections are transmitted to patients either by hospital personnel or other patients, or may arise from endogenous flora. Modes of acquisition include surgical procedures, indwelling intravenous or bladder catheters, endotracheal tubes, intravenous fluids, and equipment used for respiratory support. Of particular importance is the seemingly innocuous acquisition of pathogenic or opportunistic organisms into the pool of normal flora, predisposing compromised individuals to subsequent invasion by their own flora. This is particularly true of postoperative patients and individuals treated with antibiotics, immunosuppressants, or antineoplastic agents. These organisms include the usual pathogens such as *S. aureus* and *E. coli*, as well as other enterobacteria, *Pseudomonas*, opportunistic fungi, and viruses.

Infection committees with surveillance programs must be an integral part of the modern hospital. Nosocomial infections occur in approximately 5 percent of all patients admitted. These rates vary depending upon the type of hospital (e.g., acute vs. extended care facilities). Over 80 percent of these infections involve the urinary and respiratory tracts and surgical wounds. The use of antibiotics in hospitals predisposes to the selection of resistant organisms that may inadvertently be passed from patient to patient by the mechanisms described above. In addition, the duration of hospitalization plays a great role in the presence of pharyngeal colonization by gram-negative bacteria. Thus, physicians and nursing personnel should be constantly aware of the role of antibiotics and person-to-person transmission in the genesis of hospital-acquired infections.

Opportunistic Infections

Infections that occur as a result of abnormalities in host defenses are generally referred to as opportunistic. These infections may be caused by bona fide pathogens or by organisms of low virulence, such as those that comprise the normal body flora. Opportunistic infection may occur either as a complication of abnormal defense mechanisms or as a result of various iatrogenic or nosocomial factors. For example, diabetics and alcoholics, two groups of individuals at increased risk for gram-negative rod pneumonias, have twice the normal incidence of gram-negative rods as part of their resident pharyngeal flora. Similar increases in pharyngeal colonization may be observed after the administration of broad spectrum antibiotics such as ampicillin, cephalosporins, clindamycin, and tetracyclines. Tetracyclines are frequently associated with *Candida* vaginitis or

---

TABLE 24-2. MISCELLANEOUS CONDITIONS PREDISPOSING TO COMPROMISED HOST DEFENSES

Drugs—immunosuppressive, antibiotics, anesthetics
Alcoholism
Malnutrition
Viral infections, e.g., influenza, measles
Alterations in normal mucosal-cutaneous barriers
Iatrogenic procedures
    Prosthetic devices
    Intravenous catheters, bladder catheters
    Respiratory assist devices, i.e., inhalation therapy, nebulizers,
      respirators
    Whirlpools

TABLE 24-3. DISORDERS OF IMMUNE FUNCTION

Inherited disorders leading to an increased incidence of infection
   Complement deficiencies
   Neutrophil-killing defects
   Abnormal chemotaxis
Disorders of immune function
   Antibody deficiency syndromes
   Cell-mediated immunodeficiencies
   Combined deficiencies

thrush. Many of these same drugs may result in pseudo-membranous colitis, which is caused by an overgrowth of the colonic flora by *C. difficile.*

Other infections, such as measles or influenza, may lead to superinfections through depression of phagocytosis and chemotaxis, decreased cell-mediated immunity, or damage to the respiratory epithelium.

Hospital admission results in the acquisition of new flora, many of which may be potential pathogens. Many of the procedures performed on hospitalized patients lead to colonization and potential superinfection. These are outlined in the section on nosocomial infections and in Table 24-2.

Inherited or acquired disorders of immune function may result in abnormal antibody synthesis, absent cell-mediated immunity, or impaired neutrophil killing or chemotaxis (Table 24-3). Additionally, iatrogenic factors such as immunosuppression of cancer patients or transplant recipients produce broad disorders of normal defenses. Neutropenia induced by antitumor drugs may turn such commensals as *S. epidermidis, Propionibacterium acnes, Candida* species, bacilli, and micrococci into virulent pathogens. Examples of associations between various host defects and infecting organisms are presented in Table 24-4. With these patients, great care must be taken to prevent introduction of organisms from the environment; e.g., neutropenic patients should not be given raw fruits or vegetables, which are frequently contaminated with *P. aeruginosa.* An outbreak of aspergillosis in leukemics has been associated with the intake of air from a construction site into the ventilation system of a hospital.

Control of the resident flora has been attempted in some of these instances. Gut "sterilization" by orally administered antibiotics and placement of patients in isolation laminar flow rooms may decrease the incidence of infection. Infants with severe combined immune deficiency (total absence of cell-mediated and humoral immunity) have been kept alive in a sterile environment while awaiting bone marrow transplantation. These procedures, however, are extraordinarily expensive.

Ultimately, the successful treatment of such inherited disorders and neoplastic diseases will rest upon the ability of medical science to control and prevent infections from the body's resident flora.

FURTHER READING

Books and Reviews

Britton G, Marshall KC: Adsorption of Microorganisms to Surfaces. New York, Wiley, 1980

Drasar BS, Hill MJ: Human Intestinal Flora. New York, Academic Press, 1974

McDermott W: Conference on air borne infections. Bacteriol Rev 25:173, 1961

Noble WC: Microbiology of Human Skin. London, Lloyd-Luke Medical Books, 1981

Rosebury T: Microorganisms Indigenous to Man. New York, McGraw-Hill, 1962

Skinner FA, Carr JG: The Normal Microbial Flora of Man. New York, Academic Press, 1974

Smith T: Parasitism and Disease. Princeton, Princeton University Press, 1934

Selected Papers

Allweiss B, Dostal J, Carey KE, et al.: The role of chemotaxis in the ecology of bacterial pathogens of mucosal surfaces. Nature 266:448, 1977

Beachey, EH: Bacterial adherence: Adhesion-receptor interactions mediating the attachment of bacteria to mucosal surfaces. J Infect Dis 143:325, 1981

Cohen R, Roth FJ, Delgado E, et al.: Fungal flora of the normal human small and large intestine. N Engl J Med 280:638, 1969

Crowe CC, Sanders WE, Longley S: Bacterial interference. II. Role of the normal throat flora in prevention of colonization by group A streptococcus. J Infect Dis 128:527, 1973

Donaldson RM: Normal bacterial populations of the intestine and their relation to intestinal function. N Engl J Med 270:938, 994, 1050, 1964

Drasar BS, Shiner M, McLeod GM: Studies on the intestinal flora. I. The bacterial flora of the gastrointestinal tract in healthy and achlorhydric persons. Gastroenterology 56:71, 1969

TABLE 24-4. RELATION OF DEFECTS IN HOST DEFENSES TO INFECTION

Host Defense Defect	Diseases Precipitating Defect	Resultant Infection
Abnormal cell-mediated immunity	Hodgkin's disease, sarcoidosis	Tuberculosis, cryptococcosis
Hypogammaglobulinemia	Bruton's, common variable immunodeficiency	Pneumococcal pneumonia
Abnormal neutrophil killing	Chronic granulomatous disease	*Staphylococcus,* gram-negative
Absent complement components	C6, C7, C8 deficiency	Disseminated *Neisseria*
Drug-induced neutropenia	Acute leukemias, chemotherapy	Gram-negative, candidiasis
Abnormal mucociliary clearance	Cystic fibrosis	*Pseudomonas*

Edens CS, Eriksson B, Hanson LA: Adhesion of *Escherichia coli* to human uroepithelial cells in vitro. Infect Immun 18:767, 1977

Goldacre MF, Watt B, Loudon N, et al.: Vaginal microbial flora in normal young women. Br Med J 1:1450, 1979

Gossling J. Slack JM: Predominant gram-positive bacteria in human feces: Numbers, variety, and persistence. Infect Immun 9:719, 1974

Jennison MW: Atomizing of mouth and nose secretions into the air as revealed by high-speed photography. Aerobiology, 17:106, 1947

Mackowiak PA: The normal microbial flora. N Engl J Med 307:83, 1982

Saigh JH, Sanders CC, Sanders WE Jr: Inhibition of *Neisseria go-* *norrhoeae* by aerobic and facultatively anaerobic components of the endocervical flora: Evidence for a protective effect against infection. Infect Immun 19:704, 1978

Schimpff SC: Surveillance cultures. J Infect Dis 144:81, 1981

Schimpff SC, Green, WH, Young VM, et al.: Infection prevention in acute nonlymphocytic leukemia. Laminar air flow room reverse isolation with oral, nonabsorbable antibiotic prophylaxis. Annu Intern Med 73:351, 1975

Sen P, Kapila R, Chmel H, et at.: Superinfection: Another look. Am J Med 73:707, 1982

Sprunt K, Leidy GA, Redman W: Prevention of bacterial overgrowth. J Infect Dis 123:1, 1971

CHAPTER 25

Staphylococcus

Staphylococci are responsible for over 80 percent of the suppurative diseases encountered in medical practice. They cause most suppurative infections of the skin but may also invade and produce severe infections in any part of the body. The primary natural habitat of staphylococci is mammalian skin, where the organisms are found in large numbers. In their adaptation to parasitism, staphylococci have been among the most versatile and successful of the pathogenic bacteria. Although numerous antistaphylococcal antibiotics have been introduced during the past 40 years, the ability of the staphylococcus to develop resistance, especially in the hospital environment, continues to provide a major medical threat. Of greatest concern has been the appearance of resistance to penicillin G and, more recently, to methicillin.

At present, most of the serious staphylococcal infections are seen in patients whose normal host defenses are severely impaired. Hospitalized, debilitated patients who have serious underlying diseases or who have undergone extensive surgery are especially susceptible to infection with staphylococci. Such problems stem from the compromised status of the host rather than the virulence of the organism.

The Genus *Staphylococcus*

Staphylococcus is the only genus of medical importance in the family Micrococcaceae (Table 25-1). It contains gram-positive cocci that are facultatively anaerobic and grow in irregular clusters. Properties distinguishing staphylococci from members of the genus *Micrococcus* that are also often present in soil, water, and on the skin of humans are shown in Table 25-2. The name "staphylococcus," derived

from the Greek noun *Staphyle* (a bunch of grapes) and *coccus* (a grain or berry), was introduced to describe the organisms seen by early investigators in pus from surgical infections. Since most strains freshly isolated from staphylococcal infections produced a golden yellow pigment, the organism was named *Staphylococcus aureus* to distinguish these strains from the less pathogenic staphylococci that usually produce white colonies. Pigment production, however, is a variable trait of staphylococci, and its correlation with pathogenicity is unreliable. Its use has been superseded by coagulase production, which is the most useful single criterion for the recognition of *S. aureus*. A staphylococcus that produces coagulase is *S. aureus* irrespective of colony pigmentation.

Three species of *Staphylococcus* are currently recognized: *S. aureus*, *Staphylococcus epidermidis*, and *Staphylococcus saprophyticus*. Useful properties for distinguishing the three species are shown in Table 25-3. Although *S. aureus* is the most significant pathogen for man, there has been increased awareness that coagulase-negative staphylococci can cause infections. *S. epidermidis*, although relatively avirulent, has been associated with an increasing number of hospital-acquired infections, especially in patients whose susceptibility is increased and where there is a nidus of foreign material, such as a prosthesis or plastic catheter. *S. saprophyticus* can cause urinary tract infections in females (p. 459).

Staphylococcus aureus

Morphology

S. aureus is a nonmotile coccus, 0.8 to 1.0 μm in diameter, which divides in two planes to form irregular grapelike

TABLE 25-1. DIFFERENTIAL PROPERTIES OF THE GENERA OF THE FAMILY MICROCOCCACEAE

	Micrococcus	*Staphylococcus*	*Planococcus*
Cells: spherical, gram-positive	+	+	+
Arrangement			
Irregular clusters	+	+	−
Tetrads	v	−	+
Glucose fermentation*	−	+	−
Cytochromes	+	+	+
Catalases			
Heme	+	+	+
Nonheme	−	−	−
Hydrogen peroxide formation	−	−	−
Motility	−	−	+
Yellow-brown pigment	−	−	+
G + C content of DNA (mol%)	66–75	30–40	39–52

Adapted from Baird-Parker: In Buchanan and Gibbons (eds): Bergey's Manual of Determinative Bacteriology, 8th ed, 1974. Baltimore, Williams & Wilkins.

+, most (90% or more) strains positive; −, most (90% or more strains negative; v, inconstant—in one strain may sometimes be positive, sometimes negative.

*Growth and acid production anaerobically from glucose, with exception of strains of *Staphylococcus saprophyticus*, which only weakly ferment glucose.

TABLE 25-2. CHARACTERISTICS DISTINGUISHING MEMBERS OF THE GENERA *STAPHYLOCOCCUS* AND *MICROCOCCUS*

	Staphylococcus	*Micrococcus*
Anaerobic growth, fermentation of glucose	+	−
Cell wall:		
Glycine-containing penta- or hexapeptide cross-bridges	+	−
Ribitol or glycerol teichoic acids	+	−
DNA: G + C content (mol%)	30–40	66–75

From Baird-Parker: Ann NY Acad Sci 236:8, 1974.
+, 90% or more strains positive; −, 90% or more strains negative.

clusters of cells (Fig. 25-1). In smears from pus, the cocci appear singly, in pairs, in clusters, or in short chains. The irregular clusters are found characteristically in smears from cultures grown on solid media. In broth cultures, short chains and diplococcal forms are common. A few strains produce a capsule or slime layer that enhances the virulence of the organisms. *S. aureus* is gram-positive, but old cells and phagocytized organisms stain gram-negative.

Ultrastructure and Cell Composition

The architecture of a staphylococcus is similar to that of other gram-positive organisms. Thin sections of log phase cells reveal nucleoids, mesosomes, and a trilaminar cytoplasmic membrane that is separated from the cell wall by a periplasmic region. The thickness of the wall of young cells is 18 to 25 nm. In encapsulated strains, a loose fimbriate or capsular layer also may be seen.

The cell wall of *S. aureus* consists of three major components: peptidoglycan, teichoic acids, and protein A. The composition of these materials has been useful in distinguishing *Staphylococcus* from *Micrococcus* and *S. aureus* from *S. epidermidis* (Tables 25-2 and 25-3). The peptidoglycan comprises 40 to 60 percent of the weight of the cell wall; the amounts of the other major components vary.

Peptidoglycan. The primary stucture of the staphylococcal peptidoglycan is distinctive for the species (Figs. 6-2B and 6-3B). As in most bacteria, the glycan portion of the molecule consists of alternating N-acetylglucosamine and N-acetylmuramic acid residues joined through β-1,4 glycosidic linkages. In staphylococci, however, all of the N-acetylmuramic acid residues carry tetrapeptide chains that are crosslinked by pentaglycine bridges. The extensive crosslinking of the peptide moiety gives the staphylococcal wall a tight structure that aids the cell in its quest for survival in the hostile environment of the host tissues. Antibodies are produced to the peptidoglycan.

Teichoic Acid. In *S. aureus*, the wall teichoic acid is of the ribitol phosphate type (Fig. 6-5). The walls of *S. epidermidis* contain a glycerol teichoic acid, whereas in micrococci there is another type of teichoic acid or, usually, no teichoic acid at all. Teichoic acid is an essential component of the phage receptor of *S. aureus*. It also plays an important role in the maintenance of normal physiologic functions. By regulating the cationic environment of the bacterial cell, it controls the activity of autolytic enzymes that function in growth of the cell wall and separation of the daughter cells. Although mutants completely lacking teichoic acid do exist, showing that the polymer is not es-

TABLE 25-3. CHARACTERISTICS DISTINGUISHING MAJOR SPECIES OF THE GENUS *STAPHYLOCOCCUS*

	S. aureus	*S. epidermidis*	*S. saprophyticus*
Coagulase	+	−	−
Anaerobic growth and fermentation of glucose	+	+	−
Mannitol			
Acid aerobically	+	v	v
Acid anaerobically	+	−	−
α-Toxin	+	−	−
Heat-resistant endonucleases	+	−	−
Biotin required for growth	−	+	NT
Cell wall			
Ribitol	+	−	+
Glycerol	−	+	v
Protein A	+	−	−
Novobiocin sensitivity[†]	S	S	R

From Baird-Parker: Ann NY Acad Sci 236:9, 1974.
+, 90% or more strains positive; −, 90% or more strains negative; v, some strains positive, some negative; NT, not tested.
[†]R, MIC > 2.0 μg/ml; S, MIC < 0.6 μg/ml.

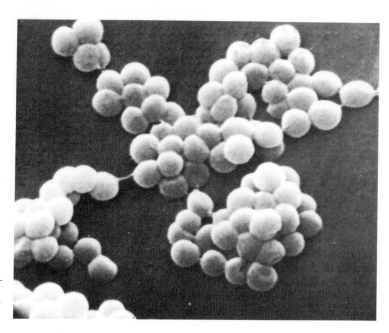

Figure 25-1. Scanning electron photomicrograph of *S. aureus* in serum-salts broth. *(From Watanakunakorn: Infect Immun 4:73, 1971.)*

sential for viability, such mutants are phage-resistant, grow more slowly than wild-type organisms, and produce large bizarre nonseparating cells with abnormal crosswall structure (p. 99).

Protein A. The major protein component of the cell wall of *S. aureus* is protein A, about one third of which is released into the medium during cell growth. This is a group antigen, specific for most strains of *S. aureus*, and is not found in other staphylococci or in the micrococci. Purified protein A contains a preponderance of basic amino acids, and its removal from the bacterial cell increases the negative charge of the cell surface. In the cell it is covalently linked to the peptidoglycan structure and uniformly distributed in the whole cell wall. There has been considerable interest in protein A because of its unique property of interacting nonspecifically with immunoglobulins. A discussion of the immunologic importance of protein A is found on page 449.

Physiology

Cultural Characteristics

The staphylococcus is a facultative anaerobe, but growth is more abundant under aerobic conditions. Some strains also require an increased CO_2 tension. Growth occurs over a wide temperature range, from 6.5 to 46C, with an optimum for *S. aureus* of 30 to 37C. The pH optimum is 7.0 to 7.5, with growth occurring over a range of pH 4.2 to 9.3. For growth on chemically defined media, staphylococci require a number of amino acids and vitamins; under anaerobic conditions, uracil and a fermentable carbon source are also required. In spite of their complex nutritional requirements, staphylococci grow well on most routine laboratory media, such as nutrient agar or trypticase soy agar.

Sheep blood agar is recommended for primary isolation from clinical materials. Since human blood contains nonspecific inhibitors or antibodies, it should not be used in the preparation of blood agar.

On agar plates, colonies are smooth, opaque, round, low-convex, 1 to 4 mm in diameter. Most strains of *S. aureus* produce golden yellow colonies upon primary isolation. The color is due to carotenoid pigments and is extremely variable, ranging from deep orange to pale yellow. Pigment production is dependent on growth conditions and is not a valid criterion for the separation of *S. aureus* from *S. epidermidis*. Pigment production is best observed by growth on agar plates at 37C for 24 hours followed by incubation at room temperature for an additional 24 to 48 hours. No pigment is produced under anaerobic conditions or in liquid medium.

On blood agar, a zone of β hemolysis surrounds colonies of organisms that produce soluble hemolysins. Although primarily associated with *S. aureus*, β hemolysis also may be produced by strains of *S. epidermidis* and, as with pigmentation, is a variable property of the staphylococcus.

Metabolism

Energy is obtained via both respiratory and fermentative pathways. Intact pathways for glycolysis, the pentose phosphate pathway and citric acid cycle, are operative under appropriate growth conditions. The ability of the staphylococcus to exist under conditions of both high and low oxidation-reduction potential is an obvious advantage to the organism in its battle for survival in its natural habitat on mucosal surfaces and in competition with other bacterial species in the mixed microflora at the site of infection.

Catalase is produced by aerobically grown cells. In testing for this enzyme in blood agar cultures, precaution must be taken to avoid carryover of blood cells with the organisms. Catalase is present in red blood cells, and any red blood cells present in the bacterial mixture may lead to a false positive reaction.

A wide range of sugars and other carbohydrates is utilized by staphylococci. Under aerobic conditions the major product of glucose dissimilation is acetic acid, with small amounts of CO_2. Under anaerobic conditions, lactic acid is the principal product; acetoin also is usually produced. The fermentation of mannitol by most strains of *S. aureus* is helpful in its differentiation from *S. epidermidis*.

Identification

Major characteristics for distinguishing *S. aureus* from other staphylococci are shown in Table 25-3. Of these, the most convenient and reliable property for diagnostic purposes is the production of coagulases, enzymes that cause the coagulation of plasma. Approximately 97 percent of staphylococci isolated from pathologic processes elaborate these enzymes. In testing for coagulase, the test tube method should be employed. The slide test, although useful for screening purposes and usually correlating well with test tube results, detects a clumping factor on the surface of the organism that is distinct from the free coagulase. It is less reliable than the test tube method. In interpreting a coagulase test, it is important to remember that certain bacteria other than *S. aureus* may produce coagulase and that false positive reactions may be elicited by citrate-utilizing bacteria, such as enterococci and *Pseudomonas*.

Classification

Bacteriophage Typing. Most strains of *S. aureus* are lysogenic: they carry phages to which they themselves are immune but that will lyse some of the other members of the species. Susceptibility of *S. aureus* strains to the various temperate bacteriophages provides the basis for a phage-typing system that has been useful in epidemiologic studies. The system is based on patterns of sensitivity shown by each strain to various phages. The phage patterns of different strains fall essentially into three broad groups, phage groups I, II, and III. The term "group" refers to strains of *S. aureus* with related phage patterns as well as to corresponding groups of phages with host range for these strains. The phages within a group are unrelated and possess different morphologic and serologic properties (groups A through L). The grouping of strains appears to be less fortuitous, i.e., group II strains of staphylococci are often associated with skin infections, such as impetigo and pemphigus of the newborn, and the production of enterotoxin is confined primarily to phage group III. Resistance to penicillin and methicillin first appeared in group III before it developed in other strains. Hospital epidemics started with strains of a few phage patterns (phages 75, 77) in group III and group I (phage 80) but soon shifted to resistant strains of the 52/52A/80/81 complex (group I), to be

followed by strains lysed by phage 83A (group III) and strains of the 83A/84/85 complex. Strains untypeable by the typing scheme currently in use are now being isolated with increasing frequency. Although incompletely understood at present, the basis of phage patterns appears to lie in the strain-dependent restriction-modification systems on which lysogenic immunity and phage-dependent restriction are superimposed (Chap. 8).

Techniques currently in use for staphylococcal phage typing are rigidly controlled by the Subcommittee on Phage Typing of the International Committee on Nomenclature of Bacteria. The 23 phages that now constitute the basic set of typing phages are shown in Table 25-4. Only coagulase-positive staphylococci may be typed with the basic set of phages.

For typing, each specific phage of the basic set is grown on its homologous propagating strain of staphylococcus and separated from the bacterial cells by centrifugation and filtration. After proper dilution, a single drop is placed on separate squares of an agar plate previously seeded with a young broth culture of the organism to be typed. The plate is air-dried and incubated at 30C overnight. Phage typing results are recorded by listing only the phages that exhibit strong lysis (i.e., a 2+ reaction indicating more than 50 plaques).

Serotyping. Procedures are available for the classification of staphylococcal isolates into specific serologic types, but at present the systems are complex and empirical, and unsuitable for routine diagnostic use. Serotyping, however, has certain advantages over other marker systems and promises to be a useful tool in the future.

Speciation. In animals, variants of *S. aureus* have arisen as a result of adaptation to a particular host. Schemes that recognize such biotypes or ecotypes have been useful for epidemiologic and ecologic studies (Table 25-5). They have also provided the basis for the more modern taxonomic schemes in which new species and subspecies are recognized. Although such schemes are of primary interest at present to the research systematist, they focus on an emerging new classification of the staphylococcus based on molecular approaches (Table 25-6).

Genetics

The extreme flexibility of the staphylococcus has always made difficult the characterization of a typical *S. aureus*. The medical implications of this variability were not fully

TABLE 25-4. LYTIC GROUPS OF *STAPHYLOCOCCUS* TYPING PHAGES IN THE BASIC SET OF TYPING PHAGES

Lytic Group	Phages in group				
I	29	52	52A	79	80
II	3A	3C	55	71	
III	6	42E	47	53	54 75 77 83A 84 85
Unassigned	81	94	95	96	

TABLE 25-5. SUBDIVISION OF *STAPHYLOCOCCUS AUREUS* INTO BIOTYPES

	Biotypes					
	A	B	C	D	E	F
Origin of biotype	Human	Pigs, poultry	Cattle, sheep	Hares	Dogs	Pigeons
Fibrinolysin	+	−	−	−	−	−
Pigment	+	+	+	v	−	−
Coagulation of						
Human plasma	+	+	+	+	v	+
Bovine plasma	−	−	+	−	+	+
α-Hemolysin	+	v	v	−	−	v
β-Hemolysin	v	v	+	+	+	+
Reduction of tellurite	+	+	+	+	W	−
Clumping factor	+	+	+	+	v	−
Growth on crystal violet agar	−	v	v	−	+	+
Typed by adapted phages	H	H	H,B	H	C	−

Data from Hájek and Maršálek: From Baird-Parker: Ann NY Acad Sci 236:10, 1974

+, more than 80% of strains positive; −, more than 80% of strains negative; v, some strains positive, some negative; W, weak reaction; H, basic international human phage set; B, bovine phage set; C, canine phages.

realized, however, until the spectacular emergence of anti-biotic-resistant strains, first to penicillin and then successively to each antibiotic included in the therapeutic regimen. The serious epidemiologic and therapeutic problems created by these drug-resistant strains prompted genetic studies on the staphylococcus similar to those carried out with the enteric organisms. It now appears that most antibiotic resistance in *S. aureus,* as in the Enterobacteriaceae, is plasmid mediated and that the genetics of the staphylococcus is analogous to the genetics of *Escherichia coli.* In *S. aureus,* however, transfer of plasmids between cells is mediated exclusively by bacteriophages, since conjugation does not occur.

Most clinical isolates of *S. aureus* harbor one or more prophages that presumably are integrated into the bacteri-

TABLE 25-6. MAJOR NATURAL HOSTS OF *STAPHYLOCOCCUS* SPECIES

Species	Host(s)
S. hominis, S. epidermidis, S. capitus	Humans
S. haemolyticus, S. warneri, S. aureus, S. saprophyticus, S. simulans	Humans and nonhuman primates
S. cohnii	Humans, nonhuman primates, tree shrews
S. simians	Nonhuman primates
S. xylosus	Primates, Carnivora, Artiodactyla, Perissodactyla, Rodentia, Marsupialia
S. intermedius	Carnivora
S. hyicus	Artiodactyla
S. sciuri	Mammalia
S. lentus	Artiodactyla

Data from Kloos and Schleifer: In Starr et al. (eds): The Prokaryotes, 1981, vol. II. New York, Springer-Verlag.

al chromosome. Of the large number of staphylococcal phages, however, only those of serologic group B are transducing and thus potentially capable of the transfer of plasmids. In nature, transduction is of the generalized type. Almost any character can be transduced at a frequency of about 10^{-4} to 10^{-10} per plaque-forming unit of phage.

About 10 percent of the total cell DNA in naturally occurring organisms is plasmid DNA. Since these genetic elements have the capacity to evolve rapidly, they impart to the population of cells carrying them a better ability to survive under changing environmental conditions than cells containing a uniform DNA content.

Two classes of staphylococcal plasmid have been demonstrated, the relatively larger penicillinase plasmids that appear to replicate under stringent control and the relatively smaller multicopy plasmids that encode one resistance determinant each (Table 25-7). Four different types of penicillinase plasmids have been recognized, but no naturally occurring strain carries more than one type. Some penicilinase plasmids also carry markers conferring resistance to erythromycin and certain metal ions, but with the exception of plasmids displaying resistance to both penicillin and erythromycin, there is no tendency to build up mutiple antibiotic-resistance plasmids. In *S. aureus,* the plasmids carrying resistance determinants for tetracycline, chloramphenicol, and neomycin/kanamycin are independent entities.

Some strains of *S. aureus* carry plasmid genes for bacteriocin production. These genes are analogous in many ways to the colicinogenic factors of the enteric bacteria. The production of the bacteriocin, staphylococcin, is limited to phage group II strains of *S. aureus.* Staphylococcin is a heat-stable protein distinct from other extracellular products of *S. aureus* and is a specific phage product. Its spectrum is wide and includes β-hemolytic streptococci, pneumococci, other staphylococci, corynebacteria, and *Ba-*

TABLE 25-7. TYPICAL *STAPHYLOCOCCUS AUREUS* PLASMIDS

| Plasmid | Incompatibility | | Copies per Cell | Genotype* |
	Type	MW		
pI258	Inc 1	18×10^6	2.7	$pen^+ \; asa^+ \; asi^+ \; ant^+ \; ero^+ \; inc \; 1^+ \; mer^+ \; bis^+ \; cad^+ \; lea^+$
pII147	Inc 2	21×10^6	2.7	$pen^+ \; asa^+ \; inc \; 2^+ \; cad \; B^+ \; bis^+ \; lea^+ \; mer^+ \; cad \; A^+$
pT169	Inc 3	2.7×10^6	~30	tet
pC221	Inc 4	3.0×10^6	>20	cml

Adapted from Novick et al.: In Schlessinger (ed): Microbiology—1974, 1975. Washington DC, American Society for Microbiology.
*The following genotype abbreviations are used for loci: *asa, asi, ant, bis, lea, cad, mer, pen, ero, tet, cml* for response to arsenate, arsenite, antimony, bismuth, lead, cadmium, and mercuric ions and to penicillin, erythromycin, tetracycline, and chloramphenicol; *inc* is for incompatibility specificity.

cillus species. Gram-negative bacteria and producer strains are resistant to its action.

Although plasmids have been implicated in the synthesis of a number of virulence factors of the staphylococcus, their precise role as structural or regulatory genes has not been defined. In the case of alpha toxin, either there is more than one genetic determinant for toxin production, or the gene for toxin production is associated with a transposon. There are both chromosomal and plasmid genes for the synthesis of exfoliative toxin, each producing a distinct species of toxin.

Transformation and transfection also occur in some strains of *S. aureus*, but only in the presence of a high concentration of Ca^{2+} and a helper bacteriophage. The competence-conferring activity of the bacteriophage appears to be due to an interaction of a unique morphogenic precursor of the phage organelle with the surface of the cell.

Resistance

Staphylococci are more resistant to adverse environmental conditions than are most nonsporulating bacteria. They survive for weeks in dried pus and sputum and, on sealed agar slants, remain viable for several months. Most strains are relatively heat resistant and require for killing a temperature of 60C for 1 hour. Staphylococci are also more resistant than most bacteria to the common chemical disinfectants, such as the phenols and mercuric chloride, but like other gram-positive organisms, they are sensitive to concentrations of unsaturated fatty acids and basic dyes that do not inhibit most gram-negative organisms. Advantage is taken of this differential susceptibility to dyes in the designing of selective media for the culture of the enteric bacteria from fecal specimens.

Antigenic Structure

The phagocytic response of the host is a crucial factor in determining the initiation and the outcome of staphylococcal infections. In this process of host recognition and immunity, the cellular antigens of the staphylococcal cell, especially the more superficial surface ones, are major determinants. The antigenic structure of *S. aureus* is very complex, and of the more than 30 antigens observed, the biologic and chemical properties of only a few have been well characterized.

Polysaccharide A

A major antigenic determinant of all strains of *S. aureus* is the group-specific polysaccharide A of the cell wall. The serologic determinant of this polysaccharide is the *N*-acetyl glucosaminyl ribitol unit of teichoic acid; specificity resides in the alpha or beta configuration of the glucosaminyl substituents. In the cell wall, polysaccharide A is associated with the peptidoglycan in an insoluble state and requires lytic enzymes for release. Most adults have a cutaneous hypersensitivity reaction of the immediate type to polysaccharide A, and low levels of precipitating antibodies are found in their sera. This group-specific polysaccharide A is not found in *S. epidermidis*, which contains instead glycerol teichoic acid with glycosyl residues rather than ribitol teichoic acid. In *S. epidermidis* the group-specific antigen is referred to as polysaccharide B.

Protein A

Protein A is a group-specific antigen unique to *S. aureus* strains. Ninety percent of protein A is found in the cell wall covalently linked to the peptidoglycan. During cell growth, protein A is also released into the culture medium, where it comprises about one third of the total protein A produced by the organism.

Protein A consists of a single polypeptide chain with a molecular weight of 42,000. Four tyrosine residues fully exposed on the surface are responsible for biologic activity. The uniqueness of protein A is centered on its ability to interact with normal IgG of almost all mammals. Within a species the interaction may be restricted to certain subgroups of IgG. Although similar to an ordinary antigen-antibody reaction, binding involves not the Fab fragment but the Fc portion of the immunoglobulin. Protein A consists of five regions: four highly homologous domains are Fc-binding, whereas the fifth, C-terminus domain, is bound to the cell wall and does not bind Fc (Fig. 25-2).

Protein A provokes a variety of biologic effects. It is chemotactic, anticomplementary, antiphagocytic, and elicits

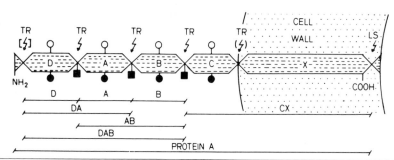

Figure 25-2. Proposed structure of protein A. Arrows indicate points of enzymic cleavage. TR, trypsin; LS, lysostaphin. Arrow in round brackets indicates no cleavage of cell wall-bound protein; arrow in square brackets indicates site may constitute the N-terminus of the protein. Fc receptor (O), structures evoking precipitation against rabbit antiprotein-A serum: (●) present in all Fc-binding regions, (■) only demonstrated in the polyvalent fragments. These structures are not necessarily true antigenic determinants but may be related to the Fc receptors. *(From Sjödahl: Eur J Biochem 73:343, 1977.)*

hypersensitivity reactions and platelet injury. It is mitogenic and potentiates natural killer activity of human lymphocytes. Although there is good correlation between protein A production and coagulase activity, there is no correlation between the absence or presence of protein A and any pathogenic property. Its ability to bind to the Fc region of IgG has led to numerous applications in immunochemical and cell surface structural studies.

Capsular Antigens

Although staphylococci are rarely encapsulated, a few strains have been isolated that carry immunologically significant surface antigens. It is probable that encapsulation in vivo is not a rare phenomenon and that the capsular material produced under these conditions imparts to the organism an antiphagocytic advantage. The capsular antigen is found only in mucoid-untypeable strains of *S. aureus* that lack detectable bound coagulase (clumping factor). A further discussion of this antigen is found below.

Determinants of Pathogenicity

One of the essential attributes of a successful parasite is the ability to survive in the animal host. In this respect the staphylococcus has exhibited exceptional adaptive potential. Much of the organism's versatility may no doubt be attributed to the possession of hydrolytic enzymes for a wide range of substrates, including native animal proteins. Proteases, lipases, esterases, and lyases are among the more important enzymes facilitating establishment of the organism on the skin and mucous membranes of the host. In order to survive in a hostile environment, however, the successful parasite must also counteract host defenses. For more than half a century, a wide array of extracellular enzymes, toxins, and cellular components of the staphylococcus have been examined in detail in an attempt to define a specific virulence factor. Unfortunately, although much has

been learned about a wide variety of factors that are often present in the more pathogenic strains, at present, no single factor can be equated with virulence. It is quite probable that this most sought-after factor may be produced only in vivo and cannot be demonstrated by our current in vitro approaches. It has been emphasized and ably expressed by Abramson that although it is not always possible to correlate the various enzymes and toxins with pathogenicity, their significance is not decreased. Each of these "may constitute a link in a complex chain of innumerable, presently undetermined, pathogenic factors that combine only in vivo to precipitate the staphylococcal phenomenon as it is currently understood."*

Surface Antigens

Surface components that possess antiphagocytic activity are of obvious advantage to the staphylococcus in its initial establishment in the host. Although capsule formation is apparently limited to a few strains, these strains are more virulent for mice, a finding that has been correlated with the resistance of such strains to phagocytosis. Vaccines prepared with them stimulate the production of protective antibodies, and the purified surface polysaccharide elicits the production of opsonic and skin-sensitizing antibodies. During infection, it is likely that most strains of staphylococci elaborate in the in vivo environment surface antigens similar to or identical with the polysaccharide surface antigens of the prototype encapsulated Smith strain.

Staphylococci bind to fibronectin, a high molecular weight glycoprotein important in many biologic systems. Although the specific binding site for fibronectin in the staphylococcal cell wall has not been fully characterized, a fibronectin-binding protein has been purified. A correlation has been demonstrated between binding and the cellular protein A content of a specific strain. Fibronectin

---

*Abramson C: Staphylococcal enzymes. In Cohen JO (ed): The Staphylococci, 1972, p. 235. New York, Wiley-Interscience.

binding to protein A-containing organisms may play a role in the pathogenicity of the staphylococcus by promoting their attachment to host tissues.

Extracellular Enzymes

Coagulases. Although the correlation between coagulase production and pathogenicity provides a convenient virulence marker, there is no definite evidence that coagulase is directly involved in pathogenicity, and there is no evidence that antibodies to any of the antigenically distinct coagulases are important in acquired resistance to staphylococcal infections.

The action of coagulase in the clotting of plasma is similar to the thrombin-catalyzed conversion of fibrinogen to fibrin. For full enzymatic activity, coagulase requires a plasma component, either prothrombin or a prothrombin derivative, referred to as coagulase-reacting factor (CRF). The coagulase-thrombin product (CT) not only causes fibrinogen clotting but also possesses proteolytic and esterolytic activity similar to that of thrombin. The fibrinopeptides released are indistinguishable from thrombin-induced fibrinopeptides, some of which possess pharmacologic activity comparable to that of bradykinin on smooth muscle.

Lipases. Staphylococci produce several lipid hydrolyzing enzymes collectively referred to as lipases. The lipases are active on a variety of substrates, including plasma and the fats and oils that accumulate on the surface areas of the body. The utilization of these materials is of survival value to the organism and explains the intense colonization of staphylococci in the sebaceous areas of greatest activity. The production of lipase apparently is essential in the invasion of healthy cutaneous and subcutaneous tissues. In primary human isolates, there is a close correlation between in vitro production of lipase and the ability to produce boils. The decreased virulence of hospital staphylococci observed during the last 20 to 30 years parallels a decrease in staphylococcal isolates that produce large amounts of the enzyme. The decrease apparently is due to the presence of a prophage that blocks lipase production.

Hyaluronidase. Over 90 percent of *S. aureus* strains produce hyaluronidase. This enzyme hydrolyzes the hyaluronic acid present in the intracellular ground substance of connective tissue, thereby facilitating spread of the infection. Since inflammation antagonizes the spreading action by hyaluronidase, its importance in staphylococcal infections is limited to the very early stages of infection. The staphylococcal enzyme is similar to the hyaluronidase extracted from pneumococci, streptococci, and *Clostridium perfringens* and consists of several enzymatically active components.

Staphylokinase (Fibrinolysin). One of the proteolytic enzymes of staphylococci has fibrinolytic activity but is antigenically and enzymatically distinct from the streptoki-nase of the streptococcus. In the dissolution of clots by the staphylococcal enzyme, the proenzyme plasminogen is converted to the fibrinolytic enzyme plasmin. Staphylokinase in some strains is controlled by genes on the bacterial chromosome, while in others it is mediated by lysogenic conversion. Although produced by most strains of *S. aureus*, there is little evidence that it is a major factor in pathogenicity.

Nuclease. The elaboration of a heat-resistant nuclease appears to be uniquely associated with *S. aureus*. The enzyme, which is present in, at, or near the cell surface, is a compact globular protein consisting of a single polypeptide chain. Heating at 65C causes structural disruptions, but the changes are rapidly and completely reversible. The nuclease is a phosphodiesterase with both endo- and exonucleolytic properties that can cleave either DNA or RNA to produce 3′-phosphomononucleotides. Antisera produced in rabbits against the staphylococcal nuclease inhibit completely its enzymatic activity. It does not precipitate streptococcal nuclease.

Toxins

A variety of toxic manifestations are associated with certain of the extracellular proteins produced by staphylococci. Three major groups of staphylococcal toxins have been defined on the basis of their biologic activity: cytolytic toxins (hemolysins and leukocidin), enterotoxins, and epidermolytic toxin.

Cytolytic Toxins. A number of bacteria produce toxins that cause physical dissolution of mammalian or other cells in vitro. Most of these are proteins, are extracellular, and induce the formation of neutralizing antibodies. There is considerable diversity, however, in the manner in which the various cytolytic toxins interact with the cell surface. The hemolysins and leukocidin elaborated by *S. aureus* are among the best defined of the cytolytic toxins, a group of toxins that also includes streptolysin O and S and various toxins of *Clostridium*. The staphylococcal cytolytic toxins are antigenically distinct from other staphylococcal toxins.

ALPHA TOXIN (α-HEMOLYSIN). This toxin exhibits a wide range of biologic activities, including the hemolytic, lethal, and dermonecrotic effects observed following the injection of broth culture filtrates of certain strains. No other bacterial toxin is so versatile in its effects. Alpha toxin disrupts lysosomes and is cytotoxic for a variety of tissue culture cells. Human macrophages and platelets are damaged, but monocytes are resistant. There is injury to the circulatory system, muscle tissue, and tissue of the renal cortex. Although not the sole virulence factor for the staphylococcus, together with the other virulence factors described above, the alpha toxin contributes significantly to pathogenicity by producing tissue damage after the establishment of a focus of infection.

Pure alpha toxin has a sedimentation coefficient ($S_{20,w}$) of 3.0 and a molecular weight of about 28,000. It consists of four different conformational forms, separable by electrophoresis. Rapid interconversion of these forms occurs upon storage. The alpha toxin polymerizes and aggregates reversibly to form both soluble and insoluble 12S products. These forms, which are biologically inactive, are referred to as "toxoids." Toxoiding of a 3S toxin with formaldehyde, however, does not change the molecular size, possibly because the reagent reacts with active groups essential for both toxicity and polymerization.

The reaction with erythrocytes involves two sequential steps: (1) an initial interaction between toxin and cells that results in the prelytic release of K^+, followed by (2) the actual lysis of the cell and release of hemoglobin. The precise mechanism of membrane damage has not been established. The formation of ring structures upon addition of alpha toxin to liposomes and to mammalian cell membranes suggests that lytic activity may be due to the capacity of alpha toxin to penetrate and disrupt hydrophobic regions of membranes. There is evidence, however, that an enzymic action also is involved and that alpha toxin is converted from an inactive protease precursor to an active protease by an enzyme in the membrane. Current data on the control of alpha toxin production indicates that there is either more than one genetic determinant or that the gene for toxin production is associated with a transposon.

BETA TOXIN (STAPHYLOCOCCAL SPHINGOMYELINASE). The most striking activity of this toxin is its ability to produce a hot–cold lysis, i.e., an enhanced hemolytic activity if incubation at 37C is followed by a period at 4C or at room temperature. The toxin is an enzyme with substrate specificity for sphingomyelin (and lysophosphatides). Sphingomyelin degradation is the membrane lesion that leads to hemolysis when the cells are chilled:

$$\text{Sphingomyelin} + H_2O \xrightarrow[Mg^{2+}]{\beta\text{-toxin}}$$

$$N\text{-acylsphingosine} + \text{phosphorylcholine}$$

Erythrocytes from different animal species exhibit impressive differences in their sensitivity to beta toxin. A correlation exists between toxin sensitivity and content of sphingomyelin, most of which is located in the outer leaflet of the lipid bilayer of the erythrocytic membrane and thus accessible to exogenous toxin.

DELTA TOXIN. This is a relatively thermostable surface-active toxin whose strong detergent-like properties are responsible for its damaging effects on membranes. Delta toxin exhibits a high degree of aggregation and is electrophoretically heterogeneous. It has a high content of hydrophobic amino acids, which, if localized in one area, could make the molecule amphipathic and strongly surface active. The membrane receptor site is thought to be a straight-chain fatty acid with 13 to 19 carbons. Delta toxin

exhibits a broad spectrum of biologic activity and displays no pronounced specificity for cells of a particular species. Erythrocytes, macrophages, lymphocytes, neutrophils, and platelets are all damaged by delta toxin as are spheroplasts and protoplasts of other bacteria. Although delta toxin-specific antibodies are elicited in response to immunization with delta toxin incorporated in complete Freund adjuvant, the antibodies do not neutralize the toxin's hemolytic activity.

In addition to the gross cytolytic effects of delta toxin, a number of more subtle responses can be induced at very low concentrations. Delta toxin has been shown to inhibit water absorption by the ileum, to stimulate AMP accumulation, and to alter ion permeability to guinea pig ileum. It also stimulates the release of insulin from isolated islets of Langerhans, the mechanism of which is attributable to the facilitation of Ca^{2+} entry into the islet cells mediated directly by an ionophoretic mechanism.

GAMMA TOXIN. This is a cytolytic toxin with pronounced hemolytic activity, but its precise mode of action is unknown. It consists of two separate protein components that act synergistically, both being essential for hemolysis and toxicity. Detailed information on the chemistry and biologic effects of gamma toxin is lacking. The finding of elevated levels of specific neutralizing antibodies to gamma toxin in human staphylococcal bone disease suggests a possible role of this toxin in the disease state.

LEUKOCIDIN. The Panton–Valentine leukocidin produced by most strains of S. aureus attacks polymorphonuclear leukocytes and macrophages but no other cell type. The toxin is composed of two proteins that are electrophoretically separable, the F (fast) and the S (slow) components. They are inactive alone but act synergistically to induce cytolysis. The unique response of leukocytes to leukocidin is an altered permeability to cations. Other changes that occur are secondary to this initial event. The membrane-leukocidin interaction stimulates potassium-sensitive acylphosphatase activity, accompanied by an increased permeability to cations and an accumulation in the cytoplasm of orthophosphate at the expense of ATP. In the presence of calcium, large amounts of protein derived from the cytoplasmic granules are secreted. This degranulation may be observed microscopically. Both components of leukocidin are highly antigenic and have been toxoided. Although leukocidin alone is not responsible for the pathogenicity of the staphylococcus, it enhances staphylococcal invasiveness by allowing the organism to resist phagocytosis.

Enterotoxins. Approximately one third of all clinical isolates of S. aureus produce exotoxins that cause diarrhea and emesis in man. These toxins are a major cause of bacterial food poisoning and also have been implicated in pseudomembraneous enterocolitis seen in patients following antibiotic therapy.

The staphylococcal enterotoxins consist of a group of simple single-chain globular proteins with molecular weights ranging from 28,000 to 35,000. Six types (A, B, C$_1$, C$_2$, D, E), of which five are noncross-reacting, have been identified. Enterotoxins A and D are most frequently associated with staphylococcal food poisoning; enterotoxin B is the toxin most likely to be associated with hospital infection. A new staphylococcal enterotoxin, enterotoxin F, has recently been isolated from toxic shock syndrome *S. aureus* strains. This enterotoxin is believed to be responsible for some of the signs and symptoms of this recently recognized syndrome. The genetic locus of staphylococcal enterotoxins has not been clearly defined. Although a close association of enterotoxin B production with methicillin resistance has been observed, proof of genetic linkage is lacking.

The mechanism of action of staphylococcal enterotoxin is unknown. The major hindrance to an understanding of its pathogenesis and mode of action has been the lack of a practical and sensitive assay system. Except for humans, the only reliable experimental animal for testing enterotoxin activity is the monkey. The emetic receptor site for staphylococcal enterotoxin is the abdominal viscera, from which site the sensory stimulus reaches the vomiting center via the vagus and sympathetic nerves. Enterotoxin-induced diarrhea has been attributed to inhibition of water absorption from the lumen of the intestine and to increased transmucosal fluid flux into the lumen. The enterotoxin is a potent mitogen for lymphocytes. It is pyrogenic and enhances gram-negative lethality. Potent toxoids to the enterotoxins may be produced by treatment with formaldehyde, which results in extensive crosslinking of the toxin into large polymers. Polymer size correlates directly with the ability of the toxoid to stimulate production of specific antibody.

Exfoliative Toxin (Epidermolytic Toxin). This toxin, elaborated primarily by phage group II staphylococci, is capable of splitting adjacent cell layers within the epidermis, causing the various skin manifestations of the staphylococcal scalded skin syndrome. The isolation and characterization of exfoliative toxin was made possible by the use of newborn mice for bioassay. In mice, the toxin produces a severe form of the disease indistinguishable histologically and clinically from the human disease.

The purified toxin is protein in nature, with a molecular weight of 24,000. It causes lysis of the intracellular attachment between cells of the granular layer of the epidermis but does not primarily cause cell death and does not elicit an inflammatory response. The mechanism of action is unknown. Toxin synthesis may be under the control of either chromosomal or extrachromosomal determinants. Some group II *S. aureus* strains express only plasmid control, some express only chromosomal control, while others express both. In addition to its toxin effects, exfoliative toxin is a potent mitogen, primarily of T cells.

Pyrogenic Exotoxins. Staphylococcal pyrogenic exotoxins have recently been described that are thought to play a role in syndromes similar to classic group A strepto-coccal scarlet fever and the toxic shock syndrome. Three of these toxins (A, B, C) have been characterized biochemically and biologically. They have been distinguished from other staphylococcal toxins and are serologically distinct from, but share many important properties with, the group A streptococcal pyrogenic exotoxin. Staphylococcal pyrogenic exotoxin type C is produced by strains of *S. aureus* from patients with toxic shock syndrome. It possesses properties similar to those of both pyrogenic exotoxins and enterotoxins. In addition, this toxin has profound effects on the immune system, including nonspecific T cell mitogenic activity and enhancement of acquired hypersensitivity. It has been proposed that the type C pyrogenic exotoxin may act in concert with host-derived endotoxin to trigger the onset of the toxic shock syndrome in susceptible individuals.

Clinical Infection

Epidemiology

The staphylococcus is a normal component of human indigenous microflora and is carried asymptomatically in a number of body sites. Its transmission from these sites causes both endemic and epidemic disease.

A better understanding of the origin and epidemiology of staphylococcal disease was provided during the 1950s following the appearance of epidemics caused by antibiotic-resistant strains of the 80/81 complex. The appearance of epidemics at that particular time apparently resulted from a set of new circumstances that had evolved as a result of medical progress. When first introduced, penicillin was dramatically effective in the treatment of staphylococcal infections. By 1946, however, an increasing number of penicillin-resistant strains were isolated from hospital infections. As penicillin became less useful clinically and as other antibiotics were introduced, resistance to these agents also rapidly appeared. Resistance to the new antibiotics was associated almost exclusively with penicillinase production and the development of multiple resistance in a few strains, which then became established endemically in the hospitals. The increasing number of highly susceptible individuals congregated in hospitals contributed to the epidemic appearance of staphylococcal disease. Group III phage types and strains within the 80/81 complex are most often incriminated in outbreaks of infection among newborn infants, older surgical and medical patients, and hospital personnel.

The source of staphylococcal infection is a patient or a member of the hospital personnel with a staphylococcal lesion. Patients with lesions draining pus externally are dangerous to others because of their ability to disseminate the organisms by contamination of the environment. Direct contact via the hands is the single most important route of transmission. There are documented cases of hospital personnel with mild staphylococcal lesions, such as furuncles, paronychia, or styes, who have initiated epidemics. An infected surgeon is a common source of infection in surgical patients.

The acquisition and carriage of *S. aureus* is a complex problem that is incompletely understood. Colonization of the infant with staphylococci occurs within a few days after birth, but because of antibodies passively received through the placenta, the carrier rate drops during the first 2 years of life. By the age of 6 the child has acquired an adult carrier rate of approximately 30 percent. Some individuals who harbor staphylococci are chronic or persistent carriers, but most are intermittent carriers harboring the organism for only a few weeks. *S. aureus* is found in the asymptomatic carrier in a number of body sites, but the anterior nares is the major reservoir of infection and source of disease. The perineum is also an important carriage site. The frequency of isolation from the skin reflects the density of colonization of the nose and rectum.

The carrier problem is an especially serious one in the newborn nursery. The umbilicus and the groin are usually the sites of primary colonization. By maintaining sterility of the umbilical stump, the nasal carrier rate can be markedly reduced. The carriage rate is determined by the presence or absence in the nursery of an epidemic strain. If such a strain is present, most of the infants will be colonized, but if a number of different strains are present, less than 20 percent will be colonized. Many of the staphylococcal lesions that develop during early infancy are thus due to nursery-acquired strains. Staphylococci are disseminated from these newborns with lesions to other infants and nursery personnel and to their families. Since lesions may not develop until after hospital discharge, newborns and patients with postoperative wound infections may transmit hospital strains of staphylococci into the community.

The carrier state is a potentially serious problem in chronic users of parenteral drugs, in whom an increased carrier rate of *S. aureus* has been found. It is probable that the high frequency of staphylococcal endocarditis in the narcotic addict is attributable to the increased opportunity for the introduction of staphylococci into the circulation as the skin barrier is penetrated by the injections.

Bacterial Interference. Clinical and epidemiologic observations suggest that colonization by one strain of coagulase-positive staphylococci prevents colonization with a second strain. These observations have been examined experimentally by inoculating infants and medical personnel with a selected strain of *S. aureus* 502A, a coagulase-positive organism of low virulence, sensitive to penicillin, and unable to produce β-lactamase. An impressive correlation was observed between the prior presence of *S. aureus* in the nose and failure to implant the 502A strain. Coagulase-negative staphylococci were less effective in this respect than coagulase-positive strains. The principle of bacterial interference has been used clinically in a number of select nursery situations where the staphylococcal colonization rate and disease rate were high. The data obtained from these studies support the conclusion that

this approach is an effective and safe method of curtailing epidemics provided reasonable precautions are taken.

Pathogenesis

In the typical staphylococcal skin infection, the organisms penetrate a sebaceous gland or hair shaft where they find an environment nutritionally suitable for growth. The defense mechanisms of the host and the size and virulence of the infective dose determine the likelihood of development of a staphylococcal infection. Although benign skin infections are common, serious staphylococcal disease is infrequent, emphasizing the excellent protective barrier provided by the skin and mucous membranes. Any condition that destroys the integrity of these surface areas predisposes the individual to infection. Third-degree burns, traumatic wounds, surgical incisions, decubitus or trophic ulcers, and certain viral infections are among the many precipitating causes of staphylococcal disease. Foreign bodies, such as intravascular prostheses and intravenous plastic catheters, provide a medium for vascular infection and bacteremia. Bacteremia is most commonly seen in individuals with diabetes mellitus, cardiovascular disease, granulocyte disorders, and immunologic deficiency. In fact, it is uncommon to observe staphylococcal bacteremia in individuals who do not have an associated disease that predisposes to infection. Alteration of the normal flora of the patient by the administration of antibiotics may also provide the proper setting for increased proliferation of staphylococci. An example of this is the development of staphylococcal enterocolitis following the oral administration of tetracycline and subsequent replacement of indigenous organisms by antibiotic-resistant enterotoxigenic staphylococci.

Phagocytosis of Staphylococci

The granulocyte is primarily responsible for resistance to staphylococcal infections. Once the organisms have penetrated the skin or mucous membranes, mobile phagocytes migrate into the area in response to the stimulus of chemotactic factors. Chemotactic activity is generated by a number of different mechanisms, each of which is significant at a different stage of the infection. Early in the infection, staphylococcal proteases generate their own chemotactically active fragments of complement components. Later in the course of infection or with repeated antigenic challenge, specific antibody may generate chemotactic activity through the classic complement pathway (Chap. 13). By virtue of their ability to activate complement, all major cell wall components of *S. aureus* contribute to the generation of chemotactic factors.

The inflammatory reaction induced following the accumulation of phagocytes at the site of the invading bacteria facilitates contact between organisms and phagocytic cells. This interaction of staphylococci with phagocytic cells plays a central role in the critical early stages of infection. Phagocytosis of nonencapsulated strains of staphylococci is promoted by either complement or antibody. For efficient phagocytosis of the more resistant encapsulated strains, however, both antibody and complement are required. In most normal individuals, as well as in many with staphylococcal infections, complement is the primary source of opsonic factors. Although specific IgG antibodies are present in the serum of most individuals as the result of subclinical infection with staphylococci, the titer of these antibodies is relatively low. In the absence of antibody or a functional classic complement pathway, staphylococcal opsonization may proceed by activation of the alternate complement pathway, which is mediated exclusively by peptidoglycan. Activation of complement by either the classic or alternative complement pathway leads to the deposition of opsonic C3b on the bacterial surface and engulfment of the organism via C3b receptors of the leukocyte.

Once they are phagocytized, most staphylococci are rapidly killed and degraded within the phagocytic vacuoles. The intracellular killing is mediated by both oxygen-dependent and oxygen-independent bactericidal mechanisms. In their response to phagocytosis, leukocytes exhibit a dramatic burst in metabolic activity, characterized by a marked increase in oxygen consumption and shift to use of the hexose monophosphate shunt pathway. Accompanying these events is an increased production of highly toxic one- and two-electron reduction products of oxygen, O_2^-, and H_2O_2. The bactericidal potential of H_2O_2 is greatly increased by myeloperoxidase and oxidizable cofactors, such as iodide or chloride. The peroxidase is present in high concentration in cytoplasmic granules and is released into the phagocytic vacuole during degranulation. Deficient bactericidal capacity against staphylococci has been demonstrated in the phagocytic cells of patients with chronic granulomatous disease of childhood. Leukocytes of these patients do not exhibit the normal metabolic response to phagocytosis that results in hydrogen peroxide accumulation. As a result, engulfed staphylococci remain viable within phagocytic vacuoles. Catalase-negative organisms, however, such as streptococci and pneumococci, which are unable to break down the hydrogen peroxide produced by their own metabolism, accumulate it and are therefore readily killed by chronic granulomatous disease leukocytes because they provide the reagent for the myeloperoxidase-halide-hydrogen peroxide bacterial system.

Oxygen-independent staphylocidal systems that are operative in the phagocytes include the low pH within the vacuole, lysozyme, lactoferrin, and granular cationic proteins.

Delayed Hypersensitivity

A small number of organisms may survive for prolonged periods within the phagocytic cell. These survivors may outlive the phagocytic cell, which could explain the occurrence of chronic, latent, or smoldering infections. An important aspect of this intimate host–parasite relationship is the development of delayed hypersensitivity to staphylococcal antigens, which has been demonstrated both experimentally and in patients with various types of recurrent *S. aureus* infections. The exaggerated hypersensitivity response may impair local resistance and increase tissue destruction. Both cell wall peptidoglycan and membrane proteins are thought to contain the major determinants for delayed hypersensitivity. The success that has been encountered by many clinicians in the use of autogenous vaccines for the treatment of patients with recurrent boils has been attributed to a hyposensitization of the patient who has an excessive amount of delayed hypersensitivity to staphylococcal products.

Clinical Manifestations

The characteristic feature of staphylococcal infection is abscess formation. This can occur in any part of the body, but in each area the basic lesion consists of inflammation, leukocyte infiltration, and tissue necrosis. In a fully developed lesion there is a central necrotic core filled with dead leukocytes and bacteria separated from the surrounding tissue by a relatively avascular fibroblastic wall.

Cutaneous Infections

FURUNCLES AND CARBUNCLES. Staphylococcal infection of the skin is the most common bacterial infection in humans. The most superficial of these is folliculitis, in which there is infection of the hair follicle. An extension into the subcutaneous tissue results in the formation of a focal suppurative lesion, the boil or furuncle. A carbuncle is similar to a furuncle but has multiple foci and extends into the deeper layers of fibrous tissue. Carbuncles are limited to the neck and upper back, where the skin is thick and elastic. In children, cutaneous lesions are less well localized than in adults.

No pain is associated with folliculitis, but as the infection penetrates into the subcutaneous tissues and inflammation progresses, the overlying skin becomes thin, stretched, and shiny, and exquisite tenderness appears. Most furuncles evolve in 3 to 5 days, followed by spontaneous drainage, relief of pain, and onset of healing. About 20 percent of patients with furuncles have one or more recurrences during the ensuing year. A small number have chronic recurrent furunculosis for months or years. Dissemination from cutaneous lesions occurs by contiguous extension or hematogenous spread.

IMPETIGO. In the newborn infant, pustules or impetigenous lesions are the most frequent staphylococcal skin manifestation. Staphylococcal impetigo also is common in young children, often occurring around the nose. It is characterized by the formation of encrusted pustules on the superficial layers of the skin. When crusts are removed, a red weeping denuded surface is exposed. The disease is highly contagious and, when introduced into a nursery or school, spreads in an epidemic manner (Fig. 25-3). In the United States, staphylococci, often phage

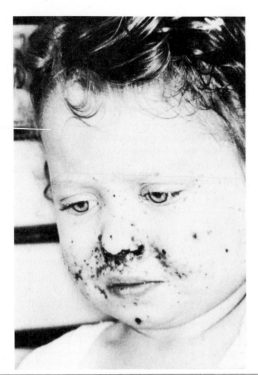

Figure 25-3. Facial impetigo in a 3-year-old child caused by group A β-streptococci and *S. aureus*. (Courtesy of Dr. Leon J. LeBeau, University of Illinois Medical Center.)

type 71, appear to participate with streptococci in most common impetiginous lesions.

SCALDED SKIN SYNDROME. The scalded skin syndrome comprises a spectrum of dermatologic disease with a common etiology, the staphylococcal exfoliative toxin. All of the clinically recognizable features of this syndrome are attributable to this toxin, the effects of which are separable from the effects of the staphylococcal infection itself (Fig. 25-4). The scalded skin syndrome comprises three distinct but related clinical entities:

1. Generalized exfoliative dermatitis (Ritter's disease, staphylococcal toxic epidermal necrolysis) is the most severe form. It is characterized by generalized painful erythema and dramatic bullous desquamation of large areas of skin. The focus of infection may be at a distant site.
2. Bullous impetigo is a localized form of the syndrome in which the infection occurs at the site of the lesion.
3. Staphylococcal scarlet fever is a mild generalized form of the scalded skin syndrome, clinically similar to streptococcal scarlet fever. A localized infection from which staphylococci may be isolated is the usual focus.

The scalded skin syndrome primarily afflicts neonates and children under 4 years of age. Its relatively rare occurrence in adults, except in the immunologically compro-

mised patient, suggests the presence of neutralizing antibodies in the majority of the population.

Pneumonia. Staphylococcal pneumonia is a very important disease because of its high mortality rate (up to 50 percent). It may be a fulminant process in all age groups but is relatively rare except during epidemic periods of influenza. Infants less than 1 year of age appear to be the most susceptible and account for about 75 percent of the cases. Primary staphylococcal pneumonia is most often seen in patients with impaired host defense: children with cystic fibrosis or measles, influenza patients, or debilitated, hospitalized persons being treated with antimicrobials, steroids, cancer chemotherapy, or immunosuppressants. Necrosis, with formation of multiple abscesses, is characteristic of the infection. The pneumonia usually is patchy and focal in nature.

Staphylococcal bacteremia from a focus elsewhere may result in hematogenous or secondary pneumonia. In recent years this has been seen most often among heroin addicts with endocarditis, predominantly in the adolescent–young adult age group.

Osteomyelitis. *S. aureus* is the cause of most cases of primary osteomyelitis. This disease occurs primarily in male children under the age of 12 years and, in most cases, follows hematogenous spread from a primary focus, usually a wound or furuncle. The organisms localize at the diaphysis of long bones, probably because the arterial circulation in this area consists primarily of terminal capillary loops. As the infection progresses, pus accumulates and emerges to the surface of the bone, raising the periosteum and producing a subperiosteal abscess. Clinical symptoms of acute osteomyelitis include fever, chills, pain over the bone, and muscle spasm around the area of involvement. When the infection occurs near a joint, staphylococcal pyoarthrosis is a common complication.

Secondary staphylococcal osteomyelitis is associated with a penetrating trauma or surgery and is frequent in patients with diabetes mellitus or peripheral vascular disease. Two increasingly common forms of osteomyelitis are vertebral osteomyelitis, seen in adults, especially intravenous drug users, and clavicular osteomyelitis, a complication of subclavian catheter usage.

Pyoarthrosis. Approximately 50 percent of all cases of bacterial arthritis are caused by *S. aureus*. Staphylococcal pyoarthrosis may occur following orthopedic surgery, in conjunction with osteomyelitis or local skin infections, or by direct inoculation of staphylococci into the joint during intraarticular injections, especially in patients with rheumatoid arthritis receiving corticosteroids. Staphylococcal joint infection destroys the articular cartilage and may result in permanent joint deformity.

Bacteremia and Endocarditis. Bacteremia may occur with any localized staphylococcal infection, but infections of the skin, respiratory, or genitourinary tract provide the

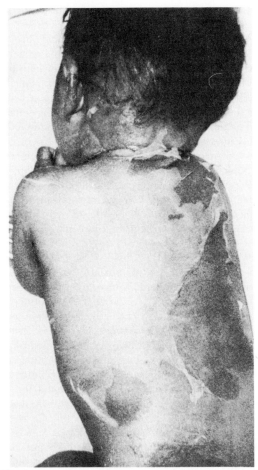

Figure 25-4. Scalded-skin syndrome produced by the epidermolytic toxin of *S. aureus*. In the generalized exfoliative form, the epidermis separates and peels off, leaving rolled skin at the margins and revealing a moist red glistening area. *(From Melish et al.: Zentralbl Bakteriol I Abt. Suppl 5:473, 1976.)*

primary focus for most of these lesions. Catheters and other foreign bodies, trauma, and debilitating diseases predispose to the seeding of the bloodstream with staphylococci. Approximately 50 percent of staphylococcal septicemias are hospital acquired. Fever, shaking chills, and systemic toxicity are usually associated with staphylococcal bacteremia. A frequent complication is endocarditis, which is usually acute and malignant, with heart valve destruction within a few days.

Metastatic Staphylococcal Infections. One of the characteristic features of *S. aureus* bacteremia is the production of metastatic abscesses. The most frequent sites of the metastatic abscesses are the skin, subcutaneous tissues, and lungs. Internal abscesses of the kidneys, brain, and spinal cord are not uncommon.

Food Poisoning. In the United States, staphylococcal food poisoning is the most common form of bacterial food poisoning. It is caused by the ingestion of food that con-

tains the preformed toxin elaborated by enterotoxin-producing strains (p. 452). The food is usually contaminated by food handlers who have the organisms on their hands. The foods most commonly involved are improperly refrigerated custard or cream-filled bakery products. Ham, processed meats, ice cream, cottage cheese, hollandaise sauce, and chicken salad are foods that are often implicated. Foods containing the enterotoxin are normal in odor, appearance, and taste. Sufficient toxin is produced in 4 to 6 hours at 86F, but not at refrigerator temperatures, to produce symptoms of food poisoning.

Symptoms, which appear abruptly 2 to 6 hours after ingestion of the food, consist of severe cramping, abdominal pain, nausea, vomiting, and diarrhea. Sweating and headache are seen, but fever is not a common feature. Recovery is usually rapid, within 6 to 8 hours.

Staphylococcal Enterocolitis. This infection is an iatrogenic acute colitis clinically distinguishable from staphylococcal food poisoning. It is observed primarily in hospitalized patients whose normal bowel flora has been suppressed by the oral administration of wide-spectrum antibiotics that selectively permit overgrowth by drug-resistant enterotoxin-producing strains of staphylococci. The finding of large numbers of gram-positive cocci in sheets or clumps in the feces of these patients suggests the diagnosis of staphylococcal enterocolitis. Clinical manifestations include abdominal cramps, copious diarrhea, fever, dehydration, and electrolytic imbalance. *S. aureus* is, however, responsible for only a small proportion of cases of antibiotic-associated enterocolitis, the majority of which are caused by toxigenic *Clostridium difficile*, especially when there is no involvement of the small intestine.

Toxic Shock Syndrome (TSS). Toxin-producing strains of *S. aureus* have been implicated in this multisystem disease that primarily afflicts young women. Onset of illness usually occurs during menses and is related to the use of tampons. Clinical features include fever, marked hypotension, diarrhea, conjunctivitis, myalgias, and a scarlatiniform rash followed by fine desquamation. There appears to be no significant correlation between the outcome of the acute illness and the type of antimicrobial therapy. However, the use of β-lactamase-resistant drugs does reduce the rate of recurrence, probably by reducing colonization with toxin-producing *S. aureus* strains (p. 453).

Immunity

Humans, as are all animals, are highly resistant to infection by the staphylococcus. Billions of organisms must be introduced in order to elicit an observable response. How much of this resistance is of a natural (inherited) type and how much is acquired in response to repeated natural exposure to organisms that constitute part of our natural flora have not been clearly defined. Once infection is established, however, an intact polymorphonuclear response is essential for containment of infection. Complement is the major opsonin for the staphylococcus. This is demonstrated clinically by the finding that patients with deficiencies in complement components have repeated and severe staphylococcal infection, while agammaglobulinemia patients have severe problems with organisms other than the staphylococcus. For efficient phagocytosis, antibodies to the surface antigens are also required, but the relative protective value of antibodies to each specific cell wall component or staphylococcal antigen is unknown. Most adults possess serum antibodies to a number of staphylococcal antigens, but high titers have not always protected from disease. Antibody levels (IgM and IgG) to both peptidoglycan and teichoic acid are usually elevated in the more serious staphylococcal infections of deep tissue sites. Immunization as a means of enhancing resistance to staphylococci has encountered little success and is not recommended.

Laboratory Diagnosis

Because of the widespread distribution of staphylococci, meticulous care must be taken in the collection of specimens. If the material for culture requires aspiration, the skin in the area must be properly sterilized. The finding of typical irregular clusters of gram-positive cocci upon direct microscopic examination of purulent material is presumptive evidence of the presence of staphylococci, but definitive identification requires laboratory isolation.

Pus, purulent fluids, sputum, and urine specimens should be streaked directly on a blood agar plate and inoculated into a tube of thioglycollate broth. Specimens from patients receiving penicillin should be treated with penicillinase prior to culture. For blood cultures, 10 ml of venous blood should be inoculated into 50 ml of tryptose-phosphate broth. Identification of staphylococci and differentiation of *S. aureus* from *S. epidermidis* is based on colonial and microscopic morphology, catalase and coagulase production, and mannitol and glucose fermentation. Staphylococci and micrococci are catalase positive; pneumococci and streptococci are catalase negative. By definition, all strains of *S. aureus* are coagulase positive.

Antibiotic sensitivity testing of the isolate is useful for clinical and epidemiologic purposes. Bacteriophage typing and serologic testing are not practical procedures for the routine diagnostic laboratory.

Treatment

In the management of localized staphylococcal infections, the basic principle of therapy is adequate drainage. Foreign bodies at the site of infection should be removed. Although antibacterial agents may control spread of the organisms from the abscess, they are less effective upon bacteria within the abscess and do not facilitate its resolution.

Antibiotic sensitivity testing is important in the selection of the appropriate antibiotic and in evaluating its effectiveness during the course of the infection. Although many antibiotics are now available for the treatment of

staphylococcal infections, the unpredictable sensitivity of a particular isolate narrows the initial choice. Unless the patient is allergic, bactericidal penicillin analogs are recommended. The initial choice should be limited to penicillinase-resistant drugs, since most isolates from both hospital and community infections are resistant to penicillin G, penicillin V, and ampicillin. If, however, upon sensitivity testing the staphylococcal isolate proves to be sensitive to penicillin, this is the drug of choice for continued treatment because it is more active and less expensive.

For cutaneous infections, oral therapy with a semisynthetic penicillin, such as cloxacillin or dicloxacillin, is usually efficacious. Oxacillin and nafcillin are not recommended for oral therapy because their absorption is too unpredictable. If the patient is allergic to penicillin, the cephalosporins, erythromycin, or clindamycin may be used orally.

For serious systemic staphylococcal disease, parenteral administration of nafcillin, methicillin, oxacillin, or a cephalosporin is recommended. Vancomycin, cephalosporins, and clindamycin are suitable parenteral substitutes in the allergic patient. Staphylococcal infections often respond slowly and relapses often occur if therapy is terminated too early. The development of resistance to penicillin during the course of treatment is unusual, but with erythromycin, it is not uncommon. Treatment of most serious staphylococcal infections should be continued for 4 to 6 weeks to prevent the later emergence of metastatic abscesses.

Although methicillin-resistant strains of *S. aureus* have not been a common problem in the United States, they are now being seen with increased frequency. Most significant outbreaks have occurred in hospitals, but methicillin-resistant infection may also arise in the community and has the potential to disseminate in both settings. Drug abusers and debilitated patients represent major sources of methicillin-resistant *S. aureus*. These organisms usually show resistance to other semisynthetic penicillins, such as nafcillin, and to the cephalosporins. They may also be resistant to gentamicin, tobramycin, and clindamycin. Vancomycin, alone or in combination with rifampin, is the recommended treatment of infections caused by these methicillin-resistant strains. Rifampin should only be used in combination with another antibiotic in order to prevent the emergence of rifampin resistance.

Prevention

Staphylococcal infection will never be completely controlled because of the carrier state in humans. In the home as well as in the hospital, spread of infection can be limited only by proper hygienic care and disposal of contaminated materials. Because of the large number and severity of hospital-acquired staphylococcal infections, prevention and control of hospital infection has been the primary focus. In the hospital setting, one is more likely to encounter a more virulent organism as well as a very susceptible patient population. Persons with staphylococcal lesions should be segregated from newborns and from highly susceptible adults. Indiscriminate use of antibiotics should be avoided in order to prevent establishment and spread of resistant strains throughout the hospital. All surgical procedures and instrumentation should be performed with maximal attention to aseptic techniques. In the newborn, proper care should be given the umbilical stump, and personnel in the nursery should be screened for staphylococcal carriers. Infection committees that have been set up in hospitals to control nosocomial infections should provide effective surveillance and follow-through of problems that are uncovered.

Other Medically Significant Staphylococci

Staphylococcus epidermidis

Identification. *S. epidermidis*, previously classified as *S. albus*, is a coagulase-negative staphylococcus that characteristically forms white colonies on blood agar. It may be distinguished from *S. aureus* by a number of important properties, some of which are listed in Table 25-3. At least 10 coagulase-negative species have been recognized, but of these only *S. epidermidis* and *S. saprophyticus* are of clinical importance.

Epidemiology. *S. epidermidis* appears to be host-specific for humans. All individuals carry the organisms on their skin. Most frequent sites include the axillae, head, arms, nares, and legs. Humans thus serves both as an exogenous source of contamination for infection to others and as an endogenous source to ourselves. Apparently, pathogenic strains of *S. epidermidis* are not present in the patient preoperatively. Biotyping and phage-typing techniques have been devised for *S. epidermidis* to trace specific isolates from the environment to the patient, but as a general epidemiologic tool these have proved inadequate. Plasmid pattern analysis, currently being evaluated, shows promise as an epidemiologic marker for clinically important isolates.

Clinical Infection. The coagulase-negative *S. epidermidis* can no longer be considered a harmless saprophyte. It is now apparent that under appropriate conditions, it can assume a pathogenic role when alterations in the integument allow it to gain entry into the body. Most of these infections are associated with foreign bodies, i.e., prosthetic valvular endocarditis, and infections of cerebrospinal fluid shunts, joint prostheses, and vascular prostheses. In addition, *S. epidermidis* can cause urinary tract infections, especially in elderly hospitalized males, and occasionally natural valve endocarditis in intravenous drug abusers.

Treatment. Multiple antibiotic resistance, including resistance to methicillin, is a common feature of disease-producing strains of *S. epidermidis*. There is no single recurring

pattern of antibiotic resistance. Multiple antibiotic resistance not only complicates therapy but also provides a reservoir of genetic antibiotic resistance for the more virulent *S. aureus*. The choice of appropriate therapy should be based on the local antibiogram to *S. epidermidis*. In the absence of this, an initial regime should include an aminoglycoside (gentamicin or tobramycin) with cephalothin, or rifampin, or vancomycin alone.

Staphylococcus saprophyticus

This coagulase-negative staphylococcus can be distinguished from *S. epidermidis* by its resistance to novobiocin and by its failure to ferment glucose anaerobically. Colonies have either an intense yellow or a white pigmentation.

Normally *S. saprophyticus* is found only occasionally on the human skin and is not commonly found in the indigenous flora of the rectum, vagina, or periurethral area. It can, however, assume the role of a primary pathogen and produce urinary tract infections in a narrowly defined population group—the adolescent female. *S. saprophyticus* is the second most common cause of urinary tract infections in young, sexually active females.

FURTHER READING

Books and Reviews

Arbuthnott JP: Staphylococcal toxins. In Schlessinger D (ed): Microbiology—1975. Washington, DC, American Society for Microbiology, 1975, p 267

Bernheimer AW: Interactions between membranes and cytolytic bacterial toxins. Biochim Biophys Acta 344:37, 1974

Buchanan RE, Gibbons NE (eds): Bergey's Manual of Determinative Bacteriology, 8th ed. Baltimore, Williams & Wilkins, 1974

Cohen JO (ed): The Staphylococci. New York, Wiley-Interscience, 1972

Fekety FR: Staphylococcal infections. In Wyngaarden JB, Smith LH Jr (eds): Textbook of Medicine, 16th ed. Philadelphia, Saunders, 1982, vol II, pp 1466–1473

Forsgren A: Immunological aspects of protein A. In Schlessinger D (ed): Microbiology—1977. Washington, DC, American Society for Microbiology, 1977, p 353

Freer JH, Arbuthnott JP: Biochemical and morphologic alterations of membranes by bacterial toxins. In Bernheimer AW (ed): Mechanisms in Bacterial Toxinology. New York, Wiley, 1976, p 169

Kloos WE: Natural populations of the genus *Staphylococcus*. Annu Rev Microbiol 34:559, 1980

Kloos WE, Schleifer K-H. The Genus *Staphylococcus*. In Starr MP, Stolp H, Trüper HG, et al. (eds): The Prokaryotes. New York, Springer-Verlag, 1981, vol II, p 1548

Lacey RW: Antibiotic resistance plasmids of *Staphylococcus aureus* and their clinical importance. Bacteriol Rev 39:1, 1975

Mudd S: A successful parasite: parasite-host interaction in infection by *Staphylococcus aureus*. In Mudd S (ed): Infectious Agents and Host Reactions. Philadelphia, Saunders, 1970, p 197

Novick R, Wyman L, Bouanchaud D, et al.: Plasmid life cycles in *Staphylococcus aureus*. In Schlessinger D (ed): Microbiology—

1974. Washington, DC, American Society for Microbiology, 1975, p 115

Rogolsky M: Nonenteric toxins. Microbiol Rev 43:320, 1979

Schleifer K-H, Kloos WE, Kocur M: The genus *Micrococcus*. In Starr MP, Stolp H, Trüper HG, Balows A, Schlegel HG (eds): The Prokaryotes. New York, Springer-Verlag, 1981, vol II, p 1539

Sheehy RJ, Novick RP: Penicillinase plasmid replication in *Staphylococcus aureus*. In Schlessinger D (ed): Microbiology—1974. Washington, DC, American Society for Microbiology, 1975, p 130

Verhoef J, Verbrugh HA: Host determinants in staphylococcal disease. Annu Rev Med 32:107, 1981

Wiseman GM: The hemolysins of *Staphylococcus aureus*. Bacteriol Rev 39:317, 1975

Selected Papers

Aly R, Maibach HI, Shinefield HR, et al.: Bacterial interference among strains of *Staphylococcus aureus* in man. J Infect Dis 129:720, 1974

Bailey RR: Significance of coagulase-negative *Staphylococcus* in urine. J Infect Dis 127:179, 1973

Baughn RE, Bonventre PF: Acquired cellular resistance following transfer of lymphocytes from mice infected repeatedly with *Staphylococcus aureus*. Cell Immunol 27:287, 1976

Bergdoll MS, Crass BA, Reiser RF, et al.: A new staphylococcal enterotoxin, enterotoxin F, associated with toxic-shock syndrome *Staphylococcus aureus* isolates. Lancet I:1017, 1981

Buxser S, Bonventre PF, Archer DL: Specific receptor binding of staphylococcal enterotoxins by murine splenic lymphocytes. Infect Immun 33:827, 1981

Carney DN, Fossieck BE Jr, Parker RH, et al.: Bacteremia due to *Staphylococcus aureus* in patients with cancer: report on 45 cases in adults and review of the literature. Rev Infect Dis 4:1, 1982

Chesney PJ, Davis JP, Purdy WK, et al.: Clinical manifestations of toxic shock syndrome. JAMA 246:741, 1981

Christensen GD, Bisno AL, Parisi JT, et al.: Nosocomial septicemia due to multiply antibiotic-resistant *Staphylococcus epidermidis*. Ann Intern Med 96:1, 1982

Crossley K, Landesman B, Zaske D: An outbreak of infections caused by strains of *Staphylococcus aureus* resistant to methicillin and aminoglycosides. II. Epidemiologic studies. J Infect Dis 139:280, 1979

Davis JP, Osterholm MT, Helms CM, et al.: Tri-state toxic-shock syndrome study. II. Clinical and laboratory findings. J Infect Dis 145:441, 1982

Doran JE, Raynor RH: Fibronectin binding to protein A-containing staphylococci. Infect Immun 33:683, 1981

Elias PM, Fritsch P, Tappeiner G, et al.: Experimental staphylococcal toxic epidermal necrolysis (TEN) in adult humans and mice. J Lab Clin Med 84:414, 1974

Eng RHK, Wang C, Person A, et al.: Species identification of coagulase-negative staphylococcal isolates from blood cultures. J Clin Microbiol 15:439, 1982

Espersen F, Clemmensen I: Isolation of a fibronectin-binding protein from *Staphylococcus aureus*. Infect Immun 37:526, 1982

Finland M, Barnes MW: Changing ecology of acute bacterial empyema: Occurrence and mortality at Boston City Hospital during 12 selected years from 1935 to 1972. J Infect Dis 137:274, 1978

Gezon HM, Thompson DJ, Rogers KD, et al.: Control of staphylococcal infections and disease in the newborn through the use of hexachlorophene bathing. Pediatrics 51:331, 1973

Grinstead J, Lacey RW: Ecological and genetic implications of pigmentation in *Staphylococcus aureus*. J Gen Microbiol 75:259, 1973

Hill HR, Williams PB, Krueger GG, et al.: Recurrent staphylococcal abscesses associated with defective neutrophil chemotaxis and allergic rhinitis. Ann Intern Med 85:39, 1976

Jordan PA, Iravani A, Richard GA, et al.: Urinary tract infection caused by *Staphylococcus saprophyticus*. J Infect Dis 142:510, 1980

Karakawa WW, Kave JA: Immunochemistry of an acidic antigen isolated from a *Staphylococcus aureus*. J Immunol 114:310, 1975

Kloos WE, Orban BS, Walker DD: Plasmid composition of *Staphylococcus* species. Can J Microbiol 27:271, 1981

Kohashi O, Pearson CM, Watanabe Y, et al.: Structural requirements for arthritogenicity of peptidoglycans from *Staphylococcus aureus* and *Lactobacillus plantarum* and analogous synthetic compounds. J Immunol 16:1635, 1976

Kondo I, Fujise K: Serotype B staphylococcal bacteriophage singly converting staphylokinase. Infect Immun 18:266, 1977

Leijh PCJ, van den Barselaar M Th, Daha MR, et al.: Stimulation of the intracellular killing of *Staphylococcus aureus* by monocytes: Regulation by immunoglobulin G and complement components C3/C3b and B/Bb. J Immunol 129:332, 1982

Lillibridge CB, Melish ME, Glasgow LA: Site of action of exfoliative toxin in the staphylococcal scalded-skin syndrome. Pediatrics 50:728, 1972

Melish ME, Glasgow LA: The staphylococcal scalded-skin syndrome: Development of an experimental model. N Engl J Med 282:1114, 1970

Morgan NG, Montague W: Studies on the interaction of staphylococcal δ-haemolysin with isolated islets of Langerhans. Biochem J 204:111, 1982

Morvitz J: A study of the biosynthesis of protein A in *Staphylococcus aureus*. Eur J Biochem 48:131, 1974

Nordstrom K, Forsgren A: Effect of protein A on adsorption of bacteriophages to *Staphylococcus aureus*. J Virol 14:198, 1974

Okabayashi K, Mizuno D: Surface-bound nuclease of *Staphylococcus aureus*: Localization of the enzyme. J Bacteriol 117:215, 1974

Osterholm MT, Davis JP, Gibson RW, et al.: Tri-state toxic-shock syndrome study. I. Epidemiologic findings. J Infect Dis 145:431, 1982

Peacock JE, Moorman DR, Wenzel RP, et al.: Methicillin-resistant *Staphylococcus aureus*: Microbiologic characteristics, antimicrobial susceptibilities, and assessment of virulence of an epidemic strain. J Infect Dis 144:575, 1981

Peterson PK, Wilkinson BJ, Kim Y, et al.: Influence of encapsulation on staphylococcal opsonization and phagocytosis by human polymorphonuclear leukocytes. Infect Immun 19:943, 1978

Rees PJ, Fry BA: The morphology of staphylococcal bacteriophage K and DNA metabolism in infected *Staphylococcus aureus*. J Gen Virol 53:293, 1981

Ruby C, Novick RP: Plasmid interactions in *Staphylococcus aureus*: Nonadditivity of compatible plasmid DNA pools. Proc Natl Acad Sci USA 72:5031, 1975

Saravolatz LD, Markowitz N, Arking L, et al.: Methicillin-resistant *Staphylococcus aureus*. Epidemiologic observations during a community-acquired outbreak. Ann Intern Med 96:11, 1982

Schlievert PM: Enhancement of host susceptibility to lethal endotoxin shock by staphylococcal pyrogenic exotoxin type C. Infect Immun 36:123, 1982

Schlievert PM, Shands KN, Dan BB, et al.: Identification and characterization of an exotoxin from *Staphylococcus aureus* associated with toxic-shock syndrome. J Infect Dis 143:509, 1981

Schmeling DJ, Gemmell CG, Craddock PR, et al.: Effect of staphylococcalα-toxin on neutrophil migration and adhesiveness. Inflammation 5:313, 1981

Sewell CM, Clarridge JE, Young EJ, et al.: Clinical significance of coagulase-negative staphylococci. J Clin Microbiol 16:236, 1982

Sjodahl J: Repetitive sequences in protein A from *Staphylococcus aureus*. Eur J Biochem 73:343, 1977

Tuazon CU, Sheagren JN, Choa MS, et al.: *Staphylococcus aureus* bacteremia: Relationship between formation of antibodies to teichoic acid and development of metastatic abscesses. J Infect Dis 137:57, 1978

Ubelaker MH, Rosenblum ED: DNA replication in bacteriophage-infected *Staphylococcus aureus.* J Virol 24:768, 1977

Verbrugh HA, Peters R, Rozenberg-Arska M, et al.: Antibodies to cell wall peptidoglycan of *Staphylococcus aureus* in patients with serious staphylococcal infections. J Infect Dis 144:1, 1981

Warren JR, Spero L, Metzger JF, et al.: Immunogenicity of formaldehyde-inactivated enterotoxins A and C_1 of *Staphylococcus aureus.* J Infect Dis 131:535, 1975

Weinstein RA, Kabins SA, Nathan C, et al.: Gentamicin-resistant staphylococci as hospital flora: epidemiology and resistance plasmids. J Infect Dis 145:374, 1982

Wheat LJ, Kohler RB, Tabbarah ZA, et al.: IgM antibody response to staphylococcal infection. J Infect Dis 144:307, 1981

Wilkinson BJ, Kim Y, Peterson PK, et al.: Activation of complement by cell surface components of *Staphylococcus aureus*. Infect Immun 20:388, 1978

CHAPTER 26

Streptococcus

The genus *Streptococcus* comprises many species pathogenic for humans. Among the major diseases caused by these organisms are streptococcal pharyngitis, scarlet fever, impetigo, neonatal sepsis and meningitis, and bacterial endocarditis. In addition, infection caused by group A streptococci may lead to the postinfectious syndromes of acute rheumatic fever, rheumatic heart disease, and acute glomerulonephritis.

Streptococci were first identified by Pasteur in the latter part of the 19th century, but many years passed before the diverse nature of the various species within this genus was fully understood. In the 1930s, Rebecca Lancefield introduced a method for the classification of streptococci into serologic groups (A, B, D . . . S). The system was based upon the antigen composition of cell wall carbohydrates and provided a mechanism for the unraveling of the spectrum of streptococcal infections and their nonsuppurative complications.

The genus *Streptococcus* is the only one of the five genera in the family Streptococcaceae that contains organisms pathogenic for humans. The most important of these pathogens are *Streptococcus pyogenes* (group A), *Streptococcus agalactiae* (group B), *Streptococcus faecalis* (group D), *Streptococcus pneumoniae* (Chap. 27), and the viridans group.

Properties of the Genus. Streptococci are gram-positive organisms, spherical to ovoid in shape, and less than 2 μm in diameter. Cell division occurs in one plane, resulting in pairs or chains. All members of the genus are homolactic fermentors and are catalase negative.

One of the most useful schemes for the preliminary classification of streptococci is based upon the type of hemolysis produced on blood agar plates. β-Hemolytic streptococci produce a clear zone of hemolysis around the colony as a result of the complete lysis of red blood cells by various hemolysins. α-Hemolytic organisms produce a zone of incomplete lysis and greenish discoloration of the medium. Many streptococci are nonhemolytic and produce no hemolysis (gamma reaction) on blood agar.

Group A Streptococci
(*Streptococcus pyogenes*)

Morphology

Group A streptococci are spherical to ovoid microorganisms, 0.6 to 1.0 μm in diameter (Fig. 26-1). Many strains form long chains, especially when grown in liquid media. Although usually staining gram-positive, organisms may become gram-variable or gram-negative with age.

The ultrastructure of the group A streptococcus is typical of other gram-positive bacteria in that there is a rigid cell wall, an inner plasma membrane with mesosomal vesicles, cytoplasmic ribosomes, and nucleoid. In addition, external to the cell wall are surface, fimbria-like append-

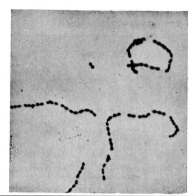

Figure 26-1. *Streptococcus pyogenes.* Gram stain. $\times$ 1200.

ages that contain the type-specific M protein (Fig. 26-2).

Some strains produce a capsule of hyaluronic acid, which may be demonstrable during the first 2 to 4 hours of growth. Since many strains also produce the enzyme hyaluronidase later during the growth cycle, capsules may not be seen in older cultures. With the exception of M protein, the ultrastructure of the group A streptococcus is identical to that of other groups of streptococci.

L Forms. Protoplasts, induced by exposure to penicillin or lytic enzymes, may be propagated on hypertonic media, giving rise to L colonies and an L form of growth. Removal of penicillin usually results in reversion of these forms to the parent strain. The role of L forms in disease states or in persistence of streptococci in tissues is unclear.

Physiology

All members of the genus *Streptococcus* are facultative anaerobes. Their metabolism is fermentative, the principal product being lactic acid. They are catalase negative, oxidase negative, and do not contain any heme compounds. The minimal nutritional requirements of the streptococcus are complex because of the organism's inability to synthesize many of its required amino acids, purines, pyrimidines, and vitamins. Group A streptococci are killed in 30 minutes at 60C, in contrast to certain other streptococci, e.g., group D, that are more heat resistant.

Cultural Characteristics. Growth of *S. pyogenes* is optimal at pH 7.4 to 7.6 at 37C. Primary isolation media usually contain blood or blood products, and growth may be enhanced by culture at a reduced oxygen tension or increased level of CO_2. Most group A streptococci are β hemolytic on sheep blood agar, although small amounts of fermentable carbohydrate (0.05 percent glucose) may decrease this reaction (Fig. 26-3). Under anaerobic conditions, hemolysis is enhanced. Sheep blood is preferred for primary isolation, since it is inhibitory to the growth of *Haemophilus haemolyticus*, an organism whose colonial morphology and β-hemolytic reaction may cause it to be con-

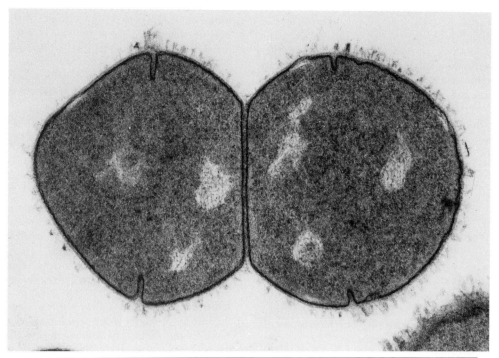

Figure 26-2. Electron micrograph of group A streptococcus, M type 23, showing external fuzz with M protein and cell wall with electron-dense inner layer and closely adherent plasma membrane. The cytosol shows numerous homogeneous ribosomes and lighter nucleoid areas. This pair of organisms demonstrates one complete crosswall and the beginning of secondary septations. Glutaraldehyde-osmium fixation. × 84,000. (Courtesy of Dr. Roger M. Cole.)

fused with the hemolytic streptococci. Human blood should not be used unless it is known to be free of inhibitory substances.

Laboratory Identification. For primary culture, clinical specimens may be processed by both pour and streak plate techniques. Characteristically, after 18 to 24 hours of growth on blood agar, *S. pyogenes* colonies are domed, grayish to opalescent, and approximately 0.5 mm in diameter. They are surrounded by a zone of β hemolysis sever-

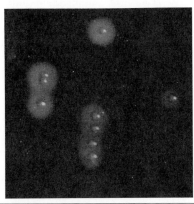

Figure 26-3. Hemolytic streptococcus. Yeast blood agar. × 3. Note β-type hemolysis. (From Li and Koibong: J Bacteriol 69:326, 1955.)

al times greater than the diameter of the colony. Beta hemolysis serves as the marker for primary isolation and may be enhanced by subsurface inoculation or incubation at reduced oxygen tension. Group A streptococci must be distinguished from other β-hemolytic streptococci (primarily groups B, C, and G) that are often present in the pharynx and other tissue sites. The serologic group may be determined by a variety of techniques, such as Lancefield extraction and precipitation, fluorescent antibody, or coagglutination. The bacitracin test, used primarily for pharyngeal cultures, may be useful for presumptively distinguishing between β-hemolytic streptococci. This test is based on the sensitivity of group A streptococci to bacitracin and predicts with 95 percent accuracy the presumptive identification of pharyngeal isolates.

Streptococci are differentiated from staphylococci on the basis of cellular and colonial morphology and the catalase test. Streptococci are catalase negative, while staphylococci are catalase positive.

Lysogeny

Lysogeny is very common among all groups of streptococci. In group A, estimates of lysogeny range as high as 90 to 100 percent. Some of these bacteriophages associated with the lysogenized state may play an important role in directing the synthesis of various group A streptococcal enzymes and toxins. This relationship has been especially well established with the pyrogenic exotoxin (erythrogenic toxin). The role of lysogeny in the induction of post-streptococcal syndromes is unclear (p. 470). However, many patients with these diseases possess phage-neutralizing antibodies. Phage-associated muralysins are produced by both groups A and C virulent phages during infection of streptococcal cultures. The group C lysin has been particularly useful in studies on the cell wall structure of group A streptococci. This enzyme is an *N*-acetylmuramyl-L-alanine amidase that is capable of lysing streptococci of many groups. This enzyme has also been useful in the production of L forms and in the preparation of purified membranes and M protein.

Antigenic Structure

Carbohydrates. The work of Rebecca Lancefield in 1933 laid the groundwork for the serologic classification of streptococci. Antigens (either wall polysaccharides or teichoic acids) may be extracted from whole cells by such procedures as dilute hydrochloric or nitrous acid, formamide, autoclaving, or lytic enzymes. The extracts thus obtained are then tested against group-specific antisera by capillary tube precipitation reactions. This is the most accurate method employed for defining the various serologic groups.

The group-specific antigen, the C carbohydrate, is composed of a branched polymer of L-rhamnose and *N*-acetyl-D-glucosamine in a 2:1 ratio, the latter being the antigenic determinant. This carbohydrate is linked (possibly by phosphodiester bonds) to the peptidoglycan, which consists of *N*-acetyl-D-glucosamine, *N*-acetyl-D-muramic acid, D-glutamic acid, L-lysine, and D- and L-alanine.

Proteins. Group A streptococci produce three surface protein antigens (M, T, and R) that are useful in serologic typing. Over 90 percent of strains may be classified by the use of M and T antigens. M protein is the major virulence factor of group A streptococci, and over 60 serotypes have been described.

Organisms producing M protein are resistant to phagocytosis in the absence of type-specific antibody. M protein is acid and heat stable and trypsin sensitive. Removal from the cell wall does not alter the viability of the organism. M typing is performed by capillary tube precipitin tests using type-specific antisera and hydrochloric acid extracts. Electron microscopy using ferritin-labeled antibody has shown M proteins to be associated with the fuzz on the outer surface of the cell wall (Fig. 26-2). M protein may be destroyed during growth by a proteinase if the pH of the medium is allowed to fall below 6.5. Some strains lack detectable M protein on initial isolation or lose it with repeated subculture. On rare occasions, two different M proteins may be detected in the same strain.

T antigens are resistant to pepsin and trypsin but are acid and heat stable. T typing is done by a slide agglutination test using trypsin-treated whole streptococci. Some T antigens are restricted to a single M type, while others may be shared by several M types (Table 26-1). T antigens are not associated with surface fuzz or with virulence. Antibodies to T antigens are not protective.

Typing systems employing the R surface antigens are not commonly used. These antigens are destroyed by pepsin but not by trypsin.

Another antigen, M-associated protein, is found in all M protein-containing group A streptococci and some strains of groups C and G but not in M-negative strains. Antibody responses to M-associated proteins are usually highest in patients with acute rheumatic fever.

Capsular Antigen. Many group A streptococci produce a hyaluronic acid capsule. However, this capsule is less important as a virulence factor than the M protein (p. 464).

Determinants of Pathogenicity

Lipoteichoic Acid. In order for an organism to gain entry into its intended host, it must be able to gain a foothold on the surface of cells at the portal of entry. It has been shown by several authors that adherence to buccal epithelial cells is mediated by the lipoteichoic acid of group A streptococci. It has also been shown by others that adherence to pharyngeal epithelial cells is much greater in those strains that contain M protein as opposed to those that do not. Regardless of the major mechanism of attachment, group A streptococci have been shown to adhere avidly to these surfaces.

M Protein. Once adherence has taken place, those strains that are able to resist phagocytosis and intracellular killing by leukocytes, i.e., those organisms rich in M protein, proliferate and begin to invade local tissues. Local pharyngeal or cutaneous infection may ensue, or the organism may invade contiguous tissues or distant tissues through the bloodstream. Once an antibody response is induced, organisms may be rapidly engulfed and killed by phagocytes. The cell walls of group A streptococci have been shown to react with immunoglobulins in a non-immune manner similar to staphylococcal protein A. The cell walls are also potent activators of the alternative complement pathway. It is interesting that the presence of M protein on the surface of the cell wall prevents these reac-

TABLE 26-1. RELATION OF T PATTERNS TO M TYPES

T Complex	M Types Bearing T Complex
1	1
2	2
3/13/B3264	3,13,33,39,41,43,52,53,56
8/25/Imp. 19	2,8,25,55,57,58
5/11/12/27/44	5,11,12,27,44,59,61
14/49	14,49
15/17/19/23/47	15,17,19,23,30,47,54

tions from recurring and, hence, may explain the rapid recognition and phagocytosis of M-negative strains.

Pharyngeal infection may induce long-lasting type-specific immunity to M protein; however, protective antibodies are rarely detected following impetigo. Other factors enhancing virulence are discussed below. Antibodies to several of these proteins (streptolysin O, NADase, DNase B, streptokinase, and hyaluronidase) are of practical importance in the serologic confirmation of streptococcal infection. Some of these proteins or toxins have been postulated to play a role in the induction of organic lesions at distant sites (streptolysin O, pyrogenic exotoxin, cardiohepatic toxin).

Hemolysins

STREPTOLYSIN O. Streptolysin O (SLO) is produced by most strains of group A streptococci and many strains of groups C and G. SLO and streptolysin S are responsible for the β-hemolytic reaction on blood agar. SLO is inactivated by oxygen. However, this can be reversed by reducing agents, such as cysteine or beta-mercaptoethanol. It is irreversibly inactivated by cholesterol. This hemolysin has immunologic cross-reactivity and properties similar to the oxygen-labile hemolysins of pneumococci, clostridia, and bacilli.

SLO is toxic for red and white blood cells and various cell types in tissue culture. Intravenous injection in animals may cause sudden death. Following pharyngeal or systemic infections, SLO induces a brisk antibody response, usually within 10 to 14 days. A more rapid anamnestic response occurs following repeated infection. Antibody titers of 300 to 500 may be seen normally in pediatric populations, where these organisms are frequently encountered, but are considerably lower in adults or protected children populations. Immune responses to this antigen following skin infection are considerably lower, possibly secondary to local inactivation of SLO by lipids.

STREPTOLYSIN S. Streptolysin S (SLS) is an oxygen-stable, nonantigenic toxin that is extractable from streptococcal cells by albumin, RNA, or detergents. This hemolysin has been shown to be produced near the cell membrane. SLS produces hemolysis by direct cell-to-cell contact or by transmission via the carrier molecules mentioned above. The molecular weight is probably considerably less than 20,000, which may account for its lack of antigenicity. It is lytic for red and white blood cells and for bacterial protoplasts and L forms. SLS is responsible for the surface hemolysis seen on blood agar, and those occasional strains that lack SLS may appear nonhemolytic on surface growth. SLS is inhibited by phosphatidylcholine and phosphatidylethanolamine.

Pyrogenic Exotoxins (Erythrogenic Toxins).

A majority of the strains of group A streptococci produce pyrogenic exotoxins. These toxins, as is the case with diphtheria toxin, appear to be synthesized as the consequence of infection with temperate bacteriophages. Classi-

cally, much attention has been paid to the scarlet fever rash produced by the toxins. However, pyrogenicity appears to be the primary result of the toxins. Dermal reactivity is, at least in part, secondary to hypersensitivity. In addition to pyrogenicity, these exotoxins also increase the susceptibility of rabbits to lethal endotoxic shock, cause reticuloendothelial blockade, act as specific and nonspecific mitogens, produce myocardial and hepatic necrosis in rabbits, and cause a decrease in antibody synthesis. There are at least three different serotypes (A, B, and C), which have molecular weights of 8000, 17,500, and 13,200, respectively. They are heat labile but are stable to acid, alkali, and pepsin.

The type C toxin causes increased permeability of the blood-brain barrier to endotoxin and bacteria and exerts its pyretic effect by direct action on the hypothalamus. Classically, it has been thought that these toxins cause a red reaction in the skin of nonimmune individuals (positive Dick test) and no reaction in individuals with immunity (negative Dick test). Antitoxin injected into the skin of a patient with scarlet fever causes localized blanching due to neutralization of erythrogenic toxin (Schultz–Charlton reaction). Recently, it has been proposed, however, that the rash in some individuals may be more related to hypersensitivity than to lack of immunity, and the occurrence of rash may depend on an interplay between cellular and humoral factors.

DNases A, B, C, and D.

These extracellular enzymes, which presumably assist in the generation of substrates for growth, are produced by most group A streptococci. Nucleases A and C have only DNase activity, while B and D also possess RNase activity. All have molecular weights of 25,000 to 30,000 and require calcium and magnesium for optimal activity. Antibody titers to DNase B are of great value in the serodiagnosis of pharyngeal or skin infection, especially the latter where the SLO response may be blunted.

Other Enzymes.

The group A streptococcus releases a large number of proteins into its external environment during growth. The role of these enzymes and toxins in pathogenicity versus generation of amino acids or nucleic acid substrates for growth is unclear. Two different streptokinases are produced by group A streptococci. These enzymes are antigenically distinct from the streptokinase of the group C streptococcus (which is the source for the commercial production of streptokinase for use in humans as a thrombolytic agent). Streptokinase forms a complex with plasminogen activator and catalyzes the conversion of plasminogen to plasmin, thus leading to the digestion of fibrin. This reaction also leads to the cleavage of the third component of complement into the chemotactic factor C3a. Streptococcal hyaluronidase hydrolyzes hyaluronic acid, both that present in the streptococcal capsule and that found in animal tissues. Both streptokinase and hyaluronidase are antigenic and may thus be of value in serodiagnosis. Many strains of streptococci also produce a proteinase (especially as the environmental pH falls during growth), NADase, ATPase, phosphatase, esterases, amylase, N-acetylglucosaminidase, neuraminidase, lipoproteinase, and a cardiohepatic toxin, possibly distinct from pyrogenic exotoxin.

Clinical Infection

Historically, streptococci have been a major human pathogen. In the preantibiotic era, streptococci were among the most frequent pathogens and produced significant mortality. Since the introduction of antibiotics, streptococcal diseases have been well controlled, and deaths are uncommon. The major pathogen for humans is the group A streptococcus (*S. pyogenes*), which is also responsible for the nonsuppurative postinfectious sequelae of rheumatic fever and glomerulonephritis.

Acute Streptococcal Infection

Epidemiology. Pharyngitis and impetigo are the most common streptococcal infections. The true incidence is unknown, but it is unlikely that a child would reach the age of 10 years without having encountered such an infection. Surveys of school children for antibodies to streptolysin O or other streptococcal exoproducts have shown that the majority have significant titers, indicating infection within the preceding 3 to 6 months.

Upper respiratory infections occur most frequently during the winter months, when asymptomatic nasal and pharyngeal carriage is also increased. Group A streptococci are primarily transmitted by droplets from the respiratory secretions. Transmission in milk and milk products has been largely controlled by pasteurization. However, explosive common source epidemics may follow contamination of foodstuffs by carriers or infected individuals. Hospital-acquired infections are occasionally caused by medical personnel with minimal infections.

Streptococcal pyoderma (impetigo) is predominantly a disease of temperate climates and occurs with highest frequency in late summer and early fall. The exact mode of transmission is unknown. Organisms first colonize normal skin and secondarily appear in the pharynx. Insect vectors, such as mosquitoes and flies, have been suspected but are unproved agents in transmission.

Pathogenesis. Group A streptococci possess numerous factors that enhance virulence and allow the organisms to establish themselves in the host. Many strains have a predilection for the upper respiratory tract, as opposed to the skin, but the mechanisms responsible for tissue specificity are unclear. Recently, however, it has been shown that pharyngeal or cutaneous strains may bind selectively to epithelial cells at these sites. These phenomena have been discussed above in the section on Determinants of Pathogenicity.

Streptococci are rapidly killed following ingestion, and disintegration of most organisms occurs within 1 to 4 hours. The cell wall, however, is resistant to lysozyme and lysosomal enzymes and may persist in cells or tissues indefinitely. The cell walls and peptidoglycan of group A streptococci may produce chronic inflammatory lesions in animal tissues and induce cutaneous nodules and myocarditis in rabbits. In addition, cell walls and peptidoglycan activate complement in vitro, with generation of chemotactic factors capable of inducing inflammation. The role of these phenomena in the induction of poststreptococcal diseases has been postulated but unproved.

Clinical Manifestations

PHARYNGITIS AND SCARLET FEVER. Streptococcal pharyngitis is usually associated with group A organisms, although sporadic cases and epidemics have been reported with groups C and G. When infection is caused by a lysogenized strain, scarlet fever may ensue. In the preantibiotic era, streptococcal pharyngitis was frequently associated with suppurative (tonsillar abscess, otitis, mastoiditis, septicemia, osteomyelitis) as well as nonsuppurative (acute rheumatic fever or glomerulonephritis) complications. Whether associated with the earlier administration of antibiotics or with changes in virulence, the incidence of serious sequelae has progressively declined in the United States and many European countries. Rheumatic fever and glomerulonephritis remain as important problems in developing countries.

Pharyngeal infection may be asymptomatic or may be associated with all gradations of the syndrome of sore throat, fever, chills, headache, malaise, nausea, and vomiting. Occasionally, abdominal pain is seen in children and may be confused with appendicitis. The pharynx may be mildly erythematous or beefy red with grayish yellow exudates that may bleed when swabbed for culture. Anterior cervical adenopathy and leukocytosis are usually present. Clinical syndromes indistinguishable from streptococcal pharyngitis may be seen with diphtheria, infectious mononucleosis, gonorrhea, and infections with many respiratory viruses, such as adenoviruses, Coxsackie, rhinoviruses, and herpes simplex virus.

The association of a scarlatinal rash is almost diagnostic of streptococcal infection. This rash characteristically blanches on pressure and begins on the trunk and neck, with spread to the extremities. Desquamation may occur during convalescence. Individuals may have several episodes of scarlet fever, since there are at least three different erythrogenic (pyrogenic) exotoxins.

Immunity. Pharyngeal infection probably confers lifelong type-specific immunity. Early treatment, however, may prevent or modify the immune response.

SKIN INFECTION. Group A streptococci produce the syndromes of impetigo, cellulitis, erysipelas, wound infection, or gangrene.

Impetigo is a very superficial infection that usually begins as small vesicles, progressing to weeping lesions with amber crust and slightly cloudy purulent exudate. The early lesions usually contain only streptococci but may be superinfected with staphylococci later in the course. Many serotypes have been associated primarily with impetigo and occasionally with outbreaks of nephritis (e.g., M types 2, 49, 55, and 57) (Table 26-2). In addition, many isolates either lack detectable M protein or, more likely, represent as yet undescribed M types. Types 59 through 63 have recently been designated, and other new serotypes are continuously being described. T agglutination patterns have been of use in defining the epidemiology of non-M typable strains (Table 26-1).

Cellulitis with lymphangitis and lymphadenitis may occur following deeper invasion by streptococci. Systemic symptoms, such as fever, chills, and malaise, may be seen, occasionally complicated by bloodstream invasion. Erysipelas is an infection of the skin and subcutaneous tissues that usually occurs on the face or the lower extremities. This lesion is characterized by erythema, edema, and induration, which usually has a distinct advancing border. Streptococci may be isolated from the subcutaneous tissues and occasionally from the blood. Some individuals are prone to recurrences, usually in the same site. Superficial cellulitis may spread to cause gangrene, especially in patients with peripheral vascular disease or diabetes. In addition to group A, groups B, C, and D and anaerobic streptococci may be involved. Sporadic outbreaks of omphalitis occasionally are seen in newborn infants. Acute glomerulonephritis may follow any of these infections when caused by group A streptococci. However, rheumatic fever occurs only following respiratory infection.

Immunity. The development of type-specific immunity is not well documented in streptococcal impetigo. Antibody to M protein has been demonstrated in less than 10 percent of the infections.

PUERPERAL SEPSIS. Puerperal sepsis, or childbed fever, was a common cause of maternal mortality in the preantibiotic era. Streptococci may be normal vaginal flora or may be introduced during delivery, occasionally by the attending physician or nurse. Outbreaks still occur occasionally, even in hospitals, and some may be traced to respiratory or anal carriers of streptococci. In recent years, with better obstetric techniques, puerperal infections are uncommon. The usual syndrome is characterized by chills, fever, facial flushing, abdominal distention with pelvic tenderness, and serosanguineous vaginal discharge.

Group A streptococci are frequently isolated from the uterus or blood. Mortality has been significantly reduced with antibiotics, but even with therapy, recovery may be complicated and prolonged.

Laboratory Diagnosis. The definitive diagnosis of streptococcal pharyngitis can be made only by direct culture of the posterior pharynx and tonsils. Swabs may be inoculated into broth and examined by fluorescent antibody or onto blood agar, with subsequent grouping by bacitracin disc sensitivity or serologic techniques.

Streptococci are best recovered from impetigo lesions early in the infection. Vesicular or pustular fluid inoculated onto blood agar may reveal a pure growth of streptococci. As lesions become older, streptococci and staphylococci may be isolated concomitantly. Cellulitis and erysipelas are best cultured by needle aspiration of tissue fluids, especially from the advancing border in erysipelas, or by subcutaneous injection of a small amount of sterile saline followed by reaspiration. Streptococci may also be isolated from the blood in patients with deeper infections.

Treatment. Streptococcal pharyngitis is frequently a self-limited disease that may resolve without complication or antibiotics. Therapy is directed primarily at the prevention of suppurative complications and the late sequelae of rheumatic fever and of decreasing the incidence of glomerulonephritis. The treatments of choice are intramuscular benzathine penicillin given in a single dose or oral penicillin V for 10 days. Alternatives in the penicillin-allergic patient include erythromycin, clindamycin, and cephalexin (Table 26-3). Tetracyclines and sulfonamides are contraindicated because of increased resistance of streptococci and lack of prevention of rheumatic fever. Recurrences of streptococcal infection in patients with previous rheumatic fever may be prevented by monthly injections of benzathine penicillin, by oral penicillin V, or by sulfonamides (Table 26-4). Sulfonamides are effective in prevention but not in therapy.

TABLE 26-2. ASSOCIATION OF CERTAIN SEROTYPES WITH ACUTE NEPHRITIS

M Type	Pharyngitis-associated	Pyoderma-associated
1	+++	
2	0	+++
3	++	±
4	+++	±
12	++++	±
25	++	±
49	++	++++
55	0	+++
57	0	++

From Wannamaker: N Engl J Med 282:78, 1970.
++++, strong evidence of association; ± or 0, questionable or no evidence.

TABLE 26-3. TREATMENT OF STREPTOCOCCAL PHARYNGITIS OR IMPETIGO

Parenteral benzathine penicillin	Children: 600,000 to 900,000 units Adults: 1,200,000 units
Oral penicillin V, erythromycin, cephalexin, or clindamycin	15 mg/kg/day in 4 divided doses

TABLE 26-4. PROPHYLACTIC REGIMENS AND RECURRENCE RATES OF RHEUMATIC FEVER

	Parenteral Benzathine Penicillin	Oral Penicillin	Oral Sulfadiazine
Number of patient years	560	545	576
Rate of recurrence*	0.4	5.5	2.8

From Wood et al.: Ann Intern Med 60:31, 1964.
*Rate per 100 patient years (patient year = number of patients × number of years followed).

Impetigo is best treated with parenteral benzathine penicillin or by oral penicillin V, erythromycin, clindamycin, or cephalexin (Table 26-3). However, parenteral high-dose therapy may be required for deeper, more invasive infections. Surgical debridement and even amputation may be required in severe infections, especially when complicated by peripheral vascular disease.

Prevention. Infection may be prevented by prompt therapeutic intervention during epidemics or prophylactic therapy given to individuals at high risk, such as military recruits or patients with rheumatic heart disease. While immunization with M protein vaccines has been shown to be effective, the use of such vaccines in selected populations remains to be explored.

Impetigo is commonly observed in hot, humid climates, especially where crowded living conditions exist. A majority of these infections could be prevented by improved skin hygiene. Epidemics are best halted by improving hygiene and treatment of all cases with effective antibiotic regimens.

Puerperal sepsis may be prevented, for the most part, by strict attention to aseptic techniques during deliveries.

Sequelae of Acute Streptococcal Infection

Acute Rheumatic Fever (ARF). Rheumatic fever is a nonsuppurative inflammatory reaction that is epidemiologically and serologically related to antecedent group A streptococcal infection. It is manifested by arthritis, carditis, chorea, erythema marginatum, or subcutaneous nodules. This constellation of symptoms usually occurs within 2 to 3 weeks after the onset of streptococcal infection, although chorea and erythema marginatum may be seen as late as 6 months. The diversity of the manifestations of rheumatic fever, the delay in onset after infection, and the inability prior to the 1930s to classify streptococci led to considerable difficulty in understanding this illness. The clinical experience of T. Duckett Jones led to the establishment of the Jones criteria for the diagnosis of acute rheumatic fever (Table 26-5).

In addition to the clinical criteria, the documentation of recent streptococcal infection by culture or serology is of utmost importance. Since the causative streptococcal in-

TABLE 26-5. MODIFIED JONES CRITERIA FOR THE DIAGNOSIS OF RHEUMATIC FEVER

Major Manifestations*	Minor Manifestations
Carditis	Fever
Arthritis	Arthralgia
Chorea	Elevated sedimentation rate or C-reactive protein
Erythema marginatum	Electrocardiographic changes
Subcutaneous nodules	History of previous rheumatic fever or rheumatic heart disease

Plus evidence of preceding streptococcal infection (scarlet fever, culture-proven group A streptococcal pharyngitis, or elevated streptococcal antibody test).

*Two major or one major and two minor manifestations with evidence of previous streptococcal infection indicate a high probability of rheumatic fever.

fection may have resolved or may have been asymptomatic, it is necessary to detect an increase in antibody titer to at least one of several streptococcal antigens (streptolysin O, DNase B, hyaluronidase, or streptokinase). The ability to document a change in titer depends on the length of the latent period, since many patients will have a maximal response at the time of acute illness. Rheumatic fever most commonly occurs after infection by serotypes 1, 3, 5, 6, 14, 18, 19, 24, 27, or 29.

Rheumatic fever may occur in up to as many as 3 percent of individuals during epidemics of pharyngitis. In contrast, in the nonepidemic situation, rheumatic fever will occur in as few as 1 per 1000 episodes of streptococcal pharyngitis. Milder poststreptococcal inflammatory states, characterized by fever, arthralgia without arthritis, and erythema nodosum, may also be seen but are not classified as acute rheumatic fever unless associated with major manifestations (Table 26-5). The major morbidity and mortality associated with rheumatic fever are linked to the subsequent development of rheumatic valvular heart disease. With a decline in the incidence of rheumatic fever, rheumatic heart disease probably accounts for less than 15,000 deaths per year in the United States.

PATHOGENESIS. The pathogenesis of rheumatic fever is poorly understood. Various theories have been proposed, including antigenic cross-reactivity between streptococcal antigens and heart tissue, direct toxicity due to streptococcal exotoxins, actual invasion of the heart by streptococci, or localization of antigens within damaged muscle or valvular tissues. Circulating immune complexes have been found in the serum of patients with acute rheumatic fever. Nevertheless, the streptococcus produces many potentially damaging exotoxins, and the components of its cell wall have been shown to produce inflammatory reactions in mammalian tissues. The true pathogenesis of ARF may never be elucidated because of the lack of a suitable experimental model.

TREATMENT. The duration and vigor of treatment depend upon the severity of the illness. Prolonged bed rest is no longer recommended unless it is necessary to control congestive heart failure. Salicylates and corticosteroids are of equal benefit in reduction of acute symptoms and control of long-term sequelae. Corticosteroids are usually administered to patients with moderate to severe heart failure. Penicillin does not alter the course of ARF but usually is administered once the diagnosis is definite or if group A streptococci are cultured from the pharynx.

PREVENTION. The prevention of streptococcal infections will prevent rheumatic fever. In addition, prompt treatment of patients with streptococcal pharyngitis within 10 days of onset greatly reduces the incidence of rheumatic fever. Patients who have had previous episodes of rheumatic fever should be placed on continuous antibiotic prophylaxis. The therapy of streptococcal pharyngitis and the prophylaxis of streptococcal infection are outlined in Tables 26-3 and 26-4.

Acute Poststreptococcal Glomerulonephritis (AGN).

AGN is similar to rheumatic fever, a postinfectious complication of group A streptococcal infection. In contrast to rheumatic fever, which occurs only following pharyngitis, AGN may be seen after either pharyngeal or cutaneous infection. It is primarily associated with a well-defined group of serotypes (1, 2, 3, 4, 12, 25, 49, 55, 57, 59, 60, and 61). Type 12 has been most frequently associated with AGN following pharyngitis, with the majority of strains being associated with pyoderma. The incidence of AGN in epidemics or sporadic streptococcal infections may vary from less than 1 percent to as high as 10 to 15 percent.

AGN is most often seen in children and may be associated with the acute onset of edema, oliguria, hypertension, congestive heart failure, or seizures. Laboratory findings include dark or smokey urine with red blood cells, red blood cell casts, white blood cells, proteinuria, depressed serum complement, decreased glomerular filtration rate, and serologic evidence of recent streptococcal infection. In addition, less severe cases frequently occur and may be associated with minimal urinary sediment changes or depressed serum complement without symptoms. These changes are frequently seen in the siblings of patients with AGN.

The latent period between streptococcal infection and the development of nephritis is 1 to 2 weeks following pharyngitis and 2 to 3 weeks following skin infection. Hematuria may occur during the latent period both in patients who do and in those who do not develop clinical AGN. In order to establish a streptococcal etiology, it is necessary to document previous or concurrent streptococcal infection or the immune response to streptococcal products. A great majority of patients will show a serologic response either to SLO or to the DNase B. It is important that the latter be sought in patients with impetigo, as the SLO response is poor following skin infection.

Renal biopsy in the typical patient with AGN shows edema and hypercellularity of the glomerular tuft, with red blood cells in Bowman's space or in the tubular lumens. Immunofluorescent examination may show complement components with or without immunoglobulins in a granular pattern. Electron microscopic examination shows subepithelial deposits on the glomerular basement membrane.

PATHOGENESIS. Granular accumulations of immunoglobulin, commonly seen by immunofluorescent staining, correspond to the subepithelial deposits seen by electron microscopy. The inflammatory response is probably due to deposition of immune complexes within the kidney, although the localization of streptococcal cellular components or exotoxins may also play a role. Circulating immune complexes have been found in the serum of patients with acute poststreptococcal glomerulonephritis. Streptococcal components are almost certainly present within the glomerulus, although controversy exists over the exact nature of these materials.

TREATMENT. Therapy is directed at the secondary phenomena of volume excess, hypertension, and seizures. This consists primarily of sodium restriction, diuretics, and anticonvulsants. Recovery is usually complete in children, although fatalities may occur during the acute phase. AGN in adults may be associated with a poorer prognosis and a higher incidence of chronic renal disease.

PREVENTION. Treatment of streptococcal pharyngitis perhaps lowers the subsequent incidence of AGN. Therapy of impetigo, however, has no effect on the prevention of AGN. Therapy, therefore, is directed toward the prevention of transmission of infection, especially when known nephritogenic strains are involved. Effective regimens are presented in Table 26-3.

Other Streptococci Pathogenic for Humans

Group B Streptococci
(*Streptococcus agalactiae*)

Morphology and Physiology. These organisms cannot be distinguished on a morphologic basis from other β-hemolytic streptococci. In liquid media, however, they tend to grow as diplococci or in short chains in contrast to the longer chains usually seen with groups A, C, and G. β Hemolysis is produced by a hemolysin distinct from those of the group A streptococcus. Its properties are more similar to those of streptolysin S than those of streptolysin O. A double zone of hemolysis on rabbit blood agar may be observed when the blood is refrigerated after initial incubation. Colonies are usually large and mucoid (1 to 2 mm), with a relatively small zone of β hemolysis. As many as 5 to 15 percent of isolates may be nonhemolytic. Ninety-seven percent of strains produce yellow, red, or orange pigment on appropriate media. These pigments (carotenoids) are associated with the cell membrane fraction, and pigment production is suppressed by the addition of glucose to the medium.

This species may be presumptively identified by its ability to hydrolyze sodium hippurate and a positive CAMP test. The latter phenomenon is characterized by an accentuated zone of hemolysis when the group B streptococcus is inoculated perpendicular to a streak of colonies of *Staphylococcus aureus*. A majority of strains can also grow in 6.5 percent sodium chloride, and a few may grow in the presence of 40 percent bile. Esculin is not hydrolyzed, a feature that distinguishes group B streptococci from group D streptococci. A small percentage are bacitracin sensitive and may be falsely identified as group A by this screening procedure.

Antigenic Structure. The group-specific carbohydrate of *S. agalactiae* is composed of D-glucosamine, D-galactose, and L-rhamnose, with the last being the major antigenic determinant. There are at least five serotypes (Ia, Ib, Ic, II, and III). Some strains may possess combinations of these antigens. The frequency of isolations of various serotypes varies from locale to locale, with II and III usually being most common.

Clinical Infection

EPIDEMIOLOGY. Group B streptococci are commonly found among the flora of the pharynx, gastrointestinal tract, and vagina. Approximately 15 to 20 percent of pregnant women may be vaginal carriers. The exact rates of transmission of group B streptococci to the newborn infant are variable but may be as high as 50 to 60 percent in infants born of maternal carriers. The incidence of disease in colonized infants is low but may have disastrous consequences.

CLINICAL MANIFESTATIONS. Group B streptococci cause skin infection, endocarditis, puerperal infection, neonatal septicemia, and meningitis. Wound infections are most common in patients with diabetes mellitus and peripheral vascular disease. In the last 20 years, group B streptococci have been noted with increasing frequency as causes of neonatal septicemia and meningitis. The organism is acquired from the mother during delivery, and clinical disease may be higher following obstetric complications, such as prolonged labor, premature rupture of the membranes, or obstetric manipulation. Mortality rates in neonates may be as high as 50 percent and are higher when the onset of infection is within 10 days of delivery.

TREATMENT. Penicillin G is the treatment of choice for group B streptococcal infections. A majority of strains are also sensitive to erythromycin, chloramphenicol, cephalosporins, lincomycin, and clindamycin. The prevention of neonatal disease by the prophylactic administration of penicillin or by immunization has been considered but is as yet controversial.

Group C Streptococci (*Streptococcus equisimilis, Streptococcus zooepidemicus, Streptococcus equi, Streptococcus dysgalactiae*)

Morphology and Physiology. The morphology of group C streptococci is similar to that of group A organisms. All species of group C are β hemolytic, with the exception of *S. dysgalactiae*, which may be α hemolytic or nonhemolytic. *S. equisimilis* produces streptolysin O, streptokinase (antigenically distinct from that of the group A streptococcus), and other extracellular products. Rises in antibody titers may thus be seen following infection with this organism. *S. equisimilis* serves as the source of streptokinase used in thrombolytic therapy in humans.

Antigenic Structure. The group-specific carbohydrate is a polymer of L-rhamnose and N-acetyl-D-galactosamine, the latter being the major antigenic determinant. Serotypes within species may be conferred by surface protein antigens that are similar to M protein. Group C streptococci differ from group A streptococci primarily in the substitution of N-acetyl-D-galactosamine for N-acetyl-D-glucosamine, yet the former are not associated with rheumatic fever, perhaps because of absence of virulence factors as important as the M proteins of the group A streptococcus.

Clinical Infection. Streptococci of group C are important causes of disease in a wide variety of animal species. The species of major interest in human infections is *S. equisimilis*, which has been isolated from the upper respiratory tracts of normal and diseased swine, cows, horses, and humans. It has been implicated as a cause of pharyngitis, puerperal sepsis, endocarditis, bacteremia, osteomyelitis, brain abscess, postoperative wound infection, and pneumonia. With the exception of one epidemic of pharyngitis that was associated with acute glomerulonephritis, poststreptococcal sequelae do not occur. Treatment is similar to that used for group A streptococcal infections.

Group D Streptococci (*Streptococcus faecalis, Streptococcus faecium, Streptococcus durans, Streptococcus bovis, Streptococcus equinus*)

Morphology and Physiology. Group D streptococci commonly grow as diplococci or short chains. Rare motile strains are encountered. The species of this serologic group are divided into enterococci (*S. faecalis, S. faecium,* and *S. durans*) and nonenterococci (*S. equinus* and *S. bovis*) based upon the ability of the former to grow in the presence of 6.5 percent sodium chloride. Group D streptococci differ from most other streptococci in their capacity to grow at 45C and to withstand temperatures above 60C. In addition, they grow in the presence of 40 percent bile and hydrolyze esculin. All give the alpha or gamma reaction on blood agar, with the exception of *S. faecalis* var *zymogenes,* which is β hemolytic.

The species within this group may be separated on the basis of their biochemical reactions, which are given in Table 26-6. An occasional nongroup D viridans streptococcus (*Streptococcus mutans*) may grow on bile and hydrolyze esculin but will not grow in the presence of 6.5 percent sodium chloride. Such strains may be separated from *S. bovis* and *S. equinus* by their inability to hydrolyze starch and the absence of group D antigen.

Antigenic Structure. In contrast to most of the other groups of streptococci, the group D antigen is not a wall carbohydrate but is a glycerol teichoic acid containing glucose and D-alanine. This antigen appears to be associated with the cytoplasm or plasma membrane. The cell wall polysaccharides in the species serve as type-specific antigens. Variations in peptidoglycan structure exist between the species of group D, specifically as additions to the peptide portion: *S. faecalis* contains only glutamic acid, lysine, and alanine, while *S. faecium* and *S. durans* also contain aspartic acid, and *S. bovis* and *S. equinus* contain threonine.

Clinical Infection. Group D streptococci commonly inhabit the skin and upper respiratory, gastrointestinal, and genitourinary tracts. A majority of infections apparently result from invasion by these normal flora. Person-to-person transmission is not of documented importance.

Group D streptococci, most commonly *S. faecalis,* are frequently associated with urinary and biliary tract infections, septicemia, endocarditis, wound infection, and intra-abdominal abscess complicating diverticulitis, appendicitis, and other diseases that alter the integrity of the gastrointestinal tract. Group D streptococci are a frequent cause of bacterial endocarditis, especially in the elderly or in patients with underlying valvular heart disease who undergo manipulation of the genitourinary or gastrointestinal tracts. A recent association has been made between endocarditis or bacteremias due to *S. bovis* and underlying gastrointestinal tumors. It has been suggested that isolation of this organism from the blood should alert the clinician to a possible occult tumor.

TREATMENT. The proper identification of group D streptococci is of practical importance in that the enterococcal strains are generally resistant to penicillin G and the penicillinase-resistant penicillins. However, many infections may show a synergistic response when treated with penicillin and an aminoglycoside, such as gentamicin or streptomycin. Among the penicillins, ampicillin is the single most effective agent. Other antibiotics useful in the penicillin-allergic patients are vancomycin and erythromycin. The nonenterococcal strains (*S. bovis* and *S. equinus*) are generally sensitive to penicillin G.

Group G Streptococci (*Streptococcus anginosus, Streptococcus* sp)

The strains containing the group G antigen, for the most part, do not have species designations. Group G contains the minute or small colony and small cell variants of the species *S. anginosus* and other strains that tend to produce larger β-hemolytic colonies. Group G streptococci produce streptolysin O, streptokinase, NADase, DNase, and hyaluronidase and may elicit antibody rises, especially to streptolysin O, following infection.

The group G carbohydrate is composed of galactose, galactosamine, and rhamnose, with the last being the major antigenic determinant. There are several serotypes within the group, but the antigens are poorly studied. Strains have been isolated that contain an antigen similar to, or identical with, group A type 12 M protein.

Other Groupable Streptococci

Other species of streptococci bearing carbohydrate group antigens have been described but are of lesser clinical importance. These consist of group E, group F (*S. anginosus* —alpha, beta, or gamma reaction), group H (*Streptococcus sanguis*—alpha reaction), group K (*Streptococcus salivarius, Streptococcus* sp—alpha reaction), and group N (*Streptococcus lactis* and *Streptococcus cremoris*), whose group antigen is a glycerol-alanine-galactose teichoic acid.

Viridans Streptococci

Morphology and Physiology. The designation "viridans group" has been assigned to a large number of alpha-reacting streptococci that resist classification by group-specific carbohydrates. This heterogeneous collection of streptococci, generally found in the mouth or up-

TABLE 26-6. BIOCHEMICAL AND GROWTH CHARACTERISTICS OF GROUP D STREPTOCOCCI

	S. faecalis	S. faecium	S. durans	S. bovis	S. equinus
Bile esculin	+	+	+	+	+
6.5% NaCl	+	+	+	−	−
Sorbitol	+	−	−	±	−
Mannitol	+	+	−	±	−
Lactose	+	+	+	+	−
Starch	−	−	−	+	+

±, variable within species.
S. faecalis var *zymogenes* is β hemolytic and *S. faecalis* var *liquefaciens* liquefies gelatin.

per respiratory tract, includes the following species: *Streptococcus mitior, Streptococcus milleri, S. sanguis,* and *S. mutans.* They are usually classified by fermentation patterns, cell wall sugar composition, and production of dextrans (glucose 1-6 polymers) or levans (fructose 2-6 polymers) from sucrose. Strict criteria for speciation have yet to be evolved. However, physiologic schemes, such as that presented in Table 26-7, are useful in characterizing species. *S. mutans,* an organism important in the formation of dental plaque, possesses a cellular and extracellular dextransucrase that produces an insoluble dextran polymer from sucrose. This polymer is an important component of dental plaque, which, in addition, may contain up to 10^{11} streptococci per gram net weight.

Antigenic Structure. This group of organisms has been notoriously resistant to serologic characterization. While many strains may possess antigens characteristic of groups F or K (and others), physiologic heterogeneity within these groups makes this phenomenon seem of less importance. Physiologic schemes, such as the one described in Table 26-7, would, therefore, seem to better characterize a species.

Clinical Infection. The viridans streptococci are frequently found in the nasopharynx, mouth, gingival crevices, gastrointestinal tract, female genital tract, and occasionally on the skin. From these sites the organisms may invade the bloodstream following chewing, dental manipulation, or gastrointestinal or genitourinary instrumentation. In addition, cellulitis or wound infection, meningitis, sinusitis, biliary or intraabdominal infection, or endocarditis may occur. It has recently been noted that *S. milleri (S. intermedius*-MG, *S. anginosus-constellatus*) has an unusual predilection, among this group of organisms, to produce abscesses in tissues, such as the brain or the liver. *S. sanguis,* on the other hand, is the most frequent single species causing bacterial endocarditis (see below).

DENTAL CARIES. The role of the viridans streptococci, especially of *S. mutans,* in the production of dental plaque and dental caries is a complex phenomenon and will be discussed in detail in Chapter 49.

Other Streptococcal Infections

Infective Endocarditis. Infective endocarditis may be defined as implantation of bacteria or fungi on the endocardial surface of the heart. Endocarditis is most frequently encountered on damaged heart valves or in congenital heart disease. The majority of cases are caused by streptococci (viridans, pneumococci, group D) and staphylococci. Endocarditis may follow a fulminant course (acute) or be a prolonged insidious illness (subacute). Streptococci are most commonly associated with the latter presentation. Endocarditis may complicate infection at other sites, such as pneumonia, abscesses, or urinary tract infection, or may be a result of transient bacteremia, such as that associated with dental manipulation.

The symptoms and signs of endocarditis are fever, weight loss, anemia, heart murmur, splenomegaly, and pe-

TABLE 26-7. SCHEMA FOR BIOCHEMICAL SPECIATION OF VIRIDANS STREPTOCOCCI*

	Mannitol[†]	Lactose[†]	Sucrose[†]	Raffinose[†]	Inulin[†]	Esculin[‡]	Hippurate[‡]	Arginine[‡]	Litmus[§]	40% Bile[ǁ]	Glucan[¶] Agar	Glucan[¶] Broth
S. mutans	+	+	+	+	+	+	−ǁ	−	+	vǁ	+	+
S. uberis	+	+	+	+	v	+	v	v	+	+	−	−
S. sanguis 1	−	+	+	v	+	v	−	v	+	v	+	v
S. sanguis 2	−	+	+	+	−	−	−	v	+	v	v	v
S. salivarius	−	+	+	+	+	+	−	−	+	v	v	−
S. mitis (mitior)	−	+	+	−	−	−	−	v	+	v	v	−
*S. intermedius (milleri-*MG)	−	+	+	v	−	+	−	v	+	v	v	−
S. anginosus	−	−	+	−	−	v	−	v	+	v	−	−
S. morbillorum	−	−	v	−	−	−	−	−	−	−	−	−
S. acidominimus	−	−	v	−	−	+	+	v	v	+	−	−

From Facklam: J Clin Microbiol 5:184, 1977.
+, ≧80%; −, ≦10%; v, 11–79%.
*Species designations in this group of organisms remain controversial. This schema is given because analysis of cell wall carbohydrates is not required. Lancefield antisera for groups A through G should be used only for β-hemolytic isolates. All bile-esculin-positive strains should be tested serologically for presence of group D antigen (see Table 26-6). Growth in 6.5% NaCl or at 10C is rarely observed. All strains were bile insoluble and optochin resistant.
[†]Production of acid.
[‡]Hydrolysis of hippurate or arginine.
[§]Reduction of litmus milk.
[ǁ]Tolerance.
[¶]Production of gel or partial gel in 5% sucrose broth.

ripheral embolization. The diagnosis is confirmed by repeated blood culture. In most cases, over 90 percent of the blood cultures taken are positive.

The duration of therapy and choice of antibiotic are determined by the sensitivity of the isolated organism to various antimicrobial agents. Parenteral therapy is given for 2 to 6 weeks. Many cases of endocarditis may be prevented by prophylactic administration of antibiotics to patients with underlying heart disease who undergo procedures associated with transient bacteremia (dental work or genitourinary manipulation).

Miscellaneous. Streptococci of all groups and those that are not identified by species or group may cause pneumonia, septic arthritis, biliary or intraabdominal infection, urinary tract infection, cellulitis or wound infection, meningitis, osteomyelitis, or sinusitis. Recently, with improved techniques of anaerobic bacteriology, microaerophilic and anaerobic streptococci (Peptococcaceae) are frequently isolated from patients with lung abscess, septic abortion or puerperal infection, endocarditis, empyema, and intraabdominal infection (Chap. 47).

REFERENCES

Books and Reviews

Bisno AL: Treatment of Infective Endocarditis. New York, Grune & Stratton, 1981

Breese BB, Hall CB (eds): Beta Hemolytic Streptococcal Disease. Boston, Houghton Mifflin, 1978

Holm SE, Christensen P (eds): Basic Concepts of Streptococci and Streptococcal Disease. Surrey, Reedbooks, 1982

Mandell GL, Douglas RG, Bennett JE (eds): Principles and Practice of Infectious Diseases. New York, Wiley, 1979, pp 1559-1588

Markowitz M, Gordis L: Rheumatic Fever. Philadelphia, Saunders, 1972

Starr MP, Stolp H, Trüper et al.: The Prokaryotes: A Handbook on Habitats, Isolation and Identification of Bacteria. New York, Springer-Verlag, 1981, vol II, pp 1572-1650

Uhr JW: The Streptococcus, Rheumatic Fever, and Glomerulonephritis. Baltimore, Williams & Williams, 1964

Wannamaker LW, Matsen JM: Streptococci and Streptococcal Diseases. New York, Academic Press, 1972

Selected Papers

Duma RJ, Weinber AN, Medrek JR, et al.: Streptococcal infections: bacteriologic and clinical study of streptococcal bacteremia. Medicine 48:87, 1969

Facklam RR: Physiological differentiation of viridans streptococci. J Clin Microbiol 5:184, 1977

Fox EN, Waldman RH, Wittner MK, et al.: Protective study with a group A streptococcal M protein vaccine. J Clin Invest 52:1885, 1973

Freedman P, Meisler HR, Lee HJ, et al.: The renal response to streptococcal infection. Medicine 49:433, 1970

Ginsburg I: Mechanisms of cell and tissue injury induced by group A streptococci: relation to poststreptococcal sequelae. J Infect Dis 126:294, 419, 1972

Langone JJ: Protein A of *Staphylococcus aureus* and related immunoglobulin receptors produced by streptococci and pneumococci. Adv Immunol 32:157, 1982

McCarty M: An adventure in the pathogenetic maze of rheumatic fever. J Infect Dis 143:375, 1981

Stollerman GH: Rheumatogenic and nephritogenic streptococci. Circulation 43:915, 1971

Wannamaker LW: Differences between streptococcal infections of the throat and of the skin. N Engl J Med 282:23, 78, 1970

Weinstein L, Schlesinger JJ: Pathoanatomic, pathophysiologic and clinical correlations in endocarditis. N Engl J Med 291:832, 1122, 1974

Zabriskie J: The role of streptococci in human glomerulonephritis. J Exp Med 130:180s, 1971

Streptococcus pneumoniae

In spite of modern antimicrobial agents, *Streptococcus pneumoniae* remains a leading cause of morbidity and mortality in persons of all ages. It is the most common cause of bacterial pneumonia and an important cause of otitis media, meningitis, and septicemia. Pneumococcal pneumonia is the classic prototype from which our present concepts of the pathogenesis of pneumonia have evolved.

The pneumococcus has a long and fascinating history. Its initial isolation was made in 1881 in independent stud-ies by Sternberg and Pasteur, who recovered the organism from human saliva. Its association with acute lobar pneumonia was demonstrated the following year by Fried-lander, and within the following 10-year period the range of pneumococcal infection was elucidated with remarkable speed.

Some of the most important achievements in both biology and medicine have resulted from studies on the pneumococcus. The recognition of different types of pneumococci based on serologic differences in capsular material provided the basis for specific serum therapy and for the classic studies of Avery and his collaborators at the Rockefeller Institute on the immunologic and chemical properties of the capsular polysaccharides. These studies also formed the basis for the development of a polyvalent polysaccharide vaccine currently used in high-risk patients for bacteremic pneumococcal pneumonia.

Of the many contributions originating from the study of the pneumococcus, the most revolutionary in its implications was the elucidation of the mechanism of transformation of pneumococcal types. The discovery that DNA is the transforming material provided the cornerstone of the modern discipline of molecular genetics.

Pneumococcal pneumonia is no longer the "captain of the men of death" described by Sir William Osler in his famous textbook before the advent of the sulfonamides and the introduction of penicillin. Although many factors, such as antimicrobial therapy, increasing age of our hospitalized population, and modern immunosuppressive and tumor chemotherapy, have altered its clinical and epidemiologic manifestations, pneumococcal pneumonia remains an important cause of morbidity and mortality. There are an estimated half-million cases per year. Pneumonia is the only infectious disease among the 10 most common causes of death in the United States today. This continued frequency and severity of pneumococcal infections, coupled with the emergence of strains resistant to most antimicrobial agents, underscore the need for a better understanding of the pathogenesis of these infections.

Morphology

The pneumococcus, formerly referred to in the United States as *Diplococcus pneumoniae,* is now officially classified in the genus *Streptococcus* because of its genetic relatedness to other members of the genus. It is an encapsulated, gram-positive coccus, oval or spherical in shape, and 0.5 to 1.25 μm in diameter. Characteristically, the organism is lancet shaped, and as observed in direct smears of sputum and body fluids, it occurs singly, in pairs, and in short chains (Figs. 27-1, 27-2). Continued laboratory cultivation, especially on unfavorable media, leads to the formation of longer chains. Pneumococci are very sensitive to the products of their fermentative metabolism, resulting in a gram-negative staining reaction as the culture ages. Capsules may be readily demonstrated by examination of wet mounts of virulent organisms in India ink or by the use of homologous type-specific antibody in the quellung reaction (p. 479). The size of the capsule varies considerably with the pneumococcal type and is especially large in types 3, 8, and 37.

Ultrastructure and Cell Composition

The outermost boundary of the pneumococcus is a typical gram-positive cell wall composed of peptidoglycan and the phosphorus-containing polysaccharide, teichoic acid. The

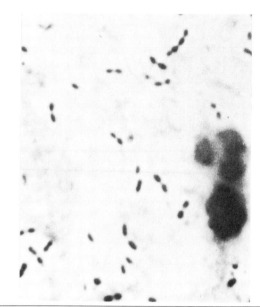

Figure 27-1. *Streptococcus pneumoniae* in sputum smear. Gram stain. *(Courtesy of Dr. Leon J. LeBeau, University of Illinois Medical Center.)*

cell wall teichoic acid contains the determinant for C polysaccharide antigenic activity. At least some of the teichoic acid units are located on the outer surface of the cell wall, since live pneumococci can be agglutinated with antisera against the C polysaccharide. An important structural component of teichoic acid is the amino alcohol, choline, which is unique to the pneumococcal wall and is thought to function as a regulatory ligand.

The plasma membrane is a typical bilayer that may give rise to invaginations varying in number and complexity from cell to cell. Essentially all of the lipid extractable from pneumococci is present in the plasma membrane. Also found in the membrane is a choline-containing teichoic acid, similar to that present in the wall but covalently bound to fatty acids. This membrane lipoteichoic acid is the carrier of the F antigen, an immunologic determinant that cross-reacts with the Forssman series of mammalian surface antigens. It contains about 15 percent of the cell's choline. Lipoteichoic acid is also a potent inhibitor of the homologous autolytic enzyme, an *N*-acetylmuramyl-L-alanine amidase, and thereby regulates murine hydrolase activity.

Physiology

S. pneumoniae is a facultative anaerobe that can use a fairly wide range of fermentable carbohydrates. Its energy-yielding metabolism is primarily of the lactic acid type, but the amount of acid accumulating is small unless the culture is periodically neutralized. Under aerobic conditions a significant amount of hydrogen peroxide is formed, along with acetic and formic acids. Since *S. pneumoniae* does not produce catalase or peroxidase, the accumulation of hydrogen

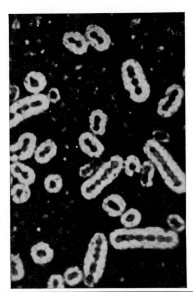

Figure 27-2. Pneumococcus from spinal fluid. Preparation stained by P. Bruce White method to show capsules. × 1800. *(Courtesy of Dr. Josephine Bittner and Dr. C.F. Robinow.)*

peroxide kills the organism unless catalase is provided by the addition of red blood cells to the culture medium.

Cultural Characteristics

The pneumococcus has complex nutritional requirements, and although chemically defined synthetic media are available for primary isolation and routine culture, such media as brain-heart infusion agar or trypticase soy agar and broth enriched with 5 percent defibrinated blood are employed. The optimum pH for growth is 7.4 to 7.8. Since 5 to 10 percent of all pneumococcal strains require an increased CO_2 concentration for primary isolation on solid media, a candle jar or CO_2 incubator should be used.

On blood agar plates, young cultures of encapsulated pneumococci produce circular, glistening, dome-shaped colonies about 1 mm in diameter. Colonies produced by type 3 organisms are usually larger and more mucoid than those produced by the other types, a reflection of the greater size of its capsule. As pneumococcus colonies on blood agar become older, autolytic changes result in a collapse of the center of the colony, giving it an umbilicate appearance. Unencapsulated strains produce small rough colonies. Colonies incubated aerobically are surrounded by a zone of α hemolysis similar to the greenish discoloration observed with the viridans streptococci. Under anaerobic conditions, however, a zone of β hemolysis is produced around the colony by an oxygen-labile pneumolysin O (p. 482).

Choline Requirement. Unlike other streptococci, *S. pneumoniae* has an absolute nutritional requirement for choline. If ethanolamine is substituted for choline during growth, a number of physiologic defects are observed. Among these are resistance to autolysin, aberrant cell division, incompetence in transformation, and phage resis-

tance. Most of these defects can be directly attributed to changes in the choline-containing teichoic acid of the cell surface. About 85 percent of the cell's choline is found in the teichoic acid of the cell wall. The remaining 25 percent is localized in the membrane lipoteichoic acid.

Laboratory Identification

Optochin Sensitivity. The procedures used in the laboratory for the identification of *S. pneumoniae* are primarily designed to distinguish it from the viridans streptococci, since they both produce α hemolysis on blood agar. The test most widely used at present for the presumptive identification of the pneumococcus is the optochin disc sensitivity test. Optochin (ethyl hydrocuprein hydrochloride) is a quinine derivative that inhibits the growth of pneumococci but not of viridans streptococci. For testing, a filter paper disc impregnated with drug is applied to the surface of a blood agar plate streaked with a lawn of a pure culture.

Bile Solubility. Another useful test for identifying pneumococci is based on the presence in pneumococci, but not in viridans streptococci, of an autolytic amidase that cleaves the bond between alanine and muramic acid in the peptidoglycan. The amidase is activated by surface-active agents, such as bile or bile salts, resulting in lysis of the organisms. For testing, a neutral pH, 10 percent desoxycholate, and viable young organisms should be used.

Quellung Reaction. The most useful and rapid method for the identification of *S. pneumoniae* is the Neufeld quellung or capsular precipitation reaction. The test not only identifies an organism as a pneumococcus but also specifies its type. It can be used directly for the identification of pneumococci in sputum, spinal fluid, exudates, or culture. The test is performed by mixing on a slide a loopful of emulsified sputum (or other clinical material) with a loopful of antipneumococcal serum and methylene blue and examining under the oil-immersion lens. In a positive reaction, which occurs when pneumococci are brought into contact with homologous capsular antiserum, the capsule becomes more refractile and greatly swollen in appearance.

At the present time, the only source of pneumococcal antisera is the State Serum Institute, Copenhagen, Denmark. Three types of antisera are commercially available: (1) omniserum, a highly concentrated polyvalent antiserum that reacts with all 83 capsular types, (2) polyvalent pooled antisera (A through I) that permit a capsular reaction of each of the 83 types in one of the nine pools that constitute the sera pool, and (3) type or group sera that react with a single pneumococcal type or group. Because the prevalence of different types varies with time and geographical area, typing surveillance is necessary to ensure the optimal composition of pneumococcal polysaccharide vaccines for use in different areas. The typing of isolates from vaccinees must also be continued in order to determine whether vaccine usage leads to a predominance of different types.

Animal Inoculation.

Animal Inoculation. The mouse is exquisitely sensitive to most types of pneumococci. When injected intraperitoneally with sputum containing pneumococci, the mouse succumbs to fatal infection within 16 to 48 hours. Other organisms that may be present in the sputum are usually eliminated, and pneumococci may be isolated in pure culture from the heart blood. However, several types of pneumococci, such as the commonly encountered type 14, are essentially avirulent for mice and do not produce fatal infections within 4 days. Although the mouse virulence test is rarely used for diagnostic purposes at the present time because it is expensive and time consuming, it remains an excellent experimental model in the research laboratory in testing the virulence of a particular isolate or type.

Genetic Variation

The possession of a capsule by the pneumococcus provides an easily recognized marker for genetic studies. When cultured on the surface of solid media, encapsulated organisms form characteristic glistening colonies that are mucoid or smooth and contain S organisms. If such organisms are cultured in the presence of homologous antiserum, they produce rough, granular colonies composed of nonencapsulated rough or R cells. In such a culture, the antiserum selects for R mutants present in any pneumococcal culture. Conversely, when broth cultures of R pneumococci are inoculated into mice, the R organisms are replaced by S pneumococci of the same serologic type as that from which the R strain was derived. In the animal, the less virulent organisms are destroyed by the host, and the more virulent encapsulated forms survive and are specifically selected. The exquisite sensitivity of the mouse to most pneumococcal capsular types permits detection of small numbers of S organisms that arise from back-mutations and that would be undetectable unless cultured in a selective environment. The conventional method of maintaining the maximum virulence of a culture by several passages through a mouse is based on this principle.

Another type of genetic variation that occurs in *S. pneumoniae* is transformation. This phenomenon was first observed in 1928 by Griffith and further studied by Avery, MacLeod, and McCarty in classic experiments that demonstrated conclusively the genetic role of DNA. In Griffith's original experiment, it was observed that mice injected with unencapsulated avirulent type 2 *S. pneumoniae*, together with heat-killed cells from a virulent encapsulated type 3 strain, frequently succumbed to the infection. From these infected animals living organisms of the virulent encapsulated type 3 could be isolated. The active principle in the heat-killed organisms responsible for transforming the avirulent organisms to virulent ones was later identified by Avery's group as DNA. The transformation in pneumococci of a number of additional characteristics other than capsular type has also been demonstrated, as has been the transformation of genetic markers between pneumococci and streptococci. The study of transformation in the pneumococcus continues to be an intriguing area of interest for the molecular geneticist who is currently examining binding and uptake of DNA by the cell and its integration into the chromosome. The recent discovery of plasmids in *S. pneumoniae* and the designing of successful plasmid transformation systems have also provided an accessible model for the cloning of recombinant DNA in this species.

Pneumococcal bacteriophages also offer promise as new genetic tools for studies of the pneumococcus. Both lytic and temperate phages have been isolated from a variety of geographic areas and are thus thought to be of widespread occurrence (Fig. 27-3). These phages vary in their morphology and serology and the ability of their DNA to transfect pneumococci. The presence of capsular polysaccharide appears to protect pneumococci against infection by all lytic phages isolated to date. However, a high frequency of lysogeny has been demonstrated among

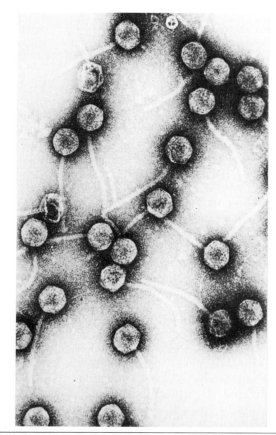

Figure 27-3. Electron micrograph of purified phage ω2, a pneumococcal bacteriophage isolated from a throat swab of an adult in the infirmary at the Massachusetts Institute of Technology. From 62 different throat swabs collected during a single week, 43 gave plaques on the standard laboratory pneumococcus strain 662. The hexagonal heads of ω2 are one-third the size of coliphage T4. *(From Tiraby: Virology 68:566, 1975.)*

pathogenic strains of the capsular types that most frequently cause infections.

Antigenic Structure

Capsular Antigens

Pneumococcal capsules consist of complex polysaccharides that form hydrophilic gels on the surface of the organism. These polysaccharides are antigenic and form the basis for the separation of pneumococci into 84 different serotypes. Some of the serotypes form serogroups, i.e., types carrying the same number but different capital letters, such as serogroup 7 consisting of serotypes 7F, 7A, 7B, and 7C. The pneumococcal polysaccharides vary in their mono- and disaccharide components, but at present the chemical structure of the polysaccharide of only a few types is completely known, e.g., type 3 consists of repeating units of cellobiuronic acid (Fig. 27-4). The pathway for the biosynthesis of type 3 has also been defined.

Some pneumococcal types exhibit no or very few cross-reactions, while other types have antigens in common with several pneumococcal types. Antisera to some of the types cross-react with polysaccharides from a number of other bacteria including *Klebsiella*, *Rhizobium*, *Salmonella*, *Escherichia coli*, type b *Haemophilus influenzae*, and viridans streptococci. In addition, type 14 polysaccharide cross-reacts with human ABO blood group isoantigens. This cross-reaction is probably attributable to *N*-acetyl-D-glucosamine, which is the common terminal end group shared by these polysaccharides. The polysaccharide capsule is essential for the pathogenicity of the pneumococcus and stimulates the production of antibodies that are protective against subsequent infection with pneumococci of the homologous type.

Somatic Antigens

C Polysaccharide. This species-specific carbohydrate is a major structural component of the cell wall of all pneumococci. It is a teichoic acid polymer containing phosphocholine as a major antigenic determinant. Phosphocholine is responsible for the agglutination of pneumococci by certain myeloma proteins and for the interaction of the polysaccharide C substance with a serum β-globulin in the presence of calcium. This β-globulin, referred to as C-reactive protein (CRP), is not an antibody but is a protein that is present in low concentrations in normal blood but is elevated in patients with acute inflammatory diseases. The binding of CRP to C polysaccharide can activate complement and mediate phagocytosis.

In addition to phosphocholine, polysaccharide C contains galactosamine, glucose, phosphate, ribitol, and a trideoxydiaminohexose. The complete structure of the repeating unit has been elucidated and has been shown to be identical in polysaccharide C derived from a number of pneumococcal types (Fig. 27-5).

F Antigen. Another major antigenic component of the pneumococcus is the F or Forssman antigen, a determinant that cross-reacts with the Forssman series of mammalian cell surface antigens. The F antigen is a lipoteichoic acid and, like the C polysaccharide antigen, contains choline as a constituent of its teichoic acid. Unlike the C antigen, however, it is localized at the outer surface of the cell membrane where it is strongly bound to fatty acids. The F antigen is a powerful inhibitor of the *N*-acetylmuramyl-L-alanine amidase, suggesting a specific physiologic role for lipoteichoic acids in the in vivo regulation of murein hydrolase activity in the organism.

M Protein. Type-specific protein antigens, analogous to the M protein of *Streptococcus pyogenes* but immunologically distinct, are present in pneumococci. No correlation has been shown between the presence of a specific type of M protein and the type of organism based on capsular polysaccharide. Antibodies to the pneumococcal M protein do not inhibit phagocytosis and are therefore not protective.

Determinants of Pathogenicity

Polysaccharide Capsule. The pneumococcus is an excellent example of an extracellular parasite that damages the tissues of the host only as long as it remains outside the phagocytic cell. Protection against phagocytosis is provided by the capsule, which exerts an antiphagocytic effect. This can be readily demonstrated by comparing the behavior in mice of an encapsulated S strain with that of a nonencapsulated R strain. The S organism is highly virulent for the mouse, whereas the R strain is avirulent and rapidly phagocytized. Removal of the capsule by treatment with

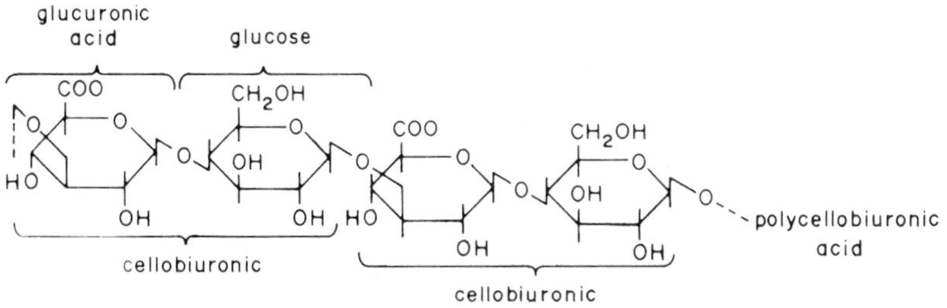

Figure 27-4. Structure of type 3 capsular polysaccharide. The basic unit of the polymer is the disaccharide, cellobiuronic acid, which consists of D-glucuronic acid and D-glucose connected by a β-1,4 glycosidic bond.

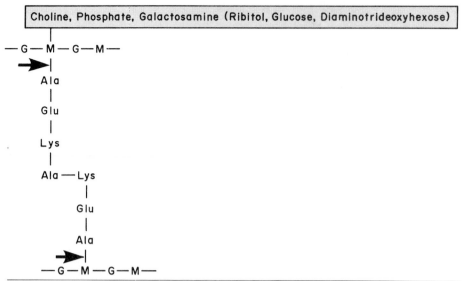

Figure 27-5. Schematic diagram of proposed cell wall structure of *S. pneumoniae*. The basic element is the *N*-acetylglucosamine *N*-acetylmuramic acid backbone of a typical peptidoglycan, which has two kinds of substituents attached to it: (1) the usual tetrapeptides (2 alanine, 1 glutamic acid, 1 lysine), which are further crosslinked, and (2) chains of a teichoic acid polymer. The arrow indicates site of action of autolysin. *(Adapted from Mosser and Tomasz: J Biol Chem 245:287, 1970.)*

an enzyme specific for the polysaccharide renders the organism nonpathogenic and readily susceptible to phagocytosis. Antibodies against the capsular polysaccharide combine specifically with it, rendering the organism susceptible to phagocytosis.

Many aspects of the pathogenesis of pneumococcal infection remain ill defined. The capsular polysaccharide is relatively nontoxic but is present in a soluble form in the body fluids of infected individuals. High levels in the serum or urine are associated with severe infections accompanied by bacteremia, empyema, and a high mortality rate. Excessive amounts of free polysaccharide neutralize antibody and make it inaccessible to the invading organisms.

Neuraminidase. A number of organisms that colonize the respiratory tract produce neuraminidase, a glycosidic enzyme of low molecular weight. The production of this enzyme by log phase cells from fresh clinical isolates of *S. pneumoniae* has been detected. In its attack on the glycoprotein and glycolipid components of the cell membrane, neuraminidase cleaves a terminal *N*-acetylneuraminic acid from an adjacent sugar. Although a specific role for the pneumococcal enzyme in disease has not been demonstrated, the organism's ability to grow in the human nasopharynx and in the mucous secretions within the bronchial tree requires special metabolic capacities. Neuraminidase is only one of the factors contributing to the invasiveness of the organism.

Adherence. Adherence to epithelial cells may also be important for pneumococci colonizing the nasopharynx or inducing otitis media. In studies on the adherence of *S.*

pneumoniae to human pharyngeal cells, the differences observed between the adhesive capacity of the various strains could be correlated with the clinical origin of the strain. Otitis strains belonging to the capsular types often associated with otitis media adhered in high numbers. The capsular polysaccharide, however, appears not to determine the adhesive capacity.

Toxins

PNEUMOLYSIN O. Pneumococci produce a hemolysin, pneumolysin O, with properties similar to those of the oxygen-labile hemolysin of *S. pyogenes,* streptolysin O. It is an SH-activated cholesterol-sensitive cytolytic toxin that disrupts cell membranes. In addition to its hemolytic activity, it is dermatoxic and lethal. The role of pneumolysin in the pathogenesis of human pneumococcal infections is unknown. In rabbits, however, pneumococcal septicemia is accompanied by a spherocytic hemolytic anemia. Prior immunization with the pneumolysin protects against a challenge dose.

PURPURA-PRODUCING PRINCIPLE. This substance produces purpura and dermal hemorrhage in experimental animals. Its activity is associated with the peptidoglycan of the cell wall solubilized by the organism's autolysin. Purpurogenic activity requires the intact β-1,4 glucosidic linkages of peptidoglycan. There is as yet no conclusive evidence for a role of this toxic principle in the pathogenesis of human infections.

The basic question that remains unanswered and that, if answered, might prevent morbidity and mortality from pneumococcal infections is the identity of the factor or factors responsible for lethality. At present, pneumococci are believed to produce disease solely through their capacity to multiply in the tissues. The possibility remains, however, that a toxin is elaborated under in vivo conditions of growth and that new techniques will be required for its demonstration. Complex multifaceted alterations occur in the metabolism of the host following infection, but it is as yet unknown whether any of the observed biochemical alterations are responsible for the lethal action of the organism.

Clinical Infection

Epidemiology

Incidence. Pneumococcal pneumonia is the most common form of bacterial pneumonia. Although it is not a reportable disease in the United States, the estimated attack rate is between 2 and 5 per 1000 population, or between 440,000 and 1,100,000 cases annually. The incidence is three to four times greater in patients over 40 years of age, where occurrence is often conditioned by underlying chronic obstructive pulmonary disease. It is increased in closed population groups, such as schools, the military, and the institutionalized chronically ill. Pneumococcal infections are usually more numerous during the winter months when frequent viral infections of the upper respiratory tract predispose to infection and spread of the organisms.

The Carrier State. Pneumococci are carried in the nasopharynx of healthy contact carriers, who constitute the major reservoir for pneumococcal infections. Multiple infections within family units, however, indicate that infections may also result from contact with another case. The carrier rate varies with age, environment, and the presence of upper respiratory infections. Carriage rates are highest in children of preschool age (25 to 50 percent) and, with increasing age, tend to decrease to about 18 percent in the adult. Adults having no contact with children have a carriage rate of only 5 percent. In military installations, where the incidence of pneumococcal infections is also very high, rates may be as high as 60 percent.

Pneumococcal Types. Although 84 pneumococcal capsular serotypes have been identified, not all pneumococcal types are equally invasive. In a surveillance study beginning in 1967 of more than 4000 isolates from municipal hospitals in several geographic areas throughout the United States, epidemiologic data have been provided on the types most often associated with infection. Types or groups 8, 4, 3, 14, 7, 12, 9, 1, 18, 19, 6, and 23, in the order given, are the 12 most common, accounting for about 80 percent of the infections. In the pediatric age group, the same types are associated with most bacteremic infections, except for the rare occurrence of type 12 in this age group. Between 50 and 60 percent of the isolates from in-

fants and children are of types or groups 6, 14, 19, and 23.

Although there is a remarkable similarity and constancy in the types responsible for infections, the rank order of frequency of different types does tend to shift slowly, and an invasive type may with time assume lesser clinical importance in a particular area. During the past 40 years a shift has been observed away from the predominance of types 1, 2, and 3 infections previously observed to a more balanced distribution. Except for type 3, which is a common inhabitant of the normal pharynx, pneumococci of the higher-numbered types are usually associated with the carrier state more often than the more virulent lower-numbered types.

Pathogenesis

Pneumococcal pneumonia is rarely a primary infection and results only when the normal defense barriers of the respiratory tract are disturbed. Chilling, anesthesia, morphine, and alcoholic intoxication commonly predispose to pneumococcal disease. By slowing epiglottal reflex, these factors facilitate aspiration of infected secretions from the upper respiratory tract. Viral infections of the upper respiratory tract are a major contributory cause of pneumococcal pneumonia and often precede its abrupt onset. Pneumococci present in the nasopharynx proliferate in the virus-modified environment and are carried down into the alveoli by the thin bronchial secretions. A number of additional clinical conditions also predispose to acute pneumococcal pneumonia: congestive heart failure, noxious gases, pulmonary stasis resulting from prolonged bed rest. In all of these cases, fluid accumulates in the alveoli, providing an excellent culture medium for the organism.

The Pneumococcal Lesion. Invasion of alveolar tissue by pneumococci results in an outpouring of edema fluid that facilitates rapid multiplication and spread of the organisms to other alveoli. Polymorphonuclear leukocytes and red blood cells accumulate in the infected alveoli, leading to complete consolidation of the lobe or segment. Crowding of leukocytes in the alveoli promotes phagocytosis and destruction of the invading organisms. Macrophages participate in the final stages of resolution, and in most of the less serious cases recovery is complete and the lung parenchyma is restored to its normal state. Effective present-day antibiotic therapy, however, frequently alters or halts the classic inflammatory response so that the distinguishing histologic features of the spreading pneumonic lesion are obscured.

In the adult, pneumococcal pneumonia characteristically involves one or more complete lobes. In infants, young children, and the aged, however, the lesions may be more patchy in their distribution and localized around the bronchi.

From the primary lesion in the lung pneumococci may invade the pleural cavity and pericardium, with the formation of extensive purulent foci (empyema). In fulminating

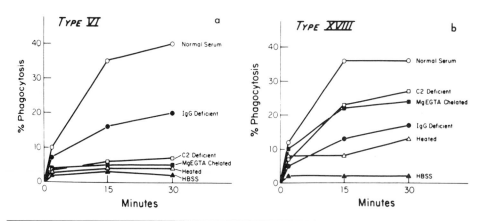

Figure 27-6. Phagocytosis of *S. pneumoniae* types 6 (**a**) and 18 (**b**) after opsonization in 10 percent normal serum, heat-inactivated normal serum, IgG-deficient serum, C2-deficient serum, and Mg EGTA (magnesium dichloride ethyleneglycoltetraacetic)-chelated serum. Bacteria were opsonized for 30 minutes before addition of PMN leukocytes, and samples were obtained thereafter at 1, 15, and 30 minutes of incubation. *(From Giebink et al.: Infect Immun 18:291, 1977.)*

infections, bacteremia is also common and may lead to infections of the meninges, heart valves, or joints.

Phagocytic Defense. The phagocytic cells in the lungs provide the major line of defense against pneumococcal infections. Before phagocytosis can proceed, the encapsulated pneumococcus must be opsonized by serum components. Complement plays a critical role, especially during the early preantibody stage of infection (Fig. 27-6). In the nonimmune individual, opsonization occurs mainly through the alternative (properdin) complement pathway. In the immune individual, however, the full effect of type-specific antibody requires both an intact alternative and classic pathway of complement activation. Pneumococcal types vary in their ability to activate the properdin pathway, an observation that may in part explain the predominance of certain types in human infection. The unusually high frequency of severe pneumococcal infections in children with sickle cell disease is associated with a deficiency in a component of the properdin pathway.

A single specific mechanism of the intracellular killing of pneumococci by phagocytes has not been defined. There is, however, strong evidence that bacterial killing is at least partially dependent upon an intact peroxide-generating system. During phagocytosis peroxidation of pneumococcal lipids is associated with the generation of peroxide and probably contributes significantly to effective phagocytic killing. An evaluation of the microbicidal capacity of the serum can be made by measuring the chemiluminescent response generated subsequent to the oxidation of the ingested organism by singlet molecular oxygen.

Clinical Manifestations

Pneumonia. Classic pneumonia strikes suddenly with a single violent shaking chill and fever that ranges between 102 and 106F (38.8 and 41.1C). The patient usually presents with a history of a mild upper respiratory infection preceding the acute onset by a few days. Severe pleuritic pain is often present, and a cough developing during the course of the disease is productive of rusty mucopurulent sputum. In untreated cases recovery may be as dramatic as the onset, with fever terminating abruptly by crisis 5 to 10 days after onset. In other cases, fever subsides more gradually by lysis. A dramatic crisis often occurs within 24 hours in patients receiving effective antibiotic therapy.

The classic presentation of acute pneumococcal pneumonia is unmistakable. Atypical cases often occur, however, in which a diagnosis is less obvious. This is especially true in the case of the alcoholic, the elderly, or the debilitated patient, in whom symptoms may be less dramatic or overshadowed by symptoms of severe prostration, confusion, or delirium.

COMPLICATIONS. The most common complication of pneumococcal pneumonia is pleural effusion, which results from inflammation of the pleura overlying the parenchymal lesion of the lung. If bacteria gain access to the effusion, the leukocyte response is greatly increased and empyema results. Empyema, meningitis, pericarditis, and endocarditis are serious complications associated with an increased risk of death. Bacteremia occurs in approximately 25 to 30 percent of the patients with pneumococcal pneumonia and carries with it a two-fold increase in mortality. An especially fulminant clinical course is seen in asplenic patients with pneumococcal bacteremia.

PROGNOSIS. The case fatality rate of untreated pneumococcal pneumonia is about 30 percent. With specific therapy the overall fatality rate is about 5 percent. However, in adults with bacteremic pneumococcal disease, there remains a high case fatality rate (approximately 25 percent) despite antibiotic treatment and shift in prevalent capsular types (Table 27-1). Prognosis is influenced adversely by increasing age, an extrapulmonary site of infection, the presence of cirrhosis or diabetes mellitus, immunodeficiency

**TABLE 27-1. CASE FATALITY RATE BY AGE FROM PNEUMOCOCCAL BACTER-
EMIA AMONG ADULTS WITH PNEUMONIA OR EXTRAPULMONARY DISEASE**

Category	Age in Years	Number in Group	Deaths Number	Deaths %
Bacteremic pneumococcal	14–29	30	1	3
pneumonia	30–49	130	20	15
	50–69	87	33	38
	70+	15	8	53
TOTAL		262	62	24
Extrapulmonary disease with	14–29	9	1	20
or without pneumonia	30–49	26	9	35
	50–69	24	12	50
	70+	8	6	75
TOTAL		63	28	44

From Mufson et al.: Arch Intern Med 134:505, 1974.

disease, and infection with certain capsular types, especially type 3. The case fatality rate in type 3 bacteremic pneumococcal pneumonia is over 50 percent even with antibiotic therapy.

Upper Respiratory Tract Infections. *S. pneumoniae* is a leading bacterial pathogen in infants and young children, accounting for a large number of infections of the respiratory tract and adjacent structures, e.g., the middle ear. By the age of 3 years more than two thirds of children have had at least one attack of otitis media. Many children have recurrent attacks, with persistent middle ear effusions and hearing loss. *S. pneumoniae* is the most frequent cause, accounting for 35 to 50 percent of all cases.

Prevention of pneumococcal otitis media by use of a polyvalent pneumococcal polysaccharide vaccine has been attempted, but its efficacy appears to be limited by the impaired immune response in young children to many polysaccharide antigens, including those contained in the vaccine.

Extrapulmonary Infections

MENINGITIS. The pneumococcus is the most common cause of bacterial meningitis in adults and of recurrent meningitis in all age groups (Table 27-2). It is the most serious pneumococcal infection, with a case fatality rate of 40 percent. Half of the cases are in young children aged 1 month to 4 years. Pneumococcal meningitis is usually preceded by pulmonary infection or by symptomatic primary infection of the upper respiratory tract and contiguous structures, i.e., sinusitis, mastoiditis, or otitis media. Alcoholism, head trauma, sickle cell disease, multiple myeloma, and general debility predispose to pneumococcal meningitis.

Immunity

Naturally Acquired Immunity. In the normal adult, natural host resistance is high, and even without any form of treatment, 7 of every 10 patients with pneumococcal pneumonia will recover. Spontaneous recovery is depen-

TABLE 27-2. BACTERIAL CAUSES OF MENINGITIS

	Neonates (<1 Month) (%)	Children (1 Month–15 Years) (%)	Adults (>15 Years) (%)
Streptococcus pneumoniae	0–5	10–20	30–50
Neisseria meningitidis	0–1	25–40	10–35
Haemophilus influenzae	0–3	40–60	1–3
Streptococcus, groups A, B	20–40 *	2–4	5
Staphylococcus aureus	5	1–2	5–15
Listeria monocytogenes	2–10	1–2	5
Gram-negative bacilli	50–60 †	1–2	1–10

From Swartz: In Wyngaarden and Smith (eds): Cecil Textbook of Medicine, 16th ed., vol II, 1982. Philadelphia, Saunders.
*Almost all isolates from neonatal meningitis are group B streptococci.
†Of all cases of neonatal meningitis, *E. coli* accounts for about 40 percent and *Klebsiella–Enterobacter* for about 8 percent.

dent on the production of type-specific antibodies to the capsular polysaccharide. These are first demonstrable in the serum 5 to 6 days after onset of the disease. In vivo, the primary role of protective antibody is opsonic, but the precise mode of clearance in most naturally occurring pneumococcal infections is unknown. The development of measurable type-specific antibody is delayed in patients with capsular polysaccharide antigenemia, but it is not known whether this delay is due to decreased antibody production or to the neutralization of antibody by circulating antigen (Fig. 27-7).

The bactericidal power of the blood of normal individuals varies with the type of pneumococcus and with the age of the individual. Blood of newborn babies has the same killing power as that of their mothers, but this is lost within 3 to 5 weeks. After that, it is observed in an increasing number of persons as they become older until the age of 55 years, when the bactericidal capacity decreases.

Type-specific immunity to pneumococcal infections is long lasting. Recurrent attacks of pneumococcal infection are usually caused by pneumococci of a different serologic type. The persistence of pneumococci in sputum and throat cultures from patients with pneumococcal pneumonia may be attributed to new types of pneumococci that are acquired during the course of infection. Multiple types are more frequently encountered in patients with chronic respiratory tract infections in whom pneumococci persist longer than in uncomplicated cases. Persistence is not related to the patient's inability to develop antibodies to specific pneumococcal types.

Artificially Acquired Immunity. Immunity to specific pneumococcal types may be induced by immunization with a polyvalent vaccine consisting of capsular polysaccharides obtained from several of the most prevalent or invasive pneumococcal types. The vaccine currently employed is composed of 14 capsular polysaccharides and will provide protection against 80 percent of the pneumococcal infections acquired in the United States and in Europe at the present time (p. 488). The preventive efficacy of the vaccine in healthy young men in epidemic conditions is 80 to 95 percent (Table 27-3). Similar responsiveness has been demonstrated for the elderly and for patients with sickle cell disease and diabetes.

Protection against pneumococcal infection, however, is complex. Such protection depends on the host's ability to produce both opsonizing antibody and complement, the presence of a functioning spleen, and leukocytes able to phagocytize the opsonized pneumococci. Any disease state, therapy, or condition that seriously impairs the host's defense mechanisms may reduce vaccine effectiveness in that patient. These same patients are those at greatest risk of significant morbidity and mortality from pneumococcal infection. The vaccine has yet to be fully assessed in all high-risk groups, but studies to date have shown that persons with asplenia, chronic renal failure, chronic pulmonary disease, and nephrotic syndromes have normal or near-normal antibody responses to vaccination. Some

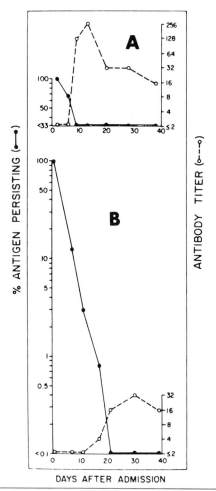

Figure 27-7. Clearance of circulating pneumococcal capsular polysaccharide associated with the development of measurable type-specific hemagglutinating antibody. A. Patient with type 12 pneumococcal pneumonia. The peak antigen level was 0.15 μg/ml. B. Patient with type 7 pneumococcal pneumonia. The peak antigen level was 51.2 μg/ml. *(From Coonrod and Drennan : Ann Intern Med 84:254, 1976.)*

TABLE 27-3. BACTEREMIC INFECTION CAUSED BY TYPES OF *STREPTOCOCCUS PNEUMONIAE* WHOSE CAPSULAR POLYSACCHARIDES WERE INCLUDED IN VACCINES

Group	Number at Risk	Number with Bacteremia*
Vaccinees	3,975	10
Controls	8,035	113
TOTAL	11,992	123

From Austrian: J Infect Dis 136[Suppl]:S38, 1977.

*Infection occurred in vaccinees and in controls later than 2 weeks after inoculation.

Note: These data show a protection rate of 82.3% (x^2 = 34.759; $P <$ 0.0001).

immunosuppressed patients, such as renal transplant recipients, exhibit partial antibody responses. However, other immunosuppressed patients with multiple myeloma and with Hodgkin's disease who have been splenectomized and have undergone radiation and chemotherapy show poor serologic responses to the vaccine.

Laboratory Diagnosis

Although a diagnosis of classic acute lobar pneumonia can usually be made on the basis of physical examination and roentgenographic findings, atypical cases, nonpulmonary infections, and pneumonia caused by other organisms are more difficult to diagnose. In addition, a number of noninfectious diseases, such as pulmonary infarction, congestive heart failure, and atelectasis, may simulate pneumonia and require specific laboratory studies for their differential diagnosis.

The proper collection of sputum is the physician's responsibility and demands that the specimens be mucus expectorated from the lungs rather than samples of saliva. When pneumococcal pneumonia is suspected, blood should be drawn by venipuncture prior to the administration of antibiotics. The pneumococcus is a delicate organism and does not survive for long periods on dry swabs or in physiologic saline. For culture, swabs should be placed immediately in sterile nutrient broth for transport to the laboratory.

Direct Examination of Sputum. A direct examination should be made of smears stained by the gram method to determine the probable etiology of the infection. If positive for gram-positive lancet-shaped diplococci, a presumptive diagnosis of pneumococcal pneumonia may be made. Cultural identification, however, is required to distinguish the pneumococcus definitely from certain other gram-positive cocci. If typing sera are available, the most simple, rapid, and accurate method for the identification of pneumococci by direct examination is the quellung reaction (p. 479).

Culture. Specimens for culture should be planted immediately (1) on an enriched medium, such as brain-heart infusion or trypticase soy agar and broth containing 5 percent blood, and (2) in thioglycollate broth. For blood culture, 5 to 15 ml of blood should be inoculated into trypticase soy broth and thioglycollate broth, maintaining a ratio of approximately 1:10 between blood and medium. Subcultures should be made by streaking the surface of a blood agar plate. Pour plates with samples of blood provide valuable information on the magnitude of the bacteremia and prognosis of the infection. For culture of body fluids, blood agar plates should be streaked, and 0.5 to 1 ml of the specimen should be inoculated into blood broth. Presumptive identification of pneumococci is based on the appearance of α-hemolytic colonies containing organisms that are bile soluble and optochin sensitive, ferment inulin, and have a positive quellung reaction (p. 479).

Serologic Diagnosis

DETECTION OF PNEUMOCOCCAL ANTIBODIES. A number of techniques have been employed for the demonstration of an immunologic response to pneumococcal infection, including agglutination, quantitative precipitation, mouse protection tests, and bactericidal tests with whole blood. The complement-fixation test cannot be used, since human antibodies to pneumococcal capsular polysaccharides do not fix complement. More recently, indirect hemagglutination, immunoelectrophoresis, indirect fluorescent antibody, and radioimmunoassay techniques have been employed. The latter test, employing as antigens pneumococcal polysaccharides labeled intrinsically with ^{14}C have been especially useful in the detection not only of an immunologic response to infection but also of a response following the administration of pneumococcal vaccines. The method is exquisitely sensitive, is capable of detecting specific capsular antibody in nanogram amounts, and requires very small amounts of serum for the assay procedure.

DETECTION OF CAPSULAR POLYSACCHARIDE. Capsular polysaccharide appears in the serum and body fluids of patients with pneumococcal infection. The presence of large quantities of the soluble polysaccharide is associated with severe infection accompanied by bacteremia, empyema, and a high mortality rate. In some patients, especially those with slowly resolving infections, capsular polysaccharide may be detected in the urine for several months. Counterimmunoelectrophoresis has been utilized to demonstrate pneumococcal polysaccharide in the blood, urine, and spinal fluid of patients and to establish a diagnosis of pneumococcal infection. The pneumococcal capsular antigen present in serum or pleural or cerebrospinal fluids is similar both physically and immunologically to purified pneumococcal polysaccharide (PPP). The polysaccharide in urine, however, is a smaller molecule and has only partial immunologic identity with the PPP. It is not known at present whether the polysaccharide found in the sera and body fluids of patients with severe infections has a deleterious effect on the host or host defense mechanisms. Also unknown is the mechanism by which polysaccharide is ultimately eliminated from the body, since there is no evidence that mammalian cells can degrade pneumococcal polysaccharides.

Treatment

Penicillin is the drug of choice for all types of pneumococcal infection. Antimicrobial therapy should be started immediately after specimens are obtained for culture and should not be withheld until culture results are available. The choice of drug in the treatment of bacterial pneumonia is based on the results of gram-stained smears of sputum and is usually directed against the pneumococcus. Delay in the administration of specific antipneumococcal therapy probably accounts for the high death rate that is still observed in patients with bacteremic pneumococcal pneumonia.

Penicillin G, given intramuscularly, is the drug of choice in the treatment of uncomplicated pneumococcal pneumonia. Oral penicillin may also be used effectively for outpatients with mild symptoms, but its absorption from the gastrointestinal tract is less predictable, especially in acutely ill patients. For patients in shock or with pneumonia plus meningitis, endocarditis, or arthritis, aqueous crystalline penicillin should be given intravenously. Patients allergic to penicillin may be given a cephalosporin or erythromycin for pneumonia and chloramphenicol for meningitis.

Response to penicillin therapy is usually very dramatic, and bacteremia, if present, will clear in a few hours. Initial response to treatment, followed by relapse, is usually due to the presence of a mixed infection. The relatively recent reports of clinical isolates with increased resistance to penicillin, however, emphasize the fact that failure to respond may sometimes be attributed to the presence of penicillin-resistant pneumococci.

Penicillin-resistant organisms, although still unusual, have been recovered from patients in many parts of the world. Most of these strains are of intermediate resistance with minimal inhibitory concentrations (MICs) for penicillin 10 to 100 times greater than those for susceptible strains. More recently, however, strains have been isolated from patients with MICs of penicillin 100 to 10,000 times greater than those for susceptible strains. Particularly alarming has been the development in South Africa of multiple resistant pneumococci with decreased susceptibility to all penicillins and cephalosporins, aminoglycosides, chloramphenicol, clindamycin, erythromycin, sulfonamides, tetracycline, and rifampin. All such reported resistant clinical isolates were of type 19A and were found both in hospitalized patients younger than 2 years of age and in the hospital personnel. Type 6A pneumococci were also isolated, but this type was present only in carriers. These reports provide a serious warning that the blind use of penicillin alone in the management of pneumococcal infection is no longer justified. Clinical isolates should always be routinely tested for antibiotic susceptibility.

Prevention

In the United States, the case fatality rate of 25 percent in bacteremic pneumococcal pneumonia has remained unacceptably high and essentially unchanged by four decades of antibiotic therapy (Table 27-1). Attempts to reduce this rate have focused on the development of a pneumococcal vaccine for use in certain high-risk groups. The composition of the vaccine, which was licensed in this country in 1977, is based on epidemiologic evidence collected during the past four decades. The vaccine is composed of capsular polysaccharide antigens of 14 different serotypes of *S. pneumoniae* that cause 80 percent of the documented pneumococcal disease in this country. The present vaccine consists of types 1, 2, 3, 4, 6A, 7F, 8, 9N, 12F, 14, 18C, 19F, 23F, and 25. The antigen content of the vaccine is 50 μg of each serotype of polysaccharide per dose. Adverse reac-

tions are usually mild and consist principally of local erythema and induration at the injection site. Immunity lasts for at least 5 years after vaccination.

Clinical indications for the administration of the vaccine include all patients at an increased risk of developing pneumococcal pneumonia and who possess the immunologic ability to respond to polysaccharide antigen challenge with the production of adequate homotypic antibody. Such patients include those with chronic pulmonary disease, asplenia, congestive heart failure, cirrhosis, diabetes mellitus, nephrotic syndrome, and renal failure. The vaccine should also be offered to the elderly, especially those in chronic care facilities. Renal transplant patients may derive some benefit, but vaccination is not likely to be useful within 6 months of transplantation. Other immunosuppressed patients, such as those with multiple myeloma and Hodgkin's disease, are unlikely to respond to immunization. The vaccine should not be given to children under 2 years of age.

FURTHER READING

Books and Reviews

Finland M: Excursions into epidemiology: selected studies during the past four decades at Boston City Hospital. J Infect Dis 128:76, 1973

Geiseler PJ, Nelson KE, Levin S, et al.: Community-acquired purulent meningitis: A review of 1,316 cases during the antibiotic era, 1954–1976. Rev Infect Dis 2:725, 1980

Kass EH, Green GM, Goldstein E: Mechanisms of antibacterial action in the respiratory tract. Bacteriol Rev 30:488, 1966

Kass EH (ed): Assessment of the pneumococcal polysaccharide vaccine. Rev Infect Dis 3 [supp] S1–S197, 1981

Quie PG, Giebink GS, Winkelstein JA (Guest eds): The pneumococcus. A symposium. Rev Infect Dis 3:183, 1981

Schwartz JS: Pneumococcal vaccine: Clinical efficacy and effectiveness. Ann Intern Med 96:208, 1982

Swartz MN: Bacterial meningitis. In Wyngaarden JB, Smith LH Jr (eds): Cecil Textbook of Medicine, 16th ed. Philadelphia, Saunders, 1982, vol II, pp 1473–1478

Ward J, Koornhof: Antibiotic-resistant pneumococci. In Remington JS, Swartz MN (eds): Current Clinical Topics in Infectious Diseases. New York, McGraw-Hill, 1980, vol 1, pp 265–287

Selected Papers

Bernheimer HP: Lysogeny in pneumococci freshly isolated from man. Science 195:66, 1977

Briles DE, Nahm M, Schroer K, et al.: Antiphosphocholine antibodies found in normal mouse serum are protective against intravenous infection with type 3 *Streptococcus pneumoniae*. J Exp Med 153:694, 1981

Briles EB, Tomasz A: Physiological studies on the pneumococcal Forssman antigen: A choline-containing lipoteichoic acid. J Gen Microbiol 86:267, 1975

Centers for Disease Control, Immunization Practices Advisory Committee: Pneumococcal polysaccharide vaccine. Ann Intern Med 96:203, 1982

Claverys JP, Lefevre JC, Sicard AM: Transformation of *Streptococ-*

cus pneumoniae with *S. pneumoniae*-phage hybrid DNA: Induction of deletions. Proc Natl Acad Sci USA 77:3534, 1980

Coonrod JD, Drennan DP: Pneumococcal pneumonia: Capsular polysaccharide antigenemia and antibody responses. Ann Intern Med 84:254, 1976

Dhingra RK, Williams RC Jr, Reed WP: Effects of pneumococcal mucopeptide and capsular polysaccharide on phagocytosis. Infect Immun 15:169, 1977

Finland M, Barnes MW: Changes in occurrence of capsular serotypes of *Streptococcus pneumoniae* at Boston City Hospital during selected years between 1935 and 1974. J Clin Microbiol 5:154, 1977

Giebink GS, Verhoef J, Peterson PK, et al.: Opsonic requirements for phagocytosis of *Streptococcus pneumoniae* types VI, XVIII, XXIII, and XXV. Infect Immun 18:291, 1977

Gray BM, Converse GM III, Dillon HC Jr: Epidemiologic studies of *Streptococcus pneumoniae* in infants: Acquisition, carriage, and infection during the first 24 months of life. J Infect Dis 142:923, 1980

Guckian JC, Christensen GD, Fine DP: The role of opsonins in recovery from experimental pneumococcal pneumonia. J Infect Dis 142:175, 1980

Heidelberg M, Nimmich W: Additional immunochemical relationships of capsular polysaccharides of klebsiella and pneumococci. J Immunol 109:1337, 1972

Höltze JV, Tomasz A: Lipoteichoic acid: A specific inhibitor of autolysin activity in pneumococcus. Proc Natl Acad Sci USA 72:1690, 1975

Hosea SW, Brown EJ, Frank MM: The critical role of complement in experimental pneumococcal sepsis. J Infect Dis 142:903, 1980

McDonnell M, Ronda-Lain C, Tomasz A: "Diplophage": A bacteriophage of *Diplococcus pneumoniae*. Virology 63:577, 1975

Mosser JL, Tomasz A: Choline-containing teichoic acid as a structural component of pneumococcal wall and its role in sensitivity to lysis by an autolytic enzyme. J Biol Chem 245:287, 1970

Mufson MA, Kruss DM, Wasil RE, et al.: Capsular types and outcome of bacteremic pneumococcal disease in the antibiotic era. Arch Intern Med 134:505, 1974

Mylotte JM, Beam TR Jr: Comparison of community-acquired and nosocomial pneumococcal bacteremia. Am Rev Respir Dis 123:265, 1981

Page MI, Lunn JS: Pneumococcal serotypes associated with acute pneumonia. Am J Epidemiol 98:255, 1973

Reed WP, Davidson MS, Williams RC Jr: Complement system in pneumococcal infections. Infect Immun 13:1120, 1976

Ronda C, Lopez R, Garciá E: Isolation and characterization of a new bacteriophage, Cp-1, infecting *Streptococcus pneumoniae*. J Virol 40:551, 1981

Ronda C, Lopez R, Tomasz A, et al.: Transfection of *Streptococcus pneumoniae* with bacteriophage DNA. J Virol 26:221, 1978

Ronda-Lain C, Lopez R, Tapia A, et al.: Role of the pneumococcal autolysin (murein hydrolase) in the release of progeny bacteriophage and in the bacteriophage-induced lysis of the host cells. J Virol 21:366, 1977

Savage DG, Lindenbaum J, Garrett TJ: Biphasic pattern of bacterial infection in multiple myeloma. Ann Intern Med 96:47, 1982

Seto H, Tomasz A: Early stages in DNA binding and uptake during genetic transformation on pneumococci. Proc Natl Acad Sci USA 71:1493, 1974

Shohet SB, Pitt J, Baehner RL, et al.: Lipid peroxidation in the killing of phagocytized pneumococci. Infect Immun 10:1321, 1974

Stassi DL, Lopez P, Espinosa M, et al.: Cloning of chromosomal genes in *Streptococcus pneumoniae*. Proc Natl Acad Sci USA 78:7028, 1981

Stephens CG, Reed WP, Kronvall G, et al.: Reactions between certain strains of pneumococci and Fc of IgG. J Immunol 112:1955, 1974

Tiraby JG, Tiraby E, Fox MS: Pneumococcal bacteriophages. Virology 68:566, 1975

Wilson D, Braley-Mullen H: Antigen requirements for priming of type III pneumococcal polysaccharide-specific IgG memory responses: Suppression of memory with the T-independent form of antigen. Cell Immunol 64:177, 1981

Winkelstein JA, Abramovitz AS, Tomasz A: Activation of C3 via the alternative complement pathway results in fixation of C3b to the pneumococcal cell wall. J Immunol 124:2502, 1980

Winkelstein JA, Shin HS: The role of immunoglobulin in the interaction of pneumococci and the properdin pathway: Evidence for its specificity and lack of requirement for the Fc portion of the molecule. J Immunol 112:1635, 1974

CHAPTER 28
Neisseria

Two species of the genus *Neisseria* are of major medical importance, *Neisseria meningitidis* and *Neisseria gonorrhoeae*. The organisms are genetically very closely related, but the clinical manifestations of the diseases they produce are quite different. Discovered in 1885 by Neisser, for whom the genus is named, *N. gonorrhoeae* is the etiologic agent of gonorrhea, the most prevalent of the classic venereal diseases. The seventh and eighth decades of the 20th century find a worldwide epidemic of gonorrhea in progress.

N. meningitidis is the causative agent of meningococcal meningitis. This disease also has the potential for occurring in epidemic form. Because of the tendency of this illness to occur in clusters, there is a great deal of concern when a patient with meningococcal infection is identified. Within the last decade, the identification and purification

of capsular antigens of several types of *N. meningitidis* have resulted in the preparation and commercial availability of vaccines for use in epidemic situations.

Humans are the only known reservoir of the members of the genus *Neisseria*, which includes, in addition to *N. meningitis* and *N. gonorrhoeae*, nonpathogenic organisms that inhabit the upper respiratory tract and other mucosal surfaces of the body. In these positions as resident flora, the other *Neisseria* can be confused with *N. gonorrhoeae* and *N. meningitidis*. Unusual situations occur in which certain of the other species may be responsible for invasive disease in the human host. Species within the genus are shown in Table 28-1.

The genus *Neisseria* is one of four genera included in the family Neisseriaceae. Other genera in the family in-

TABLE 28-1. CHARACTERISTICS DIFFERENTIATING THE SPECIES OF GENUS *NEISSERIA*

	N. gonorrhoeae	*N. meningitidis*	*N. sicca*	*N. subflava*	*N. flavescens*	*N. mucosa*	*N. lactamicus*
Acid from:							
Glucose	+	+	+	+	−	+	+
Maltose	−	+	+	+	−	+	+
Sucrose	−	−	+	±	−	+	−
Lactose	−	−	−	−	−	−	+
Polysaccharide produced from 5% sucrose	0	0	+	±	+	+	0
Reduction of:							
Nitrate	−	−	−	−	−	+	−
Nitrite	−	±	+	+	+	+	+
Pigment	−	−	±	+	+	−	−
Extra CO_2 for growth	+	+	−	−	−	−	−

clude *Branhamella, Moraxella,* and *Acinetobacter.* The taxonomic classification and an outline of the distinguishing characteristics of the four genera of the family Neisseriaceae are given in Table 28-2.

Genus *Neisseria*

Morphology

Neisseria are gram-negative cocci, 0.6 to 1.0 μm in diameter. The organisms are usually seen in pairs with adjacent sides flattened. Fresh isolates of most *N. meningitidis* serogroups are encapsulated, whereas this is less constant with *N. gonorrhoeae.* Recent laboratory observations have confirmed the presence of the *N. gonorrhoeae* capsule and demonstrated its loss with in vitro passage. Pili are present on virulent *N. gonorrhoeae,* and although frequently present on *N. meningitidis* isolates, there is no correlation with virulence. *Neisseria* are not motile.

The *Neisseria* are structurally like other gram-negative bacteria. The ultrastructure of the cytoplasm and the cell wall of the meningococcus and the gonococcus are similar. The cell envelope is composed of three major elements: the cytoplasmic membrane; the rigid peptidoglycan layer; and the outer membrane, which contains lipopolysaccharide, phospholipid, and proteins that are immunologically significant.

Physiology

Neisseria are aerobic or facultatively anaerobic organisms. Most strains of *N. meningitidis* and *N. gonorrhoeae* utilize glucose, but the acid produced arises primarily from an oxidative pathway rather than by fermentation, which explains the weak reaction that is usually observed. All *Neisseria* produce catalase and cytochrome oxidase.

Members of the genus are very susceptible to adverse environmental conditions, such as drying, chilling, and exposure to unfavorable pH or to sunlight. They should be handled in the laboratory with minimal delay.

TABLE 28-2. DIFFERENTIAL PROPERTIES OF THE GENERA OF THE FAMILY NEISSERIACEAE

Genus	Morphology	Fermentation of Glucose	Oxidase	Penicillin	G + C Moles %
*Neisseria**	Gram-negative cocci	+	+	Sensitive	47–52
Branhamella[†]	Gram-negative cocci	−	+	Sensitive	40–45
Moraxella[‡]	Short gram-negative rods	−	+	Sensitive	40–46
Acinetobacter[§]	Short gram-negative rods	±	−	Sensitive	39–47

*Species: *N. gonorrhoeae, N. meningitidis, N. sicca, N. subflava* (includes *perflava*), *N. flavescens, N. mucosa,* and *N. lactamicus* (species *incertae sedis*).
[†]Single species in genus: *B. catarrhalis* (previously *N. catarrhalis*).
[‡]Species: *M. lacunata, M. bovis, M. nonliquefaciens, M. phenylpyruvica,* and *M. osloensis.*
[§]Single species in genus: *A. calcoaceticus.* Includes two groups: (1) Ferment glucose—includes organisms known formerly as *Herellea vaginicola, Bacterium anitratum, Achromobacter anitratum,* and *Acinetobacter anitratum;* and (2) Does not ferment glucose—includes organisms known formerly as *Mima polymorpha, Acinetobacter lwoffi,* and *Achromobacter hemolyticus* var. *alcaligenes.*
G+C, guanine plus cytosine in the DNA.

Cultural Characteristics

N. meningitidis and *N. gonorrhoeae* are fastidious organisms with complex nutritional growth requirements. Iron is required for growth and the ability of the organism to compete for transferrin-bound iron is discussed in the section, Determinants of Pathogenicity (p. 494). Starch, cholesterol, or albumin should be added to the media to neutralize the inhibitory effects of fatty acids.

Cultures derived from normally sterile sites, such as cerebrospinal fluid, blood, or synovial fluid, can be inoculated on nonselective media, such as chocolate agar. Growth of primary isolates is enhanced by incubation in the presence of 2 to 8 percent CO_2. Some of the apparently beneficial effects of the CO_2 atmosphere may be a result of the increased moisture present in the incubator or candle jar used for culture.

The Thayer–Martin selective medium permits recognition of *N. meningitidis* and *N. gonorrhoeae* from materials contaminated with other bacterial flora. The medium contains chocolate agar modified by the addition of vancomycin to inhibit gram-positive bacteria; colistin for the inhibition of gram-negative enteric flora; and nystatin for the inhibition of yeast. Most nonpathogenic *Neisseria* species also fail to grow on this medium. Recently, *N. gonorrhoeae* sensitive to vancomycin have been recognized and their true incidence is unknown. Only rarely are meningococci inhibited.

Laboratory Identification

A number of biochemical reactions are useful in the differentiation of *N. meningitidis* and *N. gonorrhoeae* from other species that are present in clinical material (Table 28-1). Colonies of *Neisseria* species may be recognized by use of the oxidase test that employs the indicator dye tetramethyl-p-phenylenediamine dihydrochloride. When exposed to this dye, colonies turn dark purple within seconds. As all *Neisseria* are oxidase positive, however, the finding of oxidase-positive, gram-negative diplococci in a clinical specimen requires additional tests for confirmation and identification of species. Because organisms are rapidly killed by the oxidase reagent, subculture to chocolate agar of the unused portion of the colony should be made immediately.

The distinction between *N. gonorrhoeae* and *N. meningitidis* is usually based upon the metabolism of carbohydrates. *N. meningitidis* produces acid from both glucose and maltose, whereas *N. gonorrhoeae* produces acid from glucose only. These differential tests of pure cultures can be complicated by some of the growth characteristics of these organisms. For example, the production of acid from carbohydrate sources may be masked by the alkaline products of enzymatic peptone degradation. Supplementary methods of identification are available and may need to be utilized.

A valuable but not fully appreciated differential diagnostic test is based on the synthesis from sucrose of an iodine-reacting polysaccharide by *Neisseria* species other than *N. gonorrhoeae*, *N. meningitidis*, and *Neisseria lactamicus*. The test is done simply by incubating a streaked culture on 5 percent sucrose agar for 48 hours and then treating the culture with modified Gram's iodine. A positive test shows a darkening of the colonies to a red–blue or blue–black color.

Other techniques that are not generally available in service laboratories may become practical as the methodology is standardized. One such test is based upon the nutritional requirements of the organisms in question. Chemically defined agar media have been developed that support the growth of *N. meningitidis*, *N. gonorrhoeae*, and *N. lactamicus*. The distinct nutritional profile of each species can serve to differentiate and identify it. Gas–liquid chromatography is a second such test. This method provides a rapid and accurate means of determining the composition of bacteria and identifying their metabolic products.

In summary, the identification of isolates as *Neisseria* species is based upon the growth of gram-negative, oxidase-positive, catalase-positive diplococci that produce characteristic colonies on blood and chocolate agar plates. The differential utilization of carbohydrates forms the basis of initial speciation. Immunoserologic diagnosis of *N. meningitidis* or recognition of antigen in specimens can be accomplished employing specific antisera as discussed on page 497.

Neisseria meningitidis

Antigenic Structure

Nine serotypes of *N. meningitidis*, designated A, B, C, D, X, Y, Z, W135, and 29E, have been identified on the basis of immunologic specificity of capsular polysaccharides. Organisms in groups A, B, and C are responsible for the great majority of clinically recognized disease. The group A capsular antigen consists of N-acetyl-O-acetyl mannosamine phosphate, the B polysaccharide is $2 \rightarrow 8$-α-linked N-acetyl neuraminic acid, and the C polysaccharide is $2 \rightarrow 9$-α-linked-acetyl, O-acetyl neuraminic acid. The structural repeating unit of the capsular polysaccharides of the other meningococcal groups that have been characterized are shown in Table 28-3. Identification and purification of the groups A, B, C, Y, and W135 polysaccharide antigens have resulted in the production and licensure of effective vaccines for A and C serotypes. The type Y and W135 polysaccharides are immunogenic and currently being tested for safety and efficacy. The type B polysaccharide is nonimmunogenic.

Antigenic similarities to meningococci have been found in unrelated bacteria. Thus, *Escherichia coli* isolated from cerebrospinal fluids of newborn infants with meningitis have a K_1 capsular polysaccharide antigen that is immunologically identical to that of group B meningococcus and correlates with the apparent invasiveness of the organism in the neonate. This antigen is easily degraded in the host and is also a very poor immunogen in humans.

Surface antigens other than the group polysaccharides may stimulate bactericidal antibodies. Groups B and C *N. meningitidis* have been subdivided into 15 to 18 different serotypes based on the outer membrane proteins. Humans produce antibodies to these serotype proteins of groups B and C meningococci in response to both nasopharyngeal carriage and to systemic infection. These antibodies may be specific for the serotype proteins and/or lipopolysaccharide (LPS) antigens. Fortunately, a few serotypes seem to be responsible for the majority of illness. For example, serotype 2 has been reported to account for 65 percent of group B and 78 percent of group C disease in Canada and Belgium. Thus persons may lack anticapsular antibody and yet have demonstrable bactericidal antibody against a specific protein serotype. It may therefore be possible to utilize selected outer membrane protein antigens to induce a protective response against the strains of both the B and C groups having this particular antigen. As noted above, the group B polysaccharide is nonimmunogenic in humans and natural infection with these organisms does not seem to result in a strong antibody response to the group B polysaccharides. Membrane protein that has been detergent treated to remove LPS is a very soluble immunogen in combination with high molecular weight group B polysaccharide. This combination is currently being evaluated in animal models for potential use as a human vaccine.

In addition to the surface capsular polysaccharide antigens and outer membrane protein antigens, there are somatic antigens, which include a nucleoprotein fraction and a carbohydrate antigen. These have not been chemically defined but appear to be common to the *Neisseria* within a specific serogroup. These antigens probably account, at least in part, for the cross-reactivity observed in antibody testing.

Determinants of Pathogenicity

The capsular polysaccharides contribute to the invasive properties of the meningococci by inhibiting phagocytosis. In the presence of specific antibody, organisms are readily destroyed by the phagocytic leukocytes. Organisms visualized on gram stain may appear to be in an intracellular location, but, in fact, they are adherent to the surface of the neutrophil. There is no evidence that these organisms survive within phagocytes.

The endotoxins of the meningococci are basically similar to those of other gram-negative bacteria. These organisms produce large amounts of LPS-containing outer membrane and during division, vesicles of this material are released extracellularly. The LPS from meningococci is a more potent inducer of the dermal Shwartzman phenomenon than that from *E. coli* or *Salmonella typhimurium*. Thus it is thought that as organisms invade and multiply, they release large quantities of LPS, as do the neutrophil-

ingesting organisms. These materials released in cells of the vascular endothelium cause vascular necrosis and induce an inflammatory response. Thus endotoxin is implicated in the vascular damage, especially that visualized in the characteristic skin lesions that are a varying component of the disease that is produced. Antibody produced to a mutant strain of *E. coli* (J5 mutant of 0111) deficient in uridine 5'-diphosphate (UDP)-galactose epimerase protects animals against endotoxemia due to *N. meningitidis*, *Klebsiella*, *E. coli*, and *Pseudomonas*. The antibodies are directed against the core glycolipid because of the absence of complete O side chains in the mutant. This suggests a similarity among the endotoxins from diverse bacteria. In patients with meningitis, circulating antigen–antibody complexes have been demonstrated in the weeks following initiation of therapy. A decreased serum complement has also been seen and the late manifestations of infection such as arthritis may be attributable to these immune complexes.

All strains of every *N. meningitidis* serogroup produce an IgA protease that is excreted into the extracellular environment. All pathogenic *Neisseria* species have this capacity. These enzymes are neutral endopeptidases with a substrate specificity for human IgA_1. The proteases cleave the heavy chain of IgA_1 in the hinge region. Two enzymes, type 1 and 2 from *N. meningitidis*, have been shown to cleave a prolyl-seryl or prolyl-threonine bond, producing intact Fc-α and Fab-α fragments, respectively. The role of these enzymes in the pathogenesis of human infection remains to be characterized, but the limitation of these enzymes to pathogenic *Neisseria* species suggests that they may contribute to the ability of the organism to produce disease.

N. meningitidis is exclusively a human pathogen and must therefore be able to multiply in a host who sequesters iron either intracellularly or in association with high affinity iron-binding proteins. Iron is essential for growth of the organism and in vitro experiments have demonstrated conclusively that *N. meningitidis* can utilize transferrin-bound iron as the sole iron source. Organisms in an iron-depleted environment for a few hours acquire the ability to obtain iron from transferrin by a saturable nonenergy-requiring cell surface mechanism. There is no evidence that *N. meningitidis* has a siderophore. Uptake of iron from the bacterial cell surface, and its subsequent metabolism, is accomplished by an unknown energy-dependent mechanism. Nonpathogenic *Neisseria* species lack the ability to use iron from transferrin.

Clinical Infection

The first recognition of disease caused by *N. meningitidis* occurred in 1805 with the description of an epidemic of meningitis in Geneva, Switzerland. One year later, an outbreak in Medfield, Massachusetts, marked the first recognized outbreak in North America. The causative organism, however, was not identified until 1887, when Weichselbaum described the gram-negative diplococci in the spinal fluid of patients.

Epidemiology

Meningococcal disease is worldwide in distribution and varies from sporadic cases observed in a community to epidemics of infection. In 1980, cases were reported in every state of the United States except Nebraska and the District of Columbia. Although the reported number of cases of disease in the United States from 1948 to 1980 has been less than 3 per 100,000 population, in 1945 an outbreak of group A meningococcal disease occurred with a reported case rate of 14 per 100,000 population (Fig. 28-1). The potential for a meningococcal epidemic is illustrated by the urban epidemic that began in São Paulo, Brazil, in June 1971. The disease outbreak continued for 3 years, with an attack rate of 65 per 100,000 population per month. In February and March of 1974, the predominant strain of *N. meningitidis* changed from serotype C to serotype A. Thirteen thousand suspected cases of meningococcal disease were admitted to the hospitals in that city in July and August of 1974.

The adult nasopharyngeal carrier is important in the transmission of meningococci and provides a reservoir of infection from which the organisms are introduced into a household. The median duration of carriage in a nonepidemic setting is 10 months. The carrier rate is higher in members of the household of a patient with me-

ningococcal disease. In community epidemic situations, carriage rates of 15 percent have been observed in households with identified disease, as compared with 3.6 percent in households without disease. An investigation of an extended family with several infected children revealed a 44 percent carriage rate, whereas a rate of 3 percent was found in samples of unrelated population in the same geographic area.

The peak occurrence of disease is in children from 6 to 24 months of age. A similar peak in incidence occurs in the 10- to 20-year age group and is related primarily to outbreaks of disease in the military population. Meningococcal disease in military populations is associated with nasopharyngeal carriage rates as high as 90 percent. The intimate contact provided by army barracks increases exposure of susceptible recruits from a variety of geographic areas to carriers harboring the organism.

It is of practical importance to emphasize that susceptible intimate contacts, primarily those within a household, and potentially those in a day-care setting, provide the greatest risk for acquiring meningococcal disease. Attention should be focused on young children within the intimate setting of recognized cases, and the fear of acquisition of disease placed in the proper perspective.

Immunity

Purification and use of the A, B, C, Y, and W135 polysaccharides have extended our information concerning immu-

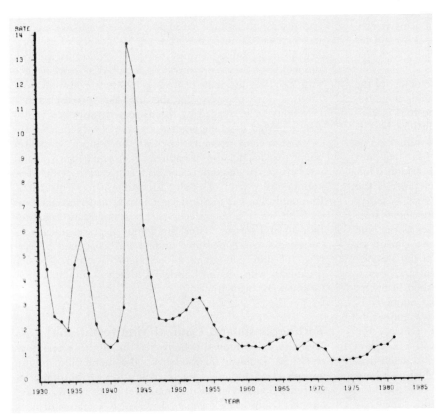

Figure 28-1. Meningococcal Infections. Reported cases per 100,000 population by year, U.S., 1930–1981. For 1981, 3525 cases were reported. The last major epidemic of meningococcal disease in this country occurred among infants less than 1 year of age. (From MMWR—Annual Summary, 1981. 30(54):58, 1982.)

nity to meningococci. The A and C polysaccharides in humans induce group-specific antibodies measured by a passive hemagglutination test, bactericidal assay, and a radioactive antigen-binding assay. The radioactive antigen-binding assay measures total anticapsular antibody, whereas the bactericidal antibody assay measures the ability of antibody to initiate immune lysis. The antibodies that induce lysis are thought to be protective and, therefore, bactericidal antibody induction is used as a means to assess efficacy of the meningococcal polysaccharides. The bactericidal activity is complement dependent. In the laboratory, it has been shown that the terminal complement sequence is essential for this antibody-associated serum bactericidal activity. The terminal complement sequence does not enhance either phagocytosis of these organisms or intracellular killing. Supportive evidence for these observations is provided by the severe, chronic, or recurrent *Neisseria* infections that have been documented in persons naturally deficient in C5, C6, C7, or C8.

Antibody-dependent, complement-mediated immune lysis or bactericidal antibodies restrict the meningococcus to mucosal surfaces. The clinical correlations with existing data corroborate the protective role of antibody. Antibodies as measured by the bactericidal assay are present in the blood of very young infants. These antibodies, detectable at birth and for the subsequent few months of life, are presumably transplacentally acquired. The lowest antibody titer is present in infants between 6 and 24 months of age, which correlates well with the peak incidence of sporadic meningococcal disease. Thus there is an inverse correlation between age-related incidence of disease and age-specific prevalence of serum bactericidal activity. Prospective longitudinal studies of army recruits also confirms that the absence of bactericidal titers correlates with susceptibility to infection.

Antibodies to the meningococcus may be of any of the three major classes (IgG, IgM, and IgA) of immunoglobulin. However, their ability to induce immune lysis is variable. IgM is always bactericidal, whereas IgG at best is only one half as effective because doublet formation is necessary for activation of complement. In fact, IgG_2 and IgG_4 poorly activate complement. IgA does not induce immune lysis, and probably of even greater relevance is the observation that IgA can block initiation of lysis by either IgM of IgG. This blocking effect is antigen specific; that is, IgA must compete for the same antigenic sites as the lytic antibody. Clinically, this observation has been linked to disseminated disease by the demonstration of the absence of bactericidal activity for the infecting strain of *N. meningitidis*, which can be restored if IgA is absorbed from the patient's serum. The uncovered bactericidal antibody is IgM. On the other hand, IgA antibody to the meningococcal capsule has been demonstrated to be necessary for complement-independent opsonophagocytosis of organisms by circulating monocytes. It has been postulated that this is a mechanism capable of destroying small numbers of invading organisms.

In the United States and several other developed na-

tions, there has not been a nationwide epidemic of meningococcal infection since the 1940s. Recently, a hypothetical immunoepidemiologic model was proposed to explain the epidemic occurrence of meningococcal infection. Enteric bacteria with surface antigens that are immunologically cross-reactive with *N. meningitidis* capsular polysaccharides have been identified, for example, group C *N. meningitidis* with *E. coli* K92 and *Streptococcus fecalis* with *N. meningitidis* group A. These colonizing organisms would not cause recognizable infection, are fecally to orally transmissible, and could stimulate the gastrointestinal tract-associated lymphoid tissue to produce IgA. Ultimately, serum IgA would be detectable. Variations in the IgA could be anticipated depending upon the stimulus and the ability of the liver to transport circulating IgA to bile. If an individual had high circulating IgA levels to *N. meningitidis* at the time this organism colonized the respiratory tract, any IgM antibodies might be blocked by the circulating IgA. Immune lysis would thus be transiently blocked and the person would be susceptible to invasive disease.

The basic premise of this hypothesis, is that an epidemic of invasive disease is realized only when the *N. meningitidis* cocirculates with a nonpathogenic enteric organism, elaborating a cross-reactive surface antigen. This hypothesis provides potential explanations for unanswered questions. Outbreaks of disease could not be correlated with carrier rates in military recruits in basic training, that is, carriage of *N. meningitidis* occurred at comparable rates in various training camps but epidemic disease occurred only sporadically in selected camps. Similar rates of susceptibity existed in various recruit camps and yet not all had disease. No single serotype of *N. meningitidis* could be incriminated as an epidemic strain, as carriage of identical organisms in similar settings occurred without disease. Additionally, as the seropositivity is acquired at a predictable rate and the majority of young adults are seropositive, there is no explanation for the apparent increase in susceptibility as evidenced by the increased incidence of disease reported in young adults.

Commensal colonization with an enteric organism might explain the above features of *N. meningitidis* epidemics. It is consistent with observed features such as the slow contiguous spread of epidemics and involvement that is frequently limited to socioeconomically deprived persons. As with many other infections, poor sanitation enhances fecal to oral spread. Natural immunity to the meningococcus seems to be identified and to be broadened by carriage of different strains of meningococci and of other organisms with identical antigens, which occurs at irregular intervals throughout life.

Pathogenesis and Clinical Manifestations

Meningococci enter the body via the upper respiratory tract and establish themselves on the membranes of the nasopharynx. Nasopharyngeal acquisition of the organism precedes hematogenous dissemination by an indeterminate length of time. The incubation period is a matter of days and is usually less than 1 week. Dissemination of meningococci via the bloodstream results in metastatic lesions in various areas of the body, such as the skin, meninges, joints, eyes, and lungs. The clinical manifestations vary depending on the site of localization.

The spectrum of illness includes a mild febrile disease, which may be accompanied by pharyngitis but is without other specific manifestations of meningococcal infection. Systemic disease characterized by fever and prostration is more readily identified. Infrequently an erythematous macular rash is observed, which usually is superseded by the appearance of a petechial eruption rapidly developing into large areas of ecchymosis. This vasculitic purpura is initiated by emboli of meningococci and is considered the hallmark of meningococcal disease. It is characteristic of the more severe fulminant disease. Meningococcemia may be accompanied by meningitis, arthritis, pericarditis, and involvement of virtually any organ system. Disseminated intravascular coagulation and gram-negative shock may be present. Hemorrhage into adrenal tissue with the resultant hypoadrenergic state is referred to as the Waterhouse–Friderichsen syndrome. The patient may survive meningococcal disease with no detectable sequelae or with direct residua of the infection that are evident for the remainder of his or her life. Such sequelae include eighth nerve deafness and central nervous system damage, and may include necrosis of large areas of the skin or tissue secondary to vascular thrombosis. These lesions may require skin grafting or amputation of necrotic digits or of an even larger portion of an extremity. The characteristic petechial eruption usually permits an accurate presumptive diagnosis and allows appropriate initial therapy of the illness.

Laboratory Diagnosis

Meningococcal infection is specifically diagnosed by the identification of N. meningitidis in materials obtained from the patient. If inflammatory exudates, such as spinal fluid, are available, a rapid presumptive diagnosis may be made by finding the characteristic gram-negative diplococci in stained smears. The organisms also may be occasionally demonstrated in gram stains of petechial lesions. In cases of overwhelming septicemia, the meningococci have been demonstrated in buffy coat smears from peripheral blood or, rarely, in a drop of blood obtained from an ear lobe or even a fingertip for routine differential white count.

The materials submitted to the laboratory for culture vary with the illness of the patient. Blood, cerebrospinal fluid, material from petechial skin lesions, synovial fluid, and a nasopharyngeal or throat swab may yield positive cultures. Thayer–Martin selective medium is used for the culture of materials expected to yield a mixture of organisms. Specimens of blood, spinal fluid, or other normally sterile materials are inoculated into blood culture bottles of trypticase soy broth with increased CO_2, and on the surface of a chocolate agar plate.

Detection of meningococcal polysaccharide in the cerebrospinal fluid, synovial fluid, and urine is possible by such techniques as countercurrent immunoelectrophoresis (CIE), latex agglutination, or coagglutination employing staphylococci with protein A. Specific antisera are the key to sensitive and specific antigen detection with any of these methods. Coagglutination employs staphylococci coated with the antibody which adheres via the Fc portion of the molecule to protein A. The Fab portion of the molecule will bind to the meningococcal antigen if present, resulting in visible coagglutination of the staphylococci. These antigen detection systems are more widely used and reproducible than immunofluorescence.

The research methods for demonstrating the serologic response of a patient with meningococcal infection are discussed on page 493. The radioactive antigen-binding assay is the most sensitive method but antibodies cannot be detected until several days after symptoms of disease have appeared.

Treatment

Penicillin remains the drug of choice for therapy of meningococcal infections. N. meningitidis is exquisitely sensitive to penicillin, with minimal inhibitory concentrations usually in the range of 0.3 μg/ml. Therapy consists of high-dosage aqueous penicillin G using the intravenous route. In penicillin-sensitive individuals, chloramphenicol is an effective alternative form of therapy. In addition to the essential specific antimicrobial therapy, supportive measures for possible complications, such as gram-negative shock or disseminated intravascular coagulation, are important aspects of the care of such patients.

Prevention

Prophylaxis. Antimicrobial prophylaxis of exposed individuals remains a controversial issue. Prior to the emergence of resistance to sulfonamide therapy, this drug was used efficiently to eradicate the organisms from the nasopharynx of individuals. It has been an interesting puzzle to microbiologists and clinicians that similar use of penicillin, to which the organism is sensitive, fails to eradicate the carrier state. Currently, when prophylaxis seems advisable, a choice may be made between two chemotherapeutic agents, rifampin and minocycline. Treatment with rifampin for a short period eliminates N. meningitidis from the nasopharynx, but during subsequent weeks, rifampin-resistant strains may recolonize the nasopharynx. Minocycline also eradicates the carrier state, but a recognized side effect of the drug—vestibular dysfunction with resultant disturbance of equilibrium—has limited its use. The combined use of two drugs probably has the greatest efficacy but is a practical impossibility because of the high incidence of side effects in patients receiving the combination.

Decisions concerning the use of such prophylaxis should be made with a complete awareness of the individuals at greatest risk. These usually include: (1) children, primarily those less than 6 years of age who live in the same household or who have a household-type of intimate contact with the index case; and (2) recruits in the setting of an army camp. If prophylaxis is given, this does not obviate the need for close observation of contact persons. Meningococcal meningitis has been reported in a patient who received rifampin prophylaxis. Similarly, penicillin at the prophylactic dosage employed does not appear to prevent meningococcal disease.

Within the hospital setting, it is not usually necessary to use prophylactic therapy for personnel exposed to meningococcal disease. The occurrence of secondary disease among this group is exceedingly rare. In this setting, recognition of disease and prompt therapy of the patient with appropriate antimicrobials tend to decrease the spread of illness. Personnel who have intimate contact with an untreated patient may take prophylaxis depending upon the specific situation.

Immunization. Group A and group C meningococcal vaccines are licensed and available. Group Y and W165 vaccines are approaching licensure. The vaccines consist of purified group-specific meningococcal polysaccharide and are administered as a single dose of 50 μg to adults and children older than 2 years. Large-scale field trials in the United States Army with group C vaccine demonstrated 90 percent efficacy in the prevention of group C disease and a decrease in the number of meningococcal carriers. Subsequent field trials of groups A and C polysaccharides in Brazil, Egypt, the Sudan, and Finland have established the safety and efficacy of these vaccines in persons older than 2 years. These trials suggest that a measurable concentration of antibody protein of approximately 2 μg/ml is necessary for protection.

The immunogenicity of these polysaccharides is age dependent. Infants can respond to the group A polysaccharide as early as 3 months of age and will show an anamnestic rise in serum antibody when given a second dose of the vaccine 3 to 4 months later. The level of antibody achieved is 2 to 3 μg/ml. A large field trial in Finland demonstrated complete protection against meningococcal group A disease in 130,000 children from 3 months to 5 years of age. Two 25-μg doses of vaccine were administered at 3-month intervals to children 3 months to 18 months of age, and a single 50-μg dose was administered to children 1.5 to 5 years of age.

In contrast, group C vaccine does not induce protective antibody levels when given to children less than 18 months of age. A booster dose does not produce an anamnestic response, and the group C vaccine did not protect infants from 6 to 24 months of age during the epidemic in São Paulo, Brazil.

In summary, group A vaccine offers effective protection for patients of all ages, and control of group A epidemics is feasible. Group C vaccine will protect persons over 2 years of age and therefore offers significant control of epidemics because it is useful for 60 to 80 percent of expected cases. These vaccines are available for use in military populations where epidemic disease is likely, and by

specific request for control of an outbreak of infection. At the present time these vaccines are not recommended for routine use in childhood.

The development of type-specific vaccines for groups A and C meningococci constitutes a significant contribution to preventive medicine. The group B organisms continue to pose a problem because the polysaccharide is a very poor immunogen (p. 493). The theoretic possibility that immunization with the group A and/or group C polysaccharide will prevent disease with organisms within these serogroups but will allow other serogroups of organisms to emerge as epidemiologically significant awaits the test of experience with the present vaccines. The persistence of antibody and duration of protection are being assessed longitudinally.

Neisseria gonorrhoeae

Antigenic Structure

The antigenic composition of *N. gonorrhoeae* is complex. Associated with the surface layers are at least three major classes of antigens: (1) pilus protein antigen; (2) polysaccharide component of the cell wall lipopolysaccharide; and (3) the outer membrane protein constituents.

Typing of *N. gonorrhoeae*. Several methods for typing gonococcal strains are currently under investigation. These methods are being applied to sharpen epidemiologic studies of gonococcal disease, and to enhance an understanding of clinical syndromes. These methods, which are summarized in Table 28-3, include the following:

1. An enzyme-linked immunosorbent assay (ELISA) for the principal outer membrane protein (Protein

TABLE 28-3. SYSTEMS FOR TYPING *NEISSERIA GONORRHOEAE* STRAINS

System	Types	Comment
ELISA to Protein I	1–9	DGI strains are types 1 and 2
Coagglutination after cross-absorption using *Staphylococcus aureus* Protein A	WI–WIII	Results correlate with MIF methods
Specific microimmuno-fluorescent antibody to whole formalinized organisms	A–C with 8 subtypes	Consort pairs are identical
Auxotyping based upon nutritional requirements	1–13	DGI strains are A⁻H⁻U⁻
Pilus typing	α, β, γ, δ	Pilus variation occurs in vivo

DGI, disseminated gonococcal infection; MIF, microimmunofluorescence; A⁻H⁻U⁻, auxotrophic for arginine, hypoxanthine, uracil.

I). Nine types may be distinguished with this assay. The majority of strains are from women with salpingitis, and almost all strains from persons with disseminated gonococcal infection are of type 1.

2. Coagglutination of strain-specific antigonococcal antibody which has been complexed with staphylococcal protein A. The sera are cross-absorbed to achieve strain specificity. Three major strains may be distinguished.

3. Microimmunofluorescence with antibody derived from immunization of mice with whole formalized organisms and cross-absorption of the antisera with heterologous strains. This system identifies three main types (A, B, and C) and at least eight subtypes (A$_1$, A$_2$, etc.). In this system, pairs from sexual consorts were found to be of identical type and subtype.

4. Auxotyping based upon the nutritional requirement of a strain for eight growth factors, which include arginine, hypoxanthine, and uracil (A, H, and U). The system identifies 13 types, of which 6 are common. It is notable that the majority of strains derived from persons with disseminated gonococcal infection require AHU and are termed A$^-$H$^-$U$^-$.

5. Pilus typing has identified four pilus types—α, β, γ and δ—by serologic methods. Pilus variation occurs in vivo, as does loss of pili. The possible use of pili for vaccine purposes renders an understanding of pilus immunology and genetics especially important.

The gonococcus shares a number of antigens with *N. meningitidis*, other *Neisseria*, *Branhamella*, and *Staphylococcus*. Antigonococcal conjugates for use in the fluorescent antibody procedure for the detection of gonococci in exudates must be absorbed with meningococci and staphylococci in order to eliminate cross-reactivity.

Pilus Protein Antigen. Gonococcal pili are serologically heterogeneous and are composed of helical aggregations of tube-like structures. They are approximately 7 nm in diameter, 2 μm in length, and each subunit has a molecular weight of about 23,000. Pili contain a common N-terminal hydrophobic core, but the C-terminal region is highly variable. At least four separate antigenic types have been identified. There is serologic cross-reactivity between some of the strains, and some, but not all, strains induce cross-reacting antibody when used as a vaccine.

Antipilus antibody appears to be responsible for immune-enhanced phagocytosis in vitro. Following use of parenteral pilus vaccine in humans, mucosal IgA immunoglobulin is produced, and is capable of blocking attachment to epithelial cells of homologous types of gonococci. There is also some protection against mucosal cell attachment by heterologous strains. Organisms readily lose the ability to produce pili, and may thereby evade the effects of local antibody production.

Lipopolysaccharide. The cell wall of *N. gonorrhoeae* contains surface components that are important in the typing, immunogenicity, and pathogenicity of strains. The lipopolysaccharide of *N. gonorrhoeae* resembles that of other gram-negative organisms. It contains lipid A but the core polysaccharide does not appear to include strain-specific, O-antigenic side chains.

Outer Membrane Proteins. The cell wall also contains several types of outer membrane proteins. The predominant type, as defined by polyacrylamide gel electrophoresis, is protein I (PI), or principal outer membrane protein (POMP). The proteins appear to be porins that form aqueous channels traversing the outer membrane. They are 36 to 39 k daltons in size. Protein I of isolates from patients with disseminated disease or from rectal cultures of male homosexuals may be partially characterized by the PI size. In general, strains with higher molecular weight PI are more resistant to serum bactericidal effects.

Protein II (PII), which occurs in lesser amounts than PI, is heat sensitive, 27 to 29.5 k daltons in size, and occurs in the outer leaflet of the outer membrane. Its presence in strains is associated with adherence to mucosal cells, opaque colony formation on agar media, and sensitivity to serum bactericidal activity. Loss of PII occurs at a high frequency, and is associated with increased invasive potential. Such organisms are recovered from the fallopian tubes of women with salpingitis and from the blood and joints of persons with disseminated gonococcal infection. Strains lacking PII are less sticky, are transparent on agar media, and are more resistant to serum bactericidal activity.

Determinants of Pathogenicity

Current concepts of virulence factors of gonococci are partially derived from observations that have been made of the five major colony types. The colony forms, designated T1 through T5, reflect differences in the surface antigens of the organisms in the colony. Differences in colony form may be distinguished by indirect microscopic examination. Types 1 and 2 are small and dense, and types 3, 4, and 5 are appreciably larger and more granular. Types 1 and 2 are predominant in primary isolates, but nonselective subculture leads to the rapid emergence of types 3, 4, and 5. Continued propagation of types 1 and 2 can be acheived, however, if individual type 1 and 2 colonies are selected by direct microscopic examination for subculture.

Colony types 1 and 2 are virulent for humans, and colony types 3, 4, and 5 are avirulent. One of the properties of types 1 and 2 colonies contributing to their ability to produce disease is piliation. The cells of types 1 and 2 are piliated, whereas types 3, 4, and 5 contain few or no pili (Fig. 28-2). Pili confer to cell walls an enhanced ability to adhere to host cells and to each other. This stickiness may be demonstrated in vitro, where organisms from colony types 1 and 2 adhere to cultured human amnion cells more avidly than do type 4 organisms. The same phenom-

Figure 28-2. Gonococci and gonococcal pili. **a.** Freeze-fracture, freeze-etch preparation of gonococci with pili on surface. × 80,000. **b.** Negatively stained gonococcus with pili radiating from surface. Uranyl acetate. ×70,000. (From Buchanan et al.: J Clin Invest 52:2896, 1973.)

enon may also occur in the human urethra, permitting attachment in spite of the flow of urine.

Organisms from the piliated colony types are relatively resistant to phagocytosis. Although gonococci of all colony types are readily destroyed following phagocytosis, ingestion of piliated gonococci by macrophages is very inefficient compared with that of nonpiliated strains. Piliated gonococci adhere to polymorphonuclear leukocytes. The polymorphonuclear leukocyte that is in contact with an adherent gonococcus discharges the contents of the phagocytic granules against the adherent organism. The attack upon the gonococcus is usually unsuccessful, and the polymorphonuclear leukocyte may then be parasitized by the gonococcus, leading to the viable intracellular organisms that characterize the gram stain of the discharge from a patient with gonorrhea.

Small spherical structures found on the cell surface are often seen in association with pili. They occur less frequently in colony type 4 organisms and may be associated with the damage done to the host during gonococcal infection.

All gonococci produce IgA$_1$ protease, an enzyme that cleaves IgA$_1$ at the hinge region of the antibody molecule. This enzyme presumably allows gonococci to adhere to mucosal surfaces even in the presence of a secretory antibody response.

For many years, evidence for encapsulation of *N. gonorrhoeae* has been sought because of the association in *N. meningitidis* of virulence with capsular antigen. Only recently, however, has a capsule been demonstrated in *N. gonorrhoeae*. Encapsulation has now been demonstrated by the India ink technique, quellung reaction, and electron microscopy in the presence of hyperimmune rabbit serum. Further studies are needed, however, to determine the role of capsular material in the virulence of gonococci.

Increased clinical virulence of some strains is suggested by the observation of a microepidemic of gonor-

rhea involving one asymptomatically infected male and nine female contacts. Seven of these women were symptomatically infected, and four experienced disseminated infection—an unusually high rate of severe disease. Characterization of strains of *N. gonorrhoeae* obtained from patients with disseminated disease has revealed the following properties:

1. Isolates of *N. gonorrhoeae* from patients with disseminated gonococcal infection are not susceptible to the bactericidal activity of sera from patients with uncomplicated gonorrhea, even though sera from those patients are usually bactericidal for other strains. This resistance to serum bactericidal activity may be due to a change in the outer membrane of the organism and may be transferred to a sensitive strain by transformation, using DNA from a resistant strain.
2. The isolates are auxotrophic for arginine, uracil, and hypoxanthine, nutrients not usually required for growth of strains from patients with asymptomatic or genitourinary disease.
3. A high degree of sensitivity to penicillin is characteristic of strains from patients with disseminated disease. Penicillin sensitivity and serum resistance appear to be independent variables.

Clinical Infection

The term gonorrhea, meaning "flow of seed," was introduced by Galen in 130 AD. Although ancient writings refer to medical conditions characterized by a urethral discharge, it was not until the 13th century that physicians were definitely applying the term to a venereally transmittable disease similar to gonorrhea as we know it. Syphilis and gonorrhea often were acquired simultaneously, and descriptions of the two diseases were intermingled. In 1767, the great physician John Hunter acquired both syphilis and gonorrhea during an autoinoculation experiment using urethral exudate of a patient erroneously thought to have only gonorrhea. Hunter ascribed his subsequent syphilitic symptoms to gonorrhea, and the confusion of the two diseases became complete. They were not effectively differentiated until the middle of the 19th century.

Epidemiology

Gonorrhea is the most common of the classic venereal diseases. Most areas of the world are now affected by the current pandemic. Since the beginning of the 20th century, when rates of gonorrhea first were recorded, increases and decreases in incidence have been associated with major social changes and with disruptions caused by warfare. Prior to the present period, the highest rates in the United States occurred during and just after World War II. As shown in Figure 28-3, the rate decreased through the late 1940s and early 1950s and was followed by a period of quiescence until the early 1960s, when the current pan-

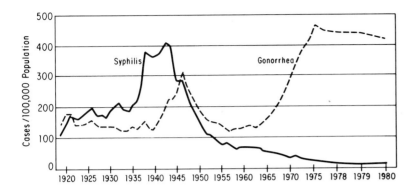

Figure 28-3. Reported cases of gonorrhea and syphilis in the U.S. per 100,000 population, 1919–1981. Recent data suggest an arrest of the annual increase in gonorrhea that occurred through the middle 1970s. (Data from the Centers for Disease Control, Atlanta, Ga.)

demic rapidly spread. In the United States, over 1 million cases were reported to the Centers for Disease Control during 1981. Even this does not represent the true incidence, as underreporting is common. However, the national rate of disease in 1980 was 443 per 100,000 population, which was slightly less than the peak of 473 per 100,000 population in 1975.

Seasonal peaks of incidence of gonorrhea occur from July through September, in both the southern and the northern areas of the United States. The prevalence of gonorrhea in the United States is markedly affected by age. In both males and females, the disease is most common in persons 20 to 24 years of age (1780 per 100,000 population) and is only slightly less common among 15 to 19 and 25 to 30 year olds. The yearly increases in incidence of gonorrhea during the 1960s were due primarily to increases among persons 15 to 30 years old. Most cases are casually acquired. Prostitution is not at present a major factor in its spread in most areas of the United States. There is a general trend toward increased prevalence in nonwhite individuals, persons of low socioeconomic groups, and urban dwellers.

Pathogenesis

Autopsy material of persons who have died of intercurrent disease during the acute phase of gonorrhea have provided some information on the histopathology of acute gonorrhea. The primary infection usually begins at the columnar epithelium of the urethra and periurethral ducts and glands of either sex. Cervical, conjunctival, and rectal mucosa also may serve as the portal of entry. Within less than 1 hour following contact with the mucosal surface, the infection is established, and the bacteria are anchored by pili to surface urethral cells. Penetration occurs through intercellular spaces, and organisms reach the subepithelial connective tissue by the third day. The resulting inflammatory response consists of a dense infiltration of polymorphonuclear leukocytes. Obstruction of ducts and glands by this exudate results in retention cysts and abscesses. Spread to other areas often occurs by direct extension through lymphatic vessels and, less commonly, by blood vessels.

In vitro perfusion of human fallopian tubes has provided a suitable model for study by electron microscopy of the disease process. Initial contact between the organism and the host is at the microvilli. Gonococcal pili extend over the membrane of the epithelial cells as the attached cocci penetrate the mucosal lining of the fallopian tube. Foci of infection develop in the subepithelial connective tissue, resulting in disorganization of collagen connective tissue and local extension of disease. Destruction of ciliated epithelial cells by gonococci depletes the mucosal lining, thereby permitting more rapid penetration by other gonococci.

Host Response to Infection

In infected males, there is a significant immunologic response to gonococcal infection that may be measured by several systems:

1. Humoral antibody of the IgG classes usually can be detected by the time the patient has clinically apparent disease. Few patients have detectable specific IgM antibodies in the early period of disease.
2. Local secretory IgA in the urethral secretions of infected males can be detected early.
3. In patients with uncomplicated gonococcal urethritis, cell-mediated immunity is activated. Lymphocyte blastogenesis may be induced by gonococcal antigens in some patients with their first infection and in most patients who have had multiple infections.

Humoral IgA and IgG antibody response to disease may block the bactericidal effects of IgM antibody. Persons who are unable to produce any of the terminal components of the complement system are susceptible to disseminated gonococcal disease and to recurrences.

Clinical Manifestations

Disease in the Male. When compared with many other infectious diseases, gonorrhea is not highly contagious. An unprotected male has approximately 22 percent chance of acquiring gonorrhea from intercourse with an infected female, and the risk is considerably reduced by use of a condom. Acute gonorrhea in the male has an incubation period of 2 to 8 days, with most cases occurring within 4 days of infection. The patient presents with burning on urination and a yellow purulent urethral discharge that signifies acute anterior urethritis. The patient may be febrile and have a leukocytosis, but systemic signs are generally lacking. The infection may be asymptomatic in approximately 10 percent of cases, although the patient retains the capacity to transmit disease. In the preantibiotic era, most males had resolution of disease within a month. However, approximately 1 percent of males develop complications, the most common being urethral strictures, epididymitis, or prostatitis. Less common are septicemia, peritonitis, and meningitis. Another frequent sequela of gonorrhea in the male is the subsequent development of nongonococcal urethritis (nonspecific urethritis) (Chap. 53).

Disease in the Female. Screening of some groups of asymptomatic women has shown prevalence rates of gonorrhea between 1 and 8 percent. Lower rates of infection are observed from such groups as private obstetric practices and higher rates from such groups as vocational training programs, neighborhood clinics, and venereal disease clinics. The risk to a female from intercourse with an infected partner is not definitely known but is probably higher than that for the male.

Between 20 and 80 percent of women with gonorrhea are asymptomatic, depending upon the population studied. Signs of disease in those who are symptomatic include burning or frequency of urination, vaginal discharge, fever, and abdominal pain. The major complication of gonorrhea in women is the development of pelvic inflammatory disease (PID) by gonococcal infection of the fallopian tubes. This disease affects approximately 15 percent of women with gonorrhea and has two important consequences: (1) Gonococcal PID is a major cause of sterility and ectopic pregnancies because the scars from the infection may block the passage of ova through the fallopian tubes; and (2) scar formation also blocks the normal flow of fluid through the fallopian tubes. In areas with fluid accumulation, infection by other bacteria, often anaerobic, may develop. This leads to chronic PID, a very debilitating and painful disease without satisfactory forms of therapy. Other complications occasionally encountered are infectious perihepatitis and generalized peritonitis.

Approximately 50 percent of females with gonorrhea have concomitant rectal colonization, and proctitis occasionally develops. In 10 percent of women, the rectal site is the only area colonized. Fewer heterosexual males have rectal colonization, but in male homosexuals, it is very common.

The other major site of extragenital colonization in both males and females is the pharynx. In about 5 percent of persons who practice fellatio it is the only site of infection. Pharyngeal gonococcal infection is most often asymptomatic, but in some instances it is associated with clinically apparent pharyngitis.

Disseminated Gonococcal Disease. The gonococcal arthritis–dermatitis syndrome is the most common manifestation of disseminated gonococcal disease and is the result of gonococcal bacteremia. Male and female patients in whom this syndrome occurs usually have had asymptomatic genitourinary infection. Although more cases have been reported in women, males with asymptomatic gonorrhea are at risk. The incidence of gonococcal bacteremia is approximately 1 percent of all persons with gonorrhea.

The acute form of the gonococcal dermatitis syndrome is heralded by fever, chills, malaise, intermittent bacteremia, polyarticular arthritis or tenosynovitis, and the development of typical skin lesions. The small distal joints are the predominant sites of involvement, and there is usually a paucity of synovial effusion. Joint fluid, if obtained, is most often sterile. The characteristic skin lesions are few in number, occur on the distal dorsal surfaces of the wrists, elbows, and ankles, and usually begin as small petechial or papular lesions. Suppuration, bullous formation, and central necrosis also are common.

If therapy is not received during this stage, which usually lasts about 3 days, the patient may progress to the septic joint form of disease. Blood cultures rarely yield *N. gonorrhoeae*, symptoms of septicemia such as fever and chills cease, and the disease becomes prominent in a single joint as overt arthritis. The synovial fluid is characteristic of pyarthrosis, with decreased sugar, poor mucin clot formation, and a pronounced granulocytic response.

Other rare forms of gonococcal disease that may follow bacteremic spread include subacute bacterial endocarditis and meningitis. The organisms responsible for disseminated gonococcal infection are usually highly sensitive to penicillin.

Disease in Children. Although gonorrhea is most commonly acquired during sexual contact between adults, a significant number of cases each year occur in infants and children. Gonorrhea in this age group may be a result of sexual abuse, but in infancy it usually results from contamination during passage through an infected birth canal.

In the perinatal period, infection of the eye is the most common manifestation of gonorrhea. Prior to the use of silver nitrate for ophthalmic prophylaxis, gonococcal ophthalmia was the cause of blindness in approximately half of the children admitted to schools for the blind. The disease is now a rare cause of blindness, but gonococcal ophthalmia neonatorum continues to occur. Prevention of disease may be based upon the diagnosis and treatment of the mother prior to birth or prophylactic treatment of the eyes of the newborn after birth. Following birth, all states require prophylactic care of the eyes of the newborn: 1 percent silver nitrate is the most satisfactory agent. Failure of silver nitrate prophylaxis, however, does occur and is more common in premature infants or after prolonged rupture of membranes. Silver nitrate instillation is inadequate treatment for established infection, and some of the failures may be due to the presence of active disease by the time of birth.

TABLE 28-4. OUTCOME OF PREGNANCY IN MOTHERS WHO WERE INFECTED WITH *NEISSERIA GONORRHOEAE* AT DELIVERY*

Outcome	Number of Cases (%)			
	Sarrell	Israel	Amstey	Edwards
Total number in study	37	39	222	19
Normal or term infant	13(35)	30(77)	142(64)	7(37)
Aborted	13(35)	1	24(11)	—
Perinatal death	3(8)	1	15(7)	2(11)
Premature	6(17)	5(13)	49(22)	8(42)

*Data from Sarrell and Pruett: Obstet Gynecol 32:670, 1968. Israel et al.: Clin Obstet Gynecol 18:143, 1975. Amstey and Steadman: J Am Vener Dis Assoc 3:14, 1976. Edwards et al.: Am J Obstet Gynecol 132:637, 1978.

Relatively little is known of the hazard to the fetus of maternal gonorrhea, but there are indications of increased rates of premature birth and of perinatal morbidity. In addition, the mother may suffer complications of delivery. For these reasons, good obstetric care now includes repeated gonococcal screening during pregnancy. Table 28-4 summarizes four separate studies on the outcome of pregnancy in women who had gonococcal disease during gestation.

Neonatal gonococcal arthritis is a highly destructive form of infectious arthritis. The organism usually is acquired from the infected mother at the time of birth. Infection of any mucous membrane may enable dissemination to occur. In several instances in which neonatal gonococcal arthritis has occurred, the mother has also had disseminated disease, a finding consistent with the observation that certain strains have particular invasive capacity.

Gonorrheal vulvovaginitis usually occurs in girls 2 to 8 years of age. The alkaline pH of the prepubescent vagina is cited as one factor favoring the establishment of gonococcal disease in this age group. The disease usually is self-limited, but occasionally progresses to invasion of the fallopian tubes or peritonitis.

An area of rapidly increasing interest and understanding concerns the role of sexual abuse of children in the acquisition of childhood gonococcal infections, as well as other sexually transmittable infections. Pediatricians are beginning to recognize sexual abuse as one aspect of the larger spectrum of child abuse and neglect.

Laboratory Diagnosis
Microscopic Examination. Gram stain of purulent materials, including urethral discharge from an infected male, conjunctival discharge, and purulent synovial fluid, will reveal many polymorphonuclear leukocytes and intracellular gram-negative diplococci. Gram stains of cervical materials from the female are often misleading because of morphologically similar saprophytic organisms. Examination of scrapings from skin lesions also may yield organisms on gram stain or by fluorescent antibody staining.

Culture. A definitive diagnosis is established by isolating the gonococcus in the laboratory. The materials submitted include urethral exudate and endocervical secretions. The pharynx and rectum also may be colonized in infected individuals and may provide evidence of infection when cultured. Blood cultures often are positive in disseminated disease. In the presence of pyarthrosis, synovial fluid usually yields the gonococcus. In the neonate, cultures of gastric aspirate and the conjunctivae are also helpful. The laboratory processing and identification of materials are discussed on page 493 of this chapter.

Persons suspected of having either meningococcal or gonococcal infection should have complete identification of the isolates. *N. meningitidis* has been recovered from all the sites in which *N. gonorrhoeae* is commonly found.

Serologic Tests. Because of the difficulty in obtaining an adequate endocervical culture from women and because a single endocervical culture from women with asymptomatic gonorrhea fails to provide a diagnosis in 20 percent of cases, a serologic method for the diagnosis of gonorrhea would be useful. Two methods for detecting antigonococcal antibody that show promise of clinical usefulness are the radioimmunoassay for pilus antibody and the ELISA test for antipilus antibody. Experience with these tests remains limited, and neither is licensed for use outside of research settings.

Treatment
Standards for the treatment of all forms of gonorrhea are regularly published by the Centers for Disease Control. As modifications of these recommendations are needed, they are published in the weekly issue of *Morbidity and Mortality Report* and should be consulted by physicians treating any patient with gonorrhea. The current recommended regimen for uncomplicated gonococcal infections in men and women is intramuscular aqueous procaine penicillin G together with oral probenecid given just before the injection. Alternative regimens are ampicillin by mouth together with probenecid. For the patient who is allergic to the penicillins or to probenecid, oral tetracycline or intramuscular spectinomycin may be used. Treatment with recommended doses of intramuscular penicillin provides adequate therapy for early stages of syphilis. This cannot be assumed if the patient is treated for gonorrhea with alternative regimens, and therefore the physician should follow-up with assessment for possible syphilis.

Penicillinase-producing *N. gonorrhoeae* (PPNG) became prevalent in the Far East during the middle 1970s and was first recognized in the United States in 1976, when the incidence was quite low. Civilian to civilian transmission was apparently infrequent and most cases were directly associated with travel in the Far East. However, in 1980 the number of cases increased dramatically and represented approximately 0.1 percent of all cases of gonorrhea at that time. Spread among heterosexual persons became apparent, but not among homosexual males. In some areas,

PPNG now represents as much as 5 percent of cases of gonorrhea. In 1981, approximately 2700 cases were reported to the Centers for Disease Control.

Penicillin-resistant strains from the Far East and most, but not all, U.S. strains are characterized by having a 4.4 md plasmid that mediates β-lactamase production. In addition to being resistant to penicillin, they are also insensitive to tetracycline. Treatment regimens suggested, but for which clinical experience is not necessarily available, include spectinomycin, cefoxitin, and trimethoprimsulphamethoxazole. Frequent changes in recommendations can be expected because of the problems surrounding the management of PPNG.

Control

Each case of gonorrhea should be reported to the local public health department. In some instances, contact tracing by the private physician may supplement public health efforts to control disease.

A number of factors contribute to the difficulties encountered in the control of gonorrhea:

1. Gonococcal infection has a very short incubation period, making it possible for secondary and tertiary cases to transmit disease before recognition and treatment of the primary case.
2. The disease is frequently asymptomatic, and the diagnosis is made only with appropriate cultures.
3. Social acceptance of sexual activity with multiple partners provides the opportunity for wide and rapid dissemination of gonorrhea and other venereally transmittable diseases.
4. Gonococcal isolates have undergone a progressive increase in the mean minimal inhibitory concentrations for penicillin during the past decade, making use of single-injection therapy increasingly difficult.

FURTHER READING

Books and Reviews

Henriksen SD: *Moraxella, Acinetobacter* and the mimae. Bacteriol Rev 37:522, 1973

Selected Papers

NEISSERIA MENINGITIDIS

Archibald FS, DeVoe IW: Iron acquisition by *Neisseria meningitidis* in vitro. Infect Immun 27:322, 1980

Artenstein MS, Gold R, Zimmerly JG, et al.: Prevention of meningococcal disease by group C polysaccharide vaccine. N Engl J Med 282:417, 1970

Davis CE, Ziegler EJ, Arnold KF: Neutralization of meningococcal endotoxin by antibody to core glycolipid. J Exper Med 147:1007, 1978

DeVoe IW: The meningococcus and mechanisms of pathogenicity. Microbiol Rev 46:162–190, 1982

Gold R, Lepow ML, Goldschneider I, et al.: Clinical evaluation of group A and group C meningococcal polysaccharide vaccine in infants. J Clin Invest 56:1536, 1975

Goldschneider I, Gotschlich EC, Artenstein MS: Human immunity to the meningococcus. I. The role of humoral antibodies. J Exp Med 129:1307, 1969

Gotschlich EC, Liu TY, Artenstein MS: Human immunity to the meningococcus. III. Preparation and immunochemical properties of the group A, group B, and group C meningococcal polysaccharides. J Exp Med 129:1349, 1969

Gotschlich EC, Rey M, Triau R, et al.: Quantitative determination of the human immune response to immunization with meningococcal vaccines. J Clin Immunol 51:89, 1972

Greenfield S, Feldman HA: Familial carriers and meningococcal meningitis. N Engl J Med 277:497, 1967

Greenfield S, Sheehe PR, Feldman HA: Meningococcal carriage in a population of "normal" families. J Infect Dis 123:67, 1971

Griffiss JM: Epidemic meningococcal disease: Synthesis of a hypothetical immunoepidemiologic model. Rev Infect Dis 4:159, 1982

Lowell GH, Smith LF, Griffiss JM, et al.: IgA-dependent, monocyte-mediated, antibacterial activity. J Exp Med 152:452, 1980

Lowell GH, Smith LF, Griffiss JM, et al.: Antibody-dependent mononuclear cell-mediated antimeningococcal activity. J Clin Invest 66:260, 1980

Munford RS, Taunay A de E, Marais JS, et al.: Spread of meningococcal infection within households. Lancet 1:1275, 1974

Munford RS, Sussuarana de Vasuncelos ZJ, Phillips CJ, et al.: Eradication of carriage of *Neisseria meningitidis* in families: A study of Brazil. J Infect Dis 129:644, 1974

Nicholson A, Lepow IH: Host defense against *Neisseria meningitidis* requires a complement-dependent bactericidal activity. Science 25:298, 1979

Weidmer CE, Dunkel TB, Pettyjohn FS, et al.: Effectiveness of rifampin in eradicating the meningococcal carrier state in relatively closed populations. Emergence of resistant strains. J Infect Dis 124:172, 1971

Wyle FA, Artenstein MS, Brandt BL, et al.: Immunologic response of man to group B meningococcal polysaccharide vaccines. J Infect Dis 126:514, 1972

Zollinger WD, Mandrell RE, Altier P, et al.: Safety and immunogenicity of a *Neisseria meningitidis* type 2 protein vaccine in animals and humans. J Infect Dis 137:728, 1978

NEISSERIA GONORRHOEAE

Blake MS, Gotschlich EG: Purification and partial characterization of the major outer membrane protein of *Neisseria gonorrhoeae*. Infect Immun 36:277, 1982

Cooperman MB: Gonococcus arthritis in infancy. Am J Dis Child 33:923, 1927

Handsfield HH, Lipman JO, Harnisch JP, et al.: Asymptomatic gonorrhea in men. Diagnosis, natural course, prevalence and significance. N Engl J Med 290:117, 1974

Handsfield HH, Sandstrom EG, Knapp JS, et al.: Epidemiology of penicillinase-producing *Neisseria gonorrhoeae* infections. N Engl J Med 306:950, 1982

Holmes KK, Counts GW, Beaty HN: Disseminated gonococcal infection. Ann Intern Med 79:979, 1971

Ingram DL, White ST, Durfee MF, et al.: Sexual contact in children with gonorrhea. Am J Dis Child 136:994, 1982

Melly MA, Gregg CR, McGee ZA: Studies of toxicity of *Neisseria gonorrhoeae* for human fallopian tube mucosa. J Infect Dis 143:423, 1981

Morse SA, Lysko PG, McFarland L, et al.: Gonococcal strains from homosexual men have outer membranes with reduced permeability to hydrophobic molecules. Infect Immun 37:432, 1982

Rice PA, Kasper DL: Characterization of gonococcal antigens responsible for induction of bactericidal antibody in disseminated infection. The role of gonococcal endotoxins. J Clin Invest 60:1149, 1977

Rice PA, Kasper DL: Characterization of serum resistance of *Neisseria gonorrhoeae* that disseminate. Roles of blocking antibody and gonococcal outer membrane proteins. J Clin Invest 70:157, 1982

Sandstrom EG, Chen KCS, Buchanan TM: Serology of *Neisseria gonorrhoeae*: Coagglutination serogroups WI and WII/III correspond to different outer membrane protein I molecules. Infect Immun 38:462, 1982

Sandstrom EG, Knapp JS, Buchanan TB: Serology of *Neisseria gonorrhoeae*: W-antigen serogrouping by coagglutination and protein I antigens. Infect Immun 35:229, 1982

Schoolnik GK, Buchanan TM, Holmes KK: Gonococci causing disseminated gonococcal infections are resistant to the bactericidal action of normal human sera. J Clin Invest 58:1163, 1976

Swanson J, Zeligs B: Studies on gonococcus infection. VI. Electron microscopic study on in vitro phagocytosis of gonococci by human leukocytes. Infect Immun 10:645, 1974

Wang S-P, Holmes KK, Knapp JS, et al.: Immunologic classification of *Neisseria gonorrhoeae* with microimmunofluorescence. J Immunol 119:795, 1977

CHAPTER 29

Haemophilus

The genus *Haemophilus* comprises a group of small, pleomorphic, gram-negative rods with fastidious growth requirements. The name of the genus comes from the requirement by these organisms for accessory growth factors found in blood, i.e., *haemo* (Greek for blood) and *philos* (Greek for loving).

Members of the genus are strict parasites for humans and other vertebrates. They exhibit a pronounced host-specificity, and with few exceptions, each species is exclusively associated with one specific host. In humans, they form a part of the indigenous flora of the upper respiratory tract and, sometimes, the vagina. Some species are associated almost exclusively with the disease state.

Historical and Clinical Perspective. Among the *Haemophilus* species, *Haemophilus influenzae* is the most important in human medicine. Although not the cause of epidemic influenza, it is responsible for a number of severe infections in humans. In infants and young children it causes acute bacterial meningitis and several other serious pediatric diseases. In adults it is primarily associated with chronic pulmonary disease.

The organism was first isolated by Pfeiffer during the 1892 influenza pandemic. The frequency of its presence in the nasopharynx of influenza patients and postmortem lung cultures led to the erroneous assumption that it was the etiologic agent of influenza—thus the designation "the influenza bacillus." As was later shown, influenza is caused by a virus. The role of *H. influenzae* during the pandemics of 1890 and 1918 was apparently that of a secondary invader. This type of synergistic interaction was subsequently demonstrated for the influenza virus of swine and a closely related organism, *Haemophilus suis.* However, as *H. influenzae* has not played a similar role in more recent influenza pandemics, the importance of viral–bacterial synergism in human disease is unknown.

Pittman discovered that the strains of *H. influenzae* that cause meningitis and other acute infections differ from those strains found in the respiratory tract of healthy individuals in that they possess capsules. This has provided the basis for our current understanding of *H. influenzae* disease. The experience of the last 50 years has shown that of the six capsular serotypes demonstrated by Pittman, serotype b organisms are responsible for virtually all acute infections caused by *H. influenzae.* These include such invasive diseases as meningitis, pyarthrosis, cellulitis, pneumonia, and acute epiglottitis. Unencapsulated *H. influenzae,* however, are commonly associated with chronic respiratory disease, principally in adults, but also cause systemic infections in immunocompromised patients.

Organisms of Medical Importance. *H. influenzae* type b is the most common cause of acute bacterial meningitis in infancy and childhood, being responsible in the United States for approximately 8000 cases annually in children less than 5 years of age. There is also an increasing awareness of the importance of unencapsulated *H. influenzae* in pulmonary disease.

Other species of *Haemophilus* that may be encountered in clinical material are listed in Table 29-1. Some of these are also known by eponyms: the influenza bacillus of Pfeiffer (*H. influenzae*), the tropical Koch–Weeks bacillus (*Haemophilus aegyptius*), the organism isolated from chancroid by Ducrey (*Haemophilus ducreyi*) and the swine bacillus of Shope (*H. suis*).

Haemophilus influenzae

Morphology

H. influenzae, as all members of the genus, is pleomorphic in appearance. In spinal fluid, joint fluid, or primary cultures of these materials on an enriched medium, the organisms are predominantly coccobacillary and uniform in shape, measuring 0.2 to 0.3 by 0.5 to 0.8 μm (Fig. 29-1). Faint refractile capsules, demonstrable by quellung reactions with type-specific antisera, may be present. *H. influenzae* is gram-negative, but may appear gram-variable unless the staining procedure is very carefully carried out. Unencapsulated organisms from sputum or ear aspirates are often more elongated and may exhibit bipolar staining with gram stain, leading to an erroneous diagnosis of *Streptococcus pneumoniae.* Organisms from rough colonies are very pleomorphic, often appearing as long threads and filaments (Fig. 29-2). The ultrastructure of *H. influenzae* is similar to that of other gram-negative organisms.

Physiology

Cultural Characteristics

Chocolate agar is the most generally used culture medium for the isolation of *Haemophilus* species. In its preparation, blood is added to an agar base and heated at 80C until a brown color appears. The mild heat releases the X and V factors from the blood cells, and also inactivates V factor splitting enzymes without destroying the V factor itself. Also useful for the culture of clinical specimens are Levinthal and Fildes enriched agar media. Because it is colorless and transparent, Levinthal agar is especially useful for differentiating between encapsulated and nonencapsulated strains. Ordinary blood agar allows growth of *H. influenzae* only when the agar plate is cross-inoculated with

TABLE 29-1. PRACTICAL DIFFERENTIATION OF *HAEMOPHILUS* SPECIES

Species	Blood or CSF	Wound or Pus	Sputum or Ear	Eye	Chancroid	Urine	Animal
H. influenzae (encapsulated)	+	+	–	–	–	–	–
H. influenzae (nontypeable)	–	–	+	–	–	±	–
H. aegyptius	–	–	–	+	–	–	–
H. haemolyticus	–	–	+	–	–	–	–
H. ducreyi	–	–	–	–	+	–	–
H. aphrophilus	+	+	–	–	–	–	–
H. parainfluenzae	±	–	+	–	–	–	+
H. parahaemolyticus	±	–	+	–	–	–	+
H. paraphrophilus	±	–	+	–	–	±	–
H. suis	–	–	–	–	–	–	+
H. gallinarium	–	–	–	–	–	–	+

CSF, cerebrospinal fluid; ±, uncommon or unconfirmed association.

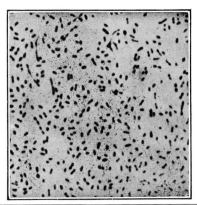

Figure 29-1. Primary growth of *H. influenzae* on chocolate agar.

a *Staphylococcus* or other V factor excreting microorganism. Blood agar is very useful for the detection of *Haemophilus* growing as satellite colonies in cultures of clinical material. Fresh human or sheep blood plates should not be used, however, because they contain heat-labile V factor inhibitors. Maximum growth occurs at 37C and pH 7.4 to 7.8 under aerobic conditions. For primary isolation, incubation in the presence of 10 percent CO_2 is recommended because of its enhancing effect on the growth of some strains.

Growth on semisolid agar is apparent 18 to 24 hours after inoculation. Because a majority of clinical isolates are from respiratory specimens and are nonencapsulated, colonies are usually very small and dew-like, and develop a coarse or rough appearance. These colonies are usually 0.5 to 1.5 mm in diameter. Encapsulated *H. influenzae* from invasive disease (usually type b from blood or cerebrospinal fluid) produces mucoid, glistening colonies that are characteristically iridescent on Levinthal agar. These may reach 3 or 4 mm in diameter on enriched medium. During

Figure 29-2. Unencapsulated *H. influenzae* from rough colony.

the second 24 hours a central umbilication develops due to the excretion of capsular polymer. Mucoid colonies often spontaneously convert to rough colonies due to loss of encapsulation (p. 510).

Cultures of *H. influenzae* are difficult to maintain in the laboratory because of their tendency to autolyze. Proper maintenance of viable virulent organisms requires frequent transfers on chocolate agar or other enriched media. Preservation is best accomplished by lyophilization.

Nutritional Requirements

All species of *Haemophilus* require either one or both of the two growth factors present in blood designated as X and V (Table 29-2). The heat-stable X factor is hemin, the prosthetic group in the iron-containing cytochromes and in heme enzymes such as catalase and peroxidase. Hemin-independent species such as *Haemophilus parainfluenzae* synthesize hemin by the common tetrapyrrole pathway, but hemin-dependent organisms have lost the ability to convert δ-aminolevulinic acid to protoporphyrin. Hemin-dependent species appear to lack all of the enzymes in tetrapyrrole synthesis except the ferrochelatase or heme synthase, which is variably present and catalyzes the final insertion of Fe^{2+}/Fe^{3+} into the protoporphyrin ring.

$$\text{Succinyl CoA} + \text{Glycine} \rightarrow \delta\text{-Aminolevulinate} \rightarrow \text{Porphobilinogen} \rightarrow$$

$$\text{Uroporphyrinogen} \rightarrow \text{Coproporphyrinogen} \rightarrow$$

$$\text{Protoporphyrin IX} \xrightarrow{Fe^{2+} \text{ or } Fe^{3+}} \text{Heme or Hemin}$$

When grown anaerobically, *H. influenzae* shifts to an anaerobic metabolism, and does not produce cytochromes. The hemin requirement is, therefore, greatly reduced if not completely absent. *H. parainfluenzae* has no requirement for an exogenous source of iron, but is metabolically less flexible, and under anaerobic conditions continues to produce and utilize the cytochrome system (p. 52).

The heat-labile V factor is nicotinamide adenine dinucleotide (NAD or NADP), which functions as a coenzyme for pyridine-linked dehydrogenases. Species of *Haemophilus* designated *para-* have a requirement only for the V factor, a differential property used in the laboratory identification of species by the use of discs impregnated with the X and V factors.

Metabolism

Haemophilus is aerobic but facultatively anaerobic. In an oxygen-free environment, nitrate is used as the final electron acceptor. The sole quinone produced by *H. influenzae* is demethylmenaquinone (DMK), but some species also produce ubiquinone. Whereas ubiquinone is used in electron transport systems leading only to high potential acceptors such as oxygen or nitrate, demethylmenaquinone may be used for both aerobic and anaerobic electron transport. The possession of DMK thus enables *H. influenzae* to produce energy both by substrate-level phosphorylation and oxidative phosphorylation.

TABLE 29-2. DIFFERENTIAL PROPERTIES OF SPECIES OF *HAEMOPHILUS**

Species	V Factor Requirement	X Factor Requirement	Increased CO_2 Requirement	Hemolysis
H. influenzae	+	+	−	−
H. aegyptius	+	+	−	−
H. haemolyticus	+	+	−	+
H. ducreyi	−	+	+	(+)
H. aphrophilus	−	+	+	−
H. parainfluenzae	+	−	−	−
H. parahaemolyticus	+	−	−	+
H. paraphrophilus	+	−	+	−
H. suis	+	+	−	−
H. gallinarium	+	+	+	−

*See also Zinnemann and Biberstein: In Buchanan and Gibbons (eds): Bergey's Manual of Determinative Bacteriology, 8th ed. Baltimore, Williams & Wilkins, 1974, p. 365.
(+), delayed.

Biotyping. Strains of *H. influenzae* may be separated into subgroups or biotypes by a battery of biochemical tests based on phenotypic properties of the organism (Table 29-3). The primary use of such schemes is in biotyping individual isolates. In a study of 130 isolates from *H. influenzae* meningitis, 93 percent belonged to biotype II, whereas this biotype is uncommon in the respiratory tract.

Genetics

In *H. influenzae,* a striking correlation exists among the colony type, antigenic structure, and virulence of the organism. Mucoid to rough transitions in colony morphology, as observed when clinical isolates from invasive disease are subcultured on artificial media, reflect loss of specific capsular polysaccharide synthesis by mutation. The spontaneous mutation rate for this property is relatively high and may be increased by suboptimal culture conditions or by the presence of type specific antisera.

The elaboration of a type-specific polysaccharide capsule by *H. influenzae* provides a convenient marker for genetic studies dealing with the transfer of DNA from one organism to another. Transformation and transfection mediated by DNA from the different serotypes and other species of *Haemophilus* have been extensively studied and the enzymology and genetics of these restriction and modification systems characterized. In *H. influenzae,* strain-specific restriction endonucleases recognize DNA from other

strains as foreign. Protection against restriction is provided by a modification methylase that methylates a limited number of adenine or cytosine residues in the recognized nucleotide sequence. Each of the serologic types of *H. influenzae* carries different DNA restriction and modification systems. These systems are powerful tools for the molecular biologist who has successfully used *Haemophilus*-derived enzymes to analyze the structure and function of specific genome fragments.

In *H. influenzae,* plasmids have been implicated in the transfer of antibiotic resistance, significantly affecting the clinical treatment of acute *Haemophilus* infections. The sudden emergence in 1974 of ampicillin resistance in clinical isolates of *H. influenzae* after a decade of ampicillin susceptibility led to the finding of plasmid DNA in resistant strains. Two types of plasmids were found: a large conjugative 30 megadalton (Mdal) plasmid; and a small nonconjugative plasmid with a 3 Mdal molecular weight. Both of these plasmids contain a transposon (TnA) that codes for β-lactamase. This transposon resembles that of the transposable element Tn3 originally found in the Enterobacteriaceae. The cores of the conjugative *H. influenzae* R plasmids isolated in different parts of the world have similar base sequences, but are not identical. Although the small 3 Mdal plasmids are unable to mediate their own conjugal transfer, nonconjugative plasmids may be conjugally transferred if a self-transmissible plasmid is also present in the cell. Such an indigenous *H. influenzae*

TABLE 29-3. BIOCHEMICAL PROPERTIES OF *HAEMOPHILUS INFLUENZAE*

Biotype	Indole	Urease	Ornithine Decarboxylase	Glucose (acid only)	Nitrate Reduction
I	+	+	+	+	+
II	+	+	−	+	+
III	−	+	−	+	+
IV	−	+	+	+	+
V	+	−	+	+	+

Adapted from Kilian: In Starr, Stolp, Trüper, et al. (eds): The Prokaryotes, Vol. II, 1981. New York, Springer-Verlag.

plasmid has been found, supporting the hypothesis that the *H. influenzae* R plasmids could have arisen simultaneously as a result of the integration of the β-lactamase transposon into different but closely related indigenous *H. influenzae* plasmids in various parts of the world. In *H. ducreyi*, a similar plasmid transfer system that will cross species and generic lines has also been demonstrated.

Resistance to chloramphenicol and tetracycline is also plasmid-mediated in *H. influenzae*. Conjugative R plasmids for these drugs are closely related to the ampicillin plasmids, and have most of their base sequences in common, as would be expected from the hypothesized origin of *H. influenzae* plasmids.

Antigenic Structure

H. influenzae contains three major classes of surface antigens: the capsular polysaccharide, lipopolysaccharide, and outer membrane proteins.

Capsular Antigens

The major antigenic determinant of encapsulated *H. influenzae* is the capsular polysaccharide. This polysaccharide confers type specificity on the organism and is the basis for the grouping of the organism into six serotypes, designated a through f. Virtually all of the strains associated with invasive disease belong to serotype b.

The capsular antigens produced by types a, b, c, and f are of the teichoic acid type, whereas those from types d and e are polysaccharides. The distinct chemical specificity of each type is shown in Table 29-4. The type b capsular polymer is unique in that it contains pentose sugars, ribose, and ribitol phosphate, instead of hexoses or hexosamines as found in the other serotypes.

Serologic Capsular Reactions.
Encapsulated *H. influenzae* may be typed by the quellung (capsular swelling) (p. 479) reaction using type-specific antisera. Typing may be done directly on fresh clinical specimens or on primary isolates. Agglutination and fluorescent antibody techniques may also be employed but must be interpreted very carefully, and confirmed with another procedure.

A modification of the capsular antigen precipitation test has been useful as a screening technique and for the identification of bacteria cross reacting with the type b polysaccharide. The technique, as illustrated in Figure 29-3, depends upon the production of immunoprecipitin halos around colonies on agar plates containing antiserum. Using this screening technique a number of organisms in other genera have been shown to have antigens cross-reactive with the type b polysaccharide of *H. influenzae*. Cross-reactions are known to occur between *H. influenzae* type b and a diverse group of gram-negative and gram-positive organisms, including certain strains of pneumococci, streptococci, *Escherichia coli* and *Staphylococcus*, and species of *Lactobacillus* and *Bacillus*. Cross-reactivity between *E. coli*

TABLE 29-4. CAPSULAR POLYSACCHARIDES OF *HAEMOPHILUS INFLUENZAE*

Type	Sugar	PO₄	Acetyl
a	Glucose	+	−
b	Ribose and ribitol	+	−
c	Galactose	+	−
d	Hexose	−	−
e	Hexosamine	−	+
f	Galactosamine	+	+

and *H. influenzae* type b has been attributed to shared specificities of acidic polysaccharides or K antigens, whereas in gram-positive species the cross-reacting antigens are cell wall polyribitol phosphate teichoic acids.

The serotype of free capsular polysaccharide antigen, which is found in culture filtrates and body fluids, may be identified serologically by countercurrent immunoelectrophoresis, latex particle agglutination, hemagglutination, or precipitation test (p. 514). As type b organisms cause over 95 percent of invasive disease, detection of the type b capsular antigen is especially important (p. 514).

Somatic Antigens

The cell envelope of *H. influenzae* consists of an outer and an inner membrane containing protein and lipopolysaccharide (LPS) antigens. The major antigenic component of LPS is its nontoxic polysaccharide fraction. Antibodies to this somatic polysaccharide are age-independent in contrast to the age-dependent responses to the capsular polysaccharide.

Of the two to three dozen proteins detected in the outer membrane of *H. influenzae*, a small number account for most of the protein content of the cell surface. Some

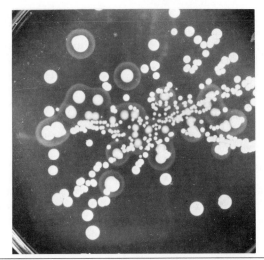

Figure 29-3. Halo of specific immune precipitin surrounding *H. influenzae* type b colonies on agar containing type b antiserum. *(Courtesy of Richard M. Michaels.)*

are consistently present in all strains of *Haemophilus,* whereas others vary with the serotype of capsulated and noncapsulated strains. A system for subclassification of isolates of *H. influenzae* type b from invasive disease based on eight consistent protein patterns has been devised. This scheme has proved useful in preliminary epidemiologic studies and in the identification of invasive strains.

The failure of the existing *H. influenzae* type b capsular vaccine to confer protective immunity against systemic disease on children less than 14 months of age has stimulated study on the potential of the noncapsular immunogens as effective immunizing agents. Especially promising are type b polysaccharide–protein conjugates, which are apparently thymic-dependent and induce an enhanced anti-type b antibody response.

Determinants of Pathogenicity

Capsule

The phosphoribosylribitol phosphate (PRRP) capsule of *H. influenzae* type b plays a critical role in the pathogenesis of invasive disease caused by this organism. Systemic infections are always caused by encapsulated strains and virtually always by those elaborating the type b capsular polysaccharide. In humans, susceptibility to systemic infections is highly correlated with the absence of serum antibodies to the type b capsule. Antibodies to the type b capsule also effectively promote phagocytosis of *H. influenzae* type b in vitro. The reason for the proclivity of type b for bacteremic infections is unknown, but is believed to be attributable to the unique structure of its capsular polysaccharide, which in the absence of specific antibody imparts to the organism an effective resistance to complement (p. 514).

Other Virulence Factors

Outer Membrane Components. Although the type b capsular polysaccharide is the critical determinant of virulence, it is likely that in *Haemophilus* virulence is multifactorial and that other at present ill-defined factors contribute to virulence in different time–site sequences. Among the components most likely to influence the disease process are outer membrane proteins and LPS. Different surface molecules may each be responsible for the various functions associated with virulence such as attachment, invasiveness, and resistance to phagocytosis. The differences that have been found in the outer membrane proteins of type b organisms may reflect a modulation of some of these functions. The importance of noncapsular somatic antigens in the pathogenesis of *H. influenzae* is emphasized by a number of studies in both human and experimental animal systems showing that antibody directed against these antigens may also be important in immunity to *H. influenzae* disease.

Surface somatic antigens of nonencapsulated *H. influenzae* also contribute to the pathogenesis of chronic non-specific lung diseases such as asmatic bronchitis. *H. influenzae* LPS exerts a paralyzing action on the ciliated respiratory epithelium and promotes proliferation of the organism in the bronchial tree. The chemical composition of the lipid A component of *H. influenzae* LPS is similar to that of enterobacterial LPS, but free lipid A does not exhibit all of the classic biologic activities of other lipid A preparations.

Adherence. Little is known about the role of adherence of *H. influenzae* to epithelial surfaces in the pathogenesis of infection. Studies indicate, however, that most (more than 90 percent) nontypeable strains are adherent to human buccal epithelial cells, whereas only a few (5 percent) of type b strains are adherent. These differences may contribute to the differences in colonization between type b and nontypeable strains and may explain the tendency for nontypeable strains to cause localized infection whereas type b strains are associated with invasive disease.

IgA Proteases. *H. influenzae* is one of five bacterial species known to produce IgA proteases—enzymes that have the unique ability to hydrolyze the human IgA_1 heavy chain as their only known substrate. The IgA proteases are neutral endopeptidases, distinguishable from other microbial enzymes in that their cleavage fragments, Fab-α and Fc-α, do not undergo a secondary degradation. As *H. influenzae* primarily infects human mucosal surfaces, whereas host defense is mediated by secretory IgA, cleavage of IgA may contribute to the organism's virulence potential. *H. influenzae* is the only member of the genus that produces this enzyme. Strains of *H. influenzae* produce three distinct types of IgA proteases that cleave different peptide bonds within the IgA_1 hinge region. The type of protease produced correlates with the serotype of the isolate. Each nontypeable strain also produces one of the three protease types.

Clinical Infection

Epidemiology

Unencapsulated *H. influenzae* are commonly carried in the nasopharynx of asymptomatic individuals. Rates of carriage are about 60 to 90 percent for healthy young children and 35 percent for adults. Of the isolates from children, about 5 percent are encapsulated, one half of which are type b. In adults, only 0.4 percent of the isolates are type b.

The frequency of invasive infections is inversely related to age; only a small percentage occur in adults and older children. Infections in the first 2 months of life are rare, probably because of transplacental transfer of maternal antibody (p. 514). Most cases of meningitis, pyarthrosis, and cellulitis occur in children under 2 years of age, and the mean age of children with epiglottitis is 3 to 5 years. Systemic *H. influenzae* in the adult, previously thought to be very uncommon, is now being recognized with increasing frequency. Also, the overall incidence of invasive *Haemophilus* disease has increased four-fold over the last 2 decades, possibly due to improved laboratory techniques for identification.

H. influenzae diseases occur worldwide and are, for the most part, endemic in nature. However, an increased incidence of secondary cases occurs among those susceptible in families and day-care centers who are exposed to an index case. Systemic infections occur more commonly among the poor and the black population. Host factors that appear to contribute to increased susceptibility include immunoglobulin deficiencies, sickle cell disease, splenectomized state, and chronic pulmonary infections. In adults, alcoholism increases the risk of *H. influenzae* pneumonia.

Pathogenesis

Respiratory Portal of Entry.
Infection with *H. influenzae* occurs following the inhalation of infected droplets from clinically active cases, convalescent patients, and carriers. Nonhuman reservoirs of *H. influenzae* are not known. The natural history of infections in children is poorly understood, but clinical experience suggests that the organisms initially colonize the nasopharynx. Both rough and smooth variants of *H. influenzae* are carried at this site. These relatively common asymptomatic infections occasionally develop into symptomatic disease, which may spread contiguously to the sinuses, middle ear, or bronchi. The organisms are established in the tracheobronchial tree, perhaps in synergy with a virus or by paralysis of normal cilia clearance functions. The relevance, however, of viral synergy, toxic bacterial products, and local immunity in producing clinical disease remains to be proved. The strains usually associated with chronic respiratory disease are nonencapsulated and probably do not invade tissues. The presence of *H. influenzae* type b in sputum or ear aspirates may indicate tissue invasion. Respiratory infections are important sources for the seeding of the blood to produce metastatic disease in the meninges or joints, and for the invasion of local tissues to cause epiglottitis, pneumonia, or cellulitis.

Bloodstream Invasion.
The critical pathogenic event for most serious diseases caused by type b organisms is invasion of the organisms from the respiratory mucous membranes and their survival in the blood. Although the precise mechanisms mediating this invasion are ill-defined at present, both bacterial virulence determinants and host resistance factors are involved. The type b capsule is essential for invasiveness. The presence of a type b capsular polysaccharide enables the organism to resist the action of complement, permitting longer survival and eventual multiplication in the blood. The increased incidence of invasive disease in nonimmune and genetically susceptible individuals emphasizes the critical role of host factors in *H. influenzae* infections. Viral synergy may also play an important role in the host–parasite interaction.

Following implantation within the tissues, *H. influenzae* characteristically evokes a nonspecific acute neutrophilic exudation that is rich in fibrin. The heavy plastic nature of the exudate may be important in protecting the organisms against the host defenses.

Clinical Manifestations

In one survey of hospitalized children, *H. influenzae* type b was the most common cause of bacteremic disease. Of those with *H. influenzae* disease, meningitis was the most common manifestation (54 percent), followed by pneumonia (14 percent), bacteremia without focus (11 percent), cellulitis (11 percent), epiglottitis (10 percent), and pericarditis (4 percent).

Meningitis.
The most serious of the diseases produced by *H. influenzae* is acute bacterial meningitis. *H. influenzae* meningitis occurs rarely in infants under the age of 3 months and is uncommon in children over the age of 6 years. Cases have been reported, however, both in neonates and in adults. The distribution of disease is equal in males and females and in races, except for the increased susceptibility associated with sickle cell anemia. Patients with humoral immunodeficiency are especially susceptible. The incidence of disease is approximately 5 per 100,000 population and is reported to be increasing in recent years. Overt symptoms, cerebrospinal fluid pleiocytosis, and positive cultures are often preceded by several days of respiratory symptoms, during which invasion presumably occurs. Clinical and laboratory findings are typical of a pyogenic infection. Therapy can prevent mortality in 90 to 97 percent of cases, but residual central nervous system deficits are demonstrable in one third of patients.

Acute Bacterial Epiglottitis.
This disease, rarely caused by organisms other than *H. influenzae* type b, has an acute onset and a dramatically rapid course. It occurs in children who are older than meningitis patients and even occurs occasionally in adults. The genetic makeup of children with epiglottitis appears to be different from those with *H. influenzae* meningitis. There is a striking predominance of occurrence in whites. The infected epiglottis has microabscesses, and the marked edema may cause complete airway obstruction, requiring emergency tracheotomy within 12 hours of onset. Severe septicemia is often present in this too-often fatal illness.

Cellulitis.
H. influenzae causes cellulitis in children below the age of 2 years and, rarely, in older adults. The most commonly involved site is the cheek, but cellulitis may occur in the periorbital area and other locations, especially the upper extremities. Classically, *H. influenzae* cellulitis of the cheek has an acute onset, and develops rapidly within a few hours accompanied by pain and edema. A distinctive bluish–purple color occurs late in the infection. Because cellulitis due to *H. influenzae* is usually a bacteremic disease, metastatic infection may result.

Bacteremia Without Local Disease. *H. influenzae* is responsible for about 20 percent of the bacteremias that occur in febrile children without any evidence of local disease. Children aged 6 to 36 months with sickle cell disease or previous splenectomy are particularly susceptible. Unsuspected *H. influenzae* bacteremia also occurs in adults with neoplastic disease undergoing chemotherapy. The clinical course may progress to septic shock, often with death, within hours of the initial medical evaluation.

Other Infections. *H. influenzae* type b is a common cause of childhood pyarthrosis, and is associated with pericarditis, usually as a part of a pneumonic episode. Most cases of *H. influenzae* pneumonia are due to type b and occur in very young children. In adults pneumonia occurs more frequently in the elderly in the setting of chronic lung disease, alcoholism, or immunologic deficiency, but it may also develop in previously healthy individuals.

H. influenzae is second in frequency to *S. pneumoniae* as the cause of otitis media. The strains isolated are usually nontypeable, but in 10 percent of the cases type b organisms are associated with the infection. Nontypeable strains often cause acute sinusitis in adults and are frequently associated with purulent sputum and clinical exacerbations of chronic bronchitis.

Immunity

One of the most striking features of *H. influenzae* disease is the relationship between age and susceptibility. The frequency of meningitis is inversely related to the bactericidal activity of the blood, whether passively acquired from the mother or actively formed. Invasive disease occurs during the age of relative humoral immunodeficiency—3 months to 3 years. Although it is now apparent that immunity to *H. influenzae* type b is mediated by antibody of multiple specificities, anticapsular antibody is the factor in serum that correlates with protection. The anticapsular antibody titer of serum varies with age in the same manner as the bactericidal activity. Anticapsular antibody is required for both complement-mediated phagocytosis and bacteriolysis. However, it is unlikely that bacteriolysis occurs in vivo. It is more likely that phagocytosis is the major host defense mechanism.

Antibodies to the type b capsular polysaccharide can be generated by infection with bacteria possessing cross-reacting surface antigens (p. 511). These antibodies are bacteriolytic in vitro and protective in experimental disease. It has been suggested that the incidence of *H. influenzae* type b carriage or disease is too low to account for the rapid and extensive age-related acquisition of anticapsular antibodies, and that cross-reacting organisms may serve as the primary immunogen. Antibodies to *H. influenzae* outer membrane antigens also appear to play a role in immunity. They promote complement-mediated phagocytosis and bacteriolysis and are protective in model systems. In certain adults and older children, resistance to clinical infection has correlated with antibody titers to somatic antigens in the absence of anticapsular type b activity. Therefore, natural resistance to *H. influenzae* is based on a multifactorial antibody response to *H. influenzae* itself as well as to heterologous antigens cross-reactive with the type b capsular polysaccharide.

The antibody response of patients recovering from systemic *H. influenzae* type b disease is age related. Infants respond infrequently and with low antibody levels. Older children and adults develop high titers. This same type of response is observed following immunization with capsular type b polysaccharide, severely limiting the effectiveness of this antigen for the vaccination of children below 18 months of age. Failure to develop a consistent antibody response to natural *H. influenzae* type b exposure or to vaccination with type b capsular polysaccharide is probably related to the young child's immunologic immaturity in processing carbohydrate antigen.

Laboratory Diagnosis

Accurate identification of the causative agent is a prerequisite for the proper management of *H. influenzae* disease.

Direct Examination. Gram-stained smears of clinical specimens are useful in providing a rapid presumptive identification. Specimens suitable for direct examination include cerebrospinal fluid, arthrocentesis, thoracentesis, middle ear aspirates, and sputum samples. However, because of the organism's tendency to retain the gram stain, extreme care must be taken during the staining procedure to decolorize sufficiently or the organism's coccobacillary forms may be erroneously interpreted as pneumococci. The use of carbol fuchsin as the counterstain in the gram stain is recommended.

Culture. *H. influenzae* is very fastidious and dies quickly in clinical materials at room temperature; specimens should thus be planted immediately. They should be streaked directly on the surface of a chocolate agar or other suitable media, and incubated aerobically in an atmosphere of 10 percent CO_2 (p. 508). In addition, blood cultures should be done on every patient believed to have meningitis or other invasive disease.

Antigen Detection. The detection of specific polysaccharide antigen in body fluids is also a valuable diagnostic aid and provides a presumptive diagnosis of *H. influenzae* infection even in the absence of a positive culture. The two techniques that have proved most useful are countercurrent immunoelectrophoresis (CIE) and latex particle agglutination, both of which provide rapid and semiquantitative results. Using CIE, rapid diagnosis can be made in 90 percent of confirmed *H. influenzae* type b meningitis and in a majority of other infections caused by this organism. For patients who have received antibiotic therapy prior to lumbar puncture—and who often pose a more complicated diagnostic challenge because gram stain and culture may be negative—the CIE may be very helpful.

Treatment

Treatment of invasive *H. influenzae* disease, especially meningitis and epiglottitis, is a medical emergency. Any unnecessary delay in starting treatment may be the difference between life and death, or, in the case of meningitis, between a normal and a brain-damaged child or adult.

About 8 percent of all strains of *H. influenzae* isolated in this country from systemic infections are now resistant to ampicillin (p. 198). The incidence of ampicillin resistance among nonencapsulated strains is even greater. Given this prevalence of resistant strains, all systemic illnesses suspected of being *H. influenzae* in origin should be treated with chloramphenicol alone or in combination with ampicillin until the etiologic agent is proved to be sensitive to ampicillin.

Ampicillin or amoxicillin, each of which is active against *S. pneumoniae* and most strains of *H. influenzae* is the drug of choice for initial treatment of otitis media in children. Alternatives include trimethoprim-sulfamethoxazole, the combination of penicillin (or erythromycin) with a sulfonamide, or cefaclor. As ampicillin-resistant strains rarely cause *H. influenzae* pneumonia in the adult, initial treatment of these infections should be with ampicillin. Ampicillin is also a suitable initial antibiotic choice for acute sinusitis.

Passive Immunotherapy. Monoclonal antibody to the capsular polysaccharide of *H. influenzae* type b has been produced by the technique of somatic cell hybridization. The antibody is of the IgM class and is bactericidal in vitro with complement. Although still in the experimental stage, results in animal models suggest that passive immunotherapy with monoclonal antibody as an adjunct to antibiotics, will have a future role in the treatment of *H. influenzae* disease.

Prevention

During the past 2 decades there has been intensive investigation into the development of a vaccine against *H. influenzae* type b. Because absence of antibody to the capsular polysaccharide polyribosylribitol phosphate (PRP) correlates clinically with susceptibility to the disease, PRP vaccines have been prepared and extensively tested. The PRP vaccine is well tolerated, produces a good antibody response, and is protective against systemic type b infections in older children. However, for children below the age of 18 months—when incidence of *H. influenzae* meningitis is greatest—the PRP vaccine is a poor immunogen and offers no protection. New approaches have been initiated to develop an effective vaccine for this younger age group. Vaccines in which PRP is coupled to protein or combined with pertussis vaccine are currently being tested.

The significant increased rate of secondary cases of invasive *H. influenzae* diseases among young children who have intimate contacts with primary cases indicates the need for an effective prophylactic program. No specific antibiotic regime, however, has proved reliably effective.

Monoclonal antibody to the capsular polysaccharide has the potential of providing protection against invasive disease in this high-risk group of children.

Other Haemophilus Species

Human disease caused by species of *Haemophilus* other than *H. influenzae* have been considered rare. Because of recent interest in these organisms and improved techniques for their isolation and identification, it is apparent that they are more frequent causes of infection than previously thought.

Although the species listed in Table 29-2 include those currently accepted at the time of publication of the Eighth Edition of Bergey's Manual, extension and refinement of taxonomic criteria has led to considerable uncertainty as to the validity of certain *Haemophilus* species. *H. aegypticus* is indistinguishable from biotype III of *H. influenzae*, and *Haemophilus parahemolyticus* has been reassigned to the *H. parainfluenzae* species.

Haemophilus aegyptius (Koch–Weeks Bacillus)

This species is associated with a communicable purulent conjunctivitis, especially in children in hot climates. Because of difficulties in differentiating *H. aegyptius* from *H. influenzae*, which also may cause conjunctivitis, the natural history of *H. aegypticus* infections is poorly understood. However, *H. aegypticus* appears to be associated with a more acute form of conjunctivitis and, in contrast to *H. influenzae*, can colonize eyes without any predisposing condition being present.

H. aegyptius has been differentiated from *H. influenzae* on the basis of its hemagglutinating ability, but this property is not universally characteristic of all strains causing acute conjunctivitis. Also, authentic and freshly isolated *H. aegyptius* strains are notably more fastidious than *H. influenzae*, and differ from it in a number of additional properties such as the failure to produce indole. It can be differentiated serologically. The conjunctivitis usually responds to topically applied sulfonamides.

Haemophilus parainfluenzae

This species is part of the normal flora of the mouth and nasopharynx. Clinical infection is the result of local or bloodstream invasion from these sites usually following dental disease, dental procedures, or other oral trauma. Other predisposing factors include respiratory tract infections, alcoholism, and other conditions that compromise host defense mechanisms. The most common *H. parainfluenzae* infection is endocarditis. Approximately 5 percent of all infective endocarditis is caused by this organism. Known preexisting cardiac disease such as congenital or rheumatic heart disease is present in about 50 percent of the cases. Most cases are young or middle-aged adults. The recommended treatment is ampicillin alone or in combination with gentamicin. *H. parainfluenzae* is also a rare cause of meningitis, epiglottitis, otitis media, bacteremia, brain abscess, and pneumonia in the adult. Ampicillin is the drug of choice unless resistance requires use of an alternate regimen employing chloramphenicol.

Haemophilus aphrophilus

This species is a part of the normal gingival flora, and an infrequent cause of disease. Most of the strains associated with infection have been cultured from the blood of patients with a damaged endocardium, congenital heart disease, secondary brain abscess, or otherwise compromised defenses. These infections often follow oropharyngeal foci of infection or trauma. Many strains are resistant to ampicillin, but are sensitive to chloramphenicol and gentamicin. For endocarditis, ampicillin in combination with gentamicin is recommended.

H. aphrophilus is a very fastidious microaerophilic organism, requiring an increased level of CO_2 for growth. It is very similar to *Actinobacillus actinomycetemcomitans*, but the relationship of the two organisms has not been sufficiently established to require reclassification.

Haemophilus ducreyi

This organism is the cause of chancroid, a sexually transmitted disease worldwide in distribution. In temperate climates, *H. ducreyi* may be responsible for up to 10 percent of venereal disease in civilian populations, but during periods of war chancroid may be nearly as great a problem as gonorrhea. In civilian populations, it is most common among nonwhite men, and is usually associated with poor socioeconomic and hygienic conditions.

Following exposure, there is a 2- to 14-day incubation period before the appearance of a single or multiple lesions that develop into sharply circumscribed, nonindurated, painful ulcers. These are usually confined to the genitalia and perianal areas, and are rarely accompanied by systemic symptoms. Suppurative inguinal buboes are characteristic and develop in about one half of the patients. Oral sulfonamide is the treatment of choice. When resistant organisms are encountered, either streptomycin or kanamycin may be used.

A laboratory diagnosis is made by finding *H. ducreyi* in gram-stained smears of ulcer exudate or bubo aspirate. The organisms are pleomorphic and may occur both extra- and intracellularly. The classic microscopic appearance is that of a school of red fish, but interpretation of smears may be very difficult because organisms in fresh smears may appear gram-positive, or other organisms may be mistaken for *H. ducreyi*. Culture of material from the ulcer or bubo should be attempted. Although primary isolation may be difficult, a positive culture provides a definitive diagnosis. The best culture results have been obtained with a chocolate agar medium containing vancomycin.

FURTHER READING

Books and Reviews

Geiseler PJ, Nelson KE, Levin S, et al.: Community-acquired purulent meningitis: A review of 1,316 cases during the antibiotic era, 1954–1976. Rev Infect Dis 2:725, 1980

Hammond GW, Slutchuk M, Scatliff J, et al.: Epidemiologic, clinical, laboratory, and therapeutic features of an urban outbreak of chancroid in North America. Rev Infect Dis 2:867, 1980

Kilian M, Frederiksen W, Biberstein EL (eds): *Haemophilus, Pasteurella, and Actinobacillus.* Proceedings of an International Symposium, Copenhagen, Denmark. New York, Academic Press, 1981

Robbins JB, Schneerson R, Argaman M, et al.: *Haemophilus influenzae* type b: Disease and immunity in humans. Ann Intern Med 78:259, 1973

Sell SHW, Karzon DT (eds): *Haemophilus influenzae.* Proceedings of a Conference on Antigen–Antibody Systems, Epidemiology and Immunoprophylaxis. Nashville, Vanderbilt University Press, 1973

Turk DC, May RF: *Haemophilus influenzae*: Its clinical importance. London, English Universities Press, 1967

Wallace RJ Jr, Baker CJ, Quinones FJ, et al.: Nontypable *Haemophilus influenzae* (biotype 4) as a neonatal, maternal, and genital pathogen. Rev Infect Dis 5:123, 1983

Selected Papers

Alexander HE, Ellis C, Leidy G: Treatment of type-specific *Haemophilus influenzae* infections in infancy and childhood. J Pediatr 20:673, 1942

Barenkamp SJ, Munson RS Jr, Granoff DM: Subtyping isolates of *Haemophilus influenzae* type b by outer membrane protein profiles. J Infect Dis 143:668, 1981

Berk SL, Holtsclaw SA, Wiener SL, et al.: Nontypeable *Haemophilus influenzae* in the elderly. Arch Intern Med 142:537, 1982

Bradshaw M, Schneerson R, Parke JC, et al.: Bacterial antigens cross reactive with the capsular polysaccharide of *Haemophilus influenzae* type b. Lancet 1:1095, 1971

Crisel RM, Baker RS, Dorman DE: Capsular polymer of *Haemophilus influenzae*, type b. 1. Structural characterization of capsular polymer of strain eagan. J Biol Chem 250:4926, 1975

Degré M, Solber LA: Synergistic effect in viral-bacterial infection. Acta Pathol Microbiol Scand B 79:129, 1971

Deneer HG, Slaney L, MacLean IW, et al.: Mobilization of nonconjugative antibiotic resistance plasmids in *Haemophilus ducreyi.* J Bacteriol 149:726, 1982

Denny F: Effect of a toxin produced by *Haemophilus influenzae* on ciliated epithelium. J Infect Dis 219:93, 1974

Edwards EA, Huehl PM, Pechinpaugh RO: Diagnosis of bacterial meningitis by counter-immunoelectrophoresis. J Lab Clin Med 80:449, 1972

Elwell LP, Roberts M, Falkow S: Common β-lactamase-specifying R plasmid isolated from the genera *Haemophilus* and *Neisseria.* In Schlessinger D (ed): Microbiology 1978. Washington, D.C., American Society for Microbiology, 1978, p 255

Fothergill LD, Wright J: Influenzal meningitis: The relationship of age incidence to the bactericidal power of blood against the causal organism. J Immunol 24:273, 1933

Gigliotti F, Insel RA: Protection from infection with *Haemophilus influenzae* type b by monoclonal antibody to the capsule. J Infect Dis 146:249, 1982

Granoff DM, Rockwell R: Experimental *Haemophilus influenzae* type b meningitis: Immunological investigation of the infant rat model. Infect Immun 20:705, 1978

Guenounou M, Raichvarg D, Hatat D, et al.: In vitro immunological activities of the polysaccharide fraction from *Haemophilus influenzae* type a endotoxin. Infect Immun 36:603, 1982

Hunter KW Jr, Hemming VG, Fischer GW, et al.: Antibacterial activity of a human monoclonal antibody to *Haemophilus influenzae* type b capsular polysaccharide. Lancet II:798, 1982

Insel RA, Anderson PW Jr: Cross-reactivity with *Escherichia coli* K100 in the human serum anticapsular antibody response to *Haemophilus influenzae* type b. J Immunol 128:1267, 1982

Insel RA, Anderson P, Loeb MR, et al.: A polysaccharide-protein complex from *Haemophilus influenzae* type b. II. Human antibodies to its somatic components. J Infect Dis 144:521, 1981

King SD, Wynter H, Ramlal A, et al.: Safety and immunogenicity of a new *Haemophilus influenzae* type b vaccine in infants under one year of age. Lancet II:705, 1981

Lampe RM, Mason EO Jr, Kaplan CL, et al.: Adherence of *Haemophilus influenzae* to buccal epithelial cells. Infect Immun 35:166, 1982

Lee CJ, Malik FG, Robbins JB: The regulation of the immune response of mice to *Haemophilus influenzae* type b capsular polysaccharide. Immunology 34:149, 1978

Laufs R, Riess F-C, Jahn G, et al.: Origin of *Haemophilus influenzae* R factors. J Bacteriol 147:563, 1981

Loeb MR, Smith DH: Outer membrane protein composition in disease isolates of *Haemophilus influenzae*: Pathogenic and epidemiological implications. Infect Immun 30:709, 1980

Loeb MR, Smith DH: Human antibody response to individual outer membrane proteins of *Haemophilus influenzae* type b. Infect Immun 37:1032, 1982

Moxon ER, Murphy PA: *Haemophilus influenzae* bacteremia and meningitis resulting from survival of a single organism. Proc Natl Acad Sci USA 75:1534, 1978

Moxon ER, Vaughn KA: The type b capsular polysaccharide as a virulence determinant of *Haemophilus influenzae*: Studies using clinical isolates and laboratory transformants. J Infect Dis 143:517, 1981

Mulks MH, Kornfeld SJ, Frangione B, et al.: Relationship between the specificity of IgA proteases and serotypes in *Haemophilus influenzae*. J Infect Dis 146:266, 1982

Peter G, Smith DH: *Haemophilus influenzae* meningitis at the Children's Hospital Center in Boston, 1958–1973. Pediatrics 55:523, 1975

Pincus DJ, Morrison D, Andrews C, et al.: Age-related response to two *Haemophilus influenzae* type b vaccines. J Pediatr 100:197, 1982

Pittman M: Variation and type specificity in the bacterial species *Haemophilus influenzae*. J Bacteriol 59:413, 1950

Raichvarg D, Guenounou M, Brossard C, et al.: Characteristics of a lipid preparation (lipid A) from *Haemophilus influenzae* type a lipopolysaccharide. Infect Immun 33:49, 1981

Robbins JB: Acquisition of "natural" and immunization-induced immunity to *Haemophilus influenzae* type b disease. In Schlessinger D (ed): Microbiology 1975. Washington, D.C., American Society for Microbiology, 1975, p 400

Robbins JB, Parke JC, Schneerson R, et al.: Quantitative measurement of "natural" and immunization-induced *Haemophilus influenzae* type b capsular polysaccharide antibodies. Pediatr Res 7:103, 1973

Scheifele SW, Fussell SJ: Frequency of ampicillin-resistant *Haemophilus parainfluenzae* in children. J Infect Dis 143:495, 1981

Schneerson R, Barrera O, Sutton A, et al.: Preparation, characterization, and immunogenicity of *Haemophilus influenzae* type b polysaccharide–protein conjugates. J Exp Med 152:361, 1980

Setlow JK, Notani NK, McCarthy D, et al.: Transformation of *Haemophilus influenzae* by plasmid RSFO885 containing a cloned segment of chromosomal deoxyribonucleic acid. J Bacteriol 148:804, 1981

Shenep JL, Munson RS Jr, Granoff DM: Human antibody responses to lipopolysaccharide after meningitis due to *Haemophilus influenzae* type b. J Infect Dis 145:181, 1982

Shope RE: The influenza of swine and man. Harvey Lect 36:183, 1935

Sutter VL, Finegold SM: *Haemophilus aphrophilus* infections: Clinical and bacteriological studies. Ann NY Acad Sci 174:468, 1970

Sutton A, Schneerson R, Kendall-Morris S, et al.: Differential complement resistance mediates virulence of *Haemophilus influenzae* type b. Infect Immun 35:95, 1982

Tarr PI, Hosea SW, Brown EJ, et al.: The requirement of specific anticapsular IgG for killing of *Haemophilus influenzae* by the alternative pathway of complement activation. J Immunol 128:1772, 1982

Zamenhof S, Leidy G, Fitzgerald PL, et al.: Polyribosephosphate, the type-specific substance of *Haemophilus influenzae* type b. J Biol Chem 203:695, 1953

Bordetella

The clinical syndrome of whooping cough or pertussis has been traced to a classic description given in the latter part of the sixteenth century. Paroxysmal coughing has been the hallmark of this acute bacterial infection of the respiratory tract. The severity of the illness prompted early investigative work, and the causative organism, *Bordetella pertussis*, was first isolated in 1906 by Bordet and Gengou. Subsequently, occasional instances of similar illness were attributed to *Bordetella parapertussis* and *Bordetella bronchiseptica*. In the United States, widespread use of standardized vaccine has resulted in a dramatic decline in the incidence of disease and its attendant morbidity and mortality. As pertussis became a rare disease, investigative interest in the organism and the pathogenesis of clinical disease declined concomitantly. A resurgence of laboratory research and renewed interest in the development of less reactogenic vaccine for immunization has occurred with the technological advances of the past decade.

Morphology

The three members of the genus *Bordetella* are *B. pertussis*, *B. parapertussis*, and *B. bronchiseptica*. The organisms are small, gram-negative coccobacilli measuring 0.2 to 0.3 μm by 0.5 to 1.0 μm that appear singly, in pairs, and in small clusters. Upon primary isolation, cells are uniform in size, but in subcultures they become quite pleomorphic, and filamentous and thick bacillary forms are common. Bipolar metachromatic staining may be demonstrated with toluidine blue. The only motile member of the genus is *B. bronchiseptica*, which possesses lateral flagella. Capsules are produced but can be demonstrated only by special stains and not by capsular swelling.

Physiology

Bordetella organisms are strict aerobes with a metabolism that is respiratory, never fermentative. They do not produce H_2S, indole, or acetylmethylcarbinol. The characteristics presently used to differentiate the three species of *Bordetella* are summarized in Table 30-1, and the antigenic differences are discussed on page 520. Recently, however, it has been suggested that the three species of *Bordetella* might be more accurately classified as a single organism. This concept is supported by DNA–DNA reassociation reactions, showing a very close genetic relationship between *B. pertussis* and each of the other two organisms.

    Unlike *Haemophilus* species, *Bordetella* organisms have no specific growth requirement for hemin (X factor) and coenzyme I (V factor). Primary isolation does require, however, the addition of charcoal, ion-exchange resins, or 15 to 20 percent blood to neutralize the growth-inhibiting

TABLE 30-1. DIFFERENTIAL CHARACTERISTICS OF SPECIES OF GENUS *BORDETELLA*

	B. pertussis	*B. parapertussis*	*B. bronchiseptica*
Motility	−	−	+
Reduces nitrate	−	−	+
Utilizes citrate	−	+	+
Produces urease	−	+	+
Growth on peptone agar	−	+	+
Browning of peptone agar	−	+	−
Growth on Bordet–Gengou agar	3–4 days	1–2 days	1–2 days
Litmus milk alkaline	−, (12–14)*	1–4 days	1–4 days
G + C content†	61	61	66

Modified from Buchanan and Gibbons (eds.): Bergey's Manual of Determinative Bacteriology, 8th ed, 1974, p 283. Baltimore, Md, Williams & Wilkins.
+, all strains positive; −, all strains negative.
*Modulated phase II to IV organisms (+) in 12 to 14 days.
†Mole percentage of guanine plus cytosine, Tm.

effects of such substances as unsaturated fatty acids, colloidal sulfur, sulfides, or peroxides. Modified Bordet–Gengou medium (potato-glycerol-blood agar) is recommended for this purpose. Colonies of *B. pertussis* on this medium are smooth, convex, glistening, almost transparent, and pearl-like in appearance. All three species produce a zone of hemolysis that varies with cultural conditions.

B. pertussis freshly isolated from patients in the catarrhal stage of pertussis are smooth colony-forming organisms (phase I or X mode). Adaptation by passage to other media, such as blood or chocolate agar, results in irreversible transition through intermediate forms (phases II and III) to the rough colony-producing form (phase IV). This phase variation is presumably a result of genetic alteration of the organism. Recent studies of the physiologic and biochemical changes accompanying phase transition are listed in Table 30-2.

Although some properties of phase I and phase IV organisms remain indistinguishable, several characteristics have been lost by phase IV organisms. These are lost in parallel with the ability to induce sensitivity to histamine, protective antibodies in animals, and lymphocytosis. The loss of these properties correlates with the loss of virulence for animals and thus provides additional markers for strain degradation. The mechanisms of phase transition or degradation are not understood but plasmid functions do not appear to be involved. It is theoretically possible that some outer membrane proteins form pores and their absence in phase IV cells provides the observed resistance to toxic substances in the medium.

The term cultural or antigenic modulation has been used to describe a change in phenotype of the organism that occurs in almost all members of a population of *B. pertussis* as a result of environmental features, for example, medium containing a high level of $MgSO_4$ (Table 30-2). The modulation is readily reversible, and a single colony may undergo the transition from phase I or X mode to C mode. Phenotypically, the C mode organisms are like phase IV organisms and have lost the properties enumerated above that are correlated with virulence for animals. Both phase variation and cultural modulation result in altered antigenicity as measured by the agglutinogen.

TABLE 30-2. BIOLOGIC PROPERTIES OF PHASE I AND IV *BORDETELLA PERTUSSIS*

Unaltered by Phase Transition I to IV	Altered by Phase Transition I to IV
DNA homology	*Diminished to absent outer membrane proteins a, b, d1, and d2
Plasmid components	*Absent cytochrome d-629
Fatty acids	Increased resistance to antibiotics including penicillin, erythromycin, and tetracycline

Enzymes
  Superoxide dismutase (high levels)
  Catalase (variable)
  Peroxidase (none detected)
	No adenylate cyclase synthesis

Adapted from Dobrogosz et al.: International Symposium on Pertussis. Bethesda Md, 1978, pp 86–93.
*Conversion of phase I to C mode with 20 mM $MgSO_4$ produces the same alterations.

Antigenic Structure

The single, heat-stable surface O antigen common to smooth strains of *B. pertussis*, *B. parapertussis*, and *B. bronchiseptica*, and to rough strains of *B. pertussis* and *B. bronchiseptica*, is a protein easily extractable from cells. It is found in the supernatant fluids of cell cultures, but does not confer protection against infection.

The antigenic differences among species and among strains of each of the species are determined by the heat-labile or capsular antigens of Kauffmann. The serotype is often indicated by numbers, e.g., *B. pertussis* 1.2.4. The existence of 14 K antigens, designated as factors, has been demonstrated on the basis of agglutinin absorption tests (Table 30-3). This scheme explains most of the observed serologic relationships. Factors 1 through 6 are found only in strains of *B. pertussis*. Factor 7 is common to all strains of the three species of *Bordetella* organisms. Factor 14 is specific for *B. parapertussis* and factor 12 is specific for *B. bronchiseptica*.

Although the serotypes per se of *B. pertussis* are not significant determinants of the severity of disease or of protection against infection, these antigens have been essential in vaccine production in providing a method for the assay of alterations occurring with phase variation and cultural modulation of strains. Factor 1 antigen is present in all strains of *B. pertussis*, and is probably the agglutinating antigen (agglutinogen) of the organism. Isolated agglutinogen is nontoxic but does not protect animals against *B. pertussis* infection. Agglutinins or antibodies to agglutinogen are a measurable response in immunized persons, or in those sustaining natural infection. These antibodies are a crude and indirect assessment of immunity because alone they are not protective and only titers of 1:320 or greater correlate with protection from disease.

Other antigenic components of *Bordetella* organisms that are probably responsible for inducing the protective antibody response are discussed in the following section.

Determinants of Pathogenicity

B. pertussis is a pathogenic organism with unique properties. The antigenicity and multiple biologic activities of the bacterium have long been recognized but only the recent resurgence of investigation has begun to elucidate the pathogenesis of disease. Pittman proposed that *B. pertussis* invasion of the respiratory tract be considered the essential prerequisite to "exotoxin" formation and disease production by this organism. Pertussis "toxin," analogous to other bacterial exotoxins, would then be responsible for the systemic manifestations of disease. Other antigens, including the O and K (discussed above), hemagglutinating, heat-labile toxin, and lipopolysaccharide (endotoxin), are also involved in disease production and/or immunity.

Pertussis Toxin: Histamine-sensitizing Factor, Lymphocytosis-promoting Factor, and Islet-activating Protein. The histamine-sensitizing factor (HSF) has been known for years, but only with the purification of a single protein from the *B. pertussis* envelope has it become clear that this one protein with a molecular weight of approximately 73,000 to 77,000 is also the lymphocytosis-promoting factor (LPF) and the islet-activating protein (IAP). The protein diffuses into the culture medium and, when treated with formaldehyde, loses its biologic activity but not its antigenicity. The homogeneous protein is thermostable and is composed of four polypeptide subunits. The purified protein can induce histamine sensitization, hypoglycemia, inability to respond to epinephrine, and leukocytosis.

In an experimental mouse model, the toxin protein produces a dose-dependent primary sensitization to histamine that is reproducible and uniform under standard conditions. There is no generalized β-receptor blockade by the protein.

This protein can also measurably alter, for periods longer than 1 month, the insulin secretory response of rats to subsequent glucose loading. Treatment with 8 M urea dissociates the protein into three subunits (F_1, F_2, and F_3), with molecular weights of 44,000, 20,000, and 11,000, respectively. Biologic activity is present only when F_3 is combined with either F_1 or F_2, but not with any of the three fragments alone or in combination with both F_1 and F_2. The contribution of altered carbohydrate metabolism to the clinical manifestations of *B. pertussis* infection has not been determined.

The striking lymphocytosis observed in association with clinical pertussis has been duplicated in the mouse with the HSF–LPF–IAP protein. It has been proposed that lymphocyte migration from small vessels is hindered by the absorption of the protein onto lymphocyte surfaces. The entrapment of lymphocytes in the vascular and lymphatic compartments creates the lymphocytosis. Also, LPF is known to be a polyclonal activator of human T lymphocytes. It is a potent T cell mitogen in vitro at concentrations in excess of those producing lymphocytosis in vivo. Experiments with mice indicate that the population of T cells stimulated is not the same as that stimulated by phytohemagglutinin. The significance in human infection of the mitogenicity of LPF is unknown.

Specific antiserum can prevent or block all three of these biologic activities. Clinical and experimental data demonstrate that after toxin binds to cells, the biologic activities are not affected by antiserum. In vitro, the biologic activities of this toxin are associated with the transfer of an adenosine diphosphate (ADP)-ribose moiety from nicotinamide adenine dinucleotide (NAD) to a target protein in the membrane of the host cell in a reaction analogous to that of diphtheria and cholera toxins, but with a different target protein and different effects.

Antibody to this protein can protect mice against lethal challenge with *B. pertussis* intranasally, intraperitoneally, and probably intracerebrally. Because of the toxicity of this protein, its use as an immunogen would require toxoid preparation. Many of the nonrespiratory tract symptoms of *B. pertussis* infection will undoubtedly be found to be attributable to this toxin and it is reasonable to predict that circulating antibody will neutralize toxin or inhibit its attachment to cells in humans, as it has been found to do in the animal model.

TABLE 30-3. SUMMARY OF HEAT-LABILE K ANTIGENS* OF *BORDETELLA*

Species	Common Antigen	Species-Specific Antigen	Other Antigens Present
B. pertussis	7	1	2–6
B. parapertussis	7	14	8–10
B. bronchiseptica	7	12	8–11

Adapted from Eldering et al.: J Bacteriol 74:135, 1957.
*Antigens often referred to as factors.

Hemagglutinins. *B. pertussis* has two hemagglutinins, one of which is a filamentous protein with an estimated molecular weight of 130,000. This hemagglutinin (F-HA) is derived from fimbriae on the surface of the organism. The host cell receptor for this hemagglutinin is thought to be the cholesterol of the cell membrane. The organisms adhere to the cilia of respiratory epithelial cells, to ependymal cells in the mouse, as well as to erythrocytes. Antibody to F-HA protects mice against lethal aerosol or intracerebral challenge with *B. pertussis*. This assay is the single best predictor of *B. pertussis* vaccine efficacy in children. Antibody directed against F-HA should provide protection against infection by preventing the adherence of these bacteria to respiratory epithelial cells. Although local or respiratory tract antibody should be more important in providing this protection than humoral antibody, infant mice receiving only passive humoral antibody have been protected from subsequent intranasal challenge.

The second hemagglutinin, HSF–LPF–IAP–HA, or toxin HA, is a round molecule and has only one twentieth of the hemagglutinating activity of the F-HA. There is no recognized role in clinical disease of the hemagglutinating properties of this protein. This hemagglutinin adheres to sialic acid-containing receptors and is expressed only by phase I organisms. The hemagglutinin portion of the molecule may therefore be important in the adherence of toxin to cells. Sialoproteins including haptoglobin and ceruloplasmin can compete for the toxin HA, thereby inhibiting attachment to cell receptors in the respiratory tract.

Heat-Labile Toxin. The heat-labile toxin (HLT) is considered to be a cytoplasmic protein and may occur in bacteria as a precursor requiring activation to induce toxicity. A homogeneous protein form of HLT has not been isolated. Released by cell lysis, HLT is destroyed when heated to 56C for 15 minutes. It is dermonecrotic and when given intraperitoneally or intravenously is lethal for mice. It is a poor antigen unless converted to toxoid by formaldehyde treatment of lysed cells (not intact cells). Toxoid-stimulated antibody does not protect mice against intracerebral challenge or children against infection. HLT is not known to stimulate antibody production in humans and its role in the pathogenesis of human illness is unknown.

Lipopolysaccharide (Heat-Stable Toxin). The lipopolysaccharide (LPS) or endotoxin of the cell wall is heat stable and basically similar to the endotoxins of Enterobacteriaceae except for differences in macromolecular structure. It consists of two different polysaccharides, each terminated by a molecule of 3-deoxy-2-octulosonic acid. Two distinct lipid fragments, lipid A and lipid X, are present and contain glucosamine, fatty acids, and esterified phosphate in similar proportions. Lipid X, which is the minor lipid, has 2-methyl, 3-hydroxydecanoic, and tetradecanoic acids that are absent from lipid A. Lipid X appears to be responsible for the acute toxicity of this endotoxin. The LPS does not induce the formation of antibodies with protective activity.

Antibody to LPS in the presence of complement determines the bactericidal activity of serum against *B. pertussis*. Bactericidal activity, however, is not correlated with protection against intracerebral challenge of mice with *B. pertussis*.

Adenylate Cyclase. *Bordetella* species possess an extracytoplasmic adenylate cyclase that is activated by calmodulin, the eucaryotic cell calcium-dependent regulatory protein. Its role in pathogenesis of disease remains to be delineated.

Experimental In Vitro and In Vivo Models. Some of the most useful synthesis of current data regarding the pathogenesis of *B. pertussis* infection comes from ultrastructural analysis of an in vivo model consisting of hamster tracheal organ culture. Phase I organisms, but not phase IV organisms, selectively adhere to the ciliated epithelial cells (Fig. 30-1). Bacterial attachment is essential for production of diminished ciliary activity, which is the first demonstrable effect of *B. pertussis* infection. The ciliated cells are then extruded from the epithelial surface, with subsequent necrosis. The fimbrial hemagglutinin is implicated in the attachment of organisms to the ciliated cells. The subsequent cell damage is not a result of cell invasion by *B. pertussis* and may be attributed to exotoxin release and adherence to cells.

Mice provide an in vivo model that allows study of the host response to infection. The pathology of experimental *B. pertussis* respiratory infection in mice is similar to that of infants, as is the duration of excretion of bacteria, the lymphocytosis, and the higher mortality in infant mice. Moreover, the interval after infection to the onset of histamine sensitization and the persistence of sensitization parallel the catarrhal and paroxysmal stages in the child. Mice that recover from respiratory infection are then resistant to intracerebral challenge. Sera obtained from mice or infants more than 4 weeks after onset of illness can provide protection to the homologous species.

Clinical Infection

Epidemiology

Several features of the epidemiology of pertussis have intrigued students of this disease for a number of years. Pertussis has not usually been a disease with marked seasonal variations in incidence, in contrast with several other childhood infections. The disease is worldwide in its distribution, and the number of reported cases in the United States has declined from approximately 120,000 in 1950 to 1730 in 1980 (Fig. 30-2). Reported deaths declined from approximately 1100 in 1950 to 12 in 1970. Twenty-six percent of cases and most deaths occur in infants under 1 year of age (Fig. 30-3), and the decline in fatalities over the past several decades has been associated with a gradual increase in the age of patients sustaining infection.

The illness is highly communicable, as evidenced by attack rates of 90 percent in unimmunized household con-

Figure 30-1. Scanning electron micrograph of *B. pertussis*-infected hamster tracheal organ culture. Unciliated cells covered with microvilli are adjacent to parasitized ciliated cells. Rod-shaped bacteria are attached to the cilia. × 6000. (*Courtesy of Dr. Kenneth E. Muse.*)

tacts of persons with pertussis. Humans are the only known source of *B. pertussis,* and excretion of organisms is limited almost entirely to persons with active infection. Immunization seems to have altered the epidemiology somewhat. In recent epidemics, persons with modification or absence of clinical illness have been shown to excrete *B. pertussis.* These persons are thought to be partially immune as a result of prior immunization, suggesting that asymptomatic carriers may be more common in the vaccine era. Prolonged presence of organisms during convalescence is extremely rare.

Pathogenesis of Pertussis

Following inhalation of infected droplets, the organisms colonize the respiratory tract. The specificity of attachment of *B. pertussis* to ciliated respiratory epithelial cells is attributable to the F-HA. Such adherence is essential for production of disease, as specific antibody can prevent damage to ciliated cells in vitro or prevent disease in an animal model. In the mouse, development of local antibody against the F-HA has been shown to occur after natural infection. Also, parenteral administration of F-HA induces systemic anti–F-HA antibodies that can protect mice against disease following lethal aerosol challenge. It is presumed that the incubation period and initial mild symptoms of rhinitis, cough, sneezing, and sometimes conjunctivitis are caused by local multiplication of the organisms in the respiratory tract. Diminished ciliary activity, as observed in vitro, would result in poorer clearance of bacteria and secretions with their resulting accumulation in the respiratory tract. Multiplication of organisms and local toxin production would be facilitated and toxin would

then contribute to necrosis and sloughing of ciliated cells. It is also tempting to speculate that the lack of bacteremia and invasion of other tissues by *B. pertussis* is related to the lack of receptors for the organism on other cell types. It is known that bacterial multiplication in other tissue does not occur, although the reason for this has not been established. The systemic manifestations of disease are most likely due to circulating HSF–LPF–IAP or toxin. The paroxysmal coughing, central nervous system manifestations, and even the rarely observed hypoglycemia, as well as the leukocytosis and lymphocytosis, may be attributable to the effects of exotoxin. The persistence of cough and lymphocytosis is a result of the fixation of toxin to cells. Additionally, bacteria are less readily detectable in the respiratory tract during the paroxysmal stage of the disease. Experimentally, the leukocytosis is clearly reproduced by purified toxin. Therefore it is conceptually easy to attribute the multiple systemic manifestations of disease to a circulating exotoxin with its array of defined biologic activities. Antibody to the toxin prevents the lymphocytosis, and also provides protection against disease in the mouse. It is thus highly likely that whooping cough results from local bacterial colonization of the respiratory tract and subsequent systemic circulation of bacterial exotoxin, although the mechanism(s) by which the toxin exerts its effects and the correlation of the in vitro biologic activities to human illness remain to be elucidated.

Clinical Manifestations

The clinical syndrome of pertussis is readily defined in the presence of the paroxysmal cough and associated whoop, but the illness is of variable severity, and the milder respi-

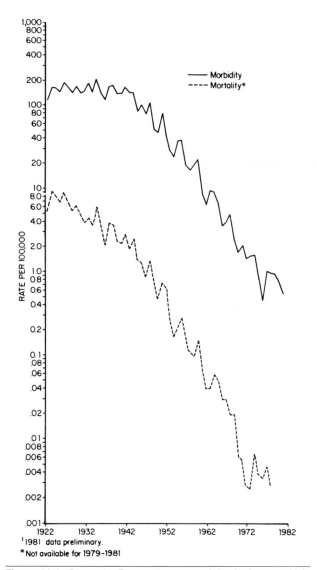

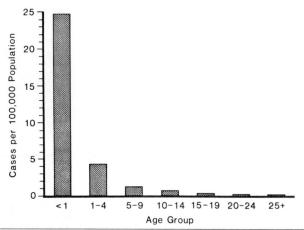

Figure 30-3. Pertussis. Reported cases per 100,000 population by age, United States, 1977–1980 average. (*From the Centers for Disease Control, Atlanta, Ga., and the U.S. Public Health Service.*)

Figure 30-2. Pertussis. Reported cases and deaths by year, United States, 1922–1981. (*From the Centers for Disease Control, Atlanta, Ga., and National Center for Health Statistics, U.S. Department of Health, Education and Welfare.*)

ratory syndromes caused by *B. pertussis* are impossible to distinguish on clinical grounds alone. As many as 20 percent of pertussis infections have been estimated to be atypical illnesses, and those patients are infectious to others.

Following inhalation of infected droplets, the organisms colonize the respiratory tract. Symptoms almost always begin within 10 days after exposure to pertussis, although the incubation period can vary from 5 to 21 days. The clinical illness is divided into three separate stages for descriptive purposes. The catarrhal or prodromal stage lasts from 1 to 2 weeks. During this period, the child exhibits only mild symptoms of an uncomplicated upper respiratory infection. Physical examination does not reveal any serious objective findings.

The second stage usually lasts from 1 to 6 weeks and is characterized by progression to a paroxysmal cough. A characteristic paroxysm is one in which 5 to 20 forcible hacking coughs are produced in 15 to 20 seconds, often terminating with production of mucus or associated vomiting. There is no time for breathing between coughs, and the paroxysm may be sufficiently prolonged to induce anoxia. The final inspiratory breath takes place through the narrowed glottis and produces the characteristic whoop. These early stages of illness are frequently associated with leukocytosis of 12,000 to 200,000 mm^3, with a lymphocytosis of 60 percent.

The third stage of illness is that of convalescence. Coughing may persist for several months after the initial onset of illness. An understanding of the pathogenesis of the cough is of potential therapeutic importance, as the clinical course of the disease and the morbidity are not appreciably altered by administration of specific antimicrobial agents.

The morbidity and mortality associated with the disease have resulted primarily from compromise of the central nervous system during the acute illness and from secondary bacterial infection, usually involving the ears, sinuses, or lower respiratory tract. The neuropathology of infants dying with pertussis and central nervous system involvement is nonspecific and indistinguishable from changes produced by anoxia. Central nervous system compromise is clearly not a result of actual invasion by *B. pertussis*.

Laboratory Diagnosis

Definitive diagnosis depends upon isolation of *B. pertussis* (or less commonly, *B. parapertussis* or *B. bronchiseptica*) from the patient. The isolation rate of the organism from the respiratory tract is greatest during the catarrhal stage, and organisms are not usually detectable for longer than the first 4 weeks of illness. Appropriate specimens for cultivation from patients are either a nasopharyngeal swab or a cough plate.

Isolation of *B. pertussis* from clinical specimens is dependent upon careful transport and efficient processing of the materials obtained for culture. If the specimen will not be planted for 1 to 2 hours, the swab should be placed in 0.25 to 0.50 ml of casamino acids solution with a pH of 7.2 to prevent drying of the swab. When the specimen is shipped to another laboratory or the holding time exceeds 2 hours, other organisms may overgrow *B. pertussis*. Swabs should therefore be placed in modified Stuart's medium (SBL) or Mishulow's charcoal agar. These media are better able to maintain the viability of organisms and to support growth under the conditions of transport, but there is a decreased recovery rate of *B. pertussis* from transport media as compared with direct inoculation. Modified Bordet–Gengou agar is recommended for primary isolation of the organism. The addition of 0.25 to 0.5 unit per ml of penicillin to a second plate is useful in inhibiting the growth of the gram-positive flora of the respiratory tract without affecting the growth of *B. pertussis*.

In addition to the biochemical reactions listed in Table 30-1, serologic identification of *B. pertussis* confirms its isolation. A slide agglutination test can be performed with a standard inoculum of organisms and specific antiserum, which is available commercially.

Fluorescent antibody (FA) staining has been used for the identification of *B. pertussis* in direct smears from nasopharyngeal swabs and for identification of organisms growing on Bordet–Gengou plates. The FA examination of nasopharyngeal swab material, however, is often unreliable, even in experienced hands. Although the FA procedure cannot substitute for cultural isolation of the organism, it can offer the advantage of more rapid laboratory identification of organisms after isolation. The Analytical Bacteriology Section of the Centers for Disease Control and many state bacteriology laboratories are prepared to culture and/or examine secretions by FA techniques for *B. pertussis*.

At present, assessment of antibodies in the serum is accomplished by measuring agglutinins. There are few laboratories in the United States prepared to perform tests for *B. pertussis* agglutinin titers. The test is not well standardized, and the agglutinin titers are not necessarily correlated with the immune status. After infection, there may be only a slight rise in agglutinins, and it tends to occur weeks into the illness. An acute and convalescent pair of sera are needed to define an antibody rise that is indicative of recent contact with antigen. A generally available, specific, and reliable assessment of the humoral response(s) to infection or immunization with *B. pertussis* is badly needed. Assays of antibody to F-HA and HSF–LPF–IAP have been developed and data are currently being obtained in association with immunization or natural disease. An enzyme-linked immunosorbent assay (ELISA) measuring IgM, IgG, or IgA antibodies to F-HA has shown 96 percent sensitivity to culture-positive patients and has identified additional patients who were culture-negative. It

is thus anticipated that these antibodies will reflect immunity in a more accurate manner.

Treatment

Erythromycin is currently the drug of choice for therapy of pertussis infection. The organism is sensitive to this drug in vitro, and administration eliminates the organism from the nasopharynx, thereby shortening the period of communicability. There is some evidence that if erythromycin is administered early during the catarrhal stage, the paroxysmal manifestations may be shortened. Tetracycline or chloramphenicol are considered adequate alternative antimicrobial agents. Secondary bacterial infection may necessitate additional therapy directed at the responsible pathogen. Supportive measures, such as careful suction to remove tenacious secretions, hydration, nutrition, and electrolyte balance are of great importance. Oxygen therapy with increased humidity appears to be beneficial. Some researchers feel that the administration of human pertussis immune serum globulin is a useful adjunct to therapy. The efficacy of this treatment has not been established.

Prevention

It has been recommended that children under 4 years of age who were previously immunized against pertussis should receive a booster dose of vaccine upon contact with an infected individual. They should also receive erythromycin, as immunity conferred by vaccine is not absolute. Furthermore, immunized persons may asymptomatically harbor organisms and thus treatment may curtail spread of infection. Unimmunized contacts should receive chemoprophylaxis with erythromycin for approximately 10 days after contact with the patient has ceased. Human pertussis immune serum globulin may be administered to exposed infants under the age of 2 years who have not been immunized, but protection afforded by this measure is not reliable. The best protective measures for young infants are adequate immunization and avoidance of contact with pertussis.

Active Immunization. Protection of the young infant against pertussis is important because the greatest number of severe complications and the highest morbidity occur in this age group. Passive protection is not afforded by the quantity of antibody that traverses the placenta. Routine primary immunization is begun at about 2 months of age, unless pertussis is prevalent in the community, in which case immunization should be begun earlier. A total of 12 protective units of pertussis vaccine is recommended, and this is divided into three equal doses given 4 to 8 weeks apart. The vaccine is usually given in a combined preparation containing absorbed diphtheria and tetanus toxoids and pertussis vaccine (DPT). These are depot antigens and appear to be more immunogenic and less reactive than a similar plain antigen product without adjuvant. A booster injection is given 12 to 18 months after primary immunization or prior to school entry.

Immunization has been successful in the prevention of disease, and widespread use of vaccine has been associated with the continued decline of reported cases of pertussis in countries employing mandatory immunization.

Additional evidence for vaccine efficacy has unfortunately accrued as a result of diminished immunization in several countries, including Denmark and the United Kingdom. In Denmark, an upswing in the number of cases from a few hundred to thousands per year has occurred several years following alterations in immunization requirements, including the use of smaller quantities of antigen. In the United Kingdom, diminished public acceptance of immunization since 1974 has resulted in pertussis reaching epidemic proportions, beginning in 1977. It is estimated that vaccine acceptance has declined from a rate of 70 to 80 percent prior to 1974 to less than 40 percent in 1982. The resulting epidemic is occurring in younger children who have not received pertussis vaccine. The epidemic in 1982 is the largest since 1957, with 47,508 cases reported from January to September. Studies during the current epidemics suggest that current vaccine is 90 percent effective in the prevention of pertussis.

The effectiveness of the vaccine in young children temporarily discouraged the development of purified immunogens. Increased concern with the reputed reactogenicity of the formalinized whole organism vaccine has renewed the study of the organism and pathogenesis of infection. The factor(s) that produce toxicity or contribute to postvaccination encephalopathy have not been defined, and thus it is impossible to test vaccines for this activity. To date, the mouse weight gain test has been employed in the United States as the most accurate animal assessment of potential toxicity of vaccine for humans. More importantly, the definition of F-HA and pertussis toxin have greatly advanced understanding of the antigens of B. pertussis necessary to induce protection. Thus purified proteins can be administered without other extraneous material from the organism.

Studies conducted in Japan with partially purified vaccine containing F-HA and pertussis toxin are far less reactogenic in children. Thus far, only those children older than 6 to 12 months have been tested. Efficacy data on vaccinees who have had household contact with B. pertussis suggest that the more purified vaccine does provide protection. It seems probable that a safer, more effective vaccine is on the horizon.

Vaccine-associated encephalopathy is estimated to occur once in 5 to 10 million injections in the United States. A prospective study is currently underway in the United Kingdom to better define vaccine-related encephalopathy. The estimated incidence of central nervous system complications of natural disease has ranged from 1.5 percent to 14 percent in hospitalized patients. One third of these individuals recover, one third have varying neurologic sequelae, and one third die or have severe deficits. The morbidity of prolonged illness and necessity for hospitalization with natural disease make it clear that the risks of immunization are far less than those associated with the natural disease.

FURTHER READING

Books and Reviews

Manclark CR, Hill JC: International Symposium on Pertussis. US Department of Health, Education and Welfare, Public Health Service, National Institutes of Health. DHEW publication no. (NIH) 79-1830, 1978

Lapin JH: Whooping Cough. Springfield. Il, Thomas, 1943

Pittman M: Pertussis toxin: The case of the harmful effects of prolonged immunity of whooping cough. A hypothesis. Rev Infect Dis 1:401, 1979

Rowatt E: The growth of Bordetella pertussis: A review. J Gen Microbiol 17:297, 1957

Selected Papers

Arai H, Sato Y: Separation and characterization of two distinct hemagglutinins contained in purified leukocytosis-promoting factor from Bordetella pertussis. Biochim Biophys Acta 444:765, 1976

Bass JW, Klenck EL, Kothermel JB, et al.: Antimicrobial treatment of pertussis. J Pediatr 75:768, 1969

Eldering G, Kendrick P: Bacillus para-pertussis. A species resembling both Bacillus pertussis and Bacillus bronchisepticus but identical with neither. J Bacteriol 35:561, 1938

Goldman WE, Klapper DG, Baseman JB: Detection, isolation, and analysis of a released Bordetella pertussis product toxic to cultured tracheal cells. Infect Immun 36:782, 1982

Granström M, Granström G, Lindförs A, et al.: Serologic diagnosis of whooping cough by an enzyme-linked immunosorbent assay using fimbrial hemagglutinin as antigen. J Infect Dis 146:741, 1982

Katada I, Ui M: Perfusion of the pancreas isolated from pertussis-sensitized rats: Potentiation of insulin secretory responses due to adrenergic stimulation. Endocrinology 101:1247, 1977

LeDur A, Caroff M, Chaby R, Szabo L: A novel type of endotoxin structure present in Bordetella pertussis isolation of two different polysaccharides bound to lipid A. Eur J Biochem 34:579, 1978

Miller JJ, Silverberg RJ, Saito TM, et al.: An agglutinative reaction for Haemophilus pertussis. II. Its relation to clinical immunity. J Pediatr 22:644, 1943

Morse SI, Morse JH: Isolation and properties of leucocytosis and lymphocytosis promoting factor of Bordetella pertussis. J Exp Med 143:1483, 1976

Munoz, JJ, Arai H, Cole RL: Mouse-protecting and histamine-sensitizing activities of pertussigen and fimbrial hemagglutinin from Bordetella pertussis. Infect Immun 32:243, 1981

Muse KE, Collier AM, Baseman JB: Scanning electron microscopic study of hamster tracheal organ cultures infected with Bordetella pertussis. J Infect Dis 136:768, 1977

Sato Y, Arai H, Suzuki K: Leukocytosis promoting factor of Bordetella pertussis. II. Biological properties. Infect Immun 7:992, 1973

Sato Y, Izumiya K, Sato H, et al.: Role of antibody to leukocytosis-promoting factor hemagglutinin and a filamentous hemagglutinin immunity to pertussis. Infect Immun 31:1223, 1981

Sumi T, Ui M: Potentiation of the adrenergic beta-receptor-mediated insulin secretion in pertussis-sensitized rats. Endocrinology 97:352, 1975

CHAPTER 31

Listeria and Erysipelothrix

Medical Significance. Only two species in the genera *Listeria* and *Erysipelothrix* are of medical importance, *Listeria monocytogenes* and *Erysipelothrix rhusiopathiae*. *L. monocytogenes* is widely distributed in nature and in a variety of animal reservoirs. In humans it produces infection with protean manifestations. Meningitis is most frequent, but the most unique of its many clinical forms is infection of the genital tract of the gravid female and infection of the offspring either before birth or during delivery (Table 31-1). Although persons in apparent good health may contract *Listeria* infections, in the adult the majority of cases occur in immunocompromised patients with underlying diseases.

    E. rhusiopathiae is the cause of erysipeloid in humans, an acute self-limited infection of the skin occurring primarily in occupational groups that handle animals and animal products.

    Morphologically, the *Listeria* and *Erysipelothrix* organisms resemble the corynebacteria, and until the Eighth Edition of Bergey's Manual they were classified in the family Corynebacteriaceae. Other properties, however, are more similar to those of the Lactobacillaceae, the group with which they are now tentatively associated. *Listeria* and *Erysipelothrix* are unlike the coryneform organisms in that they do not contain arabinogalactan in their cell wall, but contain the sugar rhamnose instead. They also differ markedly in the guanine plus cytosine (G + C) ratio of their DNA (Table 31-2).

Listeria monocytogenes

Morphology and Physiology

Morphology. The *Listeria* are small gram-positive coccobacilli that have a tendency to occur in short chains of three to five organisms. In stained preparations they often assume a typical diphtheroid palisade arrangement, a property that was responsible for their previous incorrect classification with the corynebacteria. *L. monocytogenes* is 0.4 to 0.5 by 0.5 to 2.0 μm in size. In cultures incubated for 3 to 6 hours at 37C, the bacillary forms predominate, but thereafter the prevalent form is coccoid. In cultures 3 to 5 days old, long filamentous structures 6 to 20 μm or more in length often occur, especially in rough strains. At tem-

TABLE 31-1. PRIMARY CLINICAL MANIFESTATIONS OF HUMAN LISTERIOSIS

	Cases
Meningitis, meningoencephalitis, or encephalitis	496
Septicemia:	
Neonates	51
Others	59
Pregnant women:	
Prepartum flu-like symptoms	5
Abortion	12
Postpartum, of infected infant	17
Endocarditis	7
Abscess	5
Pneumonia	5
Conjunctivitis	2
Infectious mononucleosis	2
Pharyngitis	2
Cutaneous papules and pustules	1
Persistent headache	1
Fever	2
No disease, routine culture	1
TOTAL CASES	641

From Killinger and Schubert: Proceedings of the Third International Symposium on Listeriosis, 1966. Bilthoven, The Netherlands.

peratures of 20C to 25C, *L. monocytogenes* is actively motile by means of four peritrichous flagella, but at 37C only one polar flagellum is formed. The motility of *Listeria* is useful in their differentiation from *Erysipelothrix* and the corynebacteria.

Cultural Characteristics. The optimum temperature for growth is 37C, but growth occurs over a wide range of temperature, down to 2.5C. This ability to grow at low temperatures is the basis for the cold-enrichment technique used in the clinical laboratory for the isolation of *Listeria* from specimens containing a mixed flora (p. 531).

Listeria are not fastidious organisms and grow well on tryptose agar and sheep blood agar media. On the clear colorless tryptose agar, colonies are translucent and easily

recognized by their characteristic blue–green color when viewed with oblique light. On sheep blood agar, colonies resemble *Streptococcus* colonies and are 0.5 to 1.5 mm in diameter. All strains that have been isolated from pathologic specimens of human and animal origin produce β hemolysis on blood agar. All strains of *L. monocytogenes* produce a very distinct narrow band of hemolysis except serotype 5, which produces a much more pronounced zone of hemolysis. In contrast with the β-hemolytic pathogenic isolates of *L. monocytogenes* from clinical infections, a number of isolates from the feces of healthy humans and animals are nonhemolytic and nonpathogenic. For these nonpathogenic strains, the new species *Listeria innocua* has been proposed.

Metabolism. *L. monocytogenes* is aerobic to microaerophilic, but growth is improved when cultures are incubated under reduced oxygen and a 5 to 10 percent concentration of CO_2. Catalase is produced, a property that is useful in the organisms's differentiation from streptococci. *L. monocytogenes* ferments a number of sugars with the formation of acid only (Table 31-2). Most strains from clinical material do not ferment mannitol, a useful property in separating them from the nonpathogenic species *Murraya grayi* (*Listeria grayi* and *Listeria murrayi*), which does ferment mannitol (Table 31-3). All strains hydrolyze polysorbate 80 (Tween 80), and some produce a phosphomonoesterase that causes opacity in egg yolk media.

Antigenic Structure

Based on their O (somatic) and H (flagellar) antigens, strains of *L. monocytogenes* have been separated into four major serologic groups with one to several serotypes (serovars) in each of these groups. No correlation has been detected between the various serotypes of *L. monocytogenes* and any particular clinical syndrome or specific host. There is, however, a striking difference in the geographic distribution of the various serotypes and a tendency for shifts in the prevalent serotype. In the United

TABLE 31-2. DISTINGUISHING PROPERTIES OF CERTAIN NONSPORULATING GRAM-POSITIVE BACTERIA

	Listeria monocytogenes	*Erysipelothrix rhusiopathiae*	*Streptococcus pyogenes*	*Streptococcus faecalis*	*Corynebacterium* sp.	*Lactobacillus* sp.
Morphology	Rod	Rod	Coccus	Coccus	Rod	Rod
β Hemolysis	+	−	+	−	±	−
Catalase	+	−	−	−	+	−
Motility	+	−	−	−	−	−
Acid from:						
Glucose	+	+	+	+	±	+
Mannitol	−	−	−	+	±	±
Keratoconjunctivitis	+	−	−	−	−	−
G + C ratio*	38	36	34–38	33–38	57–60	34–50

Adapted from Buchner and Schneierson: Am J Med 45:904, 1968.
*Mole fraction of guanine plus cytosine in the DNA.

TABLE 31-3. DISTINGUISHING PROPERTIES OF *LISTERIA* SPECIES AND RELATED ORGANISMS

	β Hemolysis	Mannitol Fermentation	Pathogenicity
L. monocytogenes	+	−	+
*L. innocula**	−	−	−
Murrayi grayi†	−	+	−

*New species proposed for the nonhemolytic, nonpathogenic strains of *L. monocytogenes*.
†Proposed new designation to encompass the nonpathogenic *Listeria* species, *L. grayi* and *L. murrayi*.

States and Canada, serotype 4b is the predominant strain. At the present time, three strains (1/2a, 1/2b, 4b) cause at least 90 percent of all clinical *Listeria* infections throughout the world.

Significant antigenic differences between the pathogenic *L. monocytogenes* and the nonpathogenic species *L. grayi* and *L. murrayi* support the proposal that a new genus *Murraya* be designated to encompass these nonpathogenic organisms.*

Determinants of Pathogenicity

Soluble Products. The virulence of *L. monocytogenes* is multifaceted and apparently due both to antiphagocytic components that are present at the cell surface of the organism and to soluble products that are excreted during bacterial growth. The best characterized of the soluble products is an hemolysin that appears to play an important role in the pathogenesis of the infection. It is similar in many ways to the oxygen-labile cytolytic toxins of a number of other bacterial species. The hemolysin, which is elaborated into the culture medium during growth, is nondialyzable, heat labile, and antigenic. It is sensitive to oxidative inactivation and can be reactivated by reducing substances such as cysteine.

Listeria hemolysin may function during listeric infection by disrupting membranes, especially those of the phagocytic vacuole and the lysosomes. Hydrolytic enzymes of lysosomes are solubilized and peritoneal monocytes are degranulated. When injected intravenously, hemolysin is lethal for the mouse. Electrocardiograms of these animals show serious alterations in heart rate and rhythm indicative of damage to cardiac tissue. The lethal effect following hemolysin treatment probably results from toxic injury to contractile and pacemaker myocardial tissue.

L. monocytogenes produces a soluble antigen(s) with lipolytic activity, but the precise nature of this substance is unclear. However, a correlation exists among hemolysin production, lipolytic activity, and virulence. All avirulent and most nonhemolysin-producing strains show either di-

minished or no lipolytic activity. The lipolytic antigen and the hemolytic antigen appear to be two distinct antigens rather than a single antigen with both hemolytic and lipolytic activity.

Surface Components. *L. monocytogenes* is an intracellular parasite whose virulence depends in large measure upon its successful parasitism of the cells of the mononuclear phagocyte system. A variety of antiphagocytic factors has been described, but only one—an endotoxin-like material—has been well characterized. The chemical, physical, and biologic properties of the purified cell wall component are strikingly similar to those of the classic lipopolysaccharide (LPS) endotoxins of the gram-negative organisms (Table 31-4). The listeral LPS is believed to be responsible for the transient cold agglutinin syndrome observed in some patients with septicemic listeric infections. In these patients, antibodies are induced that can, under appropriate conditions, react with the host's own erythrocytes, causing a complement-mediated in vivo lysis. The antibodies responsible are of the IgM class and appear to be directed against the blood group I system.

Clinical Infection

Epidemiology

L. monocytogenes is worldwide in its distribution. It has been isolated from humans with disease, from healthy carriers, and from a wide range of other mammals, birds, fish, ticks, and crustacea. There also is a high incidence in plant and soil samples, and in animal feces. The basic question of whether *L. monocytogenes* is primarily soil borne or originates from animals excreting the organisms in their feces

TABLE 31-4. PROPERTIES OF LIPOPOLYSACCHARIDES

Property	Lipopolysaccharides*	Listerial Lipopolysaccharides
Pyrogenicity in rabbits	+	+
Lethality in rabbits	+	+
Localized Shwartzman reaction	+	±
Amphipathic (sensitizes erythrocytes)	+	+
Limulus lysate gelatin	+	+
B cell mitogenicity	+	+
Adjuvanticity	+	+
Carbocyanine dye assay (spectral shift)	+	+
Presence of acylated hydroxy fatty acid	+	+
Presence of KDO†, heptose	+	+

Adapted from Wexler and Oppeheim: In Agarwal (ed): Bacterial Endotoxins and Host Response, 1980. New York, Elsevier/North-Holland Biomedical Press.
*From gram-negative organism.
†KDO, 2-keto-3-deoxyoctonic acid.

---

* Recommendation of the *Listeria* subcommittee of the International Committee on Systematic Bacteriology.

has not been resolved. It is currently believed, however, that *L. monocytogenes* is a saprophytic organism that lives in a plant–soil environment and can thus be contracted by humans and animals from many sources via many possible routes. The oral route of infection is probably most common.

Although direct transmission has been reported in high-risk occupational groups such as veterinarians and farm workers, most infections in the United States occur in urban dwellers with no history of direct contact with animals or potentially infected materials. This has led to the belief that asymptomatic carriers, especially fecal excretors, may be the major reservoir for transmission, and that symptoms develop only in those who are immunocompromised in some manner. Several clusters of cases of listeriosis caused by a single serotype have been described, suggesting a person-to-person or common-source spread. In one recent report, infection by *L. monocytogenes* serotype b occurred in six immunosuppressed renal transplant patients within a 10-week period. Four of these were close contacts. The source of infection and route of spread could not be identified. The potential for nosocomial spread of listeriosis is also supported by small outbreaks in newborn nurseries. The only proved example, however, of human-to-human transmission is infection of the fetus transplacentally through the umbilical vein with production of septicemia. Infection may also be acquired during delivery through contact with infective secretions.

The incidence of human listeriosis in the United States appears to be increasing. Some of this apparent increase may be due to greater awareness and improved methods for identification, but some is attributable to the increasing numbers of patients with immunosuppressive disorders who are exposed to infectious complications. The age distribution of listeriosis is uneven, with most cases occurring among neonates and the elderly. Between these extremes, it occurs primarily in immunocompromised patients. Renal transplant recipients are now one of the largest groups infected by *Listeria,* approximating in frequency that of malignancy associated with *Listeria* disease.

Pathogenesis

The gastrointestinal tract is probably the portal of entry for extrauterine *Listeria* infections. Studies in experimental animals demonstrate infection via this route and stress the role of the normal microbial flora in modulating *Listeria* colonization of the gut, and subsequent penetration of the intestinal epithelial barrier. Determinants necessary to establish mucosal colonization, or to explain the infectivity for humans of a limited number of serotypes, have not been defined. Intraluminal bacteria are presumed to gain entrance to the circulatory system when the host's defense is altered, followed by seeding of various sites.

In humans, the usual histologic response is a polymorphonuclear leukocytosis with microabscess formation, although a monocytic reaction may occur. *L. monocytogenes* exhibits a striking tropism for the fetus and placenta of most animals, and for the central nervous system of monkeys and humans. Listeriosis of the newborn, contracted intrapartum, usually is localized to the central nervous system. In prepartum infection, *Listeria* from the infected placenta are widely disseminated, resulting in granulomatous foci in many organs, including the liver, spleen, lungs, and central nervous system.

Clinical Manifestations

A wide variety of clinical syndromes is caused by *L. monocytogenes* ranging from a mild influenza-like illness to fulminant neonatal listeriosis associated with mortality rates of 54 to 90 percent. In the adult, the major infections are meningitis (55 percent), primary bacteremia (25 percent), endocarditis (7 percent), and nonmeningitic central nervous system infection (6 percent). More than half of these patients have underlying disorders such as malignancy, alcoholism, cirrhosis, diabetes, or vasculitis, or are receiving immunosuppressive drugs.

Neonatal Infections. Genital tract infection in pregnant women with infection of their offspring is the most distinctive of infections caused by *L. monocytogenes.* Usually the mother has no symptomatic illness, or only a history of a very benign and self-limited influenza-like illness during the last trimester of pregnancy. Infection in the infant may take one of two forms. The early type, granulomatis infantiseptica, results from infection in utero and presents itself within 2 days of birth. Abortion, premature birth, stillbirth, or death within a short period after birth may occur. If the infant is born alive, symptoms of septicemia develop within a few hours, often followed by fetal distress, pneumonia, diarrhea, seizures, and maculopapular skin lesions on the legs and trunk. This form of the disease has a very high mortality rate, apparently largely due to failure to diagnose in the early stages.

The second or late form of neonatal listeriosis appears after the fifth day of life, with meningitis as the usual presentation. Infection is believed to be acquired either during or after birth rather than in utero. The mother is almost always asymptomatic. About 10 percent of neonatal meningitis is due to *L. monocytogenes.*

Adult Infections. Meningitis is the most commonly recognized form of listeriosis in the adult, and a leading cause of bacterial meningitis in cancer patients and renal transplant recipients. About 30 percent of meningitis patients, however, have no preceding disease. Early in the disease the predominant cells are polymorphonuclear leukocytes, but later there may be more mononuclear cells. Clinically, meningitis caused by *L. monocytogenes* cannot be distinguished from meningitis caused by other bacteria. The overall mortality rate among patients with meningitis is 30 percent, but among cancer patients with *Listeria* meningitis the mortality rate is 60 percent.

Primary *Listeria* bacteremia usually affects patients under 50 years of age. Pregnant individuals and those with underlying diseases are particularly prone to develop this form of the disease. As with other listerial infections, diagnosis depends upon laboratory identification of the organism, as there are no specific clinical features. Endocarditis and pneumonia are among the rare focal infections that may result from the bloodstream invasion.

Immunity

L. monocytogenes is an intracellular parasite. It is readily phagocytosed by normal macrophages following activation of the alternative complement pathway by cell wall components. In addition to the C3b generated, natural antibody is also believed to contribute to opsonization. Once within the phagocytes, *Listeria* can survive and multiply, their eventual elimination depending on a cell-mediated immune response. The early stages of resistance apparently require a bone marrow-derived cell, while later acquired immunity rests on thymus cell activation and recruitment of macrophages and monocytes. Humoral antibody plays no role in immunity.

The genetics of resistance to *Listeria* infections has been examined in inbred strains of mice which exhibit differences in responsiveness to infection. The resistance trait is apparently controlled by a single, autosomal, dominant non-H-2 linked gene termed Lr. The gene is expressed phenotypically in the enhanced response of the mononuclear phagocyte system to infection. In the *Listeria*-resistant host, the Lr gene product appears to promote the early arrival of immature macrophages that develop potent antibacterial activity.

As is the case for several other bacteria, *L. monocytogenes* possesses an array of cell surface components that are capable of interacting in a variety of ways with participants of the immune response. One of the most unique of these is a factor with monocytosis-producing activity. In some animals, monocytosis is a hallmark of *Listeria* infection, and the property for which the organism was named. Various cell wall fractions have been obtained that are mitogenic for bone marrow-derived cells, act as an adjuvant, and are immunosuppressive.

No second clinical infections with *Listeria* have been observed in patients or animals cured of proven listeriosis.

Laboratory Diagnosis
Culture. The diagnosis of listeriosis is based on the isolation of *L. monocytogenes* from the appropriate clinical materials, depending upon the syndrome. The usual materials include blood, cerebrospinal fluid, amniotic fluid, and genital tract secretions. The key to the diagnosis of listeriosis is awareness. Laboratory personnel should be alerted when *Listeria* infection is suspected, as in a setting of diminished host defenses. A frequent error made by the clinical laboratory is to assume all diphtheroid isolates are contaminants, and thus are unimportant. Gram stains should be made of the infected material and examined for the presence of typical pleomorphic gram-positive bacilli. A positive finding is extremely useful, but in 60 percent of patients with *Listeria* meningitis, organisms are not seen in a gram stain of the spinal fluid. Also, the organisms may sometimes be mistaken for streptococci, or when poorly stained may resemble *H. influenzae*.

When the organisms are numerous and not mixed with other bacteria, *Listeria* is not difficult to grow from infected material unless the patient has received antibiotics. A 2 percent tryptose agar medium is excellent for cultivation and propagation. When the organisms are to be grown from tissues, successful isolation requires homogenization to release the organisms that are incarcerated often intracellularly, in the focal lesions.

If the primary culture fails to reveal *L. monocytogenes* from a suspected *Listeria* infection, the clinical specimen should be held at 4C for several weeks and replated after 6 weeks' storage, and again after 3 months' storage. The combination of the cold-enrichment technique and the use of oblique lighting for the examination of colonies is basic for the isolation of *Listeria* from specimens where the organisms are sparse or are mixed with other bacteria. Accurate identification of an isolate as *L. monocytogenes* is based on the demonstration of β hemolysis, catalase production, and tumbling motility at room temperature. Useful procedures for the differentiation of *L. monocytogenes* from diphtheroids and other organisms with which it may be confused are listed in Table 31-2. Cultures identified as *L. monocytogenes* should be sent to the Centers for Disease Control for complete serologic identification.

Although antibodies appear in the serum of patients during and after *Listeria* infections, serologic diagnosis of listeriosis is unreliable because *L. monocytogenes* cross-reacts with a number of other gram-positive organisms, including staphylococci. Also, *Listeria* agglutinins are frequently found in the sera of healthy individuals without a history of listeriosis.

Animal Inoculation. Animal pathogenicity testing is also useful in differentiating *L. monocytogenes* from gram-positive organisms possessing similar morphologic properties. The development of histologic changes in experimental listeriosis of small animals, especially rabbits, is remarkably rapid and strong. The instillation of *L. monocytogenes* into the conjunctival sac of a young rabbit produces a purulent keratoconjunctivitis within 24 to 36 hours (Anton test). The inoculation of the marginal ear vein of the rabbit produces a marked monocytosis, and the intraperitoneal injection of mice results in necrotic foci in the liver and death of the animal.

Treatment
Penicillin G or ampicillin is the recommended treatment for *Listeria* infections. Erythromycin and the tetracyclines are also effective, but cephalosporins should not be used because of their variable activity and limited penetration of the meninges. For high-risk groups, such as neonates and

immunosuppressed patients, either penicillin or ampicillin and an aminoglycoside is recommended as initial therapy.

Prevention

There is no vaccine for the prevention of listeriosis. Effective control of *Listeria* infections is hampered by difficulties in recognition of human and animal infection. In the newborn, listeriosis is preventable by early recognition and prompt treatment of the mother. Prevention should center on elimination of animal reservoirs, and contact with infected animals or animal products should be avoided. Careful attention should be given to the handling of infected infants in a neonatal unit in an attempt to prevent nosocomial transmission, and patients with immunosuppressive disorders or those receiving immunosuppressive drugs should be protected from contact with cases of listeriosis.

Erysipelothrix rhusiopathiae

Morphology and Physiology. *E. rhusiopathiae* is a nonsporogenous, nonmotile, nonencapsulated gram-positive rod. In smooth cultures, organisms are short, slender, straight or slightly curved, and measure 0.2 to 0.4 μm by 0.5 to 2.5 μm. In rough colonies, long filamentous structures and chains up to 60 μm or more are present. In smears from acute forms of erysipelas disease, organisms of the smooth-colony form are seen, whereas in chronic cases the rough form appears. *E. rhusiopathiae* has a strong tendency to dissociate from the S to the R form, resulting in changes in virulence and antigenic properties.

E. rhusiopathiae is microaerophilic when first isolated and grows in a band a few millimeters below the surface of a semisolid tube of agar. Heart infusion agar containing rabbit blood and incubation at 37C in the presence of 5 percent CO_2 is suitable for primary isolation. On blood agar plates, colonies are small, round, and grayish-white, somewhat similar to the colonies of *Streptococcus viridans*. α Hemolysis is produced, although a slight but definite clearing occurs around the colonies upon prolonged incubation. Black colonies are produced on tellurite media. Properties useful in the identification and differentiation of *Erysipelothrix* from *Listeria* and corynebacteria are listed in Table 31-2. The former's resistance to neomycin also provides a very useful rapid test for differentiating *E. rhusiopathiae* from *L. monocytogenes*.

Most strains produce hyaluronidase. There appears to be a correlation between antigenic structure, virulence, and hyaluronidase production. Most of the virulent strains belong to serogroup A, and are good hyaluronidase producers. Neuraminidase also appears to contribute to the pathogenic potential of some strains.

Clinical Infection. *Erysipelothrix* is widely distributed in nature in all parts of the world. Many wild and domestic

animals, fish, and birds harbor the organism, usually as a commensal. It occurs in the surface slime of fresh and salt water fish and is found in sewage effluent from abbatoirs and in the feces of infected animals. Disease in swine is the most common animal infection, sometimes occurring in epizootics and causing considerable economic loss. Infections also occur in horses, sheep, cows, and a number of other animals. Humans acquire the infection by contact with animals or animal products. The disease is more prevalent in the male, especially abattoir employees, butchers, and those handling fish, animal hides, and bones. Most infections are related to skin injury.

The disease in humans, erysipeloid, is characterized by a nonsuppurative purplish red lesion at the site of inoculation, which is usually on the hand or fingers. The lesions burn and itch but usually there is no pain, no systemic symptoms, nor lymphangitis. The disease remains localized, and characteristically lasts only a few days. Rarely, erysipelothrix infections may be disseminated, causing infective endocarditis and septic arthritis.

A laboratory diagnosis can be made by culture of the organism from aspirated material or biopsy specimen taken from the margin of the local lesion. Blood cultures are necessary for a diagnosis of endocarditis.

Penicillin is the drug of choice. In patients sensitive to penicillin, erythromycin may be used.

FURTHER READING

Books and Reviews

Albritton WL, Wiggins GL, DeWitt WE, et al.: In Lennette EH, Balows A, Hausler WJ Jr, et al. (eds): Manual of Clinical Microbiology, 3rd ed. Washington, DC., American Society for Microbiology, 1980, chap 11

Armstrong D: *Listeria monocytogenes*. In Mandell GL, Douglas RG Jr, Bennett JE (eds): Principles and Practice of Infectious Diseases. New York, John Wiley & Sons, 1979, chap 165

Ewald FW: The Genus *Erysipelothrix*. In Starr MP, Stolp H, Trüper HG, et al. (eds): The Prokaryotes. New York, Springer-Verlag, 1981, vol II, chap 133

Gray ML, Killinger AH: *Listeria monocytogenes* and listeria infections. Bacteriol Rev 30:309, 1966

Grieco MH, Sheldon C: *Erysipelothrix rhusiopathiae*. Ann NY Acad Sci 174:523, 1970

Hahn H, Kaufmann SHE: The role of cell-mediated immunity in bacterial infections. Rev Infect Dis 3:1221, 1981

Nieman RE, Lorber B: Listeriosis in adults: A changing pattern. Report of eight cases and review of the literature, 1968–1978. Rev Infect Dis 2:207, 1980

Proceedings of the Third International Symposium on Listeriosis, Bilthoven, The Netherlands. July 13–16, 1966

Seelinger HPR: Listeriosis. New York, Hafner, 1961

Seelinger HPR, Höhne K: Serotyping of *Listeria monocytogenes* and related species. In Bergan T, Norris JR (eds): Methods in Microbiology. New York, Academic Press, 1979, vol 13, chap II

Stamm AM, Dismukes WE, Simmons BP, et al.: Listeriosis in renal transplant recipients: Report of an outbreak and review of 102 cases. Rev Infect Dis 4:665, 1982

Sword CP, Kingdon GC: *Listeria monocytogenes* toxin. In Kadis S,

Montie TC, Ajl SJ (eds): Microbial Toxins. New York, Academic Press, 1971, pp 357–377

Welshimer HJ: The genus *Listeria* and related organisms. In Starr MP, Stolp H, Trüper HG, et al. (eds): The Prokaryotes. New York, Springer-Verlag, 1981, vol II, chap 132

Selected Papers

Baker LA, Campbell PA: *Listeria monocytogenes* cell walls induce decreased resistance to infection. Infect Immun 20:99, 1978

Chan YY, Cheers C: Mechanism of depletion of T lymphocytes from the spleen of mice infected with *Listeria monocytogenes*. Infect Immun 38:686, 1982

Cheers C, McKenzie IFC, Mandel TE, et al.: A single gene (*Lr*) controlling natural resistance to murine listeriosis. In Skamene E, Kongshavn PAL, Landy M (eds): Genetic Control of Natural Resistance to Infection and Malignancy. New York, Academic Press, 1980, p 141

Cohen JJ, Rodriguez GE, Kind PD, et al.: Listeria cell wall fraction: A B cell mitogen. J Immunol 114:132, 1975

Filice GA, Cantrell HF, Smith AB, et al.: *Listeria monocytogenes* infection in neonates: Investigation of an epidemic. J Infect Dis 138:17, 1978

Galsworthy SB, Gurofsky SM, Murray RGE: Purification of a monocytosis-producing activity from *Listeria monocytogenes*. Infect Immun 15:500, 1977

Jungi TW, Jungi R: Genetic control of cell-mediated immunity in rats: Involvement of RT1.B locus determinants in the proliferative response of T lymphocytes to *Listeria* antigens. Infect Immun 38:521, 1982

Killinger AH, Schubert JH: Proceedings of the Third International Symposium on Listeriosis, Bilthoven, The Netherlands. July 13–16, 1966, p 317

Kongshavn PAL, Sadarangani C, Skamene E: Cellular mechanisms of genetically determined resistance to *Listeria monocytogenes*. In Skamene E, Kongshavn PAL, Landy M (eds): Genetic Control of Natural Resistance to Infection and Malignancy. New York, Academic Press, 1980, p 149

McCallum RE, Sword CP: Mechanisms of pathogenesis in *Listeria monocytogenes* infection. Infect Immun 5:863 1972

Mitsuyama M, Nomoto K, Takeya K: Direct correlation between delayed footpad reaction and resistance to local bacterial infection. Infect Immun 36:72, 1982

Ratzan KR, Musher DM, Keusch GT, et al.: Correlation of increased metabolic activity, resistance to infection, enhanced phagocytosis, and inhibition of bacterial growth by macrophages from *Listeria*- and BCG-infected mice. Infect Immun 5:499, 1972

Saiki I, Kamisango K, Tanio Y, et al.: Adjuvant activity of purified peptidoglycan of *Listeria monocytogenes* in mice and guinea pigs. Infect Immun 38:58, 1982

Schuffler C, Campbell PA: Listeria cell wall fraction. Characterization of *in vitro* adjuvant activity. Immunology 31:323, 1976

Srivastava KK, Siddique IH: Quantitative chemical composition of peptidoglycan of *Listeria monocytogenes*. Infect Immun 7:700, 1973

Stuart MR, Pease PE: A numerical study on the relationships of *Listeria* and *Erysipelothrix*. J Gen Microbiol 73:551, 1972

Wexler H, Oppenheim JD: Isolation, characterization, and biological properties of an endotoxin-like material from the gram-positive organism *Listeria monocytogenes*. Infect Immun 23:845, 1979

Wexler H, Oppenheim JD: Listerial LPS: An endotoxin from a gram-positive bacterium. In Agarwal (ed): Bacterial Endotoxins and Host Response. New York, Elsevier/North-Holland Biomedical Press, 1980 p 27

Wilkinson BJ, Jones D: A numerical taxonomic survey of *Listeria* and related bacteria. J Gen Microbiol 98:399, 1977

Zinkernagel RM, Althage A, Adler B, et al.: H-2 restriction of cell-mediated immunity to an intracellular bacterium. Effector T cells are specific for *Listeria* antigen in association with H-2I region-coded self-markers. J Exp Med 145:1353, 1977

CHAPTER 32

Corynebacterium

Corynebacterium diphtheriae

Diphtheria is the prototype of a toxigenic disease. It is an acute infection caused by strains of *Corynebacterium diphtheriae* that are infected with a bacteriophage carrying the structural gene for diphtheria toxin. In diphtheria, the primary lesion usually occurs in the throat or nasopharynx, and is characterized by the presence of a spreading grayish pseudomembranous growth. As the organisms multiply at this site, they elaborate a potent exotoxin that is transported by the blood to remote tissues of the body causing hemorrhagic and necrotic damage in various organs.

The history of diphtheria is a fascinating account of the successful study and conquest of an infectious disease. It is an account that emphasizes the importance of basic research in providing practical solutions to clinical problems. Diphtheria was first established as a specific clinical entity in 1826 following publication of a classic monograph by Pierre Bretonneau, but its bacterial etiology was not fully established until 1888. Klebs had earlier described the characteristic bacilli in pseudomembranes from diphtheritic throats, and Löffler had isolated the organism in pure culture, but complete understanding of the pathogenesis of the infection was provided only with the discovery by Roux and Yersin of a soluble exotoxin in the filtrates of cultures. This finding opened the door to immunologic studies that resulted in the discovery of antitoxin and toxoid, the two biologicals that have been so successfully employed in passive and active immunization against the disease.

C. diphtheriae is the only major human pathogen of the corynebacteria group or organisms, a group that also contains a number of harmless, poorly described saprophytes frequently found on the surfaces of mucous membranes. The corynebacteria are taxonomically related to the mycobacteria and nocardia because of similarities in their cell wall composition, and they exhibit cross-reactivity with them. Peptidoglycans of the three genera contain *meso-α*, ε-diaminopimelic acid, and major sugars of their wall polysaccharide are arabinose and galactose. Corynebacteria also contain considerable amounts of mycolic acids in the lipids associated with their outer envelope. Mycolic acids found in the corynebacteria are similar to the large saturated, α-branched, β-hydroxy fatty acids of the mycobacteria but contain fewer carbon atoms (p. 100).

Morphology

C. diphtheriae is a slender, gram-positive, rod-shaped organism that is nonacid-fast and does not form spores. Cells are 1.5 to 5 μm in length, and 0.5 to 1.0 μm in width. In stained smears, they characteristically appear in palisades, or as individual cells lying at sharp angles to each other in V and L formations. These Chinese character-like formations are caused by the "snapping" movement involved when two cells divide. When grown on nutritionally complete media, diphtheria bacilli are uniform in shape, but when grown on suboptimal media such as Löffler's coagulated serum or Pai's coagulated egg medium, the cells are pleomorphic and stain irregularly with methylene blue or toluidine blue. Club-shaped swellings and beaded and barred forms are common. The metachromatic (Babes–Ernst) granules that are responsible for the beaded appearance represent accumulation of polymerized polyphosphates.

Physiology

Cultural Characteristics
C. diphtheriae is an aerobic and facultatively anaerobic organism, but grows best under aerobic conditions. Complex media are required for primary isolation and characterization. Most strains grow as a waxy pellicle on the surface of liquid media. On Löffler's coagulated serum medium, which is useful for the primary isolation of the organism,

minute grayish-white glistening colonies appear after 12 to 24 hours incubation at 37C. Löffler's medium is also useful because it does not support growth of streptococci and pneumococci that may be present in the clinical specimen.

The addition of tellurite salts to media used for primary isolation also reduces the number of contaminants. On tellurite media, colonies of diphtheria organisms assume a characteristic gray or black color, and may be differentiated into three major colonial types: gravis, mitis, and intermedius. Colonies of gravis strains are large, flat, and gray to black with a dull surface; mitis organisms produce medium-sized colonies that are smaller, blacker, glossy, and more convex; and colonies of intermedius strains are very small and either smooth or rough. The tellurite ion passes through the cell membrane into the cytoplasm, where it is reduced to the metal tellurium and precipitated. No constant relationship exists between the severity of the disease and the three colony types.

Resistance
C. diphtheriae is more resistant to the action of light, desiccation, and freezing than are most nonsporeforming bacilli. On dried fragments of pseudomembranes, organisms survive for at least 14 weeks. They are readily killed, however, by a 1 minute exposure to 100C or 10 minutes at 58C. They are susceptible to most of the routinely used disinfectants.

Antigenic Structure

All diphtheria toxins are immunologically identical. The organism *C. diphtheriae*, however, is an antigenically heterogeneous species. Agglutination tests with whole cell suspensions show a large number of serologic types. The three major colonial types—gravis, mitis, and intermedius—reflect differences in the cell surface and constitute the major biotypes of the organism. Within each of these biotypes is a more or less separate group of agglutinating serotypes. Additional differences in cell surface components have also been detected by bacteriophage typing and bacteriocin production.

K Antigen. The antigens responsible for the type specificity of *C. diphtheriae* strains are heat-labile proteins, the K antigens, localized in the superficial layers of the wall. These antigens play an important role in antibacterial immunity and hypersensitivity separate from antitoxic immunity. The occurrence of different antigenic types of *C. diphtheriae* probably explains the occurrence of diphtheria in immunized individuals who show a detectable level of circulating antitoxin. The K antigens on the surface, together with the glycolipid cord factor (see below), are major determinants of invasiveness and virulence in diphtheria bacilli.

O Antigen. The heat-stable O antigen of *C. diphtheriae* is a group antigen common to the corynebacteria parasitic for humans and animals. It is a polysaccharide containing arabinogalactans and is the antigen responsible for the cross-reactivity with mycobacteria and nocardia. Corynebacterial cells and their subcellular components are excellent antigens. When administered to animals with immunizing agents they also function as adjuvants (p. 554).

Determinants of Pathogenicity

Invasiveness

As both toxigenic and nontoxigenic strains of *C. diphtheriae* are capable of colonizing mucous membranes, factors other than toxin production contribute to the organism's invasiveness and ability to establish and maintain itself in the human host. The precise relationship of these traits to the pathogenesis of the disease, however, is ill defined. In addition to the surface K antigens, the organisms contain a cord factor that is considered to be a necessary adjunct of virulence. The cord factor, a toxic glycolipid, is a 6-6' diester of trehalose containing the mycolic acids characteristic of *C. diphtheriae*, corynemycolic acid ($C_{32}H_{62}O_3$), and corynemycolenic acid ($C_{32}H_{64}O_3$)(see p. 100).

The pharmacologic activity of the cord factor of *C. diphtheriae* is similar to that of the cord factor from *M. tuberculosis*. In the mouse, it causes a disruption of mitochondria, reduction of respiration and phosphorylation, and death.

Other factors that probably contribute to the invasive potential of *C. diphtheriae* are neuraminidase and *N*-acetylneuraminate lyase. By degrading the *N*-acetylneuraminic acid residues cleaved from its mucinous environment, these enzymes could provide a readily available source of energy for bacteria inhabiting the mucous membranes.

Exotoxin

Most naturally occurring diseases are too complex at the cellular level to dissect successfully and define definitively the primary biochemical lesion. In diphtheria, however, the exotoxin produced by *C. diphtheriae* is the major biochemical determinant in the pathogenesis of the infection and accounts for essentially all of the pathologic effects. The study of the molecular biology of its production and mode of action provides a model system for the use of modern tools and concepts in the study of a disease process.

Lysogeny and Toxin Production. Toxin is produced only by strains of *C. diphtheriae* infected with a temperate bacteriophage carrying the structural gene for toxin production. Nontoxigenic strains may be converted to the lysogenic, toxigenic state by infection with a suitable tox^+ corynephage. Conversion to toxigenicity, however, is not an obligatory property of corynephages. Although most of the studies on toxin production have been conducted with β corynephage, the *tox* gene occurs in a number of corynephages that differ both genetically and seriologically.

The production of toxin by a lysogenic strain does not require lytic growth of the phage. The *tox* gene can be expressed when phage β is present in *C. diphtheriae* as vegetatively replicating phage; as prophage; or as a superinfecting, nonreplicating exogenote in immune lysogenic cells. Under laboratory conditions, the toxin appears to serve no essential viral function. However, under natural conditions in the nasopharynx of humans, the *tox* gene imparts survival value both to the phage and to *C. diphtheriae*, its lysogenized host.

The β corynephage is a medium-sized phage containing double-stranded DNA, approximately 2.3×10^7 daltons in length. Its life cycle is similar to that of coliphage λ, growing productively in most infected cells but lysogenizing a small number. The availability of phage mutants that code for antigenically similar but nontoxic proteins (crm proteins) has made it possible to construct a map of the vegetative and prophage genomes of the β phage. The prophage map appears to be a cyclic permutation of the vegetative map with the *tox* gene at one end of the prophage map next to the attachment site on the host chromosome, and the immunity gene at the other end of the map (Fig. 32-1). Insertion into the bacterial chromosome

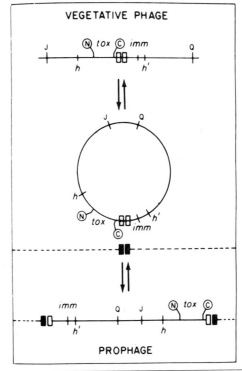

Figure 32-1. Orientation of the *tox* gene in the β bacteriophage and prophage. The figure shows the interconversion of the prophage and vegetative phage states. The *tox* gene has been enlarged, and its boundaries are defined by N and C, the locations of the codons for the N-terminal and C-terminal amino acids. Phage chromosome (—): bacterial chromosome (- - -). The open and closed rectangles represent phage and bacterial attachment sites, respectively. (*From Laird and Groman*: J Virol 19:228, 1976.)

occurs by a mechanism similar to that found in phage λ. The position of the integrated *tox* gene suggests that it may have evolved from a bacterial gene.

Regulation of *Tox* Gene Expression.

Regulation of the synthesis of diphtheria toxin encompasses both genetic and physiologic factors, and involves both the bacterium and the bacteriophage. Different strains of *C. diphtheriae* vary greatly in their capacity for toxin production when infected by specific *tox*⁺ corynephages. Also, some phages may be nontoxigenic for some host strains but *tox*⁺ for other *C. diphtheriae*. The γ corynephage, a nonconverting phage closely related to β phage, carries the *tox* gene in an inactive form. In the γ phage, a complete phage genome is present plus an adjacent small loop of bacterial DNA. It is probable that this extra DNA represents an insertion element or a transposon, and that the *tox* gene of γ phage has been inactivated by its insertion.

The yield of toxin is markedly influenced by growth conditions, especially the inorganic iron content of the medium. Diphtheria toxin is produced at maximal levels only when iron is growth-rate limiting. The addition of iron to iron-starved cultures of lysogenic *C. diphtheriae* inhibits the production of toxin almost immediately. The successful use for many years of the Park-Williams no. 8 strain for the commercial production of toxin is linked to its ability to grow in media containing very low levels of iron. Under such conditions toxin may account for approximately 5 percent of the total bacterial protein.

At the molecular level, regulation of diphtheria toxin production occurs independently of other phage functions and is directed at the level of transcription. According to the proposed model, *C. diphtheriae*, irrespective of its lysogenic state, carries a gene coding for the synthesis of the diphtheria *tox* aporepressor. In the presence of iron, a repressor-iron complex forms that binds specifically at the phage β *tox* operator locus. Under conditions of iron limitation, the repressor-iron complex dissociates and the diphtheria *tox* gene is derepressed (Fig. 32-2). Nothing is known at present concerning the identity and function in the bacterial host of the repressor protein.

Properties.

The toxin molecule is formed by *C. diphtheriae* in association with the cell membrane, and is secreted rapidly as a single polypeptide chain of 62,000 molecular weight (Fig. 32-3). When released, the native toxin molecule is nontoxic until exposure of the active enzymatic site by mild trypsin treatment. The activated molecule consists of two functionally distinct fragments, A and B, linked together by a disulfide bridge. Both of these fragments are essential for cytotoxicity. The C-terminal fragment B (40,700 daltons) binds to specific eucaryotic cell membrane receptors and mediates the entry of the enzymically active N-terminal fragment A (21,150 daltons) into the cytoplasm, where the A fragment catalytically inhibits protein synthesis.

Mode of Action.

Upon reaching the cytoplasm, fragment A disrupts protein synthesis by catalyzing transfer of the adenosine-diphosphoribose (ADPR) moiety of nicotinamide adenine dinucleotide (NAD) to the eucaryotic peptidyl-tRNA translocase, elongation factor 2 (EF2). This adenosine diphosphate (ADP)-ribosylation of EF2 is the primary target of diphtheria toxin.

$$NAD + EF2_{free} \rightleftharpoons ADPR—EF2_{free} + nicotinamide + H^+$$

Although the reaction is reversible, its equilibrium at physiologic pH lies far to the right. Only soluble EF2 can serve as substrate, and once it is fixed to ribosomes, it cannot be ADP-ribosylated.

The specific site in EF2 to which ADPR becomes covalently linked is a unique amino acid (diphthamide) that results from a novel posttranslational modification of a histidine residue. Diphthamide has not been found in any other eucaryotic protein. Toxin has no effect on polypeptide chain elongation in procaryotic systems, or in mitochondria where a different protein, elongation factor G (EFG), replaces EF2.

Uptake by Eucaryotic Cells.

In order for diphtheria toxin to express its toxicity, fragment A must cross the lipid bilayer of the cell to reach the cytosol. This is facilitated by fragment B, with which the hydrophilic enzymically ac-

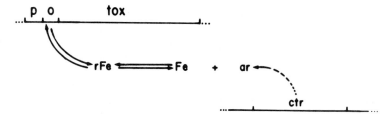

Figure 32-2. Hypothetic model of the corynebacterial regulation of the corynephage β *tox* gene. *C. diphtheriae* carries the structural information (ctr) for the synthesis of the diphtheria tox aporepressor (ar). In the presence of iron, a repressor-iron complex would form and bind to the corynephage β *tox* operator locus. Under conditions of iron limitation, the equilibrium would shift to the right, and the diphtheria *tox* gene would become depressed. (*From Murphy and Bacha: Microbiology— 1979. Washington, D.C., American Society for Microbiology.*)

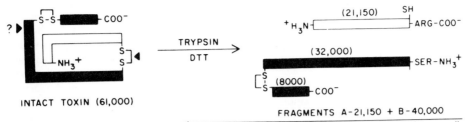

Figure 32-3. Diagram of diphtheria toxin and its fragments. Toxin is released into the culture medium as a single polypeptide chain. The intact molecule contains two disulfide bridges. The peptide loop contained within the first disulfide bridge is extraordinarily sensitive to proteolytic cleavage and is opened following mild digestion by trypsin. Subsequent reduction of the disulfide results in splitting of the nicked toxin into two fragments, A and B. (*From Pappenheimer: Microbiology—1979. Washington, D.C., American Society for Microbiology.*)

tive fragment A is associated. Fragment B recognizes and interacts with specific surface receptors that are present on the plasma membrane. The specific membrane receptor has not been identified, nor has the precise mechanism for entry of fragment A into the cell. Theories to explain the transport process suggest either that fragment B forms a channel in the plasma membrane through which fragment A passes, or that the toxin is first endocytosed and then somehow escapes to the cytoplasm from an intracellular vesicle. At some point before or during the translocation, disulfide reduction and cleavage of one or more peptide bonds are required to release fragment A from fragment B.

Host Susceptibility. Animals vary greatly in their susceptibility to diphtheria toxin. Doses as low as 160 ng/kg of body weight are lethal for humans, rabbits, guinea pigs, and birds. Rats and mice, however, are highly resistant unless the toxin is administered intracerebrally.

Antitoxin. Both A and B moieties of toxin contain a number of antigenic determinants. As a consequence, antitoxin consists of a heterogeneous mixture of antibodies specific for different domains of the toxin molecule. In native toxin or toxoid, most of the antigenic determinants in fragment A are deeply buried and are not available either to stimulate antibody production or to participate in the precipitation of antibody. Also, although anti-A antibody inhibits the enzymic activity of toxin, it does not protect animals or cells against the lethal action of toxin. Antibodies directed against fragment B, however, neutralize toxin with great efficiency, supporting the theory that antitoxin acts by competing with toxin for surface receptors on sensitive eucaryotic cells, and that fragment B is required for the initial attachment.

Active antitoxic immunization against diphtheria toxin by use of a synthetic oligopeptide has recently been accomplished. The synthetic antigen is a tetradecapeptide consisting of the loop of 14 amino acids subtended by the disulfide bridge nearer the N-terminus of the molecule. This tetradecapeptide, when linked covalently with a carrier, will elicit in guinea pigs antibodies that not only bind specifically with the toxin but neutralize its dermonecrotic

and lethal effects. This represents the first example of successful active immunization against a lethal bacterial toxin using a synthetic antigen.

Toxoid. The addition of low concentrations of formaldehyde to diphtheria toxin destroys its toxicity and converts it to toxoid. Detoxification by formaldehyde alters both the enzymatic activity of fragment A and the binding of fragment B to cells. The alteration may be due either to the chemical modification of essential residues in the toxin, or to intramolecular crosslinking of lysine and tyrosine via methylene bonds. The increased stability and resistance to proteolysis of toxoid is attributed to the crosslinking effects. Toxoid cannot be cleaved into A and B fragments. It possesses no ADP-ribosylating activity and does not bind to cell membranes of sensitive cells.

Clinical Infection

Epidemiology

Diphtheria has a worldwide distribution but is now relatively uncommon in the United States and Western Europe. The marked decrease in incidence is attributable to successful programs for the active immunization of preschool children. Even though the annual number of cases is small (about 56 cases per year since 1976) the problem remains a serious one in certain population clusters. The highest annual attack rates are in the South in the 1- to 9-year age group. The attack rate for unimmunized children is 70 times that of children having received three or more doses of toxoid. Because of the low immunization status of adults, diphtheria occurs in the older age groups at about the same low frequency as in the past, with adults 15 years of age and older now accounting for approximately 25 percent of all cases in the United States. Severe complications are also more prevalent in the elderly. During the last 10 to 15 years, the diphtheria problem has become one primarily of urban populations. Those mainly affected are of the lower socioeconomic groups, most of whom live in crowded slum areas with limited access to health care facilities. Diphtheria may occur and give rise to epidemics during any month of the year, but crowding and close interpersonal contact during the winter results in a higher frequency, especially in the South, from September through January.

Humans are the only natural hosts of *C. diphtheriae* and, thus, the only significant reservoir of infection. Asymptomatic carriers and persons in the incubation stage of the disease are the major sources of most infections. The primary habitat of *C. diphtheriae* is the upper respiratory tract, from which site they are transmitted from person to person either directly or indirectly. Transmission via droplet infection is the major mechanism of transfer in respiratory disease. Discharges from extrarespiratory sites, such as skin ulcers, also provide a source of pharyngeal as well as cutaneous disease. In tropical areas, the skin provides a major reservoir of *C. diphtheriae* infection.

Immunity

In diphtheria, immunity against clinical disease is dependent upon the presence of antitoxin in the blood formed in response to either clinical or subclinical infection, or as a result of artificial active immunization with toxoid. Under the age of 6 months, infants are passively protected from diphtheria by the transplacental passage of antitoxin from immune mothers. The antibodies present in antitoxin are of the IgG and IgA classes.

The widespread immunization of infants and preschool children with toxoid has materially reduced the incidence of diphtheria in children and, as a result, has decreased the carrier rate and opportunity for natural reinforcement of immunity by subclinical infection. As a result, an increasing number of adults are without protective antitoxic immunity. Also, a much lower percentage of newborns are immune during the first months of life. The immune status of an individual may be assessed by the determination of serum antitoxin levels or by the Schick test.

Schick Test. This test is performed by injecting into the skin of the forearm 0.1 ml of highly purified toxin.* When administering the Schick test, a control consisting of a similar amount of toxin heated to 60C for 30 minutes should be injected into the other arm. A positive reaction, indicating the absence of immunity to diphtheria, is characterized by a local inflammatory reaction that reaches a maximal intensity within 4 to 7 days and fades gradually. A negative Schick reaction signifies that the antitoxin level is greater than 0.03 U/ml blood, and that the individual is immune

under ordinary conditions of exposure. Allergic reactions are sometimes observed in adults and older children, especially in endemic areas. These reactions probably result from previous infections with corynebacteria or from artificial immunization with toxoid. If such individuals are immune, they will give a pseudoreaction characterized by erythema at both test and control sites. Such reactions reach maximum intensity within 24 to 36 hours, but fade and disappear completely within the next 72 hours. If the individuals are allergic but have no antitoxin or a low level of antitoxin, they will give a combined reaction—the reaction in the control arm subsides by the fifth or sixth day, whereas that in the test arm reaches its maximum on the fifth day and persists for several days.

Pathogenesis

Following exposure to *C. diphtheriae*, there is an incubation period of 1 to 7 days during which the organism establishes itself at the infected site. The initial lesion usually occurs on the tonsils and oropharynx, and from this site may spread to the nasopharynx, larynx, and trachea. The organisms multiply rapidly on epithelial cells in the local lesion, producing an exotoxin that causes necrosis of cells in the area. An inflammatory reaction results, accompanied by the outpouring of a fibrinous exudate. At first patchy in appearance, as the local exudative lesions coalesce, a very tough adherent pseudomembrane forms. This is grayish to black, and composed of fibrin, necrotic epithelial cells, lymphocytes, polymorphonuclear leukocytes, erythrocytes, and diphtheria bacilli. The pseudomembrane adheres very tenaciously to the underlying tissues, and if attempts are made to forcibly remove it, a raw bleeding surface is exposed. Edema of the soft tissues beneath the membrane may be pronounced.

The growth of *C. diphtheriae* is restricted to the mucosal epithelium and rarely, if ever, do the organisms invade the deeper tissues and produce lesions in other parts of the body. The reasons for this high degree of tissue specificity are unknown. The absorption of the toxin into the general circulation, however, results in degenerative lesions in a number of organs. The most serious of these are usually associated with nasopharyngeal diphtheria and involve the heart, nervous system, and kidneys.

Clinical Manifestations

Respiratory Disease. The clinical manifestations vary depending on the virulence of the organism, host resistance, and anatomic location of the lesion. In tonsillar diphtheria, the most common clinical presentation, there is an abrupt onset characterized by low grade fever, malaise, and a mild sore throat. The cervical lymph nodes become edematous and tender, especially when there is involvement of the nasopharynx. Swelling may be so pronounced that the classic bull-neck appearance results. Extension from the nasopharynx to the larynx and trachea results in a very severe form of the disease in which mechanical ob-

---

* The Schick test dose (STD) was previously defined as 1/50 MLD (minimal lethal dose) for the guinea pig. Because of the difficulty of standardizing the lethal activity of the toxin for the guinea pig, a change was made to the use of the animal erythema potency test. The STD as now defined is that amount of standard toxin that, when mixed with 0.001 unit of the U.S. standard diphtheria antitoxin and injected intradermally in a guinea pig, will induce an erythematous reaction 10 mm in diameter.

struction of the airway by the membrane and accompanying edema introduce the risk of suffocation. Death ensues unless the airway is restored by tracheotomy or intubation.

Extrarespiratory Disease. Although diphtheria is usually a disease of the upper respiratory tract, primary or secondary lesions may occur in other parts of the body. The most common extrarespiratory site is the skin. In tropical areas, cutaneous diphtheria is relatively common, especially as a secondary infection of septic skin lesions. Although less common in temperate zones, there have been an increased number of outbreaks of cutaneous diphtheria in the United States and Canada, primarily in indigent and derelict groups. In skin diphtheria, lesions usually appear at the site of minor abrasions as chronic, spreading, nonhealing ulcers covered by a grayish membrane. The etiology of cutaneous diphtheria is complex. In addition to *C. diphtheriae*, either *Streptococcus pyogenes* or *Staphylococcus aureus* is usually present in the lesion, which fails to heal until appropriate therapy for both organisms is instituted. Mitis strains are usually associated with such infections, which are usually milder and with much less systemic illness.

Rarely, diphtherial infections of the conjunctiva, cornea, vagina, and ear occur. These are almost always secondary to pharyngeal or skin infection.

Complications. The most serious complications of diphtheria are those affecting the cardiovascular and nervous systems. Cardiac abnormalities that appear after the second week of the disease are seen in approximately 20 percent of diphtheria patients and are responsible for more than half of the case fatalities. Diphtheria toxin causes fatty myocardial degeneration, resulting in cardiac dysfunction and circulatory collapse. The myocardial damage is reversible, and if the patient survives, recovery is usually complete.

When neurologic symptoms occur, they usually appear in the third to fifth week of the disease. Involvement of the cranial nerves characteristically leads to paralysis of the soft palate, with resultant difficulty in swallowing, and nasal regurgitation of fluids. The most common manifestation of peripheral nerve involvement is a polyneuritis of the lower extremities, varying in severity from a mild weakness to paralysis of certain muscle groups. Recovery from both cranial and peripheral nerve dysfunction is usually complete.

Laboratory Diagnosis
Culture. Isolation of the organism and proof of toxigenicity is necessary for the microbiologic diagnosis of diphtheria. Because early administration of antitoxin is of paramount importance, the clinician should institute therapy immediately on the basis of clinical findings without waiting for laboratory confirmation. Streptococcal pharyngitis and Vincent's infection may be confused with diphtheria and should always be considered when making a

laboratory diagnosis. Swabs containing material from both the nose and throat should be transported promptly to the laboratory and personnel alerted to the presumptive diagnosis of diphtheria. A Löffler or Pai slant, a tellurite plate, and a blood agar plate should be inoculated immediately, and a smear stained with gentian violet should be examined to rule out infection with Vincent's fusospirochetal group of organisms. No attempt should be made to identify *C. diphtheriae* directly from smears of clinical material. Löffler's slants should be examined after 16 to 24 hours by staining with methylene blue and looking for the typical pleomorphic forms (p. 536). Blood agar plates should be examined for β-hemolytic streptococci, and any characteristic grayish or black colonies on tellurite media should be transferred to Löffler slants. Confirmatory fermentation reactions may be carried out, and the isolate on Löffler's medium tested for pathogenicity either by the in vivo virulence test or the in vitro gel diffusion technique.

In Vivo Test. Animal inoculation is recommended for laboratories that seldom isolate *C. diphtheriae*. Either guinea pigs or rabbits may be used, and one animal is sufficient for both test and control. The test is performed by injecting intracutaneously into the shaved animal 0.2 ml of a 48-hour infusion broth culture of the test organism. Five hours later, 500 units of diphtheria antitoxin is injected intraperitoneally into the guinea pig (or into the ear vein of the rabbit), and after 30 minutes, a second 0.2 ml broth sample is injected intracutaneously into the control area opposite the test site. Preliminary readings are made at 24 and 48 hours. If a toxigenic strain is present, a necrotic area appears at the site of the test injection after 48 to 72 hours. At the control site, only a pinkish nodule develops but does not proceed to ulceration because of the prior administration of antitoxin.

In Vitro Test. The gel diffusion test for determining pathogenicity is more rapid than the in vivo method, but unless plates are carefully prepared, false negative results may be obtained. In the test, antitoxin-soaked strips of filter paper are placed on the surface of a serum agar medium and the plate is inoculated heavily by streaking a line of inoculum perpendicular to the strip. The plates should be read daily for 3 days. If the organism is toxigenic, a white line of toxin-antitoxin precipitate appears, extending out at a 45° angle from the intersection of the line of inoculum and the front of antitoxin diffusing from the filter paper (Fig. 32-4).

Treatment
Antitoxin. Diphtheria antitoxin in adequate amounts is the only specific and effective treatment for diphtheria. It should be administered immediately as soon as a presumptive diagnosis of diphtheria is made clinically, without waiting for a laboratory confirmation. In severe cases, the prognosis depends to a great extent upon how early in the course of the infection antitoxin therapy is initiated. The

Figure 32-4. Elik gel diffusion test for the detection of toxin-producing *C. diphtheriae*. The paper strip is saturated with antitoxin, and a heavy inoculum is streaked perpendicular to the strip. The presence of a toxin-antitoxin precipitate after 24 to 48 hours indicates a toxigenic strain. *(Courtesy of Dr. Leon J. LeBeau, University of Illinois Medical Center.)*

toxin binds rapidly to susceptible tissue cells. The initial binding is reversible, but is quickly followed by an irreversible phase as fragment A gains entrance into the cell. At this stage toxin cannot be displaced by antitoxin. The role of antitoxin is to prevent any further binding to undamaged cells of free toxin circulating in the blood. It does this by binding to determinants on the toxin molecule at the C-terminal (17,000 dalton) end, thereby preventing attachment of fragment B to the tissue cell (p. 539).

Diphtheria antitoxin should be administered intramuscularly or intravenously in a single dose. The amount of antitoxin given depends upon the severity of the infection, but there is a general lack of agreement as to what constitutes adequate therapy. One conservative scheme specifies 30,000 to 50,000 units intramuscularly for mild or moderate cases, and 60,000 to 100,000 units intravenously for severe cases.*

Because commercial diphtheria antitoxin is usually

---

* Diphtheria antitoxin is standardized in units by comparing its ability to neutralize toxin with that of the official standard unit of antitoxin, maintained in vacuo at the State Serum Institute, Copenhagen. In terms of protective units, the standard unit of antitoxin will neutralize 100 MLD of toxin.

produced in horses, the patient should be tested for hypersensitivity to horse serum before therapy is started to avoid the possible occurrence of anaphylaxis. The test utilizes a 1:10 dilution of horse serum and may be conjunctival, in which one drop of the serum is instilled into the conjunctiva of one eye; or an intracutaneous skin test utilizing 0.1 ml of the diluted antitoxin. If the patient is hypersensitive to horse serum, the conjuctiva will develop redness in 15 to 30 minutes. In the skin test, a wheal with pseudopodia surrounded by an area of erythema appears at the site of the injection. If a positive reaction is obtained by either method, desensitization with small subcutaneous doses of antitoxin should be carried out. A syringe filled with epinephrine should be available at the bedside before antitoxin is administered by any route. Although anaphylaxis is rare, allergic reactions are frequent, and complications occur in about 10 percent of patients.

Chemotherapy. *C. diphtheriae* is susceptible to a number of antimicrobials that should be used as an adjunct to, not a substitution for, antitoxin in the diphtheria patient. Penicillin G is the drug of choice, and erythromycin is effective in the penicillin-allergic patient. Antibiotics are especially useful in the prevention of secondary infections and in treatment of chronic carriers. Without chemotherapy, from 1 to 15 percent of persons who recover from diphtheria become carriers, harboring *C. diphtheriae* for weeks or months following infection. Carriage of the organism is greater following nasal infection.

Prevention
Active Immunization. Active immunization is the key to control and prevention of clinical diphtheria. For this purpose, toxoid is administered either as fluid toxoid or as alum-precipitated toxoid. Toxoid is prepared by treating diphtheria toxin with 0.3 percent formalin at 37C until the product is completely nontoxic. The addition of alum to fluid toxoid precipitates the toxin, yielding a partially purified preparation with increased antigenic efficiency because of the local stimulatory effect of the alum.

The primary course of immunization should be started in infancy during the first 6 to 8 weeks of life. It usually consists of three doses of fluid toxoid combined with tetanus toxoid and pertussis vaccine (DPT) given intramuscularly at 4- to 8-week intervals. A booster dose should be given 1 year later, and a second booster when the child enters school.

Young children rarely exhibit either local or general hypersensitivity reactions to the toxoid, but in older children and adults one must be alert to the possibility of a sensitivity reaction, especially with alum-precipitated toxoid. This may be detected by administration of a skin test in which 0.1 ml of a 1:10 dilution of fluid toxoid is given intracutaneously. A positive reaction indicates sensitivity to the proteins of *C. diphtheriae* or other corynebacteria, or to the toxin itself. Approximately 50 percent of individuals over the age of 15 give positive reactions to the test. Such reactions present a problem in the immunization of adult population groups, and require that the toxoid be administered cautiously in multiple, small doses. For persons 7 years of age and older, a combined preparation of tetanus and diphtheria toxoids (Td) is recommended. This product contains a smaller amount of diphtheria toxoid than is present in DPT. A booster dose of Td should be given every 10 years because immunity following the original immunization series wanes. Often diphtheria infection does not confer adequate immunity against subsequent infection. Therefore, active immunization should be initiated, or the primary course completed during convalescence.

Prophylaxis for Case Contacts. All household contacts and persons with intimate exposure to a patient with respiratory diphtheria should be placed immediately under careful surveillance so that antitoxin may be administered at the first sign of illness. In addition, contacts should receive an injection of toxoid, and the unimmunized or inadequately immunized individuals should also receive a chemoprophylactic regime of penicillin or erythromycin.

Passive immunization by the use of 500 to 10,000 units of antitoxin may be employed for the protection of nonimmunized individuals heavily exposed to toxigenic organisms. Because such protection is of short duration and introduces the risk of inducing sensitization or of eliciting an anaphylactic reaction in a previously sensitized individual, its use should be limited to high-risk situations.

Other Corynebacteria

As currently constituted, the genus *Corynebacterium* includes three major groups: (1) human and animal parasites and pathogens; (2) plant pathogenic corynebacteria; and (3) nonpathogenic corynebacteria. Anaerobic diphtheroids such as the previouly classified species *Corynebacterium acnes* and *Corynebacterium parvum*, are now excluded from the genus *Corynebacterium* and reclassified in the genus *Propionibacterium*. The tendency in the past to classify many organisms as corynebacteria solely on the basis of morphology has resulted in the placing of a number of unrelated or distantly related organisms in the genus. The taxonomy and classification of this group of organisms is still unsatisfactory and presents many unresolved problems.

Species currently accepted as human and animal pathogens are listed in Table 32-1.

Corynebacterium pseudotuberculosis. Except for *C. diphtheriae*, *Corynebacterium pseudotuberculosis* is the only species producing an exotoxin. This exotoxin is, however, antigenically distinct from that of *C. diphtheriae*. Closely related to *C. diphtheriae*, *C. pseudotuberculosis* is susceptible to some of the bacteriophages used in typing the diphtheria bacillus, and when lysogenized with a *tox*+ phage, synthesizes diphtheria toxin. *C. pseudotuberculosis* causes ulcerative infections in sheep, horses, cows, and occasionally in humans.

Corynebacterium ulcerans, an organism intermediate between *C. diphtheriae* and *C. pseudotuberculosis*, has been isolated from the nasopharynx of healthy individuals and from patients with a diphtheria-like disease.

Corynebacterium pseudodiphtheriticum **and** ***Corynebacterium xerosis.*** These are the two nonpathogenic species most frequently cultured from clinical materials. *Corynebacterium pseudodiphtheriticum* is found in the nasopharynx of humans. It is a short, rather uniform rod that stains evenly except for a transverse medial unstained septum. Metachromatic granules and club forms are usually absent. *Corynebacterium xerosis* inhabits the skin and mucous membranes of humans, especially the conjunctiva. Differentiating properties are listed in Table 32-1.

TABLE 32-1. *CORYNEBACTERIUM* SPECIES PARASITIC AND PATHOGENIC FOR HUMANS AND ANIMALS

Species	Hemolysis	Sucrose Fermentation	Nitrate Reduction	Urease	Other Properties
C. diphtheriae	+	−	+	−	Human pathogen, specific exotoxin
C. pseudodiphtheriticum	−	−	+	+	Nonpathogenic, nasopharyngeal mucosa of humans, no acid from any carbohydrate
C. xerosis	−	+	+	−	Nonpathogenic, skin and mucous membranes of humans
C. pseudotuberculosis	+	v*	v*	v*	Primarily animal pathogen, specific exotoxin
C. renale	−	−	−	+	Animal pathogen
C. kutscheri	−	+	−	+	Parasite of mice and rats
C. equi	−	−	+	−	Primarily animal pathogen
C. bovis	−	−	−	+	Parasite on cow's udder, may cause mastitis

From Rogosa et al.: In Buchanan and Gibbons (eds): Bergey's Manual of Determinative Bacteriology, 8th ed., 1974. Baltimore, Md, Williams & Wilkins, p. 601.
*Variable reaction.

ERYTHRASMA. This disease occurs in the intertriginous areas and is characterized by the presence of scaly plaques that fluoresce coral red under Wood's light at 365 nm. The causative agent, *Corynebacterium minutissimum*, is nutritionally exacting and probably related to *C. xerosis*. When grown aerobically on a solid medium containing tissue culture medium base and 20 percent bovine fetal serum, porphyrins are produced that give the colonies a coral red to orange fluorescence similar to that of the skin lesions. Colonies grown on blood agar do not produce this characteristic fluorescence.

A large number of lipophilic corynebacteria have also been isolated from the skin. These strains have not been well characterized, but all require lipid for growth. Except for this property, they resemble *C. xerosis*.

Corynebacterium pyogenes and Corynebacterium haemolyticum.

These organisms are a common cause of pyogenic infections in domestic animals. They are not included in Table 32-1, however, because they represent a well-defined group, distinct from the coryneform bacteria. Unlike the corynebacteria, they lack arabinose in their cell wall, but share a cell wall polysaccharide antigen with Lancefield group G streptococci. Their cultural and biochemical properties are also unlike those of the corynebacteria.

Medical Significance. Traditionally, with the exception of *C. diphtheriae*, corynebacteria have been considered unimportant as disease producers for humans. Because of the presence of such organisms as common contaminants of clinical material, their recognition and acceptance as the cause of infection has been made with reservation. There has been, however, an increasing awareness that in the proper setting, especially in the immunocompromised host, diphtheroids may assume the role of opportunistic invaders. A number of cases of endocarditis following cardiac surgery, meningitis, and osteomyelitis have been attributed a diphtheroid etiology. Therapy with the penicillins or erythromycin has usually proved effective for such infections, although there has been an increase in the prevalence of resistant strains.

Gardnerella vaginalis (Corynebacterium vaginale, Haemophilus vaginalis)

The taxonomic position of this ogranism has for some time remained unresolved. In the Eighth Edition of Bergey's Manual, it is listed as species incertae sedis, *Haemophilus vaginalis*, but with the editorial comment, "This species does not belong in the genus *Haemophilus*." Recently, however, it has been designated a new species, *Gardnerella vaginalis*.

Morphologically, the organism appears as small bacilli and coccobacilli, measuring 0.3 to 0.6 by 1 to 2 μm, and often showing club formation and metachromatic granules. It is gram variable with retention of gram stain being more pronounced in young cultures 8 to 12 hours old. On optimal media, it stains uniformly gram-positive. It is a facultative anaerobe, and requires an enriched medium for growth. Unlike *Haemophilus* species, it requires neither hemin nor nicotinamide adenine nucleotides.

The clinical significance of *G. vaginalis* remains controversial but it appears to act in concert with certain anaerobes to cause nonspecific vaginitis. It is present in the vagina of 40 percent of asymptomatic women, but is found in large numbers in over 95 percent of patients with vaginitis. Masses of bacteria may be found on the surface of epithelial cells in the discharge. These cells observed in wet mounts are referred to as "clue cells" and are considered characteristic of *G. vaginalis* infection. Oral metronidazole has been used successfully to eliminate both organisms in the infection.

G. vaginalis has also been associated with septic abortion, puerperal fever with bacteremia, and neonatal bacteremia.

FURTHER READING

Books and Reviews

Barksdale L: *Corynebacterium diphtheriae* and its relatives. Bacteriol Rev 34:378, 1970

Barksdale L, Arden SB: Persisting bacteriophage infections, lysogeny, and phage conversions. Annu Rev Microbiol 28:265, 1974

Collier RJ: Diphtheria toxin: Mode of action and structure. Bacteriol Rev 39:54, 1975

Lipsky BA, Goldberger AC, Tompkins LS, et al.: Infections caused by nondiphtheria corynebacteria. Rev Infect Dis 4:1220, 1982

Murphy JR: Structure activity relationships of diphtheria toxin. In Bernheimer AW (ed): Mechanisms in Bacterial Toxinology. New York, John Wiley & Sons, 1976, pp 31–51

Pappenheimer AM Jr: Diphtheria toxin. Annu Rev Biochem 46:69, 1977

Singer RA: Lysogeny and toxinogeny in *Corynebacterium diphtheriae*. In Bernheimer AW (ed): Mechanisms in Bacterial Toxinology. New York, John Wiley & Sons, 1976, pp 1–30

Washington JA II: Bacteriology, clinical spectrum of disease, and therapeutic aspects in coryneform bacterial infection. In Remington JS, Swartz MN (eds): Current Clinical Topics in Infectious Diseases. New York, McGraw-Hill, 1981, vol 2, pp 68–88

Selected Papers

Audibert F, Jolivet M, Chedid L, et al.: Active antitoxic immunization by a diphtheria toxin synthetic oligopeptide. Nature 289:593, 1981

Barbieri JT, Carroll SF, Collier RJ, et al.: An endogenous dinucleotide bound to diphtheria toxin. J Biol Chem 256:12,247, 1981

Barile MF, Kolb RW, Pittman M: United States standard diphtheria toxin for the Schick test and the erythema potency assay for the Schick test dose. Infect Immun 4:295, 1971

Bezjak V, Farsey SJ: *Corynebacterium diphtheriae* in skin lesions in Ugandan children. Bull WHO 43:643, 1970

Boquet P, Pappenheimer AM Jr: Interaction of diphtheria toxin with mammalian cell membranes. J Biol Chem 251:5770, 1976

Brooks GF, Bennett JV, Feldman RA: Diphtheria in the United States, 1959–1970. J Infect Dis 129:172, 1974

Buck GA, Groman NB: Genetic elements novel for *Corynebacterium diphtheriae*: Specialized transducing elements and transposons. J Bacteriol 148:143, 1981

Centers for Disease Control: Immunization Practices Advisory Committee: Diphtheria, tetanus, and pertussis: Guidelines for vaccine prophylaxis and other preventive measures. Ann Intern Med 95:723, 1981

Chang T-m, Neville DM Jr: Demonstration of diphtheria toxin receptors on surface membranes from both toxin-sensitive and toxin-resistant species. J Biol Chem 253:6866, 1978

Collier RJ, Kandel J: Structure and activity of diphtheria toxin. I. Thiol-dependent dissociation of a fraction of toxin into enzymatically active and inactive fragments. J Biol Chem 246:1496, 1971

Creagan RP: Genetic analysis of the cell surface: Association of human chromosome 5 with sensitivity to diphtheria toxin in mouse-human somatic cell hybrids. Proc Natl Acad Sci USA 72:2237, 1975

Donovan JJ, Simon MI, Draper RK, et al.: Diphtheria toxin forms transmembrane channels in planar lipid bilayers. Proc Natl Acad Sci USA 78:172, 1981

Drazin R, Kandel J, Collier RJ: Structure and activity of diphtheria toxin. II. Attack by trypsin at a specific site within the intact toxin molecule. J Biol Chem 246:1504, 1971

Gerry JL, Greenough WB III: Diphtheroid endocarditis. Report of nine cases and review of the literature. Johns Hopkins Med J 139:61, 1976

Gibson LF, Colman G: Diphthericin types, bacteriophage types and serotypes of *Corynebacterium diphtheriae* strains isolated in Australia. J Hyg (Camb) 71:679, 1973

Holmes RK: Characterization and genetic mapping of nontoxinogenic (*tox*) mutants of corynebacteriophage beta. J Virol 19:195, 1976

Honjo T, Nishizuka Y, Kato I, et al.: Adenosine diphosphate ribosylation of aminoacyl transferase II and inhibition of protein synthesis by diphtheria toxin. J Biol Chem 246:4251, 1971

Kagan BL, Finkelstein A, Colombini M: Diphtheria toxin fragment forms large pores in phospholipid bilayer membranes. Proc Natl Acad Sci USA 78:4950, 1981

Kandel J, Collier J, Chung DW: Interaction of fragment A from diphtheria toxin with nicotinamide adenine dinucleotide. J Biol Chem 249:2088, 1974

Keen JH, Maxfield FR, Hardegree MC, et al.: Receptor-mediated endocytosis of diphtheria toxin by cells in culture. Proc Natl Acad Sci 79:2912, 1982

Laird W, Groman N: Orientation of the *tox* gene in the prophage of corynebacteriophage beta. J Virol 19:228, 1976

Lory S, Carroll SF, Collier RJ: Ligand interactions of diphtheria toxin. J Biol Chem 255:12,016, 1980

McCloskey RV, Saragea A, Maximescu P: Phage typing in diphtheria outbreaks in the southwestern United States, 1968–1971. J Infect Dis 126:196, 1972

McCloskey RV, Eller JJ, Green M, et al.: The 1970 epidemic of diphtheria in San Antonio. Ann Intern Med 75:495, 1971

Michel JL, Rappuoli R, Murphy JR, et al.: Restriction endonuclease map of the nontoxigenic corynephage γc and its relationship to the toxigenic corynephage βc. J Virol 42:510, 1982

Miller LW, Bickham S, Jones WL, et al.: Diphtheria carriers and effect of erythromycin therapy. Antimicrob Agents Chemother 6:166, 1974

Moynihan MR, Pappenheimer AM Jr: Kinetics of adenosinediphosphoribosylation of elongation factor 2 in cells exposed to diphtheria toxin. Infec Immun 32:575, 1981

Murphy JR, Bacha P: Regulation of diphtheria toxin production. In Schlessinger (ed): Microbiology—1979. Washington, D.C., American Society for Microbiology, 1979, p 181

Murphy JR, Skiver J, McBride G: Isolation and partial characterization of a corynebacteriophage β *tox* operator constitutive-like mutant lysogen of *Corynebacterium diphtheriae*. J Virol 18:235, 1976

Pappenheimer AM Jr: Interaction of protein toxins with mammalian cell membranes. In Schlessinger (ed): Microbiology—1979. Washington, D.C., American Society for Microbiology, 1979, p 187

Pappenheimer AM Jr, Harper AA, Moynihan M, et al.: Diphtheria toxin and related proteins: Effect of route of injection on toxicity and the determination of cytotoxicity for various cultured cells. J Infect Dis 145:94, 1982

Pappenheimer AM Jr, Moynihan MR: Diphtheria toxin: A model for translocation of polypeptides across the plasma membrane. In Middlebrook JL, Kohn LD (eds): Receptor-Mediated Binding and Internalization of Toxins and Hormones. New York, Academic Press, 1981, p 31

Platts-Mills TAE, Ishizaka K: IgG and IgA diphtheria antitoxin responses from human tonsil lymphocytes. J Immunol 114:1058, 1975

Rittenberg MB, Pinney CT Jr, Iglewski BH: Antigenic relationships on the diphtheria toxin molecule: antitoxin versus antitoxoid. Infect Immun 14:122, 1976

Van Ness BG, Howard JB, Bodley JW: ADP-ribosylation of elongation factor 2 by diphtheria toxin. J Biol Chem 255:10,710, 1980

Welkos SL, Holmes RK: Regulation of toxinogenesis in *Corynebacterium diphtheriae*. J Virol 37:936, 1981

CHAPTER 33
Mycobacterium

Mycobacterium

The most distinctive property of organisms within the genus *Mycobacterium* is their characteristic staining. They stain with difficulty, but once stained they are resistant to decolorization with acid alcohol. For this reason they are often referred to as acid-fast bacilli. The genus contains a wide range of nutritional types, including saprophytic species that are present in the soil as well as parasitic organisms that have not been cultured in vitro. Within the genus are species responsible for two of the most dreaded diseases in the history of mankind, tuberculosis and leprosy. Both remain major public health priorities of many developing countries. A number of species that usually exist as environmental saprophytes are also causes of human infections, and are now being seen with increasing frequency.

The mycobacteria are aerobic, slightly curved or straight rods, 0.2 to 0.6 by 1.0 to 10 μm in size. The most distinctive structure of the mycobacterial cell is its cell wall, a multilayered structure containing an abundance of complex lipids, some of which are unique to the mycobacteria and exhibit profound biologic effects in the host.

In the Bergey classification scheme, mycobacteria are grouped with the Actinomycetes and Related Organisms in Part 17. Also included in this group are the corynebacteria and nocardias that have a number of features in common with the mycobacteria (p. 536).

Mycobacterium tuberculosis

Tuberculosis is an ancient disease, recognizable in skeletons from the Stone Age and in bones from some of the early Egyptian mummies. Although the infectious nature of tuberculosis was established by Villemin around 1865, the protean nature of its clinical manifestations delayed understanding of the disease until Koch's discovery of the causative agent in 1882. Koch always found the organism associated with the clinical disease, isolated it in pure culture, reproduced the disease in animals, and recovered the bacillus in pure culture from the experimentally infected animals. For the last century, these rigid requirements—referred to as Koch's postulates—have provided criteria considered essential for the complete acceptance of a particular microorganism as the cause of a specific infectious disease.

Morphology

Mycobacterium tuberculosis is a slender, straight, or slightly curved rod with rounded ends. The organisms vary in width from 0.2 to 0.5 μm and in length from 1 to 4 μm. True branching, occasionally seen in old cultures and in smears from caseous lymph nodes, may also be produced in vitro under specific cultural conditions.

The bacilli are acid-fast, nonsporogenous, and nonencapsulated. The Ziehl–Neilsen acid-fast stain is useful in staining organisms either from cultures or from clinical material. With this stain, the bacilli appear as brilliantly staining red rods against a blue background (Fig. 33-1). Organisms in tissue and sputum smears often stain irregularly and have a beaded appearance, presumably because of their vacuoles and polyphosphate content (Fig. 33-2). Although the acid-fastness of mycobacteria is attributable to their lipid content, the physical integrity of the cell is also essential. The best explanation of the acid-fastness of mycobacteria is based on a lipid–barrier principle, according to which an increased hydrophobicity of the surface layers follows the complexing of dye with mycolic acid residues that are present in the cell wall. This prevents exit of carbolfuchsin that has become trapped in the interior of the cell.

Tubercle bacilli are difficult to stain with the gram stain, and although they are usually considered to be gram-positive, staining is rather poor and irregular because of failure of the dye to penetrate the cell wall. Gram stains of clinical material are thus invalid for the identification of mycobacteria.

Mycobacterial Cell Wall

Electron micrographs of thin sections show a thick wall composed of three layers enclosing a plasma membrane that is also a three-layered structure. Chemically, the wall is very complex and unlike that of either gram-positive or gram-negative organisms. It contains an abundance of very complex lipophilic macromolecules, many of which are unique to the organism and are biologically very active. Lipids account for approximately 60 percent of the dry weight of the wall and confer on the organism properties that enable it to resist adverse environmental conditions.

The backbone of the mycobacterial cell wall is a covalent structure consisting of two polymers covalently linked by phosphodiester bonds, a peptidoglycan and an arabinogalactan. To this covalent structure are attached in a rather ill-defined manner a large number of other complex materials (Chap. 6, p. 100). Three features distinguish the peptidoglycans of mycobacteria: (1) the presence of *N*-glycolylmuramic acid instead of the usual *N*-acetyl derivative; (2) the presence of two amide groups, on both glutamate and *meso*-diaminopimelic acid (DAP) in the peptides of the repeating subunit; and (3) the presence of two kinds of interpeptide linkages: D-ala—*meso*-DAP and *meso*-DAP—DAP. As much as 70 percent of the crosslinking in the peptidoglycan consists of interpeptide bridges between

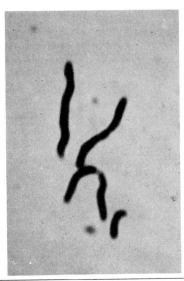

Figure 33-1. Tubercle bacilli stained uniformly by the Ziehl–Neelsen method. ✕ 3600 *(From Yegian and Kurung: Am Rev Tuberc 56:36, 1947.)*

meso-DAP residues. Interpeptide bridges of this type appear to occur only in the mycobacteria. A unit of the mycobacterial peptidoglycan is shown in Figure 33-3.

The peptidoglycan is linked to the arabinogalactan polymer by phosphodiester linkages between muramic acid residues and an arabinose of the arabinogalactan. About 1 in 10 of the arabinose residues of the polymer is esterified by a molecule of mycolic acid. The terminal branches of the arabinogalactan are linear oligosaccharides that constitute the main immunogenic determinant of the molecule. An oligomer of the cell wall, wax D, is of special interest because of its immunoadjuvant activity (p. 554).

A large number of other materials are also associated with the mycolate-arabinogalactan-peptidoglycan complex. Crude cell walls contain large amounts of most protein amino acids, which are probably present in the wall as li-

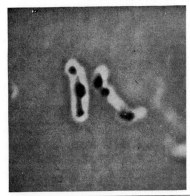

Figure 33-2. Tubercle bacilli. The protoplasm of the bacillus has been stained by the Ziehl–Neelsen method and the unstained cell wall outlined by the addition of nigrosin. ✕ 3600 *(From Yegian and Vanderlinde: J Bacteriol 54:777, 1947.)*

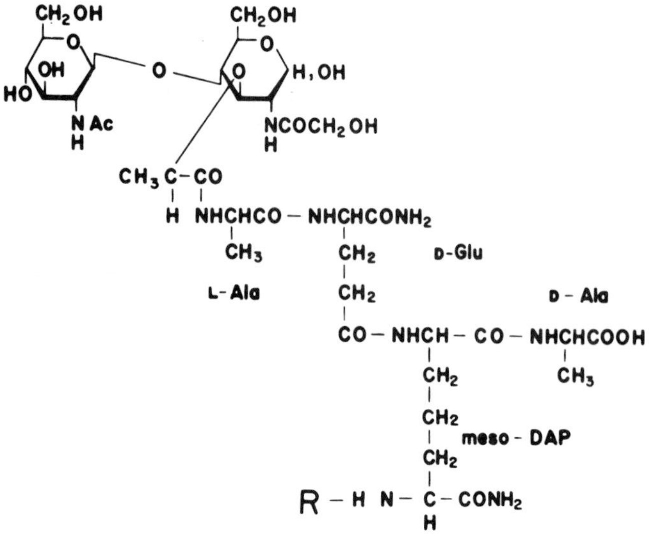

Figure 33-3. Mycobacterial peptidoglycan unit consisting of *N*-acetylglucosamine, muramic acid, L-alanine, D-glutamic acid, *meso*-diaminopimelic acid, and D-alanine. R = another unit beginning with an interpeptide bridge between the *meso*-DAP (shown) and a second *meso*-DAP linked in turn to D-Glu, L-Ala, and additional amino acids. *(From Barksdale and Kim: Bacteriol Rev 41: 217, 1977.)*

poproteins or glycolipoproteins. These peptide-containing constituents are essential for tuberculin activity of wall preparations. One of the most abundant polypeptides, a partly amidated poly-α-L-glutamic acid, is of especial interest because it is present in human and bovine tubercle bacilli, but is lacking in saprophytic species. In some strains, it accounts for up to 8 percent of the total weight of the wall.

In addition to the glycolipids bound to the peptidoglycan, other important lipid substances are present on the cell surface. The three most important of these are cord factor (trehalose 6, 6'-dimycolate), sulfatides, and mycosides, all of which have specific biologic activity (p. 102 and p. 554).

Physiology

Cultural Characteristics

M. tuberculosis is an obligate aerobe, and will not grow in the absence of oxygen. Even a small reduction in the oxygen tension results in an appreciable decrease in the rate of growth. Tubercle bacilli will grow on a very simple synthetic medium, but for primary isolation from clinical material, a more complex medium containing either an egg–potato base or serum–agar base is required. They are very slow growing even under the most optimal growth conditions, and require 10 to 20 days of incubation at 37C before growth can be visualized. Colonies are small, dry, and scaly in appearance. In synthetic liquid media, the hydrophobic properties of the organism's cell surface result in a pellicle growth confined to the surface of the medium. The addition to broth media of the nonionic detergent polysorbate 80 (Tween 80, polyoxyethylene derivative of sorbitan mono-oleate) deters aggregation of cells, and permits them to grow diffusely and to be assayed turbidimet-

rically. Aeration of cultures by rotary shaking markedly increases the growth rate and shortens the lag phase of growth. Growth is also enhanced by an increased CO_2 tension. The optimal pH for growth is 7.0, but a pH range of 6.0 to 7.6 will permit growth. For *M. tuberculosis* the optimal temperature is 37C, but for a number of other mycobacteria species the optimal temperature corresponds to the body temperature of their specific natural host, a property that must be taken into account in their isolation from clinical material.

Metabolism

The mycobacteria are strictly aerobic organisms that fulfill their energy requirements by the complete oxidation of glucose or glycerol to carbon dioxide and water. Glycerol is the preferred carbon and energy source. Although both Embden–Meyerhof (EMP) and pentose phosphate pathways are present, the EMP pathway is utilized predominantly. The tricarboxylic acid (TCA) and glyoxylate cycles are operative, and three distinct respiratory chains utilizing O_2 as terminal electron acceptor have been described. Catalase and peroxidase are present in all mycobacteria for the disposal of hydrogen peroxide generated in the final reaction of the terminal respiratory chain. Different catalases with respect to their heat stability, have been identified in various species of the mycobacteria (p. 565).

Growth Rate. Under most culture conditions, the doubling time of tubercle bacilli is 15 to 20 hours. This is also the in vivo generation time calculated from experimentally infected animals. A satisfactory explanation for the slow growth rate of tubercle bacilli is not available. However, recent findings provide evidence that the rate is limited by the DNA-dependent RNA polymerase of tubercle bacilli, and that the defect in the enzyme may be a function of limited recognition of initiation sites on the bacterial DNA.

Oxygen Requirement. The presence of oxygen is critical to the growth of *M. tuberculosis* both in vitro and in vivo. This is demonstrated very clearly by a comparison of growth curves of polysorbate 80-grown organisms in aerated vs. nonaerated cultures. In cultures well aerated by constant shaking, the growth pattern is exponential. In nonaerated cultures, an initial period of exponential growth is followed by an arithmetic linear growth period in which the bacilli replicate in the upper oxygen-rich portion of the medium at a rate that is just balanced by the rate at which the bacilli settle toward the bottom of the tube. During the settling process, the organisms adapt to survival under anaerobic conditions. When resuspended in oxygen-rich medium, they exhibit synchronous replication. The resting bacilli in this system may be analogous to the condition of tubercle bacilli lying quiescent in the host but retaining the potential of proliferating and producing overt disease after years of latency. In the in vitro system, adaptation to dormant survival under anaerobic conditions is accompanied by a marked shift to the anaplerotic glyoxylate bypass, as oxygen becomes limiting. Also brought into play is a unique glycine dehydrogenase that catalyzes reductive deamination in *M. tuberculosis*. Diversion of glycine and TCA cycle intermediates to glyoxylate synthesis serves mainly to provide a substrate for the regeneration of nicotinamide adenine dinucleotide (NAD) that may be required for the orderly shutdown before oxygen limitation stops growth completely.

Nutritional Requirements. Tubercle bacilli are prototrophic for all the amino acids, purine and pyrimidine bases, and β-complex vitamins. After primary isolation, they adapt readily to growth on simple salt solutions with ammonium ion as a source of nitrogen and glucose as a source of carbon. The most important source of carbon is glycerol, and asparagine is the preferred nitrogen source.

Iron Assimilation. Trace metal deficiences have observable consequences on the structure and metabolism of the mycobacteria. Especially pronounced are those caused by a deficiency of iron. The insoluble nature of iron, at physiologic pH values has resulted in the evolution of systems for its transport into the cell. Because of the thick lipid-rich cell wall of the mycobacteria, the system employs two iron-chelating components: (1) an extracellular water-soluble compound, exochelin, which can rob iron from ferritin (storage form of iron in the mammalian cell); and (2) mycobactin, a lipophilic molecule located in discrete regions of the cell envelope close to the cytoplasmic membrane. In the model proposed, ferric iron in the extracellular milieu is solubilized by exochelins, which then act in concert with mycobactin to transport the iron through the wall. Mycobactin, originally isolated from *M. phlei* as a growth factor for *M. paratuberculosis*, is present in all mycobacteria and appears to be confined to this genus (Chap. 5, p. 72). The basic mycobactin molecule varies only slightly from species to species.

Genetics

In spite of their medical importance, genetic studies on the mycobacteria are still in their infancy. The existence of genetic variation within members of the group is well recognized, but only in the case of drug resistance have the observed phenomena been subjected to critical genetic analysis. Fluctuation analysis has been applied to the calculation of mutation rates in *M. tuberculosis* and a number of other mycobacteria species. In these organisms, mutations occur at very low frequencies compared with those of other bacteria.

Another commonly observed class of spontaneous mutations is that causing alterations in colony morphology. Different colonial types are observable between strains as well as within a species. Smooth (S) and rough (R) variants of tubercle bacilli have been described. With few exceptions, virulent organisms produce R colonies (Fig. 33-4), but avirulent strains may also produce colonies of the R type (Fig. 33-5). The best example of this is the H37 strain of *M. tuberculosis*, which, by manipulation of culture conditions, has been dissociated into the rough virulent (Rv) and rough avirulent (Ra) variants. Although both variants are rough, their colonial morphology on both egg and liquid media is different and characteristic. These classic strains have been used extensively in experimental studies for more than 50 years.

Mycobacteriophages

Numerous phages with activities on many mycobacterial species, including *M. tuberculosis*, have been isolated from both clinical and environmental sources. Soil, human excreta, polluted water, and biopsy specimens of sarcoid and lung cancer patients are among the sources that have

Figure 33-4. Rough colonies of the human virulent strain H37 Rv. *(Grown by William Steenken, photographed by Joseph Kurung.)*

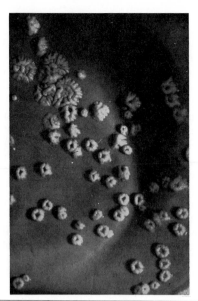

Figure 33-5. Rough colonies of the human avirulent strain H37 Ra. Note the large, flatter, intermediate colonies. *(Grown by William Steenken, photographed by Joseph Kurung.)*

yielded phages on each of several occasions. Mycobacteriophages are double-stranded DNA phages with a guanine plus cytosine (G + C) content similar to that of the host bacterium. Most of them possess an hexagonal or oval head and a long noncontractile tail (Fig. 33-6).

In addition to protein, some mycobacteriophages, including a phage grown on the host strain *M. tuberculosis* H37 Rv, also contain lipid. A high percentage of the lipid in the H37 Rv phage is phospholipid, and appears to be newly synthesized. The mycobacteriophage lipids appear to be present as a bilayer membrane-like structure essential for maintaining the structural integrity and infectivity of the phage.

In *Mycobacterium smegmatis,* the peptidoglycolipid mycoside C serves as a phage receptor, but in *M. tuberculosis* the specific phage receptor has not been chemically characterized. Although basic information on the mycobacterial host–virus relationship is still extremely limited, mycobacteriophages behave in many respects like the phages of other bacterial genera, i.e., they may either multiply within their hosts thereby causing cell death by lysis, or establish a nonlytic lysogenic relationship with the host. Pseudolysogeny is also particularly common in the mycobacteria. The establishment of lysogeny induces changes in various biologic properties of mycobacteria: colony morphology, growth rate, enzymic activity, and antigenic composition. Virulence, however, does not appear to be conditioned by the lysogenic state. Although lysogenic strains of several environmental mycobacteria (*Mycobacterium chelonei, Mycobacterium fortuitum, Mycobacterium kansasii,* and others) have been isolated from clinical material, these strains do not appear to be any more virulent than the nonlysogenic strains.

Bacteriophage Typing. Phage typing as a tool for the classification of mycobacteria is of limited value because the host range of many mycobacteriophages is not restricted to the species of the propagator strain. In the species *M. tuberculosis,* however, which is comprised of a single homogeneous serotype, bacteriophage typing offers a potentially valuable tool for genetic and epidemiologic studies. *M. tuberculosis* is divisible into several types by their susceptibility to lysis by bacteriophages. The original scheme separated all freshly isolated strains into three major and one intermediate phage types (A, B, C, and I). The geographic distribution of the phage types varies considerably, suggesting that the major phage types form distinct groups within *M. tuberculosis* rather than minor mutational variants. Type A predominates in West and Central Africa, Uganda, Japan, and Hong Kong. Types A and B are common in Europe and North America, and type C occurs predominantly in the United States. Type I, in addition to type A, is common in India but is rare elsewhere. Strains of phage type I are of especial interest because they differ from the other types in a number of properties, including virulence, lipid content, and susceptibility to various antibacterial agents. Phage types are very stable, the mutation rate to resistance to a single phage being less than 1.3×10^{-8}. Mutation to drug resistance in *M. tuberculosis* is not associated with a change in phage type.

Useful epidemiologic information on the spread of tubercle bacilli from person to person may be provided by phage typing. One of the most interesting findings is that an individual may be concurrently infected with more than one phage type. An examination of sputum from a large number of Eskimo patients showed that 14 percent contained more than one phage type of organism. Although members of the *M. tuberculosis* complex (*M. tuberculosis, Mycobacterium bovis, Mycobacterium africanum,* and *Mycobacterium microti*) cannot be distinguished by phage typing, one phage, 33D, has the useful property of distinguishing the bovine strain, bacillus of Calmette and Guerin (BCG) from other strains of tubercle bacilli by its inability to lyse BCG. Because BCG is used as an immunostimulant in tumor therapy, and in vaccination against tuberculosis, its rapid identification would be helpful to the clinician.

Unfortunately, there is at present no single internationally accepted scheme for designating phagovars (phage type). The recent proposal of a new numbering system for designating major phage types recommends the use of four major phages to designate eight phagovars of *M. tuberculosis.* This scheme promises to fulfill the need for a sound, workable, standardized method.

Resistance

Tubercle bacilli are highly resistant to drying. Cultures maintained at 37C have been found both viable and virulent after storage for 12 years. The environment in which bacilli are found is an important factor in their viability. When exposed to direct sunlight, organisms from cultures are killed in 2 hours, but bacilli contained in sputum re-

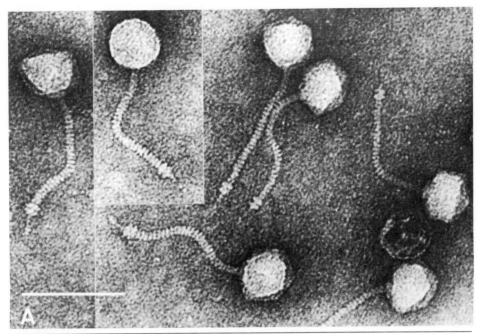

Figure 33-6. Mycobacteriophage MC-1. As with most mycobacteriophages, the symmetrical hexagonal head suggests an icosahedron. The long, noncontractile tail exhibits distinct cross-striations and ends in a single fiber extending from the tail plates. (× 270,000. Bar = 0.1 µm). *(From Barksdale and Kim: Bacteriol Rev 41:217, 1977.)*

quire an exposure of 20 to 30 hours. Bacilli remain viable for as long as 6 to 8 months in dried sputum protected from direct sunlight. The tubercle bacillus is generally more resistant to chemical disinfection than other nonsporeformers, especially when present in sputum. However, they possess no greater resistance to moist heat than other bacteria and are killed by pasteurization temperatures.

Antigenic Structure

Mycobacteria contain many unique immunoreactive substances, most of which are components of the cell wall. Cytoplasmic mycobacterial proteins also have antigenic potential. To date, however, very few of the antigens that compose this complex mosaic have been obtained in chemically pure form to permit full evaluation of their immunologic potential.

Culture Filtrate Antigens
Old Tuberculin. Old tuberculin (OT) is the original test reagent for the tuberculin test, a diagnostic skin test for tuberculosis infection. First described by Koch in 1881, OT is an antigenically crude extract prepared from 6-week-old broth cultures by boiling the culture, filtering off the organisms, and concentrating it 10-fold by steaming. The active component of this preparation is a heat-stable protein.

Purified Protein Derivative. Purified protein derivative (PPD) is a partially purified preparation of OT prepared by ammonium sulfate fractionation. The product consists primarily of a mixture of small tuberculoproteins (2000 and 9000 mol wt, respectively) in contrast with the 32,000 molecular weight of the native immunogenic tuberculoprotein. One large batch of PPD, made by Seibert in 1939, has been designated PPD-S and adopted by the World Health Organization as the International PPD–Tuberculin. Currently, PPD is the test reagent used for tuberculin skin testing (p. 560).

United States–Japan Reference System. Using unheated and thus less denatured culture filtrates of *M. tuberculosis* H37 Rv, 11 major antigens have been identified by simple one-dimensional electrophoresis, and a reference system has been established. United States–Japan antigens 1 and 2 are cell wall polysaccharides—arabinomannan and arabinogalactan, respectively—and are broadly distributed among the mycobacteria. Antigen 3 is a high molecular weight glucan, and antigens 4 through 11 are principally polypeptide proteins. There is antigenic similarity between *M. bovis* BCG and *M. tuberculosis*, and a complex sharing of antigens with other mycobacteria species.

Antigen 5. This antigen, purified from unheated culture filtrates of *M. tuberculosis* by immunoabsorbent affinity chromatography, is a cytoplasmic protein limited in species distribution to *M. tuberculosis* and *M. bovis*. Preliminary trials in human subjects using graded doses of antigen 5 have been very encouraging. The availability of a highly purified monospecific tuberculoprotein would be extremely valuable for tuberculin testing.

Polysaccharides. The protein-free polysaccharides (arabinogalactans and arabinomannans) are immunogenic and serologically active. They elicit immediate skin reactions in sensitized guinea pigs but do not elicit delayed-type hypersensitivity. Polysaccharides give precipitin reactions with antisera, and are active in complement fixation and hemagglutination reactions. The significance of these humoral antibodies, however, has not been established.

Phosphatidyl Inositol Mannosides

The phosphatidyl inositol mannosides (PIMs) are amphipathic polar lipids present in the plasma membrane of mycobacteria and related organisms. They serve an important structural role, providing a noncovalent link between the membrane and cell wall. Although purified PIMs are primarily haptenic, in the host they are undoubtedly presented in an immunogenic form, and are serologically active. The major interest in this group of glycolipids has focused on their use as potential serodiagnostic agents.

Other Immunoreactive Components

Wax D and Muramyldipeptide. The waxes D are a heterogeneous group of peptidoglycolipids composed of all of the components of the cell wall (Chap. 6, p. 100). When extracted from *M. tuberculosis*, they possess unique adjuvant activity in that they not only enhance antibody production against a protein antigen incorporated in a wax D water-in-oil emulsion, but also induce a cell-mediated immune response against the protein. Wax D duplicates the adjuvant activity of Freund's complete adjuvant, a water-in-oil emulsion of dead mycobacterial cells widely used in immunologic studies.

The least common denominator of adjuvant activity of wax D is muramyldipeptide (MDP), *N*-acetyl-muramyl-L-alanyl-D-isoglutamine (p. 549). In recent years, adjuvant research has been very active because of the immunotherapeutic effects of *M. bovis* BCG in producing regression of certain tumors.

Trehalose Glycolipids. Cord factor (trehalose-6, 6'-dimycolate), so named because of its association with the cording tendency of virulent tubercle bacilli (Chap. 6, p. 101), has a number of immunoreactive properties including adjuvant activity. It elicits extensive pulmonary granulomas that are of the foreign body type, but which are also associated with some level of protection against challenge with virulent *M. tuberculosis*. Cord factor activates the alternative complement pathway, and is endowed with demonstrable antitumor properties. A further discussion of cord factor is found below.

Sulfatides. These sulfur-containing lipids, often referred to as sulfolipids, are trehalose 2'-sulfates esterified with long-chain fatty acids (Chap. 6, p. 101). Primarily of interest because of their association with neutral red reactivity of virulent *M. tuberculosis* strains, the sulfatides, although not adjuvant active, can effectively replace cord factor as a component of an oil-BCG cell wall or endotoxin preparation causing tumor regression. Although intralesional injection of tumors with mutant *Salmonella* endotoxin admixed with sulfatide gives high cure rates, its specific role in immunostimulation may be a minor one attributable to its amphipathic properties. For further discussion of the sulfatides see below.

Determinants of Pathogenicity

M. tuberculosis produces neither exotoxins nor endotoxins. No single structure, antigen, or mechanism can explain the virulence of the organism. There is likewise no simple in vitro test based on colony morphology or serologic differences that can distinguish a virulent tubercle bacillus from its avirulent variant, a distinction that can be provided only by virulence testing in animals. A number of properties, however, are usually associated with the capacity of virulent strains of *M. tuberculosis* to produce progressive disease, and although none of these either singly or together can account completely for virulence, each undoubtedly plays a crucial role in the pathogenesis of infection.

Cord Factor. There is a high correlation between the virulence of strains of tubercle bacilli and their morphologic appearance in culture in the form of serpentine cords consisting of bacilli in close parallel arrangements (Fig. 33-7). Attenuated and avirulent forms grow in a random brush-heap pattern without this characteristic orientation (Fig. 33-8). Growth in cords can be correlated with the presence of a glycolipid, trehalose 6, 6'-dimycolate, peripherally located in the organism (p. 550 and Chap. 6, p. 101). A number of biologic responses to mycobacterial infection can be duplicated by treatment with this material. It has a peculiar and characteristic toxicity for mice; inhibits migration of polymorphonuclear leukocytes; elicits granuloma formation; and stimulates protection against virulent infection. Cord factor also attacks mitochondrial membranes, causing functional damage to respiration and oxidative phosphorylation. In spite of these varied activities of cord factor, its specific role in the pathogenesis of tuberculosis is unknown. A discussion of the adjuvant activity of cord factor is found on page 101.

Sulfatides. These peripherally located glycolipids are responsible for the neutral red reactivity associated with virulent strains of *M. tuberculosis* (p. 101). A significant correlation has also been demonstrated between the elaboration of sulfatides in culture and the rank order of virulence for the guinea pig among a series of wild-type strains spanning a broad spectrum of virulence. Although sulfatides are not toxic themselves, when administered simultaneously with cord factor, they potentiate synergistically the toxicity of cord factor. Sulfatides are readily

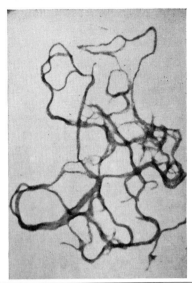

Figure 33-7. Cord growth of virulent H37 Rv. *(From Yegian and Kurung: Am Rev Tuberc 65:181, 1952.)*

endocytosed by macrophages in culture, and from within secondary lysosomes inhibit fusion of these organelles with phagosomes. Thus, as *M. tuberculosis* is an intracellular parasite, it may promote its own survival within the host by acting from within phagosomes to prevent phagolysosome formation, thus avoiding exposure to the lysosomal hydrolases.

Clinical Infection

Epidemiology
Tuberculosis is a global problem. Although effective methods for its control are available and have been applied suc-

cessfully for several decades, in some regions the prevalence of tuberculosis is still inordinately high, and a plateau in reduction of incidence is evident. Tuberculosis also remains the leading cause of death among notifiable infectious diseases.

There are an estimated 15 to 20 million infectious cases of tuberculosis in the world at the present time. Whereas the annual incidence rate in the United States has steadily declined for the past 100 years to a current rate of about 14 per 100,000 population, the rates in many developing countries, especially in some areas of Africa, Asia, and Oceania, often exceed 300 per 100,000 population. The large groups of refugees entering the United States from the developing countries are bringing with them rates of tuberculosis similar to those seen in the country from which they came. The prevalence of tuberculosis among Haitians recently entering southern Florida is 650 cases per 100,000 population. The rate for Indochinese refugees is even higher (926 per 100,000 population). These newcomers pose a serious public health problem.

The tuberculosis case rates are influenced by the race, sex, and age of the population group (Fig. 33-9). Fifty years ago the death rate from tuberculosis in the United States was exceedingly high in infants, adolescents, and young adults, and relatively low in late middle age and in old age. With improved standards of living and the adoption of control measures, a high percentage of the open carriers of tubercle bacilli have been identified and treated. This has resulted in a sharp reduction of death in infancy, adolescence, and the early years of adult life. Fifty years ago the death rate for females was higher than that of males in all age groups and all races. This is no longer true. The highest death rates now occur in nonwhite and white males after the age of 30 years, and in nonwhite and white females after the age of 60 years. There is considerable variation in the new case rate by states. Highest case rates are concentrated in a broad band across the southeastern United States, and in states with Indian reservations. The number of newly reported cases of tuberculosis each year has not declined proportionally to the decrease in deaths, and the influence of puberty on the activation of clinical disease is still apparent. Socioeconomic status is another factor closely related to tuberculosis rates. Poverty and the tubercle bacillus have always been close allies.

Tuberculosis spreads in an epidemic wave similar to that seen with other infectious diseases, but with a time course measured in decades instead of weeks. Morbidity and mortality rates rise steeply, peak, and then show a more gradual decline. The present worldwide tuberculosis epidemic began in England in the 16th century. Since then, large epidemics have been precipitated when tubercle bacilli were introduced for the first time into groups or races that had never been exposed to the infection. The Alaskan Indians and Eskimos are only now emerging from a major epidemic that has persisted on a high level for about 100 years.

Transmission. Tuberculous infection is primarily acquired by the inhalation of dried residues of droplets con-

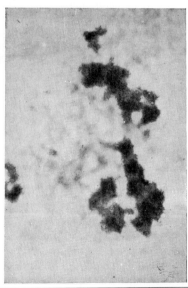

Figure 33-8. Absence of cord growth with avirulent H37 Ra. *(From Yegian and Kurung: Am Rev Tuberc 65:181, 1952.)*

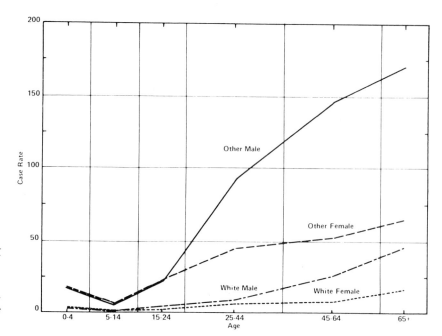

Figure 33-9. Tuberculosis case rates per 100,000 population for white and other races, males and females by age, United States, 1976. *(From 1976—Tuberculosis in the United States. Washington, D.C., U.S. Public Health Service, 1978.)*

taining tubercle bacilli that have been expelled in an aerosol created in coughing, sneezing, or talking. These droplet nuclei remain suspended in the air for prolonged periods, and those particles 1 to 10 μm in diameter are sufficiently small to reach the alveoli and initiate infection. The most important source is the infectious person with cavitary tuberculosis prior to diagnosis and the initiation of treatment. Patients on effective chemotherapy rapidly lose their infectiousness for other individuals.

Although still a problem in some parts of the world, bovine tuberculosis caused by *M. bovis* is an uncommon disease in the United States today. *M. bovis* infection results from the drinking of contaminated milk, and can be prevented by milk pasteurization and the slaughter of all infected cattle. Tuberculous infection may also be acquired by direct inoculation of abraded skin by pathologists and laboratory personnel handling contaminated tissues.

Risk of Tuberculous Infection vs. Tuberculous Disease. Humans are very susceptible to tuberculous infection but remarkably resistant to tuberculous disease. Although the prevalence of tuberculous infection, as estimated by tuberculin testing of sample populations, has demonstrated that both geographic and socioeconomic characteristics affect the risk of infection, in all cases the risk factors appear to be related to the likelihood of coming into contact with an infectious case. The degree of infectiousness of the source case is the chief factor that determines whether or not a contact will become infected.

Tuberculosis is less infectious than the common communicable diseases of childhood. Frequent and fairly prolonged association with an infectious patient is usually required for infection. Recent reports indicate that at present only 22 percent of household associates of all ages are

infected. Although the household contacts of an infected adult or adolescent patient are at greatest risk, small epidemics within certain closed environments have been attributed to schoolteachers, students, or bus drivers who were unusually good spreaders of infection. In one well-studied epidemic in Denmark, a schoolteacher infected 70 of 105 tuberculin-negative students. After infection, 41 of the 105 developed primary infections demonstrable by x-ray and cultures, and 15 of the students developed postprimary progressive tuberculosis during the next 12 years. A summary of the epidemic is shown in Figure 33-10.

It has been estimated that for all persons newly infected with tubercle bacilli, about 5 percent will develop clinical disease within a year of their infection. The remainder carry a lifelong risk of potential disease. Risk factors for development of tuberculous disease after infection are primarily intrinsic characteristics of the individual such as age, sex, body build, and genetic susceptibility. In addition, the following special risk factors predispose to disease among infected persons: use of adrenocorticosteroids and other immunosuppressive agents, hematologic and reticuloendothelial diseases that suppress cellular immunity, diabetes mellitus, and silicosis.

The risk of clinical disease is also directly related to the size of the tuberculin reaction. Reactions larger than 20 mm in diameter are almost always the result of infection by *M. tuberculosis*. As the reaction size decreases, a larger proportion of the reactions are caused by other species of mycobacteria that rarely cause disease. This correlation of the degree of sensitivity to tuberculin and risk of subsequent tuberculosis has been well documented in many diverse populations. In a study of U.S. Navy recruits, the risk of developing clinical tuberculosis was 10 times

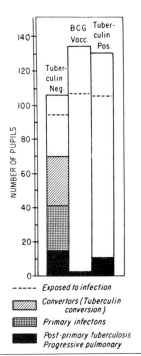

Figure 33-10. A 12-year follow-up on children infected by one teacher. *(From Hyge: Dan Med Bull 4:13, 1957.)*

greater in persons with large tuberculin reactions than in those with small reactions.

Pathogenesis

The response of an individual following exposure to virulent tubercle bacilli depends upon an interplay of two major immunologic responses: acquired cellular immunity and delayed hypersensitivity. The emergence of hypersensitivity to the proteins of the tubercle bacillus is responsible for the tissue destruction characteristic of the disease. Sensitization appears about 3 to 4 weeks after infection, and is detected by the tuberculin test. Once an individual converts to a tuberculin-positive reaction, the tuberculin-positive reaction usually remains for the rest of the individual's life (see page 560).

Individual responses following exposure are thus determined by the prior immunologic experience with the tubercle bacillus. The initial infection with *M. tuberculosis* is referred to as primary tuberculosis. Subsequent disease in a previously sensitized individual, either from an exogenous source or by reactivation of a primary infection, is known as secondary or reinfection tuberculosis.

Primary Infection. Following inhalation of virulent tubercle bacilli in droplet nuclei, the organisms reach the alveolar spaces where they are phagocytosed by alveolar macrophages. Within the macrophages, bacterial multiplication proceeds with minimal reaction, spreads to the regional nodes in the hilum of the lung, and then into the bloodstream, with a seeding of bacteria in almost all parts of the body. Asymtomatic lymphohematogenous dissemination of the primary infection occurs before the acquisition of tuberculin hypersensitivity, and sets the stage for later "reactivation" to present clinically as pulmonary or extrapulmonary disease. Circulating bacilli are efficiently cleared from the bloodstream by reticuloendothelial organs, but in the apices of the lungs, and to a lesser extent in the kidneys, vascular skeletal areas, and lymph nodes, bacterial multiplication continues. The high oxygen tensions in the lung apices provide a favorable environment for the organisms, and probably account for their predilection for these areas.

About 3 to 4 weeks after infection, the development of cellular immunity and tuberculin hypersensitivity greatly alters the course of infection. Activated macrophages limit further bacterial growth and reduce the number of organisms in both primary and metastatic foci.

TUBERCLE FORMATION. The appearance of hypersensitivity to tuberculin provokes a dramatic change in the host's response to the organisms. The nonspecific inflammatory response evoked on first exposure to tubercle bacilli becomes granulomatous, evoking the formation of tubercles. The tubercle comprises an organized aggregation of enlarged macrophages that, because they resemble epithelial cells, are referred to as epithelioid cells. A peripheral collar of fibroblasts, macrophages, and lymphocytes surrounds the granuloma. Frequently, the central region of epithelioid cells undergoes a characteristic caseous necrosis to produce a "soft" tubercle, the most characteristic hallmark of tuberculosis. When the antigen load at the initial infection site and regional lymph nodes is large, caseation necrosis may develop and later calcify. These calcified primary site lesions are referred to as the Ghon complex.

After the development of hypersensitivity, the infection becomes quiescent and asymtomatic in the majority of patients (about 90 percent). In some, however, especially the very young and adults who are immunocompromised or who have other predisposing illnesses, the primary infection may evolve into clinical disease. The progression may be local at the site of the primary lesion, or it may be at one or more distant sites where bacilli have arrived during the early hematogenous spread.

Secondary or Reinfection Tuberculosis. In a small number of those whose initial tuberculous infection subsides, secondary disease occurs in spite of the presence of acquired cellular immunity. The question of whether reinfection tuberculosis results from the breakdown of quiescent foci (endogenous) or from acquisition of new infection from an active case has long been a controversial issue; however, current opinion favors the endogenous source. In this phase of the disease, lesions are usually localized in the apices of the lungs. In about 5 percent of patients, apical pulmonary tuberculosis manifests itself within 2 years of the primary infection. In others, however, clinical disease may evolve many decades later, when resistance is lowered. Quiescent foci that harbor viable organisms thus remain a potential hazard throughout the lifetime of an individual (Fig. 33–11).

Because of the acquired cellular immunity, bacilli are more promptly phagocytized and destroyed by the activated macrophages. As a result, in secondary tuberculosis lesions remain localized and dissemination of organisms via the lymphatics is usually prevented. Hypersensitivity promotes a more rapid caseation and fibrotic walling-off of the focus. Histologically the reaction is characteristic of tubercle formation, manifest by a local accumulation of lymphocytes and macrophages. T-lymphocytes and their chemotactic lymphokines play a major role in the development of tuberculous granulomas. The chronicity of these lesions appears to be due to the persistence within them of wax D components of tubercle bacilli.

Immunity

Infections caused by *M. tuberculosis* produce a range of immunologic reactions, but since its first demonstration by Koch, immunity to tuberculosis has remained an elusive concept. Two immunologic responses, antituberculous immunity and tuberculin hypersensitivity, develop simultaneously in the naturally infected host. Koch made the initial observation of an accelerated inflammatory and healing response in guinea pigs superinfected with tubercle bacilli. In animals injected subcutaneously, the primary infection site was associated with a slowly progressive granulomatous ulcer with extensive lymph node involvement and fatal outcome. If an intradermal injection of organisms was given at a different site during this infection,

an early localized indurative response occurred, followed by rapid healing and no lymph node involvement. This observation, referred to as Koch's phenomenon, demonstrates acquired increased resistance in the infected animal, but the level of immunity is inadequate to protect the animal against death from the initial infection. Koch subsequently demonstrated that a similar localized allergic response could be induced in the skin of tuberculous animals or individuals by the injection of culture filtrates of the tubercle bacillus (tuberculin reaction). Since these early observations of Koch it has become generally recognized that neither response can be transferred to naive recipients by means of hyperimmune serum, but that an infusion of lymphoid cells from tuberculin hypersensitive donors is required. Although humoral antibodies are produced in response to naturally occurring tuberculous infection, they appear to play no beneficial role in host defense.

Acquired antituberculous immunity is the prototype of cell-mediated immunity invoked by facultative intracellular bacteria. Its development requires the cooperation of two cell types: specific T lymphocytes, which act as specific inducers; and mononuclear phagocytes, which serve as nonspecific effector cells. The dual role of the T cells is to recruit mononuclear phagocytes for the formation of granulomatous lesions, and to activate the phagocytes for enhanced bactericidal activity within the lesions. The expression of immunity ultimately depends upon the performance of the nonspecific effector cells, the activated macrophages.

Activation of macrophages is mediated by lymphokines, biologically active substances released by immuno-

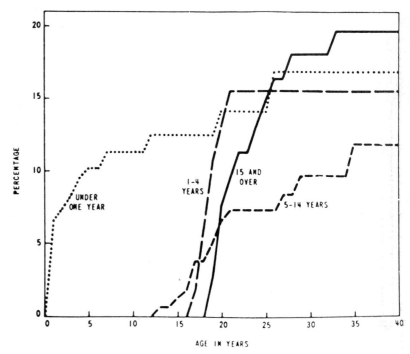

Figure 33-11. Influence of age at time of infection on the age when disease develops. *(From Zeidberg et al.: Am Rev Tuberc 70:1009, 1954.)*

competent T cells when they contact processed antigens of the tubercle bacillus. Within the developing tuberculous lesion, lymphokines with chemotactic, migration-inhibitory, and mitogenic properties cause macrophage and lymphocyte infiltration, macrophage activation, and macrophage and lymphocyte division. Activated macrophages exhibit a number of morphologic and physiologic properties that endow them with enhanced phagocytosis and bactericidal potential. Once the enhanced microbicidal activity of activated macrophages is established, it is nonspecific and can be directed against a number of unrelated microbial species, viruses, and tumor cells. This activity is the basis for the use of BCG vaccine in tumor therapy.

Macrophage activation is most pronounced within granulomas, which provide a specific focusing mechanism. Within these lesions, cells are tightly packed, facilitating cell-to-cell interactions and effective containment of the infection. Macrophages near the center of granulomas appear to be the richest in lysosomal enzymes and most capable of destroying tubercle bacilli. The alveolar macrophage is thus the key cell that determines the outcome for the host when virulent tubercle bacilli are implanted on the alveolar surface. The macrophage cytoplasm of some mammalian species is probably not favorable for growth of the organisms, especially after acquired resistance is superimposed on the basic native resistance.

The delayed hypersensitivity that develops concomitant with infection is a beneficial reaction when low amounts of bacillary antigens are involved. When large doses of antigen are present, however, the hypersensitivity reaction itself causes cell death and tissue destruction, and almost all the tissue damage characteristic of the disease.

Whether a causal relationship exists between delayed hypersensitivity and cell-mediated immunity to tubercle bacilli remains a controversial issue. A precise definition of the underlying pathologic mechanisms, specific antigens, and T cell subsets involved is required for a clear understanding of these two immunologic phenomena. However, it would appear that certain forms of delayed hypersensitivity may be mediated by the same mechanisms as protection, whereas other forms may not.

Both experimental and clinical studies have demonstrated that quiescent tuberculosis shows a high incidence of recrudescence if there is little or no response to the tuberculin reaction, whereas moderate responders show a low incidence of relapse. A high incidence of relapse is also seen in patients with a very pronounced tuberculin reaction. The magnitude of the response is thus important in assessing protective and nonprotective sensitivity.

Failure of patients to react to tuberculin—termed anergy—is considered an ominous sign in clinical medicine. It occurs in patients with miliary tuberculosis and may also be relatively common among patients with pulmonary tuberculosis, if the patients are tested while acutely ill before their disease has become chronic or responded to therapy. Studies suggest that suppressor T cells and macrophages are responsible for the increasing degree of immunosuppression that develops in severe progressive disease in humans.

Clinical Manifestations

M. tuberculosis can cause disease in any organ of the body. In the United States, however, pulmonary tuberculosis accounts for 85 percent of the cases. Infection at other sites usually results from a dissemination of organisms from the primary lesion in the lungs. Most patients with tuberculosis experience nonspecific symptoms such as malaise, fatigue, and low grade fever. The development of more specific symptoms depends on the organs involved and the extent of infection.

Pulmonary Tuberculosis

PRIMARY INFECTION. Approximately 50 percent of patients recently infected with tuberculosis develop some nonspecific clinical symptoms at the time of tuberculin conversion. Although undetectable clinically, the hallmark of initial tuberculosis infection is the prominence of hilar adenopathy compared with the relatively insignificant size of the initial focus in the lung.

In the majority of patients, the primary lesion heals completely, leaving no clinical evidence of prior infection except hypersensitivity to tuberculin. In some patients, however, the primary infection progresses directly, evolving into a pneumonic caseous process as the organisms spread through the bronchi or when a tuberculous node ruptures into a bronchus. Contiguous spread can cause infection in the pleural and pericardial spaces. In fact, pleurisy and effusion are important and not uncommon complications occurring soon after infection. The onset of pleurisy is usually abrupt, resembling bacterial pneumonia, with fever, chest pain, and shortness of breath.

CHRONIC PULMONARY TUBERCULOSIS. In adults, pulmonary tuberculosis is usually caused by organisms hematogenously seeded in the apices of the lungs during the preallergic stage of the primary infection. These metastatic foci may evolve fairly soon after seeding or after a long period of quiescence. Small patches of pneumonia develop around the foci, which become caseous as inflammation increases. The center then liquefies and empties into an adjacent bronchus, creating a cavity in open communication with inspired air from which organisms can be further disseminated to other parts of the lung and to the outside environment. As the disease progresses, there is an insidious onset and development of nonspecific constitutional symptoms such as fever, fatigue, anorexia, night sweats, and wasting. Cough and sputum are variable both in degree and time of onset, but denote more advanced disease. Hemoptysis and chest pain may also be pronounced in late chronic disease.

EXTRAPULMONARY TUBERCULOSIS. Miliary tuberculosis occurs when tubercle bacilli gain access to the lymphatics and bloodstream and seed distant organs. The term miliary is descriptive of the small barely visible foci, which resemble millet seeds, in the sites of localization. Miliary lesions may develop in almost any organ of the body, but the most favored sites for progressive tuberculous infection that develops in the absence of an adequate immunologic response are bones and joints, genitourinary tract, meninges, lymph nodes, and peritoneum. The primary infection is especially liable to develop in very young children and, together with the associated meningitis, is responsible for most tuberculosis deaths in this age group.

Tuberculin Test

The tuberculin skin test plays an essential role in the control of tuberculosis. It identifies recent or past tuberculosis infection with or without disease. The test is based on the fact that individuals infected with tubercle bacilli develop hypersensitivity to the proteins of the organism.

Purified protein derivative is the skin test reagent primarily used to detect hypersensitivity in these individuals. In order that comparable reactions may be obtained when using different batches of PPD, accurate standardization in humans against the designated standard PPD-S has been arbitrarily designated as containing 50,000 tuberculin units (TU) per mg of protein. Table 33-1 gives the relationship between the various doses of OT and PPD.

Administration of Test. The standard dose for tuberculin testing is 0.1 ml of PPD biologically equivalent to 5 TU of PPD-S (p. 553). The test is performed by intracutaneous injection (Mantoux method) and, when properly performed, produces a discrete pale elevation of the skin (wheal). To minimize reduction in potency caused by adsorption, skin tests should be given immediately after the syringe has been filled. When testing children or persons suspected of having ocular involvement, a 1 TU dose should be used. The 250 TU dose should be used only for assessing the immunologic status of patients who are negative to the 5 TU test dose.

Multiple puncture methods are also in general use for the administration of the tuberculin test. They are not recommended for diagnostic study, but because they are cheaper and easier to apply than the Mantoux test, they are widely used for screening and survey purposes. The exact amount introduced in a test cannot be measured. Multiple puncture devices introduce concentrated tuberculin into the skin either by puncture with an applicator with points coated with dried tuberculin, or by puncturing the skin through a film of concentrated liquid tuberculin. Although these methods may be practical for the busy physician in an office or clinic practice, doubtful reactions should be confirmed by the Mantoux test.

Reading and Interpretation of Test. Tuberculin tests should be read 48 to 72 hours after injection. The reading is based on the presence or absence of induration, which may be determined visually and by palpation. The diameter of induration should be measured transversely to the long axis of the forearm and recorded in millimeters. A 5 TU dose that produces 10 mm or more of induration is considered virtually diagnostic of infection with *M. tuberculosis*. A reaction that is larger than 5 mm but less than 10 mm in diameter is of doubtful significance because it may be due to other mycobacterial infections. If there is erythema without induration, or induration of less than 5 mm, the reaction should be considered negative, and the patient retested with 250 TU. If a positive reaction is obtained to this second-strength dosage, it is of doubtful significance and may reflect cross-reactions to other mycobacteria. However, a negative reaction in nonfebrile individuals who are in relatively good physical condition, and in whom anergy can be ruled out, is strong evidence against infection with *M. tuberculosis* or other mycobacteria species.

The use of PPD in the recommended skin test doses does not induce delayed hypersensitivity in an individual, even when administered over a period of months or years. Neither does tuberculin testing activate a quiescent infection. Occasionally, exquisitely hypersensitive individuals respond to tuberculin testing with vesicular and ulcerating cutaneous reactions, but only rarely does a febrile or constitutional reaction follow.

FALSE NEGATIVE REACTIONS. There are a number of potential causes of false negative reactions. Most of these are the result of injecting the tuberculin into the deeper layers of the skin, where it drains away from the local area through the lymphatics. Improper storage and handling of the tuberculin used also accounts for some errors.

About 10 percent or more of patients critically ill with tuberculosis may fail to react to the 5 TU dose of tubercu-

TABLE 33-1. COMPARABLE DOSES OF OLD TUBERCULIN AND PURIFIED PROTEIN DERIVATIVE

Dilution of Old Tuberculin	Tuberculin Injected (mg)*	Purified Protein Derivative Injected (mg)†	Tuberculin Units‡	Strength
1:100,000	0.001		0.1	
1:10,000	0.01	.00002	1.0	First
1:2000	0.05	.0001	5.0	Intermediate
1:1000	0.1		10.0	
1:100	1.0	.005	250.0	Second

From Smith: Am Rev Resp Dis 99:820, 1969.
*Based on 1 ml of concentrated OT = 1000 mg.
†Based on milligrams of protein.
‡One milligram of PPD-S contains 50,000 tuberculin units (TU).

lin or react only to the 250 TU dose. In most of these patients, the test becomes positive after a few weeks of therapy. The intensity of a tuberculin reaction may also be diminished by a number of associated illnesses and conditions: acute viral exanthems or vaccinations with live virus vaccines; immunosuppression by disease, drugs, or steroid hormones; or a state of general anergy such as that associated with sarcoidosis or malignant disease, especially lymphoma.

Fifty years ago, when almost everyone over 20 years of age had a positive tuberculin test, it was assumed that once acquired, tuberculin sensitivity would persist for the remainder of the individual's life. This is true at present only when tubercle bacilli persist in quiescent foci. If the organisms are completely eliminated, however, the tuberculin reaction is slowly diminished and finally disappears with advancing age or if the infection is treated in its earliest stages. This also occurs in individuals vaccinated with BCG but not superinfected with virulent tubercle bacilli.

FALSE POSITIVE REACTIONS. The major cause of false positive reactions is hypersensitivity to mycobacteria other than *M. tuberculosis* (Table 33–2). Although these cross-reactions tend to be smaller than reactions caused by tuberculous infections, there is no definitive point of separation. The specificity of the tuberculin test varies geographically according to the prevalence of other mycobacterial infections. In geographic areas where nontuberculous mycobacteria are prevalent in the environment, false positive reactions to 5 TU of PPD-S, particularly 4 to 12 mm reactions, are also very common.

BOOSTER EFFECT. Tuberculin test reactions that have waned with time below the level of positivity may be boosted by the stimulus of a retest, sometimes causing an apparent conversion or development of sensitivity. Although it may occur at any age, it is most frequently encountered among persons over 55 years old. The booster effect can be seen following a second test done as early as 1 week after the initial stimulating test, and can persist for a year or more. The booster effect is especially clinically important in programs of yearly tuberculin testing of hospital personnel, in whom a positive reaction on the second annual test actually caused by the booster effect may be interpreted as new infection and lead to inappropriate chemoprophylaxis. To avoid the booster phenomenon in serial tuberculin testing, it has been recommended that a repeat test be given to negative reactors 1 week later, and that positive reactions on the second test be classified as boosters rather than new infections.

Laboratory Diagnosis

Many species of mycobacteria, both saprophytes and potential pathogens, may be isolated from humans. For the individual patient, a clear-cut separation of pathogen from

TABLE 33-2. SKIN TEST REACTIONS OF MEDICAL AND NURSING STUDENTS TO 5 TU AND 100 TU OF TUBERCULIN

Student	PPD-S (5 TU) mm*	PPD-B (5 TU) mm	PPD-Y (5 TU) mm	PPD-Scot (5 TU) mm	PPD-F (5 TU) mm	OT (1:100) mm	PPD-S (100 TU) mm	PPD-A (100 TU) mm
Medical 41	0	0	0	0	—	0	0	0
Nurses 30	0	0	0	0	—	0	0	0
Medical 44	0	0	0	0	—	5.7	5.3	11.3
Nurses 24	0	0	0	0	—	—	7.6	26.0
Medical 29	0	—	—	—	0	—	5.3	12.6
Medical 8	0	—	—	—	6.6	—	2.7	12.5
Medical 39	0	0	0	10.7	—	8.3	7.1	13.1
Nurses 8	0	0	0	10.3	—	—	14.7	18.5
Medical 4	0	9.5	0	0	—	13.3	11	20
Medical 4	39	12	19	17.5	—	—	—	—
Medical 33	17	9.7	11.1	14.7	—	—	—	—

From Smith and Johnston: Am Rev Resp Dis 90:902, 1964.
TU = tuberculin units; PPD-S, human PPD; PPD-B, Battey PPD; PPD-Y, photochromogen PPD; PPD-Scot, scotochromogen PPD; PPD-F, *M. fortuitum*; OT, old tuberculin; PPD-A, *M. avium*.
*Of induration.

saprophyte is not always possible. If an isolate proves to be something other than *M. tuberculosis,* the diagnostic laboratory should be able to provide a precise species identification by use of a few in vitro tests (p. 565).

Collection of Specimens. Specimens submitted for culture should be collected before antituberculous drug therapy is started. They should be collected in sterile containers, preferably 50 ml plastic tubes with screw caps, and sent immediately to the laboratory for processing. A series of three to five single, early-morning, 5 to 10 ml samples of sputum is recommended. Because tuberculosis and other mycobacterial diseases may affect almost any organ of the body, a wide variety of other specimens such as pus, cerebrospinal fluid, urine, gastric lavage, and fluids from inflamed serous cavities are also suitable for culture.

Microscopic Examination. The detection of acid-fast bacilli in stained smears is the easiest and most rapid procedure for evaluating a clinical specimen. Because most patients with symptomatic tuberculosis demonstrate acid-fast bacilli in the sputum, such examinations play an important role in tuberculosis control programs, as transmission of the disease is due primarily to patients whose sputum contains so many organisms that they are detectable by direct microscopy.

In making the smear, small caseous areas of the sputum should be selected, spread in a thin layer on a new slide, and stained with the Ziehl–Neelsen or Kinyoun stain. Where facilities are available, fluorescent staining with auramine O or rhodamine facilitates more rapid scanning of sputum smears, and is being used increasingly. A recommended method for examining the smear microscopically is by making three longitudinal sweeps of the stained area, parallel to the length of the slide. A report from the laboratory should provide an estimate of the number of acid-fast bacilli detected. The following method is recommended by the American Lung Association:

Number of Bacilli	Report
0	No acid-fast bacilli seen
1–2 per slide	Report number found and request repeat specimen
3–9 per slide	Rare or +
10 or more per slide	Few or + +
1 or more per field	Numerous or + + + +

Digestion and Decontamination of Specimens. As most clinical specimens contain an abundance of contaminants that grow much faster than the mycobacteria, they must be decontaminated before culture. Also, because organisms are usually trapped within cellular and organic debris, exudates must be liquefied before cultures are made. The usefulness of most of the digestion–decontamination procedures depends on the greater resistance of acid-fast bacilli to strong alkaline or acidic solutions. These solutions, however, are also toxic for the mycobacteria, and

overdigestion causes a marked reduction in the numbers of survivors. The currently recommended technique used at the Centers for Disease Control employs a mixture of the mucolytic agent, *N*-acetyl-L-cysteine, and sodium hydroxide.

Culture. A number of very sensitive culture media are available that are capable of detecting as few as 10 bacteria per mililiter of digested concentrated material. They are of two types: egg-potato-base media (e.g., Lowenstein–Jensen), and agar-base media (e.g., Middlebrook 7H10). Both types should be inoculated and the culture incubated in an atmosphere of 5 to 10 percent carbon dioxide. The time from the laboratory's receipt of the specimen to the clinician's receipt of the culture report is usually 3 to 6 weeks.

Treatment

The availability of effective antituberculosis drugs has radically changed the management of patients with active disease. Surgical therapy is rarely needed and sanatoria have almost vanished. At present, the majority of patients with pulmonary tuberculosis in the United States are being treated in public health clinics, but the family physician and internist have increased responsibilities in the diagnosis, treatment, and follow-up evaluation of tuberculous patients.

Four first-line antituberculosis drugs—isoniazid (INH), rifampin, streptomycin, and ethambutol—and a number of second-line agents—pyrazinamide, cycloserine, ethionamide, p-aminosalicylic acid, viomycin, and capreomycin—are available for use in a variety of combinations (Chap. 9).

Treatment of tuberculosis is based on the use of two or more drugs in concert to prevent the emergence of resistant mutants. There are several drug regimens that can achieve a high rate of success within a treatment period of 9 months and achieve actual sterilization of the tuberculous lesion. The initial choice of regimen depends on the patient population. For newly diagnosed tuberculosis patients in whom the risk of initial isoniazid resistance is small, the treatment of choice consists of the combination of the two bactericidal drugs, isoniazid and rifampin, for a 9-month period. This two-drug bactericidal therapy for 9 months also appears adequate for the treatment of extrapulmonary tuberculosis. When there is reason, however, to suspect isoniazid resistance, two-drug bactericidal therapy with only isoniazid and rifampin is hazardous. For such patients, the use of four bactericidal drugs—streptomycin, rifampin, isoniazid, and pyrazinamide—gives uniformly favorable results. Patients who may harbor isoniazid-resistant bacilli include those who (1) have received previous chemotherapy; (2) have received isoniazid-preventive therapy; (3) have disease acquired through contact with an isoniazid-resistant patient; or (4) have infection probably acquired in countries with high prevalence of isoniazid resistance, such as Southeast Asia, Africa, and Latin America.

At present, the role of bacteriostatic drugs (ethambutol, ethionamide, and cycloserine) is limited to their use in situations where drug toxicity or the presence of multiple drug resistance precludes the use of two effective bactericidal drugs. The principal use of the previously standard chemotherapeutic regimen of isoniazid and ethambutol for 18 to 24 months is in patients who have hepatitis or impending hepatic failure.

Prevention

The prevention of tuberculosis involves either prevention of infection or, if infection has already occurred, the elimination of viable populations of organisms within the host. There are two relatively effective methods for preventing clinical tuberculosis, isoniazid prophylaxis and BCG vaccination. These methods should be considered as complementary and not competitive. The BCG vaccination is useless after the individual has been infected with tubercle bacilli, and isoniazid prophylaxis affords no protection to the uninfected individual after treatment is stopped.

Isoniazid Prophylaxis. Preventive therapy with isoniazid has become a frequently used and well-established procedure, especially in the treatment of the recent tuberculin converter. Chemotherapy with isoniazid for 1 year has been shown to reduce the risk of the evolution of a dormant infection into tuberculous disease by approximately 75 percent. Isoniazid prophylaxis is recommended for all household contacts of newly diagnosed active cases. Its greatest use is in an individual whose tuberculin conversion has occurred within the previous 2 years. Treatment of children under 5 years of age who have a positive reaction to a 5 TU dose of tuberculin is recommended without exception. Persons under the age of 20 years with a positive tuberculin reaction of unknown duration or history are generally treated, but for patients above the age of 20 years and especially those beyond the age of 35 years, there is an increased risk of isoniazid-induced hepatitis. In addition, isoniazid is recommended for patients whose health and defenses may be compromised by diabetes, alcoholism, gastrectomy, silicosis, malignancy, or prolonged corticosteroid therapy.

BCG Prophylaxis. After years of investigation, Calmette and Guerin obtained a strain of the bovine tubercle bacillus *M. bovis* with a low and relatively fixed degree of virulence (p. 564). This attenuated organism, known as the bacillus of Calmette and Guerin, or BCG, has been used to vaccinate approximately 10 million individuals. The vaccine is harmless when properly prepared and administered, but gives a relative, rather than absolute, immunity. It does reduce the immediate complications of infection stemming from lymphatic or lymphohematogenous spread, especially miliary tuberculosis and tuberculous meningitis.

The value of BCG vaccination depends on the infection rate in the population to be vaccinated, and the proportion of the population that is uninfected. The BCG vaccination can be recommended for special groups in which the morbidity rates are high and the factors favoring rapid transmission of the organisms temporarily uncontrollable. Such groups include American Indians residing on reservations; inhabitants of certain slum areas in the large cities; Naval recruits and other military personnel confined to crowded quarters and exposed to uncontrolled infection; and lastly, nurses, medical students, and hospital attendants whose professional duties necessitate almost constant exposure to infection.

The vaccine should be administered only to those individuals who have a negative tuberculin reaction to 100 TU of PPD. If a positive reaction to 100 TU dose does not develop by the end of the third month following vaccination, the procedure may be repeated. Positive tuberculin reactions usually are obtained in 92 to 100 percent of individuals receiving the vaccine, and the hypersensitivity persists for 3 to 4 years or longer. The accidental vaccination of a tuberculin-positive individual results in the rapid development, at the site of inoculation, of a superficial ulceration that persists for a few weeks but does not injure the patient. Administration by the intracutaneous method is recommended, but in the United States a vaccine for transcutaneous administration is also available.

Mycobacterium bovis

Although most cases of tuberculosis are caused by *M. tuberculosis*, two additional species, *M. bovis* and *M. africanum* also cause tuberculosis in humans. *M. africanum* has only been isolated in certain parts of Africa, but there remains a substantial residue of infection with *M. bovis* in many countries where raw milk is ingested.

Sixty years ago, dairy herds in the United States were heavily infected with bovine tubercle bacilli, and the milk from these animals provided a common source of infection in humans. Tuberculin testing of cows and slaughter of all positive reactors have dramatically reduced the incidence of infection in cows in the United States to less than 1 percent. Infection in humans has been almost eliminated in countries where pasteurized milk is consumed. However, until it is completely eradicated as a source of disease, *M. bovis* must be included in the differential diagnosis.

Morphologic and Cultural Characteristics. This species is often shorter and plumper than the human tubercle bacillus, and its primary isolation is usually more difficult. Because glycerol selectively inhibits growth of the bovine species, glycerol should not be included in the Lowenstein–Jensen or 7H10 culture media for primary isolation. The addition of 0.4 percent sodium pyruvate is stimulatory, especially for the growth of primary isolates. Strains of *M. bovis* grow more slowly, and the colonies are smaller than most clinical isolates of the human species.

In general, *M. bovis* is less aerotolerant than *M. tuberculosis*, but is more pathogenic for experimental animals. In the laboratory, the most useful single test for differentiating *M. bovis* from *M. tuberculosis* is the niacin test, a test based on the difference between the amount of free nicotinic acid produced by the two species (Table 33-3).

Pathogenesis. *M. bovis* produces spontaneous tuberculosis in a wide range of animals, including cats, dogs, and primates. In humans, the portal of entry is usually the gastrointestinal tract. Extrapulmonary lesions are prevalent, especially of the cervical and mesenteric lymph nodes, bones, and joints. When inhaled, *M. bovis* can also cause pulmonary tuberculosis indistinguishable from that caused by *M. tuberculosis.*

Bacillus of Calmette and Guerin. The bacillus of Calmette and Guerin is an attenuated mutant of *M. bovis* isolated in 1908 by the French workers Calmette and Guerin by repeated subculture on a glycerol-potato-bile medium. Subcultures of the original isolate are maintained as *M. bovis* strain BCG, and are used as an immunizing agent against tuberculosis and in cancer immunotherapy. Because BCG organisms are often isolated from abscesses arising after BCG vaccination or from the lymph nodes draining the vaccination site, the laboratory must ensure proper identification of the isolate.

Mycobacteria Associated with Nontuberculous Infections

The existence of mycobacteria having cultural characteristics different from *M. tuberculosis* had been recognized for many years, but their wide distribution in nature and lack of virulence for the guinea pig led workers to conclude that they were strictly saprophytic species. Their occasional isolation from clinical material was assumed to be completely fortuitous. Only within the last 30 years has there been a full awareness of the clinical significance of a number of these organisms. Often erroneously referred to as "atypical" mycobacteria, these organisms can and do produce severe and even fatal disease in humans.

Identification. Organisms in this group grow well on Lowenstein–Jensen, 7H10, and other media commonly used for the culture of *M. tuberculosis.* The first useful system for their classification was Runyon's scheme, in which organisms were subdivided into four groups (I to IV) on the basis of their rate of growth and pigment production. Groups I, II, and III were the slow growers, requiring 7 days or more to yield visible colonies, and group IV included the rapid growers that produced visible growth in less than 7 days. Group I was photochromogenic, producing pigmented colonies only on exposure to light. Group

II was scotochromogenic, being pigmented without light exposure, and group III was nonphotochromogenic, either nonpigmented or a very light yellow color that did not change with exposure to light. A number of organisms within each group have now been speciated and their biochemical, epidemiologic, and clinical characteristics better defined (Table 33-3). Although the number of species that has been isolated from clinical specimens is somewhat bewildering to the clinician, only a relatively few of these are known pathogens of humans.

Epidemiology. The mycobacteria in these groups are ubiquitous and have been found all over the world. They appear to be endemic in certain geographic areas, as has been impressively demonstrated by extensive skin testing (Fig. 33-12). The organisms have no known primary animal host but apparently occur in the soil. The species distribution depends upon locale, type of soil, and other climatic and environmental factors.

There is no evidence at the present time that these organisms can be transmitted directly from human to human. Epidemiologic aspects of mycobacterial pulmonary disease point strongly to the probability that most disease of this type occurs only in individuals whose lungs have already sustained damage. The current view is that mycobacterial disease caused by these organisms results from two coinciding events: (1) colonization by large numbers of mycobacteria; and (2) a localized or generalized impairment of the body's mechanisms of defense.

Clinical Manifestations. The diseases produced by this group of mycobacteria have roentgenologic, pathologic, and, to some extent, clinical similarities to tuberculosis, but there are important differences in virulence, treatment, and prognosis. Species designation of a particular infection is desirable because of differences in treatment and prognosis. Clinically these mycobacterioses may be grouped according to organ involvement as pulmonary disease, localized lymphadenitis, cutaneous disease, and, rarely, disseminated disease. The most common manifestation in the United States is pulmonary disease, which occurs mainly in older white men with chronic bronchitis and emphysema. *M. kansasii* and organisms of the *Mycobacterium avium* and *M. fortuitum* complexes are usually associated with this form of the infection. Lymphadenitis, which occurs primarily in children, is usually caused by *Mycobacterium scrofulaceum*, but any of the slowly growing species may occasionally be the causative agent. Superficial skin diseases are restricted to occupational or recreational activities involving fish. *Mycobacterium marinum* is responsible for most skin infections in the United States, whereas *Mycobacterium ulcerans* is restricted to specific areas of Africa and the Southeast Pacific. Other infections with which these mycobacteria have been associated include a large number of injection abscesses and cardiac surgical infections. Most of these have been caused by the saprophytes *M. fortuitum* and *M. chelonei.* Fatal cases of disseminated infection have

TABLE 33-3. DIFFERENTIAL PROPERTIES OF CLINICALLY SIGNIFICANT SPECIES OF *MYCOBACTERIUM*

Species	Rate of Growth	Niacin Production	Nitrate Reduction	Catalase >45 mm	Catalase pH 7, 68C	Pigment Dark	Pigment Light	Tween 80* Hydrolysis (5 days)	Tellurite Reduction	5% NaCl Tolerance	Arylsulfatase (3 days)	MacConkey Agar	Clinical Significance
M. tuberculosis	S	+	+	−	−	−	−	−	−	−	−	−	+
M. africanum	S	+		−	−	−	−	−	−	−	−	−	+
M. bovis	S	−	−	−	−	−	−	−	−	−	−	−	+
M. ulcerans	S	−	−	+	+	−	−	−			−	−	+
Group I													
M. kansasii	S	−	+	+	+	−	+	+	−	−	−	−	+
M. marinum	S	−	−	−	±	−	+	+	−	−	−	−	+
M. simiae	S	+	−	+	+	−	+	−			−	−	+
Group II													
M. scrofulaceum	S	−	−	+	+	+†	−†	−		−			+
M. szulgai	S	−	+	+	+	±†	−†	−			−	−	+
M. gordonae	S	−	−	+	+	+	−†	+			−	−	−
M. flavescens	S	−	+	+	+	+	−	+		+	−	−	−
M. xenopi	S	−	−	−	+	+†	−	−	−	−	±	−	+
Group III													
M. avium–M. intracellulare	S	−	−	−	+	−	−	−	+				+
M. gastri	S	−	−	−	−	−	−	+	−	−	−	−	−
M. terrae complex	S	−	+	+	+	−	−	+		−	−	±	+
M. triviale	S	−	+	+	+	−	−	+	−	+	−	−	−
Group IV													
M. fortuitum	R	−	+	+	+	−	−	±	±	+	+	+	+
M. chelonei	R	−	−	+	+	−	−	−	±	±	+	+	+
M. phlei	R	−	+	+	+	+	−	+	+	+	−	−	−
M. smegmatis	R	−	+	+	+	−	−	+	+	+	−	−	−
M. vaccae	R	−	+	+	+	±†	+	+	+	+	−	−	−

Date provided by George P. Kubica.

R, rapid; S, slow; +, more than 75 percent of strains positive; −, more than 75 percent of strains negative; ±, high degree of variability (40 to 60 percent with positive reactions); blank spaces, insufficient data.

*Polysorbate.

†Pigment increases either with age or after prolonged (2 weeks) exposure to light.

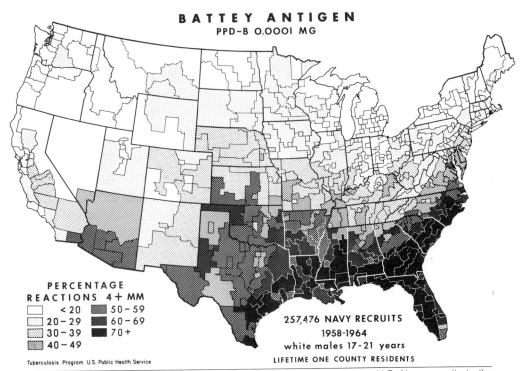

Figure 33-12. Geographic variations in the frequency of reactors among U.S. Navy recruits to the Battey antigen, the purified protein derivative prepared from a strain of *M. intracellulare. (From Palmer and Edwards: Tuberkuloza 18:193, 1966.)*

been reported in young children and immunologically deficient adults. Many different mycobacteria have been implicated, including *M. kansasii* and organisms of the *M. avium* and *M. fortuitum* complexes. In the recently described acquired immunedeficiency syndrome (AIDS) in homosexual males, however, patients have an immunologic defect that selectively favors the *M. avium-Mycobacterium intracellulare* complex over other species of mycobacteria (p. 334).

Laboratory Diagnosis. Diagnosis of the specific organism is crucial in determining therapy. Simply grouping the organisms into one of Runyon's groups is inadequate. Diagnosis depends completely on bacteriologic identification of the organism and its repeated demonstration in patient secretions in the absence of other potential pathogens. Skin testing is not generally helpful in nontuberculous mycobacterial disease. A positive acid-fast smear with a negative skin test to PPD-S, however, may lead one to suspect the diagnosis.

Treatment. Only *M. kansasii* shows significant susceptibility to the antituberculosis drugs. Organisms of the *M. avium* and *M. fortuitum* complex and all other saprophytic species of mycobacteria are highly resistant in vitro. In spite of this, various combinations of three or more antituberculosis drugs are used by many physicians, some of whom report improvement.

Slowly Growing Mycobacteria

Mycobacterium kansasii

This organism is usually longer and wider than tubercle bacilli and, with the acid-fast stain, characteristically stains unevenly to give a barred or beaded appearance. The organisms are usually arranged in curving strands.

M. kansasii is photochromogenic. Colonies, which are demonstrable after 1 to 2 weeks of incubation in the dark on glycerol egg slants, are usually smooth and ivory in color. If grown in the light, the colonies are lemon-yellow, becoming orange or reddish-orange with age.

In terms of common antigens, *M. kansasii* is closer to *M. tuberculosis* than any of the other mycobacteria species. Also, in a comparison of skin test reactions to PPDs from *M. tuberculosis* and *M. kansasii* in patients with known disease, patients reacted with complete cross-over. Experimentally, *M. kansasii* produces almost as much protection against a challenge with *M. tuberculosis* as does BCG, but other mycobacteria are less effective.

Pathogenesis. Pathogenicity for animals is extremely limited, and in most experimental animals only local lesions are produced. However, of relevance to the pathogenesis of human disease is the demonstration that guinea pigs exposed to a carbon aerosol or to a mixed carbon dust-silica dust aerosol developed progressive pulmonary disease when exposed to *M. kansasii*, whereas control animals receiving only *M. kansasii* developed minimal or no disease. Preexisting lung disease is present in a large number of persons who acquire clinically significant *M. kansasii* infections.

The pulmonary changes seen at surgery or autopsy are indistinguishable from those caused by *M. tuberculosis.* Skin lesions and lymph node lesions due to *M. kansasii* may be granulomatous or suppurative.

Clinical Infection

EPIDEMIOLOGY. Subclinical infection with *M. kansasii*, as shown by positive skin tests, indicates a definite geographic distribution. In the United States, *M. kansasii* infections are more frequent in the areas of Chicago, Louisiana, and Texas. There is also a high incidence of isolates in Northeast London, where *M. kansasii* has been isolated from the water supply. It has also been cultured from milk and animals, but its natural habitat remains a mystery. In contrast to *M. avium-M. intracellulare, M. kansasii* appears to occur most frequently in urban areas.

There is no evidence that the organism spreads directly from human to human or that children become infected when a sputum-positive member remains in a family for months to years. In the Suburban Cook County Tuberculosis Hospital-Sanatorium, 7 percent of patients admitted were infected with *M. kansasii,* but no infection has been observed in the children of these patients. In Oklahoma, another area of high infectivity, children develop positive skin tests much earlier to PPDs made from *M. kansasii* than to PPDs from *M. tuberculosis* or from other mycobacteria species. The positive skin test rate of 50 percent by the age of 17 does not increase very much after that age.

CLINICAL MANIFESTATIONS. Pulmonary disease is the most common clinical form of *M. kansasii* infection. It occurs primarily in middle-aged or elderly white males, most of whom have some preexisting form of lung disease. Clinically it resembles tuberculosis except that symptoms, when present, tend to be mild. Usually there is a very gradual progression of the disease over a period of many years, with multiple thin-walled cavity formation. As none of the opportunistic infections are contagious, isolation is unnecessary.

M. kansasii may also occasionally cause infections of the cervical lymph nodes, penetrating wound infections, and granulomatous synovitis. Although uncommon, dissemination does occur, especially in patients with conditions known to impair cell-mediated immunity.

TREATMENT. Isolates of *M. kansasii* vary in their susceptibility to the various antituberculosis agents, and in general require higher doses than *M. tuberculosis.* One recommended regimen that has resulted in healing in 90 to 95 percent of cases is isoniazid, ethambutol, and streptomycin for 18 months. Rifampin is reserved for the few cases requiring retreatment.

Mycobacterium marinum

Morphologic and Cultural Characteristics. This is a photochromogenic species implicated in a granulomatous skin disease commonly known as swimming pool granuloma. In tissues, the organisms may appear in clumps as short, thick, and uniformly staining rods, or as long, thin, beaded and barred bacilli scattered throughout the tissue. *M. marinum* grows readily at 30C to 32C on any medium commonly employed for mycobacteria. On Lowenstein–Jensen medium, colonies are grayish-white with pale yellow streaks, but upon exposure to light at room temperature, develop an intense orange-yellow pigmentation that eventually turns red.

Strains of *M. marinum* form a discrete cluster in numerical taxonomy and exhibit a unique agglutinating serovar.

Clinical Infection. Since the first isolation of *M. marinum* from granulomatous lesions of marine fish, it has been found widely distributed in nature, occurring in soil, water, and also freshwater fish. Its association with human infection was first described during an epidemic in 1954, when it was isolated both from the skin lesions of patients and from the rough cement of swimming pools. Since that time, epidemics of granulomatous skin disease have been traced to infected swimming pools, freshwater lakes, and beaches of several of the Hawaiian Islands. Sporadic cases have also been reported, a number of which were associated with the cleaning of tropical fish aquariums. Epidemic cases are readily recognized but the sporadic single case is usually misdiagnosed.

Lesions occur at the site of minor abrasions, especially on the elbows, but also on the knees, toes, fingers, and dorsum of the feet (Fig. 33-13). The infection usually begins 2 to 3 weeks after exposure as a small papule that slowly increases in size, ulcerates, and discharges pus containing the acid-fast organisms (Fig. 33-14).

In most cases, the lesions heal spontaneously after several months, but in some patients healing is prolonged, requiring 2 or more years. Various drug regimens have been used but it is difficult to assess their effectiveness. Tetracycline has been recommended as an initial conservative approach, to be replaced by rifampin and ethambutol if there is no response.

Mycobacterium simiae

Mycobacterium simiae was first isolated in 1965 during a bacteriologic survey of a colony of *Macacus rhesus* monkeys in which disease was prevalent. Since then it has been found associated with human disease in Europe, Cuba, and the United States. It has also been isolated from an environmental water supply.

Cultural Characteristics. The organism is photochromogenic, producing small dysgonic colonies that are initially buff in color, but gradually turn yellow on exposure

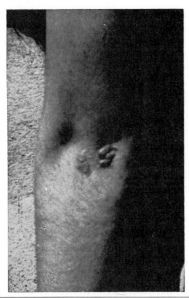

Figure 33-13. Granulomatous lesion of the elbow of 6-months' duration, showing granulations and some satellite lesions. *(Photographed by Mary S. Romer. Courtesy of Norma Johannis, Colorado State Department of Health.)*

to light. Its most distinctive property is the production of niacin, a trait not seen in the other nontuberculous mycobacteria.

Clinical Infection. Disease in humans due to *M. simiae* has not been well defined and there is relatively little epidemiologic information. It is quite probable, however, that infected *Macacus* might transmit this mycobacterium to hu-

Figure 33-14. Verrucous lesion of the toe suggesting tuberculous verrucosa cutis. *(Photographed by James A. Philpott, Jr. Courtesy of Norma Johannis, Colorado State Department of Health.)*

mans. Disease in some patients has been marked by rapid progression and extensive pulmonary cavitation, but in most patients the clinical picture has been confused by the simultaneous use of corticosteroids or by the presence of other underlying disease.

M. simiae is highly resistant to the antituberculosis drugs. An in vitro evaluation, however, should include pyrazinamide, cycloserine, capreomycin, and kanamycin, as well as the tetracyclines and erythromycin.

Mycobacterium scrofulaceum

As its name suggests, this organism is most frequently associated with a scrofula-like disease involving the cervical lymph nodes. It has been isolated from many environmental sources, from sputum, and from lesions in pigs.

Morphologic and Cultural Characteristics. Microscopically, *M. scrofulaceum* is longer, thicker, and more coarsely beaded than *M. tuberculosis*. In liquid media organisms are randomly distributed and display no evidence of cording. It is scotochromogenic, producing compact, domed colonies that are yellow to orange when grown in the dark, but that develop a reddish-orange pigmentation upon continuous exposure to light.

Because of the occasional appearance in the clinical laboratory of strains of mycobacteria that exhibit patterns of properties intermediate between those of *M. avium* and *M. scrofulaceum*, some workers prefer to consider them together as a *M. avium-M. intracellulare-M. scrofulaceum* (MAIS) complex. Taxonomically, however, *M. scrofulaceum* is sufficiently distinct to justify its designation as a separate species. *M. scrofulaceum* can also be differentiated from *M. avium* and *M. intracellulare* by skin testing and agglutination serotyping. Where identification is necessary for source-tracing, agglutination serotyping should be employed rather than further speciation based on biochemical properties.

Clinical Infection. Cervical lymphadenitis in children is the major clinical manifestation of *M. scrofulaceum* infection. The organism is very widespread in its occurrence and has been isolated from cervical lymph nodes in various parts of the world. Reports from a number of countries indicate that 75 percent of granulomatous cervical adenitis suggestive histologically of tuberculosis is actually caused by mycobacteria other than *M. tuberculosis*. In some areas, especially the Great Lakes region, Canada, and Japan, *M. scrofulaceum* appears to be the predominant cause of mycobacterial lymphadenitis, but this is not true for all regions. The widespread incidence of infection with this organism is also emphasized by the demonstration of a reaction rate of 49 percent in 30,000 Navy recruits skin tested with PPD prepared from a scotochromogenic organism.

The portal of entry in children with mycobacteriosis of the cervical nodes appears to be the oropharynx. Throat swabs and tonsils of healthy children reveal a high incidence of mycobacteria of this type, but only a relatively few children in the same area develop manifest disease. As a rule children who have developed mycobacteria scrofula have been healthy, and have not exhibited any obvious abnormalities or undue susceptibility to other infections. In some reports, however, the disease may have been preceded by a bacterial or viral pharyngitis that permitted mycobacterial invasion.

In adults, chronic pulmonary disease in association with scotochromogenic organisms has been reported, but is very difficult to assess. In some of these cases, *M. scrofulaceum* has colonized old tuberculosis cavities of the lung, and is thus suggestive of secondary invasion. Skin ulcers and abscesses, bone infections, and generalized dissemination also occur, but are very uncommon. Disseminated disease is almost invariably associated with some other serious disease.

The treatment of mycobacterial lymphadenitis consists of complete excision of the node. On the basis of in vitro reports and clinical experience, treatment of severe and threatening disease should consist of a combination of isoniazid, streptomycin, cycloserine, and rifampin until sensitivity tests on the specific strain are available.

Mycobacterium szulgai

This organism superficially resembles *M. scrofulaceum* and the nonpathogenic *Mycobacterium gordonae*, but it differs from them both biochemically and serologically. *Mycobacterium szulgai* is widespread in its distribution. Most clinical isolates to date have been from patients with pulmonary disease, rarely from lymphadenitis and bursitis. The recommended treatment is a combination of rifampin, streptomycin, and either ethambutol or isoniazid. The organism must be differentiated from *M. gordonae*, a slow-growing scotochrome that is frequently isolated from the sputum of patients who do not have mycobacterial infections.

Mycobacterium xenopi

Mycobacterium xenopi is probably the most easily recognized of potential mycobacterial pathogens. Although it was first isolated from a cold-blooded animal, the toad *Xenopus laevis*, it is unique among mycobacteria in that it grows poorly at 37C, preferring 42C to 45C. For primary isolation, 3 to 4 weeks or longer are required. Colonies are characteristically very small and granular, and microscopically show a peripheral network of hyphae. A yellow pigmentation gradually develops upon prolonged incubation. In stained smears, the bacilli are unusually long and thin and are often arranged in typical arching patterns.

M. xenopi has been isolated from both hot and cold tap water, and from granulomatous lesions in swine. It is more limited in its geographic distribution than other opportunistic mycobacterial pathogens, with more reported clinical isolations coming from northwestern Europe than from America. Most of these have occurred not far from the coast or tidal estuaries, suggesting an ecologic association with the sea or seabirds. *M. xenopi* produces a chronic slowly progressive pulmonary disease that is clinically and radiologically similar to tuberculosis. Accurate identification of the organism is important because it is more amenable to drug therapy than *M. scrofulaceum* or *M. avium-M. intracellulare*, with which it may be confused. Initial treatment should consist of isoniazid, rifampin, and ethambutol. The organism is also generally sensitive to the other antituberculosis drugs.

Mycobacterium avium–Mycobacterium intracellulare

There is considerable overlap between the properties of the two major nonphotochromogenic pathogens, *M. avium* and *M. intracellulare*, making speciation of strains extremely difficult. Because of what some mycobacteriologists consider as needless concern over the separation of the two species, the term *M. avium-M. intracellulare* complex (MAC) is used when referring to these mycobacteria.

Organisms falling within the *M. avium-M. intracellulare* and the *M. avium-M. intracellulare-M. scrofulaceum* (p. 565) complexes are too heterogeneous for successful speciation by biochemical traits. The method that has provided the best working order for the group, and which at the present time appears to be the most acceptable, is serotyping by agglutination and agglutinin absorption. *M. avium* is composed of three distinct agglutination serotypes, whereas *M. intracellulare* contains over 20 serotypes and *M. scrofulaceum* consists of five. Neither the temperature at which a serotype grows best nor pathogenicity for birds is adequate for distinguishing strains. The basis for serologic specificity is a unique class of C-mycoside glycopeptidolipids, which are present in the organism as a superficial sheath.

Morphologic and Cultural Characteristics. In clinical materials, the organisms are pleomorphic, but on culture media they usually appear as short rods with bipolar acid-fast granules (Fig. 33-15). Most virulent avian strains grow better at 44C than at 37C, while most strains of *M. intracellulare* prefer the lower temperature. Upon repeated subculture in the laboratory, however, strains of avirulent variants of avian strains may adapt to good growth at 37C, and are then indistinguishable by the usual tests from avirulent *M. intracellulare* strains. Colonies of primary isolates are predominantly thin, translucent, and smooth, but a few rough colonies are also often produced (Fig. 33-16). Upon subculture, colonies become more opaque and domed. Correlated with these colony variations are changes in virulence and other properties.

Clinical Infection

EPIDEMIOLOGY. Organisms of this group are worldwide in distribution. They are ubiquitous in the environment and have been isolated from water, soil, dairy products as well as from the tissues of both birds and mammals. Avian serotypes produce spontaneous disease in domestic fowl

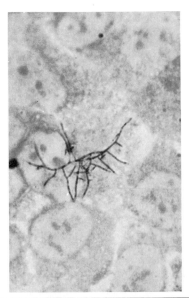

Figure 33-15. Battey strain of *M. intracellulare* from human pulmonary disease after 3 days' growth in HeLa cells. Note branching. × 1200 *(From Brosbe et al.: J Bacteriol 84:1282, 1962.)*

and other birds, and can spread from these sources to cows, swine, and humans. *M. intracellulare* serotypes, however, are associated with insects, sawdust, soil, and environmental sources. There is no accurate estimate of the prevalence of *M. avium-M. intracellulare* infections, although skin test surveys indicate that inapparent infection is common, especially in the southeastern United States. It ap-

pears to be more frequent than infection with *M. tuberculosis* but less frequent than with *M. scrofulaceum* (Fig. 33-12).

CLINICAL MANIFESTATIONS. Chronic pulmonary disease caused by *M. avium-M. intracellulare* is clinically and pathologically indistinguishable from tuberculosis. The incidence is highest among middle-aged to elderly white men, most of whom have preexisting pulmonary disease. Extrapulmonary involvement is rare, although in some geographic areas *M. avium-M. intracellulare* is associated with cervical adenitis. Until 1982, only 14 cases of disseminated infection in adults had been described, all in immunosuppressed persons. Suddenly, however, there has been an unusual clustering of cases of disseminated *M. avium-M. intracellulare* infection in homosexual men dying of AIDS. At present, this syndrome is poorly delineated, but is limited almost completely to young homosexual men, drug abusers, and Haitians who have *Pneumocystis carinii* pneumonia and Kaposi's sarcoma. These patients commonly have evidence of cytomegalovirus (CMV) infection, as well as infections with other opportunistic pathogens. They have well-defined T lymphocyte subset abnormalities and cutaneous anergy. An explanation for the sudden and selective occurrence of *M. avium–M. intracellulare* disease in this particular immunosuppressed group is currently lacking.

TREATMENT. *M. avium–M. intracellulare* are generally highly resistant to most antituberculous drugs. Clinically ill patients with progressive pulmonary disease or disseminated disease should be treated with four- to six-drug regimens including isoniazid, rifampin, ethambutol, streptomycin, cycloserine, and ethionamide.

Mycobacterium malmoense
Mycobacterium haemophilum

These are recently described nonphotochromogenic species with biochemical properties similar to those of *M. avium*. *Mycobacterium malmoense* appears to elicit a distinct agglutinating serotype, and is associated with pulmonary disease. *Mycobacterium haemophilum* is unique among slow-growing mycobacteria in its inability to grow in the absence of hemin. Also, even with hemin, it grows only at 30C but not at 37C. It is probable that this organism is sometimes not isolated because of failure to use the appropriate media. *M. haemophilum* was first isolated from granulomatous skin lesions of a patient with Hodgkin's disease.

Rapidly Growing Mycobacteria

Mycobacterium fortuitum
Mycobacterium chelonei

The rapidly growing mycobacteria are widely distributed in both aquatic and terrestrial habitats. Most of these are

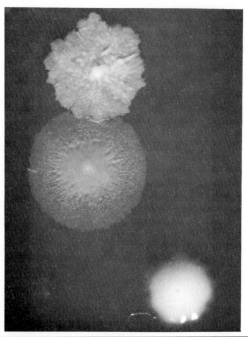

Figure 33-16. Rough, transparent, and opaque colonies of *M. avium*, serotype 2, on oleic acid-albumin agar. Original magnification × 15. *(From Schaefer et al.: Am Rev Resp Dis 102:499, 1970.)*

purely environmental saprophytes, but two species, *M. fortuitum* and *M. chelonei,* are occasional pathogens of humans, birds, and poikilothermic animals.

Morphologic and Cultural Characteristics. Although *M. fortuitum* and *M. chelonei* are now well-defined species, they share a number of similar metabolic characteristics and are not infrequently found in the same types of infection; therefore they are sometimes referred to as the *M. fortuitum-M. chelonei* complex. Organisms of this group are pleomorphic and exhibit various degrees of acid fastness. Long, filamentous forms are seen in pus, sometimes with definite branching. Colonies that appear after 72 hours incubation at 37C on the 7H10 medium, are unpigmented and are usually large and rough, but soft waxy butyrous colonies may also be seen on some media. Many strains of *M. fortuitum* may be incorrectly diagnosed as *M. tuberculosis* if the cultures are not inspected for 3 to 4 weeks after inoculation. At that time *M. fortuitum* presents an unpigmented, somewhat rough growth that may closely resemble that seen with the human tubercle bacillus. The key to the identification of *M. fortuitum* is the inspection of cultures at 4 to 14 days and selection of the colorless, rapid growers for further study.

M. fortuitum is characterized by its rapid growth on ordinary media at room temperature as well as at 37C, and its uniform resistance to isoniazid, streptomycin, and p-aminosalicylic acid. Strains exhibit marked variability in their biochemical properties, especially upon primary isolation, making speciation of strains extremely tedious and of questionable clinical importance. The species contains three major biotypes, which also differ antigenically. Almost all of the strains isolated from lesions in humans and animals are of the biochemically least active type A. *M. fortuitum* and *M. chelonei* can be distinguished by growth and physiologic properties, and by antigenic structure.

Clinical Infection. Organisms of the *M. fortuitum-M. chelonei* complex have been found in 30 to 78 percent of soil samples from various areas of the United States. They have been isolated from the sputum and saliva of healthy persons, and from scrub sinks in operating rooms.

In skin tests, *M. fortuitum* gives a specific reaction to PPD-F and exhibits very slight cross-reactions with other tuberculins. In a large study of Navy recruits, only 7.7 percent gave skin reactions to 5 TU of PPD-F, a purified protein antigen prepared from *M. fortuitum* (Table 33-4).

The most common clinical manifestation is an abscess appearing at the site of trauma, usually at the injection site of supposedly sterile products. Less frequent are corneal ulcers following some type of penetrating injury, and pulmonary infection. Pulmonary cases of *M. fortuitum* infection cannot be distinguished by x-ray from typical tuberculosis. In most of these infections, there has been evidence of preexisting disease, concomitant injury by other material, or suppression of immunity that permitted invasion by an organism that is rarely pathogenic. Serious infections have developed in wounds following open-heart surgery; venous stripping, and in renal homograft recipients. A *M. chelonei*-like organism has also been responsible for outbreaks of peritonitis in persons with renal disease associated with the use of intermittent chronic peritoneal dialysis.

M. fortuitum-M. chelonei are highly resistant in vitro to antituberculous drugs. In spite of this, some success has been achieved using multiple drug regimens that include four to six drugs. If drug therapy fails, surgical resection may be necessary.

Saprophytic Species

Although there is no evidence that these organisms can cause disease, they often occur as contaminants in clinical material. Two species, *M. smegmatis* and *Mycobacterium phlei,* have been used extensively in biochemical and genetic studies because they grow rapidly on simple media and constitute no health hazard in the laboratory. All studies on the structure and function of the mycobacterial genome and all successful attempts at genetic transfers have employed *M. smegmatis,* whereas *M. phlei* has been used in studies on mutagenesis. *M. smegmatis* is usually nonchromogenic, while the pigmentation of *M. phlei* varies with the medium upon which it is grown. *M. phlei* can grow at temperatures up to 52C and, unlike *M. smegmatis,* can survive for 4 hours at 60C. Both organisms are present in the soil. *M. phlei* occurs only occasionally in the sputum of patients with pulmonary cavities or bronchiectasis, while *M. smegmatis* is often present in the smegma around the genitalia.

Multiple Infections with Mycobacterial Species

Because of the ubiquitousness of many species of mycobacteria, some individuals will experience primary infections with two or more species. If such infections precede infection with *M. tuberculosis,* they might have an immunizing effect against clinical disease with *M. tuberculosis.* Experimental work has shown that *M. kansasii* gives animals almost as good immunity as BCG. Avian and Battey (intracellulare) strains give significant immunity, while scotochromes give less, although definitely measurable, immunity.

The importance of infections with atypical mycobacteria in protection from clinical tuberculosis was demonstrated in a study of Naval recruits who were given simultaneously 5 TU doses of PPD-S from the human tubercle bacillus and of PPD-B from the Battey bacillus. When the PPD-B reaction was 2+ mm greater than the PPD-S reaction, the morbidity rate due to *M. tuberculosis* was only 17 per 100,000 population. When the reactions were 10 mm or more and within 2 mm of each other, the morbidity rate was 95 per 100,000 population. However, when the PPD-S was definitely greater than the PPD-B, the Battey reaction represented a cross-reaction, and not a double infection, and the morbidity rate was 289 per 100,000 population.

TABLE 33-4. FREQUENCY AND SIZE OF REACTIONS AMONG NAVY RECRUITS TO 0.0001 MG OF PURIFIED PROTEIN DERIVATIVE ANTIGENS PREPARED FROM VARIOUS SPECIES OF *MYCOBACTERIUM*

Purified Protein Derivative Antigen Prepared From	Number Tested	Reactions of 2 mm or More	
		Percentage	Mean Size (mm)
M. tuberculosis	212,462	8.6	10.3
M. fortuitum	3415	7.7	4.8
Unclassified (Group III)	3729	12.0	5.8
M. kansasii	13,913	13.1	6.2
Unclassified (Group III)	9473	17.5	7.0
M. smegmatis	14,239	18.3	5.7
M. phlei	15,229	23.1	6.4
Unclassified (Group II)	10,060	28.4	9.0
M. avium	10,769	30.5	6.7
Unclassified (Battey type)	212,462	35.1	7.7
Unclassified (Group III)	8402	39.0	7.2
Unclassified (Group III)	29,540	48.7	10.3

Modified from Edwards: Annu NY Acad Sci 106:36, 1963.

Mycobacterium ulcerans

This organism produces a destructive, primarily tropical skin disease that, if not treated early, produces chronic ulcers with necrotic centers. The disease was first described in a small group of patients in Victoria, Australia, but has since been reported from a number of tropical countries in Asia, Africa, and South America.

Morphologic and Cultural Characteristics. In culture preparations, the bacilli are 0.5 μm wide and 1.5 to 3.0 μm long, but in tissue sections they are usually larger and are beaded in appearance. The optimal temperature for growth is 30C to 33C, with no growth occurring at either 25C or 37C. It is very slow-growing, requiring 6 to 12 weeks on primary isolation. Rough domed colonies, lemon-yellow in color, are produced on Lowenstein–Jensen medium. *M. ulcerans* is biochemically unreactive.

Pathogenesis. Pathologic changes in the skin of patients infected with *M. ulcerans* do not resemble tuberculosis. No caseation and only rare granulomas are seen. The base of the ulcer usually displays a rapid, undermining necrosis.

Most laboratory animals are resistant to infection, but mice may be infected by the injection of *M. ulcerans* into their footpads, a site with a low body temperature.

Clinical Infection. *M. ulcerans* infection has been most frequently encountered in isolated pockets in Australia, Africa, and Mexico. This is the third most important of the mycobacterial diseases. Proximity to rivers or swampy areas appears to be an important factor, but no studies have demonstrated the existence of the organism outside the human body.

Lesions usually occur on the legs or arms, beginning as a single subcutaneous nodular lesion that gradually breaks down to form a chronic necrotizing, undermining skin ulcer. The lesion is somewhat similar to leprosy, but more superficial and without nerve involvement. Acid-fast bacilli may be demonstrated in the lesions.

The most important form of treatment is wide surgical excision and skin grafting when lesions are extensive. In general, drug regimens have not proved effective.

Mycobacterium leprae

Leprosy is an ancient disease; its origin is shrouded in antiquity. As early as 1400 BC reference to it as an old disease in India may be found in the sacred Hindu writings of the Veda. Many accounts of leprosy may also be found in the ancient Hebrew writings, and although some of the skin lesions considered to be leprosy in the Old Testament of the Bible were probably not leprosy, many undoubtedly were.

Hansen, in 1878, described the presence of myriads of bacilli in the lesions of leprosy patients. Although this was the first description of a microorganism as the cause of human disease, the organism is still an enigma. All attempts to culture *Mycobacterium leprae* in vitro have failed. It is difficult to propagate and transmit to experimental animals, and its slow growth in both animals and patients has drastically hampered investigative efforts.

Stimulated in large part by the World Health Organization, there has been renewed interest in leprosy in recent years, in an attempt to gain a better understanding of the disease that continues to threaten the quality of life for over 12 million people in all parts of the world. The movement of the world's growing population and today's means of rapid transportation have resulted in increased contact between susceptible travelers and the millions of patients who have leprosy. The current concern in leprosy is thus of global importance.

Morphology and Physiology

When stained by the Ziehl–Neelsen method, the leprosy bacilli are seen as acid-fast rods predominantly in modified mononuclear or epithelioid structures called lepra cells. Organisms are found singly or in large masses termed globi. Large numbers of bacilli may be packed in the cells in an arrangement that suggests packets of cigars. The individual rods vary in length from 1 to 7 μm and in width from 0.2 to 0.5 μm. The rods are usually straight or slightly curved and may stain uniformly or show granules and beads that are slightly larger than the average diameter of the cell. Bacilli uniformly staining acid-fast are healthy viable cells, whereas bacilli showing beading are probably nonviable. Acid-fastness of *M. leprae* may be removed by preliminary extraction with pyridine, a useful property in distinguishing it from most other mycobacteria.

In thin sections of *M. leprae* in lepromatous nodules, the organism's structure resembles that of *M. tuberculosis* with a three-layered cell wall and a complex intracytoplasmic membrane system that connects with the plasma membrane.

All well-controlled attempts to cultivate *M. leprae* have met with failure, including attempts to grow the organism in cultures of various types of human cells. Up to the present time, no published paper has produced proof of multiplication of *M. leprae* either in the presence or absence of cultured cells. It can be grown experimentally only in animals.

The presence of a phenolase in *M. leprae* obtained from lepromatous skin nodules provides a simple test for separating *M. leprae* from other mycobacteria and from nocardias in which activity has never been detected. The phenolase converts 3,4-dihydroxyphenylalanine (dopa) to a colored product having an absorption peak at 540 nm.

Experimental Disease in Animals

M. leprae can be grown only in mice, rats, armadillos, and hedgehogs. The most frequently used animal is the normal mouse infected in the footpad. In the mouse, infections can be initiated with as few as 1 to 10 bacilli. Footpad temperatures of 30C, obtained by controlling air temperatures at 20C to 25C, are the most favorable, and are the secret of the footpad success. The intravenous injection of *M. leprae* into mice results in lesions in the nose and front feet. In humans, the bacilli that are shed from the nose have been growing at that site at a temperature of approximately 30C.

In the footpad of the mouse, the multiplying *M. leprae* produce a growth curve and histology sufficiently distinctive to permit its differentiation from other mycobacteria. During its logarithmic phase of growth, it has a generation time of approximately 12 days. Multiplication at this rate continues for 150 to 180 days after inoculation until the number of bacilli in the footpad reaches a level of about 1 $\times$ 10^6 organisms. At this time, multiplication stops, apparently because of the triggering of cell-mediated immunity. Lymphocytes infiltrate the area, and macrophages in the center of the lesion enlarge. The mouse model has been especially useful for drug screening and vaccination experiments. The armadillo has also been used extensively as an experimental model. In this animal, the disease becomes disseminated to all organs, in contrast to the disease in humans. The armadillo has been especially useful in providing a system for the study of immunologic factors that control development of the disease. It is also useful in providing large numbers of *M. leprae* for laboratory investigations. Approximately 200 g of *M. leprae* (1 $\times$ 10^{12} organisms) can be obtained from one animal 15 months after inoculation. The potential usefulness of the armadillo model, however, has also been threatened by the finding of a naturally occurring leprosy-like disease among wild armadillos. Until the armadillo can be bred in captivity and pathogen-free animals are available, its full-scale use as an experimental model is hampered. Humans are highly resistant to experimental infection. A number of attempts have been made to infect humans experimentally, but most have ended in failure.

Antigenic Structure

Different levels of antibodies to *M. leprae* are associated with different forms of leprosy, but at present there is little definitive information on the molecular structure of the specific antigens against which the antibodies are directed. As many as 20 different antigens can be detected by immunodiffusion. Some of these are antigens common to all mycobacteria. Only one determinant specific to *M. leprae* has been identified. Recently, using monoclonal antibody technology for the antigenic analysis of *M. leprae*, 11 monoclonal antibodies have been produced, two of which react only with *M. leprae*. These antibodies have many potential uses in serologic diagnostic tests and immunochemical analysis.

Clinical Infection

Epidemiology

There are approximately 15 million persons with leprosy in the world at the present time. The disease is most prevalent in tropical areas, especially in Africa, South and Southeast Asia, and parts of South America. In endemic areas of Africa, 20 to 50 persons in every 1000 may be infected. Within the United States, leprosy is endemic in Hawaii and small areas of Texas, California, Louisiana, and Florida. In recent years there has been a continuous increase in the number of new cases reported each year, most of whom have been foreign-born immigrants from leprosy-endemic areas, especially Mexico, the Phillipines, Southeast Asia, and Cuba.

Humans appear to be the only natural host for *M. leprae*. Infection is acquired by contact with patients with lepromatous leprosy who shed large numbers of the organisms in their nasal secretions and ulcer exudates, but the precise modes of transmission have not been established. The most probable portals of entry are the respiratory tract and cutaneous route through excoriations in the skin.

Although humans have traditionally been considered the sole natural host of *M. leprae*, an organism indistinguishable from the leprosy bacillus has been recovered from wild armadillos captured for use in experimental research. Up to 10 percent of the animals in certain regions of Louisiana are infected with a disseminated disease resembling lepromatous leprosy. This finding, coupled with observations that show that certain arthropods may play a role in the transmission of infection, provide additional parameters in the currently ill-defined epidemiology of this disease.

Both environmental and host factors determine susceptibility to infection with *M. leprae*. Geographic, ethnic, and socioeconomic factors contribute to the spread of leprosy by affecting the number of untreated or ineffectively treated lepromatous cases, and the opportunities for exposure. Persons in household contact with lepromatous cases have a 5- to 10-fold increase in risk compared with individuals with no known household contact.

Genetic factors have been shown to contribute considerably to the striking individual differences in susceptibility and type of response to infection with *M. leprae*. The most convincing evidence to date has been provided by studies designed to detect linkage by comparing HLA segregation patterns with the distribution of leprosy within sibships. There is evidence for the existence of a gene predisposing to tuberculoid leprosy that is linked to HLA. In the multiple-case families, the HLA-linked susceptibility gene appears to be either DR2 itself, or a gene associated with DR2. Genetic differences among hosts in response to infection with *M. leprae* reflect differences in specific immunologic competence among individuals, and this finding is consistent with the variation in pathologic types that is manifested by patients with leprosy, along the spectrum of disease.

Pathogenesis

The spectrum of disease activity in leprosy is very broad, characterized by pronounced variations in clinical, histopathologic, and immunologic findings. On the basis of these properties, Ridley and Jopling have established a classification scheme consisting of five forms of leprosy: tuberculoid (TT), borderline tuberculoid (BT), borderline (BB), borderline lepromatous (BL), and lepromatous (LL). In this spectrum, which is summarized in Table 33-5, only two forms are stable, TT and LL. The others are unstable, especially BT, which can regress in the absence of treatment to BB or BL.

M. leprae is an obligate intracellular parasite which multiplies very slowly within the mononuclear phagocytes, especially the histiocytes of the skin and Schwann cells of the nerves. It has an especially strong predilection for nerves.

Tuberculoid Leprosy. In TT, skin biopsies show mature granuloma formation in the dermis consisting of epithelioid cells, giant cells, and rather extensive infiltration of lymphocytes. Acid-fast bacilli usually cannot be demonstrated. The organisms invade the nerves, and selectively colonize the Schwann cells. The cutaneous nerve twigs are obliterated and the larger nerves are swollen and destroyed by granulomas. The nerve damage is nonspecific, and arises as the consequence of a cell-mediated immune response.

Lepromatous Leprosy. The histopathology of LL is strikingly different. Epithelioid and giant cells are absent, and lymphocytes are rare and diffusely distributed. The inflammatory infiltrate consists largely of histiocytes with a unique foamy appearance resulting from the accumulation of bacterial lipids. Massive numbers of acid-fast bacilli are found within the macrophages. Skin biopsy specimens may contain up to 10^9 bacilli per gram of tissue. *M. leprae* tends to invade vascular channels, resulting in a continuous bacteremia in lepromatous patients, and consistent involvement of the reticuloendothelial system. The nerves are also infected and numerous bacilli can be seen within the Schwann cells. Damage to the nerve structure, however, is less than in TT.

Immunity

Leprosy is a disease of low infectivity (Fig. 33-17). Most individuals never develop clinical manifestations of disease, whereas many others develop a localized lesion that heals spontaneously. This implies that those individuals who develop disease are in some way immunologically defective with respect to *M. leprae*. Also, in leprosy there is a close correlation between the various clinical forms and the cell-mediated immune response of the host. This is shown in Figure 33-18, which illustrates the inverse relationship between the intensity of the delayed hypersensitivity response to *M. leprae* and the humoral response throughout the clinical spectrum of leprosy. Patients with tuberculoid leprosy exhibit a strong delayed-type hypersensitivity to lepromin, and the histology of lesions is that of hypersensitivity granulomas. As the disease progresses across the leprosy spectrum there is a progressive loss of this hypersensitivity, and the appearance in the patient with LL of an anergic state. A concomitant loss of cell-mediated immunity parallels the decline of delayed hypersensitivity to *M. leprae* antigens. Conversely, a high serologic response characterizes LL, and polyclonal hypergammaglobulinemia is a characteristic feature. Antibodies to *M. leprae* that cross-react with other mycobacteria may be detected in the sera of 75 to 95 percent of the patients with the lepromatous

TABLE 33-5. CLINICAL HISTOLOGIC, BACTERIOLOGIC, AND IMMUNOLOGIC FEATURES OF THE DIFFERENT TYPES OF LEPROSY

Feature	Types of Leprosy				
	Tuberculoid	Borderline Tuberculoid	Borderline	Borderline Lepromatous	Lepromatous
Skin lesions					
Numbers	1 to 3	Few to moderate	Moderate	Many	Very many
Symmetry	Very asymmetrical	Asymmetrical	Asymmetrical	Slightly asymmetrical	Symmetrical
Anesthesia	Very marked	Marked	Moderate	Slight	None
Nerve enlargement*					
Cutaneous sensory	Common	May occur	0	0	0
Peripheral nerves	0 to 1	Common, asymmetrical	Common, asymmetrical	Moderately asymmetrical	Symmetrical
Skin histology					
Granuloma cell	Epithelioid	Epithelioid	Epithelioid	Histiocyte	Foamy histiocyte
Lymphocytes	+++	+++	+	± or ++	±†
Dermal nerves	Destroyed	Mostly destroyed	Some visible	Visible	Easily visible
Bacilli numbers (routine examination)	0	0, +, or ++	+, ++, or +++	++++	+++++
Lymph nodes					
Paracortical infiltrate	Nil, immunoblasts	Sarcoid-like	Diffuse epithelioid	Diffuse histiocytes	Massive infiltrate with foamy histiocytes and Virchow cells
Germinal centers	Normal	Normal	Normal	Some hypertrophy	Gross hypertrophy
Lepromin test	+++	++	± or 0	0	0
Reactions					
ENL	0	0	0	Rare	Very common
Lepra	?	Common	Very common	Very common	(Rare)‡

From Grove, Warren, Mahmoud: J Infect Dis 134:205, 1976.
*Nerves of predilection, i.e., ulnar, median, lateral popliteal, facial, great auricular, and posterior tibial.
†In lepromatous leprosy, the peripheral blood shows an absolute decrease in T and an absolute increase in B lymphocytes.
‡Lepra reactions are occasionally seen in treated lepromatous patients who have developed from borderline forms in the absence of treatment.

form of the disease. Antibodies play no protective role, however, in immune defense.

In patients with LL, erythema nodosum sometimes occurs either spontaneously, or is precipitated by treatment. It results from immune-complex deposition in the tissues. A number of abnormal serologic activities are also associated with LL, including a biologic false positive reaction in routine serologic tests for syphilis (Table 33-6).

Lepromin Test

Skin testing is not diagnostically useful, but is of value in determining the position of the patient on the immunologic spectrum. The skin test material, lepromin, consists of a heat-killed suspension of *M. leprae* prepared from lepromatous nodules. When injected intracutaneously, two types of reactions occur. The first, an early reaction (Fernandez reaction), resembles the tuberculin reaction and appears in 24 to 48 hours. The late reaction (Mitsuda reaction) is the development of an indurated nodule after 3 to 4 weeks. This corresponds to the formation of an immunologic granuloma. If the intact organisms are removed or if more purified preparations of lepromin are employed, the late

lepromin reaction can be reduced or eliminated, and the 48-hour reaction intensified. The early reaction is an indication of a delayed hypersensitivity to soluble *M. leprae* antigens by a previously sensitized individual. The late granulomatous reaction corresponds to the ability of the individual, sensitized or nonsensitized, to produce an immunologic granuloma in the presence of whole bacteria. Patients with tuberculoid leprosy usually exhibit both early and late lepromin reactions, but lepromatous patients never show these reactions because of complete anergy to the antigens of *M. leprae*. This anergy is very persistent in spite of long-term therapy.

Lepromin lacks specificity. Positive lepromin tests are elicited in patients with tuberculosis in areas of the world where no leprosy exists. Also, a positive lepromin test can be induced in normal, healthy children by vaccination with BCG.

Clinical Manifestations

The incubation period is long, usually varying from about 2 to 5 years, although periods as short as 3 months and up to 40 years have been reported. The presence of bacilli in

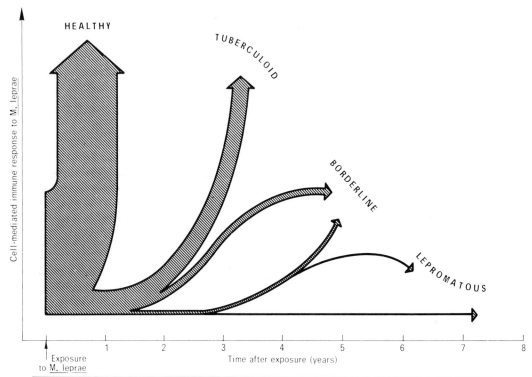

Figure 33-17. A schematic representation of the hypothesis of how the development of subclinical infection and various types of leprosy is related to time of onset of cell-mediated immune response to *M. leprae* antigens after the initial exposure. The thickness of the lines indicates the proportion of individuals from the exposed population that is likely to fall into each category. The indicated incubation periods, 2 to 3 years in tuberculoid and 6 to 7 years in lepromatous leprosy, probably represent the shortest incubation periods. They are often considerably longer. *(From Godal et al.: Bull Inst Pasteur 72:273, 1974.)*

the skin can often be demonstrated before recognition of clinical symptoms. In one reported case, leprosy bacilli were seen in smears from an apparently normal ear lobe 2 years before the patient developed cutaneous and neural leprosy of the forearms and legs.

The earliest symptom of leprosy is usually an asymptomatic, slightly hypopigmented macule several centime-

ters in diameter, which usually occurs on the trunk or distal portion of the extremities. Approximately three fourths of all patients with an early solitary lesion heal spontaneously (Fig. 33-17). In some patients, however, the infection progresses to one of a wide spectrum of patterns that vary markedly in histopathology and clinical manifestations. The immunologic status of the patient determines

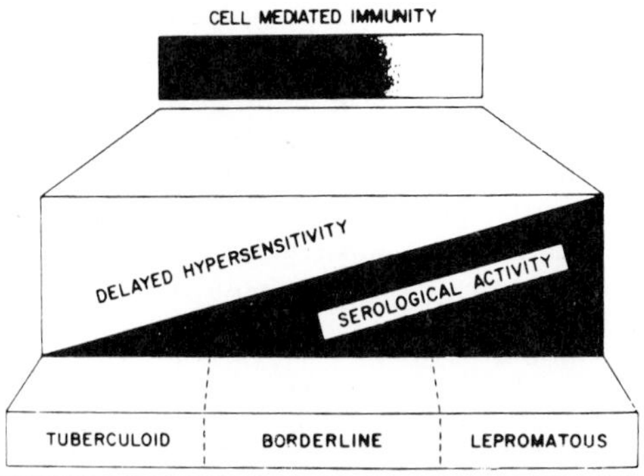

Figure 33-18. A schematic representation of the relationship between antibody production, delayed hypersensitivity, and cell-mediated immunity as related to severity of intracellular infections. *(From Bullock: J Infect Dis 137:341, 1978, as modified from Adv Int Med 21:149, 1976.)*

TABLE 33-6. ABNORMAL SEROLOGIC ACTIVITIES ASSOCIATED WITH LEPROMATOUS LEPROSY

Abnormality	Prevalence*
False positive VDRL test	> 10%
Autoantibodies	
To testicular germinal cells	†
To thyroglobulin	†
Antinuclear factors	0–30%
Rheumatoid factor	0–50%
Amyloid-related serum component protein	> 50%
Elevated level of C-reactive protein	†
Circulatory suppressor factor(s)	†
Elevated level of chemotactic factor inactivator	†
Cryoglobulinemia	> 30%
Elevated level of Clq reacting substances	†

From Bullock: J Infect Dis 137:341, 1978.
*Approximate prevalence based on published literature.
†Prevalence not established.

the prognosis. At one end of the spectrum is TT, a relatively benign form characterized by skin lesions, 3 to 30 cm in diameter, and few in number. Intermediate in its position in the spectrum, as well as in its severity of infection, is BB, in which skin lesions are more numerous and nerve involvement is more severe. Patients with BB may move towards either end of the spectrum. Such movements are accompanied by serious hypersensitivity reactions, especially as the clinical status of the patient improves, typically from a BL to a BB classification. Such reactions are regarded as an upgrading of the host's cellular immune response.

In LL, which is the most severe and extensive form of the disease, the lesions are multiple and distributed bilaterally (Fig. 33-19). As the disease progresses, the lesions coalesce and there is a marked folding of the skin, especially of the forehead, eyebrows, nose, and ear lobes, resulting in the classical leonine facies. The eyebrows and eyelashes are often lost, and the gradual destruction of the small peripheral nerves leads to trauma and secondary infection. Erythema nodosum leprosum (ENL) is a serious reactional state that occurs in more than 50 percent of patients within the first year of effective antileprosy treatment. It is believed that ENL, triggered by antimycobacterial therapy, is precipitated by the release of antigenic material following the degradation of large numbers of leprosy bacilli. Antibodies to *M. leprae* are presumed to complex with these antigens, thereby inducing severe constitutional reactions. In untreated cases, death results from respiratory obstruction, renal failure, or secondary infection.

Laboratory Diagnosis

A diagnosis of leprosy may be suspected from the symptoms, the type and distribution of the lesions, and from a history of having lived in an endemic area. Diagnosis is made by the demonstration of acid-fast bacilli in smears of

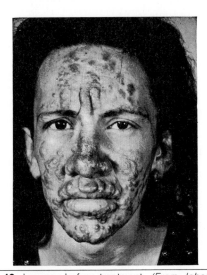

Figure 33-19. Leprosy before treatment. *(From Johansen et al.: Public Health Rep 65:204, 1950.)*

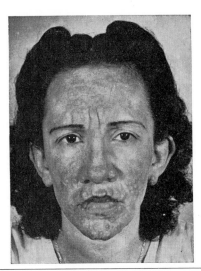

Figure 33-20. Leprosy after 12-month treatment with promacetin. *(From Johansen et al.: Public Health Rep 65:204, 1950.)*

skin lesions, nasal scrapings, ear lobes, and tissue secretions. The bacilli are very numerous in lepromatous leprosy, but in tuberculoid leprosy they are very difficult and often impossible to detect. The characteristic histologic response in biopsy material is helpful in such cases, and is essential for accurate classification of the disease within the disease spectrum. As *M. leprae* in paraffinized tissues is frequently not acid-fast when stained by the Ziehl–Neelsen method, interpretation of tissue sections can be improved by use of the Wade–Fite technique, which restores acid-fastness.

Treatment

For many years the mainstay of treatment for leprosy has been 4, 4'-diaminodiphenylsulfone (DDS, or dapsone). Since its first wide-scale use in 1950, almost all treatment has been carried out with dapsone alone (Fig. 33-20). Now, however, a marked increase in the prevalence of both primary and secondary resistance to dapsone threatens the future usefulness of the drug. Rates of resistance as high as 40 percent have recently been reported in some areas. In response to this threat, the World Health Organization has recommended a combined regimen of dapsone, rifampin, and clofazimine for all patients with LL, and one of dapsone and rifampin for all those with TT. Treatment should be continued until skin smears are free of bacilli. This may permit treatment to be stopped after only 2 years, in contrast with the traditional dapsone regimen of life-long chemotherapy.

Prevention

The control of leprosy, and perhaps its eventual elimination from the world, should be possible. Unfortunately, however, progress in this direction has been slow, primarily because of ignorance, superstition, poverty, and overpopulation in the endemic areas of the world that account for most of the cases. Only by the provision of a stronger and more secure financial base can the ultimate goal of complete eradication of leprosy be realized.

The early detection and rigid isolation of all acute lepromatous and indeterminate types of leprosy is important. This should be followed by prophylactic chemotherapy for individuals in close contact with the patient. The use of BCG for the active immunization of children in villages where lepromatous leprosy is highly endemic has been recommended as a possible control measure. Unfortunately, the results obtained to date have been widely divergent, and at present have not justified a recommendation from the International Leprosy Association for the use of BCG in the prevention of leprosy.

Mycobacteria Associated with Animal Diseases

Mycobacterium lepraemurium

Mycobacterium lepraemurium causes rat leprosy, a chronic disease that occurs spontaneously among house rats in many parts of the world. The disease is probably transmitted naturally from rat to rat by fleas. It is a disease characterized by subcutaneous indurations, swelling of lymph nodes, emaciation, and sometimes ulceration and loss of hair. Acid-fast bacilli resembling *M. leprae* are found in large numbers in the mononuclear cells of the subcutaneous tissues, lymph nodes, and nodules in the liver and lungs. Because of the similarity of the disease to human leprosy, it was thought at one time that rats could be a potential source of the human disease. Its geographic distribution, however, does not correspond with the distribution of human leprosy. At present it appears that *M. lepraemurium* and *M. leprae* are not related species, but that there is a relatedness between *M. lepraemaurium* and *M. avium*.

M. lepraemurium can be maintained for months in tissue cultures of monocytes, where it has a generation time of about 7 days (Fig. 33-21). It has also been cultured in rat fibroblasts, and in vitro in a cysteine-containing medium. *M. lepraemurium* has provided a useful model system for studying host–parasite relationships of an intracellular parasite.

Mycobacterium paratuberculosis

Mycobacterium paratuberculosis, often referred to as the Johne's bacillus, produces a chronic enteritis in ruminants. The disease is of major economic concern in cattle and sheep, where whole herds may become infected through contact with infected feces. The lesions, which are confined to the intestinal tract, are proliferative and granulomatous, and contain enormous numbers of acid-fast bacilli within the monocytes. The disease is invariably fatal.

M. paratuberculosis can be grown in the laboratory only by supplementing the medium with killed mycobacteria of readily cultivable species or with mycobactin, an iron-chelating compound unique to the mycobacteria and required

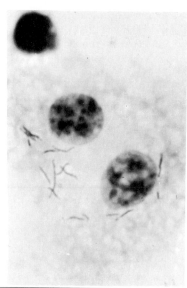

Figure 33-21. Macrophage with *M. lepraemurium* in a 20-day-old culture showing elongation of the organisms. × 2000. *(From Chang et al.: J Bacteriol 93:1119, 1967.)*

for iron transport (Chap. 5, p. 72). Taxonomically, *M. paratuberculosis* bears a strong resemblance to members of the *M. avium* complex. Strong skin test cross-reactivity has also been observed between *M. avium* and *M. paratuberculosis*.

Mycobacterium microti

Mycobacterium microti causes generalized tuberculosis in the vole. Commonly referred to as the vole bacillus, it is somewhat longer and thinner in culture than other mammalian species. Irregular S-shaped, hook-shaped, semicircular, and circular forms have been seen in tissues of infected voles. Primary growth does not occur on media containing glycerol. The organism grows more slowly than the human and bovine species, requiring 4 to 8 weeks for the appearance of minute colonies.

M. microti is immunologically closely related to *M. tuberculosis* and *M. bovis*. For this reason it was included in pilot studies with BCG to test its effectiveness when compared with BCG in the immunization of humans against tuberculosis. A level of protection was provided after 7.5 to 10 years that was almost identical to that provided by BCG.

FURTHER READING

Books and Reviews

MYCOBACTERIUM

Asselineau C, Asselineau J: Lipides specifiques des mycobacteries. Annu Microbiol (Inst Pasteur) 129A:49, 1978

Barksdale L, Kim K-S: *Mycobacterium.* Bacteriol Rev 41:217, 1977

Chaparas SD: The immunology of mycobacterial infections. CRC Crit Rev Microbiol 12:139, 1982

Daniel TM, Janicki BW: Mycobacterial antigens: A review of their isolation, and immunological properties. Micobiol Rev 42:84, 1978

Goren MB: Mycobacterial lipids. Bacteriol Rev 36:33, 1972

Grange, JM: Mycobacterial Disease. New York, Elsevier/North Holland, 1980

Hahn H, Kaufmann SHE: The role of cell-mediated immunity in bacterial infections. Rev Infect Dis 3:1221, 1981

Kubica GP, Good RC: The genus *Mycobacterium.* In Starr MP, Stolp H, Truper HG, et al. (eds): The Prokaryotes. New York, Springer-Verlag, 1981, vol II, pp 1962–1984

Lichtenstein IH, MacGregor RR: Mycobacterial infections in renal transplant recipients: Report of five cases and review of the literature. Rev Infect Dis 5:216, 1983

Ratledge C, Stanford J: The Biology of the Mycobacteria: Physiology, Identification and Classification. New York, Academic Press, 1982, vol 1

Runyon EH, Karlson AG, Kubica GP, Wayne LG: *Mycobacterium.* In Lennette EH, Balows A, Hausler WJ Jr, et al. (eds): Manual of Clinical Microbiology, 3rd ed. Washington, D.C., American Society for Microbiology, 1980, pp 150–179

Stewart-Tull DES: The immunological activities of bacterial peptidoglycans. Annu Rev Microbiol 34:311, 1980

United States Department of Health, Education, and Welfare: Procedures for the Isolation and Identification of Mycobacteria. HEW Publication No. (CDC) 77–8230, 1977

MYCOBACTERIUM TUBERCULOSIS

American Lung Association. Diagnostic Standards and Classification of Tuberculosis and other Mycobacterial Diseases. New York, 1974

American Lung Association. The Tuberculin Test. New York, 1974

Chamberlayne EC (ed): Immunization in Tuberculosis. Fogarty Intern Center Proceedings No. 14. Washington, D.C., U.S. Dept HEW Public Health Service National Institutes of Health, 1971

Daniel TM, Mahmoud AAF, Warren KS: Algorithms in the diagnosis and management of exotic diseases. XVI. Tuberculosis. J Infect Dis 134:417, 1976

Dannenberg AM Jr: Cellular hypersensitivity and cellular immunity. Bacteriol Rev 32:85, 1968

Des Prez R: Tuberculosis. In Wyngaarden JB, Smith LH Jr (eds): Cecil Textbook of Medicine, 16th ed. Philadelphia, WB Saunders, 1982, vol 2, pp 1538–1554

Eickhoff TC: The current status of BCG immunization against tuberculosis. Annu Rev Med 28:411, 1977

Luri MB: Resistance to Tuberculosis: Experimental Studies in Native and Acquired Defensive Mechanisms. Cambridge, Harvard University Press, 1964

Redmond WB, Bates JH, Engel HWB: Methods for bacteriophage typing of mycobacteria. In Bergan T, Norris JR (eds): Methods in Microbiology. New York, Academic Press, 1979, vol 13, pp. 345–375

United States Department of Health, Education, and Welfare: BCG Vaccines. Recommendations of the Public Health Service Advisory Committee on Immunization Practices. Morbidity and Mortality Weekly Report 28(21):242, 1979

United States Department of Health, Education, and Welfare: 1976—Tuberculosis in the United States. HEW Publication No. (CDC) 78-8322, 1978

United States Department of Health, Education, and Welfare: Tuberculosis in the World. HEW Publication No. (CDC) 76-8317, 1976

White RG: The adjuvant effect of microbial products on the immune response. Annu Rev Microbiol 30:579–600, 1976

Youmans GP: Tuberculosis. Philadelphia, WB Saunders, 1979

NONTUBERCULOUS MYCOBACTERIA

Chapman JS: The Atypical Mycobacterial and Human Mycobacterioses. New York, Plenum, 1977

Davidson PT (ed): International Conference on Atypcial Mycobacteria. Rev Infect Dis 3:813, 1981

Schaefer WB: Serological identification of atypical mycobacteria. In Bergan T, Norris JR (eds): Methods in Microbiology. New York, Academic Press, 1979, vol 13, pp 323–343

Sutker WL, Lankford LL, Tompsett R: Granulomatous synovitis. The role of atypical mycobacteria. Rev Infect Dis 1:729, 1979

Tellis CJ, Putnam JS: Pulmonary disease caused by nontuberculosis mycobacteria. Med Clin North Am 64:433, 1980

MYCOBACTERIUM LEPRAE

Bullock WE: Anergy and infection. Adv Intern Med 21:149, 1976

Bullock WE: Leprosy (Hansen's Disease). In Wyngaarden JB, Smith LH Jr (eds): Cecil Textbook of Medicine, 16th ed. Philadelphia, WB Saunders, 1982, vol 2, pp 1556–1561

Fine PEM: Leprosy: The epidemiology of a slow bacterium. Epidem Rev 4:161, 1982

Godal T: Immunological aspects of leprosy—present status. Prog Allergy 25:211, 1978

Hill GH: Leprosy in Five Young Men. Boulder, Colorado, Colorado Associated University Press, 1970

Pan American Health Organization Proceedings: The Armadillo as an Experimental Model in Biomedical Research. Washington, D.C., WHO, 1977

Pan American Health Organization Proceedings: Leprosy: Cultivation of the etiologic agent, immunology, animal models. Washington, D.C., WHO, 1977

Sansonetti P, Lagrange PH: The immunology of leprosy: Speculations on the leprosy spectrum. Rev Infect Dis 3:422, 1981

Shepard CC: *Mycobacterium leprae.* In Starr MP, Stolp H, Truper HG, et al. (eds): The Prokaryotes. New York, Springer-Verlag, 1981, vol II

Selected Papers

MYCOBACTERIUM TUBERCULOSIS

Bass JB Jr, Serio RA: The use of repeat skin tests to eliminate the booster phenomenon in serial tuberculin testing. Am Rev Resp Dis 123:394, 1981

Bates JH: Tuberculosis: Susceptibility and resistance. Am Rev Resp Dis 125[3 Suppl]:20, 1982

Catanzaro A: Nosocomial tuberculosis. Am Rev Resp Dis 125:559, 1982

Chaparas SD, Maloney CJ: An analysis of cross reactions among mycobacteria by in vivo and in vitro assays of cellular hypersensitivity. Am Rev Resp Dis 117:897, 1978

Chaparas SD, Brown TM, Hyman IS: Antigenic relationships of various mycobacterial species with *Mycobacterium tuberculosis.* Am Rev Resp Dis 117:1091, 1978

Chase MW: The cellular transfer of cutaneous hypersensitivity to tuberculins. Proc Soc Exp Biol Med 59:134, 1945

Chusid EL, Shah R, Siltzbach LE: Tuberculin tests during the course of sarcoidosis in 350 patients. Am Rev Resp Dis 104:13, 1971

Collins FM: The immunology of tuberculosis. Am Rev Resp Dis 125[3 Suppl]:42, 1982

Comstock GW: Epidemiology of tuberculosis. Am Rev Resp Dis 125[3 Suppl]:8, 1982

Comstock GW, Ferebee SH, Hammes LM: A control trail of communitywide isoniazid prophylaxis in Alaska. Am Rev Resp Dis 95:935, 1967

Crawford JT, Bates JH: Isolation of plasmids from mycobacteria. Infect Immun 24:979, 1979

Crowle AJ, May M: Preliminary demonstration of human tuberculoimmunity in vitro. Infect Immun 31:453, 1981

Daniel TM, Balestrino EA, Balestrino OC, et al.: The tuberculin specificity in humans of Mycobacterium tuberculosis antigen 5. Am Rev Resp Dis 126:600, 1982

Daniel TM, Oxtoby MJ, Pinto ME, et al.: The immune spectrum in patients with pulmonary tuberculosis. Am Rev Resp Dis 123:556, 1981

Daniel TM, Van der Kuyp F, Anderson PA: Initial clinical trial of Mycobacterium tuberculosis antigen 5 in tuberculin-positive human subjects. Am Rev Resp Dis 123:517, 1981

Dannenberg AM Jr: Pathogenesis of pulmonary tuberculosis. Am Rev Resp Dis 125[3 Suppl]:25, 1982

Edwards LB, Livesay VT, Acquaviva FA, et al.: Height, weight, tuberculous infection and tuberculous disease. Arch Environ Health 22:106, 1971

Edwards PQ: Tuberculosis, now and the future: Short-term therapy, preventive therapy, and Bacillus Calmette-Guerin. Bull NY Acad Med 53(6):526, 1977

Ellner JJ, Spagnuolo PJ, Schachter BZ: Augmentation of selective monocyte functions in tuberculosis. J Infect Dis 144:391, 1981

Engel HWB: Mycobacteriophages and phage typing. Annu Microbiol 129:75, 1978

Farer LS: Chemoprophylaxis. Am Rev Resp Dis 125[3 Suppl]:102, 1982

Goren MB: Cord factor revisited: A tribute to the late Dr. Hubert Block. Tubercle 56:65, 1975

Goren MB: Immunoreactive substances of mycobacteria. Am Rev Resp Dis 125[3 Suppl]:50, 1982

Goren MB, Cernich M, Blokl O: Some observations on mycobacterial acid-fastness. Am Rev Resp Dis 118:151, 1978

Goren MB, Hart PD, Young MR, et al.: Prevention of phagosome-lysosome fusion in cultured macrophages by sulfatides of Mycobacterium tuberculosis. Proc Nat Acad Sci USA 73:2510, 1976

Grange JM: The genetics of mycobacteria and mycobacteriophages—A review. Tubercle 56:227, 1975

Grange JM, Aber VR, Allen BW, et al.: The correlation of bacteriophage types of Mycobacterium tuberculosis with guinea-pig virulence and in vitro-indicators of virulence. J Gen Microbiol 108:1, 1978

Guld, J. Waaler H, Sundaksan TK, et al.: The duration of BCG-induced tuberculin sensitivity in children, and its irrelevance for revaccination. Bull WHO 39:829, 1968

Harboe M: Antigens of PPD, old tuberculin, and autoclaved Mycobacterium bovis BCG studied by crossed immunoelectrophoresis. Am Rev Resp Dis 124:80, 1981

Hardy MA, Schmidek HH: Epidemiology of tuberculosis aboard a ship. JAMA 203:175, 1968

Higuchi S, Suga M, Dannenberg AM Jr, et al.: Persistence of protein, carbohydrate and wax components of tubercle bacilli in dermal BCG lesions. AM Rev Resp Dis 123:397, 1981

Hsu HK: Isoniazid in the prevention and treatment of tuberculosis—A 20-year study of the effectiveness in children. JAMA 229:528, 1974

Hyde L: Clinical significance of the tuberculin test. Am Rev Resp Dis 105:453, 1972

Jackett PS, Aber VR, Lowrie DB: The susceptibility of strains of Mycobacterium tuberculosis to catalase-mediated peroxidative killing. J Gen Microbiol 121:381, 1980

Jones WD Jr, Good RC, Thompson NJ, et al.: Bacteriophage types of Mycobacterium tuberculosis in the United States. Am Rev Resp Dis 125:640, 1982

Kaplan MH, Chase MW: Antibodies to mycobacteria in human tuberculosis. I. Development of antibodies before and after antimicrobial therapy. J Infect Dis 142:825, 1980

Katz P, Goldstein RA, Fauci AS: Immunoregulation in infection caused by Mycobacteriun tuberculosis: The presence of suppressor monocytes and the alteration of subpopulations of T lymphocytes. J Infect Dis 140:12, 1979

Kearns TJ, Russo PK: The control and eradication of tuberculosis. N Engl J Med 303:812, 1980

Kent DC, Schwartz R: Active pulmonary tuberculosis with negative tuberculin skin tests. Am Rev Resp Dis 95:411, 1967

Kolbel HK: Anatomy of the mycobacterial cell. Annu Microbiol 129:29, 1978

Kuwabara S: Purification and properties of tuberculin-active protein from Mycobacterium tuberculosis. J Biol Chem 250:2556, 1975

Lederer E: Structure de constituants mycobactériens: Relation avec l'activite immunologique. Annu Microbiol 129:91, 1978

Lincoln EM: Epidemics of tuberculosis. Arch Environ Health 14:473, 1967

Lind A: Mycobacterial antigens. Annu Microbiol 129:99, 1978

Lowrie DB, Aber VR, Jackett PS: Phagosome-lysosome fusion and cyclic adenosine 3′,5′-monophosphate in macrophages infected with Mycobacterium microti, Mycobacterium bovis BCG or Mycobacterium lepraemurium. J Gen Microbiol 110:431, 1979

Magnus K, Edwards LB: The effect of repeated tuberculin testing on post-vaccination allergy. Lancet 2:643, 1955

Morse DL, Hansen RE, Swalbach WG, et al.: High rate of tuberculin conversion in Indochinese refugees. JAMA 248:2983, 1982

McLaugilin CA, Parker R, Hadlow WJ, et al.: Moieties of mycobacterial mycolates required for inducing granulomatous reactions. Cell Immunol 38:14, 1978

Mohr JA, Killebrew L, Muchmore HG: Transfer of delayed hypersensitivity by blood transfusions in man. JAMA 207:517, 1969

Oort J, Turk JL: The fate of (I^{131}) labeled antigens in the skin of normal guinea pigs and those with delayed hypersensitivity. Immunology 6:148, 1963

Palmer CE, Long MW: Effects of infection with atypical mycobacteria on vaccination and tuberculosis. Am Rev Resp Dis 94:553, 1966

Rado TA, Bates, JH, Engel HWB, et al.: World Health Organization studies on bacteriophage typing of mycobacteria. Am Rev Resp Dis 111:459, 1975

Ramanathan VD, Curtis J, Turk JL: Activation of the alternative pathway of complement by mycobacteria and cord factor. Infect Immun 29:30, 1980

Ratledge C, Patel PV, Mundy J: Iron transport in Mycobacterium smegmatis: The location of mycobactin by electron microscopy. J Gen Microbiol 128:1559, 1982

Retzinger GS, Meredith SC, Takayama K, et al.: The role of surface in the biological activities of trehalose 6, 6′-dimycolate. J Biol Chem 256:8208, 1981

Rideout VK, Hiltz TE: Epidemic in a high school in Nova Scotia. Can J Public Health 60:22, 1969

Riley RL: Disease transmission and contagion control. Am Rev Resp Dis 125[3 Suppl]:16, 1982

Scott DM, McConnell I, Agomo P, et al.: Purification of antigen-dependent macrophage migration inhibition factor (MIF) from lymph draining a tuberculin reaction. Immunology 34:591, 1978

Smith DT: The tuberculin unit. Am Rev Resp Dis 99:820, 1969

Smith DT: The diagnostic and prognostic value of the second strength dose of PPD (5 micrograms). Am Rev Resp Dis 101:317, 1970

Snider DE Jr: The tuberculin skin test. Am Rev Resp Dis 125[3 Suppl]:108, 1982

Stead WW: Pathogenesis of first episode of chronic pulmonary tuberculosis in man: Recrudescence of residuals of primary infection on exogenous reinfection. Am Rev Resp Dis 95:729, 1967

Stead WW, Bates JH: Evidence of "silent" bacillemia in primary tuberculosis. Annu Intern Med 74:559, 1971

Valentine FT, Lawrence HS: Lymphocyte stimulation: Transfer of cellular hypersensitivity to antigen in vitro. Science 165:1014, 1969

Wayne LG: Microbiology of tubercle bacilli. Am Rev Resp Dis 125[3 Suppl]:31, 1982

Wayne LG, Lin K-Y: Glyoxylate metabolism and adaptation of Mycobacterium tuberculosis to survival under anaerobic conditions. Infect Immun 37:1042, 1982

NONTUBERCULOUS MYCOBACTERIA

Adams RM, Remington JS, Steinberg J, et al.: Tropical fish aquariums: A source of Mycobacterium marinum infection resembling sporotrichosis. JAMA 211:457, 1970

Band JD, Ward JI, Fraser DW, et al.: Peritonitis due to a Mycobacterium chelonei-like organism associated with intermittent chronic peritoneal dialysis. J Infect Dis 145:9, 1982

Barrow WW, Brennan PJ: Immunogenicity of type-specific C-mycoside glycopeptidolipids of mycobacteria. Infect Immun 36:678, 1982

Collins FM, Cunningham DS: Systemic Mycobacterium kansasii infection and regulation of the alloantigenic response. Infect Immun 32:614, 1981

Conner DH, Lunn HF: Buruli ulceration: A clinicopathologic study of 38 Ugandans with Mycobacterium ulcerans. Arch Pathol 81:183, 1966

Edwards LB, Acguaviva FA, Livesay VT, et al.: An atlas of sensitivity to tuberculin PPD-B and histoplasmin in the United States. Am Rev Resp Dis 99:1, 1969

Edwards LB, Palmer CE: Isolation of "atypical" mycobacteria from healthy persons. Am Rev Resp Dis 80:747, 1959

Edwards ML, Goodrich JM, Muller D: Infection with Mycobacterium avium-intracellulare and the protective effects of bacille Calmette-Guérin. J Infect Dis 145:733, 1982

Feldman RA, Long MW, David HL: Mycobacterium marinum: A leisure time pathogen. J Infect Dis 129:618, 1974

Foz A, Roy C, Turado J, et al.: Mycobacterium chelonei iatrogenic infections. J Clin Microbiol 7:319, 1978

Good RC, Snider DE Jr: Isolation of nontuberculous mycobacteria in the United States, 1980. J Infect Dis 146:829, 1982

Greene JB, Sidhu GS, Lewin S, et al.: Mycobacterium avium-intracellulare: A cause of disseminated life-threatening infection in homosexuals and drug abusers. Annu Intern Med 97:539, 1982

Gruff H, Henning HG: Pulmonary mycobacteriosis due to rapidly growing acid-fast bacillus, Mycobacterium chelonei. Am Rev Resp Dis 105:618, 1972

Hoffman PC, Fraser DW, Robicsek F, et al.: Two outbreaks of sternal wound infections due to organisms of the Mycobacterium fortuitum complex. J Infect Dis 143:533, 1981

Johnston WW, Smith DT, Vandiviere HM III: Simultaneous or sequential infection with different mycobacteria. Arch Environ Health 11:37, 1965

Karlson AG, Carr DT: Tuberculosis caused by Mycobacterium bovis. Annu Intern Med 73:979, 1970

Kubica GP: The clinically significant mycobacteria: Their identification . . . By whom? Infect Dis Rev 4:29, 1976

Lincoln EM, Gilbert LA: Disease in children from other than Mycobacterium tuberculosis. Am Rev Resp Dis 105:683, 1972

Mollohan CS, Romer MS: Public health significance of swimming pool granuloma. Am J Public Health 51:883, 1961

Palmer CE, Long MW: Effects of infection with atypical mycobacteria on BCG vaccination and tuberculosis. Am Rev Resp Dis 94:553, 1966

Raucher C, Kerby G, Ruth WF: A ten-year clinical experience with Mycobacterium kansasii. Chest 66:17, 1974

Schaefer WB, Davis CL, Cohn ML: Pathogenicity of transparent, opaque, and rough variants of Mycobacterium avium in chickens and mice. Am Rev Resp Dis 102:499, 1970

Zakowski P, Fligiel S, Berlin GW, et al.: Disseminated Mycobacterium avium-intracellulare infection in homosexual men dying of acquired immunodeficiency. JAMA 248:2980, 1982

MYCOBACTERIUM LEPRAE

Beeching NJ, Ellis CJ: Leprosy and its chemotherapy. J Antimicrobiol Chemother 10:81, 1982

Bullock WE: Leprosy: A model of immunological perturbation in chronic infection. J Infect Dis 137:341, 1978

Golden GS, McCormick JB, Fraser DW: Leprosy in the United States, 1971–1973. J Infect Dis 135:120, 1977

Grove DI, Warren KS, Mahmoud AAF: Algorithms in the diagnosis and management of exotic diseases. XV. Leprosy. J Infect Dis 134:205, 1976

Harboe M, Closs O, Bjorvatn B, et al.: Antibody response in rabbits to immunization with Mycobacterium leprae. Infect Immun 18:792, 1977

Poulter LW, Lefford MJ: Relationship between delayed-type hypersensitivity and the progression of Mycobacterium lepraemurium infection. Infect Immun 20:530, 1978

Prabhakaran J, Kircheimer WF: Use of 3,4-dihydroxyphenylalanine oxidation in the identification of Mycobacterium leprae. J Bacteriol 92:1267, 1966

Ridley DS: Histological classification and the immunological spectrum of leprosy. Bull WHO 51:451, 1974

Ridley DS, Jopling WH: Classification of leprosy according to immunity. A five-group system. Int J Lepr 34:255, 1966

Shankara MK, Narayanan E, Kasturi G, et al.: Non-cultivable mycobacteria in some field collected arthropods. Lepr India 45(4):231, 1973

Shepard CC, McRae DH: Mycobacterium leprae in mice: Minimal infectious dose, relation between staining quality and infectivity, and effect of cortisone. J Bacteriol 80:365, 1965

Van Eden W, deVries RRP, Mehra NK, et al.: HLA segregation of tuberculoid leprosy: Confirmation of the DR2 marker. J Infect Dis 141:693, 1980

Van Voorhis WC, Kaplan G, Sarno EN, et al.: The cutaneous infiltrates of leprosy. Cellular characteristics and the predominant T-cell phenotypes. N Engl J Med 26:1593, 1982

Yamagami A, Chang YT: Growth of Mycobacterium lepraemurium in cultures of macrophages obtained from various sources. Infect Immun 17:531, 1977

CHAPTER 34

Actinomycetes

Actinomycetous Bacteria

The actinomycetous bacteria are associated with a number of diseases. In addition to tuberculosis and other mycobacterial infections discussed in Chapter 33, three major diseases are caused by members of other genera within the order Actinomycetales—actinomycosis, nocardiosis, and actinomycetoma. These infections will be discussed in this chapter.

The Eighth Edition of *Bergey's Manual of Determinative Bacteriology* places the actinomycetes in Part 17 within the order Actinomycetales. Also included in Part 17 are the corynebacteria and other coryneform-shaped organisms. The unifying characteristic of all actinomycetes is their tendency, varying in degree, to form filaments. They are gram-positive rods. To a varying extent, as the bacilli grow they may fail to separate following cell division and, as a consequence, they form elongated chains of cells about 1 μm in width.

TABLE 34-1. MAJOR CELL WALL CONSTITUENTS OF ACTINOMYCETES

Genus	DL-DAP	LL-DAP	Glycine	Arabinose	Galactose
Actinomyces	−	−	−(+)	−	±
Arachnia	−	+	+	−	+
Bifidobacterium	−	−	−(+)	−	−
Mycobacterium	+	−	−	+	+
Dermatophilus	+	−	−	−	−
Nocardia	+	−	−	+	+
Actinomadura	+	−	−	−	−
Streptomyces	−	+	+	−	−

+, present in cell wall; −, not found in cell wall; −(+) , present in the cell walls of only a few species or strains; DL-DAP, DL-diaminopimelic acid; LL-DAP, LL-diaminopimelic acid.

Taxonomists have evaluated many characteristics of the members of this group, including such physiologic properties as proteolytic enzyme activity, temperature and oxygen requirements, and fermentation products. DNA content (G + C ratios and homology studies), antigenic determinants, and cell wall composition have also been examined in detail.

Cell Wall. One of the most useful taxonomic aids in the classification of the genera within the order Actinomycetales is the cell wall composition. As expected, all of the cell walls contain the basic components of the peptidoglycan cell wall material—*N*-acetylglucosamine, muramic acid, alanine, and glutamic acid—but the genera can be characterized by the presence or absence of certain other amino acids and sugars. From chromatographic analysis of cell walls, four chemotypes have been defined on the basis of the content of arabinose, galactose, glycine, DL- or LL-diaminopimelic acid. Table 34-1 indicates the cell wall chemotype of some representative genera.

Several genera have the same chemotype; for example, as shown in Table 34-1, *Mycobacterium* and *Nocardia* have identical patterns. To further distinguish such genera, a subset of four additional chemotypes has been devised, based upon the presence or absence of arabinose, galactose, xylose, and madurose in whole cells.

Streptomycetes

The streptomycetes are members of the family Streptomycetaceae, a large family containing four genera and more than 500 species. The most distinctive morphologic property of this group is the formation of extensively branching aerial and substrate filaments. Streptomycetes are found naturally a few inches below the surface in soil; in water; and on organic debris. Only a few species may, on rare occasions, be pathogenic. Most are ubiquitous, benign bacteria that probably serve as essential ecologic function in the soil.

Much of the pioneering and continuing knowledge of streptomycetes and other soil microorganisms has been obtained at the Rutgers University Institute of Microbiology. The inspiration for the Institute was Selman Waksman, in whose laboratory the antibiotic streptomycin was discovered in 1944.

Morphology and Physiology

The streptomycetes are aerobic, filamentous bacteria. The substrate filaments that are produced are long, highly branched, and nonfragmenting. Aerial filaments may be rudimentary or extensive, and may be embellished with spirals, coils, or multiple branching. Many species produce spores, often in chains, from the aerial filaments.

Cultural Characteristics. Colonies grow relatively slowly, requiring several days to become visible. They are tenacious and adhere to the agar as the substrate filaments penetrate the medium and anchor the colony. After 7 to 10 days, most colonies become lusterless as aerial sporebearing filaments proliferate. The 463 species of the genus *Streptomyces* have been organized into 7 groups based on spore color. Further division of each group is based on whether the aerial spore chains are straight or curled. In stationary broth cultures, the organisms grow on the surface as a mat until they sink under their own weight. In aerated, shaken liquid cultures, streptomycetes grow in spherical microcolonies and do not produce aerial filaments.

In addition to spore morphology and pigment production, other taxonomic approaches have included serologic relationships, chemical composition, and anabolic activity.

Ecologic and Medical Importance. In their natural habitat in the soil, streptomycetes decompose organic matter. They may also provide texture in some soils, as their filaments render clay more adherent. The odor of soil is attributable to the presence of streptomycetes. Indeed, cultures often elaborate the aroma of aged straw or freshly turned earth.

The major medical importance of the streptomycetes

is the production of antibiotics. Collectively, the streptomycetes are responsible for about 85 percent of the known antibiotics, the remaining being derived from fungi and other bacteria. Among the most important antibiotics synthesized by streptomycetes are streptomycin, chloramphenicol, tetracycline, neomycin, erythromycin, kanamycin, cycloheximide, amphotericin B, and nystatin. In addition to these antibacterial and antifungal antibiotics, antiviral, antiparasitic, and anticancer agents have been produced by various *Streptomyces*. Although the medically useful potential of this group of bacteria has long been exploited by pharmaceutical companies, the natural influence or purpose of antibiotic production in the maintenance of microbial balance or evolution is unknown.

Actinomycosis

Actinomyces israelii, Actinomyces naeslundii, and Arachnia propionica

Actinomycosis is a chronic, suppurative, and granulomatous infection. It is characterized by pyogenic lesions, with interconnecting sinus tracts containing granules composed of microcolonies of the bacterial pathogen embedded in tissue elements. Actinomycosis may be caused by any of several closely related species of actinomycetes, all of which are members of the normal flora of the mouth and gastrointestinal tract. Based on the site of involvement, three clinical forms of actinomycosis are recognized: cervicofacial, thoracic, and abdominal. A fourth entity, genital actinomycosis, has been described in women with intrauterine devices.

The etiologic agents of actinomycosis, in decreasing order of frequency, are *Actinomyces israelii*, *Arachnia propionica*, and *Actinomyces naeslundii*. A single case has been

attributed to *Bifidobacterium eriksonii*. All of these bacteria are found normally in the healthy mucous membranes of the mouth and gastrointestinal tract of all humans. Two additional *Actinomyces* species, *Actinomyces viscosus* and *Actinomyces odontolyticus*, are found among the oral flora. These species have not been implicated as agents of actinomycosis, but as possible causes of or contributors to dental caries. In cattle, the predominant agent of actinomycosis is *Actinomyces bovis*. Although the clinical disease in cattle (lumpy jaw) is very similar to human actinomycosis, *A. bovis* is not associated with humans either as a commensal or pathogen.

Morphology and Physiology

Although classified as three separate genera, *Actinomyces*, *Arachnia*, and *Bifidobacterium* are similar in their morphology and physiology. On an enriched solid medium (e.g., brain–heart infusion agar) or thioglycollate broth at 37C, they produce initial microcolonies composed of branching filaments that after 24 to 48 hours fragment into diphtheroids, short chains, and coccobacillary forms (Figs. 34-1 and 34-2). They develop neither aerial filaments nor spores. They are not acid-fast and their cell walls do not contain diaminopimelic acid, arabinose, or mycolic acid.

Most strains of the five species of *Actinomyces* referred to above are facultative anaerobes that grow best in the presence of carbon dioxide. *A. propionica* is also a facultative anaerobe, but its growth is not affected by carbon dioxide. *B. eriksonii* is strictly anaerobic. Table 34-2 indicates the biochemical and physiologic characteristics that are most helpful in speciating these actinomycetes. *A. bovis* is included for comparison.

Clinical Infection

Epidemiology. Actinomycosis has been observed in persons of all ages, but the disease is rare in children under 10 years old. Most cases occur between the ages of 15 and

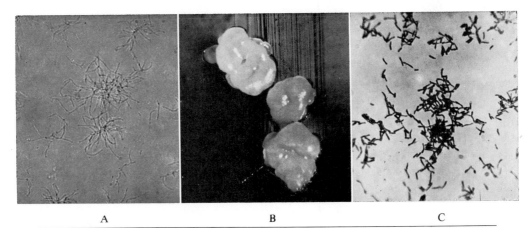

A   B   C

Figure 34-1. A. *Actinomyces israelii.* Spidery colony on brain–heart infusion agar plate, 24 hours. × 500. **B.** *Actinomyces israelii.* Molar tooth colony on brain–heart infusion agar plate, 15 days. **C.** *Actinomyces israelii.* Gram stain of smear from rough colony showing diphtheroid forms. × 1200. *(Courtesy of Mycology Unit, Centers for Disease Control, Atlanta, Ga.)*

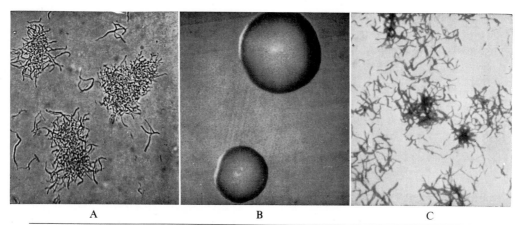

Figure 34-2. A. *Actinomyces naeslundii.* Dense tangled mycelial colony on brain–heart infusion agar plate, 24 hours. × 475. **B.** *Actinomyces naeslundii.* Smooth colony on brain–heart infusion agar plate. 7 days. **C.** *Actinomyces naeslundii.* Gram stain smear from smooth colony showing branching diphtheroid forms. × 900. *(Courtesy of Mycology Unit, Centers for Disease Control, Atlanta, Ga.)*

35 years. The infection rate in males is approximately twice as great as in females.

Pathogenesis. Although two of the four etiologic agents listed in Table 34-3, *A. israelii* and *A. propionica*, cause most cases of actinomycosis, the pathology and pathogenesis are the same regardless of the agent or location of the infection. In most cases the organisms cross the mucosal or epithelial surface of the mouth or lower gastrointestinal tract following a traumatic episode. Pulmonary or thoracic actinomycosis may be caused by aspiration of the organisms. In deeper tissue the organisms grow anaerobically, attracting a cellular infiltrate, and spread by expansion to contiguous tissue. Multiple pyogenic lesions become

joined by interconnecting sinuses. If unchecked, this process continues and draining sinus tracts may erupt to the skin surface. The organisms are contained within the pyogenic, or subsequently granulomatous lesions, and within granules in the sinus channels. The granules (grains, or sulfur granules) are composed of tissue elements and the organisms.

Clinical Manifestations

CERVICOFACIAL ACTINOMYCOSIS. Over half the cases of actinomycosis are probably of this type. The lower jaw region usually becomes involved following dental caries or gingival disease. With growth of the organisms, the area becomes swollen and discolored. Patients experience dis-

TABLE 34-2. DIFFERENTIAL CHARACTERISTICS OF *ACTINOMYCES* SPECIES, *BIFIDOBACTERIUM ERIKSONII*, AND *ARACHNIA PROPIONICA*

Characteristic	A. bovis	A. israelii	A. naeslundii	A. odontolyticus	A. viscosus	B. eriksonii	A. propionica
Oxygen requirement	A	A	F	F	F	A	F
Catalase	−	−	−	−	+	−	−
Nitrate reduction	−	V	+	+	+	−	+
Gelatin hydrolysis	−	−	−	−	−	−	−
Esculin hydrolysis	+	+	+	V	+	V	−
Fermentation (acid production)							
Arabinose	−	V	−	V	−	+	−
Glucose	+	+	+	+	+	+	+
Lactose	+	+	V	V	V	+	+
Mannitol	−	V	−	−	−	+	+
Raffinose	−	V	+	V	+	+	+
Sucrose	V	+	+	+	+	+	+
Trehalose	V	+	V	V	V	+	+
Xylose	V	+	−	V	−	+	−

F, facultative anaerobe; A, anaerobe; +, 90% or more of strains tested positive; −, 90% or more strains tested negative; V, variable reactions, strains positive or negative.

TABLE 34-3. GRANULES IN ACTINOMYCETOUS INFECTIONS

Disease	Agents	Granule Presence	Granule Color	Granule Size	Peripheral Clubs
Actinomycosis	*Actinomyces israelii*	+	White to yellow	Small (0.1–0.3 mm)	+
	Actinomyces naeslundii	Rare	White to yellow	Small (0.1–0.3 mm)	+
	Arachnia propionica	+	White to yellow	Small (0.1–0.3 mm)	+
	Bifidobacterium dentium	−			
Nocardiosis	*Nocardia asteroides*	−			
	Nocardia brasiliensis	+	White	Small (1 mm)	±
	Nocardia caviae	Rare	White to yellow	Small (1 mm)	±
Actinomycetoma	*Actinomadura madurae*	+	White, yellow, or pink	Large (1–10 mm)	+
	Actinomadura pelletieri	+	Red	Small (1 mm)	−
	A. israelii	+	White to yellow	Small (0.1–0.3 mm)	+
	N. asteroides	Rare	White	Small (1 mm)	−
	N. brasiliensis	+	White	Small (1 mm)	±
	N. caviae	+	White to yellow	Small (1 mm)	±
	Streptomyces somaliensis	+	Yellow to brown	Large (1–2 mm)	−
	Streptomyces paraguayensis	+	Black	Small (0.5 mm)	+

+, granule production; −, granule not normally formed; ±, clubs may or may not be seen.

comfort but little pain, and the prognosis is good if detected early. Pyogenic abscesses may develop, with the formation of interconnecting sinuses containing granules. The potential for bone involvement is very high because the organisms can penetrate the cortex of any bone and produce osteomyelitis.

THORACIC ACTINOMYCOSIS. The pulmonary infection may develop by extension of the cervicofacial form or by aspiration of the agent. The symptoms, which often resemble those of subacute pulmonary infection, are mild fever, cough, and purulent sputum. Eventually, lung tissue is destroyed, sinus tracts develop, and invasion of the ribs or vertebrae may occur.

ABDOMINAL ACTINOMYCOSIS. Abdominal infection is usually initiated by traumatic perforation of the intestinal mucosa, most often following a ruptured appendix or ulcer. The symptomatology reflects the location of the lesion and organs involved. The disease progresses slowly and insidiously. Involvement of the liver may produce jaundice, and extension to the urinary tract can cause cystitis or pyelonephritis. The vertebrae may become infected and sinuses may erupt through the abdominal wall. Radiography often reveals an indistinct tumor-like mass.

GENITAL ACTINOMYCOSIS. Recently, genital actinomycosis has been described in women using intrauterine devices. In most cases, diagnosis is based on histologic evidence of infection with actinomycetous bacteria, and granules are infrequent. Subclinical colonization of intrauterine devices with actinomycetes may occur more frequently.

Laboratory Diagnosis

DIRECT EXAMINATION. The presence of granules should be looked for in sputum, pus collected from draining sinuses, tissue sections, and cervical exudates. The presence and color of actinomycetous granules is indicated in Table 34-3. Actinomycotic granules are up to 1 millimeter in size, lobulated, and composed of tissue cells and bacterial filaments 1 μm in width (Fig. 34-3). These filaments often appear club-shaped at the periphery of the granule.

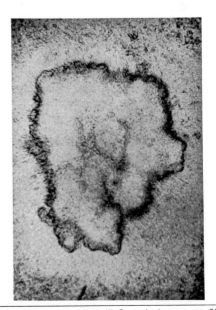

Figure 34-3. *Actinomyces israelii.* Granule in pus. × 350. *(From Conant et al.: Manual of Clinical Mycology, 3rd ed, 1971. Philadelphia, W.B. Saunders Co.)*

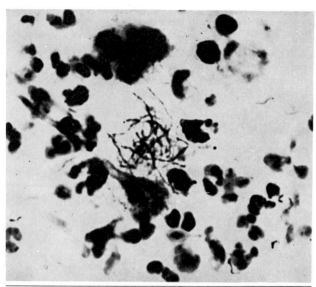

Figure 34-4. *Actinomyces israelii.* Gram stain of crushed granule, showing gram-positive branching filaments. × 1300 *(From Conant et al.: Manual of Clinical Mycology, 3rd ed, 1971. Philadelphia, W.B. Saunders Co.)*

The yellowish appearance of these sulfur granules is probably due to the abundance of macrophages containing lipid vacuoles. Crushed granules should be examined for gram-positive, nonacid-fast diphtheroids (Fig. 34-4).

CULTURE. Specimens of tissue or exudate should be cultivated on an enriched medium and incubated anaerobically as indicated above. In thioglycolate broth, *A. israelii* produces a hard colony, whereas colonies of the other *Actinomyces* species are more easily disrupted. With the exception of *A. viscosus* (Table 34-2), the catalase test can be used to distinguish these agents from *Corynebacterium, Mycobacterium,* and *Propionibacterium,* which are catalase positive.

Treatment. Penicillin is the drug of choice for the treatment of actinomycosis, and is administered parenterally or orally for a period of several weeks. Surgical excision and drainage may also be required. A variety of other antibiotics have also been used successfully, including tetracycline, clindamycin, and sulfonamide derivatives.

Nocardiosis

Nocardiosis is caused by infection with *Nocardia asteroides* or, in rare cases, *Nocardia brasiliensis* or *Nocardia caviae.* An older species, *Nocardia farcinica,* is now considered a subgroup of *N. asteroides.* All species of *Nocardia* are found in nature in the soil and aquatic environments. Nocardiosis is initiated by inhalation of the organisms. The infection rarely is acute, but usually runs a chronic course. It may be subclinical and confined to the lung, or hematogenous dissemination to any organ may occur, with the central nervous system being the preferred site.

Morphology and Staining

Nocardia are aerobic, gram-positive, partially acid-fast, nonsporeforming bacilli. Within the genus, filamentation and branching are highly developed. *Nocardia* develop both substrate and aerial filaments about 1-μm wide. Cells of the substrate filaments may separate into beaded forms, and aerial filaments may undergo fragmentation to yield unicellular, spore-like cells that are easily dispersed and aerosolized (Fig. 34-5).

As with *Mycobacterium* and *Corynebacterium,* the cell walls of *Nocardia* possess mycolic acids, called nocardic acids. Mycolic acids are long α-branched, β-hydroxy fatty acids, usually saturated or mono-unsaturated. As discussed in Chapter 6, the acid-fastness of *Mycobacterium* species is attributed to the mycolic acids with chain lengths of 50 to 90 carbon atoms. Nocardic acids are about 50 carbons in length, and corynemycolic acids are 32- to 36-carbons long. *Corynebacterium* species are nonacid-fast and *Nocardia* are weakly or partially acid-fast when stained with carbolfuchsin according to Kinyoun's method. Most *Nocardia,* if decolorized with 1 percent sulfuric acid instead of the stronger, acid-alcohol decolorant, will stain acid-fast. Some isolates, however, never exhibit acid-fastness; others lose this property after prolonged culture in the laboratory.

The cell walls of *Nocardia* species, like the mycobacteria, have been shown to enhance the antitumor and antimicrobial activities of macrophages. The immunoadjuvant and anticancer potential of these cell wall preparations is currently under investigation.

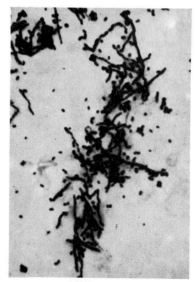

Figure 34-5. *Nocardia asteroides.* Gram-stained smear of culture on Sabouraud's glucose agar. × 1000.

Figure 34-6. *Nocardia asteroides.* Culture on Sabouraud's glucose agar at room temperature for 12 days.

Physiology

Cultural Characteristics. *Nocardia* grow readily on a variety of laboratory media. On agar media, colonies appear within 3 days. After 7 to 10 days, colonies are heaped, irregular, waxy, shiny, and several millimeters in diameter (Fig. 34-6). As aerial filaments are formed, the colony surface becomes dull and fuzzy. Colonies of *N. asteroides* vary considerably in pigmentation, and may be yellow, orange, red, or mixtures of these hues.

Differential Metabolic Properties. *Nocardia* can be distinguished from similar genera by its ability to decompose and utilize paraffin as a source of carbon and energy. This property permits *Nocardia* to be selectively isolated from mixed cultures. One method of paraffin-baiting for *Nocardia* involves the use of paraffin-coated glass rods, prepared by immersing a glass rod in melted paraffin. When the rod is removed and cooled, a thin film of paraffin wax coats the glass. The rods can be placed in natural soil or water and *Nocardia* species, if present, will grow around the paraffin-coated glass rod. Using this technique, pathogenic and saprophytic *Nocardia* species have been isolated from a wide variety of soils, fresh water, and sea water worldwide. It has been estimated that the quantity of *Nocardia* in soil exceeds 10^3 per gram of dry weight of soil. Ecologic studies tend to correlate the prevalence of a particular strain of *N. asteroides* in the environment with its incidence of infection in the indigenous population.

In addition to paraffin digestion, *Nocardia* species produce catalase and urease. The genus *Actinomadura* is morphologically similar to *Nocardia*, but differentiated on the basis of nonacid-fastness and inability to digest paraffin or urea. The three etiologic agents of noncardiosis—*N. asteroides*, *N. brasiliensis*, and *N. caviae*—can be speciated by testing the ability of each to decompose casein, gelatin, hypoxanthine, tyrosine, and xanthine. Media are prepared with coagulated or crystalline forms of these substrates and, if the *Nocardia* species being tested can utilize the substrate, a clear zone develops in the medium around the colony. As indicated in Table 34-4, *N. asteroides* is uniformly negative, whereas *N. brasiliensis* digests all but xanthine, which distinguishes it from *N. caviae*. Because of the high variability, colony pigmentation is not a reliable aid to identification.

Determinants of Pathogenicity

Studies on host–microbial interactions in the laboratory of Beaman have shown that logarithmically growing cells of *N. asteroides* are more virulent than stationary phase cells and are more resistant to phagocytosis and intracellular killing by alveolar macrophages. This increased virulence is associated with differences in the nocardial cell wall and inhibition of the fusion of macrophage lysosomes and phagosomes. In vitro experiments have also demonstrated that virulent *N. asteroides* grows out of and destroys macrophages, whereas avirulent organisms are able to survive as L-forms within macrophages. Activated macrophages are better able to withstand challenge with virulent *N. asteroides*, but maximal host resistance is associated with intact cell-mediated immunity.

Clinical Infection

Epidemiology. Approximately half of the patients who develop nocardiosis are in some way compromised. Thirty percent of patients have a history of receiving steroids or immunosuppressive therapy, or both, but chronic and pulmonary infections and cancer, with or without its attendant therapies, also provide the setting for opportunistic nocardiosis. In those cases where no underlying condition prevails, the prognosis is considerably better. The frequency of nocardiosis in the United States has been estimated to be 500 to 1000 cases per year.

Seventy-five percent of all cases of nocardiosis, whether pulmonary only or disseminated, occur in males. According to clinical reviews, the fatality rates for pulmonary and disseminated nocardiosis are approximately 40 percent and 80 percent, respectively. However, with earlier diagnosis and treatment, the fatality rate is much lower.

In a retrospective study of a specific group of patients at risk at the Stanford University Medical Center between 1968 and 1978, 13 percent of the cardiac transplant recipients developed opportunistic nocardiosis. In most cases, infection involved the lung, symptoms were nonspecific, and onset occurred between 1.5 months and 3 years following the surgery. No particular risk factor or source of infection was identified. Prognosis was excellent following rapid diagnosis and long-term treatment with sulfisoxazole. One fourth of these patients, however, subsequently developed nontuberculous mycobacterial infections following resolution of their nocardiosis.

Clinical Manifestations. Nocardiosis begins as a chronic lobar pneumonia following inhalation of the etiologic agent, which in 80 to 90 percent of the cases, is *N. asteroi-*

des. Seventy-five percent of reported cases have this initial presentation, but a variety of signs and symptoms may be observed, including fever, weight loss, or chest pain. The presenting complaints are indistinctive and mimic tuberculosis and other infections. Pulmonary consolidations may develop but granuloma formation and caseation are rare.

From the lung, metastatic foci of infection may appear anywhere, but most frequently the central nervous system becomes involved. Abscess formation occurs in the brain. The meninges are usually not inflamed enough to cause diagnostic changes in the spinal fluid. The onset of central nervous system lesions can be sudden or gradual, and patients may present with headache, lethargy, convulsions, or more severe symptoms of central nervous system dysfunction.

Another frequently involved site is the kidney, where either the cortex or medulla may be involved. Localized lesions develop in the skin and subcutaneous tissue in over half the cases caused by *N. brasiliensis.*

Regardless of the tissue infected or the specific agent, the pathology of nocardiosis is the same, namely, abscess formation. Multiple abscesses, characterized by central necrosis and dense infiltration of neutrophils, are indistinguishable from infections caused by pyogenic bacteria. Although nocardiosis resembles tuberculosis in chronicity and symptomatology, the pathology is not granulomatous but pyogenic.

Laboratory Diagnosis

DIRECT EXAMINATION. Specimens of sputum, skin lesions, tissue biopsies, or surgical material should be examined microscopically and cultured. The direct microscopic examination of positive sputum or tissue smears reveals delicate, multiply branched and beaded filaments that are gram-positive and partially acid-fast (Fig. 34-7). The organisms are very difficult to see in tissue stained with hematoxylin and eosin. Granules are not usually observed with systemic lesions.

CULTURE. *Nocardia* will grow on most laboratory media under aerobic conditions. Colonies, which appear within 3 to 7 days, are composed of branching filaments that usually stain acid-fast when decolorized with 1 percent H_2SO_4 instead of acid alcohol. Because of the formation of an abundance of substrate filaments, colonies cling tenaciously to the agar. Both substrate and aerial filaments may fragment to form bead-like chains. The genus and species identification are confirmed by the production of urease and the tests indicated in Table 34-4. On Lowenstein–Jensen medium, *N. asteroides* may be identical macroscopically to atypical mycobacteria, but *Nocardia* can be differentiated from *Mycobacterium* because the latter are strongly acid-fast and exhibit minimal branching. If *N. asteroides* is suspected, cultures of nonsterile specimens can be incubated at 40C to 50C to inhibit growth of most of the other bacteria.

Treatment. In spite of the susceptibility of *N. asteroides* to a number of newer antibiotics, the sulfonamides remain the treatment of choice for nocardiosis. Sulfa blood levels of 10 to 20 mg/mL should be maintained during treatment. The sulfamethoxazole–trimethaprim combination has also been used with success.

Actinomycetoma

Actinomycetoma (Madura foot) is a chronic suppurative and granulomatous infection of the subcutaneous tissue. The lesions are characterized by formation of abscesses, tumefaction, draining sinuses, and granules. Actinomyce-

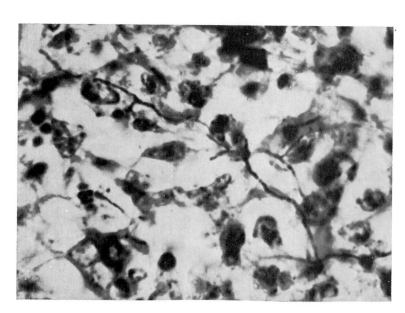

Figure 34-7. *Nocardia asteroides.* Gram-stained section of brain abscess showing gram-positive branching filaments. × 1300. *(From Conant et al.: Manual of Clinical Mycology, 3rd ed, 1971. Philadelphia, W.B. Saunders Co.)*

TABLE 34-4. IDENTIFICATION OF PATHOGENIC *NOCARDIA* SPECIES

	N. asteroides	N. brasiliensis	N. caviae
Colony pigment	Yellow to orange red	Colorless or orange red	Colorless or white
Substrates hydrolyzed			
Casein	−	+	−
Gelatin	−	+	−
Hypoxanthine	−	+	−
Tyrosine	−	+	−
Xanthine	−	−	+

toma may be caused by any one of several actinomycetes within the genera *Actinomadura, Nocardia,* and *Streptomyces.* Mycetoma may also be caused by fungal pathogens (Chap. 88) and, rarely, by *A. israelii.*

Morphology and Cultural Characteristics
Some of the agents associated with actinomycetoma are listed in Table 34-3; all except *A. israelii* grow aerobically on most bacteriologic media.

Actinomadura. The genus *Actinomadura* closely resembles *Nocardia* in culture and microscopic appearance (p. 588). Both genera are gram-positive and produce catalase, but *Actinomadura* cannot break down urea or paraffin and is nonacid-fast. Furthermore, their cell wall chemotypes are different (Table 34-1) and the filaments of *Actinomadura* do not fragment. The colony of *Actinomdura madurae* is whitish, smooth, and waxy; *Actinomadura pelletierii* produces a pink to red colony and granule. Table 34-3 compares the granules produced by pathogenic actinomycetes.

Streptomyces. Streptomyces paraguayensis develops a white to brown colony on different media and black granules. Colonies of *Streptomyces somaliensis* are initially white, but become brown to black. Granules, however, are yellow to brown. Filaments of these *Streptomyces* species do not fragment.

Clinical Infection
Epidemiology. Most of the agents of actinomyetoma are ubiquitous soil saprophytes. The agents and infection are more common in the subtropical and tropical regions. Ecologic studies of the prevalence of specific agents tend to reflect the incidence of infection. The likelihood of walking barefoot and the risk of traumatic inoculation is also greater in warmer climates. In the United States, *A. madurae* is the most frequently isolated agent.

Clinical Manifestations. Regardless of the actinomycete involved, the infection is established by contamination of the traumatized subcutaneous surface by soil containing the organism. Exposed areas of the extremities are most often involved, especially the foot or hand, although any body surface may become infected. The infection progresses slowly. An initial subcutaneous nodule develops, enlarges, and ulcerates, and satellite lesions develop that become connected by sinus tracts. If unchecked, the infection progresses to cause deformation of the involved tissue with destruction of underlying muscle and bone.

Laboratory Diagnosis
DIRECT EXAMINATION. Direct examination of pus or exudate expressed from lesions for granules can be very helpful. The shape, color, size, and microscopic appearance of the granule can usually define the bacterial etiology. Figure 34-8 depicts a typical actinomycotic granule with peripheral clubs.

CULTURE. These bacteria can be isolated on most routine bacteriologic media and identified on the basis of morphology, physiologic characteristics, and cell wall composition.

Treatment. Long-term antibiotic therapy is usually required to completely eradicate the actinomycotic infection. The treatment of *Actinomyces* and *Nocardia* infections is indicated on pages 588 and 590. Streptomycin is recommended as the first choice against *Actinomadura* and *Streptomyces.* Chemotherapy is often augmented with surgical management.

Other Actinomycetous Diseases

Dermatophilosis
Dermatophilosis, or streptothricosis, is a skin infection caused by *Dermatophilus congolensis.* The infection is most common among cattle, sheep, and similar animals. The lesion causes a pustular, exudative dermatitis that eventually becomes crusty and flakes off. Human infection is very rare. The organisms are short, branched, irregular filaments in the cutaneous tissue. Topical treatment with metallic compounds is recommended.

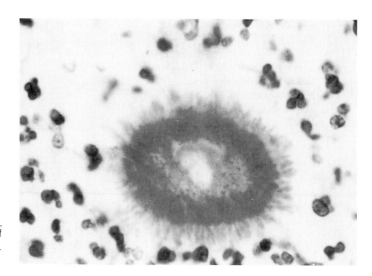

Figure 34-8. *Nocardia brasiliensis.* Tissue section of actinomycetoma granule showing peripheral clubs. × 212. *(Courtesy of Dr. L. Linares.)*

Farmer's Lung

This condition is an allergic disease induced by inhalation of thermophilic actinomycetes. The usual agents are *Micropolyspora faeni* or *Thermoactinomyces vulgaris*. They thrive in haystacks, composts, and grain storage silos at temperatures of 45C to 60C. The disease is apparently mediated by type I or type III hypersensitivity reactions to the surface antigens. In dry places, these bacteria are easily aerosolized and inhalation of the bacterial allergens induces an asthmatic reaction in sensitive individuals. As the name implies, farmer's lung is an occupational disease of persons in frequent contact with stored grain and similar reservoirs for these bacteria. This condition is treated with antiallergic drugs and by avoiding exposure.

FURTHER READING

Books and Reviews

Bradley SG: Significance of nucleic acid hybridization to systematics of actinomycetes. Adv Appl Microbiol 19:59, 1975

Bronner M, Bronner M: Actinomycosis, 2nd ed. Bristol, John Wright & Sons Ltd., 1971

Buchanan RE, Gibbons NE (eds): Bergey's Manual of Determinative Bacteriology, 8th ed. Baltimore, Williams & Wilkins, 1974

Georg LK: Agents of Human Actinomycosis. In Balows A, Dehaan RM, Guze LB, et al. (eds): Anaerobic Bacteria. Springfield, Ill, Thomas, 1974, p 237

Goodfellow M, Brownell GM, Servano JR (eds): The Biology of the Nocardiae. London, Academic Press, 1976

Goodfellow M, Minnikin DE: Nocardioform bacteria. Annu Rev Microbiol 31:159, 1977

Holdeman LV, Moore WEC (eds): Anaerobe Laboratory Manual, 3rd ed. Blacksburg, Virginia Polytechnic Institute and State University, 1975

Kalakoutskii LV, Agre NS: Comparative aspects of development and differentiation in actinomycetes. Bacteriol Rev 40:469, 1976

Lechevalier HA, Lechevalier MP: A critical evaluation of the genera of aerobic actinomycetes. In Prauser (ed): The Actinomycetales. Jena, Fischer, 1968, p 393

Lloyd DH, Sellers KC (eds): Dermatophilus Infection in Animals and Man. London, Academic Press, 1976

Mariat F, Destombes P, Segretain G: The mycetomas: Clinical features, pathology, etiology and epidemiology. Contr Microbiol Immunol 4:1, 1977

Slack JM, Gerencser MA: Actinomyces, Filamentous Bacteria. Biology and Pathogenicity. Minneapolis, Burgess, 1975

Sykes A, Skinner FA: Actinomycetales: Characteristics and Practical Importance. New York, Academic Press, 1973

Waksman SA: Actinomycetes. A Summary of Current Knowledge. New York, Ronald Press, 1967

Selected Papers

Bach MC, Adler JS, Breman J, et al.: Influence of rejection therapy on fungal and nocardial infections in renal-transplant recipients. Lancet 1:180, 1973

Bach MC, Sabath LD, Finland M: Susceptibility of *Nocardia asteroides* to 45 antimicrobial agents in vitro. Antimicrob Agents Chemother 3:1, 1973

Beaman BL, Burnside J, Edwards B, et al.: Nocardial infections in the United States, 1972–1974. J Infect Dis 134:286, 1976

Beaman BL, Smathers M: Interaction of *Nocardia asteroides* with cultured rabbit alveolar macrophages. Infect Immun 13:1126, 1976

Berardi RS: Abdominal actinomycosis. Surg Gynecol Obstet 149:257, 1979

Berd D: Laboratory identification of clinically important aerobic actinomycetes. Appl Microbiol 25:665, 1973

Berd D: *Nocardia brasiliensis* infections in the United States: A report of 9 cases and a review of the literature. Am J Pathol 59:254, 1973

Brock DW, Georg LK, Brown JM, et al.: Actinomycosis caused by *Arachnia propionica*. Report of 11 cases. Am J Clin Pathol 59:66, 1973

Brown JR: Human actinomycosis. A study of 181 subjects. Human Pathol 4:319, 1973

Causey WA, Siezer B: Systemic nocardiosis caused by *Nocardia brasiliensis*. Am Rev Resp Dis 109:134, 1974

Coleman RM, Georg LK, Rozzell AR: *Actinomyces naeslundii* as an agent of human actinomycosis. Appl Microbiol 18:420, 1969

Davis-Scibienski C, Beaman BL: Interaction of *Nocardia asteroides* with rabbit alveolar macrophages: Effect of growth phase and viability on phagosome–lysosome fusion. Infect Immun 29:24, 1980

Davis-Scibienski C, Beaman BL: Interaction of alveolar macrophages with *Nocardia asteroides*: Immunological enhancement of phagocytosis, phagosome–lysosome fusion, and microbicidal activity. Infect Immun 30:578, 1980

Dowell VR, Sonnenwirth AL: Gram-positive, nonsporeforming anaerobic bacteria. In Lennett BH, Spaulding BH, Truant JP (eds): Manual of Clinical Microbiology, 2nd ed. Washington, American Society for Microbiology, 1974, p 396

Eastridge CE, Pratter JR, Hughes FA, et al.: Actinomycosis: A 24-year experience. South Med J 65:839, 1972

Fass RJ et al.: Clindamycin in the treatment of serious anaerobic infections. Annu Intern Med 78:853, 1973

Frazier AR, Rosenow EE, Roberts GD: Nocardiosis. A review of 25 cases occurring during 24 months. Mayo Clin Proc 50:657, 1975

Goodman NL: The epidemiology of nocardiosis. In Al-Doory (ed): The Epidemiology of Human Mycotic Diseases. Springfield, Ill, Chas C Thomas, 1975, p 40

Gordon RE, Barnett DA, Handerham JE, Pang CH-N: *Nocardia coeliaca, Nocardia autotrophica,* and the nocardin strain. Int J Syst Bacteriol 24:54, 1974

Gordon RE, Mihm JM: A comparison of *N. asteroides* and *N. brasiliensis.* J Gen Microbiol 20:129, 1959

Grossman CB, Biazz DG, Armstrong D: Roentgen manifestations of pulmonary nocardiosis. Radiology 96:325, 1970

Gupta PK, Hollander DH, Frost JK: Actinomycetes in cervicovaginal smears: An association with IUD usage. Acta Cytologica 20:295, 1976

Kruk VA, Stinson EB, Remington JS: *Nocardia* infection in heart transplant patients. Annu Intern Med 82:18, 1975

Kurup PV, Randhawa HS, Mishra SK: Use of paraffin bait technique in the isolation of *Nocardia asteroides* from sputum. Mycopathol Mycol Appl 40:363, 1970

Lomax CW, Harbert GM, Thornton WN: Actinomycosis of the female genital tract. Obstet Gynecol 49:341, 1976

Mishra SK, Gordon RE, Barnett DA: Identification of nocardial and streptomycetes of medical importance. J Clin Microbiol 11:728, 1980

Nichols DR: Actinomycosis: Results in therapy in 156 patients. Progr Antimicrob Anticancer Chemother 11:8, 1970

Ortiz-Ortiz L, Bojalil LF: Delayed skin reactions to cytoplasmic extracts of *Nocardia* organisms as a means of diagnosis and epidemiological study of *Nocardia* infection. Clin Exp Immunol 12:225, 1972

Ortiz-Ortiz L, Contreras MF, Bojalil LF: The assay of delayed hypersensitivity to ribosomal proteins from *Nocardia.* Sabouraudia 10:147, 1972

Palmer DL, Harvey RL, Wheeler JK: Diagnostic and therapeutic considerations in *Nocardia asteroides* infection. Medicine 53:391, 1974

Pine L, Georg L: Reclassification of *Actinomyces propionicus.* Int J Syst Bacteriol 19:267, 1969

Ruebush TK, Goodman JS: *Nocardia asteroides* bacteremia in an immunosuppressed renal transplant patient. Am J Clin Pathol 64:537, 1975

Schiffer MD, Clguezebal A, Sultana M, et al.: Actinomycosis infections associated with intrauterine contraceptive devices. Obstet Gynecol 45:67, 1975

Simpson GL, Stinson EB, Egger MJ, et al.: Nocardial infections in the immunocompromised host: A detailed study in a defined population. Rev Infect Dis 3:492, 1981

Spilsbury BW, Johnstone FRC: The clinical course of actinomycotic infections: A report of 14 cases. Can J Surg 5:33, 1962

Stenhouse D, MacDonald DG, MacFarlane TW: Cervico-facial and intra-oral actinomycosis: A 5-year retrospective study. Br J Oral Surg 13:172, 1975

Tight RR, Bartlett MS: Actinomycetoma in the United States. Rev Infect Dis 3:1139, 1981

Wallace RJ, Septimus EJ, Williams TW, et al.: Use of trimethoprim-sulfamethoxazole for treatment of infections due to *Nocardia.* Rev Infect Dis 4:315, 1982

Weese WC, Smith IM: A study of 57 cases of actinomycosis over a 36-year period. Arch Intern Med 135:1562, 1975

Young LS, Armstrong D, Blevins A, et al.: *Nocardia asteroides* infection complicating neoplastic disease. Am J Med 50:356, 1971

Zaias N, Taplin D, Rebell G: Mycetoma. Arch Dermatol 99:215, 1969

CHAPTER 35

Enterobacteriaceae: General Characteristics

The family Enterobacteriaceae is composed of a large number of closely related bacterial species that inhabit the large bowel of humans and animals, and soil, water, and decaying matter. Because of their normal habitat, they have often been referred to as the enteric bacilli, or enterics. Included in this group of organisms are some of the most important intestinal pathogens of humans, i.e., the agents of typhoid fever and bacillary dysentery. Most enterics do not cause disease when confined to the intestinal tract of a normal host, but given an altered host or an opportunity to invade other body sites, many have the capability to produce disease in any tissue. In fact, the organisms of this family are responsible for the majority of nosocomial (hospital-acquired) infections, causing urinary tract and wound infections, pneumonia, meningitis, and septicemia. The medical and economic importance of such infections becomes apparent when one considers that an estimated 2 million patients per year in the United States (5 to 10 percent of the total hospital population) acquire infections while hospitalized. The seriousness of the problem is further complicated by the fact that many of the enterics isolated from nosocomial infections are resistant to multiple antimicrobial agents.

The Enterobacteriaceae have also played a very important role in the field of molecular biology. Most of the information contained in Chapters 7 and 8 concerning microbial genetics and regulatory mechanisms has been derived from experimentation with these organisms.

Taxonomy

As a result of the intense genetic analysis of the enterics and the need for accurate epidemiologic information on enteric infections, the speciation and characterization of this family is more detailed than that of other microorganisms. At the time of this writing there is no universally accepted classification scheme for this group. However, because of the role that the Centers for Disease Control (CDC) plays in the field of clinical microbiology in this country, the CDC-proposed taxonomy and nomenclature is used here (Table 35-1). The reader is referred to the Eighth Edition of *Bergey's Manual of Determinative Bacteriology* for other classifications used for this family.

    The CDC scheme is based on the work of Brenner and co-workers using DNA homology techniques to deter-

TABLE 35-1. CLASSIFICATION OF THE FAMILY ENTEROBACTERIACEAE AND CLINICAL SIGNIFICANCE OF SPECIES

Tribe	Genus	Species	Type of Infection
I. Escherichieae	*Escherichia*	*coli*	Opportunistic,* enteropathogenic
	Shigella	*dysenteriae*	Bacillary dysentery
		flexneri	Bacillary dysentery
		boydii	Bacillary dysentery
		sonnei	Bacillary dysentery
II. Edwardsielleae	*Edwardsiella*	*tarda*	Opportunistic, enteropathogenic (?)
III. Salmonelleae	*Salmonella*	*typhi*	Typhoid fever
		cholerae-suis	Septicemia
		enteriditis	Gastroenteritis
	Arizona	*hinshawii*	Enteropathogenic, opportunistic
	Citrobacter	*freundii*	Opportunistic
		diversus	Opportunistic
		amalonaticus	Opportunistic
IV. Klebsielleae	*Klebsiella*	*pneumoniae*	Opportunistic
		oxytoca	Opportunistic
		ozaenae	Opportunistic, atrophic rhinitis
		rhinoscleromatis	Granulomatous rhinitis
	Enterobacter	*cloacae*	Opportunistic
		aerogenes	Opportunistic
		agglomerans	Opportunistic
		sakazakii	Opportunistic
		gergoviae	Opportunistic
	Hafnia	*alvei*	Opportunistic
	Serratia	*marcescens*	Opportunistic, nosocomial
		liquifaciens	Opportunistic, nosocomial
		rubidaea	Opportunistic, nosocomial
V. Proteeae	*Proteus*	*mirabilis*	Opportunistic
		vulgaris	Opportunistic
	Morganella	*morganii*	Opportunistic
	Providencia	*rettgeri*	Opportunistic, nosocomial
		stuartii	Opportunistic, nosocomial
		alcalifaciens	Opportunistic, nosocomial
VI. Yersinieae	*Yersinia*	*pestis*	Bubonic plague
		pseudotuberculosis	Opportunistic, gastroenteritis
		enterocolitica	Opportunistic, gastroenteritis
		intermedia	Opportunistic, gastroenteritis
		fredriksenii	Opportunistic, gastroenteritis
		ruckeri	Nonhuman pathogen
VII. Erwinieae	*Erwinia*		Plant pathogen
	Pectobacterium		Plant pathogen

*Opportunistic–extraintestinal infections including septicemia, urinary tract and wound infections, meningitis, and pneumonia.

mine the degree of relatedness observed among groups of organisms and is an extension of the earlier work of Edwards and Ewing. A review of the DNA homology techniques and their application to the Enterobacteriaceae is listed in the Further Reading section of this chapter. However, a large number of organisms have characteristics of the family and have been isolated from a variety of clinical and environmental sources, but have not been studied in enough detail to be included in present classification schemes. As more isolates are obtained and studied, newer taxonomy and nomenclature will develop. For example, several new genera (*Cedecea*, *Kluyvera*, and *Tatumella*) and a new species (*Escherichia hermanii*) have been recently described.

The family is divided into seven tribes consisting of

one or more closely related genera. Six of the seven tribes are important in human infections, while the seventh, Erwineae, consists of plant pathogens. Although the tribe Yersinieae rightly belongs in the family Enterobacteriaceae, this group of organisms will be discussed separately in Chapter 40. This is because of the epidemiologic differences and historic importance of one species, *Yersinia pestis*, the cause of bubonic plague.

Morphology

The Enterobacteriaceae are small (0.5 μm by 3.0 μm), gram-negative, non-sporeforming rods. They may be motile or nonmotile. When motile, movement is by means of peritrichous flagella, a property that aids in differentiating the enterics from the polar flagellated bacteria of the families Pseudomonadaceae and Vibrionaceae. Two genera, *Shigella* and *Klebsiella*, are characteristically nonmotile. Enteric bacilli may possess a well-defined capsule, as in the genus *Klebsiella*; a loose, ill-defined coating, referred to as a slime layer, as seen in some *Escherichia*; or may be lacking in either structure. Fimbriae (pili) are present in most species and may play a role in attachment to other bacteria, phage, and host cells, and in the transfer of genetic material.

The Enterobacteriaceae possess complex cell walls composed of murein, lipoprotein, phospholipid, protein, and lipopolysaccharide arranged in layers. The murein–lipoprotein layer constitutes approximately 20 percent of the total cell wall and is responsible for cellular rigidity. The remaining 80 percent of the cell wall is joined to the lipoprotein at the lipid end of the molecule, forming a lipid bilayer. This portion of the cell wall can be removed by various treatments, such as boiling in 4 percent dodecylsulfate. A major constituent of this portion of the cell wall is the lipopolysaccharide (LPS). This molecule contains the specific polysaccharide side chains that determine the antigenicity of various enteric bacilli (p. 103), and is the portion of the cell responsible for endotoxic activity (p. 598).

Physiology

Biochemical Properties. Enterobacteriaceae are facultative organisms that are biochemically diverse and complex. Under anaerobic or low oxygen conditions, they attack carbohydrates fermentatively, but given sufficient oxygen, they utilize the tricarboxylic acid cycle and electron transport system for energy production. Various species differ in the carbohydrates they ferment. These differences, together with the variations in end-product production and substrate utilization, form the basis for speciation within this family. The important biochemical reactions of the various genera are discussed in the chapters dealing with the individual organisms and in the section on Laboratory Diagnosis (p. 600). By definition, however, all enterics ferment glucose with the production of acid and reduce nitrates to nitrites, but do not produce oxidase or liquify alginate. Most enterics ferment glucose by the mixed acid pathway, but the tribe Klebsielleae is distinguished by its ability to utilize the butanediol fermentative pathway (Chap. 4). Formation of gas, specifically hydrogen and carbon dioxide, during glucose fermentation varies with the species and represents a useful tool in preliminary identification of the organism. For example, most *Shigella* and *Salmonella typhi*, important enteropathogens, characteristically do not produce gas.

Cultural Characteristics. On nondifferential or nonselective media, such as blood agar or brain–heart infusion agar, the various genera of Enterobacteriaceae cannot be distinguished. Most species appear as moist, smooth, gray colonies. Smooth to rough variations can occur. Hemolysis, which may or may not occur, is usually of the β type. A variety of special differential and selective media are utilized in the isolation and preliminary characterization of different groups of enterics (Table 35-2). When grown in broth culture, the enterics produce diffuse growth, which reflects their facultative nature.

Genetic Interaction. The enterobacteria are very useful tools to the geneticist. They are the organisms being used in the developing recombinant DNA industry and most of the information concerning genetic mapping and gene transfer has been delineated using these bacteria (Chaps. 7 and 8). These experiments have demonstrated that genetic information can be transferred among distantly related as well as closely related enterobacteria. The transfer of this genetic material, through either transduction or conjugation, gives rise to hybrids with altered biochemical and/or structural properties. Such hybrids are not merely products of laboratory experimentation, but can be isolated in natural settings, particularly the hospital environment. Changes in structural and biochemical properties cause difficulty in proper laboratory speciation, necessitating detailed microbiologic identification procedures. Changes in antimicrobial susceptibility can also occur when a resistant organism possessing the resistance transfer factor (RTF) acts as a male cell and transfers the genes coding for resistance to a female cell via conjugation (Chap. 8). Although such hybrids do occur, it is fortunate that the majority of enterics isolated from clinical material conform to normal identification and sensitivity patterns.

Bacteriocins. A number of Enterobacteriaceae harbor plasmids that code for the production of antibacterial substances known as bacteriocins. Different bacteriocins attack different molecular sites, such as nucleic acid or protein synthesis, or adenosine triphosphate (ATP) formation. Bacteriocins produced by certain organisms within a species are generally only active against susceptible organisms

TABLE 35-2. MEDIA USED FOR ISOLATION OF ENTERICS

Differential media that permit most enterics to grow
  MacConkey's agar
  Eosin methylene blue agar
  Desoxycholate agars

More highly selective media used for cultivation of intestinal pathogens
  Salmonella–Shigella agar
  Desoxycholate citrate agar
  Hektoen–enteric agar
  Xylose–lysine desoxycholate agar

Enrichment broths used to select for growth of intestinal pathogens
  Selenite broth
  G-N broth
  Tetrathionate broth

of the same species. This selectivity is useful as an epidemiologic tool and provides a means of typing or subdividing a species more definitively. For example, if one wished to determine if all *Proteus mirabilis* isolated from wound infections in a particular hospital were the same, and thus originating from the same source, one would expose the isolates to a battery of bacteriocins. If all isolates had the same bacteriocin susceptibility pattern, the likelihood of a common source is greater.

Resistance. The enteric bacilli produce no spores and are thus relatively easily destroyed by heat and low concentrations of common germicides and disinfectants. Phenol, formaldehyde, β-glutaraldehyde, and halogen compounds are bactericidal for this group of organisms. Quaternary ammonium compounds may be only bacteriostatic, depending upon the particular formulation and situation. The use of chlorine in water has been helpful in controlling the dissemination of these organisms, particularly the agents of typhoid fever and other intestinal diseases. The enterics tolerate bile salts and bacteriostatic dyes to a varying degree, a fact that is useful in the development of primary isolation media.

These organisms are also relatively sensitive to drying but can survive for extended periods of time when provided adequate moisture. Moisture-laden respiratory care equipment and anesthesia equipment have served as sources of infections in the hospital environment. Enterics tolerate cold for extended periods of time. Isolated from snow and ice after several months, they provide a mechanism for the contamination of water supplies during the spring thaws. Contaminated ice machines are also sources of infection. Control of enterics in foods can be achieved through pasteurization, thorough cooking, and proper refrigeration.

Antigenic Structure

With certain species of enterobacteria, the antigenic characteristics play an important role in epidemiology and classification. This is particularly true with the intestinal pathogens of the genera *Salmonella* and *Shigella*. The O antigens, H antigens, and K antigens are the major components used in the serologic typing of the Enterobacteriaceae. In addition to these primary antigens, the Enterobacteriaceae share a common antigen (ECA) that is present on the outer surface of the bacteria and may be involved in reactions of the bacteria with their animal host. This antigen may also prove to be useful in further taxonomic and, possibly, epidemiologic studies.

K Antigens. The term K antigen comes from the German term *Kapsule* and was used to describe polysaccharide capsular antigen of the enteric bacilli. Most K antigens are polysaccharides and some genera, such as *Klebsiella*, can be typed using a quellung reaction, as is done with *Streptococcus pneumoniae* (Chap. 27). In other genera, the K anti-

gen resembles an amorphous slime layer surrounding the bacterial cell. Recent studies with the K antigens of *Escherichia coli* have shown that certain K antigens are proteins and form the fimbriae seen on certain strains. The K88 and K99 fimbriae antigens of *E. coli* have been shown to be important for the attachment of these strains to the intestinal wall of animals. Similar colonizing factor antigens (CFA) for *E. coli* pathogenic to humans have also been described.

Flagellar (H) Antigens. The H or flagellar antigens are proteins. Antigenic variation in flagella is thought to be due to the difference in amino acid sequence of the particular flagellar type. With two genera, *Salmonella* and *Arizona*, the flagella for a particular bacterium may exist in one of two phases: phase 1 (specific), which is not shared by many organisms; or phase 2 (nonspecific), which is found in many organisms. The Kauffman–White scheme of classification uses the variation in phase type and differences in flagellar antigen to aid in the serologic speciation of these genera.

Somatic (O) Antigens. The LPS of the cell wall is composed of three distinct regions. The O-specific or cell wall antigen is contained in region I and is a polymer of repeating oligosaccharide units of three or four monosaccharides. Within certain genera, such as *Escherichia*, *Salmonella*, and *Shigella*, these polymers vary with different isolates allowing for serologic subgrouping of the genera. Attached to the O antigen is region II, consisting of a core polysaccharide that appears to be constant within a particular genus of enterobacteria but differs between genera. The lipid A moiety, or region III, is attached to the core polysaccharide and serves to join the LPS to the murein–lipoprotein layer of the cell wall. Besides being useful as a serologic marker, the O antigen may serve as an important virulence factor in that it is toxic (see Endotoxin section, below), and it may serve as a means of binding the bacteria to certain tissue targets during infection. Further discussion of the LPS molecule of the Enterobacteriaceae and other gram-negative organisms can be found in Chapters 3 and 6.

Determinants of Pathogenicity

Endotoxin. The LPS portion of the cell wall has been referred to as endotoxin. The toxicity of LPS appears to reside in the lipid A portion. These toxins can be extracted from the bacterial cell wall by a number of agents, such as phenol-water, trichloracetic acid, or ethylene-diamine-tetraacetate. Upon injection into animals, endotoxins produce a variety of effects, such as fever, fatal shock, leukocytic alterations, regression of tumors, cytotoxicity, alterations in host response to infection, Sanarelli–Shwartzman reaction, and various metabolic changes. The cellular targets of endotoxin are varied and the exact mechanism of action of endotoxin has not been clearly delineated. (For further details of the various cellular targets of endotoxin, see the review by Bradley.)

The role of endotoxin in such diseases as urinary tract or wound infection remains unclear. However, when the enteric bacilli enter the bloodstream, endotoxic shock plays an important role. Approximately 30 percent of patients with enteric bacteremia will develop shock, with a mortality of 40 to 90 percent. The number of deaths attributed to gram-negative sepsis in the United States has ranged from 18,000 to 100,000 persons per year. The cause of shock is apparently endotoxin, as experiments with animals injected with endotoxin or with enteric organisms produce some of the reactions seen in humans. The exact mechanism and nature of this syndrome have not been determined. The chief defect in endotoxin shock is a pooling of blood in the microcirculation, resulting in inadequate blood supply to vital organs and causing cellular hypoxia and metabolic failure. Survival is directly proportional to the length of time needed to recognize the bacteremia and to institute adequate treatment of the infection as well as the shock syndrome.

Enterotoxins. Enterotoxins are bacterial substances that exert their toxic effect in the small intestine, causing a transduction of fluid into the lumen. Enterotoxin production has been found in enteropathogenic *E. coli, Klebsiella pneumoniae, Enterobacter cloacae, Shigella dysenteriae, Shigella flexneri, Shigella sonnei,* and a number of *Salmonella* species. Enterotoxin producing *E. coli* are the major causes of traveler's diarrhea, whereas enterotoxin producing *K. pneumoniae* and *E. cloacae* are rare and have been isolated primarily in countries such as India where diarrhea is extremely prevalent. The importance of enterotoxins in *Shigella* and *Salmonella* is still being determined, since other factors such as tissue penetration are also necessary for the production of disease by these organisms.

The lack of an experimental animal delayed discovery of these toxins. The discovery of *Vibrio cholerae* enterotoxin (Chap. 38) using ileal loops in rabbits provided the necessary animal model for testing enterotoxin production. Subsequent studies have demonstrated that suckling mice, infant rabbits, rabbit skin, and certain tissue cultures respond to some types of enterotoxin.

There are two types of *E. coli* toxins produced: a heat-labile enterotoxin (LT), and a heat-stable enterotoxin (ST). Production of both toxins is plasmid mediated. The LT of *E. coli* is very similar to the cholera enterotoxin (Chap. 38) and acts by increasing intracellular cyclic AMP (cAMP). This increase of cAMP causes a net secretion of electrolytes and fluid into the small intestine. The ST of *E. coli* causes a net increase in cyclic guanosine monophosphate (cGMP) with subsequent fluid and electrolyte loss. The enterotoxins of *K. pneumoniae* and *E. cloacae* appear to be related to those of *E. coli,* whereas the enterotoxins of *Shigella* and *Salmonella* do not. The enterotoxin of *S. dysenteriae* corresponds to the neurotoxin described in earlier literature. A more detailed discussion of the enterotoxins can be found in the chapters dealing with specific organisms.

Other Factors. Members of the genus *Shigella* and some strains of enteropathogenic *E. coli* have been shown to penetrate the epithelial lining of the intestinal tract. The nature of this penetration has not been determined, but it appears to be an important feature of *Shigella* infections and even occurs in enterotoxin-producing strains. Whether these organisms produce a toxin that is quickly absorbed by the intestinal lining is presently unknown.

The surface properties of the bacterial cells of certain enterics appear to play an important role in the etiology of their infections. The capsule of *K. pneumoniae* functions in a similar manner to the pneumococcal capsule and prevents phagocytosis. The Vi antigen of *S. typhi* may function in a protective manner and prevent intracellular destruction of the organism. Variation in the O antigen of *Salmonella typhimurium* can cause changes in the virulence of the organism. Losses of O-specific side chains have been associated with an increase of the LD_{50} in mice. In addition, the surface antigens of bacteria may play a significant role in the establishment of the bacteria in a particular host site, a crucial factor in the initiation of a disease state. Experiments with enteropathogenic *E. coli* of porcine origin demonstrate that the K88 fimbriae antigen is vital for attachment to the mucosa of the small intestine. The K99 antigen of bovine strains of *E. coli* appears to play a similar role, as do the CFA antigens of human isolates.

A number of enterics produce additional factors, such as hemolysins or various enzymes, but their precise role in disease is unknown. This is especially true in urinary tract and wound infections.

Clinical Infection

The Enterobacteriaceae can cause urinary tract and wound infections, pneumonia, meningitis, septicemia, and various gastrointestinal disorders. The nature of these diseases and their epidemiology is discussed in the chapters dealing with individual organisms. It is useful, however, to think of infections caused by organisms other than *Salmonella, Shigella, Yersinia,* and enteropathogenic *E. coli* as opportunistic or secondary infections—that is, these organisms require some alteration in the host by either mechanical, physiologic, or infectious processes before they can cause disease. These opportunistic infections usually occur outside the large intestine. *Salmonella, Shigella,* and *Yersinia,* however, are considered to be true enteric pathogens, and their isolation from intestinal sources of an individual implies either a diseased or a carrier state. Whereas *E. coli* can be either opportunistic or a true intestinal pathogen, most infections seen in adults in the United States are opportunistic. Table 35-1 lists the types of infections caused by the various species.

Regardless of the site of initial infection, one potential danger with the enterobacteria is the development of bacteremia and associated shock. It has been estimated that as many as 100,000 deaths in the United States are due to gram-negative bacillemia. The enteric bacilli are the most frequent isolates from these cases.

Laboratory Diagnosis

Clinical Material, Transport, and Culture. Specimens submitted for isolation of the enterobacteria include sputum, tissue, pus, body fluids, rectal swabs, and feces. To prevent overgrowth of these organisms and to obtain an accurate picture of the microbial flora, these specimens should be cultured immediately or placed in an appropriate transport medium, such as Stuart's or Amies' medium.

The method of handling the specimen in the laboratory depends upon the specimen source. With all nonfecal specimens, any enteric isolate could be a pathogen. Therefore, specimens are planted on a medium that allows the growth of most enterics but inhibits the growth of gram-positive and, possibly, other nonenteric gram-negative rods.

With fecal specimens, the laboratory is seeking to isolate only the intestinal pathogens, i.e., *Salmonella, Shigella,* and *Yersinia.* Because fecal specimens contain the opportunistic organisms in larger quantities than the pathogens, media that suppress the growth of the opportunists but favor isolation of the pathogens are required. Although some researchers have used a variety of immunologic procedures to detect the LT of *E. coli,* there are no methods readily available to the hospital laboratory for detection of enterotoxigenic bacteria at this time.

Most of the enteric media, regardless of purpose, contain various carbohydrates and acid-base indicators to demonstrate carbohydrate fermentation. Lactose is the carbohydrate most frequently used. This is because the majority of the organisms in the genera *Escherichia, Enterobacter,* and *Klebsiella*—the enteric organisms most frequently isolated and present in the greatest numbers in fecal material—ferment this carbohydrate, whereas intestinal pathogens usually do not. This provides the laboratory

with a convenient method of selecting different types of organisms from a specimen. It must be stressed, however, that because of strain variation within species, this is only a preliminary aid, and additional biochemical tests are needed for accurate identification. Some media also contain iron salts for the detection of H_2S production to aid in the identification of potential *Salmonella* colonies. Table 35-2 lists several of the common enteric isolation media and their uses.

Laboratory Identification. Colonies selected from the primary isolation media are speciated by various biochemical tests. The reactions used to divide the family into the tribes are presented in Table 35-3. Major reactions of the individual organisms are covered in the following chapters. The choice of additional tests for speciation varies from laboratory to laboratory. More detailed biochemical reactions are covered in the publications of Ewing and the *Manual of Clinical Microbiology.*

In addition to biochemical reactions, the clinical laboratory also uses serologic typing to characterize certain enteric isolates. When dealing with *Salmonella* and *Shigella,* speciation is based on serologic tests. With other enterics, serologic grouping may be useful in characterizing isolates in epidemics.

Treatment

Table 35-4 lists the various agents useful in the treatment of enterobacterial infections. Despite the discovery of newer antimicrobial agents, adequate treatment of enterobacterial infections remains a major therapeutic problem. Several factors contribute to this problem. One of the most important factors is the underlying disease of the patient. Studies on patients with gram-negative bacteremia show that, despite appropriate antibiotic therapy, patients with rapidly fatal disease (death within 1 year) have an 85 percent mortality rate; those with ultimately fatal disease (within 5 years) have a 42 percent mortality rate; and

TABLE 35-3. DIFFERENTIATION OF THE TRIBES OF ENTEROBACTERIACEAE PATHOGENIC FOR HUMANS

Test or Substrate	Tribe					
	Escherichieae	Edwardsielleae	Salmonelleae	Klebsielleae	Proteeae	Yersinieae
Hydrogen sulfide (TSI)	−	+	+	−	+ or −	−
Urease	−	−	−	− or (+)	+ or −	+ or −
Indol	+ or −	+	−	−	+ or −	− or +
Methyl red	+	+	+	−	+	+
Voges–Proskauer	−	−	−	+	−	− or +*
Citrate (Simmons)	−	−	+	+	d	−
KCN	−	−	− or +	+	+	−
Phenylalanine deaminase	−	−	−	−	+	−
Mucate	d	−	d	+ or −	−	−

From Lennette et al. (eds): Manual of Clinical Microbiology, 3rd ed., 1980. Washington, D.C., American Society for Microbiology.
Yersinia enterocolitica + at 25C.
(+), delayed positive (3 or more days); d, different biochemical reactions; + or −, most cultures positive, some negative; − or +, most strains negative, some positive; − or (+), most cultures negative, some positive delayed.

TABLE 35-4. ANTIMICROBIALS USEFUL IN TREATMENT OF ENTEROBACTERIAL DISEASES*

Penicillins	Polymyxins
Ampicillin	Colistin (polymyxin E)
Carbenicillin	Polymyxin B
Cephalosporins	Other antimicrobials
Aminoglycosides	Chloramphenicol
Amikacin	Sulfonamides
Gentamicin	Tetracyclines
Kanamycin	Trimethoprim-sulfamethoxazole
Tobramycin	

*See text and Chapters 36 and 37 for applicability to various species.

those with nonfatal disease have 10 percent mortality. In addition, the indiscriminate use of antimicrobials results in the selection of resistant strains of organisms that have the potential of transferring their resistance factors to previously sensitive organisms. A number of reports regarding outbreaks of such resistant forms are found in the literature. The biochemical nature and genetics of the resistance factors and their transfer are discussed in Chapters 7 and 8. Because of the potential variability in sensitivity of the organisms and harmful side effects of some antimicrobials, careful monitoring of the antimicrobial susceptibility of individual enterobacterial isolates is required.

Patients who develop endotoxin shock require prompt, vigorous treatment of both the shock and the infection. Treatment of shock centers on the cardiovascular system and includes the restoration of intravascular volume, digitalization, and administration of isoproterenol. Other agents such as steroids, pressor amines, and norepinephrine may be required.

FURTHER READING

Bauernfeind A, Petermüller C, Schneider R: Bacteriocins as tools in analysis of nosocomial *Klebsiella pneumoniae* infections. J Clin Microbiol 14:15, 1981

Beachey EH: Bacterial adherence: Adhesin-receptor interactions mediating the attachment of bacteria to mucosal surfaces. J Infect Dis 143:325, 1981

Berk SL, McCabe WR: Meningitis caused by gram-negative bacilli. Ann Intern Med 93:253, 1980

Boxerbaum B: Antimicrobial drugs for treatment of infections caused by gram-negative bacilli. Med Clin North Am 58:519, 1974

Brachman PS, Eichoff TC (eds): Proceedings of the International Conference on Nosocomial Infections. Baltimore, Centers for Disease Control, 1970

Bradley SG: Cellular and molecular mechanisms of action in bacterial endotoxins. Annu Rev Microbiol 33:67, 1979

Brenner DJ: Characterization and clinical identification of Enterobacteriaceae by DNA hybridization. Prog Clin Pathol 7:71, 1978

Brenner DJ, Davis BR, Steigerwalt AG, et al.: Atypical biogroups of *Escherichia coli* found in clinical specimens and description of *Escherichia hermanii* sp. nov. J Clin Microbiol 15:703, 1982

Cohen PS, Maguire JH, Weinstein L: Infective endocarditis caused by gram-negative bacteria: A review of the literature 1945–1977. Prog Cardiovasc Dis 22:205, 1980

Cowan ST: Family I. Enterobacteriaceae. In Buchanan RE, Gibbons NE (eds): Bergey's Manual of Determinative Bacteriology, 8th ed. Baltimore, Williams & Wilkins, 1974, p 290

DuPont HL: Enteropathogenic organisms. New etiologic agents and concepts of disease. Med Clin North Am 62:945, 1978

Edwards PR, Ewing WH: Identification of Enterobacteriaceae, 3rd ed. Minneapolis, Burgess, 1972

Ewing WH: Differentiation of Enterobacteriaceae by biochemical reactions, revised. Atlanta, Ga, HEW Publication No (CDC) 74-8270, 1973

Farmer JJ III, Fanning GR, Huntley-Carter GP, et al.: *Kluyvera*, a new (redefined) genus in the family Enterobacteriaceae: Identification of *Kluyvera ascorbata* sp. nov. and *Kluyvera cyrocrescens* sp. nov., in clinical specimens. J Clin Microbiol 13:919, 1981

Grimont PAD, Grimont F, Farmer JJ III, et al.: *Cedecea davisae* gen. nov., sp. nov. and *Cedecea lapagei* sp. nov., new Enterobacteriaceae from clinical specimens. Int J Sys Bacteriol 31:317, 1981

Hermans PE: General principles of antimicrobial therapy. Mayo Clin Proc 52:603, 1977

Hoeprich P, Boggs DR: Manifestation of infectious diseases. In Hoeprich PD (ed): Infectious Disease, 2nd ed. Hagerstown, Md, Harper & Row, 1977, p 63

Hollis DG, Hickman FW, Fanning GR, et al.: *Tatumella ptyseos* gen. nov. sp. nov., a member of the family Enterobacteriaceae found in clinical specimens. J Clin Microbiol 14:79, 1981

Kass EH, Wolff SM (eds): Bacterial lipopolysaccharides: Chemistry, biology and clinical significance of endotoxin. J Infect Dis 128 (Suppl):July, 1973, S1–S305

Konisky J: Colicins and other bacterocins with established modes of action. Annu Rev Microbiol 36:125, 1982

Mäkelä PH, Mayer H: Enterobacterial common antigen. Bacteriol Rev 40:591, 1976

Martin WJ, Washington JA II: Enterobacteriaceae. In Lennette EH, Balows A, Hausler WJ Jr, et al. (eds): Manual of Clinical Microbiology, 3rd ed. Washington, D.C., American Society for Microbiology, 1980, p 195

McHenry MC, Hawk WA: Bacteremia caused by gram-negative bacilli. Med Clin North Am 58:623, 1974

Morrison DC, Ulevitch RJ: The effects of bacterial endotoxins on host mediation system. A review. Am J Pathol 93:527, 1978

Noone P, Rogers BT: Pneumonia caused by coliforms and *Pseudomonas aeruginosa*. J Clin Pathol 29:652, 1976

Ottow JCG: Ecology, physiology and genetics of fimbriae and pili. Annu Rev Microbiol 29:79, 1975

Orskov I, Orskov F, Jann B, et al.: Serology, chemistry, and genetics of O and K antigens of *Escherichia coli*. Bacteriol Rev 41:667, 1977

Sanford JP: Pathogenic mechanisms in opportunistic gram-negative bacillary infections: Epidemiological and host factors. In Schlessinger D (ed): Microbiology—1975. Washington, D.C., American Society for Microbiology, 1975, p 302

Schlessinger D (ed): Bacterial antigens and host response. In Microbiology—1977. Washington, D.C., American Society for Microbiology, 1977, p 219

Shands JW Jr: Endotoxin as a pathogenic mediator of gram-negative infections. In Schlessinger D (ed): Microbiology—1975. Washington, D.C., American Society for Microbiology, 1975, p 330

von Graevenitz A: The role of opportunistic bacteria in human disease. Annu Rev Microbiol 31:447, 1977

CHAPTER 36

Opportunistic Enterobacteriaceae

Clinically, it has been useful to divide the family Enterobacteriaceae into the intestinal pathogens and the opportunistic pathogens. Traditionally, the intestinal pathogens have been the members of the genera *Salmonella, Shigella,* and *Yersinia,* and the opportunists were all the other Enterobacteriaceae. Whereas recent studies on diarrheal diseases have made this distinction less clear-cut, such an artificial distinction is nonetheless useful. All of the opportunistic enteric bacilli are capable of producing similar diseases, but the epidemiology, frequency, severity, and treatment of these diseases may vary with the different species. In fact, differences in antimicrobial susceptibility and epidemiology of the same species may be seen in different institutional settings. Discussion of the overall medi-

cal importance of these opportunists is found in the preceding chapter.

Tribe Escherichieae

Genus *Escherichia*

Two closely related genera, *Shigella* and *Escherichia,* comprise the tribe Escherichieae. All *Shigella* are intestinal pathogens and as such will be discussed in Chapter 37. The genus *Escherichia* contains two species, *Escherichia coli* and *Escherichia hermanii*. *E. hermanii* is the name given to a group of rarely isolated organisms formerly known as enteric group 11. As *E. hermanii* has been so infrequently isolated, it will not be discussed further. *E. coli*, however, is the most predominant facultative organism found in the large bowel of humans and is the most frequently isolated enteric bacillus in the clinical laboratory. It is unique among other opportunistic pathogens in that it has been associated with human gastrointestinal disease, particularly in children and travelers to developing nations. Enteropathogenic *E. coli* are also major causes of death in newborn pigs and calves. In addition, *E. coli* has been the subject of more experimental research than any other microorganism, especially in the field of molecular genetics.

Biochemical and Cultural Characteristics

E. coli grows well on most commonly used media. On enteric isolation media (Chap. 35), most strains appear as fermenting colonies. Some strains are β-hemolytic when grown on blood agar. Most strains are nonpigmented, motile, produce lysine decarboxylase, and utilize acetate as the sole source of carbon. These characteristics help to distinguish *E. coli* from other members of the tribe. Other important diagnostic tests are listed in Table 36-1.

Antigenic Structure

Serologic typing of the O antigen, K antigen, and H antigen of *E. coli* is a useful epidemiologic tool. Presently, at least 150 O antigenic types, 90 K antigenic types, and 50 H antigenic types have been described. The K antigens are further divided on the basis of physical behavior into three main types: L, A, and B. The serologic type of an *E. coli* isolate is given in the following format: O type:K type:H type; for example, *E. coli* O111:B4:H2.

Determinants of Pathogenicity

Surface Antigens. *E. coli* produces at least two different types of fimbriae: mannose-sensitive, or common pili; and mannose-resistant. Both types have been shown to be important for colonization of host tissues. The mannose-resistant colonizing factor antigens (CFAs) I and II bind

TABLE 36-1. SELECTED BIOCHEMICAL REACTIONS OF *ESCHERICHIA COLI*

Test	Reaction
Indol	+
Lysine decarboxylase	±
Mucate	+
Acetate	+
Gas from glucose	+
Lactose fermentation	+
Motility	±
Yellow pigment	−

enteropathogenic *E. coli* to human intestine, whereas the K88 and K99 fimbriae act in a similar manner in animals. Both mannose-sensitive and mannose-resistant fimbriae have been demonstrated on urinary tract isolates and are thought to be necessary for the binding of organisms to uroepithelial cells.

The K_1 capsular antigen is frequently found in *E. coli* isolates obtained from patients with bacteremia and neonatal meningitis. Its role may be to interfere with the phagocytosis of the organism by leukocytes.

The precise role of other surface antigens is not known. However, it is now apparent that although any *E. coli* serotype can be induced to produce enterotoxins by acquiring the necessary plasmids, certain serotypes have a higher probability of acquiring and retaining the enterotoxin plasmids (see below). For example, nearly all O78:H11 and O78:H12 serotypes of *E. coli* are enterotoxigenic, whereas *E. coli* O78 possessing other H antigens are less likely to be toxin producers.

Enterotoxins. Two enterotoxins have been isolated from *E. coli*—a heat-labile enterotoxin (LT), and a heat-stable enterotoxin (ST). The ability to produce these enterotoxins is associated with two transferable plasmids, one coding for both enterotoxins, and the second coding for the ST only. The LT is similar in many respects to the enterotoxin of *Vibrio cholerae* (Chap. 38). Both the LT and cholera enterotoxin stimulate the adenyl cyclase in epithelial cells of the small intestinal mucosa. This stimulation of enzyme activity increases the permeability of the intestinal lining, resulting in fluid loss, and thus, diarrhea. The LT and cholera enterotoxin also share immunologic reactivity, are cytopathic for Y-1 adrenal tumor and Chinese hamster ovary cells, and increase capillary permeability of rabbit skin at the site of injection. The structure of the two toxins is also similar in that the B chains consists of five subunits that are responsible for binding the toxin to the GM-1 ganglioside of intestinal cells. The A chains of both toxins can be split into two fragments, with fragment A_1 being the enzyme that catalyzes the transfer of adenosine diphosphate-(ADP)-ribose from nicotinamide adenine dinucleotide (NAD) to the regulatory subunit of adenyl cyclase, thus increasing cyclic AMP (cAMP) in the target cells. However, there are some antigenic differences between the cholera enterotoxin and *E. coli* LT, and the cholera toxin is more effective and produces a quicker response in Y-1 adrenal tumor cells. In addition, the potency of *E. coli* LT in animal models is about 100-fold less than that of cholera toxin.

Unlike the LT and the cholera enterotoxin, the ST does not stimulate adenyl cyclase activity and is not reactive in the rabbit skin test. When injected into ileal loops of rabbits, the ST will produce its maximal response in 4 hours, rather than the 10 hours required for the LT. The suckling mouse is unique in its ability to detect ST and is positive within 4 hours of inoculation. The ST of an *E. coli* pathogenic for humans has recently been purified 13,000-fold. It has a molecular weight of 1970 daltons, contains no basic amino acids, and has one or more disulfide bonds that account for its pH and heat stability. Human ST differs from porcine ST in molecular weight, in the number of amino acid residues, and in its terminal amino acid. The ST appears to activate guanylate cyclase to produce cyclic guanosine monophosphate, impairing net chloride and sodium absorption. The ST also appears to decrease the motility of the small bowel.

Other Factors. Hemolysin production by *E. coli* is mediated by the presence of a 41-megadalton plasmid. Loss of this plasmid is associated with a loss of nephropathogenicity. Also avirulent, nonhemolytic *E. coli* can become virulent by acquiring this plasmid. The exact role that the hemolysin of *E. coli* plays in infection is not clear, but it is cytotoxic for tissue culture cells, and hemolytic *E. coli* are more pathogenic in various animal models than nonhemolytic strains.

A number of *E. coli* strains are closely related to *Shigella* and are capable of producing gastrointestinal disease by penetrating the epithelial lining of the large intestine (Chap. 37). Whether these invasive *E. coli* induce penetration by production of low molecular weight substances, as do the *Shigella*, is not yet known. However, *E. coli* are less virulent than *Shigella* and require 10^7 organisms to produce the same effects seen with 200 *Shigella* organisms.

Enteropathogenic *E. coli* may produce diarrhea by a third method in addition to invasion and enterotoxin production. Recently, Cantey and Blake described an *E. coli* diarrhea in adult rabbits that is characterized by destruction of microvilli. A human case with similar pathology has also been described.

Clinical Infection

Clinical Manifestations. *E. coli* is the most common cause of urinary tract infections in humans. Over 85 percent of all community-acquired urinary tract infections are caused by this organism. Diseases may range from cystitis to pyelonephitis. Females are more likely to have urinary tract infections at a younger age, with factors like sexual maturation, childbirth, and tumors being most important. After the age of 45 years, the male with prostatic hypertrophy is more likely to have urinary tract infections. Exposure to instrumentation, diabetes, obstruction, and pregnancy are additional factors that predispose individuals to urinary tract infections by *E. coli* and other microorganisms.

E. coli can also cause pneumonia. In some hospitals, *E. coli* has caused as much as 50 percent of the primary nosocomial pneumonia. In other institutions the incidence is as low as 12 percent. The mean age of patients in one study was 53 years, and most patients had one or more chronic underlying diseases. Endogenous aspiration of oral flora containing *E. coli* appears to be the main cause of pneumonia, although patients with *E. coli* bacteremia may seed the lung with septic emboli. The pneumonia presents as a patchy bronchopneumonia, often of the lower lobe.

Empyema can occur, especially in patients who have the disease for more than 6 days.

E. coli is a major cause of neonatal meningitis but rarely causes meningitis in older populations. The mortality of *E. coli* meningitis is between 40 and 80 percent in newborns, and among survivors, a majority have subsequent neurologic or developmental abnormalities.

E. coli can be isolated from wound infections, particularly those occurring in the abdomen. Peritonitis caused by *E. coli* and other bowel organisms is a frequent complication of ruptured appendices. *E. coli* can also invade the bloodstream during any of the above infections. It is the most frequent cause of gram-negative sepsis and, in fact, is the leading blood culture isolate in most hospitals.

The role of *E. coli* in diarrheal disease in the United States is not fully understood. One estimate puts the incidence of *E. coli* diarrhea in children at 4 percent. No estimates are available for the adult population. It is clear, however, that enteropathogenic *E. coli* are important causes of diarrhea in children of developing countries, where personal hygiene and public sanitation is deficient. In addition, enterotoxigenic *E. coli* are a major cause of the traveler's diarrhea that occurs when persons from developed countries travel to an endemic region. In one prospective study of 73 physicians and 48 family members attending a medical conference in Mexico, 49 percent developed traveler's diarrhea. Enterotoxigenic *E. coli* was the major cause of disease in this group. Another study of students from the United States that had recently arrived in Mexico indicated that 40 percent of the diarrhea observed was due to enterotoxigenic *E. coli*.

Laboratory Diagnosis. Isolation and identification of *E. coli* from clinical specimens using the methods and media discussed in Chapter 35 and Table 36-1 constitute the major means of diagnosing *E. coli* infections. However, the mere isolation of *E. coli* from contaminated specimens such as sputa and wounds does not determine the diagnosis. Careful consideration must be given to patient factors, other organisms isolated, and stains made from clinical material.

The role of *E. coli* in intestinal disease requires that, in the absence of known pathogens, the enteropathogenic *E. coli* be considered in the differential diagnosis of diarrhea. However, enterotoxin detection requires costly animal or tissue culture studies, and the only test for detecting invasive *E. coli*, the Sereny test, requires injection of the isolate into the eye of a guinea pig. Therefore, most routine clinical laboratories do not presently have the means for detecting these organisms.

Treatment. *E. coli* isolated from community-acquired infections are usually sensitive to most antimicrobials used to treat gram-negative organisms (Chap. 35). Resistant forms, however, can and do appear, especially in patients with a history of prior antibiotic therapy. The best treatment of diarrhea appears to be management of fluid and electrolyte balance. Although infantile diarrhea has been controlled by a number of antibiotics, the rapid emergence of resistant forms requires close surveillance of sensitivity patterns. Prophylactic treatment with antibiotics such as trimethoprim-sulfamethoxazole has reduced the incidence of traveler's diarrhea. However, certain clinicians feel that this treatment may only serve to select for resistant bacteria and may serve to potentiate the carrier state in those individuals who may be infected with *Salmonella* (Chap. 37) rather than *E. coli*.

Tribe Klebsielleae

The tribe Klebsielleae consists of four genera, *Klebsiella, Enterobacter, Hafnia,* and *Serratia*. As with other enterics, these organisms normally inhabit the large bowel of humans and animals; however, a number of organisms in this tribe are found in water and soil. Some species can cause disease in plants and insects as well as in humans, and others may even play a beneficial role in nitrogen fixation. The Klebsielleae have become the third leading cause of nosocomial infections in the United States and, in some institutions, have replaced *E. coli* as the most frequent isolate in these types of infections. The epidemiology of the individual genera of this tribe has been difficult to assess because of the numerous taxonomic changes that have occurred and the lack of definitive methods of speciation. During the past 15 years, reliable procedures have become available, permitting the acquisition of more accurate epidemiologic data. Table 36-2 correlates the present classification scheme with some of the former nomenclature used to describe this tribe.

Genus *Klebsiella*
The most commonly isolated member of the tribe Klebsielleae is *Klebsiella pneumoniae* (Friedlander's bacillus). As its name implies, it can cause pneumonia and was origi-

nally thought to be the cause of the classic lobar pneumonia, the true agent of which is *Streptococcus pneumoniae*. Like all other opportunistic enterics, *K. pneumoniae* can cause infection of other body sites besides the respiratory tract. Other species of *Klebsiella* cause similar diseases but are isolated less frequently. Two species of *Klebsiella, Klebsiella ozaenae* and *Klebsiella rhinoscleromatis*, are also causative agents of chronic infections of the nasal mucosa and pharynx.

Biochemical and Cultural Characteristics. Speciation of the *Klebsiella* can be achieved by the tests shown in Table 36-3. All species are characteristically nonmotile, which is useful in separating them from other members of the tribe. With the exception of *K. rhinoscleromatis*, most isolates ferment lactose. The presence of a large capsule causes colonies of *Klebsiella* growing on agar to appear large, moist, and mucoid.

Antigenic Structure. *Klebsiella* possess O antigens and K antigens, of which the polysaccharide K antigens have proved to be the most useful for serologic typing. Seventy-two different K antigens have been described. All species of *Klebsiella* share common antigens and thus may be typed with the same set of antisera. The majority of *K. ozaenae* strains belong to type 4, whereas most *K. rhinoscleromatis* strains possess type 3 antigen.

No single serologic type of *K. pneumoniae* is more virulent, nor is one type found to be more frequently associated with specific infections. Nevertheless, serologic typing provides a useful marker for investigating apparent epidemics caused by this organism.

Determinants of Pathogenicity. The capsule of *K. pneumoniae* prevents phagocytosis and aids in establishment of the organisms in the respiratory tract. With the exception of endotoxin, no other toxin has been identified as

TABLE 36-2. COMPARISON OF PRESENT NOMENCLATURE OF KLEBSIELLEAE WITH OLDER USAGE

	Present Taxonomy	Older Synonyms
Genus	*Klebsiella*	
Species	*K. pneumoniae*	Same but includes nonmotile *Aerobacter aerogenes*, Friedländer's bacillus
	K. ozaenae	Same
	K. rhinoscleromatis	Same
	*K. oxytoca**	*Klebsiella pneumoniae*, indol positive
Genus	*Enterobacter*	*Aerobacter*
Species	*E. cloacae*	*Aerobacter cloacae, Aerobacter* A, *Cloacae* A
	E. aerogenes	*Aerobacter aerogenes, Aerobacter* B, *Cloacae* B
	E. hafniae (Hafnia alvei)†	*Hafnia* group, *E. alvei, Bacterium cadaveria*
	E. agglomerans	*Erwinia herbicola, Escherichia adecarboxylata, Bacterium typhiflavum*, and others
Genus	*Serratia*	
Species	*S. marcescens*	Same
	S. liquefaciens	*Enterobacter liquefaciens*
	S. rubidaea	*Bacterium rubidaeum*

*Designation used by Centers for Disease Control as of October 1977.
†Designation used by Centers for Disease Control and *Bergey's Manual for Determinative Bacteriology*, 8th ed.

TABLE 36-3. SPECIATION OF GENUS *KLEBSIELLA*

Test	*K. pneumoniae*	*K. oxytoca*	*K. ozaenae*	*K. rhinoscleromatis*
Lactose fermentation	+	+	±	−
Voges–Proskauer	+	+	−	−
Methyl red	−	−	+	+
Citrate (Simmons)	+	+	±	−
Lysine decarboxylase	+	+	±	−
Malonate	+	+	−	+
Indol	−	+	−	−

playing a role in opportunistic *Klebsiella* infections. Klipstein has isolated enterotoxin-producing strains of *K. pneumoniae* from patients with tropical sprue. These enterotoxins are very similar, if not identical, to the ST and LT of *E. coli*, and like *E. coli*, enterotoxin production in *K. pneumoniae* is plasmid mediated.

Clinical Infection. *K. pneumoniae* can cause a primary pneumonia. Generally this occurs in middle-aged and older men who have underlying medical problems, such as alcoholism, chronic bronchopulmonary disease, or diabetes mellitus. A thick nonputrid, bloody sputum is produced in approximately 25 to 75 percent of patients with pneumonia caused by *Klebsiella*. Abscess formation and necrosis are more likely to occur with *K. pneumoniae* than in other bacterial pneumonias, and blood cultures are positive in about 25 percent of the patients. Some authors report a mortality rate of 50 percent despite adequate antimicrobial therapy, with mortality correlating closely with the occurrence of bacteremia.

In addition to primary pneumonia, *K. pneumoniae* has been associated with urinary tract and wound infections, bacteremia, and meningitis. At some institutions, *K. pneumoniae* has replaced *E. coli* as the leading cause of bacteremia. A 5-year review of meningitis in one hospital showed that *Klebsiella* species were responsible for 15 percent of 61 cases of gram-negative bacillary meningitis. Most isolates were from neurosurgical patients and appeared secondary to infections at other sites.

The role of enterotoxigenic *K. pneumoniae* in diarrhea is difficult to assess. There have been no major systematic studies looking for these organisms in diarrhea, and most isolates have been obtained in tropical countries where diarrhea is a chronic problem.

Klebsiella oxytoca resembles *K. pneumoniae* in disease spectrum and from a clinical viewpoint can be considered as the same organism. *K. ozaenae* causes a chronic atrophic rhinitis characterized by a fetid odor. Similarly, *K. rhinoscleromatis* infects the nose and pharynx producing granulomatous destruction. Both *K. ozaenae* and *K. rhinoscleromatis* infections of the nose and pharynx are rare in the United States and are primarily seen in immigrants from endemic regions. The major role of these organisms in the United States is similar to other opportunistic enteric bacilli and the majority of isolates are obtained from urinary tract and soft tissue infections and bacteremia.

TREATMENT. The majority of *K. pneumoniae* isolates are resistant to ampicillin and carbenicillin. Cephalosporins are usually effective, a property that is useful in differentiating *Klebsiella* from organisms of the genus *Enterobacter*. Other antibiotics listed in Chapter 35 are also usually effective against this organism, although strains isolated from nosocomial infections are often resistant to multiple antimicrobials. In a prospective survey of *Klebsiella* isolates from the community and hospital, Davis and Matsen found that 18 percent of isolates from patients who had remained in the hospital 15 days or longer were resistant to several antimicrobial agents, whereas only 4 percent of isolates obtained in the community displayed multiple resistance.

Genus *Enterobacter* and Genus *Hafnia*

The genus *Enterobacter* (formerly *Aerobacter*) contains a large number of species that inhabit soil and water, as well as the large intestine of humans and animals. The clinical importance of the genus *Enterobacter* as a separate entity was not greatly appreciated until the 1960s. Prior to that time, separation of the *Enterobacter* from the *Klebsiella* was not routinely attempted, and many infections were reported as being caused by the *Klebsiella–Aerobacter* group. How many of these diseases were in reality caused by *Enterobacter* is not known. Recent studies indicate that *Enterobacter* infections occur less frequently than those caused by *Klebsiella*. The Centers for Disease Control and *Bergey's Manual of Determinative Bacteriology* presently classify *Enterobacter hafniae* as *Hafnia alvei*, but for the purposes of this chapter, it is discussed with the *Enterobacter*.

Biochemical and Cultural Characteristics. The genus *Enterobacter* can be divided into five species on the basis of the biochemical tests shown in Table 36-4. These organisms are differentiated from *Klebsiella* by the fact that all are motile, and, with the exception of *Enterobacter agglomerans*, all decarboxylate ornithine. With *H. alvei*, more biochemical tests are positive at 25C than at 37C.

Antigenic Structure. All *Enterobacter* possess O antigens and H antigens, whereas only a portion of the strains possess K antigens. The antigens for the entire genus have not been as completely characterized as those for other enteric bacilli.

TABLE 36-4. SPECIATION OF THE GENERA *ENTEROBACTER* AND *HAFNIA*

Test	E. cloacae	E. aerogenes	E. agglomerans	E. gergoviae	E. sakazakii	H. alvei (E. hafniae)
Yellow pigment	−	−	±	−	+	−
Lysine decarboxylase	−	+	−	+	−	+
Ornithine decarboxylase	+	+	−	+	+	+
Arginine dihydrolase	+	−	−	−	+	−
Urease	±	−	±	+	−	−
Sucrose fermentation	+	+	±	+	+	−

Clinical Infection. *Enterobacter*, like most Enterobacteriaceae, are capable of producing disease in any body tissue, but have been most frequently isolated from urinary tract infections. *Enterobacter cloacae* accounts for the majority of clinical isolates of this genus, but all species have been isolated from clinical specimens.

Two species, *E. agglomerans* (formerly classified as *Erwinia*) and *E. cloacae*, were associated with a nationwide epidemic involving contaminated intravenous fluids. These species were isolated from 8 hospitals in 7 states and were responsible for 150 bacteremias and 9 deaths. Enterotoxin-producing *E. cloacae* have been isolated from the jejunal aspirates of patients with tropical sprue. These strains produced toxins similar to the toxins of *E. coli*. As with the enterotoxigenic *Klebsiella* isolates, the exact role of *Enterobacter* in intestinal disease has not been fully explored.

Isolation of other *Enterobacter* species from clinical material does occur, but with less frequency than with *E. cloacae*. *Enterobacter sakazakii* was first isolated from an infant with neonatal meningitis. *Enterobacter gergoviae* has been implicated in nosocomial urinary tract infections and has been isolated from wounds, sputum, and blood. *Enterobacter aerogenes* and *H. alvei* cause similar infections.

TREATMENT. With the exception of ampicillin and cephalosporins, most of the antimicrobials discussed in Chapter 35 are useful in the treatment of *Enterobacter* infections. As with all enterics, resistance patterns may vary with individual isolates.

Genus *Serratia*

The organisms of the genus *Serratia* were at one time thought to be harmless, pigmented saphrophytes and were used as markers to demonstrate air current patterns in both the environment and in hospitals. These organisms, however, have emerged as major entities in nosocomial infections. Almost all *Serratia* infections are associated with underlying disease, changing physiologic patterns, immu-

nosuppressive treatment, or mechanical manipulations of the patient. In nature, *Serratia* is found in soil, water, plants, and animals. They are frequently isolated from insects and may play a role in diseases of certain insects. Three species, *Serratia marcescens*, *Serratia liquifaciens*, and *Serratia rubidaea*, constitute the major human pathogens of this genus.

Biochemical and Cultural Characteristics. Organisms of the genus *Serratia* can be differentiated from other enterobacteria by the production of an extracellular DNase. Speciation can be achieved with the decarboxylase and fermentation tests listed in Table 36-5. The majority of *S. rubidaea* isolates form a pink to red pigment, as do some strains of *S. marcescens*. Pigment production by *S. marcescens* is enhanced by incubation at room temperature.

Clinical Infection. *S. marcescens* is the most frequent isolate of the genus *Serratia* and has been associated with a number of nosocomial epidemics involving pneumonia, septicemia, urinary tract infections, and wound infections. Urinary tract infections are associated with underlying abnormalities or instrumentation. Pneumonia with this organism has been transmitted by contaminated respiratory care equipment and is similar to pneumonia caused by *Klebsiella*, except that necrosis and abscess formation are less likely to occur. Infection with pigmented *Serratia* may cause sputum to be tinged with red, thus giving the false impression of hemoptysis. *S. liquefaciens* and *S. rubidaea* have been found in sputa, but their role in pulmonary disease has not been clearly defined.

TREATMENT. Aminoglycoside antibiotics, chloramphenicol, and trimethoprim-sulfamethoxazole are effective against most isolates of *S. marcescens*. However, there can be considerable strain-to-strain variation, requiring close monitoring of susceptibility of individual isolates.

TABLE 36-5. SPECIATION OF THE GENUS *SERRATIA*

Test	S. marcescens	S. liquefaciens	S. rubidaea
Ornithine decarboxylase	+	+	−
Arabinose fermentation	−	+	+
DNase	+	+	+

Tribe Proteeae

The organisms of the tribe Proteeae are differentiated from other Enterobacteriaceae by their ability to deaminate phenylalanine. Ewing divided the family into two genera, *Proteus* and *Providencia,* while other taxonomists classify the *Providencia* as *Proteus inconstans.* Recent investigations on DNA relatedness have prompted Brenner and co-workers to divide this tribe into three genera: *Proteus, Providencia,* and *Morganella.*

The majority of clinical isolates of this tribe are from urine, although infections at other sites frequently occur. Some studies indicate that the Proteeae may be responsible for 10 to 15 percent of the nosocomial infections in the United States.

Genus *Proteus*

The organisms of the genus *Proteus* are found in soil, water, sewage, and decaying animal matter, as well as in the human intestinal tract. Brenner's classification scheme recognizes two clinically important species in the genus *Proteus*—*Proteus mirabilis* and *Proteus vulgaris.* A third species, *Proteus myxofaciens,* has been isolated only from gypsy moth larvae and is not be considered in this text.

Biochemical and Cultural Characteristics. *P. mirabilis* and *P. vulgaris* are actively motile at 37C, producing a thin translucent sheet of growth on nonselective agars. This phenomenon is referred to as swarming. On certain enteric isolation media, the production of hydrogen sulfide by *P. mirabilis* and *P. vulgaris* may cause the colonies of these organisms to be confused with those of the enteric pathogens of the genus *Salmonella.* Both species of *Proteus* produce a powerful urease that hydrolyzes urea to ammonia and carbon dioxide. In addition, they differ from other enteric bacilli in thir ability to grow at an alkaline pH. Differentiation between the two species can be made on the basis of indol production, ornithine decarboxylase reaction, and other tests listed in Table 36-6.

Antigenic Structure. Several studies have attempted to group *Proteus* strains on the basis of the O antigen, H antigen, and K antigen, but these studies have not been sufficently correlated to be useful as an epidemiologic tool. The most important use of the *Proteus* antigens is in the diagnosis of rickettsial disease. Certain *P. vulgaris* strains (OX-19, OX-K, and OX-2) share antigens with the *Rickettsia.* This cross-reactivity allows these organisms to serve as the antigens for the detection of rickettsial antibodies in the Weil–Felix test (Chap. 51).

Clinical Infection. *P. mirabilis* accounts for the majority of *Proteus* infections in humans, causing community-acquired as well as hospital-acquired urinary tract infections. Wound infections, pneumonia, and septicemia can also occur. The frequency of infections by other species is difficult to assess because all indol-positive *Proteus* species are lumped into one group. This group included *P. vulgaris,* *Proteus morganii (Morganella morganii),* and *Proteus rettgeri (Providencia rettgeri).* All of these organisms, however, are capable of causing the infections seen with *P. mirabilis.* *Proteus* pneumonias are similar to those of *Klebsiella.* Bacteremias are common in debilitated patients and are associated with high mortality rates.

When *Proteus* infects the urinary tract, the hydrolysis of urea to ammonia raises the pH of the urine, resulting in precipitation of calcium and magnesium salts and the formation of calculi. Urease production also correlates with the severity of the disease. When urease activity is absent, the number of organisms in the kidney and the extent of renal damage are less than when the enzyme is present. Furthermore, alkalization of urine by urease causes increased damage to renal epithelium.

TREATMENT. *P. mirabilis* differs from the other Proteeae in being sensitive to ampicillin and cephalosporins. Most isolates of all species are sensitive to aminoglycoside antibiotics and to the combination of trimethoprim-sulfamethoxazole. Resistant forms do occur, particularly in patients previously given antimicrobial therapy.

Genus *Morganella*

Brenner and co-workers using DNA homology tests found that *P. morganii* was 20 percent or less related to other Proteeae and proposed the creation of a new genus,

TABLE 36-6. IDENTIFICATION OF THE TRIBE PROTEEAE

Test	*Proteus mirabilis*	*Proteus vulgaris*	*Morganella morganii* (*Proteus morganii*)	*Providencia rettgeri* (*Proteus rettgeri*)	*Providencia stuartii*	*Providencia alcalifaciens*
Phenylalanine deaminase	+	+	+	+	+	+
Urease	+	+	+	+	±	−
Indol	−	+	+	+	+	+
H₂S	+	+	−	−	−	−
Ornithine decarboxylase	+	−	+	−	−	−
Adonitol fermentation	−	−	−	+	−	+
Inositol fermentation	−	−	−	−	+	−

Morganella. The biochemical characteristics of this organism are listed in Table 36-6. Infections caused by *M. morganii* are similar to those of other members of the tribe. Treatment is similar to that of *P. vulgaris.*

Genus *Providencia*

The organisms of the genus *Providencia* are often associated with nosocomial infections involving the urinary tract, blood, respiratory tract, and wounds. *Providencia stuartii* and *P. rettgeri* are resistant to multiple antimicrobial agents. Treatment depends upon the susceptibility results of individual isolates.

Biochemically, these organisms can be differentiated from other Proteeae by the reactions indicated in Table 36-6. The work of Penner and co-workers on the biogroups of *P. rettgeri*, together with their work on DNA homology, prompted Brenner and colleagues to classify the *rettgeri* group in the genus *Providencia*, rather than the genus *Proteus.*

Tribe Edwardsielleae

Edwardsiella tarda is the only species presently classified in the tribe Edwardsielleae. Formerly, it was classified in the Asakusa group and the Bartholomew group of organisms. A number of isolates from animal sources, primarily cold-blooded types, have been reported. In humans *Edwardsiella* infections are rare, but the organism has been isolated from wound infections, sepsis, and meningitis. The organism also has been isolated from human feces in cases of gastroenteritis. An extensive survey of 14,000 specimens obtained from human and animal sources in Panama yielded 50 human isolates of *Edwardsiella*. Ten patients had diarrheal symptoms, while 20 were asymptomatic intestinal carriers. Interestingly, all isolates were obtained in the rural population, whereas none were found in urban dwellers. Presumably the rural population had the opportunity to come in contact with the wild fauna where the organism was frequently found. There were two fatal cases; one with liver abscess, the other involving bacteremia.

E. tarda produces indol, H_2S, lysine, and ornithine de-carboxylase. Lactose is not fermented, and acetate cannot be used as the sole carbon source. The production of H_2S may cause colonies of these organisms to be confused with the enteric pathogens of the genus *Salmonella.*

Most *E. tarda* isolates are sensitive to the antimicrobials used in the treatment of enterobacterial infections.

Opportunists of Tribe Salmonelleae

In addition to the intestinal pathogens of the genus *Salmonella* (Chap. 37), the tribe Salmonelleae contains two additional genera, *Citrobacter* and *Arizona,* which are isolated from human infections. Table 36-7 lists the reactions used for the differentiation of these organisms.

Genus *Citrobacter*

This genus contains three species, *Citrobacter freundii, Citrobacter diversus,* and *Citrobacter amalonaticus.* These organisms can be isolated from a variety of clinical specimens, with the majority of isolates coming from urinary tract and respiratory infections. In one review, approximately 75 percent of patients with clinically significant *Citrobacter* infections had underlying disease or predisposing factors. *C. diversus* comprised 42 percent of all isolates; *C. freundii* comprised 29 percent; and 29 percent were identified as *Citrobacter* species. Most *Citrobacter* isolates are sensitive to the aminoglycoside antibiotics, tetracycline, nitrofurantoin, and chloramphenicol.

Genus *Arizona*

This genus is associated with reptiles and birds. One species is described, *Arizona hinshawii.* This organism was formerly referred to as *Paracolobactrum arizonae.* Because the genus *Arizona* shares many biochemical and serologic relationships with the genus *Salmonella,* many taxonomists classify this organism as *Salmonella arizonae.* Human isolates are rare, but the spectrum of disease is similar to *Salmonella* and includes gastroenteritis, bacteremia, pyelonephritis, osteomyelitis, and otitis media. Its importance in gastrointestinal disease is difficult to assess, as a large number of strains ferment lactose, and many laboratory workers have

TABLE 36-7. DIFFERENTIATION OF THE TRIBE SALMONELLEAE

Biochemical Test	*Salmonella* species	*Citrobacter diversus**	*Citrobacter freundii*	*Citrobacter amalonaticus*[†]	*Arizona hinshawii*[‡]
H_2S production	+	−	+	−	+
Indol production	−	+	−	+	−
Malonate production	−	+	±	−	+
Lysine decarboxylase	+	−	−	−	+
Growth in KCN	−	−	+	+	−

*Previously classified as aberrant *Enterobacter cloacae, Levinea, Citrobacter intermedius* biotype b, and *Citrobacter koseri.*
[†]Previously *Levinea amalonatica, Citrobacter intermedius* biotype a.
[‡]Previously *Paracolobactrum arizonae.*

been trained to disregard lactose-fermenting organisms as normal fecal flora. It appears, however, that humans are an accidental host in most infections with this organism. Ampicillin and chloramphenicol appear to be the drugs of choice, with most isolates being sensitive to the other antimicrobials used for the enterics.

FURTHER READING

Books and Reviews

Arroyo JC, Sonnenwirth AC, Liebhaber H: *Proteus rettgeri* infections: A review. J Urol 117:115, 1977

Black RE: The prophylaxis and therapy of secretory diarrhea. Med Clin North Am 66:611, 1982

Brenner DJ, Farmer JJ III, Hickman FW, et al.: Taxonomic and nomenclature changes in Enterobacteriaceae. Atlanta, Ga, HEW Publication No. (CDC) 78-8356, 1977

Carpenter CCJ: The pathophysiology of secretory diarrheas. Med Clin North Am 66:597, 1982

DuPont HL, Pickering LK: Infections of the gastrointestinal tract microbiology, pathophysiology, and clinical features. New York, Plenum, 1980, pp 61, 129, 195

Gaastra W, DeGraaf FK: Host-specific fimbrial adhesions of noninvasive enterotoxigenic *Escherichia coli* strains. Microbiol Rev 46:129, 1982

Gorbach SL, Hoskins DW: Travelers' diarrhea. Diagnostic Medium Oct 1980

Grimont PAD, Grimont F: The genus *Serratia.* Annu Rev Microbiol 32:221, 1978

John JF Jr, Sharbaugh RJ, Bannister ER: *Enterobacter cloacae* bacteremia, epidemiology and antibiotic resistance. Rev Infect Dis 4:13, 1982

Johnson RH, Latwick GA, Huntley GA, et al.: *Arizona hinshawii* infections. New cases, antimicrobial sensitivities and literature review. Annu Intern Med 85:587, 1976

Lerner AM: The gram-negative bacillary pneumonias Diagnostic Medium Nov 1980

Mangi RJ, Quintiliani R, Andriole VT: Gram-negative bacillary meningitis. Am J Med 59:829, 1975

Martin WJ, Washington JA III: Enterobacteriaceae. In Lennette EH, Balows A, Hausler WJ Jr, et al. (eds): Manual of Clinical Microbiology, 3rd ed. Washington, D.C., American Society of Microbiology, 1980, p 195

Richards KL, Douglas SD: Pathophysiological effects of *Vibrio cholerae* and enterotoxigenic *Escherichia coli* and their exotoxins on eucaryotic cells. Microbiol Rev 42:592, 1978

Sack RB: Enterotoxigenic *Escherichia coli*: Identification and characterization. J Infect Dis 142:279, 1980

Wilfert CM: *E. coli* meningitis: K₁ antigen and virulence. Annu Rev Med 29:129, 1978

Yu VL: *Serratia marcescens*—Historical perspective and clinical review. N Engl J Med 300:887, 1979

Selected Papers

Adler JL, Burke JP, Martin DF, et al.: *Proteus* infections in a general hospital. I. Biochemical characteristics and antibiotic susceptibility of the organism. Annu Intern Med 75:517, 1971

Bäch E, Möllby R, Kaijser B, et al.: Enterotoxigenic *Escherichia coli* and other gram-negative bacteria of infantile diarrhea: Surface antigens, hemagglutinins, colonization factor antigens and loss of enterotoxigenicity. J Infect Dis 142:318, 1980

Black RE, Merson MH, Rowe B, et al.: Enterotoxigenic *Escherichia coli* diarrhoea: Acquired immunity and transmission in an endemic area. Bull WHO 59:263, 1981

Brenner DJ, Davis BR, Steigerwalt AG, et al.: Atypical biogroups of *Escherichia coli* found in clinical specimens and description of *Escherichia hermanii* sp. nov. J Clin Microbiol 15:703, 1982

Brown MR, DuPont HL, Sullivan PS: Effect of duration of exposure on diarrhea due to enterotoxigenic *Escherichia coli* in travelers from United States to Mexico. J Infect Dis 145:582, 1982

Cantey JR, O'Hanley PD, Blake RK: A rabbit model of diarrhea due to invasive *Escherichia coli.* J Infect Dis 136:640, 1977

Cavalieri SJ, Snyder IS: Cytotoxic activity of partially purified *Escherichia coli* alpha haemolysin. J Med Microbiol 15:11, 1982

Clarridge JE, Mosher DM, Fainstein V, et al.: Extraintestinal human infection caused by *Edwardsiella tarda.* J Clin Microbiol 11:511, 1980

Davis TJ, Matsen JM: Prevalence and characteristics of *Klebsiella* species: Relation to association with hospital environment. J Infect Dis 130:402, 1974

Deb M, Bhujwala RA, Singh S, et al.: *Klebsiella pneumoniae* as the possible cause of an outbreak of diarrhoea in a neonatal special care unit. Indian J Med Res 71:359, 1980

Donta ST, Poindexter NJ, Ginsberg BH: Comparison of the binding of cholera and *Escherichia coli* enterotoxins to Y-1 adrenal cells. Biochemistry 21:660, 1982

Edwards LD, Cross A, Levin S, et al.: Outbreak of a nosocomial infection with a strain of *Proteus rettgeri* resistant to many antimicrobials. Am J Clin Pathol 6:41, 1974

Evans DJ Jr, Evans DG, Hohne C, et al.: Hemolysin and K antigens in relation to serotype and hemagglutination type of *Escherichia coli* isolated from extraintestinal infections. J Clin Microbiol 13:171, 1981

Gill DM, Clements JD, Robertson DC, et al.: Subunit number and arrangement in *Escherichia coli* heat-labile enterotoxin. Infect Immun 33:677, 1981

Goldstein EJC, Lewis RP, Martin WJ, et al.: Infections caused by *Klebsiella ozaenae*: A changing disease spectrum. J Clin Microbiol 8:413, 1978

Graham DR, Band JD: *Citrobacter* diversus brain abscess and meningitis in neonates. JAMA 245:1923, 1981

Guandalini S, Rao MC, Smith PL, et al.: CGMP modulation of ileal ion transport: In vitro effects of *Escherichia coli* heat-stable enterotoxin. Am J Physiol 243:G36, 1982

Hodges GR, Degener CE, Barnes YWG: Clinical significance of *Citrobacter* isolates. Am J Clin Pathol 70:37, 1978

Klipstein FA, Engert RF: Purification and properties of *Klebsiella pneumoniae* heat-stable enterotoxin. Infect Immun 13:373, 1976

Klipstein FA, Engert RF: Partial purification and properties of *Enterobacter cloacae* heat-stable enterotoxin. Infect Immun 13:1307, 1976

Klipstein FA, Engert RF, Short HB: Enterotoxigenicity of colonizing coliform bacteria in tropical sprue and blind loop syndrome. Lancet 2:342, 1978

Kourany M, Vasquez MA, Soenz R: *Edwardsiella* in man and animals in Panama. Clinical and epidemiological characteristics. Am J Trop Med Hyg 26:1183, 1977

Maki DG, Hennekens CG, Phillips CW, et al.: Nosocomial urinary tract infections with *Serratia marcescens*: An epidemiological study. J Infect Dis 128:579, 1973

Malowany MS, Chester B, Allerhand J: Isolation and microbiological differentiation of *Klebsiella rhinoscleromatis* and *Klebsiella ozaenae* in cases of chronic rhinitis. Am J Clin Pathol 58:550, 1972

Mathias JR, Nogueira J, Martin JL, et al.: *Escherichia coli* heat-stable toxin: Its effect on motility of small intestine. Am J Physiol 242:G360, 1982

Mayhall CG, Lamb VA, Gayle WE Jr, Haynes BW Jr: *Enterobacter cloacae* septicemia in a burn center: Epidemiology and control of an outbreak. J Infect Dis 139:166, 1979

Merson MH, Orskov F, Orskov I, et al.: Relationship between enterotoxin production and serotype in enterotoxigenic *Escherichia coli*. Infect Immun 23:325, 1979

Miller RH, Shulman JB, Canalis RF, Ward PH: *Klebsiella rhinoscleromatis*: A clinical and pathogenic enigma. Otolaryngol Head Neck Surg 87:212, 1979

Montgomerie JZ, Ota JK: *Klebsiella* bacteremia. Arch Int Med 140:525, 1980

Overturf GD, Wilkins J, Ressler R: Emergence of resistance of *Providencia stuartii* to multiple antibiotics: Speciation and biochemical characteristics of *Providencia*. J Infect Dis 129:353, 1974

Selden R, Lee S, Wang WLL, et al.: Nosocomial *Klebsiella* infections: Intestinal colonization as a reservoir. Annu Intern Med 74:657, 1971

Solberg CO, Matsen JM: Infections with Providence bacilli. Am J Med 50:241, 1971

Staples SJ, Asher SE, Giannella RA: Purification and characterization of heat-stable enterotoxin produced by a strain of *E. coli* pathogenic for man. J Biol Chem 255:4716, 1980

Ulshen MH, Rollo JL: Pathogenesis of *Escherichia coli* gastroenteritis in man—Another mechanism. N Engl J Med 302:99, 1980

Washington JA III, Birk RJ, Ritts RE Jr: Bacteriologic and epidemiologic characteristics of *Enterobacter hafniae* and *Enterobacter liquefaciens*. J Infect Dis 124:379, 1971

Welch RA, Dellinger EP, Minshew B, et al.: Haemolysin contributes to virulence of extra-intestinal *E. coli* infections. Nature 294:665, 1981

CHAPTER 37

Enterobacteriaceae: Salmonella and Shigella, Intestinal Pathogens

For the adult population of well-developed countries, diarrhea generally represents, at most, an inconvenience. However, among the very young, the old, the malnourished, and those living in marginal conditions, diarrhea represents a serious, crippling, life-threatening situation. As many as one third of pediatric deaths in developing countries are attributed to diarrhea and resulting dehydration. Diarrhea has even altered the course of military history by incapacitating large numbers of men, making them unfit for battle. The family Enterobacteriaceae contains two genera, *Salmonella* and *Shigella*, which are among the leading causes of bacterial diarrhea. Although they are classified in the same family and infect the same organ system, they differ greatly in their microbiologic, epidemiologic, and pathologic properties.

Shigella

Shigella species are the major causes of bacillary dysentery, a disease characterized by frequent, painful passage of low volume stools containing blood and mucus. Humans are the natural reservoir for *Shigella* and, with the exception of higher primates, these organisms do not naturally infect animals. Most disease is seen in the pediatric age group, with most infections occurring in children from 1 to 10 years of age. In the United States *Shigella* have been estimated to cause 15 percent of pediatric diarrhea, whereas in developing countries it is a leading cause of infant mortality. Most cases are a result of close contact with infected individuals and large scale outbreaks are rare. However, in

1969 and 1970 a major outbreak of shigellosis occurred in Central America and Mexico, with over 13,000 deaths reported.

Taxonomy. *Shigella* are classified in the tribe Escherichieae and are very closely related to the members of the genus *Escherichia*. There are four species of *Shigella*: *Shigella dysenteriae, Shigella flexneri, Shigella boydii,* and *Shigella sonnei*. Speciation is based upon serologic and biochemical reactions (Table 37-1). All four species can cause bacillary dysentery, but the severity of disease, mortality, and epidemiology differ for each species.

Physiology

Biochemical Properties. *Shigella* appear as nonlactose-fermenting colonies on the differential media used to isolate the enteric bacilli (Chap. 35). All *Shigella* are nonmotile, do not produce H_2S, and, except for certain types of *S. flexneri*, do not produce gas during carbohydrate fermentation. These factors distinguish them from most *Salmonella*. In contrast with *Escherichia coli*, they do not produce lysine decarboxylase, utilize acetate as a carbon source, or ferment lactose rapidly. *S. sonnei* will ferment lactose upon extended incubation. Other biochemical tests useful in speciation of the genus are found in Table 37-1. (For detailed biochemical information, see the publications of Ewing and co-workers.)

Resistance to Physical and Chemical Agents. *Shigella* are less resistant than *Salmonella* and other enterics to a variety of physical and chemical agents; most common disinfectants are lethal for *Shigella*. High concentrations of acids are detrimental, necessitating the use of well-buffered media for transport of specimens and for culture of the organisms. High concentrations of bile are inhibitory to some strains, making certain enteric media, such as Salmonella–Shigella agar, unsuitable for isolation of *Shigella* from clinical specimens. They tolerate low temperatures if adequate moisture is available, and they can survive for over 6 months in water at room temperature.

Antigenic Structure

All *Shigella* possess O antigens, and some possess K antigens. The K antigen is not significant in the serologic typing of *Shigella* but, when present, interferes with the determination of the O antigen type. This interference can usually be removed by boiling the cell suspension. The *Shigella* are divided into four major O antigen groups, designated A, B, C, and D, which correspond to the species *S. dysenteriae, S. flexneri, S. boydii,* and *S. sonnei*, respectively. Each major group or species is also subdivided into types on the basis of the O antigen. These subgroups are designated by arabic numbers. At the present time, 10 serologic types of *S. dysenteriae*, 8 of *S. flexneri*, 15 of *S. boydii*, and 1 of *S. sonnei* have been described. *S. sonnei* can be further differentiated into 15 bacteriocin types.

Determinants of Pathogenicity

Bacillary dysentery follows attachment to and penetration of the epithelial cells of the mucosal surfaces of the terminal ileum and the colon by *Shigella*. Local inflammation occurs, followed by cell death and sloughing of the lining, resulting in shallow ulcers. Virulent *Shigella* possess three factors that contribute to these events: (1) smooth lipopolysaccharide (LPS) structure; (2) invasiveness; and (3) toxin production.

Smooth Lipopolysaccharide. The importance of smooth LPS structure for the virulence of *Shigella* has been best demonstrated with *S. sonnei* and *S. flexneri*. Sansonetti and co-workers have shown that virulent *S. sonnei* possess a large 120-megadalton plasmid, which codes for production of specific O side chains of 2-amino-2-deoxy-L-altruronic acid, resulting in a phase I or smooth colony formation. Loss of this plasmid results in loss of these side chains and is evidenced by phase II or rough colony production. Phase II variants are avirulent, whereas phase I colonies are virulent. Hybrids of *S. flexneri* and *E. coli* have been produced that are virulent only when the *S. flexneri* O antigens are expressed. Although not proved, the side chains may provide a means of attaching the organism to specific host cell receptors.

Invasiveness. Virulent *Shigella* penetrate the epithelial cells of the colon in an uneven manner. The organisms rarely penetrate to the lamina propria, and exist intracellulary in cytoplasmic vacuoles. Invasion of cells depends upon the metabolic state of both the host and bacterial cells. Divalent cations such as calcium, magnesium, and iron aid in this process. After attaching to the epithelial cells, the adherent organisms produce a low molecular weight substance that induces a ruffling movement of host cell membranes and initiates pinocytosis in normally

TABLE 37-1. TESTS USEFUL IN SPECIATION OF *SHIGELLA*

Test	S. dysenteriae	S. flexneri	S. boydii	S. sonnei
Antigen group	A	B	C	D
Indol	±	±	±	−
Jordan's tartrate	±	−	−	+
Mannitol fermentation	−	+	+	+
Ornithine decarboxylase	−	−	−	+

nonphagocytic cells. Once inside the cell, *Shigella* induce a number of degenerative changes in ultrastructure, indicating cell injury. Cellular damage may be related to the production of toxin. Epithelial cell invasiveness is mediated by a chromosomally linked gene that is located near the purine E and lactose–galactose regions of the genome.

Toxin. The early literature described the production of a potent neurotoxin (shiga toxin) by *S. dysenteriae* that was lethal for rabbits. It was thought to be responsible for the convulsions seen in some children infected with *S. dysenteriae* and responsible for the higher mortality seen with this species. Subsequent research has shown that this toxin is cytotoxic for various tissue cultures and has enterotoxic activity, producing fluid accumulation in ligated rabbit ileal loops. Similar cell-free toxins have been described for *S. flexneri* and *S. sonnei*, and antibodies to these toxins can neutralize the shiga toxin of *S. dysenteriae*. The enterotoxic activity does not appear to be due to an increase in adenyl cyclase activity as seen with *E. coli* heat-labile or cholera enterotoxin (Chaps. 36 and 38), nor is the toxin immunologically related to these toxins. The mechanism by which fluid secretion is induced is not known; however, toxin-injected ileal loops produce marked histologic changes of shortened villi, cuboidal epithelial cells, and increased inflammatory cells.

The native toxin (mol wt 68,000), like other bacterial toxins, exists as a proenzyme and is composed of A and B chains. There are six or seven B chains, and one A chain in each molecule of toxin. The A chain consists of two subunits, A_1 (mol wt 30,500) and A_2 (mol wt 3000), connected by a disulfide bond. Isolated subunits are not toxic for cells, but intact A_1 is less active than isolated A_1 in cell-free systems. The B chain may serve as a binding factor, whereas the A_1 subunit inhibits protein synthesis in cell-free systems. Cells treated with toxin have inhibited protein synthesis at the level of peptide elongation. In contrast with *Pseudomonas aeruginosa* exotoxin A and diphtheria toxin, shiga toxin A_1 subunit enzymatically inactivites the 60S ribosomal unit, causes irreversible damage, and does not require nicotinamide adenine nucleotide (NAD). Attachment of the toxin appears to occur at the oligomeric β 1-4 linked *N*-acetylglucosamine of the host cell surface.

The toxin may play two roles in the pathogenesis of *Shigella* infections. First, it may act as an enterotoxin in the jejunum, producing the watery diarrhea associated with early disease. Second, its ability to inhibit protein synthesis and cytotoxicity may account for the cellular death and production of colonic ulcers. These theories are supported by investigative studies with rhesus monkeys infected intracecally with *S. flexneri*, which show that passage through the small intestine is necessary for the production of watery diarrhea, whereas dysentery occurs in animals inoculated intracecally.

A second toxin that produces morphologic changes in Chinese hamster ovary cells has been isolated, but the role of this toxin in the pathogensis of *Shigella*, if any, is not yet known.

Clinical Infection

Epidemiology. During the period of 1964 to 1973, 105,832 cases of shigellosis were reported to the Centers for Disease Control in Atlanta. Of these isolates, 73 percent were *S. sonnei*, 26 percent were *S. flexneri*, 0.7 percent were *S. boydii*, and 0.6 percent were *S. dysenteriae*. Recent data shows the same distribution of isolates. In developing countries where hygiene is poor, the isolation pattern is reversed: *S. dysenteriae* and *S. boydii* are the most frequent isolates, followed by *S. flexneri* and then *S. sonnei*. *S. dysenteriae* infections in the United States are limited to persons traveling to endemic areas.

Because animal hosts are lacking, the spread of *Shigella* is from human to human via the fecal–oral route, and the reservoirs are carriers who shed the organisms in their feces. The carrier state usually lasts for 1 to 4 weeks; however, long-term carriers have been described in patients in confined environments. From these carriers, the *Shigella* can be spread via flies, fingers, food, or feces. *Shigella* can be isolated from clothing, toilet seats, or water contaminated by infected individuals. Children under the age of 5 years account for almost one half of all isolates reported, with two thirds of all reported cases occurring in children under 10 years old.

Outbreaks involving many people occur in closed groups, such as families, mental hospitals, Indian reservations, daycare nurseries, prisoner-of-war camps, or cruise ships. Secondary transmission is very high, with children under the age of 1 year being the most susceptible and having a 60 percent infection rate, as compared with a 20 percent rate for other ages. The high communicability of these organisms is attributed to the low infective dose needed to produce disease. Studies in healthy volunteers indicate that as few as 200 bacilli are needed to produce disease in some individuals. The percentage of affected individuals increases as the number of infecting organisms increases.

Pathogenesis. As with most pathogens, the spectrum of disease varies from asymptomatic infection to severe bacillary dysentery, with high fever; chills; convulsions; abdominal cramps; tenesmus; and frequent, bloody stools. The typical case presents as watery diarrhea and fever, changing on the second day to frequent small volume stools with blood and mucus. This clinical course coincides with observations in human volunteer studies. The organisms multiply initially in the small bowel to a concentration of 10^8 organisms per milliliter. During this stage, abdominal cramps and fever are common. After 1 to 3 days, the organisms can no longer be cultured from the small intestine, but are isolated only from the colon. The temperature drops at this time and the symptoms of dysentery become obvious.

The penetration of epithelial cells by the organisms causes inflammation, sloughing of cells, and superficial ulceration. Rarely do the organisms penetrate the intestinal wall and spread to other parts of the body.

Eleven incidences of *Shigella* septicemia were reported

in 569 South African children with shigellosis. Of the 11 children, nine were suffering from kwashiorkor or marasmus. Five of the 11 patients died. The rate of *Shigella* septicemia in noncompromised individuals is unknown. In previously healthy adults, spontaneous cure can occur within 2 to 7 days. In the very young and old, however, and in malnourished individuals, the disease is longer lasting, and the mortality due to dehydration and electrolyte imbalance is higher. Rectal prolapse also can occur but is rare. Death is most likely to occur in the pediatric population and when *S. dysenteriae* is the causative organism.

Laboratory Diagnosis. The best specimen for diagnosis of shigellosis is a rectal swab of an ulcer taken by sigmoidoscopic examination. Feces can also be used, but because the *Shigella* are sensitive to the acids present in fecal material, the time interval between the collection of the specimen and the inoculation of culture media is important. Specimens that cannot be planted immediately should be placed in transport media or buffered glycerol preparations. Clinical specimens should be placed on the media listed in Table 35-2. Serologic typing and biochemical characterization of isolates are required for complete epidemiologic determination and identification (Table 37-1).

Treatment. As with other diarrheal diseases, the immediate concern is the patient's state of dehydration. In severe cases of dehydration, intravenous fluids are "pushed" (20 to 30 ml of isotonic fluid per kg) rapidly over 1 hour. Oral glucose–electrolyte solutions are used after adequate rehydration has been achieved. Oral solutions are also useful, particularly in infants, when mild or moderate dehydration occurs.

In contrast to *Salmonella* gastroenteritis, *Shigella* infections have been shown to respond to antibiotic treatment with a decrease in fever, diarrhea, and duration of the carrier state. As humans are the only source of the organism, reducing the length of time during which organisms are shed significantly reduces secondary cases. Ampicillin is the drug of choice for sensitive strains. Amoxicillin, an analog of ampicillin, is not effective. Trimethoprim-sulfamethoxazole is the drug of choice when the organism's sensitivity is unknown or the patient is allergic to pencillin-type drugs.

Control. Because humans represent the major source of organisms, adequate sanitation and detection and treatment of carriers are the only effective control methods. If at all possible, persons with disease should be kept on enteric isolation until cultures are negative. Carriers should be treated and not allowed to handle foods. Proper sewage disposal and water chlorination are important measures required for controlling all gram-negative enteropathogens.

Several oral vaccines using *Shigella–E. coli* hybrids have been shown to be effective in human volunteer studies. Other oral vaccines using mutant *Shigella* strains have been

effective in offering serotype-specific immunity for 6 months to 1 year. However, these vaccines do not alter the asymptomatic carrier rate and require at least four doses of the vaccine given in bicarbonate solution. Formal and co-workers have succeeded in forming a hybrid of *S. sonnei–Salmonella typhi* Ty 21a. An oral vaccine consisting of *S. typhi* Ty 21a has been successfully used to reduce the incidence of typhoid fever in children living in an endemic typhoid area. The usefulness of this unique hybrid vaccine has yet to be determined.

Salmonella

In contrast to the genus *Shigella*, the genus *Salmonella* is composed of a more biochemically complex and serologically diverse group of organisms. They infect many animal species besides humans and are capable of invading extraintestinal tissues and causing enteric fevers, the most severe of which is typhoid fever.

Taxonomy. The Kauffmann–White antigenic scheme for the genus *Salmonella* gives species status to each antigenic type. Ewing and co-workers have proposed that there are only three species of *Salmonella: Salmonella cholerae-suis, Salmonella typhi,* and *Salmonella enteritidis,* with the other antigenic types being serotypes of *S. enteritidis.* A comparison of the two systems is shown in Table 37-2. Further discussion of these schemes is found in the section on antigenic structure. The Ewing scheme currently is being used by the Centers for Disease Control.

Physiology
Biochemical and Cultural Characteristics. *Salmonella,* with the exception of a rare isolate, do not ferment lactose. Most strains are motile and produce H_2S from thiosulfate, and gas from glucose fermentation. *S. typhi,* however, does not produce gas from glucose fermentation, and H_2S production may be very slight. Biochemical tests used to differentiate this genus from the other genera of the tribe Salmonelleae are described in a preceding chapter (Table 35-3). The characteristics used to speciate the genus *Salmonella* are shown in Table 37-3.

TABLE 37-2. COMPARISON OF NOMENCLATURE OF KAUFFMANN–WHITE SYSTEM WITH THAT OF EWING FOR GENUS *SALMONELLA*

Kauffmann–White	Ewing
S. typhi	*S. typhi*
S. cholerae-suis	*S. cholerae-suis*
S. typhimurium	*S. enteritidis* sero typhimurium
S. derby	*S. enteritidis* sero derby

TABLE 37-3. BIOCHEMICAL SPECIATION OF *SALMONELLA*

Test	S. typhi	S. cholerae-suis	S. enteritidis*
Ornithine decarboxylase	−	+	+
Gas from glucose fermentation	−	+	+
Trehalose fermentation	+	−	+

*Commonly isolated bioserotypes.

Resistance to Physical and Chemical Agents.
Salmonella are capable of tolerating relatively large concentrations of bile, a property that is used in the designing of media for the isolation of these organisms. The members of this genus are typical of the enterics in their resistance to other physical and chemical agents. One species, *S. cholerae-suis*, is used as the standard test organism for phenolic preparations.

Antigenic Structure
The O and H antigens are the major antigens used to type the *Salmonella*. The O antigens are similar to those of other enterobacteria, but the H antigens of *Salmonella* are diphasic; i.e., the H antigens can exist in either of two major phases—phase 1, or specific phase; and phase 2, or nonspecific phase. The phase 1 antigens are shared by only a few organisms and react only with homologous antisera, whereas the phase 2 antigens are shared by many organisms and will cross-react with heterologous antisera. The numerous antigenic types of *Salmonella* were organized by Kauffmann and White to form a logical classification system that is immensely important for epidemiologic work. According to this scheme, the *Salmonella* are grouped into major groups based on common O antigens. These groups are designated by capital letters A through I. Subdivision of the major groups into species or serotypes is then accomplished by determination of the remaining O antigens and H antigens (both phases 1 and 2). Ewing utilizes the same Kauffmann–White antigenic types in his classification scheme. The only difference between the Ewing and Kauffmann–White systems is that of taxonomy. The Kauffmann–White method designates each antigenic type as a species, whereas the Ewing system designates the same antigenic type as a serotype of *S. enteritidis*.

Because of the large number of possible serotypes, only large reference centers are capable of complete serotyping of *Salmonella*. However, of the 2000 serologic types of *Salmonella* described, 38 percent of the serotypes account for 95 percent of all clinical isolates. This greatly simplifies the number of antisera required to identify clinical isolates. In a study of 500,000 *Salmonella* isolated in the United States, group B was responsible for 47 percent of all isolates. Groups C_1, C_2, D, and E_2 accounted for 13, 7, 24, and 4.4 percent, respectively, of the remaining isolates.

The capsular antigens play a minor role in the serologic classification of *Salmonella*, but may have important pathogenic significance. The *S. typhi* capsular antigen, the Vi (virulence) antigen, may play a role in preventing intracellular destruction of the organism. This antigen may be rarely found in other *Salmonella* or other enterics, such as *Citrobacter* or *Escherichia*.

Determinants of Pathogenicity
Salmonella are complex organisms that produce a number of virulence factors. These include: (1) surface antigens; (2) invasiveness; (3) endotoxin; and (4) enterotoxins.

Surface Antigens.
The ability of *Salmonella* to attach to host receptor cells and to survive intracellularly may be due to surface O antigens or, in the case of *S. typhi*, the presence of the Vi antigen. Studies in human volunteers show that those organisms containing Vi antigen are clearly more virulent than those lacking it. Non-Vi strains are capable of producing disease in volunteers, albeit at a lower rate. The Vi antigen may serve to protect the O antigen from antiserum and prevent phagocytosis. As with the *Shigella*, rough colonial variants that are deficient in O-specific side chains in LPS are avirulent, whereas smooth colonial variants are virulent. The precise role of polysaccharide side chains in facilitating attachment to host cells has not been clearly delineated.

Invasiveness.
Like the invasive *Shigella*, virulent *Salmonella* penetrate the epithelial lining of the small bowel. However, unlike *Shigella*, the *Salmonella* do not merely reside in the epithelial lining but pass directly through the epithelial cells into the subepithelial tissue. The biochemical mechanism of penetration is not known, but the process appears to be similar to phagocytosis. As the bacteria approach the epithelium, the brush border begins to degenerate and the bacteria enter the cell. They are then surrounded by inverted cytoplasmic membranes similar to phagocytic vacuoles. The *Salmonella* pass through the epithelial cells into the lamina propria. Occasionally epithelial penetration occurs at the intercellular junction. After penetration, the organisms multiply and may extend to other body sites. Epithelial destruction occurs during later stages of the disease, but the mechanism of this destruction is not known.

Endotoxin.
As with all enteric bacilli, endotoxin may play a role in the pathogenesis of *Salmonella* infection, especially during the bacteremic stages of typhoid fever and other enteric fevers. Presumably, endotoxin could be responsible for the fever seen during these diseases. Fever could be produced by endotoxin acting either directly or indirectly through the release of endogenous pyrogens from leukocytes. Endotoxin activation of the chemotactic

properties of the complement system may cause the localization of leukocytes in the classic enteric lesions seen in typhoid fever. However, the exact role of endotoxin remains unclear because endotoxin-tolerant volunteers infected with *S. typhi* still display the classic symptoms of typhoid fever.

Enterotoxin. Enterotoxin activity has been reported for a number of *Salmonella* species. These enterotoxins have the properties of both heat-stable and heat-labile enterotoxins of *E. coli* (Chap. 36). The enterotoxic activity appears to be closely associated with the cell wall or outer membrane of the bacterial cell. This close association with the bacterial cell might explain why tissue penetration is required for the induction of diarrhea. However, as with *Shigella,* the role of enterotoxin in gastroenteritis has not been completely determined.

Clinical Infection

Epidemiology
Typhoid Fever. Of all the *Salmonella* species, *S. typhi* is uniquely adapted to humans, and human carriers represent the sole natural reservoir of these organisms. These carriers can be either convalescent carriers who excrete the organism for a short period of time or chronic carriers who shed the organism for longer than 1 year. The chronic carrier state occurs in about 3 percent of patients who develop typhoid fever. Generally, the chronic carrier is an older woman with gallbladder disease. The organism resides in stones or scars in the biliary tree and is excreted in large numbers. The carriers contaminate food and water, spreading the infection to other individuals. The epidemiology of typhoid fever in a migrant worker camp in Florida serves as an example of the spread of *Salmonella* via contaminated water. This outbreak of typhoid fever involved 225 individuals (no deaths), and was the largest outbreak in the United States since 1939. The source was a typhoid carrier, and the means of transmission was drinking water from a well. A faulty sewage system coupled with poor well design led to contamination of the well water. A nonfunctioning chlorinator failed to purify the water and contributed to the epidemic. *S. typhi* can also be spread via food, as evidenced by the infamous Typhoid Mary incidents at the turn of the century, which involved a carrier who was a cook. Recent large-scale outbreaks in Aberdeen, Scotland (515 cases), and Germany (344 cases) involved contaminated corned beef and potato salad, respectively.

The control of carriers, proper sewage disposal, and water chlorination have reduced the incidence of typhoid fever in the United States from over 5000 cases in 1942 to slightly over 500 cases in 1980. Over one half of all cases seen in the United States today are acquired during travel in endemic regions. In developing countries without adequate control measures, typhoid fever remains a major health problem involving thousands of people. In these endemic areas, contaminated water represents a major source of infection.

Other *Salmonella* Infections. Salmonellosis probably represents the largest single communicable bacterial disease problem in the United States today. Approximately 30,000 cases of salmonellosis, excluding typhoid fever, are reported in the United States each year. However, as most infections are not reported, the actual rate of infection has been estimated at about 2 million cases per year.

As with typhoid fever, contaminated food and water are the mechanisms of transmission for all other *Salmonella*. In contrast to typhoid fever, where the human carrier is the sole source of the organism, contaminated animals and animal products represent the major source of other salmonelloses. In countries like the United States, where intensive animal husbandry practices are carried out, 50 percent of a herd or flock may harbor *Salmonella*. A number of these animal *Salmonella* are also resistant to many antimicrobial agents, because antibiotics are indiscriminately used in animal feeds as growth factors. Many workers feel that these resistant organisms represent a major health hazard and advocate that use of antibiotics in animal feeds be curtailed.

Poultry and beef products represent the largest sources of nontyphoid *Salmonella* in the United States. Meat products are contaminated during slaughter, and improper cooking or refrigeration may allow the *Salmonella* to proliferate to the infective dose. Eggs dried by processes that do not reach a killing temperature have contaminated cake mixes and other products.

Dogs and other pets harbor *Salmonella* for long periods. Cold-blooded animals are also efficient carriers of *Salmonella* and have been implicated as sources of human infections. Prior to restriction of the sale of pet turtles, these animals were a significant source of *Salmonella* infections in children.

The human carrier also plays a role in nontyphoid salmonellosis. Patients with *Salmonella* gastroenteritis may shed the organisms in their stool for several weeks. In contrast to *Shigella* carriers, antibiotic treatment of this infection prolongs the carrier state. In the United States, carriers have been the source of a number of food and water-borne illnesses, but generally these outbreaks were confined to a small group such as those eating in a particular restaurant or at a picnic. The general level of hygiene and sanitary measures usually taken in this country limit the role of the human carrier to these groups. In developing countries, however, where the food handling practices and water standards are lower, human carriers represent a major source of *Salmonella* infection.

Clinical Manifestations
The actual disease process may present as many as any of three distinct clinical entities: a gastroenteritis; a septicemia with focal lesions; or an enteric fever, such as typhoid fever.

Gastroenteritis. *Salmonella* gastroenteritis, like shigellosis, represents an actual infection of the bowel and usually occurs about 18 hours after ingestion of the organism. The disease is characterized by diarrhea, fever, and abdominal pain that is usually self-limiting and lasts for 2 to 5 days. In extreme cases, the symptoms may last for several weeks. In most cases, affected individuals do not seek medical attention. Dehydration and electrolyte imbalance constitute the major threat to the very young and old. While the organism can be isolated from feces for several weeks, the occurrence of chronic carriers who continue to shed the organisms after 1 year is rare. Any species of *Salmonella* can produce gastroenteritis, but the most common cause is *S. enteritidis* serotype typhimurium.

Septicemia. *Salmonella* septicemia is prolonged and characterized by fever, chills, anorexia, and anemia. Focal lesions may develop in any tissue, producing osteomyelitis, pneumonia, pulmonary abscesses, meningitis, or endocarditis. Gastroenteritis is minor or even absent, and the organism rarely is isolated from the feces. *S. cholerae-suis* is a frequent isolate from this type of disease. Osteomyelitis in persons with sickle cell trait is most frequently caused by *S. cholerae-suis*.

A chronic *Salmonella* bacteremic stage has been described in patients with schistosomiasis. The *Schistosoma* may serve as the source of the *Salmonella* because cure of the parasitic disease also produces cure of the bacteremia.

Typhoid Fever and Other Enteric Fevers. The prototype and most severe enteric fever is typhoid fever, the causative agent of which is *S. typhi*. Other *Salmonella*, particularly serotypes paratyphi A and paratyphi B, also can cause enteric fevers, but the symptoms are milder and the mortality rate is lower. Humans are the only known host for *S. typhi*, and transmission is via food and water contaminated by diseased individuals or carriers. The number of organisms present in food and water is important. Volunteer studies show that approximately 25 percent of people become infected upon ingestion of 10^5 viable organisms. The rate of infection increases to 95 percent when the number of organisms is increased to 10^9.

During the first week of infection, the symptoms—consisting of fever, lethargy, malaise, and general aches and pains—can be confused with a variety of other illnesses. Constipation, rather than diarrhea, is the rule. During this time, the organism is penetrating the intestinal wall, infecting the regional lymphatics. Some organisms are also carried by the bloodstream to other parts of the reticuloendothelial system. At both sites, they are phagocytized by monocytes but not killed. After intracellular multiplication, they reenter the bloodstream, causing a prolonged bacteremia. This occurs during the second week of illness. Infection of the biliary system and other tissue also occurs at this time. The patient is severely ill, with fever sustained at about 104F, and is often delirious. The abdomen is very tender and may have rose-colored spots. Diarrhea begins in most patients. At this time, the

organisms are reinfecting the intestinal tract from the gallbladder and may cause necrosis of Peyer's patches. By the third week, patients are exhausted and still febrile, but begin to show improvement if no complications occur. Complications include intestinal perforation, severe bleeding, thrombophlebitis, cholecystitis, pneumonia, or abscess formation. The death rate varies from 2 to 10 percent. The lower mortality rate usually occurs where adequate supportive therapy is available. Relapse occurs in about 20 percent of patients.

Laboratory Diagnosis

Isolation of *Salmonella* constitutes a positive laboratory diagnosis of salmonellosis. During the acute stages of gastroenteritis, the number of *Salmonella* in the feces is large, and the stool represents the specimen of choice. Although *Salmonella* are not as sensitive to acid conditions as *Shigella*, it is still appropriate to buffer the feces if there will be a delay in placing the material in proper culture media.

Blood cultures represent the best specimen for the detection of septicemia and enteric fevers during the first 2 weeks of illness. During this period, positive blood cultures occur in 80 percent of patients with typhoid fever, and decrease to 25 percent or less by the fourth week. Bone marrow cultures may be positive for *S. typhi* after the blood becomes negative. Positive stools are found in only about 25 percent of typhoid patients during the first week, but are found in about 85 percent of patients by the third week. Urine may be positive in about 25 percent of patients with typhoid fever. Any other appropriate specimen, such as sputum in the case of pulmonary abscesses, may yield *Salmonella*.

Treatment

Antibiotic treatment of uncomplicated gastroenteritis only serves to prolong the carrier state and may help to promote drug resistance in the *Salmonella*. Treatment should center around supportive therapy and prevention of dehydration and electrolyte imbalance.

In cases of enteric fever or septicemia, ampicillin or chloramphenicol is the drug of choice. In 1972, however, an epidemic in Mexico was caused by a chloramphenicol-resistant strain of *S. typhi*. Ampicillin resistance also occurred in some isolates. The combination of trimethopim-sulfamethoxazole proved effective in the treatment of these infections. About 50 cases involving these strains were imported into the United States. Oral amoxicillin, an analog of ampicillin, is also effective in the treatment of typhoid fever. In one study, patients treated with oral amoxicillin had a lower relapse rate than those treated with chloramphenicol.

Ampicillin—not chloramphenicol—is the drug of choice in treatment of chronic *S. typhi* carriers without gallbladder disease. Cholecystectomy and ampicillin therapy produce an 85 percent cure rate of the chronic carrier state when gallstones or gallbladder disease are present.

Control

Prevention of salmonellosis requires that water standards be observed and that all food be properly cooked and/or refrigerated. Temperatures below 40F halt *Salmonella* proliferation in foods, whereas those above 140F kill the organisms. It is critical that these temperature limits be achieved in the center of the food being processed, as the *Salmonella* could multiply in the center and later spread throughout the food.

Detection and treatment of carriers, particularly of *S. typhi*, constitutes a major control mechanism. Persons shedding organisms should not be allowed to handle food.

Various vaccines have been developed for the control of typhoid fever. The most promising is an oral vaccine consisting of multiple doses of an attenuated *S. typhi* Ty 21a administered in bicarbonate solution. Following a successful trial in volunteers, a large-scale field trial involving over 32,000 children was conducted in Alexandria, Egypt. The infection rate was 0 per 100,000 in the vaccinated group, as compared with 126 per 100,000 and 133 per 100,000 in the placebo and nontrial groups, respectively. Further studies with this vaccine and others involving different *S. typhi* antigens are being conducted.

FURTHER READING

Books and Reviews

Burney DP, Fisher RD, Schaffner W: *Salmonella* empyema: A review. South Med J 70:375, 1977

Chiodini RJ, Sundberg JP: Salmonellosis in reptiles a review. Am J Epidemiol 113:494, 1981

DuPont HL, Pickering LK: Infections of the gastrointestinal tract microbiology, pathophysiology and clinical features. New York, Plenum, 1980, pp 61, 83

Gemski P Jr, Formal SF: Shigellosis: An invasive infection of the gastrointestinal tract. In Schlessinger D (ed): Microbiology—1975. Washington, D.C., American Society for Microbiology, 1975, p 165

Giannella RA: Pathogenesis of *Salmonella* enteritis and diarrhea. In Schlessinger D (ed): Microbiology—1975, Washington, D.C., American Society for Microbiology, 1975, p 170

Levine MM: Bacillary dysentery mechanisms and treatment. Med Clin North Am 66:623, 1982

Martin WJ, Washington JA II: Enterobacteriaceae. In Lennette EH, Balows A, Hausler WJ Jr, et al. (eds): Manual of Clinical Microbiology, 3rd ed. Washington, D.C., American Society for Microbiology, 1980, p 195

Rosenberg ML, Weissman JB, Gangarosa EJ, et al.: Shigellosis in the United States: Ten year review of nationwide surveillance, 1964–1973. Am J Epidemiol 104:543, 1976

Takeuchi A: Electron microscope observation on penetration of the gut epithelial barrier by *Salmonella typhimurium*. In Schlessinger D (ed): Microbiology—1975. Washington, D.C., American Society for Microbiology, 1975, p 174

WHO Scientific Working Group: Enteric infections due to *Campylobacter, Yersinia, Salmonella* and *Shigella*. Bull WHO 58:519, 1980

Selected Papers

Anonymous. Aberdeen's typhoid bacillus. Lancet 1:645, 1973

Black RE, Craun GF, Blake PA: Epidemiology of common-source outbreaks of shigellosis in the United States, 1961–1975. Am J Epidemiol 108:47, 1978

Blaser MJ, Feldman RA: *Salmonella* bacteremia: Reports to the Centers for Disease Control, 1968–1979. J Infect Dis 143:743, 1981

Brown JE, Rothman SW, Doctor BP: Inhibition of protein synthesis in intact HeLa cells by *Shigella dysenteriae* 1 toxin. Infect Immun 29:98, 1980

Butler T, Mahmoud AAF, Warren KS: Algorithms in the diagnosis and management of exotic diseases. XXIII. Typhoid fever. J Infect Dis 135:1017, 1977

Butler T, Mahmoud AAF, Warren KS: Algorithms in the diagnosis and management of exotic diseases. XXVII. Shigellosis. J Infect Dis 136:465, 1977

Caprioli A, D'Agnola G, Roda FV, et al.: Isolation of *Salmonella wien* heat-labile enterotoxin. Microbiologica 5:1, 1982

Ewing WH: The nomenclature of *Salmonella*, its usage, and definitions for the three species. Can J Microbiol 18:1629, 1972

Feldman RE, Baine WB, Nitzkin JL, et al.: Epidemiology of *Salmonella typhi* infection in migrant labor camp in Dade County, Florida. J Infect Dis 130:334, 1974

Formal SB, Baron LS, Kopecko DJ, et al.: Construction of a potential bivalent vaccine strain: Introduction of *Shigella sonnei* form 1 antigen genes into the gal E *Salmonella typhi* Ty 21a typhoid vaccine strain. Infect Immun 34:746, 1981

Gillman RH, Hornick RB, Woodward WE, et al.: Evaluation of a UDP-glucose-4-epimeraseless mutant of *Salmonella typhi* as a live oral vaccine. J Infect Dis 136:717, 1977

Jiwa SFH: Probing for entertoxigenicity among *Salmonella*: An evaluation of biological assays. J Clin Microbiol 14:463, 1981

Keusch GT, Jacewicz M: The pathogenesis of *Shigella* diarrhea. VI. Toxin and antitoxin in *Shigella flexneri* and *Shigella sonnei* infections in humans. J Infect Dis 135:552, 1977

Keusch GT, Jacewicz M: Pathogenesis of *Shigella* diarrhea VII. Evidence for a cell membrane toxin receptor involving β 1-4 linked N-acetyl-d-glucosamine oligmers. J Exp Med 146:535, 1977

Koupal LR, Diebel RH: Assay, characterization and localization of an enterotoxin produced by *Salmonella*. Infect Immun 11:14, 1975

Levine MM, DuPont HL, Formal SB, et al.: Pathogenesis of *Shigella dysenteriae* (shiga) dysentery. J Infect Dis 127:261, 1973

Martin WJ, Ewing WH: Prevalence of serotypes of *Salmonella*. Appl Microbiol 17:111, 1969

McIver J, Grady GF, Keusch GT: Production and characterization of exotoxin(s) of *Shigella dysenteriae* type 1. J Infect Dis 131:559, 1975

Meals RA: Paratyphoid fever: A report of 62 cases with several unusual findings and a review of the literature. Arch Intern Med 136:1422, 1976

O'Brien AD, Thompson MR, Gemski P, et al.: Biological properties of *Shigella flexneri* 2A toxin and its serological relationship to *Shigella dysenteriae* toxin. Infect Immun 15:796, 1977

Olsnes S, Reisbig R, Eiklid K: Subunit structure of *Shigella* cytotoxin. J Biol Chem 256:8732, 1981

Reisbig R, Olsnes S, Eiklid K: The cytotoxin activity of *Shigella* toxin evidence for catalytic inactivation of the 60S ribosome subunit. J Biol Chem 256:8739, 1981

Rosenberg ML, Hazlet KK, Schaefer J, et al.: Shigllosis from swimming. JAMA 236:1849, 1976

Ryder RW, Blake PA: Typhoid fever in United States, 1975 and 1976. J Infect Dis 139:124, 1979

Ryder RW, Merson MJ, Pollard RA, et al.: Salmonellosis in the United States, 1968–1974. J Infect Dis 133:483, 1976

Sansonetti PJ, Kopecko DJ, Formal SB: *Shigella sonnei* plasmids: Evidence that a large plasmid is necessary for virulence. Infect Immun 34:75, 1981

Scragg JN, Rubidge CJ, Appelbaum PC: *Shigella* infection in African and Indian children with special reference to *Shigella* septicemia. J Pediatr 93:796, 1978

Sedlock DM, Deibel RH: Detection of *Salmonella* enterotoxin using rabbit leal loops. Can J Microbiol 24:268, 1978

Wahdan MH, Serie CH, Germanier R, et al.: A controlled field trial of live oral typhoid vaccine. Bull WHO 58:469, 1980

Vibrionaceae

The family Vibrionaceae includes three genera that have clinical importance for humans: *Vibrio, Aeromonas,* and *Plesiomonas.* The latter two genera rarely cause disease in humans, but have been isolated from cases of diarrhea. *Aeromonas* has also been isolated from extraintestinal sources. The majority of clinical infections caused by *Vibrio* species are enteric in nature, ranging from epidemic cholera to isolated sporadic cases of diarrhea.

All members of this family are gram-negative, facultative organisms that do not have exacting nutritional requirements. Their natural habitat appears to be water. The Vibrionaceae are oxidase-positive and, when motile, movement is by means of polar flagella. Metabolism is both respiratory and fermentative. Additional identification characteristics may be found in the sections dealing with the individual organisms and in Table 38-1.

Vibrio

The genus *Vibrio* contains some of the most important intestinal pathogens of humans, including *Vibrio cholerae,* the cause of epidemic Asiatic cholera. Another intestinal pathogen, *Vibrio parahaemolyticus,* is a leading cause of diarrhea in Japan and, as laboratories become more familiar with its characteristics, is being isolated more frequently in

TABLE 38-1. BIOCHEMICAL PROPERTIES OF THE PATHOGENIC MEMBERS OF THE FAMILY VIBRIONACEAE

Test	*Vibrio cholerae*	*Vibrio parahaemolyticus*	*Vibrio alginolyticus*	*Vibrio vulnificus*	EF-6 (*Vibrio fluvialis*)	*Aeromonas hydrophila*	*Plesiomonas shigelloides*
Sensitivity to 0/129 (150 μg)	+	+	+	+	+	−	+
Oxidase	+	+	+	+	+	+	+
Arginine dihydrolase	−	−	−	−	+	+	+
Lysine decarboxylase	+	+	+	+	−	−	+
Ornithine decarboxylase	+	+	d	d	−	−	+
Voges–Proskauer	d	−	+	−	−	d	−
ONPG	+	−	−	+	+	+	−
Growth in NaCl							
0%	+	−	−	−	d	+	+
3%	+	+	+	+	+	+	+
8%	−	+	+	−	d	−	−
10%	−	−	+	−	−	−	−
Growth on thiosulfate citrate bile salts agar	+	+	+	+	+	−	−

d, different reactions; ONPG, O-nitrophenyl-β-D-galactopyranoside.

other parts of the world. Other species of *Vibrio* have been isolated from sporadic cases of diarrhea and several species have been associated with localized and sometimes systemic infections.

Vibrio cholerae

No infection except plague arouses such panic as cholera. The disease is endemic in the Bengal region of India and Bangladesh, and from this region it has spread in a wave of pandemics. Since 1817 there have been seven pandemics, the most recent occurring from the early 1960s through the 1970s and involving Africa, western Europe, the Philippines, and other areas of Southeast Asia. This pandemic provided the impetus for research efforts that have been successful in elucidating the pathogenic processes of the disease, as well as providing significant advancements in treatment. In addition, the work on cholera has provided useful methods and approaches for research into other causes of diarrhea.

Taxonomy

There is some confusion concerning the taxonomy of *V. cholerae*, as this name is used to describe three distinct entities. Classical epidemic Asiatic cholera is caused by organisms that agglutinate in antisera directed against the O-1 antigen, and that produce disease by means of an enterotoxin. These organisms can be further distinguished biochemically into the cholerae and el tor biotypes. A second group of biochemically similar organisms type with the O-1 antisera, but do not produce the classic cholera enterotoxin. A third group do not agglutinate in the O-1 antisera, but are indistinguishable from the O-1 group both biochemically and genetically. This third group has been referred to as the nonagglutinating vibrios (NAG) or noncholera vibrios (NCV). This chapter follows the guide-

lines of the WHO Scientific Working Group of the Diarrhoeal Diseases Programme and refers to these organisms as *V. cholerae* O-1, atypical *V. cholerae* O-1, and non-O-1 *V. cholerae*, respectively.

Morphology

The cholera vibrios are short (0.5 μm by 1.5 to 3.0 μm), gram-negative rods that upon initial isolation appear to be comma-shaped. In fact, Koch initially named his isolates the *Kommabacillus*. Upon serial transfer in the laboratory, the organisms revert to straight forms. Motility is by means of a single, thick, polar flagellum with, as revealed in electron micrographs, an inner core and outer sheath.

Physiology

V. cholerae is a facultatively anaerobic organism with an optimum temperature ranging from 18C to 37C. Its metabolism is both respiratory and fermentative. Cholera vibrios will grow on simple media that provide a source of carbohydrate, inorganic nitrogen, sulfur, phosphorus, minerals, and adequate buffering. They grow best at pH 7.0, but can tolerate alkaline conditions to pH 9.5—a property used in the design of isolation media. They are extremely sensitive to an acid pH, and a pH of 6.0 or less will sterilize cultures. When grown on meat extract agar, fresh isolates of the organism develop a translucent colony with an iridescent green to red-bronze color when viewed at low magnification with oblique lighting. Older cultures, especially those transferred on laboratory media, become opaque and corrugated (rugose variant) or rough. Most strains will grow on MacConkey's agar. However, a variety of media have been developed to aid in primary isolation of the cholera vibrios from clinical specimens. These include tellurite taurocholate gelatin agar (TTGA) and thiosulfate citrate bile salts agar (TCBS). The tests currently of value in distinguishing these organisms from other members of the family are listed in Table 38-1. As with most members of the genus, the cholera vibrios are sensitive to 0/129 (2,4 diamine-6,7-diisopropyl pteridine), a trait that aids in their separation from other oxidase-positive, gram-negative rods. Biochemical differentiation of *V. cholerae* O-1 into the biotypes cholerae and el tor is epidemiologically useful and can be accomplished with the tests listed in Table 38-2.

Antigenic Structure

The O antigens, or somatic antigens, constitute the major antigens important in the serologic grouping of the cholera vibrios. All cholera vibrios appear to share the same H antigen. The majority of strains are classified into six antigenic O groups. Serogroup O type 1 (O-1) contains the biotypes cholerae and el tor. Three antigenic factors, A, B, and C, are used to subdivide the O-1 into the serotypes: ogawa, inaba, and hikojima. The biotypes el tor and cholerae of serogroup O-1 are causative agents of classic cholera, and the isolation of this serogroup from feces has epidemiologic significance. The other serogroups are associated with milder forms of disease and have apparently limited epidemic potential.

Conversion among the serotypes ogawa, inaba, and hikojima can occur in both experimental animals and in natural infection. Serologic conversion appears to be related to the appearance of agglutinating antibody in the serum.

Immunochemical studies on the lipopolysaccharide (LPS) reveal an absence of 2-keto-3-deoxyoctonate (KDO), showing a fundamental difference between the families Vibrionaceae and Enterobacteriaceae.

Like the LPS from other organisms, *V. cholerae* LPS contains lipid A, a core region of polysaccharides, and O-specific side chains of polysaccharides, which are responsible for the antigenic differences within the species.

Determinants of Pathogenicity

Enterotoxin.   The important clinical features of cholera are the result of host reaction to an extracellular enterotoxin. De and Chatterje were the first to describe the experimental model using ligated intestinal loops of rabbits, which led to the discovery of the enterotoxin and opened the way for research into the pathogenesis of other gastrointestinal diseases.

The cholera enterotoxin, or choleragen, is a complex molecule with a molecular weight of approximately 84,000 daltons. Choleragen is predominantly protein (98 percent), with approximately 1 percent lipid and 1 percent carbohydrate. It is composed of two major subunits: A, which is responsible for biologic activity; and B, which is responsible for the binding of the toxin to cell membranes. Subunit A consists of 2 unequal peptides linked together by a single disulfide bond. The toxic activity resides in A_1 (mol wt 23,000), whereas A_2 (mol wt 5000) serves as the link to

the subunit B. Subunit B consists of five identical peptides, each with a molecular weight of 11,500 daltons. The subunit B binds rapidly and irreversibly to the GM-1 monosialoganglioside molecules in the small intestine. After binding, subunit A disassociates from subunit B and enters the cell membrane. Activation of the A_1 peptide occurs with reduction of the disulfide bond. The A_1 peptide is an enzyme that transfers the adenosine diphosphate (ADP)-ribose from nicotinamide adenine dinucleotide (NAD) to a guanosine triphosphate (GTP)-binding protein that regulates adenyl cyclase activity. This binding inhibits the GTP "turnoff" mechanism and causes an increase in adenyl cyclase activity, thus increasing the level of intracellular cyclic AMP (cAMP). The increase in cAMP leads to the rapid secretion of electrolytes into the lumen of the small bowel. It is thought that this increase is caused by increased sodium-dependent chloride secretion and prevention of sodium and chloride absorption across the brush border via the sodium chloride cotransport mechanism. The result is the secretion of an isotonic fluid with a bicarbonate twice the concentration of normal plasma and a potassium concentration four to eight times that of plasma. Fluid losses may be as high as 1 liter per hour. There is a lag period of 3 to 4 hours before maximal fluid secretion is reached, and it persists for 8 to 12 hours. All effects seen in patients can be attributed to this fluid loss.

Adherence.   In addition to enterotoxin production, virulent *V. cholerae* must be able to adhere to the intestinal surfaces. Studies on adherence show that virulent cells penetrate the intestinal mucus and attach to the microvilli at the brush border of the epithelial cells. Motility may be involved in the adherence of *V. cholerae*, as nonmotile varieties that produce toxin are unable to produce disease. Chemotaxis may also be an important factor, because strains that are motile and respond to chemotaxic stimuli are better able to survive in the intestine of animals than motile, nonchemotaxic strains. Virulent *V. cholerae* also produce a mucinase that may enable the organism to penetrate the mucus lining of the small intestine. Attachment to the intestine can be inhibited by antibody directed to the LPS, indicating that somatic antigens or "adhesions" are important in disease.

Clinical Infection
Vibrio cholerae O-1

EPIDEMIOLOGY. Organisms of the serotype O-1 (biotype cholerae and el tor) are capable of causing widespread disease involving large numbers of people. A 1947 epidemic in Egypt involved 33,000 people, with 20,000 deaths. The most recent pandemic (p. 624) spread to 60 countries and affected 171,329 persons.

The human carrier serves as the source of new cases of cholera. Large numbers of organisms are shed in the carrier's feces, which contaminate water and food supplies. Two types of carriers exist, the convalescent and the chronic carrier. The convalescent carrier or individual re-

TABLE 38-2. BIOCHEMICAL DIFFERENTIATION OF *VIBRIO CHOLERAE* O-1

Test	Biotype	
	Cholerae	El tor
Voges–Proskauer test at 22C	−	+
Chicken erythrocyte agglutination	−	+
Polymyxin B sensitivity 50 IU	+	−
Group IV cholera phage sensitivity	+	−

covering from the disease is usually under 50 years of age and sheds the organism for several months to 1 year after the illness. The chronic carrier, on the other hand, is usually over 50 years of age and more difficult to detect. The chronic carrier appears to carry the organisms in the gallbladder and only sheds them intermittently. The shedding occurs during natural purging that may result from noncholera intestinal infection. The purging action apparently lowers the natural antagonism of the large bowel for the vibrios and allows them to survive passage to the outside. Detection of such carriers requires careful epidemiologic follow-up on the index cases because most chronic carriers are members of the household of the index case. Daily rectal cultures or induced purging may be required to detect the carrier. The carrier rate in endemic regions can vary from less than 1 percent to 20 percent.

The spread of the organism from the endemic areas to new areas is greatly facilitated by modern international travel and the presence of carriers. During a Bangladesh epidemic involving both el tor and cholerae biotypes, the clinical case rate to infection rate was 1:36 for the el tor, and 1:4 with the cholerae biotype. In addition to revealing a higher potential for disease with the cholerae biotype, this study demonstrated how the disease can spread via nonsymptomatic carriers. Idiopathic tropical hypochlorhydria may be a major factor in endemic areas of cholera. People with cholera in endemic areas frequently have achlorhydria or hypochlorhydria, and their gastric juices frequently do not kill the organisms. Experiments with volunteers show that the infective dose necessary to produce disease is lowered at least five-fold when the stomach acidity is reduced by administration of sodium bicarbonate.

The role of animals in the spread of cholera has been minimized. As reported recently, serotype O-1 organisms have been found in domesticated animals during periods of human infection. It has been suggested that, in the absence of human carriers, intermittent excretion of vibrios from cows and chickens may serve as a source of new infections. In addition, contaminated shellfish have served as sources of new cases.

During the past 10 years there have been several cases caused by indigenous *V. cholerae* O-1 strains in persons in the United States. One was a cause of severe diarrheal disease in Texas. Also, during September and October of 1978, *V. cholerae* was isolated from 11 symptomatic and 3 asymptomatic persons living along the Gulf coast of Louisiana. Most of the cases involved people who had eaten ei-

ther steamed or inadequately boiled crabs. Cholera organisms have been isolated from canal water near White Lake, Louisiana, and from sewage in the town of Gueydan. The organism appears to reside in the estuaries of a number of sites along the Mid-Atlantic and Southeast seacoasts, and disease from these sources is a possibility when raw or improperly cooked seafood is eaten.

PATHOGENESIS AND CLINICAL MANIFESTATIONS. Classic Asiatic cholera is one of the most devastating diseases known to humans. The incubation period may be hours or days, with a mean of 2 to 3 days. The onset is abrupt, with vomiting and diarrhea. Fluid loss in severe cases approaches 15 to 20 liters per day. The voided fluid is watery without traces of odor or enteric organisms. Hypovolemic shock and metabolic acidosis are consequences of this fluid loss. By the time the patient reaches the hospital, the eyes and cheeks are sunken, skin turgor is diminished, and hands have a washerwoman appearance. Usually the voice is low and hoarse. The untreated case fatality rate is over 60 percent, and higher attack rates are seen in children. There are also significant differences in the incidence of hospitalized cases, compared with milder forms of the disease, with respect to the biotype of the infecting agent—a more severe disease is associated with the cholerae biotype.

The organisms remain in the intestinal tract and the epithelial lining of the intestine appears to remain intact. Earlier studies that described desquamated or sloughed intestinal epithelium probably resulted from a delay in postmortem examination of victims residing in tropical climates.

LABORATORY DIAGNOSIS. The laboratory diagnosis of cholera depends upon the isolation and identification of *V. cholerae*. Immunologic studies using acute and convalescent sera are useful in retrospective epidemiologic studies. Because *V. cholerae* are susceptible to desiccation and acidic conditions, vomitus, stool, and rectal swabs should be cultured quickly or placed into a suitable transport medium. Either the Amies or Cary–Blair modification of Stuart's transport medium makes an excellent holding medium for the preservation of the sample, as does feces-soaked blotter paper stored in airtight plastic bags.

Upon arrival in the laboratory, the specimen should be placed on a nonselective medium, a selective medium, and an enrichment broth. Nutrient agar or taurocholate gelatin agar (TGA) are excellent nonselective media and are useful for visualization of the iridescent qualities of *V. cholerae* colonies illuminated with transmitted oblique lighting. *V. cholerae* colonies are surrounded by cloudy zones of hydrolyzed gelatin on TGA. Also, TCBS is an excellent medium for the isolation of *V. cholerae*. Yellow colonies (sucrose-fermenting) that are oxidase-positive should be subjected to identification procedures. However, suspected *V. cholerae* colonies growing on TCBS should be subcultured to a nonselective medium before serologic procedures are performed. This eliminates the dangers of misinterpretation that can occur using TCBS. Alkaline peptone broth (pH 8.5), incubated 6 to 8 hours, is a good enrichment broth for the isolation of *V. cholerae*. Overnight incubation of the broth is also acceptable if the 6- to 8-hour incubation cannot be accomplished. Subcultures of the broth are made using the same medium as the initial stool culture.

During recent cholera outbreaks in Calcutta and the Philippines, direct fluorescent antibody procedures on stools of patients with acute diarrhea provided over 90 percent correlation with cultural methods. Such procedures, however, are not useful in the detection of carriers.

Although prompt recognition of *V. cholerae* is important and serologic procedures are of prime importance in endemic countries, laboratories in the United States are more likely to rely on biochemical tests for initial identification. Some of the biochemical tests that distinguish *V. cholerae* from related forms are shown in Table 38-1. Serogrouping of all isolates is necessary for epidemiologic purposes. A variety of tests, such as the string test and darkfield motility, are useful in the hands of experienced field workers, but may cause confusion and misinterpretation in the hands of less experienced laboratory personnel.

TREATMENT. Recent advances in the treatment of cholera have resulted in a marked drop in mortality rate to less than 1 percent. Prompt replacement of fluid and electrolyte losses causes a rapid response and reversal of the patient's condition within a matter of hours. Initial shock symptoms are treated with intravenous fluids that provide sodium, chloride, potassium, and bicarbonate. In severe cases of dehydration and shock, fluid replacement should be rapid (2 liters in first 30 minutes). After initial recovery, fluid and electrolyte balance can be maintained with an oral solution of glucose and electrolytes. Oral therapy can also be the sole treatment in cases with mild dehydration. Assessment of the degree of dehydration is the key to the management of the disease. Oral therapy eliminates the need for large volumes of pyrogen-free solutions and is a direct result of research that showed that the absorptive powers of the colon remain intact during the disease. Oral therapy can be administered by paramedical personnel, a significant advantage in the rural areas of undeveloped countries. Tetracycline, although not directly affecting the enterotoxin, lowers the number of infecting organisms and, thereby, reduces the fluid loss by almost 60 percent. Tetracycline also aids in eliminating the carrier state because of its concentration in the bile. Tetracycline is useful in reducing subsequent cases in household contacts of an index case. However, tetracycline resistant forms have emerged in Africa, raising questions concerning the usefulness of the drug as a means of mass prophylaxis in epidemic situations.

PREVENTION AND CONTROL. The primary defense in the control of cholera is the maintenance of adequate sewage treatment and water purification systems, together with the prompt detection and treatment of patients and carriers.

In countries with adequate sanitation, cholera is limited to imported or sporadic cases. Paradoxically, the first case of cholera in the United States since 1911 occurred in a Texas man who had not left this country or knowingly had contact with a carrier. Extensive epidemiologic studies failed to reveal a source of this infection, although it is now apparent that there is an endemic source of *V. cholerae* in the area. Studies during the recent Louisiana outbreak involving crabs show that at least 10 minutes boiling of shellfish is required to kill *V. cholerae*. The present recommendations for cooking shellfish involve immersion in vigorously boiling water for at least 15 minutes and discontinuation of the practice of steaming crabs.

Travelers to countries with known cholera are cautioned against eating uncooked vegetables, unpeeled fruits, and raw seafood, and drinking unbottled beverages. Swimming should take place only at beaches not contaminated with human sewage.

The last pandemic has illustrated that the present vaccines do not afford significant protection against disease, especially if large numbers of organisms are ingested. A number of newer vaccines using toxoid, L-forms, and parenteral and enteral administration routes are currently being tested. Results of these studies are pending. At present, travelers from the United States to foreign countries are not required to have valid cholera vaccine certification.

The role of quarantine measures also has been questioned, especially in view of the failure of such measures to control the spread of the recent pandemic. The usefulness of tetracycline prophylaxis is limited and not recommended. In endemic areas, the major control measure remains the prompt detection and treatment of asymptomatic carriers.

Atypical *Vibrio cholerae* O-1. Atypical *V. cholerae* O-1 have been isolated primarily from environmental sources worldwide. Some of these sites are thought to be free from contamination with human sewage, giving rise to the suggestion that these are free-living organisms. No human infections with these organisms have been documented to date, nor were strains isolated in Brazil capable of producing disease in human volunteers given bicarbonate. These organisms also fail to give protection against subsequent challenge with typical *V. cholerae* O-1. These latter results indicate that there may be somatic antigen differences between the atypical and typical *V. cholerae* O-1 that are not detected by the antisera currently available. Some strains also show some biochemical divergence. The exact role of these organisms in the epidemiology of human infection remains unclear.

Non-O-1 *Vibrio cholerae*. The nonagglutinable vibrios, or non-O-1 *V. cholerae*, can cause isolated as well as focal outbreaks of diarrhea, but the volume of fluid loss does not approach that of classic cholera, and the disease is usually self-limiting. In a report on 26 patients with non-agglutinable vibrios isolated in the United States between

1972 and 1975, 17 of the 26 isolates were obtained from gastrointestinal sites. Thirteen of the 17 had diarrhea, and 46 percent of this group had to be hospitalized and given parenteral fluids. Unlike the *V. cholerae* O-1, which do not infect other body sites, non-O-1 *V. cholerae* have been isolated from other body sites. Septicemia has occurred in patients with underlying disease and otitis media caused by these organisms has been described.

These free-living organisms are found widely distributed in the environment including sewage, contamined water, estuaries, seafood, and animals. Their role in human disease is not known, as human disease may be limited to certain strains.

There appear to be four patterns of pathogenicity seen with the non-O-1 *V. cholerae*: (1) no toxicity in any model; (2) production of a cholera-like enterotoxin; (3) production of a heat-stable toxin; and (4) enteritis following infection with no evidence of toxin production.

More research on the epidemiology, serologic typing, and virulence of this group of organisms is needed to obtain a complete understanding of their role in human disease.

Vibrio parahaemolyticus

V. parahaemolyticus is a marine organism that inhabits estuaries throughout the world. Its role in human disease was not recognized until 1951. Since that time, efforts—primarily by the Japanese—have revealed that *V. parahaemolyticus* is a major cause of gastroenteritis involving seafood. In addition, *V. parahaemolyticus* has been isolated from extraintestinal infections.

Morphology and Physiology. *V. parahaemolyticus* resembles the other *Vibrio* species in its structural and staining characteristics. The metabolism of the organism is both fermentative and respiratory, with no gas produced during fermentation.

V. parahaemolyticus has simple nutritional requirements and preference for an alkaline environment. Optimum growth occurs between pH 7.6 and 9.0. This permits the use of the same selective media as used for the isolation of *V. cholerae*. Unlike the cholera vibrios, however, members of this species are halophilic (salt-loving) and require at least 2 percent NaCl for growth. Salt requirements can be satisfied by the addition to the medium of two or three drops of a sterile 20 to 30 percent solution of NaCl. When provided with the appropriate conditions for growth, the generation time of this species is 9 to 15 minutes, a property that may be of importance in the epidemiology of gastroenteritis.

On TCBS media, *V. parahaemolyticus* produces a large, green (nonsucrose-fermenting), smooth colony. Typical colonies giving oxidase-positive reactions should be inoculated into the media listed in Table 38-1 for identification.

Antigenic Structure. The O antigens and K antigens are useful for the serologic typing of *V. parahaemolyticus.* Currently, 12 O types and 59 K types are recognized. No particular serotype appears to be more prevalent or more virulent than any other serotype.

Determinants of Pathogenicity. The exact method by which *V. parahaemolyticus* causes disease has not been determined. Most (96 percent) of the organisms isolated from patients with diarrhea are hemolytic (Kanawaga-positive), whereas less than 1 percent of environmental isolates and isolates obtained from extraintestinal sites have this trait. The hemolytic action is caused by a heat-stable hemolysin, which is cytotoxic in tissue culture and cardiotoxic in mice. Although early reports indicated that concentrated culture filtrates of Kanawaga-positive isolates produced fluid accumulation in the rabbit ileal loop model, later reports attribute the fluid accumulation to the high salinity of the concentrates rather than to the hemolysin. Intestinal infection with Kanawaga-positive isolates, however, do produce dilation and inflammation of the small bowel and there is evidence of invasion of intestinal tissue in humans. Adherence to intestinal tissue may be important because Kanawaga-positive strains adhere more rapidly to tissue culture cells than do Kanawaga-negative strains.

Clinical Infection

EPIDEMIOLOGY. *V. parahaemolyticus* is found worldwide in estuaries and coastal waters. It can adsorb onto chitin and copepods, and in most waters is associated with the zooplankton. In temperate climates, the organism appears to be confined to the sediment in winter, releasing in spring and proliferating in water as the temperature increases. Most diarrheal cases are attributed to ingestion of raw or improperly handled seafood. Extraintestinal infections usually result from wounds or tissue being contaminated with seawater. *V. parahaemolyticus* is responsible for approximately 25 percent of the reported cases of diarrhea seen in Japan. In other parts of Southeast Asia, the incidence varies from 2 to 11 percent. There have been a number of outbreaks in the United States. Two of these outbreaks were caused by raw seafood inadequately refrigerated. Seafood cooked for too short a time followed by inadequate refrigeration and cooked seafood cross-contaminated with raw seafood accounted for the other outbreaks. In Japan, the disease is most commonly associated with the eating of sushi, a vinegared riceball topped with raw fish.

The organism has a generation time of 9 minutes and, under ideal conditions, can reach infective doses rapidly in mishandled foods. Studies in volunteers indicate that between 10^5 to 10^7 organisms must reach the intestinal tract before disease is seen.

CLINICAL MANIFESTATIONS. Gastroenteritis ranges from a self-limiting diarrhea to a cholera-like illness. Diarrhea is explosive and watery, with no blood or mucus, although a dysentery syndrome has been reported in India, Japan, and Australia. Headache, abdominal cramps, nausea, vomiting, and fever may be present. Symptoms may persist for as long as 10 days, but the median is 72 hours, and a few individuals require hospitalization. In contrast to *V. cholerae*, there is a fatty infiltration and cloudy swelling in livers of patients infected with *V. parahaemolyticus*. Localized infections may be produced in persons such as boat workers, seafood cooks, or swimmers who have contact with a marine environment. Septicemia has developed in a patient with underlying liver disease.

LABORATORY DIAGNOSIS. Feces and rectal swabs should be cultured quickly or placed in a suitable transport medium (Cary–Blair or Amies). Specimens should be inoculated into TCBS and alkaline peptone broth (pH 8.5), supplemented with 3 percent NaCl. After overnight incubation, alkaline peptone broth is subcultured to TCBS and final identification is made with tests outlined in Table 38-1.

TREATMENT. Because the gastroenteritis produced by this organism usually is self-limiting, most cases are not treated. In severe cases, however, fluid and electrolyte replacement should be given and antibiotics administered. The organisms are usually sensitive to chloramphenicol, kanamycin, tetracycline, and the cephalosporins.

CONTROL. The ubiquitous nature of *V. parahaemolyticus* prevents its elimination from the environment. Control measures are aimed at keeping the number of organisms present in seafood below the minimal infective dose. Refrigeration of seafood eaten raw is absolutely essential. Cooked seafood that is served chilled, such as crab or shrimp, should be refrigerated promptly and precautions taken to avoid recontamination. An outbreak involving crabs resulted when cooked crabs were returned to their original containers and stored in a basement until the following day. Similarly, shrimp cooked and stored at ambient temperature resulted in a major outbreak in Louisiana.

Other *Vibrio* Species

Vibrio alginolyticus
Vibrio alginolyticus is a marine *Vibrio* that has been isolated from extraintestinal sites. Most cases have either been from wounds or ears of persons with injuries or contact with seawater. Septicemia occurred in a person who was severely burned in a boating mishap. In contrast to other vibrios, no intestinal disease has been described. *V. alginolyticus* is a true halophilic organism and cannot grow in the absence of NaCl. Distinguishing features are listed in Table 38-1.

Vibrio vulnificus
Vibrio vulnificus is a lactose-fermenting marine *Vibrio* that has been associated with human disease. Two distinct syndromes are attributed to this organism. One is a wound infection characterized by swelling and erythema, which may progress with formation of vesicles and bullae. The second syndrome is a "primary septicemia" characterized by malaise, fever, chills, and prostration. Vomiting, diarrhea, and hypotension are seen in 20 to 30 percent of patients, whereas metastatic cutaneous lesions, which may progress to necrosis, appear in 75 percent within the first 2 days. The organism has been isolated from the blood of these patients. From one half of the patients, the organism has also been isolated from the lesions. Approximately one half of the patients died, with most deaths occurring within 1 day of the initiation of therapy. Preexisting liver disease was noted in 75 percent of patients with this syndrome, and five of six additional patients also had underlying disease. Patients with wound infections had previous contact with seawater or crabs, whereas many patients with primary septicemia appear to have been infected by consuming raw oysters. The incubation period appears to be less than 1 day for both forms of the infection. Diagnosis is made by isolation and identification of the organism. A heat-labile, cytolytic toxin has been described that produces effects in guinea pig skin and is lethal for mice, but the role of this toxin in *V. vulnificus* infections remains to be defined.

Group F (EF-6, *Vibrio fluvialis*)
Group F (EF-6, *Vibrio fluvialis*) is a group of *Vibrio*-like organisms that have been isolated from cases of diarrhea in Bangladesh. Most cases occurred in children under 5 years old with symptoms similar to cholera, but some patients had fever, abdominal pain, and blood and mucus in their stool. These organisms, like other vibrios, are found in estuaries and other marine environments throughout the world. The true incidence of human disease caused by them has yet to be determined.

Aeromonas

The genus *Aeromonas* contains several species, all of which are found free-living in water. Most are pathogens for cold-blooded animals and have not been incriminated in human disease. One species, however, *Aeromonas hydrophila*, is a well-documented pathogen of humans, causing septicemia, osteomyelitis, and wound infections. It has also been isolated from cases of diarrhea and from the urine of asymptomatic individuals. Most infections are seen in persons with debilitating disease, particularly neoplasms.

A. hydrophila grows quite readily on media used for the Enterobacteriaceae and is often misidentified as an enteric bacillus. In contrast to the enteric bacilli, *A. hydrophila* is oxidase-positive and only possesses a single polar flagellum. The distinguishing biochemical reactions are listed in Table 38-1. Most isolates are sensitive to gentamicin, tetracycline, and chloramphenicol and are resistant to ampicillin and the cephalosporins.

Plesiomonas

Plesiomonas shigelloides is classified by some taxonomists as an *Aeromonas* and in some medical literature is referred to as *Aeromonas shigelloides.* In humans, this organism has been isolated from the blood and spinal fluid, but it primarily causes gastroenteritis. Like *A. hydrophila,* this organism can be confused with the Enterobacteriaceae. Its biochemical characteristics are listed in Table 38-1.

FURTHER READING

Books and Reviews

Black RE: Prophylaxis and therapy of secretory diarrhea. Med Clin North Am 66:611, 1982

Blake PA, Weaver RE, Hollis DG: Diseases of humans (other than cholera) caused by vibrios. Annu Rev Microbiol 34:341, 1980

Carpenter CCJ: The pathophysiology of secretory diarrheas. Med Clin North Am 66:597, 1982

Craig JP, Benenson AS, Hardegree MC, et al. (guest eds): The structure and functions of enterotoxins. J Infect Dis 133:S1, 1976

DuPont HL, Pickering LK: Infections of the gastrointestinal tract microbiology, pathophysiology and clinical features. New York, Plenum, 1980, p 129

Finklestein RA: Cholera. CRC Crit Rev Microbiol 2:533, 1973

Richards KL, Douglas SD: Pathophysiological effects of *Vibrio cholerae* and enterotoxinogenic *Escherichia coli* and their exotoxins on eucaryotic cells. Microbiol Rev 42:592, 1978

Rodrick GE, Hood MA, Blake NJ: Human *Vibrio* gastroenteritis. Med Clin North Am 66:665, 1982

Sanyal SC, Sen PC: Human volunteer studies on pathogenicity of *Vibrio parahaemolyticus.* International Symposium on *Vibrio parahaemolyticus.* Tokyo, Japan, Saukon, 1973, p 227

von Graevenitz A: *Aeromonas* and *Plesiomonas.* In Lennette EH, Balows A, Hausler WJ Jr, et al. (eds): Manual of Clinical Microbiology, 3rd ed. Washington, D.C., American Society for Microbiology, 1980, p 220

Wachsmuth IK, Morris GK, Feeley JC: *Vibrio.* In Lennette EH, Balows A, Hausler WJ Jr, et al. (eds): Manual of Clinical Microbiology, 3rd ed. Washington, D.C., American Society for Microbiology, 1980, p 226

WHO Scientific Working Group: Cholera and other vibrio-associated diarrhoeas. Bull WHO 58:353, 1980

Selected Papers

Agarwal RK, Parija SC, Sanyal SC: Taxonomic studies on *Vibrio* and related genera. Indian J Med Res 71:340, 1980

Blake PA, Merson MH, Weaver RE, et al.: Disease caused by a marine *Vibrio* clinical characteristics and epidemiology. N Engl J Med 300:1, 1979

Blake PA, Allegra DT, Snyder JD, et al.: Cholera—A possible epidemic focus in the United States. N Engl J Med 302:305, 1980

Carpenter CCJ, Mahmoud AAF, Warren KS: Algorithms in the diagnosis and management of exotic diseases. XXVI. Cholera. J Infect Dis 136:461, 1977

Cash RA, Music SI, Libonati JP, et al.: Response of man to infection with *Vibrio cholerae.* II. Protection from illness afforded by previous disease and vaccine. J Infect Dis 130:333, 1974

Chitnis DS, Sharma KD, Kamat RS: Role of bacterial adhesion in the pathogenesis of cholera. J Med Microbiol 15:43, 1982

Chitnis DS, Sharma KD, Kamat RS: Role of somatic antigen of *Vibrio cholerae* in adhesion to intestinal mucosa. J Med Microbiol 15:53, 1982

Craig JP, Yamanoto K, Takeda Y, et al.: Production of cholera-like enterotoxin by a *Vibrio cholerae* non-O-1 strain isolated from the environment. Infect Immun 34:90, 1981

De SN, Chatterge DN: An experimental study on mechanism of action of *Vibrio cholerae* on the intestinal mucous membrane. J Pathol Bacteriol 66:559, 1953

Freter R, O'Brien PCM, Macsai MS: Role of chemotaxis in the association of motile bacteria with intestinal mucosa: In vivo studies. Infect Immun 34:234, 1981

Gill DM: The arrangements of subunits in cholera toxin. Biochemistry 15:1242, 1976

Hughes JM, Hollis DG, Gangarosa EJ, et al.: Noncholera vibrio infections in the United States. Clinical, epidemiologic and laboratory features. Annu Intern Med 88:602, 1978

Huq MI, Alam AKMJ, Brenner DJ, et al.: Isolation of *Vibrio*-like group, EF 6, from patients with diarrhea. J Clin Microbiol 11:621, 1980

Iijima Y, Yamada H, Shinoda S: Adherence of *Vibrio parahaemolyticus* and its relation to pathogenicity. Can J Microbiol 27:1252, 1981

Kreger A, Lockwood D: Detection of extracellular toxin(s) produced by *Vibrio vulnificus.* Infect Immun 33:583, 1981

Lee JV, Shread AL, Furniss AL, et al.: Taxonomy and description of *Vibrio fulvialis* sp. nov. (synonym group F vibrios, group EF-6). J Appl Bacteriol 50:73, 1981

Miyamoto Y, Obara Y, Nikkawa T, et al.: Simplified purification and biophysiochemical characteristics of Kanagawa phenomenon-associated hemolysin of *Vibrio parahaemolyticus.* Infect Immun 28:567, 1980

Morris JG, Wilson R, Davis BR, et al: Non-O group 1 *Vibrio cholerae* gastroenteritis in United States clinical, epidemiologic and laboratory characteristics of sporadic cases. Annu Intern Med 94:656, 1981

Pitarangsi C, Escheverria P, Whitmire R, et al.: Enteropathogenicity of *Aeromonas hydrophilia* and *Plesiomonas shigelloides*: Prevalence among individuals with and without diarrhea in Thailand. Infect Immun 35:666, 1982

Schmidt U, Chmel H, Cobbs C: *Vibrio alginolyticus* infections in humans. J Clin Microbiol 10:666, 1979

Schneider DR, Parker CD: Purification and characterization of mucinase of *Vibrio cholerae.* J Infect Dis 145:474, 1982

Smith MR: *Vibrio parahaemolyticus.* Clin Med 78:22, 1971

Srivastava R, Sinha VB, Srivastava BS: Events in pathogenesis of experimental cholera: Role of bacterial adherence and multiplication. J Med Microbiol 13:1, 1980

Zen-Yoji H, LeClair RA, Ohta K, Montague TS: Comparison of *Vibrio parahaemolyticus* cultures isolated in the United States with those isolated in Japan. J Infect Dis 127:237, 1973

CHAPTER 39

Pseudomonas

The genus *Pseudomonas* is composed of a large number of nonfermentative, aerobic, gram-negative rods that inhabit soil and water. In their natural habitat these widely distributed organisms play an important role in the decomposition of organic matter. Several species are major plant pathogens, whereas others can infect animals. A few are pathogenic for both plants and animals. While most *Pseudomonas* species do not infect man, some are important opportunistic pathogens that infect individuals with impaired host defenses. Such human infections are usually severe, difficult to treat, and hospital acquired.

Pseudomonas aeruginosa

The *Pseudomonas* species most frequently associated with human disease is *Pseudomonas aeruginosa*. In some hospitals, this organism causes 10 to 20 percent of the nosocomial infections. It has replaced *Staphylococcus aureus* as the major pathogen of cystic fibrosis patients and is frequently isolated from individuals with neoplastic disease or severe burns.

Morphology and Ultrastructure

P. aeruginosa is a gram-negative rod, measuring 0.5 to 1.0 μm by 3.0 to 4.0 μm. It usually possesses a single polar flagellum, but occasionally two or three flagella may be present. An extracellular polysaccharide slime layer, similar to a capsule, is produced. In mucoid strains isolated from cystic fibrosis patients, the slime layer is quite thick and is composed of ordered fibers of polyuronic acids.

The cell wall structure of *P. aeruginosa* is similar to that of the Enterobacteriaceae. The inner core of *P. aeruginosa* lipopolysaccharide (LPS) contains 2-keto-3-deoxyoctonic acid (KDO) and lipid A, as do the enteric bacilli. However, lipid A of *P. aeruginosa* LPS lacks the β-hydroxymyristic acid found in the enterobacteria. The polysaccharides of the *Pseudomonas* LPS can be divided into core polysaccharides common to all strains and side chain polysaccharides, which are strain specific.

Isolates obtained from clinical specimens frequently possess pili that promote attachment to cell surfaces.

Physiology
Biochemical and Cultural Characteristics. *P. aeruginosa* is an extremely adaptable organism that can utilize over 80 organic compounds for growth. It will grow on media used for the isolation of the enterobacteria, and its ability to tolerate alkaline conditions also permits it to grow on *Vibrio* isolation media. Although an aerobic organism, *P. aeruginosa* can utilize nitrate and arginine as electron acceptors and grow anaerobically. A temperature of 35C is optimal for growth, but growth can occur at 42C.

Clinical isolates grown on blood agar are frequently β-hemolytic.

P. aeruginosa is the only gram-negative rod that produces pyocyanin, a phenazine pigment, but not all strains produce this pigment. The use of specialized media (*Pseudomonas* P agar) enhances pigment production. In addition to pyocyanin, a number of fluorescent pigments may also be produced and can be detected in the tissues of patients, as well as in culture, by the use of a Wood's light. A few strains produce a red pigment.

Energy obtained from the utilization of carbohydrates is derived from oxidative rather than fermentative metabolism. Because the acid produced by the oxidative pathway is less than that produced by using a fermentative pathway, carbohydrate utilization by *P. aeruginosa* must be tested for in a special medium such as the O-F medium of Hugh and Leifson. The carbohydrate reactions listed in Table 39-1 are based upon reactions in this basal medium. Also listed in this table are biochemical reactions useful for the identification of *P. aeruginosa* and other medically important *Pseudomonas* species.

Resistance. *P. aeruginosa* is more resistant to chemical disinfection than are other vegetative bacteria. When adequate moisture is provided, it can survive in a variety of places, such as respiratory care equipment, cold water humidifiers, instruments, bedpans, floors, baths, and water faucets. Most of the commonly used antibiotics and antimicrobials are ineffective against *Pseudomonas* (p. 634). It has been isolated from certain types of quaternary ammonium compounds and from hexachlorophene soaps. Phenolics and β-glutaraldehyde are usually effective disinfectants (Chap. 10). Boiling water kills the organism, as does complete desiccation.

Genetics. Gene transfer between strains of *P. aeruginosa* can occur through conjugation and transduction (Chap. 8). Resistance to at least one antibiotic, carbenicillin, can be genetically transferred via an R factor. Strain differences

can be detected by serologic typing of the O antigen, phage typing, and pyocin (bacteriocin) typing.

Antigenic Structure

The O antigen, or somatic antigen, of *P. aeruginosa* has been used to group various strains for epidemiologic purposes. A number of antigenic schemes have been proposed; however, a standardized scheme—the IATS—has been established by a subcommittee of the International Association of Microbiological Societies. This scheme includes 17 serotypes, 12 of which were introduced by Habs in 1957, and five additional serotypes proposed by several other investigators. Serologic typing of the O antigen provides a less cumbersome and less variable system of epidemiologic characterization than does pyocin or phage typing. Bacteriophage and pyocin typing, however, may be necessary for complete characterization of *Pseudomonas* strains isolated during an epidemic.

The slime layer of *Pseudomonas* is also immunogenic and may play a role in protecting the bacterial cell from phagocytosis. Active and passive immunization against this slime material can protect mice from the toxic and lethal effects of challenge with live bacteria.

Determinants of Pathogenicity

A complex organism, *P. aeruginosa* possesses anatomic features that aid in the colonization of a host by the organism and produce a variety of extracellular products to induce pathologic changes in various host systems. The role of each of these factors in disease probably differs with the site of infection and the predisposing factors of the patient.

The first criterion for disease is the colonization of the appropriate site by the pathogen. *P. aeruginosa* has two factors that may be important for colonization—pili and the slime layer. Studies using tissue cultures of human buccal epithelial cells have shown that pili are the most

TABLE 39-1. SOME CHARACTERISTICS OF *PSEUDOMONAS* SPECIES ENCOUNTERED IN CLINICAL SPECIMENS

Test or Substrate	*P. aeruginosa*	*P. cepacia*	*P. maltophilia*	*P. mallei*	*P. pseudomallei*	*P. putida*	*P. stutzeri*	*P. alcaligenes*	*P. fluorescens*
Indophenol oxidase	+	+	−	d	+	+	+	+	+
Glucose oxidation*	+	+	+	+	+	+	+	−	+
Maltose oxidation*	−	+	+	+	+	d	+	−	d
Lactose oxidation*	−	+	+	+	+	d	−	−	d
Mannitol oxidation*	d	+	−	d	+	d	d	−	d
2-ketogluconate	d	d	−	−	−	d	−	−	d
Nitrate to gas	d	−	−	−	+	−	+	−	−
Pyocanin production	d	−	−	−	−	−	−	−	−
Lysine decarboxylase	−	+	+	−	−	−	−	−	−
Ornithine decarboxylase	−	d	−	−	−	−	−	−	−
Arginine dihydrolase	+	−	−	+	+	+	−	−	+
Growth at 42C	+	d	d	−	+	−	d	d	−

*Organisms grown on the O-F medium of Hugh and Leifson at 30C.
d, different strain reactions.

important factor in binding the bacteria to the tissue culture cells. Investigative studies also show that adherence does not occur on normal basal cell layers or regenerating epithelium, but only on cells of the exposed basement membrane damaged by endotracheal intubation or influenza infection. In theory, the slime layer may prevent phagocytosis, particularily in cystic fibrosis patients where very mucoid *P. aeruginosa* are the major colonizers. However, studies of mucoid *P. aeruginosa* in a guinea pig lung model failed to show decreased phagocytosis, opsonin antibody activity, or increased mortality as compared with control animals. Mucoid colonies, however, show increased clumping of bacterial cells and adherence to tissue cells. This binding permits the mucoid cells to exist as a microcolony that is better able to resist the defenses of the lung.

Two hemolysins, one of which is a phospholipase and the other a glycolipid, are produced. They have no lethal activity, but in *P. aeruginosa* pneumonia the phospholipase may contribute to the invasiveness of the organism by destroying the pulmonary surfactant and attacking the pulmonary tissue to produce atelectasis and necrosis.

At least two types of proteases are produced by *P. aeruginosa*. They may be responsible for the hemorrhagic skin lesions observed in some infections. Furthermore, one of the purified proteases, an elastase, produces the pathologic features seen in experimental *P. aeruginosa* infections of the cornea. There is also evidence that elastase production is important in lung infections, enhancing the effects seen with exotoxin A.

Exotoxin A is an enzyme produced by *P. aeruginosa*. It is lethal for animals and produces a number of pathologic effects in a variety of animal models. There is indirect evidence that exotoxin A is also produced during human infections. Elevated antibody titers to exotoxin A are seen in patients infected with *P. aeruginosa* and this antibody appears to be important for the survival of immunocompromised patients. Furthermore, patients bacteremic with exotoxin A-producing organisms have a higher mortality than patients bacteremic with non-exotoxin A-producing organisms. Exotoxin A has been purified and its biochemical mechanism delineated. Its similarity to diphtheria toxin (Chap. 32) is very striking. Like diphtheria toxin, exotoxin A is a polypeptide (mol wt 71,500) that exists as a proenzyme consisting of fragment A, the active enzyme site; and fragment B, the portion of the molecule responsible for cell binding. In addition, like the diphtheria toxin, the active enzyme possesses adenosine diphosphate (ADP)-ribose transferase activity, causing inactivation of the elongation factor 2 (EF2) of eucaryotic cells, thus halting protein synthesis.

Despite these similiarities, exotoxin A differs from diphtheria toxin in a number of ways. The major target organ of exotoxin A appears to be the liver, instead of the heart, as seen with diphtheria toxin. The nature of the proenzyme activation may also differ, and with exotoxin A it can occur in either of two ways—first, by proteolytic hydrolysis into fragments A and B; or second, by an unfolding of the molecule following splitting of the disulfide bonds. Furthermore, the receptor sites for the two toxins differ. Exotoxin A is less toxic and there is no serologic cross-reactivity with the antibody developed against the toxins. Also, unlike diphtheria, the elaboration of exotoxin A by *P. aeruginosa* is not the only mechanism of pathogenesis, as disease can be seen with non-toxin-producing strains. Antibody to LPS, as well as antibody to exotoxin A, appears to be important for survival.

A second exotoxin, exotoxin S, has been described. It also catalyses the transfer of the ADP-ribose moiety of nicotinamide adenine dinucleotide (NAD) to a number of different proteins in eucaryotic cells. It is produced by 38 percent of clinical isolates and has been detected in the burned tissue of a mouse model. The overall role of this toxin in human disease has yet to be determined.

P. aeruginosa also produces an enterotoxin that may be responsible for diarrhea associated with *P. aeruginosa* intestinal infection.

Clinical Infection

Epidemiology. *P. aeruginosa* infections occur in individuals with altered host defenses. These include burn patients, persons with malignant or metabolic disease, or those who have had prior instrumentation or manipulation. The frequency of urinary tract infections is higher in individuals of advanced age. Prolonged treatment with immunosuppressive or antimicrobial drugs and radiation therapy also predisposes individuals to *Pseudomonas* infections.

The ubiquitous nature of *P. aeruginosa* enhances its spread. It is present not only in soil and water, but also in approximately 10 percent of normal stools and on the skin of some normal individuals. The intestinal colonization rate of hospitalized patients increases and may approach 30 to 60 percent. Almost any site in the hospital environment may harbor the organism, especially if moisture is present. Contaminated respiratory equipment, catheters, instruments, intravenous fluids, and even soap are vehicles for its spread. Transmission from patient to patient via hospital personnel is more significant in spreading the organism throughout a hospital unit than is airborne spread. A nationwide epidemiologic surveillance of community hospital infections has showed that *P. aeruginosa* is responsible for 10 percent of all nosocomial infections, 11 percent of all blood isolates, and about 4 percent of nosocomial epidemics. In specialized units, such as burn or cancer centers, *P. aeruginosa* may cause 30 percent of all infections.

Pathogenesis and Clinical Manifestations. *P. aeruginosa* can infect almost any tissue or body site. Localized lesions occur at the site of burns or wounds, in corneal tissue, the urinary tract, or the lungs. Bacterial endocarditis and gastroenteritis also can be caused by *P. aeruginosa*. Infection of corneal tissue may result in loss of the eye. From a localized infection, the organisms may spread via the hematogenous route, producing septicemia and focal lesions in other tissues. With septicemia, the mortality rate may reach 80 percent. In *Pseudomonas* pneumonias, patients present with toxicity, confusion, and progressive cyanosis. Empyema is common. X-rays of the lower lobes reveal infiltrates and nodules, which may necrose with abscess formation. The mortality is high in *Pseudomonas* pneumonia. The major body defense against *Pseudomonas* infection appears to be a functioning phagocytic system. In patients with leukemia, mortality is highest when patients become severely leukopenic.

Laboratory Diagnosis A diagnosis is made by isolation of the organism. *P. aeruginosa* will grow on almost any of the laboratory media in use today. Isolation from properly collected and transported clinical specimens does not require any unusual procedures. Properties useful in the identification of the organism are listed in Table 39-1.

Treatment. Most antimicrobial agents are ineffective in the treatment of *P. aeruginosa* infections. Prevention of colonization and control of the primary disease are major factors in the survival of patients with *P. aeruginosa* infections. A majority of the strains are susceptible to the aminoglycosides (amikacin, gentamicin, and tobramycin) and to colistin, but resistant forms develop, especially during long-term treatment. Approximately 50 percent of *P. aeruginosa* are sensitive to carbenicillin and its relative, ticarcillin. Aminoglycosides and carbenicillin appear to act synergistically in vivo, and combination therapy with these drugs is recommended in life-threatening situations. The antibiotics should not be mixed prior to injection because in vitro inactivation occurs. A number of new third generation cephalosporins (cefoperazone, moxalactam, and cefotaxime) and semisynthetic penicillins (piperacillin, mezlocillin, and azlocillin) with in vitro antipseudomonal activity have been recently marketed. The overall effectiveness of these agents in human infections is still being evaluated.

When applied topically, the sulfonamide derivative sulfamylon limits the bacterial density in burns and prevents spread of the organisms to other body sites.

Another approach to treatment of *P. aeruginosa* infections has been the use of vaccines. One of these, a heptavalent vaccine, reduces the mortality in burn patients from 15 to 3 percent. However, multiple injections are required and side effects are frequent. Another vaccine, PEV-01, has also been used in a controlled prospective study involving burned patients in two hospitals. In the vaccinated group in this study, there were no bacteremic patients compared with a 20 percent bacteremia in the control group. Mortality in the vaccinated group was approximately 5 percent, compared with 21 percent in the control group. Further studies in other patient populations are needed to further delineate the place for these vaccines in the general hospital.

Hyperimmune gamma globulin may also prove to be useful in treatment of patients with neoplastic disease and *P. aeruginosa* infections. Granulocyte transfusion has also been used with some success in this patient population.

Control. The spread of *P. aeruginosa* is enhanced by failure to observe proper isolation procedures, hand-washing techniques, disinfection, and guidelines for care of catheters and respiratory equipment. Hospital surveillance and access to strain typing are useful measures in locating sources of infections and preventing their spread. Because the organism is killed by drying, attempts should be made to eliminate sources of moisture. Disinfection of sinks and hand-washing areas has also been shown to be useful in controlling the spread of this organism. Because *P. aeruginosa* is found on fresh vegetables, several authors have recommended that the diet of patients with neoplastic disease be low in these items.

Other *Pseudomonas* Species

A number of other *Pseudomonas* species are isolated from clinical specimens and the hospital environment. When these organisms are isolated, careful evaluation of the clinical setting must be made, as these organisms may represent contaminants rather than etiologic agents of disease. Most of them are common inhabitants of the soil, but *Pseudomonas mallei* appears to be a specialized animal pathogen. Characteristics that are useful for the laboratory identification of these organisms are listed in Table 39-1.

Pseudomonas cepacia. *Pseudomonas cepacia* was formerly classified as EO-1, *Pseudomonas multivorans*, and *Pseudomonas kingii*. It is frequently isolated from the hospital environment and clinical specimens. It has been associated with endocarditis, septicemia, wound infections, and urinary tract infections. Most patients in whom *P. cepacia* has been associated with disease were debilitated or had prior instrumentation or manipulation. One hospital infection involving pressure transducers contaminated with *P. cepacia* occurred in the United States. Bacterial endocarditis in heroin addicts has been documented. A number of isolates of this species are resistant to most antibiotics, including gentamicin. However, most are sensitive to chloramphenicol or to the combination of trimethoprim–sulfamethoxazole.

Pseudomonas maltophilia. This pseudomonad is frequently isolated from the oropharynx and sputum of normal adults, as well as from many sites in the environment. The significance of laboratory isolation of this organism requires close clinical evaluation. Documented cases of *Pseudomonas maltophilia* infection of wounds, the urinary tract, and blood have occurred. Nosocomial infections may also occur. One hospital reported colonization of 63 patients as a result of using contaminated disinfectant. Of the 63 patients, 7 adults had proven urinary tract infections, and 2 infants had probable infections. Antimicrobial susceptibility varies, but most isolates are susceptible to chloramphenicol, colistin, tetracycline, and sulfisoxazole.

Pseudomonas mallei. This organism is the cause of the disease glanders in horses and donkeys. Humans are infected by direct contact through skin abrasions and inhalation. Because equine glanders has been eliminated in the United States and Canada, the disease is rarely seen in these countries.

Pseudomonas pseudomallei. This organism is a common inhabitant of the soil in Southeast Asia. It causes melioidosis, a glanders-like disease in humans. The organism gains entrance into the body by inhalation or through abraded skin. In the body, it may not cause any immediate problems but follows the course of a benign pulmonary disease, mimicking primary tuberculosis or fungal disease. Melioidosis can also present as an acute fulminating septicemia that is rapidly fatal. Reactivation of quiescent disease may occur after many years, giving the disease the nickname, Vietnamese time bomb. The organism can be isolated from sputum, urine, pus, or blood. Most strains are sensitive to tetracycline, chloramphenicol, and sulfadiazine; however, the combination of trimethoprim–sulfamethoxazole appears to be the treatment of choice.

Other Species. *Pseudomonas stutzeri, Pseudomonas putida, Pseudomonas alcaligenes, Pseudomonas acidovorans,* and other *Pseudomonas* species are isolated from clinical material, but frequently are not the etiologic agents of disease. Rarely, however, they may be significant causes of wound, pleural, and urinary tract infections. *Pseudomonas fluorescens* is frequently isolated from the hospital environment or blood products. As this organism grows poorly at 37C, the symptoms seen in humans, such as fever, may be caused by endotoxin rather than by an infectious process.

FURTHER READING

Books and Reviews

Artenstein MS, Sanford JP (eds): Symposium on *Pseudomonas aeruginosa.* J Infect Dis 130(Suppl), Nov 1974

Bergan T, Forsgren A, Norrby R (eds): Symposium on intensive therapy in *Pseudomonas* infections. Scand J Infect Dis Suppl 29, 1981

Clarke PH, Richmond MH (eds): Genetics and Biochemistry of Pseudomonas. London, Wiley, 1975

Doggett RG: *Pseudomonas aeruginosa* clinical manifestations of infection and current therapy. New York, Academic Press, 1979

Gilardi GL: Identification of *Pseudomonas* and related bacteria. In Gilardi GL (ed): CRC glucose nonfermenting gram-negative bacteria in clinical microbiology. West Palm Beach, CRC Press, 1978, p 15

Homma JY: Roles of exoenzymes and exotoxin in the pathogenicity of *Pseudomonas aeruginosa* and the development of a new vaccine. Jap J Exp Med 50:149, 1980

Howe C, Sampath A, Spontnitz M: The pseudomallei group: A review. J Infect Dis 124:598, 1971

Hugh R, Gilardi GL: *Pseudomonas.* In Lennette EH, Balows A, Hausler WJ Jr, et al. (eds): Manual of Clinical Microbiology, 3rd ed. Washington, D.C., American Society for Microbiology, 1980, p 288

von Graevenitz A: Clinical role of infrequently encountered nonfermenters. In Gilardi GL (ed): Glucose nonfermenting gram-negative bacteria in clinical microbiology. West Palm Beach, CRC Press, 1978, p 119

Selected Papers

Ashdown LR: Nosocomial infection due to *Pseudomonas pseudomallei*: Two cases and epidemiologic study. Rev Infect Dis 1:891, 1979

Baker NR: Role of exotoxin A and proteases of *Pseudomonas aeruginosa* in respiratory tract infections. Can J Microbiol 28:248, 1982

Baltch AL, Hammer M, Smith RP, et al.: *Pseudomonas aeruginosa* bacteremia susceptibility of 100 blood cultures to seven antimicrobial agents and its clinical significance. J Lab Clin Med 94:201, 1979

Blackwood LL, Pennington JE: Influence of mucoid coating on clearance of *Pseudomonas aeruginosa* from lungs. Infect Immun 32:443, 1981

Cross AS, Sadoff JC, Iglewski BH, et al.: Evidence for the role of toxin A in the pathogenesis of infections with *Pseudomonas aeruginosa* in humans. J Infect Dis 142:538, 1980

Fisher MC, Long SS, Roberts EM, et al.: *Pseudomonas maltophilia* bacteremia in children undergoing open heart surgery. JAMA 246:1571, 1981

Fisher MW: Polyvalent vaccine and human globulin for controlling *Pseudomonas aeruginosa* infections. In Schlessinger D (ed): Microbiology—1975. Washington, D.C., American Society for Microbiology, 1975, p 416

Flick MR, Cluff SE: *Pseudomonas* bacteremia. Am J Med 60:501, 1976

Iglewski BH, Sadoff J, Bjorn MJ, et al.: *Pseudomonas aeruginosa* exoenzyme S: An adenosine diphosphate ribosyltransferase distinct from toxin A. Proc Natl Acad Sci USA 75:3211, 1978

Jones RJ, Roe EA, Gupta JL: Controlled trials of a polyvalent *Pseudomonas* vaccine in burns. Lancet 2:977, 1979

Kreger AS, Gray LD: Purification of *Pseudomonas aeruginosa* proteases and microscopic characterization of pseudomonal protease-induced rabbit corneal damage. Infect Immun 19:630, 1978

Kulczski LL, Murphy TM, Bellanti JA: *Pseudomonas* colonization in cystic fibrosis a study of 160 patients. JAMA 240:30, 1978

Lebek G, Häfliger W: Verification of nosocomial infection chains with *Pseudomonas aeruginosa.* Abl Bake Hyg I Abt Orig B 169:530, 1979

Leppla SH, Martin OC, Muehl LA: The exotoxin of *P. aeruginosa*: A proenzyme having an unusual mode of action. Biochem Biophys Res Comm 81:532, 1978

Martone WJ, Osterman CA, Fisher KA, et al.: *Pseudomonas cepacia*: Implications and control of epidemic nosocomial colonization. Rev Infect Dis 3:708, 1981

Pavaloskis OP, Iglewski BH, Pollack M: Mechanism of action of *Pseudomonas aeruginosa* exotoxin A in experimental mouse infections: Adenosine diphosphate ribosylation of elongation factor 2. Infect Immun 19:29, 1978

Pedersen MM, Manso E, Pickett MJ: Nonfermentative bacilli associated with man. III. Pathogenicity and antibiotic susceptibility. Am J Clin Pathol 54:178, 1970

Pennington JE: Lipopolysaccharide pseudomonas vaccine: Efficacy against pulmonary infection with *Pseudomonas aeruginosa.* J Infect Dis 140:72, 1979

Pollack M: Editorial, *Pseudomonas aeruginosa* exotoxin A. N Engl J Med 302:1360, 1980

Pollack M, Young LS: Protective activity of antibodies to exotoxin A and lipopolysaccharide at onset of *Pseudomonas aeruginosa* septicemia in man. J Clin Invest 63:276, 1979

Ramphal R, Small PM, Shands JW Jr, et al.: Adherence of *Pseudomonas aeruginosa* to tracheal cells injured by influenza infection or endotracheal entubation. Infect Immun 27:614, 1980

Saelinger CB, Snell K, Holder IA: Experimental studies on the pathogenesis of infections due to *Pseudomonas aeruginosa*: Direct evidence for toxin production during pseudomonas infection of burned skin tissues. J Infect Dis 136:555, 1977

Sokol PA, Iglewski BH, Hager TA, et al.: Production of exoenzyme S by clinical isolates of *Pseudomonas aeruginosa*. Infect Immun 34:147, 1981

Thomas ET, Jones LF, Simano E, et al.: Epidemiology of *Pseudomonas aeruginosa* in a general hospital: A four year study. J Clin Microbiol 2:397, 1975

Vasil ML, Iglewski BH: Comparative toxicities of diptheria toxin and *Pseudomonas aeruginosa* exotoxin A: Evidence for different cell receptors. J Gen Microbiol 108:333, 1979

Woods DE, Cryz SJ, Friedman RL, Iglewski BH: Contribution of toxin A and elastase to virulence of *Pseudomonas aeruginosa* in chronic lung infections in rats. Infect Immun 36:1223, 1982

Woods DE, Strauss DC, Johnson WJ Jr, et al.: Role of pili in adherence of *Pseudomonas aeruginosa* to mammalian buccal epithelial cells. Infect Immun 29:1146, 1980

Young LS: The role of exotoxins in pathogenesis of *Pseudomonas aeruginosa* infections. J Infect Dis 142:626, 1980

CHAPTER 40

Yersinia

The three species in the genus *Yersinia* are primarily animal pathogens but also produce human disease. *Yersinia pestis* is the cause of plague, and *Yersinia pseudotuberculosis* and *Yersinia enterocolitica* are most commonly associated in humans with gastrointestinal disease and involvement of the mesenteric lymphatics. *Yersinia* were previously classified in the genus *Pasteurella*. They are now a separate genus in the family Enterobacteriaceae.

Yersinia pestis

Occasional fragments of early writing suggest that plague was present in the western world over 2000 years ago, but the extent of the disease is unknown. The first document-ed pandemic occurred during the reign of the Byzantine emperor Justinian I, in 542 AD. It probably began in Egypt or Ethiopia, spread widely, and lasted 60 years. Approximately 100 million persons died of the infection, and towns were completely decimated.

The second pandemic, known as the Black Death, started in the 14th century. The disease originated in Central Asia and became rampant throughout Europe, the Near East, India, and China. Both rats and infectious droplets from pneumonic victims played a prominent role in transmission of the disease. In Europe alone, 25 million persons died, one fourth of the entire population. Following the second pandemic, the disease became endemic among urban rat populations in many affected areas, and periodic smaller epidemics continued to occur through the

17th century. From that time until 1894, a general decline of the disease occurred, the causes of which are incompletely understood.

In 1855, however, warfare facilitated spread of a Burmese focus of disease, and slow migration brought infected persons to Canton and Hong Kong in 1894. Modern transportation facilitated rapid spread of the disease and precipitated the current and third pandemic. Virtually the entire world was affected, including, for the first time, the United States. During the third pandemic, foci of plague were firmly established among wild commensal rodents in large areas. These foci currently give rise to sporadic cases of plague that have a potential for dissemination even in countries with high standards of public health. The third pandemic, however, appears to be approaching quiescence. Between 1958 and 1979, approximately 47,000 cases were reported worldwide.

Morphology and Physiology

The causative agent, *Y. pestis,* is a gram-negative nonmotile coccobacillus (Fig. 40-1). It shows marked bipolar staining, especially in tissue impressions, bubo aspirates, and pus stained with Wayson's stain. The cells have a safety-pin appearance, with the polar bodies staining blue and the remainder staining light blue to reddish. Freshly isolated virulent organisms are encapsulated.

Yersinia are facultative anaerobes. They are anaerogenic and usually do not ferment lactose. They are oxidase negative and produce catalase.

Cultural Characteristics. *Y. pestis* can grow over a wide temperature range, from 0C to 43C, the optimal tempera-

ture for growth being 28C. It can grow on ordinary laboratory media even from small inocula. On nutrient agar plates, small mucoid colonies appear in 1 to 2 days. On desoxycholate agar, very small red colonies may be seen on the second day of incubation. No hemolysis is produced on blood agar. Growth may be slow and turbidity in broth may be minimal.

Strain Identification. Three biotypes have been identified on the basis of their ability to reduce nitrates to nitrites and to ferment glycerol. These biotypes have been designated orientalis, mediaevalis, and antigua, and are characterized by differences in their geographic distribution. Orientalis is the usual biotype of western North America. *Y. pestis* strains are also characterized by quantitative differences in their antigens, as described below.

Antigenic Structure

At least 20 different antigens have been detected in *Y. pestis* by gel diffusion and biochemical analysis; 15 of these are shared with *Y. pseudotuberculosis.* Most of these antigens have received an alphabetical designation. Fraction 1 (F-1) antigen, murine toxin, and D antigen are unique to *Y. pestis;* M and N antigens are found only in *Y. pseudotuberculosis.* Quantitative differences occur in the content of the F-1 antigen and the murine toxin, which vary independently from one isolate to another. Variations also exist in the protein patterns of isolates from various areas and may be of use epidemiologically in the future.

Determinants of Virulence

The availability of a number of mutants of *Y. pestis* has provided information on phenotypic properties of the organism that contribute to its virulence. Five determinants of virulence have been defined:

1. V and W antigens
2. F-1 antigen (envelope antigen)
3. pesticin, coagulase, and fibrinolysin formation
4. ability to absorb certain pigments
5. purine synthesis.

The importance of each of these factors is difficult to assess separately. The V and W antigens appear to confer on *Y. pestis* the ability of small numbers of bacilli to establish infection in animals. Once infection is established, the F-1 (envelope) antigen, pesticin, coagulase, and fibrinolysin contribute to the disease process.

Determinants for the synthesis of purines, F-1 antigen, and pigment receptors are controlled by chromosomal genes, whereas plasmid DNA controls the expression of V and W antigens and of pesticin and its attendant invasive enzymes.

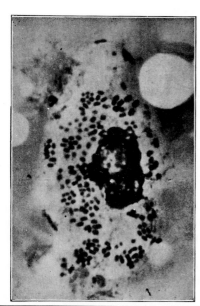

Figure 40-1. *Yersinia pestis* in monocyte from mouse lung. Giemsa stain, × 1500. *(From Meyer: J Immunol 64:139, 1950.)*

V and W Antigens. These antigens are always produced together. The V antigen consists of a 38,000 dalton peptide, and the W antigen is a 145,000 dalton lipoprotein fraction. These antigens enable the organism to resist phagocytosis by the polymorphonuclear leukocyte. They are selectively produced during periods of stasis of bacterial growth. Factors that promote bacterial growth, such as the proper ionic conditions, repress V and W formation. Synthesis is enhanced by 20 to 40 mM Mg^{2+} and the essential absence of Ca^{2+}, concentrations that characterize mammalian intracellular fluid. Synthesis of V and W occurs at 37C, but is prevented at this temperature if Ca^{2+} is present at concentrations similar to the concentration in mammalian plasma (2.5 mM). Their expression may thus be restricted to the environmental conditions that the parasite would find in the cytoplasm of the host cell and that are thought to be necessary for intracellular survival. Only yersiniae harboring a 45-megadalton plasmid produce the V and W antigens.

Envelope Antigen. F-1 antigen, or envelope antigen, is a soluble antigen contained within the bacterial envelope. It consists of two immunologically identical complexes: a protein complexed with polysaccharides (fraction 1A), which contains *N*-acetylglucosamine and hexuronic acid; and a protein alone (fraction 1B). F-1 apparently consists of a series of serologically identical molecular aggregates. Maximal production occurs at 37C, and none is produced at very low temperatures. It is essential for virulence in the guinea pig but not in the mouse. Its role in human infections is unknown. It is highly immunogenic, however, and may constitute as much as 7 percent of the dry weight of the organism. Antibody to F-1 appears to be protective in both human and experimental animals. F-1 is not present in flea-adapted organisms but develops in regional lymph nodes during the incubation period in mononuclear cells.

Pesticin I, Coagulase, and Fibrinolysin. The production of these components is always correlated. Pesticin I is a bacteriocin produced by *Y. pestis* that inhibits the growth of *Y. pseudotuberculosis* as well as some strains of *Escherichia coli* and *Y. enterocolitica*. Pesticin is a monomeric pure protein with a molecular weight of 65,000. It acts by converting sensitive bacteria to nonviable osmotically stable spheroplasts. The production of coagulase and fibrinolysin is apparently a function of bacteriocinogenic conversion. Strains of *Y. pestis* lacking these enzymes are fully infectious for the mouse or guinea pig, but lethality is significantly attenuated.

Pigment Absorption. An interesting relationship exists in *Y. pestis* between pigmentation and virulence. In virulent strains, an unidentified surface component is present that results in the absorption of hemin and basic aromatic dyes to form colored colonies. In mice, avirulent nonpigmented organisms may be restored to their original expression of virulence by providing an excess of free serum iron. Also, the in vivo virulence of pigmented *Y. pestis* is enhanced by an injection of Fe^{2+} sufficient to saturate serum transferrin. This observation, however, is not unique to *Y. pestis* but also occurs following experimental infection with *Pasteurella multocida*, *Y. pseudotuberculosis*, and many other organisms.

Purine Synthesis and Other Virulence-associated Factors. Loss of the ability to complete the synthesis de

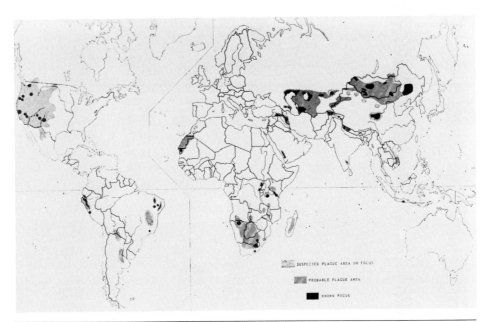

Figure 40-2. Locations of known and probable foci and areas of plague in 1969. *(From WHO Technical Report Series 447:6, 1970.)*

novo of purine ribotides is correlated in *Y. pestis* with a loss of virulence. This is also observed with other members of the Enterobacteriaceae family.

Two additional factors that may be associated with the expression of virulence of *Y. pestis* are murine toxin and endotoxin. Although the LD_{50} of murine toxin for rats and mice is less than 1 μg, it is relatively atoxic for other animals—hence the name murine toxin. It is a protein that in highly purified preparations inhibits the respiration of muscle mitochondria of sensitive animals by preventing reduction of coenzyme Q. The effect is most pronounced on the peripheral vascular system, resulting in shock and fatty degeneration and necrosis of the liver.

The role of *Y. pestis* endotoxin is ill defined. Endotoxin shock is produced in sensitive animals. A biphasic febrile response, induced tolerance, and both localized and generalized Shwartzman reactions are produced by the lipopolysaccharide of the cell wall.

Clinical Infection

Epidemiology

At the present time, over 90 percent of the total world incidence of plague occurs in Southeast Asia, especially in South Vietnam, Burma, Nepal, and Indonesia. Another major active focus is in Brazil. Outbreaks in the present pandemic, however, have been less extensive than in the past (Fig. 40-2).

Plague was introduced into the United States from China in 1900, when the first human case of the disease was reported in San Francisco. Within the next decade, studies showed the presence of infected wild rodents, especially ground squirrels, in wide areas south of San Francisco. In 1907 to 1908, a major epidemic of 167 cases occurred in San Francisco. Permanent foci of plague now exist that involve at least 57 wild rodent species and their fleas. These extend as far east as Kansas, Oklahoma, and Texas and to approximal areas of Canada and Mexico. The disease does not have natural foci in North America east of these areas. During the current pandemic, infected fleas are the major mechanism for transmission.

Plague is perpetuated by three cycles: (1) natural foci among commensal rodents with transmission by fleas (sylvatic plague, wild plague); (2) urban rat plague, which is transmitted by the rat flea (domestic plague, urban plague); and (3) human plague, which may be acquired by contact with either of the former cycles and which may be transmitted by pneumonic spread or, rarely, by the bite of a human flea (Fig. 40-3).

Sylvatic Plague. This is the only form prevalent in the United States. Wild rodent plague in the area west of the 100th meridian is one of the largest world reservoirs. A major factor in the restriction of plague to the western United States is the presence of dense colonies of rodents, such as prairie dogs, in the western areas.

FLEA-RELATED FACTORS. In nature, the flea is essential for the perpetuation of plague. At least four flea-related factors influence the epidemic potential of plague:

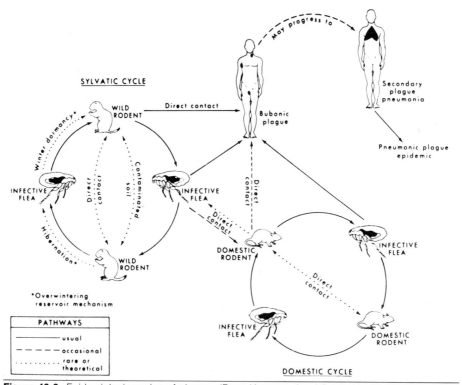

Figure 40-3. Epidemiologic cycles of plague. *(From Kartman: Am J Public Health 56:1554, 1966.)*

1. Fleas vary greatly in their vector efficiency. Most wild rodent fleas are relatively inefficient in the transmission of disease to humans. However, the oriental rat flea, *Xenopsylla cheopis*, is highly efficient and has been the classic vector in urban rat-borne epidemics.

2. The restricted feeding habits of most wild rodent fleas limit their threat to humans. However, the spread of infection within rodent populations, especially among different species that commingle, is facilitated by the transfer of fleas from one rodent host to another. Humans occasionally have been infected when wild rodent deaths during an epizootic left hungry fleas in search of a new host. Dog and cat fleas are very poor vectors and have not been associated with human plague outbreaks.

3. Some infective wild rodent fleas survive in burrows for long periods of time even after the rodent hosts have died. Survival of fleas for as long as 15 months has been shown.

4. The development of dichloro-diphenyl-trichloroethane- (DDT-) resistant fleas in some areas may influence the epidemic potential of plague. This has occurred in some instances concomitant with widespread DDT spraying during malaria control programs.

Transmission of plague by fleas may occur in several ways. The most efficient involves ingestion of the organism by the flea during a blood meal from a bacteremic host. In the flea stomach, the infected blood is coagulated by coagulase that is produced by *Y. pestis* in the presence of an enzyme from the flea stomach. Bacteria are thus trapped in a matrix of fibrin, which fixes them to the spines of the flea's proventriculus. As the bacteria multiply, the proventriculus is occluded, causing blockage. The time between ingestion and blockage is the extrinsic incubation time, and usually for *X. cheopis* it is about 2 weeks. During subsequent attempts to obtain a blood meal, regurgitation of infected material results in infection of the new host. The hungry flea also becomes less fastidious about his host and will readily attack humans. Hot, dry weather adversely affects all stages in the life cycle of the flea, explaining the subsidence of many epidemics at the beginning of a hot and dry season. Blockage is enhanced at temperatures below 26C. Above 27C the fibrinolytic factor of *Y. pestis* and the trypsin-like activity of the flea stomach enzyme are activated, destroying the fibrin meshwork needed for blockage. Decreased blockage results in decreased vector efficiency.

Mechanical transmission via contaminated mouth parts of the flea is also important in the transmission of plague, especially in wild plague. The fleas of many wild rodents do not become blocked or may be poor vectors even when blocked. Mechanical transmission by such fleas, which are highly prevalent, may be the primary means by which enzootics and epizootics occur.

WILD RODENT-RELATED FACTORS. The essential components of a natural focus for wild plague include the presence of *Y. pestis* organisms of sufficient virulence to cause infection, a dense population of rodents that may develop bacteremia when infected, and a high index and infestation rate of fleas that are capable of transmitting the infection. Over 200 species of rodents and other small animals may be infected during an enzootic. Ten rodent and two rabbit genera are especially important, and include squirrels, field mice, prairie dogs, chipmunks, and voles. Populations of wild rodents vary widely in their susceptibility to plague. The introduction of disease into a population may cause a catastrophic dying off of almost the entire population, or may cause extensive infection without apparent illness. This variation in susceptibility to plague is a function of the experience of the population with former outbreaks of plague and is genetically determined. Animals that are genetically highly susceptible are unlikely to survive the disease to manifest an immune response. Inherently resistant animals, however, may survive and demonstrate serologic evidence of exposure.

RESERVOIRS OF INFECTION. The maintenance of plague foci is enhanced by a commingling of rodents or other mammals of differing susceptibility. The relatively resistant species act as a reservoir of the infection. Although they do not succumb, they develop bacteremia that serves to perpetuate a population of infected fleas. The relatively susceptible species augment the level of infectivity to the local flea population, and deaths among this species may bring them into contact with humans or domestic animals.

The hibernating habits of some wild rodents also facilitate survival of plague during the winter months. Animals are much more resistant to plague during hibernation, and an infected animal may survive through the winter and transmit disease again in the spring.

Attenuated *Y. pestis* organisms have been isolated from natural foci. These organisms may produce infection without causing death and provide another mechanism for perpetuation of the disease. It is not known at present whether these attenuated strains reacquire virulence. Perpetuation of the infection is also facilitated by the ability of *Y. pestis* to survive for long periods in the soil of animal burrows.

HUMAN INFECTION. The transmission of infection from wild plague foci to humans is an accidental occurrence. From 1908 to 1968, 120 cases were attributed to wild sources, and in 1980 there were 18 cases reported. Thirteen of the cases occurred in New Mexico and there were five deaths. Disease is usually sporadic and most often bubonic in character. Cases often occur in areas where there is extensive mortality among the rodent populations. Most of the cases occur from June through August, and the patients are primarily children. Other victims are hunters and other persons who come into professional or avocational contact with wild animals.

A particularly high-risk group are the American Indians of the southwest, especially the Navajos, whose cultural patterns bring them into close contact with prairie dogs. In 1965, an epidemic among the Navajos was associated with an extensive epizootic of the local prairie dog colonies. Dogs associated with these outbreaks developed antibody to F-1 antigen without experiencing serious disease. Domestic dogs and cats bring infected wild rodents and their fleas into the area of human habitations, increasing the danger of infection among domestic rodents.

Urban Plague. Urban plague has been the principal cause of the massive epidemics of plague in recent history. The characteristics of this form of plague are quite distinct from those of wild plague, although there has been extensive interaction between the two forms. Urban plague is characterized by an accompanying epizootic of plague among the rodents, particularly the black rat that lives in close proximity to human habitations. An epizootic occurs when there is a sufficiently dense population of susceptible domestic rats, a high index of parasitism with fleas capable of transmitting disease (the flea index), adequate climatic conditions, and introduction of disease. Centers for commerce, especially seaports, have been frequent sites of explosive disease. During the eras of epidemic human plague, the natural history of the epidemic included a gradual onset of fatal disease among the domestic rodents. As the infected fleas moved from the dead or dying rodents to other living rodents, the rate of death among the rodents increased. Human cases could be expected when approximately 10 percent of the rats were infected. Human cases usually began as bubonic disease, and usually there was only one or a few cases per family. The disease was seasonal, and the onset of adverse climatic conditions would curtail an epidemic. The disease was also cyclical, with severe disease following seasons of no disease or mild disease.

Initial epizootics characteristically caused the death of essentially the entire population of domestic rodents. Subsequent generations of rodents, however, were relatively more resistant to plague. After several such epizootics, the rodents developed herd resistance and did not die of the disease, with the result that the severity of subsequent epidemics gradually declined. This probably accounts in large part for the fact that epidemic urban plague during the current pandemic has been declining for several decades. Repopulation of an urban area by a susceptible species of rodent, however, as occurred in Bombay in 1948 with susceptible bandicoots, may again lead to extensive epizootic and epidemic disease.

Rural Plague. Urban plague has been commonly disseminated to rural areas through lines of commerce. Grain shipment in Vietnam has probably played a role in the establishment of new foci. In rural areas, the disease first affects domestic rats. It is then spread to commensal wild rodents, thereby establishing a focus of sylvatic plague. In contrast to urban plague, sylvatic plague causes small numbers of sporadic human cases. There is little tendency for rural plague to disappear because the constant influx of susceptible rodents from wild sources precludes the development of effective herd immunity.

Interhuman Transmission. Pneumonic plague is the clinical form of the disease with the greatest potential for rapid dissemination. It occurs following close contact with a victim of bubonic plague who has developed secondary pulmonary involvement and exhales the organism in droplets. This form of disease may be widespread and rapid in onset, and is the only form of plague that is directly contagious. Epidemiologic characteristics of pneumonic plague show that familial spread is most common, and disease is most frequent in areas of overcrowding. Cold weather with high humidity fosters the disease. Although most epidemics are predominantly either bubonic or pneumonic in clinical presentation, recent epidemics in Vietnam have been mixed. Factors that determine whether an epidemic will be predominantly bubonic or pneumonic are not well understood.

Pathogenesis and Clinical Manifestations

The various clinical forms of plague overlap but may be grouped as predominantly bubonic, septicemic, or pneumonic.

Bubonic plague is essentially the only form of disease that occurs in humans following infection from a wild focus. The bubo represents the infected regional lymph node, which drains the area of skin through which the organism was introduced. The groin is the region most commonly affected, with axillary and cervical nodes less frequently. Buboes in more than one site are extremely rare. The incubation period of bubonic plague is usually less than a week, and the bubo may be preceded by prodromata of chills, fever, malaise, confusion, nausea, and pains in the limbs and back. Onset of disease is usually sudden. Patients may experience pain at the site of the future bubo before it is palpable. The lesion itself is tender, the node is enlarged and may suppurate, and erythema of the surrounding tissues is common. The bubo of plague cannot be distinguished clinically from other causes of acute lymphadenitis. Bacteremia is usually present even in mild cases—approximately one half of early blood cultures will yield the organism. The level of bacteremia is, however, usually low, and organisms are seldom seen on direct observation of stained buffy coat smears.

Pulmonary involvement is common in all severe cases of plague and is often the immediate cause of death. Hemorrhagic and edematous effusions predominate, probably associated with emboli that may be septic and originate from the bubo. Endotoxin-mediated effects are observed. Another primary manifestation of bubonic plague is congestion of the vessels of the conjunctivae. The three clinical findings that traditionally have been most useful in the diagnosis of plague are a rapid rise in temperature, regional buboes, and conjunctivitis. In very mild plague, the only clinical finding may be vesiculation at the site of inoculation. Slightly more active disease, with local buboes but without systemic signs of disease, is termed pestis minor.

At the opposite extreme is septicemic plague. In the United States, approximately 20 percent of cases are septicemic; the fatality rate is very high in this form of plague. In this form of disease, the patient experiences a very high level of bacteremia early in the course of disease before local buboes evolve. The mortality rate, treated and untreated, is high, with rapid peripheral vascular collapse. A prominent finding in this form, as well as in other forms of plague, is disseminated intravascular coagulation with a generalized Shwartzman phenomenon. Purpuric lesions with intravascular thrombi occur in all areas of the body. This manifestation of disease is more common in children than in adults.

Plague meningitis is an infrequent complication. Clinical evidence suggests that it is most common in persons who experience an attenuated form of infection, as occurs in the partially immune or following inadequate treatment. The clinical findings are those of an acute bacterial meningitis.

Pneumonic plague usually arises from septic embolization to the lungs. Patients also may acquire the pneumonic form following pharyngeal plague and direct extension into the lung from the cervical or tonsillar buboes. Inhalation of organisms in droplet nuclei dispersed by another person with plague also provides a mechanism for direct inoculation of the lung parenchyma. This was probably the etiology of the highly fatal fulminant epidemics in China in the late 1890s. The average length of time in untreated patients with pneumonic plague from the first appearance of symptoms to death is less than 2 days. The disease is highly contagious. Marked central nervous system abnormalities, including convulsions, incoordination, stuper, and delirium, usually accompany the disease.

Laboratory Diagnosis

Clinical materials containing Y. pestis may be hazardous. Suspicious cultures or specimens should be sent immediately to a laboratory with facilities for making a rapid definitive identification. Aspirates from buboes, pus from the area of the flea bite, sputum, throat swabs, or blood should be carefully collected and placed in Cary–Blair transport media for transfer to the laboratory.

Serologic Diagnosis. Antibodies to the F-1 antigen (p. 638) may be detected by use of the agglutination test or the complement-fixation test. The complement-fixation test also may be used for detecting the F-1 antigen either in tissue extracts or the organism itself. Complement-fixing antibodies decrease rapidly following recovery from plague. The passive hemagglutination test, which uses tanned erythrocytes coated with F-1 antigen or murine toxin, is also a sensitive indicator of antigen or antibody. A

serologic response is apparent by day 5, with a peak by day 14. This antibody may persist for several years following recovery from plague and is a sensitive test for identifying a quiescent plague focus.

Precipitin tests are useful in the detection of F-1 antigen in dried and decomposed carcasses of animals. An immunofluorescent test using F-1 antibody is a rapid and generally accurate method of identifying Y. pestis, although cross-reaction with Y. pseudotuberculosis may occur in a small percentage of cases.

Treatment

Streptomycin is bactericidal and highly effective for most strains. Resistant organisms have been observed in the Far East but not in the United States. Alternative treatment is tetracycline or chloramphenicol. Kanamycin appears to be equally effective as streptomycin and to show no cross-resistance. Penicillin is inadequate, and the sulfonamides are not uniformly effective. Sulfamethoxazole-trimethoprim has been used successfully in the treatment of a small number of patients, but further evaluation is necessary.

Prevention

A vaccine for immunization against plague is available. In spite of widespread plague among Vietnamese civilians, by the end of 1969, only 8 cases had occurred in U.S. military personnel. Considerable experience was gained in the past with the use of live attenuated vaccines, but these are no longer used in the United States. The plague vaccine that is licensed for use in the United States is inactivated with formaldehyde, and its efficacy appears to be directly proportional to the content of F-1 antigen. The primary series consists of three doses of vaccine and a booster 6 months thereafter. Severe reactions to the vaccine have not been common, although immediate and generalized urticaria and anaphylaxis may occur.

Immunization is recommended for persons who (1) are working in the laboratory or field with Y. pestis strains that are resistant to antimicrobials; (2) are engaged in aerosol experiments with Y. pestis; and (3) are engaged in field operations in areas with enzootic plague where prevention of exposure cannot be achieved. There is no evidence that vaccine protects against pneumonic plague, and the efficacy in prevention of bubonic disease under field conditions is also uncertain. Therefore, if a person has had a definite exposure, prophylactic antibiotics may be indicated even if the person has received vaccine.

Control

Plague is one of the internationally quarantinable diseases, and reporting of cases is mandatory. Public health authorities may institute enforced quarantine, and/or disinfection of persons, ships, and aircraft arriving with known or suspected infected persons or animals. Efforts to prevent plague have been directed toward preventing transportation of rats, especially by ships and airplanes. Current methods of shipping and docking make the classic importation of shipboard rats into western countries unlikely, but the possibility of spreading diseased rats by container shipping may still exist.

Control of urban plague has proceeded along the principles of:

1. flea control, which should precede rat control
2. rodent extermination
3. treatment and/or quarantine of cases
4. quarantine of contacts of pneumonic plague
5. restriction of movement in highly infected areas
6. thorough garbage disposal
7. application of good personal hygiene.

Yersinia pseudotuberculosis and Yersinia enterocolitica

The term yersiniosis denotes infection with *Yersinia* species other than *Y. pestis*, namely, *Y. pseudotuberculosis* and *Y. enterocolitica*. These are zoonotic diseases, and human infection appears to be acquired accidentally from disease cycles of wild and domestic animals.

Morphology and Physiology

Unlike the plague bacillus, *Y. pseudotuberculosis* and *Y. enterocolitica* are motile. The flagella are paripolar or peritrichous in location and are produced during growth at 22C but not at 37C. A microscopically visible capsule is not produced.

Both species may be isolated from clinical material by culture on blood agar and the usual media used for the enteric bacteria. They are characterized by urease production, an acid slant and butt on TSI agar, and no hydrogen sulfide production. Growth requirements and metabolic pathways are similar to those of the other Enterobac-

teriaceae. Differentiation of *Y. enterocolitica* and *Y. pseudotuberculosis* is based on biochemical differences between the species (Table 40-1). The results of many of these tests are markedly affected by temperature, a property that should be noted in the interpretation of results.

The taxonomic status of organisms currently defined as *Y. enterocolitica* has not been clearly defined. The genus contains organisms that are quite biochemically variable, especially concerning sucrose, rhamnose, indole, and trehalose. Some workers refer to nonhuman isolates that are biochemically unlike human strains as *Y. enterocolitica*-like. Four distinct DNA relatedness groups (biogroups 1 through 4) have been defined, with typical *Y. enterocolitica* strains that are sucrose positive and rhamnose negative being in biogroup 1. Some workers have proposed new species, including *Yersinia intermedia*, *Yersinia frederiksenii*, and *Yersinia kristensenii*, for strains biochemically unlike the human strains.

Antigenic Structure

Yersinia pseudotuberculosis. There are six serotypes of *Y. pseudotuberculosis*, each of which is characterized by type-specific O and H antigens. The O-antigen specificity is conferred by 3,6-dideoxyhexoses, some of which also are present in the cell walls of *Salmonella* groups B and D and are responsible for cross-reactions with these organisms. Serotyping can be performed by agglutination or hemagglutination methods.

The V and W antigens of *Y. pestis* are also present in *Y. pseudotuberculosis*.

Yersinia enterocolitica. Organisms currently regarded as *Y. enterocolitica* are serologically heterogeneous. Twenty-seven serotypes have been identified on the basis of their O and H antigens. Certain serologic types are consistently associated with human infection. *Y. enterocolitica* serotypes 0:3 and 0:9 are the most common cause of human infection in Europe, Japan, and Canada, whereas serotype 0:8 strains are responsible for infections in the United States.

TABLE 40-1. DISTINGUISHING PROPERTIES OF *YERSINIA* SPECIES

	Y. pseudotuberculosis	Y. enterocolitica	Y. pestis
Oxidase	−	−	−
β-galactosidase	+	+(d)	+
Indole	−	−(d)	−
Rhamnose	+	−	−
Melibiose	+	−	−
Cellobiose	−	+	−
Sucrose	−	+(d)	−
Ornithine decarboxylation	−	+(d)	−
Salicin	+	−(d)	+
Esculin hydrolysis	+	−(d)	+
Urease	+	+	−
Motility 25C	+	+	−

From Nilehn: Acta Pathol Microbiol Scand [Suppl] 206:20, 1969.
+, positive; −, negative; d, different biochemical types.
The signs denote the behavior of the majority of strains within a species. Rare exceptions are omitted.

Y. enterocolitica bears little antigenic relationship to other *Yersinia*, but cross-reacts with *Brucella*. Most species of *Brucella* show complete cross-reaction with *Y. enterocolitica* serotype 0:9. Differentialton between the two organisms, however, can be made by means of a quantitative rose bengal plate. For diagnostic purposes, it should be kept in mind that a positive *Brucella* agglutination titer may represent a *Y. enterocolitica* 0:9 infection. The antigenic determinant responsible for this cross-reactivity is probably cell wall lipopolysaccharide. There is also antigenic similarity between some *Y. enterocolitica* strains and *Vibrio cholerae* serotype Inaba.

Determinants of Virulence

Yersinia pseudotuberculosis. Although *Y. pseudotuberculosis* possesses the V and W antigens of *Y. pestis*, it does not produce coagulase, fibrinolysin, pesticin, or murine toxin. F-1 is usually absent from *Y. pseudotuberculosis*, but a few strains produce a similar antigen. Pigmentation of *Y. pseudotuberculosis* is less pronounced than in *Y. pestis*, but as in *Y. pestis*, virulence is greatly enhanced by the presence of free iron compounds. Potential determinants of virulence that are common to both organisms are an exotoxin that is similar to the plague murine toxin and an endotoxin. Pesticin of *Y. pestis* inhibits some strains of *Y. pseudotuberculosis*.

Yersinia enterocolitica. Major virulence factors that have been identified for *Y. enterocolitica* include:

1. V and W antigens that are immunologically identical to those produced by *Y. pestis*
2. adherence to human epithelial cells in tissue culture
3. a positive guinea pig corneal ulcer test (the Sereny test)
4. heat-stable enterotoxin production
5. pesticin I sensitivity.

In strains of *Y. enterocolitica* involved in an outbreak of human enteric disease among 218 school children in New York following the consumption of contaminated milk, the presence of a 41-megadalton plasmid was found to be absolutely correlated with tissue invasiveness as measured by the Sereny test. The elaboration of heat-stable enterotoxin by these strains was not associated with any particular plasmid but appears to be encoded by chromosomal genes. The 41-megadalton plasmid is similar to plasmids found in virulent strains of *Y. pestis* and *Y. pseudotuberculosis* and thus appears to be a common virulence determinant among yersiniae.

Clinical Infection

Epidemiology. The yersinioses have been recognized in wild and domestic mammals, birds, invertebrates, and amphibians. Clinically apparent disease with both species has occurred in humans in all areas of the world, but the majority of cases have come from northern Europe, especially France, Germany, and the Scandinavian countries.

Sources of infection in humans are poorly defined. Direct contamination of food or water by infected animals may account for cases in which identical strains have been obtained from an owner and pet or domestic animal, frequently a pig, in a *Y. enterocolitica*-related disease. Most reports are of individual cases or small family outbreaks. Secondary cases are common, as is a high attack rate. A single school outbreak of *Y. enterocolitica* enteritis in Japan involved 20 percent of the entire student body. Person-to-person spread within related families and between personnel on hospital wards has demonstrated the potential for rapid transmission under appropriate conditions. It most commonly occurs in persons from rural areas.

Contamination of water supplies by *Y. enterocolitica* has been demonstrated. Survival of both organisms in various types of water is shortest in the spring and summer and longest in the fall and winter, a finding probably attributable to the ability of the organisms to grow at the lower temperature. The majority of human infections occur during the winter and early spring. The incidence of infection with both organisms is the same for males and females. *Y. enterocolitica* is primarily a disease of the very young, including infants, whereas *Y. pseudotuberculosis* more commonly affects persons 10 to 20 years of age. Large outbreaks have been traced to contaminated tofer, chocolate milk, and other chocolate milk products.

In the United States, *Y. pseudotuberculosis* has been identified in six species of domestic mammals and several wild mammals, including deer, rabbits, and rodents. Wild birds also are reservoirs of *Y. pseudotuberculosis*. Fecal-oral spread of both *Y. enterocolitica* and *Y. pseudotuberculosis* appears to be the major natural method of transmission.

The relative clinical importance of *Y. enterocolitica* and *Y. pseudotuberculosis* as causes of gastrointestinal disease in the United States is considerable. Surveys of routine stool cultures suggest that rarely are the organisms found in normal persons. In some areas these *Yersinia* species are as common a cause of serious gastrointestinal disease as is shigellosis. It is expected that as physicians and laboratories in the United States become more familiar with these organisms, their importance will be more frequently recognized.

Pathogenesis. The primary lesion results from invasion of the wall of the small intestine, usually in the area of the ileum. Ulcers of the intestinal mucosa at the site of lymphoid tissue may develop and lead to extensive loss of blood and fluid, strongly resembling the intestinal findings in typhoid fever. The mesenteric nodes usually are the most extensively involved structures. Enlarged nodes may become confluent. Histopathologic differentiation of these lesions from gastrointestinal infection with *Francisella tularensis*, *Salmonella* species, and cat-scratch fever occasionally may be difficult.

Although usually restricted to the gastrointestinal tract, invasion of the portal system, leading to liver involvement, and generalized septicemia with colonization in other parts of the body may occur.

Clinical Manifestations. A short prodromal period of approximately 1 day precedes symptoms of gastrointestinal disease. The majority of naturally acquired human cases present primarily with gastrointestinal symptoms, including diarrhea and mesenteric lymphatic involvement. Systemic symptoms usually accompany the focal gastrointestinal complaints and consist of headache, which may be severe, malaise, and fever associated with convulsions. Both organisms produce severe abdominal pain, which together with enlarged mesenteric nodes has resulted, in many instances, in exploratory surgery in the expectation of appendicitis. The uncomplicated case of gastroenteritis caused by either *Y. enterocolitica* or *Y. pseudotuberculosis* is not clinically distinguishable from that caused by *Salmonella* or *Shigella*.

Complications consisting of septicemia and hepatic abscesses may occur in a small number of patients, most of whom have preexisting liver disease, are diabetics, or are receiving corticosteroids. An arthritis-erythema nodosum syndrome also has been reported extensively by Scandinavian physicians, but is infrequently observed in the United States or Canada. Patients with this syndrome are predominantly females in the 15- to 45-year age group. The arthritis, which usually is preceded by abdominal pain and diarrhea and often affects multiple large joints sequentially, occurs primarily in person who are HLA-B27 positive.

Laboratory Diagnosis. A definitive diagnosis can be made only by culture of the organism. The organisms can be isolated from mesenteric lymph nodes, feces, blood (in generalized septicemia), effusions from serous cavities, and organ specimens. For selective enrichment and holding, the specimen should be placed in isotonic saline with or without potassium tellurite and promptly refrigerated.

Treatment. The susceptibility of *Y. enterocolitica* and *Y. pseudotuberculosis* to ampicillin, tetracycline, and other commonly used antibiotics is variable, necessitating the sensitivity testing of each isolate. Most strains are sensitive to the aminoglycosides and to trimethoprim-sulfamethoxazole. Other forms of supportive care, such as maintenance of fluid and electrolyte balance, are essential in the care of the severely ill patient.

FURTHER READING

Books and Reviews

Ahvonen P: Human yersiniosis I and II. Annu Clin Res 40:30,39, 1972

Brubaker RR: The genus *Yersinia*: Biochemistry and genetics of virulence. Curr Top Microbiol Immunol 57:111, 1972

Chen TH: The Immunoserology of Plague. In: Kwapinski JGB (ed): Research in Immunochemistry and Immunobiology. Baltimore, University Park Press, 1972, vol 1, p 233

Pollitzer R: Plague. WHO Monograph Series, No 22. Geneva, World Health Organization, 1954

Plague vaccine. Morbidity and Mortality Weekly Report 31:301, 1982

Trends in research on plague immunization. J Infect Dis 129: (Suppl) S1, 1974

Selected Papers

Bolin I, Norlander L, Wolf-Watz H: Temperature-inducible outer membrane protein of *Yersinia pseudotuberculosis* and *Yersinia enterocolitica* is associated with the virulence plasmid. Infect Immun 37:506, 1982

Butler T: A clinical study of bubonic plague. Am J Med 53:268, 1972

Butler T, Hudson BW: The serological response to *Yersinia pestis* infection. Bull WHO 55:39, 1977

Carter PB, Zahorchak RJ, Brubaker RR: Plague virulence antigens from *Yersinia enterocolitica*. Infect Immun 28:638, 1980

Cavanaugh DC: Specific effect of temperature upon transmission of the plague bacillus by the oriental rat flea, *Xenopsylla cheopis*. Am J Trop Med Hyg 20:264, 1971

Gordon JE, Kniws PT: Flea versus rat control in human plague. Am J Med Sci 213:362, 1947

Gutman LT, Ottesen EA, Quan T T, et al.: An interfamilial outbreak of *Yersinia entercolitica* enteritis. N Engl J Med 288:1372, 1973

Hubbert WT: Yersiniosis in mammals and birds in the United States. Case reports and review. Am J Trop Med Hyg 21:458, 1972

Hudson BW, Quan SF, Goldenberg MI: Serum antibody responses in a population of *Microtus californicus* and associated rodent species during and after *P. pestis* epizootics in the San Francisco Bay area. Zoonoses Res 3:15, 1964

Kartman L: Historical and ecological observation on plague in the United States. Trop Geogr Med 22:257, 1970

Knapp W, Thal E: A simplified antigenic scheme for *Yersinia enterolitica* ("Pasteurella X") based on biochemical characteristics. Zentralbl Bakteriol (Orig A) 223:88, 1973

Legters LJ, Cottingham AJ, Hunter DH: Clinical and epidemiologic notes on defined outbreak of plague in Viet Nam. Am J Trop Med Hyg 19:639, 1970

Mair NS: Yersiniosis in wildlife and its public health implications. J Wildl Dis 9:64, 1973

Nilehn B: Studies on *Yersinia enterolitica*. Acta Pathol Microbiol Scand (B) (Suppl) 206:1, 1969

Okamoto K, Inoue T, Shimizu, et al.: Further purification and characterization of heat-stable enterotoxin produced by *Yersinia enterocolitica*. Infect Immun 35:958, 1982

Ratnam S, Mercer E, Picco B: A nosocomial outbreak of diarrheal disease due to *Yersinia enterocolitica* serotype 0:5, biotype 1. J Infect Dis 145:242, 1982

Reed WP, Palmer DL, Williams RC, Kisch AL: Bubonic plague in the Southwestern United States. Medicine 49:465, 1970

Straley SC, Brubaker RR: Cytoplasmic and membrane proteins of yersiniae cultivated under conditions simulating mammalian intracellular environment. Proc Natl Acad Sci USA 78:1224, 1981

Straley SC, Brubaker RR: Localization in *Yersinia pestis* of peptides associated with virulence. Infect Immun 36:129, 1982

Surgella MJ, Beesley EC, Albizo JM: Practical applications of new laboratory methods of plague investigations. Bull WHO 42:993, 1970

Toivanen P, Toivanen A, Olkkonen L, et al.: Hospital outbreak of *Yersinia enterocolitica* infection. Lancet 1:801, 1973

von Reyn CF, Barnes AM, Weber NS, et al.: Bubonic plague from exposure to a rabbit: A documented case and a review of rabbit-associated plague cases in the United States. Am J Epidemiol 104:81, 1976

Zink DL, Feeley JC, Wells JG, et al.: Plasmid-mediated tissue invasiveness in *Yersinia enterocolitica*. Nature 283:224, 1980

Leisure-Time Reading

Burnet M, White DO: Natural History of Infectious Diseases, 4th ed. Cambridge, Cambridge University Press, 1972

Cravens G, Marr JS: The Black Death. New York, Ballantine Books, 1977

CHAPTER 41

Francisella

Tularemia is a major zoonotic disease indigenous to many areas of the United States. It is caused by *Francisella tularensis*, which is transmitted to humans by insect vectors or by the handling or ingestion of infected animals or animal products. Human disease, often referred to as deerfly fever or rabbit fever, is characterized by a focal ulcer at the site of entry of the organisms and enlargement of the regional lymph nodes.

Most of the early work on the etiology and epidemiology of tularemia was carried out by epidemiologists in the United States. The organism was first isolated in 1911 by McCoy and Chapin from ground squirrels in Tulare County, California; hence the name tularemia. These workers found animals in the area to be infected with a plague-like organism that caused disease and produced lesions resembling those of plague. Human cases were recognized under circumstances that implicated rabbits as the source of infection. Extensive studies by Francis and his colleagues of the United States Public Health Service resulted in a classic description of the human disease, definition of the zoonotic nature of the disease, implication of several vectors and means of transmission, and bacteriologic description of the organism.

The disease was subsequently recognized in Europe and the Far East. Widespread enzootic disease with associated human cases has occurred in Russia and Scandinavia. In some of these areas, endemic disease of herds of animals has had serious economic consequences.

Previously classified as *Bacterium tularense*, *Brucella tularense*, and *Pasteurella tularensis*, the organism is currently classified in the genus *Francisella*, which is a genus of uncertain affiliation. *Francisella novicida*, isolated from water, is the only other member of the genus and is not known to infect humans.

Francisella tularensis

Morphology and Physiology

F. tularensis is a small, poorly staining, gram-negative coccobacillus, approximately 0.5 by 0.2 μm in size. It is nonmotile, and displays bipolar staining. Capsules are rare or absent. Young cultures may be relatively uniform in appearance, but older cultures are characterized by extreme pleomorphism (Fig. 41-1). *F. tularensis* is an obligate aerobe and is catalase negative. It produces a small amount of acid but no gas from glucose, maltose, fructose, dextrin, and mannose. H_2S is produced on media containing cysteine, and growth is slight on litmus milk. Biochemical characterization, however, is of little value in identification.

Cultural Characteristics. Growth occurs over a temperature range of 24C to 39C, with an optimal temperature of 35C to 37C. It survives best at low temperatures. The ge-

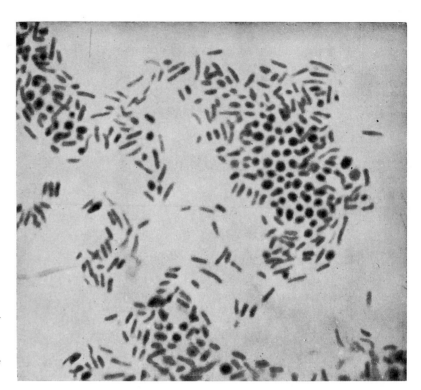

Figure 41-1. *Francisella tularensis.* From culture on glucose cystine agar, showing coccoid and bacillary forms in the same field. Approx. × 5000. *(From the U.S. Army Medical Museum. Courtesy of Edward Francis, U.S. Public Health Service.)*

nus is characterized by a growth requirement for cysteine or cystine. No growth is obtained on ordinary culture media, but slow growth is obtained on semisolid media, such as gelatinized egg yolk and media containing cysteine, glucose, and defibrinated rabbit blood or serum. Growth from small inocula is greatly enhanced by the addition of a low molecular weight cellular component that has not been completely characterized. This substance forms complexes with iron and copper ions and may be partially replaced by iron salts and sideramines. On blood media, colonies of *F. tularensis* may produce greening directly under the colony but no true hemolysis. Colonies are nonpigmented.

Determinants of Pathogenicity

A general correlation exists between a smooth colonial morphology, high degree of virulence for experimental animals, acriflavin reaction, acid agglutination, and staining with crystal violet. Fresh isolates of the organism produce smooth colonies, but repeated passage on artificial media results in a change from the smooth to the rough form and loss of a surface antigen that is a major determinant of pathogenicity.

In addition to changes in virulence that accompany colonial variation, inherent differences exist among the wild strains of *F. tularensis*. Strains with high virulence for

humans are most often associated with tick-borne tularemia of rabbits, exhibit citrulline ureidase activity, and ferment glycerol. These strains are the major type in the United States and are termed Jellison type A, or *F. tularensis* var tularensis. Strains of lesser virulence for humans are associated with water-borne disease of rodents, seldom ferment glycerol, and do not exhibit citrulline ureidase activity (Jellison type B, or *F. tularensis* var palaearctica). These strains also are found in the United States but are the predominant type in Europe, Russia, and the Far East. In some areas, both types are present. Factors conferring virulence to the organism are poorly understood. No exotoxin has been identified.

Antigenic Structure

Different immunologic types of *F. tularensis* have not been detected. Three major antigens have been obtained from all strains tested: (1) a polysaccharide antigen that produces an immediate wheal and erythematous skin reaction in patients recovering from tularemia; (2) cell wall and envelope antigens that apparently contain the immunizing antigen and are responsible for endotoxin activity; and (3) a protein antigen that is responsible for a delayed type of hypersensitivity reaction in patients with the disease. A common protein antigen is shared by *F. tularensis* and members of the genus *Brucella.*

Clinical Infection

Epidemiology

Tularemia is a reportable disease in the United States. Since 1939, when 2291 cases were reported, there has been a steady decline in the annual number of reported cases (Fig. 41-2). In 1980, there were 234 cases of tularemia. The majority of the cases are from rural areas and are predominantly adult males. Women and children commonly acquire the infection while skinning a rabbit. Seasonal peaks occur during the winter and summer months (Fig. 41-3).

Tularemia is enzootic in all areas of the continental United States, as well as most other areas of the world that are north of the equator, except for the British Isles. Rodents and rabbits are the major reservoirs of infection. Other wild and domestic mammals susceptible to tularemia include the deer, fox, mink, raccoon, opossum, beaver, mouse, rat, mole, dog, cat, sheep, and horse. Many species of birds also are probably naturally infected.

Two strains of tularemia may be recognized in the United States. Type A, which accounts for approximately 80 percent of recognized human cases in North America, is highly virulent for humans and usually tick borne and rabbit associated. A second natural complex of virulent tularemia, type B, occurs among sheep kids and may be transmitted to humans by mosquitoes and/or ticks. Type B is less virulent for humans and usually is associated with disease of water-dwelling rodents, such as muskrats, and with associated contamination of streams. The source of the disease in muskrats and beavers may be an epizootic of tularemia in a neighboring highly susceptible rodent population, leading to the rodents dying off in the area adjacent to the water, contamination of water, and disease of the larger rodents. Apparent disease from tularemia among rodent species ranges from asymptomatic disease, to bacteremia with survival, to the rapid death of the entire colony.

F. tularensis has been recovered from over 54 arthropod species, half of which are known to have transmitted the disease to humans. Although ticks, especially *Dermacentor andersoni, Dermacentor variabilis,* and *Amblyomma americanum,* are the most common arthropod vectors, other bloodsucking arthropods may be involved, including deerflies, mites, blackflies, mosquitoes, and occasionally lice. *F. tularensis* may be transmitted directly from an infected female tick to her offspring, an example of transovarial passage of infection.

There are many methods by which humans may become infected with *F. tularensis* other than by contact with infected animals or through the bite of an arthropod vector. These include ingestion of inadequately cooked

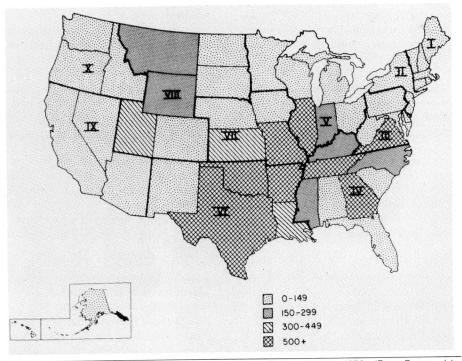

Figure 41-2. Reported cases of tularemia in the United States, 1950 to 1973. *(From Boyce: J Infect Dis 131:97, 1975.)*

0-149
150-299
300-449
500+

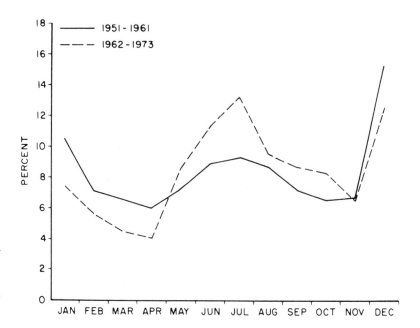

Figure 41-3. Seasonal distribution of tularemia in the United States, 1951 to 1973. Presented as the percentage of annual number of reported cases of tularemia that occurred during each month. *(From Boyce: J Infect Dis 131:97, 1975.)*

infected meat, usually rabbit, leading to primary cervical or gastrointestinal tularemia, which presents without a lesion of the skin. Ingestion of contaminated water and inhalation of contaminated aerosols may lead to disease. Two examples of this latter mode of transmission are: (1) laboratory-acquired infections, which are common and from which a great deal has been learned concerning the natural history of the disease; and (2) inhalation of dust that has been contaminated with infected vole feces during an epizootic. Other methods for acquiring the disease include direct inoculation into the skin from an infected animal bite or scratch and, very rarely, direct spread from an infected person. This is known to have occurred only in the case of a pathologist who accidentally contaminated himself while handling infected material. Under normal circumstances, person-to-person transfer does not occur. Congenital infection has been reported.

Tularemia in human beings is usually a sporadic disease, but it may be epidemic in circumstances that favor the arthropod vector, such as the deerfly, and when water or food supplies are contaminated. Regional differences in the ecology of the reservoirs of tularemia markedly affect its epidemiologic characteristics, which must be determined for each area if a clinical or veterinary problem exists. No other disease is more varied in its reservoirs, vectors, and ecology.

Pathogenesis

F. tularensis is a facultative intracellular parasite. The infective dose for highly susceptible animals, which include mice and guinea pigs, is one organism. The infective dose for humans is 10 organisms, and 10 to 50 organisms will produce human disease after exposure by aerosolization. Macrophages of the fixed reticuloendothelial system and circulating mononuclear phagocytes ingest and harbor the organism. Peripheral polymorphonuclear leukocytes, however, phagocytose *F. tularensis* only in the presence of immune serum. Infected tissue is characterized by invasion of macrophages, necrosis, and granuloma formation.

In the original description of tularemia, Francis recognized at least four clinical forms: glandular, ulceroglandular, oculoglandular, and typhoidal. Tularemic meningitis, gastrointestinal disease, bacterial endocarditis, pharyngitis, and pneumonia were subsequently described. These variations in presentation depend largely upon the method by which the person has acquired the infection, the virulence of the organism, and the individual degree of resistance to tularemia.

Most infections in the United States are acquired from the bite of a contaminated arthropod, usually a tick, deerfly, or mosquito, and lead to an ulceroglandular from of disease. Ulceration at the site of inoculation and regional lymphadenitis are similar following the arthropod bite or direct inoculation, such as occurs during the dressing of an infected rabbit. At the site of inoculation, a single punched-out ulcer develops within 1 to 3 days. Lymphangitic spread from the ulcer to the draining nodes may be apparent as induration and erythema. Painful and swollen regional nodes may be the first sign of disease that the patient notices.

Clinical Manifestations

Systemic signs of disease that occur suddenly and early in the course of illness include back pain, anorexia, headache, chills and fever, sweating, and prostration. Nonproductive cough, nausea, vomiting, and abdominal pain also are common. Over 10 percent of patients develop a rash that may be macular, papular, or blotchy, is painless, is not pruritic and lasts a week. In the Scandinavian countries, a third of the patients with tularemia develop erythema nodosum, but this is rare in other parts of the world.

In untreated cases, the primary ulcer often suppurates and heals with scarring in 4 to 7 weeks. Lymphadenopathy is often of long duration, and convalescence usually extends from 3 to 6 months. Suppuration sometimes continues for years, and relapses are frequent. Before specific therapy became available, the mortality rate was about 10 percent, but it is now less than 1 percent. In areas of the world where less virulent disease is prevalent, such as Russia and Japan, low mortality rates have always prevailed.

Oculoglandular disease accounts for less than 5 percent of the cases and is similar to ulceroglandular except that the conjunctival sac is the primary site of inoculation. Infection commonly occurs during the skinning of an animal. In the early stages, a papule may be found in the conjunctiva. The histopathology is a granulomatous conjunctivitis that may suppurate. Involved lymph nodes are the preauricular, parotid, submaxillary, and cervical. Although blindness is uncommon, the disease often is accompanied by severe systemic signs of disease that may be fatal.

The two forms of the disease that do not present with a primary ulcer are glandular and typhoidal tularemia. Glandular disease may resemble sporotrichosis. Typhoidal disease is characterized by general and severe systemic signs and must be differentiated from typhoid fever, brucellosis, tuberculosis, and other generalized bacterial diseases. Blood and sputum cultures most frequently yield *F. tularensis*. Hepatosplenomegaly is common, and the mortality rate is particularly high. Diagnosis is more difficult because there is no history of contact with an appropriate vector or source.

Pulmonary disease commonly accompanies tularemia and may be the dominant feature. Roentgenographic findings, which are not specific for tularemia, include bronchopneumonia, pleural effusion, hilar adenopathy, nodular infiltrations, and peribronchial thickening. The etiology of these changes probably includes hematogenous spread to the lung parenchyma, extension from involved hilar lymph nodes, and, occasionally, inhalation of bacteria. Most of the 42 cases of laboratory-acquired tularemia included in Overholt's report were assumed to have resulted from inhaled infected particles. Persons with tularemic pulmonary involvement, however, seldom disseminate disease to other persons, and secondary spread is extremely rare. Other rare forms of tularemia have included meningitis following septicemia, gastrointestinal disease following ingestion of infected food, and congenital infection.

Immunity

Agglutinating antibody is usually found in the serum by the second to third week of primary illness and may persist for years. Both IgM- and IgG-agglutinating antibody is formed, and both typically persist indefinitely. Therefore, the presence of IgM antibody to *F. tularensis* does not nec-

essarily indicate that the infection is recent. The agglutinating and precipitating antibodies to *F. tularensis* probably confer little protective immunity to virulent strains, as persons and animals with high titers of these antibodies have subsequently acquired the disease.

The major factor in acquired resistance appears to be a cellular immune mechanism involving an activated macrophage population. Evidence for this comes from several observations:

1. Resistance of experimental animals to tularemia is conferred by immunization with viable, attenuated strains of organisms but not with killed organisms.
2. The presence of circulating antibody, either passively or actively acquired, does not necessarily enhance the ability to resist challenge.
3. Peritoneal macrophages from immune animals have shown enhanced destruction of *F. tularensis* and prolonged cell viability after phagocytosis as compared with cells from nonimmune animals. This phenomenon is little influenced by specific antibody.
4. Protective immunity to *F. tularensis* may be conferred by the transfer of macrophages from immune animals.
5. There is no direct relationship between the titer of agglutinating antibody and the presence or absence of a population of lymphocytes that are reactive to *F. tularensis* antigen.

A recently recognized aspect of host resistance to arthropod-borne tularemia is that host resistance to tick infestation alters transmission of *F. tularensis*. Sheep and other animals that have been sensitized to ticks have a greatly decreased death rate when exposed to tularemia-infected ticks, as compared with animals that have not been sensitized to ticks.

Skin Test. The preceding findings are consistent with the concept that *F. tularensis* is a facultative intracellular parasite of the reticuloendothelial system and that development of protective immunity is coincident with the development of cell-mediated immunity. A skin test using phenolized and diluted vaccine gives a delayed reaction within 7 days of the onset of disease in approximately 90 percent of persons with known tularemia. This skin test has been useful in epidemiologic studies on the prevalence of tularemia in endemic areas and in hazardous occupations. A reactive status persists for many years after infection or vaccination, long after circulating antibody has greatly decreased. The tularemia skin test does not elicit a response in persons previously infected with *Brucella* species without *F. tularensis* infection. The skin test also is more sensitive than most serologic tests, being reactive both earlier and longer. In one study, only 72 percent of persons with delayed hypersensitivity had demonstrable circulating antibody. This skin test antigen, however, is now very difficult to obtain.

Laboratory Diagnosis

Culture. Diagnosis of tularemia by culture of *F. tularensis* is uncommon because of the requirement for special media and frequent overgrowth by other organisms. Overgrowth may be partially controlled by the use of selective inhibitors, such as penicillin G and cycloheximide. Another obstacle in the recovery of *F. tularensis* from clinical specimens is the reluctance of many laboratories to attempt isolation because of the high rate of laboratory-acquired disease. This is especially true when animal inoculation is used. Even in laboratories well equipped to attempt isolation, however, the proportion of isolates to suspected cases is low when compared with most bacterial diseases. In patients with pulmonary foci of disease, pharyngeal secretions and morning gastric aspirates may yield an isolate, as may the primary ulcer and regional node.

Serologic Tests. Fluorescent antibody identification of *F. tularensis* organisms or antigen in tissues of naturally infected animals has shown good correlation with histopathologic findings and may be useful in situations where the organisms are nonviable, as in highly decomposed tissue. Under such circumstances, a thermostable antigen may be extracted from organisms or infected tissue and identified by a precipitin reaction with appropriate antibody.

The agglutination test is the most commonly used serologic method for the diagnosis of tularemia. It usually becomes positive by the second to fourth week of disease, at which time the titer rises sharply to a maximum of over 1:1280 in 4 to 8 weeks, and persists for a variable time. The skin test is usually reactive before the agglutination test.

The antibody response to infection or to vaccination with attenuated or killed *F. tularensis* includes production of agglutinins that cross-react with *Brucella abortus* and *Brucella melitensis*. The agglutinating titer to *Brucella* that is stimulated by *F. tularensis* is usually two to four dilutions lower than to *F. tularensis*. *Brucella* fail to absorb anti-*Francisella* agglutinins. Similarly, agglutinating antibody to *Brucella* cross-reacts with *F. tularensis*, but the titer to *Brucella* is not ablated by absorption with *F. tularensis*. Cross-reaction with fluorescent antibody appears to be less common than with agglutination.

Treatment

Streptomycin is the recommended treatment for tularemia. Bactericidal concentrations are readily achieved, and there is little clinical tendency to relapse. In contrast, clinical studies with both chloramphenicol and tetracyclines have demonstrated recurrent disease, and failure of these drugs to eradicate subcutaneous deposits of bacteria has been shown in volunteers. Treatment during early infection is more likely to result in eradication of bacteria than is treatment of chronic tularemia. Strains naturally resistant to streptomycin, chloramphenicol, or tetracycline are exceedingly rare. Penicillin is inefficacious.

Prevention

Prevention involves avoidance of animals likely to be infected, especially rabbits; protection from biting arthropods; and provision of clean water supplies. The natural foci of disease appear to be stable and are not likely to be eradicated. Rabbits sick with tularemia are more easily caught by dogs or cats and may therefore be brought into contact with human beings. Persons at risk include rabbit hunters; all persons who handle wet skins of potentially infected animals, such as muskrats and beavers; sheepshearers; and persons in endemic areas whose field work brings them into contact with rodents. Precautionary measures may include wearing clothing with secure ankles and wrists to protect against attachment of ticks.

At high risk are all laboratory personnel handling cultures of *F. tularensis* and infected laboratory animals. Persons likely to be exposed in this way, including the animal handlers, should consider receiving live vaccine.

Immunization. Attenuated live tularemia vaccine is available through the United States Army Biological Laboratories and Centers for Disease Control. Experimental and clinical experience with the use of this vaccine has been acquired in Russia, Japan, and the United States. The tularemia vaccine and the vaccine of bacillus-Calmette and Guerin (BCG) for tuberculosis are the only currently licensed live bacterial vaccines. Use of the tularemia vaccine should be restricted to persons whose risk of exposure is high, such as sheepherders, sheepshearers, trappers, and laboratory personnel.

The vaccine usually is administered by scarification, producing a papular-vesicular lesion at the site of the take, similar to that in smallpox vaccination. Reactions to vaccine have not been severe, but regional lymphadenitis does occur. Vaccinated persons promptly develop cutaneous delayed hypersensitivity. Most vaccinees show an antibody response that reaches a peak within 1 month for half of the vaccinees, and within 2 months in the remainder. The infective dose of *F. tularensis* in vaccinated persons is several logarithms higher than it is for nonimmune persons, which is estimated to be 10 to 50 virulent organisms. When clinical disease does occur in previously vaccinated individuals, it is modified.

FURTHER READING

Books and Reviews

Reilly JR: Tularemia. In Davis JW (ed): Infectious Diseases of Wild Mammals. Ames, Iowa, Iowa State University Press, 1970, p 175

Selected Papers

Bell JF, Stewart SJ, Wikel SK: Resistance to tick-borne *Francisella tularensis* by tick-sensitized rabbits: Allergic flendusity. Am J Trop Med Hyg 28:876, 1979

Bell TF: Ecology of tularemia in North America. J Jinsen Med 11:34, 1965

Buchanan TM, Brooks GF, Brachman PS: The tularemia skin test: 325 skin tests in 210 persons: Serologic correlation and review of the literature. Annu Intern Med 74:336, 1971

Claflin JL, Larson CL: Infection-immunity in tularemia: Specificity of cellular immunity. Infect Immun 5:311, 1972

Dahlstrand S, Ringertz O, Zetterberg B: Airborne tularemia in Sweden. Scand J Infect Dis 3:7, 1971

Francis E: Symptoms, diagnosis and pathology of tularemia. JAMA 91:1155, 1928

Jellison WL, Owen CR, Bell JF: Tularemia and animal populations: Ecology and epizoology. Wildl Dis 17:22, 1961

Koskela P, Herva E: Cell-mediated immunity against *Francisella tularensis* after natural infection. Scand J Infect Dis 12:281, 1980

Overhold EL, Tigertt WD, Kadull PJ, et al.: An analysis of 42 cases of laboratory acquired tularemia. Treatment with broad spectrum antibiotics. Am J Med 30:785, 1961

Young LS, Bicknell DS, Archer BG, et al.: Tularemia epidemic: Vermont, 1968. Forty-seven cases linked to contact with muskrats. N Engl J Med 280:1253, 1969

CHAPTER 42

Pasteurella, Actinobacillus, Streptobacillus, and Calymmatobacterium

Pasteurella

Organisms of the genus *Pasteurella* have a wide host spectrum and cause epidemic and septicemic diseases of domestic animals and birds. Humans also may be infected. The genus comprises an extremely heterogeneous group of organisms, but only four species are currently recognized: *Pasteurella multocida, Pasteurella pneumotropica, Pasteurella haemolytica,* and *Pasteurella ureae. P. multocida,* the major human pathogen of the genus, derives its name from its wide range of hosts, i.e., "many killing," and now includes as minor biotypes those organisms previously classified as different species because of their isolation from different animal hosts.

Pasteurella multocida

Morphology and Physiology

Pasteurella are small, nonmotile, ovoid, or rod-shaped organisms, approximately 1.4 by 0.4 μm in size. They are gram-negative and show bipolar staining, especially in preparations from infected tissue. From clinical material, they may be pleomorphic and occur singly or in clusters or chains. Virulent *P. multocida* organisms are encapsulated.

Media containing blood or hematin should be used for culture. *P. multocida* is nonhemolytic but may produce a brownish discoloration of the medium in areas of confluent growth. The optimum temperature for growth is 37C, with growth occurring between 25C and 40C. *Pasteurella* does not grow in bile-containing media and does not require the X or V factors.

Pasteurella species are facultatively anaerobic. They are catalase-positive and usually oxidase-positive. Their metabolism is fermentative, and acid is produced by most strains from glucose, mannitol, and sucrose. Distinguishing properties of *P. multocida* are shown in Table 42-1. Biotyping of strains based upon fermentation patterns of maltose, mannitol, xylose, sorbitol, and trehalose has yielded 11 biotypes. Cat-related strains are primarily types A and B, whereas dog-related strains are very heterogeneous.

Colony Morphology
Four principal colony variations occur: mucoid, smooth (iridescent), smooth (noniridescent), and rough. Most mucoid colonies are serotype A. There is a relationship among colony morphology, acriflavin reaction, and hemagglutination. Flocculation with acriflavin is associated with rough or noniridescent colonies that fail to hemagglutinate and are deficient in capsular material. Cells that remain suspended in acriflavin contain capsular material that is typeable by indirect hemagglutination, and are mucoid or smooth (iridescent).

Determinants of Pathogenicity
Encapsulated organisms are usually pathogenic for mice, but other ill-defined factors, including the somatic antigens, play an important role in pathogenicity. Virulence is greatly enhanced in vivo by the provision of free iron in the form of ferric ammonium citrate, hematin, lysed mouse erythrocytes, and purified hemoglobin. Conversely, one serum component that participates in the bacteriostasis of *P. multocida* contains transferrin. Interference with the iron metabolism of the organism is believed to be a significant factor in host resistance to infection.

The cell walls of both *P. multocida* and *P. haemolytica* contain significant endotoxin activity. No exotoxin has been demonstrated. Some strains produce hyaluronidase, especially those isolates that come from animals with hemorrhagic septicemia of animals.

Antigenic Structure
Five major antigenic types of *P. multocida* have been identified and designated A to E on the basis of their capsular or surface polysaccharide, using an indirect hemagglutina-

tion assay. In this scheme, types A and D are most common among human isolates. Typing by use of standard agglutination procedures is usually complicated by the presence of common somatic antigens, and, in mucoid strains, by capsular hyaluronic acid. Somatic, or O antigen typing, has also been achieved using a precipitation technique. According to this scheme, which contains 13 serotypes, human strains are most commonly types 1, 3, 4, 6, 12, and 13. *P. multocida* strains may be characterized by their somatic O antigen and capsular components, and the various serotypes assigned formulas such as 1:A, 2:A, 1:B, where the numeral designates the somatic type and the letter designates the capsular type. There is a broad relationship between serologic type and host distribution when both capsular and somatic type are identified.

Clinical Infection
Epidemiology. The *Pasteurella* species are present as normal flora in many domestic animals. *P. multocida* occupies an ecologic niche in the nasopharynx of the cat similar to that of the α-hemolytic *Streptococcus* in humans. It survives poorly in soil and water and is transmitted most commonly by direct contact, usually a bite. The tonsils of dogs are a site commonly colonized with *P. multocida*. The organism is more common in young male dogs, and the rate of colonization is higher in the cold seasons. Rates of colonization for the dog vary, but the majority of cats harbor *P. multocida*, as do many other domestic animals. Colonized animals are generally asymptomatic, and the disease usually occurs during stressful situations, such as the shipping of cattle, when highly virulent organisms may be recovered from diseased animals. *P. multocida* is responsible for outbreaks of cholera of domestic or wild fowl, hemorrhagic septicemia of cattle, and primary and secondary pneumonias.

P. multocida types A and D are widely distributed in nature, and most respiratory tract disease in humans is caused by type A. Types B and C have been recovered primarily from cattle, bison, and buffalo. Organisms recovered from human disease following dog or cat bite or scratch are usually nontypeable and rough, and lack pathogenicity for mice.

Pathogenesis and Clinical Manifestations. Human disease caused by *P. multocida*, or by other members of the genus that are rarely pathogenic for humans, may be of

TABLE 42-1. PROPERTIES OF THE SPECIES OF GENUS *PASTEURELLA*

Characteristic	*P. multocida*	*P. pneumotropica*	*P. haemolytica*	*P. ureae*
Hemolysis on blood agar	−	−	+	+
Growth on MacConkey's agar	−	−	+	−
Indole	+	+	−	−
Urease	−	+	−	+
Mannitol	+	−	+	+

−, 90 percent or more strains negative; +, 90 percent or more strains positive.

three types: (1) infection via bites or scratches, (2) superinfection of a chronically diseased lung, and (3) other foci of disease that are secondary to septicemia.

Animal bites are common and frequently require medical attention. *P. multocida* can be recovered from approximately one half of infected animal bites. In addition, half of the wounds that initially are colonized with *P. multocida* develop frank infection. Wounds that have been sutured are particularly prone to develop infection with the organism. Complications of bites infected with *P. multocida* are common. Cat bites may progress to pyarthrosis, necrotizing synovitis, and osteomyelitis of the underlying bone, presumably due to the depth of the bite and associated trauma of adjacent tissue. Regional lymphadenitis, with severe local pain, swelling, and discoloration, follows the often sudden onset of signs of infection.

Septicemia occurs primarily in persons with underlying disease that impairs reticuloendothelial function, such as cirrhosis of the liver and rheumatoid arthritis, but also has been reported in apparently normal persons.

The second most common form of human disease with *P. multocida* is infection of the lung in patients with preexisting chronic pulmonary disease. These patients are usually middle-aged or older. Lower respiratory tract diseases associated with recovery of *P. multocida* are bronchiectasis, bronchogenic carcinoma, chronic bronchitis, emphysema, pulmonary abscess, and pneumonia, including sinusitis, mastoiditis, and chronic otitis media. *P. multocida* also has been recovered from upper respiratory tract infections. Disease with *P. multocida* in these settings of previously chronic disease may present with asymptomatic colonization, insidious progression to apparent pulmonary disease, or with an acute or fulminant onset.

Other sites of infection with *P. multocida* are unusual and represent either hematogenous dissemination during bacteremia or local extension. Examples are meningitis, chorioamnionitis with premature delivery, cerebellar abscess, and infectious endocarditis.

Laboratory Diagnosis. Identification of the causative agent requires culture of appropriate specimens from the patient, depending on the area of involvement. Early morning sputum, bronchial washing, nasal swabs from respiratory tract infections, purulent exudate from animal bites, spinal fluid, and repeated blood cultures are appropriate for this purpose.

Treatment. Among the gram-negative rods, the *Pasteurella* species are unusual in their uniform sensitivity to penicillin. The mean inhibitory concentrations of tetracycline, ampicillin, and benzyl penicillin are less than 0.5 μg/ml. As with other pyogenic infections, localized abscesses must be drained. Following an animal bite, the wound should be meticulously cleaned, and suturing should be avoided. Initiation of treatment with penicillin or tetracycline at the time of the bite is recommended by many physicians.

Prevention. A number of vaccines and antisera preparations have been used in attempts to control veterinary disease due to these organisms. These have lacked efficacy, however, and are no longer used. More precise knowledge concerning the antigenic structure of strains causing particular veterinary diseases may make possible the production of more useful products.

Actinobacillus

Species in the genus *Actinobacillus* cause acute septicemia or granulomatous lesions in cattle and sheep. One ill-defined species now associated with this genus as species incertae sedis is *Actinobacillus actinomycetemcomitans*, a human pathogen. This is a small, nonmotile, nonencapsulated, gram-negative coccobacillus. It grows best on serum or blood agar in an atmosphere of 10 percent carbon dioxide. Growth is optimal at 37C. Colonies on agar are about 1 mm in diameter after 2 or 3 days, are star-like, and are adherent to the agar. Growth in broth is granular.

Clinical Infection. Most clinical isolates of *A. actinomycetemcomitans* have been from infected blood and bone. Although some isolates have been obtained from lesions also infected with *Actinomyces* species, from which the organism's name is derived, most isolates have been obtained in the absence of *Actinomyces*. The organism is a normal inhabitant of the human mouth. The apparent etiology of some cases of human *Actinobacillus* disease is related to impaired host defenses in such diseases as malignant lymphoma and leukemias. A number of cases of subacute bacterial endocarditis due to *A. actinomycetemcomitans* also have been described.

Streptobacillus

The primary human pathogen of this genus is *Streptobacillus moniliformis*, the cause of one type of rat-bite fever and of a milk-borne disease known as Haverhill fever. *Spirillum minus* (Chap. 48) also causes rat-bite fever, and characteristics distinguishing these organisms and their clinical diseases are summarized in Table 42-2.

Morphology and Physiology. *S. moniliformis* is a nonencapsulated, nonmotile, gram-negative bacillus, 0.3 to 0.7 μm by 1 to 5 μm in length. The organism frequently occurs in chains and filaments 10 to 150 μm long. It is often pleomorphic and may produce a series of bulbous swellings (Fig. 42-1). It is a facultative anaerobe and requires CO_2 and moisture for primary isolation. The most distinguishing characteristic of the organism is the sponta-

TABLE 42-2. DISTINGUISHING CHARACTERISTICS OF TWO CAUSES OF RAT-BITE FEVER IN HUMANS

Characteristic	Streptobacillus moniliformis	Spirillum minor
Bacteriology	Microaerophilic gram-negative pleomorphic bacillus	Gram-negative spiral organism
Diagnostic serology	Agglutinins	Not available
Incubation period	Usually 1 to 3 days	Usually more than 7 days
Source of human disease	Animal bite from wild or laboratory rat, animal bite from mouse, cat, dog, contaminated food	Animal bite from rat or cat
Rash	Morbilliform, maculopapular, or petechial on palms, soles, extremities	Violaceous macules become confluent on palms, soles, extremities
Fever	Usually septic	Usually relapsing, with fever lasting 3 to 4 days occurring at irregular intervals
Epidemiology	Poor sanitation with heavy rodent infestation, occupational exposure	Same as *S. moniliformis*
Treatment	Penicillin or tetracycline	Same as *S. moniliformis*

neous development of L-phase variants during in vitro culture (Fig. 42-2). The designation L-phase was first used by Klieneberger–Nobel to refer to the pleuropneumonia-like organism associated with *S. moniliformis* in cultures. It is now known that most bacteria and fungi are capable of similar changes, but this organism remains the prototype of this phenomenon.

Culture of these organisms requires media supplemented with serum, ascitic fluid, or blood. In liquid media, typical puffballs of growth will be present in the bottom of the tube. Growth is usually slow and may take several days to become apparent. Fermentation patterns are variable, but often include glucose, maltose, galactose, and salicin.

Clinical Infection. Human disease caused by *S. moniliformis* is uncommon and is related to poor living conditions. Only 70 cases were reported in the United States through 1965. *S. moniliformis* is a common inhabitant of the nasopharynx of wild and laboratory rats. Although the disease usually is acquired by the bite of a rat or other rodent, it also may be transmitted by means of milk, water, and food. This disease is frequently recognized in laboratory workers who have been bitten by a rat or other ro-

dent, and in children who live in conditions allowing them to be bitten by rats.

The disease begins 1 to 5 days after introduction of the organism. The onset is abrupt, with chills, fever, vomiting, headache, and severe pain in the joints. A maculopapular rash develops within the first 48 hours, and one or more joints become swollen and painful. The rash characteristically is morbilliform, maculopapular, or petechial and involves the palms, soles, and extremities. Acute arthritis is a characteristic and persistent symptom. Multiple abscesses in many organ systems characterize severe forms of the disease.

LABORATORY DIAGNOSIS. Diagnosis is made by isolation of the organism from the blood or from joint fluids and pus. Agglutinins appear in the patient's serum within 10 days and reach a maximum in 3 or 4 weeks. A titer of 1:80 is considered diagnostic, and a four-fold rise in titer is significant.

TREATMENT. Penicillin, tetracycline, and streptomycin have been successfully used in treatment, and penicillin appears to be the drug of choice. If the disease is not treated, the mortality rate may reach 10 percent or higher.

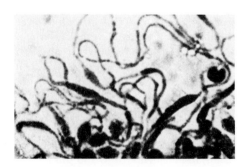

Figure 42-1. Edge of a colony of *Streptobacillus moniliformis*. × 3000. *(From Sharp: The Role of Mycoplasmas and L Forms of Bacteria in Disease, 1970. Springfield, Ill, Charles C Thomas.)*

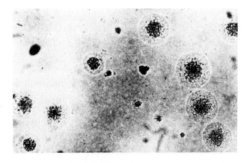

Figure 42-2. L-form colony of *Streptobacillus moniliformis* produced spontaneously. × 100. *(From Sharp: The Role of Mycoplasmas and L Forms of Bacteria in Disease, 1970. Springfield, Ill, Charles C Thomas.)*

Calymmatobacterium

This genus contains a single species, *Calymmatobacterium granulomatis*, the etiologic agent of granuloma inguinale. This is a chronic mildly infectious disease characterized by ulcerating lesions of the skin and mucosa of the genital and inguinal areas.

Morphologic and Cultural Characteristics. The generic name of the organism is derived from the Greek word *calymma,* meaning sheath or mantle. *C. granulomatis* is a pleomorphic, gram-negative, nonmotile rod, 1 to 2 μm in length. It is usually heavily encapsulated and exhibits single or bipolar condensation of chromatin, a property that gives rise to the characteristic appearance of a safety pin when gram stained. After primary isolation on fresh egg yolk medium, the organism may be adapted to growth on artificial culture media. *C. granulomatis* is antigenically very similar to, but not identical with, *Klebsiella* species.

Within the host *C. granulomatis* replicates intracellularly. The pathognomonic cell is a large, histiocytic endothelial cell containing numerous encapsulated bacilli, the Donovan bodies. These were first described in 1905 by Donovan, who observed them in epithelial cells of the skin of infected persons in India.

Clinical Infection. The disease associated with *C. granulomatis*, granuloma inguinale, is usually classified as one of the minor venereally transmitted diseases. However, evidence of direct transmission by sexual contact is poor. It is an inhabitant of the intestinal tract, and it is possible that skin disease is contracted through poor hygiene and abrasions of the skin in anal and perineal regions. It is primarily a disease of tropical and subtropical areas, but is occasionally diagnosed in the southern United States. Communicability is apparently not high, as the disease is rarely transmitted to sexual partners. It occurs primarily in persons who are promiscuous and have had other venereally transmissible diseases. Although experimental infection in animals has not been achieved, infection in humans has been obtained by inoculation of diseased tissue or exudate.

The initial lesions of granuloma inguinale may occur in any of the pubic areas. They begin as painless papules that develop into spreading ulcerating lesions, which may bleed and become secondarily infected. The lesions spread by direct extension, or by contact of one skin area with another, such as between the scrotum and thigh. The patient is unlikely to have constitutional symptoms and remains afebrile. However, the infection may progress very slowly for long periods of time, and heal with scarring. The scarring may lead to genital elephantiasis if treatment is not provided. Metastatic lesions to joints, bones, and liver have been reported.

Diagnosis depends upon the demonstration of Donovan bodies by direct examination of biopsy material that has been stained with the Wright or Giemsa stain. Culture may be attempted, but initial isolation is difficult.

The infection readily responds to antimicrobial therapy. In most instances streptomycin, tetracycline, chloramphenicol, gentamicin, and cotrimoxazole have been shown to be effective.

Other Gram-negative Bacilli

A number of other facultative gram-negative organisms may be encountered in the clinical microbiology laboratory, often from compromised patients or from patients with hospital-acquired infections. These organisms constitute a heterogenous group whose taxonomy is currently in a state of flux and whose isolation from clinical materials frequently causes problems of identification.

Cardiobacterium. The only species within this genus is *Cardiobacterium hominis*, a gram-negative rod, 0.5-μm wide and 1- to 3-μm long, arranged singly, in pairs, short chains, or clusters. On blood agar containing yeast extract, organisms stain homogeneously and are uniform in size, but in the absence of yeast extract, pleomorphic forms with enlarged ends or rosette clusters may occur.

C. hominis grows well on enriched media such as trypticase soy agar, chocolate agar, or heart infusion agar at 35C and in an atmosphere of 3 to 5 percent carbon dioxide. A high humidity requirement is growth limiting. Growth is characteristically very slow, with minute punctiform colonies visible after 24 hours and attaining a maximum size of 1 to 2 mm in 3 to 4 days. No hemolysis is observed, but in areas of heavy growth a slight greening of the blood agar may develop. *C. hominis* does not grow on MacConkey's agar and other selective media used for the isolation of the Enterobacteriaceae. *C. hominis* is oxidase positive, catalase negative, and nonmotile. It is fermentative and produces acid throughout triple sugar iron (TSI) medium. Other biochemical properties are listed in Table 42-3.

C. hominis is a part of the normal flora of the human upper respiratory tract and can be detected in the feces by immunofluorescence. Infection occurs primarily in compromised individuals and has been described most frequently in association with endocarditis and bacteremia. In cases of endocarditis, the infection has usually been chronic and has manifested itself in patients with damaged heart valves or prosthetic heart valves. Other diseased sites from which *C. hominis* is being cultured with increasing frequency include the cervix, vagina, empyema fluid, spinal fluid, mandible, and sputum. Penicillin and ampicillin have been used successfully in the treatment of *C. hominis* infection.

Chromobacterium. Organisms in this genus are gram-negative motile rods resembling pseudomonads, which produce violet colonies on suitable media. On blood agar, the colonies may appear to be black. Two species are recognized, *Chromobacterium violaceum*, a facultative anaerobe, and *Chromobacterium lividum*, a strict anaerobe. *C. violaceum* grows on MacConkey's agar. Most strains can be readily identified by pigmentation and the fermentative utilization of carbohydrates. Biochemical properties are listed in Table 42-3.

C. violaceum is an inhabitant of soil and water and is especially common in tropical countries. Although usually considered to be nonpathogenic, it has been shown to cause severe pyogenic or septicemic infections of mammals, including humans. Infection with this organism has been confined primarily to tropical and subtropical regions. In the United States, all cases have occurred in the southeast. The organism is usually introduced through injury to the skin resulting in localized abscesses and a very slowly evolving systemic disease. Infection is usually characterized by liver abscess, sepsis, and death. *C. violaceum* may also gain entry to the body via the gastrointestinal tract and is associated with diarrhea. The organism is sensitive to aminoglycosides, chloramphenicol, and tetracycline.

Capnocytophaga. These recently characterized oral organisms are fastidious gram-negative flexible rods, fusiform in shape, and with a gliding motility. They have a fermentative metabolism, are oxidase- and catalase-negative, and grow under anaerobic or aerobic conditions in the presence of 5 to 10 percent CO_2. Increased CO_2 is essential for aerobic growth. Growth is characteristically slow. On blood agar at 37C, colonies that are not visible to the unaided eye at 24 hours become 2 to 3 mm in diameter after 2 to 4 days. The genus has been divided into 3 distinct species based on physiologic and morphologic properties: *Capnocytophaga ochracea* (formerly *Bacteroides ochracea*), *Capnocytophaga sputigena*, and *Capnocytophaga gingivalis*. Group-specific and type-specific antigens have been identified, and various cell envelope components have been shown to have immunomodulating activity.

Considerable interest has recently been focused on this organism because it has been shown to be a predominant cultivable organism in the advancing front of periodontal lesions, and has been shown to be capable of causing periodontal disease, including bone loss, when implanted into gnotobiotic rats (p. 753). *Capnocytophaga* has also been isolated from a variety of clinical specimens in association with systemic disease in compromised hosts.

Eikenella corrodens. This is a small (0.5 by 1 to 3 μm) nonmotile, gram-negative, facultatively anaerobic rod that has only recently been taxonomically separated from the obligate anaerobe, *Bacteroides corrodens*. It does not grow on MacConkey's agar or other similar selective media, but grows on hemin-containing media such as blood or chocolate agars. Growth, which is enhanced in an atmosphere of 3 to 10 percent CO_2, is characteristically slow, with small pinpoint colonies usually visible after 24 hours of aerobic incubation. A distinctive feature observed with about 45 percent of the isolates is a pitting of the agar under the colony after several days of growth. *E. corrodens* is oxidase positive but is biochemically inactive, lacking both oxidative and fermentative capabilities (Table 42-4).

E. corrodens is a part of the indigenous flora of the mouth and upper respiratory tract, as well as other mucosal surfaces of the human body. It is isolated most commonly from respiratory specimens but may also be isolated from a variety of other materials such as blood, spinal fluid, abscesses, or joint aspirates. Although *E. corrodens* is usually associated with polymicrobic infections, it may also be the sole cause of infection. Infections are usually the result of predisposing factors that compromise the body's host defense mechanisms and permit the organism to penetrate surrounding tissue. Hematogenous spread from this primary infection and establishment of foci in other body sites may then occur. *E. corrodens* is sensitive to penicillin, tetracycline, colistin, and chloramphicol but is resistant to clindamycin.

TABLE 42-3. PROPERTIES OF *CAPNOCYTOPHAGA* SPECIES, *CARDIOBACTERIUM HOMINIS*, AND *CHROMOBACTERIUM VIOLACEUM*

Property	Capnocytophaga species	Cardiobacterium hominis	Chromobacterium violaceum*
β Hemolysis	−	−	v
Motility	v	−	+
Acid from			
Glucose	+	+ or (+)	+
Lactose	v	−	−
Sucrose	+	+ or (+)	v
Catalase	−	−	+
Oxidase	−	+	v
Growth on MacConkey's agar	−	−	+

From Weaver and Hollis: In Lennette et al. (eds): Manual of Clinical Microbiology, 3rd ed., 1980. Washington, D.C., American Society for Microbiology.
*92 percent produce a violet color. +, 90 percent or more positive; −, no reaction; v, more than 10 percent and less than 90 percent; (+), some strains positive after 3 or more days.

TABLE 42-4. BIOCHEMICAL PROPERTIES OF *EIKENELLA CORRODENS*

Test or Substrate	Reaction
Oxidase	+
Catalase	−
Growth on	
MacConkey's agar	−
Salmonella–Shigella agar	−
Oxidation-fermentation test	I
Urease	−
Indole	−
Motility	−
Nitrate to nitrite only	+
Pigment (pale yellow)	+

From Rubin, et al.: In Lennette, et al. (eds): Manual of Clinical Microbiology, 3rd ed., 1980. Washington D.C., American Society for Microbiology. +, 90 percent or more positive; −, no reaction in 90 percent or more; I, inactive, 100 percent.

Flavobacterium. The genus *Flavobacterium* is a taxonomically heterogenous group, but as defined by the current Centers for Disease Control classification schema, only four members are included in the genus: *Flavobacterium meningosepticum, Flavobacterium odoratum, Flavobacterium breve,* and an unnamed species of *Flavobacterium* designated group IIb. *F. meningosepticum* is the species most commonly associated with human infection. Flavobacteria are long thin nonmotile gram-negative rods with slightly swollen ends. They are oxidase-positive, proteolytic, and very weakly fermentative. After 18 to 24 hours of incubation on blood agar, the organisms produce colonies 1 mm in diameter and with a distinctive lavender–green discoloration of the red cells. *F. meningosepticum* colonies usually have a slight yellow pigment, which can be intensified by growth on nutrient agar at room temperature. The other flavobacteria have a more intense yellow color.

The flavobacteria are widely distributed in nature in soil and water, and normally are not a part of the indigenous human flora. They are often found in the hospital environment and have been isolated from a number of reservoirs such as humidifiers, ice machines, water baths, saline solutions, respiratory equipment, incubators, and indwelling catheters, all of which may serve as a source for nosocomial infections. Most of these infections are caused by *F. meningosepticum,* and occur primarily in neonates. Meningitis is the primary manifestation and the mortality rate is high. Infection in adults is generally mild, and usually follows operative or manipulative procedures. Isolates of *F. meningosepticum* vary in their antibiotic sensitivity patterns and are usually resistant to most antimicrobial agents.

FURTHER READING

PASTEURELLA

Carter GR: Pasteurellosis: *Pasteurella multocida* and *Pasteurella haemolytica*. Adv Vet Sci Comp Med 11:321, 1967

Carter GR, Chengappa MM: Hyalunonidase produced by type B *Pasteurella multocida* from cases of hemorrhagic septicemia. J Clin Microbiol 11:94, 1980

Heddleston KL, Wessman G: Characteristics of *Pasteurella multocida* of human origin. J Clin Microbiol 1:377, 1975

Holloway WJ, Scott EG, Adams YB: *Pasteurella multocida* infection in man. Am J Clin Pathol 51:705, 1969

Holmes MA, Brandon G: *Pasteurella multocida* infections in 16 persons in Oregon. Public Health Rep 80:1107, 1965

Hubbert WT, Rosen MN: I. *Pasteurella multocida* infection due to animal bite. II. *Pasteurella multocida* infection in man unrelated to animal bite. Am J Public Health 60:1103, 1109, 1970

Lee MLH, Buhr AJ: Dog bites and local infection with *Pasteurella septica*. Br Med J 1:169, 1960

Murata M, Horiuchi T, Namioka S: Studies on the pathogenicity of *Pasteurella multocida* for mice and chickens on the basis of O-groups. Cornell Vet 54:294, 1964

Oberhofer TR: Characteristics and biotypes of *Pasteurella multocida* isolates from humans. J Clin Microbiol 13:566, 1981

Smith JE: Studies on *Pasteurella septica*. I. The occurrence in the nose and tonsils of dogs. J Comp Pathol 65:239, 1955

Syuto B and Matsumoto M: Purification of protective antigen from a saline extract of *Pasteurella multocida*. Infect Immun 37:1218, 1982

Tindall JP, Harrison CM: *Pasteurella multocida* infections following animal injuries, especially cat bites. Arch Dermatol 105:412, 1972

ACTINOBACILLUS

Affias SA, West A, Stewart J, et al. *Actinobacillus actinomycetemcomitans* endocarditis. Can Med Assn J 118:1256, 1978

Baehni P, Tsai C-C, McArthur WP, et al.: Interaction of inflammatory cells and oral microorganisms. VIII. Detection of leukotoxic activity of a plaque-derived gram-negative microorganism. Infect Immun 24:233, 1979

Ellner JJ, Rosenthal MS, Lerner PI, et al.: Infective endocarditis caused by slow-growing, fastidious, gram-negative bacteria. Medicine 58:145, 1979

Holt SC, Tanner ACR, Socransky SS: Morphology and ultrastructure of oral strains of *Actinobacillus actinomycetemcomitans* and *Haemophilus aphrophilus*. Infect Immun 30:588, 1980

King EO, Tatum HW: *Actinobacillus actinomycetemcomitans* and *Haemophilus aphrophilus*. J Infect Dis 111:85, 1962

Stevens RH, Hammond BF, Lai CH: Characterization of an inducible bacteriophage from a leukotoxic strain of *Actinobacillus actinomycetemcomitans*. Infect Immun 35:343, 1982

Taichman NS, Dean RT, Sanderson CJ: Biochemical and morphological characterization of the killing of human monocytes by a leukotoxin derived from *Actinobacillus actinomycetemcomitans*. Infect Immun 28:258, 1980

Vandepitte J, DeGeest H, Jousten P: Subacute bacterial endocarditis due to *Actinobacillus actinomycetemcomitans*. Report of a case with review of the literature. J Clin Pathol 30:842, 1977

STREPTOBACILLUS

Cole JS, Stoll RW, Bulger RJ: Rat-bite fever: Report of three cases. Ann Intern Med 71:979, 1969

Holden FA, MacKay JC: Rat-bite fever—an occupational hazard. Can Med Assoc J 91:78, 1964

Raffin BJ, Freemark M: Streptobacillary rat-bite fever: A pediatric problem. Pediatrics 64:214, 1979

Roughgarden JW: Antimicrobial therapy of rat-bite fever. Arch Intern Med 116:39, 1965

CALYMMATOBACTERIUM

Anderson K: The cultivation from granuloma inguinale of a microorganism having the characteristics of Donovan bodies in the yolk sac of chick embryos. Science 97:560, 1943

Davis CM: Gramuloma inguinale. A clinical, histological and ultrastructural study. JAMA 211:632, 1970

Kuberski T, Papadimitriou JM, Phillips P: Ultrastructure of *Calymmatobacterium granulomatis* in lesions of granuloma inguinale. J Infect Dis 142:744, 1980

Kuberski T: Granuloma inguinale (Donovanosis). Sex Transmit Dis 7:29, 1980

OTHER GRAM-NEGATIVE BACILLI

Cabrera HA, Davis GH: Epidemic meningitis of the newborn caused by flavobacteria. I. Epidemiology and bacteriology. Am J Dis Child 101:289, 1961

Centers for Disease Control: National nosocomial infections study quarterly report, third and fourth quarters—1973. Atlanta, Centers for Disease Control, 1975

Centers for Disease Control: Multiple abscesses and death due to *Chromobacterium violaceum*-Florida. Morbidity and Mortality Weekly Report 23:387, 1974

Coyle-Gilchrist MM, Crewe P, Roberts G: *Flavobacterium meningosepticum* in the hospital environment. J Clin Pathol 29:824, 1976

Dorff GJ, Jackson LJ, Rytel MW: Infections with *Eikenella corrodens*: Newly recognized human pathogen. Ann Intern Med 80:305, 1974

Geraci JE, Greipp PR, Wilkowske CJ, et al.: *Cardiobacterium hominis* endocarditis: Four cases with clinical and laboratory observations. Mayo Clin Proc 53:49, 1978

Jackson FL, Goodman YE: Transfer of the facultatively anaerobic organism *Bacteroides corrodens* Eiken to a new genus, *Eikenella*. Int J Syst Bacteriol 22:73, 1972

Johnson WM, DiSalvo AV, Steuer PR: Fatal *Chromobacterium violaceum* septicemia. Am J Clin Pathol 56:400, 1971

Leadbetter ER, Holt SC, Socransky SS: *Capnocytophaga*: New genus of gram-negative gliding bacteria. I. General charactistics, taxonomic considerations and significance. Arch Microbiol 122:9, 1979

Murayama Y, Muranishi K, Okada H, et al.: Immunological activities of *Capnocytophaga* cellular components. Infect Immun 36:876, 1982

Newman MG, Sutter VL, Pickett MJ, et al.: Detection, identification, and comparison of *Capnocytophaga*, *Bacteroides ochraceus*, and DF-1. J Clin Microbiol 10:557, 1979

Olsen H, Frederiksen WC, Sibbon KE: *Flavobacterium meningosepticum* in 8 nonfatal cases of postoperative bacteremia. Lancet 1:1294, 1965

Rubenstein JE, Leiberman MF, Gadoth N: Central nervous system infection with *Eikenella corrodens*: Report of two cases. Pediatrics 57:264, 1976

Rubin SJ, Granato PA, Wasilauskas BL: In Lennette EH, Balows A, Hausler WJ Jr, et al. (eds): Manual of Clinical Microbiology, 3rd ed. Washington D.C., American Society for Microbiology, 1980, p 263

Savage DD, Kagan RL, Young NA, et al.: *Cardiobacterium hominis* endocarditis: Description of two patients and characterization of the organism. J Clin Microbiol 5:75, 1977

Slotnick IJ, Dougherty M: Further characterization of an unclassified group of bacteria causing endocarditis in man: *Cardiobacterium hominis*. Antonie van Leeuwenhoek 30:261, 1964

Stamm WE, Colella JJ, Anderson RL, et al.: Indwelling arterial catheters as a source of nosocomial bacteremia. An outbreak caused by *Flavobacterium* species. N Engl J Med 292:1099, 1975

Stevens RH, Hammond BF, Lai CH: Group and type antigens of *Capnocytophaga*. Infect Immun 23:532, 1979

Weaver RE, Hollis DG: In Lennette EH, Balows A, Hausler WJ Jr, et al. (eds): Manual of Clinical Microbiology, 3rd ed. Washington D.C., American Society for Microbiology, 1980, p 242

Zinner SH, Daly AK, McCormack WM: Isolation of *Eikenella corrodens* in a general hospital. Appl Microbiol 24:705, 1973

Brucella

Malaise, anorexia, fever, and profound muscular weakness characterized a debilitating illness first recognized by Marston in 1861 as "gastric remittent fever." The responsible organism, *Micrococcus melitensis*, was isolated in 1887 by Sir David Bruce. The organism derived its species name from Melita, the Roman name for the Isle of Malta, where the disease was recognized. The classic description of the clinical illness by Hughes in 1897 altered its designation to the more frequently used term, undulant fever. The subsequent, rapid acquisition of knowledge of brucellosis was a direct result of the morbidity sustained by the armed forces of Great Britian, who used Malta as a major military site in the first part of the 20th century.

Recognition of infections with other members of the genus *Brucella* occurred independently. Nocard, in 1862, first recognized the presence of bacteria between the fetal membranes and the wall of the uterus of the pregnant cow, but it remained for Bang, a Danish veterinarian, to isolate the organism, *Brucella abortus*. In a report of his findings in 1897, Bang linked the organisms to infectious abortions in animals. The third member of the genus was identified in 1914 when Traum isolated *Brucella suis* from a premature pig.

The relationship of these three organisms was unknown until Alice Evans, working for the Dairy Division of the Bureau of Animal Industry, noted the close bacteriologic and serologic relationships between *M. melitensis* and *B. abortus*. As a result of her observations, the genus *Brucella* was recognized and named in honor of Sir David Bruce.

Brucellae are mammalian parasites and pathogens with a relatively wide host range. They are facultatively intracellular. Six species are currently recognized: *Brucella melitensis*, *B. abortus*, *B. suis*, *Brucella neotomae*, *Brucella ovis*, and *Brucella canis.*

Morphology and Physiology

Brucella are small, nonmotile rods, usually coccobacillary but with a size range of 0.5 to 0.7 μm by 0.6 to 1.5 μm. Organisms occur singly or in groups, and capsules, if present, are small. *Brucella* are gram-negative but frequently take the counterstain poorly and require a minimum of 3 minutes for good definition.

The brucellae are strict aerobes. They grow slowly and require complex media for primary isolation. Such media as serum-dextrose agar or trypticase agar are satisfactory for this use.

Colonies of *Brucella* are spheroidal in shape, 2 to 7 mm in diameter. The colony morphology may be altered by the conditions of growth, but usually colonies are moist, translucent, and slightly opalescent.

The members of this genus comprise a closely knit genetic group as defined by DNA hybridization studies. The differentiation of the three most common *Brucella* species is based upon quantitative differences in several physiologic tests (Table 43-1). The need for increased CO_2 for growth is characteristic of *B. abortus*. The ability to pro-

TABLE 43-1. DIFFERENTIAL CHARACTERISTICS OF SPECIES AND BIOTYPES IN THE GENUS *BRUCELLA*

Species	Biotypes	CO_2 Required	H_2S Produced	Growth on Dye Media* Basic Fuchsin 1:100,000	Thionine 1:25,000	Thionine 1:100,000	Agglutination in Monospecific Sera Abortus	Melitensis
B. melitensis	1	−	−	+	−	+	−	+
	2	−	−	+	−	+	+	−
	3	−	−	+	−	+	+	+
B. abortus	1	±	+	+	−	−	+	−
	2	+	+	−	−	−	+	−
	3	±	+	+	+	+	+	−
	4	±	+	+	−	−	−	+
	5	−	−	+	−	+	−	+
	6	−	±	+	−	+	−	+
	7	−	±	+	−	+	+	−
	8	+	−	+	−	+	+	+
	9	±	+	+	−	+	−	+
B. suis	1	−	+	−	+	+	+	−
	2	−	−	−	−	+	+	−
	3	−	−	+	+	+	+	−
	4	−	−	+	+	+	+	+
B. neotomae		−	+	−	−	+	+	−
B. ovis		+	−	+	+	+	−	−
B. canis		−	−	−	+	+	−	−

Modified from WHO Technical Report Series No. 464:71, 1971.

*Species differentiation is obtained on albimi or tryptose agar with graded concentrations of dyes. Interpretation should be controlled with the reference strains of each species.

duce H_2S for a period of 4 to 5 days is more typical of *B. abortus* or *B. suis*. *B. melitensis* usually grows in the presence of both basic fuchsin and thionin, whereas thionin inhibits *B. abortus*, and basic fuchsin inhibits *B. suis*.

Within each of these three species of *Brucella*, a number of strains or biotypes have been recognized based on these and additional biochemical properties (Table 43-1).

Antigenic Structure

B. abortus, B. melitensis, and *B. suis* occur in nature in the smooth phase. Serial propagation in the laboratory results in a change in antigenicity of the organism, with visible alterations in colony morphology and a reduction in virulence for laboratory animals. For these reasons, propagation of the organisms in the laboratory prior to identification or for use as antigens in serologic testing requires rigorous monitoring of the cultures. The present hypothesis of the antigenic structure of these three species recognizes two antigenic determinants: A (*abortus*) and M (*melitensis*), present on the lipopolysaccharide (LPS) protein complex.

Quantitative agglutinin-absorption tests differentiate among the smooth phase antigens of the three species. Monospecific antisera are produced by absorption of heterologous antisera. *B. melitensis* absorption leaves serum re-

active only with *B. abortus* and *B. suis*. *B. abortus* and *B. suis* are indistinguishable by agglutination tests. Absorption of a second aliquot with *B. abortus* leaves serum specific for *B. melitensis*. *B. canis* exists only in the rough form and differentiation from other *Brucella* species is a problem only when the other species are also present in rough form. Recently, gas–liquid chromatography has demonstrated that *B. canis* lacks a 19-carbon cyclopropane acid that is common to all other species, thus providing a rapid means of identification of *B. canis*. There is some antigenic cross-reaction of *Brucella* with other organisms, such as *Yersinia enterocolitica, Francisella tularensis,* and *Vibrio cholerae*.

Determinants of Pathogenicity

The intracellular survival and multiplication of *Brucella* is a property associated with virulence. It is an essential determinant of the organism's ability to gain access to nodes and other tissues, but the mechanism by which this is achieved is not understood. It has been shown that in vitro smooth organisms are injested somewhat less rapidly by macrophages than rough organisms but do not kill the host cell and can subsequently multiply. The bacteria are partially protected in the phagolysosome and neither rough nor smooth *Brucella* organisms cause hexosemonophosphate shunt stimulation upon ingestion. It is unknown whether they fail to stimulate, or whether they actively inhibit the normal sequence of events leading to intracellular killing. Early experiments attributed the survival in phagocytes to the cell walls of brucellae organisms. It has been shown that mononuclear cells from an immune animal can kill ingested organisms so that the host defenses can ultimately eliminate infection.

Recent in vitro work suggests that rough strains of *Brucella* have antigens accessible to antibody, whereas smooth strains do not. Also, rough strains appear to bind immunoglobulin IgG and other serum proteins nonspecifically, whereas smooth strains do not. Both observations are consistent with the hypothesis that as the organisms become rough, outer membrane proteins are exposed. It is possible that these surface changes contribute to the loss of virulence by increasing the accessibility of the organisms to specific and nonspecific IgG.

Current investigation of outer membrane proteins suggests that groups 2 and 3 proteins of *B. abortus* are analogous to matrix porins and Omp A of *E. coli*, respectively. The group 2 proteins are antigenically identical in all strains examined. The LPS determinants are tightly associated with these proteins and are inducers of antibody in cattle. A study is in progress to determine any protective role of antibody to these proteins and their potential usefulness in separating natural infection from vaccine-induced immunity in cattle.

Another potential virulence factor of *Brucella* is a cell wall carbohydrate that is responsible for binding to human B lymphocytes. Although all *Brucella* can bind to lymphocytes, species pathogenic for humans exhibit greater activity. The binding appears to result from an interaction between a lectin on the lymphocyte and a specific carbohydrate on the bacterial cell wall.

Studies of cell envelope components of smooth *B. abortus* and attenuated rough organisms do not show any ultramicroscopically detectable differences. The LPS composition, however, varies as shown by the presence of both phenol- and water-soluble LPS fractions in the smooth organisms, but absence of phenol-soluble LPS in rough organisms. The fatty acid composition of *Brucella* LPS is distinct from that of enterobacterial LPS. β-OH myristic acid is lacking, and 85 percent of all fatty acids have a chain length of 16 or greater. A native hapten (NH), polysaccharide in nature, is present in endotoxin preparations from smooth *Brucella*. Poly B, a nontoxic, nonimmunogenic polysaccharide extracted from rough organisms, is immunologically identical to NH by immunodiffusion analysis. It has thus been proposed that the situation in *Brucella* is analogous to that in *Escherichia coli* and *Salmonella*, where the nontoxic hapten is thought to be an incompletely assembled cytoplasmic precursor of the O-polysaccharide chain of the smooth LPS.

Animals naturally or experimentally infected with *B. abortus* exhibit a characteristic predilection for fetal bovine tissues. Quantitation of the tissue distribution of organisms has shown that 60 to 85 percent of the organisms extracted from the tissues of infected animals are present in fetal cotyledons, 1 to 25 percent are present in the chorion, and 2 to 8 percent are present in fetal fluids (allantoic and amniotic). Also present in these same tissues of the pregnant cow, sheep, and goat, but not in the human placenta, is the 4-carbon polyhydric alcohol erythritol (OH CH$_2$ CHOH CHOH CH$_2$ OH). Erythritol seems to be a fetal product measurable in amniotic and allantoic fluids from normal bovine fetuses. This alcohol functions efficiently as a carbohydrate source in a basal medium for virulent *B. abortus*, but not for attenuated or rough organisms. It also enhances the intracellular growth of the organisms in an in vitro system employing phagocytes and *B. abortus*. There is, therefore, a significant correlation of the organotropism in cattle, sheep, and goats with the presence of erythritol. The absence of such tissue localization in human disease correlates with the absence of large amounts of erythritol in these organs. The similar intracellular localization of the bacteria in human illness, however, indicates that the role of other contributory factors to the pathogenesis of brucellosis, including those mentioned above, needs to be elucidated.

Clinical Infection

Epidemiology
Brucella organisms are distributed throughout the world, and the epidemiology is intricately related to animal infections. Human infection is a direct result of contact with infected animals, and for this reason it is necessary to consider animal disease.

Infection in Animals. The pathogenesis of infection in the various animal species is similar. Under natural conditions, goats harbor *B. melitensis*; cattle harbor *B. abortus*; swine harbor *B. suis*; and sheep harbor *B. melitensis* or, less frequently, *B. abortus*. Infrequently, dogs may be reservoirs of any of these three species if they come in contact with infected animals. *B. canis*, however, has been identified as a significant pathogen in dogs, particularly in kennels of beagles. Other animals that may be infected with brucellae include buffalo and the Bactrian camel (*B. abortus*), as well as reindeer and caribou (*B. suis*). Other farm animals, such as horses and poultry, may become infected with brucellae in very unusual situations, but they do not constitute a large reservoir or significant source of human infections.

Infection of animals occurs through the gastrointestinal tract, skin, and mucous membranes, including the conjunctivae. Animal food substances may come in contact with *Brucella*-infected materials, and when ingested by the animal, result in infection. *B. ovis*, *B. suis*, and *B. canis* are also transmitted with some frequency from an infected male to a female at the time of breeding.

Infection of lymph nodes nearest the portal of entry is followed by bacteremia, which in the pregnant or lactating animal can lead to massive multiplication of the organism in the uterus and mammary glands. Brucellae localize in chorionic epithelial cells and cause necrosis of placental cotyledons. The animal fetus may become infected or may be aborted because of asphyxia alone.

Animals usually recover spontaneously but excrete the bacteria for varying intervals of time in vaginal secretions, urine, and milk, which are infectious. *B. melitensis* is excreted for months in the milk of infected goats, as originally demonstrated by Zammit on the Isle of Malta. Sheep tend to excrete the organisms for a shorter interval of time. The organisms are extremely long lived under the proper environmental conditions. Survival is altered by pH and temperature, and exposure to sunlight will kill the organisms after a few hours. *B. melitensis* has been shown to survive in damp soil for as long as 72 days, in milk for 17 days, and in sea water for 25 days. It is clear that potential communicability can represent a threat to an entire herd of animals and persons in contact with the animals and infected materials.

Pathogenesis

The disease is similar in its pathogenesis in humans. Humans acquire the organisms through contact with infected materials. Within the United States, 90 percent of brucellosis is now due to contact with infected materials rather than ingestion of contaminated fresh milk and milk products. Fifty-six percent of the 2189 reported cases of brucellosis from 1967 to 1976 was abattoir-related illness. The occurrence of disease principally in males between the ages of 20 and 50 years reflects the occupational hazard of people in the meat-processing industry. Adequate means of detecting infected meat and prevention, except by elimination of infection in cattle and swine, are not available. Movement of animals allows the organisms to be reintroduced into uninfected herds. Other workers at risk because of necessary animal contact include veterinarians, livestock producers, farmers, dairy workers, and laboratory personnel working with the organism.

Organisms can enter through abraded skin, where they gain access to lymphatics and lymph nodes. There is often local lymphadenopathy and subsequent bloodstream invasion secondary to the bacterial multiplication and dissemination from the primary node. The subsequent localization of the organisms occurs particularly in the reticuloendothelial system. Brucellae have been observed inside the phagocytes, including mononuclear cells and macrophages. Intracellular organisms are protected from host defense mechanisms including antibody, as well as from antibiotics. The response in infected tissues is granulomatous and, in some situations, may go on to abscess formation and even caseation. Such sites as the spleen are particularly heavily involved, and the bone marrow frequently has detectable granulomas. The liver tends to be less involved but has been a good source of biopsy-positive materials.

Clinical Manifestations

The incubation period of disease may be as short as 3 days but is sometimes several months in duration. More commonly, there is a time period of approximately 3 weeks after known exposure to organisms before the onset of symptoms. Weakness is seen in the vast majority of patients and is the most outstanding complaint. Fatigue, especially that occurring late in the day, results in the inability to perform normal activities. Chills, sweats, and anorexia are seen in approximately three fourths of patients with acute illness, and over one half of patients report generalized muscle aching, headache, and backache. These nonspecific symptoms may be accompanied by associated mental depression and increased nervousness.

The findings on physical examination are minimal in contrast to the multiple complaints. Over 90 percent of patients have fever, but only 10 to 20 percent have palpable splenomegaly or lymphadenopathy. This organomegaly is more common in children. The fever tends to be intermittent, with characteristic diurnal variation. Disease may begin either insidiously or with a rather abrupt onset. The majority of patients with infection due to *B. abortus* have a self-limited disease. *B. melitensis* is the most invasive species in humans, and illness due to *B. melitensis* or *B. suis* may be more severe or chronic. Chronic brucellosis is difficult to diagnose or define but is usually nonbacteremic and occurs without demonstrable localized foci or infection.

Complications resulting from *Brucella* infections are usually attributable to the granulomatous lesions that occur in various organs and tissues. Analysis of a large series of patients by Spink showed that at least 10 percent of patients were suffering from debilitating neuropsychiatric disorders. The complaints may be nonspecific, or there may be localized specific dysfunction. The second most common complication is *Brucella* infection of a bone or joint. The vertebral column is involved most frequently, and the invasion of the intervertebral disc and adjacent vertebral bodies constitutes an osteomyelitis. In addition, a few cases of endocarditis due to *Brucella* have been recorded, and the viscera, such as spleen, liver, and bone marrow, may have evidence of infection for a significant period of time. The liver may be enlarged or tender, but only on rare occasions is there jaundice, hepatic failure, or subsequent cirrhosis.

Immunity

Following natural infection in both humans and animals there is an initial IgM antibody response, followed by an IgG antibody response. The agglutination test measures antibody directed at *Brucella* LPS antigens. Macrophages from immune animals more efficiently destroy the intracellular organisms and undoubtedly contribute to eradication of this infection by the host.

Laboratory Diagnosis

Analysis of the clinical illness described above reveals that the findings in this group of patients are nonspecific when considered individually. Therefore, evaluation of patients with these symptoms often includes a number of tests dictated by the differential diagnosis. Every patient suspected of having brucellosis should have at least one blood culture taken. Other materials can also be examined, including the cerebrospinal fluid in the presence of central nervous system symptoms; bone marrow; and tissues (e.g., lymph nodes and liver). The intracellular localization of *Brucella* organisms, particularly within reticuloendothelial cells, may be responsible for the positive cultures obtained from bone marrow aspirates at a time when blood cultures from the same patient are negative. Isolation of *Brucella* organisms provides the definitive diagnosis. Isolation of organisms from tissues of infected animals may also be important.

Culture. In vitro growth of *Brucella* organisms from patient specimens requires careful and informed laboratory processing of materials. For primary culture, direct inoculation of materials onto solid media is recommended to facilitate the recognition and isolation of the developing colonies and to limit the establishment of nonsmooth mutant organisms. Such media as serum-dextrose agar, serum-potato-infusion agar, trypticase agar, *Brucella* agar with serum, and 5 percent sheep blood agar are satisfactory for this use. Duplicate plates should be incubated under atmospheric conditions and in the presence of 10 percent CO_2, as this is often essential for growth of *B. abortus* on primary isolation. The commonly used method for attempted isolation from blood and body fluids employs the Castenada bottle containing both a solid and a liquid medium. All cultures should be kept for 21 to 35 days before discarding as negative. Transfers or subcultures should be made from the original flask every 4 to 5 days to fresh medium, and these subsequent subcultures should also be observed for 21 to 35 days. The conventional cultural methods employed in most hospital and laboratory situations are adequate to identify the positive cultures encountered from patients or in animal surveillance studies. The Food and Agriculture Organization (FAO) of the United Nations and World Health Organization (WHO) Brucellosis Reference Centers will perform the metabolic testing that may be necessary for identification of atypical cultures of epidemiologic significance.

Laboratory processing of materials, such as animal tissue or milk that may be heavily contaminated with other microorganisms, employs either animal inoculation or the use of selective media containing antibiotics to inhibit bacteria other than *Brucella*.

Serologic Diagnosis. The serologic diagnosis of infection accomplished by the standard tube agglutination test is very sensitive and yields the highest degree of reproducibility. The success of the agglutination test depends largely upon the selection and standardization of the antigen. A single antigen will accurately diagnose disease with any one of the three commonly encountered species of brucellae. Cross-reactivity of the serum agglutinins to *Brucella* with *F. tularensis*, *V. cholerae*, and organisms of the genus *Yersinia* demonstrates the slight cross-antigenicity of these organisms. The homologous titer is considerably higher than the heterologous titer, affording help in evaluating the presence of agglutinins. The card agglutination test

recognizes more than 90 percent of patients with positive tube agglutination tests. This test, which can be performed and interpreted in 4 minutes, may become a useful screening test because it is superior to the slide agglutination test. Sera should, however, also be examined by the standard tube agglutination test, particularly during the first week of illness.

It has been long observed that there is a prozone phenomenon in the measurement of agglutinating antibodies to *Brucella* organisms. This phenomenon is apparently due to the presence of blocking antibodies. During the acute phase of illness, when IgM agglutinating antibodies predominate, it is easy to detect the agglutination reaction. As IgG antibodies are formed during the course of the infection, some of them bind with antigen, thus preventing its agglutination by the large IgM molecule. A modified Coombs (antiglobulin) test has been used to increase the efficiency of serologic diagnosis. With the use of antiglobulin to bind the incomplete antibody and antigen, agglutination can again be detected.

Extensive data gathered by Spink and co-workers indicate a relatively high incidence of *Brucella* agglutinins in the serum of normal individuals. The titer varies with the geographic location and exposure to the organisms, but is usually less than 1:100. More recent comparative serologic studies by Buchanan indicate that a single titer of greater than 1:160 by the standard tube agglutination test is presumptive evidence of current or recent infection with *Brucella* organisms. A four-fold rise in agglutinins is seen in the first 3 months of infection in more than 90 percent of patients with cultures positive for brucellosis. To date, there is no evidence that antibiotic therapy alters the appearance or persistence of antibodies to the *Brucella* organisms.

Other serologic determinations include the enzyme-linked immunoabsorbent assay (ELISA), complement-fixation (CF), and the 2-mercaptoethanol (2 ME) agglutination test. The ELISA has been tested for use in surveillance for *Brucella* antibodies in animal herds. This test is safe, sensitive, and can be fully automated for large-scale application. The ELISA tests employing commercially available reagents for measurement of total, IgG, and IgM anti-*Brucella* antibodies are being tested on human sera. These tests eliminate the prozone phenomena and will replace previous serologic tests. Complement fixation requires rigid adherence to the standardization of the test conditions and reagents. The 2 ME test has been reliably utilized to demonstrates IgG (2 ME-resistant) agglutinins. The absence of such agglutinins mitigates against active disease, which is helpful in evaluation for chronic brucellosis. Maintenance of IgM agglutinating antibody titers for years by patients without active infection confuses serologic evaluation. Thus, 2 ME resistant agglutinating antibodies are the best indicator of cure as titers decline with time.

Skin Test. The *Brucella* skin test is a measurement of sensitization to the antigens of the organism at an undetermined time. It is not used for diagnosis of acute infection because it may remain positive for years after infection, even after the agglutination test becomes negative. The most commonly used antigens are those producing a delayed type of dermal reaction. One of the more standardized products is Brucellergen, a commercially available nucleoprotein fraction of *Brucella* cells. Because a positive intradermal test can result in a rise in serum agglutinins, serologic studies should be done prior to application of intradermal skin tests.

Treatment

Tetracycline remains the first drug of choice for therapy of illness and is usually continued for a minimum of 6 weeks. Streptomycin has been used in combination with tetracycline, and recent analysis of a large series of patients indicates that antibiotic combinations including streptomycin produced a significantly lower relapse rate than combinations without streptomycin. Investigations with experimental brucellosis in animals have demonstrated the efficacy of rifampin for *B. melitensis* and *B. abortus* infection. Encouraging preliminary reports of therapy of human disease with rifampin have also appeared. Although other antimicrobial agents have been effective in vitro, none of these agents has been shown to have greater therapeutic efficacy than tetracycline and streptomycin. The intracellular position of many of these organisms contributes to the continued recovery of bacteria from the blood despite antibiotic therapy. Intracellular killing of *Brucella* is essential for the final eradication of the bacteria and is dependent upon the normal mechanisms of the phagocyte. Rifampin penetrates cells, and thus theoretically would assist the host in eradication of organisms. Relapses occur in less than 5 percent of cases when therapy is instituted early and maintained for 6 weeks. Such relapses usually occur within 3 to 10 months and respond to a second course of therapy.

Prevention

Control of Animal Disease. There has been a decline in the reported cases of brucellosis in the United States, from 3510 in 1950 to 183 in 1980 (Fig. 43-1). The mean annual occurrence for the period 1976 to 1980 was 221 cases. The continued decline of recognized illness is a result of controls exerted on the animal reservoirs of infection. Regulations governing the pasteurization of milk were effective in reducing the infection of milk and dairy products. In addition, the U.S. Livestock Sanitary Association (now the U.S. Animal Health Association) and the Bureau of Animal Industry of the U.S. Department of Agriculture formulated the State Federal Cooperative Brucellosis Eradication Program. Their recommendations were reinforced by the U.S. Public Health Service 1953 Milk Ordinance and Code. The majority of human cases of disease in the United States are associated with the meat-processing industry. Thus, this infection is no longer primarily food-borne, but is an occupational hazard of those in contact with infected cattle or swine.

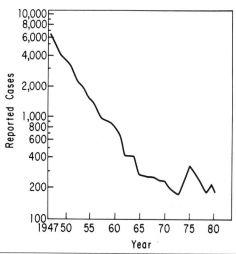

Figure 43-1. Reported human brucellosis in the United States, 1947 to 1980. *(Adapted from Annual Summary of Brucellosis 1976. Washington, D.C., U.S. Department Health, Education, and Welfare, Public Health Service, 1977, vol 26, p 51; Brucellosis Surveillance. From MMWR Annual Summary 1980. Vol 30, p 29, 1981.)*

An effective vaccine is available for animal immunization. In animals immunized during the first 6 to 8 months of life, abortion is prevented, organisms are not excreted in the milk, and the animal has a permanent immunity against natural infection. The strains employed for this purpose are the attenuated *B. abortus* strain 19 for cattle and the *B. melitensis* strain Rev I for sheep and goats.

Prophylaxis in Humans. Prevention of human brucellosis is primarily dependent upon control of the animal sources of infection. Modifications of milk and dairy-product processing, as well as animal surveillance and animal immunization, have greatly reduced the dangers of this disease within the United States. The population at risk consists almost exclusively of those persons in contact with animals or their contaminated products. Available vaccines are suitable only for animals. The disease remains one of economic importance in many countries of the world where control of infected animal herds has not been readily accomplished.

FURTHER READING

Books and Reviews

Alton CG, Jones LM: Laboratory techniques in brucellosis. In Buchanan RE, Gibbons NE (eds): Geneva, WHO, 1967

Annual Summary of Brucellosis 1976. Atlanta, Ga, U.S. Department HEW, Public Health Service, Centers for Disease Control, 1977

Thimm BH: Brucellosis. Distribution in man, domestic and wild animals. New York, Springer-Verlag, 1982

Selected Papers

Buchanan TM, Baber LC, Feldman RA: Brucellosis in the United States, 1960–72. An abattoir-associated disease. Part I. Clinical features and therapy. Medicine 53:403, 1974

Buchanan TM, Sulzer CR, Frix MK, et al.: Brucellosis in the United States, 1960–72. An abattoir-associated disease. Part II. Diagnostic aspects. Medicine 53:415, 1974

Buchanan TM, Hendricks SL, Patton CM, et al.: Brucellosis in the United States, 1960–72. An abattoir-associated disease. Part III. Epidemiology and evidence for acquired immunity. Medicine 53:427, 1974

Elberg SS: Immunity to *Brucella* infection. Medicine 52:339, 1973

Kreutzer DL, Dreyfus LA, Robertson DC: Interaction of polymorphonuclear leucocytes with smooth and rough strains of *B. abortus.* Infect Immun 23:737, 1979

Moreno E, Speth SL, Jones LM, et al.: Immunochemical characterization of *Brucella* lipopolysaccharides and polysaccharides. Infect Immun 31:214, 1981

Smith H, Keppie J, Pearce JH, et al.: The chemical basis of the virulence of *Br. abortus.* I. Isolation of *Br. abortus* from bovine foetal tissue. Br J Exp Pathol 42:631, 1961

Smith H, Keppie J, Pearce JH, et al.: Erythritol. A constituent of bovine foetal fluids which stimulates the growth of *Br. abortus* in bovine phagocytes. Br J Exp Pathol 43:31, 1962

Smith H, Keppie J, Pearce JH, et al.: III. Foetal erythritol a cause of the localization of *Br. abortus* in pregnant cows. Br J Exp Pathol 43:530, 1962

Verstreate DR, Creasy MT, Caverney NT, et al.: Outer membrane proteins of *B. abortis*: Isolation and characterization. Infect Immun 35:979, 1982

Wise RI: Brucellosis in the United States. Past, present and future. JAMA 244:2318, 1980

CHAPTER 44

Bacillus

The Genus *Bacillus*

The organisms within the genus *Bacillus* are large, gram-positive rods characterized by their ability to produce heat-resistant endospores. The genus is a very large one comprised of both strict aerobes and facultative anaerobes. Bacilli are widely distributed in nature and are found in large numbers in most soil and water samples. Many of their most distinctive properties are attributable to their quest for survival under adverse conditions encountered in their native habitat.

Historically, considerable attention has focused on the genus *Bacillus*. The ubiquity of its members and their propensity for producing endospores delayed the ultimate refutation of the theory of spontaneous generation. In the area of infectious diseases, anthrax—which is caused by a member of the genus, *Bacillus anthracis*—was one of the first bacterial infections whose etiology was established. In studies on *B. anthracis*, Koch demonstrated for the first time a set of criteria or postulates that must be satisfied before an organism can be credited as the etiologic agent of a specific infection. He also introduced the use of living attenuated organisms as a means of immunizing against the disease. Certain members of the genus *Bacillus* have assumed importance as producers of antibiotics, some of which have been clinically useful (polymyxin and bacitracin). At present, species of *Bacillus* are encountered primarily in clinical microbiology laboratories as contaminants or as reagents for testing the effectiveness of methods of sterilization. However, the recent recognition of their etiologic role in food poisoning, as a cause of sepsis in heroin addicts, and as a significant pathogen in the compromised host should make the clinical microbiologist wary of labeling *Bacillus* isolates as contaminants, or of not reporting them at all.

In the Eighth Edition of *Bergey's Manual,* 48 different species of *Bacillus* are recognized. One of the distinctive features of the genus, in contrast with most bacterial genera, is the wide range of the G + C content of the DNA of the various species—32 to 62 mole percent. This reflects the tremendous heterogenicity in the properties of the organisms within the genus. There is diversity of metabolic type, nutritional requirements, and composition and structure of the vegetative cell walls. Psychrophiles, mesophiles, and thermophiles, as well as alkalophilic, neutrophilic, and acidophilic species are included in the genus. Also, virtually all of the currently recognized species secrete a variety of soluble extracellular enzymes, which reflect the diversity of the parental habitats.

The only major pathogen in this genus is *B. anthracis.* Other organisms within the genus, however, have been of considerable significance because of their use in the industrial production of solvents, alcohols, enzymes, vitamins, and antibiotics. Partly as a result of these industrial applications, there has now accumulated a wealth of basic knowledge on the genetics and physiology of the *Bacillus* cell. Virulent and temperate bacteriophages, as well as plasmids, exist in bacilli, and the demonstrations of transformation, transduction, and transfection will undoubtedly play an important role in the future development of the fermentation industry.

Bacillus anthracis

Anthrax is primarily a disease of herbivorous animals, particularly sheep and cattle and, to a lesser extent, horses, hogs, and goats. It is caused by *B. anthracis,* a gram-positive, aerobic, sporeforming bacillus that was first isolated by Robert Koch in 1877. Humans accidentally encounter this disease in an agricultural setting, usually with the development of a local skin infection that may become generalized. The disease also may be acquired in an industrial setting, during the processing of hides or animal hair with resultant inhalation anthrax and the production of a virulent type of pneumonia. Rarely, the disease may be acquired by ingestion. Widespread immunization of animals has markedly diminished outbreaks of anthrax in herds, with the result that it is now a very rare illness in humans.

Morphology and Physiology

B. anthracis is a straight rod, 3- to 5-μm long and 1- to 1.2-μm wide. When examined in smears from the blood or tissues of an infected animal, the organisms usually are found singly or in pairs. Their ends appear square and the corners are often so sharp that the bacilli in the chains are in contact at these points, leaving an oval opening between the organisms. Unlike most members of the genus, *B. anthracis* is nonmotile. Bacilli are encapsulated during growth in the infected animal, but capsules cannot be demonstrated in vitro unless the organisms are cultured on a bicarbonate-containing medium in the presence of 5 percent CO_2. Spores are formed in culture, in the soil, and in the tissues and exudates of dead animals, but not in the blood or tissues of living animals.

Cultural Characteristics. The organisms grow well on most common laboratory media, but for demonstration of characteristic colonial morphology, specimens should be inoculated on 5 percent blood agar plates, prepared with blood free of antibiotics. Maximal growth of the organism is obtained at pH 7.0 to 7.4 under aerobic conditions, but growth does occur, although sparsely, in the absence of oxygen. The optimal temperature for maximal growth is 37C, but growth occurs over a wide temperature range of 12C to 45C. By continued cultivation, the organisms may become adapted to either a low or high temperature and eventually attain luxuriant growth. After 24 hours of incubation on simple laboratory media, the organisms produce large, raised, opaque, grayish-white, plumose colonies, 2 to 3 mm in diameter, and possessing an irregular, fringe-like edge. Tangled masses of long hair-like curls can be seen with a colony microscope. The colony is membranous in consistency and emulsifies with difficulty. No hemolysis is produced. Selective media have been described that permit growth of *B. anthracis* while inhibiting common contaminants, enteric organisms, and even closely related sporeformers, such as *Bacillus cereus.* Although selective media are useful for monitoring purposes, nonselective media are recommended for culture of clinical materials.

Laboratory Identification. Differential properties useful in the identification of *B. anthracis* are listed in Table 44-1. Neither morphology nor the usual cultural characteristics will differentiate *B. anthracis* from nonmotile strains of *B. cereus,* the organism most easily mistaken for *B. anthracis.* Virulent strains of *B. anthracis,* however, are the only organisms that produce rough colonies when grown in the absence of increased CO_2, and that produce mucoid

TABLE 44-1. DIFFERENTIAL CHARACTERISTICS OF *BACILLUS ANTHRACIS* AND *BACILLUS CEREUS*

Characteristic	B. anthracis	B. cereus
Blood agar colony	Rough, flat, usually many comma-shaped outgrowths	Rough, flat, no or few comma-shaped outgrowths
Hemolysis	None or very weak	Usually β-hemolytic
Tenacity*	Positive	Negative
Bicarbonate medium (CO_2)	White, round, raised, glistening, mucoid	Flat, dull
Fluorescent-antibody test	Positive	Negative
Gammaphage	Susceptible	Resistant
Animal pathogenicity	Positive	Negative
Litmus milk	Not reduced or slowly reduced and peptonized	Usually reduced in 2 to 3 days
Methylene blue	Not reduced or slightly reduced in 24 hr	Usually reduced in 24 hr
Motility	Negative	Usually positive

Adapted from Lennette, Spaulding, Truant (eds): Manual of Clinical Microbiology, 2nd ed, 1974. Washington, D.C., American Society for Microbiology.
*Refers to appearance of colony when pushed gently; resembles beaten egg whites.

colonies when grown on sodium bicarbonate medium in an atmosphere of 5 percent CO_2.

The string of pearls reaction also clearly separates virulent and avirulent *B. anthracis* from *B. cereus* and other aerobic sporeformers. Most dramatic, the string of pearls reaction can be demonstrated following a 3- to 6-hour incubation of *B. anthracis* on the surface of a solid medium containing 0.05 to 0.5 units of penicillin G per milliliter. The cells become large and spherical and occur in chains, which, as seen on the surface of agar, resemble a string of pearls. Another useful test for differentiating *B. anthracis* and *B. cereus* is based on the susceptibility of *B. anthracis* to a variant bacteriophage, gamma-phage; no lysis of *B. cereus* occurs. Confirmation of the identity of an isolate may be obtained by the inoculation of a mouse with a suspension of organisms from an agar plate. Death from anthrax infection usually occurs within 2 to 5 days and organisms can be recovered from the heart blood.

Resistance. Because of its ability to produce spores, the anthrax *Bacillus* is extremely resistant to adverse chemical and physical environments. A temperature of 120C for 15 minutes is usually adequate for inactivating the spore. The vegetative cell is comparable in resistance to other nonsporeforming bacteria, and is destroyed by a temperature of 54C in 30 minutes. Spores remain viable for years in contaminated pastures and remain a source of infection for long periods of time.

Antigenic Structure

Three antigens of *B. anthracis* have been partially characterized: (1) the capsular polypeptide; (2) a polysaccharide somatic antigen; and (3) a complex protein toxin. Unlike most bacterial capsules, the capsule of *B. anthracis* is a polypeptide of high molecular weight consisting exclusively of D-glutamic acid. There appears to be a single antigenic capsular type. The somatic polysaccharide antigen is a component of the cell wall and contains equimolar amounts of *N*-acetylglucosamine and D-galactose. It cross-reacts with human blood group A material and with type 14 pneumococcus polysaccharide. Antibodies to this antigen are not protective.

Anthrax toxin, derived from the thoracic and peritoneal exudates of infected animals, is a complex toxin consisting of three components: protective antigen (PA), lethal factor (LF), and edema factor (EF). All of the components appear to be nondialyzable proteins or lipoproteins, are highly thermolabile, and display evidence of molecular heterogeneity. The components are serologically active and distinct, and are also immunogenic.

Determinants of Pathogenicity

Only strains of *B. anthracis* that produce both a capsule and a toxin are fully virulent. The glutamyl polypeptide capsule interferes with phagocytosis and appears to be a major factor in the organism's pathogenesis, especially during the early stages of infection. Antibodies against the capsular antigen are produced but are not protective against the disease.

The signs and symptoms of anthrax are attributable to a toxin that gradually accumulates in the infected animal, with a maximum accumulation at the time of death. The pathophysiology of the bacillary disease and the toxemia from sterile toxin are very similar. In both cases and in all animal hosts tested, respiratory failure and anoxia result from action of the toxin on the central nervous system.

Maximum toxicity occurs only when all components of the toxin are present. A combination of PA and LF are required for lethality; EF and LF combined have no biologic activity. Although toxoiding of the toxin or its components has been demonstrated, data in this area are limited.

Clinical Infection

Epidemiology

IN ANIMALS. Spontaneous disease occurs in herbivorous animals that acquire the infection by ingestion of spores that probably enter the body through microscopic cuts or abrasions of the oral or intestinal mucosa. When a pasture has been contaminated with anthrax spores, it may remain a source of infection for 20 to 30 years. Although it is impossible to determine the precise time of infection in a case of spontaneous anthrax, it is certain that the duration of the disease is only a few days. The infected animal remains asymptomatic until a few hours before death. Mortality in herbivorous animals is usually about 80 percent. Sporadic cases continue to occur in the United States; in 1968 a total of 165 cases of anthrax occurred on 34 farms in California. Many species of animals acquire the natural disease, and epizootics continue to occur in wildlife sanctuaries in Africa.

IN HUMANS. The most common form of human anthrax is industrial anthrax, which results from contact with animal products such as wool, hide, goat hair, skin, and bones imported from Africa, the Middle East, and Asia. Less commonly, anthrax is acquired in an agricultural setting from working with infected animals. Among the recent examples of anthrax in the United States are cases acquired from imported bongo drums covered with animal hide containing the spores. Cutaneous anthrax also has resulted from contact with finished products such as shaving brushes made with animal bristles, ivory piano keys, and wool products.

In 1978, six cases, all occupationally acquired, were reported to the Centers for Disease Control. Four of these were acquired by unvaccinated employees in textile and felt mills, where they were exposed to imported goat hair. The two agricultural anthrax cases, which occurred in North Dakota and Idaho, were associated with anthrax in cattle and were acquired during postmortem examinations of cattle that had died suddenly. The widespread use of vaccines has markedly reduced morbidity in mill workers. However, the only method of completely eliminating potential exposure is to discontinue the use of imported goat hair. Although an effective vaccine for anthrax in cattle is available, the sporadic occurrence of bovine anthrax fails to provide ranchers with incentive to vaccinate their livestock routinely.

Pathogenesis. Humans become infected by one of three mechanisms:

1. The organisms can gain access through small abrasions or cuts and multiply locally with a fairly dramatic inflammatory response.
2. They also may gain access by inhalation, where they multiply in the lung and are swept to the draining hilar lymph nodes, where marked hemorrhagic necrosis may occur.
3. A rare method of infection is ingestion of infected meat, with resultant invasion and ulceration of the gastrointestinal mucosa.

From all three surface areas, invasion of the bloodstream may occur with profound toxemia. Metastatic infections such as meningitis may complicate the primary process.

Clinical Manifestations. Anthrax presents in one of three ways, depending on the mode of infection. Cutaneous anthrax begins 2 to 5 days after infection as a small papule that develops within a few days into a vesicle filled with dark bluish-black fluid. Rupture of the vesicle reveals a black eschar at the base, with a very prominent inflammatory ring of reaction around the eschar. This is sometimes referred to as a malignant pustule. The lesion is classically found on the hands, forearms, or head. It is rarely found on the trunk or lower extremities. The pulmonary infection, known as wool-sorter's disease, occurs in patients who handle raw wool, hides, or horse hair and acquire the disease by the inhalation of spores. The patient's symptoms are typically those of a respiratory infection with fever, malaise, myalgia, and unproductive cough. Within several days, however, it rapidly becomes a very severe infection with marked respiratory distress and cyanosis. With the sudden worsening of the illness, death usually occurs within 24 hours. Infection of the gastrointestinal tract, which occurs rarely, is associated with nausea, vomiting, and diarrhea. Occasionally there is loss of blood either through hematemesis or in the stools. This is associated with profound prostration with eventual shock and death. In all three of these surface infections, there may be invasion of the bloodstream and localization in the meninges, with a resultant fatal meningitis.

Anthrax infection in humans provides permanent immunity; second attacks are extremely rare.

Laboratory Diagnosis. Specimens for culture should be obtained from either a malignant pustule, the sputum, or blood. A gram stain and fluorescent-antibody stain are useful in making a presumptive diagnosis. The organism will grow readily on most laboratory media. However, the greatest problems encountered in establishing a diagnosis are the frequency with which nonpathogenic species of bacilli, such as *B. cereus,* are confused with *B. anthracis,* and the fact that most laboratory personnel have never seen *B. anthracis.*

Acute and convalescent sera should be obtained because antibodies to the organism can be demonstrated by agar-gel diffusion, complement-fixation, and hemagglutination procedures. These procedures are available at the Centers for Disease Control. Acute and convalescent sera of suspect cases may be submitted to this laboratory.

Treatment. *B. anthracis* is quite susceptible to penicillin, which is curative when used in the course of the illness. The major difficulty is the lack of clinical suspicion of anthrax because of its rarity. For this reason, cutaneous anthrax may be given another diagnosis and the inappropriate antibiotic for anthrax given. With pulmonary anthrax the diagnosis is usually made postmortem, as is the case with gastrointestinal anthrax. If the diagnosis of pulmonary anthrax is made in sufficient time, large intravenous doses of penicillin should be instituted as quickly as possible. In patients allergic to penicillin, tetracycline may be used. If a skin lesion is mistakingly identified as a staphylococcal infection, incision and drainage may be attempted. This can lead to disastrous results because of widespread dissemination of the organism.

Prevention. Animals with known or suspected anthrax should be handled with care and their carcasses buried deeply to prevent the spread of spores to new pastures. Wool, horse hair, and hides coming from areas where epidemic anthrax is present should be gas sterilized. A vaccine is available for outbreaks of human anthrax in an industrial setting.

IMMUNIZATION. Active immunization is the only known method of preventing anthrax in herbivorous animals in areas where the pasture land is already contaminated with spores. Pasteur's famous attenuated-living anthrax vaccine was effective but difficult to maintain at a desired level of virulence. Pasteur's vaccine has been superseded by a living spore vaccine made from a nonencapsulated strain of *B. anthracis.* In a comparison of this vaccine with an alum-precipitated protective antigen vaccine, it was shown that the living vaccine gave good protection but caused some local disease in the animals. The precipitated antigen vaccine provided 100 percent protection a month after vaccination, but by the end of 3.5 months, only 52 percent protection was provided. The simultaneous use of both materials provides a marked increase in the level of resistance and, when available, is the recommended procedure for animal immunization. The widespread use of a living spore vaccine in South Africa has reduced the incidence of anthrax in the cattle of this area by over 99 percent. The alum-precipitated protective antigen has been used in industrial plants to protect workers in high-risk situations. This antigen appears to be quite effective and no harmful side effects are produced.

Other Aerobic Sporeforming Bacilli

Bacillus cereus

This species is an infrequently recognized cause of food-borne illness in the United States. It is similar to *B. anthracis* in cellular morphology, but unlike *B. anthracis,* it is usually motile, β-hemolytic, and is not susceptible to gamma-phage (Table 44-1).

B. cereus food poisoning can cause two clinical syndromes. The first has a short incubation period of 4 hours. It is characterized clinically by severe nausea and vomiting and is frequently mistaken for staphylococcal food poisoning. Epidemics have been described following the ingestion of such foods as fried rice in which extensive multiplication of the organisms had occurred. The second syndrome has a longer incubation period (17 hours) and is characterized by abdominal cramping and diarrhea. It is commonly confused with clostridial food poisoning. The basic mechanism of *B. cereus* food poisoning relates to the fact that the spore forms survive cooking and that the food is allowed to reach temperatures that permit germination of the spore and elaboration of the enterotoxin. Strains of *B. cereus* elaborate at least two enterotoxins that act differently in experimental animals, depending on the nature of

the outbreak from which the strains were initially isolated. Proof of the cause of *B. cereus* food poisoning usually depends upon the isolation of the same type of organism from the food and the stools of infected patients.

In addition to food poisoning, *B. cereus* has also been implicated in serious infections associated with impairment of host defense mechanisms primarily by foreign bodies, prosthetic devices, or restricted blood supply. Patients with serious underlying diseases such as acute leukemia or who are immunosuppressed because of transplantation surgery can develop overwhelming bacteremia, endocarditis, or meningitis. Also, because these organisms are resistant to the β-lactam antibiotics, they may be selected for by the prior use of antibiotics for therapeutic or prophylactic purposes.

Antibiotic sensitivities for *B. cereus* are not applicable, but usually the organism is susceptible to chloramphenicol, aminoglycosides, vancomycin, and clindamycin.

Other *Bacillus* Species

Bacillus subtilis is present in the air, dust, brackish water, and infusion of vegetable matter. It is usually a common laboratory contaminant, but, like *B. cereus,* is capable of producing infection in the compromised host. It has also been seen in overwhelming bacteremias and eye infections in heroin addicts and has been cultured from street heroin. *B. subtilis* infections usually respond to therapy with the β-lactam antibiotics. Another species of interest is *Bacillus stearothermophilus,* the spores of which are used to evaluate the efficacy of autoclaving and other sterilization proce-

TABLE 44-2. DIFFERENTIAL TESTS FOR *BACILLUS ANTHRACIS* AND SOME COMMONLY RECOGNIZED SPECIES

Species	Spore				Differential Test	
	Swells Rods	Centrally Located	Xylose, Arabinose	Anaerobic Growth	Result	Test
B. anthracis	−	+	−	+	+	Phage*
B. cereus	−	+	−	+	−	Phage*
B. megaterium	−	+	D	−	−	
B. thuringiensis	−	+	−	+	+	Protein bodies†
B. subtilis	−	+	+	−	+	Nitrate to nitrite
B. licheniformis	−	+	+	+	+	Nitrate to nitrite
B. pumilis	−	+	+	−	−	
B. firmis	−	+	D	−	+	Starch hydrolysis‡
B. laterosporus	+	+	−	+	−	Starch hydrolysis‡
B. brevis	+	D	−	−	−	
B. sphaericus	+	−	−	−	−	
B. macerans	+	−	+G	+	−	Dihydroxyacetone
B. polymyxa	+	D	+G	+	+	Dihydroxyacetone
B. circulans	+	D	+	D	−	
B. alvei	+	D	−	+	+	Indole
B. coagulans	D	D	D	+	−	Indole
B. stearothermophilus	D	−	D	−	−	

From Lennette, Balows, Hausler, et al. (eds): Manual of Clinical Microbiology, 3rd ed, 1980. Washington, D.C., American Society for Microbiology.
+, positive for ≥90 percent of strains; −, negative for ≥90 percent of strains; D, reactions differ, positive for 11 to 89 percent of strains; G, gas.
*Other special tests positive for *B. anthracis* are encapsulation on bicarbonate agar, string of pearls, and direct fluorescent-antibody staining.
†*B. thuringiensis* contains crystalline protein bodies (see text).
‡*B. laterosporus* spores have laterally attached, spindle-shaped bodies that are easily stained.

dures. Several species produce disease in insects and in some cases have been used in insect control. The best studied of these is *Bacillus thuringiensis*, which is pathogenic for the larvae of *Lepidoptera*. This species is distinguished from *B. cereus* by the production of a crystalline protein body, or rarely, two or three bodies, in the cell during sporulation. This body separates readily from the liberated spore, and toxin is released from the crystal by enzymatic action in the larval gut. Differential properties of the most commonly recognized *Bacillus* species are shown in Table 44-2.

FURTHER READING

Books and Reviews

Bonde GJ: The Genus Bacillus. Dan Med Bull 22:41–61, 1975

Lincoln RE, Fish DC: Anthrax toxin. In Montie TC, Kadis S, Ajl SJ (eds): Microbial Toxins. New York, Academic Press, 1970, vol 3, pp 361–413

Schlessinger D (ed): Bacilli: Biochemical genetics, physiology and industrial applications. In Microbiology—1976, Washington, D. C., American Society for Microbiology, 1976, pp 5–449

Turnbull PCB: *Bacillus cereus* toxins. Pharmacol Thera 13:453–55, 1981

Selected Papers

Berke E, Collins WF, Von Graevenitz A, et al.: Fulminant postsurgical *Bacillus cereus* meningitis. J Neurosurg 55:637, 1981

Brachman PS: Anthrax. Ann NY Acad Sci 174:577, 1970

Brachman PS: Inhalation anthrax. Ann NY Acad Sci 353:83, 1980

Burdon KL, Wende RD: On the differentiation of anthrax bacilli from *Bacillus cereus*. J Infect Dis 107:224, 1960

Coonrod JD, Leadley PJ, Eickhoff TC: Antibiotic susceptibility of *Bacillus* species. J Infect Dis 123:102, 1971

Ellar DJ, Lundgren DB: Ordered substructure in the cell wall of *Bacillus cereus*. J Bacteriol 94:1778, 1967

Fish DC, Mahlandt BG, Dobbs JP, et al.: Purification and properties of in vitro-produced anthrax toxin components. J Bacteriol 95:907, 1968

Fitz-James PC, Young IE: Comparison of species and varieties of the genus *Bacillus*. J Bacteriol 78:743, 755, 765, 1959

Gianella RA, Brasile L: A hospital food-borne outbreak of diarrhea caused by *Bacillus cereus*: Clinical, epidemiologic and microbiologic studies. J Infect Dis 139:366, 1979

Gold H: Treatment of anthrax. Fed Proc 26:1563, 1967

Gordon MA, Moody MD, Barton AM, et al.: Industrial air sampling for anthrax bacteria. Arch Indust Hyg Occup Med 10:16, 1954

Ihde DC, Armstrong D: Clinical spectrum of infection due to *Bacillus* species. Am J Med 55:839, 1973

Jones WI Jr, Klein F, Walker JS, et al.: Growth of anthrax bacilli in resistant, susceptible, and immunized hosts. J Bacteriol 94:600, 1967

Klein F, DeArmon IA Jr, Lincoln RE, et al.: Immunity against *Bacillus anthracis* from protective antigen live vaccine. J Immunol 88:15, 1962

MMWR: Animal Anthrax in California. 17:279, 1968

MMWR: Anthrax in Humans—United States, 1978. 28:160, 1979

Nungester WJ: Proceedings of the conference on progress in the understanding of anthrax. Fed Proc 26:1491, 1967

Tuazon CU, Murray HW, Levy C, et al.: Serious infections from Bacillus species. JAMA 241:1137, 1979

Turnbill PCB, Jorgensen K, Kramer JM: Severe clinical infections associated with *Bacillus cereus* and the apparent involvement of exotoxin. J Clin Pathol (London) 32:289, 1979

Weinstein L, Colburn CG: *Bacillus subtilis* meningitis and bacteremia. Arch Intern Med 86:585, 1950

Wright GG: Anthrax toxin. In Schlessinger D (ed): Microbiology —1975. Washington, D.C., American Society for Microbiology, 1975, p 292

CHAPTER 45

Introduction to the Anaerobic Bacteria: Non-Sporeforming Anaerobes

Anaerobic Bacteria

Recognition of the anaerobic nature of certain species of microorganisms is credited to Pasteur, who noted in 1863 that motility of certain bacteria was lost upon exposure to air. By 1900, a variety of bacteria had been isolated that would grow only in gaseous environments having substantially reduced oxygen tensions. The study of anaerobic bacteria lagged, however, because of their sensitivity to oxygen, their generally fastidious growth requirements, and their frequent occurrence in complex mixtures of anaerobic and facultative species. Provision of anaerobic environments for culture was technically difficult, making it almost impossible to obtain pure cultures for study. With the improved anaerobic technology that has been available within the last 10 to 15 years, however, a high incidence of anaerobic organisms has been demonstrated in clinical specimens and a better understanding of these organisms has evolved.

The anaerobic bacteria are widespread in nature. They constitute the predominant part of our normal indigenous flora on mucocutaneous surfaces and outnumber facultatively anaerobic bacteria in the gut by a factor of 1000:1. On the skin, mouth, upper respiratory tract, and female genitourinary tracts, they outnumber facultatively anaerobic bacteria by a factor of 5 to 10:1. Many of these anaerobic organisms, previously considered to be harmless commensals of our indigenous flora, are now recognized as opportunistic pathogens that may produce disease when the host's resistance is reduced.

Types of Anaerobic Organisms

The anaerobic bacteria include many different types, both gram-positive and gram-negative. In the Eighth Edition of *Bergey's Manual,* they are classified in different parts on the basis of their gram stain reactions, cellular morphology, and intolerance to oxygen. Although DNA homology and other techniques have increased our understanding of relatedness among anaerobes, a need exists for further work on classification and simplification of the procedures required for cultivation and identification. The same organism may have a variety of names coined over the years, and reference to authoritative manuals on anaerobic bacteriology is helpful in reading some of the older literature. The nomenclature used in this chapter is based on the *Anaerobe Laboratory Manual* of the Virginia Polytechnic Institute and State University, but also includes more recently published changes in nomenclature.

On the basis of differences in the types of diseases produced, anaerobic bacteria may be conveniently divided into: (1) the clostridia, which form spores; and (2) the non-sporeforming anaerobic bacteria. For this reason, the pathogenic exotoxin-producing clostridia and other species are covered separately in Chapter 46. Certain of these clostridia, however, can also occur in combination with the non-sporeforming anaerobes in the types of infections, which are discussed in the present chapter. The general aspects of morphology, physiology, laboratory culture, and identification of anaerobes discussed in the first part of this chapter pertain to the clostridia as well as to the non-sporeforming anaerobic bacteria obtained from clinical sources. The pathogenic mechanisms and clinical infections especially associated with non-sporeforming anaerobes, and the particular genera and species with a recognized role in human infection, are discussed in the remainder of the chapter. Additional information on anaerobic organisms is found in Chapters 34, 47, 48, and 49 on the *Actinomyces, Treponema, Borrelia, Campylobacter,* and Oral Microbiology, respectively.

Morphology

The anaerobic bacteria include a variety of morphologic types, including bacilli, cocci, comma-shaped organisms, and spirochetes (Table 45-1). Although the clostridia gen-

erally stain boldly with the gram stain, many of the non-sporeforming anaerobic bacteria stain poorly, are pale in appearance, and are gram-variable. Better definition may be obtained by the use of Kopeloff's modification of the gram stain. Observation of certain of the anaerobes, including *Campylobacter, Treponema,* and *Borrelia,* may require phase contrast or darkfield microscopy.

As a group, anaerobic bacteria are more pleomorphic in appearance than most aerobic or facultatively anaerobic species, a property that may be useful in their recognition from clinical material or in culture (Figs. 45-1 through 45-4). Pleomorphism, however, is quite variable depending on the chemical environment present in infected material or in artificial culture, and thus is not always apparent.

Colony morphology of anaerobic bacteria on solid and in liquid media is helpful in recognition of certain anaerobes, but for many species colonies are not sufficiently unique to aid appreciably in identification. Characteristics helpful in some instances include production of turbid, granular, or flocculent growth in liquid culture and the size, shape, color, and consistency of colonies on solid media. Hemolysis does not aid in the identification of these organisms to the same extent that it does with certain facultatively anaerobic bacteria. Colony morphology, like cellular morphology, is extremely dependent on the cultural environment.

Physiology

Anaerobiosis. Complete understanding of oxygen intolerance among anaerobic bacteria remains unresolved at present. Various factors play a role, but no single mechanism has received total acceptance by investigators. Among the proposals are:

1. O_2 has a direct toxic effect.
2. O_2 is indirectly toxic via specific mediators, such as H_2O_2 or free radicals.
3. An appropriately low oxidation-reduction potential that appears to be required for many anaerobic bacteria is unachievable in the presence of normal O_2 tensions.
4. Essential sulfhydryl-containing enzymes are oxidized and therefore inactivated by O_2.
5. O_2 inhibits metabolism by reaction with flavoproteins and reduced nicotinamide adenine dinucleotide (NADH) oxidases, thereby critically lowering the reducing power of the cell.

The most popular theories advanced to explain the toxicity of O_2 for anaerobic bacteria have been based on their lack of the enzymes catalase (or peroxidase) and superoxide dismutase, which thereby allows accumulation of toxic levels of H_2O_2 or superoxide ions, respectively. Although undoubtedly important, experimentation has not substantiated the hypothesis that the absence of either of these enzymes is solely responsible for the toxicity of oxygen. Also, catalase and/or superoxide dismutase have been detected in certain species, especially in the more aero-

TABLE 45-1. GENUS IDENTIFICATION OF ANAEROBIC BACTERIA

Characteristics	Genus
I. Rods	
A. Form spores (sometimes difficult to demonstrate)	*Clostridium**
B. Do not form spores	
1. Gram-positive cells present (Kopeloff's modification of gram stain)	
a. Propionic and acetic acids as the major volatile acid products	*Propionibacterium**, *Arachnia**
b. Acetic and lactic acids (1+ to 1)	*Bifidobacterium**
c. Lactic acid sole major product (see also *Actinomyces*)	*Lactobacillus**
d. Moderate acetic, ± formic, and with (1) major succinic; (2) major succinic and lactic [1 acetic to 2+ lactic]; or (3) major lactic	*Actinomyces**
e. Other: butyric plus others, acetic and formic; or no major acids	*Eubacterium**, *Lachnospira*
2. Only gram-negative cells present (Kopeloff's modification of gram stain)	
a. Peritrichous flagella or no flagella	
i. Produce butyric (without much isobutyric and isovaleric acid)	*Fusobacterium**
ii. Produce only lactic acid	*Leptotrichia buccalis**
iii. Produce acetic acid and hydrogen sulfide; reduce sulfate	*Desulfomonas*
iv. Not as in i, ii, or iii	
a. Long, thin cells, growth in air-CO_2 (5 to 10 percent)	*Capnocytophaga**
b. Obligately anaerobic	*Bacteroides**
b. Polar flagella	
i. Fermentative	
a. Produce butyric acid	*Butyrivibrio*
b. Produce succinic acid	
1. Spiral-shaped cells	*Succinivibrio*
2. Ovoid cells	*Succinimonas*
c. Produce propionic and acetic acids	*Selenomonas*, *Anaerovibrio*
ii. Nonfermentative	
a. Produce succinic acid from fumarate	*Wolinella**
b. Do not produce succinic acid from fumarate	*Campylobacter**[†]
c. Spiral-shaped cells with axial filaments	*Treponema**, *Borrelia**
II. Cocci	
A. Gram-positive	
1. Occur in packets	*Sarcina*
2. Pairs and chains	
a. Require fermentable carbohydrate	
i. Produce butyric acid	*Coprococcus*
ii. Do not produce butyric acid	*Ruminococcus*
b. Do not require fermentable carbohydrate	
i. Produce lactic acid as sole major fermentation product	*Streptococcus**
ii. Not as in i	*Peptostreptococcus**, *Peptococcus**
B. Gram negative[‡]	
1. Produce propionic and acetic acids	*Veillonella**
2. Produce butyric and other acids	
a. Require fermentable carbohydrate and produce budding cells	*Gemmiger*
b. Do not ferment carbohydrate	*Acidaminococcus**
3. Large cells and produce complex mixture of fermentation acids	*Megasphaera*

Modified from Holdeman, Cato, Moore: Anaerobe Laboratory Manual, 4th ed, 1977. Blacksburg, Va, Virginia Polytechnic Institute and State University.
*Medically significant; these genera are commonly isolated from clinical specimens (frequencies vary from high to low). (Other listed genera may rarely occur in infections.)
[†]Microaerophilic; some strains will grow in 10 percent oxygen, but not aerobically or anaerobically in usual media.
[‡]Some cells may stain gram-positive.

tolerant clinical isolates. The simultaneous presence of both enzymes in sufficient quantities might be required. Indeed, a single mechanism may not be applicable to all anaerobes, and different mechanisms may also apply under different environmental conditions.

Anaerobes are not equally intolerant of O_2. Maximum growth occurs at a pO_2 equal to or less than 0.5 percent for strict anaerobes, and equal to or less than 3 percent

for moderate anaerobes. Thus, most anaerobic bacteria will not grow on the surface of blood agar plates exposed to air. However, it has been common practice to include among anaerobic classification schemes (Table 45-1) genera that are microaerophilic (*Campylobacter*); strains that grow in the presence of nearly ambient pO_2 if the atmosphere is CO_2-enriched (e.g., *Streptococcus*, *Actinomyces*); and strains that grow sparsely under aerobic conditions (e.g.,

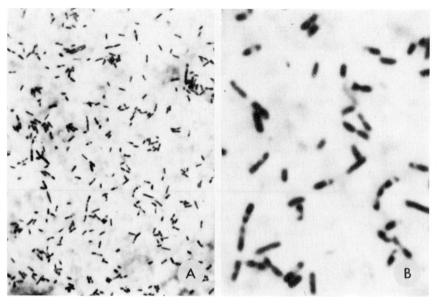

Figure 45-1. *Bacteroides fragilis* from supplemented thioglycolate. Pleomorphism, which may be considerably more or less apparent than shown here, is highly dependent on the cultural environment and on strain differences. **A.** × 1000. **B.** × 3000. Note vacuoles present within cells.

certain *Clostridium* species). Many of these strains show the most luxuriant growth and/or are isolated initially only under anaerobic conditions with the usual clinical laboratory procedures. Anaerobic gaseous environments commonly include 5 to 10 percent CO_2, which is stimulatory or a requirement for many anaerobic bacteria.

Growth Requirements. In addition to the provision of a CO_2-enriched anaerobic atmosphere, anaerobes generally are nutritionally very fastidious and require an enriched medium for growth. Growth factor requirements usually may be met by the addition of yeast extract, blood (for solid media), serum or ascites fluid (for liquid medium), vitamin K, hemin, and a fermentable carbohydrate to the basal medium, although other additives such as cystine or

arginine may be required for some strains. An enriched base such as Brucella agar, brain-heart infusion, or Schaedler is preferable; cooked chopped meat medium containing carbohydrate and thioglycolate also are suitable as broth media. Selective plates are helpful for isolation and aid in presumptive identification of anaerobes from clinical specimens, which often include multiple species of facultative and anaerobic bacteria.

Solid and liquid media should be either freshly prepared, prestored in an anaerobic environment overnight or longer, or sterilized and maintained under anaerobic conditions to reduce quantities of oxygen or oxidizable components, thereby shortening the lag phase and providing for more optimal growth of anaerobes. Further detail on media, cultivation, and identification of anaerobes can

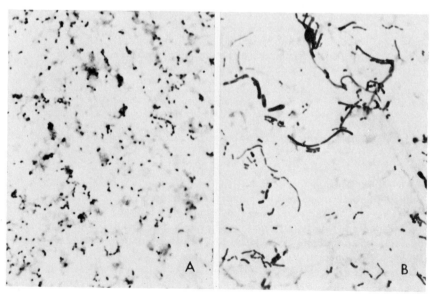

Figure 45-2. A. *Bacteroides intermedius* from supplemented thioglycolate, 24 hours. As shown here, the organism often appears as small coccobacilli. × 1000. **B.** *Bacteroides intermedius* from chopped meat medium with carbohydrate, 24 hours. × 1000. The same strain is shown in **A** and **B** to demonstrate variation in morphology dependent on the culture medium. Pleomorphism is also observed within a single culture, as demonstrated in **B**.

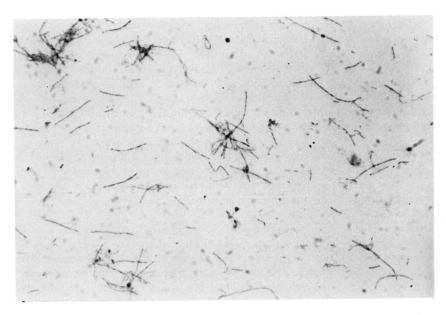

Figure 45-3. *Fusobacterium nucleatum* from sheep's blood agar plate. As shown here, this organism may appear as thin bacilli, often with very tapered ends, or may form long thin filaments. It stains palely gram-negative. × 1000.

be found in the various manuals on anaerobic bacteriology listed in the Further Reading section of this chapter.

Laboratory Culture and Identification

Anaerobic Culture Systems

At present, three primary types of anaerobic culture systems are available. As detailed below, they provide an anaerobic atmosphere during incubation and, in certain cases, during primary inoculation, subculture, and examination steps.

Anaerobic Jar. This method employs vented or unvented jars for holding culture plates or tubes. These containers have a leak-proof closure and contain a catalyst, palladium, that reacts with residual oxygen in the presence of hydrogen to form water. They are filled with an anaerobic gas mixture provided either by a gas generator envelope, which generates the proper volume of H_2 and CO_2 when water is added to the packet; or by a mechanical evacuation-replacement system, consisting of a vacuum source, manometer, and gas cylinder (mixture containing 10 percent H_2, and 5 to 10 percent CO_2 and N_2). These jars are incubated in a standard incubator. A similar system utilizes a small, O_2-impermeable, plastic bag that can hold two agar plates.

Roll Tube. This system uses a medium that has been prepared, sterilized, and stored under O_2-free gas in order

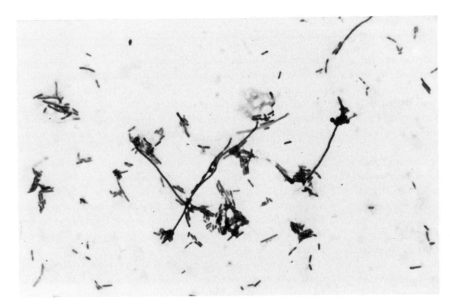

Figure 45-4. *Fusobacterium necrophorum* from chopped meat medium with carbohydrate. This organism may be highly pleomorphic, appearing as short, long, or filamentous bacilli, often with bulbous swellings and round bodies. × 1000.

to maintain a low oxidation-reduction potential and prevent oxidative changes. The medium is kept anaerobic during inoculation either by needle injection through the O_2-impermeable stopper without introducing any air, or by passing a gentle stream of O_2-free gas into the tube via a sterile cannula whenever the stopper is removed. Thus, each tube has its own anaerobic atmosphere and can be incubated in a standard incubator and observed for growth at any time.

Anaerobic Glove Box. This method consists of a closed chamber made of rigid or flexible clear plastic and fitted with gloves for manipulations within the enclosed space. An entry lock, which can be evacuated and filled with O_2-free gas, is used to pass material in and out of the chamber, which is filled with 10 percent H_2, and 5 to 10 percent CO_2 and N_2. Any oxygen introduced into the system is removed by palladium catalyst. The glove box contains all the necessary equipment for bacteriologic work, including incubators, so that standard methods of inoculation, incubation, and isolation can be accomplished completely within an anaerobic environment.

All of these anaerobic systems have certain advantages and disadvantages. Although the roll tube and anaerobic glove box systems yield a higher number of isolates of the more oxygen-sensitive species, the anaerobic bacteria associated with clinical infections usually are less oxygen-sensitive than many other anaerobic members of the normal flora, and anaerobic jar techniques work well.

Isolation and Identification

When specimens arrive in the laboratory for culture, an immediate gram stain of the material is often valuable in the choice of appropriate media and methods for culture and as a quality control for the types of bacteria that laboratory culture should reveal. Specimens should be cultured as soon as possible, with the aim of minimizing exposure to oxygen. Also, many anaerobic infections are polymicrobic, and the nutritive material present in most clinical specimens will tend to support growth of the least fastidious, most rapidly growing bacteria at the expense of other strains that may be present in high proportions in the original material. Delay in specimen transport to the laboratory or in inoculation of media, and initial culture only in broth (without directly streaking plates) may significantly alter the relative concentrations of species as they exist in the infection, thus giving misleading results.

It is generally characteristic of anaerobes to grow more slowly than facultative or aerobic species due to the lower energy yield of their fermentative metabolism. Ideally, plates should be checked at 18 to 24 hours for faster-growing species (e.g., *Clostridium perfringens* and *Bacteroides fragilis*) and daily thereafter up to 5 to 7 days in order to isolate strains that grow slowly (e.g., certain *Actinomyces*, *Eubacterium*, and *Propionibacterium* species).

Once an organism has been isolated in pure culture and is determined to be anaerobic rather than facultative,

a number of procedures are available for identification (see manuals listed in Further Reading section). A commonly used scheme of classification and identification to the genus level (Virginia Polytechnic Institute's *Anaerobe Laboratory Manual*) is primarily based on gram stain reaction, cellular morphology, motility, and gas–liquid chromatography of volatile fatty acids, nonvolatile acids, and alcohols produced as end-products of metabolism (see Table 45-1). Species determination is based on gas–liquid chromatography, fermentation of various sugars, and other biochemical reactions similar to those used for aerobic or facultative bacteria. Commercial, miniaturized identification kits are available that are suitable for identification of clinical isolates.

Immunofluorescence appears promising as a technique for rapid identification of anaerobic bacteria directly in clinical material or in pure culture. Fluorescent-antibody conjugates have been used for specific staining of species of *Bacteroides*, *Fusobacterium*, *Clostridium*, *Actinomyces*, and *Propionibacterium*.

Determinants of Pathogenicity

Specific Virulence Factors. Information on the mechanisms of pathogenicity among non-sporeforming anaerobic bacteria is limited because of the delay in recognition of their medical importance. No toxins are produced by the non-sporeforming anaerobes that are comparable to the potent toxins of certain clostridia. Yet, the existence and importance of virulence factors is apparent from the unequal pathogenicity observable with various species or strains of anaerobes. For example, certain organisms such as *B. fragilis* (Fig. 45-1) are frequently isolated in infection, but are found in the indigenous flora in much lower numbers than other prevalent endogenous organisms that are rarely associated with infection.

Lipopolysaccharide has been demonstrated in strains of *Fusobacterium*, *Bacteroides*, and *Veillonella*. However, the endotoxin of strains of *Bacteroides* and most, but not all, of *Fusobacterium* tested lack heptone and 2-keto-3-deoxyoctonate (KDO), and the biologic activity of this lipopolysaccharide is weak compared with the endotoxin of aerobic and facultative gram-negative bacteria. However, enhanced coagulation (decreased clotting time) has been demonstrated in mice injected with whole bacteria, lipopolysaccharide, or lipid A from strains of *Bacteroides* and *Fusobacterium*. This may be related to thromboembolic disease, which can be observed with clinical *Bacteroides* infections.

A polysaccharide capsule has been demonstrated among strains of *B. fragilis*, *Bacteroides asaccharolyticus*, and a few strains of other species. Studies (see section on *B. fragilis*) have shown the capsule to be an important virulence factor.

A variety of enzymes that may serve as virulence factors have been detected in different strains of anaerobic bacteria. These include collagenase, heparinase, hyaluronidase, fibrinolysin, gelatinase and other proteolytic enzymes, lecithinase, lipase, chondroitin sulfatase, deoxyribonuclease, phosphatase, neuraminidase, and elastase. Although a role in virulence is hypothetical for many, some enzymes have been demonstrated to be active in disease production. Examples are the collagenase of a strain formerly referred to as *Bacteroides melaninogenicus* (present nomenclature, *Bacteroides gingivalis*) and probably other former *B. melaninogenicus* subspecies (Fig. 45-2); the heparinase of certain *Bacteroides* species, which may be an alternate mechanism to that of endotoxin for the development of thrombophlebitis and septic emboli; and various hemolysins and lipases associated with strains of *Fusobacterium necrophorum*. In a general sense, oxygen-protective enzymes such as superoxide dismutase, peroxidase, and catalase can be thought of as virulence factors because they enhance survival of anaerobes in tissue environments prior to establishment of a low oxygen tension and oxidation-reduction potential.

Synergy. The endogenous anaerobic bacteria are opportunists, producing disease when factors combine to produce an environment in tissue that promotes their growth. Individual strains of anaerobic bacteria generally lack the full complement of virulence factors that provides for invasion of tissue, resistance to host defense mechanisms, growth in tissue, and injury to host tissue. The deficiencies in individual strains can be compensated for by other species, however, such that, collectively, a mixture of organisms possesses a full complement of virulence factors. Thus, although infections do occur with a single species, mixed infection either with a variety of anaerobic species or with a combination of facultative and anaerobic species is most common. Deficiencies in individual bacteria may also be compensated for in compromised host defenses, as discussed below.

Well-documented early studies pioneered work on the polymicrobic nature of many anaerobic infections and the combined pathologic activities of organisms in synergistic mixtures. Intratracheal inoculation of pyorrhea exudate from humans containing a fusospirochetal mixture of oral anaerobic bacteria—spirochetes, fusiform bacilli, streptococci, and usually vibrios—was shown to produce aspiration pneumonia and lung abscess in laboratory animals, as a model of human disease following aspiration. Subsequent studies in a variety of laboratory animals have demonstrated production of infection by synergistic mixtures of two or more anaerobic species or, more commonly, by facultative and anaerobic species where the species were noninfective individually. These have included models of veterinary and human infections such as appendicitis peritonitis, genital infections, soft tissue infections, peridontal disease, skin necrosis, and intrahepatic and intraabdominal abscesses. However, the specific contributions of each strain essential to the infective mixture has rarely been determined. Examples of specific virulence determinants that were essential in the particular model system studied are: (1) the requirement for a gram-positive species such as a

diphtheroid, which produced a naphthaquinone necessary for the growth of *B. melaninogenicus* (presently, *B. gingivalis*), which was itself required for infectivity of a mixture of strains from peridontal disease, (2) the ability of *B. fragilis* strains possessing a polysaccharide capsule to produce abscesses without the presence of a facultative species that was required for unencapsulated *B. fragilis;* and (3) the ability of succinate to replace the requirement of *Klebsiella pneumoniae* for infectivity of *B. asaccharolyticus* (presently, *B. gingivalis*).

Other interactions among species in a mixture include the apparent inhibition of phagocytosis of facultative species in the presence of anaerobes, by competition for opsonins or by other mechanism(s), the protection of susceptible strains in mixtures via destruction of penicillin or cephalosporins by β-lactamase-producing *Bacteroides*, and the utilization of oxygen by facultative species that aids in producing a suitable environment for growth of anaerobes.

Clinical Infection

A number of features under discussion characterize the majority of infections involving the non-sporeforming anaerobic bacteria (Table 45-2) and distinguish them, in particular, from certain more classic infectious diseases that are consistent with Koch's postulates. Most important, perhaps, is that most of the anaerobic bacteria that cause infection are members of our normal indigenous flora. Mucocutaneous surfaces of the skin and the upper respiratory tract, the gastrointestinal tract, and the genitourinary tracts are populated with a varied and great abundance of anaerobic organisms (Table 45-3). Other features also are frequently present:

1. Infections often develop slowly and may become chronic, although acute episodes such as gram-negative septicemia or necrotizing fasciitis occur.
2. Anaerobic bacteria frequently produce in infected material or in vitro culture a putrid odor, due to certain end-products of their metabolism.
3. Gas may be present in tissue or in loculations, although it may occur also with certain facultative bacteria.

TABLE 45-2. FEATURES ASSOCIATED WITH INFECTIONS INVOLVING ANAEROBES

Caused by endogenous opportunistic pathogens, usually not transmissible

Occur in settings of compromised host defense, particularly local reduction in tissue pO_2

Usually polymicrobic, synergistic mixtures of aerobes and anaerobes or exclusively anaerobes

Abscess formation, tissue necrosis

Broad spectrum antimicrobial therapy generally required

TABLE 45-3. INCIDENCE OF ANAEROBIC BACTERIA AS NORMAL FLORA IN HUMANS

| | Cocci | | Bacilli | | | | | | | | |
| | | | Gram-positive | | | | | | Gram-negative | | |
Anatomic Site	Gram-positive	Gram-negative	Clostridium	Actinomyces	Bifidobacterium	Eubacterium	Lactobacillus†	Propionibacterium	Bacteroides	Fusobacterium	Campylobacter
Skin	1	0	0	0	0	U	0	2	0	0	0
Upper respiratory tract*	1	1	0	1	0	±	0	1	1	1	1
Mouth	2	2	±	1	1	1	1	±	2	2	1
Intestine	2	1	2	±	2	2	1	±	2	1	±
External genitalia	1	0	0	0	0	U	0	U	1	1	0
Urethra	±	±	±	0	0	U	±	±	1	1	±
Vagina	2	1	±	0	±	±	2	±	1	±	±

Modified from Sutter, Vargo, Finegold: Wadsworth Anaerobic Bacteriology Manual, 2nd ed, 1975. Los Angeles, University of California.
*, includes nasal passages, nasopharynx, oropharynx and tonsils; †, includes anaerobic, microaerophilic, and facultative strains; U, unknown; 0, not found or rare; ±, irregular; 1, usually present; 2, usually present in large numbers.

Pathogenesis

The pathogenesis of infection with non-sporeforming anaerobes is incompletely understood, although the clinical settings that predispose to invasion by endogenous flora are clear. A change from the normal, apparently benign host–parasite relationship occurs most often in settings that reduce the oxidation-reduction potential in tissues that normally are well oxygenated and resistant to invasion by anaerobes.

Reduction of the normal oxidation-reduction potential of tissue (approximately +120 mV) and a lowered oxygen tension may stem from an impaired blood supply, necrosis of tissue, or growth of facultative bacteria. These conditions may be associated with vascular disease, trauma, surgery, presence of foreign bodies, malignancy, radiation therapy, injection of vasoconstrictive agents such as epinephrine, shock, cold, or edema. As with other opportunists, a generalized compromise of host defenses that accompanies administration of immunosuppressive drugs, steroids, and cytotoxic agents or that occurs in the presence of diseases such as diabetes mellitus may also predispose to infection with anaerobes. In these settings, anaerobic bacteria may initiate infection by direct extension from mucocutaneous surfaces into normally sterile adjacent tissue or be carried to more distant sites by hematogenous or other mechanisms of spread. The introduction of very high numbers of anaerobic bacteria already in a milieu of low oxidation-reduction potential, such as is true for fecal material spilled into the peritoneal cavity, can overwhelm even normal host defenses and result in infection without direct tissue damage. Also, the low oxygen tension present in anaerobic infections may interfere with oxygen-dependent pathways of phagocytic killing. Thus, disturbance of the natural balance between tissue resistance and endogenous flora is basic in the pathogenesis of these infections. Once introduced into the affected site, the production of toxins, enzymes, and other virulence factors is obviously significant, but their specific role in the overall disease process requires further clarification in many instances, as discussed previously.

Aerobic or facultative bacteria most often are present in infections involving anaerobes and usually are derived from the same endogenous source. Their presence is not required so long as conditions in tissue are appropriate for anaerobic growth and the pure or mixed species of anaerobes possess the necessary complement of virulence factors. Nevertheless, mixed aerobic-anaerobic species may be more efficient in initiating and maintaining infection; there are numerous examples of how one may complement the activity of the other.

Clinical Manifestations

The types of infections produced and the particular anaerobic species involved are related to the endogenous habitat on the skin and in the upper respiratory, gastrointestinal, and genitourinary tracts (Table 45-3). Of the tremendous variety of indigenous species, only relatively few are common as opportunistic pathogens (Table 45-4). Many species are ubiquitous as to location, but some are most typically etiologic in particular body sites. For example, *B. fragilis*, which is not a usual member of the upper respiratory tract and mouth flora but is present in the gastrointestinal tract, is very common in abdominal and, to a lesser degree, pelvic infections. *Bacteroides bivius*, which is a prevalent species in the vaginal flora, is common in genital tract infections, but considerably less common in other sites. The facultative organisms that most often accompany anaerobes in infections with anaerobes are *Escherichia coli*, *Klebsiella*, *Enterobacter*, *Proteus*, *Streptococcus*, *Staphylococcus*

TABLE 45-4. MAJOR NON-SPOREFORMING ANAEROBIC* BACTERIA OF CLINICAL SIGNIFICANCE

Gram-negative bacilli
  Bacteroides
    fragilis ⎫
    thetaiotaomicron ⎪
    distasonis ⎬ B. fragilis group
    vulgatus ⎪ (included in former
    ovatus ⎪ subspecies)
    uniformis ⎭
    asaccharolyticus ⎫
    gingivalis ⎪
    intermedius ⎪
    corporis ⎪
    melaninogenicus ⎬ Former
    denticola ⎪ B. melaninogenicus
    loescheii ⎭
    bivius
    disiens
    oris
    buccae
    ureolyticus
  Fusobacterium
    nucleatum
    necrophorum
    mortiferum
  Wolinella succinogenes
  Capnocytophaga ochracea
Gram-positive bacilli
  Actinomyces israelii
  Eubacterium
    lentum
    alactolyticum
    nodatum
  Propionibacterium acnes
Gram-positive cocci
  Peptostreptococcus anaerobius
  Peptococcus
    magnus
    prevotii
    asaccharolyticus
  Gaffkya anaerobia
  Streptococcus intermedius
Gram-negative cocci
  Veillonella parvula

*Includes microaerophilic species. Also, numerous species of *Clostridium*, histotoxic and others, can occur with the non-sporeformers (see Chap. 46).

(usually not *Staphylococcus aureus*), *Lactobacillus*, and diphtheroids, depending also on the body site.

The incidence of anaerobic bacteria in the infections they are most commonly associated with is shown in Table 45-5. The major types of infections are briefly discussed below.

Intraabdominal Infections. Infectious complications generally derive from spillage of fecal matter into the peritoneal cavity in settings such as penetrating abdominal trauma, surgery, appendicitis, diverticulitis, inflammatory bowel disease, or cancer. The first clinical manifestation is generally peritonitis, which is followed in survivors by abscess formation and localization of the infection. Infectious complications following disruption of the integrity of the lower bowel and colon are more frequent than of the upper bowel due to the higher concentrations of bacteria, especially anaerobes (10^{11}/g feces) in the colon. On the average, five species—three anaerobic and two facultative—are isolated from abdominal infections. Species of *Bacteroides*, particularly *B. fragilis* and related organisms, *Clostridium*, *Fusobacterium*, and anaerobic gram-positive cocci are the most frequent anaerobic isolates.

Obstetric and Gynecologic Infections. Anaerobes play a prime role in salpingitis, tuboovarian and pelvic abscesses, vaginal cuff infections following hysterectomy, pelvic cellulitis, endometritis, postabortal sepsis, chorioamnionitis, and other infections. Especially associated with infection are premature rupture of membranes, prolonged labor, extensive manipulations and hemorrhage during delivery, nonelective caesarean section and other surgery, spontaneous or induced abortion, malignancy, gonococcal salpingitis, and intrauterine contraceptive devices. Endometritis may be limited or spread to produce tuboovarian infection, peritonitis, pelvic abscess, and septicemia. Chronic pelvic infections may subsequently develop. Likewise, salpingitis may extend and produce generalized pelvic infections with sequelae such as chronic pelvic pain, infertility, ectopic pregnancy, and recurrent infection. Multiple anaerobic and facultative species are common, as in abdominal infections. *Bacteroides* and anaerobic gram-positive cocci are frequently isolated and have largely replaced group A streptococci in puerperal sepsis.

Pleuropulmonary Infections. Anaerobic bacteria are important causes of pneumonitis (pulmonary infiltrate without cavity formation), lung abscess, necrotizing pneumonia, and empyema, but not of lobar pneumonia or chronic bronchitis. Aspiration of mouth flora, as may occur in alcoholism or general anesthesia, generally is the inciting event for these diverse pleuropulmonary infections. Other common underlying conditions are dental infections or other extrapulmonary anaerobic disease, bronchogenic carcinoma, pulmonary embolus with infarction, and bronchiectasis. As in other anaerobic infections, pleuropulmonary foci yield mixed species of bacteria, usually three or four, but ranging to over 10 species per specimen. In these settings, the common anaerobic isolates are *Fusobacterium nucleatum* (Fig. 45-3), former subspecies of *B. melaninogenicus* (see section on black-pigmenting *Bacteroides* for present nomenclature), and gram-positive cocci. *B. fragilis* is isolated in approximately 20 percent of these infections, even though it is not present in the normal flora of the oral cavity.

Upper Respiratory Tract Infections. Chronic forms of a wide variety of infections of the upper respiratory tract are associated with anaerobes. These include periodontal disease, various fusospirochetal diseases, actino-

TABLE 45-5. INFECTIONS TYPICALLY INVOLVING NON-SPOREFORMING ANAEROBIC BACTERIA

Site	Anaerobes Present (Percent)	Approximate Percentage of Anaerobe-Positive Sites With Anaerobes Exclusively
Bacteremia	10–20	80
Central nervous system—brain abscess	89	66
Ear, nose, throat, and dental—chronic otitis media; chronic sinusitis; dental and oral infections	52	80
Thoracic		
Aspiration pneumonia	93	50*
Lung abscess	93	66
Bronchiectasis		
Empyema (nonsurgical)	76	50
Intraabdominal		
Intraabdominal infection (general)	86	10
Liver abscess (pyogenic)	50–100	66
Appendicitis with peritonitis	96	< 10
Other intraabdominal infection (postsurgery)	93	20
Obstetric–gynecologic		
Vulvovaginal abscess	74	50
Salpingitis and pelvic peritonitis	29–75	20
Tuboovarian and pelvic abscess	92	50
Septic abortion and endometritis	73	20
Postoperative wound infection	67	25
Other sites, soft tissue, or abscess		
Necrotizing fasciitis	81	< 10
Cellulitis, perirectal abscess, breast abscess	92	10

Adapted from Finegold: Anaerobic Bacteria in Human Disease, 1977. New York, Academic Press.
*Rate for solely community-acquired pneumonia is higher (66%).
Note: Numerous species of *Clostridium,* histotoxic and others, may accompany the non-sporeformers, particularly in abdominal infections, necrotizing fasciitis, and cellulitis.

mycosis, peritonsillar abscesses, otitis media, mastoiditis, and sinusitis. Former subspecies of *B. melaninogenicus* contribute significantly to the pathogenesis of many of these infections.

The association of anaerobic bacteria with chronic otitis media, mastoiditis, and sinusitis doubtlessly accounts for the contiguous spread of these organisms into the central nervous system and their high incidence in nontraumatic brain abscesses. Hematogenous spread from anaerobic pleuropulmonary infections and sepsis following dental extractions are also sources of brain abscess.

Soft Tissue Infections. Anaerobic infections of the skin and soft tissues usually evolve from traumatic injury, surgery, or ischemia associated with vascular disease or diabetes mellitus. The specific anaerobes involved depend largely on the site of infection or source of the infecting bacteria. For example, human bites that can develop into serious infections usually involve normal flora of the mouth. These diverse anaerobic infections usually produce extensive tissue necrosis with extension along subcutaneous and fascial planes, gas, and a foul odor. The better

characterized of these infections include:

1. progressive bacterial synergistic gangrene
2. chronic undermining ulcer of Meleney
3. synergistic necrotizing cellulitis
4. necrotizing fasciitis
5. streptococcal gangrene, usually caused by *Streptococcus pyogenes* and less frequently by anaerobic gram-positive cocci.

Septicemia. Abdominal and pelvic infections account for the majority of anaerobic bacteremias in a general hospital. Many anaerobic bacteremias involve gram-negative anaerobes, particularly *B. fragilis,* and may be associated with jaundice, septic thrombophlebitis, and suppuration at distant sites from metastases. The current overall mortality in anaerobic septicemia is 25 to 35 percent, but large differences are found relative to the patient population. Bacteremia stemming from abdominal infections, particularly with underlying diseases, is associated with a high mortality and poorer prognosis than bacteremia in obstetric patients. Reports of the incidence and clinical outcome of bacteroides sepsis in neonates have varied.

Laboratory Diagnosis

Because of their endogenous nature, the bacteriologic diagnosis of infections involving anaerobic bacteria differs from those with exogenously derived aerobic or facultative pathogens (Table 45-3). Thus, isolation of anaerobic bacteria may be clinically meaningless or uninterpretable unless the specimen is derived from a closed loculation or a site that is normally sterile (Table 45-6). The following specimens are generally unacceptable for anaerobic culture because the collection sites are colonized or because contamination with indigenous flora often occurs: expectorated sputum, throat swabs, nasotracheal or bronchoscopy aspirates, gastrointestinal contents, feces (except for *Campylobacter* or *Clostridium difficile*), vaginal or endocervical secretions, midstream urine, and skin or superficial wound swabs.

Aspirated or tissue specimens are preferable to swabs whenever feasible, because better survival of pathogens, greater quantity of specimen, and less contamination with extraneous organisms are often achieved. Optimally, specimens should be immediately delivered (within 20 minutes) to the laboratory for culture. This is particularly important if specimens are not placed in an anaerobic transport device. These devices are tubes or vials containing an anaerobic gas mixture substituted for air, which protects the organisms from O_2-inactivation and drying during transport to the laboratory.

Culture reports for anaerobic bacteria generally require more time than for facultative organisms, as many anaerobes grow more slowly and usually must be isolated from complex mixtures of species. A direct gram stain of the specimen read by a trained individual can be a significant aid to the clinician in determining the types of organisms involved and thus in the choice of initial antibiotic therapy.

Communication between clinician and microbiologist is important in securing appropriate specimens and in the reporting of anaerobic cultures. Most combinations of species isolated/specimen source are readily interpretable, but other situations are highly dependent on numerous clinical parameters. For example, *Propionibacterium acnes* is usually considered a contaminant (of blood, spinal fluid, or other cultures) because it is part of the normal flora of the skin. However, in certain clinical settings it may assume the role of a pathogen.

Routine antimicrobial susceptibility testing of anaerobic isolates has not been performed in many clinical laboratories because of the somewhat predictable patterns of susceptibility for many species, delayed testing and reporting due to slower growth, and the necessity for special methods. Increasing resistance among anaerobic bacteria, particularly the gram-negative bacilli, and the current availability of a wider variety of antimicrobials for which susceptibility is less predictable require more routine testing. Susceptibility testing of anaerobes should be performed in the following settings: life-threatening infections, critical sources such as blood, cerebrospinal fluid, and surgical osteomyelitis specimens, lack of response to antimicrobial therapy, and whenever serious sequelae are likely. Testing procedures appropriate for anaerobes (e.g., broth-disc method) are published in the various anaerobic bacteriology manuals and in selected publications. A reference standard agar dilution test for determination of the minimum inhibitory concentration of antimicrobials has been developed under the auspices of the National Committee for Clinical Laboratory Standards (USA).

Treatment

In many circumstances, surgical drainage and resection of necrotic tissue are therapeutic mainstays for anaerobic infections. Even where surgery is appropriate, however, effective concurrent antimicrobial therapy is important. In nonsurgical settings, antimicrobial selection may be critical because the most common clinical isolate, *B. fragilis*, and strains of many other *Bacteroides* are resistant to some of the more frequently used antibiotics. In anaerobic septicemias, especially those involving *Bacteroides*, mortality among patients treated with antimicrobials active against the infecting anaerobes is 12 to 16 percent, compared with 60 percent mortality among patients who receive inappropriate drugs. Parenteral therapy usually is required in an effort to attain adequate drug levels in necrotic tissue and abscesses; chronic infections such as lung or liver abscesses must be treated for prolonged periods to prevent relapse.

The antimicrobials with the highest activities against anaerobic bacteria, indicated both by in vitro susceptibility tests and clinical efficacy, are primarily clindamycin, metronidazole, and chloramphenicol, followed by cefoxitin, moxalactam, carbenicillin, ticarcillin, and the newer penicillins, mezlocillin and piperacillin. Clindamycin has high activity against most anaerobic bacteria, including *B. fragilis*. It has proved clinical effectiveness in body sites other than the central nervous system, where it does not penetrate. Limited instances of plasmid-mediated resistance of

TABLE 45-6. APPROPRIATE SPECIMENS FOR ANAEROBIC CULTURE

Normally sterile tissues
Body fluids—bile, pleural, sinus, joint, pericardial, peritoneal
Abscess contents
Blood cultures
Deep aspirates of wounds
Transtracheal aspirates
Bronchoscopy aspirates*
Amniotic fluid at caesarean section, amniocentesis, amniotomy
Endometrium (lochia)*
Culdocentesis fluid
Urine, catheterized or suprapubic aspirate (only in complicated cases)
Cerebrospinal fluid (only in complicated cases)

*When collected with a device that protects the sampling portion (e.g., swab, brush, aspirator) from contamination with endogenous flora en route to the normally sterile area being cultured.

B. fragilis to clindamycin have been reported. Metronidazole has specific activity for anaerobes and is mostly inactive against aerobic and facultative bacteria. It is bactericidal for most obligately anaerobic strains, which includes the *Bacteroides,* but aerotolerant or microaerophilic strains of *Streptococcus, Actinomyces, Propionibacterium,* and certain other gram-positive bacilli are often resistant. It penetrates the blood–brain barrier and is effective in anaerobic brain abscess. Chloramphenicol is widely active in vitro against anaerobic bacteria, including *B. fragilis,* and was previously the major drug available for anaerobic brain abscess. Occasional failure of the drug in vivo may relate to its decreased penetration of abscesses compared with clindamycin and metronidazole or inactivation of the drug by high concentrations of *Bacteroides* and *Clostridium.* The serious side effects of the drug, although quite infrequent, have generally restricted its use to severe or life-threatening infection.

A key element regarding a drug's performance is activity against the *B. fragilis* group of organisms and many other *Bacteroides* that constitutively produce β-lactamases that can inactivate penicillins and the older cephalosporin antibiotics. Certain of the newer cephalosporins and penicillins are more active against these resistant *Bacteroides* because of increased resistance to β-lactamase degradation and/or the higher blood levels achievable. Thus, cefoxitin and moxalactam have increased activity against *B. fragilis* and other *Bacteroides,* as do carbenicillin, ticarcillin, mezlocillin, and piperacillin, although these latter agents are still susceptible to β-lactamases. This susceptibility may be important in light of the high concentration of anaerobic bacteria in abscesses.

Members of the *B. fragilis* group are resistant to the usual clinical levels of penicillin G, ampicillin, nafcillin, methicillin, and cephalothin. Despite previous generalizations, resistance to these drugs has not been limited to the *B. fragilis* group. However, resistance also is apparently increasing among many species, and other *Bacteroides,* notably *B. bivius, Bacteroides disiens,* certain of the former subspecies of *B. melaninogenicus, Bacteroides capillosus,* and *Bacteroides oris* are frequently resistant or only moderately susceptible. The high association of *B. fragilis* and other resistant *Bacteroides* with anaerobic infections of the abdomen and pelvis precludes widespread use of penicillin or the other drugs noted above for serious anaerobic infections in these sites. Although penicillin G has been the drug of choice for pleuropulmonary infections, it recently has been demonstrated to be less effective than clindamycin in treatment of anaerobic lung abscess. Penicillin G, ampicillin, and cephalothin are highly active, however, against a variety of other anaerobes, including certain other *Bacteroides,* fusobacteria, and anaerobic gram-positive cocci and bacilli.

Anaerobic bacteria are uniformly resistant to the aminoglycoside antibiotics, making such drugs as streptomycin, neomycin, kanamycin, gentamicin, tobramycin, and amikacin useless against anaerobes in infection. Many anaerobes were sensitive to tetracycline in the past, but increasing resistance has been noted and presently about two thirds of clinical isolates of *B. fragilis* are resistant. Two tetracycline derivatives, doxycycline and minocycline, are active against 65 to 75 percent of *B. fragilis* strains, in addition to many other anaerobes.

Genetic transfer of drug resistance among *Bacteroides* apparently has a limited clinical impact at present, but is worrisome because of the high frequency of these organisms in infection and high concentrations in our endogenous flora. Resistance transfer, most intensively studied with *B. fragilis* strains, has been observed for tetracycline, clindamycin, erythromycin and high-level penicillin and ampicillin resistance. Certain resistance determinants are also transferable to *E. coli.*

Knowledge of the usual pathogens in specific sites of infection and a direct gram stain of the specimen aid in the choice of initial therapy, but changes may be necessary when the laboratory obtains the isolates from culture. Because mixed infections containing both anaerobic and facultative species frequently occur, two or even three antibiotics may be required for coverage of known pathogens unless one of the more broad spectrum agents, which also covers the facultatively anaerobic species, is appropriate. For example, clindamycin and metronidazole are not active against the facultative gram-negative rods, and an aminoglycoside or another antibiotic is usually necessary to cover organisms such as *E. coli,* especially in pelvic and abdominal infection.

Hyperbaric oxygen theoretically could be an alternate mode of therapy for anaerobic infections, but it has been primarily evaluated only for clostridial myonecrosis (Chap. 46). The bacteriostatic and bactericidal effects of high oxygen tensions on anaerobic bacteria, achievable by hyperbaric oxygenation, might be useful clinically as an adjunct to conventional antimicrobial and surgical therapy or as an alternate to surgery for inoperable patients.

Clinically Significant Non-Sporeforming Anaerobic Bacteria

This section presents a brief description of some of the species of clinically significant non-sporeforming anaerobic bacteria. The following genera, listed in approximately decreasing order of incidence, are isolated from the types of infections discussed in this chapter: *Bacteroides, Peptococcus, Peptostreptococcus, Propionibacterium, Clostridium, Fusobacterium,* and *Streptococcus,* primarily; followed by *Eubacterium, Lactobacillus, Veillonella, Actinomyces, Bifidobacterium, Arachnia,* and *Acidaminococcus. Treponema, Campylobacter,* and *Borrelia,* not included in this list, are covered in Chapters 47 and 48.

Anaerobic Gram-negative Bacilli

Bacteroides

Bacteroides fragilis. Previously, *B. fragilis* was divided into five subspecies: fragilis, thetaiotaomicron, distasonis, vulgatus, and ovatus. These now have been designated as separate species based on DNA homology studies, but are frequently referred to collectively as the *B. fragilis* group. These species, when identified by phenotypic characteristics, also have been found to contain additional homology groups, but most have not yet been named.

Among this group of saccharolytic intestinal *Bacteroides*, *B. fragilis* is by far the most common isolate from infections, yet is present in relatively lower concentrations in the normal fecal flora. *Bacteroides thetaiotaomicron* is the next most frequent from clinical sources. These *Bacteroides* primarily are associated with intraabdominal infections or septicemias derived from this site. Although *B. fragilis* and the other former subspecies are infrequently present in the vagina and cervix of healthy women, they are isolated from genital tract infections where they probably gain access from the perineal region. All of these former subspecies are generally resistant to ordinary doses of penicillin and older cephalosporin antibiotics and share many characteristics. However, because *B. fragilis* is clinically most important and possesses some unique characteristics, it will be discussed separately in the remainder of this section.

MORPHOLOGY AND PHYSIOLOGY. *B. fragilis* may appear as pleomorphic bacilli with vacuoles and swellings, which are particularly apparent when the organisms are stained from broth containing fermentable carbohydrate (Fig. 45-1). Colonies of *B. fragilis* are low convex, white to gray, semiopaque, and glistening, and some strains may be hemolytic. The organisms grow more rapidly than most nonsporeforming anaerobes, and growth is stimulated by bile. *B. fragilis* is a moderate anaerobe, growing maximally in pO_2 less than 3 percent, but is capable of surviving prolonged exposures to oxygen, particularly in the presence of blood. The organism produces superoxide dismutase and also a catalase (in the presence of hemin).

ANTIGENIC STRUCTURE. Thermolabile protein and thermostable lipopolysaccharide antigens have provided a basis for serologic classification of *B. fragilis*. Strains of the five former subspecies of *B. fragilis* can be divided into corresponding serotypes on the basis of agglutination, gel diffusion, and fluorescent-antibody assays. Different serotypes among *B. fragilis* strains have required the use of pooled antisera for identification of the species. Commercial pooled antisera are available for fluorescent-antibody detection of members of the *B. fragilis* group (all former subspecies).

More recently, a species-specific capsular polysaccharide antigen has been demonstrated for strains of *B. fragilis*. A polysaccharide capsule is rarely present among other members of the *B. fragilis* group, but is characteristic of clinical strains of *B. fragilis*.

The antibody response of patients with various infections, including septicemia, soft tissue infections, or abscesses due to members of the Bacteroidaceae, has been investigated to study the pathologic significance of bacteroides in mixed infections and possibly to devise a clinically useful method of detecting bacteroides infections. Precipitin and agglutination techniques have detected antibodies to *B. fragilis*, other *Bacteroides*, and fusobacteria in infected patients' sera that are absent in control sera. More sensitive techniques such as passive hemagglutination have detected antibodies, primarily of the IgM class, to *B. fragilis* in healthy individuals. Using a sensitive radioactive antigen-binding assay, antibody response to *B. fragilis* capsular polysaccharide has been demonstrated in experimental animals with abscesses induced with an encapsulated strain of *B. fragilis*, and in women with pelvic infection.

DETERMINANTS OF PATHOGENICITY. As previously mentioned, the lipopolysaccharide of the outer membrane of *B. fragilis* lacks certain characteristics of classic endotoxin and has much less biologic activity. The polysaccharide capsule of *B. fragilis* appears to confer added virulence to this species, which is indicated indirectly by the disproportionately higher association of *B. fragilis* in human infection. Direct evidence from experimental studies in animals includes the following:

1. Encapsulated strains of *B. fragilis* were more virulent than unencapsulated strains;
2. Heat-killed encapsulated *B. fragilis* were capable of producing abscesses; and
3. Prior immunization with capsular polysaccharide conferred protection against challenge with encapsulated *B. fragilis* in a model of intraabdominal infection.

Mechanism(s) by which the capsule enhances virulence include interference with phagocytosis and opsonophagocytic killing by neutrophils and possibly also greater adherence to rat peritoneal mesothelium than the unencapsulated strains of other former subspecies, which may also affect clearance of the organism. As with *Streptococcus pneumoniae*, laboratory passage of *B. fragilis* reduces the amount of capsular polysaccharide resulting in a diminution of virulence.

Black-pigmenting *Bacteroides*. All *Bacteroides* that produce tan to black pigment on blood agar previously were classified as subspecies of *B. melaninogenicus*. By DNA homology, DNA base composition, and other studies, these organisms have been shown to be heterogeneous even within the former subspecies groups; recently the better characterized homology groups have been elevated to species rank. Organisms previously classified as *B. melaninogenicus* subspecies asaccharolyticus presently are separated into *B. asaccharolyticus,* which is most often found in nonoral sites of infection, and *B. gingivalis,* which occurs primarily in oral sites and is associated with forms of severe periodontal disease. Organisms formerly identified as *B. melaninogenicus* subspecies melaninogenicus presently are separated into *Bacteroides denticola, B. melaninogenicus,* and *Bacteroides loescheii;* all are isolated from predominately oral, but also from other body sites. Former *B. melaninogenicus* subspecies intermedius strains now are classified as *Bacteroides corporis,* which commonly occur in gastrointestinal and urogenital sites, and *Bacteroides intermedius* (Fig. 45-2), which are isolated from oral and other body sites. Former subspecies levii has also been elevated to species rank, *Bacteroides levii.*

The cellular morphology of these *Bacteroides* differs somewhat according to species, but strains often appear as small coccobacilli, usually with longer rod forms also present (see Fig. 45-2). Colonies on blood agar usually are convex, smooth, circular, often β-hemolytic, and pigmented, becoming tan to black in 2 to 14 days. Vitamin K and hemin are required or are highly stimulatory for the growth of most strains. These species can be separated on the basis of serologic tests, and a polysaccharide capsular antigen that is species-specific for *B. gingivalis* has been isolated. As in *B. fragilis,* the lipopolysaccharide present in the outer membrane of these *Bacteroides* species is biochemically distinct from that of facultative gram-negative organisms, and its biologic potency is significantly less. Collagenase and a polysaccharide capsule apparently are virulence factors for certain strains. These organisms are important agents in oral, pulmonary, obstetric, and gynecologic infections. Clinical isolates may produce β-lactamase and have demonstrated increasing resistance to penicillin and the older cephalosporin antibiotics.

Bacteroides bivius. *B. bivius* usually appears as a small coccobacillus, often occurring in pairs or short chains. Hemin is required for growth. *B. bivius* is a prevalent anaerobic gram-negative rod in the vaginal flora and is most commonly isolated from genital tract infections, particularly obstetric infections, although it also can be pathogenic in other body sites. Many strains produce β-lactamase and are resistant to penicillin G and the older cephalosporins.

Fusobacterium

F. nucleatum is the most common of the fusobacteria isolated from infections. *F. nucleatum* characteristically is thin with pointed ends, and may resemble scattered wheat straw or appear as very long, thin filaments (Fig. 45-3). Colonies sometimes are α-hemolytic and may be convex and translucent, with internal flecking or mottling, or more umbonate, heaped, dull, and opaque. *F. nucleatum* is present in the normal flora of the mouth and, infrequently, in the urogenital tract. It is an important agent in oral infections, lung abscess, and other pleuropulmonary infections. Another species, *F. necrophorum,* is an important animal pathogen and is found in a variety of human infections, particularly abdominal infections and liver abscesses. These bacilli are generally broad, usually with rounded ends, and may be short, long, or filamentous forms, often with bulbous swellings and round bodies (Fig. 45-4). Colonies may be α- or β-hemolytic. A lipase and a partially characterized leukocidal toxin are produced. Its normal habitat is probably the gastrointestinal tract. Most of the fusobacteria are susceptible to penicillin G and the older cephalosporins.

Anaerobic Gram-positive Bacilli

Eubacterium, Propionibacterium, Lactobacillus, Actinomyces, Arachnia, **and** *Bifidobacterium*

Many of these organisms are slow growing in contrast to *Clostridium,* the sporeforming anaerobic bacilli. Strains within *Propionibacterium, Lactobacillus, Bifidobacterium,* and *Actinomyces* may show sparse to good growth in a CO_2 incubator or aerobically, while other strains are obligately anaerobic. Many gram-positive rods are isolated that cannot be identified by present schemes.

Eubacterium lentum is the most common of the eubacteria isolated from nonoral clinical specimens. It is a coccobacillus that often is found with *B. fragilis,* but little is known of its pathogenic role, if any. It is part of the normal flora of the gastrointestinal tract. Three newly described species of *Eubacterium, Eubacterium brachy, Eubacterium timidum,* and *Eubacterium nodatum,* are prevalent in periodontitis. Their relatively slow and minimal growth on artificial media has probably retarded their recognition from dental and other body sites. *E. nodatum* resembles *Actinomyces* in cellular morphology.

P. acnes and *Propionibacterium granulosum* are normal inhabitants of the gastrointestinal tract and, primarily, of the skin. Consequently, they occur most frequently as contaminants in cultures of blood and cerebrospinal fluid, but rarely they appear to be causally associated with infections. They were previously classified as anaerobic members of the genus *Corynebacterium. Propionibacterium* may closely resemble *Actinomyces* or *Arachnia propionica,* as the cells are pleomorphic and may be branched and/or diphtheroidal.

Strains of *Lactobacillus* are normal flora in the mouth and gastrointestinal tract and, in some women, are the predominant flora in the vagina. Most species apparently have minimal pathogenic potential, but *Lactobacillus catenaforme* is especially associated with pleuropulmonary infections.

The *Actinomyces* are discussed separately in Chapter 34. Previously, their characteristically slow growth resulted in their affiliation with diagnostic mycology rather than bacteriology laboratories. *A. propionica,* which can also cause actinomycosis, is closely related to the *Actinomyces* and to *Propionibacterium,* but because of certain metabolic differences it has been placed in a separate genus.

Bifidobacteria are infrequently involved in infection. *Bifidobacterium dentium* (previously *Actinomyces eriksonii* or *Bifidobacterium eriksonii*) has been isolated from a variety of sites. Various bifidobacteria are normal flora in the mouth and urogenital tract and occur in high numbers in the gastrointestinal tract.

Anaerobic Gram-positive Cocci

Peptococcus, Peptostreptococcus, and *Streptococcus*

The most common clinical isolate of *Peptostreptococcus* is *Peptostreptococcus anaerobius.* Among *Peptococcus, Peptococcus magnus, Peptococcus asaccharolyticus,* and *Peptococcus prevotii* are most common. Certain species previously classified among *Peptococcus* and *Peptostreptococcus* produce lactic acid as their major metabolic end-product, and include strains that will grow in a CO_2 incubator or, rarely, in air, usually after initial anaerobic isolation and subculture but occasionally also on primary isolation. These organisms have been reclassified as *Streptococcus,* and *Streptococcus intermedius* is the most frequently isolated from infections. The relative importance of cellular morphology (clumps or chains), the presence of catalase, and metabolic patterns in the classification of *Peptococcus* and *Peptostreptococcus* is controversial. Corresponding to the facultative genera, *Staphylococcus* and *Streptococcus,* peptococci generally occur as singles, pairs, tetrads, and irregular clumps, and peptostreptococci may be in singles, pairs, or chains. Many of these gram-positive anaerobic cocci are similar in colony appearance. These species of cocci are normal flora in the mouth, urogenital tract, and gastrointestinal tract. They are prevalent in a wide variety of human infection but are particularly important in pleuropulmonary disease, brain abscess, and obstetric and gynecologic infections. Another organism, temporarily called *Gaffkya anaerobia,* occurs in irregular clumps, tetrads, and pairs; *Gaffkya* is particularly common from obstetric and gynecologic infections.

Anaerobic Gram-negative Cocci

Veillonella and *Acidaminococcus*

Veillonella parvula is isolated from clinical specimens, but little is known of its role in the production of infection. *Veillonella* are small cocci occurring in pairs, short chains, and clumps. They are present in the normal flora of the mouth, in particular, and the gastrointestinal and urogenital tracts. *Acidaminococcus fermentans* is isolated very infrequently. The cocci are larger than *Veillonella,* and cells may stain partially gram-positive. This species has been isolated as normal flora from the gastrointestinal and urogenital tracts.

FURTHER READING

Books, Manuals, and Reviews

Balows A, DeHaan RM, Dowell VR Jr, et al. (eds): Anaerobic Bacteria: Role in Disease. Springfield, Ill, Thomas, 1974

Buchanan RE, Gibbons NE (eds): Bergey's Manual of Determinative Bacteriology, 8th ed. Baltimore, Williams & Wilkins, 1974

Dowell VR Jr, Allen SD: Anaerobic bacterial infections. In Balows A, Hausler WJ (eds): Diagnostic Procedures for Bacterial, Mycotic and Parasitic Infections, 6th ed. New York, American Public Health Association, Inc, 1981, p 171

Dowell VR Jr, Hawkins TM: Laboratory Methods in Anaerobic Bacteriology. Centers for Disease Control Laboratory Manual. Washington, D.C., US Government Printing Office. DHEW Publication No (CDC) 74-8272, 1974

Dowell VR Jr, Lombard GL: Pathogenic members of the genus *Bacteroides.* In Starr MP, Stolp H, Trüper HG, et al. (eds): The Prokaryotes. A Handbook on Habitats, Isolation, and Identification of Bacteria. New York, Springer-Verlag, 1981, p 1425

Dowell VR Jr, Lombard GL, Thompson FS, et al.: Media for Isolation, Characterization, and Identification of Obligately Anaerobic Bacteria. Atlanta, DHEW Public Health Service, Centers for Disease Control, 1977

Finegold SM: Anaerobic Bacteria in Human Disease. New York, Academic Press, 1977

Finegold SM, Shepherd WE, Spaulding EH: Practical Anaerobic Bacteriology, Cumitech 5. Washington, D.C., American Society for Microbiology, 1977

George WL, Kirby BD, Sutter VL, et al.: Gram-negative anaerobic bacilli: Their role in infection and patterns of susceptibility to antimicrobial agents. II. Little-known *Fusobacterium* species and miscellaneous genera. Rev Infect Dis 3:599, 1981

Gorbach SL, Bartlett JG: Anaerobic infections. N Engl J Med 290:1177, 1237, 1289, 1974

Hill GB: The anaerobic cocci. In Starr MP, Stolp H, Trüper HG, et al. (eds): The Prokaryotes. A Handbook on Habitats, Isolation, and Identification of Bacteria. New York, Springer-Verlag, 1981, p 1631

Holdeman LV, Cato EP, Moore WEC: Anaerobe Laboratory Manual, 4th ed. Blacksburg, Virginia Polytechnic Institute and State University, 1977

Kirby BD, George WL, Sutter VL, et al.: Gram-negative anaerobic bacilli: Their role in infection and patterns of susceptibility to antimicrobial agents. I. Little-known *Bacteroides* species. Rev Infect Dis 2:914, 1980

Lennette EH, Balows A, Hausler WJ Jr, et al. (eds): Manual of Clinical Microbiology, 3rd ed. Washington, D.C., American Society for Microbiology, 1980

Miraglia GJ: Pathogenic anaerobic bacteria. CRC Crit Rev Microbiol 3:161, 1974

Moore WEC, Cato EP, Holdeman LV: Anaerobic bacteria of the gastrointestinal flora and their occurrence in clinical infections. J Infect Dis 119:641, 1969

National Committee for Clinical Laboratory Standards: Proposed Reference Dilution Procedure for Antimicrobic Susceptibility Testing of Anaerobic Bacteria. Villanova, Pa, 1979

Prevot AR, Fredett V (trans): Manual for the Classification and Determination of the Anaerobic Bacteria. Philadelphia, Lea & Febiger, 1965

Sabbaj J, Sutter VL, Finegold SM: Anaerobic pyogenic liver abscess. Ann Intern Med 77:629, 1972

Silver S: Anaerobic Bacteriology for the Clinical Laboratory. St. Louis, Mosby, 1980

Smith LDS: The Pathogenic Anaerobic Bacteria. Springfield, Ill, Thomas, 1975

Sutter VL, Vargo VL, Finegold SM: Wadsworth Anaerobic Bacteriology Manual, 3rd ed. St. Louis, Mosby, 1980

Selected Papers

Altemeier WA: The pathogenicity of the bacteria of appendicitis peritonitis. Surgery 11:374, 1942

Aranki A, Freter R: Use of anaerobic glove boxes for the cultivation of strictly anaerobic bacteria. Am J Clin Nutr 25:1329, 1972

Bartlett JG: Anti-anaerobic antibacterial agents. Lancet ii:478, 1982

Bartlett JG, Finegold SM: Anaerobic infections of the lung and pleural space. Am Rev Respir Dis 110:56, 1974

Brown WJ, Waatti PE: Susceptibility testing of clinically isolated anaerobic bacteria by an agar dilution technique. Antimicrob Agents Chemother 17:629, 1980

Carter B, Jones CP, Alter RL, et al.: *Bacteroides* infections in obstetrics and gynecology. Obstet Gynecol 1:491, 1953

Chow AW, Guze LB: Bacteroidaceae bacteremia: Clinical experience with 112 patients. Medicine 53:93, 1974

Coykendall AL, Kaczmarek FS, Slots J: Genetic heterogeneity in *Bacteroides asaccharolyticus* (Holdeman and Moore 1970) Finegold and Barnes 1977 (approved lists, 1980) and proposal of *Bacteroides gingivalis* sp. nov. and *Bacteroides macacae* (Slots and Genco) comb. nov. Int J Syst Bacteriol 30:559, 1980

Cuchural G, Jacobus N, Gorbach SL, et al.: A survey of *Bacteroides* susceptibility in the United States. J Antimicrob Chemother 8(Suppl D):27, 1981

Dowell VR Jr: Comparison of techniques for isolation and identification of anaerobic bacteria. Am J Clin Nutr 25:1335, 1972

Finegold SM, Bartlett JG, Chow AN, et al.: Management of anaerobic infections. Ann Intern Med 83:375, 1975

Gorbach SL, McGowan K: Comparative clinical trials in treatment of intra-abdominal sepsis. J Antimicrob Chemother 8(Suppl D):95, 1981

Gregory EM, Kowalski JB, Holdeman LV: Production and some properties of catalase and superoxide dismutase from the anaerobe *Bacteroides distasonis*. J Bacteriol 129:1298, 1977

Hill GB: Anaerobic flora of the female genital tract. In Lambe DW Jr, Genco RJ, Mayberry-Carson KJ (eds): Anaerobic Bacteria: Selected Topics. New York, Plenum, 1980, p 39

Hill GB, Osterhout S, Pratt PC: Liver abscess production by nonsporeforming anaerobic bacteria in a mouse model. Infect Immun 9:599, 1974

Holdeman LV, Cato EP, Burmeister JA, et al.: Descriptions of *Eubacterium timidum* sp. nov., *Eubacterium brachy* sp. nov., and *Eubacterium nodatum* sp. nov. isolated from human periodontitis. Int J Syst Bacteriol 30:163, 1980

Holdeman LV, Cato EP, Moore WEC: Recent changes in identification and classification of some anaerobes. In Lambe DW Jr, Genco RJ, Mayberry-Carson KJ (eds): Anaerobic Bacteria: Selected Topics. New York, Plenum, 1980, p 25

Holdeman LV, Johnson JL: *Bacteroides disiens* sp. nov. and *Bacteroides bivius* sp. nov. from human clinical infections. Int J Syst Bacteriol 27:337, 1977

Holdeman LV, Johnson JL: Description of *Bacteroides loescheii* sp. nov. and emendation of the descriptions of *Bacteroides melaninogenicus* (Oliver and Wherry) Roy and Kelly 1939 and *Bacteroides denticola* Shah and Collins 1981. Int J Syst Bacteriol 32:399, 1982

Holdeman LV, Moore WEC, Churn PJ, et al.: *Bacteroides oris* and *Bacteroides buccae,* new species from human periodontitis and other human infections. Int J Syst Bacteriol 32:125, 1982

Holland JW, Hill EO, Altemeier WA: Numbers and types of anaerobic bacteria isolated from clinical specimens since 1960. J Clin Microbiol 5:20, 1977

Ingham HR, Sisson PR, Middleton RL, et al.: Phagocytosis and killing of bacteria in aerobic and anaerobic conditions. J Med Microbiol 14:391, 1981

Ingham HR, Sisson PR, Selkon JB: Current concepts of the pathogenetic mechanisms of non-sporing anaerobes: Chemotherapeutic implications. J Antimicrob Chemother 6:173, 1980

Johnson JL: Taxonomy of the Bacteroides I. Deoxyribonucleic acid homologies among *Bacteroides fragilis* and other saccharolytic *Bacteroides* species. Int J Syst Bacteriol 28:245, 1978

Johnson JL, Holdeman LV: *Bacteroides intermedius* comb. nov. and descriptions of *Bacteroides corporis* sp. nov. and *Bacteroides levii* sp. nov. Int J Syst Bacteriol 33:15, 1983

Kasper DL, Onderdonk AB, Polk BF, et al.: Surface antigens as virulence factors in infection with *Bacteroides fragilis*. Rev Infect Dis 1:278, 1979

Leadbetter ER, Holt SC, Socransky SS: *Capnocytophaga*: New genus of gram-negative gliding bacteria I. General characteristics, taxonomic considerations and significance. Arch Microbiol 122:9, 1979

Loesche WJ: Oxygen sensitivity of various anaerobic bacteria. Appl Microbiol 18:723, 1969

Mansheim BJ, Onderdonk AB, Kasper DL: Immunochemical characterization of surface antigens of *Bacteroides melaninogenicus*. Rev Infect Dis 1:263, 1979

Martin WJ: Isolation and identification of anaerobic bacteria in the clinical laboratory. Mayo Clin Proc 49:300, 1974

Mayrand D, McBride BC: Ecological relationships of bacteria involved in a simple, mixed anaerobic infection. Infect Immun 27:44, 1980

McCord JM, Keele BB Jr, Fridovich I: An enzyme-based theory of obligate anaerobiosis: The physiological function of superoxide dismutase. Proc Natl Acad Sci USA 68:1024, 1971

McDonald JB, Sutton RM, Knoll ML, et al.: The pathogenic components of an experimental fusospirochetal infection. J Infect Dis 98:15, 1956

McGowan K, Gorbach SL: Anaerobes in mixed infections. J Infect Dis 144:181, 1981

Meleney FL: Bacterial synergism in disease process. Ann Surg 94:961, 1931

Nichols RL, Smith JW, Fossedal EN, et al.: Efficacy of parenteral antibiotics in the treatment of experimentally induced intraabdominal sepsis. Rev Infect Dis 1:302, 1979

Nord CE, Olsson B, Dornbusch K: β-lactamases in *Bacteroides*. Scand J Infect Dis (Suppl) 13:27, 1978

O'Keefe JP, Tally FP, Barza M, et al.: Inactivation of penicillin G during experimental infection with *Bacteroides fragilis*. J Infect Dis 137:437, 1978

Onderdonk AB, Kasper DL, Cisneros RL, et al.: The capsular polysaccharide of *Bacteroides fragilis* as a virulence factor: Comparison of the pathogenic potential of encapsulated and unencapsulated strains. J Infect Dis 136:82, 1977

Onderdonk AB, Kasper DL, Mansheim BJ, et al.: Experimental animal models for anaerobic infections. Rev Infect Dis 1:291, 1979

Roberts DS: Editorial. Synergistic mechanisms in certain mixed infections. J Infect Dis 120:720, 1969

Rolfe RD, Finegold SM: Comparative in vitro activity of new beta-lactam antibiotics against anaerobic bacteria. Antimicrob Agents Chemother 20:600, 1981

Rudek W, Haque R: Extracellular enzymes of the genus *Bacteroides*. J Clin Microbiol 4:458, 1976

Salyers AA, Wong J, Wilkins TD: Beta-lactamase activity in strains of *Bacteroides melaninogenicus* and *Bacteroides oralis*. Antimicrob Agents Chemother 11:142, 1977

Simon GL, Klempner MS, Kasper DL, et al.: Alterations in opsonophagocytic killing by neutrophils of *Bacteroides fragilis* associated with animal and laboratory passage: Effect of capsular polysaccharide. J Infect Dis 145:72, 1982

Smith DT: Experimental aspiratory abscess. Arch Surg 14:231, 1927

Socransky SS, Gibbons RJ: Required role of *Bacteroides melaninogenicus* in mixed anaerobic infections. J Infect Dis 115:247, 1965

Sonnenwirth AC: Antibody response to anaerobic bacteria. Rev Infect Dis 1:337, 1979

Steffen EK, Hentges DJ: Hydrolytic enzymes of anaerobic bacteria isolated from human infections. J Clin Microbiol 14:153, 1981

Sutter VL, Finegold SM: Susceptibility of anaerobic bacteria to 23 antimicrobial agents. Antimicrob Agents Chemother 10:736, 1976

Sweet RL: Treatment of mixed aerobic-anaerobic infections of the female genital tract. J Antimicrob Chemother 8(Suppl D):105, 1981

Tally FP, Goldin BR, Jacobus NV, et al.: Superoxide dismutase in anaerobic bacteria of clinical significance. Infect Immun 16:20, 1977

Tanner ACR, Badger S, Lai CH, et al.: *Wolinella* gen. nov., *Wolinella succinogenes* (*Vibrio succinogenes* Wolin et al.) comb. nov., and description of *Bacteroides gracilis* sp. nov., *Wolinella recta* sp. nov., *Campylobacter concisus* sp. nov., and *Eikenella corrodens* from humans with periodontal disease. Int J Syst Bacteriol 31:432, 1981

Thadepalli H, Gorbach SL, Broido P, et al.: A prospective study of infections in penetrating abdominal trauma. Am J Clin Nutr 25:1405, 1972

Tofte RW, Peterson PK, Schmeling D, et al.: Opsonization of four *Bacteroides* species: Role of the classical complement pathway and immunoglobulin. Infect Immun 27:784, 1980

Weinstein WM, Onderdonk AB, Bartlett JG, et al.: Experimental intra-abdominal abscesses in rats: Development of an experimental model. Infect Immun 10:1250, 1974

Weinstein WM, Onderdonk AB, Bartlett JG, et al.: Antimicrobial therapy of experimental intraabdominal sepsis. J Infect Dis 132:282, 1975

CHAPTER 46
Clostridium

Clostridium

The clostridia are anaerobic, sporeforming bacilli that usually stain gram-positive. Most species are obligate anaerobes, but a few species are aerotolerant and will grow minimally in air at atmospheric pressure. The pathogenic species produce soluble toxins, some of which are extremely potent. Some species are saccharolytic, producing acid and gas from carbohydrates; many are proteolytic. The clostridia are widely distributed in nature, and are present in soil and in the intestinal tract of humans and animals.

The pathogenic clostridia can be divided into four major groups, according to the types of diseases they produce.

1. The histotoxic clostridia characteristically cause a variety of tissue infections, usually subsequent to wounding or other types of traumatic injury.
2. The enterotoxigenic clostridia produce food poisoning and more severe forms of gastrointestinal disease.
3. *Clostridium tetani*, the causative agent of tetanus, produces disease through a potent exotoxin that is produced during limited growth within tissue.
4. *Clostridium botulinum* is the etiologic agent of botulism which results from the ingestion of a powerful exotoxin previously formed by the organism in contaminated food.

Histotoxic Clostridia

The histotoxic clostridia cause a severe infection of muscle, clostridial myonecrosis. Older and frequently used synonyms for this infection are gas gangrene and clostridial myositis. The term gas gangrene, however, is misleading, as the presence of gas in the infected tissues may be a late or variable manifestation of the disease, and clostridial myositis suggests muscle inflammation rather than the actual pathologic condition, necrosis. The most important histotoxic clostridia are *Clostridium perfringens*, *Clostridium novyi*, and *Clostridium septicum*. Three other organisms of lesser importance also are capable alone of producing clostridial myonecrosis: *Clostridium histolyticum*, *Clostridium sordellii*, and *Clostridium fallax*. All of these histotoxic clostridia produce a variety of toxins of different potencies, and for each species, toxins are designated by Greek letters in order of importance or discovery. Thus, the α toxins of different species are not identical. None of these histotoxic clostridia is a highly invasive pathogen; each plays an opportunistic role that requires a special set of conditions within tissue in order to initiate infection. A spectrum of clinical involvement is seen in clostridial wound infections, ranging from simple contamination of wounds to the most serious type of infection, myonecrosis.

Because the clostridia are so widely distributed in nature, contamination of wounds with these bacteria is very common. Often more than one clostridia are present, including both saprophytic and histotoxic species. An average of 2.6 species of clostridia were isolated from cases of clostridial myonecrosis during World War II; higher numbers would probably have been demonstrated with the improved anaerobic culture techniques currently available. Reported figures for clostridial contamination of wounds in civilian life range up to 39 percent, and contamination during warfare is considerably higher. Only a small proportion of wounds contaminated with *C. perfringens* or other histotoxic clostridia, however, evolve into true clostridial myonecrosis. The incidence of the disease in civilian life is difficult to establish, with considerable variation according to the precipitating incident and geographic location. Its incidence during warfare is from 10 to 100 times greater than during peacetime, occurring in 0.2 to 1 percent of war casualties. An important corollary to the high rate of clostridial contamination of wounds is that the isolation of histotoxic clostridia from wounds or drainage material does *not* by itself indicate clostridial myonecrosis. Diagnosis of clostridial myonecrosis must be made on clinical grounds; the bacteriology laboratory contributes to a careful differential diagnosis by demonstrating histotoxic clostridia or other bacteria that are associated with diseases of similar symptomatology.

Clostridium perfringens

C. perfringens is cultured from 60 to 90 percent of cases of clostridial myonecrosis. There are five types of *C. perfringens, A to E*, separated according to their production of four major lethal toxins (Table 46-1). *C. perfringens* type A

TABLE 46-1. TOXINS AND SOLUBLE ANTIGENS OF *CLOSTRIDIUM PERFRINGENS*

Type	Group	Disease	Major Lethal Toxins*				Minor Antigens[†]				
			α	β	ϵ	τ	δ	θ	κ	λ	μ
A		Gas gangrene in humans and animals	++	–	–	–	–	++	++	–	++
		Food poisoning in humans[‡]									
B	1	Lamb dysentery									
		Enterotoxemia of foals	++	++	++	–	–	++	–	++	++
	2	Enterotoxemia of sheep and goats (Iran)	++	++	++	–	–	++	++	–	–
C	1	Enterotoxemia (struck) of sheep	++	++	–	–	++	++	++	–	–
	2	Enterotoxemia of calves and lambs (Colorado)	++	++	–	–	–	++	++	–	–
	3	Enterotoxemia of piglets	++	++	–	–	–	++	+	–	+
	4	Necrotic enteritis (pig-bel) of humans (New Guinea)	++	++	–	–	–	++	+	–	++
	5	Necrotic enteritis of humans and fowl (Germany)[‡]	++	++	–	–	–	–	–	–	–
D		Enterotoxemia of sheep, lambs, goats and cattle	++	–	++	–	–	++	++	+	++
E		Isolated from sheep and cattle; pathogenicity doubtful	++	–	–	++	–	++	++	++	+

Modified from Kadis, Montie, Ajl (eds): Microbiol Toxins, vol 2A. Bacterial Protein Toxins, 1971. New York, Academic Press.
++, produced by all or most strains; +, produced by less than 50 percent of strains; –, not produced.
*Lethal antigens primarily responsible for pathogenicity and type designation.
[†]Lower order of toxicity; some may be involved in pathogenicity.
[‡]Some strains produce heat-resistant spores.

is the organism primarily responsible for diseases in humans: clostridial myonecrosis, less severe wound infections, and a common form of food poisoning. *C. perfringens* type A has been found in the intestinal tract of almost every animal that has ever been cultured for this organism, but is a less common cause of disease in animals than in humans. In contrast, types B, C, D, and E, which occur only in the intestinal tracts of animals and occasionally humans, produce a variety of naturally occurring diseases of domestic animals.

Morphology

C. perfringens usually appears as a short, plump rod, strongly staining gram-positive (Fig. 46-1). The organisms are uniform in appearance, 2 to 4 μm in length and 1 to 1.5 μm in width. The length varies according to the stage of growth, as well as the nutritional and ionic composition of the medium. Rapidly growing organisms may appear almost coccoid or cubical, whereas more elongated cells occur in older cultures. Unlike the other pathogenic clostridia, *C. perfringens* is nonmotile. It does not produce spores in ordinary media; special media must normally be used to demonstrate sporulation. Capsules may be observed by direct examination of smears from wounds, but are not uniformly demonstrable in culture.

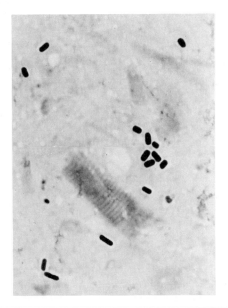

Figure 46-1. *Clostridium perfringens* directly smeared from infected muscle in clostridial myonecrosis, demonstrating short, plump bacilli that lack spores. A variety of other bacteria including other clostridia may also be present, but polymorphonuclear leukocytes are characteristically absent. Note the disintegrated muscle tissue. ×1000.

Physiology

C. perfringens is an aerotolerant anaerobe and will survive and even grow in oxygen tensions that are inhibitory to most other anaerobes, especially when streaked on blood agar plates. Some strains produce superoxide dismutase. Anaerobic culture methods suitable for *C. perfringens* and other clostridia are discussed in Chapter 45. It will grow over a pH range of 5.5 to 8.0 and a temperature range of 20C to 50C. Although *C. perfringens* is usually grown at 37C, a temperature of 45C is optimal for many strains. The generation time at this temperature may be as short as 10 minutes.

Cultural Characteristics. Surface colonies that are produced on blood agar after 24 hours incubation are circular and smooth, 2 to 4 mm in diameter, but as the colonies increase in size with age, the periphery often loses symmetry and assumes the appearance of a colony of motile bacteria showing swarming. Variation in colonial morphology occurs, depending on the degree of encapsulation and smooth to rough transition.

Laboratory Identification

The rapid growth of *C. perfringens* in chopped meat media at 45C can be utilized to isolate it from mixtures of bacteria. As *C. perfringens* will outgrow most other organisms during the first 4 to 6 hours of incubation, blood agar plates streaked after that time and incubated at 37C will have proportionally higher numbers of *C. perfringens*. Although heat treatment (80C to 100C) of mixed cultures is an aid in the isolation of many clostridial species that sporulate well, this method is not recommended for *C. perfringens*. Cultures of clinical isolates of this organism usually contain few spores, and the heat resistance of the spores appears to be inversely related to the toxigenicity of the vegetative forms.

A few easily observable characteristics aid in the identification of *C. perfringens*. In chopped meat glucose media there is abundant growth with gas formation; in vivo toxicity testing may be carried out with supernates from this medium. *C. perfringens* also produces a characteristic pattern of hemolysis on blood agar plates, precipitation in serum or egg yolk media, and "stormy" fermentation in milk media. After overnight incubation on rabbit, sheep, ox, or human blood agar, colonies of most strains demonstrate a characteristic target hemolysis resulting from a narrow zone of complete hemolysis due to the θ toxin and a much wider zone of incomplete hemolysis due to the α toxin. This double-zone pattern of hemolysis may fade with longer incubation.

A dense opalescence in human serum is produced by growing organisms or by the supernatant fluid from an overnight culture. This reaction, the Nagler reaction, is caused by the α toxin (a lecithinase C) and is specifically inhibited by *C. perfringens* antitoxin. A similar and more easily observable reaction occurs with egg yolk agar. This medium can be inoculated directly with wound specimens for screening purposes, and is helpful in the identification of pure cultures of *C. perfringens* and other clostridia that produce a lecithinase or a lipase. Presumptive identification of *C. perfringens* can be made by streaking organisms on both sides of an egg yolk agar plate, one half of which has been covered with *C. perfringens* antitoxin. The inhibition of opalescence on the antitoxin-treated portion, however, is not totally specific for *C. perfringens*. A number of other organisms produce a lecithinase, and although most are inhibited by the specific antitoxin, the lecithinases of the *Clostridium bifermentans-C. sordellii* group and *Clostridium barati* are antigenically similar and are partially inhibited by antitoxin to *C. perfringens* α toxin. These organisms can be separated by other tests.

In milk media, most strains of *C. perfringens* produce "stormy" fermentation, in which the fermentation of the lactose in milk produces a large amount of acid, causing the protein (casein) to coagulate. This acid clot is then disrupted and torn apart by the large volume of gas formed from the lactose fermentation. This action in milk media is useful in the identification of *C. perfringens*, but when used alone is not diagnostic because the reaction also may be produced by a number of other clostridial species, including *C. septicum*. Use of the test requires a pure culture of the organisms. Fermentation reactions and other biochemical tests used in the identification of *C. perfringens* are listed in Table 46-2 and are covered in greater detail in the reference manuals listed in the Further Reading section of this chapter.

Antigenic Structure

Strains of *C. perfringens* produce at least 12 different soluble substances or toxins, all of which are protein in nature and antigenic (Table 46-1). Of the four major lethal antigens, α, β, ϵ, and ι toxins, the most important is the α toxin, which is produced by all five types of *C. perfringens*. All of the toxins are exotoxins.

Many of the other soluble substances or minor antigens are enzymes with defined substrates. These substances are nonlethal and should not be referred to as toxins as has been customary in the past. Examples of these substances are collagenase (κ antigen), deoxyribonuclease (ν antigen), and hyaluronidase (μ antigen).

In general, serotyping with somatic antigens has not been practical in the further subdivision of *C. perfringens*, although a large number of serologic types exist. One useful application, however, is in epidemiologic studies of outbreaks of food poisoning, where a comparison can be made between the serotypes of heat-resistant (100C for 1 hour) *C. perfringens* type A isolated from the feces of patients and the serotype of the incriminated food.

Determinants of Pathogenicity

The toxin of primary importance in the pathogenesis of clostridial myonecrosis is the α toxin, which initially was described in terms of its lethal, dermonecrotic, and hemo-

TABLE 46-2. CHARACTERISTICS OF FREQUENTLY ENCOUNTERED CLOSTRIDIA

Species	Egg Yolk Agar		Cooked Meat Medium			Gelatin Hydrolysis	Indole	Carbohydrate Fermentation				Principal Fermentation Products
	Lecithinase	Lipase	Spores	Digestion	Milk			Glucose	Maltose	Lactose	Sucrose	
Toxigenic, pathogenic for humans												
C. perfringens	+	−	ST	−	CG	+	−	+	+	+	+	A,(P),B
C. novyi A*	+	+	ST	−	C(G)	+	−	+	+	−	−	A,P,B,V
C. novyi B*	+	−	ST	V	C(G)	+	V	+	−	−	−	A,P,B,V
C. septicum	−	−	ST	−	CG	+	−	+	+	+	−	A,B
C. sordellii	+	−	ST	+	CD	+	+	+	+	−	−	A,F,P,IB,IV,IC
C. histolyticum†	−	−	ST	+	CD	+	−	−	−	−	−	A,L
C. tetani	−	−	T	−	−	+	V	−	−	−	−	A,P,B
C. botulinum‡												
Group I*	−	+	ST	+	(C)(D)	+	−	+	+	−	−	A,P,IB,B,IV,V,IC
Group II*	−	+	ST	−	(C)	+	−	+	+	−	−	A,B
Group III*	V	+	ST	−	(C)(D)	+	−	+	V	−	−	A,P,B
Nontoxigenic, uncertain pathogenicity for humans												
C. bifermentans	+	−	ST	+	CD	+	+	+	+	−	−	A,F,P,IB,(B),IV,IC
C. ramosum	−	−	T	−	(C)(G)	−	−	+	+	+	+	A,F,L,S
C. sporogenes	−	+	ST	+	CD	+	−	+	+	−	+	A,P,IB,B,IV,IC
C. innocuum	−	−	T	−	−	−	−	+	V	−	V	A,(F),B,L
C. paraputrificum	−	−	T	−	C(G)	−	−	+	+	+	+	A,F,B,L
C. subterminale	−	−	ST	+	CD	+	−	−	−	−	−	A,(P),IB,B,IV
C. cadaveris	−	−	T	+	CG	+	−	+	−	−	−	A,(IB),B,IV
C. butyricum	−	−	ST	−	CG	−	−	+	+	+	+	A,F,B
C. tertium†	−	−	T	−	CG	−	−	+	+	+	+	A,B
C. limosum	+	−	ST	+	CD	+	−	−	−	−	−	A
C. cochlearium	−	−	ST	−	−	−	−	−	−	−	−	A,(P),B

+, positive reactions for 90 to 100 percent of strains; −, negative reaction for 90 to 100 percent of strains; ST, subterminal; T, terminal; ( ), variable; C, curd; D, digestion; G, gas; Fermentation products by gas liquid chromatography; A, acetic; F, formic; P, propionic; IB, isobutyric; B, butyric; IV, isovaleric; IC, isocaproic; L, lactic; S, succinic.

*Toxin neutralization test required for identification.

†Growth on aerobically incubated blood agar.

‡Group I contains proteolytic strains (types A, B, F, G); group II, types C and D; group III, nonproteolytic strains (types B, E, F).

lytic activities. The toxin is a lecithinase C (or phospholipase C), which splits lecithin to phosphorylcholine and a diglyceride. The toxin is activated by Ca^{2+} and Mg^{2+} ions; it also hydrolyzes sphingomyelin. Titration of α toxin can be performed using in vivo lethality testing and in vitro procedures, which are dependent upon its enzymatic action against lecithin-containing substrates such as egg yolk emulsion, human serum, or erythrocytes of certain animal species. α Toxin is an excellent antigen; in vivo protection or therapy of animals is dependent entirely on the α antitoxin titer. The in vivo action of α toxin is apparently on lecithin-containing lipoprotein complexes in the cell membrane and probably on mitochondria. Disruption or leakage of cell membranes alone can explain the lysis of erythrocytes, destruction of tissue, and edema observed in this disease. Its local activity in the muscle lesion is obvious, but the basis for the generalized toxicity or systemic manifestations and death seen in clostridial myonecrosis is not fully explained. Other substances such as θ, κ, or μ antigens apparently exert an ancillary role in abetting the local spread of the infection through tissue and providing nutrients for the proliferation of the organism.

Clinical Infection
Wound and Soft Tissue Infection

EPIDEMIOLOGY. *C. perfringens* is unbiquitous; type A strains are commonly found in the intestinal tract of humans and animals and are numerous in the soil both in the vegetative and spore forms. The natural habitat of types B through E, which cause a variety of infections in animals, appears to be the intestinal tract of animals and occasionally humans; these types however, do not permanently inhabit the soil, as does type A. Infection may be due to endogenous or exogenous clostridia. In traumatic injuries, either accidental or during warfare, the source of clostridia is usually soil carried into the tissues; the incidence of contamination and infection depends on the concentration of *C. perfringens* in the soil, which varies with the geographic location. Endogenous infections stem from fecal flora present on the skin or on particles of clothing carried into the wound, or from clostridia escaping from the bowel when its integrity is disrupted by disease, traumatic injury, or surgery.

One of the essential factors predisposing to clostridial myonecrosis is trauma associated with deep and lacerated or crush wounds of muscle and vascular damage of major vessels and capillary beds. If ischemia and necrosis are present deep within the muscle, however, the trauma may not be necessarily severe. Such a setting occurs in infections associated with injections of vasoconstrictive agents such as epinephrine. The basis for the requirement of trauma with ischemic or necrotic areas is the anaerobic nature of the clostridia, which require a reduced oxygen tension and oxidation-reduction potential for growth (see also Chap. 45). Clostridia are unable to initiate infection in healthy tissues in which the oxidation-reduction potential is normal. Even with the high frequency of pathogenic

clostridia in wounds, the incidence of gas gangrene remains relatively low because of these growth restrictions.

The principle settings for infections of this type occur during periods of war, when massive wounds of muscle contaminated with soil, clothing, and metal fragments are common. Prior to the early 1950s evacuation and medical care of the wounded following injury was delayed, providing optimal conditions for the histotoxic clostridia to initiate infection. More recently, however, rapid evacuation of the wounded and early medical care have drastically reduced the wartime incidence of clostridial myonecrosis. In civilian life, settings that may lead to this disease include automobile and motorcycle accidents, gunshot wounds, compound fractures, industrial accidents, surgical complications, septic abortion, and injections of medications such as epinephrine. A reduced blood supply stemming from edema, cold, or shock, and the presence in the wound of facultative organisms also predisposes to clostridial infection.

PATHOGENESIS. When *C. perfringens* is introduced into tissue, the primary requirement for initiation of infection is a lowered oxidation-reduction potential. In areas of reduced oxygen tension the pyruvate of muscle is incompletely oxidized and lactic acid accumulates, causing a drop of pH. The combination of lowered oxidation-reduction potential and a drop in pH may activate endogenous proteolytic enzymes, resulting in tissue autolysis. This release of nutrients and the lowered oxidation-reduction potential combine to produce conditions suitable for growth of anaerobic organisms.

Proliferation of the organisms is accompanied by the production of soluble toxins. In true clostridial myonecrosis these toxins diffuse from the initial site of growth and attack healthy muscle and surrounding tissues. These tissues are in turn destroyed by the toxins, thereby permitting spread of the infection into new necrotic areas. The edema fluid and gas accumulated from the metabolism of the organism also increase the pressure within muscle bundles so that circulation is impaired, and the oxidation-reduction potential and pH decrease, providing new areas within muscle suitable for the extension of growth of the clostridia. The disease progresses in this manner, with the organisms moving into new areas behind the destructive action of their toxins. The local infection and its extension into healthy tissues is well understood. Still unknown, however, are the cause of the generalized systemic toxicity and the immediate cause of death in clostridial myonecrosis, although the α toxin is recognized as an essential element in its pathogenesis.

CLINICAL MANIFESTATIONS. Wound infections can be divided into three categories of increasing levels of severity: (1) simple wound contamination, (2) anaerobic cellulitis; and (3) clostridial myonecrosis. Two additional clinical settings, uterine infections and clostridial septicemia, are special types of wound and soft tissue infections with certain unique features. The symptomatology of the three categories, however, may overlap, and an infection may evolve from a cellulitis into true clostridial myonecrosis. The essential clinical features of wound infections are given in Table 46-3.

Simple Wound Contamination. In simple wound contamination one or more histotoxic clostridia may be present without an obvious pathologic process. Either the clostridia present may be nontoxigenic or the environmental conditions in the wound may be unsuitable for toxin production and the initiation of a progressive infection by toxigenic strains.

Anaerobic Cellulitis. This is a more serious form of wound infection, in which the clostridia infect necrotic tissue already dead as a result of ischemia or direct trauma. The organisms in this case spread through subcutaneous tissue and along fascial planes between muscles, but do not invade healthy, intact muscle. Growth of *C. perfringens* within the necrotic tissue is extensive, and gas is normally a prominent feature. Patients, however, are not extremely toxic and the overall prognosis is considerably better than for clostridial myonecrosis. Careful distinction between this level of infection and true clostridial myonecrosis is necessary in order to avoid the sometimes extreme surgical measures that are unnecessary for anaerobic cellulitis but are required for treatment of clostridial myonecrosis.

Clostridial Myonecrosis. This term should be limited to use in characteristic anaerobic infections of muscle in which organisms are *invasive* and the infection is associated with profound toxemia, extensive local edema, variable amounts of gas, massive tissue damage, and death in untreated cases.

Following injury there is an incubation period usually of 12 to 48 hours before symptoms suddenly appear. The characteristic initial symptom is pain in the affected area, which increases in severity as the infection spreads. There is local edema and a thin blood-stained exudate. The pulse rate rises disproportionately higher than the temperature. If the disease remains untreated, the process advances rapidly, with increasing toxemia and extension of the infection. With increased exudation from the area, gas usually becomes obvious, but it is a variable symptom. Changes occur in skin color, which finally may become black. Necrosis of large muscle masses is associated with severe shock and prostration. Infrequently, intravascular hemolysis occurs, producing hemoglobinemia, hemoglobinuria, and renal failure. Death occurs rapidly in untreated cases.

Uterine Infection. These are a special type of clostridial myonecrosis, usually involving the gravid uterus. Prior to legalized abortion many cases followed illegal attempts at mechanically induced abortion by nonmedical practitioners. They may occasionally occur as puerperal infections. The source of *C. perfringens* may be exogenous or endogenous. As in wound infections, different levels of clinical involvement occur. In contrast to clostridial myonecrosis from wounds, in uterine myonecrosis septicemia and intravascular hemolysis are common and lead to secondary renal failure. The disease progresses rapidly and has a high mortality.

TABLE 46-3. DIFFERENTIATION OF GASSY INFECTIONS OF SOFT TISSUES AND WOUNDS

Criterion	Infected Vascular Gangrene	Anaerobic Cellulitis	Clostridial Myonecrosis	Streptococcal Myonecrosis
Incubation	Over 5 days, usually longer	Almost always over 3 days	Usually under 3 days	3 to 4 days
Onset	Gradual	Gradual	Acute	Subacute or insidious
Toxemia	None or minimal	None or slight	Very severe	Severe only after some time
Pain	Variable	Absent	Severe	Variable, usually fairly severe
Swelling	Often marked	None or slight	Marked	Marked
Skin	Discolored, often black and desiccated	Little change	Tense, often very white	Tense, often with coppery tinge
Exudate	None	None or slight	Variable, may be profuse, serous, and blood-stained	Very profuse; seropurulent
Gas	Abundant	Abundant	Rarely pronounced, except terminally	Very slight
Smell	Foul	Foul	Variable, may be slight, often sweetish	Very slight, often sour
Muscle	Dead	No change	Marked change	Little change at first, except edema

Modified from MacLennan: Bacterial Rev 26:177, 1962.

Clostridial Septicemia. Invasion of the bloodstream may occur in association with malignancy and may involve a localized myonecrosis in addition to a fulminating clostridial septicemia. There usually is no history of external trauma; the source of the organisms appears to be endogenous and to develop from an alteration of the patient's intestinal tract as a consequence of the malignant process. Septicemia also may follow biliary tract or gastrointestinal surgery. *C. septicum* or *C. perfringens* is usually the etiologic agent. Rapid diagnosis and treatment are essential because death may occur in less than 24 hours after the onset of symptoms in untreated cases.

LABORATORY DIAGNOSIS. An early diagnosis of clostridial myonecrosis is essential and, as previously stated, must be made on clinical grounds. Bacteriologic confirmation of the organisms present in the infection is important, however, and valuable information can be gained from a direct smear and gram stain of material from deep within the wound (Fig. 46-1). More than one clostridial species is usually present; other organisms that are commonly found include *Staphylococcus, Streptococcus, Escherichia, Proteus, Bacillus, Bacteroides,* and other anaerobes. Gram stain and culture assist in differentiating clostridial myonecrosis from rare cases of anaerobic streptococcal myonecrosis or other mixed anaerobic infections (Table 46-3).

TREATMENT. Simple wound infections with *C. perfringens* can be treated by removal of necrotic tissue and by cleansing. Administration of antibiotics is rarely required. Anaerobic cellulitis, which is a more serious infection, usually can be treated by opening the wounded area, removing all necrotic tissue, cleansing thoroughly, and administering antibiotics. These infections must be carefully monitored.

Mortality rates of clostridial myonecrosis vary from approximately 15 to 30 percent and are highly dependent on the anatomic location of the infection. Intensive and immediate therapy is indicated. The surgical removal of all infected and necrotic tissue is of prime importance. Intensive antibiotic and antitoxin therapy is partially effective, but if all infected muscles are not excised because of their anatomic location or for other reasons, the results obtained with antibiotic and antitoxic therapy alone are poor. Patients who survive this infection usually require extensive surgery and amputation. Not only is clostridial myonecrosis more common in areas of the body that have large masses of muscle, such as the buttock, thigh, and shoulder, but mortality from infections in these sites is also higher. In addition, treatment is more difficult because of spread of the infection to the trunk and involvement of areas that cannot be excised.

Although antibiotic therapy alone is ineffective in treating clostridial myonecrosis, there is general agreement on its therapeutic effectiveness as an ancillary agent. Penicillin is the drug of choice, but clindamycin or metronidazole can be substituted for patients with penicillin allergy. When used, high doses of penicillin G should be administered in any case, but it is worrisome that *C. perfringens* strains are apparently less susceptible than previously. Secondary infection with facultative gram negative organisms such as *Escherichia coli* must be treated with gentamicin or tobramycin. Clindamycin or metronidazole is indicated when *Bacteroides* are suspected or demonstrated. Although the value of antitoxin, which is available commercially as a polyvalent serum, has been disputed, it is probably a valuable therapeutic adjunct to surgery when used in adequate dosage. In spite of the risk of hypersensitivity reactions, antitoxin therapy is generally employed in a setting of extreme toxemia, particularly with intravascular hemolysis.

Hyperbaric oxygen was introduced in 1961 as an adjunct in the therapy of clostridial myonecrosis. Its effectiveness is supported by clinical use and by experimental observations in animals, but at present no controlled studies are available on its efficacy in the reduction of mortality in humans. When patients with clostridial myonecrosis involving the trunk are treated with hyperbaric oxygen in addition to the standard modes of therapy, however, the increased survival of approximately 50 percent suggests the effectiveness of this adjunct. The administration of hyperbaric oxygen also appears to have eliminated the necessity for early radical amputation of limbs or excision of tissue in an attempt to check surgically the irreversible spread of infection. Initial surgery can be limited to debridement and removal of frankly necrotic tissue, with hyperbaric oxygen used to halt further spread of the infection and improve oxygenation of marginally viable tissue.

Hyperbaric oxygen usually is administered by giving the patient five to seven intermittent exposures to breathing 100 percent O_2 in a chamber pressurized to 3 atmospheres absolute pressure. The first treatment is given as soon as possible after diagnosis. Further treatments are scheduled to complete a total of three treatments during the first 24 hours, with intervals of 6 to 8 hours between treatments; a fourth treatment 6 to 12 hours later; and usually three final treatments at intervals of 6 to 14 hours. Hyperbaric oxygen exerts a direct inhibitory action on the organism and on toxin production. It also exhibits indirect effects by raising the oxidation-reduction potential of the tissues surrounding the infection and preventing spread of the organisms. Preformed α toxin is not inactivated by hyperbaric oxygen.

Clostridial septicemia complicating malignancy is amenable to antibiotic therapy with penicillin if initiated early. If localized myonecrosis develops, additional appropriate treatment is required.

PREVENTION. The most important preventive measure against clostridial myonecrosis is early and adequate wound debridement. The incidence of the disease markedly increases with delay in debridement. Adequate cleansing, removal of necrotic tissue, delay in primary closure of large, ragged wounds, maintenance of drainage, and avoidance of tight packing are all of prime importance in prevention. Administration of prophylactic antibiotic (penicillin) probably reduces the risk of an anaerobic infection, particularly if administered shortly after wounding.

Food Poisoning. This mild form of food poisoning has been recognized with increasing frequency since its association with *C. perfringens* was first demonstrated in 1945. The organisms usually involved are strains of type A that produce heat-resistant spores and minimal amounts of θ toxin, although more typical type A strains also may cause the disease. Between 8 to 24 hours following ingestion of contaminated food, patients develop acute abdominal pain and diarrhea. Nausea may occur, but vomiting is uncommon, as are other signs of infection such as fever and headache. Symptoms normally last for 12 to 18 hours, and recovery is usually complete except for rare fatalities in elderly or debilitated patients.

The symptoms are attributable to an enterotoxin that is synthesized during sporulation of the organism. Properties of enterotoxin are erythema after intracutaneous injection, fluid accumulation in intestinal loops, and lethality for mice. Clinical symptoms are probably due to the action of the enterotoxin on the intestinal mucosa. Repeated attacks occur, indicating absence of immunity.

This type of food poisoning usually results from the ingestion of meat dishes such as roasts, poultry, fish, and stews that are heavily contaminated with *C. perfringens*. Contamination of food may occur at any time, as this organism is widespread in the environment. Raw meat may be contaminated at slaughter, through handling during preparation, or by exposure to flies and dust. The initial heating or cooking of the food may produce germination of heat-resistant spores, or food may become contaminated after cooking. The clostridia multiply during cooling of the meat or during a storage period, and will produce food poisoning if the food is served cool or is inadequately reheated. Symptoms occur only if the organisms multiply to a concentration of 10^6 to 10^7 viable cells per gram of food, so that 10^8 to 10^9 viable bacteria are ingested. A specific diagnosis is made by isolation of *C. perfringens* in higher than normal numbers from the feces of infected patients and, if possible, from samples of the ingested food.

Enteritis Necroticans (Necrotizing Jejunitis, Necrotic Enteritis). This disease is caused by type C strains of *C. perfringens* and is a more severe disease than *C. perfringens* type A food poisoning. Following an incubation period of less than 24 hours, the onset is sudden, with severe abdominal pain, diarrhea, and in some patients loss of intestinal mucosa with bleeding into the stool. The disease may be fatal, with peripheral circulatory collapse or intestinal obstruction and peritonitis. Although strains causing this disease were originally designated as a new type of *C. perfringens*, type F, they are now considered to be an atypical type C strain producing heat-resistant spores (Table 48-1). In addition to sporadic cases, major outbreaks have been reported from New Guinea, where it is

associated with the ingestion of contaminated and inadequately cooked pork (pig feasting), and the disease is called pig-bel. In New Guinea the disease occurs in four forms, varying in severity and degree of toxicity but having an overall mortality rate of 35 to 40 percent. The β toxin produced by *C. perfringens* type C is responsible for the symptomatology; the administration of *C. perfringens* type C antitoxin to patients with enteritis necroticans significantly reduces the mortality rate.

Other Histotoxic Clostridia

Clostridium septicum

C. septicum is closely related to *Clostridium chauvoei*, both of which are widely distributed in nature and in the intestinal tract of humans and animals. *C. chauvoei* is pathogenic for animals only, but *C. septicum* is pathogenic for humans and other animals. Because this organism can escape from the intestinal tract of humans and animals and invade tissues shortly after death, the presence of *C. septicum* in pathologic specimens must be interpreted with caution. *C. septicum* is 0.8 μm in diameter and 3 to 5 μm in length; it is motile by means of peritrichous flagella. Colonies on blood agar are surrounded by zones of complete hemolysis, and swarming across the surface of plates may be marked. The major toxin produced by *C. septicum* is α toxin, which is lethal, necrotizing, hemolytic, and possibly leukocidic. The organism also produces deoxyribonuclease, hyaluronidase, and an oxygen-labile hemolysin. The percentage of *C. septicum* isolates from cases of clostridial myonecrosis ranges from 5 to 20 percent, according to different reports. Endogenous *C. septicum* from the patient's own intestinal tract may produce septicemia and occasionally localized myonecrosis in patients with underlying carcinoma.

Clostridium novyi

C. novyi has been differentiated into three major types, A, B, and C, on the basis of the soluble antigens present in toxic filtrates of the organism. These bacteria are found in the soil (especially type A) and in the livers (types A and B) of a variety of apparently healthy animals. *C. novyi* type A is 4 to 8 μm in length and 1 μm in width; type B organisms are even larger. These bacteria have oval subterminal spores; they are motile by peritrichous flagella, producing swarming on the surface of blood agar plates.

C. novyi type A, rather than type B, causes most of the clostridial myonecrosis and other wound infections in humans. Both types produce an α toxin that is necrotizing, lethal, and the most potent toxic substance in filtrates of *C. novyi* cultures. The α toxin apparently increase capillary permeability and produces the intense gelatinous edema in muscle tissue that is characteristic of clostridial myonecrosis caused by *C. novyi*. β (Type B) and γ (type A) toxins are lecithinase C enzymes, which are hemolytic, necrotizing, and in the case of β toxin, lethal. These toxins produce lecithinase reactions on egg yolk agar plates. Type A also produces a lipase (ϵ antigen) that gives a pearly layer effect (similar to oil on water) on and around colonies on egg yolk agar plates.

The production of the lethal α toxin by *C. novyi* appears to be bacteriophage-dependent. Recent studies have demonstrated that curing a toxigenic *C. botulinum* type C of its prophage produces a nontoxigenic strain. Infecting this organism with another specific bacteriophage then can convert the nontoxigenic strain into *C. novyi* type A, which produces the lethal α toxin. Continued toxigenicity and interconversion of species of toxigenic *C. botulinum* type C and *C. novyi* type A thus depends upon the presence of specific bacteriophages.

The percentage of clostridial myonecrosis due to *C. novyi* varies in different reported series, but was approximately 42 percent during World War II. As this organism, especially type B, is extremely fastidious and oxygen sensitive, its true occurrence may be greater than reported. Clostridial myonecrosis with *C. novyi* generally is characterized by a high mortality rate and large amounts of edema fluid, with little or no observable gas in the infected tissue.

Clostridium histolyticum

C. histolyticum has been isolated from the gastrointestinal tract of humans and from the soil. It is an aerotolerant species and produces limited growth on blood agar plates incubated under aerobic conditions, although improved growth is produced in an anaerobic environment. The organism is markedly proteolytic, digesting a variety of native proteins. Several soluble antigens are produced, of which the most important are the α and β toxins that are lethal and necrotizing. The β toxin is a collagenase that causes the destruction of collagen fibers and marked disruption of tissues observed in cases of clostridial myonecrosis caused by this organism. The incidence of *C. histolyticum* in cases of clostridial myonecrosis during World War II was between 3 and 6 percent.

Clostridium sordellii

Although *C. bifermentans* and *C. sordellii* are very similar and for a time were considered a single species, there are sufficient serologic and physiologic differences to justify their separation into different species. *C. sordellii* consists of both pathogenic and nonpathogenic strains and produces urease; *C. bifermentans* is nonpathogenic and urease-negative. Both organisms are found in the soil and as part of the normal intestinal flora of humans and other animals. Both species produce proteolytic enzymes and a lecithinase that is serologically related to the α toxin of *C. perfringens* and is therefore partially inhibited by *C. perfringens* antitoxin. *C. sordellii* also produces a lethal toxin. Clostridial myonecrosis involving *C. sordellii* is characterized by large amounts of edema and thus may resemble infection with *C. novyi*. The incidence of *C. sordellii* in clostridial myonecrosis is about 4 percent.

Clostridium fallax

Although MacLennan considers *C. fallax* capable by itself of causing clostridial myonecrosis, there are few cases on record, and the organism is rarely encountered. *C. fallax* is a strict anaerobe that rapidly loses virulence after isolation and artificial cultivation. Its natural habitat is unknown.

Other Clostridia In Tissue Infection

There are a variety of other species of clostridia commonly encountered in soft tissue infections, abscesses, wound infections, anaerobic cellulitis, and clostridial myonecrosis. These organisms are usually considered to be nonpathogenic, although there are unanswered questions regarding their possible role in the development of infection.

There is a correlation between the incidence in wound infections of the various species of clostridia, both pathogenic and nonpathogenic, and their occurrence in the soil, but the source also may be endogenous organisms on the skin and clothing. Alteration in the integrity of the intestinal wall by disease, surgery, or trauma may also release a variety of endogenous bacteria, including these so-called nonpathogenic clostridial species that may then be isolated from subsequent infections such as peritonitis and intraabdominal abscess. It is important to remember that *C. tetani* and *C. botulinum* may also be present in a wound in addition to the histotoxic and other clostridial species.

Of the clostridia currently labeled as nonpathogenic, *Clostridium sporogenes* is frequently encountered in wound infection, along with *C. bifermentans* and *Clostridium tertium*. Additional organisms that may be isolated from these sources are *Clostridium innocuum*, *Clostridium ramosum*, *Clostridium subterminale*, *Clostridium limosum*, *Clostridium butyricum*, *Clostridium cochlearium*, *Clostridium cadaveris*, and *Clostridium paraputrificum*. Although *C. sporogenes* does not appear to produce toxins, there is some evidence that the organism, when associated with frank pathogens such as *C. perfringens* or *C. novyi*, may play a synergistic role in clostridial myonecrosis. Its presence with either of these organisms is correlated with a high mortality rate.

Clostridium difficile and Antibiotic-associated Colitis

Pseudomembranous enterocolitis is a severe, potentially lethal gastrointestinal tract disease characterized by exudative plaques with underlying necrosis of the mucosal surface of the intestine. The plaques may coalesce, forming large pseudomembranes, and sloughing may occur. This pathologic entity had been described in a variety of clinical settings, particularly with gastrointestinal tract surgery, prior to the antibiotic era, but an increased incidence and a higher predilection for involvement of the colon became apparent with antibiotic-associated disease. Studies in guinea pigs and hamsters of fatal hemorrhagic cecitis that can be induced by administration of penicillin, clindamycin, and other antibiotics significantly aided in the elucidation of the etiologic agent, *Clostridium difficile*.

C. difficile is an obligate anaerobe that is saccharolytic and weakly proteolytic, producing a complex array of acid fermentation products detectable by gas–liquid chromatography. A selective agar medium containing cefoxitin, cycloserine, fructose, and egg yolk significantly aids in its isolation from feces. The organism has been isolated from both human and animal feces. Infants may acquire *C. difficile* early in the neonatal period, with isolation rates as high as 64 percent reported for infants up to 8 months of age, but colonization decreases to a rate of approximately 3 percent in healthy adults. The source of the organism in neonates has not been conclusively demonstrated, although acquisition during vaginal delivery has been suggested.

C. difficile produces two major protein toxins that are important in antibiotic-associated colitis and are antigenically distinct. Toxin B, or cytotoxin, produces a cytopathic effect on tissue culture cells, a procedure that has been used for the detection of toxin from feces of patients and from in vitro cultures. The cytopathic effect of this toxin can be inhibited by the *C. sordellii* antitoxin component of polyvalent gas gangrene antitoxin due to antigenic cross-reactivity. It was assumed initially that this toxic moiety was entirely responsible for enterotoxic activity. Purification studies, however, led to the detection of a second toxic component, toxin A, which was significantly more active in some model biologic assay systems, but had less cytotoxic activity for tissue culture cells than toxin B. Toxigenic strains of *C. difficile* apparently produce both toxins and protection experiments in animals immunized against either toxin alone or both toxins indicate that both moieties are involved in intestinal pathology.

Pseudomembranous colitis, as confirmed by gross appearance at sigmoidoscopy or by biopsy and that occurs in association with antibiotic administration, is highly correlated with the presence of *C. difficile* by culture and detection of the cytotoxin (96 percent) from feces. Diarrhea, which usually spontaneously resolves, is a more frequent adverse reaction following administration of antibiotics than is colitis; approximately one third of patients with antibiotic-associated diarrhea have been found to be positive for the cytotoxin. Cytotoxin has been detected, although infrequently, from feces of patients who have received antibiotics but did not have diarrhea and also from patients with miscellaneous diarrheal conditions unrelated to antibiotic administration. The high frequency of *C. difficile* in neonates has confounded understanding of the role of this organism in gastrointestinal disorders in this population. The organism and cytotoxin have been detected in infants with necrotizing enterocolitis but also in a significant pro-

portion of healthy neonates. Thus, even high cytotoxin levels in feces do not always correlate with the presence of diarrhea or colitis.

A wide variety of antibiotics has been associated with the development of diarrhea or colitis, although the drugs most frequently linked to colitis are the penicillins, clindamycin, and the cephalosporins. The *C. difficile* strain isolated from a patient with pseudomembranous colitis may be resistant or susceptible to the implicated antimicrobial(s), so it has become apparent that the mechanism of disease production is not simply overgrowth by resistant strains of *C. difficile*. Yet, alteration of the normal microbial ecology of the gastrointestinal tract and suppression of prevalent endogenous organisms is obviously important in allowing *C. difficile* to colonize or in predisposing to its proliferation and toxin production. Gastrointestinal procedures, in particular, may precipitate the disease without prior antimicrobial exposure, although much less frequently. *C. difficile* is generally very susceptible to penicillin, ampicillin, metronidazole, and vancomycin, whereas strains usually are resistant to the cephalosporins and may be resistant to clindamycin and the tetracyclines.

Patients may develop diarrhea or colitis when antibiotics are being administered, or onset can occur even weeks after antibiotics have been discontinued. An awareness of the potential for mild to severe gastrointestinal disease associated with administration of antimicrobials is important in prevention. Documentation of five or more loose stools per day without other apparent cause in patients receiving antimicrobials has been commonly used as an indication to discontinue antibiotic therapy, if appropriate, or to substitute alternate antimicrobial agents believed to be less frequently associated with diarrhea and colitis sequelae. Pseudomembranous colitis can usually be effectively treated with oral vancomycin, although relapse may occasionally occur and require retreatment.

Clostridium tetani

C. tetani is the causative agent of tetanus, a disease now relatively rare in well-developed countries. In developing countries, however, where many unimmunized mothers give birth to children with neglected umbilical cord care, neonatal tetanus has a significant impact on overall mortality. In the adult, disease classically follows a puncture wound and is characterized by severe muscle spasms, the most characteristic being that of the jaw; thus the term trismus or lock-jaw. Despite many advances in treatment, the mortality rate is quite high especially in the very young and very old.

The anaerobic nature of *C. tetani* was in part responsible for its delayed isolation and discovery by Kitasato in 1889. The clinical as well as experimental recognition that a small local infection produced by this organism could result in a profound toxemia with neuromuscular manifestations led to the discovery of tetanus toxin and the detection shortly after of its specific antitoxin. Antitoxin proved to be quite effective when administered prophylactically but less so when used therapeutically. The discovery, however, that the toxin can be converted to a toxoid that is an excellent immunizing agent, provided a remarkably effective method for the prevention of this disease. Widespread use of the toxoid has resulted in a marked reduction in its incidence.

Morphology and Physiology

Morphology. The tetanus bacillus is quite long and thin when compared with other pathogenic clostridia. Individual bacilli range from 2 to 5 μm in width and 3 to 8 μm in length. Young cultures of the organism usually stain grampositive, but in older cultures and in smears made from wounds the organisms frequently are gram-negative. Under appropriate cultural conditions the organism produces a spore terminally located and of considerably greater diameter than the vegetative cell, giving the characteristic drumstick appearance. The spore does not take up the gram stain and appears as a colorless round structure. With prolonged incubation the vegetative cells autolyze, leaving behind either the spore with a portion of the vegetative cell attached or free spores. Most isolates of C. tetani possess numerous peritrichous flagella that convey active motility to the organism.

Cultural Characteristics. C. tetani is an obligate anaerobe, moderately fastidious in its requirement for anaerobiosis. The optimal temperature for growth is 37C and the optimal pH is 7.4. Nutritional requirements of C. tetani, like those of other clostridia, are complex, and include a number of amino acids and vitamins. These requirements, however, can be readily met by blood agar or cooked meat broth. Because swarming of the organisms occurs on blood agar plates, the isolation of surface colonies is difficult. The edge of the colony appears as a translucent finely granular sheet with a delicate filamentous advancing edge. This pronounced motility, especially in the presence of condensed moisture, has been used to advantage in isolating the organism from mixed cultures containing bacteria that are less motile than C. tetani. Where isolated colonies can be obtained, faint β hemolysis is observed. In cooked meat broth a small amount of growth can be detected in 48 hours; no digestion of the meat is noted. The organism does not ferment any carbohydrates. It does not usually liquify gelatin in 48 hours and produces very little change in litmus milk.

Resistance. The spore of C. tetani conveys to the organism considerable resistance to various disinfectants and to heat. It is not destroyed by boiling for 20 minutes. For practical purposes, autoclaving at 120C for 15 minutes is the best method for sterilizing contaminated materials.

Laboratory Identification. Clinical materials for culture should be transported to the laboratory in vessels containing carbon dioxide. They should be planted immediately both on prereduced solid media and on anaerobic liquid culture media such as chopped meat, and incubated under anaerobic conditions. Sometimes isolation of the organism is difficult because of the presence of other organisms in the mixture. In such cases heating the culture at 80C for 20 minutes after an initial 24-hour incubation period will kill non-sporeforming organisms and permit recovery of C. tetani. The rapid motility of the organism may also be useful in its isolation. One half of a culture plate is inoculated with a culture, and the remaining uninoculated half is examined after 24 hours for a thin film of the motile C. tetani. Final proof of the isolation of a toxin-producing C. tetani rests on the in vivo demonstration of toxin production when injected into mice and its neutralization in mice previously inoculated with antitoxin.

Antigenic Structure

Flagella (H), somatic (O), and spore antigens have been demonstrated in C. tetani. The spore antigens are different from the H and O antigens of the somatic cell. Strains of the organism have been differentiated into 10 types on the basis of their flagellar antigens. There is a single somatic agglutination group for all strains that permits identification of the organism by use of fluorescein-labeled antisera. Of tremendous practical importance, however, is the production of a single antigenic type of toxin by all strains of C. tetani, and its neutralization by a single antitoxin.

Determinants of Pathogenicity

Very little is known about the conditions that allow the tetanus bacillus to survive within the human host. C. tetani has little invasive ability, and when present alone rarely produces an invasive cellulitis. Frequently, however, it is found in association with other bacteria that play a more significant role in the local infection and that lower the oxidation-reduction potential at the site of injury.

Tetanus Toxin. All of the symptoms in tetanus are attributable to an extremely toxic neurotoxin, tetanospasmin, which is an intracellular toxin released by cellular autolysis.

PROPERTIES. The purification of intracellular toxin is best accomplished by its extraction from cells in the late exponential phase of growth before there is appreciable toxin detectable in the culture medium. The toxin is a heat-labile protein that may be inactivated by heating for 20 minutes at 60C. The intracellular molecular form consists of a single polypeptide chain with a molecular weight of 150,000. Release of the toxin from the bacterial cell is accompanied by a nicking of the toxin molecule, but the precise mechanism of release is unknown. The nicked toxin (extracellular toxin) consists of two nonidentical polypeptide chains having molecular weights of 100,000 and 50,000, respectively, and held together by disulfide bridges. The ganglioside-binding site on the toxin molecule is located on the heavy chain, whereas the light chain appears to be responsible for the toxic activity.

Tetanus toxin is one of the most poisonous substances known; only botulinum toxin and *Shigella* dysentery toxin are comparable in toxicity. There is no simple in vitro test for determining its activity; toxicity must be assayed by observing its lethal effect on an experimental animal. In quantitating the toxin, the most meaningful dose is the LD_{50} because this level of toxin lies on the steepest part of the dose-response curve. Animals vary in their susceptibility to the toxin. Humans and horses are probably the most susceptible, whereas birds and cold-blooded animals are usually quite resistant. In mice, pure toxin preparations have a potency of about 30 million minimal lethal doses per milligram of protein. Although toxin constitutes about 5 to 10 percent of the bacterial weight, the physiology of its production and function in the parent organism is largely unknown. Attempts to correlate the toxigenicity of strains with the presence of a converting bacteriophage have been unsuccessful. Recent evidence, however, implicates a large plasmid in toxin production.

Although toxin forms nontoxic dimers spontaneously, toxoiding with formaldehyde increases the degree of polymerization and produces a more stable and reliable product. This material is useful in immunization against the disease. Toxoid is nontoxic but retains the single antigenic determinant of tetanus toxin that gives rise to antitoxin antibody.

MODE OF ACTION. The molecular basis for the action of tetanus toxin is unknown. The toxin binds with gangliosides in synaptic membranes. These gangliosides are water-soluble mucolipids, ceramidyloligosaccharides, containing residues of stearic acid, sphingosine, glucose, galactose, *N*-acetylgalactosamine, and sialic acid. The toxin is transmitted intraaxonally against the flow (retrograde intraaxonal transport). When it reaches the region of the nucleus, it is transported to the inhibitory interneurons, where it inhibits the release of inhibitory transmitters. Small amounts of the toxin injected locally in experimental animals result in local tetanospasm followed by ascending tetanus, the development of increasing muscle spasticity above the site of the injection. Injection of the tetanus toxin intravenously produces descending tetanus characterized by spasticity in the head and neck, spreading to the back and limbs followed by generalized tetanic convulsions. Localization of radio-labeled tetanus toxin injected intravenously suggests that generalized tetanus results from the summation of innumerable forms of local tetanus. The inhibition of the release of inhibitory transmitter substances allows the more powerful muscles to prevail. In the human, this can be seen in the form of muscle spasms

of the masseter muscles with trismus flexion of the upper extremities and extension of the lower extremities with arching of the back (opisthotonus).

Clinical Infection

Epidemiology. In the United States, tetanus is a sporadic disease that is seen most frequently in the southern, southeastern, and midwestern states. It has become increasingly a disease of older persons because of the failure to be immunized with tetanus toxoid during childhood or military service. Recently, addiction to heroin has been associated with increased numbers of cases in large urban centers.

Pathogenesis. The spores of *C. tetani* are ubiquitous. They have been found in 20 to 64 percent of soil samples taken for culture, and are present in even higher yields in cultivated lands. They are present in the gastrointestinal tracts of humans and other animals. If one carefully cultures traumatic wounds, *C. tetani* can also be demonstrated fairly frequently. It is quite uncommon, however, for tetanus to develop from these wounds. The most significant feature of the pathogenesis of tetanus is the setting of the wound where the oxidation-reduction potential must be properly poised to permit multiplication of the organism and toxigenesis. Classically, the wounds seen in practice are the simple puncture wound from a nail, splinter, or thorn. Other settings, however, such as compound fractures, "skin popping" by drug addicts, decubitus and varicose ulcers, external otitis, and dental extractions also provide the proper conditions. The most feared form of tetanus, tetanus neonatorum, is a very significant cause of morbidity and mortality in developing nations. This form of tetanus usually results from cutting the umbilical cord with unsterile instruments or from improper care of the umbilical stump. In the United States most neonatal cases have resulted from unattended home deliveries. Tetanus may also follow operative procedures, but with modern hospital facilities this is an extremely rare event.

The lowering of the oxidation-reduction potential is associated with tissue necrosis following traumatic injuries or injection of necrotizing substances. An important contributing factor is the presence of aerobic bacteria that will grow to the point of removing oxygen and then continue to grow facultatively as anaerobes. This growth will effectively reduce the oxidation-reduction potential to the point where tetanus spores may germinate. Following the germination of the spores, toxin is elaborated and gains entrance to the central nervous system. The infection with *C. tetani* remains localized and inconspicuous with minimal reaction unless other organisms are present. In any collection of cases of tetanus, there are always a few cases that give no previous history of injury.

Clinical Manifestations. Following implantation of spores into an appropriate site, there is an incubation period of 4 to 10 days. Tetanus may occur rarely in a localized form, developing in muscles adjacent to the site of inoculation. More frequently, however, it is generalized in nature. The earliest manifestation is muscle stiffness followed by spasm of the masseter muscles—trismus or lock-jaw. This is the classic symptom of tetanus. As the disease progresses, tetanospasms cause clenching of the jaw, producing a grimace referred to as risus sardonicus, arching of the back (opisthotonos), flexion of the arms, and extension of the lower extremities. These tetanospasms are relatively brief in duration but may be frequent and exhausting. Recently, attention has been brought to the sympathetic affects of tetanus toxin-hypotension-hypertension, tachycardia, and cardiac disturbance. Respiratory complications such as aspiration pneumonia and atelectesis are common. Occasionally the spasms will be of sufficient intensity to produce bone fractures.

The disease takes several weeks to run its course; death may occur during one of these spasms. Poor prognosis is associated with a short incubation period between injury and seizure, rapid development from muscle spasm to tetanospasms, injury close to the head, extremes of age, and frequency and severity of convulsions. Patients who recover from this disease usually return to a completely normal state after a variable period of stiffness. Except for possible damage to the lungs from pulmonary complications or bone fracture, tetanus leaves no permanent residua.

Immunity. There is no evidence that the natural disease confers immunity against subsequent tetanus infection. The tetanus toxin is so toxic that an amount sufficient to cause clinical tetanus is too small to be immunogenic. Recurrent attacks are not uncommon, and for this reason patients who recover from the disease should be actively immunized with toxoid to prevent possible exogenous reinfection or recurrence of infection from spores of *C. tetani* retained within the body. Adequate immunization of pregnant patients is extremely important to insure passive immunity for the newborn, thereby diminishing the likelihood of tetanus neonatorum. This maternally transmitted immunity will last until active immunization has been started during the first year of life.

Laboratory Diagnosis. The diagnosis of tetanus is made on clinical grounds because the isolation of the organism can occur in the absence of disease; also, it is possible to have tetanus and never isolate the organism. If the local lesion can be detected and gram stained, one can occasionally demonstrate thin gram-positive or gram-negative rods, and sometimes spores, with varying amounts of the vegetative cell attached. Most such attempts to demonstrate the organisms directly, however, have been unsuccessful. Material from a known wound should be transported to the microbiology laboratory in transport vessels that have been filled with carbon dioxide for culture under anaerobic conditions.

Treatment. The treatment of tetanus varies with the severity of the disease. In general, however, it is designed to prevent the further elaboration and absorption of toxin.

Antitoxin is administered, and because of the immediate and delayed complications from antitoxin prepared in a horse or sheep, human antitoxin from pooled hyperimmune donors is recommended. The use of intrathecal human globulin is currently experimental, but reports concerning its efficacy have been enthusiastic when it is used early. Also under investigation is the intrathecal use of $F(ab')_2$ from human immune globulin. In addition, debridement of the wound and removal of any foreign bodies is recommended unless the extent or location precludes such a surgical approach. Large doses of penicillin should be given; if the patient is allergic to penicillin, tetracycline or clindamycin may be considered. Mild tetanospasm may be controlled with barbiturates and diazepan. With severe tetanospasms, however, a curare-like agent may be employed to completely paralyze the patient's muscles so that the respiratory function may be maintained by positive pressure breathing apparatus. Tracheostomy should be performed after the onset of the first tetanospasm in order to minimize respiratory complications. To minimize the frequency and severity of the tetanospasms, good supportive care of the patient should also include careful control of the environment to reduce auditory and visual stimuli. With improved control of tetanospasms and respiratory complications, greater attention has been paid to controlling the sympathetic effects of the toxin. Morphine and labetalol have been reported to be useful.

Prevention. Tetanus is almost completely preventable by properly applied active or passive immunization. During World War II, only 12 cases of tetanus occurred in 2,735,000 hospital admissions for wounds and injuries in soldiers who had been previously immunized. This successful experience resulted in the passage by most state legislatures in the United States of laws making admission to primary school contingent upon adequate immunization with tetanus toxoid. Compliance has resulted in the steady decline in the incidence of tetanus. Infants and children, the pregnant patient, and the elderly should be of primary concern for receipt of immunization.

ACTIVE IMMUNITY. Routine immunization with tetanus toxoid should begin at 1 to 3 months of age, using a combination of tetanus and diphtheria toxoid and pertussis vaccine (DPT). Three doses of DPT should be given at intervals of 3 or 4 weeks, with booster doses 1 and 4 years later. Immunity to tetanus can be maintained by a single booster dose of toxoid every 10 years. Because young children are very prone to lacerations and puncture wounds, in the past they were exposed repeatedly to booster shots when brought to emergency rooms. Also, requirements of schools, camps, and the armed forces have led to an inordinate exposure to tetanus toxoid. For this reason, patients presenting to an emergency room with a history of the basic immunizing series and a history of a booster injection within a 4-year period probably do not need to receive a booster injection at the time of injury. Individuals who present in the emergency room with no history of immunization or with partial immunization should receive in one arm the human immune globulin and in the other arm the first of a series of toxoid injections in order to prevent future tetanus (Tables 46-4 and 46-5).

PASSIVE IMMUNITY. Passive immunity may be conferred by the administration of antitoxin. This form of immunity was developed during World War I, when it was recognized that a small dose of tetanus antitoxin prepared in a horse was impressively protective when administered to humans at the time of an injury. Because of the risks involved in the use of a foreign serum in a sensitized recipient, human hyperimmune antitetanus globulin is recommended for passive immunization. The prophylactic administration of a single 250-unit dose of antitoxin should be reserved for patients with tetanus-prone wounds who have no record of immunization, or have received only one dose of tetanus toxoid, or who are not seen until 48 hours after the injury. Penicillin or tetracycline also should be given along with appropriate surgical care. Such persons receiving antitoxin should be given at the same time alum-absorbed toxin administered at a different site, and a second dose of toxoid 1 month later.

Clostridium botulinum

C. botulinum produces the most potent exotoxin known. The toxin, which is a neurotoxin, is the cause of botulism, a severe neuroparalytic disease characterized by sudden onset and swiftness of course, terminating in profound paralysis and pulmonary arrest. Although disease caused by botulinum toxin is rare in humans, it is much more common in animals. Unlike tetanus toxin, there are eight serologically distinct botulinum toxins, which are designated A, B, C_1, C_2, D, E, F, and G. Table 46-6 gives the animal species affected in outbreaks caused by the various types.

The most common form of botulism is food-borne botulism, an intoxication caused by the ingestion of preformed botulinum toxin in contaminated food. The disease botulism received its name from the Latin *botulus* (sausage). The term was introduced in 1870 to describe a fatal food poisoning syndrome associated with the eating of sausage. Although of historical interest, the name has lost much of its significance because fish and other animal proteins also transmit the disease, and because plant rather than animal products are more common vehicles in the United States. The canning industry's use of autoclaving at temperatures sufficient to kill the spores has reduced the

TABLE 46-4. ROUTINE DIPHTHERIA, TETANUS, AND PERTUSSIS IMMUNIZATION SCHEDULE SUMMARY FOR CHILDREN LESS THAN 7 YEARS OF AGE

Dose	Age/Interval	Product
Primary 1	6 weeks of age or older	DTP[†]
Primary 2*	4 to 8 weeks after first dose	DTP
Primary 3*	4 to 8 weeks after second dose	DTP
Primary 4*	Approximately 1 year after third dose	DTP
Booster	4 to 6 years of age, prior to entering kindergarten or elementary school (not necessary if fourth primary immunizing dose administered after fourth birthday)	DTP
Additional boosters	Every 10 years after last dose	Td

From Immunization Practices Advisory Committee, Centers for Disease Control: Ann Intern Med 95:723, 1981.
DTP, diphtheria and tetanus toxoids and pertussis vaccine, adsorbed; Td, tetanus and diphtheria toxoids adsorbed, (for adult use).
*Prolonging the interval does not require restarting series.
[†]DT, if pertussis vaccine is contraindicated.

TABLE 46-5. SUMMARY GUIDE TO TETANUS PROPHYLAXIS IN ROUTINE WOUND MANAGEMENT

History of Tetanus Immunization (Doses)	Clean, Minor Wounds		All Other Wounds	
	Tetanus and Diphtheria Toxoids (for Adult Use)*	Human Tetanus Immune Globulin	Tetanus and Diphtheria Toxoids (for Adult Use)*	Human Tetanus Immune Globulin
Uncertain	Yes	No	Yes	Yes
0–1	Yes	No	Yes	Yes
2	Yes	No	Yes	No[†]
3 or more	No[‡]	No	No[§]	No

From Immunization Practices Advisory Committee, Centers for Disease Control: Ann Intern Med 95:723, 1981.
*For children less than 7 years of age, diphtheria and tetanus toxoids and pertussis vaccine adsorbed (or diphtheria and tetanus toxoids adsorbed [for pediatric use], if pertussis vaccine is contraindicated) is preferred to tetanus toxoid alone. For persons 7 years of age and older, tetanus and diphtheria toxoids (for adult use) is preferred to tetanus toxoid alone.
[†]Yes, if wound more than 24-hours old.
[‡]Yes, if more than 10 years since last dose.
[§]Yes, if more than 5 years since last dose. (More frequent boosters not necessary and can accentuate side effects.)

TABLE 46-6. ANIMAL SPECIES SUSCEPTIBLE TO
***CLOSTRIDIUM BOTULINUM* TYPES**

Type	Species	Sites of Outbreaks
A	Human	United States, Soviet Union
B	Human, horse	United States, Northern Europe, Soviet Union
C_α*	Birds, turtles	Worldwide
C_β†	Cattle, sheep, horses	Worldwide
D	Cattle, sheep	Australia, South Africa
E	Human, birds	Northern Europe, Canada, United States, Japan, Soviet Union
F	Human	Denmark, United States
G	No outbreaks have been recognized	

From Smith: Botulism, 1977. Springfield, Ill, Chas C Thomas.
*Major toxin produced by type C_α is C_1 (see below).
†Major toxin produced by type C_β is C_2.

TABLE 46-7. SOME CULTURAL CHARACTERISTICS OF
CLOSTRIDIUM BOTULINUM

Characteristic	Cultural Group			
	I	II	III	IV
Digestion of coagulated protein	+	−	−*	+
Fermentation of glucose	+	+	+	−
Fermentation of mannose	−	+	+	−
Hydrolysis of gelatin	+	+	+	+
Formation of lipase	+	+	+	+
Production of indol	−	−	−	−
Reduction of nitrate	−	−	−	−
Fermentation products	A,P,IB, B,IV	A,B	A,P,B	A,P,IB, B,IV

From Smith: Botulism, 1977. Springfield, Ill, Chas C Thomas.
*Weak proteolysis by some strains.
A, acetic acid; B, butyric; IB isobutyric; IV, isovaleric; P, propionic.

relative importance of commercially canned food as a source of disease, except where procedural errors have occurred. Because the toxin is destroyed by heat, the routine cooking of home-canned food limits the frequency of this type of food poisoning.

In addition to food-borne botulism, the disease also occurs when toxin is produced by *C. botulinum* organisms contaminating traumatic wounds (wound botulism), and when toxin is elaborated within the gastrointestinal tract of infants (infant botulism).

Morphology and Physiology

Morphology. *C. botulinum* is a straight to slightly curved, gram-positive rod with rounded ends. Although exhibiting marked variation in size depending on cultural conditions and serologic type, the size falls within the range of 3.4 to 8.6 μm by 0.5 to 1.3 μm. Involution forms on artificial media are frequently observed. *C. botulinum* is motile with peritrichous flagella. It produces heat-resistant spores that are oval, subterminal, and tend to distend the bacillus. These are produced more consistently when the organism is grown on alkaline glucose media at 20C to 25C; spores usually are not produced at higher temperatures.

Cultural Characteristics. *C. botulinum* is a strict anaerobe, easily cultured in an anaerobic environment on routine media. On blood agar, all strains except those of type G are β-hemolytic. The nutritional requirements of the organism are complex, especially those of the nonproteolytic strains. Although originally classified into two groups, proteolytic and nonproteolytic, the species have subsequently been divided into four groups. Some of the cultural characteristics used to separate the four groups are depicted in Table 46-7.

Resistance. The heat resistance of the spores of *C. botulinum* is greater than that of any other anaerobe; the degree

of resistance to various physical and chemical factors depends on the specific strain and serologic type of the organism. Type A is more resistant than types B, C, and D; type E is the least heat-resistant but variants of this type that are exquisitely heat-resistant have been obtained. In general, the spores may survive several hours at 100C and up to 10 minutes at 120C. The spores are also resistant to irradiation and can survive temperatures of −190C.

Antigenic Structure

Exotoxin. The species *C. botulinum* includes a very heterogeneous group of strains that has been divided into eight serologically distinct types—A, B, C_α, C_β, D, E, F, and G—on the basis of the type of toxin produced. Immunologic differences among these types are constant and clearcut, and of epidemiologic significance. Except for types C and D strains, a single toxin is produced by each type. Types C and D strains are complex and produce three different toxins. Strains that produce predominantly C_1 and C_2 toxins are designated C_α and C_β, respectively; type D strains produce D toxin in the greatest amount.

Somatic Antigens. The antigenic composition of vegetative cells of *C. botulinum* is very complex and has not been completely defined for most of the types. Studies on types A and B have permitted their division into six subgroups on the basis of their heat-labile antigens. These strains of *C. botulinum* share one heat-stable antigen with *C. tetani, C. histolyticum,* and *C. sporogenes.* Using the fluorescent-antibody technique, cross-reactions have been demonstrated among the strains of A, B, and F, and between strains of C and D. There are different heat-stable and heat-labile agglutinating antigens among strains of type E. These strains, however, appear to be homogeneous and distinct and do not cross-react with strains of other types. For each of the types, spore antigens appear to be more specific than antigens of the vegetative cells.

Determinants of Pathogenicity

Properties of Botulinum Toxin. The clinical manifestations of botulism are attributable to the toxin of *C. botulinum* present in the ingested food, the gastrointestinal tract of infants, or in wounds. Botulinum toxin is one of the most potent toxins known. One microgram of the purified toxin contains about 200,000 minimal lethal doses for a 20-g mouse.

The optimal temperature for toxin production varies greatly both within and across types. Best results have been obtained over a temperature range of 30C to 38C, except for *C. botulinum* type E, which grows best at 25C to 28C. Initiation of growth and toxin production occurs only over a narrow pH range of 7.0 to 7.3. *C. botulinum* can be grown on a completely defined synthetic medium, but toxin production is less than that obtained on media containing casein hydrolysate or corn steep liquor.

Although usually classified as an exotoxin because of its high potency and antigenicity, botulinum toxin differs from a classic exotoxin in that it is not released during the life of the organism. Instead, it is produced intracellularly and appears in the medium only upon death and autolysis of the cell. The role played by the toxin in the metabolism of *C. botulinum* is unknown; the organism may be rendered nontoxigenic without any discernible effect on the growth rate. Original reports of a large molecular weight for the toxin (900,000) were modified when it was discovered that this original toxin contained both a neurotoxin and a hemagglutinin. Subsequent work has shown that the toxin is formed within the bacterial cell as a relatively low toxicity progenitor toxin. In the course of release of this progenitor molecule, proteolytic enzymes activate the toxin. The activation can be reproduced with trypsin. In addition, it has been noted that there is a cleavage or nicking of the molecule, resulting in a 150,000 molecular weight polypeptide composed of a 100,000 molecular weight heavy chain and a 50,000 molecular weight light chain joined by disulfide bridges. It is of interest that type E toxigenic *C. botulinum* is nonproteolytic and lacks the enzymes necessary to activate the toxin. In order to overcome this deficiency, it has been suggested that the tryptic activity within the gastrointestinal tract activates the toxin. In studying the oral toxicity of botulinum toxin, it was noted that the larger molecular weight aggregates are more toxic. It has been suggested that the larger aggregates are more stable within the gastrointestinal tract against prolonged exposure to gastrointestinal enzymes that reduce their ultimate toxicity.

The observation that there were many isolates of *C. botulinum* that were nontoxigenic and that most *C. botulinum* carry bacteriophages suggested that the production of toxins by *C. botulinum* might be mediated by bacteriophages. It has now been well documented that this is indeed the case. It is possible to cure *C. botulinum* of toxigenic bacteriophages and thereby render them nontoxigenic. It is also possible to convert one type of toxin-producing *C. botulinum* to another by reinfecting such cured *C. botulinum* strains with a different bacteriophage. In the course of this work, it was discovered that bacteriophage infection of certain strains of *C. botulinum* could convert this organism to *C. novyi*, a cause of gas gangrene.

Site of Action of Botulinum Toxin. Toxin is absorbed largely from the small intestine and appears in the lymphatics draining the intestine before it is found in the bloodstream. Much of the ingested toxin may not be absorbed from the small intestine, as it may be destroyed by proteolytic enzymes. After an incubation period that is inversely related to dose, botulinum intoxication is associated with a functional disturbance of the peripheral nervous system resulting from the inhibition of the release of acetylcholine. The toxin acts at the myoneural junction to produce complete paralysis of the cholinergic nerve fibers at the point of release of acetylcholine. Botulinum toxin affects both sets of cholinergic transmission points in the autonomic system, the synaptic ganglia and the parasympathetic motor end plates peripherally located in the junction between the nerve and cell fibers. The process of intoxication begins with a binding step. Although no specific neuroreceptor site has been identified as in tetanus toxin, there is indirect evidence that certain gangliosides are very effective inactivators of the toxin. The next step is internalization of the toxin by either pinocytosis or a protein carrier. The final step—the lytic step—is the blocking of excitation-secretion coupling. Botulinum toxin antagonizes the transmitter-releasing effects of calcium.

Inactivation of Toxin. The susceptibility of toxin to inactivation by various chemical and physical agents is of practical importance because of its role in food poisoning. The resistance of botulinum toxin to various deleterious agents is markedly dependent upon the serologic type of toxin, temperature, pH, and the presence of certain other materials in the medium in which the toxin is tested. In general, thermostability depends upon the substance in which the toxin is dissolved. All toxin types are completely destroyed by boiling for 1 minute, or by heating at 75C to 85C for 5 to 10 minutes. At room temperature toxin persists for several days in tap water, an important observation suggesting the possibility of prolonged contamination of water supplies. Toxin is destroyed by direct sunlight within 5 days unless protected from air, in which case toxin inactivation proceeds at a slower rate. A low pH of 3.5 to 6.8 favors preservation of the toxin, while an alkaline pH favors detoxification. Factors usually found in putrifying canned foods apparently have no affect on the toxin.

Laboratory Detection of Toxin. The incrimination of *C. botulinum* as the cause of food poisoning is based on the demonstration of toxin in the food or in the sera or gastric contents from the patient. The most commonly used method is the mouse toxicity and neutralization test. The sensitivity can detect as little as 10 μg (10^{-8} mg) of toxic protein. Other methods such as the enzyme-linked immunoabsorbent assay (ELISA) are available, although their sensitivity is not as great.

Clinical Intoxication

Ecology. *C. botulinum* is the causative agent of a very lethal type of food poisoning that occurs in several species of animals. In general, food poisoning in a particular animal species is usually associated with certain types of the organism. The reason for this specificity is unknown. Humans are susceptible to types A, B, E, and F; birds primarily to A and C; ruminants to C and D; and mink to A, B, C, and E (See Table 46-6). *C. botulinum* has been isolated from the sediments of lakes and rivers; from virgin and cultivated terrestrial soil; from the intestinal tract of fish; and from the intestinal tract, spleen, and liver of a variety of animals. In general, *C. botulinum* is isolated more frequently from soil containing silt and is easier to culture from manured than from nonmanured land.

Knowledge of the distribution of the various types of *C. botulinum* throughout the world is based on reports of disease in animals and humans caused by the different types, and upon isolation of the organisms from animals and the soil in various parts of the world. Table 46-6 shows the major geographic areas associated with each type. Although there is a prevalence of certain types of *C. botulinum* food poisoning in various localities, this does not rule out the possibility of spores of other types in the same locality. In the United States, types A and B are widely distributed and have been associated with most outbreaks of human botulism. Type A is the predominant type in the Pacific Coast states, in the Rocky Mountains, and in Maine, New York, and Pennsylvania. The strains found in the soil of the Mississippi River Valley, the Great Lakes region, and in New Jersey, Delaware, Maryland, Georgia, and South Carolina are predominantly type B. In recent years, type E has been isolated in several parts of the United States, but is especially prevalent in the Great Lakes area.

Disease in Animals. A number of characteristic paralytic diseases of birds and mammals are caused by the ingestion of botulinum toxin in the animals' food. The best known examples include grass or fodder sickness of horses, silage disease in cattle, limberneck in chickens, lamziekte of cattle, and dust sickness in wild birds. Outbreaks of botulism in animals reflect the geographic distribution of the organism, susceptibility of the animal species to the different toxins, and other predisposing factors. Thus, lamziekte in cattle is restricted to areas in which the soil and herbage are markedly deficient in phosphorus. In such areas the cattle are prone to eating putrid bones and carcasses of small animals that are often toxic as a result of *C. botulinum* in the intestinal tract. Botulism in sheep results from bone chewing and is associated with periods of drought or with over-grazed ranges. Forage poisoning in cattle and horses occurs in many parts of the world as a result of the ingestion of toxic hay or silage. The toxin in the hay originates from an animal carcass. Carcasses of small animals, especially cats, found in a herd's food or bedding, have been shown to contain as much as 3000 minimal lethal doses of toxin per gram. This toxin diffuses out into the hay or silage.

Type C botulinum toxin is the cause of large epidemics of botulism in aquatic and shore birds. Wild ducks are most frequently involved in these epidemics, which may affect thousands of birds. The disease is the major natural cause of death of ducks in the western United States. Outbreaks have also occurred in domestic ducks, gulls, loons, and sandpipers. Such outbreaks are probably initiated by strong winds that tear aquatic plants from their moorings on the shore, leaving long rows of uprooted vegetation to decay. Aquatic invertebrates present in the decaying vegetation die because of lack of oxygen, and *C. botulinum* proliferates in their bodies. In searching for food among the masses of decaying vegetation, ducks ingest the toxic bodies of the invertebrates that contain *C. botulinum* organisms. After the duck's death the carcass is invaded by the organisms; the carcass itself becomes toxic and serves to perpetuate the outbreak. The carcass becomes flyblown, and the fly larvae pick up a considerable amount of toxin both in their exterior slimy coating and by ingestion. The ingestion by ducks of only a few of these fly larvae results in death; the dead ducks are invaded by *C. botulinum*, and become flyblown, thus furnishing toxic larvae to poison more ducks. Botulism in pheasants follows a similar pattern.

Carrion eaters such as the vulture are almost completely resistant to botulinum toxin. The mechanism of this resistance is unknown, but apparently it cannot be attributed to the presence of antitoxin in the animal's blood.

Disease in Humans

TYPES OF BOTULISM. In the United States, cases of botulism are now classified into four categories:

1. Food-borne botulism, which is the most common, is a lethal type of food poisoning resulting from the ingestion of the neurotoxin in incompletely processed food contaminated with the organisms.
2. Infant botulism, which is a recently described syndrome in infants, is presumably related to the ingestion of *C. botulinum* spores, the multiplication of organisms within the gastrointestinal tract, and subsequent absorption of toxin.
3. Wound botulism, which is the least common, is a neuroparalytic illness associated with wounds showing little clinical evidence of active infection.
4. Unclassified botulism includes persons over the age of 1 year who have symptoms of clinical botulism but with no identifiable vehicle of transmission.

EPIDEMIOLOGY. The geographic distribution of botulism is primarily in the northern hemisphere, between 30 and 65 degrees north latitude—north of the Gulf of Mexico, the Mediterranean Sea, the Persian Gulf, and the Bay of Bengal. Most of the outbreaks occur in seven countries: Canada, France, Germany, Japan, Poland, the Soviet Union, and the United States. From 1950 through 1979, 215 outbreaks of food-borne botulism were reported in the United States, with 566 persons affected. Although improved industrial canning methods have resulted in a gradual decline in the incidence of the disease, outbreaks still occur, especially as a consequence of the ingestion of home-canned foods. Ninety-two percent of the food-borne botulism was caused by food preserved at home. Although botulism is now a comparative rarity, its potential must always be kept in mind.

At the present time, most cases of food-borne botulism occur in relatively circumscribed outbreaks following the consumption of home-preserved food. Sterilization procedures employed by commercial canneries utilize pressure apparatus in which canned products are held at a temperature of 121.1C for 30 minutes. As a result, outbreaks of botulism are rarely associated with canned food. In recent years, however, there have been outbreaks from canned tuna fish and from vichyssoise, as well as from smoked fish. The home-canned food most often incriminated as the source of botulism is green beans, which may produce only a slight sharp taste, prompting the homemaker to rinse the contents of the jar of beans and serve them in a salad. Highly acidic foods, such as tomatoes and citrus fruits, are rarely the source because the organisms will not grow and release toxin at the low pH's encountered, about 4.6. Because of increased use of home freezers, meat has become an extremely rare source of *C. botulinum* in the United States. Although the early descriptions of botulism followed the ingestion of contaminated sausage and meat products, botulism of this type is now uncommon. Specialized foods eaten by certain ethnic groups are often responsible for the prevalence of a particular type of botulism poisoning in a given locality. In Japan, sushi (fermented raw fish salad) is often implicated; among the Indians of the northwest Pacific Coast, salmon egg cheese or stink eggs are responsible. Among the Alaskan Eskimos, muklak is prepared by soaking beluga flippers in seal oil for an extended period before eating. In all of these foods, conditions are suitable for *C. botulinum* growth and toxin production.

Pathogenesis. Food-borne botulism results from the ingestion of preformed botulinum toxin in contaminated foods. After ingestion the toxin is absorbed primarily from the upper gastrointestinal tract, but toxin reaching the lower small intestine and colon may be slowly absorbed, perhaps accounting for the delayed onset and prolonged duration of symptoms seen in many patients. The toxin gains access to the peripheral nervous system and, as discussed on page 713, blocks release of acetylcholine.

In several reported cases of wound botulism, *C. botulinum* contaminated a traumatic wound. The rarity of this form of botulism is probably due to failure of *C. botulinum* spores to germinate readily in tissues. Also, within the setting of the experimental laboratory, the inhalation of aerosolized toxin has resulted in three cases of clinical botulism.

Less clear is the role of infection in the gastrointestinal tract as a source of botulism. Since the recognition in 1976 of infant botulism as a disease entity, it has been recognized with increased frequency. There appears to be no seasonality of its occurrence or of toxin type. In a significant number of cases the disease followed the ingestion of honey that was contaminated with botulinum spores, strongly supporting this as one method of intoxication. Botulinum toxin was recently isolated from the serum and stool of a 6-week-old boy with infant botulism, the first time that the toxin has been isolated from the serum of an infant with the disease.

Clinical Manifestations. The incubation period and clinical manifestations are similar for all types of botulinum toxin. Because the duration of the incubation period can be related to the dose of toxin, the shorter the incubation period, the poorer the prognosis. Characteristically, symptoms begin 12 to 36 hours following ingestion of the contaminated food, or as late as 8 days after. Type E botulism appears to have a shorter incubation period than do types A and B. Severe nausea and vomiting are frequently observed with type E intoxication. Weakness, lassitude, and dizziness are often early complaints. There is usually no diarrhea, but constipation is common. The early symptoms of botulism would rarely bring a patient to a physician's attention. Cranial nerve palsies are usually the presenting symptom: classically diplopia (double vision), dysphagia (difficulty in swallowing), and dysphonia (difficulty in speaking). The pupils are dilated, and the tongue is very dry and furry. In type E intoxication, abdominal distention is especially common, leading to a mistaken diagnosis of acute abdomen. Fever is rarely observed, and the mental processes remain intact. As the disease progresses, weakness of muscle groups, particularly of the neck, proximal extremities, and respiratory musculature, is often observed, leading ultimately to sudden respiratory paralysis, airway obstruction, and death. The mortality rate is affected by the type of toxin consumed, distribution of toxin in the food, and speed with which the disease is diagnosed and antitoxin therapy is initiated. Recent mortality is about 32 percent for type A toxin, 17 percent for type B toxin, and 40 percent for type E toxin.

Clinically, infant botulism is an acute flaccid paralysis that manifests as weakness of head, face, and throat musculature, that then extends symmetrically to involve the muscles of the trunk and extremities. Death results either from paralyzed tongue or pharangeal muscles occluding the airway, from paralysis of the diaphragm and intercostal muscles, or from secondary complications. Fulminant forms may resemble the sudden infant death syndrome (SIDS, or crib death). The age distribution rises rapidly after the first week of life, peaking at 1 to 2 months and gradually subsiding with only a small number of cases occurring beyond the age of 6 months.

The neurologic diseases most frequently confused with botulism are myasthenia gravis, Guillain-Barré syndrome, and cerebrovascular accidents.

Laboratory Diagnosis. The rapid diagnosis and establishment of the type of botulinum toxin affecting the patient is of crucial importance. Unfortunately, however, the disease is difficult to diagnose because at its onset the symptoms of botulism are often confused with the symptoms of other diseases and because few physicians are familiar with it. Diagnosis in an isolated case may be extremely difficult, and by time the nature of the disease is apparent, it is usually too late for therapy.

As soon as a diagnosis of botulism is suspected on clinical grounds, a specimen of the patient's blood should be drawn immediately and allowed to clot. Specimens of stool and gastric washings should also be obtained if possible. The Centers for Disease Control in Atlanta, Georgia, should be called to make arrangements for the laboratory diagnosis.* Collected specimens should be refrigerated until arrangements are made to send them by air express service to Atlanta (p. 713). If at all possible, the original food specimen should be obtained for similar studies, but all too frequently the original ingested food has been discarded. Because it is imperative to begin specific treatment as quickly as possible, the diagnosis must be made clinically and then confirmed by laboratory methods.

Infant botulism has presented a problem in diagnosis because the toxin is usually not detectable in the patient's serum. Diagnosis rests upon the identification of the toxin or the isolation of the organism from the stool. The isolation of the organism has been made easier by the recognition that *C. botulinum* is resistant to cycloserine, sulphamethoxisole, and trimethoprim. Addition of these antibiotics to an egg yolk agar medium has provided a selective tool for the isolation of this organism.

Electromyography may also provide supporting information, as it can demonstrate a characteristic finding of abundant brief, small-amplitude motor unit potentials.

Treatment. The therapy of botulism leaves much to be desired. Immediate administration of antitoxin has been the cornerstone of the therapy of adult botulism, but its efficacy, especially in well-developed cases of neuroparalytic disease, has been questioned. Antitoxin in the type E syndrome, however, has received enthusiastic reports. The presently used antitoxin (polyvalent A, B, E, available through the Centers for Disease Control), is of equine origin and thus carries the risk of potential side effects of immediate anaphylaxis and serum sickness. Skin testing for hypersensitivity to horse serum must be performed prior to administration of the antitoxin. In addition to antitoxin, the physiologic support of the patient is critical; this calls for management in an intensive care unit with the available support of respiratory, cardiovascular, and renal services. Saline enemas have been recommended. Other therapeutic considerations include guanidine hydrochloride, which enhances acetylcholine release. Variable success has been encountered with the use of this agent, and caution is required in its use because of potentially serious side effects. Experimentally, 4-aminopyridine has given interesting results in animal models, and may provide a clinically useful drug in the future. Unless there are infectious

complications, antibiotics are not recommended. There is one report of aminoglycoside potentiation of the paralytic effects of botulism.

Prevention. Although existing preventive measures for the control of botulism are simple and effective when properly carried out, the fact that a number of outbreaks still occur each year indicates the need for improved methods of control. The homemaker should be alerted to the necessity of using sterilized containers and pressure cookers in the canning of all foods in order to kill any *C. botulinum* spores that may be on the food. Also, before eating any home-canned food it should be boiled for 1 minute or heated at 80C for 5 minutes to destroy any toxin that might have been produced in the anaerobic environment provided.

The safety record of the canning industry during the past 40 years has been very impressive. Unfortunately, however, minor breaks in technique do occur and unless rigid controls are constantly monitored there will continue to be sporadic outbreaks. A bulging or defective can should be immediately discarded.

Because botulism is a relatively rare disease, immunization of the entire population is impractical. An effective toxoid is available, however, for laboratory workers in high-risk situations. The recommended schedule for the establishment of active immunity in humans is two injections of toxoid, either absorbed on aluminum sulfate or mixed with an equal volume of Freund's adjuvant, given at 0 and at 10 weeks, with a booster injection 1 year later.

Economically, botulism in animals is a very important disease, causing the death of many thousands of animals and birds each year. For range cattle, immunization with toxoid is the most practical method for prevention. In addition, lamziekte in cattle may be controlled by keeping the feeding area free of carcasses and by providing a diet adequate in phosphorus. Botulism in sheep can be prevented by supplementing the diet with carbohydrates and protein. Forage poisoning in cattle and horses may be prevented by keeping food and bedding free of carcasses of small animals.

FURTHER READING

Books and Reviews

HISTOTOXIC CLOSTRIDIA

Balows A, DeHaan RM, Dowell VR Jr, et al. (eds): Anaerobic Bacteria—Role in Disease. Springfield, Ill, Thomas, 1974

Buchanan RE, Gibbons NE (eds): Bergey's Manual of Determinative Bacteriology, 8th ed. Baltimore, Williams & Wilkins, 1974

Dowell VR Jr, Hawkins TM: Laboratory Methods in Anaerobic Bacteriology. CDC Laboratory Manual. Washington D.C., Government Printing Office, HEW Pub No (CDC) 74-8272, 1974

George WL, Sutter VL, Finegold SM: Antimicrobial agent-induced diarrhea—A bacterial disease. J Infect Dis 136:822, 1977

Hill EO: The genus *Clostridium* (Medical aspects). In Starr MP, Stolp H, Truper HG, et al. (eds): The Prokaryotes. A Handbook

---

*Centers for Disease Control, Atlanta, Georgia 30333. Telephone: Day (404) 329-3753 Nights, holidays, and weekends (404) 329-3644.

on Habitats, Isolation, and Identification of Bacteria. New York, Springer-Verlag, 1981, p 1756

Holdeman LV, Moore WEC: Anaerobic Bacteriology Manual, 4th ed. Anaerobe Laboratory, Blacksburg, Virginia Polytechnic Institute and State University, 1977

Kadis S, Montie TC, Ajl SJ (eds): Microbial Toxins. Bacterial Protein Toxins. New York, Academic Press, 1971, vol IIA

Lennette EH, Balows A, Hausler WJ Jr, et al.: Manual of Clinical Microbiology, 3rd ed. Washington, D.C., American Society for Microbiology, 1980

MacLennan JD: The histotoxic clostridial infections of man. Bacteriol Rev 26:177, 1962

Smith LDS: The Pathogenic Anaerobic Bacteria. Springfield, Ill, Thomas, 1975

Sutter VL, Vargo VL, Finegold SM: Wadsworth Anaerobic Bacteriology Manual, 3rd ed. St. Louis, Mosby, 1980

Willis AT: Clostridia of Wound Infection. London, Butterworths, 1969

CLOSTRIDIUM TETANI

Bizzini B: Tetanus toxin. Bacteriol Rev 43:224, 1979

Kerr J: Current topics in tetanus. Inten Care Med 5:105, 1979

Smith LD, Holdeman LV: The Pathogenic Anaerobic Bacteria. Springfield, Ill, Thomas, 1968

Van Heyningens: Tetanus toxin. Pharm Ther 11:141, 1980

Veronesi R: Tetanus—Important New Concepts. Amsterdam, Excerpta Medica, 1981

Willis AT: Clostridia of Wound Infection. London, Butterworths, 1969

CLOSTRIDIUM BOTULINUM

Arnon SS, Daus K, Chin J: Infant botulism: Epidemiology and relation to sudden infant death syndrome. Epidemiol Rev 3:45, 1981

Centers for Disease Control: Botulism in the United States, 1899–1973. Handbook for Epidemiologists, Clinicians and Laboratory Workers. Atlanta, Ga, Centers for Disease Control, 1979

Feldman RA (ed): A seminar on infant botulism. Rev Infect Dis 1:607, 1979

Lewis GE Jr (ed): Biomedical Aspects of Botulism. New York, Academic Press, 1981

Simpson LI: The origin, structure and pharmacological activity of botulinum toxin. Pharm Rev 33:155, 1981

Smith LDS: Botulism. Springfield, Ill, Thomas, 1977

Sugiyama H: *Clostridium botulinum* neurotoxin. Microbiol Rev 44:419, 1980

Selected Papers

HISTOTOXIC CLOSTRIDIA

Alpern BJ, Dowell VR Jr: *Clostridium septicum* infections and malignancy. JAMA 209:385, 1969

Altemeier WA, Fuller WD: Prevention and treatment of gas gangrene. JAMA 217:806, 1971

Aronsson B, Mollby R, Nord CE: Occurrence of toxin-producing *Clostridium difficile* in antibiotic-associated diarrhea in Sweden. Med Microbial Immunol 170:27, 1981

Bartlett JG, Gorbach SL: Pseudomembranous enterocolitis (antibiotic-related colitis). Adv Intern Med 22:455, 1977

Brown RA, Fekety R Jr, Silva J Jr, et al.: The protective effect of vancomycin on clindamycin-induced colitis in hamsters. Johns Hopkins Med J 141:183, 1977

Cato EP, Holdeman LC, Moore WEC: *Clostridium perenne* and *Clostridium paraperfringens*: Later subjective synonyms of *Clostridium barati*. Int J Syst Bacteriol 32:77, 1982

Eklung MW, Poysky FT, Meyers JA, et al.: Interspecies conversion of *Clostridium botulinum* type C to *Clostridium novyi* type A by bacteriophage. Science 186:456, 1974

George WL, Rolfe RD, Finegold SM: *Clostridium difficile* and its cytotoxin in feces of patients with antimicrobial agent-associated diarrhea and miscellaneous conditions. J Clin Microbiol 15:1049, 1982

George WL, Sutter VL, Citron D, et al.: Selective and differential medium for isolation of *Clostridium difficile*. J Clin Microbiol 9:214, 1979

Hauschild AHW, Nolo L, Dorward WJ: The role of enterotoxin in *Clostridium perfringens* type A enteritis. Can J Microbiol 17:987, 1971

Hobbs BC, Smith ME, Oakley CL, et al.: *Clostridium welchii* food poisoning. J Hyg 51:75, 1953

Holland JA, Hill GB, Wolfe WG, et al.: Experimental and clinical experience with hyperbaric oxygen in the treatment of clostridial myonecrosis. Surgery 77:75, 1975

Holst E, Helin I, Mardh PA: Recovery of *Clostridium difficile* from children. Scand J Infect Dis 13:41, 1981

Kim KH, Fekety R, Batts DH, et al.: Isolation of *Clostridium difficile* from the environment and contacts of patients with antibiotic-associated colitis. J Infect Dis 143:42, 1981

Libby JM, Jortner BS, Wilkins TD: Effects of the two toxins of *Clostridium difficile* in antibiotic-associated cecitis in hamsters. Infect Immun 36:822, 1982

Lyerly DM, Lockwood DE, Richardson SH, et al.: Biological activities of toxins A and B of *Clostridium difficile*. Infect Immun 35:1147, 1982

Lyerly DM, Sullivan NM, Wilkins TD: Enzyme-linked immunosorbent assay for *Clostridium difficile* toxin A. J Clin Microbiol 17:72, 1983

Macfarlane MG: On the biochemical mechanism of action of gas gangrene toxins. Sym Soc Gen Microbiol 5:57, 1955

Macfarlane MG, Knight BCJG: The biochemistry of bacterial toxins. I. The lecithinase activity of *C. welchii* toxins. Biochem J 35:884, 1941

Marrie TJ, Haldane EV, Swantee CA, et al.: Susceptibility of anaerobic bacteria to nine antimicrobial agents and demonstration of decreased susceptibility of *Clostridium perfringens* to penicillin. Antimicrob Agents Chemother 19:51, 1981

Pierce PF Jr, Wilson R, Silva J Jr, et al.: Antibiotic-associated pseudomembranous colitis: An epidemiologic investigation of a cluster of cases. J Infect Dis 145:269, 1982

Rolfe RD, Helebian S, Finegold SM: Bacterial interference between *Clostridium difficile* and normal fecal flora. J Infect Dis 143:470, 1981

Shertz RJ, Sarubbi FA: The prevalence of *Clostridium difficile* and toxin in a nursery population: A comparison between patients with necrotizing enterocolitis and an asymptomatic group. J Pediatr 100:435, 1982

Stern M, Warrock GH: The types of *Clostridium perfringens*. J Pathol Bacteriol 88:279, 1964

Taylor NS, Thorne GM, Bartlett JG: Comparison of two toxins produced by *Clostridium difficile*. Infect Immun 34:1036, 1981

Viscidi R, Willey S, Bartlett JG: Isolation rates and toxigenic potential of *Clostridium difficile* isolates from various patient populations. Gastroenterology 81:5, 1981

Walker PD, Murrell TGC, Nagy LK: Scanning electronmicroscopy of the jejunum in enteritis necroticans. J Med Microbiol 13:445, 1980

CLOSTRIDIUM TETANI

Black GF, Buchanan TM, Bennett JV: Tetanus toxoid immunization of adults: A continuing need. Ann Intern Med 73:603, 1970

Black RE, Huber DH, Curlin GT: Reduction in neonatal tetanus by mass immunization of non-pregnant women: Duration of protection provided by one and two doses of aluminum adsorbed tetanus toxoid. Bull WHO 58:927, 1980

Buchanan N, Crane RD, Wolfson G, et al.: Autonomic dysfunction in tetanus: The effects of a variety of therapeutic agents with special reference to morphine. Inten Care Med 5:65, 1978

Edsall G, Elliott MW, Peebles TC, et al.: Excessive use of toxoid boosters. JAMA 202:17, 1967

Erdman G, Hanauske A, Wellhouer HH: Intraspinal distribution and reaction in grey matter with tetanus toxin of intracisternally injected anti-tetanus toxin F(ab')2 fragments. Brain Res 211:367, 1980

Gupta PS, Goyal S, Kapoor R, et al.: Intrathecal human tetanus immunoglobulin in early tetanus. Lancet 2:439, 1980

Helting TB, Zwisler O: Structure of tetanus toxin. I. Breakdown of the toxin molecule and discrimination between polypeptide fragments. J Biol Chem 252:187, 1977

Helting TB, Zwisler O, Wiegandt H: Structure of tetanus toxin. II. Toxin binding to ganglioside. J Biol Chem 252:194, 1977

Hortnagel H, Brucke T, Hackl JM: The involvement of the sympathetic nervous system in tetanus. Klin Wochenschr 57:383, 1979

Immunization Practices Advisory Committee, Centers for Disease Control: Diphtheria, tetanus and pertussis—Guidelines for vaccine prophylaxis and other preventive measures. Ann Int Med 95:723, 1981

Johnson DM: Fatal tetanus after prophylaxis with human tetanus globulin. JAMA 207:1519, 1969

Kozbor D, Roder TC: Requirements for the establishment of high-titered human monoclonal antibodies against tetanus toxoid using the Epstein–Barr virus technique. J Immunol 127:1275, 1981

Laird WJ, Aaronson W, Silver RP, et al.: Plasmid associated toxigenicity in Clostridium tetani. J Infect Dis 142:623, 1980

MacLennan JD: The serological identification of C. tetani. Br J Exp Pathol 20:371, 1939

Matsuda M, Yoneda M: Antigenic substructure of tetanus neurotoxin. Biochem Biophys Res Comm 77:268, 1977

Mellanby J, Green J: How does tetanus toxin act? Neuroscience 6:281, 1981

Murphy SG, Miller KD: Tetanus toxin and antigenic derivatives. I. Purification of the biologically active monomer. J Bacteriol 94:580,586 1967

Nakamura S, Okado I, Tomo A, et al.: Taxonomy of Clostridium tetani and related species. J Gen Microbiol 113:29, 1979

Pascale LR, Wallyn RJ, Goldfein S, et al.: Treatment of tetanus by hyperbaric oxygenation. JAMA 189:408, 1964

Price DL, Griffin J, Young A, et al.: Tetanus toxin: Direct evidence for retrograde intraaxonal transport. Science 188:945, 1975

Robinson JP, Picklesimer JB, Puett D: Tetanus toxin-effect of chemical modifications on toxicity, immunogenicity, and conformation. J Biol Chem 250:7435, 1975

Ruben FL, Nagel J, Fireman P: Antitoxin response in the elderly to tetanus-diphtheria (Td) immunization. Ann J Pub Health 108:145, 1978

Suri JC, Rubbo SD: Immunization against tetanus. J Hyg (Camb) 59:29, 1961

Van Heyningen S: Binding of ganglioside by the chains of tetanus toxin. FEBS Letters 68:5, 1976

Wessler S, Avioli LA: Tetanus. JAMA 207:123, 1969

Wigley FM, Wood SH, Waldaman RH: Aerosol immunization of humans with tetanus toxoid. J Immunol 103:1096, 1969

Young LS, LaForce FM, Bennett JV: An evaluation of serologic and antimicrobial therapy in the treatment of tetanus in the United States. J Infect Dis 120:153, 1969

Zacks SI, Sheff MF: Tetanus toxin: Fine structure localization of binding sites in striated muscle. Science 159:643, 1968

CLOSTRIDIUM BOTULINUM

Arnon SS, Midura TF, Damus K, et al.: Honey and other environmental factors for infant botulism. J Pediatr 94:331, 1979

Boroff DA, Reilly JRV: Prophylactic immunization of pheasants and ducks against avian botulism. J Bacteriol 77:142, 1959

Boroff DA, Shu-Chen G: Radioimmunoassay for type A toxin of Clostridium botulinum. Appl Microbiol 25:545, 1973

Boroff DA, Nyberg S, Hoglund S: Electron microscopy of the toxin and hemagglutinin of type A Clostridium botulinum. Infect Immun 6:1003, 1972

Bott TL, Johnson J Jr, Foster EM, et al.: Possible origin of high incidence of Clostridium botulinum type E in an inland bay (Green Bay of Lake Michigan). J Bacteriol 95:1542, 1968

Brown GW Jr, King G, Sugiyama H: Penicillin-lysozyme conversion of Clostridium botulinum types A and E into protoplasts and their stabilization as L-form cultures. J Bacteriol 104:1325, 1970

Cardella MA, Duff JT, Wingfield BH, et al.: VI. Purification and detoxification of type D toxin and immunologic response to toxoid. J Bacteriol 79:372, 1960

Craig JM, Pilcher KS: Clostridium botulinum type F isolated from salmon from the Columbia River. Science 153:311, 1966

DasGupta BR, Sugiyama H: Molecular forms of neurotoxins in proteolytic C. botulinum type B cultures. Infect Immun 14:680, 1976

Dezfulian M, McCroskey LM, Hatheway CL, et al.: Selective medium for isolation of Clostridium botulinum from human feces. J Clin Micro 13:526, 1981

Dezfulian M, Dowell VR: Cultural and physiological characteristics and antimicrobial susceptibility of Clostridium botulinum isolates from food borne and infant botulism cases. J Clin Micro 11:604, 1980

Eklund MW, Poysky FT: Interconversion of type C and D strains of Clostridium botulinum by specific bacteriophages. Appl Microbiol 27:251, 1974

Eklund MW, Poysky FT, Reed SM, et al.: Bacteriophage and the toxigenicity Clostridium botulinum type C. Science 172:480, 1971

Fiock MA, Varinsky A, Duff JT: VII. Purification and detoxication of trypsin-activated type E toxin. J Bacteriol 82:66, 1961

Kitamura M, Iwamori M, Nagai Y: Interaction of Clostridium botulinum neurotoxin and gangliosides. Biochem Biophys Acta 628:328, 1980

Kozaki S, Miyazaki S, Sakaguchi G: Development of antitoxin with each of two complementary fragments of C. botulinum type B derivative toxin. Infect Immun 18:761, 1977

Mandia JW: Serological group II of the proteolytic clostridia. J Immunol 67:49, 1951

Merson MH, Dowell VR: Epidemiologic, clinical and laboratory aspects of wound botulism. N Engl J Med 289:1005, 1973

Miyazaki S, Iwasaki M, Sakaguchi G: *Clostridium botulinum* type D toxin: Purification, molecular structure, and some immunological properties. Infect Immun 17:395, 1977

Oguma K: Stability of toxigenicity in *Clostridium botulinum* type C and type D. J Gen Microbiol 92:67, 1976

Oguma K, Iida H, Shiozaki M, et al.: Antigenicity of converting phages obtained from *C. botulinum* types C and D. Infect Immun 13:855, 1976

Ohishi I, Sakaguchi G: Activation of botulinum toxins in the absence of nicking. Infect Immun 17:402, 1977

Pickett J, Berg B, Chaplain E, et al.: Syndrome of botulism in infancy: Clinical and electrophysiologic study. N Engl J Med 295:770, 1976

Scott AB: Botulinum toxin injection into extraocular muscles as an alternative to strabismus surgery. Ophthalmology 87:1044, 1980

Sugh S, Sakaguchi G: Molecular construction of *Clostridium botulinum* type A toxins. Infect Immun 12:1262, 1975

Swenson JM, Thornsberry L, McCroskey LM, et al.: Susceptibility of *Clostridium botulinum* to thirteen antimicrobial agents. Antimicrob Agents Chemother 18:13, 1980

Takagi A, Kawata T, Yamamoto S: Electron microscope studies on ultrathin sections of spores of the clostridium group with special reference to the sporulation and germination process. J Bacteriol 80:37, 1960

Terranova W, Breman JG, Lacey RP, et al.: Botulism type B: Epidemiologic aspects of an extensive outbreak. Am J Epidemiol 108:150, 1978

Williams-Walls NJ: Type E botulism isolated from fish and crabs. Science 162:375, 1968

Wonnacott S, Marchbanks RM: Inhibition by botulinum toxin of depolarization-evoked release of (^{14}C)acetylcholine from synaptosomes in vitro. Biochem J 156:701, 1976

CHAPTER 47

Treponema, Borrelia, and Leptospira

Spirochaetaceae

The family Spirochaetaceae contains motile, slender, helically coiled, flexible organisms with one or more complete turns in the helix. There are five genera in the family Spirochaetaceae, of which only *Treponema*, *Borrelia*, and *Leptospira* contain species that cause major human illness. Differentiation among the genera of the family is based primarily on morphology (Table 47-1). Members of the family Spirochaetaceae are 3 to 500 μm long and 0.2 to 0.75 μm wide. Multiplication is by transverse fission. Cellular motility includes rapid rotation around the long axis, flexation of cells, and locomotion along a helical path. Spirochaetaceae may be aerobic, facultatively anaerobic, or anaerobic. Many are best recognized by darkfield microscopy, since they may be below the resolution of light microscopy.

Spirochaetaceae are gram-negative organisms characterized by the presence of an axial fibril (also referred to as an axial filament or endoflagellum). The basic cellular components of Spirochaetaceae are the axial fibril, the outer sheath that encompasses the cell, the protoplasmic cylinder that includes the cell wall and cell membrane, and the cellular cytoplasm (Fig. 47-1).

The outer sheath appears to be a unit membrane that may be separated from the cell for examination by electron microscopy. The internal structure of the outer membrane varies from one species to another and may also be altered by fixation techniques. Although the precise chemical composition is uncertain, carbohydrates, proteins, and prospholipids are present. The function of the outer sheath is not known, but the outer sheath of *Leptospira interrogans canicola* is immunogenic, and antibodies to the membrane may be protective.

The axial fibril is morphologically similar to a flagellum. In spite of this similarity, however, an unequivocal role in motility has not been demonstrated. The axial fibril consists of a shaft and its covering sheath and an insertion apparatus. The shaft resembles a bacterial flagellum in substructure, being either filamentous or globular and composed entirely of protein. It lies between the outer sheath and the outermost layer of the protoplasmic cylin-

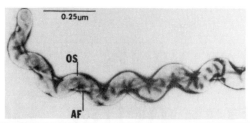

Figure 47-1. *Leptospira interrogans*, showing regular, tight coils and hooked end. The outer sheath (OS) and axial fibrils (AF) are indicated. *(From Holt: Microbiol Rev 42:114, 1978.)*

der and is, therefore, an internal structure of the bacterium. One or more axial fibrils wind around the protoplasmic cylinder and may or may not overlap at the center of the cell. The diameter of the axial fibril in most species of Spirochaetaceae is approximately 15 μm to 20 μm. The sheath covering the shaft of the axial filament often appears to have a striated substructure (Fig. 47-2).

The insertion apparatus of the axial fibril consists of a proximal hook and insertion discs. The proximal hook is an extension of the shaft and bends sharply toward and into the protoplasmic cylinder. The insertion discs are plate-like structures that are inserted into a hole or depression near the end of the cell. The number of insertion discs varies depending on the genus, *Leptospira* having three to five, *Borrelia* having two, and *Treponema* having one. They are approximately 20 to 40 μm in diameter (Fig. 47-3).

The protoplasmic cylinder lies directly beneath the outer sheath and consists of the cytoplasmic membrane and cell wall. The cell wall and cell membrane are similar to that of other gram-negative bacteria, containing two electron-dense layers.

The three *Treponema* species that are pathogenic for humans have not been cultured in vitro. *Treponema pallidum* is the cause of venereal and endemic syphilis and is the type species. *Treponema pertenue* is the cause of yaws, and *Treponema carateum* is the cause of pinta. The differentiation of these pathogenic *Treponema* species is based solely upon differences in the sites of lesions produced in several types of experimental animals (Table 47-2). These distinguish-

TABLE 47-1. PROPERTIES OF GENERA OF THE FAMILY SPIROCHAETACEAE

Type Species	Characteristics of Genus	Human Disease Produced
Spirochaeta plicatilis	Free-living, regular coils, 2 axial fibrils, anaerobic	None
Cristispira pectinis	Free-living, 2–10 loose coils, ovoid intracellular inclusions, over 100 axial fibrils	None
Treponema pallidum	Tight, regular coils, 3–8 axial fibrils, microaerophilic	Syphilis, yaws, pinta, bejel
Borrelia anserina	Coarse, irregular coils, 15–20 axial fibrils, anaerobic	Relapsing fever
Leptospira interrogans	Tight, regular coils, bent or hooked ends, 2 axial fibrils, aerobic	Leptospirosis

AF   Axial Fibril
OS   Outer Sheath
IP   Insertion Pore
LL   Lipoprotein Layer
NR   Nuclear Region

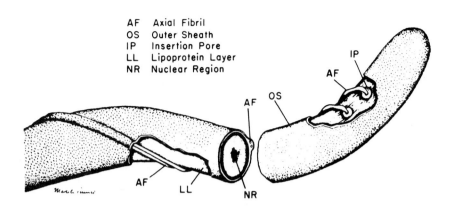

Figure 47-2. Diagrammatic representation of a typical spirochete as interpreted from electron micrographs. An outer sheath envelops the cell. The axial fibrils are between the outer sheath and the layers of the protoplasmic cylinder (seen here as lipoprotein layers) and insert into the cylinder by way of an insertion pore. *(From Holt: Microbiol Rev 42:114, 1978.)*

ing characteristics are not accepted by all investigators, some of whom consider the three organisms to be minor variants of a single species. There are seven other species of *Treponema*, most of which may be cultivated in vitro. They are seldom major human pathogens but may cause diseases of the oral cavity (Chap. 49)

Borrelia species cause relapsing fever in humans. No adequate method of speciation of *Borrelia* is available. The organisms currently are identified on the basis of the arthropod vector with which they are associated. Many have been cultured on complex media. Identification is based upon the coarse, uneven coils, the presence of 30 to 40 parallel fibrils, and atypical motion.

Leptospira species cause human and animal leptospirosis and are characterized by their motion, hooking or bending of one or both ends, and the presence of two fibrils that do not cross at the midsection. There are two recognized species, *Leptospira interrogans* and *Leptospira biflexa*. *L. interrogans* includes known human and animal pathogens, while *L. biflexa* includes free-living saprophytes. Serologic characteristics identify complexes by serotype and serogroup, but these complexes have no taxonomic standing.

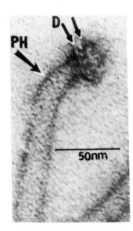

Figure 47-3. Spirochetal axial fibrils that are seen to be continuous with the proximal hook (PH), which terminate with the insertion discs (D). *(From Holt: Microbiol Rev 42:114, 1978.)*

Treponema pallidum

Morphology and Physiology

The name *Treponema* is derived from the Greek words meaning "turning thread." Individual organisms are 5 to 20 μm in length and 0.09 to 0.5 μm in diameter; the ends are finely tapered. Whole cells appear to have a flat wave with one or more planes per cell, giving it the appearance of a helical coil. There are 8 to 14 evenly distributed waves per cell. Its motility is sluggish, with a drifting motion and graceful flexuous movements; it rarely rotates.

The structure of *T. pallidum* is in general similar to that of the other Spirochaetaceae and consists of a multilayer cytoplasmic membrane, flagella-like fibrils that lie between the cell wall and the cytoplasmic membrane, the cell wall, and outer envelope. Pathogenic *T. pallidum* have also been shown to have a capsule-like outer coat that is not present in the nonpathogenic species. Some *Treponema* species contain intracytoplasmic microtubules that have not been demonstrated in other Spirochaetaceae. Pathogenic *Treponema* have a tapered end, which is oriented toward host cell surfaces during attachment. Hyaluronidase is also produced by pathogenic species of *Treponema*.

Treponemal organisms have a high content of glycolipid, and the axial filament consists of amino acids. The composition of the outer envelope and other cell components is currently under study, but to date no single com-

TABLE 47-2. TREPONEMAL SPECIES PATHOGENIC FOR HUMANS

Organism	Human Disease	Differentiating Characteristics
T. pallidum	Syphilis	Cutaneous lesions in rabbits
	Endemic syphilis	No cutaneous lesions in hamsters or guinea pigs
T. pertenue	Yaws	Cutaneous lesions in rabbits and hamsters, no cutaneous lesions in guinea pigs
T. carateum	Pinta	No cutaneous lesions in rabbits, hamsters, or guinea pigs

ponent of *T. pallidum* has been found to evoke protective antibody.

Until recently, virulent strains of *T. pallidum* could not be grown in vitro, although they could be maintained for 4 to 7 days at 25C in an anaerobic medium containing albumin, sodium bicarbonate, pyruvate, cysteine, and a bovine serum ultrafiltrate. Tissue culture cultivation of virulent *Treponema* species on rabbit epithelial cells in an atmosphere of reduced O_2 tension has now been achieved. In this system, virulence has been maintained. Virulent strains (e.g., the Nichols strain) are also propagated by intratesticular inoculation of rabbits. The division time of organisms in experimental chancres in rabbits is about 30 hours, and division is by transverse fission. Treponemes posses a functional cytochrome oxidase system and should, therefore, be capable of aerobic respiration.

Clinical Infection

History. Syphilis was first recognized in Europe at the end of the fifteenth century, when the disease first appeared in the Mediterranean areas and rapidly reached epidemic proportions at that time. One theory concerning the origin of syphilis is that it is of New World origin and that Columbus's crew acquired syphilis while in the West Indies and introduced it into Spain upon their return. Alternatively, the disease that had been endemic for centuries in Africa may have been transported to Europe at that time during the migration of armies and civilian populations. The relatively benign African diseases, yaws and bejel, may have been transformed in the susceptible population of Europe into a highly virulent disease with high mortality rates.

Syphilis initially was called the "Italian disease," the "French disease," and the "great pox" as distinguished from smallpox. Its venereal transmission was not recognized until the eighteenth century. Delineation of the characteristics of syphilis was hindered by confusion of its symptoms with those of gonorrhea. In 1767, John Hunter, a great English experimental biologist and physician, inoculated himself with urethral exudate from a patient with gonorrhea. Unfortunately, the patient also had syphilis, and the subsequent symptoms experienced by Hunter convinced two generations of physicians of the unity of gonorrhea and syphilis. The separate nature of gonorrhea and syphilis was demonstrated in 1838 by Ricord, who reported his observations on more than 2500 human inoculations. Recognition of the stages of syphilis followed, and in 1905 Schaudinn and Hoffman discovered the causative agent. The following year Wassermann introduced the diagnostic serologic test that bears his name.

Epidemiology

Syphilis is not a highly contagious disease; a person who has had sexual contact with an infected partner has approximately 1 chance in 10 of acquiring disease. The rate of primary and secondary syphilis in the United States in 1980 was 12.1 per 100,000 population, and over 27,000 cases were reported. As is typical of patients with other venereal diseases, persons who acquire syphilis are often promiscuous and have had sexual contact with an average of five other persons during the incubation period. They are also characteristically young. In the United States in 1979, the rate of infectious syphilis for persons aged 15 to 19 years was 16.2 per 100,000; that or persons aged 20 to 24 years was 32.1 per 100,000.

In each of the four years 1976–1980 the rates of primary and secondary syphilis increased, from 9.6 per 100,000 in 1976 to 12.1 per 100,000 in 1980. During this period an increasing proportion of cases were attributed to males who had contracted the disease from other males. By 1980 this accounted for almost half of all early cases (Fig. 47-4). Prostitutes traditionally have been prominent in dissemination of syphilis, but in most developed countries organized prostitution is no longer a major source of disease. Replacing their role in urban areas are casual partners and part-time prostitutes who also pursue other occupations.

An increase in the rate of other venereal diseases also has occurred in the same period. For example, in one area of the United States, 8 percent of persons with gonorrhea also had concomitant syphilis. Since many of these persons with dual diseases are treated in the preprimary stage of syphilis, this infection may never become manifest clinically or serologically.

Transmission

T. pallidum has the capacity to invade the intact mucous membranes or skin in areas of abrasions. Direct inoculation from contact with an infected person is necessary for infection, since survival of the organism outside the host is very limited. Sexual contact is the common method of transmission, and the site of inoculation usually is on the genital organs, the vagina or cervix in females and the penis in males. Other sites include lips, when infected by

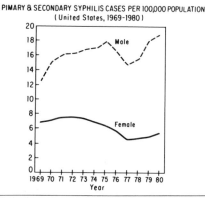

Figure 47-4. Incidence of primary and secondary syphilis in the United States by sex, 1969–1980. *(Courtesy of the Centers for Disease Control, Atlanta, Ga.)*

kissing, and other areas of the skin, when infected through abrasions. Examining physicians or pathologists may be infected in this way if appropriate barrier protection is not provided.

Pathogenesis

Syphilis is a disease of blood vessels and of the perivascular areas. After invasion, the organisms undergo rapid multiplication and are widely disseminated. Spread through the perivascular lymphatics and then the systemic circulation occurs before development of the primary lesion. Ten to ninety days later, but usually within three to four weeks, the patient manifests an inflammatory response to the infection at the site of the inoculation. The resulting lesion, the chancre, is characterized by profuse shedding of spirochetes, accumulation of mononuclear leukocytes, lymphocytes, and plasma cells, and swelling of capillary endothelia. The regional lymph nodes are enlarged, and the cellular infiltrate resembles that of the primary lesion. Resolution of the primary lesion is by fibrosis.

In experimental systems, pathogenic *T. pallidum* may be shown both in vitro and in vivo to attach avidly to a wide range of cell lines and tissue specimens. *Treponema* attached to cultivated cell lines in vitro have a prolonged survival time. The organisms attach by their tapered ends, as shown in Figure 47-5. They also appear to be able to penetrate through the hyaluronic acid-containing ground substance that joins capillary endothelial cells. Following attachment of the treponemes, an alteration of host cell membrane properties serves to block attachment of additional organisms. Specific host membrane ligands appear to function as mediators of attachment. Attachment is blocked by immune serum. Avirulent strains of *Treponema* species are unable to attach to cultured cells.

Secondary lesions develop when tissue of ectodermal origin, such as skin, mucous membranes, and central nervous system, participate in an inflammatory response. Mucous patches in the mouth are due to local vasculitis. The cellular infiltrate resembles that of the primary lesion, with a predominance of plasma cells. There is little or no necrosis, and healing is without scarring but may include pigmentary changes.

Tertiary syphilis may involve any organ system and is often asymmetrical. Gummas are lesions typified by extensive necrosis, few giant cells, and paucity of organisms. They commonly occur in internal organs, bone, and skin. The other major form of tertiary lesion, a diffuse chronic inflammation with plasma cells and lymphocytes but without caseation, may result in aneurysm of the aorta, paralytic dementia, or tabes dorsalis. Chronic swelling of the capillary endothelium and fibrosis result in the characteristic tissue changes.

Clinical Manifestations

Primary Disease. The chancre of primary syphilis is typically a single lesion, nontender, and firm, with a clean surface, raised border, and reddish color. It may be overlooked by women, in whom it is frequently situated on the cervix or vaginal wall. Systemic signs or symptoms are absent, but the draining lymph nodes are frequently enlarged and nontender.

Secondary Disease. Two to ten weeks after the primary lesion, the patient may experience secondary disease (Fig. 47-6). Prominent findings include fever, sore throat, generalized lymphadenopathy, headache, and rash. Involvement of the palms and soles is common, in contradistinction to many other dermatologic conditions. On mucous membranes the lesions may appear as white mucous patches. Condylomata lata occur around moist areas, such as the anus and vagina. All secondary lesions of the skin and mucous membranes are highly infectious.

Other signs of this stage of disease may be secondary to the generalized immunologic response. Nephrotic syndrome with immune complex nephritis results from deposition of antigen–antibody complexes within the glomerular basement membrane. Arthritis and arthralgias may have a similar etiology. Involvement of other systems also occurs.

Following the last episode of secondary disease, the patient enters the stage of latent disease, the first 4 years of which are considered early latent and the subsequent period late latent. By definition, persons in the late latent stage of disease have no signs or symptoms of active syphilis but remain seroreactive. If therapy for syphilis is first given during this stage, the patient is unlikely to show regression of nontreponemal antibody determinations. Approximately 60 percent of untreated patients in the late latent stage continue to have an asymptomatic course, while 40 percent develop symptoms of late disease. Progression of disease from late latent to late symptomatic syphilis is usually prevented if appropriate antimicrobial therapy is given at this stage.

Figure 47-5. Transmission electron photomicrograph. Specific attachment of *T. pallidum* to rabbit testicular cell membrane via terminal organelle. Axial fibrils are seen. *(From Hayes et al.: Infect Immun 17:174, 1977.)*

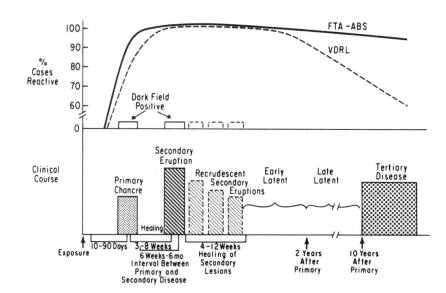

Figure 47-6. The course of untreated syphilis.

Tertiary Disease

GUMMAS. Three to ten years following the last evidence of secondary disease, the patient may develop non-progressive, localized lesions of the dermal elements or supporting structures of the body, which are called "gummas." Since these lesions are relatively quiescent, the term "benign tertiary syphilis" often is used. Spirochetes are extremely sparse or absent. The gummatous reaction is primarily a pronounced immunologic reaction of the host.

NEUROSYPHILIS. During the early stages of syphilis, approximately one third of all patients have involvement of the central nervous system, but only half of these, if untreated, develop late neurosyphilis. The interval between primary disease and late neurosyphilis usually is more than 5 years. Late neurosyphilis may present in a variety of ways. Classic presentations include paralytic dementia, tabes dorsalis, amyotropic lateral sclerosis, meningovascular syphilis, seizures, optic atrophy, and gummatous changes of the cord. Neurosyphilis may resemble virtually any other neurologic disease.

CARDIOVASCULAR SYPHILIS. Approximately 10 to 40 years after primary syphilis, the untreated patient may develop signs of cardiovascular involvement. The most commonly involved organs are the great vessels of the heart, where syphilitic aortic and pulmonary arteritis develop. The inflammatory reaction may also cause stenosis, with resulting angina, myocardial insufficiency, and death.

Congenital Syphilis.

Congenital syphilis results from transplacental infection of the developing fetus and is often a very severe and mutilating form of the disease. In spite of widespread programs to examine all pregnant women, in 1980 there were 111 reported cases of early congenital syphilis in the United States. There are approximately 2 cases of early congenital syphilis per 100 cases of primary or secondary syphilis in women of childbearing age.

At the onset of congenital syphilis, *T. pallidum* is liberated directly into the circulation of the fetus, resulting in spirochetemia with widespread dissemination. The mortality rate of untreated congenital syphilis is approximately 25 percent, and an additional 40 percent of children suffer from late stigmata (Table 47-3). Abortion because of congenital syphilis usually occurs during the second trimester of pregnancy, and histopathologic reactions to *T. pallidum* rarely are found in fetal tissue prior to that time. It has been generally thought that the fetus is protected from congenital infection until the sixteenth week of pregnancy,

TABLE 47-3. STIGMATA OF LATE CONGENITAL SYPHILIS*

Stigmata	Percentage of Total Patients
Frontal bossae of Parrot	87
Short maxilla	84
High palatal arch	76
Hutchinson's triad	75
Hutchinson's teeth	63
Interstitial keratitis	9
Eighth nerve deafness	3
Saddle nose	73
Mulberry molars	65
Higoumenakis' sign	39
Relative protuberance of mandible	26
Rhagades	7
Saber shin	4
Scaphoid scapulae	0.7
Clutton's joint	0.3

Adapted from Fiumara: Arch Dermatol 102:78, 1970.
*An analysis of 271 patients.

when the Langhans layer of the chorion atrophies. There is also evidence, however, that infection of the fetus may occur earlier but that the typical inflammatory response that results in tissue injury and fetal death does not occur until the fetus becomes immunologically competent. Syphilitic pregnant women who have not been treated may transmit the infection to their fetus at any clinical stage of their disease. In general, the greater the time that has elapsed since the women's primary or secondary infection, the less likely she is to transmit disease to the fetus. Almost all pregnant women with untreated primary syphilis, 90 percent of women with secondary syphilis, and approximately 30 percent of women with early latent syphilis may infect their fetuses.

The manifestations of congenital syphilis are highly variable in both signs and intensity. Especially prominent early symptoms include hepatosplenomegaly, jaundice, hemolytic anemia, pneumonia, and multiple long bone involvement. Snuffles, skin lesions, and testicular masses are common.

Late manifestations of congenital syphilis result both from scars of the active disease and the progression of active disease (Table 47-3). Some changes may be prevented by early treatment, but others often progress despite therapy.

Immunity

Immune Response. In spite of a great deal of interest and research into the nature of the response of the host to syphilis, the fundamental question of how the organism is capable of establishing persistent infection that lasts for decades remains unanswered. There is a vigorous immunologic response by the host to the infection, and yet the infection is neither fully controlled nor eradicated. During the initial infection with *T. pallidum*, humoral IgG and IgM antibody is detectable by the time the chancre appears. Thereafter, both IgG and IgM antibodies persist for long periods in the untreated patient. If the patient is adequately treated, IgM antibody declines during the next 1 to 2 years, but IgG antibody usually persists throughout the lifetime of the patient. If the patient is not treated, the stages of syphilis will evolve in spite of humoral antibody response.

The outer layers of virulent treponomal strains possess a dense coat that appears to protect the organism against the effects of specific antibody attachment. This coat of surface-associated protein is composed at least in part of host protein. It is strongly adherent to the treponemal envelope and may be removed only with vigorous trypsinization. Host transferrin and ceruloplasmin may be found in this host protein outer coating. The function of this coat is unknown, but it could serve to mask the host response.

Polymorphonuclear leukocytes are attracted to pathogenic *Treponema* and ingest them. After phagocytosis, the organisms are enclosed within a phagocytic vacuole, degranulation occurs, and the *Treponema* are digested. The entire phagocytic cycle thus appears to be functional and intact. There are preliminary indications that small numbers of organisms may evade polymorphonuclear leukocyte detection.

Inhibition of cell-mediated immunity occurs in early syphilis. Lymphocytes from syphilitic persons show reduced or absent response specifically to treponemal antigens. Paracortical areas of syphilitic lymph nodes in early stages of disease are correspondingly depleted of lymphocytes. There is a particularly marked depression of natural killer cell activity in patients with secondary syphilis. The effect appears to be mediated primarily by autologous serum and is found in the immunoglobulin fraction. The suppressive factor may be immune complexes. There is a correlation between poor lymphocyte transformation of lymphocytes from patients with syphilis and the presence of elevated levels of immune complexes. These alterations in cellular immune response abate during the disease. Persons with late secondary syphilis and tertiary syphilis exhibit cell-mediated immunity to treponemal antigen. In addition, there is experimental evidence of nonspecific activation of macrophages several weeks after infection with *T. pallidum.*

Natural Immunity. The progressive decline in severity of syphilis between its introduction into Europe at the end of the fifteenth century and the present time indicates that there has occurred a change in the virulence of the organism, the development of relative resistance by affected human populations, or both. Natural humoral or cell-mediated immunity sufficient to protect against disease has not been demonstrated, and the ID_{50} for humans in an experimental situation has been estimated to be as few as 57 organisms.

Acquired Immunity. Persons with untreated syphilis have a relative resistance to reinfection, so that the development of a chancre with second infection is unusual and probably depends upon the challenge inoculum. Following reexposure, untreated persons may develop an increased humoral antibody level.

In persons who have been treated for syphilis, especially if treatment was given during the secondary or earlier stages, the protective effect of prior disease is minor, and active disease following reinfection is common. This applies to persons who maintain a reactive nontreponemal antibody test (serofast) as well as to those who are serononreactive. In summary, although active or prior syphilis modifies the response of the patient to subsequent reinfection, protection is only relative and is unreliable.

Serologic Tests

There are two basic types of serologic tests for syphilis, the nontreponemal antigen tests and the treponemal antigen tests (Table 47-4). Although the latter tests indicate experience with a treponemal infection, they cross-react with antigens other than those of *T. pallidum;* hence no test

TABLE 47-4. SEROLOGIC TESTS FOR SYPHILIS

Antigen	Antigen Source	Tests	Percent Reactivity During		
			Primary Stage	Secondary Stage	Tertiary Stage
Nontreponemal reagin	Extracts of tissue (cardio-lipin–lecithin–cholesterol)	Complement fixation (Was-sermann, Kolmer)			
		Flocculation (VDRL, Hinton, Kahn)	78	97	77
Treponemal	*T. pallidum* Reiter strain	RPCF	61	85	72
	T. pallidum	TPI	56	94	92
		FTA–ABS	85	99	95
		IgM–FTA–ABS			

is specific for syphilis. However, since yaws and pinta are rare diseases in the United States, the treponemal tests generally provide a reliable indication of syphilitic infection.

Nontreponemal Tests. The original test for syphilis, as described by Wassermann, used syphilitic tissue as a complement-fixing antigen for the detection of antibody (reagin) that is induced by *T. pallidum*. Extracts of normal tissue, however, such as beef heart, have similar properties, and the purification and standardization of these materials led to the use as antigen of a preparation containing cardiolipin and lecithin in cholesterol.

Two types of tests use cardiolipin-lecithin as antigen: (1) complement-fixation tests, including the Wassermann and Kolmer tests, and (2) flocculation tests, including the VDRL (Venereal Disease Research Laboratory), Hinton, and rapid reagin tests. These tests provide similar clinical information and have similar advantages. They are inexpensive to perform and demonstrate rising and falling antibody titers and the clinical status of a patient. Disadvantages include a relatively high proportion of biologic acute and chronic false positive reactions and an increasing proportion of false negative reactions in the later stages of untreated syphilis. The technical difficulties include a negative reaction due to the prozone phenomenon when only undiluted serum is tested.

Treponemal Tests

TREPONEMA PALLIDUM IMMOBILIZATION (TPI). This test is based on the capacity of reaginic antibody and complement to immobilize a suspension of living and motile treponemes maintained in rabbit testes. The effect of the test serum on the motility of the spirochetes is determined by darkfield microscopy. The test is difficult and expensive, requires living organisms, and is also positive in the nonvenereal treponematoses, bejel, yaws, and pinta. The TPI test is now performed in only a few research laboratories, primarily for comparison with and evaluation of other tests. It also retains a useful clinical role in distinguishing between syphilis and biologic false positive reactions in patients who have collagen vascular disease with abnormal serum globulins.

REITER PROTEIN COMPLEMENT FIXATION. Antigen for this test is an extract from a nonvirulent treponeme, the Reiter strain, which may be cultured in vitro. The test detects group antigen; therefore, both false positive and false negative results are not uncommon. Nonvirulent treponemal organisms in the oral cavity may stimulate the production of cross-reacting antibody. The test is frequently nonreactive in late stages of syphilis.

FLUORESCENT ANTIBODY TESTS. The most significant development of the past two decades in the serology of syphilis is the detection of treponemal antibody by fluorescein-labeled antihuman antibody. The tests are used to confirm the validity of a positive reaginic test, to diagnose congenital syphilis, and to diagnose late stages of syphilis. The tests are both sensitive and reliable.

Fluorescent treponemal antibody (FTA) tests use lypophilized Nichols strain organisms as antigen. Antigen is fixed to a slide, and the test serum is applied, allowing reaction of antitreponemal antibody with antigen. The slide is layered with fluorescein isothiocyanate-labeled antihuman gamma globulin, and the presence or absence of antibody is determined by fluorescence microscopy.

The currently used modification of this method is the FTA-ABS (fluorescent treponemal antibody-absorption) test in which test sera are preabsorbed with sorbent* to eliminate group antibody. The test is thus rendered relatively specific for disease with virulent treponemal species, usually *T. pallidum*.

The FTA-ABS test is expensive and time-consuming. It is therefore recommended not for general screening but for confirmation of positive nontreponemal tests and diagnosis of later stages of syphilis in which the results of nontreponemal tests are frequently falsely negative.

HEMAGGLUTINATION TESTS. A hemagglutination method for serodiagnosis of syphilis has been automated and is both technically easy to perform and inexpensive. It is as sensitive as the FTA-ABS tests, except in primary syphilis, and is highly specific. Further experience with this method

---

*Sorbent originally consisted of sonicate of Reiter treponemes; other substances may be used.

will determine its usefulness as a primary test for confirming syphilis. Like the FTA-ABS test, it is unlikely to revert to a nonreactive state following treatment of the patient unless treatment is given very early. The test usually is also reactive in persons with nonsyphilitic treponematoses.

IgM-FTA-ABS TEST. In the diagnosis of congenital syphilis it is necessary to differentiate between passive transplacental transfer of maternal antibody to the fetus and production by the fetus of endogenous antitreponemal antibody. Since antibodies of the IgG but not of the IgM class cross the placenta, detection of specific IgM antibody in the fetal circulation usually indicates antibody production by the fetus due to active fetal infection. The FTA-ABS test will detect immunoglobulin of both the IgG and IgM classes and hence will not distinguish between active infection and passive transfer. This problem stimulated the development of a fluorescent antihuman antibody that is specific for IgM class antitreponemal antibody, the IgM-FTA-ABS test. A reactive test with infant blood is strong evidence of active congenital disease. However, the test may be nonreactive in infants with congenital syphilis if the disease was transmitted to the infant late in pregnancy, as is often the case. Furthermore, there are indications that a reactive test does not always absolutely confirm the diagnosis of congenital syphilis. Therefore, the test is not highly reliable and has been abandoned by many laboratories.

False Positive Reactions. All of the available serologic tests for syphilis produce occasional reaction results in patients for whom there is no other evidence of syphilitic infection. These reactions are usually called "biologic false positive" (BFP), as distinct from positive reactions due to technical errors. The majority of BFP reactions occur with nontreponemal tests; approximately 1 percent of normal adults will have a BFP reaction by nontreponemal antigen tests. Reaginic antibody is reactive with at least 200 antigens other than those of *T. pallidum*, and although the specific stimulus for this antibody in syphilis as well as other diseases is unknown, it may represent antibody to cellular lipoidal antigens of the host that are liberated during various diseases. For clinical purposes, BFP reactions may be classified as acute, in which the reactivity resolves within 6 months, or chronic, in which reactivity is persistent.

ACUTE BFP. Most BFP reactions are detected by nontreponemal tests and occur in patients with other acute illnesses, especially pneumonia, hepatitis, vaccinations, and viral exanthematous disease. The prognosis for the patient's health is not affected by the finding. The titer of antibody usually is low, less than 1:8, and in most instances the FTA-ABS is nonreactive. Approximately two thirds of patients with BFP reactions have acute reactions, and reactivity subsides in 6 months or less.

CHRONIC BFP. Drug addiction, chronic hepatitis, old age, leprosy, and collagen vascular disease are highly associated

with chronic BFP reactions. The antibody detected by the VDRL test in chronic BFP reactions is predominantly IgM, whereas in syphilis it is mainly IgG. Patients with chronic BFP reactions and systemic lupus erythematosus commonly also have a reactive FTA-ABS. The TPI test may be helpful in the differential diagnosis in these instances.

Laboratory Diagnosis

Efforts to diagnose infectious syphilis suffer from the lack of a method to culture the organism on laboratory media. Three methods are useful in the diagnosis of syphilis: (1) direct visualization of the organism by darkfield microscopy, fluorescent antibody technique, or by special stains of infected tissue, (2) animal inoculation, and (3) demonstration of serologic reactions typical of syphilis.

Syphilis in patients with a primary chancre as well as with active secondary lesions may be diagnosed by darkfield microscopy. Since this depends upon direct visualization of motile spirochetes, the organisms must be active and viable. Prior use of many antibiotics rapidly destroys the motility of the organisms, as do many topical disinfectants. Serous fluid from the base of the lesion should be collected for darkfield examination. Syphilitic lesions of the mouth may harbor indigenous treponemes whose morphologic similarity to pathogenic species can confuse the interpretation of findings. The technique is, however, particularly helpful in making a diagnosis early in the disease prior to the development of seroreactivity. If darkfield microscopy is unavailable, a direct fluorescent antibody stain for *T. pallidum* may be made. Exudate is collected in capillary tubes or on slides and stained with specific antibody. Syphilis that has progressed beyond the primary stage is diagnosed in most patients by serologic methods.

Treatment

Since *T. pallidum* cannot be grown in vitro, estimates of the sensitivity of strains to antimicrobial agents depend upon the results of treatment of experimental animals, especially rabbits. The minimal inhibitory concentration of penicillin for *T. pallidum* is approximately 0.004 U/ml, making it one of the most sensitive of human pathogens. There is no evidence that the resistance of the organism to penicillin has increased during the past three decades of penicillin use. For these reasons, penicillin has remained the single most widely and successfully used antimicrobial agent for treatment of all stages of syphilis.

Essential requirements for effective therapy include maintenance of at least 0.03 units of penicillin per milliliter of serum for 7 to 10 days in early syphilis and avoidance of penicillin-free intervals during therapy. Provision of treatment for an individual patient may require frequent injections of short-acting penicillin preparations or the use of long-acting preparations. If the disease is beyond the early stages, the patient should receive adequate doses of penicillin for at least 21 days.

Successful eradication of active disease also has been

achieved with erythromycin, tetracyclines, and cephaloridine. However, infected women have delivered syphilitic infants after treatment with these forms of therapy, possibly because of the relatively poor passage of erythromycin and tetracycline into the fetal circulation.

In most patients receiving appropriate therapy during the primary or secondary stage, active disease is totally and permanently arrested. Persistent seroreactivity as measured by FTA-ABS may be avoided if treatment is given during the preprimary stage but seldom thereafter. Nevertheless, progression to tertiary disease seldom, if ever, occurs. Similarly, therapy during early or late latent syphilis averts the development of symptomatic tertiary disease. Antimicrobial therapy for symptomatic neurosyphilis, optic neuritis, and cardiovascular syphilis may not be followed by significant clinical improvement, and established damage to vital organs may fail to resolve.

Jarisch–Herxheimer Reaction. Two to twelve hours following the treatment of active syphilis with either heavy metals or penicillins, a variable proportion of patients develop an acute focal and systemic reaction usually consisting of headache, malaise, and fever to 38C or above. The reaction is most commonly observed in the early stages of syphilis and does not affect the course of recovery. Most reactions in late syphilis are clinically insignificant, but an occasional reaction may produce damage to the central nervous system or the cardiovascular system.

Prevention

Methods to control the spread of syphilis have relied extensively on treatment of case contacts. Persons with acute syphilis are interviewed to identify all sexual contacts that may have occurred during the incubation period. The contacts are examined, and if they are not infectious, they receive treatment appropriate for primary syphilis. Advantage is thus taken of the long incubation period of syphilis by preventing disease in contacts before they themselves can transmit infection.

Other Treponemal Diseases

Yaws (Frambesia)

Yaws is a spirochetal disease of the tropics caused by *T. pertenue*, an organism very closely related to *T. pallidum* (Table 47-2). The two organisms are serologically and morphologically indistinguishable and are differentiated by the type of lesions produced in experimental animals. No serologic test distinguishes human yaws from syphilis.

Yaws is endemic in tropical forest regions of Africa, parts of South America, India, and Indonesia, and many of the Pacific Islands. In these areas it is most commonly acquired in childhood and by direct contact other than sexual contact. The disease very rarely occurs congenitally, since most infected children have passed the early stages of disease by the age of sexual maturity.

The course of yaws resembles that of syphilis. The initial lesion is called the "mother yaw" or "framboise" and occurs about a month after the primary infection. It is a painless erythematous papule that heals during the subsequent 1 or 2 months. Secondary lesions that resemble the primary lesion occur 6 weeks to 3 months later. Recurrent disease may continue to occur for several years. Tertiary lesions are most likely to involve the skin and bones with gummatous ulcerations. Infection of the feet causes a crippling form of disease, called "crab yaws."

Yaws is readily treated with penicillin. Eradication of yaws has accompanied the general improvement in sanitation and standard of living in most areas of the world.

Pinta

Pinta is a disease of tropical areas of Central and South America, caused by *T. carateum*. This organism is serologically and morphologically similar to both *T. pallidum* and *T. pertenue* and is distinguished by failure to produce cutaneous lesions in rabbits, hamsters, or guinea pigs (Table 47-2). Chimpanzees, however, may be experimentally infected.

Human pinta is acquired by person-to-person contact and rarely by sexual intercourse. The primary and secondary lesions are flat, erythematous, and nonulcerating. The healing lesion first becomes hyperpigmented and later, as scarring occurs, will be depigmented. The lesions most commonly occur on the hands, feet, and scalp. Tertiary disease, such as occurs in syphilis, is uncommon in pinta. Treatment with penicillin is highly efficacious.

Bejel

Bejel is a disease that closely resembles yaws both epidemiologically and in its clinical manifestations. It is considered to be a form of endemic syphilis and occurs in areas of the Middle East. Poor hygienic conditions are important in perpetuating these infections, which are decreasing in incidence in most areas. Bejel is transmitted by direct contact, usually during early childhood, and results uniformly in serologic reactions that are indistinguishable from those of syphilis.

Leptospira

Leptospirosis, an acute illness associated with febrile jaundice and nephritis, was first recognized by Weil in 1886 as a clinical entity distinct from other icteric fevers. Commonly referred to since that time as "Weil's disease," the infection is caused by a leptospira transmitted to humans from infected rodents.

By 1948, over 300 cases of human leptospirosis had been reported, most of which were clinically severe and accompanied by jaundice. Recognition of other forms of leptospirosis, however, was exceedingly slow in spite of indications in Europe and other parts of the world that clinically milder and nonicteric forms of leptospirosis were common and that animal reservoirs were not restricted to rodents. In 1938, Meyer described canicola fever in dogs and humans. Infection in cattle was reported in 1948, and shortly thereafter human infection by this same strain (serovar* *pomona*) was recognized in Georgia. Human leptospirosis is now known to be caused by infection with a family of organisms. These organisms may be classified into multiple serogroups and serovars. The various serogroups of *Leptospira* cause diseases that are extremely varied in their clinical presentations.

Evidence of widespread leptospiral disease among cattle, swine, horses, and other livestock led to appreciation of the economic losses attributable to these infections as well as their threat to human health. By the early 1950s, several public health laboratories were capable of evaluating serologic evidence of infection with an increasing number of leptospiral serogroups, and syndromes, such as pretibial fever (Fort Bragg fever), aseptic meningitis, and other mild febrile illnesses, were attributed to leptospiral infection. Commonly used terms for leptospirosis include swineherds' disease, Fort Bragg fever, pretibial fever, Weil's disease, canicola fever, and autumnal fever.

Morphology

The genus *Leptospira* is characterized by fine coiling of the primary spirals. The name is derived from the Greek word *lepto*, meaning "thin" or "fine" spiral. *Leptospira* are helicoidal organisms, usually 6 to 20 μm in length and 0.1 μm in diameter. The coils are 0.2 to 0.3 μm in overall diameter and 0.5 μm in pitch. In liquid media, one or both ends are usually hooked. In the living state, the organisms are clearly visible by darkfield and much less clearly by phase contrast microscopy.

Ultrastructure. *Leptospira* consist of a helicoidal protoplasmic cylinder, two axial filaments, and an outer envelope. The outer envelope is composed of three to five layers and surrounds the whole organism. Located between the outer envelope and the cytoplasmic membrane are two independent axial filaments, each of which is inserted by one end subterminally at opposite ends of the protoplasmic cylinder. The free ends are directed toward the center of the cell, where they usually do not overlap. During cellular reproduction, septal wall formation occurs at the middle region of the organism, leading to transverse division.

Lipids comprise 18 to 28 percent of the dry weight of the leptospiral cell and are composed of approximately 70 percent phospholipid and 30 percent free fatty acids. The composition of fatty acids is a reflection of those present in the culture medium, since with few exceptions *Leptospira* can neither synthesize fatty acids de novo nor elongate chains.

The major compounds of the leptospiral cell wall are polysaccharide and peptidoglycan. Alanine, glutamic acid, diaminopimelic acid, glucosamine, and muramic acid are the predominant amino acids and sugars. The diaminopimelic acid content of *Leptospira* serves to differentiate these organisms from treponemes and members of the genus *Spirochaeta*, which instead contain ornithine.

Physiology

Leptospira are aerobic in their culture requirements. Their metabolism is respiratory, with oxygen utilized as the final electron acceptor. They grow well at pH 7.2 to 7.4 in rabbit serum or Tween 80 albumin media. The generation time of pathogenic *Leptospira* cultivated in laboratory media is 12 to 16 hours and 4 to 8 hours in inoculated animals. Long-chain unsaturated fatty acids serve as the major source of carbon and energy and are required by the parasitic strains. *Leptospira* can use inorganic ammonium salts as a source of nitrogen.

Characterization of Species

Two species, *L. interrogans* and *L. biflexa* were proposed by the *Leptospira* subcommittee in 1973. *L. interrogans* includes pathogenic organisms, whereas *L. biflexa* includes saprophytic or water *Leptospira* that commonly occur in fresh, surface waters. Distinguishing characteristics of the two species, other than their ability to infect animals, include the inhibitory effect of 8-azaguanine, bivalent copper ions, serologic characteristics, and ability to grow at low temperatures (5C to 10C). The two species also are distinguishable genetically and share no nucleotide sequences as determined by DNA-DNA annealing tests. Each species can be further separated into three genetic groups of strains that have partial DNA homologies. This taxonomy remains provisional, and accurate speciation will depend upon further advances in the basic microbiology of these complex organisms.

Antigenic Structure

Each species of *Leptospira* includes a large number of serologically distinct serogroups as determined by cross-agglutination and microscopic agglutinin-absorption tests. Strains that share major agglutinogens are arbitrarily assembled into serogroups. The serogroup is not a recognized taxon and serves primarily serodiagnostic purposes. The basic taxon is the serovar. Serovars have been characterized by the use of factor sera, in which cross-absorption with related strains has yielded a serum with a narrow range of agglutination. Attempts to identify serovars have also included use of DNA base composition analysis, immunodiffusion analysis of axial filament antigen, enzymatic characteristics, and restriction endonuclease analysis. The

---

* The term "serovar" has been adopted in lieu of serotype by the International Committee on Systematic Bacteriology (*Leptospira* subcommittee), 1973.

TABLE 47-5. *LEPTOSPIRA* **SEROGROUPS AND SEROVARS COMMONLY ISOLATED FROM HUMANS AND DOMESTIC ANIMALS IN THE UNITED STATES**

Serogroup	Serovar	Host
Autumnalis	*fort-bragg*	Human
Canicola	*canicola*	Human, cattle, dog, swine
Grippotyphosa	*grippotyphosa*	Cattle, swine
Hebdomadis	*szwajizak*	Cattle
	georgia	Human
	hardjo	Cattle
Icterohaemorrhagiae	*icterohaemorrhagiae*	Human, cattle, dog, swine
	copenhageni	Human, dog
Illini	*illini*	Cattle
Pomona	*pomona*	Human, cattle, dog, goat, horse, swine
Australis	*bratislava*	Dog, raccoon, fox
Ballum	*ballum*	Mouse, pig, skunk

standard method of characterizing serogroups and serovars of *Leptospira* has remained, however, the difficult microscopic agglutination assay.

Among the parasitic leptospiras, over 150 serovars are now recognized and classified into approximately 18 serogroups, all of which are characterized by very wide distribution both in variety of animal species affected and in geographic occurrence. Table 47-5 indicates serogroups and serovars isolated from humans and domestic animals in the United States. Between 1974 and 1978, of the 498 cases of human leptospirosis in the United States, 332 were found to have been caused by one of 17 serogroups.

Determinants of Pathogenicity

Although mechanisms of virulence remain uncertain, a number of biologic properties characterize the pathogenic strains of *Leptospira.*

1. Avirulent strains are relatively more sensitive to the leptospirocidal effect of immune serum and complement than are virulent strains. The outer sheath of *Leptospira* is the primary site of action of antibody and complement and has immunogenic properties. Guinea pig macrophages phagocytize both virulent and avirulent strains. However, macrophages are not bactericidal for virulent leptospira unless in the presence of homologous antibody. In the absence of homologous antibody, macrophages both ingest and kill saprophytic leptospiras.
2. Some virulent strains of *Leptospira* produce a soluble hemolysin that appears to be important in the manifestations of leptospirosis in a number of animal species. The hemolysin is thermolabile and probably protein in nature. Previous infection with a hemolytic serotype confers immunity to subsequent hemolytic disease.
3. Some of the clinical manifestations of leptospirosis, such as conjunctival irritation and iritis, are probably caused by cell-mediated sensitivity to leptospiral antigen.

4. Some strains of *Leptospira* appear to contain small amounts of endotoxin. Findings in animals with leptospirosis suggest the presence of endotoxemia.

Clinical Infection

Epidemiology. Leptospirosis is a zoonotic disease with a wide range of host reservoirs. The predominant natural reservoirs of pathogenic *Leptospira* are wild mammals, although other vertebrates occasionally are infected. Domestic animals, such as dogs, cattle, swine, sheep, goats, and horses, also may be major sources of human infections. The improved ability of regional laboratories to group *Leptospira* has resulted in the recognition of the large number of serovars endemic in the United States, as well as the extent of infections in a variety of animal species. Nevertheless, it is an infrequently diagnosed human disease. In the period 1977–1982, there were 60 to 110 cases per year reported to the Centers for Disease Control.

The major mode of transmission between animals and humans is by indirect contact with urine infected with virulent *Leptospira* from an animal with leptospiruria. *Leptospira* from infected soil, food, and water enter the body through a break in the skin and through mucous membranes. Survival of *Leptospira* outside the host is fostered by a temperature of 22C or above, moisture, and a neutral to slightly alkaline environment. *Leptospira* are readily killed by temperatures above 60C, detergents, desiccation, and acidity.

Because of its prevalence in rodents and domestic animals, leptospirosis has been primarily a disease of persons in occupations heavily exposed to animals and animal products, such as sewer workers, swineherders, veterinarians, abattoir workers, and farmers. Also at risk are persons living in rodent-infested housing, such as urban slums. The convoluted renal tubules of animal reservoirs harbor viable *Leptospira,* which are passed in the urine, and the duration of asymptomatic urinary shedding varies with the animal species. There is a higher incidence of disease in men. At present, the majority of cases occur in the summer and fall in teenagers and young adults. Avocational exposure is now increasingly common.

Dogs are becoming an increasingly recognized reservoir for bringing humans into contact with leptospirosis. A sizable (15 to 40 percent) proportion of dogs are infected, and the majority of human cases are associated with intimate contact with a dog. Immunization of a dog to leptospirosis may fail to prevent renal shedding.

Common source outbreaks attributed to contaminated ponds or slowly moving streams are numerous; over 14 instances have been reported in the United States since 1939. A high attack rate, summer season, young age group, and the proximity of animals to the water typify most of these outbreaks. In some areas of the world, the runoff during flooding also is highly infectious.

Forms of transmission other than direct and indirect contact with contaminated urine are rare. Lactating animals shed *Leptospira* in the milk, but whole milk is leptospirocidal after a few hours, and no known human cases have occurred in this manner. *Leptospira* are not shed in saliva, and animal bites are therefore not a direct source of infection. Person-to-person transmission has not been reported and is probably rare. Humans rarely shed *Leptospira* for more than a few months.

Pathogenesis. The organism probably invades the human through small breaks in the skin or intact mucosa. The initial sites of multiplication are unknown. Nonspecific host defenses fail to contain *Leptospira* to any significant extent, and leptospiremia occurs rapidly after infection and continues through the initial acute illness. A local lesion at the site of entry does not develop.

Leptospira usually infect the kidneys. The major renal lesion, common to all forms of leptospirosis and present even in patients with normal renal function, is an interstitial nephritis with associated glomerular swelling and hyperplasia. Studies of infections in experimental mice have demonstrated that the earliest lesion of the kidney is interstitial edema, which occurs by the second day after infection. This is followed by a thickening of the basement membrane of the proximal tubules. By the tenth day *Leptospira* can be identified in areas adjoining tubular epithelial cells. The glomerulus is apparently not involved. Late manifestations of this disease may be caused by the host immunologic response to the infection.

Clinical Manifestations. The severity of human leptospirosis varies greatly and is determined to a large extent by the infecting strain and by the general health of the host. Severe icteric disease with a high fatality rate occurs in a small proportion of patients and is frequently associated with serogroup Ictohemorrhagiae serovars. Less severe and anicteric disease is far more prevalent and is commonly caused by serovars of serogroups Australis and Pyrogenes; disease due to those of Canicola, Ballum, and Pomona is often mild. There is, however, no absolute correlation between severity of disease or clinical syndrome with infecting serogroups. Leptospirosis is a disease that is unusually protean in its clinical manifestations.

The incubation period is usually 10 to 12 days but ranges from 3 to 30 days after inoculation. Prominent presenting signs include an abrupt onset of fever, chills, headache, conjunctival suffusion, myalgias, and gastrointestinal complaints. The clinical presentations of leptospirosis often suggest other disease processes, most commonly hepatitis, viral meningitis, fever of unknown etiology, and encephalitis.

Clinical illness is biphasic, the first leptospiremic stage lasting approximately 7 days in most instances. The appearance of humoral antibody coincides with the termination of fever and leptospiremia. A few days after the initial defervescence, a second and shorter febrile period may occur. Routine laboratory studies do not usually aid in the diagnosis.

Infection of the kidneys results in the excretion of organisms in the urine. Renal failure in Weil's disease is not rare and is the cause of death in most fatal cases. With the availability of extracorporeal dialysis, however, the mortality rate is very low, and there is complete return of renal function following recovery.

Hepatic injury with hepatocellular disease is common in leptospirosis. The pathogenesis of the liver disease is not certain but may be due to the vasculitis that is generally present. Jaundice may be extensive and may be due to both conjugated and unconjugated bilirubin. Electron microscopic changes in hepatocytes during leptospirosis include increases in smooth endoplasmic reticulum, destruction of mitochondria, and abnormalities in cell wall structure.

Certain presentations and complications of leptospirosis require attention. Meningeal irritation is common and probably a frequent cause of undiagnosed aseptic meningitis. Approximately half the patients examined during the second week of illness may have a cerebrospinal fluid lymphocytosis associated with a moderate elevation of cerebrospinal fluid protein. *Leptospira* may be isolated from the cerebrospinal fluid early in the disease. The later onset of symptoms of central nervous system involvement may reflect an untoward antigen–antibody reaction. Permanent neurologic sequelae are exceedingly rare.

An infectious agent from patients with a syndrome named Fort Bragg fever (pretibial fever), first described at Fort Bragg, North Carolina, in 1943, was identified in 1952 as pathogenic *Leptospira* (serovar *fort bragg*) in the Autumnalis serogroup. Sporadic cases have subsequently been reported from other parts of the United States, including the Pacific Northwest. Clinical characteristics of this syndrome included an unusual symmetrical rash limited to the pretibial areas. The lesions resembled erythema nodosum but were urticarial in a few cases. Fever, headache, a palpable spleen, and leukopenia predominated. A similar syndrome may occur with other *Leptospira* serovars.

Immunity. In all patients with leptospiral bacteremia, homologous agglutinating antibodies develop. During the initial immunologic response the antibody is of the IgM class. It is detectable within a week after onset of disease and may persist in high titer for many months. Some patients, but not all, may also develop IgG antibodies a month or more after onset of illness. Human convalescent serum contains protective and agglutinating antibodies that persist in a patient's serum for many years. The capacity of sera to protect against disease is best correlated with titer of agglutinating antibody and may be of either the IgG or IgM class.

Laboratory Diagnosis

CULTURE. During the acute phase of the disease, *Leptospira* can be readily cultured from the blood or cerebrospinal fluid. After the first week of disease and for several months thereafter, *Leptospira* may be shed intermittently in the urine by a large proportion of patients and may be demonstrated by cultural means.

Isolation of *Leptospira* may be accomplished by direct inoculation into laboratory media or by animal inoculation. Commonly used media include Fletcher's and Stewart's media, which contain rabbit serum. Ellinghausen's medium is semisolid and also provides an effective method of recovery. Isolation of *Leptospira*-contaminated specimens may be accomplished by the use of a selective inhibitor, such as 5-fluorouracil, or by intraperitoneal inoculation of young hamsters or guinea pigs.

DIRECT EXAMINATION. The direct demonstration of *Leptospira* by darkfield microscopy, fluorescent antibody silver impregnation, or staining with aniline dyes is successful in only a small portion of cases. It is not recommended as a single diagnostic procedure because of the frequently mistaken identification of artifacts as *Leptospira*.

SEROLOGIC TESTS. The macroscopic slide agglutination test, which employs formalized antigen, is a safe and rapid screening test for the detection of leptospiral antibody. Determination of serovar-specific antibody, however, is accomplished with the very sensitive microscopic agglutination test employing live organisms. The microscopic agglutination test (agglutination-lysis) was the original method for determining antibody response to leptospirosis and remains the reference method. Use of living organisms gives the most specific reaction, with highest titer and fewer cross-reactions. Formalinized antigen also may be used. Results are read by low-power, darkfield microscopy. These tests require maintenance of appropriate living or formalinized antigen, may be dangerous and arduous to perform, and are available only in reference laboratories. Since agglutinating antibodies persist for long periods after the acute episode, these tests are useful in determining the past experience of a community with leptospirosis. Various complement-fixation, hemagglutination, hemolytic, and fluorescent antibody tests have been proposed and advantageously used in some laboratories for the diagnosis of human cases. These tests, however, although less difficult, may lack the prerequisite sensitivity for detecting antibodies in animals or in individuals tested retrospectively for serologic surveys. In the hemolytic assay, antibody to the polysaccharide that is extracted as the erythrocyte-sen-

sitizing substance is of the IgM class even late in the disease.

It should be noted that even though animals are shedding virulent *Leptospira* in their urine, they may fail to demonstrate serologic evidence of leptospirosis. The absence of a positive serologic test should thus not be interpreted as having proven an animal to be free of leptospirosis.

Treatment and Prevention. Penicillin, streptomycin, tetracycline, and the macrolide antibiotics are active against *Leptospira* in vitro and in experimentally infected animals. Recovery of human cases appears to be hastened if therapy is initiated during the first 2 days after onset. When therapy is initiated after the fourth day of illness, the course of the disease usually may not be altered. A placebo-controlled study with penicillin, begun within 4 days of onset of illness, indicated that the illness was shortened in treated patients.

Vaccines have been effectively used in veterinary medicine for humans in endemic areas. Protection is serovar specific.

Borrelia

Spirochetes of the genus *Borrelia* cause the disease in humans known as "relapsing fever." This is an acute infection characterized by febrile episodes that subside spontaneously but tend to recur over a period of weeks. The organisms are transmitted by ticks or by the human body louse. Other terms used to describe these diseases are tick fever, borreliosis, and famine fever.

Relapsing fever has been known to the Western world since the time of Hippocrates. In recent times it has been associated with poverty, crowding, and warfare. Following World War I, louseborne relapsing fever was disseminated through large areas of Europe, carried by louse-infested dislocated civilians, soldiers, and prisoners. A high mortality rate occurred in these debilitated populations who often were also experiencing epidemic typhus.

Separation of the genus *Borrelia* from other members of the Spirochaetaceae is based on their characteristic morphology as revealed by the electron microscope (Table 47-1). The current speciation of *Borrelia* is based on the arthropod vector.

Morphology and Physiology

Borrelia are helical organisms 0.2 to 0.5 μm wide and 3 to 20 μm in length, with 3 to 20 uneven coils. Spirals are coarser and more irregular than those of the treponemes or leptospires and can be seen with light microscopy in preparations stained with aniline dyes, such as Wright or Giemsa stains. Borreliosis is the only disease in which spirochetes may be demonstrated by direct stain in the peripheral blood, and the presence of morphologically typical forms is adequate for diagnosis. In fresh blood the organisms are actively motile; they move in forward and backward waves and in a corkscrew-like motion. Observed variations in morphology depend upon the parasitized host and on the stage of the disease.

Borrelia are microaerophilic. Special culture media containing natural animal proteins are available, and propagation of cultures of several species has been accomplished. Little is known of the nutritional requirements of the organisms other than the fact that long-chain fatty acids are required for growth. Very little is known about the endotoxin content of the organisms. The optimum temperature for growth is 28C to 30C, and the generation time is about 18 hours.

In vitro and in vivo culture for diagnostic purposes is difficult and not always successful. If inoculation of experimental animals is attempted, great care should be taken to insure that the animals are free from preexisting borreliosis. Suckling or 21-day-old mice may be inoculated subcutaneously or intraperitoneally and smears of peripheral blood subsequently examined for *Borrelia*. Chick embryo cultures have been irregularly successful, as have tissue culture techniques.

Speciation is based upon two considerations: (1) the species responsible for louseborne disease is designated *Borrelia recurrentis*, as opposed to all tickborne strains, and (2) for tickborne strains the close vector–strain relationship has led to the definition of most species by the tick vector. For example, *Borrelia hermsii* is associated with the tick *Ornithodoros hermsi* (Fig. 47-7). The type species is *Borrelia anserina*, which is the cause of avian spirochetosis. There are 18 recognized species of *Borrelia*, 10 of which cause relapsing fever in humans and all of which have an arthropod vector. The principal species of *Borrelia* in North America are *B. hermsii*, *Borrelia parkeri*, and *Borrelia turicatae.*

Antigenic Structure

The most striking property of relapsing fever is the capacity of *Borrelia* to undergo several antigenically distinct variations within a given host during the course of a single infection. Early studies of experimental infection of rats with *B. hermsii* showed the presence of four major serotypes. Antigenic shifts were observed to occur in a regular sequence, were most readily determined by immunofluorescent methods, and were accompanied by appropriate antibody responses in the host. The organisms disappeared from the peripheral blood coincident with appearance of specific antibody and reappeared after antigenic variation had occurred.

Recent studies have shown that a single strain of *B. hermsii* can give rise to progeny that represent at least 24 separate serotypes when studied by indirect immunofluorescent methods. These strain variations occur both in vitro and in vivo and are not dependent upon the selective pressure of the host immunologic response. During relapses, the host typically demonstrates spirochetemia with several different serotypes of organisms at a time. In ani-

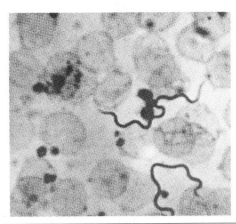

Figure 47-7. *Borrelia hermsii* from *Ornithodoros hermsi*, collected at Broune Mountain. Giemsa-stained smear of mouse blood. × 2,300. *(From Thompson: JAMA 210:1045, 1969.)*

mals, conversions from one serotype to another occur constantly and are independent of relapses.

The protein constituents of several serologic strains of *B. hermsii* have been studied. Of the four separate strains studied, all contained one identical major protein. In addition, in each of the four strains, a second protein was found that had a different molecular weight for each serotype. This protein appears to be the serotype-specific antigen, and alterations in it may account for the antigenic shifts that occur during infection with *Borrelia.*

In human relapsing fever, most patients develop immobilizing antibody to both autologous and heterologous strains. Studies have also demonstrated that a low level of protection against subsequent relapse strains is afforded by antigenic stimulation with the initial strains of the infection and that there is a frequent lack of cross-protection.

The technical problems associated with antigenic shifts and the recognition of the coexistence of mixed populations of *Borrelia* have precluded the development of a reliable serologic test for borreliosis. Although several experimental systems are being evaluated, all lack adequate standards. Agglutination of *Proteus* OX-K at moderate titers has been observed in the majority of patients with louseborne relapsing fever and in one study of an outbreak of tickborne disease. The test is nonspecific but may be marginally helpful.

Clinical Infection
Epidemiology

TICKBORNE DISEASE. The vectors of tickborne borreliosis are ticks of the genus *Ornithodoros*, which comprise the soft ticks (argasid ticks).

In the United States, tickborne disease may be transmitted by *Ornithodoros turicatae, Ornithodoros parkeri,* and *O. hermsi.* Throughout the world, over 15 species of *Ornithodoros* ticks have been found to transmit borreliosis. *Ornithodo-*

ros species feed exclusively on blood, and they often feed at night. They usually have a painless bite and feed for a short time (usually less than an hour), after which they spontaneously leave the host. An individual is therefore frequently unaware of having received a bite.

When the tick bites a borrelemic host, the borrelia penetrate the tick's coelomic cavity. There is a predilection for the coxal and salivary glands and gonads, enabling transovarian passage to occur. The infected tick may survive for years without food in environments of low humidity. The infection is transmitted by ticks both by contamination of the bite with coxal fluid and by the salivary fluid. The life span of the tick is not shortened by carrying borrelia. Many *Ornithodoros* species will feed on a variety of hosts.

B. hermsii and its vector are found primarily at elevations above 3000 feet and are associated with tree squirrels and chipmunks, which may carry the ticks into cabins where they become established. *O. turicatae* parasitizes goats, sheep, and rodents and is found in caves and animal burrows mainly in Florida and Mexico. *O. parkeri* inhabits the homes of ground squirrels and prairie dogs at lower elevations than *O. hermsi* and is widely distributed geographically. Numerous other small mammals also serve as reservoirs for tickborne *Borrelia,* including rats, mice, rabbits, opossums, and hedgehogs. Birds have not been implicated. Once infected, the tick may harbor the disease for many years. However, tickborne disease is not rapidly spread and, in the United States, is responsible only for sporadic cases.

In the United States, relapsing fever is a disease limited to persons who have come into contact with infected ticks. This most commonly results from vacationing in a tick-infested summer cottage. In one recent outbreak, 11 of 42 members of a Boy Scout troop contracted relapsing fever. Most of the infected scouts had slept in a rodent-infested cabin, while the scouts who were younger had slept in tents and did not become infected. The spring and summer distribution of disease coincides with the season of maximal tick activity and the avocational invasion of humans into tick-infested areas. In the United States, foci of tickborne disease occur mainly in the Western states, particularly Oklahoma, California, New Mexico, Colorado, Washington, Texas, and Kansas, which may reflect the distribution of *Ornithodoros* ticks.

In some areas of Africa, inhabitation of the home by *Ornithodoros moubata* is considered to be good luck and has resulted in introduction of disease.

LOUSEBORNE DISEASE. The human body and head lice, *Pediculus humanus corporis* and *Pediculus humanus capitis,* are the vectors of epidemic relapsing fever caused by *B. recurrentis,* although there is some evidence that bedbugs may occasionally also transmit the disease. After the louse ingests *Borrelia,* the organisms pass exclusively into the hemolymph and central ganglion. Since other organs are not invaded, there occurs neither transovarian transmission nor direct infection during the feeding by an intact louse.

Borrelia escape the louse to infect the host only when the louse is injured, as may occur during scratching. A single louse can, therefore, infect only one person. The infected louse remains infectious for its life span, which is approximately 10 to 60 days. Lice may rapidly and widely disseminate disease. Epidemics usually occur in the cold seasons, among the crowded and poor, and in homes with inadequate hygiene. Because of a narrow temperature preference, the louse typically leaves a febrile patient in search of a new host, potentiating rapid spread of an epidemic. Although *B. recurrentis* is considered to be the louseborne species, tickborne *Borrelia* may also be transmitted by lice. No natural animal reservoir of *B. recurrentis* is known. Currently, *B. recurrentis* is endemic, primarily in areas of Ethiopia.

Relapsing fever may occasionally be acquired by means other than louse or tick infestation. For example, transplacental transmission has caused congenital disease, and infected blood may be the cause of laboratory accidents leading to infection.

Pathogenesis. During the entire course of borreliosis, there is a constant spirochetemia, which worsens during febrile periods and wanes between recurrences. Specific pathogenic factors are ill defined. However, the organisms appear to contain a heat-stable pyrogen that is not endotoxin. Skin biopsies of infected persons have shown that there is no inflammatory response around the spirochetes that are within the dermis and that the dermal vessels show no thrombosis or other evidence of vasculitis, as would be expected if endotoxin were produced.

Borrelia are actively phagocytized by polymorphonuclear leukocytes of humans. Immune serum both enhances phagocytosis and exerts a direct effect on *B. hermsii,* causing decreased motility and viability and increased agglutination.

Clinical Manifestations. Prior to the development of effective antimicrobial agents, fever induction was used in the therapy of tertiary syphilis. Induced infection with *Borrelia* was often selected for this purpose, and much of our present knowledge concerning prodromata, incubation period, natural history, and complications stems from these experiences.

The symptoms and severity of relapsing fever depend upon the immune status of the host, geographic location, strain of *Borrelia,* and phase of the epidemic. There may also be consistent differences between some characteristics of louseborne disease and tickborne disease, but both forms will be described together (Table 47-6).

The natural history of a course of relapsing fever includes the incubation period, the primary attack, the afebrile interval, and subsequent attacks. In epidemic, endemic, and therapeutically induced disease, few prodromata have been noted. The incubation period is approximately 6 days, with a range of 2 to 14 days. Late in the incubation period the patient may experience chills. The onset is usually very sudden and accompanied by fever,

TABLE 47-6. CLINICAL MANIFESTATIONS IN RELAPSING FEVER

Manifestation	Mean Value of Incidence	
	Tickborne Disease*	Louseborne Disease[†]
Incubation period	Approx. 7 days	
Duration of primary febrile attack	3.1 days	5.5 days
Duration of afebrile interval	6.8 days	9.25 days
Duration of relapses	2.5 days	1.9 days
Number of relapses	3	1.0
Maximum temperature (primary attack)	Approx. 105F (40.5C)	
Splenomegaly	41%	77%
Hepatomegaly	17–18%	66%
Jaundice	7%	36%
Rash	28%	8%
Respiratory symptoms	16%	34%
CNS involvement	8–9%	30%

Adapted from Southern and Sanford: Medicine 48:129, 1969.
*Based on a review of 1105 reported cases.
[†]Based on review of 2073 reported cases.

headache, tachycardia, and muscle pain. The initial attack usually lasts 3 to 7 days, may be longer for louseborne than for tickborne disease, and ends by crisis. The fever is usually continuous.

A macular rash is seen in varying numbers of patients and usually appears near the end of the first paroxysm. Hepatosplenomegaly, jaundice, nausea, and vomiting are common. Bronchitis and bronchopneumonia are frequent in the United States. Meningeal signs with and without encephalitic disease may affect up to 30 percent of some groups of patients, and ocular disease is common.

The crisis is coincidental with the immune response, which is associated with lysis of the organisms. Occasionally, the crisis is associated with shock. Usually, the temperature returns to normal, and the patient is asymptomatic until the subsequent attack. The interval between initial and subsequent attacks is usually shorter with louseborne disease, 5 to 9 days, than with tickborne disease, which is approximately 14 days.

Data from the period 1921 to 1941 described the course of untreated tickborne disease as follows: no relapse, 16 percent; 1 relapse, 20 percent; 2 relapses, 27 percent; 3 relapses, 17 percent; 4 or more relapses, 18 percent of cases. A similar distribution of relapses has occurred in some outbreaks of louseborne disease.

Subsequent attacks are usually shorter in duration, less severe, and with increasingly shorter apyrexial periods between attacks but are otherwise clinically similar to the initial episode. Most physicians fail to diagnose relapsing fever until one or more relapses have occurred.

Treatment and Prevention. Treatment of relapsing fever includes general supportive measures, such as fluid and electrolyte therapy. Evaluation of efficacy of antimicrobial therapy has been inhibited by the lack of information on in vitro sensitivity. The most clinically effective antimicrobial agents appear to be tetracyclines and chloramphen-

icol. Streptomycin has also been found to modify the disease, although it may fail to prevent relapses.

Prevention of relapsing fever is dependent upon control of exposure to the arthropod vectors. In tickborne borreliosis, this includes wearing protective clothing and careful cleaning of rodent-infested cabins, followed by spraying with appropriate insecticides, such as aldrin, benzene hexachloride, or malathion. Louseborne relapsing fever is controlled by the application of good personal and public standards of hygiene.

Lyme Disease

In the middle 1970s several hundred cases of a newly described syndrome, Lyme disease, were reported in the United States. The disease was named for the village in Connecticut in which the epidemic nature of the disease was first recognized. Lyme disease is characterized by a particular skin eruption, erythema chronicum migrans, which may be followed weeks or months later by migratory polyarthritis. Approximately 85 percent of recognized cases have the rash and/or arthritis. Other prominent manifestations include carditis and a variety of neurologic abnormalities.

The disease usually begins during the summer and is geographically distributed primarily in the Northeast, West, and areas of Wisconsin, California, and Oregon. The skin lesion, which frequently begins at the site of a tickbite, is a slowly expanding, annular, erythematous single lesion. Recent studies have implicated the tick *Ixodes dammini* as the most probable vector in the Northeast and Midwest and *Ixodes pacificus* in the Western United States.

A large proportion of ticks from known endemic foci for Lyme disease have been found to carry spirochetal organisms, primarily in their midgut. The organisms resemble *Treponema* species with electron microscopy. Persons who have had Lyme disease have sharply rising IgM antibody titers to the tick-related organisms when measured by an indirect fluorescent antibody technique. A peak in IgM titer is reached at the third to sixth week of illness and is followed by an elevated IgG titer.

The identification of these organisms and further proof that they are the causative agent in Lyme disease await further study.

FURTHER READING

Books and Reviews

TREPONEMA PALLIDUM

Canale-Parole E: Physiology and evolution of spirochetes. Bacteriol Rev 41:181, 1977

Clark EG, Danbolt N: The Oslo study of the natural course of untreated syphilis. Med Clin North Am 48:613, 1964

Holt SC: Anatomy and chemistry of spirochetes. Microbiol Rev 42:114, 1978

Johnson RC (ed): The Biology of Parasitic Spirochetes. New York, Academic Press, 1976

Musher D, Schell R (eds): The Immunology of Treponemal Infection. New York, Marcel Dekker Press, 1982

Nabarro D: Congenital Syphilis. London, Edward Arnold, 1954

Pavia CS, Folds JD, Baseman JB: Cell-mediated immunity during syphilis. A review. Br J Vener Dis 54:144, 1978

Selected Papers

TREPONEMA PALLIDUM

Alderette JF, Baseman JB: Surface-associated host proteins on virulent *Treponema pallidum.* Infect Immun 26:1048, 1979

Alderette JF, Baseman JB: Surface characterization of virulent *Treponema pallidum.* Infect Immun 30:814, 1980

Fieldsteel AH, Cox DL, Moeckli RA: Cultivation of virulent *Treponema pallidum* in tissue culture. Infect Immun 32:908, 1982

Fitzgerald TJ, Johnson RC, Miller JN, et al.: Characterization of the attachment of *Treponema pallidum* (Nichols strain) to cultured mammalian cells and the potential relationships of attachment to pathogenicity. Infect Immun 18:467, 1977

Gjestland T: The Oslo study of untreated syphilis—An epidemiologic investigation of the natural course of untreated syphilis based on a restudy of the Boeck-Bruusgaard Material. Acta Derm Venerol [Suppl] 1955

Hayes NS, Muse KE, Collier AM, Baseman JB: Parasitism by virulent *Treponema pallidum* of host cell surfaces. Infect Immun 17:174, 1977

Idsoe O, Guthe T, Willcox RR: Penicillin in the treatment of syphilis. The experience of three decades. Bull WHO [Suppl] 48:1, 1972

Lukehart SA: Activation of macrophages by products of lymphocytes from normal and syphilitic rabbits. Infect Immun 37:64, 1982

Musher DM, Hague-Park M, Gyorkey F, et al.: The interaction between *Treponema pallidum* and human polymorphonuclear leukocytes. J Infect Dis 147:77, 1983

Robertson SM, Kettman JR, Miller JN, et al.: Murine monoclonal

antibodies specific for virulent *Treponema pallidum* (Nichols). Infect Immun 36:1076, 1982

Zeigler JA, Jones AM, Jones RH, et al.: Demonstrations of extracellular material at the surface of pathogenic *T. pallidum* cells. Br J Vener Dis 52:1, 1976

LEPTOSPIRA

Adachi Y, Yanagawa R: Inhibition of leptospiral agglutination by the type-specific antigens of leptospirosis. Infect Immun 17:466, 1977

Adler B, Fain S: The antibodies involved in the human immune response to leptospiral infection. J Med Microbiol 11:387, 1978

Banfi E, Cinco M, Bellini M, et al.: The role of antibodies and serum complement in the interaction between macrophages and leptospiras. J Gen Microbiol 128:813, 1982

Berman SJ, Tsai CC, Holmes KK, et al.: Sporadic anicteric leptospirosis in South Vietnam. A study in 150 patients. Ann Intern Med 79:167, 1973

Marshall RB, Wilton BE, Robinson AJ: Identification of *Leptospira* serovars by restriction-endonuclease analysis. J Med Microbiol 14:163, 1981

Martone WJ, Kaufman AF: Leptospirosis in humans in the United States, 1974–1978. J Infect Dis 140:1020, 1979

Pertzelan A, Pruzanski W: *Leptospira canicola* infection: Report of 81 cases and review of the literature. Am J Trop Med Hyg 12:75, 1963

Turner LH: Leptospirosis I, II, and III. Trans R Soc Trop Med Hyg 61:842, 1967; 62:880, 1968; 64:623, 1970

Wong ML, Kaplan S, Dunkle LM, et al.: Leptospirosis. A childhood disease. J Pediatr 90:532, 1977

BORRELIA

Barbour AG, Tessier SL, Stoenner HG. Variable proteins of *Borrelia hermsii.* J Exp Med 156:1312, 1982

Boyer KM, Munford RS, Maupin GO, et al.: Tickborne relapsing fever: An interstate outbreak originating at Grand Canyon National Park. Am J Epidemiol 105:469, 1977

Bryceson ADM, Parry EHO, Perine PL, et al.: Louse-borne relapsing fever. A clinical and laboratory study of 62 cases in Ethiopia and a reconsideration of the literature. Quart J Med 39:129, 1970

Butler T, Jones PK, Wallace CK: *Borrelia recurrentis* infection: Single-dose antibiotic regimens and management of Jarisch–Herxheimer reaction. J Infect Dis 137:573, 1978

Coffey EM, Eveland WC: Experimental relapsing fever initiated by *Borrelia hermsii.* I. Identification of major serotypes in the rat. J Infect Dis 117:23, 29, 1971

Pickett J, Kelly R: Lipid catabolism of relapsing fever borreliae. Infect Immun 9:279, 1974

Southern PM, Sanford JP: Relapsing fever. A clinical and microbiological review. Medicine 48:129, 1969

Spagnuolo PJ, Butler T, Bloch EH, et al.: Opsonic requirements for phagocytosis of *Borrelia hermsii* by human polymorphonuclear leukocytes. J Infect Dis 145:358, 1982

Stoenner HG, Dodd T, Larsen C: Antigenic variation of *Borrelia hermsii.* J Exp Med 156:1297, 1982

Thompson RS, Burgdorfer W, Russell R, et al.: Outbreak of tickborne relapsing fever in Spokane County, Washington. JAMA 210:1045, 1969

Warrell DA, Pope HM, Parry EHO, et al.: Cardiorespiratory disturbances associated with infective fever in man: Studies of Ethiopian louseborne relapsing fever. Clin Sci 39:123, 1970

LYME DISEASE

Burgdorfer W, Barbour AG, Hayes SF, et al.: Lyme disease—A tick-borne spirochetosis? Science 216:1317, 1982

Schrock CG: Lyme disease: Additional evidence of widespread distribution. Am J Med 72:700, 1982

Steere AC, Malawista SE: Cases of Lyme disease in the United States: Locations correlated with distribution of *Ixodes dammini.* Ann Intern Med 91:730, 1979

Steere AC, Grodzick RL, Kornldatt AN, et al.: The spirochetal etiology of Lyme disease. New Engl J Med 308:733, 1983

CHAPTER 48

Spirillum and Campylobacter

The family Spirillaceae is composed of the two genera, *Spirillum* and *Campylobacter*. Both genera contain organisms that are pathogenic for humans. *Spirillum minor* is the cause of one form of rat-bite fever and is the only human pathogen of the genus. *Campylobacter fetus* and *Campylobacter jejuni* may cause neonatal septicemia, diarrhea, and a variety of other infections. Their association with human infections has only recently been appreciated as a result of improvements in anaerobic culture techniques.

The Spirillaceae are rigid, helically curved rods with a variable number of turns. They are motile by means of flagella and move in a corkscrew motion. Most of the organisms in the family are free living in fresh or salt water. Others are saprophytic or parasitic and human or animal pathogens. *Spirillum* species are polytrichous, with flagella at both poles, and are strict aerobes. *Campylobacter* have a single polar flagellum at one or both poles and are microaerophilic to anaerobic.

Spirillum minor

Morphology. *S. minor* is a short, thick organism with tapering ends, 0.2 to 0.5 μm by 3 to 5 μm in size. It has two or three windings that are thick, regular, and spiral. It is gram-negative but can best be visualized in blood smears with Giemsa or Wright's stain. Silver impregnation methods, such as that of Fontana–Tribondeau, stain the polytrichous polar flagella. The outer membrane lipopolysaccharide layer forms a sheath along the entire length of the flagella. This membrane appears to be important in the protection of *Spirillum* sp. from *Bdellovibrio* predation. Darkfield illumination of a drop of blood containing the organism is the best method for demonstrating its rapid motility, spiral structure, and flagella.

Laboratory Identification. *S. minor* has not been cultured on artificial media. Proof that the organism produces rat-bite fever has been obtained by experimental inoculation of man with blood containing the organism. The diagnosis of rat-bite fever is based upon the demonstration of the organisms in inoculated animals. The primary method is inoculation of white mice and guinea pigs with the patient's blood, exudate from the initial lesion, serum expressed from exanthematous patches, material aspirated from lymph nodes, or ground-up pieces of tissue excised from lesions. Since mice often harbor this organism, it is necessary to ensure that animals are free from spirilla before inoculations are made. Alternately, diagnosis may be made by examination of blood and exudate from lesions by darkfield illumination and stains. The organism rarely has been detected with certainty in the blood of man but may be found in material from the lesions.

Clinical Infection

Rat-bite fever is an acute bacteremic infection caused by *S. minor* or by *Streptobacillus moniliformis*, both of which are present in the normal oropharyngeal flora of rodents. Differences in clinical manifestations between the two forms, however, permit differentiation of the two diseases. Rat-bite fever caused by *S. minor* is commonly referred to as "Sodoku fever." (See Table 42-2, p. 660.)

Epidemiology. Rat-bite fever is primarily a disease of wild rats that is transmissible to rats, various other animals, and humans by the bite of an infected animal. Fleas and other insects are not vectors, and there is no record of transmission of the disease from human to human by contact, excreta, or fomites. Cases attributed to the bites of cats, ferrets, and weasels have been reported.

The infecting organisms are carried into the wound of the bite by the rat's teeth. Spirilla have not been found in the saliva of the rats. They may get into the mouth and on the teeth in blood from injured gums, lesions in the mouth, infectious conjunctival exudate that drains through the lacrimal ducts, or exudate from pulmonary lesions. When several persons are bitten by an infected rat, often only the first victim will contract the disease.

Pathogenesis. Rat-bite fever begins as a wound that may be infected with organisms other than *S. minor*. A variety of cocci, bacilli, and actinomycetes have been found in these conditions.

There are few recorded autopsies of cases of rat-bite fever. The local lesion, which is a granuloma without suppuration, shows necrosis of the epithelium and dense round cell infiltration of the corium. Similar round cell infiltration with dilated vessels occurs in the lesions of the skin eruption.

Clinical Manifestations. In a case uncomplicated by mixed or secondary infection, the wound of the bite heals promptly. After an incubation period of 5 to 14 days, the site of the wound swells and becomes purplish and painful. A chancre-like indurated ulcer with a black crust may develop at this site and may reach a diameter of 5 to 10 cm. The regional lymphatics are inflamed, and the adjacent lymph nodes become enlarged and tender. The development of the local lesion is accompanied by malaise and headache and a sharp rise in temperature, usually with a chill. After this, periods of fever alternate with afebrile periods. The temperature rises abruptly, remains elevated for 24 to 48 hours, and falls rapidly to normal within about 36 hours. The intervening afebrile periods last from 3 to 9 days. In untreated cases this relapsing type of fever may continue for weeks or months, gradually subsiding.

Within the first week of the beginning of the fever, a characteristic purplish maculopapular eruption of the skin of the arms, legs, and trunk, and occasionally on the face and scalp usually appears. The skin lesions do not ulcerate. They fade somewhat during the afebrile periods but reappear, with new patches of eruption, during the parox-

ysms of fever. The major serious complication of this infection is subacute bacterial endocarditis. In the preantibiotic era, the mortality rate was estimated to be about 10 percent and was usually due to secondary pyogenic infection. Some patients develop a false positive reaction to treponemal antigens.

Treatment. Penicillin is the drug of choice. Streptomycin also has been used successfully.

Campylobacter

Morphology. The genus name *Campylobacter* is derived from the Greek word *campylo* meaning "curved." Organisms in this genus are gram-negative, curved, microaerophilic, and 1.5 to 5 μm long (Fig. 48-1). They are characteristically comma-shaped when seen in infected tissue but are filamentous or coccoid following laboratory isolation. The organisms are motile, with a single unipolar or bipolar flagellum, and the motility can best be observed by phase microscopy.

Taxonomy. The taxonomy of these organisms has undergone several recent changes, which is a frequent cause of confusion. The initial isolation of this genus, made in 1909, was classified as *Vibrio fetus*. In 1947 human infection was observed, and subsequently, isolations from blood and cerebrospinal fluid were occasionally made. In 1957, the term "related vibrios" was given to strains that had certain biochemical characteristics and grew best at 42C. In 1973, the genus and species *C. fetus* was proposed, with three

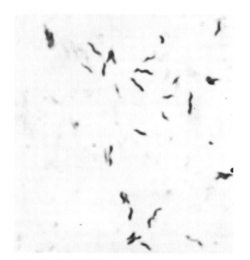

Figure 48-1. *Campylobacter jejuni.* Note curved and serpentine forms. Gram stain. ×2500.

subspecies: *fetus, intestinalis,* and *jejuni.* Current nomenclature now recognizes two human pathogens in this group: *C. jejuni* and *C. fetus* subsp. *fetus.* Organisms formerly termed *C. fetus* subsp. *intestinalis* are now termed *C. fetus* subsp. *fetus.* Organisms formerly termed "related vibrios" are now classified as *C. jejuni. C. fetus* subsp. *fetus* is the type species of this genus. Other species and subspecies are thought to be commensals or causes of diseases of animals.

Campylobacter fetus and *Campylobacter jejuni*

Laboratory Identification

Isolation of *C. jejuni* from rectal and stool specimens has been facilitated by the development of selective media, without which isolation is infrequent. Appropriate selective media include Butzler's medium, Skirrow's medium, and campy-BAP. These media contain various combinations of vancomycin, trimethoprim, polymyxin B, bacitracin, novobiocin, colistin, or cephalothin to inhibit overgrowth of competing rectal flora. *C. jejuni* grows at 42C as well as at 37C, and a candle jar provides an adequate atmosphere. Colonies are fully developed at 24 to 48 hours. *C. jejuni* is oxidase, catalase, and nitrate positive. It is sensitive to naladixic acid and is usually hippurate positive.

C. fetus subsp. *fetus* is an opportunistic pathogen in humans that is morphologically similar to *C. jejuni.* Growth is best at 25C and 37C and is very poor at 42C. It is oxidase, catalase, and nitrate positive but differs from *C. jejuni* in being resistant to nalidixic acid and hippurate negative.

Antigenic Structure

The major antigen of *Campylobacter* is the lipopolysaccharide of the outer membrane. In *C. fetus* subsp. *fetus,* the polysaccharide O-antigen chains are of intermediate length, while those of *C. jejuni* are of low molecular weight. The outer membrane of *C. jejuni* contains a single major polypeptide, while *C. fetus* subsp. *fetus* contains at least two polypeptide fractions. The protein appears to traverse the entire outer membrane.

Serotyping techniques for the identification of strains have not been well standardized. Among the most promising under study is a passive hemagglutination technique, apparently dependent on O antigens, that distinguishes 23 serotypes. A reliable typing system would contribute greatly to a better understanding of the epidemiology of these diseases.

Patients who have been infected with *C. jejuni* develop strain-specific antibody responses that may be measured by agglutination, killing, and complement-fixation techniques. Complement fixation is the least strain specific, with some cross-reactivity to noninfecting strains, while agglutinating antibody responses are usually specific for the infecting strain. High titer bactericidal antibody responses usually develop in the convalescent phase of illness and are strain specific.

Clinical Infection

Epidemiology. *Campylobacter* infections in animals are a major source of interest in veterinary medicine. Serious economic losses to farmers result from abortions and infertility of infected cattle and sheep. Infection of animals is transmitted venereally and may be harbored asymptomatically for long periods of time in the genitourinary and intestinal tracts. An extensive literature on this subject may be found in veterinary journals.

C. fetus subsp. *fetus* appears to cause an opportunistic infection in persons who are debilitated and in pregnant women. It causes a frequently fatal septicemic infection of newborns and in debilitated persons. The sources of these infections are unknown, although a venereal transmission has been suggested. Few persons with bacteremic disease due to *C. fetus* subsp. *fetus* have a known exposure history.

C. jejuni causes mainly an enteric infection of all age groups. Animal reservoirs may lead to human disease, as it is a commensal of cattle, dogs, and fowl. Contaminated milk, water, and food have led to large outbreaks. Human-to-human spread through a fecal-oral route has been demonstrated, and the care of small children with *C. jejuni* diarrhea carries a risk of disease to adult caretakers. Outbreaks in daycare centers may be spread in these ways.

The incidence of *C. jejuni* in persons with diarrhea has been similar to that with *Yersinia enterocolitica* and is about 5 percent of cases. The infecting dose is not known, but ingestion of 500 organisms with milk has resulted in diarrhea.

Pathogenesis. The mechanisms of pathogenesis of *C. jejuni* are poorly understood. Enteric infection with *C. jejuni* results in a mucosal invasion characterized by ulceration of the mucosal surface, crypt abscesses, and hemorrhagic necrosis of the ileum and jejuneum. Examination of the stools during infection reveals bloody diarrhea in a large proportion of cases, and fecal leukocytes are usually present, even in young children. Tests for heat-stable toxin, heat-labile toxin, and invasion by use of the Sereney test are all negative, although there is cell association and penetration by HeLa cell assay.

Clinical Manifestations. *C. fetus* subsp. *fetus* is usually recognized as an acute febrile bacteremic disease. The fatality rate in newborns is very high, but it is a rare disease in that age group. Bacteremic disease in older men with cirrhosis, malignancies, and cardiovascular disease may lead to localized infection of the meninges, pleural space, lungs, joints, pericardium, and peritoneum. Women in the third trimester of pregnancy are also susceptible to bacteremic disease.

C. jejuni characteristically causes an enteritis, usually in normal persons, with fever, abdominal pain, blood in the stool, and headache. It is also associated with inflammatory proctitis in male homosexuals. Persons who are HLA-B27 genotype and develop *C. jejuni* diarrhea are at considerable risk of developing a reactive arthritis.

Treatment. Most *Campylobacter* strains are sensitive in vitro to erythromycin, aminoglycosides, tetracycline, and chloramphenicol, but no controlled studies of the treatment results have been reported. Relapses following cessation of therapy are not rare.

FURTHER READING

SPIRILLUM

Babudieri B: Experimental infections by *Spirilla*. In Eichler O (ed): Handbuch der experimentellen Pharmakologie, New Series. Berlin, New York, Springer Verlag, 1973, vol 17, IIB, p 43

Bayne-Jones S: Rat-bite fever in the United States. Internat Clin [41st ser]3:235, 1931

Gilbert GL, Cassidy JF, Bennett N McK: Rat-bite fever. Med J Aust 2:1131, 1971

Roughgarden JW: Antimicrobial therapy of rat-bite fever. Arch Intern Med 116:39, 1965

Watkins CG: Rat-bite fever. J Pediatr 28:429, 1946

CAMPYLOBACTER

Blaser MJ, Reller LB: Campylobacter enteritis. N Engl J Med 305:1444, 1981

Bokkenheuser V: *Vibrio fetus* infection in man. I. Ten cases and some epidemiologic observations. Am J Epidemiol 91:400, 1970

Bokkenheuser V: *Vibrio fetus* infection in man: A serological test. Infect Immun 5:222, 1972

Chow, AW, Patten V, Bednorz D: Susceptibility of *Campylobacter fetus* to twenty-two antimicrobial agents. Antimicrob Agents Chemother 13:416, 1978

Duffy MC, Benson JB, Rubin JJ: Mucosal invasion in campylobacter enteritis. Am J Clin Pathol 73:706, 1980

Gribble MJ, Salit IE, Isaac-Renton J, et al.: Campylobacter infections in pregnancy. Am J Obstet Gynecol 140:423, 1981

King EO: The laboratory recognition of *Vibrio fetus* and a closely related vibrio isolated from cases of human vibriosis. Ann NY Acad Sci 98:700, 1962

Kosunen TU, Danielsson D, Kiellander J: Serology of *Campylobacter fetus* subsp. *jejuni* ("related" campylobacters). Acta Path Microbiol Scand [B]88:207, 1980

Logan SM, Trust TJ: Outer membrane characteristics of *Campylobacter jejuni*. Infect Immun 38:898, 1982

McCoy EC, Doyle D, Bunda K, et al.: Superficial antigens of *Campylobacter (Vibrio) fetus*: Characterization of an antiphagocytic component. Infect Immun 11:517, 1975

Patton CM, Mitchell SW, Potter ME, et al.: Comparison of selective media for primary isolation of *Campylobacter fetus* subsp. *jejuni*. J Clin Microbiol 13:326, 1981

Penner JL, Hennessy JN: Passive hemagglutination technique for serotyping *Campylobacter fetus* subsp. *jejuni* on the basis of soluble heat-stable antigens. J Clin Microbiol 12:732, 1980

Oral Microbiology

An important responsibility of physicians is to recognize oral diseases in their patients and to refer them for proper care as needed, just as the same responsibility exists for dentists with regard to systemic diseases in their patients.

The most common oral infections are dental decay (dental caries) and diseases of the gingiva and alveolar bone, which support the teeth (periodontal disease). Although these diseases are not normally life threatening, virtually the entire population is affected at some time by one or both conditions. Other infections are endodontic infections that involve the pulp of the tooth following trauma or carious exposure and periapical infections resulting from extension of bacteria from infected pulp through the apex of the tooth. Abscesses may form from periapical in-

fections or from deep periodontal pockets. Mild to severe pyogenic infections, which are potentially life threatening, can result from all of these sources or from heavy contamination of tissues during surgery.

Of additional significance to systemic health is the potential for oral bacteria to induce endocarditis in susceptible heart valves or for chronic oral infection to affect the course of some systemic diseases, e.g., diabetes. Conversely, systemic conditions (e.g., pregnancy, diabetes) enhance oral inflammation and destructive processes of periodontal disease. Patients undergoing radiation therapy that affects the salivary glands often experience severe oral complications, including rampant tooth decay.

Other infections of the tongue and oral mucosa include those caused by herpes simplex types I and II and various fungi. *Candida albicans* most commonly infects the tongue and mucosa following antibiotic therapy, especially tetracyclines. Rarely, chronic granulomatous lesions are caused by *Cryptococcus neoformans, Histoplasma capsulatum, Coccidioides immitis,* and oral *Actinomyces* species. These may be encountered about as frequently in patients as oral carcinoma and must be differentiated. These agents are described in Chapters 84 and 87.

Additionally, oral mucosa often exhibits lesions reflective of systemic infections, e.g., measles, herpes zoster. The reader is referred to more comprehensive texts referenced for detail on these and other infections that are not described in this chapter.

Oral Microbial Flora

Large masses of bacteria develop in different ecologic niches within the mouth, on the epithelial surfaces of the tongue and cheek, and on the teeth (especially in occlusal fissures, proximal areas, and surfaces along the gingival margin.) Bacterial plaque accumulations contain more than 10^{11} microorganisms per gram wet weight, and saliva contains approximately 10^8 bacteria per ml (Table 49-1). The organisms present in saliva do not represent a resident population but are organisms that have been dislodged from the oral surfaces, especially the tongue.

The oral cavity is usually sterile at birth, but within the first day, possibly coinciding with the first feeding, resident flora begin to colonize the mouth. Within a week *Streptococcus salivarius* and *Streptococcus mitior* are detected, probably acquired from the parents or attendants, and *Veillonella* and other anaerobic species appear shortly thereafter. With the emergence of teeth and the provision of a more suitable anaerobic environment, an increase occurs in the number of anaerobic organisms, such as *Fusobacterium* and *Bacteroides.* Also detectable following the eruption of teeth are certain facultative organisms distinctive of the adult flora, *Streptococcus mutans* and *Streptococcus sanguis.* By the end of the first year, the overall flora of infants is similar to that of adults. Two important groups, however, the oral spirochetes and *Bacteroides melaninogenicus,* which colonize the gingival sulcus, usually are not present in the oral cavity until the time of puberty.

Flora of Different Parts of the Mouth

Within the oral cavity are several diverse ecosystems, each with its characteristic collection of bacteria. *S. salivarius,* for example, preferentially colonizes the dorsal surface of the tongue, whereas *S. sanguis* and *S. mutans* colonize the teeth. *S. mitior* is the predominant organism in the buccal mucosa, whereas *B. melaninogenicus* and oral spirochetes prefer the gingival crevice area.

Ecologic and Environmental Influences

The ecology and intraoral distribution of the various bacterial types indigenous to the mouth are regulated by a number of physical, chemical, and mechanical factors that only recently have become accessible to study.

Physical and Chemical Influences. A high proportion of the microflora of the oral cavity is anaerobic. Until 1970, only 20 to 25 percent of the bacteria counted microscopically could be cultured using conventional anaerobic techniques. Methods now available permit investigators to sample and culture approximately 70 percent of the bacteria from many oral sites. The oxidation-reduction potential at certain anatomic sites in the mouth discriminates against certain of the organisms that gain access to the oral cavity. The teeth and associated gingival crevice area are essential for the growth of anaerobic forms, such as *Fusobacterium* and *Bacteroides,* presumably because the reduced conditions in the gingival crevice area are required for colonization. When teeth are absent, spirochetes are not found.

The importance of E_h in the development of dental plaque has been emphasized in studies in which colonization of a mechanically cleaned tooth surface was followed by use of a scanning electron microscope. Facultative streptococci were found to play a prominent role in the initial colonization of the tooth surface. Other initial colonizers included *Neisseria* and *Nocardia,* which were present in relatively high levels in 1-day old plaque. By the end of 9 days, however, there was a marked decrease in the incidence of these aerobic species and a corresponding increase in anaerobic genera, such as *Actinomyces, Veillonella,* and *Fusobacterium.* Direct measurements of plaque E_h also document the shift from a positive to a negative oxidation-reduction potential. The E_h of a clean tooth surface is about $+140$ mV; by the seventh day it has dropped to val-

---

TABLE 49-1. COUNTS OF BACTERIA ON ORAL SURFACES AND IN SALIVA

	Coronal Plaque	Gingival Plaque	Saliva	Tongue Epithelial Cells	Cheek Epithelial Cells
Direct microscopic count	2.5×10^{11}/g	1.7×10^{11}/g		100/cell	10–20/cell
Total cultivable count—anaerobic incubation	4.6×10^{10}/g	4.0×10^{10}/g	1.1×10^8/ml		
Total cultivable count—aerobic incubation	2.4×10^{10}/g	1.6×10^{10}/g	4.0×10^7/ml		

From Gibbons and van Houte: In Shaw et al. (eds): Textbook of Oral Biology, 1978. Philadelphia, Saunders.

ues of −140 to −200 mV, prompting the influx of more anaerobic organisms.

The pH of the microbial niche also exerts a selective force, permitting acid-tolerant organisms to survive in oral sites when pH levels drop. When sugar is consumed, the production of lactic acid causes the in vivo plaque pH values to fall rapidly within 5 to 10 minutes to pH minima approaching pH 5.0. Aciduric bacteria, such as the lactobacilli and *S. mutans*, benefit from this drop in pH.

Nutritional Factors. The oral cavity is nutrient-rich, providing an ideal environment for the most fastidious species. Nutrients may be provided by the host in the form of salivary components, crevicular fluid, and desquamated cells, all of which may be broken down by degradative enzymes elaborated by bacteria colonizing the mouth. Bacteria in the mouth may also derive nutrients from the host's diet. The significance of the diet in dental caries has been clearly established. Especially important is the role of sucrose, which enhances the colonization of *S. mutans* on the teeth (p. 749). Restriction in the ingestion of carbohydrate results in a reduction in the numbers of lactobacilli and *S. mutans* on the teeth.

Bacterial Adherence. If colonization is to occur, the organism must become firmly attached to a surface or find a protective niche to avoid being washed away by the fluid flow to which these environments are subjected. Within the oral cavity, the bacterial biomass doubles only two or four times a day, a rate that is insufficient for the bacteria to maintain themselves in the oral secretions. Bacterial adherence thus becomes a powerful ecologic determinant, correlating positively with the relative proportions in which that species is found in its natural habitat. The initial adsorption of bacteria to oral surfaces appears to be reversible, and only a small number of those cells that do adsorb resist the mechanical cleansing action in the mouth.

Colonization of oral surfaces is influenced by the initial concentration of bacteria to which the teeth or epithelial cells are exposed. The critical concentration needed for *S. sanguis* to adhere to teeth is approximately 10^3 per milliliter. For *S. mutans* the minimum concentration is 10-fold higher.

S. sanguis has an affinity for saliva, which strongly adheres to the enamel surface. *S. mutans* has no affinity for saliva but has receptors that can attach directly to enamel, and capsular glucans then link the *S. mutans* cells together. Although *S. salivarius* has little affinity for enamel, it possesses pili that have a strong affinity for epithelial cells. This accounts for the localization of *S. sanguis* and *S. mutans* on teeth and *S. salivarius* on the mucosa.

Desquamation is an extremely important host defense mechanism effective in preventing the development of large microbial accumulations on mucosal surfaces. The rate of desquamation is apparently related to the microbial burden. The nondesquamating nature of the tooth surface permits the accumulation of large masses of bacteria, and,

without appropriate hygiene, the ultimate development of dental disease.

Dental Plaque

Formation of Dental Plaque. The colonization of teeth results in the formation of dental plaque, a layer of bacteria on the erupted surfaces of teeth leading to caries and periodontal disease. Between the bacterial plaque and the enamel surface is a thin amorphous film less than 1 μ in thickness, the acquired pellicle. This layer consists primarily of salivary constituents that have adsorbed selectively to the enamel surface. Because of its affinity for saliva, *S. sanguis* sorbs to the acquired pellicle within minutes after the teeth are vigorously cleaned. Within 1 to 2 days macroscopically visible colonies appear. Continued bacterial growth results in a confluent microbial layer composed of organisms held together by a matrix consisting of bacterial and salivary polymers.

Bacterial colonization of plaque is a selective process dependent upon the specific ability to attach to pellicle, matrix, or other bacteria. A strong mechanism of adherence in plaque is the synthesis of extracellular polysaccharide polymers by some bacteria, e.g., glucans produced by *S. mutans* and fructans by *Actinomyces viscosus*. Somewhat unique to *S. mutans* is its dextran-binding capacity, which permits it to bind to matrix components.

There is a gradual trend in plaque development from primarily gram-positive facultative streptococci and rods toward an abundance of gram-negative anaerobic rods, curved motile forms, and spirochetes. Synergistic interactions have been demonstrated involving the production of peroxidase by facultative streptococci, catalase by *A. viscosus*, and superoxide dismutase by most facultative species. These enzymes would eliminate toxic peroxides formed from oxygen, thereby permitting anaerobic species to grow in the plaque. Lactic, formic, and isobutyric acids and polyamines serve as nutrients for various anaerobic gram-negative bacteria, such as *Veillonella* species, *Eikenella corrodens*, and vibrio and spirochetal forms. Vitamin K, which is required by gram-negative anaerobic *B. melaninogenicus*, is produced by early appearing gram-positive facultative plaque forms. Levan or other surface polysaccharides produced by *A. viscosus* will aggregate *Veillonella* and certain strains of *S. sanguis* and appear to be responsible for the formation of corncob structures seen in well-developed plaque (Fig. 49-1).

In sections of human dental plaque the microcolonies appear as parallel columns perpendicular to the tooth surface. The composition of plaque is quite varied, depending on the particular tooth and plaque site, diet, and plaque age. Plaque is most abundant in the most stagnant sites on the surface, especially in areas where adjacent teeth are in contact. The firm approximation of gum to tooth usually prevents the invasion of bacteria into the crevice between the gum and tooth and sharply delineates the plaque border in this region if plaque is regularly removed.

Figure 49-1. Two-day plaque. Some of the filaments are covered by cocci, resembling corncob formation, in an area demonstrating an increasing degree of structural complexity. × 4,000 SEM. *(From Lie: J Periodont Res 12:85, 1977.)*

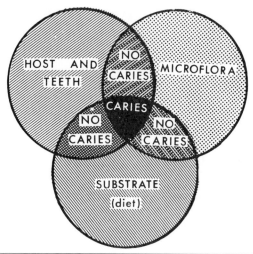

Figure 49-2. Overlapping circles depicting factors responsible for caries activity. *(Courtesy of Dr. P. H. Keyes, National Institute for Dental Research.)*

Plaque Diseases of the Oral Cavity

Dental Caries

During the past decade, we have witnessed a drastic decrease in caries incidence primarily due to impressive progress in understanding etiologic factors and in the widespread application of effective preventive measures. Potential vaccination against dental caries has been extensively studied and is now a real though not imminent possibility.

Etiology. Caries is a multifactorial disease that results only when interdependent factors coincide, as illustrated in Figure 49-2. A tooth susceptible to acid demineralization is the first requirement. Susceptibility is related to low fluoride content and to surface irregularities, such as pits and fissures, that entrap food and plaque.

The role of dietary sucrose as the major substrate for bacterial acid production has been demonstrated repeatedly. Its role in caries is further enhanced by bacteria that utilize sucrose to synthesize adherent polysaccharides (Fig. 49-3). These polysaccharides significantly thicken plaque and consequently prolong the maintenance of a low pH on the tooth surface. The length of time that this acidic environment is maintained has a significant effect on caries occurrence. Consequently, the *frequency* and *form* of sucrose in the diet have greater importance in caries formation than does the quantity of sucrose.

Recent research has indicated that other carbohydrates, e.g., starches, can also contribute to acid production and are of concern. Other studies have determined that *combinations* of foods and *sequence* of eating foods influence the acidogenic potential from sucrose and other carbohydrates. These areas, as well as the use of noncariogenic sweeteners, will probably receive the greatest attention in research that relates diet to dental caries in the next decade.

In addition to the above factors, the quantity and quality of saliva significantly influence the occurrence and progression of dental caries. Protection imparted by a normal flow rate becomes most apparent when that rate is pathologically reduced and a dramatic increase in caries occurs. This is due to reduction in mechanical cleansing and buffering capacity. The role of various natural and acquired defense factors in saliva has been speculated but not proven. Saliva is known to contain lysozyme, lactoferrin, lactoperoxidase, and secretory immunoglobulin, all of which may contribute to caries inhibition by virtue of their antimicrobial capabilities.

Enamel Caries

By far the most common types of carious lesions are those that initiate on enamel surfaces of teeth. These may be further classified as *occlusal caries*, which occur on the biting surfaces of posterior teeth, and *smooth-surface caries*, which occur on the other enamel surfaces.

BACTERIOLOGIC BASIS. Evidence is overwhelming that *S. mutans* is the primary etiologic agent in enamel caries. There is evidence, however, that other acidogenic bacteria, especially *Lactobacillus* species, play an important secondary role in the caries process, particularly on occlusal surfaces (Chaps. 26 and 45).

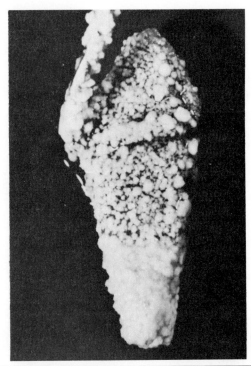

Figure 49-3. Demonstration of extensive bacterial plaque-like matts or deposits that can form on a tooth incubated in sucrose broth culture of *Streptococcus mutans* for 24 hours and transferred to fresh broth medium daily for 4 or more days. *(Courtesy of Dr. P.H. Keyes, National Institute of Dental Research.)*

Streptococcus mutans. In humans the occurrence of caries correlates very highly with the presence of *S. mutans.* Although this organism comprises only a small proportion of pooled plaque samples, the proportions are much greater on carious or precarious surfaces than on adjacent sound surfaces of the same tooth. Various animal models have been useful in proving conclusively that *S. mutans* alone can induce dental caries that is transmissible to other animals.

Extensive research, both in vitro and in vivo, has been devoted to characterizing the properties of specific strains of *S. mutans* that are associated with cariogenicity. Important to the virulence of *S. mutans* are its acidogenic potential and aciduric nature. Although it is not as acidogenic or aciduric as are lactobacilli, it more rapidly produces lactic acid from sucrose and other carbohydrates than do other oral bacteria.

A property that is highly correlated with cariogenicity is the synthesis of high molecular weight insoluble glucans (Fig. 49-4). This is a sucrose-dependent reaction that involves, first, the cleavage of sucrose by a cell surface enzyme, glucosyltransferase, followed by the subsequent polymerization of the glucose moiety into branched glucans with primarily $\alpha\ 1 \rightarrow 3$ and $\alpha\ 1 \rightarrow 6$ linkages. *S. mutans* cells also possess lectin-like cell surface receptor sites that enable them to bind to the tooth surface. Receptor sites for glucans are also present that promote the ag-

gregation of *S. mutans* and the establishment of a thick insoluble plaque. This plaque acts as a barrier, separating buffering agents in saliva from acids on the tooth surface. The combination of acidogenic/aciduric ability and the production of insoluble glucans enables *S. mutans* to maintain low pH levels approaching 5.0, which are sufficient to induce enamel demineralization.

Additional capabilities that contribute to the cariogenicity of *S. mutans* relate primarily to the capacity to utilize a variety of nutrient sources other than dietary carbohydrates. Of special importance is the ability of cariogenic strains of *S. mutans* to convert excess sucrose to intracellular storage polysaccharides (Fig. 49-4) and to metabolize them when exogenous energy sources are limited. In addition, *S. mutans* may produce glycosidic hydrolases, which can extract carbohydrate from saliva as an energy source. They often metabolize soluble levans or dextrans previously formed by themselves or other plaque bacteria. In brief, *S. mutans* has a multiplicity of nutrient resources that contribute to their stability in the oral environment and to their continual production of at least low levels of acids. The extent to which all of these capabilities are exhibited by a specific strain of *S. mutans* relates directly to the cariogenicity of that strain.

Other Bacteria. Prior to the emergence of interest in *S. mutans* as a cariogenic agent, research efforts were devoted to the *Lactobacillus* species. Observations indicated that the extent of dental caries could be correlated with the numbers of lactobacilli present. Their strong acidogenic and aciduric properties and consistent association with dental caries made the lactobacilli likely candidates as primary etiologic agents. It is now felt that lactobacilli are secondary agents in dental caries and are of particular importance in occlusal caries and in carious lesions that have progressed into the dentin of the tooth. Results of clinical caries activity tests that are often incorporated into caries prevention programs in essence reflect the numbers of lactobacilli present in plaque and saliva. These tests are based on the ability of lactobacilli to initiate growth below pH 5.0 in a glucose medium and, if sufficiently abundant, to reduce the pH to 3.0 to 4.0 within 72 hours. This causes a visible color change in the pH indicator in the medium, as described by Snyder and Grainger. The terminal pH and the time required to reach that pH are indicators of cariogenic potential and sucrose intake. Clinical methods are not readily available for bacteriologic detection of *S. mutans.*

Another group of acidogenic bacteria in the oral cavity are other oral streptococci, e.g., *S. sanguis, S. mitior* and *Streptococcus milleri.* Although none of these organisms possess the pathogenic potential that *S. mutans* does, they almost certainly contribute to acid demineralization, particularly in occlusal caries. The physiologic classification of oral streptococci is summarized in Table 49-2.

CONTROL AND PREVENTION. Efforts to prevent dental caries have been directed to control of one or more com-

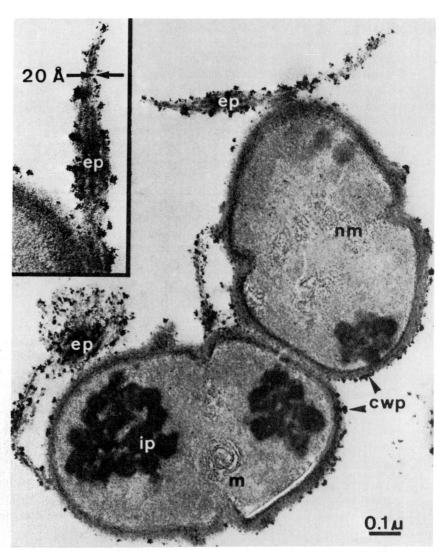

Figure 49-4. Sucrose-grown cells of strepto-cocci contrasted with osmium black. Heavily electron-dense granules of intracellular poly-saccharides (ip), mesosomes (m), and nucle-ar material (nm) are shown in detail. The cell wall has a dense middle layer; its outer sur-face carries strongly dense osmiophilic parti-cles labeling sites of polysaccharides (cwp, arrows). In direct continuation with the cell wall there are extracellular polysaccharides (ep). This material consists of very fine proto-fibrils approximately 20 nm thick (inset). ✕ 99,000. Inset, ✕ 165,000. *(From Guggenheim and Schroeder: Helv Odont Acta 11:131, 1967.)*

ponents of the etiologic triad, i.e., incorporation of fluo-ride in communal water supplies, topical fluoride appli-cation, dietary carbohydrate control, and daily removal of dental plaque. Although the impact of such preventive approaches is apparent, optimal levels of control have not been attained. Currently, efforts are being concentrated on the identification of safe and effective antimicrobial approaches that do not require such intense individual compliance. A variety of plaque-inhibiting agents have been identified, e.g., chlorhexidine and amine fluorides, but have not yet been deemed acceptable for widespread marketing in the United States.

The development and application of an effective anticaries vaccine have been demonstrated in rat, hamster, and primate animal models. These experimental vaccines have utilized almost exclusively whole cells or constituents of *S. mutans* as the immunogen. Immunization with

glucosyltransferase, for example, has been found to effec-tively diminish the synthesis of glucans and to decrease caries, though not necessarily reduce the numbers of *S. mutans*. Generally, both serum and salivary antibodies were induced, and evidence suggests that both systems can im-part protection. Results of attempts to demonstrate natu-rally acquired immunity in humans have been somewhat contradictory. There is evidence that some individuals are better protected against dental caries than others, but it is not clear whether this protection is related more to serum or to secretory immunity. In one study, adult human vol-unteers ingested capsules of formalin-killed *S. mutans* cells and developed significant specific salivary IgG antibody levels with no change in serum antibody levels. This meth-od of immunization, previously used in an experimental rat model, produced the same secretory IgG response and significantly protected the rats against dental caries. This

TABLE 49-2. BIOCHEMICAL TESTS FOR IDENTIFICATION OF ORAL STREPTOCOCCI

Test	Streptococcus				
	S. mutans	S. sanguis	S. mitior	S. milleri	S. salivarius
Mannitol fermentation	+	−	−	−	−
Sorbitol fermentation	+*	−	−	−	−
Arginine hydrolysis	−†	+	−	+	−
Esculin hydrolysis	+*	+	−	+	+*
Acetoin production	+	−	−§	+	V
Dextran production	+	+	+*	−	−
H₂O₂ production	−‡	+	+	−	−

From Hardie and Bowden: J Dent Res 55 [Special Issue A]: A166, 1976.
Results based on studies with oral isolates.
V, variable reaction.
*Some strains negative.
†Serotype *b* strains positive.
‡Some strains weakly positive.
§Some strains positive.

suggests that oral immunization may be effective in humans and avoids some of the concerns for safety that are inherent with other routes of immunization.

However, the ingestion of large numbers of *S. mutans* cells could possibly produce serum antibody if the organisms gain access to the blood supply, e.g., via ulcers. Concern results from an observation that antibodies to some *S. mutans* cell wall components cross-react with heart tissue antigens. This finding has impeded further human trials.

It is realistic to expect a breakthrough in this area in the near future. Employing genetic engineering techniques, avirulent *Escherichia coli* can be endowed with genetic information from *S. mutans*. The potential for using *E. coli* to stimulate secretory immunity against cariogenic factors of *S. mutans* is a promising endeavor that would avoid the risk of stimulating the production of serum antibodies to *S. mutans* that could cross-react with heart tissue.

Caries in Nursing Infants. An occasional occurrence in infants is nursing bottle caries, a condition that will confront the physician more often than the dentist. This condition involves a rapid carious breakdown particularly of incisors and is associated with an infant's prolonged use of a nursing bottle while going to sleep. This results in the pooling on the teeth of milk or juice that contains a fermentable carbohydrate. This extended exposure to sugar results in the accumulation of high numbers of *S. mutans* and lactobacilli in plaque on the affected teeth and subsequent carious breakdown. Diagnosis of this condition, however, is clinical rather than bacteriologic. Obviously, this is not a major problem prior to 6 months of age when the first teeth usually erupt, but thereafter, the problem becomes more severe as additional teeth erupt. Timely weaning from the bottle and not permitting the bottle to remain in the sleeping baby's mouth are the best measures for preventing nursing bottle caries. Use of sucrose-free liquids will reduce the likelihood of occurrence of this type

of caries. Therapy involves the use of these measures and improved hygiene. Other antimicrobial approaches are not indicated.

Root Surface Caries
The discussion thus far has related to dental caries that initiate in enamel and progress into dentin. A problem that is not unusual in older individuals is decay of exposed cementum surfaces. Exposure of root surfaces is most often associated with gingival recession related to periodontal pathology but may also result from mechanical abrasion of gingival tissues, usually from improper toothbrushing. Cementum is much less mineralized than enamel and is consequently susceptible to decay under less stringent requirements than enamel. Although *S. mutans* has been implicated in root surface caries, this type of caries is more frequently correlated with the presence of *Actinomyces* species, particularly *A. viscosus*. *A. viscosus* has a predilection for colonizing the cervicoradicular area of a tooth, is mildly acidogenic, and produces soluble extracellular polysaccharides. These capabilities, although inadequate to decay enamel surfaces, are often sufficient to initiate caries on the more susceptible root or cementum surface.

Prevention of root caries is not unlike that of enamel caries but usually requires a more intensified application, i.e., avoidance of sucrose and soft foods, meticulous oral hygiene, and regular fluoride exposure. Again, diagnosis and therapy is symptomatic rather than bacteriologic.

Radiation Caries
Of particular concern to medical personnel is the potential for rampant and severe decay in patients suffering from xerostomia, a condition resulting from radiation therapy that affects the salivary glands. This severe reduction in salivary flow results in a multiplicity of oral problems, thus prior consultation with a knowledgeable dentist is critical to the patient's oral health and comfort after radiation

therapy is initiated. Frequently, teeth are extracted to prevent problems of discomfort or infection that are likely to ensue. If appropriate preventive measures are not instituted, severe decay may affect all tooth surfaces, even those not normally susceptible, e.g., incisal edges of anterior teeth. The reduction of salivary flow results in a drastic increase in acidity in the oral environment. There is an increase in acidogenic/aciduric bacteria, such as *S. mutans, Lactobacillus,* and also *Candida* and *Staphylococcus* species, with a concurrent decrease in *S. sanguis, Neisseria,* and *Fusobacterium* species, which are normal constituents of the oral cavity. Thus, a major shift in microbial balance occurs that predisposes the patient to multiple problems. The risk of oral facial infections, particularly osteomyelitis, is significantly increased, especially if the individual has dental disease at the outset.

As mentioned earlier, prevention may involve extraction of compromised teeth prior to radiation therapy. Diagnosis, prevention, and therapy for radiation caries are based on symptoms rather than bacteriologic testing. If teeth are retained, a rigorous program of fluoride therapy, including daily self-application, is essential for the prevention of radiation caries. This should be accompanied by good oral hygiene, dietary sucrose restriction, and use of artifical saliva.

Periodontal Diseases

Periodontal diseases affect the vast majority of adults. By destroying the tissues that surround and support the teeth, these diseases are the major cause of loss of teeth in adults. Most result from the interaction of the body and its defense mechanisms with products from bacterial plaque deposits on the teeth.

Bacterial plaque is a dynamic ecosystem in which colonies of facultative cocci, rod forms, anaerobic cocci, and fusobacteria adhere to teeth in close proximity to the gingival tissues. If this ecosystem is allowed to accumulate relatively undisturbed, it will provide a suitable site for the proliferation of anaerobic motile rods, vibrios, and spirochetes (fusospirochetal flora). These latter organisms appear to be important in the production of gingival inflammation.

Histologic and Clinical Changes

Plaque accumulation at the gingival margin causes minimal changes in gingival tissues of young children, except for mild erythema. In adults, however, gingival tissues show histologic changes within 2 to 3 days after initial plaque accumulation. Within 2 weeks, a situation termed the "established lesion" is usually seen, characterized by plasma cell infiltrate and dissolution of subepithelial connective tissue. This is clinically recognizable as gingivitis. In some persons the disease remains at this level indefinitely. However, the majority of cases progress with time, and resorption of supporting alveolar bone and formation of soft tissue pockets between the gingiva and the teeth occur. Histologically, this is termed the "advanced lesion," and it is characterized by an infiltrate of plasma cells, lymphocytes, and macrophages. Clinically, the entity is termed "periodontitis." This is associated with persistence of bacteria on the tooth surface and extension of their growth into the gingival space (sulcus).

Pathogenesis

The pathogenesis of periodontal inflammation appears to be multifactorial. Except for acute necrotizing ulcerative gingivitis (ANUG) (i.e., trench mouth, "Vincent's infection), the bacteria found in periodontal conditions do not usually penetrate the gingival tissues, although there is ulceration of the epithelium within the sulcus. Thus, these destructive responses are dependent upon bacterial products and host responses to those products (Table 49-3). Potential mechanisms of periodontal destruction include direct effects of bacterial products, such as collagenase from *B. melaninogenicus,* endotoxins, and bacterial cell wall peptidoglycans. Other destructive mechanisms are attributable to host-mediated responses induced by bacteria or their products. Polymorphonuclear leukocytes are attracted to the area by a highly chemotactic polypeptide released within the sulcus by bacteria. Subsequent cytolysis or stimulation of the PMNs results in the release of relatively large amounts of lysosomal enzymes that are potentially more destructive than bacterial enzymes. Complement may be activated by endotoxin or by immune complexes and may induce a variety of destructive biologic responses. There is evidence that immunopathologic responses, both humoral and cellular, are operative in some periodontal disease states. The disease process is likely to be related to the cumulative effects of all these mechanisms. It is possi-

TABLE 49-3. POSSIBLE MICROBIAL MECHANISMS OF TISSUE DESTRUCTION IN PERIODONTAL DISEASE

Direct Effects	Indirect Effects
Bacterial enzymes	Host enzymes
Collagenase	Collagenase
Hyaluronidase	Hyaluronidase
Chondroitin sulfatase	Antibody-mediated
Proteases	pathogenic effects
DNase	Anaphylactic
RNase	Cytotoxic
Neuraminidase	Immune complex
Others	Delayed hypersensitivity
Cytotoxic agents	Lymphokines
Endotoxin*	Activation of complement
Peptidoglycans†	Chemotaxis of PMNs
Ammonia	
Hydrogen sulfide	
Toxic amines	
Organic acids	

* Gram-negative organisms.

† Gram-positive organisms.

ble that the importance of their roles may vary from individual to individual, with the resulting disease symptomatology being similar.

Bacteriology

Experimental Early Gingivitis in Adults. A model of the potential for plaque bacteria to harm the gingiva of adults was first established in 1965 by Loe and associates. When young adults are taught to scrupulously clean their teeth free from plaque bacteria, they can achieve an ideal degree of gingival health with no evidence of inflammation. When brushing and flossing are then suspended for a month, classic flora changes are usually observed. After only 1 day, exposed surfaces of teeth just above the gingiva begin to colonize with a film of cocci (mainly *S. sanguis*) roughly 20 cells thick. Extensive numbers of filamentous forms and fusobacteria accumulate in 4 to 7 days, followed by a proliferation of vibrios and spirochetes at the gingival margin in about 2 weeks. Marginal inflammation and easily bleeding gingiva are observed at about 2 to 3 weeks, within a few days after spirochetes become abundant. This plaque has been found to be 100 to 300 bacteria thick. Histologic evaluations indicate the presence of subclinical inflammation after about 4 days. Restoring daily brushing and flossing to disrupt and remove the plaque permits restoration of gingival health within 2 to 3 days. Bacteriologic evaluations in experimental early gingivitis demonstrate that associated plaques are composed almost entirely of gram-positive bacteria, about 50 percent being *Actinomyces* species.

Long-standing Gingivitis. In long-standing gingivitis found in dental patients, the flora is similar, but about 25 percent of the bacteria may be gram-negative, including species of *Campylobacter*, *Veillonella*, and *Fusobacterium*. The total plaque has begun to migrate into the gingival crevice (subgingivally), and gram-negative forms are found on the subgingival filamentous plaque surface, next to the gingival tissues. Erythematous, bleeding, and edematous gingiva are characteristic. Unless there are modifying systemic factors, this is a chronic inflammation that causes only slight or no discomfort to the patient.

Clinical Disease

Periodontitis. The term "periodontitis" is used when resorption of supporting bone results in loss of gingival attachment to teeth, forming pockets around the teeth. Ulcerations occur at the bottoms of the pockets at the point of tissue attachment, and tissues in the pockets bleed easily upon probing. Surface gingiva may appear healthy if pockets are deep and supragingival plaque (plaque on the tooth above the gingiva) is well controlled. Plaque that persists on teeth within the pockets can perpetuate chronic inflammation. A distinct layer of motile gram-negative rods, vibrios, and spirochetes lies next to the epithelium covering the attached filamentous plaque and extends to

TABLE 49-4. PREDOMINANT CULTIVABLE ORGANISMS PRESENT IN SUBGINGIVAL PLAQUE OF EIGHT PATIENTS WITH ADVANCED PERIODONTITIS

	Percent Cultivable Organisms in Subgingival Plaque of 8 Patients
Gram-negative organisms	
Facultative cocci	< 1
Anaerobic cocci	0.6
Facultative rods	< 1
Anaerobic rods	74
Bacteroides melaninogenicus	32
Other prominent gram-negative rods include: *Fusobacterium*, *Eikenella*, *Selenomonas*, and *Campylobacter* species	
Gram-positive organisms	
Facultative cocci	6
Anaerobic cocci	< 1
Facultative rods	4
Anaerobic rods	15
Actinomyces species	12

From Slots: Scand J Dent Res 85:114, 1977.

the bottom of the pocket. Subgingival flora differs from supragingival plaque. Subgingival flora is predominantly anaerobic and gram-negative (Table 49-4). In some instances *Bacteroides gingivalis*, *Bacteroides intermedius*, *E. corrodens*, *Fusobacterium nucleatum*, *Selenomonas*, and *Campylobacter* are found in abundance (Chaps. 45 and 48). Here, as in marginal experimental gingivitis, bacteria do not actually penetrate into the tissues, as is seen in ANUG.

Periodontal therapy and hygiene, including deep cleaning and scaling, reduce the inflammation and create a more favorable ecosystem. Antibiotics, especially tetracycline, have been used to help control progressive periodontitis.

Localized Juvenile Periodontitis (LJP). This is a progressive disease seen in adolescents who often appear to have fair to good oral hygiene with light supragingival plaque, yet characterized by deep pockets and rapid bone resorption, usually limited to incisors and first molars. In 90 percent of patients, cultures reveal *Capnocytophaga* species (similar to *Bacteroides ochraceus*) and *Actinobacillus actinomycetemcomitans;* 80 percent of patients have IgG antibody to the latter organism. These two species have been found together with other subgingival species common to rapidly progressive periodontitis. The disease has been associated with reduced phagocytic ability and chemotactic activity of the PMN leukocyte related to host deficiency or to leukotoxins produced by the above bacteria. Normal leukocyte function is sometimes restored when the disease is overcome. Antibiotics, such as chlortetracycline, have been used to treat juvenile periodontitis, with some success in halting the destructive processes.

Acute Necrotizing Ulcerative Gingivitis (ANUG).
ANUG (trench mouth, Vincent's gingivitis) is characterized by the sudden onset of a painful, inflammatory, pseudo-membranous condition of the gingiva in young adults. The interdental papilla characteristically terminate in ulcerative lesions with a punched-out appearance. The condition usually occurs in individuals under stress who also have deficient oral hygiene.

In ANUG, bacteria actively penetrate and infect the gingiva. Electron micrographic studies of diseased gingival tissues have revealed the following zones: (1) a surface zone rich in motile fusobacteria and vibrio forms as well as large, gently curved *Borrelia vincenti* and intermediate-sized spirochetes with multiple axial filaments, (2) a zone rich in neutrophils, (3) a necrotic area filled with spirochetes, PMN leukocytes, and mononuclear cells, and (4) a zone where only multifilamented spirochetes are observed actively penetrating the tissues. The fetid odor of skatoles and H_2S produced by anaerobic species is usually pronounced.

Although ANUG is not difficult to recognize clinically, microscopic confirmation of this fusospirochetal disease is helpful. If a fever is present, culture for herpes simplex is also especially desirable to distinguish a combined viral–bacterial infection from simple bacterial ANUG, which is usually afebrile. Herpes virus is probably the essential predisposing factor in apparent fusospirochetal gingivitis of children.

Pharyngitis (Vincent's angina) caused by fusospirochetal bacteria is uncommon but may resemble diphtheria. Phase contrast or darkfield examination of a wet mount preparation of a fresh swab can identify it by the presence of a fusospirochetal flora. Anaerobic cultures are not helpful.

Penicillin is the drug of choice for the pharyngitis, but treatment of ANUG preferably consists of warm saline rinses, debridement, restoring good oral hygiene, rest, and a good diet. Since penicillin therapy only temporarily suppresses the gingival infection without eliminating the cause, without other treatment recurrence with more extensive tissue damage would be likely.

Systemic Spread of Periodontal Disease Microflora. Bacteremias following extraction of periodontally diseased teeth include the facultative streptococci most often associated with endocarditis as well as most of the other bacteria found in plaque. Of particular concern is the possible infection of damaged valves of rheumatic heart patients, joint prostheses, and endocardial implants by bacteria derived from chronic periodontal disease or following dental treatments that cause bleeding. Current American Heart Association guidelines for antibiotic prophylaxis should be consulted and followed for premedicating all such high-risk patients and patients who are receiving cytotoxic drugs or immunosuppressive therapy. Antibiotic prophylaxis for such patients who are just learning to floss and brush their teeth is also important during the first few days until gingival health is established. Patients capable of establishing good oral health are no more likely to have endocarditis than are patients whose teeth are replaced with dentures.

Laboratory Diagnosis

Periodontal diseases are diagnosed primarily by clinical evaluation. However, culture evaluation of gingival crevice samples can be helpful in diagnosing juvenile periodontitis by detecting *Capnocytophaga* species (*B. ochraceus*) and *A. actinomycetemcomitans.*

Cultivation and characterization of the indigenous plaque flora in most periodontal conditions are complex, time-consuming, and unrewarding tasks except for investigational purposes. However, phase contrast microscopic demonstration of plaque flora has been useful in patient education and in evaluating the effectiveness of a patient's oral hygiene. Phase contrast or darkfield examination of subgingival plaque samples collected from gingival pockets now holds promise for helping periodontists evaluate the disease status and treatment progress of periodontally diseased patients. A nonmotile coccal flora is related to controlled plaque; motile flora is associated more with disease. Phase microscopy is helpful in diagnosing ANUG and Vincent's angina.

Following periodontal surgery to eliminate pockets, rare secondary infections involving coliform or mixed anaerobic bacteria have been identified. In these instances, culture evaluations are helpful in guiding specific therapy against the predominating pathogens.

Pyogenic Orofacial Infections

The complexity and severity of many pyogenic infections of orofacial tissues present a serious and difficult challenge to the physician, dentist, and clinical microbiologist. Approximately 10 percent of outpatients treated in surgical dental clinics may exhibit moderate to severe infections of this type. Nearly half are posttreatment infections. These are often progressive, associated with malaise, fever, lymphadenopathy, or elevated blood neutrophil count, and require immediate therapy. Mild, acute, or chronic infections, such as periodontal abscesses, periapical infections, or simple pericoronitis, may be painful but remain self-limiting, without systemic involvement. Clinical treatment without antibacterial chemotherapy is generally effective.

Source of Infection. Infections result from displacement or extension of plaque flora into tissues. Bacteria can spread into tissues from infected dental pulps, from periodontal pockets, from infected pericoronal tissues, or from traumatic injuries. Plaque may also be forced into tissues during invasive dental treatments, such as tooth extraction.

Pathogenesis of resulting infections involves the same mechanisms as those proposed for other mixed anaerobic infections.

Extension of Orofacial Infections

Progressive pyogenic orofacial infections are of serious concern because they can precipitously extend to periocular tissues or to vital organs. *B. melaninogenicus,* one of the most common of the agents in these mixed infections, produces heparinase that can induce thrombi in small vessels. In maxillary infections, thrombi can be carried by venous return to the vasculature surrounding the brain. Resulting brain abscesses are detected by neurologic symptoms. Cavernous sinus thrombophlebitis, a potential consequence, causes a rare but usually fatal occlusion of the major blood sinus beneath the brain. Mandibular infections can spread to the neck to produce suffocation (Ludwig's angina) or to the thorax to cause fatal pericarditis. Because of these possible ramifications, it is essential to provide immediate aggressive clinical and antimicrobial therapy for all orofacial infections that appear to be progressive and/or produce signs of systemic involvement.

Management. Management of severe or progressive pyogenic orofacial infections involves several critical steps: (1) collection of specimen for culture, drainage of exudate, and removal of necrotic bone and tissue debris, (2) immediate antibiotic therapy, (3) treatment of source of infection, e.g., removal or treatment of offending tooth, and (4) provision of supportive care and pain control as needed. Because of the critical nature of some orofacial infections, hospitalization should be considered early for debilitated or medically compromised patients.

Laboratory Diagnosis

Specimen Collection. Methods for specimen collection and transport are much the same as those routinely used in hospitals for anaerobic infections elsewhere in the body (Chap. 45). Except for research purposes, samples need not be collected under a continuous flow of oxygen-free gas. Pathogens of concern readily survive ordinary collection procedures if they can be immediately introduced into an anaerobic environment or be cultured within a few minutes after collection.

Exudates (0.5 to 1 ml) are best collected with a syringe and large needle before the lesion is incised and carefully transferred to anaerobic tubes or vials, avoiding oxygen contamination. Aerobic as well as anaerobic species will survive transport in a reduced atmosphere. Swabs of exudate may be collected from incised intraoral lesions if saliva contamination is kept to a minimum. An adequate amount of exudate can be collected, since counts of bacteria in exudates often equal or exceed those in saliva. Sam-

ples should be transported to the laboratory immediately and should be called to the attention of the technologist who will culture them, explaining their importance and the kinds of information needed. The better the communication, the better and more useful the results are likely to be. This may avoid the useless report: "Only normal oral flora detected."

Culture. The major value of culturing acute, severe infections is to confirm the effectiveness of the antibiotic in use or to guide any needed changes in therapy in 48 to 72 hours if the infection is not responding. Time is, therefore, crucial. Exudates are inoculated on blood agar plates under conditions suitable for cultivation of *Bacteroides* species and incubated anaerobically. Gram-stained smears of exudates are examined primarily to confirm detection of all major forms visible in such smears.

In the treatment of acute, progressive orofacial infection, it is critical that descriptions of the predominant organisms be reported within the first 24 to 48 hours and that antibiotic sensitivity tests be conducted for reporting after the next 24 hours. If indicated, other slower growing organisms can be identified later. The number of strains and antibiotics tested should be determined by consultation between technologist and clinician.

Microbiology of Pyogenic Infections. As seen in Table 49-5, most of the bacteria found in odontogenic infections are facultative and anaerobic species similar to those found in plaque. Mixtures in which two to six species usually predominate are frequently encountered. The most common mixtures include *Streptococcus faecalis* or other indigenous streptococci and the anaerobic *Bacteroides, Fusobacterium, Veillonella,* or *Propionibacterium* species. Predominating pathogens are usually those that grow within 48 hours. The finding of slow-growing *Actinomyces* species in acute pyogenic infections of a few days duration is of no clinical importance.

Among the anaerobic gram-negative rods, *B. melaninogenicus* is one of the most common isolates. *Bacteroides fragilis* may be detected less frequently. It is more common in patients with severe, persistent infections that require hospitalization. Facultative pathogens are also sometimes important. *Staphylococcus aureus* is found in less than 10 percent of all oral infections; *Streptococcus pyogenes* is found in less than 1 percent. Coliform bacteria, especially *Enterobacter* species, may be found as secondary invaders in about 14 percent of all infections.

Orofacial infections are difficult diagnostic problems for the dentist, physician, and microbiologist. Since they are mixed infections involving both anaerobic and facultative species, nearly all of the species detectable in a patient's plaque or saliva may also be present in some orofacial infections. Laboratory identification, therefore, is usually limited to the predominating two to four species. An additional obstacle is that laboratory personnel usually fail to attribute importance to facultative streptococci and

TABLE 49-5. PREDOMINANT ORGANISMS IN ACUTE OROFACIAL ABSCESSES

Anaerobic Organism	No. Lesions (%) n = 46	Aerobic Organism	No. Lesions (%) n = 46
Bacteroides melaninogenicus	16 (35)	*Streptococcus mitior*	1 (2)
Bacteroides melaninogenicus		*Streptococcus viridans*	26 (57)
subsp. *melaninogenicus**	3 (7)	*Streptococcus faecalis*	1 (2)
subsp. *intermedius†*	1 (2)	*Streptococcus* group B	1 (2)
		Staphylococcus aureus	1 (2)
Bacteroides asaccharolyticus§	2 (4)	*Klebsiella pneumoniae*	3 (7)
Bacteroides uniformis	1 (2)		
Bacteriodes sp.	19 (41)		
Fusobacterium nucleatum	2 (4)		
Fusobacterium sp.	1 (2)		
Veillonella parvula	3 (7)		
Streptococcus intermedius	2 (4)		
Peptostreptococcus parvula	1 (2)		
Peptostreptococcus anaerobius	1 (2)		
Peptostreptococcus micros	1 (2)		
Peptostreptococcus sp.	1 (2)		
Anaerobic gram-positive rods	2 (4)		

n, Total number of infections.
*Presently separated into *B. denticola, B. melaninogenicus,* and *B. loescheii* (Chap. 45, p. 692).
†Presently separated into *B. corporis* and *B. intermedius* (p. 692).
§See page 692.

consider them contaminants, just as they are in throat cultures. Nevertheless, *S. mitior, S. sanguis, S. mutans, S. faecalis,* and possibly even *S. salivarius* are significant when obtained in abundance from exudates or blood where they should not exist. Routinely, these need not be identified to species. However, the antibiotic susceptibility of the most abundant strains must be determined and reported as rapidly as possible, together with that of other rapidly growing anaerobic species, in order to effectively guide therapy.

Treatment

Penicillin and erythromycin are the initial drugs of choice for outpatient therapy. For hospitalized patients, some clinicians prefer to use more broad-spectrum antibiotics, including clindamycin, being mindful of drug side effects (e.g., pseudomembranous ulcerative colitis). Other antimicrobial agents, such as second- or possibly third-generation cephalosporins or metronidazole, have been used in resistant infections. In practice, however, and with the aid of drug susceptibility tests, an acceptable dose of a common antibiotic or elevated dose of penicillin G usually provides adequate therapy against all major species present and can help provide rapid control of severe orofacial infections. Commonly used drugs include penicillin and its derivatives, erythromycin, cephalosporins, vancomycin, doxycycline, and streptomycin.

Prevention

Early treatment of decayed, periodontally diseased, or nonvital teeth, good training of all patients in plaque control, especially those with conditions that compromise their resistance, and removal of plaque before dental and, especially, surgical treatments will all help to reduce risks of orofacial infections that can become life threatening under certain circumstances.

FURTHER READING

Books and Reviews

Crawford JJ: Periapical infections and infections of oral facial tissues. In McGhee JR, Michalek SM, Cassell GH (eds): Dental Microbiology. Philadelphia, Harper, 1982, pp. 786–814

Crawford JJ, Fine J: Infection control in hospital dentistry. In Hooley JR, Daun LG (eds): Hospital Dental Practice. St. Louis, Mosby, 1980, pp 119–154

Genco RJ: Antibiotics in the treatment of periodontal diseases. J Periodontol 52:545, 1981

Genco RJ, Mergenhagen SE: Proceedings: Host–parasite Interactions in Periodontal Diseases. American Society for Microbiology, 1982

Gibbons RJ, van Houte J: Bacterial adherence in oral microbial ecology. Ann Rev Microbiol 29:19, 1975

Hamada S, et al.: Biology, immunology, and cariogenicity of *Streptococcus mutans.* Microbiol Rev 44:331, 1980

Harvard Conference: Current research concepts fundamental to the improvement of periodontal care: Usefulness and limitations of model systems in studying periodontal disease; inflammation and repair, supplement. J Dent Res 50:236, 1971

Ikeda T: Sugar substitutes: Reasons and indications for their use. Int Dent J 32:33, 1982

Kleizberg I, Ellison SA, Mandel ID: Proceedings: Saliva and Dental Caries. Microbiology Abstracts [Spec Suppl] 1979

Lobene RR: Clinical studies of plaque control agents: An overview. J Dent Res 58:2381, 1979

McGhee JR, Michalek SM: Immunobiology of dental caries: Microbial aspects and local immunity. Ann Rev Microbiol 35:595, 1981

McGhee JR, Michalek SM, Cassell GH (eds): Dental Microbiology. Hagerstown, Md, Harper, 1982

Menaker L (ed): The Biologic Basis of Dental Caries. Hagerstown, Md, Harper, 1980

Microbial Aspects of Dental Caries. Special Supplement to Microbiology Abstracts, Proceedings of a Conference, 1976, vol I–II

Newbrun E: Dietary carbohydrates: Their role in cariogenicity. Med Clin North Am 63:1069, 1979

Newbrun E: Sucrose in the dynamics of the carious process. Int Dent J 32:13, 1982

Newman HN: Update on plaque and periodontal disease. J Clin Periodontol 7:251, 1980

Newman MG: The role of *Bacteroides melaninogenicus* and other anaerobes in periodontal infections. Rev Infect Dis 1:313, 1979

Roitt IM, Lehner T: Immunology of Oral Diseases. London, Blackwell, 1980

Rowe NH (ed): Proceedings of a Symposium on Diet, Nutrition, and Dental Caries. Ann Arbor, Mich, 1978

Shaw JH, Sweeney EA, Cappuccino CC, et al. (eds): Textbook of Oral Biology. Philadelphia, Saunders, 1978

Tanzer JM (ed): Animal Models in Cariology. Special Supplement to Microbiology Abstracts—Bacteriology, Proceedings of a Symposium & Workshop, Sturbridge, Mass, 1980

van Houte J: Bacterial specificity in the etiology of dental caries. Int Dent J 30:305, 1980

Virginia Polytechnic Institute Anaerobe Laboratory: Outline of Clinical Methods in Anaerobic Bacteriology. Blacksburg, Va, Virginia Polytechnic Inst and State Univ of Virginia, 1971

Selected Papers

Bowen WH, Amsbaugh SM, Monell-Torrens S, et al.: A method to assess the cariogenic potential of foodstuffs. J Am Dent Assoc 100:677, 1980

Brown LR, Dreizen S, Handler S, et al.: Effect of radiation-induced xerostomia on human oral microflora. J Dent Res 54:741, 1975

Chassy BM, Beall JR, Bielawski RM, et al.: Occurrence and distribution of sucrose-metabolizing enzymes in oral streptococci. Infect Immun 14:408, 1976

Coykendall AL: Four types of *Streptococcus mutans* based on their genetic, antigenic and biochemical characteristics. J Gen Microbiol 83:327, 1974

Crawford J, Sconyers J, Moriarty J, et al.: Bacteremia after tooth extractions studied with the aid of prereduced anaerobically sterilized culture media. Appl Microbiol 27:927, 1974

Germaine GR, Schachtele CF: *Streptococcus mutans* dextransucrase: Mode of interaction with high-molecular-weight dextran and role in cellular aggregation. Infect Immun 13:365, 1976

Gibbons RJ: Adherence of bacteria to host tissue. In Schlessinger D (ed): Microbiology—1977. Washington, D.C., American Society for Microbiology, 1977, p 395

Holt RG, Abiko Y, Saito S, et al.: *Streptococcus mutans* genes that code for extracellular proteins in *Escherichia coli* K-12. Infect Immun 38:147, 1982

Keene HJ, et al.: Dental caries and *Streptococcus mutans* prevalence in cancer patients with irradiation-induced xerostomia: 1–13 years after radiotherapy. Caries Res 15:416, 1981

Lee H, Theilade E, Jensen SB: Experimental gingivitis in man. J Periodontol 36:177, 1965

Meiers JC, Wirthlin MR, Shklair IL: A microbiological analysis of human early carious and non-carious fissures. J Dent Res 61:460, 1982

Page RC, Schroeder HE: Pathogenesis of inflammatory periodontal disease. A summary of current work. Lab Invest 33:235, 1976

Shklair IL, Gaugher RW: Glucan synthesis by the oral bacterium *Streptococcus mutans* from caries-active and caries-free naval recruits. Arch Oral Biol 26:683, 1981

Singletary MM, Crawford JJ, Simpson DM: Darkfield microscopic monitoring of subgingival bacteria during periodontal therapy. J Periodontol 53:671, 1982

Slots J: The predominant cultivable microflora of advanced periodontitis. Scand J Dent Res 85:114, 1977

Smith DJ, Taubman MA, Ebersole JL: Effect of oral administration of glucosyltransferase antigens on experimental dental caries. Infect Immun 26:82, 1979

Syed SA, Loesche WJ, Pape HL Jr, et al.: Predominant cultivable flora isolated from human root surface caries plaque. Infect Immun 11:727, 1975

Tanzer JM: Essential dependence of smooth surface caries on, and augmentation of fissure caries by sucrose and *Streptococcus mutans* infection. Infect Immun 25:526, 1979

van Houte J, Gibbs G, Butera C: Oral flora of children with "nursing bottle caries." J Dent Res 61:382, 1982

CHAPTER 50

Legionellaceae

During the summer of 1976 an outbreak of a fulminant pneumonia occurred in persons attending a state American Legion convention in Philadelphia. The causative agent of this outbreak was *Legionella pneumophila,* a previously undescribed bacterium genetically unrelated to any known human pathogen. Analyses of sera and tissue saved from earlier undiagnosed outbreaks of respiratory illness have revealed that *L. pneumophila* was responsible for at least three major outbreaks of respiratory diseases prior to 1976. In addition to epidemic pneumonia, *L. pneumophila* has been shown to cause sporadic pneumonic disease.

Subsequent investigations have demonstrated that there are at least six serotypes of *L. pneumophila* and 23 additional species of *Legionella* that have been either implicated in human disease or isolated from environmental sources. The additional species have been placed in the same genus as *L. pneumophila* by workers at the Center for Disease Control (CDC). However, other investigators have proposed that these additional organisms be placed in separate genera within the family Legionellaceae (Table 50-1).

Members of the family Legionellaceae appear to be widely distributed in nature, inhabiting the water of lakes, cooling towers, and the water supplies of hospitals and hotels. In nature, they have been found in association with various algae. Infection in humans appears to result from contact with these environmental sources. The importance of these organisms in human disease has not been fully delineated, but several studies have indicated that they are common causes of pneumonia and may account for 10 to 30 percent of nosocomial pneumonias in some hospitals.

Legionella pneumophila

The major human pathogen of the family Legionellaceae is *L. pneumophila*. Two distinct forms of illness are caused by this organism: the first, and most commonly recognized, is an acute pneumonia, whereas the second entity, known as Pontiac fever, is a mild upper respiratory tract infection with no known fatalities.

Morphology

L. pneumophila is a rod-shaped organism 0.3 to 0.9 μm in width and 2 to 20 μm long. Motility is observed in fresh isolates and occurs by means of flagella. The organism is difficult to stain and consequently may be missed with conventional examination of clinical material. It does not stain with hematoxylin and eosin, and with gram stain, organisms either fail to stain or will stain very faintly gram-negative if the safranin is left on for 30 to 60 seconds. The

TABLE 50-1. FAMILY LEGIONELLACEAE

Nomenclature		
CDC	Other	Isolation Source
Legionella pneumophila		Human respiratory tract, blood, water
Legionella micdadei	Tatlockia micdadei, Legionella pittsburgensis	Human respiratory tract
Legionella bozemanii	Fluoribacter bozemanae	Human respiratory tract
Legionella gormanii	Fluoribacter gormanii	Soil
Legionella dumoffii	Fluoribacter dumoffii	Human respiratory tract, water
Legionella longbeachae		Human respiratory tract
Legionella jordanis		Water, sewage

CDC, Centers for Disease Control.

organism is not acid-fast. Special stains, such as the Dieterle silver impregnation method, are useful in identifying the organism in tissue. With this stain the bacilli appear intensely brown to black against a pale yellow background. The organisms are consistently seen in areas of pneumonia and are both intracellular and extracellular. Organisms grown on artificial media have inclusions that appear to be poly-β-hydroxybutyrate.

Electron microscopy (Fig. 50-1) reveals typical procaryotic cells resembling known gram-negative bacteria. The organism is surrounded by two triple-layer unit membranes separated by the periplasmic space. Lipid vacuoles are present.

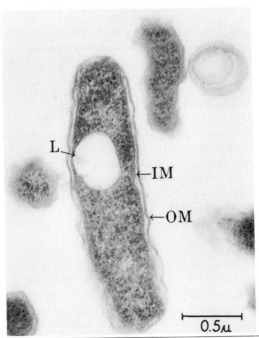

Figure 50-1. Transmission electron micrograph of *Legionella pneumophila* grown on charcoal yeast extract medium. IM, inner membrane; L, probable lipid vacuole; OM, outer membrane. *(Courtesy of Dr. John Shelburne, Duke University.)*

Physiology

Biochemical and Cultural Characteristics. *L. pneumophila* was first grown on artificial culture media using Mueller–Hinton agar supplemented with hemoglobin and Isovitalex in an atmosphere of 5 percent carbon dioxide. The hemoglobin provided a source of soluble iron and the Isovitalex furnished the L-cysteine required for growth of this organism. Charcoal yeast extract agar (CYE) buffered with *N*-(2-acetamido)-2-aminoethane sulfonic acid has subsequently proven to be a superior medium for the isolation of all members of the family Legionellaceae. Growth is slow, requiring 3 to 5 days incubation, and is best achieved in an atmosphere of 2.5 percent carbon dioxide. *L. pneumophila* produces a dark brown pigment that is more readily visible on media without charcoal, such as Feeley–Gorman agar (F–G) or yeast extract agar supplemented with additional tyrosine. Colonies of *L. pneumophila* may be distinguished from other species of Legionellaceae by incorporating 0.001 percent bromcresol and 0.001 percent bromthymol blue in CYE agar. On this medium, *L. pneumophila* grows as a relatively flat, pale green colony, whereas other members of the family are blue or bright green.

L. pneumophila is weakly oxidase positive, produces catalase, and hydrolyzes starch, gelatin, and hippurate. This latter characteristic is useful in differentiating *L. pneumophila* from other members of the family. Carbohydrate fermentation tests are negative.

The cellular fatty acid composition of *L. pneumophila* is unlike that of any other bacterial group. It contains at least 80 to 90 percent branched-chain fatty acids, 14 to 20 carbon atoms in length.

Resistance. *L. pneumophila* remains viable in broth cultures for at least 112 days at 25C and 150 days at 4C. In stream water the bacilli are able to survive for at least 415 days. In growth experiments using unsterile tap water, *L. pneumophila* grew at 32, 37, and 42C, with best growth occurring at 37 and 42C. Little or no growth occurred at 25C. The organism is rapidly killed by a 1-minute exposure to 1 percent formalin, 70 percent ethyl alcohol, 10 ppm iodine, 0.125 percent glutaraldehyde, 1:8000 quaternary ammonium solution, and 0.0021 percent phenolics.

Preliminary studies on the ability of various chemical disinfectants to inhibit growth of *L. pneumophila* in tap water have indicated the effectiveness of a compound containing 50 percent didecylimethylammonium chloride, 20 percent isopropanol, and 30 percent inert ingredients. Calcium hypochlorite also appears to be effective and of potential value in decontaminating evaporative condensers and cooling towers implicated in the transmission of *L. pneumophila.*

Antigenic Structure

There are currently six distinct serogroups of *L. pneumophila,* as defined by direct immunofluorescent staining of whole bacterial cells. There is evidence of common antigens shared among several of the serogroups. Persons infected with bacteria of one serogroup may also show elevated titers to several additional serogroups. The majority of clinical disease seen to date has been caused by members of serogroup 1. Flagellar antigens have also been detected but have not been studied in detail.

Clinical Infection

Epidemiology. Legionnaires' disease is widespread in occurrence, with cases reported worldwide. The disease occurs sporadically as well as in well-defined epidemic clusters. Although cases occur throughout the year, there appears to be a peaking of cases in the late summer months from July to October. The male to female ratio in several studies is approximately 3 to 1. Smokers appear to be more susceptible to infection, as are patients who are immunosuppressed. The majority of cases are seen in persons in the fifth decade of life or older, but cases have been seen in most age groups.

In the epidemic outbreaks caused by *L. pneumophila,* the pattern has been that of a continuing common-source airborne exposure. Person-to-person spread has not been documented, suggesting an enviromental source. The source of several epidemics has been an air-conditioning cooling tower or evaporative condenser at the site of the outbreak. A malfunctioning air-conditioning unit was also proven to be the source of the Pontiac fever outbreak in Pontiac, Michigan, in 1968. In the 1965 Washington, D.C. outbreak, windblown dust from excavations was implicated as the source of infection. A high exposure to dust from construction sites has been implicated as the source of sporadic cases of Legionnaires' disease. The water systems of hospitals with endemic Legionnaires' disease may be the main source of nosocomial Legionellosis in these institutions.

Clinical Manifestations. The overall clinical spectrum of *L. pneumophila* pneumonia varies in severity from a mild pneumonia to an adult respiratory distress syndrome accompanied by major extrapulmonary manifestations. The illness, which is usually not associated with a preceding upper respiratory infection, presents with symptoms of malaise, myalgia, and mild headache. Within 1 to 2 days the patient develops a moderate, nonproductive cough with associated chest pain and becomes febrile (39 to 41C), often with shaking chills. Additional early symptoms may include systemic manifestations, such as vomiting, diarrhea, abdominal pain, impaired renal function, confusion, and/or delirium. Progression of pneumonia can lead to respiratory failure.

The disease may be accompanied by a moderate leukocytosis, elevated sedimentation rate, elevation of liver enzymes, hematuria, hyponatremia and proteinuria. Chest x-rays show unilateral, patchy infiltrates that may progress to consolidation. Pleural effusions are seen in approximately one half of the patients. Fatalities may be high, particularly in the immunosuppressed population.

The syndrome of Pontiac fever is a self-limited illness characterized by fever, myalgia, malaise, and headache but no respiratory manifestations or pneumonia. No fatalities have been associated with this syndrome.

Evidence is accumulating to indicate that Legionnaires' disease may involve systems other than the respiratory tract. The organism has been isolated from blood and has been demonstrated in blood vessels of the kidney, liver, and spleen of at least two patients at autopsy. *L. pneumophila* has also been reported to be the cause of hemodialysis fistula infections in two patients.

Pathogenesis. Although some reports suggest that the disease involves organ systems other than the lung, the extent of this involvement is not completely understood. Autopsy findings demonstrate an acute fibrinopurulent pneumonia with exudation of neutrophils, macrophages, and large amounts of fibrin in the alveolar spaces. The inflammatory process does not involve blood vessel walls or large bronchi, and interstitial infiltration is minimal.

L. pneumophila appears to be a facultative intracellular parasite and is capable of growing in macrophages and a variety of protozoa. Wong and co-workers have shown that avirulent *L. pneumophila* could become virulent when grown in tissue cultures of human lung fibroblasts. Growth was intracellular, and no soluble toxin was produced during cell culture. However, other workers have discovered a small peptide cytotoxin that decreases hexose monophosphate and oxygen consumption in polymorphonuclear leukocytes during phagocytosis. This heat stable cytotoxin renders cells incapable of killing bacteria and may be responsible for intracellular survival of *L. pneumophila.*

Laboratory Diagnosis. Laboratory confirmation of Legionnaires' disease can be made by (1) direct demonstration of the organism in clinical specimens, (2) culture of the organism, and/or (3) detection of specific antibodies in patient's serum.

DIRECT EXAMINATION. *L. pneumophila* can be detected in clinical specimens by use of the modified Dieterle silver stain (p. 760). A direct immunofluorescence (IF) test is also available that offers a rapid, convenient, and specific method for diagnosis of Legionnaires' disease. Direct IF is useful for demonstrating *L. pneumophila* in sputa, transtrachael aspirates, biopsy specimens, and bronchial lavage specimens. Direct IF testing should include antisera directed at all six serogroups.

CULTURE. Methods for culture of the organism on artificial media are now available using buffered CYE media (p. 760). Clinical materials should be planted on at least two plates. On one plate a spot inoculum should be made without streaking, while on the second plate the specimen is streaked. Growth usually occurs at the site of the original inoculation, where the concentration of organisms is heaviest. Normal respiratory flora is inhibitory to *L. pneumophila*, hindering isolation from sputa. A semiselective medium using buffered CYE with 2-ketoglutarate, cefamandole, polymyxin B, and anisomycin has been developed for the isolation of *L. pneumophila* from contaminated sources. A biphasic CYE blood culture bottle has also been developed for the isolation of the organism from blood.

INDIRECT IMMUNOFLUORESCENCE. Despite the availability of direct IF and culture, diagnosis has been mainly epidemiologic or retrospective and has been based primarily on the use of an indirect immunofluorescent antibody test. The demonstration of a four-fold change in titer of convalescent-phase serum relative to acute-phase serum, or a titer of $\geq$ 1:256 against *L. pneumophila* in a single serum specimen is indicative of infection. As with the direct IF test, a complete indirect IF test must look for antibody to all six serogroups. The greatest obstacle, however, in the usefulness of the indirect IF test is that it requires 3 to 6 weeks for the demonstration of a significant rise in titer and, thus, is not helpful to the physician in the management of the infection.

OTHER TESTS. A rapid radioimmunoassay for the detection of *L. pneumophila* serogroup I antigens in urine of patients with Legionnaires' disease has been developed. In preliminary testing, the test was positive in 9 of 9 known positive patients and negative in 214 control subjects. The overall usefulness of this method requires further testing.

Treatment. *L. pneumophila* is sensitive to erythromycin, which is currently the treatment of choice for Legionnaires' disease. The organism is also very susceptible to rifampin. Its production of a β-lactamase that is more active on the cephalosporins than on penicillin renders these drugs ineffective and probably accounts for the approximately 48 percent mortality seen in patients treated with cephalosporin during the 1976 outbreak. For the more seriously ill patients, respiratory therapy is also required.

Other *Legionella*

The techniques used to discover and characterize *L. pneumophila* have led to the establishment of 23 additional species within the genus *Legionella*. These organisms have been isolated from both human and environmental sources. The diseases caused by these organisms are similar to that caused by *L. pneumophila*. Like *L. pneumophila*, the organisms can be isolated on artificial media only with difficulty. The direct immunofluorescence test on clinical specimens and the measurement of antibody by indirect immunofluorescence remain the major methods of diagnosing disease caused by these organisms.

Legionella micdadei. *Legionella micdadei* was isolated in 1979 from lung tissue of two renal transplant recipients who had acute purulent pneumonia. It was designated the "Pittsburgh pneumonia agent." This organism was subsequently proven to be the same as both the TATLOCK organism, which Tatlock had isolated in 1943 from the blood of a soldier with Fort Bragg fever, and the HEBA agent isolated by Bozeman in 1959 from the blood of a patient with pityriasis rosea. These organisms are phenotypically similar to *L. pneumophila* but have been shown to be genetically distinct by DNA homology studies. The name *L. micdadei* has been proposed for these agents to honor McDade who first isolated *L. pneumophila* from materials obtained during the 1976 Philadelphia epidemic. Other proposed names are listed in Table 50-1.

Like other members of the family, *L. micdadei* will grow on CYE agar. However, it will not grow on F–G agar on primary isolation, but it will grow on this medium after several passages first on CYE agar. In addition, *L. micdadei* is similar to *L. pneumophila* except that *L. micdadei* does not hydrolyze hippurate nor does it produce a β-lactamase. Disease caused by *L. micdadei* is clinically indistinguishable from that caused by *L. pneumophila* except that all cases to date have been in immunosuppressed patients. Erythromycin therapy appears to be the treatment of choice.

Legionella bozemanii. In 1959, Bozeman and co-workers isolated an agent, termed WIGA, from lung tissue of a patient who died of pneumonia. After the discovery of *L. pneumophila*, other legionella-like organisms were isolated from similarly infected patients. These organisms and WIGA were found to be genetically identical to each other and, although related to, were distinct from *L. pneumophila*. The name *Legionella bozemanii* has been proposed for these organisms. Colonies grown on CYE agar fluoresce a characteristic blue-white color when exposed to long-wave UV light. These organisms can be further distinguished from other members of the family by a negative oxidase test and the production of bright green colonies on dye containing CYE agar. Disease caused by *L. bozemanii* is typical of that caused by other *Legionella* species, and the treatment of choice is presumed to be erythromycin.

Other Species. *Legionella dumoffii* and *Legionella longbeachae* are two additional proven human pathogens in the family Legionellaceae. They are similar to other members of the family and were isolated from lung tissue. In addition, both have been isolated from environmental sources.

Legionella gormanii and *Legionella jordanis* have only been isolated from environmental sources, but serologic studies indicate that human infection may also occur. Other *Legionella*-like organisms have been isolated and are being classified as rapidly as genetic studies are completed.

FURTHER READING

Books and Reviews

Balows A, Fraser DW (eds): International Symposium on Legionnaires' Disease. Atlanta, Center for Disease Control, Nov 13–15, 1978. Ann Intern Med 90:489, 1979

Jones GL, Herbert GA (eds): "Legionnaires," the Disease, the Bacterium and Methodology. Atlanta, Ga, US Dept HEW, PHS, Centers for Disease Control, 1979

Winn WC Jr, Myerowitz RL: The pathology of *Legionella* pneumonias, a review of 74 cases and the literature. Hum Pathol 12:401, 1981

Selected Papers

Band JD, LaVenture M, Davis JP, et al.: Epidemic Legionnaires' disease—airborne transmission down a chimney. JAMA 245:2404, 1981

Berendt RF: Influence of blue-green algae (Cyanobacteria) on survival of *Legionella pneumophila* in aerosols. Infect Immun 32:690, 1981

Bock BV, Kirby BD, Edelstein PH, et al.: Legionnaires' disease in renal transplant recipients. Lancet 1:410, 1978

Brenner DJ, Steigerwalt AG, Gorman GW, et al.: *Legionella bozemanii* sp. nov. and *Legionella dumoffii* sp nov: Classification of two additional species of *Legionella* associated with human pneumonia. Curr Microbiol 4:111, 1980

Broome CV, Cherry WB, Winn WC, et al.: Rapid diagnosis of Legionnaires' disease by direct immunofluorescent staining. Ann Intern Med 90:1, 1979

Centers for Disease Control: Legionnaires' disease: Diagnosis and management. Ann Intern Med 88:363, 1978

Chandler FW, Blackmon JA, Hicklin MD, et al.: Ultrastructure of the agent of Legionnaires' disease in tissue. N Engl J Med 297:1218, 1977

Cherry WB, Gorman GW, Orrison LH, et al.: *Legionella jordanis*: A new species of *Legionella* isolated from water and sewage. J Clin Microbiol 15:290, 1982

Edelstein PH, Meyer RD, Finegold SM: Isolation of *Legionella pneumophila* from blood. Lancet 1:750, 1979

England AC III, Fraser DW: Sporadic and epidemic legionellosis in the United States. Epidemiologic features. Am J Med 70:707, 1981

England AC, Fraser DW, Plikaytis BD, et al.: Sporadic legionellosis in the United States: The first thousand cases. Ann Intern Med 94:164, 1981

England AC, McKinney RM, Skaliy P, et al.: A fifth serogroup of *Legionella pneumophila*. Ann Intern Med 93:58, 1980

Feeley JC, Gorman GW: *Legionella*. In Lennette EH, Balows A, Hausler WJ Jr, et al. (eds): Manual of Clinical Microbiology, 3rd ed. Washington, American Society for Microbiology, 1980, p 318

Fraser DW, Tsai TF, Orenstein W, et al.: Legionnaires' disease: Description of an epidemic of pneumonia. N Engl J Med 297:1189, 1977

Friedman RL, Lochner JE, Bigley RH, et al.: The effect of *Legionella pneumophila* toxin in oxidative process and bacterial

killing of human polymorphonuclear leukocytes. J Infect Dis 146:328, 1982

Glick TH, Gregg MB, Berman B, et al.: Pontiac fever—epidemic of unknown etiology in a health department. 1. Clinical and epidemiological findings. Am J Epidemiol 107:149, 1978

Harris PP, Aufdemorte T, Ewing EP, et al.: Fluorescent-antibody detection of *Legionella dumoffii* in a fatal case of pneumonia. J Clin Microbiol 13:778, 1981

Herbert GA: Hippurate hydrolysis by *Legionella pneumophila*. J Clin Microbiol 13:240, 1981

Herbert GA, Steigerwalt AG, Brenner DJ: *Legionella micdadei* species nova: Classification of a third species of *Legionella* associated with human pneumonia. Curr Microbiol 3:225, 1980

Hicklin MD, Thomason BM, Chandler FW, et al.: Pathogenesis of acute Legionnaires' disease pneumonia. Immunofluorescent microscopic study. Am J Clin Pathol 73:480, 1980

Kalweit WH, Winn WC Jr, Rocco TA Jr, et al.: Hemodialysis fistula infections caused by *Legionella pneumophila*. Ann Intern Med 96:173, 1982

Kohler RB, Zimmerman SE, Wilson E, et al.: Rapid radioimmunoassay diagnosis of Legionnaires' disease. Detection and partial characterization of urinary antigen. Ann Intern Med 94:601, 1981

McDade JE, Shepard CC, Fraser DW, et al.: Legionnaires' disease: Isolation of a bacterium and demonstration of its role in other respiratory disease. N Engl J Med 297:1197, 1977

McKinney RM, Porschen RK, Edelstein PH, et al.: *Legionella longbeachae* species nova, another etiologic agent of human pneumonia. Ann Intern Med 94:739, 1981

McKinney RM, Wilkinson HW, Sommers HM, et al.: *Legionella pneumophila* serogroup six: Isolation from cases of legionellosis, identification by immunofluorescence staining and immunological response to infection. J Clin Microbiol 12:395, 1980

Meyer RD, Edelstein PH, Kirby BD, et al.: Legionnaires' disease: Unusual clinical and laboratory features. Ann Intern Med 93:240, 1980

Morris GK, Steigerwalt P, Feeley JC, et al.: *Legionella gormanii* sp nov. J Clin Microbiol 12:718, 1980

Moss CW, Weaver RE, Dees SB, et al.: Cellular fatty acid composition of isolates from Legionnaires' disease. J Clin Microbiol 6:140, 1977

Myerowitz RL, Pasculle AW, Dowling JN, et al.: Opportunistic lung infection due to "Pittsburgh pneumonia agent." N Engl J Med 301:953, 1979

Pasculle AW, Myerowitz RL, Rinaldo CR Jr: New bacterial agent of pneumonia isolated from renal-transplant recipients. Lancet 2:58, 1979

Skaliy P, Thompson TA, Gorman GW, et al.: Laboratory studies of disinfectants against *Legionella pneumophila*. Appl Environ Microbiol 40:697, 1980

Stout J, Yu VL, Vickers RM, et al.: Ubiquitousness of *Legionella pneumophila* in the water supply of a hospital with endemic Legionnaires' disease. N Engl J Med 306:406, 1982

Thacker SB, Bennett JV, Tsai TF, et al.: An outbreak in 1965 of severe respiratory illness caused by the Legionnaires' disease bacterium. J Infect Dis 138:512, 1978

Thornsberry C, Kirven LA: β-lactamase of the Legionnaires' bacterium. Curr Microbiol 1:51, 1978

Thornsberry C, Baker CM, Kirven LA: In vitro activity of antimicrobial agents on Legionnaires' disease bacterium. Antimicrob Agents Chemother 13:78, 1978

White HJ, Felton WW, Sun CN: Extrapulmonary histopathologic

manifestations of Legionnaires' disease. Evidence for myocarditis and bacteremia. Arch Pathol Lab Med 104:287, 1980

Wong MC, Peacock WL Jr, McKinney RM, et al.: *Legionella pneumophila*: Avirulent to virulent conversion through passage in cultural human embryonic lung fibroblasts. Curr Microbiol 5:31, 1981

Yee RB, Wadowsky RM: Multiplication of *Legionella pneumophila* in unsterilized tap water. Appl Environ Microbiol 43:1330, 1982

Yu VL, Kroboth FJ, Shonnard J, et al.: Legionnaires disease: New clinical perspective from a prospective pneumonia study. Am J Med 73:357, 1982

Rickettsiae

Rickettsiae are small, pleomorphic, gram-negative coccobacilli that have adapted to intracellular multiplication in arthropods—lice, fleas, and ticks. Because of the blood-sucking habits of these arthropods, the rickettsiae, when released from the intracellular environment of the tick, have developed a complex life cycle involving infection of vertebrate hosts from which the arthropods obtain a blood meal. This infection, usually asymptomatic, produces a rickettsemia from which other arthropods are infected. The distribution of the rickettsiae is thus related to the distribution of the particular arthropod-rodent reservoir. In recent years, the human infections due to this group of agents have resulted from accidental intrusion in this life cycle, with the production of such diseases as Rocky Mountain spotted fever (and other spotted fevers), rickettsialpox, murine and scrub typhus, and Q fever. Historically, the most significant rickettsial infection resulted from an adaptation of a rickettsia to a human body louse–human infectious cycle resulting in epidemic typhus. This last infection had a tremendous impact on history. War, with its attendant human suffering, was frequently associated with a marked increase in the prevalence of body lice, with the result that epidemics of this infection occurred. Frequently, the impact of this widespread infection was most significant in the outcome of battles, such as Charles I's plans to march on London in 1653, the fall of Prague to the French army in 1741, and Napoleon's retreat from Moscow.

The requirement for living cells to support growth resulted in the early failure of usual bacteriologic techniques to isolate the causative agents of these diseases. It was not until Dr. Howard T. Ricketts took blood from a patient with Rocky Mountain spotted fever and inoculated it into a guinea pig that the etiologic agent was transmitted experimentally. Fortunately, the guinea pig developed a disease that was clinically remarkably similar to the human illness. Microscopic examination of these ticks and later of autopsy material identified the etiologic agent of this disease. Subsequently, application of these techniques to other rickettsial diseases and their expansion to the embryonated chicken egg and tissue culture have led to the isolation and identification of the agents responsible for the above-mentioned diseases. Serologic procedures were developed for more precise identification as well as clinical diagnosis. The discovery that these agents were usually susceptible to drugs, such as tetracycline and chloramphenicol, led to effective chemotherapy when the disease was recognized early in its course. Finally, the development of vaccines, as well as curbing arthropod vectors, has decreased the incidence of some of these illnesses.

This family of agents has been given the name Rickettsiaceae in honor of Dr. Ricketts, who subsequently succumbed to typhus during his investigation of this illness. Other organisms capable of causing human disease and sharing some of the characteristics of rickettsiae include *Rochalimaea quintana*, a gram-negative coccobacillus indistinguishable morphologically from rickettsiae but differing from rickettsiae in that it is quite capable of growing extracellularly in the lumen of the body louse as well as in suitably enriched artificial media. It has been established as the cause of trench fever, a common febrile illness among World War I combatants. Another agent, *Coxiella burnetii*, also shares some of the above-mentioned rickettsial properties, such as arthropod–vertebrate cycle, morphology, and requirement for intracellular growth. It possesses, however, some striking properties, such as a marked resistance to environmental factors (perhaps associated with endospore production), antigenic composition, DNA base composition, and aerosol transmission to humans with the production of a flu-like illness and frequent development of pneumonia. These differences have resulted in its separation from the rickettsial group.

Rickettsiaceae

Morphology and Physiology

Rickettsiae are pleomorphic, rod-shaped to coccoid organisms that range in size from 0.3 to 0.6 μm in width to 0.8 to 2.0 μm in length. They stain poorly with gram stain but can be visualized using both the Giemsa and Macchiavello methods. With the Gimenez modification of the Macchiavello method, the organisms stain pink or red and are visualized by counterstaining with malachite green.

The ultrastructure and chemical composition of rickettsiae resemble that of gram-negative bacteria. A characteristic three-layered cell wall, a trilaminar plasma membrane, and nuclear material have been observed by electron microscopy (Fig. 51-1). Ribosome-like particles and intracytoplasmic organelles are present in the granular cytoplasm, and in *Rickettsia prowazekii* an amorphous capsule surrounding the organism has been detected.

Biochemistry. Rickettsiae possess both RNA and DNA, have respiratory and synthetic capabilities independent of the cell, and divide by binary fission. The optimum temperature for growth is 32 to 35C. Pyruvate is the major energy source for *Coxiella*, whereas members of the genus *Rickettsia* utilize glutamate as their major carbon and energy source. Rickettsiae possess enzymatic mechanisms for the breakdown of carbohydrates, the formation of high-energy phosphate bonds, and the synthesis of lipids and proteins. Studies of the G + C content, genome size, and DNA hybridization have demonstrated wide differences among *Coxiella*, the typhus, spotted fever, and scrub typhus groups, although marked similarities exist among species of the same group. There is a very close relationship between *R. prowazekii* and *Rickettsia typhi*.

Cultural Characteristics. Except for *R. quintana* all rickettsiae require living cells for growth. They can be propagated in embryonated eggs, various tissue culture systems, laboratory animals (guinea pigs, mice, meadow voles), and

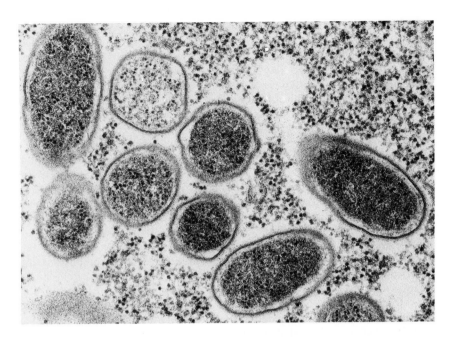

Figure 51-1. *Rickettsia rickettsii,* the causative agent of Rocky Mountain spotted fever, in ovarian tissue of *Dermacentor andersoni* × 66,000. *(Courtesy of Dr. Lyle Brinton, Rocky Mountain Laboratory, USPHS, Hamilton, Montana.)*

in certain arthropods. *R. rickettsii* has been successfully isolated and propagated in primary bone marrow cells and circulating monocyte cultures from experimentally infected laboratory animals. Because of the dangers of infection of laboratory personnel, cultural procedures should be undertaken only in specially equipped facilities.

Species differ in the location of intracellular multiplication both in vitro and in vivo. For example (with the exception of *Rickettsia canada*), typhus group rickettsiae, *C. burnetii* and *Rickettsia tsutsugamushi* characteristically grow only in the cytoplasm of infected cells, whereas spotted fever group organisms grow both in the cytoplasm and the nucleus. *C. burnetii* grows within the phagolysosomes. *R. quintana,* the agent of trench fever, is unique among the Rickettsiae in that it grows extracellularly in its arthropod vector (lice) and can be grown in vitro on cell-free media.

Resistance. Members of the genus *Rickettsia* are unstable extracellularly under ordinary environmental conditions, and careful technique is required in the successful isolation of the organisms from body fluids and tissue. In contrast, *C. burnetii* is very resistant to heat and drying, a characteristic that is important for an understanding of its ecology and epidemiology. The recent discovery of an endospore structure in *C. burnetii* may explain this resistance. Rickettsiae remain viable for long periods when stored at −70C or lyophilized from appropriate media.

Antigenic Structure

Differences in the antigenic composition of the rickettsiae have made possible their classification into genera, groups, and species. There is no common antigen for all members

of the family Rickettsiaceae or for the tribe that contains the three genera of medical interest: *Rickettsia, Rochalimaea,* and *Coxiella*. Antigenically they are completely unrelated.

Two major kinds of antigens have been detected in rickettsiae: (1) ether-soluble group-specific antigens that represent stripped-off capsular material and (2) type-specific antigens that are associated with the cell wall. On the basis of antigenic analysis and the biologic properties discussed above, the genera *Rochalimaea* and *Coxiella* are each comprised of a single species: *R. quintana* (the agent of trench fever) and *C. burnetii* (the agent of Q fever). Strains comprising the genus *Rickettsia* fall into three distinct groups or biotypes: typhus, spotted fever, and scrub typhus. Antigenic differences have permitted further speciation within the first two groups. Additional methods of separating various biotypes include cross-protection and toxin neutralization tests in laboratory animals and, more recently, DNA base ratio analysis.

Phase variation analogous to S → R variation in other bacteria has been observed in *C. burnetii* but in none of the other rickettsiae. Organisms isolated from natural infections of humans and animals are in phase I, but repeated passage in the yolk sac elicits the emergence of phase II and changes in its antigenic makeup. Phase I can usually be reestablished by passage in an animal. Phase I activity is attributable to a surface polysaccharide that is much more potent than phase II as an immunogen.

Serologic Diagnosis

Specific antibodies develop in response to rickettsial infection, and the demonstration of an immune response during convalescence is the most widely used method of confirming clinical diagnosis. In fulminant cases with fatal outcome, antibody may not appear. The separation of antibody into IgM and IgG may on a single specimen help in elucidating whether the infection is recent or remote. Complement-fixing antibodies are useful in identifying infection due to different genera of rickettsiae into groups. Highly purified reagents can minimize the confusion resulting from cross-reactions. More recently, a wide array of serologic procedures has been devised that includes microagglutination, indirect hemagglutination, and fluorescent antibody techniques. In Rocky Mountain spotted fever, fluorescent antibody identification of rickettsiae intracellularly in punch biopsies of the skin has yielded the earliest diagnosis. Early antibiotic treatment can blunt or abolish the complement-fixation antibody rise in titer.

Weil–Felix Reaction

The Weil–Felix reaction is based upon the cross-reactions between antigens in the rickettsial organism and the *Proteus* polysaccharide O antigen. Thus, antibodies to the rickettsial organism will agglutinate certain nonmotile strains of *Proteus* (*Proteus* OX-19, *Proteus* OX-2, *Proteus* OX-K). The test is very simple to perform and inexpensive, with the result that it is widely used. However, the specificity of the test leaves something to be desired. When more specific serologic tests are performed using antigens obtained from rickettsiae, as few as 17 percent have been confirmed with the Weil–Felix reaction when a documented paired serum antibody rise to the rickettsial antigens has been demonstrated. In addition, the *Proteus* agglutinins do not usually appear until after a week of illness, limiting their usefulness in early diagnosis. Some rickettsial diseases (e.g., rickettsialpox and Q fever) are not associated with Weil–Felix antibody rises, and *Proteus* bacterial infections, such as urinary tract, bacteremia, and wounds, can give rise to false positive results.

Determinants of Pathogenicity

In rickettsial diseases the microscopic pathology is characteristic. Multiplication of the organisms in the endothelial cells lining the small blood vessels causes endothelial proliferation and perivascular infiltration, resulting in leakage and thrombosis. The result is a widespread infectious vasculitis. Chick embryos, tissue cultures, and small laboratory animals have been successfully used as experimental models and have provided useful information on the penetration and multiplication of rickettsiae in host cells. Little is known, however, on how the organisms damage the host cell.

Only viable rickettsiae are capable of penetrating host cells. Organisms inactivated by heat, formalin, or UV irradiation lose their infectivity for mice. Rickettsial infection of an individual host cell is a two-stage process. Adsorption to cholesterol-containing receptors precedes penetration into the cell. Both steps are dependent upon energy produced by the infecting rickettsia. Once inside the host

cell, the rickettsiae cause little detectable damage to the parasitized cell until the cell ruptures. However, when large numbers of viable rickettsiae are injected into a mouse, the animal dies of acute toxemia within 8 hours. This effect can be blocked by the administration of type-specific antibody but not by the antirickettsial drugs, tetracycline or chloramphenicol. In order to be effective, the immune serum must be administered before infection, suggesting that the attachment of the toxin to its primary binding site of action occurs very rapidly. The isolation and chemical characterization of a toxin from rickettsiae has not been accomplished, but there is evidence that at least one component possesses endotoxic activity. In addition, the typhus fever rickettsiae are capable of hemolyzing erythrocytes of several animal species. This hemolytic activity is correlated with infectivity of the rickettsiae and with their metabolic activities. The role of these toxins and hemolytic activities in human rickettsial disease remains conjectural.

Different isolates of the same rickettsial species vary in virulence. More than 40 years ago, Spencer and Parker showed that when ticks infected with a virulent strain of *Rickettsia rickettsii* are refrigerated for several months, the rickettsiae lose their virulence for guinea pigs, although they immunize the animal against challenge with a virulent strain. This phenomenon is not due to a spontaneous mutation or to a difference in the number of rickettsiae in the tick. Despite refrigeration, *R. rickettsii* retains its virulence for chick embryos, and a single egg passage reestablishes virulence for the guinea pig. Results of studies of Rocky Mountain spotted fever in monkeys suggest that the duration of the incubation period and severity of the disease may be related to the size of the inoculum of *R. rickettsii*.

The mechanism of variations in virulence in naturally occurring human infections has been incompletely studied. Rhesus monkeys infected with *R. rickettsii* develop a clinical illness similar to that in humans. This animal model is commonly used to study the pathogenesis of human *R. rickettsii* infection.

Host Defenses

The relative importance of various host defense mechanisms in humans infected with rickettsiae is uncertain. Delayed hypersensitivity to typhus group antigens develops after human *R. prowazekii* infection, and lymphocyte-mediated hypersensitivity also can be demonstrated after vaccination or infection with *C. burnetti*. Lymphocytes collected from humans previously infected with *R. rickettsii* undergo blast transformation after in vitro exposure to spotted fever group antigens. An interaction between humoral antibody (opsonins) and macrophages is required for effective killing of *R. typhi* in tissue culture systems. Similarly, antibody-treated *R. rickettsii* are phagocytized and destroyed by guinea pig peritoneal macrophages, whereas untreated rickettsiae replicate and destroy peritoneal phagocytic cells.

Clinical Infection

Among the diverse clinical illnesses produced by rickettsiae are primary pneumonia (Q fever), fulminant vasculitis (Rocky Mountain spotted fever), a febrile illness associated with a vesicular rash (rickettsialpox), asymptomatic infection (trench fever), a recrudescent infection appearing many years after primary infection (Brill–Zinsser disease), and endocarditis (Q fever). Endothelial damage secondary to angiitis is a common finding in rickettsial infections, particularly in spotted fever and typhus.

The Spotted Fever Group

The basic pathologic process in the spotted fever group is a widespread vasculitis involving skin, with the production of a rash. More severe disease can lead to disseminated intravascular coagulopathy with petechial and purpuric skin manifestations. Vasculitis involving viscera, such as the brain, heart, and kidneys, can produce significant clinical complications. Glucose 6-P-dehydrogenase deficiency may make the disease more severe. Extensive dilatation of endoplasmic reticulum has been noted in experimentally infected cells.

Rocky Mountain Spotted Fever (American Spotted Fever, Tickborne Typhus)
Epidemiology

PREVALENCE. Although Rocky Mountain spotted fever got its name from the obvious geographic distribution, subsequent studies have shown that the incidence has dropped markedly in the Rocky Mountain area and is, instead, much more common in the Piedmont region of the southeastern United States (Fig. 51-2). However, cases continue to be reported in small numbers from almost every state in the United States. Rocky Mountain spotted fever has been recognized as a distinct clinical entity for

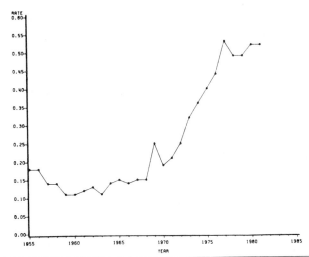

Figure 51-3. Rocky Mountain spotted fever. Reported cases per 100,000 population by year, United States, 1955–1981. The 1192 cases reported for 1981 were the highest number ever reported to the Centers for Disease Control. Of these 834 (70 percent) of the 1981 cases were reported from North Carolina, South Carolina, Virginia, Oklahoma, Tennessee, Georgia, and Maryland. *(From Morbidity and Mortality, CDC Annual Summary 30(54):98, 1982.)*

more than 75 years. Over 1100 cases of Rocky Mountain spotted fever in the United States were reported to the Centers for Disease Control during 1981, and many other cases undoubtedly were either not reported or not diagnosed (Fig. 51-3). Rocky Mountain spotted fever accounts for over 95 percent of the reported rickettsial disease in humans in the United States.

Most cases of Rocky Mountain spotted fever in the eastern and the southern United States occur in children and adolescents, whereas adult men are more commonly affected in the Rocky Mountain region. The seasonal distribution of Rocky Mountain spotted fever is related to the activities of the tick vector, with the result that the disease makes its appearance in April and continues through August. This seasonal distribution should be stressed, since

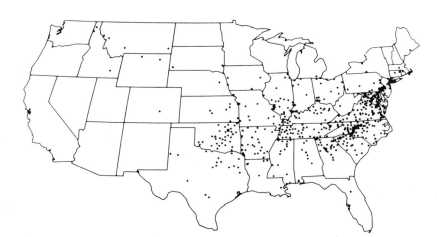

Figure 51-2. Rocky Mountain spotted fever. Reported cases by county, United States. 1977. *(From Morbidity and Mortality. CDC Ann Summary 26(53):65, 1978.)*

any febrile illness occurring during these months in endemic areas is suspect for Rocky Mountain spotted fever.
In addition, cases of this disease have been reported during the winter months, although rarely. People living in
rural and suburban locations are at a higher risk of infection than are urban dwellers. With increasing suburbaniza-
tion and popularity of outdoor recreational activities,
further increases in the incidence of spotted fever may occur.

ECOLOGY. *R. rickettsii,* the etiologic agent of Rocky Mountain spotted fever, is cycled in nature through ticks, small
rodents, and larger wild and domestic animals. Humans
are only accidentally infected. The ecology and epidemiology of spotted fever is directly related to the life cycle of
four species of ixodid (hard) ticks that are indigenous to
the United States: the Rocky Mountain wood tick (*Dermacentor andersoni*), the American dog tick (*Dermacentor variabilis*), the Lone Star tick (*Amblyomma americanum*), and the
rabbit tick (*Haemaphysalis leporispalustris*). *D. andersoni* is the
major vector in the Rocky Mountain region, whereas the
American dog tick is the major vector in the eastern and
southeastern United States. The rabbit tick is found
throughout the continental United States and rarely
attaches to humans but is thought to be important in
maintaining a reservoir of infection in nature.

Ixodid ticks pass through four stages in their life cycle
and can become infected by feeding on a rickettsemic animal; these ticks may in turn pass their infection to their
progeny transovarially. Thus, infected ticks function both
as a reservoir and as a vector of spotted fever rickettsiae.

A number of authors have speculated that strains of
R. rickettsii vary in virulence. Prior to the introduction of
specific antimicrobial therapy, for instance, the mortality
rate of Rocky Mountain spotted fever was known to be 70
to 80 percent in the Bitterroot Valley of Montana but less
than 5 percent in the nearby Idaho Snake River Valley.

TRANSMISSION. Most patients with spotted fever, but by
no means all of them, give a history of a recent tick bite.
Infection may also occur through contamination of fingers
while removing ticks that are feeding on animals or on another person. Except for rare reports of infections associated with needle-stick injections and blood transfusions,
human transmission of spotted fever does not occur. Acquisition of infection by laboratory workers handling *R.
rickettsii* is a well-known phenomenon.

Pathogenesis. *R. rickettsii* infection produces widespread
endothelial damage, resulting in occlusion of small vessels,
microthrombi, microhemorrhages, secondary fluid and
electrolyte changes, and in severe cases, necrosis, shock,
and death. However, the precise mechanism by which rickettsial infection produces endothelial damage is still largely unknown. Activation of the kallikrein-kinin system has
been observed in humans with Rocky Mountain spotted fever. It has been suggested that kinins play a role in the
pathophysiology of the vasculitis and disseminated intra-

vascular coagulation sometimes present in patients with severe infection.

Clinical Manifestations. The incubation period ranges
from 3 to 12 days. The disease typically begins abruptly
with headache, fever, and malaise. Chills may occur, and
diffuse myalgias are common. Usually the rash appears 2
to 4 days after the onset and, in many cases, first appears
on the ankles and wrists and then becomes generalized.
The rash is usually maculopapular early in the disease but
may later become petechial or hemorrhagic. Involvement
of the palms and soles is common.

Gastrointestinal complaints, arthralgias, conjunctivitis,
stiff neck, and periorbital edema are present in some
cases. Splenomegaly is detected in approximately 25 percent of cases. Hyponatremia and thrombocytopenia are
common, and overt disseminated intravascular coagulation
may also occur.

Reported fatality rates have declined from approximately 20 percent in the 1940s to 5 to 10 percent in the
past few years. Factors that increase the chance of fatality
are increasing age, increasing length of time from onset to
institution of effective chemotherapy and, perhaps, virulence of the infecting agents.

Laboratory Diagnosis. It should be emphasized that
the initial diagnosis and treatment of spotted fever should
be made on clinical grounds and not delayed until laboratory confirmation is obtained. An appropriate constellation
of symptoms in a patient with a history of recent tick attachment is enough justification to begin treatment.

Isolation of rickettsiae from blood is possible but expensive, since it requires trained personnel and specialized
facilities available in only a few locations in the United
States.

Confirmation of infection is usually based on serologic
tests. The Weil–Felix test depends upon antigenic cross-
reactions between antirickettsial antibodies and polysaccharide antigens from various strains of *Proteus vulgaris.*
Proteus OX-2 and/or OX-19 agglutinins usually appear 5 to
8 days after infection with *R. rickettsii.* A four-fold antibody
rise is suggestive of recent infection. Type-specific complement-fixation tests (available through most state health departments) may not become positive until 2 to 3 weeks
after onset. Paired sera should be tested in order to demonstrate a significant (four-fold) antibody rise. Either one
or both serologic tests may remain negative in a small percentage of clinically diagnosed cases. Early antibody treatment may blunt the complement-fixation or Weil–Felix
antibody response.

Microagglutination, indirect hemagglutination, and indirect fluorescent antibody tests have been developed.
Rickettsial mouse-toxin neutralization tests and tests to detect erythrocyte-sensitizing substances have been described
but require specialized laboratories and remain experimental tools.

Lymphocyte blast transformation occurs when lymphocytes from a patient with Rocky Mountain spotted fe-

ver are exposed to killed *R. rickettsii* in vitro. The relative importance of different mechanisms of immunity (cell-mediated and humoral) in recovery from *R. rickettsii* infection is unknown.

Most of the above serologic methods suffer from a delay in diagnosis in that they do not become positive until the patient is well into the illness. The first major advance in early diagnosis has been the application of fluorescent antibody identification of rickettsiae in tissue sections, usually a skin biopsy during the very early phases of the illness. Unfortunately, only about 50 percent of patients with this disease can be so diagnosed. Prior chemotherapy or the absence of a rash may reduce the sensitivity of the procedure. Finally, this test is limited to a relatively few laboratories that are properly equipped.

Treatment. Without treatment, the mortality rate is approximately 20 percent. The first drug shown to be effective in rickettsial diseases was chloramphenicol. The subsequent recognition of rare but serious side effects of this agent plus the recognition that the tetracyclines were quite effective has made tetracycline therapy an attractive alternative.

Because the clinical presentation and seasonal incidence can overlap in cases of meningococcemia and Rocky Mountain spotted fever, chloramphenicol is the drug of choice where these two diseases cannot be clinically differentiated. In addition, intensive care is necessary to combat the problems of marked fluid loss secondary to the widespread vasculitis and associated hypotension. However, with too vigorous replacement of fluids, the not uncommon association of kidney and myocardial involvement can result in congestive failure.

Prevention. Rocky Mountain spotted fever is much easier to treat than to prevent. Traditional measures of vector control are generally not practical. Protective clothing, such as boots, leggings, and tightly buttoned shirts, may be helpful in preventing tick attachment in persons exposed to heavily infested environments.

Individuals who are exposed to ticks and tick-infested environments should inspect their bodies regularly and remove attached ticks carefully. Experimental studies have shown that a poorly understood period of reactivation is required from the time an infected tick attaches and begins feeding until it transmits the pathogenic rickettsiae. Therefore, ticks removed within several hours of attachment may not transmit rickettsiae. For removal of ticks, forceps and gentle traction are recommended. The tick can also be removed by hand, using paper or cloth to protect the fingers. Care should be taken during removal to prevent crushing the tick and contaminating the fingers, because both tick tissues and tick feces are highly infectious.

The most widely used Rocky Mountain spotted fever vaccine is made from rickettsiae grown in eggs, extracted with ether, and killed with formalin. It is not completely effective in preventing spotted fever in humans. There is a poor correlation between the presence of antibodies in the serum and protection in volunteers given the vaccine and then exposed to virulent rickettsiae. A newer vaccine prepared from chick embryo cell cultures is more immunogenic and contains less contaminating protein material. However, this product is not yet commercially available.

In geographic areas where the disease occurs frequently, both the public and physicians should be taught and periodically reminded of the signs, symptoms, and epidemiologic features of the disease.

Rickettsialpox

Rickettsialpox is characterized by a local eschar, a papulovesicular rash, and a benign clinical course. It is caused by *Rickettsia akari*, a member of the spotted fever group that cross-reacts serologically with *R. rickettsii*.

Epidemiology. Rickettsialpox was first recognized in New York City in 1946. Since then cases have been recognized in Boston, Cleveland, and Philadelphia. However, it is possible that the true distribution of rickettsialpox in the United States is wider, as the mite vector of the disease (*Allodermanyssus sanguineus*) is found throughout a large part of the country. Because rickettsialpox is typically a mild disease that is not usually reported to the health departments, exact morbidity figures are not available.

An organism identical to *R. akari* has been isolated in urban areas in the Russian Ukraine, where the epidemiology and clinical features of the disease appear to be similar to rickettsialpox in the United States. *R. akari* has also been isolated from a wild Korean rodent (*Microtus fortis pelliccus*), which implies occurrence of a rural cycle of this disease.

ECOLOGY. *R. akari* infects the mite (*A. sanguineus*), which in turn is an ectoparasite of the common house mouse. The mite infects its progeny transovarially. As in several other rickettsial diseases, humans only enter this cycle of infection accidentally. Human infections are more apt to occur when the rodent population is suddenly reduced by vermin control programs. When murine hosts are scarce, *A. sanguineus* readily attacks humans. Most reported cases of rickettsialpox have been from urban areas where the density of both the murine mites and persons is higher than in rural areas.

Pathogenesis. Little is known concerning the pathogenesis of *R. akari* infection. Microscopically, the maculopapular rash in rickettsialpox shows a mononuclear perivascular infiltrate and necrosis of epithelial cells resulting in intraepidermal vesicles.

Clinical Manifestations. The clinical hallmarks of rickettsialpox are a vesicular rash and a local eschar that is often associated with regional lymphadenopathy. The incubation period is not precisely known, since patients typically are unaware of a mite bite. The first sign of disease is a local erythematous papule that evolves first into a vesicle and then into an eschar. Approximately 3 to 7 days after the appearance of the eschar, chills and fever begin abruptly and may be associated with headache, malaise, and myalgia. Within 72 hours of the appearance of fever, a generalized maculopapular rash becomes apparent and soon evolves into a vesicular eruption. Differentiation of the rash of rickettsialpox from chickenpox is important and is based upon the following observations concerning the rash of rickettsialpox: (1) it occurs more often in adults, (2) it is associated with a primary eschar, and (3) the cutaneous vesicles are surrounded by papular rings. In contrast, the rash of chickenpox occurs most often in children, is entirely vesicular, and lacks a primary lesion. Smallpox does not have a primary eschar, is associated with an eruption that evolves into pustular lesions, and usually is a more severe illness.

No fatalities due to rickettsialpox have been reported. Although a small scar may occur at the site of the primary lesion, the vesiculopapular eruption heals without scarring.

Laboratory Diagnosis. Weil-Felix antibodies do not appear after infection with *R. akari*. However, complement-fixing antibodies can be measured 1 to 2 months after the onset of illness.

R. akari can be isolated from both the blood and the fluid from vesicular lesions of infected persons. Such isolations require the technical facilities of specially equipped laboratories and are accomplished by the inoculation of infected specimens into laboratory animals or embryonated hens' eggs.

Treatment and Prevention. Both tetracycline and chloramphenicol produce rapid defervescence and clinical improvement.

Measures aimed at controlling both rodent populations and their mite ectoparasites will prevent transmission of *R. akari* to humans.

Other Tickborne Diseases

Other species of *Rickettsia* cause tickborne diseases that in many respects resemble Rocky Mountain spotted fever. They are found on several different continents and cause sporadic cases of a mild clinical pattern. *Rickettsia sibirica*, the agent of North Asian tick typhus, *Rickettsia australis*, the agent of Queensland tick typhus, and *Rickettsia conorii*, the agent of boutonneuse fever, are very similar to one another but, by the use of cross-immunity and mouse-toxin neutralization tests, have been shown to be separate organisms. North Asian tick typhus occurs in central Asia, Mongolia, and the Siberian region of the USSR. Queensland tick typhus occurs in Australia, and boutonneuse fever occurs in the Mediterranean region, Africa, and India. Boutonneuse fever has also been called South African tick bite fever, Kenya tick typhus, and Indian tick typhus.

All three of these rickettsiae are maintained in nature in both ixodid ticks and wild animals. Humans only accidentally enter their natural cycle of infection and are not important in the maintenance of the rickettsiae in nature.

Diseases caused by these rickettsiae are characterized by local eschars or skin lesions at the site of tick attachment. All three produce diseases that are typically milder than Rocky Mountain spotted fever, although they have some of the same symptoms—namely, fever, headache, myalgia, malaise, and maculopapular eruptions that may become petechial.

Infection with any of the three rickettsiae is followed by the production of group-specific complement-fixing antibodies and inconsistently by *Proteus* OX-19 or OX-2 agglutinating antibodies. Type-specific complement-fixing antibodies can be measured after infection, although some serologic cross-reactivity with other members of the spotted fever group occurs. All three rickettsiae are sensitive to both chloramphenicol and tetracycline.

Rickettsia Not Associated with Human Disease

Rickettsia parkeri, *Rickettsia montana*, and *Rickettsia rhipicephali* are species serologically related to the spotted fever group but whose disease potential in humans is currently unknown. In 1939 an isolation of a rickettsia was made from Gulf Coast ticks (*Amblyomma maculatum*) removed from cattle in eastern Texas. Isolations of this rickettsia were also made from the same tick species in Mississippi and Georgia. The name "maculatum disease" was given to a syndrome observed in experimentally inoculated guinea pigs, and the causal rickettsia was called the maculatum agent. In 1965, following extensive laboratory characterization of this organism, the name *R. parkeri* was proposed in honor of R.R. Parker, the first director of the Rocky Mountain Laboratory.

The name *R. montana* has been proposed for a rickettsia isolated from ticks in eastern Montana. This organism is nonpathogenic for guinea pigs and, by using serologic techniques, can be distinguished from other members of the spotted fever group. Another unique rickettsial strain related to the spotted fever group has been isolated from the tick *Ixodes pacificus* collected in Oregon.

An additional rickettsia of the spotted fever group has been isolated from brown dog ticks (*Rhipicephalus sanguineus*) in several southeastern states. Although practically nonpathogenic for guinea pigs, it has the ability to reduce immunity to challenge with virulent *R. rickettsii* in laboratory animals. Using serologic methods and mouse-toxin neutralization tests, the organism has been shown to be distinct from other spotted fever group rickettsiae. It appears that this organism is widespread and common. Its role in the epidemiology of Rocky Mountain spotted fever is as yet unknown.

The Typhus Group

These rickettsiae cause epidemic typhus (and its recrudescent infection, Brill–Zinsser disease) and murine typhus. Typhus group organisms are characterized by intracytoplasmic growth and a common, soluble, group-specific, complement-fixing antigen.

Epidemic Typhus (Louseborne Typhus)

Epidemic typhus, a louseborne disease caused by *R. prowazekii* (named after a Polish investigator, von Prowazek, who died of typhus contracted in the course of his studies), has had great impact upon the history of man. According to Zinsser, Napoleon's retreat from Moscow "was started by a louse." In World War I, typhus was responsible for the death of over 150,000 Serbians and over 3,000,000 Russians and caused nonfatal illness in many additional millions.

Epidemiology. Epidemic typhus was once worldwide in occurrence, but epidemics have disappeared in areas with high standards of living. *R. prowazekii* infection occasionally occurs in its recrudescent form (Brill–Zinsser disease). There has not been an outbreak of epidemic typhus in the United States since 1922. At present, the most important foci of epidemic typhus are in Africa. A few minor foci of epidemic typhus also exist in Central and South America.

ECOLOGY AND TRANSMISSION. *R. prowazekii*, the etiologic agent of epidemic typhus, can infect both the human body louse (*Pediculus humanus corporis*) and the head louse (*Pediculus humanus capitis*), the former being the more significant vector. The body louse feeds only on humans, and all three stages of its life cycle (egg, nymph, and adult) can occur on the same host.

Lice become infected after taking a blood meal from a rickettsemic human. Several days later the ingested rickettsiae have multiplied sufficiently in the louse, and infective rickettsiae appear in the arthropod's feces. If the louse encounters a susceptible human at this point, transmission of *R. prowazekii* may occur.

During each blood meal the louse defecates. The feeding process is irritating, and scratching by the host produces minor excoriations that function as portals of entry for the rickettsiae from the louse feces. Lice do not transmit *R. prowazekii* to their progeny but succumb to their infection within 1 to 3 weeks.

Because louse–human–louse transmission thrives under conditions in which individuals wear the same clothes continuously in crowded environments, it is not surprising that major epidemics have occurred in association with war, poverty, and famine. Persons in cold climates are more likely to acquire typhus infections, particularly if they are forced by poverty or unusually hard circumstances to wear the same clothes for long periods of time. Lice actively seek out locations where the temperature is approximately 20C, a temperature often found in the folds of clothing. Lice will abandon a host with a body temperature of 40C or greater, as well as the body of a dead person.

Recent association of sporadic cases of epidemic typhus in the United States associated with the flying squirrel has raised a question as to another reservoir of this rickettsial agent. The exact mechanism of transmission between the rodent and the human is poorly understood.

Clinical Manifestations. The incubation period typically ranges from 10 to 14 days. Prodromal symptoms of headache, malaise, and minimal temperature elevations sometimes occur, but usually the onset is abrupt, with generalized myalgias, chills or chilliness, fever, and headache. Headache is characteristically frontal, severe, and unremitting. Other less specific symptoms are often present, including gastrointestinal complaints, weakness, and cough. Splenomegaly may be present, as may meningismus. The spinal fluid is typically normal.

A skin rash usually occurs from 4 to 7 days after the onset of illness. It may first appear as a patchy cutaneous erythema and progress to maculopapular, petechial, or hemorrhagic forms. In contrast to Rocky Mountain spotted fever, the rash in typhus usually spares the palms, soles, and face and characteristically appears first on the trunk and later spreads to the extremities.

A wide variety of complications may occur in severe cases, including mental changes (stupor and delirium), hypotension, oliguria and azotemia, and even gangrene of the skin, genitalia, and digits.

Untreated, the disease may last up to 3 weeks. Mortality has varied from 10 to 40 percent in different outbreaks. Case fatality ratios have been shown to increase with increasing age. Survivors of epidemic typhus are generally immune for years following their primary infection, although mild recurrences of illness (Brill–Zinsser disease) may occur years later.

Laboratory Diagnosis. Once a clinical diagnosis is made, treatment should be instituted prior to laboratory confirmation. Substantiation of a clinical diagnosis can be obtained either by isolation of *R. prowazekii* or by serologic means. The former is difficult, potentially dangerous, expensive, and involves specialized personnel and equipment.

As in Rocky Mountain spotted fever, patients convalescent from typhus produce antibodies that agglutinate *Proteus vulgaris* OX polysaccharide antigens. These agglutinins usually appear in the second week after onset. Generally, agglutination is maximal with OX-19 strains, although strongly positive reactions with OX-2 antigens sometimes occur. Serial serum specimens should be tested rather than a single convalescent sample. A four-fold rise in agglutinating titer is suggestive of recent infection.

Antibodies against group-specific complement-fixing antigens (prepared from yolk-sac-grown rickettsiae) typically appear in the third week after onset. Microagglutination and fluorescent antibody tests are also available through specialized laboratories.

Treatment. Both chloramphenicol and tetracycline produce prompt defervescence and clinical improvement when given early in the course of the illness. Patients who develop circulatory and renal complications before receiving either antibiotic may die despite therapy.

Prevention and Control. It is possible to interrupt epidemic louse–human–louse transmission of *R. prowazekii* by mass application of insecticides to human beings and their clothing. Some populations of lice, especially in Africa, have become increasingly resistant to insecticides (including DDT and malathion). Once free of lice, patients with typhus are not infectious. Typhus vaccine prepared from infected yolk sacs is also an effective control measure. Although controlled studies in humans are lacking, it is generally accepted that typhus vaccine lessens the severity and shortens the course of clinical disease. Two doses of vaccine 4 weeks apart are necessary for primary immunization. Booster doses are recommended every 6 to 12 months during periods of exposure.

Brill–Zinsser Disease

Individuals who previously have had epidemic typhus may develop recrudescent infection many years later. This illness was named after Nathan Brill, who first recognized and described the clinical features, and Hans Zinsser, who first suggested in 1934 that the disease was a relapse of a prior epidemic typhus infection. Epidemiologic, clinical, and experimental evidence has since been published confirming Zinsser's hypothesis.

Epidemiology. In the United States, recrudescent typhus occurs primarily in immigrants from previously endemic areas, such as eastern Europe. The disease may occur in an individual living in a louse-free environment, and many years may have passed since the patient's initial infection with *R. prowazekii*. However, lice that feed on a patient with recrudescent typhus can become infected, and if local conditions are favorable for louse–human–louse transmission, an outbreak of epidemic typhus may result. Latent human infection thus represents an interepidemic reservoir for *R. prowazekii*.

Clinical Manifestations. Brill–Zinsser disease is a milder illness than classic epidemic typhus; skin rash is rarely seen, and the duration of disease is shorter (less than 2 weeks). Fever may be erratic instead of sustained. As in epidemic typhus and other rickettsial diseases, headache, malaise, and myalgias are common symptoms. Complications and fatalities are rare.

Laboratory Diagnosis. In the United States, Brill–Zinsser disease should be suspected when fever of obscure origin occurs in a foreign-born person from an area where epidemic typhus has occurred and who complains of an intense headache and develops a maculopapular skin rash on the fourth to sixth day of his or her illness.

Weil–Felix agglutinins often do not develop in patients with Brill–Zinsser disease. As a general rule, the sooner recrudescent infection occurs after primary infection, the less likely are Weil–Felix antibodies to be present.

Complement-fixing antibodies are found in the sec-
ond week after onset, which is earlier than in patients with epidemic typhus. Since some typhus patients may have detectable complement-fixing antibodies many years after primary infection, an isolated convalescent serum sample may yield confusing results. Therefore, a four-fold complement-fixing antibody rise should be sought in patients suspected of having recrudescent typhus.

Consistent with Zinsser's hypothesis, it has been found that patients with epidemic typhus initially have an IgM followed by an IgG antibody response, whereas patients with Brill–Zinsser disease initially have an anamnestic IgG antibody response. The microimmunofluorescence test is the most sensitive and reliable method to differentiate between Brill–Zinsser disease and primary epidemic typhus.

The clinical history, epidemiologic setting, and dynamics of the antibody response are helpful in distinguishing among epidemic, recrudescent, and murine typhus.

Treatment and Prevention. As in epidemic typhus, tetracycline and chloramphenicol are both effective in treatment. The ultimate prevention of Brill–Zinsser disease necessarily is dependent upon the prevention of epidemic typhus. If recrudescent typhus occurs in an environment where lice rarely infest humans, no special precautionary public health measures are required. In areas where the potential for louse–human–louse transmission is high, delousing of the patient and his or her contacts may be necessary to prevent an outbreak of epidemic typhus.

Murine Typhus (Endemic Typhus, Fleaborne Typhus, Rat Typhus)

Murine typhus is a fleaborne illness caused by *R. typhi*, a member of the typhus group. Typically, murine typhus is a mild illness characterized by fever, headache, and often by a generalized skin rash.

Epidemiology

PREVALENCE. Murine typhus is endemic in many countries, including the United States, where it occurs primarily in the Southeast and Gulf Coast region. It is also endemic in parts of Central America and Mexico. Investigations of murine typhus in Alabama and Florida completed over 45 years ago revealed that most cases occurred in individuals who worked in rat-infested shipyards and harbors. More recently, cases have been reported from inland rural locations, presumably because infected rats and mice may occur in large numbers in areas where grains and feeds are stored. In the past decade, over half of all reported cases in the United States have occurred in Texas.

Although murine typhus is a reportable disease in the United States, considerable numbers of cases may be neither diagnosed nor reported. Despite this problem, it appears that the incidence of murine typhus has gradually decreased in the past two decades (Fig. 51-4).

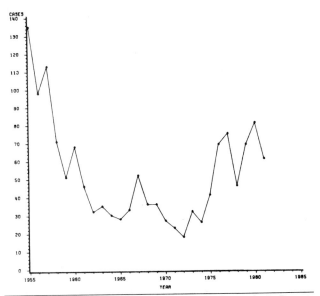

Figure 51-4. Typhus fever, fleaborne (endemic, murine). Reported cases, by year, United States, 1955–1981. For 1981, a total of 61 cases was reported from eight states, with Texas reporting 50 cases. *(From Morbidity and Mortality, CDC Annual Summary 30(54):97, 1982.)*

ECOLOGY. *R. typhi* is cycled in nature by the rat and two of its ectoparasites, the rat flea (*Xenopsylla cheopis*) and the rat louse (*Polyplax spinulosus*). The former is the more important vector. As in Rocky Mountain spotted fever, humans enter this arthropod–vertebrate–arthropod cycle only accidentally. *X. cheopis* acquires *R. typhi* infection by feeding upon a rickettsemic mouse or rat. Once infected, the flea may infect other susceptible rodents and thus a natural cycle of flea–rodent–flea infection may become established. Rodents infected with *R. typhi* do not succumb to their infection, despite the fact that viable rickettsiae can be demonstrated in rodent brains for periods up to several months. Fleas do not transmit *R. typhi* transovarially.

TRANSMISSION. Transmission of *R. typhi* to humans occasionally occurs. When infected fleas taking a blood meal defecate on the host, the host rubs the infected feces into small excoriations during scratching. Flea feces are also infective if accidently transmitted to mucosal surfaces, such as the conjunctiva.

Clinical Manifestations. Murine typhus is usually a mild illness with a mortality rate of less than 2 percent. The incubation period ranges from 1 to 2 weeks. The hallmarks of the disease are abrupt onset of fever, headache, malaise, and myalgias and, in most cases, a macular to maculopapular, nonpruritic skin rash that begins on the third to fifth day on the trunk and spreads to the extremities. As in epidemic typhus, involvement of the palms, soles, and face is rare. The rash may be fleeting or absent in some cases and may be unapparent in blacks without careful inspection. Chills or chilliness, cough, nausea, vom-

iting, arthralgias, weakness, and extreme prostration may be associated symptoms. Untreated, the illness may last up to 2 weeks. Defervescence may occur by either abrupt crisis or gradual lysis.

Fatalities are more likely to occur in the old and infirm. Such fatality may be heralded by peripheral vascular collapse and evidence of central nervous system involvement, such as stupor and coma.

Confusion between murine typhus and Rocky Mountain spotted fever may occur, since both diseases are associated with rising titers against *Proteus* OX-19 (and sometimes OX-2) antigens. Spotted fever is usually a more severe illness and is often associated with an antecedent tick bite. The rash of murine typhus begins first on the trunk and spreads to the extremities, while the opposite evolution occurs in Rocky Mountain spotted fever. Since precise information on the evolution of the rash is often unavailable, the differential point is often not helpful. In older, immigrant patients, confusion between murine typhus and Brill–Zinsser disease may also occur (p. 774).

Laboratory Diagnosis. Agglutinins to *Proteus* OX-19 and, less commonly, to OX-2 appear in the second week of infection. Complement-fixing antibodies against *R. typhi* appear slightly later. With the complement-fixation test, serologic cross-reactions among members of the typhus group are common, but in patients with murine typhus much higher antibody titers are obtained against *R. typhi* than against *R. prowazekii*. The use of specific antigens in the complement-fixation test also permits differentiation of *R. typhi* and *R. prowazekii*. An indirect fluorescent antibody test using IgM and IgG fluorescent-labeled conjugates has been developed.

Intraperitoneal inoculation of blood from patients with endemic typhus into a male guinea pig produces severe testicular lesions and scrotal swelling, in contrast to the very mild disease produced by the inoculation of blood from epidemic typhus patients.

Treatment and Prevention. Both tetracycline and chloramphenicol are effective rickettsiostatic agents. Patients with laboratory-acquired murine typhus treated with chloramphenicol 2 to 4 days after onset have been known to experience clinical relapses despite the presence of antirickettsial antibodies. These relapses responded to reinstitution of the same antimicrobials. Insecticides and rodenticides are both effective in reducing rat–flea–human transmission in endemic areas.

Rickettsia canada Infections

Rickettsia canada was first isolated from rabbit ticks collected in Ontario, Canada, in 1967. Antigenically, *R. canada* belongs to the typhus group biotype. However, it grows in both the cytoplasm and the nuclei of infected cells, a characteristic of the spotted fever group.

The clinical spectrum of human infection with *R. canada* is still largely unknown. The organism has not yet been isolated from human sources, although serologic evidence of *R. canada* infection has been reported in four patients with symptoms typical of Rocky Mountain spotted fever. In one instance a patient with a clinical history consistent with spotted fever had a negative complement-fixation test against spotted fever group antigens but an antibody titer of 1:256 against *R. canada* antigen.

In complement-fixation tests, there are strong cross-reactions among *R. canada, R. prowazekii,* and *R. typhi* antigens. In experimental infections in guinea pigs, *R. canada* produces fever but no scrotal reaction.

Scrub Typhus (Chiggerborne Typhus), Tsutsugamushi Disease (Miteborne Typhus), Japanese River Fever, Rural Fever, Tropical Typhus

R. tsutsugamushi (also called *Rickettsia orientalis*) is actually a group of rickettsiae that produce one clinical illness in humans despite the fact that different strains possess markedly different surface antigens. The name *tsutsugamushi* is derived from two Japanese words: *tsutsuga* (something small and dangerous) and *mushi* (creature). The appelation "scrub typhus" is derived from the fact that infection commonly occurs in endemic areas after exposure to terrain with secondary (scrub) vegetation. In the past several decades, it has been found that *R. tsutsugamushi* infection also occurs in a variety of habitats that cannot be described as scrub (including sandy beaches, mountain deserts, and equatorial rain forests). Therefore, it has been suggested that scrub typhus is a misnomer and that the term "chiggerborne typhus" is more descriptive.

Three major antigenic types of *R. tsutsugamushi* (Karp, Gilliam, and Kato) have been recognized.

Epidemiology

PREVALENCE. Chiggerborne typhus is endemic in a triangular geographic area of over 5,000,000 square miles, including Australia, Japan, Korea, India, and Vietnam. In World War II, it caused appreciable morbidity and mortality in both Japanese and American soldiers. In peacetime it occurs predominantly as a sporadic endemic illness. Epidemics may occur, however, when groups of people are brought into endemic mite-infested areas. Chiggerborne typhus occurred sporadically in American troops in Vietnam.

ECOLOGY. The vectors of chiggerborne typhus are several species of trombiculid mites. These mites have four-stage life cycles (egg, larva, nymph, and adult). The larva (chigger) is the only stage that feeds on vertebrates. After engorgement on a vertebrate host, chiggers detach and metamorphose into eight-legged nymphs and subsequently into adults. The latter two stages are free-living in the soil. Transmission of *R. tsutsugamushi* occurs both transstadially (from larva to nymph to adult) and transovarially (from adult to egg). Thus, as do several ixodid ticks in the ecology of Rocky Mountain spotted fever, trombiculid mites function as both vector and reservoir of chiggerborne rickettsiosis.

A natural cycle of *R. tsutsugamushi* transmission occurs between chiggers and small mammals (e.g., field mice and rats) in endemic areas. Ground-feeding birds may also be important in the ecology of chiggerborne rickettsioses, especially as transporters of the disease agent over long distances. In chiggerborne typhus, as in rickettsialpox, Rocky Mountain spotted fever, and murine typhus, humans only accidentally enter a natural cycle of rickettsial infection. Areas such as savannas, forest clearings, riverbanks, grassy fields, and gardens may provide the proper conditions that allow infected mites to thrive and thus produce small focal geographic areas of high risk to humans. Such mite-infested areas, which may be as small as a few meters in diameter, have been called "scrub typhus islands." The term "zoonotic tetrad" has been suggested to describe the coexistence and intimate relationship among *R. tsutsugamushi*, chiggers, rats, and secondary or transitional forms of vegetation. These four factors are essential components for the establishment of a microfocus of infection.

The seasonal incidence of the disease varies with the climate in different countries, occurring more frequently during the rainy season in the summer or fall.

Person-to-person transmission of infection has not been reported.

Clinical Manifestations. Approximately 1 to 3 weeks after being bitten by an infected chigger, humans abruptly develop chills, fever, and headache. Any or all of the following additional symptoms may also occur: cough, nausea, vomiting, myalgia, abdominal pain, and sore throat. The skin rash of scrub typhus is classically heralded by a local cutaneous lesion, which evolves from a small indurated or vesicular lesion into an ulcerated lesion present at the time of onset of symptoms, when it is covered by a black scab (eschar). Lymphadenopathy may be prominent in the area proximal to the eschar. Five to eight days after the onset of fever, a macular or maculopapular eruption may appear on the trunk and later become generalized.

Although it is said that eschars are commonly found in Caucasians and less commonly in Asians, only 46 percent of American servicemen in Vietnam with scrub typhus had an identifiable eschar, and no more than 34 percent had a skin rash. Other signs sometimes occurring with scrub typhus include splenomegaly, conjunctivitis, and pharyngitis. During the first week of illness, the pulse may be slow in relation to the height of fever. In severe cases, deterioration in mental status (stupor, delirium), pneumonia, and/or circulatory failure may occur. Some patients may experience a second attack of scrub typhus, since infection with one strain of *R. tsutsugamushi* does not confer protection against another strain.

Fatality ratios in epidemics have varied from 0 to 50 percent. With prompt recognition and appropriate treatment, fatality rates are almost nil.

Laboratory Diagnosis. Complement-fixation tests are of less value in the diagnosis of scrub typhus than in the diagnosis of infections due to rickettsiae of the spotted fever and typhus groups, since different strains have different surface antigens that do not cross-react with each other. Thus, if a complement-fixation test is to be used, a minimum of three different antigens (representing the Karp, Gilliam, and Kato strains) must be employed. Agglutinins to *Proteus* OX-K antigens (but not OX-2 and OX-19) appear in the convalescent sera of many but not all patients with scrub typhus. These cross-reacting antibodies usually are detectable in the second week of illness, peak in the third to fourth weeks, and then disappear by about the fifth week after onset. As in other rickettsial serologic tests, paired sera should be collected in order to demonstrate a significant (four-fold) titer rise. A fluorescent antibody test employing pooled conjugates made from the three major strains of *R. tsutsugamushi* has been a useful technique in some investigators' hands.

R. *tsutsugamushi* can be isolated by inoculating blood from patients into white mice.

Treatment. As in other rickettsial diseases, both tetracycline and chloramphenicol are effective rickettsiostatic agents that usually produce prompt defervescence and clinical improvement. Relapses may occur when antibiotics are discontinued too quickly in patients treated within the first few days after onset. In patients first treated in their second week of illness, tetracycline or chloramphenicol can be discontinued 1 to 2 days after defervescence occurs. Fluorescent antibody levels are lower in those patients given early antibiotic treatment.

Prevention. Because of strain variations, an effective scrub typhus vaccine has not been prepared. In endemic or hyperendemic areas, measures to prevent chigger bites (protective clothing, insect repellents) and to control mite populations (insecticides, clearing of vegetation, and chemical treatment of the soil) may be used to prevent chigger–human–chigger transmission.

Trench Fever (Shinbone Fever, Five-Day Fever, Quintana Fever)

Trench fever is caused by *Rochalimaea quintana* and is transmitted by the body louse (*Pediculus humanus corporis*) in a human–louse–human cycle of infection. *R. quintana* is the only rickettsia that can be grown on cell-free media.

Trench fever was first recognized in World War I as a 5- to 6-day febrile illness, associated with pains in the shins, that caused large epidemics in both German and Allied armies. By the end of the war a wider spectrum of clinical illness and the louseborne transmission of infection had been recognized. The disease disappeared during the next two decades only to reappear in epidemic form in the armies on the Eastern Front during World War II. It has been rarely reported since 1945.

Epidemiology. Trench fever has occurred in England, France, Yugoslavia, Italy, Russia, Germany, and several other countries in eastern Europe. *R. quintana* has been isolated from lice in Mexico, an area where clinical trench fever has not yet been recognized.

ECOLOGY. The body louse acquires infection by feeding on a rickettsemic human. Once infected, the louse excretes *R. quintana* in its feces for the remainder of its life, which is not shortened by the rickettsial infection. Unlike all other members of the tribe Rickettsieae, *R. quintana* proliferates in an extracellular environment in the arthropod host, in the lumen of the gut rather than within the intestinal epithelial cells. Transovarial transmission of *R. quintana* to the louse's offspring does not occur.

No animal reservoir other than humans has been identified. Thus, like epidemic typhus, a louse–human–louse cycle of infection occurs. Little is known about the prevalence and distribution of *R. quintana* in nonepidemic intervals. *R. quintana* has been isolated from apparently healthy patients years after their original attack.

Clinical Manifestations. The incubation period of trench fever in volunteers given intradermal inoculations of *R. quintana* ranges from 8 to 18 days. The clinical manifestations of the disease are highly variable and range from a mild afebrile disease to a moderately severe febrile disease with multiple relapses. The onset may be gradual or abrupt. Symptoms of acute disease include headache, malaise, fever, chilliness, myalgias, and bone pain (especially in the tibial region). Fever curves vary widely among different patients. A macular rash resembling the rose spots of typhoid fever may occur.

Laboratory Diagnosis. Laboratory animals, such as guinea pigs, rabbits, or mice, are not suitable for the isolation of *R. quintana*. Instead, either xenodiagnosis (the feeding of uninfected lice on an infected patient and the later demonstration of rickettsia in the louse tissue) or primary isolation on enriched blood-agar media are used for diagnosis. Like other members of the tribe Rickettsieae, the organism can also be grown in yolk sacs of embryonated hens' eggs. Complement-fixation and indirect immunofluorescent antibody tests have been developed and appear to be promising clinical and epidemiologic tools. A passive hemagglutination antibody test has also been described.

Treatment. Data on the efficacy of tetracycline and chloramphenicol in the treatment of trench fever are not available. Based, however, on in vitro sensitivity testing and the universal susceptibility of other Rickettsieae to these agents, both drugs are likely to be effective.

Prevention and Control. Measures to control human louse infestation will control the transmission of trench fever.

Q Fever

The name Q fever (the Q is the initial of "query") was first used in 1937 to describe an unusual febrile illness in Australian packinghouse workers. Several years later, the causal agent was isolated from several of the infected workers and was identified as a rickettsia. At about the same time, the same organism was identified in wood ticks collected in Montana. Cox, an American, and Burnet, an Australian, were honored for their early contributions to the study of this organism by the selection of the name *Coxiella burnetii* for the etiologic agent of Q fever. Along with trench fever and epidemic typhus, Q fever caused epidemics in the armies fighting in Europe in World War II. In the past three decades, *C burnetii* has been shown to have a worldwide distribution and a complex ecology and epidemiology.

Epidemiology

PREVALENCE. Exact morbidity figures on the incidence of Q fever are not available, although surveys have shown that many people throughout the world have serologic evidence of past infection with *C. burnetii*. In such surveys, the incidence of positive serologic tests far exceeds the incidence of clinical Q fever. The disease has been recognized in over 50 countries on 5 continents. In the United States, outbreaks of Q fever, all associated with livestock or livestock products, have occurred in California, Texas, and Illinois.

ECOLOGY. There are two cycles of infection with *C. burnetii* in nature. One involves arthropods (especially ticks) and a variety of vertebrates. The other is maintained among domestic animals. The significance of arthropod–vertebrate transmission is conjectural. Although humans are not directly infected by ticks, these arthropods may transmit infection to domestic animals, especially sheep and cattle. Domestic animals have unapparent infections but may shed large quantities of infectious organisms in their urine, milk, feces, and especially their placental products. Because *C. burnetii* is unique among Rickettsieae in its resistance to desiccation and exposure to light or temperature extremes, infectious organisms in placental products of domestic animals may become aerosolized after parturition and cause widespread outbreaks in humans and other animals over a distance of several miles from their place of origin. Dust in sheep or cattle sheds may become heavily contaminated and function as a source of infection for susceptible humans and animals. Once established, animal-to-animal spread is maintained primarily through airborne transmission.

Outbreaks of Q fever in humans have been traced to consumption of infected milk, to handling contaminated wool or hides, to soil contaminated by infected animal feces, to infected straw, and even to dusty clothing. Q fever may be an occupational risk in abattoir workers, sheep shearers, dairy and other farm workers, workers in tanneries, wool, and felt plants, and technicians handling the organism in the laboratory, especially those using sheep and goats for experimental purposes. *C. burnetii* may enter the body through the skin (e.g., a contaminated minor abrasion), through the lungs (e.g., inhalation of infectious aerosols), through mucous membranes (e.g., conjunctival contact with infectious materials), or through the gastrointestinal tract (e.g., ingestion of contaminated raw milk). Human-to-human transmission of Q fever has been reported but is probably rare. Asymptomatic recrudescence of infection may occur during pregnancy. There is immunologic evidence that placental transfer of *C. burnetii* may result in human fetal infection.

Clinical Manifestations. *C. burnetii* is capable of causing unapparent infection, an influenza-like illness, pneumonia, prolonged fever, endocarditis, and hepatitis. Many mild or subclinical cases probably occur but are not diagnosed. Like most other rickettsial diseases, clinically recognized Q fever usually begins abruptly with fever, chills or chilliness, headache, malaise, and myalgia. However, unlike most other rickettsial diseases, skin rash is not a part of the clinical syndrome, although evanescent macular rashes have been reported in a few cases. Patients may also complain of nonspecific gastrointestinal symptoms, sore throat, chest pain, nonproductive cough, and painful eyes.

Physical findings include hepatosplenomegaly, auscultatory evidence of pneumonic infiltration or consolidation, and a relative bradycardia. Chest x-rays may show unilateral or bilateral lower lobe infiltrates similar to infiltrates seen in viral and mycoplasma pneumonia and in psittacosis. Liver function tests are often minimally abnormal, but granulomatous hepatic lesions have been documented, with only mild liver function test abnormalities.

Q fever endocarditis may occur months or years after the acute attack and should be suspected when routine blood cultures are negative and unresponsiveness to antimicrobial therapy occurs in a clinical setting strongly suggestive of subacute bacterial endocarditis. Q fever endocarditis usually occurs on a previously damaged heart valve, often the aortic valve, and until recently was almost invariably fatal.

Laboratory Diagnosis. The most definitive diagnostic procedure, isolation of *C. burnetii* from clinical specimens, can be accomplished by intraperitoneal inoculation of guinea pigs or mice or embryonated hens' eggs. However, unless there are available experienced personnel working in specialized facilities with stringent safeguards to prevent infection of laboratory personnel and other people in the vicinity of the laboratory, such primary isolations should not be attempted.

Complement-fixation tests utilizing purified antigens are available through some state health departments. Since *C. burnetii* exists in two phases, complement-fixing reagents prepared using phase I and phase II antigens can be useful in distinguishing acute from chronic (e.g., endocarditis) or past infection. A microagglutination test has also been developed and appears to be a useful epidemiologic and clinical tool. Both the complement-fixation and micro-agglutination tests are sensitive and specific. Weil–Felix antibodies do not appear in response to infection with *C. burnetii*.

Treatment. Many patients with mild or subclinical illnesses recover without antimicrobial therapy. However, all clinically diagnosed cases should be treated. Even though tetracycline and chloramphenicol are both active against *C. burnetii*, patients with Q fever treated with these drugs do not respond as uniformly and as quickly as do patients with other rickettsial diseases. Tetracycline, which is less toxic and as effective as chloramphenicol, is the preferred drug for treatment. Therapy should be continued for 5 to 7 days after defervescence.

Successful treatment of Q fever endocarditis has been reported after prolonged (10 months) therapy with tetracycline and with combined treatment using tetracycline and trimethoprim-sulfamethoxazole. Surgical replacement of the infected valve may be necessary.

Prevention and Control. A completely satisfactory vaccine has not yet been developed. Both live and killed Q fever vaccines elicit the production of complement-fixing antibodies and have been shown effective in guinea pigs. Their use in humans has been hampered by the high rate of reactions (fever, pain, and swelling at the site of injection).

Measures to identify and decontaminate infected areas and to vaccinate domestic animal populations are difficult, expensive, and not generally done. Milkborne transmission, however, can be prevented by pasteurization.

FURTHER READING

Books and Reviews

Brettman LR, Lewin S, Holzman RS, et al.: Rickettsialpox: Report of an outbreak and a contemporary review. Medicine 60:363, 1981

Burgdorfer W, Anacker RL: Rickettsia and Rickettsial Diseases. New York, Academic Press, 1981

Kishimoto RA, Stockman RN, Redmond CL: Q fever: Diagnosis, therapy and immunoprophylaxis. Milit Med 144:183, 1979

Marchette J: Ecologic Relationships and Evolution of the Rickettsiae. Cleveland, CRC Press Inc., 1982, vols I, II

Riley HD Jr: Rickettsial diseases and Rocky Mountain spotted fever. Part I and II. Curr Prob Pediatr 11: No. 5, p 1, No. 6, p 1, 1981

Walker DH: Rickettsial diseases: An update. Current Topics in Inflammation and Infection. Baltimore, William & Wilkins, 1982, chap 12, p 188

Zdrodovski P, Golinevich R: The Rickettsial Diseases. New York, Pergamon Press, 1960

Zinsser H: Rats, Lice and History. Boston, Little, Brown, 1935

Selected Papers

Anacker RL, McCaul TI, Burgdorfer W, et al.: Properties of selected rickettsiae of the spotted fever group. Infect Immun 27:468, 1980

Bozeman FM, Elisberg BL, Humphries JW, et al.: Serologic evidence of *Rickettsia canada* infection of man. J Infect Dis 121:367, 1970

Duma RJ, et al.: Epidemic typhus in the United States associated with flying squirrels. JAMA 245:2318, 1981

Hackstadt T, Williams JC: Biochemical stratagem for obligate parasitism of eukaryotic cells by *Coxiella burnetii*. Proc Nat Acad Sci USA 78:3240, 1981

Harell GT: Rocky Mountain spotted fever. Medicine 28:333, 1949

Hart RJ: The epidemiology of Q fever. Postgrad Med J 49:535, 1973

Hechemy KE, et al.: Discrepancies in Weil–Felix and micro-immunofluorescence test results for Rocky Mountain spotted fever. J Clin Microbiol 9:292, 1979

Kimbrough RC, Ormshee RA, Peacock M, et al.: Q fever endocarditis in the United States. Ann Intern Med 91:400, 1979

Maxcy KF: Clinical observations on endemic typhus (Brill's disease) in the United States. Public Health Rep 41:1213, 1926

McCaul TF, Williams JC: Developmental cycle of *Coxiella burnetii*: Structure and morphogenesis of vegetative and sporogenic differentiation. J Bacteriol 147:1063, 1981

Meiklejohn G, Reimer LG, Graves PS, et al.: Cryptic epidemic of Q fever in a medical school. J Infect Dis 144:107, 1981

Murray ES, Baehr G, Shwartzman G, et al.: Brill's disease. I. Clinical and laboratory diagnosis. JAMA 142:1059, 1950

Murray ES, Gaon JA, O'Connor JM, et al.: Serologic studies of primary endemic typhus and recrudescent typhus. J Immunol 94:723, 1965

Newhouse VF, Shepard CC, Redus MD, et al.: A comparison of the complement-fixation, indirect fluorescent antibody and microagglutination test for the serologic diagnosis of rickettsial diseases. Am J Trop Med Hyg 28:387, 1979

Parker RR: Rocky Mountain spotted fever. JAMA 110:1185, 1273, 1938

Silverman DJ, Wisseman CL: In vitro studies of Rickettsia–host cell interactions: Ultrastructural changes induced by *Rickettsia rickettsii* infection of chicken embryo fibroblasts. Infect Immun 26:714, 1979

Silverman DJ, Wisseman CL, Waddell AD, et al.: External layers of *Rickettsia prowazekii* and *Rickettsia rickettsii*: occurrence of a slime layer. Infect Immun 22:233, 1978

Stuart BM, Pullen RM: Endemic murine typhus fever: Clinical observations. Ann Intern Med 23:520, 1945

Traub R, Wisseman CL Jr: The ecology of chiggerborne rickettsiosis (scrub typhus). J Med Entomol 11:237, 1974

Walker DH, Cain BG: A method for specific diagnosis of Rocky Mountain spotted fever on fixed, paraffin-embedded tissue by immunofluorescence. J Infect Dis 137:206, 1978

Woodward TE: A historical account of rickettsial diseases with a discussion of unsolved problems. J Infect Dis 127:583, 1973

Yamada T, Harber P, Pettit GW, et al.: Activation of the kallikrein-kinin system in Rocky Mountain spotted fever. Ann Intern Med 88:764, 1978

CHAPTER 52

Bartonella

Bartonellosis

Oroya Fever and Verruga Peruana

If there is truth in the notion that fact is stranger than fiction, it is to be found in bartonellosis or Carrion's disease. This exclusively human disease and its causative agent *Bartonella bacilliformis* are unique from nearly every point of view. The organism is an arthropodborne bacterium that causes two extraordinary and entirely different human diseases: Oroya fever, a rapidly progressive, febrile, highly fatal, hemolytic anemia, and verruga peruana, a skin disease characterized by the eruption of bright red, angiomatous, wart-like lesions. Transmission is sharply confined geographically to certain areas on the western aspects of the Andes in Peru, Ecuador, and Colombia.

Historical Perspective. Verruga peruana probably was present in pre-Columbian Peru in essentially the same areas in which it may be found today. Oroya fever, however, was not clearly described until the mid-1800s. The verrucous form of the disease developed in some of the survivors of the hemolytic disease. This gave rise to the suspicion that the two conditions might be etiologically related. It was soon after this, on August 27, 1885, that Daniel Carrion, a Peruvian medical student, inoculated himself with verrucous material in a quest for a better clinical definition of the earliest signs and symptoms of the skin disease. Just before his death from Oroya fever, 39 days later, he clearly stated his conviction, ". . . that Oroya fever and the verruga have the same origin. . . ." While he soon became a national hero, his conclusion on the unity of causation of the two diseases was debated for many years.

In 1905, Alberto Barton described the organism in erythrocytes in cases of Oroya fever. An expedition from Harvard in 1913, led by Richard Pearson Strong, confirmed and extended Barton's observations but failed to find the organism in histopathologic sections of verrugas. Strong concluded that the two diseases were unrelated and honored Barton by naming the organism *Bartonella bacilliformis*. In 1926, Hideyo Noguchi, working in New York, isolated the organisms from specimens sent from Peru. Of the major works for which this flamboyant microbiologist received worldwide adulation in the first decades of this century, only his work on bartonellosis, proving Carrion to be correct, has stood the test of time. He isolated identical organisms from blood specimens from Oroya fever patients and from verrugas excised from patients with the eruptive form of the disease. With organisms cultured from either source, he was able consistently to produce verrugas in monkeys and to reisolate the organism in pure culture from the monkey lesions. His work has been confirmed repeatedly.

In 1912, convincing epidemiologic evidence began to accumulate indicating that the organism is transmitted by certain species of flies of the genus *Phlebotomus*. In the 1940s, the findings that the organism is susceptible to penicillin and that DDT is highly effective against *Phlebotomus* flies provided tools that are as effective today as when they were first used.

Other erythrocyte-associated organisms in the families Bartonellaceae and Anaplasmataceae are found in animals. There are scattered reports of human disease associated with as yet unidentified hemotropic organisms that exhibit similarities to various members of this diverse group.

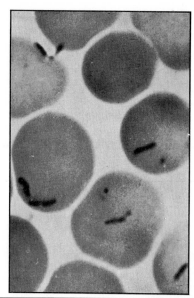

Figure 52-1. *Bartonella bacilliformis.* Human blood stained with Giemsa stain. × 3000. *(From Wigand et al.: Z Tropenmed Parasitol 4:539, 1953.)*

Morphology and Physiology

B. bacilliformis is the single species of the genus *Bartonella* and the only organism of medical significance in the family Bartonellaceae. Like other members of the family, however, it parasitizes red blood cells, possesses a cell wall, and has been cultured in vitro.

Morphology and Staining. Bartonella are small, exceedingly polymorphic, motile, gram-negative bacteria. They range in shape from small coccoid and ring-shaped structures to long angular forms in chains and clusters. In erythrocytes they usually appear as short rods ranging from 1 to 3 μm in length by 0.25 to 0.5 μm in width. (Fig. 52-1). The organisms stain weakly with aniline dyes but appear bright red to purple with Wright's or Giemsa's stain. They are not acid-alcohol fast. The cultured organisms possess up to 10 terminal flagella (Fig. 52-2), but flagella have not been seen in fresh preparations of clinical specimens from humans.

Under natural conditions *B. bacilliformis* is found in or on erythrocytes and in the cytoplasm of reticuloendothelial and vascular endothelial cells. In phlebotomus flies they are found in the lumen of the digestive tract.

Cultural Characteristics. Growth and maintenance in serial passage in the laboratory may be achieved in cell-free medium containing agar and fresh serum and hemoglobin from a number of species, including rabbits, horses, and humans. Other culture systems include yolk and chorioallantoic fluids in the embryonated hen's egg and a variety of tissue culture systems. In the latter, the organisms grow in the cytoplasm and extracellularly. In semisolid agar, colonies are 1 to 5 mm puffs of white that appear 1 to 2 weeks after inoculation. Temperatures around 30C favor growth and longevity. Cultures remain viable for long periods when stored at −70C. Nutritional requirements for growth are satisfied in semisolid nutrient agar containing 10 percent rabbit serum and 0.5 percent rabbit hemo-

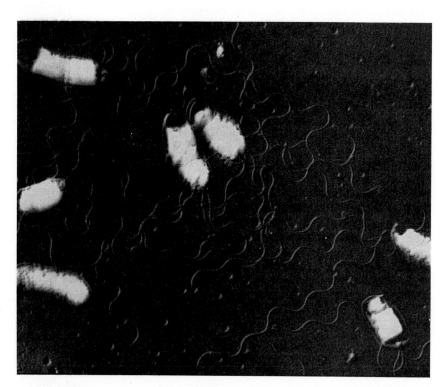

Figure 52-2. Electron micrograph of 7-day culture of *B. bacilliformis.* Note cell wall and terminal flagella apparently originating from protoplasts. × 15,000. *(From Peters and Wigand: Z Tropenmed Parasitol 3:313, 1952.)*

globin. A number of specific nutrients, such as glutathione and ascorbic acid, have been identified.

No hemolysin has been demonstrated in vitro; neither acid nor gas is produced in media containing a wide variety of sugars. The organisms are obligate aerobes. L forms have been shown to develop in hyperosmolar medium containing penicillin.

All strains appear to be similar in respect to morphology, growth characteristics, and antigenic reactivity. Organisms are agglutinated and complement is fixed by immune serum derived either from naturally infected humans or from inoculated laboratory animals.

Clinical Infection

Epidemiology. Outbreaks of Oroya fever have been associated with the intrusion of nonimmune persons into sharply demarcated areas on the western slopes of the Andes, where insect transmission occurs. In 1871, such an episode provided the first reported cases of the hemolytic disease, when expatriate laborers were building a railroad from Lima into the mountains to the city of Oroya. Hundreds of cases occurred, and the case fatality reached 40 percent. The eruptive form of the disease subsequently appeared in some of the survivors and also in individuals who were not observed to have had the fever. Subsequent epidemics have been similar. Foci of endemic activity are notable for their stability over many years.

Transmission of bartonellosis is restricted by the habits and ecology of the *Phlebotomus* fly vector. The endemic area extends from 2° north of the equator to 13° south lat-

itude, a distance of approximately 100 miles. It is further confined to a rather narrow band between 2500 and 8000 feet above sea level, generally less than 100 miles in width, on the western slopes of the Andes in Peru, Colombia, and Ecuador. Major ecologic limitations include conditions of temperature and humidity that are inimical to the fly. Transmission is at night when female flies take their blood meal. Even before insect transmission was seriously considered, disease was prevented in susceptible railroad workers by removing them from known endemic areas before nightfall.

Humans are the only known vertebrate reservoir, in spite of intensive search in many species of animals and plants. In some endemic areas, bacteremia may be detected in about 5 percent of apparently well individuals, and the duration of the bacteremia has been shown to be as long as a year, thus providing strong evidence for the suitability of humans as the major reservoir.

There are many unsolved problems in the epidemiology of bartonellosis. Experimental transmission probably has been achieved from flies caught in endemic areas, transported to nonendemic areas, and fed on laboratory monkeys. More conclusive and complete transmission experiments have not been successful because of the difficulty of colonizing *Phlebotomus* in the laboratory and the incomplete expression of bartonella infection in laboratory animals.

Pathogenesis. In patients with Oroya fever, the organism is found in large numbers of erythrocytes. The infected red blood cells are destroyed by an unknown

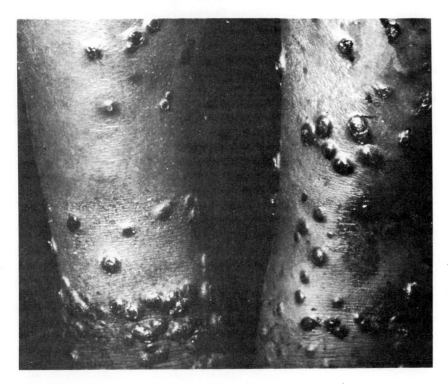

Figure 52-3. One form of skin lesions seen in verruga peruana, a clinical manifestation of bartonellosis. Compare with the smaller, more generalized lesions in Figure 52-4. *(By Arias-Stella. From Dooley: Bartonellosis in Pathology of Tropical and Extraordinary Diseases. Washington, D.C., Armed Forces Institute of Pathology, 1976.)*

Figure 52-4. One form of skin lesions seen in verruga peruana, a clinical manifestation of bartonellosis. *(By Arias-Stella. From Dooley: Bartonellosis in Pathology of Tropical and Extraordinary Diseases. Washington, D.C., Armed Forces Institute of Pathology, 1976.)*

mechanism. The organisms are also found in large clusters distending the cytoplasm of the endothelium of blood and lymph capillaries. In patients with verruga, the organisms are less easily found but are demonstrable in properly fixed sections of veruga tinted with Giemsa's stain. The proliferative response of the vascular endothelium in these lesions is so intense and disorganized as to suggest sarcoma. The properties of insect transmission and invasion of and growth within erythrocytes and capillary endothelial cells are more often associated with protozoa and viruses than with bacteria.

Clinical Manifestations. Oroya fever is a highly fatal illness characterized clinically by fever, diffuse and severe bone and muscle pains, and anemia. Most of the signs and symptoms of the disease are directly attributable to the rapidly progressive hemolysis and the resultant profound anemia. An incubation period of 2 to 5 weeks, hepatosplenomegaly, and terminal secondary infection with salmonella further typify the disease.

Verruga peruana is a chronic nonfatal illness that develops either in those who have recovered from Oroya fever or in persons with no prior clinical evidence of bartonellosis. Verruga peruana is best characterized by the presence of either localized or generalized angiomatous warts that vary in their size and degree of superficiality

(Figs. 52-3 and 52-4). Because of their histology, the more superficial ones may appear bright red. They reach the size of an egg. Systemic signs of fever, generalized pains, and malaise occur, although less frequently than in Oroya fever. The eruption lasts from a month to 2 years and averages 4 to 6 months.

Infection results in an immunologic response that includes the production of complement-fixing antibodies and varying degrees of resistance to subsequent disease and infection. Oroya fever is believed to occur in the fully susceptible individual, while verruga peruana probably signifies a state of partial immunity.

Laboratory Diagnosis. A laboratory diagnosis can be made by demonstration of the organism in erythrocytes in Giemsa-stained films of peripheral blood or by blood culture. Serologic tests demonstrate the development of complement-fixing and agglutinating antibodies but are not major diagnostic tools.

Treatment. Penicillin, streptomycin, tetracyclines, and chloramphenicol have each been reported to reverse dramatically the downhill course in patients with Oroya fever. The bacteremia, however, may not be eradicated. Appropriate transfusion therapy often is useful. Antibiotics are reported to be less beneficial in verruga.

Control. *Phlebotomus* flies are exquisitely sensitive to DDT and can be controlled by its use. Antivector measures employing this chemical, the susceptibility of *B. bacilliformis* to a variety of antibiotics, plus unknown epidemiologic factors have all collaborated to limit a more extensive distribution of what has at time been a local major public health problem. Vaccine development ceased in the 1940s when it became evident that antibiotics were curative and that the vector could be locally controlled by DDT.

FURTHER READING

Colichon HF, DeBedon C: Enfermedad de Carrion II. Nutrientes utillizables para el crecimiento de la *Bartonella bacilliformis.* Rev Lat Am Microbiol 15:75, 1973

Dooley JR: Bartonellosis. In Binford CH, Conner DH (eds): Pathology of Tropical and Extraordinary Diseases. Washington, D.C., Armed Forces Institute of Pathology, 1976, p 192

Dooley JR: Haemotropic bacteria in man. Lancet 2:1237, 1980

Kreier JP, Dominquez N, Krampitz HE, et al.: The hemotropic bacteria: The families Bartonellaceae and Anaplasmataceae. In Starr MP, Stolp H, Truper HG, et al. (eds): The Prokaryotes— A Handbook on Habitats, Isolation and Identification of Bacteria. New York, Springer-Verlag, 1981

Ristic M, Kreir JP: Hemotropic bacteria. N Engl J Med 301:937, 1979

Schultz MG: A history of bartonellosis (Carrion's disease). Am J Trop Med Hyg 17:503, 1968

Weinman D: Bartonellosis. In Weinman D, Ristic M (eds): Infectious Blood Diseases of Man and Animals. New York, Academic Press, 1968, Chap 15

CHAPTER 53
Chlamydia

Chlamydiaceae is a family of obligate intracellular bacterial parasites, characterized by a unique developmental cycle that is common to all members of the family.

Chlamydia infect a wide spectrum of vertebrate hosts within three major ecologic niches: birds, mammals, and humans. Human diseases commonly caused by these organisms include trachoma, inclusion conjunctivitis, lymphogranuloma venereum, and psittacosis. Humans are also occasionally infected with *Chlamydia* that are normally associated with disease of other animals, such as feline pneumonitis.

Only two distinct species of *Chlamydia* are recognized: (1) *Chlamydia trachomatis,* which is inhibited by sulfonamides and produces iodine-staining cytoplasmic inclusions, and

(2) *Chlamydia psittaci* which is not inhibited by sulfonamides and does not produce iodine-staining inclusions in cytoplasmic vesicles (Table 53-1).

Chlamydia

Morphology and Developmental Cycle

The unusual developmental cycle of *Chlamydia* is similar for both members of the genus. It begins with the infection of a cell by the elementary body, the extracellular, rigid, metabolically dormant phase of the organism. The cycle consists of three major phases: (1) attachment and

TABLE 53-1. CHARACTERISTICS DISTINGUISHING
CHLAMYDIA **SPECIES**

C. trachomatis	C. psittaci
Sensitive to sulfadiazine	Insensitive to sulfadiazine
Sensitive to D-cycloserine	Insensitive to D-cycloserine
Form compact microcolonies within cytoplasmic vesicles	Organisms dispersed throughout host cell cytoplasm
Iodine-staining carbohydrate and lipid produced by microcolony	Iodine-staining carbohydrate and lipid not produced
Guanine-plus-cytosine content of DNA 44.4%	Guanine-plus-cytosine content of DNA 41.2%
Less than 10% of DNA sequences of two species are homologous	

penetration of the elementary body into the host cell cytoplasm, (2) development of the elementary body into a reticulate body, and (3) maturation of reticulate bodies and formation of elementary body progeny.

Elementary Bodies. The elementary body is the minimal infecting unit and is capable of survival outside the host. It is a small, dense, spherical body 0.2 to 0.4 μm in diameter, surrounded by a rigid cell wall similar in appearance and composition to that of gram-negative bacteria. Elementary bodies contain both DNA and RNA. The DNA is compactly organized in a central nucleoid structure and, in both *C. trachomatis* and *C. psittaci*, is in a closed circular molecule with a molecular weight of 660×10^6. This size molecule could provide information for about 600 different proteins, which is about one fourth the amount provided by the *Escherichia coli* genome. The elementary bodies of *Chlamydia* contain three species of RNA with sedimentation coefficients of 21S, 16S, and 4S. The ribosomal RNA thus resembles bacterial RNA rather than that of the mammalian host cell.

Developmental Cycle. *Chlamydia* elementary bodies attach to host cells and induce phagocytosis by nonprofessional phagocytic cells. The organisms attach by specific receptors, and phagocytosis is parasite specific. Following phagocytosis by the host cell, the elementary body is surrounded by the invaginated host cell membrane to form the inclusion body. The morphology of the developing inclusion can be followed by electron microscopy. During the first 12 hours postinfection, the elementary body gradually enlarges to form a structure 0.7 to 1.0 μm in diameter. This is a reticulate body. This body is surrounded by a double membrane and contains numerous ribosomes but no nucleoid. It is metabolically active and incapable of survival outside the host. At 20 hours following infection the host cell nucleus has been displaced by reticulate bodies that fill the cytoplasm. Some of these bodies can be seen in the process of binary fission. At 30 hours, elementary body progeny can be seen within the inclusion, some of

which are mature and others in an intermediate stage of development. Multiplication continues for several hours, during which time large numbers of elementary bodies are produced. Intracellular inclusion bodies thus contain both large noninfectious reticulate bodies and elementary bodies at different stages of their development. They represent intracellular bacterial microcolonies. Upon rupture of the host cell containing an inclusion body, elementary bodies are released to begin a new infectious cycle (Fig. 53-1).

Chlamydia survive intracellularly within phagosomes. The elementary body cell wall appears to inhibit phagolysosome fusion, and following development of the intracellular inclusion, the host cell membrane incorporates elementary body antigen into its surface.

MOLECULAR ASPECTS OF THE DEVELOPMENTAL CYCLE. Marked changes in the metabolism of both chlamydiae and host cell accompany infection. The elementary body loses its dense nucleoid structure preparatory to transcription and replication. The cell wall also disappears. The initiation of development depends on the presence of functioning mitochondria in the infected cells and on active nuclear functions. Chlamydiae also utilize the total acid-soluble purine and pyrimidine pools of its host for biosynthesis of its own RNA.

Enzymatic activities of *C. trachomatis* elementary bodies include a DNA-dependent RNA polymerase, which is inhibited by rifampin. Although chlamydiae possess enzymes for the metabolism of glucose after the glucose-6-phosphate stage, hexokinase is a host enzyme. The chlamydiae are unable to synthesize ATP and other high-energy phosphates. Reticulate bodies acquire their nucleotide precursors and high-energy phosphates from host metabolic sources and for this reason are sometimes referred to as "energy parasites."

During the development of the elementary bodies into reticulate bodies there is active biosynthesis of DNA at an increasing rate. The RNA content of *C. psittaci*-infected cells gradually increases and becomes maximum at 20 hours. Chlamydiae-coded enzymes are involved in the biosynthesis of the membrane of the reticulate body. In *C. trachomatis* but not in *C. psittaci*, glycogen is synthesized for a 10-hour period beginning at 20 hours postinfection. This also is a chlamydiae-coded function. With the synthesis of the cell wall, elementary bodies are produced and the cycle completed. Treatment with penicillin inhibits the development of reticulate bodies, and infectious particles are not produced.

Laboratory Identification

Isolation of *Chlamydia* may be accomplished by inoculation of infected material into embryonated eggs, into selected tissue culture cell lines, or into experimental animals.

The original isolation procedure consisted of the inoculation of material into 6- to 8-day-old chick embryo yolk sacs. All known strains of *Chlamydia* will infect the chick

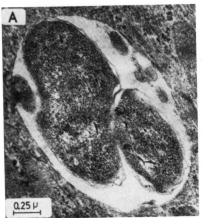
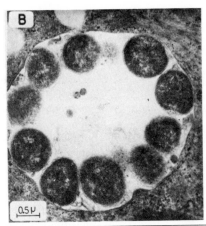
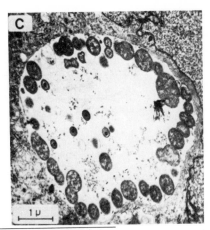

Figure 53-1. The developmental cycle of *C. trachomatis* (TE55 strain) in FL cells. **A.** An initial step in the formation of the inclusion body. **B.** An inclusion body with initial bodies. **C.** A developed inclusion body with initial bodies, dividing bodies, and elementary bodies. *(From Becker: Monogr Virol 7:31, 1974.)*

embryo, and group-specific antigen and characteristic inclusions are found in yolk sac material from infected embryos. Disadvantages in the use of the chick embryo yolk sac for diagnostic isolation included a long delay in confirmation of a result, inconvenience in culturing large numbers of specimens, and susceptibility of the embryo to superinfection with other bacteria.

Tissue cell culture methods for isolating *Chlamydia* have been improved by the use of centrifugation to enhance adsorption of *Chlamydia* to cells, by the use of new cell lines, and by the pretreatment of cells to increase detection of inclusions. Cell lines commonly used to isolate *Chlamydia* include irradiated McCoy and HeLa cells. Isolation of *Chlamydia* from clinical sources is more frequently achieved in tissue cultures than from egg yolk sac inoculations (Fig. 53-2).

Some strains of *Chlamydia* will infect mice, and a strain is partially characterized by the route of inoculation by which the infection may be established. Lymphogranuloma venereum (LGV) strains will usually infect mice when inoculated intracerebrally, while *Chlamydia* causing trachoma and inclusion conjunctivitis will not infect mice by any route of injection. They will, however, cause a rapid toxic death in mice if injected intravenously. Mice may be protected against toxic death by prior immunization with the same strain. This phenomenon was utilized in designing a typing system, the mouse toxicity prevention test.

Antigenic Structure

All *Chlamydia* possess a common heat-stable, group-specific antigen. This antigen is associated with the cell wall and is a carbohydrate–lipoprotein complex. The immunodominant group is a 2-keto-3-deoxyoctanoic acid. Since it is group specific, its presence cannot be used for identification of chlamydial strains or species. This group antigen may be detected by complement-fixation tests.

Species-specific and strain-specific antigens have been

demonstrated. Wang has typed chlamydial isolates from ocular and genital sources by the indirect immunofluorescence test using homologous mouse antisera (Table 53-2). At least 15 types have been identified, corresponding closely with the results of the mouse toxicity prevention test. This test is itself a function of the development of strain-specific antibody. Another method of species identification is by infectivity neutralization using hyperimmune sera.

Chlamydial elementary bodies contain a major outer membrane protein, which is from 38 to 42 Kdal in size

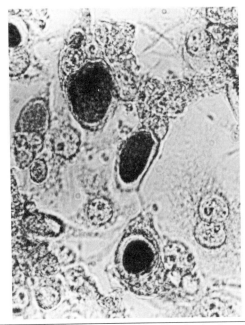

Figure 53-2. McCoy cells infected with *C. trachomatis* and stained with iodine demonstrating cytoplasmic inclusions. They are mahogany in color and discrete.

TABLE 53-2. MICROIMMUNOFLUORESCENT TYPES OF
***CHLAMYDIA TRACHOMATIS* IN CLINICAL DISEASE**

Microimmuno-fluorescent Type	Usual Disease	Geographic Distribution
A, B, C	Classic ocular trachoma	Primarily endemic in Asia and Africa
D, E, F, G, H, I, J, K, L, M	Inclusion conjunctivitis Genital trachoma Infant pneumonitis	Worldwide, sporadic
LGV I, II, and III	Lymphogranuloma venereum Genital trachoma	Worldwide, sporadic

and is immunogenic. There is considerable cross-reactivity between the outer membrane protein antibody of the various *C. trachomatis* types but very little between strains of *C. trachomatis* and *C. psittaci.* Research on diagnostic uses of chlamydial outer membrane protein and antibody is currently active.

Frei Test. The Frei test is an intradermal skin test that has been used since 1925, primarily for the diagnosis of lymphogranuloma venereum (LGV). The test initially used boiled lymph node material excised from a patient with LGV; at present it employs Lygranum antigen prepared from infected chick embryo yolk sac material. At 48 and at 72 hours the site of the skin test injection is examined for the formation of a subcutaneous nodule. The test is not specific for LGV infection, because group-specific antigen is involved in the reaction, and patients with other chlamydial infections may exhibit a positive Frei test. Similarly, intradermal tests using psittacosis antigen may be positive in patients with LGV. A modification of several of the antigens by acid extraction has resulted in a test of greater specificity, but because of difficulties in production it has not come into general use. The Frei test is also much less sensitive than is the complement-fixation test. Schachter found that four of six patients, all of whom were known to have LGV by recovery of *Chlamydia* from node biopsy, had a negative Frei test, but all developed complement-fixing antibody.

Antigenic studies of *C. trachomatis,* based upon immunofluorescence techniques, have revealed that there are 15 or more distinguishable types. Types A, B, and C have been isolated primarily from the eyes of persons with trachoma in trachoma-endemic areas. Isolation of these types from genital sites is rare. Types D, E, F, G, H, I, J, K, L, and M have been isolated from the eyes of persons in areas where trachoma is nonendemic and most frequently from the genital tracts of adults. Included among strains D through M are isolates from infants who were born to mothers with cervical infection and who developed inclusion conjunctivitis, and isolates from infants with the infant pneumonitis syndrome.

Chlamydia trachomatis

Human infections caused by *C. trachomatis* include trachoma, urethritis, cervicitis, salpingitis, inclusion conjunctivitis, infant pneumonitis syndrome, and lymphogranuloma venereum. A characteristic of the species is the development of compact, clearly defined, glycogen-containing, intracellular microcolonies or inclusions known as Halberstadter–Prowazek bodies. These are found in infected yolk sac preparations, infected animal tissue, inoculated tissue culture cells, and in conjunctival scrapings of persons with active trachoma or inclusion conjunctivitis. They are basophilic and gram-negative; they stain mahogany with iodine stain and may also be stained by fluorescent antibody or Giemsa stain. *C. trachomatis* is inhibited by sulfadiazine and usually by D-cycloserine.

Clinical Infection

Ocular Trachoma
The name trachoma, derived from the Greek word which means "rough," refers to the pebbled appearance of the infected conjunctiva. Naturally occurring trachoma is limited to humans and continues to be a leading cause of blindness in underdeveloped areas. Trachoma is a disease of poverty. In the United States, American Indians are the group most frequently infected. Trachoma is a disease optionally reported to the Centers for Disease Control.

Pathogenesis and Clinical Manifestations. The MacCallan classification is internationally accepted for describing trachoma. It describes four major stages of disease. Stage I is incipient trachoma. It may be relatively asymptomatic, with little if any conjunctival exudate. Minimal keratitis is usually present. Stage II is established trachoma, with follicular and papillary hypertrophy. Trachomatous pannus accompanies corneal infiltration. Stage III includes cicatricial complications, with scarring of the conjunctiva; trichiasis, entropion, and further pannus develop. Stage IV represents healed trachoma without evidence of active inflammation. If no complications of trachoma develop during active infection, this stage may be asymptomatic.

Long periods of latent infection occur, and superinfection with other bacteria contributes to more advanced forms of the disease. Trachomatous persons living in hygienic conditions experience a mild course or clinical resolution of infection. Repeated exposure to ocular trachoma infection is associated with an increased incidence of marginal infiltration and neovascularization. This may have a late onset after inclusions are no longer detectable in conjunctival scrapings and represents host response to chlamydial antigen, which then contributes to the severe ocular damage.

During the acute stage the ocular exudate contains primarily polymorphonuclear leukocytes, although the subepithelial infiltrate is mainly mononuclear. Limbal follicles, neovascularization with pannus, interstitial keratitis, cicatrization of the tarsal conjunctiva, and corneal ulcers lead to impairment of vision. Recurrences often occur after apparent healing; the clinical source is variable.

Immunity. Protective immunity conferred by a prior attack of trachoma is of a low order. Persons with ocular trachoma often develop neither complement-fixing antibody to the group antigen nor delayed sensitivity to the Frei antigen. However, detection of specific antibody in eye secretions and/or serum may be achieved with fluorescent antibody techniques.

Laboratory Diagnosis

SEROLOGIC TESTS. The microimmunofluorescence technique has been modified to determine the type-specific antibody response to *C. trachomatis.* In the majority of patients for whom the identity of the infecting strain is known there is a type-specific antibody response with a titer of 1:8 or greater. Early antibody formation is of the IgM class and persists for approximately 1 month before being replaced by IgG. In the few instances in which serial antibody determinations have been made, type-specific antibody has been observed to decrease fairly rapidly after primary infection, often within 1 or 2 months.

IDENTIFICATION OF *CHLAMYDIA TRACHOMATIS.* Demonstration of the causative agent may be made by everting the tarsal plate, removing the exudate, and gently scraping epithelial cells from the surface. The inclusions in scrapings may stain with fluorescent antibody or Giemsa stain or may be cultured by inoculation into yolk sac or tissue cell culture preparations.

Treatment. Antibiotics may be used topically and systematically. In hyperendemic areas, their primary effect is to limit coexisting bacterial conjunctivitis, but in the United States this is probably not a significant factor. Studies of American Indians at a boarding school showed that although systemically administered tetracyclines and sulfonamides caused a regression of clinical trachomatous activity, the agent persisted in conjunctival scrapings, and subclinical disease among the children continued to spread. Nevertheless, treatment may limit complications and should be administered.

Prevention. Systemically administered vaccines may exacerbate the disease. Since the protection afforded by most vaccines is of a relatively low order and duration, vaccines are usually not used in programs to control trachoma. Current research may provide a more satisfactory vaccine. Good standards of hygiene are essential in the control of trachoma.

Inclusion Conjunctivitis and Genital Trachoma

In the Infant. Inclusion conjunctivitis is often a disease of the newborn eye that is derived from infection of the maternal genital tract. It is caused by an agent almost indistinguishable from that causing trachoma; the term "TRIC agent" historically encompassed both (Table 53-2).

The incidence of chlamydial infection of the eyes of infants depends upon the prevalence of cervical infection in the mothers. In an English survey, 0.5 percent of infants had symptomatic disease of the eyes caused by these strains, while in South Africa, 26 percent were infected.

The disease in the newborn usually becomes clinically apparent between 5 and 12 days after birth. It is characterized by a sticky exudate and conjunctivitis and may be unilateral. Vulvovaginitis, ear infection, and mucopurulent rhinitis may accompany ocular disease. Many children ascertained to have neonatal inclusion conjunctivitis are premature, but whether this represents increased susceptibility or longer observation of the infant due to hospitalization is unknown.

Inclusion conjunctivitis of the newborn eye had been considered benign and self-limited until recent studies showed a high incidence of micropannus, conjunctival scars, and late recurrence, which were prevented by local application of tetracycline before the twelfth day of life.

Neonatal Pneumonia. *C. trachomatis* is a prominent cause of pneumonitis in infants. The children characteristically become ill at 4 to 16 weeks of age, have prominent respiratory symptoms of wheezing and cough, and lack systemic findings of fever or toxicity. They may be eosinophilic and have an elevated serum IgG and IgM and a very pronounced titer to the specific infecting strain by microimmunofluorescent typing. Chlamydial neonatal conjunctivitis often precedes the onset of the pneumonia. This disease is the subject of intense current research and interest.

In the Adult. Inclusion conjunctivitis in the adult is usually sporadic but may be epidemic following contamination of unchlorinated swimming pools. Inclusion conjunctivitis must be differentiated from epidemic keratoconjunctivitis, which is a viral disease.

The concept that these infections are harbored and transmitted sexually has received attention since the 1960s, although the association between neonatal inclusion conjunctivitis and maternal cervical disease was recognized much earlier. Jones demonstrated that adult ocular disease was associated with concomitant urethritis, cervicitis, and multiple sexual partners. Dunlop reported studies of patients whose infants had inclusion conjunctivitis. Salpingitis, cervicitis with cervical follicles, cervical discharge, prostatitis, proctitis, and nongonococcal urethritis afflicted their parents.

Adult Genital Tract Infection. Recovery of these agents and serologic evidence of infection with *Chlamydia* occurs in 20 percent of males with nongonococcal urethritis and in 45 percent of their sexual contacts. It occurs in men with a history of other venereal diseases and promiscuity and frequently in the consorts of women with cervical chlamydial infection.

Infections in adult women include chronic cervicitis and urethritis. Postpartum fever in infected women is common. In addition, maternal cervical infection is associated with an increased rate of premature delivery and perinatal morbidity. Ectopic pregnancies, secondary to chlamydial salpingitis, have increased in incidence in the past decade.

Lymphogranuloma Venereum (LGV)

Lymphogranuloma inguinale, climatic bubo, tropical bubo, and esthiomene are synonyms of a venereal disease that appeared from time to time throughout the eighteenth century. The distribution of this disease is now worldwide. It occurs more commonly in blacks than in whites and is recognized more frequently in males than in females. Humans are the sole natural hosts. LGV should not be confused with granuloma inguinale, which is caused by *Calymmatobacterium granulomatis.*

Etiology. *Chlamydia* isolated from patients with LGV differ from other chlamydial types in that they fail to cause typical follicular conjunctivitis in the monkey eye, they cause death in mice that are infected intracerebrally, and they sometimes cause rapid death of egg embryos. LGV isolates are divided by microimmunofluorescence into three antigenic types, LGV I, II, and III, which show cross-reactions with genital *Chlamydia* types E and D.

Clinical Infection

EPIDEMIOLOGY. LGV is a venereally transmittable disease. Although in the United States the incidence of syphilis and gonorrhea has been increasing during the past 5 years, the incidence of LGV has remained stable at about 500 reported cases per year. This may not remain the case, however, since recent reports suggest that importation of disease from Southeast Asia has occurred. In the United States, male homosexuals are especially likely to experience LGV infection and constitute another major reservoir of disease. Patients with LGV commonly have other concomitant venereal diseases, especially syphilis. All patients with LGV should be thoroughly examined for evidence of other venereal disease.

Little is known about the infectivity of LGV or the duration of infection when untreated. The frequency of relapse and the observation that a man may infect a new sexual partner many years after his initial infection indicate that it may be a very long, indolent, and chronically active disease.

CLINICAL MANIFESTATIONS. The usual incubation period of LGV is 1 to 4 weeks. Early constitutional symptoms, such as fever, headache, and myalgia, are common. The primary lesion is painless, small, inconspicuous, and vesicular and often escapes notice. Characteristically the presenting complaint concerns the enlarged matted inguinal and femoral lymph nodes. They are moderately painful, firm, and may become fluctuant. Aspiration of fluctuant nodes may be therapeutic and provide diagnostic material.

Women commonly experience proctitis, presumably because the lymphatic drainage from the vagina is perirectal. Infection may cause diarrhea, purulent drainage, tenesmus, anemia, abdominal pain, and the formation of infected sinuses. Rectal stricture and rectal perforation are recognized late sequelae to LGV proctitis.

The course of the disease is variable. It may cause progressive destruction of the vulva and urethra. Lymphatic obstruction in women can lead to elephantiasis of the vulva, called "esthiomene." Vulvar carcinoma is reported to be more common in women who have had LGV. An unknown percentage of persons have asymptomatic infections or heal without complications. Serologic evidence of experience with *Chlamydia* at some time has been reported to be very common, especially among persons attending venereal disease clinics. These data do not indicate which strain of *Chlamydia* was present, or when. Until more appropriate precise tests for routine use become available the extent of subclinical LGV will remain unknown.

PATHOLOGY. Autopsy of patients with chronic LGV infection has revealed lesions of the lymph nodes composed of aggregations of large mononuclear cells forming abscesses surrounded by epithelioid cells. A few giant cells of the Langhans type may be found. Numerous plasma cells may invade the granuloma formation. Occasionally, there are necrotic lesions with few or no granulocytes but also surrounded by giant cells, usually in disease of long duration. Varying degrees of fibrosis occur, with bands of granulation and connective tissue and thickened capsule.

Hyperglobulinemia is common early after infection, and a positive reaction for rheumatoid factor and cryoglobulins is frequent. A specific increase in IgA has been reported.

LABORATORY DIAGNOSIS. Diagnosis depends upon (1) a compatible clinical presentation, (2) recovery of the organism from the site of infection and its identification as *C. trachomatis*, (3) demonstration of rising LGV complement-fixation test (LGV-CFT) titer, microimmunofluorescent antibody titer rise, or (4) a reactive intradermal Frei test. In the interpretation of these parameters it must be recognized that the LGV-CFT is not specific for LGV and that the Frei test is both nonspecific and relatively insensitive. Recovery of the organism is the most satisfactory diagnostic aid, although the culture may fail to yield growth in situations that are in every other way typical of LGV, probably because of lack of sensitivity of the culture methods.

TREATMENT. Treatment may include sulfadiazine or tetracycline; penicillin has been effective when other drugs have failed. Meticulous follow-up for relapse or the development of complications is essential. A decrease in the LGV-CFT titer and reversion of the intradermal Frei test from positive to negative may follow treatment.

Chlamydia psittaci

A severe febrile disease obviously contracted from parrots was recognized in Switzerland, France, and Germany in the last 2 decades of the nineteenth century. Worldwide interest in this infection dates from 1929 to 1930, when over 700 cases were found in 12 different countries, including the United States. Psittacosis caused endemic disease in parrots and parakeets and in a wide range of other birds, including ducks, chickens, and turkeys. Infection of flocks of turkeys in the United States caused considerable human disease in the 1950s. Cattle and other animals may also experience endemic and epidemic psittacosis. Human disease transmitted from psittacine birds is referred to as "psittacosis" and that transmitted from nonpsittacine birds is "ornithosis." The diseases, however, are clinically indistinguishable.

Etiology

Four characteristics distinguish *C. psittaci* from *C. trachomatis*: (1) The intracellular microcolonies contain little glycogen and do not stain recognizably with iodine. (2) The inclusions are not compact but are more diffuse and irregular in shape. They are called LCL (Levinthal–Cole–Lillie) bodies, and stain with Giemsa or Macchiavello methods. Fluorescent microscopy has not been adapted to the clinical diagnosis of psittacosis in humans. (3) The development of inclusions is not inhibited by sulfadiazine or cycloserine. (4) The DNA base composition differs from that of strains of *C. trachomatis*, and the degree of homology is low. However, few strains have been studied (Table 53-1).

Clinical Infection

Epidemiology. The general prevalence of this disease is unknown, but a study in Wisconsin of sera referred for diagnostic studies on patients with respiratory disease found 2.8 percent positive for psittacosis. Relatively more cases occur in the autumn than in other seasons. Although only about 50 cases are currently reported annually, this reflects only the more severe cases and disease in bird handlers.

Between 1945 and 1951 in the United States an average of 28 cases of psittacosis per year were reported, after which there was a rapid increase to a peak of 568 in 1956. Some of the increase may be attributed to the relaxation of quarantine regulations for psittacine birds and some to the increased incidence of infections from turkeys. Since 1956 there has been a steady decline in the number of cases of psittacosis in the United States, and since 1980 there have been 120 to 140 cases per year.

The respiratory tract is the main portal of entry, and the usual source is inhalation of organisms from infected birds and their droppings. Many patients, but not all, give a clear history of exposure to psittacine birds. Since pigeons, turkeys, chickens, and wild birds may harbor the disease, a patient may not recognize a possible exposure. In addition, person-to-person transmission occurs. Exposure to patients who will die of psittacosis in the next 1 or 2 days is especially likely to propagate a very severe or fatal secondary infection.

SPONTANEOUS DISEASE IN ANIMALS. In parrots the naturally acquired disease is characterized by apathy, shivering, weakness, diarrhea, and respiratory symptoms. At necropsy, multiple areas of necrosis are found in the liver and spleen and occasionally in the lungs. Unapparent or subclinical infections occur in birds of the psittacine group and even more frequently in nonpsittacine birds. Small birds are often healthy carriers of the organism. Sheep in Colorado have a type of polyarthritis from which *Chlamydia* organisms have been isolated. Many other mammals experience arthritis, abortion, encephalitis, and conjunctivitis when infected with these agents.

Clinical Manifestations. The clinical disease was originally believed to be very severe, with a 20 percent mortality rate. It is now recognized that the signs, symptoms, and severity vary greatly. For example, 10 percent of persons believed to have influenza in a British chest clinic were found probably to have psittacosis. Most cases are heralded by constitutional signs of fever, myalgia, and often a severe frontal headache. This precedes the pulmonary signs of the disease, which include nonproductive cough, rales, and consolidation. Radiologic examination of the chest may suggest bronchopneumonia or primary atypical pneumonia, and inadequately treated patients may suffer repeated episodes of pneumonia.

The second most frequently involved organ system is the central nervous system. Symptoms are usually no more pronounced than a severe headache, but encephalitis, coma, convulsions, and death may occur. The etiology is usually thought to be a toxic encephalitis rather than direct invasion of the central nervous system by *Chlamydia*, although LCL bodies have been identified in the meninges of involved cases.

Patients with psittacosis may also develop carditis, subacute bacterial endocarditis, hepatitis, with or without formation of hepatic granulomata, erythema nodosum, and follicular keratoconjunctivitis. During the early phase of the illness, an acute biologic false positive test for syphilis may develop in a third of the patients, and sera may be anticomplementary.

Pathology. The pathology of psittacosis includes focal areas of necrosis of the liver and spleen, with predominance of mononuclear cells. In the lungs, consolidation is characterized by the thickening of alveolar walls, infiltration of mononuclear cells, and a gelatinous alveolar exudate also containing mononuclear cells.

Laboratory Diagnosis. *C. psittaci* may be isolated from infected material by methods similar to those used for other *Chlamydia*, including mouse inoculation, intracerebrally and subcutaneously, egg inoculation, and tissue culture.

Treatment. Results of treatment of psittacosis are imperfect. Tetracycline may be used with some success, and a good response has been achieved by the use of erythromycin. In half the patients radiologic evidence of pulmonary infiltration persists for over 6 weeks. Fall of the peak complement-fixation titer may not occur for over a year. Asymptomatic persistence of infection in psittacosis has not been well studied, but one patient is known to have shed the organism in his sputum for 12 years, both before and after penicillin treatment. However, no secondary cases were attributed to him.

Prevention. Prophylactic treatment of psittacine birds with antibiotic-supplemented feed reduces the risk of disease in bird handlers. The recognition, however, that *C. psittaci* may infect many avian and mammalian species, sometimes causing subclinical communicable disease with occasional outbreaks involving human beings, widens the need for careful epidemiologic investigation of each case. Workers in poultry-processing plants should have excellent environmental protection, as they are in particular risk of heavy exposure to this infection.

Miscellaneous Diseases

Cat-scratch fever is a human disease of unknown etiology. It is characterized by contact with a kitten, development of a primary papule at the site of a scratch or injury, and subsequent lymphadenopathy central to the papule. Approximately 25 percent of persons with this disease exhibit an antibody response to chlamydial group antigens, and 70 percent exhibit a significant titer by the immunofluorescence method to specific strains of *Chlamydia*. However, *Chlamydia* species have not been recovered from excised tissue of infected persons; hence the suggestion that the disease is caused by a *Chlamydia* species is unproven.

Reiter's Syndrome. Patients with Reiter's syndrome characteristically exhibit a triad of recurring signs, including conjunctivitis or iridocyclitis, polyarthritis, and nonbacterial urethritis. The disease usually occurs in young white males and is frequently preceded by dysentery or gonococcal urethritis. Many organisms have been etiologically associated with Reiter's syndrome. Studies implicating *Chlamydia* strains have included isolation of *C. psittaci* from synovial fluid from a patient and the demonstration that a higher proportion of Reiter's syndrome patients have antibody to *Chlamydia* group antigen than do comparable patients with gonorrhea or nongonococcal urethritis. The question of whether Reiter's syndrome patients react more vigorously to *Chlamydia* antigen or whether the disease is caused directly by *Chlamydia* infection has not been answered.

FURTHER READING

Becker Y: The agent of trachoma. Monogr Virol 7:31, 1974

Becker Y: The *Chlamydia*: Molecular biology of procaryotic obligate parasites of eucaryocytes. Microbiol Rev 42:274, 1978

Beem MO, Saxon EM: Respiratory tract colonization and distinctive pneumonia syndrome in infants infected with *Chlamydia trachomatis*. N Engl J Med 296:306, 1977

Brunham RC, Kuo CC, Cles L, et al.: Correlation of host immune response with quantitative recovery of *Chlamydia trachomatis* from the human endocervix. Infect Immun 39:1491, 1983

Caldwell HD, Schachter J: Antigenic analysis of the major outer membrane protein of *Chlamydia* sp. Infect Immun 35:1024, 1982

Eissenberg LG, Wyrick PB, Davis CH, et al.: *Chlamydia psittaci* elementary body envelopes: Ingestion and inhibition of phagolysosome fusion. Infect Immun 40:741, 1983

Grayston JT, Wang SP: New knowledge of chlamydiae and the diseases they cause. J Infect Dis 132:87, 1975

Jansson E: Ornithosis in Helsinki and some other localities in Finland. Ann Med Exp Biol Fenn 38 [Suppl 4]:1, 1960

Martin DH, Koutsky L, Eschenbach DA, et al.: Prematurity and perinatal mortality in pregnancies complicated by maternal *Chlamydia trachomatis* infection. JAMA 247:1585, 1982

Oriel JD, Ridgway GL: Epidemiology of chlamydial infection of the human genital tract: Evidence for the existence of latent infections. Eur J Clin Microbiol 1:69, 1982

Richmond SJ, Stirling P: Localization of chlamydial group antigen in McCoy cell monolayers infected with *Chlamydia trachomatis* or *Chlamydia psittaci*. Infect Immun 34:561, 1981

Schachter J: Chlamydial infections. N Engl J Med 298:428, 490, 540, 1978

Schachter J, Sugg N, Sung M: Psittacosis: The reservoir persists. J Infect Dis 137:44, 1978

CHAPTER 54

Mycoplasma

The Mycoplasmas

The mycoplasmas are the smallest and simplest procaryotes capable of self-replication. The first isolation of one of these organisms was from bovine pleuropneumonia, thus its name, the pleuropneumonia organism (PPO). Similar organisms subsequently isolated from a variety of animals both from the carrier and the disease state, were referred to as pleuropneumonia-like organisms (PPLO) because of their similarity to the orginal isolate.

The unique property that characterizes the mycoplasmas is the absence of a cell wall. They belong to the class Mollicutes (*mollis*, soft; *kutis*, skin), a taxon that contains small procaryotic organisms bounded by a single trilaminar cell membrane. The taxonomy and properties of members of this class are listed in Table 54-1. One of the major differentiating properties of the Mycoplasmataceae is the requirement for cholesterol that is incorporated into the cell membrane.

Although known for some time to be a part of the normal flora of the human respiratory and genitourinary tracts, only recently was a mycoplasma shown to cause a specific disease of humans—a pneumonia that because of its dissimilarities with bacterial pneumonia was orginally referred to as primary atypical pneumonia. Three important observations helped to define the illness clinically: (1) the discovery that cold agglutinins developed in convalescent sera of infected patients, (2) the availability of penicillin, which had no effect on this particular infection, and (3) the frequency of the disease in military personnel during World War II. The causative organism, initially isolated in hamsters and referred to as the Eaton agent, was subsequently isolated in embryonated eggs and artificial laboratory media. It was given the name *Mycoplasma pneumoniae*. With the development of specific serologic tests, it was demonstrated that pneumonia is an uncommon complication of infection. The discovery that the isolated agent could be inhibited by antibiotics led to the clinical observations that tetracycline and erythromycin could alter the natural history of the infection. Although an experimental vaccine has been tried, present efforts are being directed toward a more effective immunogen.

Morphology

Mycoplasmas are very small in size, ranging in diameter from 200 to 300 nm. They are bounded by a trilaminar membrane, 8 to 10 nm thick. Because they lack a rigid cell wall, they assume a number of morphologic forms ranging from cocci, cocci and chains, cocci with tubules, filamentous cells with branching, filamentous cells with terminal structures, and pear-shaped cells (Fig. 54-1). The mode of reproduction of mycoplasma is essentially by binary fission.

Some *Mycoplasma*, including *M. pneumoniae*, exhibit a gliding motility on liquid-covered surfaces. In these species, specialized structures have been differentiated to

TABLE 54-1. TAXONOMY AND PROPERTIES OF ORGANISMS INCLUDED IN THE CLASS MOLLICUTES

Classification	Current Number of Recognized Species	Genome		Cholesterol Requirement	NADH Oxidase Localization	Characteristic Properties	Habitat
		Size ($\times 10^8$ daltons)	G + C Content (%)				
Mycoplasmataceae							
Mycoplasma	ca. 50	5	23–41	+	Cytoplasm		Animals
Ureaplasma	1	5	28	+	ND	Urease activity	Animals
Acholeplasmataceae							
Acholeplasma	6	10	29–35	−	Membrane		Animals
Spiroplasmataceae							
Spiroplasma	1	10	26	+	Cytoplasm	Helical filaments	Insects and plants
Genera of uncertain taxonomic position							
Anaeroplasma	2	ND	29–34	Some + Some −	ND	Anaerobic, some digest bacteria	Rumens of cattle and sheep
Thermoplasma	1	10	46	−	Membrane	Thermophilic (optimum 59C) and acidophilic (optimum pH 1.0–2.0)	Burning coal, refuse piles

From Razin: Microbiol Rev. 42:414, 1978.
G + C, guanine plus cytosine; NADH, reduced nicotinamide adenine dinucleotide; ND, not determined.

serve both as an attachment site and as the leading edge of the cell during movement. In the helical spiroplasmas, there is a rapid rotary motion, and the flectional movements resemble those of spirochetes. The isolation of an actin-like protein from this group of organisms suggests the presence of contractile proteins in mycoplasmas.

Physiology

The mycoplasma genome consists of a circular double-stranded DNA molecule, which is the smallest for any known self-replicating procaryote. This is consistent with the organism's limited biosynthetic capabilities and requirement for highly complex culture media. The mycoplasma genome also differs from that of other procaryotes in its low guanine-plus-cytosine content (Table 54-1), and the lower G + C content of the ribosomes is well below that of other procaryotic rRNAs. DNA annealing techniques have not supported the theory that mycoplasma are related to L forms of existing cell wall-containing bacteria. However, ribosomal transfer-RNA analysis suggests a relationship in the distant past to bacteria of the genus *Clostridium.*

Cultural Characteristics. Most mycoplasmas have a unique requirement among procaryotes for cholesterol and related sterols for membrane synthesis. They also lack the enzymatic pathways for the synthesis of purines and pyrimidines. Complex culture media, such as beef-heart infusion, supplemented with horse serum, yeast extract, and nucleic acids, are thus required for in vitro culture. The purine and pyrimidine requirement for growth provides

the basis of a useful technique for detecting mycoplasma-infected tissue culture systems, since purines and pyrimidines are not taken up by tissue culture cells.

When grown on solid media, mycoplasma slowly form a dome-shaped colony on the surface of the agar. The central part of the colony grows down into the agar, producing a more dense central core. When viewed from above, the mycoplasma colony resembles a "fried egg" in appearance. The colonies are extremely small and often require a dissecting microscope for visualization. (Fig. 54-2) Although the mycoplasma culture follows a bacterial growth curve pattern, its doubling time of 1 to 6 hours means that up to 3 weeks may be required before colony formation is visible.

Metabolism. Most species are facultative anaerobes, although growth is better in an aerobic environment. Mycoplasmas can be divided into two broad physiologic groups, the fermentative and the nonfermentative. In the fermentative organisms, ATP is derived from sugars via glycolysis, whereas in nonfermentative mycoplasmas, the arginine dihydrolase pathway has been proposed as the major source for ATP. ATPase has been found associated with the cell membrane of every mycoplasma examined. Most of the enzymes involved in membrane lipid biosynthesis in mycoplasmas are also membrane bound.

Determinants of Pathogenicity

Mycoplasmas produce a number of extracellular products that contribute to their disease-inciting ability. Except for the neurotoxin produced by *Mycoplasma neurolyticum*, at-

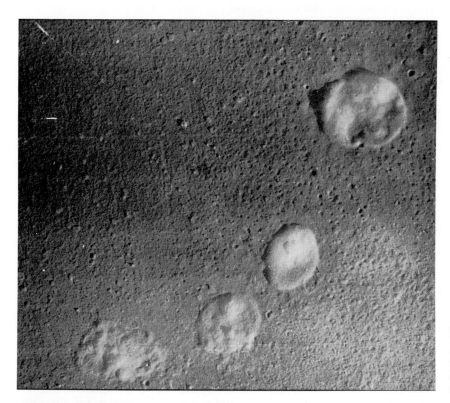

Figure 54-1. Electron micrograph of *Mycoplasma.* Note the obvious plasticity of the organism and the lack of a definite cell wall. × 20,000. *(From Morton et al.: J Bacteriol 68:697, 1954.)*

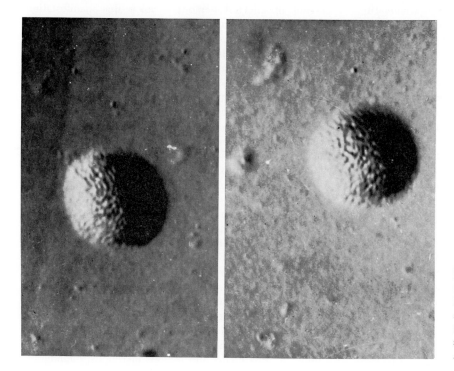

Figure 54-2. Colonies of *M. pneumoniae.* Note the granular appearance. The organisms grow down into the agar beneath the surface colony. *M. pneumoniae,* on initial isolation, frequently does not have surrounding surface growth. *(From Chanock et al.: Proc Natl Acad Sci USA 48:41, 1962.)*

tempts to detect toxins produced by the other species have been unsuccessful. Although neurotoxic symptoms similar to those caused by the *M. neurolyticum* exotoxin are also observed in turkeys inoculated intravenously with *Mycoplasma gallisepticum*, in this case, the neurotoxic effects are associated with viable mycoplasma cells and not with a soluble exotoxin. *M. gallisepticum* possesses a marked tropism for arterial walls in the brain, where it multiplies and damages the capillary endothelium.

Killed *Mycoplasma fermentans* cells or membranes are toxic to animals, producing symptoms of a gram-negative endotoxemia. Toxicity is attributable to mycoplasmal lipopolysaccharides, but these are totally unrelated to the lipopolysaccharides of gram-negative bacteria, and their specific role is at present ill defined.

Since most mycoplasmas produce surface infections and rarely invade the bloodstream and tissues, the ability to adhere to cell surfaces is important for colonization of the epithelial linings of the respiratory and urogenital tracts. Intimate contact of the mycoplasmas with their host cells is required in order to furnish nutrients and specific growth factors, especially nucleic acid precursors, which mycoplasmas are unable to synthesize.

Mycoplasmas attach to the mucous membranes of the respiratory, gastrointestinal, and genitourinary tracts. Specific receptor proteins have been identified, and a specific receptor tip on some mycoplasmas has been identified. Fusion of the membranes has been suggested as an explanation for the observed exchange of membrane antigens. The alteration of host antigens, as well as the similarity between mycoplasma antigens and host antigens, raises the question of the role of autoimmunity in the production of disease. The firm adherence of mycoplasmas to the host cell results primarily in surface infections, with invasion of the bloodstream being an extremely rare event.

Mycoplasmas have been shown to adhere to a variety of surfaces, including erythrocytes, HeLa cells, spermatozoa, tracheal epithelial cells, tracheal organ cultures, and such inert substances as glass. Avirulent strains of *M. pneumoniae* fail to adhere to cell surfaces. Tracheal organ cultures infected with *M. pneumoniae* exhibit tissue damage characterized by a ciliostatic effect and histologic changes observed by light and electron microscopy. Local accumulation of metabolites, such as H_2O_2 and ammonia, probably contribute to tissue damage. The cytopathic effects are reflected in changes in the metabolism of the infected organ culture. Among the most prominent biochemical changes are a depletion of arginine and alterations in host nucleic acid metabolism. The finding that mycoplasmas are infected with bacteriophages raises additional possibilities concerning mechanisms of pathogenicity.

Mycoplasma pneumoniae

Morphology and Physiology
M. pneumoniae is a short filament, 2 to 5 μm in length. Near one end of the filament is a bulbous enlargement with a differentiated tip structure. *M. pneumoniae* can attach to glass surfaces, red cells, and respiratory epithelial cells by this differentiated pole. When so attached, the organisms can show a gliding motility.

The organism grows more slowly than most other mycoplasmas, with colonies appearing 5 to 10 days after inoculation. The *M. pneumoniae* colony is more compact than that of most *Mycoplasma* colonies and has a minimal skirt, giving more of a mulberry colony appearance.

Antigenic Structure
The major antigenic determinants of *M. pneumoniae* are present in lipid extracts of membranes. The organism is antigenically distinct from other species of human origin, as determined by agglutination, complement-fixation, and other serologic procedures.

Clinical Infection
Epidemiology. *M. pneumoniae* is widely distributed in all parts of the world. Infection is usually endemic, but an increased incidence is observed during the colder months. Transmission is by aerosol droplets created by the cough or sneeze of the infected patient. Spread is usually quite slow, and true epidemics are rare except in confined populations of persons, such as in school, family, or barracks. It is important to stress that pneumonia is not the usual manifestation of infection, since only an estimated 3 to 10 percent of infected persons develop this manifestation. The observation that infection was common in young children but illness was rare until the ages of 5 to 20 has led to the speculation that part of the clinical illness may result from the host's reaction to the pathogen.

Clinical Manifestations. After an incubation period of 2 to 3 weeks, there is usually a gradual onset of fever, throbbing headache, and a persistent nonproductive hacking cough. Physical examination in the early stages of the illness frequently does not reveal many chest findings in contrast to an x-ray that sometimes shows bilateral feathery infiltrates. As the disease progresses, loud rales become prominent, with little physical findings of consolidation. The peripheral white count during the early stages of the illness is usually within normal limits, although as the disease progresses it may become elevated (Fig. 54-3). Microscopic examination and culture of the sputum yield only normal flora. Without treatment, the disease may last up to 3 weeks but is usually of much shorter duration. The cold agglutinins commonly found in convalescent sera are occasionally associated with hemolytic anemia or cold acrocyanosis. A fatal outcome is extremely rare.

Rarely, other organ systems have been implicated, resulting in otitis media and bullous myringitis, meningoencephalitis, or myocarditis. Pulmonary infections are occasionally complicated by a bullous eruption of the skin and mucous membranes, the Stevens–Johnson syndrome.

The illness must be differentiated from a large number of other atypical pneumonias caused by a variety of

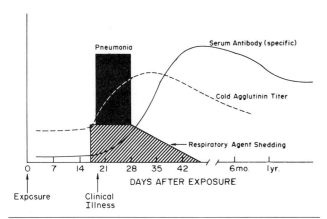

Figure 54-3. The course of *Mycoplasma pneumoniae* infection.

different agents: virus infection (e.g., influenza, adenovirus, respiratory syncytial virus), psittacosis, Q fever, and legionellosis.

Laboratory Diagnosis. In the early stages of infection, the diagnosis must be made on clinical grounds. Although fluorescence microscopy using known antibody has been used to localize mycoplasma in experimental infections, its use for direct demonstration of the agent in sputum is not available for diagnostic purposes.

CULTURAL TECHNIQUES. Culture of *M. pneumoniae* on a special enriched medium containing serum and yeast extracts may require 2 to 3 weeks for colonies to appear. The addition of penicillin and thallium to the culture medium is required to inhibit bacterial growth.

The ability of *M. pneumoniae* to ferment glucose is utilized in the preparation of a differential medium into which glucose and a pH indicator are incorporated. In the laboratory, advantage is also taken of the organism's ability to hemolyze red blood cells. If a previously inoculated agar medium is overlaid with agar containing red blood cells, a small zone of hemolysis will develop over colonies of *M. pneumoniae,* permitting easier identification.

SEROLOGIC TESTS. During the acute phase of the illness, no diagnostic serologic tests are available. A frequently used nonspecific serologic test that has been helpful in establishing diagnosis is the cold agglutination reaction. Cold agglutinins are measured in both the acute and convalescent phases of the illness in order to demonstrate a rising titer. This test is easily performed by mixing dilutions of the patient's serum with a standard concentration of washed human type O cells and incubating overnight at 4C. The test is based on the development of hemagglutination, which can be reversed by placing the tubes at 37C.

More informative and specific, however, is the complement-fixation test utilizing antigen from *M. pneumoniae.* A test based on the ability of antibodies from patients' convalescent sera to inhibit the growth of mycoplasma in vitro has also been devised. In this test, growth inhibition is a function of the inhibition of glucose fermentation or tetrazolium reduction. Other tests that have been used are the indirect hemagglutination, latex fixation, indirect immunofluorescence, and complement-mediated mycoplasmicidal and radioimmunoprecipitation tests.

Treatment. Treatment is usually instituted on the basis of a clinical impression. Penicillin has no effect on the natural history of the disease. The tetracyclines and erythromycin reduce the clinical course of the illness in terms of fever, number of hospital days, and x-ray resolution. Interestingly, however, the drugs do not alter the shedding of the agent from the respiratory tree. Erythromycin is sometimes preferred in older patients because of its effectiveness against both *Mycoplasma* and *Legionella pneumophila.* In the pediatric age group, it is unusual for *M. pneumoniae* infections to present as a pneumonia. More commonly the infection is seen in upper respiratory tract syndromes and is usually treated symptomatically.

Prevention. There is no commonly accepted method available for preventing *M. pneumoniae* infection other than avoiding close contact with acutely ill patients. One approach to prophylaxis is the use of antibiotics in persons at high risk of infection. Use of prophylactic tetracycline in members of families with an index case of infection has been reported to cause a significant reduction in clinical illness, although the incidence of infection is only minimally reduced.

Another approach to the prevention of infection has been through the development of vaccines. Trials with inactivated *M. pneumoniae* vaccines have shown them to be effective against both natural and artificially induced infection. In one study, however, illness appeared to be more severe in persons who failed to develop detectable antibody following vaccination than in controls. Strains of *M. pneumoniae* that have become attenuated by repeated passage on artificial media are effective vaccines but continue to produce illness in a portion of the vaccinees. Temperature-sensitive mutants restricted to growth in the upper respiratory tract also offer promise as effective immunizing agents.

Other Mycoplasmas

Increasing evidence is accumulating that other mycoplasmas are capable of producing disease. *Mycoplasma hominis* has been implicated in postpartum fever, postabortal fever, pelvic inflammatory disease, and pylonephritis. *Ureaplasma ureolyticum* probably contributes to a small percentage of nongonococcal urethritis. In addition, ureaplasmas may play a role in perinatal mortality. Although mycoplasmas have been demonstrated as the unequivocal etiologic agents of arthritis in animals, there is not as yet convincing evidence that they are the etiologic agent in human illness.

FURTHER READING

Books and Reviews

Archer DB: The structure and function of the *Mycoplasma* membrane. Int Rev Cytol 69:1, 1981

Barile MF, Razin S (eds): The Mycoplasma. Cell Biology. New York, Academic Press, 1979, vol I

Cassell GH, Cole BC: Mycoplasmas as agents of human disease. N Engl J Med 304:80, 1981

Denny FW: Atypical pneumonia and the Armed Forces Epidemiological Board. J Infect Dis 143:305, 1981

Razin S (ed): Mycoplasma infections. Isr J Med Sci 17:509, 1981

Taylor-Robinson D, McCormack WM: The genital mycoplasmas. N Engl J Med 302:1003, 1063, 1980

Tully JG, Whitcomb RF (eds): The mycoplasmas. Human and Animal Mycoplasmas. New York, Academic Press, 1979, vol II

Whitcomb RF, Tully JG (eds): The Mycoplasmas. Plant and Insect Mycoplasmas. Academic Press, New York, 1979, vol III

Selected Papers

Hu PC, Cole RM, Huang YS, et al.: *Mycoplasma pneumoniae* infection: Role of a surface protein in the attachment organelle. Science 216:313, 1982

Kundsin RB, Driscoll S, Pelletier PA: *Ureaplasma ureolyticum* incriminated in perinatal morbidity and mortality. Science 213:474, 1981

Stopler T, Gerichter CB, Branski D: Antibiotic resistant mutants of *Mycoplasma pneumoniae.* Isr J Med Sci 16:169, 1980

Thomsen AC, Lindskov HO: Diagnosis of *Mycoplasma hominis* pyelonephritis by demonstration of antibodies in urine. J Clin Microbiol 9:681, 1979

Tully JG, Pose DZ, Whitcomb RF, et al.: Enhanced isolation of *Mycoplasma pneumoniae* from throat washings with a newly modified culture medium. J Infect Dis 139:478, 1979

Woese CR, Maniloff J, Zablen LB: Phylogenetic analysis of the mycoplasmas. Proc Natl Acad Sci USA 77:494, 1980

SECTION IV
BASIC VIROLOGY

CHAPTER 55

The Nature, Isolation, and Measurement of Animal Viruses

Many important infectious diseases that afflict mankind are caused by viruses. Some are important because they are often fatal; among such are rabies, smallpox, poliomyelitis, hepatitis, several hemorrhagic fevers, and some encephalitic diseases. Others are important because they are very contagious and cause acute discomfort; among such are influenza, the common cold, measles, mumps, and chickenpox, as well as respiratory–gastrointestinal disorders. Still other viruses, such as rubella and cytomegalovirus, can cause congenital abnormalities, and finally there are viruses that can cause tumors and cancer in animals and also in humans.

There is little that can be done to interfere with the growth of viruses, since they multiply within cells, using the cells' synthetic capabilities. Only a limited number of highly specialized reactions are under their own control. It is hoped that their selective inhibition will form the basis for a rational system of antiviral chemotherapy, thereby permitting virus diseases to be brought under effective control, just as antibiotics have brought most bacterial diseases under control.

In addition to their medical importance, viruses provide the simplest model systems for many basic problems in biology. The reason is that viruses are essentially small segments of genetic material encased in protective shells. Since the information encoded in viral genomes differs from that in host cell genomes, viruses afford unrivaled opportunities for the study of the mechanisms that control the replication and expression of genetic material. Knowledge of these mechanisms is fundamental to an understanding of the development and operation of differentiated functions in higher organisms and is, therefore, directly applicable to the practice of medicine and the improvement of human welfare.

Historical Background

There are three major classes of viruses: animal viruses, plant viruses, and bacterial viruses. Since knowledge concerning each of these classes has accumulated along distinctive lines, extensive specialization has developed. Bacterial viruses are, therefore, dealt with only briefly in this book, and plant viruses are not considered at all. Yet discoveries made concerning each of these classes of viruses have influenced profoundly our understanding of the nature of each of the others.

The existence of viruses became evident during the closing years of the 19th century when, as the result of newly acquired expertise in the handling of bacteria, the infectious agents of numerous diseases were being isolated. For some infectious diseases this proved to be an elusive task until it was realized that the agents causing them were smaller than bacteria. Iwanowski, in 1892, was probably the first to record the transmission of an infection (tobacco mosaic disease) by a suspension filtered through a bacteria-proof filter. This was followed in 1898 by similar experiments of Löffler and Frosch concerning foot-and-mouth disease of cattle. Beijerinck (1898) considered the infectious agents in bacteria-free filtrates to be living but fluid—that is, nonparticulate—and introduced the term "virus" (Latin, poison) to describe them. It quickly became clear, however, that viruses were particulate, and the term "virus" became the operational definition of infectious agents smaller than bacteria and unable to multiply outside living cells. In 1911 Rous discovered a virus that produced malignant tumors in chickens, and during World War I Twort and d'Herelle independently discovered the viruses that multiply in bacteria, the bacteriophages.

During the next 25 years the experimental approaches in the three areas of virology diverged. Plant viruses proved easy to obtain in large amounts, thus permitting extensive chemical and physical studies. This work first led to the demonstration that plant viruses consist only of nucleic acid and protein and culminated in the crystallization of tobacco mosaic virus by Stanley in 1935. This feat evoked great astonishment, since it cut across preconceived ideas concerning the attributes of living organisms and demonstrated that agents able to reproduce in living cells behaved under certain conditions as typical macromolecules.

Work with bacteriophages concentrated on their clinical application. It was hoped that bacteria could be destroyed inside the body by injecting appropriate bacteriophages. Their activity in vivo, however, never matched their activity in vitro, most probably because they are eliminated efficiently from the bloodstream.

Work with animal viruses concentrated on the pathogenesis of virus infections and on epidemiology. Throughout this period, fundamental studies on animal cell–virus interactions were severely hampered by the absence of rapid and efficient techniques for quantitating viruses. The only method then available was the expensive and time-consuming serial end point dilution method, using animals (p. 808).

Around the year 1940 came several breakthroughs. First, the advent of electron microscopy permitted visualization of viruses for the first time. As will become evident,

not only is morphology an important criterion of virus classification, but the study of the morphology of viruses has also had a profound impact on our understanding of their behavior and function. Second, techniques for purifying certain animal viruses were being perfected, and a group of workers at the Rockefeller Institute headed by Rivers carried out some excellent chemical studies on vaccinia virus. Third, Hirst discovered that influenza virus agglutinates chicken red cells. This phenomenon, hemagglutination, was rapidly developed into an accurate method for quantitating myxoviruses, as a result of which this group of viruses became in the 1940s the most intensively investigated group of animal viruses. Finally, this period marked the beginning of the modern era of bacterial virology. Until then the interaction of bacteriophages with bacteria had been analyzed principally in terms of populations rather than at the level of a single virus particle interacting with a single cell. This conceptual block was removed by Ellis and Delbrück's study of the one-step growth cycle, as a result of which the bacteriophage–bacterium system became extraordinarily amenable to experimentation. Indeed, during the past three decades, many of the major advances in molecular biology have resulted from work in the bacteriophage field. Among these are the demonstration that initiation of virus infection involves the separation of viral nucleic acid and protein, the demonstration that the virus genome can become integrated into the genome of the host cell, the discovery of messenger RNA, and elucidation of the factors that control initiation and termination of both transcription and translation of genetic information.

In animal virology, rapid advances followed the development, in the late 1940s, of techniques for growing animal cells in vitro. Strains of many types of mammalian cells can now be grown in media of defined composition. As a result, animal cell–virus interactions can now be analyzed with the same techniques that have proved so powerful in the case of bacteriophages.

The Nature of Viruses

Viruses are a heterogeneous class of agents. They vary in size and morphology; they vary in chemical composition; they vary in host range and in the effect that they have on their hosts. There are certain characteristics, however, that are shared by all viruses:

1. Viruses consist of a genome, either RNA or DNA, that is surrounded by a protective protein shell. Frequently this shell is itself enclosed within an envelope that contains both protein and lipid.
2. Viruses multiply only inside cells. They are absolutely dependent on the host cells' synthetic and energy-yielding apparatus. They are parasites at the genetic level.

3. The multiplication of viruses involves as an initial step the separation of either their genomes or their nucleocapsids from their protective shells (see below).

In essence, therefore, viruses are nucleic acid molecules that can enter cells, replicate in them, and code for proteins capable of forming protective shells around them.

Given this definition of viruses, are they to be regarded as living organisms or as lifeless arrangements of molecules? The answer to this question depends on whether one is concerned with viruses as extracellular suspensions of particles or as infectious agents. Isolated virus particles are arrangements of nucleic and protein molecules with no metabolism of their own; they are no more active than isolated chromosomes. Within cells, however, virus particles are capable of reproducing their own kind manyfold by virtue of precisely regulated sequences of reactions. Considered in this light, viruses may indeed be said to possess at least some of the attributes of life. Such terms as "organism" and "living," however, are not really applicable to viruses. It is preferable to refer to viruses as being functionally active or inactive, rather than living or dead.

The Origin of Viruses

The question of the origin of viruses poses a fascinating problem. The two likeliest hypotheses are (1) viruses are the products of regressive evolution of free-living cells. An evolutionary pathway of this type has been suggested for mitochondria, which still retain vestiges of cellular organization as well as a mechanism for replicating, transcribing, and translating genetic information. The largest animal viruses, the poxviruses, are so complex that one could imagine them also to be derived from a cellular ancestor. (2) Viruses are derived from cellular genetic material that has acquired the capacity to exist and function independently. Nowadays the latter hypothesis is considered much more likely for all viruses (with the possible exception of poxviruses).

The Characteristics of Cultured Animal Cells

The medical practitioner should understand not only how viruses affect the patient as a whole but also how viruses interact with cells. This understanding can be acquired far more readily by studying isolated infected cells than by examining infected cells in the intact organism. Animal virology provided the main impetus for the development of tissue culture—the technique of growing cells in vitro. Tissue culture is now used extensively in many fields of bi-

ologic and biomedical research. Since knowledge concerning the normal cell is crucial to an understanding of the virus–cell interaction, we will first examine briefly the characteristics of animal cells cultured in vitro.

The Establishment of Animal Cell Strains

Cells of many organs can be grown in vitro. As a rule, small pieces of the tissue in question are dissociated into single cells by treatment with a dilute solution of trypsin, and a suspension of the cells is then placed into a flask, bottle, or Petri dish. There the cells attach to the flat surface, and provided that they are supplied with a growth medium, they multiply. The essential constituents of a growth medium are physiologic amounts of 13 essential amino acids and 9 vitamins, salts, glucose, and a buffering system that generally consists of bicarbonate in equilibrium with an atmosphere containing about 5 percent carbon dioxide. This medium is supplemented to the extent of about 5 percent with serum, the source of which is not predicated by the species from which the cells were derived; calf and fetal calf serum are the two most commonly employed. Antibiotics, such as penicillin and streptomycin, are also usually added in order to minimize the growth of bacterial contaminants, and a dye, such as phenol red, is generally included as a pH indicator. This medium, or more complex versions of it, will permit most cell types to multiply with a division time of 24 to 48 hours.

When cells are brought into contact with a surface, they generally attach firmly and flatten so as to occupy the maximum surface area. The only time when they are not maximally extended is during mitosis, when they become round and are therefore easily dislodged from the substratum. Cells multiply until they occupy all available surface area—that is, until they are confluent—but no further. The reason for this is that cells cease dividing when they make contact with neighboring cells, a phenomenon known as "contact inhibition" (Chap. 62).

Animal cells can be cloned just like bacterial cells, although the efficiency of cloning is frequently less than 100 percent. Numerous genetically pure cell strains are now available. They fall into two morphologic categories, epithelial cells with a polygonal outline and fibroblasts with a narrow spindle-like shape (Fig. 55-1).

The first cultures after tissue dispersion are known as "primary cultures." When such cultures are confluent they are passaged by dislodgment from the surface by treatment with trypsin or the chelating agent ethylene diamine tetraacetate (EDTA) and reseeding into several new containers, in which they form secondary cultures. Passaging can be continued in this manner, provided that an adequate supply of growth medium is supplied at regular intervals.

The overall properties of cell strains are generally stable on continuous culturing. However, mutations occur constantly, so that one particular mutant, or variant, usually emerges as the dominant population component under

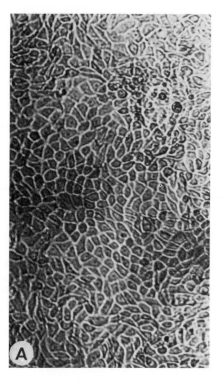

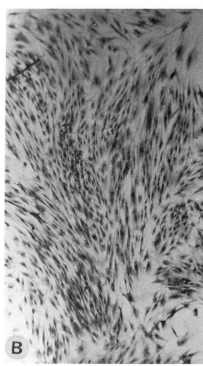

Figure 55-1. Cultured mammalian cells. **A.** Unstained monkey kidney cells, which exhibit a typical epithelioid morphology. **B.** Chick embryo fibroblasts (Giemsa stain). Note characteristic spindle shape and orderly alignment. (**A** from Eagle and Foley: Cancer Res 18: 1017, 1958; **B** courtesy of Dr. R.E. Smith.)

any given set of conditions. As a result, the same cell strain cultured in two different laboratories may exhibit detectable phenotypic differences.

The Multiplication Cycle

The multiplication of each individual cell conforms to a regular pattern, which can be thought of as a cycle (Fig. 55-2). According to this scheme, the interval between successive mitoses is divided into three periods: the G1 period that precedes DNA replication, the S period during which DNA replicates, and the G2 period during which the cell prepares for the next mitosis. RNA and protein are not synthesized while mitosis proceeds—that is, during metaphase—but are otherwise synthesized throughout the multiplication cycle. Nongrowing cells are usually arrested in the G1 period; the resting state is often referred to as G0 (G zero).

Under conditions of normal growth, the individual cells of a growing culture pass through this multiplication cycle in an unsynchronized fashion, so that cells at all stages of the cycle are always present. It is, however, possible to synchronize cells so that they multiply in step for several generations. Synchronized cell cultures are useful in studies of the reactions that are essential for progression through the multiplication cycle.

The Aging of Cell Strains

Cells derived from normal tissues cannot be passaged indefinitely. Instead, after about 50 passages, which generally occupy about 1 year, their growth rate inevitably begins to slow. The amount of time that they spend in G0 following each mitosis gradually increases, fewer and fewer cells enter the S period, and the cells' karyotype (that is, their chromosomal complement) changes from the euploid (diploid) pattern characteristic of normal cells to an aneuploid one, characterized by the appearance of supernumerary chromosomes, chromosome fragments, and chromosomal aberrations (that is, changes in the structure of individual chromosomes). Finally, the cell strain dies out. Loss of cell strains in this manner is generally guarded against by growing large numbers of cells during the early passages and storing them at −196C, the boiling point of liquid nitrogen.

Continuous Cell Lines

While cells derived from normal tissues have the properties described thus far, malignant tissues give rise to aneuploid cell lines that have an infinite life span and are referred to as "established cell lines." Infrequently such

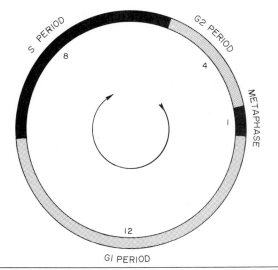

Figure 55-2. The multiplication cycle of mammalian cells. The duration of the cycle illustrated here is 25 hours; the average lengths (in hours) of the individual periods is indicated by the numbers inside the cycle.

cell lines seem to arise from euploid cell strains, but the possibility that malignant or premalignant cells were not present originally is difficult to rule out. In addition to being aneuploid and immortal, such cell lines usually have two other significant properties: they form tumors when transplanted into animals, and they can grow in suspension culture like bacteria. Cells growing in suspension are used extensively for studies of virus multiplication, since they are easier to handle experimentally than cells growing as monolayers.

Patterns of Macromolecular Biosynthesis

Since virus multiplication consists essentially of nucleic acid and protein synthesis, a brief description of the patterns of macromolecular synthesis in animal cells is relevant. The essential feature of the animal cell is its compartmentalization. The DNA of the animal cell is restricted to the nucleus at all stages of the cell cycle except during metaphase, when no nucleus exists. All RNA is synthesized in the nucleus. Most of it remains there, but messenger RNA and transfer RNA migrate to the cytoplasm. Ribosomal RNA is synthesized in the nucleolus; the two ribosomal subunits are assembled partly in the nucleolus and partly in the nucleus, and then they also migrate to the cytoplasm. All protein synthesis proceeds in the cytoplasm. The only exception to this brief summary concerns the mitochondria, which contain DNA-synthesizing, RNA-synthesizing, and protein-synthesizing systems of their own and which are located only in the cytoplasm.

The Detection of Animal Viruses

The presence of viruses is recognized by the manifestation of some abnormality in host organisms or host cells. In the organism, symptoms of virus infection vary widely, from inapparent infections (detectable only by the formation of antibody), the development of local lesions, or mild disease characterized by light febrile response to progressively more severe disease culminating in death. In cells, the symptoms of virus infection vary from changes in morphology and growth patterns to cytopathic effects, such as rounding, breakdown of cell organelles, the development of inclusion bodies, and general necrotic reactions, finally resulting in complete disintegration.

The Isolation of Animal Viruses

Many techniques have been developed for isolating viruses. The source of virus may be excreted or secreted material, the bloodstream, or some tissue. Samples are collected and, unless processed immediately, are sheltered from heat, preferably by storage at −70C, the temperature of dry ice. If necessary, a suspension is then prepared by grinding or sonicating in the presence of cold buffer solution, and this is then centrifuged in order to remove large debris and contaminating microorganisms.

This suspension is then tested for the presence of virus in several ways. First, it is injected back into the original host species in order to determine whether the first noted abnormality is produced. Second, the suspension is injected into other animals in order to establish whether there exist more susceptible hosts in which the disease develops more rapidly, more severely, or in a more easily recognizable manner. Newborn or suckling animals (often mice or hamsters) or developing chick embryos are hosts that permit many viruses to multiply more extensively than do adult animals and are accordingly widely used for virus isolation. Third, a search is conducted for a cultured animal cell strain or line in which the virus will multiply well, so that it may be isolated and characterized. Finally, cells are needed in which the virus rapidly elicits readily observable cytopathic effects and which can therefore be used to assay it.

The final stage of the isolation procedure is passage at limiting dilution in order to ensure that only a single unique virus is being isolated. This may be accomplished either by limiting serial dilution, when the virus suspension is diluted to such an extent that only one out of several aliquots inoculated gives a positive response, or by plaque isolation (p. 807). The latter is preferable, since plaques originate from single virus particles, just as bacterial colonies originate from single bacterial cells.

While virus isolation from severely diseased hosts may present no difficulty, it may be a formidable task if the original source is merely suspected of containing a small amount of virus. As a result, no symptoms may result when the initial virus suspension is inoculated into the various test systems. In such cases one generally resorts to so-called blind passaging, in the hope that gradual enrichment of virus will occur. In this procedure, cells are disrupted several days after inoculation even if they appear healthy and unaltered, and an extract of them is inoculated into fresh cells. This is repeated several times until symptoms appear. It is important that this procedure be adequately controlled by passaging extracts of uninfected cells under the same conditions, since animal cells are known to harbor latent viruses that may be induced to multiply and that may then be mistaken for the etiologic agent of whatever condition is under study.

Adaptation and Virulence

During the isolation of viruses, there may emerge variants capable of multiplying more efficiently in the host cells used for this purpose than the original wild-type virus. This phenomenon, which is known as "adaptation," has as its basis the selection of spontaneous mutants that constantly arise during virus multiplication. These mutants multiply more efficiently in the cells used for isolating the virus than in the cells of the original infected tissue. Such variants damage the original host less severely than the wild-type virus and are therefore said to be less "virulent." Viruses are often purposely adapted in order to alter growth and virulence characteristics. An example is provided by the attenuated vaccine virus strains, which are obtained by repeated passaging of virus virulent for one

host in some different host, until virus strains with decreased virulence for the original host are selected.

The Measurement of Animal Viruses

Viruses are measured by several methods that can be divided into two categories. First, viruses may be measured as infectious units, that is, in terms of their ability to infect, multiply, and produce progeny. Second, viruses may be measured in terms of the total number of virus particles, irrespective of their function as infectious agents.

Measurement of Viruses as Infectious Units

Measurement of the amount of virus in terms of the number of infectious units per unit volume is known as titration. There are several ways of determining the titer of a virus suspension, all of them involving infection of host or target cells in such a way that each particle that causes productive infection elicits a recognizable response.

Plaque Formation

In this method a series of monolayers of susceptible cells are inoculated with small aliquots of serial dilutions of the virus suspension to be titrated. Wherever virus particles infect cells, progeny virus particles are produced and released and then immediately infect adjoining cells. This process is repeated until, after a period ranging from 2 to 12 days or more, there develop areas of infected cells that

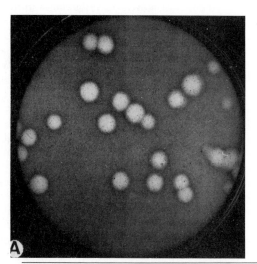 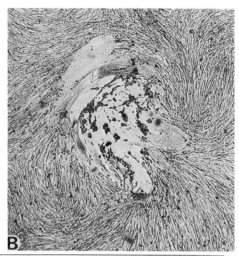

Figure 55-3. Virus plaques. **A.** Plaques of influenza virus on monolayers of chick embryo cells, 4 days after inoculation. The monolayers were stained with neutral red on day 3. **B.** Photograph showing the microanatomy of a herpesvirus plaque on BHK 21 cells. (**A** *courtesy of Dr. G. Appleyard,* **B** *courtesy of Dr. S. Moira Brown.*)

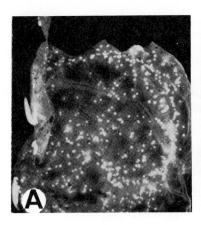 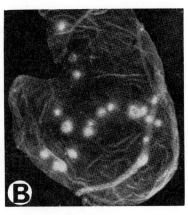

Figure 55-4. Pocks on the chorioallantoic membrane of the developing chick embryo. The membrane is cut out 2 or 3 days after inoculation, washed, and spread on a flat surface. **A.** Variola virus. **B.** Vaccinia virus. *(From Kempe: Fed Proc 14:468, 1955.)*

can be seen with the naked eye. These are called "plaques." In order to ensure that progeny virus particles liberated into the medium do not diffuse away and initiate separate (or secondary) plaques, agar is frequently incorporated into the medium.

The fundamental prerequisite for this method of enumerating infectious units is that the infected cells must differ in some way from noninfected cells; for example, they must either be completely destroyed, become detached from the surface on which they grow, or possess staining properties different from those of normal cells. In practice, the most common method of visualizing plaques is to apply the vital stain neutral red to infected cell monolayers after a certain number of days and to count the number of areas that do not stain (Fig. 55-3). Titers are expressed in terms of numbers of plaque-forming units (PFU) per milliliter.

There is a linear relationship between the amount of virus and the number of plaques produced; that is, the dose–response curve is linear. This indicates that each plaque is caused by a single virus particle. The virus progeny in each plaque therefore are clones, and virus stocks derived from single plaques are said to be "plaque purified." Plaque purification is an important technique for isolating genetically pure virus strains.

Plaque formation is often the most desirable method of titrating viruses. It is economical of cells and virus, as well as technically simple. However, not all viruses can be measured in this way, because there may be no cells that develop the desired cytopathic effects. For these viruses, alternative titration methods must be used.

Pock Formation

Many viruses cause macroscopically recognizable foci of infection or lesions on the choriollantoic membrane of the developing chick embryo. This membrane may be used in a manner similar to the cell monolayers employed for plaque assay. The main advantage is ready availability, wide virus susceptibility, and ease of handling. The main disadvantage is variation in virus susceptibility among different eggs of even the same hatch, so that a larger num-

ber of eggs than cultured cell monolayers is necessary to attain the same level of statistical significance. The lesions caused by viruses are known as pocks and are generally recognizable as opaque white or red areas caused by cell disintegration, migration, and proliferation, as well as edema and hemorrhage (in the case of red pocks) (Fig. 55-4). The actual titration is carried out as described for plaques, with enumeration of pocks taking the place of counting plaques.

Focus Formation

Many tumor viruses do not destroy the cells in which they multiply and, therefore, produce no plaques. However, they cause cells to change morphology and to multiply at a faster rate than uninfected cells. As a result, transformed cells develop into foci that gradually become large enough to be visible to the naked eye (Fig. 55-5). Assay by focus

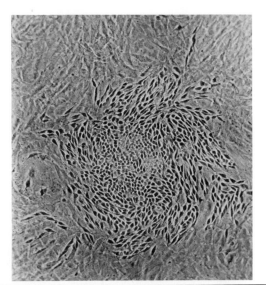

Figure 55-5. Focus of NRK (normal rat kidney) cells transformed by Kirsten murine sarcoma virus. × 200. *(Courtesy of Dr. S.A. Aaronson.)*

formation (counting the number of focus-forming units, or FFU) is analogous to assay by plaque and pock formation.

The Serial Dilution End Point Method

Although many viruses destroy cells, they do not produce the type of cytopathic effects necessary for visible plaque formation. Such viruses may be titrated by means of the serial dilution end point method. In this method serial dilutions of virus suspensions are inoculated into cell monolayers, which are then incubated until the cell sheet shows clear signs of cell destruction (Fig. 55-6). The end point is that dilution that gives a positive (cell-destroying) reaction, and the titer is calculated assuming that the last positive

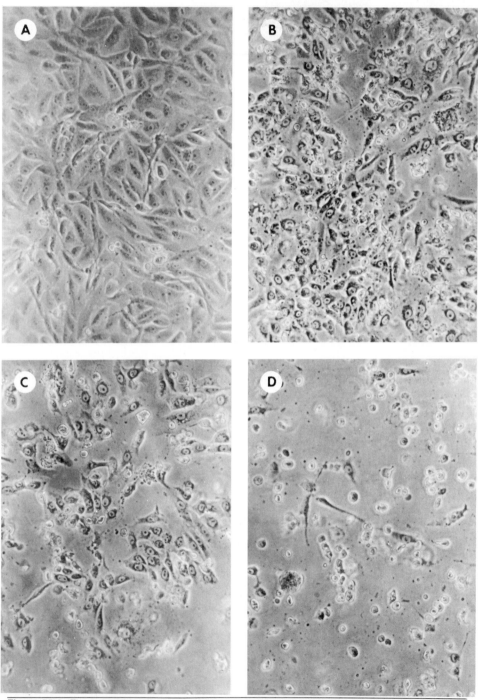

Figure 55-6. The cytopathic effects caused by reovirus serotype 3 in Vero monkey kidney cells. **A.** Normal cell sheet. **B.** Partial cell destruction at 20 hours after infection. **C.** 36 hours after infection. **D.** 48 hours after infection. × 125. *(Courtesy of Dr. E.C. Hayes.)*

dilution originally contained at least one infectious unit. Considerable accuracy can be attained by the use of statistical methods of treating results.

The dilution end point method is also employed when virus is titrated in laboratory animals. Examples are the titration of togaviruses (arboviruses) and Group A Coxsackie viruses in the brains of suckling mice, with death as the end point.

Enumeration of the Total Number of Virus Particles

It is universally true for animal viruses that even though one virus particle is capable of causing infection, not all particles in a population actually do so. The total number of virus particles in a given preparation can be determined by either direct or indirect methods.

Counting by Means of Electron Microscopy

Direct counting of virus particles by means of electron microscopic examination is carried out according to either of two methods. The first involves mixing virus preparations with suspensions of latex spheres of similar size and known concentration and spraying the mixture onto coated electron microscope grids. The number of virus particles and spheres in individual spray droplets is then counted; knowing the concentration of the spheres, the number of virus particles can be calculated (Fig. 55-7). The second method involves centrifuging virus preparations onto electron microscope grids and counting the virus particles; knowing what volume of the virus suspension was centrifuged, the virus concentration can be calculated.

Measurement of Optical Density

The concentration of highly purified virus preparations can be routinely determined by very simple methods once

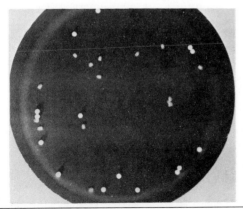

Figure 55-7. A spray droplet containing 15 latex particles (spheres) and 14 vaccinia virus particles (slightly smaller brick-shaped particles). × 6500. *(From Dumbell et al.: Virology 4:467, 1957.)*

they have been standardized by electron microscopy. One of these methods is measurement of the optical density. For example, 1 ml of a suspension of reovirus particles that absorbs 90 percent of incident light at a wavelength of 260 nm (that is, 1 optical density unit or 1 $OD_{260 \text{ nm}}$) contains 2.1×10^{12} virus particles, and 1 $OD_{260 \text{ nm}}$ of vaccinia virus corresponds to 1.2×10^{10} virus particles.

The Hemagglutination Assay

The most common indirect method of measuring the number of virus particles is the hemagglutination assay. Many animal viruses adsorb to the red blood cells of various animal species. Each virus particle is multivalent in this regard; that is, it can adsorb to more than one cell at a time. In practice, the maximum number of cells with which any particular virus particle can combine is two, because red cells are far bigger than viruses. In a virus-cell mixture in which the number of cells exceeds the number of virus particles, the small number of cell dimers that may be formed is generally not detectable, but if the number of virus particles exceeds the number of cells, a lattice of cells is formed that settles out in a highly characteristic manner readily distinguishable from the settling pattern exhibited by unagglutinated cells.

The hemagglutination assay is performed by determining the virus dilution that will just hemagglutinate a given number of red cells (Fig. 55-8). Since the number of virus particles necessary for this is readily calculated, hemagglutination serves as a highly accurate and rapid method of quantitating virus particles. It was and still is particularly useful in studies with myxoviruses, particularly influenza virus, and many others.

The Significance of the Infectious Unit: Virus Particle Ratio

For all animal viruses the number of virus particles in any given preparation exceeds the number of demonstrably infectious units; usually the ratio of infectious units to particles is in the range of 1:10 to 1:1000 or even less. There are two possible explanations for this. The first is that virus preparations contain a majority of noninfectious particles. Although this may be so sometimes, it is unlikely to be the general rule. It is more likely that although all virus particles in a given preparation are capable of causing productive infection, only a small proportion of them are actually successful in doing so. Two lines of evidence support this view. The first is that the titer of a given virus preparation varies markedly, depending on the nature of the assay system. For example, the titer often differs with the route of inoculation if the virus is assayed in whole animals and with the type of cell if it is assayed in cultured cells. Second, before a virus particle can manifest itself (e.g. as a plaque, pock, or focus), it must initiate several productive infection cycles that require numerous reactions, many of which have a low probability of occurring (Chap. 58). Therefore, the number of infectious units can-

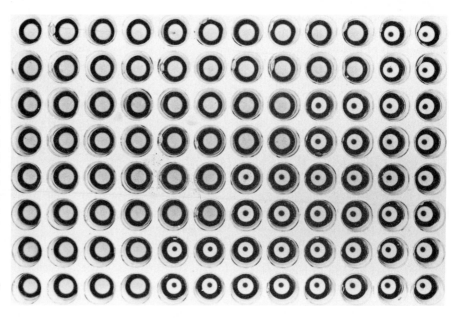

Figure 55-8. Hemagglutination titration of influenza virus. In the top two rows a sample of influenza virus was diluted in serial twofold steps from left to right; in the next two rows the amount of virus in the first well was the same as in the third well in the top row, and so on down. The same number of red blood cells was then added to all wells, and after mixing, the tray was placed at 4C for 2 hours. Unagglutinated cells form a dark button; where the virus has agglutinated cells, the resulting lattice has prevented button formation. The pattern developed in this tray attests to the reproducibility of the technique. *(Courtesy of Dr. E.C. Hayes.)*

not equal the total number of virus particles. The ratio of the number of infectious units to the total number of virus particles may generally be regarded as a measure of the probability with which virus particles achieve productive infection.

FURTHER READING

Books

VIRUSES

Andrews C, Pereira HG, Wildy P: Viruses of Vertebrates, 4th ed. London, Bailliere Tindall, 1978

Fenner FJ, McAuslan BR, Mims CA, et al.: Animal Viruses, 2nd ed. New York, Academic Press, 1974

Fenner FJ, White DO: Medical Virology, 2nd ed. New York, Academic Press, 1976

Fraenkel-Conrat H, Wagner RR (eds): Methods Used in the Study of Viruses, Comprehensive Virology. New York, Plenum, 1981, vol 17

Howard CR (ed): New Developments in Practical Virology. New York, Liss, 1982

Hughes SS: The Virus: A History of the Concept. New York, Neale Watson Academic Publication, 1977

Luria SE, Darnell JE Jr, Baltimore D, et al.: General Virology, 3rd ed. New York, Wiley, 1978

CELLS

Alberts B, Bray D, Lewis J, et al.: Molecular Biology of the Cell. New York, Garland Publishing, 1983

Barigozzi C (ed): Original and Natural History of Cell Lines. New York, Liss, 1978

Clarkson B, Marks PA, Till JE (eds): Differentiation of Normal and Neoplastic Hematopoietic Cells. New York, Cold Spring Harbor Laboratory, 1979

Crow R, Ozer H, Rifkin D (eds): Experiments with Normal and Transformed Cells. New York, Cold Spring Harbor Laboratory, 1979

Dingle JT, Gordon JL (eds): Cellular Interactions. Amsterdam/New York, Elsevier, 1981

Frederick JF (ed): Origins and Evolution of Eukaryotic Intracellular Organelles. New York, New York Academy of Sciences, 1980

Gall JG, Porter KR, Siekevitz P (eds): Discovery in Cell Biology. New York, Rockefeller University Press, 1981

Glick JL (ed): Fundamentals of Human Lymphoid Cell Culture. New York, Marcel Dekker, 1980

Goldstein L, Prescott DM (eds): The Structure and Replication of Genetic Material. Cell Biology. New York, Academic Press, 1979, vol 2

Hoffman JF, Giebisch GH, Bolis L (eds): Membranes in Growth and Development. New York, Liss, 1981

Jeter JR, Cameron IL, Padilla GM, et al. (eds): Cell Cycle Regulation. New York, Academic Press, 1978

Lerner RA, Bergsma D (eds): The Molecular Basis of Cell–Cell Interaction. New York, Liss, 1978

Lloyd CW, Rees DA (eds): Cellular Controls in Differentiation. New York, Academic Press, 1981

Lloyd D, Poole RK, Edwards SW (eds): The Cell Division Cycle. New York, Academic Press, 1982

Maul GG (ed): The Nuclear Envelope and the Nuclear Matrix. New York, Liss, 1981

Nicolini C (ed): Cell Growth. New York, Plenum, 1982

Pollack R (ed): Readings in Mammalian Cell Culture. New York, Cold Spring Harbor Laboratory, 1981

Poste G, Nicholson GL (eds): Cytoskeletal Elements and Plasma Membrane Organization. New York, Elsevier, 1981

Salmon SE (ed): Cloning of Human Tumor Stem Cells. New York, Liss, 1980

Sato G (ed): Functionally Differentiated Cell Lines. New York, Liss, 1981

Sato GH, Pardee AB, Sirbasku DA (eds): Growth of Cells in Hormonally Defined Media. New York, Cold Spring Harbor Laboratory, 1982

Whitson GL (ed): Nuclear–Cytoplasmic Interactions in the Cell Cycle. New York, Academic Press, 1980

Zimmerman AM, Forer A (eds): Mitosis/Cytokinesis. New York, Academic Press, 1982

Selected Reviews and Papers

Amsterdam A, Jamieson JT: Techniques for dissociating pancreatic exocrine cells. J Cell Biol 63:1037, 1974

Barnes D, Sato G: Serum-free culture. Cell 12:649, 1980

Branton D, Cohen CM, Tyler J: Interaction of cytoskeletal proteins on the human erythrocyte membrane. Cell 24:24, 1981

Conrad GW, Hart GW, Chen Y: Differences in vitro between fibroblast-like cells from cornea, heart and skin of embryonic chicks. J Cell Sci 26:119, 1977

Dustin P: Microtubules. Sci Am 243:66, 1980

Edelman GM: Cell adhesion molecules. Science 219:450, 1983

Mullinger AM, Johnson RT: Packing DNA into chromosomes. J Cell Sci 46:61, 1981

Osborn M, Weber K: Intermediate filaments: Cell-type-specific markers in differentiation and pathology. Cell 31:303, 1982

Pastan IH, Willingham MC: Journey to the center of the cell: Role of the receptosome. Science 214:504, 1981

Porter KR, Tucker JB: The ground substance of the living cell. Sci Am 244:56, 1981

Rothman JE: The Golgi apparatus. Science 213:1212, 1981

Smith JR, Hayflick L: Variation in the life span of clones derived from human diploid strains. J Cell Biol 62:48, 1974

Stack SN, Brown DB, Dewey WC: Visualization of interphase chromosomes. J Cell Sci 26:281, 1977

Watt FM, Harris H: Microtubule organizing centers in mammalian cells in culture. J Cell Sci 44:103, 1980

Wishnow RN, Steinfeld JL: The conquest of the major infectious diseases in the United States: A bicentennial retrospect. Ann Rev Microbiol 30:427, 1976

CHAPTER 56

The Structure, Components, and Classification of Viruses

The Morphology of Animal Viruses

Although animal viruses differ widely in shape and size, they are nevertheless constructed according to certain common principles. Basically, viruses consist of nucleic acid and protein. The nucleic acid is the genome that contains the information necessary for virus multiplication; the protein is arranged around the genome in the form of a layer or shell that is termed the capsid. The structure consisting of shell plus nucleic acid is the nucleocapsid. Many animal virus particles consist of naked nucleocapsids, whereas others possess an additional envelope that is usually acquired as the nucleocapsid buds from the host cell. The complete virus particle is known as the virion, a term that denotes both intactness of structure and the property of infectiousness.

Capsids

The essential feature of capsids is that they are composed of numerous repeating subunits—identical or belonging to only a few different species—arranged in precisely defined patterns. The simplest subunits are single protein molecules; more complex forms are morphologic subunits termed capsomers that can be seen with the electron microscope and that consist of several either identical or different protein molecules. The use of only a few types of subunits for capsid construction has two noteworthy consequences: (1) It minimizes the amount of genetic informa-

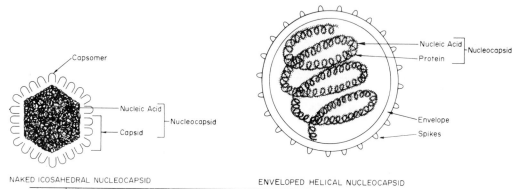

NAKED ICOSAHEDRAL NUCLEOCAPSID   ENVELOPED HELICAL NUCLEOCAPSID

Figure 56-1. The two basic patterns of animal virus structure. **Left:** The condensed genome is enclosed by a shell of capsomers arranged so as to display 5:3:2 rotational symmetry. **Right:** The extended genome is enclosed by protein molecules arranged so as to display helical symmetry. The resultant structure, the nucleocapsid, is enclosed in an envelope to whose outer surface glycoprotein spikes are attached.

tion necessary to specify capsids, and (2) it assures that they will be assembled efficiently. Capsid proteins exhibit a strong tendency to bind to one another, and much of the information necessary for the morphogenesis of nucleocapsids is inherent in their amino acid sequence.

Capsids (and envelopes) have a dual function. The first is to protect viral genomes from potentially destructive agents in the extracellular environment, such as enzymes. All viral genomes are protected from nucleases by virtue of the capsids that enclose them. Animal viruses are generally resistant to attack by the proteases of higher animals, such as pepsin, trypsin, and chymotrypsin. Some, such as the enteroviruses, are completely resistant, whereas others, such as the poxviruses and reoviruses, possess susceptible outer shells but resistant cores. Glycoprotein spikes of enveloped viruses (see below) can generally be removed by treatment with proteolytic enzymes, and the resulting "bald" particles lack infectivity. Phospholipases usually inactivate enveloped viruses by hydrolyzing their phospholipids.

The second function of viral capsids is to introduce viral genomes into host cells. The need for this latter function stems from the fact that viral nucleic acids are often longer than cell diameters and cannot penetrate into cells by themselves. Capsids (and envelopes), on the other hand, adsorb readily to cell surfaces and can enter cells by several mechanisms (see Chap. 58).

Envelopes

Only six families of animal viruses exist as naked nucleocapsids. In all others the nucleocapsids are enclosed by membrane-containing envelopes that are acquired as the nucleocapsids bud through special patches of cell membrane—either in outer plasma cell membranes or in vacuolar membranes—on their way to the exterior of the cell. The membrane patches through which nucleocapsids bud are virus-modified; usually the cell-specified proteins in them are completely replaced by virus-specified proteins,

and virus-specified glycoprotein spikes are attached to their outer surface. However, there are exceptions. The envelopes of herpesviruses and RNA tumor viruses probably still contain some host-coded proteins, and although herpesvirus envelopes contain glycoproteins, they do not possess obvious spikes.

Viral envelopes generally lack rigidity. As a consequence they usually appear heterogeneous in shape and size when fixed for electron microscopy, and enveloped

0   100 Å

Figure 56-2. Schematic representation of tobacco mosaic virus. As can be seen in the cutaway section, the ribonucleic acid helix is associated with protein molecules in the ratio of three nucleotides per protein molecule. *(From Klug and Caspar: Adv Virus Res 7:225, 1960.)*

viruses are therefore often said to be pleomorphic. There is little doubt, however, that in their native state most enveloped viruses are spherical, enclosing either icosahedral nucleocapsids or spherically coiled helical nucleocapsids. However, two types of enveloped viruses are not spherical. These are the rhabdoviruses, which possess a highly characteristic bullet-like shape, rounded at one end and flat at the other, and certain strains of influenza virus, whose helical nucleocapsids become enveloped not in a coiled but in an extended configuration, which causes the enveloped virus particles also to be long and filamentous.

Nucleocapsids

Viral nucleocapsids are constructed according to a small number of basic patterns. Two of these have been studied in great detail at both the structural and the molecular level. In one, the nucleic acid is extended; in the other, it is condensed (Fig. 56-1). Superimposed on these two patterns are variations dictated both by the size of the genome and by the nature of the capsid proteins.

Nucleocapsids with Helical Symmetry. The prototype of nucleocapsids in which the nucleic acid occurs as an extended filament is a plant virus, tobacco mosaic virus

(TMV), whose structure has been studied extensively by x-ray diffraction. In this virus, the extended nucleic acid molecule is surrounded by protein molecules arranged helically so as to yield a structure with a single rotational axis (Fig. 56-2). The ortho- and paramyxoviruses and rhabdoviruses possess nucleocapsids constructed in this manner, each with its own characteristic length, width, periodicity, flexibility, and stability (Fig. 56-3). It should be noted that these nucleocapsids are not the complete virus particles; the particles of these virus families consist of the nucleocapsids coiled more or less tightly inside envelopes (Figs. 56-4 and 56-5).

Nucleocapsids with Icosahedral Symmetry. In the second pattern of virus structure, the nucleic acid is condensed and forms the central portion of a quasispheric nucleocapsid. Here the capsid consists of a shell of protein molecules that are clustered into small groups called capsomers, with the bonds between molecules within capsomers being stronger than those between capsomers. Capsomers are morphologic units that can often be seen with the electron microscope; they vary in size and shape from virus to virus.

X-ray diffraction analysis indicates that in this type of nucleocapsid the capsomers are arranged very precisely ac-

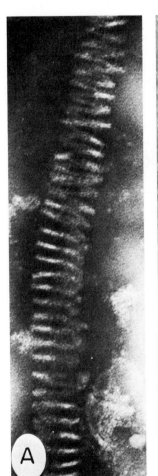
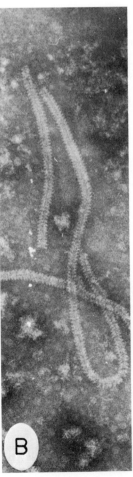
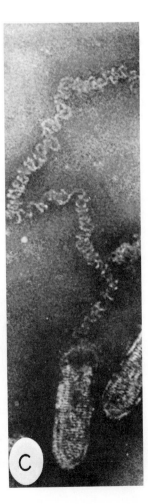

Figure 56-3. The nucleocapsids of (**A**) influenza virus strain PR8 (× 225,000); (**B**) measles virus (× 150,000); and (**C**) vesicular stomatitis virus (VSV) (× 160,000). The latter is emerging from a damaged virus particle. (**A,** from Almeida and Waterson: In Barry and Mahy (eds): The Biology of Large RNA Viruses, 1970. New York, Academic Press; **B,** from Finch and Gibbs: J Gen Virol 6:144, 1970; **C,** from Simpson and Hauser: Virology 29:660, 1966.)

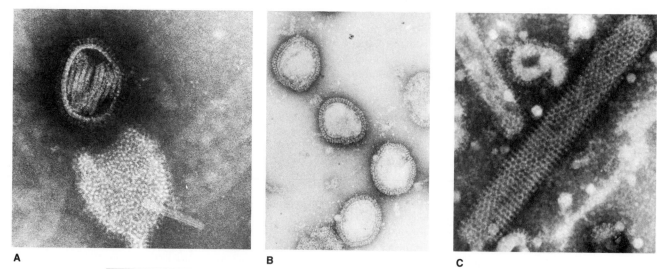

Figure 56-4. The structure of influenza virus. **A.** Influenza virus A_2, stained with phosphotungstate. One particle is penetrated by the stain, thereby revealing the arrangement of the internal nucleocapsid. × 155,000. *(Courtesy of Dr. M. V. Nermut.)* **B.** Influenza virus A_0/WSN, stained with phosphotungstate, revealing the characteristic arrangement of spikes on the particle surface. There are two types of spikes; hemagglutinin spikes are about six times more numerous than neuraminidase spikes. × 135,000. *(Courtesy of Dr. I. T. Schulze.)* **C.** A filamentous particle of an influenza C strain. Note the regular subunit surface pattern. × 115,000. *(From Apostolov and Flewett: J Gen Virol 4:366, 1969.)*

cording to icosahedral patterns characterized by 5:3:2-fold rotational symmetry (Fig. 56-6). Two such patterns are found among animal viruses. The first is exhibited most clearly by adenoviruses. The adenovirus capsid is constructed in the shape of an icosahedron with six capsomers along each edge and 252 capsomers altogether (Fig. 56-7). Of these, 240 are spherical and are situated along the edges and on the faces of the icosahedron; each has six nearest neighbors and is known as a hexon or hexamer. The remaining 12 are situated at the 12 vertices of the icosahedron and have five nearest neighbors; these are known as pentons or pentamers. They have a highly characteristic shape, consisting of a spherical base and a long fiber that may serve as the cell attachment organ.

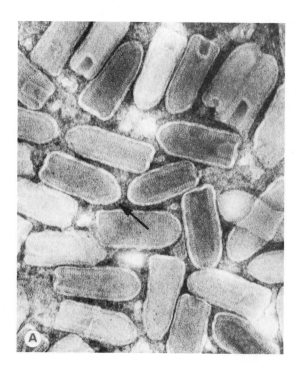

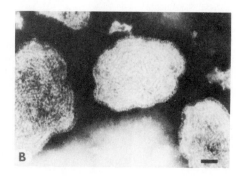

Figure 56-5. A. Highly characteristic bullet-shaped vesicular stomatitis virus (VSV) particles, some penetrated by stain, revealing the tightly coiled nucleocapsid. Note glycoprotein spikes (arrow). *(Courtesy of Dr. Erskine Palmer).* **B.** Sendai virus, a paramyxovirus. Note the tightly coiled nucleocapsid. × 73,500. *(From Maeno et al.: J Virol 6:492, 1970.)*

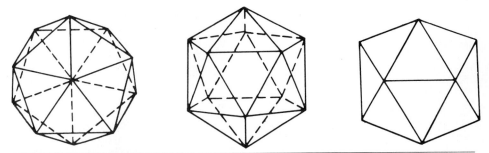

Figure 56-6. The icosahedron viewed normal to five-, three-, and two-fold rotational axes. Edges of the upper and lower surfaces are drawn in solid and broken lines respectively. The five-fold rotational axes pass through the vertices (**left**); the three-fold rotational axes pass through the centers of the triangular faces (**center**); and the two-fold rotational axes pass through the edges (**right**). In this view, the edges on the upper and lower surfaces coincide. Note that the icosahedron possesses 12 vertices, 20 triangular faces, and 30 edges.

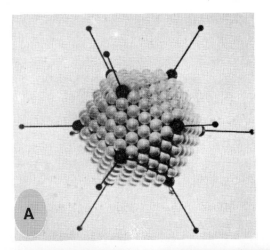

Figure 56-7. A. Model of the adenovirus particle constructed by R. C. Valentine. **B.** Adenovirus freeze-dried and shadowed with platinum. × 400,000. *(Courtesy of Dr. M.V. Nermut.)*

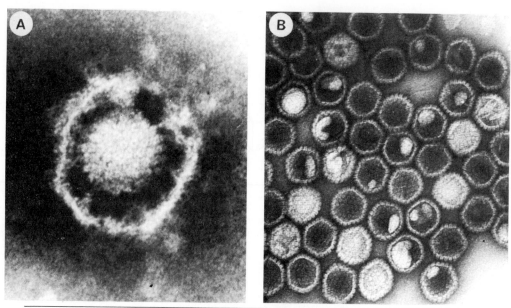

Figure 56-8. The morphology of herpesviruses. **A.** Enveloped equine abortion virus (EAV) particle. × 125,000. **B.** EAV particles from which the envelope has been removed by treatment with detergent. × 75,000. *(From Abodeely et al.: J Virol 5:513, 1970.)*

The capsids of four other virus families are constructed similarly. Iridovirus capsids possess 10 capsomers along each edge, and there are 1112 capsomers altogether—1100 hexons and 12 pentons; herpesvirus capsids possess five capsomers along each edge and 162 capsomer altogether—150 prism-shaped hexons and 12 pentons (Fig. 56-8); papovavirus capsids consist of 72 capsomers—60 hexons and 12 pentons (Fig. 56-9); and parvovirus capsids appear to be made up of 32 capsomers—20 hexons and 12 pentons (Fig. 56-9).

The second pattern is exhibited most clearly by picornaviruses. Here 60 identical capsomers, each composed of four different proteins, are situated equidistantly from a common center, which results in a spherical capsid (Fig. 56-10). However, instead of being bonded equally strongly to each other, these capsomers are bonded into groups of five, 12 of which make up the capsid. Reoviruses are unique in possessing two capsids shells (Fig. 56-11). Both possess icosahedral symmetry, but it has so far proved impossible to discern either the total number of capsomers or the precise manner in which they are arranged. The capsid shell of some members of this family,

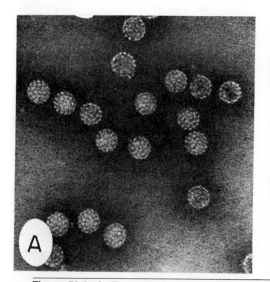
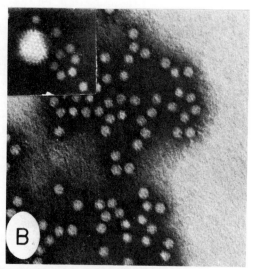

Figure 56-9. A. The papovavirus SV40 (× 160,000). **B.** The parvovirus adeno-associated virus (AAV) type 4 (× 150,000). **Insert:** A particle of the simian adenovirus SV15, which enabled the AAV to multiply (see Chap. 60). *(Courtesy of Dr. Heather Mayor.)*

Figure 56-10. Model of the picornavirus capsid. There are 60 capsomers, each consisting of three protein molecules, represented here by white, gray, and black balls. (In fact, in the mature virion, one of these protein molecules is cleaved into a larger and a smaller fragment, both of which remain in position.) Groups of five capsomers (as outlined) form units that are intermediates during morphogenesis; the capsid is composed of 12 of these "groups of five." *(Adapted from Johnston and Martin: J Gen Virol 11:77, 1971.)*

such as the orbiviruses and the rotaviruses, appears to be composed of 32 large ring-shaped capsomers. It is more likely however that this type of capsid is composed of numerous small subunits arranged in ring-shaped (or hexagonal) patterns and that many of these subunits are shared by adjacent rings so that what is visible is 32 holes, rather

than 32 capsomers. Reovirus capsids are probably structured similarly, although here the rings are much less prominent.

In the case of the picornaviruses, caliciviruses, adenoviruses, papovaviruses, parvoviruses, and reoviruses, the virus particles are the naked nucleocapsids. In the case of the herpesviruses and the sole iridovirus that is a mammalian virus, however, the naked nucleocapsids themselves are relatively noninfectious; here the virus particles consist of enveloped icosahedral nucleocapsids.

Structure of Togaviruses, Bunyaviruses, Arenaviruses, and Coronaviruses

These viruses are all enveloped particles to whose outer surface glycoprotein spikes are attached; the spikes are particularly prominent in coronaviruses (Fig. 56-12), where they are large and club-shaped and surround the virus particles like a corona (hence the name). The nucleocapsids of these viruses exhibit a variety of morphologic patterns. In the case of the togaviruses, the alphaviruses—which comprise the old group A arboviruses—contain condensed RNA molecules intimately associated with protein to form nucleocapsids that possess distinct icosahedral symmetry (Fig. 56-13); but the nucleocapsids of the slightly smaller flaviviruses—the old group B arboviruses, the name deriving from the prototype yellow fever virus—possess no obvious symmetry elements. The bunyaviruses have a diameter that is about twice that of togaviruses and possess three coiled circular nucleocapsids that possess helical symmetry (Fig. 56-14). The arenaviruses (Fig. 56-15)

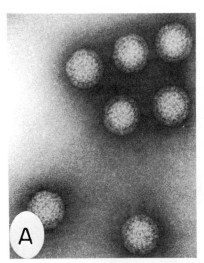
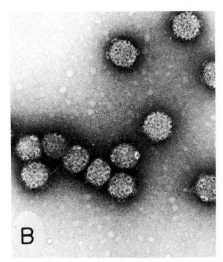
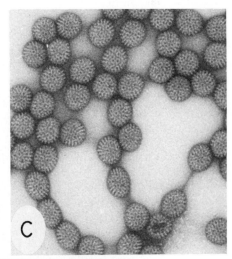

Figure 56-11. Morphology of reovirus and human rotavirus. **A.** Reovirus. Note double capsid shell. The arrangement of capsomers is clearly discernible at the periphery. × 120,000. **B.** Reovirus cores are derived from reovirus particles by digesting their outer capsid shell with chymotrypsin. Note the large spikes; there are 12, located as if situated on the 12 vertices of an icosahedron. **C.** Human rotavirus. Its structure is similar to, but not identical with, that of orbiviruses. × 136,000. *(**A, B,** courtesy of Drs. R.B. Luftig and W.K. Joklik; **C,** courtesy of Dr. Erskine Palmer, Centers for Disease Control, Atlanta.)*

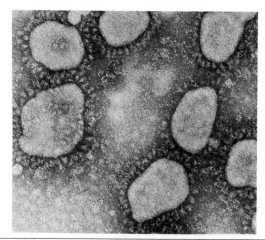

Figure 56-12. Coronavirus particles. Note pleomorphic envelopes studded with characteristic widely spaced club- or pear-shaped surface projections. × 144,000. *(From Kapikian: In Lennette and Schmidt (eds): Diagnostic Procedures for Viral and Rickettsial Infections, 4th ed, 1969. New York, American Public Health Assoc.)*

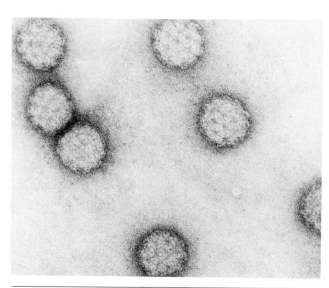

Figure 56-13. The togavirus Sindbis virus, stained with uranyl acetate. × 240,000. *(Courtesy of Dr. P.J. Enzmann.)*

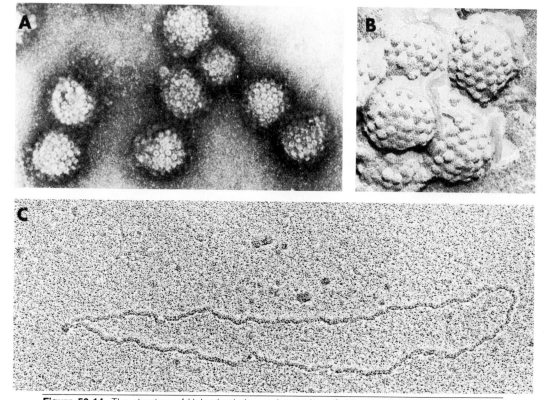

Figure 56-14. The structure of Uukuniemi virus, a bunyavirus. **A.** Virus particles fixed with glutaraldehyde and negatively stained with uranyl acetate. × 100,000. **B.** Freeze-etched, glutaraldehyde-fixed virus particles: a group of particles showing the icosahedral arrangement of the surface projections. × 180,000. **C.** Circular nucleocapsid of this virus, shadowed with platinum. Nucleocapsids are released from virus particles by treatment with the nonionic detergent Triton X-100. × 60,000. *(From von Bonsdorff: J Virol 16:1296, 1975.)*

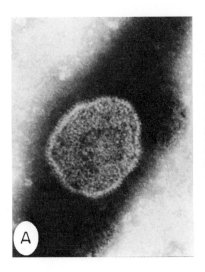

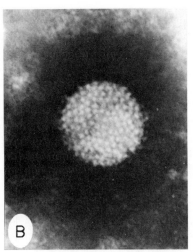

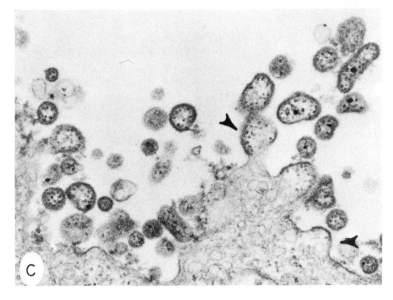

Figure 56-15. Tacaribe virus, an arenavirus. **A, B.** Two virus particles, one of which has been partially penetrated by negative contrast medium, showing the glycoprotein spikes that cover their surface. × 135,000 and × 235,000, respectively. **C.** Thin section of Paraña virus particles budding from the plasma membrane of Vero African green monkey kidney cells. Note the characteristic dense granules, which are ribosomes. × 45,000. *(From Murphy et al.: J Virol 6:507, 1970.)*

are slightly larger than bunyaviruses and contain two circular coiled nucleocapsids; they also contain several highly characteristic granules (Latin *arenosus,* sandy) that have been shown to be ribosomes (which, however, are not required for virus multiplication). Finally, the coronaviruses (Fig. 56-12), which are about the same size as bunyaviruses, possess linear helical nucleocapsids that resemble those of myxoviruses.

Structure of RNA Tumor Viruses, Poxviruses, and Some Miscellaneous Viruses

RNA tumor viruses have a more complex structure; they consist of a concentrically coiled nucleocapsid that possesses icosahedral symmetry and is closely associated with

an "inner coat" that is itself bounded by a membrane bearing more, less prominent, glycoprotein spikes (Fig. 56-16) (see Chap. 62).

Poxviruses are the largest and most complex of all animal viruses. Morphologically, there are two classes of poxviruses. Most poxvirus particles are prolate ellipsoids that are often brick-shaped when fixed for electron microscopy. They are covered on their outer surface with tubules or filaments arranged in a characteristic whorled or mulberry pattern (Fig. 56-17). Within this outer layer there is a protein coat that contains two lateral bodies of unknown composition and function, and a DNA-containing nucleoid or core bounded by a layer of well-defined protein subunits.

Parapoxviruses are somewhat smaller, ovoid rather than brick-shaped, and covered on their outer surface by tubules or filaments similar to those that cover the poxvirus particles just described, except that they are arranged

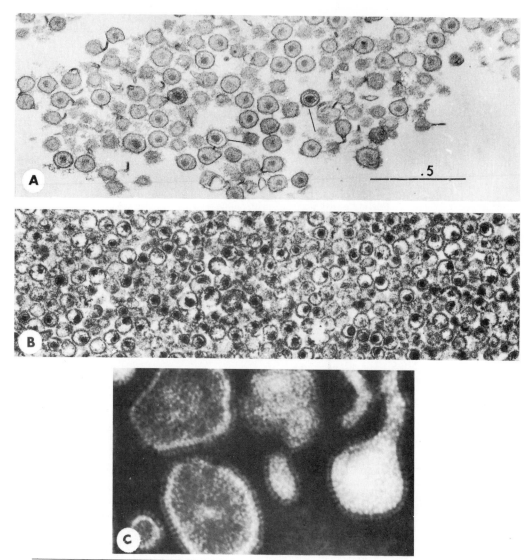

Figure 56-16. The structure of RNA tumor viruses. **A.** Thin section of Rous sarcoma virus, a C-type virus. Outer and inner membranes as well as nucleoids are clearly visible (arrows); note the central location of the nucleoid. × 52,000. **B.** Thin section of mouse mammary tumor virus, a B-type virus. Note the eccentric location of the nucleoid. **C.** Mouse mammary tumor virus stained with phosphotungstate. Note the prominent glycoprotein spikes. **(A,** from Courington and Vogt: J Virol 1:400, 1967; **B, C,** courtesy of Dr. D. Moore.)

in a highly regular, crisscross pattern, which is probably caused by one continuous filament wound round each particle in 12 to 15 left-handed turns (Fig. 56-18). The internal components of parapoxvirus particles are similar to those of the poxvirus particles described above.

Finally, there are several viruses that have not yet been assigned to any virus family. Among these are hepatitis B virus (serum hepatitis virus), which has a diameter of 42 nm (Dane particles, see Chaps. 62 and 76); and the chronic infectious neuropathic agents.

The relative sizes of some important animal viruses are illustrated in Fig. 56-19. Their morphology is summarized in Table 56-1.

The Nature of the Components of Animal Viruses

Purification of Viruses

Little serious work on the properties, composition, and molecular biology of viruses is possible unless pure virus is available. The starting material for virus purification may be any material that contains a sufficiently large amount of virus. This may be cellular material such as infected organs or cultured cells, or it may be extracellular material such as plasma, allantoic fluid, or cell culture medium. If the virus concentration in such material is not high

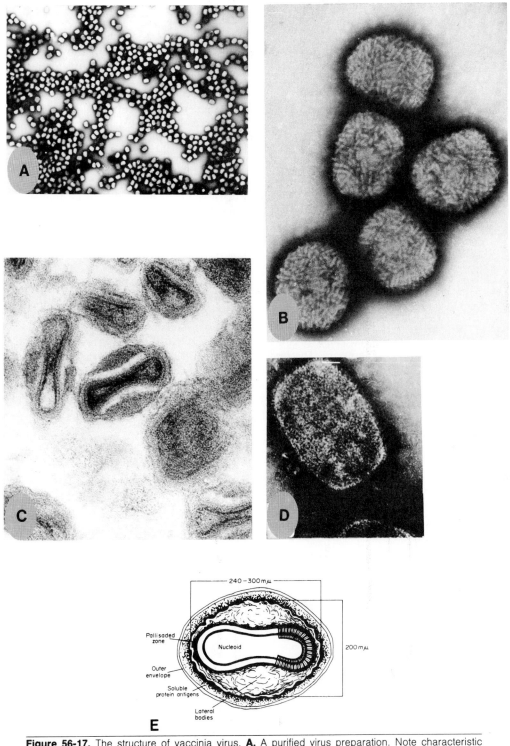

Figure 56-17. The structure of vaccinia virus. **A.** A purified virus preparation. Note characteristic brick shape. × 6,000. **B.** Vaccinia virus particles stained with phosphotungstate to reveal surface structure. Note the characteristic arrangement of rodlets or tubules. × 60,000. **C.** Cross-section of vaccinia virus particle. × 70,000. **D.** An isolated core, showing regular surface elements. × 90,000. **E.** A model of the vaccinia virus particle. *(A, B, C, courtesy of Dr. Samuel Dales; D, from Easterbrook: J Ultrastr Res 14:484, 1966; E, adapted from Westwood et al.: J Gen Microbiol 34: 67, 1964.)*

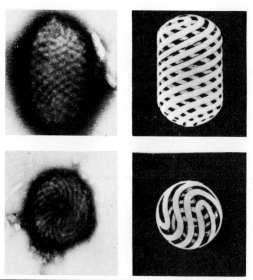

Figure 56-18. Parapoxvirus particles. These are particles of contagious pustular dermatitis virus (ORF), negatively stained so as to reveal the crisscross arrangement of surface strands or tubules. × 90,000. *(From Buttner et al.: Arch Ges Virusforsch 14: 657, 1964.)*

gation or by adsorption to and elution from red blood cells (in the case of myxoviruses), or by passage through columns that contain materials capable of separating viral and cellular components. For the final purification step, fractionation by means of density gradient centrifugation is almost always used. There are two modes of employing this technique (Fig. 56-20). In velocity density gradient or rate zonal centrifugation, the virus suspension is layered onto a density gradient; that is, a solution of either sucrose or glycerol or some salt of gradually increasing density, the maximum density being such that virus particles would migrate to the bottom of the tube if centrifuged long enough. If centrifuged for shorter periods, particles with the same sedimentation coefficient, which depends on size, density, and shape, sediment as homogeneous bands that may be collected. This step eliminates all impurities except those with the same sedimentation coefficient as virus particles. In order to eliminate the remaining impurities, the particles in the virus-containing band recovered from the first density gradient are then centrifuged to equilibrium in a second density gradient composed of solutions of higher density. Here particles form bands where the density of the medium is identical to their own buoyant density. Because contaminants with both the same sedimentation coefficient and the same density as virus particles are rare, virus purified by two such density gradient centrifugation steps is generally considered to be essentially pure (that is, at least 97 to 99 percent free of nonviral material).

enough, an initial concentration step that employs either precipitation or centrifugation may be necessary. The next step is then generally designed to achieve a preliminary purification by removing the bulk of nonviral material. This may be achieved by treatment with detergents or emulsification with organic solvents followed by centrifu-

The purification of icosahedral viruses generally presents no difficulties. Some enveloped viruses, however, are

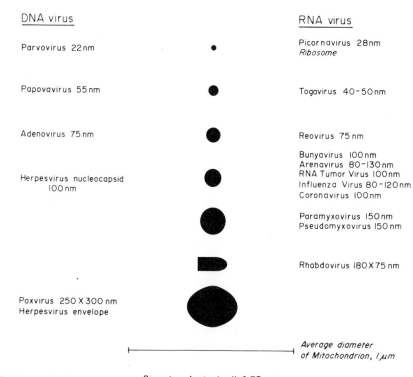

Figure 56-19. The relative sizes of the principal families of animal viruses. Unless otherwise indicated, the scale is the same for all.

DNA virus

Parvovirus 22 nm

Papovavirus 55 nm

Adenovirus 75 nm

Herpesvirus nucleocapsid 100 nm

Poxvirus 250 × 300 nm
Herpesvirus envelope

RNA virus

Picornavirus 28 nm
Ribosome

Togavirus 40-50 nm

Reovirus 75 nm

Bunyavirus 100 nm
Arenavirus 80-130 nm
RNA Tumor Virus 100 nm
Influenza Virus 80-120 nm
Coronavirus 100 nm

Paramyxovirus 150 nm
Pseudomyxovirus 150 nm

Rhabdovirus 180 × 75 nm

Average diameter of Mitochondrion, 1 μm

Diameter of animal cell, 0.75 meter

Length of DNA in poxvirus particle, 7.5 meters

TABLE 56-1. THE MORPHOLOGY OF ANIMAL VIRUSES

Virus	Morphology
DNA Virus	
Poxvirus	Complex
Iridovirus	Enveloped icosahedral nucleocapsid
Herpesvirus	Enveloped icosahedral nucleocapsid
Adenovirus	Naked icosahedral nucleocapsid
Papovavirus	Naked icosahedral nucleocapsid
Parvovirus	Naked icosahedral nucleocapsid
RNA Viruses	
Picornavirus	Naked icosahedral nucleocapsid
Togavirus	Enveloped icosahedral nucleocapsid
Bunyavirus	Enveloped helical circular nucleocapsids
Reovirus	Naked double-shelled icosahedral nucleocapsid
Orthomyxovirus	Enveloped helical nucleocapsid
Paramyxovirus	Enveloped helical nucleocapsid
Rhabdovirus	Enveloped helical nucleocapsid
RNA tumor virus	Enveloped icosahedral nucleocapsid
Arenavirus	Enveloped helical nucleocapsid
Coronavirus	Enveloped helical nucleocapsid

not easily purified because the amount of envelope per virus particle may be variable, which causes them to be heterogeneous with respect to both size and density.

It is almost impossible to establish absolute criteria of purity for viruses, mainly because small amounts of cellular constituents tend to absorb to them. Absence of particulate impurities is best assessed by electron microscopic examination.

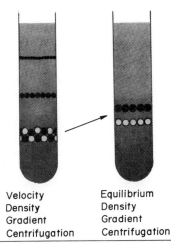

Velocity Density Gradient Centrifugation   Equilibrium Density Gradient Centrifugation

Figure 56-20. Application of the technique of density gradient centrifugation to virus purification. A partially purified virus preparation is layered onto the left-hand density gradient, which is centrifuged so that particles with the same sedimentation coefficient form distinct bands. Three such bands are shown, one of which contains virus (open spheres). This band is then centrifuged to equilibrium in the right-hand density gradient, in which particles are separated according to their buoyant density. See text for further details.

Viral Nucleic Acids

The nucleic acids of animal viruses are astonishingly diverse. Some are DNA, others RNA; some are double-stranded, others single-stranded; some are linear, others circular; some have plus polarity, others minus polarity. Information concerning these and other properties of viral nucleic acids is essential for an understanding of the key reactions during virus multiplication cycles.

Size of Viral Nucleic Acids

The nucleic acid content of animal viruses varies within wide limits. At the lower end of the scale, only about 2 percent of influenza virus particles are RNA, and only about 5 percent of poxvirus particles are DNA; at the upper end, about 25 percent of picornavirus particles are RNA. However, the proportion of nucleic acid in virus particles is not as significant as its absolute amount, which is the factor that determines the amount of genetic information that it contains. The smallest animal virus genomes are those of the picornaviruses and parvoviruses, whose molecular weights range from 1.5 to 3×10^6. Because the coding ratio (the ratio of the molecular weight of single-stranded nucleic acid to the molecular weight of the protein for which it can code) is about nine, these viral genomes can code for protein with a total molecular weight of from 150,000 to 300,000, which is equivalent to four to eight average-sized proteins. The largest animal virus genomes, those of the poxviruses and herpesviruses, are about 100 times larger; their molecular weights range from 100 to almost 200×10^6 (Fig. 56-21). The molecular weights of viral genomes are listed in Table 56-2.

Structure of Viral Nucleic Acids

Strandedness. Both double-stranded and single-stranded DNA as well as RNA can act as the genome of animal viruses (Table 56-2).

Terminal Redundancy. The nucleic acids of several animal viruses are terminally redundant or repetitious, that is, their base composition may be represented as A, B, C. . . . X, Y, Z, A, where A, B, C, and so on are nucleotide sequences. This is most readily demonstrated, in the case of double-stranded DNAs, by treating them briefly with bacteriophage λ exonuclease, which digests DNA strands from their 5'-phosphate-containing termini. In this way, one end of one strand and the other end of the other strand are digested. On melting and reannealing, DNA digested in this manner circularizes, indicating that the two single-stranded regions at the two ends are complementary, hence that their sequences must be repetitious (Fig. 56-22). In the case of herpesvirus DNA, the length of the repeated sequences is about 400 nucleotide base pairs, which is about 0.25 percent of its total length. Terminal redundancy of this type is also exhibited by the RNA of RNA tumor viruses.

Figure 56-21. Electron micrograph of an intact herpesvirus DNA molecule. It is about 44-μm long. The small circular molecules are intact DNA molecules of the bacteriophage φX 174 (see Chap. 63), which were added to provide a size marker. *(Courtesy of Dr. Edward K. Wagner.)*

Adenovirus, parvovirus, and poxvirus DNAs, as well as rhabdovirus, bunyavirus, arenavirus RNA, are also terminally redundant, but here the repeated sequences are the inverted or reversed complements of each other; that is, their structure is of the form 5'-GATCAT. . . . ATGATC-3'. This is demonstrated by the fact that here single-stranded DNA circles form upon melting and reannealing, without first having to be digested with nuclease (Fig. 56-23). The length of the repeated sequence in adenovirus DNA is about 100 base pairs (0.3 percent of its length); for adeno-associated virus and poxvirus DNAs, the corresponding figures are 145 bases (about 3 percent) and 10,000 base pairs (about 5 percent), respectively. For rhabdovirus RNA the repeated inverted complementary sequence is about 20 residues long. Bunyavirus and arenavirus RNA exist in circular form in the native state, but not following denaturation. The reason for this appears to be the presence of inverted complementary base sequences some 10 to 30 nucleotides long at their ends, which causes them to be sticky or cohesive, that is, able to hybridize and therefore circularize.

The significance of the various forms of terminal redundancy is no doubt related to the mode of replication and expression of these nucleic acids. Examples of situations in which the reasons for terminal redundancy seem clear are described in Chapters 62 and 63.

Crosslinking. The DNAs of poxviruses are unique in being crosslinked covalently at their ends. This is shown by the fact that when these DNAs are melted and reannealed, single-stranded circles are generated whose circumference is twice the length of the linear double-stranded poxvirus DNA molecules. The structure of the ends of vaccinia virus DNA is shown in Fig. 56-24.

Covalent Linkage With Protein. Several viral nucleic acids are linked covalently to protein molecules. Thus the RNAs of picornaviruses like poliovirus are liked at their 5'-

TABLE 56-2. CHARACTERISTICS OF VIRAL NUCLEIC ACIDS

Virus	Nature of Nucleic Acid	Mol Wt ($\times 10^6$)	Strandedness	Structure	Number of Segments	Polarity	Infectivity of Naked Nucleic Acid
Poxvirus	DNA	125–185	Double	Linear crosslinked	1		
Herpesvirus	DNA	100, 150*	Double	Linear	1		+
Adenovirus	DNA	23	Double	Linear	1		+
Papovavirus	DNA	3, 5†	Double	Supercoiled circular	1		+
Parvovirus	DNA	2	Single	Linear	1	+ and − or −§	+ −
Hepatitis B virus	DNA	2	Partially double	Linear, cohesive ends	1		−
Picornavirus	RNA	2–3	Single	Linear	1	+	+
Calicivirus	RNA	2–3	Single	Linear	1	+	+
Togavirus	RNA	4.5	Single	Linear	1	+	+
Bunyavirus	RNA	5	Single	Linear, cohesive ends	3	−	−
Reovirus	RNA	15	Double	Linear	10, 11‖		−
Orthomyxovirus	RNA	4	Single	Linear	8	−	−
Paramyxovirus	RNA	6	Single	Linear	1	−	−
Rhabdovirus	RNA	3–4	Single	Linear	1	−	−
RNA tumor virus	RNA	5–8	Single	Linear	2‡	+	−
Arenavirus	RNA	5	Single	Linear, cohesive ends	2	−	−
Coronavirus	RNA	5.5	Single	Linear	1	+	+

*100×10^6 for α- and γ-herpesviruses; 150×10^6 for β-herpesviruses.
†5×10^6 for papilloma viruses; 3×10^6 for all others.
‡The genome is not really segmented; the two molecules are identical.
§Adeno-associated viruses, + and −; most others, −.
‖ Orthoreoviruses and orbiviruses, 10 segments; rotaviruses, 11 segments.

termini to a 22 amino acid-long protein (through a tyrosine residue), and each 5'-terminus of the double-stranded DNA of adenoviruses is linked to a protein with a molecular weight of about 55,000. The 5'-terminus of calicivirus RNA and of the minus strand of hepatitis B virus DNA is also linked to a protein. These proteins probably function as primers in the replication of these nucleic acids.

Circularity and Supercoiling. Most viral nucleic acids are linear, but papovavirus DNA exists in the form of double-stranded circles, circles that are supercoiled. The reason why papovavirus DNA circles are supercoiled is as follows. Papovavirus DNAs, like SV40 DNA, exist in virus particles in the form of minichromosomes, that is, chains of beads or nucleosomes (Fig. 56-25) that resemble in structure the chromatin of eucaryotic cells. Nucleosomes consist of pairs of the four histones H3, H4, H2A, and H2B, around which the viral DNA is wound. There are about 24 nucleosomes per SV40 DNA molecule. Because it is about 5200 base pairs long, there are about 220 base

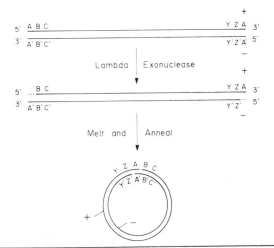

Figure 56-22. Demonstration of terminal redundancy in double-stranded DNA. Note that the A-A' sequences that are paired in the circularized molecule were originally located at opposite ends of the linear molecule.

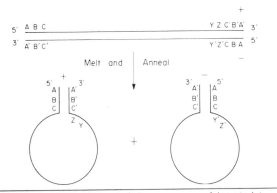

Figure 56-23. Demonstration of the presence of inverted terminally redundant sequences in double-stranded DNA.

S

```
        40              30              20              10          C A
A-A-G-T-T-A-G-T-A-A-A-T-T-A-T-A-T-A-T-A-T-A-A-T-T-T-T-A-T-A-A-T-T-A-A-T-T-T-A-A-T-T-T-T-A-T-A-T-T-T-T-A-T-T-T-A-G-T-G-T-3'
I  . .  . .  . . .  .  .  . . . .  . . .  . . . .  .  . .  . .  . . . .  .  . . .  .  I  I
A-T-C-A-A-T-C-A-T-T-T-A-A-T-A-T-A-T-A-T-T-A-A-A-A-T-A-T-T-A-A-T-T-A-A-A-A-T-T-A-A-A-A-T-A-T-A-A-A-A-T-A-A-A-T-C-A-C-A-5'
        T            T  T        C      C    G          A        T-T
   50           60          70           80         90           100
```

F

```
        50              40              30          20              10
              A              A  A            G      G  C        A      T          A-A
T-A-G-T-T-A-G-T-A-A-A-T-T-A-T-A-T-A-T-A-T-A-A-T-T-T-T-A-T-A-A-T-T-A-A-T-T-T-A-A-T-T-T-T-A-T-A-T-T-T-T-A-T-T-T-A-G-T-G-T-3'
I  . .  . . .  . . .  . . . .  . . .  .  .  . . . .  . .  . . .  . .  . .  . . . .  I
T-T-C-A-A-T-C-A-T-T-T-A-A-T-A-T-A-T-A-T-A-T-T-A-A-A-A-T-A-T-T-A-A-T-T-A-A-A-T-T-A-A-A-A-T-A-T-A-A-A-A-T-A-A-A-T-C-A-C-A-5'
        70              80          90              100
                                                      G  T
```

Figure 56-24. The structure of the ends of the DNA of vaccinia virus strain WR. Its plus and minus strands are crosslinked by normal phosphodiester bonds. Note the slight deviations from perfect double-strandedness near the crosslinks, which means that the DNA can assume a variety of configurations. Two of the most highly base-paired configurations are illustrated. The sequences of the two strands are identical from base pair 105 on into the interior of the molecule, which is toward the right. Because vaccinia virus DNA possess inverted terminal repeats (see text), its two ends are identical. *(From Baroudy et al.: Cell 28:315, 1982; Pickup et al.: Virology 124:215, 1983.)*

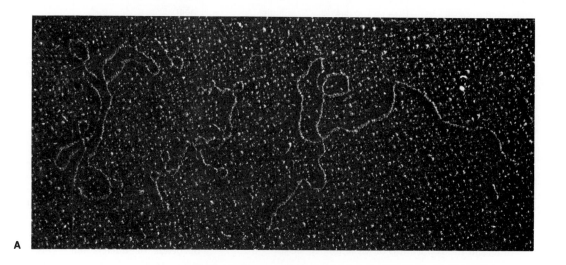

A

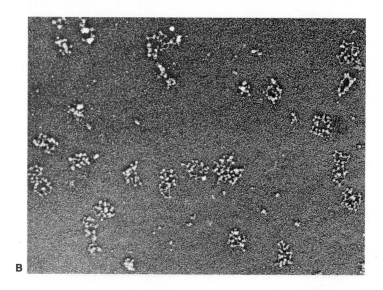

B

Figure 56-25. A. The three forms of bovine papilloma virus DNA. At the top, a supercoiled twisted circular molecule, which is the form in which the DNA exists within the virus. To the left of it is a "relaxed" circular molecule, in which one strand has been nicked, or broken by treatment with deoxyribonuclease, thereby relieving the supercoiling by permitting free rotation of the remaining intact strand. On the right, there is a linear molecule, generated by the introduction of nicks close to one another in both strands. × 66,000. *(Courtesy of Dr. H. J. Bujard.)* **B.** Electron micrograph of the SV40 DNA nucleoprotein complex in the chromatin-like, beads-on-a-string nucleosome conformation. × 50,000. *(From Keller and Mueller: Science 201:406, 1978.)*

pairs per nucleosome, 180 of which are initimately associated with each nucleosome, whereas 40 are present in the space between adjacent nucleosomes. Coiling the circular DNA molecule around each nucleosome requires a reduction in the number of superhelix turns if the resulting structure is to be devoid of strain. Therefore, when the nucleosomes are removed and the naked DNA is examined in pure form, it is supercoiled, the supercoiling having been caused by the deficiency in superhelical turns. For the SV40 DNA molecule, the deficiency is about 24 superhelical turns. The negative supercoils are introduced into the DNA by the enzyme DNA gyrase, which is a topoisomerase, a class of enzymes that catalyze adenosine triphosphate-dependent changes in the linking number of circular DNA; that is, they coil the DNA helix axis itself by permitting DNA strands to move or pass "through" DNA strands by transiently breaking such strands. Supercoiling is relieved by the action of widespread untwisting enzymes, which are also topoisomerases (DNA gyrase being the only topoisomerase that introduces supercoils), or by nicking one strand with a nuclease, or by the intercalation of substances such as ethidium bromide (Fig. 56-26).

The significance of circularity is not known. It is not essential for infectivity, as the DNA in poxviruses, herpesviruses, and adenoviruses is linear. Conceivably, circularity is a prerequisite for integration into the host genome (see Chap. 62), and the DNAs of adenoviruses and herpesviruses, which, like papovaviruses, can transform cells (see Chap. 62), circularize before being integrated.

Circular Permutation. As discussed in Chapter 63, bacteriophage DNAs are frequently circularly permuted. The DNAs of mammalian animal viruses are generally not circularly permuted. However recent evidence indicates that the DNA of a vertebrate iridovirus is circularly permuted

and terminally redundant, and that the genome of the porcine herpesvirus pseudorabies virus possesses features that can best be explained on the basis of circular permutation of some of its sequences.

Segmentation. For a long time it was assumed that viral genomes consist of unbroken strands of nucleic acid. However, this is not always so. The genomes of reoviruses and rotaviruses consist of 10 and 11 segments of double-stranded RNA, respectively (Fig. 56-27); the genomes of influenza viruses, bunyaviruses, and arenaviruses consist of eight, three, and two single-stranded RNA molecules, respectively; and the genome of RNA tumor viruses consist

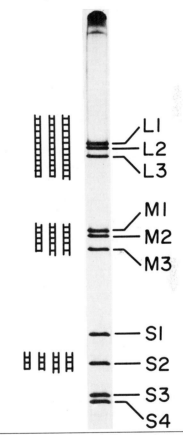

Figure 56-27. The structure of the reovirus genome. RNA extracted from reovirus particles by treatment with the detergent sodium dodecyl sulfate (SDS) was subjected to electrophoresis in a polyacrylamide gel. In such gels, molecules of various sizes migrate as discrete bands, with the smallest ones moving fastest. The direction of migration in the gel shown here was from top to bottom. Bands of RNA were visualized by autoradiography. Each band corresponds to a reovirus gene. The reovirus genome is seen to consist of 10 molecular species that fall into three size classes, designated L, M, and S. Because the rate of migration is inversely proportional to the square of the molecular weight, estimates of relative molecular weights can be made by measuring the distances traveled by the various bands. The relative sizes of the 10 reovirus genes are indicated at the left of the gel.

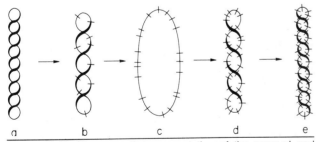

Figure 56-26. Diagrammatic representation of the removal and reversal of supercoiling turns by intercalation of ethidium bromide. The Watson–Crick helix is here represented as a single continuous line. The number of supercoiling turns in the original molecule (**a**) decreases as ethidium bromide molecules (represented by bars perpendicular to the helix axis) bind to the DNA and intercalate between base pairs (**b**) (see Chap. 7). At equivalence, the accumulated untwisting due to the number of intercalated drug molecules balances the initial number of supercoiling turns (**c**). Intercalation of further drug molecules (**d**) leads to the introduction of supercoiling turns in the opposite sense (**e**). *(From Crawford and Waring: J Gen Virol 1:387, 1967.)*

of two single-stranded RNA molecules that are identical (so that the genome is not really segmented [see Chap. 62]). One of the consequences of segmentation is highly efficient genetic recombination caused by random reassortment of segments in multiply infected cells (see Chap. 60).

Polarity of Viral Nucleic Acids

The single-stranded RNA molecules present in picornavirus, calicivirus, togavirus, coronavirus, and RNA tumor virus particles can be translated by ribosomes and serve as messenger RNA. All the genetic information necessary for the formation of progeny virus is translated directly from them. This is not the case for the RNA molecules present in ortho- and paramyxovirus, rhabdovirus, bunyavirus, and arenavirus particles. These RNAs must first be transcribed into RNA strands of opposite polarity, and it is these transcripts that are then translated by ribosomes. Because the polarity of messenger RNA is generally designated as plus, the polarity of the RNA in the former group of viruses is plus, and that of the RNA in the latter group is minus.

The adeno-associated satellite viruses, which belong to the parvovirus family and contain single-stranded DNA, present a unique situation. There are two kinds of these virus particles; one contains plus strands, the other minus strands. These two kinds of particles are produced in equal amounts. When the DNA is extracted from them it hybridizes rapidly, thus giving the illusion that it is double-stranded. It is only when the DNA is extracted under conditions where the plus and the minus strands are prevented from hyridizing with each other that the true situation is revealed. All other parvoviruses (except the insect Densonucleosis viruses) contain single-stranded DNA of one polarity only, namely minus.

Genetic Relatedness of Viral Nucleic Acids

The genetic relatedness of animal virus nucleic acids is of interest for its taxonomic and evolutionary significance. For example, there are many viruses that are poxviruses, according to a variety of criteria. The question naturally arises as to how closely related they are.

The most definitive measure of genetic relatedness is determination of similarity of nucleic acid-base sequence by direct sequence analysis. Very efficient techniques for sequencing both RNA and DNA have been perfected, and the genomes of many animal viruses have been sequenced either completely or in part.

Another, less laborious, way to assess genetic relatedness, which is still sensitive enough to be useful taxonomically, is measurement of how extensively nucleic acids can hybridize with each other. If the nucleic acids under investigation are double-stranded, the two molecules whose relatedness is to be determined are denatured to the single-stranded state, mixed, and allowed to reanneal. Single strands derived from the two genomes are thus presented with the opportunity of pairing with each other, and conditions are readily arranged so that such pairing can be quantitated. If the two genomes are very closely related,

pairing will occur extensively; if there is no relatedness, no pairing will occur. For example, the genomes of highly oncogenic adenoviruses hybridize with each other to the extent of 80 percent or more; in other words, they share over 80 percent of their base sequences. However, they share only about 25 percent of base sequences with the genomes of the nononcogenic adenoviruses, which, on the other hand, share over 80 percent of base sequences among themselves. Another example is provided by strains of the three serotypes of reovirus. The RNAs of strains of serotypes 1 and 3 hybridize with each other to the extent of about 70 percent, but both hybridize only to the extent of about 10 percent with the RNAs of reovirus strains of serotype 2.

If the viruses that are to be compared possess single-stranded nucleic acid, the strategy is to hybridize the viral nucleic acid in question to the double-stranded, or replicative form (see Chap. 58) of the nucleic acid of the virus strain whose genetic relatedness is to be assessed.

Another interesting application of this methodology is heteroduplex analysis, in which the hybrid molecules are examined in the electron microscope. Regions of genomes that are identical or nearly identical, that is, homologous, are then seen to be double-stranded, whereas regions that are so dissimilar that they will not base-pair with each other, or regions that are present in only one genome and not in the other (either insertions or deletions) appear as single-stranded regions. An example of heteroduplex analysis is shown in Figure 58-12.

Another measure of genetic relatedness employs comparison of proteins rather than of nucleic acids. Similarity of nucleic acid-base sequences signifies similarity of the amino acid sequences of the proteins encoded by them. This in turn implies antigenic similarity; that is, ability of proteins to react with each other's antibodies. Ability of nucleic acids to hybridize, and of the proteins coded by them to cross-react immunologically, are thus measures of the same parameter. As an example of this approach, Figure 56-28 illustrates the relatedness of five herpesviruses, based upon analysis of their antigenic similarity.

Infectivity of Animal Virus Nucleic Acids

Viral nucleic acids contain all the information necessary for the formation of virus particles. This was first shown by Hershey and Chase in 1952, when they found that the injection of bacteriophage DNA into the host cell causes the formation of progeny virus particles. Later, in 1956, Gierer and Schramm showed that the same was true for RNA when they found that RNA extracted from tobacco mosaic virus particles was infectious.

The nucleic acids of several groups of animal viruses, such as that of the picornaviruses, caliciviruses, togaviruses, coronaviruses, papovaviruses, adenoviruses, and herpesviruses, are also infectious (see Table 56-2). These are all nucleic acids that either can act as messenger RNA themselves or that are transcribed into messenger RNA by host-coded RNA polymerases. Viral nucleic acids that are transcribed into messenger RNA by virus-coded polymer-

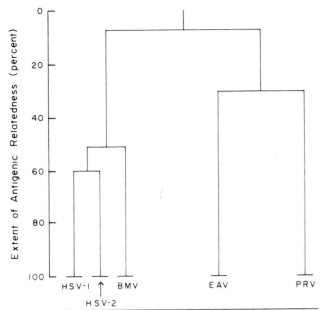

Figure 56-28. Dendrogram illustrating the serologic relatedness of five herpesviruses, herpes simplex virus type 1 (HSV-1), herpes simplex virus type 2 (HSV-2), bovine mammillitis virus (BMV), equine abortion virus (EAV), and pseudorabies virus (PRV). The extent of antigenic relatedness was assessed by counting the number of precipitin bands formed in agar gel immunodiffusion reactions, using antisera to the various herpesviruses and the virus-coded proteins in extracts of cells infected with homologous and heterologous herpesviruses. *(Adapted from Honess and Watson: J Gen Virol 37:15, 1977.)*

ases, such as the minus-stranded RNAs of ortho- and paramyxoviruses, rhabdoviruses, bunyaviruses, and arenaviruses; the double-stranded RNA of reovirus; and the double-stranded DNA of poxviruses, are not infectious, as they cannot express themselves in the cell.

In all cases, the naked nucleic acids are less infectious by factors varying from 10^3 to 10^6 than the virus particles from which they were extracted. There are two principal reasons for this. First, naked viral nucleic acids are quickly degraded by nucleases, which are generally present in extracellular fluids, as well as on outer cell membranes; second, naked nucleic acids are taken up very poorly by cells. Uptake can be increased significantly by (1) treating cells briefly with concentrated salt solutions, which promotes pinocytosis, thereby facilitating nucleic acid entry; or (2) complexing the naked nucleic acids with polycations such as protamine of DEAE-dextran; or (3) adsorbing them to precipitated calcium phosphate, which is taken up well by cells. The inefficiency with which naked viral nucleic acids penetrate into cells emphasizes the role of the viral capsid in this vital function.

The host range of naked viral nucleic acids is very much broader than that of the respective virus particles. This stems from the fact that the host range of a virus particle is restricted by the specificity of the interaction between capsid and cell surface receptors (see Chap. 58). This is not the case for naked viral nucleic acids. For ex-

ample, whereas poliovirus can infect and therefore multiply only in cells of human or primate origin, poliovirus RNA can also infect chick cells and mouse cells.

Infectious viral nucleic acid can be extracted not only from infectious virus particles but from all virus particles that contain undamaged nucleic acid. Among such are virus particles inactivated by heat, proteolytic enzymes, and detergents.

Presence of Host Cell Nucleic Acids in Virus Particles

As a rule, capsids enclose viral nucleic acids. However, sometimes segments of host nucleic acid become encapsidated instead. For example, particles exist that contain, within a papovavirus capsid, a linear piece of host DNA roughly the same size as papovavirus DNA. Each such particle, known as a pseudovirion, contains a different segment of host DNA. Pseudovirions usually make up only a small fraction of the yield, but in some cell lines the majority of the particles formed as a result of infection with polyoma virus are pseudovirions. Another example is provided by virus particles that are formed when papovaviruses are passaged repeatedly at high multiplicity; that is, in cells infected with many virus particles. These virus particles contain DNA molecules that consist partly of viral and partly of host cell sequences. They are formed when the viral genomes, which become integrated into host cell DNA prior to replication (see Chaps. 62 and 63), are imperfectly excised (cut out of the host genome).

Viral Proteins

The principal constituent of all animal viruses is protein. Proteins are the sole component of capsids, the major component of envelopes, and are also intimately associated with the nucleic acids of many icosahedral viruses as internal or core proteins. All these proteins are referred to as structural proteins, as their primary function is to serve as virus particle building blocks. They are almost always encoded by the viral genome.

Viral proteins vary widely in size, from less than 10,000 to more than 150,000 daltons. They also vary in number, with some virus particles containing as few as three species, and others containing more than 50. Viral proteins are characterized most conveniently by dissociating highly purified virus preparations with detergent and subjecting the resulting protein mixtures to electrophoresis in polyacrylamide gels. The most commonly used detergent is sodium dodecyl sulfate (SDS), which not only destroys the secondary structure of proteins but also forms complexes with them. These complexes carry numerous strong negative charges, so that upon electrophoresis in polyacrylamide gels they migrate strictly according to size. As a result, SDS-polyacrylamide gel electrophoresis provides a very convenient method for determining not only the number of different protein species that make up virus particles, but also their sizes. The polyacrylamide gel electrophoresis profile of vaccinia virus proteins is shown in Figure 56-29.

All members of the same virus family display the same or almost the same highly characteristic electrophoretic protein patterns. The patterns of reovirus, rhinovirus, murine encephalomyelitis (ME) virus, and Sendai virus (a paramyxovirus) proteins are illustrated in Figure 56-30.

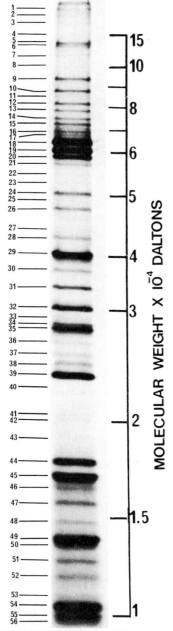

Figure 56-29. Polyacrylamide gel of vaccinia virus structural proteins. Vaccinia virus labeled with ^{35}S-methionine was dissolved in a buffer containing sodium dodecyl sulfate, and the resulting solution was electrophoresed in an 11 percent polyacrylamide gel. An autoradiogram was then prepared from it by exposing x-ray film to it. The direction of electrophoresis was from top to bottom. Over 50 proteins are clearly visible. There is evidence that the vaccinia virus particle may comprise as many as 100 different species of proteins. *(From Dales: Virology 95:355, 1979.)*

Glycoproteins

Viral envelopes often contain glycoproteins in the form of spikes or projections. Their carbohydrate moieties consist of oligosaccharides comprising 10 to 15 monosaccharide units, which are linked to the polypeptide backbone through N- and O-glycosidic bonds involving asparagine and serine or threonine, respectively. Their principal components are generally galactose and galactosamine, glucose and glucosamine, fucose, mannose, and neuraminic acid, which always occupies a terminal position.

It was long thought that the structure of the oligosaccharides of viral glycoproteins is specified solely by the nature and relative abundance of the various host cell glycosyl transferases that assemble them from their monosaccharide components. This is certainly a very important factor, but it now appears that the nature of the protein that is being glycosylated also influences oligosaccharide structure, particularly near the chain termini. As a result, the oligosaccharides of viral glycoproteins probably differ from those of cells. Further, the oligosaccharides of glycoproteins of different viruses grown in the same cell are probably not identical, and the oligosaccharides of the glycoproteins of the same virus grown in different cells, although closely related, are probably different. The structure of the oligosaccharide of the spike glycoprotein of vesicular stomatitis virus is shown in Figure 56-31.

Viral Proteins with Specialized Functions

Some viral proteins have specialized properties and functions. Among them are the following:

Hemagglutinins. Many animal viruses, both naked and enveloped (such as picornaviruses, togaviruses, reoviruses, ortho- and paramyxoviruses, adenoviruses, and papovaviruses), agglutinate the red blood cells of certain animal species. This property, which is called hemagglutination, reflects the fact that these red blood cells possess recep-

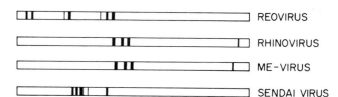

Figure 56-30. Drawings of polyacrylamide gel patterns of the proteins of reovirus, rhinovirus, murine encephalomyelitis (ME) virus, and Sendai virus. Highly purified preparations of these viruses were dissolved in buffer containing sodium dodecyl sulfate (SDS); the illustration shows gels in which the SDS complexes of the proteins of these four viruses had been electrophoresed from left to right. The location of each band is a measure of its molecular weight; thus the largest protein shown is the left-most protein of reovirus (mol wt 150,000) and the smallest is the right-most protein of rhinovirus (mol wt about 8000). The relative amount of each protein is indicated by the thickness of the band representing it. It is clear from the gel patterns shown that the protein complements of rhinovirus and ME virus, both picornaviruses, are very similar, and that they are quite different from those of reovirus and Sendai virus.

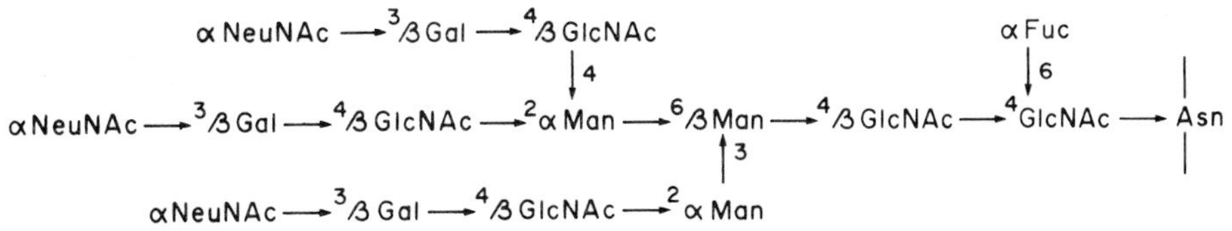

$$\alpha\,NeuNAc \xrightarrow{\ 3\ } \beta\,Gal \xrightarrow{\ 4\ } \beta\,GlcNAc \qquad\qquad\qquad\qquad \alpha\,Fuc$$

Figure 56-31. The structure of the oligosaccharide unit of the vesicular stomatitis virus spike glycoprotein. There are two of these units per protein molecule. Asn, asparagine; GlcNAc, N-acetylglucosamine; Man, manose; Gal, galactose; NeuNAc, N-acetyl-neuraminic acid; Fuc, fucose. *(From Reading, et al.: J Biol Chem 253:5600, 1978.)*

tors for certain surface components of virus particles. In the case of ortho- and paramyxoviruses, these viral surface components, called hemagglutinins, are glycoprotein spikes present on the virus particle surface (see Fig. 56-4). The ability to hemagglutinate can be used to quantitate virus (see Chap. 55).

Enzymes. Virus particles often contain enzymes (Table 56-3). Among them are the following:

1. Orthomyxovirus and paramyxovirus articles contain an enzyme, neuraminidase, that hydrolyzes the galactose-N-acetylneuraminic acid bond at the ends of the oligosaccharide chains of glycoproteins and glycolipids, thereby liberating N-acetylneuraminic acid. Like the hemagglutinin, this enzyme is located on glycoprotein spikes. Orthomyxoviruses possess two types of spikes, one with hemagglutinin and the other with neuraminidase activity (See Fig. 56-4), whereas paramyxoviruses also possess two types of spikes, on one of which both activities are located. The primary function of neuraminidase appears to be to release virus particles from the cells in which they were formed. The reason why the release of these viruses requires a special enzyme is that their hemagglutinins possess strong affinity for the galactose-N-acetylneuraminic acid (sialic acid) residues at the ends of oligosaccharide chains of glycoproteins. By hydrolyzing them, the enzyme permits the liberation of virus particles that would otherwise not be released or that would readsorb immediately after being liberated.

2. Many virus particles contain RNA polymerases. The necessity for such enzymes stems from the fact that the nucleic acid present in virus particles must be able to express itself once it has gained access to the interior of the host cell. If this nucleic acid itself can act as messenger RNA, like picornavirus or togavirus RNA, there is no problem. If it cannot, either because it is double-stranded or because it is RNA with minus polarity, RNA with plus polarity must first be synthesized. There are two possible sources for the enzymes necessary for this purpose. The first is enzymes existing in the host cell. Herpesvirus, adenovirus, and papovavirus DNAs are transcribed into messenger RNA by DNA-dependent RNA polymerases specified by the host cell. The alternative is that the virus particle itself contains the enzyme. Examples of this type are the poxviruses that possess a DNA-dependent RNA polymerase in their cores, reoviruses that possess an RNA-dependent RNA polymerase in their cores, and negative RNA-stranded viruses that contain RNA polymerases that synthesize plus RNA strands from the minus RNA strands present within them. These enzymes are usually not active

TABLE 56-3. ENZYMES IN ANIMAL VIRUSES

Virus	Enzyme
DNA Viruses	
Poxvirus	DNA-dependent RNA polymerase
	Messenger RNA capping enzymes
	Poly (A) polymerase
	Nucleases (nicking enzymes)
	DNA-dependent nucleotide phosphohydrolase
	Topoisomerase
	Protein kinase
Herpesvirus	None
Adenovirus	Protein kinase
Papovavirus	None
Parvovirus	DNA polymerase
Hepatitis virus	DNA polymerase
RNA Viruses	
Picornavirus	None
Calicivirus	None
Togavirus	None
Bunyavirus	RNA polymerase
Reovirus	RNA-dependent RNA polymerase
	Nucleotide phosphohydrolase
	Messenger RNA capping enzymes
Orthomyxovirus	Neuraminidase
	RNA polymerase
Paramyxovirus	Neuraminidase
	RNA polymerase
	Messenger RNA capping enzymes
	Poly (A) polymerase
Rhabdovirus	RNA polymerase
	Messenger RNA capping enzymes
	Poly (A) polymerase
	Protein kinase
RNA tumor virus	RNA-dependent DNA polymerase (reverse transcriptase)
	Ribonuclease H
	Protein-cleaving enzyme
	Protein kinase
Arenavirus	RNA polymerase
	Poly (A) polymerase
Coronavirus	None

in intact virus particles but are activated when the envelope or capsid is partially degraded, which generally occurs very soon after infection.

3. RNA tumor viruses possess a DNA polymerase, the so-called reverse transcriptase, which transcribes their single-stranded RNA into double-stranded DNA, which is then integrated into the genome of the host cell (see Chap. 62). Furthermore, hepatitis B virus contains a DNA polymerase that appears to be able to transcribe both DNA and RNA into DNA (see Chap. 62).

All these enzymes are virus-specified. The evidence for this conclusion is provided by the existence of virus strains that contain variant enzymes and the fact that the enzymes in question often simply do not exist in uninfected cells.

Viruses also often contain other enzymes. Among them are the enzymes that modify both ends of the messenger RNA molecules synthesized by these polymerases, notably the capping enzymes (that is, the enzyme that guanylylates their 5'-termini and the methylases that methylate both the terminal G as well as the ribose residue that was N-terminal in the uncapped molecule), and poly (A) polymerases that polyadenylate the 3'-termini of messenger RNA molecules. Protein kinases, the function of which is unknown, are also often present, and the very complex poxviruses also contain deoxyribonucleases, DNA-dependent phosphohydrolases, and topoisomerases that may function in DNA replication.

Viral Lipids

Viral envelopes contain complex mixtures of neutral lipids, phospholipids, and glycolipids. As a rule, the composition of these mixtures resembles that of the membranes of the host cells in which the virus multiplied. As the lipid composition of membranes varies markedly from one cell strain to another and even for the same cell strain depending on the composition of the medium, this means that the same virus grown in different cell strains may have widely differing neutral lipid, phospholipid, and glycolipid complements, even though its biologic properties, including infectivity, are identical.

The nature and extent of the differences in composition that are encountered are best illustrated by the example of the paramyxovirus SV5. On the average, this virus contains about 20 percent total lipid, of which roughly 25 percent and 5 percent are cholesterol and triglyceride, respectively; 55 percent are phospholipid; and 15 percent are glycolipid. If grown in cells the membranes of which have a molar ratio of cholesterol to phospholipid of 0.81, the ratio for the viral envelope is 0.89; if the cellular ratio is 0.51, that for the viral envelope is 0.60. The same holds for the relative amounts of several phospholipids. If the ratio of phosphatidylcholine to phosphatidylethanolamine in host cell membranes is 3.55, the ratio in the viral envelope is 2.55; if the ratio in host cell membranes is 0.8, the ratio in the viral envelope is 0.6. However, for one cell the

ratio is 1.6, whereas that in the viral envelope derived from it is 0.6, which suggests that a limited degree of lipid selection is possible.

The Classification of Animal Viruses

Attempts to devise a system of classification for animal viruses began almost as soon as their importance as pathogens became apparent. However, the criteria on which to base such a system have changed with advances in our knowledge of their nature and properties. For example, host range cannot be such a criterion; not only is each animal species subject to infection by a wide variety of viral agents, but also numerous viruses infect several different animal species. Similarly, the pattern of pathogenesis of the disease that is caused is an unreliable foundation on which to base a system of classification.

It has become more and more apparent that a system of classification must be based on the physical and chemical properties of the virus particles themselves.

Criteria for Classification

Morphology
The primary criterion for classification is morphology. It is easy to apply because it does not require purified virus. Virus particles can be examined either within cells, that is, in thin sections of infected tissue, or in their extracellular state. Morphologic detail is usually brought out by shadowing with thin films of some heavy metal, such as uranium or tungsten; by staining with osmic acid or uranyl acetate; or by negative staining with phosphotungstic acid.

Physical and Chemical Nature of Viral Components
Morphologic similarity correlates closely with similarity of viral components. For example, all viruses with the morphology of adenoviruses contain double-stranded DNA genomes with a molecular weight of about 23 million; all papovaviruses contain circular supercoiled DNA genomes with molecular weights of 3 to 5 million, and all reoviruses contain segmented double-stranded RNA genomes. In fact, a system of virus classification based on the structure and size of viral genomes yields the same grouping as one based on morphology. Similarly, viruses with similar morphology are composed of similar populations of proteins; and a classification system based on polyacrylamide gel patterns, such as illustrated in Figures 56-29 and 56-30, again yields the same grouping.

Genetic Relatedness
Although the members of the major virus families presumably derived from a common ancestor, genetic relationships among them are often no longer discernible. For example, several groups of mammalian poxviruses exist (Table 56-4) whose members are unrelated genetically as

TABLE 56-4. THE MAJOR FAMILIES OF ANIMAL VIRUSES

DNA-containing Viruses

Classification	Description
I. POXVIRIDAE	All genera except *Parapoxvirus*: Brick-shaped complex particles, whorled surface filament pattern Dimensions: 225 × 300 nm *Parapoxvirus*: Ovoid complex particles, regular surface filament pattern Dimensions: 150 × 200 nm

Strains pathogenic for many animal species exist. There are seven genera; the members of each are closely related antigenically but share only one antigen with members of the others. Poxviruses multiply in the cytoplasm.

Subfamily	Host	Symptoms in humans
Chorodopoxvirinae (poxviruses of vertebrates)		
A. Genus *Orthopoxvirus*		
Variola major	Humans	Smallpox
Variola minor	Humans	Alastrim
Monkeypox	Monkey, humans	Smallpox-like disease
Vaccinia	Cattle	—
Cowpox	Cattle, humans	Vesicular eruption of the skin
Buffalopox	Buffalo	—
Rabbitpox	Rabbit	—
Ectromelia (mousepox)	Mouse	—
B. Genus *Leporipoxvirus*		
Myxoma	Rabbit	—
Fibroma	Rabbit	—
C. Genus *Avipoxvirus*		
Fowlpox	Chicken	—
Other birdpox viruses	Various birds	—
D. Genus *Capripoxvirus*		
Sheeppox	Sheep	—
Goatpox	Goat	—
E. Genus *Suipoxvirus*		
Swinepox	Pig	—
F. Genus *Parapoxvirus*		
Orf (contagious postular dermatitis [CPD])	Sheep, goats, humans	Nodules on hands
Pseudocowpox (milker's nodule virus)	Cattle, humans	Nodules on hands
Bovine papular stomatitis	Cattle	—
G. Ungrouped		
Molluscum contagiosum	Humans	Benign epidermal tumors
Yaba monkey tumor virus	Monkeys, humans	Benign subcutaneous tumors
Tanapoxvirus	Monkeys, humans	Benign subcutaneous tumors
Entomopoxvirinae (poxviruses of insects)		
Various strains	Coleoptera, lepidoptera, diptera	—

Classification
II. IRIDOVIRIDAE

These are the icosahedral cytoplasmic deoxyviruses. Many iridoviruses are insect viruses and nonenveloped. The sole mammalian iridovirus is enveloped. The size of its DNA is about 150×10^6.

	Host	Symptoms in humans
African swine fever virus	Swine	—
Frog virus 2	Amphibia	—
Frog virus 3		
Lymphocystis virus of fish	Fish	—
Tipula iridescent viruses	Insects	—
Iridescent viruses of	Molluscs, annelids, protozoa	—

(continued)

TABLE 56-4. (cont.)

DNA-containing Viruses	
Classification	*Description*
III. HERPESVIRIDAE	Enveloped icosahedral nucleocapsids Diameter of enveloped virions: 180–250 nm Diameter of naked nucleocapsids: 100 nm

The most logical classification of herpesviruses is based on a combination of biologic properties and genome structure. α-Herpesviruses have a variable host range (from very wide to very narrow); most, but not all, have a relatively short replication cycle; most, but not all, are highly cytopathic in cultured cells; they frequently establish latent infections in sensory ganglia; and their DNA has a molecular weight of about 100×10^6. β-Herpesviruses have a narrow host range and a relatively long replication cycle; are less cytopathic; frequently establish latent infections in the salivary gland and in other tissues; and their DNA has a molecular weight of about 150×10^6. γ-Herpesviruses have a narrow host range; a predilection for lymphoblastoid cells which they can transform (that is, they can cause tumors); and their DNA has a molecular weight of about 100×10^6. Few herpesviruses are closely related to each other as judged by DNA–DNA hybridization, but most show some relatedness (2 to 10 percent cross-hybridization). Almost all herpesviruses possess some common antigenic determinants. Herpesviruses cause type A nuclear inclusions (single large acidophilic inclusion bodies separated by a nonstaining halo from basophilic marginated chromatin).

	Host	*Symptoms in humans*
A. *α-Herpesviruses*		
Herpes simplex virus type 1 (human (α-)herpesvirus 1)	Humans	Stomatitis, upper respiratory infections, generalized systemic disease, severe and generally fatal encephalitis
Herpes simplex virus type 2 (human (α-)herpesvirus 2)	Humans	Genital infections
Varicella-zoster (human (α-)herpesvirus 3)	Humans	Chickenpox, herpes zoster
B virus	Monkey, humans	Fatal encephalitis in humans
Equine abortion virus (EAV) (equine herpesvirus 1)	Horse	—
LK virus (equine herpesvirus 2)	Horse	—
Coital exanthema virus (equine herpesvirus 3)	Horse	—
Pseudorabies virus (suid herpesvirus)	Pig	—
Infectious bovine rhinotracheitis (IBR)	Cattle	—
Infectious bovine keratoconjunctivitis (IBKC)	Cattle	—
Feline rhinotracheitis virus	Cat	—
Infectious laryngotracheitis virus (ILT)	Chicken	—
B. *β-Herpesviruses*		
Human cytomegalovirus (human (β-)herpesvirus 5)	Humans	Jaundice, hepatosplenomegaly, brain damage, death
Other Cytomegaloviruses	Monkey	—
	Rodents	—
	Swine	—
C. *γ-Herpesviruses*—All these viruses either cause, or are strongly suspected of causing, tumors		
Epstein–Barr virus (human (γ-)herpesvirus 4)	Humans	Burkitt lymphoma Nasopharyngeal carcinoma Infectious mononucleosis
Marek's disease virus	Chicken	—
Herpesvirus saimiri	Squirrel monkey	—
Herpesvirus ateles	Spider monkey	—
Herpesvirus sylvilagus	Rabbit	—
Guinea pig herpesvirus	Guinea pig	—
Lucke' tumor virus	Frog	—

(continued)

TABLE 56-4. (cont.)

DNA-containing Viruses	
Classification	*Description*
IV. ADENOVIRIDAE	Naked icosahedral nucleocapsids Diameter: 70 nm

Adenoviruses have been isolated from many species of animals; their host range is generally rather narrow. Adenoviruses share common group-specific complement-fixing antigenic determinants (on hexons) and, in addition, possess type-specific determinants (on pentons and fibers). Host-specific adenoviruses are often represented by numerous serotypes. For example, there are 35 human adenovirus serotypes that can be grouped into subgroups on the basis of antigenic cross-reactivity, DNA hybridization characteristics, ability to transform cells of various animal species, and ability to agglutinate rhesus monkey and rat erythrocytes. More specifically, viruses of subgroup A possess 48 percent (G + C) and regularly cause tumors in newborn hamsters (high tumorigenicity); those in subgroup B possess 51 percent (G + C) and sometimes cause tumors in newborn hamsters (low tumorigenicity); and members of subgroups C, D, and E possess 58 percent (G + C) and do not cause tumors, but still transform cultured cells. Members of subgroup A exhibit about 60 percent homology with each other. For members of the other subgroups, this value is 90 to 100 percent. The extent of homology between subgroups is 10 to 20 percent. Recombination occurs only within, not among, subgroups. Members of the various subgroups exhibit different pathogenicities. Adenoviruses produce intranuclear type B inclusions (basophilic masses sometimes connected to the nuclear periphery by strands of chromatin).

	Host	*Symptoms in humans*

A. Genus *Mastadenovirus*
  Human adenoviruses
    35 serotypes
	Humans	

Subgroup A Serotypes 12,18,31		No known pathogenicity; regularly isolated from feces of apparently healthy individuals; high incidence of antibodies
Subgroup B Serotypes 3,7,11,14,16,21,34		Acute respiratory disease (7, also 3,14, and 21); pharyngitis (3,7); acute hemorrhagic cystitis in children (11,21); low incidence of antibodies
Subgroup C Serotypes 1,2,5,6		Mild infections of the respiratory tract, especially in infants and children; latent infections in lymphoid tissue
Subgroup D Serotypes 8,9,10,13,15,17,19,20,22–30,32,33		Epidemic keratoconjunctivitis (8, also 19); no known pathogenicity for others; low incidence of antibodies
Subgroup E Serotype 4		Acute respiratory disease
Ungrouped Serotypes 31,35		
Simian adenoviruses 23 serotypes	Monkey	—
Canine adenoviruses (infectious canine hepatitis [ICH])	Dog	—
Adenoviruses of	Cattle, pigs, sheep, frogs, and many other species	—
B. Genus *Aviadenovirus* Avian adenoviruses (CELO, chicken-embryo-lethal-orphan; GAL, gallus-adeno-like)	Chicken, quail, other birds	—
C. Noncultivatable adenoviruses (Enteric adenoviruses [EA])	Humans	Enteritis-associated enteric infections

(continued)

TABLE 56-4. (cont.)

DNA-containing Viruses

Classification	Description
V. PAPOVAVIRIDAE	Naked icosahedral nucleocapsids Diameter: 55 nm (papilloma viruses) 45 nm (all other papovaviruses)

Papovaviruses (papilloma-polyoma-simian vacuolating agent) fall into two distinct groups on the basis of size; members of the genus papillomavirus possess both larger capsids and larger genomes than members of the genus polyomavirus. All except K virus produce tumors in animals; all produce latent and chronic infections in their natural hosts. Most adults possess antibodies against BK virus and JC virus. Polyomaviruses agglutinate erythrocytes of certain animal species.

	Host	Symptoms in humans
A. Genus *Papillomavirus*		
Human papilloma viruses (HPV) 1–15	Humans	
HPV 1 and 4		Plantar warts
HPV 2		Common warts (verruca vulgaris)
HPV 3,5,8,9,10,12,14, and 15		Flat warts (verruca plana) and/or epidermodysplasia verruciformis (EV)
HPV 6 and 11		Anogenital warts (condylomata acuminata), otolaryngeal warts
HPV 7		Meat-handlers' warts
HPV 13		Oral focal hyperplasia
Bovine papilloma viruses 1–5	Cattle	—
Shope rabbit papilloma virus	Rabbit	—
Various viruses pathogenic for other animal species		—
B. Genus *Polyomavirus*		
Polyoma virus	Mouse	—
Simian vacuolating agent (SV-40)	Monkey, humans(?)	—
Rabbit vacuolating agent (RKV)	Rabbit	—
K virus	Mouse	—
BK virus	Humans	Isolated from the urine of renal transplant patients
JC virus	Humans	Isolated from brains of patients with progressive multifocal leucoencephalopathy (PML)
Other primate polyomaviruses [Lymphotrophic papovavirus (LPV)]	African green monkey	Multiplies only in monkey or human B lymphoblasts. About 30 percent of humans have antibody against it

Classification	Description
VI. PARVOVIRIDAE	Naked icosahedral nucleocapsids Diameter: 22 nm

These viruses are grouped together on the basis of morphology and nucleic acid structure (single-stranded DNA). All multiply in the nucleus. Most are antigenically unrelated. Adeno-associated viruses (AAVs) multiply only in cells simultaneously infected with adenoviruses (Chap. 60); rat virus (RV) multiplies only in cells that are themselves multiplying actively. Aleutian mink diseases virus causes a slow disease characterized by hypergammaglobulinemia, systemic proliferation of plasma cells, glomerulonephritis, and hepatitis, and is invariably fatal. The disease condition is caused by activation of the immune response; it is exacerbated by administration of inactivated virus or passive antibody. Although it causes a slow disease, Aleutian mink disease virus replicates as rapidly in cultured cells as viruses that cause acute disease.

	Host	Symptoms in humans
A. Genus *Adeno-associated virus*		
Human AAV 4 serotypes	Humans	No known symptoms
Other AAVs	Cattle, dog, chicken	

(continued)

TABLE 56-4. (cont.)

DNA-containing Viruses		
	Host	Symptoms in humans
B. Genus *Parvovirus*		
Rat virus (Kilham)	Rat	Latent viruses isolated from various hosts including human tumors; none is oncogenic or capable of transforming cultured cells. They produce a mongoloid osteolytic deformity in newborn hamsters
Hamster osteolytic viruses (H-1, H-3, X-14, etc.)		
Minute virus of mice (MVM)	Mouse	—
Other parvoviruses (porcine parvovirus, feline panleukopenia virus, bovine parvovirus, canine parvovirus, etc.)	Pig, cattle, cat, dog	—
Aleutian mink disease virus	Mink	—
C. Genus *Densovirus*		
Densonucleosis viruses	Insects	—

RNA-containing Viruses	
Classification	Description
I. PICORNAVIRIDAE	Naked icosahedral nucleocapsids Diameter: 25–30 nm

Picornaviruses comprise a large number of virus strains pathogenic for many animal species. They are subdivided into four genera: *Enterovirus* and *Cardiovirus,* whose members are acid-stable, and *Rhinovirus* and *Aphthovirus,* whose members are acid-labile.

	Host	Symptoms in humans
A. Genus *Enterovirus*		
Human enteroviruses		
Poliovirus 3 serotypes	Humans, monkey	Poliomyelitis
Coxsackie virus A 23 serotypes	Humans, mouse	Differentiated from Group B Coxsackie viruses primarily on the basis of selective tissue damage: Group A, primarily general striated muscle damage; Group B, primarily fatty tissue and central nervous tissue damage. Group A viruses are associated with herpangina, aseptic meningitis, paralysis, and the common cold syndrome
Coxsackie virus B 6 serotypes	Humans, mouse	Pleurodynia (Bornholm disease), aseptic meningitis, paralysis, severe systemic illness of newborns
ECHO viruses (enteric cytopathogenic human orphan) 32 serotypes	Humans	Paralysis, diarrhea, aseptic meningitis
Human enterovirus 72 (hepatitis A virus)	Humans	Infectious hepatitis, jaundice
Simian enteroviruses 18 serotypes	Monkey	—

(continued)

TABLE 56-4. (cont.)

RNA-containing Viruses		

	Host	*Symptoms in humans*
Murine encephalomyelitis (ME) viruses	Mouse	—
Poliovirus muris		
(Theiler's virus)		
GDVII strain and others		
Bovine enteroviruses	Cattle	—
7 serotypes		
Porcine enteroviruses	Swine	—
8 serotypes		
B. Genus *Cardiovirus*		
Encephalomyocarditis virus (EMC)	Various species including humans	Mild febrile illness
Mengovirus	Mouse	—
ME virus	Mouse	—
C. Genus *Rhinovirus*		
Human rhinoviruses	Humans	Common cold, bronchitis, croup, bronchopneumonia
113 serotypes		
Other rhinoviruses	Strains pathogenic for horses, cattle	—
D. Genus *Aphthovirus*		
Foot-and-mouth disease virus (FMDV)	Cattle, swine, sheep, goats	—
7 serotypes		

Classification	*Description*
II. CALICIVIRIDAE	Naked icosahedral nucleocapsids
	Diameter: 35–40 nm

Caliciviruses differ significantly from picornaviruses in size and structure, but the primary reason they have been constituted a separate family is that the strategy of genome expression during the multiplication cycles of picornaviruses and caliciviruses is quite different. Vesicular exanthema (VE) virus and San Miguel sealion virus (SMSV) are very closely related; feline picornaviruses are related to them to the extent of about 10 percent as judged by RNA hybridization analysis.

	Host	*Symptoms in humans*
VE of swine virus	Swine	—
SMSV	Seals	—
Feline picornaviruses	Cat	—
Probable calicivirus	Humans	Gastroenteritis
Norwalk virus		

Classification	*Description*
III. TOGAVIRIDAE	Enveloped icosahedral nucleocapsids
	Diameter: 60–70 nm (alphaviruses and rubiviruses)
	45–55 nm (flaviviruses and pestiviruses)

Togaviruses include many of the viruses previously known as arboviruses (arthropodborne). They multiply in bloodsucking insects as well as in vertebrates; in their natural environment they alternate between an insect vector (usually a mosquito or tick) and a vertebrate reservoir, rarely producing disease in either. Many cause subclinical infections in humans, particularly in the tropics, but several are among the most virulent and lethal of all viruses. They are commonly named for the geographic site where they were isolated. They are divided into four genera, primarily on the basis of antigenic relationships (neutralization, complement fixation and hemagglutination inhibition). The alphaviruses and flaviviruses are the old arbovirus groups A and B.

	Reservoir	*Symptoms in humans*
A. Genus *Alphavirus*		
(mosquito-borne)		
Eastern equine encephalitis (EEE)	Birds	Encephalitis: frequently fatal
Semliki forest virus	Monkey	Undifferentiated febrile illness
Sindbis	Monkey	None
Chikungunya	Monkey	Myositis-arthritis
O'Nyong-Nyong	?	Fever, arthralgia, rash

(continued)

TABLE 56-4. (cont.)

RNA-containing Viruses		
	Reservoir	*Symptoms in humans*
Ross river virus	Mammals	Fever, rash, arthralgia
Venezuelan equine encephalitis (VEE)	Rodents	Encephalitis
Western equine encephalitis (WEE)	Birds	Encephalitis
B. Genus *Flavivirus*		
1. Mosquito-borne		
Yellow fever	Monkey	Hemorrhagic fever, hepatitis, nephritis, often fatal
Dengue (4 serotypes)	Humans	Fever, arthralgia, rash
Japanese encephalitis	Birds	Encephalitis: frequently fatal
St. Louis encephalitis	Birds	Encephalitis
Murray Valley encephalitis	Birds	Encephalitis
West Nile	Birds	Fever, arthralgia, rash
Kunjin	Birds	—
2. Tickborne		
Central European tickborne encephalitis (biphasic meningoencephalitis)	Rodents, hedgehog	Encephalitis
Far Eastern tickborne encephalitis [Russian spring–summer encephalitis (RSSE)]	Rodents	Encephalitis
Kyasanur forest	Rodents	Hemorrhagic fever
Louping III	Sheep	Encephalitis
Powassan	Rodents	Encephalitis
Omsk hemorrhagic fever	Mammals	Hemorrhagic fever
C. Genus *Rubivirus*		
Rubella virus	Humans	Severe deformities of fetuses in first trimester of pregnancy
D. Genus *Pestivirus*		
Mucosal disease virus (bovine virus diarrhea virus)	Cattle	—
Hog cholera virus (European swine fever)	Pig	—
Border disease virus	Sheep	—
E. Also included among the togaviruses are several as yet unclassified viruses. Among them are:		
Riley's lactic dehydrogenase elevating virus (LDHV)	Mouse	Produces lifelong chronic viremia in mice; elevates lactic dehydrogenase levels by decreasing the rate of enzyme clearance)
Equine arteritis virus	Horse	—
Simian hemorrhagic fever	Monkey	—

Classification	*Description*
IV. BUNYAVIRIDAE	Enveloped nucleocapsids Diameter: about 100 nm

Bunyaviruses include all former arbovirus group C viruses, as well as previously ungrouped arboviruses.

	Host	*Symptoms in humans*
A. Genus *Bunyavirus*		
Bunyamwera and related viruses	Mammals	—
California encephalitis group including La Crosse, Lumbo, and Snowshoe hare virus)	Mammals	Encephalitis
B. Genus *Phlebovirus*		
Sandfly fever Sicilian virus	Sandfly, mammals	Facial erythema
Rift Valley fever virus	Humans, sheep, cattle	Fever, arthralgia, retinitis
C. Genus *Nairovirus*		
Crimean–Congo hemorrhagic fever (CCHF) (2 serotypes)	Mammals	Hemorrhagic fever

(continued)

TABLE 56-4. (cont.)

RNA-containing Viruses		
	Host	*Symptoms in humans*
Five other serotypes including Nairobi sheep disease virus	Mammals	
D. Genus *Uukuvirus*		
Uukuniemi and related viruses	Birds, mammals	—

Classification	*Description*
V. REOVIRIDAE	Naked nucleocapsids possessing two capsid shells (except cytoplasmic polyhedrosis virus), each with icosahedral symmetry Diameter: 75 nm

The name is an acronym based on respiratory-enteric-orphan (because of lack of association with any disease in humans). The primary criterion for inclusion in this family is possession of a genome consisting of 10, 11, or 12 segments of double-stranded RNA. There are six genera with widely differing host ranges and somewhat differing morphologies. The vertebrate reoviruses possess two clearly defined capsid shells; the orbiviruses (many of which are transmitted by arthropods and are functionally arboviruses) possess a structurally featureless outer shell and an inner shell composed of 32 large ring-shaped capsomers (hence the name, Latin *orbis,* ring; however, see p. 815). The cytoplasmic polyhedrosis viruses possess only one capsid shell with clearly defined icosahedral symmetry. Phytoreoviruses closely resemble the vertebrate reoviruses, while Fijiviruses possess a structure more reminiscent of cytoplasmic polyhedrosis viruses. Rotaviruses present a wheel-like appearance (see Fig. 56-11), hence the name. The members of the six genera are not related antigenically.

	Host	*Symptoms in humans*
A. Genus *Orthoreovirus*		
Mammalian reoviruses 3 serotypes	Humans, other mammals	Pathogenicity not established
Avian reoviruses 5 serotypes	Chicken, duck	—
B. Genus *Orbivirus*		
Bluetongue virus	Culicoides, sheep	—
Eugenangee virus	Mosquitoes	—
Kemerovo	Ticks	—
African horse sickness virus	Culicoides, horse	—
Colorado tick fever virus	Ticks, humans	Encephalitis
C. Genus *Cypovirus*		
Cytoplasmic polyhedrosis virus Numerous strains	*Bombyx mori* (silkworm) and other Lepidoptera, Diptera, and Hymenoptera	—
D. Genus *Phytoreovirus*		
Wound tumor virus	Plants, leaf hoppers	—
Rice dwarf virus	Plants, leaf hoppers	
	—	
E. Genus *Fijivirus*		
Maize rough dwarf virus	Plants, leaf hoppers	—
Fiji disease virus	Plants, leaf hoppers	—
F. Genus *Rotavirus*		
Human rotavirus	Humans	Diarrhea in infants
Calf rotavirus (Nebraska calf diarrhea virus)	Calf	—
Murine rotavirus (Epizootic diarrhea of infant mice [EDIM]	Mouse	—
Simian rotavirus (SA11)	Monkey	—
Bovine or ovine rotavirus ("O" agent)	Cattle or sheep	—
Numerous other rotaviruses	Guinea pig, goat, horse, deer, antelope, rabbit, dog, duck	—

Classification	*Description*
VI. ORTHOMYXOVIRIDAE	Enveloped helical nucleocapsids Diameter: 80–120 nm

(continued)

TABLE 56-4. (cont.)

RNA-containing Viruses

The term myxovirus was coined to denote the unique affinity of influenza viruses for glycoproteins. Nowadays, members of this family are characterized by possession of nucleocapsids with helical symmetry that reside within envelopes, to whose outer surface are attached glycoprotein spikes of two types: One is the hemagglutinin, the other the neuraminidase. Influenza virus strains of type C differ from those of type A and type B in that (1) their buoyant density is lower, (2) the receptors for their hemagglutinins do not appear to contain sialic acid, and (3) their receptor-destroying enzyme seems to cleave some bond other than the -gal-neuraminic acid bond.

	Host	Symptoms in humans
A. Genus *Influenza virus*		
Influenza virus type A		
Human subtypes	Human	Acute respiratory disease
A_0 1933–1947*		Acute respiratory disease
A_1 1947–1957		Acute respiratory disease
A_2 1957–1964 (Asian)		Acute respiratory disease
A_2 1968 (Hong Kong)		Acute respiratory disease
Swine influenza virus	Swine	Acute respiratory disease
Avian subtypes		
Fowl plague virus and numerous other strains	Chicken, duck, turkey and others	—
Equine subtypes	Horse	—
Influenza virus type B		
Human subtypes		
B_0 1940–1945	Human	Acute respiratory disease
B_1 1945–1955		Acute respiratory disease
B_2 1962–1964		Acute respiratory disease
B_3 1962 (Taiwan)		Acute respiratory disease
Influenza virus type C	Humans	Acute respiratory disease
(possible separate genus)		

Classification	Description
VII. PARAMYXOVIRIDAE	Enveloped helical nucleocapsids
	Diameter: about 150 nm

Members of this family were until recently grouped with the orthomyxoviruses in the family Myxoviridae. They have been placed in a separate family because they differ from orthomyxoviruses in that their genomes are not segmented and their hemagglutinin and neuraminidase are located on the same glycoprotein spike. The other type of spike is responsible for the cell-fusing and hemolyzing activities (genus *Paramyxovirus*). Members of the *Morbillivirus* and *Pneumovirus* genera possess no neuraminidase, and members of the genus *Pneumovirus* possess no hemagglutinin either.

	Host	Symptoms in humans
A. Genus *Paramyxovirus*		
Parainfluenza virus type 1		
Sendai virus	Human, pig, mouse	Croup, common cold syndrome
(hemagglutinating virus of Japan [HVJ])		
HA-2 (hemadsorption virus)	Human	Mild respiratory disease
Parainfluenza viruses types 2 to 5		
Numerous strains including HA-1, SV5	Human and other animals	Respiratory tract infections
Newcastle disease virus (NDV)	Chicken	—
Mumps	Human	Parotitis, orchitis, meningoencephalitis
B. Genus *Morbillivirus*		
Measles	Human	Measles
Subacute sclerosing panencephalitis (SSPE)	Human	Chronic degeneration of the central nervous system
Distemper	Dog	—
Rinderpest	Cattle	—

(continued)

TABLE 56-4. (cont.)

RNA-containing Viruses		
	Host	**Symptoms in humans**
C. Genus *Pneumovirus*		
Respiratory syncytial virus (RSV)	Human	Pneumonia and bronchiolitis in infants and children, common cold syndrome
Bovine respiratory syncytial virus	Cattle	—
Pneumonia virus of mice (PVM)	Mouse	—

Classification	Description
VIII.　RHABDOVIRIDAE	Bullet-shaped, enveloped helical nucleocapsids Dimensions: 180 × 75 nm

This group comprises all viruses with the unique bullet-shaped morphology, as well as some that are bacilliform (rounded at both ends). It includes vesicular stomatis virus (VSV), rabies, and some viruses isolated from insects that do not appear to cause disease in vertebrates, but antibodies to them are found in birds and mammals, including humans. Some members of this family multiply in arthropods as well as in vertebrates.

	Host	**Symptoms in humans**
A. Genus *Vesiculovirus*		
VSV	Cattle, horse, swine	—
Chandipura virus	Isolated from humans	—
Flanders-Hart Park virus	Mosquitoes, birds	—
Kern Canyon virus	Bats	—
B. Genus *Lyssavirus*		
Rabies	All warm-blooded animals	Encephalitis, almost invariably fatal
C. Fish Rhabdoviruses (Two genera)	Fish	—
D. Other Rhabdoviruses		
Drosophila sigmavirus	Drosophila	—
Lettuce necrotic yellow virus	Plants	—
Other plant rhabdoviruses		—
E. Possible Rhabdoviruses		
Marburg virus	Monkey	Hemorrhagic fever, frequently fatal
Ebola hemorrhagic fever	?	Acute hemorrhagic fever, almost 90 percent case mortality

Classification	Description
IX.　RETROVIRIDAE (RNA tumor viruses)	Enveloped particles containing a coiled nucleocapsid with an icosahedral core shell Diameter: about 100 nm

The RNA tumor virus family comprises a large group of viruses characterized by possession of an RNA genome that comprises two identical molecules, a common morphology, and reverse transcriptase. There are three subfamilies. The first, the Oncovirinae, comprises the C-, B-, and D-type RNA tumor viruses. The second subfamily, the Lentivirinae, comprises the Visna group of viruses. They resemble the Oncovirinae with respect to morphology, nature of the genome, possession of a DNA polymerase, and ability to transform cultured cells in vitro, but have not yet been shown to possess oncogenic potential. The third subfamily, the Spumavirinae, comprises the foamy viruses, which are found in spontaneously degenerating kidney (and other) cell cultures, causing the formation of multinucleated vacuolated giant cells that have a highly characteristic appearance. Spumavirinae resemble Oncovirinae in morphology and in certain key characteristics of their mode of replication.

Subfamily	Host	Symptoms in humans
Oncovirinae		
A. Genus *Oncornavirus C*		
Subgenus Oncornavirus C avian		
Endogenous leukemia/leukosis viruses (RAV-0, RAV-1, RAV-2, etc.)	Chicken	—

(continued)

TABLE 56-4. (cont.)

RNA-containing Viruses		
Subfamily	*Host*	*Symptoms in humans*
Avian sarcoma viruses (ASV)	Chicken	—
Nondefective		
Rous Sarcoma virus (RSV)		
Sarcoma/acute leukemia viruses, defective	Chicken and other birds	—
Fujinami sarcoma virus (FSV)		
Y73 Sarcoma virus and		
Esh sarcoma virus		
UR-2 virus		
Avian myeloblastosis virus (AMV)		
Avian erythroblastosis virus (AEV)		
Avian myelocytomatosis virus (MC29)		
Reticuloendotheliosis virus (REV)	Duck, chicken	—
Lymphoproliferative disease virus of turkeys (LPDV)	Turkey	—
Subgenus Oncornavirus C mammalian		
Endogenous leukemia viruses, numerous strains	Mammals	—
Sarcoma/acute leukemia viruses, defective	Rodents	
Abelson murine leukemia virus		
Murine sarcoma viruses (Harvey, Kirsten, Moloney, Rasheed)		
Feline sarcoma viruses (Snyder–Theilen, Gardner–Arnstein, McDonough)	Cats	—
Simian sarcoma virus	Woolly monkey	—
Subgenus Oncornavirus C Reptilian	Reptiles	—
Reptilian Type C retroviruses		
B. Genus *Oncornavirus B*		
Mouse mammary tumor virus (Bittner virus (milk factor))	Mouse	—
Viruses of guinea pigs, baboons and other mammals	Mammals	—
C. Genus *Oncornavirus D*		
Mason–Pfizer monkey virus (MPMV)	Rhesus monkey	—
Viruses from primates	Primates	?
Guinea pig virus	Guinea pig	—
↶ Lentivirinae		
Visna	Sheep	—
Maedi	Sheep	—
Progressive pneumonia virus	Mice	—
↶ Spumavirinae		
Human foamy virus	Human cells	—
Simian foamy viruses (9 serotypes)	Monkey kidney cells	—
Canine foamy virus	Dog kidney cells	—
Bovine syncytial virus	Bovine kidney cells	—
Feline syncytial virus	Feline cells	—
Hamster syncytial virus	Hamster cells	—
Probable Retrovirus		
Equine infectious anemia virus	Horse	—

Classification	*Description*
X. ARENAVIRIDAE	Enveloped coiled nucleocapsids Diameter: 80–130 nm

This family comprises viruses characterized by well-defined envelopes that bear closely spaced projections and enclose an unstructured interior containing a variable number of characteristic electron-dense granules about 25 nm in diameter that have been shown to be ribosomes. They share a group-specific antigen, but antisera do not cross-neutralize.

(continued)

TABLE 56-4. (cont.)

RNA-containing Viruses		
	Host	*Symptoms in humans*
Lymphocytic choriomeningitis virus (LCM)	Mouse	Latent infection in mice, may produce fatal meningitis in many other species, including humans
Tacaribe virus complex		
Several viruses including Argentinian (Junin) and Bolivian (Machupo) hemorrhagic fever	Isolated from insects and rodents	Hemorrhagic fever, Machupo frequently fatal
Lassa virus	?	Hemorrhagic fever, frequently fatal

Classification	*Description*
XI. CORONAVIRIDAE	Enveloped helical nucleocapsids Diameter: about 100 nm

Nucleocapsids helical, characteristic large club-shaped projections (spikes). There is some antigenic relationship between certain human and murine strains.

	Host	*Symptoms in humans*
Infectious bronchitis virus (IBV)		
Avian strains	Chicken	—
Human strains	Human	Acute upper respiratory disease
Mouse hepatitis virus	Mouse	—
Probable coronaviruses		
Human enteric coronavirus	Humans	Intestinal disorders
Feline infectious peritonitis (feline coronavirus)	Cats	—

XII.   MISCELLANEOUS VIRUSES

Several viruses do not fit into any of the families listed so far. The most important are:

A. Hepatitis B Virus. Clinically, two types of viral hepatitis are distinguished. One is characterized by a short incubation period (infectious hepatitis, or epidemic jaundice); the other is characterized by a long incubation period and usually requires parenteral transmission (serum hepatitis). The etiologic agent of the former, hepatitis virus A, is a picornavirus. The etiologic agent of the latter, hepatitis B virus (HBV), appears to infect only humans in nature, and experimental infection has only been achieved in a few additional primates. Its infectious form appears to be a particle about 40 nm in diameter, the Dane particle, that contains a partially double-stranded circular DNA molecule consisting of a complete minus strand and a plus strand that is about two thirds as long. Evidence is accumulating that HBV is implicated in the etiology of hepatomas. Recently, several viruses have been discovered that closely resemble HBV, namely the woodchuck hepatitis virus, the ground squirrel hepatitis virus, and the Pekin duck hepatitis virus. These viruses are used as models for studies on HBV.

B. Chronic infectious neuropathic agents (CHINA viruses). These agents have a preclinical period lasting months to several years, succeeded by a slowly progressing, usually fatal disease. Most of them affect the central nervous system. Degenerative diseases of other organs and tissues may be caused by similar agents. They include the agents that cause Kuru and Creutzfeldt–Jakob disease in humans, scrapie in sheep, and transmissible mink encephalopathy in mink. All are slow degenerative disorders of the central nervous system, marked by ataxia and wasting, and end in death. The etiologic agents have been transmitted, but not yet isolated or even visualized.

*In 1971 a WHO Study Group adopted a new nomenclature for influenza type A viruses. In this system virus strains are described in terms of both the hemagglutinin (HA) and neuraminidase (NA) antigens. A_0 strains are now designated as HON1, A_1 strains as H1N1, the Asian type A_2 of 1957 as H2N2, and the Hong Kong type A_2 of 1968 as H3N2.

judged by nucleic acid hybridization, immunologic cross-reactivity, or ability to recombine (see Chap. 60). The same applies to members of the herpesvirus family; the human, simian, canine, and avian adenoviruses; and so on. These groups form the basis for subdividing the viruses within families into genera.

Major Families of Animal Viruses

Table 56-4 presents a summary of the distinguishing characteristics of the major families of animal viruses, together with a list of the most important animal and human patho-

gens in each. It is based in large part on recommendations made by the International Committee on Taxonomy of Viruses.

FURTHER READING

Books and Reviews

Bishop DHL: Virion polymerases. In Fraenkel-Conrat H, Wagner RR (eds): Comprehensive Virology. New York, Plenum, 1977, vol 10, p 117

Caspar DLD, Klug A: Physical principles in the construction of regular viruses. Cold Spring Harbor Symp Quant Biol 27:1, 1962

Compans RW, Klenk H: Viral membranes. In Fraenkel-Conrat H, Wagner RR (eds): Comprehensive Virology. New York, Plenum, 1979, vol 13, p 293

Diener TO: Viroids and their interactions with host cells. Ann Rev Microbiol 36:239, 1982

Tooze J (ed): DNA tumor viruses. New York, Cold Spring Harbor Laboratory, 1980

Wimmer E: Genome-linked proteins of viruses. Cell 29:199, 1982

POXVIRUSES

Baxby D: Jenner's Smallpox Vaccine: The Riddle of Vaccinia Virus and Its Origin. London, Heineman, 1981

HERPESVIRUSES

Roizman B: The structure and isomerization of herpes simplex virus genomes. Cell 16:481, 1979

Roizman B (ed): The Herpesviruses. New York, Plenum, 1982

ADENOVIRUSES

Nermut MV: The architecture of adenoviruses: Recent views and problems. Arch Virol 64:175, 1980

PARVOVIRUSES

Berns KI, Hauswirth WW: Adeno-associated viruses. Adv Virus Res 25:407, 1980

PICORNAVIRUSES

Perez-Bercoff R (ed): The Molecular Biology of Picornaviruses. New York, Plenum, 1978

Putnak JR, Phillips BA: Picornaviral structure and assembly. Microbiol Rev 45:287, 1981

TOGAVIRUSES

Schlesinger RW (ed): The Togaviruses. New York, Academic Press, 1980

BUNYAVIRUSES

Calisher CH, Thompson WH (eds): California Serogroup Viruses. New York, Alan R. Liss, 1983

REOVIRUSES

Joklik WK (ed): The Reoviridae. New York, Plenum, 1983

RHABDOVIRUSES

Bishop DHL (ed): Rhabdoviruses. Boca Raton, Fla, CRC Press, 1979

RETROVIRUSES

Weiss R, Teich N, Varmus H, et al. (eds): RNA Tumor Viruses. New York, Cold Spring Harbor Laboratory, 1982

ARENAVIRUSES

Howard CR, Simpson DIH: The biology of the Arenaviruses. J Gen Virol 51:1, 1981

CORONAVIRUSES

MacNaughton MR, Davies, HA: Human enteric Coronaviruses. Arch Virol 70:301, 1981

Selected Papers

POXVIRUSES

Baroudy BM, Venkatesan S, Moss B: Incompletely base-paired flip-flop terminal loops link the two DNA strands of the vaccinia virus genome into one uninterrupted polynucleotide chain. Cell 28:315, 1982

Essani K, Dales S: Biogenesis of vaccinia: Evidence for more than 100 polypeptides in the virion. Virology 95:385, 1979

Geshelin P, Berns KI: Characterization and localization of the naturally occurring cross-links in vaccinia virus DNA. J Mol Biol 88:785, 1974

Kates JR, McAuslan BR: Messenger RNA synthesis by a "coated" viral genome. Proc Natl Acad Sci USA 57:314, 1967

Nevins JR, Joklik WK: Isolation and properties of the vaccinia virus DNA-dependent RNA polymerase. J Biol Chem 252:6930, 1977

Oie M, Ichihashi Y: Characterization of vaccinia polypeptides. Virology 113:263, 1981

Pickup DJ, Bastia D, Stone HO, et al.: Sequence of terminal regions of cowpoxvirus DNA: arrangement of repeated and unique sequence elements. Proc Natl Acad Sci USA 79:7112, 1982

Preston VG, Davison AJ, Garon CF, et al.: Visualization of an inverted terminal repetition in vaccinia virus DNA. Proc Natl Acad Sci USA 75:4863, 1978

Sarov I, Joklik WK: Studies on the nature and location of the capsid polypeptides of vaccinia virions. Virology 50:579, 1972

IRIDOVIRUSES

Lee MH, Willis DB: Restriction endonuclease mapping of the frog virus 3 genome. Virology 126:317, 1983

HERPESVIRUSES

Graham FL, Veldjuisen G, Wilkie NM: Infectious herpesvirus DNA. Nature [New Biol] 245:265, 1973

Heine JW, Honess RW, Cassai E, et al.: Proteins specified by herpes simplex virus. XII. The virion polypeptides of type 1 strains. J Virol 14:640, 1974

Palmer EL, Martin ML, Gary GW: The ultrastructure of disrupted herpesvirus nucleocapsids. Virology 65:260, 1975

ADENOVIRUSES

Brown DP, Westphal M, Burlingham BT, et al.: Structure and composition of the adenovirus type 2 core. J Virol 16:366, 1975

Everitt E, Lutter L, Philipson L: Structural proteins of adenoviruses. Virology 67:197, 1975

Nermut MV: Fine structure of adenovirus type 5. I. Virus capsid. Virology 65:480, 1975

Rekosh DMK, Russell WC, Bellett AJD: Identification of a protein linked to the ends of adenovirus DNA. Cell 11:283, 1977

Roberts RJ, Arrand JR, Keller W: The length of the terminal repetition in adenovirus-2 DNA. Proc Natl Acad Sci USA 71:3829, 1974

PAPOVAVIRUSES

Fiers W, Conteras R, Haegeman G, et al.: Complete nucleotide sequence of SV40 DNA. Nature 273:113, 1978

Finch JT: The surface structure of polyoma virus. J Gen Virol 24:359, 1974

Gibson W: Polyoma virus proteins: A description of the structural proteins of the virion based on polyacrylamide gel electrophoresis and peptide analysis. Virology 62:319, 1974

Müller U, Zentgraf H, Eicken I, et al.: Higher order structure of simian virus 40 chromatin. Science 201:406, 1978

Reddy VB, Dhar R, Weissman SM: Nucleotide sequence of the genes for the simian virus 40 proteins VP2 and VP3. J Biol Chem 253:621, 1978

Reddy VB, Thimmappaya B, Dhar R, et al.: The genome of simian virus 40. Science 200:494, 1978

PARVOVIRUSES

Berns KI, Kelley TJ: Visualization of the inverted terminal repetition of adeno-associated DNA. J Mol Biol 82:267, 1974

Chesebro B, Bloom M, Hadlow W, et al.: Purification and ultrastructure of Aleutian disease virus of mink. Nature 254:456, 1975

McPherson RA, Rose JA: Structural proteins of adenovirus-associated virus: Subspecies and their relatedness. J Virol 46:523, 1983

Mayor HD, Torikai K, Melnick JL: Plus and minus single-stranded DNA separately encapsidated in adeno-associated satellite virions. Science 166:1280, 1969

Spear IS, Fife KH, Hauswirth WW, et al.: Evidence for two nucleotide sequence orientations within the terminal repetition of adeno-associated virus DNA. J Virol 24:627, 1977

Türler H: Interactions of polyoma and mouse DNAs. III. Mechanism of polyoma pseudovirion formation. J Virol 15:1158, 1975

PICORNAVIRUSES

Ambros V, Baltimore D: Protein is linked to the 5′ end of poliovirus RNA by a phosphodiester linkage to tyrosine. J Biol Chem 253:5263, 1978

Coulepis AG, Locarnini SA, Westaway EG, et al.: Biophysical and biochemical characterization of hepatitis A virus. Intervirology 18:107, 1982

Golini F, Nomoto A, Wimmer E: The genome-linked protein of picornaviruses. Virology 89:112, 1978

Lund GA, Ziola BR, Salmi A, et al.: Structure of the mengo virion. V. Distribution of the capsid polypeptides with respect to the surface of the virus particle. Virology 78:35, 1977

Medappa KC, McLean C, Rueckert RR: On the structure of rhinovirus 1A. Virology 44:259, 1971

CALICIVIRUSES

Black DN, Brown F: A major difference in the strategy of the calici- and picornaviruses and its significance in classification. Intervirology 6:57, 1975/1976

Burroughs JN, Doel TR, Smale CJ, et al.: A model for vesicular exanthema virus, the prototype of the calicivirus group. J Gen Virol 40:161, 1978

TOGAVIRUSES

Aliperti G, Schlesinger MJ: Evidence for an autoprotease activity of Sindbis virus capsid protein. Virology 90:366, 1978

Bell JR, Bond MW, Hunkapillar MW, et al.: Structural proteins of western equine encephalitis virus: Amino acid compositions and N-terminal sequences. J Virol 45:708, 1983

Burke D, Keegstra K: Carbohydrate structure of Sindbis virus glycoprotein E2 from virus grown in hamster and chicken cells. J Virol 29:546, 1979

Enzmann PJ, Weiland F: Studies on the morphology of the alphaviruses. Virology 95:501, 1979

de Madrid AT, Porterfield JS: The flaviviruses (group B arboviruses): A cross-neutralization study. J Gen Virol 23:91, 1974

Pedersen CE, Eddy GA: Separation, isolation, and immunological studies of the structural proteins of Venezuelan equine encephalomyelitis virus. J Virol 14:740, 1974

von Bonsdorff CH, Harrison SC: Hexagonal glycoprotein arrays from Sindbis virus membranes. J Virol 28:578, 1978

BUNYAVIRUSES

Dahlberg JE, Obijeski JS, Korb J: Electron microscopy of the segmented RNA genome of La Crosse virus: Absence of circular molecules. J Virol 22:203, 1977

Hewlett MJ, Petterson RF, Baltimore D: Circular forms of Uukuniemi virion RNA: An electron microscopic study. J Virol 21:1085, 1977

Murphy FA, Harrison AK, Whitefield SG: Bunyaviridae: Morphologic and morphogenetic similarities of Bunyamwera serologic supergroup viruses and several other arthropod-borne viruses. Intervirology 1:297, 1973

Petterson RF, von Bornsdorff CH: Ribonucleoproteins of Uukuniemi virus are circular. J Virol 15:386, 1975

Smith JF, Pifat DY: Morphogenesis of Sandfly fever viruses (Bunyaviridae family). Virology 121:61, 1982

REOVIRUSES

Esparza J, Gil F: A study on the ultrastructure of human rotavirus. Virology 91:141, 1978

Luftig RB, Kilham SS, Hay AJ, et al.: An ultrastructural study of virions and cores of reovirus type 3. Virology 48:170, 1972

Nicolas JC, Cohen J, Fortier B, et al.: Isolation of a human pararotavirus. Virology 124:181, 1983

Palmer EL, Martin ML: The fine structure of the capsid of reovirus type 3. Virology 76:109, 1977

Palmer E, Martin ML: Further observations on the ultrastructure of human rotavirus. J Gen Virol 62:105, 1982

Shatkin AJ, Sipe JD, Loh P: Separation of ten reovirus genome segments by polyacrylamide gel electrophoresis. J Virol 2:986, 1968

Smith RE, Zweerink HJ, Joklik WK: Polypeptide components of virions, top component and cores of reovirus type 3. Virology 39:791, 1969

Urasawa S, Urasawa T, Taniguchi K: Three human rotavirus serotypes demonstrated by plaque neutralization of isolated strains. Infect Immun 38:781, 1982

ORTHOMYXOVIRUSES

Desselberger U, Palese P: Molecular weights of RNA segments on influenza A and B viruses. Virology 88:394, 1978

Inglis SC, McGeoch DJ, Mahy BWJ: Polypeptides specified by the influenza virus genome. II. Assignment of protein coding functions to individual genome segments by in vitro translation. Virology 78:522, 1977

Murti KG, Bean WJ Jr, Webster RG: Helical ribonucleoproteins of influenza virus: An electron microscopic analysis. Virology 104:227, 1980

Racaniello VR, Palese P: Influenza B virus genome: Assignment of viral polypeptides to RNA segments. J Virol 29:361, 1979

Schwartz RT, Klenk H: Carbohydrates of influenza virus. IV. Strain-dependent variations. Virology 113:584, 1981

PARAMYXOVIRUSES

Orvell C: Structural polypeptides of mumps virus. J Gen Virol 41:527, 1978

Scheid A, Choppin PW: Identification of biological activities of paramyxovirus glycoproteins. Activation of cell fusion, hemolysis, and infectivity by proteolytic cleavage of an inactive precursor protein of Sendai virus. Virology 57:475, 1974

Shimizu K, Shimizu YK, Kohama T, et al.: Isolation and characterization of two distinct types of HVJ (Sendai virus) spikes. Virology 62:90, 1974

RHABDOVIRUSES

Emerson SU, Wagner RR: L protein requirement for in vitro

RNA synthesis by vesicular stomatitis virus. J Virol 12:1325, 1973

Reading CL, Penhoet EE, Ballou CE: Carbohydrate structure of vesicular stomatitis virus glycoprotein. J Biol Chem 253:5600, 1978

Tabas I, Schlesinger S, Kornfeld S: Processing of high mannose oligosaccharides to form complex type oligosaccharides on the newly synthesized polypeptides of the vesicular stomatitis virus G protein and the IgG heavy chain. J Biol Chem 253:716, 1978

Wagner RR, Prevec L, Brown F, et al.: Classification of rhabdovirus proteins: A proposal. J Virol 10: 1228, 1972

ARENAVIRUSES

Palmer EL, Obijeski JF, Webb PA, et al.: The circular, segmented nucleocapsid of an arenavirus—Tacaribe virus. J Gen Virol 36:541, 1977

Young PR, Howard CR: Fine structure analysis of Pichinde virus nucleocapsids. J Gen Virol 64:833, 1983

CORONAVIRUSES

Guy JS, Brian DA: Bovine coronavirus genome. J Virol 29:293, 1979

King B, Brian BA: Bovine coronavirus structural proteins. J Virol 42:700, 1982

MacNaughton MR, Davies HA, Nermut MV: Ribonucleoprotein-like structures from coronavirus particles. J Gen Virol 39:545, 1978

MacNaughton MR, Madge MH: The genome of human coronavirus strain 229 E. J Gen Virol 39:497, 1978

HEPATITIS B VIRUS

Barker LF, Almeida JD, Hoofnagle JG, et al.: Hepatitis B core antigen: Immunology and electron microscopy. J Virol 14:1552, 1974

Krugman S, Gregg MB, Kabat EA, et al.: Nomenclature of antigens associated with viral hepatitis B. Intervirol 2:134, 1974

Robinson WS, Greenman RL: DNA polymerase in the core of human hepatitis B virus. J Virol 13:1231, 1974

VIROIDS

Dickson E, Diener TO, Robertson HD: Potato spindle tuber and citrus exocortis viroids undergo no major sequence changes during replication in two different hosts. Proc Natl Acad Sci USA 75:951, 1978

Owens RA, Erbe E, Hadidi A, et al.: Separation and infectivity of circular and linear forms of potato spindle tuber viroids. Proc Natl Acad Sci USA 74:3859, 1977

CHAPTER 57

Viruses and Viral Proteins as Antigens

Interaction of Virus with Neutralizing Antibody

Complement-fixing Antigen and Antibody

Hemagglutination Inhibition

Gel Immunodiffusion and Immunoelectrophoresis

Visualization of Viral Antigens Using Tagged Antibody

Detection of Minute Amounts of Viral Antigens and Antibodies

Virus-coded Cell Surface Antigens

Immune Response to Virus Infection

Many proteins coded by viruses are good antigens. This is of vital significance both medically and scientifically. Although great strides have been made during the past two decades in defining the biochemistry and molecular biology of the multiplication cycles of animal viruses, the chemotherapeutic control of viral diseases is not yet practical. In fact, with very few exceptions, there is no way in which virus infections can be controlled; in almost all cases one relies on the natural ability of the host to form antibody to the invading virus. When the spread of virus infection to essential organs is too rapid, or when for some reason antibody formation does not take place early enough, the patient may succumb. Gammaglobulin from hyperimmune sera is sometimes administered as a last resort; even then, one must rely on antibodies, not on drugs. By the same token, the only presently practical form of antiviral prophylaxis is provided by antibodies produced in response to vaccines (Chap. 61).

Structural viral proteins stimulate the formation of antibodies not only as components of virus particles, but also as components of virus particle subunits, such as capsomers or nucleocapsids, and also in the free state. The principal antigenic determinants are often the same in all three forms, but extra antigenic sites are sometimes generated as individual proteins become part of more complex structures. For example, adenovirus hexons exhibit antigenic determinants not expressed by free hexon proteins, and the capsids of picornaviruses possess antigenic specificities that are not exhibited by their component structural proteins.

The range of antiviral antibodies formed under conditions when viruses can and cannot multiply differs greatly. If a virus cannot multiply, either because it has been inactivated or because the host is not susceptible, only antibodies to surface components of the virus particle are usually formed. However, if the virus can multiply, not only is far more antibody formed because progeny virus will also act as antigen, but also the range of antibodies produced is much wider. This is because antibodies are then also formed to the unassembled and partially assembled viral components that are synthesized as a result of virus multiplication, as well as to nonstructural virus-coded proteins. For example, antisera to inactivated vaccinia virus contain only a few species of antibody directed against its surface components; but antisera from animals in which vaccinia virus has multiplied contain antibodies against at least 30 different viral proteins in readily detectable amounts.

The antigenic determinants of virus-coded proteins are of great importance for virus classification. Viruses that belong to the same genus or major group share common determinants, or group-specific antigens, that are generally located on internal virus components. The most specific, or individual, viral antigenic determinants are usually located on surface components of virus particles, which are the type-specific antigens that identify individual virus strains.

Possession of group- and type-specific antigenic determinants provides an important tool for epidemiology. Newly isolated virus strains are usually characterized on the basis of their reaction with antisera of known specificity, which reveals the nature of their group- and type-specific antigens. Furthermore, it is possible to determine whether a given human or animal population has been exposed to a particular virus strain by testing serum samples for the presence of antiviral antibody.

The interaction of viruses and viral proteins with antibodies can be recognized and measured in several ways. The four most important ways are described below.

Interaction of Virus with Neutralizing Antibody

Antibodies to viral surface components neutralize infectivity; these are the neutralizing antibodies that protect against disease. They usually persist in the body for many years, and even when their level drops, a secondary or anamnestic response to virus generally boosts their titers to very high levels, so that no second cycle of infection ensues. This explains the fact that animals generally contract any particular virus disease only once. Exceptions, such as the common cold and influenza, are due to special circumstances. The reason for the frequent recurrence of the common cold syndrome is that it is elicited by a large group of viruses, among which are rhinoviruses, enteroviruses, adenoviruses, ortho- and paramyxoviruses, and coronaviruses. The reason for recurrent epidemics caused by influenza virus is that new antigenic variants arise readily, and are readily selected in humans (Chap. 60).

The reaction between neutralizing antibody and virus follows first-order kinetics, which indicates that one antibody molecule can inactivate one virus particle. It does so by interfering with the initial events of the virus multiplication cycle. There are two likely mechanisms. The first is steric hindrance—virus particles combined with antibody molecules are unlikely to be able to react with their receptors on cell surfaces. The second mechanism involves interference with uncoating. Under conditions of normal infection, viral genomes are liberated into the interior of the cell, ready to start multiplication (see below); but virus–antibody complexes are apparently engulfed by and inactivated in phagocytic vacuoles, so that no intact viral genomes are able to reach the interior of the cell (Chap. 58).

Complement-fixing Antigen and Antibody

The virus–antibody interaction can also be measured by taking advantage of the fact that complexes of viral protein and antibody often fix complement. As sensitive methods for titrating complement are available, this provides a convenient and accurate method of measuring either the amount of viral antigens (complement-fixing antigens or CFA), or of antibody to such antigens. The chief advantage of this method of detecting viral antigens is that any virus-coded protein may be a complement-fixing antigen; both structural and nonstructural virus-coded proteins, as well as viral subunits and virus particles themselves, may form complexes with antibody that fix complement. This method of quantitating viral proteins is particularly useful for detecting abortive virus infections when only part of the genetic information present in the viral genome is expressed and no virus particles are produced (Chap. 59). It is also of great importance in epidemiology, as it is often far easier to identify newly isolated virus strains by determining whether extracts of cells infected with them fix complement with antisera of known specificity than by measuring the ability of such antisera to neutralize infectivity.

Hemagglutination Inhibition

As described in Chapter 55, many viruses can agglutinate red blood cells. Virus particles coated with antibody molecules cannot do so, and antisera that contain virus-specific antibodies therefore inhibit hemagglutination. The antigens that are involved here are viral surface components, which, as pointed out above, possess the most type-specific antigenic determinants; hemagglutination-inhibition therefore provides a highly specific characterization of viruses. Because it is also very easy to perform, this technique is very useful in virus identification and epidemiology.

Gel Immunodiffusion and Immunoelectrophoresis

Under appropriate conditions of antigen–antibody equivalence, antigen–antibody complexes are insoluble. This property is used in gel immunodiffusion and gel immunoelectrophoresis, techniques that are widely employed for resolving mixtures of viral antigens such as occur in extracts of infected cells.

The most widely used version of gel immunodiffusion is the Ouchterlony method, which employs Petri plates containing agar into which are cut a number of wells, one being situated centrally and the others equidistantly from it and from each other. Antiserum or antibody is placed into the center well, antigen is placed into the outer ones, and diffusion is then allowed to proceed. Where the concentration of antigen–antibody complexes exceeds their solubility product, precipitin lines form, the location of which depends on the relative diffusion rates, and therefore on the relative sizes, of antigen and antibody. Identity of antigens is revealed by the familiar fusion of precipitin lines (Fig. 57-1) (Chap. 13).

In the gel immunoelectrophoresis technique, antigens are not separated by free diffusion but by electrophoresis in agar slabs, after which antiserum is applied in a trough cut parallel to the direction of electrophoresis. After diffusion, precipitin lines form as above (Fig. 57-2). The concentrations of antibody and antigen used in both these techniques are generally adjusted so that the precipitin lines are very thin, thus permitting great resolution.

A modification of this technique is rocket immunoelectrophoresis, in which proteins are electrophoresed into agarose gels that contain antibodies to them. If the pH of such gels is adjusted so that antibody molecules remain stationary (about pH 8.6), rocket-shaped zones of precipitation result that are demonstrable with protein stains and the area of which is proportional to the quantity of antigen that is applied (see Chap. 13). This technique is very versatile because it permits not only the quantitation of single antigens in complex mixtures, provided that the appropriate monospecific antibody is available, but is also applicable to mixtures of antigens. The technique is then known as crossed immunoelectrophoresis, and the antigens are first separated by electrophoresis in a standard agarose gel before being electrophoresed at right angles into the gel that contains the appropriate mixture of antibodies. An ex-

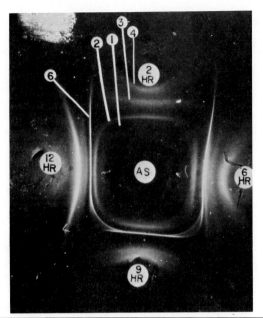

Figure 57-1. The use of gel diffusion analysis to detect virus-specified proteins in extracts of infected cells. Antiserum to vaccinia virus was placed into the center well (AS) of a Petri dish containing a layer of agar. Extracts of HeLa cells infected for 2, 6, 9, and 12 hours, respectively, with vaccinia virus were placed into the other four wells, and the antibodies and antigens were allowed to diffuse toward each other. Precipitin lines, formed as described in the text, were then stained with Poinceau S. The pattern becomes increasingly complex with increasing time after infection, as more virus-specified proteins are synthesized. The advantage of this method is that virus-specified proteins are revealed without the necessity for purification. *(From Salzman and Sebring: J Virol 1:16, 1967.)*

ample of this technique is to use extracts of infected cells as the source of virus-coded antigens, and antiserum from an infected rabbit as the source of the gamut of antibodies to all these antigens. This combination yields a profile of the type shown in Fig. 12–25, which permits the simultaneous quantitation of numerous virus-coded proteins.

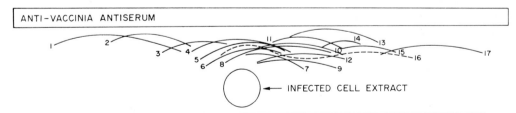

Figure 57-2. Diagrammatic representation of an immunoelectrophoretic pattern of the virus-specified proteins present in an extract of chick cells infected with vaccinia virus. In this technique, the cell extract is placed into a circular well cut into an agar slab and an electric field is applied, causing antigens to migrate at rates governed by their charge and size. After electrophoresis, antiserum to vaccinia virus is placed into a trough cut parallel to the direction of electrophoresis and allowed to diffuse toward the separated cell extract components. Virus-specified proteins able to react with antibodies in the antiserum form precipitin lines. Seventeen such proteins can be detected in the extract shown here. *(From Rodriguez-Burgos et al.: Virol 30:569, 1966.)*

Visualization of Viral Antigens Using Tagged Antibody

There are many occasions when direct visual localization of viral antigens is desired. This can be achieved by using antibody that is tagged, conjugated, or labeled with some material that can be visualized with either the light microscope or the electron microscope.

Antibody labeled with the dye fluorescein fluoresces brightly when viewed with a microscope equipped with a source of ultraviolet light (Fig. 57-3). Such antibody is a sensitive research tool in viral pathogenesis—that is, in studies of the route of infection and the spread of virus within the organism—because it can reveal a small number of infected cells. It is also useful for measuring the proportion of infected cells in a variety of experimental situations, and it can serve as a rapid diagnostic tool, as minute amounts of infected biopsy material can be treated with fluorescein-labeled antibodies to several suspected viruses, one of which will cause the infected cells to fluo-

resce. Finally, as fluorescein-labeled antibody can also reveal the pattern of viral antigen distribution within infected cells, and because this pattern is often highly characteristic (nuclear or cytoplasmic, diffuse or highly localized), it can also serve as a useful adjunct to virus identification.

Antibody molecules can also be tagged with large molecules or particles that can be seen with the electron microscope. Among these are ferritin, a large iron-containing protein, bacteriophage or virus particles with characteristic shapes, and latex spheres. Use of such antibody permits exquisitely detailed observation of the distribution of viral antigens in infected cells. It is therefore invaluable in studies aimed at establishing the exact location of the sites of synthesis, accumulation, and assembly of viral protein components (Fig. 57-4).

Detection of Minute Amounts of Viral Antigens and Antibodies

Appropriate patient management frequently depends on early detection of virus infections or on the recognition of persistent virus infections. In both situations, diagnosis requires the correct and rapid identification of minute amounts of viral antigens and antibodies, and in recent years great strides have been made in the development of extremely sensitive techniques for this purpose. Two such techniques are as follows. The first is radioimmunoassay, such as is described in Chapter 12. Not only is this technique capable of detecting extremely small amounts of virus-coded proteins, but it is also very useful for assessing serologic relatedness, as one can measure the efficiency with which unknown antigens can prevent antibody from precipitating the standard labeled test antigen—the closer the serologic relationship, the more efficient the competition.

The second technique is the enzyme-linked immunosorbent assay (ELISA), in which antigen is immobilized on some surface, such as the wells of a plastic microtiter plate (see Fig. 55-8), and specific antiserum coupled with alkaline phosphatase or peroxidase is then added. Following extensive washing, substrate is added, and the reaction product that is formed by the antibody-linked enzyme bound to the antigen is measured colorimetrically. This technique can be made to be over 100 times more sensitive than complement fixation assays.

A further dimension has recently been added to these techniques by the use of the "protein A" that is present in the cell walls of certain strains of *Staphylococcus*, which possesses high affinity for the Fc portion of most mammalian IgGs. Cells of such strains, or isolated A protein coupled to inert particles such as agarose, can be used to adsorb minute amounts of antigen–antibody complexes from complex mixtures, and the antigens in them can then be analyzed.

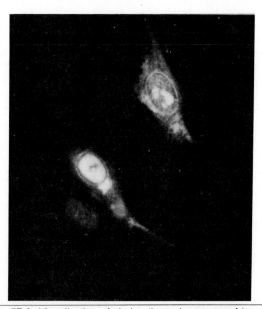

Figure 57-3. Visualization of viral antigens by means of immunofluorescence. Cells infected with herpes simplex virus were washed, fixed in acetone, and allowed to react either with herpesvirus antibody conjugated with fluorescein isothiocyanate (direct immunofluorescence), or with herpesvirus antibody prepared in rabbits, followed by antirabbit globulin conjugated with fluorescein isothiocyanate (indirect immunofluorescence). Cells were then examined under ultraviolet light. The cells shown here were stained by the indirect method. The top cell shows fluorescent nuclear patches as well as fluorescence at the nuclear membrane and some diffuse cytoplasmic fluorescence. The other cell shows bright fluorescence of practically the whole nucleus, as well as cytoplasmic fluorescence. × 550. *(From Ross et al.: J Gen Virol 2:115, 1968.)*

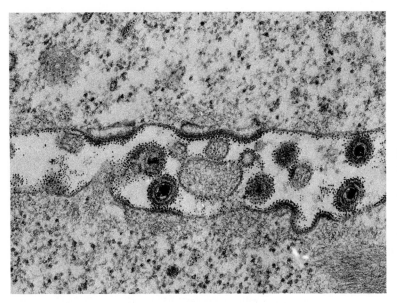

Figure 57-4. Visualization of viral antigens by means of ferritin-conjugated antibody. This electron micrograph reveals the localization of virus-specified proteins on the surfaces of cells infected with herpes simplex virus. Infected cells were allowed to react with herpesvirus antibody conjugated with ferritin; the cells were then washed, fixed, embedded, and sectioned. The surfaces of two adjacent cells are seen, both with intensely labeled patches. Budding virus particles and detached cytoplasmic fragments are also labeled. × 48,000. *(From Nii et al.: J Virol 2: 1172, 1968.)*

Virus-coded Cell Surface Antigens

As described in Chapters 58 and 59, new antigenic determinants frequently appear on the surfaces (outer plasma membranes) of infected cells. These new determinants are on virus-coded proteins. In the case of enveloped viruses, these proteins become, in due course, part of the viral envelope. In the case of nonenveloped viruses, they are nonstructural proteins that are synthesized early during the infection cycle. In either case, virus-coded cell surface proteins provide a clear signal to the immune mechanism that cells are infected; antibodies are formed against them and destroy the cells.

Immune Response to Virus Infection

During virus infection, antibodies are formed against all classes of virus-coded antigens. Those that are most important for eliminating virus from the body are those directed against virus particle components and against virus-coded cell surface antigens. The former include the neutralizing and complement-fixing antibodies that eliminate virus particles as infectious units (p. 852). As for the latter, their combination with virus-coded antigens on cell surfaces renders the cells subject to destruction by at least two mechanisms: combination with complement followed by lysis, and attack by cytolytic T lymphocytes. In either case, the infected cell is eliminated as a source of progeny virus.

Usually the mechanisms for destroying infectious virus and infected cells are beneficial to the host. However, it is now recognized that sometimes these mechanisms may be harmful. Let us consider first the destruction of infected cells. As a rule, the number of cells that are destroyed is not large enough to cause serious problems for the host organism, but there are exceptions. An example is provided by lymphocytic choriomeningitis virus (LCM), which causes encephalitis in mice and also in humans. LCM, an enveloped virus, is not a very "lytic" virus; cells infected with it are not severely damaged and may survive for long periods of time. In mice, LCM produces no overt disease if the immune mechanism is not operative (in tolerant or immunosuppressed animals). However, in immunologically competent mice, LCM causes a fatal meningitis within a week, that is, as soon as antibody begins to be formed, death being due to the destruction of infected cells by activated macrophages. Thus the disease is not caused by the destruction of the host's cells by the virus, but by the destruction of infected cells by the host's immune mechanism. A similar interaction between immune lymphocytes and virus-coded cell surface antigens may account for the symptoms associated with some viral diseases of humans, such as hepatitis.

The same is true for the second mechanism for destroying infected cells—namely, combination with antibody and complement, which will destroy infected cells long before cells break down as a direct result of virus infection. This mechanism also, though no doubt generally very valuable as a defense against infection, may sometimes cause severe damage to the host. For example, it appears that the sometimes fatal hemorrhagic shock syndrome associated with dengue fever is caused by sudden increases in vascular permeability that may be triggered by the interaction of immune complexes with the complement and clotting systems.

Although virus–antibody complexes are usually eliminated from the body without difficulty either before or after combination with complement, they may cause diseases quite unrelated to those caused by viruses alone. This realization has come from studies of several virus infections in animals, particularly LCM and lactic dehydrogenase elevating virus (LDHV). Infection with both these viruses results in the presence in the bloodstream of large amounts of virus–antibody complexes; it is also characterized by the development of glomerulonephritis and the presence in kidney capillaries of large amounts of virus–antibody–complement complexes. Similar observations have been made with respect to Aleutian mink disease and equine infectious anemia, in which the inflammatory changes are not confined to the kidneys but also involve the blood vessels (with the development of arteritis) and other parts of the body. Some forms of human glomerulonephritis may also be caused by virus–antibody complexes.

Finally, it is now suspected that autoimmune diseases, such as rheumatoid arthritis and lupus erythematosus, are also caused by the interaction of viruses with the immune system.

FURTHER READING

Books and Reviews

Cooper NR: Humoral immunity to viruses. In Fraenkel-Conrat H, Wagner RR (eds): Comprehensive Virology. New York, Plenum, 1979, vol 15, p 123

Della-Porta AJ, Westaway EG: A multi-hit model for the neutralization of animal viruses. J Gen Virol 38:1, 1977

Haller O (ed): Natural resistance to tumors and viruses. Curr Top Microbiol Immunol 92, 1981

Mandel B: Interaction of viruses with neutralizing antibodies. In Fraenkel-Conrat H, Wagner RR (eds): Comprehensive Virology. New York, Plenum, 1979, vol 15, p 37

Mogensen SC: Role of macrophages in natural resistance to virus infections. Microbiol Rev 43:1, 1979

Norrild B: Immunochemistry of herpes simplex virus glycoproteins. Curr Top Microbiol Immunol 90:67, 1980

Notkins A (ed): Viral Immunology and Immunopathology. New York, Academic Press, 1976

Oldstone MBA: Virus neutralization and virus-induced immune complex disease. Prog Med Virol 19:85, 1975

Oldstone MBA: Immune responses, immune tolerance, and viruses. In Fraenkel-Conrat H, Wagner RR (eds): Comprehensive Virology. New York, Plenum, 1979, vol 15, p 1

Oldstone MBA, Fujinami RS, Lampert PW: Membrane and cytoplasmic changes in virus-infected cells induced by interactions of antiviral antibody with surface viral antigen. Prog Med Virol 26:45, 1980

Sissons JGP, Oldstone MBA: Killing of virus-infected cells: The role of antiviral antibody and complement in limiting virus infection. J Infect Dis 142:442, 1980

Selected Papers

Babiuk LA, Acres SD, Rouse BT: Solid-phase radioimmunoassay for detecting bovine (neonatal calf diarrhea) rotavirus antibody. J Clin Microbiol 6:10, 1977

Ghose LH, Schnagl RD, Holmes IH: Comparison of an enzyme-linked immunosorbent assay for quantitation of rotavirus antibodies with complement fixation in an epidemiological survey. J Clin Microbiol 8:268, 1978

Halstead SB, O'Rourke EJ: Antibody enhanced dengue virus infection in primate leukocytes. Nature 265:739, 1977

Harmon SA, Summers DF: Characterization of monospecific antisera against all five vesicular stomatitis virus-specific proteins: Anti-L and anti-NS inhibit transcription in vitro. Virology 120:194, 1982

Icenogele J, Gilbert SF, Grieves J, et al.: A neutralizing monoclonal antibody against poliovirus and its reaction with related antigens. Virology 115:211, 1981

Minor PD, Schild GC, Bootman J, et al.: Location and primary structure of a major antigenic site for poliovirus neutralization. Nature 301:674, 1983

Oldstone MBA: Immune complexes in cancer: Demonstration of complexes in mice bearing neuroblastomas. J Natl Cancer Inst 54:223, 1975

Stollar V: Immune lysis of Sindbis virus. Virology 66:620, 1975

Zinkernagel RM, Oldstone MBA: Cells that express viral antigens but lack H-2 determinants are not lysed by immune thymus-derived lymphocytes but are lysed by other antiviral immune attack mechanisms. Proc Natl Acad Sci USA 72:3666, 1976

CHAPTER 58

The Virus Multiplication Cycle

Virus particles represent the static or inert form of viruses. The very existence of viruses is recognizable only in terms of their interaction with cells, which is the central theme of virology.

The interaction of virus and cell generates a novel entity, the virus–cell complex, the fate of which varies widely, as it depends both on the nature of the cell and on the nature of the virus. The two most commonly observed virus–cell interactions are: (1) the lytic interaction, which results in virus multiplication and lysis of the host cell; and (2)

the transforming interaction, which results in the integration of the viral genome into the host genome and the permanent transformation or alteration of the host cell with respect to morphology, growth habit, and the manner in which it interacts with other cells with which it comes into contact.

In studying the virus–host interaction, one can focus primarily either on the fate and functioning of the invading virus particle and on the production of virus progeny, or on the reaction of the host cell to virus infection. Both approaches are of fundamental importance to the medical practitioner. The former is particularly relevant to the development of a rational approach to antiviral chemotherapy; the latter to an understanding of chronic virus infection and cancer. This chapter focuses on the invading virus particle; the next chapter focuses on the response of the cell.

Lytic Virus–Cell Interaction

The lytic virus–cell interaction is best thought of in terms of a cycle—the infection or multiplication cycle—during which the virus enters cells, multiplies, and is released. This cycle is repeated many times when a virus particle infects an organism, until, for one reason or another, further

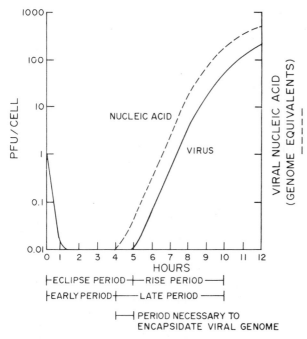

Figure 58-1. The one-step growth cycle. Its essential features are: Following adsorption, infectivity is abolished or "eclipsed"; this is caused by uncoating of the infecting virus particles. During the eclipse or early period, which can last from a few minutes to many hours, the stage is set for viral nucleic acid replication. The appearance of the first progeny genome marks the beginning of the late period, and the appearance of the first mature progeny virus particle marks the beginning of the rise period. It should be noted that the interval between the beginning of the late and rise periods represents the average time necessary for virus maturation; that is, the incorporation of a free nucleic acid molecule into a mature virus particle. The lengths of all periods, as well as the extent of virus multiplication, vary greatly for different viruses and cells.

multiplication is arrested, or the host dies.

One of the principal goals of virology is to define in molecular terms all the various reactions that proceed during the virus multiplication cycle. As an example, when a poliovirus particle infects a cell, one RNA molecule and about 200 protein molecules are introduced into it. How does this RNA molecule replicate? What are the proteins for which it codes, and what are their functions? How are mature virus particles assembled? What is the fate of the 200 parental protein molecules? What effect do they have on the host cell? What are the reactions that cause the host cell to die?

It is impossible to answer these and many other questions by studies in the intact organism. Instead, simple experimental systems that can be manipulated at will are required. Such systems are provided by cloned strains of cultured animal cells that grow in vitro and can be infected under any desired set of conditions with pure (plaque-purified) strains of virus. Among their many advantages is that they permit focusing on one multiplication cycle, rather than on many repeated cycles, which is achieved by infecting all cells at the same time. In fact, one of the major conceptual breakthroughs in virology occurred about 40 years ago when Ellis and Delbrück demonstrated how very much simpler the analysis of the one-step growth cycle is than that of numerous successive unsynchronized cycles. In populations of cultured cells infected at high multiplicity—that is, with many virus particles per cell—so as to ensure that infection commences at the same time in all cells, the various reactions that together comprise virus multiplication proceed synchronously according to a regulated progressive pattern that is amenable to study by the techniques of biochemistry, biophysics, and cellular and molecular biology.

One-Step Growth Cycle: General Aspects

The virus multiplication cycle can be divided into several periods, using events of critical importance as markers (Fig. 58-1). As discussed below, infectivity of virus particles is destroyed or eclipsed when they adsorb; the initial period of the cycle is therefore often referred to as the eclipse period. This period ends with the formation of the first mature progeny virus particle, which marks the beginning of the rise period. Alternatively, the synthesis of the first progeny genome is often regarded as dividing the multiplication cycle into the early and late periods. The eclipse and early periods and the rise and late periods overlap substantially, with the interval between the beginning of the late and rise periods representing the time necessary for the incorporation of a viral genome into a mature virus particle.

Eclipse Period

Adsorption

The first step of the virus–cell interaction is adsorption, which can itself be separated into several stages. The first of these is ionic attraction. Both cells and virus particles are negatively charged at pH 7, and positive ions are therefore required as counter-ions. As a rule this requirement is met most efficiently by magnesium ions. The second stage involves the interaction of virus particles with specific receptor molecules. Indirect evidence for the existence of such specific virus receptor molecules has been available for some time. For example, poliovirus adsorbs only to cells of human or primate origin. In fact, in the body, poliovirus adsorbs only to cells of the central nervous system and to cells lining the intestinal tract; other human and primate cells develop the ability to adsorb poliovirus only after being cultured in vitro, which causes unmasking of receptors (the gene for which is located on human chromosome 19). Recently, techniques have been devised for isolating such specific receptor molecules in the form of virus–receptor complexes. The fact that attachment of human rhinovirus type 2 and of poliovirus type 2 to human cells is inhibited by concanavalin A, a lectin that reacts with α-D-mannose residues, suggests that some of these receptors are glycoproteins.

The time course of virus adsorption follows first order kinetics. The rate of adsorption is independent of temperature if suitable corrections are made for changes in the viscosity of the medium, but the rate is directly proportional to the amount of surface to which virus can adsorb, that is, to the cell concentration. The kinetics of adsorption are described by the relation

$$\frac{V_t}{V_o} = e^{-Ktc},$$

where V_o and V_t are the concentrations of free virus at time 0 and after t minutes, respectively; c is the cell concentration; t is time in minutes; and K is the adsorption rate constant.

The number of virus particles or infectious units adsorbed per cell is referred to as the multiplicity of infection (moi). Animal cells are generally capable of adsorbing very large amounts of virus. It has been shown, for example, that cells contain about 100,000 receptor sites for Sindbis virus (Fig. 58-2).

Penetration and Uncoating

The second stage of the virus multiplication cycle involves penetration and uncoating, which are considered together because, although they are separated both temporally and spatially for some viruses, they occur simultaneously for others. Penetration, sometimes referred to as viropexis, concerns the entry into the cytoplasm of either the whole virus particle or that part of it that contains the genome. It may be observed directly by means of the electron microscope or indirectly by measuring the loss of ability of antiviral antiserum to arrest initiation of virus multiplication. The reason for this is that as long as virus particles remain outside the cell, combination with antibody significantly decreases their ability to cause productive infection; but once the particles are within the cell, they are no longer accessible to antibody.

Uncoating signifies the physical separation of viral nucleic acid from viral protein or, in the case of negative-stranded RNA viruses, the disruption of virus particles with resultant liberation of nucleocapsids. Double-stranded RNA-containing viruses also are not uncoated completely, but only to subviral particles (see below). Uncoating is of taxonomic significance because viruses are the only intracellular infectious agents or parasites for which this is an obligatory step of the multiplication cycle.

Uncoating is best assessed by measuring physical and chemical changes in the adsorbed virus particles. Among these changes are progressive labilization of the capsid structure as judged by loss of its ability to shield the viral genome from hydrolysis by nucleases, development of susceptibility to reagents such as urea to which intact virus particles are resistant, loss of antigenic determinants, and progressive loss of capsid protein. The total time from adsorption to final uncoating ranges from several minutes to several hours.

The actual pathways of penetration and uncoating of the various types of virus particles differ markedly—a fact not surprising in view of their diversity. The essential features of this process are as follows. Virus particles usually bind to the surface of microvilli, elongated projections extending from the cell surface. They then move down the microvillus shaft toward the body of the cell, where they encounter coated pits. These are specialized regions of the cell membrane that are formed continually and that are coated on their cytoplasmic, or inner, side by clathrin, a large protein. These coated pits, with the virus particles attached to them, then fold inward, pinch off, and move into the interior of the cell as coated vesicles, or phagocytic vacuoles. After the clathrin, which is now on their outer surface, is removed, they fuse with lysosomes. It is within the resulting vesicles, which now contain the lysosomal proteolytic enzymes, that virus particles are uncoated or that the uncoating process begins. In the case of adenoviruses, the fibers and pentons are removed and partially uncoated virus particles are released into the interior of the cell (Fig. 58-3). In the case of reoviruses, the outer capsid shell is partially digested and subviral particles, consisting of the inner capsid shell or core to which about one half of the components of the outer capsid shell are still attached, are released into the cytoplasm, where they remain throughout the multiplication cycle, free reovirus RNA never being liberated (see p. 879). In the case of poxviruses, the outer regions of the virus particles are removed and viral cores are liberated into the cytoplasm. The degradation of these cores, which results in the uncoating of the viral DNA, requires the synthesis of a special virus-encoded uncoating

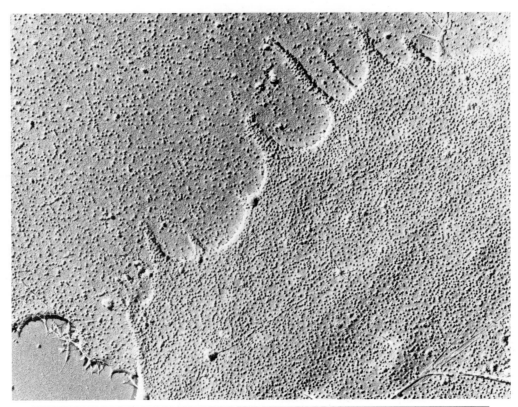

Figure 58-2. Surface replica of Sindbis virus adsorbed to the surface of two chick cells. × 9,300. *(From Birdwell and Strauss: J Virol 14:672, 1974.)*

protein (Fig. 58-4). In the case of most enveloped viruses except paramyxoviruses, the very acidic vesicle milieu causes the viral envelope to fuse with the vesicle membrane, thereby liberating the helical nucleocapsid into the interior of the cell.

The only viruses that are not taken up and uncoated by this mechanism are the paramyxoviruses. Here the viral envelope—probably through the mediation of the glycoprotein F spikes, which possess powerful cell-fusing activity—fuses directly with the plasma cell membrane, thereby liberating the nucleocapsid directly into the cytoplasm. There are several lines of evidence for this pathway, including direct morphologic evidence and the fact that cells infected with such viruses are killed by antiviral antibody plus complement (because fusion of the viral envelope with the host cell membrane causes viral antigens to be incorporated into it).

Eclipse

Adsorption, penetration, and uncoating result in loss of infectivity, which is referred to as eclipse. The only residual infectivity is that due to the viral nucleic acid itself (or to the nucleocapsid, as the case may be), which, however, is never more than a small fraction of that of the virus particles themselves.

The first three stages of infection are usually inefficient processes. Virus particles often adsorb to portions of the cell surface at which penetration will not proceed; viral genomes may be damaged by ribonuclease, which is frequently associated with outer (or plasma) cell membranes; and virus particles may fail to be released from the phagocytic vacuoles in which they have become engulfed. All these inefficiencies account in large part for the fact that the ratio of infectious to total animal virus particles is almost always far less than 1 (Chap. 55).

Synthetic Phase of the Virus Multiplication Cycle

Once the viral genome is uncoated, the synthetic phase of the virus growth cycle commences. In essence, this encompasses, in a precisely regulated program, the replication of the viral genome, the synthesis of viral proteins, and the formation of progeny virus particles.

The location of viral genome replication is characteristic for each virus (Table 58-1). There is no correlation between this location and any other property, such as chemical nature or size of genome. Viral protein is always synthesized in the cytoplasm on polyribosomes composed of viral messenger RNA, host cell ribosomes, and host cell

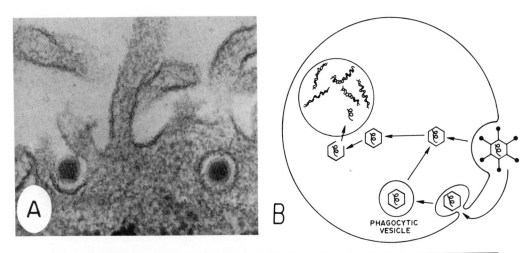

Figure 58-3. The penetration and uncoating of adenovirus. **A.** One virus particle about to be engulfed in a coated pit; another inside a coated vesicle. Virus particles that have lost their fibers and pentons are liberated from such vesicles. These particles are then transported to the nucleus, where uncoating is completed near the nuclear pore complex; only viral DNA enters the nucleus. *(From Chardonnet and Dales: Virology 40:462, 1970.)* **B.** Diagrammatic representation of the uptake and uncoating of adenovirus particles.

transfer RNA. In the case of RNA-containing viruses most, if not all, their genetic information is expressed soon after uncoating. In the case of the double-stranded DNA-containing viruses, however, the multiplication cycle can be divided into clearly defined early and late periods, with the onset of viral DNA replication marking the beginning of the late period (see Fig. 58-1).

Early Period

The early period of the synthetic phase is devoted primarily to the activation of reactions that are prerequisites for the initiation of viral genome replication. This activation proceeds as the result of viruses exercising certain early functions. Among them are: (1) inhibition of host DNA, RNA, and protein synthesis—which may involve the syn-

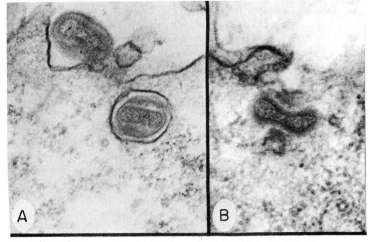

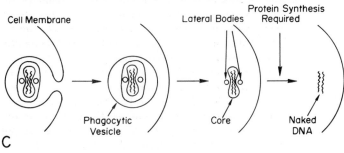

Figure 58-4. The penetration and uncoating of vaccinia virus. Three stages are shown. **A.** One of the two virus particles is in a coated pit; the other is already in the cytoplasm within a coated vesicle. **B.** The vesicle has broken down, as has the virus particle's outer protein coat. The core is now free, and the two lateral bodies have moved some distance away. The final stage of uncoating is the breakdown of the core, which results in the liberation of the DNA. This step is not achieved if protein synthesis is inhibited. This indicates that the synthesis of a special uncoating protein is required. Poxviruses are the only viruses that require the synthesis of a special protein for uncoating. The whole process is depicted diagrammatically in **C. A** and **B,** × 80,000. *(A, B, from Dales: J Cell Biol 18: 63, 1963; C, modified from Joklik: J Mol Biol 8:277, 1964.)*

TABLE 58-1. UNCOATING, VIRAL GENOME REPLICATION, AND VIRUS MATURATION

Virus	Uncoated to	Genome Replication	Virion Maturation
Poxviruses	Free DNA	Cytoplasm	Cytoplasm
Herpesviruses	Free DNA	Nucleus	At nuclear membrane
Adenoviruses	Free DNA	Nucleus	Nucleus
Papovaviruses	Free DNA	Nucleus	Nucleus
Picarnaviruses	Free RNA	Cytoplasm	Cytoplasm
Togaviruses	Free RNA	Cytoplasm	At membranes
Bunyaviruses	Nucleocapsid	Cytoplasm	At membranes
Reoviruses	Subviral particle	Cytoplasm	Cytoplasm
Orthomyxoviruses	Nucleocapsid	Nucleus	At membranes
Paramyxoviruses	Nucleocapsid	Cytoplasm	At membranes
Rhabdoviruses	Nucleocapsid	Cytoplasm	At membranes
RNA tumor viruses	Free RNA	Nucleus	At membranes
Arenaviruses	?	Cytoplasm	At membranes
Coronaviruses	Free RNA	Cytoplasm	At membranes

thesis of virus-coded proteins that either inhibit or alter the specificities of the DNA-replicating, RNA-transcribing, and protein-synthesizing systems—so that viral rather than host cell genetic information is processed; (2) synthesis of proteins that form the matrix of inclusions, either in the nucleus or in the cytoplasm, within which viral nucleic acids replicate and viral morphogenesis proceeds; and (3) synthesis of enzymes that synthesize viral DNA and RNA.

The extent to which early functions are expressed varies greatly from virus to virus. Some viruses possess so little genetic information that only very few early functions are expressed; others possess so much that they may express from 30 to 50 early functions. Early functions are expressed through (early) virus-coded proteins that are transcribed from early viral messenger RNA species. Viral genomes that are plus-stranded RNA serve directly as messenger RNA. For all other viruses, plus-stranded messenger RNA must first be transcribed from infecting parental genomes by means of polymerases either associated with them or preexisting in the host cell.

Late Period

During the late period, the late viral functions are expressed. The late viral proteins are primarily virus particle components and enzymes and other nonstructural proteins that function during viral morphogenesis; they are encoded by late viral messenger RNA molecules that are transcribed from different regions of the viral genome than early ones. Furthermore, late messenger RNA molecules are transcribed from progeny genomes; and because there are always more progeny than parental genomes, many more late messenger RNA molecules are always formed than early ones. Therefore, the amount of late proteins that is synthesized always greatly exceeds that of

early ones. Activation of the regions of the viral genome that code for late functions may or may not be accompanied by deactivation of the regions that code for early functions. In either case, a mechanism exists that specifies that one set of genes (the early set) is transcribed from parental genomes, whereas another set (the late set) is transcribed only from progeny genomes. The basis for this mechanism lies in the specificity of the enzyme(s) that transcribes DNA. Indeed, work with certain bacteriophages has indicated that the host-specified DNA-dependent RNA polymerase is modified early during infection to a form that can transcribe those sites on the phage genome that code for early functions, and that at the beginning of the late period its specificity is altered again so as to enable it to respond to those signals that specify late functions (Chap. 63). Similar mechanisms may also operate in the case of animal viruses.

During the late period, the newly formed virus genomes and capsid proteins are assembled into progeny virus particles, a process that is known as morphogenesis. This is a spontaneously occurring process, as most of the information for virus assembly resides in the amino acid sequences of the capsid proteins. Nucleic acid performs no essential function during morphogenesis, a fact demonstrated by the occurrence among the yield of most icosahedral viruses of empty virus particles—that is, virus particles containing no nucleic acid—that are morphologically indistinguishable from mature virus particles.

The duration of the late period is generally limited by the ability of the host cell to supply energy for macromolecular synthesis. This is a critical factor because infection with lytic viruses invariably interferes with the functioning of the host cell by multiple mechanisms that are discussed in Chapter 59. As a result, the synthesis of viral nucleic acids and proteins slows down progressively, thereby limiting the amount of viral progeny.

Release of Progeny Virus

The final step of the infection cycle is the release of progeny virus. There is no special mechanism for the release of unenveloped viruses and poxviruses; infected cells simply disintegrate more or less rapidly, liberating the viral progeny that has accumulated within them. The amount of cell-associated virus therefore exceeds the amount of released virus until the very last phase of the multiplication cycle (Fig. 58-5). A special mechanism does, however, exist for the enveloped viruses, for which release is the final stage of morphogenesis. Here virus-coded envelope proteins are incorporated into certain areas of host cell membranes while nucleocapsids are being synthesized. The nucleocapsids then bud through modified membrane patches and become enveloped by them (p. 884). Budding occurs both at the outer plasma cell membrane and at the membranes lining intracytoplasmic vacuoles, which then transport the virus to the exterior of the cell. These viruses do not exist in mature infectious form within cells, and the amount of extracellular virus, therefore, greatly exceeds the amount of cell-associated virus at all stages of the multiplication cycle (Fig. 58-5).

The duration of the phases of the virus multiplication cycle varies greatly, depending on the nature of the virus and on the nature of the host cell. Table 58-2 lists the minimum lengths of the eclipse periods and of the complete multiplication cycles of some well-studied viruses.

Multiplication Cycles of Several Important Viruses

Of the many facets of virus multiplication, the two that are central are: (1) the nature of the information encoded in viral genomes; and (2) the manner in which this information is expressed. Not only does description of a virus in this manner define it in its most fundamental terms, but it also provides the framework of knowledge essential for a rational approach to antiviral chemotherapy. A discussion of the strategy used by several important and intensively studied viruses to encode and express their information content follows.

Multiplication Cycles of Double-stranded DNA-containing Viruses

Knowledge concerning the multiplication of double-stranded DNA-containing viruses has expanded greatly in recent years owing to the advent of techniques for mapping their genomes by the use of bacterial restriction endonucleases (see Chap. 7). These enzymes, which are components of the restriction–modification systems in bacteria, cleave DNA at palindromic sequences that are highly specific for each enzyme. These sequences, which possess two-fold rotational symmetry, may be as simple as

$$\downarrow$$
$$5' \dots GGCC \dots$$
$$\dots CCGG \dots 5',$$
$$\uparrow$$

or as complex as

$$\downarrow$$
$$5' \dots GCCNNNNNGGC \dots$$
$$\dots CGGNNNNNCCG \dots 5'.$$
$$\uparrow$$

Clearly, the simpler the recognition site, the more often it will occur and the more often the DNA will be cut. The fragments that result may be ordered or mapped by

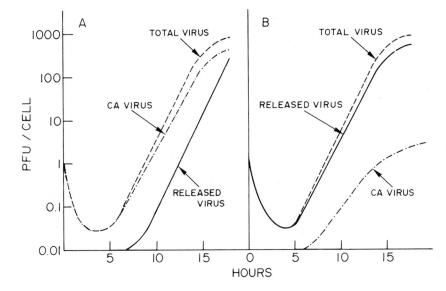

Figure 58-5. The relationship between virus multiplication and release. **A.** This graph refers to viruses with icosahedral nucleocapsids. Such viruses are not released readily from cells. Viral progeny accumulates within cells, so that for much of the rise period the amount of cell-associated (CA) virus greatly exceeds the amount of released virus. Release of virus occurs only when cells break down at the end of the rise period. **B.** This graph refers to all enveloped viruses. Such viruses mature only in the process of being released, as it is only then that they acquire their envelope. The amount of liberated virus therefore always greatly exceeds the amount of cell-associated virus. The only cell-associated virus particles are those in the process of budding from the plasma membrane and those that bud into intracytoplasmic vacuoles.

TABLE 58-2. APPROXIMATE DURATION OF ECLIPSE PERIOD AND OF ENTIRE MULTIPLICATION CYCLE

Virus	Eclipse Period (hours)	Total Multiplication Cycle (hours)
Poxvirus: vaccinia virus	4	24
Herpesvirus: herpes simplex virus	3–5	24–36
Adenovirus	8–10	48
Papovavirus: polyoma virus	12–14	48
Poliovirus	1–2	6–8
Togavirus: Sindbis virus	2	10
Reovirus	4	15
Orthomyxovirus: influenza virus	3–5	18–36
Rhabdovirus: vesicular stomatitis virus	2	8–10

digesting DNA under conditions when it is only partially cleaved, and analyzing the incompletely hydrolyzed pieces for neighboring fragments. Figure 58-6 shows the 6 *Hind* III restriction endonuclease fragments of SV40 DNA, and Figure 58-7 shows the order in which they are arranged, that is, the *Hind*III map. Several other restriction endonuclease maps of SV40 DNA are also shown. Given such maps—and almost 100 restriction endonucleases are now available—this type of analysis is capable of defining precisely the position in the viral genome of even short DNA sequences. It also permits one to determine to which endonuclease fragment the messenger RNA molecules that are transcribed at any time during the multiplication cycle will hybridize, which can provide information concerning when and how frequently each portion of the viral genome is transcribed during the multiplication cycle. This technique is known as physical mapping because here the viral DNA is treated as a sequence of nucleic acid bases. Physical maps can be correlated with genetic maps if they are available. In that case it becomes possible to determine the order of the various genes that comprise the viral genome and the extent to which they are expressed. The ultimate aim of this type of analysis is to define exactly where each gene is located in the viral genome, when it is transcribed during the virus multiplication cycle, how frequently it is

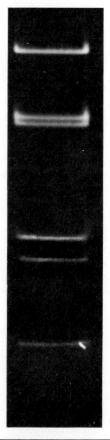

Figure 58-6. Autoradiogram of a polyacrylamide gel in which the six fragments were electrophoresed that result when circular ^{32}p-labeled SV40 DNA is digested with restriction endonuclease *Hind* III. *(Courtesy of Drs. J.K. Li and C. Huang.)*

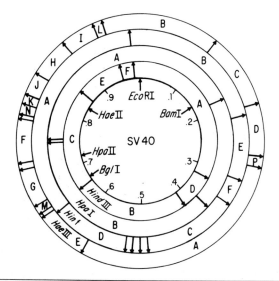

Figure 58-7. Restriction endonuclease cleavage maps of SV40 DNA, using *Hind*III, *Hpa*I, *Hin*f and *Hae*III. *Eco*RI, *Bam*I, *Bgl* I, *Hpa*II and *Hae*II cleave only once. The origin of DNA replication is at a palindromic sequence, part of which forms the recognition sequence for *Bgl*I.

transcribed, and, finally, what controls or regulates the transcription program.

Papovaviruses

Papovaviruses are the smallest double-stranded DNA-containing viruses. Although practically nothing is known of the multiplication cycles of papillomaviruses (because they do not grow in cultured cells), a great deal is known about how polyoma virus and SV40 multiply. It should be noted that two human viruses, namely BK virus and JC virus (see Chap. 62), are more or less closely related to SV40.

When these viruses infect certain cells of their natural hosts, viral DNA replicates, viral proteins are made, progeny virus particles are formed, and the cells lyse; in other words, they initiate productive infection. These host cells are known as permissive cells. In contract, many other types of cells are more or less nonpermissive—that is, they do not support efficient multiplication. Instead, the viral DNA becomes integrated into that of the host cells, thereby creating new, genetically different cells, namely, virus-transformed cells. Such cells form tumors in animals. The transforming papovavirus–host cell interaction is discussed in Chapter 62. The present chapter considers the nature of their lytic multiplication cycles.

The basic feature of the multiplication cycle of polyoma virus and SV40, like that of all double-stranded DNA-containing viruses, is that it can be divided into well-defined early and late phases (see Fig. 58-1). During each phase, different portions of the viral genome are transcribed into messenger RNAs, which are then processed and translated into proteins by the cellular protein-synthesizing system.

Early Period
The multiplication cycle of SV40 is best understood by reference to the genetic and transcription maps of its genome, a circular molecule of 5226 nucleotide base pairs that has been completely sequenced (Fig. 58-8). During the early phase, which lasts a surprisingly long time (14 to 18 hours), about 50 percent of the viral DNA, namely the region that extends from map position 0.65 (relative to the *Eco*RI restriction endonuclease cleavage site) to map position 0.17, is transcribed into RNA. This region comprises a single gene, known as the A gene; however the transcripts of the region are processed into not one, but two species of messenger RNA. The first is about 2230-nucleotides long and is composed of two parts, about 330- and 1900-nucleotides long, that map from map position 0.65 to 0.60 and from 0.54 to 0.17, respectively. RNA corresponding to the DNA between map position 0.60 and 0.54 is not present in this messenger RNA; this region is spliced out of the original transcript by a mechanism that is not yet clear. Splicing-out of regions of RNA transcripts is a common phenomenon even among the transcripts of cellular genes; numerous cellular messenger RNAs are now known to comprise sequences from widely separated

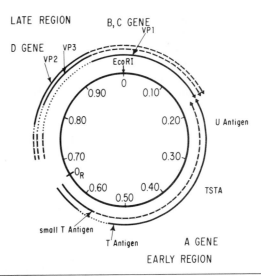

Figure 58-8. The genetic and transcription map of SV40 DNA. It is located so that the restriction endonuclease *Eco*RI cleavage site is at the 12 o'clock position. O_R is the origin of replication of SV40 DNA. The coding sequences of messenger RNAs are indicated by solid lines, noncoding sequences by dashed ones. Sequences spliced out during the processing of transcripts to messenger RNAs are indicated by dotted lines.

regions of DNA. A well-known example is the messenger RNA of the β-globin gene of the mouse. This messenger RNA is derived from the transcript of a DNA sequence that is about 1700-base pairs long, which is processed so that two sequences of 780 and 125 nucleotides, respectively, are spliced out. Thus the β-globin messenger RNA consists of three sequences of about 480, 205, and 155 nucleotides, respectively, that are covalently joined but were originally separated. Regions that are spliced out are known as introns (inserted sequences); regions that are not spliced out are known as exons. Splicing-out is a very widespread feature of the mechanism by which RNA transcripts of eucaryotic DNA are processed to messenger RNAs; other features of this mechanism are capping of their 5′-termini and polyadenylation of their 3′-termini.

The second species of early SV40 messenger RNA is 2500-nucleotides long and is also composed of two exons, 630- and 1900-nucleotides long, respectively, that map from map position 0.65 to 0.55 and from 0.54 to 0.17, respectively.

The two messenger RNAs code for two proteins. First, the 2230-nucleotide long messenger RNA codes for the T antigen, a protein with a molecular weight of about 95,000 that is located in the nucleus, binds to SV40 DNA near the origin of replication (see below) and may function in initiating or facilitating the initiation of its replication. The T antigen may also be responsible for inducing the synthesis of substantial amounts of cellular DNA (generally it replicates at least once) and of several enzymes involved in DNA synthesis during the initial stages of the multiplication cycle. These must be cellular enzymes because SV40 DNA is too small to code for them. Finally, the T antigen is also responsible for the maintenance of the transformed state when SV40 transforms cells (see Chap. 62).

The second species of early messenger RNA, that which is 2500-nucleotides long, encodes a protein that is known as the small t antigen, whose molecular weight is about 17,000 and which is essential for initiating the transformed cell state. Whereas the T antigen is encoded by the sequences that extend from map position 0.65 to 0.60 and from 0.54 to 0.17, small t antigen is encoded by the sequence from 0.65 to 0.55. T antigen and small t antigen therefore share amino acid sequences and they cross react immunologically. The reason small t antigen is so short is that there are several termination codons near map position 0.55. These condons are spliced out of the messenger RNA species that codes for T antigen.

In polyoma virus-infected cells there is yet a third protein, middle T antigen (mol wt 55,000), which shares antigenic determinants with both the T and the small t antigens. Its amino-terminal portion is the same as that of the T and small t antigens, but its carboxyl-terminal end is different, as it is encoded by the portion of the messenger RNA that follows the splice junction; that is, by the second exon, but in a different reading frame from that used for the T antigen. Like T antigen and small t antigen, middle t antigen plays an essential role in transformation, but cannot transform cells completely by itself.

Cells infected with SV40 also contain a protein, the U antigen (mol wt 28,000), which corresponds to the carboxyl-terminal portion of the T antigen. Its existence was first detected in cells infected with virus particles that are SV40-adenovirus hybrids (Ad2$^+$ND$_1$) (see Chap. 60) and that only contain SV40 DNA corresponding to map positions 0.28 to 0.11.

The SV40 72-bp Repeat. A very interesting feature of this early SV40 transcription unit is that about 150-base pairs (bp) upstream from the transcription start site two identical 72-bp sequences are located directly adjacent to each other. This 72-bp repeat acts as a powerful promoter element. It potentiates, activates, or enhances transcription by eucaryotic RNA polymerase II, even if it is thousands of base pairs away in either direction. It appears to act as a bidirectional super entry site for RNA polymerase II. This element has been positioned near a variety of genes by genetic engineering techniques, and always greatly enhances transcription. Its action is similar to that of the long terminal repeat (LTR) of retroviruses (see Chap. 62). The recognition and characterization of such transcriptional super promoters, which no doubt also exist in eucaryotic genomes including the human genome, are extremely important "fall-out" by-products of molecular virology.

Late Period

During the late period, the A gene continues to be transcribed, and in addition, the remainder of the SV40 genome is also transcribed. This region contains three genes

—known as the B, C, and D genes—which overlap, as shown in Fig. 58-8; it is transcribed in the opposite direction, that is, from the opposite strand, as the A gene. Its transcripts are processed into three messenger RNAs that correspond to map positions 0.76 to 0.17, 0.83 to 0.17, and 0.95 to 0.17, respectively. In addition, all three messenger RNAs possess leader sequences that are about 120-nucleotides long and that are transcribed from the region between map positions 0.72 to 0.76. (The region that is spliced out in the first messenger RNA species is very short, and except for it, this messenger RNA species is a transcript of the entire late region of the SV40 genome.) These messenger RNAs are translated into three proteins. The map position 0.95 to 0.17 RNA species, which is the most abundant, is transcribed into VP1, a 46,000-dalton protein that is the major component of the SV40 capsid. The map position 0.83 to 0.17 messenger RNA species is translated only for part of its length—namely, the portion that corresponds to the region from map position 0.83 to about 0.97. This portion yields VP3, a 25,000 dalton protein that is a minor capsid component. Finally, the map position 0.76 to 0.17 RNA species also is translated only for part of its length—namely, the portion that corresponds to the region between map positions 0.76 and 0.97 —and its translation product is VP2, a 35,000-dalton protein that is also a minor capsid component. The fascinating feature of this mode of gene expression is that the gene for VP3 lies entirely within the gene for VP2; their messenger RNAs are read in the same phase, and translation of both is terminated by the same termination codon, which is a long way from the 3′-terminus of their messenger RNAs. Furthermore, the initiation signal for VP1 lies about 120 nucleotides within the genes for VP2 and VP3, and VP1 is read in a different phase from VP2 and VP3.

Replication of SV40 DNA

The replication of SV40 DNA starts at map position 0.67, the origin of replication, where there is a remarkable palindromic sequence 27-base pairs long. Replication proceeds bidirectionally until the two forks meet again, with one DNA strand being synthesized in a continuous manner, and the other in the form of Okazaki fragments that are then joined. Newly replicated SV40 DNA does not exist in free form within infected cells, but is complexed with histones into a chromatin-like structure. This structure differs from that of the minichromosomes present in SV40 particles (see Chap. 56) in being more condensed. The reason for this is that the intracellular SV40 chromatin also contains histone H1 (in addition to histones H3, H4, H2A, and H2B), which causes the intranucleosomal distance to be reduced. Upon removal of histone H1 by the addition of salt, this structure unfolds into the beads-on-a-string conformation. SV40 DNA cannot be transcribed while in the compact chromatin configuration, but can be transcribed in the beads-on-a-string conformation.

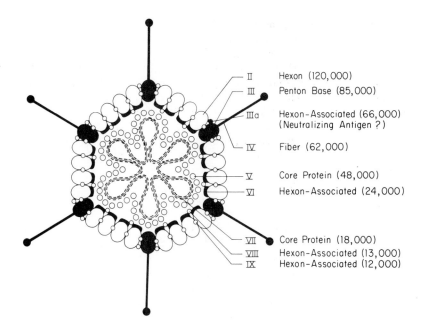

II   Hexon (120,000)
III   Penton Base (85,000)
IIIa Hexon-Associated (66,000)
      (Neutralizing Antigen ?)
IV   Fiber (62,000)
V   Core Protein (48,000)
VI   Hexon-Associated (24,000)
VII   Core Protein (18,000)
VIII Hexon-Associated (13,000)
IX   Hexon-Associated (12,000)

Figure 58-9. Model of adenovirus showing the inner DNA protein core and the outer icosahedral capsid, as well as the location and sizes of the various structural proteins. The fiber is the viral cell attachment organ, as well as the hemagglutinin. Proteins VI, VII, and VIII are synthesized in the form of precursors that are cleaved to the actual virus particle components. *(Adapted from Brown et al.: J Virol 16:366, 1975.)*

General Observations on the Strategy of Papovavirus Gene Expression

The preceding discussion of the multiplication cycle of SV40 and polyoma virus highlights the remarkable strategy according to which these viruses express the information encoded in their genomes. Most astonishing is that many of their coding sequences are used twice, and even three times. One obvious advantage of this strategy is that it minimizes the amount of DNA necessary to encode their proteins. It is also noteworthy that the functions of the early and late proteins are quite different. All early proteins have regulatory functions, and at least one, the T antigen, appears to act in a pleiotropic manner—that is, it seems to be multifunctional. In contrast, the late proteins are the structural virus particle components. Finally, it is remarkable that the DNA of these viruses exists in the form of a complex that has many of the properties of cellular chromatin, and that it uses cellular histones for this purpose.

Adenoviruses

The multiplication cycle of adenoviruses (Fig. 58-9, Table 58-3), whose genome (mol wt 23×10^6; 36,500 base pairs) is six to seven times the size of SV40 or polyoma virus, lasts about 36 to 48 hours. It can be divided into an early period of about 8 hours, during which about 30 percent of the viral genome is transcribed, and a late period during which the remainder of the information that it encodes is expressed. Viral DNA replication begins at 6 to 8 hours. Host macromolecular biosynthesis (synthesis of

TABLE 58-3. STRUCTURAL PROTEINS OF ADENOVIRUS (SEROTYPE 2)

Protein	Designation	Number per Virus Particle	Size (Mol Wt)
Hexon	II	720 (3 per hexon)	120,000
Penton	III	60 (5 per penton)	85,000
Fiber	IV	36 (3 per fiber)	62,000
Major core protein	VII	1000	18,000
Minor core protein	V	200	48,000
Hexon-associated	VI	450	24,000
Hexon-associated	VIII	?	13,000
Hexon-associated (peripentonal)	IIIa	?	66,000
Hexon-associated	IX	?	12,000

Additional structural proteins IVa1, IVa2, X, XI, and XII (mol wt 60,000, 56,000, 6500, 6000, and 5000, respectively), have been described, but not studied in detail.

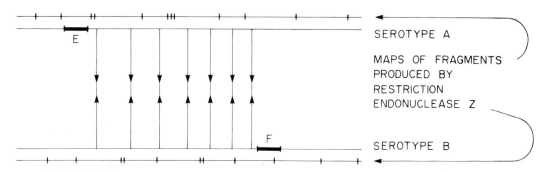

Figure 58-10. Mapping of genes by restriction endonuclease analysis of recombinant DNA. Two temperature-sensitive mutants (see Chap. 60) are taken, one in gene E of serotype A, the other in gene F of serotype B. (Note that there are 35 serotypes of human adenovovirus; see Table 56-4.) Wild-type recombinants (that is, recombinants able to grow at the normal temperature) are then isolated and their restriction endonuclease cleavage pattern profiles compared with those of the two mutants. Restriction endonuclease Z is used here as an example. Note that all recombination events must occur in the sequence between genes E and F. Clearly the DNAs of recombinants will have profiles partially characteristic of one parent and partially of the other. Fixing the distance between the two genes is thus a matter of defining the limits of the DNA segment between them; the more recombinants that are examined, and the more restriction endonucleases that are used, the more accurately can this distance be determined.

host protein, RNA, and DNA) is not shut off until the late period has commenced. Late proteins are often synthesized in very large amounts, greatly in excess over what is assembled into progeny virus particles; much of this protein may be deposited in infected cells in the form of quasicrystalline arrays (see below). Usually about 20 virus-coded proteins can be detected in infected cells, about half of which are structural components of adenovirus particles. But this is no more than about one half of the total number of adenovirus-coded proteins, because at least 45 virus-specific messenger RNA species have been identified.

Adenovirus Genetic Map

The location of many adenovirus genes has been determined by a variety of techniques, all of which take advantage of the fact that completely ordered, detailed restriction endonuclease cleavage maps are available for the DNAs of adenovirus strains of several serotypes. In one such method, the mixture of virus-specific messenger RNAs from infected cells is hybridized to individual restriction endonuclease fragments. The messenger RNA species that hybridize with each fragment, and are therefore transcribed from it, are then dissociated from the DNA and translated in cell-free protein-synthesizing systems, and the proteins that are formed are identified by reference to authentic samples of adenovirus-coded proteins. As the position on the viral genome of each restriction endonuclease cleavage fragment is known, the location of the genes coding for the proteins that are translated is also known. In a variation of this technique, known as the hybridization-arrested translation method, the nature of the proteins that can be translated in vitro from the messenger RNAs isolated from infected cells is com-

pared with the proteins that are translated from such messenger RNA mixtures after they have been hybridized to individual restriction endonuclease fragments. Such hybridization prevents individual species of messenger RNA from being translated, which results in the absence of specific proteins in the mixtures of translated proteins. Another method involves analysis of wild-type recombinants between mutants of viruses of different serotypes. The use of mutants permits the selection of recombinants; because the DNAs of viruses that differ in serotype have different restriction endonuclease patterns, it is possible to map the position of the crossover points on the viral genome (Fig. 58-10); and because some of the proteins encoded by the different serotypes differ detectably in size (as determined by electrophoresis in sodium dodecyl sulfate-containing polyacrylamide gels), it is possible to assign their genes to specific regions on the adenovirus genetic and physical maps.

The location of several genes on the adenovirus genome is shown in Figure 58-11. Interestingly, not all sequences of the adenovirus genome are essential for infectivity. For example, a deletion mutant that lacks the sequences between map positions 79 to 85, which include the gene that codes for the 21K glycoprotein that is present on the surface of infected cells, grows well; these sequences are therefore not necessary for growth.

Nature of Adenovirus Transcripts and Messenger RNAs

Just as in the case of SV40 described above, the primary adenovirus genome transcripts are not messenger RNAs; rather they must be processed to messenger RNAs. Processing involves three major modifications. First, the 5'-termini of transcripts must be capped (which involves

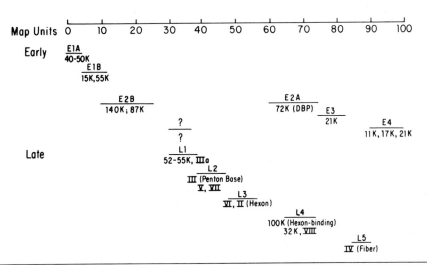

Figure 58-11. Location of the genes for various adenovirus proteins. The 55K protein (region E1B) is probably the adenovirus T antigen; the 140K protein and the 87K protein (region E2B) are, respectively, the adenovirus-coded DNA polymerase and the precursor of the 55K terminal protein that is covalently linked to the 5′-termini of adenovirus DNA; the 72K protein (region E2A) is a DNA-binding protein; the 100K protein (region L4) binds strongly to hexon protein and functions during morphogenesis; and the 21K protein (region E3) is a glycoprotein that is present on the surface of infected cells. Other adenovirus-specific proteins have been detected in infected cells. Some may be cleavage products of other adenovirus-specified proteins.

guanylylation and methylation); second, the 3′-termini of transcripts must be polyadenylated; and third, the transcripts must be spliced. The purpose of capping is most probably to provide strong ribosome binding sites to ensure efficient translation (although some messenger RNAs are known that are translated very efficiently, yet are not capped; and vice versa). The purpose of polyadenylation is not known; again, some messenger RNAs are known that are translated very efficiently, yet are not polyadenylated. As for splicing, the discovery of spliced adenovirus RNAs in the spring of 1977, and the demonstration shortly thereafter that splicing is a general phenomenon that applies not only to viral but also to most eucaryotic cellular messenger RNAs, was one of the major discoveries in biology in recent times—one that changed fundamentally our ideas and concepts concerning the nature and arrangement of eucaryotic genetic material.

The basic discovery was that a single messenger RNA species hybridizes to (and therefore is transcribed from) several widely separated regions of the adenovirus genome. One of the techniques for showing this was the so-called R-loop technique in which adenovirus DNA and RNA transcripts are mixed and partially melted; the DNA–DNA strands dissociate and more stable RNA–DNA hybrids are formed instead. When these hybrids are examined with the electron microscope, the regions where RNA and DNA have hybridized are seen as loops (R-loops), one arm of which is seen to be single stranded (a DNA strand), and the other double stranded (the RNA–DNA hybrid) (Fig. 58-12). This technique is extremely powerful for defining the precise locations of transcribed regions, as not only the length of R-loops, but also their distance from the ends of DNA molecules can be measured very precisely.

Using this technique it was shown that most late adenovirus messenger RNA molecules begin with a leader sequence that is derived from three locations at map positions 16.7, 19.7, and 26.7, the combined length of which is about 200 nucleotides (Fig. 58-13). Presumably the original transcripts encompass not only the leader sequences and the coding sequences, but also the intervening sequences, the introns, that are then spliced out as part of the processing mechanism.

Adenovirus Transcription Program

Two transcription programs control the expression of adenovirus genes, one during the early period, and the other during the late period. During the early period, four widely separated regions of the genome are transcribed (Fig. 58-14). The first of these, region E1A and E1B at the left-hand end of the adenovirus genome, controls the earliest events in the multiplication cycle and the functions involved in cell transformation. The second region, which controls primarily DNA replication, consists of region E2A, which is transcribed into four messenger RNAs, all from different promoters; and region E2B, which gives rise to three messenger RNAs that use a promoter near map position 76. Finally, there is region E3, which is nonessential for virus growth, and region E4, the function of which is not known. Less than half the proteins coded by these early messenger RNA species have been identified.

Adenovirus late proteins are translated from about 20 messenger RNAs that are all processed from a transcript that covers the entire region between map positions 30 and 100. This was documented by measuring their relative

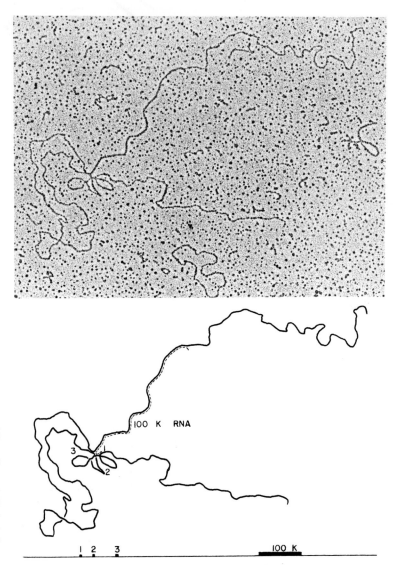

Figure 58-12. R-loops. An electron micrograph of a heteroduplex of human adenovirus serotype 2 (Ad2) RNA and single-stranded Ad2 DNA. The RNA was extracted from HeLa cells at a late time after infection. The DNA is constrained into a series of deletion loops corresponding to the intervening sequences removed from the spliced RNA. The first, second, and third leader segments and the main coding body of the messenger RNA are clearly visible. *(From Chow and Broker: Cell 15:497, 1978.)*

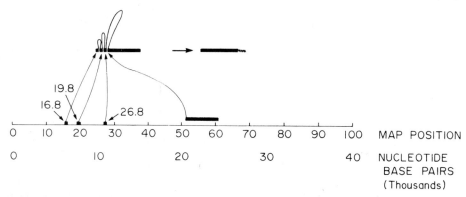

Figure 58-13. Structure of the adenovirus hexon messenger RNA. It consists of five portions: (1) a leader sequence about 50-nucleotides long, transcribed at map position 16.7; (2) a leader sequence about 80-nucleotides long, transcribed at map position 19.7; (3) a leader sequence about 75-nucleotides long, transcribed at map position 26.7; (4) a coding sequence about 3800-nucleotides long, transcribed from a genome sequence beginning at map position 51.7; and (5) a poly (A) sequence about 200-nucleotides long. The intervening sequences between map position 16.7 and 19.7, 19.7 and 26.7, and 26.7 and 51.7, which are present in the primary transcript, are spliced out.

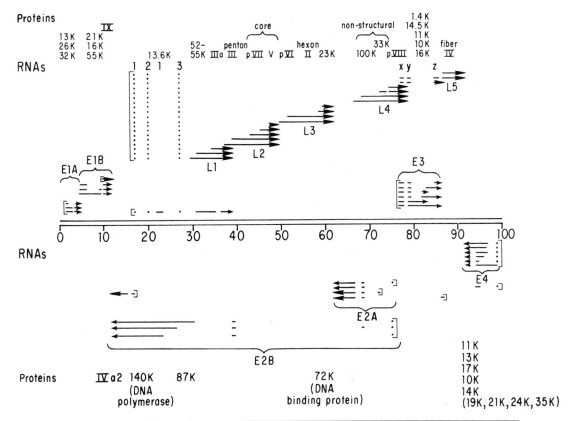

Figure 58-14. The adenovirus transcription map. Some transcripts are transcribed from one strand, others from the other. For each messenger RNA the following are indicated: the transcription region; the location of the promoters ([or]) and of the exons; and the proteins encoded by various messenger RNAs.

resistance to ultraviolet irradiation. Cells infected with adenovirus were irradiated with increasing doses of ultraviolet light, and the amount of each messenger RNA that accumulated was measured. It was found that the sensitivity to ultraviolet irradiation of each transcript of the four early regions was proportional to its physical size; but that the sensitivity of accumulation of each late messenger RNA species was proportional to the distance of the sequence from which it was transcribed from map position 30 (see Fig. 58-14). This indicated the existence of several discrete early region transcription units, but of only one late transcription unit. This late transcript is about 25,000-nucleotides long and is processed into the individual late messenger RNA species, all of which possess the same tripartite leader transcribed from the short sequences at map positions 16, 19, and 27, and falling into five groups that share common 3'-termini (which are polyadenylated) at map positions 39, 49, 61, 78, or 91. Their coding sequences vary, as shown in Figure 58-14. Neither the signals for splicing, nor the signals for polyadenylation, are known. Presumably, the nature of the cleavage/polyadenylation signals controls how many messenger RNA molecules of each type are made. (Clearly, many of some and few of other species are required, depending on how many of the corresponding protein molecules are needed.)

Whereas the preceding describes the overall features of the adenovirus transcription program, there are exceptions. For example, protein IX, a small hexon-associated protein (see Fig. 58-9) is a late protein, but is transcribed from a region around map unit 10, which is in early region E1B. Another messenger RNA uses the late messenger RNA promoter, possesses the late tripartite leader sequence, and maps in late region L1, yet is transcribed early. Clearly, the transcription program is fine-tuned by additional subtle features that remain to be defined.

In summary, the adenovirus transcription program provides an extremely sophisticated system for regulating both when each gene is transcribed and the extent to which it is transcribed (Fig. 58-15). The controlling features of this system are the existence of promoters where transcription is initiated, and the transcript processing mechanism that provides the means, through splicing and cleaving/polyadenylation, for specifying both the untranslated and the translated regions of messenger RNAs. This system permits any given region of DNA to be used over and over again, given the fact that there are three reading frames and that there is flexibility as to where coding regions start and finish. In other words, many adenovirus genes overlap, both on the same strand and on the complementary strand.

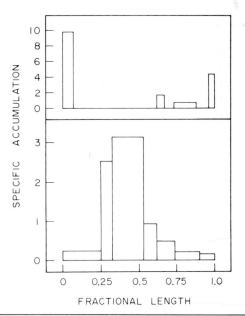

Figure 58-15. Frequence of accumulation of messenger RNAs transcribed from various regions of the adenovirus genome during the early and late periods of the multiplication cycle. Upper panel, early RNA (5.5 to 7.5 hours after infection); lower panel, late RNA (24 to 26 hours after infection). *(From Smiley and Mak: J Virol 28:227, 1978.)*

Adenovirus DNA Replication

Adenovirus DNA replication commences 6 to 8 hours after infection and reaches its peak rate 6 to 10 hours later. Adenovirus codes for a new DNA polymerase, but host cell DNA polymerase α may also function in adenovirus DNA replication. Replication begins at each end of the DNA molecule (Fig. 58-16) and proceeds in the 5′ to 3′ direction by displacement and complementary strand synthesis, the products being two double-stranded and two single-stranded molecules. The latter then probably cyclize by virtue of their inverted terminally repeated sequences, which would again permit complementary strand synthesis by 5′ to 3′ displacement, the products being two more double-stranded DNA molecules. This replication mechanism, parts of which are known to occur while others still lack rigorous proof, is unique because it does not involve the formation of Okazaki fragments and because the transcription of the two parental strands does not proceed simultaneously, but sequentially. Presumably, the two 5′-terminally linked protein molecules play some critical role in adenovirus DNA replication, possibly as primers.

There is some evidence that adenovirus DNA is integrated into host DNA during the lytic multiplication cycle. The significance of this phenomenon is not yet clear. It would seem that such integration is not essential for productive infection.

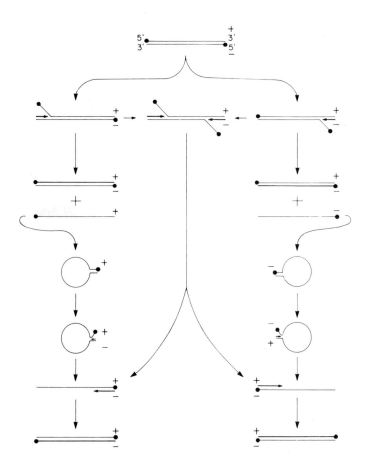

Figure 58-16. Proposed model for adenovirus DNA replication. Replication commences at either end and causes the displacement of either the plus or the minus strand. The resulting single strands then cyclize via their inverted terminally repeated sequences (see Chap. 56) and are transcribed again, yielding double-stranded molecules. Occasionally replication starts simultaneously at both ends, which yields replication intermediates of the type indicated. ●, the protein (mol wt 55,000) that is linked covalently to the 5′-termini of both DNA strands. *(Adapted from Lechner and Kelly: Cell 12:1007, 1977.)*

Adenovirus Morphogenesis

The morphogenesis of adenovirus, like that of papovaviruses and herpesviruses, proceeds in the nucleus, where paracrystalline arrays of mature virus particles, of incomplete empty particles, and even of the structural proteins that are often synthesized in great excess (see above) may be formed (Fig. 58-17).

Herpesviruses

Herpesvirus Particle

Herpesvirus particles possess four distinct morphologic elements: (1) an electron-opaque core; (2) an icosahedral capsid that surrounds the core; (3) electron-dense amorphous material called the tegument, which surrounds the capsid; and (4) an envelope whose outer surface is studded with small glycoprotein spikes. The tegument, which has a fibrous structure, is unique to herpesviruses. Its amount varies greatly from strain to strain, and even from virus particle to virus particle.

Herpesvirus particles contain many (probably at least 30) proteins; the exact number is difficult to determine, as herpesviruses are not easily purified because the amount of tegument and envelope material per virus particle is not constant, and because many proteins are present in small amounts only. At least six proteins are present in the nucleocapsid, including the major component, the hexon (mol wt 155,000); and at least five glycoproteins are located on the outer surface of the envelope. Little is known concerning the protein composition of the tegument.

The DNA of herpesvirus (mol wt 96×10^6; 160,000 base pairs) is 30 times larger than that of SV40 and polyoma virus, and four times larger than that of adenovirus; the DNA of cytomegalovirus is 50 percent larger still. It consists of two regions, L and S, which comprise 82 and 18 percent of the viral DNA, respectively. Each component consists of largely unique sequences, U_L and U_S, bracketed by inverted repeated sequences (Fig. 58-18). Herpesvirus DNA extracted from virus particles consists of four isomers in which the orientations of the L and S components are inverted relative to each other; the herpesvirus ge-

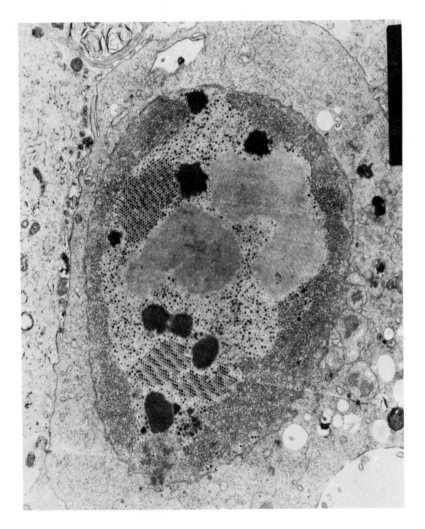

Figure 58-17. The nucleus of a Vero African green monkey kidney cell, 70 hours after infection with adenovirus type 2. Paracrystalline arrays of virus particles, crystals of vertex or core proteins, and intranuclear inclusions (the densely staining masses) are visible. × 10,000. *(From Henry et al.: Virology 44:215, 1972).*

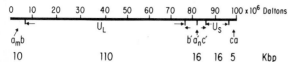

Figure 58-18. The sequence arrangement in herpes simplex virus DNA. U_L and U_S are unique regions flanked by inverted repeats. Regions *a, a'*, etc., are inverted forms of the same sequence. U_L is flanked by *b* sequences (about 8000 base pairs) and a variable number of *a* sequences (about 400 base pairs); U_S is flanked by *c* sequences (about 5000 base pairs) and *a* sequences. The *a* sequence is itself composed of five smaller sequence elements (from 12- to 64-base pairs long), two of which are unique, with the others present in 2, 2 to 3, and 19 to 22 copies, respectively. Note that the *a* sequence is present in the same orientation at the ends of the molecule and in the inverted orientation at the L–S junction. During the course of infection, the L and S components invert relative to each other so that the progeny DNA consists of equimolar amounts of four isomers that differ from each other solely in the relative orientation of the two components. The signal that directs the inversion of L and S components relative to each other is contained solely in the *a* sequence; in fact, insertion of *a* sequences into the U_L region causes additional inversions to occur. As a rule, the segment that inverts is flanked by *a'* sequences; segments that are flanked by *a* sequences do not invert. The inversion itself results from intramolecular recombination between terminal and inverted *a* sequences.

nome therefore exists in four configurations. The mechanism that causes this state of affairs is intramolecular recombination within a small region of the DNA (see Fig. 58-18). The significance of the extraordinary sequence anatomy of herpes simplex virus DNA is not clear, especially because it is modified or absent in other herpesviruses. Thus, in the genome of pseudorabies virus, a porcine herpesvirus, only the S segment is flanked by inverted repeated sequences and only the S segment inverts relative to the L segment, so that there are only two sequence isomers; and the genome of Epstein–Barr virus exists in only one configuration (see Chap. 62). Also, herpes simplex virus containing DNA that has been modified by recombinant DNA technology so that it can exist in only one configuration is infectious.

Herpesvirus Multiplication Cycle
The manner in which herpesvirus particles penetrate into cells is still a matter for debate; the evidence favors fusion of the viral membrane with the plasma cell membrane and liberation of naked nucleocapsids into the cytoplasm. Certainly, the herpesvirus envelope, most probably through the activity of one or more of the glycoproteins on its surface, causes cell membranes to fuse very readily, resulting in the formation of giant syncytia (see Chap. 59).

Once the parental DNA is uncoated, it moves to the nucleus where it begins to replicate after about 4 hours. The peak rate of herpesvirus DNA replication occurs at 6 to 8 hours, but the DNA continues to replicate at a gradually decreasing rate until the end of the multiplication cy-

cle at about 36 hours. Herpesvirus DNA replication is mediated by a virus-coded DNA polymerase, which is an early protein, and probably proceeds via a rolling circle mechanism.

Substantial progress has recently been made in defining the herpesvirus genetic map. Using the techniques described above for adenovirus, more than 50 genes have so far been mapped on the herpesvirus physical map.

Herpesvirus Transcription Program
The herpesvirus transcription program has been intensively investigated by the techniques described above for SV40 and adenovirus. Like the transcription programs of these viruses, it can be divided into an early and a late period. The early period can be divided into two stages. During the first stage, which proceeds in the absence of de novo protein synthesis, about one third of the viral genome is transcribed by the host cell DNA-dependent RNA polymerase II into a series of apparently large, extensively processed transcripts; only about one third of their sequences reaches the cytoplasm in the form of messenger RNA molecules. The nature of the processing is not known; some herpesvirus messenger RNAs are spliced, but the majority appear to be unspliced. Only 4 of the 10 to 20 proteins that could be translated from these messenger RNA molecules have been detected and mapped on the herpesvirus genome; the others are probably formed in small amounts only. The proteins that are formed during this period are known as the immediate early or α proteins. Their rate of formation reaches a peak between 2 and 4 hours after infection and declines rapidly thereafter.

The second stage of the early period proceeds only when the α proteins have been synthesized. This stage is termed the β or true early stage. During this stage, which, like the α stage, proceeds even if herpesvirus DNA replication is inhibited, about 40 percent of the viral genome is transcribed, and the messenger RNAs that are translated in the cytoplasm correspond to about a quarter of the information encoded in herpesvirus DNA. Presumably, transcription is now initiated at new promoters, which are recognized by a new DNA-dependent RNA polymerase, perhaps the host cell RNA polymerase II modified by an α protein.

Once viral DNA replication has commenced, the transcription program changes again. The late period of the herpesvirus multiplication cycle can also be divided into two stages, the $\beta\gamma$, or leaky, late stage, and the γ late stage; $\beta\gamma$ genes are those that are transcribed marginally before DNA replication, and more strongly thereafter. During the late period more than one half of the viral DNA is transcribed.

The α, β, and γ herpesvirus proteins are rapidly being characterized (Fig. 58-19). Four α proteins are known; they seem to be phosphoproteins, and presumably are all regulatory proteins. About 10 β proteins have been identified, among them a DNA polymerase, a deoxypyrimidine kinase, and a DNA-binding protein. A deoxyribonuclease,

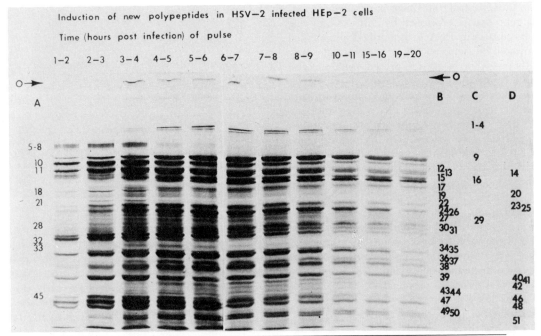

Figure 58-19. Autoradiogram of a polyacrylamide gel in which the proteins synthesized at various stages of the herpesvirus infection cycle (1 to 2 hours, 2 to 3 hours, 3 to 4 hours, and so on) had been electrophoresed. The direction of electrophoresis was from O (origin) downward. A, B, C, and D refer to kinetic groups to which the individual numbered proteins can be assigned, and which correspond roughly to the products of the α, β, and γ genes. *(Courtesy of Dr. Richard J. Courtney.)*

a ribonucleotide reductase, and a deoxycytidylic acid deaminase also appear to be early proteins. Another interesting early herpesvirus protein is an Fc receptor that begins to appear on the surface of infected cells at about this time. These virus-induced Fc receptors have affinity for normal immunoglobulin Gs and appear to be operationally equivalent to the Fc receptors expressed on certain cells of the immune system. The herpesvirus Fc receptor activity is associated with glycoprotein E, which also appears to be present in the envelopes of herpesvirus particles. The role of Fc-binding receptors in herpesvirus replication and biology is not understood. It is conceivable that the Fc-binding activity is fortuitous and that the receptor serves some entirely different function; but it has also been suggested that the binding of normal immunoglobulin or antiviral immunoglobulin to the Fc receptors can somehow interfere with cytotoxic immune reactions, thus sparing the infected cells and perhaps favoring the establishment of latency. Aggregated immunoglobulin G can in fact protect herpesvirus-infected cells from immune cytolysis in vitro. The binding of immunoglobulin to the Fc receptor may also repress the expression of viral functions in the cell, again favoring the establishment of latency. The other side of the coin, of course, is that cells infected with herpesvirus may be lysed nonspecifically by Fc receptor-positive killer cells; this has actually been observed.

As for the late proteins, they are predominantly structural components of virus particles.

Herpesvirus Morphogenesis and Release

Little is known concerning these processes except that assembly of herpesvirus nucleocapsids takes place in the nucleus, the capsid being formed first, followed by insertion of the viral DNA, and that the major site of envelopment is the inner nuclear membrane. Thus the envelopment of herpesvirus nucleocapsids differs sharply from the two important processes by which RNA-containing nucleocapsids are enveloped (budding through the plasma membrane or budding into vacuoles, see below). Instead, enveloped herpesvirus particles are transported through the cytoplasm by an as yet unidentified mechanism and are released from the plasma cell membrane without acquiring significant extra material in the process.

Poxviruses

Poxviruses are unique among animal viruses in that they require a newly synthesized protein for uncoating. Vaccinia virus particles are taken up into cells via phagocytic vacuoles (coated vesicles, see above) from which they are liberated into the cytoplasm in the form of cores. In the presence of inhibitors of protein synthesis, these cores are not uncoated; they do not liberate the DNA that they contain. Instead, uncoating of cores requires a protein that is encoded by the DNA that they contain and that is transcribed by the DNA-dependent RNA polymerase that is present within them. In fact, this uncoating protein gene is not the only gene that is expressed by cores. About one third of the vaccinia virus DNA is transcribed into capped and polyadenylated messenger RNA—the immediate early messenger RNAs—while it is still present within cores, that is, before it is uncoated.

Once vaccinia virus DNA is fully uncoated, it initiates the formation of inclusions, or factories, in the cytoplasm, within which it is transcribed further, replicates, and is encapsidated into progeny virions. The inclusions, which are easily visible with light microscopy, are composed of fibrillar material and may be located anywhere in the cytoplasm. Their number per cell is proportional to the multiplicity of infection, which suggests that each infecting virus particle initiates its own factory (Fig. 58-20).

During the early period of the multiplication cycle, almost one half of the vaccinia virus genome is transcribed into messenger RNA. Among the early proteins are several enzymes, such as a thymidine kinase and a DNA polymerase, as well as several structural proteins.

Vaccinia virus DNA replication commences at about 1.5 hours after infection and is complete by about 5 hours (Fig. 58-21). Its onset heralds the commencement of the

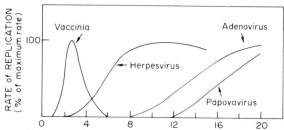

Figure 58-21. The replication of vaccinia virus, herpesvirus, adenovirus, and papovavirus DNA. Vaccinia virus DNA is atypical in replicating only during a brief period early in the multiplication cycle. Progeny DNA molecules form a pool from which individual molecules are selected at random for incorporation into progeny virus particles.

late period of the multiplication cycle, which is characterized by new patterns of viral gene expression. Synthesis of at least some of the early structural proteins continues into the late period, but synthesis of the early enzymes ceases. Cessation of their synthesis is due to the so-called switch-off phenomenon (Fig. 58-22), which is of interest because it provides one of the few well-documented examples of

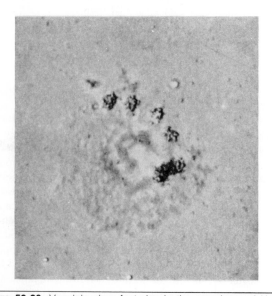

Figure 58-20. Vaccinia virus factories in the cytoplasm of HeLa cell. Cells growing on a cover slip were infected at a multiplicity of 6 plaque forming units (PFU) per cell. At 6 hours after infection tritiated thymidine was added, and at 7 hours the cells were fixed. Autoradiographic stripping film was then applied, and the slide stored for 2 weeks. On developing, the picture shown here was obtained. There are no grains (indicative of thymidine incorporation and, therefore, DNA replication) over the nucleus, but there are five labeled areas or factories in the cytoplasm (one actually composed of two coalesced areas)—this is where viral DNA is being synthesized. This cell has been stained with antibody to vaccinia virus coupled to fluorescein before autoradiography, and it was thereby demonstrated that the only areas in the cell that contained appreciable amounts of vaccinia virus antigens were the factories. Both viral DNA replication and viral morphogenesis therefore proceed within the factories. *(From Cairns: Virology 11:603, 1960.)*

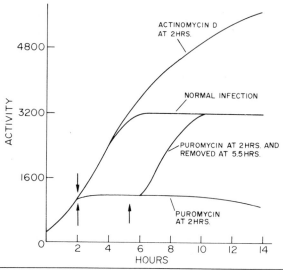

Figure 58-22. The switch-off phenomenon. This graph depicts the synthesis of early enzymes, such as thymidine kinase and DNA polymerase, during vaccinia virus infection. Under normal conditions these enzymes begin to be formed soon after infection, and their synthesis is "switched-off" at about 4 hours. If actinomycin D, which inhibits messenger RNA formation, is added at 2 hours, switch-off does not occur. This demonstrates (1) that the messenger RNAs from which these enzymes are translated are very stable, and (2) that switch-off itself requires the synthesis of some other messenger RNA. If protein synthesis is inhibited with puromycin at 2 hours, enzyme synthesis immediately ceases; if puromycin is removed at 5.5 hours, enzyme synthesis resumes and is again switched off after a time interval equivalent to that between the addition of puromycin and the onset of normal switch-off. This indicates that switch-off is due to the accumulation of a certain amount of some specific protein. *(Modified from McAuslan: Virology 21:383, 1963.)*

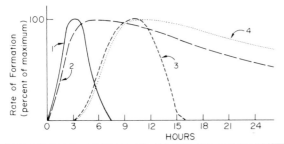

Figure 58-23. The synthesis of four classes of vaccinia-specified proteins. First, there are some early proteins whose synthesis is switched off when DNA replication commences. Early enzymes and several structural vaccinia virus proteins belong to this class. Second, there are those early proteins whose synthesis continues throughout the multiplication cycle. One of the major components of immature virus particles belongs to this class. Third, there are late proteins that are synthesized for some time following the onset of DNA replication and that are then switched off; finally, there are late proteins that are synthesized throughout the entire late period of the multiplication cycle. Most structural vaccinia virus proteins belong to this class. *(Modified from Holowczak, and Joklik: Virology 33:726, 1967.)*

regulation of gene expression at the level of translation. In essence, cessation of the synthesis of early enzymes is not due to inhibition of the transcription of their genes, to instability of their messenger RNA species, or to instability of the enzymes themselves. Rather it is due to a suddenly developing inability of the messenger RNA species that code for early enzymes to be translated. If viral DNA replication and protein synthesis are inhibited, this inability does not develop. It is thought, therefore, that one of the first late proteins to be synthesized specifically prevents the translation of the messenger RNA molecules that code for early enzymes. This mechanism of controlling protein synthesis is obviously highly selective, as many other viral messenger RNA molecules—for example, those that code for structural proteins—continue to be translated. It is of potential significance for antiviral chemotherapy because it implies the existence of a chemical difference between those messenger RNA molecules that continue to be translated and those that are switched off. In may prove possible to expoit this difference (see Chap. 61).

Another class of vaccinia proteins whose synthesis is switched off comprises certain late proteins that are formed for only several hours. Together with the early and late proteins whose synthesis is never switched off, there are thus at least four different vaccinia virus protein translation patterns, and there may be many more (Fig. 58-23).

Somewhat more than one half of vaccinia virus DNA codes for late proteins. By far the quantitatively most important of these are the structural viral proteins, most of which, like DNA, are usually formed in great excess. The vaccinia virus-coded proteins that are synthesized as the multiplication cycle progresses are shown in Figure 58-24; it is readily seen that different proteins are synthesized at

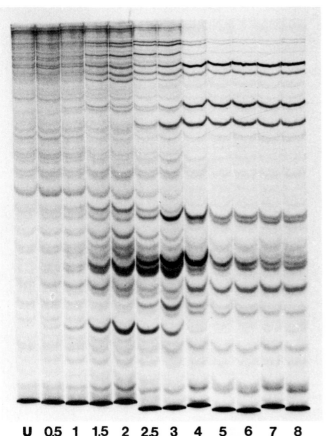

U 0.5 1 1.5 2 2.5 3 4 5 6 7 8

— 100

— 60

— 40   **Mol. wt. x 10⁻³**

— 25

Figure 58-24. Autoradiogram of a 9 percent polyacrylamide slab gel in which the proteins synthesized at various stages of the vaccinia virus multiplication cycle had been electrophoresed. At the times indicated (hours after infection) the cells were labeled for 15 minutes with [¹⁴C]-protein hydrolysate. At the right is a molecular weight scale as determined by electrophoresing proteins of known size under identical conditions. *(Courtesy of Dr. T.H. Pennington.)*

different times. It should be noted that very few host-coded proteins are formed even at the earliest times; like herpesvirus, vaccinia virus shuts down host–cell protein synthesis very efficiently (see Chap. 59).

Another way of analyzing the vaccinia virus transcription program is to hybridize early and late vaccinia messenger RNA populations to a series of mapped restriction endonuclease fragments of vaccinia virus DNA, and to translate the messenger RNAs that hybridize to each in an in vitro cell-free protein-synthesizing system. The results of such an analysis are summarized in Table 58-4. Clearly, numerous proteins are translated during periods of the multiplication cycle, and some portions of the vaccinia virus genome are transcribed predominantly in the early period, others in the late period.

Progeny genomes of DNA-containing viruses generally replicate faster than they are incorporated into virus particles. They therefore accumulate to form pools from which individual genomes are withdrawn at random for encapsidation. Whereas some DNA molecules are withdrawn very soon after they are formed, others may remain naked for long periods of time. This is true particularly in the case of the vaccinia virus multiplication cycle, where DNA replication ceases at about 5 hours after infection, whereas viri-

TABLE 58-4. TRANSLATION MAP OF THE VACCINIA VIRUS GENOME

Hind III Fragment	C*			N+M+K	F	E	O+I	G	L+J	H	D	A	B	Total
Size (mol wt × 10⁻⁶)	5.9	4.4	3.8	5.3	8.9	9.9	5.3	6.0	5.8	5.6	10.3	30	19	120
Early Proteins (mol wt × 10⁻³)	42	60	38	54	68	95	80	54	41	40	86	58	42	
	19	21	32	53	62	67	79		21	39	84	45	35	
	8	19	21	46	59	64	33		17	14	79	41	35	
		13	15	40	54	62	32			11	52	40	31	
		6	14	30	45	55	25				34	39	27	
			12	23	39	36					28	37	19	
				20	35	34					27	35	8	
				16	33	32					24	31		
				11	27	30					17	27		
				9	17	26					14	24		
					16	22					12	23		
					16	17						20		
					10	15						18		
												16		
Medium–Early Proteins			23					33	110					
								30						
Late Proteins	—	22	40	—	52	12	46	65	44	55	70	96		
					40		35	44	37	46	35	84		
					15		11	33	36	36	17	45		
								30	33	30	14	22		
								14	30	28				
								12	28	18				
								9	28	17				
									19	16				
									19	13				
									18					
									14					

Adapted from Isle et al.: Virology 112:306, 1981.
*Extracts were prepared of HeLa cells infected with vaccinia virus under three sets of conditions: (1) In the presence of cycloheximide, an inhibitor of protein synthesis, when vaccinia virus cores are not uncoated and only early messenger RNA is made; (2) in the presence of cytosine arabinoside, an inhibitor of DNA replication, when both early and medium–early messenger RNAs are made; and (3) during the late period of the multiplication cycle, when some early and medium–early, but also all late messenger RNAs are made. The viral messenger RNAs in these extracts were hybridized to cloned Hind III restriction endonuclease fragments of vaccinia virus DNA, and those that hybridized were translated in a cell-free protein synthesizing system prepared from reticulocytes in order to determine which proteins they encoded. The Hind III C fragment was further divided into its EcoRI fragments, A, B, C, and D; because they are small, C and D are combined in this map. For the same reason, Hind III fragments N, M, and K, O and I, and L and J are also combined. Hind III fragments A and B are too large to be cloned readily and were used in uncloned form. Note the sizes of the proteins that were obtained, which range from over 100,000 to less than 10,000; and the fact that the early and late proteins are clustered on the vaccinia virus genome, late proteins being encoded primarily in the regions defined by Hind III fragments G, L, J, and H.

on morphogenesis continues for about 25 hours. The time necessary for a complete virus particle to be assembled around a vaccinia DNA molecule is about 1 hour.

Vaccinia Virus Morphogenesis. The morphogenesis of vaccinia virus proceeds in the cytoplasm, where mature virus particles accumulate (Fig. 58-25). Most virus particles are released only when infected cells disintegrate. However, mature vaccinia virus particles can also migrate to the

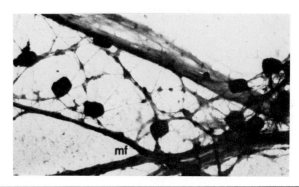

Figure 58-26. Vaccinia virus particles associated with elements of the cytoskeleton. Infected cells, extracted with 0.5 percent triton X-100. Uranyl acetate staining. Magnification, × 30,000. For details see text. *(From Hiller et al.: Virology 98:142, 1979.)*

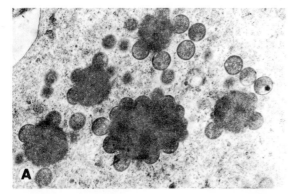

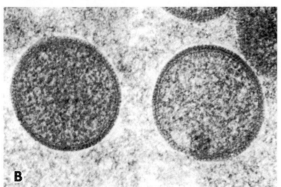

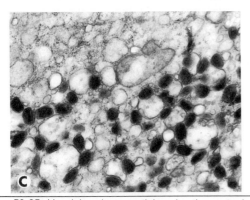

Figure 58-25. Vaccinia virus particles in the cytoplasm of infected cells. **A.** Immature virus particles developing from intracytoplasmic inclusions 1 hour after reversal of vaccinia virus morphogenesis arrest by rifampicin (see Chap. 61). × 9,000. **B.** Characteristic structure of immature vaccinia virus particles. × 48,000. **C.** Mature vaccinia virus particles in the cytoplasm of infected cells. × 12,000. *(A, B, courtesy of Dr. T.H. Pennington; C, from Dales and Siminovitch: J Biochem Biophys Cytol 10:475, 1961.)*

plasma cell membrane in association with the actin-containing microfilaments of the cytoskeleton (Fig. 58-26). Late in infection, numerous specialized microvilli that contain predominantly F-actin and F-actin–binding proteins (α-actinin, filamin, and fimbrin) are formed on the surface of infected cells, and these microvilli usually contain a vaccinia virus particle at their tip. They may provide a second mechanism for vaccinia virus release and dissemination.

Multiplication Cycle of Double-stranded RNA-containing Viruses

When reovirus infects cells, its outer capsid shell is partially degraded to form an entity known as a subviral particle (SVP) (Fig. 58-27). This process activates an RNA polymerase (known as the transcriptase) within them that transcribes all 10 reovirus genes into capped, but unspliced and not polyadenylated, messenger RNA molecules. The parental double-stranded RNA never escapes from the SVP, which persists in infected cells throughout the multiplication cycle.

The messenger RNA molecules are translated for several hours and then begin to associate with viral proteins to form complexes within which they are transcribed once —and once only—into minus strands with which they remain associated, thereby giving rise to double-stranded RNA molecules, the progeny genes. These immature progeny virus particles then transcribe more messenger RNA molecules, which are translated into more proteins, and so on. At the same time, immature virus particles associate with the proteins of the outer capsid shell to form mature virus particles. It should be noted that the stages of the multiplication cycle when the plus and minus strands of progeny double-stranded RNA are synthesized are separated by an interval of several hours. Double-stranded RNA is thus synthesized by a mechanism that is quite different from that by which double-stranded DNA is formed.

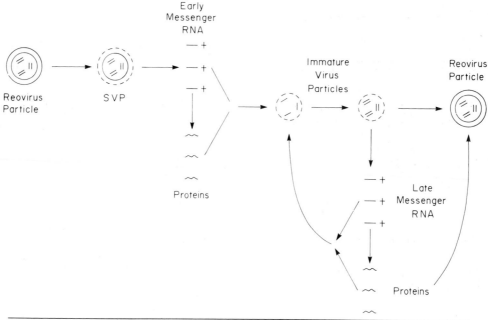

Figure 58-27. The reovirus multiplication cycle. For details, see text. Note that here early and late messenger RNA and protein molecules are identical. Because many immature virus particles are formed from each parental subviral particle, most viral messenger RNA molecules are transcribed from progeny templates.

Multiplication Cycles of Single-stranded RNA-containing Viruses

The principles involved in the multiplication of single-stranded RNA-containing viruses differ in several aspects from those described for the double-stranded DNA-containing viruses. Most importantly, their multiplication cycles cannot be divided into clearly defined early and late periods, as can those of the double-stranded DNA-containing viruses. A brief description of the multiplication cycles of several families of single-stranded RNA-containing viruses follows.

Multiplication Cycles of Plus-stranded RNA-containing Viruses

Picornaviruses

The multiplication cycle of poliovirus is considered here, as it has been investigated in most detail. The strategy of the poliovirus multiplication cycle is summarized schematically in Fig. 58-28.

After uncoating, the poliovirus RNA, a 7441-nucleotide long molecule that has been completely sequenced, is translated into one very large protein with a molecular weight of about 240,000. This protein, the polyprotein, is then cleaved in a precisely defined sequence of reactions into about 20 smaller proteins, among which are the four structural proteins, the RNA-linked protein VPg, an RNA polymerase, and a protease that carries out this cleavage program (Fig. 58-29). Little is known concerning the nature of the intermediate cleavage products or whether any of them exercise a specific function during the poliovirus multiplication cycle. They clearly possess the potential for doing so, and if they did, this would be another example of the multiple use of viral genetic information. In this respect, it is of interest that posttranslational cleavage of viral proteins occurs frequently; there are numerous examples of vaccinia virus, adenovirus, myxovirus, togavirus, retrovirus, and other viral capsid proteins being synthesized in the form of precursors larger than themselves. The reason in most cases is not expansion of the pool of genetic information, but rather one of the following:

1. Viral capsid proteins may be too insoluble to be transported from the site where they are synthesized to the site of morphogenesis unless their solubility is increased by the presence of removable amino acid sequences.
2. Precursors may be the forms in which capsid proteins must exist prior to assembly to enable them to interact correctly with other capsid proteins.
3. Cleavage of precursors to capsid proteins may provide at least part of the energy for morphogenesis.

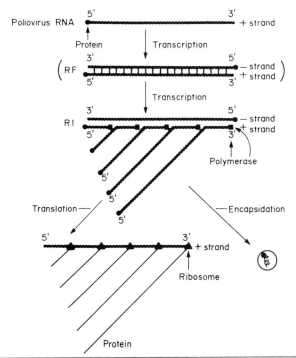

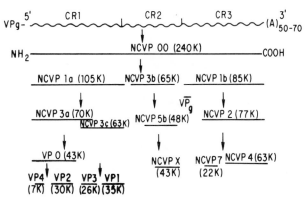

Figure 58-29. Synthesis and processing of the poliovirus precursor polyprotein NCVP 00 (noncapsid viral protein 00). Poliovirus RNA can be subdivided into three coding regions; CR1, which encodes the structural proteins; CR2, which encodes the protease, and CR3, which encodes the RNA polymerase. Some precursor proteins can be cleaved at more than one location. The small 5'-terminal genome-linked protein VPg has been mapped to a location at or near the amino terminus of the replicase precursor protein NCVP 1b. Cleavages that occur before the putative proteases, NCVP X and NCVP 7, have been formed probably occur autocatalytically.

Figure 58-28. Replication and functioning of poliovirus RNA. Parental RNA strands are translated into the polyprotein, and the polymerase derived from the polyprotein then transcribes them into strands of minus polarity, to yield double-stranded replicative forms (RF), which probably have a very short half-life. Progeny plus strands are then transcribed repeatedly from the minus-strand templates by a peeling-off type mechanism. The structure consisting of minus-stranded template and several plus-stranded transcripts at various stages of completion is known as the replicative intermediate (RI). Early during the infection cycle, the number of RFs, and therefore of RIs, increases, so that the rate of formation of progeny plus strands first increases and then becomes constant. Progeny plus strands are either translated or encapsidated.

The enzyme that cleaves the precursors is not known with certainty but appears to be either NCVP X, NCVP 7, or both. All cleavages occur between glutamine and glycine residues. The only exception is the cleavage of VP 0 to VP 4 and VP 2, which occurs only during morphogenesis. Proteins of other viruses with protease function are capsid protein C of togaviruses and p15 of retroviruses (see Chap. 62).

Poliovirus RNA replication, which starts within an hour of infection and appears to be catalyzed by the poliovirus polymerase NCVP 4 in concert with a cellular protein, takes place in two stages: First, the parental plus strand is transcribed into a minus strand; and then, the minus strand serves as the template for repeated transcription into progeny plus strands. Some progeny plus strands are then again transcribed into minus strands, but this is not a common occurrence. The total number of minus strands in the infected cell probably does not exceed 1000, whereas up to a million plus strands may be formed. RNA

strands of both plus and minus polarity are covalently linked at their 5'-termini to a tyrosine residue of VPg, a protein with 22 amino acids that probably serves as a primer for transcription. This protein is also present on the RNA that is present in virus particles, but is not present on poliovirus RNA molecules that serve as messenger RNA. Thus, progeny RNA molecules that are encapsidated have never served as messenger RNAs, and poliovirus RNA molecules that serve as messengers do not become encapsidated.

Whereas translation of the poliovirus genome into a single precursor protein has many advantages, there are also potential disadvantages. In particular, it implies that all portions of the viral genome are expressed with equal frequency. This is very wasteful, as many more capsid protein molecules are required than, for example, polymerase molecules. In practice, the various processed poliovirus-specified proteins are not formed in equimolar amounts, especially during the later stages of the multiplication cycle when most of the proteins that are formed are indeed capsid components. The mechanism by which such translational control is achieved is probably as follows. It is known that the capsid proteins are encoded in the 5'-terminal region of the poliovirus RNA molecule. This means that in the event of premature termination of translation, which is not unlikely for a long messenger RNA such as poliovirus RNA, the formation of capsid proteins would be favored over those like the polymerase that are encoded by sequences in its 3'-terminal half.

As is the case for other icosahedral viruses, poliovirus progeny often accumulates in the form of large, intracytoplasmic, paracrystalline arrays (Fig. 58-30).

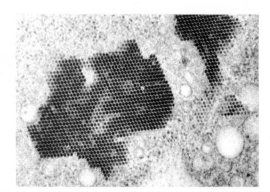

Figure 58-30. A large crystal of poliovirus in the cytoplasm of a HeLa cell infected for 7 hours. × 50,000. *(From Dales et al.: Virology 26:379, 1965.)*

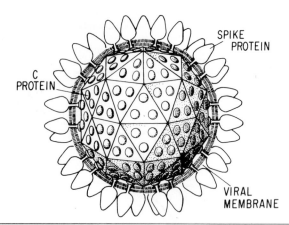

Figure 58-31. Structure of the alphavirus Semliki Forest virus. One hundred eighty molecules of the C protein (mol wt 30,000) form the icosahedral nucleocapsid; and 180 spikes are inserted into the lipid bilayer of the envelope. *(Adapted from Simons et al.: Sci Am 246:58, 1982.)*

Togaviruses

Togavirus particles, as exemplified by the alphavirus Semliki Forest virus, consist of an icosahedral nucleocapsid made up of 180 molecules of a single protein, the C protein, surrounded by an envelope that consists of a lipid bilayer into which 180 spikes are inserted (Fig. 58-31). Each spike consists of three glycoproteins, E1, E2, and E3, two of which span the lipid bilayer, and is associated with a C protein molecule (Fig. 58-32). Interestingly, proteins E1 and E2 are not only glycosylated, but also acylated; that is, several molecules of fatty acid are linked to them covalently via ester bonds, presumably in the regions that are most intimately associated with the lipid bilayer.

The RNA of alphaviruses, which form one of the togavirus genera, is a plus stranded RNA molecule about 12,700-nucleotides long (mol wt 4.3×10^6), with a sedimentation coefficient of 42S. The strategy of the alphavirus multiplication cycle (Fig. 58-33) is quite different from that of picornaviruses. When the alphavirus genome is uncoated, about two thirds of it, starting from the 5′-termini, is translated into one large protein that is cleaved into four nonstructural proteins. They are, from the amino termini, ns70-ns26-ns78-ns60 (the numbers being their molecular weights), and two of them form the RNA polymerase. This enzyme transcribes the parental plus strands into minus strands, which then act as templates for plus strand synthesis. The crucial point is that two types of plus strands are transcribed: (1) full-length 42S plus strands that, like the progeny plus strands of poliovirus RNA, can function either as messenger RNA, act as templates for the synthesis of additional minus strand synthesis, or be encapsidated into progeny virus particles; and (2) strands that are only about one third as long and possess a sedimentation coefficient of 26S (mol wt 1.6×10^6, about 4800-nucleotides long). These RNA molecules represent the 3′-terminal one third of the 42S RNA molecules, and they code for the four structural alphavirus proteins that are translated from it in the form of a 130,000-dalton precursor (Fig. 58-34).

In essence, then, the alphavirus genome expresses itself via two messenger RNAs instead of via only one. The special 26S messenger RNA is transcribed about three times as frequently as the 42S RNA, which acts as the messenger RNA only for the nonstructural protein precursor. These relative transcription frequencies are controlled by at least two proteins: one of the nonstructural proteins described above, which promotes transcription of 26S RNA; and the C protein, which expresses it.

The morphogenesis of alphaviruses involves the formation of cores or nucleocapsids that consist of 42S plus-stranded RNA and C protein, and budding through

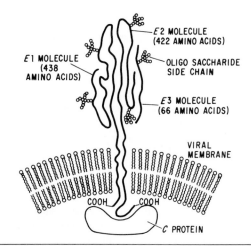

Figure 58-32. The togavirus spike. Each spike consists of three glycoproteins, E1 (envelope protein 1) (mol wt 48,000), E2 (mol wt 46,000), and E3 (mol wt 7,000). E1 and E2 possess two carbohydrate chains; E3 possesses only 1. The two large glycoproteins are anchored in the lipid bilayer by spanning it; thirty-one amino acids of E2 and two amino acids of E1 penetrate through it into the interior of the particle where they make contact with C protein molecules. *(Adapted from Simons et al.: Sci Am 246:58, 1982.)*

patches of plasma cell membrane into which the glycosylated envelope proteins have been inserted. This process is discusses in more detail below.

Members of the *Flavivirus* genus, which are considerably smaller than alphaviruses (50 nm in diameter, rather than 60 to 65 nm), use a different strategy of genome expression and maturation. Here the viral RNA itself serves as messenger RNA for all virus-coded proteins. There is good evidence that this RNA—which in contrast with that of alphaviruses, is not polyadenylated—possesses multiple internal protein synthesis initiation sites, a unique property for eucaryotic messenger RNAs. Furthermore, flaviviruses tend to mature in association with cytoplasmic membranes, bud into cytoplasmic vacuoles, and are liberated by exocytosis and cell lysis, instead of budding through plasma cell membranes (see below).

Multiplication Cycles of Negative-stranded RNA-containing Viruses

Rhabdoviruses

One of the most intensively investigated negative-stranded RNA-containing viruses is the rhabdovirus vesicular stomatitis virus (VSV). The VSV particle consists of a helical nucleocapsid composed of a minus-stranded RNA molecule about 11,300-nucleotides long, and 2300 molecules of N protein (mol wt 52,000), 230 molecules of NS protein (mol wt 32,000), and 60 molecules of L protein (mol wt

175,000). The latter two proteins constitute the RNA polymerase. This nucleocapsid is surrounded by an envelope that consists of a layer of M protein (mol wt 25,000; 4700 molecules) and a lipid bilayer, to which about 700 spikes consisting of trimers of the G protein (mol wt 67,000; 2100 molecules) are attached. The G protein is a transmembrane protein; about 30 amino acids at its carboxyl-terminal end penetrate through the lipid bilayer into the interior of the particle, where they contact not only M protein molecules, but also the nucleocapsid.

These five structural proteins are the only proteins encoded by VSV RNA. The order of their genes, from the 3'-end of the RNA in virus particles, is N-NS-M-G-L. They are separated by short 13-residue long sequences of almost identical structure.

VSV enters the cell via invagination into coated vesicles in which its envelope is removed, and from which the nucleocapsid is liberated into the cytoplasm. Removal of the envelope activates the RNA polymerase in the nucleocapsid, which then transcribes the minus-stranded RNA within it into plus-stranded RNA of two types, using two quite different mechanisms. The first consists of the faithful transcription of the minus-strands into plus-strands, which in turn serve as templates for the synthesis of numerous progeny minus-strands. This is the process of minus-strand replication. The second mechanism involves the transcription of the minus-stranded RNA into plus-stranded RNA molecules that can serve as the messenger RNA through which the VSV genome expresses itself. For this purpose a different type of transcript is synthesized. In its simplest terms, this type of transcription process can be visualized as follows. The first gene to be transcribed is

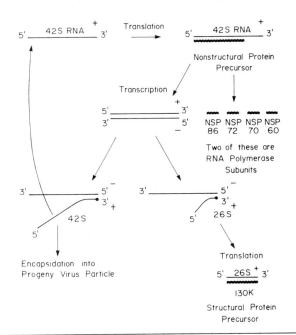

Figure 58-33. The strategy of the togavirus multiplication cycle. For details see text. For fate of 130K structural protein precursor, see Figure 58-34.

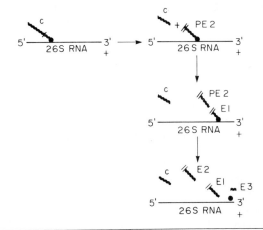

Figure 58-34. The processing of the togavirus 130K structural protein precursor. The precursor only accumulates to a significant extent within cells if its processing is inhibited by protease inhibitors. Under normal conditions, the C or nucleocapsid protein is cleaved off the amino-terminal end of the growing precursor chain, and at the same time the newly generated amino terminus of the E2 precursor (PE2) is inserted into membrane. PE2 is then cleaved off the growing precursor chain, and the amino terminus of E1 is inserted into membrane in similar manner. E3 is cleaved from PE2 during virus maturation. The protease activity is located on the C protein.

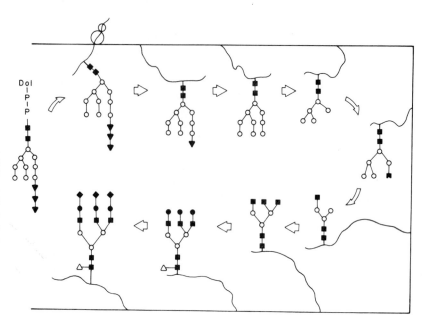

Figure 58-35. Proposed sequence of reactions for the synthesis of the oligosaccharide subunit of the G spike protein of vesicular stomatitis virus. The first step is the transfer of a mannose-rich precursor from dolichol pyrophosphate to an asparagine residue of the G protein; this precursor is then processed to yield the final oligosaccharide unit. Dol, dolichol; ■, N-acetylglucosamine; ○, mannose; ▲, glucose; ●, galactose; ◆, sialic acid (N-acetylglucosamine acid-galactose); △, fucose. *(From Kornfeld et al.: J Biol Chem 253:7771, 1978.)*

the N gene. At the end of its coding region, the intragenic sequence is encountered—the first 11 residues of which act as a polyadenylation signal. Accordingly, about 250 adenine residues are added by a process such as reiterative copying or slippage polymerization. The polymerase then resumes transcription of the next gene, the preceding sequence is cleaved off at the end of the poly (A) sequence, and this process is repeated until messenger RNAs corresponding to all five genes have been synthesized. Interestingly, there is a sixth transcript: the extreme 3′-terminal region of VSV RNA, preceding gene N, is transcribed into a 48-nucleotide long RNA, the so-called leader sequence, which migrates to the nucleus of infected cells where it may help shut down host RNA and DNA synthesis.

Transcription of parental minus strands into plus-stranded RNA molecules is referred to as primary transcription; because the nucleocapsid contains the polymerase, it proceeds without requiring newly synthesized protein. Transcription of plus strands into progeny minus strands, and of progeny minus strands into additional genome-length plus strands or messenger RNAs is referred to as secondary transcription and requires the synthesis of newly synthesized nucleocapsid proteins (N, L, and NS proteins). Nothing is known of the mechanism that determines whether a minus strand is to be transcribed into a genome-length plus strand, or into messenger RNAs.

The five VSV proteins are synthesized in about the same molar ratios as those in which they exist in virus particles; for example, about 30 times as many N as L protein molecules are synthesized. These ratios appear to be controlled at the level of transcription rather than translation. This presents a paradox, for the manner in which VSV RNA is transcribed implies that equal numbers of all five species of messenger RNA molecules should be formed. It seems, however, that there is a polarity effect (attenua-

tion); that is, as the polymerase transcribes the RNA it tends to "fall off," so that the further a gene is situated from the origin of transcription, the less likely it is to be transcribed. Indeed, the gene order 3′-N-NS-M-G-L-5′ in the VSV RNA minus strand parallels the frequency, in a decreasing sense, of the transcription of each gene into messenger RNA.

Like other enveloped viruses, VSV has a special mechanism for being liberated from cells, namely, budding. Enveloped viruses code for two types of structural proteins, nucleocapsid proteins and envelope proteins. The nucleocapsid proteins (the N, NS, and L proteins in the case of VSV) are assembled into nucleocapsids either in the nucleus (in the case of orthomyxoviruses) or in the cytoplasm (in all other enveloped RNA viruses), where they are often found in large numbers. The envelope proteins (the membrane-associated M protein and the G glycoprotein, in the case of VSV) follow a different pathway. The glycoproteins are synthesized on ribosomes that either are attached to membranes, or become attached to membranes very soon after they begin to be translated. Like other proteins that must pass through membranes, viral glycoproteins possess hydrophobic amino acid leader sequences some 15- to 20-amino acids long at their amino-terminal ends, which are cleaved off as soon as they have penetrated the lipid bilayer and have led and fed the rest of the protein chain through it. Glycoproteins are usually anchored in the lipid bilayer of the membrane via their carboxyl-terminus sequences, which do not pass through it.

As they are being synthesized, glycoproteins are glycosylated. Glycosylation proceeds in several stages, the first of which is the transfer of a mannose-rich precursor oligosaccharide from dolichol pyrophosphate to specific asparagine residues. These precursors are then processed to the final prosthetic group in a series of reactions depicted in Figure 58-35. As discussed in Chapter 56, the precise nature of the final oligosaccharide is determined in part by the nature and relative amounts of the various cellular glycosyl transferases, and in part, particularly beyond the mannose fork, by the nature of the protein that is being glycosylated. While glycosylation is proceeding, the glycoproteins are transported through the cell membrane system to patches on the plasma membrane and membranes lining intracytoplasmic vacuoles, where they replace host-specified membrane proteins. Nucleocapsids then migrate to these modified areas, which they recognize with great specificity; align themselves with them; and then bud through them, becoming coated by them in the process (Figs. 58-36 and 58-37). The M protein is incorporated into virus particles immediately prior to budding.

Interestingly, actin-containing microfilaments may play a role in the release of budding viruses, because treatment of cells with cytochalasin B, a drug that disrupts such filaments, inhibits the release of viruses such as measles virus.

Whereas budding is very efficient in some strains of cells, it is very inefficient in others. This sometimes leads to the accumulation in the cytoplasm of very large numbers of nucleocapsids (Fig. 58-38).

The budding process itself does not harm the host cell significantly. Many cells persistently infected with enveloped viruses (see Chap. 59) remain normal in appearance and continue to multiply for many generations while viruses bud from their surfaces. This is not to say that many enveloped viruses are not cytopathic; it merely says that budding per se is not a factor in cytopathogenicity.

Paramyxoviruses

Paramyxovirus RNA (mol wt 5.4×10^6) encodes six proteins, all of which have counterparts among the rhabdovirus structural proteins; there are three nucleocapsid proteins (NP, P, and L), and three envelope proteins (the M [matrix] protein and two glycoproteins, F and HN). Together, these proteins account for most of the coding capacity of the paramyxovirus genome. Their order, from the 3'-terminus of the minus strand, is NP-P-F-M-HN-L.

Paramyxoviruses enter cells by fusing with plasma cell membranes, with resultant liberation of their nucleocap-

sids into the cytoplasm. The very pronounced ability of the envelopes of these viruses to fuse with cell membranes is a function of one of the two paramyxovirus glycoproteins spikes, namely, that which is made up of the F (fusion) protein (see Chap. 56).

The strategy of the paramyxovirus multiplication cycle resembles that of the rhabdovirus multiplication cycle.

A remarkable feature of the two paramyxovirus glycoproteins, F and HN, which form the two types of spikes on the virus particle surface—the former being responsible for the fusion and hemolytic, and the latter for the hemagglutinin and neuraminidase activities (see Chap. 56)—is that they must be cleaved once, and once only, for virus particles to be infectious. The resultant fragments are not lost, but remain covalently bound via -SS- bonds to the other fragment of each molecule, which is anchored in the lipid bilayer like the VSV G glycoprotein. The cleavage is accomplished by a protease that is present in some cells but not in others, which therefore produce noninfectious virus particles. Cleavage can also be accomplished by certain proteases in vitro, which results in the activation of noninfectious virus particles.

Orthomyxoviruses—Influenza Virus

The influenza viruses are enveloped helical nucleocapsids. Their envelope consists of a lipid bilayer derived from the plasma membrane of the host cell, to which two types of spikes are attached. One is the hemagglutinin spike, of which there are about 350 per virus particle, each composed of a trimer of two -SS-linked proteins, HA1 and HA2, that are formed by cleavage of the hemagglutinin protein. Both are glycosylated, and HA2 is also esterified with fatty acid in a hydrophobic region at its carboxyl terminus, which is anchored in the lipid bilayer of the envelope, with about 10 amino acids penetrating into the interior of the virus particle.

The second type of spike is the neuraminidase. There are about 50 of these per virus particle, each made up of a tetramer of the neuraminidase protein, which is also glycosylated and may also be acylated. It is anchored in the lipid bilayer not by its carboxyl terminus, but by its amino terminus.

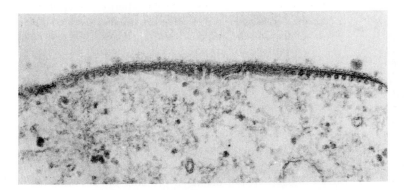

Figure 58-36. Modification of the plasma membrane of a monkey kidney cell infected with SV5. A layer of dense material resembling the spikes present on viral envelopes is present on the outer surface of the membrane. Nucleocapsid strands, many seen in cross-section, are aligned immediately beneath modified patches of the cell membrane. In due course they will bud through these membrane patches, becoming enveloped by them in the process. × 76,000. *(From Compans et al.: Virology 30:411, 1966.)*

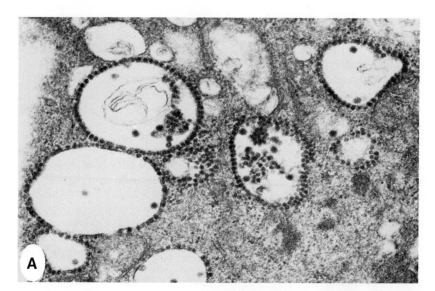

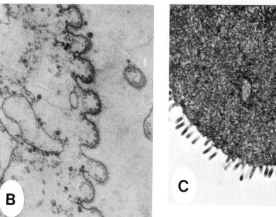

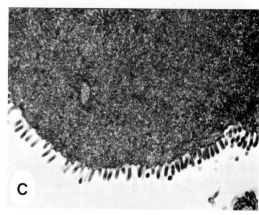

Figure 58-37. Budding of enveloped viruses. **A.** A chick embryo cell infected with the togavirus Semliki Forest virus (SFV). Numerous nucleocapsids lining cytoplasmic vacuoles prior to budding into them can be seen. × 22,000. **B.** A row of SV5 particles budding from the plasma membrane of a monkey kidney cell, showing many cross-sections of nucleocapsids. × 50,000. **C.** Vesicular stomatitis virus (VSV) budding from the plasma membrane of a mouse L cell. In L cells, the majority of VSV particles bud from the plasma membrane; in other cells, such as chick embryo fibroblasts and pig kidney cells, VSV buds mostly into cytoplasmic vacuoles. × 21,500. (**A,** from Grimley et al.: J Virol 2: 1326, 1968; **B,** from Compans et al.: Virology 30:411, 1966; **C,** from Zee et al.: J Gen Virol 7:95, 1970.)

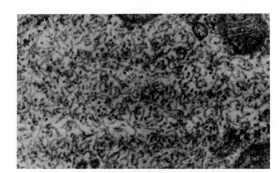

Figure 58-38. Accumulation of SV5 nucleocapsids in the cytoplasm of a BHK-21 cell. Such accumulation does not occur in monkey kidney cells, from which nucleocapsids bud as rapidly as they are synthesized. × 34,000. (From Compans et al.: Virology 30:411, 1966.)

Within the lipid envelope lies the M (matrix) protein, and within the matrix shell are eight negative-stranded RNA molecules, the influenza virus genes, associated in helical nucleocapsid structure with the NP (nucleocapsid) protein, and several molecules of three large proteins, namely P1, P2, and P3. The RNA accounts for about 2.5 percent of the virus particle mass.

Table 58-5 summarizes current information concerning the eight genes of influenza virus and the proteins encoded by them, and Figure 58-39 shows electropherograms of the proteins encoded by the genes of two influenza virus A strains. The three large proteins are present in few copy numbers only and appear to function in RNA transcription and replication: P1 and P3 initiate transcription and elongation, respectively, whereas P2 appears to be the cap-binding protein that causes short capped 5′-terminal sequences from host cell messenger RNA molecules to be used as primers to initiate the transcription of influenza virus genes into messenger RNAs (see below).

The two smallest genes are fascinating. Each codes not for one protein, but for two—each encodes a major protein in about three quarters of its sequences, and a smaller one in about one third. These smaller proteins are encoded by messenger RNAs that are either spliced or the transcription of which is initiated at an internal site, and whose coding sequences not only overlap those of the messenger RNA of the major proteins, but are also read in a different reading frame. The two minor proteins, M2

TABLE 58-5. THE EIGHT INFLUENZA VIRUS GENES AND THE PROTEINS THEY ENCODE

Gene	Size of Gene		Protein	Size of Protein Mol Wt	Approx. No. of Protein Molecules per Virus Particle	Approx. Percent of Virus Particle Protein	Function of Protein
	Mol Wt	No. of Nucleotides					
1	750,000	2341	P1	96,000	16	1	Initiation of transcription
2	730,000	2300	P2	87,000	16	1	Cap-binding protein
3	715,000	2233	P3	85,000	16	1	Elongation of transcription
4	565,000	1765	HA ⟨ HA1 / HA2	36,000 / 27,000	750 / 750	25	Hemagglutinin spike. The size of the carbohydrate groups on HA1 and HA2 are 11,500 and 1400 daltons, respectively
5	500,000	1565	NP	56,000	1000	30	Structural protein of helical nucleocapsid
6	450,000	1413	NA	50,000	160	4	Neuraminidase
7	330,000	1027	M1	27,000	3000	38	Matrix protein
			M2	11,000	0		Nonstructural protein of unknown function
8	285,000	890	NS1	26,000	0		Nonstructural protein of unknown function
			NS2	12,000	0		Nonstructural protein of unknown function

and NS2, are both nonstructural proteins of unknown function.

Influenza viruses are classified into three types, A, B, and C, on the basis of their antigenic properties. Protein NP and M possess antigenic determinants common to all strains of influenza virus. Type A influenza virus strains are subdivided into subtypes based on the antigenic characteristics of their surface antigens, namely, the hemagglutinin and the neuraminidase, each of which possesses four major antigenic determinants. Thirteen distinct hemagglutinin subtypes and nine neuraminidase subtypes are currently recognized. Human influenza A strains contain hemagglutinins of three subtypes (H1, H2, or H3), and neuraminidases of two subtypes (N1 and N2); the other hemagglutinin and neuraminidase subtypes are present in influenza virus strains of other animal species (horses, swine, birds, etc.). The nature of the extensive antigenic variations of influenza virus that cause the emergence of the virus strains responsible for successive pandemics is discussed in Chapter 60.

Influenza Virus Multiplication Cycle

Influenza virus probably enters cells via coated vesicles. It has recently been found that the hemagglutinin possesses cell membrane-fusing activity, and this activity may also play a role in influenza virus penetration and uptake. The liberated nucleocapsids do not remain in the cytoplasm, but migrate to the nucleus, where they are transcribed and replicate. The overall strategy of the influenza virus multiplication cycle resembles that of rhabdoviruses and paramyxoviruses, with primary transcription catalyzed by parental nucleocapsids leading to the formation of plus-stranded templates for the transcription of progeny minus

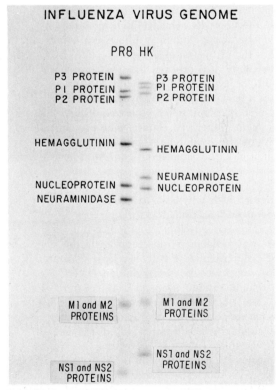

Figure 58-39. Electropherogram of a polyacrylamide gel of the eight negative-stranded genes of influenza A (strains PR8 and HK), together with the proteins that they encode. Proteins P1, P2, and P3 function in RNA replication and transcription; the nucleoprotein NP is the nucleocapsid protein; the M1 protein is associated with the internal side of the envelope membrane; the hemagglutinin (HA) and neuraminidase (NA) are the two spike glycoproteins; and MS2, NS1, and NS2 are nonstructural proteins of unknown function. *(From Ritchey et al.: J Virol 20:307, 1976.)*

strands, followed by secondary transcription of both plus and minus strands catalyzed by newly synthesized polymerase. A remarkable feature of influenza virus RNA transcription is that it is primer dependent; and the primers that are used are 10- to 15-residues long, capped oligonucleotides from the 5'-ends of host transcripts. Apparently these 5'-terminus sequences are recognized by P2, cleaved by a virus-encoded endonuclease, and then transcription of influenza virus RNA is initiated on them. As a result, all influenza transcripts possess a heterogeneous assortment of host-derived sequences at their 5'-end. These transcripts are of two types. First, there are the messenger RNAs. These are incomplete transcripts because transcription of them stops about 40-nucleotide residues before the polymerase reaches the 5'-ends of the minus-strand templates being transcribed. All messenger RNA species are polyadenylated at their 3'-termini. The second type of transcripts, formed mostly during the later stages of the multiplication cycle, are faithful plus-stranded nonpolyadenylated copies of the negative-stranded genes. They, in turn, serve as the templates for the synthesis of progeny minus strands.

The morphogenesis of influenza virus proceeds as described above for VSV, but there are some unique features. The first is that in order for influenza progeny virus to be infectious, its hemagglutinin must be cleaved three times. The hemagglutinin of one typical strain is synthesized in the cytoplasm as one polypeptide chain, 566-amino acids long. It is processed as a glycoprotein and passes through the plasma cell membrane by virtue of the presence of a 16-amino acid long hydrophobic leader sequence at its amino termini, which is cleaved off by a cellular protease on the outer cell surface. The protein remains anchored to the lipid bilayer by virtue of a hydrophobic region at its carboxyl terminus end. As in many other viral glycoproteins, such as paramyxovirus, rhabdovirus, togavirus, and coronavirus glycoproteins, the carboxyl terminus is esterified with fatty acid. At about the time when influenza virus nucleocapsids bud through what has by now become the viral envelope, the hemagglutinin protein is cleaved by a cellular trypsin-like protease, and a virus-associated carboxypeptidase B (either loosely associated or an intrinsic activity of the hemagglutinin/HA1/HA2) removes one arginine residue from the carboxyl terminus of the distal (HA1) fragment. The result is that the hemagglutinin is converted to HA1 (328 amino acids) plus HA2 (221 amino acids), which remain linked covalently by -SS-bonds. In some cells, the protease is inactive and virus with uncleaved hemagglutinin is released. Such virus is noninfectious, but can be activated by treatment in vitro with trypsin, the arginine being removed immediately thereafter by an enzyme activity intrinsic to the virus.

A second unique feature concerns the function of the influenza virus neuraminidase. It seems that the primary function of this enzyme is to remove neuraminic acid from the oligosaccharide moieties of the hemagglutinin protein. If this function is prevented (either by the addition of inhibitors or by the use of appropriate mutants), the neuraminic acid-containing hemagglutinin serves as the receptor for virus about to bud from the cell surface, with resultant inhibition of virus liberation. It is also known that hemagglutinin cannot be cleaved to HA1 plus HA2, unless neuraminic acid is first removed from its obligosaccharides.

FURTHER READING

Books and Reviews

GENERAL

Dimmock NJ: Initial stages in infection with animal viruses. J Gen Virol 59:1, 1982

Simons K, Garoff H: The budding mechanisms of enveloped animal viruses. J Gen Virol 50:1, 1980

Lenard J, Miller DK: Uncoating of enveloped viruses. Cell 28:5, 1983

Doerfler W: DNA methylation—A regulatory signal in eukaryotic gene expression. J Gen Virol 57:1, 1981

Shatkin AJ (ed): Initiation signals in viral gene expression. Curr Top Microbiol Immunol 93, 1981

Broker TR: Animal virus RNA processing. In Apirion D (ed): RNA Processing. Boca Raton, Fla, CRC Press, 1983

Challberg MD, Kelly TJ: Eukaryotic DNA replication: Viral and plasmid model systems. Ann Rev Biochem 51:901, 1982

Klenk H, Rott R: Cotranslational and posttranslational processing of viral glycoproteins. Curr Top Microbiol Immunol 90:19, 1980

Simons K, Garoff H, Helenius A: How an animal virus gets into and out of its host cell. Sci Am 246/2:58, 1982

Ward DC, Tattersall P (eds): Replication of Mammalian Parvoviruses. New York, Cold Spring Harbor Laboratory, 1978

Becht H: Infectious bursal disease virus. Curr Top Microbiol Immunol 90:107, 1980

Mahy BWJ, Barry RD (eds): Negative Strand Viruses and the Host Cell. London, Academic Press, 1978

Pedersen IR: Structural components and replication of arenaviruses. Adv Virus Res 24:277, 1979

Howard CR, Simpson DIH: The biology of the arenaviruses. J Gen Virol 51:1, 1980

Siddell S, Wege H, ter Meulen V: The biology of coronaviruses. J Gen Virol 64:761, 1983

Siddell S, Wege H, ter Meulen V: The structure and replication of coronaviruses. Curr Top Microbiol Immunol 100:131, 1982

PAPOVAVIRUSES

Cremisi C: Chromatin replication revealed by studies of animal cells and papovaviruses (simian virus 40 and polyoma virus). Microbiol Rev 43:297, 1979

Elder JT, Spritz RA, Weissman SM: Simian virus 40 as a eukaryotic cloning vehicle. Annu Rev Genet 15:295, 1981

Ziff EB: Transcription and RNA processing by the DNA tumor viruses. Nature 287:491, 1980

ADENOVIRUSES

Logan JS, Shenk T: Transcriptional and translational control of adenovirus gene expression. Microbiol Rev 46:377, 1982

Nevins JR: Adenovirus gene expression: Control at multiple steps in RNA biogenesis. Cell 28:1, 1982

Nevins JR, Chen-Kiang S: Processing of adenovirus nuclear RNA to mRNA. Adv Virus Res 26:1, 1981

Persson H, Philipson L: Regulation of adenovirus gene expression. Curr Top Microbiol Immunol 97:157, 1982

HERPESVIRUSES

Roizman B: The structure and isomerization of herpes simplex virus genomes. Cell 16:481, 1979

Ben-Porat T: Replication of herpesvirus DNA. Curr Top Microbiol Immunol 91:81, 1981

POXVIRUSES

Moss B: Poxviruses. In Nayak DP (ed): The Molecular Biology of Animal Viruses, New York, Marcel Dekker, 1978, vol 2, p 849

Holowczak JA: Poxvirus DNA. Curr Top Microbiol Immunol 97:27, 1982

REOVIRUSES

Joklik WK: Structure and function of the reovirus genome. Microbiol Rev 45:483, 1981

PICORNAVIRUSES

Crowell RL, Landau DJ: Receptors in the initiation of picornavirus infections. In Fraenkel-Conrat H, Wagner RR (eds): Comprehensive Virology. New York, Plenum, vol 18, p 1

Sangar DV: The replication of picornaviruses. J Gen Virol 45:1, 1979

Putnak RJ, Phillips BA: Picornaviral structure and assembly. Microbiol Rev 45:287, 1981

MacNaughton MR: The structure and replication of rhinoviruses. Curr Top Microbiol Immunol 97:1, 1982

TOGAVIRUSES

Kaariainen L, Soderlund H: Structure and replication of alphaviruses. Curr Top Microbiol Immunol 82:15, 1978

Garoff H, Kondor-Koch C, Riedel H: Structure and assembly of alphaviruses. Curr Top Microbiol Immunol 100:1, 1982

Schlesinger RW (ed): The Togaviruses: Biology, Structure, and Replication. New York, Academic Press, 1980

BUNYAVIRUSES

Bishop DHL: Genetic potential of bunyaviruses. Curr Top Microbiol Immunol 86:1, 1979

RHABDOVIRUSES

Bishop DHL (ed): The Rhabdoviruses. Boca Raton, Fla, CRC Press, 1979

PARAMYXOVIRUSES

Morgan EM, Rapp F: Measles virus and its associated diseases. Bacteriol Rev 41:636, 1977

Wechler SL, Meissner HC: Measles and SSPE viruses: Similarities and differences. Prog Med Virol 28:65, 1982

MYXOVIRUSES

Choppin PW, Richardson CD, Merz DC, et al.: The functions and inhibition of the membrane glycoproteins of paramyxoviruses and myxoviruses and the role of the measles virus M protein in subacute sclerosing parencephalitis. J Infect Dis 143:352, 1981

Hinshaw VS, Webster RG, Rodriguez RJ: Influenza A viruses: Combinations of hemagglutinin and neuraminidase subtypes isolated from animals and other sources. Arch Virol 67:191, 1981

Lamb RA, Choppin PW: The gene structure and replication of influenza viruses. Ann Rev Biochem 52:467, 1983

Skehel JJ, Hay AJ: Influenza virus transcription. J Gen Virol 39:1, 1978

Ward CW: Structure of the influenza virus hemagglutinin. Curr Top Microbiol Immunol 94/95:1, 1981

Selected papers

PAPOVAVIRUSES

Reddy VB, Thimmappaya B, Dhar R, et al.: The genome of Simian virus 40. Science 200:494, 1978

Wasylyk B, Wasylyk C, Augereau P, et al.: The SV40 72bp repeat preferentially potentiates transcription starting from proximal natural or substitute promoter elements. Cell 32:503, 1983

Laimonis LA, Khoury G, Gorman C, et al.: Host specific activation of transcription by tandem repeats from simian virus 40 and Moloney sarcoma virus. Proc Natl Acad Sci USA 79:6453, 1982

Frisque RJ: Nucleotide sequence of the region encompassing the JCV origin of DNA replication. J Virol 46:170, 1983

ADENOVIRUSES

Chow LT, Gelinas RE, Broker TR, et al.: An amazing sequence arrangement at the 5′ ends of adenovirus 2 messenger RNAs. Cell 12:1, 1977

Chow LT, Lewis JE, Broker TR: RNA transcription and splicing at early and intermediate times after adenovirus-2 infection. Cold Spring Harbor Symp Quant Biol 44:401, 1980

Nevins JR, Darnell JE Jr: Steps in the processing of ad2 mRNA: poly (A)+ nuclear sequences are conserved and poly (A) addition precedes splicing. Cell 15:1477, 1978

Wilson MC, Fraser NW, Darnell JE: Mapping of RNA initiation sites by high doses of UV irradiation: Evidence for three independent promoters within the left 11% of the Ad-2 genome. Virology 94:175, 1979

Shaw AR, Ziff EB: Transcripts from the adenovirus-2 major late promoter yield a single early family of 3′ coterminal mRNAs and five late families. Cell 22:905, 1980

Stillman BW, Lewis JB, Chow LT, et al.: Identification of the gene and mRNA for the adenovirus terminal protein precursor. Cell 23:497, 1981

Friefeld BR, Lichy JH, Hurwitz J, et al.: Evidence for an altered adenovirus DNA polymerase in cells infected with the mutant H5ts149. Proc Natl Acad Sci USA 80:1589, 1983

Challberg MD, Desiderio SV, Kelly TJ Jr: Adenovirus DNA replication in vitro: Characterization of a protein covalently linked to nascent DNA strands. Proc Natl Acad Sci USA 77:5105, 1980

Cepko CL, Sharp PA: Assembly of adenovirus major capsid protein is mediated by a nonvirion protein. Cell 31:407, 1982

HERPESVIRUSES

Parris DS, Dixon RAF, Schaffer PA: Physical mapping of herpes simplex virus type 1 ts mutants by marker rescue: Correlation of the physical and genetic maps. Virology 100:275, 1980

DeLuca N, Zik DJ, Bond VC, et al.: Nucleotide sequences of herpes simplex virus 1 (HSV-1) affecting virus entry, cell fusion, and production of glycoprotein gB (VP7). Virology 122:411, 1982

Honess RW, Roizman B: Regulation of herpesvirus macromolecular synthesis: Sequential transition of polypeptide synthesis requires functional viral polypeptides. Proc Natl Acad Sci USA 72:1276, 1975

Mackem S, Roizman B: Differentiation between α promoter and regulator regions of herpes simplex virus 1: The functional domains and sequence of a movable α regulator. Proc Natl Acad Sci USA 79:4917, 1982

Poffenberger KL, Tabares E, Roizman B: Characterization of a viable, noninverting herpes simplex virus 1 genome derived by insertion and deletion of sequences at the junction of components L and S. Proc Natl Acad Sci USA 80:2690, 1983

Post LE, Norrild B, Simpson T, et al.: Chicken ovalbumin gene fused to a herpes simplex virus α promoter and linked to a thymidine kinase gene is regulated like a viral gene. Mol Cell Biol 2:233, 1982

Lee GT-Y, Para MF, Spear PG: Location of the structural genes for glycoproteins gD and gE and for other polypeptides in the S component of herpes simplex virus type 1 DNA. J Virol 43:41, 1982

Stringer JR, Holland LE, Wagner EK: Mapping early transcripts of herpes simplex virus type 1 by electron microscopy. J Virol 27:56, 1978

Mocarski ES, Roizman B: Structure and role of the herpes simplex virus DNA termini in inversion, circularization and generation of virion DNA. Cell 31:89, 1982

POXVIRUSES

Ichihashi Y, Oie M: Proteolytic activation of vaccinia virus for the penetrationphase of infection. Virology 116:297, 1982

Oda K, Joklik WK: Hybridization and sedimentation studies on "early" and "late" vaccinia messenger RNA. J Mol Biol 27:395, 1967

Cooper JA, Moss B: In vitro translation of immediate early, early, and late classes of RNA from vaccinia virus-infected cells. Virology 96:368, 1979

Isle HB, Venkatesan S, Moss B: Cell-free translation of early and late mRNAs selected by hybridization to cloned DNA fragments derived from the left 14 million to 72 million daltons of the vaccinia virus genome. Virology 112:306, 1981

Weir JP, Moss B: Nucleotide sequence of the vaccinia virus thymidine kinase gene and the nature of spontaneous frameshift mutations. J Virol 46:530, 1983

Hiller G, Weber K, Schneider L, et al.: Interaction of assembled progeny pox viruses with the cellular cytoskeleton. Virology 98:142, 1979

Krempien U, Schneider L, Hiller G, et al.: Conditions for pox virus-specific microvilli formation studied during synchronized virus assembly. Virology 113:556, 1981

REOVIRUSES

Schonberg M, Silverstein SC, Levin DH, et al.: Asynchronous synthesis of the complementary strands of the reovirus genome. Proc Natl Acad Sci USA 68:505, 1971

Silverstein SC, Schonberg M, Levin DH, et al.: The reovirus replicative cycle: Conservation of parental RNA and protein. Proc Natl Acad Sci USA 67:275, 1970

Zweerink HJ, McDowell MJ, Joklik WK: Essential and nonessential noncapsid reovirus proteins. Virology 45:716, 1971

McCrae MA, Joklik WK: The nature of the polypeptide encoded by each of the ten double-stranded RNA segments of reovirus type 3. Virology 89:578, 1978

Huismans H, Joklik WK: Reovirus-coded polypeptides in infected cells: Isolation of two native monomeric polypeptides with affinity for single-stranded and double-stranded RNA, respectively. Virology 70:411, 1976

Antczak JB, Chmelo R, Pickup DJ, et al.: Sequences at both termini of the 10 genes of reovirus serotype 3 (strain Dearing). Virology 121:307, 1982

Cashdollar LW, Esparza J, Hudson GR, et al.: Cloning the double-stranded RNA genes of reovirus: Sequence of the cloned S2 gene. Proc Natl Acad Sci USA 79:7644, 1982

Lee PWK, Hayes EC, Joklik WK: Characterization of anti-reovirus immunoglobulins secreted by cloned hybridoma cell lines. Virology 108:134, 1981

PICORNAVIRUSES

Nomoto A, Omata T, Toyoda H, et al.: Complete nucleotide sequence of the attenuated poliovirus Sabin 1 strain genome. Proc Natl Acad Sci USA 79:5793, 1982

Adler CJ, Elzinga M, Wimmer E: The genome-linked protein of picornaviruses. VIII. Complete amino acid sequence of poliovirus VPg and carboxy-terminal analysis of its precursor, P3-9. J Gen Virol 64:349, 1983

Summers DF, Maizel JV, Darnell JE: Evidence for virus-specific noncapsid proteins in poliovirus-infected HeLa cells. Proc Natl Acad Sci USA 56:505, 1965

Larsen GR, Anderson CW, Dorner AJ, et al.: Cleavage sites within the poliovirus capsid protein precursors. J Virol 41:340, 1982

Korant B, Chow N, Lively M, et al.: Virus specified protease in poliovirus infected HeLa cells. Proc Natl Acad Sci USA 76:2992, 1979

Palmenberg AC, Rueckert RR: Evidence for intramolecular self-cleavage of picornaviral replicase precursors. J Virol 41:244, 1982

Dasgupta A, Baron MH, Baltimore D: Poliovirus replicase: A soluble enzyme able to initiate copying of poliovirus RNA. Proc Natl Acad Sci USA 76:2679, 1979

TOGAVIRUSES

Garoff H, Frischauf A-M, Simons K, et al.: Nucleotide sequence of cDNA coding for Semliki Forest virus membrane glycoproteins. Nature 288:236, 1980

Rice CM, Strauss JH: Nucleotide sequence of the 26 S mRNA of Sindbis virus and deduced sequence of the encoded virus structural proteins. Proc Natl Acad Sci USA 78:2062, 1981

Fuller FJ, Marcus PI: Sindbis virus. 1. Gene order of translation in vivo. Virology 107:441, 1980

Huggins JW, Jahrling PB, Rill W, Linden CD: Characterization of the binding of the TC-83 strain of Venezuelan equine encephalitis virus to BW-J-M, a mouse macrophage-like cell line. J Gen Virol 64:149, 1983

RHABDOVIRUSES

Schlegel R, Tralka TS, Willingham ME, et al.: Inhibition of VSV binding and infectivity by phosphatidylserine: Is phosphatidylserine a VSV-binding site? Cell 32:639, 1983

Ball LA, White CN: Order of transcription of genes of vesicular stomatitis virus. Proc Natl Acad Sci USA 73:442, 1976

Emerson SU: Reconstitution studies detect a single polymerase entry site on the vesicular stomatitis virus genome. Cell 31:635, 1982

Herman RC, Schubert M, Keene JD, et al.: Polycistronic vesicular stomatitis virus RNA transcripts. Proc Natl Acad Sci USA 77:4662, 1980

Schubert M, Keene JD, Herman RC, et al.: Site on the vesicular stomatitis virus genome specifying polyadenylation and the end of the L gene mRNA. J Virol 34:550, 1980

Keene JD, Schubert M, Lazzarini RA: Intervening sequence between the leader region and the nucleocapsid gene of vesicular stomatitis virus RNA. J Virol 33:789, 1980

Villarreal LP, Breindl M, Holland JJ: Determination of molar ratios of vesicular stomatitis virus-induced RNA species in BHK$_{21}$ cells. Biochemistry 15:1663, 1976

Toneguzzo F, Ghosh HP: In vitro synthesis of vesicular stomatitis virus membrane glycoprotein and insertion into membranes. Proc Natl Acad Sci USA 75:715, 1978

Jacobs BL, Penhoet EE: Assembly of vesicular stomatitis virus: Distribution of the glycoprotein on the surface of infected cells. J Virol 44:1047, 1982

Rose JK, Welch WJ, Sefton BM, et al.: Vesicular stomatitis virus glycoprotein is anchored in the viral membrane by a hydrophobic domain near the COOH terminus. Proc Natl Acad Sci USA 77:3884, 1980

Schlesinger MJ, Malfer C: Cerulenin blocks fatty acid acylation of glycoproteins and inhibits vesicular stomatitis and Sindbis virus particle formation. J Biol Chem 257:9887, 1982

Birdwell CR, Strauss JH: Maturation of vesicular stomatitis virus: Electron microscopy of surface replicas of infected cells. Virology 59:587, 1974

Zee YC, Hackett AJ, Talens L: Vesicular stomatitis virus maturation sites in six different host cells. J Gen Virol 7:95, 1970

PARAMYXOVIRUSES

Markwell MA, Svennerholm L, Paulson JC: Specific gangliosides function as host cell receptors for Sendai virus. Proc Natl Acad Sci USA 78:5406, 1981

Peeples M, Levine S: Respiratory syncytial virus polypeptides: Their location in the virion. Virology 95:137, 1979

Collins PL, Hightower LE, Ball LA: Transcriptional map for Newcastle disease virus. J Virol 35:682, 1980

Glazier K, Raghow R, Kingsbury DW: Regulation of Sendai virus transcription: Evidence for a single promoter in vivo. J Virol 21:863, 1977

Huang YT, Wertz GW: Respiratory syncytial virus mRNA coding assignments. J Virol 46:667, 1983

Stallcup KC, Raine CS, Fields BN: Cytochalasin B inhibits maturation of measles virus. Virology 124:59, 1983

MYXOVIRUSES

Jones KL, Huddlestone JA, Brownlee GG: The sequence of RNA segment 1 of influenza virus A/NT/60/68 and its comparison with the corresponding segment of strains A/PR/8/34 and A/WSN/33. Nucleic Acids Res 11:1555, 1983

Mowshowitz SL: P1 is required for initiation of cRNA synthesis in WSN influenza virus. Virology 91:493, 1978

Blaas D, Patzelt E, Kuechler E: Cap-recognizing protein of influenza virus. Virology 116:339, 1982

Kaptein N, Nayak DP: Complete nucleotide sequence of the polymerase 3 gene of human influenza virus A/WSN/33. J Virol 42:55, 1982

Palese P, Ritchey MB, Schulman JL: P1 and P3 proteins of influenza virus are required for complementary RNA synthesis. J Virol 21:1187, 1977

Allen H, McCauley J, Waterfield M, et al.: Influenza virus RNA segment 7 has the coding capacity for two polypeptides. Virology 107:548, 1980

Briedis DJ, Lamb RA, Choppin, PW: Influenza B virus RNA segment 8 codes for two nonstructural proteins. Virology 112:417, 1981

Bouloy M, Plotch SJ, Krug RM: Globin mRNAs are primers for the transcription of influenza viral RNA in vitro. Proc Natl Acad Sci USA 75:4886, 1978

Garten W, Bosch FX, Linder D, et al.: Proteolytic activation of the influenza virus hemagglutinin: The structure of the cleavage site and the enzymes involved in cleavage. Virology 115:361, 1981

Varghese JN, Laver WG, Colman PM: Structure of the influenza virus glycoprotein antigen neuraminidase at 2.9 Å resolution. Nature 303:35, 1983

CORONAVIRUSES

Lai MMC, Brayton PR, Armen RC, et al.: Mouse hepatitis virus A59: Messenger RNA structure and genetic localization of the sequence divergence from hepatotropic strain MHV-3. J Virol 39:823, 1981

Siddell S: Coronavirus JHM: Coding assignments of subgenomic mRNAs. J Gen Virol 64:113, 1983

GENERAL

Darnell JE Jr: Implications of RNA : RNA splicing in evolution of eukaryotic cells. Science 202:1257, 1978

Fan DP, Sefton BM: The entry into host cells of Sindbis virus, vesicular stomatitis virus and Sendai virus. Cell 15:985, 1978

Schmidt MFG: Acylation of viral spike glycoproteins: A feature of enveloped RNA viruses. Virology 116:327, 1982

CHAPTER 59

The Effect of Virus Infection on the Host Cell

Cytopathic Effects

Inhibition of Host Macromolecular Biosynthesis

Changes in the Regulation of Gene Expression

Appearance of New Antigenic Determinants
on the Cell Surface

Cell Fusion

Abortive Infection

Persistent Infections

Latency

Modification of Cellular Permissiveness

The effect of virus infection on host cells is far more difficult to study in molecular terms than is the process of virus multiplication. Study of the multiplication process
requires merely the ability to recognize and measure virus-
specified macromolecules; study of the effect on host cells
requires a detailed knowledge of the functioning of the
normal host cell. In fact, studies that have focused attention on the effect of virus infection on host cells have
broadened our knowledge of the functioning of uninfected
cells, and the acquisition of such knowledge has been one
of the important spinoffs of the study of virus–cell interactions.

Whatever the reason that lytic viruses destroy their
host cells, several causes can be ruled out. One is that virus synthesis creates an excessive demand for protein and
nucleic acid precursors, so that competition causes a short-
age of building blocks that prevents synthesis of host cell
macromolecules. This is unlikely, since the amount of viral
material that is synthesized rarely exceeds 10 percent of
total host cell material. Considerations of this nature also
rule out the necessity for breakdown of host cell material
in order to provide precursors for the synthesis of viral
macromolecules.

Since both nucleic acid and protein synthesis are absolutely dependent on the supply of energy, the largest virus yields would be expected in cells that are damaged
least for the longest periods of time. Many highly lytic
viruses are, however, very successful. This is primarily because they take over the host cell's synthetic apparatus
very rapidly and multiply extensively in the brief period of
time for which the apparatus can function.

Cytopathic Effects

The most easily detected effects of infection with lytic
viruses are the cytopathic effects, which can be observed
both macroscopically and microscopically. Plaque formation is due to the cytopathic effect of viruses; viruses kill
the cells in which they multiply, and plaques are the areas
of killed cells. The light microscope, as well as the electron microscope, often reveals changes in a variety of cell
organelles soon after infection. Frequently the nucleus is
affected first, with pyknosis, changes in nucleolar structure, and margination of the chromatin. Changes in the
cell membrane usually follow; cells gradually lose their
ability to adhere to supporting surfaces and therefore
round up and sometimes develop a strong tendency to
fuse with one another (see below). This is then often
followed by the appearance, either in the nucleus or in the
cytoplasm, of distinct spreading foci that are generally
composed of fibrillar material. These are the classic inclusion bodies that have long been described by cytologists,
and they represent the sites of virus-directed biosynthesis

and morphogenesis. Finally, at about the time when structural viral protein synthesis proceeds at its maximum rate, necrotic and grossly degradative changes become noticeable. They may be attributed to at least three causes. First, by this time interference with host cell macromolecular biosynthesis is generally complete (see below). It is known that all host cell macromolecules turn over to a greater or lesser extent—that is, they are continually broken down and resynthesized by a mechanism that operates no matter whether the cells grow or not. Inhibition of resynthesis in the presence of continuing breakdown could clearly lead to structural and functional failure. Second, plasma membrane function declines, probably because host protein and lipid synthesis ceases and because patches of host membrane protein are replaced by virus-coded ones. Certainly, permeability increases soon after infection, and before long loss of plasma membrane function results in failure to maintain the proper intracellular ionic environment and in diminished transport of essential nutrients into the cell and of waste products out of it. Third, the membranes lining lysosomes also begin to fail, and as a result the degradative hydrolytic enzymes that they contain begin to leak out into the cytoplasm, thereby exacerbating the effects caused by the other mechanisms.

The net result of cell necrosis is the release of those viruses that do not bud from the cell membrane. In general, the smaller the virus, the more readily it is released. Large viruses, such as poxviruses, are often retained in the ghosts of infected cells for considerable periods of time. In the body, the situation may be different, for damaged cells may become phagocytized, thereby providing an additional mechanism for the dissemination of viral progeny.

Whereas the reasons for the necrotic and degradative changes that occur during the late phases of the virus multiplication cycle are reasonably well understood, it is not as obvious why some viruses cause cytopathic effects soon after infection. For some viruses, it is likely that early cytopathic effects are caused by early viral proteins—that is, proteins that are synthesized after infection; for others, it seems that such effects are caused by components of the invading virus particles. It is interesting in this regard that the adenovirus fiber is known to be cytotoxic, that the surface tubule protein of vaccinia virus inhibits protein synthesis, and that the vesicular stomatis virus (VSV) glycoprotein inhibits cellular DNA and RNA synthesis. On the other hand, it is also known that empty reovirus particles—that is, particles that contain no RNA—cause no cytopathic effects, even at very high multiplicities.

Inhibition of Host Macromolecular Biosynthesis

The induction of cytopathic effects early after infection may be intimately related to the more or less rapidly developing inhibition of host protein, DNA and RNA synthe-

sis, which is a fundamentally important factor in the lytic virus–cell interaction.

Host protein synthesis is inhibited first. This effect is often missed if only the overall rate of protein synthesis is measured, since viral protein synthesis generally takes over as host protein synthesis declines. Inhibition of host protein synthesis then quickly causes inhibition of host DNA replication, which is known to depend on the activity of several short-lived proteins. Host RNA synthesis also soon ceases, probably not as a result of any direct effect on DNA-dependent RNA polymerases but because of changes in the physical state of the DNA (margination of chromatin, changes in nucleolar structure—see above) that prevent it from acting as a template for transcription. As a result, not only is progressively less and less host cell messenger RNA synthesized, but also the supply of new ribosomes is quickly interrupted. In fact, the only ribosomes that are usually available for protein synthesis during the virus multiplication cycle are those that are present in the cytoplasm at the time of infection.

The reasons why infection with viruses inhibits host protein synthesis have been studied extensively but are not well understood. Sometimes the reason may be that less and less host cell messenger RNA reaches the cytoplasm. However, it is clear that in many virus infections the rate of host protein synthesis decreases far more rapidly than would be expected on the basis of the decay rate of messenger RNA in uninfected cells; for example, infection with high multiplicities of vaccinia virus inhibits host protein synthesis by well over 90 percent within 2 hours (Fig. 59-1). This is not a general toxic effect on the host cell, since viral proteins are already being synthesized vigorously at this time; nor is inhibition of host protein synthesis usually due to destruction of host cell messenger RNA, since the continued existence of undiminished amounts of fully functional messenger RNA can often be demonstrat-

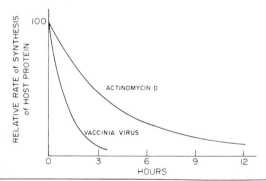

Figure 59-1. Effect of infection with vaccinia virus, as well as of actinomycin D, on cellular protein synthesis in mouse L fibroblasts. Actinomycin D inhibits RNA transcription; the decrease in the rate of host protein synthesis in its presence, therefore, reflects the stability of cellular messenger RNA, whose average half-life is about 3 hours. Infection with vaccinia virus inhibits host protein synthesis much more rapidly. Presumably the infection either causes host messenger RNA to be inactivated, or it actively prevents host messenger RNA from being translated.

ed by its ability to be translated in in vitro protein-synthesizing systems prepared from uninfected cells. Rather, shortly after infection a mechanism apparently develops that actively prevents from being translated the cellular messenger RNA that is present in the cytoplasm at the time of infection.

There is probably not one but several such mechanisms, depending on the nature of the virus. As pointed out above, the plasma cell membrane is often damaged soon after infection, and as a result the Na^+–K^+ gradient collapses, which causes the Na^+ concentration to rise and the K^+ concentration to fall. There is evidence that these changes inhibit the translation of host cell messenger RNAs more than that of viral messenger RNAs. Further, soon after infection with some viruses such as poliovirus, cap binding protein, essential for the translation of capped messenger RNAs such as most cellular messenger RNAs, is inactivated, while translation of uncapped poliovirus RNA is of course not affected. However, it is known that not *all* uncapped messenger RNAs can be translated in poliovirus-infected cells. This finding is in line with earlier observations that suggested that viral messenger RNAs may, in general, be more efficient messenger RNAs than host cell messenger RNAs; thus, in vitro protein-synthesizing systems presented with mixtures of host and viral messenger RNAs usually translate the latter preferentially. The most important reason that viral messenger RNAs are translated almost exclusively once the late period has begun is simply that by that time viral messenger RNA molecules greatly outnumber host cell messenger RNA molecules; by that time transcription of host messenger RNAs is greatly reduced for the reasons discussed above, and large amounts of viral messenger RNAs are being transcribed. Even if virus messenger RNAs were not translated preferentially, simple competition would cause host protein synthesis to be almost completely shut down.

While the hypothesis that inhibition of host DNA and RNA synthesis is secondary to inhibition of host protein synthesis is probably correct for most virus infections, several exceptions have been reported. Thus VSV and Sindbis virus rapidly shut off host RNA and DNA synthesis, respectively, by mechanisms that do not appear to involve protein-synthesis inhibition; a component of parental poxvirus particles has been reported to inhibit host DNA synthesis; and the isolated G glycoprotein of VSV inhibits cell DNA, but not protein synthesis.

Changes in the Regulation of Gene Expression

Virus infection may also affect the regulation of host genome expression. Thus, the activity of enzymes on the pathway of nucleic acid biosynthesis often increases after infection; for example, infection with the papovaviruses polyoma and SV40 causes increases in the activity of at least six enzymes. One of these enzymes is a deoxypyrimidine kinase that phosphorylates both deoxyuridine and deoxycytidine. Cells possess two forms of this enzyme, one of which is synthesized in resting cells, the other in growing cells. When resting cells are infected with SV40 or adenovirus, it is the latter form of the enzyme whose formation is induced. Further, infection with almost all viruses leads to the synthesis of a new protein, interferon (Chap. 61). All this evidence suggests that infection may upset the mechanisms that usually prevent the expression of a large portion of cellular DNA and that during the early part of the infection cycle more of the cell's genome may be expressed in infected than in uninfected cells.

Appearance of New Antigenic Determinants on the Cell Surface

Sooner or later following infection, the outer cell membrane is modified. This manifests itself in a variety of ways: the cells' morphology changes, they become more agglutinable by the lectin concanavalin A (Chap. 62), permeability increases, and new antigenic determinants appear on the cell surface. When the infecting virus is an enveloped virus, these new determinants are likely to be viral envelope proteins that have become incorporated into the cell membrane, but new antigenic determinants also appear on the surfaces of cells that are infected with nonenveloped viruses. The presence of these determinants serves to alert the immune mechanism to the fact that the cells are infected and should be eliminated (Fig. 59-2). These new antigens are discussed further in Chapter 62 in relation to tumor and transplantation antigens.

Cell Fusion

Several viruses, particularly paramyxoviruses such as Sendai virus and herpesviruses, cause cells to fuse with one another, which results in the formation of giant syncytia, masses of cytoplasm bounded by one membrane that may contain hundreds and even thousands of nuclei. Fusion seems to be caused by changes induced in cell membranes as a result of interaction with certain glycoproteins in viral envelopes (glycoproteins gB and gD in herpesvirus envelopes and the F spike glycoprotein of paramyxoviruses). It is therefore caused not only by active but also by inactivated virus particles. Further, it can be induced not only among identical but also among different cells. The products of fusion are either heterokaryons—cells containing several nuclei of different types (Fig. 59-3)—or hybrid cells that contain the fused nuclei of the parents. Hybrid cells are frequently viable and have become widely used

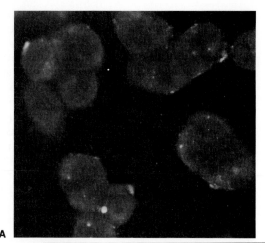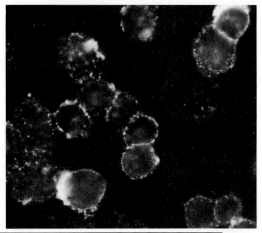

Figure 59-2. Virus-specified surface antigen on HeLa cells infected with vaccinia virus. **A.** Uninfected cells. **B.** Cells infected for 8 hours. The cells were stained with rabbit antiserum to rabbit cells infected with vaccinia virus. *(Courtesy of Dr. Yoskiaki Ueda.)*

for studies in somatic cell genetics. They are produced by fusing mixtures of the two cell lines to be hybridized with inactivated Sendai virus or, more commonly nowadays, by treatment with polyethylene glycol, and are then cloned. They are particularly useful for determining the chromosomal location of specific genes; as hybrid cells multiply, they often lose chromosomes, and one can readily correlate the loss of a particular gene function with loss of a specific chromosome. In recent years, many human genes have been assigned to chromosomes in this manner, using human–mouse hybrid cells that lose human chromosomes. The genetic factors that control susceptibility to virus infection and the expression of tumorigenicity can also be studied with the aid of somatic cell hybrids. The technique has also become very important in immunology for the production of hybridomas that produce individual species of antibodies (monoclonal antibodies) (Chap. 12).

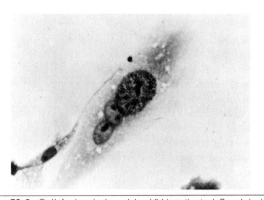

Figure 59-3. Cell fusion induced by UV-inactivated Sendai virus. Three chick embryo fibroblasts have fused to yield the heterokaryon shown here, which contains three nuclei. The two small ones are normal chick nuclei; the large one is from a chick cell transformed with Rous sarcoma virus. × 400. *(From Svoboda and Dourmashkin: J Gen Virol 4:523, 1969.)*

Abortive Infection

Most of the changes described so far relate to the lytic multiplication cycle under conditions of productive infection, that is, when a virus multiplies to high titer in permissive cells. However, viruses can also infect cells that are not fully permissive and even cells that are nonpermissive. In such cells, viruses cannot multiply because some essential step of the multiplication cycle cannot proceed. Examples of this type are the abortive infection of HeLa cells by influenza virus, of dog kidney cells by herpes simplex virus, of pig kidney cells by certain mutants of rabbitpox virus, of monkey cells by human adenovirus, and many others. The infection of permissive cells in the presence of antiviral agents (Chap. 61) is also abortive. In almost all such cases the viral genome begins to express itself, the alterations in the host cell that were described above occur, and the cells die.

Persistent Infections

As a rule, the infection of permissive cells with lytic viruses leads to productive infection and cell death. Occasionally, however, cell cultures are observed that, while multiplying more or less normally, nevertheless release significant amounts of virus. Such cultures are said to be persistently infected. Stable relationships between cell growth and virus multiplication, which occur not only in vitro but also in vivo, are of two kinds.

The first involves infection by viruses that cause a minimum of the type of cell damage discussed above. A good example is provided by the paramyxovirus SV5. This virus interacts with many cells by means of a lytic, cytocidal interaction. However, when it infects monkey kidney cells, it causes practically no cell damage and permits the infected cells to grow freely while multiplying itself. In such cultures, all cells multiply, all cells are infected, and all cells produce virus. Infection of new cells plays no role in this situation; consequently, this type of infection cannot be cured by the addition of neutralizing antibody.

The second situation is quite different. Here the virus–cell interaction is lytic—that is, the cells that are infected always die—but the extent of virus multiplication is limited, so that the yield is small. In addition, various factors reduce the probability of reinfection, so that the proportion of infected cells in the total cell population is kept small and constant. The factors that induce and maintain this type of persistent infection are as follows: First, this type of infection occurs most readily in cells that are almost, but not quite, nonpermissive and that, therefore, produce only small amounts of virus. Second, it occurs in situations where factors in the medium, such as antibody or interferon, prevent the majority of released progeny virus from infecting new cells. The role of small amounts of interferon is sometimes not readily apparent but becomes obvious if antibody to interferon is added, when persistent infections become lytic infections. Third, many of the virus strains that have been isolated from persistently infected cells and animals are less virulent, less cytopathic, and, interestingly enough, temperature-sensitive with respect to their ability to multiply. This has now been found for foot-and-mouth disease virus, Coxsackie virus, Sindbis virus and WEE, influenza virus, Newcastle disease virus, Sendai virus, mumps and measles virus, VSV and herpesvirus, and others. Attempts to isolate viruses from the tissues of patients with suspected chronic and persistent virus infections should take account of the possibility that they may grow better at 31C to 33C than at 37C.

The fact that virus strains recovered from persistently infected cells are frequently temperature-sensitive has aroused considerable interest, and, indeed, persistent virus infection can be established readily by infecting cells with temperature-sensitive mutants. However, the important fact is not that virus strains present in persistently infected cells are temperature-sensitive but rather that they are not cytopathogenic. Although temperature-sensitive virus strains are usually much less cytopathogenic than wild-type virus, presumably because much less viral protein causing cytopathogenicity is synthesized in cells infected with them (Chap. 60), virus strains do not have to be temperature-sensitive to cause persistent infection. For example, a strain of VSV, isolated from persistently infected cells, is not temperature-sensitive (except above 40C) but transcribes much less of its minus strand into messenger RNAs relative to complete plus strands than does wild-type virus (Chap. 58). This, in turn, results in less viral protein being synthesized and the virus strain being relatively noncytopathic.

Finally, virus yields from persistently infected cells sometimes contain, in addition to infectious virus, deletion mutants, that is, virus particles that lack a portion of their genome (Chap. 60). These virus particles, which are unable to multiply on their own, are nevertheless capable of interfering with the multiplication of infectious virus and are, therefore, known as defective interfering or DI particles. Their presence dampens the effect of virus infection, that it, it reduces both cell damage and the amount of infectious virus that is produced. Their role in initiating and maintaining persistent infection is readily demonstrated as follows: if DI particles of VSV, the so-called T particles (Chap. 60), are inoculated into animals together with normal, infectious VSV, an otherwise fatal disease is converted to a slowly progressing persistent infection; and if cultured cells are infected with wild-type virus together with DI particles, persistent infection is readily established. The formation of DI particles has now been implicated in persistent and chronic infections with influenza virus, Sendai virus, NDV and measles, lymphocytic choriomeningitis virus, Sindbis virus, Semliki Forest virus, WEE, and rubella, as well as VSV. The nature of the deletions in the DI particles of various viruses and the nature of the mechanisms that cause such particles to inhibit the multiplication of wild-type virus is discussed in Chapter 60.

In summary, the replacement of virulent by less virulent virus strains and the inhibition of virus multiplication by DI particles are major factors in the establishment and maintenance of persistent virus infections. Persistent infections of this class are important in the intact organism, where the presence of antiviral antibody and interferon in amounts too low to eliminate virus may provide conditions favoring low-level persistent infections. It should be noted that this kind of persistent infection can be cured by the addition of large amounts of neutralizing antibody.

Latency

We have so far considered only persistently infected cell populations that release significant amounts of virus. However, there are also persistently infected cells that release virus only rarely. In such cells virus multiplication can start but is arrested at some stage or other. The best known infections of this type are caused by herpesviruses. Most herpesviruses have a pronounced tendency to infect cells in which their development is normally blocked but from which they can be released in infectious form if conditions are right. The released virus particles can then initiate productive infection in some other type of cell. This is the phenomenon of latency, the best known example of which is the latent herpes simplex virus type 1 infection of the neurons of sensory ganglia, from which the virus emerges to cause infections of the skin, such as fever blisters (Chap. 67). The form in which herpesvirus persists in ganglion cells is not known. An interesting recent observation that may explain why viral DNA is transcribed minimally, if at all, in latent infections is that latent herpesvirus DNA is heavily hypermethylated, and hypermethylation is known to reduce gene expression very significantly.

This type of persistent infection, whether by herpes simplex virus or varicella-zoster virus, generally does not cause progressive disease; that is, the virus multiplies for some time, then becomes latent, and then reinitiates productive infection. Other persistent infections of this type tend to cause progressive disease. Examples are subacute sclerosing panencephalitis (SSPE), an infection of the central nervous system by a virus closely related to measles virus, in which large amounts of viral nucleocapsids are present in infected cells (Chap. 58); and progressive multifocal leukoencephalopathy (PML), a papovavirus infection (Chap. 69).

Modification of Cellular Permissiveness

Occasionally, virus infection can alter to an amazing degree the permissiveness of cells for completely unrelated viruses. The effects are always highly specific. Thus, adeno-associated virus (AAV) can multiply only in cells infected with adenovirus; SV40 enables human adenovirus to grow in monkey cells; and poxviruses, such as vaccinia and fibroma virus, enable VSV to multiply in rabbit cells. The helper functions appear always to involve translation of messenger RNA, that is, protein synthesis. For example, in rabbit cornea cells infected with VSV, viral messenger RNA is made and forms polyribosomes, but no VSV proteins are synthesized; VSV protein synthesis is inhibited at the stage of elongation. If such cells are also infected with vaccinia virus, VSV multiplies normally. Apparently vaccinia virus modifies the protein-synthesizing mechanism of these cells in such a way that VSV messenger RNA can be translated normally. The ability of SV40 to correct the inability of human adenoviruses to multiply in monkey cells appears to have a similar basis. However, other mechanisms may also be involved. For example, the reason why VSV cannot multiply in rabbit kidney cells appears to be the fact that it is not adsorbed, that is, an absence of VSV receptors. If such cells are first infected with Shope fibroma virus, the plasma membrane is modified, VSV adsorbs normally and, as a result, is able to replicate.

FURTHER READING

Books and Reviews

GENERAL

Ehrenfeld E: Poliovirus-induced inhibition of host-cell protein synthesis. Cell 28:435, 1982

Holland JJ, Kennedy SIT, Semler BL, et al.: Defective interfering RNA viruses and the host-cell response. In Fraenkel-Conrat H, Wagner RR (eds): Comprehensive Virology. New York, Plenum, 1980, vol 16, p 137

Maltzman W, Levine AJ: Viruses as probes for development and differentiation. Adv Virus Res 26:65, 1981

PATHOGENESIS

Ennis A: Some newly recognized aspects of resistance against and recovery from influenza. Arch Virol 73:207, 1982

Fields BN: Molecular basis of reovirus virulence. Arch Virol 71:95, 1982

Johnson RT (ed): Viral Infections of the Nervous System. New York, Raven, 1982

Klein RJ: The pathogenesis of acute, latent and recurrent herpes simplex virus infections. Arch Virol 72:143, 1982

Rott R: Molecular basis of infectivity and pathogenicity of myxovirus. Arch Virol 59:285, 1979

Sever JL, Madden DL (eds): Polyomaviruses and Human Neurological Disease. New York, Liss, 1983

ter Meulen V, Stephenson JR, Kreth HW: Subacute sclerosing panencephalitis. In Fraenkel-Conrat H, Wagner RR (eds): Comprehensive Virology. New York, Plenum, 1983, vol 18, p 105

Wege H, Siddell S, ter Meulen V: The biology and pathogenesis of coronaviruses. Curr Top Microbiol Immunol 100:165, 1982

Wolinsky JS, Johnson RT: Role of viruses in chronic neurological diseases. In Fraenkel-Conrat H, Wagner RR (eds): Comprehensive Virology. New York, Plenum, 1980, vol 16, p 257

PERSISTENT INFECTIONS

Norkin LC: Papovaviral persistent infections. Microbiol Rev 46:384, 1982

Porter DD, Cho HJ: Aleutian disease of mink: A model for persistent infection. In Fraenkel-Conrat H, Wagner RR (eds): Comprehensive Virology. New York, Plenum, 1980, vol 16, p 233

Rapp F: Persistence and transmission of cytomegaloviruses. In Fraenkel-Conrat H, Wagner RR (eds): Comprehensive Virology. New York, Plenum, 1980, vol 16, p 193

Stevens JG, Todaro DJ, Fox CF (eds): Persistent Viruses. New York, Academic Press, 1978

Streissle G: Persistent viral infections as models for research in virus chemotherapy. Adv Virus Res 26:37, 1981

Stroop WG, Baringer JR: Persistent, slow and latent viral infections. Prog Med Virol 28:1, 1982

Youngner JS, Preble OT: Viral persistence: Evolution of viral populations. In Fraenkel-Conrat H, Wagner RR (eds): Comprehensive Virology. New York, Plenum, 1980, vol 16, p 73

Selected Papers

INHIBITION OF MACROMOLECULAR BIOSYNTHESIS

Abreu SL, Lucas-Lenard J: Cellular protein synthesis shutoff by mengovirus translation of nonviral and viral mRNAs in extracts from uninfected and infected Ehrlich ascites tumor cells. J Virol 18:182, 1976

Bablanian R, Esteban M, Baxt B, et al.: Studies on the mechanisms of vaccinia virus cytopathic effects. I. Inhibition of protein synthesis in infected cells is associated with virus-induced RNA synthesis. J Gen Virol 39:391, 1978

Brown D, Jones CL, Brown BA, et al.: Translation of capped and uncapped VSV mRNAs in the presence of initiation factors from poliovirus-infected cells. Virology 123:60, 1982

Castiglia CL, Flint SJ: Effects of adenovirus infection on rRNA synthesis and maturation in HeLa cells. Mol Cell Biol 3:662, 1983

Coppola G, Bablanian R: Discriminatory inhibition of protein synthesis in cell-free systems by vaccinia virus transcripts. Proc Natl Acad Sci USA 80:75, 1983

Ebina T, Satake M, Ishida N: Involvement of microtubules in cytopathic effects of animal viruses: Early proteins of adenovirus and herpesvirus inhibit formation of microtubular paracrystals in HeLa-S3 cells. J Gen Virol 38:535, 1978

Eggleton KH, Norkin LC: Cell killing by simian virus 40: The sequence of ultrastructural alterations leading to cellular degeneration and death. Virology 110:73, 1981

Ehrenfeld E, Lund H: Untranslated vesicular stomatitis virus messenger RNA after poliovirus infection. Virology 80:297, 1977

Etchison D, Milburn SC, Eldery I, et al.: Inhibition of HeLa cell protein synthesis following poliovirus infection correlates with the proteolysis of a 220,000-dalton polypeptide associated with eucaryotic initiation factor 3 and a cap binding protein complex. J Biol Chem 257:14, 806, 1982

Fields BN, Greene MI: Genetic and molecular mechanisms of viral pathogenesis: Implications for prevention and treatment. Nature 300:19, 1982

Golini F, Thach SS, Birge CH, et al.: Competition between cellular and viral mRNAs in vitro is regulated by a messenger discriminatory initiation factor. Proc Natl Acad Sci USA 73:3040, 1976

Jaye MC, Godchaux W III, Lucas-Lenard J: Further studies on the inhibition of cellular protein synthesis by vesicular stomatitis virus. Virology 116:148, 1982

Jen G, Detjen BM, Thatch RE: Shutoff of HeLa cell protein synthesis by encephalomyocarditis virus and poliovirus: A comparative study. J Virol 35:150, 1980

Kauffman RS, Wolf JL, Finberg R, et al.: The σ1 protein determines the extent of spread of reovirus from the gastrointestinal tract of mice. Virology 124:403, 1983

Koizumi S, Simizu B, Hashimoto K, et al.: Inhibition of DNA synthesis in BHK cells infected with Western equine encephalitis virus. 1. Induction of an inhibitory factor of cellular DNA polymerase activity. Virology 94:314, 1979

Marcus PI, Sekellick MJ: Cell killing by viruses. 1. Comparison of cell-killing, plaque-forming, and defective-interfering particles of vesicular stomatitis virus. Virology 57:321, 1974

Mbuy GN, Morris RE, Bubel HC: Inhibition of cellular protein synthesis by vaccinia virus surface tubules. Virology 116:137, 1982

McSharry JJ, Choppin PW: Biological properties of the VSV glycoprotein. I. Effects of the isolated glycoprotein or host macromolecular synthesis. Virology 84:172, 1978

Nishioka Y, Jones G, Silverstein S: Inhibition by vesicular stomatitis virus of herpes simplex virus-directed protein synthesis. Virology 124:238, 1983

Ray BK, Brendler TG, Adya S, et al.: Role of mRNA competition in regulating translation: Further characterization of mRNA discriminatory initiation factors. Proc Natl Acad Sci USA 80:663, 1983

Rose JK, Trachsel H, Leong K, et al.: Inhibition of translation by poliovirus: Inactivation of a specific initiation factor. Proc Natl Acad Sci USA 75:2732, 1978

Schaefer A, Kuhne J, Zibirre R, et al.: Poliovirus-induced alterations in HeLa cell membrane functions. J Virol 44:444, 1982

Schnitzlein WM, O'Banion MK, Poirot MK, et al.: Effect of intracellular vesicular stomatitis virus mRNA concentration on the inhibition of host cell protein synthesis. J Virol 45:206, 1983

Sharpe AH, Fields BN: Reovirus inhibition of cellular RNA and protein synthesis: The role of the S4 gene. Virology 122:381, 1982

Stenberg RM, Pizer LI: Herpes simplex virus-induced changes in cellular and adenovirus RNA metabolism in an adenovirus type 5-transformed human cell line. J Virol 42:474, 1982

Thomas JR, Wagner RR: Inhibition of translation in lysates of mouse L cells infected with vesicular stomatitis virus: Presence of a defective ribosome-associated factor. Biochemistry 22:1540, 1983

CELL FUSION

Ege T, Krondahl U, Ringertz NR: Introduction of nuclei and micronuclei into cells and enucleated cytoplasms by Sendai virus induced fusion. Exp Cell Res 88:428, 1974

Lee GT-Y, Spear PG: Viral and cellular factors that influence cell fusion induced by herpes simplex virus. Virology 107:402, 1980

Manservigi R, Spear PG, Buchan A: Cell fusion induced by herpes simplex virus is promoted and suppressed by different viral glycoproteins. Proc Natl Acad Sci USA 74:3913, 1977

Tedesco TA, Diamond R, Orkwiszewski KG, et al.: Assignment of the human gene for hexose-1-phosphate uridylyltransferase to chromosome 3. Proc Natl Acad Sci USA 71:3483, 1974

PERSISTENT AND LATENT INFECTIONS

Barrett PN, Atkins GJ: Establishment of persistent infection in mouse cells by Sindbis virus and its temperature-sensitive mutants. J Gen Virol 54:57, 1981

Frey TK, Youngner JS: Novel phenotype of RNA synthesis expressed by vesicular stomatitis virus isolated from persistent infection. J Virol 44:167, 1982

Galloway DA, Fenoglio C, Shevchuk M, et al.: Detection of herpes simplex RNA in human sensory ganglia. Virology 95:265, 1979

Gerdes JC, Marsden HS, Cook ML, et al.: Acute infection of differentiated neuroblastoma cells by latency-positive and latency-negative herpes simplex virus ts mutants. Virology 94:430, 1979

Newton SE, Short NJ, Dalgarno L: Bunyamwera virus replication in cultured *Aedes aldopictus* (mosquito) cells: Establishment of a persistent viral infection. J Virol 38:1015, 1981

Preble OT, Youngner JS: Temperature-sensitive defects of mutants isolated from L cells persistently infected with Newcastle disease virus. J Virol 12:472, 1973

Price RW, Schmitz J: Route of infection, systemic host resistance, and integrity of ganglionic axons influence acute and latent herpes simplex virus infection of the superior cervical ganglion. Infect Immun 23:373, 1979

Rock DL, Fraser NW: Detection of HSV-1 genome in central nervous system of latently infected mice. Nature 302:523, 1983

Thacore HR, Youngner JS: Persistence of vesicular stomatitis virus in interferon-treated cell cultures. Virology 63:345, 1975

Williams MP, Brawner TA, Riggs HG Jr, et al.: Characteristics of a persistent rubella infection in a human cell line. J Gen Virol 52:321, 1981

Yoshida T, Hamaguchi M, Naruse H, et al.: Persistent infection by a temperature-sensitive mutant isolated from a Sendai virus (HVJ) carrier culture: Its initiation and maintenance without aid of defective interfering particles. Virology 120:329, 1982

Youngner JS, Preble OT, Jones EV: Persistent infection of L cells with vesicular stomatitis virus: Evolution of virus populations. J Virol 28:6, 1978

PERMISSIVENESS

Chen C-Y, Crouch NA: Shope fibroma virus-induced facilitation of vesicular stomatitis virus adsorption and replication in nonpermissive cells. Virology 85:43, 1978

Jones EV, Whitaker-Dowling PA, Youngner JS: Restriction of vesicular stomatitis virus in a nonpermissive rabbit cell line is at the level of protein synthesis. Virology 121:20, 1982

The Genetics of Animal Viruses

The genetic approach to studying the virus–cell interaction in general and the virus multiplication cycle in particular has long been recognized as being extremely powerful in the bacteriophage field (Chap. 63). During the last two decades it has become very useful in animal virology as well. The principal techniques of viral genetics are the isolation of mutants in the various genes, their characterization, and identification of the proteins that they encode. These techniques permit dissection of the complex series of interactions that is initiated when viruses infect cells into its individual components.

Types of Virus Mutants

Viruses are encapsidated segments of genetic material, and like other genetic systems, viral genomes are not invariate but are subject to change by mutation. Spontaneous mutations occur constantly in the course of virus multiplication, and while many mutations are lethal, others are not. Virus populations may thus be regarded as genetically heterogeneous, since they are likely to contain mutants, at an average rate of 10^{-6}, at each locus. This value refers to DNA-containing viruses. It is sometimes suggested that the error rate for RNA-containing viruses is much higher, since RNA polymerases appear to lack the proofreading mechanism that some DNA polymerases possess. However, RNA-containing viruses are not inherently unstable genetic systems. In fact, many genes of RNA-containing viruses exhibit sequence conservation as extensive as those of DNA-containing viruses.

Mutant virus strains can also be generated in the laboratory as a result of mutagenesis, and it is from mutagenized virus populations that mutant isolation is usually undertaken. Among the procedures for mutagenizing RNA-containing viruses are treatment of virus particles with nitrous acid, hydroxylamine, N-methyl-N-nitro-N-nitrosoguanidine (NTG) or ethane methane sulfonic acid, and propagation in the presence of 5-fluorouracil, 5-azacytidine, or proflavine (for the double-stranded RNA-containing reovirus). For DNA-containing viruses, treatment of virus particles with nitrous acid, hydroxylamine, or ultraviolet irradiation, and growth in the presence of 5-bromodeoxyuridine, NTG, or proflavine are used most commonly.

There are two principal types of virus mutants: point mutants, in which there is a change in a single nucleotide base, and deletion mutants, in which a whole sequence or region of nucleic acid has been deleted. Deletion mutants will be considered below when defective virus particles are discussed. The most important point mutants are the conditional lethal mutants.

Conditional Lethal Mutants

Conditional lethal mutants, as the name implies, can multiply under some conditions but not under others that will, however, still permit the multiplication of wild-type virus. There are two classes of such mutants. The first comprises mutants that are temperature-sensitive (ts) with respect to their ability to multiply. Wild-type animal viruses can generally multiply over a temperature range that extends from a lower limit of about 20C to 24C to an upper limit of about 39.5C for mammalian viruses and 40C to 41C for avian ones. In ts mutants there is a nucleic acid base substitution that causes an amino acid replacement in some virus-coded protein, as a result of which it cannot assume or maintain the structural conformation necessary for activity at elevated or nonpermissive (restrictive) temperatures, though still able to do so at lower or permissive temperatures. The typical ts mutation causes the formation of an enzyme or a structural protein that cannot function in a temperature range (typically from about 36C to 41C) where the corresponding protein of the wild-type strain can function.

The second class of conditional lethal mutants is the host-dependent mutants, which have been extremely useful in the bacterium/bacteriophage fields. In these mutants, the codon for some amino acid is changed to a termination codon (UAG, UAA or UGA), and, as a result, proteins are formed that are shorter than those specified by wild-type virus and therefore cannot function. However, mutant bacterial strains exist that contain mutated tRNA molecules that recognize these termination codons as the codon for some amino acid (generally different from the original one), which is therefore inserted into the amino acid sequence, thereby again permitting a full-length protein to be formed. If the amino acid change is such that the altered protein can still function, the effect of the original nonsense mutation is therefore suppressed. The mutated tRNA molecules are known as suppressor tRNAs and the mutant bacterial strains as suppressor strains. The mutations that give rise to UAG, UAA, and UGA are known as *amber*, *ochre*, and *opal* respectively. Recent work has shown that this type of mutation also exists in mammalian cells/viruses.

The reason that conditional lethal mutants are so important in studies seeking to define the reactions essential for virus multiplication is that they permit study of the virus multiplication cycle with one, and only one, reaction unable to proceed. Use of such mutants permits both the assessment of the role of any particular known reaction during the course of virus multiplication and also the detection of hitherto unknown functions.

Mutants with Other Commonly Observed Phenotypes

In addition to temperature-sensitive and host-dependent mutants, several other mutant phenotypes are commonly observed, primarily because selection for them is easy and they are therefore readily observed and isolated.

Plaque-size Mutants

Many virus strains give rise to spontaneous mutants that form smaller plaques than wild-type virus because their adsorption is inhibited by sulfated polysaccharides present in agar (Fig. 60-1). Large plaque mutants are also known, and in this case the ability of wild-type virus to adsorb is inhibited by the polysaccharides, whereas that of the mutants is not. In either case, the site of the mutation is in a capsid protein that functions in adsorption.

Host Range Mutants

Host range mutants of several animal viruses are known. The best characterized are those of adenoviruses, papovaviruses, and poxviruses. Adenovirus *hr* mutants map in the

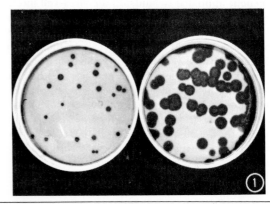

Figure 60-1. Comparative sizes of two Mengovirus plaque size variants, S-Mengo (left) and L-Mengo (right). The plaques are 48 hours old. The L mutant is more virulent in animals and more cytopathic in cell cultures. *(From Amako and Dales: Virology 32: 184, 1967.)*

leftmost portion of the viral genome, the region that also contains the genetic information required for transformation. Not surprisingly, these adenovirus *hr* mutants tend to be defective in their ability to transform cells.

Drug-resistant Mutants

Drugs capable of inhibiting the multiplication of certain viruses are known (Chap. 61), and mutants exist that are resistant to these drugs. Examples are poliovirus mutants resistant to guanidine, herpesvirus mutants resistant to phosphonoacetic acid, and vaccinia virus mutants resistant to rifampicin and IBT. Poliovirus mutants dependent on guanidine and vaccinia virus mutants dependent on IBT also exist.

Enzyme-deficient Mutants

Viruses code for several enzymes essential for virus multiplication, and mutations that result in loss of this ability are obviously lethal. Some viruses also code for enzymes that are not essential, and mutants lacking the ability to code for them are viable. For example, poxviruses and herpesviruses code for enzymes that phosphorylate thymidine (thymidine kinases). Virus mutants that are deficient in the ability to induce the synthesis of these enzymes are known. They multiply well, which suggests that the survival advantage conferred by the ability to code for them is small.

Hot Mutants

These are mutants that can grow well at temperatures higher than can wild-type virus. For example, whereas 41C is near the upper limit of the temperature growth range of wild-type poliovirus, mutant strains exist that grow as well at 41C as at 37C, or even better. Not surprisingly, such strains are very virulent, since they can multiply rapidly in patients with high fever, when the multiplication of wild-type virus is at least partially inhibited.

Interactions Among Viruses

Under conditions of multiple infection, cells may become infected with two or more virus particles with different genomes. If they are sufficiently closely related—that is, if they belong to the same genus or virus family—they can even interact. There are several types of such interactions.

Recombination

The detection of recombination between two virus strains depends on the availability of techniques that permit recombinants to be detected in the presence of a large excess of the two parents. If the two parents are single-step mutants of some wild-type strain, each differing from it in some recognizable manner, some of the recombinants will have the wild-type genotype and phenotype and be easily detectable. The detection of other recombinants, such as the reciprocal recombinants or recombinants between viruses that differ in several loci, is usually more difficult.

Viruses differ greatly in the ease with which they undergo recombination, the principal relevant factor being the nature of their genomes. With the exception of poliovirus and foot-and-mouth-disease virus, viruses that possess a single molecule of single-stranded RNA do not recombine. By contrast, viruses that contain double-stranded DNA recombine efficiently, most probably by a mechanism that is analogous to that by which the genomes of bacteria and higher organisms recombine.

The important point here is that viruses must be reasonably closely related for recombination to occur: only viruses in the same genus recombine, and ability to recombine may well become a useful taxonomic criterion. (Note, however, the infectious adenovirus-SV40 recombinants, which present a very special case, see below.)

Genome Segment Reassortment

The most efficiently recombining viruses are those whose genomes consist of several nucleic acid segments. In such cases recombination proceeds not by classic recombination involving breakage and reformation of covalent bonds but by simple reassortment of segments into new sets (Fig. 60-2). Both single-stranded and double-stranded RNA segments participate in this type of recombination, as shown by the fact that pairs of reovirus and influenza virus mutants with mutations in different genome segments generate wild-type virus with high frequency. For reoviruses and rotaviruses, bunyaviruses, and arenaviruses, genome segment reassortment does not appear to play a major role under natural conditions: although reassortants can readily be detected among the viruses that circulate in populations infected with two virus strains, such reassortants tend to die out, perhaps because some of their structural proteins are specified by one virus strain, others by the other, and the resulting heterologous capsids are less stable than homologous capsids. However, genome segment reassortment is very important in generating new influenza virus strains. It is very likely that such reassortment is the cause of the major antigenic shifts that have occurred in human influenza virus during the last half-century. The influenza viruses that exist in nature each possess one of 13 different kinds (subtypes) of hemagglutinin (H1 to H13) and one of 9 different subtypes of neuraminidase (N1 to N9). These different subtypes differ markedly in amino acid sequence and possess quite different antigenic determinants. Before 1957 the influenza virus strains that circulated in humans had the constitution H1N1; in 1957 they were replaced by Asian influenza, which was H2N2; in 1968 the

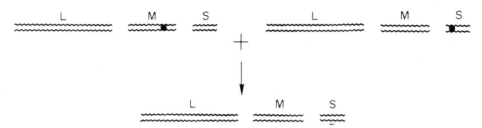

Figure 60-2. Generation of wild-type genomes by reassortment of damaged genome segments. Two genomes, each consisting of three segments of double-stranded nucleic acid, are shown. One carries a mutation in an M segment, the other in an S segment. When they are introduced into the same cell, sets of undamaged segments are generated by reassortment. This type of mechanism can account for the generation of new genotypes among reoviruses, influenza viruses, bunyaviruses, and arenaviruses.

Hong Kong virus appeared, which is H3N2 and is still circulating; and in 1977 H1N1 human influenza virus reappeared after an absence of 20 years. All these major antigenic shifts occurred in China. The most likely mechanism for the appearance and reappearance of these new subtypes is the introduction of new genome segments into human influenza virus from some animal influenza virus by genome segment reassortment. For example, the Hong Kong influenza virus strain of 1968 most probably derived its H3 hemagglutinin gene from an equine or duck influenza virus.

Multiplicity Reactivation

Viruses that contain double-stranded nucleic acid frequently exhibit multiplicity reactivation after being subjected to ultraviolet radiation. Ultraviolet radiation damages nucleic acids. In particular, it causes the formation of covalent bonds between adjacent pyrimidines, thereby giving rise to cyclobutane derivatives. In DNA, the most commonly formed pyrimidine dimers are those between adjacent thymine rings. In RNA, dimers are formed between any adjacent uracil and cytosine rings. Dimer formation inactivates viral genomes by preventing replication and probably also transcription and translation.

Dimers are removed from DNA by several mechanisms. These have been studied mainly in bacteria, but there is evidence that they operate in animal cells as well. One involves an enzyme system that utilizes radiation of longer wavelengths, particularly those of visible light, for dissociating dimers. This is the so-called photoreactivating repair system. Another mechanism (the dark or excision repair mechanism) involves nucleases that recognize dimers and excise them, the gaps then being repaired by a DNA polymerase acting in conjunction with polynucleotide ligase. This is the host cell reactivating system.

Ultraviolet radiation also induces crosslinking of the two strands of double-stranded DNA by a mechanism that is not clear. No doubt this also contributes to virus inactivation.

Ultraviolet radiation also causes the addition of water molecules across the C5—C6 double bond of pyrimidines in both DNA and RNA, which results in the formation of photohydrates (6-hydroxy-5,6-dihydro derivatives). These photohydrates represent a major portion of the lethal damage caused by ultraviolet light in many RNA-containing viruses.

The most radiation-sensitive property of a virus is its infectivity. The reason is that infectivity requires expression of the genome's entire information content and thus presents the largest target. Sometimes virus particles that have lost the ability to reproduce can still express some special function or group of functions that originate from cistrons that have not sustained radiation damage. Examples of such functions are the ability to synthesize early enzymes and the ability to transform cells. At very high radiation doses, damage to capsid proteins becomes important. This causes loss of the ability to interfere with the multiplication of related viruses, loss of ability to hemagglutinate, and loss of antigenicity.

Multiplicity reactivation is recognized by the fact that the frequency of viral survivors increases sharply with multiplicities of infection above 1. It is due to cooperation between viral genomes that have been damaged by the radiation and that can, therefore, no longer multiply on their own. The nature of the cooperation is recombination, that is, the damaged genomes recombine until an intact genome arises, which can then replicate and form progeny.

Complementation

Viral genomes can also interact indirectly by means of complementation. A typical example of this type of interaction is provided by infection of cells at the restrictive temperature with two virus mutants that bear temperature-sensitive mutations in different cistrons and neither of which can multiply alone. If complementation occurs, progeny comprising both mutants is produced. The explanation of this phenomenon is that each mutant produces functional gene products of all cistrons except that bearing the temperature-sensitive mutation, so that in cells infected with both mutants all gene products necessary for virus multiplication are formed, and both mutants can therefore multiply. Complementation plays a major role in permitting the survival of viruses with genomes that contain damaged or nonfunctional genes.

Phenotypic Mixing and Phenotypic Masking

The dual phenomena of phenotypic mixing and phenotypic masking represent a special case of complementation. When two closely related viruses, for example, poliovirus type 1 and poliovirus type 3, infect the same cell, the two types of progeny genomes may become encapsidated not only by their own capsids but also by hybrid capsids—that is, capsids composed of proteins encoded by both genomes (phenotypic mixing)—or even by capsids entirely specified by the other genome (phenotypic masking or transcapsidation) (Fig. 60-3). This situation is mostly readily detected by antigenic analysis, for the former class of virus particles is neutralized by antiserum to either parent, whereas virus particles of the latter class are neutralized by antiserum to one of the parents, while their progeny is neutralized by antiserum to the other.

A similar phenomenon occurs among enveloped viruses, but here it involves not only viruses that are related but also viruses that are completely unrelated. In particular, the nucleocapsid of the rhabdovirus vesicular stomatitis virus (VSV) possesses a remarkable ability to become encapsidated in envelopes that are only partially, or not at all, specified by it. For example, among the yield from cells simultaneously infected with both VSV and the paramyxovirus SV5, there are bullet-shaped particles that contain VSV nucleocapsids encased in envelopes that bear not only VSV-specified glycoprotein spikes but also both

types of SV5-specified spikes. Another example is provided by VSV nucleocapsids completely encased in RNA tumor virus envelopes. Since such particles are easily and rapidly quantitated (because VSV causes plaques rapidly on cell monolayers), they are useful for studies on RNA tumor virus host range (which is specified by the envelope) and for detecting the presence of RNA tumor virus-specified envelope proteins, particularly in connection with the search for human tumor viruses (Chap. 62). Finally, VSV nucleocapsids can even be encased in herpesvirus envelopes. Nucleocapsids of one virus enclosed in envelopes specified by another are known as "pseudotypes." Viruses differ in their propensity to form pseudotypes; those that do so most readily are VSV and RNA tumor viruses.

Defective Virus Particles

There are several types of virus particles that cannot multiply on their own but can multiply in cells simultaneously infected with some infectious "helper" virus. They can be subdivided into two classes; those that interfere extensively with the multiplication of their helper virus, and those that do not.

Defective Interfering (DI) Virus Particles

When viruses are passaged repeatedly at high multiplicity, the progeny frequently includes, in addition to mature virus particles, defective virus particles that are capable of interfering with the multiplication of homologous virus. Such virus particles have the following properties: (1) they contain the normal structural capsid proteins, (2) they contain only a part of the viral genome—that is, they are de-

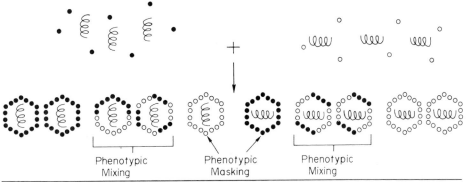

Figure 60-3. Phenotypic mixing and phenotypic masking. Simultaneous infection with two related viruses is illustrated. Either genome can be encapsidated in capsids that are composed exclusively of homologous capsomers or mixed capsomers (phenotypic mixing) or of exclusively heterologous capsomers (phenotypic masking). The method of detecting the latter two classes of particles is described in the text.

letion mutants, (3) they can reproduce only in cells infected with homologous virus, which acts as helper, (4) although unable to reproduce on their own, they can nevertheless express a variety of functions in the absence of helper, such as inhibition of host biosynthesis, transformation of cells, and synthesis of viral proteins, and (5) they specifically interfere with the multiplication of homologous virus.

The following are some examples of defective interfering virus particles that have been characterized in some detail.

Defective Interfering Influenza Virus Particles

Under conditions of repeated passaging at high multiplicity, the infectivity of successive yields of influenza virus gradually decreases a million-fold or even more, though the total number of virus particles that is produced remains roughly the same. In other words, noninfectious, defective virus particles gradually replace virions in the yields. This phenomenon was first described in 1952 by von Magnus and bears his name. Defective particles are not formed if influenza virus is passaged at low multiplicity, and they are readily eliminated from virus stocks by passaging at a multiplicity of less than 1, which shows that defective particles cannot multiply on their own. The ability of the defective particles to inhibit the multiplication of

infectious virus is demonstrated by the fact that the addition of defective particles to influenza virus preparations free of them immediately reduces the yield of infectious virus in most types of cells.

The essential difference between infectious and defective influenza virus particles is that the latter have lost most of their three large RNA segments (Chap. 58) and acquired instead several new RNA segments that are smaller than the smallest segment of wild-type virus and that are derived from the three largest segments. The loss of infectivity is a consequence of the deletions; the ability to interfere with the multiplication of wild-type virus is due to the presence of the new small RNA segments. The nature of several small influenza interfering RNA species is shown in Figure 60-4. The mechanism of the interference and the reason that these deletion mutants compete successfully with wild-type virus are discussed below.

Defective Interfering Particles of Other Viruses

DI particles of such rhabdoviruses as VSV contain RNA that is about one-third as long as genome RNA. As a result, they are about one-third as long as infectious virus particles, which makes them very nearly spherical (Fig. 60-5). The interfering RNA molecules are, like those of influenza virus, the result of aberrant transcription: for rea-

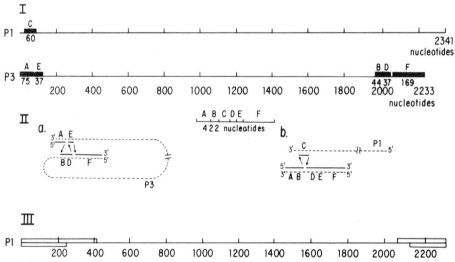

Figure 60-4. I. Origin of one of the two small interfering RNA molecules of influenza virus strain A/NT/60/68. It is 422 nucleotides long, a little less than one-half as long as the smallest influenza gene (see Table 58-5). It is derived from five separate regions, four on the *P3* gene, including both ends, and one on the *P1* gene. The size of each region is indicated. **II. a** and **b.** Models for the origin of the small interfering RNA molecule. Template switching occurs when the RNA polymerase pauses at U-rich sequences and reinitiates synthesis at another site. **III.** The A/WSN/33 strain of influenza virus gives rise to two different interfering RNA molecules; each is derived from gene *P1* via a single large deletion. The larger and smaller RNA species (composed of the two terminal sequences indicated above and below the gene, respectively) are 683 and 441 nucleotides long, respectively. Note that all small interfering influenza RNA molecules contain both ends of influenza virus genes. *(From Fields and Winter: Cell 28:303, 1982; Nayak et al.: Proc Natl Acad Sci USA 79: 2216, 1982.)*

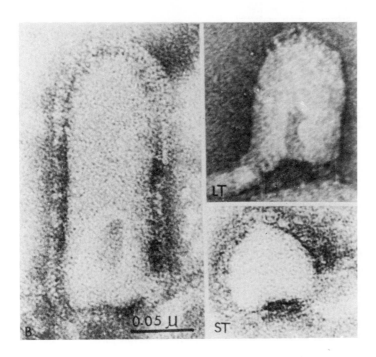

Figure 60-5. Defective interfering particles of VSV. B, a bullet-shaped normal virus particle; LT, a long truncated particle; ST, a short truncated particle. LT particles are about half as long as normal virus particles; ST particles, which are round, are about one-third as long. The defective particles contain RNA molecules that are proportionately shorter than the RNA molecules in normal virus particles. *(Courtesy of Dr C.Y. Kang.)*

sons that are poorly understood, the RNA polymerase suddenly switches transcription from one template sequence to another, causing deletions if reinitiation occurs downstream on the same template, or the synthesis of sequences complementary to portions of the strand being synthesized if reinitiation occurs on it. As a result, interfering VSV RNA molecules may possess either both ends of normal viral RNA or complementary copies of one end. The nature of small interfering VSV RNA molecules is shown in Figure 60-6. Small interfering RNA molecules of paramyxoviruses, such as Sendai virus, are quite similar.

DI particles of togaviruses, such as Semliki Forest virus and Sindbis virus, contain small interfering RNA molecules one-sixth to one-half as long as full-length RNA. They are generated by a mechanism in which template switching is carried to the nth degree. They tend to comprise a short region derived from one end of the genome, two or three tandemly arranged copies of regions 500 to 800 nucleotides long composed of regions transcribed from several portions of the viral genome, regions that may themselves be repeated, and a unique sequence derived from the other end of the viral genome. All regions from the 3'-terminal one-third of the viral genome, which comprises the structural togavirus genes (Chap. 58), are usually completely deleted. The generation of these molecules requires very extensive template switching, with the RNA polymerase disengaging from the template it is transcribing and, while remaining attached to the transcript that it is in the process of synthesizing, reinitiating transcription elsewhere on the same strand, either upsteam or downstream, or on a different strand, which may be the transcript to which it is attached.

DI particles also occur in high passage yields of polioviruses, reoviruses, bunyaviruses, herpesviruses such

as pseudorabies virus, and probably most other viruses. All are deletion mutants that require the presence of wild-type virus as helpers but outgrow them quickly. DI polioviruses tend to be simple deletion mutants, the deletions being only 4 to 15 percent of the genome, and DI mutants of reovirus simply lack one or two genes.

General Observations on the Significance of DI Particles

The basic attributes of DI particles are that they outgrow infectious virus particles and that they interfere very successfully with the multiplication of wild-type virus. No doubt these two properties are connected. Small interfering RNA molecules are heterogeneous populations, but they must all possess two regions: the recognition site(s) for RNA polymerase and the recognition site(s) for encapsidation. Since these recognition sites are at the termini of viral RNAs, all small interfering RNAs contain at least one end of intact viral RNA. All available evidence indicates that small interfering RNAs are encapsidated and serve as template for RNA polymerase at least as efficiently as intact RNA, if not more so. Since, in addition, they are formed much more rapidly than intact RNA because they are smaller, very soon most RNA polymerase and encapsidation recognition sites will be on small interfering RNA molecules, not on intact RNA molecules. Interestingly, the host cell can influence profoundly both the extent to which DI particles are formed and their nature. Thus, a single clonal isolate of influenza virus or VSV will produce different types of small interfering RNA molecules in different cell types and will produce DI particles in different cell types after different numbers of passages (that is, at different passage levels). Sometimes DI particles are not

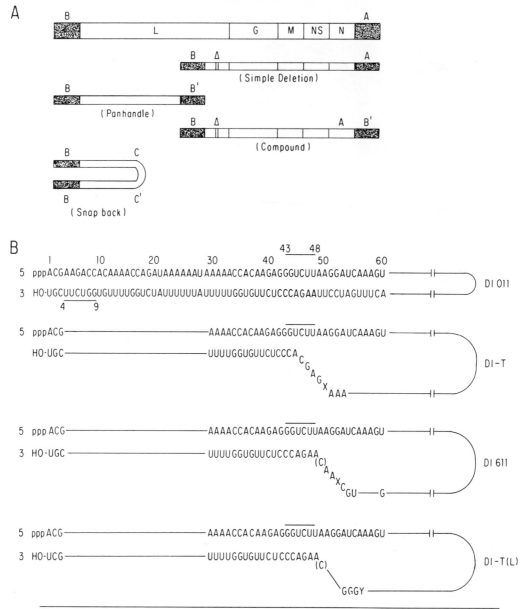

Figure 60-6. A. The nature of four types of small interfering RNAs of VSV. The structure of intact VSV RNA is shown at the top; *A* and *B* serve to identify the sequences at its two ends, and *L, G, M, NS,* and *N,* refer to the five VSV genes. The first interfering RNA is a simple deletion mutant, with most of the *L* gene deleted. Small interfering Sendai virus RNA molecules have been described in which genes *other* than the *L* gene are deleted (although some of the *L* gene is deleted also). To generate the second small interfering RNA, the RNA polymerase transcribes part of the *L* gene, then stops, reinitiates on the transcript that it is in the process of synthesizing about 45 nucleotides from the end, and transcribes it back to the end. B′ and B are complementary to each other. This process is described in more detail in **B.** Many panhandle molecules of this type are generated in which the length of the *L* gene sequences that are transcribed are all different, but the polymerase always starts transcribing backwards at the same position, where there is the sequence UUCUGG (overlined), which is also present near the 3′-end (underlined) and which is probably an RNA polymerase recognition sequence. Thus all RNA molecules of this type contain an identical panhandle and varying lengths of *L* gene sequences. The third type of small interfering RNA is a combination of the first and the second, while in the fourth the polymerase transcribes the *L* gene for a certain distance and then immediately transcribes it backwards, probably using as a template not the strand it is synthesizing but the strand that is being transcribed by the polymerase molecule immediately in front of it (note that one template can be transcribed by many RNA polymerase molecules at the same time). *(From Schubert et al.: Cell 18:749, 1979; Lazzarini et al.: Cell 26:145, 1981.)*

formed at all; for example, VSV will not form DI particles in cells treated with actinomycin D, which inhibits host cell transcription, or in cells simultaneously infected with fibroma virus. All this evidence suggests that host factors are required for the generation of DI particles, but whether they function in the *generation* of small interfering RNAs (by decreasing fidelity of transcription and/or promoting template switching) or by increasing the efficiency with which DI particles are formed is not known.

Finally, a word about the role of DI particles in the establishment and maintenance of persistent infections in nature. There is no question that persistent infections can be established readily in animals by infecting them with mixtures of infectious and DI particles and that DI particles are often present in persistently infected cells. However, it seems that, in general, persistent infections are caused by viral variants that are less cytopathogenic than wild-type virus. Often such variants are also temperature-sensitive with respect to the ability to grow, but they need not be so (Chap. 59). Perhaps the generation and propagation of DI particles is a mechanism for ensuring the survival of viruses by permitting them to establish persistent infections. If this is true, DI RNA should not only be smaller than wild-type RNA, possess higher affinity for RNA polymerase, and replicate more rapidly but should also preferably not code for structural viral proteins (so as to minimize cytopathic effects). Small interfering RNAs appear to fulfill all of these criteria.

Adenovirus and Papovavirus Deletion Mutants

When repeatedly passaged at high multiplicity, adenoviruses and papovaviruses give rise to a variety of deletion mutants, among which are the following:

1. Viable adenovirus deletion mutants. A sizable region of the adenovirus genome centered around map position 83 (in early transcription region 3) is not essential for the ability to replicate. Mutants that lack more than 2200 base pairs in this region have been isolated and are viable.
2. Adenovirus and papovavirus particles that contain DNA molecules that are almost the same size as the viral genome but comprise only a small fraction of it, the remainder being cellular DNA. These particles are called "evolutionary variants" because they evolve over the course of repeated passages from a variety of simple deletions or substitutions to more complex reiterations. Usually they acquire a selective growth advantage (but do not interfere significantly with the multiplication of wild-type virus) and become a significant fraction of the virus yield, and they require infectious virus as helper. Like the nucleic acids in DI particles, their DNA must contain recognition sites for the initiation of viral DNA replication and encapsida-

tion. Extreme examples of adenovirus evolutionary variants are

(a) one that contains only the leftmost 3 percent of the viral DNA and host DNA corresponding to 93 percent of the viral DNA and another that contains about 6 percent of each end of viral DNA and host DNA corresponding to 80 percent of the viral DNA, and
(b) one that contains a deletion comprising only about 10 percent of the viral DNA that is partially substituted to the extent of about 3 percent by host DNA.

Among extreme SV40 evolutionary variants is one with a DNA molecule that comprises a sequence of about 150 base pairs that includes the SV40 origin of replication and is repeated tandemly three times, linked to about 1500 base pairs of host cell DNA.

Adenovirus-SV40 Hybrid Particles

Human adenovirus cannot multiply in monkey cells but will do so in the presence of the simian papovavirus SV40, which performs some helper function related to the ability of viral messenger RNA to be translated. The progeny of such mixed infections sometimes includes particles that contain, within an adenovirus capsid, a hybrid DNA molecule that contains both adenovirus and SV40 DNA sequences but does not contain all the sequences of either. One well-known example of such particles is the E46+ strain of adenovirus serotype 7, which was generated during attempts to adapt human adenovirus to grow in monkey kidney cells for vaccine production. It turned out that the monkey kidney cells contained SV40, which recombined with the adenovirus type 7 to produce genomes that contain covalently linked sequences of both adenovirus and SV40 DNA. These genomes lack the adenovirus DNA sequences between map positions 5 and 21 and contain instead two SV40 DNA sequences joined end to end—namely, those between map positions 0.50 and 0.71 and between 0.11 and 0.66. These particles cannot, of course, multiply on their own, since they contain neither a complete adenovirus genome nor a complete SV40 genome, but they can perform that function that enables human adenoviruses to multiply in monkey cells. They are, therefore, called "PARA" (particles aiding the replication of adenovirus). Further, they can themselves multiply in the presence of the human adenovirus, which presumably supplies the function(s) coded by the piece of adenovirus DNA that is missing in the hybrid particle DNA. Thus, here are two types of virus particles, human adenovirus and hybrid PARA particle, neither of which can multiply in monkey cells by itself but both of which multiply if they infect monkey cells simultaneously. The significance of this system is that viruses that can multiply only in cells also harboring other viruses may be of great importance in causing human diseases of as yet undefined etiology. These situations are very difficult to detect, but detailed studies such as those described above may provide valuable clues in the search for others of clinical relevance. Another interesting class of adenovirus-SV40 hybrids is the infectious adenovirus type 2-SV40 hybrids, five of which have been characterized in detail (Table 60-1). In these hybrids, a portion of the adenovirus genome is deleted from a region that is not essential for virus multiplication. This region comprises 4 to 7 percent of the adenovirus genome and ends at map position 86 in all hybrids (see above). In its place is inserted a portion of SV40 DNA that codes for early functions. This portion varies from 7 to 43 percent of the SV40 genome and starts at map position 0.11 in all five hybrids. The largest piece (43 percent) of SV40 DNA expresses all early SV40 functions, while the smallest piece (7 percent) can code for no more than 10,000 to 15,000 daltons of protein. Yet this is sufficient to endow even this hybrid with the ability to multiply in monkey cells.

Interference Between Viruses

It has been known for a long time that when two different viruses infect the same cell, they may interfere with each other and diminish each other's yield. Although the precise nature of the inhibition is usually not known, it appears that there are two primary causes of interference. First, the first virus may inhibit the ability of the second virus to adsorb, either by blocking its receptors (certain pairs of enteroviruses) or by destroying its receptors (certain pairs of myxoviruses). Second, one virus may prevent the messenger RNA of the second virus from being translated. Thus, just as poliovirus inhibits the translation of host-cell messenger RNA by inactivating the cap-dependent mRNA recognition mechanism (Chap. 59), so does it inhibit the translation of VSV messenger RNA and interfere with the ability of VSV to multiply. Similarly, the translation of vaccinia virus messenger RNA is prevented

in cells infected with adenoviruses. The ability of Sindbis virus to interfere with the multiplication of VSV and NDV and of rubella virus to interfere with the multiplication of NDV may also be due to interference with the ability of viral messenger RNA to be translated. This mechanism of interference would account for the marked specificity of such inhibitory effects. Thus, whereas cells infected with rubella virus become resistant to NDV, they remain susceptible to a variety of other viruses.

Yet another mechanism may operate in cells infected with VSV and herpesvirus. VSV inhibits host cell protein synthesis by inhibiting transcription; herpesvirus inhibits host cell protein synthesis by causing polyribosomes to dissociate (an effect of a capsid protein) and host cell messenger RNA to be degraded (a function of an early herpesvirus protein). In cells infected with both these viruses, VSV is dominant, as it inhibits herpesvirus transcription. In cells infected with both herpesvirus and adenovirus, however, transcription of adenovirus messenger RNAs is inhibited, a function, apparently, of yet another early herpesvirus protein.

Studies that are directed specifically at determining how certain pairs of viruses interfere with each other's ability to multiply may possess great potential for defining how viruses interfere with host cell protein and RNA and DNA synthesis (Chap. 59).

Evolution of Viruses

The advent of recombinant DNA technology and the development of rapid nucleic acid sequencing techniques during the last decade have yielded a rapidly increasing body of information concerning the genetic relationship between and the relatedness of virus strains that circulate in natural host populations. They have also provided insight into the process of viral evolution. For example, there are three serotypes of mammalian reovirus, two of which, serotypes 1 and 3, are related to the extent of about 70 percent, while serotype 2 is related to the other

TABLE 60-1. NONDEFECTIVE ADENOVIRUS TYPE 2-SV40 HYBRIDS

Hybrid	Portion of Adenovirus DNA Deleted		Portion of SV40 DNA Inserted	
	Map Position*	Mol Wt	Map Position	Mol Wt
Ad2$^+$ND$_1$	80.6–86.0	1.24×10^6	0.11–0.39	0.58×10^6
Ad2$^+$ND$_2$	79.9–86.0	1.40×10^6	0.11–0.43	1.02×10^6
Ad2$^+$ND$_3$	80.7–86.0	1.22×10^6	0.11–0.18	0.22×10^6
Ad2$^+$ND$_4$	81.5–86.0	1.04×10^6	0.11–0.54	1.38×10^6
Ad2$^+$ND$_5$	78.9–86.0	1.63×10^6	0.11–0.39	0.90×10^6

*According to convention, the adenovirus and SV40 maps are referred to as being 100 and 1.00 map units long, respectively.

two to the extent of only about 10 percent. Interestingly, the antigenic determinants on the proteins specified by most of their genes have been highly conserved during evolution, as demonstrated by the fact that antisera against any one serotype cannot readily differentiate between the proteins encoded by any of the three serotypes, even the distantly related (in genetic terms) serotype 2. The only exception is provided by the proteins encoded by the *S1* gene; the three *S1* genes are related only about one-quarter as closely as the other genes. Thus, we have an ancestral set of genes that diverged along three pathways to yield three gene pools, two closely and one by now only slightly related, and nine of the ten genes evolved more or less together, while the tenth diverged markedly.

This type of analysis is also applicable to viruses of other families, particularly influenza viruses. Influenza viruses, particularly influenza A viruses, exhibit a large degree of antigenic variation. Minor antigenic variants cause new epidemics every 1 or 2 years, and major antigenic variants cause new pandemics every 10 to 15 years. The marked degree of antigenic variation of influenza viruses, that is their rapid evolution, cannot be explained by an enhanced capacity to produce mutant viruses, for the frequencies with which antigenic variants in Sendai virus, VSV, and influenza A virus arise are about the same, namely, $10^{-4.5}$ to $10^{-4.7}$. Rather, it is due to the fact that many antigenic variants of the two influenza spike proteins, the neuraminidase and especially the hemagglutinin, are functional, presumably because they are not structural proteins in the sense of having to fit into a capsid structure. It is also due to the fact that influenza A virus is a ubiquitous virus to which humans are constantly exposed, so that new antigenic variants are constantly being selected by an immune system that has multiple prior experiences with previously encountered variants.

Antigenic variation of influenza A virus is of two types. The first is antigenic drift, minor antigenic changes that occur continuously in partially immune host populations during intrapandemic periods and that sporadically lead to the emergence of new epidemic virus strains. These minor antigenic changes are caused by the selective growth advantage that new antigenic variants possess in the presence of antibodies capable of neutralizing the previous, parental virus strains. The second type of antigenic variation is antigenic shift, the emergence of new major antigenic variants, primarily in the hemagglutinin (H) but also to a lesser extent in the neuraminidase (N). These new antigenic variants are genetic reassortants that are generated when an *H* or an *N* gene from an influenza virus of some animal species is introduced into the gene set of an influenza virus pathogenic for humans. For example, the *H3* gene of the Hong Kong strain of human influenza A virus that emerged in 1968 and that caused the most recent human pandemic was probably derived from an equine or a duck influenza virus strain. Both equine and duck influenza virus A strains contain *H3* genes that are very closely related to that now present in human H3N2 influenza virus strains (Chap. 59).

Mapping of Genomes of Animal Viruses

The basic aims of virologists are (1) to characterize the various functions involved in virus multiplication and (2) to identify the portions of the viral genome that encode these functions. The first of these aims depends primarily on the availability of mutants. As pointed out above, mutants permit the virus multiplication cycle to be examined when one of its component reactions fails to function. These reactions can, therefore, be identified and characterized one by one.

Great strides have recently been made in mapping animal virus genomes, that is, in identifying the regions or genes that encode specific proteins. A brief summary follows of the techniques that are most useful for mapping the genes of DNA- and RNA-containing viruses.

DNA-containing Viruses

1. Recombination analysis. This has been described on page 868.
2. Restriction endonuclease analysis of the DNA of recombinants of pairs of mutants in different genes, using related viruses. This was described in Figure 58-10.
3. Marker rescue. Cells are infected with a ts mutant together with individual restriction endonuclease fragments of wild-type virus DNA. The fragments will recombine with the mutant genome, generating a wild-type recombinant if they contain the gene, or portion of the gene, that contained the original mutation. If the location on the genome of the restriction endonuclease fragment is known, the location of the mutated gene is known.
4. Virus-coded messenger RNAs can be isolated from infected cells and translated in cell-free protein-synthesizing systems, and the proteins for which they code identified. Since the origin of these messenger RNAs can be specified by determining with which restriction endonuclease fragment they can hybridize, the regions of the viral genomes that code for specific viral proteins can be identified. Another version of this analysis is described on page 867.

RNA-containing Viruses

The gene order on viral genomes that consist of RNA can be identified as follows.

UV-transcriptional Mapping

UV irradiation damages RNA (page 904) and inactivates it as a template for transcription. Since transcription of such viral RNA as paramyxovirus and rhabdovirus RNA is initiated from a single site (at its 3' terminus), the effect of irradiation is to interfere differentially with the transcription of the various genes; transcription of those farthest from the transcription initiation site is inhibited most, while transcription of those closest to it is inhibited least. Since the relative amount of each species of messenger RNA that is transcribed can be determined readily by measuring the amount of protein that is translated from it, the gene order can be determined (Chap. 58).

Pactamycin Translation Inhibition Mapping

This technique is capable of mapping the gene order of viral genomes that act as messenger RNAs (such as the genomes of picornaviruses). The technique depends on the fact that these genomes are multigenic but monocistronic, that is, there is a single site (at their 5' terminus) where translation is initiated, but it comprises several cistrons (genes), the RNA being translated into a large precursor protein, the polyprotein, which is then cleaved into the various functional proteins (Chap. 58). Pactamycin inhibits initiation of translation but not elongation. Thus, if pactamycin is added to infected cells and a radioactively labeled amino acid is then added at various times thereafter, its incorporation into the region of polyprotein closest to the 5' end will be inhibited first, and its incorporation into the region of the polyprotein that is furthest from the 5' end will be inhibited last. It has been found in this way that the 5' terminal half of poliovirus RNA codes for the structural proteins, while the 3' terminal half codes for nonstructural proteins, such as polymerase.

Identification of Genes in Segmented Genomes

Identification of the genes of reovirus and influenza virus can be accomplished in two ways. The first is the direct one: it is possible to translate the individual RNA species in cell-free protein-synthesizing systems and to identify the proteins for which they code. The second way takes advantage of the fact that the individual RNA and protein species of different serotypes of reovirus and influenza virus can be identified because they possess different migration rates when electrophoresced in polyacrylamide gels (Fig. 60-7). By crossing pairs of virus strains that belong to different serotypes and examining the RNA and protein patterns of the reassortants that are produced, one can determine which RNA segment codes for each individual protein. In the simplest case, one would look for recombinants in which all RNA segments except one are derived from one parent. Then there would be one, and only one, protein that would be characteristic of the other parent. This technique also permits ts mutations to be assigned to RNA segments. To do this, ts mutants of one serotype are crossed with the wild-type strain of another serotype, progeny are then plaqued at the restrictive (normal) temperature, and the pattern of RNA segments of several

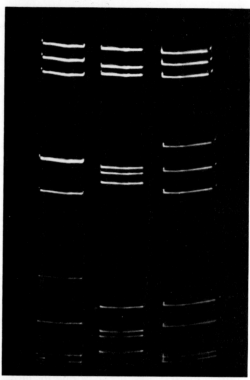

Figure 60-7. Separation by electrophoresis in polyacrylamide gels (Maizel Tris-glycine system) of the 10 double-stranded RNA-containing genes of reovirus serotype 3, 2, and 1. Note that two genes in the reovirus serotype 3 genome overlap in this system (but can be readily separated in others). The cognate segments differ slightly but significantly in electrophoretic mobility and can therefore be identified. Similar patterns are obtained if the protein species of these strains are compared by electrophoresis in SDS-containing polyacrylamide gels. *(Courtesy of Dr. R. Gaillard.)*

plaque isolates is then examined by polyacrylamide gel electrophoresis. Since recombination in reovirus and influenza virus occurs by random assortment of RNA segments (p. 903), one would expect that all RNA segments would be randomly derived from either one parent or the other, except one—namely, that which bears the ts lesion. That segment will always be derived from the wild-type parent.

FURTHER READING

Books and Reviews

Boettiger D: Animal virus pseudotypes. Prog Med Virol 25:37, 1979

Bratt MA, Hightower LE: Genetics and paragenetic phenomena of paramyxoviruses. In Fraenkel-Conrat H, Wagner RR (eds): Comprehensive Virology. New York, Plenum, 1977, vol 9, p 457

Cooper PD: Genetics of picornaviruses. In Fraenkel-Conrat H, Wagner RR (eds): Comprehensive Virology. New York, Plenum, 1977, vol 9, p 133

Fields BN: Genetics of reovirus. Curr Top Microbiol Immunol 91:1, 1981

Gorman BM: Variation in orbiviruses. J Gen Virol 44:1, 1979

Halliburton IW: Intertypic recombinants of herpes simplex viruses. J Gen Virol 48:1, 1980

Holland J. Spindler J, Horodyski F, et al.: Rapid evolution of RNA genomes. Science 215:1577, 1982

Lazzarini RA, Keene JD, Schubert M: The origins of defective interfering particles of negative-strand RNA viruses. Cell 26:145, 1981

Nayak DP: Defective interfering influenza viruses. Annu Rev Microbiol 34:619, 1980

Palese P, Young JF: Variation of influenza virus A, B and C viruses. Science 215:1468, 1982

Patch CT, Levine AS, Lewis AM Jr: The adenovirus-SV40 hybrid viruses. In Fraenkel-Conrat H, Wagner RR (eds): Comprehensive Virology. New York, Plenum, 1979, vol 13, p 459

Pfefferkorn ER: Genetics of togaviruses. In Fraenkel-Conrat H, Wagner RR (eds): Comprehensive Virology. New York, Plenum, 1977, vol 9, p 209

Pringle CR: The genetics of vesiculoviruses. Arch Virol 72:1, 1982

Reanney DC: The evolution of RNA viruses. Annu Rev Microbiol 36:47, 1982

Sambrook J, Sleigh M, Engler JA, et al.: The evolution of the adenoviral genome. Ann NY Acad Sci 354:426, 1980

Webster RG, Laver WG, Air GM, et al.: Molecular mechanisms of variation in influenza viruses. Nature 296:115, 1982

Selected Papers

MUTANTS, VARIANTS, AND RECOMBINANTS

Desselberger U, Nakajima K, Alfina P, et al.: Biochemical evidence that "new" influenza virus strains in nature may arise by recombination (reassortment). Proc Natl Acad Sci USA 75:3341, 1978

Drayna A, Fields BN: Genetic and biochemical studies on the mechanism of chemical and physical inactivation of reovirus. J Gen Virol 63:149, 161, 1982

Dubbs DR, Kit S: Isolation and properties of vaccinia mutants deficient in thymidine kinase inducing activity. Virology 22:214, 1964

Gaillard RK Jr, Joklik WK: Quantitation of the relatedness of reovirus serotypes 1, 2, and 3 at the gene level. Virology 123:152, 1982

Hinshaw VS, Webster RG, Bean WJ: Swine influenza-like viruses in turkeys: Potential source of virus for humans? Science 220:206, 1983

Klessig DF, Hassel JA: Characterization of a variant of human adenovirus type 2 which multiplies efficiently in simian cells. J Virol 28:945, 1978

Larsen SH: Evolutionary variants of mouse adenovirus containing cellular DNA sequences. Virology 116:573, 1982

Laver WG, Air GM, Webster RG, et al.: Amino acid sequence changes in antigenic variants of type A influenza virus N2 neuraminidase. Virology 122:450, 1982

Lee TNH, Nathans D: Evolutionary variants of simian virus 40: Replication and encapsidation of variant DNA. Virology 92:291, 1979

Portner A, Webster RG, Bean WJ: Similar frequencies of antigenic variants in Sendai, vesicular stomatitis, and influenza A viruses. Virology 104:235, 1980

Ramig RF, Ahmeed R, Fields BN: A genetic map of reovirus: Assignment of the newly defined mutant groups H, I, and J to genome segments. Virology 125:299, 1983

Sambrook JF, Padgett BL, Tomkins JKN: Conditional lethal mutants of rabbitpox virus. 1. Isolation of host-cell dependent and temperature-dependent mutants. Virology 25:592, 1966

Schuerch AR, Matsuhisa T, Joklik WK: Temperature-sensitive mutants of reovirus. VI. Mutant ts447 and ts556 particles that lack either one or two genome RNA segments. Intervirology 3:36, 1974

Smith GL, Mackett M, Moss B: Infectious vaccinia virus recombinants that express hepatitis B virus surface antigen. Nature 302:490, 1983

Werner G, zur Hausen H: Deletions and insertions in adenovirus type 12 DNA after viral replication in vero cells. Virology 86:66, 1978

NONGENETIC INTERACTIONS AMONG VIRUSES

Chen C, Crouch NA: Shope fibroma virus induced facilitation of vesicular stomatitis virus adsorption and replication in nonpermissive cells. Virology 85:43, 1978

Choppin PW, Compans RW: Phenotypic mixing of envelope proteins of the parainfluenza virus SV5 and vesicular stomatitis virus. J Virol 5:609, 1970

Huang AS, Palma El, Hewlett N, et al.: Pseudotype formation between enveloped RNA and DNA viruses. Nature 252:743, 1974

Joklik WK, Abel P, Holmes IH: Reactivation of poxviruses by a nongenetic mechanism. Nature 186:992, 1960

Joklik WK, Woodroofe GM, Holmes IH, et al.: The reactivation of poxviruses. I. The demonstration of the phenomenon and techniques of assay. Virology 11:168, 1960

Trautman R, Sutmoller P: Detection and properties of a genomic masked viral particle consisting of foot-and-mouth disease virus nucleic acid in bovine enterovirus protein capsid. Virology 44:537, 1971

Zavadova Z, Zavada J: Unilateral phenotypic mixing of envelope antigens between togaviruses and vesicular stomatitis virus or avian RNA tumor virus. J Gen Virol 37:557, 1977

GENE MAPPING

Ball A: Transcriptional mapping of vesicular stomatitis virus in vivo. J Virol 21:411, 1977

Frost E, Williams J: Mapping temperature-sensitive and host-range mutations of adenovirus type 5 by marker rescue. Virology 91:39, 1978

Greenberg HB, Flores J, Kalica AR, et al.: Gene coding assignments for growth restriction, neutralization and subgroup specificities of the W and DS-1 strains of human rotavirus. J Gen Virol 64:313, 1983

Halliburton IW, Morse LS, Roizman B, et al.: Mapping of the thymidine kinase genes of type 1 and type 2 herpes simplex viruses using intertypic recombinants. J Gen Virol 49:235, 1980

Mustoe TA, Ramig RF, Sharpe AH, et al.: A Genetic map of reovirus. III. Assignment of the double-stranded RNA-positive mutant groups A, B, and G to genome segments. Virology 85:545, 1978

Palese P, Ritchey MB, Schulman JL: Mapping of the influenza virus genome. II. Identification of the P1, P2, and P3 genes. Virology 76:114, 1977

Ramig, RF, Fields BN: Revertants of temperature-sensitive mutants of reovirus: Evidence for frequent extragenic suppression. Virology 92:155, 1979

Wilson MC, Fraser NW, Darnell JE Jr: Mapping of RNA initiation sites by high doses of UV irradiation: Evidence for three independent promoters within the left 11% of the Ad-2-genome. Virology 94:175, 1979

DEFECTIVE AND DEFECTIVE INTERFERING VIRUS PARTICLES

Amesse LS, Pridgen CL, Kingsbury DW: Sendai virus DI RNA species with conserved virus genome termini and extensive internal deletions. Virology 118:17, 1982

Ben-Porat T, Demarchi JM, Kaplan AS: Characterization of defective interfering viral particles present in a population of pseudorabies virions. Virology 61:29, 1974

Fields S, Winter G: Nucleotide sequences of influenza virus segments 1 and 3 reveal mosaic structure of a small viral RNA segment. Cell 28:303, 1982

Huang AS, Greenawalt JW, Wagner RR: Defective T particles of vesicular stomatitis virus. Virology 30:161, 173, 1966

Huang AS, Baltimore D: Defective viral particles and viral disease processes. Nature 226:325, 1970

Lehtovaara P, Söderlund H, Keränen S, et al.: Extreme ends of the genome are conserved and rearranged in the defective interfering RNAs of Semliki Forest virus. J Mol Biol 156:731, 1982

McClure MA, Holland JJ, Perrault J: Generation of defective interfering particles in picornaviruses. Virology 100:408, 1980

Nayak DP, Sivasubramanian N, Davis AR, et al.: Complete sequence analyses show that two defective interfering influenza viral RNAs contain a single internal deletion of a polymerase gene. Proc Natl Acad Sci USA 79:2216, 1982

Nomoto A, Jacobson A, Lee YF, et al.: Defective interfering particles of poliovirus; mapping of the deletion and evidence that the deletions in the genomes of DI(1), (2), and (3) are located in the same region. J Mol Biol 128:179, 1979

Schubert M, Keene JD, Lazzarini RA: A specific internal RNA polymerase recognition site of VSV RNA is involved in the generation of DI particles. Cell 18:749, 1979

Sivasubramanian N, Nayak DP: Defective interfering influenza RNAs of polymerase gene 3 contain single as well as multiple internal deletions. Virology 124:232, 1983

Winship TR, Thacore HR: Inhibition of vesicular stomatitis virus-defective interfering particle synthesis by Shope fibroma virus. Virology 93:515, 1979

INTERFERENCE BETWEEN VIRUSES

Aubertin A, Guir J, Kirn A: The inhibition of vaccinia virus DNA synthesis in KB cells infected with frog virus 3. J Gen Virol 8:105, 1970

Dubovi EJ, Youngner JS: Inhibition of pseudorabies virus replication by vesicular stomatitis viruses. J Virol 18:526, 1976

Ehrenfeld E, Lund H: Untranslated vesicular stomatitis virus messenger RNA after poliovirus infection. Virology 80:297, 1977

Hunt JM, Marcus PI: Mechanism of Sindbis virus-induced intrinsic interference with vesicular stomatitis virus replication. J Virol 14:99, 1974

Marcus PI, Carver DH: Intrinsic interference; a new type of viral interference. J Virol 1:334, 1967

CHAPTER 61

Antiviral Chemotherapy, Interferon, and Vaccines

Rational Approach to Antiviral Chemotherapy

Virus multiplication consists of the synthesis of viral nucleic acids and proteins and their assembly into virus particles. A rational approach to antiviral chemotherapy should examine whether and where these processes can best be interrupted without detriment to the host. The following is a relevant brief analysis.

Replication of Viral Nucleic Acids

The replication of the nucleic acids of many viruses is catalyzed by enzymes that do not exist in uninfected cells. This is true for all RNA-containing viruses, as well as for poxviruses, herpesviruses, and adenoviruses. It should be possible to isolate and characterize these enzymes and find specific inhibitors for them.

Synthesis of Viral Proteins

Several lines of evidence suggest that viral messenger RNAs differ in some fundamental way from host messenger RNAs. Viral messenger RNAs are generally translated in preference to host cell messenger RNAs. Furthermore, translation to host messenger RNAs often ceases entirely several hours after infection, when translation of viral messenger RNAs proceeds rapidly and extensively. This suggests that viral messenger RNAs differ in some recognizable manner from host messenger RNAs, and this difference should be exploitable. In fact, two antiviral agents, isatin-β-thiosemicarbazone and interferon, act by preventing viral messenger RNAs from being translated. A promising avenue of approach, therefore, is analysis of the features in which viral messenger RNAs differ from host cell messenger RNAs.

Viral Morphogenesis

Viral morphogenesis proceeds at several levels. Many viral capsid proteins are now known to be synthesized in the form of precursors that are cleaved to furnish the actual proteins used for the formation of virus particles. The nature of the enzymes that cleave these precursors is not known. It may be fruitful to characterize them and design inhibitors for them. In fact, recent experiments indicate that peptide analogs of amino acid sequences in the regions where viral proteins are cleaved possess antiviral activity.

No means of specifically inhibiting the assembly of virus particles exists. It may be possible to prevent budding when more is known about the properties of cell membranes, but it seems that cell membranes are such important organs that it may be unwise to attempt to prevent virus particle formation at this stage. By the same token, inhibition of the uptake and penetration of virus particles at the beginning of the multiplication cycle has not been attempted seriously. As will be discussed below, the drug α-adamantanamine does inhibit the uptake of certain myxoviruses, but it does not do so by directly affecting the functioning of the cell membrane. In summary, the two most promising avenues of approach for specifically inhibiting virus multiplication are inhibition of viral genome replicases and exploitation of the differences between viral and host cell messenger RNAs.

Mode of Action of Certain Antiviral Chemotherapeutic Agents

An appreciable number of compounds inhibit virus multiplication very efficiently in cultured cells and have no adverse effect on the growth of uninfected cells. Unfortunately, many of these compounds either inhibit virus multiplication in the body much less efficiently or have unacceptable toxic side effects. They are therefore not suitable for antiviral chemotherapy in man. However, clinical trials currently underway have revealed several compounds of great promise. Most of them act via one of the mechanisms discussed above.

Before considering individually the mode of action of successful and promising antiviral agents, it is well to examine which viral pathogens of humans present the most suitable targets for chemotherapy at the present time. It is reassuring to note that the incidence of many severe virus-caused diseases, such as smallpox, poliomyelitis, yellow fever, and measles, has decreased dramatically in recent decades, owing to the success of vaccination campaigns. Other important virus-caused diseases fall into two classes.

First, there are those that, although severe and life-threatening, are rare in most parts of the world; such diseases are rabies, the encephalitides, and the African and South American hemorrhagic fevers (Marburg disease, Ebola fever, Lassa fever, and Machupo). Second, there are diseases that are usually no more than mild to moderate but may become life-threatening; such diseases are caused by influenza virus and the various herpesviruses. In fact, herpesviruses may well be regarded as the primary targets for antiviral chemotherapy at this time, closely followed by influenza virus and the rare severe diseases mentioned above.

Isatin-β-thiosemicarbazone (IBT)

IBT is a potent inhibitor of poxvirus multiplication (Fig. 61-1). At a concentration of 3 mg per liter, it inhibits vaccinia virus multiplication in cultured cells by over 90 percent without having any detectable effect on the host cells themselves. In the presence of IBT, the early period of the poxvirus multiplication cycle, viral DNA replication, and transcription of late messenger RNA all proceed normally. However, the translation of late messenger RNA is inhibited. The synthesis of late proteins, which include most of the viral capsid proteins, is therefore prevented, and no progeny virus particles are formed.

A derivative of IBT, N-methyl-IBT (Marburan), was administered by mouth to known smallpox contacts in field trials in India and Pakistan, and beneficial results of this prophylactic treatment were observed.

2-Hydroxybenzylbenzimidazole (HBB) and Guanidine

These two reagents (Fig. 61-2) inhibit the multiplication of many picornaviruses, such as poliovirus, echoviruses, Coxsackie viruses, and foot-and-mouth disease virus. They are examples of reagents that interfere with the replication of viral RNA, either by preventing initiation of the synthesis of progeny plus strands or by preventing progeny plus strands from separating from the replicative form-replicase complex (Chap. 58). The precise manner in which this is accomplished is not known. Although both drug-resistant and drug-dependent virus mutants quickly emerge, HBB has been used with some success in controlling enterovirus infections in animals.

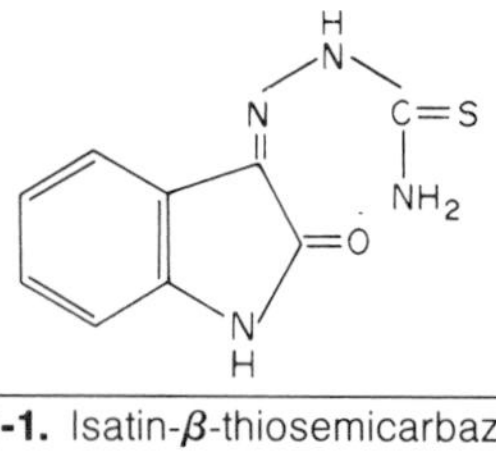

Figure 61-1. Isatin-β-thiosemicarbazone (IBT).

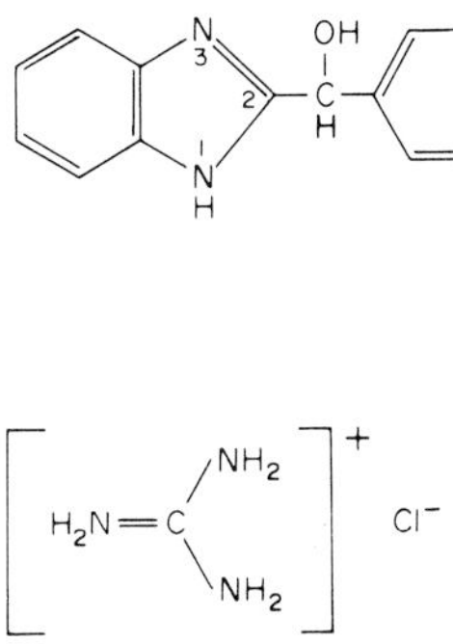

Figure 61-2. 2-Hydroxybenzylbenzimidazole (HBB) and guanidine hydrochloride.

Rifampin

Rifampin (see Fig. 9-24) and related rifamycin derivatives bind to bacterial RNA polymerases, thereby preventing the initiation of transcription. Rifampin does not bind to animal RNA polymerases, but it does inhibit the multiplication of poxviruses and adenoviruses. The mechanism by which it achieves this has been studied most intensively in vaccinia virus-infected cells. Inhibition of viral RNA polymerase is not involved, since both early and late messenger RNAs are transcribed normally. Rather, the mechanism involves some event in viral morphogenesis, for in the presence of the drug immature virus particles of the type illustrated in Figure 58-25 accumulate. The specific nature of the step that is inhibited is not known. However, it is clear that it involves a diffusible product, most probably a protein, since wild-type virus sensitive to rifampin matures normally in its presence in cells simultaneously infected with mutants resistant to it. Presumably the mutants code for a protein that is unaffected by the drug and can thus function when the normal wild-type protein cannot. As a result, resistant mutants rescue sensitive virus.

Although rifampin does not inhibit the vaccinia DNA-dependent RNA polymerase, the RNA-dependent DNA polymerase of RNA tumor viruses (the reverse transcriptase, Chap. 62) is very sensitive to it and to certain of its derivatives.

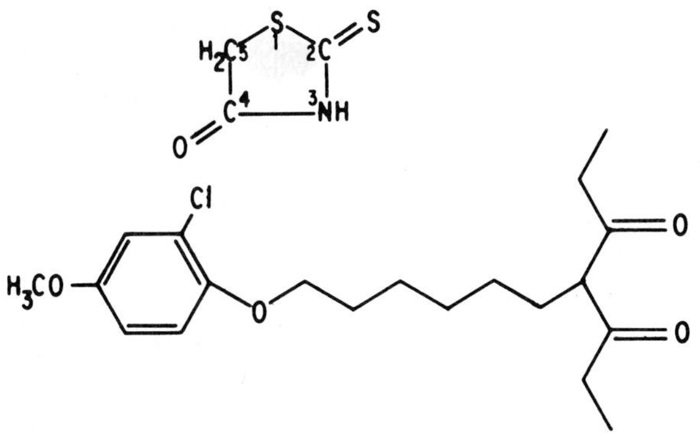

Figure 61-3. Top. Rhodanine (2-thio-4-oxothiazol idine). **Bottom.** Arildone (4-[6-(2-chloro-4-methoxyphenoxyl)hexyl]-3,5-heptanedione).

Arildone and Rhodanine

Arildone (4-[6-(2-chloro-4-methoxyphenoxy)hexyl]-3,5-heptanedione) and rhodanine (2-thio-4-oxothiazolidine) (Fig. 61-3) inhibit the uncoating of poliovirus and Echovirus 12, respectively. Neither drug affects adsorption or penetration, but both interact with the capsids of these viruses, increasing their stability to such an extent that active virus can be recovered from cells 4 hours after infection with treated virus. The mechanism by which arildone and rhodanine increase the stability of poliovirus and Echovirus capsids is not known.

α-Adamantanamine

α-Adamantanamine (Amantadine, Symmetrel) (Fig. 61-4), a substance with a remarkably rigid structure, inhibits an early event in the multiplication cycle of influenza virus (and also of arenaviruses). It does not inhibit adsorption and penetration but completely inhibits primary transcription. It is thought, therefore, that it inhibits uncoating. Drug-resistant mutants exist, which are mutants in the M protein gene. It is conceivable, therefore, that the drug interacts with the M protein.

On the basis of large-scale trials that demonstrated that it had a significant prophylactic effect, α-adamantanamine has been recommended for the prevention of disease caused by influenza virus type A. It has been used extensively in the Soviet Union and is presently the only drug for virus respiratory disease that is licensed in the United States. However, it is not used extensively in this country because it seems impractical to control, through chemoprophylaxis, a disease that is generally mild and that the individual patient has a good chance of avoiding anyway. As for the problem of protecting individuals at high risk and in whom the disease may be potentially dangerous, there is a choice between this drug and the influenza vaccine (p. 923). At least at this time, the vaccine would seem to be preferred.

Phosphonoacetic Acid (PAA)

Phosphonoacetic acid (Fig. 61-5) is a potent inhibitor of herpesvirus-coded DNA polymerase but not of cellular DNA polymerases. Unfortunately it has undesirable side effects, and drug-resistant mutants emerge rapidly. It is conceivable that modification of the structure of this drug will yield compounds to which resistance will develop less readily and that will be less toxic. The closely related

Figure 61-4. α-Adamantanamine (Amantadine).

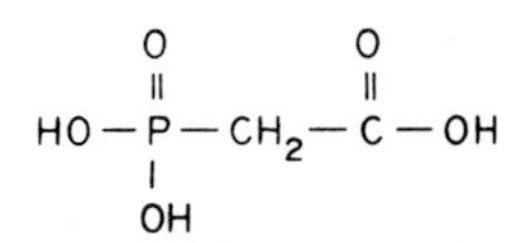

Figure 61-5. Phosphonoacetic acid (PAA).

phosphonoformic acid is also an inhibitor of herpesvirus DNA polymerase and is effective in treating cutaneous herpes simplex virus infections in guinea pigs.

Analogs of Ribonucleosides and Deoxyribonucleosides

The agents that have been discussed so far inhibit virus multiplication very well in cultured cells, but only IBT and α-adamantanamine are suitable antiviral agents in humans. The class of compounds that appears to have the greatest potential for antiviral chemotherapy in humans is the analogs of ribo- and deoxyribonucleosides. These substances consist of nucleic acid bases or derivatives of them, linked to either ribose or analogs of ribose.

The mechanism of action of these compounds is twofold. First, all become phosphorylated to mono-, di-, and triphosphates, the final precursors of nucleic acids. The analog-containing phosphorylated nucleosides may inhibit to a greater or lesser extent the various enzymes on the nucleic acid biosynthetic pathway and interfere competitively with the metabolism of normal phosphorylated nucleosides; indeed, many of the analogs of ribo- and deoxyribonucleosides with antiviral activity were designed and first tested as anticancer agents. Their importance as antiviral agents lies in the fact that if the virus that is to be inhibited codes for its own enzyme(s) on the nucleic acid biosynthetic pathway, such as a deoxyribonucleoside kinase or a DNA polymerase, this enzyme may be inhibited to a greater extent by the analog-containing phosphorylated nucleosides than the corresponding host cell enzyme, thus causing a specific antiviral effect. Further, the analog-containing nucleoside triphosphates are incorporated into nucleic acids and may then interfere with the functioning of such nucleic acids. For example, if the analog is 5'-iodouracil (IU), which is an analog of thymine, IU does not pair with adenine as faithfully as thymine, and mismatching occurs during both the replication and the transcription of the substituted DNA. This causes the formation of defective progeny DNA strands and defective messenger RNAs and inhibits virus multiplication.

Among the most promising analogs that have been studied are the following:

1. 5'-Iododeoxyuridine (IDU) and trifluorothymidine (F_3T) (Fig. 61-6), both analogs of thymidine. Both inhibit the multiplication of herpesviruses. IDU, applied topically to the eye, has been used successfully in the treatment of herpetic conjunctivitis. Although IDU is most probably incorporated into host cell as well as into viral DNA, it reaches so few host cells in the eye that no serious damage is

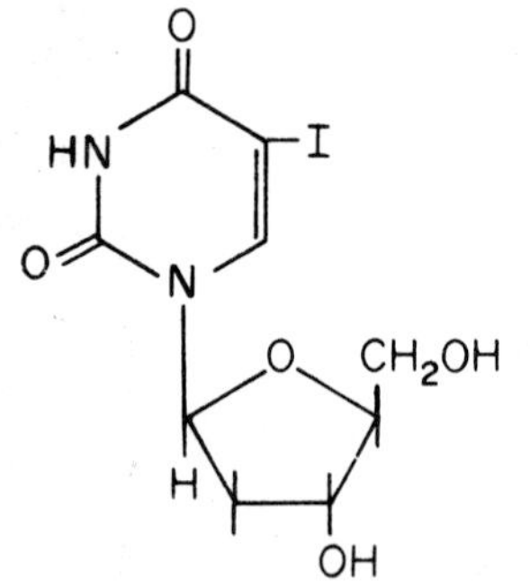

Figure 61-6. 5'-Iododeoxyuridine (IDU). In trifluorothymidine (F_3T), I is replaced by CF_3.

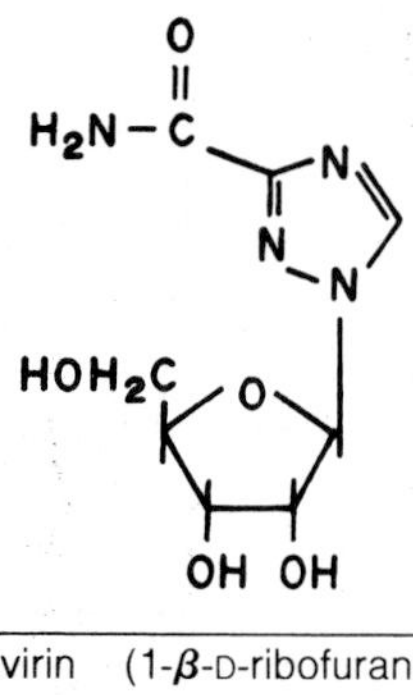

caused. At worst, damaged corneal cells regenerate after virus multiplication has been inhibited and infected cells have been eliminated.

2. Cytosine arabinoside (ara-C), adenosine arabinoside (ara-A or vidarabine), and ara-AMP. All these substances have had more or less extensive clinical trials as antiherpesvirus agents. They are inhibitors of DNA polymerases, and one of their effects is to act as chain terminators. The reason that they are antiviral agents is that herpesvirus DNA polymerase is more sensitive to them than are host cell DNA polymerases.

3. Ribavirin or virazole (Fig. 61-7), an analog of the purine precursor 5'-aminoimidazole-4-carboxamide. This substance inhibits the biosynthesis of guanine nucleotides. Its antiviral target is probably virus-coded enzymes that have a higher affinity for its phosphorylated derivatives than have normal cellular enzymes. It has a wide antiviral spectrum and is effective against RNA- as well as DNA-containing viruses.

4. (E)-5-(2-bromovinyl)-2'-deoxyuridine (BVdU) and E-5-(2-iodovinyl)-2'-deoxyuridine (IVdU); 5-vinyl-2'-deoxyuridine and 5-vinyl-2'-deoxycytidine; 1-(2'-deoxy-2'-F-β-D-arabinofuranosyl)-5'-iodocytosine (FIAC); and 9-[(1,3-dihydroxy-2-propoxy)-methyl]-guanine (DHPG) (Fig. 61-8). These compounds are the most recent and powerful antiherpesvirus compounds. They possess low cytotoxicity. All inhibit

Figure 61-7. Ribavirin (1-β-D-ribofuranosyl-1,2,4-triazole-3-carboxamide).

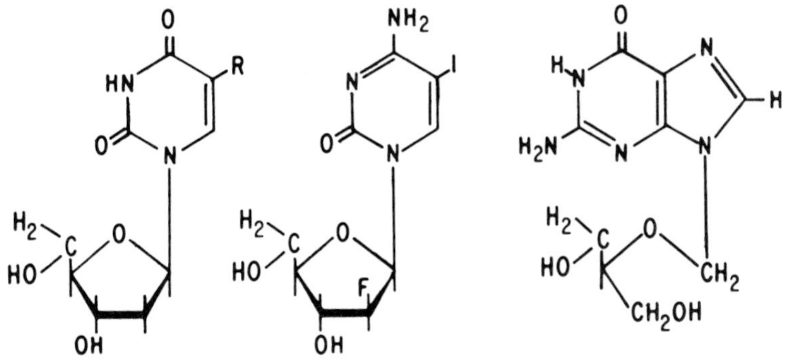

Figure 61-8. Left. R = —CH = CH$_2$, 5-vinyl-2'-deoxyuridine; R = —CH=CHBr, (E)-5-(2-bromovinyl)-2'-deoxyuridine. **Center**. FIAC. **Right**. DHPG.

the herpesvirus DNA polymerase much more powerfully than do cellular DNA polymerases (see also below).

Target Cell Approach

It has long been realized that the most successful strategy in antiviral chemotherapy would be to use drugs that could enter, or would be activated in, only virus-infected cells—in other words, a strategy that focused on the infected cell as the target cell.

Several approaches to the target cell strategy are as follows. The first takes advantage of the fact that one of the early herpesvirus-coded enzymes is a deoxypyrimidine kinase capable of phosphorylating compounds that cannot be phosphorylated by any cellular enzyme. It is, therefore, possible to use as highly specific antiherpesvirus agents analogs of nucleosides that are phosphorylated only in cells infected with herpesviruses. Examples of such compounds are acycloguanosine or acyclovir [9-(2-hydroxyethoxymethyl)-guanine] (Fig. 61-9), IVdU, and DHPG (Fig. 61-8). These drugs are converted to their triphosphates and incorporated into DNA much more extensively in herpesvirus-infected cells (by the herpesvirus-encoded deoxypyrimidine kinase) than in uninfected cells. They are now undergoing clinical trials against several herpesviruses.

The second approach takes advantage of the fact that infection with viruses often changes the permeability properties of cell membranes (Chap. 59). This provides the opportunity for introducing into infected cells antiviral agents that would not be taken up by uninfected cells. In fact, this approach should permit the administration of drugs that are quite toxic to normal cells.

The third approach is to couple powerful antiviral agents that may be so toxic that they could not be used on their own to monoclonal antibodies directed against virus-coded antigens on the surfaces of infected cells (Chap. 59). The monoclonal antibodies would, in effect, be used as homing devices. Once the coupled antibodies reached their target cells, they would presumably be internalized, and the antiviral drug would inhibit virus multiplication. Like the second, this approach still requires a great deal of research before becoming a practical reality.

Interferon

The antiviral agent on which interest is currently focused most intensely is one that is elaborated by living cells themselves. Animal cells infected with viruses very often produce a protein that, when added to uninfected cells, protects them against virus infection or, more precisely, greatly decreases the chance that subsequent virus infection will initiate a productive multiplication cycle. This substance is called "interferon." Interferon provides the first line of defense against virus infection: in animals treated with antiserum to interferon, virus infections progress much more rapidly and cause much more severe disease than in untreated animals.

Interferons are host-coded proteins. Normal cells do not as a rule contain detectable amounts of interferon. The formation of interferon is induced by a variety of agents, the best known of which is infection with a virus (see below). Since it is a cellular protein, interferon is host cell species-specific but not virus-specific. Sometimes this specificity is very narrow. For example, chick and duck interferons exhibit little if any cross-protective activity, nor do mouse and rat interferons. However, not unexpectedly, there are exceptions; for example, human interferon protects bovine cells even better than does bovine interferon.

There is not one type of human interferon but three: IFN-α and IFN-β, produced predominantly by leukocytes and fibroblasts, respectively, in response to virus infection, and IFN-γ, formerly called "immune" or type 2 interferon, which is produced by unsensitized lymphoid cells in re-

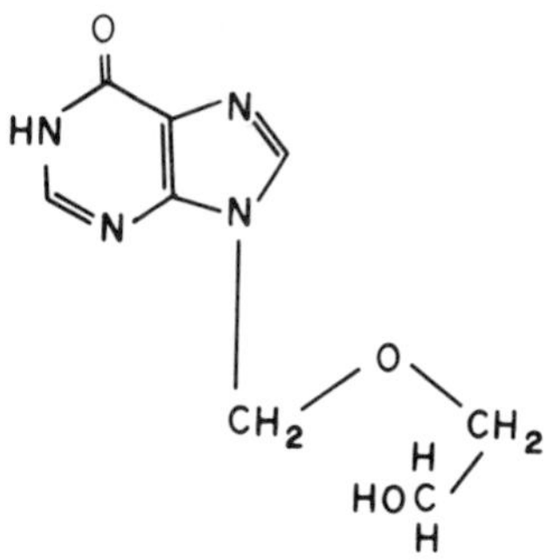

Figure 61-9. Acycloguanosine.

TABLE 61-1. CHARACTERISTICS OF HUMAN INTERFERONS

Type	Alpha (Leukocyte)	Beta (Fibroblast)	Gamma (Immune or class 2)
Produced by	Peripheral leukocytes	Fibroblasts	Lymphocytes
Inducing agent	Virus infection, dsRNA	Virus infection, dsRNA	Mitogens (nonsensitized lymphocytes), specific antigen (sensitized lymphocytes)
Number of genes	14	At least 2, possibly 5	1
Presence of introns	No	No	Yes
Chromosomal location	9	9,2,5	12
Size of primary protein (number of amino acids)	166	166	166
Length of signal sequence (number of amino acids)	23	21	20
Size of actual protein (number of amino acids)	143	145	146
Molecular weight	17,000	17,000	17,000
Glycoprotein	No	Yes	Yes
Stability at pH 2	Yes	Yes	No
Activity in the presence of SDS	Yes	Yes	No

sponse to mitogens and by sensitized lymphocytes when stimulated with specific antigens (Table 61-1).

Nature of Interferon Genes

Recombinant DNA technology (cloning of some species of interferon, and the use of cloned genes to search through human gene libraries) has demonstrated that there are at least 14 different human IFN-α genes, from 2 to 5 IFN-β genes, and 1 IFN-γ gene. All these genes encode proteins 166 amino acids long with 20 to 23 amino acid long N-terminal signal sequences, which are necessary because interferons are secreted proteins. These signal sequences are not present in normal interferon molecules.

The various IFN-α genes are 80 to 90 percent related to each other. The coding sequences of the IFN-α and IFN-β gene families are about 30 percent related to each other, which indicates that they diverged at least 500 million years ago, before the emergence of vertebrates. The IFN-γ gene exhibits essentially no homology to the other interferon genes but codes for a similar sized protein in which certain single amino acids occupy the same positions as in α and β interferons.

Most if not all IFN-α genes are arranged tandemly on chromosome 9. IFN-β genes are dispersed throughout the genome, one also being located on chromosome 9 and the others on chromosomes 2 and 5, and the IFN-γ gene is located on chromosome 12. The latter is the only interferon gene that possesses introns.

In addition to all these genes, the human genome also contains at least six interferon pseudogenes whose sequences are very similar but that contain alterations or deletions that prevent the expression of full length interferon proteins.

Induction and Production of Interferons

The formation of interferons can be induced by a variety of conditions and agents other than infectious virus. Among these are:

1. Viruses inactivated by heat or ultraviolet irradiation. Among such viruses are influenza virus, NDV, reovirus, rotavirus, and bovine enterovirus.
2. Double-stranded RNA, such as reovirus RNA, and synthetic double-stranded polyribonucleotides, such as poly (I):poly (C). The amount of interferon that is induced by poly (I):poly (C) can often be increased greatly by *superinduction*. This is a phenomenon that is elicited when cells are treated with poly (I):poly (C) together with the protein synthesis inhibitor, cycloheximide. If, after 5 hours, actinomycin D is added, followed 1 hour later by reversal of the inhibition of protein synthesis, about 50 times more interferon is produced than if cycloheximide had not been added initially. The basis of this phenomenon is thought to be the existence of an unstable regulator of the expression of the interferon gene. During the first 5-hour period, both interferon and regulator messenger RNA would accumulate. After the addition of actinomycin D, both interferon and regulator messenger RNA would decay, the latter more rapidly than the former. Upon the release of the protein synthesis block, only interferon would then be formed in the presence of greatly reduced levels of regulator.
3. Single-stranded RNA and single-stranded or double-stranded DNA are inactive as interferon inducers, as are double-stranded RNA-DNA hybrid molecules. However, certain synthetic polycarbo-

xylic acids and pyran copolymers are inducers of interferon formation at relatively high concentrations, as is bacterial endotoxin. The feature common to all these substances, including the polyribonucleotides, is that they are polyanions.

4. The preceding agents induce IFN-α and IFN-β. IFN-γ is induced by stimulation of lymphocytes, as described above.

The mechanism by which interferon formation is induced is not clear, nor is it known why substances as diverse as double-stranded RNA and endotoxin can induce interferon, whereas a substance as similar to double-stranded RNA as double-stranded DNA cannot.

In practice, IFN-α is usually prepared from leukocytes, obtained as the buffy coat layer by low-speed centrifugation of human blood. Since IFN-α genes are located close to each other, they are generally induced coordinately. IFN-β species are prepared from fibroblasts grown in monolayer culture, induced with poly (I):poly (C) and superinduced with inhibitors of RNA and protein synthesis (see above). Since IFN-β genes are dispersed, they tend to be expressed in a noncoordinate manner, that is, different strains of diploid fibroblasts express different relative amounts of the various IFN-β genes.

The disadvantages of these methods of producing interferon are (1) interferon is produced in small amounts only (typically about 1 mg interferon per 10 liters of tissue culture supernatant medium), and (2) they are destructive, that is, the induced cells produce interferon for 48 to 72 hours and then die.

Recently two advances have enormously improved methods of producing and purifying interferon. First, the cloned IFN-α, IFN-β, and IFN-γ genes have been inserted into expression vectors by genetic engineering technology, and interferons can now be produced in large amounts in either procaryotic or eucaryotic cells that synthesize them constitutively. Second, monoclonal antibodies to all classes of interferon have been obtained and are now being used to purify interferons efficiently and rapidly by means of immunoaffinity chromatography. The biologic properties of the cloned interferons are the same as those of naturally occurring interferons.

Assay of Interferons

Interferons are commonly assayed by exposing cells to serially diluted interferon preparations for about 12 hours and then infecting them with a standard amount of virus known to produce a certain number of plaques. The titer of an interferon preparation is the reciprocal of that dilution that reduces this number by 50 percent. The specific activity of pure interferons is 5 to 10 $\times$ 10^8 PRD_{50} (50 percent plaque reduction doses) per milliliter. This astonishingly high biologic activity means that in some very sensitive cell-virus systems interferon activity is detectable at concentrations as low as 10^{-14} M.

Nature of Interferons

The IFN-β and IFN-γ species are glycoproteins; the IFN-α species are not. IFN-α and IFN-β species, but not IFN-γ species, are remarkably resistant to low pH—both are quite stable at pH 2 at 4C—and retain activity in the presence of sodium dodecylsulfate, an unusual property that is very useful in the final stages of purification.

There is evidence for two highly conserved domains in the interferon molecule: one, in the aminoterminal half of the molecule, probably contains the site that binds to the cell surface receptor, and the other, in the other half of the molecule, appears to modulate such binding and mediate other biologic functions.

As isolated from cells and organs, interferons sometimes appear to be larger than the 17,000 dalton molecules specified by their genes. There are several possible reasons for this. First, some interferons are glycoproteins; second, some naturally occurring interferon species may be dimers; and third, under certain conditions interferons may associate or complex with other proteins.

The three classes of human interferon are immunologically quite distinct, that is, antisera raised against any one class do not inactivate heterologous interferons.

Finally, interferons possess two types of biologic activity: antiviral activity and anticellular activity. Both types of activity vary markedly with the nature of the cell. The antiviral activity of all three classes of interferon is comparable, but the anticellular activity of IFN-γ seems to be far greater than that of IFN-α and IFN-β.

Antiviral Activity of Interferons

Interferons interfere with virus multiplication; this is the property for which they are named. Interferons themselves are not the proteins that actually inhibit virus multiplication, but rather, interferons are inducers that cause cells to synthesize proteins that are the actual effectors of the antiviral state. In fact, interferons do not induce the antiviral state if messenger RNA or protein synthesis is inhibited. The antiviral state usually lasts for several days and then decays. Expression of the antiviral state is controlled by a gene on chromosome 21 that probably codes for a specific interferon receptor on the cell surface. Interferons do not enter cells; they express themselves via second messengers that are activated at the cell surface. Interferon injected into cells fails to induce the antiviral state.

It seems clear that the basic mechanism by which interferons inhibit virus multiplication involves interference with the ability of parental or early viral messenger RNA molecules to be translated. As a result, no virus-specified proteins are synthesized, no progeny viral genomes are formed, and infection is aborted.

The precise mechanism by which interferon inhibits viral protein synthesis is under intense investigation. It has been found that treatment with interferon—and all three classes of interferon act similarly—induces the synthesis of several cellular proteins. One is an enzyme, 2',5'-oligo (A)

TABLE 61-2. EFFECT OF INTERFERONS ON CELL FUNCTIONS

Effect	Dose Required (PRD$_{50}$/ml)	Description
Inhibition of cell growth	0.2–1000	Most cell types are sensitive but differ widely in sensitivity; tumor cells may be somewhat more sensitive than normal cells; molecular basis not known
Enhancement of differentiation	1–1000	Seen most strongly in functioning effector cells of the immune system
Inhibition of differentiation	100–1000	Observed when induced RNA and protein synthesis is required for expression of differentiation
Effects on the immune system	10–1000	Inhibition of proliferative responses, such as primary and secondary antibody response, mitogen- and antigen-induced lymphocyte blastogenesis, and delayed-type hypersensitivity
		Enhancement of differentiated responses, such as a natural killer cell activation, T cell cytotoxicity, macrophage phagocytosis and cytotoxicity, and IgE-mediated histamine release
Increased expression of histocompatibility antigens (HLA-A, B, and C) and β microglobulin	1–1000	May lead to increased susceptibility of infected and malignant cells to cell killing by effector cells that use these antigens for cell recognition; this effect is far stronger for IFN-γ than for IFN-α or IFN-β

synthetase, which, in the presence of double-stranded RNA, converts ATP to 2′-5′-linked oligo (A) molecules up to 15 residues long, the trimer being the most abundant. This 2′,5′-oligo (A) then activates an endoribonuclease, RNase L, that is present in normal cells in widely varying amounts and the level of which increases in some types of cells in response to interferon treatment. It is hypothesized that this enzyme hydrolyzes viral messenger RNAs, thereby inhibiting viral protein synthesis.

The principal problems with this explanation of the mechanism of interferon action are: (1) What is the source of the double-stranded RNA (especially for DNA-containing viruses)? (2) Can it account for the selectivity of interferon action? (Note that interferon prevents the synthesis of viral proteins but not of host cell proteins, yet RNase L hydrolyzes both viral and host cell RNAs.), (3) In certain cells the antiviral state is established without 2′,5′-oligo (A) synthetase being induced, while in others the synthesis of 2′,5′-oligo (A) synthetase is induced, but the antiviral state is not established.

Another effect of interferon is that it causes the phosphorylation of a 67,000 dalton membrane-associated protein that is itself a protein kinase, and in the presence of double-stranded RNA, phosphorylates the α subunit of protein synthesis initiation factor eIF-2, thereby inactivating it. The problems with this explanation of how interferon inhibits viral protein synthesis are similar to those enumerated above. It remains to be seen whether either or neither of these mechanisms is the basis of the antiviral state.

Anticellular Activity of Interferons

In addition to their antiviral activity, interferons also affect a variety of cell functions, summarized in Table 61-2. It is conceivable that all these functions are mediated by messenger RNAs that share certain sequences and that the primary function of interferon is to regulate the expression of these *cellular* messenger RNAs rather than those of viral

messenger RNAs (which may be inhibited because they are structurally similar). Thus interferon may have evolved as a cellular regulator rather than as an antiviral defense mechanism.

Their remarkable and varied effects on immunoregulatory mechanisms have recently focused intense attention on interferons as lymphokines, soluble effectors of immune reactions. This, coupled with their cell multiplication inhibitory activity (CMI), the molecular basis of which is not known and which is often greater against tumor than against normal cells, has given rise to hopes that interferons may be useful as anticancer agents in humans, and numerous clinical trials are currently underway to test this possibility. All classes of interferons are under investigation, but expectations are highest for IFN-γ, for which the ratio of anticellular to antiviral activity is much higher than for IFN-α and IFN-β. Such clinical trials were impossible until recently because of the very large amounts of interferon that are required. Daily injections of 1 to 5 million PRD$_{50}$ are generally regarded as necessary for obtaining a therapeutic effect. Such large amounts of interferon could not be prepared from leukocytes or cultured fibroblasts except for only a very few patients. However, the situation is now changing radically as large amounts of cloned interferons are becoming available.

Prospects for Clinical Use

Extensive clinical trials are currently underway to evaluate interferons as anticancer agents (for the treatment of breast cancer, lung cancer, brain tumors, osteogenic sarcoma, myeloma, lymphoma, hepatoma, and others) and as antiviral agents (in such life-threatening infections as rabies, hemorrhagic fevers, and herpesvirus-caused encephalitis, as well as in persistent infections, such as hepatitis B, herpesvirus zoster, wart viruses, and cytomegalovirus infections). Interferon treatment does cause some undesirable side effects, such as fever, hypotension, myalgia, tachycardia, and impaired liver function, but such side effects may be acceptable in the context of life-threatening disease. The primary need at this time is large-scale well-controlled clinical investigations. There is no doubt that the availability of large amounts of cloned interferon species will usher in a new era in the employment of interferon as an antiviral and as an anticancer agent. In particular, as the characteristics of the functional domains on interferon molecules become defined, recombinant DNA technology may permit the construction of nontoxic interferon molecules tailor-made for the treatment of specific infections or tumors.

Vaccines

The only currently feasible means of preventing diseases caused by viruses is through mobilization of the immune mechanism by means of vaccines. There are two types of antiviral vaccines: inactivated virus and attenuated active virus.

Inactivated Virus Vaccines

The primary requirements for an effective vaccine of this type are complete inactivation of infectivity coupled with minimum loss of antigenicity. These requirements are not easily met simultaneously, since few reagents are available that inactivate viral genomes, the source of infectivity, without also affecting viral protein, the source of antigenicity. Ultraviolet irradiation could accomplish this best but is inapplicable because virus inactivated in this manner is capable not only of expressing the function of those genes that have not received a lethal hit but also of undergoing multiplicity reactivation (Chap. 60). Photodynamic inactivation, in which virus is treated with such dyes as neutral red dyes that are capable of intercalating between adjacent nucleic acid bases, and is then irradiated with white light, inactivates viral nucleic acids efficiently and irreversibly without damaging viral proteins, but suffers from similar drawbacks in that genome segments that have not received a hit remain potentially functional. For example, herpesvirus inactivated in this manner can still transform cultured cells in vitro and may be tumorigenic. Beta-propiolactone is a potentially useful inactivating agent but is a potent carcinogen. The best reagent for inactivating viral nucleic acid without compromising antigenicity is formaldehyde, which destroys infectivity primarily by reacting with those amino groups of adenine, guanine, and cytosine that are not involved in hydrogen bond formation. Viruses that contain single-stranded nucleic acid are therefore inactivated readily, while those that contain double-stranded nucleic acid are resistant to formaldehyde. It also reacts with amino groups of proteins, forming addition compounds of the Schiff's base type and crosslinking polypeptide chains without, however, significantly disturbing protein conformation. Reaction with protein is most prob-

ably responsible for the occasional generation of a formaldehyde-resistant infectious virus fraction, which appears to be caused by such extensive crosslinking of capsid proteins that formaldehyde cannot reach and inactivate the viral nucleic acid. Careful control of reaction conditions and rigorous checks for residual infectious virus are mandatory for the preparation of formaldehyde-inactivated virus vaccines.

Attenuated Active Virus Vaccines

The second method of immunizing against viral pathogens is by administering attenuated virus strains, antibody to which is capable of neutralizing the pathogen (Chap. 57). This is the principle on which Jenner's vaccination procedure against smallpox in 1798 was based. He inoculated with cowpox virus to induce antibodies against the highly virulent smallpox virus. Since then many attenuated virus vaccine strains have been developed, among them Theiler's yellow fever virus vaccine strain, the attenuated Sabin poliovirus vaccine strains, and attenuated measles and rubella virus strains. The most commonly used method of producing such attenuated virus strains is by repeated passage of the human pathogen in other host species, which results in the selection of multistep variants with drastically reduced virulence for humans.

Attenuated virus vaccines are effective in very small amounts, since the attenuated virus can multiply. This provides a powerful amplification effect: the viral progeny, rather than the virus in the inoculum, acts as the antigen. The attenuated vaccines also possess the advantage of stimulating the formation of all the correct types of antibody molecules (i.e., IgA, etc., as well as IgG).

Immunization with Chemically Synthesized Antigenic Determinants

An entirely new perspective for the development of antiviral vaccines has opened recently as a result of recombinant DNA technology. Numerous viral genomes are now being cloned and sequenced, and as a result the amino acid sequences of many proteins, including those of the proteins that elicit neutralizing antibodies, have become known. Antibodies are made against relatively short sequences at the antigenic determinant sites of proteins. Therefore, knowing the amino acid sequence of a type-specific, neutralizing- antibody-eliciting viral antigen, peptides about 10 amino acids long that correspond to its antigenic determinants can be synthesized chemically, and these peptides will elicit the formation of fully functional, high-affinity neutralizing antibodies. Such synthetic peptide vaccines should be economical to produce and safe. They are the ideal vaccines, presenting the minimum amount of antigen. Several such vaccines are under development at this time and appear to be effective in containing, limiting, and eliminating viral infections. No doubt such vaccines will soon be developed against influenza virus and paramyxoviruses, hepatitis B, herpesviruses, rotaviruses, arena-viruses, togaviruses, and others.

FURTHER READING

Books and Reviews

CHEMOTHERAPEUTIC AGENTS

Gauri KK (ed): Anti-herpes Virus Chemotherapy: Experimental and Clinical Aspects. New York, Karger, 1979

Grunert RR: Search for antiviral agents. Annu Rev Microbiol 33:335, 1979

Hann FE (ed): Virus Chemotherapy. New York, Karger, 1980

INTERFERON

Billiau A: Interferon therapy: Pharmacokinetic and pharmacological aspects. Arch Virol 67:121, 1981

De Maeyer E, De Maeyer-Guignard J: Interferons. In Fraenkel-Conrat H, Wagner RR (eds): Comprehensive Virology. New York, Plenum, 1979, vol 15, p 205

Gordon J, Minks MA: The interferon renaissance: Molecular aspects of induction and action. Microbiol Rev 45:244, 1981

Gresser I (ed): Interferon 1979. New York, Academic Press, 1979

Gresser I, Tovey MG: Antitumor effect of interferon. Biochim Biophys Acta 516:231, 1978

Hayes TG: Differences between human α (leukocyte) and β (fibroblast) interferons. Arch Virol 67:267, 1981

Lengyel P: Biochemistry of interferons. Annu Rev Biochem 51:251, 1982

Rubinstein M: The structure of human interferons. Biochim Biophys Acta 695:5, 1982

Sehgal PB: The interferon genes. Biochim Biophys Acta 695:17, 1982

Stewart WE II: The Interferon System. New York, Springer-Verlag, 1979

Taylor-Papadimitriou J, Balkwill FR: Implications for clinical application of new developments in interferon research. Biochim Biophys Acta 695:49, 1982

VACCINES

Arnon R: Chemically defined antiviral vaccines. Annu Rev Microbiol 34:593, 1980

Chanock RM: Strategy for development of respiratory and gastrointestinal tract viral vaccines in the 1980s. J Infect Dis 143:364, 1981

Lerner RA, Sutcliffe JG, Shinnick TM: Antibodies to chemically synthesized peptides predicted from DNA sequences as probes of gene expression. Cell 23:309, 1981

Melnick JL: Viral vaccines. Prog Med Virol 23:158, 1977

Sutcliffe JG, Shinnick JM, Green N, et al.: Antibodies that react with predetermined sites on proteins. Science 219:660, 1983

Selected Papers

CHEMOTHERAPEUTIC AGENTS

Allaudeen HS, Descamps J, Sehgal PK, et al.: Selected inhibition of DNA replication in herpes simplex virus infected cells by 1-(2'-deoxy-2'-fluoro-β-D-arabinofuranosyl)-5-iodocytosine. J Biol Chem 257:11879, 1982

Benedetto A, Rossi GB, Amici C, et al.: Inhibition of animal virus production by means of translation inhibitors unable to penetrate normal cells. Virology 106:123, 1980

Carrasco L: Membrane leakiness after viral infection and a new approach to the development of antiviral agents. Nature 272:694, 1978

Cheng Y-C, Huang E-H, Lin J-C, et al.: Unique spectrum of activity of 9-[(1,3-dihydroxy-2-propoxy)methyl]-guanine against herpesviruses in vitro and its mode of action against herpes simplex virus type 1. Proc Natl Acad Sci USA 80:2767, 1983

Descamps J, DeClercq E: Specific phosphorylation of E-5-(2-iodovinyl)-2'-deoxyuridine by herpes simplex virus-infected cells. J Biol Chem 256:597, 1981

Katz E, Moss B: Formation of a vaccinia virus structural polypeptide from a higher molecular weight precursor: Inhibition by rifampicin. Proc Natl Acad Sci USA 66:677, 1970

Korant BD: Poliovirus coat protein as the site of guanidine action. Virology 81:25, 1977

Lubeck MD, Schulman JL, Palese P: Susceptibility of influenza A viruses to amantadine as influenced by the gene coding for the M protein. J Virol 28:710, 1978

Richardson CD, Scheid A, Choppin PW: Specific inhibition of paramyxovirus and myxovirus replication by oligopeptides with amino acid sequences similar to those at the N-terminal of the F_1 or HA_2 viral polypeptides. Virology 105:205, 1980

Schlegel R, Dickson RB, Willingham ME, et al.: Amantadine and dansylcadaverine inhibit vesicular stomatitis virus uptake and receptor-mediated endocytosis of α_2-macroglobulin. Proc Natl Acad Sci USA 79:2291, 1982

Whitley RJ, Alford CA: Developmental aspects of selected antiviral chemotherapeutic agents. Annu Rev Microbiol 32:285, 1978

Willis RC, Carson DA, Seegmiller JE: Adenosine kinase initiates the major route of ribavirin activation in a cultured human cell line. Proc Natl Acad Sci USA 75:3042, 1978

Woodson B, Joklik WK: The inhibition of vaccinia virus multiplication by isatin -β-thiosemicarbazone. Proc Natl Acad Sci USA 54:946, 1965

INTERFERON

Creasey AA, Eppstein DA, March YV, et al.: Growth regulation of melanoma cells by interferon and (2'-5')oligoadenylate synthetase. Mole Cell Biol 3:780, 1983

Friedman RM, Sonnabend JA: Inhibition of interferon action by p-fluorophenylalanine. Nature 203:366, 1964

Jacobsen H, Czarnieck CW, Krause D, et al.: Interferon-induced synthesis of 2-5A-dependent RNase in JLS-V9R cells. Virology 125:496, 1983

Joklik WK, Merigan TC: Concerning the mechanism of action of interferon. Proc Natl Acad Sci USA 56:558, 1966

Kerr IM, Brown RE: pppA2'p5'A2'p5'A: An inhibitor of protein synthesis synthesized with an enzyme fraction from interferon-treated cells. Proc Natl Acad Sci USA 75:256, 1978

Krust B, Rivière Y, Hovanessian AG: p67K kinase in different tissues and plasma of control and interferon-treated mice. Virology 119:240, 1982

Le J, Vilcek J, Saxinger C, et al.: Human T cells hybridomas secreting immune interferon. Proc Natl Acad Sci USA 79:7857, 1982

Miyata T, Hayashida H: Recent divergence from a common ancestor of human IFN-α genes. Nature 295:165, 1982

Rehbert E, Kelder B, Hoal EG, et al.: Specific molecular activities of recombinant and hybrid leukocyte interferons. J Biol Chem 257:11497, 1982

Samuel CE, Knutson GS: Mechanism of interferon action. J Biol Chem 257:11791, 1982

Shaw GD, Boll W, Taira H, et al.: Structure and expression of cloned mouse IFN-α genes. Nucl Acid Res 11:555, 1983

Slate DL, D'Eustachio P, Pravtcheva D, et al.: Chromosomal location of a human α interferon gene family. J Exp Med 155:1019, 1982

Stewart WE II, Gosser LB, Lockhart RZ: Priming: A nonantiviral function of interferon. J Virol 7:792, 1971

Stewart WE II, de Somer P, Edy VG, et al.: Human interferons: Requirements for stabilization and reactivation of human leukocyte and fibroblast interferon. J Gen Virol 26:327, 1975

Tanaka S, Oshima T, Ohsuye K, et al.: Expression in *Escherichia coli* of chemically synthesized gene for the human immune interferon. Nucl Acid Res 11:1707, 1983

Virelizier J, Gresser I: Role of interferon in the pathogenesis of viral diseases of mice as demonstrated by the use of anti-interferon serum. J Immunol 120:1616, 1978

Wallach D, Fellous M, Revel M: Preferential effect of γ-interferon on the synthesis of HLA antigens and their mRNAs in human cells. Nature 299:833, 1982

Wiebe ME, Joklik WK: The mechanism of inhibition of reovirus replication by interferon. Virology 66:229, 1975

Zilberstein A, Kimichi A, Schmidt A, et al.: Isolation of two interferon-induced translational inhibitors: a protein kinase and an oligo-isoadenylate synthetase. Proc Natl Acad Sci USA 75:4734, 1978

VACCINES

Bittle JL, Houghten RA, Alexander H, et al.: Protection against foot-and-mouth disease by immunization with a chemically synthesized peptide predicted from the viral nucleotide sequence. Nature 298:30, 1982

Neurath AR, Kent SBH, Strick N: Specificity of antibodies elicited by a synthetic peptide having a sequence in common with a fragment of a virus protein, the hepatitis B surface antigen. Proc Natl Acad Sci USA 79:7871, 1982

CHAPTER 62

Tumor Viruses

Viruses are at the forefront of a large effort that is being devoted to discovering how normal cells are turned into cancer cells; they are at the cutting edge of studies of the mechanisms of tumorigenesis. There are two reasons for this. First, infection with tumor viruses transforms *all* cells in a population into potential tumor cells; with all other agents the frequency of transformation is far lower. As a result, studies at the biochemical and molecular level are possible on virus-induced transformation but are not possible on transformation caused by other agents. Second, mutants of tumor viruses exist that are incapable of transforming cells or maintaining the transformed state at nonpermissive temperatures (typically 38C or above), but capable of doing so at permissive temperatures (typically 35C or below). Transformation of cells is therefore a function of a *single* virus-coded protein. Clearly, it should be possible to isolate such proteins, determine what their function is, identify their targets, and thereby gain insight into the basic reactions that cause normal cells to be converted into tumor cells. This is impossible with any other carcinogenic agent.

Both RNA-containing and DNA-containing viruses can cause various types of neoplasms in animals, and most probably also in humans. Awareness of the principles of tumor virology is, therefore, important for the medical practitioner.

Origins of Tumor Virology

Tumor etiology, the study of the causes of tumors, has brought to light a bewildering variety of tumorigenic agents that fall into three classes: chemical substances, physical stimuli, and biologic agents. The chemical substances include compounds of the most diverse constitution, ranging from polycyclic hydrocarbons, such as methylcholanthrene, benzo(a)pyrene, and dimethylbenzanthracene, on the one hand, to such multifunctional compounds as dimethylnitrosamine, nitrosomethylurea, and 4-nitroquinoline-1-oxide, on the other. The physical agents include both x-rays and ultraviolet irradiation, and the most important biologic agents are viruses.

The concept that infectious agents might be involved etiologically in the cancer process was advanced as early as 1908 by Ellerman and Bang, who observed that the mode of transmission of leukemia in the fowl was similar to that of an infectious disease. Shortly thereafter, Rous demonstrated that the infectious agent in avian sarcomas could pass through filters that would not permit passage of bacteria. For many years this discovery remained an isolated finding, until in 1932 Shope discovered in wild cottontail rabbits a viral agent that transmitted a wart-like growth not only to cottontails but also to domestic rabbits. While most warts in cottontails remained benign or regressed, those in domestic rabbits sometimes developed into highly malignant carcinomas. Then, in 1938, Bittner discovered, in mammary gland tumors of mice, the mammary tumor

virus or milk factor, a virus that is passed from mother to offspring through suckling. The finding that, more than any other, elicited the upsurge of interest in viruses as carcinogenic agents was the discovery by Gross in 1951 of a virus that induced leukemia in mice, a disease that is remarkably similar to leukemia in humans. Efforts to develop animal models applicable to the human disease have resulted in the isolation of a large number of viruses that cause many kinds of cancers in every major group of animals.

Characteristics of Virus-transformed Cells

Discussion of the virus–cell interaction up to this point has been focused on the lytic interaction, which involves multiplication of the virus and destruction of the host cell. However, certain DNA viruses can interact with cells not only by means of the lytic interaction but also by means of an interaction in which virus multiplication is repressed and the host cell is not destroyed. In this type of interaction the viral DNA is either integrated into the genome of the host cell, or it replicates as a plasmid. In either case a new entity, the transformed or tumor cell, is created that can multiply indefinitely.

In addition, a family of RNA-containing viruses, the RNA tumor viruses (oncornaviruses or retroviruses), can also transform cells and cause tumors. At first, the problem of how viral RNA could be integrated into the host genome presented a major conceptual hurdle. This has now been overcome by the discovery in RNA tumor viruses of a DNA polymerase that can transcribe RNA into DNA; this is the so-called reverse transcriptase. It is the DNA transcribed from viral RNA by this enzyme that is integrated into the cellular genome.

Transformed or tumor cells differ from normal cells in several important respects. Since these changes are at the genetic level, they are passed on to the cells' descendants. The principal properties in which virus-transformed cells differ from normal ones are as follows.

Possession of the Viral Genome

Transformed cells contain either the whole or part of the genome of the virus that causes the transformation. More often than not, this genome is integrated into the cell DNA, but in some transformed cell nuclei it may exist as a free plasmid-like entity (Chap. 7). The extent to which viral genetic information is expressed in transformed cells varies widely, from full expression, with resultant formation of progeny virus particles, to complete silence, as judged by the absence of messenger RNA transcribed from it or proteins encoded by it.

Tumorigenicity

Transformed cells generally give rise to tumors when injected into animals, particularly immunosuppressed or immunologically deficient animals, such as athymic nude mice. Like naturally occurring tumors, different lines of virus-transformed cells exhibit wide variation in invasiveness. Even large numbers of some types of transformed cells produce only benign tumors at the site of infection, while even small numbers of others give rise to highly invasive cancers.

Morphology

Normal and transformed cells differ in morphology. There are two major differences. First, transformed cells are usually more rounded and refractile than are normal cells. Second, normal and transformed cells differ in the orientation of cells relative to each other. Normal cells usually arrange themselves in regular patterns, while transformed cells tend to orient themselves randomly (Fig. 62-1). These morphologic changes are often virus-specific. There are variants of Rous sarcoma virus and polyoma virus that give rise to transformed cells with morphologies that differ from those induced by the corresponding wild-type viruses.

Changes in Growth Patterns

Most types of untransformed cells grow in vitro to a certain cell density. Cell division stops (or almost stops) when a monolayer of uniformly spread cells has formed. Under the same conditions, transformed cells continue to divide so that very much higher cell densities are attained. This alteration of growth patterns is due primarily to two factors.

Loss of Contact Inhibition. When an untransformed cell comes into contact with another cell, there is cessation of the rapid movement of pseudopodia (ruffles) that are constantly extended and retracted, and forward movement is arrested. At the same time the cell stops dividing. This dual phenomenon is known as "contact inhibition." The ability to respond to contact with other cells ensures that any given cell grows only in its appropriate location within the complete organism and does not proliferate unless it receives a signal that more of its kind are needed for the organism's orderly growth or maintenance.

While cell strains differ appreciably in their susceptibility to contact inhibition, transformed cells are always much less susceptible than their untransformed counterparts. This loss of contact inhibition represents release from one of the normal controls over multiplication, and it may account in part for their unregulated growth in the organism. Some transformed cells, such as cells of benign tumors, have lost only the ability to respond to contact with each other but are still inhibited by contact with other cell types. Others, such as those of metastasizing tumors,

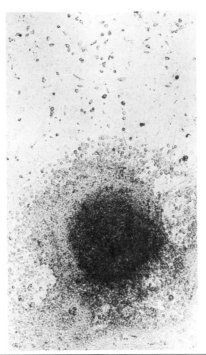

Figure 62-1. Focus of chick embryo fibroblasts transformed with Rous sarcoma virus. Note drastically altered morphology and growth pattern of the transformed cells, which are densely piled on top of one another in the center of the focus. *(Courtesy of Dr. P.K. Vogt.)*

respond to contact neither with each other nor with other types of cells.

Reduction in Requirement for Serum. Contact inhibition is not the only factor that limits cell growth. The addition of serum to a nondividing contact-inhibited monolayer of untransformed cells results in further rounds of division, so that cells may actually pile up on top of one another. Virus transformation markedly reduces cellular serum dependence, and transformed cells require much less serum to initiate division than do untransformed cells. For example, 3T3 cells, a line of mouse fibroblasts, will not grow optimally unless the serum concentration is greater than 5 percent. By contrast, 3T3 cells transformed with SV40 can divide to a small but significant extent in serum concentrations as low as 0.5 percent. The nature of the serum factors that are required by cells and how these factors act is not known. It is hoped that studying them will hasten the deciphering of the signals that guide cells through their growth cycles.

Ability to Form Colonies in Soft Agar Suspension. Untransformed cells of fibroblast origin must attach to a solid surface before they can divide; this requirement is known as "anchorage dependence of multiplication." By contrast, transformed cells will divide in suspension culture. In particular, ability to grow and form colonies when

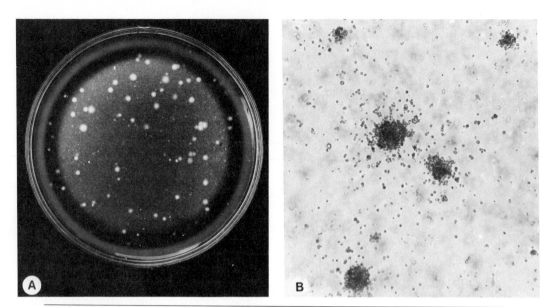

Figure 62-2. Suspension colonies in soft agar of chick embryo fibroblasts transformed with Rous sarcoma virus. **A.** Appearance of a Petri dish with numerous colonies (14 days after infection and plating). **B.** Enlargement of several colonies (8 days after infection and plating). *(Courtesy of Dr. Thomas Graf.)*

suspended in soft (0.5 percent) agar (Fig. 62-2) provides not only a very useful test for the stably transformed state but also the basis for a selective procedure that permits the isolation of transformed cells from populations of predominantly untransformed ones.

Changes in Membrane Transport Properties

Simple sugars and other nutrients are transported several times more rapidly across the plasma membrane of transformed than of untransformed cells. This is generally the earliest observable change following transformation. It is conceivable that increased sugar transport into transformed cells is responsible, at least in part, for the increased rate of glycolysis that is generally observed in transformed cells.

Acquisition of New Surface Antigens

The surfaces of virus-transformed cells possess antigenic determinants not present on untransformed cells. These new surface antigens can be detected by immunofluorescence (Fig. 62-3) or by tests for cytotoxic and cell-mediated immunologic response. Indeed, their presence causes transformed cells to be recognized as foreign and therefore subject to immunologic surveillance. Most of these new antigenic determinants are virus-specified; some may be coded by the host genome. Among the mechanisms that may cause new host-specified antigens to manifest themselves following transformation are derepression of

genes that do not ordinarily express themselves and unmasking or exposure of determinants that are not normally apparent.

Increased Agglutinability by Lectins

Many animal cells are agglutinated by certain glycoproteins and proteins that are present in plant seeds, snails, crabs, and some fish and that possess affinity for sugars in general and glycoproteins in particular. They are collectively known as lectins (or agglutinins). Some of these lectins preferentially agglutinate tumor cells. The two lectins that best discriminate between normal and transformed cells are the jack bean agglutinin concanavalin A (con A) and wheat germ agglutinin (WGA). At first it was thought that transformed cells are agglutinated more readily than nontransformed ones because they bind more lectin molecules. However, it is now known that this is not so. Rather the reason seems to lie in the arrangement of the lectin-binding sites, which are dispersed in untransformed cells and clustered in transformed ones. Further, it is now known that both con A and WGA can agglutinate untransformed cells under certain conditions—they will agglutinate cells in metaphase and cells exposed briefly to proteases, such as trypsin. Interestingly, contact-inhibited protease-treated cells escape from growth control for brief periods of time, probably until the proteins that are removed by protease are regenerated. The reason why treatment with proteases renders cells more agglutinable is not known. One explanation is that proteases split off surface

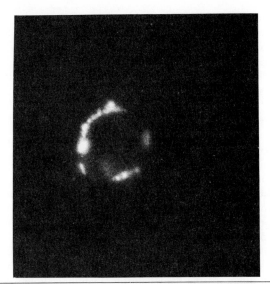

Figure 62-3. Demonstration, by means of immunofluorescence, of the tumor-specific transplantation antigen (TSTA) on a cell derived from a tumor induced in hamsters with adenovirus type 12. The cell was treated first with a hamster antiserum to such cells and then with rabbit antihamster immunoglobulin conjugated with fluorescein isothiocyanate. The hamster antiserum was prepared by repeated injection of adenovirus type 12, followed by one injection of tumor cells. Note the annular pattern of specific fluorescence on the cell membrane. TSTA on cells transformed with papovaviruses can be demonstrated by analogous procedures. *(From Vascencelos-Costa: J Gen Virol 8:69, 1970.)*

glycopeptides, thereby increasing membrane fluidity and facilitating clustering of lectin receptor sites.

Changes in Chemical Composition of Plasma Cell Membrane Components

Whereas the protein composition of plasma membranes appears to be fairly constant, their lipid, glycolipid, and glycoprotein carbohydrate components are notoriously variable. They are exquisitely sensitive to changes in the medium, changes in the cell growth rate, and changes in growth conditions in general. It is difficult, therefore, to pick out changes attributable to transformation, but the following appear to be significant.

1. When untransformed cells become density (contact) inhibited, their glycolipids become larger and more complex. Apparently, glycosyl transferases that are not expressed in rapidly growing cells are expressed in contact-inhibited cells. Transformed cells do not exhibit this cell density-dependent glycolipid extension response. Fucose-containing, galactose-containing, and NANA(N-acetylneuramic acid)-containing glycolipids are present in transformed cells in smaller amounts than in untransformed cells and are smaller in size.

2. Strangely, changes of a similar but opposite nature occur in the oligosaccharides of glycoproteins. In transformed cells these groups are larger than in untransformed cells, and they contain more NANA (which always occupies a terminal position).

3. One of the major components of plasma membranes of normal cells is fibronectin, a large polymorphic glycoprotein (subunit mol wt 200,000 to 250,000). Fibronectin possesses affinity for collagen, fibrin, heparin, and cell surfaces. Transformed cells usually, but not always, lack this protein, which is therefore also known as LETS (large external transformation-sensitive) protein. Its absence may decrease cell-cell and cell-tissue matrix interactions and may also increase membrane fluidity, thereby facilitating access to nutrients (thus lowering the requirement for serum) as well as lateral movement of membrane components (which would explain the increased agglutinability by lectins).

Decreased Levels of Cyclic AMP

In normal cells the level of cyclic AMP increases as they reach confluency. Lowering the concentration of serum in the medium has a similar effect. Transformed cells generally possess lower levels of cyclic AMP than normal ones, and lowering the serum concentration fails to elevate these levels.

Increased Secretion of Plasminogen Activator

Most cells produce small amounts of a protease, commonly known as plasminogen activator, that converts plasminogen to plasmin, the protease that digests fibrin. Many, but not all, transformed cells produce very much more plasminogen activator than their normal counterparts. The significance of this increased level of protease production is not clear. Since normal cells produce less plasminogen activator when contact-inhibited than when growing and since tumor cells are less subject to contact inhibition than normal cells, production of high levels of plasminogen activator may have a role in cell growth stimulation.

Chromosomal Changes

Transformation generally results in changes in karyotype. Among the most common changes are deletion of portions of some chromosomes, duplication/amplification of portions of others, translocation of material from one chromosome to another, and duplication of entire chromosomes.

Members of six virus families are known to be tumorigenic and capable of transforming cells in culture. The manner in which transformation by these viruses is established, maintained, transmitted, and detected will now be described.

DNA Tumor Viruses

Papillomaviruses

Papillomaviruses cause both benign and malignant tumors in a wide variety of animals. There are good reasons for considering human papillomaviruses (HPV) as candidate etiologic agents of cancer in humans:

1. Human papillomaviruses are ubiquitous.
2. They are epitheliotropic, which is important because most human cancers are carcinomas.
3. HPVs are the only viruses known to induce tumors (warts or papillomas) naturally in humans.
4. Papillomaviruses are the natural causes of cancer in animals. For example, Shope papilloma virus induces skin warts in cottontail rabbits, and these warts transform into squamous cell carcinomas in about 25 percent of animals. Further, bovine papilloma viruses induce upper alimentary tract papillomas in cattle that convert to carcinomas, as well as adenomas and adenocarcinomas of the intestine, urinary bladder tumors, meningiomas, fibropapillomas, penile and esophageal papillomas, and urinary bladder papillomas. Bovine papilloma viruses may also be involved in naturally occurring equine connective tissue tumors.

In humans, at least 15 distinct types of HPV cause a variety of lesions (see Table 56-4). Many HPVs cause benign warts, but in about 25 percent of patients with epidermodysplasia verruciformis, a rare hereditary disease involving generalized degeneration of flat warts, malignant transformation occurs. Human genital warts (condylomata acuminata) tend to convert to squamous cell carcinomas of the penis, vulva, and the perianal region, and recurrent laryngeal warts also sometimes convert to malignant tumors.

The various HPVs are surprisingly diverse. Their genomes possess quite distinct restriction endonuclease maps, and most of them exhibit very little homology; for example, the DNA of HPV-1 hybridizes to the DNAs of HPV-2 and HPV-7 only to the extent of about 6 and 0.6 percent, respectively. Not surprisingly, the proteins encoded by the various types of HPV are serologically quite distinct. As a result, individual patients may be simultaneously infected with numerous HPV types.

Little is known concerning the interaction of HPV with cells because they are extremely fastidious and have not yet been grown in cells cultured in vitro. It is clear, however, that HPV DNA tends to persist in the cells that it infects and transforms in stable episomal, rather than in integrated, form. From 5 to 200 copies of it per cell have been detected. Further, there is evidence that papilloma virus DNA in transformed cells is heavily hypermethylated. Active gene sequences are generally undermethylated, while inactive genes show a high degree of methylation. The hypermethylation of papilloma virus DNA would,

therefore, suggest that it is transcribed sparingly in infected cells, which agrees with the finding that transcripts corresponding to only about 10 percent of the papilloma virus genome are present in infected cells, and these only in small amounts. Thus papilloma virus genomes express themselves only minimally in transformed cells.

Polyomaviruses

Polyoma virus was isolated from mouse cell extracts used for the transmission of leukemia. It is only rarely responsible for tumors in nature. When infected into newborn mice or hamsters, it produces a wide variety of histologically distinguishable tumors, hence its name.

SV40 was first isolated from apparently normal cultures of monkey kidney cells. The only host in which it causes tumors is the baby hamster. Lymphocytic leukemia, lymphosarcoma, reticulum cell sarcoma, and osteogenic sarcoma are all produced.

There are also two human polyomaviruses, BK virus and JC virus (BK and JC are the initials of the persons from whom they were isolated in 1971). Both are more closely related to SV40 than to polyoma virus. For example, the amino acid sequences of the proteins encoded by SV40 and BKV are about 80 percent homologous, and the antigenic determinants on their T antigens are almost completely cross-reactive. However, BKV and JCV are quite distinct from each other; and there are only weak immunologic cross-reactions between them.

BKV was isolated from the urine of an immunosuppressed renal transplant patient, while JCV was isolated from the brain of a patient with progressive multifocal leukoencephalopathy (PML). Both viruses are very widespread in humans. Infection occurs in childhood, and most adults have antibodies to them. Apparently primary infection is asymptomatic and is followed by a low-grade lifelong persistent infection that is activated when the immune system is compromised. The viruses then establish silent foci of infection in the urinary tract, which causes them to be excreted in the urine, and in a very few individuals JCV reaches oligodendrocytes in the brain and causes PML, a chronic, fatal demyelinating disease. Interestingly, a very similar disease occurs in rhesus monkeys, the natural host of SV40; and two human PML cases have yielded SV40, rather than JCV.

Both BKV and JCV possess oncogenic potential. Both transform cultured cells and cause tumors in a variety of animals. Both are markedly neurotropic. Thus, JCV induces brain tumors in owl monkeys, and most of the tumors caused by BKV in Syrian hamsters are ependymomas, while JCV shows a predilection for neuroectodermal cells. However, neither virus is associated with any form of cancer in humans.

SV40 and polyoma virus are moderately closely related. Their proteins share some antigenic determinants, and their amino acid sequences are about 40 percent homologous (that is, they possess the same amino acids in equivalent positions).

Transformation by Polyomaviruses

Polyomaviruses not only interact with cells by means of the lytic pathway (Chap. 58) but also transform cells. Cells are transformed if they are nonpermissive—that is, if the virus cannot multiply in them—or if they are permissive, they can be transformed if they are infected with defective virus particles, such as the deletion mutants that arise upon repeated passage at high multiplicity (Chap. 60), or if they are infected with inactivated virus particles (for example, by photodynamic inactivation). In other words, if a polyomavirus can multiply in a cell, that cell will generally not be transformed; if it cannot multiply, the cell will often be transformed. The establishment of the transforming papovavirus-cell interaction requires high multiplicities of infection (10^6 to 10^7 particles per cell). There is evidence that the frequency of transformation is enhanced, by some as yet unknown mechanism, by treating cells with chemical carcinogens, such as 4-nitroquinoline-1-oxide. The essential feature of the transforming polyomavirus-cell interaction is that viral DNA becomes stably integrated into the DNA of the host cell. As a result it cannot multiply, except as a cellular gene, and no progeny virus is produced. Not all infected cells are transformed in this manner; the majority are transformed abortively—that is, they escape from growth control for a few cell generations but then revert to the normal uninfected state for reasons that are not understood.

Polyomavirus DNA can integrate at many sites in the cellular genome. The number of integrated viral genomes generally varies from 1 to 3 but can be as high as 20 to 50. Transformation by one polyomavirus does not preclude transformation by another nor by unrelated viruses. Thus, double transformants of SV40 and polyoma, as well as of SV40 and adenovirus and SV40 and various RNA tumor viruses, have been isolated and studied.

Function of Polyomavirus Tumor Antigens

Although viral DNA in transformed cells does not replicate except as part of the host genome, it does express itself. In fact, substantial amounts of messenger RNA are generally transcribed from it by the α-amanitin-sensitive cellular RNA polymerase II. However, not the entire viral genome is transcribed but only that portion of it that encodes the early functions, that is, the various tumor antigens [large T and small t in the case of SV40, and large T, middle t, and small t antigens in the case of polyoma virus (see Chap. 58)]. It is the presence of these proteins that is responsible for the induction and maintenance of the transformed state, and a very large amount of work has been done to define their functions. Clearly, knowledge of these presumably complementary functions would provide fundamentally important clues concerning the mechanisms by which normal cells are transformed into cancer cells. The most important approach has been examination of the characteristics of cells infected with SV40 and polyoma virus mutants that contain well-defined deletions—in other words, cells in which either large T, middle t, or small t antigen cannot be made. This work has permitted the following conclusions:

1. Large T antigen (Fig. 62-4) is responsible for stimulating host cell DNA synthesis, a function that is inseparable from ability to transform. It is also responsible for the ability to form colonies in soft agar and proliferate in the presence of low serum concentrations. Further, at least a portion of it is exposed on the transformed cell surface as the *transplantation antigen*, the antigen that elicits antibodies and sensitizes lymphocytes that protect against subsequent challenge with polyomavirus-induced tumor cells or with polyomavirus particles.

2. Small t antigen complements the functions of large T antigen. The presence of functional small t antigen is also usually necessary for the ability to form colonies in soft agar, produce plasminogen activator (p. 931), and cause dissolution of the intracel-

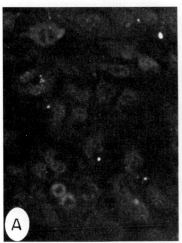
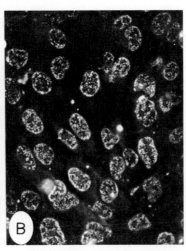

Figure 62-4. Demonstration of the SV40 T antigen by means of immunofluorescence. Cells from a tumor induced by SV40 in a hamster were exposed to fluorescein-conjugated serum from a nontumor-bearing hamster (**A**) and a tumor-bearing hamster (**B**). × 280. *(From Rapp, et al.: Proc Soc Exp Biol Med 116:1131, 1964.)*

lular actin cable network, absence of which is a
hallmark of fully transformed cells. Maintenance of
the transformed state and tumor progression are
also dependent on the continued presence of small
t antigen.

3. In cells transformed with polyoma virus, middle t
antigen plays an additional role in enhancing full
expression of the transformed phenotype and
transformed cell growth in general and membrane
alterations in particular. It is located in the plasma
cell membrane and is associated with a protein kinase activity that phosphorylates tyrosine residues
in middle t antigen itself. Phosphorylation of tyrosine residues in proteins also occurs in cells
transformed with RNA tumor viruses (see below),
and a great deal of research is currently directed at
determining the significance of tyrosine phosphorylation in the cancer process. The reason for the
excitement is that most phosphoproteins in normal
cells are phosphorylated in serine and threonine
residues, not in tyrosine residues.

4. SV40 does not code for a middle t antigen. Instead, intracellular large T antigen is tightly associated with a cellular phosphoprotein, the p53
protein, whose formation it stimulates some 25- to
50-fold over normal cell levels. The p53 protein is
also present in cells transformed by other agents,
such as chemicals or irradiation, and in spontaneously arising carcinoma cells. It is also induced
in nondividing lymphocytes when stimulated with
mitogens, which suggests that it functions somehow in cell division.

Rescue of Transforming Genome

Whereas viral DNA in transformed cells never expresses itself in its entirety, it is nevertheless the entire viral DNA
that is integrated, not portions or segments of it, as can be
proved by several types of experiments. For example, the
genome of SV40 can be rescued from transformed cells by
fusing them with uninfected permissive cells. Fusion may
be effected by simple cocultivation or, much more efficiently, by treatment with inactivated Sendai virus (Chap.
59). Apparently only the cytoplasm of permissive cells is
necessary, since fusion with enucleated cells has been
shown to rescue SV40. This rescue of virus can be effected
from cells that have been transformed for many generations and many years. Presumably the mechanism involves
excision and/or recombination to generate free circular viral genomes.

Transformed cells can be shown to harbor the entire
SV40 genome because infectious SV40 DNA can be
extracted from them.

Revertants of Transformed Cells

Numerous attempts have been made to isolate revertants
of transformed cells. Among the techniques that have led
to the selection of revertant cell lines that exhibit growth

control comparable to that of normal cells are selection
for serum dependence, passage at high dilution, resistance
to nucleic acid-base analogs at high cell density (i.e., inability to multiply at high cell density), and resistance to
con A. All these methods yield cells that display contact inhibition of growth and resemble normal cells morphologically. However, almost always they still synthesize the
tumor antigens, display unaltered virus-specific transcription patterns, and contain the entire viral genome. Therefore, reversion is due to a change in the cellular rather
than in the viral genome. Revertants cannot be retransformed with the virus whose genome they already harbor,
but they can be retransformed with other tumor viruses.
Finally, revertant cell lines often contain more chromosomes than their transformed antecedents. This has
suggested a hypothesis that the cell phenotype is modified
as a result of changes in the relative amounts of chromosomes with genes that promote and genes that oppose
transformation.

Adenoviruses

Adenoviruses also can transform animal cells in vitro, and
some but not all cause tumors in animals.

Human adenoviruses can be divided into three groups
on the basis of their oncogenicity. Serotypes 12, 18, and
31 are highly oncogenic: when injected into newborn hamsters, they cause tumors rapidly and with high frequency.
Serotypes 3, 7, 14, 16, and 21 are weakly oncogenic: they
produce tumors in newborn hamsters with low frequency
and after a long latent period. Most of the remaining serotypes, exemplified by serotypes 2 and 5, are nononcogenic, but even they can transform rodent cells in culture.

Adenovirus-transformed cells display many of the
properties described above for cells transformed by
polyomaviruses, but there are also important differences.
For example, in contrast to integrated SV40 and polyoma
virus DNA, adenovirus DNA integrated into the genomes
of transformed cells cannot be rescued by fusion with permissive cells. This suggests either that only a portion of
adenovirus DNA is integrated or that it is integrated in the
form of fragments. Analysis of the nature of viral DNA sequences present in a variety of cells transformed by several
adenoviruses has shown that (1) none of them contained
the complete adenovirus genome, (2) all contained at least
the left-hand 8 percent of the viral genome (which agrees
with the finding that cells can be transformed with its
leftmost restriction endonuclease fragments), and (3) some
contain most of the viral genome but in fragmented form.
To take specific examples, one rat embryo cell line
transformed by adenovirus type 2 contains three copies of
the sequence comprising the leftmost 17 percent of the
adenovirus genome, another contains about 13 copies of
the leftmost 70 percent of the viral DNA, 3 of which are
linked covalently to the sequence between map units 98
and 96.5 *inverted* with respect to its normal orientation in
the viral DNA, and a hamster cell line transformed by ade-

novirus type 12 contains about 8 copies of fragments that together comprise virtually the entire viral genome.

Virus-specific proteins were first detected in cells transformed by adenoviruses by immunologic means, such as immunofluorescence or complement fixation, using antisera from hamsters bearing tumors induced by adenoviruses. Such proteins were designated "adenovirus T antigen." It is clear now that various adenovirus-transformed cell lines contain very different amounts of viral DNA and that a variety of viral proteins, all early region proteins, are formed in them. The term "adenovirus T antigen" should be reserved for those proteins that are encoded by the minimal segment of adenovirus DNA capable of transforming cells, that is, its leftmost 4.5 percent, which corresponds to early region E1A (Chap. 58). This region codes for about four proteins, one of which has recently been highly purified and appears to be a true T antigen that is present in all adenovirus-transformed cells and is therefore diagnostic of adenovirus transformation.

The demonstrated oncogenicity of human adenoviruses in rodents raises the possibility that they may cause neoplasia in humans. This has been investigated using exquisitely sensitive hybridization techniques that have been refined to the extent that one part of viral DNA can be detected in the presence of 10^6 parts of host DNA, which corresponds to less than one adenovirus genome per cell. So far no human neoplasms have been found that contain adenovirus DNA.

Herpesviruses

Numerous herpesviruses either are oncogenic, are associated with tumors, or can transform cells in vitro. Eight are oncogenic in animals (Table 62-1); they include herpesviruses of primates and other mammals, birds, and amphibians. In humans, no fewer than three herpesviruses may be involved in tumorigenesis: Epstein–Barr virus (EBV), the genome of which is often present in cultured Burkitt's lymphoma and nasopharyngeal carcinoma cells and which can transform human B lymphocytes; herpes simplex virus type 2 (HSV 2), which seroepidemiologic evidence suggests may be associated with carcinoma of the cervix and which can transform a variety of cells, especially when inactivated with ultraviolet irradiation so as to preclude the productive or lytic virus-cell interaction; and human cytomegalovirus (HCMV), which may play a role in the etiology of Kaposi's sarcoma and which, after exposure to ultraviolet irradiation, can also transform cultured cells.

Epstein–Barr Virus

Burkitt's lymphoma is a tumor that is relatively common in East Africa and New Guinea but rare in other parts of the world. It occurs chiefly in children 5 to 12 years of age. In the body, the tumor cells contain no virus, but a small proportion of lymphoblastoid cells established in vitro from tumor tissue produce a herpesvirus that is antigenically distinct from all known herpesviruses and is known as Epstein–Barr virus (EBV) after its discoverers. EBV is very widely distributed; 80 percent of adults worldwide possess antibodies to it. Thus, even in regions where Burkitt's lymphoma is endemic, only a very small proportion of infected individuals develop the tumor. Interestingly, a small proportion of African Burkitt's lymphomas and most American and European Burkitt's lymphomas contain no trace of EBV infection.

EBV is one of three tumorigenic B lymphotropic herpesviruses of humans and primates, the others being herpesvirus papio (baboons) and herpesvirus pan (chimpanzees), the DNAs of which share about 40 percent sequence homology. All cause lymphoproliferative disease in primates, such as marmosets. Their host range is very narrow; only B lymphocytes possess receptors for them, and they transform these cells into lymphoblastoid cells with an infinite life span. All cells infected with EBV or cell lines isolated from EBV-induced tumors contain EBV DNA and a tumor antigen known as EBNA (Epstein–Barr nuclear antigen), but productive virus multiplication occurs in only a very small proportion of them, possibly because EBV DNA is unusually hypermethylated (hypermethylation being an attribute of genetically silent genome regions). The proportion of cells in which EBV multiplies can be in-

TABLE 62-1. ONCOGENIC HERPESVIRUSES

Virus	Host	Host Cell	Malignancy
Epstein–Barr virus	Humans	B lymphocytes	Burkitt's lymphoma, nasopharyngeal carcinoma
Herpesvirus pan	Primates	B lymphocytes	Lymphoma
Herpesvirus papio	Primates	B lymphocytes	Lymphoma
Marek's disease virus	Chicken	T lymphocytes	Lymphoma
Herpesvirus saimiri	Primates	T lymphocytes	Lymphoma
Herpesvirus ateles	Primates	T lymphocytes	Lymphoma
Herpesvirus sylvilagus	Rabbits	B lymphocytes (?)	Lymphoma
Cytomegalovirus	Humans		Kaposi's sarcoma (?)
Lucké frog herpesvirus	Frogs		Adenocarcinoma of the kidney

creased by arginine deprivation or treatment with bromodeoxyuridine, TPA (12-0-tetradecanoylphorbol-13-acetate, a tumor promoter), 5-azacytidine, or dimethylsulf-oxide. The last three reagents are demethylating agents. More than 80 messenger RNAs and 50 proteins have been characterized in cells infected with EBV.

The number of EBV DNA molecules (Fig. 62-5) present in transformed or tumor cells ranges from 2 to more than 200, mostly as free episomal closed circles. Only a few are integrated into the host genome. It is present not only in Burkitt's lymphoma cells but also in the cells of nasopharyngeal (postnasal) carcinoma, which occurs with rather high frequency among Southern Chinese and, to a lesser extent, among Algerians.

It is impossible to say, at present, whether the relationship between EBV and Burkitt's lymphoma and nasopharyngeal carcinoma is casual or causal. On the one hand, there is no doubt that EBV possesses oncogenic potential in vivo and in vitro, its target cells being cells of the B lymphocytic lineage. Further, the cells of most of the Burkitt's tumors that arise in Africa and some of the tumors that arise elsewhere contain EBV DNA and EBV antigens. However, some such tumors do not, and if EBV has an etiologic role in Burkitt's lymphoma or nasopharyngeal carcinoma, other cofactors must also be involved.

EBV and Infectious Mononucleosis. The majority of the normal adult population of the United States possesses antibodies against EBV. Conversion from the seronegative to the seropositive state occurs during the acute phase of infectious mononucleosis. Apparently, therefore, EBV is the etiologic agent of infectious mononucleosis. Presumably, most infections with EBV are subclinical, that is, so mild as to go unnoticed. Some can cause overt infectious

Figure 62-5. Schematic representation of the DNA of EBV. Its size is similar to that of herpes simplex virus [about 100 × 10^6 or 170,000 base pairs (bp)].

TR: terminal repeat, about 500 bp, repeated 6 to 12 times
IR1: internal repeat 1, about 3000 bp, repeated 6 to 12 times
IR2: internal repeat 2, about 123 bp, repeated 8 to 12 times
IR3: internal repeat 3, about 123 bp, repeated about 7 times
IR4: internal repeat 4, about 103 bp, repeated 6 to 12 times
D$_L$ and D$_R$: partially homologous regions about 2000 bp long
The various regions are not drawn strictly to scale. About three quarters of the sequences of EBV DNA are distributed in 5 unique regions, while about one quarter are present in the form of repeated sequence elements. IR1 is transcribed very extensively into messenger RNA, but the proteins that it encodes have not yet been identified. IR2 appears to contain the origin of replication. IR3 is also transcribed very actively and appears to encode the EBNA, a 380 amino acid long protein with a remarkable composition: more than two thirds of its amino acids are alanine and glycine, which no doubt causes EBNA to possess a unique structure.

mononucleosis, and, very rarely in most populations but more frequently in others where special factors may be operative, Burkitt's lymphoma or nasapharyngeal carcinoma ensues. The nature of these factors and their mode of action are not known.

Herpes Simplex Virus

Two types of herpes simplex virus infect humans: type 1, which causes primarily oral lesions, and type 2, which is associated with genital infections. Several lines of epidemiologic and serologic evidence suggest that infection with herpes simplex virus type 2 is associated with carcinoma of the cervix. In particular, women with genital herpetic infections have a higher than average incidence of cervical carcinoma, and women with cervical carcinoma have a higher than average incidence of antibodies to herpes simplex virus type 2.

Although suggestive, this evidence is not sufficient to establish a definitive causal relationship between herpes simplex virus and malignant disease in humans. In fact, such a relationship will be difficult to establish because herpesviruses are ubiquitous and tend to establish latent and persistent infections. Since experimentation with humans is not feasible, the most direct evidence is likely to be provided by a vaccine that markedly reduces the incidence of some form of cancer.

Poxviruses

Two poxviruses are tumorigenic: fibroma virus, including the classic Shope fibroma virus that is pathogenic for rabbits, as well as several closely related viruses pathogenic for deer, squirrels, and hares, and Yaba monkey tumor virus, which is pathogenic for several species of monkeys and for humans. Both produce benign tumors that soon regress.

The interest in poxvirus tumorigenicity derives from the fact that poxviruses are DNA-containing viruses that have always been thought to multiply exclusively in the cytoplasm. Recently acquired evidence suggests, however, that while vaccinia virus does not appear to have an obligatory nuclear phase in its multiplication cycle, some viral DNA and RNA are often present in the nucleus. Poxvirus DNA may therefore have the opportunity to become integrated into the host cell genome or to become otherwise stably associated with it, and when that happens, cell transformation may result. When fibroma virus transforms cells, host DNA replication is first arrested, and viral DNA replication is initiated. However, it quickly ceases, host DNA replication recommences, and the cells become transformed. They then multiply more rapidly than before, exhibit an altered morphology, are less sensitive to contact inhibition, and display new antigens on their surface. While the precise nature of the association of fibroma DNA and the host cell genome is not known, it is clear that the association is not a stable one, since the tumors often regress and contain fibromavirus particles when they do so.

Hepatitis B Virus

The only known hosts of hepatitis B virus are humans. About 200 million humans are currently infected with this virus, which causes hepatitis B or serum hepatitis. Many of them are carriers in whom the virus established a chronic or persistent infection manifested by the circulation for months or years of relatively high concentrations of HBsAg particles, defective or incomplete viral forms that consist predominantly of hepatitis B surface antigen, as well as small amounts of infectious virus. In other individuals the virus establishes a highly virulent, often fatal disease (Chap. 79). What has recently caused great excitement is evidence that infection with this virus is strongly associated not only with cirrhosis of the liver but also with the most common fatal cancer of humans, primary hepatic carcinoma.

The complete hepatitis B virus particle, known as the "Dane" particle, consists of an outer layer, the so-called HBsAg, and an inner core, the HBcAg. In the serum of patients and carriers, most HBsAg and HBcAg exists in free form, not in the form of intact virus particles. Both consist of several protein species. Hepatitis B DNA in Dane particles is *partially* double-stranded; it consists of a complete 3200 nucleotide long minus strand, but the plus strand is on the average only about two thirds of full length. The virus contains a DNA polymerase, and if it is incubated in vitro with nucleoside triphosphates, plus strand synthesis is completed.

Efforts to understand its life cycle have been hindered by the failure of hepatitis B virus, or three closely related viruses recently discovered in woodchucks, ground squirrels, and domestic ducks, to grow in cultured cells. Current evidence suggests that hepatitis B virus multiplies as depicted in Figure 62-6. It appears that soon after infection the minus strand of its genome is transcribed into plus-stranded RNA, which after being translated becomes encapsidated in hepatitis B core-like particles and is then reverse-transcribed into DNA in maturing progeny virus particles. The fascinating feature of this sequence of reactions is that it is a temporally permuted version of the retrovirus multiplication cycle (see below), where infectious virus particles contain RNA that is reverse-transcribed into double-stranded DNA soon after infection and, after integration into the host genome, is used as the template for the repeated synthesis of genome RNA later in infection.

As pointed out above, chronic hepatitis B infection is associated with increased risk of primary hepatic carcinoma. The question then arises as to whether hepatitis B virus causes cancer the same way retroviruses do, namely, by integrating their DNA into that of the cell. Little is known yet concerning the integration of hepatitis B DNA into the human genome, but preliminary evidence indi-

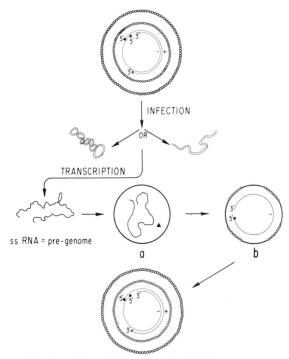

Figure 62-6. The multiplication cycle of hepatitis B virus. Following infection and uncoating, the viral DNA is transcribed into plus-stranded RNA—the pre-genome. After being translated, the pre-genome is encapsidated into immature cores (**a**) and reverse-transcribed into negative-stranded DNA, while the RNA pre-genome is degraded (**b**). The outer virus particle coat is then added, the minus DNA strand is partially transcribed into its plus strand, and the virus is liberated.

cates that it does occur and that the DNA is not integrated intact, as is the DNA provirus of retroviruses, but that integration is accompanied by complex rearrangements within multimeric inserts and loss of genetic material. It seems, however, that one particular region of the hepatitis B genome is always retained, and it is conceivable that it contains a super promoter like polyoma and SV40 DNA (Chap. 58) and the LTR of retroviruses (p. 942). As a result, promoter insertion may play a role in transformation by hepatitis B virus, just as it does for certain retroviruses.

RNA Tumor Viruses

The most intensively investigated tumor viruses are the RNA tumor viruses. There are three primary reasons for this. First, in many animal species they cause leukemias and solid tumors that are often strikingly similar to human neoplasms. This raises the question of their role in the genesis of cancer in humans. Second, the genomes of RNA tumor viruses are integrated into the genomes of many species of animals ranging from reptiles to higher apes. Why not also into that of humans? Third, the transforming genes of RNA tumor viruses are very closely related to genes that are present in the genomes of all animals, genes that may function during embryonic development. Thus RNA tumor viruses seem to provide the best chance for solving the basic secret of cancer, namely, why genes are switched on inappropriately and out of context and how they are switched on. In other words, since cancer undoubtedly has a genetic basis, our best hope of unraveling its molecular mechanisms lies in studying the nature of the interaction of tumor virus genomes with those of the host cells, and RNA tumor viruses provide a far wider spectrum of viruses than any of the DNA tumor virus families.

Since the term "RNA tumor virus" is cumbersome and since viruses exist that, while exhibiting the same strategy of genome structure and expression, are not tumorigenic (Table 58-4), attempts have been made to devise a more concise and/or generally applicable designation. Among the proposed names are oncornavirus (oncoRNA), oncovirus, and retrovirus. The latter term has found wide acceptance and is used in the discussion that follows.

Nature of Retrovirus Particles

Retrovirus particles possess an electron-dense nucleoid or core surrounded by two shells (Figs. 62-7 and 62-8). The inner shell, which appears to possess icosahedral symmetry, is the nucleocapsid shell, while the outer shell consists of a lipid bilayer membrane to which glycoprotein spikes are attached. Retrovirus particles comprise seven structural proteins (Table 62-2). Two are glycoproteins that make up the envelope spikes, the larger forming the knobs, the smaller the stalks. They are linked by covalent-SS-bonds. The larger, glycoprotein gp85 in the case of avian retroviruses and gp70 in the case of mammalian ones, is the antigen that gives rise to neutralizing antibodies and is responsible for host range specificity. These envelope proteins are encoded by the *env* gene. The other structural proteins, p27, p19, p15, p12, and p10 in the case of avian retroviruses, and p30, p15, p12, p12E and p10 in the case of mammalian ones, are nucleocapsid components. The p27 and p30 proteins make up the nucleocapsid shells and are the principal virus particle components; p12 and p10 in avian and mammalian viruses, respectively, are intimately associated with the RNA. All these nucleocapsid proteins are derived by cleavage of a precursor that is the primary protein product of one of the three retrovirus genes, the so-called *gag* protein gene (see below). The order of these proteins in the precursors is NH_2-p19-p10-p27-p12-p15-COOH and NH_2-p15-p12-p30-p10-COOH for avian and mammalian retroviruses, respectively. There is evidence that in the avian system the protease that effects the cleavages is p15. Retrovirus particles also contain

about 10 molecules of an RNA-dependent DNA polymerase, the reverse transcriptase, which is coded by the *pol* gene (see below).

All these proteins exhibit a multiplicity of antigenic determinants. There are type-specific antigenic determinants private to individual virus strains, group-specific determinants common to groups of viruses with common hosts, and interspecies-specific determinants that are common to all mammalian or avian retroviruses.

Three Morphologic Types of Retrovirus Particles

There are three morphologic types of retrovirus particles: C-type particles characteristic of most leukemia and sarcoma viruses; B-type particles characteristic of mammary tumor viruses that differ from C-type particles in the eccentric rather than central location of their nucleocapsid in thin sections, their more prominent glycoprotein spikes, and in the fact that their nucleoids can often be seen in the cytoplasm of infected cells (these are the so-called *intracytoplasmic* A-type of particles); and D-type particles isolated from a variety of primates, the properties of which are intermediate between those of C-type and B-type retrovirus particles.

Retrovirus Genome

The retrovirus genome consists of two identical plus-stranded RNA molecules that are 5,000 to 10,000 nucleotides long (mol wt 1.5 to 3×10^6) and hydrogen-bonded to each other via palindromic sequences some 15 nucleotides long that are centered about 60 nucleotides from their 5′ ends (Fig. 62-9).

The essential features of the basic retrovirus genome are shown in Figure 62-10. This is the genome of nondefective avian and mammalian (particularly murine) leukosis or leukemia viruses (ALV and MLV), also sometimes called lymphoid leukosis viruses (LLV). It possesses sequences with regulatory features at each end flanking the coding sequences of three intronless genes: the *gag* gene that encodes the four or five nucleocapsid proteins, the *pol* gene that encodes the RNA-dependent DNA polymerase (the so-called reverse transcriptase), and the *env* gene that encodes the glycoprotein spike components. Other important features are the presence of direct terminal repeats (the R regions), the PB (-) region where a tRNA molecule binds to act as primer for the transcription of the RNA into DNA, the fact that the 5′ terminus untranslated region [R-U5-PB(-)-L] serves as a leader for subgenomic mRNA species (for example, the *env* proteins are transcribed from a messenger RNA that is derived from a complete genomic plus strand by splicing out the *gag-pol* coding sequences), and the presence of a strong promoter of transcription in the U3 region.

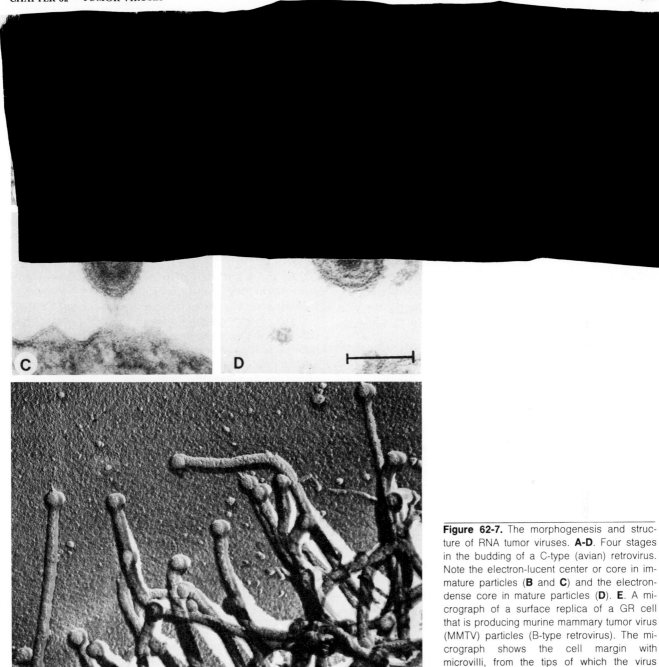

Figure 62-7. The morphogenesis and structure of RNA tumor viruses. **A-D**. Four stages in the budding of a C-type (avian) retrovirus. Note the electron-lucent center or core in immature particles (**B** and **C**) and the electron-dense core in mature particles (**D**). **E**. A micrograph of a surface replica of a GR cell that is producing murine mammary tumor virus (MMTV) particles (B-type retrovirus). The micrograph shows the cell margin with microvilli, from the tips of which the virus buds. (**A** through **D**, courtesy of Dr. Heinz Bauer; **E**, courtesy of Dr. J.B. Sheffield.)

Molecular Biology of Retroviruses

There has long existed a conceptual difficulty concerning the role of retroviruses as oncogenic agents. Transformation represents a heritable alteration of cellular biosynthetic patterns that is most plausibly accounted for by a change in genetic capabilities. Conceivably there are mechanisms whereby a viral DNA genome might cause such changes after being integrated, but how could a viral RNA

genome achieve this? Furthermore, many cells transformed by retroviruses produce virus for generation after generation. What type of mechanism could ensure that host and viral genome replication would proceed in step? In 1964 Temin advanced a revolutionary hypothesis: he proposed that upon infection the viral RNA genome is transcribed into DNA by a reversal of the usual flow of information transfer, that this DNA is then integrated into the host genome, and that progeny viral genomes are transcribed from it just as all other RNA is normally transcribed. In essence, the hypothesis implied that the RNA present in retroviruses is not the genome but rather the messenger RNA of a genome that exists only intracellularly.

The discovery of reverse transcriptase independently by Temin and Baltimore in 1970 proved Temin's hypothesis correct.

Formation and Integration of Proviral DNA

Following adsorption and penetration into the cell, the RNA in parental nucleocapsids is transcribed by the reverse transcriptase into linear double-stranded DNA, the proviral DNA. The exact mechanism by which transcription is accomplished is not known, but it probably proceeds approximately as depicted in Figure 62-11. Linear proviral DNA then moves to the nucleus, circularizes to form supercoiled circles, and is integrated into the genome of the host cell.

The structure of the integrated provirus is shown in Figure 62-12. As a consequence of the transcription process described in Figure 62-11, its coding sequences are flanked by long terminal repeats (LTRs) composed of the

U3, R, and U5 regions of retrovirus RNA. Their significance lies in the fact that they contain strong promoters of transcription. The integrated provirus is flanked by short inverted repeats of viral sequences (IR) and direct repeats (DR) of host cell sequences. This suggests that the mechanism of provirus integration is formally analogous to the insertion into bacterial DNA of transposable elements: proviruses are effectively transposons.

All available evidence indicates that integration can occur anywhere into cellular DNA. There are no preferred insertion loci. However, once integrated, a provirus is a stable genetic element; it is a gene. The number of integrated proviruses per haploid host genome may be as low as 1 or as high as 100, depending on the virus and on the cell.

Transcription of Integrated Proviral DNA

Integrated proviral DNA is transcribed by cellular RNA polymerase II into transcripts that are capped at their 5′ termini and polyadenylated at their 3′ termini. They are used in one of three ways. They may be encapsidated into progeny virus particles, or they may be translated either into the gag gene product, a protein with a molecular weight of

TABLE 62-2. RETROVIRUS STRUCTURAL PROTEINS*

Protein[†]		Percent of Total Protein[‡]	Coded by	Location and Properties
Avian	Murine			
p10	p12E	5,9	gag	Envelope
p12	p10	14,6	gag	Core, RNA-associated, phosphoprotein
p15		6	gag	Between core and envelope, protein-cleaving activity
p19		20	gag	Between core and envelope phosphoprotein, hydrophobic
	p15	14	gag	Between core and envelope, hydrophobic
	p12	8	gag	Between core and envelope, phosphoprotein
p27	p30	36,50	gag	Core shell, major virus particle component
gp85	gp70	13,11	env	Major envelope protein, spike knob
gp37	gp15E	6,2	env	Minor envelope protein, spike stalk

*The structure of B-type retroviruses is similar to that of the C-type viruses. The major B-type internal gag-coded protein is p28; the major surface env-coded glycoprotein is gp52.

[†]p, protein; gp, glycoprotein. The number indicates the approximate size of daltons $\times 10^{-3}$. In addition to the seven proteins listed, each virus particle also contains about 10 molecules of the RNA-dependent DNA polymerase (reverse transcriptase).

[‡]Approximate percentages in avain and murine retroviruses, respectively.

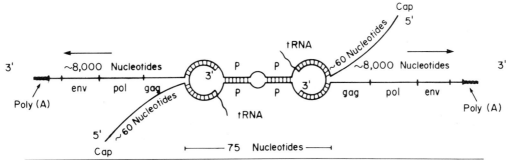

Figure 62-9. The avian retrovirus genome. Its two RNA subunits are held together by base pairing between the palindromic sequences P and by base pairing with tRNA molecules. *(Adapted from Haseltine et al.: Proc Natl Acad Sci USA 74:989, 1977.)*

about 70,000 that is cleaved into the 4 or 5 *gag* proteins, or, infrequently, into a *gag-pol* readthrough product, a large protein, P180, that is cleaved to the *pol* gene product (the reverse transcriptase) as well as the *gag* proteins. This mode of synthesizing the *pol* gene product (that is, without a messenger RNA of its own and by infrequent readthrough from another gene) provides a convenient mechanism for ensuring that it will not be synthesized in excessive amounts (clearly far more *gag* gene products are required for viral morphogenesis than *pol* gene product). Finally, provirus transcripts may have the *gag* and *pol* gene sequences spliced out, thereby providing messenger RNA molecules that are translated into the *env* gene product. If a transforming gene is also present in the provirus (see below), the *env* gene sequences may also be spliced out to provide messenger RNA for the transforming gene product.

Proviruses as Transposable Elements

All available evidence suggests that there are no specific integration sites for proviruses in animal genomes, but that they can be integrated in very many sites, perhaps anywhere. Since an RNA transcribed from one proviral site can, after being encapsidated, be transferred via a virus particle to another cell only to be reverse-transcribed and inserted into another site, proviruses are transposable elements. The essential features of this aspect of retroviruses are (1) the fact that their DNA genomes are transcribed into RNA, in which form their genetic information is transported, and (2) the fact that transcription of proviral DNA into RNA and RNA into proviral DNA generates unique structures at the provirus termini, the LTRs —these are the structures that permit integration. LTRs possess two essential functions: they are necessary for integration, and they possess strong promoters for transcription (see below).

Endogenous Retroviruses: Leukemia and Leukosis Viruses

Retroviruses enjoy an unusually intimate relationship with their hosts. They exist in provirus form in numerous, perhaps all, animal species. These proviruses are transmitted

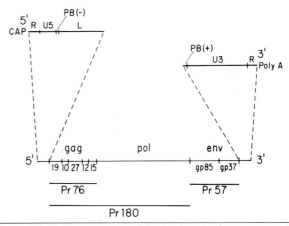

Figure 62-10. The structure of the basic retrovirus genome: the genome of nondefective avian or murine leukemia/leukosis viruses. Note that the RNA is capped at its 5′ terminus and polyadenylated at its 3′ terminus.

R: 20 to 80 nucleotides repeated at both ends

U5: 80 to 100 nucleotide unique sequence

PB(−): A sequence complementary to the 16-19 3′ terminal residues of $tRNA_{trp}$, $tRNA_{pro}$, or $tRNA_{lys}$ in avian leukosis viruses (ALVs), murine leukemia viruses (MLVs) and murine mammary tumor viruses (MMTVs), respectively. The tRNAs bind at these primer binding (hence PB) sites to serve as primers for the transcription of the RNA into complementary DNA (cDNA).

L: An untranslated region about 250 nucleotides long that precedes the coding sequences of the *gag* gene.

gag, pol, and *env*: The genes that encode the nucleocapsid (*gag*) proteins, the RNA-dependent DNA *pol*ymerase (the so-called reverse transcriptase), and the *env*elope proteins or spike components.

Pr76: The precursor of the ALV *gag* proteins. The precursor of MLV *gag* proteins is a Pr65.

Pr180: The infrequently formed readthrough produce of the *gag* and *pol* genes, cleavage of which provides the RNA-dependent DNA polymerase.

Pr57: The precursor of gp85 and gp37, the knob and stalk proteins of the spikes of ALVs. The corresponding MLV precursor protein is a Pr80.

PB(+): A short, highly conserved sequence upstream from the initiation site for *positive* strand DNA synthesis. Its primer is not known.

U3: A unique region 200 to 1,000 nucleotides long. It is thought to contain a powerful transcriptional promoter.

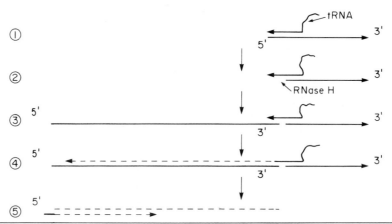

Figure 62-11. Scheme for the transcription of retrovirus RNA into DNA. Step 1. Using tRNA as primer, the 100 or so nucleotides at the 5′ terminus are transcribed. Step 2. The terminally redundant (R) RNA sequence is removed by the action of ribonuclease H, another enzymic activity of the reverse transcriptase. Step 3. The DNA transcript region of the R sequence pairs with an R sequence at the 3′ terminus of another RNA subunit, or it could also be on the same subunit that has just been transcribed at its 5′ terminus. Step 4. The new RNA subunit is transcribed into minus-strand DNA. There may be another strand switch at the 5′ end. Step 5. Transcription of minus-strand into plus-strand DNA proceeds, perhaps using viral RNA as primer, to yield a double-stranded DNA transcript of RNA. *(Adapted from Bishop: Annu Rev Biochem 47:35, 1978.)*

vertically as dominant genetic traits, like cellular genes, and their expression is controlled by individual specific regulatory controls that no doubt involve repression/derepression at the DNA transcription level. These are the endogenous retroviruses. In feral animal populations the expression of these proviruses as virus particles is greatly repressed. However, breeding programs for mice and chickens in particular have produced many strains of inbred animals in which the mechanisms that control endogenous provirus expression have been dissociated from the proviruses themselves, with the result that in these animals RNA tumor viruses are produced continuously and often throughout the animal's life. For example, there are mice, such as the albino AK and the C58BL mice, that have large numbers of leukemia virus particles in their bloodstream and develop leukemia between 6 and 12 months of age with high frequency. These mice are high-incidence-of-leukemia mouse strains. In other mice, such as in BALB/c and DBA strain mice, endogenous proviruses express themselves during embryonic life, even to the extent of virus particle formation, but then become repressed, so that the virus is generally undetectable in young adults. Later in life, however, the proviruses again begin to express themselves and leukemia may result, albeit with low

frequency. Mice that harbor such endogenous proviruses are known as low-incidence-of-leukemia mouse strains. Finally, some mice, such as NIH Swiss and NZB mice, do not develop leukemia at all, because the endogenous proviruses capable of producing spontaneous leukemia have been bred out of them.

In addition to the spontaneously inducible endogenous viruses described above, which generally exist in the animal genome in the form of 1 to about 100 copies, mice and chickens, as well as other rodents, cats, monkeys, and primates, contain endogenous viruses that do not normally express themselves but can be induced to do so by treatment with a variety of substances, including inhibitors of protein synthesis, inhibitors of nucleic acid synthesis, such as halogenated deoxyribonucleosides, inhibitors of glycosylation, and B lymphocyte mitogens. Different inducers induce different viruses, which indicates that the nature of the cellular controls that regulate their ability to express themselves are diverse. That proviruses are indeed under the control of *cis*-acting control elements is suggested by transfection experiments. Proviral DNA is infectious, even when integrated, but proviruses have been found that are not infectious unless the cellular DNA in which they reside is reduced in size to approximate provirus size, a proce-

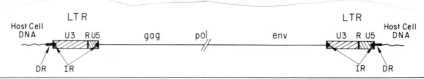

Figure 62-12. Structure of the integrated provirus. The coding sequences of the *gag, pol,* and *env* genes are flanked by long terminal repeats (LTRs) from 250 to 1200 bp long, depending on the virus, each composed of the U3, R, and U5 regions (Fig. 62-10), which are themselves flanked by short (5 to 22 bp) inverted repeat regions (IR). The entire provirus is flanked by direct repeats (DR) 4 to 6 bp long that are of host cell origin.

dure that would be expected to remove control elements linked to proviruses.

Ecotropic and Xenotropic Endogenous Viruses

There are two groups of endogenous viruses that differ in host range, interference patterns, and antigenic determinants. The first group comprises viruses that multiply readily in cells of the species in which they were induced; these are the ecotropic viruses. The second group comprises viruses that cannot infect cells of the species in which they were induced but can infect cells of other species; these are the xenotropic viruses. The host range is determined by the *env* gene.

Many ecotropic and xenotropic viruses have now been isolated from numerous animal species, including primates. For example, chickens contain at least 16 proviruses that have been mapped and that differ with respect to how readily they express themselves (some do not express themselves at all and are noninducible, others only express their *env* gene, still others form infectious virus particles), as well as in the structure of their LTRs. Mice also contain numerous proviruses in their genomes. It is probably conservative to estimate that wild-type mice, that is, mice that have not been inbred for proviral repression or derepression, harbor as many as 10 different proviruses of both types, each present to the extent of an average of 10 copies.

Most of these viruses are related to each other, as judged by nucleotide sequence analysis, and some fascinating evolutionary relationships have emerged from studies to determine whether proviruses in germline cells of various animal species are more closely related to each other than are their genes. If they are, this may indicate horizontal transmission and insertion of virus into germline cells. For example, the avian reticuloendotheliosis virus (REV) is much more closely related to murine viruses than to endogenous or exogenous (see below) chicken viruses, which suggests that a murine provirus was transferred from mice to birds many millions of years ago. The exogenous primate simian sarcoma virus and gibbon ape leukemia virus are derived from a murine xenotropic virus, and the cat xenotropic virus RD114 was acquired in the Pliocene from a monkey. Further, the diversity of integration sites (see above) suggests that vertebrate species experience a continual accretion of new endogenous proviruses in their germlines. This is the case for the high-incidence-of-leukemia AKR mouse strain in which new germline provirus integrations are accumulating at a rate of about 1 per 30 generations.

Expression and Significance of Endogenous Retroviruses

The expression of endogenous viruses appears to be under controls similar to those of developmentally regulated genes. Activation of an endogenous virus in a given tissue apparently reflects transcriptional activity of the insertion site. This may explain the occasional expression of endogenous viruses during early development and in some tissues of the adult, which has led to speculation that they may play a role in cellular differentiation. However, this is unlikely. Not only is provirus gene expression idiosyncratic with respect to tissue and age in different mouse strains, which is not compatible with provirus expression playing an essential functional role in ontogeny, but healthy chickens that lack all endogenous proviruses have been bred. However, it is conceivable that possession of endogenous proviruses may confer some form of survival advantage under natural conditions, such as protection against infectious retrovirus particles (see below). Beyond that, it is clear that endogenous proviruses represent a mode of persistence developed by retroviruses.

Retroviruses as Exogenous Viruses

Retroviruses not only exist as endogenous viruses that are passed on vertically from generation to generation, occasionally causing tumors in the animals in which they are activated, but also they can infect animals horizontally, as any other infectious agent. In fact, several natural populations suffer from retrovirus infections that cause tumors: examples are household cats infected with feline leukemia virus, cattle infected with bovine leukemia virus, and feral mice infected with murine leukemia viruses. The ability of endogenous viruses to act as exogenous viruses is demonstrated best by using, as hosts, animals such as AKR strain mice that have been bred to remove many of the alleles that protect normal mice against virus-induced leukemia.

There is also another class of retroviruses that only exist as exogenous viruses. Endogenous proviruses occasionally incorporate cellular genes that seem to play a role in development and differentiation into their sequences. As a result, they acquire the ability to transform cells and cause cancer. However, these chimeric viruses, the sarcoma/acute leukemia viruses (see below), are not integrated into germline cells, and they are therefore not endogenous viruses. Nor are they transmitted horizontally in nature with any frequency because most of them cannot multiply on their own, the acquired cellular gene having replaced one or other of the *gag, pol,* or *env* genes. The only exception is the group of avian sarcoma viruses whose prototype is Rous sarcoma virus (see below). As a result, these viruses arise occasionally in viremic animals and can be isolated and propagated in the laboratory but very probably die out in nature with the afflicted host.

Viruses with and without Transforming Genes: Nondefective and Defective Viruses

C-type retroviruses can be divided into two groups—those that can transform fibroblasts as well as certain other cells in vitro and rapidly cause tumors in animals, and those that cannot do so. It turns out that the difference lies in whether the viruses encode transforming genes or not.

Leukemia Viruses

Leukemia viruses possess the three genes, *gag, pol,* and *env,* that are essential for their multiplication; they are nondefective. However, they do not transform fibroblasts or any other cells in vitro, and while they are widely distributed and the animals that they infect are generally viremic throughout life, they cause tumors (lymphocytic, myeloid, and erythroblastic leukemias, osteopetrosis, nephroblastoma, and thymic and generalized lymphosarcomas) only late in life. When injected in large amounts into animals, they cause neoplasms only after a prolonged latent period.

Sarcoma/Acute Leukemia Viruses

The viruses in this group possess high oncogenic potential. They transform fibroblasts and hematopoietic cells in vitro and rapidly cause a wide variety of tumors of the hematopoietic system as well as sarcomas, endotheliomas, neurolymphomatosis, and splenomegaly. The reason for this is that their genomes contain an additional gene, a transforming gene or oncogene, that encodes a protein that triggers the transformation of normal cells into cancer cells. The first viruses of this group to be discovered and studied were several avian sarcoma viruses, the prototype of which is Rous sarcoma virus. These viruses possess four genes: *gag, pol, env,* and *src* (for sarcoma), which encodes a phosphoprotein, pp60^{src}, that possesses protein kinase activity. The *src* gene is the transforming gene, since viral mutants with temperature-sensitive lesions are temperature-sensitive with respect to ability to transform or maintain transformation. These viruses are also nondefective viruses, since they possess all three genes necessary for multiplication.

The other sarcoma/acute leukemia viruses include avian sarcoma viruses such as the Fujinami sarcoma virus, avian leukemia viruses such as MC29 (avian myelocytomatosis virus), avian erythroblastosis virus (AEV), and avian myeloblastosis virus (AMV), murine leukemia viruses such as Abelson murine leukemia virus, murine sarcoma viruses such as the Moloney, Kirsten, and Harvey murine sarcoma viruses, feline sarcoma viruses, and simian sarcoma virus. All these viruses possess genomes that are deficient in the genes that encode their structural proteins. Many lack part of the *gag* and *pol* genes, and some lack part of the *env* gene. They are therefore defective. They can only multiply in the presence of a helper virus, typically a leukemia virus, that encodes the missing nucleocapsid, polymerase, or envelope proteins. Strains of these viruses therefore consist of two types of particles, the helper virus particles and particles specified by the helper virus that contain the genome of the defective sarcoma/acute leukemia virus. These particles are known as "pseudotype particles" and are denoted thus: Mo-MSV(Mo-MLV) for Moloney murine

sarcoma virus, the virus in parentheses denoting the helper virus that is generally present in large excess. Usually any one of numerous leukemia viruses can serve as helper.

The part of the viral genome that is missing in these viruses is replaced by sequences like the *src* gene, sequences that are unrelated to normal retrovirus genetic information. These sequences are specific for each defective sarcoma/acute leukemia virus and represent their transforming genes. Usually, the amount of genetic material that is deleted exceeds that which replaces it, so that the genomes of the defective sarcoma/acute leukemia viruses are generally smaller than, and sometimes only one-half the size of, standard leukemia virus genomes.

Transforming Genes

The reason that sarcoma/acute leukemia viruses transform cells readily in vitro and rapidly induce cancer in animals is that they possess special genes, transforming genes or oncogenes. About 20 transforming genes have been discovered thus far on the basis of the presence in retrovirus genomes of specific nucleic acid sequences and the detection in cells transformed by them of proteins of specific size, possessing specific antigenic determinants and specialized functions. Ten of the best characterized of these oncogenes are identified in Table 62-3, and their physical maps are illustrated in Figure 62-13. Some are encoded only by specific transforming genes, such as pp60$^{v\text{-}src}$ of Rous sarcoma virus (RSV), P45$^{v\text{-}myb}$ of avian myeloblastosis virus (AMV), P45$^{v\text{-}erb\text{-}B}$ of avian erythroblastosis virus (AEV), P37$^{v\text{-}mos}$ of Moloney murine sarcoma virus (Mo-MSV), pp21$^{v\text{-}ras}$ of Harvey and Kirsten murine sarcoma virus (Ha-MSV and Ki-MSV), and P28$^{v\text{-}sis}$ of simian sarcoma virus (SiSV). Others are fusion proteins in which the N-terminal portion of the transforming gene product is the N-terminal portion of the *gag*-encoded protein, while the remainder is encoded by the specific oncogene.

Recently, transforming genes have also been detected in human cancer cells. This finding is of enormous potential significance as it affords yet another avenue for studying the basic mechanisms underlying tumorigenesis. The basic finding is that when DNA from a variety of human cancers is added to NIH 3T3 cells, an established strain of mouse cells, the cells become transformed and form readily identifiable foci. Table 62-4 lists some human tumors that have been shown to contain dominant transforming or cancer genes by this transfection technique, and the list is growing. There are several such genes, as can be shown by testing their sensitivity to various restriction endonucleases. All except one are related to the *ras* oncogene in Harvey and Kirsten murine sarcoma virus; that in human breast cancer is closely related to the oncogene in mammary carcinomas induced by mouse mammary tumor virus (MMTV). Interestingly, *c-Ki-ras* is the oncogene that is frequently activated in fibrosarcomas and skin carcinomas induced in the mouse by chemical carcinogens.

TABLE 62-3. VIRAL ONCOGENES

Oncogene	Virus	Type of Tumor	Animal	Protein Product	Cellular Protooncogene
v-src	Rous sarcoma virus, B77 ASV	Sarcoma	Chicken	pp60$^{v\text{-}src}$	pp60$^{c\text{-}src}$
v-myb	Avian myeloblastosis virus (AMV)	Leukemia	Chicken	P45$^{v\text{-}myb}$	P75$^{c\text{-}myb}$
v-myc	Avian myelocytomatosis virus (MC29)	Carcinoma, sarcoma, leukemia	Chicken	P110$^{v\text{-}gag\text{-}myc}$	c-myc
v-erb-B	Avian erythroblastosis virus (AEV)	Leukemia, sarcoma	Chicken	P45$^{v\text{-}erb\text{-}B}$	c-erb
v-fps	Fujinami sarcoma virus (FSV)	Sarcoma	Chicken	pp140$^{v\text{-}gag\text{-}fps}$	P98$^{c\text{-}fps}$
v-mos	Moloney murine sarcoma virus (Mo-MSV)	Sarcoma	Mouse	P37$^{v\text{-}mos}$	c-mos
v-ras	Kirsten murine sarcoma virus	Sarcoma	Rat	pp21$^{v\text{-}ras}$	pp21$^{c\text{-}ras}$
	Harvey murine sarcoma virus	Leukemia	Rat		
v-abl	Abelson murine leukemia virus	Leukemia	Mouse	pp120$^{v\text{-}gag\text{-}abl}$	pp150$^{c\text{-}abl}$
v-fes	Feline sarcoma virus (FeSV)	Sarcoma	Cat	pp85$^{v\text{-}gag\text{-}fes}$	pp92$^{c\text{-}fes}$
v-sis	Simian sarcoma virus	Sarcoma	Wooly monkey	P28$^{v\text{-}sis}$	c-sis

Protooncogenes

The fact that human cancers harbor transforming genes and that some of these transforming genes are very similar to genes in sarcoma/acute leukemia viruses, although not predictable, was not totally unexpected since a most surprising discovery had been made shortly before: namely that retrovirus transforming genes are closely related to genes present in the genomes of most animals. As indicated in Table 62-3, each of the viral oncogenes has a cellular counterpart, or homolog. The degree of relatedness of the viral oncogenes with their cellular counterparts is often surprisingly close; for example, v-myb and c-myb differ in only 15 nucleotides out of 1197. The major difference between the viral oncogenes and their cellular counterparts is that the latter possess introns, whereas the viral oncogenes do not.

These transforming genes, or protooncogenes, are extraordinarily widespread in animals. They are present not only in all vertebrates but probably also in all metazoan genomes. The presence of src sequences is, for example, easily recognizable in the Drosophila genome. Clearly, these genes have survived more than a thousand million years; they exhibit extraordinary evolutionary conservation. Presumably the cellular genes arose first, since they are distributed throughout innumerable animal genomes, whereas the viral oncogenes are restricted to a very few sarcoma/acute leukemia viruses. The fact that retroviruses do indeed possess the ability to acquire these cellular protooncogenes has been demonstrated conclusively as follows. When mutants of RSV with large deletions in their src gene are injected into chickens, virus can be recovered that can again cause tumors (which the deletion mutant could not do) because the src gene has been reconstituted in them by pirating the chicken c-src gene. This indirect evidence that cellular protooncogenes possess, under appropriate conditions, the same oncogenic potential as viral oncogenes was soon supplemented by direct and conclusive proof. It was shown that whereas cellular protooncogenes do not transform cells in transfection experiments of the type described above (p. 944), at least some of them do so when linked by genetic engineering to LTR regions of proviruses, which, as pointed out above, contain strong promoters (p. 940).

Exactly how protooncogenes are incorporated into retrovirus genomes is not known. Presumably the protooncogene first recombines with and is inserted into the provirus, and its introns are lost when the transcript of this new recombinant provirus is spliced to yield the RNA genome of the new transforming virus. This insertion of cellular material into the retrovirus genome apparently provides the opportunity for further recombination with homologous regions within the host cell genome. Thus, the provirus of a murine sarcoma virus has been sequenced, and it is clear that it contains, in addition to portions of the gag gene, a ras gene flanked by sequences derived from both mouse and rat DNA, which suggests that it was generated as a result of numerous recombination events.

Functions of Oncogenes/Protooncogenes

The first retrovirus transforming gene to be studied in detail was the src gene of RSV. The protein encoded by this gene, pp60$^{v\text{-}src}$, is a phosphoprotein that either possesses protein kinase activity itself or is very closely associated with a protein kinase. Phosphorylation of proteins has for some time been recognized as a mechanism for modifying protein function. The excitement occasioned by the discovery of the protein kinase activity of pp60$^{v\text{-}src}$ was that it phosphorylates tyrosine residues rather than serine or threonine residues, which are the amino acids phosphorylated by most other protein kinases. Further, it is itself phosphorylated on a serine and on a tyrosine residue. Phosphorylation of tyrosine residues in proteins is much more common in tumor cells than in normal cells, although it does occur in normal cells. For example, it oc-

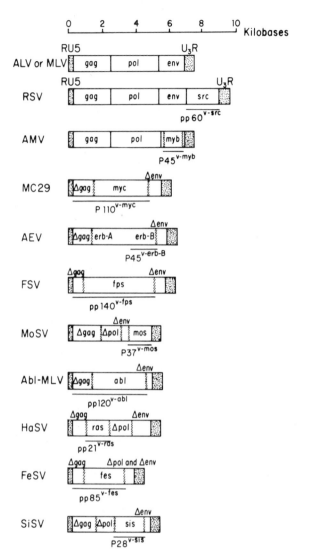

Figure 62-13. The genomes, transforming genes, and transforming gene products of acute leukemia/sarcoma viruses. The genomes of avian and murine leukemia viruses (ALV and MLV) and of Rous sarcoma virus (RSV), which are nondefective, are shown for comparison. The transforming proteins are indicated beneath the RNA sequences that encode them. Phosphoproteins are denoted by the prefix pp, other proteins by the prefix P. Note that some transforming proteins are fusion proteins with portions of *gag* proteins (the v-myc, v-fps, v-abl, and v-fes gene products), while others (the v-src, v-myb, v-erb-B, v-mos, v-ras, and v-sis gene products) are not. Note that in all cases, except RSV, the transforming gene replaces a very substantial portion of the normal retrovirus genome, leaving only partially deleted genes (Δ*gag*, Δ*pol,* and Δ*env*), which are not large enough to encode functional proteins/enzymes. ALV and MLV, avian and murine leukemia virus, respectively; RSV, Rous sarcoma virus; AMV, avian myeloblastosis virus; MC29 is a strain of avian myelocytomatosis virus; AEV, avian erythroblastosis virus; FSV, Fujinami sarcoma virus; MoSV, Moloney murine sarcoma virus; Abl-MLV, Albelson murine leukemia virus; HaSV, Harvey murine sarcoma virus; FeSV, feline sarcoma virus; SiSV, simian sarcoma virus.

TABLE 62-4. PROTOONCOGENES/ONCOGENES IN HUMAN TUMORS DEMONSTRATED WITH THE NIH 3T3 CELL TRANSFECTION ASSAY

Tumor	Gene
Bladder carcinoma	c-Ha-ras
Lung cancer	c-Ki-ras
Colon cancer	c-Ki-ras
Neuroblastoma	c-N-ras*
Pancreatic carcinoma	As yet unidentified
Wilms' tumor	As yet unidentified
Testicular tumors	As yet unidentified
Breast cancer	As yet unidentified

*c-N-ras is related to both c-Ha-ras and c-Ki-ras but is not identical with them.

curs when cells are stimulated with growth factors, such as epidermal growth factor (EGF) whose binding to the surface of cells stimulates DNA synthesis and cell division.

The protein pp60[v-src] appears to be localized mainly in adhesion plaques, specialized regions of the plasma membrane that adhere to solid surfaces and that are dismantled in cancer cells. Interestingly, the protein kinase activity of pp60[v-src] phosphorylates a tyrosine residue in vinculin, a protein that is a constituent of adhesion plaques and becomes dispersed throughout the cell following transformation by RSV. It seems reasonable to suggest that the phosphorylation of vinculin precipitates the dismantling of adhesion plaques, but the actual importance of this event in cell transformation remains to be established, especially since other cellular proteins, in particular the glycolytic enzymes enolase, phosphoglycerate mutase, and lactic dehydrogenase, as well as several other as yet unidentified proteins, also become phosphorylated very soon after infection with RSV.

Three other retrovirus oncogenes also possess or are very closely associated with tyrosine-phosphorylating protein kinase activity (Table 62-5), but other viral oncogenes do not. For example, pp21[v-ras] is also a protein kinase, but it phosphorylates threonine residues; P45[v-erb-B] is a glycoprotein that becomes associated with dense intracytoplasmic membrane; P110[v-gag-myc] is a DNA-binding protein that localizes in the nucleus, which is where P45[v-myb] is also found; and P37[v-mos] is a soluble protein in the cytoplasm.

The retrovirus oncogenes, and therefore also their cellular protooncogene homologs, are therefore a heterogeneous group of proteins with a plethora of functions. It is interesting to compare them with the tumor antigens that are expressed by DNA-containing tumor viruses: the middle t antigen of polyoma virus, which also possesses tyrosine-phosphorylating protein kinase activity, the large T antigens of SV40, polyoma virus, and adenovirus, which are nuclear DNA-binding proteins, as is the EBNA of EBV, and the SV40 and adenovirus T antigens, which possess strong affinity for p53, a cellular nuclear protein with a possible function in cell proliferation.

The picture that emerges is that all vertebrate cells

TABLE 62-5. FUNCTIONS OF ONCOGENES/PROTOONCOGENES

Gene	Species of Origin	Protein Kinase	GTP-binding Activity	Glyco-protein	DNA-binding Protein	Intracellular Location
v-src	Chicken	+ tyr				Plasma membrane
v-myb	Chicken	−				Nucleus
v-myc	Chicken	−			+	Nucleus
v-erb-B	Chicken	−		+		Intracytoplasmic membrane
v-fps	Chicken	+ tyr				Plasma membrane
v-mos	Mouse	−				Cytoplasm
v-ras	Rat	+ thre	+			Plasma membrane
v-abl	Mouse	+ tyr				Plasma membrane
v-fes	Cat	+ tyr				?
v-sis	Monkey	?				?

contain at least one copy of about 20 distinct genes, the protooncogenes, the messenger RNAs of which may on occasion be incorporated into retrovirus RNA. The chromosomal location of many of these protooncogenes in the human genome is known: thus c-myb is located on chromosome 6, c-myc and c-mos are on chromosome 8, c-ras is on chromosomes 6, 11, 12, and X (the human genome contains two c-Ha-ras genes and two c-Ki-ras genes, genes that are incorporated into the Harvey and Kirsten strains of MSV, respectively), c-N-ras (Table 62-4) is on chromosome 1, c-fes is on chromosome 15, and c-sis is on chromosome 22.

Most of these protooncogenes express themselves to a certain extent in normal cells and presumably fulfill specific essential functions. The nature of these functions is not clear, though clues are provided by examining the functions of their viral homologs. It is known that c-ras is transcribed extensively during liver regeneration, that c-abl and c-ras are transcribed during prenatal and early postnatal development of the mouse, and that c-myc, c-erb and c-src are transcribed in a variety of chick cells and tissues. Further, c-myb is transcribed in immature lymphoid and erythroid hematopoietic cells, and c-ras is transcribed at elevated levels in a human hematopoietic precursor cell line.

Interestingly, the various protooncogenes are related to a most surprising degree. It is clear, for example, partly from studies of their nucleic acid sequences and partly from studies of the amino acid sequences of their gene products, that src, fps, abl, fes, mos, and erb, of which some are tyrosine phosphorylating protein kinases while others are not, are all quite closely related and that they exhibit homology not only with certain cellular serine phosphorylating cyclic nucleotide-dependent protein kinases but also with the β-subunit of mitochondrial and bacterial ATP-synthetase, primarily in the region that is believed to contribute to nucleotide binding. It may be noted, in this regard, that pp21^{v-ras} is also a GTP-binding protein (see Table 62-5). It seems that all these protooncogenes are derived from a common ancestor that arose before the emergence of metazoan phyla.

Whereas in normal cells several oncogenes are usually transcribed simultaneously to greater or lesser extent, only one protooncogene is transcribed in large amounts in cancer cells. For example, the HL-60 cell line, derived from a patient with acute promyelocytic leukemia, expresses high levels of c-myc, the gene that is also expressed at high levels in most lymphomas; a human bladder cancer cell line expresses high levels of c-Ha-ras; avian erythroleukemia cells express high levels of c-erb; and a murine plasmacytoma cell line expresses high levels of c-mos. Thus, each type of tumor is associated with one specific oncogene. This target cell specificity is not due to the host range of the virus that carries the oncogene. Thus MC29 virus transforms both fibroblasts and macrophages in culture but not erythroblasts, though the myc gene expresses itself in such cells, and RSV infects both macrophages and fibroblasts and pp60^{src} is synthesized in both, but only fibroblasts are transformed. Rather, target cell specificity is a property of the oncogene/protooncogene itself, and these genes appear to be tissue-specific regulators of cellular growth and development. This conclusion is supported by the finding that different transforming genes are activated in neoplasms representative of different stages of normal B and T cell differentiation, which suggests that an excess of protooncogene products inhibits the maturation of the various hematopoietic precursor cells from one compartment to the next.

How Do Retroviruses that Lack Oncogenes Transform Cells?

This brings us to the two most central questions concerning the role of retroviruses and oncogenes/protooncogenes in transforming cells and causing cancer. There is no conceptual hurdle concerning how sarcoma/acute leukemia viruses accomplish this, since they possess oncogenes, and following provirus integration, these genes are presumably brought under the control of provirus LTRs and express themselves vigorously so that large amounts of the proteins encoded by them are formed. But how do leukemia/leukosis viruses that do not possess

TABLE 62-6. CELLULAR GENES IMPLICATED IN NONVIRAL HUMAN AND MURINE TUMORS BY TRANSLOCATION OR AMPLIFICATION

Gene	Tumor	Alteration
c-myc	Burkitt's lymphoma	Translocation
c-mos	Plasmacytoma	Translocation
c-myc	Myeloid leukemia	Amplification
c-abl	Myeloid leukemia, erythroid leukemia	Translocation, amplification

oncogenes cause cancer, as they all do late in life, and why do protooncogenes, most of which are expressed all the time, suddenly transform cells?

There may be several mechanisms (Table 62-6).

Insertional Mutagenesis

Leukemia/leukosis provirus can probably become integrated anywhere in the animal genome, and in most cases no noticeable change will result. If, however, insertion occurs in the vicinity of a protooncogene, it will come under the influence of the proviral LTR and be stimulated to be expressed vigorously, with resultant cell transformation. Indeed, it is known that the c-myc gene is expressed to a high degree in many of the lymphomas caused by infection of chickens with avian leukosis viruses and that it takes no more than a 5-fold increase in the amount of v-src to transform a cell.

Gene Amplification

Many tumor cells display abnormalities in their chromosomes due to the presence of excess amounts of amplified DNA. The general progression is as follows. A region of a chromosome, usually encompassing many genes, begins to replicate excessively, up to 1000-fold and beyond, for reasons that are not known, probably because of a rare spontaneous event. Sooner or later, the amplified DNA departs the chromosome and exists in the form of a ring chromosome and then in the form of double minute chromosomes. Finally, it reenters chromosomes as homogeneously staining regions. These homogeneously staining regions have been shown to contain protooncogenes, c-myc and c-Ki-ras among them in different tumor cells, which therefore exist in such cells in the form of far more copies than in normal cells, that is, in highly amplified form. They then give rise to very large amounts of messenger RNA and of the proteins that they encode. Once again, cell transformation is associated with enhanced protooncogene expression.

Mutation

The two preceding mechanisms invoke increased *dosage* of protooncogene products as the cause of transformation. Another mechanism involves a *qualitative alteration* of the protooncogene product. Such an alteration could be caused by a point mutation in the protooncogene. Indeed, it has been found that the c-Ha-ras protooncogene that is activated in a human bladder carcinoma to the point that it becomes able to transform NIH 3T3 cells in the transfection assay differs in one nucleotide residue from the c-Ha-ras protooncogene that is present in normal human cells. This change causes a single amino acid substitution in the $pp21^{ras}$ protein that is expressed in the cancer cell. It is conceivable that it is this change that triggers cell transformation.

Translocation

Finally, many cancer cells display chromosomal abnormalities caused by translocations, where one portion of a chromosome breaks away and exchanges position with another portion of another chromosome. Some of these translocations involve DNA sequences that contain protooncogenes. For example, in Burkitt's lymphoma a portion of chromosome 8 is transferred to chromosome 14, and the transferred DNA includes c-myc. Murine plasmacytomas also often exhibit translocations in which the translocated chromosome portion includes c-myc. Such translocations may have two effects: they may affect the extent to which a protooncogene is transcribed, and altered protooncogenes may be produced as a result of recombination, the gene products of which, like that of the mutated gene discussed above, may trigger cell transformation.

Naturally, we are only considering here the very first of a series of steps that culminate in the generation of fully fledged cancer cells. It is conceivable, and indeed likely, that various oncogenes may act sequentially in the same cell. It is known, for example, that in B cell lymphomas of chickens induced by infection with avian leukemia viruses (ALVs), not only is the expression of the c-myc gene enhanced but another gene is also activated, a gene that can transform NIH 3T3 cells in the transfection assay described above. This second gene has been cloned and found to encode a protein that is homologous to the N-terminal portion of transferrin (although it is much smaller than transferrin). Interestingly, a related gene is also activated in Burkitt's lymphoma cells. This second gene may well act during a later stage in tumorigenesis. However this may be, it is clear that recognition of the relationship between viral oncogenes and cellular protooncogenes has provided a most powerful tool with which to probe the reactions fundamental to the initiation of the cancer process.

Recombinant Endogenous Viruses: MCF Viruses

As pointed out above, mice, and no doubt other vertebrates as well, harbor retrovirus proviruses in their genomes that cause leukemia at some stage of their life. In the numerous inbred mouse strains that have been developed over the years, the ability of these proviruses to cause leukemia has been separated from some of the elements that control their ability to express themselves. Very likely, some of the factors that cause leukemia early in life or late in life are those that were discussed in the preceding section, that is, protooncogene activation or protooncogene alteration. However, it appears that other factors may also be involved. It has been found, for example, that the development of spontaneous leukemia tends to be associated with the generation of a new class of retroviruses. Such new viruses have been isolated from the thymuses of late preleukemic or leukemic AKR and preleukemic C58 mice and from the lymphomas that develop in NIH Swiss mice. These viruses are recombinants in the *env* gene between ecotropic and xenotropic endogenous viruses. They possess the *gag* and *pol* genes of ecotropic viruses, but their *env* genes comprise sequences from both ecotropic and xenotropic endogenous viruses. They multiply not only in mouse cells but also in cells of other species and are, therefore, said to be dual-tropic. They also reduce focal morphologic alterations in a mink lung cell line and are therefore often referred to as mink cell focus (MCF)-inducing viruses. They are *not* endogenous viruses, since they are not present in germline cells.

The emergence of MCF viruses from lymphoid tissue in the preleukemic and leukemic state indicates that they may be involved in the pathogenesis of leukemia. It remains to be established whether the recombinant *env* glycoproteins themselves are the transforming proteins or whether their role is host range modification/expansion.

Expression of *env* Gene of Endogenous Viruses

While endogenous viruses usually do not express themselves to the extent of virus particle formation, they do frequently express their *env* gene. The first indication in this direction was the discovery that a strain of RSV whose RNA contains a deletion in the *env* gene (the Bryan strain) is incapable of growing in some chick cells but is capable of growing in others, which were therefore said to possess a "chick-helper-factor" (chf). This factor turned out to be the spontaneously expressed glycoprotein encoded by the *env* gene of the endogenous chick RNA tumor virus RAV-O. Thus RAV-O does not express itself at all in chick cells that are chf⁻; it expresses its *env* gene in cells that are chf⁺; and there are one or two inbred strains of chickens in which it expresses itself completely, to the extent of infectious virus particle formation (and causing leukemia infrequently and late in life).

The situation is similar in the mouse. It has long been known that a differentiation alloantigen designated G_{IX} (since it appeared to be specified by a gene in linkage group IX of the mouse) is present in the thymocytes of certain mouse strains. It is also present in the serum of some mouse strains (particularly strain 12A), in epithelia associated with the digestive tract, and in the epithelial lining of the male reproductive tract and, therefore, in semi-

nal fluid. It was then found that the G_{IX} antigen could be induced in rats by infection with murine leukemia virus and that it represents the gp70 glycoprotein specified by the *env* gene of a provirus that expresses itself irrespective of *gag* and *pol* gene expression, just as does chf. Mice harbor numerous proviruses, all of which may express their *env* genes independently of each other in the form of gp70 molecules located on the surface of cells of various organs. Thus in mouse strains such as the 129/J mouse, from which no C-type virus has ever been isolated, the gp70s of four different endogenous viruses are expressed in different developmental compartments—one in the serum, one on thymus and spleen cells, one on bone marrow cells, and one on cells lining the male genital tract. This polymorphism may benefit the mouse through the various gp70s acting as differentiation antigens.

Mouse Mammary Tumor Viruses (MMTV)

Mouse mammary tumor virus, the prototype of B-type RNA tumor viruses, exists in the form of numerous endogenous proviruses that express themselves to varying degrees in different strains of mice. The most familiar is the milk factor of Bittner, a virus strain that is resident in C3H and A strain mice (high-incidence strains), expresses itself in all cells of its hosts, and is present in particularly large amounts in lactating mammary tissue and, therefore, in milk. As a result, it is passed on readily to progeny animals, in which it produces mammary adenocarcinomas with high frequency early in life (generally between age 6 and 12 months). It is also passed on to animals of low-incidence strains when they are reared by foster mothers of the C3H and A strains and, with high frequency, produces tumors in them also.

The other MMTV strains express themselves less readily. They are not present in large amounts in milk and are transmitted vertically through the gametes. They are also much less oncogenic and cause tumors with low frequency late in life.

These various endogenous proviruses express themselves to very different degrees in different mouse tissues. In virus-producing tumors and lactating mammary tissue of high-incidence mouse strains there are from 100 to 1000 genome equivalents of MMTV RNA per cell; in tumors of low-incidence mouse strains only 1 to 2; and in livers and spleens from either high- or low-incidence mouse strains (that is, in cells that never become transformed and never produce MMTV), there is only 0.1 to 1 genome equivalent of viral RNA per cell.

The factors that regulate the transcription of MMTV proviruses are both genetic and hormonal. Estrogens enhance the development of mammary cancer (which normally occurs only in female mice but can also be induced in male mice of high-incidence strains); and glucocorticoids greatly increase the degree of MMTV formation in lactating mammary tissue cells and mammary tumor cells, but do not effect significant increases in other mouse cells.

The hormone-sensitive control element for MMTV provirus expression appears to reside in its LTR sequences, which are about five times as long as those of C-type retroviruses.

A transforming gene active in the NIH 3T3 cell transfection assay has been detected in the DNA of MCF7 cells, a human breast cancer cell line, and this gene retains its hormone inducibility. This human gene is very closely related to the transforming gene of MMTV. Thus, in both mice and humans, breast cancer appears to involve the activation and functioning of an oncogene/protooncogene. Whether the human gene is also part of a retrovirus provirus remains to be determined.

Retrovirus-like DNA Sequences and Particles

Three types of DNA sequences have been detected in animal genomes that share certain basic features with the proviruses of retroviruses. They are:

Intracisternal A-type Particles (IAP)

Mouse cells often contain particles some 60 to 80 nm in diameter that consist of toroidal (ring-shaped) nucleoids enclosed by a membrane (Fig. 62-14). These particles are found budding abundantly into the cisternae of the rough endoplasmic reticulum in all murine plasmacytomas, as well as in a variety of other murine tumors (rhabdomyosarcoma and neuroblastoma) and in normal murine embryos at certain stages of development but only rarely in normal adult murine tissues. Usually they remain attached to the rough endoplasmic reticulum membrane, but in many myelomas they are released in extracellular form.

Like retrovirus particles, IAP contain two identical molecules of RNA about 7000 nucleotides long (mol wt about 8×10^6) and possess an RNA-dependent DNA poly-

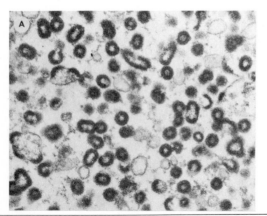

Figure 62-14. Intracisternal A-type particles (IAP) extracted from mouse plasma cell tumor MOPC 104E and partially purified by sucrose density gradient centrifugation. × 70,000. *(Courtesy of Dr. N.A. Wivell.)*

merase. They possess no infectivity, but their major structural protein is encoded by their RNA. Their RNA is not related to the RNA of any retrovirus of the common mouse *Mus musculus* but is closely related (about 25 percent homology and similarly situated restriction endonuclease sites) to the RNA of M432, an endogenous retrovirus of the Asian mouse *Mus cervicolor*. Amazingly, the genome of *M. musculus* contains about 1000 dispersed copies of the IAP provirus, while the genome of *M. cervicolor* contains only about 20 copies. These two mouse species diverged about 5 million years ago. Clearly, the evolution of *M. musculus* was accompanied by a striking amplification of IAP proviral DNA, which now makes up about 0.3 percent of the total cellular DNA. Recently, IAP-related sequences have also been detected in moderate copy number in the hamster and in low copy number in other rodents. The extent of sequence divergence indicates that they have been resident in the hamster genome for millions of years.

Examination of mouse gene libraries has revealed that the 1000 or so IAP proviruses in the mouse genome are a strikingly heterogeneous population of sequences. About a quarter of the IAP provirus, near one end, is highly conserved, but varying lengths of its central portion are often deleted; and the other end is highly variable. Thus, while there is a common provirus pattern, there is also very extensive sequence switching, which is no doubt the result of recombination and sequence deletion. One feature common to all IAP proviruses is possession of LTRs with the same structure as retrovirus LTRs and flanking short direct repeats of cellular DNA. IAP proviruses are thus integrated into cellular DNA just as are retrovirus proviruses and, like them, they are transposable elements.

The relatedness of IAP to retrovirus M432 indicates that they have evolved from a C-type retrovirus. They are nonfectious, which is not surprising in view of the obviously degenerate form of their RNA; and although they appear regularly in early mouse embryos, they have no as yet defined physiologic role. In view of the fact that IAP proviruses are so abundant and widely dispersed in the *M. musculus* genome, they may have assumed in this species some function beyond coding for the particles themselves. It is conceivable, for example, that they can activate cellular genes and possess regulatory functions controlling other genetic elements of the mouse.

VL30 DNA Sequences

Recently, two fascinating extensions of this work have come to light. First, it has been found that many species of mice contain yet another set of dispersed reiterated DNA sequences about the same size as IAP or retrovirus proviruses, the so-called VL30 (for virus-like) sequences. They possess the following properties: (1) some mouse species contain only 1 to 2 copies of them, whereas others, such as *M. musculus*, contain about 200; (2) like retrovirus and IAP proviruses, VL30 sequences are flanked by LTRs, attributes of transposable elements; (3) in certain mouse cells they are transcribed extensively into 30S polyadenylated RNA that tends to exist in dimeric form like retrovirus RNA and is efficiently encapsidated into noninfectious C-type retrovirus particles that provide a means for transferring them from cell to cell; (4) they are not related to any retrovirus or IAP provirus; (5) certain portions of VL30 sequences appear to exist in the mouse genome in a form that is not part of intact VL30 sequences; and (6) more or less related sequences are also present in the DNAs of other animals, such as rats and humans.

VL30 sequences clearly have a long evolutionary history, part of which they may share with IAP and retroviruses. If their LTRs contain promoters of transcription like the LTRs of retroviruses, they may function as portable promoters and play a significant role in differentiation and cellular regulation.

The *copia* Element of *Drosophila*

The *copia* element is a transposable element found throughout the *Drosophila* genome. Usually 100 to 200 DNA sequences are present, and they constitute about 5 percent of the total DNA. Like retrovirus and IAP proviruses and VL30 sequences, *copia* sequences are flanked by LTRs several hundred base pairs long, short inverted repeats, and very short direct repeats of cellular DNA (p. 942), and they are transcribed into long polyadenylated RNA molecules. Virus-like particles have now been found in *Drosophila* that contain *copia* RNA sequences and RNA-dependent DNA polymerase activity. Like IAPs they are not infectious, and, further, like retrovirus and IAP RNA but unlike VL30 RNA, *copia* RNA encodes what appears to be the major structural component of the *copia* virus-like particles.

The existence of yet another class of these large transposable genetic elements intensifies still further the question of their biologic role, which may be fundamentally important in view of the extraordinary extent to which they have been conserved during evolution. It may also be asked, are *copia* elements evolving toward retroviruses, or are they descended from them?

Do Viruses Cause Cancer in Humans?

Many viruses transform cells, and many viruses cause cancer in animals. But which viruses cause cancer in humans?

In most cases, it is impossible to demonstrate a direct causal relationship; but there are four groups of viruses that seem to play a very important role in carcinogenesis in humans.

Human Papillomaviruses. As described on page 932 and in Chapter 69, several types of HPV cause lesions that progress to malignant carcinomas.

EBV. This virus plays a very important role in the genesis of Burkitt's lymphoma and nasopharyngeal carcinoma. The role of human herpes simplex virus type 2 in cervical cancer is less well defined.

Hepatitis B Virus. The DNA of this virus is known to be integrated into the human genome, where its presence correlates strongly with the incidence of hepatoma.

Retroviruses. An enormous amount of work has been devoted to attempts to find and isolate a human retrovirus. Only one, HTLV, has so far been isolated (see below). This is remarkable for two reasons. First, most animal species harbor endogenous retroviruses that express themselves more or less frequently during the animal's life. This includes the closest relatives of humans, namely, gorillas, orangutans, chimpanzees, and baboons. Second, all animals contain protooncogenes, cellular genes that can be incorporated into retrovirus genomes and thereby transferred from cell to cell. The human genome also contains these protooncogenes, and, as described above, they are known to be activated in human cancer cells. Extensive searches for human retroviruses during the last decade or so, which took the form of searches for RNA, DNA, and protein sequences in normal and cancerous human tissues that were related to primate, feline, or murine retroviruses, yielded results that were sometimes tantalizingly close to being convincingly positive. Yet they were never sufficiently regular, frequent, and reproducible to provide conclusive evidence for the existence of human retroviruses. These results can now be ascribed to the presence in the human genome of protooncogenes and their activation in human cancers.

Human T Cell Leukemia Virus (HTLV)

Recently a human retrovirus has been isolated that does indeed appear to play a role in a human cancer. This is *human T cell leukemia virus (HTLV)*, which has been isolated from peripheral blood T lymphocytes of patients with adult T cell leukemia (ATL). It is not an endogenous retrovirus, and its provirus is not present in human germline cells. Not surprisingly, the distribution of ATL is that of a disease caused by an infectious agent: it occurs most frequently in Southwest Japan and in the Caribbean basin (where the disease is called T cell lymphosarcoma cell leukemia). About 10 percent of the population in Southwest Japan possess antibodies to the major components of HTLV particles. Such antibodies are absent in humans in parts of the world where the disease is not endemic. Recent evidence suggests that there may be at least two types of HTLV that are related but substantially different serologically. HTLV is not closely related to any known retrovirus, but its major core protein, p24 (though not its reverse transcriptase), displays some relatedness to the p24 of bovine leukemia virus. However, the degree of homology observed is not sufficient for detectable immunologic cross-reactivity.

The evidence suggests, therefore, that HTLV is a unique novel retrovirus, exogenous for humans and of unknown origin, that is capable of causing T cell leukemia in humans. It is the only retrovirus so far associated with any human neoplasm, though protooncogene activation occurs in most human cancers.

FURTHER READING

Books and Reviews

GENERAL

Bishop JM: Cancer genes come of age. Cell 32:1018, 1983

Cairns J: The origins of human cancers. Nature 289:353, 1981

Doerfler W: DNA methylation and gene activity. Annu Rev Biochem 52:93, 1983

Ponder BAJ: Genetics and cancer. Biochim Biophys Acta 605:369, 1980

DNA TUMOR VIRUSES

Tooze J: DNA Tumor Viruses. New York, Cold Spring Harbor Laboratory, 1980

Klein G (ed): DNA-virus Oncogenes and their Action. New York, Raven P, 1983

Ziff EBl: Transcription and RNA processing by the DNA tumor viruses. Nature 287:491, 1980

Doerfler W: Uptake, fixation and expression of foreign DNA in mammalian cells: The organization of integrated adenovirus DNA sequences. Curr Top Microbiol Immunol 101:127, 1982

Papovaviruses

Lancaster WD, Olson C: Animal papillomaviruses. Microbiol Rev 46:191, 1982

Benjamin TL: The hr-t gene of polyoma virus. Biochim Biophys Acta 695:69, 1982

Hand R: Functions of T antigens of SV40 and polyomavirus. Biochim Biophys Acta 651:1, 1981

Levine AJ: The nature of the host range restriction of SV40 and polyoma viruses in embryonal carcinoma cells. Curr Top Microbiol Immunol 101:1, 1982

Klein G (ed): The Transformation-associated Cellular p53 Protein. New York, Raven, 1982

Norkin LC: Papovaviral persistent infections. Microbiol Rev 46:384, 1982

Walker DL, Padgett BL: Progressive multifocal leukoencephalopathy. In Fraenkel-Conrat H, Wagner RR (eds): Comprehensive Virology. New York, Plenum, 1983, vol 18, p 161

Herpesviruses

Fleckenstein B: Oncogenic herpesviruses of non-human primates. Biochim Biophys Acta 560:301, 1979

Kieff E, Danbaugh T, Heller M, et al.: The biology and chemistry of Epstein–Barr virus. J Infect Dis 146:506, 1982

Knipe DM: Cell growth transformation by herpes simplex virus. Prog Med Virol 28:114, 1982

Nazerian K: Marek's disease lymphoma of chicken and its causative herpesvirus. Biochim Biophys Acta 560:375, 1979

Vonka V, Hirsch I: Epstein–Barr-virus nuclear antigen. Prog Med Virol 28:145, 1982

zurHausen H, Fresen K: Heterogeneity of Epstein–Barr virus. Biochim Biophys Acta 560:343, 1979

Hepatitis B Virus

Maupas P, Melnick JL (eds): Hepatitis B virus and primary hepatocellular carcinoma. Prog Med Virol 27, 1980 and 1981

Tiollais P, Charney P, Vyas GN: Biology of hepatitis B virus. Science 213:406, 1981

RNA TUMOR VIRUSES

General

Weiss R, Teich N, Varmus H, et al.: RNA Tumor Viruses. New York, Cold Spring Harbor Laboratory, 1982

Bishop JM: Cellular oncogenes and retroviruses. Annu Rev Biochem 52:301, 1983

Coffin JM: Structure, replication and recombination of retrovirus genomes: Some unifying hypotheses. J Gen Virol 42:1, 1979

Eisenman RN, Vogt VM: The biosynthesis of oncovirus proteins. Biochim Biophys Acta 473:188, 1978

Gilden RV, Rabin H: Mechanisms of viral tumorigenesis. Adv Virus Res 27:281, 1982

Ihle JN, Enjuanes L, Lee JC, et al.: The immune response to C-type viruses and its potential role in leukemogenesis. Curr Top Microbiol Immunol 101:31, 1982

Jaenisch R: Endogenous retroviruses. Cell 32:5, 1983

Levy JA: Xenotropic type C viruses. Curr Top Microbiol Immunol 79:111, 1978

Montelaro RC, Bolognesi DP: Structure and morphogenesis of type C retroviruses. Adv Cancer Res 28:63, 1978

Nooter K, Bentvelzen P: Primate type-C oncoviruses. Biochim Biophys Acta 695:461, 1980

O'Connor TE, Rauscher FJ Jr (eds): Oncogenes and Retroviruses: Evaluation of Basic Findings and Clinical Potential. New York, Alan R. Liss, 1983

Risser R: The pathogenesis of Abelson virus lymphomas of the mouse. Biochim Biophys Acta 651:213, 1981

Scolnick EM: Hyperplastic and neoplastic erythroproliferative diseases induced by oncogenic murine retroviruses. Biochim Biophys Acta 651:273, 1982

Stephenson JR (ed): Molecular Biology of RNA Tumor Viruses. New York, Academic Press, 1980

Varmus H (ed): Readings in Tumor Virology. New York, Cold Spring Harbor Laboratory, 1983

Temin HM: Origin of retroviruses from cellular moveable genetic elements. Cell 21:599, 1980

Troxler DH, Ruscetti Sk, Scolnick EM: The molecular biology of Friend virus. Biochim Biophys Acta 605:305, 1980

Weinberg RA: Origins and roles of endogenous retroviruses. Cell 22:643, 1980

Oncogenes/Protooncogenes

Bishop JM: Enemies within: The genesis of retrovirus oncogenes. Cell 23:5, 1981

Fine D, Schochetman G: Type D primate retroviruses: A review. Cancer Res 38:3123, 1978

Dalla-Favera R, Martinotti S, Gallo RC: Translocation and rearrangements of the c-myc oncogene locus in human undifferentiated B-cell lymphomas. Science 219:963, 1983

Erikson RL: The transforming protein of avian sarcoma viruses and its homologue in normal cells. Curr Top Microbiol Immunol 92:25, 1981

Hayman MJ: Transforming proteins of avian retroviruses. J Gen Virol 52:1, 1981

Hayward WS, Neel DG: Retroviral gene expression. Curr Top Microbiol Immunol 91:217, 1981

Huebner RJ, Todaro GJ: Oncogenes of RNA tumor viruses as determinants of cancer. Proc Natl Acad Sci USA 64:1087, 1969

Graf T, Stéhelin D: Avian leukemia viruses: Oncogenes and genome structure. Biochim Biophys Acta 651:245, 1982

Klein G (ed): Oncogene Studies. New York, Raven, 1982

Klein G: Specific chromosomal translocations and the genesis of B-cell-derived tumors in mice and men. Cell 32:311, 1983

Temin HM: Function of the retrovirus long terminal repeat. Cell 28:3, 1982

Varmus HE: Form and function of retroviral proviruses. Science 216:812, 1982

Weinberg RA: Use of transfection to analyze genetic information and malignant transformation. Biochim Biophys Acta 651:25, 1981

Weinberg RA: Fewer and fewer oncogenes. Cell 30:3, 1982

Weinberg RA, Steffen DL: Regulation of expression of the integrated retrovirus genome. J Gen Virol 54:1, 1981

Mammary Tumor Viruses

Bentvelzen P: Interaction between host and viral genomes in mouse mammary tumors. Ann Rev Genet 16:173, 1982

Hynes NE, Groner B: Mammary tumor formation and hormonal control of mouse mammary tumor virus expression. Curr Top Microbiol Immunol 101:51, 1982

Ringold GM: Glucocorticoid regulation of mouse mammary tumor virus gene expression. Biochim Biophys Acta 560:487, 1979

Selected Papers

DNA TUMOR VIRUSES

Papovaviruses

Carmichael CG, Schaffhausen BS, Dorsky DI, et al.: Carboxy terminus of polyoma middle-sized tumor antigen is required for attachment to membranes, associated protein kinase activities, and cell transformation. Proc Natl Acad Sci USA 79:3579, 1982

Chandrasekaran K, McFarland VW, Simmons DT, et al.: Quantitation and characterization of a species-specific and embryo stage-dependent 55-kilodalton phosphoprotein also present in cells transformed by simian virus 40. Proc Natl Acad Sci USA 78:6953, 1981

Eckhart W, Hutchinson MA, Hunter T: An activity phosphorylating tyrosine in polyoma T-antigen immunoprecipitates. Cell 18:925, 1979

Gissmann L, Wolnik L, Ikenberg H, et al.: Human papillomavirus types 6 and 11 DNA sequences in genital and laryngeal papillomas and in some cervical cancers. Proc Natl Acad Sci USA 80:560, 1983

Ito Y, Spurr N: Polyoma virus T-antigens expressed in transformed cells: Significance of middle T-antigen in transformation. Cold Spring Harbor Symp Quant Biol 44:149, 1980

Kahn P, Topp WC, Shin S-I: Tumorigenicity of SV40-transformed human and monkey cells in immunodeficient mice. Virology 126:348, 1983

Kremsdorf D, Jablonska S, Favera M, et al.: Biochemical characterization of two types of human papillomaviruses associated with epidermodysplasia verruciformis. J Virol 43:436, 1982

Oren M, Reich NC, Levine AJ: Regulation of the cellular p53 tumor antigen in teratocarcinoma cells and their differentiated progeny. Mol Cell Biol 2:443, 1982

Rubin H, Figge J, Bladon MT, et al.: Role of small t antigen in the acute transforming activity of SV40. Cell 30:469, 1982

Topp WC, Rifkin DB: The small-t protein of SV40 is required for loss of actin cable networks and plasminogen activator synthesis in transformed rat cells. Virology 106:282, 1980

Volckart G, van de Voorde A, Fiers W: Nucleotide sequence of the simian virus 40 small-t gene. Proc Natl Acad Sci USA 75:2160, 1978

Adenoviruses

Gallimore PH, Sharp PA, Sambrook J: Viral DNA in transformed cells. II. A study of the sequences of adenovirus 2 DNA in nine lines of transformed rat cells using specific fragments of the viral genome. J Mol Biol 80:49, 1974

Houweling A, van den Elsen PJ, van der EB AJ: Partial transformation of primary rat cells by the leftmost 4.5% fragment of adenovirus 5 DNA. Virology 105:337, 1980

Sarnow P, Ho YS, Williams J, et al.: Adenovirus Elb-58kd tumor antigen and SV40 large tumor antigen are physically associated with the same 54 kd cellular protein in transformed cells. Cell 28:387, 1982

Herpesviruses

Duff R, Rapp F: Oncogenic transformation of hamster embryo cells after exposure to inactivated herpes simplex virus type 1. J Virol 12:209, 1973

Henle G, Henle W, Diehl V: Relation of Burkitt's tumor-associated herpes-type virus to infectious mononucleosis. Proc Natl Acad Sci USA 59:94, 1968

Hummel M, Kieff E: Mapping of polypeptides encoded by the Epstein–Barr virus genome in productive infection. Proc Natl Acad Sci USA 79:5698, 1982

Klein G, Giovanella BV, Lindahl T, et al.: Direct evidence for the presence of Epstein–Barr virus DNA and nuclear antigen in malignant epithelial cells from patients with poorly differentiated carcinoma of the nasopharynx. Proc Natl Acad Sci USA 71:4737, 1974

Hepatitis B Virus

Summers J, Mason WS: Replication of the genome of a hepatitis B-like virus by reverse transcription of an RNA intermediate. Cell 29:403, 1982

RNA TUMOR VIRUSES

Endogenous Retroviruses

Elder JH, Gautsch JW, Jensen FC, et al.: Multigene family of endogenous retroviruses: Recombinant origin of diversity. J Natl Cancer Inst 61:625, 1978

Gardner MB: Type C viruses of wild mice: Characterization and natural history of amphotropic, ecotropic and xenotropic MuLV. Curr Top Microbiol Immunol 79:215, 1978

Herr W, Schwarz D, Gilbert W: Isolation and mapping of DNA hybridization probes specific for ecotropic and nonecotropic murine leukemia proviruses. Virology 125:139, 1983

Igel HJ, Heubner RJ, Turner RC, et al.: Mouse leukemia virus activation by chemical carcinogens. Science 166:1624, 1969

Weiss RA, Friis PR, Vogt PK: Induction of avian tumor viruses in normal cells by physical and chemical carcinogens. Virology 46:920, 1971

Recombinants between Endogenous Viruses

Bryant ML, Roy-Burman P, Gardner MB, et al.: Genetic relationship of wild mouse amphotropic virus to murine ecotropic and xenotropic viruses. Virology 88:389, 1978

Chattopadhyay SK, Cloyd MW, Linemeyer DR, et al.: Cellular origin and role of mink cell focus-forming viruses in murine thymic lymphomas. Nature 295:25, 1982

Chien Y-H, Verma IM, Shih TY, et al.: Heteroduplex analysis of the sequence relations between the RNAs of mink cell focus-inducing and murine leukemia viruses. J Virol 25:352, 1978

Cloyd MW: Characterization of target cells for MCF viruses in AKR mice. Cell 32:217, 1983

Coffin JM, Tsichlis PN, Conklin KF, et al.: Genomes of endogenous and exogenous avian retroviruses. Virology 126:151, 1983

Hartley NW, Wolford NK, Old LJ, et al.: A new class of murine leukemia virus associated with development of spontaneous lymphomas. Proc Natl Acad Sci USA 74:789, 1977

Hoffman PM, Davidson WF, Ruscetti SK, et al.: Wild mouse ecotropic murine leukemia virus infection of inbred mice: Dual-tropic virus expression precedes the onset of paralysis and lymphoma. J Virol 39:597, 1981

Rapp UR, Todaro GJ: Generation of oncogenic type C viruses: Rapidly leukemogenic viruses derived from C3H mouse cells in vivo and in vitro. Proc Natl Acad Sci USA 75:2468, 1978

Thomas CY, Coffin JM: Genetic alterations of RNA leukemia viruses associated with the development of spontaneous thymic leukemia in AKR/J mice. J Virol 43:416, 1982

Troxler DH, Lowy D, Howk R, et al.: Friend strain of spleen focus-forming virus is a recombinant between ecotropic murine type C virus and the *env* gene region of xenotropic type C virus. Proc Natl Acad Sci USA 74:467, 1977

Acute Leukemia/Sarcoma Viruses

Devare SD, Reddy EP, Law JD, et al.: Nucleotide sequence of the simian sarcoma virus genome: Demonstration that its acquired cellular sequences encode the transforming gene product p28^{sis} Proc Natl Acad Sci USA 80:731, 1983

Duesberg PH, Bister K, Vogt PK: The RNA of avian acute leukemia virus MC29. Proc Natl Acad Sci USA 74:4320, 1977

East JL, Amesse LS, Kingsbury DW, et al.: Sequence relationships between Kirsten retrovirus genomes and the genomes of other murine retroviruses. Virology 126:126, 1983

Goff SP, Gilboa E, Witte ON, et al.: Structure of the Abelson murine leukemia virus genome and the homologous cellular gene: Studies with cloned viral DNA. Cell 22:777, 1980

Graf T, Beug H: Avian leukemia viruses: Interaction with their target cells in vivo and in vitro. Biochim Biophys Acta 516:269, 1978

Hu SSF, Vogt PK: Avian oncovirus MH2 is defective in gag, pol, and env. Virology 92:278, 1979

Reddy EP, Smith MJ, Aaronson SA: Complete nucleotide sequence and organization of the Moloney sarcoma virus genome. Science 214:445, 1981

Schwartz DE, Tizard R, Gilbert W: Nucleotide sequence of Rous sarcoma virus. Cell 32:853, 1983

Shibuya M, Hanafusa H: Nucleotide sequence of Fujinami sarcoma virus: Evolutionary relationship of its transforming gene with transforming genes of other sarcoma viruses. Cell 30:787, 1982

Wong TC, Lai MMC, Hu SSF, et al.: Class II defective avian sarcoma viruses: Comparative analysis of genome structure. Virology 120:453, 1982

Retrovirus Provirus

Barbacid M, Stephenson JR, Aaronson SA: Evolutionary relationships between gag gene-coded proteins of murine and primate endogenous type C RNA viruses. Cell 10:641, 1977

Hizi A, Joklik WK: RNA-dependent DNA polymerase of avian sarcoma virus B77. I. Isolation and partial characterization of the α, β_2, and $\alpha\beta$ forms of the enzyme. J Biol Chem 252:2281, 1977

McLellan WL, Ihle JN: Purification and characterization of a murine tumor cell surface glycoprotein of 75,000 daltons that is related to the major envelope glycoprotein of murine leukemia virus. Virology 89:547, 1978

Vogt VM, Wight A, Eisenman R: In vitro cleavage of avian retrovirus gag proteins by viral protease p15. Virology 98:154, 1979

Nature of Oncogenes/Protooncogenes

Alitalo K, Bishop JM, Smith DH, et al.: Nucleotide sequence of the v-myc oncogene of avian retrovirus MC29. Proc Natl Acad Sci USA 80:100, 1983

Capon DJ, Chen EY, Levinson AD, et al.: Complete nucleotide sequence of the T24 human bladder carcinoma oncogene and its normal homolog. Nature 302:33, 1983

Collett MS, Brugge JS, Erikson RL: Characterization of a normal avian cell protein related to the avian sarcoma virus transforming gene product. Cell 15:1363, 1978

Cooper JA, Reiss NA, Schwarz RJ, et al.: Three glycolytic enzymes are phosphorylated at tyrosine in cells transformed by Rous sarcoma virus. Nature 302:218, 1983

Dhar R, Ellis RW, Shih TY, et al.: Nucleotide sequence of the p21 transforming protein of Harvey sarcoma virus. Science 217:934, 1982

Fung YT, Crittenden LB, Fadley AM, et al.: Tumor induction by direct injection of cloned v-src DNA into chickens. Proc Natl Acad Sci USA 80:353, 1983

Gay NJ, Walker JE: Homology between human bladder carcinoma oncogene product and mitochondrial ATP-synthase. Nature 301:262, 1983

Gondo TJ, Bishop JM: Structure and transcription of the cellular homolog (c-myb) of the avian myeloblastosis virus transforming gene (v-myb). J Virol 46:212, 1983

Groffen J, Heisterkamp N, Shibuya M, et al.: Transforming genes of avian (v-fps) and mammalian (v-fes) retroviruses correspond to a common cellular locus. Virology 125:480, 1983

Hayman MJ, Ramsay GM, Graf T, et al.: Identification and characterization of the avian erythroblastosis virus erbB gene product as a membrane glycoprotein. Cell 32:579, 1983

Hirano A, Neil JC, Vogt PK: ts transformation mutants of avian sarcoma virus PRCII: Lack of strict correlation between transforming ability and properties of the P105-associated kinase. Virology 125:219, 1983

Iwashita S, Kitamura N, Yoshida M: Molecular events leading to fusiform morphological transformation by partial src deletion mutant of Rous sarcoma virus. Virology 125:419, 1983

Kloetzer WS, Maxwell SA, Arlinghaus RB: P85$^{gag-mos}$ encoded by ts110 Moloney murine sarcoma virus has an associated protein kinase activity. Proc Natl Acad Sci USA 80:412, 1983

Langbeheim H, Shih TY, Scolnick EM: Identification of a normal vertebrate cell protein related to the p21 src of Harvey murine sarcoma virus. Virology 106:292, 1980

Levinson AD, Opermann H, Levintow L, et al.: Evidence that the transforming gene of avian sarcoma virus encodes a protein kinase associated with a phosphoprotein. Cell 15:561, 1978

Maness PF, Levy BT: Highly purified pp60^{src} induces the actin transformation in microinjected cells and phosphorylates selected cytoskeletal proteins in vitro. Mol Cell Biol 3:102, 1983

Parker RC, Varmus HE, Bishop JM: Cellular homolog (c-src) of the transforming gene of Rous sarcoma virus: Isolation, mapping, and transcriptional analysis of c-src and flanking regions. Proc Natl Acad Sci USA 78:5842

Shi TY, Papageorge AG, Stokes PE, et al.: Guanine nucleotide-binding and autophosphorylating activities associated with the p21^{src} protein of Harvey murine sarcoma virus. Nature 287:686, 1980

Shimizu K, Goldbarb M, Suard Y, et al.: Three human transforming genes are related to the viral ras oncogene. Proc Natl Acad Sci USA 80:2112, 193

Snyder MA, Bishop JM, Colby WW, et al.: Phosphorylation of tyrosine-416 is not required for the transforming properties and kinase activity of pp60^{v-src}. Cell 32:891, 1983

Takeya T, Hanafusa H: Structure and sequence of the cellular gene homologous to the RSV src gene and the mechanism for generating the transforming virus. Cell 32:881, 1983

Vennstrom B, Sheiness D, Zapielski J, et al.: Isolation and characterization of c-myc, a cellular homolog of the oncogene (v-myc) of avian myelocytomatosis virus strain 29. J Virol 42:773, 1982

Watson R, Oskarsson M, Vande Woode GF: Human DNA sequence homologous to the transforming gene (mos) of Moloney murine sarcoma virus. Proc Natl Acad Sci USA 79:4078, 1982

Transfection of Oncogenes

Balmain A, Pragnell ID: Mouse skin carcinomas induced in vivo by chemical carcinogens have a transforming Harvey-ras oncogene. Nature 303:72, 1983

Chang EH, Furth ME, Scolnick EM, et al.: Tumorigenic transformation of mammalian cells induced by a normal gene homologous to the oncogene of Harvey murine sarcoma virus. Nature 297:479, 1982

Der CJ, Cooper GM: Altered gene products are associated with activation of cellular ras^k genes in human lung and colon carcinomas. Cell 32:201, 1983

Der CJ, Krontiris TG, Cooper GM: Transforming genes of human bladder and lung carcinoma cell lines are homologous to the ras genes of Harvey and Kirsten sarcoma viruses. Proc Natl Acad Sci USA 79:3637, 1982

Parada LF, Tabin CJ, Shih C, et al.: Human EJ bladder carcinoma oncogene is homologue of Harvey sarcoma virus ras gene. Nature 297:474, 1982

Activation of Oncogenes

PROMOTER INSERTION

Hayward WS, Neel BG, Astrin SM: Activation of a cellular onc gene by promoter insertion in ALV-induced lymphoid leukosis. Nature 290:475, 1981

Joyner A, Yamamoto Y, Bernstein A: Retrovirus long terminal repeats activate expression of coding sequences for the herpes simplex virus thymidine kinase gene. Proc Natl Acad Sci USA 79:1573, 1982

Kuff EL, Feenstra A, Lueders K, et al.: Homology between an endogenous viral LTR and sequences inserted in activated cellular oncogene. Nature 302:547, 1983

AMPLIFICATION

Alitalo K, Schwab M, Lin CC, et al.: Homogeneously staining chromosomal regions contain amplified copies of an abundantly expressed cellular oncogene (c-myc) in malignant neuroendocrine cells from a human colon carcinoma. Proc Natl Acad Sci USA 80:1707, 1983

Collins S, Groudine M: Amplification of endogenous myc-related DNA sequences in a human myeloid leukemia cell line. Nature 298:679, 1982

Favera RD, Wong-Staal F, Gallo RC: Onc gene amplification in promyelocytic leukemia cel line HL-60 and primary leukemic cells of the same patient. Nature 299:61, 1982

TRANSLOCATION

Erikson J, Ar-Rushdi A, Drwinga HL, et al.: Transcriptional activation of the translocated c-myc oncogene in Burkitt lymphoma. Proc Natl Acad Sci USA 80:820, 1983

Mushinski JF, Bauer SR, Potter M, et al.: Increased expression of myc-related oncogene mRNA characterizes most BALB/c plasmacytomas induced by pristane or Abelson murine leukemia virus. Proc Natl Acad Sci USA 80:1073, 1983

MUTATIONS

Reddy EP, Reynolds RK, Santos E, et al.: A point mutation is responsible for the acquisition of transforming properties by the T24 human bladder carcinoma oncogene. Nature 300:149, 1982

STAGE-SPECIFIC ACTIVATION

Groffen J, Heisterkamp N, Blennerhassett G, et al.: Regulation of viral and cellular oncogene expression by cytosine methylation. Virology 126:213, 1983

Lane M-A, Sainten A, Cooper GM: Stage-specific transforming genes of human and mouse B- and T-lymphocyte neoplasms. Cell 28:873, 1983

Muller R, Slamon DJ, Tremblay JM, et al.: Differential expression of cellular oncogenes during pre- and postnatal development of the mouse. Nature 299:640, 1982

Chromosomal Location

Chang EH, Gonda MA, Ellis RM, et al.: Human genome contains four genes homologous to transforming genes of Harvey and Kirsten murine sarcoma viruses. Proc Natl Acad Sci USA 79:4848, 1982

Favera RD, Gallo RC, Giallongo A, et al.: Chromosomal localization of the human homolog (c-sis) of the simian sarcoma virus onc gene. Science 218:686, 1982

Neel BG, Jhanwar SC, Chaganti RSK, et al.: Two human c-onc genes are located on the long arm of chromosome 8. Proc Natl Acad Sci USA 79:7842, 1982

Retrovirus env Gene Products as Differentration Antigens

del Villano BC, Lerner RA: Relationship between the oncornavirus gene product gp70 and a major protein secretion of the mouse genital tract. Nature 259:497, 1976

Elder JH, Jensen FC, Bryant ML, et al.: Polymorphism of the major envelope glycoprotein (gp70) of murine C-type viruses: Virion associated and differentiation antigens encoded by a multigene family. Nature 267:23, 1977

Hesselink WG, van der Kemp ACM, Bloemers HPJ: Moloney cell surface antigen (MCSA) has properties of the env gene products of the Moloney strain of murine leukemia virus. Virology 110:375, 1981

Mammary Tumor Viruses

Calahan R, Drohan W, Tronick S, et al.: Detection and cloning of human DNA sequences related to the mouse mammary tumor virus genome. Proc Natl Acad Sci USA 79:5503, 1982

Diggelman H, Vessaz AL, Buetti E: Cloned endogenous mouse mammary tumor virus DNA is biologically active in transfected mouse cells and its expression is stimulated by glucocorticoid hormones. Virology 122:332, 1982

Huang AL, Ostrowski MC, Berard D, et al.: Glucocorticoid regulation of the Ha-MuSV p21 gene conferred by sequences from mouse mammary tumor virus. Cell 27:245, 1981

Morrios VL, Medeiros E, Ringold GM, et al.: Comparison of mouse mammary tumor virus-specific DNA in inbred, wild and Asian mice, and in tumors and normal organs from inbred mice. J Mol Biol 114:73, 1977

Parks WP, Ransom JC, Young HA, et al.: Mammary tumor virus induction by glucocorticoids. Characterization of specific transcriptional regulation. J Biol Chem 250:3330, 1975

Ringold GM, Yamamoto KR, Bishop JM, et al.: Glucocorticoid-stimulated accumulation of mouse mammary tumor virus RNA: Increased rate of synthesis of viral RNA. Proc Natl Acad Sci USA 74:2879, 1977

Ucker DS, Firestone GL, Yamamoto K: Glucocorticoids and chromosomal position modulate murine mammary tumor virus transcription by affecting efficiency of promoter utilization. Mol Cell Biol 3:551, 1983

Human T Cell Leukemia Virus (HTLV)

Gallo RC, Mann D, Broder S, et al.: Human T-cell leukemia-lymphoma virus (HTLV) is in T but not B lymphocytes from a patient with cutaneous T-cell lymphoma. Proc Natl Acad Sci USA 79:5680, 1982

Haynes BF, Miller SE, Palker TJ, et al.: Identification of human T cell leukemia virus in a Japanese patient with adult T cell leukemia and cutaneous lymphomatous vasculitis. Proc Natl Acad Sci USA 80:2054, 1983

Popovic M, Sarin PS, Robert-Guroff M, et al.: Isolation and transmissin of human retrovirus (human T-cell leukemia virus). Science 219:856, 1983

Intracisternal A-type Particles

Kuff EL, Lueders KK, Scolnick EM: Nucleotide sequence relationship between intracisternal type A particles of *Mus musculus* and an endogenous retrovirus (M432) of *Mus cervicolor*. J Virol 28:66, 1978

Kuff EL, Feenstra A, Lueders K, et al.: Intracisternal A-particle genomes as movable elements in the mouse genome. Proc Natl Acad Sci USA 80:1992, 1983

Lueders KK, Kuff EL: Sequences associated with intracisternal A particles are reiterated in the mouse genome. Cell 12:963, 1977

Ono M, Cole MD, White AT, et al.: Sequence organizatin of cloned intracisternal A particle genes. Cell 21:465, 1980

Shen-Ong GLC, Cole MD: Differing populations of intracisternal A-particle genes in myeloma tumors and mouse subspecies. J Virol 42:411, 1982

VL30 Sequences

Giri OP, Hodgson CP, Elder PK, et al.: Discrete regions of sequence homology between cloned rodent VL30 genetic elements and AKV-related MuLV provirus genomes. Nucl Acid Res 11:305, 1983

Itin A, Rotman G, Keshet E: Conservation patterns of mouse "viruslike" (VL30) DNA sequences. Virology 127:374, 1983

copia Sequences of *Drosophila*

Hoffman-Falk H, Einat P, Shilo BZ, et al.: *Drosophila melanogaster* DNA clones homologous to vertebrate oncogenes: Evidence for a common ancestor to the src and abl cellular genes. Cell 32:589, 1983

Shiba T, Saigo K: Retrovirus-like particles containing RNA homologous to the transposable element *copia* in *Drosophila melanogaster*. Nature 302:119, 1983

CHAPTER 63

The Bacteriophages

In 1915 Twort published a note describing the infectious destruction of micrococcal colonies and offered three possible explanations for this phenomenon: (1) the destroyed colonies represented a stage in the bacterial life cycle that induced normal colonies to undergo a glassy transformation; (2) the lytic agent was an enzyme that caused both cell destruction and the production of more enzyme; and (3) the causative agent was a virus that grew in bacteria and destroyed them. All of these explanations could be reconciled with the original findings that the agent would not grow on any medium, passed through bacteria-proof filters, and was inactivated by heating to 60C for 1 hour. d'Hérelle, who discovered this phenomenon independently, soon demonstrated the particulate nature of what he called "bacteriophages." The elegant pioneering work of Burnet and of Schlesinger in the 1930s confirmed the viral nature of the Twort-d'Hérelle agents.

Early hopes of using the action of phage on susceptible bacteria as a means of preventing and treating infectious diseases were not fulfilled (Chap. 55). However, in the early 1940s, Delbrück and a group of investigators around him realized that the availability of viruses multiplying in cloned populations of rapidly growing host cells provided an ideal tool for gaining an insight into the mechanisms of biologic self-replication. Their expectation was amply borne out. The bacteriophage-bacterium system proved to be highly amenable to experimentation, and intensive investigation of it over the last four decades has provided many of the fundamental concepts concerning molecular genetics, nucleic acid replication, and the tran-

TABLE 63-1. CHARACTERISTICS OF SOME WELL-STUDIED BACTERIOPHAGES

Phage	Host	Particle Dimensions (nm)			Nucleic Acid			
		Head	Tail Length	Structure	Type	Strandedness	Mol Wt × 10⁶	Structure
T1	E. coli	50	150	Hexagonal head, simple tail	DNA	DS	27	Linear
T2, T4, T6	E. coli	80 × 110	110	Prolate icosahedral head, complex tail with fibers	DNA	DS	112	Linear, contains glycosylated 5-hydroxymethylcytosine
T3, T7	E. coli	60	15	Hexagonal head, short tail	DNA	DS	25	Linear
T5	E. coli	65	170	Hexagonal head, simple tail	DNA	DS	80	Linear, one strand segmented
N4	E. coli	70	15	Hexagonal head, short tail	DNA	DS	40	Contains U, contains RNA polymerase
Lambda	E. coli	54	140	Hexagonal head, simple tail	DNA	DS	31	Linear, cohesive ends
P22	S. typhimurium	55	20	Hexagonal head, complex tail	DNA	DS	27	Linear, circularly permuted, terminally redundant
SPO1, SP82	B. subtilis	100	210	Hexagonal head, complex tail	DNA	DS	100	Linear, contains 5-hydroxymethyluracil
φ 29	B. amyloliquefaciens	30 × 40	30	Prolate icosahedral head with attached fibers, complex tail with collars and neck appendages	DNA	DS	11	Linear
PM2	Pseudomonas BAL-31	60	None	Hexagonal head, envelope contains lipid	DNA	DS	6	Circular
φX 174, S13, M12, G4	E. coli	27	None	Icosahedral	DNA	SS	1.6	Circular
fl, fd, M13, φ 6	E. coli	5–10 × 800	None	Filamentous	DNA	SS	1.3	Circular
	Pseudomonas phaseolica	65	None	Polyhedral head, envelope contains lipid	RNA	DS	9.5	3 linear segments (2.2, 2.8 and 4.5 × 10⁶)
MS2, f2, fr, Qβ	E. coli	24	None	Icosahedral	RNA	SS	1.2	Linear

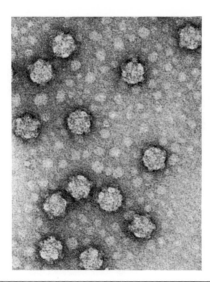

Figure 63-1. Bacteriophage φX 174. × 225,000. *(Electron micrograph by Dr. R.B. Luftig.)*

scription and translation of genetic information. These concepts are applicable not only to bacterial viruses but also to animal viruses and to cells in general. Thus, although bacteriophages were a failure as therapeutic agents, they have proved invaluable for the elucidation of the reactions most basic to life.

Structure of Bacteriophages

There are bacteriophages for every bacterial species. Very few have been investigated in detail, but intense concentration on a small number of them has permitted rapid progress. Many of the well-studied phages are active on *Escherichia coli* or on *Bacillus* or *Pseudomonas* species (Table 63-1).

Structure of Phage Capsids

The structure of bacteriophages is governed by the same principles as are described in Chapter 56. Some of the smaller phages, such as φX 174 and MS2, have icosahedral capsids (Fig. 63-1); others, such as phage fl, are filamentous and possess helical symmetry (Fig. 63-2). Larger phages generally consist of a head that comprises the phage genome enclosed within a single capsid shell that is usually hexagonal and may or may not be elongated, and a tail that serves both as the cell attachment organ and as a tube through which phage DNA passes into the host cell. The complexity of this tail varies greatly from phage to phage, but in most phages it conforms to one of three morphologic patterns. In the first, exemplified by coliphages T3 and T7, the tail is very short (Fig. 63-3). In the second, exemplified by coliphages T1 and lambda, the tail is long but rather simple in construction. It consists essentially of a noncontractile flexible tube that may or may not possess a knob or one or several spikes or fibers at its distal end (Fig. 63-4). The third morphologic tail pattern, exhibited by the coliphages T2, T4, and T6, the so-called T-even phages, is staggeringly complex (Fig. 63-5). These tails consist of a hollow core that is attached at one end to the head and bears at its distal end a hexagonal base plate to which are attached six pins and six long tail fibers bent in the middle. A thin collar with several (possibly six) whiskers is attached to the core close to the head, and it is surrounded for most of its length by a sheath composed of 24 rings of helically arranged capsomers. These tails, which consist of more than 20 different protein species, serve as syringes by means of which phage DNA is injected into the cell (p. 962).

Very few phages are enveloped. Among them are the *Pseudomonas* phages PM2 (Fig. 63-6) and φ6.

Structure of Phage Nucleic Acids

Bacteriophage genomes take several forms (Table 63-1). Most of them consist of single, linear, double-stranded

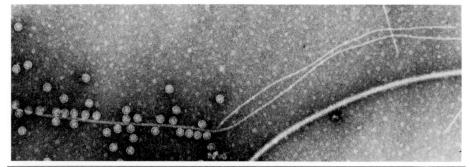

Figure 63-2. An F-pilus with two types of male-specific phage attached: icosahedral RNA-containing MS2, and filamentous DNA-containing fl. The former are attached along the entire length of the F-pilus; the latter are adsorbed by their ends to the tip of the pilus. *(From Caro and Schnos: Proc Natl Acad Sci USA 56:128, 1968.)*

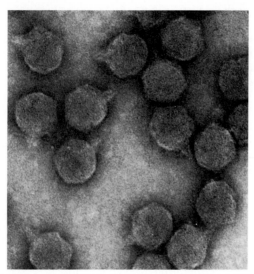

Figure 63-3. Bacteriophage T7. × 225,000. *(Electron micrograph by Dr. R.B. Luftig.)*

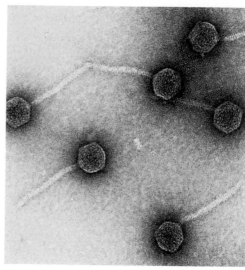

Figure 63-4. Bacteriophage lambda. × 135,000. *(Electron micrograph by Dr. R.B. Luftig.)*

DNA molecules that vary in molecular weight from about 1 million to more than 100 million. Some of them, in particular those of the T-even phages, exist in a form that is both circularly permuted and terminally redundant. In them the nucleotide sequence of some genomes is A, B, C, D, . . . W, X, Y, Z, A, B; of others B, C, D, . . . W, X, Y, Z, A, B, C; of others still C, D, . . . W, X, Y, Z, A, B, C, D; and so on. In other words, all contain the same sequences, but the terminal sequences differ in individual genomes and some nucleotides are reiterated at the ends of the genomes. This is because their replication proceeds via replicative intermediates that consist of many genomes linked covalently end-to-end, which are then cut at random so as to yield molecules of the same size (p. 968). The reason for the terminal redundancy is to avoid loss of the gene that would otherwise be cut at each end.

Other phage genomes, such as that of PM2, consist of circular double-stranded DNA, others, such as those of phages φX 174, G4, and fd, consist of single-stranded circular DNA with plus polarity, and still others are linear single-stranded nonpermuted RNA molecules with plus polarity, analogous to poliovirus RNA. There is also a phage of *Pseudomonas phaseolica*, φ6, that contains double-stranded RNA. Like the double-stranded RNA-containing genomes of animal viruses, φ6 RNA is segmented. It consists of three unique segments whose molecular weights are 2.2, 2.9, and 4.8 million.

Phage nucleic acids are infectious, just as are the nucleic acids of animal viruses. The bacterial cell wall, however, presents a more formidable barrier to the entry of nucleic acids than is presented by the plasma membrane of animal cells, and infection by naked phage nucleic acids, termed transfection, can only be demonstrated under special circumstances. Transfection of whole bacterial cells is rare. Cells of *Bacillus subtilis* that are competent for transformation (Chap. 3) can be transfected, and so can cells of *E. coli* treated with Ca²⁺ (only with lambda DNA).

The most readily transfectible bacterial forms are spheroplasts (Chap. 8), and frozen-thawed cells that can be transfected not only with small single-stranded phage RNAs and DNAs but also with large double-stranded phage DNAs.

Unusual Nucleic Acid Bases. Whereas the genomes of animal viruses contain only the normal nucleic acid bases, adenine, guanine, cytosine, and uracil or thymine, phage genomes sometimes contain unusual or substituted bases. For example, the DNA of the T-even bacteriophages contains not cytosine but 5-hydroxymethylcytosine and glycosylated 5'-hydroxymethylcytosine; the DNA of certain *B. subtilis* phages contain not thymine but uracil, 5-hydroxymethyluracil, glycosylated and phosphorylated 5-(4', 5'-dihydroxypentyl) uracil, or α-glutamyl thymine; and in the DNA of a *Pseudomonas acidovorus* phage, thymine is partially replaced by 5-(4'-aminobutylaminomethyl) uracil, also known as α-putrescinylthymine.

Glycosylation and Methylation. The bases of bacteriophage DNA are often glycosylated and/or methylated in highly characteristic patterns. Both 5-hydroxymethylcytosine and 5-hydroxymethyluracil are often glycosylated by phage-coded enzymes, that is, they are linked to either one or two glucose residues via α or β bonds. In the T-even phages the situation is as shown in Table 63-2. In addition, a small proportion of the cytosine and adenine residues in phage DNAs is generally methylated by host-specified enzymes.

The presence of modified bases alters the properties of phage DNAs in two major ways. First, both glycosylation and, especially, methylation confer resistance to degradation by nucleases. As discussed in Chapter 8, bacteria guard themselves against foreign DNAs by degrading them with highly specific restriction endonucleases, and in order to prevent their own DNA from being destroyed,

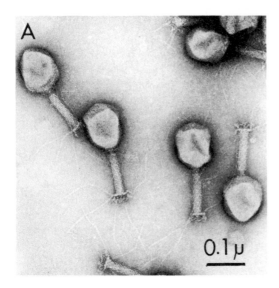

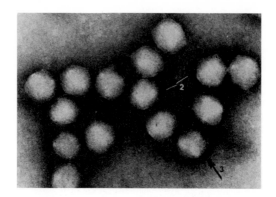

Figure 63-6. The enveloped phage PM2 fixed with glutaraldehyde and negatively stained with phosphotungstate. The numbered arrows indicate axes of two-fold and three-fold symmetry. × 120,000. *(From Silbert, et al.: Virology 39:666, 1969.)*

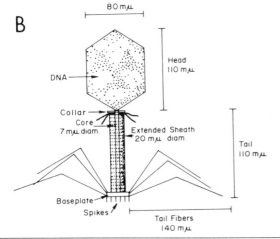

Figure 63-5. A. Bacteriophage T4. × 100,000. **B.** Model of phage T4. *(Electron micrograph by Dr. R.B. Luftig.)*

degraded by the restriction endonucleases of other bacteria. However, any DNA molecules that escape degradation in such hosts can replicate, and their progeny will then be modified by the methylating enzymes of the new hosts, acquiring new specificities in the process.

Second, modification of the apolar character of pyrimidine 5'-methyl substituents alters the stability of the DNA double helix. In general, uncharged and negatively charged substituents decrease helix stability, and positively charged substituents increase it (by neutralizing some of the phosphate negative charges). Such modified bases could further protect DNAs from host and phage nucleases, serve as signals for transcription and replication, and facilitate packaging of DNAs on the one hand, and their injection into cells on the other (see below). However, although modified bases are clearly advantageous under certain specific sets of conditions, their presence has not conferred sufficient survival advantage during evolution for them to become used widely in nature.

Physiology of Lytic Phage Infection

Like animal viruses, some bacteriophages lyse their hosts, while others integrate their genomes into the host genome. The former are known as virulent phages, the latter

they cause methyl groups to be attached to their own DNA in specific locations. Since phage nucleic acids also become methylated, highly specific restriction patterns, known as "host-induced modifications," exist among bacteriophages. For example, the DNA of a phage grown in host A will be methylated by that host's enzymes in such a manner that it will be resistant to that host's restriction endonucleases, but most of the DNA of this phage will be

TABLE 63-2. EXTENT OF GLYCOSYLATION OF 5-HYDROXYMETHYLCYTOSINE IN THE DNA OF T-EVEN PHAGES*

Phage	Not Gylcosylated	α-Glucosyl	β-Glucosyl	β-1 : 6-Glucosyl-α-Glucosyl
T2	25	70	0	5
T4	0	70	30	0
T6	25	3	0	72

From Lehman: J Biol Chem 235:3254, 1960.
*Figures represent percentages.

as temperate or lysogenic phages. We will consider first the lytic phage one-step multiplication cycle, which is formally analogous to that of animal viruses (Chap. 58) but proceeds much more rapidly. For example, whereas the length of the latent period (the interval between infection and release of progeny virus) is at least 4 hours for poliovirus and 12 hours for vaccinia virus, the latent period is 13 minutes for phages T1, T7, and ϕX 174, and 21 to 25 minutes for the T-even phages and MS2.

Adsorption

The initial step of the phage multiplication cycle illustrates with particular clarity the existence of virus-specific receptors on host cells. Phages are usually highly specific for a limited number of bacterial host strains. The basis of this specificity resides in the complementarity of the molecular configurations on phage attachment organs on the one hand, and on receptor molecules on the bacterial surface on the other. The nature of the phage attachment organ varies from phage to phage: small icosahedral phages, such as ϕX 174 and MS2, have multiple attachment sites; filamentous phages, such as f1, adsorb with their ends (Fig. 63-2); tailed phages adsorb with the knobs, spikes, or fibers that are located at the tips of their tails. Phage receptors on the bacterial surface are sometimes on lipopolysaccharide, at other times on lipoprotein, and sometimes on F pili (Chap. 3), which causes the phages that adsorb to them, typically single-stranded DNA-containing and RNA-containing phages, to be male-specific (Fig. 63-2).

Mutations in bacteria that destroy the complementarity of the phage-receptor interaction result in resistance. Phage populations usually contain mutants that are themselves altered in such a way as to restore the necessary complementarity. These are known as "host-range mutants" and can grow in the mutated bacteria.

Bacteria capable of adsorbing the same phage are often antigenically related. This observation has found diagnostic application in the practice of phage typing. In this method the sensitivity/resistance patterns of bacterial strains to a series of bacteriophages are determined. Since these patterns are both readily determined and highly characteristic, they are useful tests for identification. For example, some phages of *Salmonella typhi* are specific for strains possessing the Vi antigen, which is characteristic of virulent strains (Chap. 37), and these phages are used for the identification of virulent strains of typhoid bacilli. Several other phage-typing systems exist, the best known being that for staphylococci.

Injection and Uncoating

The events immediately following phage adsorption are involved with the injection of phage nucleic acid into the cell. These events provide an excellent illustration of the general principle that uncoating of viral genomes involves the physical separation of genome and capsid. This was first shown in 1952 by Hershey and Chase in an experiment that represents one of the milestones of virology. They infected bacteria with phage T2 labeled in the protein with the radioisotope ^{35}S. Following an incubation period of several minutes, the mixture of phage and bacteria was sheared by blending in a Waring blender. After blending, only a small amount of radioactive label was associated with the bacteria. When this experiment was repeated with phage in which the DNA had been labeled with the radioisotope ^{32}P, the converse was true, that is, the radioactive label was now associated with the cells. This experiment was correctly interpreted as signifying that the DNA had passed into the interior of the cell, while the protein coat had remained attached to the outer cell surface from which it could be dislodged by shearing forces. Since the bacteria from which phage coats were removed by shearing yielded normal phage progeny, this experiment clearly showed that viral DNA itself contained all the information necessary for phage multiplication.

The actual infection process has been best studied for the T-even phages. Following adsorption by means of the tail fiber-receptor interaction, the six tail pins make contact with the host surface and firmly anchor the phage tail plate to it. The conformation of the tail plate (an extremely complex structure consisting of at least 14 different protein species) then changes. The reason for this change is apparently the fact that several of the tail plate components are enzymes. The six wedges that constitute the major portion of the base plate possess dihydrofolate reductase activity, and the central tail plug possesses both thymidylate synthetase and folyl polyglutamate cleaving activity. It is thought that the configuration of these components changes as a result of their interaction with substrates that leak from cells in response to infection. These conformational changes apparently trigger a change in the manner in which the protein subunits of the tail sheath are arranged, as a result of which the sheath contracts, thereby driving the tail core about 12 nm into the bacterial cell (Fig. 63-7). The phage DNA then passes through the tail core into the bacterial cell (Fig. 63-8). The holes created in the cell membrane by the phage tail cores are normally quickly sealed by cell wall material newly synthesized under the control of a phage gene called "spackle." Clearly, it would not be to the advantage of the phage to permit these holes to remain, since cell contents would leak out through them. In fact, if the multiplicity of infection is too high, excessive leakage does occur, resulting in a phenomenon known as "lysis from without," and the multiplication cycle is aborted.

As pointed out previously, most phages do not possess tails with contractile sheaths, and still others possess no tails at all. Nevertheless, here also the phage protein coat remains on the outside, and only the nucleic acid is introduced into the cell.

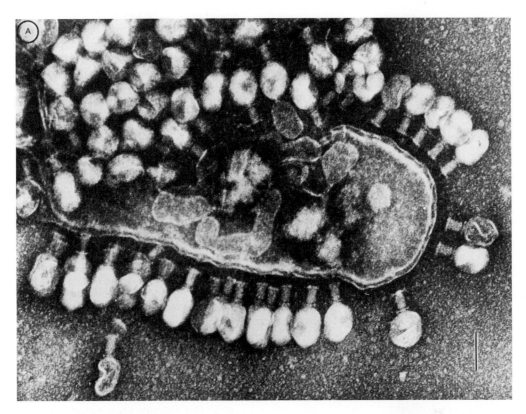

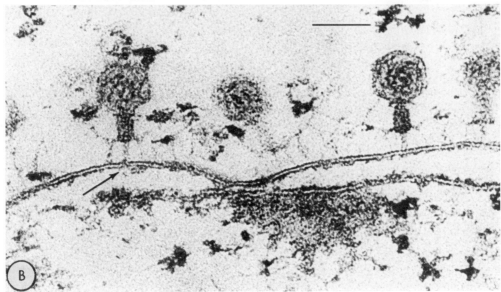

Figure 63-7. Bacteriophage T4 adsorbed to *Escherichia coli.* **A.** The phages' sheaths are con-tracted, and their base plates are 30 to 40 nm from the cell wall. **B.** The phages visible in this sec-tion are seen to be attached to the cell wall by the tail fibers. The cell wall appears as two continuous electron-dense lines separated by an electron-lucent space. The arrow indicates where the tail core of an adsorbed phage may have penetrated the cell wall. *(From Simon and Anderson: Virology 32:279, 1967.)*

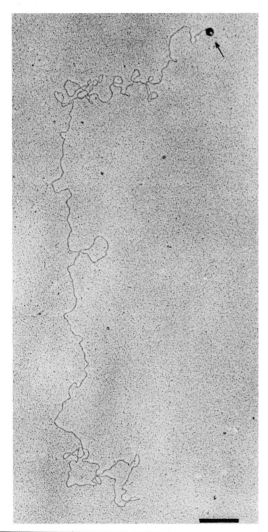

Figure 63-8. A T7 phage particle ejecting its DNA after being treated with formamide. Almost the entire DNA molecule (12 μm long) has been ejected. The bar represents 0.3 μm. *(Courtesy of Dr. Kaoru Saigo.)*

Multiplication Cycle of Phages Containing Double-stranded DNA

Nature of Information Encoded in Phage Genome

During the phage multiplication cycle, the genetic information encoded in the viral genome expresses itself according to an exquisitely refined transcription program that has been studied intensively along the lines discussed in Chapter 58. In fact, many of the techniques and methods that are now being applied to the analysis of the multiplication cycles of animal viruses were first designed and developed for studying the multiplication of bacteriophages. These techniques include, first and foremost, the extensive

use of conditional lethal mutants, both temperature-sensitive and missense, such as amber (Chap. 60).

Thousands of phage mutants have by now been isolated and characterized with respect to the function in which they are defective. For example, there are mutants that fail to synthesize early proteins, mutants that fail to replicate phage DNA, mutants that fail to form mature phage particles, and so on. Over 100 such functions, each corresponding to a specific gene, have so far been discovered for bacteriophage T4, which is probably about one half of the total number of T4 genes. For smaller, less complicated phages, this proportion is considerably higher. Most of the known T4 genes have been mapped. All lie on a circular linkage map, a finding that, together with the fact that T4 DNA is a linear molecule, indicates that it is circularly permuted (Fig. 63-9). Similar maps have been prepared for several other phages, notably T3 and T7, lambda and P22, SP01 and SP82, and the small phages that contain single-stranded nucleic acid.

The proteins specified by the genes of the large complex phages, such as the T-even phages, have a variety of functions. Most of them can be grouped into eight classes:

1. Proteins that repair the bacterial cell membrane early in infection
2. Enzymes that degrade host DNA
3. Enzymes that synthesize nucleic acid precursors
4. Proteins that function in phage DNA replication
5. Proteins that program transcription of the phage genome
6. Proteins that are structural phage components
7. Proteins that function catalytically in phage morphogenesis
8. Enzymes that degrade the bacterial cell wall and cell membrane during the late stages of the multiplication cycle

In addition, the T-even phages code for eight species of transfer RNA, the genes for which are clustered in a small region of the phage genome where they may form a single transcription unit or operon. Mutants lacking these genes can multiply, but their burst size, or yield, is smaller than that of wild-type phage. This suggests that the phage-coded transfer RNAs supplement the reading capacity of those codons that are used more commonly by the virus than by the host, thereby ensuring optimal rates of phage protein synthesis.

We will now consider each of the eight categories of proteins and examine their role in phage multiplication.

Functions of Various Categories of Phage Gene Products

Proteins that Repair Bacterial Cell Membrane Early in Infection

Infection causes the bacterial cell membrane to become leaky, so that cellular contents begin to escape; but within several minutes sealing occurs, and the preinfection integ-

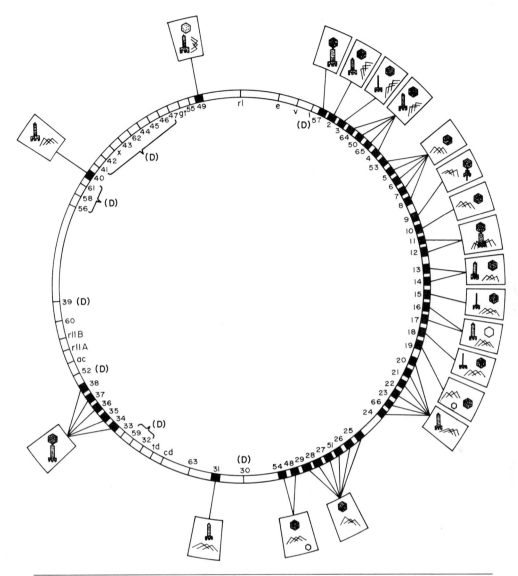

Figure 63-9. Simplified genetic map of T4. The solid segments indicate late genes with morphogenetic functions, while the short bars represent early genes with enzymatic or regulatory functions. Genes with functions in DNA replication are identified by (D). The boxes indicate viral components, such as heads, tails, and tail fibers, that are present in extracts of cells infected with conditional lethal mutants under restrictive conditions. *(Adapted from Edgar and Wood: Proc Natl Acad Sci USA 55:498, 1966.)*

rity of the membrane is restored. The precise mechanism by which this is achieved is not clear, but it is known that at least five phage-coded proteins are synthesized very early during the infection cycle and incorporated into the cell membrane.

Enzymes that Degrade Host DNA

Among the earliest effects of phage T4 infection is the breakdown of host DNA (Fig. 63-10), which has several consequences. First, it provides a source of nucleic acid precursors, which may be limiting under certain condi-

tions. More important, transcription of host-specified messenger RNAs ceases abruptly. Since most bacterial messenger RNAs are short-lived, this not only leads to the sudden cessation of host-specified protein synthesis but also quickly provides ribosomes for the translation of phage-specified messenger RNA. The following is a graphic example of the consequences of the cessation of host protein synthesis. If a culture of *E. coli* grown in glucose-containing medium and infected with T4 is transferred to medium containing lactose, phage infection is aborted. The reason is that lactose utilization requires the induction of the synthesis of certain proteins, among them a perme-

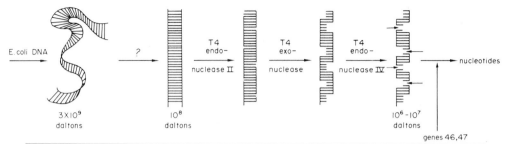

Figure 63-10. A tentative scheme to account for the degradation of *E. coli* DNA after infection with phage T4. *E. coli* DNA is first degraded to fragments about one-tenth its size, which are then nicked by endonuclease II. The nicks are then widened by an exonuclease, and the single-stranded stretches so produced are cut by endonuclease IV. The products are duplex DNA molecules that are more than 100 times smaller than the original *E. coli* genome. Since fragments of this size accumulate in cells infected with phage mutants carrying lesions in genes 46 and 47 but not in cells infected with wild-type phage, these genes presumably code for enzymes that function in their further degradation. *(Adapted from Warner et al.: J Virol 5:700, 1970.)*

ase and the enzyme β-galactosidase. T4 infection halts host protein synthesis and therefore prevents these enzymes from being synthesized. Thus, lactose cannot be metabolized, energy production is prevented, and no phage progeny can be formed.

Enzymes the Synthesize Nucleic Acid Precursors

Numerous enzymes function in the synthesis of nucleic acid precursors, particularly when the phage DNA contains both unusual bases and substituted bases. For example, in the case of T4, which contains glycosylated 5-hydroxymethylcytosine instead of cytosine, the following enzymes are required: (1) enzymes that destroy deoxycytidine triphosphate (so that cytosine is not incorporated into phage DNA), (2) an enzyme that hydroxymethylates deoxycytidylic acid, and (3) enzymes that glycosylate 5-hydroxymethylcytosine residues in polymerized DNA. In addition, numerous enzymes, among them enzymes that function in the synthesis of the other three deoxyribonucleoside triphosphates, are specified by the T-even genome. They are not absolutely essential but increase the yield of progeny phage.

Proteins that Function in Phage DNA Replication

The replication mechanism of double-stranded phage DNA is exceedingly complex, and although enormous effort has been devoted to its elucidation, no clear picture has yet emerged (Chap. 7). The following facts are pertinent:

1. Phage DNA replicates by a semiconservative mechanism.
2. During the early stages of the replication of some phage DNAs, extensive recombination occurs. This

results in the dispersal of the parental genome among numerous progeny genomes.

3. Phage DNA most probably replicates while associated with membrane in some as yet unspecified manner.
4. The replication of some double-stranded DNA genomes, notably those of the T-even phages, is followed by extensive recombination that leads to the formation of giant molecules that consist of up to 100 repeats of the phage genome.
5. Progeny phage genomes are cut from this intermediate in headful units that consist of circularly permuted and terminally redundant genomes (see above).

Among the enzymes known to function in the replication of T4 DNA are a DNA-unwinding protein (coded by gene 32), DNA polymerase (coded by gene 43), a polynucleotide ligase (coded by gene 30), several DNA-binding proteins of unknown function (for example, the proteins encoded by genes 44, 62, and 45), and several nicking enzymes (nucleases).

Proteins that Function in Transcription Programming

Phage genomes, like all other genomes, express themselves via messenger RNA. In fact, it is worth noting that messenger RNA was discovered in phage-infected cells. In 1956 Volkin and Astrachan noted that the base composition of newly synthesized RNA in T2-infected *E. coli* resembled that of T2 DNA rather than that of host DNA, and in 1959 Brenner, Jacob, and Meselson firmly established the concept of messenger RNA by showing that this RNA combines with ribosomes and is responsible for the synthesis of phage proteins.

Phage-specified proteins, like those specified by animal virus genomes (Chap. 58), are synthesized in a strictly ordered sequence. The necessity for this is obvious; for example, enzymes breaking down host DNA should be synthesized first, and viral structural proteins should be synthesized last. There are two ways of achieving programmed protein synthesis: (1) either the entire viral genome is transcribed continuously and translation is programmed, or (2) transcription is programmed, and messenger RNAs are translated as they are formed. Although the former mechanism may apply to a few species of messenger RNA, it is primarily transcription that is programmed in phage-infected and animal virus-infected cells. This is shown most readily by hybridization experiments of the following type. At various times after infection, cells are incubated for brief periods, e.g., for 2 minutes, with RNA precursors labeled with some radioisotope. On extraction, such cells yield labeled messenger RNA populations synthesized from 0 to 2, 2 to 4, 4 to 6, and so on, minutes after infection. These messenger RNA populations are then allowed to hybridize with denatured phage DNA in the presence of excess unlabeled messenger RNA extracted from cells at various stages of infection. If the labeled and unlabeled RNA populations are identical, the large excess of unlabeled RNA will, by competition, prevent the labeled RNA from hybridizing. If the two populations are different, the presence of unlabeled RNA will have no effect on the ability of labeled RNA to hybridize to DNA.

Another type of analysis that can be carried out is to hybridize the messenger RNA molecules that are formed at various times after infection to the series of fragments that are generated when phage DNA is digested with restriction endonucleases. As described in Chapter 58, such fragments can be ordered to provide restriction endonuclease fragment maps. With this technique, it can readily be demonstrated that different regions of the phage genome are transcribed at different stages of the multiplication cycle. The situation here is quite analogous to that described in Chapter 58 concerning the sequential expression of viral genetic information during the multiplication of the double-stranded DNA-containing poxviruses and herpesviruses. There are "immediate early" T4 genes, "early" genes, several classes of "middle" genes, and finally the truly "late" genes that code for most of the T4 structural proteins.

How is such a complex transcription program managed? The complete answer is not yet evident, but one of the factors of critical importance is certainly the specificity of the transcribing enzymes. Since phages do not contain RNA polymerases, the first phage genes are always transcribed by the host RNA polymerase. As discussed in Chapter 7, the *E. coli* RNA polymerase is a complex enzyme that consists of four subunits—two α subunits, one β subunit, and one β' subunit—and has associated with it a protein molecule, the σ factor, which controls its specificity. The portion of the phage genome that is transcribed by this enzyme is usually not large and is specified by the presence of initiation and termination signals that it can recognize. The messenger RNAs transcribed by it are known as early or immediate early RNAs (in the case of very complex programs).

The remainder of the transcription program is managed by proteins specified by early phage messenger RNAs. There are two principal means by which this is accomplished. The first, employed by the simpler phages, such as T7, involves the synthesis of a new RNA polymerase, the specificity of which is such that it transcribes its homologous DNA more efficiently than host DNA. The second means, employed by the more complex phages, such as T4, involves the synthesis of a series of proteins that modify the host RNA polymerase in a series of successive steps. These steps are as follows. Within 30 seconds, one of the α subunits is ADP-ribosylated and the σ factor is partially phosphorylated. During the next 5 minutes both the β and the β' subunits are altered in some as yet unspecified manner that changes their antigenicity and tryptic peptide maps, and between 5 and 10 minutes later, the β' subunit is further modified, as evidenced by an alteration in its electrophoretic mobility. Finally, after 10 minutes, three phage-coded proteins, the proteins encoded by genes 55, 33, and 45, become associated with the modified enzyme. If the ability to recognize and react with promoters and transcription termination signals of each of these modified RNA polymerase species is different, these modifications could readily effect transcription programs such as those described above.

Structural Phage Components

The messenger RNAs that code for the proteins described so far are, with few exceptions, transcribed only from parental DNA. Those that code for the proteins to be described now are usually transcribed only from progeny DNA. These proteins are known as the "late" proteins, and the quantitatively most important late proteins are the structural components of progeny phage particles (compare Chap. 58). Tailed phage particles are complex structures composed of a large number of protein species; for example, T4 contains at least 30 protein species, and even simple phages, such as T7 and lambda, contain 10 to 15 different species.

Proteins that Function Catalytically in Morphogenesis

Phage genomes usually code for several proteins that function in a catalytic capacity during morphogenesis. For example, in addition to the 30-odd structural protein species, T4 DNA codes for at least 17 additional proteins that are also essential for the formation of mature phage particles. Their nature and the precise manner in which they function is largely unknown (but see below for the scaffolding protein). The genes for tail proteins are clustered into three sets in the T4 genome (Fig. 63-9): one for proteins of which only very few copies are needed, one for proteins of which intermediate numbers of copies are needed, and one for proteins of which very many copies per cell are needed. The genetic apparatus for tail formation is thus very highly refined. The situation seems to be similar for several other phages.

Enzymes Necessary for Progeny Phage Liberation: Cell Lysis

Most virulent phages are liberated from their hosts by a mechanism that differs fundamentally from any employed by animal viruses (Chap. 58): the bacterial cell suddenly lyses, thereby liberating the entire progeny. This is known as cell lysis, and, in the case of the T-even phages, is the result of the function of two late phage-coded proteins. One is a lipase encoded by the *t* gene, which attacks the cell membrane, thereby halting metabolic processes; the other is the enzyme lysozyme, encoded by the *e* gene, which then hydrolyzes the cell wall.

Phage Morphogenesis

Phage morphogenesis has been investigated most intensively for coliphages T4, T3, and T7, and lambda, and for the *Salmonella* phage P22. The most successful approach has been to study the products of infection of a large number of conditional lethal mutants, both temperature-sensitive and amber, under restrictive conditions. Cells infected with structural or catalytic gene mutants under restrictive conditions accumulate not mature phage particles but phage components, such as immature and mature proheads, mature heads, tail fibers, tail cores, tail sheaths, partially assembled tails, and so on. As a result, it has been possible to link defects in both structural and catalytic proteins to specific genes whose location on the genetic map of the phage DNA could be identified.

Phage morphogenesis often involves the processing of capsid proteins by means of proteolytic cleavage (during T4 and P2 head assembly, as well as during T5 and lambda tail assembly). Possible reasons why viral capsid proteins are synthesized in the form of precursors are discussed in Chapter 58.

As for the study of the morphogenetic pathway itself, one of the most useful approaches has been to examine the ability of extracts of cells infected with phage mutants to complement each other. In other words, pairs of such extracts are mixed, and one determines whether infectious phage particles are formed from the components present in each. By such means, the scheme for T4 assembly shown in Figure 63-11 has been worked out. Heads, tails, and tail fibers are assembled independently of each other and then combine to form complete virus particles. Heads begin as complex core-containing immature proheads that are assembled on the cytoplasmic membrane of the bacterial cell. They are then proteolytically processed to proheads, which are released from the membrane and filled with DNA, a process that requires the function of several gene products.

Detailed insight into the mechanism by which DNA is inserted into phage heads was provided by studies on the *Salmonella* phage P22. It appears that P22 proheads consist

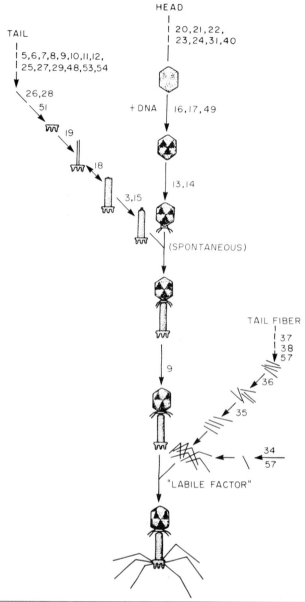

Figure 63-11. The morphogenetic pathway of T4. There are three principal branches leading independently to the formation of heads, tails, and tail fibers. The numbers refer to the gene product(s) involved in each step. *(Adapted from Wood and Edgar: Sci Am 217:74, 1967.)*

of 200 to 300 molecules of a catalytic head assembly or scaffolding protein around which the actual phage head components are assembled. During DNA packaging the scaffolding protein exits and is replaced by DNA. Possibly the exit of the scaffolding protein exposes charged sites that serve to collapse the DNA. The actual amount of DNA that is inserted into the phage head is determined by the so-called headful cutting mechanism, that is, the DNA, in the form of long multimers, is inserted until the head is full, and the DNA is then cut by special endonucleases.

As for tail assembly, this is an extremely complex process that requires the functioning of at least 25 genes (including the genes encoding the tail structural components). One of the proteins involved in tail assembly is a fibrous core component that may act as a ruler that specifies the length of the T4 tail.

Multiplication Cycle of Phages Containing Single-stranded DNA

Several *E. coli* phages contain not double-stranded but single-stranded DNA (Table 63-1). The most completely studied of these phages is φX 174, whose genome is a plus-stranded single-stranded DNA molecule 5375 nucleotides long (mol wt 1.7×10^6), which was sequenced in 1976. It contains 10 genes that have been mapped (Table 63-3). The φX 174 genome was the first shown to contain overlapping genes, both in the same and in different reading frames, and genes within genes. Three of the proteins specified by φX 174 DNA are structural proteins, four are involved in DNA synthesis, one causes cell lysis, and the function of one is not clear.

The main problems posed by the discovery of these phages were (1) how does single-stranded DNA replicate and still obey the rules of base pairing, and (2) since messenger RNAs are transcribed from double-stranded DNA, how can single-stranded DNA serve as their template? These questions were answered when it was found that upon entering the host cell, single-stranded DNA is converted to a double-stranded replicative form.

A scheme describing the mode of replication of φX 174 DNA is presented in Figure 63-12. Although most of the steps depicted are no doubt oversimplified, the scheme's key elements have been demonstrated experimentally. The parental DNA (V) first attaches to a host cell membrane component (M), and a complementary strand is synthesized so as to yield a supercoiled circular double-stranded replicative form, known as RF1. This is then nicked by an endonuclease to yield RF2, a relaxed circular molecule (step 2). In step 3 the V strand is elongated using the C strand as template, and a new C strand is synthesized, with the elongated portion of the V strand serving as its template. This process is a modification of the rolling circle mode of DNA replication (Chap. 7). In step 4 this intermediate is split into a membrane-attached RF1 identical to the original one and a free RF1. The original RF1 then repeats steps 2 to 4, forming new RFs, while the free RF1 has two options. If membrane sites are available, it can attach to them (step 5) and then also generate progeny RFs, or it may be converted to RF2 (step 6), on which the V strand is again elongated (step 7). This intermediate is then split to yield another RF2 (possibly via RF1) and a V strand, which is encapsidated (step 8). The net result is that double-stranded RF molecules first repli-

cate and then serve as templates for the synthesis of large numbers of single-stranded progeny DNA molecules. It is worth noting that infectious φX 174 DNA has been synthesized in vitro with purified enzymes. It was the first functioning DNA genome to be synthesized outside a living cell.

The enzymology of φX 174 DNA replication is extraordinarily complex. It requires at least 12 *E. coli* proteins, including DNA polymerase III, DNA polymerase I, ss DNA binding protein, DNA ligase, DNA gyrase, the products of genes dnaC and dnaB as well as of several others involved in priming, and the φX 174 A protein, which is the only protein encoded by φX 174 DNA that is directly involved in its replication (the functions of the A* protein and of the B, C, and D proteins not being known with certainty).

Messenger RNAs are transcribed from membrane-bound RFs. Among the proteins into which they are translated are the structural phage components, one species of which attaches to the loose end of progeny V strands that arise in step 7, thereby initiating encapsidation.

The mutiplication cycle of the filamentous phages f1, fd, and M13 differs somewhat from that of φX 174. Whereas nascent φX 174 DNA is immediately encapsidated, forming progeny that accumulates within the cell and is eventually liberated when cells lyse, large amounts of free viral DNA accumulate intracellularly in the case of the filamentous phages. These molecules are then encapsidated at the cell membrane, and virus particles are released without cell lysis in a process that is in part analogous to that by which enveloped animal viruses are released from their host cells.

TABLE 63-3. GENES AND GENE FUNCTIONS OF THE SINGLE-STRANDED DNA-CONTAINING BACTERIOPHAGE φX 174

Gene Designation	Molecular Weight of Protein Product	Function
A	56,000	RF replication, ss DNA synthesis
A*	35,000	DNA synthesis
B	14,000	SS DNA synthesis
C	7,000	SS DNA synthesis
D	17,000	SS DNA synthesis
E	10,000	Lysis
J	4,000	Structural
F	46,000	Major capsid protein
G	19,000	Spike (vertex) protein
H	36,000	Spike (vertex) protein

Adapted from Denhardt: In Fraenkel-Conrat and Wagner (eds): Comprehensive Virology, vol. 7, p 1, 1977. New York, Plenum.
The coding regions for the A* protein and A protein overlap; the coding region for the A* protein is the righthand (C terminal) three fifths of the A gene in the same reading frame. The coding region for the B protein is entirely within that of the A/A* protein(s) in a different reading frame. Similarly, the coding region of the E protein is entirely within that of the D protein in a different reading frame. The gene order, starting from the junction of the A and the H gene, is A(A*) (B)-C-D(E)-J-F-G-H.

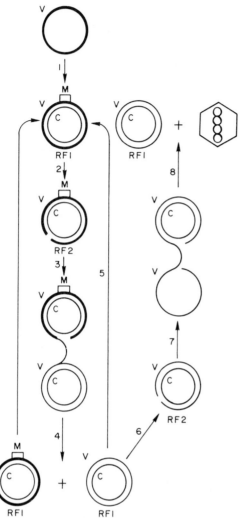

Figure 63-12. A scheme to account for the replication of φX 174 DNA. V represents the DNA strand present in phage (the plus strand); C represents the complementary strand. The heavy V strand is that which was present in the parental virus particle. To replicate the RF1, the A protein nicks the V strand at the origin of replication that lies within the *A* gene and is a 30 nucleotide-long AT-rich sequence; it is therefore a region where the two strands can be readily separated. This AT-rich sequence contains three symmetrical (palindromic) sequences (-CAACTTG-, -TATTATA-, and -CTATAG-); such sequences are often present in protein binding sites. The A protein becomes covalently attached to the 5' end of the nicked strand and then initiates DNA synthesis in a rolling circle mode. It also linearizes and circularizes the displaced plus strand after each round of replication.

Multiplication Cycle of Phages Containing Single-stranded RNA

Single-stranded RNA-containing phages are small icosahedral viruses that resemble picornaviruses. Like them, their capsids consist of 180 protein molecules, but whereas there are 60 molecules each of three different protein species present in picornaviruses, all 180 protein molecules are identical in the case of the RNA phages. In addition, each phage particle contains one molecule of an additional protein, the maturation or A protein, which is necessary both for proper assembly and for ability to adsorb. The RNA of these phages has a molecular weight of just over 1 million, which is less than one-half the size of the picornavirus genome.

Most of the known RNA phages are coliphages. They fall into three or four serologic groups. Most of the well-studied RNA phages (f2, fr, MS2, R17, and so on) belong to one group, while one, Qβ, belongs to another. All RNA phages are male-specific because their receptor sites are on F-pili.

Study of the multiplication cycle of these phages has provided important clues for investigations of animal viruses that contain RNA. In addition, their extraordinary simplicity—they are the simplest viruses known—has made them fascinating objects for the elucidation of fundamental biologic processes.

The genomes of RNA phages comprise only three genes. These are, from the 5'-terminus of the RNA, the genes for the A protein, the coat protein, and the RNA polymerase. The latter protein is actually only one of four proteins that together make up the functional RNA polymerase, the others being three bacterial proteins that are normally located on ribosomes, namely, factor i, elongation factor Tu, and elongation factor Ts. The precise function of these host-coded components of the viral RNA polymerase is not known, but it has been suggested that they somehow control specificity and affinity for initiation sites and that they are essential for maintaining the enzyme complex in its active conformation.

The genomes of RNA phages possess plus polarity and start functioning as messenger RNA very soon after they enter cells. After coding for the RNA polymerase, they replicate by a mechanism similar to that described for poliovirus RNA (Fig. 58-28). This involves, first, the formation of complementary minus strands and then their successive transcription into progeny plus strands. Coat protein then begins to be synthesized in large amounts together with some A protein. As soon as their concentrations are sufficiently high, encapsidation of RNA commences and mature progeny phage accumulates (Fig. 63-13). These events are quite analogous to those that occur during the picornavirus multiplication cycle (Chap. 58).

Programming of RNA Phage Multiplication Cycle

The multiplication cycle of RNA phages, like that of other viruses, is programmed, with the various reactions necessary for morphopoiesis taking place in a regularly progressing sequence. Yet, the genome of RNA phage is too small to encode proteins with a purely regulatory function. How then does it happen that coat protein, which is present in virus particles in 180-fold excess over A protein, is in fact

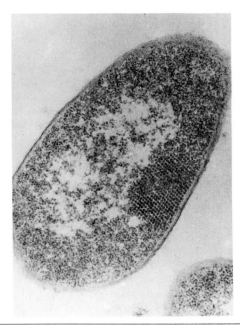

Figure 63-13. An *E. coli* cell 50 minutes after infection with MS2. Note the paracrystalline virus particle array. The virus is slightly larger than the ribosomes that are present throughout the cell. This cell is about to lyse. × 37,500. *(Courtesy of Dr. M. Van Montagu.)*

synthesized in vastly greater amounts than A protein? What mechanism causes the RNA polymerase to be synthesized early during the infection cycle, when it is needed, and not throughout the cycle? And since parental RNA genomes must be both translated and transcribed, what mechanism prevents collisions between ribosomes, which traverse messenger RNA from the 5'- to the 3'-terminus, and RNA polymerase molecules, which traverse templates in the opposite direction?

The genomes of RNA phages are polycistronic messenger RNAs. There are three ribosome attachment sites, one at the start of each cistron. Like other RNAs, they possess a pronounced secondary structure that is specified by base pairing between complementary nucleotide sequences. There is a marked difference in how accessible ribosome-binding sites at the 5'-termini of the three genes are to ribosomes. In normal phage RNA, the most accessible ribosome binding site is at the beginning of the cistron that codes for the coat protein, which would therefore be formed in the greatest amount. However, attachment of ribosomes there renders the site at the beginning of the polymerase cistron also accessible to ribosomes, so that during the initial stages of infection both coat protein and polymerase are formed freely. Coat protein itself, however, can bind to the ribosome-binding sites at the beginning of the A protein and polymerase cistrons, thereby preventing ribosome attachment, so that as coat protein accumulates, it gradually inhibits translation of these two cistrons more and more strongly. The net result is that the A protein cis-

tron is translated only infrequently throughout most of the infection cycle, and the polymerase cistron is translated frequently during the early stages but only very infrequently during the later stages of the infection cycle.

The collision problem is solved in the following manner. When a polymerase molecule attaches to the 3'-terminus of the RNA strand, it also binds to the coat protein cistron ribosome-binding site, thereby preventing further ribosome attachment. When ribosomes have cleared from the region between these two points, the polymerase molecule can progress along the RNA strand unimpeded.

It is clear therefore that, although it is very simple and small, the RNA phage genome regulates a sophisticated transcription and translation program. This is because several of the proteins that it encodes possess dual functions.

Lysogeny

Not all phage infections result in lysis of the host cell. There are some phages that, upon entering a sensitive cell, either elicit a lytic response, that is, a typical multiplication cycle controlled by a transcription program that involves expression of the entire phage genome, or a second type of response in which all phage genes except one are completely repressed and in which the phage genome is either inserted into the host genome or exists in the bacterial cell in free plasmid-like form. The phage genome, the prophage, behaves from then on like any other portion of the bacterial chromosome, and the bacteria that harbor prophage appear perfectly normal. Occasionally, however, the integrated prophage is excised from bacterial DNA, and phage multiplication and cell lysis ensue. This phenomenon is therefore known as "lysogeny." Bacteria that harbor such prophages are known as lysogens, and phages capable of eliciting the lysogenic pathway are known as lysogenic, or temperate, phages.

When it was discovered that some animal viruses could enter into a similar relationship with their hosts (Chap. 62), the analogy between them and temperate phages was quickly recognized. Although there are differences in the manner in which tumor viruses and temperate phages interact with their host cells, the concepts that evolved from studies of temperate phages have profoundly influenced our thinking concerning tumor viruses.

Prophage

The frequency of lysogenization varies both with the phage-host system and with the multiplicity of infection. It is greater the higher the multiplicity. This suggests that each phage genome has a finite probability of lysogenization. The number of any particular species of prophage per lysogen is generally the same as the number of bacterial genomes, which suggests that the prophage attachment site is restricted to a particular locus on the bacterial chromosome. That a unique location generally exists can be demonstrated by genetic studies, such as the interrupted mating technique (Chap. 8). However, occasionally there is more than one attachment site for the same prophage, and all may be occupied simultaneously. Superinfection may result in the integration of two prophages at the same site, inserted either into one another or adjacent to each other. Further, bacterial chromosomes generally contain prophage attachment sites for several different unrelated prophages, and all of these may be occupied at the same time. Finally, one temperate phage, phage mu, has no specific integration site but can insert its genome anywhere in the bacterial chromosome. However, it is not essential that the prophage be integrated. If prophage cannot integrate (because of mutations, for example, in the attachment site or in the enzyme system responsible for integration), it can exist and be carried in the lysogen in the form of supercoiled circular plasmid-like molecules.

Concept of Immunity

Under normal conditions prophage is stable. Termination of the lysogenic state occurs spontaneously with low frequency (of the order of 10^{-5}). The phage genome is then excised from the host chromosome (see below), and a normal lytic multiplication cycle ensues. Cultures of lysogenic bacteria therefore contain low titers of infectious virus, but they are not lysed by it: they are immune. This is not because the phage cannot adsorb and inject its DNA but because the injected DNA cannot express itself and replicate. The reason for this is that one of the prophage genes codes for a repressor that prevents transcription of all other phage genes, both those on prophage and those on superinfecting phage genomes. Immunity is highly specific; bacteria lysogenic for one phage are only immune to it and to closely related phages. It is ability to effect immunity, that is, possession of a repressor gene, that is the essential attribute of a lysogenic phage, not ability to integrate its DNA into that of the host cell.

Nature of Phage Lambda

The best known of the many temperate bacteriophages is the coliphage lambda, probably the most intensively studied of all viruses. The exquisite and intricate mechanisms that control its interaction with the host and its multiplication represent the system par excellence with which to study biologic control. We will, therefore, discuss the mechanisms that operate in lysogeny in terms of the lambda system.

Phage lambda possesses a hexagonal head and a simple noncontractile tail with one terminal tail fiber; it comprises about 10 protein species. Its DNA, which contains no unusual bases, is linear and has a molecular weight of 31×10^6, representing about 50 genes. Figure 63-14 shows a simplified genetic map of lambda. About 20 genes code for its structural proteins and for proteins that function in its morphogenesis. These genes, which include genes *A* to *J*, are in the left arm of the DNA, which contains 55 percent guanine plus cytosine. Some 30 other genes have been identified as regulating in one way or another the interaction of the lambda genome with that of the host cell and its transcription and translation. These genes are located in the right arm of the DNA, which contains 46 percent guanine plus cytosine. Not all genes are transcribed from the same DNA strand. As shown in Figure 63-14, some genes are transcribed from one strand, while others are transcribed in the opposite direction from the other strand.

Events Leading to Lysogeny: Integration

When lambda DNA enters the cell, it cyclizes and forms covalently closed circles. Within a brief critical period of time the decision is then made as to whether a lytic multiplication cycle is to be initiated or whether the phage genome is to be repressed. The decision is influenced principally by how rapidly several viral gene products reach critical concentrations and begin to function.

The critical genes and control elements are illustrated in Figure 63-15, which is an enlargement of one portion of the lambda genetic map. The key element is the operator O_R, a sequence located at map position 78.5. It is about 50 nucleotides long and contains three 9 bp long sequences to which *E. coli* DNA-dependent RNA polymerase and two types of phage lambda-coded repressor proteins can bind. At the very beginning of infection, only RNA polymerase binds there and transcribes both genes that flank O_R, one from one strand and the second from the other. The gene on the left is the *cI* gene that encodes the lambda repressor, a 236 amino acid long protein (mol wt 26,000) with two domains, an N-terminal domain that rec-

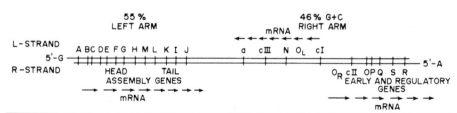

Figure 63-14. A simplified genetic map of phage lambda. *(Adapted from Taylor et al.: Proc Natl Acad Sci USA 57:1618, 1967.)*

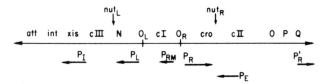

Figure 63-15. Portion of the lambda genetic map centered around operator O_R. It illustrates the various genes and control elements that determine whether the lytic or lysogenic pathway will be followed. For details, see text.

ognizes operator sites and a C-terminal domain that provides contacts for dimerization, dimers being the form in which it binds to O_R. The gene on the right is the *cro* gene, which encodes the *cro* repressor, a 66 amino acid long protein (mol wt 8000), the dimeric form of which also binds to O_R. The nature of lambda infection depends on whether more *cI* or more *cro* gene product is synthesized. If more *cro* repressor is made, it binds to O_R, preventing lambda repressor from binding there, and prevents the synthesis of the *cII* and *cIII* gene products that greatly increase lambda repressor synthesis (see below). In essence, it greatly reduces lambda repressor synthesis. As a result, a second operator, O_L, is not blocked by lambda repressor, and RNA polymerase binds there and proceeds to transcribe gene N. Expression of this gene is essential for the transcription of most of the lambda genome. The reason is that the early lambda genes, except *cI* and *cro*, contain transcriptional terminators, short sequences that cause RNA polymerase to cease transcription. The N gene product, a very basic protein 107 amino acids long (mol wt 12,000), is an antiterminator. It prevents, via a very complex mechanism that involves at least five host proteins, termination of transcription by attaching to so-called *nut* sites (N protein *ut*ilization sites) that are located just upstream from these terminators. If *N* gene product is bound to *nut* sites, RNA polymerase reads through the transcriptional terminators into adjacent genes, and in particular into gene *O*, the product of which is essential for lambda DNA replication, and into gene *Q*, the product of which is an antiterminator for lambda late genes. As a result, a normal lytic lambda multiplication cycle ensues, just like that described above for T4.

If, on the other hand, more *cI* product, that is, lambda repressor, is made than *cro* repressor, then the lambda repressor binds very tightly to operator O_R. This turns off transcription from promoter P_R (and therefore prevents synthesis of *cro* repressor) and turns on transcription from promoter P_{RM}. In fact, transcription from P_{RM} is much more efficient if O_R is occupied by lambda repressor than if it is not occupied. Lambda repressor is thus a positive regulator of its own synthesis.

The interaction of lambda repressor and *cro* repressor with the operator O_R thus forms an exquisitely sensitive, efficient, and sophisticated molecular switch capable of switching between two physiologic states in response to a transient signal. It represents the most intensively studied and best understood interaction between protein and nucleic acid.

In addition to lambda repressor positively controlling its own synthesis, two other early proteins augment *cI* transcription. These are the proteins encoded by genes *cII* and *cIII*, both of which promote transcription from P_E, a promoter that appears to lie within gene *cII*, from which transcription of *cI* can be initiated. Not only that, but the *cII* protein also positively regulates transcription, from promoter P_I, of the *int* gene, which codes for the integrase, the enzyme necessary for the integration of lambda DNA into that of the host cell. Thus, if lambda repressor is synthesized in excess over *cro* repressor, it binds to O_R and promotes its own transcription from P_{RM}, and the *cII* and *cIII* proteins additionally stimulate its transcription from P_E. All this causes a rapid buildup of lambda repressor. The *cII* protein also causes the formation of large amounts of integrase, and then, no more *cII*, *cIII*, and integrase proteins being required, the lambda repressor rapidly and with extraordinary efficiency binds to both O_R and O_L, thus blocking all transcription from the lambda genome except from gene *cI*. That is, it prevents the synthesis of all lambda proteins except its own.

Once a cell contains sufficient lambda suppressor and also adequate amounts of the *int* gene product, the lambda genome is integrated into the host genome by a mechanism that is illustrated schematically in Figure 63-16. The phage DNA region aa' between genes *J* and *cIII* aligns itself with region bb', the prophage attachment site on the bacterial chromosome, which is situated between the galactose and biotin genes. Enzyme(s) encoded by the phage *int* system then make scissions in all four nucleotide strands, the phage DNA is inserted by means of double crossing over, and covalent bonds are then reformed. The lambda genome is now integrated. If the aa' and bb' regions do not match, usually because of a deletion, the phage genome cannot become integrated. It then multiplies in the host cell as an episome (Chap. 8). In either event, the only lambda gene that transcribed is gene *cI*.

Release from Repression: Induction

The termination of the lysogenic state, or induction, requires both release from repression and excision of the phage DNA from the host DNA. The cellular signal for release from repression is inhibition of host DNA synthesis, such as starvation of thymine-requiring mutants of thymine, treatment with inhibitors of DNA replication, such as mitomycin C, or irradiation with ultraviolet (UV) light. All these treatments initiate a multifunctional cellular response termed the SOS pathway, which causes the cellular *recA* protein, a protein that normally functions in genetic recombination, to be activated as a protease. This protease then cleaves the lambda repressor between amino acids 111 and 112, thereby inactivating it. As a result, RNA polymerase can again bind to operators O_L and O_R and initiates transcription there. As a result enzymes specified by the *int* and *xis* phage genes are synthesized, and these enzymes excise the phage DNA from the host DNA. Excision proceeds essentially via reversal of the integration

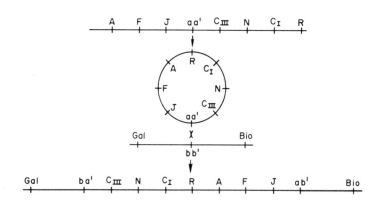

Figure 63-16. The Campbell model for prophage integration. After circularization of lambda DNA by joining of cohesive ends, reciprocal recombination between phage aa' and bacterial bb' recognition sites provides for linear insertion of the viral DNA into that of the host. *(Adapted from Echols and Joyner: In Fraenkel-Conrat (ed): Molecular Basis of Virology, 1968. New York, Reinhold Book Corp.)*

process and yields a circular phage DNA molecule that, active lambda repressor no longer being present, replicates in the normal manner characteristic of virulent lambda infection. Under appropriate conditions, treatment with mitomycin C or UV irradiation induces all cells in cultures of lysogenic bacteria at the same time, with lysis and progeny phage liberation occurring some 20 minutes later.

The juxtaposition of bacterial and phage genes in the chromosome of the lysogen sometimes leads to excision of some bacterial DNA together with phage DNA. This provides an opportunity for the transfer of bacterial genetic material from one bacterium to another via phage; this is known as transduction. Transduction is discussed in Chapter 8.

Significance of Lysogeny

Every bacterium may well carry one or more prophages. There are numerous defective or cryptic prophages, that is, prophages that lack the complete set of information necessary for phage multiplication, so that even when induced and therefore excised, no infectious phage particles are formed. Bacteria harboring such prophages are analogous in some respects to cells transformed by defective SV40 (Chap. 61). Even if they possess complete genomes, lysogenic phage cannot be detected unless there are available nonimmune bacterial indicator strains that lack prophage attachment sites and that can be infected by the lytic pathway.

Lysogenic Phage Conversion
Although in most cases the phenotype of bacteria that carry prophage is indistinguishable from that of bacteria that do not, there are several interesting examples where there is a clear difference. These bacteria exhibit the phenomenon known as lysogenic conversion.

One case of lysogenic conversion involves the phage-mediated conversion of the somatic O antigen of *Salmonella*. This antigen forms part of the lipopolysaccharide structure of the cell wall and is composed of highly esterified lipid (lipid A) linked to a polysaccharide core to which are attached side chains consisting of repeating sequences of a

variety of sugars (Chap. 6). The antigenic specificity resides in the sugars and can be modified by altering their nature.

A wide variety of temperature phages of *Salmonella* can modify the antigenic properties of the somatic O antigen. For example, the O antigen of *Salmonella anatum* normally has the antigenic formula 3,10 (based on its response pattern to various antisera) and carries terminal O-acetyl-D-galactosyl groups linked by means of α-1:6 linkages to D-mannosyl residues. When lysogenized by the temperature phage ϵ^{15}, antigenic specificity 10 is lost, and specificity 15 is acquired. This is due to loss of the O-acetyl group and replacement of the α-1:6 bond by a β-1:6 bond. Cells lysogenized by phage ϵ^{15} can themselves be lysogenized by phage ϵ^{34}, which causes a further change in antigenic specificity, this time due to the addition of D-glucosyl residues by means of β-1:4 linkages. There are numerous other examples.

There is no doubt that the chemical alterations in *Salmonella* polysaccharides are phage-induced, since ϵ^{15} causes repression or inhibition of the α-1:6-galactosyl transferase and acetylase and encodes a β-1:6-galactosyl transferase. These converting functions are, however, not essential to the phage, since mutants defective in them grow well. On the other hand, it is significant that ϵ^{15} cannot adsorb to cells with antigenic specificity 3,15, that is, cells carrying ϵ^{15} cannot be infected by it. It is likely that the significance of this type of lysogenic conversion lies in the specification of host range patterns, perhaps to augment phage immunity systems.

A second example of lysogenic conversion concerns *Corynebacterium diphtheriae*. Toxigenic strains carry the prophage β, and nontoxigenic strains can be made toxigenic by lysogenization. It is clear that the structural gene for toxin is located on the phage genome, since phage mutants coding for defective toxin have been isolated. However, it is not clear what advantage the phage gains by coding for a protein with the remarkable biologic and enzymatic properties of the toxin (Chap. 32) Most probably, the advantage relates not to the phage but to the bacterial lysogen. In other words, corynebacteria that can synthesize toxin most probably have a survival advantage over those that cannot do so.

A third example involves *Clostridium botulinum* types C and D, which produce toxin only when demonstrably infected with (and yielding) phages CE β and DE β, respectively. However, with types C and D, the relation between host cell and phage may not be true lysogeny, since infected *C. botulinum* becomes nontoxigenic (and nonvirogenic) when treated with antiserum to phage.

FURTHER READING

Books and Reviews

DOUBLE-STRANDED DNA-CONTAINING PHAGES

Anderson TF: Reflections on phage genetics. Annu Rev Genet 15:405, 1981

Cairns J, Stent GS, Watson JD: Phage and the origins of molecular biology. New York, Cold Spring Harbor Laboratory, 1966

Casjens S, King J: Virus assembly. Annu Rev Biochem 44:555, 1975

Earnshaw WC, Casjens SR: DNA packaging by the double-stranded DNA bacteriophages. Cell 21:319, 1980

Eiserling FA: Bacteriophage structure. In Fraenkel-Conrat H, Wagner RR (eds): Comprehensive Virology. New York, Plenum, 1979, vol 13, p 543

Hausmann R: Bacteriophage T7 genetics. Curr Top Microbiol Immunol 75:77, 1976

Koerner JF, Snustad DP: Shutoff of host macromolecular synthesis after T-even bacteriophage infection. Microbiol Rev 43:199, 1979

Kruger DH, Schroeder C: Bacteriophage T3 and bacteriophage T7 virus-host cell interactions. Microbiol Rev 45:9, 1981

Mathews CK: Reproduction of large virulent bacteriophages. In Fraenkel-Conrat H, Wagner RR (eds): Comprehensive Virology. New York, Plenum, 1977, vol 8, p 179

Murialdo H, Becker A: Head morphogenesis of double-stranded DNA containing bacteriophages. Microbiol Rev 42:529, 1978

Reanney DC, Ackermann H: Comparative biology and evolution of bacteriophages. Adv Virus Res 27:206, 1982

Wagner ES, Auer B, Schweiger M: *Escherichia coli* virus T1: Genetic controls during virus infection. Curr Top Microbiol Immunol 102:131, 1983

Warren RAJ: Modified bases in bacteriophage DNAs. Annu Rev Microbiol 34:137, 1980

Wood WB, King J: Genetic control of complex bacteriophage assembly. In Fraenkel-Conrat H, Wagner RR (eds): Comprehensive Virology. New York, Plenum, 1979, vol 13, p 581

PHAGE LAMBDA: LYSOGENY

Calendar R, Geisseloder J, Sunshine MG, et al.: The P2-P4 transactivation system. In Fraenkel-Conrat H, Wagner RR (eds): Comprehensive Virology. New York, Plenum, 1977, vol 8, p 329

Campbell A: Defective bacteriophages and incomplete prophages. In Fraenkel-Conrat H, Wagner RR (eds): Comprehensive Virology. New York, Plenum, 1977, vol 8, p 259

Echols H, Murialdo H: A genetic map of lambda. Microbiol Rev 42:577, 1978

Gottesman S: Lambda site-specific recombination: The att site. Cell 25:585, 1981

Greenblatt J: Regulation of transcription termination by the N gene protein of bacteriophage lambda. Cell 24:8, 1981

Herskowitz I, Hagen D: The lysis-lysogeny decision of lambda: Explicit programming and responsiveness. Annu Rev Genet 14:402, 1980

Johnson AD, Poteete AR, Lauer G, et al.: Lambda repressor and cro—components of an efficient molecular switch. Nature 294:217, 1981

Nash HA: Integration and excision of bacteriophage lambda: The mechanism of conservative site-specific recombination. Annu Rev Genet 15:143, 1981

Scott JR: Immunity and repression in bacteriophage P1 and P7. Curr Top Microbiol Immunol 90:49, 1980

Shimatake H, Rosenberg M: Purified lambda regulatory protein cII positively activates promoters for lysogenic development. Nature 292:128, 1981

Ward DF, Gottesman ME: Suppression of transcription termination by phage lambda. Science 216:946, 1982

Weisberg RA, Gottesman S, Gottesman ME: Bacteriophage lambda: The lysogenic pathway. In Fraenkel-Conrat H, Wagner RR (eds): Comprehensive Virology. New York, Plenum, 1977, vol 8, p 197

SINGLE-STRANDED RNA AND DNA-CONTAINING PHAGES

Denhardt DT, Dressler D, Ray DS (eds): The Single-stranded DNA Phages. New York, Cold Spring Harbor Laboratory, 1978

Fiers W: Structure and function of RNA bacteriophages. In Fraenkel-Conrat H, Wagner RR (eds): Comprehensive Virology. New York, Plenum, 1979, vol 13, p 69

Ray DS: Replication of filamentous bacteriophages. In Fraenkel-Conrat H, Wagner RR (eds): Comprehensive Virology. New York, Plenum, 1977, vol 8, p 105

Zinder ND: Portraits of viruses: RNA phage. Intervirology 13:257, 1980

Selected Papers

THE LYTIC PHAGE MULTIPLICATION CYCLE

Black LW, Silverman DJ: Model for DNA packaging into bacteriophage T4 heads. J Virol 28:643, 1978

Brennan SM, Chelm BK, Romeo JM, et al.: A transcriptional map of the bacteriophage SP01 genome: II. The major early transcription units. Virology 111:604, 1981

Coombs DH, Eiserling FA: Studies on the structure, protein composition and assembly of the neck of bacteriophage T4. J Mol Biol 116:375, 1977

Crowther RA, Lenk EV, Kirkuchi Y, et al.: Molecular reorganization in the hexagon to star transition of the base plate of T4. J Mol Biol 116:489, 1977

Fuller MT, King J: Purification of the coat and scaffolding proteins from procapsids of bacteriophage P22. Virology 112:529, 1981

von Gabain A, Bujard H: Interaction of *Escherichia coli* RNA polymerase with promoters of several coliphage and plasmid DNAs. Proc Natl Acad Sci USA 76:189, 1979

Heintz N, Shub DA: Transcriptional regulation of bacteriophage SP01 protein synthesis in vivo and in vitro. J Virol 42:951, 1982

Kozloff LM, Lute M, Crosby LK: Bacteriophage T4 virion baseplate thymidylate synthetase and dihydrofolate reductase. J Virol 23:637, 1977

Majumder HK, Bishayee S, Chakraborty PR, et al.: Ribonuclease

III cleavage of bacteriophage T3 RNA polymerase transcripts to late T3 mRNAs. Proc Natl Acad Sci USA 74:4891, 1977

Matsuo-Kato H, Fujisawa H, Minagawa T: Structure and assembly of bacteriophage T3 tails. Virology 109:157, 1981

McCorquodale DJ, Chen CW, Joseph MK, et al.: Modification of RNA polymerase from *Escherichi coli* by pre-early gene products of bacteriophage T5. J Virol 40:958, 1981

Milhausen MJ, Whiteley HR: In vitro synthesis of a peptide which modifies the transcriptional specificity of *Bacillus subtilis* RNA polymerase. Biochem Biophys Res Commun 99:900, 1981

Minagawa T, Murakami A, Ryo Y, et al.: Structural features of very fast sedimenting DNA formed by gene 49 defective T4. Virology 126:183, 1983

Nordstrom B, Randahl H, Slaby I, et al.: Characterization of bacteriophage T7 DNA polymerase purified to homogeneity by antithioredoxin immunoadsorbent chromatography. J Biol Chem 256:3112, 1981

Spiegelman GB, Whiteley HR: Bacteriophage SP82 induced modifications of *Bacillus subtilis* RNA polymerase result in the recognition of additional RNA synthesis initiation site on phage DNA. Biochem Biophys Res Commun 81:1058, 1978

Talkington C, Pero J: Restriction fragment analysis of the temporal program of bacteriophage SP01 transcription and its control by phage-modified RNA polymerases. Virology 83:365, 1977

PHAGE LAMBDA: LYSOGENY

Ho Y-S, Lewis M, Rosenberg M: Purification and properties of a transcriptional activator: The cII protein of phage lambda. J Biol Chem 257:9128, 1982

Hochschild A, Irwin N, Ptashne M: Repressor structure and the mechanism of positive control. Cell 32:319, 1983

Hoyt MA, Knight DM, Das A, et al.: Control of phage lambda development by stability and synthesis of cII protein: role of the viral cIII and host hfla, himA and himD genes. Cell 31:565, 1982

Kikuchi Y, Nash HA: The bacteriophage lambda int gene product. J Biol Chem 253:7149, 1978

Sanger F, Coulson AR, Hong GF, et al.: Nucleotide sequence of bacteriophage lambda DNA. J Mol Biol 162:729, 1982

Sauer RT, Ross MJ, Ptashne M: Cleavage of the lambda and P22 repressors by recA protein. J Biol Chem 257:4458, 1982

Shaw JE, Jones BB, Pearson ML: Identification of the N gene protein of bacteriophage lambda. Proc Natl Acad Sci USA 72:2225, 1978

Takeda Y, Folkmanis A, Echols H: Cro regulatory protein specified by bacteriophage lambda: Structure, DNA-binding and repression of RNA synthesis. J Biol Chem 252:6177, 1977

SINGLE-STRANDED NUCLEIC ACID-CONTAINING PHAGES

Arai K-I, Kornberg A: Mechanism of dnaB protein action. IV. General priming of DNA replication by dnaB protein and primase compared with RNA polymerase. J Biol Chem 256:5267, 1981

Grant RA, Lin T-C, Konigsberg W, et al.: Structure of the filamentous bacteriophage f1. J Biol Chem 256:539, 1981

Meyer TF, Geider K: Enzymatic synthesis of bacteriophage fd viral DNA. Nature 296:828, 1982

Sanger F, Coulson AR, Friedmann T, et al.: The nucleotide sequence of bacteriophage φX174. J Mol Biol 125:225, 1978

Shlomai J, Polder L, Arai K-I, et al.: Replication of φX174 DNA with purified enzymes. I. Conversion of viral DNA to a supercoiled, biologically active duplex. J Biol Chem 256:5233, 1981

DIPHTHERIA TOXIN PRODUCTION

Uchida T, Kanei C, Yoneda M: Mutations in corynephage β that affect the yield of diphtheria toxin. Virology 77:876, 1977

SECTION V
CLINICAL VIROLOGY

CHAPTER 64

Host-Virus Interactions

Invasion of Host
Multiplication in Host Cell Tissues
Antibody
Complement

Adverse Consequences of Altered Virus–Host
  Interactions
Cell-mediated Immunity

As available knowledge has increased, discussion of virus infections in humans has become increasingly detailed and concepts of virus–host interaction have been expanded. Initially, only acute self-limited illnesses, such as a respiratory tract infection or measles, were included. The number of viruses identified that cause specific illnesses with a short course and attendant morbidity and mortality has greatly increased with the availability of cell cultures.

A second virus–host interaction is that of teratogenesis. Viral agents, such as rubella and cytomegalovirus (CMV), are well appreciated causes of intrauterine infection. The consequences of such infection are dependent upon such factors as the time of infection, the agent causing the infection, and the response of the host. One virus may cause alterations in organogenesis, as is the case with rubella virus, or another, such as CMV, may cause destruction of poorly differentiated tissue. The immature fetus does not mimic the adult immune response to infection, and thus it has been shown with rubella and CMV that infectious viruses may replicate and be present for months or years. The effects of virus replication may be compounded by the additional consequences of the host's response to the virus, for example, circulating immune complexes are demonstrable in infants with congenital CMV.

Virus latency and recurrent infection with the same agent are well documented in the case of the herpes group of viruses. Herpes simplex virus is notorious for its ability to cause intermittent overt infection and has been demonstrated to be latent in neural tissue. CMV may be intermittently excreted by the infected individual, and when appropriate studies have been done, such as restriction endonuclease analysis, it appears that the infecting virus is a single strain of virus. Similarly, it has long been appreciat-

ed that varicella virus infection induces a primary generalized illness and, in subsequent years, a localized infection in the form of zoster with the same virus. Epstein–Barr virus, the etiologic agent of infectious mononucleosis, would certainly appear to cause latent infection of circulating lymphocytes and perhaps of other cells. The herpes group of viruses exemplifies a type of virus–host interaction where an initial infection is followed by an indeterminate period of time when no demonstrable signs of infection are present and subsequent recurrent active infection occurs.

At the present time it is apparent that some human viruses may cause persistent infection of the host. The most salient example of this is hepatitis B virus, which may be present in large quantities for years. The results of this persistent infection can be asymtomatic infection or symptomatic aggressive liver disease, with the possible outcomes of cirrhosis and hepatocellular carcinoma linked to this infection. The central nervous system infections with enteroviruses occurring in patients with agammaglobulinemia are a second example of persistent infection but one that is dependent on an abnormal host response. Virus may be readily detectable for years in the central nervous system, and the clinical manifestations are progressive and ultimately fatal. Serologic studies indicate that human papovaviruses (BK and JC) asymptomatically infect the majority of persons by adulthood, yet disease becomes apparent only with immunosuppression. Relentlessly progressive central nervous system disease, such as Kuru or Jacob–Creutzfeldt disease, is due to transmissible agents called chronic infectious neuropathic agents, or CHINA viruses. These agents have not been visualized or isolated.

A final example of the spectrum of illness is oncogenesis. It now seems quite certain that several human viruses are linked to the ability to form tumors in people. Hepatitis B virus is associated with hepatocellular carcinoma. Epstein–Barr virus has been associated with Burkitt's lymphoma and a polyclonal malignant lymphoma, as well as nasopharyngeal carcinoma. Finally, the recently recognized human leukemia viruses must be considered in any discussion relating viruses to human malignancy. Thus, in general terms, our appreciation of the virus–host interaction has extended across the spectrum from acute, temporally limited illness through persistent infection to latency and recurrent infection, as well as teratogenesis and oncogenesis.

The virus–host interaction is likely to differ for viral agents of different groups. The age of the infected individual can influence the outcome of the host–virus interaction by virtue of variations in the immune response. Many other specific features of the virus–host interaction will determine which tissues of the host are infected and, therefore, where the detectable effects of infection will be.

Invasion of Host

The virus must first gain access to the host, and most often this is via mucous membranes of the respiratory, gastrointestinal, or genitourinary tracts. Aerosolization of virus affords entry into respiratory tract, whereas direct inoculation of virus occurs with sexual transmission and fecal–oral spread. Less frequently, the virus may gain entry through the skin or by direct inoculation into the bloodstream, as occurs with insect vectors or transplacental transmission of agents. Intact skin poses obvious problems for viral entry with the layers of keratinized nonmetabolizing cells. Mucous membrane surfaces are bathed in an array of secretions, including saliva and mucus, and these materials contain both nonspecific and specific defenses against invasion by viral pathogens.

Adherence of the virus to the host cells is essential to invasion of the cell. This is mediated by specific receptors of the virion and host. For polioviruses, the neuronal affinity is mediated by specific receptors, as is the species susceptibility. The specific cell receptors are known to be coded for by human chromosome 19. The specific viral receptor is unknown.

The ability of viruses to survive transit through the gastrointestinal tract may be partially dependent upon their acid stability, as with the enteroviruses. A protease resistant outer capsid protein (μIC) enables reovirus type 1 to survive transit through the gastrointestinal tract, adsorb to M epithelial cells, and subsequently replicate. There are mucoproteins in host secretions containing sialic acid that may be the receptor recognized by several viruses of the myxo/paramyxovirus families. There may be competition by the mucoproteins for the virus, which would diminish its infectivity by preventing contact with cells.

Influential features of the outcome of the virus–host contact include many factors that are not completely understood. The age of the host is clearly important when

such agents as herpes simplex virus, rotavirus, hepatitis B virus, and adenoviruses are considered. In the case of herpes simplex virus, an overwhelming disseminated infection may occur in the unfortunate newborn infant who acquires infection during delivery. In the absence of immunocompromise this is an almost unheard of event in older individuals. In the case of rotaviruses, asymptomatic excretion of virus appears to be the rule in the newborn nursery when such infection is documented. On the other hand, these viruses are the leading cause of diarrhea in infants between the ages of 6 months and 18 months. Although hepatitis B virus can infect and ultimately cause serious disease in any age individual, it is far more likely to become established as a persistently infecting agent if an infant acquires this virus in the newborn period. It has been estimated that 90 percent of infants who contact this virus in the perinatal period become chronic carriers of hepatitis B virus. This is clearly not the case for older children or adults who acquire infection with this agent. Adenoviruses can cause a severe and overwhelming pneumonia in infants, and this is very unusual in older individuals unless immunosuppression or immunodeficiency exists.

Another such factor influencing the outcome of the infection is the nutritional status of the host. Elaborate studies have been done to define the changes in the host's immune status with malnutrition. The complex changes in immune function that are secondary to protein calorie malnutrition contribute to the severity of the infection.

Multiplication in Host Cell Tissues

Virus replication must occur when the virus gains access to the host if infection is to occur. Assuming the virus gains access to the relatively few permissive cells, other features of the environment in the host can limit virus replication. The paramyxoviruses have a surface fusion protein that is essential for induction of infection and is responsible for cell-to-cell spread of virus. This protein must be cleaved by a host cell protease for the virus to be infectious. Thus, the tissue tropism may be determined by the presence of an appropriate host cell protease capable of cleaving the target protein.

As virus replication must occur intracellularly, viruses gain access to both the professional phagocytes and host cells that do not normally phagocytize. Their intracellular replication may provide a protected environment, as they are clearly inaccessible to such elements of the host immune system as antibody. In certain instances, for example, lymphocytic choriomeningitis in animals, it is easy to demonstrate that the virus actually infects the effector cells of the immune system. This too will have a distinct effect on the outcome of the infection if virus can continue its replication within these cells and alter their function or destroy them.

As intracellular pathogens, viruses may spread by lysing the host cell, in which case the virus will spread via the extracellular environment and ultimately be accessible to all elements in the host immune system. Enteroviruses are examples of agents that are generally thought to lyse cells. Secondly, viruses may spread from cell to cell, as do those viruses having a fusion protein responsible for the joining of cellular membranes. Measles virus is a good example of an agent that can spread by fusion of adjacent cell membranes. Intact infectious virus is also released to the extracellular environment by budding from the host cell. Finally, some viruses may be passed directly from the parent cell to progeny cells without being liberated.

Children with immunodeficiencies have provided insight into the normal host–virus interaction. Several children with combined humoral and cellular immunodeficiencies have been observed to have chronic pulmonary infection with a parainfluenza virus because they are not able to eradicate viruses from their respiratory tract. The deficient cellular immunity seems more important, as children with only antibody deficiency have not had prolonged respiratory tract infection with these viruses. These cellular immunodeficient children have also been observed to have virus detectable in tissues other than the respiratory tract. It is not yet clear whether these viruses are mutants whose cleavage requirements have been altered or whether this is simply related to the persistent infection and the inability to prevent virus dissemination. The paramyxoviruses are only one group of agents that are known to cause difficulty in these children. The herpes agents and measles virus produce lethal infections. These children have also been demonstrated to have noncultivatable adenoviruses and/or rotaviruses present in the gastrointestinal tract for prolonged periods of time with active gastrointestinal disease.

It is possible to speculate that those viruses that insert viral proteins into host cell membrane by fusion during cell entry or cause de novo expression of viral antigens on the cell surface prior to formation of intact infectious virus might conceivably be rapidly eliminated if host defense mechanisms, antibody, complement, and sensitized lymphocytes, can destroy the cell in which the virus is replicating. It is also possible that the infection will actually be temporarily amplified if the infected cell is identified and lysed by the host defense mechanisms at a time when infectious progeny are contained within the cytoplasm.

Antibody

The host sees the virus as multiple antigens. Whether this is as multiple determinants on a single protein and/or as multiple proteins, the host develops numerous specific responses to these antigens. Generally speaking, neutralization of the virus involves the interaction of antibody with surface protein(s). In the case of reoviruses the sigma 1 capsid protein stimulates the production of antibody that is capable of neutralizing the virus. The sigma 3 protein, an outer capsid antigen, also stimulates antibody that is capable of neutralizing the virus. The use of monoclonal antibodies has allowed more intricate dissection of the interaction of such proteins with specific determinants of the virus. Thus, with polioviruses, at least two different antigenic sites on a single protein (VP1) have been shown to be capable of neutralizing the virus. Neutralization of infectivity may not of itself indicate destruction of virus. The interaction of virus and antibody occurs rapidly, and one antibody molecule can neutralize one virus particle. Presumably, the entire capsid of the virus is altered conformationally, or additional combining sites are obscured. Initial combination of antibody with the virus results in loss of infectivity, but release of antibody from the virus particle restores infectivity. When antibody–virus mixtures are prepared in vitro, there is almost always a small unneutralizable fraction of virus as a result of virus–antibody complexes, with subsequent dissociation of infectious virus from antibody. The presence of these complexes can be demonstrated by neutralization of all residual infectivity by anti-immunoglobulin serum.

There is general agreement that the most significant role of antibody is to prevent reinfections, although antibody may contribute to the termination of primary infection by limiting the spread of extracellular virus. These generalizations may also be applicable to virus replication in the cells of mucous membranes, although when the primary site of replication is the mucosa, IgA antibody becomes more important than IgG antibody.

Much of what we know about the host–virus interaction is a result of the unfortunate natural occurrence of immunodeficient patients. These patients, who are unable to synthesize antibodies normally, provide useful data through the observation of their interaction with viruses. Such patients generally receive replacement globulin, which provides them only with humoral IgG. Nevertheless, even before this human deficiency was recognized, many of the common viral infections did not appear to be a major problem for these patients. They were able to cope normally with vaccinia immunization and such illnesses as varicella and measles. They have not had recognized problems with such viruses as RSV, influenza, herpes simplex, or CMV. On the other hand, inability to limit replication of viruses in the central nervous system is now recognized to be a consequence of humoral antibody deficiency. Enteroviruses may replicate specifically in neuronal cells, as the polioviruses do, or generally in central nervous system tissue, as other enteroviruses seem to do. These unfortunate patients sustain persistent and chronic infections with these viruses in this sequestered location, and antibody therefore is important in the termination of these infections.

Virus–antibody complexes may circulate and be responsible for pathology within the host. Infants congenitally infected with CMV have been shown to have circulating complexes of virus and antibody. Such complexes have been shown to be deposited in the kidney. Similarly, patients with hepatitis B infection have also been shown to have circulating immune complexes and have had resultant pathology, such as arthritis and glomerulonephritis, as a result of these circulating complexes. In the experimental animal, perhaps the most extensively studied system is that of lymphocytic choriomeningitis (LCM). In utero and neonatal infection of mice with this virus produces an initially asymptomatic infection. Persistent lifetime viremia is established, and the blood contains virus–antibody complexes. Study of the animal has demonstrated that antibody to each of the polypeptides of the virus is present, yet virus is not cleared from the animal. In fact, the infectious virus that circulates is protected by being bound to antibody. These complexes may lead to glomerulonephritis, choroiditis, and vasculitis, with virus subsequently multiplying in the tissues where the complexes are deposited. It is not clear in this animal model why cytotoxic T cells are not detectable. However, some fascinating observations have been made, demonstrating that initially the viral antigens are expressed on the surface of infected cells as well as in the cytoplasm. Sequentially, the surface antigen diminishes, which may be related to the capping and subsequent stripping of the antigen from the cell by antibody. The viral nucleoprotein in the cytoplasm persists, but the cell is then no longer susceptible to specific lysis, as there are no expressed viral antigens on its surface. Simultaneously, in both in vivo and in vitro infection, defective interfering LCM virus is present. It has been demonstrated that defective virus can block infection with wild-type LCM virus. Details of these interactions are not clearly delineated, but in this model the host can be immunologically intact and yet the virus is able to escape immune surveillance and successfully establish persistent infection.

The concept that these persistently infected cells manage to survive despite intracellular virus replication is important. Some functions of the cells may be lost. For example, infection of the pituitary has resulted in absence or diminution of growth hormone because of the presence of LCM virus. There are no hallmarks of tissue destruction or inflammation in this setting, and by light microscopy the tissue morphology appears normal. With the use of appropriate antisera, however, viral nucleoprotein can be demonstrated within the cytoplasm of the cells. This infection would go undetected without knowledge of the infectious agent and use of specific antisera.

Complement

The alteration of the cell membrane by such enveloped viruses as measles or respiratory syncytial virus is apparent, with the resulting fusion of cell membranes producing multinucleated giant cells. Thus, with adsorption of the virus to the cell, the cell membrane is altered and viral proteins can be inserted in the host cell membrane without de novo synthesis of viral antigens. The presence of complement may initiate lysis of either cell free virus or virally infected cells in the presence of antibody. The expression of viral antigens on the cell surface has been more carefully described for enveloped viruses, but this is also true for reovirus (nonenveloped)-infected cells, which are replicating virus. For some of the other viruses, the expression of their proteins on the cell surface has yet to be delineated.

Thus, in the presence of complement, antibody may result in lysis of the infected cell and/or the virus particle.

Adverse Consequences of Altered Virus–Host Interactions

Atypical measles is an example of an adverse effect that may be precipitated by the combination of virus infection and host response. The atypical measles syndrome occurred in persons who had received inactivated measles virus vaccine and subsequently, some months to years later, contacted wild-type measles virus. It has been learned that the formalin-inactivated measles virus vaccine had lost the fusion, or F, protein. Therefore, the vaccine recipients never formed antibody to the F protein but did form antibody to the hemagglutinin and all other proteins of measles virus. Introduction of wild-type virus into such a host permitted its replication and its cell-to-cell spread. An anamnestic response in antibody occurred with recognition of the expressed hemagglutinin, as well as a cell-mediated immune response to the other viral proteins. Virus replication and cell-to-cell spread could not be eliminated until antibody to the fusion protein formed. It is also possible that infected cells expressing antigen could exhibit capping and liberate antigen–antibody complexes. Thus, a different result of infection with measles virus was produced by the antecedent immunization with an incomplete antigen. It is assumed that the unusual manifestations of illness, including the prominent pulmonary disease, a vasculitic type rash, arthritis, and hepatic dysfunction, are more likely a result of the host response to the virally infected cells than a result of virus replication alone, since these manifestations differ from those of natural measles.

Cell-mediated Immunity

Antigens are processed by macrophages and T cells. Clearly, macrophages phagocytize virus particles and, in some instances, destroy the virus. In other instances, the virus may replicate and destroy the macrophage. Sensitized T cells have been demonstrated to an array of human viruses. Such cells must recognize more than the viral antigens, that is, they must share the H2 antigens with the target cells. Interactions of various subgroups of T cells are complex. If the sensitized lymphocyte can eliminate the infected cell prior to the production of infectious progeny, the infection may be aborted.

Again, using measles virus as an example, this virus has been isolated from circulating lymphocytes by cocultivation of these cells in vitro. It is not yet known whether the inability to produce infectious virus in lymphocytes is related to absence of a host cell protease capable of cleaving the fusion protein. Experimentally, infection of lymphocytes can occur in vitro without any demonstrable expression of antigens on the cell surface or in the cytoplasm until a mitogenic stimulus is applied for the cells. The infected lymphocytes appeared morphologically normal but did not function normally as effectors in antibody-dependent cell-mediated cytotoxicity assays. Thus, this is another potential example of infected cells having an altered specific function but otherwise appearing normal. The phenomenon of loss of intradermal sensitivity to tuberculin protein has long been recognized with acute measles infection. Is this because effector cells are now infected with measles virus and have lost their effector functions?

Finally, it has been suggested that inflammation might nonspecifically destroy contiguous infected cells. All of the contributing features to the inflammatory response, such as the lymphokines, including migration inhibition factor and interferon, may contribute nonspecifically to the control of infection. Further details of the host response to infection are given in the following chapters.

FURTHER READING

Buchmeier MJ, Welsh RM, Dutkof J: The virology and immunobiology of LCMV infection. Adv Immun 30:275, 1980

Choppin PW, Richardson CD, Merz DC: The functions and inhibition of the membrane glycoproteins of paramyxoviruses and myxoviruses and the role of the measles virus M protein in subacute sclerosing panencephalitis. J Infect Dis 143:352, 1981

Fishaut M, Tubergen D, McIntosh K: Cellular response to respiratory viruses, with particular reference to children with disorders of cell-mediated immunity. J Pediatr 96:179, 1980

Fraenkel-Conrat H, Wagner RR (eds): Virus–host interactions. Comprehensive Virology. New York, Plenum, 1979, vol 15

Oldstone MBA: Immunopathology of persistent viral infections. Hosp Pract 17:61, 1982

Zinkernagel RM: Major transplantation antigens in host responses to infection. Hosp Pract 13:83, 1978

CHAPTER 65

Diagnostic Virology

Treatment of Specimens
Identification of Virus
Antigen Detection

Electron Microscopic Examination
Tissue Pathology

Specific diagnosis of virus infection is useful to the practicing physician by delineating an etiologic agent for the observed illness. This information may alter unnecessary antimicrobial therapy, allow earlier discharge from the hospital, provide a more accurate prognosis, or generate informative research data. Additionally, some diagnoses may prevent nosocomial transmission or alert the physician to possible transmission in the community. New viruses or unique manifestations of old viruses may be identified. Finally, a specific diagnosis may allow appropriate use of prophylactic agents or vaccines.

The time required for viral diagnosis by standard inoculation of tissue culture, embryonated egg, or animals often exceeds the duration of the illness in question. Increasingly rapid diagnosis will provide far more information for the responsible physician. Antigen detection rather than isolation of virus will ultimately replace the classic, more cumbersome isolation techniques. Development of specific antiviral drugs will produce a need for more rapid specific diagnosis if the drug is to be administered effectively.

Viral diagnostic facilities vary greatly, and the clinician must determine what specific resources are available in his or her hospital or community. The facilities of the municipal or state health departments are accessible without direct patient costs, and through these laboratories referral of clinical materials can be made to the Centers for Disease Control in Atlanta, Georgia. In addition, laboratories within community hospitals and associated with medical schools are accessible in some areas. More recently, commercial laboratories have incorporated some viral diagnostic skills into their available procedures. Generally, these utilize serologic diagnosis or consist of commercially available kits containing reagents for detection of a few specific antigens, e.g., HBsAg or rotavirus.

There are viral illnesses where the accurate diagnosis is of extreme importance to the patient, the patient's family, or his or her community. The correct diagnosis of an initial case of paralytic poliomyelitis can allow effective immunization to be instituted in a susceptible community, thereby preventing a sizable outbreak of disease. This requires isolation, identification, and typing of the virus in the laboratory. When rubella infection is suspected in the first trimester of pregnancy, the currently available accurate serologic means of diagnosis are mandatory to confirm the diagnosis prior to any consideration of pregnancy interruption. The serologic assessment of rubella immunity should become an integral part of the female premarital examination and is required by law in many states, since attenuated rubella vaccine offers a means of establishing immunity in susceptible women prior to pregnancy. Introduction of varicella into the hospital setting of high-risk patients is a frequent problem. Ascertainment of the susceptibility of exposed persons and protection of those at high risk with prophylactic zoster immune globulin requires laboratory definition of the serologic status of many persons.

Finally, within the framework of any program responsible for training physicians, it is important that these physicians become familiar with viral illnesses, institute appropriate control measures when necessary, and provide appropriate prophylaxis for others when this is available. Laboratory documentation of viral illness confirming the clinical impression or, in some cases, providing an unsuspected diagnosis will improve the judgment of these physicians providing patient care in the years to come.

Treatment of Specimens

A basic understanding of the laboratory procedures is necessary in order to utilize optimally the available virus laboratory facilities. Whenever a specimen is submitted, adequate information must accompany the request. This information is essential, as specimens are often processed in the laboratory according to their source and the physician's provisional diagnosis. Commonly needed information includes patient identification and age, the clinical diagnosis, the referring physician, the source of the specimens, and the date they were obtained. In addition, it is often helpful to know whether the patient has recently received viral vaccines. The responsible laboratory processing these materials should be consulted whenever questions arise, as they will be able to specify the optimal handling of the materials.

Patient materials are cultured for the presence of viruses to detect a specific etiologic agent. Tissue culture systems are usually more readily available, less cumbersome, and less expensive than embryonated eggs or animals. The cells utilized are dictated in part by the facilities available in the particular laboratory and by the suspected type of infection. It is usually advisable to collect the specimens as early as possible in the course of a clinical illness. Since viral agents are not affected by antimicrobial agents, cultures may be obtained even if therapy for potential bacterial pathogens has previously been initiated. Such mate-

TABLE 65-1. SUGGESTED SPECIMENS TO BE SUBMITTED FOR VIRAL DIAGNOSIS

Clinical Illness	Suspected Virus	Clinical Specimens	Postmortem Materials	Notes
Respiratory Pharyngitis, croup, bronchiolitis, pneumonia	Influenza, parainfluenza, respiratory syncytial virus, reovirus, adenovirus	N-P secretions	Lung, trachea, blood Stool, eye	Immunofluorescence requires intact cells in secretions, immunocompromised patients may have virus in tissue other than respiratory tract
Gastrointestinal Diarrhea	Rotavirus, adenovirus, Norwalk-like, miscellaneous others	Stool	Intestinal contents	Examination by EM or testing for antigen more productive than attempted cell culture isolation, if parotitis present, culture throat and urine
CNS Meningitis/encephalitis	Enteroviruses, mumps, herpes simplex virus, togaviruses	CSF, throat, stool Brain tissue, blood, brain	Brain, liver, intestinal contents Brain tissue, other viscera	No routine serology available for enteroviruses except polioviruses; many enteroviral syndromes are without symptoms Togaviruses present in CNS tissue and blood; serologic diagnosis more practical
Gingivostomatitis Keratoconjunctivitis Genital vesicular lesions Vesicular lesions of lip Eczema vaccinatum Chickenpox Zoster	Herpes simplex Varicella	Vesicle fluid, throat swab or mouth wash, vaginal swab		
Mononucleosis	EBV CMV	Buffy coat, urine, throat swab	Lung, kidney, salivary gland	Serologic assessment of antibodies to EBV include IgM, IgG, anti-VCA, EBNA, EA
Hepatitis	HBV HAV NANB	Serum, liver, stool	Liver, blood	Serologic assessment allows detection of HBV antigens and antibodies (HBsAg, HBeAg, αHBsAg, αHBeAg, αHBeAg); IgM anti-HAV provides estimate of recent infection
Measles	Measles	Respiratory secretions, blood, urine	Brain, lung	
SSPE		CSF	Brain, lung	CSF antibodies for measles virus
Intrauterine or perinatal infection	CMV Rubella Enteroviruses Herpes simplex virus	Buffy coat, urine Throat, urine, blood, Throat, stool, blood, CSF Throat, stool, blood, CSF	Liver, kidney, intestinal contents	

rials are obtained from the same site(s) as specimens for bacterial cultures (Table 65-1). Respiratory tract secretions, stool, urine, cerebrospinal fluid, pleural fluid, pericardial fluid, blood, bone marrow, vesicle fluid, and skin scrapings may be utilized in attempted virus isolations. Tissue specimens from biopsies or postmortem examinations are suitable for attempted virus isolation if fresh or frozen but not in formalin.

The principles adhered to in the collection of bacterial specimens are applicable to specimens submitted for virus culture. Containers must be clean and sterile and handled accordingly. The more stable viral agents include vaccinia, the enteroviruses, and the adenoviruses. As it is rarely possible to know the particular etiologic agent until the virus is isolated, all specimens should be treated as though the agent is potentially labile. Many viruses tend to be unstable at an acid pH or at temperatures above 4C. All patient specimens should be transported as rapidly as possible to the laboratory. If a delay of even 1 hour occurs, the material should be refrigerated at 4C or transported on ice. Freezing materials at −20C for brief periods of time may inactivate many of the more labile enveloped viruses, and keeping specimens at 4C is preferred. Delays of several days prior to inoculation usually necessitate freezing of the specimens, preferably at −70C.

In addition to their heat and acid lability, many viruses do not withstand drying. Swabs of mucosal surfaces, skin scrapings, and tissues are placed in a transport medium, which provides stability with a protein source such as 0.5 percent gelatin or antibody-free serum and a buffered salt solution essential to preserve infective virus. Several such media are commercially available or easily made. Veal infusion broth or Stuart's transport medium is employed by bacteriology laboratories. A 1 percent solution of skimmed milk in distilled water, Hanks' balanced salt solution with 10 percent fetal calf serum, or 0.5 percent bovine albumin or minimal essential medium and 40 percent sucrose have been used successfully. All of these media have antibiotics (e.g., penicillin, 50 units/ml, amphotericin, 1 μg/ml; streptomycin, 5 μg/ml) added to prevent overgrowth by resident host bacterial flora and to decrease exogenous contamination of the specimen.

Identification of Virus

Isolation of an agent from a patient, especially from a mucosal surface, does not alway prove an etiologic relationship to his illness. For this reason, patients being thoroughly evaluated for viral disease should have both acute and convalescent sera drawn for antibody determinations. The acute specimen of blood is obtained from the patient as early as possible in the illness, and the convalescent specimen is drawn 2 to 3 weeks later. The serum is separated and kept at refrigerator or freezer temperature. Whole blood should not be frozen, as the erythrocytes lyse and the subsequent assay of hemolyzed serum is difficult and inaccurate.

The traditional techniques to assay viral antibodies include complement fixation (CF), hemagglutination inhibition (HI), and neutralization of the viral cytopathic effect. The commonly accepted serologic confirmation of acute infection is a four-fold or greater rise in antibody titer when serial two-fold dilutions of serum are employed in these traditional test procedures. Complement fixation offers a relatively rapid, efficient way of screening sera for a number of viral antibodies. It is particularly useful in determining that recent infection has occurred. As a general rule, the antibodies thus detected are group-specific and not type-specific. For example, adenovirus antibody can be distinguished from poliovirus antibody, but antibody to type 1 adenovirus cannot be distinguished from any other adenovirus CF antibody. Complement-fixation antibodies usually decline more rapidly than do specific neutralizing antibodies, so that their absence does not necessarily denote susceptibility.

Hemagglutination-inhibition studies are limited by the fact that not all viruses possess a hemagglutinin. Agents with recognized hemagglutinins are influenza, mumps, measles, the parainfluenzae, rubella, vaccinia, variola, the arboviruses, reoviruses, and some of the adenoviruses and echoviruses. The limitations of the test itself are primarily technical, as it is often difficult to eliminate nonspecific inhibitors of hemagglutination from the serum being tested. The HI antibodies are type-specific, and the method is quite sensitive, correlating well with the antibodies observed in the neutralization test.

Neutralizing antibodies are highly specific for virus type, appear early in the course of illness, and persist for a long time thereafter. The determination of neutralizing antibody requires the growth of an agent in a cell culture or an animal system. The serum in question is incubated with the suspected etiologic agent and is then placed into the appropriate culture system. The results are commonly unavailable for at least a week, as the appearance of the cytopathic effects in the cell culture or illness in the animal determines the length of time necessary for the performance of the test.

Technologic innovations are changing traditional serologic methods. Enzyme-linked immunosorbent assay (ELISA) tests, radioimmunoassays (RIAs), and quantitative immunofluorescence determinations are reporting results with numerical values that have no obvious correlation with traditional values that are reported as the reciprocal of two-fold dilutions. Each test brings with it a new set of values, each of which must be compared to acceptable standards. Each method demands individual interpretation. Currently, such serologic procedures are coming into general use. Many laboratories are dependent upon the manufacturers' standards and are not aware of the limitations of the pre-licensure testing or of the variability of the procedure itself and cannot relate the information obtained to older methods. Thus, there are many predictable changes occurring in laboratory serology, and there is already attendant confusion concerning interpretation of the data. The utilization of these methods will increase the availability of tests, and ultimately will increase the sensitivity and efficiency of testing.

Regardless of the method employed, serologic demonstration of an antibody response may provide evidence of infection in the absence of virus isolation from cultures. Unfortunately, documentation of infection is delayed because of the need to assess convalescent serum for antibodies and is limited by the multiplicity of viruses infecting humans. Serology may also be used to assay the immune status of an individual following previous infection or vaccination.

Antigen Detection

In recent years the development and application of new technology has begun to be applied to detection of viral antigen in clinical materials. The site of virus multiplication, the complex composition of patient specimens, and the differences in amount of virus present influence the usefulness of these techniques. Far better serologic reagents, such as monoclonal antibodies, have the potential of increasing the sensitivity and specificity of antigen detection. Methods, such as immunoprecipitation and countercurrent immunoelectrophoresis, have been utilized for specific viruses, such as varicella and hepatitis. These tests have provided more rapid recognition of the viral antigens than cell culture isolation of virus and have made tests available on a larger scale but have been superseded by more sensitive methods such as ELISAs or RIAs. No other human virus has been as easily demonstrable in serum as HBV and its antigens because of the quantity that is present.

ELISAs are being increasingly applied to the detection of viral antigens. Reagents are less costly than radiolabeled materials, and expensive equipment is not a necessity. ELISAs are designed with a specific enzyme linked to antibody. The enzyme (e.g., alkaline phosphatase) can produce a color change when provided with the right substrate (e.g., p-nitrophenyl phosphate). The visible color change is appreciated with the eye and can be easily quantitated by a spectrophotometer. In practice, antigen is adsorbed to a surface (microtiter plate, metal beads, and so on), and the test serum is added. Any specific antibody will attach to antigen. Enzyme-labeled antiglobulin is then added and attaches to the bound antigen–antibody complex. After exhaustive washing, addition of the substrate results in the visible color change. If no specific antibody is present, the enzyme-labeled antiglobulin will not adhere to the plate, and no hydrolysis of substrate can occur.

Immunofluorescence has been utilized for rapid viral diagnosis of respiratory tract infections, vesicular exanthems, and examination of tissues. Respiratory epithelial cells obtained from nasal secretions may be examined by indirect immunofluorescence for the presence of many of the common respiratory viruses. A large successful experience with diagnosis of respiratory syncytial virus, parainfluenza, influenza, and adenovirus infections has been accumulated in European laboratories. Meticulous preparation of materials is required, and antisera, which are not generally available, must be both sensitive and specific. The basic techniques can be standardized but still require an experienced person to interpret the results. The method offers possible etiologic diagnosis within one working day, which is a distinct advantage to the clinician. Standardization of reagents and increased availability will render immunofluorescence a more broadly available method.

Experimental detection of viral nucleic acids and proteins has reached a level of sensitivity and practicality that has begun to permit investigators to apply the technology to clinical materials. Immobilization of the DNA, RNA, or protein on nitrocellulose paper precedes the use of an enzyme or radiolabeled probe that specifically binds to the nucleic acid or protein and is detected by the reaction product or autoradiography. Epstein–Barr virus and cytomegalovirus DNA have been reproducibly demonstrated in peripheral lymphocytes by placing lymphocytes on nitrocellulose paper and performing in situ hybridization with appropriately labeled probes.

The sensitivity and specificity make this approach to detection of a known virus attractive. The obvious disadvantage is the multiplicity of agents infecting humans and the need for multiple specific probes. Clearly, the physician's knowledge of the clinical illness and epidemiology must be utilized when a patient is having materials submitted for analysis. Ideally, definition of group-reactive antigen or nucleoprotein would occur initially, so that, for example, definition of an influenza virus versus a parainfluenza virus would precede subsequent, more specific identification. In many instances, the rapid detection of an etiologic virus without type specificity would provide extremely useful information. To identify influenza A viral antigens in secretions provides the physician with the opportunity to use antiviral therapy for the patient, prophylaxis of contacts at risk, and administration of vaccine in the setting of an outbreak among high-risk persons. Knowing the specific serotype would not add appreciably to the initial decisions.

Electron Microscopic Examination

The electron microscope has been utilized recently to assist in viral diagnosis. Negative staining by heavy metal salts allows electron microscope visualization and morphologic characterization of the virus group. Examination of materials from vesicular lesions by experienced personnel can readily distinguish a viral agent of the herpes group from one of the poxvirus group. Identification of the etiologic agent of hepatitis B was accomplished by electron microscope visualization of the virus. Another example of the use of the electron microscope has been the detection of previously unrecognizable virus particles in stool specimens. Ten percent suspensions of stool examined directly will reveal the presence of virus at a concentration of 10^5 particles/g of stool. These observations have led to the identification of agents, such as rotaviruses and Norwalk-like agents, now considered causative in acute gastrointestinal disease. The human rotaviruses are not yet easily grown in any cell culture system, so antigen detection by electron microscopy or ELISA is the most practical diagnostic method at the present time. Other refinements of methodology have broadened the diagnostic applications of the electron microscope. The use of specific antiserum causes adherence of virions to antibody and results in clumps of virus particles. This clumping permits detection of smaller numbers of virus particles. Such immune electron microscopy provides direct visualization of the antigen–antibody reaction. Although such work is primarily investigational at the present time, new insights into gastrointestinal disease have come from work with the electron microscope. The future will see electron microscope diagnosis supplanted by more practical methods.

Tissue Pathology

Any laboratory where tissue pathology is done can examine tissues by microscopy, which may facilitate a viral diagnosis. Histologic examination of pertinent clinical specimens with hematoxylin and eosin or Giemsa stains may disclose pathognomonic features of virus infection, such as inclusion bodies of multinucleate giant cells. If specific antisera are available, immunofluorescence may also be used.

Measles infection, either that of natural disease or of attenuated virus, may produce multinucleate giant cells in the urine, as well as intranuclear and intracytoplasmic inclusions. Such findings have also been described in the sputum or throat scrapings of patients with giant cell pneumonia or in malnourished patients with measles. Rubella, mumps, Coxsackie A virus, and variola have been reported to produce eosinophilic cytoplasmic inclusions in the cells of the urine sediment. Agents, such as cytomegalovirus and herpes simplex, have been observed to produce intranuclear inclusions in cells in the urine. Such positive findings in the urine, sputum, or tissues will point toward the probability of virus infection, but without culturing the virus, the laboratory cannot identify the specific agent involved. The absence of such findings does not exclude viral disease, and it is very important to recognize

that many agents (for example, the enteroviruses) do not cause specific hallmarks of virus infection.

Cutaneous lesions, such as vesicles, macules, or pustules, can also be examined for the presence of multinucleate giant cells and inclusion-bearing cells. Materials obtained from such lesions can be examined easily and quickly. The vesicle is sponged with alcohol (if no virus cultures are being taken from the lesion), and the roof of the vesicle is reflected. The fluid is blotted, the base of the lesion is scraped, avoiding gross bleeding when possible, and the cellular material is spread on a glass slide and air dried. It can be fixed with methyl alcohol and stained with Wright or Giemsa stain. Giant cells are readily apparent, and inclusions may be seen with carefully made preparations. The presence of such giant cells will exclude vaccinia and variola from the differential diagnosis but will not distinguish between herpes simplex and varicella-zoster. Similarly, the diagnosis of herpes simplex infection of the female genital tract can be approached by obtaining a swab of the cervix and examining by Papanicolaou stain for giant cells and/or intranuclear inclusions. Again, immunofluorescence may provide specific etiologic information in some of these situations.

The information provided in this brief discussion can be applied in the diagnosis of the virus infections discussed in the subsequent clinical virology chapters.

FURTHER READING

Gardner PS, McQuillin J: Rapid Virus Diagnosis. Application of Immunofluorescence. London, Butterworth, 1974

Hsiung GD: Diagnostic Virology, 3rd ed. New Haven, Yale University Press, 1982

Lennette EH, Schmidt NJ: Diagnostic Procedures for Viral, Rickettsial and Chlamydial infections, 5th ed. Washington, D.C., Am Public Health Assoc, 1979

McIntosh K, Wilfert C, Chernesky M, et al.: Summary of a workshop on new and useful techniques in rapid viral diagnosis. J Infect Dis 142:793, 1980

Rapid Laboratory Techniques for the Diagnosis of Viral Infections. Report of a WHO scientific group. Tech Rep Series 661. Geneva, WHO, 1981

Yolken RH: Enzyme immunoassays for the detection of infectious antigens in body fluids: Current limitations and future prospects. Rev Infect Dis 4:35, 1982

CHAPTER 66

Poxviruses

The family Poxviridae includes 20 or more mammalian viruses in five genera, a group of bird viruses (genus *Avipoxvirus*), and nearly two dozen insect species (*Entomopoxvirus*). They are large, brick-shaped, enveloped, containing double-stranded DNA, two lateral bodies, and a central core or nucleoid. They possess a common internal nucleoprotein antigen and share the clinical manifestation of vesicular skin lesions in the susceptible host. Until recently, smallpox (variola), an *Orthopoxvirus*, received most attention because of its responsibility for epidemic disease throughout the world for many centuries. With smallpox eradication in 1977, poxvirus research and surveillance programs have been emphasized to investigate the relationships of other mammalian poxviruses and to explore the possiblity of mutations with reemergence of variola.

Smallpox (Variola)

Smallpox is a member of the genus *Orthopoxvirus*, which includes also vaccinia, monkeypox, buffalopox, camelpox, cowpox, rabbitpox, and mousepox (ectromelia). There is only one serotype, which has apparently persisted for many years. The disease was recognized as a major health problem for at least 3000 years. Early epidemics may have been confined to the Orient, Middle East, and Africa until spread into Europe by the Arab expeditions of the sixth century. Early 16th century Spanish conquistadors introduced smallpox to the Western Hemisphere with transmission to Aztec natives. It spread throughout the Americas. Early American colonists were familiar with the illness. A noteworthy epidemic in 1721 involved 5889 of Boston's 12,000 inhabitants. The publication in 1798 by Edward Jenner of his work with cowpox heralded the beginning of a new era in the prevention of smallpox. Despite widespread acceptance of vaccination and its use with increasing frequency throughout Europe and the United States, another century passed before major epidemics were eliminated in these nations. By 1967 there were still 33 countries, mainly in Asia and Africa, where smallpox remained endemic, claiming millions of victims annually. The World Health Organization (WHO) program for eradication was initiated that year and culminated in success, with the last naturally occurring case detected in Somalia in October 1977.

Epidemiology

Humans were the only known natural host of smallpox virus, but nonhuman primates have been successfully infected in the laboratory. Transmission of virus occurred directly from person to person. The major source of infectious material was respiratory tract secretions with droplet spread. The virus was able to persist for lengthy periods in the protein and cellular debris of the sloughed scabs, but they were of minor importance in transmission. Susceptible members of a patient's household were the most likely to become infected. Another group with a high rate of spread were unimmunized hospital personnel caring for a patient with the illness. Although respiratory tract shedding of small amounts of virus preceded the appearance of the exanthem (Fig. 66-1), the patient was most infectious with the onset of rash for 7 to 14 days.

Widespread use of smallpox vaccine was responsible for striking alterations in the geographic patterns of the disease. The last recorded case in the United States was an importation in 1949, but there had been 48,000 patients in 1930. The WHO eradication program began in 1967, when 62 nations still harbored the disease. By 1974 only 8 countries reported cases, and it was endemic in only 4 (India, Pakistan, Bangladesh, and Ethiopia). The last-known naturally acquired case was recorded in the town of Merka, Somalia, on 26 October 1977. Because there is no evidence that smallpox virus persists in the human host and there is no animal reservoir, it is likely that deep-freezes in four laboratories (U.S., U.S.S.R., South Africa, and England) now hold the only variola virus remaining in the world.

Clinical Illness

Three main forms of disease were seen in unvaccinated patients. Classic smallpox (variola major) was a severe, biphasic, febrile illness with a mortality rate of 20 to 50 percent. Variola minor (alastrim) had <1 percent mortality, while an intermediate form of smallpox ranged from 10 to 15 percent mortality rate. An initial prodromal phase was characterized by abrupt onset of high fever, chills, headache, backache, and malaise for 2 to 4 days. The patient was toxic and obtunded or delirious. This was followed by a sudden, dramatic improvement as temperature fell and respiratory symptoms developed. At this time rash began, often starting on the face and arms but progressing to involve the trunk and lower extremities. Simultaneously, an enanthem appeared with painful ulcers of the buccal, pha-

ryngeal, and bronchial mucosa. The rash started as macules, which within a few hours had become papules. Over the next several days the lesions vesiculated and then, between the seventh and ninth days evolved into frank pustules. Fever rose again, and the patient became severely ill with constitutional symptoms as well as intensely painful local lesions. By the tenth day the pustules started to rupture and then to crust, with sloughing of the scabs over the next 5 to 10 days, leaving scars in the areas of previous involvement.

There was considerable individual variation in the clinical picture, and the rash ranged from discrete to confluent lesions. A small number of patients underwent severe hemorrhagic smallpox, with a rapid, fulminant course, including hemorrhages of the skin and mucous membranes in the absence of the usual eruption. This form of the disease was almost always fatal. In contrast, a very mild, modified illness occurred in partially immune individuals. Because patients did not feel ill and remained ambulatory, they were a source of virus spread throughout the community.

Variola minor had a prolonged incubation period of 15 to 20 days, a mild prodromal phase, a discrete, rapidly evolving eruption, moderate constitutional symptoms, and a mortality rate of less than 1 percent.

Complications included pneumonia, keratitis with corneal damage, encephalitis, and peripheral neuritis. Most common were secondary bacterial infections of the skin and respiratory tract, with cellulitis, furunculosis, impetigo, and pneumonia. Bacteremia resulted in metastatic foci, including pyogenic arthritis and osteomyelitis. No specific effective antiviral chemotherapy was available, but the oral administration of methisazone (methylisatin-beta-thiosemicarbazone) very early in the incubation period diminished the attack rate in susceptible contacts. Supportive therapy with fluids, electrolyte replacement, nutritional supplementation, and antibiotic treatment of secondary bacterial complications was of importance.

Pathology and Pathogenesis

After its entry via the upper respiratory tract, smallpox virus spread to regional lymphatics and nodes, from which a primary viremia then emanated. Dissemination was widespread, with involvement of all tissues, especially epithelial cells. Lesions began in the areas surrounding infected endothelial cells of small vessels. The skin lesions evolved as foci of mononuclear cell infiltration and edema, progressing to swollen epithelial cells within which acidophilic inclusions (Guarnieri bodies) appeared in the cytoplasm. Epithelial cell necrosis and accumulations of interstitial fluid resulted in vesicle formation. A similar evolution took place in mucous membranes, where rapid unroofing of the vesicle surface initiated ulcerative lesions. Focal necrotic lesions also developed in spleen, testis, liver, lung, and other organs. Virus released from these multiple sites of replication produced a secondary viremia coinciding with the eruptive phase of the disease.

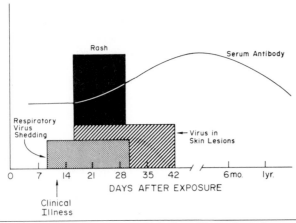

Figure 66-1. The course of infection with smallpox virus.

Those patients at the extremes of life, the young infant and the aged, had more severe disease and were more likely to die as a result. Prognosis correlated well with the severity of the prodromal phase and the type of rash. Striking toxemia and prostration in the first days of illness heralded a poor outcome. The hemorrhagic rash presaged an 80 percent or higher mortality; confluent exanthem, 50 percent; discrete lesions, less than 10 percent.

Immunity

An attack of smallpox confers active immunity of lifetime duration. Prior to the discovery of vaccination, variolation was widely practiced as a means of inducing active immunity. A mild form of illness ensued, which in most instances produced enduring immunity. However, occasional recipients developed severe smallpox from variolation, and there were deaths in 1 to 2 percent.

The introduction of vaccination in 1798 began one of the fascinating sagas of medical history. With abundant controversy, vaccination gradually achieved a widespread role as the principal prophylactic method for smallpox prevention. It was not until the WHO program began in 1967 that the final stages in disease eradication provided the culmination in 1977 of Jenner's observations 181 years earlier.

Following natural infection or immunization, virus-specific antibodies appear in the patient's serum. These will fix complement, inhibit hemagglutination, neutralize viral infectivity, and/or inhibit agar gel precipitation (AGP).

Diagnosis

When smallpox was a common disease, diagnosis was often established on the basis of (1) a clinically compatible illness, (2) a known recent exposure in an endemic area, and (3) the absence of a history of vaccination or a visible vaccination scar. When the infection had become a rarity, laboratory confirmation of any suspected case was absolutely essential because of the public health implications. A number of available techniques (Table 66-1) offer rapid and reliable laboratory diagnosis.* In actual practice those laboratories most familiar with smallpox virus employed four tests: electron microscopy (EM), AGP, and viral propagation in cell cultures and on the chorioallantoic membrane (CAM). Electron microscopic identification of a poxvirus may be accomplished within an hour; AGP requires only 1 to 4 hours. Rapid diagnosis is, therefore, possible with the skill of those working regularly with various poxviruses.

Because chickenpox (varicella), disseminated herpes hominis, vaccinia, and some noninfectious exanthemata may mimic smallpox, it is not surprising that many false

---

*Any suspicious patient or specimen material should be discussed promptly with the Centers for Disease Control, Atlanta, Georgia, telephone number (404) 633-3311, Biohazards Control Officer.

TABLE 66-1. LABORATORY METHODS FOR POXVIRUS DIAGNOSIS

A. Direct visualization of poxvirus
  1. Electron microscopy (EM)
B. Visualization of elementary bodies
  1. Light microscopy of stained smears
C. Identification of viral antigens
  1. Agar gel precipitation (AGP)
  2. Fluorescent antibody (FA)
  3. Complement fixation (CF)
D. Isolation of virus
  1. In ovo on chorioallantoic membrane (CAM)
  2. In cell cultures
    a. Primary rhesus kidney cells
    b. Infant foreskin fibroblasts (FS-32)
    c. Vero cells
E. Serologic confirmation
  1. Specific antibody rise (four-fold) in paired sera
    a. Complement fixation (CF)
    b. Virus neutralization (VN)
    c. Agar gel precipitation (AGP)
    d. Hemagglutination inhibition (HI)

---

alarms have been avoided or aborted by use of the techniques listed. Monkeypox, although indistinguishable by EM and AGP, produces distinct CAM lesions that can be identified by pock morphology. Vaccinia and variola are similarly distinguished by varying pock characteristics and growth in cell culture systems. In the 4 years from January 1978 to March 1982, 176 smallpox suspects have been investigated in 60 different countries. Aside from 2 laboratory-acquired smallpox cases in 1978 in Birmingham, England, all have been disproven.

Prevention

The appropriate use of vaccination has prevented smallpox. Vaccinia virus confers protective immunity against variola for a lengthy period of time, albeit less enduring than the active immunization produced by the disease itself. Past field experience with studies of vaccine prophylaxis under conditions of natural exposure demonstrated a 99 percent efficacy for 1 year, 95 percent for 3 years, 85 percent for 10 years, and 50 percent for 20 years. Previously vaccinated individuals, whose immunity had waned, tended to have a much milder illness and a mortality rate far lower than that of unvaccinated patients. This ameliorating effect persisted as long as 20 years.

Vaccinia

Jenner introduced vaccination to the public with a paper published in 1798 based on his studies begun in 1796. It is not possible to establish with certainty today what virus he actually employed or what has happened to it over the ensuing nearly two centuries of multiple human and animal passages. Jenner believed his strain was horsepox (grease) which had been passed in cows before transfer to humans. Laboratory markers show that current vaccinia strains differ from cowpox as well as from other orthopoxviruses. Some investigators suggest that it is a recombinant of cowpox and smallpox.

Until 1971 it was recommended that all infants in the United States receive smallpox vaccination as a routine component of the childhood immunization schedule. The absence of endemic smallpox, the absence of disease importations since 1949, and the recognition of certain common and other rare complications of vaccination led public health and pediatric authorities to discontinue this practice after 1971–1972. Until that time nearly four million primary vaccinations and eight million revaccinations had been performed annually. With careful surveillance of vaccine complications, it was possible to exclude certain candidates who were at increased risk of these problems. Patients with exfoliative dermatitis (especially eczema) were likely to develop diffuse vaccinial lesions over the involved skin. This followed accidental vaccination of patients themselves or exposure to household or other close contacts who had an active vaccinial lesion. Individuals with immunodeficiency or those being treated with immune-suppressive therapy, particularly those with compromised T cell function, were more apt to develop progressive vaccinia. It was not possible to predict the rare patient who might develop encephalitis in the 10 to 14 days after inoculation, but there seemed to be an increased risk correlated with increasing age at the time of primary vaccination. Autoinoculation was relatively common among infants who scratched the site of their primary lesion and then introduced the virus elsewhere on their uninvolved skin. This was not a source of any serious problems unless the secondary site was the eye, where keratitis was a possible outcome. Local bacterial infections with pyogenic skin flora, especially staphylococci or streptococci, occasionally precipitated a cellulitis at the lesion site.

Molluscum Contagiosum

This infection of humans is caused by a poxvirus that does not replicate in any current in vitro cell culture systems but can be visualized by electron microscopy of affected cells from papillomatous skin lesions. It may be related to another poxvirus, the Yaba monkey tumor agent. The lesions are easily distinguished by their umbilicated appearance produced by central degeneration and by the formation of satellite nodules at the periphery of the parent lesion. They are 2 to 5 mm long, pink, white, or yellow, shiny, and rounded. On pathologic examination, the epidermis shows ballooning degeneration, acanthosis, hyperplasia, and the presence of large acidophilic inclusions

(molluscum bodies) in superficial epithelial cells. The incubation period varies widely from 14 to 50 days.

In children the lesions are commonly on the face, trunk, or limbs. Adults often acquire their infection by sexual transmission and develop genital lesions. Most patients have fewer than 20 individual papules, but some immunocompromised individuals may have hundreds of lesions. Although spontaneous resolution usually occurs, this may require 6 to 36 months, so that surgical removal, cryotherapy, or other physical methods have been employed to accelerate clearing.

Orf

This old Saxon term refers to the infection of humans with the parapoxvirus of contagious pustular dermatitis (scabby mouth) of sheep and goats. In lambs the disease is characterized by the development of watery papillomatous lesions on the cornea, lips, and mouth, which usually resolve in 4 to 6 weeks without consequence. In humans, vesicles usually develop at the site of abrasions on the hand or face, which may evolve into hyperplastic nodular masses accompanied by regional lymphadenopathy. Ordinarily, lesions require no treatment. Since parapoxviruses and orthopoxviruses are only very distantly related immunologically, vaccination provides no protection. Orf is an occupational disease associated with handling of sheep, particularly in shearers.

Milker's Nodules (Pseudocowpox)

This condition is also caused by a *Parapoxvirus* (paravaccinia). It is a cutaneous disease of the udders of cow that may be transmitted to humans by contact. Lesions appear 1 to 2 weeks after contact and are similar to those produced by contagious pustular dermatitis. Recovery is complete in 4 to 8 weeks. Reinfection may occur, indicating that immunity does not last long.

Cowpox

Cowpox, an *Orthopoxvirus*, produces a vesicular eruption on the udders of cattle from which it may be transmitted to humans. Usually the illness in humans is a limited vesicular eruption of the skin, although rarely the disease develops into a widespread eruption with systemic symptoms.

Currently, reports of human cowpox are exceedingly rare, and doubts exist whether the cow is indeed the true reservoir host of this virus. Spontaneous cowpox virus infections have been detected in humans with no prior contact with cows and in some large felines (tigers) in Russian, Swedish, and English zoos. A similar, closely related virus was isolated from gerbils in the Turkmen Republic of U.S.S.R., suggesting that rodents in western Europe may be the true natural reservoir of cowpox.

Monkeypox

From 1970 through 1982, more than 70 laboratory-confirmed cases of human infection with this *Orthopoxvirus* have been reported, mainly in the tropical rain forest areas of West and Central Africa. The responsible agent had initially been recovered in the 1950s and 1960s from monkeys who became ill in captive laboratory colonies. Human illness closely resembles smallpox and, therefore, is critically important in surveillance for any recrudescence of variola infections. Monkeys are suspected to be the reservoir of this apparent zoonosis, but the epidemiology remains unclear. Antibodies specific to monkeypox virus have been detected in sera of wild monkeys trapped in West Africa. One human case has been reported in an infant bitten 12 days earlier by a chimpanzee.

Of 63 cases reported by WHO field teams, most were children under 15 years of age with no history of prior smallpox vaccination. There were 8 deaths. Although secondary spread from human to human appears unusual, 6 of 139 unvaccinated close contacts of the initial patients developed rash 7 to 20 days after exposure.

FURTHER READING

Arita I, Cromyko A: Surveillance of orthopoxvirus infections, and associated research, in the period after smallpox eradication. Bull WHO 60:367, 1982

Benenson AS: Smallpox. In Evans AS (ed): Viral Infections of Humans, 2nd ed. New York, Plenum, 1982, chap 22, pp 541–568

Dales S, Pogo BGT: Biology of poxviruses. In Kingsbury DW, Zur Hausen D (eds): Virology Monographs. New York, Springer-Verlag, 1981, vol 18

Fenner F: A successful eradication campaign. Global eradication of smallpox. Rev Infect Dis 4:916, 1982

Mutombo MW, Arita I, Jezek Z: Human monkeypox transmitted by a chimpanzee in a tropical rain-forest area of Zaire. Lancet 1:735, 1983

Ricketts TF, Byles JB: The diagnosis of smallpox (Reprinted from the 1908 London edition). US Dept of HEW, Bureau of Disease, Prevention and Environmental Control, Division of Foreign Quarantine, 1966, vols I and II

World Health Organization: Declaration of global eradication of smallpox. Weekly Epidemiol Rev 55:148, 1980

CHAPTER 67

Herpesviruses

The group of human herpesviruses includes herpes simplex virus (HSV) types 1 and 2, varicella-zoster virus (VZV), cytomegalovirus (CMV), and Epstein–Barr virus (EBV). These large, enveloped DNA viruses are responsible for a broad spectrum of clinical illness and share the unique characteristic of becoming latent in the body after a primary infection and having the potential for subsequent reactivation after varying periods of time. This persistence in the body has raised concern about potential oncogenicity, and, indeed, HSV and EBV have been associated with human cancer, and CMV has been shown to have transforming properties. Most people have antibodies to these viruses as adults, and although most of the clinical diseases are only temporarily annoying, life-threatening infections can occur.

Herpes Simplex Virus

Infections caused by herpes simplex virus have troubled us throughout history. Diseases resembling herpes were described in biblical times with the term "herpes" (from the Greek work, $\epsilon\rho\pi\epsilon\iota\nu$, to creep) first appearing around 100 AD in descriptions of herpes labialis. Genital herpes infection was described by the physician to the king of France in 1736.

In 1912 Gruter and Lowenstein first transmitted HSV infection by using material from herpes keratitis and labialis to infect rabbit corneae. Since that time, advances in viral technology have led to the development of in vitro models of infection, recognition of the two types of HSV, increasing understanding of the mechanisms of latency and reactivation, and development of the first successful antiviral agents for use against HSV.

Epidemiology

Herpes simplex viruses are found worldwide, with humans as the only natural reservoir. Since the virus cannot remain infectious for long outside the body, direct contact with infected secretions is the mode of spread. Generally, HSV type 1 is transmitted in oral secretions, while HSV type 2 is transmitted genitally. Asymptomatic excreters can transmit infection, although that amount of virus is less than that seen in active lesions and presumably less infectious. Serologic studies indicate that lower socioeconomic status is associated with greater rates of seropositivity, which is thought to be related to closer living conditions with increased chance for spread. Primary infection with HSV-1 often occurs very early in life, and by adulthood, 30 to 100 percent of populations, depending on socioeconomic status, will have antibodies. In contrast, HSV-2 antibodies do not appear until puberty, and the acquisition rate is highest from 14 to 29 years, reflecting the venereal transmission of this virus. A sharp increase in the number of recognized cases has been noted recently in the United States, making genital herpes infection one of the most common venereal diseases seen today. Other modes of infection include autoinoculation to another body site and transmission from mother to infant via an infected birth canal at the time of delivery.

Recurrent infections can occur following primary infection with either HSV-1 or HSV-2. Recurrent disease is not usually as severe as primary disease, and most recurrences are due to reactivation of endogenous virus rather than to reinfection.

Pathogenesis

Herpes simplex virus enters the body through skin and mucous membranes. Most primary infections go unrecog-

nized or are subclinical. When studied histologically, a local lytic infection of the parabasal and intermediate epithelial cells is noted, with an accompanying local mononuclear inflammatory response. Together, these make up the characteristic vesicle of HSV infection. Multinucleated cells and occasional typical Cowdry type A intranuclear inclusions are seen, along with ballooning degeneration of the cells and chromatin margination. During primary infection, involvement of the lymphatics and regional lymph nodes is common, and viremia with further dissemination can occur in patients who do not have an intact immune system.

HSV becomes latent following a primary infection by traveling up the sensory nerves to the corresponding sensory ganglia. Human autopsy studies have demonstrated the presence of latent virus in sacral, trigeminal, and vagal ganglia. The site of latency is felt to be within the neuron where viral DNA has been detected, but the exact physical nature of latent HSV is still not known. The signal for reactivation is also still a mystery, but various stimuli, such as fever, stress, sunlight, and local trauma, may be recognized by individuals as precipitating factors. Following reactivation, the virus presumably travels back down the sensory nerve pathway and again establishes infection on the skin or mucous membranes. Antibody titer is not usually affected, and lesions may or may not be clinically evident.

Both humoral and cell-mediated immune mechanisms contribute to limitation of virus spread, but the severe forms of disseminated HSV infection are more often seen in patients with depressed cell-mediated immunity. Viremia disseminates the virus in severe cases, resulting in infection that can involve the liver, lungs, kidneys, and central nervous system. Infants and malnourished or immunosuppressed patients are at highest risk for this often lethal complication.

Clinical Illness

Neonate

Herpes simplex virus infection of the newborn can be a devastating and often fatal disease, with an estimated incidence of between 1 in 3000 and 1 in 30,000 live births. Although cases of true intrauterine infection have been described, most infants seem to be infected via passage through the mother's infected birth canal or via retrograde spread of maternal genital infection shortly before birth. Since most genital herpes is due to HSV-2, most babies are also infected with this virus type. If HSV-1 is transmitted, it causes an identical disease picture. In a natural history study of neonatal HSV occurring in 56 mother–infant pairs, Whitley et al. found that 70 percent of the mothers had asymptomatic infection at the time of delivery, and nearly 50 percent of the births occurred prematurely.

Disease becomes apparent in infants from birth to 4 weeks of age, depending on the time of infection. Disseminated congenital infection is manifest at birth or soon after by seizures, irritability, jaundice, hepatosplenomegaly, bleeding disorders, chorioretinitis, pneumonitis, and skin vesicles. Progressive deterioration, with respiratory distress, circulatory collapse, and coma, often follows, and death occurs in up to 70 percent of untreated infants. Neonatal infections acquired during labor and delivery are clinically manifest around the end of the first week of life. Skin vesicles alone were the presenting sign in 70 percent of the neonates studied, but continued progression to involvement of other sites occurred in a further 70 percent of these infants. About 20 percent of infants, however, never developed skin vesicles, and viral isolation from other sites was required for diagnosis.

Disease that remains localized to the skin, eye, or mouth carries a better prognosis (mortality 30 percent) compared with localized central nervous system or disseminated disease (mortality 50 percent and 80 percent, respectively), which make up more than three quarters of all cases. Vidarabine therapy has markedly decreased the mortality of this disease, but few treated survivors of the localized central nervous system and disseminated forms of infections are neurologically normal at 1 year of age.

Oropharynx

Most primary HSV infections of the mouth are clinically inapparent. Those that are symptomatic most often present in early childhood (ages 1 to 6) as gingivostomatitis. This disease occurs year-round and begins with fever, drooling, and a sore mouth following an incubation period of 2 to 12 days. Initial pharyngeal erythema and edema progresses to vesiculation involving the pharyngeal and oral mucosa and gum tissue. Ulceration occurs quickly and may be accompanied by gingival bleeding. Lesions primarily involve the buccal mucosa, tongue, soft palate, and lips, in contrast to herpangina, caused by Coxsackie A virus, which primarily affects the posterior oropharynx. Tender submandibular and cervical adenopathy is common. Fever, pain, and toxicity may last several days, and secondary dehydration may occasionally become a problem. The illness lasts 1 to 2 weeks, and recovery is complete. Young adults may have a different presentation of their primary oral HSV-1 infection, with involvement of tonsils and the posterior pharnyx. This must be distinguished from other common causes of pharyngotonsillitis. This picture may also be seen in 10 percent of patients with primary genital HSV infections, presumably due to orogenital contact.

Herpes labialis (fever blister) is the most common disease caused by HSV, affecting 20 to 40 percent of the population. In contrast to the intraoral location of primary gingivostomatitis, recurrences usually occur at the vermilion border of the outer lip. A prodrome of pain, burning, tingling, or itching often precedes the clinical development of vesicles, which usually appear at the same location with each recurrence. Pain and viral titer decrease rapidly after 24 hours, and complete healing occurs within 7 to 10 days.

Eye

HSV infections involving the eye are usually due to type 1 virus. The disease is most often unilateral and can present as follicular conjunctivitis or blepharitis with associated preauricular adenopathy. Corneal involvement, in the form of dendritic lesions or punctate opacities on the epithelium, occurs in some patients. In the absence of stromal involvement, the eye heals without scarring in 1 to 3 weeks, but delayed healing, aggravated by the host's immune response, is sometimes seen if the infection spreads more deeply into the stroma and may be associated with opaque scars that require surgical therapy.

Recurrent eye disease occurs within 2 years in 25 to 50 percent of patients who experience recognized primary infection. Dendritic keratitis is the most common form and is easily diagnosed using fluorescein staining. Amebic ulcers (named for their shape) occasionally occur, especially in those patients treated with topical steroids, and can be a destructive and chronic problem. Unilateral keratoconjunctivitis and blepharitis may also be seen as manifestations of recurrent disease.

Genitalia

Primary genital infection with HSV may go unnoticed by the patient or run a distressing clinical course that can involve many body sites. The incubation period is 2 to 7 days. HSV-2 causes 70 to 90 percent of these infections. The 10 to 30 percent associated with HSV-1 usually result from oral–genital contact. Patients who present with symptoms have multiple painful vesicular or ulcerated lesions, vaginal or urethral discharge from the associated cervicitis or urethritis, and local itching. Tender inguinal adenopathy is common, and about 40 percent of patients will have associated systemic findings, including fever, malaise, myalgias, or photophobia. Dysuria and urinary retention secondary to urethral involvement may be seen, and some patients have an accompanying sacral radiculomyelitis that can be associated with neuralgia and urinary retention. Viral meningitis has been documented in 4 to 8 percent of cases, implying that viremia has occurred, and is accompanied by CSF pleocytosis. Cerebrospinal fluid culture for HSV is positive in this situation, in contrast to encephalitis due to HSV, where virus isolation from cerebrospinal fluid is distinctly unusual. The symptoms of primary genital disease last for 2 weeks, and healing of the lesions is usually complete at the end of 3 weeks. This disease has been seen in young children, and as with other sexually transmitted diseases presenting at a young age, the possibility of child abuse should be kept in mind.

Recurrent genital HSV disease is clinically much milder than its primary form. Approximately 70 percent of patients experience prodromal symptoms that are followed by the development of a small group of vesicles that may coalesce to form one or two ulcers. Pain and itching, if present, can last 3 to 4 days but are much milder than with primary infection. Lesions are usually healed in 7 to 10 days. In a study of 137 patients with a first symptomatic episode and 87 patients with recurrent disease, Corey et al. noted that only 14 percent of HSV-1 infections versus 60 percent of HSV-2 infections recurred during mean follow-up of 221 days. The probability of recurrence was directly related to the presence and titer of neutralizing antibody to HSV-2 in convalescent serum. Symptomatic patients with recurrent genital disease have five to eight recurrences per year, but cervical and urethral shedding of virus can occur without disease, so the true recurrence rate is unknown. The majority of recurrences are due to reactivation of endogenous virus, but reinfection with a different strain can occur.

The most important complication of this disease is the emotional stress that patients experience in dealing with a chronic illness that affects their sexual relationships. Women have the added burden of being at increased risk for abnormalities in cervical cytology and need to be monitored during pregnancy to avoid exposing their infants to herpes infection at the time of birth.

Perianal Area

Primary HSV disease of the anus and perianal area is being recognized with increasing frequency, particularly in male homosexuals. Vesicles and ulcers can be seen in the perianal and anal area and are associated with pain, itching, discharge, and tenesmus. Bilateral inguinal adenopathy is common, and systemic symptoms similar to those seen in primary genital infection have been noted.

Skin

Primary HSV infection of the skin occurs following a break in the integrity of this important host defense, and several well-recognized syndromes have been described. Young children with eczema can develop a rare but potentially fatal disease known as "eczema herpeticum." The term "Kaposi's varicelliform eruption" has also been used for this disease, but its use should be avoided since it implies a different etiologic agent. High fever and irritability are followed by the development of crops of vesicles that may continue to appear over the course of a week. Fever and toxicity subside after 7 to 10 days, and lesions are usually healed by the end of the third week unless bacterial superinfection complicates the process.

Primary infection of the finger, "herpetic whitlow," causes intense pain and itching, usually involving just one digit. Vesicles form deep under the skin and the accompanying axillary adenopathy and neuralgias can be impressive. Medical and dental personnel are at risk due to occupational exposure while examining infected patients, but this infection also occurs in the general population. Virus is readily communicated from these lesions, and medical personnel should cease all patient care while the disease is active. Lesions heal in 2 to 3 weeks and should not be incised, since bacterial superinfection is a frequent complication of this unnecessary procedure.

Superficial skin abrasions can also become infected with HSV. In wrestlers, this presentation has come to be known as "herpes gladiatorum." Burn patients are susceptible to developing HSV infections involving damaged skin areas.

Brain

Herpes simplex infection of the brain is considered to be the most common cause of fatal sporadic encephalitis occurring in the United States, and although still relatively uncommon (with an estimated incidence of 1 case/million/year), it carries a mortality rate of up to 70 percent in untreated cases. Outside of the neonatal period, disease is almost always caused by HSV-1 and can affect persons of any age. Herpes encephalitis can occur as a primary infection, but many patients have a history of prior HSV infection with detectable antibody titers in acute-phase serum. Only 8/92 encephalitis patients studied by Whitley et al. had HSV isolated from a peripheral site in addition to the brain isolate. Restriction endonuclease patterns were identical for five of the pairs but different in three. The reason for such rare involvement of the brain in HSV infections, which are otherwise so common in the general population, is still unclear.

The clinical illness can begin suddenly or after a flu-like illness. Fever, headache and stiff neck, and behavioral changes are common, and most patients will have focal neurologic signs. Findings in the cerebrospinal fluid are quite variable, but elevated protein, pleocytosis, and erythrocytosis, can all be seen. Most cases will eventually demonstrate localized abnormalities by electroencephalography, technetium brain scanning, or computerized axial tomography of the head, but these changes may take days to become detectable. Localization to the temporal lobe is characteristic, but other areas of brain involvement have also been described, and the process can even be patchy and diffuse. Since isolation of HSV from cerebrospinal fluid is rare and noninvasive techniques may not be helpful early in the disease process, brain biopsy is still generally recommended for appropriate histologic and culture diagnosis. Untreated, the disease can run a very rapid course, progressing from lethargy to coma and death. The mortality in biopsy-proven untreated cases is 60 to 80 percent, and more than 90 percent of survivors have significant neurologic sequelae.

Immunocompromised Host

Severe herpes simplex infections can complicate the course of patients with a variety of immunocompromised states. Patients who are HSV seropositive before receiving bone marrow, cardiac, or renal transplants have a reactivation rate of 35 to 85 percent. Many reactivation episodes are asymptomatic, but severe and/or chronic disease may cause significant morbidity in this population, and patients have died from HSV pneumonia. HSV reactivation in patients receiving immunosuppressive therapy for malignancy correlates with suppression of cell-mediated immunity. Similarly, patients with congenital or acquired immunodeficiency syndromes affecting cell-mediated immunity may have severe, persistent infection. Patients with severe combined immunodeficiency, nucleoside phosphorylase deficiency, common variable immunodeficiency, and the newly described acquired immune deficiency syndrome (AIDS), primarily affecting young homosexual males, fall into this category.

Oncogenic Potential

HSV-2 has been associated with squamous cell carcinoma of the cervix, although inconsistencies exist in the data used to support this contention. Evidence centers around the following findings: (1) HSV can transform cells in vitro, (2) antibodies to HSV-2 are found more frequently and in higher titer in patients with cervical carcinoma than in controls, and (3) HSV-2 DNA and mRNA have been detected in some cancer cells. Conflicting results have been reported by different laboratories and may reflect different definition of patient and control groups, different serologic techniques and reagents, and possibly different geographic factors. At this time one can say that there is an association between HSV-2 and cervical carcinoma, but it remains to be proven that a cause-and-effect relationship exists.

Diagnosis

Patients presenting with fever, localized vesicular rash (especially one that involves mucous membranes), and constitutional symptoms should be suspected of having primary HSV infection. Clinical recognition of recurrent disease is made easier if the patient can give a history of previous similar eruptions. Confirmation of the diagnosis, especially in the face of severe illness where antiviral therapy is considered, is most reliably made by virus isolation. Infection of embryonated eggs or experimental animals has become unnecessary in light of the advances made in tissue culture techniques. Several primary cell lines, including rabbit kidney, human amnion, and human embryonic kidney, can be used to cultivate HSV, in addition to continuous human diploid and monkey kidney cell strains. Because of its rapid growth, cytopathic effect due to HSV can be seen in 24 to 48 hours if a large enough inoculum is used. Specimens for culture should be obtained from fresh vesicular lesions, which contain the highest titer of virus, and inoculated onto tissue cultures as soon as possible.

A more rapid bedside procedure, known as the Tzanck preparation, is made by scraping the base of a suspected skin lesion, fixing the specimen with ethanol, and staining with Wright or Giemsa stain. Examination under the microscope will reveal multinucleated giant cells indicative of HSV or varicella-zoster infection. The Papanicolaou stain may also be used to demonstrate intranuclear inclusions, but both of these techniques are less sensitive than is actual virus isolation. Paired acute and convalescent sera may show a rise in neutralizing or complement-fixing antibodies in primary infection but are unreliable in diagnosing recurrent disease. Immunofluorescence has been used to rapidly detect HSV antigen in brain and other tis-

sue specimens but is not as sensitive as tissue culture in detecting small amounts of HSV. Electron microscopy has also been used for rapid diagnosis but suffers from the same insensitivity problem and is unable to differentiate among the herpes group viruses, since thay are morphologically similar.

Treatment

The development of effective antiviral therapy for the treatment of HSV infections has been one of the most important breakthroughs in medical virology. Although supportive care is still an important part of the management of all HSV infections, several antiviral agents have been shown to effectively treat various syndromes associated with this virus.

Idoxuridine

Idoxuridine (5-iodo-2′-deoxyuridine, IUDR) was the first agent shown to be efficacious against HSV. Developed in 1959, this deoxythymidine analog inhibits thymidine kinase and is incorporated into viral and cellular DNA. Its associated myelosuppressive effect makes it too toxic for systemic use, and its primary usefulness has thus been in the topical treatment of dendritic herpes keratitis. Development of in vitro resistance can be easily demonstrated, and lack of clinical response has been noted in several studies. Some patients develop allergic responses to application of the solution. Controlled clinical studies determined that use of idoxuridine is harmful in the treatment of encephalitis due to HSV, and its efficacy in the treatment of skin infections has not been well supported.

Vidarabine

This purine analog was first shown to have anti-HSV activity in 1964 after having been developed as an anticancer agent. Also known as ara-A and adenine arabinoside, the compound is phosphorylated within the cell and, in its triphosphate form, acts as a preferential noncompetitive inhibitor of viral DNA polymerase over host DNA polymerase. It is used topically to treat HSV keratitis and is effective against some IUDR-resistant isolates. Vidarabine has much lower toxicity than IUDR, which enables it to be administered systemically. Multicenter studies using the intravenous form of the drug have demonstrated that it clearly reduces mortality and favorably affects morbidity in patients who have HSV encephalitis or disseminated neonatal HSV infection, although many survivors are still left with significant neurologic sequelae. As with IUDR, there is no benefit from topical application to oral or genital lesions.

Trifluridine

Trifluridine (trifluorothymidine, TFT), is another deoxythymidine analog. First synthesized in 1962 as an anticancer agent, this agent is useful topically in the treatment of herpes keratitis. Its antiviral activity is thought to be a result of direct incorporation into viral DNA. Myelosuppression occurs following systemic administration, and therefore its use is also limited to topical therapy of HSV keratitis. Trifluorothymidine has better efficacy against stromal disease than either IUDR or vidarabine, and it can effectively treat disease caused by some viruses that are resistant to those agents.

Acyclovir

The latest antiherpes agent licensed for use is acyclovir (acycloguanosine, ACV), a deoxyguanosine analog first reported as an antiherpes agent in 1977 by Schaeffer and Elion. Its major advantage is a selective action on HSV-infected cells and consequent low toxicity for normal uninfected host cells. The compound itself is relatively inert, but HSV thymidine kinase, present in infected cells, preferentially phosphorylates acyclovir to its monophosphate form. Cellular thymidine kinase has a much lower affinity for the drug, so this step does not occur to any great extent in uninfected cells. Further phosphorylation to the triphosphate form is accomplished by cellular enzymes, and acyclovir-triphosphate then can act as an inhibitor of, and substrate for, the HSV DNA polymerase and effectively stop viral replication. When topical acyclovir is used for treatment of primary genital infection due to HSV, duration of viral shedding and time to crusting are shorter than in placebo-treated patient with primary disease, but only duration of viral shedding was shortened in patients with recurrent disease. The intravenous form, when used in the treatment of severe genital herpes infections and HSV disease in the compromised host, has an even more dramatic effect on the clinical and virologic course of the disease. An oral preparation is still being clinically tested but has suppressed symptomatic recurrences in immunosuppressed patients who were given the drug for up to 65 days.

Prevention

Avoidance of infection is currently the mainstay of prophylaxis attempts. Neonates and children with extensive areas of eczema should avoid contact with persons having active HSV lesions. Patients with genital HSV infection should abstain from close sexual contact during the time that active lesions are present. Cesarean section is recommended for mothers in labor who have genital lesions, or are known to be virus-positive, if the membranes have not been ruptured for more than 4 hours.

In terms of more specific prevention, prophylactic administration of acyclovir to seropositive bone marrow transplant recipients has been shown to prevent acute reactivation of HSV, although some patients did develop mild disease after cessation of the drug. Attempts at vaccine development have been discouraging to date and raise a number of questions concerning possible teratogenicity, oncogenicity, and effects on latency and recurrent disease. It must be remembered that patients with latent virus that recurs already have detectable antibodies in their serum, which, therefore, can hardly be called protective.

Varicella-Zoster Virus

Infection with varicella-zoster virus (VZV) is an experience most people have before reaching adulthood. Primary infection produces the disease, varicella (or chickenpox), a common childhood disease that is considered a normal part of growing up. Adults and those individuals who are immunosuppressed form a small group at risk of developing a much more serious illness that can be fatal. Individuals who have had prior varicella infection may later develop zoster, which is the illness produced by reactivation of latent VZV.

The distinction between chickenpox and smallpox was first made in 1802 by Heberden. Von Bohan, in 1892, was the first to propose that varicella and zoster were the same infectious agent after observing cases of chickenpox that occurred in close family contacts of zoster patients. The theory that zoster is a manifestation of reactivation of latent varicella virus was expressed by Garland in 1943. Propagation of the virus in tissue culture was accomplished by Weller and co-workers in 1952. This was followed by investigations that show that virus isolates from chickenpox and zoster patients are identical immunologically and morphologically and that the molecular weight of their DNA and restriction endonuclease patterns are the same.

Epidemiology

The only natural hosts for VZV are humans, although an animal model of experimental infection in guinea pigs has been described and may be useful in studying pathogenesis and effect of therapy. Outbreaks of varicella occur worldwide, primarily in late winter and early spring, with transmission occurring by close personal contact. It is assumed that virus is present in the respiratory tract of infected patients, since transmission of disease from an infected person prior to development of a rash has been noted, and epidemiologic studies document high communicability primarily by aerosolized virus. The incubation period is 10 to 23 days, and studies of household outbreaks indicate that 88 percent of the susceptibles will develop disease within 2 weeks of exposure and 96 percent within 1 month. Vesicular fluid from skin lesions contains easily detectable, significant amounts of virus, particularly during the first 3 days of rash, and close contact with this material is also a potential mode of transmission. Susceptible children exposed to a household member with zoster can also develop chickenpox, but the attack rate is much lower.

A much rarer form of transmission can occur during pregnancy. If a woman develops varicella during the first trimester, she may have as much as a 1 in 14 chance of transmitting the infection to the fetus. A spectrum of clinical findings has been described in these infants with congenital varicella syndrome. This is in contrast to those fetuses infected later in pregnancy, who are normal at birth but have zoster as their first manifestation of VZV disease postnatally.

In contrast with varicella, zoster occurs year-round and is generally believed to be the result of activation of latent VZV. It can affect patients of any age but occurs with increased frequency after the fourth decade and in immunosuppressed populations. There is no clear evidence to support the view that zoster can develop as a consequence of exposure to a patient with varicella or zoster. A population study reported by Ragozzino et al. has also disputed the notion that patients with zoster have an increased risk of developing cancer.

Pathogenesis

Because of the lack of good animal models, our understanding of the pathogenesis of VZV infection is based on clinical observations. After first contact with the virus, perhaps via the respiratory tract, a secondary site of replication is probably established within the body. Peripheral blood cell-associated viremia has been documented during the incubation period and during active infection in immunosuppressed children but not in normal children. Presumably the virus spreads via the bloodstream to the skin. Sequential histologic sections taken during the course of varicella infection demonstrate the presence of eosinophilic intranuclear inclusions in endothelial cells that surround small blood vessels in the corium, followed by spread to other cells and the formation of multinucleated cells. The fluid that accumulates between the prickle cell layer and outer epidermis forms the vesicle characteristic of the VZV rash. The lesion evolves as polymorphonuclear leukocytes invade the corium and appear in the vesicular fluid, which then becomes cloudy. Scab formation follows reabsorption of this fluid, and eventually the epidermis desquamates.

The sequence of events involved in the establishment of ganglionic infection is unknown. During active zoster infection, the skin pathology is the same as for varicella, although the lesions are generally confined to a single dermatome. Within the dorsal spinal ganglion corresponding to the distribution of the lesions, intense inflammation can be detected, with hemorrhagic necrosis of nerve cells and destruction of portions of the ganglion. Observation of virus particles in ganglion and perineural cells also adds credence to the theory that VZV becomes latent in the dorsal spinal ganglia following varicella infection and can be activated to travel down sensory nerves and establish infection of the skin. In contrast to HSV, however, the isolation of VZV during this proposed latent state has not been accomplished. It is possible that VZV reactivates periodically,

as suggested by a longitudinal serologic study, but is aborted before causing the full constellation of signs recognized as zoster. Support for this concept comes from the occurence of the pain syndrome typical of zoster that is reported by some patients who never develop skin lesions.

Intact cellular immunity is important in preventing reactivation. Thus, immunosuppression due to malignancy, chemotherapy, or both is associated with an increased rate of reactivation. In vitro lymphocyte stimulation responses to VZV antigen wane with advancing age and may contribute to the propensity for VZV reactivation in older individuals.

Clinical Illness

Varicella

Following the incubation period of 10 to 23 days (mean 14.3 days), uncomplicated infection becomes manifest as a vesicular eruption beginning on the scalp and trunk and spreading to the extremities. Children may have an associated fever of 38 to 39C for 2 to 3 days, which parallels the severity of the rash, but they generally feel fairly well during the illness. Macules quickly evolve to papules and superficial vesicles 2 to 3 mm in diameter, appearing in crops that are characteristically in different stages of development. The pruritis that accompanies the vesicle stage can be distressing to both the child and the parents. Mucous membrane lesions can appear in the mouth, conjunctivae, rectum, and vagina, but because of their location, they rupture easily and appear as ulcers. Lesions can continue to appear for up to 1 week and tend to cluster in areas of irritation. Total lesion counts vary considerably from a few to many hundreds, with older children and secondary cases tending to have more lesions. The vesicles differ from smallpox in their superficial nature, the distribution, and the simultaneous existence of different lesion stages.

Varicella in adolescents and adults is generally more severe, and may be preceded by a 1 to 2 day prodrome of fever, malaise, anorexia, and headache. Fever tends to be higher and the rash more profuse than in young children. Primary viral pneumonia, uncommon in children, has been documented in 14 percent of military recruits studied with varicella but was accompanied by clinical signs in only 4 percent. Nonetheless, fatal cases of varicella pneumonia in adults have been described.

Immunosuppressed patients who develop varicella may go on to suffer from progressive varicella, with continued lesion formation into the second week of illness and temperature elevations up to 40 to 41C. Lesions may be more deep-seated than usual, and visceral spread in the form of pneumonia, meningoencephalitis, and hepatitis can be seen. This form of illness occurs in approximately 30 percent of children with leukemia and carries a mortality rate of 20 percent. The overall mortality in children with leukemia who develop varicella is 7 percent.

There are several well-described complications of varicella in normal children, although it should be emphasized that uncomplicated disease with complete recovery is the rule. Acute cerebellitis is the most common neurologic manifestation of varicella and usually occurs at the end of the first week of rash, although it occasionally precedes lesion development. Fulminant encephalitis, occurring in less than 1 of every 1000 cases, has been noted to occur shortly after varicella lesions appear. Some of these cases are now felt to be examples of Reye's syndrome, which has been epidemiologically associated with varicella infection. Less common neurologic manifestations include aseptic meningitis, Guillain–Barré syndrome, and myelitis. Bleeding disorders, such as Henoch–Schönlein purpura and purpura fulminans, may occur 1 to 2 weeks into the illness. Secondary skin infections are usually due to staphylococci or streptococci and, prior to antibiotic availability, served as foci for systemic infection.

Congenital and Neonatal Varicella

Women who have varicella during the first trimester may transmit the infection to their infants. The spectrum of manifestations of the congenital varicella syndrome includes cicatricial lesions of an extremity in a dermatomal distribution, growth retardation, microphthalmia, cataracts, chorioretinitis, deafness, and cortical atrophy of the brain.

If primary maternal varicella infection occurs within 4 days prior to or after delivery, the attack rate of infection for the infant is about 20 percent. Mortality in these infants, however, is about 35 percent because they cannot control viral replication like an older child. Transmission of maternal IgG antibody across the placenta will partially protect the newborn infant, who will then develop only mild infection.

Zoster

Radicular pain is often the first manifestation of zoster, particularly in adults. The rash appears morphologically similar to varicella, but its unilateral distribution covers only one to three dermatomes. New lesions may appear for several days, and occasionally a few may be found outside the original dermatomes. Thoracic dermatome involvement is most common. Involvement of the ophthalmic division of the fifth cranial nerve becomes a particular problem in elderly populations, where it occurs more frequently than in the young. Complications such as iritis, conjunctivitis, keratitis, and ocular muscle paralysis can occur. Other reported complications include Bell's palsy and the Ramsay Hunt syndrome.

Patients receiving radiation therapy or organ transplants and those with malignancies, especially Hodgkin's disease and chronic lymphocytic leukemia, have a higher incidence of zoster. From 24 to 50 percent of the immunosuppressed population develop disseminated zoster, which is heralded by the onset of new lesions in a generalized distribution beginning about 1 week after the localized lesions appear. Visceral involvement in the form of hepatitis, pneumonitis, and occasionally encephalitis occurs in 10 to 20 percent, and the mortality rate in this population is 0 to 3 percent. Postherpetic neuralgia is a frequent and quite disabling sequela of zoster in both normal and immunosuppressed patients and may last for months following lesion healing. Studies addressing this problem have defined the term in such a variety of ways that the exact incidence of this complication is unknown. It occurs more after age 60 and can be severe enough to prevent sleep.

Diagnosis

Typical cases of varicella and zoster can usually be diagnosed on clinical grounds. An exposure history to chickenpox is often possible to elicit in patients suspected of having varicella, but patients with zoster, particularly the elderly, may not be able to remember having chickenpox in the past. Herpes simplex can occasionally present in a zosteriform distribution, and disseminated disease due to either HSV or VZV may be difficult to differentiate. In these cases, culture of vesicular fluid will correctly identify the causative agent. Human diploid fibroblasts support the growth of VZV, and isolation is best accomplished if lesions are cultured within 3 days of onset. The lability of the virus requires that samples be kept at 4C and transported promptly to the laboratory or inoculated directly onto the cell culture. Viremia may be detectable by culture of white blood cells in patients with progressive varicella or disseminated zoster. Detection of VZV antigen in vesicle fluid is also possible using counterimmunoelectrophoresis (CIE).

Acute and convalescent serum will show a rise in titer of antibody to VZV if further confirmation is needed. The complement-fixation (CF) test is adequate for this purpose but not sensitive enough to pick up low levels of antibody for determination of susceptibility. Some of the more sensitive, but not yet generally available, serologic tests include fluorescent antibody to membrane antigen (FAMA), immune adherence hemagglutination antibody (IAHA), enzyme-linked immunosorbent assay (ELISA), radioimmunoassay (RIA), immunoperoxidase antibody (IPA), and complement-enhancing neutralizing antibody.

Treatment

Uncomplicated varicella is a benign disease that requires only supportive care. Drying lotions, such as calamine, can reduce pruritis, and effects of scratching can be minimized by cutting the fingernails. The use of aspirin in varicella should be avoided because it has been epidemiologically associated with an increased risk of developing Reye's syndrome.

Several agents have been tried in the treatment of varicella in immunosuppressed patients, although none have yet been approved for this indication. High-dose interferon diminished the duration of new lesion formation and reduced the incidence of serious dissemination compared to placebo in a study of children with cancer. Vidarabine (ara-A) has also been shown to decrease new lesion formation, mean daily lesion counts, and duration of fever, and it prevented visceral disease when given within 3 days of onset. A collaborative study of the effect of acyclovir on chickenpox in immunosuppressed children indicated that administration of this antiviral agent also prevented visceral dissemination when given early in the course of the disease.

Treatment of zoster in normal patients is mainly limited to control of the accompanying pain. Steroids given early in the course have been shown to significantly reduce postherpetic neuralgia. A placebo-controlled trial of intravenous acyclovir found that the rate of healing was significantly increased and the period of pain was shortened by this agent.

Studies of immunosuppressed patients with zoster have shown that high-dose interferon decreased the time of new lesion formation, the incidence of dissemination, postherpetic neuralgia, and visceral complications. Vidarabine therapy also has been shown to decrease dissemination, visceral complications, and rate of cutaneous dissemination. The duration of post-herpetic neuralgia was diminished although the frequency of this complication was the same as that seen in the placebo group. Therapy with acyclovir has produced similar results.

Prevention

Since uncomplicated varicella is usually a mild illness, attempts at preventing it are not thought to be worthwhile. The focus has been in preventing infection with VZV in the immunocompromised patients who can suffer more significant morbidity and occasional mortality from this virus. Children incubating varicella should not be electively admitted to the hospital, where contact with immunocompromised patients may result in transmission of disease. If hospitalization is unavoidable, patients with varicella and zoster should be kept in strict isolation.

Passive immunization has been shown to be effective in modifying or preventing illness. Initial studies of immune serum globulin (ISG) showed that the severity of disease could be diminished if administration occurred within 72 hours of exposure. Testing with the high-titered, convalescent, IgG fraction of plasma from recuperating zoster patients, known as "zoster immune globulin" (ZIG), prevented disease in healthy children and modified illness in susceptible immunosuppressed patients. ZIG has effectively prevented severe varicella in the susceptible newborn whose mother developed varicella within 4 days of delivery. Outdated, high-titer normal donor plasma has

been used to make varicella-zoster immune globulin (VZIG) and is available for use in high-risk cases. It should be administered within 72 hours of exposure or delivery for optimal benefit.

Experimental live attenuated VZV vaccine has been developed in Japan, which results in no morbidity in serosusceptible normal children and adults. It is immunogenic, and antibodies can be detected for as long as 2 years. When this vaccine is given to immunosuppressed children, a few will have mild rash and will rarely transmit virus to susceptible household contacts. Protective neutralizing antibody levels develop and predictably persist in those who develop rash. A proportion of these patients lose detectable antibodies with time. Thus far, these children have been protected from severe varicella, but reinfection, as manifested by antibody rise after exposure, is fairly common. Longitudinal studies with the vaccine are in progress.

Cytomegalovirus

This member of the group of herpesviruses is another ubiquitous agent and is responsible for a broad spectrum of disease affecting humans in many different clinical settings. While the great majority of cytomegalovirus (CMV) infections are inapparent, this virus is now known to be the most common cause of congenital viral infection, an etiologic agent of the mononucleosis syndrome, and a feared pathogen in post bone marrow transplant patients.

The first description of probable congenital CMV disease was published by Rippert in 1904. In examining sections of kidney from a stillborn child, he noted large protozoan-like cells. In 1907, Lowenstein described intranuclear and intracytoplastic inclusions in the parotid glands of infants, and over the next 25 years, further case reports of apparent cytomegalic inclusion disease appeared in the literature. Isolation of murine cytomegalovirus was accomplished by Smith in 1954 and followed by the independent isolation of human CMV by Smith, Rowe, and Weller. Since that time, the development of serologic tests has further enabled investigators to define the importance of this virus as a cause of human infection.

Epidemiology

Serologic studies using the complement-fixation (CF) test have shown that the prevalence of CMV antibodies in adults is 40 to 100 percent depending on socioeconomic conditions. CMV is found worldwide and has no seasonal predilection. Age at CMV antibody acquisition is dependent on the mode of transmission. Population studies have shown that serologic conversion rates increase in early childhood, with a peak at 1 to 2 years. A second slow increase in the seroconversion rate occurs in adults from 16 to 50 years, although detection of virus shedding is not as common after the age of 30 years.

Several methods of CMV transmission have been described and result in different manifestations of infection. Intrauterine infection occurs in 0.05 to 2 percent of all live births, as documented by infant viruria shortly after birth. The neonatal infection rate following primary CMV in mothers during pregnancy is high, and 23 percent of infants born to women with seroconversion during pregnancy have stigmata of congenital CMV infection at birth. Numerically, more babies with intrauterine infection are born to seropositive mothers with secondary or reactivation infection. These infants are less likely to be symptomatic at birth than those infected during primary infection.

Perinatal CMV infection of an infant is much more common than intrauterine infection, occurring in 3 to 10 percent of babies born in various populations. In contrast to intrauterine CMV, viruria is not detectable until 4 to 8 weeks of age. Several prospective studies of pregnant women have documented an increased rate of cervical CMV excretion during the latter stages of pregnancy. Infants born to mothers who excrete virus in the third trimester or postpartum have a higher infection rate than those born to non-CMV excreters, suggesting that infection during the birth process probably occurs. Some immune mothers have also been shown to excrete CMV into breast milk. Infants who are breastfed by these mothers have a higher infection rate than their bottle-fed counterparts. A recent day-care center study has suggested that toddlers in this setting may acquire CMV infection from children who orally excrete the virus. CMV was recovered from toys mouthed by children known to be oral excreters, and the infection rate was highest in the 1 to 2 year age group, corresponding to the age where the oral excretion rate and oral contact between children was highest.

Several pieces of evidence point to venereal transmission as another mode of CMV acquisition. The virus has been isolated from semen and secretions from the female genitourinary tract. Multiple sex partners are associated with higher cervical infection rates in women, and homosexual males have a higher urinary CMV isolation rate and seropositivity rate than do heterosexual males. Cases of CMV mononucleosis transmitted by sexual contact have been reported.

Blood transfusions have also been implicated as a source of CMV. A postperfusion syndrome of fever, splenomegaly, and atypical lymphocytosis was noted in patients following extracorporeal cardiopulmonary bypass surgery for open heart surgery and linked to CMV. Susceptible infants of seronegative mothers who receive blood from seropositive donors are also at high risk and can develop severe and sometimes fatal infection. Most posttransfusion infections outside of the neonatal period are asymptomatic. The risk of infection is about 3 percent per unit transfused and greatest if donors are seropositive and recipients seronegative. CMV can infect peripheral blood leukocytes, but isolation of virus from donor blood has not been successful.

CMV infection occurs in most patients about 2 months (range 1 to 4 months) following bone marrow or renal transplantation. There is some evidence that the donor kidney may be a source of primary CMV following renal transplantation, although isolation of the virus from this organ has not been possible. Cytotoxic drug regimens, which are part of the care of transplant patients, have been associated with development of virus shedding in seropositive patients, presumably on the basis of depression of cell-mediated immunity. The use of antithymocyte globulin, cyclosporin A, and cyclophosphamide has been particularly associated with an increased incidence of viremia and CMV illness. Graft-vs.-host (GVH) reactions in bone marrow transplant recipients are also associated with a higher incidence of CMV infection. Patients with malignancy and undergoing chemotherapy form another high-risk group.

Pathogenesis

Primary infection, as mentioned above, can be acquired through the bloodstream or through presumed oral, venereal, and respiratory routes. Epithelial cells in the oropharnyx may serve as a primary focus, with spread to lymphoid tissue and leukocytes finally producing viremia. Patients with CMV mononucleosis often have virus isolated from peripheral leukocytes for up to 3 weeks following infection, and persistence for as long as 3 months has been documented.

The typical cytopathology seen with CMV infection is the presence of large giant cells with intranuclear inclusions. These inclusions have been shown to contain virions and specific viral antigens. In young children the salivary glands, particularly the parotid, are a site of localization, but this has not been demonstrated in older children and adults. Virus shedding occurs over prolonged periods of time from throat, breast milk, feces, semen, and cervix in spite of the presence of antibody to CMV. After active shedding ceases, the virus presumably enters into a latent state, possibly in the leukocytes, and can be reactivated on a periodic basis under a number of different conditions.

Clinical Illness

Congenital Cytomegalic Inclusion Disease

Twenty percent of all infants with congenital CMV infection are symptomatic at birth and may manifest hepatosplenomegaly, jaundice, petechiae, microcephaly, chorioretinitis, and cerebral calcifications. Lethargy, respiratory distress, and seizures can occur shortly after birth, often followed by death in days to a few weeks. This severe form of disease is found in 25 percent of symptomatic infants (5 percent of all infants born with congenital infection). The other 75 percent of the symptomatic infants (15 percent of the total born with congenital infection) manifest psychomotor retardation, hepatomegaly, or deafness. The remaining 80 percent, detected by viruria within 1 week after birth, are clinically asymptomatic at birth, but over time, 5 to 20 percent of this group may manifest evidence of central nervous system involvement, such as sensorineural hearing loss and school failure problems.

Perinatal CMV

The vast majority of infants who acquire CMV in the perinatal period are asymptomatic. Pneumonitis and manifestations of the mononucleosis syndrome may be seen occasionally, with onset during the first 3 months of life, but subtle long-term CNS sequelae have not yet been detected. Infection in the neonatal period via blood transfusion, however, can be serious and sometimes fatal.

CMV Mononucleosis

This syndrome may occur in children but is most commonly seen in adults. Fever is the dominant symptom, accompanied by an absolute lymphocytosis with atypical lymphocytes that are detectable within 1 to 2 weeks of onset of fever. Unlike EBV-induced mononucleosis, tonsillopharyngitis, splenomegaly, and lymphadenopathy are uncommon. Mildly elevated liver function tests are present but usually not associated with severe hepatitis or jaundice. Prospective studies of mononucleosis patients have documented that CMV infection is responsible for about 45 percent of heterophil-negative cases. The clinical picture of posttransfusion CMV mononucleosis, which occurs 3 to 4 weeks after transfusion, is the same as that described for spontaneous CMV mononucleosis. Associated findings that are occasionally observed include interstitial pneumonitis, hepatitis (with or without granulomas), meningoencephalitis, Guillain–Barré syndrome, myocarditis, hemolytic anemia, and thrombocytopenia.

CMV in Immunocompromised Patients

Primary and reactivation CMV infections are commonly manifest about 2 months posttransplantation, affecting nearly all renal transplant recipients and half of the bone marrow recipients. Most of these renal patients (50 to 60 percent) have asymptomatic infection with CMV, 40 to 50 percent will have a self-limited nonspecific syndrome, and 2 to 3 percent develop disseminated disease and death. CMV can cause renal disease, with diminished renal function. This should be carefully distinguished from rejection because immunosuppressive therapy should be decreased, not increased. Symptomatic CMV infection is more likely to occur in primarily infected patients than in those who are initially seropositive. Interstitial pneumonitis can be a rapidly fatal process in transplant recipients. In bone marrow transplant recipients, the incidence of CMV pneumonia is about 20 percent, and mortality is more than 80 percent.

Patients with leukemia or lymphoma, especially children, are at high risk for acquiring CMV infection. The rate of seroconversion has been related to the number of

chemotherapeutic agents used. Clinical syndromes in this population include pneumonia, fever and rash, gastrointestinal involvement, and myocarditis. In both the posttransplant and malignancy groups, CMV has frequently been isolated with other agents. Its role as a copathogen in the recently described acquired immune deficiency syndrome (AIDS) is also currently under study.

Diagnosis

The manifestations of CMV disease are not distinctive enough to make a clinical diagnosis without laboratory confirmation. Isolation of the virus is the most specific means of making a diagnosis, and cytopathic effect can be detected as early as 3 to 4 days postinoculation with a large inoculum, although sometimes 1 to 2 weeks are required. Histologic examination of lung biopsy tissue will often reveal the presence of cytomegalic cells, primarily in the alveolar and bronchial epithelium, in patients with CMV pneumonitis.

Complement-fixing antibodies can be detected in immunosuppressed patients undergoing clinical CMV infection but are not sensitive enough to reliably diagnose perinatal and cervical infection. Serologic tests with increased sensitivity include indirect fluorescent antibody (IFA) and anticomplement immunofluorescent (ACIF) tests.

Treatment

Significantly less progress has been made in the development of effective therapy for CMV infections than for either HSV or VZV. Many antiviral agents have been tested on the more severe forms of disease, but none has yet been found to be clinically useful. Human leukocyte interferon has been used to treat congenital CMV and infection in renal and bone marrow transplant recipients. There was no clinical benefit for any patients except for a delay of virus excretion in the renal transplant group. Initial treatment and prophylaxis studies of recombinant α-$_2$ interferon in bone marrow transplant recipients demonstrated that it did not prevent CMV infection or clinically benefit the patients. Therapy with cytosine arabinoside (ara-C), vidarabine (ara-A), idoxuridine (IUDR), and acyclovir (ACV), including combination studies of vidarabine or acyclovir with human leukocyte interferon, has been unsuccessful to date.

Prevention

Efforts at vaccine development have been targeted at preventing the most serious forms of CMV illness—primary infection in the immunosuppressed and congenital CMV infection, which is the cause of severe neurologic sequelae in more than 5000 children born each year in the United States. Since reactivation occurs in the presence of specific antibody, the theoretical effect of vaccine on already seropositive patients should be nil. Concern exists over questions of potential oncogenicity and effect of vaccine on the latent state and reactivation potential of CMV, but development has continued and human trials are currently in progress.

Two vaccine strains have been studied, the AD169 and Towne 125 strains. Both have been shown to be immunogenic as defined by seroconversion, although titers diminish over time, and there has been no evidence of persistent infection. Vaccination with Towne 125, however, failed to protect seronegative renal transplant recipients who received kidneys from seropositive donors. Further studies are currently underway.

Epstein–Barr Virus

Epstein–Barr virus (EBV), like its fellow members of the human herpesvirus group, is a common cause of infection worldwide. Infection in early childhood is largely asymptomatic, but adolescents and young adults often develop the disease, infectious mononucleosis. Patients undergoing posttransplant immunosuppression or those with various congenital immunodeficiency syndromes may develop more severe disease, and reports of the development of fatal polyclonal lymphomas associated with the presence of EBV genome and antigens in tumor tissue are increasing. EBV has now been firmly associated with African Burkitt lymphoma and nasopharyngeal carcinoma.

Although EBV is the most recently recognized member of the human herpesviruses, clinical descriptions of infectious mononucleosis date back to 1889, when Emil Pfeiffer reported the characteristic findings of Drüsenfieber, (glandular) fever, including fever, adenopathy, sore throat, and occasional hepatosplenomegaly. In 1920, Sprunt and Evans named the disease and described the characteristic mononuclear leukocytosis. Three years later, Downey and McKinley definitively described the morphologic appearance of atypical lymphocytes in infectious mononucleosis. Paul and Bunnell discovered the heterophil antibody in 1932, which, with Davidsohn's addition of guinea pig kidney and beef red blood cell absorption, enabled clinicians to make a serologic diagnosis.

In 1958 Burkitt described a malignant tumor involving the jaw and abdomen that occurred in children of central and western Africa. Because the distribution of the tumor seemed to be affected by climate, a search for an infectious agent carried by an anthropod vector was begun. Culture of the tumor cells was accomplished by Epstein and Barr in 1964, and electron microscopy of these cells revealed the presence of viral particles typical of the herpes group. Henle and Henle developed an indirect immu-

nofluorescent antibody test for EBV antibody in 1966 that helped to define epidemiologically the importance of this virus. When their technician developed infectious mononucleosis in 1968 after working with Burkitt tumor tissue, serial examination of her serum showed the development of EBV antibodies, her lymphocytes were able to be grown in tissue culture, and EBV antigen was detected in the lymphocytes. The link with EBV and infectious mononucleosis has been subsequently confirmed.

Epidemiology

Seroepidemiologic studies indicate that EBV infections occur in all populations, with age of acquisition determined by socioeconomic factors. Seroconversion has been documented shortly after birth in New Guinea, Polynesia, and Ghana, where poor hygiene and low socioeconomic conditions predominate, while the more industrialized populations, such as England, Sweden, and the United States, tend to delay their encounter with this virus until adolescence and adulthood. Increasing use of day-care centers may change this pattern, however, as seen with CMV. EBV antibodies can be found in 90 to 95 percent of most adult populations.

When primary EBV exposure occurs early in childhood, infection most often goes unrecognized. Seroconversion has been documented following nonspecific febrile illness, upper respiratory infection, pharyngotonsillitis, rash, lymphadenopathy, hepatosplenomegaly, and pneumonia. Sometimes infectious mononucleosis-like illness occurs in young children, but this presentation becomes much more common in college-age populations, where 28 to 74 percent of students with documented seroconversion have associated mononucleosis. Throat washings from acutely ill patients yield EBV, and excretion may continue intermittently for months after all signs of illness have disappeared. Cultures of throat washings from 10 to 20 percent of healthy seropositive adults who have no history of mononucleosis yield low titers of EBV, indicating that intermittent excretion occurs even after clinically inapparent infection. This helps to explain why only 6 percent of patients with mononucleosis can report exposure to another case of mononucleosis. EBV has been isolated from saliva, parotid gland secretions, and B lymphocytes but not from other sources of transmissions, such as feces, urine, or the genital tract. Intimate oral contact is thus felt to be the primary mode of transmission. Infection via blood transfusions and bone marrow has been reported but occurs much less frequently than CMV-induced posttransfusion mononucleosis. The virus does not seem to be very contagious, since follow-up studies of susceptible college roommates of mononucleosis patients found that the risk of infection was no greater than in the general population. A family study of EBV transmission in households following diagnosis of a child with infectious mononucleosis found evidence of spread in 20 percent, with an incubation period of 4 to 6 weeks.

Pathogenesis

Since saliva seems to be the probable mode of transmission, it is thought that the virus attaches to cells in the oropharynx, perhaps in the parotid glands, where it is able to replicate. Bloodstream invasion occurs at some point, with the ensuing infection of B lymphocytes and dissemination through the lymphoreticular systems. During acute clinical disease, 0.005 to 0.5 percent of circulating mononuclear cells have been found to contain EBV. The polyclonal activation of B cells by EBV is accompanied by production of antigens that are expressed by the cell. Actively proliferating T cells belonging to the B suppressor/cytotoxic subset then appear, accounting for most of the atypical lymphocytes seen on peripheral smears taken from patients with infectious mononucleosis.

Histologic examination of the enlarged lymph nodes accompanying mononucleosis reveals enlarged, active lymphoid follicles and germinal centers but an intact reticular framework. The spleen can be enlarged two to three times its normal size, and blast cells are seen throughout the red pulp. Tonsils, lung, liver, heart, kidneys, adrenals, skin, and central nervous system can also be involved with focal mononuclear infiltrates. The host's antibody response to EBV infection forms the basis of laboratory diagnosis.

EBV remains latent in the blood, lymphoid tissue, and throat of individuals who have been infected and is reactivated, as detected by intermittent oral shedding, by as yet undefined mechanisms. The importance of cellular immune mechanisms has been suggested by a higher prevalence of EBV excretion in patients with malignancy and those receiving immunosuppressive therapy.

Clinical Illness

Infectious Mononucleosis

Infectious mononucleosis is the most commonly recognized manifestation of EBV infection. This is a disease primarily seen in adolescents and young adults, although cases may occasionally be seen in young children and the elderly, and the spectrum of illness is very broad. Onset of disease can be abrupt or heralded by a prodrome of chills, sweats, anorexia, or malaise. The classic clinical triad of sore throat, which can be extremely severe, fever, and lymphadenopathy is often accompanied by hepatosplenomegaly. Fever may be as high as 40C and usually resolves in 1 to 2 weeks. Tonsillopharyngitis is accompanied by thick exudate in about half of the patients, and palatal petechiae can occasionally be seen. The lymphadenopathy occurs in over 90 percent of patients and is usually symmetrical, involving the posterior cervical chain. Anterior cervical, submandibular, axillary, and inguinal node involvement can also be present. Adenopathy usually subsides in days to weeks depending on the severity of the illness. Splenomegaly is maximal in the second week and is detectable in half of the patients with mononucleosis if it is looked for regularly during the course of illness. Hepatomegaly occurs in 10 to 15 percent of patients, with associated mild elevations in hepatic cellular enzymes, but only 5 percent of patients will develop jaundice. Various forms of skin rash may also be seen in 5 percent of patients. If, for some other reason, individuals are given ampicillin during the course of their illness, more than 90 percent will develop a pruritic maculopapular eruption. Although the exact mechanism is not known, this usually does not imply an allergic response to penicillin.

The vast majority of patients will recover uneventfully, but occasional individuals may exhibit complications. Splenic rupture, heralded by abdominal pain, most commonly occurs in the second or third week of illness. This dramatic but rare complication is often preceded by trauma and has fostered recommendations against contact sports and vigorous palpation of the spleen during the first weeks of illness. Neurologic complications occur in less than 1 percent of patients and can present as encephalitis, meningitis, Guillain–Barré syndrome, myelitis, Reye's syndrome, and cranial nerve palsies. Autoimmune hemolytic anemia is noted in 0.5 to 3 percent of patients, with hemolysis becoming apparent after the second week of illness. Thrombocytopenia is common but only rarely severe. Pericarditis and myocarditis have been reported, and serologic evidence of acute EBV infection has been found in children with pneumonia, some of whom had evidence of infection with other pathogens. Persistent infection with EBV producing illness lasting longer than 1 year has been described in several patients presenting with nonspecific malaise, fever, and weight loss.

EBV in Immunosuppressed Patients

Severe and sometimes fatal infection with EBV has been described in several groups of patients with underlying immunodeficiencies. The X-linked recessive lymphoproliferative syndrome (XLP) has been described in several kindreds and is associated with an extreme susceptibility to serious EBV infection. Fatal EBV disease occurs in 40 to 50 percent, 15 to 30 percent have acquired variable immunodeficiency, and 40 percent develop lymphoreticular malignancies. A functional T cell defect and deficiency of natural killer (NK) cell activity have been demonstrated. A non-X-linked form has been described in a family where three siblings developed severe infectious mononucleosis associated with diminished NK cell activity. Renal transplant recipients have occasionally been reported to develop polymorphic diffuse B cell hyperplasias and polymorphic B cell lymphomas. EBV genome has been detected in tumor cells of patients with serologic evidence of primary or reactivated EBV infection.

Burkitt's Lymphoma and Nasopharyngeal Carcinoma

Several lines of evidence point to the association of EBV to African Burkitt's lymphoma and nasopharyngeal carcinoma. EBV genome has been detected in the tumors, tumor cells express EBV-associated nuclear antigen (EBNA), and patients have high titers to EBV antibodies. Salivary IgA antibody to EBV viral capsid antigen is currently being used as a screening tool to detect early nasopharyngeal carcinoma in China, where the frequency of this tumor is very high.

Diagnosis

Because infection with EBV produces such a broad range of clinical illness, the diagnosis is usually based on laboratory findings. The hematologic picture is often the first clue, with 70 percent of patients having an absolute lymphocytosis at presentation. Atypical lymphocytes are not specific for EBV infection, but at the height of lymphocytosis, they account for about 30 percent of the differential count. Some patients may have none while others have up to 90 percent of their total lymphocyte count made up of these large cells with lobulated nuclei and vacuolated, basophilic cytoplasm. Neutropenia and thrombocytopenia are common.

Heterophil antibodies are present in 90 percent of adult cases at some point during the illness, usually peaking at 2 to 3 weeks. Children in the 2 to 5 year group develop detectable heterophil antibodies when they present with classic infectious mononucleosis, but since this is an unusual manifestation of disease in this age group, the majority of children will not have a positive test.

Immunofluorescent tests for EBV-specific antibodies are used in heterophil-negative and atypical cases to confirm EBV infections. Antibodies to viral capsid antigen (VCA) arise early and are detectable in virtually all patients. IgG VCA persists for life, while IgM VCA lasts only 4 to 8 weeks, making it a more specific indicator of acute infection. There are two types of antibody to early antigen (EA). Anti-D diffusely stains both the nuclei and cytoplasm of infected cells and is detected in 70 percent of patients with infectious mononucleosis. It persists for 3 to 6 months following infection and is also detected in patients with nasopharyngeal carcinoma. Anti-R stains only aggregates in the cytoplasm and is seen in atypical, protracted cases of mononucleosis and in patients with African Burkitt's lymphoma. Antibodies to Epstein–Barr nuclear antigen (EBNA) appear 3 to 4 weeks after infection and persist for life, as do antibodies to soluble complement-fixing antigen (anti-S) and neutralizing antibodies. Isolation of EBV is not routinely available because of the need for specific laboratory methods involving cultivation of lymphocytes and antigen detection. Throat washings and lymphocytes yield virus in 80 to 90 percent of patients with infectious mononucleosis in an experimental laboratory. EBV antigen detection in lymphocytes will become more practical as labeled specific antibodies or probes become available.

Treatment

Infectious mononucleosis is generally a self-limited disease, and treatment is limited to supportive measures. Bed rest during the acute stage is advocated, and contact sports should be avoided for the first 2 to 3 weeks. Fever and sore throat can usually be relieved by aspirin or acetaminophen. Patients with severe disease manifest by impending airway obstruction, hemolytic anemia, thrombocytopenia, and progressive neurologic involvement may be treated with corticosteroids, but this is not felt to be necessary in mild cases. The use of steroids has been shown to decrease the duration of fever, and response is usually prompt.

EBV is inhibited in vitro by vidarabine, phosphonoacetic acid, and acyclovir. Clinical trials of acyclovir for use in the treatment of infectious mononucleosis, the X-linked recessive lymphoproliferative syndrome, and posttransplant EBV-associated lymphomas are currently in progress.

Prevention

Patients with infectious mononucleosis need not be isolated, since intimate contact is necessary for disease transmission. Immunization has been considered, but the problem of potential oncogenicity will be difficult to overcome unless purified polypeptides with immunogenic properties can be developed.

FURTHER READING

HERPES SIMPLEX VIRUS

Allen WP, Rapp F: Concept review of genital herpes vaccines. J Infect Dis 145:413, 1982

Arvin AM, Pollard RB, Rasmussen LE, et al.: Cellular and humoral immunity in the pathogenesis of recurrent herpes viral infections in patients with lymphoma. J Clin Invest 65:869, 1980

Bastian FO, Rabson AS, Yu CL, et al.: Herpesvirus hominis: Isolation from human trigeminal ganglion. Science 178:306, 1972

Binder PA: Herpes simplex keratitis. Surv Ophthalmol 21:313, 1977

Cook ML, Bastone B, Stevens JG: Evidence that neurons harbor latent herpes simplex virus. Infect Immun 9:946, 1974

Corey L: The diagnosis and treatment of genital herpes. JAMA 248:1041, 1982

Corey L, Nahmias AJ, Guinan ME, et al.: A trial of topical acyclovir in genital herpes simplex virus infections. N Engl J Med 306:1313, 1982

Hanshaw JB, Dudgeon JA: Herpes simplex infection of the fetus and newborn. In Major Problems in Clinical Pediatrics. Viral Diseases of the Fetus and Newborn, Philadelphia, Saunders, 1978, vol. 17

Nahmias AJ, Roizman B: Infection with herpes simplex viruses 1 and 2. N Engl J Med 289:667, 719, 781, 1973

Nahmias AJ, Dowdle WR, Schinazi RF (eds): The Human Herpes Viruses: An Interdisciplinary Perspective. New York, Elsevier, 1981

NIAID Collaborative Antiviral Study: Adenine arabinoside—therapy of biopsy-proven herpes simplex encephalitis. N Engl J Med 297:289, 1977

Nilsen AE, Aasen T, Halsos AM, et al.: Efficacy of oral acyclovir in the treatment of initial and recurrent genital herpes. Lancet 2:571, 1982

Openshaw H, Puga A, Notkins AL: Herpes simplex virus infection in sensory ganglia: Immune control, latency and reactivation. Fed Proc 38:2660, 1979

Rawls WE, Garfield CH, Seth P, et al.: Serological and epidemiologic considerations of the role of herpes simplex virus type 2 in cervical cancer. Cancer Res 36:829, 1976

Reeves WC, Corey L, Adams HG, et al.: Risk of recurrence after first episodes of genital herpes. N Engl J Med 305:315, 1981

Saral R, Burns WH, Laskin OL, et al.: Acyclovir prophylaxis of herpes simplex virus infections: A randomized double-blind, controlled trial in bone marrow transplant recipients. N Engl J Med 305:63, 1981

Spruance SL, Overall JC, Kern ER, et al.: The natural history of recurrent herpes simplex labialis: Implications for antiviral therapy. N Engl J Med 297:69, 1977

Stevens JG: Latent herpes simplex virus and the nervous system. Curr Top Microbiol Immunol 70:31, 1975

Whitley RJ, Nahmias AJ, Soong S-J, et al.: Vidarabine therapy of neonatal herpes simplex virus infection. Pediatrics 66:495, 1980

Whitley RJ, Nahmias AJ, Visintine AM, et al.: The natural history of herpes simplex virus infection of mother and newborn. Pediatrics 66:489, 1980

Whitley RJ, Soong S-J, Hirsch MS, et al.: Herpes simplex encephalitis: Vidarabine therapy and diagnostic problems. N Engl J Med 304:313, 1981

VARICELLA-ZOSTER VIRUS

Arbeter AM, Starr SE, Wiebel RE, et al.: Live attenuated varicella vaccine: Immunization of healthy children with the OKA strain. J Pediatr 100:886, 1982

Arvin AM, Kushner JH, Feldman S, et al.: Human leukocyte interferon for the treatment of varicella in children with cancer. N Engl J Med 306:761, 1982

Bastian FO, Rabson AS, Yee CL, et al.: Herpesvirus varicellae: Isolated from human dorsal root ganglia. Arch Pathol 97:331, 1974

Burgoon CF, Burgoon JS, Baldridge GD: Natural history of herpes zoster. JAMA 164:265, 1975

Burke BL, Steele RW, Beard OW: Immune responses to varicella zoster in the aged. Arch Intern Med 142:291, 1982

Brunell PA: Zoster in infancy: Failure to maintain virus latency following intrauterine infection. J Pediatr 98:71, 1981

Brunell PA, Ross A, Miller LH, et al.: Prevention of varicella by varicella zoster immune globulin. N Engl J Med 280:1191, 1969

Brunell PA, Shehab Z, Geiser C, et al.: Administration of live varicella vaccine to children with leukaemia. Lancet 2:1069, 1982

Eaglstein WH, Katz R, Brown JA: The effects of early corticosteroid therapy on the skin eruption and pain of herpes zoster. JAMA 211:1681, 1970

Feldman S, Hughes WT, Daniel CB: Varicella in children with cancer—seventy-seven cases. Pediatrics 56:388, 1975

Feldman S, Hughes WT, Kim HY: Herpes zoster in children with cancer. Am J Dis Child 126:178, 1973

Gordon JE: Chickenpox: An epidemiologic review. Am J Med Sci 244:362, 1962

Merigan TC, Rand KH, Pollard RB, et al.: Human leukocyte interferon for the treatment of herpes zoster in patients with cancer. N Engl J Med 298:981, 1978

Orenstein WA, Heymann DL, Ellis RJ, et al.: Prophylaxis of varicella in high-risk children: Dose-response effect of zoster immune globulin. J Pediatr 98:368, 1981

Peterslund NA, Seyer-Hansen K, Ipsen J, et al.: Acyclovir in herpes zoster. Lancet 2:827, 1981

Ragozzino MW, Melton JL, Kurland LT, et al.: Risk of cancer after herpes zoster: A population-based study. N Engl J Med 307:393, 1982

Steele RW, Myers MG, Vincent MM: Transfer factor for the prevention of varicella zoster infection in childhood leukemia. N Engl J Med 303:355, 1980

Takahashi M, Otsuka T, Okuno Y, et al.: Live vaccine used to prevent the spread of varicella in children in hospital. Lancet 2:1288, 1974

Weller TH: Serial propagation in vitro of agents producing inclusion bodies deriving from varicella and herpes zoster. Proc Soc Exp Biol Med 83:340, 1953

Whitley RJ, Chien LT, Dolin R, et al.: Adenine arabinoside therapy of herpes zoster in the immunosuppressed. N Engl J Med 294:1193, 1976

Whitley RJ, Hilty M, Haynes R, et al.: Vidarabine therapy of varicella in immunosuppressed patients. J Pediatr 101:125, 1982

Whitley RJ, Soong S-J, Dolin R, et al.: Early vidarabine therapy to control the complications of herpes zoster in immunosuppressed patients. N Engl J Med 307:971, 1982

CYTOMEGALOVIRUS

Drew WL, Mintz L, Miner RC, et al.: Prevalence of cytomegalovirus infection in homosexual men. J Infect Dis 143:188, 1981

Evans AS (ed): The risk factors of cytomegalovirus infection transmitted by blood transfusion. Yale J Biol Med 49:3, 1976

Fiala M, Payne JE, Berne TV: Epidemiology of cytomegalovirus infections after transplantation and immunosuppression. J Infect Dis 132:421, 1975

Hanshaw JB: Congenital cytomegalovirus infection: A fifteen-year perspective. J Infect Dis 123:555, 1971

Hanshaw JB, Scheiner AP, Moxley AW, et al.: School failure and deafness after "silent" congenital cytomegalovirus infection. N Engl J Med 295:468, 1976

Ho M (ed): Cytomegalovirus, Biology and Infection. New York, Plenum, 1982

Huang E-S, Alford CA, Reynolds DW, et al.: Molecular epidemiology of cytomegalovirus infections in women and their infants. N Engl J Med 303:958, 1980

Lang DJ, Kummer JF: Cytomegalovirus in semen: Observations in selected populations. J Infect Dis 132:472, 1975

Meyers JD, McGuffin RW, Bryson YJ, et al.: Treatment of cytomegalovirus pneumonia after marrow transplant with combined vidarabine and human leukocyte interferon. J Infect Dis 146:80, 1982

Meyers JD, Spencer HC, Watts JC, et al.: Cytomegalovirus pneumonia after human marrow transplantation. Ann Intern Med 82:181, 1975

Pass RF, August AM, Dworsky M, et al.: Cytomegalovirus infection in a day-care center. N Engl J Med 307:477, 1982

Pass RF, Stagno S, Myers GS, et al.: Outcome of symptomatic congenital cytomegalovirus infection: Results of long-term longitudinal follow-up. Pediatrics 66:758, 1980

Pass RF, Whitley RJ, Diethelm AG, et al.: Cytomegalovirus infection in patients with renal transplants: Potentiation by antimyocyte globulin and an incompatible graft. J Infect Dis 142:9, 1980

Schopfer K, Lauber E, Krech U: Congenital cytomegalovirus infection in newborn infants of mothers infected before pregnancy. Arch Dis Child 53:536, 1978

Stagno S, Pass RF, Dworsky MF, et al.: Congenital cytomegalovirus infection: The relative importance of primary and recurrent maternal infection. N Engl J Med 306:945, 1982

Stagno S, Reynolds DW, Huang E-S, et al.: Congenital cytomegalovirus infection: Occurrence in an immune population. N Engl
J Med 296:1254, 1977

Stagno S, Reynolds DW, Pass RF, et al.: Breast milk and the risk
of cytomegalovirus infection. N Engl J Med 302:1073, 1980

Stagno S, Reynolds D, Tsiantos A, et al.: Comparative serial virologic and serologic studies of symptomatic and subclinical congenitally and natally acquired cytomegalovirus infection. J Infect
Dis 132:568, 1975

Starr SE, Glazer JD, Friedman HM, et al.: Specific cellular and humoral immunity after immunization with live Towne strain cytomegalovirus vaccine. J Infect Dis 143:585, 1981

Suwansirikul S, Rao N, Ho M, et al.: Clinical manifestations of
primary and secondary CMV infection after renal transplantation. Arch Intern Med 137:1026, 1977

Yeager AS, Grumet FC, Hafleigh EB, et al.: Prevention of transfu-
sion-acquired cytomegalovirus infections in newborn infants. J
Pediatr 98:281, 1981

EPSTEIN–BARR VIRUS

Cheeseman SH, Henle W, Rubin RH, et al.: Epstein–Barr virus
infection in renal transplant recipients: Effects of antithymocyte
globulin and interferon. Ann Intern Med 93:39, 1980

Epstein MA, Achong BG (ed): The Epstein–Barr Virus. New York,
Springer-Verlag, 1979

Epstein MA, Achong BG, Barr YM: Virus particles in cultured
lymphoblasts from Burkitt's lymphoma. Lancet 1:702, 1964

Fleisher G, Lennette ET, Henle G, et al.: Incidence of heterophil
antibody responses in children with infectious mononucleosis. J
Pediatr 94:723, 1979

Fleisher GR, Pasquariello PS, Warren WS, et al.: Intrafamilial
transmission of Epstein–Barr virus infections. J Pediatr 98:16,
1981

Gerber P, Nonoyama M, Lucas S, et al.: Oral excretion of Epstein–
Barr virus by healthy subjects and patients with infectious
mononucleosis. Lancet 2:988, 1972

Grose C, Henle W, Henle G, et al.: Primary Epstein–Barr virus infections in acute neurologic disease. N Engl J Med 292:392,
1975

Hallee TJ, Evans AS, Niederman JC, et al.: Infectious mononucleosis at the United States Military Academy: A prospective study
of a single class over 4 years. Yale J Biol Med 47:182, 1974

Hanto DW, Frizzera G, Gajl-Peczalska KJ, et al.: Epstein–Barr

virus-induced B cell lymphoma after renal transplantation:
Acyclovir therapy and transition from polyclonal to monoclonal
B cell proliferation. N Engl J Med 306:913, 1982

Hanto DW, Frizzera G, Purtilo DT, et al.: Clinical spectrum of
lymphoproliferative disorders in renal transplant recipients and
evidence for the role of Epstein–Barr virus. Cancer Res
41:4253, 1981

Heath CW, Brodsky AL, Potolsky AI: Infectious mononucleosis in
a general population. Am J Epidemiol 95:46, 1972

Henle W, Henle G, Hewetson J, et al.: Failure to detect heterophile antigens in Epstein–Barr virus infected cells and to demonstrate interaction of heterophile antibodies with Epstein–Barr
virus. Clin Exp Immunol 17:281, 1974

Henle W, Henle G, Ho HC, et al.: Antibodies to Epstein–Barr virus in nasopharyngeal carcinoma, other head and neck neoplasms and control groups. J Natl Cancer Inst 44:225, 1970

Kieff E, Dambaugh T, Heller M, et al.: The biology and chemistry
of Epstein–Barr virus. J Infect Dis 146:506, 1982

Porter DD, Wimberly I, Benyesh-Melnick M: Prevalence of
antibodies to EB virus and other herpesviruses. JAMA
208:1675, 1969

Purtilo DT, Bhawan J, Hutt LM, et al.: Epstein–Barr virus infections in the X-linked recessive lymphoproliferative syndrome.
Lancet 1:798, 1978

Purtilo DT, Cassel CK, Yang JPS, et al.: X-linked recessive progressive combined variable immunodeficiency (Duncan's disease). Lancet 1:935, 1975

Robinson J, Brown N, Andiman W, et al.: Diffuse polyclonal B
cell lymphoma during primary infection with Epstein–Barr virus. N Engl J Med 302:1293, 1980

Sawyer RN, Evans AS, Niederman JC, et al.: Prospective studies
of a group of Yale University freshmen. 1. Occurrence of infectious mononucleosis. J Infect Dis 123:263, 1971

Svedmyr E, Jondal M: Cytotoxic effector cells specific for B cell
lines transformed by Epstein–Barr virus are present in patients
with infectious mononucleosis. Proc Natl Acad Sci USA
72:1622, 1975

Tosato G, Magrath I, Koski I, et al.: Activation of suppressor T
cells during Epstein–Barr-virus-induced infectious mononucleosis. N Engl J Med 301:1133, 1979

Zur Hausen H, Schulte-Holthausen H, Klein G, et al.: EBV DNA
in biopsies of Burkitt's tumors and anaplastic carcinomas of the
nasopharynx. Nature 228:1056, 1970

CHAPTER 68

Adenoviruses and Adenovirus-associated Viruses

Adenoviruses

Adenoviruses were first isolated in the winter of 1952–1953 by two groups of investigators. Rowe and his colleagues at the National Institutes of Health cultured adenoids surgically removed from young children and isolated an agent from 33 of 53 adenoids that caused cytopathic effects in the epithelial cells. During the same winter, an epidemic of acute respiratory illness occurred in military recruits at Fort Leonard Wood, Missouri. Although many of the infections were caused by type A influenza, approximately 20 percent of the patients did not have myxovirus infection. The throat washings from one of these patients contained a virus that caused similar cytopathic effects in tissue culture and was related by complement-fixation and neutralization tests to the agent described by Rowe and co-workers. The recovery of these prototype adenovirus strains illustrates their ability to cause epidemic and endemic respiratory illness and also to establish latent or persistent infections in normal hosts. The adenoviruses were named as a group in 1956 by Enders et al. to serve as a reminder of the source of the original prototype strain of adenoviruses and also to designate the prominent involvement of lymphoid tissue in these infections.

Classification

There are at present 34 recognized serotypes of human adenoviruses and 5 candidate serotypes. The 34 human adenoviruses are subdivided by their biophysical, biochemical, biologic, and immunologic characteristics into 5 subgroups (A,B,C,D,E), as indicated in Table 56-4. The biophysical and biologic characterizations of these viruses and the clinical syndromes caused by the subgroups are remarkably parallel. Subgroup A adenoviruses (types 12, 18, and 31) are regularly isolated from the feces of apparently healthy humans, and the incidence of antibody to them is high. This group of adenoviruses is associated with the induction of tumors in hamsters. Subgroup B adenoviruses (types 3, 7, 11, 14, 16, 21, and 34) are most frequently isolated from the throat washings of patients during epidemics of acute respiratory illness and may be associated with severe pneumonia. This group of adenoviruses is not associated with latency. Subgroup C adenoviruses (types 1, 2, 5, and 6) are most commonly isolated from the respiratory tract and gastrointestinal tract of children with mild upper respiratory infections; they are also isolated from long-term cultures of adenoid or tonsillar tissue. Apparently, subgroup C adenoviruses are able to establish latent or masked infections in lymphoid tissue, which persist for extended periods of time. Subgroup D adenoviruses (types 8, 9, 10, 13, 15, 17, 18, 19, 20, 22, 23, 24, 25, 26, 27, 28, 29, 30, 32, and 33) are most often associated with the clinical entities of conjunctivitis and pharyngoconjunctival fever. They are not usually associated with the development of latent infections. Antibodies to these serotypes are also less commonly detected in seroepidemiologic surveys of the population. The only subgroup E adenovirus is adenovirus type 4, which has been associated with severe respiratory disease in both sporadic and epidemic form and with pharyngoconjunctival fever.

Several other adenoviruses have been isolated from patients with clinically significant illness over the past years. They are classified as new serotypes on the basis of unique serum neutralization tests and restriction endonuclease analysis of their genomes. Adenovirus 35 was first isolated in 1972 from the lung and kidney tissues of an immunosuppressed patient who died from interstitial pneumonia following kidney transplant. It has since been frequently identified in immunosuppressed patients with a variety of clinical illnesses and occasionally from nonimmunocompromised patients. Adenovirus 36 was isolated in February 1978 in a stool specimen from a 6-year-old child with enteritis. Adenovirus 37 was first identified in May 1976 in an eye swab from a patient with epidemic catarrhal conjunctivitis. It has also been associated with pharyngoconjunctival fever, and conjunctivitis and urethritis. Adenovirus 38 is the first of the adenoviruses to be termed "noncultivatable," "fastidious," or "enteric." Studies in the United States, England, and the Scandinavian countries have associated this adenovirus with approximately 5 to 15 percent of infantile gastroenteritis. Adenovirus 39 was recently isolated from a stool specimen obtained from a 2-year-old South American child hospitalized with severe respiratory illness.

Clinical Associations

Adenoviruses show a predilection for infection of conjunctival, respiratory, and intestinal epithelium, in addition to regional lymphoid tissue. Incubation periods have been 1 to 2 weeks where discernible. Latent infections, clinically asymptomatic infections, and prolonged viral shedding following clinical illness (particularly of the intestine) have all been described.

Respiratory Illness

Adenoviruses of subgroups B and C cause acute respiratory illness. Subgroup B viruses, particularly serotypes 3, 7, and 14, cause epidemics of respiratory disease in addition to sporadic cases of illness, while subgroup C serotypes 1, 2, 5, and 6 cause sporadic mild respiratory illness of infants and children and can also establish latent infections in tonsillar tissue. Characteristically, the patient with adenoviral respiratory infection is febrile, with pharyngitis,

cervical adenitis, and conjunctivitis. Coryza and cough are also frequently present. Atypical pneumonia may occur, and adenoviruses have been recovered from the lungs of fatal cases of pneumonia. A pertussis-like syndrome has been associated with adenovirus infection. Gastrointestinal symptoms can be prominent, and occasionally adenovirus-induced lymphoid hyperplasia in the gastrointestinal tract can serve as a focus, promoting intussusception.

Adenoviruses are responsible for a significant proportion of serious, acute, lower respiratory tract infections in children. Adenovirus infection may occur at any time of the year, but large outbreaks of pharyngoconjunctival fever tend to occur in the summer and are associated with swimming pools. Prolonged excretion of virus, both from the pharynx and from the intestinal tract, may follow illness. Therefore, fecal to oral spread and respiratory droplet transmission contribute to the spread of virus in the community.

Outbreaks of severe pulmonary disease due to adenovirus types 3, 7, and 21 have been well documented in children as well as in adults in closed populations. Affected children are generally young infants or toddlers less than 3 years old. Longitudinal studies of these children show that a significant proportion will develop evidence of chronic pulmonary disease. Adenovirus types 3, 4, 7, 14, and 21 have also been associated with epidemics of febrile respiratory illness in closed populations, particularly among military recruits. Syndromes include an influenza-like illness, febrile pharyngitis, and atypical pneumonia. The conjunctival mucosa is frequently involved, and there are occasionally gastrointestinal symptoms.

Conjunctivitis

Epidemic keratoconjunctivitis is most commonly caused by adenovirus type 8. Several recent outbreaks due to both adenovirus type 19 and the candidate serotype adenovirus 37 have been described. The disease is characterized by the acute onset of tearing, erythema, suffusion, lymphoid follicles beneath the conjunctiva, and preauricular adenopathy. As the conjunctivitis begins to subside after 1 to 2 weeks, discrete corneal infiltrates appear. The latter may persist for 1 to 2 years.

Gastroenteritis

Adenoviruses can often be cultured from the stools of children with and without symptomatic diarrhea or gastroenteritis. Early epidemiologic studies show that adenoviruses types 1 through 31 are not related to gastrointestinal disease. Several large epidemiologic studies have used electron microscopy to screen the stools of children admitted to the hospital with diarrhea for the presence of virus particles. These studies, which established the rotaviruses and a group of 27 nm viruses related to the prototype Norwalk agent as etiologic agents for acute infectious gastroenteritis in children, also showed that some children had adenovirus-like particles present in their stools. In approximately half the children with gastroenteritis and adenovirus particles detectable in stools by transmission electron microscopy, no adenovirus could be isolated in tissue culture. These noncultivatable, enteric adenoviruses form a unique serotype of adenoviruses, candidate serotype 38, which may be responsible for between 5 and 15 percent of endemic diarrhea in young children (Chapter 71).

Cystitis

Adenovirus 11 infection is a cause of acute hemorrhagic cystitis in children. In a study of hermorrhagic cystitis conducted in Sendai, Japan, and Chicago, Illinois, 51 percent of the Japanese children and 23 percent of the American children had adenoviruria during their illness. Shedding of virus in the urine was associated with rising levels of neutralizing type-specific antibody during convalescence and the detection of adenovirus antigens in exfoliated bladder cells by immunofluorescence, supporting the association of adenovirus infection with hemorrhagic cystitis. Hemorrhagic cystitis due to adenovirus 11 has also been reported in several renal transplant recipients. Adenovirus 21 has been infrequently associated with hemorrhagic cystitis in children.

Clinically, hemorrhagic cystitis due to adenoviruses is an acute self-limited disease lasting 4 or 5 days in the typical American population. Males are affected more commonly than females. Gross hematuria, frequency, and dysuria are common. Enuresis, fever, and suprapubic pain are infrequent symptoms.

Other Clinical Adenovirus Infections

Several other syndromes and illnesses have rarely been associated with adenovirus infection. Among them are myocarditis, hepatitis, renal disease, arthritis, acute and subacute meningoencephalitis, and general exanthems. Many of the higher adenovirus serotypes have been recovered from persons with inapparent infections, and their significance in disease is uncertain. Several of the newly defined adenovirus serotypes, particularly candidate adenovirus 35, have been isolated from immunosuppressed patients and may cause life-threatening infections, primarily interstitial pneumonitis. These infections may represent reactivation of latent virus in a manner analogous to cytomegalovirus infection in this population.

Epidemiology

Human adenoviruses spread from person to person with no other known reservoir. Serologic evidence of infection with one or more subgroup C serotypes is common by age 5. Approximately 5 to 10 percent of civilian respiratory disease appears to be due to adenovirus infection. Occasional outbreaks of subgroup D conjunctival infection have been linked to swimming pools, but most have no such association.

The method of acquisition of adenovirus appears to be important in determining disease. Attempts to artificially induce conjunctivitis in volunteers are unsuccessful unless the conjunctival surface is mildly irritated by swabbing. Dusty environments, chlorinated swimming pool water, and optical instruments, such as tonometers, can all provide the necessary conjunctival irritation and, in some cases, transmit the virus as well.

Nasopharyngeal inoculation of volunteers with adenovirus and ingestion of the virus usually results in asymptomatic infection or mild, afebrile illness. However, inhalation of adenovirus aerosols into the lower respiratory tract has resulted in the full range of clinical syndromes. It is thus suggested that the epidemic adenovirus disease seen in army recruits is the result of airborne spread, facilitated by close contact within a large group of susceptibles. The fecal shedding of virus may facilitate transmission by the fecal to oral route, thus contributing to the spread of infection, particularly between young children or where personal hygiene and sanitation are poor.

Diagnosis

The differential diagnosis of the clinical syndromes caused by adenoviruses includes infections with bacteria, chlamydia, *Mycoplasma pneumoniae*, and several other viruses. The differentiation between streptococcal and adenovirus pharyngitis and a culture for streptococci should be obtained whenever possible. The atypical pneumonia caused by adenoviruses cannot be clinically distinguished from that caused by mycoplasma, and as many as one in five persons with adenovirus atypical pneumonia have modest elevations in cold agglutinins.

Adenoviruses can be recovered in cell culture from respiratory secretions, throat swabs, urine, feces, blood, and biopsy or autopsy tissue of infected persons. Cell culture isolation remains the most sensitive means of detecting these viruses except for the fastidious enteric adenoviruses. The specimen is inoculated into susceptible cells of human origin. Human embryonic kidney (HEK) cells are the most sensitive, but continuous human cell lines, such as HeLa, Hep-2, or KB cells can also be used. Typical adenovirus cytopathic effects (cpe) are grape-like clusters of rounded, refractile cells that have intranuclear inclusions when stained with hematoxylin and eosin. Cpe may be seen in 2 to 7 days but can require several weeks; the rapidity of appearance of cpe depends on virus concentration, serotype, and host tissue. Identification of the agent as an adenovirus can be accomplished using antiserum to detect the hexon group antigen by fluorescent-antibody (FA) or complement-fixation (CF) tests. A monoclonal antibody specific for adenovirus hexon protein has recently been described. Although adenoviruses can be subgrouped according to their hemagglutination reaction with rhesus and rat blood cells, identification of serotype is usually accomplished by neutralization tests. Adenoviruses or adenoviral antigens have also been detected in clinical specimens using electron microscopy, enzyme-linked immunoabsorbent assays (ELISAs), and radioimmunoassays. These methods of adenoviral antigen detection are of primary importance in the diagnosis of adenoviral enteric disease in which virus isolation cannot be accomplished.

Because adenoviruses can be recovered from healthy persons, increases in antibody titer between acute and convalescent sera should be sought to document recent acute infection. This may be done using a complement-fixation test for the group-reactive adenovirus hexon antigen or specific neutralization or hemagglutination-inhibition tests employing type-specific antisera.

Treatment

Treatment of adenovirus infections is symptomatic and supportive. Secondary bacterial infection is not common. Interferon has been used for treatment of adenoviral keratoconjunctivitis. Although there appears to be little modification of the course of established conjunctivitis, interferon does seem to protect against severe manifestations of disease in the contralateral eye in those cases presenting with only one eye involved.

Prevention

Immunity following adenovirus infection is serotype-specific and appears to be long lasting. Because of the problem that adenovirus infections pose in military recruit camps, vaccines have been developed and tested in these populations. Experience with subcutaneous inoculation of an inactivated polyvalent vaccine containing types 3, 4, and 7 indicated that such vaccines could control infection.

More recently, infectious adenovirus vaccines of types 4, 7, and 21 have been developed for oral administration. The virus is encased in an enteric-coated capsule and is released in the intestine, where it causes an asymptomatic, nontransmissible infection. Good protection has been provided by these vaccines. However, the rate of seroconversion to adenovirus 7 and 21 decreases when several vaccine types are given simultaneously.

Currently, efforts are being directed toward the exploration of additional forms of immunoprophylaxis, including new modes of vaccine administration and the development of purified subunit vaccines. The development of new vaccines, particularly those that do not require administration of whole virus, is desirable, since adenoviruses may establish persistent infection and are capable of both including tumors in lower animals and causing transformation of cells in vitro.

Adeno-associated Viruses

The adeno-associated viruses (AAV) are small DNA viruses of the parvovirus family. AAV were first observed by electron microscopy as small particles in adenovirus prepara-

tions. They were thought to represent either precursor or breakdown products of the adenovirus virions. It was later determined that they represented a second virus that was biologically and structurally different from adenovirus. Four serotypes of AAV have been identified. AAV-1 was isolated from adenovirus preparations that had been cultured in both human and monkey cells, AAV-2 and AAV-3 were isolated from human cell lines and considered to be human serotypes, and AAV-4 has only been isolated from African green monkey cells. Since the initial demonstration of AAV, serotypes of bovine, avian, equine, and canine origin have been discovered. The host range of AAVs is not as limited as that of their helper adenoviruses: avian AAVs can productively infect man using a human adenovirus helper, and herpes simplex viruses type 1 and type 2 can provide the necessary helper functions for AAV replication.

Several seroepidemiologic studies have shown that AAV infection is common in the general population. Antibodies are generally acquired between 6 months and 2 years of age, and 50 to 80 percent of persons tested can be shown to have serologic evidence of AAV infection. AAV cryptically infects human and primate cells in vivo, and latent infection of human cells with AAV has been documented. No overt clinical syndromes have been associated with the AAV infection.

Coinfection or previous infection of cells with AAV may modify the ability of some DNA viruses to transform those cells or induce tumor formation in animal models. The ability of adenovirus 12 to induce tumors in hamsters is decreased from 44 to 18 percent by AAV infection. Perinatal transplacental infection of mice with AAV protects these mice against a lethal postnatal adenovirus infection. In a study of patients with either cervical carcinoma or carcinoma of the prostate, neoplasms for which an etiologic role for herpes simplex virus has been discussed, the incidence of antibodies to AAV 2 and AAV 3 was lower than in a comparable group of patients without carcinoma who were matched for sex and age. The suggestion implicit in these studies is that patients with AAV infection may be at decreased risk of developing either cervical or prostatic carcinomas when later exposed to herpes simplex virus. These data are provocative and not proven. Any role of AAV in human disease remains to be established.

FURTHER READING

Books and Reviews

ADENOVIRUSES

Chanock RM: Impact of adenoviruses in human disease. Prev Med 3:466, 1974

Spencer MJ, Cherry JD: Adenoviral infections. In Fergen RD, Cherry JD (eds): Textbook of Pediatric Infectious Diseases. Philadelphia, Saunders, 1981

Tullo AB: Clinical and epidemiological features of adenovirus keratoconjunctivitis. Trans Ophthalmol Soc UK 100:263, 1980

ADENO-ASSOCIATED VIRUSES

Berns KI, Hauswirth WW: Adeno-associated viruses. Adv Virus Res 25:407, 1979

Selected Papers

ADENOVIRUSES

Brandt CD, Kim HW, Vargosko AJ, et al.: Infections in 18,000 infants and children in a controlled study of respiratory tract disease. I. Adenovirus pathogenicity in relation to serologic type and illness syndrome. Am J Epidemiol 90:484, 1969

D'Angelo LJ, Hierholzer JC, Kennlyside RA, et al.: Pharyngoconjunctival fever caused by adenovirus type 4: Report of a swimming pool-related outbreak with recovery of virus from pool water. J Infect Dis 140:42, 1979

Faden H, Gallagher M: Disseminated infection due to adenovirus type 4. Clin Pediatr 19:427, 1980

Fox JP, Hall CE, Cooney MK: The Seattle virus watch. Am J Epidemiol 105:362, 1977

Harnett GB, Bucens MR, Clay SJ, Saker BM: Acute haemorrhagic cystitis caused by adenovirus type 11. Med J Aust 1: 565, 1982

Hierholzer JC, Sprague JB: Five-year analysis of adenovirus and antibody levels in an industrial community following an outbreak of keratoconjunctivitis. Am J Epidemiol 110:132, 1979

James AG, Lang WR, Liang AY, et al.: Adenovirus type 21 bronchopneumonia in infants and young children. J Pediatr 95:530, 1979

Ladisch S, Lovejoy FH, Hierholzer JC, et al.: Extrapulmonary manifestations of adenovirus type 7 pneumonia simulating Reye syndrome and the possible role of an adenovirus toxin. J Pediatr 95:348, 1979

Mackey JK, Green M, Wold WSM, et al.: Analysis of human cancer DNA for DNA sequences of human adenovirus type 4. J Natl Cancer Inst 62:23, 1979

Mufson MA, Belshe RB: A review of adenoviruses in the etiology of acute hemorrhagic cystitis. J Urol 115:191, 1976

Nicolas JC, Ingrand BF, Bricout F: A one-year virological survey of acute intussusception in childhood. J Med Virol 9:267, 1982

Sundmacher R, Wigand R, Cantell K: The value of exogenous interferon in adenovirus keratoconjunctivitis: Preliminary results. Graefe's Arch Clin Exp Ophthalmol 218:139, 1982

Takafuji ET, Gaydos JC, Allen RG, et al.: Simultaneous administration of live, enteric-coated adenovirus types 4, 7 and 21 vaccines: Safety and immunogenicity. J Infect Dis 140:48, 1979

Wenman WM, Pagtakhan RD, Reed MH, et al.: Adenovirus bronchiolitis in Manitoba. Chest 81:605, 1982

Zahradnik JM, Spencer MJ, Porter DD: Adenovirus infection in the immuno-compromised patient. Am J Med 68:725, 1980

ADENO-ASSOCIATED VIRUSES

Blacklow NR, Hoggan MD, Kapikian AZ, et al.: Epidemiology of adenovirus-associated virus infection in a nursery population. Am J Epidemiol 88:368, 1968

Buller RML, Janik JE, Sebring ED, et al.: Herpes simplex virus type 1 and 2 completely help adenovirus-associated virus replication. J Virol 40:241, 1981

Lipps BV, Mayor HD: Defective parvoviruses acquired via the transplacental route protect mice against lethal adenovirus infection. Infect Immun 37:200, 1982

Maza M, Carter BJ: Inhibition of adenovirus oncogenicity in hamsters by adeno-associated virus DNA. J Natl Cancer Inst 67:1323, 1981

Rosenbaum MJ, Edwards EA, Pierce WE, et al.: Serologic surveillance for adeno-associated satellite virus antibody in military recruits. J Immunol 106:711, 1971

Yates VJ, Dawson GJ, Pronovost AD: Serologic evidence of avian adeno-associated virus infection in an unselected human population and among poultry workers. Am J Epidemiol 113:542, 1981

CHAPTER 69

Human Papovaviruses

Papillomaviruses

Papillomaviruses are the etiologic agents causing warts in humans and papillomas in rabbits, dogs, and livestock. The inability to propagate human papillomaviruses in cell culture systems has impeded their classification and characterization based on biologic properties. Papillomaviruses have now been assigned to several different groups based on patterns of restriction endonuclease digestion of their DNA, DNA homology, and serologic cross-reactivity. A classification of human papillomaviruses is provided in Table 56-4. Each type of human papillomavirus (HPV) is associated with the production of a single variety of wart. A single clinical type of wart, however, may be caused by several HPV types.

Epidemiology and Clinical Associations

The papillomaviruses associated with common warts (common, plantar, or flat warts) are usually found in children and young adults. Natural transmission of these viruses is presumed to be through contact and minor abrasions. There are no specific predisposing factors for the development of common warts, although more extensive warty disease may be seen in individuals with primary immune deficiencies and patients treated with immunosuppressive therapy. These warts usually are caused by HPV types 1, 2, 3, and 4. HPV types 1, 2, and 3 are commonly isolated from warts of patients age 5 to 15, while HPV-4 warts oc-

cur predominantly in patients age 20 to 25. Serologic studies show corresponding serotype-specific antibody rises in these age groups. HPV-7 infection is also associated with the development of common warts. It is seen only in the warts found in meat handlers. The natural history of these lesions is spontaneous regression.

The warts and macular lesions associated with epidermodysplasia verruciformis may be seen throughout the lifetime of infected individuals. The natural route of transmission for the three HPV serotypes associated with these lesions (HPV-5, 8, and 9) is not known. Epidermodysplasia verruciformis is a familial disease that may be linked to an autosomal recessive gene. It is usually associated with depressed cell-mediated immunity, which may contribute to the lifetime persistence of disseminated wart lesions. Malignant conversion of some of the warty lesions, particularly those associated with HPV-5, is observed in approximately 25 percent of the affected patients. Those warts that undergo malignant transformation usually occur on sun-exposed areas.

Condylomata acuminata are genital warts caused by HPV-6. They generally occur in young adults, and their transmission is mainly venereal. Sexual promiscuity and changes in hormone balance (for example, in pregnancy) may be predisposing factors for their development. Condylomata acuminata may regress spontaneously but are subject to recurrence and may give rise to extensive lesions. Malignant conversion of these anogenital warts has been observed in rare instances in lesions of the vulva, penis, and anus. Laryngeal papillomas form another group of HPV-caused lesions. The juvenile type of laryngeal papillomatosis is generally found in children under 5 years of age. The HPV-6 genome has been demonstrated in some of these lesions, suggesting that the development of these papillomas follows transmission of virus at birth from mothers with condylomata acuminata to their children. Laryngeal papillomas are also seen in adults, and a small percentage of these lesions can also be shown to contain the HPV-6 genome. Laryngeal papillomas show frequent and rapid recurrence after excision, but their malignant conversion is only rarely observed. Reported transformation has usually been seen after radiation therapy of the lesions. In the absence of radiation therapy, malignant transformation may be associated with heavy smoking. A new type of human papillomavirus, HPV-11, has recently been reported to be another etiologic agent of laryngeal papillomatosis.

Epidermal infection with human papillomaviruses is associated with the development of serotype-specific antibodies and cell-mediated immunity. Both IgM and IgG humoral antibodies appear in wart patients. Detection of IgG antibodies and cell-mediated immunity correlate with resolution of the warts. The number of warty lesions and the persistence of those HPV infections is inversely related to the presence of IgG antibody. Thus, those patients without any antibodies to HPV are more likely to have a greater number of warts that persist for a longer period of time. Those patients who possess HPV-specific IgM but have not developed HPV-specific IgG have intermediate numbers of warts and intermediate healing times.

Human Papillomaviruses in Cancer

Most human papillomaviruses, including the etiologic agents of common, plantar, and flat warts, are entirely benign and are not associated with subsequent development of carcinomas. However, as indicated previously, several HPVs have recently been implicated in the malignant transformation of special groups of warty lesions to squamous cell carcinomas. HPV-6 is related to juvenile laryngeal papillomatosis and the malignant transformation of condylomata acuminata. Recently, Ostrow and his colleagues have identified HPV-5-specific nucleotide sequences in the squamous cell carcinomas evolving from the chronic wart disease, epidermodysplasia verruciformis. Not only was HPV-5 DNA present in the warty lesions of these patients, but the viral genome could also be detected in a subcutaneous metastatic tumor from one of these patients. This latter observation eliminated the possibility that the HPV-5 DNA present in the malignant tumors in these patients resulted from cross-contamination from an adjacent benign wart containing HPV-5 genome but does not establish the virus as the etiologic agent of the carcinoma.

An additional papillomavirus-associated clinical entity that may be linked to malignancy has been reported recently. It is described as both noncondylomatous cervical wart viral infection and subclinical papillomavirus infection. Lesions are associated with atypia on Papanicolaou smears and appear as abnormal mucosa with vascular surface contour and pigment changes similar to cervical intraepithelial neoplasia under colposcopy. Tissue sections from cervical biopsies taken from these lesions show pathologic changes that are similar to, but distinct from, those of cervical intraepithelial neoplasia. HPV-specific antigens have been detected in 50 percent of these dysplastic cervical lesions using a cross-reactive genus-specific antiserum. This association of HPV infection with cervical dysplasia is of particular interest, since cervical dysplasia is epidemiologically associated with carcinoma in situ and invasive cervical carcinoma. The role that HPV infection plays in cervical intraepithelial neoplasia is not yet known.

Several points may be made regarding the relationship between papillomaviruses and carcinomas. The oncogenic potential of any given papillomavirus type is unique. Therefore, the viruses of common warts probably have low oncogenic potential, while HPV-5 and HPV-6 are more strongly related to the development of malignancies. The role of cocarcinogens may be crucial to potentiating papillomavirus-induced malignant transformation. This is particularly evident in epidermodysplasia verruciformis, where malignant transformation of warty lesions occurs only in those areas exposed to UV light. Additionally, the malignant transformation associated with HPV-6 in laryngeal papillomatosis appears to occur only after radiation therapy to the lesions. The immunocompetence of the host also effects the transformation of viral papillomas. Defects in immunity in patients with epidermodysplasia verruciformis or immunosuppression (such as in renal transplant patients) are associated with both an increased number of warts and an increased incidence of cutaneous carcinomas. The state of the papillomavirus genome in the malignant transformed cell is unknown at this time.

Polyomaviruses

Classification

In 1971 two human polyomaviruses were isolated. Each was designated by the initials of the person from whom it was obtained. BKV was isolated by Gardner et al. from the urine of a patient receiving immunosuppressive therapy following kidney transplant. This urine contained transitional epithelial cells with enlarged inclusion-bearing nuclei. Electron microscopic examination of the exfoliated cells showed the presence of intranuclear virus-like particles of the size and structure typical of polyomaviruses. The virus hemagglutinates human type O and guinea pig red blood cells. In serologic tests, BKV reacts slightly with antisera specific for SV40 but not with antisera against polyomavirus or human papillomaviruses. JCV was isolated by Padgett et al. from brain tissue obtained at autopsy from a person with progressive multifocal leukoencephalopathy (PML). JCV also hemagglutinates human and guinea pig erythrocytes. Weak reactions are detected between JCV and antisera against SV40.

Other viruses serologically indistinguishable from BKV and JCV have been recovered from a wide variety of patients. Although these viruses are isolated with the highest frequency from patients immunocompromised either by virtue of inherent defects in immunity or by immunosuppression due to pregnancy or suppressive medication, they can also be isolated at a low frequency from apparently normal individuals. All JCV-related viruses tested appear to be identical to prototype JCV. Most BKV-related viruses are very similar to the prototype BKV, although some differences may be seen with DNA reassociation techniques and restriction endonuclease mapping.

Epidemiology

Several epidemiologic studies of human polyomaviruses have shown that these infections are widespread in many parts of the world. Infection with these viruses is a common occurrence during childhood, with BKV and JCV infections occurring independently. Approximately 80 percent of children age 5 to 9 can be found to have antibodies to BKV present in their serum. Antibody to JCV may be acquired somewhat later in life. Padgett and Walker found that 65 percent of children in Wisconsin were JCV-seropositive by 10 to 14 years of age. In another study, 50 percent of the children had antibodies to JCV by the age of 3 years. Seventy to one hundred percent of adults are seropositive for antibodies to both JCV and BKV. There is probably no animal reservoir for these viruses.

The high incidence of infection with these viruses implies that they are transmitted easily. However, almost nothing is known about the route of transmission. The viruses have, with rare exception, been recovered only from the urine. It may be, however, that a respiratory route of viral transmission is the source of most primary BKV and JCV in infections.

Several recent studies have addressed the possibilities of congenital infections with human polyomaviruses. For example, in a serologic study of pregnant women in the United States, four-fold or greater polyomavirus antibody rises during pregnancy occurred. These antibody titer changes were all consistent with virus reactivation. Virus-specific IgM was detected in the serum of mothers after BKV reactivation but not after JCV reactivation. No evidence of congenital transmission of either JCV or BKV has been found in either English or American studies, although some Japanese studies have turned up some positive evidence.

A high percentage of adults have antibodies to BKV and JCV detectable decades after their primary childhood infection. Many patients have significant levels of antibody against BKV prior to immunosuppression and nevertheless show four-fold or greater rises of titer thereafter. In some persons this increase has been correlated with the appearance of IgM antibodies in the serum. These data suggest that there is lifelong persistence of virus with reactivation of virus subsequent to immunosuppression.

Clinical Association

Primary Infection

A mild respiratory illness has been recorded in children at the time of appearance of antibodies to BKV. Goudsmit et al. retrospectively investigated sera from 177 children admitted to hospital for acute respiratory disease. Sera from 7 of these children showed seroconversion suggestive of primary infection with BKV. All children with BKV infection had upper respiratory tract infections, and BKV was associated with 8 percent of all upper respiratory tract infections in this series. Tonsillar tissue from children with recurrent attacks of acute respiratory disease has been found to contain BKV DNA by hybridization with cloned genomic DNA of prototype BKV, but no infectious virus could be isolated from these tonsils.

Primary infection with JCV has not been documented. The lack of data regarding primary JCV infections may be related to difficulty in obtaining primary fetal glial cells for virus isolation. In addition, the slightly older age of children when infected may make primary JCV disease less severe and preclude patients seeking medical attention.

Urinary Tract Infection

BKV was first isolated from the urine of an immunosuppressed patient who had received a kidney and ureter transplant 3 months previously. Virus-containing cells were shed from the lumen of a stenosed donor ureter. Subsequently, Coleman et al. documented the excretion of large quantities of polyomavirus-infected cells in the urine of renal transplant patients who had been given large doses of cortisol steroids for possible rejection episodes. The transplanted ureters of these patients proved to be stenosed and ulcerated, and the urothelium contained cells with large inclusion bodies containing polyomavirus particles. Hogan et al. studied infection of JCV and BKV in 61 immunosuppressed renal transplant patients. Polyomavirus excretion was detected in the urine of 12 of 61 patients (20 percent), 11 of whom excreted JCV, while 9 excreted BKV. Serologic data suggested that most JCV infections were primary, while most BKV infections resulted from virus reactivation. Urinary tract excretion of polyomavirus is associated with drug-dependent diabetes mellitus, arterial occlusive disease, and urethral stricture with loss of renal function.

Progressive Multifocal Leukoencephalopathy

The one disease closely linked to human polyomaviruses is progressive multifocal leukoencephalopathy (PML). PML is an uncommon, generally fatal, demyelinating disease that occurs in patients with altered immunocompetence. It occurs most frequently as a complication in patients with chronic lymphocytic leukemia or Hodgkin's disease but may also be seen in patients who are immunosuppressed following renal transplantation or with inherited immunodeficiency syndromes.

PML is caused by infection of oligodendrocytes by polyomavirus (most commonly JCV). Several cases of PML have been associated with an SV40-like virus designated as SV40-PML. Pathologic lesions of PML occur in both the gray and white matter and may be distributed throughout the neuraxis. These lesions are classically small discrete areas of demyelination that may become confluent. Cytologically, the outstanding features are unusual astrocytes with bizarre chromatin patterns and atypical oligodendroglia with enlarged, ill-defined inclusions. As a result of the infection of the oligodendrocytes, demyelination occurs. Presumably, this oligodendroglial degeneration is secondary to activation of previously latent JCV by immunosuppression.

Although PML is now recognized as a viral disease, diagnosis of this illness still rests on finding the characteristically altered oligodendroglia in biopsied brain tissue. Studies of serum antibody levels are not helpful because of the high background level of antibody in the general population and because affected patients often have impaired immune responses. Neither specific JCV antibodies nor other proteins characteristic of demyelinating disease have been found in the cerebrospinal fluid. In a recently reported case of PML in a male homosexual with T cell insufficiency, oligoclonal bands were documented.

There is no known treatment for PML. Human leukocyte interferon has been used for prophylaxis of BKV and JCV infections in renal transplant patients with success.

Oncogenicity of Human Polyomaviruses

The human polyomaviruses, BKV and JCV, are oncogenic in neonatal hamsters. Tables 69-1 and 69-2 document the types of tumors that are induced by BKV and JCV, respectively, in these laboratory animals. Malignant, histologically distinct brain tumors developed in two of four adult owl monkeys inoculated with JCV intercerebrally, subcutaneously, and intravenously simultaneously. Four owl monkeys inoculated similarly with BKV and four inoculated with SV40 did not develop tumors after 3 years.

TABLE 69-1. TYPES OF TUMORS INDUCED BY BK VIRUS IN HAMSTERS

Osteosarcoma
Fibrosarcoma
Undifferentiated and poorly differentiated sarcoma
Ependymoma
Choroid plexus papilloma
Reticulum cell sarcoma
Insulinoma

TABLE 69-2. TYPES OF TUMORS INDUCED BY JC VIRUS IN HAMSTERS

Medulloblastoma
Undifferentiated neuroectodermal tumor
Glioblastoma
Ependymoma
Pineocytoma
Meningioma
Neuroblastoma
Leiomyosarcoma
Hemangioma

Attempts have been made to associate BKV and JCV infection with human tumors, using both serologic and DNA hybridization techniques on hundreds of patients and tumors. All results have been uniformly negative. No data, therefore, exist linking BKV or JCV to human carcinomas.

SV40 is tumorigenic in laboratory animals. It produces subclinical infections indicated by serum antibody prevalence and virus isolation in humans. SV40-contaminated poliovirus vaccine was administered to millions of people in the early 1960s. Although it has been reported that SV40-like antigens are present in the tumors of patients with meningiomas and other brain tumors, there is no evidence that cancer trends have been affected by poliovirus vaccine containing SV40. Mortimer et al. recently reported on the 17- to 19-year follow-up of newborns exposed to SV40-contaminated poliovirus vaccine in the first 3 days of life and found no excess risk of cancer of any sort.

FURTHER READING

Books and Reviews

Adler A, Safai B: Immunity in wart resolution. J Am Acad Dermatol 1:305, 1979

Bender ME, Pass F: Papillomaviruses and cutaneous malignancy. Int J Dermatol 20:468, 1981

Briggman RA, Wheeler CE: Immunology of human warts. J Am Acad Dermatol 1:297, 1979

Dvoretzky I, Lowy DR: Infections by human papillomavirus (warts). Am J Dermatopathy 4:85, 1982

Howley PM: The human papillomaviruses. Arch Pathol Lab Med 106:429, 1982

Mantyjarvi RA: New oncogenic human papovaviruses. Med Biol 57:29, 1979

Orth G, Jablonska S, Breitburd F, et al.: The human papillomaviruses. Bull Cancer 65:151, 1978

Padgett BL, Walker DL: Natural history of human polyomavirus infections. In Stevens JG (ed): Persistent Viruses. New York, Academic Press, 1978, p 751

Shah K, Nathanson N: Human exposure to SV-40: Review and comment. Am J Epidemiol 103:1, 1976

Von Krogh G: Warts: Immunologic factors and prognostic significance. Int J Dermatol 18:195, 1979

Selected Papers

Coleman DV, et al.: A prospective study of human polyomavirus infection in pregnancy. J Infect Dis 142:1, 1980

Daniel R, Shah K, Madden D, et al.: Serological investigation of the possibility of congenital transmission of papovavirus JC. Infect Immun 33:319, 1981

Dei R, Marmo F, Corte D, et al.: Age-related changes in the prevalence of precipitating antibodies to BK virus in infants and children. J Med Microbiol 15:285, 1982

Gardner DV, Field AM, Coleman DV, et al.: New human papovavirus (BK) isolated from urine after renal transplantation. Lancet 1:1253, 1971

Goudsmit J, Baak ML, Slaterus KW, et al.: Human papovavirus isolated from urine of a child with acute tonsillitis. Br Med J 283:1363, 1981

Goudsmit J, Wertheim-van Dillen P, van Strien A, et al.: The role of BK virus in acute respiratory disease and the presence of BKV DNA in tonsils. J Med Virol 10:91, 1982

Grossi MP, et al.: Lack of association between BK virus and ependymomas, malignant tumors of pancreatic islets, osteosarcomas and other human tumors. Intervirology 15:10, 1981

Heritage J, Chesters PM, McCancer DJ: The persistence of papovavirus BK DNA sequences in normal human renal tissue. J Med Virol 8:143, 1981

Hogan TF, Borden EC, McBain JA, et al.: Human polyomavirus infections with JC virus and BK virus in renal transplant patients. Ann Intern Med 92:373, 1980

Miller JR, et al.: Progressive multifocal leukoencephalopathy in a male homosexual with T-cell immune deficiency. N Engl J Med 307:1436, 1982

Mortimer EA, Lepow ML, Gold E, et al.: Long-term follow-up of persons inadvertently inoculated with SV40 as neonates. Med Intell 305:1517, 1981

Mounts P, Shah KV, Kashima H: Viral etiology of juvenile- and adult-onset squamous papilloma of the larynx. Proc Natl Acad Sci USA 79:5425, 1982

Orth G, et al.: Identification of papillomavirus in butchers' warts. J Invest Dermatol 76:97, 1981

Ostrow RS, Bender M, Niimura M, et al.: Human papillomavirus DNA in cutaneous primary metastasized squamous cell carcinomas from patients with epidermodysplasia verruciformis. Proc Natl Acad Sci USA 70:1634, 1982

Padgett BL, Walker DL, ZuRhein GM, et al.: Cultivation of papova-like virus from human brain with progressive multifocal leucoencephalopathy. Lancet 1:1257, 1971

Pfister H, zur Hausen H: Seroepidemiological studies of human papilloma virus (HPV-1) infections. Int J Cancer 21:161, 1978

Reid R, Stanhope CR, Herschman BR, et al.: Genital warts and cervical cancer. I. Evidence of an association between subclinical papillomavirus infection and cervical malignancy. Cancer 50:377, 1982

Reid R, Laverty CR, Coppleson M, et al.: Noncondylomatous cervical wart virus infection. Obstet Gynecol 55:476, 1980

Shah K, Daniel R, Madden D, et al.: Serological investigation of BK papovavirus infection in pregnant women and their offspring. Infect Immun 30:29, 1980

Shokri-Tabibzadeh S, Koss L, Molnar J, et al.: Association of human papillomavirus with neoplastic processes in the genital tract of four women with impaired immunity. Gynecol Oncol 12:S129, 1981

Tabuchi K, Kirsch WM, Low M, et al.: Screening of human brain tumors for SV40-related T antigen. Int J Cancer 21:12, 1978

Zang KD, May G: Expression of SV40-related T antigen in cell cultures of human meningiomas. Naturwissenschaften 66:59, 1979

The Enteroviruses

The *Enterovirus* genus of the Picornaviridae family contains six major groups: polioviruses, echoviruses, Coxsackie viruses group A and group B, the new enterovirus serotypes 68, 69, 70, and 71, and hepatitis A viruses. All of these viruses infect and multiply in the gastrointestinal tract. There are three distinct poliovirus serotypes, 23 Coxsackie A serotypes, 6 Coxsackie B serotypes, 31 echovirus serotypes, 4 viruses designated as enteroviruses 68–71, and 1 serotype of hepatitis A.

Enteroviruses are 30 nm particles composed of a single-stranded RNA genome with a protein coat of icosahedral symmetry. The viruses are indistinguishable morphologically, stable at pH 3, and resist inactivation by ether. Assignment of a virus to one of these groups is based on biophysical properties, differences in growth and tissue culture systems, pathogenicity for various strains of laboratory animals, and serologic reactivity.

The RNA of enteroviruses is a positive single strand and, therefore, by definition, serves as the messenger RNA for the replication cycle of the virus. All enteroviruses code for four proteins (VP1, VP2, VP3, VP4), derived by cleavage of a precursor polyprotein, that constitute the capsid of the virus. Experiments with members of each of these groups suggest that the four capsid proteins contain the antigens that determine the serotype of the virus strain. Recent work suggests that VP2 induces specific neutralizing antibodies for several echoviruses and Coxsackie B viruses. Group-reactive antigenic determinants are found in all of the capsid proteins for Coxsackie B viruses.

In contrast, it would appear that antibody to VP1 of the polioviruses is neutralizing and, therefore, defines serotype specificity. However, antibodies to other capsid polypeptides seem to contribute to neutralization. More than one capsid protein may be required to form the antigenic determinant that elicits the formation of serotype-specific antibodies.

History

Sporadic cases of paralytic disease are as old as recorded history. A famous Egyptian stele from the period 1580 to 1350 BC depicts a priest with a flail atrophic lower limb, typical of paralytic polio. Sir Walter Scott underwent an illness in 1772 that was one of the earliest and certainly the most renowned case of polio described in the British Isles. Acute paralytic illness in children was first described by Underwood in his textbook in 1789, while the name "anterior poliomyelitis" is attributed to Kussmaul in the late nineteenth century. The term "poliomyelitis" was derived from the Greek for gray marrow of the spinal cord and the Latin (*itis*) for inflammation. The location of the involved cells in the anterior horns of the spinal cord contributed the designation "anterior." Outbreaks of illness were regularly identified in the nineteenth and twentieth centuries by the observation of paralysis among young children. The first isolations of poliovirus were achieved in 1908 by the inoculation of central nervous system tissue into susceptible monkeys via the intracerebral route. In 1949, Enders, Robbins, and Weller reported their classic experiments on the cultivation of poliovirus in tissue cultures of nonneural human cells.

The histories of the other enteroviruses are relatively recent. In 1948, Dalldorf and Sickles isolated a filterable agent from a stool of a patient from Coxsackie, New York, who had paralytic illness. Subsequently, a large group of antigenically related agents has been characterized and designated Coxsackie (A and B) viruses. The echoviruses were also isolated initially from human fecal specimens, frequently from patients without overt disease, and their name represents an acronym (*e*nteric, *c*ytopathic, *h*uman, *o*rphan). With increasing clinical study, they have been associated with a wide variety of illness, so that most no longer are "orphans." As distinct antigenic serotypes were identified, they were assigned sequential numbers. Since 1969, new enterovirus types have been assigned enterovirus type numbers rather than being subclassified as Coxsackie viruses or echoviruses, hence, the designation enteroviruses 68 through 71. Finally, hepatitis A virus now appears to be a member of the picornavirus family.

Epidemiology

The epidemiology of all human enteroviruses is quite similar. The pattern is most clearly defined for the polioviruses because paralytic disease has been so readily identifiable. As early as 1916, the clinical epidemiologic features were defined on the basis of an outbreak that occurred that year in New York City. These principles were: (1) Poliomyelitis is, in nature, exclusively a human infection, transmitted from person to person without the necessary intervention of a lower animal or insect host. (2) The infection is far more prevalent than is apparent from the incidence of clinically recognized cases, since a large majority of persons infected become carriers without clinical manifestations. It is probable that during an epidemic, such as that in New York City, a very considerable proportion of the population became infected, adults as well as children. (3) The most important agencies in disseminating infection are the unrecognized carriers and perhaps mild, abortive cases ordinarily escaping recognition. It is fairly certain that frank paralytic cases are a relatively minor factor in the spread of infection. (4) An epidemic of 1 to 3 recognized cases per 1000, or even less, immunizes the general population to such an extent that the epidemic declines spontaneously due to the depletion of susceptible persons.

Enteroviruses have a worldwide distribution, with increased prevalence during the warm months of the year in temperate climates. Within the United States, some variation in geographic distribution of infections may be due to importation of viruses, for example, from Latin America. Epidemics occur between May and October in the United States and other areas of the north temperate zone. The seasonal prevalence of these viruses is shown in Figure 70-1. Sporadic infections due to these viruses can occur at any time throughout the year.

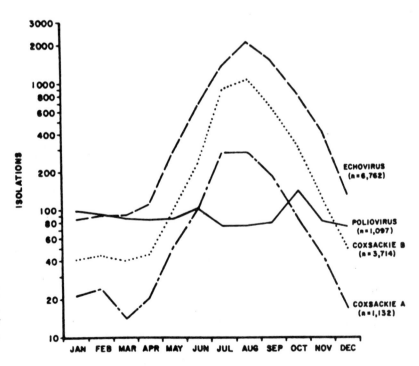

Figure 70-1. Seasonal prevalence of human enterovirus infections, United States, 1970–1979. *(From CDC Enterovirus Surveillance Summary 1970–1979. Issued November 1981, p 6. U.S. Dept. HHS, PHS.)*

The seroepidemiology of enterovirus infections, including polio and hepatitis A viruses, demonstrates an increased transmission of infection at a young age among persons of lower socioeconomic status. Crowding creates intimate living conditions, and it may also be associated with poor hygiene, which enhances the fecal to oral transmission of these agents.

Enterovirus illness and laboratory isolation of enteroviruses is most commonly reported from children in the 1- to 4-year age group. It is not clear whether such data reflect an enhanced concern over any illness at that age, whether the infections occur more frequently in infants, or whether they are more likely to be overt rather than occult. Nevertheless, when specific outbreaks occur within a community, a broad age spectrum of persons may be infected.

A 2- to 5-year periodicity has been observed with Coxsackie B infection, suggesting that a limited number of serotypes contributes to this pattern. These agents are also able to cause outbreaks of disease when a newly susceptible population is present in the community. When a specific virus circulates frequently, younger infants are more likely to be susceptible and develop disease. Perhaps the numerous serotypes of echoviruses have a broader range of age-related attack rates, since a given serotype circulates through a community sporadically. It is unlikely that any individual would encounter all serotypes of enteroviruses in childhood.

The clinical epidemiology of these infections suggests that respiratory excretion of virus is not as important a means of spread as is fecal to oral transmission. Intimate human contact is important in transmission of virus, and communicability within households is greatest between children. Diapered infants appear to be more efficient disseminators of infection than do other individuals. In the current era of day-care centers and nursery schools, it has already been shown that hepatitis A virus is transmitted in this setting and, if present, is transmitted to virtually all children in the nursery. It is likely that outbreaks of infection due to other enteroviruses are also occurring in these facilities.

Community outbreaks of enterovirus infection can spread to hospital nurseries. A newborn who acquires virus from the mother or from nursery personnel may spread it throughout the nursery. The viruses can be introduced into intensive care units by patients or personnel. Recognition of such infections imposes a need to institute isolation precautions, such as cohorting of infants and personnel, to minimize spread of infection.

Clinical Illness

The broad spectrum of clinical disease produced by the enteroviruses overlaps among groups. A listing of the various syndromes is included in Table 70-1. The more common manifestations associated with infection are discussed briefly.

TABLE 70-1. CLINICAL MANIFESTATIONS OF ENTEROVIRUS INFECTIONS

| Clinical Syndrome | Viruses Implicated | | | |
| | Polio | Coxsackie | | Echo |
		A	B	
Asymptomatic infection	X	X	X	X
Nonspecific febrile illness	X	X	X	X
Respiratory disease		X	X	X
Exanthems		X	X	X
Enanthems		X		
Pleurodynia			X	
Orchitis			X	
Myocarditis			X	
Pericarditis			X	X
Aseptic meningitis and meningoencephalitis	X	X	X	X
Disseminated neonatal infection			X	X
Transitory muscle paresis	X	X	X	X
Paralytic disease	X	X	X	X

Febrile Illness

The great majority of infections with enteroviruses produce no specific clinical hallmarks. Although the portal of entry is the gastrointestinal tract, they are not frequently responsible for gastroenteritis. In young children, undifferentiated febrile illness, nonspecific malaise, and myalgias are frequently associated with enterovirus infections. There is nothing unique about this type of clinical presentation. A recent prospective study of newborn infants in Rochester, New York, demonstrated that during a typical enterovirus season, as many as 13 percent of infants acquired infection with these viruses during the first month of life. Although 80 percent of these patients were asymptomatic, most of the symptomatic infants were admitted to the hospital because they were suspected of having bacterial sepsis. These observations allow an estimate of the frequency of enterovirus infection of very young infants. It is an impressive 7 infections per 1000 live births, which occur in the first month of life during the months of seasonal prevalence of these agents. It has come to be appreciated that enterovirus infection is as common as other neonatal infections of widespread clinical concern.

Congenital and Neonatal Infections

Transplacental and neonatal transmission have been demonstrated with Coxsackie B viruses, resulting in a serious disseminated disease that may include hepatitis, myocarditis, meningoencephalitis, and adrenal cortical involvement. Infants infected in the newborn period with echoviruses have died with disseminated disease, with the predominant feature of the infection being hepatic necrosis. Serologic studies have suggested that maternal antibody protects the infant from severe disease, although they may acquire infection, as documented by virus excretion. When poliovirus infections were common, examples of transmission from infected mother to fetus were apparent by the birth of a paralyzed infant.

Respiratory Disease

Mild upper respiratory tract illness has been associated with several of the Coxsackie and echoviruses. Very few cases of pneumonia have been attributed to Coxsackie virus infection.

Exanthems and Enanthems

Various enteroviruses (particularly echoviruses 9 and 16, and Coxsackie A9, A16, and B5) have been associated with large outbreaks of febrile rash disease. Younger children are more likely to develop exanthems, which vary widely in their characteristics. Small vesicular lesions have been described on the hands and feet in association with ulcers of the buccal mucosa (hand–foot–mouth disease), particularly with Coxsackie A16. Macular and maculopapular eruptions indistinguishable from rubella have been observed with a number of the Coxsackie and echoviruses. Petechiae have accompanied some rashes, especially with echovirus 9. The presence of virus has been demonstrated in the skin lesions themselves. Some of the syndromes include an associated enanthem. Herpangina, one of these, is most commonly associated with Coxsackie A infections. Ulcerative lesions of the mucosa appear on the posterior pharynx and soft palate. The associated discomfort, fever, and sore throat are prominent. Because many different serotypes of the echo and Coxsackie groups can cause identical clinical pictures, virus isolation is necessary to identify the specific etiologic agents.

Pleurodynia (Bornholm Disease)

This is a febrile illness with extreme myalgia, pleuritic chest pain, and headache. Most often it has been associated with Coxsackie B viruses, but other agents have occasionally been recovered from patients.

Orchitis

Although viral orchitis most often is due to mumps, it has accompanied infections with the Coxsackie B viruses.

Myocarditis and Pericarditis

Isolated myocarditis and/or pericarditis in older children and adults has been shown to result from Coxsackie B or echovirus infections. The spectrum has ranged from benign, self-limited pericarditis to severe, chronic, fatal myocardial disease.

Viral Meningitis and Meningoencephalitis

Many of the enteroviruses have been isolated from the cerebrospinal fluid (CSF) of patients suffering from aseptic meningitis and/or encephalitis. These illnesses are characterized by varying degrees of fever, headache, nuchal rigidity, malaise, and altered central nervous system function. The most severely affected patients may be obtunded or comatose. The least involved patients may complain only of mild headache and/or stiff neck. A CSF pleocytosis is detected, with usually less than 500 cells per microliter. Early in the disease, these cells are likely to be predominantly polymorphonuclear leukocytes, but after 24 to 48 hours of illness, mononuclear cells are the majority. The CSF protein may be normal or mildly elevated, but the glucose is normal. Most patients with enteroviral meningitis recover so that pathologic descriptions are based on only a few cases. There is inflammation of the meninges with perivascular inflammatory cell infiltration. The most acute processes are likely to have a predominance of polymorphonuclear cells, whereas mononuclear cells become predominant after a relatively few days. The prognosis for patients with enteroviral central nervous system infections is better than that for similar pyogenic infections. Although most of the youngest infants appear to recover completely, several longitudinal studies of such patients have demonstrated that neurologic deficits, particularly in receptive language function, are demonstrable in babies sustaining such infection during the first several months of life.

Paralytic Disease

Poliovirus infection, especially with types 1 and 3, was responsible for almost all the paralytic disease associated with the enteroviruses. Occasional cases of transient paralysis and muscle weakness have been noted with these viruses, particularly the Coxsackie B agents. With classic paralytic polio, there is a 2 to 6 day incubation period with an initial nonspecific febrile illness. This probably coincides with early replication of virus in the pharynx and gastrointestinal tract. With the subsequent hematogenous spread of virus, central nervous system involvement may result in meningitis and anterior horn cell infection. From 1 to 4 percent of susceptible patients infected with polioviruses develop central nervous system involvement. The spectrum of paralytic disease is enormously variable and may involve only an isolated muscle group or extensive paralysis of all extremities. Characteristically, the picture is one of asymmetrical distribution, the lower extremities more frequently involved than the upper. Large muscle groups are more often affected rather than the small muscles of the hands and feet. Involvement of cervical and thoracic segments of the spinal cord may result in paralysis of the muscles of respiration. Infection of cells in the medulla and the cranial nerve nuclei results in bulbar polio with compromise of the respiratory and vasomotor centers. With the return of the patient's temperature to normal, the progress of paralysis ceases, and the subsequent weeks and months reveal a varying spectrum of recovery ranging from full return of function to significant residual paralysis. Atrophy of involved muscles becomes apparent after 4 to 8 weeks. Recovery may be exceedingly slow, and its full extent cannot be judged for 6 to 18 months.

Pathology

Most Coxsackie and echovirus infections are transient and nonfatal, therefore limited histologic information is available. The pathogenic changes of poliovirus in the central nervous system are most prominent in the spinal cord, medulla, pons, and midbrain. After initial cytoplasmic alterations in the Nissl substance of the motor neurons, nuclear changes develop next, and pericellular infiltrates of polymorphonuclear and mononuclear cells accumulate. The final stage is destruction, with neuronophagocytosis and dropping out of the necrotic cells.

Fatal newborn infection with enteroviruses has shown nonspecific but extensive damage of the infected tissues. The echovirus infections are associated with hepatic necrosis and evidence of disseminated intravascular coagulation in multiple organs. If the patient has lived long enough, the liver may then show cirrhosis, as opposed to the acute process.

Immunity

Enteroviruses gain entry into the host via the mouth. The virus establishes infection in the oral pharynx and portions of the gastrointestinal tract, where multiplication subsequently occurs. Virus then gains access to adjacent lymph nodes and the bloodstream. The incubation period of enterovirus infection is usually from 1 to 5 days (Fig. 70-2). Invasion of other tissues, such as the meninges or the myocardium, typically occurs from 7 to 10 days after initial exposure to the virus, resulting in the classic biphasic illness such as occurs with poliovirus infection.

Enteroviruses induce secretory and humoral antibody responses. The humoral responses are initially predominantly IgM antibodies, followed by IgA and IgG antibodies that persist for months to years. Coproantibodies, primarily IgA, have been studied as a response to administration of poliovirus vaccines or to natural infection, e.g., by echovirus type 6. Local secretory immunoglobulin A production occurs at the site of contact of virus with lymphoid cells. The antibody formation and virus excretion for typical enterovirus infections are depicted in Figure 70-2. Development of type-specific antibody provides lifelong protection against clinical illness due to the same agent. Local reinfection of the gastrointestinal tract may recur, but this is accompanied by only an abbreviated period of virus replication without clinical illness.

In experimental poliovirus infection, specific IgG antibody-producing cells and measurable antibodies can be demonstrated in areas of the central nervous system where virus is replicating. Local central nervous system antibody production is independent of systemic humoral antibody production. Although specific central nervous system enterovirus antibodies are probably produced primarily within the central nervous system, some passive transfer of serum antibody to the CSF may occur as permeability is increased by inflammation. In viral meningitis, only minimal elevation of CSF globulin has been documented. Evidence for the extreme importance of central nervous system antibody is deduced from agammaglobulinemic patients who are unable to eliminate enteroviruses from the central nervous system. Administration of extraordinarily large quantities of parenteral globulin or plasma with specific antibody is necessary to passively achieve measurable antibody levels in the CSF.

Cellular immunity against enterovirus infection is not well defined. Circulating peripheral white blood cells have been a source of virus isolation during acute illness. Recognition of virus by lymphocytes occurs in experimental models employing Coxsackie viruses. The abnormal response of some children with immunodeficiency disease to infection with attenuated polioviruses may offer further insights to immune processes normally stimulated by enterovirus infection.

Diagnosis

The clinical illnesses in some instances may permit a presumptive diagnosis of enterovirus infection. However, as shown in Table 70-1, the spectrum of illness is wide. The time of the year may be an indication, with a predilection for summer and early autumn circulation of enteroviruses occurring in the temperate zones. Specimens for virus isolation should be obtained early in the course of illness. The CSF of patients with viral meningitis and/or meningo-

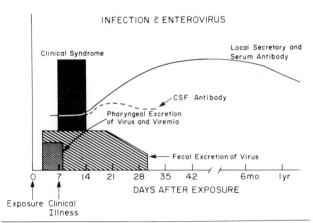

Figure 70-2. The course of typical enterovirus infection.

encephalitis has been a rich source of enteroviruses, except for the three polio types. Materials, such as pleural and pericardial fluid, should also be cultured when available.

Specific etiologic diagnosis of enteroviral disease is presently dependent upon demonstration of the enterovirus by cell culture techniques. Unfortunately, most of the Coxsackie A viruses grow poorly in cell cultures, and newborn mice provide a more reliable system for detection of these agents. The distinction between Coxsackie groups A and B has classically depended on the pathologic lesions produced in mice. Group A viruses cause generalized myositis and flaccid paralysis. Group B viruses cause focal myositis and typical lesions in the infrascapular fat pad and brain. Various primate and human cell culture systems will support replication of most of the enteroviruses, with cytopathic effects revealing their presence. The specimens submitted most often for attempted virus isolations are nasopharangeal swabs and stool specimens. An enterovirus may be excreted in fecal material for several weeks after the onset of clinical illness. Recovery of an enterovirus from the throat or stool of the patient does not in itself establish this as the etiologic agent of the illness observed. The temporal association of illness, virus recovery, and an antibody rise specific to that agent provide firmer evidence of a causative relationship.

A potentially useful and more practical approach to diagnosis would be the demonstration of viral antigen in clinical specimens, especially CSF. It would be ideal if antigen could be detected by a method capable of providing a specific diagnosis within hours. Although there are only scanty data concerning the quantitation of enterovirus present in various sites, it would appear that an efficient diagnostic test for the detection of virus/antigen in the CSF or blood should be capable of detecting as little as 1 to 100 tissue culture infectious doses of virus. Diagnostic assays, such as the enzyme-linked immunosorbent assay (ELISA) tests, presently available for antigen detection do not achieve this level of sensitivity. On the other hand, excretion of these viruses in stool is more likely to produce 10^3 to 10^7 tissue culture infectious doses of virus. The existing assays are capable of detecting this quantity of virus. Thus, at the present time virus isolation remains the most sensitive and specific means of diagnosing enterovirus infections. It is anticipated that future tests for antigen detection and body fluids will be more sensitive.

Acute and convalescent serum samples obtained from 7 to 21 days apart will help to define quantitative changes in antibody titers. Complement fixation, virus neutralization, immunoprecipitation, and, in a few instances, hemagglutination inhibition are the available techniques for assaying enterovirus antibodies. Neutralizing antibodies are type-specific, whereas complement fixation demonstrates group-reactive antibodies. In the course of a lifetime, humans sustain multiple infections, occult or overt, with a variety of enteroviruses. A specific infection elicits the production of antibody specific to that virus type but also may prompt an anamnestic response demonstrated by an increase in group-reactive antibody and by parallel rises in antibodies to serotypes of some of the other enteroviruses previously encountered. The concomitant serologic rises in antibody titer create some problems with serologic surveys, rendering the complement-fixation test inadequate to define a specific infection. The isolation of a specific virus provides the opportunity for assessment of the patient's antibody against his or her own viral agent. In the absence of the recovery of a virus, one faces the problem of seeking specific antibody rises against the whole genus of *Enterovirus*. Thus the complexity of serology makes the serologic diagnosis of these infections impractical.

Prevention

Because there currently is no specific treatment for enterovirus infections, efforts have focused on means of prevention. The multiple antigenic types and the usually benign, self-limited course of most echovirus and Coxsackie virus infections have resulted in little stimulus for the development of vaccines. The story of the poliovirus vaccines, however, has been one of the most exciting and rewarding sagas in the history of microbiology. Prior to the work of Enders and colleagues with successful tissue culture techniques for growth of the polioviruses, there had been several ill-fated vaccines prepared from emulsions of spinal cord removed from monkeys infected with wild-type poliovirus. These preparations were treated with formalin or other inactivating agents. Trials of such vaccines in 1935 proved unsuccessful and unacceptable.

Enders' tissue culture techniques lent themselves to the propagation in vitro of sufficient amounts of relatively pure poliovirus, so that controlled formaldehyde inactivation could be used to produce noninfectious virus that retained its antigenicity. Salk and his colleagues pursued this line of research and by 1954 were able to embark on a field trial that established the efficacy of an inactivated poliovirus vaccine in the prevention of paralytic disease. This was a trivalent preparation incorporating the three poliovirus types. After an initial series of two or three injections spaced several weeks to months apart, followed by a booster 6 to 12 months later, there was demonstrable serum antibody to all three polio serotypes. The vaccine was widely used in the United States during the 5 years from 1956 through 1960. The results were dramatic. Previous years had seen from 10,000 to 20,000 cases of paralytic disease reported annually. With the widespread use of Salk vaccine, this rapidly dropped to 2000 to 3000 cases annually, as increasingly large numbers of susceptible individuals were immunized (Fig. 70-3).

By the early 1960s, a second vaccine was available. Strains of poliovirus that Sabin had selected and studied in his laboratory were proven attenuated for monkey and humans. Their ingestion resulted in intestinal infection and virus excretion, so that humoral and gastrointestinal tract immunity developed without any illness. Because these could be administered more readily, by the oral route, and

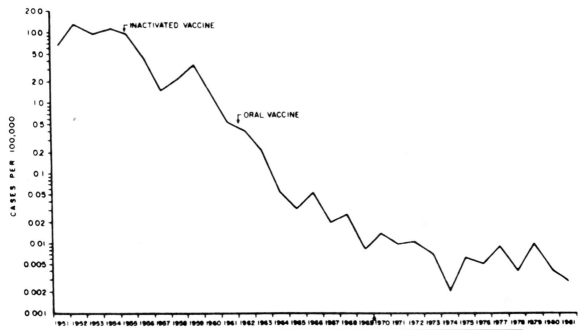

Figure 70-3. Reported paralytic poliomyelitis attack rates, United States, 1951–1981. *(From CDC Poliomyelitis Surveillance Summary 1980–1981. Issued December 1982, p 9. U.S. Dept. HHS, PHS.)*

because their multiplication in the gastrointestinal tract more closely mimicked natural infection, they offered certain selected advantages, which led to their replacing the injectable Salk vaccine. Over the first 5 years of the 1960s, more than 400 million doses of oral vaccine were distributed in the United States. At the same time, trials also were successfully conducted in European nations, Japan, and other countries. The use of oral vaccine in this country was accompanied by a further decrease in the annual reported polio cases (Fig. 70-3), so that beginning in 1966 fewer than 100 have occurred each year. Between 1969 and 1981 with continued use of the oral vaccines, 203 cases of paralytic disease were reported in the United States. In less than 20 years, a disease that had claimed thousands of victims annually and had been the source of indescribable community anxiety was reduced to a rarity!

In the complex processes of vaccine development, commercial production, and widespread utilization, a number of unexpected events took place that merit consideration. After the highly successful field trials of 1954, commercial manufacture of the Salk type vaccine was licensed. Within a few weeks of its use, paralytic disease was observed in April through June 1955 among children in California and Idaho who had received some of the first lots of commercial vaccine manufactured by the Cutter Laboratories. By the time this had been fully investigated and resolved, it was learned that there were 204 cases of vaccine-associated disease. Seventy-nine were among children who had received the vaccine, 105 were among their family contacts, 20 were in community contacts. Nearly three quarters of the cases were paralytic, and there were

11 deaths. The agent isolated from these patients was type 1 poliovirus. Laboratory tests on vaccine distributed by Cutter Laboratories revealed viable virulent type 1 poliovirus in 7 of 17 lots. Revisions of the federal regulations governing the steps in vaccine manufacture were promptly promulgated and implemented to prevent recurrence of such a tragic episode.

Manufacturers faced further difficulties in maintaining the fine balance between the complete elimination of the infectious live virus from the production process and the retention of effective antigenicity of the inactivated components. A number of lots of vaccine subsequently proved to be poorly antigenic for type 3 poliovirus. As a result, when community polio outbreaks occurred among well-immunized groups, there were breakthroughs with paralytic disease, especially due to type 3 virus, in previously immunized subjects. Such an outbreak was studied in 1959 in Massachusetts, where an analysis of polio cases revealed that 47 percent (62 of 137) of the patients had previously received three or more inoculations of inactivated vaccine.

Enthusiasm for attenuated oral vaccine was enhanced by the aforementioned unfortunate episodes involving inactivated vaccine. A final unanticipated discovery in 1960 was the detection of a simian virus, SV-40, as a contaminant of the monkey kidney cell cultures utilized in the preparation of both inactivated and oral attenuated vaccines. Once again, a revision of standards for preparing and testing the tissue cultures was necessary to exclude this previously unrecognized agent. In the United States, inactivated vaccine has been used sparingly in the past 15 years. Almost all immunization has been conducted with the oral attenuated product. A number of European countries, especially those in Scandinavia, have adhered to the use of inactivated vaccine. With successful production of fully potent antigens, their record of achievement in the control of polio has been parallel to that of this country.

The marked decrease in paralytic disease due to wild polioviruses has disclosed a small but significant number of oral vaccine recipients who have developed paralytic illness in temporal association to the ingestion of vaccine. In addition to these few cases among recipients of vaccines, there have also been paralytic episodes reported among susceptible family or community contacts of the vaccine recipients. These have been few in number and are difficult to characterize with complete clarity. A small portion of these patients have been found to be immunodeficient children, particularly those with congenital hypogammaglobulinemia. One concern has always been that the attenuated strains of virus might prove genetically unstable in human intestinal passage, so that increased neurovirulence might result from widespread dissemination. This has not been demonstrated. Monovalent oral polio vaccine (MOPV) was used extensively until 1964, when it was supplanted by trivalent vaccine (TOPV). With MOPV the risk of vaccine-associated illness in recipients was estimated to have been 0.19 per million doses distributed. With TOPV an overall figure of risk to recipients and their contacts has been calculated at 0.28 per million or 1 case per 3.6 million doses.

The achievements with poliovirus vaccination have been impressive in the United States, Canada, most of Europe, Australia, and some Asian and African nations. Polio remains an endemic disease in many tropical lands. It is premature, therefore, to relax the use of poliovirus vaccination in those parts of the world where the disease has been nearly eradicated. The possibility of the inadvertent introduction of virulent virus is omnipresent. A number of such episodes have occurred already on the Texas–Mexico border. Although polio immunization is now confined principally to infants and children, it is recommended that adult Americans traveling abroad to endemic areas receive polio vaccine prior to departure. Armed Forces recruits are routinely administered oral poliovirus vaccine. Its major use remains that for infants and children in the first 18 months of life.

FURTHER READING

Beatrice S, Katze MG, Zajac BA, et al.: Induction of neutralizing antibodies by the Coxsackie virus B3 virion polypeptide VP2. Virology 104:426, 1980

Centers for Disease Control Enterovirus Surveillance. Summary 1970–1979. Issued November 1981. USDHHS, PHS

Hughes JR, Wilfert CM, Moore M: Echovirus 14 infection associated with fatal neonal hepatic necrosis. Am J Dis Children 123:61, 1972

Kibrick S, Benirschke K: Severe generalized disease (encephalohepatomyocarditis) occurring in the newborn period and due to infection with Coxsackie virus, group B. Pediatrics 22:857, 1958

Melnick JL, Rennick V: Infectivity titers of enterovirus as found in human stools. J Med Virol 5:205, 1980

Modlin JF: Fatal echovirus 11 disease in premature neonates. Pediatrics 66:775, 1980

Modlin JF, Polk BF, Horton P, et al.: Perinatal echovirus infection: Risk of transmission during a community outbreak. N Engl J Med 305:368, 1981

Morbidity and Mortality Weekly Report. Annual Summary 1980. 30[54]:25, 103, 1982

Morens DM: Enteroviral disease in early infancy. J Pediatr 92:374, 1978

Nathanson N, Langmuir AD: The Cutter incident, I, II, II. Am J Hyg 78:16, 1963

Nightengale EO: Recommendations for a national policy on poliomyelitis vaccination. N Engl J Med 297:249, 1977

Ogra PL: Distribution of echovirus antibody in serum, nasopharynx, rectum and spinal fluid after natural infection with echovirus type 6. Infect Immun 2:150, 1977

Ogra PL, Ogra S, Al-nakeeb S, et al.: Local antibody response to experimental poliovirus infection in the central nervous system of rhesus monkeys. Infect Immun 8:931, 1973

Paul JR: A History of Poliomyelitis. New Haven, Conn., Yale University Press, 1971

CHAPTER 71

Viruses in Gastrointestinal Tract Infection

Acute infectious diarrheal illness is recognized as one of the leading causes of morbidity and mortality in developing nations. Even in developed countries, infectious gastroenteritis is cited as second only to respiratory infections as a cause of morbidity in childhood. Viral agents recognized during the past decade have been shown to be responsible for a large proportion of the diarrhea for which an etiologic agent can be defined. Acute viral gastroenteritis affects all age groups of people and may occur in either sporadic or epidemic form. Most such illnesses are self-limited, and in normal hosts recovery is complete. If severe dehydration occurs, morbidity and even mortality may be substantial.

Infection of the gastrointestinal tract with agents, such as the enteroviruses, is usually an asymptomatic infection. However, the initial virus replication may be followed by viremia, with spread of virus to other target organs and symptoms related to the infection of the additional tissues. Contrary to early suggestions, enteroviruses do not appear to be responsible for any significant proportion of gastrointestinal tract disease. It is also necessary to note briefly that several nondiarrheal diseases of the gastrointestinal tract have a postulated viral etiology. Intussusception occurs most commonly in children under the age of 2 years, and mesenteric lymph node enlargement in the area of the terminal ileum has been observed at the time of surgery. Since lymphoid hyperplasia may ocur in association with virus infection and adenoviruses have been isolated from children with intussusception, they are one of the etiologic agents suggested for this entity. Coxsackie B viruses have been epidemiologically associated with diabetes mellitus, and several case reports have associated Coxsackie B4 with the development of diabetes mellitus. Experimentally, other viruses have been shown to infect pancreatic beta cells, and reovirus type 3 and murine encephalomyocarditis virus can produce a diabetes-like syndrome in inbred strains of mice. It must be emphasized that intussusception and diabetes mellitus have not been proven to be of viral etiology. It may be that virus infection represents only one of several contributing factors to the development of subsequent disease.

The viral agents discussed in detail in this chapter are those that seem to be associated with clinical symptoms related to virus replication within the gastrointestinal tract.

Rotaviruses

In 1943 an outbreak of diarrhea occurring in infants was reported by Light and Hodes. They isolated a filterable agent from stool specimens that caused diarrhea in calves. They established the incubation period and reproduced diarrhea with serial passage of the agent. The pathology of the bowel, the development of immunity to the agent, and passive protection by the administration of immune serum were described. At the time there was no confirmation of the viral etiology of the diarrhea as no agent could be detected in available cell culture systems. It was not until 1973 when virus particles were visualized in a duodenal biopsy by electron microscopy that the etiologic agent was finally defined as a rotavirus. In the past decade, rotaviruses have come to be appreciated as the single most common agent causing epidemic diarrhea in infants from 6 to 24 months of age. Rotaviruses are known to cause diarrhea in foals, lambs, piglets, rabbits, deer, monkeys, and other species. Experimental infection of animals other than the species of origin occurs with most rotaviruses.

Rotaviruses constitute one genus of the family Reoviridae. The rotavirus genome consists of 11 segments

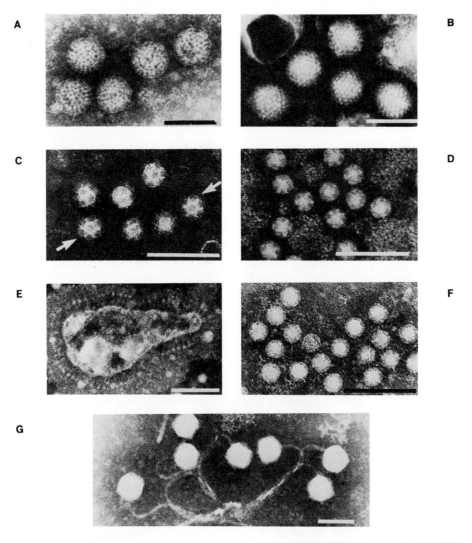

Figure 71-1. Viruses in feces from patients with gastroenteritis, negatively stained with 2 percent potassium phosphotungstate at pH 6.5. Bars present 100 nm. **A**. Rotaviruses. **B**. Adenoviruses. **C**. Calciviruses—two particles (arrowed) exhibit the characteristic morphology of a six-pointed star. **D**. Astroviruses—some particles show a five- or six-pointed star on the surface. **E**. Coronavirus—pleomorphic particle with large surface projections. **F**. Small round viruses (SRV)—particles with no obvious morphology and a diameter of 27 to 30 nm. **G**. Bacteriophages—these particles have hexagonal heads with long flexible tails attached to pieces of bacterial debris. *(Courtesy of H.A. Davies.)*

of double-stranded RNA, and rotavirus particles possess a double shell of outer and inner capsids (Fig. 71-1). The inability of these viruses to grow in the usual cell cultures has slowed research on them. It is the fourth gene that restricts ability of human rotaviruses to grow in cell culture, and if this gene is supplied by a bovine rotavirus (all animal rotaviruses *can* grow in cultured cells), the reassortant is able to grow. Gene *9* codes for the neutralizing antigen and gene *6* for the major internal structural protein responsible for type specificity. The fact that a high percentage of the strains can be typed indicates that the number of distinct serotypes may be limited. Additional information is of critical importance in defining the antigens responsible for inducing immunity and the feasibility of developing protective immunogens.

Epidemiology and Clinical Illness

Sporadic diarrhea occurring predominantly in infants and young children is associated with rotavirus infection. In fact, rotaviruses appear to be responsible for at least one half of the cases of infantile diarrhea that require hospitalization. Peak prevalence of rotavirus infection occurs during the cooler months of the year in temperate climates, but in tropical areas cases are identified throughout the year. Serum antibodies are usually acquired between the ages of 6 and 24 months. Transmission of rotavirus from person to person occurs via the fecal to oral route. The incubation period is 2 to 5 days. Adults are often contacts of symptomatc children and have infections that may be asymptomatic or associated with clinical illness. Asymptomatic or very mild infections, but not severe disease, have been reported in newborn nurseries.

Acute infection due to rotaviruses is characterized by the rather abrupt onset of severe watery diarrhea that is not characteristically associated with blood or mucus in the stool. Fever and vomiting are often present at the onset of illness. Dehydration and metabolic acidosis are observed in children hospitalized with this infection. Those most severely infected and affected are between the ages of 6 and 24 months of age.

Virus is excreted in large amounts, with as many as 10^{11} virus particles per gram of feces. Virus is excreted in feces for approximately 8 days after onset of symptoms. Virus does not seem to be shed from the respiratory tract, but aerosolization of fecally contaminated material may result in inhalation or ingestion of infectious virus.

Rotavirus particles have been visualized by electron microscopy in intestinal epithelial cells, aspirated duodenal secretions, and feces of infected persons. Morphologically, shortening and blunting of the villi of the duodenum and small intestine accompany acute illness. Virus particles are visualized in the cytoplasm and bud into the cisternae of the endoplasmic reticulum of the enterocytes. Immunofluorescence studies have also demonstrated rotavirus antigens in the cytoplasm of the villus epithelial cells but not in the cells of the crypts or lamina propria. This suggests that the specificity of the virus particle is for the mature or differentiated enterocytes located on the villi. This destruction of mature enterocytes is associated with the decrease in one or more mucosal disaccharidases. The destroyed infected cells are replaced by immature cells, which results in a deficit in glucose-facilitated sodium transport. Diarrhea then results from decreased absorption secondary to the altered ion transport in immature cells. In contrast to the toxin-mediated diarrheas, there is no increase in intestinal secretion. Complete recovery has been documented by biopsy as early as 4 weeks after the episode of diarrhea.

Host Response

Specific antibody has been demonstrable in the stool and serum. Primary infection seems to produce an initial serum IgM response, followed by an IgG response. Adults with rotavirus infection manifest an anamnestic response with elevation of IgG antibodies. Some studies in volunteers suggest that protection from infection with rotaviruses correlates better with secretory IgA of the small intestine than with serum antibody. Persons may experience more than one infection with rotaviruses. It is not yet clear whether these are always different serotypes.

Specific antibody has been demonstrated in colostrum and milk for as long as 9 months of lactation. In nurseries where rotavirus infection has been endemic, breast-fed infants seem to acquire infection less often than do formula-fed infants. Those who are infected excrete less virus and are less symptomatic. Antibodies in human colostrum and milk are capable of neutralizing rotavirus in vitro. Weaning from breast milk in developing nations is temporally associated with the onset of a diarrhea/malnutrition cycle. Despite the fact that breast milk has been shown to have preventive activity against undifferentiated infantile diarrhea and, even more specifically, has been shown to contain antibodies against some of the causative agents, breastfeeding is declining in popularity in many parts of the world with the greatest incidence of diarrheal disease. Breast milk does provide both specific and nonspecific protection.

Diagnosis

Human rotaviruses are extremely fastidious, but recent progress has been made in the cell culture propagation of these agents. Human rotavirus strains have been efficiently propagated directly from clinical specimens into cell cultures. In particular MA 104 cells, which are a line derived from embryonic rhesus monkey kidney, have been successfully used. The specimen is pretreated with trypsin (10 μg/ml), and 1 μg/ml of trypsin is added to the maintenance medium. The ability to grow these viruses in cell culture will enhance the study of these agents.

The quantity of virus excreted in fecal specimens has enabled various antigen detection systems to identify virus in the stool. Electron microscopic examination of fecal extracts obtained during the acute stage of gastroenteritis, particularly when specific antiserum is used to agglutinate virus particles, is almost invariably positive. Enzyme-linked immunosorbent assay (ELISA) and radioimmunoassay techniques have also become available for the detection of rotavirus particles in stool. All known rotaviruses share a common antigen thought to be located in the inner capsid, which makes the immunologic detection of these viruses a practical reality. Indirect assays employ an unlabeled specific antibody and subsequent measurement of the antibody/antigen combination with an enzyme-labeled antiglobulin. ELISAs are as sensitive as elecron microscopy and are more practical, since commercially marketed reagent kits are now available for detection of rotavirus in stool.

Prevention and Therapy

Therapy of viral gastroenteritis is limited to supportive measures, since there are no effective antiviral agents available for specific treatment.

There can be no doubt that prevention of rotavirus illness would be a major contribution to reduction or morbidity from gastroenteritis. For immunization to be effective, immunity within the gastrointestinal tract is probably a necessity. Various approaches to immunization that are being considered include a live attenuated human rotavirus vaccine, an attenuated reassortant rotavirus vaccine, a vaccine made from an animal rotavirus strain if such a strain is able to infect humans without causing illness, cloning of the human rotavirus genome by DNA technology, and production of synthetic vaccines derived from the specific amino acid sequences responsible for induction of protective antibodies. Reassortants have already been created, and work with human rotaviruses in cell culture has facilitated the study of these agents. The task of defining protective antigens has made the feasibility of a vaccine an accessible goal.

Norwalk Group of Viruses

Agents Associated with Epidemic Viral Gastroenteritis

Epidemic gastroenteritis occurs in a form that produces an explosive, self-limited disease lasting for 24 to 48 hours, may be communitywide, and involves school-age children, family contacts, and adults. Such as outbreak of gastroenteritis occurred in Norwalk, Ohio, in 1969. Within a matter of 48 hours gastrointestinal illness developed in half of the students and teachers in an elementary school, with the secondary symptomatic attack rate affecting approximately one third of family contacts. A bacteria-free filtrate from a stool specimen produced gastroenteritis in several volunteers, and stools from the infected individuals could be serially passaged in additional volunteers, but no agent could be isolated in the usual cell culture systems. In 1972, immune electron microscopy with serum from a symptomatic patient demonstrated small, 27-nm-diameter particles in an infectious stool filtrate. Norwalk agent particles contain RNA and only a single protein species with a molecular weight of 59,000. Among mammalian viruses in this size range, the caliciviruses are the only ones that possess a single structual protein. The Norwalk agent is, therefore, probably a calicivirus (Fig. 71-1).

Epidemiology and Clinical Infection

Roughly one third of outbreaks of gastroenteritis can be attributed to a Norwalk-like agent. This is a surprisingly high number and suggests that the antigenically unique viruses of this group may be limited in number. The settings in which such outbreaks have occurred include schools, recreational camps, cruise ships, persons using contaminated water, nursing homes, and ingestion of inadequately cooked shellfish. Outbreaks have occurred at all times of the year, and antibody surveys suggest that the agent occurs worldwide. In most undeveloped nations, infection occurs at an early age. In the United States, antibody is uncommon during childhood and develops during late adolescence and early adulthood.

Norwalk agents are transmitted by the fecal to oral route, and volunteers who have shed virus had detectable virus in their stools during the first 72 hours after the onset of illness. Although respiratory symptoms are very unusual, the rapidity of spread suggests that some aerolization of virus may occur from fecal contamination of the environment.

Infection with these agents result in delayed gastric emptying, although the gastric mucosa is morphologically normal. Microscopic broadening and blunting of the villi in the jejunum is apparent. The mucosa remains histologically intact, but there is mononuclear cell infiltration. Virus has not yet been detected in involved mucosal cells by elecron microscopy.

The majority of patients who sustain these infections have nausea, vomiting, and abdominal cramps, and about one half of them have associated diarrhea. Some have fever and chills. The symptoms last from 12 to 24 hours, and the incubation period appears to be around 48 hours. Usually the stools are not bloody and do not have mucus or leukocytes. A transient lymphopenia has been observed in volunteers challenged with these agents.

Host Response

Immunity to these viruses is a puzzling feature of the infections that they cause. Challenge of volunteers produces disease in some persons and not in others. Repeat challenge with homologous virus within several months of the original infection will not produce clinical illness. Subsequent challenge 2 to 4½ years later produces disease in the same volunteers who had symptomatic illness with the first contact with the virus. Those individuals who were asymptomatic and did not acquire infection with the first contact do not do so with subsequent challenge. Those individuals who acquire clinical infection have demonstrable antibody that wanes and then have an anticipated boost with a subsequent challenge. Volunteers who fail to develop illness have no demonstrable antibody. Measurement of intestinal antibody also showed that those individuals who develop illness have higher mean antibody titers in jejunal fluid than those who failed to develop illness. Neither serum nor local intestinal antibody correlates with resistance to a Norwalk challenge. Thus, individuals who have demonstrable serum antibody are those who are at risk for symptoms induced by infection. These findings are not understood, but it is possible that some individuals are not susceptible to these viruses by reason of genetically determined factors. It is also possible that repetitive exposure to the virus is necessary to induce an antibody response.

Diagnosis and Therapy

Immune electron microscopy is the best available procedure for identifying members of this group of agents. Specific coating of particles with antibody needs to be observed if the immune response is being assessed.

The radioimmunoassay is more efficient than immune electron microscopy because it is able to detect soluble as well as particulate antigens. This test depends on the availability of appropriate high titered serum. In research laboratories radioimmunoassay has been used to assess serum antibody status.

The illnesses associated with the Norwalk group of agents are generally self-limited and mild. No specific therapy is available. Replacement fluid therapy is essential in management of such patients. The rather simplistic approach of good handwashing and effective hygiene is the best we have to offer to prevent the spread of these viruses.

Coronaviruses

Coronaviridae are pleomorphic, enveloped, single-stranded RNA-containing viruses with widely spaced 20 nm club-shaped surface projections (Fig. 71-1). This group of viruses has been well identified as etiologic agents of respiratory and intestinal disease in animals. In 1960 the first of these agents was isolated from a human with a cold. It grew in human tracheal organ culture and was morphologically similar to infectious bronchitis virus, an avian coronavirus. A second isolate, now the prototype strain of human coronavirus (229E), was isolated in human kidney cell culture in 1962.

Human coronaviruses were first etiologically associated with upper respiratory tract disease in adults and lower respiratory tract disease in hospitalized children. It has also been found that coronaviruses may be responsible for an exacerbation of the respiratory symptomatology in hospitalized asthmatic children. Coronavirus infections tend to occur in small outbreaks that take place during the late winter and early spring. Sporadic outbreaks in other seasons can also occur. Periods of high incidence with one specific serotype have been reported to occur in cycles of 2 to 3 years. Reinfection with coronaviruses is a frequent event, as shown by the presence of infection in persons with preexisting neutralizing antibodies.

It was predicted in the mid-1970s that an association of coronaviruses with enteric disease of humans would be identified because of the known involvement of these agents with enteric disease in animals. Electron microscopic examination of stools then revealed coronavirus-like particles in human feces. In fecal specimens it has been estimated that there are as many as 10^8 coronavirus particles per gram of feces. Many of these particles are defective, which suggests that maturation may be faulty within the gastrointestinal tract.

Of interest is an apparent difference in the epidemiology of these viruses. They seem to be less likely to be responsible for diarrhea within the first year of life and are more commonly visualized in the diarrheal stools of older children and young adults. Enteric coronaviruses can be excreted for as long as a year and a half, and in some of the persistent excretors chronic gastroenteritis is present. Prolonged viral excretion complicates conclusions concerning the epidemiology because asymptomatic individuals can be excreting virus. It has been documented that endemic diarrheal disease occurs in situations where poor hygiene exists, and acquisition of virus does coincide with development of gastroenteritis.

Work with the coronaviruses isolated from humans has been difficult because they grow best in organ cultures, and those from the respiratory tract appear to require the use of ciliated epithelium. They are labile under conditions of virus purification. Thus, detailed analyses of avian and other mammalian coronaviruses have proceeded at a much more rapid rate. The successful cultivation in vitro of several human strains is beginning to advance our knowledge of these viruses. Human coronaviruses possess four major structural proteins; the surface projections that are responsible for hemagglutination and complement fixation consist of two large glycoproteins. There are at least two antigenic groups of mammalian coronaviruses. Coronaviruses do not appear to bud from the plasma membrane of infected cells but bud into the cisternae of the endoplasmic reticulum and are found in large intracytoplasmic vesicles (Chap. 58).

Enteric Adenovirus

The fastidious enteric adenoviruses were first described by Flewett in 1975. These agents have been established as a significant pathogen of diarrheal illness in children. Preliminary characterization of agents isolated from geographically distinct outbreaks indicates that they are all representatives of the adenovirus 38 candidate serotype (Fig. 71-1). These agents do not grow in the usual cell culture systems, and thus far successful in vitro cultivation has been dependent upon the use of Graham 293 cells, a human embryonic kidney cell line containing some of the genome of adenovirus type 5. In these cells the cytopathic effects produced are typical of adenoviruses.

These agents most often appear to be responsible for acute diarrhea in infants less than 1 year of age. The onset of illness is manifest by diarrhea, and the young child may also have respiratory symptoms, including cough, rhinorrhea, or wheezing. Pneumonia and conjunctivities have also been seen in association with adenovirus gastroenteritis. The illness is usually mild to moderate in severity and may occur in any month of the year. The viruses are visualized in the feces of patients for a period of 4 days to 2 weeks.

The estimated frequency of adenoviruses as a cause of diarrhea varies. If the study period is selected when rotavirus disease is not occurring, adenoviruses are responsible for a larger proportion of illness, recorded to be as high as 50 percent. If a prospective study is managed over a several year period, 4 to 8 percent of all the diarrhea observed may be associated with these fastidious adenoviruses. Serologic studies of children suggest that approximately one half of children have antibodies at an early age.

The diagnosis of these agents is also dependent upon demonstration of the virus particles or antigen in stool. Initial recognition by electron microscopy has been followed by the development of ELISAs and radioimmunoassays.

Astroviruses

Astroviruses were described in 1975 after visualization in the stool of newborn infants. They are 28-nm particles shed in the stool in very large numbers. They are present in the stools of infants both with and without acute gastroenteritis. The star shape designates the agents as astrovirus particles (Fig. 71-1). Similar morphologic structures have been identified in feces from both adults and children and in diarrheal feces from lambs and calves. These agents infect monolayers of human embryonic kidney cells without producing cytopathic effects. The evidence that they cause disease is equivocal. Early serologic studies have demonstrated seroconversion in association with demonstrable infection and fecal excretion. However, feeding of these agents to adult volunteers resulted in poor transmission of disease but antibody rises in the majority of volunteers. It may be that most adults have previously met the agents and, therefore, are protected against disease.

In lambs, these agents infect the epithelial cells and the subepithelial macrophages of the villi of the small intestine. Atrophy of the villi is demonstrable. At least in an-

imals it would appear that they are capable of causing gastroenteritis.

Other Putative Viruses

The interest in examining feces by electron microscopy has resulted in the description of numerous virus species. Some of them are clearly defined as etiologic agents of gastroenteritis, and others have yet to be assigned a definitive etiologic role. The relationship to each other or other morphologic species also remains to be determined. These as yet unidentified and unclassified types of particles are discussed below.

Small Round Viruses, Picorna-Parvovirus-like

This is a heterogeneous collection of small round viruses visualized in stools. The particles have a diameter of 27 to 30 nm, no detectable surface structure, and do not grow in vitro in routine cell culture systems (Fig. 71-1). They have been seen in stools from patients who have no clinical symptoms. More information is needed to establish their causative role in gastroenteritis.

Mini-Reovirus

These agents are 30 nm particles with a double capsid. They have been visualized in stools from children with diarrhea and in infants who have acquired diarrhea within hospital settings. Additional information is needed to establish their role as etiologic agents in gastroenteritis.

FURTHER READING

GENERAL

Tyrrell DAJ, Kapikian AZ: Virus Infections of the Gastrointestinal Tract. New York, Basel, Marcel Dekker, 1982

ROTAVIRUSES

Blacklow NR, Cukor G: Viral gastroenteritis. N Engl J Med 304:397, 1981

Steinhoff MC: Rotavirus: The first 5 years. J Pediatr 96:611, 1980

NORWALK-LIKE AGENTS

Adler I, Vicki R: Winter vomiting disease. J Infect Dis 119:668, 1969

Agus SG, Dolin R, Wyatt RG, et al.: Acute infectious nonbacterial gastroenteritis: Intestinal histopathology. Histologic and enzymatic alterations during illness produced by the Norwalk agent in man. Ann Intern Med 79:18, 1973

Blacklow MR, Cukor G, Bedigian MK, et al.: Immunoresponse and prevalence of antibody to Norwalk enteritis virus as determined by radioimmunoassay. J Clin Microbiol 10:903, 1979

Blacklow MR, Dolin R, Fedson DS, et al.: Acute infectious nonbacterial gastroenteritis: Etiology and pathogenesis. Ann Intern Med 76:993, 1972

Dolin R, Blacklow MR, DuPont H, et al.: Transmission of acute infectious nonbacterial gastroenteritis in two volunteers by oral administration of stool filtrates. J Infect Dis 123:307, 1971

Greenberg HB, Valvesuso J, Kalica AR, et al.: Proteins of Norwalk virus. J Virol 37:994, 1981

Schaffer FL: Calicivirus. In Fraenkel-Conrat H, Wagner RR (eds): Comprehensive Virology. New York, Plenum, 1978, vol 14, p 249

OTHER VIRUSES

Flewett PH, Davies H: Caliciviruses in man. Lancet 1:311, 1976

MacNaughton MR, Davies HA: Human enteric Corona viruses. Arch Virol 70:301, 1981

Madaley CR, Cosgrove BP: Viruses in infantile gastroenteritis. Lancet 2:124, 1975

Madaley CR, Cosgrove BP, Bell EJ, et al.: Stool viruses in babies in Glasgow. I. Hospital admissions with diarrhea. J Hyg 78:261, 1977

Kurtz JB, Lee TW, Pickering D: Astrovirus-associated gastroenteritis in a children's ward. J Clin Pathol 30:948, 1977

Kurtz JB, Lee TW, Craig JW, et al.: Astrovirus infection in volunteers. J Med Virol 3:221, 1979

Chiba S, Sakuma Y, Kogasaka R, et al.: Fecal shedding of virus in relation to the days of illness in infantile gastroenteritis due to calicivirus. J Infect Dis 142:247, 1980

Kapikian AZ, Wyatt RG, Greenberg HB, et al.: Approaches to immunization of infants and young children against gastroenteritis due to rotaviruses. Rev Infect Dis 2:459, 1980

CHAPTER 72

Influenza Viruses

Clinical Features
Epidemiology
Diagnosis

Treatment
Prevention

Members of this group of ubiquitous myxoviruses produce epidemic illness so regularly that all readers of this chapter will have experienced several infections by them. They have been inconstant companions of human beings for at least as long as recorded history, and the broad age range of those afflicted, as well as the broad spectrum of illness observed, was well described in English texts as early as the 16th century, when it was called the "newe acquaintance." Because of the seasonality of influenza epidemics, it was thought initially to be a disease affected by celestial movement ("malattia influenza per le stelle"), and this belief permanently baptized both the viruses and the clinical illnesses that they produce. Although several major epidemics occurred during the early part of the 20th century, the viral etiology of this disease was not demonstrated until the 1930s, and an understanding of the antigenic variability of these viruses and thus the periodic occurrence of epidemic illness had to await the more advanced virologic techniques of the 1940s and 1950s.

Clinical Features

As with most viral diseases, influenza represents a very broad spectrum of severity of clinical illness, from the occasional asymptomatic infection to the more common, bothersome, and irritating grippe, to the often fatal primary influenza pneumonia. The reasons for this wide variation are poorly understood and may be partly, but not exclusively, related to the age, health, and antigenic background of the patient, the presence of preexisting abnormalities of the pulmonary parenchyma or vasculature, and possibly to the virulence of the virus itself. The tissue tropism for influenza virus and thus the pathogenesis of in-

fection may be explained by a unique virus–host cell interaction. Hemagglutinin, one of the glycoprotein spikes present on the surface of the virus (Chap. 62), is responsible for adsorption of virus to the host cell. The host cell receptors not only must contain N-acetyl neuraminic acid residues but must also express proteases that cleave the hemagglutinin to render it infectious. There is significant homology in the cleavage region between the hemagglutinin of influenza A and B as well as the F protein of paramyxoviruses. This suggests that fusion of viral and cellular membranes may occur during penetration.

The significant histologic abnormalities induced by the virus are relatively uniform, beginning with virus replication in all superficial respiratory cells, destruction of the ciliated columnar epithelium, and subsequent denudation of the tracheobronchial tree. These anatomic changes invariably lead to impairment in respiratory function and physiology, even in patients without obvious clinical evidence of pulmonary involvement. Defects include diminished defusing capacity, diminished tracheobronchial clearance, increased bronchial reactivity (sensitivity to irritation), and variable degrees of small airway obstruction. The pharyngitis and tracheitis thus induced explain the severe throat and substernal pain, cough, and shortness of breath often seen during acute illness, as well as the paroxysms of cough that may be exacerbated by sudden exposure either to cold or environmental pollutants for weeks after the subsidence of clinical illness. Some of the functional pulmonary defects may require weeks or, occasionally, months to disappear. Furthermore, influenza virus infection not only ablates the normal removal of bacteria in the mucous blanket by destroying ciliated cells but also decreases T lymphocyte function and granulocyte chemotaxis. The tracheobronchial tree that has been denuded of its usual defense mechanisms is therefore more susceptible to invasion by bacteria. This occurs most commonly in patients with chronic bronchitis (where large numbers of bacteria may already be present in the lower respiratory tract), in aged individuals, and in others with poor cough reflexes and poor efficiency of expectoration, or in any patient in whom pulmonary mechanics or host defense mechanisms are already abnormal.

The usual clinical disease begins rather suddenly within 1 or 2 days after exposure to infected aerosol droplets (Fig. 72-1), most often in epidemic circumstances and primarily during the colder months of the year. A sudden rise in temperature, sometimes to 102F, though rarely higher in adults, is accompanied by rigors and myalgias and is then followed by the variable appearance of sore throat, nasal congestion and dryness, conjunctivitis, a nonproductive cough, and headache. Marked lassitude and moderate anorexia accompany these symptoms and are frequently so severe that even the most stalwart and ambitious individual must spend the first day or two of illness in bed. Physical examination will reveal generalized vasodilation of the dermal and mucosal surfaces of the head and neck, as well as edema and little to moderate discharge from the upper respiratory tract. On auscultation, the lungs are generally clear in individuals with otherwise healthy pulmonary function. Young children, however, are prone to develop laryngotracheobronchitis, bronchiolitis, gastrointestinal upset, febrile seizures, conjunctivitis, and the common sequela of otitis media. Laboratory studies will generally show a moderate lymphocytopenia, and blood gas analysis will often indicate a surprising level of hypoxia even in patients without physical or roentgenologic evidence of pulmonary involvement. Patchy diffuse infiltrates of small size can occasionally be noted on chest x-ray. However, these are usually evanescent and of little consequence but must be differentiated from the more severe forms of pneumonia seen during influenza epidemics and described below. Acute bronchitis, possibly bacterial, and exacerbations of chronic bronchitis are seen more commonly. The symptoms generally subside within several days, and the patient is able to return to normal daily functions shortly thereaf-

ter. However, a severe postviral aesthenia, as well as a depressed affect, may be quite striking and persist for several weeks or more after resolution of the illness. The etiology of this weakness and easy fatigability is unknown, and its perceived severity is often proportionate to the expectations of the patient, as is often the case with infectious mononucleosis occurring during adolescence.

Pneumonias are the most common serious complications of influenza and can be divided into two fairly distinct types. The more common is that produced by secondary bacterial invasion. It occurs particularly in patients with preexisting lung disease and in elderly or debilitated individuals. It usually follows a brief quiescent period after the initial viral illness, at a time when the patient feels he is beginning to recover. The most common infecting organisms are pneumococci, *Haemophilus influenzae*, and staphylococci, although any species of bacteria can cause such an infection. The pneumonias themselves differ in no way from those caused by the same organisms under any other circumstance, and mortality is generally low (less than 20 percent). As with all bacterial pneumonias, nevertheless, the precise prognosis of the patient depends on the virulence and sensitivity to antibiotics of the infecting organism, the age and general health of the patient, the expertise of the physician, and the zeal and skill with which supportive care is given.

A second type of infection caused solely by the virus itself is much less common but considerably more lethal. The patient's clinical illness gives him no respite from the explosive onset of the grippe and continues inexorably through coughing, often in paroxysms, and marked shortness of breath, finally requiring admission to hospital. The sputum is scant, with little mucus, and is occasionally tinged with blood. Hypoxia is marked, and chest x-rays often reveal more than does physical examination. Initially there are small patches and streaks of hazy infiltration, usually extending from the hilum to involve one or several lobes. These infiltrates increase in size, distribution, and density each day in spite of increasing levels of respiratory support and the desperate use of antibiotics until near total opacification of both lung fields occurs, and the patient expires from extreme hypoxia. This sequence may evolve so rapidly that occasionally a patient may die suddenly after only a few days' or hours' symptoms, whereas patients with severely depressed cell-mediated immunity may rarely manifest a much more indolent course. On pathologic examination, the lungs appear firm and red, and the cut surfaces give the appearance of raw beef. Microscopic examination reveals no bacteria but rather severe interstitial edema and hemorrhage, as well as occlusion of alveoli with proteinaceous debris and blood. While this disease is seen most commonly in patients with mitral stenosis and, in some epidemics, in women during the last trimester of pregnancy, it may occasionally be seen in otherwise healthy individuals.

It should be emphasized that many more patients die as a result of influenza that can be accounted for solely by those succumbing from pneumonia. Total excess mortality

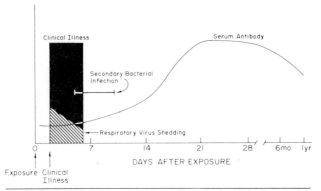

Figure 72-1. The course of influenza virus infection.

(the difference between the observed and expected number of deaths in a given time period) is generally two to three times greater during influenza epidemics than mortality from pneumonias. This can be easily understood if one accepts that influenza infection causes major derangements in many of the body's homeostatic mechanisms, and a large number of patients with marginal cardiac, metabolic, neurologic, and renal function may succumb rapidly from their primary disease when overcome by the virus-induced stress of fever and hypoxia.

A number of other less common complications or sequelae of influenza have been observed at variable frequencies, and these include myositis, myocarditis, pericarditis, orbital cellulitis, aseptic meningitis, as well as a postinfectious neuritis impossible to distinguish from the Guillain–Barré syndrome. Elegant epidemiologic studies have strongly established that von Economo's encephalitis (encephalitis lethargica) and thus most postencephalitic parkinsonism were a result of H swine-like influenza, which was epidemic or sporadic between 1918 and the late 1920s. This illness has not been observed following infection with other strains of influenza, but an increased incidence of bacterial meningitis has occasionally been noted during some influenza epidemics, confirming experimental animal studies that demonstrated that influenza increases the susceptibility to meningococcal and *H. influenzae* infections by mechanisms that have not been fully elucidated.

Epidemiology

The interpersonal spread of influenza viruses is caused by the formation of virus-laden aerosol droplets when an infected patient sneezes or coughs. Individuals in the immediate vicinity inhale these particles, the larger ones being deposited in the nares and upper nasopharynx, the smaller ones reaching the lower tracheobronchial tree. If the virus is not neutralized, as for example by local IgA-specific antihemagglutinin antibody, initial replication occurs. Further replication, and thus clinical illness, is dependent on a host of factors, including local IgA, serum IgG, and possibly various effectors of cell-mediated immunity. The efficiency of aerosol spread is exceedingly high and is in direct proportion to the number of particles generated, the density of susceptibles, their physical and temporal contiguity to the aerosols, and the degree of stagnation of the ambient atmosphere. It is not surprising, therefore, to observe high attack rates (at times approaching 90 percent) in nursing homes, classrooms, barracks, ships, and aircraft, where these conditions are at their worst. Cigarette smoking increases the susceptibility to infection and is associated with more severe clinical illness. There is little evidence that other modes of contagion play a significant part in the spread of influenza, although the role of hand contact and certain fomites, such as nasal spray applicators and drinking glasses, remains to be investigated.

Young children of preschool or gradeschool age appear to be the most effective disseminators of this disease, probably as a direct result of their limited prior experience with influenza (and thus high attack rates), as well as their primitive concepts of hygiene when they are actively excreting virus. An explanation of the appearance of epidemic influenza in the winter in temperate zones has been assiduously sought, but few positive conclusions have been reached. It does not appear to be directly related to temperature, relative humidity, or the duration of sunlight exposure. Likewise, wetting or chilling of the susceptible, so prominent a factor in avian respiratory infections, also does not appear to influence the susceptibility to or virulence of human influenza infections. It must be remembered that influenza epidemics may occur at any time in the tropics, and there have been epidemics recorded in every month of the year in the United States. Our largest recorded epidemic occurred in the late summer and early fall of 1918.

An understanding of the periodicity of influenza epidemics and pandemics requires a basic knowledge of the ultrastructure of the virus, as delineated in Chapter 62, for the surface antigens of the virus and the antibodies they induce within a population determine the epidemiology of influenza. The lipid envelope of the virus is covered by a number of glycoprotein spikes made up primarily of hemagglutinin with a smaller number of separate neuraminidase spikes. Because neuraminidase cleaves neuraminic acid from glycoproteins or glycolipids, it permits elution of virus from cell receptors or glycoproteins in secretions. Thus neuraminidase facilitates cell-to-cell spread and may even remove neuraminic acid residues from hemagglutinin, facilitating its cleavage and activation. Antibody to hemagglutinin prevents infection, either in tissue culture or in people, whereas antibody to neuraminidase does not neutralize virus or prevent infection but does limit viral spread in tissue culture and, if antihemagglutinin antibody is absent, lessens the severity of clinical influenza. Changes in the amino acid sequences of these glycoproteins lead to changes in their antigenic identity. The mechanisms that engender these changes have been a matter of great speculation for many years. It is now known, however, that influenza viruses are genetically extremely unstable as a result of either recombination/reassortment, insertions, frequent point mutations, and/or an absence of "proofreading" enzymes found in most DNA viruses. All of these factors are further multiplied by the relatively high replication rate of influenza viruses.

Generally, when a new virus with a different hemagglutinin type appears, there is little antibody against it in the population at large. The first wave of the epidemic infects a substantial proportion (25 to 60 percent) of the population. The virus also has the opportunity to undergo spontaneous, albeit limited, mutation (antigenic drift), so that a separate wave occurring 1 to 5 years later infects many of the remaining antibody-negative individuals, as well as some of the individuals who were previously infected but whose antihemagglutinin antibody response was either initially low or had waned over the intervening years. This sequence continues until the vast majority of the population has developed protective levels of antibody. At that point, a new hemagglutinin type, which shares no obvious antigenic cross-reactivity with the first, begins the cycle once again.

The grouping of these major antigenic types is based on the ability to observe serologic cross-reactivity in hemagglutination- or neuraminidase-inhibition (HI or NI) tests. To date, there have been five major types of human influenza hemagglutinin identified (H swine, H0, H1, H2, and H3) and two major types of neuraminidase (N1 and N2). Antibody to one will not be protective against infection with the other. That a greater hierarchy occurs than can be detected by serologic HI assays, however, is demonstrated by the fact that one can note antigenic priming among members of the H swine, H0, and H1 group, and similar priming among the H2 and H3 groups. Such priming is observed when prior infection with one member of the group causes an accelerated antibody response when subsequent infection or immunization occurs with another member of the group. The immunologic relatedness of the H swine, H0, and H1 group is so strong, in fact, that they have all recently been placed in the same group, designated H1.

Serologic analysis of individuals who were alive during the epidemics occurring in the latter part of the 19th century and the early part of the 20th century have clarified and greatly expanded our understanding of the periodic recurrences of influenza epidemics. Epidemics of viruses bearing the H2 hemagglutinin occurred in 1889 to 1890, and these viruses were later replaced by viruses probably bearing the H3 antigen around 1900. The H swine antigen first appeared in 1918 and gave rise to one of the greatest epidemics in recorded history. Next to appear in the 1930s was the H0 antigen, and viruses carrying the H1 antigen became epidemic in 1946. The H2 antigen then reappeared in 1957, to be supplanted by the H3 antigen in 1968, a sequence similar to that noted in 1889 and 1900. After the abortive reappearance of viruses bearing the H swine antigen in 1976, the H1 antigen reappeared in Southeast China in mid-1977 and caused widespread epidemics with apparently low mortality throughout the world in 1977 and 1978. Since that time, both H1 and H3 viruses have circulated simultaneously in the world population, with outbreaks caused by H3 viruses generally associated with more significant morbidity. The coexistence of H1 and H3 antigens had not been regularly observed in the past and defies current epidemiologic explanation.

Many recent pandemics (1957, 1968, 1977) began in China. The reasons for this are not clear, although it has been hypothesized that the density of human and swine or duck populations in the area near Canton provides an opportunity for dual infections with viruses indigenous to each species. Subsequent recombination may produce numerous new viruses, some of which may be novel or more virulent or against which there is little protective antibody in the population. Credence was given to the possibility of increased interspecies virulence when the mass die-off of harbor seals on the New England coast in 1980 was found to have been caused by an influenza virus bearing avian markers. Although spread from China to the rest of the world may take several months or more, the inexplicable simultaneous appearance of a new influenza virus in several widely separated parts of the world is well documented and was observed even before the advent of rapid intercontinental travel. Epidemics tend to sweep rapidly through individual communities and rarely last more than a month from beginning to end. Mortality rates are difficult to quantitate, but are generally quite low, often being quoted as 0.01 to 0.1 percent for the population at large, although the case fatality rate is considerably higher in the very young and particularly the very old. No valid explanation has yet been brought forth to account for the high fatality rate, especially in patients in their third and fourth decades of life, observed in the 1918 H swine epidemic.

Influenza B viruses also undergo major and minor antigenic changes, though at lower frequencies and longer intervals. Epidemics caused by influenza B viruses have occasionally been associated with excess mortality, but the significance of major antigenic changes, as occurred for example in 1973, is more obvious in younger rather than older age groups. Children are more readily infected with this virus, are more obviously symptomatic once infected, and are more likely to suffer one of the most serious sequelae of infection by this virus, Reye's syndrome. The association of this disease, characterized by hepatocerebral fatty infiltration and degeneration, with influenza B and a few other virus infections, including influenza A, is extremely strong, and over 300 cases were noted in the United States during the brief influenza B epidemic of 1973–1974. The pathomechanisms of this disease remain to be elucidated, but epidemiologic studies have shown a considerably higher risk of this complication in those patients who received aspirin rather than acetaminophen as part of their treatment for clinical influenza. Although a direct cause-and-effect relationship has yet to be defined to the satisfaction of all concerned, it would seem most prudent to use acetaminophen rather than aspirin in children with influenza B infection, at least until the matter has been fully clarified.

Diagnosis

While the influenza viruses can produce a wide variety of clinical illnesses, it must also be remembered that a broad spectrum of infectious and noninfectious agents can produce clinical syndromes that are similar to those of influenza. Patients infected by other common respiratory pathogens (RSV, adenovirus, coronavirus, rhinovirus, parainfluenza viruses) will often state that they have "the flu." Although influenza is more easily diagnosed during epidemics, a pathognomonic complex of signs and symptoms does not exist, and there is little to distinguish it on clinical grounds from other illnesses that may also occur during these periods. Recent experience in the Caribbean, for example, has shown that even epidemic dengue is difficult to distinguish from influenza unless the patient has prominent respiratory signs and symptoms.

Rapid confirmation of the clinical impression of influenza can be obtained by examination of material obtained by swabbing the posterior pharynx. Viral antigen can be detected within nasopharyngeal cells by direct or indirect immunofluorescent stains directed at any of the viral antigens. Such tests, however, are not generally available and require excellent reagents and appropriate positive and negative controls, as well as experienced observers.

The virus may also be grown from nasopharyngeal swabs after inoculation into any of several substrates. One of the more convenient and effective is continuous Madin–Darby canine kidney (MDCK) cells. Other effective substrates are primary monkey kidney cells and the allantoic or amniotic cavity of 10-day-old embryonated hen's eggs. In most instances, little viral cytopathic effect is induced, and the presence of virus must be determined by immunofluorescence, hemadsorption, or hemagglutination. Plaques can be formed in MDCK cells if the overlay contains trypsin, but this is of little use in primary isolation. Viral culture of the blood is of no use, as viremia is exceedingly rare in influenza.

It is usually not practical, necessary, or even desirable to attempt virus isolation in most clinical situations. Nonetheless, programs of surveillance that include virologic culture of select populations, are highly necessary for the general well-being. Individuals involved in such programs serve as sentinels and permit the early awareness of epidemic influenza. These surveillance systems operate best when centralized, as through the Centers for Disease Control or the World Health Organization, and when they disseminate their information rapidly and effectively. Public health authorities may then make plans for immuno- or chemoprophylaxis.

Serologic confirmation of influenza infection is rarely required. If such a confirmation is deemed necessary for scientific investigational purposes, however, several facts should be borne in mind. Complement-fixation (CF) tests depend on the interaction between antibody and internal (nucleocapsid) proteins. Since these proteins are antigenically (if not chemically) identical for all viruses within a group (A, B, or C), positive CF tests can signify infection by a member of that group but cannot more specifically identify the virus. CF titers generally rise relatively rapidly (within 1 to 2 weeks), but their decline is somewhat variable, often in terms of several weeks to several months. In some patients, however, they may persist for longer periods of time, though usually at a rather low titer. A single postillness serum sample, therefore, cannot be considered absolutely diagnostic but does become more useful if the titer is elevated ($\geq 1 : 32$). Studies of postinfection CF titers are generally of more epidemiologic than clinical value.

A more precise method of determining acute influenza infection is the quantitation of HI antibodies simultaneously on a sample obtained before or at the beginning of illness and a second sample obtained 3 or more weeks later. A four-fold rise in antibody titer is considered diagnostic of influenza if the patient has not been recently vaccinated. Additionally, because this antibody is directed against the specific surface glycoprotein, the precise identity of the originally infecting organism can generally be determined with some precision.

Since most influenza infections are self-limited and recovery can be hastened only moderately by early amantadine therapy, the major goal in diagnosis is often to determine whether other diseases causing influenzal syndromes are present or whether pneumonia caused by bacterial superinfection has occurred. The criteria for establishing the latter diagnosis include sustained pyrexia, leukocytosis, evidence of pulmonary consolidation on physical or x-ray examination, and, most importantly, the presence of substantial numbers of granulocytes and bacteria in the sputum. It cannot be overemphasized that the ability to perform a proper gram stain of the sputum and examine it with accuracy is the sine qua non of a physician competent to treat all pneumonias, including those following influenza. The proper treatment and final clinical outcome of the patient will often be dependent on the physician's ability to distinguish granulocytes from epithelial cells (and thus to distinguish a sputum from spit) and to recognize properly stained bacteria.

Treatment

General supportive measures have been prescribed for centuries to alleviate the bothersome symptoms of common influenza. There is little evidence that any of these measures shortens the duration of illness or diminishes the frequency of complications. Antipyretics such as acetaminophen for fever, fluids for thirst, analgesics for muscle pains, and bed rest for lassitude all support the comfort of the patient. Topical and systemic decongestants relieve nasal obstruction, and antihistamines promote restful sleep.

Evidence is accumulating that amantadine, a symmetrical primary amine, is effective in the treatment of clinical disease caused by influenza A viruses. This effect must be carefully differentiated from its prophylactic effect, which will be discussed below. The interpretation and performance of studies on the therapeutic effectiveness of this synthetic inhibitor of viral uncoating are difficult because influenza is a fairly rapidly evolving and resolving illness, and only a few days of clinical disease are ordinarily observable after the patient has consulted with his physician and sought therapy. Earlier studies during the 1960s demonstrated that amantadine definitely lessened the number of days of pyrexia, but this effect appeared little different from that of aspirin. More recent studies, however, have shown that amantadine often diminishes the severity of most symptoms to a significant extent and hastens clinical recovery by 1 to 4 days if therapy is begun within 48 hours of the onset of symptoms. Although this improvement may appear to be of marginal benefit, any improvement will be welcomed by most patients. Amantadine also significantly reduces the titer and duration of viral excretion, which, while of little direct benefit to the patient may be very significant in terms of limiting the spread of disease. Most importantly, it has been demonstrated that the duration of altered pulmonary function (as measured by diminished frequency-dependent compliance), which may last for several weeks after clinical influenza, can be markedly shortened following amantadine therapy. The dose of amantadine is usually 100 mg by mouth twice a day. Side effects can occur with variable frequency and are more common in the elderly. The most common effects in young adults appear to be psychologic. Insomnia, excitement, and a poor attention span or difficulty in mental concentration may be seen in up to 10 percent of recipients. Some studies have shown that rimantadine, a close analog of amantadine, is associated with fewer adverse reactions. Other studies have shown that although amantadine does not induce adverse reactions that are more common or more severe than those associated with a commonly used antihistamine, the combination of the two agents markedly increases the frequency of these reactions. In older individuals, the drug can occasionally precipitate congestive heart failure, cause more severe mental changes, or be responsible for other less common adverse effects. Amantadine has also been administered by aerosol to the upper and lower tracheobronchial tract, but the advantages and practicality of this technique require further study.

Prevention

Because the therapy of influenza is limited, major attempts have been made to prevent this disease. That these efforts have not been successful in the past is confirmed by the large number of people who die during epidemics and even during interpandemic years. The etiology of this lack of success is primarily logistical and attitudinal (as was the case for smallpox) and not because effective prophylactic measures are not available. Early attempts at preventing the disease involved growing influenza viruses in the allantoic cavity of embryonated hens' eggs and inactivating the virus in this slurry with formaldehyde. These vaccines generally induced high levels of HI antibody, which were closely correlated with protection from infection. Serum HI antibodies of $\geq 1:32$ generally are associated with solid protection from infection, whereas lesser levels of antibody (1:8 and 1:16) are associated with either moderate protection from infection or attenuation of clinical disease. These early vaccines required large volumes of allantoic fluid for each vaccine dose and were associated with endotoxin-like reactions, consisting of fever, chills, lassitude, and myalgias, which the recipient often considered to be similar to (or worse than) the early phases of clinical influenza. The etiology of these reactions was not clear. It was thought that the large amount of nonviral (egg) protein present in the vaccines caused these reactions, although

precise clinical investigations both in the 1940s and the 1970s did not support this hypothesis, and vaccines containing larger amounts of egg protein (such as the 17D yellow fever vaccine) do not induce such reactions. An equally valid hypothesis could be that the large amount of virus with its lipoprotein coat may cause these reactions. Vaccines produced in the late 1960s and early 1970s by density gradient centrifugation (Chap. 62), which eliminated most of the nonviral protein, were associated with fewer adverse reactions and equal immunogenicity, although very young individuals continued to experience at times severe and debilitating adverse reactions, including seizures. Disruption of the virus by a variety of detergents and solvents yielded vaccines that produced fewer adverse reactions and were equally immunogenic in primed populations. Single doses of such vaccines, however, are less immunogenic in individuals who have little or no prior antigenic experience with influenza. A second dose given a month or more later, however, ordinarily induces protective antibodies in over 85 percent of these recipients. The recent cloning of the hemagglutinin antigen by genetic engineering techniques is another step in the potential refinement of influenza vaccines.

Inactivated vaccines, nonetheless, are little used for a variety of reasons, the first being the memory of both the patients and doctors of older, more highly reactogenic vaccines. Second, these vaccines are effective only when the antigens in the vaccine are identical or closely related to the antigens of the epidemic virus. Since the preparation of these inactivated vaccines requires 6 to 9 months of lead time, a certain amount of guessing must occur each winter in an effort to establish which virus will be epidemic in the succeeding year. These prognostications are sometimes accurate. In addition, the immunity induced is not highly effective for much more than a year, so that annual vaccination must be given. Finally, the economics of vaccine production are difficult and marginal. Manufacturers must deal with an everchanging target, which only intermittently captures the attention of patients, public health authorities, and physicians. As a result, they tend to produce only the number of doses of vaccine they feel will be sold in the succeeding year. This has generally been in the range of 15 to 20 million doses annually in the United States. The practical problems of administering all of these doses at the optimal time (generally several weeks to a few months before the expected epidemic) can be easily imagined.

Because of these logistical problems, vaccine is currently recommended only for those who are at high risk of dying should they acquire influenza. They would include any patient with a chronic illness, particularly pulmonary, but also any chronic cardiac, metabolic, and neurologic disease, and the elderly (those over 65 years of age). Its use in vital community personnel, such as physicians, nurses, policemen, and firemen, was temporarily slowed by the occurrence of the Guillain–Barré syndrome in approximately 1 of 100,000 recipients of the A/New Jersey (swine) strain influenza vaccine. About 10 percent of these cases were fatal. Subsequent studies have shown that vaccines that do not contain the H swine antigen are not associated with Guillain–Barré syndrome, further emphasizing the unusual neuroreactivity of this virus as exemplified by the association of von Economo's encephalitis with natural infection only with this strain of influenza. It should be noted also that influenza vaccines, like other biologic response modifiers, transiently depress hepatic cytochrome p-450 enzymatic activity and thus diminish the rate of metabolism or degradation of certain drugs, such as theophyllin. This observation would have particular relevance to asthmatics receiving this drug and who are also a high-risk group requiring immunization.

As it is thought optimal by some to induce local (nasal and/or tracheobronchial) immunity to prevent respiratory disease, great enthusiasm has been generated over the past 3 decades in Eastern Europe and for the past 1 to 2 decades in the West for live influenza vaccines that may be applied topically in the nostrils. The viruses in these vaccines have been attenuated by a variety of means, including multiple egg passages, recombination with highly attenuated strains, cold adaptation, selection of temperature-sensitive mutants, or the selection of strains resistant to nonspecific inhibitors. Clinical studies with several of these types have shown that immunity can be induced at a level and duration roughly equivalent to that induced by inactivated vaccines. One advantage of these vaccines is that a much greater number of doses can be produced per embryonated egg. Fertilized eggs are often a limiting economic and supply factor in the production of influenza vaccines. A disadvantage of these vaccines, on the other hand, is that because they contain infectious virus, much more stringent testing in humans must be performed before their widespread general use in tens or hundreds of millions of recipients can be encouraged. When such testing is completed, the original epidemic strain for which the vaccine was designed has often disappeared from the population. Attempts to use nonhuman and in vitro models of immunogenicity and safety are being intensely investigated, but to date none has been sufficiently consistent and convincing to obviate or circumvent extensive and time-consuming clinical trials in humans.

Amantadine has also been shown to be effective in the prevention of influenza. Its efficacy is approximately that of inactivated vaccines. Its advantage is that its prophylactic activity is not as strain-specific as that of vaccines and therefore may be used when the virus epidemic in the population is different from that in the vaccine or when vaccines are not available. Its disadvantages are that it will induce a substantial number of side effects in patients who are otherwise healthy, and it must be taken regularly during the entire epidemic exposure season.

FURTHER READING

Barry DW, et al.: Comparative trial of influenza vaccine. I. Immunogenicity of whole virus and split product vaccine in man. Am J Epidemiol 104:34, 1976

Dolin R, Reichman RC, Madore HP, et al.: A controlled trial of amantadine and rimantadine in the prophylaxis of influenza A infection. N Engl J Med 307:580, 1982

Dowdle WR, Coleman MT, Gregg MB: Natural history of influenza type A in the United States, 1957–1972. Prog Med Virol 17:91, 1974

Hall WJ, Douglas RG, et al.: Pulmonary mechanics after uncomplicated influenza A infection. Am Rev Respir Dis 113:141, 1976

Jackson GG, Muldoon RL: Viruses causing common respiratory infections in man. V: Influenza A (Asian). J Infect Dis 131:308, 1975

Kark JD, Lebiush M, Rannon L: Cigarette smoking as a risk factor for epidemic A(H_1N_1) influenza in young men. N Engl J Med 307:1042, 1982

Kavet J: A perspective on the significance of pandemic influenza. Am J Public Health 67:1063, 1977

Kilbourne ED: The Influenza Viruses and Influenza. New York, Academic Press, 1975

Linnemann CC, et al.: Reye's syndrome: Epidemiologic and viral studies, 1963–1974. Am J Epidemiol 101:517, 1975

Rogers DE, Louria DB, Kilbourne ED: The syndrome of fatal influenza virus pneumonia. Trans Assoc Am Physicians 71:260, 1959

Selby P (ed): Influenza: Virus, Vaccines and Strategy. New York, Academic Press, 1973

Stuart Harris CH, Schild GC: Influenza: The Viruses and the Disease. Littleton, Mass, Publishing Sciences Group, 1976

Webster RG, Laver WG, et al.: Molecular mechanisms of variation in influenza viruses. Nature 296:115, 1982

CHAPTER 73

Paramyxoviruses

Parainfluenza Virus

In the quarter of a century since their initial isolation, the epidemiology, clinical illnesses, and the host response to infection associated with these agents have been described, as well as some of the molecular events associated with replication. The pathogenesis of infection has been better understood with the advent of investigations by the molecular virologists. These paramyxoviruses have a nonsegmented single-stranded RNA genome with enveloped helical nucleocapsids. They have both hemagglutination and neuraminidase activity associated with one glycoprotein and cell fusion or red blood cell hemolytic activity attributed to a second membrane glycoprotein.

Clinical Illness

Parainfluenza virus type 1 has been the predominant agent associated with laryngotracheobronchitis (croup). This agent tends to occur in biennial epidemics that peak in the fall months. Children usually acquire infection between the ages of 6 months and 3 years. The vast majority of young

adults have demonstrable serum antibodies against parainfluenza 1.

Parainfluenza type 2 is also associated with acute laryngotracheobronchitis in the fall, but it occurs in sporadic outbreaks. Parainfluenza 2 will often alternate years of occurrence with parainfluenza type 1. Although children of the age group 8 to 36 months are at greatest risk of infection, the morbidity from parainfluenza 2 infections is substantially less than that from parainfluenza type 1.

Parainfluenza virus type 3 is associated with bronchiolitis and pneumonia in infants less than 1 year of age, with croup in children 1 to 3 years old, and with tracheobronchitis in older children. This virus is similar to respiratory syncytial virus, with its predilection for causing infection in infants during the first 6 months of life, as well as through 3 years of age. The majority of primary infections in infants and children are associated with febrile illness of approximately 4 days duration, and at least one third of these illnesses are complicated by pneumonia or bronchiolitis. Serologic studies have demonstrated that approximately one half of children have been exposed to this agent by the end of their first year of life.

Parainfluenza type 4 has been recovered much less frequently from children and adults with mild respiratory illness. The laboratory isolation of this agent is more diffi-cult, which contributes to its less frequent association with illness. The common symptoms of infection with any of the parainfluenza agents in adults are nasal congestion, sore throat, and malaise. These viruses have been associat-ed with exacerbations of chronic bronchitis, pharyngitis, and tonsilitis in college-age students, and asymptomatic re-infection.

Several studies have demonstrated excretion of para-influenza virus type 3 for 2 to 5 months in a population of adult patients with chronic bronchitis and emphysema. Several of these patients had virus-specific local antibody detected in their respiratory tract prior to detection of vi-rus excretion. This persistence of parainfluenza type 3 in-fection did not produce a significant increase in local respiratory antibody. This could be contrasted to those pa-tients showing symptomatic illness, brief excretion of virus, and an associated increase in titers of serum and respirato-ry secretory antibody.

Pathogenesis of Infection

These viruses are excreted in respiratory secretions, and therefore virus is aerosolized in small and large droplets with breathing, coughing, and sneezing. Inhalation of in-fectious virus transmits infection. It is probable that hand-to-mucous membrane contact may also transmit infection, as has been shown with rhinovirus and respiratory syncy-tial virus. The incubation period is usually from 1 to 4 days. Virus has been isolated from the respiratory tract for as long as 6 days before onset of symptoms and is isolated from about two thirds of symptomatic patients during the first week of illness. The parainfluenza viruses are found in secretions as long as the second and third week after onset of symptoms more often than are respiratory syncy-tial viruses or influenza viruses.

The replication of each of these agents is normally limited to the respiratory tract. The two membrane glyco-proteins, the hemagglutinin/neuraminidase (HN) and fu-sion protein (F) are spike-like surface projections on the virion and are anchored in the viral membrane by the hy-drophobic portion of the proteins. The HN protein has re-ceptor binding activity and requires a receptor with neuraminic acid residues for adsorption of the virus to the cell. Glycoproteins and glycolipids with neuraminic acid are ubiquitous on vertebrate cells. These viruses can therefore adsorb to many cell types, and thus virus–cell re-ceptor interaction does not explain the tissue tropism of parainfluenza viruses. The fusion of virus and cell mem-branes is involved in virus penetration and, therefore, in-fection of the host cell. The F protein is essential for infectivity of these viruses and must be cleaved by a pro-teolytic enzyme provided by the host. Host-dependent proteolytic activity therefore determines tissue tropism and host range. It has even been shown that a specific amino acid sequence at the N terminus of the activated F poly-peptide is essential for biologic activity and that homology for this region exists between the proteins of various members of the paramyxovirus group. Proteolytic cleavage

occurs in the same site on the F protein of different para-influenza viruses, and oligopeptides resembling the N ter-minus have been used experimentally to inhibit fusion/penetration.

Since adsorption occurs independently of fusion, it is possible to hypothesize different outcomes of the virus–host cell interaction. If the virus with cleaved F protein in-fects cells lacking an appropriate protease, virus replica-tion may occur, but progeny virus particles will have uncleaved F protein and therefore will not be infectious. If these virus particles can adhere to cells capable of cleaving F protein during adsorption, they become infectious. If cells cannot activate F protein after adsorption, the virus is not infectious. Although details are yet to be determined, the limitation of virus replication to the human respiratory tract probably is a result of specific proteolytic cleavage of the F protein, which occurs efficiently in the respiratory tract.

In vitro, tracheal ring cultures have been utilized to describe the sequential morphologic events attendent upon parainfluenza infection. Epithelial cell cytoplasmic antigens are easily demonstrable with immunofluorescence by 24 hours after infection. Ciliary motion is lost, and the pseudostratified columnar organization of the epithelial layer disappears as some ciliated cells are lost and others fuse to form multinucleate giant cells. Such fusion is also seen in the lamina propria and cartilage, and budding vi-rus particles from these cells are visible by electron mi-croscopy. Virus replication can be sustained for 2 weeks in vitro before focal destruction of the epithelium occurs.

The available pathology of pulmonary tissue from infected children shows an increase in bronchial-associated lymphoid tissue, peribronchial infiltrates, and exudates in lamina of the bronchus. Interstitial pneumonia with associ-ated atelectasis may also be present.

Host Response

A single infection does not confer immunity to infection or symptomatic illness. Reinfections occur frequently, al-though the presence of antibody may modify the clinical manifestations. In fact, studies with parainfluenza 2 have demonstrated that protection against clinical reinfection is best correlated with the presence of detectable local neu-tralizing antibody in nasal secretions. Patients show homol-ogous IgA antibody formation in nasal secretions, usually within 7 to 10 days after onset of symptoms, which peaks in about 2 weeks. Virus neutralizing activity is also demon-strable but does not necessarily correlate with IgA. This may be because neutralization is only one measure of anti-body function, and all neutralization is not due to specific IgA antibody.

Infection causes the host to produce humoral anti-body, as measured by hemagglutination inhibition, neutral-ization, and complement fixation. The neutralizing and hemagglutination-inhibitory antibodies are directed against the HN protein. Antibody to the F protein is also critical to limitation of virus replication. Antibodies to both pro-teins are probably necessary to prevent clinical infection.

Reinfection stimulates an anamnestic response with heterotypic antibodies to other serotypes of the parainfluenza viruses.

Lymphocytic cellular host defenses seem to be important in the normal response to these viruses and may be essential for their eradication. Severe combined immunodeficient children have persistent respiratory excretion of these viruses for months or years and have had virus isolated from peripheral blood, liver, CSF, and pericardial fluid. Systemic infection is well documented, but whether these are mutant viruses that have evolved in a defective host or whether this is only due to the host's deficits remains to be determined.

Diagnosis

The overlapping spectrum of respiratory disease produced by a number of viral pathogens can be etiologically defined only by specific identification of the infecting virus. Rapid and accurate diagnosis by immunofluorescence of nasal secretions has been accomplished by several laboratories. Careful collection of materials and fluorescent staining of the exfoliated epithelial cells from the upper respiratory tract provide a diagnosis within 1 day of obtaining the specimen. The fluorescence is cytoplasmic and has been observed in the ciliated epithelial cells.

Isolation and identification of these agents in susceptible cells in tissue culture may require up to 2 weeks. The infected tissue culture can be examined by immunofluorescence with successful demonstration of viral antigens during the first several days of culture. More commonly, hemadsorption employing guinea pig red blood cells is utilized in 4 to 7 days to detect the presence of a hemadsorbing agent.

Tissue culture cells that are used for parainfluenza virus isolation include primary human embryonic kidney, human amnion, and primary monkey kidney cells. Unfortunately, the presence of hemadsorbing simian viruses must be taken into consideration when working with simian cell cultures.

The parainfluenza viruses do hemagglutinate guinea pig and chicken red blood cells. Thus, hemagglutination inhibition can be useful for identification of virus and for determining the titers of specific serum or secretory antibody against known virus strains. With primary infection, the response to the infecting agent tends to be type-specific. With each subsequent infection, there is an increased heterologous antibody response.

Treatment

The treatment of these respiratory infections is only by the provision of supportive therapy. Hydration, maintenance of an adequate airway, and therapy of bacterial superinfection are important in the management of these patients. The severity of the lower respiratory tract disease is greatest in the youngest age group.

Parainfluenza virus 3 infection has been documented to be a significant nosocomial infection, with increased morbidity to the patients who acquire these infections in the hospital. All of the viruses causing severe respiratory infections in children should be regarded as communicable within the hospital, and appropriate precautions should be taken to prevent spread of the infection within the hospital.

Prevention

As implied in the preceding paragraphs, immunity to parainfluenza viruses is only partial. That is, reinfection may occur naturally, although the severity of the illness may be considerably diminished.

Parainfluenza viruses grown in embryonated eggs or tissue culture have been administered as formalin-inactivated vaccines. Two to three injections of the individual parainfluenza types are necessary before seroconversion occurs in young infants. These materials have been tested as monovalent or trivalent immunogens. The vaccine can produce a 95 to 100 percent seroconversion rate in antibody-negative subjects, but no resulting significant protection against naturally occurring infection has been demonstrated. Parenteral administration of these inactivated products does not produce local secretory antibody. It has been learned that formalin destroys the F protein of measles and RSV. It seems likely that such treatment destroys the F glycoprotein of the parainfluenza viruses, too, and therefore no antibody to this protein was produced. Thus, the failure to achieve protection with inactivated antigen may be due to destruction of an antigen essential for protection. This explanation is important in view of suggestions that preexisting antibody might enhance disease and provide an explanation for severe disease in infants with maternal antibody. A Syrian hamster model of pulmonary infection with live virus has been used to show no enhancement of disease by the administration of passive antibody or by reinfection of an animal.

Significant morbidity results from infection with these agents in the youngest infants. Prevention of infection is a desirable goal, but the available immunogens have not been successful. Our better understanding of the surface glycoproteins may facilitate successful immunization.

Respiratory Syncytial Virus

Clinical Illness

Respiratory syncytial virus (RSV) was first described in 1956 and since that time has become recognized as the single most important respiratory tract pathogen of infants and young children. Severe disease caused by RSV is most often manifest as bronchiolitis or pneumonitis and occurs predominantly in children less than 6 months of age. Studies of hospitalized young children with acute respiratory disease reveal 20 to 30 percent of their illnesses are attributable to RSV. RSV is the only known virus of humans in the genus *Pneumovirus* of the Paramyxoviridae family. The virus is enveloped; it has a nonsegmented single-stranded RNA genome in a helical nucleocapsid. RSV has a fusion protein but lacks a hemagglutinin.

The virus comes in contact with the mucosal surfaces of the nasopharynx, and virus replication occurs. Immunofluorescence of nasal secretions provides evidence that ciliated epithelial cells and macrophages contain RSV antigen and are, therefore, sites of virus replication. Infection is normally limited to the respiratory tract, and virus is generally isolated only from respiratory tissues and secretions. The incubation period preceding clinical illness is ordinarily 1 to 4 days after exposure to virus.

A calculated risk for hospitalization due to RSV infection during the first year of life is 5 per 1000 live births based on studies in low income families. These very young children are severely ill, have fever, lethargy, and apnea, which is a prominent feature of their respiratory tract disease. Passively acquired maternal serum antibody levels correlate with acquisition and severity of illness. Infants born with higher titers of antibody acquire disease at a later age, and those who have moderate levels of antibody have milder illnesses. Bronchiolitis and/or pneumonia is manifest by labored, rapid respirations, with radiologic changes ranging from hyperaeration to consolidation. Involvement of the lungs may be sufficiently extensive to necessitate respiratory assistance and, in a small percentage of children, is fatal. Infection in the youngest infants is their primary encounter with the virus. In a day-care setting where longitudinal observations are possible, virtually all antibody-negative children develop infections, as documented by virus excretion and seroconversion after their first exposure to RSV. Clinically apparent illness, including fever, lower respiratory illness, and middle ear effusions, occurs significantly more often with primary infection. One infection will produce some resistance to reinfection, so about 75 percent of these exposed seropositive youngsters develop infection with the second exposure to virus, as evidenced by antibody rise and/or virus shedding.

Children between the ages of 1 and 3 years who have symptomatic illness are less likely to have bronchiolitis. The age-related occurrence of bronchiolitis suggests anatomic determinants of this manifestation of RSF infection. Middle ear effusion is a common concomitant finding after the age of 6 months. Older children and adults are more likely to have colds, with cough, nasal congestion, or asymptomatic infection evidenced only by serologic or cultural evidence of infection. The percentage of asymptomatic/mild illness increases from 2 percent to 50 percent from the first to the third or fourth infection, respectively. This reduction in the number of symptomatic infections is independent of age and correlates with previous infections. Even mild symptomatic illness can produce measurable morbidity for periods up to 8 weeks. There may be an increase in respiratory tract resistance, apparently due to altered reactivity of small airways.

Reinfection with RSV is the rule. Antigenic variation of RSV does not seem to explain reinfection of humans, since the measured antibody response to different isolates of virus does not detect any antigenic difference. As indicated above, 75 percent of infants will become infected with their second exposure to virus and 65 percent with their third. School-age children have a risk of infection of about 40 percent. Thus, immunity is reinforced by subsequent infections.

Host Response

The severity and frequency of RSV infections in infants have led to rather extensive study of the virus in an attempt to guide the development of preventive procedures. Serum antibody against RSV is largely IgG and, as such, crosses the placenta to the fetus as delineated above. Serum antibodies appear in the infant in approximately 10 days to 2 weeks after the onset of infection, persist for months, and then show an anamnestic response with natural or artificial challenge with virus. Antibody persists longer after reinfection as compared to primary infection. The presence of humoral antibody in the youngest infants correlates with protection against serious lower respiratory tract disease. The mechanism of protection is unknown, and hypotheses, such as diffusion of humoral antibody at the alveolar or bronchiolar level or mediation of antibody-dependent cell-mediated cytotoxicity (ADCC), have been suggested. A ^{51}Cr release assay showed ADCC in nasopharyngeal secretions collected from infants that was primarily mediated by IgG. It is possible that humoral antibody itself may not be the protective factor. These antibodies do not cross-react with any other known virus.

Antibody to RSV in nasopharyngeal secretions as demonstrated by immunofluorescence appears promptly during illness, usually within 3 days of onset. IgG and IgM are detectable with peak titers available for IgG and M from 8 to 15 days after onset. IgA is the predominant class of antibody, with peak titers attained from 14 to 28 days after onset and sustained for about 2 months in the face of waning IgG and IgM. Infants older than 6 months produce higher IgA titers in response to primary infection than do younger infants. Secondary infection stimulates an anamnestic response in all three classes of immunoglobulin in nasopharyngeal secretions, with a less rapid decline in antibody.

Studies demonstrate a better correlation with protection against RSV infection by secretory antibodies than by serum IgG antibodies. It has been shown in virus challenge studies in adults that there is an inverse correlation of the titer of nasal wash antibody with the quantity of virus excreted and the antibody rise. Those persons with high levels of specific IgA antibody challenged with RSV excrete less virus and may show no antibody rise in secretions or serum. Those adults with low or absent levels of specific nasal IgA show excretion of increased amounts of virus and do develop rises in antibody. Clinical illnesses were observed only in the volunteers with low nasal levels of IgA who received the larger of two inoculation dosages of virus.

More recently, RSV-specific IgE has been demonstrated more frequently and in higher titers in the nasopharyngeal secretions of infected children with wheezing. The degree of RSV-IgE response correlated with hypoxia and, thus, with severity of illness.

RSV-specific IgA has been measured in colostrum and milk, and because virus replicates only in the respiratory tract, it would seem that this is the site of antigen exposure to lymphocytes, suggesting a bronchomammary axis in the immune response. Any protective or ameliorating effect of specific antibody in breast milk is still to be determined, although epidemiologic evidence suggests that some protection may be provided.

RSV antigens are expressed on the surface of infected cells, and as expected, increased blastogenic responses of lymphocytes to RSV are seen in children with RSV infection. The combined immunodeficient children have prolonged shedding of this virus from their respiratory tracts. Cellular immunity is probably important in limiting infection with this enveloped respiratory virus.

Pathology

Infants with RSV infection show severe changes in their lungs. Pulmonary tissues from children with pneumonia show marked inflammation, with mononuclear cell infiltrates in interstitial tissues, alveoli, small bronchioles, and alveolar ducts. There may be demonstrated syncytia formation and intracytoplasmic inclusions, which are consistent with, but not pathognomonic of, RSV infection. Infants with bronchiolitis show less extensive interstitial and alveolar involvement but instead have moderate to marked changes in the bronchioles. Epithelial necrosis and plugs consisting of cell debris and fibrin are seen, as well as peribronchiolar lymphocytic infiltrates. Virus has been isolated from tissues of children with both types of illness.

Epidemiology

RSV infection is worldwide in distribution and tends to cause yearly outbreaks of illness alternating from midwinter to late spring in occurrence. Infection has been experimentally produced by introduction of virus into the upper respiratory tract of volunteers and animals (chimpanzees). Within the hospital setting, RSV transmission is readily demonstrable when persons have close contact with infected infants, suggesting transmission by touch or large droplets. Small particle aeresol transmission does not transmit infection easily in this setting. Experimental inoculation of adults shows the eye and nose to be efficient portals of entry, whereas the mouth is less sensitive. The increased frequency of severe disease in infants has been noted, and during community studies, attack rates of 30 to 60 percent are estimated for exposed infants less than 1 year old. RSV infection seems to be introduced into a family by an older (2 years or more) sibling. The secondary attack rate is then greatest for infants less than 1 year of age, but appreciable attack rates occur at all ages, with approximately 50 percent of family members acquiring infection. Investigations conducted simultaneously on different populations in the same geographical setting pointed out that patients seen in private practice had RSV infection at an older age than either urban or rural clinical patients. These older infants with less severe disease were seldom hospitalized.

A final epidemiologic consideration is that of nosocomial infections with RSV. The hospitalized infant excretes virus for days, and it is transmitted to ward personnel as well as to patients. The acquired illness also results in prolongation of hospitalization for infants and may produce fatal illness in those infants hospitalized with other underlying problems. Consideration should be given to patient isolation and careful handwashing in providing care for infected infants.

Diagnosis

Presumptive clinical diagnosis can be made in an infant with bronchiolitis or pneumonia with no demonstrable bacterial pathogens. A definite etiologic diagnosis of RSV infection can be made by detection of viral antigen or serologic tests. The virus can be isolated from nasal secretions utilizing tissue cultures of cells such as the Hep2 line. Longitudinal study of families has demonstrated that virus can be detected as long as 6 days before symptoms occur and in about 73 percent of infants during the first week but in only 6 percent in the second week (after onset of symptoms). Hospitalized children may excrete virus for longer periods. The mean titer of virus detected in secretions of hospitalized children is $5.0 \log_{10} TCID_{50}$. Children less than 2 years of age excrete virus longer than do older children or adults. Although this information may assist in management of severely ill infants, all too frequently it is not known until the patient has completed the natural course of illness.

The virus is extremely labile, and respiratory secretions, throat swab material, lung biopsy, or postmortem material must be kept at 4C and immediately processed by a knowledgeable laboratory utilizing specific techniques in order to preserve cells for immunofluorescence or infectious virus for tissue culture inoculation. Several laboratories have demonstrated RSV antigens in cells present in nasal secretions by fluorescent antibody techniques or ELISA assay. The indirect immunofluorescent method employing bovine anti-RSV serum and fluorescein-conjugated antibovine globulin has been successful in the few laboratories familiar with the procedure. The technique affords a means of identification of RSV infection within hours but is not generally available. Later in disease, it can also identify viral antigen despite the presence of antibody, which may neutralize the infective virus in secretions.

Serologic techniques allow the recognition of RSV antibodies by complement-fixation, neutralization, or plaque-reduction techniques, with the latter being the most sensitive method. Experimentally, radioimmune and enzyme-linked immunosorbent assays are presently being developed. Acute and convalescent sera are necessary to define the significance of measured antibody. As with other virus infections, a four-fold or greater antibody rise indicates recent infection, but a single value will indicate only previous experience with the virus at an undetermined time. Investigational use of a specific IgM radioimmunassay may prove helpful in defining acute illness.

Therapy

The therapy of RSV infection is entirely nonspecific and consists of support of the respiratory system, control of fever, adequate hydration and nutrition, and therapy of secondary bacterial infection.

Prevention

The prevalence of RSV infection has resulted in attempts to develop an effective vaccine. A formalin-inactivated, alum-precipitated virus vaccine was prepared and utilized in field trials. The vaccine produced a rise in serum antibodies as measured by neutralization in adults and children but no rise in secretory antibody. Although antibody to the F protein was not measured, it is probable that formalin destoryed this protein, rendering the immunogen incomplete. During naturally occurring outbreaks of RSV disease, the attack rate among infants immunized with inactivated vaccine was the same as that in the control (unimmunized) populations. This indicated that this immunization did not protect from subsequent infection. In addition, the clinical illness was more severe in vaccinees, and the age group suffering severe disease was extended to include older infants. It is probable that immunization was analogous to that with inactivated measles virus. The infants could identify RSV antigens except the fusion protein, and therefore virus replication and cell-to-cell spread occurred until antibody to the F protein was formed. The host response to other viral components resulted in destruction of some infected cells. Such unanticipated observations stopped the use of inactivated vaccine.

Attempts are being made to develop a live virus vaccine that will be protective. It is predicted that an effective vaccine will induce formation of respiratory tract IgA because of its recognized importance in protection against disease. The first RSV vaccine administered intranasally was a cold-adapted strain that was found to be attenuated in adults but produced mild disease and prolonged virus shedding in seronegative infants. Temperature-sensitive RSV mutants multiplying only at 33 to 35C have been isolated. It was hoped that such mutants would multiply only in the upper respiratory tract and thus reproduce the immunologic events of natural infection without causing illness in the infants. Administration of the temperature-sensitive mutant produced asymptomatic infection in seropositive children and infection with mild rhinitis in seronegative children. Using intranasal administration of 100 $TCID_{50}$, children with preexisting nasal antibody were not infected. Virus shedding was monitored, with recovery of a small population of genetically altered virus with partial loss of the temperature-sensitive property in a few instances. The vaccine virus was not transmissible to control children in intimate contact with vaccinees. Sequential studies have shown that subsequent natural RSV infection occurred in vaccinees. Additional trials with a genetically stable, temperature-sensitive mutant of RSV having a defect consistent with an altered fusion protein have been conducted. The virus was poorly infectious when administered intranasally in chimpanzees, adults, and children. A similar vaccine administered parenterally was not efficacious.

The natural history of RSV infection suggests that no immunogen will prevent multiple infections with this virus. If the initial infection with an attenuated virus could prevent severe illness, which is more likely to occur in the younger children, it would be worthwhile. Perhaps multiple immunizations will be necessary. To date, prevention of illness has never been demonstrated, and no candidate immunogen clearly modifies subsequent infections.

Mumps

Mumps virus infection was recognized as epidemic parotitis as early as the fifth century BC by Hippocrates. From its position as a universal infection of childhood, mumps has receded to an uncommon illness in nations employing vaccine. The most frequently recognized symptom of infection is parotitis, although the illness may have multiple manifestations, indicating the widespread multiplication of virus. The last decade has added far more substantive understanding to the molecular virology of mumps virus, and it is possible to incorporate some of these facts into our understanding of the clinical consequences of infection with this agent. Mumps virus is a member of the paramyxovirus family and, as such, is pleomorphic, with a diameter of approximately 150 nm. The lipid envelope surrounds a helical nucleocapsid containing a single-stranded, nonsegmented RNA genome. There are two surface glycoproteins on the envelope that are equivalent to the hemagglutinin/neuraminidase (HN) and fusion (F) proteins described for other paramyxoviruses.

Clinical Illness and Pathogenesis

The portal of entry is thought to be the upper respiratory tract, as the virus gains direct access to the mucous membranes of the mouth and nose. Virus is transmitted by the saliva or other secretions of infected persons. The virus replicates after entering the host, although the specific tissue in which primary multiplication of virus occurs is unknown. Sialic acid residues are ubiquitous on mammalian cells and may be the host cell receptors for this virus. Although the functions of the mumps glycoproteins have been incompletely studied, by analogy with other paramyxoviruses the HN protein probably determines the initial virus–host cell interaction. The HN protein induces antibody that inhibits hemagglutination and neuraminidase activity and neutralizes the infectivity of the virus.

Viremia occurs with uncomplicated infection, and mumps is therefore different from parainfluenza infections where virus replication is limited to the respiratory tract. Secondary infection of multiple tissues can occur. The fusion protein on the viral envelope is essential for cell-to-cell spread for this virus. Cleavage of the F protein by the host cell protease(s) is necessary for its activation, and it can be postulated that this protein has cleavage sites susceptible to different enzymes than the parainfluenza viruses because the tissue tropism is far more generalized. Such tissues as the salivary glands (predominantly the parotids), meninges, testes, pancreas, ovaries, thyroid, and heart can support virus replication. Virus is excreted in the urine, and transient abnormalities in renal function have been found in adult males with mumps infection, implying that virus replicates in renal tissue.

The temporal relationships of virus excretion and communicability of the illness with clinical symptoms and host–antibody response are shown in Figure 73-1. Salivary gland infection with pain, edema, and consequent swelling results in the characteristic parotid gland enlargement diagnosed as mumps. This infection may involve the parotids, submandibular, and, less often, the sublingual salivary glands. Salivary gland involvement usually precedes other clinical symptoms, lasts for 2 to 7 days, and may be unilateral or bilateral. Any of the other manifestations of mumps infection occasionally precede, coincide with, or occur in the absence of salivary gland involvement.

Mumps virus may involve the central nervous system. The vast majority of such recognized illness is transient meningitis with few sequelae. The incidence of this type of central nervous system infection has been estimated to be as high as 65 percent when lumbar punctures have been done on a group of hospitalized patients with clinical mumps. The central nervous system involvement was manifest in each case by a predominantly lymphocytic pleocy-

tosis of the cerebrospinal fluid. Only one half of these individuals evidenced clinical signs of meningitis. Signs of meningeal involvement are most often manifest 2 to 10 days after the onset of parotitis, last for 3 to 4 days, and are self-limited.

A more serious, and fortunately much less common, central nervous system manifestation of mumps is postinfectious encephalitis or encephalomeyelitis. The reported ratio of mumps encephalitis to mumps cases is 2.5/1000 cases for each year from 1968 to 1972, but because of the lack of a uniform definition, meningitis cases are probably included, giving a falsely high case ratio. The time of onset is usually later than the transient meningitis and occurs 10 to 14 days after the clinical salivary gland involvement. The patient appears severely ill, is deeply obtunded, and may succumb to the illness. Although occurring less often than the postinfectious encephalitis associated with measles or varicella, it is clinically and pathologically indistinguishable from them.

Clinicians have studied the antibody response to mumps central nervous system infection and find that specific mumps IgG becomes detectable in the cerebrospinal fluid and is thought to be synthesized within the central nervous system. Titers of antibody usually reach a peak in 4 to 10 days after the onset of meningitis and decrease thereafter (Fig. 73-1). The IgG antibodies are oligoclonal, and in several patients with prolonged pleocytosis, antibodies could be detected as long as 11 to 12 months after meningitis. Similarly, interferon activity could be measured during the peak of clinical illness and tended to parallel the evidence of inflammation in the cerebrospinal fluid. Individuals with a prolonged pleocytosis had demonstrable interferon for longer periods of time.

An observation recorded in 1967 demonstrated the development of hydrocephalus in hamsters after experimental mumps infection. Suckling hamsters were inoculated intracerebrally with mumps virus isolated from the cerebrospinal fluid of a patient and passaged only once in cell cultures. Sequential studies demonstrated virus replication and associated inflammatory changes during the first week after inoculation. Virus was localized almost entirely within ependymal cells of the ventricles and choroid plexus. The animals did not show any signs of acute illness, but hydrocephalus became evident in 3 to 6 weeks after infection. These animals had aqueductal stenosis secondary to the resolving inflammation. Thus an inapparent, acute inflammation resulted in later sequelae at a time when viral antigen and infectious virus were no longer evident. The clinical application of such observations remains to be established. Although there are several case reports recording mumps infection in children who subsequently developed aqueductal stenosis, it has not been proven that the infection caused the subsequent hydrocephalus.

An occasional sequel of mumps virus infection is deafness, which may occur even in the absence of other evidence of central nervous system involvement. The loss of hearing may be preceded by tinnitus and a sense of fullness of the ear. Such deafness is relatively uncommon but

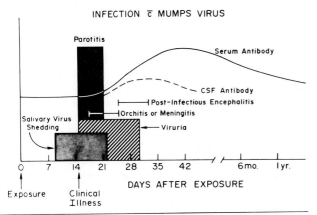

Figure 73-1. The course of mumps virus infection.

occurs suddenly during the period of parotid swelling. It is usually unilateral, but an estimated 20 percent of those so affected may have bilateral disease. Once deafness has occurred, the damage is irreversible and is apparently due to virus replication, inflammation, and subsequent destruction of the organ of Corti.

The complication of mumps infection best known by nonmedical persons is the involvement of the testes. Orchitis most commonly develops during the first week of infection and occurs predominantly in postpubertal males, with 20 to 30 percent of this age group manifesting testicular involvement during the course of the mumps infection. Bilateral disease is present in approximately 2 to 6 percent of the total number of patients with orchitis. Inflammation of the testes is extremely painful, but the fear of this affliction stems from the concern that sexual impotence and sterility will follow. It is common to have some atrophy of the involved testes, but the incidence of sterility following mumps orchitis is extremely infrequent, and experience during World War I and II failed to show that either impotence or sterility was an important consequence of this infection. Once it has occurred, therapy of orchitis is symptomatic, without any clearly demonstrated benefits from either steroid therapy or immune globulin.

Pancreatitis with typical symptoms of abdominal pain, fever, and vomiting may occur in association with mumps infection. Recent in vitro work has demonstrated by fluorescence mumps infection of beta cells and non-beta cells of human pancreas in cell culture. Much less frequently, thyroiditis, mastitis, myocarditis, and oophoritis are seen.

In the normal individual, a single infection with mumps virus confers permanent immunity against clinically evident infection. It is probable that reinfection, defined as an antibody rise after exposure to the virus, may occur, but neither virus shedding nor clinical illness has been demonstrated with such reinfection. Second attacks of parotitis have been observed in the same individual but without documentation by culture or serology of two clinical attacks of mumps virus infection. Serologic cross-reactivity with other paramyxoviruses may be responsible for the rare seropositive person who develops mumps.

Special consideration should briefly be given to persons commonly thought to be at increased risk from virus infections. The pregnant woman or her fetus is frequently considered at higher risk with regard to particular infections, such as varicella and rubella. At this time, conflicting reports relating to mumps infection during pregnancy have been published. There is no confirmed evidence of the occurrence of congenital anomalies in the human fetus as a result of mumps infection sustained by the pregnant mother and transmitted to her fetus in utero. Although there have been documented cases of mumps in mothers during pregnancy and in the immediate postpartum period, neither clinical mumps nor excretion of virus has been documented in these infants. Patients who have altered immunity, either humoral (e.g., the agammaglobulinemics) or cellular (the naturally occurring lymphopenic or the immunosuppressed patient), have not been shown to be at

increased risk with respect to mumps infection.

Host Response

The host initially responds to mumps virus infection with serum IgM antibodies and subsequently antibodies predominantly in the IgG class. Neutralizing antibodies are formed within the first week of symptoms and ordinarily persist for a lifetime. Hemagglutination-inhibiting and complement-fixing antibodies may appear from the first through the third week and usually reach a peak within 3 to 6 weeks after the onset of symptoms (Fig. 73-1). Complement-fixing antigens have been differentiated into two distinct types. The first is the soluble (S) or nucleoprotein antigen, and the second is the viral (V) or surface antigen, which includes the HN and F proteins. Complement-fixing antibodies against the S antigens are present within 2 to 3 days of onset, peak at about 10 days, and disappear in 8 to 9 months. Complement-fixing antibodies against the V antigen appear at about the tenth day of infection and persist for years. The transient nature of the S antibodies may sometimes assist in defining recent infection, and the presence of V antibodies may define the presence of previous experience with this virus.

Specific antibodies to the two surface glycoproteins HN and F are measurable after natural infection and are essential for protection from disease. Neutralization of infectivity, inhibition of hemagglutination, and the inhibition of neuraminidase activity are functions attributable to antibodies to the HN protein.

Mumps virus can replicate in vitro in human continuous lymphoblastoid cell lines with T cell characteristics or in human peripheral blood mononuclear cells. In the presence of pokeweed mitogen, replication is enhanced and occurs primarily in T lymphocytes, suggesting that these cells may be infected during natural infection. A T cell-mediated cytolytic response by human lymphocytes has been demonstrated after natural infection in humans.

Pathology

Examination of involved tissues is unusual because of the ordinarily benign nature of the illness. When salivary glands have been studied, there is no disruption of the general architecture of the gland after several days of illness. The involved salivary ducts demonstrate changes in the epithelial lining cells, ranging from swelling to complete desquamation. The ducts are dilated, and the lumen may be filled with cellular debris and polymorphonuclear cells. There is a moderate amount of periductal edema around the involved ducts, and the interstitial inflammatory cell is predominantly mononuclear.

The pathology of other involved tissues, e.g., the testes, is similar, with mononuclear infiltrates, edema, and no specific hallmarks allowing the diagnosis of mumps infection to be made solely on the basis of the observed pathology.

Epidemiology

In susceptible populations, mumps virus infection is predominantly a disease of childhood, with the majority of clinically evident infections being seen between the ages of 5 and 10 years. It has been estimated that in the prevaccine era 90 percent of the population were immune by the time they reached 15 years of age. Although mumps virus infection is contagious, it is less communicable than measles and varicella. The degree of communicability is estimated most accurately by serologic surveys of exposed individuals because as many as one fourth of the infections with mumps virus occur without clinical symptoms.

Isolation of the patient within the hospital setting or in homes, when it is attempted, has not effectively curtailed spread of disease. This is usually attributed to the period of virus shedding that occurs prior to the symptomatic onset of illness and thus precedes the recognition of infection. As previously mentioned, one fourth of patients have an asymptomatic infection, but they also excrete virus. Their infection is self-limited, and their immunity is comparable to those with symptomatic infection. To the best of our knowledge, there are no animal reservoirs or human carriers of mumps virus.

Diagnostic Approach

The work of Johnson and Goodpasture first established that mumps was caused by a filterable virus and demonstrated that rhesus monkeys could be experimentally infected. The description of the complement-fixation test and successful propagation of virus in chick embryo preceded the now generally employed standard tissue culture techniques. These methods employ monolayers of one of several cell types, including primary monkey kidney, human amnion, or human kidney and cell lines, such as HeLa. In vitro multinucleate giant cells are seen, and hemabsorption inhibition provides a practical means of identification of the virus. With these techniques, virus has been isolated from such varied sources as blood, cerebrospinal fluid, urine, saliva, salivary gland tissue, and human milk.

In many academic and large hospital settings, viral diagnostic laboratories are available, and virus isolation can be attempted from clinical materials. The responsible laboratory will provide directions for submitting materials for culture. Saliva and urine can be collected at the time of clinical central nervous system symptoms and submitted for culture. Mumps isolation in tissue culture is usually not necessary for either diagnosis or management of patients with parotitis, but the techniques and facilities are available for defining the unusual or complicated situation.

For practical reasons, many diagnostic laboratories can offer more extensive serologic diagnosis than cultural facilities for virus isolation. They will evaluate sera for the presence of antibodies to mumps virus. The serum for evaluation should be obtained as early as possible in the illness, and a convalescent specimen should be obtained after an interval of 2 to 3 weeks. A pair of sera can determine whether a specific illness is mumps infection by demonstrating an increase in antibody titer. A single serum can determine whether a person has ever had mumps infection but cannot define when it occurred. As indicated above, there are several types of antibody elicited by mumps infection.

Treatment

There is no specific therapy available for mumps infection. Symptomatic management of patients includes adequate hydration and analgesic and antipyretic therapy.

Prophylaxis and Immunization

The problem repeatedly occurs of what to do after exposure to mumps infection. Usually the person concerned is an adult without previous symptomatic mumps infection. Hyperimmune globulin or pooled serum IgG has been administered after exposure, with no proven efficacy. However, there has been a controlled study purporting to demonstrate that the administration of hyperimmune globulin after the appearance of parotitis can decrease the incidence and severity of orchitis. For this reason, hyperimmune globulin has been administered to postpubertal males who already have parotitis.

There has been limited experience in Scandinavia using a formalin-inactivated mumps vaccine, which affords some short-term protection from infection. No anti-F protein antibodies are demonstrable. It is probable that the F protein, which is sensitive to formalin, is not present. However, animal studies suggest that whole virus preparations can only induce anti-HN antibodies when the virus fails to replicate in the host. Purified F protein is an effective immunogen, and antibody to this protein is necessary to limit cell-to-cell spread of virus.

Live attenuated mumps virus vaccine was licensed in January 1968 and is available for prophylactic use. It is recommended for administration to children more than 1 year old and to young adults for induction of immunity parallel to that induced by natural infection. Vaccine should not be given to pregnant women because of the potential vulnerability of the fetus. Although no data exist that demonstrate transmission of attenuated virus to the fetus, placental infection has been documented after maternal immunization. For practical purposes there is only a single serologic strain of mumps virus, hence a single infection with either natural or attenuated virus confers immunity. The vaccine is a live attenuated virus produced in tissue cultures of chick embryo fibroblasts and is administered parenterally. Virus is not shed by the vaccinee, and immunization does not cause any side effects. The vaccine produces 95 to 100 percent serologic conversion from antibody-negative to antibody-positive in vaccinated susceptibles. The antibody levels, although considerably lower, parallel those produced by natural infection and persist for the 12 to 15 years that vaccine has been available for study. Immunized children in contact with naturally occurring mumps in their families, institutional settings, or the community have been protected against clinical illness.

The vaccine will not offer protection against mumps if someone has already been exposed to natural infection and is in the incubation period of illness. On the other hand, no harmful effects have been noted after administration of vaccine to an exposed susceptible.

Over 40 million doses of mumps vaccine have been administered since its licensure in 1967. The incidence of mumps has declined from 90 to 200/100,000 reported prior to 1967 to 2/100,000 in 1981. This is a decrease of 97 percent from 1968. Based on this reported occurrence of the disease, the vaccine has been effective in reducing disease morbidity, mortality, and the calculated costs of illness.

FURTHER READING

PARAINFLUENZA VIRUSES

Chanock RM, Parrott RH, Cook MK, et al.: Newly recognized myxoviruses from children with respiratory disease. N Engl J Med 258:207, 1958

Choppin PW, Scheid A: The role of viral glycoproteins in adsorption, penetration, and pathogenicity of viruses. Rev Infect Dis 2:40, 1980

Frank AL, Taber LH, Wells CR, et al.: Patterns of shedding of myxoviruses in children. J Infect Dis 144:433, 1981

Gardner PS, McQuillin J, McDuckin R, et al.: Observations on clinical and immunofluorescent diagnosis of parainfluenza virus infections. Br Med J 3:7, 1971

Glezen WP, Denny FW: Epidemiology of acute lower respiratory disease in children. N Engl J Med 288:498, 1973

Glezen WP, Fernald GW: Effective or passive antibody on parainfluenza type 3 pneumonia in hamsters. Infect Immun 14:212, 1976

Gross PA, Green RH, Curnen MG: Persistent infections with parainfluenza type 3 virus in man. Am Rev Respir Dis 108:894, 1973

Gross PA, Green RH, Lerner E, et al.: Immune response in persistent infection. Further studies on persistent respiratory infection in man with parainfluenza type 3 virus. Am Rev Respir Dis 110:676, 1974

Hall CB, Geiman JM, Breese BB, et al.: Parainfluenza viral infections in children: Correlation of shedding with clinical manifestations. J Pediatr 91:194, 1977

Kim HW, Canchola JG, Vargosko AJ, et al.: Immunogenicity of inactivated parainfluenza type 1, type 2, and type 3 vaccines in infants. JAMA 196:111, 1966

Klein JD, Collier AM: The pathogenesis of human parainfluenza type 3 virus infection in hamster tracheal organ culture. Infect Immun 10:883, 1974

Musson MA, Mocega HE, Krause HE: Acquisition of parainfluenza 3 virus infection by hospitalized children. I. Frequencies, rates, and temporal data. J Infect Dis 128:141, 1973

Workshop on broncholitis. Pediatr Res 11:209, 1979

Yanagihara R, McIntosh K: Secretory immunological response in infants and children to parainfluenza virus types 1 and 2. Infect Immun 30:23, 1980

RESPIRATORY SYNCYTIAL VIRUS

Belshe RB, Van Voris LP, Mufson MA: Parenteral administration of live respiratory syncytial virus vaccine: Results of a field trial. J Infect Dis 145:311, 1982

Chin J, Magoffin RL, Shearer LA, et al.: Field evaluation of a respiratory syncytial virus vaccine and a trivalent parainfluenza virus vaccine in a pediatric population. Am J Epidemiol 89:449, 1969

Fishaut M, Murphy D, Neifert M, et al.: Bronchomammary axis in the immune response to respiratory syncytial virus. J Pediatr 99:186, 1981

Fishaut M, Tubergen D, McIntosh K: Cellular response to respiratory viruses with particular reference to children with disorders of cell-mediated immunity. J Pediatr 96:179, 1980

Frank AL, Taber LH, Wells CR, et al.: Patterns of shedding of myxoviruses and paramyxoviruses in children. J Infect Dis 144:433, 1981

Fulginiti VA, Eller JJ, Sieber F, et al.: Respiratory virus immunization. I. A field trial of two inactivated respiratory virus vaccines; an aqueous trivalent parainfluenza virus vaccine and an alum-precipitated respiratory syncytial virus vaccine. Am J Epidemiol 89:435, 1969

Glezen WP, Paredes A, Allison JE, et al.: Risk of respiratory syncytial virus infection for infants from low-income families in relationship to age, sex, ethnic group, and maternal antibody level. J Pediatr 98:708, 1981

Hall CB, Douglas RG Jr, Geiman JM: Quantitative shedding patterns of RSV in infants. J Infect Dis 32:151, 1975

Hall WJ, Hall CB, Speers DM: Respiratory syncytial virus. Infection in adults. Clinical, virologic, and serial pulmonary function studies. Ann Intern Med 88:203, 1978

Henderson FW, Collier AM, Clyde WA Jr, et al.: Respiratory-syncytial-virus infections, reinfections and immunity. N Engl J Med 300:530, 1979

Kapikian AZ, Mitchell RH, Chanock RM, et al.: An epidemiologic study of altered clinical reactivity to respiratory syncytial (RS) virus infection in children previously vaccinated with an inactivated RS virus vaccine. Am J Epidemiol 89:405, 1969

Kaul TN, Welliver RC, Wong DT: Secretory antibody response to respiratory syncytial virus infection. Am J Dis Child 135:1013, 1981

Kim HW, Canchola JG, Brandt CD, et al.: Respiratory syncytial virus disease in infants despite prior administration of antigenic inactivated vaccine. Am J Epidemiol 89:422, 1969

Klein BS, Dollete FR, Yolken RH: The role of respiratory syncytial virus and other viral pathogens in acute otitis media. J Pediatr 101:16, 1982

Welliver RC, Wong DT, Sun M, et al.: The development of respiratory syncytial virus-specific IgE and the release of histamine in nasopharyngeal secretions after infection. N Engl J Med 305:842, 1981

Wright PF, Belshe RB, Kim HW, et al.: Administration of a highly attenuated, live respiratory syncytial virus vaccine to adults and children. Infect Immun 37:397, 1982

Wright PF, Shinozaki T, Fleet W, et al.: Evaluation of a live, attenuated, respiratory syncytial virus in infants. J Pediatr 88:931, 1976

MUMPS

Brunell PA, Brickman A, O'Hara D, et al.: Ineffectiveness of isolation of patients as a method of preventing the spread of mumps. N Engl J Med 279:1357, 1968

Fryden A, Link H, Norrby E: Cerebrospinal fluid and serum immunoglobulins and antibody titers in mumps meningitis and aseptic meningitis of other etiology. Infect Immun 21:852, 1978

Koplan JP, Preblud SR: A benefit-cost analysis of mumps vaccine. Am J Dis Child 136:362, 1982

Morishima T, Miyazu M, Ozaki T, et al.: Local immunity in mumps meningitis. Am J Dis Child 134:1060, 1980

Orvell C: Immunological properties of purified mumps virus glycoproteins. J Immunol 41:517, 1978

Orvell C: Structural polypeptides of mumps virus. J Immunol 41:527, 1978

Server AC, Merz DC, Waxham MN, et al.: Differentiation of mumps virus strains with monoclonal antibody to the HN glycoprotein. Infect Immun 35:179, 1982

Vandik B, Norrby E, Steen-Johnsen J, et al.: Mumps meningitis: Prolonged pleocytosis and occurrence of mumps virus-specific oligoclonal IgG in the cerebrospinal fluid. Eur Neurol 17:13, 1978

Weibel RE, Buynak EB, McLean AA, et al.: Persistence of antibody after administration of monovalent and combined live attenuated measles and rubella virus vaccines. Pediatrics 61:5, 1978

Yamauchi T, Wilson C, St. Geme JW Jr: Transmission of live attenuated mumps virus to the human placenta. N Engl J Med 290:710, 1974

CHAPTER 74

Measles and Subacute Sclerosing Panencephalitis

Measles (Rubeola)

Because of its distinctive clinical features, measles was recognized as a disease entity by Hebrew and Arabic physicians long before the demonstration of its viral etiology. Home, a Scottish physician, demonstrated in 1758 the transmissibility of the disease by scarification of susceptible individuals with blood taken from infected patients. Measles virus was first isolated successfully in tissue culture in 1954 by Enders and his associates. This permitted investigations of the virus, studies of immunity, and the development and selection of variants for vaccine evaluation.

The use of live attenuated vaccines began in 1963 and has extended to many parts of the world, altering dramatically the incidence and epidemiology of the disease. Where infant measles immunization is widely practiced, the illness has become uncommon. In other lands, especially among economically disadvantaged nations, measles persists in epidemic fashion responsible for the deaths of approximately 1.5 million infants and children each year.

Pathogenesis and Pathology

Measles is acquired as an infection of the respiratory tract, with principal damage to surface mucosal lining cells. Secondary bacterial infections may be enhanced by dimin-ished function of alveolar macrophages. The virus spreads via lymphatics and through the blood to give widespread involvement, particularly in lymphoid cells. The large, inclusion-bearing, multinucleated giant cells found in tissues are quite similar to those produced by measles virus when grown in vitro. Patients excrete large amounts of virus during the catarrhal phase. Virus can be recovered from the blood, particularly from the white cell fraction, for several days before rash appears but rarely thereafter. The urine may remain positive for virus up to 4 days after the onset of rash. The pathology of acute central nervous system involvement includes edema, congestion, and scattered petechial hemorrhages, with perivascular cuffs of round cells. There may be some perivascular demyelination in the later stages.

The exact nature of the measles rash is uncertain but involves a vasculitis. Viral microtubular aggregates have been observed by electron microscopy of nuclei and cytoplasm of skin biopsies. They are also found in the oral lesions (Koplik spots). An important host mechanism for controlling the virus infection is the recognition and lysis of measles-infected cells by cytotoxic T lymphocytes. Malnourished, immunodeficient, or immunosuppressed children often undergo more severe measles, with prolonged virus replication, giant cell pneumonia, virus dissemination to many organs, and high case-fatality rates. Depressed T cell function appears to be the liability shared by these patients. Further evidence of an interaction of measles infections and cellular immune mechanisms is the loss of delayed hypersensitivity to tuberculoprotein among tuberculin-positive patients with measles. This persists for weeks to months following acute infection. A worsening of underlying tuberculosis in children or adults who acquire measles infection has been reported.

Clinical Course

Measles has an incubation period of 10 to 14 days (Fig. 74-1). A prodromal stage marked by catarrhal symptoms of cough, coryza, and conjunctivitis is accompanied by fever, which rises steadily until the appearance of rash 2 to 4 days after onset. Preceding the rash, the pathognomonic Koplik spots appear on the lateral buccal mucosa. They are pinpoint and grayish white surrounded by bright red inflammation and found over the lateral buccal mucosa and the inner lips, occasionally involving the entire inner mouth. They also may be detected in the palpebral conjunctiva. The exanthem begins on the head, behind the ears, on the forehead, and on the neck. Discrete red macular and papular lesions progress downward to involve the trunk and upper extremities. Over a period of 3 days, the entire body is involved. When the lower extremities show discrete lesions, those on the head and neck have begun to coalesce. With rash on the lower extremities, fever recedes dramatically. The bright red rash fades, to leave a brown discoloration that does not blanch with pressure.

The respiratory tract manifestations vary in severity but include laryngitis, tracheobronchitis, bronchiolitis, and interstitial pneumonitis from the primary virus infection that damages surface mucosal cells. With defervescence, there is improvement, but secondary bacterial infections of the respiratory tract may complicate the recovery phase in 5 to 10 percent of patients. These include otitis media, sinusitis, mastoiditis, and pneumonia. Encephalomyelitis occurs in approximately 0.1 percent of cases. This follows a 3 to 4 day period of recovery from the acute illness and is marked by a sudden onset, with seizures, confusion, and

coma. The mortality rate of central nervous system involvement approaches 25 percent. Nearly half of those who survive are left with some sequelae involving impaired intellectual, motor, or emotional development. The responsibility of measles virus for a late central nervous system complication, subacute sclerosing panencephalitis (SSPE), has been firmly established. This occurs in 1 or 2 per 100,000 cases and is discussed in greater detail later in this chapter.

Epidemiology

Measles is one of the most highly communicable of all virus infections, so that nearly all susceptible children acquire the infection. In rural settings or among isolated communities, it has been possible for a population to reach adult life without exposure. Under such circumstances, the introduction of measles virus produced devastating epidemics. An epidemiologic classic is Panum's description in 1846 of such an outbreak in the Faroe Islands. He showed the persistence of lifelong immunity among individuals who had acquired the infection 6 and 7 decades previously.

In the temperate zones, measles has occurred in winter–spring epidemics in 2- or 3-year cycles, apparently related to the new groups of susceptible children born since the last outbreak. Maternal antibody is transplacentally acquired by the infant, so that infection under 6 months of age is rare. As shown in Figure 74-1, the catarrhal stage of the illness is marked by extensive respiratory virus excretion, so that infected droplet nuclei provide the usual mode of transmission within families, schoolrooms, or other crowded settings. A single attack confers lifelong immunity.

Although measles is a member of the genus *Morbillivirus*, which includes the agents of canine distemper and rinderpest of cattle, there is no evidence of natural spread from one species to another. Except for minor variations, only one distinct serotype of measles virus has been identified. Infection sustained in any part of the world confers enduring, uniform geographic protection. With the striking decline of indigenous measles in the United States (Table 74-1) an increasing proportion of cases arise as a result of disease importations from other countries. Of the 1697 cases of measles reported in the United States in 1982, 36 percent were imported or the result of transmission to a family or social contact from an imported case. A second change has been an upward shift in the age-specific incidence rate, with sporadic outbreaks on college campuses among previously unimmunized students.

Diagnosis

Diagnosis is clinical, based on the characteristic history and findings. Examination of the urinary sediment or of nasal smears will show characteristic inclusion-bearing

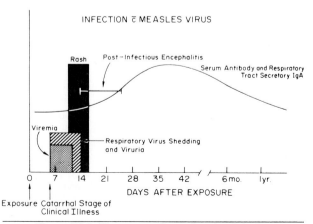

Figure 74-1. The course of measles virus infection.

TABLE 74-1. REPORTED MEASLES CASES IN THE UNITED STATES, 1963–1982

Year	Cases	Rate (per 10^5 Population)
1963	385,156	204.3
1964	458,083	239.7
1965	261,904	135.3
1966	204,136	104.4
1967	62,705	31.8
1968	22,231	11.1
1969	25,826	12.8
1970	47,351	23.2
1971	75,290	36.5
1972	32,275	15.5
1973	26,690	12.7
1974	22,094	10.5
1975	24,374	11.4
1976	41,126	19.2
1977	57,345	26.5
1978	26,871	12.3
1979	13,597	6.2
1980	13,506	6.0
1981	3,032	1.3
1982	1,697	0.7

syncytia, and there is peripheral leukopenia. Indirect immunofluorescent microscopy has been used to show measles antigen in nasopharyngeal cells. Although virus can be isolated from blood, respiratory tract secretions, conjunctival secretions, or urine, this is not ordinarily required or available.

A number of serologic tests are available, utilizing the antigens of the measles virion or its infectivity. These include virus neutralization, hemagglutination inhibition, complement fixation, enzyme-linked immunosorbent assay (ELISA), and immunofluorescence. Antibodies rise very rapidly after the appearance of rash and reach high titers in the next 30 to 60 days. A pair of sera obtained early in the course of illness and 7 to 14 days thereafter will show a marked rise in antibody titer by any of the methods described. Because of its ease and rapidity, the hemagglutination-inhibition test is most often utilized.

Treatment and Prevention

The primary viral disease is not amenable to any current therapy. Supportive measures may be employed to reduce fever, ameliorate cough, and maintain hydration. Secondary bacterial complications are treated with antibiotics selected by culture of appropriate specimens. The treatment of encephalomyelitis is also symptomatic, with careful attention to the maintenance of an airway, reduction of increased intracranial pressure, control of seizures, and provision of fluid, electrolyte, and caloric requirements. The administration of immune globulin early in the incubation period of measles may completely abort or modify the infection, depending on the amount employed. At a dose of 0.04 ml per kg body weight, immune globulin will

reliably modify measles so that a more benign course ensues, followed by lasting immunity. Using a dose of 0.2 ml per kg body weight, it is possible to abort the infection completely, and the patient remains susceptible to future infection after catabolism of the exogenous globulin.

The prevention of measles by proper use of the available attenuated active virus vaccines offers the most reliable and enduring protection. Vaccines are recommended for all healthy children shortly after 1 year of age. The virus is propagated in chick embryo fibroblast or human diploid cell cultures. It is given parenterally, with successful infection in at least 95 percent of susceptibles. The infection is usually occult but may cause fever in 15 percent of recipients. Rarely, there is moderate, transient rash following the fever. This attenuated infection is noncommunicable and results in antibody responses somewhat lower than those that follow the natural infection. In patients studied to date, antibodies have persisted for periods up to 20 years after immunization. When exposed to natural infection, immunized children remain solidly protected. From 1963 to 1967 an inactivated measles vaccine was available. It has been abandoned because a severe unusual illness followed the exposure of children to naturally occurring measles several years after the receipt of inactivated vaccine. They developed fever, pneumonitis, and an unusual rash with petechiae. It seemed to represent a hypersensitivity reaction. The formalin or detergent treatment required to prepare the inactivated vaccine apparently degraded the F (penetration, fusion, hemolytic) protein of the measles virus so that recipients lacked specific antibody to that antigen despite the presence of antibodies to the other five virion proteins. On exposure to natural measles, virus penetration occurred, with spread of newly synthesized virus by cell-to-cell fusion. These inactivated vaccines have been withdrawn from the market.

In the initial 5 years after the onset of vaccine use in the United States, reported cases of measles were reduced by 90 percent. Table 74-1 demonstrates the striking decline in reported cases of measles in the United States since vaccine licensure in 1963 and public health immunization programs began in 1966. Studies of patients involved in current small outbreaks in the United States showed a heterogeneous population with illness. The majority had never received measles vaccine and had not previously acquired natural immunity to the infection. Some had received inactivated vaccine in the past. Others had been given active attenuated vaccine prior to their first birthday, at a time when persistent maternally transmitted antibody aborted the vaccine infection. In some instances vaccine failures were traced to improper storage of the attenuated vaccine, so that heat and/or light had reduced the infectivity of the material injected.

Recent public concerns about the hazards of immunization have prompted the collection of data, such as those displayed in Table 74-2, to permit a statistical evaluation of the comparative risks of the natural disease and the attenuated vaccine infection. The higher rates of pneumonia and deaths represent figures collected in India, Nigeria, Gambia, Bangladesh, and other nations with developing

TABLE 74-2. RATES OF COMPLICATIONS OF MEASLES AND MEASLES IMMUNIZATION

Adverse Reaction	Measles per 10^5	Vaccine per 10^5
Encephalomyelitis	50–400	0.1
SSPE	0.5–2.0	0.05–0.1
Pneumonia	3800–7300	—
Seizures	500–1000	0.02–19
Deaths	10–10,000	0.01

Adapted from reports of the Centers for Disease Control and World Health Organization.

health systems. Continued attention to immunization of all children 1 year of age or older is needed in order to attain the goal of measles eradication.

Subacute Sclerosing Panencephalitis (SSPE)

This degenerative disease of the brain was first described in 1933, but it was not attributed to measles infection until 1967. Since that time abundant laboratory data have confirmed the etiologic agent as measles virus (or a defective variant) that persists in the host after initial conventional measles in infancy or early childhood.

Clinical Features

SSPE occurs almost exclusively in children and adolescents, usually between the ages of 5 and 10 years and predominantly in males (3:1). Rarely, young infants are affected, and a few cases have been reported in adults. The usual clinical course is outlined in Table 74-3. The progress of the disease may appear to be arrested for periods of years, usually when the patient is in coma, but almost invariably the patients die, often of intercurrent infection. A very few older patients with well-documented disease have recovered.

Epidemiology

Although the incidence of SSPE varies in many countries and among population groups within nations, it is estimated that 1 case follows each 100,000 cases of measles. The majority of patients give a history of uncomplicated natural measles at a younger age than average, and all remain well during the interim of approximately 5 years until the onset of their central nervous system disease. Occasionally, no prior history of measles is elicited, or live attenuated measles vaccine is the only known exposure to measles virus. No consistent immunologic defects have been demonstrated in SSPE patients.

Data have suggested that the disease may have been more common in the Southeastern United States, that the primary measles infection occurred at a younger age in such patients, and that rural rather than urban residents were at greater risk. With the widespread use of measles vaccine and the resultant decline of measles in the United States, there has been a parallel, sharp decrease in reported cases of SSPE. Prior to measles vaccination programs, 20 to 40 cases of SSPE occurred annually, whereas only 4 or 5 new patients are now reported each year.

Diagnosis and Pathogenesis

The diagnosis should be suspected on the basis of the history and the clinical findings of progressive personality changes, myoclonus, and variable focal neurologic deficits. A characteristic spike wave discharge is seen on the EEG, and the cerebrospinal fluid is usually normal apart from a first-zone colloidal gold sol curve. The diagnosis is confirmed by the finding of measles antibodies in the cerebrospinal fluid. Brain biopsy is not necessary for diagnosis, but should tissue become available, it may show perivascular round cell infiltration, neuronal degeneration, gliosis, and type A Cowdry intranuclear inclusion bodies when stained with hematoxylin and eosin and examined with the light microscope. With the electron microscope, these inclusion bodies contain microtubular filaments, corresponding in size and configuration to the nucleocapsids of measles virus. With appropriate fluorescent antibody staining, measles antigen can be demonstrated at these sites. Finally, when brain tissue is cocultivated in the laboratory with permissive cells, complete infectious virus may be re-

TABLE 74-3. SUBACUTE SCLEROSING PANENCEPHALITIS (SSPE)

Clinical		Laboratory
Personality change; declining school performance; intellectual deterioration often manifested by impaired memory, altered judgment, and inappropriate behavior; occasionally chorioretinitis; impaired motor activity, gait difficulty, speech difficulty; myoclonic jerks progressing to repetitive, often sound-sensitive myoclonic seizures; paralysis; gradual deterioration in consciousness; coma; death	EEG	Paroxysmal, synchronous spike discharges with interim suppression of electrical activity
	CSF	Usually acellular; normal total protein; increased gamma globulins (IgG); detectable measles antibody
	Blood	Markedly elevated measles antibody
	Brain	Specific immunofluorescence to measles antigen; microtubular filaments within nuclear inclusions on electron microscopy; SSPE virus recovered in tissue culture

EEG, electroencephalogram; CSF, cerebrospinal fluid.

covered. Analysis of serum and cerebrospinal fluid antibodies reveals that SSPE patients lack detectable response to the M (membrane) protein of measles virus, although they respond to all the other five major structural proteins of the virion. This observation coincides with the failure of maturation of the measles nucleocapsids detected in the neuronal and glial cells.

Although measles virus infects central nervous system cells, there is no incorporation of the newly replicated components into cell membranes to form complete new infectious virus. Direct analysis of measles-virus proteins in central nervous system tissue of SSPE patients shows a lack of synthesis of M protein. This host cell-dependent restriction is consistent with the findings of accumulated viral nucleocapsids intracellularly, failure to recover infectious measles virus except by cocultivation with known permissive cells, antibodies in serum and cerebrospinal fluid to the other five measles proteins, and the abortive persistent infection.

Treatment

Although numerous approaches to therapy for this condition have been attempted, all have failed. At present one can only offer general supportive care for the patient, attempt to control seizures, and try to provide early diagnosis and proper explanation.

FURTHER READING

MEASLES

Abu Beer Mohammed Ibn Zacariya Ar-Razi (Rhazes): A Treatise on the Smallpox and Measles. Translated from the original Arabic by WA Greenhill, Sydenham Society, London, 1848

Annunziato D, Kaplan MH, Hall WW, et al.: Atypical measles syndrome: Pathologic and serologic findings. Pediatrics 70:203, 1982

Centers for Disease Control: Measles Surveillance Report No. 11, 1977–1981, issued September 1982

Frank JA Jr, Hoffman RE, Mann JM, et al.: Imported measles: A potential control problem. JAMA 245:264, 1981

Hall WW, Choppin PW: Measles-virus proteins in the brain tissue of patients with subacute sclerosing panencephalitis. N Engl J Med 304:1152, 1981

Katz SL, Krugman S, Quinn TC: International symposium on measles immunization. Rev Infect Dis 5:389, 1983

Lucas CJ, Biddison WE, Nelson DL, el al.: Killing of measles virus-infected cells by human cytotoxic T cells. Infect Immun 38:226, 1982

Machamer CE, Hayes EC, Zweerink HJ: Cells infected with a cell-associated subacute sclerosing panencephalitis virus do not express M protein. Virology 108:515, 1981

terMeulen V, Stephenson JR, Kieth HW: Subacute sclerosing panencephalitis. In Fraenkel-Conrat H, Wagner RR (eds): Comprehensive Virology. New York, Plenum Press, 1983, vol 18, p 105

CHAPTER 75

Rubella (German Measles)

Clinical Features and Pathogenesis
Epidemiology

Diagnosis
Treatment and Prevention

From the mid-19th century until 1941, rubella was regarded as a benign childhood exanthem. When the Australian ophthalmologist, Sir Norman Gregg, reported the association of intrauterine rubella infection with congenital cataracts, this attitude changed completely. Subsequently, deafness, congenital heart disease, and other malformations were found to result from maternal rubella during the first 4 months of pregnancy. The recovery in 1962 of rubella virus in cell culture systems led to the development and, in 1969, the licensure of attenuated active vaccines that have proven safe and effective. Congenital rubella has become infrequent in the United States as a result of widespread use of these vaccines.

On the basis of its biochemical and biophysical properties, rubella is classified as a togavirus, but it has no involvement with arthropod vectors. Because many different viruses may cause a similar clinical illness with rash and fever, but only rubella is teratogenic, virus isolation techniques and specific serologic tests for antibody have provided the necessary differentiation of etiologic agents.

Clinical Features and Pathogenesis

Rubella is a mild rash disease that occurs principally in children but is seen at all ages. As shown in Figure 75-1, the incubation period is approximately 2 weeks, with minimal prodromal signs or symptoms. Most often, the first awareness of illness is mild fever and respiratory signs immediately preceding the onset of rash. The exanthem is pink macules and papules, at first on the face and then on the neck, trunk, and extremities, where they remain discrete and rarely coalesce. The rash has ordinarily disappeared by the third day. Preceding and accompanying the rash there is lymphadenopathy, which may involve the postauricular, suboccipital, and cervical nodes. Rash is ob-

served commonly among children, but infection may be occult or only a febrile pharyngitis in as many as one third of adult patients. Although major complications are rare (thrombocytopenic purpura and encephalitis), the incidence of arthralgia and arthritis is much greater than generally appreciated. The frequency of joint involvement is directly correlated with increasing age and appears also to be more common among women. In a few patients, persistence of rubella virus in synovial cells has been associated with polyarthritis and arthralgia of lengthy duration. Usually joint involvement is acute and transient without sequelae (Fig. 75-1).

The route of infection is via the respiratory tract, with spread to lymphatic tissues and then to the blood. Both viremia and respiratory tract shedding of virus may precede the rash by 1 week, and the latter may follow it for another several weeks. Because much virus excretion occurs prior to the recognition of illness, secondary infection of intimate contacts has usually transpired before the primary patient has been diagnosed. Little is known of the actual pathology of the postnatal disease because it is not a fatal one. However, the pathogenesis of congenital infection has been well studied during and since the 1964 pandemic. Maternal viremia is followed by infection of the placenta, which may lead to virus invasion of the fetus. Multiple tissues and organs support the replication of virus, which continues to multiply throughout the remainder of pregnancy and in the postnatal period. A large percentage of maternal infections that occur in the first 3 months of pregnancy result in fetal illness. There is a diminishing number in the fourth month, and it is uncertain whether any fetal infections have resulted from maternal rubella in later pregnancy. Although the exact mechanism of damage to fetal organs is not clear, rubella infection of human embryonic cells in vitro is associated with both chromosomal breakage and inhibition of normal mitosis. Infants with

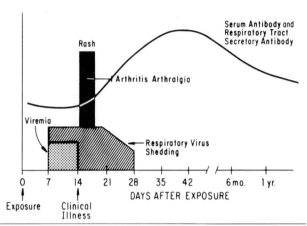

Figure 75-1. The course of postnatal rubella virus infection.

congenital rubella undergo intrauterine growth retardation and have a subnormal number of cells in some infected organs, suggesting that these same features may also occur in vivo.

Congenital rubella infection may result in a large variety of abnormalities, including deafness, congenital heart disease, eye defects (cataracts, glaucoma, retinitis, microphthalmia), growth retardation, thrombocytopenic purpura, osteitis, hepatitis, pneumonitis, encephalitis, and cerebral damage with mental retardation. In contrast to the postnatal infection, intrauterine disease is marked by a continued replication and excretion of virus, which may persist throughout the first year of life and has been demonstrated in selected tissues, such as the lens of the eye, as long as 3 or 4 years postnatally. In addition to malformations compatible with life, intrauterine rubella infection may produce fetal death, abortion, or neonatal death. Although many of the acute neonatal manifestations resolve over the first months of life, long-term sequelae result in multiple developmental handicaps. Recent studies also indicate a significant increase in diabetes mellitus, chronic pneumonitis, thyroiditis, and degenerative encephalitis among long-term survivors of intrauterine infection.

The immunity that follows naturally acquired postnatal rubella is enduring. Only one serologic type of virus has been identified, and a single attack confers lifelong immunity. The immunologic events that accompany intrauterine infection differ strikingly. The antibody response in utero is principally one of IgM rather than IgG. Although IgG specific for rubella is found in fetal circulation, it is mainly of maternal origin and transplacentally acquired. Despite the presence of specific IgM and IgG in the fetal circulation, chronic infection of cells continues. Postnatally, the infected infant synthesizes rubella-specific IgG. Virus replication diminishes gradually over ensuing months. A small number of congenitally infected infants have developed unusual forms of hypogammaglobulinemia in the first year of life. Another group have lost all their detectable rubella antibody postnatally but have nonetheless proven resistant to challenge with attenuated rubella

viruses. The explanations for a number of these immunologic paradoxes are not yet known.

Epidemiology

In most urban communities, rubella infection was acquired during early childhood, principally in the school years. Because it is not as highly communicable as measles or varicella, 15 to 20 percent of women reached childbearing age without having acquired natural immunity. In the United States, large epidemics occurred at 6- to 8-year intervals, with smaller numbers of cases in the intervening years. This cycle was interrupted in 1969 by the onset of rubella vaccine programs, and no large epidemic has developed in the United States in the 19 years since the 1964–1965 pandemic. The usual transmission is by the respiratory route, but the prolonged viruria of congenitally infected infants may be of importance in spread to close contacts. In situations where adolescents and young adults have been placed in crowded living conditions, rubella outbreaks have been observed. Examples are regularly seen among Armed Forces recruits, preparatory school and college groups, and summer camps. With widespread use of rubella vaccines, childhood outbreaks have disappeared, and sporadic clusters of cases are reported now on college campuses, in Armed Forces training camps, and among other teenage groups who have escaped both immunization and natural infection.

Diagnosis

Until 1962, the diagnosis was entirely a clinical one and unreliable. Currently, virus isolation techniques are not readily available. They involve the use of cell culture systems susceptible to the virus, with direct observation for cytopathic effect, or the use of an interfering agent as an indicator of virus replication. Because these are time consuming and fastidious, serologic tests are more commonly used. The antigens of rubella virus are responsible for the induction of a number of antibody types that can be assayed in serum. These include complement-fixing (CF), virus-neutralizing, hemagglutination-inhibiting (HI), immunodiffusion, enzyme-linked immunosorbent assay (ELISA), and immunofluorescent antibodies. The HI test has been the most commonly employed. Paired serum specimens obtained early in the course of illness and 10 to 20 days thereafter will demonstrate an antibody rise. The presence in many sera of nonspecific inhibitors of hemagglutination makes the HI test susceptible to some error unless care is taken first to treat the sera to remove inhibitors. The HI test is also valuable for screening populations to determine serosusceptibility or immunity and in the consideration of the use and efficacy of vaccines. The diagnosis of intrauterine rubella is more complicated but may be accomplished on a single serum specimen if neonatal blood is assayed for IgM antibodies specific to rubella virus. If such a technique is not available, it may be necessary to test paired specimens obtained over a period of several months in order to determine whether there is active post-natal antibody synthesis by the infant or merely a decline of transplacentally acquired antibody.

Treatment and Prevention

There is no specific therapy for rubella virus infection. In the case of documented maternal infection during the first trimester of pregnancy, therapeutic abortion is commonly employed. Even though rubella is confirmed virologically and/or serologically in the mother, it is not possible to be certain that fetal infection has occurred. It is estimated that maternal rubella in the first month of gestation carries a 30 to 50 percent incidence of fetal infection, in the second month of gestation a 25 percent risk, and in the third month a 10 percent incidence. After the third month there are insufficient numbers to quantitate the minimal risk of fetal involvement. Amniocentesis has been employed in a few cases to examine cells and fluid for evidence of rubella virus infection. However, there are insufficient data to evaluate the reliability of this method.

The use of immune serum globulin to prevent fetal rubella in an exposed pregnant woman has been ineffec-tive and misleading. In most situations, it has been diffi-cult to ascertain for how long a mother has been exposed to the virus. As depicted in Figure 75-1, she may have had a week's exposure to respiratory shedding from a contact by the time the contact develops rash disease.

Since 1969, attenuated active rubella virus vaccines have been used. These vaccines confer lasting effective im-munity. They have been prepared in cell cultures of duck embryo, canine kidney, rabbit kidney, and a human diploid line (WI-38). Their use is followed by some respiratory shedding of the vaccine virus for 7 to 28 days after vacci-nation. In contrast to that following natural infection, this is not transmissible. The human diploid cell vaccine (RA 27/3) is most widely available in this country. A serologi-cally detectable HI antibody response is produced in more than 95 percent of susceptible vaccine recipients within 4 to 6 weeks of immunization. As with naturally acquired ru-bella, vaccination may be followed by joint complaints but in a smaller percentage of recipients. Approximately 25 percent of adult female vaccinees experience arthralgia and only 1 percent arthritis. Whether the attenuated strains of virus might also be teratogenic for the fetus or

embryo is uncertain. Vaccine virus can cross the placenta to reach the products of conception. The risk of any tera-togenesis is negligible in 14 years' accumulated experience but theoretically possible. Of 215 infants born to serone-gative mothers who inadvertently received rubella vaccine in early pregnancy, none had a detectable congenital mal-formation. The use of vaccine in pregnant women remains nonetheless contraindicated. Before administering attenu-ated rubella vaccine to a susceptible woman of childbearing age, it is important to be certain she is not pregnant and that she will follow an acceptable method of pregnancy prevention for 2 to 3 months thereafter.

Reinfection with wild rubella virus has been demon-strated to occur following either naturally acquired or vac-cine-induced immunity. This usually happens in individuals whose antibody titers have fallen to low levels. It is more likely to occur in vaccinated individuals who lack detectable rubella-specific respiratory tract secretory im-munoglobulins (sIgA). The virologic events of such rein-fections are markedly abbreviated and have not been accompanied by viremia. There is a rapid secondary type of antibody response with no overt illness. Such a reinfec-tion does not threaten the fetus of a woman in early preg-nancy. Continued study will be required to answer fully all the questions raised by the vaccines. In 15 years of utiliza-tion in the United States, they have produced a striking re-duction in reported cases of intrauterine and postnatal rubella. Prospective surveillance will be required to deter-mine the long-range effects of these changes in population immunity and of the reduction in circulation of rubella virus.

FURTHER READING

Chantler JK, Ford DK, Tingle AJ: Persistent rubella infection and rubella-associated arthritis. Lancet 1:1323, 1982

Crawford GE, Gremillion DH: Epidemic measles and rubella in Air Force recruits. J Infect Dis 144:403, 1981

Gregg NM: Congenital cataract following German measles in the mother. Trans Ophthal Soc Aust 3:35, 1942

Herrmann KL, Halstead SB, Wiebenga NH: Rubella antibody per-sistence after immunization. JAMA 247:193, 1982

Hinman AR, Bart KJ, Orenstein WA, et al.: Rational strategy for rubella vaccination. Lancet 1:39, 1983

Horstmann DM: Rubella. In Evans AS (ed): Viral Infections of Humans. New York, Plenum, 1982, chap 21, p 519

Sever JL, Cleghorn BA: Rubella diagnostic tests. Postgrad Med 71:73, 1982

CHAPTER 76

Arboviruses

> Rift Valley fever really pulled a surprise act near the end of the decade (1970s) in affecting thousands of people as well as domestic animals in an invasion of Egypt.
> — *RW Chamberlain: Am J Trop Med Hyg 31:434, 1982*

Arboviruses (*ar*thropod *bo*rne) are maintained in nature usually by cyclical transmission between susceptible vertebrates and bloodsucking arthropods, with the virus multiplying in both. In vertebrates, arboviruses produce, after an incubation period, a viremia of sufficient titer and duration to permit infection of arthropods. In arthropods, virus multiplication results in the capacity to transmit virus by bite to new vertebrates after a period of time known as the "extrinsic incubation period." This biologic transmission, which depends on the multiplication of virus in the arthropod, is in contrast to mechanical transmission, which occurs as the result of rapid carriage of virus by an arthropod from one host to another. The consequences of infection of the vertebrate range from total absence of disease to major illness and death. With rare exceptions, no effect on the arthropod has been detected.

Most arboviruses are members of the families Togaviridae and Bunyaviridae; some are Reoviridae and Rhabdoviridae. A small number, generally of no significance in human disease, may be found in other families. Proof of biologic transmission by arthropods has not been obtained for many arboviruses but is inferred from epidemiologic and taxonomic characteristics. The discipline of arbovirology often deals, in addition, with some viruses, such as arenaviruses, that probably are not arthropod-borne and for which other modes of transmission have been demonstrated. This is because of a historical, initial presumption of arthropod transmission based on clinical and epidemiologic similarities to known arbovirus disease.

The year 1900 marked the beginning of arbovirology, when Walter Reed demonstrated the biologic transmission of yellow fever by the mosquito *Aedes aegypti*. The facts, so convincingly elucidated by the commission of which Reed was chairman, proved the remarkable perception of Dr. Carlos Finlay and the essentials of the basic definition of arboviruses.

During ensuing decades, other viruses, many of which are human pathogens, were recognized to possess life cycles that involve arthropod–vertebrate transmission. By the end of the 1930s, there were approximately 16 arboviruses. Since World War II, hundreds more have been discovered, so that now there are in excess of 445 known arboviruses. More than 100 may infect and produce disease in humans.

Beginning in the 1920s with the yellow fever program of its International Health Division, the Rockefeller Foundation has provided one of the major physical and intellectual bases of arbovirology. Besides developing the conceptual base, much of the basic technology, and most of the facts of arbovirology, the Foundation's conquest of yellow fever by the development of the first tissue culture-attenuated, live virus vaccine stands as a major achievement.

Classification and Nomenclature

Initially, arboviruses were named for the disease that they caused, for example, dengue and yellow fever. Later, combined names were devised that comprised both geography and diseases, such as St. Louis encephalitis. Names currently derive from the place of collection of the specimen from which the first isolation was made, especially for those agents for which no disease is recognized. Most arboviruses fall into this group.

Arboviruses are classified according to morphologic and serologic relationships based on tests employing complement fixation, hemagglutination inhibition, and infectivity neutralization. At first, serogroups were designated A, B, and C; these groups have been renamed alphaviruses, flaviviruses, and bunyaviruses, respectively. More recently, groups are given the name of the first virus isolated. Fifty-seven serogroups have been identified, and 97 viruses are as yet ungrouped.

There are several other ways of grouping arboviruses that may be useful in certain circumstances. These include classification according to vector, geographic range, type of disease, or combinations of these, such as "tickborne hemorrhagic fevers." Because this text is intended primarily for medical students, it seems appropriate to use a clinically oriented classification (Tables 76-1, 76-2, and 76-3).

Technology

The epidemiologic and clinical study of arboviruses involves two basic technologies: virus isolation and measurement of antibodies. The major culture medium for isolation is the infant mouse, in which nearly all arboviruses are pathogenic when inoculated intracerebrally. Cell cultures are the next most useful medium, especially for the study of many established isolates. In the usual search for arboviruses, serum or tissue specimens from humans or other vertebrates, as well as aqueous suspensions of ground-up arthropods, are routinely inoculated intracerebrally into each infant mouse in a litter. The mice are observed for signs of illness over the next month. Brain tissue from mice that become sick during this time is used for further passage. Serial passages are made until mouse pathogenicity is constant, at which time a large pool of infected brain tissue is stored frozen for use in biologic, physical, chemical, and serologic characterization of the agent.

Serum antibody measurements are usually essential in the diagnosis of individual patients (Fig. 76-1). Antibodies also provide a record of past infection that permits detection of the involvement of specific populations of humans and other vertebrates in the transmission cycle. These seroepidemiologic investigations are of great value to the understanding of the geographic distribution of arboviruses. Geographic distribution is also delineated through the use of sentinel animals, such as caged mice or chickens placed in the forest. Presence of virus in these animals is revealed by disease or by a rise in serum antibodies.

The techniques of virus isolation and antibody measurement may be applied to captured or colonized vertebrates and arthropods to determine their potential as natural hosts for a given virus. Experts in field and labora-

TABLE 76-1. SUMMARY OF SELECTED HUMAN ARBOVIRUS DISEASES CHARACTERIZED BY FEVER WITH OR WITHOUT RASH

Virus			Maintenance Hosts		Clinical (Other than Fever) and Epidemiologic Features
Name*	Group	Geography	Vertebrate	Arthropod	
Dengue	Flavivirus	Tropics and subtropics throughout world	Humans, monkeys	Mosquito	Headache, myalgia, rash, arthralgia, endemic and epidemic
Colorado tick	Orbivirus	Western North America	Small mammals	Tick	Headache, myalgia, rash, biphasic fever, spring and summer focal endemicity
West Nile	Flavivirus	Mediterranean, Africa, Middle East, Russia, Asia	Birds	Mosquito, ? ticks	Headache, lymphadenopathy, rash, meningoencephalitis, endemic and epidemic
Sandfly	Phlebotomus fever	Mediterranean, Asia, tropical America	Humans, small mammals	Phlebotomus flies	Myalgia, headache, gastrointestinal symptoms, seasonally epidemic and sporadically endemic
Rift Valley	Phlebotomus fever	Africa	Large mammals, humans, ? forest rodents	Mosquito	Headache, myalgia, photophobia, retinal damage may occur; sporadic from mosquito or by contact with tissue from infected animals
O'nyong-nyong	Alphavirus	Africa	Humans	Mosquito	Like dengue, epidemic
Chikungunya	Alphavirus	Africa, Asia	Humans	Mosquito	Like dengue, epidemic and endemic
Mayaro	Alphavirus	South and Central America	Monkeys	Mosquito	Like dengue, forest-associated, endemic and epidemic
Ross River	Alphavirus	Australia, South Pacific	Humans, ? marsupials	Mosquito	Like dengue, arthritis, endemic and seasonal epidemic
Oropouche	Simbu	Tropical America	Humans, monkeys, birds	Culicoides gnats	Headache, fever, myalgia, epidemic

*Disease name is generally the virus name plus fever, e.g., Colorado tick fever.

TABLE 76-2. SUMMARY OF SELECTED HUMAN ARBOVIRUS DISEASES CHARACTERIZED BY ENCEPHALITIS

Virus			Maintenance Hosts		Clinical (Other than Encephalitis) and Epidemiologic Features
Name*	Group	Geography	Vertebrate	Arthropod	Case Fatality (F) and Sequelae (S)
Western equine	Alphavirus	Western and South America	Birds	Mosquito	Summer epidemics North America, equine epizootics F = high, S = severe
Eastern equine	Alphavirus	Eastern North America, Caribbean, South America	Birds	Mosquito	Summer epidemics and equine and bird epizootics F = high, S = severe
Venezuelan equine	Alphavirus	South and Central America	Rodents, birds, horses	Mosquito	Influenza-like, CNS involvement far less common, encephalitis epizootics in equines F = low, S = moderate
St. Louis	Flavivirus	Western hemisphere, especially U.S.	Birds	Mosquito	Summer epidemics in U.S. F = low, S = moderate
Rocio	Flavivirus	Brazil	Birds	Mosquito	Seasonal epidemics F = low, S = moderate
California (La-Crosse)	Bunyavirus-California	USA	Small mammals	Mosquito	Summer outbreaks and sporadic cases F = low, S = ?
Japanese	Flavivirus	Asia, Japan	Birds, pigs	Mosquito	Summer and fall epidemics in temperate areas F = high S = severe
Murray Valley	Falvivirus	Australia and New Guinea	Bird	Mosquito	Epidemic F = high
Powassan	Flavivirus	U.S. and Canada	Small mammals	Tick	Two cases of human disease recognized, one fatal
Russian spring–summer	Flavivirus	Russia	Birds, mammals	Tick†	Diphasic pattern, bulbospinal paralysis F = high, S = severe
Central European	Flavivirus	Europe	Birds, mammals	Tick†	Diphasic pattern F = low, S = mild
Kyasanur Forest disease	Flavivirus	India	Birds, mammals	Tick†	Hemorrhagic manifestations more common F = low, S = 0
Louping ill	Flavivirus	British Isles	Birds, mammals	Tick†	Diphasic pattern F = 0, S = 0

*Disease and virus name listed plus, usually encephalitis, e.g., St. Louis encephalitis.
†Transovarially transmitted.

tory zoology and entomology are therefore essential. As a result, the arbovirus laboratory is staffed by a broad-based medical zoology group.

Ecology, Epidemiology, and the Public Health

Arboviruses are distributed throughout the world. While most exist in the tropics and subtropics, some have also been isolated from arthropods and vertebrates in climatically less hospitable areas, such as Finland, Canada, Siberia, and Alaska. Their epidemic potential continues to surprise us. Outbreaks of major and often new clinical disease in the savannahs, forests, and swamps of the tropical world provide episodes of romance and high adventure as well as tragedy and misery. Here in the United States, St. Louis and LaCrosse encephalitis viruses cause epidemics of serious disease.

The conceptual and technical base of arbovirus ecol-

ogy grew from the work in the late 1920s and 1930s on yellow fever. Yellow fever initially was thought to be entirely a problem of human-to-human transmission by the relatively domestic mosquito *A. aegypti*. When the entirely nonhuman jungle cycle was discovered and the study broadened zoologically, several arthropod-transmitted viruses were discovered that were not yellow fever. After World War II, an extraordinary global effort was made to study these agents. Arbovirus laboratories were established around the tropical world in such places as Jamaica, Trinidad, Egypt, India, Malaya, Taiwan, Brazil, Panama, Nigeria, and Uganda. The hundreds of arboviruses that we know today were discovered in this effort.

Arboviruses generally maintain themselves in nature by constant and fairly frequent transmission between an arthropod and a vertebrate host (Fig. 76-2). Many variables influence the likelihood and frequency of successful transmission. For example, the frequency of feeding of an

TABLE 76-3. HEMORRHAGIC FEVERS DUE TO ARBOVIRUSES, ARENAVIRUSES, AND OTHER (SIMILAR TO RHABDO) VIRUSES

Fever	Causative Agents	Vector(s)	Vertebrate Host(s)	Geographic Distribution	Epidemiologic Features of Involvement of Humans	Control	Remarks
Yellow fever (urban)	Flavivirus	*Aedes aegypti*	Humans	Human populations in tropics of Africa and South America	Person-to-person passage by *Aedes*	*A. aegypti* control: vaccination	Sylvan YF can spread to *A. aegypti* cycle
Yellow fever (sylvan)	Flavivirus	*Haemagogus* mosquitoes in New World, *Aedes* species in Africa	Monkeys	Forests and jungles of South America and West, Central, and East Africa	Humans infected by exposure in jungles (i.e., woodcutters, hunters, and so on)	Vaccination	Human cases epidemic or sporadic and unpredictable, monkey deaths signal epidemic in tropical America
Dengue homorrhagic fever	Flaviviruses	*Aedes aegypti*	Humans	Tropical and subtropical cities of Asia and Caribbean	Small children usually involved in cities where *A. aegypti* densities are high	*A. aegypti* control: mosquito repellent, screens, and so on	Disease probably due to immune enhancement following sequential infection with different dengue serotypes
Omsk hemorrhagic fever	Alphavirus	Ticks of genus *Dermacentor*	Small rodents and muskrats	Omsk region of U.S.S.R., northern Romania	People exposed in fields and wooded lands	Tick repellents and protective clothing	
Kyasanur forest disease	Flavivirus	Ticks of several species in genus *Haemaphysalis*	Monkeys (rhesus and langur) and small rodents and birds	Mysore State, India	People exposed in fields and wooded lands	Tick control: tick repellents and protective clothing	Monkey mortality signals epidemic
Argentinian hemorrhagic fever	Junin virus, an arenavirus	None recognized	Small rodents, *Akodon, Calomys*	Argentina, NW of Buenos Aires Province extending west to Province of Cordoba	Field workers at harvest time are particularly at risk	Rodent control in fields	Infected rodents contaminate environment with urine
Bolivian hemorrhagic fever	Machupo virus, an arenavirus	None recognized	Small rodent, *Calomys callosus*	Beni Province of Bolivia	Residents of small rodent-infested villages and homes, 1971 nosocomial outbreak in Cochabamba, Bolivia	Rodent control in fields	High mortality in humans
Lassa fever	Lassa virus, an arenavirus	None	Small rodent, *Mastomys natalensis*	West Africa; Nigeria, Liberia, Sierra Leone	Residents of small rodent-infested villages, dramatic nosocomial outbreaks	None known, possibly rodent control	High mortality in humans
Crimean hemorrhagic fever	CHF Congo virus, (bunyavirus)	Ticks of several genera	Larger domestic animals implicated, also African hedgehog	Southern U.S.S.R., Bulgaria, East and West Africa	Cowhands and field workers in U.S.S.R., nosocomial outbreaks reported	Tick control relating to livestock, full isolation in patient care	Human disease important in U.S.S.R., importance to humans in Africa not known

(continued)

TABLE 76-3. (cont.)

Fever	Causative Agents	Vector(s)	Vertebrate Host(s)	Geographic Distribution	Epidemiologic Features of Involvement of Humans	Control	Remarks
Korean hemorrhagic fever (hemorrhagic fever with renal syndrome)	Hantaan virus (bunyavirus)	Not known	Rodents	Korea, northern Eurasia to and including Scandinavia	Rural or sylvan exposure (military, forest occupations, farmers), outbreaks in laboratories related to white rats	None	Antibodies found in Norway rats worldwide
African hemorrhagic fever (Marburg disease)	Marburg and Ebola viruses, similar to rhabdoviruses	Not known	Unknown	Africa	Person-to-person	Clinical isolation techniques	Highly fatal

Adapted from Johnson: In Beeson and McDermott (eds): Textbook of Medicine, 14th ed, 1975, p 239. Philadelphia, W.B. Saunders Co.

arthropod on a vertebrate host depends on numerous innate and behavioral factors of each animal that may be environmentally determined. These factors may differ from time to time and from place to place. Arthropod longevity, a major determinant, depends in turn on climatic factors, such as humidity and temperature, availability of food, and the prevalence of predators, as well as on noxious physical, chemical, and biologic agents. Many other complex factors similarly influence transmission. Clearly, a systems analysis approach is needed to assess the effect of any change in any variable. Because of the large number of agents, the numerous nonhuman hosts, and the great gaps in our knowledge, this approach has not yet come into its own.

Although viremia is usually of sufficient intensity for transmission only for a matter of days, there is evidence that long-lasting infections in some hibernating vertebrates, such as rodents infected with Eastern equine encephalitis virus, may permit a virus to remain in an area

during a period of unsuitable climate and to be transmitted again when arthropod feeding resumes. Transovarial transmission through generations of arthropods occurs with many viruses and provides the prospect of indefinite survival of virus without cyclical transmission. The important subject of how arboviruses survive periods when cyclical transmission cannot occur has been reviewed recently by Reeves. Other than transovarially, arthropod-to-arthropod transmission of some viruses may occur venereally. Direct transmission may occur between vertebrate hosts, as in the case of Eastern equine encephalitis virus, which is passed from pheasant to pheasant under crowded conditions of commercial pheasant farms. The importance in nature of these alternate maintenance cycles is an area of active, current research.

Humans are generally incidental hosts and a dead end for virus transmission. Notable exceptions are the human–mosquito cycles of dengue, yellow fever, Ross River, and Chikungunya viruses. As a rule, arboviruses are clinically

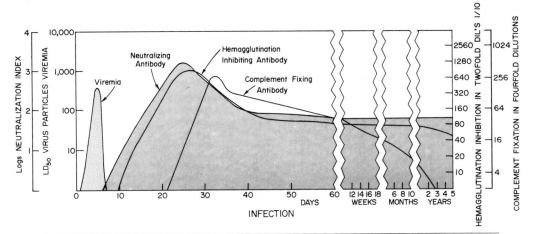

Figure 76-1. Hypothetical diagnostic virologic and serologic features of arbovirus infections of humans. *(From Work: In Hunter et al. (eds): Tropical Medicine, 5th ed, 1976. Philadelphia, W. B. Saunders Co.)*

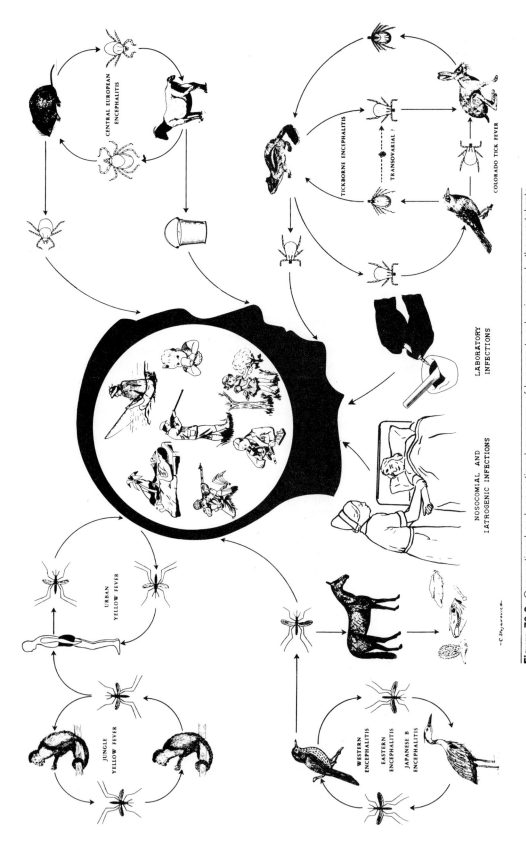

Figure 76-2. Occupational and recreational exposure of humans to arbovirus and other vertebrate virus infections. *(From Work TH et al.: In Feigin RD, Cherry JD (eds): Textbook of Pediatric Infectious Diseases, vol II, 1981. Philadelphia, Saunders.)*

silent as they cycle in nature. Disease occurs in unnatural hosts, such as humans and horses, and these hosts are of no consequence to viral transmission.

There are two basic ecologic mechanisms that ensnare humans and other animals: either humans enter the geographic area of the transmission cycle, or the cycle changes and moves closer to them. Yellow fever ecology provides examples of both. The former occurs when a person enters a forest where yellow fever virus is cycling between monkeys and the tree-top mosquitoes. Without the intrusion, these mosquitoes might never feed on humans. When the infected person returns to town, he may introduce yellow fever virus into a cycle involving humans and the peridomestic mosquito *A. aegypti*, and an urban epidemic may result. Some viruses may come in contact with humans by another mechanism, the amplification cycle. These viruses have a basic cycle away from humans that involves arthropods that feed almost exclusively on animals. With seasonal build-up of populations of different arthropods and, let us say, birds, another cycle is started that carries the virus closer to humans and arthropods that are more likely to bite humans.

From time to time an arbovirus disease appears where it has been previously unknown. Dengue hemorrhagic fever of Southeast Asia, Kyasanur Forest disease in India, Chikungunya and O'nyong-nyong fevers in East Africa, and Rift Valley Fever in Egypt are all examples of new disease situations that have been recognized within the past 2 decades. Did O'nyong-nyong fever virus become more pathogenic for humans as a result of passage in anopheline mosquitoes? This is a possibility, since all other group A arboviruses (alphaviruses) are transmitted by culicenes. Was pick-up and transmission by a certain genus of tick responsible for human pathogenicity of Kyasanur Forest disease virus? It is difficult to prove that pathogenicity for humans may be enhanced by a change in the transmission cycle, but the question is of great importance, since we persist in meddling with our environment. The possibility that the clinical consequences of infection may be affected by sensitization from prior infection by a related virus has been raised to explain the hemorrhagic fever syndromes in the Philippines and Southeast Asia caused by dengue viruses. Before the 1950s, dengue virus was known only in connection with classic dengue, a relatively benign disease characterized by fever and nonhemorrhagic rash. Thus one disease may change as a result of a change in the ecology of another infection. Whatever the reasons prove to be, we are clearly in an unstable situation, with some surprises in the future. Additional expression of some basic concepts that help to explain the behavior of arboviruses can be found in the paper by Chamberlain.

Arboviruses cause disease and death in thousands of persons each year. From year to year and decade to decade, the epidemic potential results in wide swings in incidence in a given geographic area. In the United States, the number of known and reported cases of arboviral encephalitis varies from tens to thousands each year. In the 1960s, there was an epidemic of yellow fever in remote areas of Ethiopia that is believed to have killed 10,000 persons. Dengue hemorrhagic fever in the Philippines and Southeast Asia affects thousands each year, with substantial mortality. Laboratories must not be overlooked in a discussion of epidemiology. Human curiosity about arboviruses has resulted in more than 400 reported laboratory infections, of which 18 caused the deaths of investigators.

Disease in Humans

Arboviruses cause acute febrile illnesses that range from minimal severity to highly fatal diseases with devastating sequelae for the survivors. There are three general types of disease. One type is characterized by undifferentiated fever that may or may not be accompanied by rash and arthralgia (Table 76-1). In another type, there is viral invasion of the central nervous system, productive of an often severely damaging encephalitis (Table 76-2).

In the third category are those arboviruses that cause hemorrhagic fever, as in yellow fever, hemorrhagic dengue fever, and a number of tickborne hemorrhagic fevers. Hemorrhagic fevers (Table 76-3) are also caused by a number of arenaviruses and agents similar to rhabdoviruses. Additional material on this category of diseases is found in the appropriate chapters.

The pathogenetic mechanisms of arbovirus disease are only beginning to be understood. In some diseases direct tissue destruction occurs for unknown reasons. In others, such as dengue, an immunologically modulated pathologic process seems to occur. The conditions permitting this chain of events are briefly referred to in the section below on hemorrhagic dengue.

There follow brief descriptions of representatives of each of these categories, somewhat arbitrarily selected because they occur within the United States or because of their significance to the public health elsewhere.

Arboviruses that Cause Fevers, Rash, Arthralgia

Undifferentiated arbovirus fevers with or without rash and arthralgia are generally of low mortality, and recovery is complete with rare exceptions. Classic dengue fever is the prototype, and many of the other members of the group often are described as dengue-like. Members of the group that are transmitted in the United States include Colorado tick fever and vesicular stomatitis viruses.

Dengue fever is a disease of rapid onset, with the temperature rising suddenly to high levels and persisting for several days. Accompanying the fever are headache and aching in the muscles and joints. There is an associated macular rash that blanches with pressure; later in the disease it may become papular and morbilliform, sparing only the palms and soles. Lymphadenopathy is usually present. After 5 to 10 days, the acute symptoms disappear, but generalized weakness and lassitude may delay complete recovery for weeks. In the midst of the acute disease, there may be one or several days when manifestations of illness disappear. This provides the biphasic course that characterizes many arbovirus diseases. Leukopenia is the most striking and consistent laboratory abnormality. Etiologic diagnosis is confirmed by virus isolation from the blood or by the demonstration of a rise in serum antibody titer. Death is rare, and most patients recover completely. The disease occurs throughout the tropical world. Epidemics may be widespread, affecting a very high proportion of the population. Typically, the virus cycles between humans and *Aedes* mosquitoes, especially *A. aegypti*. In some areas (Malaysia and West Africa) there may be a jungle cycle involving primates.

O'nyong-nyong, Chickungunya, Mayaro, and *Ross River* are four related, mosquitoborne alphaviruses that cause dengue-like disease in Africa, southern Asia, South and Central America, and Australia.

Colorado tick fever is characterized by fever, headache, myalgia, and lethargy. Rash and central nervous system signs and symptoms occur less commonly. The rash is maculopapular, sometimes petechial, and does not have any particular distribution. A biphasic fever pattern occurs in about half of the cases. Disease begins approximately 5 days from tick bite and lasts for about a week, although malaise may continue for several more weeks, especially in the older patient. Approximately 20 percent of patients require hospitalization. Rarely, pericarditis or orchitis occurs; very rarely, a hemorrhagic syndrome with death has been reported. Diagnosis is confirmed by laboratory methods that detect antibodies in serum, infectious virus in blood, or viral antigen in erythrocytes. Curiously, infectious virus and/or viral antigens may be detected in the blood for up to several weeks after onset of disease. There are some reports of transmission by blood transfusion.

The disease is transmitted throughout the Rocky Mountain states from March to October, with peak incidence in May and June. During 1973 and 1974, active search revealed 228 cases in Colorado alone. The arthropod host is *Dermacentor andersoni*.

Arboviruses that Cause Encephalitis

The arthropodborne viral encephalitides are a group of serious illnesses that occur sporadically or in epidemic form throughout the tropical and subtropical world, except in Africa. Infected persons are at varying risk of death or severe central nervous system damage depending on their age and on the type of virus. In the United States since 1960, there have been an average of 300 arbovirus encephalitis cases reported to the Centers for Disease Control annually, ranging from a low of 45 in 1960 to a high of 2113 cases in 1975. In recent years, California and St. Louis encephalitis viruses account for most of these cases, followed by Western, Eastern, and Venezuelan equine encephalitis viruses, with Powassan virus making its appear-

ance in human disease for the first time in single cases in 1971 and 1972. Rociovirus encephalitis emerged as an epidemic disease in southern Brazil in 1975.

In Britain, louping ill is a tick-transmitted encephalitis of sheep and humans. Central European and Russian spring–summer encephalitides are found in Europe and Russia, while Japanese encephalitis occurs in Japan, Asia, and the Pacific. Kyasanur Forest disease in India entails signs of central nervous system involvement, as well as abnormal bleeding, and is thus classified both as an encephalitis and a hemorrhagic fever. Murray Valley encephalitis is found in Australia.

Western equine encephalitis is an acute generalized illness characterized by fever, meningeal signs, somnolence, coma, convulsions, and paralysis. In infants and young children fatality is high, and permanent brain damage is common. The virus cycles in nature between a variety of small birds and the mosquito *Culex tarsalis*. Humans and equines are dead-end hosts because viremia adequate for mosquito infection does not occur. The overwintering mechanism is unknown. Diagnosis depends on demonstration of a rise in antibodies between acute and convalescent serum specimens. Virus isolation attempts from human tissue are rarely successful. The disease is confined to western North America, and it occurs in epidemic form in early and midsummer.

Eastern equine encephalitis is an acute generalized illness characterized by fever and central nervous system signs and symptoms, including those of meningeal irritation, lethargy, coma, pareses, and convulsions. Death is frequent, and in survivors residual central nervous system damage is common and severe. A striking polymorphonuclear leukocytosis occurs in blood and in cerebrospinal fluid. Etiologic diagnosis depends on virus isolation from brain tissue or the demonstration of an increase in serum antibody. Mosquitoborne epidemics and equine epizootics occur in late summer along the eastern strip of North America from Florida to Canada and the Caribbean.

California encephalitis is an acute febrile illness of 7 to 10 days' duration, characterized by headache, vomiting, lethargy, disorientation, seizures, and focal neurologic signs. Although convalescence may be prolonged, fatality and morbidity are low. The cerebrospinal fluid contains abnormally high protein and increased numbers of white blood cells during the acute phase. Etiologic diagnosis depends on demonstration of a rise of specific antibodies. The virus has been recovered from brain tissue from two fatal cases. It is a disease of children and occurs during summer months, principally in the north central United States. Mosquitoes of the genus *Aedes* transmit the virus to humans. Overwintering of the virus occurs by transovarial transmission in vector mosquitoes; the eggs survive the winter in tree holes. Vertebrate hosts include small animals.

St. Louis encephalitis is an acute febrile illness that occurs in a small percentage of those infected. It is usually characterized by fever and headache, tremors, and abnormalities of coordination and motor cranial nerve function. The disease is most common in older persons. Signs of urinary tract inflammation may occur. Fatality may reach 25 percent, and mental and emotional sequelae are common. Cerebrospinal fluid changes are consistent with virus infections. Diagnosis depends on demonstration of a serum antibody rise; virus isolation is rarely achieved from human specimens. The virus cycles between birds and *Culex* mosquitoes. Large urban and suburban outbreaks occur in the United States.

Powassan virus encephalitis has been identified as the cause of encephalitis in an occasional patient for the past 20 years in northern United States and Canada. While human infections are very uncommon, clinical consequences can be severe, with fever, headache, seizures, coma, and paralysis. Permanent neurologic damage may follow. Powassan virus is transmitted by ticks, although history of tick bite is frequently lacking.

Venezuelan equine encephalitis is characterized by fever, severe headaches, chills, and gastrointestinal symptoms. In a few children, serious encephalitis occurs, with convulsions, coma, paralysis, and abnormal reflexes. Myalgia, conjunctivitis, and sore throat commonly appear as the disease progresses. Leukopenia is typical. Diagnosis is suspected in patients who have been in appropriate geographic areas during epidemics. Demonstration of a rise in antibodies in serum samples obtained in the acute and convalescent stages of disease is necessary.

The disease occurs as equine epizootics in South and Central America, the Caribbean, Mexico, and southern United States. Strains of the virus that are responsible for endemic cases cycle in mosquitoes, small rodents, and birds. Epidemic strains cycle in equines and a variety of mosquitoes; it is this poorly understood amplification cycle that results in humans becoming infected. A live attenuated vaccine has been shown capable of immunizing humans and horses. While it may be too dangerous for widespread use in people, this vaccine can be applied on a mass basis to horses, with prompt cessation of both equine and human cases. Insecticides can be used concurrently.

Japanese B encephalitis is an acute illness that occurs in a small percentage of those infected with the virus. The usual consequence of infection is either inapparent or mild, undifferentiated disease. The encephalitis disease is similar to St. Louis encephalitis, with more severe neurologic involvement. Fatality is high, especially in young children and older persons. Diagnosis is confirmed serologically, and virus isolation from human specimens is unusual. The virus cycles annually between *Culex* mosquitoes and birds or pigs in temperate areas, including Japan, Siberia, Korea, Taiwan, Pacific islands, and in tropical areas of southeast Asia. Japanese encephalitis also produces disease in horses, and large epizootics are reported.

The tickborne flaviviruses comprise a group of related agents that are found, each in a circumscribed domain, in North America, Britain, Europe, Russia, and Asia. The most common ones are Powassan, Russian spring–sum-

mer, central European, and louping ill encephalitides, and Omsk hemorrhagic fever and Kyasanur Forest disease. The clinical spectrum includes hemorrhagic fever or encephalitis or a combination of the two. They are listed in both Tables 76-2 and 76-3. The diseases vary greatly in fatality and in the severity of permanent neurologic damage. Diagnosis is by virus isolation from blood or brain tissue and by serologic tests.

Arboviruses that Cause Hemorrhagic Fevers

The viral hemorrhagic fevers comprise 10 distinct diseases with widely differing epidemiologies and geographic loci. They are caused by viruses representing several taxonomic groups (Table 76-3). Some are mosquitoborne, others are tickborne, and the arthropod vectors of a third group are unknown or nonexistent. In addition to humans, vertebrate hosts generally include primates or small mammals or rodents. The diseases are severe, with high fatality, but are self-limited, and survivors usually recover completely. None of these diseases is transmitted within the United States, although recently, antibodies to Hantaan virus (hemorrhagic fever with renal syndrome) have been found in Norway rats in the United States.

Yellow fever and hemorrhagic dengue are described here, and others are described in Chapters 77 and 78.

Yellow Fever. In spite of the presence of a safe, effective, attenuated vaccine that has been available since the 1930s, yellow fever remains today a real or potential public health problem for many peoples of the tropical world. Clinically, the disease is marked by the sudden onset of fever associated with severe headache, myalgia, back pain, conjunctivitis, and photophobia. As the disease progresses, prostration increases, and there are signs of involvement of the liver, kidneys, and heart with hepatomegaly, albuminuria, and a striking slowing of the heart rate. A hypovolemic shock phase associated with bleeding gums, hematemesis, oliguria, and jaundice follows. Case fatality is high; recovery is complete. Diagnosis is confirmed by the isolation of the virus from the blood or from the liver or by demonstration of a specific serologic response.

The characteristic histopathology of midzonal necrosis in the liver lobule has given rise in some areas to an epidemiologic method wherein mandatory postmortem corings of the liver are obtained from all deaths and sent in fixative to a central examining point, thus permitting an assessment of the proportion of deaths that are due to yellow fever.

Yellow fever has two distinct cycles in nature. One is an *A. aegypti*–human cycle that is operative in the large urban epidemics that are now, fortunately, history in the new world and Europe. In parts of Africa, however, large epidemics with transmission by this vector still occur. The virus also cycles in the forest, involving monkeys and different genera of mosquitoes. The virus is carried from the forest cycle to the urban cycle by infected monkeys or humans. Transmission can be effectively interrupted, in the case of the urban cycle, by *A. aegypti* control measures. Such measures, used prior to the recognition of the sylvan cycle, rid the urban centers of the Western hemisphere of dreadful epidemics that affected cities as far north as Philadelphia into the first decade of the present century. These same measures permitted completion of the Panama Canal. They are however, less useful when one is dealing with zoophilic forest mosquitoes. As if in defiance of the best human effort, yellow fever epidemics occur today in areas where lack of basic logistical underpinnings of health services prevent vector control or immunization programs.

Hemorrhagic Dengue. This acute febrile illness begins abruptly like classic dengue with fever, nausea, vomiting, and cough. However, after a day or two, hemorrhagic manifestations appear, first as petechiae, followed by purpura and signs of gastrointestinal hemorrhage. The clinical picture of shock ensues. Hypoproteinemia, rising hematocrit, and hemostatic abnormalities, including thrombocytopenia, are present. During the next days the patient either dies or begins a complete recovery. Left untreated, fatality can be as high as 50 percent, but with proper medical management, the fatality rate is less than 3 percent. Diagnosis is confirmed by the demonstration of rise in antibodies or by isolation of virus from specimens of blood.

This hemorrhagic form of dengue was first seen in the 1950s in the Philippines and Southeast Asia and subsequently in India, Indonesia, and Oceania. In 1981, the disease appeared in this hemisphere in an outbreak in Cuba. Children comprise the vast majority of those affected.

Recently, substantial progress has been made in the development of data that support the hypothesis that dengue hemorrhagic fever/dengue shock syndrome (DHF/DSS) occurs in circumstances where circulating antibody to one dengue virus type from a prior infection changes the clinical expression of a second infection to include DHF/DSS. In this sequence, called "immune enhancement," antibody and virus form immune complexes that are phagocytized by monocytes. Virus multiplies within the monocyte, which in turn interacts with sensitized T cells. Factors are then released that activate a complement pathway, increase vascular permeability, and effect blood clotting systems. The resulting hypocomplementemia, thrombocytopenia with hemorrhage, and increased vascular permeability with hypovolemia and shock are the hallmarks of DHF/DSS. An impressive array of results from in vitro laboratory experiments and field epidemiologic observations support this hypothesis. The well-documented, although far less frequent, occurrences of DHF/DSS in association with primary dengue infection remain unexplained, however.

The usual vector is *A. aegypti*, as it is for classic dengue. Accordingly, the frequency of transmission is governed by climate and numerous domestic factors (such as those relating to collections of standing water) that affect breeding and feeding of this mosquito. Control measures entail all those measures that reduce the frequency of the vector mosquito feeding on potential patients. No vaccine is available.

Treatment, Control, and Prevention

There are no therapeutic agents that are specific for arboviruses. The management of patients with disease involves measures designed to restore and maintain nutrition and reasonably normal physiology. The latter might include, for example, restoration of intravascular volume during the hypovolemic shock phase of hemorrhagic fever or measures to reduce destructive degrees of cerebral edema in patients with encephalitis.

With the exception of the highly effective, safe, attenuated, tissue culture vaccine that has been in use since the 1930s for yellow fever, there are no arbovirus vaccines in general use.

Venezuelan equine virus vaccine has a restricted use for laboratory workers and can protect humans indirectly by preventing the development of a large reservoir of infected horses. Many experimental vaccines are under trial or laboratory development.

For arboviruses with limited host range and an accessible vector, there have been instances of effective prevention based on vector control. Urban yellow fever was eradicated in the early decades of this century by measures against *A. aegypti*. The task becomes infinitely more complex when one considers arboviruses that cycle silently in a variety of natural settings in birds, mammals, and arthropods. In these situations, where no ecologically acceptable vector control is available, the possibilities may be restricted to lessening human exposure to the ecosystem in question. Such a measure may fail because of anticipated and unacceptable consequences as, for example, might beset a woodcutter denied access to the forest.

FURTHER READING

Berge TO: International Catalogue of Arboviruses, 2nd ed. Dept. HEW, Washington, D.C., US Govt Printing Office, 1975. *See also* Karabatsos N (ed): Supplement to International Catalogue of Arboviruses Including Certain Other Viruses of Vertebrates. Am J Trop Med Hyg [Suppl] 27:372, 1978

Chamberlain RW: Arbovirology—then and now. Am J Trop Med Hyg 31:430, 1982

Halstead SB: The pathogenesis of dengue molecular epidemiology in infectious disease. Am J Epidemiol 114:632, 1981

Johnson KM, Monath TP, Tesh RB, et al.: Arthropod-borne viral fevers, viral encephalitides and viral hemorrhagic fevers (arboviruses and arenaviruses). In Wyngaarden JB, Smith LH (eds): Textbook of Medicine, 16th ed. Philadelphia, Saunders, 1979

Johnson KM: Nephropathia epidemica and Korean hemorrhagic fever: The veil lifted? J Infect Dis 141:135, 1980

Monath TP (ed): St. Louis Encephalitis. Washington, D.C., American Public Health Assoc, 1980

Morens DM: Dengue fever and dengue shock syndrome. Hosp Pract 17:103, 1982

Strode GK: Yellow Fever. New York, McGraw-Hill, 1951

Theiler M, Downs WG: The Arthropod-Borne Viruses of Vertebrates. New Haven, Yale University Press, 1973. *See also review of this book*: Work TH: Science 182:273, 1973

World Health Organization. Rift Valley Fever—An Emerging Human and Animal Problem. WHO Offset Publication #63. Geneva, WHO, 1982

Work TH, Monath TP, Gear JHS, et al.: Arbovirus infections and other zoonotic virus infections and diseases. In Feigin RD, Cherry JF (eds): Textbook of Pediatric Infectious Diseases. Philadelphia, Saunders, 1981, vol II

CHAPTER 77

Rhabdoviruses and Marburg and Ebola Viruses

Rabies
    Epidemiology
    Disease in Humans
    Prevention and Control

Marburg and Ebola Diseases

Vesicular Stomatitis Virus

> A rabid racoon recently cost the State of South Carolina approximately $10,000 by coming in contact with at least 25 people.
>
> — *USPHS, Centers for Disease Control Veterinary Public Health Notes, Feb. 1980, p. 1*

> On May 10, 1980, a rabid dog in Yuba County, California, bit three persons. A total of 70 persons were identified as having been exposed to the dog and were given rabies vaccine. The total cost of the episode was $105,790.
>
> — *USPHS, Centers for Disease Control MMWR 30:527, 1981*

There are more than 60 rhabdoviruses. Of these, 10 are known to cause disease in humans (Table 77-1). Two important agents, Marburg and Ebola, although previously so classified, possess distinctive physicochemical characteristics that may lead to the formation of a new group. Rabies, Marburg, and Ebola cause serious illnesses, often with very high fatality rates. Rabies is nearly 100 percent fatal, while for Marburg and Ebola, fatality rates ranging from 20 to 90 percent have been observed. Other members of the group, such as Isfahan virus, isolated in Iran in 1975, probably infect humans, as evidenced by a high prevalence of antibodies, but no disease has been observed. Rabies was the first rhabdovirus isolated, in the early years of this century, while the newest member of the group was isolated in the 1970s.

Transmission from human to human or animal to animal is clearly by arthropod with certain viruses, while others are transmitted mainly through animal-to-animal contact. For Marburg and Ebola, in addition to person-to-person transmission, humans probably acquire infection as a result of environmental contamination by chronically infected viruric rodents.

Serologic relationships and morphology define the subgroups of the Rhabdoviridae and individual members of subgroups. Arbovirologists often have been drawn into the study of rhabdoviruses either because of definite arthropod transmission or epidemiologic features that were initially highly suggestive of arthropodborne transmission. In addition, rhabdoviruses possess complex life cycles and transmission patterns, and these aspects in themselves attract the attention of medical scientists and clinicians whose interests are broadly based zoologically. Other than support of physiologic processes in the sick individual, the only protection that humans and animals have from rhabdoviruses is natural or induced immunity and environmental manipulations based on an understanding of the zoonotic details of the life cycles.

In this chapter, attention will be paid only to those agents known to cause human disease: rabies, Marburg, Ebola, and certain members of the vesicular stomatitis group.

Rabies

Rabies has been recognized for centuries. In the late 1800s, Pasteur sorted out the confusion concerning the etiologic agent and pathogenesis of the disease. He then crowned these extraordinary achievements with the development of a method of active immunization, which, although never tested in a controlled trial in humans, is generally believed to have been effective. Today, rabies

TABLE 77-1. RHABDOVIRUSES AND SIMILAR AGENTS THAT CAUSE ILLNESS IN HUMANS

| Virus | | Year of Original Isolation | Natural Geographic Distribution | Disease in Humans |
Group	Serotype			
Rabies	Rabies	1903	Worldwide	Meningoencephalitis
	Mokola	1968	Africa	Meningoencephalitis
	Duvenhage	1971	Africa	Meningoencephalitis
Marburg	Marburg	1967	Africa	Hemorrhagic fever
	Ebola	1976	Africa	Hemorrhagic fever
Vesicular stomatitis	Indiana	1925	Western hemisphere	Fever and vesicles
	New Jersey	1952	Western hemisphere	Fever and vesicles
	Alagoas	1964	Western hemisphere	Fever
	Piry	1960	Western hemisphere	Fever
	Chandipura	1965	Asia and Africa	Fever

continues to frighten human beings, and it poses a distinct and usually fatal hazard to them as well as to their domestic animals. In the 1960s the physical and chemical properties and the morphology of the rabies virus were established. Markedly improved vaccines were developed in the late 1970s.

Epidemiology

Rabies virus persists in nature by passage from animal to animal, usually during direct contact. Adequate contact generally requires that virus-containing saliva be inoculated through broken skin by bite or scratch into a susceptible animal. The infected animal's illness alters behavior in favor of transmission by making the animal deranged and aggressive and thus more likely to attack or bite more frequently and to less purpose than normally. The timing and duration of virus excretion in saliva of infected animals vary dramatically according to species of the infected animal, strain of virus, and dose of virus inoculated. For dogs, virus in the saliva has been documented as long as 2 weeks prior to onset of disease, although a few days appears to be a more probable period. Infected, unvaccinated dogs have been observed to excrete virus intermittently for months in the absence of disease. How frequently this occurs, however, is not known.

In some situations rabies virus may be transmitted by aerosol. Such situations occur in caves that contain large populations of insectivorous bats. At certain times of the year, when bats are being born and are developing and when their mothers are lactating, sufficient quantities of rabies virus may be present in the air of the cave to infect susceptible animals placed in such a fashion as to be exposed only to cave air. Beyond such observations as these, however, the ecology of bat cave rabies is poorly understood.

Some biologic determinants of transmission include the following. The saliva must contain an adequate titer of virus. In this connection, infected skunks have been found to develop the highest titers. The bitten animal must be of a susceptible species. In the laboratory, the route of inoculation is important, with intracerebral being among the most sensitive routes, whereas intraperitoneal injection is less likely to transmit infection. Transmission transplacentally or via the gastrointestinal tract by drinking milk of infected animals or by arthropod vector does not seem to occur to any significant degree in nature. The age of the receiving animal can be important; generally, the younger the animal, the more susceptible.

The site of the inoculation is clearly important. In humans, bites about the head and neck are likely to result in disease earlier than are bites at areas more peripheral to the central nervous system. This is thought to be because there is a greater density of nerve endings about the head and face, providing more neurologic pathways for the virus to travel to the central nervous system. In addition, the distance to the brain is shorter from the head and neck area.

Virus multiplies first in muscle tissue at the site of inoculation. Thereafter virus enters peripheral nerves and progresses, probably passively, in the axoplasm to the spinal cord and then to the brain. Virus spreads rapidly throughout the brain. Disease ensues after a variable incubation period, which is not explained solely by the time it takes the virus to migrate from peripheral inoculation site to brain.

Virus also migrates toward the periphery from the brain, reaching such sites as skin and cornea, where it may be detected by immunofluorescent techniques.

Sources of infection for humans and their economically important domestic animals vary to some extent geographically. For example, in North America, foxes, skunks, racoons, cattle, dogs, cats, and bats are the usual reservoir and vector hosts. In Latin America, bats of various species seem to be the most important, while in the Mediterranean area the wolf, in India the jackal, and in certain Caribbean areas, such as Cuba, Puerto Rico, and Granada, the mongoose is the most significant reservoir/vector animal.

Bat rabies may be transmitted in two ways. In South America, vampire bats transmit rabies to domestic cattle while taking their blood meal. This results in enormous economic and nutritional losses to people in these areas. As mentioned above, aerosol transmission can occur in some insectivorous cave-dwelling bats of North America. Presumably this mechanism, as well as transmission by bite, can be operative in vampire populations.

The epizootic and enzootic behavior of rabies in domestic and wild animals may display swings of incidence in one or another species and from one place to another that are often poorly understood. For example, in the United States during the past few years there has been a sharp increase in rabies in domestic cattle and wild racoons.

Spread of animal rabies into a geographic area typically involves but one species. Evidence that strains of wild rabies may differ in their pathogenicity for one species as compared to others may explain such epizootic behavior.

Density of susceptible and infected animal populations, the biology of the virus in the infected host, the degree to which rabies can become latent, and factors that influence aggressive behavior, other than the rabies infection itself, all require further understanding before the determinants of rabies transmission are completely elucidated.

Serologic surveys for the purpose of detecting rabies antibodies indicative of nonfatal, past infection has not been helpful in rabies because no adequately sensitive, specific, and inexpensive test is available. Accordingly, the question of whether latent or clinically silent infections are epidemiologically significant is not yet answered.

There is seasonal variation in rabies in animals as well as humans. In animals the greatest number of cases occur in the spring and early summer during the breeding sea-son. In humans, most cases occur in the summer, possibly as a result of more outdoor contact with rabid animals.

In the 1970s two viruses, Mokola and Duvenhage, were found in Africa that are morphologically and serologically related to rabies. For each agent there is at least one reported case of fatal central nervous system disease.

Disease in Humans

Generally, disease begins within 2 to 8 weeks of exposure, with extremes of the incubation period ranging from 9 days to more than a year, and even many years in very rare instances (Fig. 77-1). Early symptoms include loss of appetite, nausea, vomiting, and fever associated with feeling poorly, pain, and paresthesias at the site of the bite. Anxiety, depression, and agitation ensue. As the disease progresses, spasms begin to occur. When they involve the muscles of deglutition, the inability to drink liquids or to swallow one's own saliva properly gives rise to the picture of hydrophobia. Soon, a wide variety of central nervous system signs appear, including hyperactivity, confusion, delirium, hallucinations, disorders of coordination, and paralyses. Finally, coma develops, and by this time secondary infection, nutritional deficiencies, and respiratory problems bring on death.

Recovery from rabies is considered to be extremely rare. There is one definite, well-documented case of recovery reported in the literature.

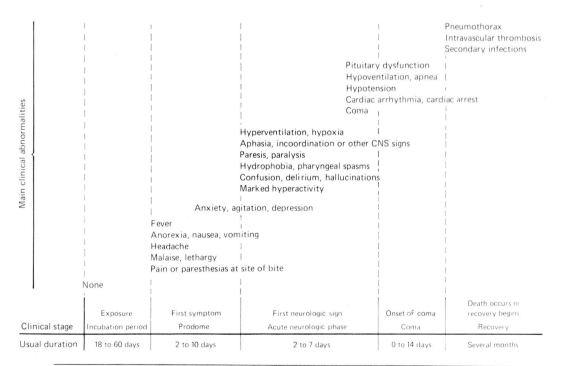

Figure 77-1. The natural history of clinical rabies in humans; hypothetical composite case. All clinical abnormalities need not be present in each case. *(From Hatwick, Gregg: In Baer (ed): The Natural History of Rabies. New York, Academic Press, 1975, vol 2, p 289.)*

Laboratory findings include an elevated white cell count of 20 to 30 thousand cells per cu mm associated with an increased percentage of polymorphonuclear leukocytes. Cerebrospinal fluid is usually normal, but a mild pleocytosis may be observed consisting of less than 100 mononuclear cells per cu mm. Protein, sugar, and hyaline casts may be seen in the urinary sediment.

The clinical features described above following a bite by a rabid animal are the major criteria for diagnosis. Virus isolation attempts from saliva, cerebrospinal fluid, tears, urine, and nasopharyngeal secretions may be successful, but they are meaningless if negative. Fluorescent antibody staining of smears of corneal epithelial cells or of sections of skin from face or neck may reveal virus antemortem after disease has begun. A rise in serum antibodies is generally not detectable by the time the patient dies. Following death, virus can be demonstrated in the central nervous system by one or more of the three methods used to detect rabies virus in animal brain specimens. One method, the most rapid, involves the detection of intracellular rabies virus antigen by the use of fluorescent antibody. A second method is the identification of typical intracytoplasmic, eosinophilic inclusions (Negri bodies, see Color Plate) in nerve cells in certain areas of the central nervous system. Third, intracerebral inoculation of mice with suspensions of brain tissue is followed by identification of rabies virus in brain tissue after a suitable period of time for virus multiplication to occur.

The principal pathologic finding in rabies is inflammatory necrosis that chiefly involves the brain. Grossly, the brain is swollen and congested. Microscopically, nerve cells are seen to be destroyed, and perivascular, mononuclear cuffing is present in addition to meningeal mononuclear cell infiltrates. Negri bodies are found. The most severe changes are found in the thalamus, hypothalamus, and cranial nerve nuclei in the floor of the fourth ventricle. The spinal cord may show similar changes.

Prevention and Control

Other than avoidance of rabid animals and their bite, prevention of human disease is based on immunization. Preexposure immunization is recommended for certain people at high risk of exposure because of their animal-handling occupation. Because of the long incubation period, active immunization to rabies *after* exposure is considered to be effective for preventing disease and has been the basic measure in the management of human rabies since it was first performed by Pasteur.

The decision to immunize depends on evidence that the biting animal is rabid. Such evidence results from examination of the animal's brain for rabies virus by one or more of the methods mentioned above. Presently in the United States, such animals as skunk, fox, coyote, racoon, or bat that carry out unprovoked attacks on humans may be presumed rabid.

Effective vaccines for human use both pre- and postexposure are available. These modern vaccines are prepared from the strain of virus originally isolated by

Pasteur in 1882. Virus is grown in cell cultures derived from human embryo lung. Physical, chemical, and immunologic studies of rabies virus reveal that the glycoprotein found in the form of spikes in the viral envelope stimulates neutralizing and protective antibody. Based on this work of the 1960s and 1970s, a number of vaccines employing either whole virion or glycoprotein split product are now in use. Antiserum derived from hyperimmunized people is used concurrently with vaccine in certain clinical situations. Details of active and passive rabies immunization procedures for animals and humans are constantly changing as products exhibiting fewer undesirable side effects and better antigenicity are developed. Groups, such as the Advisory Committee on Immunization Practices of the U.S. Public Health Service, publish their recommendations periodically in the Morbidity and Mortality Weekly Report of the Centers for Disease Control.

The control of rabies is considerably more difficult in wild animals than in domestic animals and humans. Human rabies in the United States has decreased to less than 3 cases per year for the past 19 years. Dog rabies has decreased from 8000 cases in 1947 to 216 in 1981 as a result of immunization practices. On the other hand, thousands of cases of wild animal rabies continue to be recognized. Measures for control of wild animals rabies include quarantine or slaughter. More recently, some innovative work employing baited food with live virus vaccine designed to attract specific species, such as foxes, is being tried in certain parts of the world. The Morbidity and Mortality Weekly Report should be consulted for the latest status reports and recommendations for control of animal rabies.

Marburg and Ebola Diseases

In 1967 in Marburg, Germany, and in Yugoslavia, 25 persons became ill (Table 76-3). There were 6 secondary cases among persons in direct contact with the primary 25 cases. Seven of these 31 patients died. All of the primary cases were laboratory workers who had been in contact with tissue or body fluid of green monkeys (*Cercapithicus aethiops*). The monkeys were from two shipments from Uganda. On route there was an overnight stay at Heathrow Airport in London. A new virus was isolated from the patients and subsequently named "Marburg virus." On the basis of its physical and chemical properties, Marburg virus is similar to but distinct from rhabdoviruses.

During early February 1975, a young Australian, accompanied by his girlfriend, traveled extensively in Rhodesia (Zimbabwe). He then became ill and died in Johannesburg. Marburg virus was isolated from blood and throat specimens. His girlfriend and a nurse who assisted her in caring for him during his fatal illness suffered a similar illness but recovered. Intensive epidemiologic investigations have failed to reveal the source of their infections.

Between August and November 1976, in two towns

COLOR PLATE

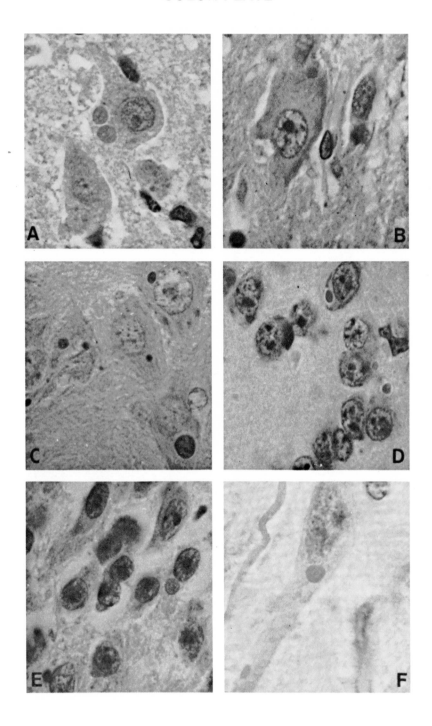

Negri bodies are shown here as the round eosinophilic cytoplasmic inclusions in brain cells from a variety of animals. **A,** Human; **B,** Cat; **C,** Dog; **D,** Mouse; **E,** Bovine; **F,** Fox. All sections are magnified approximately 1000–1400 ×. *(From Atanasiu P, Sisman J: In Baer GM (ed): The Natural History of Rabies, Vol 1. New York, Academic Press, 1975, p 374.)*

600 miles apart, one in southern Sudan and the other in northeastern Zaire, hundreds of persons became ill, and most died. From these patients a virus was isolated that was found to be antigenically distinct from but morphologically identical to Marburg virus. This agent was named "Ebola virus," for the river that runs near the location of the outbreak in Zaire. In Zaire, except for the index case, all 235 cases were linked by intimate contact or by syringes and needles used for medicinal purposes. Of these cases, 201 died, including all of those infected by syringe and needle transfer. Ebola hemorrhagic fever was again seen in Zaire in 1977–1978.

Thus, Marburg and Ebola viruses have struck at humans only occasionally but always with devastating effect. While the method of spread within a given outbreak has become evident quickly, the natural life cycle and the source of the infection for the index cases remain a mystery except in the original outbreak involving green monkeys from Uganda. But even in that instance, the epidemiologic determinants of the monkey infection are not at all clear. Investigations in Uganda failed to reveal serologic evidence in the monkeys at the time of the Marburg outbreak. Moreover, the nearly universal mortality in this species makes it unlikely that monkeys serve as reservoir hosts in a maintenance cycle. Thousands of green monkeys shipped from Uganda during the years before the Marburg incident were apparently not infected. As one prominent worker in this field has stated, "The green monkey is a red herring."

Exhaustive studies of the 1975 Rhodesian cases provide evidence suggestive of arthropod transmission of infection to the index case, but no proof could be obtained. Similarly, the sources of infection in the index cases of the two simultaneous 1976 outbreaks are unknown. Serologic surveys show that between 1 and 20 percent of persons, depending on the geographic area surveyed, may show antibodies, and it is thus likely that considerable human transmission occurs in clinical silence. Beyond that, however, routes of transmission, reservoir hosts, and all of the other facts that permit rational defense are totally lacking. Recently, Ebola antibodies were found in domesticated guinea pigs living in village huts in Zaire. The source of the infection remains a mystery.

Transmission routes within outbreaks, however, are much better understood. Close personal contact during acute illness clearly results in transmission. Handling of secretions, tissues, and body fluids from acutely ill patients is another source. In several instances, venereal transmission has occurred through infected semen long after the acute disease. In Zaire, improper use of syringes and needles spread the infection nosocomially in a uniformly fatal manner among hundreds of persons.

Clinically, Marburg/Ebola disease begins with fever, headache, nausea, vomiting, myalgia, and diarrhea. After a few days, sore throat, conjunctivitis, abdominal pain, and rash appear. The rash is fine and maculopapular and may appear initially in relation to hair follicles. It begins either on the trunk or limbs, spreading thereafter to involve both. After 5 to 7 days, severe bleeding begins, involving chiefly the gastrointestinal tract. Shock and death often follow soon thereafter.

Laboratory findings include leukopenia, thrombocytopenia, and often the findings of disseminated intravascular coagulation. Proteinuria and markedly elevated serum amylase and transaminases are also found.

The major pathologic finding is focal hepatocellular necrosis. Eosinophilic cytoplasmic inclusions are seen, and on electron microscopy numerous virus particles are seen in the liver.

Diagnosis rests on isolation of virus in Vero cell cultures or guinea pigs from blood or oropharyngeal secretions from acute-phase specimens. After approximately 2 weeks, serum antibodies may be best detected by fluorescent antibody and various other techniques.

Case fatality approaches 100 percent under some circumstances, but this has been held to 20 percent when first class intensive care facilities are available. The roles of heparin to combat intravascular coagulation, specific antibody-containing plasma, and interferon in the treatment of the acute disease are not fully defined, but all have been used.

Index cases cannot be prevented for lack of knowledge. Secondary cases resulting from person-to-person contact or the handling of patient's secretions or tissues can be controlled by rigorous isolation and protective practices. Laboratory work on Marburg viruses must be confined to maximum containment laboratories, such as the facility at the Centers for Disease Control in Atlanta.

Vesicular Stomatitis Virus

The vesicular stomatitis virus (VSV) group is generally considered to comprise seven serologically related but distinct viruses; they are the Indiana, New Jersey, Cocal, Alagoas, Piry, Chandipura, and Isfahan strains of VSV. The first of these was isolated in 1925 and the latest in 1975. All but two have been found solely in the western hemisphere, while the remainder have been found in Africa, Asia, and Iran.

As a group, the vesicular stomatitis viruses have been found in nature in a variety of animals ranging from insects to humans. Evidence for arthropod-borne transmission includes the fact that VSV has often been isolated from naturally infected insects, including sandflies, mites, blackflies, and mosquitoes. With some serotypes, insect transmission has been achieved experimentally. For other serotypes, the above information is lacking. In addition, the seasonal pattern of transmission of some of these agents strongly suggests arthropod vectors.

The disease, vesicular stomatitis, is most frequently seen in cattle, pigs, and horses, where it occurs in outbreaks of illness associated with vesicular lesions in the mouth and elsewhere on the body. Lesions may also be seen on the feet, thus mimicking foot-and-mouth disease, a much more severe condition. Although three quarters of a herd may become ill, fatality is less than 5 percent. Transmission among domestic animals seems to require contact with very sharply circumscribed pasture areas, epidemiologic conditions that are difficult to explain on the basis of arthropod transmission.

The disease in humans is an acute, self-limited, benign, febrile illness associated with general malaise, myalgia, nausea and vomiting, and headache. No deaths have been reported. In some instances, vesicles may be seen on the gums and buccal and pharyngeal mucosa. Herpes-like lesions on the lips may also be seen. Illness is generally of less than a week's duration. Many laboratory infections have occurred, and the illnesses resulting from these infections seem to be identical to those occurring as a result of natural transmission.

In the United States, small outbreaks have been detected wherein direct transmission from animal to human is the most probable route of infection. Several serologic surveys in a variety of populations throughout the world suggest that infections with these viruses may be quite widespread. No serious public health problem, however, in the form of human disease at least, has seemed to occur as a result of these widespread infections. Because of the nonspecific nature of the illness, it is possible that many cases of human disease go undetected.

No characteristic clinical laboratory or histopathologic features have been recognized. Diagnosis depends on the clinical picture and the knowledge of close contact with diseased animals. Specific, complement-fixing, and neutralizing antibodies are detectable approximately 2 weeks after the onset of illness.

Except for avoidance of contact with diseased domestic animals, no preventive measures are known. No specific treatment of the sick individual is available.

FURTHER READING

RABIES

Centers for Disease Control, U.S. Public Health Service: Morbidity and Mortality Weekly Report:
  Rabies—United States, 1980. 30:147, 1981
  Rabies prevention (ACIP). 29:265, 1980
  Supplementary statement on rabies vaccine and serologic testing (ACIP). 30:535, 1981
  Supplementary statement on pre-exposure rabies prophylaxis by the intradermal route. 31:279, 1982
  Rabies in racoons—Virginia. 30:354, 1981
  Compendium of animal rabies vaccines, 1983. 31:685, 1982
Familusi JB, Osunkoya BO, et al.: A fatal human infection with Mokola virus. Am J Trop Med Hyg 21:959, 1972
Fekadu M, Shaddock JH, Baer GM: Excretion of rabies virus in the saliva of dogs. J Infect Dis 145:715, 1982
Hattwick MAW, et al.: Recovery from rabies: A case report. Ann Intern Med 76:931, 1972
Meredith CD: An unusual case of human rabies thought to be of Chiropteran origin. South Afr Med J 45:767, 1971 (Note: This virus is known as Duvenhage.)

Mertz GJ, Nelson KE, Vithayasai V, et al.: Antibody responses to human diploid cell vaccine for rabies with and without human rabies immune globulin. J Infect Dis 145:720, 1982
Plotkin SA: Rabies vaccine prepared in human cell cultures: Progress and perspective. Rev Infect Dis 2:433, 1980
Shope RE: Rabies. In Evans AS (ed): Viral Infections of Humans, Epidemiology and Control. New York, Plenum, 1976
Shope RE, Baer G, Allen WP: Summary of workshop on the immunopathology of rabies. J Infect Dis 140:431, 1979
Sikes RK: Rabies. In Hubbert WT, et al. (eds): Diseases Transmitted from Animals to Man, 6th ed. Springfield, Ill., Thomas, 1975
Steele JH: The epidemiology and control of rabies. Scand J Infect Dis 5:299, 1973

MARBURG AND EBOLA VIRUS DISEASE

Centers for Disease Control, U.S. Public Health Service: Marburg Virus Disease, Kenya. Morbidity and Mortality Weekly Report 29:145, 1980
Conrad LT, et al.: Epidemiologic investigation of Marburg virus disease, Southern Africa, 1975. Am J Trop Med Hyg 27:1210, 1978
Editorial: After Marburg, Ebola. . . . Lancet 1:581, 1977
Editorial: Ebola virus infection. Br Med J 1:539, 1977
Esmond RTD, et al.: A case of Ebola virus infection. Br Med J 1:541, 1977
Gear JSS, et al.: Outbreak of Marburg virus disease in Johannesburg. Br Med J 4:489, 1975
Heymann DL, Weisfeld JS, Webb PA, et al.: Ebola hemorrhagic fever: Tandala Zaire, 1977–1978. J Infect Dis 142:372, 1980
Johnson KM, et al.: Isolation and partial characterization of a new virus causing acute hemorrhagic fever in Zaire. Lancet 1:569, 1977
Johnson KM, Scribner CL, McCormick JB: Etiology of Ebola virus: A first clue. J Infect Dis 143:749, 1981
Kissling RE: Marburg virus. In Hubbert WT, et al. (eds): Diseases Transmitted from Animals to Man, 6th ed. Springfield Ill., Thomas, 1975
Pattyn SR (ed): Ebola Virus Hemorrhagic Fever. Amsterdam, Elsevier, 1978
Regnery RL, Johnson KM, Kiley MP: Marburg and Ebola viruses: Possible members of a new group of negative strand viruses. In Bishop DHL, Compans RW (eds): The Replication of Negative Strand Viruses. Amsterdam, Elsevier/North Holland, 1981, p 971

VESICULAR STOMATITIS VIRUS DISEASES

Bhatt PN, Rodrigues FM: Chandipura: A new arbovirus isolated in India from patients with febrile illness. Indian J Med Res 5:1295, 1967
Fields BN, Hawkins K: Human infection with the virus of vesicular stomatitis during an epizootic. N Engl J Med 277:989, 1967
Tesh RB, Johnson KM: Vesicular stomatitis. In Hubbert WT, et al. (eds): Diseases Transmitted from Animals to Man, 6th ed. Springfield, Ill., Thomas, 1975, chap 73
Tesh RB, Saidi S, Javadian E, et al.: Isfahan virus, a new vesiculovirus infecting humans, gerbils and sandflies in Iran. Am J Trop Med Hyg 26:299, 1977
Watson WA: Vesicular diseases: Recent advances and concepts of control. Can Vet J 22:311, 1981

CHAPTER 78

Arenaviruses

Human Diseases
Lymphocytic Choriomeningitis
Hemorrhagic Fevers

Junin Virus in Argentina
Machupo Virus in Bolivia
Lassa Fever in West Africa

... the most spectacular outbreaks in Africa (in the 1970s) were of Lassa fever and African hemorrhagic fever.
—*RW Chamberlain: Am J Trop Med and Hyg 31:434, 1982*

Arenaviruses have in common a similar morphology, related antigens, and the capacity to cause chronic infections in rodents. Three of them—Junin, Machupo, and Lassa—can cause serious, highly fatal, hemorrhagic fever in humans; one—lymphocytic choriomeningitis (LCM)—causes a sporadic, relatively benign meningoencephalitis. The arenaviruses are listed in Tables 56-4, 76-3, and 78-1, together with some of their clinical and epidemiologic features. LCM virus is found worldwide. Other arenaviruses are found in the Americas and in Africa, as shown in Figures 78-1 and 78-2. Perhaps the most impressive feature of arenaviruses is their capacity to surprise us. Lassa fever outbreaks warned us in the 1970s, as Machupo and Junin did in the 1960s, of the dire consequences of our ignorance and of our constant meddling with the balance of nature.

Arenaviruses establish chronic infections in a variety of laboratory and wild rodent species. When the young animal is inoculated perinatally, a chronic, lifetime infection can result, with continuous viremia, as well as high titer viruria. It is this sort of rodent infection that is of great importance in the perpetuation of the virus in nature, as well as the main source of human infection. Inoculation of normal adult rodents, on the other hand, may cause severe, acute, often fatal disease. By varying such factors as dose, route of inoculation, strain, and age of animal, intermediate results may be produced with, for example, chronic immune complex disease being the result of the infection.

The virologic and immunologic events characterizing these three types of host response (acute disease, asymptomatic carrier state, and chronic immune complex dis-

ease) have been extensively studied for LCM virus with development of knowledge that, in many particulars, is applicable to an understanding of the pathogenesis of viral disease in general. The most striking result of these studies is the fact that both the acute and chronic forms of disease are the result of immunopathologic processes.

The events depicted in Figure 78-3 occur in the acute disease situation. It has been demonstrated repeatedly by a variety of techniques, that compromise of the infected animal's immune system results in less severe disease. Subsequent studies showed that the acute disease is mediated by specifically oriented cytotoxic T cells. Furthermore, the capacity of such T cells to damage infected cells requires, in addition to viral antigen at the cell surface, an appropriate histocompatibility complex. This additional requirement is known as the "H-2 restriction" phenomenon and was described first with LCM virus infection. The roles of interferon, natural killer cells, the increase in number and changes in morphology of macrophages, the B cell activation and immunoglobulin production in the outcome of acute disease are not clearly understood.

In the chronic disease situation, humoral antibody is produced but is not detectable unless one searches for it in immune complexes that are deposited in certain tissues, such as glomeruli. The chronic disease is a progressive glomerulonephritis and arteritis resulting from the deposited immunocomplexes. Antibody production also occurs in asymptomatic carrier animals. Due to complex formation, this antibody is not detectable by older, conventional methods, and it was thought formerly that such animals were immunologically tolerant to their LCM virus infection. The situation in chronic disease and in the carrier state with regard to T cell activation and function is not clear.

Transmission by arthropods has been demonstrated

TABLE 78-1. ARENAVIRUSES

Virus	Date and Source of Specimen Yielding Initial Isolate			Disease in Humans
Lymphocytic choriomeningitis	1934	United States	Humans	Meningoencephalitis
Tacaribe	1956	Trinidad	Bats	None
Junin	1958	Argentina	Humans	Hemorrhagic fever
Machupo	1963	Bolivia	Humans	Hemorrhagic fever
Amapari	1964	Brazil	Rodent	None
Latino	1965	Bolivia	Rodent	None
Parana	1965	Paraguay	Rodent	None
Pichinde	1965	Columbia	Rodent	None
Tamiami	1965	United States	Rodent	None
Lassa	1969	Nigeria	Humans	Hemorrhagic fever
Mozambique	1977	Mozambique	Rodent	None

experimentally for one of these agents (Junin), but such transmission does not seem to be of importance in nature.

Human Diseases

When a person is infected, one of a spectrum of consequences ensues. On the one hand, no clinical disease appears, with infection detectable only by a rise in titer of serum antibody. On the other hand, virulent disease with a 25 percent fatality rate may result. In many instances during the acute stages of infection, virus may be isolated readily from a variety of body fluids or tissues by the use of laboratory rodents or cell cultures.

Figure 78-1. Geographic locations of arenavirus isolates in the New World. All viruses with the exception of LCM virus are serologically defined as being members of the Tacaribe complex. *(From Howard and Simpson: J Gen Virol 51:1, 1980.)*

Lymphocytic Choriomeningitis

LCM virus was isolated initially in the 1930s from patients with acute aseptic meningitis. From the very beginning, it has been recognized that LCM virus causes but a fraction of a percent of the overall number of these cases.

In nature the virus is found in house mice, where, as a result of natural perinatal inoculation, chronic infection with no clinical manifestations is established in a high proportion of animals. Contact with mice or their excreta brings the virus to humans; a dead end for the virus, as person-to-person or person-to-animal spread does not seem to occur. Captive Syrian hamsters have been found to be a source of human infection. There have been sizable outbreaks in laboratory personnel who have been in contact with infected hamsters that were being used in cancer research. In one instance, the infected hamsters were widely dispersed to many cancer laboratories from a single provider's colony.

Disease begins with fever, headache, myalgia, and malaise. Conjunctival infection and cough may be present. Within a few days the disease may subside, with rapid return of well-being. In some, however, it may worsen, with the addition of confusion and somnolence. Stiff neck, vomiting, and marked increase in the severity of the headache may become prominent. Gross disorientation may be a feature. Physical examination reveals fever and meningeal and encephalitic signs. Children tend to have less severe disease than adults.

The total and differential white blood counts are often normal, as is the urine. Cerebrospinal fluid, especially from those persons showing a meningeal or encephalitic clinical picture, is generally under abnormally high pressure, with an elevated number of lymphocytes and increased protein concentration. Low spinal fluid sugar is often found.

The disease is rarely, if ever, fatal, and it generally subsides after a few days so that the acute febrile stage of the illness is usually over within 7 to 14 days. In older and more severely affected patients, convalescence may be prolonged.

Clinically, there is often little to suggest LCM viral eti-

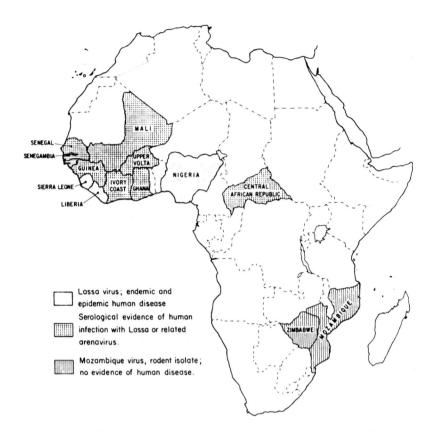

Figure 78-2 Geographic distribution of Lassa and Mozambique viruses in Africa. *(Courtesy of Dr. P. Monath.)*

Legend:

☐ Lassa virus; endemic and epidemic human disease

▦ Serological evidence of human infection with Lassa or related arenavirus.

▥ Mozambique virus, rodent isolate; no evidence of human disease.

ology as being any more probable than many of the far more usual causes of viral meningitis, such as enteroviruses. Diagnosis is made by isolation of virus and/or demonstration of high titer, complement-fixing antibodies

in serum obtained during the acute disease or by a rise in titer between samples obtained early in the acute stages and again in convalescence. Adult white mice from colonies known to be free of LCM virus are the medium of choice for isolation. Following intracerebral inoculation, the mice become ill at about 10 days, at which time antigen can be detected in brain tissue by immunofluorescent techniques.

Human deaths are very rare. Consequently, the pathology of LCM infection must be inferred from that observed in experimental animals, such as mice and monkeys, where lymphocytic infiltrates of the meninges and choroid plexus are prominent features from which the name of the virus is derived.

Treatment is supportive; there are no specific measures available.

Hemorrhagic Fevers

Argentinian and Bolivian hemorrhagic fever and Lassa fever are similar diseases.

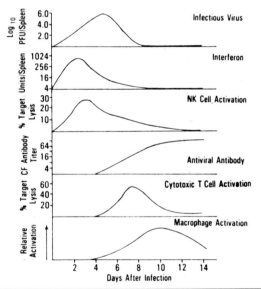

Figure 78-3. A composite diagram of host response to acute lymphocytic choriomeningitis (LCM) virus infection following intraperitoneal inoculation. CF, complement-fixing; NK, natural killer. *(From Buchmeier et al.: Adv Immunol 30:305, 1980.)*

Junin Virus in Argentina

Argentinian hemorrhagic fever occurs in certain agricultural areas of northern Argentina. The disease was recognized in the 1940s, and in the 1950s definitive epidemiologic and clinical studies culminated in the identification of Junin virus as the causative agent. The disease occurs predominantly in male field workers, who in the course of their labors contact the virus, which contaminates the environment from the excreta of chronically infected wild rodents of the genus *Calomys*. During the period from February to May when agriculture, especially of maize, is intense, thousands of cases a year may occur, with a 10 to 20 percent case fatality. Concurrent transmission of LCM virus and arboviruses may confound the clinical and epidemiologic features of the annual outbreaks.

Machupo Virus in Bolivia

Bolivian hemorrhagic fever was recognized in 1959 as a disease of agricultural workers in specific areas of Bolivia. In 1962 and 1963, the disease moved into the town of San Joaquin. Intensive investigation of this outbreak revealed Machupo virus to be the etiologic agent, and the source of human infection to be the contamination of the household environment with the excreta of chronically infected rodents of the species *Calomys*. These mice, normally found in the fields, invaded the houses of the town and thereby brought the virus to family groups in the general population. The reasons for this ecologic shift are not well understood but serve to remind us of the complexity of the balances of nature. The efficacy of domestic rodent control was established in the San Joaquin outbreak. In 1971, a small, highly fatal outbreak in an area where there are no *Calomys* was due to person-to-person spread of the infection.

Lassa Fever in West Africa

Lassa fever was first encountered in 1969 in a small outbreak of serious disease among the hospital personnel from missions in the towns of Lassa and Jos in northeastern and central Nigeria (Fig. 78-2). Subsequent to the startling events surrounding this initial outbreak, a more balanced picture of this zoonosis has emerged. Additional nosocomial outbreaks occurred in the subsequent years, and considerable concern spread to Europe and America as the result of the importation of nine cases from West Africa. Fortunately, very little spread, if any, resulted from these importations, and it can be definitively stated now that careful isolation procedures are adequate to protect hospital staff and visitors. Handling of specimens and tissues in the laboratory, however, must be done with extreme care, as fatal laboratory infections have occurred. Presently, maximum security laboratory facilities, such as the one available at the Centers for Disease Control in Atlanta, are the only laboratories where work with this virus should be carried on.

The virus, as is so far known, seems to be present in several countries of West Africa, including Liberia, Nigeria, and Sierra Leone. Studies in West Africa, guided by knowledge gained from studies of LCM, Junin, and Machupo viruses, resulted in the recognition that the rodent *Mastomys natalensis* (Fig. 78-4) carries the agent in na-

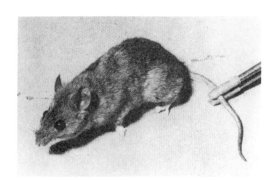

Figure 78-4. *Praomys* (*Mastomys*) *natalensis*, chronically infected reservoir host of Lassa fever virus. *(From Isaacson: Bull WHO 52: 631, 1975.)*

ture with the lifelong infection that characterizes arenaviruses.

Humans provide the only overt indication that virus is present (indicator host) and are of little significance in the perpetuation of this agent in nature. Humans probably acquire infection from contamination by these rodents of the environment with infectious urine and excreta. Human practices that entail field burning and capture of rodents fleeing therefrom contribute to the intimate human–rodent contact necessary for transmission. For all three hemorrhagic fever arenaviruses, human-to-human transmission has been documented but is relatively unusual.

A Lassa-like virus has been isolated from the other side of Africa. This agent, named Mozambique virus, was isolated from pools of rodent (*Mastomys*) organs. So far, it is not known to cause human disease.

The clinical consequences of infection with the three viruses range from no disease and simple seroconversion to a fulminant and highly fatal condition. Following an incubation period of 1 to 2 weeks, disease begins in a nonspecific manner with fever, headache, sore throat, and myalgia. Toward the end of the first week, flushing, edema, and petechial rash of the face and neck appear in association with adenopathy and worsening of the above symptoms. Mouth lesions, exudative pharyngitis, and pulmonary symptoms may predominate in Lassa fever. During the second week, serious complications relating to loss of intravascular mass through generalized capillary leakage cause a shock syndrome characterized by peripheral vasoconstriction, generalized edema, renal failure, and metabolic acidosis. Myocardial involvement may be evident. There is moderate hemorrhage in the form of petechiae and bleeding from the gastrointestinal and genitourinary tracts, which is not sufficient to produce the shock syndrome. After a few days of the shock syndrome, the patient dies or recovery begins. During convalescence, there may be loss of hair and neurosequelae, including nystagmus, tremulousness, and episodic dizziness. In spite of considerable damage to the liver seen histopathologically, no jaundice or other signs of liver failure occur. The spleen does not enlarge. Major complications in severe

Lassa fever occur in no more than 5 percent of cases and include deafness, abortion, pleural effusion, pericarditis, orchitis, and iridocyclitis.

Laboratory findings include leukopenia, decreased platelets, increased prothrombin time, and evidence of disseminated intravascular coagulation. Protein, cells, and casts are found in the urine. The cerebrospinal fluid is normal. A chest x-ray may reveal a patchy pneumonia, and there may be electrocardiographic changes consistent with diffuse myocardial damage. During the shock syndrome, hemoconcentration may be evident.

Pathologic findings include little inflammatory response other than modest monocytic infiltrates. There is widespread edema indicative of the capillary damage that permits leakage of plasma from the intravascular space. Cellular damage may be found in many organs, including heart, muscles, lungs, liver, adrenals, kidneys, spleen, lymph nodes, and brain.

Diagnosis depends on isolation of virus from either body fluids or from tissue specimens. Mice, hamsters, or cell cultures are inoculated, depending on the virus suspected. The presence of virus in the laboratory system is detectable by a variety of means, including disease, cytopathology, and immunofluorescent techniques. Antibody response in man occurs within a few weeks of onset and is generally detected in testing for a rising titer of group-specific complement-fixing antibodies or type-specific neutralizing antibodies. Immunofluorescent antibodies appear as early as 7 days after the onset of symptoms in Lassa infections. A satisfactory neutralization test for Lassa fever has not been developed.

The general theme of clinical management of arenavirus disease calls for efforts to support basic physiologic processes long enough for the body to cure itself. There are no specific therapeutic agents except humoral antibodies derived from persons who have survived natural infection. Such immunotherapy has seemed to improve dramatically a few patients who appeared to be in imminent danger of dying. No vaccine is currently available, although experimental vaccines have been prepared and tested.

Prevention of person-to-person transmission can be achieved in any hospital by standard strict isolation techniques. Outbreaks can be controlled with measures aimed at breaking the intimate contact between the excreta of certain rodents and humans. A protocol for the management and containment of imported cases of Lassa fever has been used and is available through the Center for Disease Control of the U.S. Public Health Service.

FURTHER READING

Arenaviruses in perspective. Br Med J 1:529, 1978

Buchmeier MJ, Welch RM, Dutko FJ, et al.: The virology and immunobiology of lymphocytic choriomeningitis virus infection. Adv Immunol 30:275, 1980

Buckley SM, Casals J: Pathobiology of Lassa fever. Int Rev Exp Pathol 18:97, 1978

Cooper CB, Gransen WR, et al.: A case of Lassa fever: Experience at St. Thomas's Hospital. Br Med J 285:1003, 1982

Dowdle WR: Exotic viral diseases. Yale J Biol Med 53:109, 1980

Emond RTD, Bannister B, et al.: A case of Lassa fever: Clinical and virological findings. Br Med J 285:1001, 1982

Fuller JG: Fever! The Hunt for a New Killer Virus. New York, Readers Digest Press, 1974

Howard CR, Simpson DIH: The biology of the arenaviruses. J Gen Virol 51:1, 1980

International Symposium on Arenaviral Infections of Public Health Importance, July 14–16, 1975, Atlanta, Georgia. Bull WHO 52:381, 1975

Jahrling PB, Hesse RA, Eddy GA, et al.: Lassa virus infection of rhesus monkeys: Pathogenesis and treatment with ribavirin. J Infect Dis 141:580, 1980

Johnson KM, Monath TP, Tesh RB, et al.: Arthropod-borne viral fevers, viral encepalitides and viral hemorrhagic fevers (arboviruses and arenaviruses). In Wyngaarden JB, Smith LH (eds): Textbook of Medicine, 16th ed. Philadelphia, Saunders, 1982.

Kiley MP, Tomori O, Regnery RL, et al.: Characterization of the arenaviruses Lassa and Mozambique. In Bishop DHL, Compans RW (eds): The Replication of Negative Strand Viruses. Amsterdam, Elsevier, North Holland, 1981

CHAPTER 79

The Hepatitis Viruses

Although recognized since the fifth century BC, hepatitis is only recently becoming well understood. By the 1940s and into the 1950s, viral agents were accepted as etiologic agents of hepatitis. The most notable of these were the MS-1 and MS-2 strains of hepatitis virus reported by Krugman and later identified as hepatitis A and hepatitis B, respectively. More detailed understanding of these viruses began with the discovery of the Australia antigen in 1964 by Blumberg. This chapter summarizes current concepts in viral hepatitis, and reflects largely the data acquired by many investigators since 1964.

Characterization of the Hepatitis Viruses

By definition, viral hepatitis is an infection of the liver caused by one of four or more distinctly different viruses: hepatitis A virus (HAV), hepatitis B virus (HBV), non-A, non-B hepatitis virus or viruses (NANB), and δ-associated agent. Table 79-1 summarizes some of their known characteristics.

Noted in Table 79-2 is the terminology used for the hepatitis viral antigens and antibodies, which serve as the basis for identification. Because the sequence of development and disappearance of these viral markers is generally predictable, it serves as the basis for diagnosing the specific diseases that each virus causes. Although there is a simple interpretation for each marker, as indicated in Table 79-2, it should also be noted that observation of the course of events is as useful as the presence or absence of any single marker.

Both commercial and research methods are available to test for these markers, which each have various advan-

tages and disadvantages. The sensitivity and the specificity of each method are noted in Tables 79-3 and 79-4. In these tests, each antigen is specific to a particular virus; that is, antigen has been detected that is common to more than one class of virus. Within each individual group, all strains are alike in terms of virulence, disease produced, complications, and cross-protection, except for NANB, which in all probability is composed of distinctly different virus types.

Pathogenesis and Pathology

Direct viral-induced cytopathogenicity is one plausible explanation for some of the pathologic features of the disease, but immune mechanisms (both humoral and cellular) may better account for some of the extrahepatic manifestations. In acute hepatitis, pathologic findings typically consist of an accumulation of inflammatory cells and parenchymal necrosis, which are variously distributed in the portal, periportal, and lobular areas of the liver. These findings usually disappear completely after clinical recovery. Among the four groups there are currently no accepted differences in the pathology of acute hepatitis. Chronic hepatitis, a disease peculiar to HBV and NANB, is described in the following section.

Clinical Disease

As well as being pathologically indistinguishable, the acute infections are also symptomatically indistinguishable. Patients are commonly anicteric, a finding present in from 30 to 70 percent of the reported cases. NANB hepatitis may

TABLE 79-1. CHARACTERISTICS OF THE HEPATITIS VIRUSES

Virus	Virus Family: Genus	Size	Nucleic Acid	Cloned	Grown in Tissue Culture	Identified by Electron Microscopy	Specific Serologic Tests	Infectious for Animals	Special Characteristics
HAV	Picornaviridae: *Enterovirus 72*	27 nm	Single-stranded RNA	No	Yes	Yes	Yes	Yes	—
HBV	Not yet classified	42-nm Dane particle 22-nm surface antigen	Primarily circular, double-stranded DNA	Yes	Yes*	Yes	Yes	Yes	—
NANB	Unknown	Unknown	Unknown	No	No	No	No	Yes	Probably two or more different viruses
δ-associated agent	Unknown	35–37 nm	RNA-associated	No	No	No	Yes	Yes	Requires HBV for replication

*Reported but unconfirmed.
HAV, hepatitis A virus; HBV, hepatitis B virus; NANB, non-A, non-B hepatitus virus or viruses.

be the one most likely to fall into this category. The typical illness, when it occurs, consists of lassitude, anorexia, weakness, nausea, vomiting, headache, chilliness, abdominal discomfort, fever, dark urine, and a variety of less frequent symptoms and signs. Laboratory findings reflect liver cell necrosis, such as aminotransferase abnormalities. Bilirubin and, less frequently, alkaline phosphatase may also be abnormal. Each type of abnormality may persist beyond the typical 2 to 4 weeks of illness, may appear to relapse after apparent resolution, or may resolve completely save for persistent hyperbilirubinemia, but these courses are unusual. These unusual courses are benign, like most cases of acute hepatitis, in that they resolve completely and spontaneously. The diagnostic laboratory tests

that can identify the specific disease are noted in Table 79-2.

Despite the general similarity of symptoms and pathology, each class of virus has unique features. Figure 79-1 shows schematically the usual course of acute hepatitis for each virus, and Table 79-5 outlines the sequelae associated with each. Distinctive features are as follows:

Hepatitis A

Hepatitis A is characterized by an absence of many of the features found in other forms of hepatitis. In particular, there are no extrahepatic manifestations of the acute infection, no chronic hepatitis, no long-term carrier state, and

TABLE 79-2. NOMENCLATURE OF THE HEPATITIS VIRUS ANTIGENS AND ANTIBODIES AND THEIR GENERAL INTERPRETATION

Virus	Antigens		Antibodies	
	Name	Interpretation	Name	Interpretation
HAV	HA Ag* (Major antigen of HAV)	Acute infection	Anti-HA IgG† IgM†	Immune to HAV / Recurrent or current acute infection
HBV	HBsAg† (Surface antigen of HBV)	Prior exposure to HBV	Anti-HBs†	Immune to HBV
	Subtypes adr, adw, ayw, ayr	Distinctive strains of HBV		
	HBcAg* (Core antigen of HBV)	Infectivity—acute or chronic	Anti-HBc†	Early or late convalescence or chronic hepatitis
	HBeAg† (A core-related antigen)	Infectivity—acute or chronic	Anti-HBe†	Late convalescence
NANB	—‡	—	—‡	—
δ-associated agent	δ antigen*	Acute δ-associated hepatitis	Anti-δ*	Immune to δ-associated hepatitis

*Research tools only.
†Tests currently clinically available.
‡Reportedly identified but not confirmed.
HAV, hepatitis A virus; HBV, hepatitis B virus; NANB, non-A, non-B hepatitis virus or viruses.

TABLE 79-3. METHODS FOR DETECTING HEPATITIS B s ANTIGEN (HBsAg)

Methods	Relative Sensitivity	Ease of Performance	Time Required for Completion (hours)
Immunodiffusion (ID, AGD)	1	Simple	24–72
Counterimmunoelectrophoresis (CIE)	2–10	Simple	2
Complement fixation (CF)		Moderate	18–24
Reversed passive latex agglutination		Simple	0.1–0.2
Passive hemagglutination inhibition (HAI)		Moderate	2
Immune adherence hemagglutination		Moderate	2
Radioimmunoassay (RIA)	100		
Solid phase		Moderate	4–24
Double antibody (DA-RIA, RIP)		Complex	24–72
Reversed passive hemagglutination (RPHA)		Simple	1–4
Enzyme-linked immunosorbent assay (ELISA)		Moderate	2–24

Adapted from Howard, Burrell: Structure and nature of hepatitis B antigen. Prog Med Virol 22:36, 1976.
AGD, agar gel diffusion; RIP, radioimmunoprecipitation.

no recognized association with either cirrhosis or primary hepatocellular carcinoma. A fulminant form of acute hepatitis occurs in 1 to 4 percent of patients, as it does in hepatitis B.

Hepatitis B

Hepatitis B differs from hepatitis A in a number of significant respects. First, there is a wide range of extrahepatic manifestations including arthralgias, particularly of the small joints, and a variety of immunologically mediated diseases such as arteritis, nephritis, and dermatitis.

Second, there are chronic forms of hepatitis B, defined as those illnesses lasting longer than 6 months. These fall into two general classifications: chronic persistent hepatitis and chronic aggressive hepatitis. In terms of ultimate outcome, chronic persistent hepatitis is a benign illness. This form of hepatitis B is ordinarily not symptomatic, but patients who are tested show abnormal findings on liver function tests and the persistence of many HBV markers, including hepatitis B s antigen (HBsAg), hepatitis B e antigen (HBeAg), and antihepatitis B c (HBc) antibody. Liver biopsy findings further reveal the transient abnormalities associated with acute hepatitis and, most importantly, the absence of the more advanced pathology found in the serious and second form of chronic hepatitis B, namely chronic aggressive hepatitis.

TABLE 79-4. METHODS FOR DETECTING ANTI-HBs

Method	Relative Sensitivity
Immunodiffusion	1
Counterimmunoelectrophoresis	1–4
Complement fixation	2–10
Passive hemagglutination	1000–10,000
Solid-phase radioimmunoassay	1000–10,000
Radioimmunoprecipitation	10,000–100,000

Adapted from Howard, Burrell: Structure and nature of hepatitis B antigen. Prog Med Virol 22:36, 1976.

In chronic aggressive hepatitis, the patient is clinically indistinguishable from the one with chronic persistent hepatitis. However, the anti-HBc antibody may be present in very high titer, and the liver biopsy shows widespread lymphocyte infiltration of portal tracts and lobules, piecemeal necrosis, and bridging necrosis. This pathologic picture is a recognized precursor to cirrhosis and possibly to primary hepatocellular carcinoma. As yet, it is not possible to predict the frequency with which this sequence of events occurs, but the cause-and-effect association seems certain. How long it takes to develop these very serious consequences is also not known. Once it has developed, however, most cases of chronic aggressive hepatitis seem to persist indefinitively. This is most unlike the expected course with chronic persistent hepatitis, which resolves spontaneously and completely, without progression to the more serious sequelae. As desirable as it might be to be able to predict which cases of acute hepatitis B will become chronic, no reliable markers have yet been identified to allow this prognostication. Once chronic hepatitis develops, HBeAg should be measured.

HBeAg was described by Magnius and Espmark in 1972. Distinct from either the 22-nm free HBsAg and the 42-nm Dane particle, this antigen is a soluble protein that correlates with HBcAg and DNA polymerase, is a marker of infectivity, and suggests continued activity of chronic hepatitis. In addition, HBeAg–IgG immune complexes are thought to be pathogenetic in the induction of membranous glomerulonephritis. On the other hand, antibody to HBeAg correlates with the absence of HBcAg and DNA polymerase; much less infectivity; a better prognosis for chronic hepatitis; and, possibly, omission of nephritis. Continued liver disease, however, may be observed independent of infectious virus, and immunopathologic mechanisms may continue to play a role in the pathogenesis.

The third way that hepatitis B differs from hepatitis A is that the patient with hepatitis B may continue for a long time to be a carrier of either the whole infectious virus or the HBsAg. About 0.1 to 1 percent of the adult population

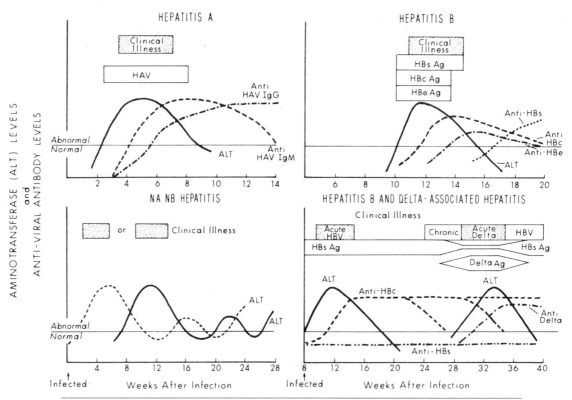

Figure 79-1. The panels illustrate the sequence of development of clinical illness, liver function abnormalities, and hepatitis antigen and antibody for each of the four currently recognized forms of viral hepatitis.

are asymptomatic carriers, and approximately 5 percent of the recognized cases of acute HBV infection become carriers of HBV. Some patients spontaneously cease to be carriers, but most persist beyond 6 months after the acute infection. Patients more likely to be long-term carriers are: (1) nonwhites; (2) males; (3) infants; and (4) those who are immunosuppressed, such as dialysis patients or patients receiving immunosuppressive drugs. The carrier may or may not be infectious. There are currently no available tests that make this distinction, but among the several types of patients more prone to being infectious carriers are: (1) those who have had a recent infection; (2)

those with a high titer of anti-HBc antibody; (3) those with detectable HBeAg; and (4) those who are immunocompromised. In contrast, those with anti-HBs or anti-HBe antibodies are less likely to be infectious. From a practical point of view, all carriers should be considered potentially infectious.

The fourth difference is that cirrhosis and primary hepatocellular carcinoma are probable sequelae of hepatitis B because the HBV genome is integrated into the DNA of tumor cells. Furthermore, epidemiologic studies have confirmed a significant association of HBV infection and the incidence of hepatocellular carcinoma.

TABLE 79-5. CLINICAL CHARACTERISTICS OF VIRAL HEPATITIS

| Virus | Anicteric: Icteric | Incubation Period | Acute | Extra-hepatic Manifestations | Fulminant/ Mortality | Chronic Hepatitis | | | Associated with Cirrhosis |
						Persistent	Aggressive	Carrier	Primary Hepatocellular Carcinoma
HAV	1:2.1	25 days	+	−	+/1–4 percent	−	−	−	−/−
HBV	1:2.1	75 days	+	+	+/1–4 percent	2–4 percent	2–4 percent	5 percent	+/+
NANB	4–6:1	50 days post-tx sporadic	+	+	+/0.1–0.2 percent	15 percent	30 percent	+(? percent)	+/−
δ	Unknown	Unknown	+	Unknown	Unknown	Unknown	Unknown	Unknown	Unknown

HAV, hepatitis A virus; HBV, hepatitis B virus; NANB, non-A, non-B hepatitis virus or viruses; post-tx, post-transfusion; +, present; −, absent.

Non-A, Non-B Hepatitis

Non-A, non-B hepatitis is a diagnosis of exclusion. It can be made only after the exclusion of HAV with an IgM anti-HAV antibody; HBV with an HBsAg and an IgM anti-HBc antibody; and a variety of other infectious and noninfectious diseases, i.e., infectious mononucleosis, cytomegalovirus, syphilis, measles, varicella, rubella, mumps, leptospirosis, bacterial sepsis, cholecystitis, ischemic necrosis, and Wilson's disease. Several investigators have reported promising antigen/antibody systems for making a positive diagnosis of NANB, but none is completely satisfactory.

Evidence is substantial that more than one virus causes NANB hepatitis. For one, there is a wide range of incubation periods. A preliminary suggestion is that there are at least two broad categories—one of rather short incubation (2 to 4 weeks), and the other of rather long incubation (16 to 24 weeks). For another, there are second episodes of NANB. Other evidence is the suspected histopathologic differences and the definite cross-challenge experiments in chimpanzees. Aside from these, three general groups of NANB have been identified based on its occurrence—the epidemic form, those cases associated with parenteral exposures, and sporadic cases.

As with HBV, NANB seems to be transmitted primarily by parenteral routes and is the leading cause of posttransfusion hepatitis (PTH). In efforts to obviate this complication, it has been found that a significant reduction in the incidence of PTH can be achieved by using donor blood that has an aminotransferase activity of less than 45 IU/liter. Unfortunately, most donors have normal alanine transferase (ALT) levels, and the logistic and financial costs of routine donor testing make this a problematic solution to the unacceptable frequency of PTH.

Both a carrier state and the incidence of chronic hepatitis have been documented with NANB. The frequency of chronic hepatitis seems to be even greater with this virus than it is with HBV. Posttransfusion chronic hepatitis may occur in up to 54 percent of cases, whereas sporadic occurrence may account for only 20 percent. Although the frequency of chronic non-A, non-B hepatitis is high, the severity is usually more benign than that accompanying HBV. Also, complete resolution without sequelae is usual, especially among the sporadic. But parenterally acquired (e.g., that involving PTH, illicit drug use, or hemodialysis) chronic hepatitis may persist and progress. Symptoms and liver function tests are not necessarily infallible indicators of continuously active liver disease.

δ-Associated Agent

Rizzetto and his colleagues described the δ/anti-δ antigen–antibody system in 1977. They had detected δ antigen in the nuclei of liver cells of only HBsAg-positive patients with chronic liver disease. Antibody to δ was later found to be associated with, but distinct from, HBV (it was present in 19.1 percent of patients with chronic hepatitis; in 2.6 percent of chronic carriers of HBsAg; and in low titers

transiently in 4.8 percent of patients with acute HBV infection), and it was not detected in the sera of HBsAg-negative controls. Patients who had been multiply transfused and were HBsAg-positive had a higher incidence of anti-δ antibody than nontransfused HBsAg carriers did. These data suggested that the δ-associated agent was transmitted both parenterally and nonparenterally by superinfection or coinfection of HBsAg carriers. Animal studies confirmed that it is transmittable and that it requires HBV as some sort of helper while it inhibits the replication of HBV. It has been further characterized biochemically and biophysically as being a protein that is encapsulated with low molecular weight RNA in a particle 35 to 37 nm in diameter. Recent studies suggest that it may be a very important determinant in the development of the carrier state and chronic liver disease.

Epidemiology

Consideration of the trends in occurrence, risk factors, and modes of transmission is extremely important for understanding viral hepatitis. Between 1970 and 1980 the number of new reported cases of hepatitis of all types has decreased. This decrease is accounted for by a reduction in the rate (cases per 100,000 population) of hepatitis A from 27.87 in 1970, to 12.84 in 1980. Hepatitis B increased from 4.08 to 8.35 over the same interval, and unspecified hepatitis increased from 3.95 to 5.39 from 1974 to 1980.

Viral hepatitis is transmitted by a wide variety of routes. Table 79-6 summarizes these routes and their relative frequency. HAV is present in the feces of infected patients during the 2 weeks preceding and several weeks following the onset of clinical disease. It may be present in the blood and other body fluids for a short period of time, but fecal–oral transmission of the HAV virus is by far the most prevalent route. The finding of an unexpectedly high incidence of HAV among male homosexuals suggests the possibility of sexual transmission, but more likely, this is due to fecal–oral transmission during oral–anal contact.

For HBV, NANB, and δ-associated diseases, blood is the major route, although other body fluids may possibly transmit virus as well. This has been especially well documented for HBV. In contrast with HAV, feces are not infectious or, if they are, it is at a very low level unless contaminated by blood. Therefore, not surprisingly, the usual mode of transmission is by parenteral exposure to viral-contaminated blood or blood products. This mechanism is facilitated because HBV and NANB may persist in the serum in carriers. It is especially important to recognize that blood and blood products from different sources have a different risk of possessing infectious virus. Most whole blood and its derivatives, which are not prepared by the pooling of units, have a relatively smaller risk of inducing infection with HBV. This is largely due to the infrequency of HBV carriage in the normal adult population, the exclusion of potentially infected donors, and perhaps

TABLE 79-6. MAJOR ROUTES OF HEPATITIS VIRUS TRANSMISSION

Virus	Major Infectious Body Fluid	Route of Transmission*			
		Parenteral	Fecal/Oral	Sexual	Vertical
HAV	Feces	−	4+	+/−	−
HBV	Blood and other body fluids	4+	+/−	2+	3+
NANB	Blood	4+	+/−	+/−	+/−
δ	Blood	4+	−	−	−

*Estimates of the relative frequency (from − to 4+).
HAV, hepatitis A virus; HBV, hepatitis B virus, NANB, non-A, non-B hepatitis virus or viruses; +, present; −, absent.

most importantly, the routine testing of donor blood for HBsAg. Preparations generally considered to be safe are serum albumin, thrombin, profibrinolysin, fibrinolysin, immune serum globulin (ISG), and all hyperimmune globulins. At higher risk are blood and blood products derived from commercial (as opposed to volunteer) donors, and those derived from pools of plasma and clotting factors I, II, VII, IX, and X. Washed and frozen human blood cells are not reliably virus free, and should also be considered a risk. Furthermore, as no tests are available to exclude NANB from blood or blood products, none of the above measures are known to eliminate this virus. As it might be expected, posttransfusion hepatitis is most often due to NANB. Hepatitis caused by either HBV or NANB may be

acquired by other parenteral routes, such as illicit drug use or accidental needle-stick, or accidental injury such as might occur during surgery.

Nonparenteral routes of transmission have also been reported for HBV, NANB, and δ-associated disease. There is strong evidence that all these viruses, but particularly HBV, may be transmitted by close personal contact, including sexual contact.

Prevention

The mainstays of viral hepatitis prevention are the correct diagnosis and reporting of new cases, attention to standard principles of cleanliness and hygiene, and specific

TABLE 79-7. RECOMMENDATIONS FOR USE OF IMMUNE GLOBULINS

Virus Exposure	Time of Evaluation	Nature of Exposure	Recommended Immune Globulin	
			Immune Serum Globulin*	Hyperimmune Hepatitis B Immune Globulin†
HAV	Preexposure	Travel to endemic area	+	−
	Postexposure	Household	+	−
		School or work	−	−
		Institution	+	−
		Primate handler	+	−
		Medical and paramedical personnel	+	−
		Common source epidemic	+	−
HBV	Preexposure	Personnel or family members attending and infected patient	+	−
	Postexposure	Parenteral		
		Transfusion of HBsAg positive blood	−	+
		Accidental	−	+
		Needle-stick, surgical injury, blood on open wound, or mucous membrane		
		Nonparenteral		
		Intimate contact	+	−
		Postnatal	−	+
NANB	Postexposure	Parenteral		
		Transfusion	+	−
		Accidental	+	−
δ-associated agent‡				

*0.01–0.02 ml/kg body weight dose within 14 days of exposure, if possible.
†0.05–0.07 ml/kg body weight dose within 7 days of exposure, and repeated 1 month later.
‡No present indications for use of globulins with δ-associated agent exposure.
HAV, hepatitis A virus; HBV, hepatitis B virus; NANB, non-A, non-B hepatitis virus or viruses; +, indicated; −, not indicated.

measures to eliminate the sources of infection. These methods are only partially successful, and additional measures have, therefore, been used.

Passive immunity through the use of immune globulin is one such approach. Table 79-7 summarizes current recommendations for the use of ISG and hyperimmune hepatitis B immune globulin. It should be noted that there is general, but not complete, agreement on these recommendations among authorities.

Active immunization is now a viable option for the prevention of HBV, for a vaccine is now available. The vaccine is a preparation of purified HBsAg from HBsAg carriers who are free of infectious virus of all known types. In studies performed on over 10,000 recipients, the vaccine has been shown to induce anti-HBS antibody in over 95 percent of normal adult recipients. Because anti-HBs antibody is considered to be the immunologic defense that confers immunity, it is not surprising that it is 85 to 95 percent effective in the prevention of naturally acquired hepatitis B. In the same studies, minor self-limited side effects were observed at a rate equal to that of the placebo control. No serious long-term side effects were observed in the period up to 7 years after immunization. It appears that this vaccine will be an extremely useful preventive in the control of this disease.

Treatment

No specific effective therapy has been identified for any of the diseases associated with the hepatitis viruses. Supportive measures are still the best therapy. Although corticosteroids have been used for a variety of the sequelae of hepatitis, there are no proved indications for this form of therapy. In fact, these agents may have adverse effects.

FURTHER READING

Barker LF, Gerety RJ, Tabor E: The immunology of the hepatitis viruses. Adv Intern Med 23:327, 1978

Beasley RP, Hwang LY, Lin CC, et al.: Hepatocellular carcinoma and hepatitis B virus. A prospective study of 22,707 men in Taiwan. Lancet 2:1129, 1981

Bernstein LH, Koff RS, Siegel ER, et al.: The hepatitis knowledge base—A prototype information transfer system. Ann Intern Med 93:165, 1980

Blumberg BS: Polymorphisms of the serum proteins and the development of iso-precipitating antibodies in transfused patients. Bull NY Acad Med 40:377, 1964

Centers for Disease Control Hepatitis Surveillance. Washington, D.C., U.S. Department of Health and Human Services, Public Health Service, Report 47, 1981

Dienstag JL: Immunopathogenesis of the extrahepatic manifestations of hepatitis B virus infection. Springer Semin Immunopathol 3:461, 1981

Francis DP, Hadler SC, Thompson SE, et al.: The prevention of hepatitis B with vaccine: Report of CDC multicenter efficacy trial in homosexual men. Ann Intern Med 97:362, 1982

Gerety RJ: Non-A, non-B hepatitis. New York, Academic Press, 1981

Howard CR, Burrell CJ: Structure and nature of hepatitis B antigen. Prog Med Virol 22:36, 1976

Ishak KG: Light microscopic morphology of viral hepatitis. Am J Clin Pathol 65:787, 1976

Koff RS: Viral hepatitis. New York, Wiley, 1978

Krugman S, Giles JP, Hammond J, et al.: Infectious hepatitis—Evidence for two distinctive, epidemiological and immunological types of infection. JAMA 200:365, 1967

Krugman S, Gocke DJ: Viral hepatitis. Philadelphia, Saunders, 1978

Levy GA, Chisari FV: A proposed role for the immune system in the pathogenesis of hepatitis B virus induced liver disease. CRC Crit Rev Clin Lab Sci 15:335, 1981

Seeff LB, Hoofnagle JH: Immunoprophylaxis of viral hepatitis. Gastroenterology 77:161, 186, 1979

Szmuness W, Stevens CE, Harley EJ, et al.: Hepatitis B vaccine—Demonstration of efficacy in a controlled clinical trial in a high-risk population in the United States. N Engl J Med 303:8323, 1980

Szmuness W, Alter HJ, Maynard JE (eds): Viral Hepatitis. Philadelphia, Franklin Institute Press, 1981

Szmuness W, Stevens CE, Harley EJ, et al.: HBV vaccine in medical staff of hemodialysis units. Efficacy and subtype cross-protection. N Engl J Med 307:1481, 1982

Vyas GN, Cohen SN, Schmid R (eds): Viral Hepatitis—A Contemporary Assessment of Etiology, Epidemiology, Pathogenesis and Prevention. Proceedings of the Second Symposium on Viral Hepatitis. Philadelphia, Franklin Institute Press, 1978

CHAPTER 80

Miscellaneous Viruses: Reoviruses and Rhinoviruses

Reoviruses

Possession of a segmented genome of 10 to 12 genes of double-stranded RNA is now the criterion for classifying viruses as Reoviridae. The family name has evolved from the original designation of human isolates as respiratory enteric orphan (REO) viruses.

Human reoviruses were first isolated in the early 1950s and initially identified as echovirus type 10 until 1959, when these agents were designated as reovirus serotype 1. A macaca monkey isolate (SV59) was designated reovirus type 2 in 1960, and a human virus pathogenic for mice, first reported in 1953, was designated reovirus type 3 in 1961.

Mammalian reoviruses are inhabitants of the human gastrointestinal tract. They have been detected in healthy persons and occasionally in persons with an array of inconsequential respiratory and/or gastrointestinal illnesses. Three different serotypes have been isolated from sporadic cases of encephalitis, hepatitis, meningitis, fatal pneumonia, and other diseases. They have no clearly defined role in these associated human illnesses. In infants, reovirus infection has been associated with mild febrile illness, diarrhea, and exanthem. The most characteristic pattern of adult infection as judged by recovery of virus is that of an afebrile coryzal illness occurring in the winter. It is clear that asymptomatic infections, as documented by virus isolation, are common. The presence of neutralizing antibody in serum seems to be relatively protective against infection with the homologous serotype and may give some degree of protection against infection with heterologous types. The reoviruses have not been of major epidemiologic importance, but infection of humans is common throughout the world.

A recent clinical serologic association of reovirus type 3 with human biliary atresia is a provocative observation in view of the ability of reovirus type 3 to infect bile duct cells and produce obstructive jaundice in a weanling mouse model.

The three serotypes of mammalian reoviruses have also been obtained from many species of animals including chimpanzees, calves, pigs, cats, and sheep. Serologic evidence of infection has been shown in virtually all of the biologic species tested.

Pathogenesis of Infection

The reoviruses have been studied in great detail in mouse models of infection. The segmented genome has facilitated their study because gene assortment occurs easily and is readily detected in recombinant progeny. Thus, the molecular virology of these viruses has been utilized to gain insight into clinical problems. It has been possible to dissect several virulence factors of these viruses for mice, thus illustrating the complexity of interaction of the host and the virus. The direct relevance of these observations to the pathogenesis of human infection is not yet established.

A summary of the interaction of the virus with the murine host follows. The virus gains entry to the host through the gastrointestinal tract. Reovirus type 1, but not type 3, successfully establishes infection in the mouse gut. Through the use of recombinant viruses, it has been demonstrated that an outer capsid protein, μ1C (a cleavage product of protein μ1), coded for by gene M1, varies in its sensitivity to proteases. The μ1C protein of reovirus type 1 confers resistance to proteolysis in the gastrointestinal tract, and enables pinocytosis of the virus by M epithelial cells, subsequent access to Peyers patches, replication in lymphoid tissue, and dissemination of virus. Thus, one facet of virulence is the ability to resist nonspecific degradation of infectious virus to noninfectious cores by the action of host proteases within the gastrointestinal tract. Specific antibody directed against μ1C does not neutralize infectivity or prevent hemagglutination, but will bind and precipitate virus particles.

Of great importance in virulence is the σ1 protein encoded by the S1 gene. Only 24 molecules of this protein are present per virion in contrast with an estimated 900 of another outer capsid protein, σ3. Protein σ1 specifies cell tropism, as it is the attachment protein of the virus that interacts with specific receptors on the cell surface. The specificity of the σ1 protein of reovirus type 1 is for ependymal cells, whereas that of reovirus type 3 is for neurons. Thus, intracerebral inoculation of reovirus type 1 produces infection of ependymal cells, an inflammatory response, and scarring of the aqueduct with resulting hydrocephalus at a later time. In contrast, intracerebral inoculation of reovirus type 3 produces infection of neurons and rapidly fatal encephalitis. Thus, a single gene, S1, encodes a protein that determines specific cell interactions and tissue tropism in the host.

The σ1 protein is also the hemagglutinin, and the first of three proteins stimulating the formation of neutralizing antibody. These functions are on separate domains of the σ1 protein. This protein is also responsible for the association of reovirus particles with microtubules in cells. Sensitized T lymphocytes, both cytolytic and suppressor cells, identify this protein, which is also expressed on the surface of infected cells.

The outer capsid shell of reovirus is largely (about 99 percent) composed of the μ1C and σ3 proteins. They have a strong affinity for each other and probably exist as complexes in the capsomers. Proteins σ1 and σ3 are removed after virus gains entry into cells. The loss of part of the outer capsid shell results in activation of enzymes essential for mRNA synthesis. Antibody to protein σ3 also neutralizes infectivity and inhibits hemagglutination. Thus, a second protein is capable of stimulating a host antibody response, which may contribute to protection of the animal.

The third protein inducing neutralizing antibodies is λ2, a core protein, present in about 60 copies per virus particle. It is the major component of the 12 core projections or spikes through which single-stranded transcripts of reovirus genes can be liberated. The fact that antibodies to this protein neutralize infectivity of virus and inhibit hemagglutination indicates that these spikes project through the outer capsid shell. It has been shown that antibody to λ2 prevents antibody to σ1 from binding to this protein. Thus, σ1 must be near λ2, and the λ2 antibody may neutralize infectivity by blocking the σ1 protein from access to specific host cell receptors. Finally, antibody to λ2 of any of the three serotypes of reovirus interacts with the λ2 protein of all three serotypes. Therefore, by definition, this protein is group reactive, which suggests common determinants are present on the λ2 protein of all three serotypes of reovirus.

As would be predicted from the above information, creating a recombinant virus with a serotype 1 M1 gene and a serotype 3 S1 gene produces a virus capable of infecting the mouse by the oral route, of disseminating infection, and of establishing a fatal encephalitis. Reovirus has two proteins defined thus far that are essential to their virulence in animals. Proteins located on the surface of the virus are involved in adsorption to host cells, stimulation of neutralizing antibodies and sensitization of lymphocytes.

Diagnosis

The mammalian reoviruses replicate and produce visible cytopathic effects in a wide variety of cell cultures. Primary rhesus cell cultures are satisfactory for routine isolation, but primary human kidney cells have been used when it was essential to exclude the possibility of endogenous infection of monkey kidney cells. The cytopathic effect consists of a granular change in the cytoplasm and the cells do not slough off the glass quickly. Often they remain attached to the surface by a single cytoplasmic process and may appear to be undergoing nonspecific degeneration. Stained preparations show intracytoplasmic eosinophilic inclusions. Isolates have been obtained primarily from fecal specimens and, less often, from the respiratory tract. They have also been seen on occasion from other sources such as urine, cerebrospinal fluid, and various tissues obtained at autopsy.

The serotypes of reoviruses have been identified by the hemagglutination inhibition technique. Human infection is associated with a demonstrable antibody rise to the homotypic serotype. The antibody rise is measured by type-specific hemagglutination inhibition or neutralization. The heterologous rise in antibodies to other serotypes is explained by the antibody to group reactive λ2 protein. The high prevalence of reovirus antibodies makes it mandatory that an antibody titer rise be demonstrated to document the presence of acute infection. The detailed studies of the antigens associated with reoviruses have made it possible to measure an antibody response to individual proteins and to use measurements other than neutralization or inhibition of hemagglutination. These methods include radioimmunoprecipitation, immunodiffusion, and immunofluorescence. For example, studies that have examined the potential role of reovirus type 3 in biliary atresia have utilized patient sera for indirect immunofluorescence of infected cells. This has demonstrated that these sera have reovirus antibodies despite the absence of neutralizing antibodies. Although the antigens (and the induced antibody) responsible for the immunofluorescence remain to be defined, it seems probable that the σ1 protein will not be one of these antigens. Thus, if this is reovirus infection, the absence of neutralizing antibodies suggests an unusual virus–host interaction.

Rhinoviruses

Over 100 serotypes constitute the genus Rhinovirus of the Picornaviridae family. These 25- to 30-nm particles have a naked icosahadral nucleocapsid containing a single-stranded RNA genome. The first isolate in 1956 was initially classified as echovirus 28, but subsequently it has been appreciated that, in contrast to other picornaviruses, the rhinoviruses are acid labile and are destroyed at a pH of less than 3. This property explains their failure to survive passage through the gastrointestinal tract. The rhinoviruses have a buoyant density of 1.38 to 1.41 g/cm^2 in cesium chloride, which is higher than that of enteroviruses. The picornaviruses all share the property of capsid assembly in the cytoplasm of infected cells, and mature virus particles contain four capsid proteins as a result of cleavage of a common precursor. Rhinoviruses (nose viruses) are important causes of acute respiratory infections, with predominant illness occurring in the airway above the larynx.

Clinical Illness

The usual symptoms of a rhinovirus common cold are nasal obstruction and discharge, sneezing, scratchy throat, mild cough, and malaise. Virus replication is limited to the respiratory tract. Human epithelial cells containing rhinovirus antigens have been demonstrated in the nasal mucus of infected individuals, but the amount of damage to the nasal epithelium appears to be minimal. Severe tracheobronchitis and even atypical pneumonia may occur rarely in adults, and rhinovirus infection has been associated in a few studies with acute exacerbation of chronic obstructive pulmonary disease. As with influenza and respiratory syncytial virus (RSV) infections, hyperreactive small airways have been demonstrated in adult volunteers with rhinovirus-caused illness. Fever and significant lower respiratory tract involvement in the form of bronchiolitis or pneumonia are more common with rhinovirus infection of children, and symptomatic infections have been temporally associated with the precipitation of asthma attacks. There is need to clarify the role of rhinoviruses in these illnesses and, although they are infectious agents contributing to the overall occurrence of respiratory disease, they are not of the importance of respiratory syncytial virus and the parainfluenza viruses.

Detectable serum antibody correlates positively with resistance to infection. Repeated infections with these agents occurs throughout life and, therefore, antibody positivity increases with age. Nasal antibody is often detected in individuals with naturally occurring serum antibody. Experimental work suggests that the quantity of serum antibody is a good indicator of type-specific resistance to rhinovirus infection, and that nasal antibody is an important component of immunity to rhinoviruses. Volunteer study has shown initial appearance of IgA and IgG in nasal secretions after infection and specific antibodies become detectable within the first several days after experimental infection. If this is the first experience with the virus, the serotype specific antibody is present after approximately 1 week of infection. Similarly, serum antibody—usually IgG, but occasionally IgM—also begins to be detectable at about 1 week into the infection, with peak levels occurring at approximately 1 month. If the person has experienced rhinovirus infection previously, an anamnestic response is observed.

Epidemiology

Rhinovirus infections have been documented in all populations studied. They occur throughout the year, but in a temperate climate there is some tendency for an increased incidence from April through October. Children frequently serve to introduce rhinoviruses into the family unit, with subsequent illnesses occurring usually after a 2- to 3-day incubation period. Occasional epidemics with a single serotype have been described, but more often multiple serotypes appear to be circulating at the same time. Rhinoviruses seldom repeat within a population from year to year. Human rhinoviruses are limited in their host range, and only the chimpanzee and gibbon have been shown to be experimentally susceptible.

Shedding of rhinoviruses in respiratory tract secretions has been amply documented. Experimental studies of transmission of these viruses have shown that transmission may occur by way of aerosols under intimate circumstances, but, surprisingly enough, a more efficient means of transmission of virus is by hand contact and self-inoculation of the mucous membranes of the eye or nasal mucosa. Comparative studies have indicated that this route of infection is more efficient than that of either large or small particle aerosols. The importance of hand contact and self-inoculation observed in experimental circumstances where the variables are well controlled raises questions as to the main mode of transmission of infection in naturally occurring infection. Viral contamination of objects in the environment frequently occurs. Experimental studies tend to support the concept that rhinoviruses will not ordinarily be transmitted in public gathering places or with relatively short exposure to friends or relatives with colds. Rather, transmission is far more likely where hand contact with viral-containing secretions occurs. The observed unpredictability of intrafamilial transmission of these viruses tends to support the concept that virus is not easily transmitted by aerosol. Finally, recent studies have successfully interrupted experimental rhinovirus transmission by the use of local application of aqueous iodine solution to the hands. Effectiveness of iodine in interrupting viral transmission in volunteers suggests that hand contamination and self-inoculation of the virus may be the most important route by which these agents are transmitted.

Diagnosis

Acute upper respiratory symptoms can be caused by multiple viruses. A tentative assignment of probable rhinovirus etiology can be made to illnesses occurring in the characteristic seasonal pattern and producing typical mild symptoms. To specifically distinguish rhinovirus infection from other etiologic agents requires virus isolation and demonstration of a rise in antibody titer between acute and convalescent sera. The assays for antibody must be done with neutralization tests against the specific rhinovirus causing the infection. This specific serologic identification of the agent requires multiple cross-neutralization tests and is impractical for general use.

Rhinoviruses are best recovered by inoculating a nasal specimen into human fibroblast cells or human embryonic kidney cell cultures. To obtain an optimal yield of these viruses, the culture should be incubated at 33C to 35C and at a neutral pH. The cytopathic effects usually become visible within the first week after inoculation and infected cells are visible as a focus of oval or refractile cells. Cell lysis and diminution in cell number occurs. At the present time, detection of antigen by immunofluorescence of respiratory secretions is impractical because the large number of rhinovirus serotypes requires multiple antisera. No commerically available sources of hyperimmune sera exist. An isolate can be identified as a rhinovirus by its physicochemical properties, especially acid lability, which will distinguish rhinoviruses from enteroviruses.

Treatment

Treatment of rhinovirus infection is generally aimed at relief of symptoms. Hydration and preventing obstruction of the airways, paranasal sinuses, and eustachian tubes are the mainstays of therapy. Aspirin has been commonly prescribed and although some diminution in symptoms may occur, increased virus shedding has been documented in persons receiving aspirin therapy. Interferon has been tested for effectiveness against the rhinoviruses. The total number of infections was unaltered, although some diminution in the frequency of symptomatic illness could be attributed to the administration of low doses of prophylactic human leukocyte interferon applied topically to the nasal mucosa. The quantity of interferon appears to be of critical importance, and thus far there is no practical or inexpensive way in which this material can be used. Finally, in apparently well-designed studies, vitamin C has not significantly decreased the number of colds.

Prevention

Resistance to reinfection with the same rhinovirus serotype can be shown following an initial infection. The major ob-

stacle to the formation of any proposed vaccine is the existence of the more than 100 antigenically distinct serotypes of rhinoviruses. Trials have suggested that production of materials of appropriate potency is difficult, and there are questions concerning whether a parenterally administered rhinovirus vaccine is protective because of the need for stimulation of nasal antibody. At the present time, vaccination appears unlikely to be a successful method of prophylaxis against rhinovirus colds.

Probably of greater importance is the possibility of diminishing the spread of infection. Recognition that contaminated hands play a role provides the opportunity to interrupt transmission with such basic measures as hand disinfection. One study has shown that iodinated paper, which is virucidal for rhinoviruses, successfully diminished the incidence of respiratory illness.

FURTHER READING

REOVIRUSES

Epstein RL, Powers ML, Weiner HL: Interaction of reovirus with cell surface receptors. III. Reovirus type 3 induces capping of viral receptors on murine lymphocytes. J Immunol 125:1800, 1981

Fontana A, Weiner HL: Interaction of reovirus with cell surface receptors. II. Generation of suppressor T cells by the hemagglutinin of reovirus type 3. J Immunol 125:2660, 1980

Morecki R, Glaser JH, Cho S, et al.: Biliary atresia and reovirus type 3 infection. N Engl J Med 307:481, 1982

Rubin DH, Fields BN: Molecular basis of reovirus virulence. Role of the M2 gene. J Exp Med 152:853, 1980

Sabin AB: Reovirus: A new group of respiratory and enteric viruses formerly classified as ECHO type 10 is described. Science 130:1387, 1959

Weiner HL, Drayna D, Averill DR, et al.: Molecular basis of reovirus virulence. Role of the S1 gene. Proc Natl Acad Sci USA 74:5744, 1977

Weiner HL, Powers ML, Fields BN: Absolute linkage of virulence and central nervous system cell tropism of reovirus to viral hemagglutinin. J Infect Dis 141:609, 1980

RHINOVIRUSES

Butler WT, Waldmann TA, Rossen RD, et al.: Changes in IgA and IgG concentrations in nasal secretions prior to the appearance of antibody during viral respiratory infection in man. J Immunol 105:584, 1970

Cate TR, Roberts JJ, Russ MA, et al.: Effects of common colds on pulmonary function. Am Rev Resp Dis 108:858, 1973

Greenberg SB, Harmon MW, Couch RB, et al.: Prophylactic effect of low doses of human leukocyte interferon against infection with rhinovirus. J Infect Dis 145:542, 1982

Gwaltney JM, Moskalski PB, Hendley JO: Hand-to-hand transmission of rhinovirus colds. Ann Intern Med 88:463, 1978

Gwaltney JM, Moskalski PB, Hendley JO: Interruption of experimental rhinovirus transmission. J Infect Dis 142:811, 1980

Stanley ED, Jackson GG, Panusarn C, et al.: Increased virus shedding with aspirin treatment of rhinovirus infection. JAMA 231:1248, 1975

Turner RB, Hendley JO, Gwaltney JM Jr: Shedding of infected ciliated epithelial cells in rhinovirus colds. J Infect Dis 145:849, 1982

CHAPTER 81

Subacute Spongiform Encephalopathies/Unconventional Viruses

The agents considered in this chapter are those causing progressive degenerative central nervous system diseases with very similar histopathologies. It is from this histologic appearance that the term subacute spongiform encephalopathy is derived. The four known naturally occurring diseases of the central nervous system caused by unconventional viruses include two in humans, kuru and Creutzfeldt–Jakob disease (CJD), and two in animals, scrapie and transmissible mink encephalopathy. The agents causing the subacute spongiform encephalopathies have been more difficult to characterize than the conventional viruses. Examination of their chemical and physical properties suggests that they are simpler and smaller than conventional viruses. Recognizable virus particles have not been visualized by electron microscopy of infected cells or in preparations concentrated by density gradient banding that have subsequently been shown to have a high infectivity. Perhaps the best working hypothesis for a model of these agents is one similar to a membrane protein, featuring a small nucleic acid capable of coding for a single polypeptide.

The illnesses caused by these agents are transmissible conditions characterized by slow evolution of the clinically evident symptomatology, which invariably involves the central nervous system. Each of these diseases has an incubation period of months to years, and a protracted, relentlessly progressive illness that is almost invariably fatal. All of these diseases have been transmissible to animals by injection of infected tissues. One of the unique features of this group of illnesses is the lack of inflammatory reaction of involved tissues (central nervous system), and the normal cerebrospinal fluid in the face of devastating illness. In accord with these findings, there has been no measurable host cellular or humoral immunity to these agents, although infected hosts have no impairment of the overall function of their immune system.

The technical difficulties of working with these agents are immense. Quantitation of virus is dependent upon infection of animals. The prolonged incubation period, as well as the species specificities (Table 81-1), make quantitation tedious, expensive, and time-consuming.

It is yet to be determined if a single "unit" causes infection or if successful infection requires many "units." Similarly, the basic question of whether these agents are truly self-replicating or whether they liberate an undefined infectious product whose properties are specified by cellular genes is not yet known. It is clear that infectivity of tissue is associated with cellular membranes.

Primary cell lines maintaining the infectious viruses of kuru, CJD, and scrapie have been derived from the brains

TABLE 81-1. HOST RANGES OF THE UNCONVENTIONAL VIRUSES

	Nonprimate	Primate
Scrapie, kuru, and Creutzfeldt–Jakob disease		Capuchin, squirrel, cynomologous macaque, rhesus, spider
Kuru and Creutzfeldt–Jakob disease	Ferret	Chimpanzee; marmoset; wooly monkey, African green monkey, mangabey, pig-tailed macaque
Kuru	Mink	Gibbon, bonnet
Creutzfeldt–Jakob disease	Cat, goat, guinea pig, hamster, mouse	Stump-tailed macaque, patas
Scrapie	Gerbil, goat, hamster, mink, mouse, rat, sheep, vole	Not susceptible: cat, chimpanzee

of patients and animals with spongiform encephalopathies. The cultures do not differ morphologically from those of uninfected tissues, and infectivity is ultimately lost after several months. The infected cultures contain amounts of virus much smaller than those from the original tissues and, thus, investigations must continue primarily to use animals as sources of virus.

Infections of Humans

Kuru

The Fore people of Papua, New Guinea's eastern highlands, used the word *kuru* to describe trembling from fear or cold. This came to be the name of a disease characterized by tremors, which was confined to this Melanesian tribe of 35,000 persons in isolated mesolithic villages. At the time of the first description of this illness by Gajdusek and Zigas in 1957, the disease was the most common cause of death among these people, and the death rate was as high as 2 to 3 percent annually.

Kuru is a distinctive neurologic syndrome, and there is an extraordinary uniformity of the clinical signs, symptoms, and course of the disease. Clinical disease is initially recognized by clumsiness in walking or inability to maintain balance. There may be associated minor changes in mood and personality. The relatively minor changes in coordination may be present for 12 to 18 months before generalized disease becomes readily apparent. After definite neurologic symptoms are recognized, the course of the disease is ordinarily fatal over the next 12 to 18 months. The initial clumsiness progresses so that the patient becomes unable to balance and finally unable to sit, stand, or perform any voluntary movement. At this stage, excessive contraction of synergistic muscles and associated movements in other limbs create exhausting ineffective muscular activity. Speech is slurred and becomes unintelligible, and chewing and swallowing are affected. Clinical examination reveals signs of cerebellar dysfunction. Changes in mood, emotional lability, depression, and epi-

sodes of confusion are common. Disability of memory and language functions are apparent. Unable to eat, aspirating secretions, and developing pneumonia and decubitus ulcers, the patients die.

Epidemiology

The incubation periods of kuru range from 4.5 years to more than 30 years. Pregnant women have been infected, but the agent was not transmitted transplacentally. Found commonly in children of both sexes and women, kuru is uncommon in adult males. The ritualistic cannibalism of the immediate family members ordinarily involved women and small children, and it is likely that the handling of the infected tissues allowed access of the agent to breaks in the skin or mucous membranes, with the consumption of infected tissue less important in transmission of disease. No one born after cannibalism ceased has acquired kuru, and no outsider in contact with patients has ever acquired the disease. Human-to-human transmission has been limited exclusively to the ritual of cannibalism.

Pathology

The pathology of this disease is as uniform as the clinical illness. Macroscopically the cerebral hemispheres appear normal, whereas the cerebellum is atrophied. Microscopically the most severe degeneration occurs in the cerebellum, where there is a loss of Purkinje and granule cells. The spongiform changes of gray matter of the cerebral cortex and cord seem to be vacuolar enlargements of neuronal processes, and are apparently of intradendritic origin. Accumulations of membranous material in these swellings are apparent. Microglial cells are seen in all stages of phagocytic activity, and there is a proliferation of astrocytes with a dense gliosis throughout the cerebellar cortex. Amyloid plaques are also apparent. There has never been any evidence of inflammatory reaction or demyelination of white matter.

Kuru has been transmitted to experimental animals and the pathology of nonhuman primates resembles that of humans. Infectivity is as high as 10^9 infectious units per gram of brain tissue, and animal inoculation studies have shown lower titers sometimes present in spleen, liver, or nodes. Infectivity has not been found in blood, cerebrospinal fluid, urine, white blood cells, milk, amniotic fluid, or placenta. Intracerebral, intravenous, intramuscular, subcutaneous, or intraperitoneal injections regularly transmit infection, but incubation periods range from 1 to 8.5 years. A single experiment successfully transmitted infection orally to a squirrel monkey, whereas many other experiments have failed. Thus, experiments suggest penetration of skin or mucosal membrane is a far more likely avenue of infection.

Creutzfeldt–Jakob Disease

Clinical and Epidemiologic Features

Creutzfeldt–Jakob disease, a progressive fatal disease of the central nervous system, occurs in middle life and affects both sexes. It was originally described as a rare form of presenile dementia in approximately 1920. Vague prodromal symptoms are followed by dementia, with subsequent rapid progression to coma and death, usually within 2 years after recognition of symptoms. The dementia can be accompanied by other neurologic disturbances, which may include focal cortical degeneration, both upper and lower motor neuron signs, extrapyramidal signs, cerebellar ataxia, and seizures. Associated myoclonus is characteristic, and there is often a typical electroencephalogram. There is great variability in the clinical syndrome.

This disease occurs throughout the world, with the rate thought to be 1 or 2 per 1 million population. Libyan Jews have a rate of over 30 times higher, and there may be a familial form of the disease in approximately 10 percent of patients, involving several generations of a family. It is not highly transmissible, as evidenced by the fact (1) that there are only two reports of conjugal cases of disease and (2) that none of the people in closest contact with CJD patients, including wives, friends, and health care professionals, appear to be at any higher risk of contracting infection than the general population. Thus far, neuropathologists, research scientists, and laboratory personnel do not appear to be at excessive risk.

The only definitive evidence of transmission has occurred with the recognition of several patients who presumably acquired infection from contaminated stereotactic electroencephalographic electrodes or from a corneal transplantation. The incubation period is estimated to be at least 15 to 20 months in these situations, where direct inoculation has occurred. The natural latency period must be much longer because of the absence of disease in children or young adults.

There can be no doubt about the transmissibility of this disease, as Gadjusek and associates have reported that 111 inoculations of 373 specimens resulted in successful transmission of disease to nonhuman primates. The results of additional inoculations are still pending. The agent has been found less regularly and in lower concentrations in lymph nodes, liver, kidney, spleen, lung, cornea, and cere-

brospinal fluid than in the brain and spinal cord. The brain has been shown to contain at least 10^8 infectious units per gram. This agent has not yet been isolated from body surfaces, secretions, or excretions. It therefore seems that acquisition of diseases occurs by tissue penetration.

Pathology

The diverse clinical manifestations are matched by the spectrum of the distribution and severity of the pathologic lesions. Brains may look normal or they may be small and atrophic. The microscopic changes always include neuronal loss, status spongiosus (less marked than in kuru), and proliferation of hypertrophic astrocytes. Brains of inoculated animals also show status spongiosus as one of the major pathologic findings, together with neuronal loss, proliferation of astrocytes, and absence of inflammation.

Disinfection for Creutzfeldt–Jakob Disease Agent

No special ward isolation procedures are warranted. In fact, such procedures may be an inconvenience to patient and personnel alike, and, if specified, may prevent optimal care of the patient or admission to a long-term facility. Exposure to breath, saliva, nasopharyngeal secretions, or urine of CJD patients should not cause special concern. Washing of hands or other exposed areas of the skin with appropriate detergent is recommended, as in contact with any other patient with potentially transmissible agents. Although a patient's blood or cerebrospinal fluid could be potentially infectious, simple washing should be sufficient to protect exposed persons. Caution must be taken to avoid accidental percutaneous exposure to blood, cerebrospinal fluid, or brain tissue. If a puncture occurs, careful washing in iodine solution, phenolic antiseptic, or 0.5 percent sodium hypochlorite is appropriate. Some of the physical and chemical properties of these agents are listed in Table 81-2. The preferred method of disinfection of equipment is either a 1-hour exposure to 5 percent sodium hypochlorite or autoclaving at 121C. Choice of method depends upon the material being treated. Surgical instruments, tonometers used on demented patients, and all instruments used in eye operations should be considered contaminated.

TABLE 81-2. SOME PHYSICAL AND CHEMICAL PROPERTIES OF UNCONVENTIONAL AGENTS

1. Sensitive (loss of infectivity) to: autoclaving (121C, 15 ψ, 60 min), phenol, sodium hypochlorite (5 percent), ether, acetone (moderate)
2. Resistant to: formaldehyde, glutaraldehyde, B-propriolactone, EDTA, 80C (partial inactivation at 100C), proteases (trypsin, pepsin, pronase, proteinase K), nucleases (DNase, RNase, micrococcal nuclease), phospholipases, nonionic or mildly ionic detergents (triton, NP-40, deoxycholate, sarkosyl)

Subacute Spongiform Encephalopathies in Animals

Scrapie

Scrapie has been known for some 200 years as a clinical entity in sheep in Europe, and for over a 100 years a chronic disease affecting the central nervous system of sheep was prevalent in a region of northern Iceland. The name of the disease in Icelandic, *rida,* means either ataxia or tremor. This is a progressive disease of adult sheep, which is inevitably fatal. The sheep fail to thrive, become frightened and excitable, and then the tremor and uncoordinated movements of the head appear. The animal becomes clumsy, with initial involvement of hind limbs. Finally, the animal is unable to walk, and despite remaining alert, is totally immobilized by tremors and spastic movements. This disease is scrapie, but without the characteristic pruritus seen in other geographic locations. Vertical transmission of infection to offspring is usual and, in nature, lateral transmission is also quite common. Hadlow first suggested that the neuropathology was similar to that of kuru, and Gajdusek then pursued the possibility that kuru was caused by an infectious agent.

Scrapie can be transmitted by inoculation to a number of animals (Table 81-1), and sheep, goats, and mice have been infected with scrapie by the oral route. The pathogenesis of scrapie has been studied in mice and, although the route of infection influences the incubation period and ultimate outcome, several general features emerge. Inoculated virus can be almost completely recovered from spleen in the first week after infection. Infectivity is no longer detectable for the subsequent several weeks. About 1 month after inoculation, replication becomes apparent, with initial reappearance of the agent in the spleen and nodes, then spreading through the lymphoid system. Replication in lymphoid tissue does not alter immune function or produce any apparent pathology. About 3 months after inoculation, infectivity is detectable in the central nervous system. It also is detectable in salivary glands, lung, intestine, and other tissues. The animals remain asymptomatic for a total of 4 to 5 months.

Transmissible Mink Encephalopathy

Transmissible mink encephalopathy is a rare sporadic disease occurring only in ranch mink. It was first noted in 1947 in Wisconsin, with additional outbreaks from 1961 through 1963. It, too, is a chronic progressive neurologic disease characterized by early behavioral changes, ataxia, and somnolence. The incubation period is approximately 8 to 9 months after oral inoculation, and 5 to 6 months following parenteral inoculation. Experimentally, the disease is transmitted to mink by various routes of inoculation using suspensions of infected brain or other tissues. The similarity to scrapie was noted in 1965. Like the other diseases discussed above, pathologic lesions are restricted to the central nervous system and the pathology is similar. Differences from the other illnesses consist only of variation in the topographic distribution of the lesions.

Experimental studies demonstrate detectable virus 16 weeks after inoculation, visible pathologic lesions 24 weeks after inoculation, and recognizable clinical disease approximately 30 weeks after inoculation. It is now thought that transmissible mink encephalopathy most likely originates from mink having been fed scrapie-infected sheep or goat tissues. Experimentally, mink are susceptible to United States sources of scrapie, but not those from the United Kingdom. It has also been suggested that the natural route of infection in mink is from bite wounds inflicted by litter mates. These observations have led to questions concerning the identities of these agents relative to each other. Is there only one such agent, i.e., scrapie that has been transmitted to mink and humans? Is kuru a result of CJD spreading in the Fore by inoculation through cannibalism? At the present time there is no evidence for epidemic spread of any of the spongiform encephalopathies.

FURTHER READING

Books and Reviews

Prusiner SB, Hadlow WJ (eds): Slow Transmissible Diseases of the Nervous System. Vol I. Clinical, Epidemiological, Genetic and Pathological Aspects of the Spongiform Encephalopathies. Vol II. Pathogenesis, Immunology, Virology, and Molecular Biology of the Spongiform Encephalopathies. New York, Academic Press, 1979

Selected Papers

Brown P, Gibbs CJ Jr, Amyx HL, et al.: Chemical disfunction of Creutzfeldt–Jakob disease virus. N Engl J Med 306:1279, 1982

Gajdusek CD, Gibbs CJ Jr, Asher DM, et al.: Precautions in medical care of, and in handling materials from patients with transmissible virus dementia (Creutzfeldt–Jakob disease). N Engl J Med 297:1253, 1977

Gibbs CJ Jr, Amyx HL, Bacote A, et al.: Oral transmission of Kuru, Creutzfeldt–Jakob disease and scrapie to non-human primates. J Infect Dis 142:205, 1980

Klatzd I, Gajdusek DC, Zigas V: Pathology of Kuru. Lab Invest 8:799, 1959

Manuelidis EE, Angelo JN, Gorgacz EJ, et al.: Experimental Creutzfeldt–Jakob disease transmitted via the eye with infected cornea. N Engl J Med 296:1334, 1977

Sigierdsson B: Observations on three slow infections of sheep. Br Vet J 110:255, 307, 341, 1954

Zigas V, Gajdusek DC: Kuru: Clinical study of a new syndrome resembling paralysis agitans in natives of the Eastern Highland of New Guinea. Med J Aust 2:765, 1957

SECTION VI
MEDICAL MYCOLOGY

CHAPTER 82

General Characteristics of Fungi

Mycology is the study of fungi. The term mycology derives from the Greek word *mykos*, meaning mushroom. Fungi are now recognized to constitute a separate kingdom, within which are approximately 100,000 species characterized by pronounced differences in structure and physiologic characteristics. Fortunately, fewer than 300 species of fungi have been implicated directly as agents of human and animal disease, and less than a dozen of these species cause about 90 percent of all fungous infections. To understand how fungi develop, how some become pathogenic, and how they can be recognized as pathogens, it is necessary to have an understanding of the characteristics of fungi in general.

All fungi are eucaryotic organisms; as such, fungous cells possess at least one nucleus, a nuclear membrane, endoplasmic reticulum, and mitochondria. Fungous cells resemble those of higher plants and animals and are quite advanced microorganisms. Most fungous cells possess a rigid cell wall, and many species produce flagellated, motile cells. Some of the higher fungi show more than rudimentary differentiation into tissues and specialized structures. Unlike members of the plant kingdom, fungi lack the property of photosynthesis.

The natural habitats for most fungi are water, soil, and decaying organic debris. All fungi are obligate or faculative aerobes. They are chemotrophic organisms, obtaining their nutrients from chemicals found in nature. Most fungi survive by secreting enzymes that degrade a wide variety of organic substrates into soluble nutrients, which are then passively absorbed or taken into the cell by active transport systems. The breakdown of natural substrates by fungi is an ecologic necessity.

Morphology

Growth Forms

Yeasts

Fungi grow in two basic morphologic forms, as yeasts or molds. Yeast refers to the unicellular growth of fungi. Yeasts are usually spheric to ellipsoidal and vary in diameter from 3 to 15 μm. Most yeasts reproduce by budding, although a few undergo division by binary fission (Fig. 82-1). The budding process is initiated when the yeast cell wall softens at a specific point as a result of localized lysis of the cell wall. The internal pressure on this area of weakened cell wall causes the wall to balloon outward. This swollen portion enlarges and, following nuclear division, a progeny nucleus migrates into the newly formed bud. The bud may then continue to enlarge. The cell wall grows to-

gether at the constricted point of attachment. Eventually the bud breaks off from the parent cell; the cycle is complete and ready to be repeated. Yeasts retain a characteristic scar on their cell walls where a bud was once attached, as shown in Figure 82-1B. Some species typically produce multiple buds before detachment occurs. If single buds fail to separate, chains of spheric yeast cells are formed. Other species of yeasts produce buds that characteristically fail to detach and become elongated; the continuation of the budding process then produces a chain of enlongated yeast cells that resemble hyphae (see below) and are called pseudohyphae. The cells that compose a stretch of pseudohyphae are characteristically constricted where they are attached to each other (see Chap. 87).

Macroscopically, yeasts produce colonies on solid media that are pasty, opaque, and generally attain a diameter of 0.5 to 3.0 mm. A few species are characteristically pigmented, but most are cream-colored. Because the mi-

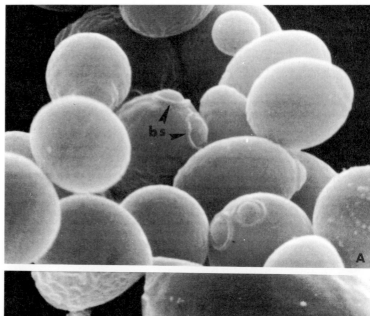

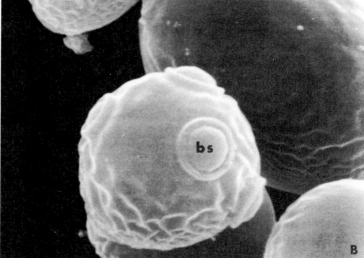

Figure 82-1. *Saccharomyces cerevisiae.* **A.** Scanning electron photomicrograph of budding yeast cells, × 9800. **B.** Bud scars, × 1800. *(Courtesy of Dr. Sara Miller.)*

croscopic and colonial morphology of yeast species differ so little, physiologic tests are necessary for their speciation (see Chaps. 84 and 87).

Molds

Growth in the mold form refers to the production of multicellular, filamentous colonies. These colonies consist basically of branching cylindric tubules varying in diameter from 2 to 10 μm and termed hyphae (singular hypha). Hyphal growth occurs by apical elongation, i.e., extension in length from the tip of a filament. Hyphal tips are densely packed with membrane-enclosed vesicles, some of which fuse with the cell membrane during active growth. The hyphal width of a given species remains relatively constant during growth. The mass of intertwined hyphae that accumulates during active growth in the mold form is called a mycelium (plural, mycelia). Some hyphae are divided into cells by septa or crosswalls, that are typically formed at regular intervals during filamentous growth. Other molds are composed of nonseptate hyphae. As hyphal septa are perforated, cytoplasmic continuity is maintained in septate as well as nonseptate mycelia.

Molds preferentially grow on the surface of natural substrates or of laboratory media. Under these conditions, hyphae that penetrate the supporting medium and absorb nutrients are called vegetative or substrate hyphae. These hyphae also serve to anchor the mycelium to its natural substrate or to laboratory agar medium. Other hyphal filaments project above the surface of the mycelium into the air, and such aerial mycelium usually bears the reproductive structures of the fungus. Identification of most filamentous fungi can be made by observation of their

morphology. Macroscopic examination of a mold isolate should include notation of such characteristics as the rate of growth, topography (e.g., glabrous, verrucose), surface texture (e.g., pasty, cottony, powdery), and any pigmentation (surface, reverse, or diffusible into the medium). Microscopically, the type of reproductive spores produced (pigment, size, shape, mode of attachment) and their ontogeny are characteristic for each species.

Dimorphism

In addition to growth as a yeast or mold, some species of fungi are dimorphic, capable of growing in more than one form under different environmental conditions. For example, some of the pathogenic fungi grow as yeasts at 37C and in the mold form at 25C.

Subcellular Structure

The fine structure of all fungi includes a unique cell wall, cell membrane, and cytoplasm containing an endoplasmic reticulum, nuclei, nucleoli, storage vacuoles, mitochondria, and other organelles (Fig. 82-2).

Capsule

Some fungi elaborate an external coating of slime or a more compact capsule. The capsule or slime layer is composed predominantly of amorphous polysaccharides that may be mucilaginous and cause the cells to adhere and clump together. The quantity, chemical composition, physical properties such as viscosity and solubility, and the antigenicity of capsular polysaccharides vary with different

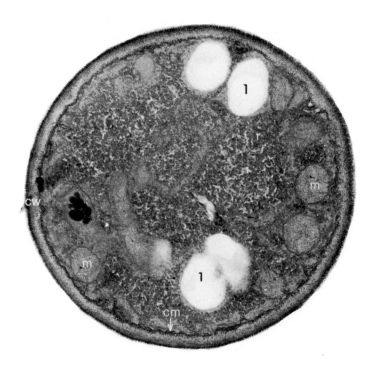

Figure 82-2. *Candida albicans.* Transmission electron photomicrograph of yeast cell showing internal structure: cw, multilayered cell wall; cm, cell membrane; m, mitochondrium; l, lipid; r, ribosomes. *(Courtesy of Dr. Sara Miller.)*

species. The capsule does not appear to affect permeability or other functions of the cell wall and membrane. However, because of its gelatinous nature, this capsular material may influence the growth of the fungus by preventing either the dissociation of buds from yeast cells, or the dispersion of yeasts in air or water. The fungal capsule may be medically significant; the best example is the capsule of *Cryptococcus neoformans*, which has both antigenic and antiphagocytic properties (Chap. 84).

Cell Wall

The cell wall is an extremely important component of a fungus, comprising approximately 90 percent of its dry weight. The cell wall provides rigidity and strength, and it protects the cell membrane from osmotic shock. As the wall determines the shape of any fungus, the process of fungal morphogenesis (e.g., sporulation or yeast—mold dimorphism) must involve changes in the cell wall. The cell wall is generally thicker in yeasts (200 to 300 nm) than in molds (<200 nm), but in both cases it is a multilayered structure that appears highly refractile under light microscopy.

Composition. Eighty to ninety percent of the cell wall is carbohydrate. Actually, a relatively small number of polysaccharides are usually found in the walls of a wide variety of species of fungi, albeit these are present in different quantities and in combination with varying amounts of other polysaccharide components that may be less common or even unique. Some of the more prevalent polysaccharides are listed in Table 82-1.

Many fungi share surface antigens as a result of the similarity among these polysaccharides. However, because the side chains of glucans vary considerably in the number, length, and linkage of their residues, many unique antigenic determinants can also be found. Some antigens are therefore widespread among fungi, whereas others are found only within certain groups or individual species. This concept is illustrated in Table 82-2, which indicates the association of certain polysaccharides with broad taxonomic groups. Antigenic analysis of the cell wall has become a powerful tool for the study of classification, evolution, and phylogenetic relatedness among fungi.

Futhermore, the detection of species specific surface antigens in solution has recently been shown to provide a rapid method for the identification of slow-growing and/or poorly sporulating pathogenic fungi (see Chaps. 83 and 84).

Approximately 10 percent of fungal cell walls consist of protein and glycoprotein. This protein includes enzymes involved in wall growth, certain extracellular enzymes, and structural proteins that cross link the polysaccharide chains. The concentration of wall protein is higher near the cell membrane. Wall proteins possess large amounts of sulfur-containing amino acids and disulfide bonds. Disulfide bonds are more prevalent in hyphal than in yeast walls, and in some species (*Candida albicans* and *Paracoccidioides brasiliensis*) the reduction of these bonds is associated with transformation from the filamentous to the yeast form of growth.

Ultrastructure. Cell wall polysaccharides are fibrillar in structure and multilayered. As may be seen in Figure 82-2, the polysaccharides appear as long microfibrils. Two to four distinct layers are usually observed but the degree of organization, whether a linear alignment or a cross-hatched pattern, varies considerably. Generally, the most compact layer is nearest the cell membrane, and the external layer(s) tend to be more amorphous, less organized, and less compact. Some fungal cell walls contain tightly interwoven microfibrils embedded in an amorphous, polysaccharide matrix.

Methods of Study. Fungal cell walls may be studied in situ by several techniques:

1. selective enzymatic digestion of exposed surface components
2. immunologic analysis of external antigens (e.g., fluorescent antibody, agglutination)
3. incorporation and localization of radiolabeled cell wall precursors (cell-associated label; autoradiography)
4. susceptibility to selective chemical degradation (e.g., solubility in hot acid or alkali)
5. histochemical stains
6. lectin-binding.

TABLE 82-1. MAJOR POLYSACCHARIDES OF FUNGAL CELL WALLS

Polymer	Monomer	Linkage and Structure
Chitin	N-acetylglucosamine	β-1,4-long, unbranched polymer
Chitosan	D-glucosamine	β-1,4-
Cellulose	D-glucose	β-1,4-
β-Glucan	D-glucose	β-1,3-linked backbone with β-1,6-linkages at branchpoints
α-Glucan	D-glucose	α-1,3 and α-1,4-
Mannan	D-mannose	α-1,6-linked backbone with frequent α-1,2- and α-1,3-linked branches of one to five residues each

TABLE 82-2 DISTRIBUTION OF MAJOR CELL WALL POLYSACCHARIDES

Fungous Group	Predominant Cell Wall Polysaccharides
Lower aquatic fungi	Cellulose
Class Zygomycetes	Chitin, chitosan
Ascomycetous yeasts	β-glucans, mannans
Basidiomycetous yeasts	Chitin, mannans
Fungi with septate hyphae	Chitin, β-glucans

From Bartnicki-Garcia: Ann Rev Microbiol 22:87, 1968.

A more precise analysis can be made by the study of isolated cell walls. As a first step in the preparation of cell walls, fungal cells are disrupted by physical means and the cytoplasmic contents are released. Cell walls vary in toughness and a number of methods differing in violence have been used. These include:

1. grinding fungal suspensions with mortar and pestle to break the cells
2. use of high pressure to crack the cells (e.g., French press, Ribi press)
3. shredding the cells in a blender
4. sonification
5. violent mixing with sand, glass beads, or steel pellets.

With all these procedures, certain precautions should be observed, such as preventing the system from overheating, and buffering with inhibitors of degradative enzymes to minimize autodigestion of macromolecules during breakage. After the cells are broken, the walls can be separated from intact cells and cytoplasmic contents by differential filtration and extensive washing. Purified cell walls are essentially membrane-free upon examination by electron microscopy.

Biochemistry. Isolated cell walls can be subjected to routine techniques of carbohydrate biochemistry. Following removal by hot alkali or acid, polysaccharides can be purified by chromatographic or electrophoretic techniques. Once purified, a polysaccharide can then be hydrolyzed and its component sugar(s) identified. If more than one sugar is present, the relative proportion of each may be determined. The linkage of an intact polysaccharide can be studied by optical rotation or x-ray crystallography. The specific structure can be analyzed by determination of reducing sugars and susceptibility to specific enzymes (e.g., β-glucosidase, endo-β-1,3-gluconase). Acetylation and methylation analyses can be used to determine exposed hydroxyl groups. These and other methods have been combined to determine the size, composition, and linkage of several cell wall polysaccharides.

Polysaccharide Biosynthesis. Chitin is a major component of the cell wall of many fungi. It is a homopolymer of N-acetylglucosamine (GlcNAc) and, like cellulose, which it structurally resembles, it is insoluble in water and forms crystalline arrays of parallel chains.

Polymerization of chitin is catalyzed by the enzyme chitin synthase, which is located in the cell membrane. The amino sugar donor, uridine-diphospho-N-acetyl-glucosamine (UDP-GlcNAc), is derived de novo from fructose-6-phosphate and glutamine. In the following general type of reaction, the enzyme is allosterically activated upon complexing with UDP-GlcNAc, and the preexisting chitin polymer is extended in length by one monomeric unit, as an aminosugar is added.

$$\text{UDP-GlcNAc} + \begin{array}{c}\text{(GlcNAc)n}\\\text{preexisting}\\\text{chitin}\end{array} \xrightarrow[\text{Mg}^{2+}]{\begin{array}{c}\text{chitin}\\\text{synthase}\end{array}}$$

$$\begin{array}{c}\text{(GlcNAc)}_{n+1}\\\text{lengthened}\\\text{chitin chain}\end{array} + \text{UDP}$$

Chitin synthase exists in a latent (zymogen) state and is activated by partial proteolysis.

The synthesis of glucans follows the same general type of reaction. The enzyme involved is glucan synthase, Mg^{2+} is required, and UDP serves as carrier for the D-glucose monomer. In the synthesis of mannan, guanosine-diphosphomannose is produced in the cytoplasm and transferred to a lipid carrier, polyprenolmannose phosphate. It is probable that mannan biosynthesis occurs on the endoplasmic reticulum and that the monomeric units are transported to the cell wall via the lipid intermediate. In the cell wall, two mannan chains are linked to a peptide. Mannan synthase requires Mn^{2+}.

Cell Wall Biosynthesis. Fungal cell walls are synthesized in situ between the membrane and existing wall. In the case of hyphal growth, it is probable that the apical vesicles contain the components of the synthetic machinery, the wall precursor material, biosynthetic enzymes, lytic enzymes, and cofactors. Lytic enzymes are believed to be transported to the site of new wall growth, where they cleave glycosidic bonds of polysaccharides. Less is known of the biosynthesis of the yeast cell wall, but undoubtedly similar substances and mechanisms are involved.

A unified theory has been advanced by Bartnicki-Garcia to explain the events involved in wall biosynthesis. The steps may occur in sequence or simultaneously. During growth, vesicles migrate to the site of wall growth, fuse with the cell membrane, and empty their contents into the space between the membrane and wall. These vesicles contain lytic enzymes that cleave the polysaccharide fibrils, the ends of which may serve as biosynthetic primers for further elongation of polymer. This cleavage weakens the wall, and the internal pressure causes it to stretch and increase in surface area. Activating factors, such as proteases and UDP-sugar carriers, are also carried in vesicles and deposited in this space. Wall growth occurs as synthases construct new polysaccharide to lengthen broken fibrils and strengthen the wall. This filling in of the increased surface area caused by the pressure-stretched wall produces growth.

While plausible, this concept still leaves a number of fundamental questions unanswered. What governs the spatial orientation of newly formed polysaccharide? Somehow, following focal disruption and biosynthesis, the integrity of the wall and arrangement of the microfibrils is maintained. What controls the balance between wall lysis and synthesis? It seems probable that the process of wall biosynthesis is the same in both yeast and molds. If so, the difference in shape between yeasts and molds may reflect the result of diffuse versus localized (apical) wall growth, respectively; indeed, dimorphism in fungi involves regulation of cell wall growth. These and a number of other questions, e.g., the molecular basis of bud formation, sporulation, and germination, are currently under active investigation. These areas of research have obvious relevance to morphogenesis and eucaryotic cell transformation and provide excellent models for study. Practical benefits may also emerge as results are applied to medically important fungi. As more becomes known about how fungal growth is controlled, it may be possible to devise ways to perturb the growth of fungi. To cite one example, a natural inhibitor of chitin synthase has been isolated. If this or a similar inhibitor were effective against pathogenic fungi and nontoxic for human tissue, it would have excellent promise as an antibiotic.

Medical Importance of Fungal Cell Walls. Fungal cell walls are potent antigens. It is not clear in all cases whether the wall immunogen is purely carbohydrate or how much, if any, protein must be present. Many wall extracts lose their immunogenicity upon fractionation into polysaccharide, protein, and glycoprotein components. It is apparent, however, that humans and animals develop specific immunity to a number of wall determinants, some of which can be identified as specific oligosaccharides. Through normal environmental exposure most adults have become immunized to the more prevalent mannans and glucans. Atopic individuals may develop a severe hypersensitivity reaction to specific cell wall determinants (see Chaps. 83, 84, and 87). In fact, the human exposure to some fungal antigens (candida, dermatophytin) is so widespread that reactivity to them constitutes one measure of normal immunologic status.

Mammalian tissues lack the enzymes to degrade the wall polysaccharides, and, as a result, fungal walls are cleared from the body very slowly. The retention of wall material following fungal infection undoubtedly contributes to the pathogenesis of fungal infection. Specific examples of this are cited in subsequent chapters.

Staining Properties. The cell wall polysaccharides have unique staining properties that have been exploited by the histopathologist. Often fungi are poorly stained or too sparse to be detected by the routine hematoxylin and eosin preparation. Two very helpful fungal wall stains are the periodic acid-Schiff reagent and the methenamine silver stain. These are true wall stains; neither the internal contents nor capsular material are stained by them. All fungal walls probably stain positive with the gram stain. This stain is usually not applied to fungi, however, because it does not help in their identification and may obscure internal structures. One occasion when the gram stain is useful is in the examination of smears for both *C. albicans* and bacteria.

Cell Membrane

Fungi possess a bilayered membrane similar in structure and composition to the cell membranes of higher eucaryotes. This membrane protects the cytoplasm, regulates the intake and secretion of solutes, and facilitates the synthesis of the cell wall and capsular material. The membrane contains several phospholipids, and their relative amounts vary with different species. Most prevalent are phosphatidylcholine and phosphatidylethanolamine, with smaller amounts of phosphatidylserine, phosphatidylinositol, and phosphatidylglycerol; there are other phospholipids in the cell membranes of some species. The phospholipid content varies not only among species, but within strains of a species, and within a given strain depending upon the conditions of growth.

Unlike bacteria (except the mycoplasmas), but similar to other eucaryotes, fungal membranes contain sterols. The principal fungal sterols are ergosterol and zymosterol; mammalian cell membranes possess cholesterol. This difference has been exploited in the successful use of the polyene antibiotics, which complex with sterols both in solution and in membranes. Their effect is to perforate intact membranes (Chap. 9, p. 204). A major antifungal, amphotericin B, is a polyene with greater affinity for ergosterol than cholesterol. The interaction of this antibiotic with sterols probably explains both its effectiveness against fungi and its toxicity for humans.

Cytoplasmic Content

Cells of fungi, both yeasts and molds, often contain several nuclei. All hyphae can be considered multicellular, as cytoplasmic continuity is maintained. Hyphal filaments with septa or crosswalls have pores that permit streaming of cytoplasmic contents and migration of organelles, including nuclei, through the hyphae.

The mitochondria of fungi resemble those of plant and animal cells. The number of mitochondria per cell varies considerably with the level of respiratory activity. For example, the sporulation process demands a large expense of energy, and is followed by a reduction in respiration, number of mitochondria, and overall metabolic activity. In tissue, fungi may also have fewer mitochondria. Spore germination is accompanied by an increase in respiratory activity. In several species, germination has been correlated with increased energy consumption, increased numbers of mitochondria, and higher ratios of mitochondrial DNA to cellular DNA. Although increased respiration seems to be required for spore germination, it is not certain that other observed changes, such as synthesis of mitochondrial DNA, are actually necessary.

Many fungi possess characteristic vacuoles (see Fig. 82-2). In yeasts, an active secretory apparatus has been characterized. Fungous viruses, plasmids, and other extrachromosomal genetic systems also have been described.

Some fungi elaborate characteristic secondary metabolites. These are usually small compounds that are not essential for viability and that have no obvious function or survival value. Secondary metabolites include a number of compounds with diverse and pronounced biologic effects, such as carcinogens (e.g., aflatoxin), toxins (e.g., amanitin), antibiotics, anticancer substances, and pharmacologically active compounds (e.g., ergotamine).

Reproduction

Asexual Spores

Fungi reproduce via asexual, sexual, and/or parasexual processes. Asexual reproduction can occur simply as growth and expansion of a mold or yeast colony. When a few yeast cells or fragments of hyphae are transferred to a fresh substrate, such as is routinely performed in the laboratory when fungi are transplanted from one culture to another, the transferred portion grows to produce a new colony. Asexual reproduction, however, usually refers to the production of spores, which are generally more resistant to adverse environmental conditions. The properties of spores that facilitate their dispersion are often essential for the dissemination and propagation of the fungus in nature. For example, spores are usually dry and easily airborne. Some spores are equipped with rough surfaces for adherence to fomites.

Conidia are the major type of asexual spores. Conidia are produced by specialized structures, conidiogenous cells, and classified according to their developmental process. Although the method of conidial formation is often not discernible with routine light microscopy, the conidia of many fungi are distinctive and can be easily identified. Two basic types of conidial ontogeny are thallic and blastic. These terms, and others applicable to medically important fungi, are described below. They represent only a portion of the diversity of conidiogenesis among the fungi.

Thallic conidia are derived from cells of the thallus or body of the fungus. That is, a hyphal cell becomes delineated by a septum and is transformed into a conidium. One example is the chlamydospore, a large, round, thickwalled, unicellular structure formed by enlargement of a hyphal cell. Chlamydospores may be produced in a position lateral, intercalary, or terminal on a hypha (Fig. 82-3B). An arthroconidium is a unicellular, thallic conidium produced by condensation of the cytoplasm and thick-

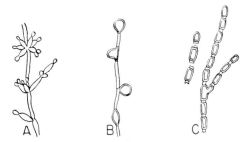

Figure 82-3. Various types of asexual spores. **A.** Blastoconidia and pseudohyphae. **B.** Chlamydospores. **C.** Arthroconidia. *(From Conant, et al.: Manual of Clinical Mycology, 3rd ed., 1971. Philadelphia, Saunders.)*

ening of the wall of a hyphal cell. Usually adjacent cells, or alternate cells, as in the case of *Coccidioides immitis*, undergo this transformation into small, dense conidia to produce a chain of rectangular arthroconidia from a hypha. With maturation, the arthroconidia become desiccated and fragment to form an easily aerosolized batch of conidia, quite uniform in size (Fig. 82-3C). The dermatophyte species (see Chap. 86) also develop thallic conidia, usually from terminal hyphal cells; they may be unicellular (microconidia) or multicellular (macroconidia). A fungus species may characteristically produce either or both microconidia and macroconidia (Fig. 82-4). Arthroconidia, chlamydospores, and other thalloconidia are not specific for any single species; a given fungus may produce none, one, or more than one type of thallic conidia.

In the formation of blastic conidia, only a portion of the conidiogenous cell contributes to the development of the conidium, and conidium formation is initiated before it is delimited by a septum. There are two subdivisions of blastic conidia, holoblastic and enteroblastic. With holoblastic conidiogenesis, all of the cell wall layers of the conidiogenous cell participate in conidium development. The buds formed by most species of yeasts, including pseudohyphae, are examples of holoblastic conidia, or blastoconidia (Fig. 82-3A). The mold genus *Cladosporium* produces chains of holoblastic conidia.

With enteroblastic conidia, only the inner cell wall layer(s) or no portion of the cell wall of the conidiogenous cell contributes to the development of the conidium. Several of the varieties or subtypes of enteroblastic conidiogenesis are important in medical mycology. Tretic conidia develop through a channel or pore in the outer wall of the conidiogenous cell, and the inner wall layer(s) participates in formation of the conidium. Phialides are flask-shaped conidiogenous cells that produce conidia through a distal opening in the cell wall, but no portion of the phialide cell wall contributes to the cell walls of the conidia (phialoconidia). Several familiar genera produce phialoconidia, including *Aspergillus*, *Penicillium*, and *Phialophora*. An annellide is another blastic conidiogenous cell; it elongates with the production of each conidium and acquires ring-shaped scars or annellations at the point of co-

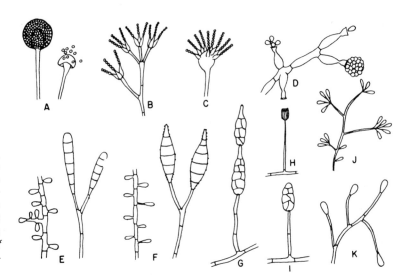

Figure 82-4. Various types of asexual spores. **A.** Sporangia and sporangiospores. **B, C, D.** Phialoconidia, *Penicillium* (**B**), *Aspergillus* (**C**), *Phialophora* (**D**). **E, F.** Thallic microconidia and macroconidia,*Trichophyton* (**E**), *Microsporum* (**F**). **G.** Tretic macroconidia, *Alternaria*. **H–K.** Other types of conidia and conidiophores. *(From Conant, et al.: Manual of Clinical Mycology, 3rd ed., 1971. Philadelphia, Saunders.)*

nidial production. Annelloconidia are produced by *Scopulariopsis, Exophiala,* and other genera. (These latter conidia are further discussed in Chapter 85).

Conidia that are not released during their production (some are forcibly discharged) are usually liberated by physical dislodgement or by the disintegration of adjacent cells. A specialized conidium-bearing hypha is called the conidiophore, which may be a simple and undistinguished stem to which the conidium is attached. Other conidiophores are branched, complexly modified, or uniquely shaped. For example, *Aspergillus* species produce at the end of the conidiophore a swollen vesicle from which short phialides project, and these in turn produce spheric, unicellular, often pigmented phialoconidia in radiating chains (Fig. 82-4C). This type of conidial formation is basipetal, as the conidia are extruded from the attached end of the chain. The newest and youngest conidium is the one nearest the point of attachment of the chain to the phialide. With other fungi (e.g., *Cladosporium*), acropetal sporulation occurs, whereby chains of blastoconidia are produced by a budding process from the youngest spore at the distal end.

Finally, asexual spores in many of the lower fungi are produced in a sac-like structure termed a sporangium (Fig. 82-4A). The sporangium develops at the tip of a sporangiophore, and the spores formed within it are called sporangiospores. Upon maturation, the sporangial wall ruptures to release its spores.

Sexual Reproduction

Sexual reproduction in fungi follows the same pattern as in other biologic forms. The process is initiated by plasmogamy, whereby two compatible, haploid nuclei are brought together in the same cell. Karyogamy is the fusion of these two nuclei to form a diploid nucleus. Karyogamy may immediately follow plasmogamy, or it may be delayed,

as in some higher fungi, with the development of a mycelium consisting of binucleate cells. Sooner or later meiosis occurs, resulting in genetic exchange, reduction, and then division to yield four haploid progeny nuclei. This sequence occurs in all the fungi for which a sexual cycle has been discovered, but many variations exist among the different kinds of fungi. Some fungi produce distinct sex organs and gametes. In others, the somatic nuclei perform the sexual function. In some species, nuclei within a single thallus are capable of fusion, and in others, sexual compatibility is genetically determined, and compatible thalli are required for mating. Nuclei from two compatible thalli may be brought together by specialized gametangia or by anastomosis of their hyphae and exchange of nuclei by migration from the hypha of one to the other. Hyphal fusion and migration of nuclei and cytoplasmic contents from one filament to another are a common phenomenon among the same species of fungi.

Because the sequence of plasmogamy, karyogamy, and meiosis is not a single continuous event for many fungi, it is helpful to describe the life cycle of a fungus as consisting of: (1) a haplophase, during which the uninuclear or multinuclear thallus contains only haploid nuclei; (2) a dikaryophase, in which two genetically distinct haploid nuclei occupy each cell of the thallus; and (3) a diplophase, referring to the diploid nucleus formed as a result of karyogamy. Most of the lower fungi lack a distinct dikaryophase, and the haplophase is the predominant state. In the more complex higher fungi, the dikaryophase assumes greater significance. For example, with mushrooms, the binucleate mycelium of the dikaryophase constitutes most of the structures and pertains much of the time. In some species both the haplophase and dikaryophase occupy significant portions of the life cycle, and asexual sporulation may serve to propagate both phases.

Following meiosis, the progeny nuclei develop into sexual spores. The basic type of sexual spore that is formed, as well as the supporting tissue, if any, that is produced, are used to define the major taxonomic groups of fungi.

Parasexual Reproduction

Parasexuality refers to a sequence of events that culminates in genetic exchange via mitotic recombination. First described by Pontecorvo, parasexual reproduction has become a laboratory tool for the genetic analysis of many imperfect fungi. It has also been shown to occur in several species of ascomycetes, basidiomycetes, and deuteromycetes (see following classification scheme).

Parasexuality is initiated by the formation of a heterokaryon, a thallus containing haploid nuclei of two different genotypes. Heterokaryons are most commonly formed by hyphal anastomosis and nuclear exchange between genetically different strains of the same species. Rarely, nuclear fusion will occur with the formation of a stable diploid heterozygote that multiplies along with the two haploid nuclei. It is during mitosis of the diploid nuclei that homologous chromosomes may pair up and permit somatic recombination to occur. The cycle is completed when the diploid undergoes haploidization to the original chromosome number, and the haploid recombinant is isolated.

The parasexual process, or mitotic recombination, provides a natural mechanism for genetic exchange among imperfect fungi. It also provides a model for genetic manipulation in other somatic systems, such as mammalian tissue cultures.

Classification

Mycologic taxonomy is constantly being revised. The following is an abridged list of taxonomic groups of special interest in medical mycology. Classification within the fungous kingdom depends primarily on the type of spore formation that follows sexual reproduction (the "perfect" stage).

Class Myxomycetes

The myxomycetes, or slime molds, are fungi whose vegetative form is an aseptate mass of protoplasm called a plasmodium, which contains many diploid nuclei.

Class Zygomycetes

The zygomycetes and several smaller classes of fungi were once called phycomycetes, a term that is no longer taxonomically legitimate. These fungi include water molds, arthropod parasites, and various types of motile, flagellated fungi. In most cases the mycelium of these fungi lack crosswalls. The class Zygomycetes contains molds that reproduce sexually by the fusion of two compatible gametes to form a zygospore, and asexually with the production of conidia or sporangiospores.

Order Mucorales. Asexual reproduction occurs by the production of sporangia. It includes the genera *Rhizopus, Absidia, Mucor, Phycomyces, Pilobolus, Mortierella, Cunninghamella,* and *Saksenaea, among others.*

Order Entomophthorales. Asexual reproduction occurs by conidia that are forcibly discharged. Some genera are *Basidiobolus, Entomophthora,* and *Conidiobolus.*

Subdivision Ascomycotina
Sexual reproduction involves a sac-like structure, an ascus, that contains the spores produced following karyogamy and meiosis.

Order Eurotiales. Asci are produced within a cleistothecium, a closed structure composed of compact specialized hyphae. Many of the medically important, imperfect fungi for which sexual phases have been discovered have now been reclassified into this order.

FAMILY GYMNOASCACEAE. This family produces spheric cleistothecia with unorganized peridial hyphae. Genera include *Arthroderma, Nannizzia, Emmonsiella,* and *Ajellomyces,* which are the perfect genera for *Trichophyton, Microsporum, Histoplasma,* and *Blastomyces,* respectively.

FAMILY EUROTIACEAE. With these molds the peridial hyphae around the cleistothecium are more compact and tissue-like. This group contains *Pseudoallescheria boydii* and many of the perfect forms of *Aspergillus* species (e.g., *Sartorya, Eurotium, Emericella*) and of *Penicillum* species (e.g., *Talaromyces, Carpenteles*).

Subdivision Basidiomycotina
Sexual reproduction results in progeny spores produced on the surface of a special structure called a basidium. This is a large and complex group of fungi, some of whose members show a high degree of differentiation. The basidiomycetes include many plant and insect pathogens, the smuts and rusts, puffballs, bracket fungi, mushrooms, and *Filobasidiella,* the perfect genus of *Cryptococcus.*

Subdivision Deuteromycotina (Fungi Imperfecti)
This group contains all the fungi for which a sexual reproductive cycle has not been discovered. The taxa within this group do not reflect phylogenetic relationships.

Class Blastomycetes. This class contains the imperfect (asexual) yeasts, some of which may produce pseudohyphae or true hyphae. The genera include *Crypto-*

coccus, Candida, Torulopsis, Trichosporon, Pityrosporum, and
Rhodotorula.

Class Hyphomycetes. These fungi produce septate mycelia that reproduce asexually by conidia which are not
borne on a highly specialized tissue stroma. Most of the
pathogenic fungi belong here. Some of the genera are *Alternaria, Aspergillus, Blastomyces, Cladosporium, Epidermophyton,
Exophiala, Geotrichum, Histoplasma, Microsporum, Penicillium,
Philaphora, Sporothrix,* and *Trichophyton.*

FURTHER READING

Books and Reviews

Ainsworth GC, Sussman AS (eds): The Fungi. An Advanced Treatise. Vol I, The Fungal Cell. London, Academic Press, 1965

Ainsworth GC, Sussman AS (eds): The Fungi. An Advanced Treatise. Vol II, The Fungal Organism. London, Academic Press,
1966

Ainsworth GC, Sussman AS (eds): The Fungi. An Advanced Treatise. Vol III, The Fungal Population. London, Academic Press,
1968

Ainsworth GC, Sparrow FK, Sussman AS (eds): The Fungi. An
Advanced Treatise. Vol IVA, Ascomycetes and Fungi Imperfec-
ti. London, Academic Press, 1973

Ainsworth GC, Sparrow FK, Sussman AS (eds): The Fungi. An
Advanced Treatise. Vol IVB, Taxonomic Review with Keys: Basidiomycetes and Lower Fungi. London, Academic Press, 1973

Alexopoulos CJ, Mims CW: Introductory Mycology, 3rd ed. New
York, Wiley, 1979

Ballou CE, Raschke WC: Polymorphism of the somatic antigen of
yeast. Science 184:127, 1974

Bartnicki-Garcia: Cell wall chemistry morphogenesis, and taxonomy of fungi. Annu Rev Microbiol 22:87, 1968

Burnett JH: Mycogenetics. London, Wiley, 1975

Burnett JH: Fundamentals of Mycology, 2nd ed. London, Arnold,
1976

Cabib E: Molecular aspects of yeast morphogenesis. Annu Rev
Microbiol 29:191, 1975

Calam CT: The evaluation of mycelial growth. Methods Microbiol
1:567, 1968

Carmichael JW, Kendrick WB, Conners IL, et al.: Genera of Hyphomycetes. Edmonton, Alberta Univ Press, 1980

Cole GT, Kendrick B (eds): Biology of Conidial Fungi. New York,
Academic Press, 1981

Deacon JW: Introduction to Modern Mycology. New York, Wiley,
1980

Farkas V: Biosynthesis of cell walls of fungi. Microbiol Rev
43:117, 1979

Gander JE: Fungal cell wall glycoproteins and peptido-polysaccha-
rides. Annu Rev Microbiol 28:103, 1974

Gilardi GL: Nutrition of systemic and subcutaneous pathogenic
fungi. Bacteriol Rev 29:406, 1965

Nickerson WJ: Symposium on the biochemical basis of morphogenesis in fungi. IV. Molecular basis of form in yeast. Bacteriol
Rev 27:305, 1963

Pontecorvo G: The parasexual cycle in fungi. Annu Rev Microbiol
10:393, 1956

Prescott DM: Methods in Cell Biology, Vol XI, XII. Yeast Cells.
New York, Academic Press, 1975

Rose AH, Harrison JS (eds): The Yeasts. Vol I, Biology of Yeasts.
New York, Academic Press, 1969

Rose AH, Harrison JS (eds): The Yeasts. Vol II, Physiology and
Biochemistry of Yeasts. New York, Academic Press, 1973

Rose AH, Harrison JS (eds): The Yeasts. Vol III, Yeast Technology. New York, Academic Press, 1973

Ross IK: Biology of the Fungi. New York, McGraw-Hill, 1979

Smith JE, Berry DR: An Introduction to Biochemistry of Fungal
Development. London, Academic Press, 1974

Smith JE, Berry DR (eds): The Filamentous Fungi. Vol II, Biosynthesis and Metabolism. London, Arnold, 1975

Smith JE, Berry DR (eds): The Filamentous Fungi. Vol III, Developmental Mycology. London, Arnold, 1978

Sypherd PS, Borgia PT, Paznokas JL: Biochemistry of dimorphism
in the fungus *Mucor.* Adv Microb Physiol 18:67, 1978

Webster J: Introduction to Fungi, 2nd ed. Cambridge, Cambridge
University Press, 1980

Selected Papers

Barkai-Golan R, Sharon N: Lectins as a tool for the study of yeast
cell walls. Exp Mycol 2:110, 1978

Cabib E, Farkas V: The control of morphogenesis: An enzymatic
mechanism for the initiation of septum formation in yeast. Proc
Natl Acad Sci USA 68:2052, 1971

Elmer GW, Nickerson WJ: Nutritional requirements for growth
and yeastlike development of *Mucor rouxii* under carbon dioxide.
J Bacteriol 101:595, 1970

Jones GR, Stewart-Tull DES: Antigenic analysis of yeast cell walls.
Sabouraudia 13:94, 1975

Lopez-Romero E, Ruiz-Herrera J, Bartnicki-Garcia S: Purification
and properties of an inhibitory protein of chitin synthetase
from *Mucor rouxii.* Biochem Biophys Acta 525:338, 1978

Novick P, Ferro S, Schekman R: Order of events in the yeast secretory pathway. Cell 25:461, 1981

Orlowski M, Ross JF: Relationship of internal cyclic AMP levels,
rates of protein synthesis and *Mucor* dimorphism. Arch
Microbiol 129:353, 1981

Ruiz-Herrera J, Lopez-Romero E, Bartnicki-Garcia S: Properties
of chitin synthetase in isolated chitosomes from yeast cells of
Mucor rouxii. J Biol Chem 252:3338, 1977

Ulane RE, Cabib E: The activating system of chitin synthetase
from *Saccharomyces cerevisiae.* J Biol Chem 251:3367, 1976

CHAPTER 83

Principles of Fungous Diseases

Fungi are ubiquitous microorganisms that, viewed collectively, represent a unique and fascinating biologic entity. They have profound effects, both good and bad, on other forms of life. The degradative enzymes produced in soil by many saprophytic fungi provide an essential function in the biologic recycling of organic matter. Cellulolytic fungi decompose vegetative debris into humus. Some of the yeasts are necessary for the processing of certain foods, such as bread, cheese, and alcoholic beverages. Higher fungi, mostly basidiomycetes, may be eaten directly. Many fungi are phytopathogens and exert tremendous economic influence on agriculture and the annual food supply.

In medicine, fungi are an important source of pharmacologically active compounds, including hallucinogens, adrenergic alkaloids, vitamins, mutagens, carcinogens, antibiotics, and potential anticancer substances. Basic researchers continue to employ fungal models to explore such fundamental areas as the control of genetic information, the cytoplasmic transmission of sensory stimuli, and the regulation of morphogenesis. The potential value of several fungous components as stimulators of various host defense mechanisms is being actively explored.

Types of Fungous Diseases

Fungi are able to cause human disease in three generalized ways:

1. Allergies may result from sensitization to specific fungal antigens.

2. Fungi may elaborate or indirectly generate substances that have a toxic effect.

3. Some fungi are able to cause infection, growing actively on an animal host.

Fungous Allergies

The respiratory tract of all animals, including humans, is constantly exposed to the aerosolized spores of many free-living fungi. These spores, or other fungous components, may contain potent allergens to which certain individuals may respond with a strong hypersensitivity reaction. The manifestation of these allergies does not require growth or even viability of the inducing fungus, although in some cases both infection and allergy may occur simultaneously.

The fungal spore count in outdoor air varies considerably with the season, time of day, geographic location, and weather conditions. The average spore concentration in normal outdoor air is approximately 10^5 spores per cubic meter. In enclosed areas with suitable growth conditions for the proliferation of fungal contaminants, such as farm buildings, the spore counts may exceed 10^9 spores per cubic meter. Individuals exposed to such heavy doses of fungous aerosols are likely to become sensitized to the fungous (and actinomycetous) antigens, which may lead to allergic reactions upon subsequent exposure. Depending upon the site of deposition of these allergens, patients may exhibit rhinitis, bronchial asthma, alveolitis, or generalized pneumonitis. Atopic individuals are more susceptible. Table 83-1 lists some of these well-characterized

TABLE 83-1. RESPIRATORY ALLERGIES (RHINITIS, BRONCHIAL ASTHMA, ALVEOLITIS) CAUSED BY FUNGI AND ACTINOMYCETES

Allergy	Source	Etiology
Cheese washer's lung	Cheese	*Penicillium casei*
Maltster's lung	Barley malt	*Aspergillus clavatus*
Maple-bark stripper's lung	Maple tree bark	*Cryptostroma corticale*
Sequoiosis	Redwood sawdust	*Aureobasidium pullulans, Graphium*
Suberosis	Cork	*Penicillium frequentans*
Wood-pulp worker's disease	Wood pulp	*Alternaria*
Farmer's lung	Stored hay	*Micropolyspora faeni, Thermoactinomyces vulgaris*
Bagassosis	Sugar cane	*Thermoactinomyces saccharii*
Humidifier lung	Humidifiers, air conditioners	*Thermoactinomyces vulgaris, Thermoactinomyces candidus*

syndromes. Farmer's lung is a classic example of extrinsic allergic alveolitis. In general, the clinical manifestations are determined by the immune responses of the host and the nature of the fungous challenge—namely, the particle size, antigenicity, and the amount of the inoculum. The clinical description, mechanisms of immunopathology, and management of these hypersensitivity diseases are covered in Chapter 18. These factors are also discussed in Chapter 87 in reference to one of the more ubiquitous genera, *Aspergillus*. In addition to specific medical treatment, these conditions are often controlled by eliminating the environmental hazard, or otherwise removing the patient from the allergen.

Mycotoxicoses

Fungi can generate substances with direct toxicity for humans and animals. Such toxins are secondary metabolites that are synthesized and secreted directly into the environment and include a variety of mycotoxins elaborated by mushrooms. Exposure to these toxins after their accidental ingestion results in a disease termed mycetismus, whose severity depends upon the amount and type of mycotoxin ingested. A recent history of mycophagy precedes the manifestation of symptoms by several hours. Heating of mycotoxins has little effect on reducing the toxicity. Table 83-2 summarizes the common clinical forms of mycetismus.

Amatoxins and Phallotoxins. The amatoxins and phallotoxins represent two important families of mycotoxins. The amatoxins are among the most potent. The phallotoxins, which are not absorbed by the gastrointestinal tract and are not a cause of mycetismus, have been studied in animals following injection. Both toxins are produced by poisonous mushrooms, such as *Amanita*, which may yield several toxins, including phalloidin, phalloin, α-, β-, and γ-amanitin. The amatoxins are cyclic octapeptides, and the phallotoxins are cyclic heptapeptides. The liver is

a target organ for both families of toxins. α-Amanitin binds to a subunit of RNA polymerase II and consequently interferes with mRNA and protein synthesis. Phalloidin binds to actin in cell membranes. Globular actin (G-actin), after binding adenosine triphosphate (ATP) and Ca^{2+}, is converted to the fibrous form (F-actin). Phalloidin binds to and stabilizes F-actin, causing vacuolization, membrane leakage, and disruption of the endoplasmic reticulum. Treatment for mushroom poisoning is largely supportive, as specific antidotes are not available. The basis for the efficacy of thioctic acid in some patients is not apparent. In addition to these and other direct toxins, fungal contamination and breakdown of grains and other foods may result in by-products that are toxic upon ingestion.

Aflatoxins and Other Tumorogenic Mycotoxins. Other fungi elaborate a variety of mutagens and carcinogens; their production is not essential for survival of the fungus. Although these toxins can be lethal or tumorogenic for animals, no direct evidence has yet linked any of the carcinogenic mycotoxins with human disease. The most potent and best characterized example is aflatoxin, of which eight varieties are produced by certain strains of *Aspergillus flavus* and other molds.

The structures of the two most potent aflatoxins, B_1 and G_1, are shown in Figure 83-1. Aflatoxins are converted in the host, e.g., by liver microsomal enzymes, into active, unstable compounds that bind to DNA, prevent base pairing, and induce frameshift mutations. Aflatoxin B_1, the most potent liver carcinogen, also induces many other molecular changes. Birds are extremely sensitive to aflatoxin. Sublethal amounts in the avian diet cause a multitude of dose-related effects, including fatty liver changes, reduced liver enzymes, anemia, decreased leukocyte chemotaxis and phagocytosis, and increased susceptibility to infection. Other mycotoxins with demonstrated carcinogenesis for experimental animals include ochratoxin, sporidesmin, zearalenone, and sterigmatocystin. As mentioned, there is no direct evidence to implicate mycotoxins in human dis-

TABLE 83-2. SUMMARY OF HUMAN MYCETISMUS (ABRIDGED)

Site of Involvement	Latent Period	Duration	Etiology	Mycotoxin	Mechanism of Action	Symptoms	Prognosis and Treatment
Gastrointestinal tract	Minutes to hours	36–72 hr	*Boletus satanas, Lactorius torminosus, Lepiota morgani, Russula emetica,* and others	Unidentified		Nausea and diarrhea; mild to severe	Spontaneous recovery
Gastrointestinal tract (choleratype) and parasympathetic nervous system	6–24 hr		*Amanita phalloides, Amanita virosa*	Amatoxin(s) (phallotoxins)	Cholinergic effect on smooth muscles and exocrine glands	Diphasic (1) violent vomiting, diarrhea, dehydration, muscle cramps; (2) renal and hepatic failure, confusion, perspiration, lacrimation, salivation, twitching, jaundice, coma	Second phase treated with atropine; often fatal; thioctic acid
	15–30 min		*Clitocybe* species, *Inocybe* species	Muscarine	Same	Violent gastrointestinal upset, perspiration, salivation; central nervous system symptoms: delirium, hallucination, or coma; high doses cause cardiac and/or respiratory failure	Same
	6–8 hr		*Helvella esculata*	Gyromitrin (volatilized with heating)	Gastrointestinal toxicity, hemolysis	Nausea, vomiting, diarrhea; hemoglobinuria, jaundice	Usually self-limiting
Central nervous system	30–60 min	5–10 hr	*Psilocybe cubensis, Psilocybe* species	Psilocybin		Hallucination	Spontaneous recovery

From Becker et al.: West J Med 125:100, 1976.

ease, but some provocative epidemiologic investigations have correlated liver damage in certain tribal populations with dietary exposure to aflatoxin. There is regular monitoring of levels of aflatoxins and other mycotoxins in grains, peanuts, and other foods frequently contaminated with the toxin-producing fungi.

Mycoses

The most common form of mycotic disease is infection—that is, actual growth of a fungus on a human or animal host. Mycosis is the term used to indicate an infection caused by a fungus. The names of many fungous infections are formed by coupling "mycosis" as a suffix to another word that designates the etiologic agent (e.g., coccidioidomycosis) or the site of involvement (e.g., otomycosis). In general, the establishment of an infection depends upon the host defenses, or lack thereof, the route of exposure to the fungus, the size of the inoculum, and the virulence of the organism. Chapters 84 through 87 provide detailed coverage of the ecology, mycology, pathogenesis, and immunology of the more prevalent infectious fungi, and the epidemiology, clinical symptoms, diagnosis, and treatment are indicated for the respective mycoses. The remainder of this chapter develops some general principles of the host–fungus interaction that pertain to exposure, infection, diagnosis, and treatment.

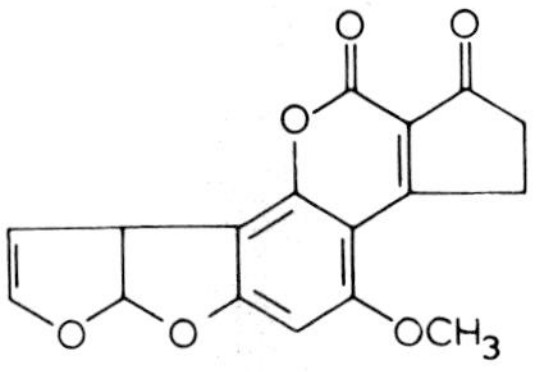

Figure 83-1. Structures of aflatoxins B₁ and G₁. *(From Fishbein and Falk: Chromatog Rev 12:42, 1970.)*

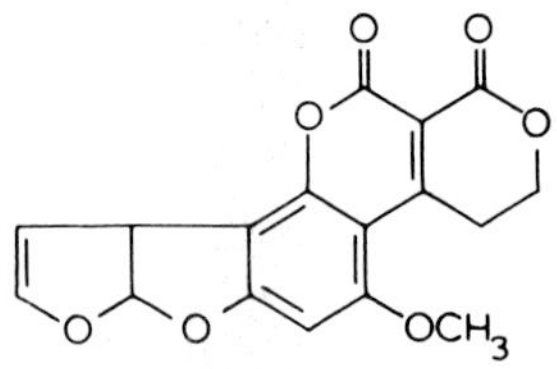

TABLE 83-3. MORTALITY RATES DUE TO MYCOSES FOR SELECTED YEARS, BASED UPON VOLUNTARY REPORTS TO THE CENTERS FOR DISEASE CONTROL*

Mycosis	1955	1960	1965	1970	1975
Aspergillosis	NR[†]	NR	NR	1.98	3.33
Blastomycosis	1.24	0.88	1.59	0.16	0.11
Candidiasis	NR	6.31	5.09	7.96	11.36
Coccidioidomycosis	4.06	3.21	2.84	2.19	3.17
Cryptococcosis	4.84	4.15	3.39	5.83	6.92
Histoplasmosis	3.47	4.67	4.05	2.92	3.12

*Number of mycotic deaths per 100,000 adult deaths. Calculated from Morbidity and Mortality Weekly Reports, Annual Summary and Vital Statistics Reports, Vols 7–26, U.S. Dept Health, Education, and Welfare.
[†]NR, Mycotic deaths not recorded by Centers for Disease Control.

Mycosis

Incidence

Because none of the mycoses is a reportable disease, the prevalence of fungal infections is unknown. An accurate census cannot be obtained either by the Centers for Disease Control in the United States or by the World Health Organization. It is apparent, however, that dermatophytoses are among the most common infectious diseases in the world. Furthermore, they have always been prevalent; the lesions are superficial, and historical descriptions of ringworm date from antiquity. Other historical accounts are compatible with many of the systemic mycoses. It has been suggested, for example, that the curse of Tutankhamen's tomb, was in fact, residual spores of *Histoplasma capsulatum*.

Several mycologists have attempted to estimate the incidence of systemic mycoses. Table 83-3 gives the annual deaths voluntarily reported to the Centers for Disease Control for selected years from 1955 to 1975 and adjusted for the total number of annual deaths. For several reasons these figures represent a gross underestimation of the prevalence of these infections. Not all deaths due to fungal infection are reported to the Public Health Service, as reporting is not a legal requirement. Many mycotic deaths are probably undiagnosed, misdiagnosed, or unspecified because they are secondary to some preexistent malady. Because the mortality rate for primary systemic infection with any of the three dimorphic fungi endemic to this country (blastomycosis, coccidioidomycosis, and histoplasmosis) is probably less than 5 percent, the number of patients with active clinical diseases in any year is at least 20 times greater than the number of deaths. Many fungal infections are chronic; the management and care of individual patients may extend for years. Furthermore, the incidence of mycotic disease is not geographically uniform. Most mycoses are caused by fungi that reside saprophyti-

cally in nature, but their distribution varies considerably. For this reason, the attack rates and incidence of specific mycoses vary widely in different areas. Despite the lack of more reliable data, it is apparent from Table 83-3 that in recent years there has been a steady increase in the overall prevalence and mortality from the opportunistic mycoses, aspergillosis, candidiasis, and cryptococcosis. Conversely, the attack rates of the primary pathogens have remained stable or declined.

Hammerman et al. have utilized data from the Commission on Professional and Hospital Activities to estimate the incidence of mycoses requiring hospitalization. Their data base consisted of vital statistics and discharge diagnoses from over 10 million hospital records obtained from participating acute-care hospitals during the year 1970. From these data, regional and national projections of incidence and mortality were made for specific diseases. For histoplasmosis and coccidioidomycosis, the prevalence of skin test reactivity in the population was used for extrapolation. The geographic distribution of histoplasmosis, coccidioidomycosis, and blastomycosis is well circumscribed. The projected hospitalized cases of histoplasmosis per 100,000 population was between 2.2 and 3.2 in the endemic area. Coccidioidomycosis is largely confined to the southwestern states, and the projected number of hospitalized cases per 100,000 population in these states varied from 1.2 to 24.7. Overall incidence of other mycoses (as the primary diagnosis) requiring hospitalization was estimated between 0.02 and 0.09 per 100,000. Table 83-4 shows the actual number of cases reported, the case fatality rate, and the projected number of mycoses that would require hospitalization throughout the United States in 1970. These estimates on incidence are conservative, as patients not requiring hospitalization have been excluded; furthermore, none of the Veterans Administration or other federal hospitals contributed to the data base. It is apparent from Table 83-4 that certain mycotic agents, such as *Coccidioides immitis* and *Blastomyces dermatitidis*, occur more often as primary pathogens, whereas others, such as *Candida*, are encountered predominantly as opportunists. In all cases, the mortality rates are double for secondary or opportunistic fungal infection in a previously compromised

TABLE 83-4. NUMBER OF CASES OF SYSTEMIC MYCOSES AND PERCENTAGE OF FATALITIES REPORTED TO COMMISSION ON PROFESSIONAL AND HOSPITAL ACTIVITIES AND PROJECTED NUMBER OF HOSPITALIZED CASES FOR THE UNITED STATES DURING 1970

Infection	Fungal Disease as Primary Diagnosis			Fungal Disease as Secondary Diagnosis		
	Cases Reported	Case Fatality Rate (percent)	Cases Projected	Cases Reported	Case Fatality Rate (percent)	Cases Projected
Aspergillosis	42	7.1	144	53	22.6	235
Blastomycosis	39	7.7	145	14	14.3	35
Candidiasis (systemic)	31	6.5	88	76	13.2	250
Coccidioidomycosis	270	1.5	1389	146	4.1	695
Cryptococcosis	39	18.0	139	38	36.8	121
Histoplasmosis	449	1.6	1462	855	2.6	2544

From Hammerman, et al.: Sabouraudia 12:33–45, 1974.

patient, and, as reflected in Table 83-3, these mycoses are on the rise. Hammerman et al. also reported that the average time of hospitalization varied from 11 days for patients with acute histoplasmosis to 30 days for those with aspergillosis.

Portal of Entry

To establish an infection, a potential fungal pathogen must first enter the body. This can occur in only a limited number of ways. First, fungi may invade the body directly through the skin. The skin presents a strong barrier to fungal penetration. Most fungi lack the enzymes or other invasive properties that would permit them to penetrate intact skin; however, if the skin surface has been abraded, burned, macerated, or its integrity otherwise compromised, fungi, as well as other microorganisms, are able to gain access to the cutaneous and subcutaneous tissue, and the opportunity for infection vastly increases. Several studies have shown that the chances of establishing an experimental cutaneous mycosis with *Candida albicans* or the dermatophytes are greatly enhanced by abrading the skin before exposure to the infectious agent. The protective defenses of the skin have not been completely defined, but it is clear that both anatomic and physiologic factors are involved. For example, dermatophytes are fungi that preferentially infect the skin, hair or nails (see Chap. 86). Certain chemicals that are often present on the skin surface (amino acids, fatty acids) and lipids found in sebum are able to inhibit the growth or spore germination of some dermatophytes. Although lysozyme is not a potent antifungal agent, the dermatophytes may be actively deterred by other factors, such as hormone-induced changes in the skin or hair chemistry, salinity, pH, and the secretion of specific growth inhibitors.

Inhalation is the most important method of initiating a fungal infection. Although throughout life the respiratory tract is exposed to a barrage of airborne fungi, the incidence of respiratory mycoses is relatively low. One of the

natural host barriers is the anatomy of the respiratory tract, which determines the depth to which particles can be inhaled. The size of the inhaled fungal cells or spores will delimit the extent of penetration. Particles 10 μm or larger may be deposited on the tracheal or nasal epithelium and are usually retained and then expelled from this area. Particles 5 to 10 μm in diameter may penetrate the bronchioli, but they are usually removed by bronchial secretions and ciliated epithelial lining of the respiratory tract. Therefore, fungal inocula greater than 5 μm in diameter usually are limited to local colonization of the nasal sinuses or bronchial tree. Alternatively, their surface antigens may induce a local allergic response; this reaction is also associated with inert allergens, such as pollen and dust particles. Particles or spores less than 5 μm in diameter may be inhaled to the alveoli.

Once in the alveolus, the microorganism is confronted by surfactant, serum elements, alveolar macrophages, and a subsequent inflammatory response. The host response may culminate in either inactivation of the fungus, after which it will be cleared slowly from the host, or protracted interaction between the fungus and the host. This transient episode is characterized by a mild or asymptomatic lung infection, pulmonary infiltration, sensitization of the host to fungal antigens, and, eventually, a stabilized, or arrested lesion (usually granulomatous) that becomes sealed off for an indefinite period. Alternatively, the host defenses may be inadequate to quell the fungal challenge, and a progressive infection ensues. Symptomatology varies considerably; the developing mycosis may be acute, subacute, or chronic; it may remain localized in the lung or metastasize. Whether the fungal challenge is successfully contained or results in clinical disease depends upon the dynamic interaction of three determinants: the host defenses (both immunologic and nonspecific), the pathogenic potential of the fungus, and the extent of exposure.

In addition to the skin and respiratory tract, the other two epithelial surfaces infrequently serve as portal of entry for fungal infection. For example, the urogenital tract is occasionally bridged by endogenous fungi such as *C. albicans* when their growth is enhanced by physiologic changes in the local mucosa. Similarly, the gastrointestinal tract may become a source of fungal infection following changes induced by age, trauma, neoplasm, or an imbalance in the normal flora. As discussed in Chapter 87, there is increasing evidence that yeasts, such as *C. albicans*, have cell wall determinants that mediate their specific attachment to receptors on the membrane surfaces of epithelial cells.

Iatrogenic inoculation—via contaminated indwelling catheters, during surgery, following antibacterial or immunosuppressive chemotherapy, administration of steroids, or radiation treatment—has dramatically increased the incidence of opportunistic mycoses in recent years. Under these conditions fungi are either introduced into the host directly, often bypassing normal defense mechanisms, or the host defenses are sufficiently suppressed to permit enhanced fungous invasion (see Chaps. 84 and 87).

Classification

No clinical classification of mycoses is entirely adequate. Fungous infections are usually organized on the basis of the general body area predominantly involved. This classification, given in Table 83-5 reflects both the portal of entry and major site of involvement. For example, most primary systemic mycoses are initially pulmonary whereas subcutaneous mycoses follow traumatic inoculation of the skin. Exceptions exist, however, as systemic or deep mycotic agents can produce cutaneous and subcutaneous lesions following dissemination from the lung as well as from primary cutaneous inoculation. Similarly, many subcutaneous mycoses may have systemic manifestations.

The term opportunistic mycoses is not precise. It generally refers to fungous infections in patients whose host defenses have been somehow compromised. However, under certain circumstances, any pathogenic fungus may be opportunistic. Furthermore, the agents generally considered to be opportunistic may also cause primary disease. Because the range of host-compromising conditions has not been fully defined, it may be that many more mycoses are opportunistic than detected as such. Opportunistic is a frequently and loosely used term, and it is probably best applied functionally and not confined to specific pathogens.

Pathogenesis

Fungous Determinants. The fungous properties required for a mycotic infection to develop can be summarized in general terms as follows. Initially, contact between the host and fungal pathogen must be established. Most fungi that can produce disease exist saprophytically in a natural reservoir and are acquired by exogenous contact. Some fungi (e.g., *C. immitis*) are confined to specific geographic regions, while others (e.g., *Aspergillus fumigatus*) are ubiquitous. The conditions of exposure (inoculum size,

route, host immunity) will determine whether or not infection ensues. A few common pathogens are endogenous and are either part of the human flora (*Candida* species) or adapted for persistent infection (certain dermatophytes).

Another important determinant of whether infection follows contact is the inherent virulence of the fungus. Tissue-reactive enzymes, irritants, attachment to host cells, antiphagocytic properties, and inflammatory components have been described for many infectious fungi, but virulence factors as clear-cut and potent as bacterial exotoxins or endotoxins have not been discovered. Although experimental infections have shown wide differences in virulence among various fungal species or strains of a single species, the explanation for these differences is poorly understood.

Fungal pathogens may produce hyphae or spheric structures (yeasts, sporangia, sclerotic cells) in vivo, and this morphology influences pathogenesis. Anatomically, hyphae tend to penetrate the lumen of vessels and lymphatics. Yeasts are less confined. They can be transported via the bloodstream and lymphatics to virtually any part of the body and therefore, in theory, their pathogenic potential is unlimited.

The concept of infection and disease must be clearly understood. Some fungi are completely superficial; they grow on the host without invasion and cause minimal irritation. Two examples are piedra (the formation of nodules on hair), and the colonization of the external ear or nasal sinuses by *Aspergillus* species. Aspergilli can also colonize the bronchial tree and pulmonary cavity. In other situations, many fungi cause asymptomatic infection in spite of tissue involvement. Some colonizing fungi, however, have the potential to cause more serious disease. Fungal disease, in most cases, refers to symptomatic infection and is generally associated with more extensive damage to host tissue.

Many fungal pathogens closely resemble each other in morphology, taxonomic relationship, antigenic determinants, cell wall composition, growth requirements, and physiologic properties. These similarities are relevant to the diagnosis and treatment of specific fungal infections.

Host Factors. From the high prevalence of fungi in the environment and the low incidence of mycotic disease, it is obvious that most humans are highly resistant to fungal invasion. During infection both humoral and cellular immune responses are stimulated. In general, the antibodies that form are not protective, although they may be useful in establishing a diagnosis and evaluating prognosis. Resistance to several mycoses (e.g., disseminated coccidioidomycosis, chronic mucocutaneous candidiasis) is clearly associated with the thymus-dependent immune system. Furthermore, patients are predisposed to the opportunistic mycoses listed in Table 83-5 by a cellular immunodeficiency resulting from various sarcomas (and their treatments), steroids, and/or immunosuppressive therapy following organ transplantation. The importance of the host immune response in controlling mycotic infections, where understood, is discussed more fully under the individual

TABLE 83-5. CLINICAL CLASSIFICATION OF MYCOTIC INFECTIONS (ABRIDGED)

Area of Predominant Involvement	Mycosis	Etiology
Superficial	Pityriasis versicolor	*Malassezia furfur*
	Tinea nigra	*Exophiala werneckii*
	White piedra	*Trichosporum beigelli*
	Black piedra	*Piedraria hortai*
Cutaneous	Dermatophytosis	*Microsporum* species, *Trichophyton* species, and *Epidemophyton floccosum*
	Candidiasis of skin, mucous membranes, and nails	*Candida albicans*, other *Candida* species
Subcutaneous	Sporotrichosis	*Sporothrix schenckii*
	Chromomycosis	*Philaphora verrucosa*, *Fonsecaea pedrosoi*, and others
	Mycetoma	*Pseudoallescheria boydii*, *Madurella mycetomatis*, and others
	Subcutaneous phycomycosis	*Basidiobolus haptosporus*
	Rhinoentomophthoromycosis	*Entomophthora coronata*
	Rhinosporidiosis	*Rhinosporidium seeberi*
	Lobomycosis	*Loboa loboi*
Systemic	Primary fungus infections	
	Coccidioidomycosis	*Coccidioides immitis*
	Histoplasmosis	*Histoplasma capsulatum*
	Blastomycosis	*Blastomyces dermatitidis*
	Cryptococcosis	*Cryptococcus neoformans*
	Paracoccidioidomycosis	*Paracoccidioides brasiliensis*
	Opportunistic fungus infections	
	Candidiasis, systemic	*Candida albicans*, other *Candida* species
	Aspergillosis	*Aspergillus fumigatus*, other *Aspergillus* species
	Mucormycosis	Species of *Rhizopus*, *Absidia*, *Mucor*, and others

mycoses. In some situations the neutrophil appears to be necessary for optimal defense.

Other host factors that have been shown to influence susceptibility to fungal infection in specific instances include age, sex, race, heredity, and physiologic condition.

Dynamics of Host–Fungus Interactions. With the possible exception of some mycoses that may be limited to the subcutaneous or mucocutaneous tissue by anatomic or temperature-dependent mechanisms, the pathogenesis of fungous infections is determined by the interplay of host defenses with the invading fungus. After a fungal cell has bridged one of the body surfaces, it confronts numerous nonspecific, immunologic, humoral, and cellular host defenses.

The nonspecific humoral defenses include inhibitors of fungal growth in serum, hormones that may inhibit growth or regulate phagocytosis, serum substances that clump yeast cells, and opsonins. Both neutrophils and macrophages are able to phagocytize most pathogenic yeasts, but the extent of phagocytosis is inversely related to the size of the yeast cells. The ability to kill ingested yeasts, though inefficient, may be adequate to ensure resistance to infection. Activated macrophages are more phago-

cytic but the killing ability of macrophages from different sources varies considerably. The antifungal activity of natural killer cells has recently been described. Specific antibodies do not appear to be directly injurious to most pathogenic fungi; however, antibodies can serve as opsonins and enhance phagocytosis. Lymphocytes from sensitive individuals may serve a protective function directly or through the mobilization of macrophages or modulation of other immunologic components.

Blocking factors and suppressor lymphocytes have also been implicated as possible mechanisms enhancing fungous disease.

Diagnosis

Three diagnostic approaches are routinely applied in clinical mycology. The direct microscopic examination of specimens can sometimes establish a diagnosis and usually narrow the etiologic possibilities. Pathogenic fungi are large enough to be observed in tissue or body fluids digested with 10 percent potassium hydroxide. If few organisms are present, material can be concentrated and/or stained with fluorescent antibody reagents or histochemical stains for fungous cell wall material.

Regardless of the results of a direct examination, specimens are cultured for the pathogenic fungi. Nonsterile specimens (e.g., skin scrapings, sputa) are planted on media containing antibiotics to inhibit bacteria and nonpathogenic fungous contaminants. The routine fungous culture medium is Sabouraud's agar, consisting of 2 or 4 percent glucose, 1 percent neopeptone, and 2 percent agar. Richer media, such as brain–heart infusion agar with sheep blood, are also used in the diagnosis of systemic mycoses. Fungous isolates are identified by appropriate morphologic, physiologic, and/or immunologic properties.

Serology is the third laboratory tool with established diagnostic and prognostic value for certain mycotic infections. Individual tests and their limitations are discussed under specific infections. Mycoserologic techniques include the measurement of specific antibodies, delayed and immediate hypersensitivity, and, in some cases, fungal antigen. Complete immunologic evaluation of patients, both in vivo and in vitro (i.e., quantity and functional capacity of immunoglobulins, subpopulations of T and B cells, and lymphokine production), yields information of prognostic value in several mycoses.

With some mycoses, additional diagnostic methods (e.g., radiology, clinical chemistry) can be helpful.

Therapy

A limited number of antibiotics are used in the treatment of fungal infections. Among these are the polyenes (Chap. 9), 5-fluorocytosine, and the imidazoles. Table 83-6 sum-

TABLE 83-6. SUMMARY OF ANTIFUNGAL ANTIBIOTICS (ABRIDGED)

Name	Indications	Adult Dosage	Side Effects	Comments
Nystatin	Oral thrush	Solution of 1–2×10^5 units taken orally daily for 1–2 weeks	Rarely diarrhea, nausea	
	Cutaneous candidiasis	Topical ointment of 10^5 units applied 2–3 times daily	Rarely hypersensitivity	
Amphotericin B	Systemic mycoses	1–5 mg/day taken intravenously, increased to 0.3–0.6 mg/kg/day in 5 percent glucose with mannitol; total dose 1–3 g (light sensitive)	Manifold; especially renal toxicity; thrombophlebitis, hypokalemia, anemia, chills, fever, headache, nausea, anorexia	Monitor BUN and creatinine levels; intrathecal administration for some meningitis cases
Flucytosine	Candidiasis Cryptococcosis Aspergillosis Chromomycosis	150 mg/kg/day, taken orally in 4 doses	Skin rash, diarrhea, nausea, hematopoietic (aplastic anemia, granulocytopenia) and hepatic toxicities	Resistant organisms emerge; check WBC and liver function weekly; patients with impaired renal function may accumulate toxic levels
Griseofulvin	Dermatophytosis	0.5–1.0 g daily taken orally after fatty meal	Some GI distress; rarely neurologic symptoms	
Tolnaftate	Dermatophytosis of skin	Topical 1 percent solution in cream base		Not effective against hair and nail infections
Miconazole	Systemic mycoses	0.2–1.2 mg taken intravenously every 8 hr for 3–6 weeks	Phlebitis, nausea, anemia, hyponatremia, pruritis, rash, cardiac arrhythmia; rarely marrow, liver toxicity	May give intrathecally
Miconazole nitrate	Dermatophytosis	Topical; 2 percent in cream base		
Ketoconazole	Chronic mucocutaneous candidiasis Superficial candidiasis Dermatophytosis Systemic mycoses, especially Paracoccidioidomycosis	0.2–1.2 g taken orally in multiple daily doses for 1–6 months or more	Nausea, vomiting, rash; rarely liver toxicity, gynecomastia, endocrine dysfunction	Monitor liver function
2-Hydroxy-stilbamidine	Blastomycosis	225 mg daily taken intravenously 30–90 min; total dose 8–16 g (light sensitive)	Skin rash, nausea, others	

BUN, blood urea nitrogen; GI, gastrointestinal; WBC, white blood cell count.

marizes the current applications of these and other chemotherapeutics. More specific information is discussed in Chapters 84 through 87.

In addition to antibiotics, other therapeutic approaches have been applied to certain mycoses, including topical chemical preparations, heat, surgery, and administration of transfer factor.

FURTHER READING

Books and Reviews

Abramovici A: Mycotoxins and abnormal fetal development. Contrib Microbiol Immunol 3:81, 1977

Ajello L: Comparative ecology of respiratory mycotic disease agents. Bacteriol Rev 31:6, 1967

Al-Doory Y (ed): The Epidemiology of Human Mycotic Disease. Springfield, Ill, Thomas, 1975

Becker CE, Tong TG, Buerner U, et al.: Diagnosis and treatment of *Amanita phalloides*-type mushroom poisoning. Use of thioctic acid. West J Med 125:100, 1976

Bennett JE: Chemotherapy of systemic mycoses. N Engl J Med 290:30, 1974

Bode FR, Pare JAP, Fraser RG: Pulmonary diseases in the compromised host. Medicine 53:255, 1974

Buechner HA (ed): Management of Fungus Diseases of the Lungs. Springfield, Ill, Thomas, 1971

Cohen J: Antifungal chemotherapy. Lancet 2:532, 1982

Conant NF, Smith DT, Baker RD, et al.: Manual of Clinical Mycology, 3rd ed. Philadelphia, Saunders, 1971

Drutz DJ: Newer antifungal agents and their use, including an update on amphotericin B and flucytosine. In Remington JS, Swartz MN (eds): Current Clinical Topics in Infectious Diseases. New York, McGraw-Hill, 1982, p 97

Emmons CW, Binford CH, Utz JP, et al.: Medical Mycology, 3rd ed. Philadelphia, Lea & Febiger, 1977

Feigin RD, Shearer WT: Opportunistic infection in children. Parts I-III. Pediatr 87:507,677,852, 1975

Hoeprich PD: Chemotherapy of systemic fungal disease. Annu Rev Pharmacol Toxicol 18:205, 1978

Howard DH: Mechanisms of resistance in the systemic mycoses. In Nahmias AJ, O'Reilly RJ (eds): Immunology of Human Infection. New York, Plenum, 1981, p 475

Kobayaski GS, Medoff G: Antifungal agents: Recent developments. Annu Rev Microbiol 31:291, 1977

Koneman EW, Roberts GD, Wright SF: Practical Laboratory Mycology, 2nd ed. Baltimore, Williams & Wilkins, 1978

Kong YM, Levine HB: Experimentally induced immunity in the mycoses. Bacteriol Rev 31:35, 1967

Lincoff G, Mitchel DJ: Toxic and Hallucinogenic Mushroom Poisoning. New York, Van Nostrand Reinhold, 1977

Litten W: The most poisonous mushrooms. Sci Am 232:90, 1975

McGinnis MR: Laboratory Handbook of Medical Mycology. New York, Academic Press, 1980

Moore GS, Jaciow DM: Mycology for the Clinical Laboratory. Reston Va, Reston Pub Co, 1979

Palmer DF, Kaufman L, Kaplan W, et al.: Serodiagnosis of Mycot-ic Diseases, Springfield, Ill, Thomas, 1977

Pepys J: Allergy. In Dick G (ed): Immunological Aspects of Infectious Diseases. Baltimore, Univ Park Press, 1979, p 215

Polak A: 5-Fluorocytosine. Contrib Microbiol Immunol 4:158, 1977

Rippon JW: Medical Mycology, 2nd ed., Philadelphia, Saunders, 1982

Rosen PP: Opportunistic fungal infections in patients with neoplastic diseases. Pathol Annu 11:255, 1976

Rumack BH, Salzman E: Mushroom Poisoning: Diagnosis and Treatment. Boca Raton, Fla, CRC Press, 1978

Salvin SB, Neta R: Immunopathology of mycotic infections. In Rose NR, Siegel BV (eds): the Reticuloendothelial System, A Comprehensive Treatise, Vol 4, Immunopathology. New York, Plenum, 1983, p 145

San-Blas G: The cell wall of fungal human pathogens: Its possible role in host-parasite relationships. A review. Mycopathologia 79:159, 1982

Speller DCE: Antifungal Chemotherapy. Chichester, Wiley, 1980

Thompson RA: Clinical features of central nervous system fungus infection. Adv Neurol 6:93, 1974

Warnock DW, Richardson MD (eds): Fungal Infection in the Compromised Patient. Chichester, Wiley, 1982

Weiland T, Faulstich H: Amatoxins, phallotoxins, phallolysin, and antamanide: The biologically active components of poisonous *Amanita* mushrooms. Crit Rev Biochem 5:185, 1978

Williams DM, Krick JA, Remington JS: Pulmonary infection in the compromised host. Part I. Am Rev Resp Dis 114:359, 1976

Wogan GN: Mycotoxins. Annu Rev Pharmacol 15:437, 1975

Wright DE: Toxins produced by fungi. Annu Rev Microbiol 22:269, 1968

Selected Papers

Fraser DW, Ward JI, Ajello L, et al.: Aspergillosis and other systemic mycoses. The growing problem. JAMA 242:1631, 1979

Haley LD, Trandel J, Coyle MB: Practical Methods for the Culture and Identification of Fungi in the Clinical Microbiology Laboratory. Cumitech 11. Washington, D.C., American Society for Microbiology, 1980

Hammerman KJ, Powell KE, Tosh FE: The incidence of hospitalized cases of systemic mycotic infections. Sabouraudia 12:33, 1974

Kaufman L, Standard P: Immuno-identification of cultures of fungi pathogenic to man. Curr Microbiol 1:135, 1978

McCormick DJ, Arbel AJ, Gibbons RB: Nonlethal mushroom poisoning. Annu Intern Med 90:332, 1979

Mote RF, Muhm RL, Gigstad DC: A staining method using acridine orange and auramine O for fungi and mycobacteria in bovine tissue. Stain Technol 50:5, 1975

Schlueter DP, Fink JN, Hensley GT: Wood-pulp workers' disease: A hypersensitivity pneumonitis caused by *Alternaria*. Ann Intern Med 77:907, 1972

Stobo JD, Paul S, Van Scoy RE, Hermans PE: Suppressor thymus-derived lymphocytes in fungal infection. J Clin Invest 57:319, 1976

Stoddert RW, Herbertson BM: The use of lectins in the detection and identification of human fungal pathogens. Biochem Soc Trans 5:233, 1977

CHAPTER 84

Systemic Mycoses

The major systemic mycoses, coccidioidomycosis, histoplasmosis, blastomycosis, paracoccidioidomycosis, and cryptococcosis, are all caused by fungi that exist in nature. Inhalation of any one of these fungous cells can lead to pulmonary infection, which may or may not be symptomatic. Dissemination may occur to other parts of the body. With the exception of cryptococcosis, which is widespread, the prevalence and geographic distribution of these mycoses are delimited. Except for a few extremely rare cases, there is no evidence of transmission among humans or animals.

The mycotic agents of coccidioidomycosis, histoplasmosis, blastomycosis, and paracoccidioidomycosis are thermally dimorphic fungi.

Coccidioidomycosis

Coccidioidomycosis is caused by *Coccidioides immitis,* a dimorphic fungus that normally lives in soil in a highly restricted geographic area. This fungus and the infection it causes are almost entirely limited to this endemic area. The organism was discovered in tissue from a fatal case and was named *Coccidioides* (*Coccidia*-like) because the tissue forms (spherules) resemble *Coccidia.* The species name, *immitis,* means not mild. Most early cases were diagnosed at autopsy, and until 1930, the disease was erroneously thought to be invariably severe and disseminated. It was recognized quite early that coccidioidomycosis is confined to the southwestern United States, contiguous regions of northern Mexico, and specific areas of Central and South America. The more common primary form is a mild, respiratory ailment, also called valley fever or San Joaquin Valley fever.

Much of the knowledge concerning coccidioidomycosis has been provided by Dr. Charles E. Smith and his students at Stanford. The ecology of *C. immitis* and the epidemiology of coccidioidomycosis were elucidated by Smith in a series of pioneering investigations that began in the 1930s and have since inspired many other mycologic studies. Smith discovered the natural reservoir of *C. immitis* by isolating it from soil samples collected throughout the Southwest and determined the environmental conditions under which the organism was propagated. From mycelial culture filtrates, he developed coccidioidin, the skin test antigen used to detect exposure to *C. immitis,* and conducted population studies of skin reactivity. His contributions have reached beyond the accumulation of information about coccidioidomycosis to the establishment of fundamental concepts of mycotic infection.

Morphologic and Cultural Characteristics
Life Cycle. As depicted in Figure 84-1, the life cycle of *C. immitis* encompasses at least four distinct morphologic structures that are produced under different conditions. In nature and in the laboratory, *C. immitis* grows as a mold (Fig. 84-1, A–E). It produces branching septate hyphae, and as the culture ages, characteristic arthroconidia are usually, but not invariably, produced in alternate hyphal cells (Figs. 84-1, E and 84-2). In older cultures, the arthroconidium-forming hypha fragments readily releasing unicellular, barrel-shaped arthroconidia, which often retain at the ends appendage-like remnants of wall material from adjacent cells. Arthroconidia are approximately 3 by 6 μm in size, are easily airborne, and are small enough to be inhaled into the alveoli. They are highly resistant to desiccation, temperature extremes, and deprivation of nutrients and may remain viable for years. Under appropriate growth conditions, the arthroconidia will germinate to recycle the saprophytic mycelial phase (Fig. 84-1, B). In the infected host, *C. immitis* exists as spherules—spherical thick-walled structures, 15 to 80 μm in diameter—that are filled with a few to several hundred endospores. The endospores, which are 2 to 5 μm in size, are freed when the spherules rupture and, in turn, enlarge to mature spherules (Fig. 84-1, F–S). Occasional reports have described hyphae as well as spherules in the tissues and sputum of patients with coccidioidal cavities of the lungs. Spherules can be produced in the laboratory on a complex medium at 39C to 40C under 20 percent CO_2. As a spherule enlarges (Fig. 84-1, H–S) the nuclei undergo mitosis, the cytoplasm partitions around these nuclei, and a cell wall forms around each developing endospore. At maturation, the spherule ruptures to release its endospores.

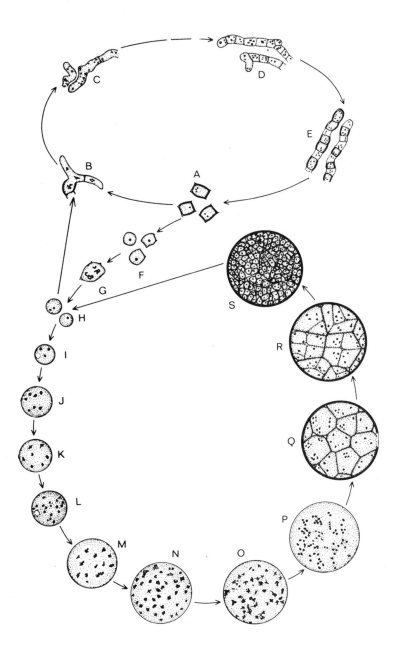

Figure 84-1. *Coccidioides immitis.* Life cycle in saprophytic (**A–E**) and parasitic (**F–S**) phases. **A.** Arthroconidia (3 × 6 μm). **B–E.** Mycelium. **H–S.** Formation of spherule containing endospores. **S.** Mature spherule (usually 30 to 60 μm diameter) containing endospores (2 to 5 μm). *(From Sun, Huppert: Sabouraudia 14:185, 1976.)*

Cultural Characteristics. On Sabouraud's agar at room temperature, different isolates produce a wide variety of colony types. Colonies may be white, gray, or brownish in color, with a powdery, woolly, or cottony texture. Because numerous infectious arthroconidia are produced in culture and are readily aerosolized in the dry state, extreme caution should be exercised when handling cultures of *C. immitis.* Tubes or plates should be tightly sealed and opened only under a safety hood that protects both the laboratory worker and the environment.

Although spherules have been obtained, in vitro growth of the tissue phase even under the most optimal conditions is seldom extensive.

Clinical Infection

Epidemiology and Ecology

In the United States, the geographic areas endemic for coccidioidomycosis and from which *C. immitis* can be isolated from the soil correspond to the Lower Sonoran life zone. These areas are characterized by a semiarid climate, alkaline soil, and characteristic indigenous desert plants and rodents. The endemic foci in Mexico, Argentina, and other scattered areas of Central and South America are ecologically very similar. Although *C. immitis* grows in the laboratory over a wide range of temperature, pH, and salt concentration and requires only glucose and ammonium

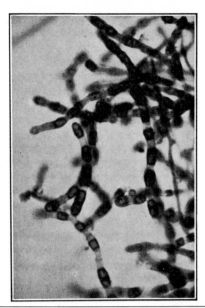

Figure 84-2. *Coccidioides immitis.* Typical arthroconidia formation in hyphae. From Sabouraud's glucose agar at room temperature. × 736.

salts to grow, it has never become established in soils outside the endemic area, in spite of transmission to other regions by infected animals and fomites. The mycelia, which can be found several inches beneath the soil surface, are recovered at the surface after the spring rains. As the weather becomes hot and dry, the mycelia convert to infectious arthroconidia. This accounts for the peak infection rate during the summer. Studies have shown that *C. immitis* is inhibited by other microorganisms, cultivated soil, or treatment with various chemicals. However, none of these factors explains its restricted habitation. In the endemic area, natural infections occur among animals (e.g., desert rodents, dogs, and cattle).

Outbreaks of primary infection have been reported among individuals simultaneously exposed to a heavy aerosol of arthroconidia. Coccidioidomycosis may thus be considered an occupational hazard for construction workers, archeology students, and others who disrupt the soil in the endemic areas. In a similar manner, many cases of acute disease developed subsequent to a severe wind storm in California in 1977, when contaminated soil was blown from the San Joaquin Valley far north and west, exposing large populations of unsensitized individuals.

Skin test surveys with coccidioidin confirm the areas of endemicity defined by isolation of *C. immitis* from soil. Within the endemic areas, which include portions of California, Arizona, New Mexico, Nevada, Utah, and Texas, the percent reactivity varies, with the highest rate being found in Phoenix and Tucson, Arizona, and Kern County, California. Isolated cases of coccidioidomycosis that occur outside the endemic areas are attributed to fomite transmission of the arthroconidia or to patient travel through the endemic area.

Virtually everyone who inhales the arthospores of *C. immitis* becomes infected and acquires a positive delayed hypersensitivity response. Approximately half the infections are benign and most of the others are symptomatic but self-limited. Approximately 1 percent of these cases will disseminate. Some individuals, however, have a much greater than average risk of developing disseminated disease following primary infection. These include persons in certain racial groups, namely, blacks, Filipinos, Latin Americans, and American Indians. This apparent racial predilection for severe disease may be confirmed by correlation with human genetic markers, such as HLA type. In addition to race, males, women in the third trimester of pregnancy, persons with cellular immunodeficiency, and individuals at the age extremes are more susceptible to severe disease.

Pathogenesis

Although arthroconidia and endospores are readily engulfed within phagosomes by alveolar macrophages, the intracellular fusion of lysosomal granules with the phagosomes is inhibited and neither form of *C. immitis* is killed. However, following activation of macrophages with either immune T cells or lymphokines, phagosome–lysosome fusion and killing are enhanced. Many of the patients who develop disseminated coccidioidomycosis have depressed cell-mediated immunity. Recent in vitro experiments by Cantanzaro indicate that certain adherent cells from patients with active dissemination are able to supress T cell activity by localized enhancement of prostaglandin production.

Clinical Manifestations

Primary Coccidioidomycosis. With the rare exception of cutaneous inoculation, the primary form follows inhalation of arthroconidia, and in most individuals, the infection is asymptomatic. Others may have fever, chest pain, cough, or weight loss. Radiographic examination often reveals discrete nodules in the lower lobes. Primary pulmonary coccidioidomycosis has an incubation period of 10 to 16 days and usually resolves without complication in 3 weeks to 3 months. A small percentage of patients retain cavities, nodules, or calcifications, but endogenous reactivation of residual pulmonary lesions is rare.

Up to 20 percent of patients with primary coccidioidomycosis manifest allergic reactions, usually erythema nodosum or multiforme. These appear with the primary symptoms, are very painful, and persist for approximately 1 week. The allergic manifestations are associated with strong immunity and a good prognosis.

Disseminated Coccidioidomycosis. Disseminated or secondary coccidioidomycosis usually develops within a few months as a complication of the primary form. The numerous forms of secondary coccidioidomycosis include chronic and progressive pulmonary disease, single or mul-

tiple extrapulmonary dissemination, or generalized systemic infection. Chronic pulmonary coccidioidomycosis usually
involves a single, thin-walled cavity. Dissemination may be
fulminant or chronic, with periods of remission and exacerbation. Extrapulmonary lesions most frequently involve
the meninges, skin, and bone. Chronic cutaneous coccidioidomycosis develops from initial lesions that usually appear on the face or neck and that, over a period of years,
evolve into thick, raised, verrucous lesions with extensive
epithelial hyperplasia. Bone involvement may accompany
generalized systemic disease. Both osteomyelitis of long
bones, vertebrae, and other bones and arthritis may develop. Draining sinus tracts may evolve from subcutaneous
and osseous lesions.

Skin Test

A crude toluene extract of a mycelial phase culture filtrate,
coccidioidin, is used for skin testing. The reaction is of the
delayed type and is considered positive if induration exceeds 5 mm in diameter. Another *C. immitis* antigen, prepared from cultured spherules and termed spherulin, is
more sensitive, but possibly less specific, than coccidioidin.
Skin testing with either antigen does not induce or boost
an immune response. The skin test becomes positive within 2 weeks after the onset of symptoms and before the appearance of precipitins and complement-fixing antibodies
and often remains positive indefinitely. A positive reaction
has no diagnostic significance without a history of conversion, but a negative test can be used to exclude coccidioidomycosis, except in patients with severe disseminated
coccidioidomycosis who may become anergic. Indeed, a
negative skin test in proved cases is associated with a
grave prognosis. Conversely, a positive skin test in healthy
subjects implies immunity to symptomatic reinfection.

Laboratory Diagnosis

Direct Examination. A definitive diagnosis of coccidioidomycosis requires the finding of spherules of *C. immitis* in
sputum, draining sinuses, or tissue specimens (Figs. 84-3
and 84-4). Clinical exudates should be examined directly
in 10 to 20 percent potassium hydroxide (KOH) preparations, and tissue obtained from biopsy should be stained
with hematoxylin and eosin or special fungal stains. Direct
microscopic examination of cutaneous or deep tissue specimens, either in KOH preparation or histologic sections,
yields positive results in approximately 85 percent of
proved cases. However, sputum specimens are positive by
direct examination or culture in less than half of the cases.

Culture. Clinical specimens are cultured on Sabouraud's
agar, a routine fungal medium composed of 4 percent glucose and 1 percent neopeptone. A modified Sabouraud's
agar, containing 2 percent glucose, has been recommended as a more effective isolation medium. For the culture of
nonsterile specimens, such as sputum or skin, antibiotics
are included in the media, usually cycloheximide and

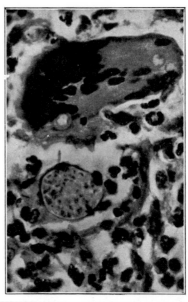

Figure 84-3. *Coccidioides immitis.* Section of lung showing mature spherule and giant cell containing immature cells. × 736.

chloramphenicol or gentamicin, to inhibit saprophytic fungi and bacterial contaminants. Colonies develop within 1
or 2 weeks and are examined microscopically for the production of characteristic arthroconidia. Microscopic preparations of mycelia should always be prepared under a
safety hood.

The identification of *C. immitis* may be confirmed by
inducing spherule production in vitro by incubation in a
complex medium at 40C with 20 percent CO_2, or by ani-

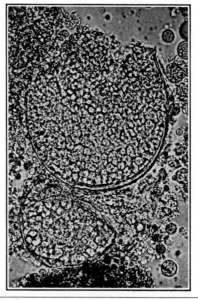

Figure 84-4. *Coccidioides immitis.* Large spherules with endospores in pus. × 315. *(From Smith: Am J Med 2:594, 1947.)*

mal inoculation (e.g., intraperitoneal injection of mice or intratesticular inoculation of guinea pigs). A newer method of confirmation involves the simple demonstration by immunodiffusion of antigens specific for *C. immitis* (or other fungi) in the concentrated supernatant fluid of a short-term broth culture of the isolate. As indicated in Table 84-1, specific precipitins are identified with reference antisera. This method is rapid and is applicable to nonsporulating cultures.

Serologic Tests. Fortunately, because of the time required for identification and the hazards involved in working with cultures of *C. immitis*, serologic tests are extremely helpful (Fig. 84-5). Table 84-2 summarizes the useful antibody responses to coccidioidomycosis and other systemic mycoses. Precipitins (IgM) are produced early and assist in the diagnosis of primary infections. They are detected by a sensitive tube test that becomes positive in 90 percent of patients within 2 weeks after the appearance of symptoms and disappears in most cases by 4 months. Therefore, a positive tube precipitin (TP) test indicates active primary (or reactivation) coccidioidomycosis. Results obtained with the original TP method correlate quite well with those obtained with the faster and more convenient latex particle agglutination test; the latter procedure is more sensitive but less specific than the TP test. The TP antigen, a component of coccidioidin, is heat stable at 60C, whereas the antigen detected in the complement-fixation (CF) test is heat labile.

The CF test for antibodies (IgG) to coccidioidin is a powerful diagnostic and prognostic tool. Because the CF test becomes positive more slowly and persists longer, the presence of CF antibodies may reflect either active infection or the recovery stage. The CF titer correlates with the severity of disease. Most patients with secondary coccidioidomycosis develop a titer of 1:16 or higher, whereas in nondisseminated cases, the titer is almost invariably lower. Therefore, a critical titer of 1:32 or higher reflects active, disseminated disease. A lower titer, however, does not exclude disseminated disease, as many patients with single extrapulmonary lesions (e.g., coccidioidal meningitis) do not develop high titers.

Most useful are multiple serum specimens, because a change in the CF titer reflects the prognosis. With recovery, the CF titer declines and eventually disappears. A ris-

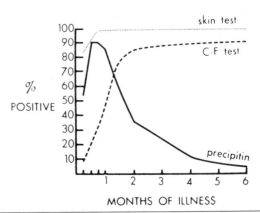

Figure 84-5. Frequency of serologic reactivity following exposure to *Coccidioides immitis*. (From Huppert: Mycopathologia 41:107, 1970.)

ing titer indicates active, uncontrolled infection and a poor prognosis. A stable or fluctuating titer often indicates the presence of a recalcitrant or stabilized lesion. An exceptional situation is coccidioidal meningitis, in which only half of the patients have a titer of 1:32 or higher. However, most of these patients will have a positive CF in their spinal fluid, which is equally valuable.

The immunodiffusion (ID) method can be used to detect both TP and CF antibodies by using reference antisera and heated (TP only) and unheated antigen. Antibodies to two specific heat-labile antigens, termed F (or CF) and HL, may be detected.

Other Methods. Some laboratories recommend direct animal inoculation of specimens as an additional method of primary isolation.

Treatment

Symptomatic treatment is usally adequate for the patient with primary coccidioidomycosis. If the primary infection is severe, however, or if there is evidence of dissemination, amphotericin B should be administered. The chronicity of the disseminated disease usually requires prolonged therapy, increasing the risk of undesirable side effects. Miconazole has also been used extensively in recent years, and many, but not all, patients respond well to it. Intrathecal chemotherapy is often required for chronic coccidioidal meningitis. Successful treatment with ketoconazole may require continuous administration for more than a year.

Histoplasmosis

Histoplasmosis, the most common pulmonary mycosis of humans and animals is caused by the thermally dimorphic fungus, *Histoplasma capsulatum (Emmonsiella capsulata)*. The infection is worldwide in its occurrence. Its incidence

TABLE 84-1. SPECIES-SPECIFIC EXOANTIGENS FOR THE IDENTIFICATION OF SYSTEMIC DIMORPHIC FUNGAL PATHOGENS

Fungus	Exoantigens*
Coccidioides immitis	TP, HL, F (CF)
Histoplasma capsulatum	h, m
Blastomyces dermatitidis	A
Paracoccidioides brasiliensis	1, 2, 3

From Kaufman and Standard: Curr Microbiol 1:135, 1978.
*Exoantigens are detected by precipitin lines of identity in immunodiffusion tests of concentrated culture supernatant fluids vs. reference antisera.

TABLE 84-2. SUMMARY OF MYCOSEROLOGY FOR SYSTEMIC MYCOSIS (ABRIDGED)

Mycosis	Test	Antigen	Sensitivity and Value		Limitations/Specificity
			Diagnosis	Prognosis	
Coccidioidomycosis	TP	C	Early primary infection; 90 percent cases positive	None	None
	CF	C	Titer ≥ 1:32 = secondary disease	Titer reflects severity	Rarely cross-reactive with H
	ID	C	More than 90 percent cases positive, i.e., F and/or HL band		More specific than CF test
Histoplasmosis	CF	H	83 percent of cases positive, i.e., titer ≥ 1:8	Four-fold change in titer	Cross-reactions in patients with blastomycosis, coccidioidomycosis, cryptococcosis, aspergillosis; titer may be boosted by skin test with H
		Y	94 percent of cases positive (titer ≥ 1:8)	Four-fold change in titer	Less cross-reactivity than with H
	ID	H(10X)	82 percent of cases positive, i.e., m or m and h bands	Loss of h	Skin test with H may boost m band; more specific than CF test
Blastomycosis	CF	By	Less than 50 percent of cases positive; reaction to homologous antigen only is diagnostic	Four-fold change in titer	Highly cross-reactive
	ID	Bcf	80 percent of cases positive, i.e., A band	Loss of A band	None; more specific and sensitive than CF test
Cryptococcosis	LA		Detect capsular antigen in serum or spinal fluid; 92 percent of cases positive (in spinal fluid)	Titer reflects severity	Proper controls eliminate false-positives due to rheumatoid factor
	TA	Cn	38 percent of cases serum positive; i.e., titer ≥ 1:2; usually positive early and/or posttreatment	May become positive with recovery	Few cross-reactions
	CPA	CP	Titer ≥ 1:4 considered positive	As above	Few cross-reactions; more sensitive, less specific than TA test
Paracoccidioidomycosis	CF	P	80 to 95 percent of cases positive (titer ≥ 1:8)	Four-fold change in titer	Some cross-reactions at low titer with aspergillosis and candidiasis sera
	ID	P	98 percent of cases positive (bands 1, 2, and/or 3)	Loss of bands	None; band 3 and band m (to H) are identical

Tests: CF, complement fixation; ID, immunodiffusion; TP, tube precipitin; LA, latex agglutination, latex particles coated with rabbit anti-*C. neoformans* globulin; CPA, charcoal particle agglutination; TA, tube agglutination.
Antigens: C, coccidioidin; H, histoplasmin; Y, yeast phase of *Histoplasma capsulatum;* By, yeast phase of *Blastomyces dermatitidis;* Bcf, culture filtrate of *B. dermatitidis* yeast phase; P, culture filtrate of *Paracoccidioides brasiliensis* yeast phase; Cn, killed yeast cells of *Cryptococcus neoformans;* CP, charcoal particles coated with capsular antigen.

varies considerably, being negligible in many parts of the world but pronounced in local regions wherein most of the population have been infected. Infection is initiated by inhalation of the fungus. Ninety-five percent of the infections, however, are inapparent and are detected only by the manifestation of residual lung calcification(s), delayed hypersensitivity, or both.

Histoplasmosis is one of the more common primary mycoses. Throughout history, its existence has been alluded to on the basis of descriptions of its pathogenesis and natural history. It has been suggested, for example, that when Lord Carnarvon's expedition unearthed the tomb of King Tutankhamen, the curse they succumbed to soon after was caused by the disturbance of the histoplasmal spores rather than by ancient Egyptian spirits.

The etiologic cause of histoplasmosis was discovered around the turn of the century by Samuel Darling, a pathologist working in the Canal Zone when he described the histopathology of infected tissue. He observed intracellular organisms that resembled encapsulated protozoa and named the organism *Histoplasma capsulatum*. Darling's accurate descriptions of the disease led to the recognition of other cases. The fungal etiology of histoplasmosis was not established until 1929 by William de Monbreun, and, even though *H. capsulatum* is neither a parasite nor encapsulated, the misleading name, by precedence and taxonomic custom, remains. Recently, following the discovery of the sexual stage by Kwon-Chung, *H. capsulatum* was transferred from the subdivision Deuteromycotina to the family Gymnoascaceae within the subdivision Ascomycotina (Chap. 82, p. 1121). With taxonomic reclassification, the organism has been renamed *Emmonsiella capsulata*—the generic name honors Dr. Chester Emmons, an eminent mycologist whose contributions included many pioneering investigations on the ecology of medically important fungi. Although taxonomic rules require that the new name eventually replace the older one, for the present, the more familiar *H. capsulatum* is retained by common usage.

The extent of inapparent *Histoplasma* infection was not realized until the 1940s. During extensive skin test surveys for tuberculosis in student nurses from various parts of the country, radiologists observed small calcifications in the lungs of many apparently healthy individuals who were nonreactive to old tuberculin (OT). Subsequent skin testing with histoplasmin, an antigen from *H. capsulatum* analogous to OT, revealed a very strong correlation between delayed skin test reactivity and pulmonary calcifications in tuberculin-negative individuals. The availability of skin test antigens for histoplasmosis has provided a very useful epidemiologic tool.

Morphologic and Cultural Characteristics

Dimorphism. *H. capsulatum* is a thermally dimorphic fungus. At temperatures below 35C, it grows as a mold, white to brown in color, but at 37C, it grows as a yeast with small, heaped, and pasty colonies. The organism is characteristically slow growing. Under optimal conditions, the mold colony develops after 1 or 2 weeks, and asexual spores are produced shortly thereafter. Cultures from clinical specimens, however, sometimes require incubation periods of 8 to 12 weeks before there is detectable growth.

Both microconidia and macroconidia are produced at temperatures below 35C. The microconidia are 1 to 5 μm in diameter and are produced singly from short conidiophores. Because of their small size, the microconidia are important in transmission of the infection. Upon dehydration, the conidia are easily dislodged by air currents and aerosolized. The macroconidia, or tuberculate chlamydospores, are very distinctive. Mature forms are large, 8 to 16 μm in diameter, spherical thick-walled structures with spikes projecting from the cell wall (Figs. 84-6 and 84-7). Young macroconidia are smooth walled and tubercles develop as the conidia mature.

At 37C, the mold converts to a budding yeast-type of growth (Fig. 84-8). Conversion is often difficult to effect in

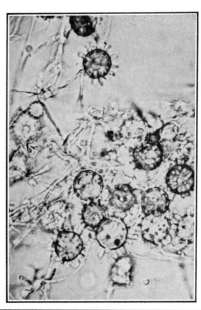

Figure 84-6. *Histoplasma capsulatum.* Typical tuberculate macroconidia from Sabouraud's glucose agar culture. × 658. *(From Smith: Am J Med 2:594, 1947.)*

vitro but is enhanced by a rich, complex medium. The yeast cells are small, ellipsoidal, approximately 1 to 3 μm by 3 to 5 μm in size, and virtually identical to the yeasts observed in vivo within phagocytes (Fig. 84-9).

Sexual Reproduction. *H. capsulatum* possesses a sexual reproductive cycle. It was the finding of a perfect stage in the organism that prompted its reclassification and change

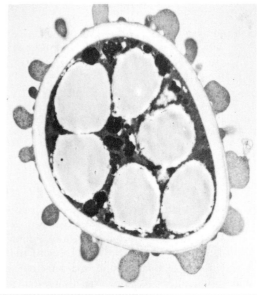

Figure 84-7. *Histoplasma capsulatum.* Transmission electron photomicrograph of tuberculate macroconidium. × 12,000. *(Courtesy of Dr. John Spahr.)*

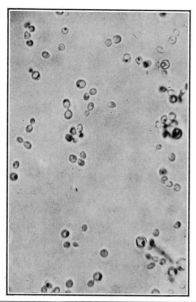

Figure 84-8. *Histoplasma capsulatum.* Yeast cells from blood agar culture at 37C. × 700.

of name to *E. capsulata. E. capsulata* is heterothallic: two opposite (compatible) mating types are necessary for its sexual reproduction. Although the mating types (plus and minus) are equally prevalent among soil isolates, almost 90 percent of infections are caused by the minus type.

Physiology

Significant differences have been detected in the chemical composition of the cell walls of the yeast and mycelial phases of *H. capsulatum.* Based on these differences in the

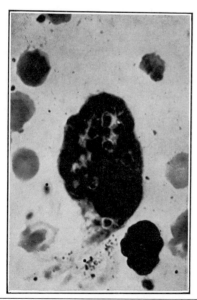

Figure 84-9. *Histoplasma capsulatum.* Yeast cells within mononuclear cell in peripheral blood smear. × 1540.

relative amounts of chitin, α-glucan, and β-glucan, cell-wall chemotypes for each phase have been defined. The three RNA polymerases in each phase have also been found to be different.

In a synthetic medium, conversion between the mycelial phase and the yeast phase is temperature dependent:

$$\text{Mold} \underset{25C}{\overset{37C}{\rightleftharpoons}} \text{Yeast.}$$

The molecular basis for morphogenesis in *H. capsulatum* involves a protein, histin, isolated from the cytoplasm of the mycelial phase. Histin inhibits RNA polymerase and may play a regulatory role in the temperature-dependent control of transcription. It is plausible that an inhibitor, such as histin, may be operative in the mycelial phase, because protein synthesis is considerably reduced during mycelial growth.

In studying the crucial mycelium to yeast conversion at 37C, the importance of cysteine has been confirmed. This conversion proceeds in three stages: (1) During the first 40 hours at 37C, respiration gradually decreases and the intracellular amino acid pools become almost depleted. (2) The cells are then dormant for 4 to 6 days. (3) Then, with commencement of the third stage, concentrations of intracellular cysteine and other amino acids increase and respiration is restored. Cysteine and cystine are essential for the transition to yeast-phase cells. These compounds and other reducing agents appear to stimulate the mitochondrial electron transport chain, perhaps via a unique cysteine oxidase. This mycelium to yeast conversion is inhibited by elevated cyclic AMP, which may also be regulated intracellularly by the action of sulfhydryl agents.

Two colonial forms of the mycelial phase of *H. capsulatum* have been described, the A or albino type and the B or brown type. Both phenotypes produce identical yeast and tissue forms. Primary isolates are often brown and become white with prolonged cultivation. If grown in the dark, the B type can be maintained. Type B strains are more pathogenic for mice and rabbits and produce more macroconidia (but fewer microconidia) than type A.

Determinants of Pathogenicity

Both mating types of *E. capsulata* can be isolated from soil with a comparable frequency. One allele, however, is predominantly recovered from clinical material and is, therefore, associated with human pathogenicity. This genotype may be linked to virulence or to increased conidiation or conidial survival in nature.

All clinical forms are believed to evolve from the same natural history. Microconidia are inhaled from an exogenous source and penetrate to the alveoli, where they convert to small, budding yeast cells. The yeasts are readily phagocytized by alveolar macrophages. At this stage, the yeast-laden macrophages may be cleared through the upper respiratory tract. They may disseminate via the bloodstream, spreading the yeasts to other reticuloendothelial organs, and/or they may invoke a tissue response in situ. The tissue reaction may involve an early infiltration of neutrophils and lymphocytes, but the pyogenic inflammatory response gives way to epithelioid cell tubercle formation. In the course of these various possible reactions, the intracellular yeasts may or may not be inactivated by the phagocytes (see below).

Clinical Infection

Epidemiology

Ecology. In nature, *H. capsulatum* grows in soil with a high nitrogen content. It also has a definite association with bat and avian habitats. The organism has been isolated many times from bird roosts, chicken houses, bat caves, and similar environments. Conidia, when dry, are easily airborne and spread by wind currents, as well as by birds and bats.

The organism is most prevalent in the natural environment in areas where the disease is most endemic, namely in the Ohio–Mississippi Valley—in Missouri, Kentucky, Tennessee, Indiana, Ohio, and southern Illinois. This area also has the highest population of starlings, which tend to congregate in large numbers. The excrement from these birds provides a superlative medium for the enrichment of *H. capsulatum*. In South America, the chief reservoir appears to be chicken coops and bat caves.

The survival of *H. capsulatum* in soil appears to depend upon strict temperature and humidity requirements. It survives best in moist soil (95 to 100 percent humidity) at temperatures of 37C and below. In dry soil, the vegetative cells rapidly dehydrate and lyse, whereas the microconidia and macroconidia will remain viable for some time. However, with prolonged desiccation, only the macroconidia survive. Similarly, only the macroconidia may be able to survive in moist soil at temperatures above 40C.

It was empirically observed, and subsequently proven in the laboratory, that soil supplemented with fecal extracts from bats and several kinds of birds (starlings, chickens, and blackbirds), provided a much better environment for the growth of *H. capsulatum* than soil alone. It was shown by Smith and associates that aged feces contains the highest amounts of growth-stimulating components, nitrogenous compounds, phosphates, and cations, and less of the toxic ammonium salts and uric acid.

Both *H. capsulatum* and a stable variant, *H. capsulatum* var *duboisii*, have been isolated from cases of histoplasmosis in Africa. African histoplasmosis, caused by *H. capsulatum* var *duboisii*, is distinguished from the usual infection by (1) a greater frequency of skin and bone lesions, (2) diminished pulmonary involvement, (3) pronounced giant cell formation, and (4) larger, thick-walled yeast cells in tissue. Although these clinical features are unique and reproducible, *H. capsulatum* var *duboisii* cannot be reliably differentiated in vitro from the type species on the basis of morphology, physiology, and antigenic composition. Indeed, Kwon-Chung has proven that the agents of both forms of histoplasmosis are the same species, as *H. capsulatum* var *duboisii* mates with *H. capsulatum*, and its sexual form is identical to *E. capsulata*.

Histoplasmin Skin Test. Much of the knowledge concerning the prevalence of histoplasmosis has been derived from extensive skin test surveys conducted since World War II.

The antigen, histoplasmin, is produced by growing the mycelial phase of *H. capsulatum* in the same asparagine broth medium used for preparing OT. The filtrate from the culture is dialyzed, the concentration is standardized, and 0.1 ml of the appropriate dilution (usually a 10^{-2} or 10^{-3} dilution of the original material) is injected intradermally. A positive reaction is indicated by induration of 5-mm diameter or greater after 48 hours.

A positive test, if specific, indicates previous sensitization to the fungus. Without a history of prior negativity, the positive test has no diagnostic significance. Histoplasmin is a crude, polyvalent mixture of antigens, only some of which are specific for *H. capsulatum*. Because some antigenic determinants are shared by other pathogenic fungi, cross-reactions can occur. For example, some individuals sensitive to *Blastomyces* or *Coccidioides* will give a false-positive reaction to histoplasmin. For this reason, epidemiologists routinely administer, along with histoplasmin, a battery of skin test antigens, including coccidioidin and blastomycin in the United States and paracoccidioidin in South America. A reaction to a single antigen is generally considered specific. Reactions to two antigens may be caused by sensitization to one or both, although the larger reaction is often considered more specific. In comparing the specificity of the systemic fungal skin test antigens, the decreasing order of specificity (i.e. least likely to be cross-reactive) is coccidioidin, paracoccidioidin, histoplasmin, and blastomycin.

Incidence. Since the 1950s, skin test surveys have been conducted all over the world. The region with the highest level of reactivity is the central United States, along the river valleys of the Ohio, Mississippi, St. Lawrence, and Rio Grande Rivers, where in some locales 80 to 90 percent of the population may be skin test positive by the age of 20 years. Foci of high reactivity exist elsewhere in the world, such as southern Mexico, Indonesia, the Philippines, and Turkey. In the United States alone, projections based on skin test surveys have led to estimations that at least 40 million people have been exposed, with 500,000 new infections every year. Of these, perhaps 55,000 to 200,000 cases will be symptomatic, 1500 to 4000 will require hospitalization annually, and 25 to 100 deaths will occur.

Transmission. In addition to humans, many animals, both wild and domestic, are susceptible to histoplasmosis. Some animals, including the bat, may act as vectors to disseminate the organism in nature.

Epidemics of acute respiratory histoplasmosis result from the simultaneous exposure of a large number of people. These epidemics are *not* caused by direct spread among humans or animals. The experience of youths on Earth Day, 1970, in Delaware, Ohio, is more ironic than most, but otherwise typical of these epidemic outbreaks. The young people gathered to reclaim an abandoned park and, in so doing, overturned several truckloads of soil, which was enriched with starling feces and contaminated with millions of histoplasmal conidia. Several cases of acute respiratory histoplasmosis followed inhalation of heavy inocula of aerosolized conidia. Many similar episodes have been documented: multiple exposure follows the sudden release of a heavy inoculum that has accumulated in a dormant environment. Silos, air-conditioning units contaminated with bird droppings, and accumulations of guano in caves, attics, or parks have all been implicated as fungal reservoirs in epidemics of this type. Perhaps the largest outbreak occurred in Indianapolis between the fall of 1978 and 1979. It is estimated that more than 100,000 persons were infected during this time, resulting in over 300 hospitalized cases and at least 15 deaths. The incidence of disseminated histoplasmosis and the fatality rate were unusually high. The environmental source of the fungus was not determined. Indeed, *H. capsulatum* was recovered from none of the soil samples collected at the most likely site, where an abandoned amusement park had been recently dismantled.

Males develop symptomatic histoplasmosis more often than do females, and approximately 75 percent of cases occur in males. Before puberty, the attack rate for males and females is identical, and the percentage of positive skin test reactors at all ages is the same for both sexes. These epidemiologic data suggest that adult males are inherently more susceptible to the disease than females. Severity of disease and mortality are greater at the age extremes, in infancy and beyond the fifth decade of life.

Pathogenesis

After being phagocytized, the yeast cells of *H. capsulatum* are killed by neutrophils, but they are able to survive and multiply within macrophages. However, macrophages from immunized animals, as well as normal macrophages activated by immune lymphocytes, restrict the growth of intracellular yeasts. By use of an experimental model of self-limited murine histoplasmosis, Artz and Bullock have demonstrated that various parameters of cell-mediated immunity are suppressed during the height of antigen (yeast) burden. Concomitant with resolution of the infection, the number of suppressor lymphocytes in the spleen diminishes and helper cells increase. These correlations of competent cell-mediated immune responses with resistance to infection are supported by the clinical data.

Clinical Manifestations

The manifestations of infection with *H. capsulatum* are protean. Several clinical classifications have been devised, but none is completely satisfactory or universally accepted. The scheme presented in Table 84-3 is among the most

TABLE 84-3. CLINICAL FORMS OF HISTOPLASMOSIS

	Acute	Chronic
Pulmonary	Asymptomatic Mild Moderate Severe	Pneumonic or cavitary (anatomic defect)
Disseminated	Benign Progressive	Progressive (mucocutaneous)

useful. The initial pulmonary episode may be acute or chronic, or dissemination may occur by hematogenous or lymphatic spread from the lung to other organs.

Most normal individuals contain the infection. The granulomata that form may undergo fibrosis, and residual scars may remain in the lungs or spleen. Resolution appears to confer some immunity to reinfection. This process occurs asymptomatically in 95 percent of all persons with acute, primary histoplasmosis, whether disseminated or confined to the lung.

Acute Pulmonary Histoplasmosis. Patients with acute pulmonary histoplasmosis manifest symptoms ranging from a mild flu-like illness that clears spontaneously to a moderate or severe type of disease. In healthy hosts, the degree of involvement and symptomatology is roughly proportional to the size of the inoculum inhaled. In the previously sensitized individual, such reinfection exposure results in a shorter and milder infection with minimal histopathology. The incubation period varies from 1 week to several weeks. A moderate disease is characterized by cough, chest pain, dyspnea, and hoarseness. In more severe cases, fever, night sweats, and weight loss also develop. Occasionally, yeast cells may be observed in the sputum. Radiologic examination may reveal multiple lesions scattered throughout the lungs, and in patients with active disease hilar lymphadenopathy is usually present.

The differential diagnosis includes tuberculosis, bacterial bronchiectasis, and lymphoblastoma. Indeed, because there are so many similarities among the different forms of histoplasmosis and the various stages of tuberculosis, it is imperative that a diagnosis be established. *Histoplasma* lesions in the lung resolve slowly. Healing may be complete or with fibrosis, but, typically, calcification occurs. An experienced radiologist can differentiate between the calcifications of histoplasmosis and tuberculosis. Calcifications produced by *H. capsulatum* are more regular, with halos, and may be found in the liver and spleen as well as in the lungs. Miliary calcifications may also occur. Calcifications are produced in children more rapidly than in the adult. Single, solitary, uncalcified lesions, known as coin lesions, are also produced and are similar to those seen in tuberculosis. As these resemble neoplasm, they are often removed surgically. Another tuberculosis-like pulmonary manifestation usually found in the adult lung is a histoplasmoma. The histoplasmoma, which may be 2 to 3 cm in diameter, contains a central necrotic area encased in a fibrotic capsule. Calcification begins in the center of the lesion and is followed by the development of concentric rings of fibrosis and calcification.

Chronic Pulmonary Histoplasmosis. This form is seen most often in adult males. It is considered to be an opportunistic complication of underlying chronic obstructive lung disease with emphysema and abnormal pulmonary spaces.

With small emphysematous air spaces, transient pneumonitis develops, and infection of large bullous spaces may result in cavitary histoplasmosis. Symptoms of the latter may be indistinguishable from those of chronic cavitary tuberculosis. The chronic form is secondary to the underlying pulmonary disease. It may develop immediately after primary inhalation or after years of apparent quiescence. Pathologic and immunologic evidence suggests that the late onset results from reactivation of an old lesion rather than exogenous reinfection. Chronic pulmonary histoplasmosis is usually apical. Patients experience a low-grade fever, productive cough, progressive weakness, and fatigue. Chest films show centrilobular or bullous emphysema.

Disseminated Histoplasmosis. The gamut of clinical forms and pathology observed in pulmonary histoplasmosis can also occur in any other part of the body. The yeast cells are probably disseminated throughout the body while inside macrophages. The most common sites of involvement, after the lung, are the reticuloendothelial tissues of the spleen, liver, lymph nodes, and bone marrow. However, lesions have been documented in almost every organ. Dissemination may be completely benign and inapparent except for the presence of calcified lesions, usually in organs of the reticuloendothelial system.

Alternatively, disseminated histoplasmosis may be acute and progressive. In such cases, the pulmonary symptoms are insignificant, and patients may have splenomegaly and hepatomegaly, weight loss, anemia, and leukopenia. Granulomatous lesions and macrophages packed with yeast cells can be observed throughout the liver, spleen, marrow, and, quite often, the adrenals. Acute progressive histoplasmosis is often fulminant and rapidly fatal—ultimately every organ can become diseased. This form of histoplasmosis is an opportunistic disease associated with compromised cell-mediated immunity, such as patients receiving immunosuppressive drugs and those with underlying lymphomatous neoplasia. In some cases, the compromising condition may have reactivated an old, dormant histoplasmal lesion that originally occurred years before. Within the endemic area, infants with histiocytosis may develop disseminated histoplasmosis that is characteristically fulminant. Chronic disseminated histoplasmosis may evolve from protraction of the acute disease. This form is progressive, with eventual involvement of every organ, especially the mucocutaneous areas around the eye, tongue, and anus.

Primary Cutaneous Histoplasmosis. Very rarely, chronic disseminated histoplasmosis develops following primary inoculation of the skin or mucocutaneous tissue.

The lung is not involved in these cases, as the organisms are typically introduced following the contamination of a traumatic wound. Such infections may be anatomically localized, as with some ocular cases, or they may chronically progress with involvement along the draining lymphatics.

Laboratory Diagnosis

Microscopic Examination. Histoplasmosis can be diagnosed upon finding the yeast cells in clinical materials. Suitable specimens include sputa, tissue material obtained from biopsy or surgery, spinal fluid, and blood. The buffy coat of a blood specimen may reveal yeast-laden macrophages. Bone marrow obtained when patients are febrile may contain the yeast cells. Smears of infected sputum, blood, marrow, or tissue that have been fixed with methanol and stained with the Wright or Geimsa stain will reveal the characteristically small, ellipsoidal yeast cells (approximately 2 by 4 μm) inside macrophages. With either stain, the larger end of the yeast cell contains an eccentric, red-staining mass (see Fig. 84-9).

Culture. Sputum specimens should be collected early in the morning, and purulent or sanguineous portions of the sputum should be selected for culture. A bronchial wash is even more likely to be positive. Nonsterile specimens (e.g., sputum, urine) should be cultured on a blood-enriched medium and Sabouraud's agar with antibiotics (cycloheximide and chloramphenicol or gentamicin) and incubated for at least 4 weeks at 25C. Because *H. capsulatum* may grow very slowly, cultures should be kept up to 12 weeks, if possible, before being discarded as negative. If a sporulating mold develops, *H. capsulatum* can be identified by the presence of its characteristic macroconidia (Fig. 84-6) and by conversion to the yeast phase by growth on an enriched medium at 37C (Fig. 84-8). Alternatively, conversion to the yeast may be effected by growth in tissue cultures, such as HeLa cells, or by animal inoculation, such as the intraperitoneal injection into mice. Occasional isolates of *H. capsulatum* will not produce conidia, but it may be possible to identify these variants by conversion to the yeast phase or by the detection of *H. capsulatum*-specific antigens (Table 84-1).

It has been recommended that, in endemic areas or in cases where histoplasmosis is suspected, specimens should be inoculated on at least four media: (1) Sabouraud's agar without antibiotics at 25C to 30C; (2) Sabouraud's agar with antiobiotics (cycloheximide and chloramphenicol, gentamicin, or penicillin and streptomycin) at 25C to 30C; (3) brain–heart infusion agar with 5 percent sheep blood and antibiotics at 25C to 30C; and (4) brain–heart infusion agar with 5 percent sheep blood without cycloheximide at 37C. The pH of these media should be near neutrality, since *H. capsulatum* is inhibited below pH 6.

Skin Test. As mentioned above, the skin test antigen, histoplasmin, is a valuable epidemiologic tool. Within 2 weeks after infection, most persons become skin test positive, and this reactivity usually persists for many years. The diagnostic value of the skin test is minimal. A negative re-action can be used to rule out active histoplasmosis in the immunocompetent subject, but patients with anergy may be falsely negative. Without a prior history of a negative skin test, a positive reaction is meaningless except in in-fants, in whom a positive test can be presumed to result from recent or current infection. With most patients, only a history of conversion from negative to positive is diag-nostic. Because of its limited diagnostic value and the pos-sibility of the skin test's obscuring the humoral antibody measurement (see below), skin testing with histoplasmin should be avoided in most patients.

Serology. Specific antibodies to histoplasmal antigens can be detected during infection. Two serologic tests are now widely accepted because of their convenience and availability: the measurement of CF antibodies and the im-munodiffusion test for precipitins. Both tests may be help-ful in the diagnosis and prognosis of histoplasmosis, provided the results are properly interpreted (see Table 84-2).

COMPLEMENT-FIXATION TEST. The CF test is routinely performed under standard conditions for measuring fixa-tion of complement by the classic pathway (Chap. 14). Sensitized sheep red cells are the indicator system, and two antigens are usually employed: histoplasmin and a standardized suspension of killed *H. capsulatum* yeast cells. Because of the possibility of cross-reactivity, patient serum is also tested at the same time against other fungal anti-gens, such as coccidioidin, spherulin, *Blastomyces dermatiti-dis*, or *Paracoccidioides brasiliensis*. Serum antibodies specific for histoplasmal antigens can be detected in the CF test 2 to 4 weeks following exposure. Most laboratories perform the CF test on two-fold dilutions of patient serum, begin-ning with a dilution of 1:8. With resolution of the infec-tion, the antibody titer gradually declines and disappears (i.e., titer < 1:8, in most cases by 9 months. The CF test with either *H. capsulatum* yeast or mycelial (histoplasmin) antigen is very sensitive, and 90 percent of patients are positive (i.e., titer ≥ 1:8). A titer of 1:32 that persists or rises over the course of several weeks indicates active dis-ease in patients with an established diagnosis of histoplas-mosis. Unfortunately, in sensitive patients, the skin test an-tigen may boost the CF antibody titer to histoplasmin, and the elevated titer may remain for as long as 3 months. Ob-viously, the CF test, which can deliver results as rapidly as the skin test, is preferable for diagnostic purposes. Howev-er, a positive CF test, even in high titer, is not by itself di-agnostic, as the results can be caused by cross-reacting antibodies. If a patient's serum is reactive to more than one fungal antigen or if it is anticomplementary, the ID test should be conducted.

IMMUNODIFFUSION. Precipitins can be detected by dou-ble diffusion of serum and antigen in agarose gel. The an-tigen is histoplasmin in 10-fold the concentration used for the CF test. The ID test becomes positive in about 80 per-cent of patients with histoplasmosis by the third or fourth week of infection. This test, while less sensitive and requir-ing a longer time to become positive, is more specific than the CF test. Precipitin lines or bands specific for *H. capsulatum* are detected by the formation of lines of identity with reference serum. Kaufman and associates defined two specific precipitin bands, m and h. The m line, which is observed more frequently, appears soon after infection and may persist in the serum up to 3 years following re-covery. The h band, which forms closer to the serum wells, is more transient. Because it disappears soon after the disease, the presence of an h band in serum denotes active infection. As with the CF titer, the m band may be boosted by the administration of the histoplasmin skin test, and the boosting effect may last up to 3 months.

Other serologic tests for antibodies to *H. capsulatum* that are in current use include the indirect fluorescent an-tibody test and counterimmunoelectrophoresis. Tests for circulating antigen are being developed that promise in the near future to be more specific and probably more in-dicative of progressive disease.

Treatment

Most cases of histoplasmosis remain undetected and re-quire no treatment. With symptomatic, progressive dis-ease, the treatment of choice is amphotericin B. The regimen is similar to that applied to other systemic myco-ses, as mentioned in Chapter 83. Recovery following treat-ment with amphotericin B is generally faster, and fewer relapses occur than are experienced with blastomycosis and coccidioidomycosis.

Arrested pulmonary lesions are often removed surgi-cally.

Blastomycosis

Blastomycosis is a chronic infection characterized by gran-ulomatous and suppurative lesions initiated by inhalation of a thermally dimorphic fungus, *B. dermatitidis (Ajellomyces dermatitidis)*. From the lung, dissemination may occur to any organ, preferentially to the skin and bones. The dis-ease was previously referred to as North American blasto-mycosis because initial cases were confined to the United States, Canada, and Central America. Although the preva-lence continues to be highest on the North American con-tinent, blastomycosis has been documented in Africa and South America. It is endemic for humans and dogs in the eastern United States.

Blastomycosis was first described in its chronic cutane-ous form in the 1890s by Gilchrist. As early case reviews included a large number of cases from Chicago, blastomy-cosis became known as Gilchrist's or Chicago disease. Many early case descriptions were undoubtedly confused with coccidioidomycosis, cryptococcosis (European blasto-mycosis), and paracoccidioidomycosis (South American blastomycosis) until Benham firmly established the distinct etiology of blastomycosis in 1934. A complete understanding of the relationship between the cutaneous and systemic manifestations of the disease was not available, however, until 1951 when Schwarz and Baum presented evidence derived from pathologic specimens that both forms originate in the lung. At present, blastomycosis is considered to be primarily a pulmonary infection characterized by secondary spread to the skin and other parts of the body. The respiratory episode, however, may be completely subclinical. Very rarely, truly primary extrapulmonary infection has been demonstrated.

Morphologic and Cultural Characteristics

Cultural Characteristics. Blastomycosis is caused by a single dimorphic species, *B. dermatitidis*. At temperatures below 35C, the organism grows as a mold, producing a filamentous colony that is composed of uniform, septate hyphae and conidia. On Sabouraud's glucose agar at 25C, different isolates of *B. dermatitidis* vary in their rate of growth, colony appearance, and degree and type of sporulation. Usually, however, colonies require at least 2 weeks for full development. Many strains produce a white, cottony mycelium that becomes tan to brown with age. On enriched media at 37C, *B. dermatitidis* produces a yeast type of growth with colonies that are folded, pasty, and moist.

Microscopic Appearance. The mycelial form produces abundant conidia from the aerial hyphae and lateral conidiophores (Fig. 84-10). The conidia are spherical, ovoid, or pyriform in shape and are 3 to 5 μm in diameter. Thick-walled chlamydospores, 7 to 18 μm in diameter, may also be observed. Because the colony and conidia of *B. dermatitidis* are indistinguishable from those of many other fungi,

identification must be confirmed by conversion to the characteristic yeast phase. This conversion can be accomplished by in vitro cultivation at 37C or by animal inoculation. Under these conditions, the organism grows as a thick-walled spherical yeast that usually produces single buds (Fig. 84-11). The bud and parent yeast have a characteristically wide base of attachment, and the bud often enlarges to a size equal to that of the parent cell before becoming detached. Yeasts normally range in size from 8 to 15 μm, although some cells reach a diameter of 30 μm.

Sexual Reproduction. The perfect or sexual phase of *B. dermatitidis* was discovered by McDonough and Lewis. As the sexual progeny are produced within an ascus, the fungus is an ascomycete, a member of the subdivision Ascomycotina (Chap. 82). It has been renamed *Ajellomyces dermatitidis* and reclassified in the family Gymnoascaceae on the basis of its sexual apparatus. *A. dermatitidis* is heterothallic, requiring two compatible (opposite) mating types for sexual reproduction. Although the two mating types possess different antigens, they are similar in many other respects, including pathogenicity. Mating compatibility has been used to confirm that a single species is responsible for blastomycosis among dogs and humans, and probably also for cases in North America and Africa.

Determinants of Pathogenicity

Strains of *B. dermatitidis* vary in their virulence for experimental animals, but an explanation for this difference is lacking. A correlation has been observed between virulence for the mouse and the ability of the alkali-soluble

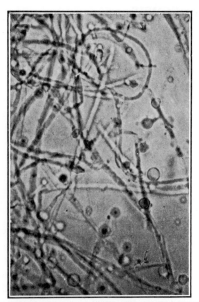

Figure 84-10. *Blastomyces dermatitidis.* Mycelium and conidia from culture of Sabouraud's glucose agar at room temperature. × 736.

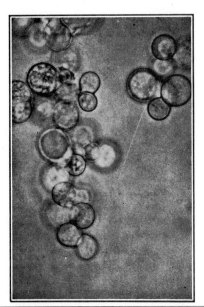

Figure 84-11. *Blastomyces dermatitidis.* Budding, yeast cells from culture on blood agar at 37C. × 700. *(From Conant, et al.: Manual of Clinical Mycology, 3rd ed, 1971. Philadelphia, Saunders.)*

cell wall fraction of the yeast cells to induce granulomata following intracutaneous injection. Also, a unique chemotactic factor for leukocytes is elaborated by yeast cells of *B. dermatitidis*. The fungal properties and host reponses that determine which organs will be most involved are unknown.

Clinical Infection

Epidemiology

Ecology. The natural habitat of *B. dermatitidis* is an enigma. If it is assumed that most cases of blastomycosis are acquired by inhalation of exogenous, infectious particles, the fungus should grow and produce airborne cells in the natural environment. However, with only four exceptions, all attempts to isolate this agent from the environment have failed. In one of these, Denton and DiSalvo isolated *B. dermatitidis* from 10 of 356 soil samples collected on two occasions almost a year apart. Subsequent samplings were negative. The positive samples were collected from a rural environment, including a chicken house, a cattle crossing, and an abandoned shack. The ecologic niche occupied by *B. dermatitidis* cannot be defined by these meager findings.

The ecology of *B. dermatitidis* must be highly specialized. There is no evidence that an animal reservoir exists to perpetuate the organism. The fungus probably exists in nature during most of the year in a protected and dormant state but is stimulated during the cooler months by suitable climate or other specific environmental condition to propagate and become airborne and infectious.

Geographic Distribution.

As *B. dermatitidis* is not readily recoverable from nature and an adequate skin test antigen is not available for conducting population surveys of exposure, the geographic distribution of blastomycosis has been estimated from reports of human and animal cases. The endemic area extends roughly east from states that border the Mississippi River. Blastomycosis is endemic in southern Canada east of Manitoba and, in the United States, in Illinois, Minnesota, Ohio, the Atlantic Coastal states, and the southeastern states, with the exception of Florida. The occurrence of cases in New England and elsewhere in the United States is rare. The incidence is highest in Arkansas, Kentucky, Louisiana, Mississippi, North Carolina, and Tennessee. Within the endemic region, cases are often clustered, indicating local pockets of high endemicity.

Clinical reports have also documented autochthonous cases in Africa, both north (Tunisia, Morocco) and south (Uganda, Tanzania, Zimbabwe), as well as India, Israel, Mexico and Venezuela. Reports of the infection occurring elsewhere are dubious either because of a questionable diagnosis or a history of patient travel to or contact with fomites from an endemic area. Blastomycosis is also a disease of dogs and may occur more frequently in them than in humans. Canine cases follow the same endemic pattern as those of humans.

Incidence. Blastomycosis occurs more frequently during middle age, in blacks, and in males. One of the most extensive case reviews was that of Furcolow and associates, which presented data from 1114 human and 247 canine case reports. Although the disease can occur at any age, in this series of cases, 60 percent of the human cases occurred in patients between the ages of 30 and 60 years. Only 3.4 percent of these patients were under 20 years of age, and only 4 (0.9 percent) were less than 10 years old. The incidence for whites and blacks was 0.5 and 1.5 cases per 100,000, respectively. Most of the male to female ratios reported in several surveys involving hundreds of patients vary from 6:1 to 15:1, but lower ratios of 3:1 to 4:1 have been reported in smaller studies. Unlike coccidioidomycosis, racial or sexual differences in either severity of disease or tissue tropism have not been clearly defined.

Socioeconomic and occupational data frequently associate blastomycosis with squalid housing, malnutrition, manual labor, agriculture, construction work, and exposure to dust and wood. Although rare, blastomycosis also occurs in immunocompromised patients.

Seven outbreaks of blastomycosis have been documented. Table 84-4 summarizes these episodes. Each consisted of a cluster of cases that occurred at approximately the same place and time. All but one area, Westmont, Illinois, are rural. Five outbreaks occurred during the fall and winter (1954, 1972, and 1975–76), and two in the summer (1976 and 1979).

Striking and unexplained exceptions to the usual age spectrum of blastomycosis occurred in two of the North Carolina outbreaks. Seven of the eleven patients in the Grifton outbreak and three of the five patients in the Enfield cluster were 16 years old or younger, and seven of the combined total of 16 cases were less than 10 years of age.

As with the other systemic mycoses caused by dimorphic, exogenous fungi, blastomycosis occurs more often in males. The explanation for this sexual difference in predisposition to blastomycosis can be attributed to differences in exposure to *B. dermatitidis* or to host susceptibility. The older theory of differential exposure based on the assumption that males have greater contact with the organism because they spend more time outdoors is becoming increasingly difficult to accept. There is some experimental evidence to support the concept of sex-linked differences in susceptibility. Landay and associates have compared the mortality of ovariectomized and castrated hamsters with control animals and those treated with hormones. The survival of untreated males and females was 7 and 52 percent, respectively. Overall, the mortality was apparently potentiated by both androgens and estrogens, although to a greater extent by the former.

Pathogenesis

Almost all cases of blastomycosis originate in the lung. In the alveoli, *B. dermatitidis* induces an inflammatory response characterized by the infiltration of both macro-

TABLE 84-4. REPORTED OUTBREAKS OF BLASTOMYCOSIS*

| Site | No. of Cases* | Dates of Onset | No. of Cases | | Diagnosis† | | | Skin Test | |
			Males	< 16 yr.	Smear	Culture	CF‡	B	H
Grifton, NC	11	October 1953–March 1954	5	7	9/10	10/10	3/8	5/10	0/10
Bigfork, MN	12	October–November 1972	7	4	4/4	4/4	4/12	10/12	1/12
Westmont, IL	5	August 1974–April 1975	3	0	?	5/5	2/5	ND	ND
Sioux Lookout, Ontario	4	October 1974–March 1975	?	?	?	4/4	?	?	?
Enfield, NC	5	November 1975–January 1976	1	3	1/1	5/5	1/1	ND	ND
Trenton, NC§	3	July 1976	2	0	3/3	3/3	ND	ND	ND
Haywood, WI	3	July 1979	?	?	?	3/3	0/6	?	?

From Furcolow et al.: Morbid Mortal Weekly Rep 25:205, 1976; Kitchen, et al.: Am Rev Respir Dis 115:1063, 1977; Smith, et al.: JAMA 158:641, 1955; Tosh, et al.: Am Rev Respir Dis 109:525, 1974; Goldthorpe, Butler: Can Dis Weekly Rep 1–13:49, 1975; Brewer, et al.: Morbid Mortal Weekly Rep 28:450, 1979; Baron, unpublished observations.

*All 43 patients had abnormal radiographic examinations.

†No. positive/No. tested. Specimens were sputum, lung tissue, bronchial or gastric washing. Smear refers to direct microscopic examination of specimens for tissue form (see Fig. 84-12).

‡A positive serum CF antibody test to yeast phase antigen of *Blastomyces dermatitidis* is defined as a titer ≥ 1:8. Only 5 of the 10 positive patients had CF titers greater than 1:8 (three had 1:16 and two had 1:32). At least four of these CF-positive patients also had a positive CF test to the yeast phase antigen of *Histoplasma capsulatum*.

§Four canine cases were also diagnosed in Trenton, NC, during July 1976.

CF, complement fixation; ND, not done; ?, data unknown.

Skin test antigens: B, *Blastomyces* vaccine (Grifton) or blastomycin lot KCB-26 (Bigfork); H, histoplasmin.

Additional skin tests: three of the Grifton cases had a positive skin test to 1:1000 Old Tuberculin, and none of nine tested reacted to coccidioidin. None of 11 Bigfork cases tested reacted to intermediate-strength Purified Protein Derivative.

phages and neutrophils and the subsequent formation of granulomata. The accumulation of neutrophils presents a suppurative component that is uncharacteristic of most mycoses and other chronic diseases. Unlike monocytes and macrophages, which inhibit the growth of *B. dermatitidis* yeast cells in vitro, neutrophils apparently stimulate their growth.

If the pathogenesis of blastomycosis is similar to that of histoplasmosis and coccidioidomycosis, most infections may be subclinical and resolve spontaneously. Without specific and sensitive skin test antigens, however, it has not been possible to determine the extent of exposure to *Blastomyces* in the general population. Also, as calcification is uncommon, there is little radiologic or histopathologic evidence of residual blastomycotic lesions. The best evidence for the existence of subclinical blastomycosis derives from the Bigfork epidemic, where specific skin test reactivity was used to document exposure to *Blastomyces* in the absence of symptoms.

An alternative theory of the pathogenesis of blastomycosis has been advanced by Furcolow and Smith. Based on the apparent scarcity of *B. dermatitidis* in nature and the marginal evidence for subclinical infections, it is postulated that blastomycosis is a rare and unusually serious disease. Studies of experimental canine blastomycosis tend to support this hypothesis. In humans, the primary pulmonary lesion may be inapparent to severe. If inapparent, dissemination to the skin and bones may follow. If the pulmonary episode is severe, the generalized systemic disease may develop, with the potential involvement of multiple organs.

Clinical Manifestations

Two basic forms of blastomycosis are recognized: pulmonary, often disseminated blastomycosis, and the chronic cutaneous form. In a Veterans Administration Cooperative Study of 198 patients, the most common symptoms that initially led patients to seek medical attention were cough, weight loss, chest pain, skin lesions, fever, hemoptysis, and localized swelling. No characteristic pattern of symptoms was apparent, as other complaints were also documented by patients, albeit less frequently.

Pulmonary Blastomycosis. The primary pulmonary infection may persist locally and/or spread to any organ. In some patients, the initial pulmonary infection causes symptoms of mild respiratory infection. In others, the pulmonary lesion heals by fibrosis and resorption. Unlike tuberculosis and histoplasmosis, blastomycotic lesions rarely caseate or calcify. In patients whose pulmonary lesions have resolved, hematogenous, lymphatic, or macrophage-borne dissemination may already have occurred, generally to the skin. Busey and associates recognized pulmonary blastomycosis in 60 percent of the patients, including dissemination in 39 percent. Of the case total, 35 percent had involvement in both lung and skin, whereas 19 percent had infection only in the skin. Alternatively, in some patients, the pulmonary focus becomes more severe and is accompanied by pleuritis. An acute to chronic lung infection may develop. The most common forms of pulmonary involvement are infiltration, cavitation, pneumonia, or nodular. There are accounts of self-limited, acute, pulmonary blastomycosis. A wide variety of symptoms, pathology, and radiographic appearance may be observed, and the extent of these manifestations tends to reflect the severity of the disease. Because of the tremendous variation in symptomatology, blastomycosis is quite often misdiagnosed as some other infection, sarcoid, or cancer. It is too often diagnosed by accident or following a process of elimination.

Chronic Cutaneous Blastomycosis. In chronic cutaneous blastomycosis, the initial skin lesion is one or more subcutaneous nodules that eventually ulcerate. Lesions are most common on exposed skin surfaces, such as the face, hands, wrist, and lower leg. Spread may occur by extension to the trunk or other areas and may require weeks or months for the ulcerative process to evolve. If untreated, elevated, granulomatous lesions with advancing borders will develop in time. The organisms can be found in microabscesses near the dermis. Extensive, often verrucous, epithelial hyperplasia overlying the yeasts may develop and resemble carcinoma. These extensive cutaneous lesions are characteristically discolored and crusty, and they tend to heal and scar in the central, older areas. The active microabscesses found at the leading edge of the lesion can be aspirated or biopsied, and the typical yeast cells of *B. dermatitidis* can be observed on direct microscopic examination (Fig. 84-12).

Disseminated Blastomycosis. Dissemination may be widespread in blastomycosis. The most frequently involved extrapulmonary sites are the skin, bones, genitourinary tract, central nervous system, and spleen. Less frequently, the liver, lymph nodes, heart, and other viscera are infected. The progressive systemic form of blastomycosis develops in patients with unresolving pulmonary infection, but the degree of pulmonary involvement is not related to the extent of dissemination. The infection may be chronic,

with few organisms present, or multiple pulmonary foci may be demonstrable at the time generalized systemic disease develops.

From the lungs, the yeasts disseminate throughout the body, with a characteristic predilection for the skin and bones. Skin lesions may be more severe than those in chronic cutaneous blastomycosis and are seen in about 75 percent of the patients. Overall skeletal involvement is observed in approximately 33 percent of cases. Osteomyelitis and, in some cases, draining sinuses to the skin, develop and should be examined for the presence of organisms. Because of the frequency of bone involvement and because almost any bone can be affected, a total body radiographic examination is advisable upon diagnosis of blastomycosis. Arthritis may develop by extension from infected bone or by direct dissemination from the lung without bone infection. In 4 to 22 percent of patients, there is involvement of the urogenital tract, especially of the male genitalia, kidney, and adrenals. Metastasis to the central nervous system, with resultant meningitis or brain abscess, occurs in up to 10 percent of patients.

Primary Cutaneous Blastomycosis. This form of blastomycosis is initiated by traumatic autoinoculation or contamination of an open wound with the infectious material. The symptomatology, pathology, and pathogenesis of this form differ considerably from the other forms of blastomycosis. The lymphatics and regional lymph nodes are involved, but the infection remains localized and often resolves without treatment.

Laboratory Diagnosis
Microscopic Examination. In wet mounts or in KOH preparations of pus, exudate, sputum, or other specimens, a diagnosis can be made by detection of the characteristic yeast cells of *B. dermatitidis*. These yeasts are large (8 to 15 μm in diameter) and typically thick-walled. The cell wall is highly refractile and often resembles a double wall. Budding usually occurs singly. The bud is attached to the parent cell by a broad base and enlarges to the size of the parent yeast before it is detached. These features, as depicted in Figure 84-12, are pathognomonic for blastomycosis and permit an immediate diagnosis to be made. Unsuspected cases of blastomycosis are occasionally diagnosed from Papanicolaou-stained sputum specimens. In tissue stained with hematoxylin and eosin, the yeast cytoplasm stains dark and the cell wall appears colorless. The cells may be multinucleated. The yeasts are often abundant in cutaneous lesions, and these specimens are often positive on direct examination. If the yeasts are sparse, fungal cell wall stains, such as periodic acid-Schiff or methenamine silver, are helpful.

Rarely, small forms of *B. dermatitidis* are seen in tissue. These cells are typical in shape and budding but only 2 to 5 μm in diameter. Their multinucleation may help to differentiate them from *Cryptococcus neoformans* and *H. capsulatum* var *duboisii.*

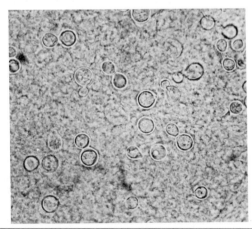

Figure 84-12. *Blastomyces dermatitidis.* Budding yeast cells in pus. $\times$ 762. *(From Conant, et al.: Manual of Clinical Mycology, 3rd ed, 1971. Philadelphia, Saunders.)*

Culture. Specimens should also be cultured on Sabouraud's agar and sheep blood-enriched media. If the specimen is not normally sterile, media with antibiotics should also be used. Cultures should be incubated at room temperature for at least 4 weeks. Colonies are white to brown, variably textured, and produce potentially infectious conidia. The identification is confirmed by detecting the A exoantigen (Table 84-1) or by subculturing on an enriched medium, such as brain–heart infusion blood agar or Kelly's medium at 37C, for subsequent conversion to the yeast phase (see Fig. 84-11).

Some laboratories employ direct animal inoculation as a method of rapidly isolating and identifying *B. dermatitidis*. Mice are injected intraperitoneally with the clinical specimen, and their peritoneums are examined for typical yeast cells 1 to 2 weeks later.

Skin Test. Delayed-type hypersensitivity to *Blastomyces* has been detected by skin tests with both whole cell and culture filtrate antigens (blastomycin). The skin test, however, has no diagnostic value. False-negative results are often observed in patients, and false-positive cross-reactions occur in many individuals sensitized to other fungi. Reactivity is transient and disappears with time. In individuals doubly reactive to blastomycin and histoplasmin, the blastomycin reaction is considered specific if it is equal to or larger than the response to histoplasmin. In the Bigfork epidemic (see Table 84-4), blastomycin was extremely valuable, as a significant number of the infections were detected by monospecific reactions, whereas the background level of reactivity in the population was negligible.

Serology. Measurement of CF antibodies to various antigen preparations of *B. dermatitidis* has not proved reliable. Sera from patients with blastomycosis may react with higher CF titer to heterologous antigens, especially histoplasmin, than to blastomycin. The yeast phase of *B. dermatitidis* provides a more specific antigen, but only 30 to 50 percent of patients are positive.

The most useful serologic procedure is an ID test for specific precipitins. As indicated in Table 84-2, the ID test is more sensitive and more specific than the CF test. Using a yeast phase filtrate antigen and positive reference sera, antibodies to a specific precipitin line, designated A, can be detected. Although other precipitin lines may occur, apparently only sera from patients with blastomycosis develop antibodies to antigen A, which do not cross react with the specific precipitin lines formed to antigens of other systemic fungal pathogens (Table 84-1). It is not known how soon after infection the ID test becomes positive, but within a few months after successful treatment the precipitin lines disappear.

Treatment

The first effective agent for treatment of blastomycosis was the aromatic diamidine, 2-hydroxystilbamidine isethionate. Adults are treated with 2-hydroxystilbamidine by daily in-travenous administration of 225 mg in saline for a total dose of 8 g. The same regimen may be repeated following an intervening rest period of 3 to 4 months. Although the relapse rate is about 30 percent and the side effects may be severe, this therapy continues to be judiciously recommended by many clinicians, especially for patients who may tolerate it better than they do amphotericin B. Side effects associated with 2-hydroxystilbamidine may include elevated serum levels of glutamic oxaloacetic transaminase, bilirubin, urea nitrogen, or hemoglobin.

Amphotericin B is also quite effective against blastomycosis. A total dose of at least 2 g is required to eradicate all the organisms, as the relapse rate is significant if 1.5 g or less is administered. The protocol for administration of amphotericin B and monitoring renal function are the same as its application for other mycoses (see Table 83-6). In a comparison of 2-hydroxystilbamidine and amphotericin B, amphotericin B has been shown to be more effective in the more severe cases of blastomycosis, i.e., those patients with extensive or cavitary pulmonary lesions or dissemination to organs other than the lung and skin. Noncavitary pulmonary blastomycosis and cases involving only lung and/or skin respond well to either drug. Corrective surgery may be recommended during chemotherapeutic treatment. To date, clinical results with ketoconazole in the treatment of blastomycosis are mixed.

Cryptococcosis

Cryptococcosis is caused by infection with an encapsulated yeast, *C. neoformans*. Outdated synonyms for cryptococcosis include torulosis, European blastomycosis, and torula meningitis. The natural reservoir for *C. neoformans* is the soil and avian feces, and infection follows airborne exposure and inhalation of the yeasts. The organisms may metastasize from the lungs to virtually any organ in the body, but they preferentially invade the central nervous system. Although the organism is ubiquitous in nature, the incidence of cryptococcosis is relatively low. Therefore, it is likely that many more persons inhale the yeast than become ill. The actual extent of subclinical exposure is unknown because a good skin test antigen has not been developed for population surveys. The yeast was recently shown by Kwon-Chung to represent the imperfect form of the basidiomycetous species, *Filobasidiella neoformans*.

Morphology and Physiology
Cultural Characteristics. Visible colonies of *C. neoformans* develop on routine laboratory media within 36 to 72 hours. They are white to cream colored, opaque, and may be several millimeters in diameter. Colonies may develop sectors that differ in pigmentation. As seen in Figure 84-13, the colonies are typically mucoid in appearance, and, indeed, the amount of capsule produced can be

Figure 84-13. *Cryptococcus neoformans.* Colony on Sabouraud's glucose agar at room temperature for 10 days.

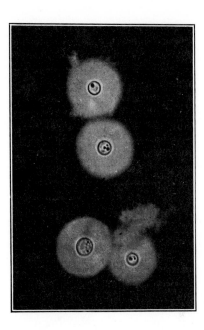

Figure 84-14. *Cryptococcus neoformans.* India ink preparation of spinal fluid. × 736.

judged by the degree of colony wetness. Highly encapsulated colonies will actually run down a slant to pool in the bottom of the tube, or they may drip off the medium of inverted plates.

Microscopic Appearance. Although filamentous variants have been isolated, most clinical isolates are uniformly spherical, budding, encapsulated yeast cells in both tissue and culture. Rarely, short hyphal forms are also observed. The yeast cells vary in size from 5 to 10 μm in diameter and exhibit both single and multiple budding. The hallmark of *C. neoformans* is its capsule, which may be up to two or three times the width of the cell (Fig. 84-14). Considerable variation exists in the size of the capsule, which is determined both by inherent strain differences and conditions of growth. Some strains characteristically produce large capsules, whereas in others the capsule is minimal. Most strains, however, even those that are consistently small capsuled in vitro, develop large capsules during infection. In a few rare cases, nonencapsulated cells of *C. neoformans* have been found in tissue. Figure 84-15 depicts the sexual life cycle of *F. neoformans*. Among clinical and natural isolates of *C. neoformans* (which are fertile and heterothallic), more than 95 percent belong to only one of the two mating types of *F. neoformans*.

PROPERTIES OF THE CAPSULE. The capsular material can be removed from yeast cells by chemical or enzymatic techniques. During growth, it is also solubilized and can be precipitated from the culture supernatant fluid. The purified capsular material is a high molecular weight polysaccharide. Complete hydrolysis yields mannose, xylose, and glucuronic acid in ratios of 3:2:1, respectively. Recent evidence supports an α-1,3-linked polymannose backbone with monomeric branches of xylose and glucuronic acid. The capsular polysaccharide is a large hapten; it is poorly immunogenic by itself, but if it is conjugated to a protein carrier, antibodies are formed to the polysaccharide determinants. Antibodies to the capsule are formed naturally during infection or immunization with whole yeast cells. Using rabbit antisera, four different antigenic determinants have been detected in the capsular polysaccharide of *C. neoformans* and designated A through D. Kwon-Chung has shown that isolates of serotypes A and D and isolates of serotypes B and C represent two distinct varieties of *C. neoformans*. These capsular polysaccharide serotypes can also be correlated with differences in ecology and prevalence of disease. Strains of serotype A can be isolated worldwide and are the most frequent cause of cryptococcosis. Type D is more common in Europe. Serotypes A and D have been isolated from soil and avian, especially pigeon, feces. Serotyes B and C have only rarely been isolated from natural environments, and infections caused by strains of these two serotypes are concentrated in southern California.

Physiology. Of the 19 species of *Cryptococcus*, only *C. neoformans* and some strains of other species are able to grow at 37C. *C. neoformans* is inhibited or killed at 41C. This temperature restriction may be an important determinant of its pathogenicity. *C. neoformans* produces a unique phe-

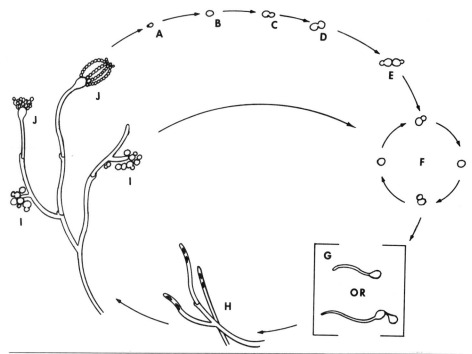

Figure 84-15. Life cycle of *Filobasidiella neoformans*. **A–E,** swelling of haploid basidiospore and subsequent budding; **F,** asexual yeast cell cycle (*Cryptococcus neoformans*); **G,** initiation of hyphal formation by sexually compatible yeast cells; **H,** dikaryotic hyphae with clamp connections; **I,** budding yeast cells; **J,** basidia and basidiospores formed at hyphal tips. *(From Erke: J Bacteriol 128:445, 1976.)*

nol-oxidase(s) that converts a variety of hydroxybenzoic substrates (including 3,4-dihydroxyphenylalanine) into brown or black pigments, which impart a dark color to the cells and/or the medium. This reaction, which was first observed empirically by Staib when the organism was cultured on a medium containing crushed seeds of *Guizotia abyssinica*, has been exploited for the rapid identification of *C. neoformans*.

All *Cryptococcus* species are nonfermentative, hydrolyze starch, assimilate inositol, and produce urease. These characteristics distinguish them from other clinically important yeasts. Table 84-5 summarizes some useful physiologic reactions for the most frequently isolated species of *Cryptococcus* and compares them with other common yeasts.

Physiologic tests are being developed to readily distinguish the two serotypic varieties of *C. neoformans*. On appropriate media, more than 85 percent of isolates of serotypes B and C assimilate glycine and malate and are resistant to concentrations of cycloheximide or canavanine, which inhibit isolates of serotypes A and D. Most isolates of the latter pair of serotypes are unable to utilize malate for carbon, or glycine as a sole carbon and nitrogen source. Both varieties utilize creatinine and are induced to synthesize creatinine deaminase, but this enzyme is repressed by the ammonia by-product in isolates of A and D only. Several selective media containing creatinine, glucose, and pH indicators have been devised that exploit this difference to differentiate between the two varieties.

Determinants of Pathogenicity

Different strains of *C. neoformans* serotype A have been shown to vary considerably in both murine virulence and in vitro capsule size. These two properites, however, are not directly correlated. Nevertheless, the capsule is probably a crucial virulence factor. The purified capsular polysaccharide has been shown, under the appropriate conditions, to specifically inhibit both phagocytosis of the yeast and the production of antibody to it.

Recent genetic analyses support the association of virulence with diphenol oxidase production. This enzyme can utilize natural catecholamines (e.g., norepinephrine, dihydroxylphenyl-alanine) as substrates, but its role in pathogenicity remains unclear.

Clinical Infection

Epidemiology

Cryptococcosis is a sporadic infection with a worldwide distribution. The causative agent is ubiquitous in the soil and in avian fecal material, such as pigeon droppings, which apparently provides a reservoir of organisms. Although spontaneous infections in animals occur, no spread of infection among animals or humans has been reported. Despite its infrequent isolation from the respiratory tract of healthy individuals, *C. neoformans* is not considered a part of the normal flora of humans or animals. Approxi-

TABLE 84-5. CHARACTERISTICS OF *CRYPTOCOCCUS* AND OTHER YEAST SPECIES FREQUENTLY ISOLATED FROM CLINICAL SPECIMENS

Species	Growth at 37C	Pseudo/True Hyphae	Capsule	Assimilations Glucose	Maltose	Sucrose	Lactose	Galactose	Inositol	Xylose	Dulcitol	Glucose Fermentation	Urease Activity	KNO₃ Utilization	Diphenol Oxidase
Cryptococcus neoformans	+	_*	+	+	+	+	_	+	+	+	+	_	+	_	+
Cryptococcus albidus var *albidus*	_*	_*	+	+	+	+	+	+*	+	+	+*	_	+	+	_
Cryptococcus albidus var *diffluens*	+*	_	+	+	+	+	_	_*	+	+	+*	_	+	+	_
Cryptococcus gastricus	_	_	+	+	+	+*	_	+	+	+	_	_	+	_	_
Cryptococcus laurentii	+*	_	+	+	+	+	+	+	+	+	+	_	+	_	_
Cryptococcus luteolus	_	_	_	+	+	+	_	+	+	+	+	_	+	_	_
Cryptococcus terreus	_*	_	+	+	+*	_	+*	+*	+	+	_*	_	+	+	_
Cryptococcus uniguttulatus	_	_	+	+	+	+	_	_*	+	+	_	_	+	_	_
Rhodotorula glutinis	+	_	_*	+	+	+	_	+*	_	+	_	_	+	+	_
Rhodotorula rubra	+	_	_*	+	+	+	_	+	_	+	_	_	+	_	_
Candida albicans	+	+	_	+	+	+	_	+	_	+	_	F	_	_	_
Saccharomyces cerevisiae	+	_*	_	+	+	+	_	+	_	_	_	F	_	_	_
Trichosporon beigelii	+*	+	_	+	+	+	+	+*	+*	+	+*	_	+*	_	_
Trichosporon pullulans	+*	+	_	+	+	+	+*	+	+	+*	_	_	+	+	_
Geotrichum candidum	_*	+	_	+	_	_	_	+	_	+	_	_	_	_	_

From Silva-Hutner, Cooper: In Lennette et al. (eds): Manual of Clinical Microbiology, 3rd ed, 1980. Washington, D.C., American Society for Microbiology.

*, strain variation; +, growth greater than negative control; —, growth not greater than negative control; F, sugar is fermented.

mately half of the cases are primary infections, and half are secondary to some other existing condition. The underlying conditions that most frequently predispose to opportunistic cryptococcosis are leukemia, lymphoma, Hodgkins's disease, sarcoidosis, systemic lupus erythematosus, and immunosuppression. The risk factor having the highest correlation with opportunistic cryptococcosis is treatment with steroids.

Patients with acquired immune deficiency syndrome (AIDS) have a marked depression of cell-mediated immunity, and, as might be predicted, are highly susceptible to cryptococcosis, as well as other opportunistic infections. Many of the AIDS patients with cryptococcal meningitis have extremely high antigen titers in both spinal fluid and serum. The microscopic examination of spinal fluid often reveals large numbers of yeast cells and few, if any leukocytes. Cryptococcosis and esophageal candidiasis are the most common mycotic infections among patients with AIDS (see Chap. 87).

Cryptococcosis occurs equally in both sexes. It is curious that despite larger numbers of children at risk, the occurrence of cryptococcosis in patients below the age of puberty is unaccountably rare.

Pathogenesis

The frequency of subclinical infection is unknown because there is not an adequate skin test antigen with which to assess exposure, and the primary pulmonary infection does not usually calcify or leave marked residua. However, because of the prevalence of *C. neoformans* in nature and the relatively low frequency of disease, many persons are probably exposed without symptoms developing. The few reports of human skin testing with experimental antigens indicate a high level of reactivity.

Systematic cryptococcosis is initiated in the lung following inhalation of yeast cells of *C. neoformans*. The isolation of small (less than 5 μm in diameter), minimally encapsulated, yeast cells of *C. neoformans* from air and soil supports this hypothesis. The predominance of yeasts of a single mating type among both clinical and natural isolates, including airborne isolates, almost certainly excludes the basidiospore (2 by 3 μm) as a routine source of infection. In the alveolar spaces, the yeast cells are initially confronted by the alveolar macrophages. Whether or not active infection and disease follows this interaction depends upon the competence of the host cellular defenses and the numbers and virulence of the yeast cells inhaled. The mechanisms responsible for normal host defense, as well as the role of cryptococcal polysaccharide in escaping or subverting those defenses, are presently being investigated.

Clinical Manifestations

Pulmonary Cryptococcosis. The primary pulmonary infection may evolve in any portion of the lung, mimic an influenza-type respiratory infection, and then resolve spontaneously. Pulmonary cryptococcosis is rarely fulminant, and hilar lymphadenopathy, calcification, and cavitation are seldom observed. Patients may have no symptoms, or a minority may experience cough, sputum production, weight loss, or fever. The tissue response is usually minimal or granulomatous.

Disseminated Cryptococcosis. *C. neoformans* is neurotropic and disseminates to the central nervous system. Meningitis may be acute or chronic. The usual progression of symptoms, i.e., fever, headache, stiff neck, and disorientation, are accompanied by spinal fluid that is typically clear, increased opening pressure, presence of cells (predominantly mononuclear), elevated protein, and normal or reduced chloride and sugar. In addition to the central nervous system, dissemination may occur to the skin and viscera.

Laboratory Diagnosis
Microscopic Examination. Spinal fluid, aspirates from skin lesions, sputum, tissue, and other appropriate specimens should be examined directly in an India ink preparation for the presence of yeast cells with capsules (Fig. 84-14). Encapsulated yeasts in tissue sections appear to be surrounded by large empty spaces because of the poor staining of the capsular polysaccharide and distortion resulting from sectioning. The capsule can be stained with mucicarmine. In histologic sections, the cells of *C. neoformans* often appear collapsed.

Culture. *C. neoformans* can be isolated on most laboratory media, but Sabouraud's medium containing cycloheximide should not be used because the organism is sensitive to this agent. Most *Cryptococcus* species are encapsulated, similar in microscopic and colony morphology, produce extracellular starch and urease, and all are nonfermentative. Several unique physiologic reactions permit laboratory identification of *C. neoformans*. *C. neoformans* is unique in its pathogenicity, its ability to grow at 37C, and its production of diphenol oxidase, or a brown pigment when grown on Staib's birdseed agar or caffeic acid medium. Spinal fluid cultures are frequently positive, but blood cultures are negative except in very severe cases.

Serologic Tests. The mycoserology for the diagnosis of cryptococcosis is most specific and sensitive. During infection, capsular material is solubilized in the body fluids, and, being an antigen, it can be titered with a specific rabbit anti-*C. neoformans* antiserum. The most commonly employed method is a latex agglutination test (see Table 84-2). Latex particles are coated with the specific rabbit immunoglobulin and mixed with dilutions of patient serum or spinal fluid. Controls include latex coated with normal rabbit globulin to check for nonspecific agglutination (e.g., rheumatoid factor). A positive agglutination at any serum or spinal fluid dilution is diagnostic for cryptococcosis. The serum or spinal fluid titer can be used prognostically. A rising or constant titer indicates a poor prognosis due to active or stabilized infection. With recovery, the antigen ti-

ter drops and disappears. Diagnosis by antigen detection is more sensitive than either India ink or culture.

Antibodies to *C. neoformans* are usually not detectable in the active disease state. There is evidence that the free polysaccharide in the serum may combine with circulating antibody or inhibit its synthesis. With recovery, however, the serum may be positive for antibody.

Treatment
Cryptococcosis responds to amphotericin B and flucytosine, and a combined therapy protocol is generally recommended. Resistance to flucytosine occurs.

Paracoccidioidomycosis

Paracoccidioidomycosis or South American blastomycosis is a systemic mycotic infection caused by a thermally dimorphic fungus, *P. brasiliensis*. The infection is confined to Central and South America. It is a chronic, granulomatous disease that begins with a primary, pulmonary, usually inapparent infection that disseminates to produce ulcerative granulomata in the mucosal surfaces of the nose, mouth, and gastrointestinal tract. Internal organs, as well as the skin and lymph nodes, may become infected.

Morphologic and Cultural Characteristics
On Sabouraud's glucose agar at room temperature, colonies of *P. brasiliensis* grow very slowly, reaching a diameter of 1 to 2 cm after 2 or 3 weeks' incubation. The gross characteristics are variable and nonspecific.

Various asexual spores are produced by *P. brasiliensis*, including chlamydospores, arthroconidia, and singly borne conidia. In the absence of conidia, which may not be produced for 10 weeks, the mycelial phase may be indistinguishable from *B. dermatitidis* or many saprophytes. By growth on a rich medium at 35C to 37C, the yeast form can be induced. The yeasts are readily identifiable by their unique appearance. As seen in Figure 84-16, they produce multiple buds, and each is attached by a narrow base. The yeasts are large, up to 30 μm in diameter, and have thinner walls than the yeasts of *B. dermatitidis*. These forms have been called pilot wheels.

Of potential importance is the observation that yeast cells of *P. brasiliensis* die after approximately 2 weeks in broth cultures. Death is attributed to the accumulation of a toxic phenolic metabolite in the medium.

Clinical Infection
Epidemiology. The natural habitat of *P. brasiliensis* is not known, but is presumed to be soil. The organism has been recovered only sporadically and irreproducibly from soil in Venezuela and Argentina. Like *B. dermatitidis*, its natural

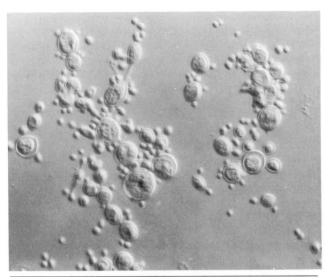

Figure 84-16. *Paracoccidioides brasiliensis.* Multiply budding yeast cells from 2-week culture on brain–heart infusion blood agar at 37C. Nomarski. × 400 *(From McGinnis: Laboratory Handbook of Medical Mycology, 1979. New York, Academic.)*

existence and life cycle in nature are unknown. There is no evidence for its transmission or an animal vector. Natural infections do not occur in either wild or domestic animals, although experimental infections can be established in mice and bats following inhalation of an infectious inoculum. Infections are presumed to follow exposure to the organisms from an exogenous source.

Several thousand cases has been reported from Brazil and lesser numbers from Venezuela, Colombia, Argentina, Ecuador, and other South and Central American countries, with the exception of Chile and the Caribbean nations. Discrete endemic foci exist within this broad area of geographic distribution. However, all cases are isolated, and epidemics have not been observed. The endemic zones are associated with moderate temperatures (14C to 30C) and rainfall, elevation of 500 to 6500 feet, subtropical forests, and river valleys, but not all areas fitting this description have paracoccidioidomycosis.

Skin test surveys have been conducted with various antigens derived from *P. brasiliensis.* These paracoccidioidins exhibit cross-reactivity with histoplasmin, and it is difficult to interpret double reactions of equal size in the same individual. As with the skin test antigens of the other dimorphic, systemic pathogens, paracoccidioidin elicits a delayed, indurative reaction indicative of previous exposure. The percentage of reactivity in endemic areas varies from 5 to 25 percent and occurs equally in both men and women. The highest percentage of reactors occurs among workers in agriculture, specifically coffee, sugar cane, and cattle workers.

Most patients are males (more than 90 percent), agricultural workers, often malnourished, and are usually 30 to 60 years of age.

Clinical Manifestations. The clinical classification of paracoccidioidomycosis is presented in Table 84-6. This organization, recently devised by Restrepo and associates, relates the clinical forms of infection to the natural history. The initial form refers to the first contact with *P. brasiliensis* via inhalation. The episode is inapparent, and the organism becomes quiescent for an indefinite period, which may be several decades in some individuals, or the lesion may resolve, perhaps with scarring. This asymptomatic infection results in conversion to a positive delayed skin test reaction to paracoccidioidin. The eventual development of symptomatic disease depends upon the host–fungal interaction, namely, the integrity of the host's cell-mediated immunity, environmental conditions (e.g., temperature, nutrients), host conditions (e.g., age, sex, state of nutrition), and the virulence of the strain of *P. brasiliensis.*

JUVENILE TYPE. Patients under 30 years of age may develop an acute, progressive infection characterized by lymphonodular lesions in the lung. This form is rare. The yeasts may disseminate to the reticuloendothelial tissue, lymph nodes, liver, spleen, skin, bone, joints, or other organs. The severity and duration of the illness depend upon the extent of organ involvement, but it may be fatal within a period of several weeks or months.

ADULT TYPE. More than 90 percent of cases are of this type and develop from the latent form, usually after several years. Lesions may be localized in the lung, or metastasis may occur from the lung to other organs, particularly the skin and mucocutaneous tissue, lymph nodes, spleen, liver, adrenals, and combinations thereof. Mucocutaneous, often petechial, lesions frequently develop on the corners of the mouth, lips, gingiva, or tongue. Pulmonary lesions are granulomatous nodules that may cavitate but rarely calcify.

The initial symptomatic form includes patients who develop disease shortly after infection without a pronounced latent period. Rare cases of primary cutaneous inoculation are also placed in this category.

Laboratory Diagnosis
MICROSCOPIC EXAMINATION. Sputum, tissue, or scrapings of mucocutaneous lesions may reveal the multiply

TABLE 84-6. CLINICAL FORMS OF PARACOCCIDIOIDOMYCOSIS

Asymptomatic infection
  Initial form
  Latent or residual form
Symptomatic disease
  Juvenile type or acute/subacute progressive form
  Adult type or chronic progressive form
    Pulmonary
    Disseminated
  Initial symptomatic form

Adapted from Restrepo et al.: Am J Med 61:33, 1976.

budding yeast form that is pathognomonic for *P. brasiliensis* (Fig. 84-16).

CULTURE. Specimens should be cultured at 25C to 30C on Sabouraud's agar with antibiotics, on Sabouraud's agar without cycloheximide, and on brain–heart infusion blood agar at 35C to 37C.

SEROLOGIC TESTS. The ID test is extremely useful. As indicated in Table 84-2, nearly 100 percent of patients have at least one of three specific precipitin lines (designated 1, 2, and 3) detected by identity with reference serum. The ID test also has prognostic value, as the bands disappear with clearing of the infection, and the number of bands is somewhat proportional to the severity of the disease.

The CF test is useful for assessing prognosis, but cross-reactions occur with other fungi.

Treatment. Ketoconazole has been highly successful in clinical trials and is now the drug of choice. Amphotericin B is very effective against paracoccidioidomycosis. The total dose usually required is 2 g or less. Nonhospitalized patients may also be given sulfa drugs, such as sulfamethoxypyridazine, at 1 g per day for 3 to 6 months and 0.5 g per day thereafter. Serologies are checked every few months to monitor the effectiveness of treatment. Successful cure requires treatment for approximately 2 years.

FURTHER READING

COCCIDIOIDOMYCOSIS

Ajello L (ed): Coccidioidomycosis, Proceedings of the Second Symposium. Tucson, Arizona University Press, 1967

Ajello L: Coccidioidomycosis and histoplasmosis. A review of their epidemiology and geographical distribution. Mycopathologia 45:221, 1971

Bayer AS, Yoshikawa TT, Calpin JE, et al.: Unusual syndromes of coccidioidomycosis: Diagnostic and therapeutic considerations. Medicine 55:131, 1976

Beaman L, Banjamini E, Pappagianis D: Activation of macrophages by lymphokines: Enhancement of phagosome–lysosome fusion and killing of *Coccidioides immitis*. Infect Immun 39:1201, 1983

Bouza E, Dreyer JS, Hewitt WL, et al.: Coccidioidal meningitis— An analysis of 31 cases and review of the literature. Medicine 60:139, 1981

Buechner HA, Seabury JH, Campbell CC, et al.: Committee recommendations. The current status of serologic, immunologic and skin tests in the diagnosis of pulmonary mycoses. Report of the Committee on fungus disease and subcommittee on criteria for clinical diagnosis—American College of Chest Physicians. Chest 63:259, 1973

Catanzaro A: Suppressor cells in coccidioidomycosis. Cell Immunol 64:235, 1981

Catanzaro A, Spitler L, Moser KM: Immunotherapy of coccidioidomycosis. J Clin Invest 54:690, 1974

Catanzaro A, Spitler LE, Moser KM: Cellular immune response in coccidioidomycosis. Cell Immunol 15:360, 1975

Cox RA, Vivas JR: Spectrum of in vivo and in vitro cell-mediated immune responses in coccidioidomycosis. Cell Immunol 31:130, 1977

DeFelice R, Galgiani JN, Campbell SC, et al.: Ketoconazole treatment of nonprimary coccidioidomycosis. Evaluation of 60 patients during three years of study. Am J Med 72:681, 1982

Deresinski SC, Stevens DA: Coccidioidomycosis in compromised hosts. Experience at Stanford University Hospital. Medicine 54:377, 1975

Deresinski SC, Galgiani JN, Stevens DA: Miconazole treatment of human coccidioidomycosis: Status report. In Ajello L (ed): Coccidioidomycosis. Current Clinical and Diagnostic Status. Miami, Symposia Specialists, 1977, p 267

Deresinski SC, Applegate RJ, Levine HB, Stevens DA: Cellular immunity to *Coccidioides immitis*: In vitro lymphocyte response to spherules, arthrospores, and endospores. Cell Immunol 32:110, 1977

Diamond RD, Bennett JE: A subcutaneous reservoir for intrathecal therapy of fungal meningitis. N Engl J Med 288:186, 1973

Drutz DJ, Catanzaro A: Coccidioidomycosis, Parts I and II. Am Rev Respir Dis 117:559,727, 1978

Edwards PQ, Palmer CE: Prevalence of sensitivity to coccidioidin, with special reference to specific and nonspecific reactions to coccidioidin and to histoplasmin. Dis Chest 31:35, 1957

Einstein HE: Coccidioidomycosis of the central nervous system. Adv Neurol 6:101, 1974

Flynn NM, Hoeprich PD, Kawachi MM, et al.: An unusual outbreak of windborne coccidioidomycosis. N Engl J Med 301:358, 1979

Huppert M: Serology of coccidioidomycosis. Mycopathologia 41:107, 1970

Huppert M, Krasnow I, Vukovich KR, et al.: Comparison of coccidioidin and spherulin in complement-fixation test for coccidioidomycosis. J Clin Microbiol 6:33, 1977

Levine HB, Gonzales-Ochoa A, Ten Eyck DR: Dermal sensitivity to *Coccodioides immitis*. A comparison of responses elicited in man by spherulin and coccidioidin. Am Rev Respir Dis 107:379, 1973

Levine HB, Restrepo A, Ten Eyck DR, et al.: Spherulin and coccidioidin: Cross-reactions in dermal sensitivity to histoplasmin and paracoccidioidin. Am J Epidemiol 101:512, 1975

Nelson AR: The surgical treatment of pulmonary coccidioidomycosis. Curr Probl Surg 11:1, 1974

Pappagianis D: Epidemiological aspects of respiratory mycotic infections. Bacteriol Rev 31:25, 1967

Rhea TH, Einstein H, Levan NE: Dinitrochlorobenzene responsivity in disseminated coccidioidomycosis: An inverse correlation with complement-fixing antibody titers. J Invest Dermatol 66:34, 1976

Smith CE: Coccidioidomycosis. Med Clin North Am 27:790, 1943

Smith CE, Saito MT, Beard RR, et al.: Serological tests in the diagnosis and prognosis of coccidioidomycosis. Am J Hyg 52:1, 1950

Smith CE, Whiting EG, Baker EE, et al.: The use of coccidioidin. Am Rev Tuberc 57:330, 1948

Standard PG, Kaufman L: Immunological procedure for the rapid and specific identification of *Coccidioides immitis* cultures. J Clin Microbiol 5:149, 1977

Stevens DA (ed): Coccidioidomycosis. A Text. New York, Plenum, 1980

Stobo JD, Paul S, Van Scoy RE, et al.: Suppressor thymus-derived lymphocytes in fungal infection. J Clin Invest 57:319, 1976

Wheat RW, SuChung KS, Ornellas EP, et al.: Extraction of skin test activity from *Coccidioides immitis* mycelia by water, perchloric acid, and aqueous phenol extraction. Infect Immun 19:152, 1978

HISTOPLASMOSIS

Ajello L, Chick EW, Furcolow ML (eds): Histoplasmosis. Springfield, Ill, Thomas, 1971

Artz RP, Bullock WE: Immunoregulatory responses in experimental disseminated histoplasmosis: Depression of T-cell dependent and T-effector responses by activation of splenic suppressor cells. Infect Immun 23:893, 1979

Berliner MD, Biundo N: Effects of continuous light and total darkness on cultures of *Histoplasma capsulatum.* Sabouraudia 11:48, 1973

Boguslawski G, Kobayashi GS, Schlessinger D, et al.: Characterization of an inhibitor of ribonucleic acid polymerase from the mycelial phase of *Histoplasma capsulatum.* J Bacteriol 122:532, 1975

Brodsky AL, Gregg MB, Lowenstein MS, et al.: Outbreak of histoplasmosis associated with the 1970 Earth Day activities. Am J Med 54:333, 1973

Christie A, Peterson JC: Pulmonary calcification in negative reactors to tuberculin. Am J Public Health 35:1131, 1945

Darling STA: A protozoan general infection producing pseudotuberules in the lungs and focal necrosis in the liver, spleen, and lymph nodes. JAMA 46:1283, 1906

deMonbreun WA: The cultivation and cultural characteristics of Darling's *Histoplasma capsulatum.* Am J Trop Med 14:93, 1934

Domer JE: Monosaccharide and chitin content of cell walls of *Histoplasma capsulatum* and *Blastomyces dermatitidis.* J Bacteriol 107:870, 1971

Domer JE, Moser SA: Histoplasmosis—A review. Rev Med Vet Mycol 15:159,1980

Goodman NL, Larsh HW: Environmental factors and growth of *Histoplasma capsulatum* in soil. Mycopath Mycol Appl 33:145, 1967

Goodwin RA, Des Prez RM: Histoplasmosis. Am Rev Respir Dis 117:929, 1978

Howard D: Effect of temperature on the intracellular growth of *Histoplasma capsulatum.* J Bacteriol 93:438, 1967

Kanetsuma F, Carbonell LM, Gil F, Azana I: Chemical and ultrastructural studies on the cell walls of *Histoplasma capsulatum.* Mycopathologia 54:1, 1974

Kaufman L, Terry AT, Schubert JH, et al.: Effects of a single histoplasmic skin test in the serological diagnosis of histoplasmosis. J Bacteriol 94:798, 1967

Kumar BV, McMillian R, Gutwein M, et al.: Comparison of the ribonucleic acid polymerases from both phases of *Histoplasma capsulatum.* Biochemistry 19:1080, 1980

Kwon-Chung KJ: Sexual stage of *Histoplasma capsulatum.* Science 175:326, 1972

Kwon-Chung KJ: Perfect state (*Emmonsiella capsulata*) of the fungus using large-form African histoplasmosis. Mycologia 67:980, 1975

Larsh HW: The epidemiology of histoplasmosis. In Al-Doory Y (ed): The epidemiology of human mycotic diseases. Springfield, Ill, Thomas, 1975, p 52

Maresca B, Lambowitz AM, Kumar VB, et al.: Role of cysteine in regulating morphogenesis and mitochondrial activity in the dimorphic fungus *Histoplasma capsulatum.* Proc Natl Acad Sci USA 78:4596, 1981

Marcesca B, Schlessinger D, Medoff G, et al.: Regulation of di-

morphism in the pathogenic fungus *Histoplasma capsulatum.* Nature (London) 266:447, 1977

McVeigh I, Morton K: Nutritional studies of *Histoplasma capsulatum.* Mycopathologia 25:294, 1965

Palmer CE: Nontuberculous pulmonary calcification and sensitivity to histoplasmin. Public Health Rep 60:513, 1945

Picardi JL, Kauffman CA, Schwarz J, et al.: Detection of precipitating antibodies to *Histoplasma capsulatum* by counterimmunoelectrophoresis. Am Rev Respir Dis 114:171, 1976

Schwarz J: Histoplasmosis. New York, Praeger, 1981

Tewari RP, Berkhart FJ: Comparative pathogenicity of albino and brown types of *Histoplasma capsulatum* for mice. J Infect Dis 125:504, 1972

Wheat LJ, Slama TG, Eitzen HE, et al.: A large urban outbreak of histoplasmosis: Clinical features. Ann Intern Med 94:331, 1981

BLASTOMYCOSIS

Bassett FH, Tindall JP: Blastomycosis of bone. South Med J 65:547, 1972

Benham RW: Fungi of blastomycosis and coccidioidal granuloma. Arch Dermatol 30:385, 1934

Brass C, Volkmann CM, Klein HP, et al.: Pathogen factors and host factors in murine pulmonary blastomycosis. Mycopathologia 78:129, 1982

Brewer NS, Rhodes KH, Roberts GD, et al.: Blastomycosis in canoeists—Wisconsin. Morbid Mortal Weekly Rep 28:450, 1979

Brummer E, Stevens DA: Opposite effects of human monocytes, macrophages, and polymorphonuclear neutrophils on replication of *Blastomyces dermatitidis* in vitro. Infect Immun 36:297, 1982

Busey JF: Blastomycosis. III. A comparative study of 2-hydroxystilbamidine and amphotericin B therapy. Am Rev Respir Dis 105:812, 1972

Busey JF, Baker L, et al.: Blastomycosis. I. A review of 198 collected cases in Veterans Administration hospitals. Am Rev Respir Dis 89:659, 1964

Cox RA, Mills LR, Best GK, Denton JF: Histologic reactions to cell walls of an avirulent and a virulent strain of *Blastomyces dermatitidis.* J Infect Dis 129:179, 1974

Denton JF, DiSalvo AF: Isolation of *Blastomyces dermatitidis* from natural sites of Augusta, Georgia. Am J Trop Med Hyg 13:716, 1964

Duttera M, Osterhout S: North American blastomycosis: A survey of 63 cases. South Med J 62:295, 1969

Fountain FF: Acute blastomycotic arthritis. Arch Intern Med 132:684, 1973

Furcolow ML, Chick EW, Busey JF, et al.: Prevalence and incidence studies of human and canine blastomycosis. I. Cases in the United States, 1885–1968. Am Rev Respir Dis 102:60, 1970

Furcolow ML, Smith CD: A new hypothesis on the epidemiology of blastomycosis and the ecology of *Blastomyces dermatitidis.* Trans NY Acad Sci, Ser II 34:421, 1973

Furcolow ML, Smith C, Gallis H, et al.: Blastomycosis—North Carolina. Morbid Mortal Weekly Rep 25:205, 1976

Gilchrist TC: A case of blastomycetic dermatitis in man. Johns Hopkins Hosp Reports 1:269, 1896

Goldthorpe WG, Butler KF: Blastomycosis outbreak in Sioux Lookout Zone. Can Diseases Weekly Rep 1–13:49, 1975

Kaufman L, McLaughlin DW, Clark MJ, et al.: Specific immunodiffusion test for blastomycosis. Appl Microbiol 26:244, 1973

Kitchen MS, Reiber CD, Eastin GB: An urban epidemic of North American blastomycosis. Am Rev Respir Dis 115:1063, 1977

Kunkel WM, Weed LA, McDonald JR, et al.: North American blastomycosis—Gilchrist's disease; clinicopathologic study of ninety cases. Int Abst Surg 99:1, 1954

Landay ME, Mitten J, Millar J: Disseminated blastomycosis in hamsters. II: Effect of sex on susceptibility. Mycopathologia 42:73, 1970

Martin DS, Smith DT: Blastomycosis I. A review of the literature. Am Rev Tuberc 39:275, 1939

McDonough ES: Blastomycosis. Epidemiology and biology of its etiologic agent, *Ajellomyces dermatitidis*. Mycopathologia 41:195, 1970

McDonough ES, Dubats JJ, Wisniewski JR, et al.: Soil streptomycetes and bacteria related to lysis of *Blastomyces dermatitidis*. Sabouraudia 11:244, 1973

McDonough ES, Lewis AL: *Blastomyces dermatitidis*: Production of the sexual stage. Science 156:528, 1967

Mitchell TG: Blastomycosis. In Feigin RD, Cherry JD (eds): Textbook of Pediatric Infectious Diseases. Philadelphia, Saunders, 1981, pp 1478–1488

Repine JE, Clawson CC, Rasp FL, et at.: Defective neutrophil locomotion in human blastomycosis. Evidence for a serum inhibitor. Am Rev Respir Dis 118:325, 1978

Sarosi GA, Davies SF: Blastomycosis. Am Rev Respir Dis 120:911, 1979

Sarosi GA, Hammerman KJ, Tosh FE, et al.: Clinical features of acute pulmonary blastomycosis. N Engl J Med 290:540, 1974

Schwarz J, Baum GL: Blastomycosis. Am J Clin Pathol 21:999, 1951

Sixbey JW, Fields BT, Sun CN, et al.: Interactions between human granulocytes and *Blastomyces dermatitidis*. Infect Immun 23:41, 1979

Smith JG, Harris JS, Conant NF, et al.: An epidemic of North American blastomycosis. JAMA 158:641, 1955

Snapper I, McVay LO Jr.: The treatment of North American blastomycosis with 2-hydroxystilbamidine. Am J Med 15:603, 1953

Tenenbaum MJ, Greenspan J, Kerkering TM: Blastomycosis. CRC Crit Rev Microbiol 9:139, 1982

Tosh FE, Hammerman KJ, Weeks RJ, et al.: A common source epidemic of North American blastomycosis. Am Rev Respir Dis 109:525, 1974

CRYPTOCOCCOSIS

Adamson DM, Cozad GC: Effect of antilymphocyte serum on animals experimentally infected with *Histoplasma capsulatum* or *Cryptococcus neoformans*. J Bateriol 100:1271, 1969

Armstrong D, Young LS, Meyer RD, et al.: Infectious complications of neoplastic disease. Med Clin North Am 55:729, 1971

Bennett JE, Kwon-Chung KJ, Howard DH: Epidemiologic differences among serotypes of *Cryptococcus neoformans*. Am J Epidemiol 105:582, 1977

Bhattacharjee AK, Kwon-Chung KJ, Glaudenmans CPJ: On the structure of the capsular polysaccharide from *Cryptococcus neoformans* serotype C. Immunochemistry 15:673, 1978

Bindschadler DD, Bennett JE: Serology of human cryptococcosis. Ann Intern Med 69:45, 1968

Blandamer A, Danishefsky I: Investigations on the structure of the capsular polysaccharide from *Cryptococcus neoformans* type B. Biochem Biophys Acta 117:305, 1966

Bulmer GS, Sans MD: *Cryptococcus neoformans*. III. Inhibition of phagocytosis. J Bacteriol 95:5, 1967

Chaskee S, Tyndall RL: Pigment production by *Cryptococcus neoformans* from *para*- and *ortho*-diphenols: Effect of nitrogen source. J Clin Microbiol 1:509, 1975

Dabbagh R, Conant NF, Nielsen HS, et al.: Effect of temperature on saprophytic cryptococci: Temperature-induced lysis and protoplast formation. J Gen Microbiol 85:177, 1974

Denton JF, DiSalvo AF: The prevalence of *Cryptococcus neoformans* in various natural habitats. Sabouraudia 6:213, 1968

Diamond RD: Antibody-dependent killing of *Cryptococcus neoformans* by human peripheral blood mononuclear cells. Nature 247:148, 1974

Diamond RD, Bennett JE: Prognostic factors in cryptococcal meningitis. Ann Intern Med 80:176, 1974

Diamond RD, Root RK, Bennett JE: Factors influencing killing of *Cryptococcus neoformans* by human leukocytes in vitro. J Infect Dis 125:376, 1972

Dykstra MA, Friedman L, Murphy JW: Capsule size of *Cryptococcus neoformans*: Control and relationship to virulence. Infect Immun 16:129, 1977

Erke KH: Light microscopy of basidia, basidiospores, and nuclei in spores and hyphae of *Filobasidiella neoformans* (*Cryptococcus neoformans*). J Bacteriol 128:445, 1976

Evans EE: The antigenic composition of *Cryptococcus neoformans*. I. A serologic classification by means of the capsular and agglutination reactions. J Immunol 64:423, 1950

Fromtling RA, Shadomy HJ: Immunity in cryptococcosis: An overview. Mycopathologia 77:183, 1982

Gadebush HH, Johnson AG: Natural host resistance to infection with *Cryptococcus neoformans* V. The influence of cationic tissue proteins upon phagocytosis and on circulating antibody synthesis. J Infect Dis 116:566, 1966

Gentry LO, Remington JS: Resistance against *Cryptococcus* conferred by intracellular bacteria and protozoa. J Infect Dis 123:22, 1971

Gordon MA, Lapa E: Charcoal particle agglutination test for detection of antibody to *Cryptococcus neoformans*: A preliminary report. Am J Clin Pathol 56:354, 1971

Graybill JR; Alford RH: Cell-mediated immunity in cryptococcosis. Cell Immunol 14:12, 1974

Griffin FM: Roles of macrophage Fc and C3b receptors in phagocytosis of immunologically coated *Cryptococcus neoformans*. Proc Natl Acad Sci USA 78:3853, 1981

Jong SC, Bulmer GS, Ruiz A: Serologic grouping and sexual compatibility of airborne *Cryptococcus neoformans*. Mycopathologia 79:185, 1982

Karaoui RM, Hall NK, Larsh HW: Role of macrophages in immunity and pathogenesis of experimental cryptococcosis induced by the airborne route—Part II: Phagocytosis and intracellular fate of *Cryptococcus neoformans*. Mykosen 20:409, 1977

Kerkering TM, Duma RJ, Shadomy S: The evolution of pulmonary cryptococcosis. Clinical implications from a study of 41 patients with and without compromising host factors. Ann Intern Med 94:611, 1981

Kozel TR: Non-encapsulated variant of *Cryptococcus neoformans* II. Surface receptors for cryptococcal polysaccharide and their role in inhibition of phagocytosis of polysaccharide. Infect Immun 16:99, 1977

Kozel TR: Dissociation of a hydrophobic surface from phagocytosis of encapsulated and non-encapsulated *Cryptococcus neoformans*. Infect Immun 39:1214, 1983

Kozel TR, Follete JL: Opsonization of encapsulated *Cryptococcus neoformans* by specific anticapsular antibody. Infect Immun 31:978, 1981

Kwon-Chung KJ: Description of a new genus, *Filobasidiella*, the perfect state of *Cryptococcus neoformans*. Mycologia 67:1197, 1975

Kwon-Chung KJ: A new species of *Filobasidiella*, the sexual state of

Cryptococcus neoformans B and C serotypes. Mycologia 68:942, 1976

Kwon-Chung KJ, Bennett JE: Distribution of alpha and α mating types of *Cryptococcus neoformans* among natural and clinical isolates. Am J Epidemiol 108:337, 1978

Kwon-Chung KJ, Polacheck I, Bennett JE: Improved diagnostic media for separation of *Cryptococcus neoformans* var *neoformans* (serotypes A and D) and *Cryptococcus neoformans* var *gatti* (serotypes B and C). J Clin Microbiol 15:535, 1982

Littman ML: Cryptococcosis (torulosis). Current concepts and therapy. Am J Med 27:976, 1959

Mitchell TG, Friedman L: In vitro phagocytosis and intracellular fate of variously encapsulated strains of *Cryptococcus neoformans*. Infect Immun 5:491, 1972

Polacheck I, Kwon-Chung KJ: Creatinine metabolism in *Cryptococcus neoformans* and *Cryptococcus bacillisporus.* J Bacteriol 142:15, 1980

Randall RE, Stacy WK, Torne ED, et al.: Cryptococcal pyelonephritis. N Engl J Med 279:60, 1968

Rhodes JC, Polacheck I, Kwon-Chung KJ: Phenoloxidase activity and virulence in isogenic strains of *Cryptococcus neoformans*. Infect Immun 36:1175, 1982

Schmeding KA, Jong SC, Hugh R: Sexual compatibility between serotypes of *Filobasidiella neoformans* (*Cryptococcus neoformans*). Curr Microbiol 5:133, 1981

Schupback DW, Wheeler CE, Briggaman RA, et al.: Cutaneous manifestations of disseminated cryptococcosis. Arch Dermatol 112:1734, 1976

Staib F: *Cryptococcus neoformans* and *Guizotia abyssinica*. Zentralblatt für Hygiene 148:466, 1962

Utz JP, Garrigues IL, Sande MA, et al.: Therapy of cryptococcosis with a combination of flucytosine and amphotericin B. J Infect Dis 132:368, 1975

Wilson DE, Bennett JE, Bailey JW: Serologic grouping of *Cryptococcus neoformans*. Proc Soc Exp Biol Med 127:820, 1968

Young RC, Bennett JE, Geelhoed GW, et al.: Fungemia with compromised host resistance. A study of 70 cases. Ann Intern Med 80:605, 1974

Zimmerman LE: Fatal fungus infections complicating other diseases. Am J Clin Pathol 25:46, 1955

Zimmerman LE, Rappaport H: Occurence of cryptococcosis in patients with malignant disease of the reticuloendothelial system. Am J Clin Pathol 24:1050, 1954

PARACOCCIDIOIDOMYCOSIS

Giraldo R, Restrepo A, Gutierrez F, et al.: Pathogenesis of paracoccidioidomycosis: A model proposed on the study of 46 patients. Mycopathologia 58:63, 1976

Greer D, Bolanos B: Role of bats in the ecology of *P. brasiliensis*: The survival of *P. brasiliensis* in the intestinal tract of fructivirous bats. Sabouraudia 15:273, 1977

Greer D, D'Acosta D, Agredo L: Dermal sensitivity to paracoccidioidin and histoplasmin in family members of patients with paracoccidioidomycosis. J Trop Med Hyg 23:87, 1974

Kenetsuna F, Carbonell LM, Azuma I, et al.: Biochemical studies on the thermal dimorphism of *Paracoccidioides brasiliensis*. J Bacteriol 110:208, 1972

Paracoccidioidomycosis. PAHO Publ. No. 254, 1972

Restrepo A: Paracoccidioidomycosis. Acta Med Colombiana 3:33, 1978

Restrepo A, Moncada LH: Characterization of the precipitin bands detected in the immunodiffusion test for paracoccidioidomycosis. Appl Microbiol 28:138, 1974

Restrepo A, Robledo M, Giraldo R, et al.: The gamut of paracoccidioidomycosis. Am J Med 61:33, 1976

San-Blas F, DeMarco G, San-Blas G: Isolation and partial characterization of a growth inhibitor of *Paracoccidioides brasiliensis*. Sabouraudia 20:159, 1982

San-Blas G, San-Blas F: *Paracoccidioides brasiliensis*: Cell wall structure and virulence. A review. Mycopathologia 62:77, 1977

CHAPTER 85

Subcutaneous Mycoses

Sporotrichosis

Sporotrichosis is a chronic infection of the cutaneous and subcutaneous tissues and lymphatics caused by the thermally dimorphic fungus, *Sporothrix schenckii.* The infection is initiated by traumatic inoculation of the organisms into the skin. Secondary spread may follow, with involvement of the draining lymphatics, lymph nodes, and rarely, the underlying muscle and bone. Occasionally, dissemination

occurs to internal organs, such as the lungs, central nervous system, or genitourinary tract.

Historically, the clinical forms and etiology of sporotrichosis were described in the early part of this century. A massive epidemic among gold miners in South Africa served to clarify the natural history of the infection. During a 2-year period, over 3000 cases were diagnosed. The epidemic was caused not by transmission among the infected laborers but by exposure of large numbers of susceptible miners to a common source of the infectious particles. The fungus was found growing saprophytically on the timber supporting the mine shafts and tunnels. As the miners brushed against the contaminated wood, the skin of their exposed arms and shoulders was abraded, and spores or hyphal fragments of *Sporothrix* were implanted under the skin. Once the source was discovered, the epidemic was controlled by treating the timber with fungicides to eradicate the organisms.

In nature, *S. schenckii* occurs in association with plant life and in the soil. The infection, which is acquired in the course of outdoor activities, follows traumatic implantation of the fungus into the skin. Patients frequently recall a history of trauma. The primary nodule often develops at the site of a previous wound caused by a thorn or splinter, by brushing against tree bark, or by handling of reeds and grasses. Some cases have resulted from skin injury caused by metal objects or animal bites contaminated with soil containing the organism. There is evidence that the fungus may grow on biting insects, such as ants and fleas, and inoculation may occur in this manner. Exposure and infection may possibly follow contact with patients or their contaminated dressings.

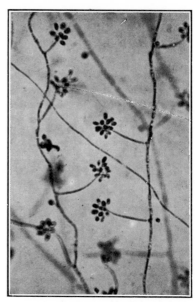

Figure 85-1. *Sporothrix schenckii.* Clusters of pyriform conidia (2 to 3 μm by 3 to 6 μm) from the tips of distally tapering conidiophores. × 650. *(From Conant, et al.: Manual of Clinical Mycology, 3rd ed, 1971. Philadelphia, Saunders.)*

Sporothrix schenckii

Morphologic and Cultural Characteristics

On Sabouraud's agar at room temperature, colonies of *S. schenckii* develop in 3 to 5 days. At first, they are usually blackish and shiny but become fuzzy with age as aerial hyphae are produced. The colony appearance of different strains varies considerably. In some strains, colonies are initially yeast-like but convert later to the mycelial phase. In some cases, colonies may be white instead of black or gray.

Microscopic examination of the colony reveals thin, branching septate hyphae and small (approximately 3 μm by 5 μm) conidia that are delicately attached to the distal, tapering ends of slender conidiophores. These conidia usually become detached in a teased preparation, but in a microculture that permits examination in situ, conidia are arranged in flower-like clusters (Fig. 85-1). Conidiation is sympodial and increases with age. In addition to the characteristic florets, some strains also produce larger, darkly pigmented conidia from the hyphae.

S. schenckii is dimorphic. Conversion to the yeast phase may be accomplished in vitro by cultivating the organisms on a rich medium, such as brain–heart blood infusion agar at 35C to 37C. Sometimes the mycelial growth persists, and the yeast cells develop only at the periphery of the colony. Yeast colonies are pasty and grayish. Yeast cells from these colonies are variable in shape but are usually fusiform, approximately 1 to 3 μm by 3 to 10 μm, with multiple buds (Fig. 85-2). As an alternative to in vitro conversion, animal inoculation can be used to induce the yeast phase.

Etiology of Sporotrichosis

The possibility that sporotrichosis may be caused by more than one species has been suggested for years, but the theory of multiple etiology is based on tenuous and circumstantial evidence. Several species of *Sporothrix* and *Ceratocystis* are phytopathogens or closely associated with vegetation. A number of these species are difficult to separate. In fact, *S. schenckii* may be morphologically indistinguishable from *Ceratocystis stenoceras*. Because these species occupy the same habitat and closely resemble each other, it has been suggested that more than one species might be pathogenic for humans. This would account for the considerable variation in some of the early morphologic descriptions. Species of *Sporothrix* and *Ceratocystis* are similar in cell wall composition and antigenicity. Sera from patients with sporotrichosis contain antibodies to antigenic determinants that are found in both genera. One of these determinants, a rhamno-mannan moiety, is also present in the streptococcal cell envelope. The presence of antibodies or delayed hypersensitivity to both *Sporothrix* and *Ceratocystis* presumably results from cross-reactivity. However, experimental infection with *Ceratocystis* has been reported.

Another observation that is compatible with the multiple agent etiology is the contrasting conditions during which the infection is most prevalent in the endemic areas. Sporotrichosis, while found worldwide, is endemic in parts of Brazil, Uruguay, South Africa, and Zimbabwe. In these areas, most cases occur during the hot, rainy season. Conversely, in Mexico, where the disease is also endemic, the

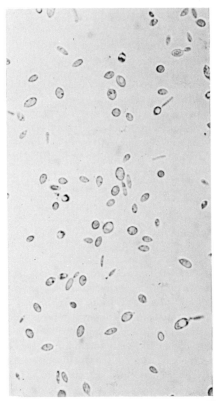

Figure 85-2. *Sporothrix schenckii.* Fusiform, round, oval cells from culture on brain–heart infusion blood agar at 37C. × 790. *(From Conant, et al.: Manual of Clinical Mycology, 3rd ed, 1971. Philadelphia, Saunders.)*

peak incidence occurs during the cool, dry season. These differences in climatic conditions during which the disease is most prevalent may reflect population differences in host defense or exposure, or possibly different etiologic strains. The epidemiology of sporotrichosis also varies in different regions (see below).

Determinants of Pathogenicity

The usual explanation for the subcutaneous nature of sporotrichosis invokes the temperature sensitivity of the organism. The temperature of skin and subcutaneous tissue is slightly lower than that of deeper tissue. In support of this theory, clinical reports describe the therapeutic effects of fever and the application of hot packs.

The maximal temperature at which most strains of *S. schenckii* will grow in vitro is 38C, and many soil isolates fail to grow at 37C. Passage through mice appears to enhance the virulence of strains, perhaps by selecting temperature-tolerant cells. For mice, temperature is a very important determinant of pathogenicity. When mice are inoculated intravenously with identical inocula of yeast cells of *S. schenckii* and subsequently housed at different temperatures, mice kept at 13C to 17C develop lesions only in the muscle tissue of the legs, whereas mice housed in the cold (ambient temperatures of 2C to 5C) develop miliary lesions in the liver as well as in the extremities. However, intrathoracic inoculation of mice with yeast cells results in a disseminated form of the disease. Isolates of *S. schenckii* from nonlymphangitic (fixed) lesions grow less well at 35C and 37C (both in vitro and in vivo) than isolates obtained from cases of lymphocutaneous sporotrichosis.

At temperatures below 37C, both *Sporothrix* and *Ceratocystis* produce neuraminidase. As neuraminic acid is only found in mammalian tissue and some microorganisms, but not in plants, this finding suggests a possible mechanism for tissue tropism and toxicity.

Clinical Infection

Epidemiology. Sporotrichosis is worldwide in its occurrence and attacks persons of all ages. Although 75 percent of the patients are males, it is not known whether this preference is attributable to sex-linked susceptibility or to increased exposure. The incidence is higher among agricultural workers and persons who live or work around vegetation. Indeed, sporotrichosis must be considered an occupational disease of forest rangers, horticulturists, and similar workers.

Animals, both wild and domestic, are susceptible to *S. schenckii* infection. The disease is similar to that which occurs in humans and has been especially well documented in horses. The highest incidence in the world is in Mexico, where sporotrichosis is the second most common fungal infection. Most cases occur among rural, often malnourished peasants in the endemic areas of Mexico, Central America, and Brazil. Infections are most pronounced in debilitated or malnourished patients, which may partially explain the high incidence in the underdeveloped countries. In developed countries, such as the United States, Canada, and France, cases of sporotrichosis are less frequent and are associated with gardening soil and the cultivation of roses. The occurrence in the United States among affluent, middle-aged, white males who may have a history of alcoholism has been described as the alcoholic rose-gardener syndrome. Occasionally, a cluster of cases occurs following exposure of several individuals to a common source at the same time, similar to the epidemic among the gold miners of South Africa. An outbreak of this kind occurred among Mississippi Forestry Commission workers, who developed sporotrichosis of the upper extremities after planting pine seedlings that were packed in sphagnum moss contaminated with *S. schenckii*. A fight among college students who threw contaminated bricks at each other also precipitated a cluster of cases.

Host factors related to nutrition, alcohol, or sex may predispose certain individuals to infection. Finally, the presence of cell-mediated immunity to *Sporothrix* appears to restrict the infection.

Clinical Manifestations

LYMPHOCUTANEOUS SPOROTRICHOSIS. Seventy-five percent of the cases of sporotrichosis are lymphocutaneous, with lesions first appearing in the cutaneous or subcutaneous tissue and progressively involving the draining lymphatics. The classic history begins with the traumatic implantation of the organisms into the skin. The incubation period is highly variable, ranging from a few days to several months, with an average of about 3 weeks. The duration probably depends upon such factors as the host defenses and the size of the inoculum. Initially, a small, movable, nontender, subcutaneous nodule develops. In some cases, if the organisms are more epidermal than subcutaneous, a small ulcer may develop. The subcutaneous nodule becomes discolored, and the overlying skin darkens to a reddish color and eventually blackens. This necrotic lesion subsequently erupts through the skin surface to form an ulcer or sporotrichotic chancre. After a few weeks, the primary lesion heals, and new ones develop nearby.

CHRONIC SPOROTRICHOSIS. Following the primary nodule, multiple subcutaneous nodules develop along the lymphatic channels, which become hard and cord-like. The sequence of development is the same: a movable nodule forms, which then becomes attached to the overlying skin, discolors due to suppuration, and ulcerates. The initial lesion is more exudative and less gummatous than the secondary ones. Untreated, sporotrichosis usually becomes chronic, but it may heal spontaneously.

FIXED SPOROTRICHOSIS. Fixed sporotrichosis refers to the presence of only one lesion. The infection is restricted and less progressive, and although the lesion may alternately wax and wane, it may also resolve completely. It is confined, and the lymphatics are not involved. The lesion may be ulcerative or appear as a plaque or rash. Because the appearance is so variable, a positive culture is required to differentiate sporotrichosis in these cases from a number of other bacterial and fungal skin lesions and carcinomas. The fixed lesion is more common in highly endemic areas, where this limited pathogenicity may be related to the overall high frequency of delayed hypersensitivity.

OTHER FORMS. Infection may spread to mucocutaneous areas, including the eyes. The most common site of dissemination is the bone, usually following a negligible, primary skin lesion. When metastasis occurs to deeper tissues, the meninges are the most common site.

In recent years, primary pulmonary sporotrichosis has been described more frequently. This form results from inhalation of the infectious particles rather than secondary dissemination from a superficial location. Pulmonary sporotrichosis of this form can be opportunistic and has been observed in compromised patients in urban hospitals. Patients with systemic sporotrichosis usually have impaired cell-mediated immunity. The disease may present a spectrum of manifestations. It can be chronic and mimic cavitary tuberculosis. It can also be acute and rapidly progressive with lymphadenopathy.

Laboratory Diagnosis

MICROSCOPIC EXAMINATION. The histopathology of sporotrichotic lesions resembles that of epithelioma. Sections of biopsy specimens yield very few organisms. The yeast cells are so sparse in most infected tissues that special stains, such as the periodic acid-Schiff or methenamine silver stain, are usually required. Fetter and Tindall recommend enzymatic digestion of tissue smears before staining to remove nonfungal polysaccharides. Fluorescent antibody can also be used to locate the rare yeast cells in tissue.

When yeasts are observed in tissue, they are usually spherical, multiply budding cells 3 to 5 μm in diameter. Some of the yeast cells may be fusiform, elongated, or cigar-shaped. They may appear to be encapsulated, but this is an artifact caused by shrinkage of the cytoplasm away from the cell wall during fixation. When present, the asteroid body is characteristic of sporotrichosis. However, asteroids are not specific for S. schenckii, as they can also be formed during infection with other fungi, especially Aspergillus, as well as bacteria. The asteroid body consists of the central, basophilic yeast cell surrounded by radiating extensions of eosinophilic material that may be as thick as 10 μm. The rays contain antigen–antibody complexes, complement, and tissue components and probably constitute an immunologic reaction to the organism. Potassium hydroxide preparations can be made of smears of biopsy material or exudates from ulcerative lesions.

CULTURE. The most reliable method of diagnosis is culture. Specimens should be streaked on Sabouraud's agar containing antibacterial antibiotics and incubated at room temperature. However, S. schenckii will grow on a variety of media. The following are suitable specimens for culture: aspirated fluid; pus; exudative material; biopsy tissue; or pulmonary infection is suspected.

SEROLOGY. Because specific antibodies are usually not present in the early stages of infection, serology is of little help in establishing the diagnosis. Perhaps the most useful of several tests that have been described is the yeast cell agglutination test. Agglutinins to the yeast cells can be used to monitor the course of infection. Patients with active sporotrichosis have an agglutinating antibody titer of 1:160 or greater. Titers up to 1:40 can be found in recovered patients, healthy individuals, or those with cross-reactive antibodies. Therefore, only titers above 1:40 are significant.

Delayed skin test reactions are elicited in sensitive persons by sporotrichin. Positive reactions are observed in many healthy individuals in the endemic regions, and the presence of delayed hypersensitivity has been associated with less severe infections.

Treatment. Neither lymphocutaneous nor fixed sporotrichosis is a debilitating infection. The lesions are chronic and may persist without treatment for years, with periods of improvement and regression. Often lesions heal spontaneously. The prognosis for disseminated sporotrichosis is invariably grave. With this form, spontaneous cures never occur, as the infection is progressive, and the organisms are numerous.

The treatment of choice for cutaneous sporotrichosis is potassium iodide (KI), discovered empirically to be effective in 1912 and recommended ever since. This can be administered topically, orally, or both. A typical regimen consists of the oral administration of a saturated solution of KI in milk. Dosage is increased daily at 0.5 to 1.0 ml increments from 1 ml three times per day to 4 to 6 ml three times per day. Adverse side effects may be indigestion, rash, cardiac arrhythmias, watering eyes, or swelling of salivary glands. Treatment is tapered off if these occur. Most patients tolerate KI, and treatment is continued for at least 4 weeks after resolution of the clinical symptoms. Surface lesions can be treated topically with 2 percent KI in 0.2 percent iodine.

The mode of action of iodide therapy is unclear. In vitro, *S. schenckii* is resistant to 10 percent KI. In vivo, iodides cause resolution of granulomata and abscesses, which is probably the basis for their therapeutic effects. After partial disruption of the tissue response, the organisms are exposed to the host's immunologic defenses.

Except in cases of severe disseminated disease, KI should always be tried first. If this fails, amphotericin B may be employed. Case reports have also supported the use of other antifungals, including dihydroxystilbamidine, griseofulvin, and 5-fluorocytosine. With ulcerative lesions, it may also be necessary to treat superinfecting bacteria with appropriate antibiotics.

Chromomycosis and Phaeohyphomycosis

Chromomycosis (chromoblastomycosis) is caused by traumatic implantation of any one of several dematiaceous fungus species into the subcutaneous tissue. The infection is chronic and characterized by the slow development of verrucous, cutaneous vegetations. Dematiaceous fungi are imperfect fungi that produce varying amounts of melanin-like pigments. These pigments are found in the conidia and/or hyphae and give the organism an olive green, brown, or black color. Taxonomic characterization of the dematiaceous fungi is controversial. The terminology used below is based on that proposed by McGinnis. Species generally recognized as agents of chromomycosis are *Fonsecaea pedrosoi*, *Phialophora verrucosa*, *Cladosporium carrioni*, and *Fonsecaea compacta*. The natural reservoir of these fungi is soil and plant debris.

In addition to chromomycosis, several other diseases are also caused by dematiaceous fungi and are summarized in Table 85-1. Tinea nigra and keratomycosis are discussed in Chapter 86, and fungus ball (aspergilloma) is described in Chapter 87. In addition to the species of dematiaceous fungi listed in Table 85-1, many others have been encountered in rare infections, but will not be discussed here.

Dematiaceous Fungi
Cultural Characteristics. The dematiaceous fungi are similar not only in their pigmentation but also in their antigenic determinants, morphology, and physiologic properties. In culture, the colonies are almost indistinguishable. They are compact, dark brown to black, and develop a fuzzy surface (see Fig. 85-5A).

Microscopic Appearance. The generic criteria are based on the type and ontogeny of conidial production. Table 85-2 presents one useful classification. *Cladosporium* species produce chains of branching conidia by acropetalous (distal) budding. The length of the chains and conidial size differ with the individual species. *Phialophora* species produce conidia from flask-shaped phialides with cup-shaped collarettes (Figs. 85-3 and 85-4). Mature conidia are extruded from the phialide and usually accumulate around it. With *Exophiala*, structures called annellides (see Chap. 82) produce conidia from a tapered, ringed tip (Fig. 85-5B). The end of this conidiogenous cell increases in length and in number of rings with the development of conidia from the tip. *Wangiella* produces conidia via phialides that lack collarettes. *Fonsecaea* is a polymorphic genus; it may exhibit *Phialophora*-type or *Cladosporium*-type sporulation. It also produces lateral or terminal conidia from a lengthening conidiogenous cell. This process is described as sympodial, and is typical of the genus *Rhinocladiella*.

Clinical Infection
Clinical Manifestations
CHROMOMYCOSIS. The fungi that cause chromomycosis, of which *F. pedrosoi* is the most common, have a low level of virulence. The localized infection, which usually occurs on the exposed lower extremities, forms a primary sore that is discolored. A mononuclear cellular infiltrate and satellite lesions develop over a period of months or years. With time, these lesions become raised 1 to 3 mm and appear scaly and dull red or grayish in color. The pathology is typically granulomatous nodules and epithelial hyperplasia. After several years, elevated (1 to 3 cm), verrucous cauliflower lesions develop. Patients experience minimal discomfort. Systemic invasion is extremely rare, but when it does occur, the organisms are neurotropic and tend to infect the central nervous system.

TABLE 85-1. MYCOSES CAUSED BY DEMATIACEOUS FUNGI

Mycosis	Etiology
Chromomycosis	*Fonsecaea pedrosoi, Phialophora verrucosa, Cladosporium carrionii, Fonsecaea compacta*
Phaeohyphomycosis	
Subcutaneous (phaeomycotic cyst)	*Exophiala jeanselmei, Wangiella dermatitidis, Exophiala spinifera, Phialophora hoffmannii, Phialophora parasitica, Phialophora repens, Phialophora richardsiae,* and others
Systemic (brain abscess)	*Cladosporium bantianum (Cladosporium trichoides)*
Mycetoma	*Exophiala jeanselmei, Madurella mycetomatis,* and others; also nondematiaceous fungi
Tinea nigra	*Exophiala werneckii*
Keratomycosis	*Phialophora hoffmannii, Phialophora verrucosa, Curvularia senegalensis,* and others; also nondematiaceous fungi
Fungus ball	*Cladosporium cladosporioides* and nondematiaceous fungi

PHAEOHYPHOMYCOSIS. This term is applied by Ajello to all infections characterized by the presence of darkly pigmented, septate hyphae in tissue (Fig. 85-6). This definition includes both cutaneous and systemic mycoses. The clinical forms include solitary encapsulated cysts in the subcutaneous tissue and brain abscess (see Table 85-1). The phaeomycotic cyst usually develops on the extremities and may enlarge to several centimeters. The skin and tissue surrounding the cyst are relatively normal.

About 30 cases of brain abscess have been reported, most of which were caused by *Cladosporium bantianum* (*trichoides*). Patients have a variety of symptoms reflecting the size and location of the lesion, and encapsulated abscesses filled with brown hyphae are found at autopsy. The fungus has been isolated from wood, but the source of infection is uncertain, as patients lack cutaneous and pulmonary lesions. This species is clearly neurotropic.

Mycetoma may be caused by a variety of fungi, including the dematiaceous species *Exophiala jeanselmei, Madurella grisea,* and *Madurella mycetomatis.* It will be discussed in the following section.

Laboratory Diagnosis. The appearance of fungal elements in potassium hydroxide (KOH) preparations or tissue specimens is characteristic for each type of dematiaceous mycosis. Chromomycotic lesions contain spherical, pigmented cells (4 to 12 μm), called sclerotic bodies, that exhibit transverse septation (muriform). The dark pigment of the sclerotic cells is evident in direct preparations and in stained tissue sections. With phaeohyphomycosis, masses of irregular dark hyphae are seen in tissue (Fig. 85-6).

TABLE 85-2. CONIDIA PRODUCTION BY PATHOGENIC GENERA OF DEMATIACEOUS FUNGI

Genus	Type of Conidiogenous Cells
Cladosporium	Blastoconidia (in chains)
Phialophora	Phialides with collarettes
Exophiala	Annellides
Wangiella	Phialides without collarettes
Fonsecaea	Sympodulae, also phialides and blastoconidia

Specimens from chromomycosis and phaeohyphomycosis should be cultured on Sabouraud's agar with and without antibiotics, as some of the agents are susceptible to cycloheximide. In general, dematiaceous saprophytes have active proteolytic enzymes capable of digesting gelatin or Loeffler's serum agar. Pathogenic dematiaceous species tend to grow more slowly than saprophytic species, and most pathogens, unlike saprophytes, grow at 37C. *C. bantianum* is capable of growth at 42C. Speciation within the pathogenic genera is based upon the mode(s) of conidiation (Table 85-2) and the size, shape, and arrangement of conidia.

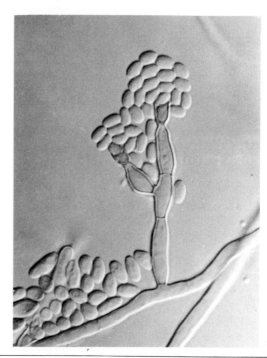

Figure 85-3. *Phialophora verrucosa.* Phialide with cup-shaped collarette and mass of conidia, from 2-week culture on potato-dextrose agar at 30C. Nomarski. × 1200. *(From McGinnis MR: Laboratory Handbook of Medical Mycology, 1979. New York, Academic.)*

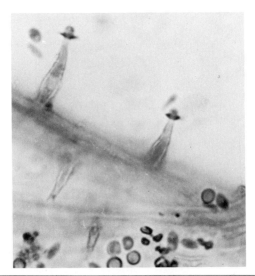

Figure 85-4. *Phialophora richardsiae.* Phialides with terminal saucer-shaped collarette. × 1387. *(From Conant, et al.: Manual of Clinical Mycology, 3rd ed, 1971. Philadelphia, Saunders.)*

Treatment. Chromomycosis can be complicated by secondary bacterial infection, which may be more threatening than the fungal infection. Advanced cases have been treated by surgical management and topical compounds. Flucy-

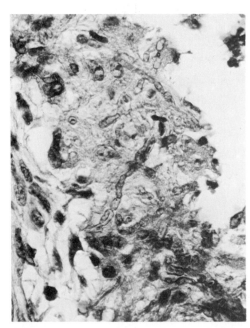

Figure 85-6. *Phialophora richardsiae.* Pigmented hyphae seen in wall of abscess.

tosine has achieved the most success. Phaeomycotic cysts are managed by excision. Even with surgical removal of the cerebral abscess, most patients with brain lesions have died within a few months.

Mycetoma

Mycetoma (Madura foot, maduromycosis) is a chronic, subcutaneous infection induced by traumatic inoculation with any of several saprophytic fungus species. A similar, but generally more severe, infection can be caused by various actinomycetes (Chap. 34). One of the common agents of mycetoma, *Pseudoallescheria boydii,* has also been documented as a rare cause of opportunistic systemic infection in compromised patients. This clinical entity is very similar to invasive aspergillosis (see Chap. 87).

Etiologic Agents
Cultural Characteristics. The fungal agents of mycetoma include *P. boydii* (*Petriellidium boydii, Allescheria boydii, Monosporium apiospermum*), *E. jeanselmei, Madurella mycetomatis, M. grisea, Acremonium falciforme* (*Cephalosporium falciforme*), *Leptosphaeria senegalensis,* and several other species. *P. boydii,* the most common cause of mycetoma in the United States, and *L. sengalensis* are homothallic ascomycetes and may produce cleistothecia or perithecia, respectively, in culture. The colony of *P. boydii* is grayish, and those of *E. jeanselmei, Madurella* species, and *L. senegalensis* are darkly pigmented. *A. falciforme* develops a white to pinkish colony.

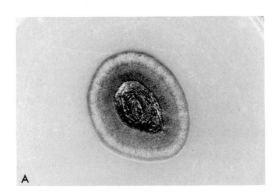

Figure 85-5. *Exophiala spinifera.* **A.** Culture on cornmeal agar at room temperature. **B.** Deeply pigmented, rigid annelides on cornmeal agar. × 450. *(From Conant, et al.: Manual of Clinical Mycology, 3rd ed, 1971. Philadelphia, Saunders.)*

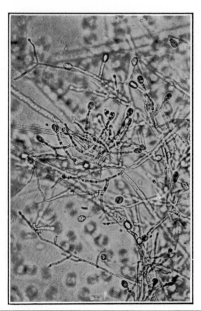

Figure 85-7. *Pseudoallescheria boydii.* Single conidia borne on ends of conidiophores. × 305. *(From Conant, et al.: Manual of Clinical Mycology, 3rd ed, 1971. Philadelphia, Saunders.)*

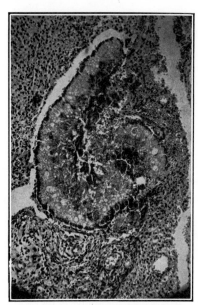

Figure 85-8. Mycetoma. Section of tissue from foot showing granule. × 112. *(From Fineberg: Am J Clin Pathol 14:239, 1944.)*

Microscopic Appearance. The identification of these fungi is often difficult. *P. boydii* produces both conidia (Fig. 85-7) and ascospores. *E. jeanselmei* sporulates via annelides. *Madurella* species may not produce any characteristic spores, but several physiologic tests are available to assist in their identification. *A. falciforme* produces characteristic crescent-shaped one- and two-celled conidia that accumulate at the tips of phialides.

Clinical Infection

Mycetoma develops after traumatic inoculation with soil contaminated with one of the etiologic agents. The feet, lower extremities, hands, and exposed areas are most often involved, but any part of the skin can become infected.

Regardless of the agent, the pathology is characterized by suppuration and abscess formation, granulomata, and the formation of draining sinuses containing granules or microcolonies of the fungi packed with tissue debris. Tumefaction and deformation of the tissue also occur. The infections are chronic, and the risk of dissemination is minimal.

The disease and most of the agents are prevalent worldwide, but the incidence is highest in hot climates (Table 85-3).

Laboratory Diagnosis. The granule, color, texture, size, and presence of hyalin or pigmented hyphae are helpful in determining the etiology (Table 85-3). As shown in Figure 85-8, a mycetomatous granule in tissue is grossly similar to actinomycotic granules (Chap. 34).

Treatment. The management of mycetoma is difficult. Amphotericin B is recommended for *Madurella* infections. Topical nystatin and potassium iodide are used in the treatment of *P. boydii* infections, and flucytosine is recommended for *E. jeanselmei* mycetoma. Surgery may be necessary in protracted cases, since the drugs frequently do not penetrate the infected tissue well enough to reach the fungal pathogens.

Rhinosporidiosis

Rhinosporidiosis is a chronic infection characterized by the development of polypoid masses of the nasal mucosa. The etiologic agent, which has never been recovered from infected tissue, has been named *Rhinosporidium seeberi.* Since the first report from Argentina by Seeber in 1900, over 2000 cases have been recognized. Approximately 90 percent of these are from India and Sri Lanka. It is more common in children and young adults, and over 90 percent of the cases occur in males.

Etiologic Agent

The etiology of rhinosporidiosis has puzzled mycologists and clinicians for years. *R. seeberi* produces large spherules in tissue, but has never been grown in vitro or recovered after inoculation of animals with infected tissue. *R. seeberi* is presumed to have a natural reservoir from which patients become infected. This habitat may be associated with water, fish, or aquatic insects, as many patients are divers. Natural infection occurs in horses, cattle, and other animals.

Clinical Infection

Lesions are most often found on the mucosa of the nose, nasopharynx, or soft palate, but many other mucocutaneous sites may become infected including the conjunctiva, skin, larynx, genitalia, or rectum. Lesions are initially flat but develop into discolored, cauliflower-type polypoid masses varying in size up to 20 g. Regardless of location, these characteristic, pedunculated lesions are formed. In the nasal area, respiration may be blocked, and there is a profuse seropurulent discharge. Sporangia may be grossly visible as small white spots on the lesion.

Laboratory Diagnosis. Histologic examination of infected tissue reveals epithelial hyperplasia and a cellular infiltrate of neutrophils, lymphocytes, plasma cells, and giant cells. Also present are the large, thick-walled sporangia. With growth, the sporangia enlarge in size to a diameter of 200 to 300 μm and are packed with thousands of endospores (6 to 7 μm) (Fig. 85-9). At maturity, the cell wall thins and a pore develops for release of the endospores which then continue the cycle. The cell wall of the spherule is multilayered and stains with mucicarmine. By contrast, the spherules of *Coccidioides immitis* rarely exceed a diameter of 100 μm. Rhinosporidial antigen is detectable in patient sera, and the frequency of antigenemia increases with the duration of infection.

Rhinosporidiosis has been treated topically, surgically, and by local injection of ethylstilbamidine.

Lobomycosis

Lobomycosis is a chronic subcutaneous infection of humans and dolphins caused by a fungus tentatively named *Loboa loboi*. Since being described in 1931 by Lobo, approximately 100 additional cases have been reported. Most

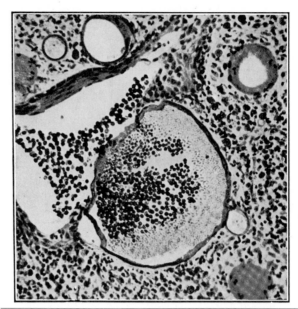

Figure 85-9. Rhinosporidiosis. Section of nasal polyp showing mature and immature sporangia. × 175. *(From Conant, et al.: Manual of Clinical Mycology, 3rd ed, 1971. Philadelphia, Saunders.)*

human cases have occurred in the region of the Amazon River basin among the Caiabi Indians and, with one exception, patients have been males and mostly adults. Natural infection has been discovered in Atlantic bottle-nose dolphins off the coasts of Florida and South America.

Clinical Infection

Clinical Manifestations. The initial lesions are small, hard subcutaneous nodules usually appearing on the extremities, face, or ear, presumably as a result of traumatic inoculation of the etiologic agent. Lymph nodes are not involved. The infection is chronic and progressive; lesions do not resolve spontaneously but continue to expand gradually over a period of years. There is one case of 50 years' duration. The lesions, which are not painful, may become verrucose or ulcerative and resemble, depending

TABLE 85-3. COMMON AGENTS OF MYCETOMA

| Species | Granule | | Predominant Geographic Distribution |
	Color	Size	
Pseudoallescheria boydii	White	<2 mm	Worldwide, tropical, North America
Madurella mycetomatis	Dark red to black	0.5–2 mm	Worldwide, tropical
Madurella grisea	Black	0.3–0.6 mm	Central and South America
Acremonium falciforme	White	0.5–1.5 mm	Worldwide, tropical
Exophiala jeanselmei	Brown to black	0.2–0.3 mm	Worldwide (very rare)
Leptosphaeria senegalensis	Black	0.5–2 mm	West Africa

upon the duration and tissue response, chromomycosis, mycetoma, or carcinoma.

Laboratory Diagnosis. Lobomycosis is diagnosed by direct, microscopic examination of skin scrapings, biopsies, or wet preparations of exudative lesions. The fungus appears in tissue as large, spherical, or oval yeasts (approximately 10 μm in diameter) that exhibit multiple budding and characteristically form short chains of three to six or more yeast cells. They are multinucleated and thick-walled. Unlike *Paracoccidioides brasiliensis*, which also produces large, multiply budding yeast cells in tissue, the buds and parent cells of *L. loboi* are the same size. Tissue sections reveal granulomatous nodules and occasional asteroid bodies. The yeasts stain with periodic acid-Schiff or methenamine silver reagents and may be seen inside macrophages. Infected tissues stained with fluorescent antibodies have demonstrated cross-reactions with antigens of many other pathogenic fungi.

The fungal agent of lobomycosis has not been successfully cultured in the laboratory or recovered by experimental infection with clinical material; neither has it been knowingly isolated from nature.

Lobomycosis has been treated with sulfa drugs and surgical excision.

Rhinoentomophthoromycosis

Rhinoentomophthoromycosis is a rare infection of the nasal mucosa caused by *Entomophthora coronata* (*Conidiobolus coronatus*). This fungus is a soil saprophyte and an insect parasite that has been isolated worldwide. The first human infection was described in 1965, and since then over 30 cases have been recognized in Nigeria, elsewhere in Africa, India, and southeast Asia.

Etiologic Agent. On Sabourand's agar, *E. coronata* produces a fast-growing colony that is flat, glabrous, and colorless or gray to yellow; radial folds and thin aerial hyphae develop. Large, spherical conidia (10 to 20 μm) with hair-like appendages are borne singly on the tips of conidiophores and forcibly ejected at maturity.

Clinical Infection. The disease is confined to the above-mentioned tropical areas. Infections are more prevalent in the young adult male. Eighty-three percent of the reported cases have occurred in males. Sixty percent of patients have been in their twenties, and only 10 percent were younger. Rhinoentomophthoromycosis also occurs in horses.

The disease begins with an initial swelling in the nasal area. Hard, subcutaneous nodules form, and an acute or chronic inflammatory response may ensue. Severe edema of the nose may block the passage of air. The swelling continues to expand and a large, disfiguring tissue mass develops.

Diagnosis is established by histologic examination of infected tissue and culture. In tissue, numerous branching hyphae approximately 4 to 10 μm in width are observed. These hyphae stain eosinophilically due to the deposition of tissue elements or immune complexes on the walls. An eosinophil infiltrate is often seen but, unlike other zygomycetous pathogens, the blood vessels are not invaded. Specimens are cultured on media without cycloheximide at 25C or 37C.

Treatment has included surgery, potassium iodide, and amphotericin B.

Subcutaneous Phycomycosis

Subcutaneous phycomycosis is a chronic self-limiting infection of the subcutaneous tissue caused by *Basidiobolus haptosporus*. Since the first two cases were described in Indonesian children in 1956, over 100 cases have been reported from Africa (Uganda, Nigeria, Kenya), India, and Indonesia.

Etiologic Agent. *B. haptosporus*, like the nonpathogen *Basidiobolus ranarum*, can be isolated from the intestinal tract of beetles, frogs, lizards, and similar creatures. The colony is colorless or brownish, thin, flat, and glabrous. Radial folds develop and become covered with short, white aerial mycelium. The hyphae are 8 to 20 μm wide and produce chlamydospores, forcibly ejected spores, and spherical, smooth-walled zygospores (30 to 50 μm). The zygospores of *B. ranarum* have an undulating wall quite distinct from *B. haptosporus*.

Clinical Infection. This infection is most prevalent in Africa. The highest incidence is among children 5 to 9 years old, and 75 percent of cases are less than 15 years of age. Seventy to 80 percent of patients are males.

Infection begins on the torso or limb with a small, firm, movable nodule in the subcutaneous tissue. The nodule enlarges, edema develops and may become massive, involving an entire leg or shoulder. The skin is generally intact but may become very tough. The lesions are not painful; they persist for several months and then resolve spontaneously.

LABORATORY DIAGNOSIS. Direct examination of tissue biopsies reveals multiple granulomata, giant cells, and eosinophils. The vasculature is not invaded. Broad, branching hyphae (5 to 18 μm wide) with infrequent septa are surrounded by a sheath of eosinophilic material. Specimens are cultured at 25C or 37C on Sabouraud's agar without cycloheximide and within 2 to 3 days colonies should appear. Iodides are an effective treatment and the prognosis is very good.

FURTHER READING

SPOROTRICHOSIS

Ajello L, Kaplan W: A new variant of *Sporothrix schenckii.* Mykosen 12:633, 1969

Altner PC, Turner RR: Sporotrichosis of bones and joints. Review of the literature and report of six cases. Clin Orthop 68:138, 1970

Auld JL, Beardmore GL: Sporotrichosis in Queensland: A review of 37 cases at the Royal Brisbane Hospital. Aust J Dermatol 20:14, 1979

Beardmore GL: Recalcitrant sporotrichosis: A report of a patient treated with various therapies including oral miconazole and 5-fluorocytosine. Aust J Dermatol 20:10, 1979

Berson SD, Brandt FA: Primary pulmonary sporotrichosis with unusual fungal morphology. Thorax 32:505, 1977

Blumer SO, Kaufman L, Kaplan W, et al.: Comparative evaluation of five serological methods for the diagnosis of Sporotrichosis. Appl Microbiol 26:4, 1973

Charoenvit Y, Taylor RL: Experimental sporotrichosis in syrian hamsters. Infect Immun 23:366, 1979

Comstock C, Wolson AH: Roentgenology of sporotrichosis. Am J Roentgenol Radium Ther Nucl Med 125:651, 1975

Cunningham KM, Bulmer GS, Rhoades ER: Phagocytosis and intracellular fate of *Sporothrix schenckii.* J Infect Dis 140:815, 1979

Dellatome DL, Latlanard A, Buckley HR, et al.: Fixed cutaneous sporotrichosis of the face. Successful treatment and review of the literature. J Am Acad Dermatol 6:97, 1982

Fetter BF, Tindall JP: Cutaneous sporotrichosis. Arch Pathol 78:613, 1964

Howard DH: Dimorphism in *Sporotrichosis schenckii.* J Bacteriol 81:484, 1961

Ishizaki H, Wheat RW, Kiel DP, et al.: Serological cross-reactivity among *Sporothrix schenckii, Ceratocystis, Europhium,* and *Graphium* species. Infect Immun 21:585, 1978

Kaplan W, Invens MS: Fluoresent antibody staining of *Sporotrichosis schenckii* in cultures and clinical material. J Invest Dermatol 35:151, 1961

Kedes KH, Siemienski J, Braude AI: The syndrome of the alcoholic rose gardener: Sporotrichosis of the radial tendon sheath. Ann Intern Med 61:1139, 1964

Kwon-Chung KJ: Comparison of isolates of *Sporothrix schenckii* obtained from fixed cutaneous lesions with isolates from other types of lesions. J Infect Dis 139:424, 1979

Latapi F: Sporotrichose an Mexique. Local Med 34:732, 1963

Mackinnon JE: The effect of temperature on the deep mycoses. In Wolstenholme GEW, Porter R (eds): Systemic Mycoses, Boston, Little, Brown, 1968

Mackinnon JE: Ecology and epidemiology of sporotrichosis. In Proceedings of the International Symposium on Mycoses. Washington, D.C., Pan American Health Organization, 1970, pp 169–181

Mariat F: The epidemiology of sporotrichosis. In Wolstenholme GEW, Porter R (eds): Systemic Mycoses, Boston, Little, Brown 1968, p 144

Mariat F: Variant, non-sexue de *Ceratocystis* sp. pathogenic pour-le-hauster. R Acad Sci 269:2329, 1969

Mariat F: Observations sur l'ecologie de *Sporothrix schenckii* et de *Ceratocystis steroceras* en corse et en Alsace; provinces Françaises indemnes de sporotrichose. Sabouraudia 13:217, 1975

Mendonca-Hagler LC, Travassos LR, Lloyd K, et al.: Deoxyribonucleic acid base composition and hybridization studies on the human pathogen *Sporothrix schenckii* and *Ceratocystis* species. Infect Immun 9:934, 1974

Mohr JA, Griffiths W, Long H: Pulmonary sporotrichosis in Oklahoma and susceptibilities in vitro. Am J Resp Dis 119:961, 1979

Müller HE: Uber das Vorkommen von Neuraminidase bei *Sporothrix schenckii* und *Ceratocystis stenoceras* und ihre Bedentang fur die Okologie und den Pathomechanismus dieser Pilze. Zentralbl Bakteriol 232:365, 1975

Plouffe JF, Silva J, Fekety R, et al.: Cell-mediated immune response in sporotrichosis. J Infect Dis 139:152, 1979

Powell KE, Taylor A, Phillips BJ, et al.: Cutaneous sporotrichosis in forestry workers. JAMA 240:232, 1978

Proceedings of the Transvaal Mine Medical Officers Association: Sporotrichosis infection on mines of the Witwatersrand, Transvaal Chamber Mines, Johannesburg, 1947

Sanders E: Cutaneous sporotrichosis: beer, bricks, and bumps. Arch Intern Med 127:482, 1971

Sethi KK: Experimental sporotrichosis in the normal and modified host. Sabouraudia 10:66, 1972

Travassos LR, Lloyd KO: *Sporothrix schenckii* and related species of *Ceratocystis.* Microbiol Rev 44:683, 1980

Wada R: Studies on mode of action of potassium iodide upon sporotrichosis. Mycopathologia 34:97, 1968

Walbaum S, Duriez T, Dujardin L, et al.: Etude d'un extrait de *Sporothrix schenckii* (forme levure). Analyse électrophoretique et immunoélectrophoretique; characterisation des activités enzymatiques. Mycopathologia 63:105, 1978

Welsh RD, Dolan CT: Sporothrix whole yeast agglutination test. Am J Clin Pathol 59:82, 1973

Wilson DE, Mann JJ, Bennett JE, et al.: Clinical features of extracutaneous sporotrichosis. Medicine (Baltimore) 46:265, 1967

CHROMOMYCOSIS AND PHAEOHYPHOMYCOSIS

Ajello L: Phaeohyphomycosis: Definition and etiology. In Proceedings of the Third International Conference on Mycoses, PAHO Sci Publ No 304, 1975, p 126

Al-Doory Y: Chromomycosis. Missoula, Mont, Mountain Press, 1972

Carrion AL: Chromoblastomycosis and related infections. Int J Dermatol 14:27, 1975

Connor DH, Gibson DW, Ziefer A: Diagnostic features of three unusual infections: Micronemiasis, pheomycotic cyst, and prototothecosis. In Majno G, Cotran RS, Kaufman N (eds): Current Topics in Inflammation and Infection. Baltimore, Williams & Wilkins, 1982, p 205

Cooper BH, Schneidau JD: A serological comparison of *Phialophora verrucosa, Fonsecaea pedrosoi,* and *Cladosporium carrionii* using immunodiffusion and immunoelectrophoresis. Sabouraudia 8:217, 1970

de Hoog GS: *Rhinocladiella* and allied genera. Stud Mycol 15:1, 1977

Dixon DM, Shadomy JH, Shadomy S: Dematiaceous fungal pathogens isolated from nature. Mycopath 70:153, 1980

Ellis MB: Dematiaceous Hyphomycetes. Kew, Surrey, UK, Commonwealth Mycology Institute, 1971

Ellis MB: More Dematiaceous Hyphomycetes. Kew, Surrey, UK, Commonwealth Mycology Institute, 1976

Emmons CW, Binford CH, Utz JP, et al.: Medical Mycology, 3rd ed, Philadelphia, Lee & Febiger, 1977

Kwon-Chung KJ, Schwartz K, Rybak BJ: A pulmonary fungus ball produced by *Cladosporium cladosporioides*. Am J Clin Pathol 64:564, 1975

McGinnis MR: Human pathogenic species of *Exophilia, Phialophora,* and *Wangiella*. In The Black and White Yeasts. Proceedings of the Fourth International Conference on Mycoses. PAHO Publ. No 356, 1978, p 37

Middleton FG, Jurgenson PF, Utz JP, et al.: Brain abscess caused by *Cladosporium trichoides*. Arch Intern Med 136:444, 1976

Silva M: Growth characteristics of the fungi of chromoblastomycosis. Ann NY Acad Sci 89:17, 1960

Vollum DI: Chromomycosis: A review. Br J Dermatol 96:454, 1977

OTHER SUBCUTANEOUS MYCOSES

Baruzzi RG, Marcopito LF, Vicente LS, et al.: Jorge Lobo's disease (Keloidal blastomycosis) and tinea imbricata in Indians from the Xingu National Park, Central Brazil. Trop Doctor 12:13, 1982

Bhawan J, Bain RW, Purtilo DT, et al.: Lobomycosis. An electron microscopic, histochemical and immunologic study. J Cutan Pathol 3:5, 1976

Bras G, Gordon CC, Emmons CW, et al.: A case of phycomycosis observed in Jamaica, infection with *Entomophthora coronata*. Am J Trop Med Hyg 14:141, 1965

Chitravel V, Sundararaj T, Subramanian S, et al.: Detection of circulating antigen in patients with rhinosporidiosis. Sabouraudia 20:185, 1982

Clark BM: The epidemiology of phycomycosis. In Wolstenholme GE, Porter R (eds): Symposium on Systemic Mycoses. Boston, Little, Brown, 1968, p 179

Greer DL, Friedman L: Studies on the genus *Basidiobolus* with reclassification of the species pathogenic for man. Sabouraudia 4:231, 1966

Hurion N, Fromentin H, Keil B: Specificity of the collagenolytic enzyme from the fungus *Entomophthora coronata*: Comparison with the bacterial collagenase from *Achromobacter iophagus*. Arch Biochem Biophys 192:438, 1979

Jaramillo D, Cortes A, Restrepo A, et al.: Lobomycosis. Report of the eighth Colombian case and review of the literature. J Cutan Pathol 3:180, 1976

Joe LK, Njo-Injo TE, Kjokronegro S, et al.: *Basidiobolus ranarum* as a cause of subcutaneous phycomycosis in Indonesia. Arch Dermatol 74:378, 1956

Lobo J: Um caso de blastomicose produzida por uma especie nora, encontrada em Recife. Rev Med Pernambuco 1:763, 1931

Lutwick LI, Galgiani JN, Johnson RH, et al.: Visceral fungal infections due to *Petriellidium boydii*. Am J Med 61:632, 1976

Mariat F, Destombes P, Segretain G: The mycetomas: Clinical features, pathology, etiology and epidemiology. Contrib Microbiol Immunol 4:1, 1977

Martinson FD, Clark BM: Rhinophycomycosis entomophthorae in Nigeria. Am J Trop Med Hyg 16:40, 1967

Seeber GR: Un neuvo esporozuario parasito del hombre. Dos casos encontrades en polipos nasales. Tesis. Univ Nat de Buenos Aires, 1900

Vanbreuseghem R: Rhinosporidiose: klinischer Aspekt, Epidemiologic und ultrastrukturelle Studien von *Rhinosporidium seeberi*. Dermatol Monatsschr 162:512, 1976

CHAPTER 86

Dermatophytosis and Other Cutaneous Mycoses

Dermatophytosis

Dermatophytosis is an infection of the skin, hair, or nails by any of a group of keratinophilic fungi called dermatophytes. Dermatophytes parasitize the nonliving, cornified integument. They secrete keratinases, proteolytic enzymes that digest keratin, the structural protein of hair, nails, and epidermis. Species of dermatophytes, of which 40 to 50 are recognized, are relatively similar in their morphology, physiology, and biochemical composition. They cause a variety of specific clinical conditions. A single species may be able to produce several types of distinctive skin diseases, and, conversely, several dermatophyte species may be etiologic agents of the same disease.

Dermatophytoses are among the most prevalent infections in the world. Though they are extremely annoying, and millions of dollars are spent annually in their treatment, with few exceptions, they are not debilitating or life threatening. The incidence varies considerably. Among military personnel in the United States and England, there is a prevalence of 17 to 24 percent, but the incidence among service personnel in the tropics increases to 60 to 80 percent. The attack rate is higher in institutions and under crowded living conditions.

Dermatophytes are highly specialized for the infection of skin. If animals are infected by the parenteral route, systemic infection does not occur. This restriction is probably related both to host defenses (e.g., serum growth inhibitor) and fungal properties (e.g., affinity for keratin, inhibition at 37C). Because dermatophytes produce visible lesions, these infections have been observed throughout history and recorded since antiquity. Dermatophyte, or ringworm, infections were the first recognized infectious diseases of humans. During the 1840s, several European physicians independently recognized that ringworm infections were caused by fungi that could be cultivated on artificial media and that these fungi could produce similar infections in healthy skin. These observations represent the first documentation of Koch's postulates some 40 years before Koch formulated them!

For more than 100 years, dermatophytes have been isolated and identified. Their speciation, geographic distribution, and clinical manifestations have been subjected to many investigations. Three genera and 40 species of dermatophytes are recognized by most taxonomists: 21 species of *Trichophyton,* 17 species of *Microsporum,* and 2 species of *Epidermophyton.*

The Dermatophytes

Biologic Properties

The biology of the dermatophytes is unique in several respects. With the exception of certain yeasts, pathogenic fungi are predominantly saprophytes that live in nature and accidently cause infection—that is, infection represents a transient alteration in the normal life cycle; transmission does not occur; and infection is not advantageous for the fungus. Dermatophytes, however, appear to be evolving toward a more permanent and benign host–parasite relationship.

Taxonomy. The genera *Trichophyton, Microsporum,* and *Epidermophyton* are all classified within the subdivision Deuteromycotina (Chap. 82). However, since 1959, the perfect or sexual reproductive stage of a number of specific dermatophytes has been discovered. Such species have subsequently been reclassified and renamed, but because of common usage, the old names have been retained. To date, all of the species of *Trichophyton* for which a sexual phase has been discovered have been reclassified within a single new genus, *Arthroderma.* Similarly, the perfect *Microsporum* species are members of the genus *Nannizzia.* Both of these genera are members of the family Gymnoascaceae in the subdivision Ascomycotina. This family also contains the etiologic agents of histoplasmosis and blastomycosis. Eleven species of *Arthroderma* and ten *Nannizzia* species have been discovered.

Sexual Reproduction. To observe the sexual phase in the laboratory, two compatible (opposite) mating types of each species are required. These mating types are designated plus or minus, and, when they are mixed together on a suitable medium (e.g., moist, sterile soil and hair), they will develop characteristic cleistothecia containing asci and ascospores. The sexual phase provides an epidemiologic marker because each species has two mating types. Furthermore, because mating occurs only within the species, sexual reproduction may be used to identify isolates.

Morphology and Physiology. As might be expected from their close taxonomic relatedness, the dermatophytes are very similar in their physiology, growth requirements, morphology, antigenicity, and infectivity. The most distinctive physiologic property of the dermatophytes is their ability to digest keratin. Only a few insects and certain other microorganisms have this capability. Differences in the nutritional requirements of the various dermatophyte species are few, but the differences that do exist, such as a requirement for thiamine or histidine, are utilized in their identification. Dermatophytes also share a number of surface antigens, both within this group and with several saprophytic fungi. Despite their marked similarities, however, subtle but important differences do exist among these fungi, such as the specificity for keratin substrates and the types of conidia produced.

Identification. The common dermatophytes are identified on the basis of their colonial appearance and microscopic morphology, as indicated in Table 86-1. The *Microsporum* species tend to produce distinctive macroconidia (Figs. 86-1 and 86-2). The *Trichophyton* species develop very similar macroconidia, but characteristic microconidia (Figs. 86-3 and 86-4). *Epidermophyton floccosum* produces only macroconidia, which are clavate (Fig. 86-5). A few nutritional tests are also useful. For example, *Trichophyton tonsurans, Trichophyton violaceum,* and *Trichophyton verrucosum* require thiamine, and most strains of *T. verrucosum* also require inositol.

Clinical Infection

Epidemiology and Ecology

The natural reservoirs of most of the dermatophytes that have been implicated in human infection are given in Table 86-2. Dermatophytes are classified as geophilic, zoophilic, or anthropophilic, depending upon whether their usual habitat is soil, animals, or humans. Many dermatophytes whose natural reservoir is soil or animals are still able to cause human infection. Evidence supports the concept that dermatophytes have evolved from geophilism to zoophilism to anthropophilism. In general, as species evolve from habitation in soil to a specific human or animal host, a decrease in both sporulation (asexual and sexual) and host toxicity occurs. Anthropophilic species tend to produce relatively mild and chronic infections in humans, whereas zoophilic dermatophytes cause infections that are more inflammatory and acute. The latter respond better to treatment and usually do not recur. Some anthropophilic species (e.g., *Microsporum audouinii*) apparently survive exclusively by transmission from one human to another. Other dermatophytes not listed in Table 86-2 have been recovered from soil and never implicated in human or animal infection. About 11 species are responsible for the majority of human dermatophytoses throughout the world, and only about six species are endemic in the United States. The distribution of dermatophyte species in specific environments and the specialization for certain host species or tissues present a fascinating balance between pathogen and host.

An understanding of the particular evolution and diversity among dermatophytes provides useful clinical information of prognostic value, such as (1) the duration of the infection, (2) the probability of recurrence, (3) the severity of the infection, and (4) the source of the dermatophyte. For example, if a child develops tinea capitis caused by the anthropophilic dermatophyte *M. audouinii,* the infection will probably be self-limiting but will spread to the child's siblings. If the agent is *Microsporum canis,* the lesion will be more inflammatory, and the family dog or cat should be treated to prevent recurrent infections.

TABLE 86-1. DISTINGUISHING MORPHOLOGIC AND PHYSIOLOGIC CHARACTERISTICS OF COMMON DERMATOPHYTES

Species	Culture*	Micromorphology
Microsporum audouinii	Slow growth; flat, silky; surface cream to brown; reverse red-brown; on rice, no growth but brown color	Conidia rare; chlamydospores
Microsporum canis	Rapid growth; surface cottony white, reverse deep yellow	Macroconidia: spindle shaped, more than 6 cells, thick, rough wall
Microsporum cookei	Rapid growth; surface flat, powdery, and white to yellow or pink; reverse red	Macroconidia: 6 to 8 cells, thick rough wall; many microconidia
Microsporum distortum	Surface white to yellow, folded or flat; good growth on rice	Macroconidia: highly distorted, thick rough wall
Microsporum ferrigineum	Slow growth; surface smooth, folded, and yellow	No conidia; distorted hyphae
Microsporum gallinae	Rapid growth; surface whitish; reverse pigment diffusible red	Macroconidia: clavate, 2 to 10 cells
Microsporum gypseum	Rapid growth; surface powdery, tan	Macroconidia: 3 to 9 cells, thin, rough wall; microconidia: rare, clavate
Microsporum nanum	Rapid growth; surface powdery white, yellow, or pinkish, reverse pink to brown	Macroconidia: 1 to 2 cells, thin, rough wall; microconidia: clavate
Microsporum vanbreuseghemii	Rapid growth; surface fluffy or powdery, tan; reverse colorless to yellow	Macroconidia: cylindrical, 5 to 12 cells, thick, rough wall; microconidia: pyriform
Trichophyton ajelloi	Rapid growth; surface flat; cream to tan; reverse colorless to red to black	Macroconidia: numerous, 5 to 12 cells, thick, smooth wall; microconidia: pyriform
Trichophyton equinum	Rapid growth; surface flat to fluffy; reverse yellow to brown, diffusible; requires nicotinic acid	Microconidia: small, pyriform; macroconidia: rare
Trichophyton megninii	Slow growth; nondiffusible red pigment; requires histidine	Microconidia: pyriform; macroconidia: rare
Trichophyton mentagrophytes	Surface flat; white to yellow, either cottony and downy (var *interdigitale*) or powdery and granular (var *mentagrophytes*)	Microconidia: numerous, single or in clusters, spherical; macroconidia: rare, 2 to 5 cells; spiral hyphae common
Trichophyton rubrum	Slow growth; surface fluffy, white; reverse deep red, nondiffusible pigment; rare isolates yellow or powdery	Microconidia: pyriform, small; macroconidia: rare, pencil-shaped, 3 to 5 cells
Trichophyton schoenleinii	Slow growth; surface compact, white to tan, folded; 37C growth	Conidia: rare; chlamydospores; swollen, knobby hyphae (chandeliers)
Trichophyton tonsurans	Slow growth; surface flat, powdery to velvety; reverse yellow to brown-red; requires thiamine	Microconidia: mostly elongate; macroconidia: rare
Trichophyton verrucosum	Slow growth; surface heaped, wrinkled; requires thiamine and inositol; 37C growth	Conidia: rare; distorted hyphae; chlamydospores in chains at 37C
Trichophyton violaceum	Slow growth; surface heaped, folded; purple; requires thiamine	Conidia: rare
Epidemophyton floccosum	Slow growth; surface fluffy to powdery to velvety, flat or folded, tan to olive-green; reverse yellow to tan	Macroconidia: clavate, 2 to 4 cells, thin, smooth walls; no microconidia

*Growth on Sabouraud's dextrose agar at 25C for 2 weeks.

Pathogenesis

Certain etiologic agents are endemic only in specific geographic areas (see Table 86-2). Geophilic and zoophilic species are obtained from contact with dermatophytes in soil or on animals, respectively. Anthropophilic species may be transmitted by direct contact or fomites.

Dermatophyte infections begin in the cutaneous tissue following contact and trauma. Various studies have shown that host susceptibility may be enhanced by moisture, warmth, specific skin chemistry, composition of sebum and perspiration, youth, heavy exposure, and genetic predisposition. The incidence is higher in hot, humid climates and under crowded living conditions. Several studies have documented poor cell-mediated immune responses to dermatophytic antigens in patients who develop chronic, noninflammatory dermatophyte infections.

Clinical Manifestations

Table 86-3 presents the clinical forms of dermatophytoses, which were erroneously termed ringworm or tinea because of the raised circular lesion. Tinea capitis is dermatophytosis of the scalp and hair, which appears as dull, gray, circular patches of alopecia, scaling, and itching. Species of

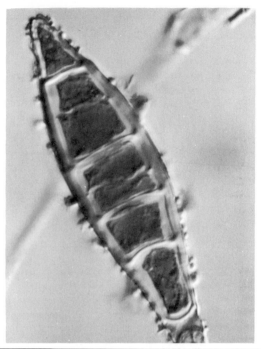

Figure 86-2. *Microsporum gypseum.* Typical rough, thin-walled macroconidia. × 1000.

Figure 86-1. *Microsporum canis.* Typical echinulate, thick-walled, spindle-shaped macroconidium. Nomarski. × 1250 *(From McGinnis: Laboratory Handbook of Medical Mycology, 1979. New York, Academic Press.)*

Microsporum or *Trichophyton* may cause tinea capitis. Zoophilic species may induce a severe combined inflammatory and hypersensitivity reaction, called a kerion. Epidemic tinea capitis in prepubescent children is self-limiting and caused by *T. tonsurans* or *M. audouinii* in the United States.

Favus is an acute inflammatory infection of the hair and follicle caused by *Trichophyton schoenleinii* and similar dermatophytes.

Tinea barbae, tinea corporis, and tinea manum are caused by *Trichophyton rubrum* (80 percent), *Trichophyton mentagrophytes* var *interdigitale* (10 percent), or *E. floccosum* (5 percent). Classic lesions on glabrous skin are annular and scaly, and they may be embellished with erythema, vesicles, or allergic reactions. The thickness and duration of the lesion and extent of the inflammatory response are determined by the nature of the dermatophyte–host interac-

TABLE 86-2. ECOLOGY OF HUMAN DERMATOPHYTE SPECIES

Anthropophilic	Zoophilic	Geophilic
Species Found Worldwide		
Epidermophyton floccosum	*Microsporum canis* (dogs, cats)	*Microsporum gypseum*
Microsporum audouinii	*Microsporum gallinae* (fowl)	*Microsporum fulvum*
Trichophyton mentagrophytes var	*Microsporum nanum* * (pigs)	*Microsporum nanum* *
interdigitale	*Microsporum persicolor* (moles)	*Trichophyton ajelloi*
Trichophyton rubrum	*Trichophyton equinum* (horses)	*Trichophyton terrestre*
Trichophyton schoenleinii	*Trichophyton mentagrophytes* var	
Trichophyton tonsurans	*mentagrophytes* (rodents)	
Trichophyton violaceum	*Trichophyton verrucosum* (cattle)	
Rare and Geographically Limited Species		
Microsporum ferrugineum, Far East,	*Microsporum distortum* (monkeys),	*Microsporum racemosum,* Europe, North
Africa, Asia, Europe	Australia, New Zealand	and South America
Trichophyton concentricum, Far East,	*Trichophyton mentagophytes* var *erinacei*	*Microsporum vanbreuseghemii,* Africa,
Central and South America	(hedgehogs), New Zealand, Europe;	Europe, North America
Trichophyton gourvillii, Africa	*T. mentagrophytes* var *quinckaenum*	
Trichophyton megninii, Europe, Asia,	(mice)	
Africa	*Trichophyton simii* (monkeys), India	
Trichophyton soudanense, Africa,		
Europe, North America		
Trichophyton yaoundei, Africa		

* *Microsporum nanum* can be isolated from human, animal (pigs), and soil habitats.

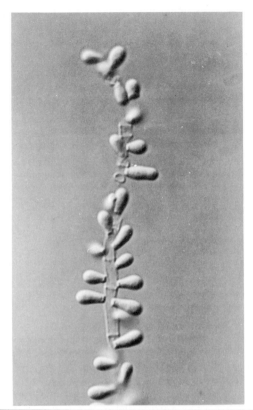

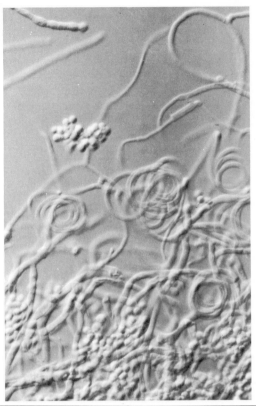

Figure 86-3. *Trichophyton rubrum.* Typical pyriform and clavate microconidia. Nomarski. × 1250. *(Courtesy of Dr. Michael R. McGinnis.)*

Figure 86-4. *Trichophyton mentagrophytes.* Clusters of spherical microconidia and coiled hyphae. Nomarski. × 500. *(Courtesy of Dr. Michael R. McGinnis.)*

tion (keratinases, toxins, antigens). Tinea cruris, or jock itch, is caused by the same three dermatophytes. Most cases (99 percent) occur in males, as a dry expanding lesion in the groin.

Tinea pedis (athlete's foot) presents as chronic involvement of the toe webs. Other forms are vesicular, ulcerative, or moccasin type, with hyperkeratosis of the sole.

Tinea unguium, or onychomycosis, is most often caused by *Trichophyton rubrum* or *T. mentagrophytes.* Nails may show white, patchy, or pitted lesions on the surface. Hyphal invasion beneath the nail results in digestion, discoloration, and deformation of the nail.

A dermatophytid, or id, reaction is an allergic response to fungal antigens. A dermatophyte infection in one area (e.g., tinea pedis) elicits an allergic reaction elsewhere (e.g., the hands).

TABLE 86-3. CLINICAL CLASSIFICATION OF THE DERMATOPHYTOSES

Clinical Name	Site of Lesions	Organisms Most Frequently Isolated
Tinea capitis, epidemic	Scalp	*Microsporum audouinii, Trichophyton tonsurans* (U.S.), *Trichophyton violaceum, Microsporum ferrugineum* (outside U.S.)
Tinea capitis, nonepidemic	Scalp	*Microsporum canis, Trichophyton verrucosum, Microsporum gypseum* (rare)
Tinea favosa (favus)	Scalp, torso	*Trichophyton schoenleinii, Trichophyton violaceum*
Tinea barbae	Beard	*Trichophyton rubrum, Trichophyton verrucosum*
Tinea corporis	Arms, legs, torso	*Trichophyton rubrum, Microsporum canis, Trichophyton mentagrophytes*
Tinea cruris	Genitocrural folds	*Trichophyton rubrum, Trichophyton mentagrophytes, Epidermophyton floccosum*
Tinea pedis and manus	Feet, hands	*Trichophyton rubrum, Trichophyton mentagrophytes*
Tinea unguium	Nails	*Trichophyton rubrum, Trichophyton mentagrophytes, E. floccosum*
Tinea imbricata	Torso	*Trichophyton concentricum*

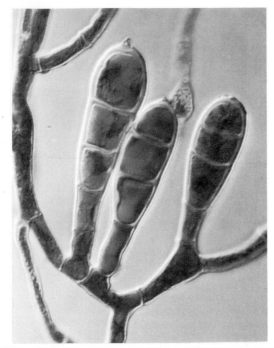

Figure 86-5. *Epidermophyton floccosum.* Typical clavate macroconidia. Nomarski. × 1250. *(From McGinnis: Laboratory Handbook of Medical Mycology, 1979. New York, Academic Press.)*

Laboratory Diagnosis

Direct Examination. Cases of suspected tinea capitis should be examined under Wood's light (365 nm), where hairs infected with *Microsporum* species or *T. schoenleinii* ex-

hibit a greenish fluorescence. Hairs can be examined directly under the microscope for endothrix involvement, the formation of arthroconidia within the hair shaft (Fig. 86-6), or ectothrix infection, a sheath of spores around the shaft (Fig. 86-7). *T. tonsurans* and *T. violaceum* produce endothrix infections, whereas other hair infections are of the ectothrix type. Favic hairs present characteristic air spaces in the hair (Fig. 86-8).

Skin and nail infections may be diagnosed by dissolving skin scrapings or nail clippings in 20 percent potassium hydroxide (KOH) and looking for hyaline, branched, septate hyphae among the squamous epithelial cells (Fig. 86-9).

The id reaction is diagnosed on the basis of a negative microscopic and cultural examination of the site and finding of dermatophytosis elsewhere on the body.

Culture. Hair, skin, or nail specimens should be cultured at room temperature on Sabouraud's medium with antibiotics. Isolates are identified on the basis of colony appearance (growth rate, surface texture, obverse and reverse pigmentation), and morphology of reproductive structures (see Table 86-1).

Treatment

Dermatophytoses may be treated with topical antibiotics, such as solutions of tolnaftate, miconazole nitrate, or clotrimazole. The most effective antibiotic is griseofulvin, which is given orally for long periods. This poorly absorbed drug is concentrated in the stratum corneum, where it inhibits hyphal growth.

Figure 86-6. Endothrix involvement of hair caused by *Trichophyton tonsurans*: arthroconidia are formed inside the hair. × 170. *(From Conant, et al.: Manual of Clinical Mycology, 3rd ed, 1971. Philadelphia, Saunders.)*

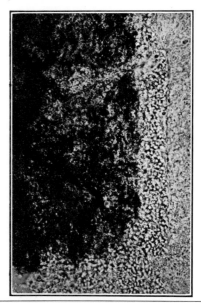

Figure 86-7. Ectothrix involvement of hair caused by *Microsporum audouinii*: conidia are around the partially digested hair shaft. × 350.

Figure 86-8. Favic involvement of hair by *Trichophyton schoenleinii*: hyphae and air spaces in hair shaft. × 170.

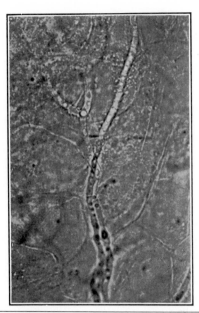

Figure 86-9. Dermatophytosis: Branching hypha seen in potassium hydroxide preparation of the skin. × 275.

Other Cutaneous Mycoses

Tinea Nigra

Tinea nigra (tinea nigra palmaris) is a superficial, chronic, and asymptomatic infection of the stratum corneum caused by *Exophiala werneckii*. This fungus, which is a ubiquitous saprophyte in nature, is dimorphic. It converts from yeast to mycelial growth with age, although both forms are usually observed.

E. werneckii is a dematiaceous fungus (see Chapter 85). The colony initially is shiny, moist, and often white to gray in color. Within a few days, the colony darkens and becomes olive to black. Later, mycelium develops and the colony appears dull and fuzzy. Microscopically, *E. werneckii* first produces budding yeasts and chains of yeast cells. Hyphae then develop and a mixture of variably pigmented hyphae and yeasts is common. Conidia are produced by annelides.

Clinical Infection. Tinea nigra is most frequently found in tropical areas. It is prevalent in Florida and other warm, coastal regions. Approximately 95 percent of cases occur in teenagers and 75 percent are in women. Lesions usually consist of an innocuous macule that is single with sharply defined margins, and spreads by expansion. The brownish color of the lesion is darkest at the advancing periphery, where most of the actively growing organisms are located. While many cases involve the palms, other areas of glabrous skin may also be infected, including the fingers and face. The lesions resemble a faded silver nitrate stain. As tinea nigra may resemble melanoma and other types of skin cancer, it is important to establish a diagnosis.

Diagnosis is easily confirmed by direct KOH examination and culture of skin scrapings from the periphery of the lesions. Microscopic examination of skin scales reveals brown-pigmented, branched, septate hyphae and budding yeast cells (1 to 5 μm in diameter). The hyphae may be distorted but the brown color excludes a diagnosis of dermatophytosis, candidiasis, or tinea versicolor. Culture of skin on Sabouraud's medium with and without antibiotics should recover *E. werneckii*.

Tinea nigra responds well to topical keratolytic solutions of sulfur, salicylic acid, or tincture of iodine. Recurrence is thought to be due to reinfection.

Pityriasis Versicolor

Pityriasis versicolor (tinea versicolor) is a chronic and nonirritating superficial infection of the stratum corneum caused by *Malassezia furfur (Pityrosporum orbiculare)*. Invasion of the cornified skin and the host responses are both minimal. Discrete, serpentine, hyper- or depigmented patches occur on the skin, usually on the chest, upper back, arms, or abdomen.

P. orbiculare and *Pityrosporum ovale* are lipophilic yeasts and part of the normal microbial flora of the skin and scalp. Originally isolated by Gordon on Sabouraud's medium supplemented with olive oil, *Pityrosporum* can utilize a variety of fatty acids. Colonies have a yeast-like consistency. Microscopically, *P. orbiculare* produces spherical cells, and *P. ovale*, ovoid or cylindrical cells. The name originally applied to the fungus observed in lesions of pityriasis versicolor was *M. furfur*. Whether or not *M. furfur* and *P. orbiculare* are identical, as suggested by indirect fluorescent antibody studies of infected skin scales, and whether or not *P. orbiculare* and *P. ovale* are indeed separate species, remain inconclusive.

Clinical Infection. The incidence of pityriasis versicolor increases where the climate is hot and humid, and is highest in the tropics. Several studies have also indicated a physiologic predisposition to this infection (e.g., excessive perspiration, corticosteroids, malnutrition, and heredity). Its occurrence has been related to the presence of certain amino acids and hydrophobic compounds on the skin, as well as to a decrease in the epithelial turnover in the stratum corneum.

The lesions present as macular patches of discolored skin that may enlarge and coalesce, but scaling, inflammation, and irritation are usually minimal. Patients seek medical care for cosmetic reasons. The affected skin does not suntan well and therefore the lesions become more pronounced in the summer.

Diagnosis is established by direct microscopic examination of skin scrapings of the infected skin, which has been digested with 10 percent KOH. The finding of short unbranched hyphae and spherical cells is diagnostic of pityriasis versicolor (Fig. 86-10). Because the clinical appearance and microscopic examination are so characteristic, cultures on a medium with lipid substrate are not required to establish the diagnosis. Indeed, the meaning of a positive culture of *Pityrosporum* is dubious, as it can also be recovered from normal skin and it has not yet been unequivocally proved as the cause of pityriasis versicolor. A golden yellow fluorescence extending beyond the periphery of the lesions may be observed under the Wood's light.

It is difficult to effect a permanent cure, but the usual treatment is topical application of 1 percent selenium sulfide for 10 minutes, three times weekly. Lesions may remain clear for a year or longer. Miconazole nitrate has been reported to be more effective.

Keratomycosis

Keratomycosis (mycotic keratitis) refers to fungous infections of the cornea. Many patients have a history of trauma leading to inoculation of the eye with a fungus. The etiologic agent is introduced from an exogenous source, and most are normally saprophytic fungi, although *Histoplasma capsulatum* and other frankly pathogenic fungi have been known to cause primary keratomycosis. Conversely, endophthalmitis usually represents the ocular manifestions of a systemic mycosis, and occurs not infrequently with systemic candidiasis, histoplasmosis, and other systemic mycoses.

Many cases of keratomycosis in the United States are caused by *Fusarium solani*. Others are attributed to other *Fusarium* species and related saprophytes, including dematiaceous fungi (e.g., *Alternaria*) and yeasts (e.g., *Candida*).

Keratomycosis occurs more often in males and individuals below the age of 50 years. Lesions appear as raised corneal ulcers with occasional satellite lesions, plaques, or hypopyon. Diagnosis is established by direct examination of corneal scrapings or surgical specimens for the presence of hyphae. *Fusarium* species grow rapidly in Sabouraud's or enriched media and most specimens yield positive culture.

The treatment of choice is pimaricin, a tetracene antibiotic, administered topically in 5 percent solution every few hours. Treatment is decreased with clearance of the lesions. Both keratoplasty and other antibiotics (e.g., nystatin, amphotericin B) have also been used.

FURTHER READING

DERMATOPHYTOSIS

Ajello L: Natural history of the dermatophytes and related fungi. Mycopathologia 53:93, 1974

Bibel DJ, Smiljanic RJ: Interactions of *Trichophyton mentagrophytes* and micrococci on skin culture. J Invest Dermatol 72:133, 1979

Cox FW, Stiller RL, South DA, et al.: Oral ketoconazole for dermatophyte infections. J Am Acad Dermatol 6:455, 1982

Dinh-Nguyen N, Hellgren L, Vincent J: A fermentation system for filamentous fungi, with special reference to dermatophytes. Mykosen 17:13, 1974

Grappel SF, Bishop CT, Blank F: Immunology of dermatophytes and dermatophytosis. Bacteriol Rev 38:222, 1974

Hashimoto T, Blumenthal HJ: Survival and resistance of *Trichophyton mentagrophytes* arthrospores. Appl Environ Microbiol 35:274, 1978

Hashimoto T, Wu CDR, Blumenthal HJ: Characterization of L-leucine-induced germination of *Trichophyton mentagrophytes* microconidia. J Bacteriol 112:967, 1972

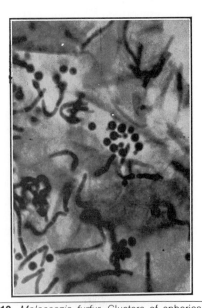

Figure 86-10. *Malassezia furfur.* Clusters of spherical cells and short hyphae in skin scrapings from pityriasis versicolor. × 736.

Hejtmanek M, Bartek J: Effects of media with different nitrogen source on nuclear ratio in heterokaryons of *Microsporum gypseum.* Acta Univ Palackianae Olomucensis 74:63, 1975

Jones HE, Reinhardt JH, Rinaldi MG: Immunologic susceptibility to chronic dermatophytosis. Arch Dermatol 110:213, 1974

Kane J, Fischer JB: The effect of sodium chloride on the growth and morphology of dermatophytes and some other keratolytic fungi. Can J Microbiol 21:764, 1975

Paveia MH: Culture medium alkalinization by dermatophytes. Mycopathologia 55:35, 1975

Philpot CM: The use of nutritional tests for the differentiation of dermatophytes. Sabouraudia 15:141, 1977

Rebell G, Taplin D: Dermatophytes. Their Recognition and Identification. Coral Gables, Fla, University of Miami Press, 1974

Rippon JW: Medical Mycology, 2nd ed. Philadelphia, Saunders, 1982, p 154

Rippon JW, Scherr GH: Induced dimorphism in dermatophytes. Mycologia 51:902, 1959

Swanson R, Stock JJ: Biochemical alterations of dermatophytes during growth. Appl Microbiol 14:438, 1966

Vincent J: The importance of fatty acids in the pathogenesis of dermatophytosis. Curr Ther Res 22:83, 1977

Walters BAJ, Breadmore GL, Halliday WJ: Specific cell mediated immunity in the laboratory diagnosis of dermatophyte infections. Br J Dermatol 94:55, 1976

TINEA NIGRA

Carrion AL: Yeastlike dematiaceous fungi infecting human skin. Arch Dermatol Syph 61:996, 1950

Castellani A: Tinea nigra. Mycopath Mycol Appl 30:193, 1966

Mok YK: Nature and identification of *Exophiala werneckii.* J Clin Microbiol 16:976, 1982

Vaffee AS: Tinea nigra palmaris resembling malignant melanoma. N Engl J Med 283:1112, 1970

Van Velso H, Singletary M: Tinea nigra. Arch Dermatol 90:59, 1964

PITYRIASIS VERSICOLOR

Belew PW, Rosenberg EW, Jennings BR: Activation of the alternative pathway of complement by *Malassezia ovalis (Pityrosporum ovale).* Mycopathologia 70:187, 1980

Burke RC: Tinea versicolor: Susceptibility factors and experimental infection in human beings. J Invest Dermatol 36:389, 1961

DaMert GI, Kirkpatrick CH, Sohnle PG: Comparison of antibody responses in chronic mucocutaneous candidiasis and tinea versicolor. Int Arch Allergy Appl Immunol 63:97, 1980

Faergemann J: Tinea versicolor and *Pityrosporum orbiculare*: Mycological investigations, experimental infections and epidemiological surveys. Acta Dermato-Venereol (Suppl)86:1, 1979

Gordon MA: The lipophilic mycoflora of the skin. I. In vitro culture of *Pityrosporum orbiculare n. sp.* Mycologia 43:524, 1951

Keddie FM, Shadomy S: Etiological significance of *Pityrosporum orbiculare* in tinea versicolor. Sabouraudia 3:21, 1963

Marinaro RE, Gershanbaum MR, Roisen FJ, et al.: Tinea versicolor: A scanning electron microscopic view. J Cutan Pathol 5:15, 1978

McGinley KJ, Leyden JJ, Marples RR, et al.: Quantitative microbiology of the scalp in nondandruff, dandruff, and seborrheic dermatitidis. J Invest Dermatol 64:401, 1975

Nazzaro Porro M, Passi S, Caprilli F, et al.: Induction of hyphae in cultures of *Pityrosporum* by cholesterol and cholesterol esters. J Invest Dermatol 69:531, 1977

Roberts SOB: *Pityrosporum orbiculare*: Incidence and distribution on clinically normal skin. Br J Dermatol 81:264, 1969

Salkin IF, Gordon MA: Polymorphism of *Malassezia furfur.* Can J Microbiol 23:471, 1977

Sohnle PG, Collins-Lech C: Relative antigenicity of *P. orbiculare* and *C. albicans.* J Invest Dermatol 75:279, 1980

Sternberg TH, Keddie FM: Immunofluorescence studies of tinea versicolor. Arch Dermatol 84:999, 1961

KERATOMYCOSIS

Bulmer C: The ocular mycoses. Contr Microbiol Immunol 4:56, 1977

DeVoe AG, Silva-Hutner M: Fungal infections of the eye. In Locatcher-Khorazo D, Seegal BC (eds): Microbiology of the Eye. St. Louis, Mosby, 1972, p 208

Forster RK, Rebell G: The diagnosis and management of keratomycosis, Parts I and II. Arch Ophthalmol 93:975, 1134, 1975

Forster RK, Rebell G, Wilson LA: Dermatiaceous fungal keratitis. Br J Ophthal 59:372, 1975

Gugnani HC, Talwar RS, Njoku-Obi ANU, et al.: Mycotic keratitis in Nigeria. A study of 21 cases. Br J Ophthalmol 60:607, 1976

Jones BR: Principles in management of oculomycosis. Trans Am Acad Ophthalmol Otolaryngol 79:15, 1975

Jones BR: Fungal keratitis. In Duane T (ed): Clinical Ophthalmology, vol IV. Hagerstown, Md, Harper & Row, 1978.

Opportunistic Mycoses

Opportunistic mycoses are associated with compromised patients. For such patients, they are life threatening and the most frequently encountered of the systemic fungous infections. During the past decade, their incidence has increased continuously and alarmingly. Persons with immunodeficiencies, endocrinopathies, organ transplants, and cancer must be considered at risk for systemic candidiasis, invasive aspergillosis, and mucormycosis. Because the etiologic agents of these mycoses are ubiquitous in the environment or are part of the normal microbial flora, it is almost impossible to avoid exposure.

In addition to their role in opportunistic disease, species of *Candida* and *Aspergillus* also cause other clinical entities.

Candidiasis

Candidiasis is an infection caused by any of several species of the yeast, *Candida.* These organisms are members of the normal flora of the skin, mucous membranes, and gastrointestinal tract. During birth, or shortly thereafter, all humans acquire and become colonized by *Candida* species. The risk of endogenous infection is, thus, clearly ever present. Indeed, candidiasis is the most common systemic mycosis and is worldwide in occurrence.

Of more than a hundred species of *Candida*, several are part of the normal flora and are potential pathogens. Most *Candida* infections are caused by *Candida albicans*, although at least seven other *Candida* species have also been encountered.

Morphology and Physiology

As indicated in Figure 87-1, *C. albicans* is capable of producing yeast cells, pseudohyphae, and true hyphae (Chap. 82). As part of the normal flora, the organism grows as a budding yeast; hyphal forms are produced only during tissue invasion. Although a number of environmental stimuli are known to trigger or block conversion in vitro from yeast to hyphal growth, regulation of morphogenesis in *C. albicans* remains inconclusive. One unquestionable stimulant is normal serum. After 90 minutes in serum at 37C, *C. albicans* begins to form hyphae. This reaction is manifested by the appearance of a germ tube, an elongated appendage, growing out from, and about half as wide and twice as long as, the yeast cell (Fig. 87-2). Germ tubes are distinct from pseudohyphae and are formed only by *C. albicans* and some isolates of *Candida stellatoidea.*

Cultural Characteristics. On most media, *Candida* species cannot be differentiated on the basis of their colony appearance. They produce within 24 to 48 hours raised, cream-colored, opaque colonies about 1 to 2 mm in diameter. After several days on agar medium, hyphae can be

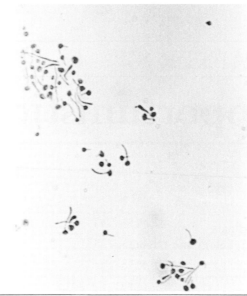

Figure 87-2. *Candida albicans.* Germ tubes formed at 37C. × 350. *(Courtesy of Dr. Marcia Manning.)*

observed penetrating the agar. Incubation of an isolate in serum at 37C provides a rapid, presumptive indentification of *C. albicans,* but speciation requires a battery of physiologic tests (Table 87-1).

Microscopic Appearance. *Candida* species produce ellipsoidal or spherical budding yeasts about 3 to 6 μm in size. Multiple buds and pseudohyphae are routinely formed on medium deficient in readily metabolizable substrates (e.g., cornmeal agar). *C. albicans,* unlike other species, characteristically produces chlamydospores on this medium (Fig. 87-3). *C. albicans* is capable of producing

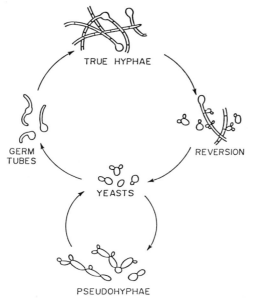

Figure 87-1. The morphogenesis of *Candida albicans. (Courtesy of Dr. Marcia Manning.)*

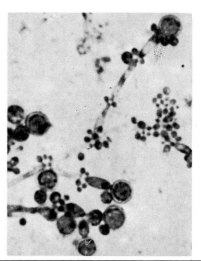

Figure 87-3. *Candida albicans.* Characteristic chlamydospores from cornmeal agar. × 790. *(From Conant: Am Rev Tuberc 61: 696, 1950.)*

TABLE 87-1. MORPHOLOGIC AND PHYSIOLOGIC CHARACTERISTICS OF *CANDIDA* AND OTHER YEAST SPECIES FREQUENTLY ISOLATED FROM CLINICAL SPECIMENS

Species	Property					Assimilations												Fermentations				
	Growth at 37C	Pseudo/True Hyphae	Chlamydospores	Germ Tubes	Capsule	Glu	Mal	Suc	Lac	Gal	Mel	Cell	Ino	Xyl	Raf	Tre	Dul	Glu	Mal	Suc	Lac	Gal
Candida albicans	+	+	+	+	−	+	+	+	−	+	−	−	−	+	−	+	−	F	F	−	−	F
Candida glabrata	+	−	−	−	−	+	−	−	−	−	−	−	−	−	−	+	−	F	−	−	−	−
Candida guillermondii	+	+	−	−	−	+	+	+	−	+	+	+	−	+	+	+	+	F	−	F	−	F*
Candida krusei	+	+	−	−	−	+	−	−	−	−	−	−	−	−	−	−	−	F	−	−	−	−
Candida parapsilosis	+	+	−	−	−	+	+	+	−	+	−	−	−	+	−	+	−	F*	−	−	−	F*
Candida pseudotropicalis	+	+	−	−	−	+	−	+	+	+	−	+	−	+*	+	−	−	F	−	−	F	F
Candida rugosa	+	+	+	−	−	+	−	−	−	+	−	−	−	+	−	−	−	−	−	−	−	−
Candida stellatoidea	+	+	+*	+	−	+	+	−	−	+	−	−	−	+	−	+*	−	F	F	F	−	−
Candida tropicalis	+	+	−*	−	−	+	+	+	−	+	−	+*	−	+	−	+	−	F	F	F	−	F
Cryptococcus neoformans	+	−*	−	−	+	+	+	+	−	+	−	+	+	+	+*	+	+	−	−	−	−	−
Rhodotorula rubra	+	−	−	−	−*	+	+	+	−	+	−	+*	+	+	+	+	+	−	−	−	−	−
Saccharomyces cerevisiae	+	−*	−	−	−	+	+	+	−	+	−	−	−	−	+	+*	−	F	F	F	−	F
Trichosporon beigelii	+*	+	−	−	−	+	+	+	+	+*	+*	+*	+*	+	+*	+*	+*	−	−	−	−	−
Trichosporon pullulans	+*	+	−	−	−	+	+	−	+*	+	+*	+	+	+*	+*	+	−	−	−	−	−	−
Geotrichum candidum	−*	+	−	−	−	+	−	−	−	+	−	−	−	+	−	−	−	−	−	−	−	−

From Silva-Hunter and Cooper: In Lennette, et al. (eds): Manual of Medical Microbiology, 3rd ed, 1980. Washington, D.C., American Society for Microbiology.
*Strain variation.
+, growth or presence of structure; −, neither growth over control nor fermentation; F, fermentation with gas production; Glu, glucose; Mal, maltose; Suc, sucrose; Lac, lactose; Gal, galactose; Mel, melibiose; Cell, cellobiose; Ino, inositol; Xyl, Xylose; Raf, raffinose; Tre, trehalose; Dul, dulcitol.

true hyphae of uniform width that grow by apical elongation and form septa at right angles with membrane-lined
pores. Pseudohyphae are formed by budding cells that
elongate and remain connected. Such cells are wider than
true hyphae and are constricted where they are attached.
These differential features of the two forms are compared
in Table 87-2. Both hyphae and pseudohyphae may revert
to a yeast growth, and it is not uncommon to see all three
forms in vitro or in tissue (Fig. 87-1).

Determinants of Pathogenicity. No distinctive virulence factors for *C. albicans* have been discovered, nor has
an explanation been provided for the variation in animal
pathogenicity that exists among strains. High doses of *C.
albicans* extracts possess endotoxin-like activity, but neither
this nor any other antigen nor enzyme extracted from the
organism correlates with pathogenetic potential. The relationship of hyphal production to virulence is currently
unresolved. However, recent studies indicate that cells of
C. albicans are able to attach to epithelial cell membranes
via a specific ligand–receptor interaction, and that germ
tubes appear to be more adhesive than yeast cells. Another provocative aspect of *C. albicans* is its immunomodulating activity. Depending upon the route and fraction of
C. albicans administered, either immunoadjuvant or immunosuppressive activity can be observed in experimental animals.

Clinical Infection

Epidemiology. There are many conditions that predispose individuals to opportunistic *Candida* infection (Table
87-3). Certain physiologic changes in otherwise healthy individuals provide the setting for opportunistic candidiasis.
In nonpregnant women, the incidence of candidal vaginitis
is between 10 and 17 percent, but this incidence approximately doubles during pregnancy. The physiologic changes
in the cervical and vaginal mucosa that result in overgrowth of *Candida* have been correlated with: (1) an increase in moisture and carbohydrate substrates on the mucosal surface; (2) a local decrease in transferrin, which
would lead to increased levels of available iron, an essential growth requirement for *Candida;* (3) increased secretion of steroids that might promote candidiasis indirectly
by reducing local host defenses, such as phagocytosis; and
(4) a decrease in the concentration of specific IgA secreto-

**TABLE 87-2. COMPARISON BETWEEN THE APPEARANCE
OF HYPHAE AND PSEUDOHYPHAE**

Feature	Hyphae	Pseudohyphae
Growth process	Apical elongation	Budding
Terminal cell	Longer; cylindrical	Shorter; spherical
Cell walls	Parallel	Constricted at septa
Septa	Straight; perpendicular	Curved or pinched
Branches	No constriction or septum at orgin	Constricted and septated at origin

**TABLE 87-3. FACTORS PREDISPOSING TO *CANDIDA*
INFECTIONS**

Physiologic
  Pregnancy
  Old age
  Infancy
Traumatic
  Maceration
  Other infection
Hematologic
  Cellular immunodeficiency
  Aplastic anemia
  Agranulocytocis
  Lymphoma, Hodgkin's disease, leukemia
  Hypogammaglobulinemia and agammaglobulinemia
Endocrine
  Diabetes mellitis
  Hypoparathyroidism
  Addison's disease
Iatrogenic
  Immunosuppression
  Transplantation
  Postoperative
  Steroid treatment
  Antibacterial antibiotics
  Birth control pill
  Catheters
  Vaccination
  Hyperalimentation
Miscellaneous
  Malignancy
  Malnutrition
  Malabsorption
  Thymoma
  Heredity

ry component, although the protective value of this antibody has not been established.

Infants are especially at risk if they are heavily exposed to *Candida* before the normal microbial flora of the
gastrointestinal tract and skin have been established. Normally, the attack rate of candidiasis among infants is approximately 4 percent, but this is increased if the mother
has candidal vaginitis. Infants usually develop oral thrush,
perianal and genital infections, gastroenteritis with severe
diarrhea, or prolonged and painful diaper rash. Nursery
epidemics have resulted from spread of the infection from
one infant to another by nursery attendants.

Although the intact adult epithelium is normally impervious to *Candida* invasion, certain conditions increase
the opportunity for superficial candidiasis. Any trauma,
burn, abrasion, or break in the epithelial integrity of skin
or gut provides an opportunity for *Candida* to invade the
cutaneous, mucosal, or subcutaneous tissue. Excessive
moisture and warmth increase the numbers of *Candida* on
the skin. Skin infections are an occupational disease of
dishwashers, bartenders, fruit pickers, and similar workers.
Warmth, moisture, and friction result in intertriginous

candidal infections of the skin folds, toe webs, groin, or under the breasts, especially in the obese. *Candida* may also secondarily invade lesions or epithelium damaged by other infections.

Many blood dyscrasias predispose patients to systemic candidiasis, as does cellular or, less frequently, humoral immunodeficiency. A decrease in numbers or functional capacity of neutrophils lowers resistance to *Candida*, resulting in recurrent systemic infections. Neutrophils from patients with leukemia or chronic granulomatous disease show subnormal in vitro phagocytosis or killing. Defective T cell immunity is evident in most patients with chronic mucocutaneous candidiasis.

Endocrinologic disturbances, such as diabetes mellitus, hypoparathyroidism, and Addison's disease result in an increased incidence of candidiasis, but a reason for the increased incidence has not been found. In many cases, *Candida* infection precedes by several years the open manifestation of diabetes. Patients with decreased parathyroid hormone secretion or adrenal insufficiency (Addison's disease) manifest developmental abnormalities that may involve the skin and mucosa and are prone to cutaneous and mucocutaneous candidiasis.

Ironically, many recent advances of modern medicine that serve to prolong life also increase the likelihood of life-threatening opportunistic infections. Patients who are treated with such medications and whose host defenses are unduly compromised are especially at risk if they are iatrogenically exposed to *Candida*. Immunosuppressive treatment, as an adjunct to transplanation or anticancer therapy, decreases resistance to *Candida*. As many as 30 percent of leukemic patients acquire systemic candidiasis. Most patients receiving organ transplants succumb to opportunistic infection, of which *Candida* is the leading mycotic agent. The probability of postoperative systemic candidiasis is related to the length of the operation and is caused not only by contamination with yeasts during surgery but also by diverse postoperative procedures, such as indwelling catheters or the use of prophylactic antibacterial antibiotics. An additional risk is imposed by prosthetic devices, such as artificial heart valves or intravenous lines, which provide a foreign body that can be colonized by *Candida* in the bloodstream. Similarly, any trauma to the heart valves may induce vegetations that can provide a nidus for *Candida*.

Many patients who develop systemic candidiasis present a history of having received corticosteroids prior to the infection. Among the many and uncertain effects of corticosteroids are depressed phagocytic activity and cell-mediated immunity. Studies have shown that cortisone-treated animals are less resistant to challenge with *Candida* and other fungi. Antibacterial antibiotics predispose to candidiasis by reducing the bacterial flora and indirectly increasing the yeast population. Contamination of intravenous lines, especially at the point of entry, is an unfortunate and too frequent source of candidemia. The skin around injection sites must be scrupulously cleaned and monitored to minimize this source of yeasts. Contamination of Foley catheters may result in infection of the urinary tract and bladder.

Among the opportunistic infections to which patients with acquired immune deficiency syndrome (AIDS) are highly susceptible are candidiasis, especially involving the mucosal surface of the esophagus, cryptococcal meningitis, and a myriad of other mycotic infections. The depression of cell-mediated immunity in these patients is often manifested as abnormally low numbers of T-helper/inducer cells and an inversion of the normal T-helper to T-suppressor/cytotoxic cell ratio. This immunologic defect in patients with AIDS is consistent with the evidence that competent cell-mediated immunity is essential for resistance to both mucocutaneous candidiasis and cryptococcosis (See Chap. 84).

Pathogenesis. The intact or physiologically normal epithelium is usually resistant to *Candida* invasion. If, however, there is a marked increase in the number of *Candida* present, or if the skin and mucosa are traumatized or are hormonally altered, these barriers are susceptible to *Candida* invasion. Iatrogenic candidemia and candiduria, induced by catheters, surgery, or hyperalimentation, are often successfully managed by the normal host defense mechanisms. However, the ability of patients with hormonal imbalances, immunodeficiencies, and malignancies to control invasion of the deeper tissues is limited.

The host defenses against candidiasis are both specific and nonspecific, cellular and humoral. Serum components, such as opsonins, complement, and transferrin, may inhibit either directly or indirectly the survival of *Candida*. Specific antibodies to *Candida* have a minimal direct effect but may (1) inhibit the normal clumping of yeasts by serum, (2) affect yeast morphogenesis or respiration, (3) function as opsonins, or (4) mediate antibody-dependent cellular cytotoxicity. Cellular host defenses against *Candida* involve neutrophils, which kill 30 to 40 percent of the ingested yeasts, and the affector and effector functions of macrophages. The host factors involved in chronic mucocutaneous candidiasis are associated with the thymus-dependent immune system.

Clinical Manifestations. Clinical candidiasis may be broadly grouped into three categories: cutaneous, systemic, and chronic mucocutaneous. Table 87-4 lists most of the forms in these categories. Allergic reactions to *Candida* antigens may also occur.

CUTANEOUS CANDIDIASIS. Infection of the skin, mucous membranes, and nails by endogenous *Candida* can be caused by conditions that result in chronic maceration of these areas, physiologic changes in the host, or a compromised immune status.

Candidiasis of the mucous membranes is often referred to as thrush. Oral thrush is most commonly seen in infants, patients with chronic mucocutaneous candidiasis, and adults undergoing treatment with steroids, cytotoxic drugs, or antibacterial antibiotics. The lesions may be sin-

TABLE 87-4. CLINICAL CLASSIFICATION OF CANDIDIASIS (ABRIDGED)

Cutaneous and subcutaneous candidiasis
　Thrush (oral, vaginal)
　Stomatitis
　Intertriginous candidiasis (groin, axillary, interdigital)
　Onychomycosis
　Esophagitis
　Severe diaper rash
　Balanitis
Systemic candidiasis
　Esophagitis
　Intestinitis
　Infant diarrhea
　Bronchopulmonary candidiasis
　Pyelonephritis
　Cystitis
　Endocarditis
　Mycocarditis
　Endophthalmitis
　Meningitis
　Arthritis
　Peritonitis
　Macronodular skin lesions
Chronic mucocutaneous candidiasis

gular, patchy, or confluent, and a whitish pseudomembrane composed of yeasts and pseudohyphae may cover the tongue, soft palate, and oral mucosa. A diagnosis can be established by direct microscopic and cultural examination of scrapings from the lesions.

Vaginal thrush occurs more often in pregnant women, diabetics, and women receiving antibacterial or hormonal treatment, including birth control pills. Patches of gray-white pseudomembrane develop on the vaginal mucosa, and a yellow-white discharge may accompany the infection. From the mucous membranes, infection and inflammation may spread to the adjacent skin. The characteristic yeasts and pseudohyphae abound in these lesions. Infants develop similar lesions in the perianal region, which may persist as a diaper rash of the genital, perianal, and groin areas. By scratching, spread may occur to other skin sites. In addition to *C. albicans, Candida tropicalis, C. stellatoidea,* and *Candida pseudotropicalis* also may cause vaginitis. *Candida* may cause chronic nail infections and paronychia. The skin around the nail becomes swollen, erythematous, and painful, unlike dermatophytosis of the nail (tinea unguim). *Candida parapsilosis, C. tropicalis,* and *Candida guilliermondi* are also associated with onychomycosis.

SYSTEMIC CANDIDIASIS. Numerous systemic manifestations of candidiasis may follow introduction of *Candida* into the bloodstream. Candidemia may result from contamination of indwelling catheters, surgical procedures, trauma to the skin or gastrointestinal tract, or aspiration. The extent and severity of the infection that follows is determined by the inoculum size, the virulence of the organisms, and, most importantly, the host defenses.

The scope of systemic candidiasis is protean. Some of the more recognized syndromes are listed in Table 87-4. Clinical indications of occult systemic candidiasis include candiduria (in the absence of catheterization and an imbalanced flora), candida endophthalmitis, and maculonodular skin lesions. Although *C. albicans* is the most common agent of candidiasis, *C. guilliermondi, C. parapsilosis,* and *C. tropicalis* are frequent causes of endocarditis. Overall, *C. tropicalis* is second to *C. albicans* in pathogenetic potential.

CHRONIC MUCOCUTANEOUS CANDIDIASIS. A unique set of predisposing conditions and clinical manifestations is associated with the entity, chronic mucocutaneous candidiasis (CMC). This condition is defined as infection, invariably with *C. albicans,* of any or all of the epithelial surfaces of the body: the skin, oral mucosa, upper respiratory tract, gastrointestinal, urinary, and genital epithelium. Invasion of the bloodstream or deeper tissues is unusual. *C. albicans* apparently penetrates the plasma membrane of viable epithelial cells, causes considerable distortion of these cells, and exists as an intracellular parasite. The onset of CMC begins early in life and often persists for a lifetime. The degree of involvement of the epithelium and mucous membranes varies with different patients and with individual patients at different times. While some children develop total skin involvement that only minimally recedes with treatment, others have limited but persistent lesions. Some patients respond temporarily to therapy, sometimes for years, but permanent cures rarely, if ever, occur. The classic lesions, as seen in Figure 87-4, are verrucous and warty, with horn-like projections growing out from the

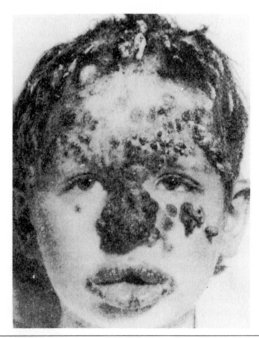

Figure 87-4. Child with chronic mucocutaneous candidiasis.

skin. These lesions appear at an early age and become chronic, with the development of extensive epithelial hyperplasia.

Several underlying conditions have been correlated with CMC. Most patients have a deficiency in their cell-mediated immunity, but the precise defect has not been defined. Immunologic analyses of CMC patients have revealed no consistent deficiency. Some patients have single, and others have multiple defects. It now seems probable that any of several T-dependent immunodeficiency syndromes may provide the setting for CMC. Individual patients may be deficient in any or all of the following ways: (1) delayed skin test anergy and lack of in vitro lymphocyte responsiveness to antigens, (2) unresponsiveness to mitogens, and (3) failure to produce lymphokines.

Many patients with CMC have normal neutrophils, immunoglobulin levels, and levels of T and B lymphocytes. Confusingly, some have defective neutrophils (chronic granulomatous disease), whereas others apparently have no immunologic dysfunction. Transfer factor therapy appears to be effective, although it has not been administered without concomitant chemotherapy. The available data would suggest defective macrophage processing, T lymphocyte responsiveness, excessive T suppressor activity, and/or humoral blocking factors.

Some patients with CMC have other dominant underlying conditions. Endocrinopathies, especially hypoparathyroidism, are present in some patients. In others, abnormalities in iron metabolism have been described, usually iron deficiency. CMC has been observed in patients with leukemia, thymoma, and other blood diseases. Montes described seven CMC cases associated with hypovitaminosis A. The mechanism of pathogenesis of CMC remains unclear. Some of the conditions mentioned could result in abnormal physiologic changes in the epithelium, and others may relate to available iron or compromised host defenses. As might be expected, an autosomal genetic predisposition to CMC and its underlying conditions has also been documented.

Three groups of CMC have been delineated by Kirkpatrick and Windhorst based on the time of onset. With early onset, the predisposing defect or defects in cell-mediated immunity are apparently inherited. These patients are subdivided into those with or without endocrinopathy. Onset between the ages of 10 and 30 years offers the best prognosis because the defects in cell-mediated immunity in these patients are often restored after therapy. Adult onset CMC occurs after the age of 40 years, and most, if not all, patients have an associated thymoma.

Laboratory Diagnosis

MICROSCOPIC EXAMINATION. The appearance in tissue of pseudohyphae or true hyphae along with budding yeast cells is pathognomonic for invasive candidiasis. The presence of hyphal forms in freshly examined skin scrapings, vaginal exudate, and specimens of sputum, centrifuged urine, spinal fluid, or joint fluid also indicates candidiasis.

Figure 87-5 shows typical forms seen in a fresh preparation.

CULTURE. Because *Candida* species are very prevalent, positive cultures are invariably suspect. Specimens from normally sterile sites can be cultured on bacteriologic media. Otherwise, Sabouraud's agar containing antibiotics should be employed. Germ tube production provides tentative identification of *C. albicans*. A germ tube-negative yeast that produces pseudohyphae and lacks a capsule and arthroconidia may be *Candida* and should be speciated according to the physiologic pattern, as detailed in Table 87-1.

Candida isolated from normally sterile specimens is significant and should be evaluated in terms of the patient's clinical history. For example, a positive blood culture in a postsurgical patient may reflect a self-limiting transient candidemia and additional specimens should be cultured. Positive cultures from sputum have no diagnostic value. Cultures from surface specimens (skin, vaginal mucosa) should confirm the etiology of clinical lesions. A census of 10^4 to 10^5 per ml from a properly obtained urine specimen without indwelling urethral catheter is considered indicative of systemic infection.

SEROLOGY. Because of the difficulty of establishing a diagnosis of systemic candidiasis by direct examination or culture, an intensive search has been underway for the past decade for a diagnostic serologic test. Different antigens and methods have been applied to the detection of diagnostic levels of antibodies. Although controversial, the detection of precipitins to cytoplasmic antigens by immunodiffusion or counterimmunoelectrophoresis is probably the most acceptable, readily available test. By itself, a positive test is not diagnostic but must be interpreted with oth-

Figure 87-5. *Candida albicans.* Gram stain of sputum smear showing budding cells and pseudohyphae. × 1175.

er clinical data. Table 87-5 summarizes the most reliable serologic aids for diagnosis of candidiasis and other opportunistic mycoses.

As everyone is exposed to *Candida*, serologic tests are limited to discriminating between normal and disease levels of antibodies. More specific tests for specific antigen are currently under development. Both *Candida* surface mannan and cytoplasmic proteins can be detected in sera by enzyme-linked immunosorbent assay, radioimmunoassay, or quantitative immunofluorescence. Mannan and arabinitol, a *Candida* metabolite, can also be measured in sera by gas–liquid chromatographic methods. The detection of circulating *Candida* antigens or metabolites appears to be diagnostic for systemic candidiasis.

Treatment. Cutaneous candidiasis can be treated with topical antibiotics (ketoconazole, nystatin, miconazole) or chemical solutions (gentian violet). For the treatment of systemic candidiasis, amphotericin B and/or flucytosine is recommended. Unfortunately, up to one half of the clinical isolates of *C. albicans* (mostly serotype A) develop resistance to flucytosine. Chronic mucocutaneous candidiasis has been treated with flucytosine, amphotericin B, miconazole, topical chemical solutions, and transfer factor. Ketoconazole has produced dramatic resolution of lesions in many patients. The response of individual patients is highly variable, but resolution of lesions is associated with improvement in immunocompetence. Both responses, however, are often only temporary.

Aspergillosis

Aspergillosis refers to a spectrum of disease conditions that may be induced by a number of *Aspergillus* species. In humans, the disease is worldwide in its occurrence, and the various species are ubiquitous in nature. Inhalation of *Aspergillus* spores or mycelial fragments may elicit in certain individuals an immediate hypersensitivity response without invasion of the body. In allergic bronchopulmonary aspergillosis, however, there is actual infection by an *Aspergillus* species, and the organism grows in the bronchial tree. Invasive aspergillosis occurs in certain types of compromised patients and involves actual invasion of the tissue. In other patients, aspergillosis is characterized by a noninvasive colonization of exposed tissue, such as the pulmonary cavity, external ear canal, or cornea. Certain secondary metabolites produced by species of *Aspergillus* are toxic and carcinogenic.

One hundred fifty different species and subspecies of *Aspergillus* have been recognized. They can be isolated from vegetation, especially nuts and grains, during either growth or storage, from decaying matter, soil, and air. Many species are pathogenic for plants, and some infect insects, birds, and domestic animals.

Aspergillus fumigatus is the most common pathogenic species for humans, although many other species have been known to produce infection. Considering the overall prevalence of pathogenic species and the constant exposure of humans and animals, the incidence of infection is relatively rare. Clearly, a high degree of natural resistance exists in the healthy host. When, however, the exposure is overwhelming, as in the occurrence of extrinsic allergic alveolitis among malt workers exposed to the spores of *Aspergillus clavatus,* or when the host defenses are compromised, as in the case of leukemic patients, aspergillosis may develop.

Morphology
Cultural Characteristics. *Aspergillus* species grow rapidly on many natural substrates and laboratory media. The abundant aerial mycelium becomes powdery and pigmented as conidia, which are characteristic of each species, are produced. Although *A. fumigatus* is the most common pathogenic species, more than a dozen additional species of *Aspergillus* have been known to cause infection. Some properties of the most frequent clinical isolates are compared in Table 87-6.

Microscopic Appearance. All species of *Aspergillus* are characterized by conidiophores, which expand into large vesicles at the end and are covered with phialides that produce long chains of conidia (Fig. 87-6). Phialides may arise directly from the vesicle (uniseriate) or from metulae, which are attached to the vesicle (biseriate). Species are identified primarily on the basis of the conidial structures, namely, the size, color, and shape of the conidiophore, conidia, and phialides (Table 87-6).

Clinical Infection
Epidemiology. Adults are infected more frequently than children, and males more often than females. Certain host factors clearly predispose some individuals to invasive aspergillosis. In general, the risk factors of invasive aspergillosis are similar to those associated with systemic candidiasis. There is a significant correlation of invasive aspergillosis with cancer (especially acute or chronic leukemia, lymphoma, or Hodgkin's disease); granulocytopenia; therapy with corticosteroids, antibacterial antibiotics, or cytotoxic drugs; or combinations of these conditions. The allergic and infectious forms occur more often in those who are exposed frequently to massive doses of the conidia. Indeed, aspergillosis is an occupational disease of agricultural workers and whiskey distillery workers exposed to numerous conidia in malting barley.

Pathogenesis. Most cases of aspergillosis develop in individuals who have structural abnormalities within the lung or who have severely impaired resistance to infection because of metastatic cancer, leukemia, lymphomatous diseases, or because of the therapy used in combating these

TABLE 87-5. SUMMARY OF MYCOSEROLOGY FOR OPPORTUNISTIC MYCOSES (ABRIDGED)

| Mycosis | Test | Antigen | Sensitivity and Value | | Limitations/Specificity |
			Diagnosis	Prognosis	
Candidiasis (systemic)	AG	Ca	60 to 75 percent of cases positive, paired sera and four-fold titer rise required	Four-fold change in titer	Many healthy persons positive
	CIE	HS or S	90 to 100 percent of cases positive; one or more precipitin bands	Titer change or loss of bands	Patients with superficial candidiasis or transient candidemia may also be positive
	ID	HS or S	88 percent of cases of systemic candidiasis positive	Loss of bands	More specific, less sensitive than CIE test
Aspergillosis (invasive)	ID	Acf	Sensitivity highly vairable; 3 to 4 bands indicative of invasive aspergillosis or aspergilloma	Number of bands reflects severity	80 to 100 percent of cases with allergic bronchopulmonary aspergillosis and more than 90 percent with aspergilloma are positive; cross-reactions with other fungi and C-reactive protein
Mucormycosis	ID	Zs	73 percent cases positive	Unknown	None

Tests: AG, yeast agglutination; CIE, counterimmunoelectrophoresis; ID, immunodiffusion.
Antigens: Ca, killed yeast cells of *Candida albicans;* HS, Hollister-Steir commercial *Monilia* antigen; S, cytoplasmic antigen of *C. albicans* yeast cells; Acf, pool of culture filtrate of 3 to 5 *Aspergillus* species; Zs, pool of cytoplasmic antigens of 11 zygomycetous species.

diseases. For example, 17 cases of precipitous, invasive aspergillosis in cancer patients inadvertantly exposed to a heavy, environmental contamination of aspergilli has recently been reported by Aisner and associates. Twelve of these patients had acute nonlymphocytic leukemia, and all 17 displayed some degree of neutropenia. In vitro studies support the clinical evidence that leukocytes appear to be essential in the host defense against aspergillosis. Healthy macrophages are able to contain conidia, and hyphae are susceptible to neutrophil and monocyte killing mechanisms.

Clinical Manifestations. Several classifications of the clinical types of aspergillosis have been proposed. Table 87-7 presents a current view and the associated immune response.

ALLERGIC FORMS. Inhalation of antigens associated with *Aspergillus* species elicits in certain atopic individuals an immediate, asthmatic reaction, mediated by reaginic antibody (IgE). Diagnosis is confirmed by patient history, positive type I skin tests (wheal and flare) to *Aspergillus* antigens, and presence of specific IgE antibody. Bronchoconstriction occurs in the absence of both fungal growth and pulmonary consolidation in radiographs of the chest.

Allergic bronchopulmonary aspergillosis is the major and well-characterized allergic form of aspergillosis. Table 87-8 lists major diagnostic criteria: (1) recurrent pulmonary densities in radiographs of the chest; (2) eosinophilia; (3) asthma; and (4) hypersensitivity to *Aspergillus* antigen, as shown by an immediate skin test reaction (type I) followed some hours later by an Arthus reaction (type III). Other criteria include positive culture of *Aspergillus* from sputum, anti-*Aspergillus* precipitins in serum, elevated serum IgE levels, and a history of recurrent pneumonia.

Nonatopic hosts can develop another form of hypersensitivity reaction to *Aspergillus* antigens, extrinsic allergic alveolitis. This entity is characterized by the presence of

TABLE 87-6. CHARACTERISTICS OF COMMON PATHOGENIC *ASPERGILLUS* SPECIES

| Species | Conidiophore Length | Vesicle Width (μm) | Phialides | Conidia | |
				Color	Diameter (μm)
A. fumigatus	Less than 300 μm	20 to 30; only top half conidiogenous	Uniseriate	Gray, green, or blue-green	2.5 to 3.0
A. flavus	Less than 1 mm	25 to 45	Uniseriate or biseriate	Yellow to green	3.5 to 4.5
A. niger	1.5 to 3.0 mm	45 to 75	Biseriate	Black	4.0 to 5.0
A. nidulans	60 to 130 μm	8 to 10	Biseriate	Dark green	3.0 to 3.5
A. terreus	100 to 250 μm	10 to 16	Biseriate	Orange to brown	1.8 to 2.4

Adapted from Raper and Fennell: The Genus *Aspergillus*, 1965. Baltimore, Williams & Wilkins.

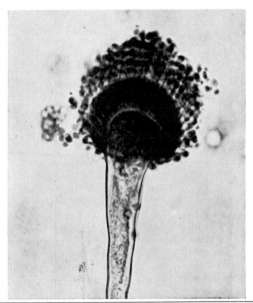

Figure 87-6. The typical conidiophore of *Aspergillus fumigatus* expands into a vesicle from which phialides arise directly (uniseriate) to produce basipetal chains of conidia. *(From Conant et al.: Manual of Clinical Mycology, 3rd ed, 1971. Philadelphia, Saunders.)*

serum precipitins to *Aspergillus* antigens and, upon skin testing, an Arthus (type III) reaction that occasionally is preceded by an immediate reaction. Inhalation of the antigen will induce fever, leukocytosis, dyspnea, nonproductive cough, myalgia, and rales after several hours. Yocum and associates demonstrated delayed (type IV) hypersensitivity and tissue inflammation characteristic of lymphocyte-mediated immunopathology in a patient with extrinsic allergic alveolitis caused by *Aspergillus* sensitivity.

ASPERGILLOMA AND EXTRAPULMONARY COLONIZATION. The conidia of *Aspergillus* species are able to germinate and colonize the surfaces of open pulmonary cavities, paranasal sinuses, and ear canals. Aspergilloma, or fungus

TABLE 87-7. CLINICAL TYPES OF ASPERGILLOSIS

Clinical Forms	Hypersensitivity Types
Allergic forms	
Atopic host	
Asthma	I only
Allergic bronchopulmonary aspergillosis	I and III
Nonatopic host	
Extrinsic allergic alveolitis	III only, or with I or IV
Noninvasive colonization	I, III, or none
Aspergilloma	
Nonpulmonary, local infection	
Invasive aspergillosis	None or any
Pulmonary	
Disseminated	

TABLE 87-8. ALLERGIC BRONCHOPULMONARY ASPERGILLOSIS: DIAGNOSTIC CRITERIA

Criteria	Percent Positive
Asthma	100
Pulmonary infiltrates	91–100
Eosinophilia (in blood)	80–100
Immune responses to *Aspergillus fumigatus* antigen	
Type I (wheal and flare)	100
Type III (Arthus)	100
Serum precipitins	84–91
Sputum	
Eosinophilia	44–100
History of plugs	74
Culture of *A. fumigatus*	46–83

From Safirstein et al.: Am Rev Respir Dis 108:450, 1973; Malo et al.: Thorax 32:269, 1977; Khan et al.: Scand J Respir Dis 57:73; 1976.

ball, refers to the colonization of a pulmonary cavity that may have been caused originally by bronchiectasis, carcinoma, histoplasmosis, malformation, sarcoidosis, or tuberculosis. This is probably the most common type of pulmonary aspergillosis, and the radiographic appearance of an air space surrounding the cavity is highly characteristic, although not diagnostic, of aspergilloma. Many patients with aspergilloma are asymptomatic. Some have productive cough and hemoptysis, but dyspnea and even pulmonary hemorrhage may also follow the appearance of a fungus ball in the chest radiograph.

Noninvasive colonization may also occur in the ears, paranasal sinuses, and nasal cavity. Symptoms of chronic otitis and sinusitis may ensue. Haley has demonstrated that both allergy and superficial trauma are necessary for the development of ear infections. Aspergillus is a major cause of otomycosis. Primary localized infection of the conjunctiva, eyelids, orbit, and intraocular structures with *Aspergillus* species have also been described. Cutaneous lesions have been reported. Onychomycosis can be caused by *Aspergillus flavus*, *Aspergillus nidulans*, and *Aspergillus glaucus*.

INVASIVE ASPERGILLOSIS. This form may be localized in the lung or generalized and disseminated. The course depends largely on the underlying illness. Most cases of invasive pulmonary aspergillosis now occur in patients with acute leukemia, granulocytopenia, lymphoma, or, less commonly, other malignancies; or in the immunosuppressed state associated with organ transplantation. In the absence of early treatment, the course is uniformly fatal. In the rare, otherwise healthy host, the prognosis is much better. Invasion of the lung parenchyma may develop immediately in a compromised patient exposed to an infectious dose of *Aspergillus* conidia, or it may follow as a rare complication of an aspergilloma. Invasion seldom occurs with allergic bronchopulmonary aspergillosis. The hyphae exhibit a propensity for invading the lumen and walls of blood vessels, causing thrombosis, infarction, and hemorrhage. Dissemination from the lungs may result in generalized aspergillosis involving a number of other organs, including, most frequently, the gastrointestinal tract, brain, liver, kidney, and many other sites (e.g., heart, skin, eye).

Laboratory Diagnosis

MICROSCOPIC EXAMINATION. Fresh sputum should be examined for the presence of branching, septate hyphae of uniform width (4 to 7 μm). Sputum may contain sparse filaments or plugs of mycelium. In the bronchi, *Aspergillus* may produce aerial hyphae and characteristic conidiophores. In tissue sections, hyphae are often seen in blood vessels forming parallel arrays with dichotomous branching at acute angles. Conidia may be formed in the air space associated with an aspergilloma. The true hyphae found in aspergillosis should not be confused with the yeast cells, hyphae, and pseudohyphae associated with invasive candidiasis (see Table 87-2).

CULTURE. *Aspergillus* species will grow on most routine media, but cycloheximide-containing media should not be used. The developing colonies produce surface mycelia within a few days and become pigmented as conidia develop. Most species are identified by their morphology (see Table 87-6 and Fig. 87-6).

As *Aspergillus* species are common contaminants, the significance of a positive culture is often questionable. Active allergic bronchopulmonary aspergillosis, aspergilloma, and other local colonizations can be diagnosed by direct examination and/or repeatedly positive, often pure cultures. Invasive aspergillosis, however, is usually negative on direct examination of sputum and culture of sputum or blood. A single positive culture in a compromised patient may, therefore, be highly significant. Aggressive diagnostic efforts in patients with leukemia, leucopenia, and pulmonary infiltrates have been recommended. These include transtracheal aspiration, bronchial brush biopsy, or open lung biopsy to obtain specimens for examination in patients who can tolerate these procedures.

SEROLOGY. The immunologic responses detected following exposure to *Aspergillus* can sometimes be used to clarify the pathogenesis, diagnosis, and prognosis. Serologic studies involving hundreds of patients have clearly associated the presence of precipitins to various antigen preparations of *Aspergillus* species, as detected by immunodiffusion, with different forms of aspergillosis (see Table 87-5). Eighty to 100 percent of patients with allergic bronchopulmonary aspergillosis or aspergilloma have one or more serum precipitins to *A. fumigatus*. Unfortunately, sera from patients with invasive aspergillosis are much less frequently positive. False-positive tests are rare but may occur with serum containing cross-reacting fungal antibodies or C-reactive protein. Approximately 30 percent or more of patients with cystic fibrosis have *Aspergillus* precipitins. In patients with allergic bronchopulmonary aspergillosis, specific IgE antibodies to *A. fumigatus* can be detected by the radioallergosorbent test.

Treatment. Allergic forms of aspergillosis have been treated with corticosteroids and antifungal therapy. Disodium chromoglycate also reduces symptoms but may not prevent recurrence.

The treatment of aspergilloma varies considerably with its severity. Asymptomatic patients may not warrant treatment, while other patients require surgical resection. Both amphotericin B and flucytosine have been recommended, and pulmonary lavage has been suggested to facilitate penetration of the cavity by the drug. The accelerated administration of amphotericin B is instituted as soon as invasive aspergillosis is diagnosed. Local, superficial aspergillosis is treated with nystatin.

Mucormycosis

Mucormycosis (phycomycosis, zygomycosis) is an opportunistic mycotic infection caused by a number of mold species classified in the order Mucorales of the class Zygomycetes (Chap. 82). These fungi are ubiquitous, thermotolerant saprophytes. Patients with acidosis, leukemias, and immunodeficiencies are particularly at risk from opportunistic mucormycosis.

The etiologic agents of mucormycosis, in approximate order of their frequency as pathogens, are *Rhizopus arrhizus, Rhizopus oryzae, Absidia corymbifera, Absidia ramosa, Mucor pusillus,* and species of *Mucor, Rhizopus, Absidia, Cunninghamella,* and other genera. They can be isolated from the air, soil, water, and hospital environments worldwide.

Morphology

Cultural Characteristics. These fungi grow rapidly and produce abundant, cottony or fluffy aerial mycelia that often fill the agar test tube or petri plate. Identification of most genera can be made on the basis of morphology, but species identification is often difficult.

Microscopic Appearance. These fungi produce broad generally nonseptate hyphae that often appear twisted and ribbon-like upon microscopic examination. They reproduce asexually by the formation of sporangia and sexually with production of a zygospore (Chap. 82). Figure 87-7 depicts the sporangia and characteristic rhizoids of *Rhizopus*.

Clinical Infection

Epidemiology. Patients with ketoacidosis resulting from diabetes mellitus, drugs, or uremia are predisposed to invasive mucormycosis. The disease is also associated with burn patients, leukemia, lymphoma, steroid treatment, and immunosuppression, either natural or induced.

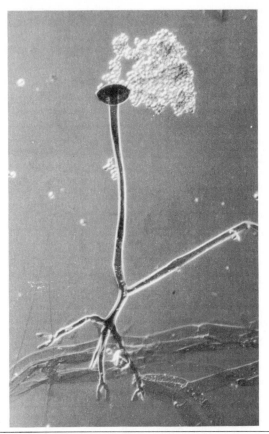

Figure 87-7. *Rhizopus.* Typical sporangium and rhizoids. Nomarski. × 200. *(From McGinnis: Laboratory Handbook of Medical Mycology, 1979. New York, Academic Press.)*

Pathogenesis. The sporangiospores apparently germinate and thrive in environments like the nasal, oropharyngeal, or respiratory mucosa of compromised patients. Hyphae invade the lumen and walls of blood vessels, causing thrombosis, infarction, and necrosis. This process is more rapid than the similar involvement with *Aspergillus* or *Candida.* A suppurative response is elicited, and local necrosis develops.

Clinical Manifestations. Invasive mucormycosis presents in two forms, defined by the site of involvement. Local, colonizing infections have also occurred in different types of patients.

RHINOCEREBRAL MUCORMYCOSIS. Invasion begins typically in the nasal region and progresses rapidly to involve the sinuses, eye, brain, and meninges. There is characteristic edema of the involved facial areas, necrosis, and a bloody exudate. The disease can be precipitous, with a terminal outcome in 1 week. Damage to the fifth and seventh cranial nerves, orbital cellulitis, and exophthalmia are frequent manifestations.

THORACIC MUCORMYCOSIS. This form begins with primary involvement in the lung following inhalation of the sporangiospores. The pathology caused by invasion of blood vessels results in profound destruction of lung parenchyma. Pulmonary lesions may be focal or diffuse. The usual course is 1 to 4 weeks from onset to death.

OTHER FORMS. Localized mucormycosis has been described in the kidney following tissue trauma. Cutaneous infection may complicate burn wounds. The application of contaminated bandages for surgical dressings has caused nosocomial infection of the skin.

Laboratory Diagnosis. In specimens of tissue, sputum, or nasal exudate, the hyphae are broad and irregular in width (10 to 15 μm) and exhibit branching, often at right angles. Hyphae are distorted and usually nonseptate (Fig. 87-8).

In systemic cases, blood cultures are invariably negative. Specimens of tissue or drainage should be cultured. If severe disease is not apparent, positive cultures may only be contaminants.

An immunodiffusion test has been developed by Jones and Kaufman (see Table 87-5).

Treatment. The mortality of invasive mucormycosis is approximately 90 percent. When diagnosis is established antemortem, amphotericin B, and surgical debridement are recommended.

Geotrichosis

Geotrichosis is caused by *Geotrichum candidum,* a fungus found saprophytically in nature and as a commensal in the mouth, gastrointestinal tract, and genitourinary tract. The

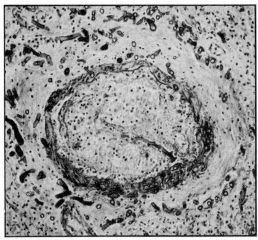

Figure 87-8. Mucormycosis. Hyphae in wall and lumen of blood vessel and in pulmonary alveoli. × 150. *(From Conant, et al.: Manual of Clinical Mycology, 3rd ed, 1971. Philadelphia, Saunders.)*

clinical forms of this rare infection are pulmonary geotrichosis, which mimics tuberculosis in its manifestations, and superficial infections of the skin, oral mucosa, and gastrointestinal tract. Diagnosis is made by culture (Table 87-1). Reported treatments include aerosolized nystatin, iodides, and amphotericin B.

Other Opportunistic Mycoses

Many individuals with compromised host defenses are susceptible to infections by other fungi. Many of the thousands of ubiquitous saprophytic fungi have caused rare infections. Such mycoses occur much less frequently than opportunistic candidiasis, aspergillosis, and mucormycosis because these rare etiologic agents have a low pathogenetic potential due to their weak virulence or high susceptibility to host defenses. In some cases, exposure may be limited by environmental factors.

Humans and other animals are constantly exposed to the hundreds of species of *Penicillium* that abound in nature. Pulmonary and systemic penicillosis has been reported sporadically. Species of another common saprophyte, *Fusarium*, have infected burn patients. Systemic infections have occurred in transplant and cancer patients. The appearances of *Penicillium, Fusarium, Aspergillus,* and other molds in tissue are indistinguishable, and cultures must be observed to establish etiology. In the patient with severely compromised host defenses, almost any fungus may become an opportunistic pathogen. To date, McGinnis has documented nearly 300 species of fungi as agents of human disease.

FURTHER READING

CANDIDIASIS

Aisner J, Schimpff SC, Sutherland JC, et al.: *Torulopsis glabrata* infections in patients with cancer. Am J Med 61:23, 1976

Araj GF, Hopfer RL, Chesnut S, et al.: Diagnostic value of the enzyme-linked immunosorbent assay for detection of *Candida albicans* cytoplasmic antigen in sera of cancer patients. J Clin Microbiol 16:46, 1982

Auger P, Dumas C, July J: A study of 666 strains of *Candida albicans*: Correlation between serotype and susceptibility to 5-fluorocytosine. J Infect Dis 139:590, 1979

Bayer AS, Blumenkrantz MJ, Montgomerie JZ, et al.: Candida peritonitis. Report of 22 cases and review of the English literature. Am J Med 61:832, 1976

Caroline L, Rosner F, Kozinn PJ: Elevated serum iron, low unbound transferrin and candidiasis in acute leukemia. Blood 34:441, 1969

Cassone A, Marconi P, Bistoni F, et al.: Immunoadjuvant effects of *Candida albicans* and its cell wall fractions in a mouse lymphoma model. Cancer Immol Immunother 10:181, 1981

Cawson RA, Rajasingham KC: Ultrastructural features of the invasive phase of *Candida albicans.* Br J Dermatol 87:435, 1972

Cutler JE: Acute systemic candidiasis in normal and congenitally thymic-deficient (nude) mice. J Reticuloendothel Soc 19:121, 1976

Cutler JE, Friedman L, Milner KC: Biological and chemical characterization of toxic substances from *Candida albicans.* Infect Immun 6:616, 1972

Cutler JE, Lloyd RA: Enhanced antibody responses induced by *Candida albicans* in mice. Infect Immun 38:1102, 1982

Dee TH, Rytel MW: Clinical application of counterimmunoelectrophoresis in detection of *Candida* serum precipitins. J Lab Clin Med 85:161, 1975

Diamond RD, Krzesicki R, Jao W: Damage to pseudohyphal forms of *Candida albicans* by neutrophils in the absence of serum in vitro. J Clin Invest 61:349, 1978

Drazin RE, Lehrer RI: Fungicidal properties of a chymotrypsinlike cationic protein from human neutrophils: Adsorption to *Candida parapsilosis.* Infect Immun 17:382, 1977

Edwards JE, Lehrer RI, Stiehm ER, et al.: Severe candidal infections. Clinical perspective, immune defense mechanisms, and current concepts of therapy. Ann Intern Med 89:91, 1978

Glew RH, Buckley HR, Rosen HM, et al.: Serologic tests in the diagnosis of systemic candidiasis. Enhanced diagnostic accuracy with crossed immunoelectrophoresis. Am J Med 64:586, 1978

Gold JWM, Wong B, Bernard BM, et al.: Serum arabinitol concentrations and arabinitol/creatinine ratios in invasive candidiasis. J Infect Dis 147:504, 1983

Greenfield RA, Jones JM: Purification and characterization of a major cytoplasmic antigen of *Candida albicans.* Infect Immun 34:469, 1981

Gresham GA, Whittle CH: Studies on the invasive, mycelial form of *Candida albicans.* Sabouraudia 1:30, 1961

Harding SA, Brody JP, Normansell DE. Antigenemia detected by enzyme-linked immunosorbent assay in rabbits with systemic candidiasis. J Lab Clin Med 95:959, 1980

Higgs JM, Wells RS: Chronic mucocutaneous candidiasis: associated abnormalities or iron metabolism. Br J Dermatol 86:88, 1972

Hughes WT: Systemic candidiasis: A study of 109 fatal cases. Pediatr Infect Dis 1:11, 1982

Jorizzo JL, Sams WM, Jegasothy BV, et al.: Cimetidine as an immunomodulator: Chronic mucocutaneous candidiasis as a model. Ann Intern Med 92:192, 1980

Kauffman CA, Tan JS: *Torulopsis glabrata* renal infection. Am J Med 57:217, 1974

Keller MA, Sellers BB, Melish ME, et al.: Systemic candidiasis in infants—Case presentation and literature review. Am J Dis Child 131:1260, 1977

Kimura LH, Pearsall NN: Relationship between germination of *Candida albicans* and increased adherence to human buccal epithelial cells. Infect Immun 28:464, 1980

Kirkpatrick CH, Rich RR, Bennett JE: Chronic mucocutaneous candidiasis: Model building in cellular immunity. Ann Intern Med 74:955, 1971

Kirkpatrick CH, Windhorst DB: Mucocutaneous candidiasis and thymoma. Am J Med 66:939, 1979

Lehmann PF, Reiss E: Detection of *Candida albicans* mannan by immunodiffusion, counterimmunoelectrophoresis, and enzyme-linked immunoassay. Mycopathologia 70:83, 1980

Lehrer RI, Cline MJ: Interaction of *Candida albicans* with human leukocytes and serum. J Bacteriol 98:996, 1969

Maisch PA, Calderone RA: Role of surface mannan in the adherence of *Candida albicans* to fibrin-platelet clots formed in vitro. Infect Immun 32:92, 1981

Manning M, Mitchell TG: Analysis of cytoplasmic antigens of the yeast and mycelial phases of *Candida albicans* by two-dimensional electrophoresis. Infect Immun 30:484, 1980

Manning M, Mitchell TG: Morphogenesis of *Candida albicans* and cytoplasmic proteins associated with differences in morphology, strain, or temperature. J Bacteriol 144:258, 1980

Meunier-Carpentier F: Significance and clinical manifestations of fungemia. In Klastersky J (ed): Infection in Cancer Patients. New York, Raven, 1982

Montes LF, Krumdieck C, Cornwell PE: Hypovitaminosis A in patients with mucocutaneous candidiasis. J Infect Dis 128:227, 1973

Munoz M, Estes G, Kilpatrick M, et al.: Purification of cytoplasmic antigens from the mycelial phase of *Candida albicans*: Possible advantages of its use in candida serology. Mycopathologia 72:45, 1980

Odds FC: *Candida* and Candidiasis. Baltimore, University Park Press, 1979

Olson VL, Hansing RL, McClary DO: The role of metabolic energy in the lethal action of basic proteins on *Candida albicans*. Can J Microbiol 23:166, 1977

Petersen EA, Alling DW, Kirkpatrick CH: Treatment of chronic mucocutaneous candidiasis with ketoconazole. A controlled clinical trial. Ann Intern Med 93:791, 1980

Piccolella E, Lombardi G, Morelli R: Generation of suppressor cells in the response of human lymphocytes to a polysaccharide from *Candida albicans*. J Immunol 126:2151, 1981

Pitchenik AE, Fischl MA, Dickinson GM, et al.: Opportunistic infections and Kaposi's sarcoma among Haitians: Evidence of a new acquired immunodeficiency state. Ann Intern Med 98:277, 1983

Poor AH, Cutler JE: Partially purified antibodies used in a solid-phase radioimmunoassay for detecting candidal antigenemia. J Clin Microbiol 9:362, 1979

Rogers TJ, Balish E: Immunity to *Candida albicans*. Microbiol Rev 44:660, 1980

Scheld WM, Brown RS, Harding SA, et al.: Detection of circulating antigen in experimental *Candida albicans* endocarditis by an enzyme-linked immunosorbent assay. J Clin Microbiol 12:679, 1980

Scherwitz C: Ultrastructure of human cutaneous candidosis. J Invest Dermatol 78:200, 1982

Segel E, Berg RA, Pizzo PA, et al.: Detection of *Candida* antigen in sera of patients with candidiasis by an enzyme-linked immunosorbent assay-inhibition technique. J Clin Microbiol 10:116, 1979

Silva-Hutner M, Cooper BH: Medically important yeast. In Lennette EH, Spaulding EH, Truant JP (eds): Manual of Clinical Microbiology, 3rd ed. Washington, D.C., American Society for Microbiology, 1980, p 562

Simonetti N, Strippoli V: Pathogenicity of the Y form as compared to the M form in experimentally induced *Candida albicans* infections. Mycopathologia 51:19, 1973

Smith JK, Louria DB: Anti-*Candida* factors in serum and their inhibitors. II. Identification of a *Candida*-clumping factor and the influence of the immune response on the morphology of *Candida* and on anti-*Candida* activity of serum in rabbits. J Infect Dis 125:115, 1972

Solomkin JS, Flohr A, Simmons RL: *Candida* infections in surgical patients. Dose requirements and toxicity of amphoterin B. Ann Surg 195:177, 1982

Stevens P, Huang S, Young LS, et al.: Detection of *Candida* antigenemia in human invasive candidiasis by a new solid phase radioimmunoassay. Infection (Suppl)8:334, 1980

Syverson RE, Buckley HR, Gibian JR: Increasing the predictive value positive of the precipitin test for the diagnosis of deep-seated candidiasis. Am J Clin Pathol 70:826, 1978

Taschdjian CL, Seelig MS, Kozinn PJ: Serological diagnosis of candidal infections. CRC Crit Rev Clin Lab Sci 4:19, 1973

Valdimarsson H, Higgs JM, Wells AS, et al.: Immune abnormalities associated with chronic mucocutaneous candidiasis. Cell Immunol 6:348, 1973

Valdivieso M, Luna M, Bodey G, et al.: Fungemia due to *Torulopsis glabrata* in the compromised host. Cancer 38:1750, 1976

Wilton JMA, Lehner T: Immunology of candidiasis. In Nahmias AJ, O'Reilly RJ (eds): Immunology of Human Infections. New York, Plenum, 1981, p 525

Yamamura M, Valdimarsson H: Participation of C3 in intracellular killing of *Candida albicans*. Scand J Immunol 6:591, 1977

ASPERGILLOSIS

Aisner J, Schimpff SC, Wiernik PH: Treatment of invasive aspergillosis: relation of early diagnosis and treatment to response. Ann Intern Med 86:539, 1977

Aisner J, Schimpff SC, Bennett JE, et al.: *Aspergillus* infections in cancer patients. Association with fireproofing material in a new hospital. JAMA 235:411, 1976

Andrews CP, Weiner MN: Immunodiagnosis of invasive pulmonary aspergillosis in rabbits. Am Rev Respir Dis 124:60, 1981

Atkinson GW, Israel HL: 5-Fluorocytosine treatment of meningeal and pulmonary aspergillosis. Am J Med 55:496, 1973

Bardana EJ, Sobti KL, Cianciulli FD, et al.: Aspergillus antibody in patients with cystic fibrosis. Am J Dis Child 129:1164, 1975

Binder RE, Faling LJ, Pugatch RD, et al.: Chronic necrotizing pulmonary aspergillosis: A discrete clinical entity. Medicine 61:109, 1982

British Thoracic and Tuberculosis Association. Aspergilloma and residual tuberculous cavities—The results of a resurvey. Tubercle 51:227, 1970

Charpin J, Ohresser PL, Boutin C, et al.: Aspergillose pulmonaire allergique. J Fr Med Chir Thor 21:517, 1967

Citron KM: Respiratory fungus allergy and infection. Proc R Soc Med 68:587, 1975

de Haller R, Suter F (eds): Aspergillosis and Farmer's Lung in Man and Animal. Bern, Hans Huber, 1974

Diamond RD, Clark RA: Damage to *Aspergillus fumigatus* and *Rhizopus oryzae* hyphae by oxidative and nonoxidative microbicidal products of human neutrophils in vitro. Infect Immun 38:487, 1982

Diamond RD, Huber E, Haudenschild CC: Mechanisms of destruction of *Aspergillus fumigatus* hyphae mediated by human monocytes. J Infect Dis 147:474, 1983

Gallant SP, Rucker RW, Groncy CE, et al.: Incidence of serum antibodies to several *Aspergillus* species and to *Candida albicans* in cystic fibrosis. Am Rev Respir Dis 114:325, 1976

Gerber JD, Jones RD: Immunologic significance of aspergillin antigens of six species of *Aspergillus* in the serodiagnosis of aspergillosis. Am Rev Respir Dis 108:1124, 1973

Golbert TM, Patterson R: Pulmonary allergic aspergillosis. Ann Intern Med 72:395, 1970

Haley LD: Etiology of otomycology. I. Mycologic flora of the ear. II. Bacterial flora of the ear. III. Observations on attempts to produce otomycosis in rabbits. IV. Clinical observations. Arch Otolaryngol 52:202, 1952

Haslam P, Lukoszek A, Longbottom JL, et al.: Lymphocyte sensitization to *Aspergillus fumigatus* antigens in pulmonary diseases in man. Clin Allergy 6:277, 1976

Khan ZU, Sandhu RS, Randhawa HS, et al.: Allergic bronchopulmonary aspergillosis: A study of 46 cases with special reference to laboratory aspects. Scand J Respir Dis 57:73, 1976

Malo JL, Hawkins R, Pepys J: Studies in chronic allergic bronchopulmonary aspergillosis. 1. Clinical and physiological findings. Thorax 32:254, 1977

Malo JL, Longbottom J, Mitchell J, et al.: Studies in chronic allergic bronchopulmonary aspergillosis. 3. Immunological findings. Thorax 32:269, 1977

Meyer RD, Young LS, Armstrong D, et al.: Aspergillosis complicating neoplastic disease. Am J Med 54:6, 1973

Nalesnik MA, Myerowitz RL, Jenkins R, et al.: Significance of *Aspergillus* species isolated from respiratory secretions in the diagnosis of invasive pulmonary aspergillosis. J Clin Microbiol 11:370, 1980

Patterson R, Fink JN, Pruzansky JJ, et al.: Serum immunoglobulin levels in pulmonary allergic aspergillosis and certain other lung diseases with special reference to immunoglobulin E. Am J Med 54:16, 1973

Pauwels R, Stevens EM, vander Straeten M: IgE antibodies in bronchopulmonary aspergillosis. Ann Allergy 37:195, 1976

Pepys J, Hutchcroft BJ: Bronchial provocation tests in etiologic diagnosis and analysis of asthma. Am Rev Respir Dis 112:829, 1975

Raper KB, Fennel DI: The Genus *Aspergillus.* Baltimore, Williams & Wilkins, 1965

Safirstein BH, D'Souza MF, Simon G, et al.: Five-year follow-up of allergic bronchopulmonary aspergillosis. Am Rev Respir Dis 108:450, 1973

Schaefer JC, Yu B, Armstrong D: An aspergillus immunodiffusion test in the early diagnosis of aspergillosis in adult leukemia patients. Am Rev Respir Dis 113:325, 1976

Schaffner A, Douglas H, Braude A: Selective protection against conidia by mononuclear and against mycelia by polymorphonuclear phagocytes in resistance to *Aspergillus*: Observations on these two lines of defense in vivo and in vitro with human and mouse phagocytes. J Clin Invest 69:617, 1982

Sidransky H, Friedman L: The effect of cortisone and antibiotic agents on experimental pulmonary aspergillosis. Am J Pathol 35:165, 1959

Slavin RG, Million L, Cherry J: Allergic bronchopulmonary aspergillosis: Characterization of antibodies and results of treatment. J Allergy 46:150, 1970

Soltanzadeh H, Wychulis AR, Fauraokh S, et al.: Surgical treatment of pulmonary aspergilloma. Ann Surg 186:13, 1977

Turner KJ, O'Mahony J, Wetherall JD, et al.: Hypersensitivity studies in asthmatic patients with bronchopulmonary aspergillosis. Clin Allergy 2:361, 1972

Varkey B, Rose HD: Pulmonary aspergilloma. A rational approach to treatment. Am J Med 61:626, 1976

Yocum MW, Saltzman AR, Strong DM, et al.: Extrinsic allergic alveolitis after *Aspergillus fumigatus* inhalation. Evidence of a type IV immunologic pathogenesis. Am J Med 61:939, 1976

Young RC, Bennett JE: Invasive aspergillosis. Absence of detectable antibody response. Am Rev Respir Dis 104:710, 1971

Young RC, Bennett JE, Vogel CL, et al.: Aspergillosis. The spectrum of the disease in 98 patients. Medicine 49:147, 1970

MUCORMYCOSIS

Abramson E, Wilson D, Arby RA: Rhinocerebral phycomycosis in association with diabetic ketoacidosis. Ann Intern Med 66:735, 1967

Carbel MJ, Eades SM: Experimental phycomycosis in mice: Examination of the role of acquired immunity in resistance to *Absidia ramosa.* J Hyg 77:221, 1976

Fisher J, Tuazon CU, Geelhoed GW: Mucormycosis in transplant patients. Am Surg 46:315, 1980

Jones KW, Kaufman L: Development and evaluation of an immunodiffusion test for diagnosis of systemic zygomycosis (mucormycosis): Preliminary report. J Clin Microbiol 7:97, 1978

Keys TF, Haldorson RN, Rhodes KM, et al.: Nosocomial outbreak of *Rhizopus* infections associated with Elastoplast wound dressings—Minnesota. MMWR 27:33, 1978

Kurrasch M, Beumer J, Kagawa T: Mucormycosis: Oral and prosthodontic implications. A report of 14 patients. J Prosthetic Dent 47:422, 1982

Langston C, Roberts DA, Porter GA, et al.: Renal phycomycosis. J Urol 109:491, 1973

Lehrer RI, Howard DH, Sypherd RS, et al.: Mucormycosis. Ann Intern Med 93:93, 1980

Marchevsky AM, Bolton EJ, Geller SA, et al.: The changing spectrum of disease, etiology, and diagnosis of mucormycosis. Hum Pathol 11:457, 1980

Meyer RD, Rosen P, Armstrong D: Phycomycosis complicating leukemia and lymphoma. Ann Intern Med 77:871, 1972

Pillsbury HC, Fischer ND: Rhinocerebral mucormycosis. Arch Otolaryngol 103:600, 1977

Reich J, Renzetti AD: Pulmonary phycomycosis. Am Rev Respir Dis 102:959, 1970

Straatsma BR, Zimmerman LE, Gass JD: Phycomycosis. Lab Invest 11:963, 1962

GEOTRICHOSIS AND OTHER OPPORTUNISTIC MYCOSES

DiSalvo A, Fickling AM, Ajello L: *Penicillium marneffei* infection in man. Description of first natural infection. Am J Clin Pathol 60:259, 1973

Fishback RS, White ML, Finegold S: Bronchopulmonary geotrichosis. Am Rev Respir Dis 108:1388, 1973

Jaqirdar J, Geller SA, Bottone EJ: *Geotrichum candidum* as a tissue invasive human pathogen. Hum Pathol 12:668, 1981

Krick JA, Remington JS: Opportunistic invasive fungal infections in patients with leukemia and lymphoma. Clin Haematol 5:249, 1976

McGinnis MR: Laboratory Handbook of Medical Mycology. New York, Academic Press, 1980

Sheehy TW, Honeycutt BK, Spencer T: *Geotrichum* septicemia. JAMA 235:1035, 1976

Wheeler MS, McGinnis MR: *Fusarium* infection in burned patients. Am J Clin Pathol 75:304, 1981

Young CN, Meyers AM: Opportunistic fungal infection by *Fusarium oxysporum* in renal transplant patients. Sabouraudia 17:219, 1979

SECTION VII
MEDICAL PARASITOLOGY

CHAPTER 88

Introduction to Medical Parasitology

Introduction
Definitions

Introduction

Many clinicians in the United States and other developed countries fail to realize the impact of diseases caused by parasitic protozoa and worms on human populations. This is due largely to the fact that people living in developed countries are reasonably well nourished, have high standards of sanitation, benefit from a temperate climate, and enjoy the absence of certain vectors of disease. On a global basis, however, and even after relatively successful campaigns against such infections as malaria, hookworm, and blood fluke in many parts of the world, parasitic infection in association with malnutrition is still the primary cause of morbidity and mortality. In fact, 15 million, or one half of the 30 million children under 5 years of age who die from all causes each year, die from a combination of malnutrition and parasitic infection. These 15 million thus represent one fourth of the total 60 million annual worldwide adult and child deaths from all causes. Recent summaries of worldwide prevalence of selected infections show the following:

Infectious Agent	Number Infected	
Ascaris	1.25	billion
Hookworm	900	million
Trichuris	700	million
Filaria	650	million
Schistosoma	275	million
Malaria	300	million

Although parasite-induced diseases are relatively uncommon, they do exist in the United States and are still far from being eradicated. Some estimates place the number of people infected with parasites in the United States at over 50 million. Most of these infections were acquired in the United States, but some, like schistosomiasis and

malaria, are constantly being imported by travelers and immigrants.

The recent increased interest in medical parasitology is due to a number of factors, including intensified travel to endemic areas, the "back-to-nature" trend within the United States, the development of drug resistance by some parasites, and the wider use of immunosuppressive therapy, which permits the exacerbation of chronic, low-level infection and a general decrease in resistance to infection. Case histories of deaths in the United States due to fulminating strongyloidiasis after kidney transplantation, due to pneumocystosis during immunosuppressive therapy for leukemia, and due to drug-resistant malaria are no longer reported only as medical curiosities in American medical journals. One also reads case histories of homosexual men dying due, somehow, to the association between pneumocystosis and Kaposi's sarcoma, as well as of rapid deaths due to brain damage caused by a normally free-living ameba found in fresh water swimming holes, lakes, and pools.

It is incumbent upon the medical profession to help dispel the notion that people in the United States are free of infection caused by protozoa and worms. This illusion results from a variety of reasons including: (1) the topic is rarely discussed by "refined" people, (2) the media are reluctant to disseminate such information, (3) poor people are usually the ones most seriously affected, and (4) travel agents and tourist bureaus are both reluctant and inadequately prepared to give travelers an accurate picture of the health hazards that may be encountered abroad. Also, many physicians are inadequately trained and fail to consider parasites as a possible cause of symptoms and disease.

Parasitology is the science or study of parasitism—

1201

that is, the relationship between parasites and the organisms (hosts) that harbor them. In the fields of medicine and public health, the subject matter of parasitology is usually limited to parasitism in humans. In a broad sense, parasitism involves the study of all organisms parasitic on or within the human body. This comprehensive consideration includes five specialized fields: medical bacteriology, virology, mycology, entomology, and parasitology. In medical parasitology, the most important parasites of humans belong to four groups: the Protozoa (eucaryotic protists, i.e., single-celled microorganisms with a true, membrane-bound nucleus); the Nematoda, or true roundworms; the Platyhelminthes, or flatworms; and the Arthropoda, which include not only the true insects, but ticks, mites, and others. Although arthropods affect the health of humans in many ways, they are considered here only where they serve as vectors.

Furthermore, only the more common parasites of the four groups selected are discussed, especially those endemic or indigenous (i.e., present and being transmitted at all times) within the United States. However, space is also given to selected exotic or imported (i.e., of foreign origin and not being transmitted in the United States) forms because of their ever-increasing prevalence, especially in travelers to, and immigrants from, endemic areas. Two forms discussed—malarial parasites and blood flukes—produce diseases now considered to be the most important infectious diseases in the world. In this global context, it is worth noting that along with the diseases produced by these two agents, there are three other parasitic diseases (trypanosomiasis, leishmaniasis, and filariasis, including onchocerciasis) singled out by the World Health Organization and the United Nations Development Program as initial target areas in a special program aimed at the acceleration of health and economic betterment in underdeveloped countries. Leprosy is the remaining disease chosen for initial attack.

In presenting the material of this section, an attempt is made to coordinate the medical and biologic approaches to the subject; that is, to give proper attention to both the host and the parasite. The life cycle of the parasite is given, followed by a brief consideration of the principal damage produced in the host, and the striking signs and symptoms resulting therefrom. After these discussions for each infection, information on the epidemiology, diagnosis, and treatment is given. By relating the important host–parasite relationships, rather than emphasizing certain factors concerned only with the host or the parasite, the reader is better able to gain an appreciation and understanding of these infections. In following this approach, however, the space customarily given to details of classification, morphology of the parasite, pathology, and so forth is necessarily limited.

Definitions

Before presenting the information outlined above, a few definitions will help the reader's understanding of the sections on protozoan and helminthic infections that follow.

Most medically oriented parasitologists consider *symbiology* to be a study of the relationship that exists between two dissimilar organisms that live together in close association. When both of these partners benefit in the association, it is spoken of as *mutualism.* When one partner benefits and the other partner neither benefits nor is harmed, the association is known as *commensalism.* However, when one partner harms the other, or in some sense lives at the expense of the other without killing it immediately, the association is known as *parasitism.* In a mutualistic relationship, both partners are called mutuals; in a commensalistic relationship, the partner that benefits is called a commensal and the other the host. In a parasitic relationship, the harmful agent is the parasite and the partner that is harmed is the host.

A parasite that lives on the surface of its host is called an *ectoparasite;* if it lives in the tissues or gastrointestinal tract, it is an *endoparasite.* A parasite that must spend all or part of its life in or on its host is called an *obligatory parasite.* An organism that is not normally parasitic but does become so when, for example, it is accidentally eaten or enters a wound or natural body orifice, is called a *facultative parasite.*

Hosts are also of different kinds. A *final* or *definitive host* is one in which the parasite reaches sexual maturity. If no sexual reproduction occurs in the life cycle of the parasite, such as with many protozoa, the organism arbitrarily deemed the most important is called the final or definitive host. An *intermediate host* is one in which some development of the parasite occurs, but sexual maturity is not reached. Finally, a *reservoir host* is an animal other than a human that is normally infected with a parasite that is also infective for humans. When a parasite can develop only in one or two species of hosts, it is spoken of as manifesting a high degree of *host specificity.* A parasite, such as that causing trichinosis, can infect almost any warm-blooded vertebrate and is spoken of as exhibiting loose host specificity.

FURTHER READING

There are many excellent references dealing with all phases of medical parasitology. The few listed below are considered among the best.

Binford CH, Connor DH (eds): Pathology of Tropical and Extraordinary Diseases: An Atlas. Washington, D.C., Armed Forces Institute of Pathology, 1976, vols 1 and 2

Brown HW, Neva FA: Basic Clinical Parasitology, 5th ed. New York, Appleton-Century-Crofts, 1982

Faust EC, Beaver PC, Jung RC: Animal Agents and Vectors of Human Disease, 4th ed. Philadelphia, Lea & Febiger, 1975

Faust EC, Russell PF, Jung RC: Craig and Faust's Clinical Parasitology, 8th ed. Philadelphia, Lea & Febiger, 1970

Hunter GW III, Swartzwelder JC, Clyde DF: Tropical Medicine, 5th ed. Philadelphia, Saunders, 1976

Manson-Bahr PEC, Apted FIC: Manson's Tropical Diseases, 18th
ed. London, Balliere Tindall, 1982

Peters W, Gilles HM: Color Atlas of Tropical Medicine and Parasitology. Chicago, Year Book, 1977

Yamaguchi T: Color Atlas of Clinical Parasitology. Philadelphia,
Lea & Febiger, 1981

Zaman V: Atlas of Medical Parasitology. Philadelphia, Lea &
Febiger, 1979

CHAPTER 89

Medical Protozoology

Protozoa are unicellular organisms composed of a nucleus, or nuclei, and cytoplasm. The cytoplasm is differentiated into an outer layer (the ectoplasm) and an inner layer (the endoplasm). Locomotion, if accomplished, is carried out by special ectoplasmic organelles. Four groups of Protozoa contain important parasites of humans: (1) Rhizopoda (amebae), (2) Mastigophora (flagellates), (3) Ciliatea (ciliates), and (4) Sporozoa. The amebae locomote by means of pseudopodia, the flagellates by flagella, the ciliates by cilia, and the sporozoans lack definite organelles of locomotion.

In the presentation of the material of this chapter, the protozoa are grouped according to their usual location within the body of man. The groups to be considered are the intestinal, urogenital, blood, and tissue protozoa.

Intestinal Protozoa

Amebae

Amebae are common in the environment and many are parasitic in invertebrate and vertebrate animals. Relatively few species parasitize humans (Table 89-1), and only two, *Entamoeba histolytica* and *Dientamoeba fragilis*, are known to cause intestinal pathology. *D. fragilis* is considered by many to be a flagellate, but because of its morphologic similarity to the amebae and its lack of flagella, it will be, for convenience, presented as an ameba.

Entamoeba histolytica

This ameba, although often living as a harmless inhabitant in the large intestine, has the capacity to colonize on and penetrate the intestinal tissues and to cause ulceration. After it penetrates the intestinal tissues, the ameba may spread by metastasis to other tissues and organs.

It is estimated that 10 percent of all people in the world are infected with *E. histolytica*; in the United States, 1 to 3 percent of the people are thought to harbor the ameba. Prevalence rates are highest in tropical areas, especially where crowding and poor sanitation exist. However, infection is cosmopolitan, being present literally from pole to pole. Fortunately, clinical disease is seen in only a small fraction of those infected. There is increasing evidence for the existence of a variety of strains and this may partially account for differing pathogenicity.

Life Cycle. The life cycle of *E. histolytica* is comparatively simple, involving the trophozoite (actively metabolizing and motile) and cyst stages (Fig. 89-1). The parasite is passed in the stools of infected individuals. If the specimen is dysenteric or diarrheic or has been obtained following purgation, trophozoites will predominate. Few cysts are observed in such specimens because, presumably, evacuation is too rapid for encystment. Therefore, cysts usually are found only in normal stools that are formed, at least in part. Cysts will remain viable only if kept moist and if other favorable conditions exist because they are readily destroyed by desiccation, sunlight, and heat.

Reinfection of the same individual or infection of a new host may occur directly via contaminated fingers

TABLE 89-1. INTESTINAL AMEBAE OF HUMANS

Name	Transmission	Pathology	Size (μm)		Remarks
			Trophozoite	Cyst	
Dientamoeba fragilis	Ingestion of trophozoite (?)	Diarrhea	5–15	None	May be transmitted inside pinworm eggs
Endolimax nana	Ingestion of cyst	None	6–15	4–14	
Entamoeba coli	Ingestion of cyst	None	10–50	10–35	
Entamoeba hartmanni	Ingestion of cyst	None	5–12	4–10	Some consider this as a small race of *E. histolytica*
Entamoeba histolytica	Ingestion of cyst; anal intercourse	Diarrhea, dysentery, abscess in extra-intestinal tissues (e.g., liver, brain)	12–60	10–20	Most often lives as a harmless commensal
Entamoeba polecki	Ingestion of cyst	None	10–20	5–10	Rare in humans; probably of animal origin
Iodamoeba bütschlii	Ingestion of cyst	None	6–25	6–20	

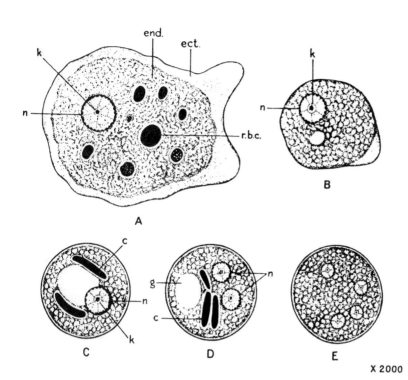

Figure 89-1. Schematic representation of *Entamoeba histolytica*. **A.** Trophozoite containing red blood cells undergoing digestion. **B.** Precystic ameba devoid of cytoplasmic inclusions. **C.** Young uninucleate cyst. **D.** Binucleate cyst. **E.** Mature quadrinucleate cyst. c, chromatoid bodies; ect., ectoplasm; end., endoplasm; g, glycogen vacuole; k, karyosome; n, nucleus; r.b.c., red blood cells. *(From Belding: Textbook of Clinical Parasitology, 2nd ed., 1952. New York, Appleton-Century-Crofts.)*

X 2000

(hand-to-mouth infection) or indirectly chiefly via contaminated food and water. Waterborne epidemics have occurred in the United States and elsewhere. Sexual transmission via anal intercourse, especially within the male population, is well known. Contaminated apparatus used in colonic irrigation is known to be a mode of transmission.

Ingested cysts pass through the stomach and excyst in the lower small intestine. An ameba containing the four cystic nuclei emerges from each cyst. Cytoplasmic divisions take place to form four small trophozoites, which feed and soon grow to full size. Hence, cysts provide for reproduction as well as transmission.

The trophozoites pass along the intestinal canal until conditions are favorable for colonization, which occurs as the result of rapid and repeated transverse binary fission. This usually occurs first at the cecal area but may take place at a lower level of the large intestine. Tissue invasion is probably accomplished by both lytic and physical means. After the trophozoites enter the tissue, lytic digestion of host cells provides food for the ameba, allowing it to advance. Under certain conditions, some of the trophozoites may metastasize to the liver and other extraintestinal sites. When trophozoites are extruded into the intestinal lumen, they begin their passage from the body. If this passage is not too rapid, they pass first through a precystic stage and occur in the stool as cysts.

It should be added that many are of the opinion that a small, nonpathogenic strain of *E. histolytica* exists. Others believe it to be a separate species and have named it *Entamoeba hartmanni*. However, it is known that geographic strains vary in pathogenicity, and that the bacterial flora of the intestine, the resistance, and the nutritional status of the host all are factors in determining whether demonstrable pathology results from a given infection. Thus, it would seem wise, until final proof has been established, to consider all strains capable of damaging the host.

Pathology and Symptomatology. Clinically, amebiasis usually presents itself as an intestinal disease or as an amebic abscess of the liver. The incubation period, after the ingestion of cysts, is generally 3 to 4 weeks in duration but may be as short as 5 days or as long as several months. The severity of the disease varies, but the majority of those infected are asymptomatic cyst passers (carriers).

The intestinal pathology of amebiasis is most commonly observed in the cecal area, followed by the sigmoidorectal area, but lesions may occur at any point from the lower ileum downward. The process begins with a small lesion produced at the site of entry into the tissue. A minute cavity is formed from lytic necrosis. As viewed from the surface, these small lesions are surrounded by a raised yellowish ring and are separated by mucosa that appears normal. These lesions usually show no evidence of inflammatory reaction, hence they fail to suggest the degree of subsurface damage, which may be considerable. As the colony of amebae increases, abscesses are formed, and often, in time, a narrow channel is formed to the base of the mucosa. There, due probably to the greater resistance of the muscularis mucosae, the lesion enlarges and forms an early, characteristic, flask-shaped or teardrop ulcer. This appears to be a critical stage, determining whether or not extensive damage will be produced, and it may, therefore, be thought of as an expression of the degree of adaptation between the parasite and the host. If the organisms are unable to penetrate this layer, repair may keep pace with the damage, and in some cases the amebae may be eliminated. If, however, the organisms erode a passage through the muscularis mucosae into the submucosa, they usually spread out radially and produce an enlarged, ragged ulcer. The mucosa surrounding this is rolled and elevated due to the undermining. Secondary infections usually are not observed in these early lesions, despite extensive necrosis, and there is essentially no tissue reaction present. Later, the surface of the ulcer sloughs off, exposing shaggy, overhanging edges. At this time, secondary infections are common, and the ulcer is infiltrated with neutrophilic leukocytes and other wandering cells, tending to thicken the overhanging edges. In severe cases, the organisms may spread from the submucosa through the muscular coats into the serosa, where they are likely to cause perforation. Amebomas—granulomatous formations that resemble carcinomas—are sometimes seen in chronic cases. The differentiation between an ameboma and cancer is difficult. Most cases are not correctly identified until after surgery.

From the intestinal wall, the amebae may enter the portal venules or, less commonly, the lymphatics and be carried to the liver and other organs. Liver involvement is most common, followed by involvement of the lungs, which usually is by direct extension from the liver abscess. Lesions outside the intestinal tract are always secondary to those in the intestine. It is important to note that immunosuppressive therapy may cause rapid progression of an *E. histolytica* infection, with resulting liver abscess and increased intestinal damage. As will be noted later, other parasites as well may produce fulminating infections in patients being treated with various immunosuppressants.

The pathology of amebic liver abscess results from the establishment and multiplication of trophozoites in that organ. The early abscess is small, with a grayish brown matrix of necrosed hepatic cells. Connective tissue does not appear to be destroyed by the lytic property of the amebae. As the abscess increases in size, the center liquefies, the wall thickens, and in most cases the contents become a viscid (creamy) yellow, gray, or chocolate-colored mass. At all stages of abscess formation, the amebae are seen invading marginal tissue. Abscesses occur more frequently in the right lobe (right upper quadrant) than in the left and tend to be multiple rather than solitary. It is important to emphasize that liver abscess may be a sequela of chronic intestinal amebiasis as well as of the acute type (amebic dysentery). In fact, liver abscess without evidence of intestinal involvement and without organisms being seen in the stool is not unusual.

Cutaneous lesions and abscesses also occur. They usually result from an extension of rectal amebiasis to perianal skin and vulva. Abdominal wall skin lesions also occur as extensions of a hepatic abscess, and penile skin infections as a result of anal intercourse are not rare.

Epidemiology. The life cycle outlined above is simple and does not involve intermediate hosts or nonhuman reservoirs. Cysts may remain viable for days if mild environmental conditions prevail. Fecal contamination of food and water occurs due to poor sewage disposal and personal hygiene. Laboratory observations indicate that a rapid transmission of parasites from human to human enhances pathogenicity, and this may partly account for the minor clinical significance of amebiasis in the United States. However, recent reports incriminate oral–anal sex practices for its alarming resurgence among the homosexual population.

Diagnosis. A variety of classification schemes for the clinical syndromes caused by *E. histolytica* have been proposed. The one adopted by the World Health Organization is presented in Table 89-2.

In intestinal amebiasis, there often is no definite pattern of symptoms. In fact, the disease may manifest itself clinically in a deceptive fashion. In acute amebiasis (amebic colitis), the individual usually is acutely ill, complains of general abdominal discomfort and tenderness, and passes numerous, malodorous stools, which in most cases are dysenteric. The symptoms usually are referable to the cecal areas and may resemble appendicitis or various other

TABLE 89-2. CLASSIFICATION OF AMEBIASIS

Classification	Characteristics
I. Asymptomatic intestinal infection	Colonization without tissue invasion
II. Symptomatic intestinal infection	Invasive infection
A. Amebic dysentery	Fuliminant ulcerative intestinal disease
B. Nondysentery colitis	Ulcerative intestinal disease
C. Ameboma	Proliferative intestinal granuloma
D. Complicated intestinal amebiasis	Perforation, hemorrhage, fistula
E. Postamebic colitis	Mechanism unknown
III. Extraintestinal amebiasis	
A. Nonspecific hepatomegaly	No demonstrable invasion accompanies intestinal infection
B. Acute nonspecific infection	Amebae in liver but without abscess
C. Amebic abscess	Focal structural lesion
D. Amebic abscess complicated	Direct extension to pleura, lung, peritoneum, pericardium
E. Amebiasis cutis	Direct extension to skin
F. Visceral amebiasis	Metastatic infection of lung, spleen, or brain

conditions. In subacute infections, the picture is similar but less striking.

Considering the number of cases involved and the difficulty of diagnosis, the major problem is chronic amebiasis. The symptomatology in this case exhibits even greater latitude. Some patients may have periodic bouts of diarrhea or, less commonly, dysentery, but longer periods of constipation are characteristic. Others are without distinctive signs and symptoms and may complain of a low fatigue threshold, moderate loss of weight, mental dullness, and the like. In many such cases, tenderness can be demonstrated in the right lower quadrant. Still others proved to be infected are entirely asymptomatic.

The symptomatology of amebic liver abscess is characterized by hepatomegaly, tenderness over the liver and referred pain around the right scapula, bulging and fixation of the right leaf of the diaphragm, moderate leukocytosis, and a low-grade, inconstant fever. Blood chemistry studies are not helpful. Liver enzymes may be mildly elevated, and serum proteins are usually depressed. Mild anemias are common. Other diseases to consider during differential diagnosis are appendicitis, diverticulitis, ulcerative colitis, idiopathic colitis, Crohn's disease, bacillary dysentery, and other parasitic diseases.

Definitive diagnosis of amebiasis depends on the demonstration of amebae in stool or in material aspirated from lesions and abscesses. Various methods are used, depending upon the type of specimen and other factors. If the specimen is dysenteric or diarrheic or has been obtained by proctoscopic swabbing or following purgation, it is likely to contain only trophozoites (12 to 60 μm). If possible, it should be examined immediately because the trophozoites may soon lose their characteristic features. A fecal smear should be prepared on a microscope slide. The smear is made by emulsifying a bit of the specimen in a drop of tepid physiologic salt solution. A coverglass is added to the preparation, and the emulsion is carefully examined under the low power and high dry power of a compound microscope. If, in suspected cases of amebiasis, such direct examinations fail to reveal the organism and especially if Charcot–Leyden crystals have been noted, it is worthwhile to culture some of the stool.

If the fecal specimen is formed, or at least semiformed, it is likely to contain mostly cysts (10 to 20 μm) rather than trophozoites. Again, direct examination should be made using a dual preparation. In this case, one smear is made in the salt solution, especially to observe the refractive cyst wall, and the other is made in D'Antoni's iodine solution, which will stain the cyst, making the internal structures visible. If direct examination fails to reveal cysts, a concentration method should be used. The correct identification of *E. histolytica* requires considerable expertise. The most common error made by diagnosticians is that host leukocytes are identified as amebae.

E. histolytica must be differentiated from the other intestinal amebae (Fig. 89-2). The single characteristic that is most diagnostic is the structure of the nucleus (Fig. 89-3). All four species of the genus *Entamoeba* have conspicuous peripheral chromatin on the inner surface of the nuclear

membrane, whereas such material is lacking or is not conspicuous in the others. In *E. histolytica*, the nuclear membrane is delicate, and the chromatin on the inner surface appears as fine beads. The karyosome (nucleolus) is small and usually central. In *Entamoeba coli*, the membrane is thicker, and the chromatin on the inner surface is in the form of coarse plaques; the karyosome is much larger than that of *E. histolytica* and is located eccentrically. The striking feature of the nucleus of *Endolimax nana* is the large karyosome located in the center or slightly off center. In the case of *Iodamoeba bütschlii*, the karyosome is also large but is surrounded by a ring of achromatic granules, giving a halo effect around the karyosome.

Proctoscopic aspirates are occasionally submitted for direct examination and culturing. These should be examined as soon as possible. In a fresh specimen, there is usually a variety of host tissue cells present, so it is important to observe the typical motility of the trophozoites before making a definitive diagnosis.

In the direct diagnosis of amebic liver abscess, the aspirated specimen should be treated with streptococcal DNase to free the trophozoites from the coagulum. After this, some of the material should be examined in unstained smears, and some should be placed in culture. Many investigators have come to rely upon the indirect hemagglutination (IHA) test as one means of diagnosis. This test seems to be more sensitive than the complement-fixation (CF) and gel-diffusion (ID) tests that are available. An enzyme-linked immunosorbent assay (ELISA) test has recently been developed, and this holds considerable promise.

Chest and abdominal x-rays are sometimes useful adjuncts in diagnosis. For hepatic abscess, the x-ray may show significant elevation of the right hemidiaphragm and hepatomegaly. Radioisotope imaging is also efficient for demonstrating liver abscess.

Laboratory diagnosis of chronic amebiasis, especially in children, is often difficult. Both the clinician and the laboratorian should be aware that the best results will be obtained from a battery of tests on a series of suitable specimens.

Treatment. Treatment depends on drug therapy, and prevention requires the proper disposal of human wastes and good personal hygiene. Sometimes combination drug therapy is necessary for the treatment of intestinal amebiasis, and even then treatment is not always effective. In contrast, extraintestinal abscesses respond quite well to a variety of individual drugs.

For the treatment of asymptomatic intestinal amebiasis, a large number of antiamebic drugs given by mouth are available. Excellent results have been obtained with diiodohydroxyquin. For symptomatic intestinal as well as hepatic infection, metronidazole plus diiodohydroxyquin is recommended. For the treatment of liver abscess, chloroquine phosphate (Aralen) is an alternative drug.

In acute amebiasis with severe dysentery, emetine HCl, an alkaloid, usually affords prompt relief of symptoms. However, since it is ineffective for eliminating the

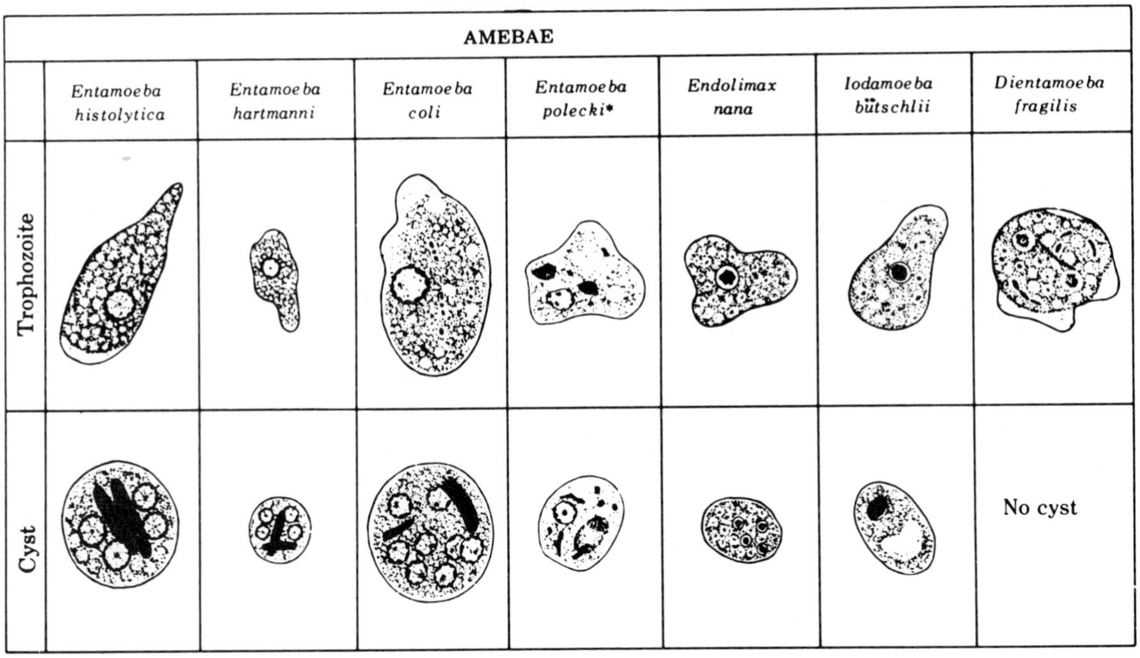

AMEBAE
Entamoeba histolytica
Entamoeba hartmanni
Entamoeba coli
Entamoeba polecki*
Endolimax nana
Iodamoeba bütschlii
Dientamoeba fragilis
Trophozoite
Cyst
No cyst

*Rare, probably of animal origin

Figure 89-2. Amebae found in human stool specimens. *(From Brooke and Melvin: Morphology of Diagnostic Stages of Intestinal Parasites of Man. Public Health Service Publication No. 1966, 1969.)*

Entamoeba histolytica
Entamoeba hartmanni
Entamoeba coli
Endolimax nana
Dientamoeba fragilis
Iodamoeba bütschlii

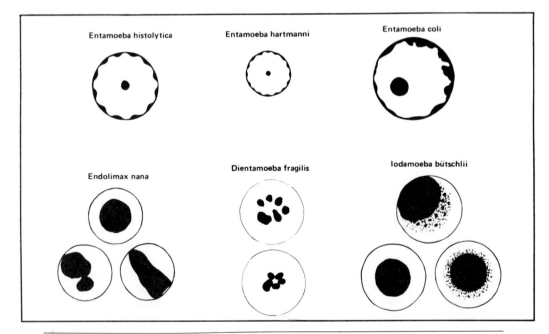

Figure 89-3. Diagrammatic representation of the various types of nuclei in the amebae of humans. *(From Brooke and Melvin: Morphology of Diagnostic Stages of Intestinal Parasites of Man. Public Health Service Bulletin No. 1966, 1969.)*

amebae, the usual treatment procedure is first to control the dysentery with emetine and then to use one of the above treatments. Since emetine may be highly toxic and is recommended only for hospitalized patients, it should be noted that an equally effective but less toxic drug, dehydroemetine, can be obtained on an investigational basis from the Parasitic Diseases Division, Centers for Disease Control (CDC), Atlanta. Another drug for treating amebiasis is available from CDC: diloxanide furoate or furamide is a luminal amebicide highly effective for treating asymptomatic and mildly symptomatic patients who are passing cysts. It, therefore, fills a void in the spectrum of amebicidal drugs available to American physicians. As will be noted in later discussions, additional drugs not on the market in the United States are recommended to be obtained from CDC. Table 89-3 lists all such recommended drugs.

Dientamoeba fragilis

D. fragilis is an ameboid parasite that inhabits the large intestine. Its taxonomic status and pathogenicity are uncertain. Although this organism is placed with the amebae in medical texts because of its appearance, careful study shows it to be a flagellate. This organism was originally considered to be a commensal, but it is now known to cause a disturbing gastroenteritis. Although generally thought to be an uncommon parasite, recent surveys show that it is worldwide in distribution; incidence rates as high as 20 percent have been reported. In the United States, a prevalence rate of over 2 percent has been proposed.

Life Cycle. D. fragilis lacks a cyst stage and, thus, differs from the true amebae. Although the mode of transmission is unknown, it has been suggested that it may be carried inside the shells of pinworm eggs. Others suggest that the portal of entry is ingestion of the trophozoite, which somehow escapes being killed by the digestive juices of the stomach.

Pathology and Symptomatology. Once the ameba is in the large intestine, it lives in the crypts of the colon where it multiplies by binary fission. This organism is not known to be invasive; it is a lumen dweller. The pathogenesis is unknown, but it is thought that increased mucus production through irritation of the mucosa is a possibility. Fibrosis of the appendix has been reported frequently.

Epidemiology. It is very unlikely that the naked trophozoite is the infective stage, as gastric juices are known to be lethal to amebae. If it is proved to be transmitted through pinworm eggs or larvae, this would help explain the lack of association between D. fragilis and the other protozoan species that are usually transmitted by fecally contaminated food and water. Infection is seen most often in children under 13 years of age. Infections seem to be most common in summer and fall months.

Diagnosis. Gastrointestinal symptoms include abdominal pain, diarrhea sometimes alternating with constipation, nausea, anorexia, and flatulence. Common systemic complaints are fever, weight loss, headache, and irritability. Occasionally blood is seen in the stool. The symptoms may continue for a month or more if untreated.

A definitive diagnosis is made upon demonstrating the unique, binucleate trophozoite (5 to 15 μm in diameter) in the stool (Figs. 89-2 and 89-3). Since no cyst stage has been seen, routine concentration techniques are of no value. Thus, direct or stained fecal smears are essential for diagnosis, and oil immersion techniques are necessary for correct identification. Of all the intestinal protozoa, D. fragilis is probably the most commonly overlooked organism.

Although it is unusual with protozoan infections, a mild eosinophilia is present in the majority of those proved to be infected. Other blood values have no diagnostic value.

Treatment. Diiodohydroxyquin or tetracycline is currently recommended for treatment. Metronidazole has also been used, as have other amebicides.

Flagellate: Giardia lamblia

This flagellate is recognized as an important cause of acute gastroenteritis in human beings. Waterborne outbreaks are being reported with increasing frequency, and backpackers have become ill after drinking water from mountain streams. Giardiasis frequently occurs among travelers to foreign countries. Person-to-person transmission occurs in mental institutions, day-care centers, and among male homosexuals. Finally, recent evidence has shown that fecally contaminated food can also be a vehicle for transmission. Infection occurs throughout the world and is most common in children. At least half of those who are infected are asymptomatic. Data show that 2 to 20 percent of the U.S. population harbors this parasite.

Life Cycle. G. lamblia is found in the small intestine; it has both trophozoite and cyst stages. The trophozoites usually are found in the intestinal crypts at the duodenal level, where they are firmly attached to the epithelial surface. At times, they are also found at lower levels of the intestine and in the common bile duct and gallbladder. As revealed in electron micrographs, they also may occur within epithelial cells. Multiplication is by longitudinal binary fission, which may result in myriads of organisms. Because of their location, trophozoites are not seen in the stool unless the individual has been given a saline cathartic or is persistently diarrheic. Therefore, cysts are usually the only stage found. These occur intermittently and may be very numerous. Under moist conditions, the cysts may remain viable for long periods. They presumably reach the mouth of the same person or others by the avenues of transmission discussed for E. histolytica. They pass unharmed through the stomach and excyst in the upper small intestine. Two trophozoites are produced by each cyst, and thus, this stage provides for reproduction as well as for transmission.

TABLE 89-3. RECOMMENDED THERAPY FOR PARASITIC INFECTIONS

Parasitic infections are now encountered throughout the world. With increasing travel, and especially with the recent large emigration from Southeast Asia, the Caribbean, and Central and South America, physicians anywhere may see infections caused by previously unfamiliar parasites. This table lists first-choice and alternative drugs. In every case, the need for treatment must be weighed against the toxicity of the drug. A decision to withhold therapy may often be correct, particularly when the drugs can cause severe adverse effects. When the first-choice drug is initially ineffective and the alternative is more hazardous, it may be advisable to try a second course of treatment with the first drug before using the alternative. Several of the drugs recommended in the table have not been approved by the U.S. Food and Drug Administration. When a physician prescribes an unapproved drug, or an approved drug for an unapproved indication, it may be advisable to inform the patient of the investigational status and possible adverse effects of the drug.

Infection	Drug of Choice	Alternative Drug
Amebiasis (*Entamoeba histolytica*)		
Asymptomatic	Diiodohydroxyquin	Diloxanide furoate and paromomycin
Mild to moderate intestinal disease	Metronidazole plus diiodohydroxyquin	Dehydroemetine plus diiodohydroxyquin, or emetine plus diiodohydroxyquin
Hepatic abscess	Metronidazole plus diiodohydroxyquin	Dehydroemetine followed by chloroquine PO_4 plus diiodohydroxyquin, or emetine followed by chloroquine PO_4 plus diiodohydroxyquin
Amebic meningoencephalitis, primary (*Naegleria* sp, *Acanthamoeba* sp)	Amphotericin B	
Ascaris lumbricoides (roundworm)	Mebendazole or pyrantel pamoate	Piperazine citrate
Balantidium coli	Tetracycline	Diiodohydroxyquin
Coccidiosis (*Isospora belli*)	Furazolidone	Trimethoprim–sulfamethoxazole
Cutaneous larva migrans (creeping eruption)	Thiabendazole	
Dientamoeba fragilis	Diiodohydroxyquin or tetracycline	
Enterobius vermicularis (pinworm)	Pyrantel pamoate or mebendazole	Piperazine citrate and pyrvinium pamoate
Filariasis		
Wuchereria bancrofti	Diethylcarbamazine	
Onchocerca volvulus	Diethylcarbamazine followed by suramin	Mebendazole
Flukes, hermaphroditic		
Clonorchis sinensis (Chinese liver fluke)	Praziquantel	
Fasciolopsis buski (intestinal fluke)	Praziquantel	Tetrachloroethylene
Paragonimus westermani (lung fluke)	Praziquantel	Bithionol
Giardiasis (*Giardia lamblia*)	Quinacrine HCl	Metronidazole, furazolidone
Hookworms (*Ancylostoma duodenale, Necator americanus*)	Mebendazole or pyrantel pamoate	Thiabendazole
Leishmaniasis		
Leishmania braziliensis (American mucocutaneous), *Leishmania mexicana* (American cutaneous)	Stibogluconate Na	Amphotericin B
Leishmania donovani (kala azar, visceral leishmaniasis)	Stibogluconate Na	Pentamidine
Leishmania tropica (oriental sore, cutaneous leishmaniasis)	Stibogluconate Na	Topical treatment
Malaria (*Plasmodium falciparum, Plasmodium ovale, Plasmodium vivax, Plasmodium malariae*)		
Suppression or chemoprophylaxis of disease while in endemic area*	Chloroquine PO_4	
Prevention of attack after departure from *P. vivax* and *P. ovale* endemic areas	Primaquine PO_4	

(continued)

TABLE 89-3. (Cont.)

Infection	Drug of Choice	Alternative Drug
Treatment of uncomplicated attack*	Chloroquine PO₄	
Treatment of severe illness*	Quinine dihydrochloride or chloroquine HCl†	
Prevention of relapses by *P. vivax* and *P. ovale* only ("radical" cure after "clinical" cure)	Primaquine PO₄	
P. falciparum (chloroquine-resistant)		
Suppression or chemoprophylaxis	Pyrimethamine plus sulfadoxine	
Treatment of uncomplicated attack	Quinine SO₄ plus pyrimethamine plus sulfadiazine	Quinine SO₄ plus tetracycline
Treatment of severe illness	Quinine dihydrochloride (parenteral dosage)	
Pneumocystis carinii	Trimethoprim–sulfamethoxazole	Pentamidine
Schistosomiasis		
Schistosoma haematobium	Metrifonate	Praziquantel
Schistosoma japonicum	Praziquantel	Niridazole
Schistosoma mansoni	Oxaminquine	Praziquantel
Strongyloides stercoralis	Thiabendazole	
Tapeworms (adult or intestinal stage)		
Diphyllobothrium latum (fish tapeworm), *Taenia saginata* (beef tapeworm), *Taenia solium* (pork tapeworm)	Niclosamide	Paromomycin
Hymenolepis nana (dwarf tapeworm)	Niclosamide or praziquantel	Paromomycin
Toxoplasmosis (*Toxoplasma gondii*)	Pyrimethamine plus trisulfapyrimidines	Spiramycin
Trichinosis (*Trichinella spiralis*)	Steroids for severe symptoms plus thiabendazole	Mebendazole
Trichomonas vaginalis	Metronidazole	
Trichuris trichiura (whipworm)	Mebendazole	
Trypanosomiasis		
Trypanosoma cruzi (South American trypanosomiasis, Chagas' disease)	Nifurtimox	
Trypanosoma gambiense, Trypanosoma rhodesiense (African trypanosomiasis, sleeping sickness)		
Hemolymphatic stage	Suramin	Pentamidine
Late disease with CNS involvement	Melarsoprol	Tryparsamide plus suramin
Visceral larva migrans	Thiabendazole	

*All *Plasmodium* except chloroquine-resistant *P. falciparum*.
†Parenteral dosage only if oral dose cannot be administered, regardless of severity.

Pathology and Symptomatology. Symptomatic giardiasis may present with any of a variety of signs and symptoms, including epigastric pain, diarrhea or loose stools, flatulence, abdominal cramps, malaise, weight loss, and steatorrhea. In more severe cases, malabsorption and a celiac-like syndrome may occur. Gallbladder irritation and blunting of villi have been noted. It has been suggested that those with severe disease may be immunodeficient with respect to IgA production.

Epidemiology. Asymptomatic infection appears to occur more frequently, and may be more important epidemiologically, than symptomatic giardiasis. Individuals with asymptomatic infections are much less likely to be detected or to seek treatment than are those with symptomatic infections and, therefore, are more likely to serve as carriers or disseminators of the disease. The *Giardia* carrier may excrete cysts for months or years. At present, giardiasis is the intestinal parasite most frequently identified in public health laboratories in the United States. Recent interest has centered on the transmission of *Giardia* cysts in drinking water, since epidemics have occurred in various cities of the United States. Chlorination of water for drinking purposes does not kill the cysts, so safe water purification procedures, including filtration, are necessary to eliminate cysts. Another problem has arisen among outdoor enthusiasts who camp in remote, uninhabited areas and drink water from mountain streams and other surface waters long assumed to be safe for human consumption. The source of outbreaks of giardiasis among members of such groups is not known, but it is likely to be cysts excreted by wild animals. *Giardia* must now be included among those organisms that may be transmitted by sexual contact. Although evidence suggestive of disease spread in this fashion has been encountered thus far principally among male homosexuals, the possibility of its spread among heterosexuals should not be overlooked.

Diagnosis. Clinically, giardiasis is characterized by severe diarrhea alternating with constipation. During bouts of diarrhea, stools are exceedingly foul-smelling and steatorrheic. Clinical signs are abdominal discomfort, weight loss, and protein malabsorption. These are also suggestive of peptic ulcer. Infection is common in those who are hypogammaglobulinemic or suffering from IgA-selective deficiency. Symptomatic patients are sometimes deficient in folic acid and fat-soluble vitamins.

Definitive diagnosis is made by finding the cysts or trophozoites in diarrheic specimens, or cysts in formed stools. The trophozoite (10 to 18 μm by 6 to 11 μm) is bilaterally symmetrical, with two nuclei and four pairs of flagella (Fig. 89-4). Living trophozoites on direct smear have a characteristic fluttery, falling-leaf type of movement. There is no oral opening, but on the ventral surface near the anterior end there is a characteristic adhesive disc. The cyst (8 to 14 μm by 6 to 10 μm) is ovoid, and the wall is relatively thickened (Fig. 89-4). When stained, two, four, or occasionally more nuclei can be seen, as well as curved fibrils.

At times no cysts are excreted in stools. Alternative methods of diagnosis include intestinal aspiration, biopsies, or the use of a string test (Enterotest). No serologic tests are routinely used, although work with an ELISA test is currently being conducted.

A number of other nonpathogenic flagellates often inhabit the human intestine. These, such as *Trichomonas hominis* and *Chilomastix mesnili* (Fig. 89-4), must be differentiated from *G. lamblia*. There is some evidence, however, that these organisms, when present alone or together in large numbers, may irritate the intestine and play a role in persistent diarrhea, since the condition often subsides upon their eradication.

Treatment. Quinacrine HCl (Atabrine) is considered the drug of choice. Metronidazole is also effective. Both drugs cause adverse side effects, and caution in their use must be exercised. It appears that none of the drugs currently available will cure all infections, and this raises the possibility of drug-resistant strains that further complicate the epidemiologic picture.

Ciliate: *Balantidium coli*

B. coli is the only ciliated protozoan infective for humans. Infection is known as balantidiasis, ciliary dysentery, or balantidial dysentery. The infection, although rare in the United States, is distributed worldwide; the general epidemiologic rule is wherever there are pigs, there are likely to be infections in humans. The precise relationship between *Balantidium* in humans and that in pigs is subject to considerable debate and needs to be clarified. In addition to pigs, rats also serve as reservoir hosts.

Life Cycle. Both trophozoites and cysts are present in the life cycle (Fig. 89-4). Humans become infected when cysts are ingested, either in contaminated water or through

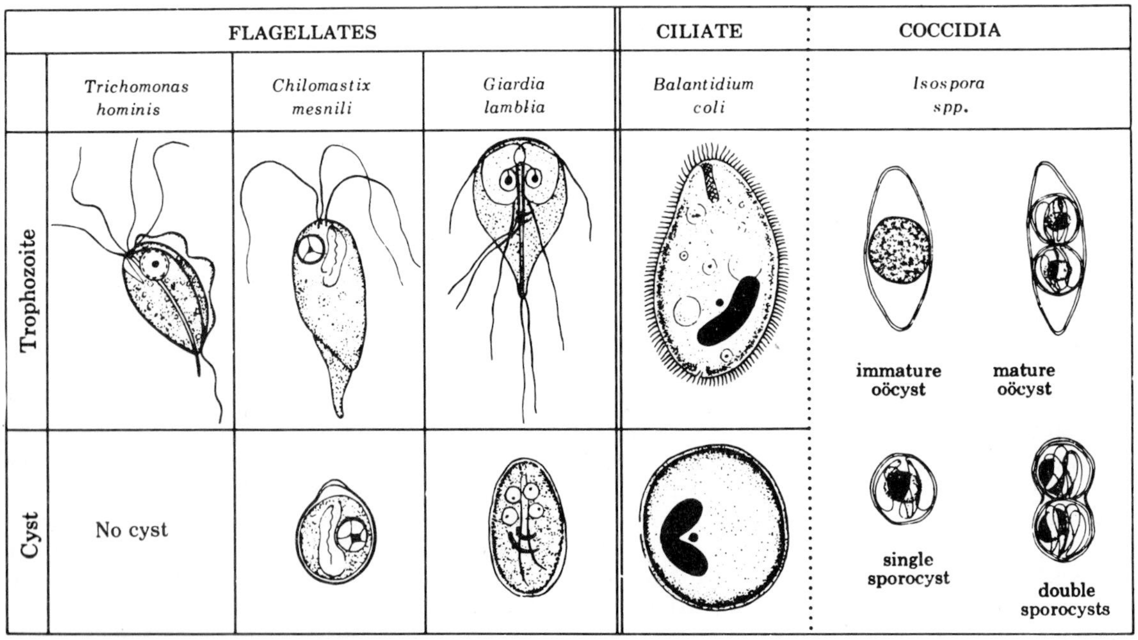

FLAGELLATES			CILIATE	COCCIDIA
Trichomonas hominis	*Chilomastix mesnili*	*Giardia lamblia*	*Balantidium coli*	*Isospora spp.*
Trophozoite				immature oöcyst / mature oöcyst
Cyst — No cyst				single sporocyst / double sporocysts

Figure 89-4. Flagellate, ciliate, and sporozoan protozoa found in stool specimens. *(From Brooke and Melvin: Morphology of Diagnostic Stages of Intestinal Parasites of Man. Public Health Service Publication No. 1966, 1969.)*

fecal–oral contamination. *B. coli* colonizes the large intestine.

Pathology and Symptomatology. This organism occurs usually in the cecal area but may be found at both higher and lower levels of the large intestine. The trophozoites feed upon bacteria and other substances in the lumen but take in host cells after they enter the tissues. They reproduce by transverse binary fission, and in certain hosts they may become numerous. The mucosal layer appears to be penetrated mainly by boring action, and because of the size of the organism, the opening is much larger than that produced by *E. histolytica*. Having gained entrance into the mucosa, the organism has little difficulty in passing through the muscularis mucosae into the submucosa, where it spreads out radially and may produce considerable destruction. Unlike *E. histolytica*, it apparently invades the muscular layers only on rare occasions, and although observed in lymphatics, it is only rarely found in extraintestinal sites. The ulcers produced may occur at all levels of the large intestine but are most common in the cecal and sigmoidorectal areas. They often resemble those produced by *E. histolytica*, especially after bacteria invade them. When this occurs, extensive inflammatory reactions are noted around the organisms, as well as a diffuse infiltrate throughout the tissue.

Epidemiology. Although rare in humans, *B. coli* is common in hogs throughout the world. Infections are most prevalent in the tropics where pigs share habitation with people. The pig is usually considered the source of human infection, but once human infection is established, person-to-person transmission may occur. This becomes significant in situations where group hygiene may be poor, such as mental hospitals and other places of confinement.

Diagnosis. Many people harboring *B. coli* are asymptomatic, but others show symptoms ranging from mild to profuse diarrhea and even fulminating, fatal dysentery. Laboratory diagnosis usually is made by finding the characteristic trophozoites in the stool. Because of their great size, they are easily detected (Fig. 89-4). The trophozoites vary considerably in size (50 to 100 μm by 40 to 70 μm or even larger), are ovoidal in shape, and are covered entirely with short, constantly moving cilia. A funnel-shaped peristome (mouth) leads into the cytostome (gullet). One or two contractile vacuoles may be seen within the cytoplasm, as well as two nuclei, a large, kidney-shaped macronucleus, and, usually within its concavity, a small micronucleus. The cyst, observed less frequently than the trophozoite, is smaller (45 to 65 μm in diameter) and is almost spherical in shape. Stained cysts reveal clearly the nuclear components.

Anemia and a mild leukocytosis may occur in some symptomatic patients. No serologic or other diagnostic tests are available.

Treatment. Balantidiasis can be treated effectively with tetracycline.

Sporozoans

Isospora species

These sporozoans are the cause of a gastroenteritis called isosporosis or coccidiosis. Taxonomists cannot agree as to the number of species that may be found in humans, so it may be best to take a conservative approach and simply diagnose such infections as being caused by *Isospora*.

Human coccidiosis is a rare disease that occurs in the warm regions of the world and has been reported from California and the southeastern United States. The life cycle is very complex. Although it is known for some of the *Isospora* infections in animals other than humans, the life cycle in humans is poorly understood. Infection is considered to begin when a person ingests the oocyst stage in fecally contaminated food or water. Inside the oocysts are eight banana-shaped forms, the sporozoites. Once released from the oocysts, the sporozoites invade epithelial cells of the intestine. Division of the parasite occurs within the epithelial cells, and the progeny of this division are released to penetrate other epithelial cells. Eventually some of these progeny become sex cells, unite, and become oocysts, which are shed in the feces. Oocysts are immature when passed and require 24 to 48 hours to become infective.

Coccidiosis is usually a mild, self-limiting infection characterized by diarrhea and colicky pains. Sometimes, however, it may present itself as a chronic infection, with acute episodes of fever, severe diarrhea, and steatorrhea. Some mortality has been reported.

Diagnosis is made by finding the oocysts (10 to 60 μm) in the stool (Fig. 89-4). Their transparent appearance may cause them to be overlooked by unskilled microscopists. Oocysts usually are present in small numbers, even in diarrheic stools. As a result, a concentration technique, such as the zinc sulfate or formalin–ether technique, should be used. No serologic tests are available. Furazolidone is the drug of choice; an alternative therapy is the use of trimethoprim–sulfamethoxazole. The former drug is currently an investigational drug and must be obtained from the CDC in Atlanta.

Cryptosporidium species

This protozoan parasite is closely related taxonomically to the coccidian parasites *Isospora* and *Toxoplasma*. Unlike these latter two genera, however, infections with *Cryptosporidium* parasites in humans are restricted to the epithelial cells of the microvillous border of the intestinal tract. Human infections, until recently, were thought to be rare and the result of opportunistic infection. Hosts normally associated with intestinal cryptosporidosis are numerous and include calves, lambs, guinea pigs, and mice; intestinal and respiratory infections occur in turkeys, chickens, and various reptiles. Since 1980, numerous reports have appeared in the literature and the concept of cryptosporidosis has changed from that of a rare and mainly asymptomatic infection to an important cause of diarrhea and enterocolitis in several species of animals, including humans.

Two events have brought human infection with this agent to the attention of the medical community. The first of these is related to acquired immune deficiency syndrome (AIDS). It has been found that one of the opportunistic pathogens is *Cryptosporidium*. The other event is the recent development of better diagnostic techniques and the demonstration of both asymptomatic and symptomatic infection in immunocompetent hosts. Most recently, it has been shown that zoonotic infection occurs, with transmission from diarrheic calves to humans.

The life cycle of *Cryptosporidium* follows in general that of other enteric coccidia. The infective stage is an oocyst containing sporozoites. Upon ingestion, the sporozoites are released from the oocyst once the sporozoites become associated with epithelial cells of the bowel, most often the small intestine. Sporozoites undergo division to form a schizont containing numerous forms known as merozoites. These are released from the host epithelial cell, become associated with other epithelial cells, and by the process of gametogenesis eventually form oocysts. The oocyst is released from the epithelial cell and is passed in an undeveloped state in the feces. It should be noted that the question of whether the parasite ought to be considered as intracellular or extracellular is still in question. Whatever the precise nature of the association, the physical location of the parasite is outside the main cell boundary; in other words, it does not develop intracytoplasmically. This fact also distinguishes *Cryptosporidium* from other enteric coccidia.

Symptomatic infection in immunocompetent hosts is usually self-limiting (3 to 14 days) and is characterized by diarrhea accompanied by vomiting, anorexia, and abdominal pain. In AIDS and other immunocompromised hosts, a prolonged and severe diarrhea that may last for years is seen. In these cases, malabsorption would be expected. Intestinal biopsies examined histologically show varying degrees of mucosal damage, including crypt lengthening, partial villous atrophy, and cellular inflammation of the lamina propria.

Sixteen species of animals are known to be susceptible to *Cryptosporidum;* 10 of these have demonstrated some degree of illness. Ruminants, in general, usually manifest diarrhea if infected at an early age. Several human cases show that the infection may be transmitted from calves and, therefore, is indeed a zoonotic disease. Serological surveys indicate that infection is not uncommon among 10 mammaliam species examined so far. Although our understanding of this infection in humans is unclear, it is suggested that immunodeficient people should not work around young animals such as calves, lambs, goats, and pigs.

Histologic demonstration of the organisms attached to the brush border of epithelial cells of the small bowel is a means of diagnosis and has been most widely used in the past. Recently, noninvasive new techniques have been developed, including the modified zinc sulfate centrifugal flotation, Giemsa-stained fecal smears, modified Ziehl-Nielson carbolfuchsin staining, and Sheather's sugar flota-

tion. Several antibiotics, antiprotozoal, and even anthelmintics have been tested against *Cryptosporidium*. None that have been tested appear to be effective in treating human disease.

Urogenital Tract Protozoan:
Trichomonas vaginalis

The human urogenital tract is, under normal conditions, a rather hostile environment to most parasites because of its acidity and mechanical barriers, such as mucus and cilia. However, when the normal environment is altered by, for example, an increase in pH, organisms such as *T. vaginalis* can reproduce and flourish.

Trichomoniasis is one of the most common, if not the most common, sexually transmitted disease. In the United States in some areas, incidence among females is reported to be 50 percent; incidence among males is considerably less.

Life Cycle. In females, the vagina is the usual habitat of the organism. In chronic infection, however, it may invade the urethra. In males, the organisms colonize the urethra and prostate gland. There is no cyst stage, and trichomoniasis is, therefore, essentially a venereal disease. Because the trophozoite can survive a few hours under warm and moist conditions, infection via contaminated towels, toilet seats, and underwear is theoretically possible. Newborns of both sexes may acquire infection from their infected mothers. The vaginal pH of the newborn is alkaline and, therefore, suitable for colonization.

Pathology and Symptomatology. In males with nongonorrheal urethral discharge found to be infected with *T. vaginalis,* the symptoms usually are mild until the infection is aggravated by secondary bacterial invasion, after which the condition becomes a purulent urethritis and prostatovesiculitis. In females, trichomoniasis does not always result in a complaint of symptoms, but the vaginal secretions are altered invariably. In typical cases of *T. vaginalis* vaginitis, the normal pH of the vagina (3.8 to 4.4 during sexual maturity) becomes more alkaline, and the glycogen stores of the vaginal mucosa, especially in the superficial layers, are reduced greatly. The normal processes of cellular destruction make the glycogen available to the Döderlein bacillus, an inhabitant of the normal vagina, which metabolizes glycogen and excretes lactic acid. This metabolism maintains the normal acid state of the vagina. In the absence of normal stores of glycogen, the numbers of this organisms are reduced, and in severe cases they may be eliminated. When these events occur, the physiologic protection offered by the normal vaginal acidity is altered, and the growth of *T. vaginalis* and other organisms is encouraged. Patients with *T. vaginalis* infection usually have a profuse, watery leukorrheic discharge that produces a chafing of the vagina, vulva, and perineum. Pruritis vaginae and vulvae are distressing.

Epidemiology. In the United States, trichomoniasis occurs in all age groups. Highest infection rates are in females 30 to 40 years of age. It is estimated that in the United States, 25 percent of all females are infected. Black females have an infection rate of 30 percent which is higher than that of white females. There is a marked decline of infection in postmenopausal females.

Diagnosis. Clinical diagnosis is based on subjective complaints and evidence of infection. The complaints of discomfort range from irritation and itching to severe vulval pruritis with intense pain. A purulent exudate from the genitalia of both sexes may be seen. Differential diagnosis includes gonorrhea, candidiasis, endometritis, staphylococcal infection, and pelvic inflammatory disease.

Diagnosis is best accomplished by the examination of a temporary microscopic slide preparation of vaginal exudate or urethral discharge in males for detection of the mobile trophozoites. These are 5 to 20 μm in length, pear-shaped, and have a jerky-type movement. Exudate from females is best collected from the vaginal canal via a vaginal speculum.

If wet preparations cannot be examined almost immediately after collection, a stained smear of the exudate should be made. Papanicolaou, Giemsa, and acridine-orange stained smears are used. The last stain, although excellent, is not often used because it requires a microscope fitted with fluorescent illumination. Culture in appropriate media is sometimes helpful in those cases where the organism cannot be found in vaginal smears and urethral discharges. Finally, *T. vaginalis* may be found in urine, especially when the urethra has been colonized.

Complement-fixation, IHA, and immunofluorescent antibody (IFA) tests have been developed. Despite the potential value of these tests for identifying the asymptomatic male and female carriers, none has been adopted for routine use in the lab.

Treatment. Metronidazole is the drug of choice, and efforts should be made to treat the sex partner(s) of those found to be infected. There is some indication that resistance is developing to the drug. Acidic douching to restore normal pH of the vaginal canal is sometimes helpful.

Blood and Tissue Protozoa

Amebae: *Naegleria* and *Acanthamoeba*

Free-living soil amebae of the genera *Naegleria* and *Acanthamoeba* are known to produce infections in humans. Among the various species of *Naegleria*, *Naegleria fowleri* is considered the species that causes an acute, rapidly progressing disease of the central nervous system, primary amebic meningoencephalitis (PAM). With only one exception, this infection has always had a fatal outcome. *Acanthamoeba*, on the other hand, usually produces subacute or chronic infections that resemble viral or mycotic disease of the brain. Investigators have termed infections with this organism amebic meningoencephalitis (AM). This separation is partially based on the fact that *Naegleria* infections are usually found in healthy, immunocompetent children and young adults, whereas *Acanthamoeba* infections are usually present in patients with other medical problems characterized by a lowering of resistance and immunity. Separation into the two disease categories is also warranted because there are strong indications that the therapy may differ.

Infections with *Naegleria* have been reported from many parts of the world that have a warm climate. In the United States, cases have occurred in California, Florida, Georgia, and Virginia. *Acanthamoeba* infection is not known to be related to any geographic factor or area. Although fewer cases of *Acanthamoeba* infection than *Naegleria* have been reported on a worldwide basis, infections of *Acanthamoeba* in the United States have been reported from Arizona, Texas, Louisiana, Florida, Virginia, Pennsylvania, and New York.

Life Cycles. Both amebae are normally free-living and ubiquitous in nature and may be found in freshwater lakes, ponds, puddles, swimming pools, and brackish water. Both have trophozoite and cyst stages. In addition, the trophic stage of *Naegleria* not only occurs as an ameba but, under certain conditions, may change into a flagellated stage. However, in humans, *Naegleria* occurs only in the amebic, nonflagellated form. Infection is thought to occur as a result of diving or swimming underwater, when water containing the amebic stage is forced into the nasal passage, allowing the amebae to invade the mucosa.

Some infections with *Acanthamoeba* probably occur in a similar manner. However, studies have shown that the amebae also exist in the cyst form in nasal secretions and in tissue. In addition to the meningoencephalitis caused by *Acanthamoeba*, upper respiratory, skin, and eye lesions are known. This indicates that the mode of infection and transmission may indeed be quite different for the two organisms.

Pathology and Symptomatology. For infections with *Naegleria*, the portal of entry is considered to be the nasal mucosa overlying the cribriform plate. The clinical course is dramatic. Prodromal symptoms of headache and fever are followed by rapid onset of nausea and vomiting, accompanied by signs and symptoms of meningitis, with involvement of the olfactory, frontal, temporal, and cerebellar areas of the brain. Involvement of the olfactory area, with disturbances in the sense of smell early in the course of the disease, is characteristic in most patients. Hematogenous spread occurs, and pulmonary edema is not uncommon. Irrational behavior, coma, and death within 5 to 16 days are usual. Spinal puncture reveals a cloudy, purulent, or sanguinopurulent fluid with high leukocyte cell counts (mostly neutrophils), and motile amebae with lobate or blunt, not filiform or spine-like, pseudopodia.

Infection with *Acanthamoeba* is thought to be spread throughout the body by the circulatory system after the ameba invades the nasal mucosa, skin, or lungs. The incubation period is thought to be about 10 days, and a more prolonged illness results. Infection may be subacute or chronic, characterized by abscesses or granulomas, or both. Thrombosis and hemorrhage are seen, which clinically resemble mycotic or chronic viral disease. *Acanthamoeba* organisms have also been found in and around corneal ulcers and in the conjunctival fluid. Whether or not the amebae play a primary or secondary role in these eye lesions is not known. Clinicians should, however, consider acanthamoebic keratitis when eye lesions fail to respond to bacterial, fungal, and viral therapy.

Epidemiology. *Naegleria* infections have been reported from many areas of the world having warm climates; *Acanthamoeba* infection is not known to be related to any geographic or climatic factor. *Naegleria* infections are primarily in healthy children, whereas *Acanthamoeba* infection is most common in people who have some deficiency of the immune or resistance mechanisms.

Cystic forms of the amebae are quite resistant, surviving both freezing temperatures and chlorination of water. Bacteria normally serve as food for the trophic forms and, thus, thermal pollution of recreational waters presents a serious potential public health hazard. There is also some evidence that *Naegleria* can survive in hospital hydrotherapy pools. Recently, it was suggested that amebae in air-conditioning cooling equipment may serve as vectors of *Legionella*. Trophozoites that have ingested *Legionella* may become encysted and then serve as a source of infection for humans via airborne particles.

Diagnosis. The onset of PAM is sudden and resembles acute meningococcal meningitis. Prior to actual neurologic involvement, fever, neck rigidity, nausea, vomiting, and headache may develop. Within 1 to 2 days after symptoms appear, the patient will become comatose. At this time the sense of taste may be lost, and the sense of smell may be altered. Death usually occurs within 1 week from cardiorespiratory failure and severe cerebral edema.

Diagnosis is very difficult because of the nonspecific symptoms and the rapidity with which the disease develops. Sediment from central nervous system fluid should be examined for amebae; phase contrast microscopy is recommended. Blood counts show a leukocytosis, and there is no eosinophilia. Fluid from the central nervous system may be hemorrhagic.

Patients having AM have signs and symptoms very much like those with PAM. However, there is no prior history of swimming. The incubation period is thought to be 10 days or more, and the illness is more prolonged than in PAM. Unlike PAM, where only the trophozoite is present in human fluids and tissues, both cysts and trophozoites can be demonstrated in cerebrospinal fluid and tissues. Both stages may also be seen in lesions and in vacuoles in

giant cells within a lesion. Differential diagnosis for both PAM and AM should include bacterial meningitis, cerebral amebiasis, toxoplasmosis, viral encephalitis, and tick fever.

Treatment. The usual drugs for the treatment of intestinal amebiasis are not effective. Amphotericin B therapy has been used in mouse studies, and there are indications that it may be beneficial in infections with both *Naegleria* and *Acanthamoeba*. Sulfadiazine, gentamicin, and pimaricin have been shown to be effective in the treatment of *Acanthamoeba* infections. It is recommended that the Parasite Diseases Division at the CDC be consulted for the treatment of any proven case.

Flagellates

These flagellates differ greatly from those discussed above. Besides having an entirely different structure, the blood and tissue flagellates are found in various tissues in humans and require a bloodsucking insect to complete their life cycles. Six species are involved, three in the genus *Trypanosoma* (*Trypanosoma brucei gambiense*, *Trypanosoma brucei rhodesiense*, and *Trypanosoma cruzi*) and three in the genus *Leishmania* (*Leishmania donovani*, *Leishmania tropica*, and *Leishmania braziliensis*). Aside from a few reported cases of *T. cruzi* and cutaneous leishmanial infections suspected of being acquired in the United States, these various species of so-called hemoflagellates have an exotic origin.

African Trypanosomiasis

Two morphologically identical species of trypanosomes cause new infections in thousands of people each year on the African continent and cause considerable mortality and morbidity. One species, *T. b. gambiense*, causes a disease known as Gambian sleeping sickness and is endemic in west and central Africa. The other species, *T. b. rhodesiense*, causes Rhodesian sleeping sickness and is endemic in east and central Africa. Gambian trypanosomiasis is primarily a disease of humans, whereas Rhodesian trypanosomiasis is a zoonotic disease, with numerous wild and domestic animals serving as potential reservoir hosts for humans. Both parasites are transmitted to humans by the bite of infected tsetse flies of the genus *Glossina*.

It is estimated that a million Africans are infected at any given time, despite control efforts and chemoprophylactic activities. Although endemic only in Africa, it is well to remember that thousands of citizens of the United States travel to and from Africa each year, and numerous Africans visit the United States for pleasure, business, and education.

Life Cycle. The transmission cycle begins when a tsetse fly feeds on blood from an infected human. Once inside the vector, the trypomastigotes (Fig. 89-5) pass through the proventriculus into the midgut, where multiplication by longitudinal binary fission results in the production of slender forms. These finally make their way into the sali-

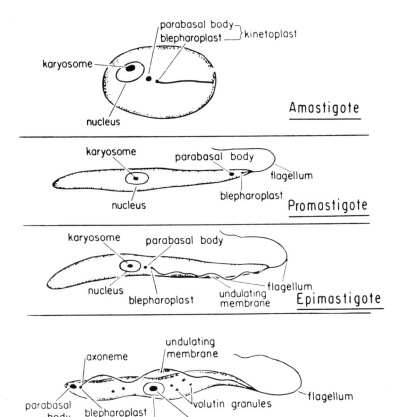

Figure 89-5. Schematic representation of morphologic forms in the *Trypanosoma* and *Leishmania* genera. *(Modified from Mackie et al.: A Manual of Tropical Medicine, 2nd ed, 1954. Philadelphia, Saunders.)*

vary glands, where multiplication results in the production of epimastigote forms (Fig. 89-5). Later, these transform into infective-stage trypomastigotes (metacyclic trypomastigotes), which accumulate in the ducts of the salivary glands of the tsetse fly. The entire development within the fly requires about 2 to 3 weeks.

Pathology and Symptomatology. Gambian trypanosomiasis is a chronic disease that eventually gives rise to the torpor and eventual coma that is characteristic of classic sleeping sickness. Untreated patients usually die as a result of malnutrition and concurrent infection. There usually are three progressive stages of tissue involvement: (1) parasitemia, when the parasites are numerous in the blood, (2) lymphadenitis, when they are concentrated in the lymph nodes, and (3) central nervous system involvement, when the parasites are numerous in the brain substance and arachnoid spaces. Lymphadenitis is most pronounced in the posterior cervical triangle (Winterbottom's sign). Physical weakness and mental lethargy follow invasion of the central nervous system. Sleepiness develops and becomes progressively pronounced until, in the advanced stage, the patient sleeps continuously. It is thought that the condition may have an autoimmune basis, since antibodies against myelin have been demonstrated. Histologically, a demyelinating encephalitis is noted in advanced cases.

Disease caused by *T. b. rhodesiense* is subacute or acute and is associated with irregular febrile paroxysms, edema, anemia, weight loss, and myocarditis. Although death may occur before the central nervous system is involved, when the disease does progress to central nervous system involvement, behavioral changes and decreased activity are noted. Once the central nervous system is involved, a patient will become progressively weaker and anemic and within 1 year may die of myocarditis or secondary infection, such as pneumonia.

Epidemiology. *Trypansoma brucei brucei,* a natural parasite of wild game animals in Africa, especially the antelope, is considered to be the ancestral form of the two human forms. Many domesticated animals are also susceptible to infection, and this leads to a disease known as nagana. Both human trypanosomiasis and nagana seriously inhibit the economic development of the continent.

Tsetse flies bite during the day. The species that transmit Rhodesian trypanosomiasis inhabit the open savannah, whereas the species of *Glossina* that transmit the Gambian form are present in riverine areas. Both sexes of tsetse flies ingest blood, are long lived, and have a long flight range.

African trypanosomes can change their antigenic make-up and thereby evade a host's immune response. This changing antigenicity, known as "antigenic variation," makes it very difficult to develop long-term protective substances, such as vaccines.

Diagnosis. The clinical picture, although having some common features, differs in the rapidity with which the diseases develop. In general, the Rhodesian form evolves more rapidly to a fatal outcome and many times does not involve the central nervous system. The Gambian form, in contrast, can take several years to develop into the typical sleeping sickness syndrome.

Trypomastigotes of both forms occur in the circulating blood during febrile episodes and sometimes can be demonstrated in thick blood smear preparations. However, lymph node aspirates contain a larger concentration of the organisms and are more useful for laboratory diagnosis. After involvement of the central nervous system, examination of spinal fluid may yield positive results. For the reason explained above, this method applies only to *T. b. gambiense* infections. If these various methods fail to establish the diagnosis, susceptible laboratory animals (guinea pigs, white rats, and others) may be inoculated intraperitoneally with some of the patient's blood and later examined at intervals for the parasites.

Both indirect fluorescent antibody and indirect hemagglutination (IHA) tests have been developed and used. Recently, enzyme-linked immunosorbent assay (ELISA) and a direct agglutination test have been developed. Although the indirect fluorescent antibody test may be helpful in diagnosing an early infection, the ELISA test in conjunction with a demonstration of increased IgM levels by radial immunodiffusion represents the best methods for determining the status of infection.

The laboratory finding most suggestive of trypanosomiasis is a marked increase in the IgM level; it may be 10 times the normal value in the Gambian form of the disease but somewhat less in the Rhodesian form. Anemia, increased erythrocyte sedimentation rate, elevated protein in the fluid of the central nervous system, as well as cells in spinal fluid are seen in most cases.

Treatment. During the hemolymphatic stage of both forms of the disease, suramin is the drug of choice. Once the central nervous system is involved, melarsoprol is recommended. Pentamidine is an alternative drug for the hemolymphatic phase, and tryparsamide plus suramin are used as an alternative for the cases where central nervous system involvement has occurred.

American Trypanosomiasis

American trypanosomiasis, or Chagas' disease, has been reported in almost all countries of the Americas, including the southern United States. Its main foci, however, are located in the poorer, rural areas of Latin America. Unfortunately, this infection can be a severely debilitating disease for which there is no known effective curative drug.

Even though suitable vectors and reservoir animals are widely distributed in the southern United States, only a few autochthonous cases have been reported. However, the clinician should pay close attention to the travel history of any symptomatic patient who has returned from an endemic area.

Life Cycle. *T. cruzi*, the flagellate causing Chagas' disease, differs from the African trypanosomes in that it has an intracellular phase involving cells not only of the lymphoid-macrophage system but also those of the myocardium, endocrine glands, and the glial cells of the brain. In this phase, the parasite is a typical amastigote (leishmanial) form (Fig. 89-5), being ovoid in shape and 1.5 to 3 μm in diameter. No true flagellum or undulating membrane is present. When freed from the host cell, this stage enters the bloodstream, where it transforms into a C-shaped trypomastigote stage, about 20 μm long, which is the stage taken by the vector. The vector, in this case, is a triatomid (cone-nosed) bug. Many species of these bugs have been found infected naturally. When taken into the bug, the trypomastigote stage of the parasite is carried into the midgut, where it transforms into the epimastigote stage. Later, following multiplication in the hindgut, the infective-stage trypomastigotes are formed, which pass out in the liquid feces of the bug when it is taking a blood meal or immediately thereafter. This usually occurs at night, and the pruritus produced by the bite often results in an infection caused by rubbing into the wound, or nearby skin or mucous membrane, the excrement containing the infective parasites. The entire cycle in the bug requires about 2 weeks.

Pathology and Symptomatology. The disease course varies according to the virulence of the strain of *T. cruzi* and host factors, such as age and immune status. The acute phase of the disease occurs during the first 6 months, and symptoms are usually severe. An infected person will manifest fever, enlargement of the lymph glands, liver, and spleen, and, most importantly, damage to the heart. Children are particularly susceptible, and death may occur in 10 percent of those infected. The chronic disease follows the development of the parasite in several visceral organs. Here they multiply as amastigotes, destroying the host cells and bringing about inflammatory reactions in the involved tissues. The principal pathologic changes are noted grossly in the heart and brain, and the predominant symptoms are those of cardiac failure. The spleen and liver are enlarged and congested in relation to the degree of cardiac failure. The brain is congested and edematous and often contains scattered petechial hemorrhages. Collections of organisms may be associated with glial nodules.

Careful study has shown that many infected people, although manifesting few if any clinical signs and symptoms, do not escape damage to the heart. There is evidence of damage to the heart muscle cells by the invading and reproducing amastigotes. Toxin-like materials affect the nerves regulating the heartbeat, and autoantibodies may also play an additional role in the destruction of heart tissue. Heart failure may follow.

Megacolon and megaesophagus are rare manifestations of chronic Chagas' disease. It is thought that the neurons serving these organs are damaged by toxin-like materials (or autoantibodies), and the loss of nerve control leads to flaccidity, loss of peristaltic action, and eventually distention of the toneless colon or esophagus.

Epidemiology. *T. cruzi* is a zoonotic infection, infecting not only humans but many reservoir hosts. Armadillos and opossums are often infected, but cats and dogs may be more important sources for domestic transmission.

Transmission is related to the environment and habits of both humans and the vectors. The disease is most common when people live in adobe mud huts, where the vectors make their homes in the cracks and crevices of the mud walls. The vectors are night feeders and are able to feed undisturbed, frequently on the conjunctiva of the eye. The fact that we have considerable sylvatic infection in the United States but little human infection attests to the importance of improved housing and living conditions.

Since *T. cruzi* can cross the placental barrier, newborn infants with advanced Chagas' disease have been reported. Infections via blood transfusion, hypodermic needle contamination, and laboratory accidents are also known.

Diagnosis. An early, almost constant sign of acute *T. cruzi* infection, especially in infants, is edema of the face and eyelids (Romana's sign). The acute form predominates in younger age groups, is usually of short duration, and has a high case fatality rate if untreated. Some patients may develop a chagoma, which is a nodular lesion containing inflammatory cells, at the site of the vector's bite. Fever, edema, diarrhea, adenopathy, hepatosplenomegaly, myocarditis, and sometimes meningoencephalitis may be noted.

T. cruzi also may be found in the circulating blood during febrile periods, but the trypomastigotes are never numerous and, therefore, are difficult to demonstrate even in thick blood film preparations. Another trypanosome, *Trypanosoma rangeli*, occurs in the blood of humans in the same geographic areas as does *T. cruzi*, and it may be misdiagnosed. This species is not pathogenic, has no extravascular stages, and is easily differentiated from *T. cruzi*, being nearly twice as long.

Laboratory animals may be inoculated with blood from suspected cases. A widely used xenodiagnostic test is available and involves the use of parasite-free, laboratory-raised triatomids, which are allowed to take blood from the patient. Examination of the feces or intestinal contents of the bugs after 8 to 10 days usually will reveal trypanosomes if they were present in the blood of the patient. The organisms can be found in stained tissue aspirates and also may be cultivated on various media, e.g., Nicolle, Novy, MacNeal (NNN) or Chang's, and in tissue cultures, all of which reveal the epimastigote and trypomastigote forms. Finally, complement-fixation, indirect fluorescent antibody, and indirect hemagglutination tests are available to assist in the diagnosis of latent and chronic cases of the disease.

A lymphocytosis may occur during the acute phase of the disease. Unlike African trypanosomiasis, there is usually no elevation of the IgM level. However, a slight increase in the IgG level may occur during the chronic phase. Heterophile antibody to rat and sheep erythrocytes may appear early in acute infection.

Treatment. Nifurtimox is used for the treatment of Chagas' disease. This eliminates the trypomastigotes from the circulating blood and will improve the clinical condition. However, it does not destroy amastigotes in the tissues, and relapses do occur. As a result, there is no satisfactory treatment for either the acute or chronic forms of the infection.

Leishmania

Although leishmania belong to the same major taxonomic group as the trypanosomes, they differ in two major respects. First, instead of being flagellated protozoans that swim freely in the blood of humans, leishmania are small (1 to 3 μm, Fig. 89-5), aflagellated organisms that live inside macrophages. Second, leishmania are transmitted to humans by the bite of small flies of the genera *Phlebotomus* and *Lutzomyia* (sandflies), instead of the larger tsetse flies and triatomids that transmit the various species of *Trypanosoma*.

The taxonomy of the genus *Leishmania* is not well understood at this time. Morphologically, all species are indistinguishable, and, as a result, speciation is based on the clinicopathologic picture, antigenic relationships, and geographic distribution. Traditionally, and based on these factors, infections have been separated into three main types, as presented in Table 89-4.

TABLE 89-4. LEISHMANIA OF HUMANS

Type	Species	Site of Lesions	Geographical Distribution
Visceral	*Leishmania donovani*	Macrophages of the deep viscera (liver, spleen, bone marrow)	India, China, Euro-Asia, Mediterranean basin, Middle East, Africa, and South America
Dermal	*Leishmania tropica* *Leishmania mexicana* *Leishmania peruviana*	Macrophages surrounding skin lesions	Asia, Middle East, India, China, Central Africa, Mediterranean basin, Mexico, and South America
Mucocutaneous	*Leishmania braziliensis*	Macrophages of the nose and pharynx	South America

Life Cycle. Several species of small biting flies called "sandflies" are mainly responsible for transmission to humans and reservoirs. The flies ingest the amastigote form directly from the infected skin or from blood or tissue juices. After being ingested, the parasites develop into the promastigote stage in the midgut (Fig. 89-5). They multiply rapidly, and within a few days many have migrated into the pharynx and mouth parts. In this location, they are ready to be injected into the skin of the next individual when the fly takes another blood meal.

Pathology and Symptomatology. *L. donovani*, agent of kala azar, causes hyperplasia of the cells of the lymphoid-macrophage system, especially of the spleen, liver, and bone marrow. Both the liver and spleen become enlarged. The usual signs and symptoms include an undulant-type fever, loss of weight (which may be masked by edema), abdominal protuberance, visible pulsation of the carotid arteries, bleeding of the gums, lips, and nares, anemia, and hemorrhage from the intestinal mucosa. The case fatality rate in some areas may be as high as 90 percent in untreated cases. *L. tropica* and *L. braziliensis* infections produce local lesions that appear first as a macule, then as a papule with a slightly raised center over a crater. Later the lesion opens at the center to discharge necrotic material. The *L. tropica* ulceration usually occurs late, after which rapid healing is the rule, always with the formation of a scar. However, secondary infections may occur, slowing the healing process and causing more extensive scarring. The lesion produced by *L. braziliensis* develops more rapidly, and ulceration and secondary infection usually occur early in the disease. After extension to the mucous membranes, destruction of tissue usually is considerable, and even if healing occurs, the scars are disfiguring to the extent of causing deformity of the face.

Epidemiology. Two different types of transmission of visceral leishmaniasis (kala azar) are seen. In an urban situation, transmission is primarily human to human. In the rural type, infection is epizootic in rodents and other wild animals, and humans are somewhat accidental hosts. Clinical variants of the infection have been described and are considered by some to be due to different strains or subspecies. These variants range from being the classic form of kala azar in India to a form in Kenya that may be manifested primarily as skin lesions.

Vaccination by inoculating serum from naturally acquired lesions into a nonimmune person is used in various countries to prevent the development of cutaneous leishmaniasis. It should be added that recently the New World forms that cause cutaneous ulcers generally have been referred to as members of the *L. braziliensis* complex, and evidence suggests strongly that the cases confirmed recently in Texas were caused by forms within this complex.

Diagnosis. The onset of visceral leishmaniasis may be sudden or, as more often the case, gradual. When development is sudden, diarrhea and fever resembling typhoid fever are seen. Hepatosplenomegaly, ascites, and lymphad-enopathy may develop during the first few months following infection. Darkening of the skin and other alterations in skin pigmentation may be observed on the forehead, mouth, chest, and legs.

The clinical manifestations of cutaneous leishmaniasis are largely determined by the site of infections. Facial lesions, especially those around the nose and mucous membranes, may be quite extensive and give rise to serious disfigurement. The differential diagnosis of this form of leishmaniasis should include syphilis, yaws, leprosy, tuberculosis, blastomycosis, and skin cancer.

In laboratory diagnosis, *L. donovani* may be demonstrated sometimes in stained blood smears. However, cultivation of the organisms by inoculating the blood into special media (NNN, Chang's, and others) and incubating at 20C to 25C is more likely to be successful. The promastigote stage occurs in cultures (Fig. 89-5). Many workers prefer to prepare stained tissue smears following removal of a specimen by biopsy. In the past, spleen and liver specimens were most often used, but today most workers favor the use of a bone marrow specimen from the iliac crest obtained by the van den Bergh technique. A portion of each sample should also be handled aseptically and cultured. If these various methods fail to reveal the amastigotes in suspected cases, hamsters may be inoculated. An immunofluorescence test and a complement-fixation test are available and of some value.

L. tropica and *L. braziliensis* may be demonstrated in stained smears made from the crater of an early lesion or, less often, from material obtained from the indurated margin of older lesions. The material can also be cultured, but care must be taken to cleanse the area before taking a sample, as the organisms seldom will grow in the presence of bacteria. A complement-fixation test and a skin test (Montenegro) are available for diagnosis of both *L. tropica* and *L. braziliensis*, but the skin test has been more widely used in clinics and in the field to diagnose *L. braziliensis*.

A very large increase in the amount of serum IgG is the most prominent laboratory finding in those infected with visceral leishmaniasis; such an increase is not noted in either the cutaneous or mucocutaneous forms. Anemia and leukopenia with a relative increase in lymphocytes are also characteristic of the visceral form.

Treatment. Stibogluconate sodium, a drug available from the Parasite Diseases Division at the Centers for Disease Control, is the drug of choice for all forms of leishmanial infection. As alternative drugs, amphotericin B can be used to treat the mucocutaneous forms, and pentamidine may be used for kala azar. After secondary bacterial infections of ulcers, the sulfa drugs or antibiotics are used to destroy these agents before specific treatment is given.

Sporozoa

Sporozoa are obligate intracellular protozoa with no organelles of locomotion. Most species produce a spore, which is the infective stage for humans. Infection is either by ingestion or by injection as a result of a biting arthropod. Sexual and asexual generations are present in each species.

Within the sporozoa are two of the most important and widespread parasites of humans, causing malaria and toxoplasmosis. *Plasmodium,* the genus of malaria, is composed of four species (one of which has developed drug-resistant strains) that are of major medical and public health importance. In addition, the mosquito that transmits this disease to humans has also developed resistance to various insecticides, and this hampers control efforts. *Toxoplasma,* the agent of toxoplasmosis, is a sporozoan that infects and multiplies in nucleated cells of almost all vertebrate animals. The domestic cat is the definitive host, and humans can become infected in a variety of ways. Although fetal death, congenital infection leading to mental retardation, and blindness are known to be caused by this parasite, it is still poorly understood and, like visceral larva migrans, may eventually be proved to be of even more importance in human morbidity.

Malaria

The malaria parasites, once endemic in the United States, are members of the genus *Plasmodium,* and their life cycles involve both an asexual phase (schizogony in humans as the intermediate host) and a sexual phase (sporogony in certain female anopheline mosquitoes as the definitive host). There are four species of human malarial parasites: (1) *Plasmodium vivax,* agent of benign, tertian malaria, is the most widely distributed and on a worldwide basis is the most prevalent species. (2) *Plasmodium falciparum,* agent of malignant, subtertian malaria, is as prevalent as *P. vivax* in subtropical and tropical regions but fails to establish itself in areas where there are long, cold seasons. (3) *Plasmodium malariae,* agent of quartan malaria, is limited almost entirely to tropical and subtropical areas, where it is considerably less prevalent than *P. vivax* and *P. falciparum.* (4) *Plasmodium ovale* is the agent of ovale malaria, a tertian-type of malaria, less common than the other three types, and has been reported sporadically from widely separated regions in Africa, South America, and Asia. It appears it has almost completely supplanted *P. vivax* on the West African coast. The parasite resembles *P. vivax* in certain characteristics and *P. malariae* in others.

Life Cycle. Although there are striking morphologic differences among the four species, the general features of their life cycles are the same. Specific differences will be considered below under laboratory diagnosis. The cycle in humans begins with the inoculation of infective sporozoites by a female anopheline mosquito. Within 40 minutes, the sporozoites disappear from the circulating blood, and no parasites can be demonstrated in red blood cells for many days. It is now known that during this negative phase, the parasite is residing in fixed tissue cells of the liver. The various stages of the parasite observed in exo-erythrocytic foci resemble those seen later within the red

blood cells, but the characteristic malarial pigment (hemozoin), derived from the breakdown of hemoglobin, is, of course, not seen. After the parasites develop in exoerythrocytic foci for many days, their density increases, and certain of the progeny enter the bloodstream and initiate erythrocytic infection. It should be added that the exoerythrocytic forms of *P. vivax* and *P. ovale* in the liver may persist long after the eradication of the erythrocytic forms and, thus, are believed to be the cause of later episodes of parasitemia and clinical relapses.

The first stage observed within the red blood cells is the trophozoite. The youngest is referred to as a "ring," which has a central vacuole and a ring of cytoplasm containing a chromatin dot. The growth of the parasite proceeds gradually, the vacuole disappears, and the cytoplasm increases in size. With the increase in volume of the cytoplasm, pigment increases in amount with increasing age of the parasite, since it is a waste product of hemoglobin metabolism. When the single nucleus of the trophozoite stage divides to form two nuclei, the schizont stage is reached. The chromatin continues to divide until the number of chromatin masses characteristic for the species is reached. At this point, the presegmenter has been produced. The segmenter or mature schizont is observed soon thereafter, when each chromatin mass has been provided, after division of the cytoplasm, with an envelope of cytoplasm to form the merozoites. By this time, the pigment, scattered previously, has become clumped, usually near the center of the parasite. The red blood cell ruptures, and some of the merozoites enter new cells to begin again the asexual cycle. The length of this cycle—from the entry of the merozoite to the rupture of the host cell—varies with the species of parasite, being 48 hours for *P. vivax* and *P. ovale,* 72 hours for *P. malariae,* and 36 to 48 hours for *P. falciparum.* The cycle of *P. falciparum,* however, has considerably less synchronization than those of the other three species. In addition to trophozoites and schizonts, a third stage, gametocyte, is seen within the red blood cells. The gametocytes, male and female sex cells, nearly fill the red blood cell and have only a single chromatin mass. It is this stage that initiates the sporogenous cycle in the female anopheline. In summary, there are only three distinct stages within the red blood cells of humans: the trophozoite, the schizont, and the gametocyte.

After a female anopheline ingests a blood meal containing ripe gametocytes of both sexes, the sexual cycle begins. In the lumen of the midgut, the gametocytes escape from the host cells and soon undergo changes preparatory to fertilization. The female gametocyte (macrogametocyte) extrudes polar-like bodies, indicating that the haploid condition is being assumed. The male gametocyte (microgametocyte), through a process known as "exflagellation," forms a number of sperm-like bodies. The fact that these changes in both cells are noted within about 20 minutes under favorable conditions, suggests that the lining up process within the nucleus of each actually begins in the human bloodstream. In any event, fertilization occurs when a microgamete enters a macrogamete, and a zygote is formed. Being motile, this form is referred to as the "ookinete." The ookinete penetrates beneath the peritrophic membrane lining the midgut and, after about 24 hours, penetrates through the cells and becomes encysted under the hemocoelic membrane on the outside wall of the midgut. Here the oocyst is formed, which, when mature, contains a large number of sporozoites. The oocyst ruptures into the body cavity after about 2 weeks, and many of the sporozoites eventually find their way into the trilobed salivary glands. After a few days under optimum conditions, the sporozoites become infective and are capable of initiating an infection in a susceptible individual after being inoculated when the mosquito punctures the skin to take a blood meal.

Pathology and Symptomatology. The most characteristic features of the pathology of malaria are anemia, pigmentation of certain organs, and hypertrophy of the liver and, especially, the spleen. The anemia—a microcytic, hypochromic type—may be produced not only by the direct loss of red blood cells as a result of their destruction by the parasite but also from interference with hematopoiesis, by increased phagocytosis of red blood cells, and as the result of capillary hemorrhages and thrombosis. In acute cases, *P. falciparum* produces the greatest degree of anemia, especially because of the marked loss of red blood cells by the growth of the parasite within them. In chronic cases and especially during malarial cachexia, the anemia is particularly outstanding. During the latter cases, leukopenia, with a 20 percent or more monocytosis, is considered diagnostic of malaria. The pigmentation noted in the tissues of malaria victims is caused by the phagocytosis of hemozoin, the true malarial pigment, released into the blood upon rupture of the host cells at the termination of each asexual cycle. Hemozoin is taken up in large amounts by cells of the lymphoid-macrophage system, especially by the macrophages in the spleen and bone marrow, and the Küpffer cells in the liver. The pigmentation constantly increases with the age of the infection, so that it may be observed grossly in autopsied cases of chronic malaria. As would be expected from the massive blood destruction, there is also deposition of pigment in the tissues. The liver is enlarged due to congestion during acute malaria, and it increases considerably in size during a chronic infection. The spleen, the organ affected most seriously in malaria, is also enlarged, first as a result of congestion (following cavernous dilatation of the sinusoids) and later as a result of a great increase in macrophage elements, especially in Billroth's cords. With repeated attacks, the enlargement becomes progressively greater, and the organ may reach considerable size, especially in width, in chronic malaria. Fibrosis of the Billroth's cords is outstanding here. Splenomegaly in malaria is so characteristic that palpation of this organ has been used for a long time as a rapid and effective means of appraising the malaria problem in communities. Changes in the bone marrow, although much less striking, are similar in character to those in the spleen.

In addition to these changes that may be expected in any malarial infection, but especially in those of long standing, another important change, capillary occlusions,

should be mentioned. These are most characteristic in *P. falciparum* infections and are most dangerous in the brain. The parasitized red blood cells become sticky (some believe as a result of a specific antibody–antigen reaction) and agglutinate. Such cells marginate at the periphery of the vessel lumen, probably as a result of centrifugal force, and later the capillary becomes occluded. Following this, hemorrhages occur about such vessels, exclusively in the subcortical and paraventricular white matter, producing a ring effect. The tissue immediately surrounding the vessel is necrotic, and the ring hemorrhage is somewhat removed from the vessel. Usually associated with ring hemorrhages are the so-called malarial granulomas, consisting essentially of a rosette of one or several layers of glial cells arranged around the necrotic zone. Anoxia with necrosis of tissue in the immediate vicinity must be the logical consequence of occluded vessels.

Blackwater fever should be mentioned, because it is associated with malarial infections, especially *P. falciparum* infections. The disease is characterized by intravascular hemolysis with hemoglobinemia and hemoglobinuria. Hemoglobinuric casts occur in the distal convoluted tubules in the kidney. Degeneration, and some regeneration, of the tubular epithelium is also seen.

The symptomatology of malaria differs between *P. vivax, P. ovale,* and *P. malariae* infections and that caused by *P. falciparum.* The typical paroxysm caused by the first three parasites involves, usually after a brief prodromal period, a cold stage (shaking chill), followed by a fever stage (a characteristic remittent-type fever quickly reaching a level of 41 to 42C and, after many hours, returning suddenly to near normal), and a marked terminal sweating stage resulting from the sudden fall in body temperature. *P. falciparum* paroxysms differ in many ways: the chill stage is less pronounced (there may be only a chilly sensation), the fever stage is more prolonged and intensified (fever tends to be a continuous or only a briefly remittent type), and because the fever fails to remit sharply, the sweating stage is usually absent. *P. falciparum* infection is more dangerous than those of the other three species, since it often is accompanied by pernicious manifestations, such as hyperpyrexia, convulsion, coma, and/or cardiac failure. *P. falciparum* parasites may localize in any organ, and those organs bearing the brunt of the attack will give rise to the most striking signs and symptoms. Therefore, the symptomatology of *P. falciparum* malaria may resemble that of many other diseases.

Epidemiology. Malaria is present in almost all tropical and subtropical areas. The Polynesian and Micronesian islands and Australia are about the only such areas free of this disease. At one time malaria was widespread in Europe and the United States; fortunately, due to control efforts it has now largely disappeared from the temperate zones. A recent report shows two autochthonous cases of *P. vivax* in California.

Malaria is a real threat to nonimmune persons who travel or reside temporarily in malarious areas. The recent appearance of drug-resistant strains of *P. falciparum* has increased this threat, and clinicians in the United States should be well aware of the steps that need to be taken to prevent their patients from contracting malaria when they travel abroad to tropical and subtropical countries.

Rainfall, temperature, and vegetation are factors that influence mosquito breeding and thereby affect seasonal transmission. Other epidemiologic factors affecting endemicity are prevalence in the indigenous host population, genetic factors of the host including those controlling the Duffy blood group and sickle cell antigens, parasite species and strains, immunity level of the population, and housing conditions. In endemic areas, children bear the brunt of malarial mortality, with an estimated 1 million deaths each year in Africa alone.

Diagnosis. *P. vivax, P. ovale,* and *P. malariae* cause an illness characterized by a high fever of a characteristic periodicity in well-established infections, anemia, and splenomegaly. Infections with these species are usually self-limiting and usually do not cause serious illness or death in untreated cases. On the other hand, *P. falciparum* ju infection is potentially life-threatening, and unless diagnosed and treated promptly it can lead to rapid death. Cerebral disease may be manifested by symptoms that range from headache to convulsions to coma. Kidney and liver dysfunctions are also noted, and, not infrequently, an attack may mimic influenza.

Definitive diagnosis is made by the demonstration of malarial parasites in blood smears. Both thick smears, for the detection of organisms, and thin smears, for speciation, are recommended. Smears should be stained with Giemsa stain as promptly as possible when malaria is suspected. Repeated blood smears should be made in the case of suspected infection when the initial smears are negative. Comparative information on the four species of malarial parasites in thin blood smears, including the most important differential diagnostic characteristics, is given in Tables 89-5 and 89-6.

Indirect fluorescent and indirect hemagglutination tests are available. However, they are used primarily for epidemiologic purposes and have little application in the clinical laboratory because they will not differentiate between an active and a past infection.

Treatment. The initial aim of treatment should be to eliminate as soon as possible an acute attack by eliminating the red blood cell stages of the parasite. At the same time, steps to combat dehydration, renal failure, hypoxia, and so on must be undertaken, especially in *P. falciparum* infection. Chloroquine is currently the drug of choice to accomplish this with all species of malaria. Chloroquine does not, however, eliminate the persisting tissue stages in the liver of those infected with either *P. vivax* or *P. ovale.* Primaquine should be administered to accomplish this in order to minimize the chance of subsequent relapse.

Quinine sulfate plus pyrimethamine are the drugs recommended for the treatment of chloroquine-resistant *P. falciparum.* Resistance is not total, and, as a result, some suppression does occur in drug-resistant strains. However, recrudescence will occur in several days to several weeks, reflecting a failure of the chloroquine to eliminate all erythrocytic parasites.

Currently, chloroquine plus fansidar (sulfadiazine plus pyrimethamine) are the recommended prophylactic treatment for those living in nonendemic areas who plan to travel to endemic areas. Table 89-3 lists the drugs and dosages currently recommended for the treatment of all types of malaria.

Toxoplasmosis

Toxoplasmosis, caused by *Toxoplasma gondii,* is a cosmopolitan and common infection afflicting nearly a third of the human race. It has an amazing lack of host specificity, being found, for example, in various primates, carnivores, rodents, birds, and ungulates. Due to the wider use of immunosuppressive treatment that permits exacerbation of latent infection as well as permitting primary infection, toxoplasmosis is now recognized as an important medical and public health problem. Fortunately, most infections are asymptomatic. When overt disease is present, however, it leaves blindness, deformity of the newborn, mental retardation, and death in its wake.

Life Cycle. Domestic house cats and other felines are the definitive hosts of *T. gondii.* In the definitive host, an asexual cycle as well as a sexual cycle occurs in the epithelial

TABLE 89-5. DIAGNOSTIC CHARACTERISTICS OF MALARIAL PARASITES

	Plasmodium vivax	Plasmodium malariae	Plasmodium falciparum	Plasmodium ovale
Other names	Benign tertian malaria	Quartan malaria	Malignant tertian estivo-autumnal malaria	Benign tertian or ovale malaria
Incubation period (days)	14 (8–27) (sometimes 7–10 months)	15–30	12 (8–25)	15 (9–17)
Erythrocytic cycle (hr)	48	72	48	48
Persistent EE stages	Yes	No	No	Yes
Parasitemia (mm³)				
Average	20,000	6,000	50,000–500,000	9,000
Maximum	50,000	20,000	Up to 2,500,000	30,000
Duration of untreated infection (yr)	1.5–4.0	1–30	0.5–2.0	Probably 1.5–4.0

TABLE 89-6. DIFFERENTIAL DIAGNOSIS OF MALARIAL PARASITES IN STAINED THIN BLOOD SMEARS

Most Striking Differences	Plasmodium vivax	Plasmodium ovale	Plasmodium malariae	Plasmodium falciparum
Abundance of parasites	More abundant than P. malariae	As for P. vivax	Least abundant	Most abundant
Stages of parasite usually observed	Trophozoites, schizonts, gametocytes	As for P. vivax	As for P. vivax	Young trophozoites, gametocytes
Changes in infected red blood cells	Enlarged (2× normal), malshaped, Schüffner's dots	Enlarged (1½ × normal), 20% and/or fimbriated	Changes not common	Changes not common
Trophozoites ring stage	Small and large (⅓ diameter of red blood cell) with vacuole and usually one chromatin dot	Much like P. vivax	Much like P. vivax	Very small (⅙ diameter of red blood cell) with vacuole, often with 2 chromatin dots, peripheral forms (accolé) common, multiple infected cells very common
Half-grown	Ameboid, irregular with vacuoles, pigment yellow-brown rods	Oval and/or fimbriated (20%) enlarged cell, compact dark brown pigment	Band forms (25%) compact, pigment coarse black granules	Not usually seen
Schizonts (mature)	12–24 merozoites	6–14 merozoites	8–12 merozoites arranged in rosette	Not usually seen
Gametocytes	Large and rounded, fills up red cell, golden brown pigment	Red cell enlarged and/or fimbriated, dark brown pigment	Parasite fills up cell, dark brown pigment	Kidney-bean shape, round or pointed ends

cells of the small intestine. The end result is the production of oocysts, which are passed in the feces. After a few days, these oocysts sporulate, so that each one contains two sporocysts, each with four sporozoites. If these mature oocysts are ingested by an intermediate host (animals other then felines, including humans), infection results, and trophozoites may be produced in many tissues. Human infection can occur, thus, not only by ingestion of cysts in undercooked meat, as has been known for years, but by ingestion of mature oocysts as well. Recent evidence of an epidemic in army troops on maneuvers in Panama indicates that the infection may be waterborne.

Pathology and Symptomatology. *T. gondii* is an obligate intracellular parasite. The individual cells, known as tachyzoites, are 4 to 7 µm long and are crescent shaped (Fig. 89-6). They may be found at times within wandering macrophages in peritoneal, pleural, and cerebral exudates and in circulating blood. The cells of the lymphoid-macrophage system are most often involved, but the parenchymal cells of the liver, lungs, brain, and other tissues may be parasitized. A cyst, 5 to 100 µm or more in size, may be observed under certain conditions. A cyst is formed by aggregates of *Toxoplasma* bradyzoites and has a delicate, delimiting argyrophilic membrane produced by the parasite. Tissue reaction is seldom associated with cysts, which are usually located in the lungs, heart, or brain. Hence, the cysts are believed to be the basis for the persistence of the organism during chronic and latent infections. The release of bradyzoites from cysts must be responsible in

most instances for the fulminating *Toxoplasma* infections reported in immunosuppressed patients. Such reports raise the question of whether suppression similarly affects other previously established protozoan infections. These organisms may give rise to at lease five types of infections:

1. a congenital infection with onset in utero
2. an acquired encephalitic infection in older children
3. an acute febrile illness, usually in adults, resembling typhus or spotted fevers and often producing pulmonary involvement (a typical diffuse interstitial pneumonitis), myocarditis, and so on
4. an infection resembling infectious mononucleosis, with lymphadenopathy, rash, fever, marked weakness, and so on
5. a latent infection, in children or adults, which usually can be recognized only be the presence of specific antibodies in the serum.

In addition, there is evidence of an association between *Toxoplasma* organisms and unilateral granulomatous uveitis and retinochoroiditis in adults. The congenital infection, with its onset in utero, occurs as a fetal or neonatal encephalomyelitis, which often is fatal soon after birth but which may remain asymptomatic until much later. Marked lesions and necrosis usually occur in the central nervous system and are associated with calcification there and in the eyes. Bilateral retinochoroiditis is common. At times, hydrocephaly or microcephaly and psychomotor disturbances are evident. In infections acquired after birth, lesions in the viscera are more common than those in the

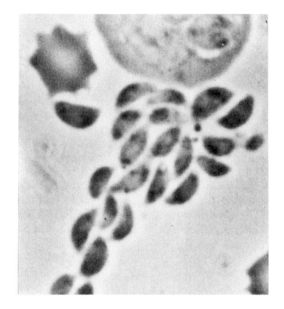

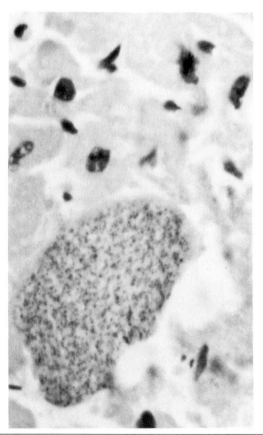

Figure 89-6. *Toxoplasma gondii.* **A.** Tachyzoites. **B.** Cyst containing bradyzoites.

central nervous system. In summary, during toxoplasmic infection, the sites most commonly attacked are the lymph nodes, brain, eyes, and lungs.

Epidemiology. Seroepidemiologic screening surveys have shown that seropositivity may range from 2 to 93 percent worldwide, depending on the population surveyed. In the United States, rates vary from 17 to 50 percent. Although toxoplasmosis is usually a latent infection, outbreaks in the United States have been reported since 1969 as a result of eating raw hamburger, inhalation or ingestion of oocysts at a riding stable, and drinking contaminated water.

Research has shown that no more than 1 percent of domestic cats in the United States may be shedding oocysts. Whether cats shed oocysts only once or several times is not known, nor is the immune status of infected cats known.

To prevent infection, which is especially important for nonimmune women who become pregnant, the following precautions are recommended. Meat should be heated to at least 66C before it is eaten, and raw meat should never be fed to cats. If possible, keep cats indoors and change cat litter in boxes daily, disposing of the litter by either burning or flushing down the toilet. When people are working in flower or vegetable gardens, places where cats frequently defecate, care should be exercised to ensure that dirt is not ingested. Finally, children's sandboxes should be kept covered when not in use.

Diagnosis. The clinical manifestations of toxoplasmosis are highly variable and often mimic other diseases. The majority of patients with acquired infection are asymptomatic. When symptoms are present, they may mimic pneumonia, myocarditis, hepatitis, and lymphadenitis. Fatigue, chills, fever, headache, and myalgia are sometimes seen. Fatal disease may occur in the immunosuppressed patient.

Ocular toxoplasmosis, either acquired or congenital, usually affects only one eye. It is estimated that 5 percent of all blindness in the United States is due to retinal involvement. Congenital involvement, which occurs in 30 to 50 percent of the cases when infection is acquired during pregnancy, may remain asymptomatic in the newborn. When damage does occur, it may be severe. It should be noted that a mother who has transmitted *T. gondii* to her fetus thereafter becomes immune, and no subsequent pregnancies will lead to transplacental infection.

With regard to the laboratory diagnosis, three approaches are available: direct examination, isolation of organisms, and examinations for a presumptive diagnosis. For direct examination under the oil-immersion objective, tissue sections and impression films of suspected tissues or fluids should be air-dried and stained with Giemsa stain. The preparations usually examined are tissues taken by biopsy, sputum, vaginal exudates, and the sediment of spinal, pleural, or peritoneal fluids. In an attempt to isolate the organisms, white mice should be inoculated intraperitoneally with fresh, untreated tissue or fluids most likely to contain the organisms. If organisms are present, a generalized infection will be produced in 5 to 10 days, at which time the organisms usually can be demonstrated easily in the extensive peritoneal exudate. It should be added that this is an excellent source of organisms to prepare antigens for serologic tests. The animals that die may be examined for *Toxoplasma* organisms by preparing films or sections of the peritoneal fluid, lungs, brain, and other tissues. A presumptive diagnosis can be made by a positive delayed skin reaction or by serologic means. Many serologic tests are available to assist in the diagnosis, including the Sabin–Feldman dye test, which is still widely used. This test depends on the fact that in the presence of specific antibodies, the cytoplasm of the living organisms loses its affinity for methylene blue. Newer tests, such as the indirect hemagglutination and indirect fluorescent antibody and ELISA tests, have been reported. Of all tests available, the ELISA appears to be best.

Treatment. Triple sulfonamides (that is, equal parts of sulfadiazine, sulfamerazine, and sulfamethazine) and pyrimethamine, which act synergistically, are effective and should be used as the treatment of choice for toxoplasmosis. A corticosteroid is suggested for treating eye involvement.

Unclassified *Pneumocystis* Organism

Pneumocystis carinii, an organism cosmopolitan in distribution, is generally accepted as a protozoan, but its taxonomic position has not been determined. The organism is present in a wide variety of animals, and, in humans, causes a diffuse pneumonia in the immunocompromised host. Even in fatal cases, the organism and the disease usually remain localized to the lung. Rarely does the disease occur in healthy people.

Life Cycle. The exact life cycle is unknown, but various structural forms have been recognized: (1) a thin-walled trophozoite, (2) a thick-walled cyst that measures 4 to 6

μm in diameter, and (3) the intracystic body, a crescent-shaped body within the cyst. As many as eight intracystic bodies may be present in a cyst (Fig. 89-7). Neither the mode of replication nor the mode of transmission is known.

Pathology and Symptomatology. Since the organism is usually found only in lung tissue, it is assumed that the likely mode of transmission is by inhalation. Studies have indicated that the organism does not become intracellular but, rather, attaches to the pneumocyte cell surface during a phase of its replicative cycle. In the vast majority of individuals, the organisms may be dormant and sparsely dispersed, with no apparent damage to host tissue. In others, such as the immunocompromised host, the organisms occur in massive numbers. There is usually a panlobular involvement of the lungs, which are enlarged, of a firm rubbery consistency, and do not collapse when the chest is opened. Fever, tachypnea, hypoxia, cyanosis, and asphyxia are common manifestations in acute cases. In debilitated infants, the alveolar septum is thickened with an interstitial plasma cell and lymphocyte infiltration. Illness may last 4 to 6 weeks, and a mortality rate of 25 to 50 percent is seen. In most overt infections, the onset is abrupt.

Epidemiology. *P. carinii* has been found in the lungs of rats, rabbits, mice, dogs, sheep, guinea pigs, horses, and a variety of lower animals. Since up to 70 percent of healthy individuals have humoral antibody to the organism, subclinical infection must be widespread.

Recently a new syndrome—acquired immune deficiency syndrome (AIDS)—has been described, with over 2500 cases occurring since June 1981. An analysis of AIDS cases shows that there is some relationship between Kaposi's sarcoma and various opportunistic infections, including *P. carinii*. Approximately 75 percent of AIDS cases have occurred among homosexual ("gay plague") or bisexual

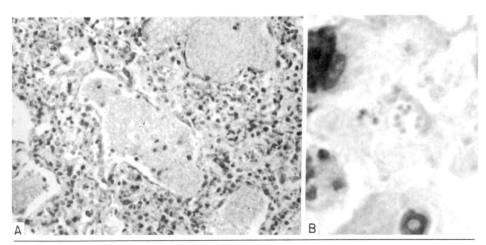

Figure 89-7. *Pneumocystis carinii* in lungs. **A.** Note ground-glass appearance of exudate. $\times$ 250. **B.** Organism showing eight intracystic bodies. *(From Brown and Neva: Basic Clinical Parasitology, 5th ed, 1983. Norwalk, Conn, Appleton-Century-Crofts.)*

men, and nearly 80 percent of all cases were concentrated in metropolitan areas on the east and west coasts of the United States. A more detailed analysis of AIDS cases shows that they may be separated into groups based on these risk factors: homosexual or bisexual males, 75 percent; intravenous drug abusers with no history of male homosexual activity, 13 percent; Haitians with neither a history of homosexuality nor a history of intravenous drug abuse, 6 percent; persons with hemophilia A who were not Haitians, homosexuals, or intravenous drug abusers, 0.3 percent; and persons (including children) in none of the other groups, 5 percent.

Diagnosis. Tachypnea and fever are consistent features of *P. carinii* pneumonia. In premature and debilitated infants, onset is subtle, with mild tachypnea. After a week, respiratory distress becomes severe, with flaring of the nasal alae and cyanosis. In the immunodeficient child or adult, onset of symptoms is rapid and quickly progresses to death if untreated.

Definitive diagnosis depends on the identification of *P. carinii* in samples taken by transbronchoscopic biopsy, open lung biopsy, or needle aspiration. Sputum, although not a reliable specimen, should be examined before invasive measures are taken. Cysts and extracystic bodies released from them can be found in smears or sections of lung tissue. Methenamine silver nitrate, toluidine blue, and Giemsa-type stains are used to detect and identify them. A diffuse bilateral alveolar disease is apparent by radiograph. Serum antibody titers are of little diagnostic value. Counterimmunoelectrophoresis techniques to detect circulating antigen may become a useful diagnostic procedure in the near future.

Treatment. Drugs available are pentamidine, trimethoprim, and sulfamethoxazole, with the latter two drugs being the drugs of choice. However, resistance is showing up with these two drugs, and, in this situation, pentamidine must be used. Supportive measures include high oxygen therapy with volume respirator assistance and digitalization to support the myocardium damaged by hypoxia.

Medical Helminthology

In addition to the Protozoa, there are numerous many-celled animals, often referred to as Metazoa, that parasitize humans. "Helminth" is a general term meaning worm and, in the present context, refers to those animals that belong to two groups within the animal kingdom, the Nematoda and the Platyhelminthes. The nematodes are represented by a single group and are commonly called roundworms. The Platyhelminthes are represented by two distinct groups of medical importance, namely, the Cestoda or tapeworms and the Trematoda or flukes. Together, the tapeworms and flukes are commonly called flatworms. In medical literature, helminths are often grouped according to the host organ in which they reside, such as blood flukes, liver flukes, intestinal roundworms, and tissue roundworms. This chapter will present the helminths of humans in the following order: nematodes, cestodes, and trematodes.

Nematodes

The adult nematodes, or roundworms, are characterized by having an elongate, cylindrical body that is round in cross-section. They are covered with an acellular cuticle and have a complete digestive tract, with mouth and anus, as well as excretory, nervous, and reproductive systems. As in all parasitic worms, sex differences are the most conspicuous. The sexes are separate, and the males are almost invariably smaller than the females (Fig. 90-1). Nematodes vary considerably in size, from forms difficult to see readily by the unaided eye to others many centimeters in length. In order to increase in size, the immature worm (larva) passes through a series of molts, or ecdyses, which are accomplished by shedding the cuticle. Complete stages

in the life cycle are: egg, four stages of larvae separated by molts, and adult male or female (Fig. 90-2).

In presenting the material of this section, nematodes will be grouped according to the usual location of the adult worms in humans. The main groups to be considered are (1) the intestinal nematodes and (2) the tissue nematodes. All of the intestinal nematodes are endemic in the United States, whereas the tissue forms are exotic in origin. As an aid to the student, the intestinal nematodes will be divided into two groups: (1) those infective for humans in the egg stage and (2) those infective for humans in a larval stage. Under each grouping, they will be presented in increasing order of the complexity of the life cycle.

Intestinal Nematodes Infective for Humans in the Egg Stage

Enterobius vermicularis
The pinworm, sometimes called the seatworm, is probably the most frequently encountered helminth parasite in the United States. Indeed, it is not unreasonable to think that nearly everyone, at one time or another, has had pinworms. Pruritus ani and occasional bouts of abdominal discomfort in children are the primary manifestations of infection. Although it is rare, most physicians are not aware of the fact that enterobiasis can cause severe, ectopic disease in various parts of the body, especially the urogenital tract of both female children and female adults.

Pinworm is cosmopolitan in distribution and is considered by most to be more prevalent in temperate than in tropical areas. Recent surveys show, however, that 67 percent of kindergarten-aged children in Shanghai, 80 percent in Central Europe, and 70 percent in West Africa are

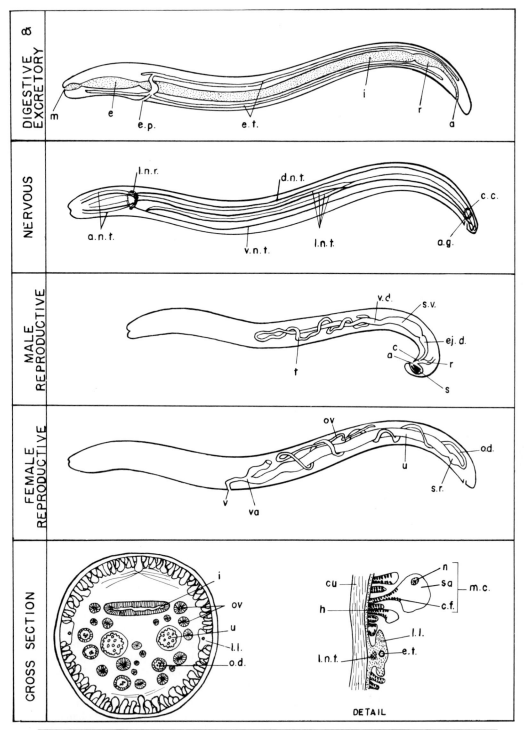

Figure 90-1. Morphology of a nematode, based on *Ascaris*. a, anus; a.g., anal ganglion; a.n.t., anterior nerve trunks; c, cloaca; cu, cuticle; c.c., circumcloacal commissure (male); c.f., contractile fibers; d.n.t., dorsal nerve trunk; e, esophagus; e.p., excretory pore; e.t., excretory tubules; ej.d., ejaculatory duct; h, hypodermis; i, intestine; l.l., lateral line; l.n.r., circumesophageal ring; l.n.t., lateral nerve trunks; m, mouth; m.c., muscle cells; n, nucleus; ov, ovary; o.d., oviduct; r, rectum; s, spicules; sa, sarcoplasm; s.r., seminal receptacle; s.v., seminal vesicle; t, testis; u, uterus; v, vulva; va, vagina; v.d., vas deferens; v.n.t., ventral nerve trunk. *(From Brown and Neva: Basic Clinical Parasitology, 5th ed, 1983. Norwalk, Conn, Appleton-Century-Crofts.)*

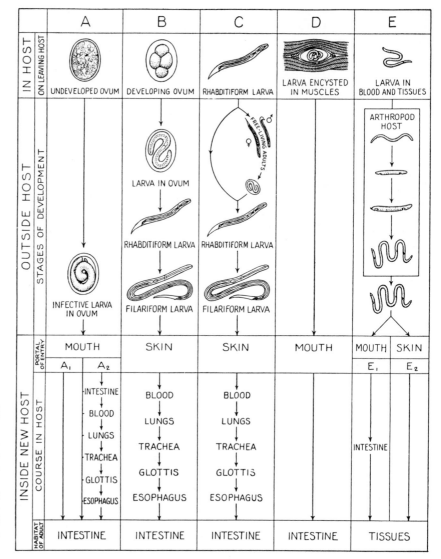

Figure 90-2. Typical life cycles of important nematodes. **A₁**. *Enterobius vermicularis, Trichuris trichiura.* **A₂**. *Ascaris lumbricoides.* **B**. *Ancylostoma duodenale, Necator americanus.* **C**. *Strongyloides stercoralis.* **D**. *Trichinella spiralis.* **E₁**. *Dracunculus medinensis.* **E₂**. *Wuchereria bancrofti, Wuchereria malayi, Loa loa, Acanthocheilonema perstans, Onchocerca volvulus.* (From Belding: Basic Clinical Parasitology, 1958. New York, Appleton-Century-Crofts.)

infected. Thus, this concept may change as more data are collected. In the United States, recent surveys show that 30 to 40 percent of all white kindergarten-aged children have pinworms; American blacks show about one third this amount. Most often, infection is found in groups of children, such as in families, schools, summer camps, day-care centers, and those who are institutionalized.

Life Cycle. Ingested eggs hatch in the small intestine, each releasing a single larva. These migrate to the large intestine, especially the cecum and colon, where they attach superficially to the mucosa. Many are found down at the base of intestinal crypts. The larvae reach adulthood within 2 to 4 weeks. Infection usually lasts 1 to 2 months, but reinfection is common. No specific immunity seems to develop, but resistance to infection does increase with increasing age of the host.

The gravid females migrate to the anus and emerge at night to extrude their eggs on the perianal and perineal skin. Each female may deposit up to 11,000 eggs before she dies. Eggs have a sticky, albuminous covering and adhere to skin, hair, clothing, and bedding. They become infective for humans in 4 to 6 hours. Occasionally, migrating females may enter the vagina, fallopian tubes, uterus, and even into the peritoneal cavity. Such migrations are rare but can cause serious disease. To a limited extent, eggs on the perineum may prematurely hatch, and the released larvae may reenter the intestinal tract (retrofection) through the anus. When this occurs, the cycle may be lengthened considerably.

Pathology and Symptomatology. Anal pruritus is outstanding, and the scratching that follows may lead to scarification, which is subject to invasion by bacteria and other infectious agents. Frequently, children manifest insomnia or fretful sleep, irritability, sometimes nausea, nail biting, and pallor. Abdominal discomfort with diarrhea has also been reported.

Lower pelvic discomfort in female patients, with or without cystitis and enuresis, has been seen and is due to the abnormal migrations of adult worms. Histologic examination has revealed pelvic peritoneal granulomas around adult worms or ova. Larger nodules, white or yellow, are sometimes recovered at surgery or necropsy and may be confused initially with *Mycobacterium tuberculosis* or metastatic carcinoma.

Mild catarrhal inflammation with small lesions in the large intestine are usually seen. In rare cases, necrosis of the mucosa occurs, and sympathetic nerve endings may be exposed. This, and the absorption of worm metabolites across the intestinal border, may be responsible for the commonly observed nervous symptoms. The worm has been found in both normal and diseased appendices but is not considered an important factor in appendicitis.

Despite the high prevalence of pinworm infection, the parasite is relatively nonpathogenic. It is estimated that a full third of those who are infected show no symptoms. It can also be said that most adults who are infected are asymptomatic, whereas infected children usually suffer from perianal itching. This writer suspects that some urinary tract infections are caused by bacteria being passively carried by migrating worms and that some cases of sterility in females are due to pinworm-initiated lesions in the fallopian tubes.

Epidemiology. The intense pruritus produced by the crawling females and, especially, the eggs, results in scratching of the affected areas. In this way, eggs are transferred to the fingertips and may become lodged beneath the nails. Reinfection follows when the fingers or food contaminated by the fingers are placed in the mouth. Aside from this common type of direct transfer of eggs—a hand-to-mouth transfer—the eggs may reach the mouth of the infected individual and others via one or more indirect routes. The eggs adhere to clothing and bed linen, and some get into air currents to be inhaled or to settle on objects. Thus, the household becomes contaminated, and all members sooner or later swallow some of the eggs. Eggs remain viable several weeks under cool and moist conditions. At a temperature above 25C and in dry air, most eggs perish within 1 to 2 days. Good personal hygiene is recommended, of course, but no matter how diligent and meticulous the family is, reinfection usually follows. In the long run, repeated treatment is the only practical solution.

Enterobius vermicularis is the only pinworm that infects humans; there are no animal reservoirs. This is important to remember because many lay people and physicians alike believe that human infection can result from similar infections in their pet cats and dogs. This does not happen, and veterinarians become justifiably disturbed when this misinformation is promulgated.

Diagnosis. Clinical diagnosis, based on the symptoms discussed above, is relatively easy. In the laboratory diagnosis, the characteristic eggs usually are not found in the stool but may be recovered easily from the perianal area by use of a simple cellophane-tape or other swab. The tape, sticky side down, is pressed onto the perianal region, then it is adhered to a microscope slide, where it is then examined. It is best to make the cellophane-tape preparation in the morning before the child has bathed or gone to the toilet. Eggs are asymmetrical, being flattened on one side, and range in size from 50 to 60 μm by 20 to 30 μm (Fig. 90-3). A larva is usually contained in each egg and many times can be seen to move when examined microscopically.

At times, and especially at night a few hours after retiring, migrating adult worms can be seen on the perianal and perineal tissues. Thus, parents may observe worms by using a flashlight while the child is asleep. Not infrequently, and especially with children still in diapers, adult worms are recovered on a stool. Adult worms are small and whitish, with females being 8 to 13 mm long, whereas males are smaller, being 2 to 5 mm long. Both have a characteristic cephalic swelling (Fig. 90-4). No serologic tests are available for diagnosis, and there are no abnormal hematologic or clinical values.

Treatment. Several drugs are effective in treating enterobiasis; mebendazole and pyrantel pamoate are the most widely used. Since the life span of the infection is less than 2 months, the major concern should be to prevent reinfection. Unfortunately, meticulous hygienic efforts are most times ineffectual, and repeated treatment is the only recourse. When infection is noted in a family, many physicians treat the entire household. In large families and institutions where pinworms are perennially present, it is now feasible to deworm the entire population several times a year.

Trichuris trichiura

This nematode is commonly known as the whipworm because of its resemblance to a buggy whip. The anterior end is long and narrow, whereas the posterior end is more robust and shorter. Females may reach a length of 50 mm; males are slightly shorter, seldom reaching 40 mm, and can be recognized by the curled tail (Fig. 90-5).

Infection with whipworm occurs extensively throughout the world but is most common in the tropics and subtropics when poor sanitation exists. The World Health Organization has estimated that some 350 to 500 million people are infected, with an 80 percent incidence in some tropical areas. In the United States, infection is not common but does occur in the warm, moist areas of the South. Areas of high endemicity include the area south of the Piedmont Plateau, in the foothills of the southern Appalachian range, and in southwest Louisiana. Multiple infections with *Ascaris* and hookworms are not uncommon. In

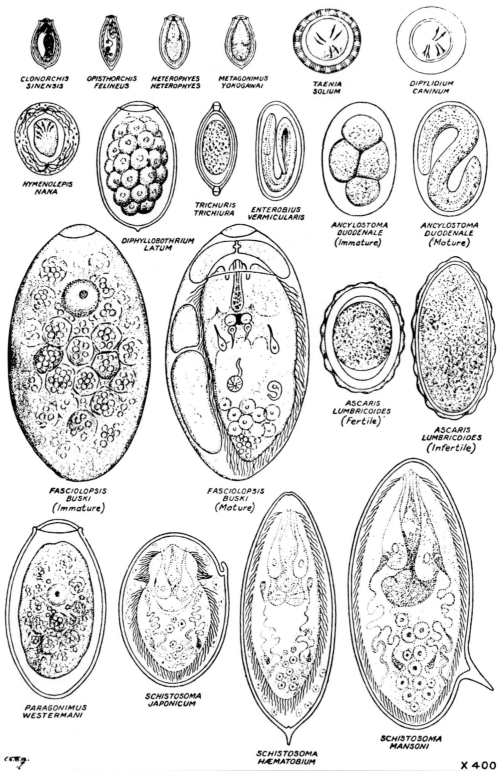

Figure 90-3. Eggs of the common helminths of humans. The size of this illustration has been increased by one fifth. *(From Belding: Textbook of Clinical Parasitology, 2nd ed, 1952. New York, Appleton-Century-Crofts.)*

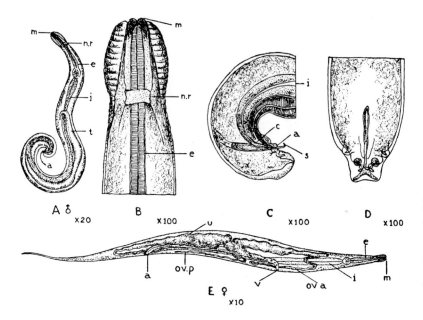

Figure 90-4. *Enterobius vermicularis.* **A.** Male. **B.** Anterior end of worm. **C.** Posterior end of male, lateral view. **D.** Posterior end of male, ventral view. **E.** Female. a, anus; c, cloaca; e, esophagus; i, intestine; m, mouth; n.r., nerve ring; ov a, anterior ovary; ov p, posterior ovary; s, spicule; t, testis; u, uterus; v, vulva. *(From Belding: Basic Clinical Parasitology, 1958. New York, Appleton-Century-Crofts.)*

temperate zones, those most heavily infected are usually inmates of institutions for the mentally retarded, especially among those who eat dirt.

Life Cycle. Infection is acquired by the ingestion of food or water that is contaminated with embryonated eggs. The incubation period is 1 to 3 months, and infection may persist as long as 6 years. When infective eggs are ingested, they hatch in the small intestine. The freed larvae enter nearby intestinal crypts and penetrate into glands and stroma. They reenter the lumen of the small intestine from these sites after about 10 days and migrate to the large intestine. Here the anterior ends of the developing adults are sewn into the mucosa of the large intestine, typically of the cecum and appendix. The robust posterior end of each worm hangs freely in the lumen of the intestine. At sexual maturity, copulation occurs and daily egg production of 2000 to 6000 eggs per female worm begins.

Eggs released by the female worms occur in the stool while in the undeveloped (noninfective) stage. Under favorable environmental conditions, which include warm temperature, shade, moisture, and sandy humus soil, eggs develop and become infective in 3 to 6 weeks. *Trichuris* eggs lack the great resistance of *Ascaris* eggs, and their survival is considered to be relatively brief, i.e., a matter of weeks.

Pathology and Symptomatology. The bloodsucking worms damage the tissues that they penetrate and may carry bacteria and other infectious agents into these sites. If the heads of the worms penetrate into blood capillaries, petechial hemorrhages are produced. The degree of damage corresponds to the number of worms involved. In light infections, relatively little damage results, but in heavy infections, the mucosa is hyperemic and eroded su-

perficially and may be inflamed extensively. Extreme irritation in the wall of the lower colon and rectum may provoke partial or complete prolapse of the rectum. Depending upon the degree of infection and reaction of the individual, the signs and symptoms vary from mild (discomfort in the right lower quadrant, flatulence, loss of appetite and weight, and so on) to severe (nausea, vomiting, mucous diarrhea or dysentery, anemia, and so on).

Epidemiology. Infections occur most often in areas where humidity is high, temperatures are warm, and soils have good moisture-holding ability. Dooryard pollution and poor sanitation lead to heavy infection. Infections usually coexist with *Ascaris* and, many times, *Entamoeba* and hookworm. Distribution may be more spotty than that of *Ascaris*, however, and this is thought to be due to the lesser resistance of the eggs of *Trichuris*. In general, the severity of infection is correlated with age, nutritional status, and worm burden. Malnourished children are the most severely infected. No animal reservoirs exist.

Diagnosis. Diagnosis based on symptoms is difficult, but parasitologic diagnosis is easily made by finding the characteristic eggs in the feces (Fig. 90-3). These are about 20 to 50 μm in size and are barrel-shaped, having a golden brown color and transparent prominences called polar plugs at each end. Eggs may be few in number and difficult to find in fecal smears, in which case it is of assistance to use a concentration method.

Adult worms are rarely seen in the stool because of the firmness with which they are fastened to the intestinal wall. Adults can be seen, however, in heavy infections when the rectum has prolapsed or as a result of sigmoidoscopy.

Anemia of the microcytic, hypochromic type, reduced hemoglobin concentrations (as low as 3 g), and an eosino-

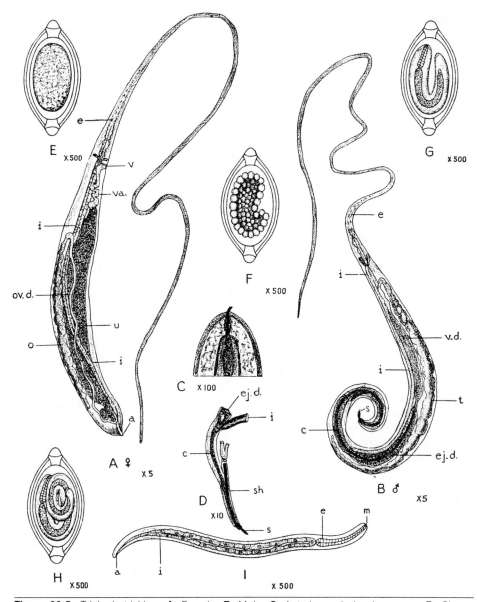

Figure 90-5. *Trichuris trichiura.* **A**. Female. **B**. Male. **C**. Anterior end showing spear. **D**. Cloaca and copulatory organs of male. **E**. Unicellular stage of egg. **F**. Multicellular stage of egg. **G**. Early larva in egg shell. **H**. Mature larva in egg shell. **I**. Newly hatched larva (**A, B, D-I** adapted from Leuckart, 1876. **C** drawn from photograph by Li, 1933.) a, anus; c, cloaca; e, esophagus; ej.d., ejaculatory duct; i, intestine; m, mouth; o, ovary; ov.d., oviduct; s, spicule; sh, sheath of spicule; t, testis; u, uterus; v, vulva; va., vagina; v.d., vas deferens. *(From Brown and Neva: Basic Clinical Parasitology, 5th ed, 1983. Norwalk, Conn, Appleton-Century-Crofts.)*

philia of 5 to 20 percent are sometimes associated with infection. No serologic tests are available.

Treatment. Mebendazole is considered the drug of choice but is not effective in all cases, especially in heavy infections. It is suggested that light, asymptomatic cases not be treated.

Ascaris lumbricoides

Next to the pinworm, infection with *Ascaris* (commonly called the large intestinal roundworm) is probably the most common nematode in humans. The World Health Organization has estimated that between 650 million and 1 billion people are infected, with most being in the tropics and subtropics. In the United States, infections are most common in the Gulf Coast states and in Appalachia. Coinfection with *Trichuris* is common.

Historically, the domestication of the pig and the bringing of the animal into the home probably account for the development of the strain of *Ascaris* for humans. The role of *Ascaris suum*, the swine ascarid, in human disease has been a controversy for many years. Experimental and laboratory accidents have shown that the swine ascarid will infect humans, but naturally occurring infections have been difficult to prove. Recently, however, a few cases of children being infected with *A. suum* have been documented. In each case, the worms recovered after anthelmintic treatment were shown to be sexually immature; furthermore, worm burdens were light, limited to no more than three worms. There is no doubt that the ingested swine ascarid eggs will hatch in human intestinal tract and cause a form of visceral larva migrans, with the lungs and liver being most affected. As a result, pediatricians and those family practitioners who practice in rural areas should be aware of the part that pigs may play in human ascariasis.

Life Cycle. Infection begins when the embryonated eggs are ingested with food or drink. The parasite then begins its life cycle by the migration of the larvae throughout various tissues of the body. The larvae eventually return to the small intestine where they mature; 8 to 10 weeks are required to reach maturity after ingestion of the eggs. Adults live for about a year and then are passed in the feces.

These large nematodes (males 15 to 31 cm and females 20 to 35 cm or more in length) usually live unattached in the lumen of the small intestine (Fig. 90-6). After the worms reach sexual maturity, copulation occurs, and the females soon thereafter begin to lay eggs. The daily egg production per female is phenomenal, averaging about 200,000. The fertilized eggs are still in the one-cell stage when they pass from the host in the feces (Fig. 90-3). Infertile eggs, differing considerably in morphologic detail from the fertilized eggs, are sometimes seen (Fig. 90-3). If the stool containing the fertilized eggs is deposited in a warm, shady, moist area, the eggs will develop. Accidental ingestion of infective eggs results in infection. The eggs hatch in the duodenum, and the emerging larvae penetrate the intestinal wall, enter the circulatory system, and are carried to the right heart and thence to the lungs. They are filtered out of the lung capillaries and later penetrate into the alveoli.

After about 2 weeks in the lungs, the larvae migrate up the respiratory tract to the epiglottis and are swallowed into the stomach. Upon reaching the small intestine the worms develop into adult males and females. It should be emphasized that the lung migration is a necessary part of the cycle.

Pathology and Symptomatology. The pathogenesis of ascariasis is related to the life cycle of the parasite. The principal damage produced by the larvae occurs in the

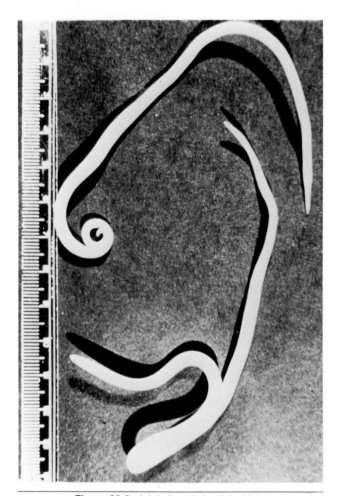

Figure 90-6. Adult *Ascaris lumbricoides.*

lungs. Petechial hemorrhages are produced following entry into the alveoli. A striking serocellular exudate, in which eosinophils are prominent, will then collect. Eosinophils are commonly associated with parasitic worms or larvae in tissues and eosinophilia is characteristic. The cardinal signs and symptoms of *Ascaris* pneumonitis consist of dyspnea, dry cough, fever, and eosinophilia. X-ray findings show scattered mottling of the lungs, suggestive of tuberculosis and other infections.

The adult worms derive most of their nourishment from the semidigested food of the host, and if the worms are abundant, they may have a detrimental effect on the host's nutrition, especially in infants and young children. However, most light infections produce little or no change in the host. Effects from the adult worm are usually noted only in children, who often show loss of appetite and weight, and intermittent intestinal colic. Indeed, colicky epigastric or periumbilical pain is the cardinal symptoms of ascariasis. The greatest danger from the adult worm results from abnormal migrations within the body. They may enter the ampulla of Vater and block the common bile duct or penetrate into the liver parenchyma or the pancreas. They also have been found on rare occasions in other abnormal sites in the body. Sometimes they perforate the intestinal wall, but much more often, especially in young children, a mass of tangled worms causes an acute obstruction of the small intestine. Finally there is evidence that absorption of the toxic and allergenic metabolites of the adult worms accounts for various referred symptoms, especially those of a neurologic nature.

Epidemiology. Incidence rates as high as 90 percent have been reported. Disease is endemic in tropical and subtropical countries, but infection and sporadic disease occur worldwide. People living in rural areas where poor sanitation prevails are most at risk. Adults usually acquire the infection by eating raw vegetables that have been contaminated by feces and human night soil. Children acquire the infection in this manner but also are infected by hand-to-mouth, by pica, or by placing contaminated toys in their mouths.

Eggs are very resistant to adverse environmental conditions. They develop best in warm, moist soil and, under such conditions, become fully embryonated 2 to 3 weeks after having been passed in human feces. Eggs are also resistant to disinfection, can survive sewage treatment, and remain viable for months or years in night soil. Pollution of wells through improper drainage of surface waters may also result in transmission. Prevention depends on the proper disposal of feces.

Diagnosis. The clinical diagnosis of ascariasis is difficult, since pneumonitis, eosinophilia, and intestinal symptoms are similar to those of other intestinal helminthic infections. Diagnosis before eggs appear in the feces and infections with male worms only are rare. Sometimes, radiologic examination reveals adult worms in the intestine, and, at times, parents will bring to a physician's office the adult worm that was either passed in the feces or actively migrated out the anus. Ascarids are occasionally vomited and can be recovered from the emesis.

A moderate eosinophilia, increased temperature, dyspnea, Charcot–Leyden crystals in the sputum, hemoptysis, coughing, and rales are characteristic of ascaris pneumonitis. Pneumonia, Löffler's syndrome, and asthma must be considered in the differential diagnosis. Patients are often hypersensitive to the worms and display urticaria and other allergic reactions, such as increased serum IgE levels.

Definitive diagnosis usually requires the demonstration of the characteristic egg in the feces. Eggs are present in large numbers and can be found in smears by examination under the low power of the microscope (Fig. 90-3). The egg is bluntly ovoid in shape, 45 to 70 μm by 35 to 50 μm, and has several layers between the single cell of the egg and the outside. The outer layer is coarsely mammillated. The various layers account for the thick shell, which is characteristic.

For serologic diagnosis, the indirect hemagglutination (IHA) test using body cavity fluid from the swine ascarid is sometimes used. Others use this test in conjunction with the bentonite flocculation test (BFT), the BFT being less sensitive than the IHA test but more specific. Recently, an enzyme-linked immunosorbent assay (ELISA) test using an antigen recovered from the embryonated *Ascaris* egg was developed, and it is reported to be superior in both sensitivity and specificity. Lack of specificity is due to marked cross-reactions with toxocariasis. Finally, complement fixation (CF), agglutination, immunofluorescence (IF), immunodiffusion (ID), and other tests have been used, but none is sufficiently sensitive and specific to be of practical value.

Treatment. Only supportive therapy is indicated during the lung migration phase of the infection. Intestinal obstruction by the adults may require intubation, surgery, and drainage. When adults are in ectopic sites due to abnormal migration (e.g., peritoneal cavity, liver), surgery is necessary. Saline enemas may be helpful in removing worms from the large bowel during chemotherapy. Many drugs are reasonably effective in treating intestinal ascariasis. Pyrantel pamoate is commonly used, as is mebendazole. It must be stressed, however, that when *Ascaris* coexists with other infectious agents, such as *Trichuris*, it is important to remove the ascarids first. Otherwise, the ascarids may become irritated and migrate into ectopic sites.

Visceral Larva Migrans

Visceral larva migrans (VLM) is a clinical term referring to human infection with the nematode larvae of *Toxocara*, an ascarid of dogs, other canids, and cats. Thus, VLM is a zoonotic disease, and humans are an accidental host. VLM is primarily a disease of preschool children, and it is estimated that 10 to 30 percent of children between the ages of 1 and 6 who have the habit of eating dirt are at risk of being infected. Although the larva of *Toxocara canis* is most often the etiologic agent of VLM, other nematodes of lower animals are known to be involved.

Life Cycle. Humans become infected when they ingest the developed eggs of the nematode. Subsequent to ingestion, the eggs hatch in the small intestine, and the freed larvae penetrate into the gut wall and enter the circulatory system. Although there are a few reports of larvae returning to the small intestine (typical ascarid life cycle) and becoming adults, it is thought that since humans are accidental hosts, the larvae wander around in the human body and come to rest in any of a variety of organs or body cavities. Reports indicate that the liver, lungs, and eyes are most often the sites of habitation. The exact incubation period is not known, but it is thought to be at least 1 week in duration. Larvae may remain alive for 2 to 3 years.

Pathology and Symptomatology. The pathogenesis of VLM is related to (1) the number of eggs ingested, (2) the number of larvae that enter the tissue, (3) the amount of larval migration within the host, (4) the location of the larvae in the host, and (5) the host's immune respone to the larvae. Fortunately, most infected children are asymptomatic or only mildly ill. Fatalities do occur, however, and pediatricians should be aware of this. When death occurs, it is due to extensive cardiac or central nervous system involvement.

Symptoms are related to the migration of larvae within tissues and are manifested as hepatomegaly, eosinophilia, and hypergammaglobulinemia. Fever, cough, and wheezing are characteristic of pneumonic involvement. Pruritic eruptions and tender nodules on the trunk and lower extremities are sometimes seen. Ocular lesions resembling retinoblastoma are fairly common and may lead to loss of vision and the eye itself. Eye lesions, although usually unilateral and painless, may lead to total retinal detachment.

Epidemiology. *T. canis* is an extremely common parasite of dogs, with over 80 percent of all puppies being infected. Surveys show that 30 to 50 percent of households in the United States include at least one dog, so that the chance for egg transmission is great. Epidemiologic studies also show that the presence of puppies less than 3 months of age and children in the same house are almost tantamount to infection. Exposure to dogs and cats outside the house (school yards, playgrounds) is not as risky. However, the increased number of infected dogs that are allowed to run free certainly contributes to the increasing incidence of VLM. Control depends on the frequent deworming of puppies and dogs, as well as the proper disposal of dog feces. Unfortunately, once soil has become contaminated with dog feces, it is almost impossible to kill the infective eggs. Thus, geophagia (dirt eating) should be prevented if at all possible.

Diagnosis. Any child manifesting hepatomegaly, hypereosinophilia, hypergammaglobulinemia, and a history of dirt eating should be suspected of having VLM. These symptoms may last for months or years. In addition, fever and respiratory distress are common, but inconsistent, features.

Laboratory data are sometimes helpful in making a diagnosis because patients with VLM usually manifest an eosinophilia of 50 to 90 percent, increased IgM levels, leukocytosis, and high isoagglutination titers. Serum glutamine oxalacetic transaminase (SGOT) levels will be elevated in 20 percent of patients. Chest x-ray reveals transient pulmonary infiltrates in many. Differential diagnosis should include a consideration of pulmonary ascariasis, asthma, retinoblastoma, trichinosis, eosinophilic leukemia, tropical eosinophilia, and collagen disease.

Stool examination is of no use in diagnosing VLM because the adult worms do not occur in humans. A definitive diagnosis is made only by the finding of larvae in tissue biopsy specimens. The liver is the preferred source of tissue for biopsy, but due to the small number of worms, the chance of finding larvae is not great. When

found, they exhibit an eosinophilic granuloma formation (Fig. 90-7).

Serologic testing for VLM disease is increasing. Although far from being completely satisfactory at this time, the IHA, BFT, indirect fluorescent antibody, and ELISA tests are used. Recent improvements in antigens and the ELISA test hold promise for a much better test in the near future. One recent study showed that the ELISA test had 90 percent sensitivity and a specificity of 91 percent in detecting ocular toxocariasis.

Treatment. Visceral larva migrans is usually a self-limiting disease, and treatment is given only in severe cases. Glucocorticoids and bronchodilators are sometimes used when pulmonary disease is severe. Glucocorticoids are also used to improve vision when endophthalmia is manifested. Diethylcarbamazine and thiabendazole have also been used by some. Chemotherapy does not, however, seem to be satisfactory.

Intestinal Nematodes Infective for Humans in the Larval Stage

Hookworms

Four species are involved in human infections: two of these, *Necator americanus* and *Ancylostoma duodenale*, are true human parasites, and two, *Ancylostoma braziliense* and *Ancylostoma caninum*, are parasites of cats and dogs. The latter two, especially *A. braziliense*, produce a dermatitis (cutaneous larva migrans or creeping eruption) in humans following penetration of the skin by the filariform larvae (see below).

All hookworms have common morphologic characteristics, the most striking being the umbrella-like copulatory bursa of males, the anterior hook or curvature of the body (Fig. 90-8), and thin-shelled eggs. Adults are approximately 1 cm long and reside in the small intestine.

Hookworm disease, once an important and common human condition in the southern United States, although still endemic, is not commonly seen in clinics. *N. americanus* is the only species of importance in human infection throughout the hookworm belt of the southern United States. It is necessary, however, to be on the alert for *A. duodenale*, which is exotic in origin and occurs in immigrants and others who were infected in endemic areas. The life cycles of these various hookworms are similar.

Life Cycle. The adult worms are usually attached to the upper levels of the small intestine. The head is anchored securely to the intestine, the tip of a villus being drawn into the mouth of the worm, and blood is sucked from the capillaries. After copulation, females lay eggs, and these are passed with the feces in an undeveloped state. Under favorable environmental conditions, the eggs will develop in 1 to 2 days, and a free-living larval stage (rhabditiform) is released. The larvae feed on bacteria and detritus in the feces and soil and eventually metamorphose into the infec-

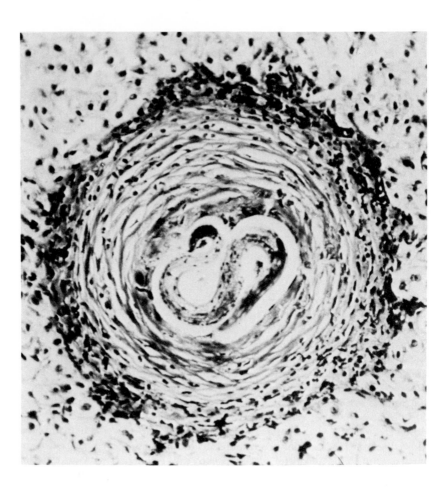

Figure 90-7. Granuloma formation around larva of *Toxocara.*

tive (filariform) larvae. Filariform larvae do not feed and have a life span of up to 2 weeks. The usual means of human infection is penetration by these larvæ into the tender skin between the toes of barefooted individuals, but they can penetrate any skin surface. After many hours, the larvae enter the cutaneous blood vessels and are carried through the right heart to the lungs. After a few days, they have succeeded in penetrating from the pulmonary capillaries into the alveoli, have ascended the respiratory tract to the epiglottis, and, after having been swallowed, have descended to the upper small intestine. After attaching to the villi, the young worms grow into sexually mature adults. Recent research suggests that a transmammary infection route is also possible for *A. duodendale.* Although some of the worms in humans may live for 10 years, most of them will have been lost from the host after 1 to 2 years. The incubation period is about 6 weeks long.

Pathology and Symptomatology. At the site of entry of the filariform larvae into the skin, most individuals experience intense itching and burning, followed by edema and erythema. A papule may appear, which transforms into a vesicle. This condition, known as "ground itch," is more serious in those sensitized by previous infections and in those developing secondary infections. The larvae reaching the lungs produce small focal hemorrhages as

they penetrate from the capillaries into the alveoli, but usually only a subclinical pneumonitis is produced. The greatest damage results from the adult worms that are attached to the small intestine. The superficial mucosa of that part of the intestine that is contained within the buccal cavity of the worm becomes denuded, and the surrounding mucosa usually shows a mild inflammatory infiltrate. More important is the blood loss of the host. Blood is sucked from the capillaries of the villi and passes through the digestive tract of the worm. Apparently, only certain products are removed, the remainder being extruded through the anus in a wasteful fashion. Adding to this blood loss, which may average more than 0.2 ml per worm daily, is that occurring after a change in site of attachment. The former wound continues to ooze blood for a time after the worm has released the tissue. Blood loss from the intestinal wall constitutes the greatest damage from hookworm infection. This loss, unless compensated adequately, will give rise to a microcytic, hypochromic anemia. Whether an individual with hookworm infection develops hookworm disease is dependent, in general, upon the number of worms harbored and the host's nutritional status. Even moderately heavy infections may fail to produce a significant anemia if the individual has been on an adequate, well-balanced diet rich in animal proteins, iron and other minerals, and vitamins. However, in heavy infections, even

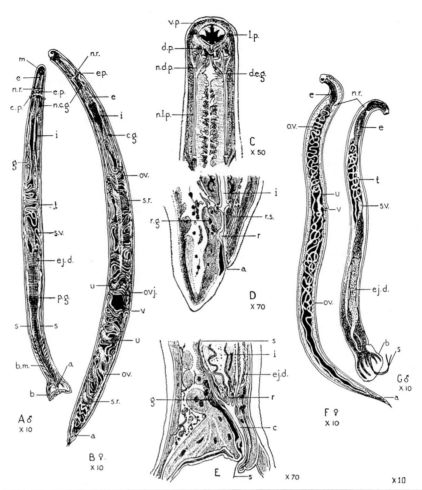

Figure 90-8. Important hookworms of humans. **A.** Adult male *Ancylostoma duodenale* from ventral side. **B.** Young adult female *A. duodenale* from right side. **C.** Anterior end of *A. duodenale* from dorsal side. **D.** Longitudinal section through end of female *A. duodenale,* somewhat diagrammatic. **E.** Longitudinal section through end of male *A. duodenale,* not quite median. **F.** Female *Necator americanus.* **G.** Male *N. americanus.* a, anus; b, bursa; b.m., bursal muscles; c, cloaca; c.g., cervical gland; c.p., cervical papilla; d.e.g., dorsal esophageal gland; d.p., dorsal papilla; e, esophagus; e.p., excretory pore; ej.d., ejaculatory duct; g, gubernaculum; i, intestine; l.p., lateral papilla; m, mouth; n.c.g., nucleus of cephalic gland; n.d.p., nerve of dorsal papilla; n.l.p., nerve of lateral papilla; n.r., nerve ring; ov., ovary; ovj., ovejector; p.g., prostatic glands; r, rectum; r.g., rectal ganglion; r.s., rectal sphincter; s, spicules; s.r., seminal receptacle; s.v., seminal vesicle; t, testis; u, uterus; v, vulva; v.p., ventral papilla. *(From Belding: Basic Clinical Parasitology, 1958. New York, Appleton-Century-Crofts.)*

with a highly fortified diet, the hematopoietic mechanism is unable to keep pace with the great loss of red blood cells. Moderate blood decompensation causes anemia, heart palpitations, and lassitude.

In some endemic areas of hookworm, the natives are infected repeatedly from early childhood. Thus, chronic hookworm disease is most characteristic. Common signs and symptoms are anemia, epigastric burning, flatulence, sallow skin, tender abdomen, irritability, alternating diarrhea and constipation, dry skin, and blurred vision. When the condition finally produces marked physical weakness, the individual is not fit for any type of manual labor. This

accounts for the early term, "lazy disease." Cardiomegaly and cardiac symptoms are evident in many, and in severe cases, death may result from cardiac failure. Physical (including sexual) stunting is characteristic in children, and many are also retarded mentally.

Epidemiology. Both species occur predominantly in the tropics and subtropics betweeen 36° north and 30° south latitudes. The World Health Organization estimated that 500 million in the world are infected. All races and ages are susceptible. Occupation plays an important role because those who come in contact with soil are more likely to be exposed. The wearing of shoes and the proper disposal of feces will markedly reduce the incidence of infection. As evidence of this, recent surveys show that in rural areas of certain parts of Asia and Africa, prevalence rates of 60 to 80 percent are not unusual, whereas in the rural areas of the southern United States, it is rare to find areas where rates exceed 10 percent.

Temperatures between 25C and 35C and a shady, sandy, or loamy soil are optimum conditions for egg development. It is doubtful if important reservoirs for human infection exist, but it is known that pigs are occasionally infected with *A. duodenale* and some nonhuman primates with *N. americanus.*

Diagnosis. Clinically, the signs and symptoms of disease usually include weakness, pallor, and fatigue. The infection is generally chronic, and the manifestations of infection are related to worm burden and nutritional status, as described above.

Ground itch, a sign of acute infection, is characterized by itching, erythema, vesiculation, and secondary infections. Coughing, sore throat, and bloody sputum are pulmonary manifestations and occur within a few days to a week after exposure. Intestinal symptoms arise 2 weeks after infection, and anemia of the microcytic, hypochromic type does not appear for 10 to 20 weeks. Differential diagnosis should include:

1. For ground itch—allergic dermatitis and fungal infections
2. For pulmonary phase—asthma and atypical pneumonia
3. For the intestinal phase—enteritis and other types of iron deficiency anemias.

Laboratory diagnosis depends on the demonstration of thin-shelled eggs (40 by 60 μm) in a stool specimen (Fig. 90-3). At times, when the stool examination has been delayed, the eggs may have developed and hatched, releasing the rhabditiform larvae. These must be differentiated from *Strongyloides stercoralis,* another intestinal nematode of humans. Stools from heavily infected patients may be viscous and tarry. These, as well as those from less heavily infected patients, often are positive for occult blood. Decreased hemoglobin and serum iron levels are common. Eosinophilia of 5 to 15 percent may occur in the early phase of infection, although normal levels of eosinophils are usually present in the chronic stage. In hypersensitive individuals, eosinophilia of 70 percent has been reported. No practical and useful serologic tests are available.

Treatment. Hookworm disease may be difficult to eradicate because reinfection usually occurs. Anthelmintics available include mebendazole and pyrantel pamoate. Most clinicians agree that if the patient is anemic, hemoglobin levels should be restored by diet and iron supplement to at least 50 percent of normal value before treatment. If *A. lumbricoides* coexists with the hookworms, the ascarids should be removed first.

Cutaneous Larva Migrans

Cutaneous larva migrans (CLM), sometimes referred to as creeping eruption, is due to the filariform larvae of dog and cat hookworms (*A. caninum* and, especially, *A. braziliense*) penetrating the human skin and subcutaneous tissues. These species of hookworms are unable to mature in humans, but the larvae are capable of living for long periods in subcutaneous tissues.

Cutaneous larva migrans is a condition seen worldwide, wherever there are cats and dogs and wherever warm, moist climatic factors prevail. In the United States, the infection occurs from Texas to New Jersey, primarily in the southern states, with Florida and Georgia reporting the most cases.

Life Cycle. The life cycles of cat and dog hookworms are the same as that of the human hookworms. It must be emphasized that cat and dog hookworm larvae, although capable of penetrating the human skin, do not mature to the adult stage in humans.

Pathology and Symptomatology. An intensely pruritic eruption caused by the migrating larvae is the primary symptom. At each point of entry into the skin, an itching, reddish papule is produced followed by a vesicle. After a few days, each larva has developed a serpiginous tunnel in the epidermis as it proceeds (usually at the rate of several millimeters a day). Edema and then a raised inflammatory tract with a crusty opening are characteristic. There also is a striking cellular infiltration, especially of eosinophilis. Over a period of weeks or months, this results in extensive skin involvement (Fig. 90-9). At times, severe systemic illness may result. These movements, limited to the skin of humans, and the resulting irritation produce an intense pruritus. This leads to scratching and often opens the lesions to pyogenic organisms.

Epidemiology. Cutaneous larva migrans is an occupational disease of plumbers, construction workers, duck hunters, and others who are exposed to infected soil under buildings, crawlways, and around hunting blinds. Children are often infected at beaches, park play areas, and in home sandboxes. Plastic sheeting to protect workers from exposure to infected soil, covering for sandboxes, and the frequent deworming of cats and dogs would help to control CLM. Limiting access of cats and dogs to public beaches and park areas would be of value.

Diagnosis. Clinicians usually make a presumptive diagnosis on the basis of patient history, signs, and symptoms. Although not routinely done, biopsy of the leading edge of a serpiginous tract may show the larva. There are no laboratory values or findings that are helpful in diagnosis. A low-grade eosinophilia might be expected in those who are repeatedly infected. No serologic tests are indicated.

Treatment. Treatment is aimed at killing the larvae and reducing the pruritis. Oral or topical treatment of the lesions with thiabendazole is effective. Lesions may be treat-

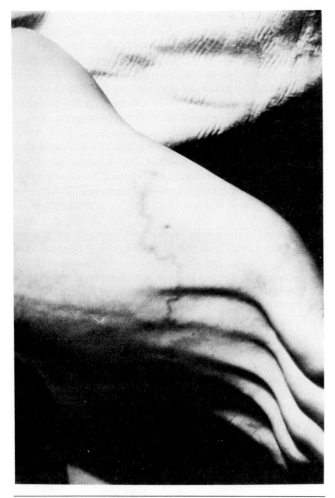

Figure 90-9. Creeping eruption.

ed by freezing the leading edge of the larva's tract with ethyl chloride.

Strongyloides stercoralis

Along with two protozoan infections of humans, toxoplasmosis and pneumocystosis, strongyloidiasis has recently received an added degree of notoriety. This is because the strongyloid worm is capable of producing an overwhelming and often fatal infection in the immunologically compromised host.

The geographic distribution is primarily tropical and subtropical but generally parallels that of hookworms, since the environmental conditions of warmth, shade, moisture, sandy–loamy soil, and indiscriminate defecation apply to both parasites. Within the temperate regions of the world, strongyloidiasis is rare in Europe, but in the United States, endemic foci in Appalachia and the southeastern states are well known. The World Health Organization estimates on a worldwide basis that 35 million people are infected. Cats and dogs are known to be sus-

ceptible to infection, but they are most likely unimportant reservoirs.

Life Cycle. The life cycle or, more accurately, life cycles are very complex and unique among the nematode parasites of humans. There are two forms of sexually mature adults: one is a parasite within the tissues of the intestine, and the other is a free-living worm that can live in the soil. *S. stercoralis* is the only nematode whose adults can increase in numbers entirely within a human host.

The direct cycle (Fig. 90-10) is the most common. The parasitic females are slender, filariform worms about 2 mm in length. They usually reside among the epithelial and glandular cells or in the tunica propria in the upper levels of the small intestine. Here, apparently in the absence of males, they lay several dozens of eggs daily. As the eggs filter through the mucosa toward the lumen, they develop and hatch, freeing the larvae. These make their way to the lumen and are passed in the stool. Under favorable conditions and after further development to the infective filariform stage, they enter exposed human skin to initiate a new infection. Their migration to the lungs and thence to the small intestine is similar to that described above for the hookworm larvae. After reaching the tissues of the small intestine, the larvae develop into mature females in about 25 days, and these worms may live for several years. They probably are parthenogenetic.

A second type of cycle, common only in the tropics, involves the interpolation of one or more free-living generations. This is the indirect cycle. In this case, the larvae in the stool develop into free-living males and females. These mate, and eventually, some of them develop into filariform larvae to begin the parasitic cycle. The intrahuman cycle is the same as that of the direct cycle.

The third type of cycle is that known as internal autoreinfection or hyperinfection. In this case, the larvae, while still in the intestinal lumen, molt into the filariform

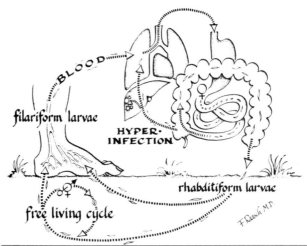

Figure 90-10. Most common life cycles of *Strongyloides stercoralis*. (*AFIP photograph No. 70-73312-2. Artist, Frank Raasch, MD.*)

larvae, which after entering the intestinal wall and blood circulation migrate through the tissues of the body. Some reach the lungs, as in the other cycles, and return to the small intestine, but many find their way into various other tissues, where considerable damage may result. Larvae in skin tunnels (larva currens) migrate rapidly, covering centimeters per day. This cycle is most common in those with severe protein-calorie malnutrition or immune deficiencies, and it may produce a fatal outcome. The fourth type of cycle results from external autoreinfection when precociously developed filariform larvae on fecally contaminated areas penetrate the skin, usually in the perianal area.

Pathology and Symptomatology. When large numbers of larvae penetrate into the skin, a pruritis or ground itch usually develops. A severe pneumonitis may be produced by the larvae in the lungs, somewhat similar to that caused by *Ascaris* larvae. Damage also is produced in the wall of the small intestine by the females, eggs, and rhabditiform larvae. Mechanical and perhaps lytic damage is produced, especially by the females, which move about considerably. Cellular infiltration, often striking, consists chiefly of eosinophils, lymphocytes, and epithelioid cells. Panmucosal duodenitis is characteristic; that is, there is involvement throughout large areas of the mucosa, and it is not, therefore, limited to areas adjacent to the worms. The affected tissue becomes increasingly nonfunctional, and at times sloughing of the tissue down to the muscularis mucosae occurs. There is evidence of systemic damage, consisting of sensitization and probably toxic reactions. As stated above, autoreinfection may produce severe damage in various tissues, with striking signs and symptoms. However, the most characteristic symptoms of the usual direct type of infection are midepigastric nonradiating pain, a watery mucous diarrhea, and eosinophilia.

Epidemiology. The pattern of transmission is similar to that of hookworms. Although coexisting with hookworm infection around the world, strongyloidiasis is more spotty in distribution and not nearly as prevalent. Autoreinfection accounts for the persistence of the parasite in an individual for as long as 40 years after having left an endemic area.

Diagnosis. Half of those infected are asymptomatic, and the majority of those who are symptomatic do not usually have severe clinical manifestations. The development of hyperinfection and its attendant severe clinical manifestations are probably due to a decreased resistance brought about by a variety of debilitating diseases, malnutrition, and immunosuppressive therapy.

Macular eruptions of the skin due to larval penetration may be noted. Perianal inflammation is sometimes seen when external autoreinfection takes place. Pulmonary signs and symptoms resemble those of hookworm infection. During the intestinal phase, there may be pain, nausea, vomiting, and diarrhea alternating with constipation. Midepigastric pain and an eosinophilia are the hallmarks

of this stage, and the symptoms mimic those of peptic or duodenal ulcers. Because of rather nonspecific signs and symptoms, differential diagnosis should include enteritis, allergies, VLM, colitis, hepatitis, cholecystitis, and malnutrition with or without malabsorption.

A definitive diagnosis depends on the demonstration of larvae in the feces (Fig. 90-11). In general, the presence of larvae in freshly passed feces strongly suggests *Strongyloides;* the presence of thin-shelled eggs suggests hookworm. The formalin–ether concentration technique is very effective in concentrating larvae for diagnosis. In cases when no larvae can be found in the stool but strongyloidiasis is still suspected, duodenal aspirates, Enterotests (string test), and sputum or urine examinations may be helpful.

The most consistent laboratory finding is an eosinophilia of 5 to 40 percent. However, many immunocompromised hosts do not show an eosinophilia and may even be eosinopenic. Serum protein and potassium levels may be abnormally low in severe infections. Levels of SGOT and SGPT may be increased. Stools vary in character, from normal to diarrheic and mucoid; some may be steatorrheic. No good serologic test is available. A CF test has been used, but its poor sensitivity and specificity do not merit its routine use.

Treatment. *Strongyloides* has been one of the more difficult helminthic infections to treat. The risk of autoreinfection and its consequences requires that all worms be eliminated, not just a reduction in worm burden. Thiabendazole is the drug of choice and is effective against both adult and larval worms. If necessary, treatment can be repeated. When autoreinfection is suspected, broad-spectrum antibiotics may be given to control bacterial infections.

Trichinella spiralis

Infection with trichina worms now has a prevalence of about 4 percent in the United States. Forty years ago, 15 to 20 percent of all Americans were infected. The cooking of garbage fed to pigs, widespread commercial and home freezing, better pig-rearing methods, and the inspection of pork for transportation across state lines has greatly reduced the chance of infection. However, undercooked or raw meat, especially pork, can still be hazardous, so thorough cooking should still be encouraged.

Currently, the highest prevalence rates are in temperate and Arctic regions. In Arctic regions, the pig is replaced by bears, seals, and walruses as the sources of human infection. Recent data show that 95 percent of all Eskimos in northern Canada are infected. Within the United States, and excluding Alaska, trichinosis remains a health hazard primarily because the infection is still enzootic among domesticated pigs. Surveillance shows that in the United States, 75 percent of infections are due to consumption of pork or pork products. Most of the remaining 25 percent are due to eating improperly cooked ground beef that has been adulterated with pork.

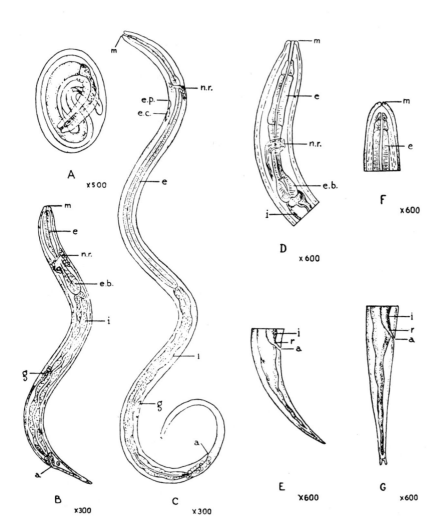

Figure 90-11. Larvae of *Strongyloides stercoralis.* **A.** Egg containing mature larva of *Strongyloides simiae.* **B.** Rhabditiform larva. **C.** Filariform larva. **D.** Anterior end of rhabditiform larva. **E.** Posterior end of rhabditiform larva. **F.** Anterior end of filariform larva. **G.** Posterior end of filariform larva. a, anus; e, esophagus; e.b., esophageal bulb; e.c., excretory cell; e.p., excretory pore; g, genital rudiment; i, intestine; m, mouth; n.r., nerve ring; r, rectum. *(From Belding: Textbook of Clinical Parasitology, 2nd ed, 1952. New York, Appleton-Century-Crofts.)*

Life Cycle. This parasite is unusual because in its life cycle it has no external phase, either as free-living larvae or in a vector. Both the adults and larvae are present in an infected host. Humans are infected by ingesting infective larvae (Fig. 90-12) that are encysted in striated muscles of a reservoir host. Gastric digestion frees most of the larvae. They soon enter the small intestine and penetrate into the mucosa. Within 2 days, they have reached the adult stage. Copulation occurs and the females (2 to 4 mm in length), after burrowing more deeply into the mucosa or even to lower levels, begin to give off larvae. The release of larvae begins about 5 days after infection, and most of the larvae occur in the circulating blood in 7 to 14 days. The total number of larvae released by each female over a period of several weeks is 500 or more. These larvae are small, about 100 by 6 μm. They enter the circulation and are distributed throughout the body. However, only those entering striated muscles are capable of further development. Here, within 3 weeks after infection, encapsulation begins. This host reaction, initiated by infiltration mainly by lymphocytes and eosinophils, results in the formation of a double-walled, adventitious capsule (Fig. 90-12). The larvae contained within grow to about 0.8 mm in length and

become tightly coiled. The larvae are infective for a new host a few days after encapsulation has begun. Some remain viable within cysts for many years, despite the fact that calcification of the cyst is usually observed after 6 months. Stray larvae that invade the myocardium and other tissues, such as the central nervous system, cause damage, but cyst formation does not occur.

Pathology and Symptomatology. Trichinosis has been confused with many other diseases, because the parasite may damage many different tissues of the body. For a clear understanding of the disease, it is necessary to consider in turn the various stages involved. During the first stage, the adult worms are becoming established in the mucosa of the small intestine. Due to extensive burrowing, tissue is destroyed, and an intense panmucosal and submucosal inflammatory reaction results. Nausea, vomiting, diarrhea, and fever may be experienced. These and other signs and symptoms resemble salmonellosis and other enteric infections, so trichinosis is not suspected. This stage usually lasts for about 10 days, and thus, it overlaps the second stage. The second stage, set arbitrarily between 7

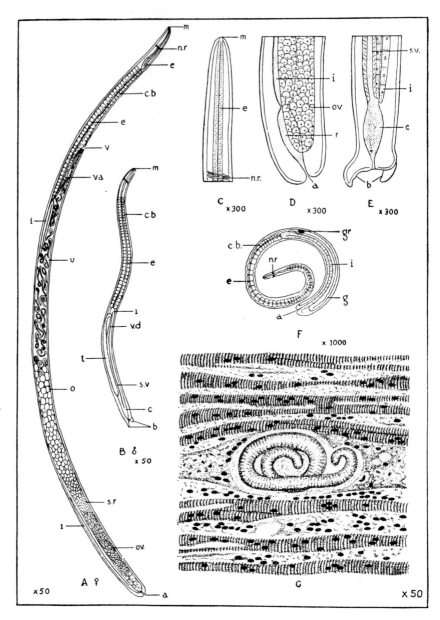

Figure 90-12. Schematic representation of *Trichinella spiralis*. **A.** Adult female. **B.** Adult male. **C.** Anterior end of worm. **D.** Posterior end of female. **E.** Posterior end of male. **F.** Young larval worm. **G.** Early encysted larva in muscle. (**C** adapted from Leuckart, 1868.) a, anus; b, bursa; c, cloaca; c.b., cell bodies; e, esophagus; g, gonads (anlage); gr, granules; i, intestine; m, mouth; n.r., nerve ring; o, ova; ov., ovary; r, rectum; s.r., seminal receptacle; s.v., seminal vesicle; t, testis; u, uterus containing larvae; v, vulva; va, vagina; v.d., vas deferens. *(From Brown and Neva: Basic Clinical Parasitology, 5th ed, 1983. Norwalk, Conn, Appleton-Century-Crofts.)*

and 14 days after infection, involves most of the larvae in the circulation. The fever usually reaches a peak (41C), and there is characteristic edema, especially around the eye(s), splinter hemorrhages under the fingernails, and eosinophilia. In addition, dyspnea, difficulty in mastication and speech, and petechial hemorrhages, especially in the conjunctivae and retinal vessels, may be noted. Myocarditis with congestive heart failure and neurologic signs and symptoms often are noted. It should be remembered that any organ may be damaged by the migrating larvae. The features of this stage are most pronounced during the second week but continue throughout the period when larvae are released. The third stage, after 14 days, is a culmination of the traumatic, allergenic, and toxic effects of the infection, and the patient may not be symptomatically cured for 5 to 8 weeks or longer. Myositis is outstanding, and

muscular pains are usually the chief complaint. The edema, especially around the eyes, persists, as does the hypereosinophilia, which reaches a peak about 21 days after infection. Cachexia may be profound. Many fatal cases show congestive heart failure due to myocardial lesions, respiratory paralysis, and anaphylaxis. It is noteworthy that eosinophilia is characteristically absent in such patients. Except during epidemics, most cases are relatively asymptomatic, due presumably to the consumption of small numbers of larvae. However, it should be kept in mind that death will result inevitably if large numbers are ingested.

The differential diagnosis should include other causes of gastroenteritis, asthma, rickettsial diseases, VLM, brucellosis, dengue fever, myositis, myocarditis, and meningoencephalitis.

Epidemiology. In the United States, trichinosis most often is seen clinically among members of ethnic groups who enjoy eating raw pork. Some recent outbreaks occurred among new immigrants who did not understand the need to either cook or freeze American pork. In the southern United States, the custom of tasting raw homemade sausage for flavor during the addition of spices is a significant cause of infection. The entire zoonotic picture of this infection is unknown because, in addition to most carnivores and omnivores, horse and mutton have also been known to be the source of infection. There is also some evidence that marine fish and crabs can serve as transport hosts.

Diagnosis. Myalgia, high eosinophilia, facial edema, and fever should lead a physician to suspect trichinosis, especially if there is a recent history of eating raw meat, mainly pork. Similar symptoms of those eating at the same place further substantiate the presumptive diagnosis.

A direct (specific) means of diagnosis is provided by biopsy of a small piece of the deltoid or other muscle. The larvae may be demonstrated by pressing the tissue firmly between two microscope slides and then examining microscopically or by recovering the larvae following digestion in artificial gastric juice. To relate the presence of larvae to the current illness, it is important to consider the stage of cyst formation. For example, the finding of calcified cysts only could not be related to an illness of recent origin. A negative biopsy, however, does not exclude the possibility of infection. Various serologic tests are available, but the bentonite flocculation test (BFT) is considered the best to detect acute infections. An intradermal test is available. Finally, the degree of and changes in the eosinophilic response (as high as 80 percent), although not specific and not always striking, are of assistance and should be charted.

Treatment. In general, only palliative and supportive measures are of value. However, cortisone and related drugs are of great assistance during the larval encystment period. Allergic reactions to and/or toxic effects from the metabolites of the worms often show amelioration after use of cortisone, and on occasion this drug has appeared to be life saving. Thiabendazole is effective against the adult worms, but the diagnosis is almost always delayed until after most, if not all, of the migrating larvae—the agents of toxic and allergic reactions—have been released by the female worms.

Tissue Nematodes

This group includes seven species of filariae. *Dracunculus medinensis*, the agent of guinea worm infection, although not a filarial worm, is also included with this group because of its morphologic similarity. All are exotic in origin (Table 90-1), and all are characterized by the following features: (1) the adults live in human tissues, (2) female worms give birth to larvae known as microfilariae that circulate in blood and lymph or are present in tissues, and (3) transmission requires the ingestion of microfilariae by a bloodsucking arthropod, which, in turn, transmits the infection to another human.

Filariasis is a major public health problem involving up to half of the adult human population in many tropical and subtropical areas. Incubation periods are long (6 months to a year), and adult worms can live for 15 or more years. Diagnosis is usually made upon recovering the microfilariae from blood or skin. Among those listed in Table 90-1, *Wuchereria bancrofti* and *Onchocerca volvulus* are of greatest medical importance and will be discussed in more detail.

TABLE 90-1. FILARIAE AND *DRANCUNCULUS MEDINENSIS*

Parasite	Disease	Location of Adults in Humans	Location of Microfilariae in Humans	Arthropod Intermediate Hosts
Wuchereria bancrofti	Bancroft's filariasis	Lymphatics	Blood	Mosquitoes, especially *Anopheles, Culex, Aedes* spp.
Brugia malayi	Malayan filariasis	Lymphatics	Blood	Mosquitoes, *Mansonia, Anopheles* spp.
Onchocerca volvulus	Onchocerciasis	Subcutaneous nodules	Skin tissues	*Simulium* spp., blackflies
Loa loa	Loasis	Subcutaneous tissues	Blood	*Chrysops* spp., deerflies
Dipetalonema perstans	Dipetalonemiasis	Body cavities, especially perirenal tissues	Blood	*Culicoides* spp., biting midges
Dipetalonema streptocerca		Skin, especially of the trunk	Skin	*Culicoides* spp., biting midges
Mansonella ozzardi	Ozzard's filariasis	Body cavities, especially mesentery fat	Blood	*Culicoides* spp., biting midges
Dracunculus medinensis	Dracontiasis	Visceral connective tissues, in subcutaneous tissues	No microfilarial stage; female discharges rhabditiform larvae	*Cyclops* spp., small crustaceans

Wuchereria bancrofti

This filarial worm, causing Bancroft's filariasis and elephantiasis, is spotty in distribution throughout the tropics and subtropics; at one time it was endemic in Charleston, South Carolina. The infection occurs primarily in coastal areas and islands with long periods of high humidity and heat. In the western hemisphere, it is found in the West Indies, along the coast of South America from Brazil to Colombia, and at a focus along the Atlantic coast of Costa Rica. In the United States, the infection is most frequently seen in immigrants, military veterans, and missionaries.

Life Cycle. The adults, usually coiled together, live within the human lymphatic system. The males are about 4 cm long, and the females are about 9 cm. Both sexes are thread-like, being no more than 0.3 mm in diameter. After copulation, the female gives birth to living microfilariae. These ultimately find their way into the bloodstream. In most strains, they circulate in the blood in largest numbers at night and are said to have nocturnal periodicity. In certain other strains, either no particular periodicity is noted, or there is a tendency for the occurrence of the largest numbers at dusk. The circulating microfilariae, 125 to 320 μm in length, develop only in certain mosquitoes. Hence, pathology will not result if blood containing them is transfused into a recipient. Female mosquitoes of various genera (especially *Culex*, *Aedes*, and *Anopheles*) serve as necessary intermediate hosts for this parasite. When the mosquitoes feed upon an infected individual, they ingest the microfilariae from the peripheral bloodstream. The microfilariae pass into the midgut and invade the intestinal wall. Within 24 hours, most of the microfilariae find their way to the thoracic muscles, where they undergo metamorphosis to infective larvae in 1 to 3 weeks. This stage is active and migrates to the tip of the proboscis sheath, from which it penetrates into the skin of the human host at the time the mosquito takes the next blood meal. The larva probably enters the wound made by the mosquito. After entering the skin, these larvae pass to the lymphatic system and grow to maturity. In most cases this probably requires at least 1 year. The adult worms may live for many years in certain individuals.

Pathology and Symptomatology. In endemic areas, most of those who are infected are more or less asymptomatic, demonstrating what appears to be an excellent host–parasite adjustment. The host will experience little or no disturbance, and the parasite is able to develop normally, which results in the release of large numbers of microfilariae. These individuals, therefore, are the main source of infection for the mosquito host, as humans are the only known definitive host. In those patients showing clinical evidence of the infection, two distinct stages are evident. In the acute stage, there is a characteristic and often profound lymphatic inflammation in response to trapped worms and/or their metabolites. Tissue changes tend eventually to constrict the wall of the lymphatic vessel or other affected parts of the system. This partial obstruction results in lymph stasis and edema. The cardinal manifestation is a recurrent lymphangitis, usually with an associated lymphadenitis. Some patients experience a low-grade fever, usually of short duration. The location of the obstruction determines the part of the body affected, the external genitalia of both sexes being the most common. In the absence of reinfection, there usually is steady improvement in the individual, each relapse being milder than the former. Thus, even without specific therapy, this condition is self-limiting and presumably will not become chronic in those acquiring the infection during a brief sojourn in an endemic area. It is important to note that microfilariae are difficult to demonstrate during the acute phase, indicating a pronounced disturbance in the host–parasite equilibrium. Sectioning of biopsied tissues reveals the worms and the characteristic reaction, consisting of a necrotic zone around the worms and a palisaded area of foreign body giant cells, epithelioid cells, and eosinophils.

There is good evidence that the advanced chronic type of filariasis, known as elephantiasis, develops only in a small percentage of highly reactive individuals infected repeatedly for many years. Following lymphatic obstruction, striking proliferative changes occur, and the worms die and are absorbed or become calcified. The edema, soft at first, becomes fibrotic following the growth of connective tissue in the area. The redundant skin, being nourished poorly, cracks and becomes fissured and often is secondarily infected with pyogenic or mycotic organisms. This resemblance to elephant skin accounts for the term "elephantiasis" being applied to this chronic, disfiguring condition. Although microfilariae may appear in the blood of chronic cases, because of new, active infection superimposed upon the older ones, they are not often demonstrated. Thus, it appears that the more reactive the host, the less likely is the development of the worms and the release of embryos.

Tropical pulmonary eosinophilia (Weingartner's syndrome) is a form of occult filariasis caused by human and nonhuman filarial parasites. It is characterized by an immunologic hyperresponsiveness with a marked increase in IgE and IgG antiparasite antibodies as well as a hypereosinophilia. Sometimes paroxysmal nocturnal coughing, breathlessness, and wheezing are characteristic of infection. At other times, lymphadenopathy and hepatosplenomegaly are seen. Microfilariae are rarely present in blood, but remnants of them are found in eosinophilic granulomas in the spleen, liver, lymph nodes, and lungs.

Various syndromes are known to coexist with filariasis or are found in filarial endemic regions and have been suggested as being manifestations of infection. These include arthritis, endomyocardial fibrosis, tenosynovitis, dermatosis, lateral popliteal nerve palsy, and others. Further study is necessary to determine the precise role that filariae play, if any, in the above syndromes.

Epidemiology. To date, no naturally infected reservoirs other than humans have been identified. Several species of monkeys can be experimentally infected, however.

Dense populations living in tropical areas where mosquitoes breed unhampered offer ideal conditions for transmission. Reduction in mosquito populations will aid in control, but some vectors have now acquired resistance to almost all of the residual insecticides. The World Health Organization estimates that some 250 million people are infected, with 400 million being at risk. No prophylactic drugs are available. Consequently, travelers to endemic areas should avoid being bitten by mosquitoes. Repellents and mosquito netting, if sleeping outdoors, are advised.

Diagnosis. The acute clinical manifestations of filariasis are characterized by recurrent attacks of fever and lymphadenitis, sometimes precipitated by hard physical exertion. While lymphadenitis commonly occurs in the inguinal regions, male genitalia are frequently involved, leading to funiculitis, epididymitis, and orchitis. The whole acute clinical course of an episode of fever and lymphadenitis may last from 3 weeks to 3 months and can result in prolonged inability to work.

The clinical course of chronic infection occurs after prolonged residence in an endemic area. Thus, when symptomatic disease occurs, patients should be reassured that the prognosis is good, and advised that elephantiasis is a rare complication that is limited to persons from endemic areas who have had constant exposure to the infected mosquitoes for years.

In the laboratory, a direct diagnosis is made by demonstrating the microfilariae, usually in thick blood smears or blood filtrates. Membrane filtration techniques are the most sensitive methods currently available. In most areas of the world, microfilariae are present in appreciable numbers only at night, usually with greatest frequency between 10 PM and 2 AM. Microfilariae are also sometimes seen in urine and in chylous and hydrocele fluids. It should be remembered that microfilariae cannot be demonstrated during the incubation period and usually not during the acute phase of the infection. In the absence of microfilariae, a presumptive diagnosis can be made on the basis of the history of exposure, clinical evidence of the disease, and positive serologic tests (IHA and BFT) and/or a positive intradermal test. Lymph node biopsy, during clinical quiescence only, often will reveal adult or immature worms. This, however, is a research tool and should not be used except in unusual circumstances and then only by adhering to strict surgical techniques.

The differential diagnosis of filariasis should include hernia, venous thrombosis, obstructive lesions of the lymphatics, multiple lipomatosis, heart failure, and gonococcal infection.

Treatment. Patients with acute disease resulting from recent primary exposure should be removed from the endemic area, if possible. Bed rest and supportive measures, such as hot and cold compresses, are of assistance in reducing the edema. Psychotherapy often is necessary, especially for young males with scrotal involvement. Although surgical excision of lymph nodes is rarely indicated, and usually not successful, surgical removal of elephantoid breast, vulva, or scrotum is successful. Administration of antibiotics for patients with secondary bacterial infections and analgesics as well as anti-inflammatory agents during the painful, acute stage is helpful.

Diethylcarbamazine is the drug of choice. It should be remembered that chemotherapy is of little value during advanced disease, since the worms may already be dead and irreversible tissue damage has already occurred.

Onchocerca volvulus

Onchocerciasis is the term used to describe an infection with this nematode. Geographically, it occurs in the tropical zone of Africa and Central and South America. A small focus has recently been discovered in the Yemen Arab Republic.

The infection has emerged as a major health problem, with 40 million people being infected and 2 million of these blinded by the worm. Depending upon the region of the world, it is known as river blindness, blinding filarial disease, gale filarienne, Roble's disease, and craw-craw.

Life Cycle. Filamentous females, 30 to 40 cm in length, and 2-cm-long dwarf males live in subcutaneous tissues. Here they are either coiled up within fibrous nodules the size of a pea or walnut or remain free. Microfilariae produced by the female worms migrate to the upper layers of the skin and other parts of the body, including the eye. It is thought that females may live for up to 20 years and microfilariae for up to 2.5 years.

Blackflies of the genus *Simulium* ingest microfilariae when they ingest blood. Within the vector, metamorphosis to the infective larval stage takes place in 2 weeks. When the infected *Simulium* feeds, the larvae enter the subcutaneous tissues and mature into the adult form in 6 to 9 months.

Pathology and Symptomatology. The disease may be arbitrarily divided into three forms: dermatologic, nodular, and ocular. The dermatologic and nodular forms usually occur simultaneously. Adult worms are most likely rather innocuous, whereas the microfilariae can cause severe pathology.

Itching and scratching, with subsequent development of dermal lesions in the form of papular rashes, are usually the first signs of infection. Altered pigmentation of the skin usually follows. Skin may be mottled ("leopard skin"), scaly and wrinkled ("lizard skin"), or hyperpigmented ("sowda"). Skin changes result because the normal architecture of the dermis is lost and is replaced by a toneless, thickened, pachyderm-like layer. In some African patients, skin and lymph node involvement in the inguinal region leads to the loss of tissue elasticity, resulting in a "hanging groin." It has been postulated that immune complexes are formed because of the antigenicity of the living and dead microfilariae. These complexes trigger a chronic inflammatory response with subsequent perivascular fibrosis and obstructive lymphadenitis. This form of the disease is similar to the elephantoid form of wuchereriasis.

Nodules are usually visible, palpable, firm, rounded, and nontender masses varying in size from 0.5 to 10 cm in diameter. Nodules may be single or may occur in clusters and resemble a tumor. Nodules tend to be located over bony prominences, such as skull, elbow, and scapula. In Africa, nodules tend to be on the lower portion of the body along the iliac crest, groin, knees, thighs, and spine. In Central America, over half of the nodules are on the head.

Ocular onchocerciasis frequently climaxes as visual impairment or blindness before the infected patient reaches adulthood. This form of the disease should be regarded as a hypersensitive type, because symptoms do not appear until the microfilariae begin to die. The other forms of the disease (nodular and dermal) may precede or occur simultaneously with the ocular form. In the early stage, microfilariae are present in the anterior chamber and cornea. Cellular accumulations occur around the dead microfilariae, causing punctate keratitis. With larger numbers of microfilariae and with greater hypersensitivity, a severe sclerotic keratitis, often with an anterior uveitis, may end in blindness. Iridocyclitis, cataracts, secondary glaucoma, and postneuritic optic atrophy may also occur.

Epidemiology. Onchocerciasis occurs in the environment containing fast-running streams where the blackflies breed. The flies have a limited flight range, and as a result, infection is most prevalent along streams. This has created "ghost villages" in Africa, when entire populations have moved inland to get away from the blindness that accompanies chronic infection. Humans are the only known host. There is some evidence, however, that the horse and other domestic animals may serve as reservoirs.

Diagnosis. The living and dying microfilariae produce an intense itching and pruritis. In time, the skin pigmentation is altered, and it becomes toneless and thickened. It has been said that onchocerciasis makes young people look old and old people look like lizards. Nodules and corneal opacities should, of course, suggest onchocerciasis in those from endemic areas.

Diagnosis is best made by obtaining superficial skin snips. Skin removed near a nodule affords the best chance of finding the microfilariae. The removed skin is teased apart in a drop of saline and examined microscopically. At times, microfilariae have been demonstrated in urine and sputum. A recent advance in diagnosis has been the membrane filter concentration technique. In this technique, skin is teased apart in saline, allowed to stand 6 to 12 hours, and then expressed through a Nucleopore or Millipore membrane of 5 μm porosity. The membrane is then removed, stained, and examined.

Ocular involvement can be detected by slit lamp examination. Histologic sections of removed skin or nodules may also demonstrate adults and microfilariae. It should

be remembered that blood examinations are of no value because the microfilariae do not circulate in the blood.

Low to moderate eosinophilia is common in onchocerciasis. Serum IgG, IgM, and IgA levels may be increased two-fold higher than normal values. Various skin and serologic tests are available. At present, the best test is the indirect fluorescent antibody test.

Treatment. Diethylcarbamazine (DEC) and suramin are used. Microfilariae die rapidly after DEC treatment, and many times the treated person will respond with a severe hypersensitive reaction (pruritis and erythema). To minimize this reaction, such anti-inflammatory agents as aspirin, antihistamines, and corticosteroids are prescribed. Nodules, especially those on the face and near the eyes, should be removed surgically before chemotherapy is started.

Cestodes

Cestodes, or tapeworms, are so named because the adult stage resembles a measuring tape. Anatomically, the adult tapeworm is composed of a joined chain (strobila) of segments (proglottids). Each mature proglottid contains a complete set of male and female reproductive organs (Fig. 90-13). No digestive tract is present, and all nutrients are absorbed through the body wall (tegument). At the most anterior end is a hold-fast or attachment organ known as a scolex that may bear hooks, suckers, or sucker-like grooves (bothria). Immediately posterior to the scolex is a short germinal center (neck) from which new proglottids arise. Maturation of proglottids proceeds toward the posterior end of the worm. Thus, groups of proglottids from anterior to posterior are termed immature, mature, and gravid (proglottids that are full of eggs). Many times, gravid proglottids will detach from the strobila and pass in the stool.

Even though some tapeworms are large (up to 10 meters in length) the adult forms in the human intestine are generally well tolerated. In most instances, major complaints are absent until the patient observes proglottids that have been passed during defecation or have emerged from the anus at other times. In general, the patient's attitude to infection with adult tapeworms is either undue alarm or indifference. Both attitudes are to be discouraged, but certainly, any intestinal tapeworm infection should be treated.

The life cycles of most tapeworms infecting humans are complicated and involve one or more intermediate hosts. The larvae in these intermediate hosts are usually cyst or bladder-like forms. In a few instances, humans can serve as hosts for the larval stage, and this can lead to severe disease and even death.

In presenting the material of this section, the cestodes will be grouped according to their common habitats in the body—the intestinal tract and various tissues. The intesti-

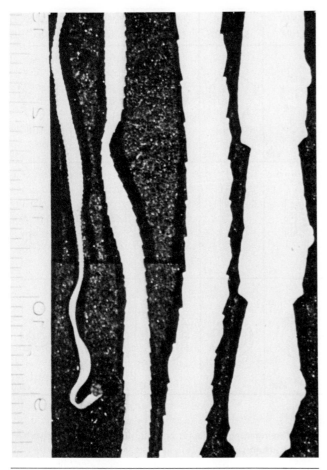

Figure 90-13. Adult tapeworm.

nal cestodes live as adults in humans, whereas the tissue cestodes occur as larvae.

Intestinal Cestodes

Only the four most common intestinal cestodes of humans will be presented: *Diphyllobothrium latum* (fish or broad tapeworm), *Taenia saginata* (beef tapeworm), *Taenia solium* (pork tapeworm), and *Hymenolepis nana* (dwarf tapeworm). All are endemic in the United States, but *T. solium* appears to be uncommon. In the adult stage, the first three are very large, ranging in length from 2 to 10 meters or more, whereas the fourth is, by comparison, a dwarf, being 45 mm or less.

Diphyllobothrium latum

This parasite is called the broad or fish tapeworm of humans: "broad" because the individual proglottids are wider than they are long, and "fish" because humans acquire the infection by ingesting raw or improperly cooked fish of certain species. *D. latum* is the largest of the human tapeworms, frequently reaching 10 meters in length.

Although worldwide in distribution, it is most common in north temperate areas where pickled or raw freshwater fish are eaten. Endemic centers include various Scandinavian countries (especially Finland), Canada, Chile, Argentina, and in the United States, the Great Lakes region as well as Alaska. The prevalence of this infection in the Great Lakes region is much lower now than in previous decades, and some consider it now a rare infection. Bears, dogs, cats, seals, and walruses are known to be infected and may serve as reservoirs for human infection.

Life Cycle. *D. latum* differs from other human tapeworms, both in structure and the complexity of its life cycle. The scolex is spatulate and is provided with median ventrally and dorsally grooved suckers (Fig. 90-14). The proglottids are broad, and the centrally located uterus has a characteristic rosette arrangement. Eggs within the fully developed uterus are discharged continuously in large numbers from the uterine pore (Figs. 90-3 and 90-14). These are undeveloped when passed in the stool and will embryonate only upon reaching cool, fresh water. A ciliated embryo (coracidium) escapes from the shell and swims about actively. When this stage is ingested by one of the first intermediate hosts (copepods), the embryo burrows into the body cavity and transforms into a mature first larva stage (procercoid). Ingestion of the first intermediate host containing these mature larvae by the second intermediate host (many different species of fish) continues the cycle. The larvae migrate into the flesh, often between the muscle fibers, of the fish and then metamorphose within several weeks into a third larva stage (plerocercoid, also known as a sparganum). Larger, edible fishes acquire the infection from eating their infected young or infected smaller species. Human consumption of fish flesh containing mature plerocercoid larvae completes the cycle. The worm, usually only one per infection in the United States, develops to maturity in the small intestine in 3 to 5 weeks. The worm may live for many years.

Pathology and Symptomatology. Only about half of those people infected exhibit symptoms. Those with symptoms usually complain of vague abdominal discomfort, nausea, diarrhea, vomiting, and weight loss. When several worms are present, their bulk can cause blockage of the intestine. In some people, and when the worm or worms are attached to the proximal portion of the small intestine, a pernicious anemia of the megaloblastic type may result. The anemia is associated with the worm's ability to compete with the host for vitamin B_{12}, which it readily absorbs from the intestinal chyme.

Epidemiology. In the United States, uncooked or pickled pike and walleyes are the usual sources of human infection. These fish are considered choice ones in the preparation of gefilte fish, and tasting for seasoning before cooking is not uncommon. The pollution of streams and lakes is a major factor in the infection cycle.

At one time, only freshwater fish were considered to

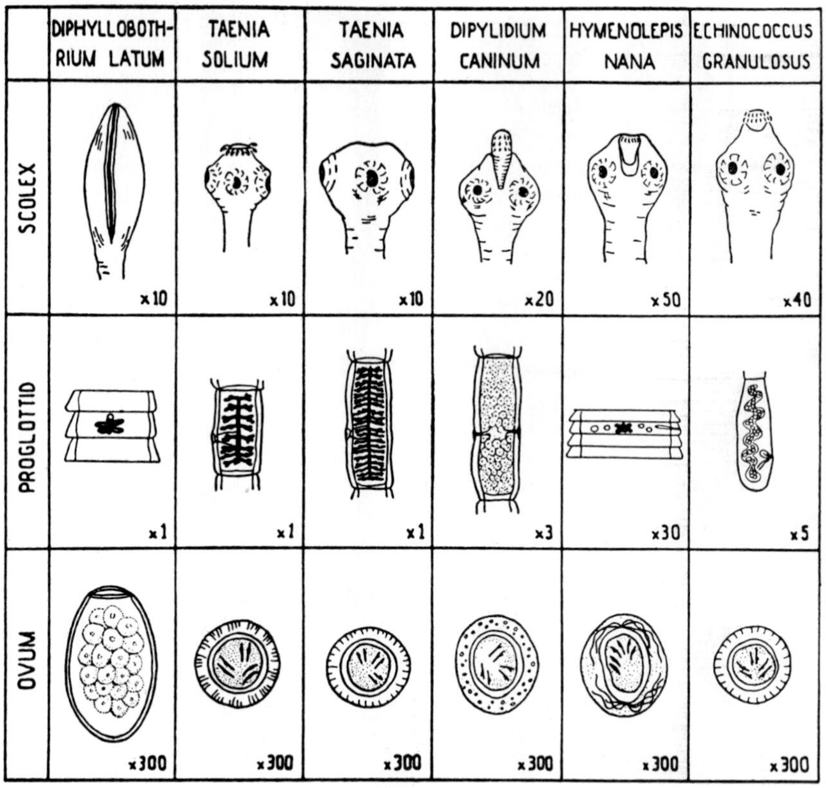

	DIPHYLLOBOTH-RIUM LATUM	TAENIA SOLIUM	TAENIA SAGINATA	DIPYLIDIUM CANINUM	HYMENOLEPIS NANA	ECHINOCOCCUS GRANULOSUS
SCOLEX	x10	x10	x10	x20	x50	x40
PROGLOTTID	x1	x1	x1	x3	x30	x5
OVUM	x300	x300	x300	x300	x300	x300

Figure 90-14. Differential characteristics of common tapeworms of humans. *(From Belding: Textbook of Clinical Parasitology, 3rd ed, 1965. New York, Appleton-Century-Crofts.)*

be suitable intermediate hosts. Recent evidence indicates, however, that marine fish also harbor the larvae that may infect humans. Thus, Japanese sushi and sashimi, Latin American ceviche, and Dutch green herring are potential sources of infection.

Diagnosis. Infection is usually benign, and many are asymptomatic. Clinically, abdominal discomfort with diarrhea, vomiting, and nausea are not uncommon. A travel history and questions concerning eating habits will help in diagnosis.

The gravid proglottids usually disintegrate prior to being passed, and consequently, only eggs are found in the stool. The eggs measure 45 by 70 μm, are thin-shelled, ovoid, and yellow-brown in color. An operculum (lid) is at one end, and a small, knob-like protuberance may be present at the other end (Figs. 90-3 and 90-14). As for all operculated eggs, concentration must be done by a sedimentation technique, such as the formalin-ether method.

Laboratory findings that usually accompany infection are a low to moderate eosinophilia, slight leukocytosis, and a low serum vitamin B_{12} level. If tapeworm-induced anemia is present, it is typical of the pernicious type. No serologic test is available.

Treatment. Yomesan (niclosamide) is the drug of choice. This drug is a nonabsorbed oxidative phosphorylation inhibitor that kills the scolex and anterior segments on contact, after which the worm is expelled. Those patients

manifesting megaloblastic anemia should also receive vitamin B_{12} therapy.

Hymenolepis nana

H. nana, the dwarf tapeworm, is the most common tapeworm in humans in the United States, being most prevalent in the southeastern states where it is estimated that 3 percent of all children under 8 years of age either are or have been infected. On a worldwide basis, it is cosmopolitan in distribution, with highest infection rates being in the tropics.

Life Cycle. Unlike the large tapeworms of humans where only a single worm is present in an infection, the adults of *H. nana* are usually numerous. An adult worm, 2 to 4 cm in length, attaches to the mucosa of the small intestine. The parasite's life cycle is unique in that eggs can be directly infective for humans, without the necessity of an intermediate host. In the direct cycle, eggs initiate infection upon ingestion. Such infections may be of a direct hand-to-mouth type or of an indirect type via contaminated foods and/or fluids. The eggs hatch in the duodenum, and the liberated embryos penetrate into nearby villi. The resulting larval stage (cysticercoid) matures in about 4 days, returns into the intestinal lumen, attaches to the mucosa, and develops in about 2 weeks into the adult worm. Adults live for months. In the indirect cycle, arthropods, especially certain beetles, serve as intermediate hosts. Accidental ingestion of those containing the mature larvae (cysticercoids) results in infection.

Pathology and Symptomatology. Variability in clinical manifestations is common in mild to moderate infections. Some infections are asymptomatic, whereas others produce diarrhea, abdominal discomfort, and anorexia. In heavy infections, symptoms are usually more pronounced and include profuse diarrhea, abdominal pain, pruritis, nervous disorders, and apathy. Generalized toxemia may develop in some heavily infected children.

Epidemiology. Though humans themselves are the most important source of human infection, rats and mice may also be involved. The accidental ingestion of rodent feces (direct cycle) that contain *H. nana* eggs is thought to be more important in contributing to the prevalence of the disease than is the ingestion of infected beetles (indirect cycle). Infection seems to be as common in urban settings as in rural areas, which is unusual for helminthic infections.

Infection is usually confined to children less than 8 years old. Autoreinfection does occur and may account for infection with hundreds of worms within a single host. Prevention is by exercising proper hygiene and practicing rodent and vermin control.

Diagnosis. Clinical symptoms are rather vague, and diagnosis depends upon laboratory demonstration of the typical egg (Figs. 90-3 and 90-14) as proof of infection. Proglottids are not usually found in or on the stool because these disintegrate in the intestine before being passed. The egg, 50 μm in diameter, is slightly oval, with a thin, colorless outer shell. Within the egg is another membrane enclosing the six-hooked larva (oncosphere). Between the outer shell and the membrane surrounding the oncosphere are 8 to 10 thread-like polar filaments. Adult worms, when recovered, can be identified on the basis of their size and the presence of a baseball bat-shaped scolex that bears four suckers and a circlet of hooks on the rostellum. About a third of those infected will have an eosinophilia of 5 percent or more. No serologic tests are available.

Treatment. On a comparative basis, *H. nana* is difficult to eradicate. Yomesan, however, has proved to be effective and is the drug of choice.

Taenia saginata

The beef tapeworm (*T. saginata*) and the pork tapeworm (*T. solium*) are large worms that may occur in humans when raw or insufficiently cooked meat is eaten. Both infections are known as taeniasis, but it is imperative that a species diagnosis be made because the pathologic consequences of *T. solium* can be severe, whereas those of the beef tapeworm are rather innocuous.

The beef tapeworm may be found wherever beef is eaten. Areas of high prevalence are Kenya, Ethiopia, Taiwan, the Philippines, and Iran. Infection in these areas is associated with eating raw beef, poor sanitation, and the practice of letting cattle graze on pastures fertilized by sewage sludge. In addition to cattle, several other ungulates, such as camels and antelope, may serve as sources of infection. Humans are the only host infected with the adult tapeworm. In the United States, infection is seen infrequently, but meat inspection records show that 10,000 to 15,000 beef animals are slaughtered each year that harbor the larva stage. As a result, there may be more human infections than are generally realized.

Life Cycle. The adult tapeworm, usually only one worm per infection in the United States, is attached to the mucosa of the small intestine by the scolex. The distalmost gravid proglottids become separated from the strobila and actively migrate out of the anus or are evacuated in the stool. If grazing cattle, or other suitable ungulates, ingest the proglottids or, more commonly, the eggs that are freed after disintegration of the proglottids on moist earth or in raw sewage, the cycle proceeds. The six-hooked embryos (oncospheres) escape from the eggs, following hatching in the duodenum, and penetrate into the intestinal tissue to reach the circulation ultimately. They are carried through the blood, and most of them reach skeletal muscles or the heart. In these sites, they transform in 60 to 75 days into a typical cysticercus larval stage (*Cysticercus bovis*), which contains a scolex, similar to that of the adult worm, invaginated into a fluid-filled bladder. Human infection is through ingestion of these larvae in beef, either raw or processed inadequately. The head of the larva attaches to the wall of the ileum, and the adult worm develops to maturity in 8 to 10 weeks, when it can reach a length of 5 to 10 meters. The worm may live for several years.

Pathology and Symptomatology. Most infections cause little discomfort and, indeed, are usually symptomless until proglottids have been passed and detected by the infected person. Some patients complain of abdominal discomfort, hunger pains, episodes of diarrhea, anorexia, and weight loss. There is some evidence to suggest that gastric secretion is reduced.

Epidemiology. The most effective control measure is the prevention of soil contamination with human feces where cattle are likely to graze. Eggs are resistant and can survive for months in soil. Chemical treatment of soil and even routine sewage plant treatment does not kill the eggs. The thorough cooking of beef and the freezing of meat before consumption will also disrupt the cycle.

Diagnosis. Clinically, signs and symptoms are either absent or very vague and indicate some kind of gastric and digestive disturbance. As previously noted, most of those infected are unaware of the infection until proglottids are noticed in the stool or, more alarmingly, crawl out of the anus and down the leg.

Definitive diagnosis depends on the recovery of the typical rhomboidal scolex, equipped with four suckers but without hooks (Fig. 90-14). The egg, about 35 μm in diameter and characterized by a striated shell containing the six-hooked oncosphere, may be detected microscopically. However, the eggs are indistinguishable from those of *T. solium.* More commonly, the gravid proglottids are used for making a diagnosis. When passed, proglottids, each approximately 1 to 2 cm long, are creamy white in color. When pressed gently between two glass slides and held in front of a bright light, 15 to 20 main lateral branches of the uterus on each side can be seen (Fig. 90-14). Abnormal laboratory values rarely occur. Sometimes a mild eosinophilia is seen, as are lymphocytosis and anemia.

Treatment. Yomesan is the drug of choice.

Taenia solium

The major differences between *T. solium*, the pork tapeworm, and the beef tapeworm are (1) infection is acquired by eating raw or improperly cooked pork, and (2) humans can serve as the intermediate host as well as the definitive host. The fact that humans can be infected with the egg of *T. solium* and can harbor the larval stage (cysticercus) makes this infection a serious and, many times, life-threatening condition.

Pork tapeworm infection occurs wherever pigs are raised, sanitation is poor, and pork is eaten raw or is insufficiently cooked. Fortunately, human infection in the United States is rare. However, it is endemic in Mexico, Latin America, tropical Africa, southeast Asia, the Philippines, and in the Indian subcontinent. Prevalence rates of 1 to 3 percent have been reported from Spain, Hungary, and Czechoslovakia.

Life Cycle. Attachment of the adult tapeworm to the mucosa of the small intestine is by means of the scolex. In addition to the four cup-shaped suckers there are 22 to 32 small hooklets on the rostellum. The length of the worm is shorter (2 to 8 meters) than that of *T. saginata*, and the structure of the proglottids differs in the two species (Fig. 90-14). The life cycles of the two species are similar, except that the hog is the usual intermediate host for *T. solium.* The scolex of the larval stage (*Cysticercus cellulosae*) is provided with four suckers and a crown of hooklets. This stage may also occur in humans and cause serious injury (see below). For this reason, *T. solium* has greater medical and public health importance than has *T. saginata.*

Pathology and Symptomatology. The clinical manifestations of a *T. solium* infection with adult worms are the same as those induced by *T. saginata.* For a discussion of the clinical manifestations due to infection with the larval stage, see below.

Epidemiology. In addition to swine, other animals, such as camels, dogs, sheep, and deer, can serve as intermediate hosts. Although humans are the only known definitive host, the incidence of human infection is rare in some areas compared to the prevalence of cysticercosis in hogs ("measly pork"). Fecally contaminated slop and the foraging of swine on fecally contaminated forage results in an incidence of cysticercosis of 25 percent in some regions. Thorough cooking of pork is, obviously, essential in prevention of infection.

Diagnosis. Clinically, symptoms due to the adult stage are the same as those of *T. saginata.* Since it is important to know whether the infected person has *T. solium* or *T. saginata,* either the recovered scolex with its row of hooks (Fig. 90-14) or the gravid proglottid must be recovered, because the eggs of both species are identical. The pork tapeworm proglottid, when pressed between two glass slides and held in front of a bright light, will show 7 to 13 main lateral branches of the median stem of the uterus (Fig. 90-14).

Treatment. Yomesan is the drug of choice. Because internal autoreinfection is possible (see below), it is suggested that a purge be done after therapy to prevent reverse peristalsis. Nausea and vomiting during therapy should also be avoided because this could lead to the backwash of eggs and possible internal autoreinfection.

Tissue Cestodes

Sparganosis

Sparganosis is a tissue infection of humans with the plerocercoid larval stage of diphyllobothrium-like tapeworms. In most cases, the genus involved is *Spirometra*, a tapeworm of cats and dogs. Although worldwide in distribution, it is most common in the Orient. Due to the immigration of large numbers of Asians into the United States within the past few years, sparganosis is now occasionally seen.

Spargana are whitish, elongate (a few mm to cm long), and have a wrinkled appearance. They may enter human tissues during treatment of wounds or sores with poultices of plerocercoid-infected frog or snake flesh. When the poultice application is near the eye, ocular sparganosis may occur. Humans may also develop sparganosis by drinking water containing infected water fleas (*Cyclops*). The procercoids within *Cyclops* invade the human gut wall and usually migrate to subcutaneous tissues. The migratory phase is usually asymptomatic.

When migration stops, a painful inflammatory reaction develops; encystation does not occur. Periorbital edema and ocular ulcers develop as a result of ocular sparganosis. A rare budding type of sparganum has been reported, and this is known to cause death.

Diagnosis is made only after surgical removal of the sparganum. Leukocytosis and eosinophilia are usually seen. There is no satisfactory drug therapy.

Cysticercosis

Cysticercosis in humans is exclusively caused by the larvae of *T. solium* (i.e., *C. cellulosae*), which have a predilection for skeletal muscles and the nervous system. The egg stage initiates human infection. A patient harboring the adult worm may become infected by transferring mature eggs from the anus to the mouth on fingertips (external autoinfection), or these may be transferred indirectly to another person (heteroinfection). The third type, internal auto-reinfection, is thought to occur when detached gravid proglottids are transferred by reverse peristalsis into more proximal parts of the small intestine, where some of the eggs become liberated. In all instances, the eggs hatch in the small intestine, and the larvae (oncospheres) are carried to the tissues via the blood circulation.

Cysticerci reach a size of 2 cm and may occur in various human tissues, but the greatest concern is the common involvement of the eyes and the brain. Ocular cysticercosis may result in uveitis, dislocation of the retina, and other conditions. Pain, flashes of light, grotesque figures in the field of vision, and other complaints have been noted. Cerebral cysticercosis usually follows involvement of the meninges, with jacksonian (rolandic) epilepsy as the most characteristic consequence. Infection of the third and fourth ventricles, with hydrocephalus, headache, and diplopia, are also noteworthy. Cysticerci may live for 3 to 5 years before they die, degenerate, and become calcified. Cysts in skeletal muscle, liver, lungs, kidneys, and heart may or may not cause symptoms. There is virtually no inflammation due to the living larvae, but dead larvae precipitate an acute cellular inflammation.

Cysticercosis is common in regions where taeniasis due to *T. solium* is endemic, such as Mexico, Thailand, and eastern Europe. Often a patient will remember a previous tapeworm infection, and physicians should be aware that symptoms of cysticercosis may not occur until several years after infection. A recent report shows neural cysticercosis in 2 percent of all autopsies carried out in Mexico.

Infections of tissues other than neural are usually asymptomatic. At the time of death of the cysticercus, however, there is the liberation of the cyst's fluid contents, which appear to be toxic and allergenic.

The diagnosis of cerebral cysticercosis is primarily clinical and may be quite difficult. Symptoms include headache, vomiting, impaired vision, and convulsions.

Various laboratory values are of limited use. Blood chemistry and hematologic data are not distinctive, although central nervous system fluid may show pleocytosis, increased protein, decreased glucose, and eosinophils. Radiologic data are especially helpful because vascular changes and calcified cysticerci are readily seen in muscle and soft tissues. Computerized axial tomography is also a useful diagnostic tool.

Until recently, there was no satisfactory chemotherapy, and surgical removal of those cysts in operable sites was the only treatment possible. However, use in clinics of praziquantel and metrifonate appears to be effective. These drugs have not yet been completely evaluated but do hold considerable promise. Epileptic-like symptoms are treated with anticonvulsant drug therapy.

Serologic tests of a variety of types are available in the United States and are strongly indicative of infection. Cross-reactions and false-negative results do occur, however, so some caution in interpretation must be exercised.

Hydatidosis

Hydatidosis is a zoonotic infection in which humans harbor the larval stage of the canine tapeworms, *Echinococcus granulosus* and *Echinococcus multilocularis*. Domestic dogs, wolves, foxes, and coyotes are the normal definitive hosts. A wide range of animals, including sheep, moose, field mice, and voles, are the normal intermediate hosts. Humans become infected by ingesting eggs of either species.

After ingestion of the eggs, the hatched oncospheres invade the mucosa of the intestine and enter the hepatic portal system. In the circulation, they may be filtered out in the liver, but if not, they may eventually come to reside in the lungs, brain, bones, spinal column, or other visceral organs. The incubation period is extended, being from several months to years. Infections are frequently asymptomatic and chronic.

Infection with the larval stage of *E. granulosus* (known as unilocular hydatid disease) is cosmopolitan in distribution. Its greatest incidence is in areas of the world where sheep raising is an important industry, such as Australia, South America, and the Mediterranean. In the United States, endemic foci are in the southwest, especially Utah, Colorado, Montana, and California. *E. multilocularis*, agent of alveolar or multilocular hydatid disease, is endemic in northern, temperate climates, including Canada and the northern United States. This is primarily a sylvatic disease, with the typical life cycle involving the fox and vole, whereas *E. granulosus* has a typical cycle involving the dog and sheep. Sheep raisers are those most often infected with larvae of *E. granulosus*, and hunters and trappers are most often infected with *E. multilocularis*.

As stated above, human infection is through inadvertent ingestion of eggs. In the case of *E. granulosus*, any organ or tissue may be involved, but the liver and lungs are the most common sites. By a remarkable process of asexual reproduction, the embryos metamorphose into hydatid cysts. These vacuolated larval cestodes are called unilocular hydatid cysts. Cysts in soft tissues, such as liver and lungs, are somewhat similar, but the type in bone tends to elongate as it flows into and erodes the bony canal. The unilocular hydatid cyst of the liver is the most common in humans. It consists typically of a central fluid-filled cavity lined with a germinative, protoplasmic layer, surrounded by a cuticular, protective layer that tends to become laminated. It usually requires many years for the cyst to reach a large size. In time, it becomes covered with a fibrous host-tissue capsule. Arising from the germinative layer, internal buds (brood capsules) are produced (Fig. 90-15). When full size, each forms vesicles along its inner margin. Typically, each vesicle (10 to 20 or more) develops into a small protoscolex, usually invaginated, which serves to protect the 20 to 30 rostellar hooklets, and a short neck region. Thus, a large mature cyst containing thousands of these protoscolices produces a heavy infection of adult worms when eaten by a definitive host.

When humans ingest the egg of *E. multilocularis*, the

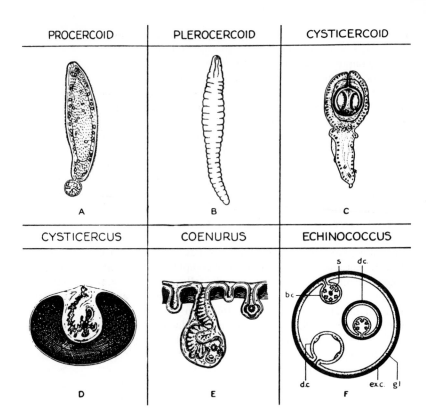

Figure 90-15. Larval forms of tapeworms. b.c., brood capsule; d.c., daughter cyst; ex.c., external laminated cuticle; g.l., germinal or inner nucleated layer; s, scolex. *(From Belding: Textbook of Clinical Parasitology, 2nd ed, 1952. New York, Appleton-Century-Crofts.)*

released larval stage penetrates the gut wall and is carried to the liver and, less often, to other visceral sites and grows by exogenous budding to form another larval stage, the alveolar hydatid cyst. The alveolar cyst in humans lacks a strong protective layer, which appears to encourage branching formations or cavities with little fluid and few, if any, protoscolices. The liver is most often involved, and the damage resembles that produced by a large amebic liver abscess. The liver disease is almost always fatal.

The hydatid cysts of *E. granulosus* are usually single, but they may be multiple. Their size and contour depend upon the site of implantation and the age of the cysts. Because of the slow growth of cysts, vital processes usually are not disturbed sufficiently to be of concern to the patient until many years after infection. Ultimately, however, there may be tissue destruction and striking signs and symptoms. The type and degree of this damage and the resulting clinical manifestations correspond to the exact location and size of the cyst(s). It should be added that systemic intoxication or sensitization often occurs in those with a unilocular hydatid cyst having a vascularized wall that permits leakage of sensitizing fluids. Sensitized individuals exhibit marked eosinophilia, and in some, urticaria or angioneurotic edema is evident. Anaphylaxis may be precipitated by the sudden release of hydatid fluid, as in the case of spontaneous rupture of a large intraabdominal cyst or rupture following a severe blow to the area.

Both infections usually are diagnosed by indirect means, such as x-ray, and intradermal and serologic tests.

Recently, ultrasonic echotomography has been used to great advantage in detecting hydatid cysts.

In treatment, surgical intervention is limited to those patients with unilocular cysts in operable sites. Meticulous care must be taken to prevent spilling the cyst contents into the operative cavity because of the probability of causing anaphylaxis. Formalin usually is injected into the cyst to kill the germinative membrane and protoscolices, since if they are released from the cyst, they can form new hydatid cysts.

Mebendazole in high doses has been used with some success. This drug appears to cause regression of cyst size, as verified by echotomography, and also destroys the cyst's contents before surgery. Mebendazole has also been used when a cyst ruptures either before or during surgery.

Trematodes

Flukes, as these parasitic flatworms are called, have unique life cycles involving sexual reproduction in vertebrate hosts and asexual reproduction in snail intermediate hosts. Adult trematodes are usually flat, elongated, and leaf-shaped. They vary considerably in size, from less than 1 mm to several centimeters. Characteristic external features include an oral and, in most species, a ventral sucker, the acetabulum. The principal internal organs include a blind, bifurcated intestinal tract, an excretory system, prominent reproductive organs, and a primitive nervous system. The arrangement, shape, and size of these various structures are characteristic for different species.

Unlike the tapeworms presented above, none of the trematodes of humans are endemic in the United States. However, due to immigration, trematode infections, especially the bloodflukes, occur in large numbers in new residents in the United States.

With the exception of the bloodflukes (schistosomes) all trematodes are hermaphroditic. Eggs that are produced are passed from humans in either a developed or an undeveloped stage. When fully developed, the larval stage within (miracidium) must take up residence within the tissues of an appropriate species of snail where prodigious asexual reproduction occurs. Eventually, cercariae are produced, and these leave the snail and either pass to the next intermediate host where they encyst and become metacercariae or actively swim and penetrate through the skin of the definitive host. As a result of ingestion of the metacercaria or the penetration of vertebrate skin by the cercaria, the larva develops to the adult worm stage after a prepatent period that is characteristic of the species.

Information concerning the most commonly found trematodes of humans in the United States and the infections they cause is listed in Table 90-2. The schistosomes, being of greatest medical and public health importance,

will be discussed in greater detail than the others. It should be remembered that there are other trematode infections of humans, but these are only rarely encountered in the United States and are, therefore, not discussed here.

Schistosomes or Bloodflukes

Schistosomes differ in many ways from the other trematodes. The important differences include the presence in schistosomes of separate sexes, nonoperculated eggs with spines, fork-tailed cercariae, and the absence of a true metacercarial stage in the life cycle. As noted in Table 90-2, there are three species that occur commonly in humans: *Schistosoma japonicum*, *Schistosoma mansoni*, and *Schistosoma haematobium*. Although the other two species lack important natural reservoirs, *S. japonicum* has many mammalian reservoirs (cats, dogs, cattle, water buffalos). However, it may be stated that humans usually are the most important source of eggs of all three species. The adult worms occur characteristically in pairs in mesenteric veins (*S. japonicum* and *S. mansoni*) and vesical veins (*S. haematobium*) on the outside wall of the intestines or urinary bladder. The male attaches to the wall of the blood vessel, holding the female in a ventral groove, the gynecophoral canal. The female is able to extend its anterior end into the smaller venules where the eggs are discharged.

The World Health Organization has stated that this

TABLE 90-2. IMPORTANT TREMATODES OF HUMANS

Parasite	Disease	Location of Adults Where Eggs Are Laid	Stage Passed from Humans	Second Intermediate Host (Certain Snails Are the First Intermediate Host)	Means of Human Infection	Laboratory Diagnosis
Fasciolopsis buski	Fasciolopsiasis	Small intestine	Undeveloped eggs in stool	Water caltrop, certain other freshwater plants	Ingestion of metacercariae	Eggs in stool (140 × 80 μm)
Clonorchis sinensis	Clonorchiasis	Distal bile ducts of liver	Developed eggs in stool	Certain freshwater fishes	Ingestion of metacercariae	Eggs in stool (30 × 16 μm)
Paragonimus westermani	Paragonimiasis	In lung capsules	Undeveloped eggs in stool and sputum	Certain freshwater crabs and crayfishes	Ingestion of metacercariae	Eggs in stool and sputum (85 × 50 μm)
Schistosoma japonicum	Schistosomiasis japonica	Venules of superior and inferior mesenteric veins	Developed eggs in stool	None	Penetration of skin by cercariae	Eggs in stool (89 × 66 μm with rudimentary lateral spine)
Schistosoma mansoni	Schistosomiasis mansoni	Venules of inferior, and at times superior, mesenteric veins	Developed eggs in stool, rarely in urine	None	Penetration of skin by cercariae	Eggs in stool or urine (150 × 60 μm, with large lateral spine)
Schistosoma haematobium	Schistosomiasis haematobia	Venules of vesical and pelvic plexuses	Developed eggs in urine, rarely in stool	None	Penetration of skin by cercariae	Eggs in urine or stool (150 × 60 μm, with large terminal spine)

infection is second only to malaria in causing morbidity and mortality in the tropics. Widespread over three continents, it affects an estimated 200 million people. *S. mansoni*, the agent of Manson's bloodfluke infection, is common throughout most of tropical Africa, especially in the Nile Delta, and in the western hemisphere in Brazil, Venezuela, the West Indies, and Puerto Rico. It is estimated that 400,000 infected persons live in the United States, many of whom are Puerto Ricans. Little risk of transmission is involved, however, since appropriate snails are lacking, and sanitation is generally good. *S. japonicum*, agent of the Oriental bloodfluke infection, is confined to the Far East. *S. haematobium*, agent of urinary bloodfluke infection, coexists with *S. mansoni* in much of tropical Africa and is also endemic in the Near East, India, and Portugal. Other species, such as *Schistosoma mekongi* (Far East), *Schistosoma intercalatum* (Zaire), and a variety of cattle bloodflukes, also occur in humans but are of minor importance when compared to the others.

Life Cycle. The adults (Fig. 90-16) of *S. japonicum* and *S. mansoni* are found normally in the tributaries of the superior and inferior mesenteric veins, respectively, whereas those of *S. haematobium* find the venous plexus of the bladder the optimum location. Although these are the usual locations, it should be remembered that the worms may be found in other sites. Perhaps most important in this connection is the well-known fact that following infection with *S. haematobium* a small percentage of the worms fails to reach the normal site and remains in the rectal vessels. Thus, some of the eggs released may occur in the stools rather than in the urine, as expected in this infection. After copulation, the female flukes give off a considerable number of eggs over a long period of time. These are undeveloped when laid but usually contain a fully developed ciliated larva (miracidium) after they have succeeded in passing through the wall of the intestine or bladder to occur in the stools or urine (Fig. 90-13). Upon reaching fresh water, the eggs hatch, and the miracidium swims about. If appropriate intermediate hosts are present in the immediate vicinity, the miracidium will penetrate into the soft tissues. The intrasnail cycle lasts for several weeks and involves three distinct stages: the first (mother) and second (daughter) generations of sporocysts and cercariae. The latter, with characteristic forked tails, escape from the snail at intervals and swim about in the water. When they contact the skin of humans or other susceptible definitive hosts, they discard their tails and penetrate the skin. These larvae, now called schistosomula, reach the bloodstream and are carried through the right heart to the lungs. Here, they pass the capillary filters and are carried to the left heart and, thence, into the large arterial vessels. From the superior mesenteric artery, where most of them are carried, they pass through the capillaries into the intrahepatic portal blood. They feed and grow in this site, and when sexual maturity approaches (after about 16 days of residence), they migrate against the portal blood flow to the areas where egg laying is to occur. *S. haematobium* is

thought to pass from the rectal veins through hemorrhoidal anastomoses into the pudendal vein to reach ultimately the vesical venous plexus. Several weeks are required for the maturation of the adult worms (4 to 5 for *S. japonicum*, 6 to 7 for *S. mansoni*, and 10 to 12 weeks after skin penetration in the case of *S. haematobium*), and they may live for many years.

Pathology and Symptomatology. For all three species, penetration of the skin by the cercariae produces small hemorrhages. After the schistosomula break out of the capillaries in the lungs, they cause an acute inflammatory reaction predominated by eosinophils. Upon their arrival in the intrahepatic portal blood, an acute hepatitis may follow, as well as systemic intoxication and sensitization, all due presumably to the toxic and/or allergenic metabolites released. Many patients exhibit toxic manifestations, such as fever and sweats, epigastric distress, and pain in the back, groin, or legs. Some develop giant urticaria and toxic diarrhea. Eosinophilia is common. These reactions may continue long after the worms have migrated to the area of oviposition. The penetration of the cercariae and the migrations of the schistosomula, and, later, the movements of the developing worms usually produce detrimental effects. However, the main agents of pathology in schistosomiasis are the eggs released from the females.

The period of egg deposition and extrusion from the body usually is referred to as the acute stage, or as Katayama fever. In the instance of the two intestinal forms, *S. japonicum* and *S. mansoni*, the events of this stage are, in general, similar. However, it is important to add that the females of *S. japonicum* release considerably more eggs, and, therefore, the damage is proportionately greater. In both species, the intestinal tissue is the first to be damaged, usually the small intestine in the case of *S. japonicum* and the colon in the case of *S. mansoni*. Considerable trauma and hemorrhage are produced by the eggs as they are filtered through the perivascular tissues into the lumen. An allergen released by the developing miracidium escapes through pores in the egg and causes a striking cell-mediated response in the affected areas. Eggs trapped in these sites are walled off, usually individually, by an eosinophilic abscess, which later transforms into a characteristic granuloma (pseudotubercle). The egg or its shell only is usually surrounded by a peripheral ring of connective tissue and then by eosinophils, plasma cells, and lymphocytes. Fibrous nodules and scarring result. The acute stage is ushered in with diarrhea or dysentery and the appearance of eggs in the stools. Daily fever, anorexia, loss of weight, severe abdominal pain, and anemia are common. Many of the eggs are swept into the intrahepatic portal vessels, where they provoke granuloma formation. Liver involvement is more rapid and severe in *S. japonicum* infections. The liver becomes tender and enlarged. Coarse bands of dense connective tissue, chiefly about the large radicals of the portal vein, have been responsible for the term "Symmer's clay pipestem fibrosis," associated with schistosomiasis japonica and mansoni. Blockage causes portal hypertension, and opening of a secondary circulatory shunt leads to varices in esophageal and gastric veins, ascites, and gross hepatosplenomegaly. It should be added that nests of *S. japonicum* eggs often occur also in ectopic sites, such as the brain and heart.

The chronic stage of schistosomiasis is one of tissue proliferation and repair. The intestinal wall becomes thickened by fibrosis, and the lumen may be reduced considerably. Anal polyps are common, as are papillomas and fistulas. Hemorrhoids may be the first indication of the infection, resulting from portal obstruction. The liver may become increasingly damaged due to extensive periportal fibrosis, and there may be a compensatory congestive enlargment of the spleen, especially in *S. japonicum* infections. Thus, in many patients, there is a rapidly developing dysfunction of the intestinal wall and periportal tissues. In the late stages of the disease in those with heavy infections, emaciation is severe, and many patients die of exhaustion or of a concurrent infection, such as salmonellosis.

In infection with *S. haematobium*, the acute stage involves mainly the wall of the urinary bladder, but the lungs also may be involved. The latter involvement is due to eggs and, at times, worms that are probably carried via the common iliac vein, the inferior vena cava, and right heart to reach the pulmonary arterioles. Here, granulomas are produced as described above, and as a result fibrosis of arterioles and pulmonary hypertension may develop. In time, heart disease (cor pulmonale) may follow. The damage produced in the wall of the urinary bladder is similar to that described above for the wall of the intestine. Hematuria usually is the first evidence of infection. As time passes, the bladder wall becomes thickened by dense fibrosis of the muscular and submucous coats, and multiple urinary polyps, papillomas, and fistulas are common. The superficial mucosa of the bladder may show metaplasia, an intense inflammatory infiltrate, and eggs (many calcified). Fever, suprapubic tenderness, and difficulty in urination are common. Bladder colic is a cardinal symptom. In addition to the bladder, other parts of the genitourinary system often become involved. Finally, it is worth noting that in some areas, there is a close association between chronic schistosomiasis of the bladder and squamous cell carcinoma, since eggs in capillaries are seen in the midst of an infiltrating carcinoma. In fact, the term "Egyptian irritation cancer" is well known.

Epidemiology. Schistosomiasis, also known as bilharziasis or snail fever, is present in many countries and afflicts at least 200 million people. Unfortunately, the infection is spreading in association with implementation of water resource projects in most developing countries. As an example of this increase, the prevalence rose from 10 to 100 percent among the inhabitants around Volta Lake, Ghana, within 5 years of water impoundment.

While it is unlikely that schistosomiasis can be eradicated, it is possible to reduce both the incidence and prevalence and the morbidity and mortality. To this end advances are being made in (1) the treatment of people so that they do not pass eggs, (2) the prevention of eggs in feces and urine from reaching water, (3) the control of snail hosts, and (4) the development of vaccines and other prophylactic treatment.

Diagnosis. Schistosomiasis usually manifests itself as a long-term, chronic illness and, in large measure, is an immunologic disease due to cell-mediated granulomatous reactions around the eggs. Acute disease is rare in the United States and is associated with a primary infection obtained while living within endemic areas, such as Puerto Rico, Africa, and the Philippines. Clinical signs and symptoms of each of the three phases—penetration, acute, and chronic—of infection were discussed above and will not be repeated. It is important to remember that a history of travel or residence in an endemic area is an important consideration in diagnosis. The differential diagnosis of acute disease should include a consideration of amebic or bacterial dysentery, hepatitis, and typhoid fever.

With regard to the laboratory diagnosis, in most cases *S. haematobium* eggs can be demonstrated in the sediment that settles out of urine. In some instances, a small bladder biopsy specimen will reveal the eggs when they cannot be demonstrated in urine. If these measures fail in suspected cases, stool examinations and/or rectal biopsies should be considered, since these worms may involve the rectum as well. The mature eggs (Fig. 90-3) range from 110 to 170 μm in length by 40 to 70 μm in width (average 150 by 60 μm). Intradermal and serologic tests, although available, usually are not needed to provide evidence of infection.

The eggs of *S. japonicum* and *S. mansoni* usually can be recovered from the stools of patients during the acute stage, but they tend to be released in clutches, making it necessary to perform repeated examinations at intervals for a period of 1 month or more before ruling out the infection. *S. japonicum* eggs are rotund, measure 70 to 100 μm by 50 to 70 μm (average 89 by 66 μm), and have a rudimentary lateral spine within a hook cavity (Fig. 90-3). Those of *S. mansoni* are rounded at both ends, measure 115 to 175 μm by 45 to 70 μm (average 150 by 60 μm), and have a conspicuous lateral spine near one pole (Fig. 90-3). The eggs of *S. japonicum* are more numerous and tend to be mixed with the feces, and, therefore, a cross-section of the fecal bolus should be used for examination. On the other hand, eggs of *S. mansoni* tend to be concentrated in the outer layer, especially in mucus or blood. In both infections, but especially in those with *S. mansoni*, simple fecal smears may fail to reveal the eggs, making it necessary to use a concentration technique. In chronic cases, rectosigmoid punch biopsy often will reveal eggs when they have not been found in many different stool specimens. Intradermal, complement-fixation, and other serologic tests with schistosome antigen are available.

Other laboratory findings are a moderate to high eosinophilia during the acute stage, elevated transaminases, abnormal liver function tests, anemia, and increased levels of IgG, IgM, and IgE. If eggs have been largely confined to the intestine, there are few abnormal values. Even in some patients with hepatosplenic involvement, blood and chemistry values may be near normal. These have been termed "compensated cases."

Treatment. The treatment of schistosomiasis is undergoing rapid change at the present time, with optimal therapy varying according to the species and geographic strain. For example, *S. mansoni* infection in the New World is treated with oxamniquine, and in Africa, hycanthone is used. Metrifonate appears to be the preferred drug for *S. haematobium* and praziquantel for *S. japonicum.* It is recommended that the current literature and the Parasitic Diseases Division of the Centers for Disease Control be consulted when cases are discovered.

Schistosome Dermatitis

Many birds and mammals are infected with their own peculiar species of schistosomes, and the life cycles are similar to those species normally parasitic in humans. As a result of birds or mammals defecating in bodies of water, the infected snails may shed cercariae that can penetrate the skin of humans who come in contact with the water. In these cases, humans are an abnormal host, and the cercariae are walled off in the skin and evoke an acute inflammatory response characterized by a leukocyte infiltration and edema. Severe itching follows, and, hence, the condition is popularly called "swimmer's itch." Papules, hemorrhagic rash, and pustules may last a week or longer, and secondary infections due to scratching are common.

Schistosome dermatitis is a plague of swimmers during the summer season in the northern lakes of the United States and southern Canada. Similar outbreaks are seen along the saltwater beaches of the eastern coasts. Some of the beaches in the Gulf States, California, and Hawaii are also involved. Topical ointments to relieve the itching and edema are recommended.

Clonorchis sinensis

Infection with this fluke is known as clonorchiasis, Oriental liverfluke infection, biliary distomiasis, and biliary trematodiasis. In addition to *C. sinensis,* various species of *Opisthorchis, Fasciola,* and *Dicrocoelium* are also liverflukes of humans. Because treatment varies for each species, it is important to determine which fluke is involved.

The World Health Organization estimates that some 20 million people are infected, with endemic centers being in the Far East. The fluke is being seen more frequently in the United States because of the recent immigration of thousands of Indochinese.

Life Cycle. The hermaphroditic adults, up to 2 cm in length, usually live in the distal bile ducts but have also been found throughout the biliary passages, gallbladder, and pancreatic duct. Eggs are shed in the feces, and a typical intrasnail cycle occurs when suitable snails ingest the eggs. Cercariae leave the infected snails and seek out suitable freshwater fish that are used as the second intermediate host. The cercariae encyst in the muscles of the fish, and humans become infected by eating raw or improperly cooked fish. After ingestion, the excysted metacercaria migrates up the ampulla of Vater and enters the biliary tract. The incubation period varies, but most worms become mature in 1 month and live for over 20 years.

Pathology and Symptomatology. Many times clonorchiasis is an asymptomatic infection. When symptoms are present, their severity will depend on worm burden, duration of infection, and the number of reinfections suffered by the host. Pathogenesis seems to be related to the mechanical actions and toxic products of the worms. Epidemiologic data indicate a possible relationship between clonorchiasis and adenocarcinoma of the liver in those in endemic areas.

Pathologic changes include biliary hyperplasia, connective tissue hyperplasia, fatty degeneration of liver parenchyma, and fibrosis, which may lead to portal cirrhosis. A pancreatitis may develop when worms have invaded the pancreatic duct.

Diagnosis. The signs and symptoms of infection may include epigastric pain, anorexia, diarrhea, ascites, and hepatomegaly. In heavy and chronic infections, progressive hepatic dysfunction is common. Malignancies, hepatitis, and cirrhosis due to other causes need to be considered in a differential diagnosis. There are few, if any, abnormal laboratory values in asymptomatic and mild infections. Moderate to heavy infections may show eosinophilia, anemia, and elevated serum alkaline phosphatase and bilirubin values. A variety of serologic tests has been developed, but none is of practical use in the clinic laboratory setting.

A definitive diagnosis depends on the demonstration of the typical small, operculated, brownish egg that measures 30 by 15 μm (Fig. 90-16). In difficult cases, duodenal aspirates are sometimes of value.

Treatment. At present, there does not appear to be any drug of real value for liverfluke infection. Aralen has been used, and it may provide some relief of symptoms, but the drug does not kill the adult worms. Recently, praziquantel has been used, and the results look promising.

Paragonimus westermani

This lungfluke is the cause of paragonimiasis, or endemic hemoptysis. Human infections are common in the Far East, especially in China, Korea, Japan, Taiwan, and the Philippines. Other endemic foci are in Africa and India, and unidentified species are reported in Mexico, Central America, Peru, and Ecuador. Recently, cases are being detected among Indochinese (Laotian) refugees in the United States.

The adult flukes, about 1 cm long, live in the parenchyma of the lung. Other tissues and organs are some-

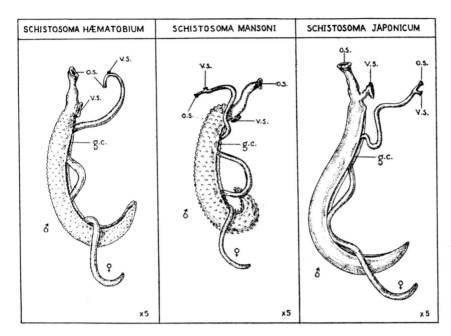

SCHISTOSOMA HÆMATOBIUM	SCHISTOSOMA MANSONI	SCHISTOSOMA JAPONICUM

Figure 90-16. Schematic representation of important schistosomes of humans. g.c., gynecophoric canal; o.s., oral sucker; v.s., ventral sucker. *(From Belding: Textbook of Clinical Parasitology, 3rd ed, 1965. New York, Appleton-Century-Crofts.)*

times involved, including the central nervous system and skin. Humans become infected by ingesting raw freshwater crabs or crayfish that harbor the infective larval stage. After ingestion, the excysted larva penetrates the intestinal wall, migrates into the peritoneal cavity, and eventually passes through the diaphragm and into the lungs.

There are usually no or few signs and symptoms during the migration phase. Once the larvae are in the lungs, hemoptysis with brown or red sputum occurs, as does pleurisy. Paragonimiasis mimics tuberculosis, and because of the high rate of tuberculosis among Indochinese refugees, diagnosis is sometimes difficult; both fluke infection and tuberculosis coexist in some patients. Hemoptysis and cough in the absence of a reaction to tuberculin should increase suspicion of paragonimiasis.

Diagnosis is based on finding the characteristic egg (Fig. 90-3) in the sputum or feces. Complement-fixation tests may aid in diagnosis but should not be the sole basis for treatment. Radiographic films may also show cysts in lung, brain, and so on. Treatment should be administered when diagnosis is confirmed. Praziquantel, available from the Parasitic Diseases Division at the Centers for Disease Control, is recommended. This drug seems to be very effective and is widely used in endemic areas.

Fasciolopsis buski

In addition to *F. buski*, other species, such as *Metagonimus yokogawai* and *Heterophyes heterophyes*, are intestinal flukes of humans. All are endemic in the Far East and have numerous reservoir hosts.

F. buski, known as the giant intestinal fluke (up to 8 cm in length), lives in the small intestine and attaches to the mucosa. Humans become infected by ingesting the encysted metacercariae that are present on water plants, such as caltrop, water bamboo, and water hyacinth. Mature flukes are present 3 months after ingestion of the infective larvae.

Adult worms cause ulceration, inflammation, and hypersecretion of mucus around attachment sites. In heavy infections, worms may inhabit the stomach and colon. Diarrhea, edema of face and legs, and marked anemia may be seen. Compared to adults, children usually are more severely affected. Adult worms may live 20 to 30 years. Death is usually attributable to cachexia and intercurrent infection.

Definitive diagnosis depends on finding the large operculated egg (Fig. 90-16). Laboratory findings include anemia and eosinophilia. Praziquantel is the drug of choice.

Index